32nd Polymer Processing Society Annual Meeting 2016 (PPS-32)

Lyon, France
25 - 29 July 2016

Volume 1 of 2

ISBN: 978-1-5108-4339-4

Printed from e-media with permission by:

Curran Associates, Inc.
57 Morehouse Lane
Red Hook, NY 12571

Some format issues inherent in the e-media version may also appear in this print version.

Printed by Curran Associates, Inc. (2017)

For permission requests, please contact Polymer Processing Society
at the address below.

Polymer Processing Society
810 E. 10th St.
Lawrence, KS 66044
USA

Phone: +1-800-627-0326
+1-785-865-9403
Fax: +1-785-843-6153

pps@allenpress.com

Additional copies of this publication are available from:

Curran Associates, Inc.
57 Morehouse Lane
Red Hook, NY 12571 USA
Phone: 845-758-0400
Fax: 845-758-2633
Email: curran@proceedings.com
Web: www.proceedings.com

TABLE OF CONTENTS

VOLUME 1

BLOW MOLDING, THERMOFORMING AND ROTOMOLDING

NANOCOMPOSITES AND FILLED POLYMERS

EXTRUSION AND EXTRUSION PROCESSES

POLYMER BLEND AND ALLOYS

POLYMER FOAMS AND LIGHTWEIGHT STRUCTURES

BIOPOLYMERS AND POLYMER FROM RENEWABLE RESOURCES

PROCESS MODELING AND SIMULATION

VOLUME 2

PROCESS MONITORING, CONTROL AND SENSORS

REACTIVE PROCESSING

RHEOLOGY AND RHEOMETRY

ELASTOMERS AND THERMOPLASTIC ELASTOMERS

FIBER, FILMS, AND MEMBRANES

INJECTION MOLDING AND MOLD

MIXING AND COMPOUNDING

MORPHOLOGY AND STRUCTURE DEVELOPMENT

RECYCLING, PLASTIC SOLUTIONS FOR ECO-CONCEPTION

HIGH PERFORMANCE COMPOSITE PROCESSING

ADDITIVE MANUFACTURING FOR PLASTIC COMPONENTS

Optimization of the IR-Heating Phase in Thermoforming of Thermoplastic Sheets: Characterization and Modelling

B. Buffel[a*], B. Van Mieghem[b], A. Van Bael[b] and F. Desplentere[a]

[a]*KU Leuven Technology Campus Oostende, Zeedijk 101, 8400 Oostende, Belgium*
[b]*KU Leuven Technology Campus Diepenbeek, Wetenschapspark 27, 3590 Diepenbeek, Belgium*

Presenting author: bart.buffel@kuleuven.be
Reference author: bart.buffel@kuleuven.be

Abstract. The thermoforming industry is currently seeking for methods to increase process monitoring and control. The present paper addresses this topic by focusing on the heating phase in thermoforming. A coupled experimental-modelling approach was developed to characterize the heating equipment on site. This allows taking local effects such as ambient air or machine specific heating efficiencies into account. The characterization results in a machine specific data set which is subsequently used to optimize the heating of the thermoplastic sheet. An optimization script was developed which proposes directly applicable machine settings in order to obtain a through-thickness temperature profile within the desired forming range.

Keywords: Thermoforming, infrared heating, finite difference.
PACS: 44.10, 44.27, 44.40

INTRODUCTION

The thermoforming process for thermoplastic sheets consists out of three different phases: (1) heating of the sheet, (2) forming of the product by means of vacuum and (3) cooling and demoulding of the product. In industry, process start up and control is still based on personal experience and trial and error methods. This leads to varying machine settings, excessive material waste and eventually high start-up times and costs. There is thus a need for a higher level of process control in the thermoforming industry [1]. Since the importance of a well heated sheet cannot be overestimated, the present paper focusses on this phase of thermoforming process. As the complexity of the heating phase increases with increasing sheet thickness, heavy gauge thermoforming of sheets with a thickness larger than 3 mm is the main area of interest. Nevertheless, the methods and results presented in this paper can be applied to thin gauge thermoforming as well.

A controlled heating phase resulting in a as uniform as possible through-thickness temperature distribution is key in thermoforming of thermoplastic sheets. In practice, the heating is done by means of infrared radiation via ceramic, quartz or halogen elements. Since this type of contact-free heating is based on a (constant) heat flux, a perfectly uniform through thickness temperature distribution is very difficult to obtain with standard industrial thermoforming equipment [2].

This paper presents a framework to characterize the heating phase of the thermoforming process based on an equipment specific dataset. This dataset is derived from experimental measurements which are used in a reverse engineering fitting step. The parameters which are used in this fitting step are: absorbed radiative power, IR penetration depth, convective heat transfer coefficient, temperature of the convective environment and a "switch-on" delay time in case of halogen heater elements. The results of the fitting step can help the machine operator to understand and optimize the heating phase of the actual thermoforming process.

MODELLING THE HEATING PHASE

As described in literature, the heating phase of the thermoforming process can be modelled by means of the standard transient heat balance equation in which conduction, convection and radiation are taken into account:

$$\frac{dQ_{in}}{dx} - \frac{dQ_{out}}{dx} = mC_p(\theta)\frac{d\theta}{dt} \qquad (1)$$

In the current paper the heat flow is considered to be uniform in plane. Therefore only the through-thickness orientation x is taken into account in equation (1). In this equation Q denotes the heat flow, m is the mass, $C_p(\theta)$ is the temperature dependent specific heat capacity, θ is the temperature in °C and t is the time variable.

Equation (1) is solved by means of the explicit finite difference method [3]. Depending on a one or two-sided heating set-up in the thermoforming process the following boundary conditions are applied to the outer surfaces:

$$Q_{in} = \alpha A(\theta_\infty - \theta_i) + q_{rad} A \tag{2}$$

In this equation α represents the convective heat transfer coefficient, θ_∞ the temperature of the environment, A the exposed surface area and q_{rad} the radiative heat flux absorbed by the thermoplastic sheet. The through-thickness transport of heat in the sheet is modelled using the Fourier's law:

$$Q_{conduction} = -\lambda A \frac{dQ}{dx} \tag{3}$$

With λ representing the thermal conductivity of the thermoplastic material.

The combined set of equations can be used to model the through-thickness temperature distribution in the heating phase as a function of time. These equations assume however a complete absorption of the incident radiative flux at the surface of the sheet. Since a thermoplastic material is semi-transparent in practice [4], this needs to be taken into account by implementing the optical penetration depth, d_p [5,6]. This parameter represents the distance measured from the irradiated surface where the intensity has decreased to $1/e$ of the intensity at the surface. The higher the optical penetration depth is, the deeper the IR radiation will penetrate into the thermoplastic sheet resulting in a more volumetric heating rather than a surface heating effect. The optical penetration is both material and process (in casu the type of heater element) dependent.

In the present paper, the method proposed by Gauthier et al. [7] has been used to model the penetration of the radiative heat flux and to determine the through-thickness radiative heat flux distribution:

$$\beta = 1 - e^{\Delta x / d_p} \tag{4}$$

$$q_i = \beta (1-\beta)^{i-2} (1-\beta) q_{rad} \tag{5}$$

Next to the integration of the penetration depth, a second effect is added to the model. This additions deals with the experimentally observed time delay present in the heating process when using halogen heater elements. In the case halogen heater elements are used, the elements require a certain period to reach the requested power setting. This delay in the heating process was modelled by implementing a time constant τ in a simple first order approach:

$$q(t) = q_{rad}\, e^{-\tau / t} \tag{6}$$

The described set of equations allows to calculate the through-thickness temperature distribution as a function of time based on one-dimensional heat transfer by means of convection, conduction and radiation taking both radiative penetration and halogen heater element start-up delay into account. This model will be used to characterize the used thermoforming heating equipment by reverse engineering the different parameters (heater power, optical penetration depth, convective environment and start-up time delay) based on experimental measurements.

EXPERIMENTAL TEMPERATURE MEASUREMENTS

In the thermoforming industry, temperature measurements are usually done by pyrometers. These devices are able to measure surface temperatures in situ and are based on IR radiation. Experimental observations have however proven that pyrometer measurements are highly influenced by reflection and transmission of the IR waves [8]. Obtaining reliable absolute temperature data is therefore not straightforward by means of a pyrometer. To overcome these difficulties, an instrumented vulcanized silicone sheet was developed. This sheet contains 15 thermocouples in total:

3 thermocouples through thickness at 5 different locations in-plane. The sheet is 3.9 mm thick and 50x50 cm wide. The sheet allows to determine the through-thickness temperature distribution as a function of time, as well as the in-plane temperature distribution. The sheet can be used for any type of heater elements and for one- or two-sided heating approaches.

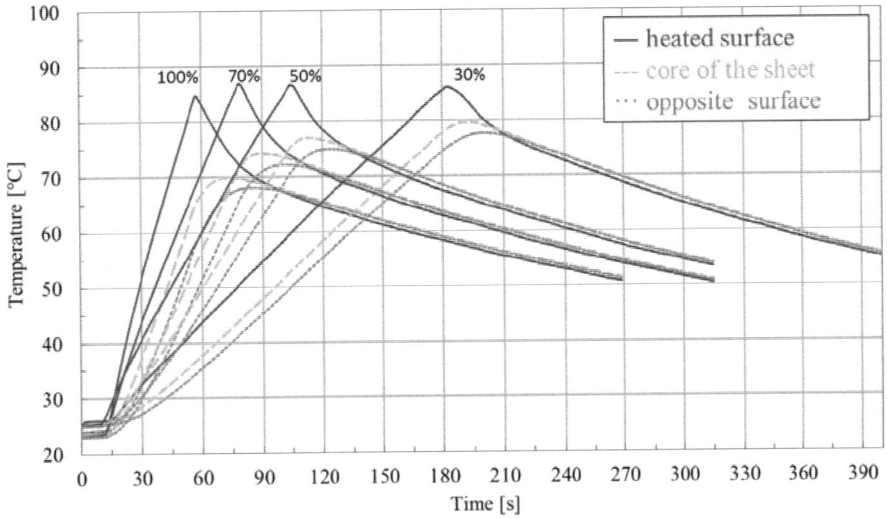

FIGURE 1. Experimental through-thickness temperature vs. time data for one-sided heating on a Geiss U8 thermoforming machine equipped with halogen heater elements. Different power settings (100-70-50-30 %) were applied with respect to the maximum heat flux of 51.5 kW/m².

The data in figure 1 clearly show that the higher the power setting of the heater bank is, the faster the sheet surface is heated. This is a desired effect in industry where a shorter heating time would result in a gain in productivity. The measurements indicate however an increasing temperature difference between the heated surface and the core or non-heated (opposite) surface of the sheet. Since a uniform through thickness temperature distribution is required for the forming step, this puts a limit on the maximum feasible heating rate and thus applied heating power setting. Since exact temperature data or heating characteristics of the heating equipment is often not known in industry, process control remains based on trial and error or on operator experience.

REVERSE ENGINEERING

In this section the developed Finite Difference model is combined with the experimental temperature measurements. Experimental temperature data of different heater settings will be used to characterize the heating equipment of the thermoforming machine. As a result of this reverse engineering step, a dataset is derived that characterizes the heating equipment. This dataset can afterwards be used to optimize the heating step. All types of heating elements can be characterized. In the case of halogen elements, experimental measurements are performed for different power settings (e.g. 100-90-80-70-60-50-40 % of the maximum heat flux), while in the case of ceramic and quartz elements different temperature settings are used (e.g. 250-300-350-400-450 °C). This approach creates the link between machine settings and virtual modelling work, which is not always included in literature.

The thermal material data of the silicone sheet are: a density of 1380 kg/m³, a thermal conductivity of 0.18 W/mK and a specific heat capacity of 1330J/kgK. The parameters which were used in the fitting procedure are summarized in table 1. All parameters in this table are determined for each heater setting in order to fully characterize the equipment. Since the temperature of halogen heater elements is an order of magnitude higher than the sheet temperature, a constant radiative heat flux was implemented. The lower operating temperature of quartz and ceramic elements does not allow this assumption.

In figure 2 the result of a single fitting procedure for a specific heating setting (ceramic heaters at 400°C) is shown. All curves in this graph are fitted using a single set of optimized heating parameters. The good correlation between the experimental and fitted curves justifies the extraction of the heating parameters (table 1). The small differences between the experimental and fitted data are explained by uncertainties in silicone material data and the exact through-thickness positions of the thermocouples.

TABLE 1. Used parameters for the fitting procedure.

Parameter	Halogen heater elements	Ceramic/Quartz heater elements
Constant radiative heat flux	V	-
Temperature of the heater elements	-	V
Radiative number (Boltzmann Law)	-	V
Convective α top side	V	V
Convective α bottom side	V	V
Environment temperature top side	V	V
Environment temperature bottom side	V	V
Optical penetration depth	V	V

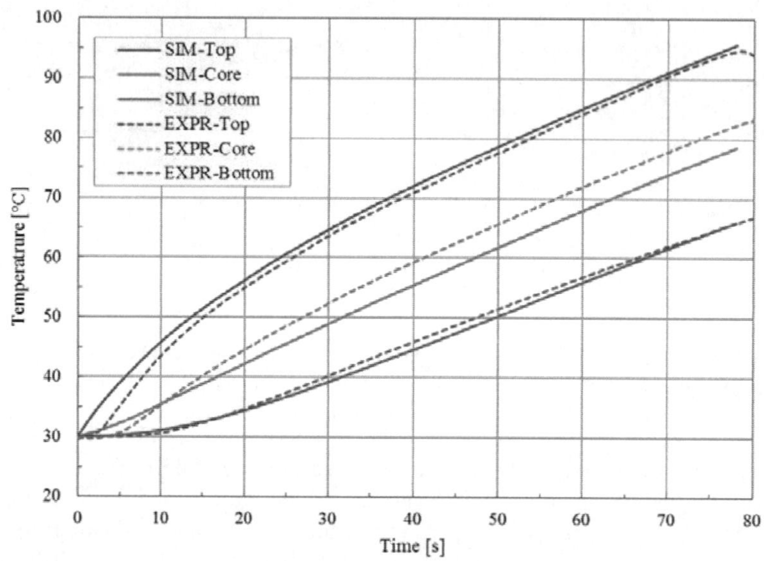

FIGURE 2. Experimental and reverse engineered time-temperature curves for one-sided heating of the silicone sheet. The sheet is heated from the top side by means of ceramic heaters set at 400°C

OPTIMIZATION PROCEDURE

In industry, an as uniform as possible through-thickness temperature distribution in the shortest possible time is desired. The theoretically fastest way of obtaining this result would be a two-step heating in which the outer surface is first heated up to the forming temperature. Secondly, the surface temperature is kept constant at the forming temperature in order to allow conductive heat transfer to reach a uniform through thickness temperature. This second step requires a fast response heating element in combination with a feedback loop coupled to an in situ accurate temperature measurement. The response times for ceramic and quartz elements are too long for this application, and the required low power settings will significantly decrease the expected lifetime of halogen elements. Despite the availability of fast responding heating methods like metal foil or ceramic surface heaters, they are seldom applied in thermoforming industry due to the high investment costs [9].

The optimization procedure presented in this paper is based on the temperature forming range of the processed thermoplastic material [2]. The procedure runs an iterative loop in which the heating power settings are adjusted. In this loop the heated surface of the sheet is heated up to the maximum allowable forming temperature. Next, the through-thickness temperature profile is checked. In case the temperatures in this profile are not completely within the specified forming range, the power setting is adjusted. This loop is repeated until the complete temperature profile fits within the forming range. As this procedure takes the fitted parameters from the reverse engineering phase into account, the result of the optimization can directly be linked to the actual machine settings.

In figure 3, illustrations of thermoformed parts have been added to demonstrate the efficiency of the proposed optimization procedure. Only when the through-thickness temperature profile was completely within the forming range, the sharp edges at the top of the part could be formed.

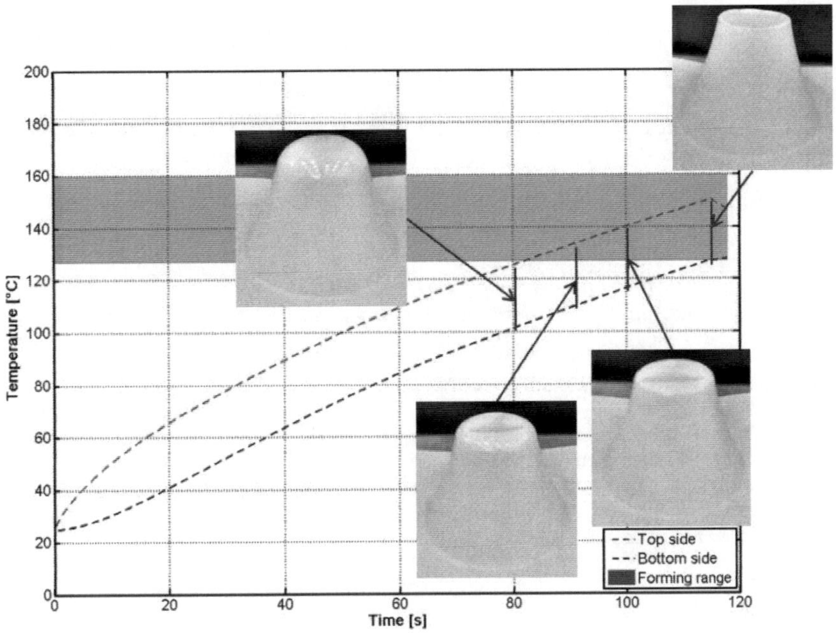

FIGURE 3. Illustration of the effects of different heating times, resulting in an accurate part when using the proposed heating optimization procedure, i.e. the shortest heating time that guarantees a through-thickness temperature profile within the forming range (green zone).

CONCLUSIONS

The current paper presents a procedure in which experimental temperature measurements are coupled to finite difference modelling in order to characterize the heating equipment of a thermoforming installation. This characterization gives the industry a better understanding and an optimization of the heating procedure. A heating strategy is proposed that guarantees through-thickness temperatures that fall within the forming range of the material at hand. Applying the proposed procedure will result in shorter start-up times, a decrease in material waste and a more controlled thermoforming process.

ACKNOWLEDGMENTS

The authors wish to thank the Flemish government for the funding in the IWT TETRA1301179 project. The results discussed in the present paper where obtained within this project.

REFERENCES

1. E. Gilham, *Lean manufacturing*, SPE- European Thermoforming Division conference, Barcelona (Sitges), 2016
2. J.L. Throne, *Technology of thermoforming*, Munich: Hanser Gardner Publications, 1996
3. J.P. Holman, *Heat transfer*, New York, McGraw-Hill, eight edition, 1997
4. J.L. Throne, *Thermoforming Quarterly* **35**,22-27 (2016)
5. J.F. Puehringer, G. Zitzenbacher and C. Spreitzer, *Intern. Polymer Processing XXVIII* **1**, 14-23 (2013)
6. J.F. Puehringer, G. Zitzenbacher and C. Spreitzer, PPS 27 Conference Proceedings, Marrakech, 2011
7. G. Gauthier, A. Mark, B. Boulet, A. Haurani, P. Girard, R. Diraddo, ANTEC Conference Proceedings, 2005, pp. 353-357
8. B. Neubig, *Determination of Temperature Profiles Across the sheet thickness during the heating phase of thermoforming processes*, SPE- European Thermoforming Division conference, Barcelona (Sitges), 2016
9. S. Bach, *Heat under control- A novel contact heating technology for thermoforming evolutions* SPE- European Thermoforming Division conference, Barcelona (Sitges), 2016

Investigation of Halloysite as Reinforcement for Polyethylenes in Rotational Moulding

Günther Höfler, Richard J.T. Lin, Krishnan Jayaraman

Centre for Advanced Composite Materials, Department of Mechanical Engineering, The University of Auckland, Private Bag 92019, Auckland Mail Centre, Auckland, 1142, New Zealand

Presenting author's email: ghoe113@aucklanduni.ac.nz
Reference author's email: rj.lin@auckland.ac.nz, k.jayaraman@auckland.ac.nz

Abstract. Rotational moulding is primarily used for the manufacture of hollow, polymer products at atmospheric pressure. The predominant material used worldwide in this industry is polyethylene of various densities, most commonly in powder form. Previous studies regarding the use of filler materials in rotational moulding often report problems with agglomerations or inward migration of the filler. In spite of the advances in machine adaptations and mould configurations to apply internal pressure, this study focusses mainly on non-pressurized composite production. Halloysite is a natural, nano-size, mineral clay with different reactivities for the internal aluminol and external siloxane surfaces. Due to its morphology and chemical nature, halloysite is easier to process compared to other fillers and achieves good particle dispersion, making it a potential candidate for reinforcing rotationally moulded products. In this study, nanoparticle-reinforced composites containing halloysite with medium density or high density polyethylene are produced by rotational moulding. The influence of halloysite on the processability, and mechanical properties - tensile, flexural and impact -, has been investigated.

Keywords: Rotational moulding, halloysite, linear medium density polyethylene, high density polyethylene, nanoparticle-reinforced composite.
PACS: 81.05.Qk, 81.20.Hy, 81.05.Lg, 81.07.Nb, 81.40.Lm, 81.70.Bt

INTRODUCTION

Rotational moulding of polymers is a closed-mould process, with low-pressure and high temperatures. The four main cycle stages are: material charging, heating, cooling and demoulding. It is mainly used for the production of large hollow parts such as water tanks, boats and playground equipment. Some of the main advantages of this process are the relatively low manufacturing and material costs, little material waste, low residual stresses, and the ability to produce large parts with complex geometries. On the other hand, limitations that need to be considered are long processing times, high mould costs for complex shapes and limited selection of materials. Polyethylene, in all its varieties, still accounts for more than 85% of raw materials used in rotational moulding, mainly due to its outstanding resistance against thermal degradation in a prolonged high temperature environment. As the industry expands and the limits of the modest mechanical properties of the available materials are reached, the introduction of reinforcement materials in rotational moulded products has moved into the focus of attention. [1, 2, 3]

Attempts to use reinforcing materials in rotational moulding have led to mixed results, with poor adhesion and mixing – segregation and agglomeration of reinforcements – being the main problems [4, 5, 6]. In order to achieve a good material flow and coalescence in rotational moulding, a maximum particle size of 500 microns is advantageous. Due to this dimensional limitation, it is difficult to attain high aspect ratios of the reinforcement material, which are essential in all plastic processes for stiffness and dimensional stability. [7] By applying little internal air pressure, or with a multi-shot process, improvements in the incorporation of reinforcement material could be reached [8, 9, 10]. But these adaptions require either expensive mould modifications or result in longer cycle times, both uneconomical. Hence, research continues on finding new ways of improving material properties of rotational moulded parts.

Halloysite, with a molecular/chemical structure of $Al_2Si_2O_5(OH)_4 \cdot H_2O$, is a tubular, platy, or spheroidal nano-sized mineral clay. In its tubular form, the internal surface is aluminol, whereas the external surface is siloxane dominated. Due to low surface charges - exfoliation or intercalation is not necessary [11, 12, 13] - halloysite is easier to process compared to other clay fillers and achieves good particle dispersion, making it a potential candidate for reinforcing rotationally moulded products. Additionally, halloysite is known to improve the thermal stability of

LLDPE [14], which is very important as rotational moulding is usually performed close to the polymer degradation temperature.

EXPERIMENTAL DETAILS

Materials Preparation and Moulding

All polyethylene grades used as matrix materials (Table 1) in this project were supplied by Vision Plastics New Zealand Ltd. Two grades of linear medium density polyethylene (LMDPE) - VP305 and VX567, and one grade of high density polyethylene (HDPE) - VX573, the latter with a new catalyst technology for superior mechanical properties in comparison to conventional HDPE. All of the polyethylenes are rotational moulding grades in powder form with an average particle size lower than 500 microns. The halloysite used in the experiments has an average particle size of below 4 μm and a density of 2.55 g/cm³.

TABLE 1. Material Properties of polyethylene grades used.

Product Name	Polyethylene Grade	Melt Flow Index [g/10min]	Density [g/cm^3]
VP305	LMDPE	3.2	0.938
VX567	LMDPE	6.0	0.935
VX573	HDPE	4.0	0.947

Rotational moulding was carried out in a laboratory scale rock-and-roll machine with a Teflon coated stainless steel mould of internal dimensions 220x210x190 mm^3. A hole at the centre of the top lid acted as a vent and provided access for a thermocouple. The oven temperature was set to 250°C for the heating cycle. Three peak internal air temperature (180°C, 190°C or 200°C) were used as an indicator to initiate the cooling to study the effects of peak internal air temperature (PIAT) on the mechanical performance of rotomoulded products. For the production of composites, the charging materials were dry blended, manually or with a high speed mixer (Fluid High Speed Mixer Type LMX5-S-VSFI by Labtech Engineering Co. Ltd.) to study the effects of mixing methods on the mechanical performance of rotomoulded products..

Materials Characterisation

The specimens for all samples were exclusively cut out from the side walls of the rotationally moulded products and tested at room temperature to determine tensile (ASTM D638), impact (ASTM D6110) and flexural (ASTM D790) properties. Each test method was performed on at least 5 specimens per sample. An Instron universal testing machine was used for tensile and flexural testing, whereas the impact tests were carried out on a Ceast Charpy impact tester.

RESULTS AND DISCUSSION

Moulding Quality

While all halloysite containing products showed small pinholes, the composites produced from manually blended charging materials exhibited more bubbles and pinholes. The surface quality of rotationally moulded product and the dispersion of halloysite showed improvements when halloysite was dried for one day at 90°C (moisture content reduction from about 6% to 2.5%) and blended with the polymer using a high speed mixer (600rpm for 5min), which led to a macroscopic more uniform distribution of halloysite within the polymer powder.

It was found that compression moulded sheets of LLDPE-halloysite composite had a reduced complex viscosity and storage modulus compared to those of pure LLDPE, especially at low shear rates. [14] Rotational moulding of plastics is a low shear rate process, experiments showed that with increasing amount of halloysite, the mouldability of HDPE-halloysite composites is reduced. The incorporation of halloysite led to cavities at the bottom of the moulded product (Fig. 1). This imperfection can be attributed to a possibly high level of melt flowability/reduced melt viscosity, which has an adverse impact on the uniform and complete adhesion along the whole mould surface.

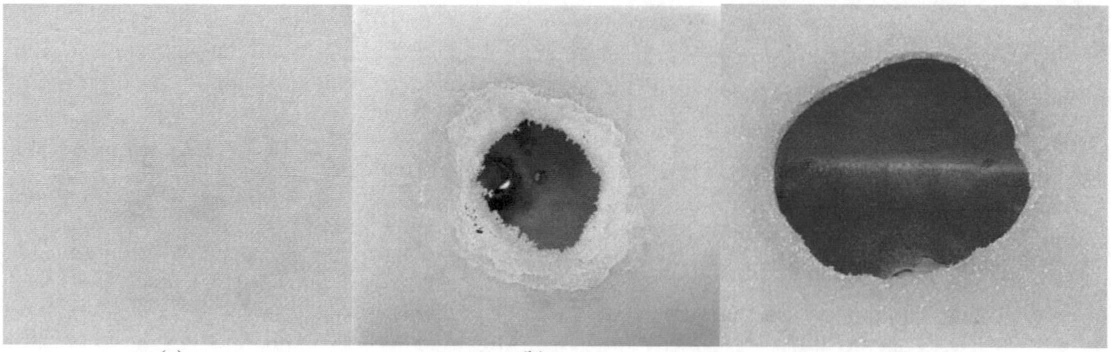

(a) (b) (c)

FIGURE 1. Influence of Halloysite content on the mouldability of HDPE-halloysite composites; a) 2 wt% halloysite, b) 5wt% halloysite, c) 10wt% halloysite

Mechanical Performance

Influence of the PIAT on the mechanical properties of PE

As cycle time has been a critical issue for rotational moulding, with heating and cooling being the longest steps in the processing cycle, the PIAT was kept in a relatively small range (180-200°C) in order to keep the overall cycle time as short as possible. On one hand, the PIAT has to be high enough to allow all the material to melt and to provide good coalescence, and on the other hand, to be low enough to avoid thermal degradation as the measured oxygen onset temperatures are between 210-220°C for the PEs used in this research. The influence of the PIAT on the stiffness of the rotationally moulded polyethylene parts is marginal for all grades used. The LMDPE grades performed very similar, whereas the HDPE is superior to them, in all conducted mechanical tests. (Fig. 2 and Fig. 3) As the property variations are within the standard deviations, and a lower temperature is more economical (due to cycle time reduction), 180 °C was chosen as PIAT for the investigation of the influence of halloysite content on the mechanical properties.

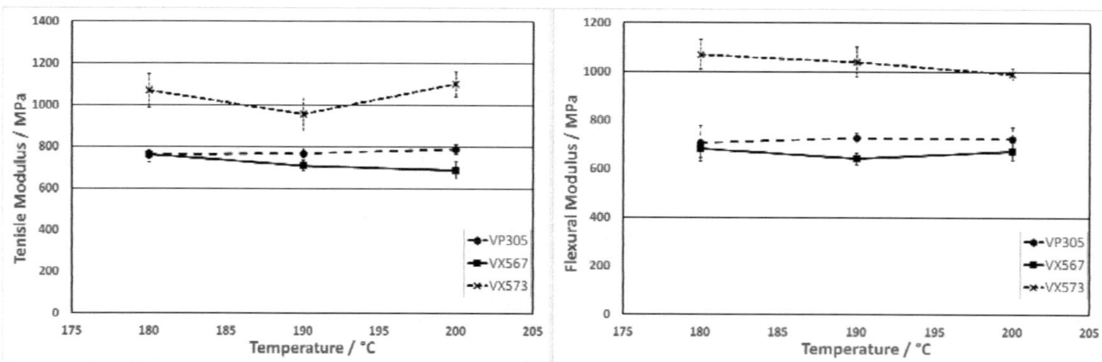

FIGURE 2. Effect of processing temperature on the tensile and flexural moduli of polyethylene

Influence of halloysite content and mixing method on mechanical properties of HDPE

The influence of mixing method was investigated on HDPE-halloysite composites. The utilization of a high speed mixer (HSM) resulted in a better visual appearance of the product, but no significant improvement in the mechanical properties could be observed. Both mixing methods, considering the standard deviation, showed the similar results. A slight increase in tensile modulus (4-8%) could be achieved by adding halloysite, whereas the flexural modulus was more stable with roughly a ± 4% variation (Fig. 4). Further, small amounts of halloysite decreased the charpy impact energy by 50-70%. (Fig. 5) The addition of 5 wt% halloysite to LMDPE led to an increase in flexural modulus of 5% and 16% for VX567 and VP305, respectively. Yet, the impact properties were cut in half for both, from 117 J/m to 68 J/m for VP305 and from 145 J/m to 67 J/m for VX567. Halloysite particles, and their agglomerations, may act as crack initiator, which lead to deteriorated impact properties of the composite part. Another possible factor for poor mechanical performance of composites is a poor interfacial filler-matrix interaction.

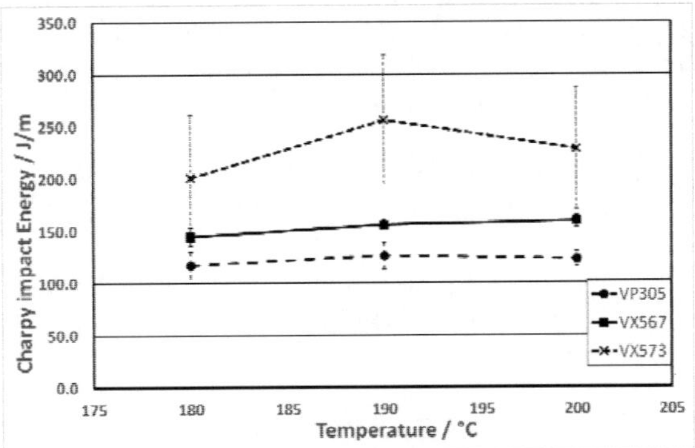

FIGURE 3. Effect of processing temperature on charpy impact energy of polyethylene

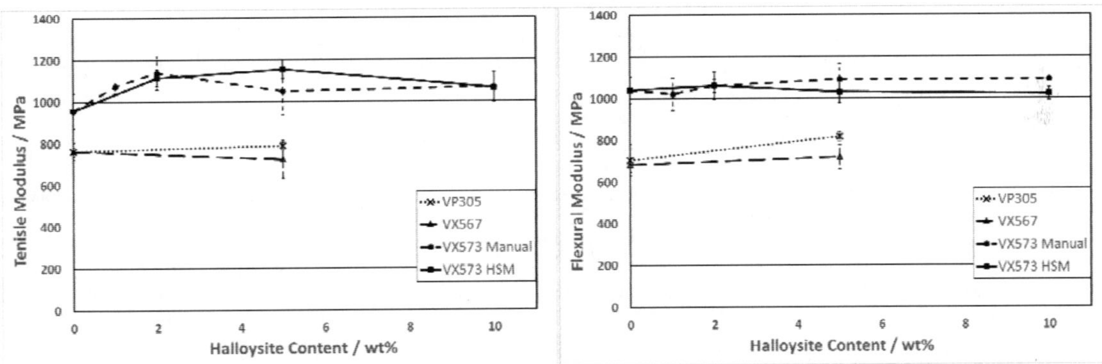

FIGURE 4. Tensile and flexural moduli of PE-halloysite composites as a function of halloysite content

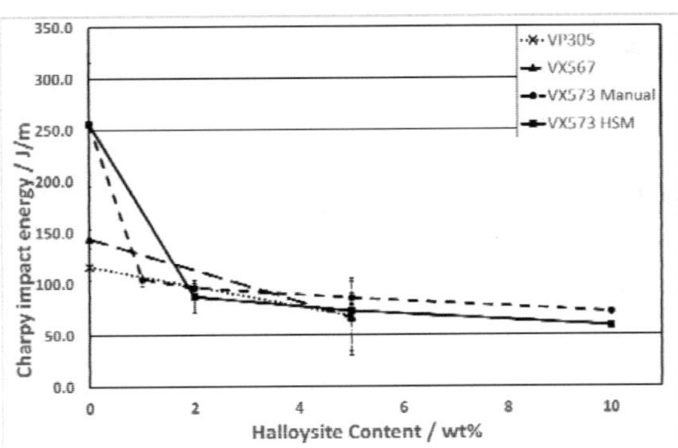

FIGURE 5. Effect of halloysite content on charpy impact energy of PE-halloysite composites

CONCLUSIONS

Within the chosen processing temperature range, the mechanical properties changed very little. Thus, a lower temperature (PIAT = 180 °C) was chosen for further experiments for economic benefits. Halloysite particles were successfully incorporated in rotational moulded polyethylene products. Dry blending of the charging materials with a high speed mixer improved the dispersion of the halloysite particles within the polymer matrix - compared to manual mixing. Additionally, HSM blending also improved the product surface quality and reduced agglomerations and pinholes of halloysite within the composites. However, no significant influence of the mixing method on the mechanical performance could be observed. In comparison to unreinforced HDPE, the tensile modulus could be increased by 20%with the addition of 5 wt% of halloysite and the use of a high speed mixer, for blending. For the LMDPE grades, 5 wt% of halloysite caused a small increase in flexural modulus of 9% (VP305) and 14% (VX567). However, the impact properties of all polyethylenes were reduced significantly with 5 wt% of halloysite addition (decrease of 64% for VX573, 54% for VX567 and 45% for VP305).

ACKNOWLEDGMENTS

Special thanks to the Ministry of Business, Innovation and Employment for financial support to the first author, and to Vision Plastics New Zealand Ltd for materials supply.

REFERENCES

1 R.J. Crawford and J.L.Throne., *Rotational Molding Technology*, Williams Andrew Publishing/Plastics Design Library, Norwich, USA, 2001
2 R.J. Crawford and M.P Kearns., *Practical Guide to Rotational Moulding*, Rapra Technology Limited, Telford, UK, 2003
3 P.E. Lescher, “Moulding of Water-Free Thermoplastic Starch Blends.” Ph.D. Thesis, The University of Auckland, 2010.
4 W. Yan, R. J. T. Lin, and D. Bhattacharyya. "Particulate reinforced rotationally moulded polyethylene composites–Mixing methods and mechanical properties." Composites science and technology 66.13, 2006, pp. 2080-2088.
5 Z. Ortega, M.D. Monzón, A.N. Benítez, M. Kearns, M. McCourt, P.R. Hornsby, "Banana and abaca fiber-reinforced plastic composites obtained by rotational molding process." Materials and Manufacturing Processes 28.8, 2013, pp. 879-883.
6 F.G. Torres, M. Aguirre. "Rotational moulding and powder processing of natural fibre reinforced thermoplastics." International Polymer Processing 18.2, 2003, pp. 204-210.
7 Lee, Stuart M. *Handbook of composite reinforcements*. John Wiley & Sons, California, USA, 1992.
8 G. Gogos. "Bubble removal in rotational molding." Polymer Engineering & Science 44.2, 2004, pp. 388-394.
9 E. Harkin-Jones, and R. J. Crawford. "Mechanical properties of rotationally molded nyrim." Polymer Engineering & Science 36.5, 1996, pp. 615-625.
10 A.H. Steinberg, F. Petruccelli, M.E. Lucas, L. Zlatkevich, "Rotational molding multilayered articles." U.S. Patent No. 4,548,779. 22 Oct. 1985.
11 E. Joussein, S. Petit, J. Churchman, B. Theng, D. Righi, B. Delvaux, "Halloysite clay minerals—a review." Clay minerals 40.4, 2005, pp. 383-426.
12 P. Yuan, D. Tan, and F. Annabi-Bergaya. "Properties and applications of halloysite nanotubes: recent research advances and future prospects." Applied Clay Science 112, 2015, pp. 75-93.
13 D. Rawtani, Y. K. Agrawal. "Multifarious applications of halloysite nanotubes: a review." Rev. Adv. Mater. Sci 30, 2012, pp. 282-295.
14 D. Pedrazzoli, A. Pegoretti, R. Thomann, J. Kristóf, J. Karger-Kocsis, "Toughening linear low-density polyethylene with halloysite nanotubes." Polymer Composites 36.5, 2015, pp. 869-883.

Effect of Molecular Weight Distribution on the Thermal and Physical-Mechanical Properties of High Density Polyethylene Blow Molding Grade

Afzali Kamal, Keshavarzian Ali*

Jam Petrochemical Company, Pars Special Economic Energy Zone, Assaluyeh, Boushehr Province, I.R.Iran.

ali_keshavarz@aut.ac.ir

Abstract. High density polyethylene (HDPE) is a thermoplastic material that is characterized as a semi-crystalline polymer. Density, molecular weight and molecular weight distribution (MWD) dominate the resin properties that influence the manufacture of the polyethylene products and the subsequent performance of them. In this research, two blow molding grade of HDPE (BL3) with the same specifications (melt flow rate (MFI) range and density) that those produced by different manufactures were compared and studied. The MWD of the BL3 samples were studied with Gel permeation chromatography (GPC) and Capillary Rheometer (CR). Differential scanning calorimetry (DSC) was used to study the thermal properties and the Notched impact test used to study Impact Strength. The GPC results have shown that both samples, had similar bimodal shaped distribution curves, but curve of sample 2, located partially in the higher range of MWD (placed in the higher quantities on the horizontal axis (molecular weight)) than the sample 1. The CR results also showed that both samples had similar distribution, but the sample 1 shows better processability in the blow molding shear rate range than the other sample. The DSC results have shown that sample 2 had higher crystallinity and Tm than the other sample. Notched impact test results also showed that the sample 2 had higher impact resistance. Results showed that sample with the higher range MWD, had better thermal and physical-mechanical properties.

Keywords: High density polyethylene, Molecular weight distribution, Characterization, Thermal properties.
PACS: 82.35.-x

INTRODUCTION

High-density polyethylene (HDPE) is a thermoplastic material composed of carbon and hydrogen atoms joined together forming high molecular weight products. The polymer chain may be 500,000 to 1,000,000 carbon units long. Short and/or long side chain molecules exist with the polymer's long main chain molecules. The longer the main chain, the greater the number of other common polyethylene (PE) materials are medium-density polyethylene (MDPE), low-density polyethylene (LDPE), Linear low-density polyethylene (LLDPE), Less common PE materials are ultra-high molecular weight polyethylene (UHMWPE) and very low density polyethylene (VLDPE). The property characteristics of polyethylene depend upon the arrangement of the molecular chains. The molecular chains are three-dimensional and lie in wavy planes. Branching off the main chains, are side chains of varying lengths. The number, size and type of these side chains determine, in large part, the properties of density, stiffness, tensile strength, flexibility, hardness, brittleness, elongation, creep characteristics and melt viscosity that are the results of the manufacturing effort and can occur during service performance of polyethylene products [1].

Polyethylene is characterized as a semi-crystalline polymer, made up of crystalline regions and amorphous regions. Crystalline regions are those of highly ordered, neatly folded, layered (in parallel) and densely packed molecular chains. These occur only when chains branching off the sides of the primary chains are small in number. Within crystalline regions, molecules have properties that are locally (within each crystal) directionally dependent. Where tangled molecular chains branching off the molecular trunk chains interfere with or inhibit the close and layered packing of the trunks, the random resulting arrangement is of lesser density, and termed amorphous. An abundance of closely packed polymer chains results in a tough material of moderate stiffness [2].

High-density polyethylene resin has a greater proportion of crystalline regions than low-density polyethylene. The size and size distribution of crystalline regions are determinants of the tensile strength and environmental stress crack resistance of the end product. HDPE, with fewer branches than MDPE or LDPE, has a greater proportion of crystals, which results in greater density and greater strength. LDPE has a structure with both long and short molecular branches. With a lesser proportion of crystals than HDPE, it has greater flexibility but less strength. LLDPE structurally differs from LDPE in that the molecular trunk has shorter branches, which serve to inhibit the polymer chains becoming too closely packed. Hypothetically, a completely crystalline polyethylene would be too brittle to be functional and a completely amorphous polyethylene would be wax like, much like paraffin. Upon heating, the ordered crystalline structure regresses to the disordered amorphous state; with cooling, the partially crystalline structure is recovered. This attribute permits thermal welding of polyethylene to polyethylene [3]. Density, molecular weight and molecular weight distribution dominate the resin properties that influence the manufacture of the polyethylene product and the subsequent performance of their [4,5].

The distribution of different sized molecules in a polyethylene polymer typically follows the bell shaped normal distribution curve described by the Gaussian probability theory. As with other populations, the bell shaped curve can reflect distributions ranging from narrow to broad [6].

A polymer containing a broad range of chain lengths is said to have a broad molecular weight distribution (MWD). Resins with this type of distribution have good Environmental Stress Crack Resistance (ESCR), good impact resistance and good processability. A polymer with a narrow MWD contains molecules that are nearly the same in molecular weight. It will crystallize at a faster, more uniform rate. This results in a product that will hold its shape. Polymers can also have a bimodal shaped distribution curve which, as the name suggests, seem to depict a blend of two different polymer populations, each with its particular average and distribution. Resins having a bimodal MWD contain both very short and very long polyethylene molecules, giving the resin excellent physical properties while maintaining good processability. MWD is dependent upon the type of process used to manufacture the particular polyethylene resin. For polymers of the same density and average molecular weight, their melt flow rates are relatively independent of MWD. Therefore, resins that have the same density and melt index (MI) can have very different molecular weight distributions [2].

The HDPE process of Basell, that used to produce samples, is based on the discovery of Karl Ziegler and co-workers in 1953, that reaction product of transition metals and organo-metallic compounds are powerful catalysts in the polymerization of ethylene even at atmospheric pressure and at moderate temperatures. Recognizing the importance of the Ziegler discovery Hoechst AG was able to implement the new process on a full commercial scale as the first company in the world. As the result of the intensive efforts about 360 t of polyethylene were already produced up to the end of 1955 and were marketed under the trade name Hostalen G. A typical combination of the original Ziegler catalyst system was titanium trichloride and aluminum-diethyl-monochloride. This type of catalyst system was still of relatively low activity [7].

In this research, two blow molding grade of HDPE (BL3) with the same specifications that those produced by different manufactures were compared and studied. The MWD of the BL3 samples were studied with Gel permeation chromatography (GPC) and Capillary Rheometer (CR). Differential scanning calorimetry (DSC) was used to study the thermal properties and the Notched impact test used to study Impact Strength.

EXPERIMENTAL

The specifications of two blow molding grade of HDPE that used in this research shown in table 1. The plants were able to produce the full product range for injection moulding, blow moulding and extrusion applications. The melt flow rate (MFR 190/5) range of < 0.1 – 23 g/10 min were covered by the product slate. Product densities were varies between 0.944 – 0.962 g/cm^3 dependent on comonomer concentration and melt flow rate. The plants were designed for an annual capacity of 300,000 metric tons.

TABLE(1). Specifications of HDPE blow molding grades (BL3)

NO.	Manufactures	MFI 190/5 (gr)	MFI 190/21.6 (gr)	Density (gr/cm^3)
Sample 1	Kermanshah Pet. Co.(KPC)	0.9-1.5	19-27	0.952-0.956
Sample 2	Jam Petro. Comp. (JPC)	0.9-1.5	19-27	0.952-0.956

Properties of samples were characterized with these test methods:
Melt Flow Index: ASTM D-1238, Density: ASTM D-1505, Gel permeation chromatography (GPC): ASTM D-6474, Capillary Rheometer (CR): ASTM D-3835, Differential scanning calorimetry (DSC): ASTM D-3418, Izod Impact: ASTM D-256 and Charpy Impact: ASTM D-6110.

RESULTS AND DISCUSSION

Figure 1 shows GPC chromatograms for HDPE samples that their Specifications observed in table 1. As seen in this figure, both samples had similar bimodal shaped distribution curves, but curve of sample 2, located partially in the higher range of MWD (placed in the higher quantities on the horizontal axis (molecular weight)) than the sample 1.

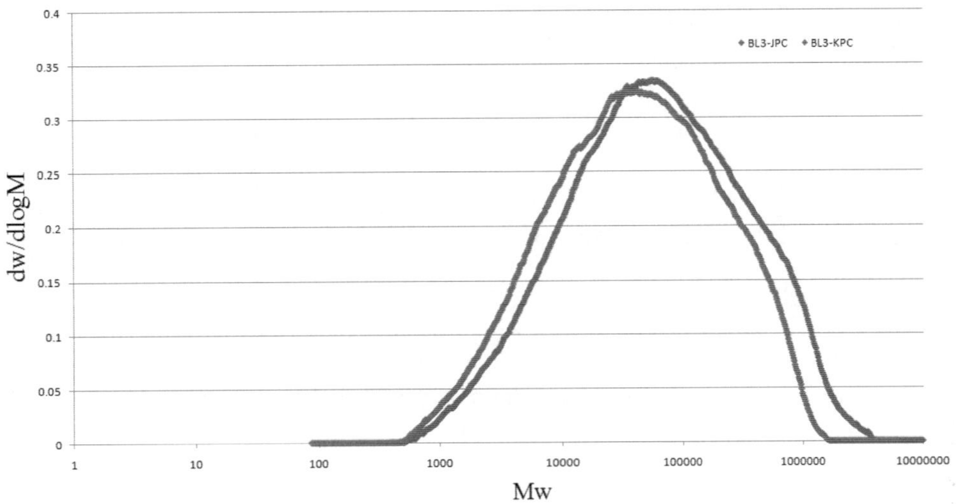

FIGURE 1. GPC graph for two HDPE sample (BL3 grade)

Figure 2 shows CR graphs for BL3 samples. The CR results also showed that both samples had similar distribution, but the sample 1 shows better processability in the blow molding shear rate range than the other sample.

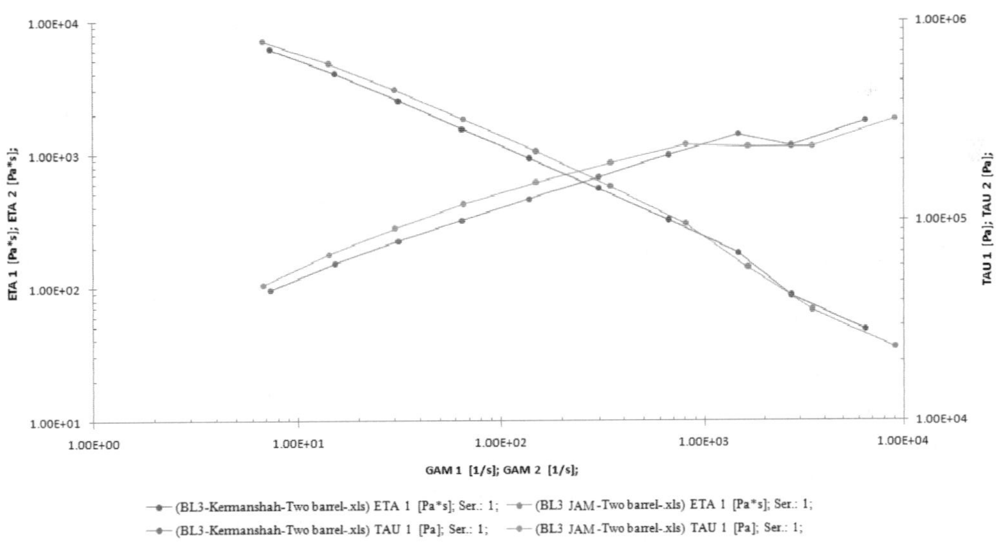

FIGURE 2. CR graph for two HDPE sample (BL3 grade)

This was observed because the sample 1 included the lower molecular weights than the sample 2.

Figure 3 shows DSC graphs for two samples. As seen in this figure, melting peak of the sample 2 located partially in the higher temperature range than the sample 1. Broad melting and crystallinity peak was due to existence of chains with different molecular weights in samples.

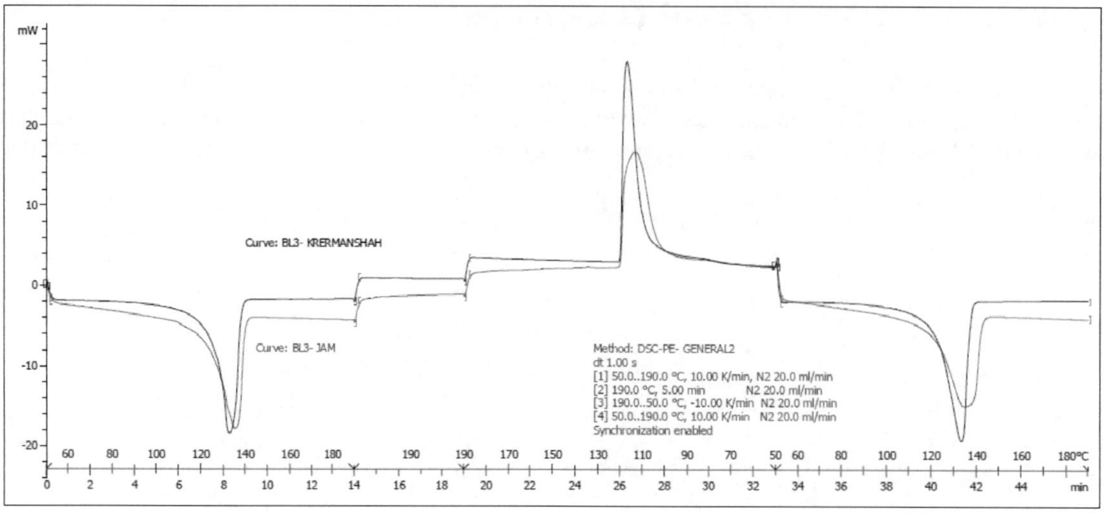

FIGURE 3. DSC curve for two HDPE sample (BL3 grade)

The DSC results have shown that sample 2 had higher crystallinity than the other sample (Figure 4).

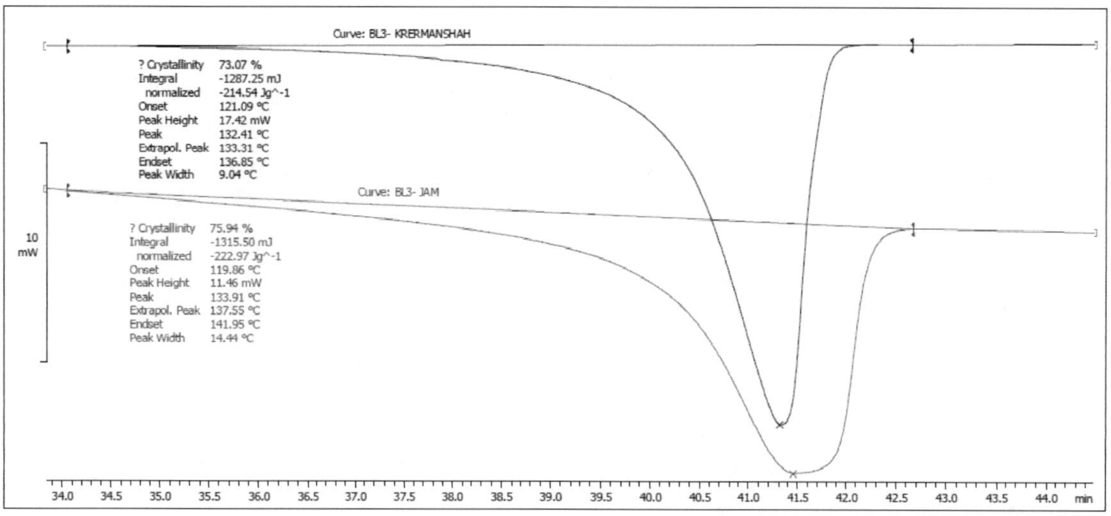

FIGURE 4. Crystallinity zone of two HDPE sample (In DSC graph)

Notched impact test results also showed that the sample 2 had higher impact resistance.

According to the DSC and impact resistance results (table 2), sample with the higher range MWD, had better thermal and physical-mechanical properties.

TABLE(2). DSC and impact resistance results for two HDPE blow molding grade sample

NO.	Tm (°C)	Crystallinity (%)	Izod Impact (J/m)	Charpy Impact (Kj/m^2)
Sample 1	132	73.07	110.2	8.55
Sample 2	134	75.94	144.4	10.66

CONCLUSION

In this research, two blow molding grade of HDPE (BL3) with the same specifications were compared and studied. The GPC results have shown that both samples, had similar bimodal shaped distribution curves, but curve of sample 2, located partially in the higher range of MWD than the sample 1. The CR results also showed that both samples had similar distribution, but the sample 1 shows better processability in the blow molding shear rate range than the other sample. The DSC results have shown that sample 2 had higher crystallinity and Tm than the other sample. Notched impact test results also showed that the sample 2 had higher impact resistance. According to these results, sample with the higher range MWD, had better thermal and physical-mechanical properties.

REFERENCES

1. L. H. Gabriel, "History and Physical Chemistry of HDPE", John Wiley & Sons, Inc., 2010, pp. 1-18.
2. O. Olabisi and K. Adewale, "Handbook of Thermoplastics", Marcel Dekker Inc., 1997, pp. 1-57.
3. M. A. Ali and E. M. Ali, *Computers & Chemical Engineering* **35**, 2480-2490 (2011).
4. X. Sun, H. Shen, B. Xie, W. Yang and M. Yang, *Polymer* **52**, 564-570 (2011).
5. J. Lee and Y. Kim, *Journal of Industrial and Engineering Chemistry* **18**, 429-432 (2012).
6. E. M. Ali and M. A. Ali, *ISA Transactions* **49**, 177-187 (2010).
7. A. J. Peacock, "Handbook of Polyethylene: Structures, Properties, and Applications", Marcel Dekker Inc., 2000, pp. 27–67.

Modelling and Prototyping of the Centrifugal Casting Process of Paraffin-Based Thermoplastic Polymers

Davide Masato (presenting author)[a], Giovanni Lucchetta (reference author)[a]

[a] *Dept. Industrial Engineering, University of Padova (Italy)*

davide.masato@dii.unipd.it
giovanni.lucchetta@unipd.it

Abstract. Centrifugal casting is a casting technique that is used for the manufacturing of thick-walled paraffin cylinders, which are difficult to cast by other means. A permanent axisymmetric mold is continuously rotated about its longitudinal axis at high speed as the molten polymer is poured. The liquid paraffin is centrifugally thrown towards the mold wall, where it solidifies after cooling. The casting is a thick-walled cylinder with an inner circular port. The technology is used to help avoid cracking and void formation inside the solid paraffin as it cools. These voids are formed because the liquid paraffin shrinks upon cooling by 15-25% depending on the grade of wax. In this work, the centrifugal casting process for the manufacturing of paraffin cylinder was modelled and prototyped. A numerical model of growth and dissolution of bubbles during centrifugal casting was developed for the prediction of the quality of the casting. The numerical solution of the model, which captures the combined effects of buoyancy, drag and diffusion, allowed the prediction of the time required to bubbles to either dissolve into the molten polymer or to leave the region of interest. Moreover, the influence of process parameters on the mechanical properties of the final casting was firstly investigated on a small scale, and then verified on a bigger scale.

Keywords: centrifugal casting, paraffin, hybrid rocket engine
PACS: 89.20.Bb (Industrial and Technological Research and Development)

INTRODUCTION

Hybrid propulsion combines the advantages of liquid and solid propulsion, thus having the potential to replace current rocket motors functionality as boost phase launch vehicle components. Hybrid rocket engines (HRE) are characterized by the possibility of multiple shut on/off, thrust modulation through oxidant mass flow regulation and higher safety. However, despite the high performances in terms of specific impulse, the development and the application of hybrid systems has been delayed by the low regression rate of the fuel.

Currently, the fuel is hydroxyl-terminated polybutadiene (HTPB), which is low cost, processes easily, and will not self-deflagrate under any conditions. However, HREs made of HTPB must have complex design, in order to guarantee the required fuel surface area. Hence, leading to manufacturing complexity and possible structural failures.

In the last decade, some researches highlighted the marked regression rate increase offered by particular solid fuels as solid paraffin and other alkanes with respect to conventional polymeric fuels, due to the entrainment of melted solid fuel droplets by the oxidizer mass flow [1]. In addition to propulsion performances, innovative solid fuel formulations suitable for HREs needs also to ensure the necessary mechanical properties that allow the grain to sustain launch loads and stresses. The achievement of good mechanical properties can be pursued by working on different strengthening strategies, as materials selection [2] and process optimization [3].

The centrifugal casting has been designed and applied to HRE manufacturing to help avoid cracking and void formation inside the solid paraffin [4]. In fact, liquid paraffin shrinks upon cooling by 15-25% depending on the grade of the wax. Generally, the castings are thick-walled paraffin cylinder with an inner circular port, which are impossible to efficiently cast by other means. The process is characterized by a permanent axisymmetric mould that is continuously rotated about its longitudinal axis at high speed as the molten polymer is poured. The liquid paraffin is centrifugally thrown towards the mould wall, where it solidifies after cooling.

In this work, a suitable paraffin was identified between three commercial paraffin considering mechanical, rheological and thermal properties. A numerical model of growth and dissolution of bubbles during centrifugal casting was then developed and applied to the prediction of the quality of the casting. The numerical solution of the model, which captures the combined effects of buoyancy, drag and diffusion, allowed the prediction of the time required by bubbles to either dissolve into the molten polymer or to leave the region of interest. Furthermore, the design and the optimization of the process were conducted considering the optimization of the mechanical properties of the casting grains.

MATERIALS AND METHODS

Material Selection

Three commercial pure paraffin were selected as candidate fuels for the manufacturing of HRE grains. For confidentiality reasons the three materials will be referred to as Paraffin A, Paraffin B and Paraffin C.

The thermal properties of the pure waxes were characterized by means of Differential Scanning Calorimetry (TA Instruments, Modulated DSC Q200), in agreement with ISO 11357-3. All tests were performed from -80 to 150 °C with a temperature ramp of 10 °C/min, using nitrogen gas (50 mL/min) as heating/cooling gas. As displayed in Fig. 1, all the paraffin exhibits a common thermal behavior, typical of microcrystalline waxes, characterized by a main melting peak and a lower temperatures transition peak [7]. Paraffin B and C showed a wider melting changing phase interval with more or less distinctive peaks and a lower specific heat flow, hence indicating that they are characterized by the presence of a secondary phase, i.e. iso-paraffin.

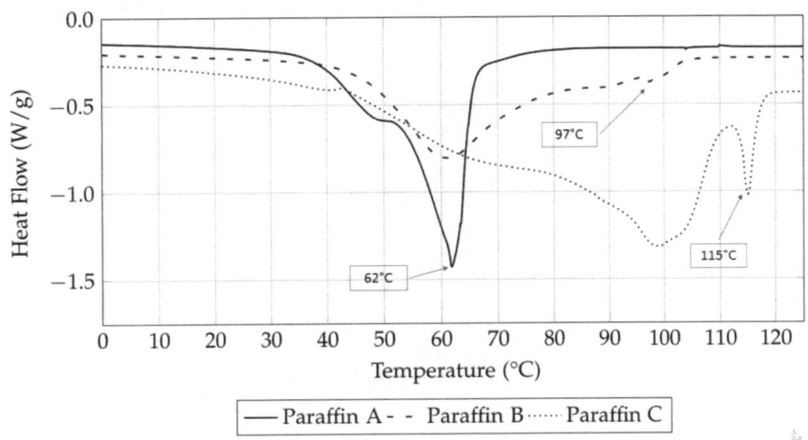

FIGURE 1. Pure paraffin DSC scan of melting peaks

Rheological investigations were performed using a parallel plate (diameter 50 mm, gap 0.5 mm) rheometer setup (TA Instruments, ARES) and tests were performed with respect to shear rate (from 5 to 2000 s^{-1}) at a constant temperature of 120 °C. The experimental tests indicated that the considered paraffins behave as Newtonian fluids. Table 1 reports the results for the three materials, indicating that Paraffin A has the lowest viscosity, hence the possibly lowest regression rate.

TABLE (1). Rheological and mechanical properties of the three paraffin.

Material	Paraffin A	Paraffin B	Paraffin C
Viscosity [Pa*s] - 50 s^{-1}	0.0032	0.0052	0.0672
Elastic Modulus [MPa]	571.3	528.0	426.8
Stress at break [MPa]	3.1	3.2	3.9
Strain at break	0.023	0.020	0.026

The mechanical behavior of the paraffin was investigated by means of uniaxial tensile tests at ambient temperature with a speed of 1 mm/s, accordingly with ISO 527-2:2012. Small specimens (type 1BA) were obtained by gravity casting the melt paraffin into a mould. A thermally controlled procedure was adopted in order to prevent the formation of internal defects due to shrinkage. Elastic modulus, stress at break and strain at break were taken into consideration for the comparison of the mechanical properties of the tested paraffin (see Table 1).

Considering the results of the characterization for the three materials, paraffin A was selected and used for the centrifugal casting of HRE grains. Indeed, its mechanical properties suits the application and its viscosity is the lowest, thus making it possible to obtain the highest regression rate.

Centrifugal Casting Machine

The main objective was the manufacturing of a 25 kg single-port paraffin grain. The centrifugal casting process was selected over gravity casting because paraffin markedly shrinks upon cooling thus making it impossible to obtain a thick-walled cylinder of regular geometry without any defects. At first the critical issues related to the centrifugal casting of paraffin were identified, then a specifically designed centrifugal casting machined was constructed (Fig. 2). A composite (carbon fiber T300 – epoxy resin) centrifugal mould was designed to control the thermal boundary conditions of the process and promote a directional solidification of the paraffin. The directional solidification will also reduce internal shrinkage porosities and made impurities to be thrown to the surface of the inner diameter, which can then be refined.

The machine allowed the possibility to mount also a smaller mould, capable of casting 2.5 kg paraffin grains. This prototypal scale of the grain was used to investigate and then set the process for the 25 kg casting.

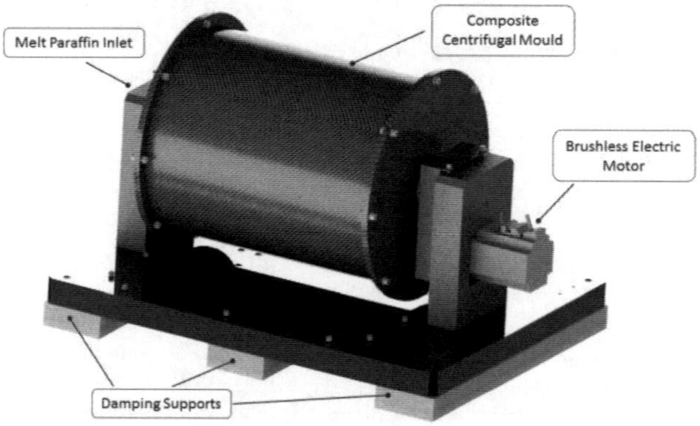

FIGURE 2. Centrifugal casting machine

Experimental

The capabilities of the centrifugal casting process to produce paraffin grains characterized by adequate structural properties were experimentally investigated using the prototypal scale version of the machine (2.5 kg paraffin grains). A design of the experiments (DoE) approach was implemented to study the effect of controllable process parameters on the mechanical properties of centrifugal casting grains. A two-level three-factor full factorial plan was designed considering the following parameters: rotational speed (800 – 1800 rpm), paraffin melt temperature (80 – 115 °C) and mould filling time (30 – 300 s). The tensile properties of each of the grains were chosen as the response variable for the analysis. The mechanical characterization was performed following the same procedure adopted for gravity casting paraffin specimens. Small tensile specimens were obtained from castings in a radial intermediate position.

The information derived from the experimental investigation on the prototypal scale were then transferred to the manufacturing of the 25 kg paraffin grain. The optimum selection of process parameters was set and the mechanical properties were then verified.

RESULTS AND DISCUSSION

Effect of Controllable Process Parameters

The results of the experimental investigation on the mechanical properties of centrifugal casting prototypal scale grains were analyzed both in terms of stress and deformation at break and compared with the basic properties of paraffin A (Table 2).

TABLE (2). Effect of centrifugal casting process parameters on paraffin grains mechanical properties.

Test	Rotational Speed [rpm]	Melt Temperature [°C]	Filling Time [s]	σ_r [Mpa] Avg. Val.	σ_r [Mpa] Std. Dev.	ε_r Avg. Val.	ε_r Std. Dev.
Paraffin A	/	/	/	3.12	0.35	0.023	0.005
a	1800	80	30	3.29	0.15	0.042	0.005
b	1800	115	300	3.15	0.12	0.043	0.005
c	800	80	30	3.08	0.54	0.037	0.014
d	800	115	30	2.86	0.16	0.039	0.003
e	800	115	300	2.72	0.26	0.044	0.013
f	800	80	300	2.91	0.28	0.028	0.013
g	1800	115	30	3.33	0.21	0.033	0.009
h	1800	80	300	3.28	0.23	0.038	0.023

The analysis of variance indicated that among investigated factors, the rotational speed is the only statistically significant parameter affecting the stress at break (p-value = 0.004). Indeed, the main effect of increasing the rotational speed from 800 to 1800 rpm is to increase the stress at break by + 13%. In general, a selection of high levels of the selected parameters resulted to have a positive effect on the deformation at break of the grains. While, considering the stress at break, the response variable is maximized by a selection of high rotational speed, low melt temperature and low filling time.

The deformation at break obtained by centrifugal casting of prototypal scale grains is always higher than that from gravity casting, while the effect of process parameters on the stress at break is not always improved.

Centrifugal Casting Optimization

The manufacturing of the 25 kg paraffin grain took into consideration the information derived from the previous experimentation. Considering the positive effect of its high value and a scale factor, the rotational speed was set at 900 rpm. The melt temperature was set at an intermediate value of 100°C, as a compromise between the positive and negative effect on the deformation and stress at break respectively.

The casting grain was characterized by sampling some small tensile specimens in two radial positions. The resulting mechanical properties were ranging from 3.21 to 3.3 MPa for the stress at break, and from 0.038 to 0.040 for the deformation at break.

ANALYTICAL MODELING

The main defects that could occur in the casting of paraffin are internal shrinkage porosities and air bubbles that are initially formed when the melt paraffin is poured into the mould. These defects can be simplified as air bubble trapped within the melt paraffin, which can degrade the mechanical properties of the final solid grain.

Centrifugal casting is a recognized method for eliminating bubbles in polymer solutions [5]. In particular, there are two mechanism contributing to the elimination of internal defects:

- diffusion, which occurs when the concentration of gas dissolved within the bubble wall is greater than the concentration of gas in the surrounding solution;
- buoyant transport, i.e. the bubbles tend to move toward the center axis of rotation.

The diffusion mechanism can be quantified by the rate of change of bubble radius, which can be derived from the work of Epstein and Plesset [6]:

$$\frac{dR}{dt}=\frac{\kappa\left(c_{s\infty}\left(c_f-1\right)-\frac{2k_D\sigma}{R}\right)}{\rho_\infty+\frac{2\tau}{3R}}\left\{\frac{1}{R}+\frac{1}{\sqrt{\pi\kappa t}}\right\} \tag{1}$$

where $\tau = 2M\sigma/(BT)$, M is the molecular weight of the gas, B the universal gas constant, k the coefficient of diffusion, ρ_∞ the density of the gas and $c_{S\infty}$, c_f, c_i are concentration parameters.

The radial velocity (v_r) of a bubble that moves is obtained balancing the buoyant and the drag forces:

$$v_r \approx \frac{\left(\rho_a-\rho_f\right)}{12\mu_f} r\omega^2 d^2 \tag{2}$$

where ω is the angular velocity of the centrifuge, d is the diameter of the bubble, r the distance of the bubble from the axis of rotation, ρ_a is the density of the air within the bubble and ρ_f, μ_f the density and the viscosity of the melt paraffin.

The final quality of the paraffin grain, in terms of presence of internal defects, was predicted using a numerical solution of first-order nonlinear equation (1) and equation (2). The time required for bubbles to either dissolve into solution or leave the region of interest, at different processing conditions, was estimated considering an isothermal process in which the mould is completely filled before spinning.

TABLE (3). Time required for bubbles removal during spinning

		Casting Temperature: 80°C				Casting Temperature: 115°C			
	Initial Dimensions [μm]	**10**	**95**	**300**	**1000**	**10**	**95**	**300**	**1000**
Rotational Speed [rpm]	250	0.32	239.79	25.42	2.84	0.30	11.38	0.93	0.10
	850	0.06	5.99	2.55	0.27	0.06	1.27	0.09	0.01
	1840	0.03	1.40	0.75	0.08	0.02	0.39	0.03	0.01
	2000	0.02	1.22	0.67	0.07	0.02	0.35	0.03	0.01

The results of the simulations, reported in Table 3, gave indications about the effectiveness of the centrifugal casting process in removing internal porosities. The model allowed also the evaluation of the effect of controllable process parameters. In particular, a positive effect on the removal time, of higher rotational speed and higher casting temperature was observed. The complete removal of bubbles from the melt paraffin occurs in times that are not relevant compared to grains solidification time; hence also corroborating the isothermal assumption.

CONCLUSIONS

The main objective of this research work was the characterization of the centrifugal casting process for the manufacturing of paraffin-based 25 kg HRE grains. The process was modelled in order to evaluate the time required for possible internal porosities to either dissolve or exit the melt paraffin during spinning. An experimental evaluation of the effect of controllable process parameters was carried out, highlighting that the rotational speed is the most important parameters affecting the mechanical properties of the grain. The comparison with the gravity casting tensile properties indicated that the centrifugal casting process can be effectively exploited for the manufacturing of paraffin-based HRE grains.

REFERENCES

1. M. A. Karabeyoglu, D. Altman and B. J. Cantwell, *Journal of Propulsion and Power* **18**, (2002).
2. M. Boiocchi, L. Merotto, F. Maggi, L. Di Landro and L. Galfetti, *EUCASS Conference*, 1199-1887 (2011).
3. J. D. DeSain, B. B. Brady, K. M. Metzler, T. J. Curtiss and T. V. Albright, *AIAA* **5115**, 1-27 (2009)
4. M. A. Karabeyoglu,, G. Zilliac, B. J. Cantwell, S. De Zilwa and P. Castellucci, *AIAA*, 1162 (2003).
5. A. D. Mazzeo, M. E. Lustrino and D. E. Hardt, *Polym. Eng. & Sci.* **52**, 80-90 (2012).
6. C. E. P. Epstein and M. S. Plesset, *J. Chem. Phys.* **18**, 1505-1509 (1950).
7. G. Meyer, *SOFW Journal* **135**, 43-50 (2009).

Optimization and Modelling of Delignification Process for Nanocrystalline Cellulose Production from Rice Husk Biomass

[a]Md. Sakinul Islam, [a]Nhol Kao, [a]Sati N. Bhattacharya* and [a]Rahul Gupta

[a]School of Engineering, Department of Chemical Engineering, RMIT University, Melbourne, VIC 3001, Australia

Presenting & Reference Author's Email: rahul.gupta@rmit.edu.au

Nanocrystalline cellulose (NCC) has been attributed as a noble material due to its superior properties for miscellaneous applications in medical science, pharmaceutical production and engineering fields. The major problems associated with the production of NCC are low production rate and inefficient process. Alkaline delignification plays an important role for the mass production of NCC from lignocellulosic biomass. In order to fulfil these research gaps a fractal kinetic model of alkaline delignification of rice husk biomass (RHB) was studied with process optimization. Fifteen kinetic experiments of alkaline delignification for NCC production from RHB, as designed by the Minitab® 2014b software using the Box-Behnken method, were performed in a 2L jacketed glass reactor under three process parameters (time, temperature and alkali concentration) with ranges of 2-10 hours, 40-100°C and 1-4M, respectively. The nucleic growth model of fractal kinetics, developed by Nguyen & Dang (2007), has been successfully applied to the data sets available from this experimental work. The process parameters estimation for the fractal kinetic model was carried out using Matlab ® 2014b software. From this study it has been found that the most efficient delignification process with first order kinetics was obtained at 15 h, 95 °C and 2.5 M alkali concentration. It has also been found that the kinetic rate coefficient of alkaline delignification is dependent on time instead of alkali concentration. The correlation coefficients R^2 for this model was found to be 0.94, which indicated that the model can be considered with high level of confidence. From this hypothesis it can be concluded that the optimised condition gave maximum cellulose pulp for producing high quality NCC.

Key words: Nanocrystalline cellulose, Fractal kinetic, Alkaline delignification, Rice husk biomass

INTRODUCTION

Cellulose is a linear polysaccharide consisting of a chain of β (1-4) linked D-glucose units. Cellulose is the major component of lignocellulosic biomass such as cotton, flax, hemp, jute, ramie and sisal fibers. Cellulose can be extracted from various biomasses by mechanical, chemical and microbial processes [1, 2]. NCC has become very demanding nano-bio-material at this time due to its noble applications in biomedical, pharmaceutical and nano-engineering industries [3]. NCC has been successfully produced from lignocellulosic biomass in small scale but the production rate is inadequate due to inefficient processes [4]. In order to scale-up the production rate of NCC from lignocellulosic biomass by chemical extraction process, delignification is a very important. The long delignification time under high temperature degrades cellulose to various degradation products (DPs) with other impurities. The most common DPs of cellulose are glucose, fructose, lactose, mannose, arabinose and furfural, etc. [5, 6]. As a result, the prolong delignification process was unable to yield high quantity of CP for NCC production. However the major focus of this paper is optimization and fractal kinetic modeling of alkaline delignification of RHB. The optimization of alkaline delignification of RHB has not widely been explored and therefore it was simultaneously investigated with modeling study. Kinetics of alkaline delignification was extensively studied for Kraft, sulfite, bisulphite and soda pulping processes [7-9]. One of the earliest kinetic model for the Kraft process is the H factor model (Eq. 01) was developed by Vroom in 1957 which combined the reaction time and temperature into a single variable [10]. The H factor is an expression of the reaction severity and is derived from the effect of temperature in the form of the Arrhenius equation (Eq. 02) on a first order rate equation whereas k_0 is the pre-exponential factor, R is the universal gas constant, E_a is the activation energy and T is the absolute temperature [11]. In 1970 the Kerr model of bulk delignification (Eq. 03) was developed to investigate the pulping kinetics and this is the good kinetic model for control system. In this following model equation k and R are the kinetic and universal gas constant, T is temperature and L is the initial lignin content. The Purdue model (Eq. 04) of Kraft delignification was developed in 1973 and it is significantly more complex than that of the H factor model. This model was applied for high and low reactive lignin where high reactive lignin depends on $[OH^-]$ and $[HS^-]$ concentrations [12]. Another typical kinetic model of delignification which is called Hatton's model [13] was developed in 1973 to predict yield or kappa number for variety of wood species (Eq. 05) where Y is the yield, K is the kappa number, α and β are the adjustable parameters, H is the H factor, and (EA) is the effective alkali. In 1983, Gustafsan model [14] was developed for three phase lignin such as initial (Eq. 06), bulk (Eq. 07) and residual (Eq. 08) where L is the respective lignin contents, k_1, k_2, k_3 and k_4 are the kinetic constant, [OH] and [HS] are the concentrations of alkali and bisulphate respectively. In 2006, the nucleic growth (NG) model of fractal kinetics (Eq. 09) was developed by Nguyen & Dang will be used in this study [15]. This model is a fairly recent method of characterising the kinetics of the alkaline and other delignification processes that deviates significantly from the above mentioned traditional kinetic models. In this model a, b, E and n are the parameters to be determined from experimental data, where X is the lignin in pulp, C is the concentration (i.e. acid, alkali, other solvent), E is the activation energy, t is the time, T is the temperature (°K), a and b are constant, and R is the universal gas constant. Although there are a lot of alkaline and Kraft delignification models [12, 14, 16-20], they are generally used to describe the delignification process of the paper industry. Nevertheless, the kinetic investigation of delignification for NCC production was not explored widely.

Therefore this study was performed in order to recover these specified research gaps for the sustainable development of NCC production. The main objective of this study is to optimize the alkaline delignification of RHB for CP production with fractal kinetic modeling.

$$H_{factor} = k_0 \int_0^t e^{-\frac{32,000}{RT(t)}} dt \qquad [01]$$

$$k = k_0 \exp(-E_a / RT) \qquad [02]$$

$$\frac{dl}{dt} = k \times e^{-32,000/RT}[OH] \times L \qquad [03]$$

$$\frac{dl}{dt} = (k_1[OH^-]^{0.5} + k_2\,[HS^+]^{0.2})(L - L_f) \qquad [04]$$

$$K \text{ or } Y = \alpha - \beta\,[\log(H) \times EA^n] \qquad [05]$$

$$\frac{dl}{dt} = k_1 L \qquad [06]$$

$$\frac{dl}{dt} = (k_2[OH^-] + k_3[OH^-]^{0.5}\,[HS^-]^{0.4})L \qquad [07]$$

$$\frac{dl}{dt} = k_4[OH^-]^{0.7}L \qquad [08]$$

$$-\frac{dX}{dt} = ae^{(\frac{E}{RT})^n}[C]^b t^{n-1} X \qquad [09]$$

METHODOLOGY

The experimental design of optimization and kinetics of alkaline delignification was determined by response surface method (RSM) followed by Box-Behnken response using Minitab® 2014b software. The proposed design of delignification kinetic modeling (DKM) and optimization was done for three variables such as time, temperature and alkali (NaOH) concentration with ranges of 3-15 hours, 40 – 95 °C and 1-4M, respectively. The resulting design and experimental parameters are given in Table 1. Figure 1 shows the experimental set-up of delignification which was conducted in 2.0L jacketed glass reactor under atmospheric condition at 400 RPM for all typical runs as shown in the Table 1. In all delignification experiments, the RHB to liquid ratio (NaOH solution) was 100g/2.0L. The average particle size of powder RHB of 700 μm was used in this work which was previously processed by washing and drying. During the experimental work all parameters were set and followed according to the design. After completing the delignification, the delignified rice husk pulp was separated from the black liquor by filtration using 45 μm sieves. The black liquor was discarded and the rice husk pulp was taken for further processing. Delignified rice husk pulp (D-RHP) was then washed using deionized distilled water (DDW) until pH 6.5-7.0 was reached. After washing, delignified pulp was then dried in the oven at 45 °C until constant weight was achieved. Finally the oven dried pulp was analyzed for optimization and DKM study.

TABLE (1). Experimental design of alkaline DKM of CP production by RSM

Ex. Run	Time (Min)	C_{NaOH} (M)	T (°C)
DKM01	540	1.0	40.0
DKM02	180	4.0	67.5
DKM03	180	1.0	67.5
DKM04	540	2.5	67.5
DKM05	900	2.5	40.0
DKM06	540	2.5	67.5
DKM07	900	4.0	67.5
DKM08	180	2.5	40.0
DKM09	540	1.0	95.0
DKM10	900	1.0	67.5
DKM11	540	4.0	40.0
DKM12	180	2.5	95.0
DKM13	900	2.5	95.0
DKM14	540	4.0	95.0
DKM15	540	2.5	67.5

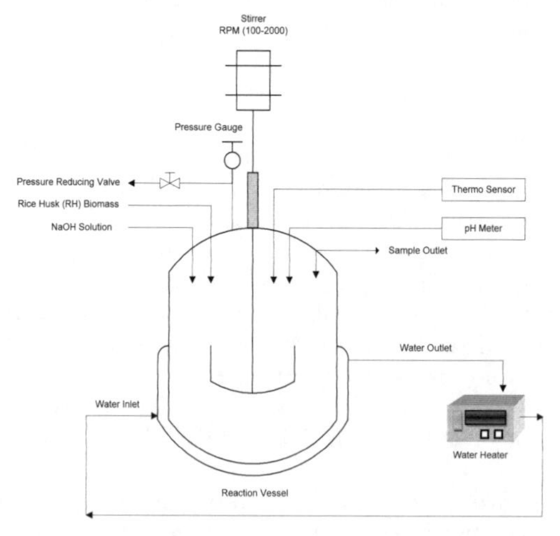

FIGURE 1. Schematic diagram of experimental set-up for DKM study using low pressure reactor (LPR)

CHARACTERIZATION AND MODELING

The chemical composition of raw RHB and D-RHP was determined in accordance to ASTM (American society for testing and materials) and TAPPI standard methods for different components of RHB and resultant products - namely T222 OS-83 for lignin, T203 OS-61 for cellulose, ASTM D 1104-56 for hemicelluloses, T245 OM-94 for silica and ASTM D 4442-92 for moisture content. These standards are commonly used by many researchers for determining biomass constituents [21, 22].

For kinetic modeling and unknown parameters prediction Matlab® 2014b software was used by selecting FMIN solver function. The integrated and linearised form of the Dang and Nguyen's fractal model [18] (Eq. 09) shown in the following relationship (Eq. 10) was used for data fitting.

$$Ln\left[-Ln\left(\frac{X}{X_i}\right)\right] = Ln\ a + bLn\ [C] - \left(\frac{nE}{RT}\right) + nLn\ t \quad \ldots\ldots\ldots\ldots\ldots\ldots \quad [10]$$

RESULTS AND DISCUSSION

This section mainly highlights the experimental results of chemical composition analysis and modeling. From the compositional analysis lignin values of RHB and D-RHP were used for NG modeling and regression analysis.

Chemical Composition Analysis

The chemical composition of raw RHB and D-RHP obtained from fifteen delignification experiments such as DKM01-DKM15 as shown in the Table (2). It is clear that the alkaline delignifications of RHB significantly increase the cellulose content with decreasing amorphous components such as lignin, hemicelluloses, pectin, waxes and silica. Delignification process is basically the substantial breakdown of lignocellulosic structure - resulting in improved defibrillation. The partial hydrolysis of hemicelluloses and depolymerization of lignin unit occurred during the delignification process as well, which gave rise to sugars and phenolic compounds that are soluble in water as reported by other authors [23]. When the lignocellulogic biomass is subjected to chemical modification in the alkaline medium the hemicellulose and lignin components present in the RHB will dissolve [24]. In this way the percentage of cellulose content of RHB increases when it undergoes delignification by the alkali treatment. From the compositional analysis it was found that DKM13 produced maximum cellulose with low impurities. On the other hand DKM 14 and DKM02 were also found to be productive than others. Based on the cellulose and remaining impurities DKM13 can be considered as the optimum condition for CP production.

TABLE (2). Chemical composition of RHB and D-RHP at various delignification experiments

Components Wt. (%)	RHB	DKM Experiments (01-15)														
		01	02	03	04	05	06	07	08	09	10	11	12	13	14	15
Lignin	18	14	8	15	12.5	13	12	9	16	10	11	13	11	7	8	13
Cellulose	35	42	50	40	43	41	43	49	44	43	44	48	42	56	52	43
Hemicellulose	26	17	11	16	15	16	15	12	14	16	11	12	14	8	11	14
Silica	13	9	6	8	10	9	10	9	9	8	7	8	9	5	7	9

Delignification Kinetic Modeling

Table 3 represents the predicted unknown parameters, activation energy of fifteen delignification experiments with error analysis. Very negligible error was observed in every case. The raw data fitting in the model predicts very stable activation energy with consistent values. The other parameters are also very consistent as shown in Table 3. From the activation energy analysis it is observed that all delignification experiments required low energy which is 5.54 kJ/mol. From these observations it can be concluded that, the kinetic of alkaline delignification in this study are first order.

TABLE (3). Delignification models [15] applied to RHB and prediction of parameters with activation energy

Experiment	a	b	n	E (J/mol)	Error
DKM01	1×10^3	1.0	493.250×10^{-3}	5.547115×10^3	1.69E-15
DKM02	1×10^3	1.0	725.289×10^{-3}	5.547115×10^3	2.56E-17
DKM03	1×10^3	1.0	858.695×10^{-3}	5.547115×10^3	6.72E-19
DKM04	1×10^3	1.0	799.903×10^{-3}	5.547115×10^3	8.59E-18
DKM05	1×10^3	1.0	480.001×10^{-3}	5.547115×10^3	1.71E-16
DKM06	1×10^3	1.0	790.336×10^{-3}	5.547115×10^3	8.22E-18
DKM07	1×10^3	1.0	740.342×10^{-3}	5.547115×10^3	4.90E-18
DKM08	1×10^3	1.0	528.287×10^{-3}	5.547115×10^3	2.59E-17
DKM09	1×10^3	1.0	1000.01×10^{-3}	5.547115×10^3	0.993371
DKM10	1×10^3	1.0	772.503×10^{-3}	5.547115×10^3	1.58E-16
DKM11	1×10^3	1.0	480.010×10^{-3}	5.547115×10^3	3.18E-17
DKM12	1×10^3	1.0	1000.01×10^{-3}	5.547115×10^3	0.995133
DKM13	1×10^3	1.0	1000.01×10^{-3}	5.547115×10^3	2.77E-26
DKM14	1×10^3	1.0	979.804×10^{-3}	5.547115×10^3	0.979804
DKM15	1×10^3	1.0	810.017×10^{-3}	5.547115×10^3	3.56E-17

The Nuclei Growth (NG) alkaline delignification model was studied by many researchers for different biomass as well. In the open literature the same kinetic study for wood, wheat straw, black cottonwood, giant red and bamboo was conducted as shown in Table 4. In order to compare our results with existing literature the predicted values of unknown parameters are given in the table. From this table it is apparent that, in our study the average activation energy RHB is lower than other biomass. The other values are very consistent and reliable as well except DKM09, DKM12 and DKM14 but these are still <1.0 error. Therefore from this overall observation, the final delignification model of RHB (Eq. 12) can be considered as a reliable model for kinetic study.

TABLE (4). A comparison of model predicted parameters between RHB and other biomasses

Biomass	References	a	b	n	E (kJ/mol)	R^2
RHB (Rice Husk Biomass)	This Study	1×10^3	1.0	0.75	5.54	0.94
Wheat Straw	[25] Siridas and Koukios (2004)	9.71E11	1.56	0.74	128	0.76
Black Cottonwood	[26] Tirtowidjojo et al. (1988)	1.09E6	n.d.	0.54	96	0.89
Bamboo	[27] Vu et al. (2004)	86	0.90	0.12	115	0.98
Giant Red	[28] Shatalov and Pereira (2005)	1177	n.d.	0.39	71	0.93

REGRESSION ANALYSIS

From the linear regression analysis of bamboo, wheat straw and cottonwood it was found that R^2 value of bamboo is 0.98 whereas wheat straw and cottonwood are 0.76 and 0.89, respectively. The slope of bamboo is 1, while the slope of cottonwood and wheat straw are 0.96 and 0.94, respectively. From this observation it can be concluded that the NG alkaline delignification model could be used for the following biomasses. From the regression analysis of RHB with NG alkaline delignification model the following responses were observed where R^2 values are 0.9499 and 0.8796. The slope of linear regression lines are 0.7082 and 0.9423. For regression analysis of RHB, two set of data was fitted in the model to confirm the model validation. Apart from this investigation it is clearly apparent that, R^2 and slope are quite similar and have the same trend with other biomasses as described above. Therefore, by considering all responses and regression analysis the alkaline delignification of RHB as conducted in this research is first order kinetic with negligible error.

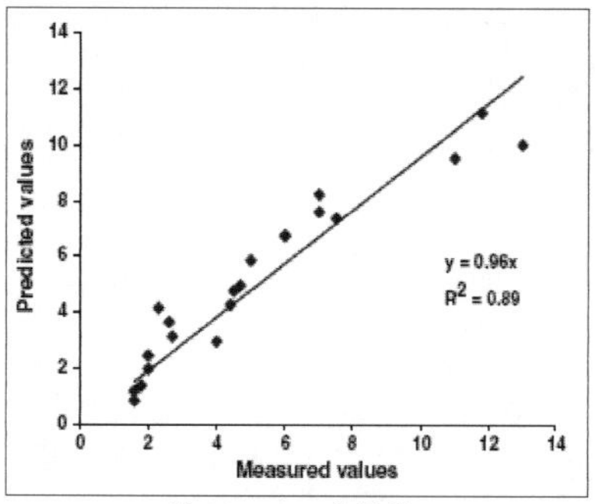

Cottonwood [26]

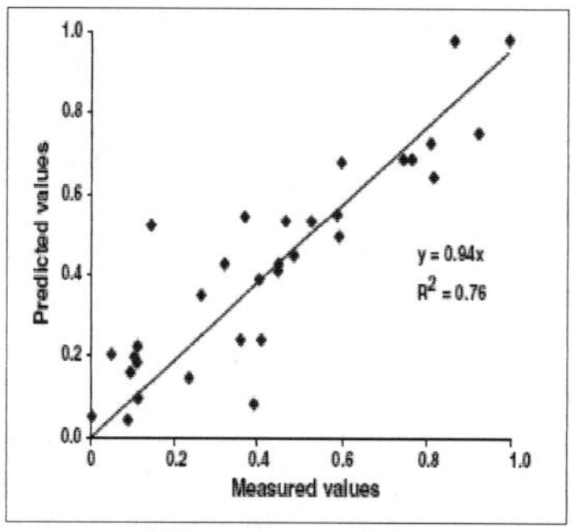

Wheat straw [25]

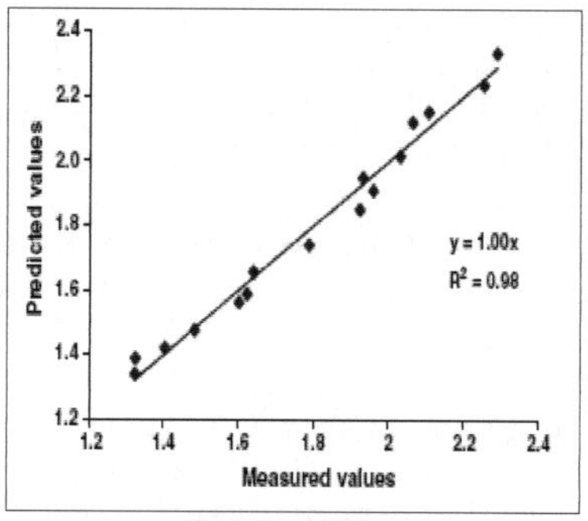

Bamboo [27]

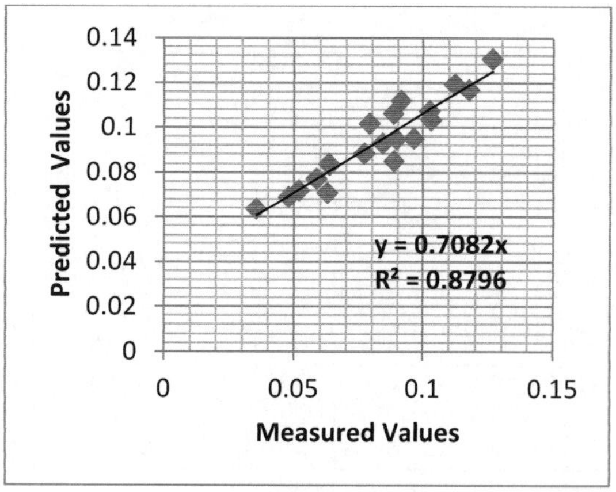

RHB (a). [this study]

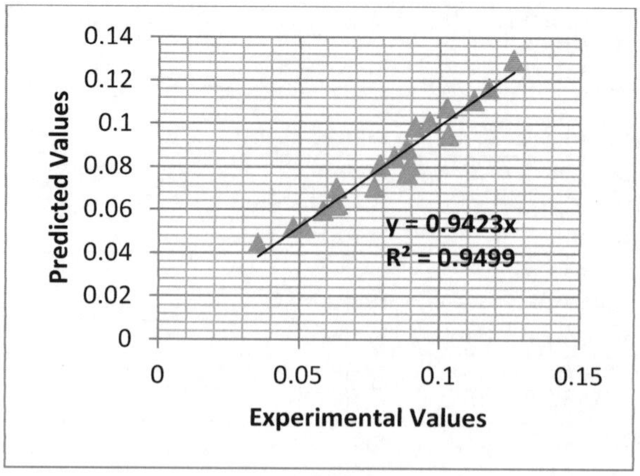

RHB (b). [this study]

CONCLUSIONS AND RECOMMENDATION

The effects of chemical concentration, temperature and time are accounted for in the Nuclei Growth (NG) model (Eq. 11). The rate equation for the delignification process has the form of a first order rate equation with a time-dependent rate coefficient. The model is statistically robust and can be used to investigate the fractal kinetics of the delignification of lignocellulosic materials. The model was applied for RHB in this study and the first order kinetic was observed with very significant values of unknown parameters. The final form of the NG model using predicted values is given in the following equation (Eq. 12). As a recommendation author would like to suggest future work on RHB for investigating the NG fractal kinetics at high temperature. For the optimization of CP production from RHB, the experiments DKM02, DKM13 and DKM14 were found more efficient. For the mass production of NCC from RHB, identified efficient runs are recommended for alkaline delignification. Nevertheless, high pressure delignification is also recommended to overcome long time delignification.

$$-\frac{dL}{dt} = a * [C]^b * \exp\left(-\frac{E}{RT}\right) * L * t^{n-1} \qquad [11]$$

$$-\frac{dL}{dt} = 1 \times 10^3 * [C]^1 * \exp\left(-\frac{5.54}{RT}\right) * L * t^{0.75-1} \qquad [12]$$

ACKNOWLEDGMENTS

The research work reported in this paper was funded by the RMIT University, Melbourne, VIC 3001, Australia. The authors would like to acknowledge all technical staff of Chemical Engineering Department and RMMF lab at RMIT for their kind support during the experimental work and analysis. One final year chemical engineering student Keegan Richard Boyd-Moss, who assisted with the lab work, is gratefully acknowledged as well.

REFERENCES

1. F. Esa et al., *Agric. Agric. Sci. Procedia*, 2, pp. 113-119, 2014. DOI: 10.1016/j.aaspro.2014.11.017
2. A. Alemdar and M. Sain. *Bioresour. Technol.* 99 (6), pp. 1664-1671, 2008a
3. X. Lv et al., *ACS Biomater. Sci. Eng.,* 2, 19–29, 2016. DOI: 10.1021/acsbiomaterials.5b00259
4. L. Brinchi et al., *Carbohydr. Polym.* 94, pp. 154-169, 2013
5. W. Qianqian et al., *Ind. Eng. Chem. Res*, 53, pp. 11007-11014, 2014.
6. N. Soltani et al., *Chem. Eng. J.* 12, pp. 889-935, 2014. DOI: 10.1016/j.cej.2014.11.056
7. J. Gierer. *Wood Sci. Technol.* 14, pp. 241-266, 1980
8. M. Dolk et al., Holzforschung. 43, pp. 91 98, 1989
9. V. L. Chiang et al., *J. Wood Chem. Technol.* 10, pp. 293-310, 1990
10. K. E. Vroom. *Pupl. Pap. Mag. Can.,* 58, pp. 228, 1957
11. H. L. Chum et al., *App. Blochern. Biotechnol.* 24-25(1), pp. 1-14, 1990
12. N. Andersson et al., *Nordic Pulp and Paper Research Journal*, 18 (2), pp. 200-209. 2003.
13. J. V. Hatton et al., TAPPI, 74, pp. 79, 1973
14. R. Gustafson et al., *I&EC Process Design and Development*, 22(1), pp. 87-96, 1983.
15. V. Q. Dang and K. L. Nguyen. Cellulose, 14, pp. 153-160, 2007
16. J. Bogrena et al., *Technical Articles*, pp. 1-11, 2008.
17. J. Castro et al., *Chem. Eng. J.*, 139, pp. 615-621, 2008.
18. V. L. Dang and K. L. Nguyen. *Bioresour. Technol.* 97, pp. 1353-1359, 2006.
19. R. Giudici and S. Won. *Ind. Eng. Chem. Res*, 35, pp. 856-863, 1996.
20. L. Yang and S. Liu. *Ind. Eng. Chem. Res,* 44, pp. 7078–7085, 2005.
21. E. Abraham et al., *Carbohydr. Polym.* 86, pp. 1468– 1475, 2011.
22. N. Johar et al., *Ind. Crops Prod.* 37, pp. 93-99, 2012.
23. C. Cara et al., *Process Biochem.* 41, pp. 423-429, 2006.
24. M. N. Islam and M. S. Islam. *J. Thermoplast. Compos. Mater.* (2013) DOI: 10.1177/0892705713503671.
25. D. Siridas and E. Koukios. *Bioresour. Technol.*, 94, pp. 91 –98, 2004
26. S. Tirtowidjojo et al., *Holzforschung* 42 (3), pp. 177 –183, 1988
27. T. H. M Vu 2004. *Ind. Crops Products* 19: 49 –57, 2004
28. A. Shatalov and H. Pereira. *Ind. Crops Products,* 21, pp. 203 –210, 2005

Polymer Carbon Nanotube Composites For Thermoelectric Applications

Jinji Luo, Petra Pötschke*, and Beate Krause

Leibniz-Institut für Polymerforschung Dresden e.V., Hohe Straße 6, 01069 Dresden, Germany

poe@ipfdd.de

Abstract. The thermoelectric (TE) performance of electrically conductive thermoplastic composites prepared by melt mixing was investigated. A cost effective and industrially widely used polymer, namely polypropylene (PP), was chosen as the matrix to fabricate the composites. Singlewalled carbon nanotubes (SWCNTs), the amount (2 wt%) of which was selected to be above the electrical percolation threshold (< 0.2 wt%), were used to form an electrical conducting network. Next to as produced SWCNTs also plasma modified tubes were employed to study the influence of the functionalization on the morphology, dispersion and TE properties of the PP composites. In addition, melt processing conditions, e.g. temperature, rotation speed and mixing time were varied. Furthermore, an ionic liquid (IL, 1-methyl-3-octylimidazolium tetrafluoroborate) was used as processing additive during melt mixing which improved the electrical conductivity of the composites. Simultaneously increases in the Seebeck coefficient were recorded, leading to much better power factor as compared to composites without IL. This melt mixing strategy opens new avenues for solvent-free large scale fabrication of polymer based TE materials.

Keywords: conductive polymer composite, thermoelectric, processing condition, ionic liquid, carbon nanotube
PACS: 82.35.Np, 82.35.Lr, 72.80.Tm, 72.20.Pa

INTRODUCTION

Nanometer sized fillers are widely embedded into polymer matrices to fabricate functional nanocomposites. These nanofillers include carbon based nanomaterials (e.g. carbon nanotubes and carbon black), metal nanoparticles, oxides and etc. In particular, composites with carbon nanomaterials are well developed as electrically conductive polymer composites (CPC) for diverse applications. For instance, carbon based nanocomposites are mainly employed for the dissipation of static electricity, floor heating elements, electronic parts such as electromagnetic interference (EMI) shielding[1]. Besides these traditional applications, CPCs are developed as thermoelectric materials to turn waste heat into electrical energy.

The thermoelectric (TE) effect (also called Seebeck effect) is described as an electrical potential (voltage ΔU) induced by a temperature difference (ΔT) between the two sides of a material. Accordingly, the Seebeck coefficient (S) of the material is defined as

$$S = -\Delta U / \Delta T \tag{1}$$

The Seebeck coefficient is an intrinsic material property and independent of the sample geometry. The free charge carriers inside the material determine the sign of Seebeck coefficient. Negative Seebeck coefficient (N-type) implies dominant electron carriers while positive Seebeck coefficient (P-type) shows holes are the dominant carriers. The TE performance of a material is evaluated by a dimensionless figure of merit, ZT, which is given as[2]

$$ZT = \sigma^2 ST / \kappa \tag{2}$$

with σ is the electrical conductivity, κ is the thermal conductivity and $\sigma^2 S$ represents the power factor (PF). An efficient TE material should have high electrical conductivity and Seebeck coefficient (resulting in high power factor). At the same time, low thermal conductivity is required.

Traditional TE materials are based on semiconductors, e.g. Bi_2Te_3. These materials are heavy and contain toxic rare earth elements (Te). Despite of their high power factor, these materials have high thermal conductivity, for instance, 1.2 to 1.6 W/(mK)for a commercial Bi_2Te_3 crystal[3]. Thereby, sophisticated and high energy demanding techniques are required to reduce the value below 1 W/(mK)[4]. Polymer materials, on the other hand, have intrinsic low thermal conductivity in the range of 0.1 to 0.6 W/(mK)[5]. Furthermore, the light weight and flexibility of polymers make them very attractive in the integration of TE materials into portable devices for power supply[6]. Nevertheless, the challenge in polymer based TE materials is to achieve high power factors. With their high electrical conductivity, intrinsic conducting polymers (conjugated polymers) have been intensively investigated. In particular, poly(3,4-ethylenedioxythiophene) (PEDOT) is reported to have a metallic electrical conductivity over 3000 S/cm by a post treatment with sulfuric acid[7,8]. However, the Seebeck coefficient of this type of polymers is usually in the range of 20 μV/K[9,10]. By reducing the oxidation level of PEDOT, the Seebeck coefficient of PEDOT based film could be increased up to 200 μV/K and high power factor of 324 $\mu W/(mK^2)$ was obtained[11]. However, the reduction in the oxidation level results in a concurrent drop in the electrical conductivity[11-14]. As a result, it is only possible to obtain an optimized power factor by the control of the oxidation level. This intercoupling behavior between Seebeck coefficient and electrical conductivity is similar to what occurs by doping in semiconductor based TE materials.

Polymer composites filled with nanoparticles provide another promising approach to improve the TE performance by the choice of different fillers. In this case, both insulating and electrical conducting polymers are applicable. These fillers could be either organic such as carbon nanotubes (CNTs)[15-18] and graphene[19] for higher electrical conductivity or inorganic semiconductors such as Bi_2Te_3 nanoparticles[20,21], Te nanowires[22] and PbTe nanoparticles[23] for higher Seebeck coefficient. These composites are typically fabricated by solution mixing, in which organic solvents are involved. Compared to solution mixing, melt mixing technique can be easily upscaled to industry level and it avoids the usage of organic solvents. Due to the low electrical conductivity of composites with insulating polymer matrix, the TE properties of melt mixed composites are rarely studied. Antar et al. reported the TE performance of multiwalled CNTs and expanded graphite filled melt mixed poly(lactic acid) composites, in which the highest Seebeck coefficient is 17 μV/K at 32 vol% expanded graphite and highest power factor is 9.6×10^{-2} at 18 vol% CNTs[24]. Our previous work was focused on the influence of the surface functionalization of multiwalled CNTs on TE properties of melt mixed polymer nanocomposites based on polycarbonate, in which the highest Seebeck coefficient was below 12 μV/K and electrical conductivity below 0.1 S/cm even at 2.5 wt% CNT inclusion[25]. Furthermore, a cyclic butylene terephthalate (CBT) oligomer was utilized to successfully improve both the processing and electrical conductivity of PC/CNT nanocomposites[26]. Nevertheless, the Seebeck coefficients of all investigated samples are still below 14 μV/K.

As all reported Seebeck coefficients of melt mixed composites are extremely low, the focus of our work was to improve the Seebeck coefficient by a selection of different types of singlewalled CNTs. At the same time, the CNT concentration and the processing conditions were investigated. Ionic liquids (IL) were previously reported to improve both the dispersion and electrical conductivity of CNTs for melt mixed polymer composites, e.g. in PMMA[27], rubber[28] and polyamide[29]. Therefore, one kind of IL was employed during the processing of the composites to study its influence on the TE properties.

EXPERIMENTS AND CHARACTERIZATIONS

All materials were used as received. PP (Moplen HP400R) was purchased from LyondellBasell Industries. Different types (A, B and mA) of singlewalled CNTs (> 75% purity, Tuball™) with diameters of 1-2 nm and length larger than 5 μm were kindly provided by OCSiAl Ltd. Type A and B represent as synthesized singlewalled CNTs prepared at different synthesis parameters. CNTs of type A were plasma oxidized to produce the CNT type mA. Ionic liquid (IL, 1-methyl-3-octylimidazolium tetrafluoroborate) (purity >97%, HPLC) was purchased from Sigma Aldrich.

A conical twin-screw microcompounder (DSM Xplore 15) with a capacity of 15 ccm was used for melt mixing. The weight percentage of CNTs was calculated regarding to the whole composite. CNTs were manually mixed with IL in an agate mortar with the ratio of CNT/IL fixed to 1:1. Afterwards, mixtures of grounded PP powders (diameter < 1mm) and defined amounts of CNTs and IL pre-shaken in glass vials were fed into the running microcompounder. The standard processing conditions were fixed to a melting temperature of 210°C, a rotation speed of 250 rpm and a mixing time of 5 min. The melting temperature and rotation speed were modified as it is specified in the sample notation to study the influence of processing condition. Afterwards, these mixed composites were pressed at 210°C under a pressure of 50 kN into circular plates with a diameter of 60 mm and a thickness of 0.3 mm using a press PW 40 EH (Paul Otto Weber GmbH, Germany). Rectangular samples were manually cut from pressed circular plates and used for the electrical conductivity and Seebeck coefficient.

For conductivity characterization, a 2-electrode test fixture combined with a Keithley electrometer E6517A was used for high resistance samples (CNT content ≤ 0.8 wt%), whereas a 4-electrode test fixture combined with Keithley multimeter DMM 2000 for samples with lower resistance (CNT content > 0.8 wt%). The reported electrical conductivities are the averaged value of eight measurements for each sample. Seebeck coefficient at room temperature was determined using a measurement platform SRX[30] and the results are the average of at least 3 measurements for each sample. For samples with CNT content < 0.8 wt%, the equipment is not able to give reliable values due to the high electrical resistance, thus, no results are presented. SEM investigations of cryo-fractured surfaces were performed using a Gemini 1530 microscope. The cryo-fractured samples were deposited with a thin layer of carbon before SEM imaging.

RESULTS AND DISCUSSION

Figure 1 shows the TE properties of melt processed composites containing plasma oxidized CNTs (type mA). With increasing amount of CNTs inside the insulating PP matrix an electrical conducting network is built up starting already at 0.2 wt%. Further addition of CNTs results in only small increase in electrical conductivity, which saturates above a content of 4 wt% CNTs. As CNTs are the only conductive parts in these composites, the Seebeck coefficients are determined by the SWCNTs. All composites (> 0.8 wt% CNT) display positive Seebeck coefficients due to the p-type characteristic of CNTs[31]. The increase in electrical conductivity with CNT concentration is attributed to the injected charge carriers by the CNTs. However, the increased charge carrier concentration is detrimental for the Seebeck coefficient, leading to a reduction tendency with CNT content. Due to the intercoupling behavior between the Seebeck coefficient and electrical conductivity, the highest power factor of 6.6×10^{-2} $\mu W/(mK^2)$ is observed for composites containing 4 wt% CNTs.

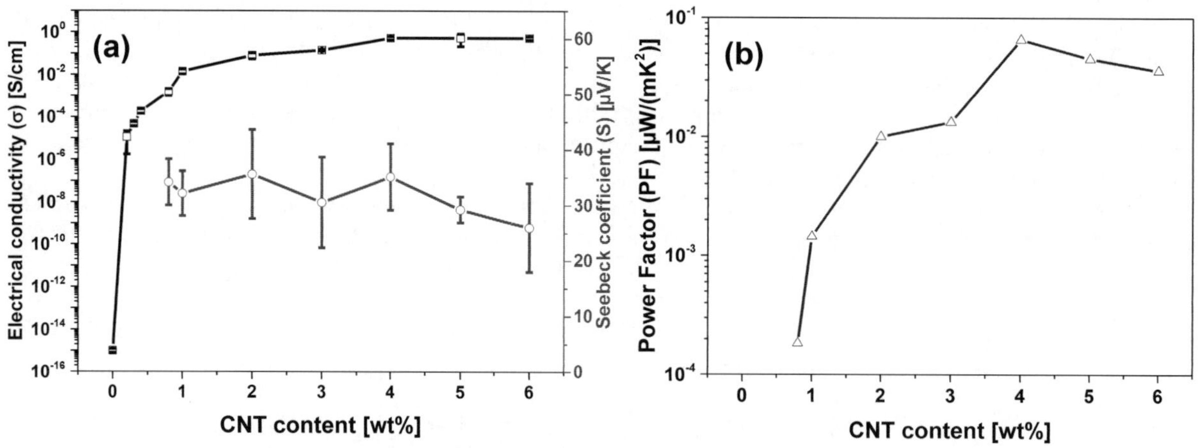

FIGURE 1. Thermoelectric properties of CPCs versus CNT concentration (type mA).

In addition, two types of unfunctionalized singlewalled CNTs (types A and B) were applied. At the same time, IL was added as dispersing agent during the melt mixing. The CNT concentration in the composites was kept to 2 wt%. As it is shown in Figure 2, the use of unfunctionalized CNTs results in higher electrical conductivity than that of plasma functionalized CNTs (type mA). However, the plasma oxidation increases the Seebeck coefficient of the composite from 23.2 μV/K (for CNTs type A) to 35.6 μV/K (for CNTs type mA). Nevertheless, this change results in only small increase in the power factor. Compared to A and mA types of CNTs, the addition of CNTs (type B) generates composites with much higher electrical conductivity and Seebeck coefficient, leading to much higher power factor (Table 1).

Simultaneous increase in the electrical conductivity and Seebeck coefficient is evidenced for all three types of CNTs when IL was added during mixing. The magnitude of enlargement in the power factor is determined by the CNT type. The addition of IL only slightly increases the power factor of the composite filled with CNTs (type B), which already had a high value without IL. On the other hand, for composites containing CNTs prepared with the same synthesis condition (types A and mA), the magnitude in the power factor enhancement by IL addition is much higher. Nevertheless, the composite with B type CNTs and IL shows the highest Seebeck coefficient (63 μV/K) and PF (0.26 $\mu W/(mK^2)$).

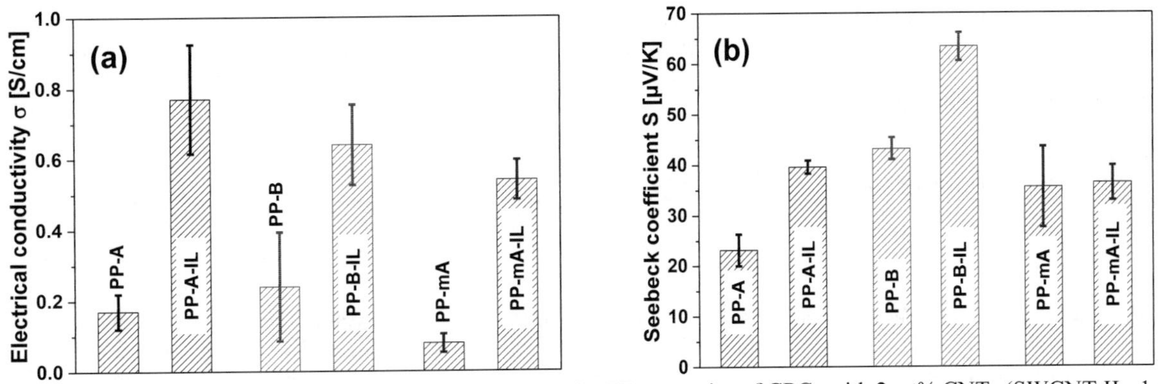

FIGURE 2. The influence of CNT type and IL addition on the TE properties of CPCs with 2 wt% CNTs (SWCNT:IL=1:1).

TABLE 1. Influence of CNT type and IL addition on the power factor of PP composites with 2 wt% SWCNTs and 2 wt% IL.

Material	Power Factor [μV/mK^2]
PP-A	0.009
PP-A-IL	0.12
PP-B	0.12
PP-B-IL	0.26
PP-mA	0.01
PP-mA-IL	0.07

As the addition of CNTs (B type) results in composites with the highest power factor, morphological investigations of corresponding composites filled without or with ionic liquid are shown in Figure 3. Long CNTs are well dispersed inside the PP matrix. Holes in the SEM images are assigned to IL as it is liquid at room temperature. IL is immiscible with the PP matrix. Nevertheless, it is difficult to distinguish the difference of the CNT dispersion state between composites with and without IL and to prove the expected better dispersion as seen in previous studies[27-29].

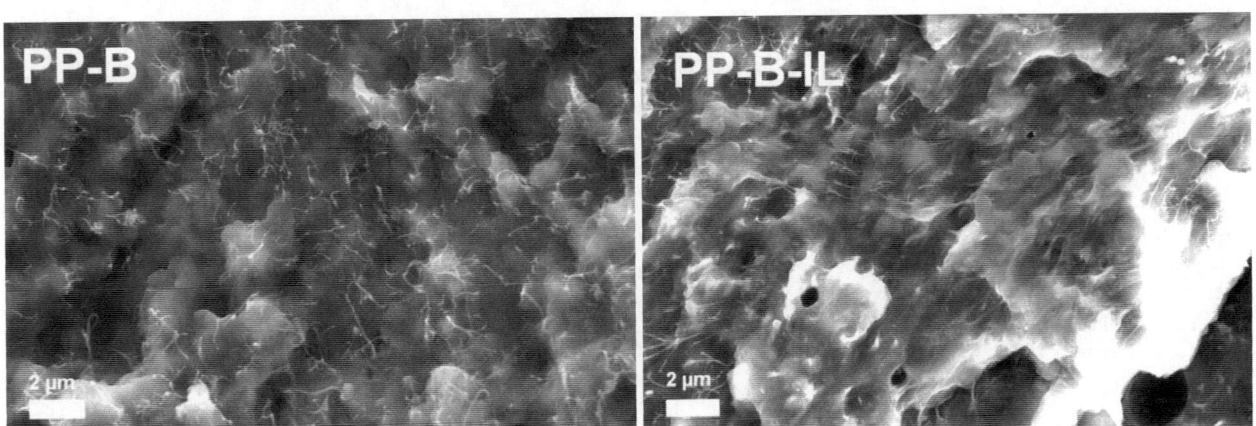

FIGURE 3. SEM of cryofractured surfaces of the composites PP-B and PP-B-IL.

The influence of processing condition was investigated using samples containing CNTs (type mA). Reduction of mixing speed from 250 rpm to 100 rpm, resulting in lower shear stresses during mixing, slightly reduced the electrical conductivity, but increased Seebeck coefficient. Increasing melt mixing temperature from 210 to 230°C showed only an effect at 100 rpm, where lower electrical conductivity, Seebeck coefficient and power factor were measured for samples mixed at the higher mixing temperature. The maximum Seebeck coefficient and power factor were obtained for the combination of low mixing temperature (210°C) and low rotation speed (100 rpm) where interestingly the electrical conductivity was not the highest among all variations. However, in summary the processing conditions do not affect much the TE properties of these composites.

TABLE 2. Influence of processing condition on the composites filled with 2 wt% CNTs type mA and 2 wt% IL.

Material	Electrical conductivity [S/cm]	Seebeck coefficient [μV/K]	Power Factor [$\mu V/mK^2$]
PP-mA-IL-250rpm-210C-5	0.54	36.4	0.072
PP-mA-IL-100rpm-210C-5	0.47	45.6	0.098
PP-mA-IL-250rpm-230C-5	0.54	36.7	0.073
PP-mA-IL-100rpm-230C-5	0.41	42.2	0.073

CONCLUSIONS

Electrically conductive melt mixed polymer composites containing singlewalled CNTs are suitable materials for the use in TE applications. Even if the achieved values for Seebeck coefficient and power factor are still far below those for traditional TE materials, the strategy of incorporating conductive carbon nanostructures is promising for future applications, especially when low price, light weight and high flexibility of polymer based composites are taken in consideration. The influence of structural properties of CNTs and their dispersion are factors which have to be studied more in detail. The use of the dispersing additive IL was shown to improve not only state of dispersion, but also conductivity, Seebeck coefficient and the corresponding power factor.

ACKNOWLEDGMENTS

We thank OCSiAl Ltd for providing the different kinds of CNTs.

REFERENCES

1. T. Tanka, G. C. Montanari, R. Mülhaupt, IEEE Transactions on Dielectics and Electrical Insulation, vol 11, 763-784 (2004)
2. "General Principles and Basic Considerations," in Thermoelectrics Handbook: Macro to Nano, edited by D. M. Rowe, Boca Raton, CRC Press, 2005, pp. 1-14
3. http://www.everredtronics.com/thermoelectric.P.N.pellets.html
4. B. Poudel, Q. Hao, Y. Ma, Y. Lan, A. Minnich, B. Yu, X. Yan, D. Wang, A. Muto, D. Vashaee, X. Chen, J. Liu, M. S. Dresselhaus, G. Chen and Z. Ren, Science 320, 634-638 (2008)
5. J. E. Mark, Physical Properties of Polymers Handbook, 2nd Edition, ISBN 978-0-387-69002-5
6. B. T. McGrail, A. Sehirlioglu, E. Pentzer, Angew. Chem. Int. Ed. 54, 1710-1723 (2015)
7. N. Kim, H. Kang, J. H. Lee, S. Kee, S. H. Lee, K. Lee, Advanced Materials 27, 2317-2323 (2015)
8. Y. Xia, K. Sun, J. Ouyang, 2012, Advanced Materials 24, 2436-2440 (2012)
9. J. H. Bahk, H. Fang, K. Yazawa, A. Shakouri, Journal of Materials Chemistry C 3, 10362-10374 (2015)
10. Q. Zhang, Y. Sun, W. Xu, D. Zhu, Advanced Materials 26, 6829-6851 (2014)
11. B. Bubnova, Z. U. Khan, A. Malti, S. Braun, M. Fahlman, M. Berggren, X. Crispin, Nature Materials 10, 429-433 (2011)
12. O. Bubnova, M. Berggren, X. Crispin, Journal of the American Chemical Society 134, 16456-16459 (2012)
13. H. Park, S. H. Lee, F. S. Kim, H. H. Choi, I. W. Cheong, J. H. Kim, Journal of Materials Chemistry A 2, 6532-6539 (2014)
14. S. H. Lee, H. Park, W. Son, H. H. Choi, J. H. Kim, Journal of Materials Chemistry A 2, 13380-13387 (2014)
15. Y. Choi, Y. Kim, S. G. Park, Y. G. Kim, B. J. Sung, S. Y. Jang, W. Kim, Organic Electronics 12, 2120-2125 (2011)
16. C. A. Hewitt, A. B. Kaiser, S. Roth, M. Craps, R. Czerw, D. L. Carroll, Nano Letters 12, 1307-1310 (2012)
17. Q. Yao, L. Chen, W. Zhang, S. Liufu, X. Chen, ACS Nano 4, 2445-2451 (2010)
18. C. Yu, K. Choi, L. Yin, J. C. Grunlan, ACS Nano 5, 7885-7892 (2012)
19. G. H. Kim, D. H. Hwang, S. I. Woo, Phys. Chem. Chem. Phys. 14, 3530-3536 (2012)
20. M. He, J. Ge, Z. Lin, X. Feng, X. Wang, H. Lu, Y. Yang, F. Qiu, Energy and Environmental Science 5, 8351-8358 (2012)
21. Y. Du, K. F. Cai, S. Chen, P. Cizek, T. Lin, ACS Applied Materials & Interfaces 6, 5735-5743 (2014)
22. N. E. Coates, S. K. Yee, B. McCulloch, K. C. See, A. Majumdar, et al., Advanced Materials 25, 1629-1633 (2013)
23. Y. Wang, K. Cai, X. Yao, ACS Applied Materials & Interfaces 3, 1163-1166 (2011)
24. Z. Antar, J. F. Feller, H. Noël, P. Glouannec, K. Elleuch, Materials Letters 67, 210-214 (2012)
25. M. Liebscher, T. Gärtner, L. Tzounis, M. Micusik, P. Pötschke, M. Stamm, G. Heinrich and B. Voit, Composites Science and Technology 101, 133-138 (2014)
26. L. Tzounis, T. Gärtner, M. Liebscher, P. Pötschke, M. Stamm, B. Voit and G. Heinrich, Polymer 55, 5381-5388 (2014)
27. L. Zhao, Y. Li, X. Cao, J. You and W. Dong, Nanotechnology 23, 255702 (2012)
28. K. Subramaniam, A. Das, K. W. Stöckelhuber and G. Heinrich, Rubber Chemistry and Technology 86, 367-400 (2013)
29. R. Socher, PA12-MWCNT-Nanokomposite: Wege zur effektiven MWCNT-Dispergierung und zu niedrigen elektrischen Perkolationsschwellen, phd thesis Technische Universität Dresden (2012), Verlag Dr. Huth, ISBN 9783843909013.
30. http://www.ipm.fraunhofer.de/en/services/contract-measurements/thermoelectric-measurements.html
31. M. Shim, A. Javey, N. W. S. Kam and H. Dai, J. Am. Chem. Soc. 123, 11512-11513 (2001)

Crystalline Nanocellulose in Biodegradable Polyester Nanocomposites Prepared by In Situ Polymerization

Ashley Clarke, Alexandros A. Vasileiou, Marianna Kontopoulou*

Department of Chemical Engineering, Queen's University, Kingston, Ontario K7L 3N6, Canada

*kontopm@queensu.ca

Abstract. In situ polymerization is implemented to disperse crystalline nanocellulose (CNC) in poly(ethylene succinate) (PESu). The addition of CNCs directly into the PESu polymerization mixture resulted in excellent nanoparticle dispersion, without visible cluster formation. The polymerization kinetics were altered with increasing CNC content, causing a decrease in the polymer molecular weight. The addition of small amounts of CNCs, up to 0.25 wt%, resulted in a slight decrease in viscosity, whereas significant increases in the zero shear viscosity were observed at 1 wt% CNC. PESu containing well-dispersed 0.25 wt% CNC maintained the tensile properties of the matrix, while demonstrating significantly improved crystallization kinetics and reduced haze due to the formation of smaller crystallites. CNC acted as a very efficient nucleating agent, while biodegradability remained unimpeded.

Keywords: In situ polymerization, crystalline nanocellulose, poly(ethylene succinate).

PACS: 82.35.Np.

INTRODUCTION

Cellulose is an abundant, naturally occurring, biodegradable, non-toxic material with widespread applications. The repeating subunits comprising the molecular structure of cellulose contain three hydroxyl groups per unit, allowing for high functionality (1). A derivative of cellulose, cellulose nanocrystals (CNCs) has the potential to create value-added materials that can be used for research and industrial applications (1). The high functionality of CNCs coupled with their high specific strength and surface area make them desirable as a nucleating agent in polymers to decrease processing time and improve crystallization (2-4).

A great deal of research has focused on how to integrate CNCs into polymeric materials. Typically, CNCs are extracted from cellulose sources and dried in a protonated form for effective shipping (4). The polar and hydrophilic nature of CNCs makes them incompatible with most polymers, since those are usually non-polar and hydrophobic. This incompatibility is further exacerbated due to the large specific surface area (or high surface energy) of the nanosized cellulose particles, which promotes aggregation and results in inefficient dispersion and weak interfacial interactions (1-5). Poor dispersion results in an uneven composition across the polymer, affecting mechanical properties and thus the usefulness in industry. Chemical modification of the surface of the nanocrystals often leads to a loss of the inherent properties of the CNC (1). Furthermore methods such as melt compounding have been met with limited success. In this work, in-situ polymerization is employed as a method to efficiently disperse CNC within a polymer matrix.

In situ polymerization requires mixing the CNCs into a liquid reactant followed by initiating the polymerization using heat (6). Ultrasonic vibrations will be used to ensure even dispersion of CNCs is achieved in the liquid reactant (7), (8), leading to a higher probability that as polymerization occurs the polymer molecules will wrap around the CNCs to create a uniform composite without large clusters of CNCs (6). In situ polymerization also avoids surface modification of the dried CNCs prior to use for industrial applications, allowing for simpler and more efficient manufacturing processes.

Polyesters are industrially relevant thermoplastic polymers, which are used in a variety of applications. Biodegradable aliphatic polyesters, such as poly(ethylene succinate) (PESu) have properties similar to polypropylene and low-density polyethylene (9), but with very low crystallization rates, hindering their processability. In this investigation, CNCs are used to create poly (ethylene succinate) (PESu)-based nanocomposites. Specifically the successful preparation of (PESu)/CNC composites containing well-dispersed CNC using the in-situ polymerization technique described above is described.

MATERIALS AND METHODS

Succinic acid (purity 99%+), 1,2-ethanediol (purum>99.8%), and titanium (IV) butoxide (≥97%) were purchased from Sigma-Aldrich Corporation and triphenyl phosphate (95%) was purchased from Fluka Analytical. The freeze-dried cellulose nanocrystals supplied by FPInnovations had a length of about 100nm and a width of about 10-20nm. All other materials and solvents used in the analytical methods were of analytical grade.

A two-stage melt polymerization was used to produce nanocomposites of poly(ethylene succinate) containing 0.25, 0.5, 0.75, and 1.0 weight percent (wt%) CNC. The composites were produced in situ; CNC were dispersed in 1,2-ethanediol using ultrasonic vibrations for 30 minutes prior to the reaction, then 1:1.1 molar ratios of succinic acid and 1,2-ethanediol were placed in a glass batch reactor degassed with argon. In the first melt stage, the mixture was heated for 3h at 200°C under an argon atmosphere and constant stirring while water was removed as a reaction by-product. Before the second melt stage, 0.3wt% triphenyl phosphate (TPP) and 1.0x10-3 mol of titanium (IV) butoxide (TBT) per mol of succinic acid were added to act as heat stabilizer and catalyst respectively. Mixing continued under a vacuum of about 1.0 Pa at a temperature of 220°C for 1.5h and at 240°C for 1h to remove reaction by-products. After the polymerization was complete, the mixture was cooled quickly to room temperature and the solid product washed with distilled water.

The dispersion of CNC particles within the polymer was gauged by searching for clusters of CNCs using optical (OM), scanning electron (SEM) and transmission electron (TEM) microscopy.

Rheological characterization was done at 120°C using a Reologica ViscoTech oscillatory rheometer equipped with 25mm parallel plate fixtures at a gap of 0.5 mm. Thermal analysis of the composites was performed using a differential scanning calorimeter (TA Instruments Q100 Series DSC) calibrated with indium and zinc standards. Samples of 10±0.5mg sealed in aluminium pans were scanned in the instrument. The cycle began with heating from 30°C to 130°C at a heating rate of 10°C/min and holding at this temperature for 2 min in order to erase any thermal history of the sample. Next, rapid cooling to -30°C and equilibration followed by reheating at a rate of 30°C/min from -30°C to 130°C and holding for 2min occurred. The rapid cooling followed by reheating at 30°C/min and 2min holding time was repeated using cooling rates of 2.5-10°C/min.

Tensile properties were measured using an Instron 3369 universal tester at crosshead speeds of 5mm/min. Dumbbell-shaped specimens were cut with a type-V die according to ASTM D 638 from sheets with average thickness of about 0.4mm. The sheets are prepared by compression moulding of the compounded samples at 140°C and 10MPa using a Carver press.

The biodegradability of the nanocomposites was studied using enzymes. Samples with a starting surface area of 2cm^2 are placed in solution with 1mg of enzyme and are held at 50°C to accelerate degradation, with the mass of each sample being checked on a weekly basis until the samples have lost >50% of their original mass.

RESULTS AND DISCUSSION

TEM imaging (Figure 1) clearly revealed the presence of well-dispersed CNC platelets within the polymer matrix, confirming the efficiency of this technique at the nanoscale level. Furthermore the absence of clusters of particles in optical microscopy and SEM images suggested that the CNCs were successfully dispersed into PESu.

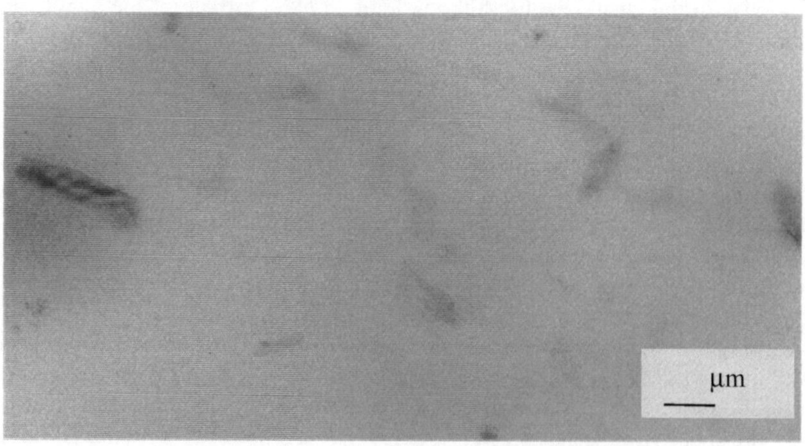

FIGURE 1. TEM image of PESu/1 wt% CNC composite

The rheological data reveal a substantial increase in the complex viscosity at low frequencies upon addition of CNC beyond 0.75 wt%. These enhancements are ascribed the very high surface area, and aspect ratio, of the dispersed CNC particles, resulting in very good interfacial interactions with the polymer matrix.

The presence of the CNCs in the polymerization mixture slowed down the polymerization kinetics, probably due to the increase in viscosity impeding heat/mass transfer during polymerization. The result was that resulting polymer matrix had lower molecular weights at higher fractions of CNCs. This is the reason for the reduction of the viscosity of the composites at high frequencies, seen in Figure 2, which are representative of the properties of the matrix.

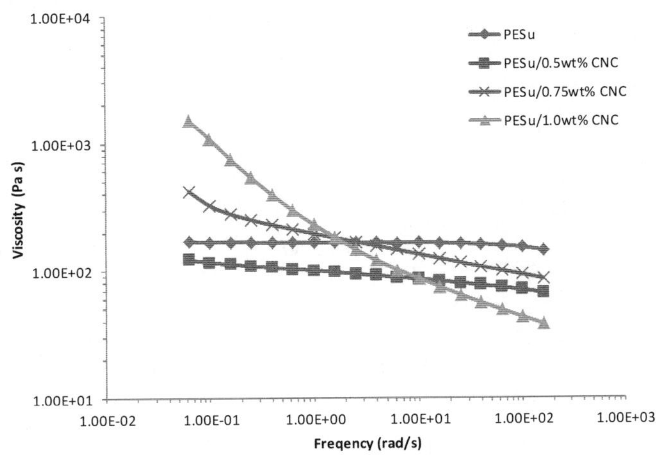

FIGURE 2. Complex viscosity as a function of frequency at 120°C of PESu and PESu/CNC composites.

Non-isothermal melt crystallization experiments revealed that CNCs are a very active nucleating agent. As shown in Figure 3 very small fractions of CNCs are capable of raising the crystallization peak temperature significantly, compared to neat PESu. Although peak temperature decreased with increasing cooling rate, the samples containing CNCs consistently crystallized at much higher temperatures than neat PESu, even at the smallest CNC concentrations.

Figure 4 reveals substantial increases in the crystallization rates. For example when using a cooling rate of 7.5°C/min, the time required for neat PESu to fully crystallize is 6-7 minutes, whereas the PESu/CNC composites take about half the time at 3-4 minutes, as seen in Figure 6. This increase in crystallization rate is important for large-scale production operations to minimize the time required to process each batch of polymer.

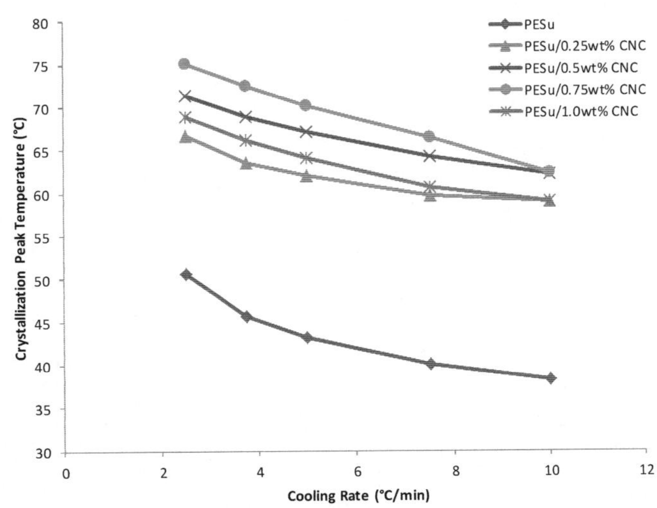

FIGURE 3. Crystallization peak temperature as a function of cooling rate for each of the nanocomposites and neat PESu.

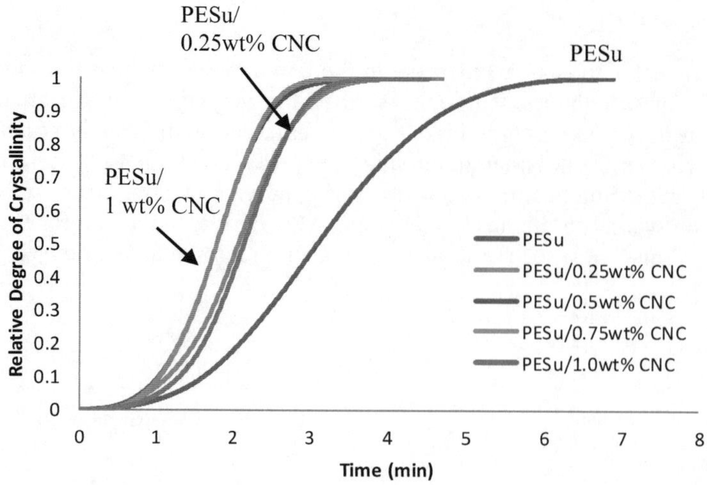

FIGURE 4. Relative degree of crystallinity as a function of time, of neat PESu and PESu/CNC composites at a cooling rate of 7.5°C/min.

Addition of up to 0.25 wt% CNC to PESu resulted in a higher Young's modulus, while maintaining the ductility of the polymer. CNC contents beyond this caused an essentially complete loss of ductility, as shown in Table 1.

TABLE 1. Summary of tensile testing of PESu and PESu/CNC composites

	Yield Stress (MPa)	**Stress at Break (MPa)**	**Elongation at Break (%)**	**Young's Modulus (MPa)**
PESu	25.0 ± 1.3	32.9 ± 1.7	716.9 ± 57.8	243.3 ± 4.7
PESu/0.25wt% CNC	29.4 ± 2.1	28.7 ± 1.8	506.2 ± 11.9	265.0 ± 25.0
PESu/0.5wt% CNC	-	28.9 ± 0.6	15.5 ± 2.9	313.3 ± 9.4
PESu/0.75wt% CNC	-	21.0 ± 2.3	11.3 ± 2.2	276.7 ± 12.5
PESu/1.0wt% CNC	-	24.6 ± 1.1	9.2 ± 1.5	340.0 ± 21.6

The biodegradability studies of neat PESu and nanocomposites revealed that the presence of CNCs affects slightly the ability of PESu to break down, as seen in Figure 5.

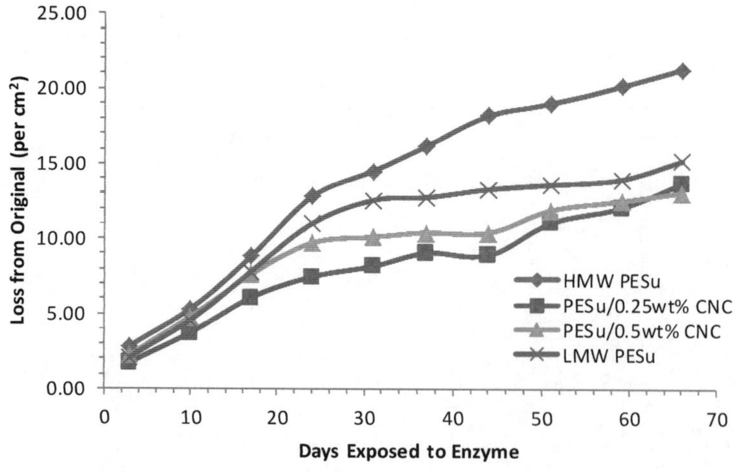

FIGURE 5. Weight loss of neat PESu and nanocomposites over time while being exposed to enzymatic degradation. A low molecular weight (LMW) PESu is included for comparison.

CONCLUSIONS

The proposed method of adding CNCs directly into the thermoplastic polyester matrix using in situ polymerization was very effective in dispersing the CNCs within the polymer matrix at the nanoscale. The addition of CNCs to PESu helps to minimize processing time by increasing the crystallization rates. This could be accomplished by using very small amounts of CNCs, as low as 0.25 wt%, thus reducing the cost of such formulations, while maintaining a fully biodegradable formulation and not compromising the mechanical properties substantially. The large improvement in crystallization kinetics of the PESu/0.25wt% CNC composite suggests further research could be conducted to pursue small amounts of CNC as a nucleating agent for other biodegradable industrially relevant polyesters. An improvement to the crystallization of PESu using CNCs could create a biodegradable "green" alternative to plastics commonly used for general-purpose biodegradable containers and food packaging.

ACKNOWLEDGMENTS

FP Innovations is acknowledged for kindly supplying the CNC used in this work. Funding was provided by the Natural Sciences and Engineering Research Council of Canada (NSERC) – Discovery Grants program and an NSERC/USRA scholarship to Ms. Clarke.

REFERENCES

1. Hubbe, M. A.; Rojas, O. J.; Lucia, L. A.; Sain, M. Cellulosic Nanocomposites: A Review. Bioresources 2008, 3 (3), 929-980.
2. Pei, A.; Zhou, Q.; Berglund, L. A. Functionalized cellulose nanocrystals as biobased nucleation agents in poly(L-lactide) (PLLA) - Crystallization and mechanical property effects. Compos. Sci. Technol. 2010, 70, 815-821.
3. Yu, H.-Y.; Qin, Z.-Y.; Liu, Y.-N.; Chen, L.; Liu, N.; Zhou, Z. Simultaneous improvement of chemical properties and thermal stability of bacterial polyester by cellulose nanocrystals. Carbohyd. Polym. 2012, 89, 971-978.
4. Peng, B. L.; Dhar, N.; Liu, H. L.; Tam, K. C. Chemistry and Applications of Nanocrystalline Cellulose and its Derivatives: A Nanotechnology Perspective. Can. J. Chem. Eng. 2011, 89, 1191-1206.
5. Beck, S.; Bouchard, J.; Berry, R. Dispersability in water of dried nanocrystalline cellulose. Biomacromolecules 2012, 13, 1486-1494.
6. Advani, S.; Hsiao, K.-T. Manufacturing Techniques for Polymer Matric Composites (PCMs); Elsevier, 2012.
7. Viet, D.; Beck-Candanedo, S.; Gray, D. G. Dispersion of cellulose nanocrystals in polar organic solvents. Cellulose 2007, 14, 109-113.
8. Park, C.; Ouanies, Z.; Watson, K. A.; Crooks, R. E.; Smith, J. . J.; Lowther, S. E.; Connell, J. W.; Siochi, E. J.; Harrison, J. S.; St. Clair, T. L. Dispersion of single wall carbon nanotubes by in situ polymerization under sonication. Chem. Phys. Lett. 2002, 364, 303-308.
9. Tokiwa, Y.; Calabia, B. P.; Ugwu, C. U.; Aiba, S. Biodegradability of plastics. Int. J. Mol. Sci. 2009, 10, 3722-3742.

INFLUENCE OF NANOCLAY ON MECHANICAL PROPERTIES OF GLASS FIBER REINFORCED POLYPROPYLENE COMPOSITES

Gabriel J. Peláez[a], José Alexandrino Sousa[a*]

[a]*Materials Engineering Department, Universidade Federal de São Carlos, São Carlos, Brazil*
* *jasousa@ufscar.br*

Abstract: Hybrid filler reinforced thermoplastics composed with short glass fiber (GF) and different mineral fillers present special interest for engineering applications, as substantial reductions in mechanical anisotropy and molding warpage are expected from these systems as compared to binary composites with GF only. However, when nanoscale fillers, such as nanoclay (NC), are incorporated in GF-reinforced thermoplastics, the mechanical strength is severely affected, even at very low NC content. Thus in this work, GF-reinforced polypropylene (PP) composites with varying total and relative concentrations of GF and organophilic Montmorillonite clay (O-MMT) along with maleated PP as compatibilizer were twin-screw extrusion compounded with three different processing protocols, in order to identify the factors leading to the observed loss in mechanical strength properties. Using tensile, flexural tests, DMTA and SEM/TEM microscopy data, it is concluded that the physical presence of NC particles at the interface contributes towards substantial reduction of shear stress transfer at the fiber-polymer interface.

Keywords: Polypropylene, glass fiber, nanoclay, hybrid ternary composites, interfacial stress transfer.
PACS: 81, 62, 68.

INTRODUCTION

Nowadays, short glass fiber (GF) reinforced polypropylene (PP) composites use fiber content in the range of 15-40 wt. %, required for enhanced mechanical performance for most engineering applications. Nevertheless, higher GF content compromises PP's ease of processing, introducing high fiber orientation in injection molded parts and, consequently, leading to shortcomings such as high mechanical properties anisotropy, warpage and drop in mechanical performance of weld lines. These deleterious effects can be minimized by the partial substitution of GF with particulate mineral fillers, such as talc or calcium carbonate[1], resulting in a hybrid composite. This concept of hybridization of fibrous and particulate reinforcements can be improved by replacing micrometric with nanometric mineral fillers, such as nanoclay. The combination of reinforcements at the micrometric and nanometric scale with different aspect ratios allows new polymer composites formulation concepts with less total filler/reinforcement content and advantages such as viscosity reduction and improved processability, which results in lighter and better surface finish parts with greater specific properties, besides reducing abrasion of the processing equipment[1,2].

Some studies[1,2], have shown that incorporating mineral filler particles into the matrix of fiber reinforced composites, results in synergistic effects in the form of higher modulus and reduced material costs, not proportional to the total content of reinforcements incorporated; nevertheless, accompanied with decreased strength and impact toughness. However, hybrid reinforced composites form a complex system and there is inadequate data available about the phenomena behind the mechanical properties changes due to the addition of particulate fillers to the fiber reinforced thermoplastic composites[2].

Previous studies carried out in our research group[3,4] showed that the incorporation of organically modified Montmorillonite (O-MMT) nanoclay in PP/GF binary systems by traditional extrusion compounding process has deleterious effects on the strength and toughness properties of PP/GF/O-MMT hybrid ternary composites. Some tentative hypotheses were formulated to explain these properties' drop. A first possible effect considered was the premature capture of the maleated-PP compatibilizer by the organophilic nanoclay (NC) particles, avoiding complete compatibilization between glass fibers and the polymer matrix. Another hypothesis was the erosive action of the NC particles on the compatibilized GF surface and thereby removing the fiber-polymer interfacial bonds created. A reduction of the GF wettability by the polymer matrix, due to increased polymer viscosity in the presence of NC particles being fed at earlier stages than the GF during the traditional extrusion compounding processing, was also considered. Finally, the physical presence of NC particles near the fiber surface is considered to have deleterious influence on the shear stress transfer mechanism at the fiber-polymer interface.

Thus, this study is an attempt to elucidate the mechanisms and the phenomenology leading to the verified drop on strength and toughness properties in PP/GF binary composites prepared by the traditional extrusion compounding process when nanometric particles are introduced into the system. In addition, this work intends to overcome the presented limitations by using different compounding protocols, so as to formulate and characterize the short-term mechanical properties and the microstructure of PP composites with hybrid GF/O-MMT reinforcement, as a material for engineering end-use applications with appropriated balance of rigidity, strength, toughness properties and improved processability.

MATERIALS AND METHODS

Materials

A mixture of isotactic homopolymer polypropylenes, supplied by Quattor/Braskem S.A. (HP648S with ρ=0.905 g/cm^3 and MFI= 40 g/10 min and grade HP550K with ρ=0.905 g/cm^3 and MFI= 3.5 g/10 min), was chosen as the matrix polymer for the investigated PP hybrid composites. The mixture of HP648S:HP550K was in the proportion 3:1 and the matrix was heat stabilized with Irganox 1010 and Irgafos 168 antioxidants. The short glass fiber (GF) was supplied by Vetrotex/Owens South America in the form of chopped roving (GF968) with ρ=2.54 g/cm^3, 13 μm fiber diameter and 4.5 mm length, pre-treated with aminosilane coupling agent and sizing compatible with PP [1]. To increase the fiber–matrix interaction, maleated-PP (PP-g-MAH - Polybond 3200) compatibilizer supplied by Crompton/Chemtura Brasil was used with 1 % nominal MAH content and MFI= 250 g/10 min. The compatibilizer's choice was made based on previous studies of this research group[3,4]. The nanoclay used was O-MMT Cloisite 20A (d_{001} = 24.2 Å) supplied by Southern Clay Company.

Composites of stabilized homopolymer i-PP with constant short GF content (25 and 37 wt. %), varying PP-g-MAH compatibilizer content (in proportion of 3:1 m-PP:NC) and varying O-MMT content (0.5, 1.0 and 2.0 wt. %) were compounded in a Werner-Pfleiderer ZSK-30 corotating twin-screw extruder (TSE) with 35 mm diameter. The extrusion profile and parameters choice was based in previous works of this group[1,3,4]. The extruder temperature profile was maintained at 200°C so as to preserve the integrity of the NC's organic treatment.

Processing

In a first mixing protocol, PP/GF binary composite and PP/GF/O-MMT hybrid ternary composites were compounded in the TSE in the proportions required to obtain the desired formulations (mentioned above), and then injection molded. It is worth noting that these composites were prepared by mixing previously extrusion compounded PP/30% GF and PP/25% NC "masterbatch" concentrates, both fed at the first feeding funnel of TSE. This set of composites was labelled with the prefix "e" that means extruded, to refer to this first compounding protocol. In a second compounding protocol, the binary composite and the hybrid ternary composites were prepared by mixing the previously prepared concentrates directly in the injection machine during the molding of tests specimen. This set of composites was labelled using prefix "d" that means directly, referring to the direct mixing step in the injection molding machine. In a third compounding protocol, binary composites were prepared by introducing GF downstream in the eighth barrel of the extruder and for the hybrid ternary composites preparation, the PP/25%NC concentrates plus the PP matrix pellets were fed at the first feeding funnel and the GF was fed at the eighth barrel of the extruder. Then the extruded composite granules were injection molded. These composites were labelled with "t" prefix. that means traditional referring to this type of process commonly used. All composites were injection molded in tensile, flexure, Izod impact and HDT ASTM standard specimens in an automatic Arburg Allrounder (270V/300-120) injection machine.

The "e" compounding protocol was formulated to ensure complete interfacial compatibilization between the GF and the PP matrix during the masterbatch preparation and then being mixed with the PP/NC concentrate, so as to avoid the hypothetical capture of the maleated-PP compatibilizer by the organophilic nanoclay particles. The "d" compounding protocol was formulated to ensure the complete compatibilization between the GF and the PP matrix the same way as in "e" mixing protocol, but with the PP/GF and PP/NC masterbatch concentrates mixture occurring in the injection machine. Using this protocol, it is expected to reduce the time and degree of erosive action by avoiding hypothetical collisions of NC lamellae on the fiber surface, thereby decreasing fiber-polymer interfacial adhesion/interactions, besides reducing the fiber breakage. The binary PP/GF composites eGF25, dGF25, tGF25 and tGF37 are considered as reference materials for the hybrid ternary composites. The two different GF concentrations were used in order to verify the influence of the higher total reinforcement concentration.

Characterization

With the short-term mechanical properties evaluation, it is intended to understand how the stress transfer from the matrix to the fibers is influenced by the presence of the NC particles. The tensile and flexural tests were carried out with an Instron universal testing machine, model 5569, according to standards ASTM D638 (crosshead displacement of 5 mm/min) and ASTM D790 (crosshead displacement of 1.3 mm/min) respectively. Scanning electron microscopy observations were carried out over the cryofractured cross-sections surfaces of pre-stressed (to 70 % of the ultimate tensile strength) tensile bars using a FEI Magellan XHR 400L FE-SEM equipment.

RESULTS AND DISCUSSION

Mechanical Properties Characterization

Figure 1 shows the PP composites' tensile strength (TS) values as a function of the NC content, where the property of the reference binary composites, for each of the three types of composite compounding protocols, decreases as the NC content increases in the systems. Similar results were also obtained for the flexural strength and Izod impact resistance properties data (data not shown due to space restrictions). For the tensile and flexural strength (FS), there is a sharper decrease in these properties with the introduction of just 0.5 % NC content, followed by a less intense property loss with further increase in NC content. The TS and FS values of "e" composites is significantly lower than those of "t" and "d" composites due to the fact that the processing protocol adopted for this formulation, with pre-mixture of concentrates in the TSE machine previous to the injection molding process, leads to greater fiber breakage, as was confirmed by fiber length measurements (data not shown here). The drop of the properties of the "e" and "d" hybrid composites prepared from PP/30%GF and PP/25%NC concentrates indicates that premature capture of m-PP compatibilizer by NC particles does not occur, as the fiber-polymer interfacial compatibilization was previously established during the PP/GF masterbatch preparation.

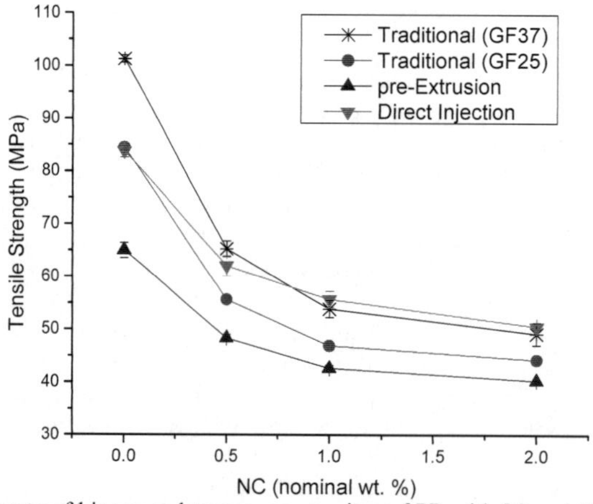

FIGURE 1. Tensile strength property of binary and ternary composites of PP with 25 and 37 wt.% of GF, as a function of NC nominal wt. content, obtained using three different compounding procedures (E, D & T= pre-extrusion and direct injection of concentrates and traditional twin-screw extrusion processing protocols respectively).

Table 1 shows the percentage loss in tensile strength values of the hybrid composites with increasing NC content, relative to their respective reference binary composites. The percentage loss in tensile strength values of eGF25 and dGF25 composites when 0.5 % NC is introduced into the binary system is nearly the same (25.6 and 26.00 respectively) in both cases. Even with very different mixing times and shear histories during the compounding of both hybrid composites, the relative loss in properties remains very similar in both cases and stays nearly equivalent for the other hybrid composites with further addition of 1.0 % and 2.0 % of NC content (Table 1). This demonstrates that there is no erosive action of NC particles on the sizing of fiber surface. From these arguments, it is possible to discard all three of the previously mentioned hypothetical reasons for the substantial drop in mechanical strength properties of the hybrid composites prepared using "e" and "d" compounding protocols, which are related to the premature capture of the compatibilizer, erosive action and poor fiber wettability, all induced by the presence of NC particles.

TABLE (1). Tensile strength reductions in hybrid ternary composites with respect to their reference binary composites.

Composite	TS of Binary Reference (MPa)	TS Property Drop in Hybrids (%)		
		0.5 NC wt. %	1.0 NC wt. %	2.0 NC wt. %
eGF25	64.98	25.6	34.3	37.9
dGF25	83.78	26.0	33.6	39.7
tGF25	84.42	34.1	44.6	47.6
tGF37	101.25	35.5	46.8	51.4

The tGF25 composite presents a sharper percentage loss in tensile strength (34.1 %) with the addition of 0.5 % NC, as can be seen from data in Table 1 and Figure 1. In this case, as in the traditional compounding protocol the NC concentrate is fed much earlier than the chopped rovings of glass fiber to the extruder, it is quite possible that some premature capture of the compatibilizer by NC particles may have induced incomplete fiber-polymer interfacial compatibilization reactions and, therefore, accounting for the higher drop in the TS values. This same property drop is also verified for hybrid tGF37 composite prepared by the same traditional compounding protocol. Similar behavior was also observed in the flexural strength data of all hybrid composites (data not shown here).

From general analysis of the TS and FS properties of all hybrid composites, with respect to their reference binary composites, it can be deduced that analyzing all the hypothetical effects contributing for the verified substantial drop in strength properties, the only common effect to all the compounding protocols used is related to the physical presence of NC particles near the GF surface leading to a deleterious effect on the shear stress transfer mechanism at the fiber-polymer interface.

Microstructural Characterization

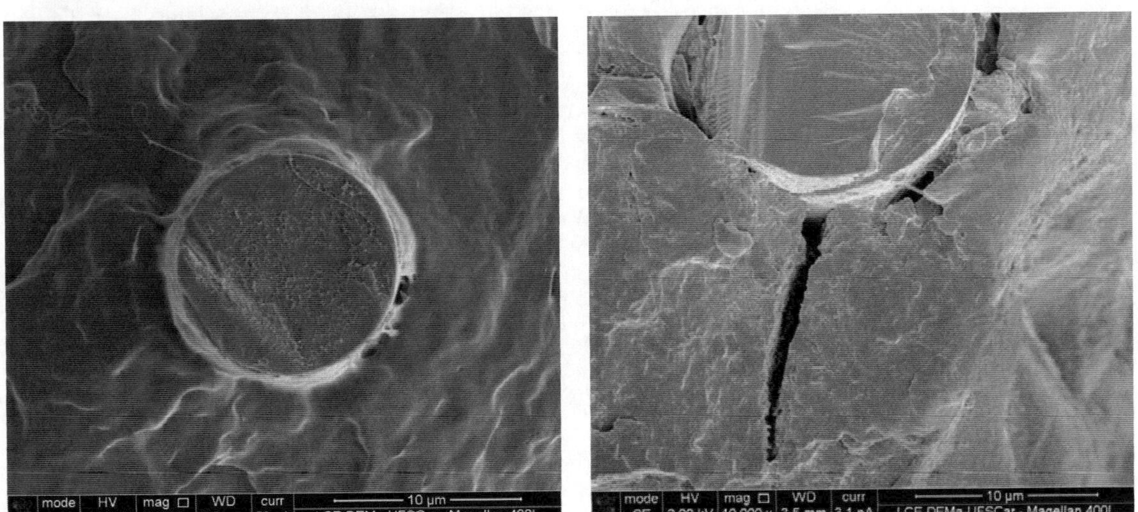

(a) (b)

FIGURE 2. SEM microphotographs of cryofractured surfaces of (a) binary dGF25 composite and (b) hybrid ternary dGF25 with 2% O-MMT content (samples were pre-stressed up to 70% of TS).

Figure 2 shows the SEM micrographs obtained for cryofractured surfaces from cross sections of tensile specimen of PP/PP-g-MAH/25%GF composites, without O-MMT (Fig. 2-a) and with 2% of O-MMT (Fig. 2-b). The tensile tests specimen were pre-stressed up to 70% of its TS in each case prior to cryofracture, in order to elucidate the fiber-polymer interfacial adhesion. SEM analysis was carried out on the binary dGF25 composite and the hybrid ternary dGF25NC2.0 composite. These formulations were used considering that both were prepared using the same PP/PP-g-MAH/30%GF masterbatch, which provided prior complete fiber-polymer interfacial compatibilization for both binary and ternary composites.

In Figure 2-a, the dGF25 composite's micrograph shows complete matrix wettability on the fiber surface, due to chemical coupling leading to strong interfacial interactions, even after the samples were submitted to a pre-stress process. In the hybrid ternary composite, as evidenced in Figure 2-b, there is extensive fiber decoupling and the polymer matrix presents a brittle fracture surface. This indicates that the presence of NC particles near the fiber surface damages the shear stress transfer at the fiber-polymer interface, in spite of prior interfacial compatibilization. In binary PP/GF composites with optimized interfacial adhesion achieved with compatibilizers, interfacial shear stress transfer occurs up to the limit of the shear strength of the matrix polymer. At these high levels of shear stress transfer at the fiber-polymer interface, the NC particles decouple, probably due to lower interfacial compatibilization with the polymer matrix, creating nanocracks that coalesce leading to greater size cracks and, therefore, contributing to premature fiber-polymer decoupling. This effect is also verified even at low (0.5%) O-MMT content, as evidenced in the data presented in Figure 1.

CONCLUSIONS

The mechanical properties of binary PP/GF and hybrid ternary PP/GF/O-MMT composites, with constant fiber concentration (25 and 37 wt. %) and variable O-MMT content (0.5, 1 and 2 wt. %) were evaluated. A substantial drop of tensile, flexural and impact strength properties of hybrid composites was verified, as a function of NC content. This drop in mechanical performance is attributed to the physical presence of NC particles at the fiber-polymer interface contributing towards substantial reduction of shear stress transfer at the fiber-polymer interface. These results are corroborated by SEM analyses of cryofractured cross section surfaces of pre-stressed tensile specimens.

ACKNOWLEDGMENTS

The authors wish to thank FAPESP for the research funding and the companies Quattor/Braskem S.A. and Vetrotex/Owens South America for kind supply of the materials used. One of the authors gratefully acknowledges his graduate research grant from CAPES.

REFERENCES

1. C. Morelli, A. S. Pouzada and J. A. Sousa, Influence of Hybridization of Glass Fiber and Talc on the Mechanical Performance of Polypropylene Composites, J. Appl. Pol. Sci. 114 (2009) 3592-3601.
2. J. Hartikainen et al, Polypropylene hybrid composites reinforced with long glass fibres and particulate filler, Comp. Sci. Tech. 65 (2005) 257-267.
3. J. Motezuki, Influência da Nanoargila no Envelhecimento Higrotérmico de Compósitos de Polipropileno / PP Maleado / Fibra de Vidro, Master's thesis, Universidade Federal de São Carlos, São Carlos, Brazil, 2009.
4. J. Motezuki, M. H. M. Dias; M. C. Branciforti and J. A. Sousa, Hygrothermal Ageing of Polypropylene/Glass Fiber Composites: Influence of Montmorillonite and Maleated PP Compatibilizer Content, Proceedings of the Polymer Processing Society, 26th Annual Meeting, Banff, Canada, 2010.

Improvement of the Mechanical and Electrical Properties of Technical Thermoplastics by Utilizing 2nd Generation Layered Silicates

Volker Altstädt[a*], Josef Breu[b]

[a] *Department of Polymer Engineering, University of Bayreuth, Universitätsstraße 30, 95447 Bayreuth, Germany*
[b] *Department of Inorganic Chemistry I, University of Bayreuth, Germany*

In this study a novel kind of reinforcing polymethylmethacrylate (PMMA) and improving the electrical properties of polyetherimide (PEI) with a 2nd generation of layered silicates is shown. Layered silicates are used as nanofillers in polymers due to their ability to increase the mechanical strength as well as to improve flame retardancy [1] and barrier properties [2], respectively. The first studies about polymeric nanocomposites with layered silicates were published in the mid eighties of the last century and lead to the development of a nylon-6-montmorillonite nanocomposite which has been the first layered silicate nanocomposite to be commercialized [3]. Since these first studies about polymer/clay nanocomposites [4], a rapid development has brought further improvement of the overall properties of these materials. However, the potential of commercially available natural layered silicates seems to have reached their limitations due to small lateral dimensions and a high heterogeneity of surface charge. Processing via melt compounding results mostly in incomplete delamination of the tactoids, which further reduces the maximal possible aspect ratio and therefore the desired properties. Also the incorporation of commercially available natural organo-clay in PMMA by melt-compounding leads to an increase of the stiffness. However, an unsatisfactory dispersion quality of the nanoclay in the PMMA matrix leads to an decrease in toughness. Therefore a newly developed kind of synthetic layered silicate was used in a transfer batch moulding process to create a PMMA-nanocomposite. With these synthetic layered silicates, which have aspect ratios of up to 600, it was possible to significantly increase the young's modulus about 55% and the fracture toughness about 70 %, without any decrease in tensile strength. Furthermore analysis of the corresponding fracture surfaces by scanning electron microscopy show in case of the novel filler an additional energy dissipating mechanisms like crack deflection, crack bridging as well as debonding effects with pull-out of the platelets leading to enhanced fracture toughness.
In addition to the improvement of the mechanical behavior, the layered silicates possess the ability to decrease the coefficient of thermal expansion (CTE) of the matrix material [5]. Therefore layered silicates provide the possibility of utilizing thermoplastic materials for applications which require a lower CTE. In electric devices e.g. substrate material for printed circuit boards has to have a CTE in the range of those of the copper (around 17 ppm/K) to avoid thermal stresses between materials. Therefore current studies are evaluating the effect of layered silicates on thermal and electrical properties of PEI.

References:
[1] M. Schütz, et.al.; Polymer; 2011; 52; pp. 3288-3294
[2] D. Kunz; et.al.; ACS Nano; 2013; 7; pp. 4275-4280
[3] Q. Nguyen, D. Baird; Advances in Polymer Technology; 2006; 25; pp. 270-285.
[4] A. Usuki, et.al.; Journal of Materials Research; 1993; 8; pp. 1179-1184.
[5] YQ. Rao, T. Blanton; Macromolecules; 2008; 41; pp. 935-941

Keywords: PMMA, PEI, nanocomposite, clay, PMMA/clay nanocomposites, PEI/clay nanocomposites, large aspect ratio filler, mechanical properties, electrical properties

***Corresponding Author**: Prof. Dr.-Ing. Volker Altstädt, altstaedt@uni-bayreuth.de

Influence Of Graphite And SEBS Addition On Thermal And Electrical Conductivity And Mechanical Properties Of Polypropylene Composites

B. Krause[a], A. Cohnen[b], P. Pötschke[a*], T. Hickmann[c], D. Koppler[d], B. Proksch[d], T. Kersting[b] and Ch. Hopmann[b]

[a]*Leibniz-Institut für Polymerforschung Dresden e.V. (IPF), Hohe Straße 6, 01069 Dresden, Germany*
[b]*Institute of Plastics Processing (IKV), Seffenter Weg 201, 52074 Aachen, Germany*
[c]*Eisenhuth GmbH & Co. KG, Friedrich Ebert Straße 203, 37520 Osterode am Harz, Germany*
[d]*Allod Werkstoff GmbH & Co. KG, Steinacher Str. 3, 91593 Burgbernheim, Germany*

krause-beate@ipfdd.de
poe@ipfdd.de

Abstract. In this study, composites based on polypropylene (PP) and different graphite fillers were melt mixed using small scale microcompounder Xplore DSM15 as well as lab-scale co-rotating twin screw extruder Coperion ZSK26Mc. The measurements of the electrical and thermal conductivity as well as mechanical properties of the composites were performed on pressed plates. It was found that the addition of graphite powders having different particle size distributions leads to different increases of the thermal conductivity. For synthetic graphite, the PP composites filled with TIMCAL Timrex® KS500 reached the highest value of thermal conductivity of 0.52 W/(m·K) at 10 vol% loading, whereas this composite was not electrical conductive.
Furthermore, the influence of a styrene-ethylene-butylene-styrene block copolymer (SEBS) based impact modifier on the mechanical properties of PP filled with 80 wt% of different synthetic graphites was investigated. For that the proportion of SEBS in the PP component was varied systematically. The conductivities were influenced by the type of graphite and the content of impact modifier. The results indicate that the impact strength of the composite containing TIMCAL Timrex® KS300-1250 can be increased by approx. 100 % when replacing 50 wt% of the PP component by SEBS.

Keywords: Polymer composites, electrical conductivity, thermal conductivity, impact modification, fuel cells
PACS: 65.80.-g; 72.80.Tm; 72.80.Le; 88.30.mj

INTRODUCTION

In the discussion of anthropogenic climate change and efforts to replace fossil fuels with renewable energies, modern storage technologies such as fuel cells or redox-flow-batteries play an important role [1, 2]. Fuel cells and redox-flow-batteries offer a high potential to convert efficiently chemical energy into electrical energy. In these systems, the costs are mainly determined by the catalyst, the membrane electrolyte unit and the bipolar plate. To replace metallic bipolar plates, which tend to corrosion and are thus restricted in their longevity, composite bipolar plates based on plastics are used increasingly. Composite bipolar plates are made of a combination of a plastic (e.g. PP) and conductive filler (e.g. graphite) prepared using conventional processing methods like hot pressing or injection molding. Compared to pure graphite, composite materials offer the possibility to produce bipolar plates with a higher flexibility in a simplified manufacturing process.

The bipolar plates must have a high electrical and thermal conductivity orthogonal to the plate dimension. Therefore, the incorporation of electrical and thermal conductive fillers in plastics is of great interest. The durability of bipolar plates based on thermoplastic composites and their production are not yet satisfactory due to the needed high proportion of the conductive filler and the resulting material embrittlement. Therefore, there is a need to reduce the proportion of the conductive filler on a high level (e.g. single or hybrid filler systems of carbon based materials) and to improve the mechanical properties by blending the plastic matrix with suitable impact modifiers.

EXPERIMENTAL PART

Materials

A polypropylene PP Sabic 579S (Sabic Deutschland GmbH, Düsseldorf, Germany) with a melt flow rate of 47 g/10 min (230 °C/ 2.16 kg) was used. A mixture of PP with a styrene-ethylene-butylene-styrene block copolymer (SEBS) based impact modifier was supplied by ALLOD Werkstoff GmbH & Co. KG, Burgbernheim, Germany.

As filler materials different kinds of graphite powders having different particle size distributions were chosen, namely synthetic graphite EP1005 and EP1200 (both from Richard Anton KG, Munich, Germany) and Timcal Timrex®KS150, KS500, KS300-1250 (all from Imerys Graphite & Carbon Switzerland Ltd., Bodio, Switzerland).

Processing and Analytics

Characterization of Graphite Materials

The particle size distribution of the graphite powders was determined by laser diffraction using a Sympatec Helos/BF particle size analyzer coupled with a RODOS dry dispersion unit and ASPIROS micro dose module. The measurement range is 4.5-875 µm. The morphology of graphite powders was observed with field emission scanning electron microscopy (FESEM) using a measuring device Sigma VP from Carl Zeiss AG (Oberkochen, Germany).

The determination of the specific surface area (BET) was done using Autosorb-1 (Quantachrome, USA) based on a volumetric principle at measuring temperature of 77 K (cooling in liquid nitrogen) and the adsorbate nitrogen.

Small-scale Melt Mixing

Melt compounding in small scale was performed using a conical twin screw microcompounder Xplore DSM15 at 210 °C, 150 rpm, and 5 min mixing time. Before processing, the graphite fillers were dried at 80 °C in a vacuum oven overnight. The fillers and the PP granules were pre-mixed in the solid state by shaking.

For the measurements of electrical resistivity and thermal conductivity the composite materials were compression molded in the required shape at 210 °C for 2 min using a hot press PW40EH (Otto-Paul-Weber GmbH, Germany). To perform electrical resistivity measurements on compression molded plates, a 4-point test fixture (gold contact wires with a distance of 16 mm between the source electrodes and 10 mm between the measuring electrodes) combined with a Keithley electrometer 6517A was applied on strips cut from pressed plates. For electrical resistivity values higher than 1 E+07 Ω cm, a Keithley 8009 Resistivity Test Fixture based on ring electrodes was used to measure the pressed circular plates with 60 mm diameter and 0.5 mm thickness.

Thermal conductivity was measured on pressed plates (diameter 12.5 mm, thickness 2 mm) using the light flash apparatus LFA 447 NanoFlash® (Netzsch-Gerätebau GmbH, Selb, Germany) at 25 °C.

Lab-scale Extrusion

The lab-scale extrusion of PP-SEBS blends was performed at ALLOD Werkstoff GmbH & Co. KG (Burgbernheim, Germany) using a co-rotating twin screw extruder ZE25 A of KraussMaffei Berstdorf GmbH (Hannover, Germany). The proportion of SEBS in PP was varied from 10 wt% to 50 wt%. The production of graphite-PP-SEBS composites was performed using a co-rotating twin screw extruder ZSK 26 Mc of Coperion GmbH (Stuttgart, Germany) with a screw diameter D of 26 mm and an extrusion length of 44 D at 200 °C, rotation speed of 150 rpm and throughput of 10 kg/h. The proportion of graphite was 80 wt% in all formulations.

The compounds were compression molded at 190-195 °C to plate geometries of 220 mm length, 200 mm width and a thickness of 4 mm in a heated press at Eisenhuth GmbH & Co. KG (Osterode am Harz, Germany).

To analyze the mechanical properties like impact strength, flexural strength and bending strain, corresponding sample shapes were prepared from these plates. The impact strength was measured in accordance with DIN ISO 179-1/1fU by Charpy impact test. The bending properties of the compounds were measured in a three-point bending test in accordance with DIN EN ISO 178 on a testing machine of type Zwick 1456 Zwick GmbH & Co. KG (Ulm, Germany).

The volume resistivity was measured on five circular samples having a diameter of 13 mm and a thickness of 3 mm at Eisenhuth GmbH & Co. KG (Osterode am Harz, Germany).

Thermal conductivity was measured as described above.

RESULTS

Small Scale Mixing of PP/Graphite Composites

The morphology of the different graphite materials is shown in Figure 1. The FESEM images illustrate that the graphites of types EP1005 and KS150 have a plate-like structure. The graphite particles of type EP1200, KS 500 and KS300-1250 show a spherical structure. EP1200 has a larger aspect ratio than KS500 and KS300-1250.

FIGURE 1. FESEM-analysis of the morphology of synthetic graphite materials: EP1005 (a), KS150 (b), EP1200 (c), KS500 (d), and KS300-1250 (e).

In Table 1 the particle size distribution and specific surface area of the different graphite types are shown. The smallest particles were measured for EP1005 having a x_{50}-value of 4.8 µm whereas also bigger particles were present indicated by a x_{90}-value of 315.8 µm. The highest particle size was found for KS300-1250 with a mean value of 327.4 µm. A broad particle size distribution was detected for KS150 having x-values between 9.9 and 178.7 µm. For EP1200 and KS300-1250 no very small particles were found illustrated by x_{10}-value of 69.9 or 79.3 µm, respectively. The BET value for EP1005 was with 17.4 m^2/g the highest, compared to the values of the other one around 1-3 m^2/g for EP1200, KS150, KS500, KS300-1250.

For screening of the different graphite fillers, PP composites filled up to 10 vol% filler content were melt mixed using the microcompounder. In Table 2 thermal conductivity and electrical volume resistivity of the PP composites are summarized. For the synthetic graphite, the highest thermal conductivity of 0.52 W/(m·K) was found for PP/10 vol% KS500. Among the PP/graphite composites only the PP/EP1005 composites starting at 5 vol% loading and the PP/10 vol% EP1200 composite showed lowered electrical resistivity values. It is assumed that the reason for the behaviour of PP/EP1005 is the significantly higher BET value when compared to the other synthetic graphites.

TABLE 1. Particle size distribution and specific surface area (BET) of different kinds of synthetic graphite

Filler type	Filler name	x_{10} [µm]	x_{50} [µm]	x_{90} [µm]	BET [m^2/g]
Synthetic graphite	EP1005	1.3	4.8	315.8	17.4
	EP1200	69.9	123.7	205.3	1.5
	KS150	9.9	54.3	178.7	3.5
	KS500	22.8	143.5	416.2	1.6
	KS300-1250	79.3	327.4	470.3	1.1

TABLE 2. Thermal conductivity and electrical volume resistivity of PP composites containing different fillers (the values for unfilled PP are 0.26 W/(m·K) and 1 E+17 Ω cm).

Filler	Filler concentration [vol%]	Thermal conductivity of composite [W/(m·K)]	Volume resistivity ρ of composite [Ω cm]
EP1005	2.5	0.27	5.7 E+14
	5	0.33	6.4 E+08
	7.5	0.42	7.3 E+08
	10	0.47	1.1 E+07
EP1200	2.5	0.26	1.1 E+17
	5	0.33	4.0 E+16
	7.5	0.38	9.9 E+14
	10	0.42	1.7 E+08
KS150	2.5	0.29	4.0 E+16
	5	0.34	4.0 E+16
	7.5	0.40	5.2 E+13
	10	0.51	1.1 E+14
KS500	2.5	0.31	4.0 E+16
	5	0.34	4.0 E+16
	7.5	0.47	4.0 E+16
	10	0.52	4.0 E+16
KS300-1250	2.5	0.36	4.0 E+16
	5	0.38	4.0 E+16
	7.5	0.44	4.0 E+16
	10	0.47	4.0 E+16

Lab-scale Preparation of PP/Graphite Composites Containing SEBS

For composite testing, the graphites EP1200, KS500 and KS300-1250 were chosen to investigate the influence of filler and SEBS on the mechanical and conductive properties. In Table 3 results are shown for the conductivity and the bending properties of the PP-SEBS composites containing 80 wt% graphite and 20 wt% PP/SEBS.

TABLE 3. PP-SEBS-composites containing different fillers and various proportions of SEBS in PP: Electrical volume resistivity and bending properties.

Filler	PP/SEBS [wt%]	ρ [Ω cm]	a [kJ/m²]	σ_{fl} [MPa]	E_{fl} [MPa]	ε_{fl} [%]
	100/0	0.23±0.01	-*	-*	-*	-*
	90/10	0.18±0.13	1.4±0.2	25.1±1,8	6101±285	0.85±0.15
EP1200	80/20	0.25±0.02	1.5±0.5	22.6±0.7	4505±37	0.89±0.02
	70/30	0.31±0.06	2.2±0.1	17.1±0.7	3649±163	1.46±0.39
	50/50	0.46±0.03	3.2±0.4	12.5±0.8	2090±75	1.97±0.31
	100/0	-*	-*	-*	-*	-*
	90/10	0.28±0.01	1.7±0.1	21.4±0.6	5225±108	0.76±0.05
KS500	80/20	0.39±0.03	1.9±0.2	16.3±0.8	3770±108	1.04±0.15
	70/30	0.34±0.02	2.3±0.3	11.7±2.0	2229±676	1.54±0.39
	50/50	0.41±0.04	2.7±0.2	8.6±0.2	1240±36	2.06±0.06
	100/0	0.29±0.04	1.5±0.2	23.5±0.5	5726±188	0.64±0.05
	90/10	0.36±0.05	1.6±0.6	18.0±1.5	4102±83	0.90±0.11
KS300-1250	80/20	0.40±0.01	2.0±0.2	16.9±0.9	3790±192	1.01±0.04
	70/30	0.43±0.07	2.2±0.2	11.5±1.2	2210±173	0.98±0.18
	50/50	0.63±0.11	2.9±0.2	8.8±1.1	1336±187	2.33±0.35

ρ – electrical volume resistivity; **a** – charpy impact strength; $\boldsymbol{\sigma_{fl}}$ – flexural strength; $\mathbf{E_{fl}}$ – flexural modulus; ε_{fl} – bending strain; *) composite was not processable due to the high brittleness

The results in Table 3 show that with increasing amount of SEBS in PP, the impact strength of highly filled composites can be improved. The addition of 50 wt% SEBS in PP leads to a maximum value of the impact strength of 3.2 kJ/m^2 using EP1200. This represents an increase in the impact strength by 128 % compared to the PP/SEBS (90/10) + EP1200 composite. EP1200 has a larger aspect ratio compared to the spherical graphites KS500 and

KS 300-1250. For fillers with a high aspect ratio, the impact energy can be dissipated better in the composite [3]. By the incorporation of SEBS, the bending properties are significantly improved. Pressed plate geometries with a proportion of 30 wt% SEBS in PP already exhibit a significantly improved flexibility compared to plate geometries without SEBS. The flexural strength of the composite decreases as expected with increasing content of SEBS in PP. The incorporation of SEBS in PP leads to a decrease in the flexural strength to 50 % for EP1200, 40 % for KS500, and 49 % for KS300-1250 compared to the PP composite containing 10 wt% SEBS. For the graphite KS300-1250 the comparison between the PP without SEBS and with 50 wt% SEBS showed a decrease to 37 %. The addition of 50 wt% SEBS leads to a minimum flexural strength of 8.6 MPa using KS500. The incorporation of SEBS in PP has a significant effect on the bending strain. The addition of 50 wt% SEBS leads to an increase of the bending strain by 132 % for EP1200 and 171 % with KS500 in comparison to composites with 10 wt% SEBS. The maximum bending strain of 2.33 % is achieved when graphite KS300-1250 was used.

However, the improvement in mechanical properties by the incorporation of impact modifier SEBS leads to an increase of the volume resistivity. It can be shown that the incorporation of 50 wt% SEBS in PP/graphite composites leads to a significant increase of the volume resistivity by approximately 100 % from 0.29 Ω cm to 0.63 Ω cm using KS300-1250 or from 0.23 Ω cm to 0.46 Ω cm using EP1200. The lowest value of the volume resistivity for composites with 50 wt% SEBS in PP was achieved by using KS500 at 0.41 Ω cm.

The thermal conductivity of PP/EP1200 composites with different contents of SEBS was determined around 13-14 W/(m·K) and was independent on the addition of SEBS to the composite. Quite higher values were measured for PP/SEBS composites containing graphite of KS type between 14 and 16 W/(m·K). However, beside the differences of the values no clear correlation between thermal conductivity and SEBS content in PP could be observed.

SUMMARY AND CONCLUSION

In small-scale microcompounder experiments, PP/graphite composites were melt mixed to screen different kinds of synthetic graphite having different particle size distribution and shape. It was found that the addition of KS500 leads to the highest thermal conductivity of 0.52 W/(m·K) at 10 vol% filler. Only for PP composites filled with up to 10 vol% EP1005 or EP1200 a decrease of volume resistivity were measured.

Composite materials based on a PP and SEBS containing 80 wt% graphite were prepared in laboratory scale in a compounding process. Graphite having different particle size distributions was used and the influence of graphite and SEBS on the composite properties was examined. By adding SEBS in PP, the impact strength and flexural properties could be significantly increased. For PP/SEBS (50/50) filled with 80 wt% graphite EP1200 a maximum impact strength of 3.2 kJ/m^2 was obtained. The highest bending strain with 2.33 % was achieved using graphite KS300-1250 in PP/SEBS (50/50) composite. However, the incorporation of SEBS into PP results in an increase of the volume resistivity. The thermal conductivity of PP/ 80 wt% graphite composites was independent on the content of SEBS and was found around 13-16 W/m·K. The development of composites with impact modification represents a compromise between electrical conductivity and mechanics. In order to improve the electrical conductivity other conductive fillers have to be investigated.

ACKNOWLEDGMENTS

The authors thank German Federal Ministry of Education and Research (BMBF) for financial support (project No. 01LY1307B). The authors thank Mr. Geißler and Mrs. Albrecht for the determination of particle size and BET surface (both IPF).

REFERENCES

1. P. Kurzweil, *Brennstoffzellentechnik: Grundlagen, Komponenten, Systeme, Anwendungen*, Wiesbaden: Springer Vieweg Verlag, 2013
2. M. Skyllas-Kazacos, M. H. Chakrabarti, S. A. Hajimolana, F. S. Mjalli and M. Saleem, *Journal of the Electrochemical Society* **158**, R55-R79 (2011).
3. G. Wypych, *Handbook of Fillers*, Toronto: ChemTec Publishing, 2010.

Microcapsules of Poly(methyl metharcylate) Containing Epoxy Resin and Multi-Walled Carbon Nanotubes

M. Galip Icduygu[a] Meltem Asilturk[b] and M. Cengiz Altan[c*]

[a] *School of Civil Aviation, Giresun University, Giresun TURKEY*
[b] *Material Science and Engineering Department, Akdeniz University, Antalya, TURKEY*
[c] *School of Aerospace and Mechanical Engineering, University of Oklahoma, Norman, OK, USA*

mehmet.icduygu@giresun.edu.tr
altan@ou.edu

Microcapsules of poly(methyl methacrylate) [PMMA] containing epoxy and multi-walled carbon nanotubes were formed by solution evaporation in an aqueous emulsion. Thin PMMA shells precipitated at the resin/water interface as the solvent was removed, thus encapsulating the dispersion of epoxy and nanotubes. Dichloromethane (DCM) was selected as the common solvent for both resin and PMMA, and slowly evaporated from the aqueous mixture agitated at 40°C. After trying various epoxy/PMMA ratios, a 1:1 w/w epoxy/PMMA ratio yielded stable, individual microcapsules with smooth PMMA shells. Four different mixing speeds (300, 500, 800, and 1000 rpm) and four carbon nanotube contents (0.25, 0.50, 1.00, 2.00 w/w %) were used during the encapsulation process. Microcapsules with 10 to 55 μm average diameter were obtained by increasing the mixing speed from 300 to 1000 rpm. Increasing the mixing speed yielded a decrease in average particle size, whereas carbon nanotube presence led to a broader capsule size distribution at any given mixing speed. In addition, roughness of the microcapsule surface increased with increasing carbon nanotube content and mixing speed. SEM images of microcapsules showed that microcapsules with various types of core-shell morphology were formed during the encapsulation process.

Keywords: Self-healing composites, PMMA microcapsules, carbon nanotube, epoxy
PACS: 81

INTRODUCTION

Micro encapsulation is a well-known technique in which solids, liquids, or gases can be enclosed in a thin-shelled, spherical capsule having a diameter between a few micrometers to millimeters. These microcapsules have been used in various industrial applications such as pharmaceuticals, adhesives, and cosmetics. More recently, microcapsules containing thermosetting resins have been successfully used in fabricating self-healing composites. Typically, externally induced damage in a composite is "healed" by mixing and curing the resin and hardener, released from the fractured microcapsules embedded in the composite. Formation of microcapsules by a wide variety of polymers and nanomaterials as shell and/or core materials have been reported in recent studies [1-5]. In two-component self-healing systems, such as thermoset (resin/hardener) or thermoplastic (monomer/initiator) systems, both components may be placed in the same capsules or in separate capsules of appropriate sizes. Nanomaterials such as carbon nanotubes have also been incorporated into microcapsules as an additional component. For example, Caruso et al. [6] synthesized microcapsules containing carbon nanotubes (CNTs) using an *in situ* emulsification polymerization of urea–formaldehyde for damage sensing. In addition, autonomic restoration of electrical conductivity using polymer-stabilized carbon nanotubes and graphene microcapsules has been studied by Odom et al. [7].

Solvent evaporation has been extensively used in encapsulation of self-healing materials. Epoxy and diamines are two example materials that can be encapsulated by using the solvent evaporation approach. In this approach, self-healing materials and a shell-forming polymer are dissolved in an organic solvent and dispersed in an aqueous phase. A hard shell is formed, encapsulating epoxy or hardener micro droplets after evaporation of the solvent. When used with epoxy and dissolved in dichloromethane (DCM), poly(methyl methacrylate) (PMMA) forms smooth shells around the epoxy droplets after DCM is removed from the solution [8-9].

The encapsulation of epoxy/carbon nanotube dispersions by PMMA through solvent evaporation technique has not been reported in the literature. The purpose of this study is to investigate the possible use of solvent evaporation technique to prepare microcapsules containing epoxy/carbon nanotube dispersion with PMMA shells that are suitable for self-healing composites.

MATERIALS AND METHODS

EPON 828 was used as the low-viscosity epoxy resin (Miller Stephenson Chemical Co.). Dichloromethane (DCM), poly(methyl methacrylate) [Mw~120000], sodium dodecyl sulfate (SDS) and multi-walled carbon nanotube (MWNT) [>90% carbon, D × L: 110-170 nm × 5-9 μm] (Sigma Aldrich) were used without any further processing or purification.

Preparation of Microcapsules

Four different mixing speeds (300, 500, 800 and 1000 rpm) and four multi-walled carbon nanotube (MWNT) contents (0.25, 0.50, 1.00, and 2.00 % wt. fiber/resin) were used to prepare the microcapsules. MWNT, EPON 828 and PMMA weighed in appropriate ratios and mixed in 60 ml DCM on a magnetic stirrer until EPON 828 and PMMA completely dissolved in DCM. This mixture was sonicated by using a Vibra-Cell high intensity ultrasonic processor for 10 minutes to enhance MWNT dispersion. Forty ml of this mixture was added to a 50-ml 1% w/V aqueous SDS solution drop-by-drop over a 90-minute period to get the organic phase/water emulsion at the desired mixing rate. After the emulsion was stabilized, the mixture was added to a 200-ml 1% w/V SDS solution and the temperature of the mixture was raised to 40°C. After 4 hours, the mixture was cooled to room temperature and filtered by using Whatman Grade 93 filter paper. Filtered microcapsules were washed with excess distilled water and dried at room temperature under the fume hood.

Characterization of Microcapsules

Chemical analyses of microcapsules were carried out by a Bruker IFS 66/S Fourier Transform Infrared Instrument. A Zeiss NEON 40 EsB Scanning Electron Microscope (SEM) was used to characterize both the surface and shell morphology of the microcapsules. Test samples were prepared on a glass slide and coated by using Hummer VI Triode Sputter Coater.

RESULTS AND DISCUSSION

FTIR Characterization

Figure 1 shows the Fourier Transform Infrared (FTIR) spectrum of the microcapsules prepared with different nanotube contents and mixing rates.

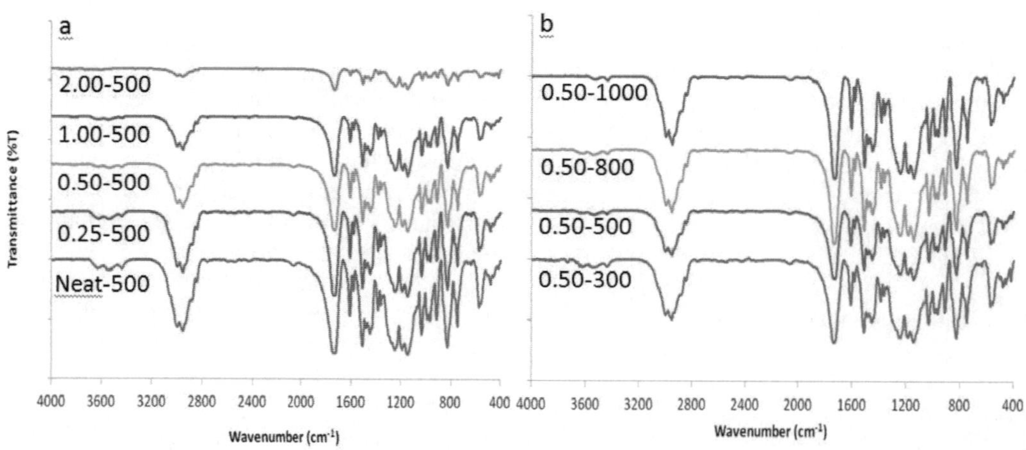

FIGURE 1. The FTIR spectrum of microcapsule prepared at (a) various nanotube/resin weight ratios at 500 rpm and (b) various mixing rates at 0.50 % wt. of MWNT/EPON 828.

The FTIR spectra of the microcapsules show characteristic PMMA absorption bands between 1,150 and 1,250 cm^{-1}, which can be attributed to the C-O-C stretching vibrations. The bands at 1,388 cm^{-1} and 754 cm^{-1} correspond to the α-methyl group vibrations. EPON 828 shows absorbing bands in the O-H stretching region (3200-3650 cm^{-1}), and the aliphatic C-H stretching bands are observed at 2860-3060 cm^{-1}. The C=C phenyl ring stretching bands appear at 1620 and 1462 cm^{-1} and the symmetric stretching band of epoxy ring can be seen at 1250 cm^{-1}. Hence, presence of EPON 828 and PMMA is evident in all microcapsule samples.

Particle Size Distribution

Particle size distribution of the microcapsules was determined by analyzing of SEM images. Multiple images taken at 50X magnification were utilized to determine the particle size distribution for each sample. Table 1 shows the mean and standard deviation values of the microcapsule diameter. As shown in Table 1, the presence of MWNT did not have a significant effect on the capsule size distribution for microcapsules prepared at 500 rpm. These microcapsules had average capsule diameters ranging between 27 and 31 μm for most cases. The mean particle size was slightly reduced when nanotube content increased to 2 % wt. The most apparent difference in the capsule size resulted from changing the mixing speed. Both the mean and the standard deviation of capsule diameter considerably decreased with increasing mixing speed.

TABLE 1. Particle size distribution of microcapsules.

Sample Codes MWNT (% wt.)-rpm	Mean Capsule Diameter (μm)	Standard Deviation (μm)
Neat-500	29.37	11.09
0.25-500	27.32	10.06
0.50-500	31.34	19.47
1.00-500	29.06	13.77
2.00-500	23.11	11.87
0.5-300	55.23	22.17
0.5-500	31.34	19.47
0.5-800	17.65	11.95
0.5-1000	10.34	9.30
Neat-1000	12.99	6.90

Microcapsule Morphology

The SEM images, depicted in Fig. 2, revealed that the microcapsules with multi-walled carbon nanotubes had a rougher surface morphology than the neat capsules. As shown in Figs. 2b and 2d, roughness was mainly caused by the presence of nanotubes, which were clearly visible at the capsule surface even at the low nanotube weight fraction of 0.5 %, covered by a thin layer of PMMA. Nanotubes appeared either as isolated, individual tubes, or as clusters. These clusters were formed by several entangled and nearly collimated nanotubes. Many of the nanotubes seemed to conform to the surface curvature of the capsule and were somewhat aligned near each other, forming a discernable but local preferred orientation. In other words, the location and alignment of the nanotubes on the surface appeared to be affected by the presence of other nanotubes within their proximity, even though they were not necessarily in physical contact or part of a larger cluster. Occasionally, nanotubes were not straight and bend with a very small radius of curvature. It is highly unlikely that these nanotubes were bent during the formation of the microcapsule, since one does not expect sufficiently high compressive stresses in the tangential direction to buckle the nanotube on the capsule surface. On the other hand, it is possible that these tubes were bent when they were initially synthesized, and retained their shape during encapsulation. It is worth noting that the nanotube clusters do not exhibit a spatially uniform and observable preferred alignment direction on the capsule surface.

Another prominent morphological feature of the surface of these capsules was several impact craters of various sizes. These craters may have been formed by the collision of microcapsules as they were being agitated at high speeds in the aqueous suspension. These collisions could have been highly plastic in nature, and thus, might have led to the formation of larger aggregates of multiple capsules adhering to each other after the collision, especially if the capsule surfaces were sufficiently soft. The mixing speed and capsule size distribution were likely to affect the formation of these aggregates as well as the adhesion of the capsules after collision. The microcapsule aggregates could be separated either after the shells were hardened or during the washing/drying steps, thus yielding a surface morphology observed in Fig. 2. This effect was especially visible at the 1000 rpm mixing rate due to the increase in both the number and strength of the collisions between the microcapsules during the encapsulation process.

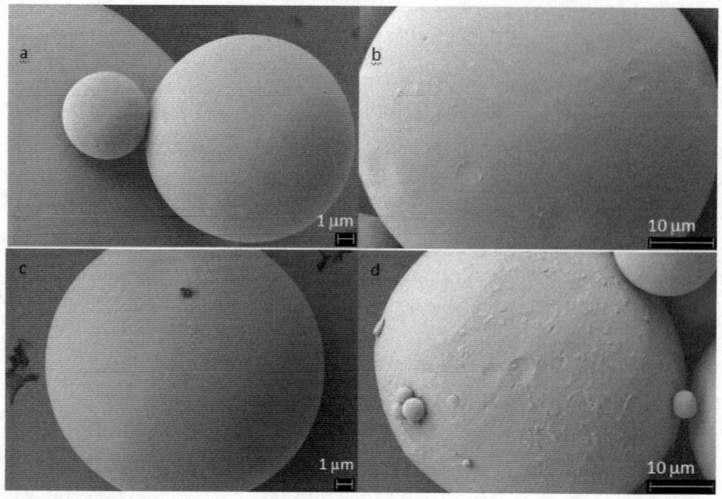

FIGURE 2. Surface morphology of microcapsules with and without MWNT:
(a) Neat PMMA at 500 rpm, (b) 0.5 % MWNT at 500 rpm, (c) Neat PMMA at 1000 rpm, and (d) 0.5 % MWNT at 1000 rpm.

When mixing speed was kept constant and the MWNT content was increased, the surface of the microcapsules started to contain more nanotubes as shown in Fig. 3. The size of the nanotube clusters increased, and they became more isotropic (Fig. 3c) as untangling and aligning a larger number of nanotubes on the capsule surface became more difficult. Increasing the nanotube content in the PMMA shell would enhance the shell strength and stiffness, but also decrease the flexibility of the capsules. Hence, at higher nanotube loadings, brittle fractures were observed on the capsules as shown in Fig. 3d. Collisions between stiffer capsules seemed to induce these eggshell-type fractures instead of leaving an impact crater on the surface as observed for capsules at lower nanotube content.

Some level of variation in shell morphology was observed even within capsules from the same sample as shown in Fig. 4. A particular sample set may exhibit a non-uniform shell thickness. In fact, in many cases, multiple sub-micron pin holes were observed on the interior surface of the PMMA as shown in Fig. 4a. Spherical particles with a homogenous, solid core were also formed during the encapsulation process (Fig. 4d). The interior shell structure showed variations from sample to sample and was either solid or contained entrapped voids across the shell wall. The presence of carbon nanotubes embedded through the shell thickness can also be seen in Fig. 4a.

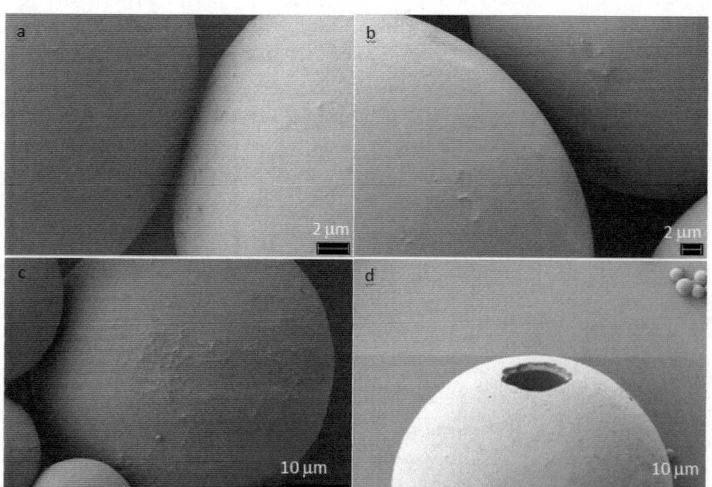

FIGURE 3. Surface morphology of microcapsules formed at 500 rpm mixing speed showing the effect of increased MWNT content: (a) 0.25 % wt., (b) 0.5 % wt., (c) 1.0 % wt., and (d) 2.0 % wt.

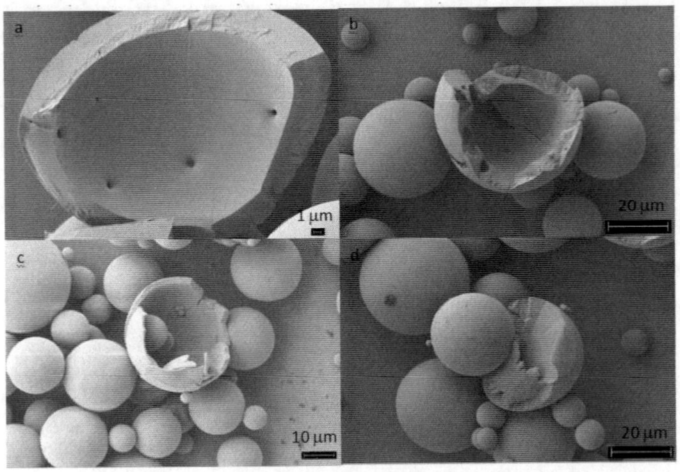

FIGURE 4. Shell and interior morphology of microcapsules: (a) 0.50 % wt. MWNT at 500 rpm, (b) Neat microcapsules at 500 rpm, (c) Neat microcapsule at 1000 rpm, and (d) Neat microcapsules at 500 rpm.

CONCLUSION

Microcapsules containing epoxy/carbon nanotube dispersion were successfully synthesized by using PMMA as a shell material. Multi-walled carbon nanotube content up to 2 % wt. was dispersed in an epoxy resin/PMMA solution dissolved in dichloromethane (DCM). Epoxy/carbon nanotube dispersion was subsequently encapsulated by PMMA via the evaporation of the solvent in an aqueous solution. It was possible to reduce the average diameter of microcapsules from 55 to 13 µm by increasing the mixing speed from 300 to 1000 rpm. Microcapsules exhibited sub-micron surface roughness due to the presence of nanotubes covered by a thin layer of PMMA at the surface. At the surface, both a number of single, isolated nanotubes and clusters comprised of several nanotubes were observed. In addition, impact craters of various sizes were often seen on the capsule surface due to collisions among microcapsules that were facilitated by agitation of the aqueous suspension. Increasing the carbon nanotube content created a more rigid shell that exhibited brittle fracture. The ability to tailor the shell stiffness and strength by changing the MWNT content could be especially beneficial in self-healing applications of composite laminates.

ACKNOWLEDGMENTS

The authors would like to thank Dr. Preston Larson from the Samuel Noble Microscopy Laboratory at the University of Oklahoma for SEM imaging of the microcapsules.

REFERENCES

1. H. Jin, C. L. Mangun, D. S. Stradley, S.M. Moore, N. R. Sottos, S. R. White, *Polymer* **53**, 581-587 (2012).
2. B. J. Blaiszik, M. M. Caruso, D. A. McIlroy, J.S. Moore, S. R. White, N. R. Sottos, *Polymer* **50**, 990-997 (2009).
3. C. Y. Zhang, X. B. Jiang, M. Z. Rong, M.Q. Zhang, *Journal Of Intelligent Material Systems and Structures* **25**, 31-39 (2014).
4. H. Wang, Y. Yuan, M. Rong, M. Zhang, *Colloid and Polymer Science* **287**, 1089-1097 (2009).
5. G. O. Wilson, J. W. Henderson, M. M. Caruso, B. J. Blaiszik, P.J. McIntire, N. R. Sottos, S. R. White, J. S. Moore, *Journal of Polymer Science Part A-Polymer Chemistry* **48**, 2698-2708 (2010).
6. M. M. Caruso, S. R. Schelkopf, A. C. Jackson, A. M. Landry, P.V. Braun, J. S. Moore, *Journal of Material Chemistry* **19**, 6093-6096 (2009).
7. S. A. Odom, T. P. Tyler, M. M. Caruso, J.A. Ritchey, M.V. Schulmerich, S. J. Robinson, R. Bhargava, N. R. Sottos, S. R. White, M. C. Hersam, J. S. Moore, *Applied Physics Letters* **101**, 043106 (2012) .
8. Q. Li, K. Siddaramaiah, H. K. Nam, D. Hui, J. H. Lee, *Composites Part B- Engineering* **55**, 79-85 (2013).
9. Q. Li, A. K. Mishra, N. H. Kim, T. Kuila, K. Lau, J. H. Lee, *Composites Part B-Engineering* **49**, 6-15 (2013).

AC Electrical Conductivity, Shielding Effectiveness and Viscoelastic Characteristics of Nanocomposites based on RTV Silicon Rubber and Nano Graphite Sheets/Carbon Black Hybrid System

Ali AsgharKatbab[a] ,JavadJeddi[a*]

[a]Department of polymer engineering, Amirkabir university of technology, Tehran, Iran

katbab@aut.ac.ir

javad.jeddi@aut.ac.ir

Abstract:

Highly flexible nanocomposites based on room temperature vulcanizing (RTV) silicon rubber, and conductive nano system comprised of high structure carbon black (CB) and graphite nano plates (GNP) were fabricated by solution processing method. The synergistic characteristic of this nano system uponthe involved electromagnetic interference shielding mechanisms of prepared composites was investigated. Electrical properties and EMI shielding efficiency of nanocomposites generated by various ratios of CB/GNP were measured and compared with the binary nanocomposites based on SR/CB and SR/GNP. The composites prepared by SR/CB showed higher electrical conductivity and lower percolation threshold than composites based on SR/GNP. However, despite of higher aspect ratio of GNP, the electrical properties of the hybrid based composites showed to be governed by the major filler phase and synergistic characteristic was not exhibited by hybrid filled nanocomposites. AC electrical conductivity and dielectric characteristic of the all prepared nanocomposites were evaluated as a function of fillers ratio within frequency range of 10^0-10^5Hz. For all binary and ternary nanocomposites ac conductivity showed increasing by frequency below percolation threshold without plateau (nonohmic conduction) , whereas above percolation threshold ac conductivity showedplateau within low frequency region, but sudden increase within high frequency region indicating nonohmic conductivity [3]. Dielectric and electrical analysis suggested that, above percolation threshold direct contact plays as the main mechanism for electric chargecurrent throughout SR phase for all nanocomposites. EMI shielding effectiveness measurement results within X-band frequency revealed that absorption loss is dominant mechanism for the attenuation of the intensity of incident electromagnetic wave in all nanocomposites generated by SR/CB, SR/GN and SR/CB/GN nano systems.

Keywords: Carbon black, Nano graphite, hybrid nanocomposites, electrical conductivity, EMI shielding

Introduction

Electrically conductive polymer nanocomposites (CPNs) have received remarkable attentions from both academic and industrial point of views[1] due to their ease of processing, shaping and engineering their microstructure [2,3]. CPNs fabricated by the incorporation of a conductive filler system onto an insulating polymer matrix. At a critical volume fraction of conductive filler, electrical conductivity of polymer shows a transition from insulating to known as percolation threshold concentration(φ_c)[2].

Recently, composites based on hybridized filler systems comprising fillers with different dimensions and aspect ratio have great attention for the preparation of the electrically conductive polymer composites with low percolation threshold and potential for manufacturing of electromagnetic interference(EMI) shielding materials [6].

In the present work, nanocomposites based on a hybrid system comprising, conductive Carbon black (CB) and Graphite nanoplates (GNP), and silicon rubber matrix were prepared. Electrical conductivity of binary and ternary composites as a function of GNP/CB ratio were studied to understand conduction mechanisms and microstructure of composites to developed new ternary composites to use in EMI shielding.

Experimental

Sample preparation

In order to prepare graphite nanoplates, as received expanded graphite(EX095200 , Jiangchem) was dispersed in a mixture of the ethanol and deionized water (65/35, v/v) and then subjected to ultrasonic sonication . Mixing method was employed to fabricate composite samples, RTV silicon rubber (Siltec, Taiwan) and compatibilizer(BYK-9076, BYK-chemie, Germany) were added n-hexane solution of graphite nano particle and

x
oc
ıp
δ

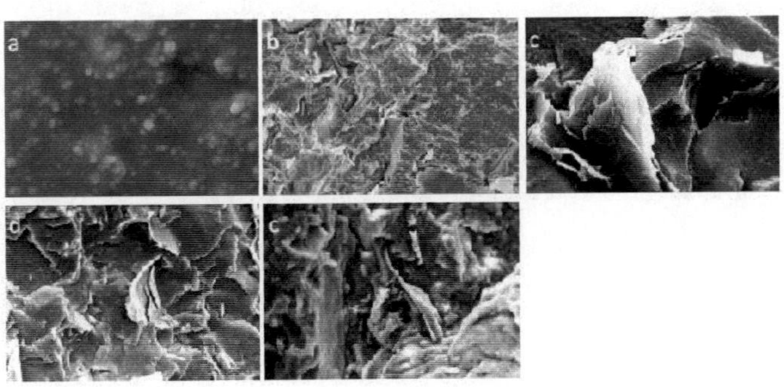

tł ısı :o
rє ɔr p:

$\sigma \quad \varphi \quad \varphi_c)^t$

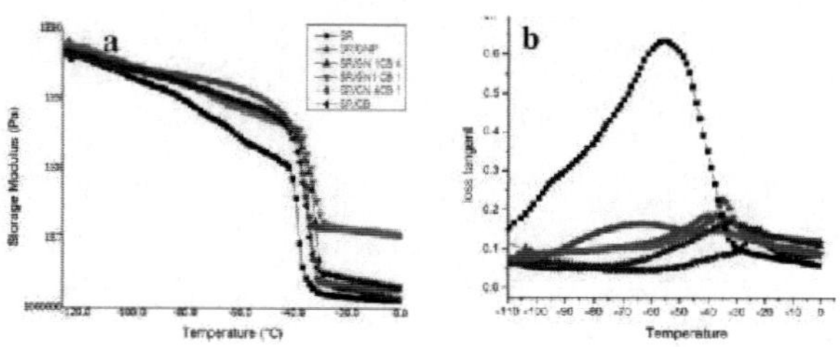

FIGURE 2. Dynamic-mechanical behavior of neat SR , and prepared nanocomposites (a) storage modulus, (b) tangent loss

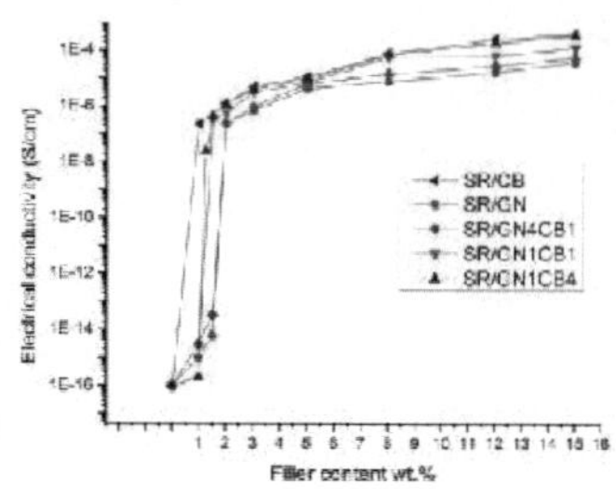

FIGURE 3. Electrical conductivity of prepared nanocomposites as function of filler content

composites:

$$\frac{m_{GNP}}{P_{SR/GN}} + \frac{m_{CB}}{P_{SR/CB}} = 1$$

A cı condu(
equatı ɔn as tollow [⁷

$$\sigma_{AC} = \sigma_{DC} + A\omega^s$$

ω

ɔc

ti

s snow

ɔad by

σ_{DC} ol

black

attenuate the incident EM waves t

$$SE_T = SE_R + SE_A + SE_M$$

R, A a M

es e l E

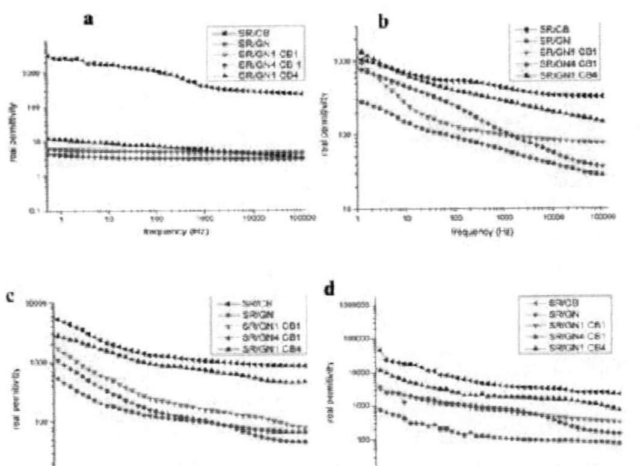

FIGURE 4. Real permitivitty of prepared nanocomposites(a) nanocomposites contained 1% filler (b) nanocomposites contained 3% filler (c) nanocomposites contained 5% filler (d) nanocomposites contained 8% filler

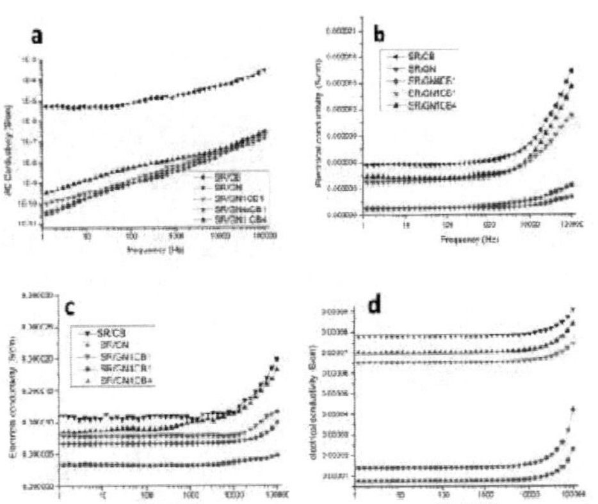

FIGURE 5. AC electrical conductivity ofprepared nanocomposites (a) nanocomposites contained 1% filler (b) nanocomposites contained 3% filler (c) nanocomposites contained 5% filler (d) nanocomposites contained 8% filler

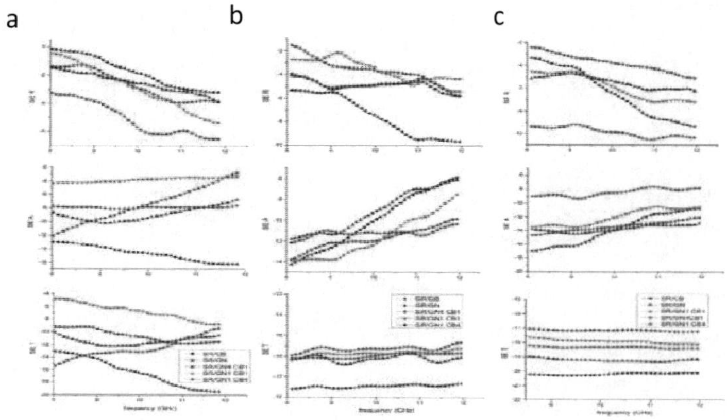

FIGURE 6.Effect fillers ratio on the reflection loss, absorption loss and total SE of (a) nanocomposites contained 3wt% fillers, (b) nanocomposites contained 5wt% fillers, (c) nanocomposites contained 8wt% fillers

Conclusion

In this study microstructure, viscoelastic, electrical behavior as well as EMI shielding efficiency of hybrid composites based on SR/GN/CB were investigated and compared to SR/GNP and SR/CB binary composites. FESEM micrograph of composites revealed that CB particle dispersed more uniformly through SR matrix but GNP particle dispersed in the form of stacked layers in large variety of particle sizes distribution. Micrograph of ternary hybrid composites showed that GNP and CB prefer to make their own network and a segregated microstructure obtained. Electrical conductivity investigation showed that SR/GN has higher percolation threshold than SR/CB, and ternary composites percolation thresholds are between SR/GN and SR/CB and no synergistic effect in electrical conductivity observed. In ternary composites SR/GN_1 CB_4 and SR/GN_4 CB_1 electrical characteristic of composites governed by major filler phase. EMI shielding investigation revealed that the composites with higher conductivity have higher SE T and dominant shielding mechanism is adsorption loss in all samples.

REFERENCES

[1].Zhong BL-W-H. Review on polymer/graphite nanoplatelet nanocomposites. J Mater Sci. 2011;46:5595–614.
[2].Balberg I. A comprehensive picture of the electrical phenomenain carbon black–polymer compositesCarbon. 2002;40:139–43.
[3].Zhang SM, Lin L, Deng H, Gao X, Bilotti E, Peijs T, et al. Synergistic effect in conductive networks constructed withcarbon nanofillers in different dimensions 6. 2012;6:159–68.

[4].Palza H, Garzón C, Arias O. Modifying the electrical behaviour of polypropylene/carbonnanotube composites by adding a second nanoparticle andby annealing processeseXPRESS Polymer Letters 2012;6,No.8 639-46.

[5].Ha-Da Bao Z-XG, Jian Yu. Effect of electrically inert particulate filler on electrical resistivity of polymer/multi-walled carbon nanotube compositesPolymer 2008;49 3826–31

[6]Maiti S, Shrivastava NK, Suin S, Khatua BB. Polystyrene/MWCNT/Graphite Nanoplate Nanocomposites: EfficientElectromagnetic Interference Shielding Material through GraphiteNanoplate–MWCNT–Graphite Nanoplate NetworkingACS Appl Mater Interfaces 2013;5:4712–24.
[7].Al-Saleh MH, Al-Anid HK, Husain YA, El-Ghanem HM, Jawad aSA. Impedance characteristics and conductivity of CNT/ABS nanocompositesJ Phys D: Appl Phys. 2013;46:385305.

Viscosity Amplification In Dilute Suspensions Of Non-Spherical Particles In A Non-Newtonian Matrix Fluid

Jan Domurath (presenting author, reference author)[a,b*], Marina Saphiannikova[a], Gilles Ausias[b], Julien Férec[b] and Gert Heinrich[a,c]

[a]*Leibniz-Institut für Polymerforschung Dresden e.V., Hohe Straße 6, 01069 Dresden, Germany*
[b]*LIMATB, Université de Bretagne-Sud, BP 92116, 56321 Lorient, France*
[b]*Technische Universität Dresden, Institut für Werkstoffwissenschaft, 01069 Dresden, Germany*

domurath@ipfdd.de

Abstract. When non-spherical particles, like rods or discs, are added to a fluid, the increase in viscosity of the mixture is often described by the Lipscomb model [1]. One important result of Lipscomb's model is that it predicts a strong increase in the viscosity of the composite with increasing aspect ratio of the filler particles. Despite the fact that this model was originally proposed for a Newtonian matrix fluid it is also applied to polymer melts filled with non-spherical particles. For example in [2,3] it is proposed to use the Lipscomb model with transient viscosity of the Giesekus viscoelastic model instead of a constant Newtonian viscosity as used in Lipscomb's derivation. This approach completely decouples the influence of the particle shape from the non-linear properties of the suspending fluid. Yet, since polymer melts often exhibit strong non-Newtonian behavior, e.g. shear thinning, it is to be expected that such a simple superposition will give a wrong prediction of the suspension viscosity.

To investigate this problem we performed a numerical study of a suspension based on a non-Newtonian matrix fluid and rigid spheroidal particles. In particular, we simulated an elongational flow of a Bird-Carreau fluid around prolate and oblate spheroidal particles and used numerical homogenization to obtain the effective viscosity of the suspension for different applied rates of deformation, different thinning exponents and different aspect ratios. To analyze the simulation results we compute the intrinsic viscosity as a function of applied rate of deformation, thinning exponent and aspect ratio. In the Newtonian regime we also compare with results from literature. In the transition region from Newtonian to non-Newtonian behavior we obtained lower values of the intrinsic viscosity. In the power-law regime of the Bird-Carreau model, i.e. at high deformation rates, we found that the intrinsic viscosity of the suspension is independent of the applied rate of deformation. Further we obtained from the simulations that the intrinsic viscosity at high deformation rates strongly depends not only on the aspect ratio of the particles but also one the thinning exponent in the Bird-Carreau model, implying that the superposition approach in fact leads to a wrong prediction of the suspension viscosity at high deformation rates.

Keywords: polymer melt, rigid filler, anisometric particles, suspension
PACS: 83.60.Fg, 83.80.Hj, 83.80.Sg

REFERENCES

1. G. Lipscomb et al. 'The flow of fiber suspensions in complex geometries'. In: Journal of Non-Newtonian Fluid Mechanics 26.3 (1988), pp. 297–325.
2. Y. Wang et al. 'Melt shear rheology of carbon nanofiber/polystyrene composites'. In: Rheologica Acta 45.6 (2006), pp. 919–941.
3. C. Kagarise et al. 'A unified model for polystyrene–nanorod and polystyrene–nanoplatelet melt composites'. In: Rheologica Acta 47.9 (2008), pp. 1061–1076.

Impact of Synthesis Temperature on Structure of Carbon Nanotubes and Morphological and Electrical Characterization of Their Polymeric Nanocomposites

Mohammad Arjmand[a*], Seyyed Alireza Mirkhani[a], Petra Pötschke[b], Beate Krause[b], and Uttandaraman Sundararaj[a*]

Department of Chemical and Petroleum Engineering, University of Calgary, Canada
Department of Functional Nanocomposites and Blends, Leibniz Institute of Polymer Research Dresden (IPF Dresden), Hohe Straße 6, D-01069 Dresden, Germany
arjmand64@yahoo.com
u.sundararaj@ucalgary.ca

Abstract. Carbon nanotubes (CNTs) were synthesized by chemical vapor deposition technique at a broad range of temperatures, i.e. 550ºC to 950ºC (at 100ºC intervals). CNTs were synthesized by flowing source and carrier gases (ethane, argon, and hydrogen) over Fe catalyst in a quartz tubular reactor. CNTs were melt mixed with a polyvinylidene fluoride (PVDF) matrix in a miniature mixer. The resulting nanocomposites were then compression molded, and electrically and morphologically characterized. Moreover, a wide range of characterization techniques were employed to obtain detailed information about the physical and morphological characteristics of CNTs. It was surprisingly observed that, despite the ascending trend of powder conductivity with the synthesis temperature, the nanocomposites made with $(CNT)_{650°C}$ had significantly lower percolation threshold (around 0.4wt.%) and higher electromagnetic interference shielding (20.3dB over the X-band for 1.1mm thickness) compared to the other temperatures. The characterization of nanofillers showed that the synthesis yield and quality of $(CNT)_{650°C}$ were highly superior to the other types of CNTs. At 850ºC and 950ºC, most of the synthesized carbonaceous materials formed graphitic structures around the sintered catalyst particles. It was also observed that the dispersion state of $(CNT)_{650°C}$ within the PVDF matrix was much better than that of CNTs synthesized at the other temperatures. Superior electrical properties of $(CNT)_{650°C}$ nanocomposites can be attributed to a combination of high synthesis yield, low diameter and decent quality of CNTs coupled with good state of dispersion within the PVDF matrix.

Keywords: Carbon Nanotube Synthesis, Chemical Vapor Deposition, Polyvinylidene Fluoride, Electrical Conductivity, Electromagnetic Interference Shielding.
PACS: 88.30.rh, 72.80.Le, 74.25.N-

INTRODUCTION

Electromagnetic interference (EMI) shielding is a protection provided to safeguard the performance of electrical and electronic equipment against EMI failure. Compulsory obedience of electromagnetic compatibility regulations is a major factor endorsing global demand for EMI solutions. Metals are currently the chief material for EMI shielding; however, compliance issues and environmental concerns over metals disposal coupled with their high density and processing difficulty are slowing their growth; this sets to make over the global market of conductive polymeric nanocomposites (CPNs). Different types of conductive nanofiller have been employed to develop CPNs, viz. carbonaceous nanofillers and metallic nanowires [1, 2], among which carbon nanotube (CNT) has appealed remarkable attention due to its large surface area and outstanding electrical and mechanical properties [3, 4]. CNTs are rolled layers of graphite coaxially arranged around a central hollow. Following the discovery of CNTs, several synthesis protocols were introduced for CNT production, among which CVD is the simplest and most economic approach to synthesize CNTs at low temperatures and ambient pressure [5]. The merits of this approach are ease of operation, low set-up cost, high production yield, and facility of scale-up.

Since the early observation of percolation-dominated electrical conductivity in a CPN by Coleman et al. [6], there have been many papers dedicated to processing and resulting electrical properties of fabricated CPNs. Nevertheless, to the best of our knowledge, there are a rare number of papers correlating the effect of CNT synthesis conditions with the physical characteristics of CNTs, quality of CNT-polymer interaction, and EMI shielding of CPNs [7, 8]. Accordingly, this study strives to cover the whole hierarchy of synthesis-processing-structure-property of CPNs. The first part of this article is devoted to a detailed characterization of CNTs synthesized at different temperatures using a CVD setup. After that, effects of synthesized carbonaceous fillers on the morphology and electrical properties of the developed CPNs, i.e. electrical conductivity and EMI shielding, are studied. Huge discrepancies between the physical properties of the generated nanofillers, and therefore, properties of the developed CPNs were observed. This provided an in-depth discernment on correlation between the physical properties of the nanofillers and final properties of the nanocomposites. Moreover, surprisingly, this study revealed an optimum synthesis temperature with electrical properties far superior to the other synthesis temperatures.

METHODOLOGY

CNTs were synthesized via chemical vapor deposition (CVD) method wherein source and carrier gases (ethane, argon, and hydrogen) were conveyed over alumina-supported iron catalyst in a quartz tubular reactor. The catalyst was prepared by incipient wetness impregnation of iron (III) nitrate nanohydrate dissolved in water on aluminum oxide support (Sasol Catalox Sba-200), followed by drying, calcination and reduction. CNTs were synthesized at a broad range of temperatures, i.e. 550ºC to 950ºC (at 100ºC intervals). Further information about the synthesis procedure is provided in our former study [9]. Synthesized CNTs were mixed with a PVDF matrix (11008/0001, 3M Canada) using a miniature melt mixer at 240°C. The samples were prepared at different CNT concentrations, i.e., 0.3, 0.5, 1.0, 2.0, 2.7 and 3.5wt%. A Carver compression molder (Carver Inc., Wabash, IN) was used to mold the samples at 220°C under 38MPa pressure for 10min.

The yield and thermal stability of synthesized CNTs were evaluated using a Thermogravimetric Analyzer (TGA), TA instruments Q500, under air atmosphere (Praxair AI INDK) from room temperature to 900ºC at a ramp rate of 10ºC/min. The electrical conductivity of CNT powders was measured using an especially designed instrument. The powder of 40–50mg was filled into a cylinder and compressed with a piston, and then the electrical conductivity of the powder was measured using a Keithley 2001 electrometer at various piston pressures, i.e. 5, 10, 20, 25, and 30MPa. Further details about this procedure are provided elsewhere [10]. In order to investigate the morphology of synthesized CNTs, a high-resolution Philips XL30 was used to obtain scanning electron microscopy (SEM) images. A transmission electron microscope (Tecnai TF20 G2 FEG-TEM (FEI, Hillsboro, Oregon, USA)) was used to take the images to measure the length and diameter of CNTs (100 individual CNTs) for each sample using the MeasureIT software (Olympus Soft Imaging Solutions GmbH).

The microdispersion state of CNTs within the PVDF matrix was quantified using light transmission microscopy (LM) on thin cuts (5µm thickness) of the compression-molded samples, prepared with a Leica microtome RM2265 (Leica Microsystems GmbH, Wetzlar, Germany). Electrical conductivity of conductive and insulative nanocomposites was measured using a Loresta GP resistivity meter (MCP-T610 model, Mitsubishi Chemical Co., Japan) and a Keithley 6517A electrometer connected to a Keithley 8009 test fixture (Keithley instruments, USA), respectively. EMI shielding measurement in the X-band frequency range (8.2-12.4GHz) was performed using an E5071C network analyzer (ENA series 300KHz - 20GHz). EMI shielding effectiveness (SE) is defined as the ratio of the incident power to the transmitted power and is expressed in dB.

RESULTS AND DISCUSSION

Figure 1 depicts TG, DTG and powder conductivity graphs of CNTs synthesized at various temperatures. These three curves provide invaluable information about the thermal stability, synthesis yield ((mass of nanofiller) / (mass of nanofiller and catalyst)) and quality of the synthesized nanofillers. The catalyst particles comprised 80wt% alumina and 20wt% metallic particles. Alumina is insulative and metallic particles have much lower surface area than synthesized CNTs, and their surface area reduced further due to sintering phenomenon at the synthesis temperatures. This explains why low yield can drastically deteriorate the electrical properties, and confirms the significance of synthesis yield on the final electrical properties. As shown in Figure 1(a) and tabulated in Table 1, $(CNT)_{550°C}$ showed by far the lowest synthesis yield about 43% and $(CNT)_{650°C}$ presented the highest synthesis yield around 85%. Carbonaceous fillers synthesized at the other three temperatures presented synthesis yields around 70%. In this regard, $(CNT)_{650°C}$ stands out in terms of the synthesis yield.

Figure 1(b) shows that $(CNT)_{550°C}$ depicts a much broader DTG curve than CNTs synthesized at the other temperatures. Moreover, the onset degradation point for $(CNT)_{550°C}$ is much lower than the other types of CNTs. This can be ascribed to low synthesis temperature, facilitating the formation of amorphous carbon [11]. $(CNT)_{650°C}$ and $(CNT)_{750°C}$ showed narrow DTG curves, denoting superior quality of CNTs. Nonetheless, it was surprisingly observed that $(CNT)_{850°C}$ and $(CNT)_{950°C}$ indicated a very different shape of DTG curve compared to the other three temperatures. The yellow box presents the portion of carbonaceous fillers synthesized at 850°C and 950°C made of CNTs. The left side of the yellow box coincides with the onset degradation of $(CNT)_{850°C}$ and $(CNT)_{950°C}$, which is far higher than the other synthesis temperatures. This shows that no amorphous carbon formed at these two synthesis temperatures, as confirmed by high inflection points defined in Table 1. The right side of the yellow box signifies the temperature in which CNTs were fully decomposed, matching the plateau in the TG curves of $(CNT)_{550°C}$, $(CNT)_{650°C}$, and $(CNT)_{750°C}$. This shows that the final material synthesized at 850°C and 950°C consisted of a thermal-resistant carbonaceous filler, CNT and catalyst. We believe that the thermal-resistant carbonaceous filler synthesized at high temperatures arose from sintering and coalescence of catalyst particles, leading to the coverage of large catalyst particles by the carbonaceous filler. Figure 1(c) shows the powder conductivity of the synthesized carbonaceous fillers as a function of piston pressure. It is vivid that the powder conductivity had an ascending trend with the synthesis temperature. Given the high conductivity of the materials synthesized at 850°C and 950°C, it can be asserted that carbonaceous fillers formed on the surface of large catalyst particles had graphitic sp^2 structure.

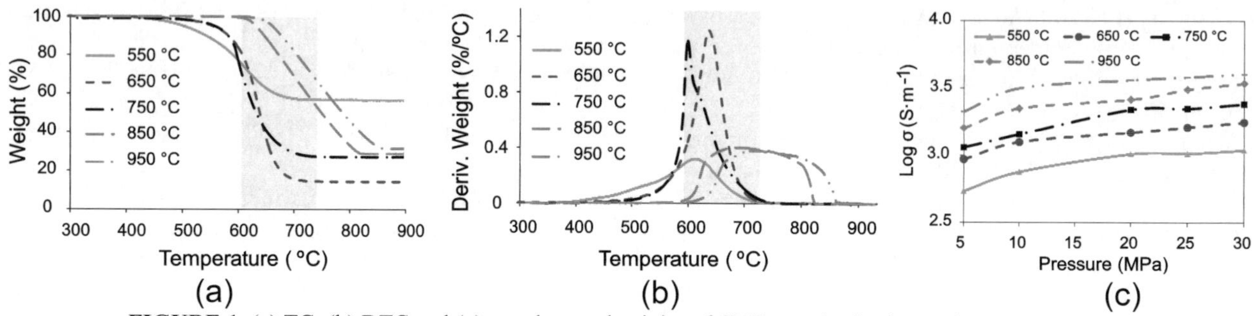

FIGURE 1. (a) TG, (b) DTG and (c) powder conductivity of CNTs synthesized at various temperatures.

Figure 2 demonstrates the SEM micrographs of CNT powders and LM micrographs of CNT/PVDF nanocomposites. Substantial differences are observable in the morphology of developed CNTs and nanocomposites depending on CNT synthesis temperature. As shown in Figure 2, $(CNT)_{550°C}$ micrograph shows a large bulk of catalyst particles covered by small-diameter CNTs. $(CNT)_{550°C}$ had also some amorphous carbon as validated by TGA analysis. The micrographs for $(CNT)_{650°C}$ and $(CNT)_{750°C}$ show the well formation of low-diameter CNTs over the catalyst particles, which are quite entangled, although the synthesis yield for 650°C was higher. For 850°C and 950°C, the diameter of observable CNTs was much higher than the lower synthesis temperatures, which is attributable to sintering and coalescence of catalyst particles at higher temperatures. Although, based on TGA analysis, we believe that some graphitic carbon covered the surface of large catalyst clusters. Measurement of the length and diameter of CNTs using the TEM images indicated that $(CNT)_{750°C}$ had the largest length, whereas $(CNT)_{650°C}$ presented the lowest diameter (largest surface area). Nonetheless, we could not find sufficient number of CNTs at 850°C and 950°C to obtain the average length and diameter.

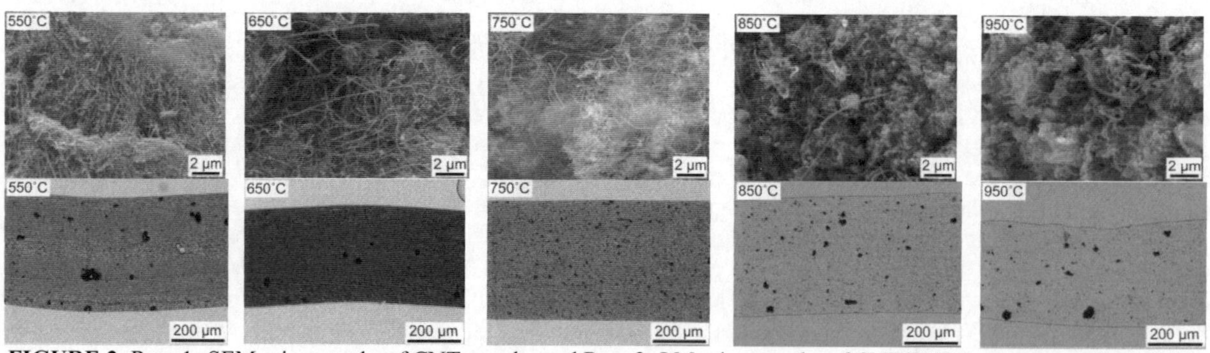

FIGURE 2. Row 1: SEM micrographs of CNT powder and Row 2: LM micrographs of CNT/PVDF nanocomposites. CNTs were synthesized at different temperatures.

TABLE 1. Physical characteristics of CNTs synthesized at different temperatures and their polymeric nanocomposites.

Synthesis Temperature	550°C	650°C	750°C	850°C	950°C
Inflection Point (°C)	612	636	599	696	711
Synthesis Yield %	43.1	85.4	72.7	71.1	68.7
Length (nm)	853	629	1696	-	-
Diameter (nm)	22	15	29	-	-
Agglomerate Area Ratio % (Nanocomposite)	3.2	0.8	4.7	2.7	1.9
Relative Transparency % (Nanocomposite)	45.0	23.3	46.2	86.9	90.6

Another nanofiller characteristic that controls the electrical properties of the resultant nanocomposites is its state of dispersion in the matrix. LM was used to assess the microdispersion state in which the nanotubes are not dispersed and appear as agglomerates. This non-dispersed part was quantified using the agglomerate area ratio. Moreover, in a smaller scale the agglomerates with sizes equal to or slightly larger than the wavelength of visible light, ca. 400-700nm, but smaller than visually recognizable agglomerates contribute to the grey appearance of the sample cut. A relative transparency ratio (transparency of sample over glass slide/Aquatex/cover glass assembly) gives a measure of the quantity of the nanotubes dispersed in this size range [7, 12]. As presented in Figure 2 and Table 1, nanocomposites with CNTs made at 650°C presented the lowest transparency, implying decent dispersion of carbon nanotubes, whereas the ones engaged with 850°C and 950°C had the highest transparency far beyond the other synthesis temperatures. Moreover, the agglomerate area ratio was the lowest and highest for 650°C and 750°C, respectively. These results clearly reveal that 650°C was the best temperature for the CNT dispersion, while 850°C and 950°C had the worst state of dispersion.

Electrical conductivity is of utmost significance for effective EMI shielding material [13]. Accordingly, prior to investigating the EMI shielding of the generated nanocomposites, their electrical conductivities were compared. Figure 3(a) shows the percolation curves of CNT/PVDF nanocomposites made at different synthesis temperatures. As vivid in Figure 3, the synthesis temperature had a tremendous impact on the electrical conductivity of the generated nanocomposites. Contrary to the ascending trend of the powder conductivity with the synthesis temperature, it was surprisingly observed that the nanocomposites made with $(CNT)_{850°C}$ and $(CNT)_{950°C}$ demonstrated an insulative behavior, spanning the whole concentration range. It was also observed that $(CNT)_{650°C}$ nanocomposites showed electrical conductivity and percolation threshold far superior to the other temperatures.

Employing the statistical percolation theory [14], $CNT_{650°C}$ nanocomposites presented a percolation threshold equal to 0.4wt%, whereas $(CNT)_{550°C}$ and $(CNT)_{750°C}$ nanocomposites presented percolation thresholds of 1.0wt% and 2.3wt%, respectively. The percolation threshold obtained for $(CNT)_{650°C}$ nanocomposites is comparable or lower than the results presented for CNT/PVDF nanocomposites in the literature [15, 16]. It was also observed that the maximum conductivity obtained for $(CNT)_{650°C}$ nanocomposites is much higher than the other temperatures. The maximum conductivity for $(CNT)_{550°C}$, $(CNT)_{650°C}$, and $(CNT)_{750°C}$ nanocomposites at 3.5wt% were 0.4, 31.2, and $0.1 S \cdot m^{-1}$, respectively. The maximum conductivities of the nanocomposites were fairly lower than the conductivities obtained for the compressed powders. This was due to the presence of polymer layers between CNTs in CPNs, highlighting the role of tunneling and hopping mechanisms besides the conduction mechanism [17].

Figure 3(b) depicts EMI shielding of the generated nanocomposites as a function of CNT content. It was interestingly observed that $(CNT)_{850°C}$ and $(CNT)_{950°C}$ nanocomposites, despite having very high powder conductivity, presented *SE* very close to pure PVDF. Among the other three synthesis temperatures, $(CNT)_{650°C}$ nanocomposites showed *SE* far away superior to the other synthesis temperatures. These results are consistent with the conductivity data and reveal the remarkable impact of synthesis temperature on EMI shielding of CPNs. *SE* of pure PVDF was 0.3dB, which increased to 8.9 and 20.2dB for $(CNT)_{650°C}$ nanocomposites with 1.0 and 3.5wt% loading, respectively. However, for instance, *SE* of $CNT_{550°C}$ nanocomposite was 1.0 and 4.2dB for 1.0 and 3.5wt% CNT content, respectively. It is worth noting that *SE* results obtained for $(CNT)_{650°C}$ nanocomposites were very close to the results acquired for commercial CNT (NC7000^{TM}, Nanocyl, Belgium) nanocomposites prepared with the same technique, endorsing the high quality of synthesized $(CNT)_{650°C}$. The insulative nature of $(CNT)_{850°C}$ and $(CNT)_{950°C}$ nanocomposites was due to insufficient formation of CNTs and poor state of filler dispersion within the PVDF matrix. As confirmed by TGA, powder conductivity and SEM, graphitic structures were formed around the sintered catalyst particulates at these two temperatures.

The observed discrepancies in the percolation curves of $(CNT)_{550^\circ C}$, $(CNT)_{650^\circ C}$, and $(CNT)_{750^\circ C}$ nanocomposites are attributable to several parameters, including length and diameter of CNTs, synthesis yield, crystallinity of CNTs, state of dispersion within the PVDF matrix, etc. It has been established well, both theoretically and experimentally, that well-dispersed CNTs with higher aspect ratio (higher length and lower diameter) demonstrate lower percolation threshold, and loftier electrical conductivity and EMI shielding [18, 19]. As a matter of fact, lower diameter signifies higher interacting surface area per mass of CNTs, and CNTs with higher length have more likelihood to be in close contiguity or contact each other. In addition, CPNs with better state of CNT dispersion present enhanced conductive network [20]. Our results show that decent state of dispersion, high synthesis yield, high quality, and low diameter of $(CNT)_{650^\circ C}$ were the contributing factors to its superior electrical behavior.

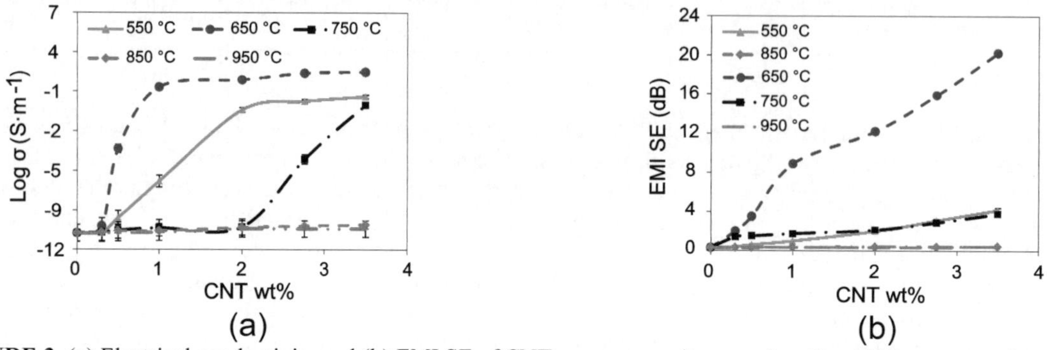

FIGURE 3. (a) Electrical conductivity and (b) EMI SE of CNT nanocomposites as a function CNT content and synthesis temperature.

ACKNOWLEDGMENTS

Financial support from the Natural Sciences and Engineering Research Council (NSERC) of Canada is highly appreciated. Dr. Mohammad Arjmand thanks IPF Dresden for granting a research stay.

REFERENCES

1. F. Qin and C. Brosseau, *J. Appl. Phys.* **111**, 061301 (2012).
2. A. B. da Silva, M. Arjmand, U. Sundararaj and R. E. S. Bretas, *Polymer* **55**, 226-234 (2014).
3. W. Bauhofer and J. Z. Kovacs, *Compos. Sci. Technol.* **69**, 1486-1498 (2009).
4. E. T. Thostenson, Z. F. Ren, and T. W. Chou, *Compos. Sci. Technol.* **61**, 1899-1912 (2001).
5. A. Magrez, W. S. Jin, R. Smajda, M. Mionić, and L. Forró, *Materials* **3**, 4871-4891 (2010).
6. J. N. Coleman, S. Curran, A. B. Dalton, A. P. Davey, B. McCarthy, and W. Blau, *Phys. Rev. B*, **58**, R7492-R7495 (1998).
7. M. Arjmand, K. Chizari, B. Krause, P. Pötschke, and U. Sundararaj, *Carbon* **98**, 358-372 (2016).
8. B. Krause, M. Ritschel, C. Täschner, S. Oswald, W. Gruner, and A. Leonhardt, *Compos. Sci. Technol.* **70**, 151-60 (2010).
9. M. Arjmand and U. Sundararaj, *ACS Appl. Mater. Inter.* **7**, 17844-17850 (2015).
10. B. Krause, R. Boldt, L. Häußler, and P. Pötschke, *Compos. Sci. Technol.* **114**, 119-125 (2015).
11. H. P. Li, N. Q. Zhao, C. N. He, C. S. Shi, X. W. Du, and H. J. Li, *Mater. Sci. Eng. A* **473**, 355-9 (2008).
12. M. Arjmand and U. Sundararaj. *Compos. Sci. Technol.* **118**, 257-263 (2015).
13. D. Markham, *Mater. Design*, **21**, 45-50 (2000).
14. M. Weber, and M. R. Kamal, *Polym. Compos.* **18**, 711-725 (1997).
15. R. Ram, M. Rahaman, and D. Khastgir, *Composites Part A* **69**, 30-39 (2015).
16. V. Eswaraiah, V. Sankaranarayanan, and S. Ramaprabhu, *Nanoscale Res. Lett.* **6**, 1-11 (2011).
17. Z. Spitalsky, D. Tasis, K. Papagelis, and C. Galiotis, *Prog. Polym. Sci.* **35**, 357-401 (2010).
18. J. Li, P. C. Ma, W. S. Chow, C. K. To, B. Z. Tang, and J. K. Kim, *Adv. Funct. Mater.* **17**, 3207-15 (2007).
19. A. Behnam, J. Guo, and A. Ural, *J. Appl. Phys.* **102**, 044313 (2007).
20. J. Li, P. C. Ma, W. S. Chow, C. K. To, B. Z. Tang, and J. K. Kim, *Adv. Funct. Mater.* **17**, 3207-3215 (2007).

Localization of MWCNTs in EOC/COC blends

A.Hosseinpour, H.Nazockdast*, A.A.Katbab, M.Mehranpour, Sina Ghiassinejad

Department of Polymer Engineering, Amirkabir University of Technology,
Tehran, Iran, ۱۵۸۷۵-۴۴۱۳
Alihah1369@gmail.com
nazdast@aut.ac.ir

Abstract. The localization of un-functionalized MWCNTs in the immiscible blends of ethylene 1-octene copolymer (EOC) and cyclic olefin copolymer (COC) was investigated. The sea-island morphology of the blends of EOC/COC at 80:20 ratio that distinguished by FESEM images, was chosen for this investigation. Theoretical prediction of wetting coefficients, rheological measurements and microscopic observations confirmed that independent of the mixing sequence, the MWNTs preferentially localized either in the COC phase or at the interface. The transfer of MWNTs from the EOC phase into the COC phase during melt mixing is explained by interfacial effects. Rheological measurements showed that, in this blend only one factor (wetting parameters, associated with surface energy difference) may dominate over the other (melt viscosity ratio or kinetic factors) in determining the localization of the filler in one of the phases.

Keywords: ethylene 1-octene copolymer, cyclic olefin copolymer, MWCNT, Nanocomposite, selective localization, Rheology.
PACS: 83.60.Bc, 83.80.Tc, 82.35.Np.

INTRODUCTION

Polymer nanocomposites containing carbon nanotubes (CNTs) have received great attention from both academia and industry, because of their unique properties, such as desired mechanical, electrical, magnetic, and thermal characteristics. [1, 2]Carbon nanotubes (CNT) have been shown to be an attractive filler for the electrical conductivity increment of polymers at much lower filler loadings compared to those conventional conductive fillers[3]. This efficient behavior is originated from the excellent electrical properties and the high aspect ratios of the CNTs. Polymeric nanocomposites containing CNTs are promising high performance conductive composites with potential applications as electromagnetic shielding coatings, electrostatic dissipative materials, electronic packaging materials, active elements in sensors and etc[4].

Polymer blending has been recognized as an effective, economical and flexible method to prepare materials with many advantages[5] due to the combination of the desirable properties of components. The ultimate properties of polymer blends are effected by different factors such as morphology, interfacial interactions, fillers and processing methods[6]. Recently, Conductive-filler-filled immiscible polymer blends have been attracted much interest. In comparison to single-polymer composites, the conductive fillers into immiscible polymer blends give a much higher potential for the enhancement of conductive composites with significantly lower filler loadings. The conductive filler can be localized either in one or both of the phases or at the interface.

Different concepts predict the localization of nanofillers in polymer blends. The filler distribution is often explained by filler affinity to each polymer forming the blend. so the position of the filler can be predicted by classical thermodynamics. The wetting coefficient ω_α generally used to express this principle that is derived from the Young's equation[7]. Other approaches to explain CNT localization in immiscible polymer blends have considered the kinetics of CNT transfer between phases during melt mixing[8]. The melt viscosity ratio also plays an important role in confining filler specifically in one of the phases. In addition the adsorption-desorption of the polymer which first come in contact with the MWNTs and the second polymer is also effective[9].

The present work deals with blends of Ethylene octene copolymer (EOC) and cyclic olefin copolymer (COC) by using carbon nanotubes as a conductive filler. At first we theoretically predict that MWCNTs preferentially locate in which phase of this blend. Also we experimentally prove our claim using a combination of microscopic examination and oscillatory rheology.

MATERIAL AND METHOD

The cyclic olefin copolymer (COC) was supplied by Ticona company (Celanese, Frankfurt, Germany) under the trade name Topas 6013, with ρ = 1.02 g/cm3 and Tg =140 °C. An ethylene–octene copolymer polyolefin elastomer (EOC) with trade name LC670, with ρ = 0.87 g/cm3 and Tm =55 °C, was supplied by LG Chem. The multiwall carbon nanotubes (Nanocyl NC7000) were purchased from Nanocyl S.A., Belgium. According to the manufacturer, the average diameter and length of MWCNTs were 9.5 nm and 1.5 µm, respectively.

Blends and nanocomposites with different compositions were prepared via melt mixing in an internal mixer (Brabender Plasticorder W50) at a temperature of 240 °C with a rotor speed of 60 rpm. The preparation of hybrid nanocomposites was carried out by the following methods; direct method: EOC, COC and MWCNT simultaneously added to the mixer; sequence method: EOC was first mixed with COC then MWCNTs added to the mixer; masterbatch method: MWCNTs first mixed with EOC or COC, then in the next step other component added to mix with it. The resulting samples were molded to sheets with 1 mm thickness for more characterizations.

Melt rheological behavior was characterized using a Universal Dynamic Spectrometer (Physica UDS 200, Austria) at 240 °C. The linear viscoelastic properties of samples were measured in oscillatory mode at frequencies varying from 0.1 to1000 rad/s. The morphology of samples as well as the localization of the CNTs were investigated using a field emission scanning electron microscope (FESEM) (Mira 3-XMU) on the cryo-fractured samples.

RESULT AND DISCUSSION

The localization of MWCNTs within immiscible EOC/COC blends is predicted by calculating the wetting coefficient w_a as given by Eq. (1)

$$\omega_\alpha = \frac{\gamma_{COC-CNT} - \gamma_{EOC-CNT}}{\gamma_{COC-EOC}} \quad (1)$$

Where $\gamma_{COC/CNT}$, $\gamma_{EOC/CNT}$ and $\gamma_{COC/EOC}$ are the interfacial energy between CNTs and COC, CNTs and EOC, COC and EOC, respectively. According to the value of ω_α obtained, three cases can be considered: $\omega_\alpha>1$, which CNTs preferentially located within the EOC phase; $\omega_\alpha<-1$, which CNTs preferentially located within the COC phase; $-1\leq\omega_\alpha<1$, which CNTs preferentially located at the interface of the polymer blend.

The values of the surface energy of polymers in the melt state were extrapolated from the literature values [10, 11]at 20 °C as summarized in Table 1. For MWCNTs, values from literature were used[12]. Interfacial tension between the different blend components and wetting coefficients determined using both harmonic and geometric mean equations see Table 2.

TABLE1a). The surface tensions values of components at room and processing temperatures.

Materials	γ (20°C)	γ^p(20°C)	γ^d(20°C)	γ^d(240°C)	γ^p(240°C)	γ (240°C)
MWCNT	27.8	10.2	17.6	-	-	-
EOC	17.5	1.9	15.6	13.840	1.686	15.53
COC	29.66	22.39	7.27	14.320	4.650	18.97

TABLE2b). The calculated interfacial tensions and wettability parameter.

Calculation Method	$\gamma_{EOC/COC}$	$\gamma_{COC/CNT}$	$\gamma_{EOC/CNT}$	ω_α
Harmonic Mean	6.552	2.411	1.398	-2.960
Geometric Mean	3.822	1.245	0.744	-3.463

It can be seen, according to the both harmonic and geometric mean equations, the thermodynamics predicts that the MWNTs should be localized preferentially at the COC phase. This result is related to the lower difference of COC surface tension relative to EOC. The higher surface energy of COC may be due to the presence of norbornen sequence in COC. On the other hand, CNTs generally prefer to localize in the phase with lower melt viscosity[12]. In this point of view, due to the lower melt viscosity of EOC, CNTs prefer to be located in this phase.

Figure 1 shows the FESEM micrographs of EOC/COC (80:20) blend. The typical two-phase structure can be seen in this micrograph, in which the dark and light parts correspond to discrete COC and continuous EOC phases, respectively. The small sizes of COC domains throughout the matrix indicated relatively good interfacial adhesion between both polymers.

Figure 2 shows the FESEM micrographs of EOC/COC/CNT (80:20) blend containing 2 wt% MWCNTs which prepared by sequence method. On the amplified images, it is seen clearly as predicted theoretically the MWCNTs are mainly localized in the COC phase and on the phase interface. The selective localization of the CNTs in some polymer blends like EOC/COC, in which the two polymer components show little difference in their affinity to the CNTs has been investigated [9, 13]. In these blend observed that the CNTs are mainly located in one polymer phase with lower interfacial tension. Therefore, in spite of unfavorable kinetics effect such lower viscosity and melting point of EOC, thermodynamically driving forces are dominated in the CNTs localization.

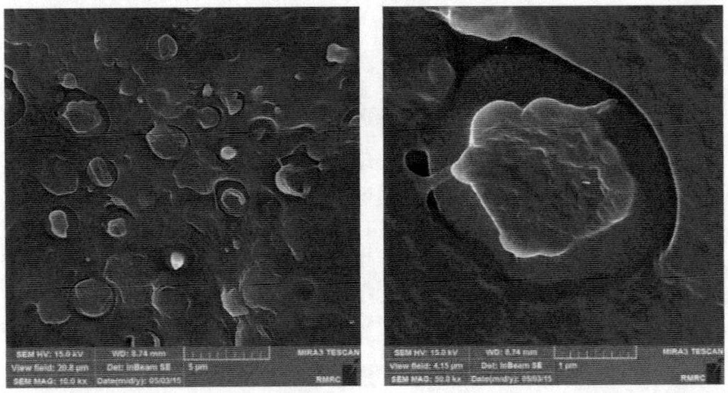

FIGURE 1. FESEM micrograph of EOC/COC (80:20) blend

It is also obvious that with addition of MWCNTs, the size of the COC domains increases remarkably and the shape of COC domains changes from spherical droplets to elongated structures which are connected to each other. This growth in the size of droplet can be related to the localization of CNTs inside the COC domains.

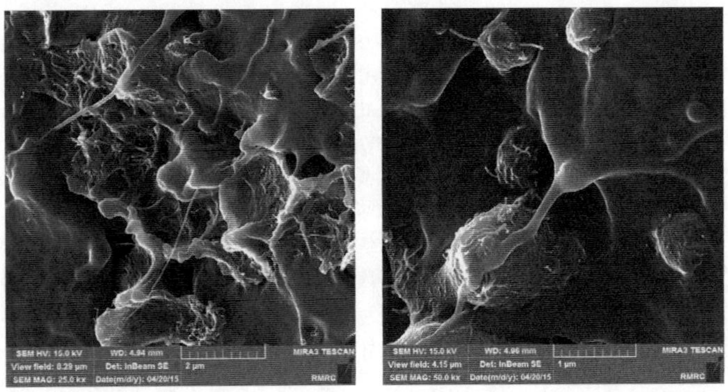

FIGURE 2. FESEM micrograph of EOC/COC/CNT (80:20:2) nanocomposite (sequence)

In the sequence order of feeding, CNTs added to the melted blend, so they are able to localize in the favorable phase. But, when EOC, COC and MWCNTs are melt-mixed in one step (direct) due to the much lower melting point of EOC, before the melting of COC phase, EOC chains which first come in contact with CNTs, be absorbed on their surfaces. Figure 2 shows the FESEM micrographs of this nanocomposite. Unexpectedly it can be seen that MWCNTs are localized in the COC phase and in its interface. This indicates the effectiveness of better affinity of CNT with COC phase compared to EOC matrix. Therefore CNTs desorbed from EOC phase, then migrated to the COC phase and adsorbed by its macromolecules.

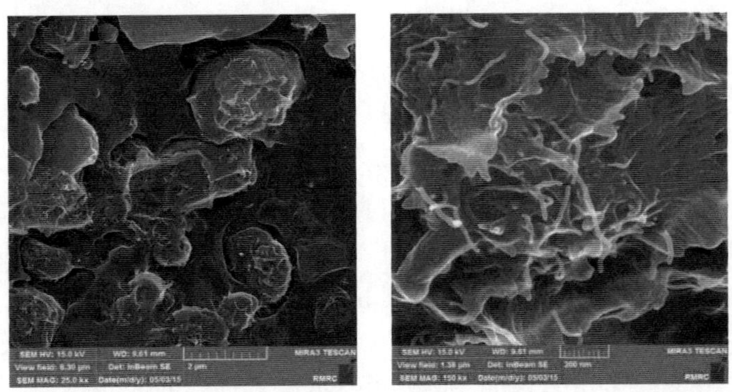

FIGURE 3. FESEM micrograph of EOC/COC/CNT (80:20:2) nanocomposite (direct)

The last investigated case, is the (EOC/CNT)/COC nanocomposite in which MWCNTs compounded with EOC at first then melt-mixed with COC. Figure3 shows the FESEM micrographs of this case. As it can be seen, most of CNTs are located exclusively in the COC phase except for a few CNTs that remain in the EOC phase. This shows the migration of CNTs from EOC to the COC phase during mixing due to better affinity of CNTs to COC phase.

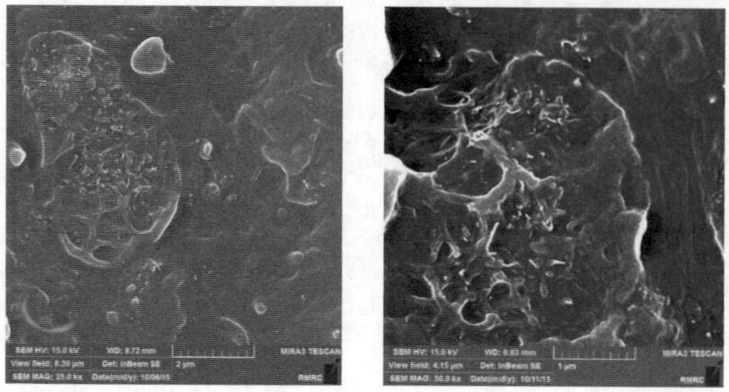

FIGURE 4. FESEM micrograph of (EOC/CNT)/COC (80:20:2) nanocomposite (masterbatch A)

The observed morphology indicates that independent of the mixing sequence, most of CNTs are localized in the COC phase and its interface. Even when three components are simultaneously mixed or when CNTs are premixed with EOC. According to Krasovitski's[14] investigation, very high aspect ratio particles such as carbon nanotubes exhibit complete penetration into the better wetting liquid even for very small differences in the wetting behavior. As Pötschke et al.[15] showed in the PC/SAN blends in which the interfacial energy of PC/CNTs was slightly lower than the one of SAN/CNTs, the CNTs were exclusively located in PC even if they were first premixed with SAN.

Figure 5 illustrates the storage modulus of COC/CNT and EOC/CNT composite containing 3wt% CNT, relative to one of their neat polymers. It is obvious that growth of storage modulus in the presence of CNTs, is much more significant in COC, especially at lower frequencies. This shows the higher affinity of CNTs to COC, which confirmed by morphological observations. However, due to no remarkable difference in the affinity of two polymers to the CNTs, both of polymers present good dispersion of the CNTs.

Figure 6a shows the storage modulus of EOC/COC/CNT (80:20:2) nanocomposites, which prepared by different order of feedings. Addition of CNTs increases the storage modulus of all samples and resulted in a quasi-plateau at low frequencies, indicating the formation of interconnected structures via elongated nanotube-rich COC droplets, with regards to the localization of CNTs in the COC discrete phase. In this Figure it is obvious that the curves of all of samples with the exception of EOC/(COC/CNT), overlapped each other; this can be related to the similar localization of CNTs and the resulted morphology. The deviation of EOC/(COC/CNT) curve could be results of the tendency of COC phase to the CNTs, as well as because of the high strength of the COC domains which caused by the presence of CNTs in those, droplet break up was more difficult, therefore the formation a completely interconnected network is not possible.

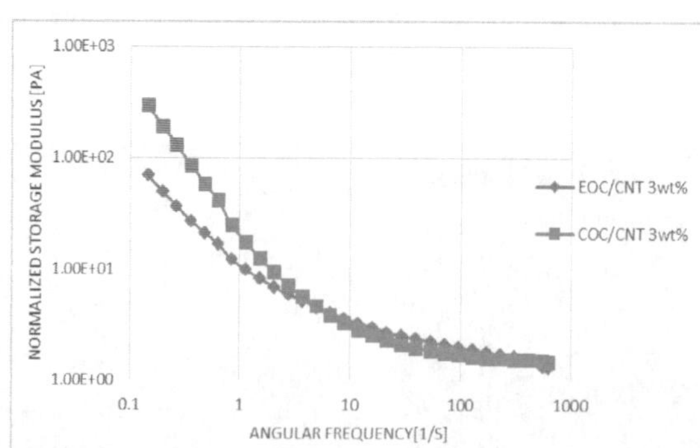

FIGURE 5. Normalized Storage modulus of EOC, COC nanocomposites containing 3wt% MWCNTs relative to their matrix

Figure 6b illustrates the storage modulus of the COC/EOC/CNT (80:20:2) nanocomposites which prepared by different order of feedings as a function frequency. The storage modulus of samples increased in the presence of nanotube, and slopes of the curves tend rapidly toward zero and become independent of the frequency for all samples except for COC/(EOC/CNT) sample in which the precompounded EOC/CNT melt mixed with COC. This plateau could also be explained by the existence of a percolated network by CNTs. The deviations of COC/(EOC/CNT) curve from the other could be related to the inadequate time for desorption of nanotubes from EOC droplet and their migration to the COC matrix. The lower viscosity and flexibility of these droplet during melt mixing also could be another reason for remaining CNTs in the EOC droplet phase.

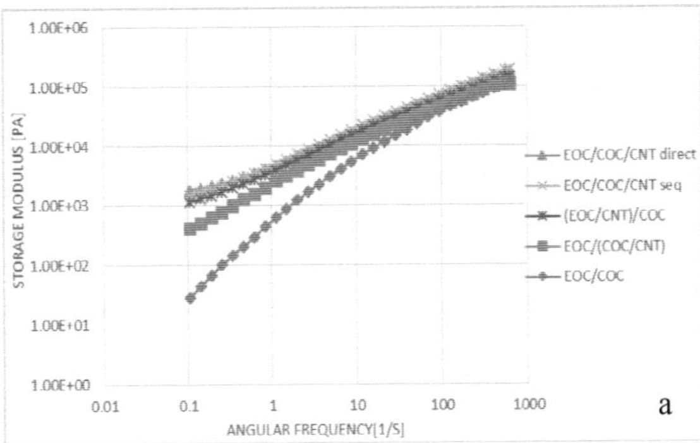

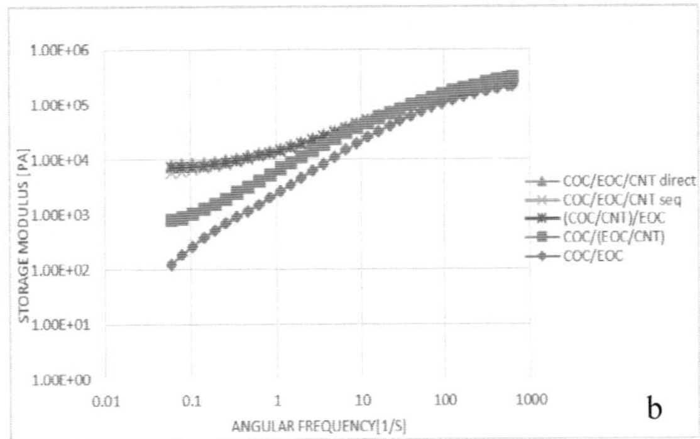

FIGURE 6. Storage modulus of EOC/COC/CNT (80:20:2) and COC/EOC/CNT (80:20:2) prepared by different order of feeding

CONCLUSION

The localization of un-functionalized MWCNTs in the immiscible blends of ethylene 1-octene copolymer (EOC) and cyclic olefin copolymer (COC) is investigated. FESEM micrographs shows the sea-island morphology of the EOC/COC blend at 80:20 weight ratio. Theoretical prediction of wetting coefficients indicates that in the processing temperature (240 °C), CNTs localized in the COC component of EOC/COC blend. Microscopic observations confirms that independent of the mixing sequence, most of CNTs preferentially localized either in the COC phase or at the interface. The transfer of MWNTs from the EOC phase into the COC phase during melt mixing is explained by interfacial effects. Rheological measurements shows that in the sequence and direct order of feeding, CNTs completely located in the COC domains; in the masterbatch methods, in which CNT premixed with EOC at first, CNTs couldn't transfer to the COC phase completely. This is related to the inadequate time for desorption and migration of CNTs. Therefore, in this case one of the factors (wetting parameters, associated with surface energy difference) may dominate over the other (melt viscosity ratio or kinetic factors) in determining the localization of the filler in one of the phases (COC).

REFERENCES

1. P. Pötschke, T. Fornes and D. Paul, *Polymer* **43**, 3247-3255 (2002).
2. M. Abdel-Goad and P. Pötschke, *Journal of non-newtonian fluid mechanics* **128**, 2-6 (2005).
3. I. Alig, T. Skipa, D. Lellinger and P. Pötschke, *Polymer* **49**, 3524-3532 (2008).
4. M. H. Al-Saleh and U. Sundararaj, *Carbon* **47**, 1738-1746 (2009).
5. S. Hobbs, M. Dekkers and V. Watkins, *Polymer* **29**, 1598-1602 (1988).
6. B. Majumdar, D. Paul and A. Oshinski, *Polymer* **38**, 1787-1808 (1997).
7. S. Wu, *Polymer interface and adhesion*, M. Dekker, 1982.
8. A. Göldel, G. R. Kasaliwal, P. Pötschke and G. Heinrich, *Polymer* **53**, 411-421 (2012).
9. L. W. Zhuo-Yue Xiong, Yao Sun, Zhao-Xia Guo, Jian Yu, *Polymer* **54**, 447-455 (2013).
10. C. Ma, M. Zhang and M. Rong, *Journal of applied polymer science* **103**, 1578-1584 (2007).
11. J. Y. Shin, J. Y. Park, C. Liu, J. He and S. C. Kim, *Pure and applied chemistry* **77**, 801-814 (2005).
12. T. M. b. Romain Cardinaud *European Polymer Journal* **49**, 1287-1297 (2013).
13. C. Li, Q. Zhao, H. Deng, C. Chen, K. Wang, Q. Zhang, F. Chen and Q. Fu, *Polymer international* **60**, 1629-1637 (2011).
14. A. Marmur and B. Krasovitski, *Langmuir* **18**, 8919-8923 (2002).
15. A. Göldel, G. Kasaliwal and P. Pötschke, *Macromolecular rapid communications* **30**, 423-429 (2009).

Influence of Filler Dispersion and Interfacial Resistance on Thermal Conductivity of Polypropylene/Carbon Nanotubes Systems

Antonella Patti[a], Domenico Acierno[b], Pietro Russo[c]

[a]*Department of Chemical, Materials Engineering and Industrial Production, University of Naples Federico II, P.le Vincenzo Tecchio 80,80125 Naples, Italy*
[b]*Centro Regionale di Competenza (CRdC Tecnologie scarl), Via Nuova Agnano 11, 80125 Naples, Italy*
[c]*Institute for Polymers, Composites and Biomaterials, National Research Council, Via Campi Flegrei 34, 80078 Pozzuoli, Naples, Italy*

Carbon nanotubes effectiveness, as thermally conductor, in polymer/CNTs nanocomposites, is mainly related to various key aspects as dispersion, matrix/CNT interfacial resistance and inter-CNT contact resistance. Features of raw materials and processing parameters for the realization of nanotube/polymer composites, can simultaneously affect these critical issues, also in adverse way. In this contribution, polypropylene based-composites, containing multi walled carbon nanotubes (MWNTs), are prepared by melt-blending and then characterized prevalently in term of thermal transport behavior, choosing different matrix grades, filler functionalization and process conditions. In more details, two commercial polypropylene (PP) grades with different melt flow rates are considered as matrices and three kind of MWNTs with the same aspect ratio, one neat and two chemically modified with carboxyl –COOH and amino $-NH_2$ groups, respectively, are used as fillers. The presence of a compatibilizer, as maleic anhydride grafted polypropylene (Ma-g-PP), and the use of various processing conditions as temperature, time, screw speed, pressure, filler drying and so on, are taken into account in order to optimize the thermal conductivity enhancement of PP/MWNTs compounds.

Keywords: Dispersion, Interfacial resistance, Thermal conductivity, Polypropylene, Carbon nanotubes.

INTRODUCTION

Recently, heat dissipation issues, increasingly important in various applications as microelectronic devices and electric motors, together with the monotonous trend to replace traditional metals with lighter polymer based materials, have generated a growing interest of the research towards the development of new thermal conductive polymer nanocomposites. These materials consist of a polymer phase in which particles with high thermal conductivity are adequately dispersed. At this regard, carbonaceous nanofillers as carbon nanotubes (CNTs) and graphene nanoparticles are widely used given their extraordinary properties including low density, high aspect ratios and high thermal conductivity and a convenient method to produce new compounds is the melt blending owing to its cost effectiveness, fast production and environmental benefits.

The improvement of thermal conductivity of nanocomposites filled with carbon nanotubes has been widely reported [1,2]. However, previous investigations has also highlighted that thermal conductivities of products are relatively low with respect to the intrinsic thermal conductivity of included carbon nanotubes. This drawback is mainly related to the large interfacial thermal resistance between CNTs and the surrounding polymer matrix, known as Kapitza resistance (R_K) [3] and to usual inter-CNTs contacts commonly leading to the formation of bundles [4]. The increase of the former thermal resistance and the reduction of the latter one hinder the transfer of phonon dominating heat conduction in polymer based systems.

On the other hand, the thermal conductivity of polymer composites depends on many factors as chemical, molecular, structural characteristics of the matrix, dispersion, functionalization, size, structural defects and amount of fillers. About the dispersion, a key role is played by processing conditions [5] and this is particularly true in case of carbon nanotubes, given their large aspect ratios and strong Van-der-Waals interactions favoring aggregations rather than dispersion phenomena, especially in high viscous polymer melts. Regarding functionalization aspects, instead, typical non-covalent and covalent approaches or the use of opportune compatibilizers, obviously alter the interfaces, affecting thermal transport ability of investigated materials [6].

In this work, many polymer/CNTs systems were prepared by melt compounding involving two commercial polypropylene (PP) resins with different melt flow rates, a common coupling agent, polypropylene-grafted-maleic anhydride (PP-g-MA), and three different multi wall carbon nanotubes, one neat (MWNT) and two functionalized by amino ($MWNT-NH_2$) and carboxyl (MWNT-COOH) groups, respectively.

Effects of the fluidity of the matrix, the type of interface and processing conditions on the thermal conductivity of products were systematically analyzed and interpreted in terms of the achieved state of carbon nanotubes dispersion, thermal resistances and average mutual distance among carbon nanotube agglomerates.

MATERIALS AND METHODS

In this work, two commercial polypropylene resins supplied under the trade codes Monsten MA524 (M_n= 32,400 g/mol, M_w=250,084; melt flow index 24,0 g/10min @ 230°C, 2,16 Kg) and Monsten MA712 (M_n= 33,519 g/mol, M_w=301,727; melt flow index 12,0 g/10min @ 230°C, 2,16 Kg) by UNIPETROL RPA, Litvínov-Czech Republic, are used as matrices. A maleic anhydride grafted polypropylene (PP-g-MA) (Polybond 3200, melt flow index 115,0 g/10min at 190°C, 2,16 Kg, maleate content 1% in wt.) was used at 10% by weight to modify the filler-matrix interface.

Not functionalized, carboxyl-functionalized and amino-modified multi-walled carbon nanotubes with trade codes MWNT-3150, MWNT-3150 and MWNT-3150, respectively, all with an average diameter equal to 9.5 nm and length approximately of 1 μm and purchased by Nanocyl S.A. (Sambreville - Belgium), are chosen as fillers.

All materials are prepared by melt mixing using a batch mixer (Brabender Plastograph EC – Brabender GmbH & Co. KG Germany). In all cases, the matrix, as received, is introduced into the preheated chamber and processed for 2 min, until its melting. Then carbon nanotubes are added and mixed with the molten polymer for a set time.

Thermal conductivity measurements are performed using a thermal conductivity meter (UNITHERM MODEL 2022 from THERMAL CONDUCTIVITY INSTRUMENT, fabricated according to the ASTM E1530 [ASTM E1530-99 "Standard Test Method for Evaluating the Resistance to Thermal Transmission of Materials by the Guarded Heat Flow Meter Technique".]) based on a classic steady-state technique. This machine allows to measure the thermal conductivity in through-plane modality. All tests are conducted at room temperature on compression moulded disc-shape specimens with a diameter of 40 mm and a thickness of 1.4 mm.

RESULTS AND DISCUSSION

About processing conditions, Fig. 1 summarizes the influence of some melt compounding conditions as processing temperature, rotor speed and mixing time on the thermal conductivity of sample containing 1.5 vol. % of MWNTs. In each case, samples are compared keeping two variable constant and varying the third one. From the picture 1 a), considered 190 °C, 220 °C and 250 °C as mixing temperatures, it is evident that the thermal conductivity assumes the highest value for samples obtained at the lowest temperature. Taking into account that increasing the mixing temperature reduces the melt viscosity, it is likely that at higher temperatures agglomeration phenomena prevail worsening carbon nanotubes dispersion and, consequently, reducing thermal conductivities of samples (~ -14%).

However the previous effect is not so pronounced as the influence of the rotor speed. In this case (see picture 1 b)), a non monotonous trend of the thermal conductivity with this parameter is detected. In details, a maximum seems to be verified in correspondence of a rotor speed equal to 60 rpm. The justification of this behavior lies in the fact that the rotational speed is obviously reflected on the viscous stress acting on included aggregates/agglomerates of MWNTs. Increasing the speed can favor dispersion mechanisms (rupture/erosion) but, beyond a certain threshold, could be responsible for carbon nanotubes shortening with a consequent deterioration of the intrinsic thermal conductivity. Analogously, even if with less pronounced effects, the influence of the mixing time is easily interpreted (Fig.1 c)).

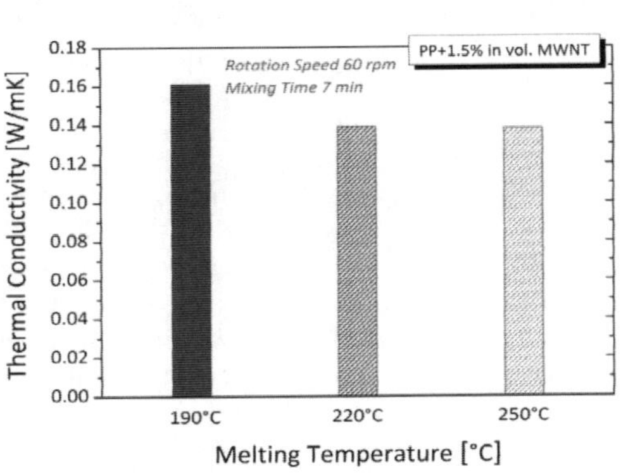

(a)

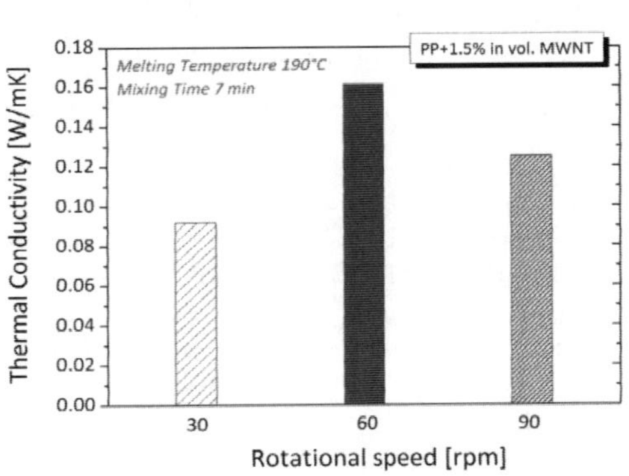

(b)

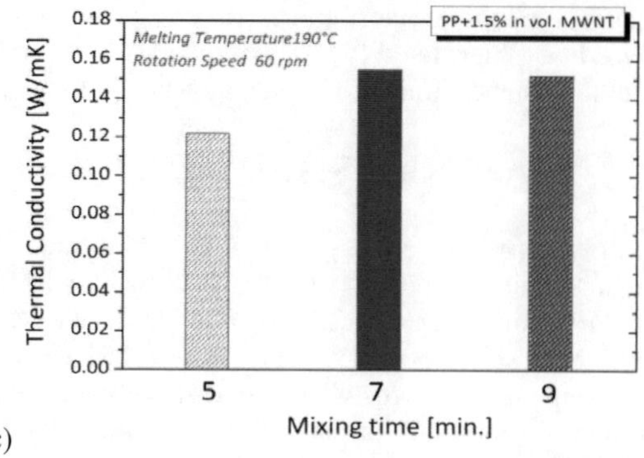

(c)

FIGURE 1- Thermal conductivity of nanocomposites prepared under different process conditions: (a) changing speed (30, 60, 90 rpm) at a fixed time (7 min) and temperature (190°C); (b) changing temperature (190, 220, 250 °C) at a fixed speed (60rpm) and time (7min); (c) changing time (5 ,7 ,9 min) at a fixed temperature (190°C) and speed (60 rpm).

The intrinsic hydrophobic nature of carbonaceous fillers is another factor that promotes agglomeration rather than dispersion mechanisms especially in viscous polymer melts and it could reduce their intrinsic thermal conductivity [7]. For this reason, composite samples consisting of PP MA524 matrix and as received or pre-dried non functionalized carbon nanotubes has been prepared and analyzed in terms of thermal conductivity. Unexpectedly,our results, collected in Fig. 2, highlight that pre-drying worsen the thermal conductivity of products. This effect is mainly due to the compaction of pre-dried MWNT entanglements, confirmed by morphological observation not included here, and which further hinders their dispersion at the expense of the thermal transport properties.

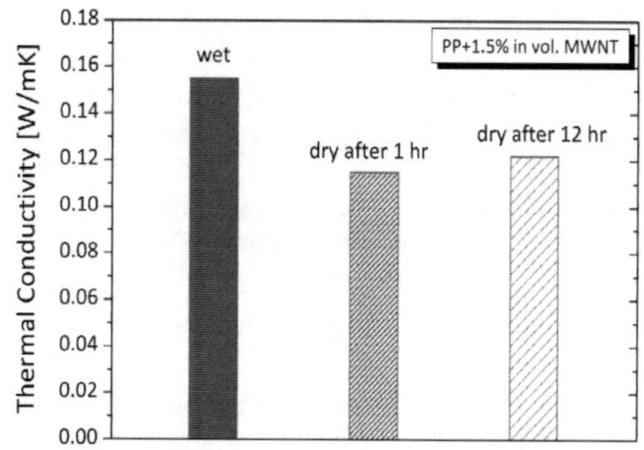

FIGURE 2- Thermal conductivity of nanocomposites based on PP MA524 and filled with 1.5% vol. of MWNT, as received and pre-dried at 80°C for 1 and 12 h, respectively.

Fig. 3 indicates thermal conductivities of the considered two polypropylene grades and composites containing 1.5 vol. % of non functionalized multi wall carbon nanotubes based on the same. Tests reveal that the thermal transport in the resin with MFI equal to 24 is slightly better than in the one with MFI equal to 12. The addition of carbon nanotubes always increases the thermal conductivity of the hosting matrix with an effect more pronounced in case of the one with higher fluidity. This evidence can be explained by assuming that a higher fluidity can facilitate the impregnation of carbon nanotubes agglomerates, reducing their strength, and favor typical dispersion mechanisms as agglomerate ruptur and erosion [8].

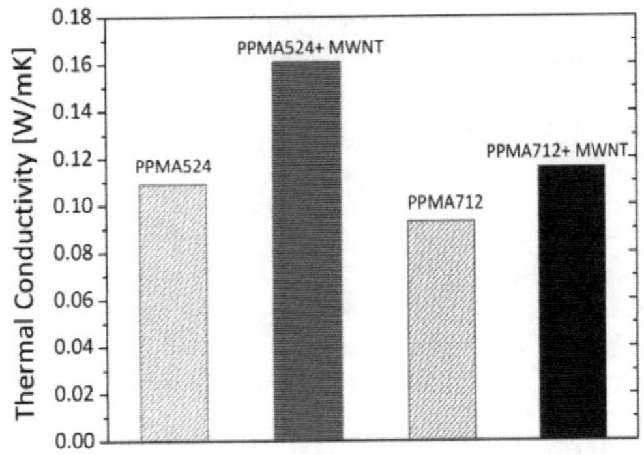

FIGURE 3 - Thermal conductivity of neat PP matrices and nanocomposites filled with 1.5% in vol. of MWNTs.

In general, compatibilizers are aften used to improve filler dispersion in a polymer, changing surface energies, improving wetting and compatibility between two phases. A common compatibilizer, added in CNTs/polyolefins nanocomposites, is maleic anhydride grafted polypropylene (Ma-g-PP), whose functionalities, reacting with amino, carboxyl or hydroxyl groups on CNTs' surface, stabilize the morphology and favor filler-matrix interfacial adhesion. [9] In our case, as cleary shown in Fig. 5 comparing the neat PPMA524 grade with the same resin modified by adding the coupling agent and composites containing 1.5 vol. % of MWNT or MWNT-NH_2 based on the same matrices, the addition of PP-g-MA does not influence the thermal transport of the neat PP but, significantly reduces the contribution of included carbon nanotubes to the ultimate thermal transport properties, from +42.2% to +19.6% in presence of MWNT and from +47.7% to +11.7% for systems containing MWNT-NH_2. Probably, MA-g-PP, wrapped around MWNT, especially in the case of amino-functionalized ones, reduces contact points among conductive particles and forms a polymeric coating on them. These conditions, improving the interfacial thermal resistance, hinder the heat transport ability in nanocomposites [10].

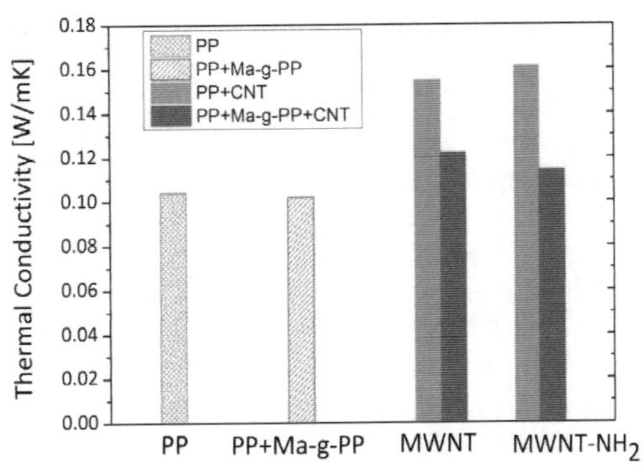

FIGURE 4- Thermal conductivity of PP MA524 nanocomposites with or without compatibilizer.

Finally, Fig. 5 reports thermal conductivity trends for nanocomposite systems based on PP MA524 by varying the amount of included carbon nanotubes. Regardless the type of carbon nanotubes, in all cases a cubic polynomial trend is observed. This unexpected behavior can be interpreted in light of the expected different average size of MWNT agglomerates. This parameter together with the nanotubes content, in turn, affect the average mutual distance of included agglomerates and, consequently, the chances for the formation of thermal conductive pathways: higher the agglomerates sizes and content, higher the average mutual distance and lower the thermal conductivity of composite samples.

In light of this considerations, it is possible to assume that, at relatively low MWNTs contents (up to 1.5 vol. %), low viscous forces developed in the mixing chamber give rise to a poor infiltration of MWNTs agglomerates to an extent essentially depending on their porosity as well as on the filler-matrix mutual affinity. This situation result in a non homogeneous filler distribution including exfoliated entanglements and bundled areas. By increasing the MWNTs content, dispersion mechanisms are favored by the greatest interfacial stress occurring in the melt state but, on the other hand, the probability the agglomerates meet to form larger ones also increases. Taking into account that these phenomena influence the thermal conductivity of samples in opposite ways, it is possible to explain the trend observed on the range of MWNTs concentrations from 1.5 to 3.7 vol. %. For higher loadings, the high probability of collisions among large agglomerates, the high effectiveness of dispersion mechanisms and the increased number of distributed aggregates, easily explain the monotonous increase of the thermal conductivity of investigated materials.

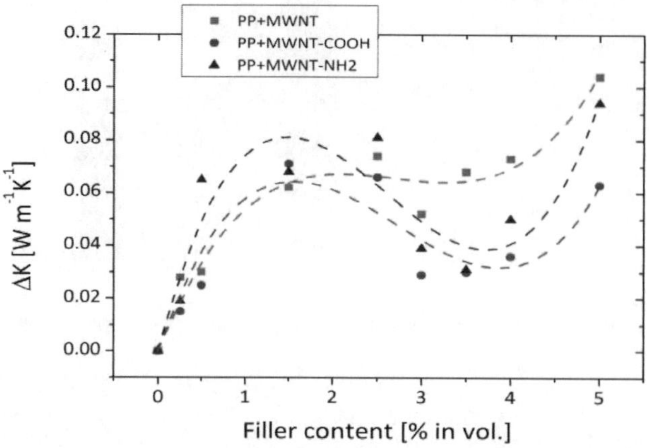

FIGURE 1-MWNT network contribution to the thermal conductivity of MWNT/PP compound as a function of MWNT volume fraction (in term of $\Delta k=k_c-k_m$; where k_c and k_m are thermal conductivity of compound and neat PP, respectively).

CONCLUSIONS

Melt compounded binary systems involving two polypropylene grades with MFI equal to 12 and 24 g/10 min, a polypropylene grafted with maleic anhydride as compatibilizer and three kind of multiwalled carbon nanotubes with the same aspect ratio but different surface chemical features were prepared in various processing conditions using a batch mixer and characterized in terms of thermal conductivity.

The research showed a greater improvement in thermal transport ability for systems based on the matrix with high melt flow index, probably for the achievement of a better filler dispersion. About the influence of processing parameters, the best results were detected for samples prepared at a processing temperature of 190 °C and using a rotor speed equal to 60 rpm. These finding, mainly interpreted by expected influences of processing parameters on both polymer melt viscosity and extent of stresses acting on included MWNT aggregates, allowed to optimize their production, at least on the lab scale. On the other hand, an adverse effect on thermal efficiency of investigated compounds was shown by adding 10 wt% of a typical coupling agent.

Finally, a cubic polynomial trend of the thermal conductivity as a function of the MWNTs content was always revealed and explained in terms of expected change of carbon nanotubes aggregate/agglomerate sizes, in turn related to various aspects as their wettability, melt viscosity and stress acting on the same during the mixing, filler content and mutual interactions. At this regard, this behavior was commented in terms of a geometrical parameter named *"Average Inter-aggregate Distance"* and simply coded as AID.

REFERENCES

[1] Haddadi M., Agoudjil B., Boudenne A. *Mat. Sci. For.* **714**, 99-113,2012.

[2] Wang S., Liang R., Wang B., Zhang C. *Carbon* **47**, 53-57, 2009.

[3] Li Q., Liu C., Fan S., *NanoLett.* **9**, 3805-3809, 2009.

[4] Wang Y., Wu J., Fei W. *Carbon* **41**, 2939-2948, 2003.

[5] Krause B., Pötschke P., Häußler L. *Comp.Sci. Tech.* **69**, 1505-1515, 2009.

[6] Tu H., Ye L. *Journ.Appl.Poly.Sci.* **116**, 2336-2342, 2010.

[7] J.A. Thomas, R.M. Iutzi, A.J.H. McGaughey *Phys. Rev. B* **81**, 045413 (2010).

[8] G.R. Kasaliwal, A. Golden, P. Potschke, G. Heinrich *Polymer* **52**, 1027-1036 (2011).

[9] Lee S.H., Cho E., Jeon S.H., Youn J.R. *Carbon* **45**, 2810-2822, 2007.

[10] Díez-Pascual A. M., Naffakh M.; Gómez, M. A., Marco C., Ellis G., González-Domínguez J. M., Ansón A, Martínez M. T., Martínez-Rubi Y., Simard, B., Ashrafi B. *NanoTechn.* **20,** 315707-315720, 2009.

Effect of OMMT on Fire Resistance of EVA-based Ceramifying Composites

Gong Xinhao[a], Shen Yucai[a] and Wang Tingwei[a*]

[a]*College of Materials Science and Engineering, Nanjing Tech University, Nanjing, 210009, China*

Presenting author's email: gxhnjtech@njtech.edu.cn
Reference author's email: wangtw@njtech.edu.cn

Abstract: EVA-based ceramifying composites, a kind of promising materials, can be widely applied in fire protection field, especially fire resistant sheaths for cables. However, the quick decomposition of the matrix has resulted in cracks and even holes in ceramic residues, which has a disastrous impact on the fire resistance of the residues. Here ceramifiable EVA composites were prepared by melt blending with wollastonite, glass frits and OMMT. Influence of OMMT on the pyrolysis behavior of the composites and properties of ceramic residues was investigated. Thermogravimetric analysis showed that the initial degradation temperature of the composite increased by 12℃ and its maximum weight loss rates at the first stage and the second stage decreased by 25% and 36% respectively, when 10phr OMMT was added. Besides, for samples heated from room temperature to 1000℃, addition of 10phr OMMT led to a 20% increase in the flexural strength of the residues. XRD showed that an intercalated structure between EVA and OMMT was formed. Therefore, it was concluded that significant improvement in the thermal stability and the flexural strength is a result of the formation of the barrier layer.

Keywords: ethylene vinyl acetate copolymer (EVA); organic montmorillonite (OMMT); ceramifying composites; intercalated structure
PACS: 81.05.Qk; 81.70.Pg; 83.80.Ab

There is considerable interest in the fire resistance of polymer based composites as the demand for improved safety in domestic, public and industrial situations is required by fire-safety standards. One method improving the fire resistance of the composites was presented by Australian researchers. They prepared a series of ceramifying composites by means of adding refractory fillers such as mica [1-4] and wollastonite [5] and fluxing agents like zinc borate [6] and glass frits [7] into silicone rubber. The composites have the properties of silicone rubber at room temperature and of ceramics at temperature increasing. Recently, to broaden the application field of ceramifying composites, attention has been shifted to ceramifying organic polymer composites. However, organic polymer especially conventional polymer, has more poor thermal stability and higher thermal decomposition rate, compared to silicone rubber [8-9]. This worsens the fire resistance of the composites. In this work, all composites were prepared on a two-roll mill. EVA-based composites, as summarized in TABLE (1), were moulded into flat sheets by compression moulding at 150℃.

TABLE (1). Composites formulations (g)

Composition code	EVA	Wollastonite	Glass frit	OMMT
A	100	100	50	
B	100	100	50	10

RESULTS

To evaluate their fire resistance, the bulk density, water absorption, dimensional change in length and flexural strength of composites heated from room temperature to 600, 700, 800, 900 and 1000℃ were tested and shown in FIGURE 1. The bulk density, water absorption and flexural strength of Sample A fired at 600 and 700℃ are too weak to be estimated. The dimensional change in length of ceramifying composites transformed expansion into shrink at 10phr OMMT loading. Besides the expansion rate of sample A declined and the shrinkage of sample B increased with increasing firing temperature, which suggests the increase of the firing temperature promotes densification via liquid phase sintering. At the same time, for composites fired at each firing temperature, addition of OMMT caused the improvement of their bulk density and flexural strength, and the reduction of amount of their open pores. And the highest flexural strength was obtained when the sample with OMMT was heated at 1000℃. The strength was 1.2 times higher than that of sample A fired at 1000℃. It can be inferred that the reduction of amount of open pores benefits the improvement of the flexural strength.

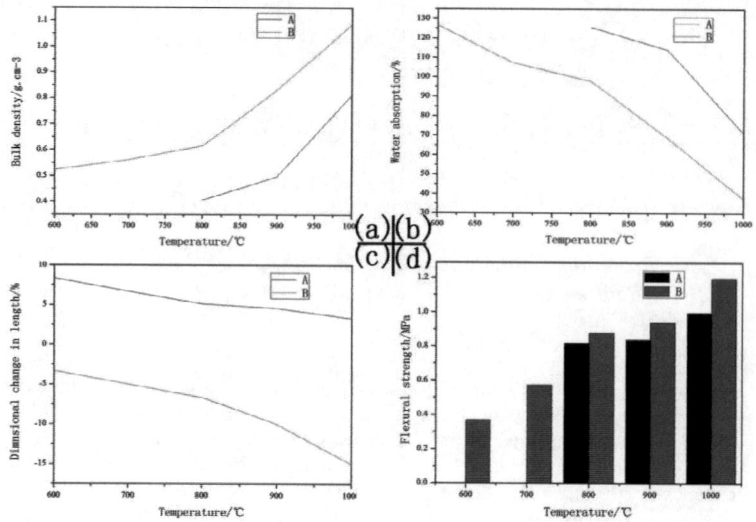

FIGURE 1. Bulk density (a), water absorption (b), dimensional change in length (c) and flexural strength (d) of sample A and B heated from room temperature to 600, 700, 800, 900 and 1000℃ respectively

The TG-DSC examination of sample A and sample B revealed two peaks of weight loss rate responded to two exothermic peaks respectively, as shown in FIGURE 2 and 3. The first peak was in line with the loss of acetic acid in the matrix and the second peak was related to the deterioration of the main chain of the matrix. The initial degradation temperature of sample B increased by 12℃, compared to that of sample A. Furthermore, OMMT added led to 25% and 36% decline in maximum weight loss rates at the first stage and the second stage.

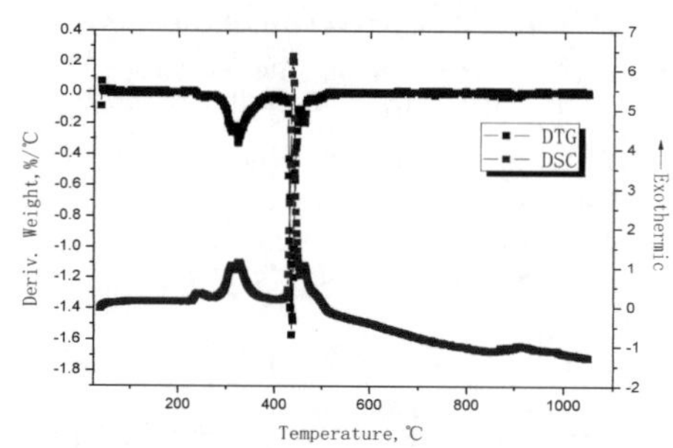

FIGURE 2. DTG and DSC curves of Sample A in air

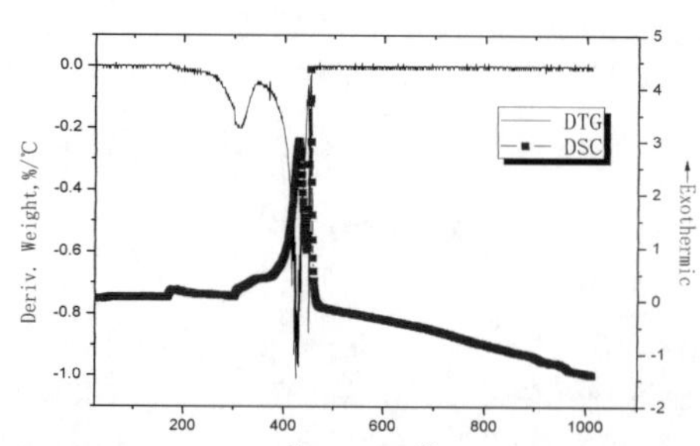

FIGURE 3. DTG and DSC curves of Sample B in air

Fig. 4 shows the wide-angle powder XRD patterns of OMMT and sample B. For sample B, there is a significant shift of the silicate layer (001) lattice diffraction peak to a lower angle from that of the corresponding OMMT. This means that an intercalated structure between EVA and OMMT in sample B has been formed.

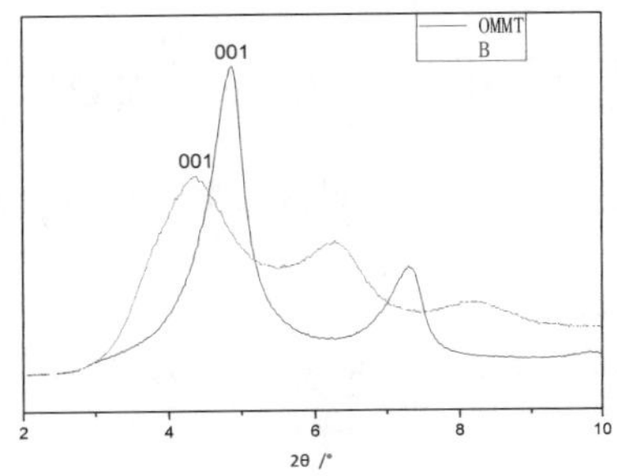

FIGURE 4. XRD patterns of OMMT and sample B

CONCLUSION

The strong interaction between EVA and OMMT delays the decomposition of the matrix, which is in favor of the densification of ceramic residue. Thus the thermal stability of composites and the flexural strength of the residue all be enhanced.

ACKNOWLEDGMENTS

This work was supported by the Innovation Foundation for Graduate Students of Jiangsu Province (KYLX15_0778) and the Priority Academic Program Development of Jaingsu Higher Education Institutions (PAPD).

REFERENCES

[1] Hanu L G, Simon G P, Mansouri J and Burford R P, J. Mater. Process. Tech, 153, 401-407 (2004).
[2] J. Mansouri, R. P. Burford, Y. B. Cheng and L. Hanu, J. Mater. Sci. 40, 5741-5749(2005).
[3] Hanu L G, Simon G P and Cheng YB, Mater. Sci. Eng. , A 398(1), 180-187(2005).
[4] Mansouri J, Burford R P and Cheng Y B, Mater. Sci. Eng. , A 425(1), 7-14(2006).
[5] Rodrigo P and Cheng YB, WO Patent No. 2010/139011A1 (2 June 2010).
[6] Z. Al-Hassany, A. Genovese and R.A. Shanks, Express Polym. Lett, 4(2), 79-93(2010).
[7] Mansouri J, Wood C A, Roberts K, Yi-Bing C and Burford R P, J. Mater. Sci. 42, 6046-6055(2007).
[8] R.R Buch, Fire safety J. 17, 1-12(1991).
[9] F.Y. Hsieh and R.R. Buch, Fire Mater. 21, 265-270(1997).

How to Efficiently Enhance the Thermal Conductivity of Poly(vinylidene fluoride)-based Composites ?

Yong Wang*, Yan-jun Xiao and Jing-hui Yang

School of Materials Science & Engineering, Key Laboratory of Advanced Technologies of Materials (Ministry of Education), Southwest Jiaotong University, Chengdu, 610031, China

E-mail: yongwang1976@163.com

Abstract. In this work, several kinds of poly(vinylidene fluoride) (PVDF) composites were prepared, including PVDF/carbon nanotubes (CNTs), PVDF/boron nitride (BN) and PVDF/graphene (GE). To efficiently enhance the thermal conductivity of these composites, a small amount of polyvinylpyrrolidone (PVP), graphene oxide (GO) or other components, were introduced into these composites, respectively. The results showed that compared with the binary PVDF-based composites, the presence of the third component largely enhanced the thermal conductivity of these composites. The dispersion of conductive fillers, the crystallization of the PVDF matrix and the interfacial interaction between PVDF and conductive fillers were systematically investigated. The denser conductive filler network structure, the improved interfacial interaction and the crystal transformation of the PVDF matrix were suggested the main mechanisms for the largely enhanced thermal conductivity of the PVDF-based composites.

Keywords: Poly(vinylidene fluoride), Carbon nanotube, Boron nitride, Graphene, dispersion, thermal conductivity.
PACS: 65.90.+i

INTRODUCTION

There are many fields that need quickly transferring the heat so that the temperature of product can be maintained at relatively low level. The related application fields include generators, heat exchangers in power generation, automobile, and microelectronic devices, etc. Now, the heat dissipation capability has become one of the key parameters that must be considered when designing the electronic devices. The traditional methods that developed to enhance the thermal conductivity of polymer materials are usually related to the use of a large amount of conductive fillers. However, adding thermal conductive fillers into polymers does not insure the great enhancement of the thermal conductivity of the composites. This is because that the thermal conductivity of polymer composites is greatly influenced by many factors, including the microstructure of polymer matrix, the interfacial interaction between thermal conductive fillers and polymer matrix, and the dispersion states of conductive fillers.[1-3] So far, there are several strategies that have been developed to enhance the thermal conductivity of polymer composites, including increasing the degree of crystallinity of polymer matrix, enhancing the interfacial interaction through surface modification of conductive fillers, and promoting the formation of an effective thermal conductive path of conductive fillers in the composites.[4-6]

Poly(vinylidene fluoride) (PVDF) is a typical semicrystalline polymer with good processing ability and exciting properties, including superior piezoelectric and pyroelectric properties. It has many applications ranging from supercapacitors, transducers, actuators to batteries. The thermal conductivity of PVDF is very small, only about 0.19 $Wm^{-1}k^{-1}$.[7] Therefore, improving the thermal conductivity of PVDF-based material becomes very significant today due to the fact that the PVDF-based electronic devices become slimmer and more integrated today.

In this work, we attempt to seek efficient but simple strategies to prepare the PVDF-based composites with high thermal conductivity at relatively low filler content. Different conductive fillers are mainly used to prepare the PVDF composites, including carbon nanotubes (CNTs), boron nitride (BN) and graphene (GE). Furthermore, a small amount of the other components, including polyvinylpyrrolidone (PVP) and graphene oxide (GO), are also introduced to tune the dispersion of fillers and the interfacial interaction between conductive fillers and PVDF matrix.

EXPERIMENTAL SECTION

The materials used in this work are commercially available. PVDF (with a trade name of Kynar 720) was purchased from Arkema (France). CNTs with carboxyl content of about 0.51 wt% were obtained from Chengdu Institute of Organic Chemistry, Chinese Academy of Science (Chengdu, China). Hexagonal BN (with a purity of 99.99% and a density of 2.3 g/cm3) was purchased from Chengdu UPAM applied materials Co., LTD (China). PVP (trade name of k30) was obtained from Kelong Chemical Reagent Co. Ltd. GE was purchased from Deyang Carbonene Science and Technology Co. LTD (China). GO was prepared in our lab according to a modified Hummer's method using graphite as the raw material.[8]

Samples were prepared through a two-step processing procedures, i.e. solution compounding using N,N-dimethylformamide (DMF) as the solvent and subsequent melt compounding processing that was conducted on a micro-extruder at melt temperature of 200 °C for about 10 min. After that, the pellets of the composites were then compression-molded to obtain the samples for microstructure and physical property measurements. The compression-molding was carried out at the melt temperature of 200 °C and the pressure of 5 MPa.

The dispersion states and microstructures of conductive fillers in the composites were characterized using a transmission electron microscope (TEM), a scanning electron microscope (SEM) and rheological measurements. The crystalline structures of the PVDF matrix were characterized using a differential scanning calorimetry (DSC) and wide angle X-ray diffraction (WAXD). Fourier transform infrared spectroscope (FTIR) was further used to investigate the crystalline structure of the PVDF matrix and the interfacial interaction between conductive fillers and the PVDF matrix. The DRL-III Thermal Conductivity Tester (Xiangke instrument manufacture, China) was applied to measure the thermal conductivity of the samples.

RESULTS AND DISCUSSION

Effect of PVP on Thermal Conductivity of PVDF/CNT Composites

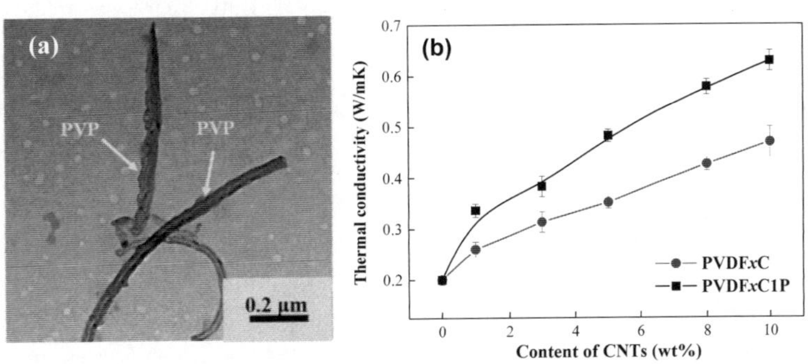

FIGURE 1. (a) TEM image showing the typical morphology of PVP-treated CNTs and (b) Variation of thermal conductivity of PVDF/CNT and PVDF/CNT/PVP samples versus the content of CNTs.

From Fig. 1 one can see that after being treated by PVP, CNTs are wrapped by a thin layer of PVP, proving that there is a special interfacial interaction between PVP and CNTs. From Fig. 1b it is seen that the thermal conductivity of all the composites increases gradually with increasing CNT content. However, it is worth noting that the PVDF/CNT@PVP composites exhibit higher thermal conductivity compared with the PVDF/CNT composites. For example, at CNT content of 10 wt%, the thermal conductivity of the PVDF10C1P sample is 0.63 $Wm^{-1}K^{-1}$, which is much higher than the 0.47 $Wm^{-1}K^{-1}$ of the PVDF10C sample, and even about 315% higher than that of the neat PVDF. Obviously, largely improved thermal conductivity of PVDF composites is achieved by adding PVP treated CNTs.

Effect of GO on Thermal Conductivity of PVDF/CNT Composites

In this section, the PVDF/CNT/GO composites were fabricated by adding very few GO (1 wt%) into PVDF/CNT composites. As shown in Fig. 2a, compared with the PVDF/CNT composites, largely enhanced thermal conductivity is achieved for the PVDF/CNT/GO composites. Further results demonstrate that the presence of GO facilitates the dispersion of CNTs and the formation of denser CNT/GO network structure on one hand (Fig. 2b). On the other hand, GO also promotes the formation of a large number of polar γ-form crystallites (Fig. 2c). Therefore, the mechanisms for the largely enhanced thermal conductivity are mainly related to the formation of the denser hybrid CNT/GO network structure and the polar γ-form crystallites.

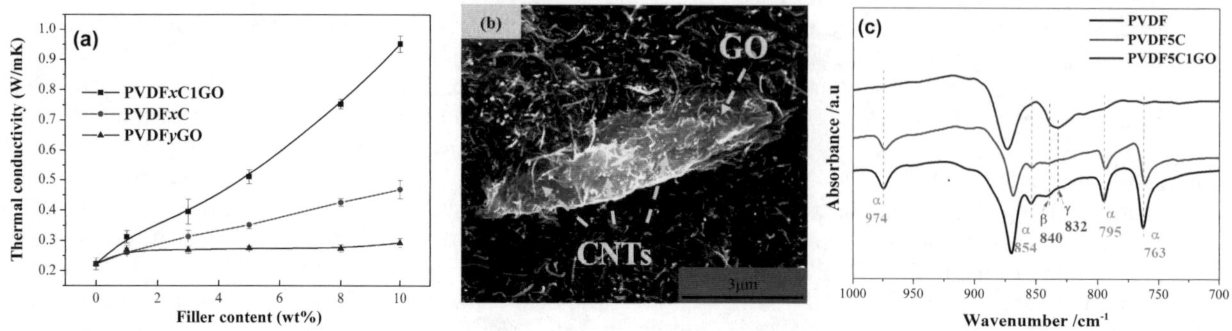

FIGURE 2: (a) Variation of thermal conductivity of the PVDF composites versus the content of fillers, (b) SEM image showing the dispersion state of fillers in the composites, and (c) FTIR spectra showing the crystal transformation of the PVDF matrix induced by GO.

Effect of CNTs on Thermal Conductivity of PVDF/BN Composites

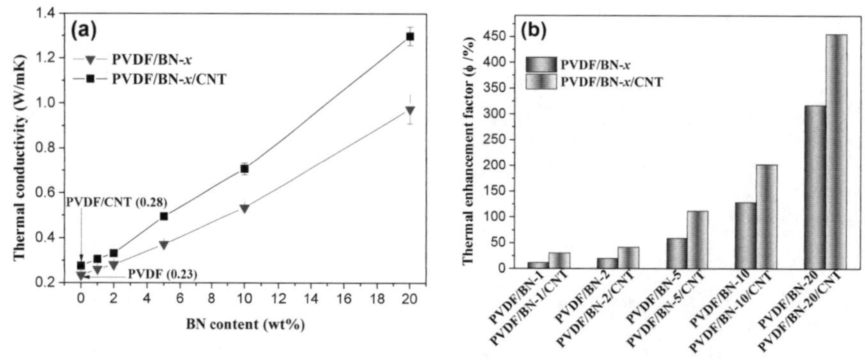

FIGURE 3: (a) variations of thermal conductivity versus the content of BN in the composites and (b) comparison of the thermal enhancement factor between PVDF/BN and PVDF/BN/CNT composites.

Fig. 3 shows that the 2 wt% CNTs exhibit the similar role to that of the 1 wt% BN particles in improving the thermal conductivity of the PVDF composites. Interestingly, Adding the same content of CNTs (2 wt%) into the PVDF/BN composites promotes the apparent enhancement of the thermal conductivity. A thermal enhancement factor is proposed to represent the degree of enhancement of thermal conductivity of composites. As shown in Fig. 3b, the thermal enhancement factor of the PVDF/BN samples increases with increasing BN content. This indicates that BN particles with high content induce more apparent enhancement in thermal conductivity. With the presence of 2 wt% CNTs, the thermal enhancement factor of the ternary PVDF/BN/CNT samples exhibits more apparent enhancement with increasing BN content. Similarly, compared with the PVDF/BN composites, largely enhanced thermal conductivity is achieved for the PVDF/BN/CNT composites.

Effect of CNTs on Thermal Conductivity of PVDF/GE Composites

To further demonstrate the effect of a small amount of the third component on the thermal conductivity of the PVDF-based composites, in this section, a small amount of CNTs (2 wt%) were introduced into the PVDF/GE composites. The results further demonstrate that compared with the binary PVDF/GE composites, the ternary composites exhibit largely enhanced thermal conductivity (Fig. 4a). CNTs form the local thermal conductive path in the composites (Fig. 4b). Rheological measurements demonstrate that the ternary composites exhibit much higher storage modulus at low frequency ranges (Fig. 4b). This indicates that denser three-dimensional network structure of fillers formed in the ternary composites, possibly mainly contributing to the largely enhanced thermal conductivity of the composites.

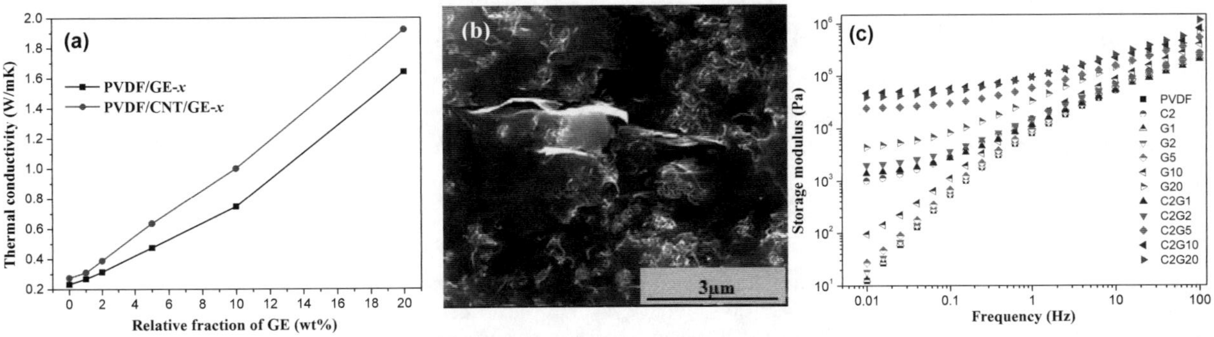

FIGURE 4: (a) variations of thermal conductivity versus the content of GE in the composites, (b) comparison of storage modulus of different samples obtained through rheological measurements, and (c) SEM image showing the dispersion states of CNTs and GE in the representative PVDF/CNT/GE-5 sample.

CONCLUSIONS

In summary, several kinds of PVDF composites containing different conductive fillers have been prepared. To efficiently enhance the conductivity of the composites, a small amount of another component, including PVP, GO and CNTs, etc. have been introduced into the PVDF/CNT, PVDF/CNT, and PVDF/BN and PVDF/GE composites, respectively. The results demonstrate that the presence of the third component greatly enhances the thermal conductivity of these composites. The interfacial interaction between PVDF matrix and conductive fillers, the crystalline structure of the PVDF matrix, and the microstructures of conductive fillers in the composites have been comparatively investigated. The mechanisms for the largely enhanced thermal conductivity of the composites are mainly related to the improved interfacial interaction induced by PVP, the crystal transformation of PVDF matrix induced by GO and the denser thermal conductive path induced by conductive fillers with the aid of the third component.

ACKNOWLEDGMENTS

Authors express their sincere thanks to the National Natural Science Foundation of China (51473137, 51203129) and the Opening Project of State Key Laboratory of Polymer Materials Engineering (Sichuan University) (Grant No. sklpme2015-4-21) for financial support.

REFERENCES

1. M. A. Stroscio, M. Dutta, D. Kahn and K. W. Kim, *Superlattices Microstruct.* 2001, **29**, 405-409 (2001).
2. M. Grujicic, G. Cao and B. Gersten, *Mater. Sci. Eng., B* **107**, 204-216 (2004).
3. C. W. Nan, Z. Shi and Y. Lin, *Chem. Phys. Lett.* **375**, 666-669 (2003).
4. J. Yu, B. Sundqvist, B. Tonpheng and O. Andersson, *Polymer* **55**, 195-200 (2014).
5. S. Shenogin, L. Xue, R. Ozisik, P. Keblinski and D. G. Cahill, *J. Appl. Phys.* **95**, 8136-8144 (2004).
6. T. Zhou, X. Wang, X. Liu and D. Xiong, *Carbon* **48**, 1171-1176 (2010).
7. Z. D. Han and A. Fina, *Prog. Polym. Sci.* **36**, 914-944 (2011).
8. J. H. Yang, C. X. Feng, J. Dai, N. Zhang, T. Huang and Y. Wang, *Polym. Int.* **62**, 1085-1093 (2013).

Determination of the Optimum Coupling Agent Content for Composites based on Hemp and High Density Polyethylene

Denis Rodrigue[a*], Azam Kavianiboroujeni[a] and Alain Cloutier[b]

[a] *Department of Chemical Engineering, Université Laval, Quebec City, G1V 0A6, Canada*
[b] *Department of Wood and Forest Sciences, Université Laval, Quebec City, G1V 0A6, Canada*

**Denis.Rodrigue@gch.ulaval.ca*

Abstract. It is well known that for polymer composites based on natural fibers, the addition of a coupling agent is necessary to improve fiber dispersion and adhesion with the matrix. Nevertheless, an optimum content must be found which is related to the total surface area created between the fibers and the matrix. But in most cases reported in the literature, a single property (like tensile strength or flexural modulus) is used to determine this optimum value. In this work, high density polyethylene (HDPE) was reinforced with hemp fibers as a typical system. In particular, the addition of a coupling agent based on maleated polyethylene (MAPE) was studied to determine its optimum content. To better detect the specific effect of the selected coupling agent, reinforcement content was limited to 10% wt., while the MAPE content was controlled at different levels (0, 5, 7, 9 and 11% wt. based on total hemp content). Compounding was performed in a twin-screw extruder and the samples were produced by compression molding. From the composites obtained, a complete set of characterization in terms of morphology (SEM), mechanical properties (tension, flexion, and impact), and density (pycnometry) was made. From all the results obtained, it can be shown that 9% wt. MAPE is the optimum content maximizing all the mechanical properties. These results indicate that any physical property can be used to determine the optimum coupling agent content.

Keywords: Polyethylene, hemp, coupling agent, optimization, mechanical properties.
PACS: 81

INTRODUCTION

Natural fibers are used mainly as filler in composites to decrease production costs, reduce final products weight, and enhance mechanical properties such as strength and stiffness. Moreover, natural fibers are renewable, recyclable, and biodegradable with minimum CO_2 footprint and low energy consumption to grow and process. They also have high electrical resistance, as well as good thermal and acoustic insulation properties. Another advantage of natural fibers is surface modification and grafting is practical and easy. But, their main drawbacks are high moisture absorption, low degradation temperature, low fire resistance, and incompatibility with most polymers.

Natural fibers, compared to thermoplastic resins, have high Young's moduli which is contributing to high composite stiffness [1]. Improving composite Young's modulus depends on several parameters such as fiber content, fiber orientation, interaction, and fiber-matrix adhesion. Mixing polar and hydrophilic fibers with nonpolar and hydrophobic matrices influences fiber dispersion in the matrix, as well as interfacial adhesion. Maximizing the adhesion between both phases can significantly affect the mechanical properties of the composites by maximizing the role of the fibers (improved interfacial stress transfer).

A low cost and effective method to improve the interfacial adhesion between the fibers and the matrix is via compatibilizing agents addition. These compounds are bonding agents (surfactants, compatibilizers, and dispersing agents) [2], making bridges and connecting the fibers to the matrix by one or more of the following mechanisms: covalent bonding, polymer chain entanglement, and strong secondary interactions such as hydrogen bonding.

Several efforts have been made in the last two decades to improve the interfacial bonding strength between fibers and matrices in a composite using different methods [1,3-13]. But for most studies, a single property was used to determine the optimum composite's composition.

So, based on the limited amount of information available on hemp/polyethylene composites, the objective of this study is to investigate the influence of compatibilizing agent content on different mechanical properties of hemp/high density polyethylene (HDPE) composites. From the results obtained, a comparison is made to determine if the optimum occurs at the same concentration for each property studied.

EXPERIMENTAL

Materials

High density polyethylene (HDPE) grade C-7525 with a melting temperature of 126°C, a melt index of 0.15 g/10 min (190°C, 2.16 kg) and a density of 0.930 g/cm^3, was supplied by Petromont (Canada). Hemp fibers were provided by the Canadian Hemp Trade Alliance (length between 1 and 1.2 mm) with a density of 1.34 g/cm^3. Figure 1 presents the state of the material used at different magnifications. It is clear that a wide range of particle diameter and length was produced for these porous fibers. Maleic anhydride-grafted polyethylene, also called MAPE (Epolene C-26), was supplied by Westlake Chemical Corporation (USA) with a density of 0.919 g/cm^3 and a melting point of 121°C.

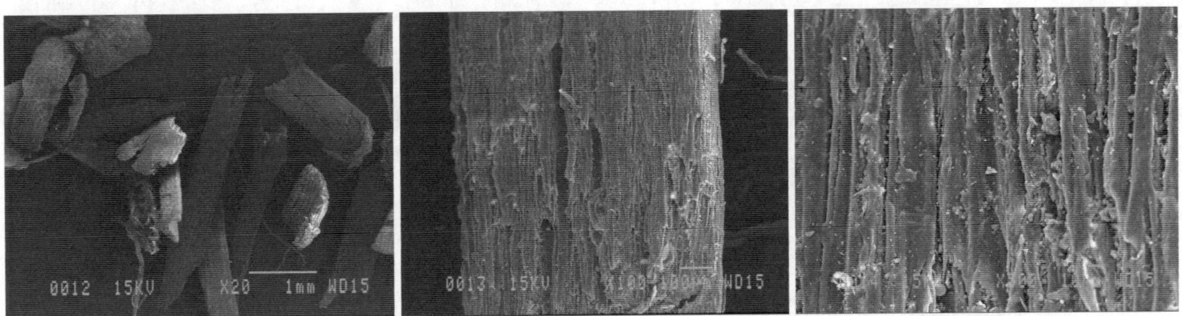

FIGURE 1. Typical SEM micrographs of the hemp fibers used at different magnifications.

Sample Preparation

Composites with 10% wt. hemp and different MAPE content (0, 5, 7, 9 and 11% wt. based on total hemp weight) were prepared. In the first step, hemp fibers were dried overnight in an oven at 80°C to remove moisture. Then, HDPE/MAPE and hemp were respectively introduced in the first and fourth zone of a twin-screw extruder (Leistritz ZSE-27, L/D = 40, 10 zones) with a screw speed of 100 rpm. The temperature profile was set at: 180/185/195/195/195/195/195/195/200/200°C from the main feeder to the die exit. Then, the extruded compounds were pelletized and dried at 80°C overnight. In the next step, compression molding was used to produce composite plates by a Carver laboratory press at 170°C under a load of 3 tons. The compounds were first preheated for 3 minutes, then pressed for 5 minutes in a mold with dimensions of 250 × 250 × 3 mm^3. More details on the materials and methods can be obtained elsewhere [14].

Characterizations

The morphology and interfacial adhesion of the fiber and matrix were studied by scanning electron micrographs (SEM). A JEOL model JSM-840A was used to study the structures exposed through cryogenic fracture and coated with a thin layer of Au/Pd prior to analysis.

Tensile properties were determined using an Instron model 5565 universal testing machine with a 500 N load cell at a strain rate of 5 mm/min and at room temperature. The specimens were cut directly from the molded plates according to ASTM D638 (type V). Young's modulus (E_t), tensile strength (σ_t), and elongation at break (ε_t) values are reported based on a minimum of five samples.

Three-point bending flexural tests (60 mm span) were conducted at room temperature on specimens with dimensions of 75 × 12.7 × 3 mm^3 according to ASTM D790. An Instron model 5565 with a 50 N load cell was run at a crosshead speed of 5 mm/min to measure the flexural modulus (E) and flexural strength (σ_{max}) of the samples. The average and standard deviation results were based on five replicates.

Charpy impact tests were conducted at room temperature according to ASTM D6110 with a Tinius Olsen testing machine Model Impact 104 with a pendulum force of 22.76 N. Rectangular samples with dimensions of 120 × 12.7

× 3 mm^3 were V-notched with an automatic sample notcher Dynisco model ASN. The average value of Charpy impact strength (I_s) with standard deviation is reported based on ten specimens.

Density was measured using a gas (nitrogen) pycnometer model ULTRAPYC 1200e (Quantachrome Instruments, USA). The reported values are the average of at least three measurements.

More details on procedures and characterization can be found in a previous document [14].

RESULTS AND DISCUSSION

Morphology

Figure 2 presents typical SEM micrographs of the hemp/HDPE composites with different MAPE content. It can be seen that the adhesion and bonding quality between both phases strongly depend on the compatibilizer content. Figure 2A shows an image for the composite with 10% wt. hemp and 5% wt. MAPE. The low compatibility between the fibers and the matrix is recognizable due to the presence of several gaps which confirms the lack of mechanical and chemical bonding across the interface. As seen in Figure 2B, the gaps between the matrix and the fibers at 7% wt. MAPE decreased considerably, but some can still be observed. Figure 2C shows that 9% completely wetted the hemp surface by the matrix. The elimination of gaps at the fiber-matrix interphase indicates some adhesion improvements at the interface. But increasing the content of compatibilizer did not affect positively the quality of the fiber-MAPE contact and clearly some voids are noticeable at the interphase. In this case (over-dosing), the compatibilizer did not act as a coupling agent, but acted more as a plasticizer at higher content which limited interfacial stress transfer [15].

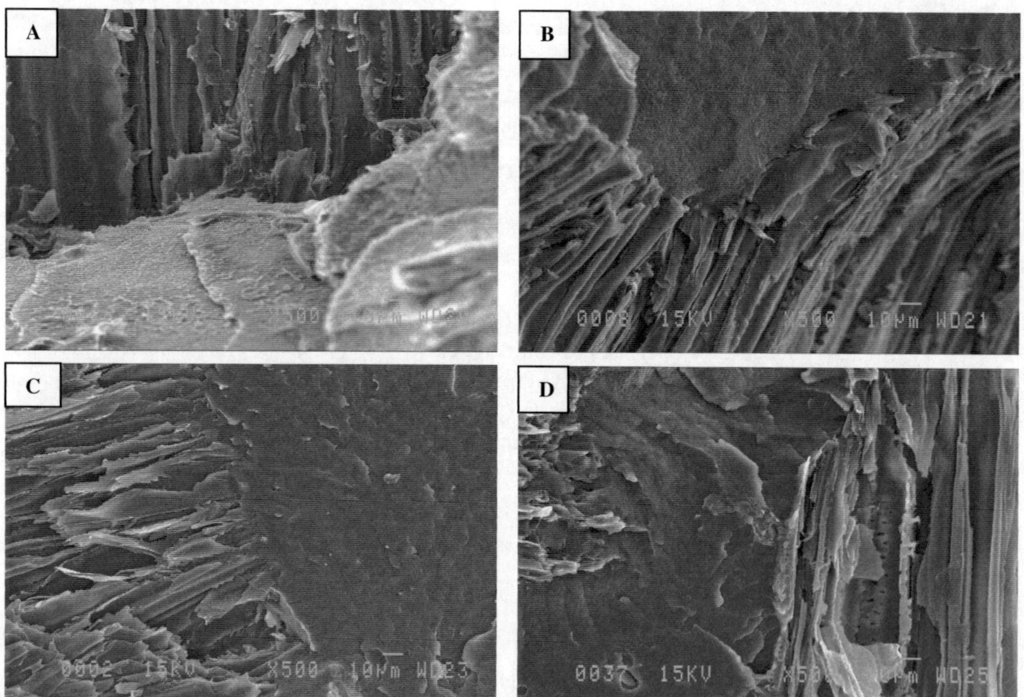

FIGURE 2. Typical SEM micrographs of the composites with 10% wt. hemp fiber and: A) 5, B) 7, C) 9, and D) 11% wt. MAPE [14].

Mechanical Properties

As shown in Table 1, tensile modulus increased with MAPE content up to 9% with a maximum of 274 MPa which represents a 7% increase over the neat composite. But increasing furthermore the MAPE content to 11% decreased the tensile modulus to 266 MPa due to the plasticizing effect of the compatibilizer. Tensile strength had a similar trend as tensile modulus, but the effect is more easily detected as the maximum value at 9% MAPE is 18.7 MPa, which is 48% higher than the neat matrix (12.6 MPa). Inclusion of the compatibilizer also enhanced the elongation at break of the composites. Again, the maximum is obtained at 9% MAPE.

Table (1). Mechanical properties of the composites with 10% wt. hemp at different MAPE content.

MAPE (% wt.)	E_t (MPa)	σ_t (MPa)	ε_t (%)	E (GPa)	σ_{max} (MPa)	I_s (J/m)	*Density* *(kg/m^3)*
0	256 (22)	12.6 (0.6)	12.5 (0.9)	1.01 (0.01)	17.8 (1.1)	108 (6)	969
5	264 (10)	15.4 (0.7)	14.1 (1.8)	1.03 (0.01)	20.7 (0.8)	135 (13)	970
7	265 (15)	16.4 (0.3)	15.0 (0.8)	1.04 (0.03)	21.3 (1.2)	148 (5)	970
9	274 (11)	18.7 (0.9)	16.4 (0.5)	1.11 (0.04)	23.4 (0.9)	154 (15)	971
11	266 (25)	15.6 (0.1)	14.9 (0.7)	1.07 (0.07)	21.3 (0.7)	160 (13)	971

Numbers in parenthesis denote standard deviations (SD). Density SD is 1 kg/m^3.

Similar results were obtained for flexural modulus and strength, but a slight difference, which might be negligible within experimental uncertainty, can be seen for impact strength where the values are very close above 7% MAPE. Nevertheless, all the mechanical results seem to confirm the results of the morphological analysis: the optimum MAPE content for this system (10% wt. hemp) is 9% wt. under the processing conditions and the materials selected. Therefore, a next step would be to prepare composites with different hemp content with 9% MAPE and characterized them to evaluate the effect of hemp content on the mechanical properties of these composites. Some results can be found in other publications [14,16].

It was found that density was not significantly influenced by MAPE concentration. Nevertheless, the quality of the fiber/matrix adhesion could be recognizable by studying density changes of the samples. The density of the sample without MAPE is 969 kg/m^3. Interestingly, increasing the MAPE content did not affect the density for the samples with 5 and 7% wt. MAPE. Although increasing the MAPE content with a density of 919 kg/m^3 from 5 to 7% wt., should decrease the density because HDPE with higher density (930 kg/m^3) was replaced with MAPE with lower density (919 kg/m^3). But it seems that the voids and gaps in the composite with 5% wt. were replaced with matrix due to the improved adhesion which led to the same final density for both composites: 970 kg/m^3. Increasing MAPE to 9%, increased the density of the composite to 971 kg/m^3, which was the maximum for all the composites with different MAPE contents.

CONCLUSION

In this study, high density polyethylene (HDPE)/hemp fiber composites were prepared by a combination of extrusion and compression molding. A special focus was made on the effect of coupling agent (maleated polyethylene, MAPE) content (0, 5, 7, 9, 11% wt.). From the samples produced, morphology, density, and mechanical properties in tension, flexion, and impact were reported.

As a first step, composites with only 10% wt. hemp were prepared with different MAPE concentrations. Based on the morphological and mechanical results, it was found that 9% wt. MAPE was the optimum content, independent of the property studied. This was attributed to increased adhesion and stress transfer at the interface between hemp and HDPE without having extra MAPE molecules not present at the interface which can act as plasticizers.

Finally, the density of the resulting composites was almost the same for all the samples produced. This behavior can be attributed to opposing effects: a density decrease by replacing a part of HDPE (930 kg/m^3) by MAPE (919 kg/m^3) compared to a density increase related to the elimination of voids and gaps due to better interfacial contact. Nevertheless, more work needs to be done to determine the best type (chemical composition) and structure (physical properties) of a coupling agent for a specific system (fiber and matrix).

ACKNOWLEDGMENTS

The authors acknowledge the financial support of the Natural Sciences and Engineering Research Council of Canada (NSERC). The technical help of Mr. Yann Giroux was also much appreciated.

REFERENCES

1. M. Zabihzadeh, F. Dastoorian and G. Ebrahimi, *J. Reinf. Plast. Comp.*, **29**, 123-131 (2008).
2. J. K. Kim and K. Pal, *Recent Advances in the Processing of Wood-Plastic Composites*, Springer Science & Business Media, 2010.
3. R. M. Rowell, A. R. Sanadi and D. F. Caulfield. *Lignocellulosic-Plastic Composites,* 1997, pp. 23-51.
4. D. Maldas, B. V. Kokta and C. Daneault, *J. Appl. Polym. Sci.*, **37**, 751-775 (1989).
5. J. Z. Lu, Q. Wu and I. Negulescu, *J. Appl. Polym. Sci.*, **96**, 93-102 (2005).
6. R. T. Woodhams, G. Thomas and D. K. Rodgers, *Polym. Eng. Sci.*, **24**, 1166-1171 (1984).
7. B. V. Kokta, D. Maldas and C. Daneault, *Polym. Plast. Tech. Eng.*, **29**, 87-118 (1990).
8. M. H. B. Snijder and H. L. Bos, *Compos. Interf.*, **7**, 69-75 (2000).
9. K. B. Adhikary, S. Pang and M. P. Staiger, *Composites Part B: Eng.*, **39**, 807-815 (2008).
10. A. L. Catto, B. V. Stefani and V. F. Ribeiro, *Mater. Res.*, **17**, 203-209 (2014).
11. T. J. Keener, R. K. Stuart and T. K. Brown, *Composites Part A: Appl. Sci. Manuf.*, **35**, 357-362 (2004).
12. A. K. Rana, A. Mandal and S. Bandyopadhyay, *Comp. Sci. Tech.*, **63**, 801-806 (2003).
13. P. Hedenberg and P. Gatenholm, *J. Appl. Polym. Sci.*, **56**, 641-651 (1995).
14. A. Kavianiboroujeni, "Mechanical Characterization of Wood Plastic Composite Sandwich Panels with Foam Core", M.Sc. thesis, Laval University, 2015.
15. S. Y. Leu, T. H. Yang and S. F. Lo, *Const. Build. Mater.*, **29**, 120-127 (2012).
16. A. Kavianiboroujeni, A. Cloutier and D. Rodrigue, *J. Sand. Struct. Mater.*, **17**, 748-765 (2015).

Fiber Length Reduction during Shearing in Polymer Processing

Elmar Moritzer[a], Gilmar Heiderich[a*]

[a]*University of Paderborn, Kunststofftechnik Paderborn (KTP), Germany*

gilmar.heiderich@ktp.upb.de
elmar.moritzer@ktp.upb.de

Abstract. In recent years a continuously rising demand of short-fiber-reinforced thermoplastics has been noted. The reason for the rising demand are the excellent weight specific mechanical properties compared to non-reinforced thermoplastics. The mechanical properties of the compound of high-strength fibers and the ductile matrix polymer is based on complex physical relations in which the fiber length is a major influencing factor. When processing short fiber-reinforced thermoplastics with screw machines (injection molding, extrusion) the fiber length is a resulting material property that is, to some extent, influenced by the process parameters. The extent of the influence and the effects of the process parameters have been carried out in many different studies, mainly with parameter variations. Nevertheless, the influence of the shearing stress on the fiber breaking is also physically modelled. However, the influences of different fiber contents are not taken into account.

To investigate the correlation between the fiber content of the thermoplastic polymer and the fiber length degradation a defined flow field is necessary. For this purpose, a test stand with a heated rotating piston and a stationary outer cylinder is used. In the gap between the piston and the outer cylinder the melted material faces shear stress depending on different rotational speeds. For the investigation different short-fiber-reinforced thermoplastics (polypropylene matrix) with different fiber content are used. By measuring the fiber length after different shear times the time-varying fiber length is determined. With these results the influences of the fiber content and shear stress on the fiber length can be identified.

Keywords: Fiber-reinforced polymers, fiber degradation, glass fibers, shear flow
PACS: 81.05.Ni

INTRODUCTION

In recent years, a continuously rising demand has been noted for densely-filled, short-fiber-reinforced polypropylene. This development is chiefly to be seen as a reflection as of the superior weight-specific mechanical properties of these materials as compared to un-reinforced thermoplastics. This improvement in mechanical properties can be seen for a fiber content as low as 10 %. As the fiber content increases further, the mechanical properties increase correspondingly, until a critical fiber content of 40 % is reached, after which point the tensile strength of the compound once again begins to decrease. The stiffness, however, can be improved still further by continuing to increase the fiber content [1].

In addition to the fiber content, fiber length also plays a significant role here; this is evident when comparing the mechanical properties of short-fiber-reinforced and long-fiber-reinforced polypropylene [2]. While short-fiber-reinforced polypropylene exhibits fiber lengths of less than a millimeter after being processed in an injection molding process, polypropylene reinforced with long glass fibers can have fiber lengths of up to 3 mm. This makes it possible to obtain improvements in mechanical properties of up to 40 % as compared to the short-fiber-reinforced polymer.

The resulting fiber lengths after the injection molding process are, not least, dependent on the initial fiber lengths in the pellets; in addition, they depend on the processing conditions, as well as the material properties of the compounding elements. Material processing initiates fiber-shortening mechanisms that are ultimately responsible for the characteristic fiber length determined at the intervals for each operational point. Three shortening mechanisms can be distinguished which occur when processing thermoplastic fiber-reinforced polymers in screw machines [3]:

- Fiber-fiber interactions
- Fiber-polymer interactions
- Fiber-machine interactions

In the past, fiber/polymer interactions due to viscous forces were primarily those investigated and described by models, such as that of Forgacs and Mason (among others) [4], which investigated the buckling of artificial fibers in thinned suspensions. Newer models for fiber/polymer interactions [5] go beyond buckling to include fiber breakage. Fiber/fiber interactions, which are especially abundant in densely filled plastics [2], represent a serious challenge for physical modelling as a consequence of the complex relationships between acting forces in such scenarios.

Because of these complex relationships and the attendant difficulty in modelling, fiber/fiber interaction has been investigated experimentally here and the experimental results used in this paper. To do so, a Couette rheometer is used to subject short-fiber-reinforced polypropylenes to defined shear loading in a drag flow.

MATERIALS

The short-fiber-reinforced materials required for the investigations are produced by employing a twin-screw compounder. For the compounding process two polypropylene homopolymers (Lyondellbasell Moplen HP500N and Moplen HP500V) with different flow properties (FIGURE 1) and e-glass fibers are used.

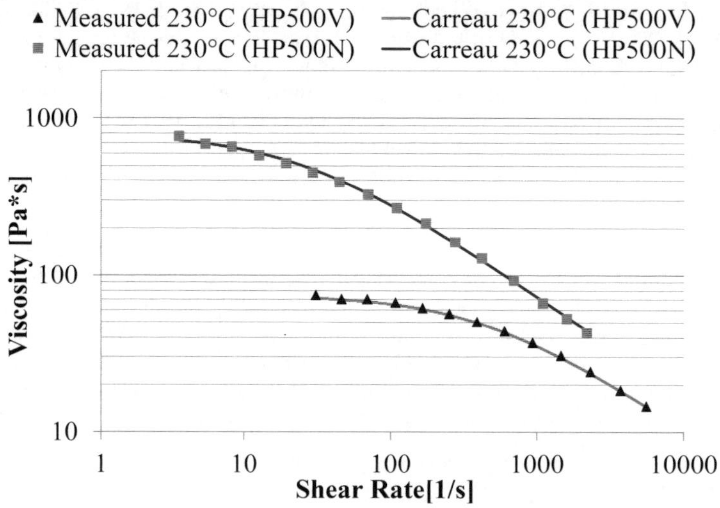

FIGURE 1. Shear Viscosity of the Moplen HP500V and the Moplen HP500N (measured and modelled)

To receive short fiber-reinforced materials with different fiber contents (20 %, 30 %, 40 %) varied proportions of cutted fibers (average fiber length 4.5 mm, average diameter 10 μm) are added during the compounding process. The resulting fiber content and the resulting fiber length in the granule are shown in TABLE (1).

TABLE (1). Produced Fiber-Reinforced Materials

	GF20-V	GF30-V	GF40-V	GF20-N	GF30-N	GF40-N
Matrix Polymer		HP500V			HP500N	
Nominal Fiber Content [%]	20	30	40	20	30	40
Actual Fiber Content [%]	20.00	30.40	39.20	19.99	29.64	39.95
Number-Average Fiber Length [μm]	855.53	631.26	594.83	663.33	510.59	466.00

EXPERIMENTAL

The apparatus used to determine the fiber degradation is a Couette rheometer. This rheometer consists of a rotating piston inside a cylinder. Between the piston and the cylinder there is a shearing gap, which allows to apply defined shear loads to the material. The shear rate that occurs during the investigations is proportional to the rotation speed of the piston. The cylinder is heated via an oil tempering system with helical tempering channels set directly into the cylinder wall (see Figure 2). The piston is heated as well with a heating cartridge. To determine the melt temperature during the investigations thermocouples are used, which are arranged axially in the cylinder and which protrude slightly into the melt.

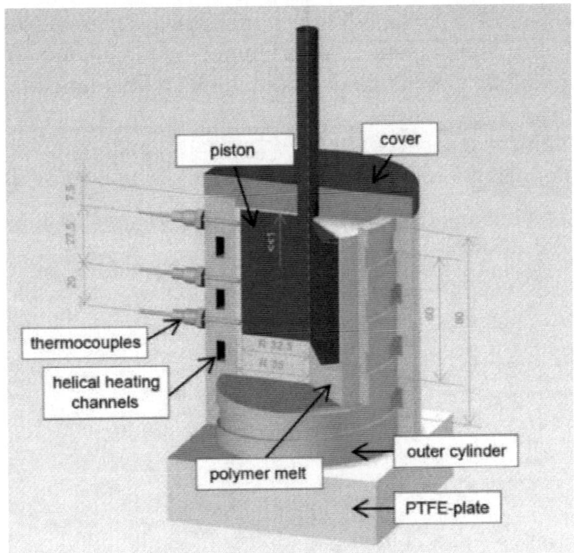

FIGURE 2. Couette Rheometer

At the beginning of the experiments the cylinder and the piston is heated. To fill the cylinder with the material the piston is pulled out axially. Subsequently, the piston is driven back into the cylinder cavity. With this movement, the excess material in the cavity is thus pressed out under low pressure and gets displaced from the piston. After that the piston is removed again to take the first samples from the piston in order to determine the fiber damage caused by the filling procedure. The samples are taken from three positions distributed along the circumference of the piston. The same sampling methodology is also used during the investigation of shearing behavior to take slightly different fiber lengths at different locations of the piston into account.

In the studies, a full factorial experimental design is used, in which the rotational velocity of the piston as well as the materials are variated to define the test points. For each test point, samples are taken after defined shear intervals. The shear intervals are dependend to the rate of the fiber length reduction, which was determined in preliminary studies. To get a high resolution of the fiber length degradation, small time steps for the shear are chosen at high rates of the fiber length degradation. The entire experimental design contains 18 test points (six materials at three speeds) for each of which samples are taken at varying shear times, meaning considerable time and effort is invested experimentally to ensure useable results. For the studies, the rheometer is heated to 230 °C. This temperature corresponds to the processing temperatures normally used for glass fiber-reinforced polypropylene materials.

MEASUREMENT OF FIBER LENGTH

To measure the fiber length of the samples the polymer first has to be pyrolitically removed in an oven at 600 °C. The fiber residue is then submerged in a water bath to separate the individual fibers, after which the separated fibers can be analyzed optically. The fibers are analyzed using electronic image processing, whereby the number-average fiber length is taken as a representative value for the respective fiber distributions during the following evaluation. The number-average length is calculated according to Equation 1.

$$L_N = \frac{\sum L_i n_i}{n} \tag{1}$$

RESULTS AND DISCUSSION

When evaluating the number-average fiber length degradation, plots of the fiber length decreasing as a function of the shear time are used. To determine the fiber-fiber interactions and their influence on the fiber length decrease materials with the same matrix polymer, but different fiber contents, are compared in one plot. Furthermore, different matrix materials are compared to demonstrate the influence of different shear viscosities on the shear time related fiber length degradation.

An increased damage to the fibers with increasing fiber content is shown in FIGURE 3. This is similar detected for the different matrix materials. The increased fiber degradation at higher fiber contents becomes already obvious by having a view at the number-average fiber length in the granulate. The granulate with a higher fiber content has a considerably lower fiber length, indicating enhanced damage in the twin screw process. The shear time related fiber length decrease is also influenced by the fiber content. The stationary fiber length, which is achieved earlier at higher fiber contents, indicates this influence.

When looking at the absolute fiber length for the different matrix materials, a general lower damage to the fibers for the low viscosity matrix materials (GF20-V, GF30-V, GF40-V) can be recognized. This is due to the lower load on the fibers caused by shear stresses.

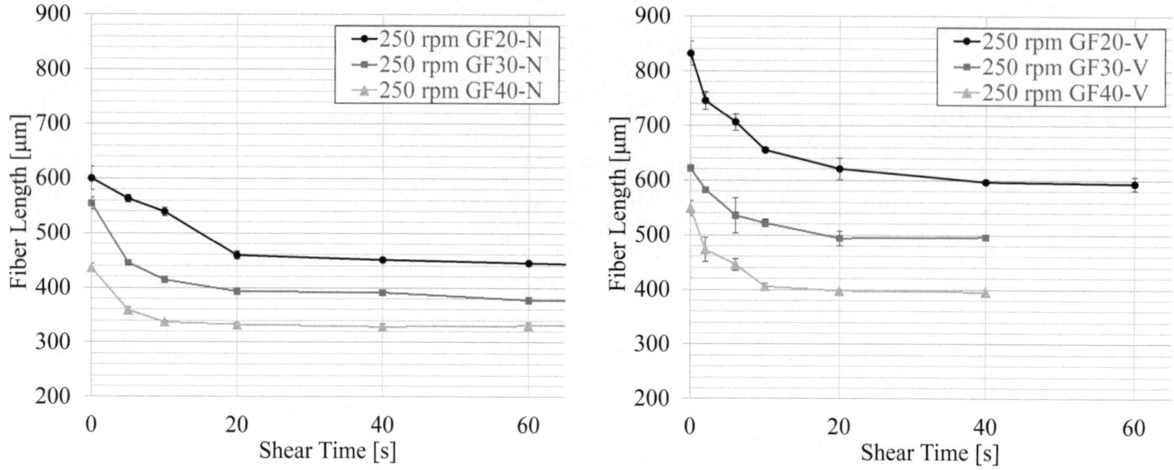

FIGURE 3. Fiber length degradation at 250 rpm

FIGURE 4 shows the shear time related fiber length degradation at a speed of 500 rpm. Compared to the fiber length degradation at 250 rpm similar trends are apparent. The fiber degradation for the low viscose matrix polymer (at all fiber contents) is likewise lower compared to the high viscose matrix polymer. Furthermore, a greater decrease of the fiber length can be detected with increased fiber content. The increase of the fiber content also leads to a earlier achievement of the steady state fiber length. In contrast to the rotational speed of 250 rpm, a lower end value of the steady fiber length is achieved. In addition to that the steady state value of the fiber length former is achieved earlier with higher rotational speed. This shows the increasing of the fiber degradation with higher rotational speed.

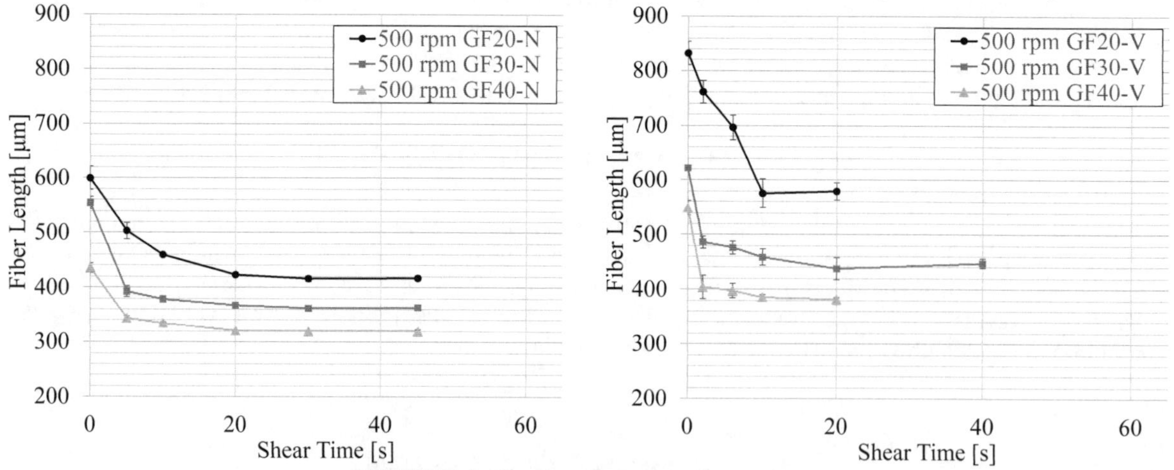

FIGURE 4. Fiber length degradation at 500 rpm

In FIGURE 5 the fiber length degradation is shown after further increasing the rotational speed to 750 rpm. The already observed influence of the fiber content on the fiber damage (250 rpm and 500 rpm) can also be seen here. A significantly faster decrease of the fiber length to a steady value however, can not be detected in spite of the high rotational speed. This is due to the poor starting performance of the test rig, which is caused by the high translation of the gear.

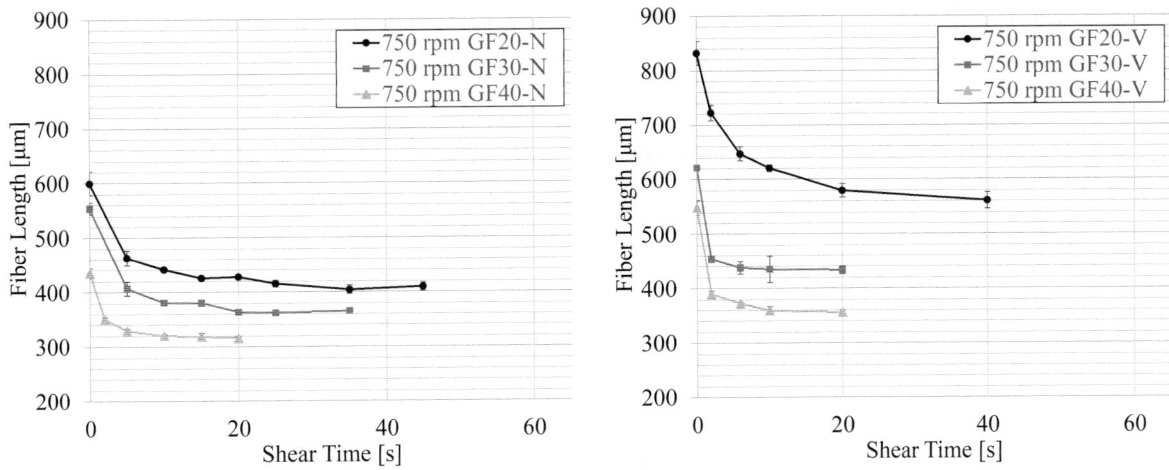

FIGURE 5. Fiber length degradation at 750 rpm

CONCLUSIONS AND OUTLOOK

The studies in this paper show the fiber length reduction of fiber-reinforced plastics with different viscous matrix materials and different fiber contents in a drag flow. Generally, a decreasing fiber length during shearing occurs until a stationary final value is reached. Strong influences of the fiber content on the fiber length degradation in the drag flow are clearly visible. An acceleration of the fiber length degradation and a lower stationary fiber length becomes visible with an increasing fiber content. These effects can be shown for two matrix materials with different viscosities. Also, this effect can be seen at different shear rates. This is caused by the increased fiber-fiber interactions at higher fiber contents. Furthermore, increased fiber damage can be seen when increasing shear rate and matrix viscosity. This reveals itself not only by the lower steady state value of the fiber length but also by an increased drop in the fiber length during the first shear period. An accurate identification of the cause can not be clearly made. In addition to the fiber-fiber interactions, the fiber-polymer interactions or the fiber-machine interactions can be identified as possible causes.

Based on the results of the studies a modeling will be constructed which takes the fiber-fiber interactions into account. This will allow predictions of the fiber length degradation for other matrix materials, shear rates and fiber contents to be made. A validation of this modeling will be performed first by examining additional parameter settings with the existing materials.

ACKNOWLEDGMENTS

We like to thank LyondellBasell Industries for providing the matrix material that is used for the experiments.

REFERENCES

1. Thomason, J. L.: The influence of fiber length and concentration on the properties of glass fiber reinforced polypropylene: 6. The properties of injection moulded long fibre PP at high fibre content. *Composites Part A: Applied Science and Manufacturing*, Volume 36, Issue 7, July 2005, Pages 995–1003.
2. Thomason, J. L.: The influence of fibre length and concentration on the properties of glass fibre reinforced polypropylene: 5. Injection moulded long and short fibre PP. *Composites Part A: Applied Science and Manufacturing*, Volume 33, Issue 12, December 2002, Pages 1641–1652.
3. Turkovich R. v.; Erwin, L.: Fiber Fracture in Reinforced Thermoplastic Processing. *Polymer Engineering And Science 23*, Issue 13, 1983, S. 743–749.
4. Forgacs, O. L.; Mason, S. G.: Particle Motions in Sheared Suspensions. IX. Spin and Deformation of Threadlike Particles. *Journal of Colloid Science (14)*, 1959, S. 457–472.
5. Phelps, J. H.; Abd El-Rahman, A. I.; Kunc, V.; Tucker, C. L.: A model for fiber length attrition in injection-molded long-fiber composites. *Composites Part A: Applied Science and Manufacturing 51*, 2013, S. 11–21.

The rheological and thermal conductivity study on PA/COC blend filled by Boron Nitride

Nikoo Ghahramani[a], Seyed Armin Seyed Esfahani[b], Milad Mehranpour[b], Hossein Nazockdast[a]

[a] *Polymer Engineering Department, Amirkabir University of Technology, Tehran, Iran*
[b] *Islamic Azad University, Science and Research Branch, Tehran, Iran*

Abstract: Thermally conductive and electrically insulating polymer/boron nitride (BN) composites are highly attractive for various applications in many thermal management fields. However, the preparation methods for polymer/BN composites with prescribed performance have usually caused difficulties in the material post processing. Therefore, using polymer blends with selective localization can be utilized as a solution for this problem. The effect of selective localization of boron nitride (BN) particles on the thermal conductivity of immiscible PA/COC blends is studied. The aim of the present work is to use PA/COC blend with co-continuous morphology in order to produce PA/COC/BN thermally conductive composites with minimum volume fraction of fillers. The efficiency of this method is evaluated through a comparison between thermal conductivity of PA/BN and PA/COC/BN systems. The rheological results and scanning electron microscopy (SEM) images reveal a greater affinity of BN with PA compared to COC phase. It is also shown that in order to form a thermal path and obtain an appropriate thermal conductivity, high concentrations of BN (≥30 wt. %) are needed. Thermal conductivity measurements show that the thermal conductivity of PA/COC/BN composites are lower than what expected based on the mixture law for polymer blends in comparison with PA/BN composites with the same concentration of BN in PA. The cooling rate is found to affect the thermal conductivity of the composite samples which is explained in terms of the crystallinity of PA.

Keywords: **Thermal conductivity, Boron Nitride, Polyamide, Cyclo Olefine Copolymer.**

Introduction: As the electronic devices become denser and powerful, the heat evolved during operation becomes more severe because the generated heat may cause degradation and subsequently reduce their reliability and performance. Therefore, the management of excess heat is an important job in developing these products. Polymer composites for electronic packaging and other applications require high thermal conductivity to dissipate heat. So far, different techniques have been employed to improve the thermal conductivity of polymeric materials. A common method used to produce a thermally conductive composite is adding high thermally conductive fillers. Ceramic powders, including aluminum nitride [1,2], boron nitride [3,4] and silicon carbide [5] have been used extensively due to their high thermal conductivity and electrical resistivity.[6,7] The conductive polymer composites and their potential applications are dependent on the dimension (size and shape) of fillers, the amount of filler loading, and the degree of dispersion. Higher filler loading and close packing are typical requirements to achieve the appropriate level of thermal conductivity. A large amount of fillers is usually needed to form a well-developed conductive network. However, a high volume fraction of fillers could result in poor mechanical properties, complex processing, high density, and high cost [8-11]. Therefore, a composite with high thermal conductivity at lower filler loading is highly desired. Recently double percolation structure has been used to improve the conductivity because it can effectively reduce the volume fraction of fillers by controlling the selective localization of fillers [12]. To achieve a double percolation structure, controlling the selective localization of the fillers is extremely crucial, many factors are affecting the selective localization of filler in polymer blends. Among them, two factors are the most important: polymer viscosity; the fillers prefer to be selectively distributed in the phase with lower viscosity and surface energy; the dispersed phase prefers to stay in the continuous phase with the closer surface energy. This work deals with blends of COC and PA by using BN as thermal conductive filler. In this study COC/PA/BN blends with different composition are prepared with masterbatch method. This work aims to reduce volume fraction of fillers by using masterbatch process to control the selective localization of BN particles.

Experimental details

Materials: PA6 (Akulon F223-D) with a density of 1.13 $g\ cm^{-3}$ was obtained from DSM, Netherlands. Cyclo Olefin Copolymer (COC 6013) with a density of 1.02 $g\ cm^{-3}$ were received from TOPAS. Boron Nitride platelets with the average particle size, 7 μm and a thermal conductivity of 150 $w\ m^{-1}\ k^{-1}$ were produced by Saint-Gobain Co.

Sample preparation: Blends and composites with different compositions were prepared via melt mixing in an internal mixer (Brabender Plasticorder W50) with a capacity of 60 cm^{-3}. All samples were mixed at a temperature of 240 °C with a rotor speed of 60 rpm for 20 min. For preparing blends, PA/BN master batches with different BN content were used. Table 1 introduces blend samples and their BN content. The ratio of PA/COC is 3/7 in all blends.

Characterization: Melt rheological behavior was characterized using a Universal Dynamic Spectrometer (Physica UDS 200, Austria) at 240 °C with parallel plate–plate geometry (plate diameter of 25 mm and gap of 1 mm). The viscoelastic properties of samples were measured under controlled shear deformation oscillatory mode. The strains were specifically chosen to be within the linear viscoelastic range.

Thermal conductivity was measured by thermal conductivity apparatus manufactured by Armfield Technical Company Limited.

Table 1. Blend samples

sample	BN mass ratio in PA/BN masterbatch	BN mass ratio in the composite
Pa/COC 3700	0	**0**
Pa/COC 3710	0.1	**0.03**
Pa/COC 3720	0.2	**0.07**
Pa/COC 3730	0.3	**0.11**
Pa/COC 3740	0.4	**0.17**

Result and Discussion

Rheology

Figure 1(a) shows the variation of storage modulus of PA6, and its composites as a function of loading frequency measured at 240°C. As it can be seen the addition of BN increased the storage modulus of both polymers and resulted in a plateau at low frequencies, indicating the formation of interconnected structures. This suggests that BN has a good affinity with PA.

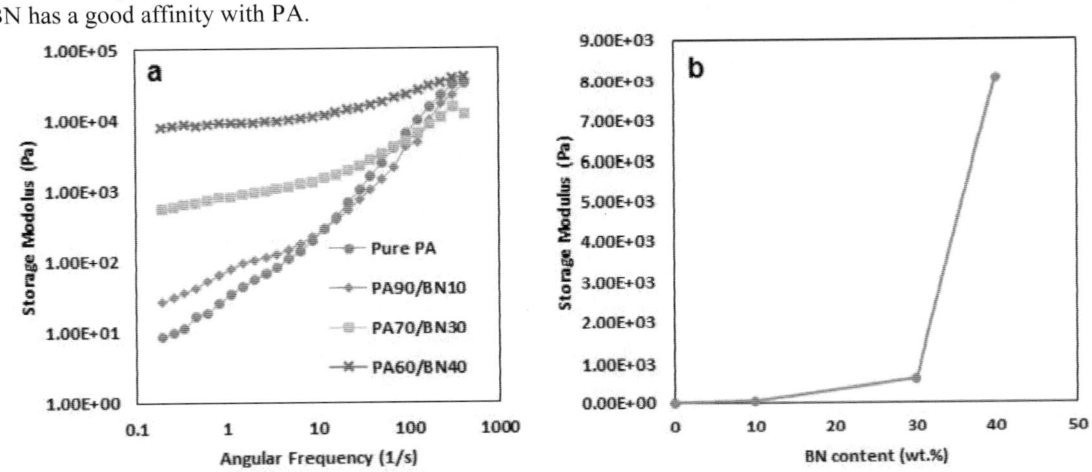

FIGURE 1. a) Storage modulus of PA and its composites. b) Variation of Storage modulus as a function of BN content at frequency=0.192 s^{-1}

Variation of storage modulus of PA/BN composites as a function of BN content at low frequencies has been showed in figure 1(b). Since the slope of the curve has changed significantly in 30 wt.% of BN, it can be easily conducted that there is a rheological percolation threshold about 30 wt.% of BN. Figure 2(a) shows variation of storage modulus of PA/COC/BN composites as a function of loading frequencies. In the figure 2(b) storage modulus of blends at law frequencies vs. BN content has been depicted. According to this figure, there is a rheological percolation at 20 wt.% of BN in the PA phase. Figure 2(c) exhibits the cole-cole diagram of blends. It can be seen that there is an upward branch in the blends which illustrates that there is a double percolated morphology in the blends. From this curves, co-continuous morphology is possible.

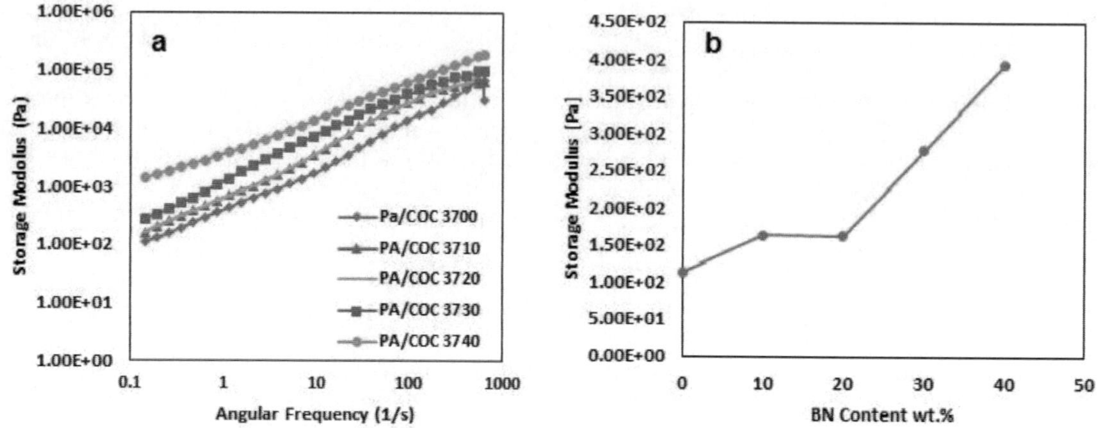

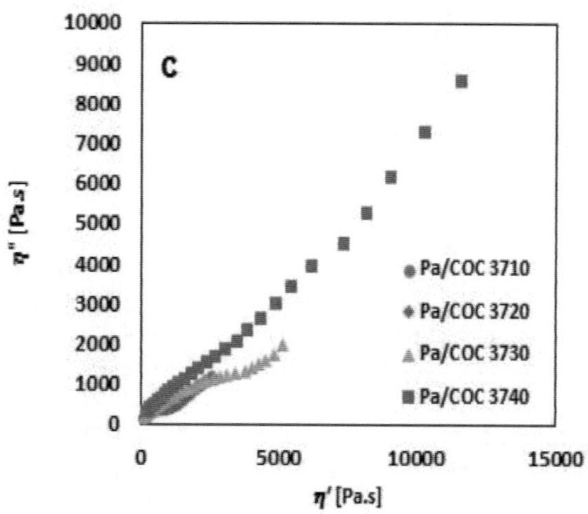

FIGURE 2. a) Storage modulus of PA/COC blends with various BN content vs. loading frequency. b) Variation of Storage modulus as a function of BN content. c) cole-cole diagram for PA/COC blends with various BN content.

Scanning Electron Microscopy

In the figure 3, the SEM image of PA/COC 3740 has been represented. It can be seen that BN fillers have been localized in PA phase, so there is a PA layer on the BN platelet and interconnected structure has been formed in the PA phase.

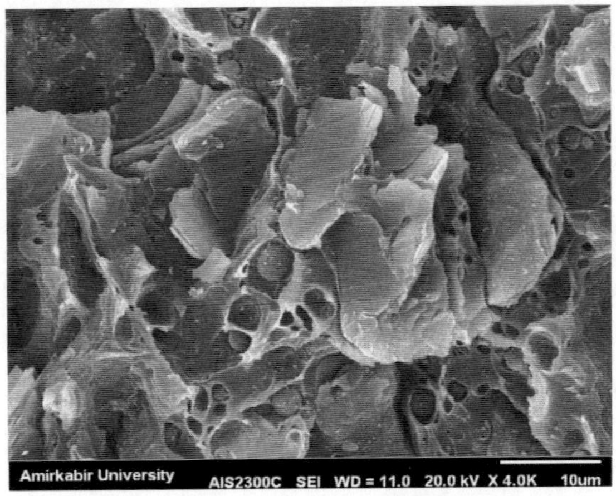

Figure 3. SEM image of PA/COC 3740.

Thermal Conductivity

Figure 4(a) illustrates the thermal conductivity of PA/COC blends with various BN content and its composites. It should be noted that in order to improve thermal conductivity, cooling rate was controlled. It can be seen that for BN≥ 30 $wt.\%$, thermal path way is formed and there is a jump in the thermal conductivity at this concentration. In order to investigate the efficiency of using blend, there is a comparison between blend and PA/BN composite with 40 wt. BN in figure 4(b). Thermal conductivity of PA60/BN40 is 1.43 $w\,m^{-1}\,k^{-1}$ and with using mixture law, thermal conductivity of PA/COC 3740 should be 0.66 $w\,m^{-1}\,k^{-1}$ while it is 0.59 $w\,m^{-1}\,k^{-1}$. This difference between expected thermal conductivity and the experimental one is due to the creation of a new surface between PA and COC phases which exert an extra phonon scattering on the blend. Another possible reason is that during blending, some of broken up fillers could migrate to COC phase, so they do not participate in the thermal conductivity which is created by thermal pathway.

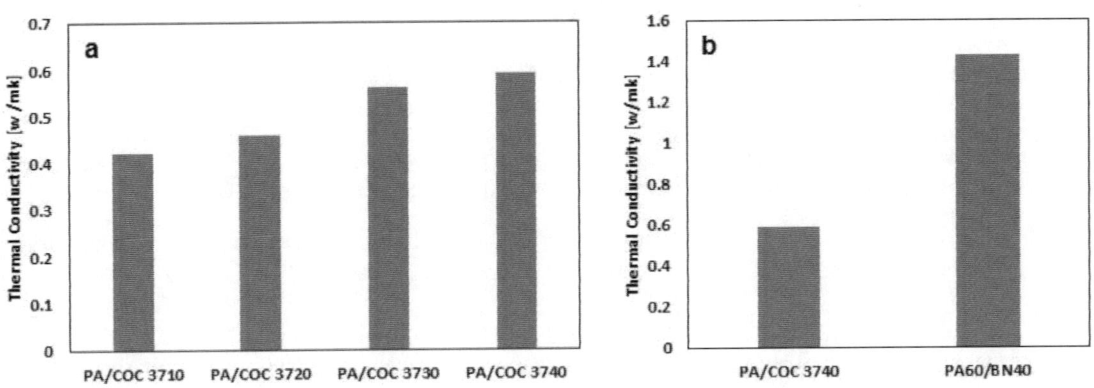

FIGURE 4. a) Thermal conductivity of PA/COC blends with various BN content. b) A comparison between thermal conductivity of blend and PA/BN composite with similar BN content in PA phase.

Conclusion

The thermal conductivity of COC/PA blends are improved by selective localizing BN particles. In order to achieve a desired thermal conductivity, high concentrations of filler were required. Thermal conductivity of the polymer blend expected to follow the rule of mixture, but it did not; by adding BN particles, an extra surface adds to the system and it increases phonon scattering on the surfaces; therefore, thermal conductivity deceased in compared with what was expected according to PA/BN composite. In addition, during the blending process, some of the BN particles broken up and become smaller particles, so they migrate form PA phase to the COC phase. These two factors decrease the effect of BN and increase phonon scattering, respectively. However, due to using blends with double percolated morphology and localization of fillers, a desired thermal conductivity was achieved.

References

[1] Y. Xu, D. D. L. Chung and C. Mroz, "Thermally Conducting Aluminum Nitride Polymer-Matrix Composites," Composites Part A: Applied Science and Manufacturing,2001, Vol. 32, No. 12, pp. 1749-1757.

[2] C. Y. Hsieh and S. L. Chung, "High thermal conductivity epoxy molding compound filled with a combustion synthesized AIN powder," Journal of Applied Polymer Science,2006, vol. 102, no. 5, pp. 4734–4740.

[3] H.Ishida and S.Rimdusit, "Very high thermal conductivity obtained by boron nitride-filled polybenzoxazine",1998,V.320, pp.177-186.

[4] W. Zhou, S. Qi, H. Li, S. Shao , "Study on insulating thermal conductivity BN/HDPE composites", Thermochim Acta, (2007), pp. 36–42.

[5] T.Zhou., X.Wang, G. U., Mingyuan, X.Liu., "Study of the thermal conduction mechanism of nano-SiC/ DGEBA/EMI-2,4 composites", Polymer,2008,Vol. 49, 4666– 4672.

[6] Z. Han and A. Fina, "Thermal conductivity of carbon nanotubes and their polymer nanocomposites: a review," Progr. Polym. Sci., vol. 36, no. 7, pp. 914-944, 2011.

[7] R. Kochetov, T. Andritsch, U. Lafont, P.H.F. Morshuis, S.J. Picken, and J.J. Smit, "Thermal behavior of epoxy resin filled with high thermal conductivity nanopowders," 2009, IEEE El. Ins. Conf., Montreal, QC, Canada, pp. 524-528.

[8] H.Im, J.Kim, "Thermal conductivity of a graphene oxide-carbon nanotube hybrid/epoxy composite", Carbon, 2012, Vol.50, pp. 5429–5440.

[9] S. R. Kim, M. Poostforush, J. H. Kim, S. G. Lee.: Thermal diffusivity of in-situ exfoliated graphite intercalated compound/polyamide and graphite/polyamide composites. Express Polymer Letters, 6, 476–484 (2012).

[10] Z.Shi, M.Radwan, S.Kirihara , Y.Miyamoto, Z.Jin, "Enhanced thermal conductivity of polymer composites filled with three-dimensional brush like AlN nano whiskers",Applied Physics letters,2009 Vol.95,224104.

[11] T.Zhou, X.Wang, Xi. Liu,D.Xiong, "Improved thermal conductivity of epoxy composites using a hybrid multi-walled carbon nanotube/micro-SiC filler",2010, Vol.48,pp.1171-1176.

[12] D.Yorifuji and Sh. Ando, "Enhanced thermal conductivity over percolation threshold in polyimide blend films containing ZnO nano-pyramidal particles: advantage of vertical double percolation structure",2011, Vol.21, pp.4402-4407.

Melt Mixing Functionalized Graphite Nanoplates Into PC/SAN Blends

M. C. Paiva[a*], E. P. Cunha[a], O. Voigt[b], M. Liebscher[b], F. Simon[b], J. Pionteck[b], P. Pötschke[b*]

[a]*Institute for Polymers and Composites/i3N, University of Minho, 4800-058 Guimarães, PORTUGAL*
[b]*Leibniz Institute of Polymer Research Dresden, Hohe Str. 6, 01069 Dresden, GERMANY*

mcpaiva@dep.uminho.pt
mcpaiva@dep.uminho.pt; poe@ipfdd.de

Abstract. The present work focusses on the analysis of the effect of chemical functionalization of graphite nanoplates (GNP) on the dispersion and localization of the GNP in polycarbonate (PC)/poly(styrene-acrylonitrile) (SAN) blends, the composite morphology, and the resulting electrical properties. For this, GNP have been chemically modified to covalently bond pyrrolidine groups at the graphene surface of the GNP. The amount and detailed structure of the functionalization has been analyzed by XPS. The dispersion of different GNP in the blends has been studied by light microscopy and the morphology of the blends and blend composites by SEM and TEM. At a composition of 60 wt.% PC and 40 wt.% SAN the blend and its composites form co-continuous structures. By means of TEM the selective localization of the GNP and f-GNP within the PC component or near the PC-SAN interface could be proven. The functionalization resulted in less agglomerated structures of the GNP in the composites. The electrical percolation concentration is not significantly influenced when comparing blend composites prepared with GNP or f-GNP. However, the conductivity in the composite blends containing f-GNP is always lower compared to the GNP composites.

Keywords: Graphite Nanoplates; Polymer blends; Composites; Morphology; Electrical properties.
PACS: 61.46.-w; 62.23.Pq; 72.80.Tm; 81.05.Qk

INTRODUCTION

The topic of polymer blends has been quite important and intensively studied along the past decades. The major aim is to attain target properties by mixing polymers with complementary characteristics, aided by the development of specific blend morphology. The inclusion of nanoparticles with special properties such as high electrical or thermal conductivity can add desired properties to the blends according to the aimed application, but certainly fillers will also induce relevant changes in the processing behavior and final composite morphology and properties. The blend composition and mixing conditions may be tailored to maximize specific properties.

The present work reports the analysis of the effect of chemical functionalization of graphite nanoplates (GNP) on the dispersion and localization of the GNP in PC/SAN blends and the resulting blend morphology. For this, blends exhibiting co-continuous blend morphology were selected, where one component is conductive due to localization and network formation of the conductive filler within this component. Even if the other blend component does not contain any conductive filler, the blend will be electrically conductive because of the continuity of the conductive component (principle of double percolation).

Such effects of selective localization in one blend component could be shown for multiwalled carbon nanotubes (MWCNT) in the same PC/SAN = 60/40 wt.-% blend system, where always selective location in PC was found [1-5] resulting in much lower resistivity as compared to PC or SAN based composites at the same filler loading [1,3]. Even if premixed with the thermodynamic less favorable SAN component, during melt blending the nanotubes migrated very fast into the PC phase [1,2,4,5]. By using a reactive component miscible with SAN it could be shown that a tuning of the nanotube localization is possible [3]. When the amount of reactive groups of the modifier and on the surface of the nanotubes was high enough, MWCNT localized within the modified SAN instead of PC. In addition, for blends containing 1 wt% GNP it was shown that there is also a preferred localization within PC [6]. However, when premixed with SAN, depending on the processing conditions and the achieved filler dispersion, parts of the filler located in SAN or at the blend interface due to stacking during the transfer process.

The aim of this work was to evaluate in which polymer the selected type of GNP localizes, whether its localization changes with functionalization, and which influence the functionalization has on blend morphology and electrical properties. Thereby, the GNP modification was performed using a mild functionalization method based on the dipolar cycloaddition reaction of azomethyne ylides to the GNP aromatic carbon surface, already successfully applied for surface modification of carbon nanotubes [7]. GNP and functionalized GNP (f-GNP) were melt mixed using a two-step procedure into PC/SAN at small scale and the composites' morphology and electrical properties were studied on compression molded samples.

EXPERIMENTAL

Materials

The graphite nanoplates (GNP) type XGnP Grade C were purchased from XG Sciences, polycarbonate (PC) Makrolon 2205 from Bayer Material Science and poly(styrene-co-acrylonitrile) (SAN) Luran 358N from BASF. The chemical modification of the GNP surface was carried out with the amino acid *N*-benzyloxycarbonylglycine (Z-Gly-OH, 99% from Aldrich) and paraformaldehyde (reagent grade) from Sigma.

Graphite Nanoplate Functionalization and Characterization

The functionalization reaction was performed according to the solvent-free procedure described previously for the functionalization of carbon nanotubes [7]. An ethanolic solution/suspension of Z-Gly-OH and paraformaldehyde (1:5 molar ratio) was prepared. A suspension of the GNP in the ethanolic solution was gently heated until the solvent was completely removed, so that a thin layer of Z-Gly-OH and paraformaldehyde was deposited on the GNP surface. The solid mixture was heated at 250 °C during 5 hours. The specific surface areas were measured by BET using nitrogen at 77 K, after drying for 3 hours at 353 K under vacuum, by means of an Autosorb-1 (Quantachrome, USA). Thermogravimetric analysis (TGA) was performed on a Q500 thermobalance from TA Instruments under N_2 (g) and a heating rate of 10 °C/min. Electrical conductivity of the powdered fillers was determined under defined pressure by a self-made two electrode apparatus equipped with a DMM 2001 multimeter (Keithley Instruments).

Preparation of Composite Blends and Characterization

The blend composite preparation was performed as two-step process. First, the filler was melt–mixed with PC. In the second mixing step the obtained PC/filler composite was melt-mixed with SAN. The compositions were chosen in such way that the final composition (in wt.-%) of the blend composites were:

PC (60-x) / SAN (40) / filler (x) (x = 1, 2, 4, 6, 8).

Mixing was performed by means of a 15 mL MicroCompounder (DSM Xplore) with ca. 12 mL filling. The conditions in the first step were T = 260 °C, r = 250 rpm, and t = 15 min., in the second step the blend composite preparation was done at T = 260 °C with r = 100 rpm and t = 5 min. Compression molding of the obtained materials to sheets with 60 mm diameter and 0.5 mm thickness was performed by a hot press PW 40 EH (Paul Otto Weber GmbH) at a pressure of 100 kN for 1 min. Polyimide foils on both sides of the mold were used to hinder sticking of the composite on the metal plates.

The quality of filler dispersion was analyzed by optical microscopy (OM, Olympus BH2 transmission light microscope equipped with DP 71 digital camera, booth Olympus Germany) on thin cuts with a thickness of 10 μm. Strands of the blends and composites were cut perpendicular to the extrusion direction by a Leica RM 2265 Microtome (Leica Microsystems) at room temperature. The morphology of the blends and blend composites was observed on smoothed cuts after dissolving the PC phase in NaOH (30 %) solution at 105 °C for 30 min. SEM images were taken by a Zeiss Neon 40 EsB (Carl Zeiss) after sputtering with Pt to hinder electrostatic charging. The localization of the filler in the blend composite was analyzed by TEM (Libra 200, Carl Zeiss) on thin cuts of 50 – 70 nm, which were prepared by a Leica UC7 Microtome (Leica Microsystems), again perpendicular to the extrusion directions of the used strands. Volume resistance measurements were carried out on a 6175A Electrometer (Keithley Instruments), coupled with an 8009 Resistivity Test Fixture with two plate electrodes located on both sides of the samples. When the resistance was below 10^7 Ω a home-made 4-

electrode test fixture for strips equipped with a DMM 2000 (Keithley Instruments) was used. Resistivity was calculated taking into account the geometric conditions.

RESULTS AND DISCUSSION

TGA analysis of the powder GNP materials showed an increase of the weight loss from 7.3 to 17.4 wt.% for pristine to functionalized GNP, respectively, at 800 °C, indicative of the organic functionalization of the f-GNP surface.

XPS analysis of the pristine and functionalized GNP demonstrates the formation of a secondary amine by chemical functionalization (in the pyrrolidine form, compound **1** in the scheme of Figure 1). This is the expected product for the 1,3-dipolar cycloaddition reaction of the azomethine ylide formed by reaction of Z-Gly-OH and formaldehyde to the GNP surface at 250 °C. The atomic ratio [N]/[C] increased from 0.005 to 0.035 after functionalization of GNP, showing a large component peak centered near 400 eV in the high resolution N1s spectrum (Figure 1d) evidencing the formation of the cyclic amine. The wide scan spectra and N1s high resolution spectra for the pristine and functionalized GNP are presented in Figure 1a, b, c and d. Under melt mixing conditions with PC (high temperature, high viscosity) reaction of the amine at the f-GNP surface and the carbonate group in PC is expected, thus forming a strong PC/GNP interface (compound **2** in the reaction scheme of Figure 1).

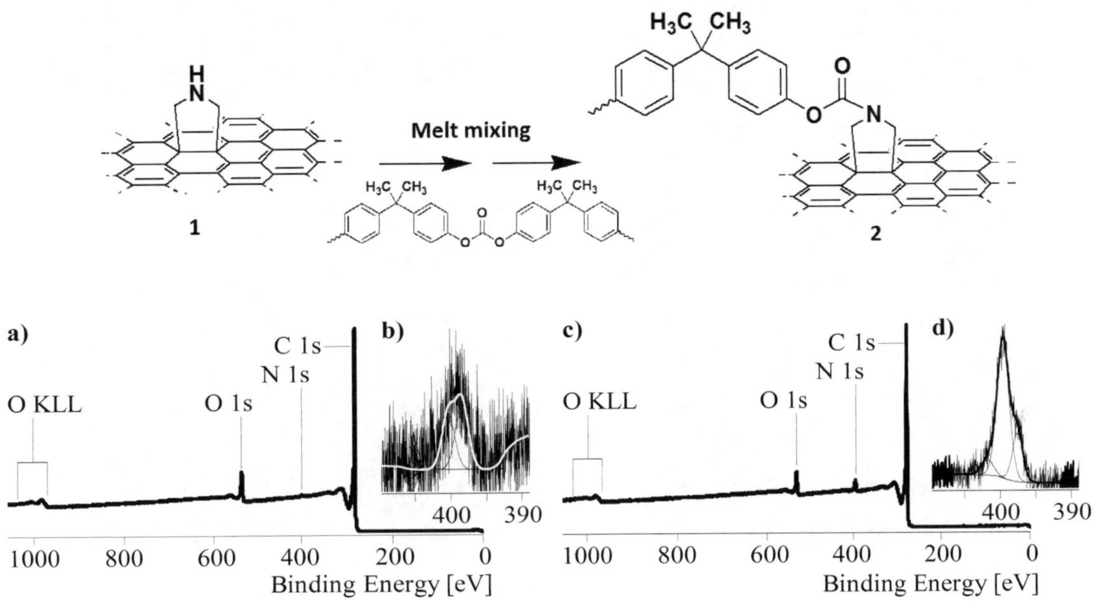

Figure 1. Top: scheme for the reaction of pyrrolidine groups at f-GNP surface and PC; bottom a) and b) wide scan XPS spectrum and high resolution N1s spectrum for the pristine GNP; c) and d) the same for the f-GNP.

The specific surface area, measured by BET, decreased from 668 to 132 m^2/g from pristine to f-GNP, which may result from the formation of larger agglomerates and higher powder density after chemical functionalization. The covalent modification of the f-GNP surface may also reduce the graphene surface available for N_2 adsorption, as compared to GNP. It is expected that the GNP powder conductivity of 7.7 S/cm (at 30 MPa) obtained for the pristine material will be reduced after functionalization, but unfortunately the apparatus [8] was not able to measure the conductivity of the f-GNP powder due to its too low compressibility.

The functionalization of GNP has a considerable effect upon the dispersability of GNP in PC. In spite of its more compact powder form, the dispersion within PC was enhanced after GNP functionalization. OM shows much smaller agglomerates in the case of f-GNP compared to GNP composites. Figure 2 illustrates the morphological characteristics of the composites with 1 wt.-% filler. All blends and blend composites studied exhibit co-continuous morphologies with only slightly changed domain sizes. More important, due to the better thermodynamic compatibility of PC and GNP and the mixing algorithm used, the fillers always stay inside the PC domains, thus enabling the desired double percolation.

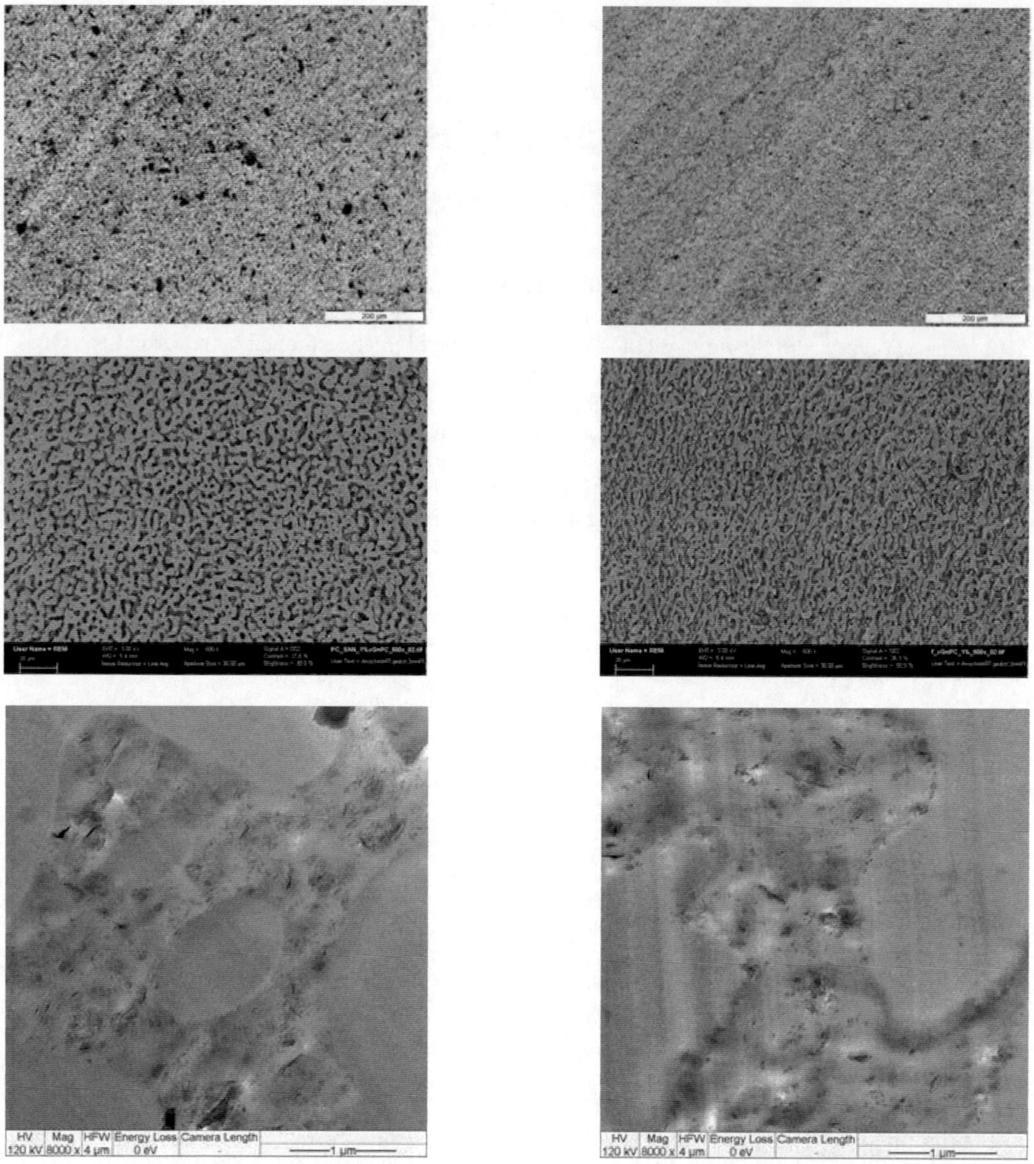

FIGURE 2: OM (top), SEM (middle) and TEM (bottom) images of co-continuous structure of blend composites containing 1 wt.-% (OM, SEM) or 8 wt.-% (TEM) GNP (left) or f-GNP (right)

The variation of the electrical resistivity with filler content is given in Figure 3. Typically, electrical percolation is detected as an abrupt decrease in resistivity within small concentration changes of electrical conductive fillers. In this system, the insulating/conductive transition is spread over a large concentration range, an observation which is not well understood yet. Already 1 wt.-% filling results in a resistivity reduction, which continues up to 6 wt.-% loading. In spite of the better dispersion of f-GNP in PC, the composites with f-GNP have slightly higher resistivity values than those containing GNP. This may be due to the enhanced interactions of the functional groups on f-GNP surface with the PC, or even chemical grafting of PC chains on f-GNP, leading to a thicker polymer layer around the f-GNP. While this may induce better dispersion in the polymer, the larger average distance between neighboring f-GNP, as compared to GNP, may hamper electron hopping and/or tunneling thus reducing the composite conductivity.

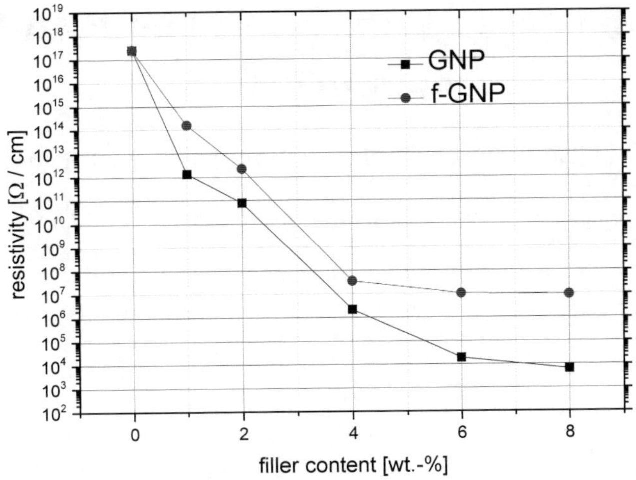

FIGURE 3: Dependence of electrical volume resistivity of PC/SAN blends on filler content and functionalization

ACKNOWLEDGMENTS

The authors gratefully acknowledge the financial support by Instituto de Polímeros e Compósitos of the University of Minho and Fundação para a Ciência e Tecnologia (FCT) through Project PEst-C/CTM/LA0025/2013 (Strategic Project e LA 25-2013-2014). M. C. Paiva acknowledges the Leibniz Institute of Polymer Research Dresden for receiving her as guest researcher, and E. Cunha acknowledges FCT for PhD grant (SFRH/BD/87214/2012).

REFERENCES

1. A. Göldel, G. Kasaliwal, P. Pötschke, *Macromolecular Rapid Communications* **30**, 423-429 (2009).
2. A. Göldel, A. Marmur, G.R. Kasaliwal, P. Pötschke, G. Heinrich, *Macromolecules* **44**, 6094-6102 (2011).
3. M. Gültner, A. Göldel, P. Pötschke, *Composites Science and Technology* **72**, 41-48 (2011).
4. A. Göldel, G.R. Kasaliwal, P. Pötschke, G. Heinrich, *Polymer* **53**, 411-423 (2012).
5. M. Liebscher, L. Tzounis, P. Pötschke, G. Heinrich, *Polymer* **54**, 6801-6808 (2013).
6. M. Liebscher, M.-L. Blais, P. Pötschke, G. Heinrich, *Polymer* **54**, 5875-5882 (2013).
7. M. C. Paiva, F. Simon, R. M. Novais, T. Ferreira, M. F. Proença, W. Xu, F. Besenbacher, *ACS Nano* **4** (12), 7379-7386 (2010).
8. B. Krause, R. Boldt, L. Häußler, P. Pötschke, *Composites Science and Technology*, **114**, 119-125 (2015).

Effect of Carbon Nano Tube (CNT) on PA6/PE base Shape memory alloy

Seyed Ali Mohammadi[a], Ismail Ghasemi[b] , Milad Mehranpour [a *]

[a] *Department of polymer ,Science and Research Branch, Islamic Azad university, Tehran, Iran*

[b]*Iran Polymer and Petrochemical institute, Tehran*

sa.mohammadi@srbiau.ac.ir
m.mehranpour@srbiau.ac.ir

Abstract. Shape memory polymers (SMPs) are new class of smart materials which their main characteristics is to recover their initial form after undergoing a large recoverable deformation via stimulation by heat , electricity , magnetic field and light . Non-conductive nature of polymers hinders the effect of stimulus such as heat and electricity to stimulate the shape recovery process.Shape memory polymers have two parts of soft and hard phase . That soft phase causes the deformation ability and other part is the harder phase and retentive initial form that called fixing phase. Having high rate of conductivity by conductive fillers such as MWCNT particles would be utilized in shape memory polymer alloys.|Adding MWCNT particles to SMPs not only dramatically boosts their conductivity and accelerates the rate of shape recovery process but also MWCNTs at hard phase of SMPs reinforces the hard phase and increases the fixity and therefor recovery rate of SMPs .At the present work the PE/PA6 alloy compatibilized by polyethelyne maleic anhydride was blended by functionalized MWCNTs through acid treatment and was used as a shape memory material. In order to have selective localization of carbon nano tubes, two mixing procedure were selected. Thermal programing for inducing of shape memory properties considered using general tensile testing machine equipped with hot oven.The localized MWCNT in the well distributed polyamide phase might result in improving conductivity and better stimulation of SMP.The rheological properties and morphology of the samples were investigated by RMS and SEM. The results showed that well mixing of two phases as well as Nano filler. Finally the SMP properties of the composite showed improvement because of better conductivity and performed morphology.

Keywords: SMP,PE/PA6/CNT,Nano Composites
PACS: 62.20.fg

INTRODUCTION

Thermal-responsive shape memory polymers (TSMPs) which have an increasing applications in various industries, a material is able to recover their initial form at the presence of the right stimulus This kind of the polymers generally consists of two phases, one is a reversible phase that could change and recover its shape. The other one can memorize original deformation that is called the''fixed phase,''which could be a chemical or physical crosslink [1,2]. For the first time, production of shape memory polymers of PA/PE alloys began in 1998, and a large group of researchers did considerable researches in order to get high rate recovering of shape memory. In order to improve the properties of immiscible PA/PE6 alloy, the researchers have employed Maleic anhydride as compatibilizer. In 2011 & 2013 researchers studied the effect of adding the functionalized multiwall carbon nanotube particles to this alloy [Because of high polarity of CNT particles and their natural tendency to diffuse into PA6 phase, different mixing methods were studied.] The aim of this work was to study the effect of adding CNT particles on the shape memory properties of this alloy. Due to Structurally MWCNTs consist of multiple layers of graphite superimposed and rolled in on themselves to form a tubular shape they can improve the thermal conductivity in the composites and its leads to rapidly stimulation of shape memory recovery.[3-7]

EXPERIMENTAL

Materials

LLDPE (LL0209AA), with melting index of 0.9 gr/10 min that used for this study was produced by Amirkabir petrochemical company(iran). Polyamide 6 (PA6,Akulon F223-D) and PE-MAH with MFI 7 gr/10min and 1.5% MAH were purchased. Functionalized multi walled carbon nano tubes (FMWCNTs) were purchased from Time Nano Chengdu company,(China). FMWCNTs Length and diameter were 10-30 μm and 2-5 μm (Inner Diameter) and <8nm (Outer Diameter) with 3.68 wt% COOH content .

Composite preparation

before using all the materials were dried at oven in 90 ℃ for 24 hours LLDPE/PA6/PE-g–MAH/FMWCNT blends were prepared by using HAAKE internal mixer at a temperature of 230 ℃. Two protocol of mixing were considered to localization FMWCNTs localize position in blend. at First protocol (P1), all the materials were mixed simultaneously. at second protocol (P2) in order to improve FMWCNTs diffusion into PA6 phase ,CNTs and PA6 were mixed together and then added to LLDPE/PE-MAH phase. Sample compositions is shown in Table.1

TABLE (1). Samples composition

Number of sample	Blend	%
1	LLDPE/ PA6	80/20
2	LLDPE/ PE-g-MAH /PA6	60/20/20
3	LLDPE/ PE-g-MAH /PA6 /0.5%CNT P1 (first protocol)	60/20/20/0.5
4	LLDPE/ PE-g-MAH /PA6 /1% CNT P1	60/20/20/1
5	LLDPE/ PE-g-MAH /PA6 /1.5%CNT P1	60/20/20/1.5
6	LLDPE/ PE-g-MAH /PA6 /2% CNT P1	60/20/20/2
7	LLDPE/ PE-g-MAH /PA6 /0.5%CNT P2 (second protocol)	60/20/20/0.5
8	LLDPE/ PE-g-MAH /PA6 /1% CNT P2	60/20/20/1
9	LLDPE/ PE-g-MAH /PA6 /1.5%CNT P2	60/20/20/1.5
10	LLDPE/ PE-g-MAH /PA6 /2 %CNT P2	60/20/20/2

Characterization

Capillary Rheometric test can give us the viscosity vs shear rate and the shear stress vs shear rate curves . so the miscibility of the blend can be observed by describing this curves . the samples were prepared to pellet form and the die length and diameter were 2.5cm and 0.07cm and the (L/D) is 34 and temperature was set at 250 ℃. The PE/PA6 blends were Immiscible and should be used some compatibilizers to improve their interaction and modified the alloy properties by increasing their miscibility. in order to miscibility observation Rheological measurements such as capillary Rheometry and RMS were used. RMS Rheometer (MCR300) at 250 ℃ was used to Rheological measurements in order to investigate the viscosity variation of the PE/PA6 alloy.The morphology of the composites were studied by using field emission scanning electron microscopy (FESEM;MIRA3-TE SCAN). The shape memory properties were investigated with a tensile test in INSTRON 6025 tensile machine with hot oven, First, two approximate lines were drawn at two sides of the central of a dumbbell specimen, the distance between the two lines was 20mm and then specimen were strengthened at 128°C with 5 mm/s stretching speed, when the sample strengthened To the 50% . the specimen was quenched by Liquid nitrogen spray to -18 °C for 2 min, to investigated shape recovery rate and velocity of recovery the samples were drop in hot oil bath at about 128 °C with the ruler in bottom of pyrex container on the heater and the camera was set on top of the system to record the samples recovery.

RESULTS & DISCUSSION

Miscibility

the effect of compatibilizer was investigated by viscosity curve. Compatibilizer makes good interaction between PA6 & PE and they compose a network with each other. when shear rates raises up, the composed network collapses and as the result, the viscosity would decrease. The network resistance against network collapse occurs when shear rate is low.[6-8]

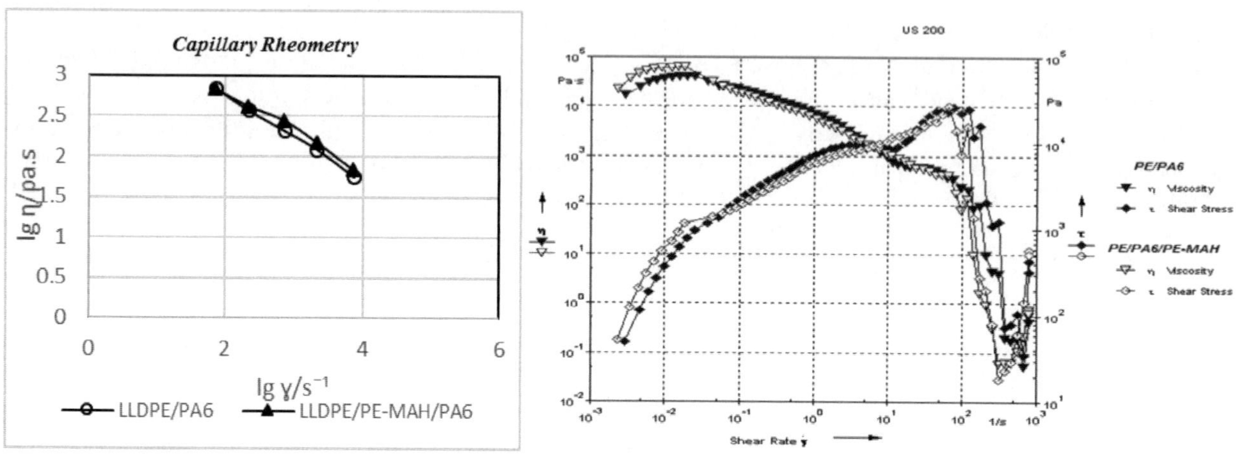

(a) (b)

FIGURE 1. Rheometry Characterization, (a)Capillary Rheometer,(b)RMS Rheometer

As shown in FIGURE 1.(a),(b) adding PE-MAH into the LLDPE/PA6 alloy improve the interaction between PA6 & PE. the matrix phase is LLDPE and the dispersed phase is PA6. when the PE-G-MAH with anhydride groups was added to alloy the anhydride was linked with Amin groups in PA6 and increased the miscibility. In lower shear rates sample No2 shows higher viscosity and shear stress against sample No 1 its related to network resistance.[5-8]

Morphology

Figure 3 shows the FESEM micrograph of PE/PA6/PE-MAH without CNTs . these results indicate that a good effect of compatibilizer leads to aggregations size reduction.[8]

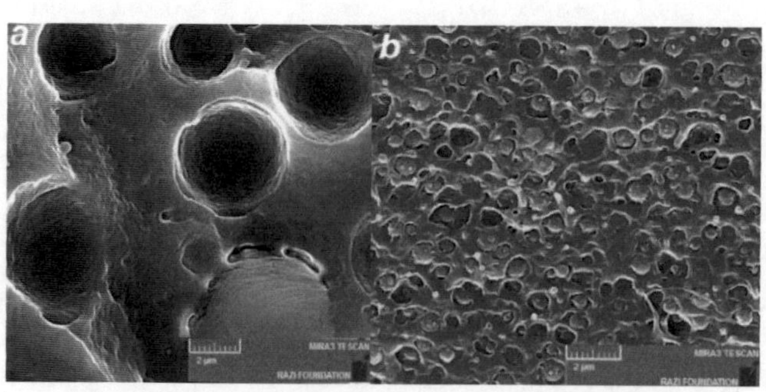

FIGURE 3. (a)PE/PA6, (b)PE/PA6/PE-G-MAH

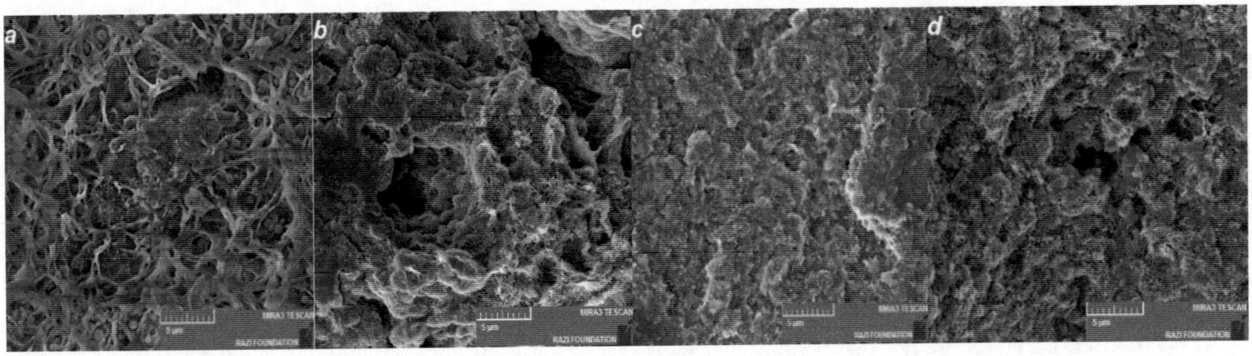

FIGURE 4.FESEM images of PE/PA6/PE-MAH/FWM CNT by simultaneously protocol mixing (P1). (a)0.5%CNT, (b)1%CT, (c)1.5%CNT, (d)2%CNT

FIGURE 5 FESEM images of PE/PA6/PE-MAH/FWMCNT by second protocol mixing (P2).
(a)0.5%CNT, (b)1%CNT, (c)1.5%CNT, (d)2%CNT

Figure 4 and 5 illustrate the effect of CNTs on PE/PA6 morphology. The CNTs localization in PE/PA6 alloy at simultaneously protocol were shown in figure 4. As illustrated, from a to c, the amount of CNTs percent in the samples has increased and enhanced the alloys morphology,alhough the degree of this effect depends on the mixing protocol . As shown in figure 5, by increasing CNTs amount from a to c the morphology changes between sea-island and co-continuous and CNTs particles can do the compatibilizer role in some conditions and in some conditions can reduced the compatibility. it depended on CNTs amount, mixing energy, mixing protocol, their polarity,etc. viscoelastic properties was more important to makes SMPs, the elongation was dependent on alloys morphology. In co-continuous condition CNTs has good relation and can affect viscoelastic properties and improve conductivity. in the other hand, when the morphology was changed to sea-island this connection was lost and the shape memory properties enhanced by this morphology changes. figure 4.b shows sea-island morphology and figure 4.c illustarates co-continuous morphology. Comparison of these morphology variation with shape memory results define a prominent relation between morphology and shape memory properties.[8-14]

Shape Memory Properties

As shown in figure 6, some of samples that include higher percentage of CNTs were broken at stretching process. For instance, sample 6 has 2 percent of nanotubes and mixed in simultaneously protocol, CNTs reduced the flexibility and increased the elongation in the alloy. in sample No 9 and 10 with 1.5 and 2 percentage of nanotubes respectively, which are mixed in second protocol, the highest elongation occur. As illustrated in Figure 2, sample No. 3 and 7 has low percentage of CNTs, approximately 0.5%, but due to different process of mixing have different mechanical properties. Sample No.3 CNTs are repartition in two phases but due to CNTs tendency to transfer into PA6 can be estimate the CNTs percentage into PA6 phase was more than in PE phase. when CNTs percentage increase, this proportion grows progressively.

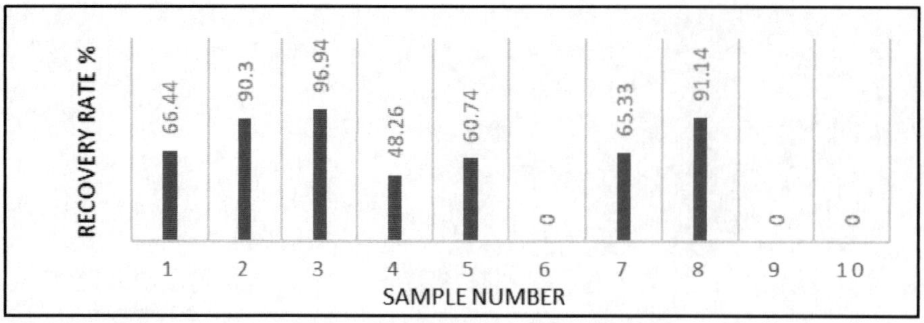

FIGURE 6.shape memory recovery percentage in mechanical shape memory test at 128 °C by 50% streching

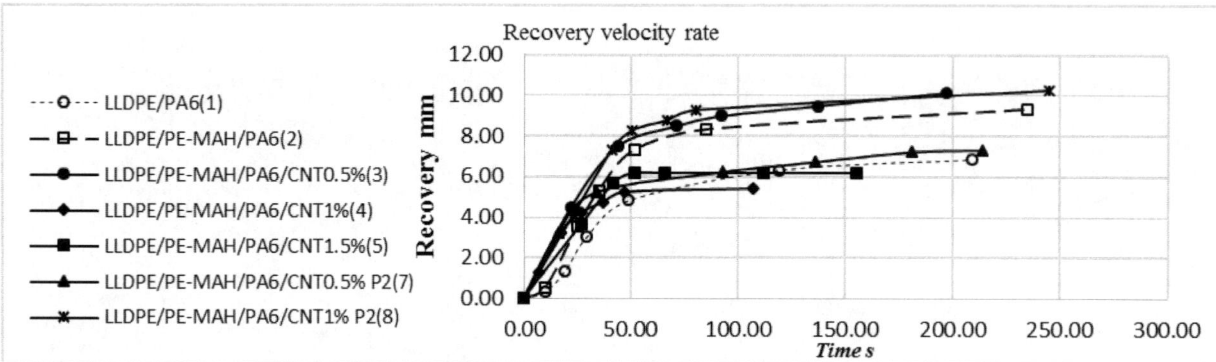

FIGURE 7. shape recovery velocity

As shown in Figure 7 it can be observed that the samples without CNTs (No1,2) unrecovered in 10 first seconds. adding CNTs into SMPs decreased their recovery Time and its happened more rapidly.

CONCLUSION

Adding CNTs into PE/PA6 alloy has lead to 96.9% shape recovery and dramatically reduced recovery time and improved shape recovery properties. better thermal conductivity leads to Slightly increase the speed of shape recovery.

REFERENCES

1. Debdatta Ratna J. Karger-Kocsis, J Mater Sci (2008) 43:254–269
2. Guo-Ming Lin, Guo-Xin Sui, Rui Yang Journal of Applied Polymer Science, Vol. 126, 350–357 (2012)
3. Fengkui Li, Yan Chen, Wei Zhu, Xian Zhang and Mao Xu Polymer Vol. 39 No. 26, pp. 6929–6934, 1998
4. Jinlian Hu, Yong Zhu, Huahua Huang, Jing Lu Progress in Polymer Science 37 (2012) 1720– 1763
5. Shu-Cai Li & Lei Tao , Polymer-Plastics Technology and Engineering, 49:2, 218-222
6. Debdatta Ratna J. Karger-Kocsis J Mater Sci (2008) 43:254–269
7. Marc Behl and Andreas Lendlein Materials today APRIL 2007 | Volume 10 | Number 4
8. Dashan Mia, Kejun Liua, Hainan Dua and Jie Zhang Polym. Adv. Technol. 2014, 25 364–371
9. F Xiang, Y Wang, Y Shi, T Huang,C Chen, Y Peng, Yong Wang.Polymer Int Volume 61, Issue 8, 1334–1343, August 2012
10. Patrick T. Mather, Xiaofan Luo, and Ingrid A. Rousseau Annu. Rev. Mater. Res. 2009.39:445-471.
11. Matricesa Petra Po¨tschke, Sven Pegel, Michael Claes, Daniel Bonduel. Rapid Commun. 2008, 29, 244–251
12. Zhong-Ming Li, Sha-Ni Li, and Xiang-Bin Xu, Polymer-Plastics Technology and Engineering, 46: 129–134, 2007
13. Xuelian Wu, Wei Min Huang, Yong Zhao, Zheng Ding , Cheng Tang and Jiliang Zhang , *Polymers* 2013, *5*, 1169-1202
14. Yuanliang Wang ,Yonggang Li ,Yanfeng Luo ,Meina Huang ,Zhiqing Liang . Materials Letters 63 (2009) 347–349

Bio-Based Epoxy Clay Nanocomposites

Andres Rigail-Cedeño (presenting author) [a] *and Daniel F. Schmidt* (reference author) [b]
[a]*Faculty of Mechanical Engineering and Production Science, ESPOL, Km 30.5 via Perimetral, Campus Gustavo Galindo, Guayaquil, Ecuador*
[b]*Department of Plastics Engineering, University of Massachusetts Lowell Lowell, MA, USA*
arigail@espol.edu.ec
Daniel_Schmidt@uml.edu

Abstract. The present research focuses on the development of a new family of bio-epoxy clay nanocomposite for coatings and other high performance applications and describes the degree of clay dispersion, curing kinetics, and mechanical properties associated with these formulations. The epoxy matrix consisted of a commercial sorbitol glycidyl ether (SGE) epoxy resin cured with one of three polyamines – two polyetheramines (one based on poly(ethylene oxide), PEO, the other based on poly(propylene oxide), PPO), and one aliphatic polyamine (triethylenetetraamine, TETA) – and reinforced with organically modified montmorillonite (OMMT) nanoclay. The degree of OMMT dispersion was assessed via x-ray diffraction (XRD). The SGE/PEO and SGE/TETA systems gave shorter gel times compared to the SGE/PPO system as observed by viscosity measurements, consistent with relative differences in amine hydrogen equivalent weight. Epoxy conversion was followed via near-IR (NIR) spectroscopy. The addition of OMMT to the SGE/TETA system gave a larger increase in mechanical properties compared to the other epoxy formulations. Based on the results of these evaluations, it is concluded that SGE/polyamine systems show promise as aliphatic epoxy nanocomposite networks for green coatings and high performance applications.
Keywords: bio-based, epoxy, clay, nanocomposite
PACS: 81.05.Qk

INTRODUCTION

Epoxy resins are among the most heavily used thermosets because of their excellent chemical and heat resistance, mechanical strength and toughness, electrical properties and anti-corrosion performance. In addition, due to their chemical compatibility, epoxies have received attention for many years as the basis for various clay nanocomposites with unique and interesting properties.

In this context, researchers have studied several variables influencing properties enhancements in epoxy / clay nanocomposites. Some authors have focused on the ability of organically modified montmorillonite clays to catalyze epoxy polymerization in the interlayer galleries, for instance [1,2]. In tandem, a range of different mixing procedures, including ultrasonic mixing and high shear mixing, have been used to produce epoxy / clay nanocomposites [3]. Most of these efforts at nanocomposite formation utilize solvents [4] and / or high temperature processing [5] to obtain clay dispersion in the epoxy-amine matrix.

In this research, a polyfunctional sorbitol glycidyl ether (SGE) epoxy resin was used. This material is free of compounds suspected of being endocrine disruptors in comparison to the traditional diglycidyl ether of bisphenol A (DGEBA) so frequently used in epoxy applications. The SGE resin is a promising material for the replacement of traditional aromatic epoxies in solvent-free coatings cured at room temperature. This is relevant in light of new environmental regulations that encourage a substantial reduction in volatile and / or potentially toxic components in coating formulations [6]. These concerns represent an issue for coating manufacturers as well in light of the hazards associated with their manufacture and coatings applicators [7]. It is with this in mind that we have focused on formulations with greater acceptability [8]. In this work, epoxy / OMMT nanocomposites based on commercial SGE with three aliphatic polyamines were studied in order to better understand the potential of this family of materials.

EXPERIMENTAL

Materials

Sorbitol glycidyl ether (SGE), Erisys GE-60, was supplied by Emerald Performance Materials. This epoxy resin has an epoxy equivalent weight (EWW) in the range of 160 to 195 g/eq. Two polyetherdiamines, a poly(propylene oxide) (PPO) diamine, Jeffamine D-230, and a poly(ethylene oxide) (PEO) diamine, Jeffamine EDR-176, were supplied by Huntsman Corporation. A polyamine, triethylenetetraamine (TETA), was supplied by Dow Chemical. D-230, EDR-176 and TETA have amine hydrogen equivalent weights (AHEWs) of 60 g/eq, 44 g/eq and 24 g/eq, respectively. A methylditallowammonium montmorillonite, Cloisite C93A (C93A), was purchased from Southern Clay Products.

Processing

Materials were prepared by dispersing 5 wt% of C93A in the SGE epoxy resin using a Flacktek DAC 400.1 FVZ speed mixer operating at 2,500 RPM for 3 minutes. A stoichiometric amount of the chosen polyamine was then added, after which mixing was repeated at 2,500 RPM for 10 seconds. The final mixture was then poured into one of a number of custom-made silicone molds in order to obtain samples of an appropriate size and shape for further evaluation via various means. For comparison purposes, all SGE/polyamine formulations were prepared with and without C93A.

Characterization

X-ray diffraction patterns were obtained using a Phillips PANalytical XPert-Pro model diffractometer equipped with Cu Kα (λ=1.54060 Å) radiation operated at 40 kV and 45 mA with a 1/8" incident slit and 1/16" detector slit, scanning from 1° to 9° 2θ with a 0.05° step size a 20 second count time per step. Extent of conversion was assessed by gathering near-IR (NIR) absorption spectra of samples in the range of 7,500-4,000 cm^{-1}. 1" square pieces of glass microscope slides were separated by a 0.044" thick, flat, U-shaped silicone rubber gasket and held together with small binder clips so as to make disposable sample cells with a known path length. SGE/polyamine and SGE/clay/polyamine mixtures were poured into these sample cells and curing was monitored for 7 days at room temperature. Gelation was monitored by viscosity measurements using a Malvern Kinexus Pro+ Rotational Rheometer at room temperature. A parallel plate (40 mm / 60 mm) arrangement with a 1 mm gap setting was used, with angular frequencies varied between 1 and 0.001 Hz for the SGE/PPO systems to avoid unacceptably high torque values and angular frequencies varied between 1 and 0.01 Hz for the SGE/PEO and SGE/TETA systems.

Evaluation

Tensile samples measuring 5.3" × 0.19'' × 0.055" were prepared by curing epoxy formulations in a silicone mold. Tensile tests were performed at room temperature at a testing rate of 0.08 in/min using a Shimadzu AG-1S 10 kN load frame. For this test, the ASTM D882 standard was followed as the experimental procedure. Specimens for hardness testing measuring 0.39" cubed were also manufactured by pouring mixtures into an appropriately shape silicone mold. Shore D Hardness testing was performed according to the ASTM D2240 standard, using Qualitest DKD-K-16501 hardness testing equipment.

RESULTS AND DISCUSSION

Figures 1 shows XRD patterns of the bottom faces of the cured nanocomposites (NCs); patterns of the top faces are omitted due to similarity, as are unfilled systems, which gave no diffraction peaks in this range. The XRD patterns exhibited broad silicate reflections at about 5° (~18 Å), 4° (~22 Å) and 6° (~14 Å) for cured PEO/NC, PPO/NC and TETA/NC, respectively. This implies a lack of intercalation, since pure C93A gives a diffraction peak at about 3.4° (~26 Å). However, it has been extensively illustrated than low molecular weight amines can readily intercalate natural and modified clays [9]. In contrast, aliphatic epoxies may be at a disadvantage vs. aromatic epoxies as far as exfoliation is concerned due to favorable interactions between aromatic rings and the clay surface [10]. In this framework, the broad diffraction patterns illustrated are proposed to arise due to deprotonation of the tertiary ammonium modifier present in C93A and its replacement by the protonated amine hardeners, consistent with the fact that all d-spacings, while lower than that in the pure C93A, track with the molecular weight of the hardener molecules studied here. This would also explain the lack of the usual harmonic series of peaks associated with well-ordered intercalated systems possessing only a single dominant interlayer spacing; these observations are more consistent with the co-existence of multiple clay phases with different interlayer spacings [11].

Table (1) shows the degree of conversion for the all networks studied, both neat and in the presence of C93A. The degree of epoxy conversion was evaluated by measuring the decrease in the peak area of the epoxy band, as observed at ~4,530 cm^{-1} in the FT-NIR spectrum [12]. This peak is due to combinations of the C-H deformation band at ~1460 cm^{-1} and the C-H fundamental stretching band at ~3050 cm^{-1}. It was normalized to the C-H stretching overtone peak at ~5,990 cm^{-1}; the epoxy conversion (α_{NIR-Ep}) is calculated according to equation 1 [13].

$$\alpha_{NIR-Ep} = 1 - \frac{[(A_{4530,t})/(A_{5990,t})]}{[(A_{4530,0})/(A_{5990,0})]} \tag{1}$$

After seven days at room temperature, the degree of epoxy conversion realized in the SGE/PEO and SGE/PPO networks was observed to exceed 95%. In contrast, the SGE/TETA networks only reached ~80% epoxy conversion during the same period of time due to their vitrification during room temperature curing. The addition of C93A results in a decrease in the epoxy conversion reached after seven days, but only when polyetheramine hardeners are used; with TETA, no change is observed. This may be related to interactions between the curing agents and the clay modifiers and / or the silicate layers themselves. Another option is that the presence of the OMMT physically disrupts network formation due to partitioning. So long as the surface of the clay layers cannot react with the polymer network, it will effectively interrupt network formation locally, resulting in some concentration of unreacted functional groups near the particle surface. The greater the exposed surface area of the filler particles in question, the more significant this effect. Additionally, the enhanced barrier properties associated with clay nanocomposite formation may hinder the diffusion of reacting species during the later stages of network formation. slowing the rate at which reactive groups find one another as a result.

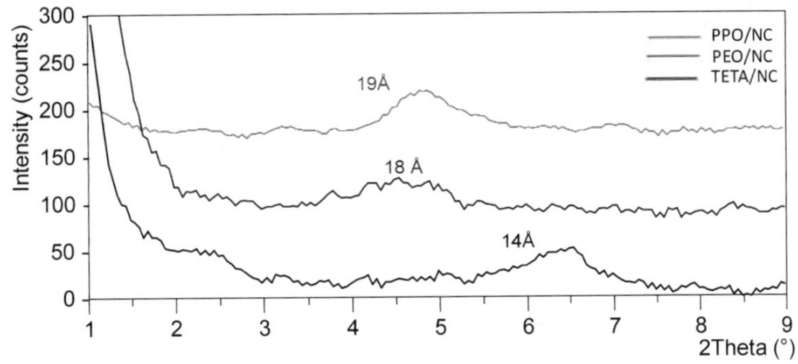

Figure 1. X-ray diffraction of SGE/clay nanocomposite samples.

Table (1). Degree of epoxy conversion by Near-IR of SGE cured with PPO (a), PEO (b) and TETA (c) hardeners.

	PEO	PEO/NC	PPO	PPO/NC	TETA	TETA/NC
Epoxy Conversion (%)	99	96	98	88	79	79

Gelation studies were conducted in a parallel plate rheometer as another means of following curing behavior; the results are presented in Figure 4. The SGE/TETA (48 min. gel time) and SGE/PEO (60 min. gel time) systems show earlier gelation compared to the SGE/PPO (300 min. gel time) system. This difference is may be explained by differences in the initial concentration of functional groups due to differences in the functionality and AHEW of these amines. Steric hindrance may also play a role in the PPO case in particular, given the presence of methyl groups immediately adjacent to the primary amine chain ends in this system in particular. There is no significant change in the gel times observed for the more reactive systems (SGE/TETA, SGE/PEO) upon incorporation of 5 wt% C93A in the system, while a slight increase in the gel time is observed in the SGE/PPO case, consistent with the largest reduction in epoxy conversion observed via NIR in the SGE/PPO NC case.

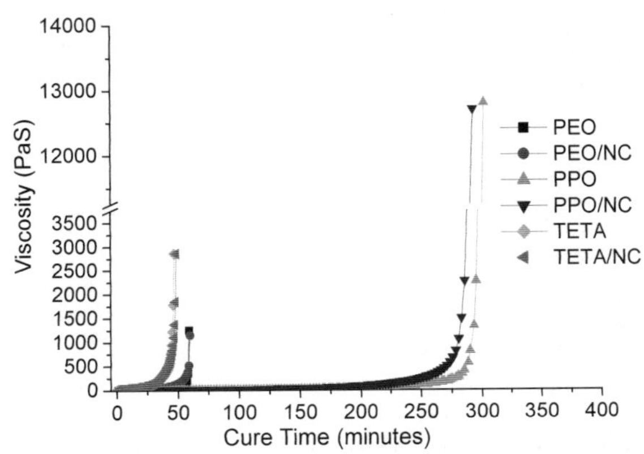

Figure 4. Viscosity data for all SGE/polyamines systems with and without 5 wt% C93A.

Table (2) illustrates the mechanical properties of these materials. The addition of C93A results in improvements in the hardness, modulus and strength of the SGE/PEO and SGE/TETA systems, as compared to the SGE/PPO system, where the observed changes are less evident. This is most likely explained by the reduction in chemical crosslink density in the latter system as a consequence of nanocomposite formation, as opposed to the other systems, where the crosslink density is minimally affected. The results of nanocomposite formation in the SGE/TETA system are particularly impressive given the observation of large enhancements in modulus, strength and energy to failure without compromising failure strain. This implies that a combination of high crosslink density and good compatibility between the network and the clay / the clay modifiers can dramatically enhance mechanical performance.

Table (2). Mechanical Properties of SGE/polyamine systems with and without 5 wt% C93A.

	PEO	PEO/NC	PPO	PPO/NC	TETA	TETA/NC
Shore D Hardness	77.5 ± 0.3	78.7 ± 0.3	76.9 ± 1.2	76.8 ± 0.4	79.9 ± 0.7	80.6 ± 0.7
Young's Modulus (GPa)	1.2 ± 0.0	1.6 ± 0.1	1.1 ± 0.1	1.3 ± 0.2	1.3 ± 0.1	1.7± 0.2
Ultimate Tensile Strength (MPa)	32.0 ± 0.2	34.8 ± 2.4	28.6 ± 5.1	28.2 ± 1.9	25.6 ± 2.1	40.5 ± 3.0
Failure Stress (Mpa)	27.8 ± 2.1	33.6 ± 1.2	25.1 ± 2.9	27.4 ± 2.3	25.5 ± 1.6	40.5 ± 3.0
Failure Strain (%)	5.7 ± 0.7	3.5 ± 0.8	6.4 ± 1.5	3.6 ± 0.7	2.8 ± 0.9	3.2 ± 0.8
Energy to Failure (J)	1.2 ± 0.2	0.8 ± 0.2	1.2 ± 0.4	0.5 ± 0.1	0.4 ± 0.3	0.7 ± 0.3

CONCLUSIONS

SGE/polyamine systems and their nanocomposites show potential as epoxy networks for green coatings and high performance applications. XRD provides information concerning the nanostructure and interactions present in these systems, and provides evidence of replacement of clay modified by the amine hardener in all cases. Near-IR spectroscopy enables us to track epoxy conversion in these systems, while rheometry provides information concerning gel times. Enhanced mechanical properties are observed upon nanocomposite formation in the SGE/PEO and SGE/TETA systems, but not in the SGE/PPO system due to a reduction in conversion. In the SGE/TETA system especially these improvements are significant and the final properties are attractive, implying a definite potential for practical application. Further studies are needed to more thoroughly evaluate the potential of SGE-based systems for protective coatings and other applications.

ACKNOWLEDGEMENTS

The authors thank the technical staff of the Laboratory of Testing Materials (LEMAT) and Center of Nanotechnology Research and Development (CIDNA) at ESPOL, as well as Prof. Xuejun Lu (UML Electrical Engineering), Dr. Mauricio Cornejo (CIDNA-ESPOL), Dr. Ana Rivas (Simon Bolivar University-Venezuela) and Dr. Schmidt's research group (UML Plastics Engineering). This research has been support by Secretary of Science and Technology (SENESCYT) of Ecuador. The authors would also like to thank Huntsman and CVC Thermosets / Emerald Performance Materials for kindly supplying the prepolymers used in this research.

REFERENCES

1. T. Lan and T Pinnavaia, *Chem. Mater*, 6, 2216-2219 (1994).
2. K. Chen and S. Yang, *Journal of Applied Polymer Science*, 86, 414-421 (2001).
3. S. McIntyre, I. Kaltzakorta, J. Liggat, R. Pethrick and I. Rhoney, *Ind. Eng. Chem. Res,* 44, 8573-8579 (2005).
4. B. Chen, J. Liu, H. Chen and J. Wu, *Chem. Mater*., 16, 4864-4866 (2004).
5. O. Becker, Y. Cheng, R. Varley and G. Simon, *Macromolecules*, 36, 1616-1625 (2003).
6. P. Lima-Neto, A. Araujo, W. Araujo and A. Correia, *Progress in Organic Coatings*, 62, 344-350 (2008).
7. R. Karshan, Cortec Corporation. (private communication).
8. (a) J.Möller, C. Kuncho, D.F. Schmidt and E. Reynaud, Bioepoxy/Glass Fiber Composites, *SPE-ANTEC Tech. Papers,* 2557-2560 (2014) (b) S. Sivasubramanian, K. Jafferji, D.F. Schmidt and E. Reynaud, *Green Materials,* 2-10 (2013) (c) R. Romano, S. Patil and D.F. Schmidt, *SPE-ANTEC Tech. Papers* , 2742-2746 (2013).
9 H.C Greenwell, M. J Harvey, P. Boulet, A. A. Bowden, P. V. Covenery, A. Whitting, Macromolecules, 38, 6189-6200 (2005)
10. J. Park and S.C.Jana, Macromolecules, 36, 8391-8397 (2003).
11. Soares BG, Silva AA, Solllymossy AP, Braga RA, Duchet J. Applied Clay Science. 83-84: 244-252 (2013)
12. L. Xu, J. Fu and J. Schlup, *Ind. Eng. Chem. Res.*,35, 963-972 (1996).
13. B. Min, Z. Stachurski, J. Hodgkin and G. Heath, *Polymer,* 34, 3620-3627 (1993).

Effect of Different Mineral Fillers on Rheological and Thermal Properties of Polylactide/Poly(Methyl Methacrylate) Blends

Mauricio Gonzalez, Shant Shahbikian, Michel A. Huneault

Department of chemical and biotechnological engineering, Université de Sherbrooke. 2500 Boulevard de l'Université, Sherbrooke, QC, J1K 2R1, Canada

michel.huneault@usherbrooke.ca

Abstract. This study presents the rheological and thermal behavior of mineral reinforced Poly(lactic acid)/Poly(methyl methacrylate) (PLA/PMMA) blends prepared by melt processing. In particular, combinations of amorphous or semi-crystalline PLA with high or low viscosity PMMA were used to prepare the blends. The effect of the addition of three commercial minerals, clay, calcium carbonate and quartz, was investigated. The glass transition, crystallization and melting of these blends were examined by differential scanning calorimetry. The blend rheology was investigated by small amplitude oscillatory shear rheology. Miscibility was inferred from the shift in glass transition temperature. A single glass transition temperature was observed for all blends indicating initial miscibility of the PMMA/ PLA blends over the entire composition range. This observation was reinforced by the fact that the mixtures remained transparent. It was however possible to induce a phase separation when placing the blends in conditions that favored PLA crystallization. The ability of PLA to crystallize was strongly restricted by the presence of the amorphous PMMA fraction but remained possible in selected conditions. All binary blends showed a typical viscoelastic behavior with a Newtonian plateau at low frequency range. However, the filled blends showed an increase in viscosity at low frequency typical of materials exhibiting a yield stress.

Keywords: Poly(lactic acid), Poly(methyl methacrylate), mineral fillers, polymers blends.
PACS: 81.05.Lg

INTRODUCTION

Commodity PLA grades have mainly been serving the packaging market where its biobased origin and its compostability are valued by consumers as an alternative to commodity petrochemical based materials. In this market, the life cycle of the material is relatively short since it is typically aimed at the protection a fresh food products. PLA has the highest tensile modulus and tensile strength of all commodity materials. Taking advantage of this great mechanical strength in longer life-cycle engineering applications would be interesting. The high tensile resistance could enable thinner and lighter parts and could be cost-competitive with some high-end petroleum based plastics. Furthermore, the expanded use of PLA would provide environmental benefits over current engineering materials since PLA production boasts lower fossil fuel consumption and lower Global Warming Gas emissions compared to most fossil based resins. Nevertheless, for engineering longer-term applications, PLA has two major deficiencies. First, it suffers from a relatively low temperature resistance when used in its amorphous form due to its low glass transition temperature, around 60 °C. Secondly, PLA can suffer from hydrolysis in hot and humid environments. Hydrolysis is the first step in the composting/biodegradation of PLA but can be seen as a disadvantage when contemplating more durable material applications. There is a whole window of opportunity in longer term engineering applications that is nearly untouched for the moment greatly due these deficiencies. Expanding the use of PLA through polymer blending technologies to tackle a wider range of engineering applications would have several key benefits for society. Properties of blends depend not only on the chemical composition of the blend but also on the compatibility or miscibility of the components [1]. There has been a tremendous effort in immiscible polymer blends involving PLA, typically with PLA as the major phase.

In this study, the focus will be placed on miscible rather than immiscible blends, namely the blend of PLA with poly(methyl methacrylate) or PMMA. The latter has good chemical resistance, a Tg around 115 °C and is also a high tensile resistance material. PLA/PMMA blends could be used as commodity materials in different kind of applications requiring high tensile strength. Early reports on PLA/PMMA blends reported miscibility but the subject has yet not been thoroughly investigated [2-4]. In this study, we will examine the miscibility window of four different polymer pairs out of 2 PLA and 2 PMMA resins and will evaluate the potential interest for highly filled version of these blends. The use of mineral fillers can be an interesting solution to reduce the global cost and to improve some specific properties such as rigidity, heat deflection temperature and processability.

METHODOLOGY

Two grades of PLA (supplied by NatureWorks LLC) were used: IngeoTM 4032D and 8302D. These differ by their L-lactic acid content and thus in their crystallization behavior. We will refer to the grades, respectively as the c-PLA and a-PLA because of their semi-crystalline and amorphous nature. For the acrylics, two different grades of PMMA (supplied by Arkema) were used: Plexiglass V826 and Plexiglas-VH100. The first one is a PMMA homopolymer while the second one is a copolymer that comprised methyl methacrylate and approximately 5% butyl acrylate. They will be referred to as PMMA-h and PMMA-c respectively. Three different reinforcing minerals were used: clay, quartz and calcium carbonate. The specific grades were Kamin35 with a particle size of 7 μm as clay, Imsil A10 with a particle size of 2.4 μm as quartz and Omyacarb UF with a particle size of 0.7 μm as calcium carbonate. Specific areas of the minerals have been measured by BET and were respectively equal to 11.2 m2/g, 5.5 m2/g and 10.3 m2/g. For processing, since PLA is prone to hydrolyze at high temperatures in presence of humidity, the PLA and PMMA were dried overnight under vacuum at 50 °C prior to processing. An internal mixer HAAKE Polylab 300P was used for melt-blending. The blending was carried out at 180 °C and 60 rpm for 10 minutes.

Differential Scanning Calorimetry (DSC) measurements were performed using a DSC TA Instrument Q-1000 under nitrogen flow. Glass transition, melting and crystallinity were evaluated using following procedure: heating at 10 °C min−1 from -10 °C to 200 °C, cooling at 10 °C min−1 to −10 °C, and heating at 10 °C min−1 to 200 °C. The first scan was to erase prior thermal history of the samples. The cooling and second heating scans were used to evaluate the ability to crystallize, the position of the glass transition and the melting peak (if present).

The rheological characterization was carried out in oscillatory shear at 180 °C and 200 °C with an Anton-Paar MCR-502 rheometer under nitrogen flow. The samples were compression molded in the form of 25 mm disk. Plate-plate geometry (25 mm diameter) and a strain of 1 % were used for all oscillatory experiments over the frequency range of 0.1 to 300 rad/s. For all measurements, samples were dried in vacuum at 50 °C for minimum of 12 h to remove humidity before testing and equilibrated for 10 min in the rheometer after loading at the desired temperature to erase the thermal history.

RESULTS AND DISCUSSION

Figure 1(a) presents the second heating scan of c-PLA/PMMA-h blends. c-PLA presents a glass transition temperature of 60.8 °C and a melting endotherm around 165 °C. The pure PMMA-h exhibited only a glass transition around 115 °C. The c-PLA/PMMA-h blends showed a single broad glass transition at intermediate temperatures between the Tg of c-PLA and PMMA-h. This broadness of the glass transition increased with PMMA content as already reported in literature [2]. The c-PLA crystallization was reduced in these blends and for PMMA content higher than 50 % PLA crystallinity was totally inhibited. Similar observations in terms of glass transition shift were observed with all polymer pairs and are summarized in Figure 1 (b). In all cases, a single transition was observed indicating that there were no observable phase separations. This aspect was reinforced by the fact that the sample remained transparent after melt mixing and compression molding.

For the purpose of demonstration, the glass transition temperature of c-PLA/PMMA-c blend is represented in Figure 2 along with the prediction from Fox and Gordon-Taylor equation [5]. The Fox and Gordon-Taylor equations can respectively be written as:

$$\frac{1}{T_g} = \frac{w_1}{T_{g1}} + \frac{1-w_1}{T_{g2}} \tag{1}$$

$$T_g = \frac{w_1 T_{g1} + k_{GT}(1-w_1)T_{g2}}{w_1 + k_{GT}(1-w_1)} \tag{2}$$

where w is the weight fraction, k is a fitting parameter, and T_{gi} denote glass transition temperature of the pure component, 1 and 2. The measured glass transition temperature exhibited a slightly negative deviation compared to the Fox equation predictions. A good agreement was obtained using the Gordon-Taylor prediction with *k*=0.26 for c-

PLA/PMMA-c blend. The k value has been interpreted as a miscibility measure that can be used to represent the interaction strength between polymers in the blend. Different values for parameter k from 0.25 to 0.52 have been reported in literature for different PLA/PMMA systems [1, 3, 6-8].

Traditionally, low values of k have been interpreted as a weak interaction between polymer chains. In this particular case, that assumption could be valid and is in agreement with broadness obtained for all PLA/PMMA systems evaluated in the present work. In fact, the data can suggest that we are dealing with systems that are close to the miscibility limit. Furthermore, regardless of polymer pairs in the blend, no significant difference was found using the a-PLA or the c-PLA with PMMA.

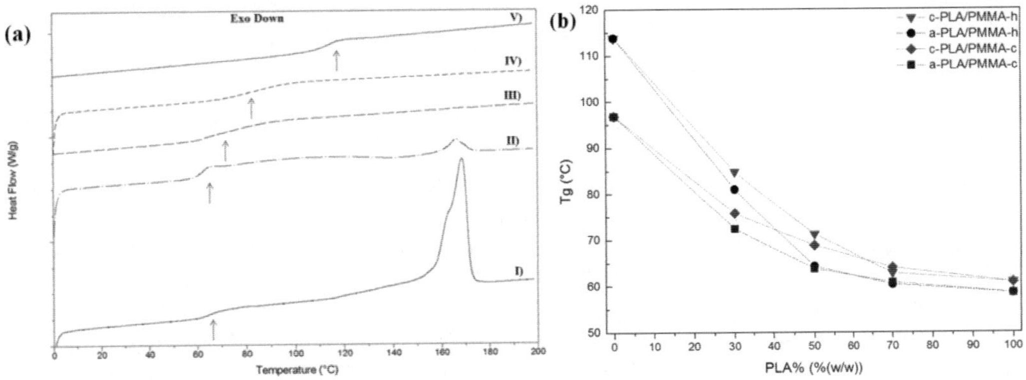

FIGURE 1. (a), DSC heating thermogram of c-PLA/PMMA-h blend I) 100/0, II) 70/30, III) 50/50, IV) 30/70, V) 0/100. (b), Glass transition temperature of unfilled PLA/PMMA blends as a function of PLA concentration.

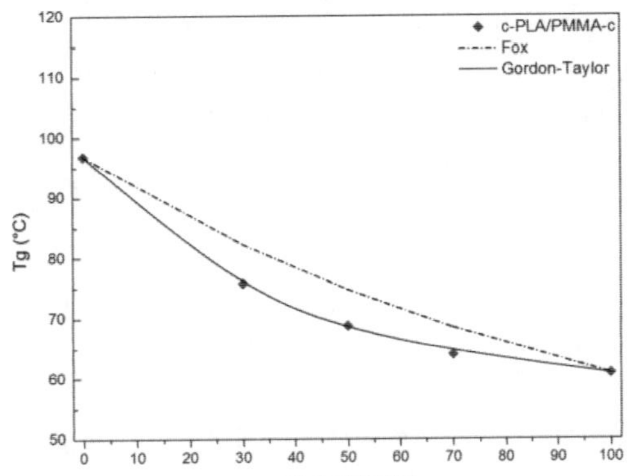

FIGURE 2. Glass transition temperature versus PLA composition. Gordon-Taylor equation prediction for k=0.26

Figure 3 display the Newtonian plateau viscosity of the same blend series as a function of the PLA concentration. Again, the data showed a slightly negative deviation compared to an additive mixing rule. There is no generally accepted mixing rule for the viscosity of miscible blends. However, some authors have reported morphological phase inversion for immiscible systems detected by an occurrence of a transition between positive and negative deviations of viscoelastic features from additive rule [9, 10]. In the current case, since just a negative deviation was observed, this could be interpreted by an increase in free volume upon mixing.

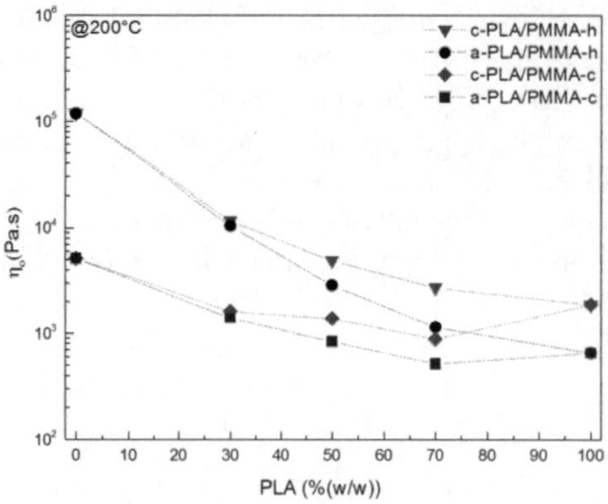

FIGURE 3. Newtonian plateau viscosity of unfilled PLA/PMMA blends as a function of PLA concentration.

Figure 4 (a) represents the complex viscosities of highly filled c-PLA-PMMA-h (50/50) blend as function of frequency. As expected from theoretical relationships such as the Krieger- Dougherty equation [11], the viscosity increased with quartz and clay concentration as a solid network developed in the material. At 53 vol%, the complex viscosity of the blend was increased by three orders of magnitude. Surprisingly, the addition of calcium carbonate leads to a significant drop in viscosity. The only valid explanation for this effect is that the effect of the solid fraction is overwhelmed by a matrix degradation effect and that for some reason; the calcium carbonate phase does not have the same tendency to form a solid network at the investigated concentration. PLA is very sensitive to hydrolysis in alkaline conditions. It is possible that at the melt-processing temperatures, the residual water or water produced by PLA hydrolysis during processing have significantly increased the pH near the surface of the calcium carbonate mineral and has led to rapid degradation of the PLA matrix. To reinforce this assumption, blends of pure c-PLA filled with calcium carbonate were also analyzed at 180 °C. The result presented in Figure 4 (b) showed a similar overall drop in complex viscosity throughout the whole frequency range compared to clay and quartz, despite the upward trend. Further investigation will help determine if this detrimental effect can be delayed by using different pre-drying conditions or by blending at lower temperatures.

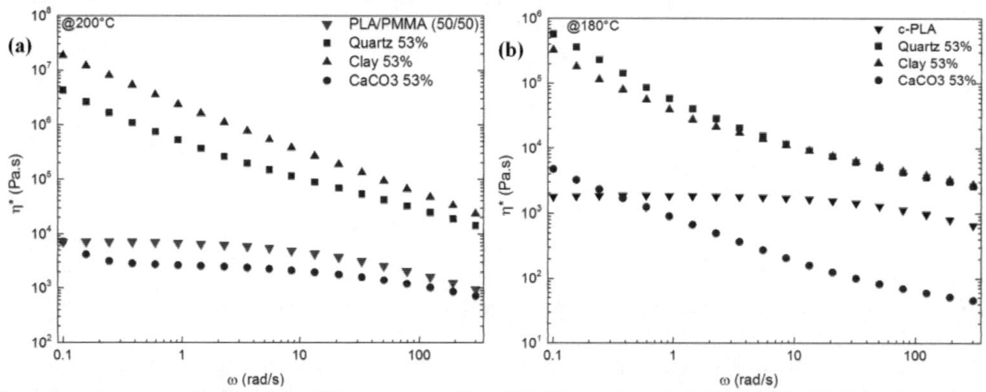

FIGURE 4. Complex viscosity as a function of frequency at 53 *vol%* filler. (a), c-PLA/PMMA-h (50/50) blend. (a), pure c-PLA.

Figure 5 shows the complex viscosity of c-PLA/PMMA-h (50/50) blend as a function of mineral concentration obtained at a fixed oscillation rate of 0,1 rad/s The viscosity increased with the mineral volume fraction but this increase was not the same for all fillers. The viscosity increase is expected to depend on the minerals' maximum packing fraction [12]. This phenomenon is closely related to the particle shape, structure (morphology) and inter-particle interactions which can be clearly observed in the upward trend of complex viscosity in the case of blends containing clay and quartz. However, in the case of calcium carbonate, the complex viscosity trend in Figure 5 is presumably over shadowed by degradation of PLA matrix.

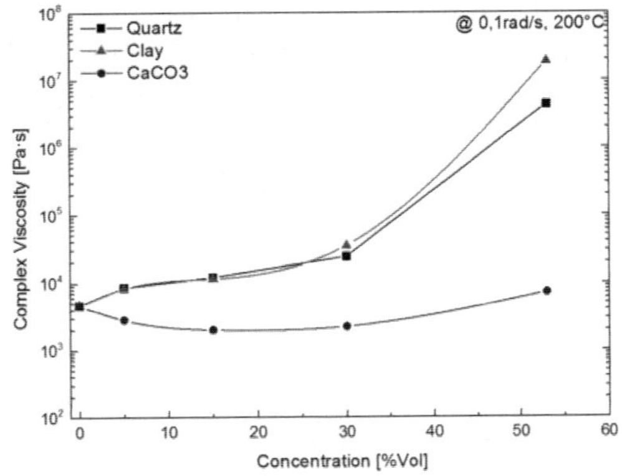

FIGURE 5. Complex viscosities at 200 °C of 50/50 c-PLA/PMMA-h blends as a function of mineral volumetric concentration.

The use of minerals opens the range of mechanical and thermal properties could be expanded to contribute to potential use of bio-based materials for longer term applications.

CONCLUSIONS

The PLA/PMMA blend is miscible even when copolymers comprising up to 10 %D-LA in PLA or around 5 % BuAr in PMMA are used. The blends exhibit a single glass transition and a blend viscosity slightly negative compared to a linear mixing rule. However, the blending of PLA and PMMA leads to a broadening of the glass transition. The Gordon-Taylor approach describes well the Tg shifts. The k parameter does not suggest different tendency between amorphous or crystalline PLA in the formation of miscible blends with PMMA. The PLA/PMMA blends are fully amorphous when PMMA is used at concentration equal or above 50 %. It is possible to reinforce the blend with large volume fractions of minerals (quartz and clay) but the use of calcium carbonate apparently leads to matrix degradation.

REFERENCES

1. Zhang, G., et al., Miscibility and phase structure of binary blends of polylactide and poly(methyl methacrylate). Journal of Polymer Science Part B: Polymer Physics, 2003. 41(1): p. 23-30.
2. Samuel, C., et al., Stereocomplexation of polylactide enhanced by poly (methyl methacrylate): improved processability and thermomechanical properties of stereocomplexable polylactide-based materials. ACS applied materials & interfaces, 2013. 5(22): p. 11797-11807.
3. Canetti, M., A. Cacciamani, and F. Bertini, Miscible blends of polylactide and poly(methyl methacrylate): Morphology, structure, and thermal behavior. Journal of Polymer Science Part B: Polymer Physics, 2014. 52(17): p. 1168-1177.
4. Samuel, C., J.-M. Raquez, and P. Dubois, PLLA/PMMA blends: A shear-induced miscibility with tunable morphologies and properties? Polymer, 2013. 54(15): p. 3931-3939.
5. Gordon, M. and J.S. Taylor, Ideal copolymers and the second-order transitions of synthetic rubbers. i. non-crystalline copolymers. Journal of Applied Chemistry, 1952. 2(9): p. 493-500.
6. Kucharczyk, P., et al. Characterization of partially biodegradable poly (L-lactic acid)/poly (methyl methacrylate) blends as potential biomaterials. in 5th International Conference on Energy and Development-Environment-Biomedicine. 2011.
7. Eguiburu, J.L., et al., Blends of amorphous and crystalline polylactides with poly(methyl methacrylate) and poly(methyl acrylate): a miscibility study. Polymer, 1998. 39(26): p. 6891-6897.
8. Shirahase, T., et al., Miscibility and hydrolytic degradation in alkaline solution of poly(l-lactide) and poly(methyl methacrylate) blends. Polymer, 2006. 47(13): p. 4839-4844.
9. RunMing, L., Y. Wei, and Z. ChiXing, Phase inversion and viscoelastic properties of phase-separated polymer blends. Polymer Bulletin, 2007. 59(4): p. 545-554.
10. Schnell, M., V. Ziegler, and B.A. Wolf, Evolution of viscosities and morphology for the two-phase system polyethylene oxide/poly(dimethylsiloxane). Rheologica Acta, 2008. 47(4): p. 469-476.
11. Krieger, I.M. and T.J. Dougherty, A Mechanism for Non-Newtonian Flow in Suspensions of Rigid Spheres. Transactions of The Society of Rheology, 1959. 3(1): p. 137-152.
12. Cassagnau, P., Linear viscoelasticity and dynamics of suspensions and molten polymers filled with nanoparticles of different aspect ratios. Polymer, 2013. 54(18): p. 4762-4775.

Non-isothermal Crystallization Kinetics of Carbonaceous Nanofiller Reinforced Poly(butylene succinate) Nanocomposites

S. Filizgök[a,b], M. Kodal[b] and G. Özkoç[b*]

[a]*Kocaeli Provincial Directorate of Environment and Urbanization, Kocaeli 41100, Turkey*
[b]*Department of Chemical Engineering, Kocaeli University, Kocaeli 41380, Turkey*

mehmet.kodal@kocaeli.edu.tr
guralp.ozkoc@kocaeli.edu.tr

Abstract: In the recent years, biodegradable products have raised great interest since the non-biodegradability of the commercially available plastics has caused many environmental problems associated with their solid waste pollution. Poly(butylene succinate) (PBS), a biodegradable biopolyester, has many applications such as food packaging, mulch films, nonwovens, fiber/textile, consumer goods, electrics and electronics, automotive interior and others. In this study, we have focused on the influence of the carbon-based nanofillers (carbon nanotubes (CNT) and fullerene (F)) on the non-isothermal crystallization kinetics of poly(butylene succinate). Biodegradable PBS composites based on carbonaceous nanofiller containing 0.5 and 1.0 wt % nanoparticles were prepared by melt compounding in an Xplore microcompounder. The data for non-isothermal crystallization kinetics studies were obtained using differential scanning calorimetry (DSC). The combined Avrami-Ozawa method could adequately describe the primary stage of non-isothermal crystallization kinetics of the neat PBS and PBS nanocomposites. The spherulitic morphology and growth were observed with hot stage polorized optical microscopy at 95 °C. The structure of nanocomposites in the nanometer scale was characterized using transmission electron microscopy (TEM) observations. The results indicated that the nanocomposite samples exhibited enhanced crystallinity and nucleation density significantly, along with smaller spherulite size. There was a strong dependence of the rate of crystallization and extent of crystallinity of PBS on nanofiller loading and its shape. CNTs were found to be more effective nucleating agent for PBS under experimental conditions.

Keywords: Poly(butylene succinate), carbonaceous nanofiller, nonisothermal crystallization kinetics

PACS: 82.35.Np

INTRODUCTION

Poly(butylene succinate) (PBS) is the most promising polymer in biodegradable aliphatic polyester family, which has many applications such as packaging materials, nets, nonwoven, tape and so on. Its mechanical properties are comparable with polypropylene (PP) and low density polyethylene (LDPE) while its crystallization behavior is similar to that of polyethylene. Additionally, it has many desirable properties including easy processability, high flexibility, thermal and chemical resistance as well as biodegradability. However, other properties of PBS such as softness, tensile and gas barrier properties are frequently insufficient for various end-use applications [1-4].

Nanocomposites are new materials with improved properties due to the incorporation of small amounts of nano-sized fillers into a polymer matrix. Nanofillers can be classified according to their morphology, such as particles that are layered, spherical or circular. Their specific geometrical dimensions, and thus aspect ratios, partly affect the final materials properties. Generally, carbon nanotubes and fullerene are widely used as nanofiller in plastics [2, 5].

Also, the crystallinity plays a significant role in the final physical properties and biodegradability of polymers. Similarly, the crystalline structure and morphology of polymers are influenced greatly by the thermal history. Thus it is very important to investigate crystallization kinetics of composites to understand the relationship between structural features and crystallization conditions [1-6].

In this study, the effects of particle size and shape of carbonaceous nanofillers on the morphology, crystallization, melting behavior, and thermal stability of PBS nanocomposite were studied. PBS reinforced with unmodifed carbon nanotube (CNT) and fullerene was prepared by melt-compounding method. The investigation of non-isothermal crystallization kinetics was performed by using differential scanning calorimetry (DSC). The spherulitic morphology and growth were observed with polarized optical microscopy (POM) while the structure of nanocomposites was characterized by using transmission electron microscopy (TEM) observations.

EXPERIMENTAL

Materials

Poly(butylene succinate) (PBS), which is a thermoplastic, aliphatic polyester, was supplied from Reverdia, The Netherlands. The carbonaceous fillers, unmodifed carbon nanotube (CNT) and fullerene used for the preparation of PBS nanocomposites, purchased from Nanocyl and Sigma Aldrich respectively.

Sample Preparation

PBS nanocomposites were prepared via melt blending technique using a twin screw microcompounder (Xplore Instruments, The Netherlands) operated at a rotor speed of 100 rpm at 170 °C for 3 min. Carbon nanotube and fullerene with loading of 0.25 - 2 wt% were used as carbon based nanoparticles for nanocomposites preparation. Then, the melts were compressed for 3 minute at 170 °C to get films with a thickness of around 0.5 mm. The samples were abbreviated with respect to the type and loading level of carbon based nanoparticles, for instance CNT0.5 represents the composite including 0.5 wt % carbon nanotube.

Characterization

Transmission electron microscopy (TEM) (JEOL 2100 electron microscope) was used to investigate the morphology of the PBS nanocomposites.

Thermal analysis was performed in nitrogen atmosphere by using a Differential Scanning Calorimeter (DSC) (Mettler Toledo DSC 1 Star). For the non-isothermal crystallization kinetic studies of pure PBS and PBS nanocomposites, the samples were first heated to 150 °C with a heating rate of 15 °C/min (first heating), kept at 150 °C for 5 min to eliminate the thermal history and cooled to 25 °C at various cooling rates of 5, 15, 25 and 50 °C /min. Then, the samples were heated to 150 °C again at 10 °C/min The melting behaviors of neat PBS and its nanocomposite were studied from the second heating section to determine the degree of crystallinity (X_c) from melting enthalpy values.

The spherulitic morphology and growth rate of neat PBS and its nanocomposites were studied with a polarizing optical microscope equipped with a hot stage. The samples were heated from 25 to 150 °C at a heating rate of 100 °C/min. In order to eliminate the thermal history, they were held at 150 °C for 5 min and 100 °C for 3 min and then they were cooled to 95 °C with a cooling rate of 1°C/min. The photomicrographs showing optical images of these samples which were isothermally crystallized at 95 °C with holding time of 1 hr were obtained under crossed polarizer with a video camera.

RESULT AND DISCUSSION

Non-Isothermal Crystallization Kinetics

The nonisothermal melt crystallization exotherms at different cooling rates for neat PBS, PBS/CNT0.5 and PBS/F0.5 nanocomposites showed that, the exothermic peak shifted to lower temperature and also become broader with increasing cooling rate. This indicated that when the samples were cooled slowly, crystallization began at higher temperature. The reason of the effect was that at lower cooling rates, polymer chains have sufficient time to overcome nucleation energy barrier and initiated crystallization at higher temperature. It was obtained that T_{cc} values (upon cooling) for all the nanocomposites were higher than that of neat PBS independently from the amount of fillers indicating that crystallization process of PBS was enhanced in the presence of carbonaceous nanofillers. However, T_{cc} values of the nanocomposites filled with CNT nanoparticles were higher than the nanocomposites including F. This observation revealed that CNT was more effective heterogeneous nucleating agent as compared to the fullerene.

In order to investigate the kinetics of nucleation and crystal growth processes under non-isothermal conditions, Avrami is probably the most widely used kinetic model to investigate the primary stage of non-isothermal crystallization. Avrami rate constant (Z_C) that provides information about growth rate while the Avrami exponent, n, gives information about nucleation type and morphology of the crystallite formed. Avrami kinetic parameters of PBS and its nanocomposites at different cooling rates are summarized in Table 1.

TABLE 1. Avrami kinetic parameters of PBS and its selected nanocomposites at different cooling rates

Sample	β (°C/min)	*n*	*Z*	Z_c	R^2
PBS	5	2.33	4.60E-03	3.41E-01	0.998
	15	2.37	7.58E-02	8.42E-01	0.997
	25	2.53	4.06E-01	9.65E-01	0.994
	50	4.33	9.03E+01	1.09E+00	0.992
PBS/CNT0.5	5	2.48	9.08E-03	3.90E-01	0.999
	15	2.23	6.12E-02	8.30E-01	0.999
	25	2.71	2.82E-01	9.51E-01	0.996
	50	4.62	2.39E+02	1.12E+00	0.996
PBS/CNT1.0	5	2.57	7.48E-03	3.76E-01	0.995
	15	3.09	3.35E-01	9.30E-01	0.997
	25	3.56	1.64E+00	1.02E+00	0.995
	50	4.00	5.86E+01	1.08E+00	0.993
PBS/F0.5	5	3.37	4.37E-02	5.30E-01	0.994
	15	3.76	3.16E+00	1.08E+00	0.998
	25	3.69	1.29E+01	1.11E+00	0.999
	50	2.75	1.32E+01	1.05E+00	0.997
PBS/F1.0	5	2.62	3.25E-02	6.90E-01	0.996
	15	3.51	2.33E+00	6.00E-02	0.994
	25	3.68	1.09E+01	1.00E-01	0.998
	50	3.63	5.30E+01	8.00E-02	0.997

Avrami exponents "*n*" were obtained from the slope of the primary crystallization stage. As shown in Table 1, the *n* values of neat-PBS were between 2.33 and 4.33 while that of PBS nanocomposites were between 2.48 and 4.62. It was observed that the changes of n values of the nanocomposites were not proportional to the cooling rate or amount of loading fillers. Avrami exponents that greater than 4 expressed conelike spherulite growth

The Z_c values for neat PBS and PBS nanocomposites increased slightly with increasing cooling rate, which implying an increase in crystallization rate. At 5 and 15 °C cooling rate, the Z_c values of PBS/F0.5 were higher than other PBS nanocomposites including 0.5 wt% nanofiller while the Z_c values of PBS/F1.0 were lower than other PBS nanocomposites including 1.0wt% nanofiller. This could be ascribed that the crystallization rate of polymer depends upon to nucleation mode, nucleation density and growth rate of individual crystallites.

The non-isothermal crystallization studies showed that the presence of fillers caused different effects on the PBS crystallization. CNT exhibited the best nucleating activity in PBS matrix as compared to F nanofillers. The results were in accordance with the TEM and POM images where the increased nucleation density, small spherulites and better dispersion were obtained.

The crystallization morphology of PBS and its nanocomposites were analyzed by polarized optical microscopy (POM). As can be seen from Figure 1, the PBS nanocomposites showed decreased spherulite size due to the higher number of nucleating sites within sample. Thus, nucleation density of PBS increased in the presence of nanoparticles It was obvious from Figure 1 that the PBS/CNT nanocomposites exhibited the smallest spherulites, a fact which lead to higher nucleation density, hence enhanced crystallization rate of PBS.

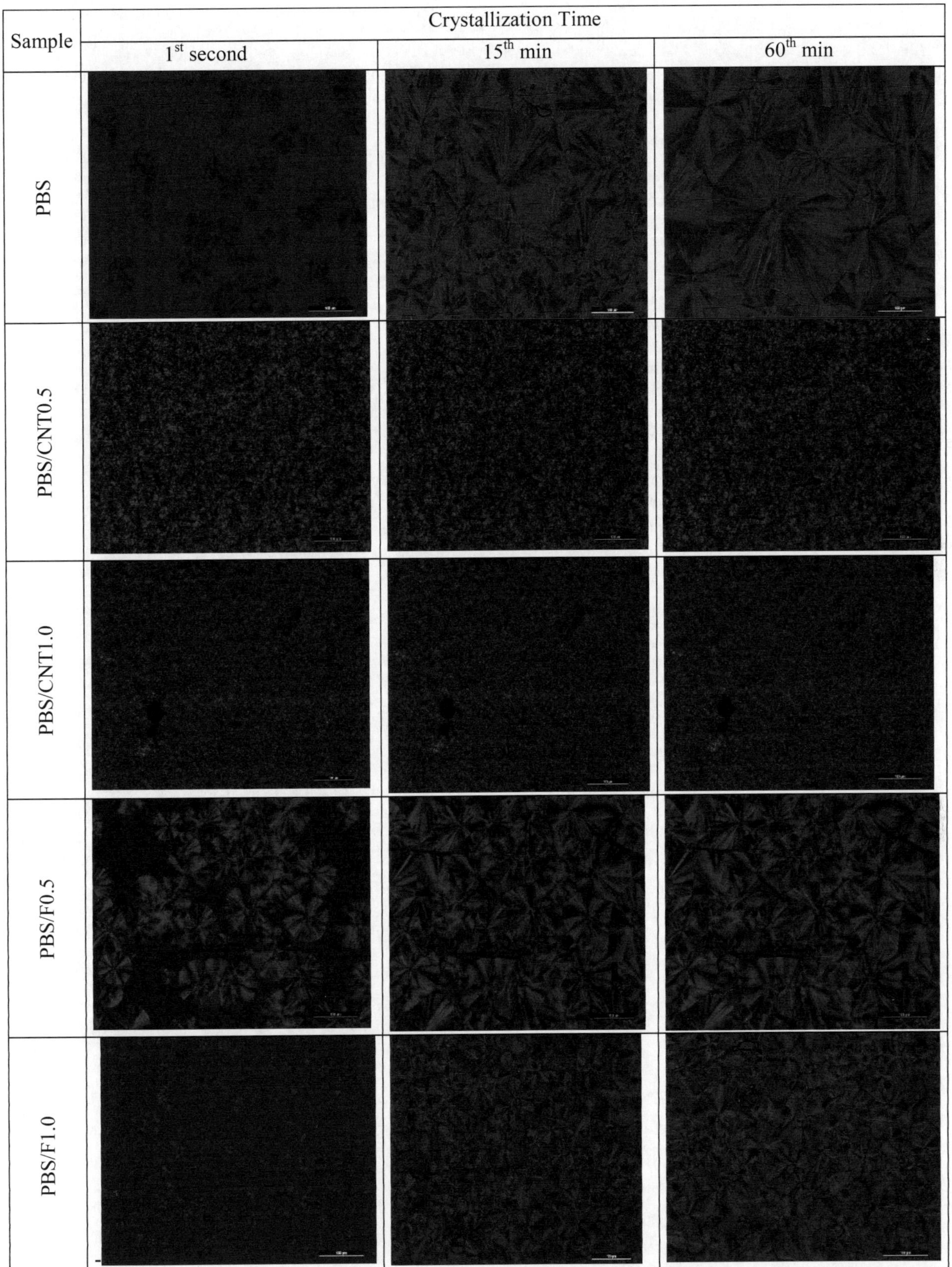

FIGURE 1. The POM pictures of neat PBS and its nanocomposites

CONCLUSIONS

Kinetic models revealed that the incorporation of nanofillers to the PBS enhanced the crystallization rate of PBS. CNT particles are found to be the most effective nucleating agent compared to fullerene.

ACKNOWLEDGEMENTS

This study is granted by the research foundation of Kocaeli University (Project No: 2014/086).

REFERENCES

1. T. Uesakaa et al., *Polymer* **41**, 8449-8454 (2000).
2. D. G. Papageorgioua et al., *Thermochimica Acta* **590**, 181–190 (2014).
3. V. Ojijo V and S. Sinha Ray, *Prog. Mater. Sci.*, **62**, 1-57 (2014).
4. W-J. Xie and X-M. Zhou, *Mater. Sci. Eng.*, **46**, 366-373 (2015).
5. Y. F. Shih et al., *Polymer*, **49** 4602-4611 (2008).
6. C. S. Lin et al., *J. Taiwan Inst. Chem. Eng.*, **43**, 322–328 (2012).

Influence of Mixing Conditions on Carbon Nanotube Shortening and Curling in Polycarbonate Composites

Beate Krause*, Jérôme Carval and Petra Pötschke

Leibniz-Institut für Polymerforschung Dresden e.V., Hohe Straße 6, 01069 Dresden, Germany

krause-beate@ipfdd.de

Abstract. Polycarbonate composites containing multiwalled carbon nanotubes (MWCNTs, 0.2-2.0 wt%) were melt mixed in small scale at different conditions of screw speed and mixing time to vary the specific mechanical energy (SME) input between 0.4 and 4.0 kWh/kg. Next the electrical properties of compression molded plates and the MWCNT macrodispersion also the nanotube length and shape were analyzed. For this, the matrix of the composites with 0.75 wt% loading was dissolved and the remaining nanotubes were investigated using TEM. It was found that with increasing SME input the number of remaining CNT agglomerates decreases. The MWCNT length decreased from initially about 1.4 micrometers towards 350 nanometers at a SME of 4 kWh/kg and the mean curling values were also reduced. The electrical percolation threshold increases with SME from about 0.4 wt% to 0.6 wt%.

Keywords: carbon nanotube length, electrical properties, polycarbonate, melt mixing
PACS: 06

INTRODUCTION

In composites of polymers with carbon nanotubes (CNTs) the electrical percolation at low filling grades is connected with the very high aspect ratio of CNTs. For this, good dispersion of as synthesized agglomerates into single tubes is needed as well as preserving the initial length of CNTs. Melt mixing is an effective method to disperse MWCNTs in thermoplastic matrices. Thereby, the melt mixing conditions influence the state of dispersion significantly. Based on the steps of dispersion of primary agglomerates of MWCNTs, melt viscosity (related to mixing temperature and shear) influences e.g. the infiltration step of polymer chains into the agglomerate structures but also the shear stresses applied on such infiltrated agglomerates in the following steps of rupture and erosion [1]. Whereas low viscosity is favorable for the infiltration step, higher viscosity favors the dispersion steps. Mixing time also plays an important role on all mechanisms involved in agglomerate dispersion. In general it was found that higher specific mechanical energy (SME) input , achieved e.g. using higher screw speed or lower temperature (or in discontinuous mixing by higher mixing time) results in better dispersion [1-3]. On the other hand, higher SME was also found to shorten the nanotube length to higher extent [3]. As low electrical percolation threshold as well as high mechanical properties requires both, good dispersion and high retained CNT length, optimization of processing conditions is needed.

In this study, polycarbonate (PC)/MWCNT composites were melt mixed in a small-scale microcompounder by varying the mixing conditions. The electrical properties and the CNT macrodispersion were studied depending on the mixing conditions by varying the CNT content. In addition, the CNT length distribution and the curling of nanotubes dissolved from the composites were examined to judge to which extent the initial length could be retained.

EXPERIMENTAL

Composites of polycarbonate (PC, Makrolon® 2600, Bayer MaterialScience AG) and 0.25-2.0 wt% MWCNTs (Nanocyl™ NC7000, carbon purity 95%) were melt mixed at 280°C in a conical twin-screw microcompounder DSM Xplore 15 by varying the mixing time (5 min, 15 min) and rotation speed (100 rpm, 250 rpm).

The specific mechanical energy (SME) input was calculated according to [2]. Electrical measurements were done on pressed plates (pressing conditions: 280°C, 1 min, hot press PW40EH). For composites with resistivities $>10^7$ Ohm cm a Keithley text fixture 8009 and for conductive samples a 4-point text fixture with gold electrodes were used combined with a Keithley electrometer E6517A. The study of CNT macrodispersion using light microscopy (LM) was performed on thin sections (10 μm) which were cut from the extruded strands at room temperature by a diamond knife. The quantification was done by calculation of agglomerate area ratio A_A (%) by division of area of agglomerates A through area of complete image A_0 [4]. For the determination of CNT length distribution and CNT curling the composites containing 0.75 wt% CNT were dissolved in chloroform, a drop of this dispersion was put on a carbon covered TEM grid and the single tubes were imaged by transmission electron microscopy (TEM) using a methodology described in [5]. The curling value was defined as the ratio of the total and the effective length of a single CNT (diagonal of the rectangular area which encased the CNT) as shown exemplarily in Fig. 1 (left).

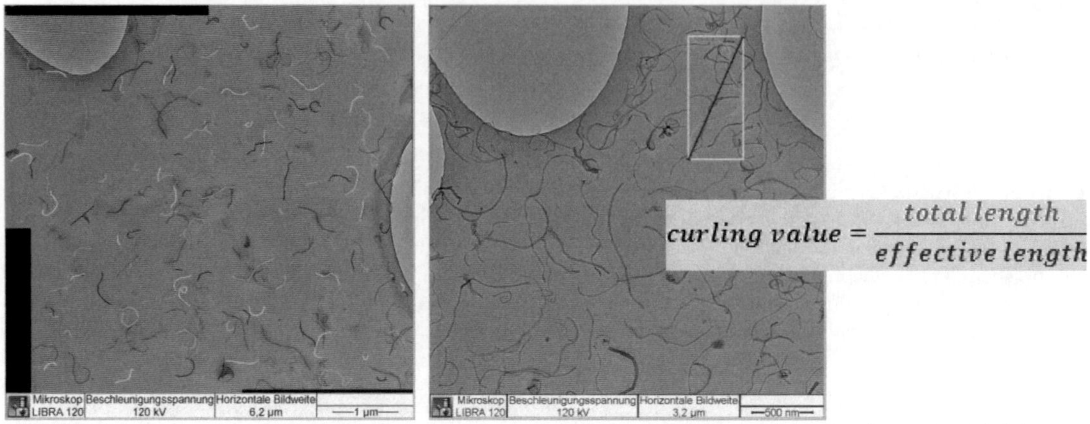

FIGURE 1. Typical TEM images with marked nanotubes (left) and the definition of curling value (right)

RESULTS AND DISCUSSION

When varying the melt mixing conditions, the MWCNT macrodispersion was influenced as illustrated on LM images of composites containing 0.75 wt% CNTs as shown ín Fig. 2.

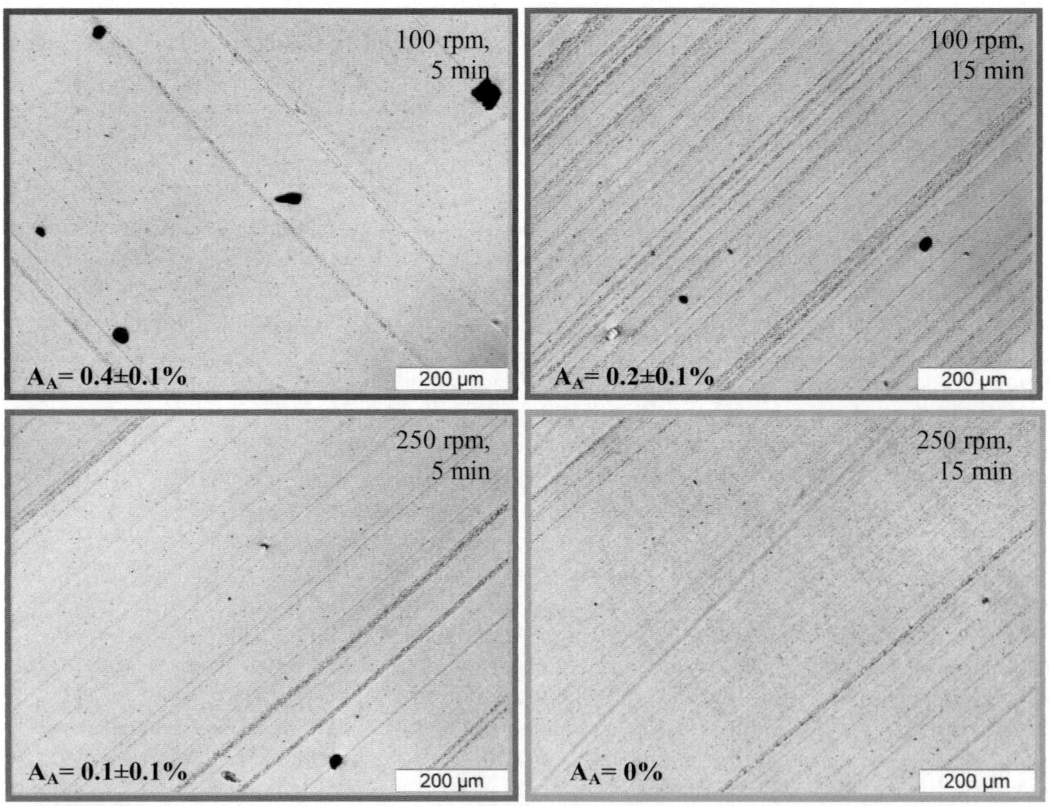

FIGURE 2. Light microscopy images and agglomerate area ratio A_A of PC/0.75 wt% MWCNT composites prepared at different mixing conditions

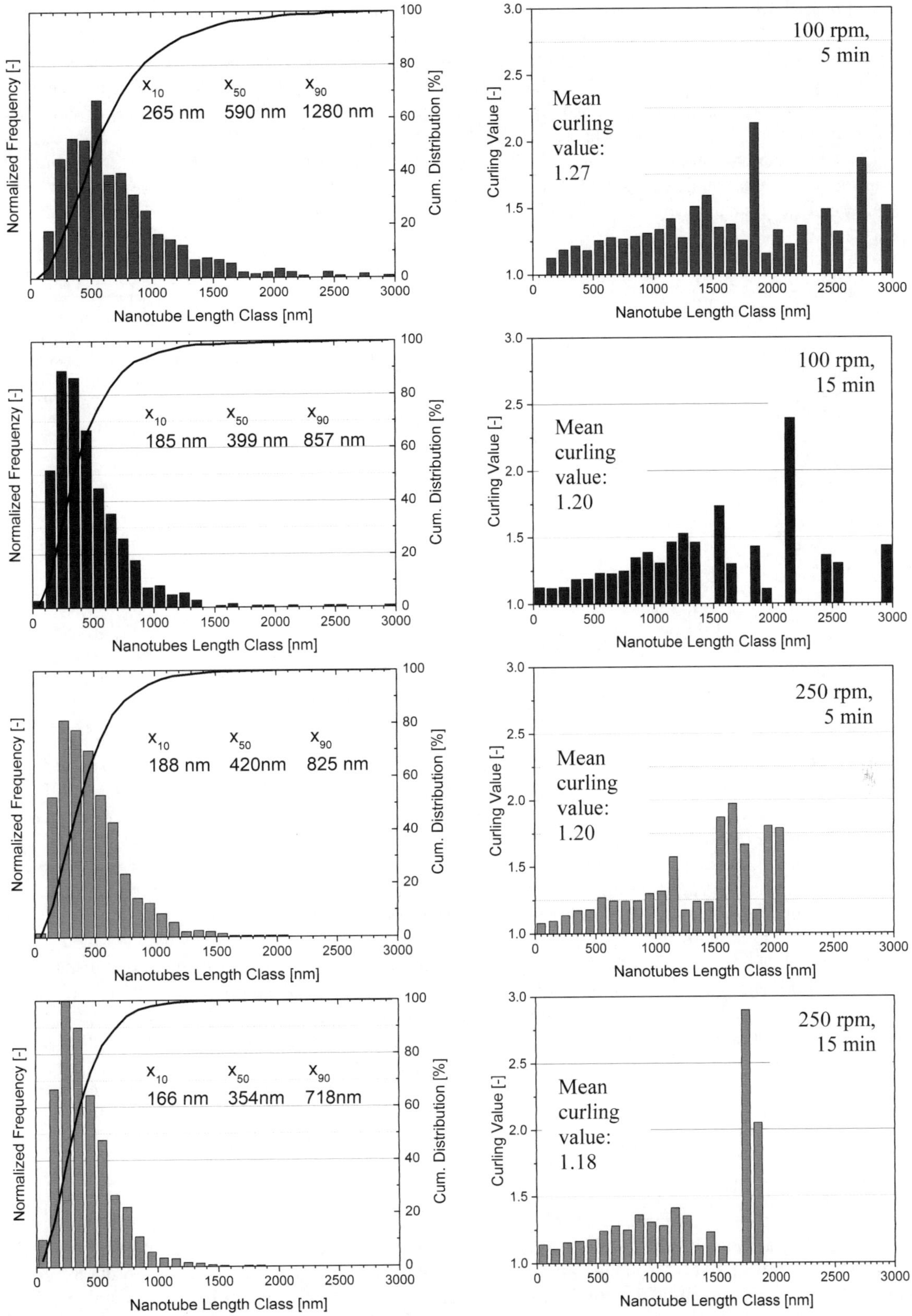

FIGURE 3. Distributions of CNT length (left) and curling value (right) for nanotubes extracted from the PC/0.75 wt% CNT composite prepared at 280°C and different conditions

Increasing SME input, achieved either by increased rotation speed or mixing time improved the state of dispersion as indicated by the decrease of agglomerate area ratio A_A. Agglomerate free composites could be prepared at the highest SME, based on higher rotation speed (250 rpm) and longer mixing time (15 min).

Additionally, a decrease of the MWCNT length after melt mixing could be measured. From the initial mean MWCNT length x_{50} of 1341 nm [5] the MWCNT length in the composites was reduced to values between 590 nm and 354 nm whereas the shortening was more pronounced at higher SME. The length histograms are shown in Fig. 3 (left) and the dependency of mean MWCNT length on SME in Fig. 4 (left). Similar relations between SME, macrodispersion and nanotube length were described before for polycaprolactone/0.5 wt% MWCNT composites prepared using the same small-scale compounder at different rotation speeds [3]. The authors found with increasing SME (rotation speed) an improvement of the MWCNT macrodispersion with simultaneous reduction of the mean MWCNT length.

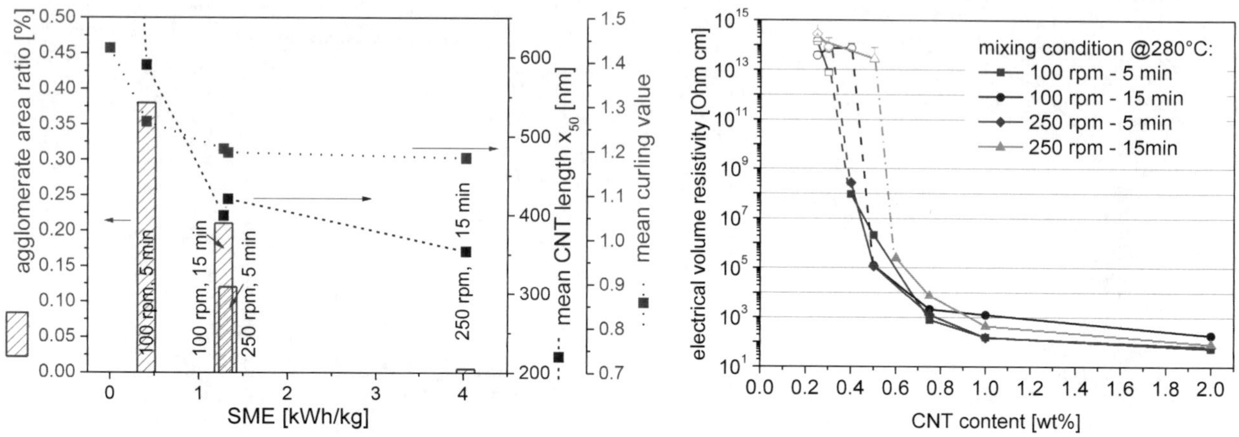

FIGURE4. Dependence of agglomerate area ratio, mean CNT length x_{50} and mean curling value on the specific mechanical energy (SME) for PC/0.75 wt% MWCNT composites prepared at different mixing conditions (left), Electrical volume resistivity versus CNT content for PC composites prepared at different mixing conditions (right)

Comparing the mean curling value at different mixing conditions, a decrease of the curling with increasing SME can be seen (Fig. 3 right, Fig. 4 left). Generally, the curling value lowered with decreasing MWCNT length. It can be assumed that the nanotubes break at the defect sites and the remaining nanotube parts have a lower defect density and are straighter and more like an ideal rod shape nanotube. A reduction of the defect sites in MWCNTs with lower length detected using RAMAN study was described before for MWCNTs which were broken using ball milling [6].

Furthermore, the electrical percolation threshold p_c increased with increasing SME input despite better nanotube dispersion which correlates well with the more pronounced MWCNT shortening (Fig. 4 right). The values for p_c increased from ~0.4 wt% for composites prepared at 100 rpm/5 min and 250 rpm/5 min (towards ~0.5 wt% for those prepared at 100 rpm/15 min and ~0.6 wt% for composites mixed at 250 rpm/15 min. When comparing the two composites with medium energy input (1.28 or 1.33 kWh/kg) interestingly the sample with the shorter mixing time but higher speed showed slightly lower shortening and lower electrical percolation threshold illustrating. This illustrates that the mixing time seems to be also a critical factor. In addition it shows that the combination of better dispersion and higher CNT length, as obtained at 250 rpm/ 5 min results in the lowest electrical resistivity values. The sample with best dispersion having the same time most pronounced shortening (250 rpm /15 min) shows the highest electrical percolation.

CONCLUSIONS

When varying mixing conditions during melt mixing of PC/MWCNT composites, increasing SME input resulted in improved MWCNT macrodispersion and decreased mean MWCNT length as well as mean curling value. With decreasing MWCNT length the curling values decrease indicating that entangled nanotubes are broken into more straight parts. This is most probably due to breakage at defect sites of the nanotubes. Electrical percolation sensitively reflects both, improved dispersion and reduced nanotube length. It seems that for electrical properties preserving the nanotube length is more critical than to obtain excellent dispersion.

ACKNOWLEDGMENTS

The main part of this work was performed in the frame of a 3 month internship of J.C. from the Université de Rennes I, St Brieuc Cedex 1, France at the Leibniz-Institut für Polymerforschung Dresden e.V. (IPF).

REFERENCES

1. I. Alig, P. Pötschke, D. Lellinger, T. Skipa, S. Pegel, G. R. Kasaliwal and T. Villmow, *Polymer* **53** (1), 4-28 (2012).
2. B. Krause, P. Pötschke and L. Häußler, *Composites Science and Technology* **69** (10), 1505-1515 (2009).
3. P. Pötschke, T. Villmow and B. Krause, *Polymer* **54** (12), 3071-3078 (2013).
4. R. Socher, B. Krause, M. T. Müller, R. Boldt and P. Pötschke, *Polymer* **53** (2), 495-504 (2012).
5. B. Krause, R. Boldt and P. Pötschke, *Carbon* **49** (4), 1243-1247 (2011).
6. B. Krause, T. Villmow, R. Boldt, M. Mende, G. Petzold and P. Pötschke, *Composites Science and Technology* **71** (8), 1145-1153 (2011).

PP/SWCNT Composites Modified With Ionic Liquid

Beate Krause[a]*, Mikhael Predtechensky[b], Evgeny Ilin[b], Petra Pötschke[a]

[a] *Leibniz-Institut für Polymerforschung Dresden e.V., Hohe Straße 6, 01069 Dresden, Germany*
[b] *OCSiAl Ltd., 630090 Novosibirsk, Russia*

krause-beate@ipfdd.de

Abstract. Polypropylene composites filled with singlewalled carbon nanotubes TUBALL® (SWCNTs) were studied with regard to the effect of ionic liquid (IL) addition in different SWCNT:IL ratios (1:0.5 -1:6). The incorporation of IL leads to a decrease of the electrical percolation threshold and already at 0.025 wt% SWCNT loading reduced resistivity values can be observed. However, the SWCNT macro dispersion, already relatively good without IL, was not affected by the IL incorporation. In addition, the nucleation effect of the SWCNT in polypropylene is not influenced when adding simultaneously IL whereas the crystallization enthalpy slightly decreases with its addition.

Keywords: polypropylene, singlewalled carbon nanotubes, ionic liquid, electrical properties, thermal properties
PACS: 73

INTRODUCTION

Carbon nanotubes (CNTs) have been shown in the last years as a versatile filler to influence several polymer properties favorably. This concerns especially the electrical conductivity of singlewalled (SWCNTs) and multiwalled CNTs (MWCNTs) which can be transferred to insulating matrices by establishing a percolated filler network. Melt mixing was shown to be a highly convenient method to produce thermoplastic composites containing CNTs. One crucial point in producing such nanocomposites is the filler dispersion, which is difficult to achieve due to strong van-der-Waals forces and entanglements between the tubes. In order to facilitate the dispersion process, the addition of dispersion agents is a common way. Recently, a new high quality SWCNT product with the trade name TUBALL™ from the company OCSiAl LTD came into market, which suitable dispersion using melt mixing in polypropylene (PP) and very low electrical percolation threshold were shown in ref. [1]. In order to further improve dispersion and lower the percolation concentration, an ionic liquid (IL) was selected as an additive introduced simultaneously with the CNTs. The suitability of this IL to disperse MWCNTs was shown in previous reports [2-4].

EXPERIMENTAL PART

In this study, polypropylene (PP) Moplen HP 400R (LyondellBasell) having a melt flow rate (230°C, 2.16 kg) of 25 g/10 min was used in powder shape (particle size <1mm). As singlewalled CNTs (SWCNTs) non-functionalized TUBALL® (OCSiAl S.a r.l., Luxembourg) with a diameter of 1-2 nm and a length of more than 5 μm were used. An ionic liquid (IL, 1-methyl-3-octylimidazolium tetrafluoroborate) (purity >97%, HPLC) purchased from Sigma Aldrich was selected as additive.

The PP composites were melt mixed using the masterbatch dilution approach using a conical twin-screw small-scale compounder Xplore DSM15 at 210°C, rotation speed of 250 rpm, and mixing time of 5 min. The SWCNT concentration in the masterbatch was varied between 1, 2 and 3 wt%, whereby a SWCNT:IL ratio of 1:1 was applied. For the selected most suitable SWCNT concentration (2 wt% SWCNT) the ratio of SWCNT to IL was varied to be 1:0.5, 1:1, 1:2, 1:3, and 1:6. Consequently, the different masterbatches were diluted under the conditions named before to lower concentrations, ranging between 0.01 and 0.2 wt% in order to study the electrical percolation concentration.

Electrical measurements were done on pressed plates (pressing conditions: 210°C, 2 min, hot press PW40EH). For composites with resistivities $>10^7$ Ohm cm a Keithley text fixture 8009 and for conductive samples a 4-point text fixture with gold electrodes were used combined with a Keithley electrometer E6517A.

The study of CNT macrodispersion was done on thin sections (5 μm) which were cut at room temperature using Leica microtom RM2265 and a diamond knife. The light microscopy investigations were performed using a microscope Olympus BH2 and a camera Olympus DP71. The quantification was done by calculation of agglomerate area ratio A_A (%) by division of area of agglomerates A through area of complete image A_0 [5]. Results are shown for 0.1 wt% loading.

Differential Scanning Calorimetry (DSC) analysis was carried out with a DSC Q 1000 (TA Instruments) between -80°C and 150°C under nitrogen atmosphere at a scan rate of 10 K/min in heating-cooling-heating cycles. The melting temperature in the second heating run $T_{m,}$ as well as onset ($T_{c,onset}$) and maximum ($T_{c,max}$) temperatures of crystallization and the PP polymer corrected crystallization enthalpy $\mathbf{\Delta H_C}$ from the cooling run were evaluated.

RESULTS

Evaluation of the influence of masterbatch concentration

To select the most suitable masterbatch concentration for the further dilution into composites, three different IL containing masterbatches were prepared having 1, 2 or 3 wt% SWCNT. The masterbatches were diluted to determine the electrical percolation threshold concentration of SWCNT. The results presented in Fig. 1 show the lowest percolation threshold when using the masterbatches containing 1 or 2 wt%. For the further work, the masterbatch with 2 wt% was selected as for this the lowest volume resistivity values were achieved.

The resistivity values achieved at 0.1 wt% with the IL containing masterbatches (1 and 2 wt%) are about 2 decades lower when compared to a composite diluted from a masterbatch (2 wt% loading) without IL addition [1], indicating the favorable effect of IL.

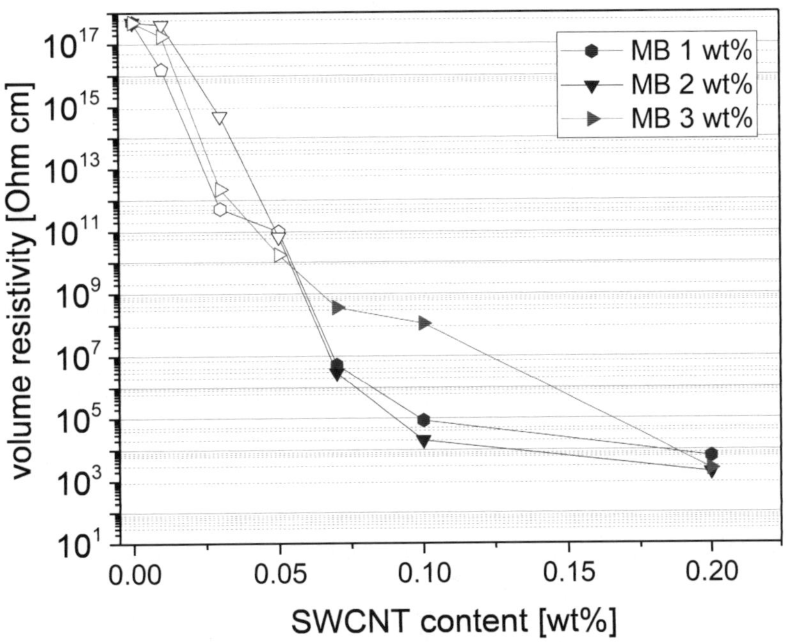

FIGURE 1. Electrical volume resistivity versus SWCNT content for PP/SWCNT composites prepared by masterbatch (MB) dilution from masterbatches with different SWCNT contents (SWCNT:IL = 1:1)

Evaluation of SWCNT: ionic liquid ratio

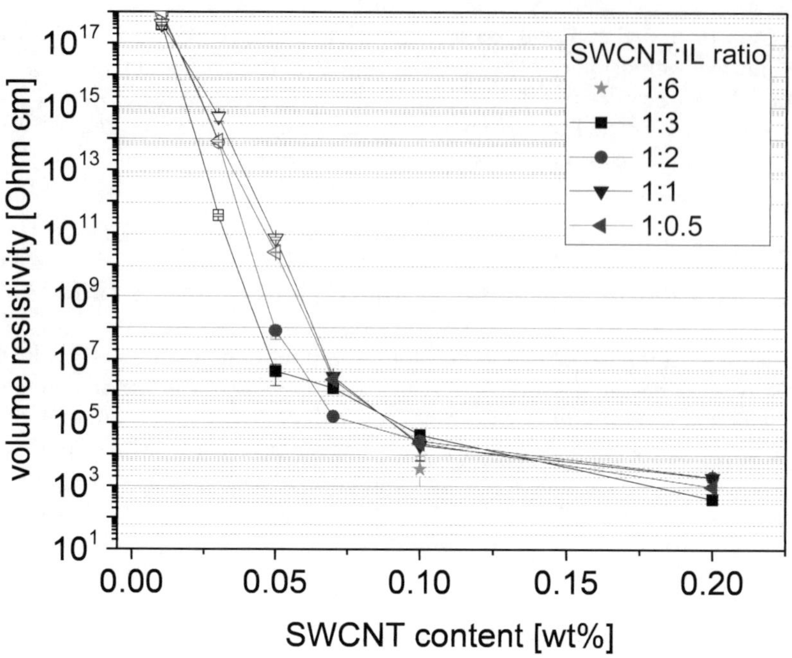

FIGURE 2. Electrical volume resistivity versus SWCNT content for PP/SWCNT composites prepared using the masterbatch with 2 wt% SWCNTs and different contents of ionic liquid

The electrical resistivity was investigated for PP composites containing different ratios of SWCNT and ionic liquid (Fig. 2). As compared to composites without the use of IL, where electrical percolation was found between 0.075 and 0.1 wt%[1], all composites with IL show lower percolation thresholds. For all SWCNT:IL ratios reduced resistivity values can be already observed at 0.025 wt% loading indicating the formation of conductive pathways. The electrical resistivity of the Tuball® filled PP composites is lower if the SWCNT:IL ratio (and with it the absolute IL content) increases, especially seen in the ratios 1:0.5 or 1:1 are compared to that of 1:2. The lowest values at 0.025 and 0.05 wt% SWCNT loading were achieved when using the SWCNT:IL ratio of 1:3. At the SWCNT:IL ratio of 1:6 (only investigated for 0.1 wt% loading) an additional decrease of volume resistivity value could observed, however this recipe was not further investigated because the ionic liquid migrated partially to the surface of the PP composite. These results indicate that the influence of the ionic liquid on the electrical properties of PP composites is significant.

Additionally, the SWCNT macrodispersion was studied and light microscopy images are shown in Fig. 3. For the PP composites without and with ionic liquid same only very small remaining agglomerates were visible and the agglomerate area ratios A_A were very low. As dispersion was already very good without IL, the influence of ionic liquid on the SWCNT macrodispersion was not significant.

Furthermore, the thermal behavior of PP/SWCNT composites was investigated to observe the influence of ionic liquid on the melting and crystallization behavior. In Table 1 the results of PP/0.1 wt% SWCNT composites filled with different IL content are summarized. It was found that the melting temperature T_m was not influenced by the addition of SWCNT and IL. The crystallization temperatures increase with the incorporation of SWCNT. However, an additional influence of the ionic liquid on the crystallization temperatures could not be observed. The value of enthalpy $\mathbf{\Delta H_C}$ (cooling) also increased with the SWCNT addition illustrating the well-known nucleation effect of SWCNTs. The comparison of ΔH_C for PP/SWCNT and PP/SWCNT/IL composites shows slightly lower values for the PP/SWCNT/IL composites. It can be assumed that the ionic liquid slightly disturbs the growth of crystals. The results in Table 2 shows PP/SWCNT/IL composites with constant SWCNT:IL ratio of 1:1. Already the concentration of 0.01 wt% SWCNT leads to an increase of the crystallization temperature from 123.7°C ($T_{c,onset}$) or 120.5°C ($T_{c,max}$) to 131.3 or 128.8°C, respectively. With further increase in the SWCNT content the crystallization temperatures also increase.

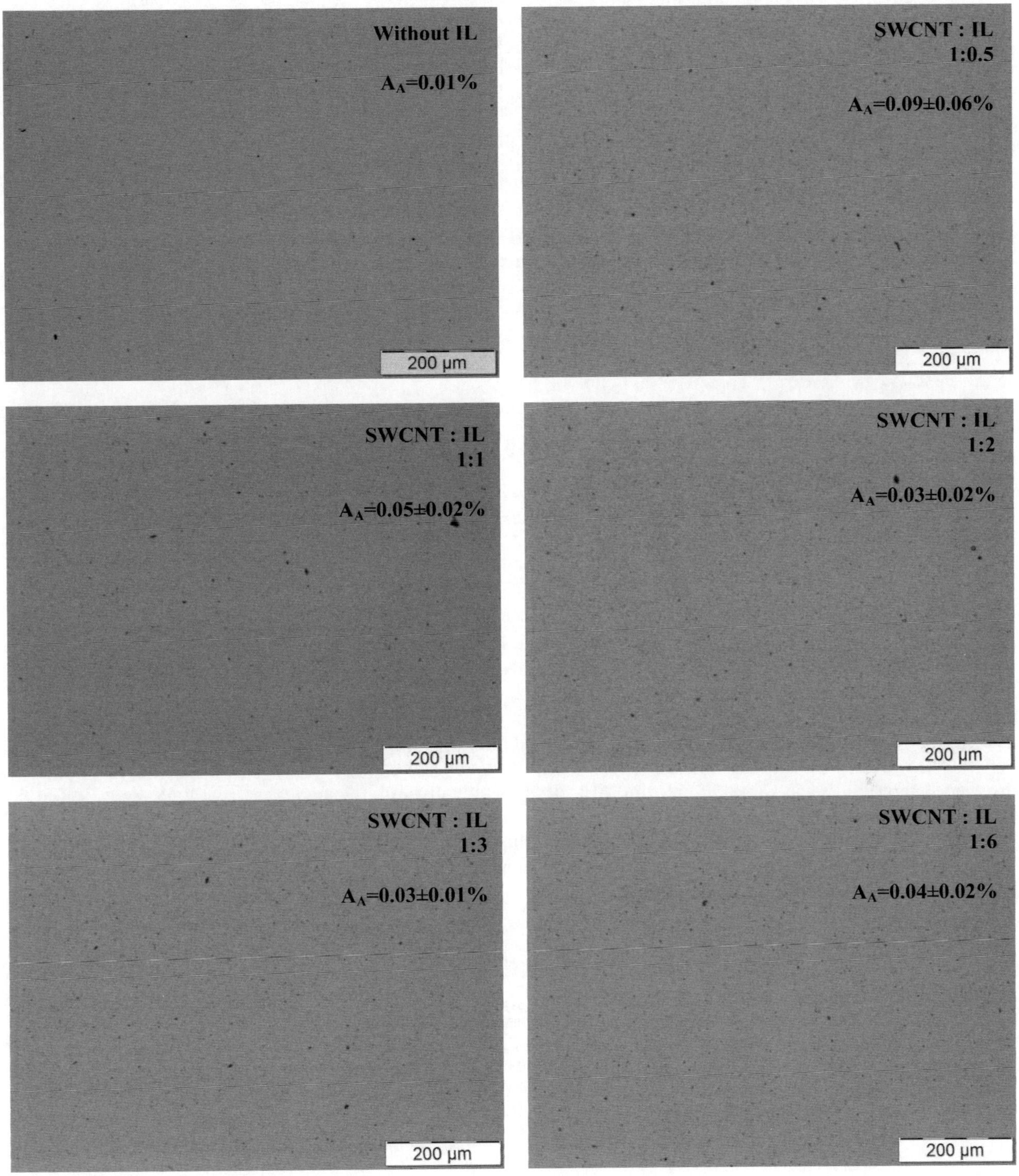

FIGURE 3. Optical microscopy images of PP/0.1 wt% SWCNT composites without ionic liquid and with different ratios of SWCNT:IL

TABLE 1. DSC data of PP (melt processed) and PP/SWCNT/IL composites containing 0.1 wt% SWCNT and different SWCNT:IL ratios.

PP composite	T_m [°C]	$T_{c,onset}$ [°C]	$T_{c,max}$ [°C]	ΔH_C [J/g $_{polymer}$]
PP processed	162.7	123.7	120.5	104.5
PP + 0.1% SWCNT	163.8	132.3	129.4	109.7
PP + 0.1% SWCNT + 0.05% IL	163.1	131.8	129.2	108.8
PP + 0.1% SWCNT + 0.1% IL	163.6	131.2	128.8	106.5
PP + 0.1% SWCNT + 0.2% IL	162.7	131.8	129.9	108.4
PP + 0.1% SWCNT + 0.3% IL	163.5	131.9	129.2	106.9
PP + 0.1% SWCNT + 0.6% IL	163.2	132.0	129.4	107.2

TABLE2. DSC data of processed PP and PP/SWCNT/IL composites at SWCNT:IL ratio of 1:1 containing different SWCNT contents.

PP composite	T_m [°C]	$T_{c,onset}$ [°C]	$T_{c,max}$ [°C]	ΔH_C [J/g $_{polymer}$]
PP processed	162.7	123.7	120.5	104.5
PP + 0.01% SWCNT + 0.01% IL	163.6	131.3	128.8	106.4
PP + 0.03% SWCNT + 0.03% IL	162.3	129.0	127.1	107.4
PP + 0.05% SWCNT + 0.05% IL	162.5	130.5	128.5	109.9
PP + 0.07% SWCNT + 0.07% IL	162.8	130.9	128.8	110.4
PP + 0.1% SWCNT + 0.1% IL	163.6	131.2	128.8	106.5
PP + 0.1% SWCNT	163.8	132.3	129.4	109.7
PP + 0.2% SWCNT + 0.2% IL	163.7	133.3	130.9	106.3

CONCLUSION

For melt mixed PP/SWCNT composites the influence of the addition of an ionic liquid on the electrical and thermal properties and SWCNT macrodispersion was investigated. It was found that the electrical percolation threshold decreases with increasing ionic liquid content. The SWCNT macrodispersion was not influenced by the IL addition which was already very good without IL addition. In DSC investigations, the melting temperature was not influenced by the addition of SWCNT and/or IL. The crystallization temperatures increase with the incorporation of SWCNT. However, an additional influence of the ionic liquid on the crystallization temperatures could not be observed. Interestingly, the values of enthalpy ΔH_C decrease slightly with the addition of the ionic liquid and it is assumed that IL disturbs the crystal growth.

In summary the results show that the selected ionic liquid is suitable to improve electrical properties of melt mixed PP/SWCNT composites and to achieve electrical percolation at low SWCNT loadings. Further work is required to study the effect of IL on other composite properties.

REFERENCES

1. B. Krause, P. Pötschke, E. Ilin and M. Predtechenskiy, *Polymer*, submitted (2016).
2. L. Zhao, Y. Li, X. Cao, J. You and W. Dong, *Nanotechnology* **23** (25), 255702 (2012).
3. K. Subramaniam, A. Das, K. W. Stöckelhuber and G. Heinrich, *Rubber Chemistry and Technology* **86** (3), 367-400 (2013).
4. R. Socher, *PA12-MWCNT-Nanokomposite: Wege zur effektiven MWCNT-Dispergierung und zu niedrigen elektrischen Perkolationsschwellen*. (München, Deutschland: Verlag Dr. Huth, 2013).
5. R. Socher, B. Krause, M. T. Müller, R. Boldt and P. Pötschke, *Polymer* **53** (2), 495-504 (2012).

Thermal Conductivity of Hybrid Filled HDPE Nanocomposites

Michael Thomas Müller[a], Beate Krause[a], Bernd Kretzschmar[a], Ivonne Jahn[b] and Petra Pötschke[a]

[a]*Leibniz-Institut für Polymerforschung Dresden e.V. (IPF), Hohe Straße 6, 01069 Dresden, Germany*
[b]*Fraunhofer Pilot Plant Center (PAZ) for Polymer Synthesis and PolymerProcessing, ValuePark A74, 06258 Schkopau, Germany*

krause-beate@ipfdd.de, poe@ipfdd.de

Abstract. In this study composite materials based on high-density polyethylene (HDPE) with fillers containing nanostructures were prepared using melt mixing. Vapour Grown Carbon Fibers (VGCF), multiwalled carbon nanotubes (MWCNT) of the types Baytubes® C150P and Nanocyl™ NC7000, anthracite powder, microsilica, organoclay and expanded graphite (EG) as well as mixtures of these fillers were used. The amount and mixing ratios of the hybrid filled systems have been varied to determine the effect on the achievable level of thermal conductivity as measured on pressed plates. The filler dispersion and phase adhesion were studied using scanning electron microscopy. When limiting the maximum filler content to 10 wt%, the highest enhancement in thermal conductivity by 166% was found for VGCF followed by a 1:1 filler combination of VGCF with EG (148%).

Keywords: Thermal conductivity, Nanocomposites, Carbon nanotubes, Vapor grown carbon fibers, Expanded graphite.
PACS: 44

INTRODUCTION

Managing heat dissipation is for example in electronic applications a great issue in context with the ongoing size reduction of electronics devices. To achieve enhanced thermal conductivity in polymer based electronic packaging currently metallic particles [1, 2] or mineral materials [3, 4] are used. As metallic particles have the disadvantage to promote thermo-oxidative polymer degradation alternatives have to be found. One approach is the use of carbon fillers like MWCNT, VGCF or expanded graphite. Such polymer nanocomposites combine improvements in heat transport with established plastics processing methods.

In this contribution, composite materials based on high-density polyethylene (HDPE) containing nanostructured fillers were prepared. To test the effects of Vapour Grown Carbon Fibers (VGCF), multiwalled carbon nanotubes (MWCNT) of the types Baytubes® C150P and Nanocyl™ NC7000, anthracite powder, microsilica, organoclay, and expanded graphite as well as mixtures of these fillers, small scale mixing using a DSM Xplore 15 twin screw compounder was performed. The amount and mixing ratios of the hybrid filled systems were varied to determine the effect on the achievable level of thermal conductivity. Due to the orientation of rod-like filler particles leading to anisotropic heat flow, the thermal conductivity measurements were carried out in the through-plane direction as this direction is most interesting for industrial applications. In order to improve filler dispersion and induce chemical interactions between the polymer and the filler, the use of maleic anhydride-grafted polyethylene and amino-functionalized MWCNT were investigated. As reported before, covalent matrix filler bonding can benefit the thermal transfer between the nanotubes and the polymer matrix [5]. The filler dispersion and phase adhesion were studied using scanning electron microscopy (SEM).

EXPERIMENTAL PART

A high-density polyethylene (HDPE) copolymer based on hexane prefilled with 2 wt% carbon black having a melt temperature range of 130– 132°C and a melt flow index of 0.3 g/10 min (ISO 1133) was used for melt mixing.

Different commercially available multiwalled carbon nanotubes were employed, namely (i) Baytubes® C150P (Bayer MaterialScience AG, Leverkusen, Germany) with diameters of 5 – 20 nm and length of 1 - >10 µm, bulk density 120-170 kg/m^3 with a specific density of 1.82 g/cm^3 (measurements at IPF), (ii) Nanocyl™ NC7000 (Nanocyl S.A., Sambreville, Belgium) with an average diameter of 9.5 nm and an average length of 1.3 µm [6], bulk density 66 kg/m^3 with a specific density of 1.85 g/cm^3 (measurements at IPF), (iii) Nanocyl™ NC3150 (Nanocyl S.A., Sambreville, Belgium) with an average diameter of 9.5 nm and an average length of < 1 µm, (iv) NH_2 functionalized Nanocyl™ NC3152 (Nanocyl S.A., Sambreville, Belgium) with an average diameter of 9.5 nm and an average length of < 1 µm with a functionality grade of <0.5%, and (v) Nanocyl™ NC9000 series (Nanocyl S.A., Sambreville, Belgium) representing an in-situ polymerised nanocomposite polyolefin powder based on NC7000 with 34.8 wt% MWCNT.

As a coupling agent between the polyethylene matrix and the nanofillers, a maleic acid anhydride (MA) functionalized (2%) LLDPE-g-MA with the designation TSPE 1112 GALL SCONA® (Kometra Kunststoff-Modifikationen und- Additiv GmbH, Germany) having a melt flow index of 1 g/10 min (MFR 190°C/2,16kg) was used. High thermal conductive Vapor Grown Carbon Fibers (VGCF-S) (SHOWA DENKO K. K., Tokyo, Japan) with a diameter of 80 nm and a length of 10 µm, bulk density 20 kg/m^3 and a specific density of 2.1 g/cm^3 were employed to study a long fiber filler. As microsilica the filler SIDISTAR® T 120 XP (Elkem AS, Materials, Norway) with a median particle size of 0.15µm and a specific density of 2.2 g/cm^3 [7]was used. In addition, a layered silicate, namely Nanofil® 5 (organically modified microsilica with C-18 n-alkyl chains, Süd-Chemie, Germany) with a mean particle size of 8 µm and a specific density of 1.8 g/cm^3 was used to promote filler dispersion. Furthermore, as expanded graphite ECOPHIT® GFG5 (EG, SGL TECHNOLOGIES GmbH, Germany) having a mean particle size of 5 -7 µm and a specific density of 2.25 g/cm^3 was used as a platelet-like filler. Finally, also an anthracite powder trial product of the deposit in Ibbenbüren (Germany) with a specific density of 1.4 g/cm^3 (measurements at IPF) was applied.

The premixed materials were melt mixed using a DSM Xplore 15 twin-screw microcompounder with a chamber volume of 15 cm3. The mixing conditions were 210°C set temperature, 150 rpm mixing speed, and 5 min mixing time. The extruded strands were compression moulded (210°C, 50 kN) into plates (27 mm diameter, 2 mm thickness) using a hot press (Model PW 40 EH, Paul Otto Weber GmbH).

The achievable level of thermal conductivity was determined on pressed plates with the NanoFlash™ LFA447 device (NETZSCH, Gemany) at a temperature of 25°C. From six measurements on three different plates mean values and standard deviation were obtained.

To study the morphology of the composites, scanning electron microscopy (SEM) was applied. SEM of the extruded composite strands was performed on cryofractured gold sputtered surfaces using an Ultra plus Field Emission Gun Scanning Electron Microscope (FEG-SEM, Carl Zeiss GmbH, Oberkochen, Germany) with an InLens detector. An excitation voltage of 20 kV was used.

RESULTS

Thermal Conductivity

Table 1 shows the through-plane thermal conductivity of composites containing 10 wt% of some of the single carbon nanofillers. As MWCNT and VGCF have a rod-like structure, the fillers may are oriented transverse to the measurement direction when performing through-plane measurements. Therefore, the thermal conductivity values may be lower than in in-plane direction. The additions of 10 wt% Baytubes® C150 P or Nanocyl™ NC7000 enhance the thermal conductivity of HDPE only by about 40% and 54%, respectively. The addition of LLDPE-g-MSA as coupling agent does not result in benefits for thermal conductivity. Even when the NH_2-modified nanotube material Nanocyl™ NC3152 was used no significant enhancement compared to the other nanotube materials was observed. Nanocyl™ NC9000 and expanded graphite showed enhancements by about 105% and 111%, respectively, as compared to pure HDPE. The long fibers of VGCF-S improved the through-plane thermal conductivity by about 166% and therefore indicated to be the best but also a quite expensive solution.

In Figure 1 left the dependence of the through-plane thermal conductivity on the VGCF-S filler volume fraction is shown. With the increase in filler content of VGCF the thermal conductivity increases linearly. With a filler content of 1.9 vol% (4 wt%) VGCF-S a thermal conductivity of 0.55 W·m^{-1}·K^{-1} was achieved which is in the same range like the results obtained by the addition of 10 wt% Baytubes® C150 P or Nanocyl™ NC7000.

To use the advantage of VGCF-S but to produce a cheaper VGCF containing composite, different filler combinations were studied. Table 2 shows through-plane thermal conductivity results for filler combinations containing 5 wt% VGCF-S as well as values for their parent materials. The highest achievable enhancement was found for the combination of 5 wt% VGCF-S with 5 wt% expanded graphite resulting in a thermal conductivity of 0.87 $W \cdot m^{-1} \cdot K^{-1}$ which is almost the same like 0.93 $W \cdot m^{-1} \cdot K^{-1}$ for the composite with 10 wt% VGCF-S. Also, the combination of 10 wt% of low heat-conductive microsilica (0.40 $W \cdot m^{-1} \cdot K^{-1}$) with 5 wt% VGCF-S generated a relatively high through-plane thermal conductivity of 0.71 $W \cdot m^{-1} \cdot K^{-1}$ in the composite.

TABLE 1. Through-plane thermal conductivity results for single fillers in HDPE.

Filler composition	**Through-plane thermal conductivity [$W \cdot m^{-1} \cdot K^{-1}$]**	**Standard deviation**
No filler (HDPE)	0.35	± 0.012
10 wt% Baytubes® C150 P	0.49	± 0.002
10 wt% NanocylTM NC7000	0.54	± 0.002
10 wt% expanded graphite	0.74	± 0.042
10 wt% NanocylTM NC9000	0.72	± 0.020
10 wt% LLDPE-g-MA +10 wt% NanocylTM NC3152	0.54	± 0.000
10 wt% LLDPE-g-MA +10 wt% NanocylTM NC3150	0.53	± 0.010
10 wt% LLDPE-g-MA +10 wt% NanocylTM NC7000	0.52	± 0.030
10 wt% VGCF-S	0.93	± 0.040

TABLE 2. Through-plane thermal conductivity results for hybrid fillers in HDPE.

Filler composition	**Through-plane thermal conductivity [$W \cdot m^{-1} \cdot K^{-1}$]**	**Standard deviation**
No filler (HDPE)	0.35	± 0.012
10 wt% VGCF-S	0.93	± 0.040
10 wt% anthracite powder	0.38	± 0.000
20 wt% anthracite powder	0.37	± 0.010
5 wt% VGCF-S + 10 wt% anthracite powder	0.36	± 0.010
5 wt% VGCF-S + 20 wt% anthracite powder	0.41	± 0.010
10 wt% microsilica	0.40	± 0.010
5 wt % VGCF-S + 10 wt% microsilica	0.71	± 0.030
5 wt % VGCF-S + 7.5 wt% Nanofil®5 + 7.5 wt% LLDPE-g-MA	0.66	± 0.030
5 wt% VGCF-S + 5 wt% expanded graphite	0.87	± 0.070

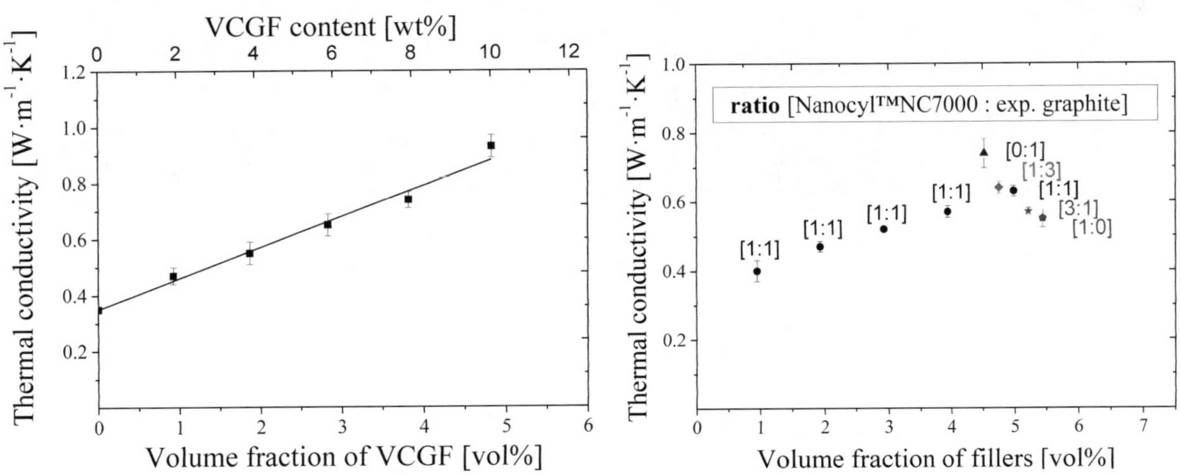

FIGURE 1. Through-plane thermal conductivity: HDPE/ VGCF-S composites in dependence on the filler volume fraction (left); mixtures of Nanocyl™ NC7000 and expanded graphite in HDPE with different ratios (right)

Additionally, nanotubes Nanocyl™ NC7000 in a filler combination with expanded graphite (EG) were studied. The results of thermal conductivity are shown in Figure 1 right. In the through-plane direction an advantage in the thermal conductivity was found when replacing NC7000 by EG. At a mixing ratio of NC7000/EG = 1:1 wt%, the thermal conductivity increased from 0.55 W·m^{-1}·K^{-1} using 10 wt% NC7000 (5.4 vol%) to 0.63 W·m^{-1}·K^{-1} by using the mixture of 5 wt% of EG and 5 wt% NC7000 (5.0 vol% filler). In that series the composite filled with 10 wt% (4.5 vol%) EG as single filler performed the highest through-plane thermal conductivity of 0.74 W·m^{-1}·K^{-1}.

Morphology

Figure 3 show the morphologies of the cryofractured surface of HDPE with some of the nanotubes and VGCF-S at 10 wt% loading. The SEM images illustrate especially the different wetting behaviour of the HDPE melt and the fillers. For Nanocyl™ NC9000 nanocomposite filler, a quite good wetting of the nanotubes can be observed, whereas for the pure nanotube materials more free nanotube surfaces illustrating worse wetting can be seen. Furthermore, it is clearly visible that the long, thick fibers of the VGCF-S filler form an excellent network structure even if wetting is not pronounced for that filler material.

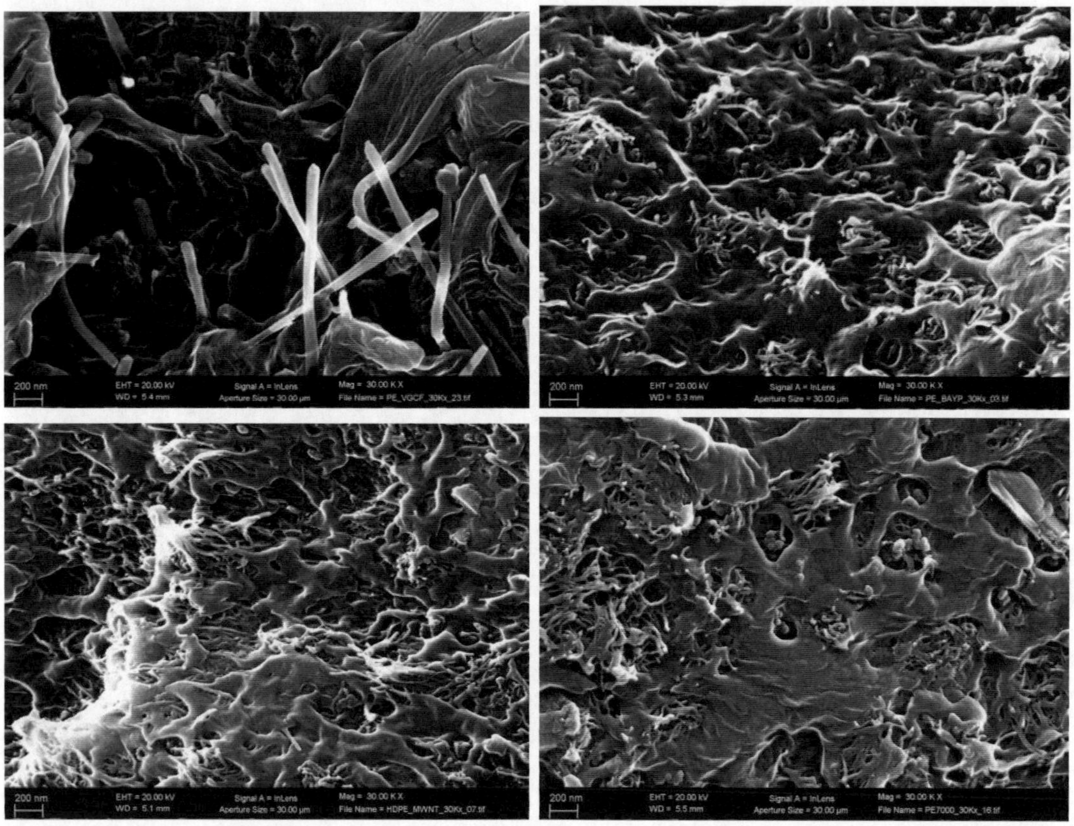

FIGURE 2. Morphologies of the cryofractured surfaces of HDPE composites with 10 wt% of selected carbon nanofillers: VGCF-S, Baytubes® C150P, Nanocyl™ NC9000, Nanocyl™ NC7000 (from left to right)

SUMMARY AND CONCLUSION

The aim of this study was to determine the effects of carbon nanofillers and their mixtures with other fillers on the thermal conductivity of melt mixed composites based on HDPE. Different kinds of MWCNT, VGCF-S, expanded graphite, microsilica, and anthracite powder were investigated. Based on a thermal conductivity of LDPE of 0.35 W·m^{-1}·K^{-1}, for single-filler systems and limiting the filler content to 10 wt%, expanded graphite and VGCF-S enhanced the through-plane thermal conductivity by 111% (0.74 W·m^{-1}·K^{-1}) and 166% (0.93 W·m^{-1}·K^{-1}), respectively.

All studied mixtures of Nanocyl™ NC7000 with expanded graphite showed no enhancement of thermal conductivity in through-plane direction as compared to the use of the expanded graphite as single filler. This may be due to the different orientation effects of spherical, platelet-like and rod-like fillers during the compression moulding process. On this issue, further investigations have to be carried out which involve the measurement of thermal conductivity also in in-plane direction.

The combination of 5 wt% VGCF-S with 5 wt% expanded graphite generated a through-plane thermal conductivity of 0.87 $W \cdot m^{-1} \cdot K^{-1}$ in the composite. The addition of 10 wt% of low heat-conductive microsilica to 5 wt% VGCF-S resulted in an enhancement in through-plane thermal conductivity from 0.61 $W \cdot m^{-1} \cdot K^{-1}$ towards 0.71 $W \cdot m^{-1} \cdot K^{-1}$. It may be concluded, that the combination of VGCF-S with conductive expanded graphite or low-conductive microsilica promotes the formation of thermal conductive pathways through the HDPE matrix.

ACKNOWLEDGMENTS

We thank the German Federal Ministry of Education and Research (BMBF) for financial support within the Innovation Alliance Inno.CNT, project CarboTube, BMBF 03X0054B.

REFERENCES

1. D. Kumlutas, I. H. Tavman and M. Turhan Çoban, *Composites Science and Technology* **63** (1), 113-117 (2003).
2. A. Boudenne, L. Ibos, M. Fois, J. C. Majesté and E. Géhin, *Composites Part A: Applied Science and Manufacturing* **36** (11), 1545-1554 (2005).
3. H. Ishida and S. Rimdusit, *Thermochimica Acta* **320** (1-2), 177-186 (1998).
4. S. Yu, P. Hing and X. Hu, *Composites Part A: Applied Science and Manufacturing* **33** (2), 289-292 (2002).
5. S. Shenogin, A. Bodapati, L. Xue, R. Ozisik and P. Keblinski, *Applied Physics Letters* **85** (12), 2229-2231 (2004).
6. B. Krause, R. Boldt and P. Pötschke, *Carbon* **49** (4), 1243-1247 (2011).
7. Elkem AS. Material, 2003.

Investigation of the Correlation between Adhesion Strength and Thermodynamic Work of Adhesion on Silicone Rubber-Kevlar/Rayon Fibers

Humeyra Sirin, Mehmet Kodal, Bagdagul Karaagac, Guralp Ozkoc*

Department of Chemical Engineering, Kocaeli University, 41380, KocaeliTurkey

* *guralp.ozkoc@kocaeli.edu.tr*

Abstract. The aim of this work is to investigate the correlation between mechanical adhesion strength and thermodynamic work of adhesion of poly(hedral oligomeric silsesquioxane) (POSS)/silicone rubber (SR) composites and Kevlar or Rayon fibers. The rubber composites were prepared by using laboratory type internal mixer at pre-optimized conditions. The POSS was used to enhance the adhesion properties of SR to reinforcing fibers, such as Kevlar and Rayon. Three different POSS types including different functional groups were selected. These functional groups were expected to provide crosslinking with SR and grafting to fiber surfaces. The POSS type and reinforcing the fiber type (Kevlar and Rayon) were handled as experimental parameters. The fiber/silicone adhesion was investigated by H-adhesion tests. In addition the thermodynamic work of adhesion (TWA) was determined by using the surface free energy components of the compounds and fibers. The correlation between adhesion strength values and the TWA was investigated to understand the adhesion mechanism. It was found that H- adhesion tests showed an increased adhesion values by the addition of POSS particles. It was seen that there is a linear correlation between H-adhesion values and TWA.

Keywords: Silicone Rubber, POSS, Reinforcing Fiber, H-Adhesion, Thermodynamic Work Adhesion.
PACS: 82.35.Gh

INTRODUCTION

Silicone has some unique properties such as good chemical stability, oxidation resistance, thermal stability, low temperature toughness, and low toxicity. Thanks to these properties, silicone rubber (SR) is applicable in different industry areas like automotive, medical, aerospace, construction, and textile. However, unfilled silicon exhibits lower mechanical properties although it has good flexibility and thermal stability in broad temperature range. To deal with this disadvantage, the reinforcing fillers can be used. POSS, organic/inorganic hybrid materials, are new generation nanofillers with the possibility of reinforcement and stabilization of the polymers [1]. The size of POSS molecule is about 1.5 nm, which makes it comparable with polymer chain dimensions [2]. The inorganic structure may provide polymers with molecular reinforcement, increased thermal stability, better flame resistance, enhanced mechanical properties and etc [3-5].

In this study, adhesion properties of SR/POSS composites with Kevlar and Rayon and the correlation between adhesion strength values and TWA were reported. The POSS types used in this study, octavinyl-POSS (O-POSS), methacryl-POSS (M-POSS) octamaleamic acid-POSS (OM-POSS), were able to react with SR during peroxide vulcanization due to the presence of double bonds in the side chains. Dicumyl peroxide (DCP) was used as the crosslinking agent. DCP and POSS concentrations in the compounds were 0.5 and 3 phr, respectively. The silica content in composites was 5 phr.

EXPERIMENTAL

High temperature vulcanization (HTV) type poly(dimethyl siloxane) SR (Elastosil R401/70) was purchased from Wacker, Germany. POSS types were purchased from Hybrid Plastics, USA. Silane modified silica (VP Coupsil® 6508) was provided by Evonik Industries AG, Germany. The DCP was obtained from Pergan, Turkey. Kevlar and Rayon were supplied from Kordsa Global, Turkey.

Compounding of SR composites was carried out in a Brabender banbury mixer at 50 rpm of rotor speed and at an average temperature of 50°C. The curing of SR composites was carried out in a hot press. The curing temperature was determined as 160°C. The curing time was determined from moving die rheometer (MDR) analysis. H-adhesion test was performed according to ASTM D4776 after compounds cured with reinforcing fibers. Topology of the fiber surfaces was investigated on a JEOL JSM 6060 scanning electron microscope (SEM). TWA approach was adopted using the experimentally determined surface energies. Contact angles were measured using an Attension Theta Lite model contact angle measurement machine. Diiodomethane was used as the probe liquid for Liftshitz-Van der Waals interactions. Ethylene glycol and deionized water were used to identify the acid-base interactions.

RESULTS and DISCUSSION

The selected results were showed for this study. H-adhesion values between matrix/fiber are shown in Figure 1 and 2. Adhesion force value between Reference and Kevlar was 17 N. O-POSS and M-POSS improved adhesion by two times while OM-POSS effected adhesion negatively. The same trend been observed for matrix/Rayon adhesion. Addition of any type of filler increased the H-adhesion values between matrix and Rayon.

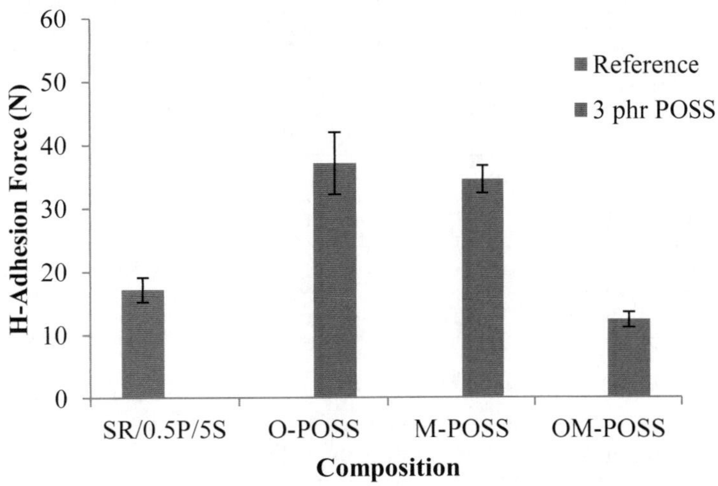

FIGURE 1. H-Adhesion values between SR/POSS composites and Kevlar

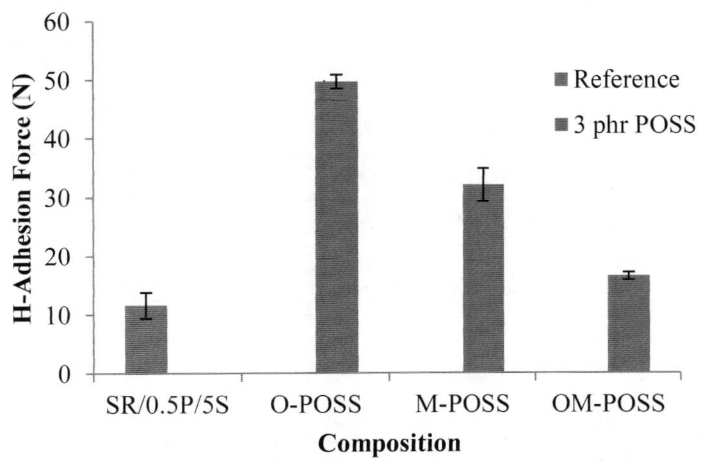

FIGURE 2. H-Adhesion values between SR/POSS composites and Rayon

Thermodynamic work of adhesion (W_{AD}^{TOT}) can be calculated by using the contact angles and surface free energy components of the compounds and fibers. It was seen that OM-POSS had highest acid-base component value due to the presence of acidic group in its structure. Thus W_{AD}^{TOT} value between SR/OM-POSS composite and Kevlar was obtained higher than that of others (Table 1. The same finding has been observed for TWA between Matrix/Rayon (Table 2).

TABLE 1. Thermodynamic work of adhesion of Matrix/Kevlar according to POSS types

Matrix/Fiber	Acid-Base Component W_{AD}^{AB} (mJ/m^2)	Dispersive Component W_{AD}^{LW} (mJ/m^2)	Work of Adhesion W_{AD}^{TOT} (mJ/m^2)
SR/0.5P/5S - Kevlar	21,7	20,3	42,0
SR/0.5P/5S/3 O-POSS – Kevlar	15,3	27,8	43,1
SR/0.5P/5S/3 M-POSS – Kevlar	13,6	28,5	42,1
SR/0.5P/5S/3 OM-POSS – Kevlar	31,5	27,7	59,2

TABLE 2. Thermodynamic work of adhesion of Matrix/Rayon according to POSS types

Matrix/Fiber	Acid-Base Component W_{AD}^{AB} (mJ/m^2)	Dispersive Component W_{AD}^{LW} (mJ/m^2)	Work of Adhesion W_{AD}^{TOT} (mJ/m^2)
SR/0.5P/5S - Rayon	7,2	40,7	48,0
SR/0.5P/5S/3 O-POSS - Rayon	4,4	57,4	61,8
SR/0.5P/5S/3 M-POSS – Rayon	8,6	55,5	64,0
SR/0.5P/5S/3 OM-POSS – Rayon	13,4	55,7	69,1

H-adhesion force and work of adhesion values were given to determine a correlation between H-adhesion and interfacial interactions (Table 3 and 4). As can be seen that a linear correlation was obtained between H-adhesion values and interfacial adhesion except for SR/OM-POSS system.

TABLE 3. Correlation between TWA and H-adhesion force for SR/POSS/Kevlar systems

Matrix/Fiber	H-Adhesion Force (N)	Work of Adhesion W_{AD}^{TOT} (mJ/m^2)
SR/0.5P/5S/3 O-POSS – Kevlar	37,0	43,1
SR/0.5P/5S/3 M-POSS – Kevlar	34,4	42,1
SR/0.5P/5S/3 OM-POSS – Kevlar	12,3	59,2

TABLE 4. Correlation between TWA and H-adhesion force for SR/POSS/Rayon systems

Matrix/Fiber	H-Adhesion Force (N)	Work of Adhesion W_{AD}^{TOT} (mJ/m^2)
SR/0.5P/5S/3 O-POSS - Rayon	49,7	61,8
SR/0.5P/5S/3 M-POSS – Rayon	32,0	64,0
SR/0.5P/5S/3 OM-POSS – Rayon	16,5	69,1

CONCLUSION

In this study, adhesion between SR/POSS composites and reinforcing fiber was investigated. Three different types of POSS were chosen since they have multiple double bonds on the side chains that allow participating in peroxide crosslinking. The usage of different POSS particles improved the adhesion between matrix and reinforcing fibers. Moreover, it was obtained that there is a correlation between interfacial strength and H-adhesion for SR/O-POSS and SR/M-POSS composites.

ACKNOWLEDGMENTS

This study is granted by The Scientific and Technological Research Council of Turkey (TUBITAK) (Project No: 113M258).

REFERENCES

1. M. Kodal, H. Sirin and G. Ozkoc, *Polym. Eng. Sci.,* **54,** 264-275 (2014).
2. J. D. Lichtenhan and J. J. Schwab, *Polym. Preprints*, **41,** 527 (2000).
3. Z. Qiu and H. Pan, *Comp. Sci. Tech.,* **70,** 1089-1094 (2010).
4. F. Baldi, F. Bignotti, A. Fina, D. Tabuani and T. Rico, *J. Appl. Polym. Sci.,* **105,** 935-943 (2007).
5. X. Wang, S. Xuan, L. Song, H. Yang, H. Lu and Y. Hu, *Journal of Macromolecular Science, Part B: Physics*, **51,** 2806-2812 (2009).

Synthesis and Properties of POSS modified PS Microbeads

Nilay Tuccar Kilic, Gizem Turgut, Seyhan Boyacioglu Aysun Yildirim, Ozge Yogurtcu, Mehmet Kodal, Guralp Ozkoc*

Department of Chemical Engineering, Kocaeli University, 41380, Kocaeli/TURKEY

nilaytuccar@yahoo.com.tr
**guralp.ozkoc@kocaeli.edu.tr*

Abstract: In this study, polystyrene (PS) microbeads having POSS nanoparticles on the surface were synthesized by using dispersion polymerization technique. Polyhedral oligomeric silsesquioxanes (POSS) with functional groups were used to obtain PS/POSS microbeads. Five different types of POSS; amino, hydroxyl, epoxy functional groups (aminopropyl isobutyl POSS (aPOSS), octa isobutyl POSS (oPOSS), trisilanol isobutyl POSS (tPOSS), glycidyl isobutyl POSS (moEpPOSS), glycidyl POSS (muEpPOSS)) were used to modify surface properties of PS microbeads. Scanning electron microscope (SEM) was used to investigate shapes and morphologies of microbeads. Energy-dispersive X-ray spectroscopy (EDX) was used to investigate whether POSS was bonded onto PS microbeads. Gel permeation chromatography (GPC) was used to determine molecular weights of PS-POSS microbeads. Thermogravimetric Analysis (TGA) was used to specify thermal stability of microbeads. It was proved that POSS nanoparticles were grafted onto the PS microbeads successively. PS-POSS microbeads have higher molecular weights and thermal stability in comparison to neat PS microbeads. PS and most of PS-POSS microbeads have monodisperse particle size distribution.

Keywords: Polymeric microbeads, dispersion polymerization, PS, POSS
PACS: 82.35.-x

INTRODUCTION

Microbeads are polymeric particles whose size is between 1-1000 microns (μm) in diameter [1]. Micrometer-size monodisperse polymer particles have a wide diversity of scientific and technological applications such as diagnosis of illness, electronics and microelectronics, fillers, catalyst carriers, coatings, information storage materials, biological sensors and colloidal crystals [2]. Dispersion polymerization is generally suitable method to generate micrometer-sized monodisperse particles. Organic/inorganic nanocomposite microbeads have attracted increasing attention because of their unique properties emerging from the combination of the advantageous properties of inorganic nanoparticles and polymers. Thus, the properties of polymers are improved such as increased modulus and strength, enhanced barrier properties, improved heat resistance, contained functional groups and decreased flammability [3]. Polyhedral oligomeric silsesquioxanes (POSS) is one of the nanoparticles are used to make nanocomposite microbeads. POSS with the empirical formula $(RSiO_{1.5})_n$, (generally, n>4) where R can be an organic functional group such as hydroxyl, amines, alcohols, thiols, epoxide etc. units. [4].

In this study, it was aimed to synthesize the microbeads, which contains functional POSS particles by using dispersion polymerization that enables to attach various materials such as ligands, proteins, etc. onto the surface of PS microbeads. Five types of POSS were used with different functional groups (aPOSS, has amino functional group, oPOSS, isobutyl side group, moEpPOSS, one epoxy functional group, muEpPOSS, multi epoxy functional group, tPOSS hydroxyl functional group) as surface modification chemicals.

EXPERIMENTAL

Materials

Styrene (liquid, 145°C boiling point) and ethyl alcohol (EtOH, liquid, 75°C boiling point) were purchased from Merck. Hydroxypropyl cellulose (HPC, white powder, M_n~10000) was purchased from Sigma-Aldrich. Benzoylperoxide (BPO, white, solid, 104 °C melting point) was purchased from Fluka. Aminopropyl isobutyl POSS (aPOSS), octa isobutyl POSS (oPOSS), glycidyl isobutyl POSS (moEpPOSS), glycidyl POSS (muEpPOSS), trisilanol isobutyl POSS (tPOSS) were purchased from Hybrid Plastics (USA). aPOSS, oPOSS, tPOSS , moEpPOSS are crystalline white powder form, muEpPOSS is viscous liquid form. All POSS types are dissolvable in THF.

Sample Preparation

The PS particles containing POSS nanoparticles were synthesized by using dispersion polymerization method. The dispersion polymerization was carried out in a three-neck flask reactor with a mechanical stirrer, a reflux condenser, a nitrogen inlet, and a temperature controller. A 100 mL three-neck flask reactor was charged with degassed ethanol as a medium. Flask was placed in a 50˚C oil bath and stirred with magnetic stirrer, then as a stabilizer HPC was dissolved in ethanol medium. HPC and ethanol medium was mixed during 30 min. At the same time, BPO as initiator and POSS nanoparticles (aPOSS, oPOSS, moEpPOSS, muEpPOSS and tPOSS) with amount of %3 wt of styrene were dissolved in styrene in 30 min. Afterwards, styrene mixture was introduced into three-neck flask containing ethanol mixture drop by drop. The final medium was mixed with a mechanical stirrer during 2 h and nitrogen inlet. The color of the medium was turned to white in 2 h. After 2 h, temperature was increased and polymerization was allowed to proceed at 125 rpm with N_2 inlet for 22 h.

Characterization

QUANTA 400F Field Emission scanning electron microscope (SEM) was used to observe microbeads' morphological characteristics. EDX analysis was carried out for surface characterizations during SEM analysis.

Molecular weight measurements were performed by gel permeation chromatography. For GPC measurements, Agilent Series 1100 equipped with refractive index and light scattering detector was used. For separation two PLgel Mixed B (300x7.5 mm) columns from Agilent were used as stationary phase. As mobile phase, tetrahydrofuran (THF) at a flow of 1 ml/min at 25 °C.

Thermal properties of samples were monitored by a thermogravimetric analyzer (TGA, METTLER TOLEDO TGA 1 STAR System). Measurements were carried out from room temperature 25° C–600° C at a heating rate 20°C/min under nitrogen gas flow of 10 ml/min.

RESULTS and DISCUSSION

Size and morphology of microbeads were observed (Figure 1) through SEM images. It is seen that the microbeads exhibiting a mono size distribution except aPOSS-PS. SEM micrographs showed that different types of POSS differently affected the roughness the surface of microbeads.

It was obtained from EDX spectra that PS-POSS microbeads exhibitted Si peaks. These peaks were so small because of low loading ratio of POSS to the system. However, neat PS microbeads did not have same peaks. Thus, it can be resulted that Si peak comes from POSS nanoparticles. It basically means that POSS nanoparticles are located on the surfaces of the PS microbeads.

In order to investigate the correlation of POSS types and molecular weight of PS microbeads, GPC was used. According to GPC results, neat PS microbeads had lowest molecular weight compared other ones. It was clearly obtained that in the presence of POSS nanoparticles, the molecular weight of PS microbeads increased possibly due to the grafted POSS molecules. As it can be seen from Table 1, muEpPOSS-PS microbeads had the highest molecular weight and neat PS microbeads had lowest.

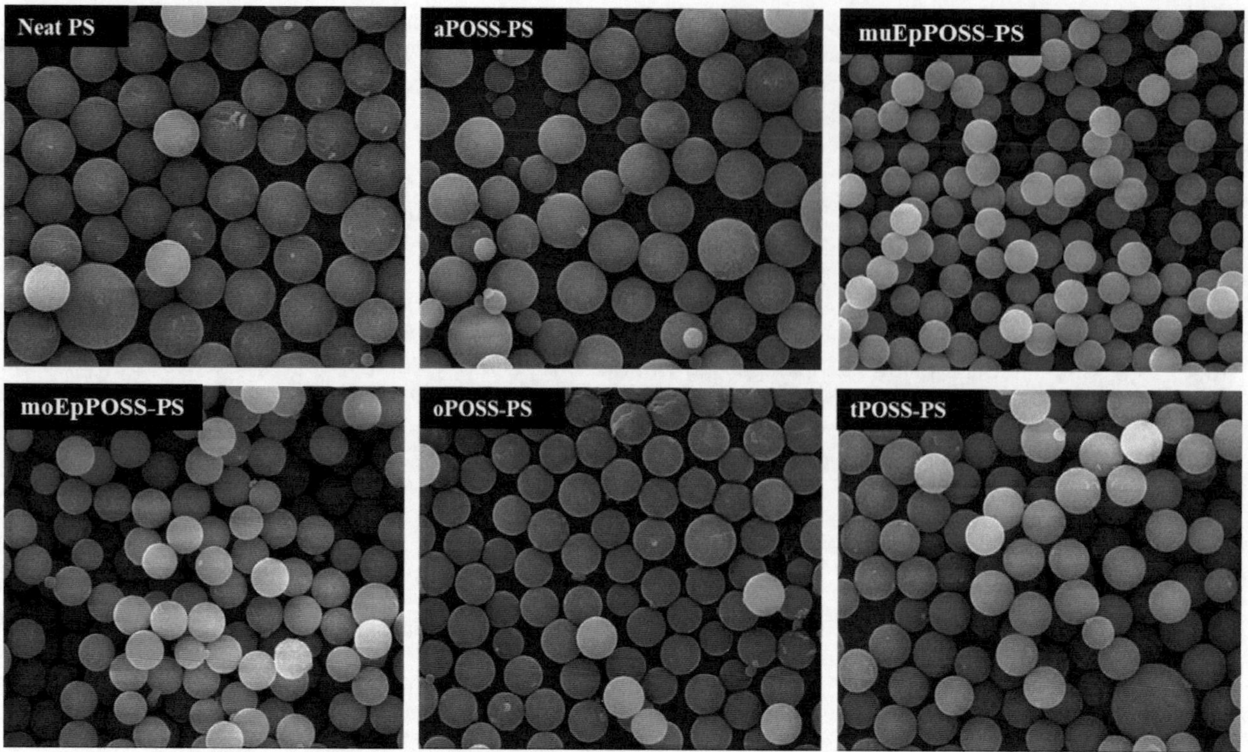

FIGURE 1. SEM images of neat PS microbead and different types of POSS-PS microbeads

FIGURE 2. EDX results of PS and POSS- PS microbeads.

TABLE 1. GPC results of neat PS microbead and different types of POSS-PS microbeads.

Sample	M_n	M_w	PDI
Neat PS	15255	29486	1,933
aPOSS-PS	20207	39781	1,969
moEpPOSS-PS	18851	35938	1,906
muEpPOSS-PS	24643	40802	1,656
oPOSS-PS	19509	33308	1,707
tPOSS-PS	14182	35305	2,489

Decomposition temperatures are seen in Table 2, and corresponding TGA thermograms in Figure 3. Neat PS microbeads had the highest decomposition temperature. The low decomposition temperatures of PS-POSS microbead are possibly due to the decomposition of side groups of POSS. The char yield in PS-POSS composites are high with respect to PS.

TABLE 2. TGA results of neat PS microbead and different types of POSS-PS microbeads.

Sample	$T_{decomposition}$ (°C)	Char Yield (%)
Neat PS	411	0.00
aPOSS-PS	407	2.84
moEpPOSS-PS	408	1.21
muEpPOSS-PS	407	2,15
oPOSS-PS	408	2,36
tPOSS-PS	406	1,31

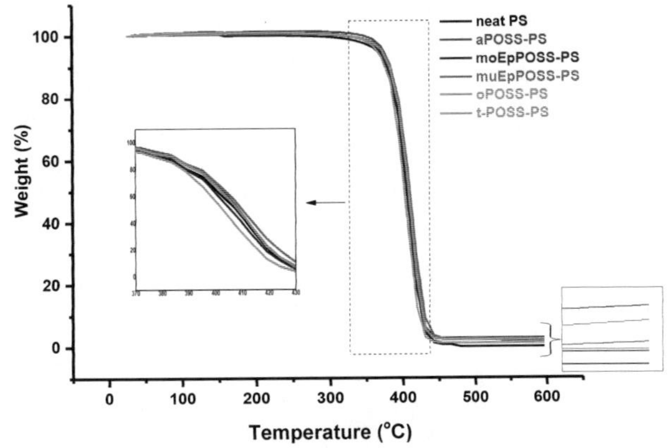

FIGURE 3. TGA thermograms.

CONCLUSIONS

Monodispersed PS-POSS microbeads were sucessfuly synthesized via dispersion polymerization. The POSS modifications of PS microbeads were confirmed by EDX analysis. Si bonds were observed on EDX spectrum, which proved the presence of POSS on the surface of microbeads. Molecular weight measurement results showed that in presence of POSS nanoparticles caused higher molecular weight than neat PS microbeads.

REFERENCES

1. Y. Lipovetskaya, *Cosmetics and Toiletries*, **125:4**, 46, (2010).
2. S. Kawaguchi and K. Ito, *Advanced in Polymer Science* **175**, 299–328 (2005).
3. J. Ugelstad and A. Berge, *Progress in Polymer Science* **17**, 87–161(1992).
4. A.P. Papavlu, V. Dinca, I. Paraico, A. Moldovan, and J.S. Stewart, *Journal of Applied Physics* **108**, 033111(2010).
5. W.H. Zhang, X.D. Fan and W.W. Fan, *eXPRESS Polymer Letters* **6,** 532–542(2012).
6. K.J. Shea and D.A. Loy, *Chemical of Materials* **13**, 3306-3319(2001).

Effects Of Nanoclay Content On Melt Flow Rate Of Polyamide 12/Clay Nanocomposites

Josip Stojšić[a*], Pero Raos[a] and Tomislav Galeta[a*]

[a] *Mechanical Engineering Faculty in Slavonski Brod, J. J. Strossmayer University of Osijek, Trg I. B. Mažuranić 2, Slavonski Brod, Croatia*

**jstojsic@sfsb.hr*

Abstract. Nanocomposites are materials in which at least one phase is in nanometer dimensions (dimensions of the particle, platelet or fibre modification are in the range of 1–100 nm). Important class of nanocomposite materials are polymer layered silicates, in which layered filler (mostly modified natural clay minerals) is dispersed in polymer matrix. Many of parameters (i.e. type of polymer matrix, type and amount of nanofiller, and a number of process parameters as melting parameters, screw speed, number and shape of mixing screws etc.) affect properties of nanocomposites. The aim of the investigation was to determine the influence of nanoclay content, on melt flow rate of PA12/clay nanocomposites

Keywords: Polymer nanocomposite, polyamide 12, montmorillonite, MFR
PACS: 81.05.Lg

INTRODUCTION

By definition, the nanocomposites are materials in which at least one phase is in nanometre dimensions, (dimensions of the particle, platelet or fibre modification are in the range of 1–100 nm), whereby is achieved higher specific interfacial area. Therefore the properties of obtained nanocomposites depend more on interactions at the phase boundary than on the phase itself. Polymer nanocomposites are composed of polymer matrix and nanofiller. Typical nanofiller / reinforce filler includes: layered filler (with nanometre layer thickness and sheet structure, fibre reinforcing filler (carbon nanotubes and nanofibres) and nanoparticles (SiO2 nanometre dimensions particles).

PRODUCTION OF POLYMER NANOCOMPOSITES BY MELT INTERCALATION PROCESS

The most common and simple method, particularly useful for thermoplastic polymers is melt intercalation. In melt intercalation, fillers are mechanically dispersed into a polymer matrix using a high temperature and high shear force mixer or compounder. This approach is simple and compatible with current industrial practices

The biggest challenge in getting nanocomposite reinforced with clay layers is the separation and dispersal of an individual layer in the polymer matrix.

Unless there is no separation of clay layers, common micro composite is obtained (Figure 1a). Nanocomposites with a few polymer molecules inserted in the interlayer between the clay sheets can occur depending on the strength of interfacial interactions (Figure 1b). In the above-mentioned dependence can also occur stratified nanocomposite where the clay layers are completely separated and evenly dispersed in the polymer matrix (Figure 1c). Stratified polymer composite is preferred because it produces the largest matrix-filler contact area, which leads to the best nanocomposite properties [1]

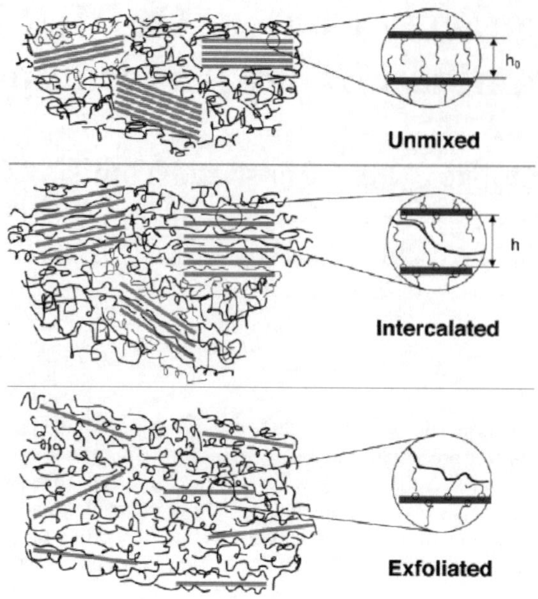

FIGURE 1. Schematic diagrams of the possible layered silicate based nanocomposite structures [2]

INFLUENCE OF CLAY CONTENT ON MELT FLOW RATE OF PA 12 / CLAY NANOCOMPOSITE

Investigation Material

Material used for the nanocomposite matrix is PA12 in the powder form (made in Eos Company), known under the trade name "PA 2200 Balance 1.0". This material is used for the production of laser-sintered fully functional products, which are replacing typical injection molding products due to their excellent mechanical properties. The main material properties are given in Table 1.

TABLE 1. PA 2200 Balance 1.0 properties [3]

Properties	Value	Unit	Test Standard
Tensile Modulus	1650	MPa	ISO 527-1/-2
Tensile Strength	48	MPa	ISO 527-1/-2
Charpy impact strength (+23°C)	53	kJ/m²	ISO 179/1eU
Shore D hardness (15s)	75	-	ISO 868
Melting temperature (10°C/min)	176	°C	ISO 11357-1/-3
Density (lasersintered)	930	kg/m³	EOS Method

Cloisite 93A of Southern Clay Products are the nanofiller used in the experiment. Based on previous studies this type of nanofiller is selected as optimal for PA 2200 [4], [5]. Cloisite 93A is a natural montmorillonite modified with a ternary ammonium salt. This filler is designed for use as an additive for plastics and rubber to improve various physical properties, such as reinforcement, HDT, CLTE, synergistic flame retardant and barrier.

Equipment

Nanocomposite compounds were prepared on Brabender extruder line, which has two-screw extruder, cooling paths and granulator. Previously weighed and mixed powder (PA 12 + Cloisite 93A) is inserted into the two-screw extruder which mixes melt, pushes it forward and after mixing extrudes it through a nozzle. Nanocomposite mixture is put down on the cooling strip, which transports it and cools it in the granulation thread.

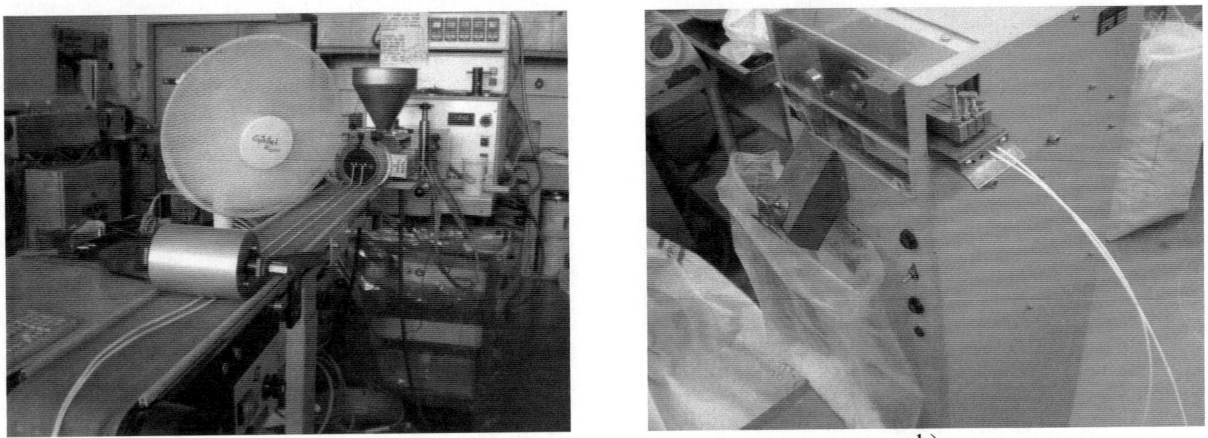

a) b)

FIGURE 2. Equipment for nanocomposite mixing (a) and granulating for MFR test (b)

Investigation Design And Results

Based on previous researches and pre-experiments the three main parameters were selected: content of nanoclay, screw rotation frequency and mixing temperature. Lower and upper limits were determinate for three selected specific factors: content of nano-clay from 0,95 to 11,05%, screw rotation frequency from 13,18 to 46,82 min^{-1} and mixing temperature from 203,18 to 236,82 °C. MFR test was made according to ISO 1133.

TABLE 2. Investigation Results

Number of samples	Factor 1: contnet of nano-clay, %	Factor 2: screw rotation frequency, min^{-1}	Factor 3: temperature, °C	$MFR_{(230\ C,\ 2{,}16kg)}$, g/10min
1	3	20	210	41,20
2	9	20	210	39,29
3	3	40	210	41,36
4	9	40	210	38,51
5	3	20	230	38,07
6	9	20	230	36,90
7	3	40	230	40,52
8	9	40	230	34,92
9	0,95	30	220	42,94
10	11,05	30	220	36,10
11	6	13,18	220	39,08
12	6	46,82	220	39,65
13	6	30	203,18	39,76
14	6	30	236,82	37,96
15	6	30	220	38,50
16	6	30	220	38,55
17	6	30	220	36,63
18	6	30	220	37,76
19	6	30	220	35,97

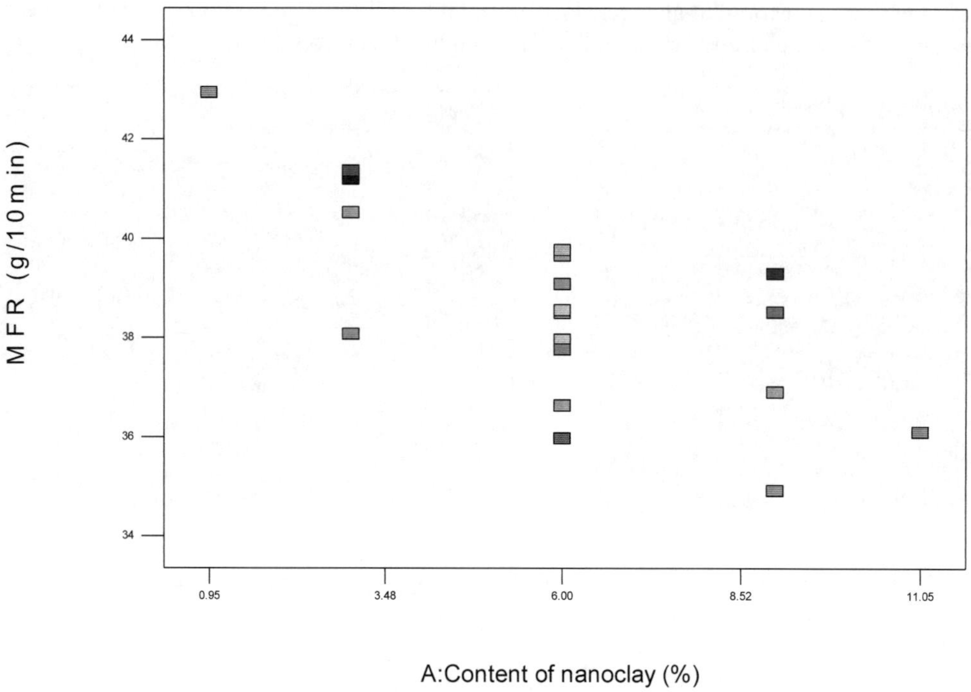

FIGURE 3. Dependence of MFR on the share of nanoclay (with the impact of remaining 2 factors)

CONCLUSION

The experimental part investigated the dependence of MFR on the mixing parameters PA12/layered clay nanocomposites. It is evident from the results that increasing the content of nanoclay from 1% to 11% (with the impact of the other two parameters) significantly decreases MFR. Further research will go in the direction of getting the appropriate mathematical model that will determine the dependence of MFR on the mixing parameters while making nanocomposite mixture of PA 12/layered clay needed to define specific mixing parameters in order to achieve desired properties

REFERENCES

1. P.Raos, J. Stojsic, M. Šercer, A. Pilipović, M. Somolanji, Mechanical properties of PA12/clay nanocomposites. TEAM 2013, 5th International scientific conference TEAM 2013, November 04-06, 2013 Prešov, Slovakia
2. R. Krishnamoorti; K. Yurekli, Rheology of polymer layered silicate nanocomposites. Current Opinion in Colloid & Interface Science 6(2001), pp. 464-470
3. EOS GmbH, Electro Optical Systems, München, http://eos.materialdatacenter.com/eo/en, 16.03.2016.
4. J. Stojsic, A. Kalendova, P.Raos, M. Somolanji, Compounding of PA12/clay nanocomposites. TEAM 2011, 3rd International scientific conference TEAM 2011, October 19-21, 2011 Trnava, Slovakia
5. J. Stojsic, A. Kalendova, P.Raos, M. Somolanji, Compounding of PA12/clay nanocomposites, 2nd International conference on manufacturing engineering & management ICMEM 2012, December 5-7, 2012 Prešov, Slovakia

Relationship between Crystallization Kinetics and Interfacial Interactions in PLA/POSS Nanocomposites

Mehmet Kodal[a], Humeyra Sirin[a] and Guralp Ozkoc[a*]

[a]*Department of Chemical Engineering, Kocaeli University, 41380, Kocaeli/Turkey*

mehmet.kodal@kocaeli.edu.tr
guralp.ozkoc@kocaeli.edu.tr

Abstract. In this study, the relationship between crystallization rate and surface chemistry of polyhedral oligomeric silsesquioxanes (POSS) in poly(lactic acid) (PLA)/POSS nanocomposites was investigated. Four different kinds of POSS particles having different chemical structures were used. The composites were processed in a Xplore micro-compounder at a screw speed of 100 rpm and 180°C barrel temperature. The POSS content was varied as 1, 3, 7 and 10 wt%. Common kinetic models such as Avrami, Ozawa and Dobreva and Gutzow were used to determine the non-isothermal crystallization behaviour of PLA/POSS nanocomposites by using the thermal data obtained from differantial scanning calorimetry (DSC). The morphology during the crystal growth was investigated by a polarized optical microscope (POM) equipped with a hot-stage. Thermodynamic work of adhesion approach was adopted using the experimentally determined surface energies to investigate the interfacial interactions between POSS particles and PLA. A strong correlation was obtained between interfacial chemistry and the nucleation rate in PLA/POSS nanocomposites. It was found that the polar interactions are the dominating factor which determines the nucleation activity of the POSS particles.

Keywords: PLA, POSS, nanocomposites, non-isothermal crystallization kinetic, interfacial interactions
PACS: 82.35.Gh

INTRODUCTION

Non-biodegradable of the common petrochemical-based polymers caused serious ecological problems [1]. To obtain the eco-friendly polymeric materials, several biodegradable polymers have been investigated during the last few decades as alternatives to non-biodegradable polymers. Among the biodegradable polymers, poly(lactic acid) (PLA) has been broadly studied due to its natural biodegradability and availability from renewable resources such as corn. Full-scale commercial production of PLA is now a reality; the cost of production has been reduced significantly and PLA is being used in packaging and automotive. However, one of the most important drawbacks of PLA is its low crystallization ability limiting its versatile applications especially in packaging and automotive applications. This disadvantage can be eliminated by incorporating nanofillers within the PLA matrix. In the literature, several potential nucleating agents such as montmorillonite, silica, silver nanoparticles, talc, carbon nanotubes and graphene were used to increase the crystallization ability of PLA [2, 3]. Alternative to the conventional nanofillers utilized as nucleating agent, poly(hedral oligomeric silsesquioxane) (POSS) nanoparticles have a high potential to increase the crystallization rate of PLA since they have flexible chemical and physical hybrid properties with a reasonably lower price.

Chain conformation and segmental mobility of polymers are influenced by substrate/polymer interfacial interaction. For a better understanding of filler/matrix interphase mechanism, it is necessary to determine the interfacial strength and possible interactions between polymer and filler. This is usually carried out by means of contact angle measurements to determine the thermodynamic energy of adhesion formed between filler and polymer [3].

In the current study, it was aimed to investigate the non-isothermal crystallization behavior of POSS reinforced PLA by differential scanning calorimeter (DSC) adapting different crystallization methods. Besides, for the first time in literature, correlation between crystallization behavior of PLA/POSS nanocomposites and thermodynamic adhesion was studied.

EXPERIMENTAL

Materials

Poly(lactic acid) (PLI005, MFI: 10-30 g/10 min at 190°C/2.16 kg) was purchased from NaturePlast Company, France. Aminopropylisobutyl-POSS (A-POSS), trisilanolisobutyl-POSS (T-POSS), glycidylisobutyl-POSS (G-POSS) and octaisobutyl-POSS (O-POSS) were purchased from Hybrid Plastics Company, USA. All the POSS types are dissolvable in THF, chloroform and hexane as PLA does, which points out a possible compatibility. They are all in crystalline white powder form. The isobutyl groups on the POSS molecules are responsible from the solubility and compatibility, and the functional groups are responsible for the physical or chemical interaction with the PLA.

Sample Preparation

PLA based nanocomposites, PLA/A-POSS, PLA/T-POSS, PLA/G-POSS and PLA/O-POSS having different POSS contents (1, 3, 7 and 10 wt %) were prepared via melt blending in a twin-screw microcompounder (15 ml Microcompounder, Xplore Instruments, The Netherlands) at a screw speed of 100 rpm. Before compounding, PLA was dried in an oven at 50 °C for 12 h under vacuum. Prior to processing, predefined amount of POSS and PLA were dry-mixed. The barrel temperature of microcompounder was set to 180 °C. The mixing time was kept constant at 3 min. Then, the melts were compressed for 3 minute at 180 °C to get films with a thickness of around 0.5 mm. The samples were abbreviated with respect to the type and loading level of POSS, for instance G-POSS1 represents the composite including 1 wt % glycidylisobutyl -POSS.

Characterization

Contact angles were measured using an Attension Theta Lite model contact angle measurement machine. Bromonaphtalene (BN) was used as the probe liquid for Liftshitz-van der Waals interactions. Ethyleneglycol (EG) and deionized water (DW) were used to identify the acid–base interactions between sample and the probe liquid. The contact angle from each image and the average contact angle for the drop were obtained after each measurement.

Differential scanning calorimeter (DSC) analysis was performed to study the non-isothermal crystallization kinetics of pure PLA and PLA/POSS nanocomposites. The DSC procedure was consisting of three segments. At first segment, the films were heated from 25 to 200 °C with a heating rate of 15 °C/min, then they were held at this temperature for 5 min to eliminate the thermal history, and then they were cooled to 25 °C at cooling rates of 1, 2, 5, 10 and 20 °C/min and held at °C for 5 min. In the last segment, they were reheated to 200 °C at a heating rate of 10 °C/min. Please note that only some selected results were given for non-isothermal crystallization kinetic studies of PLA/POSS nanocomposites due to the excess of results.

RESULT AND DISCUSSION

Non-Isothermal Crystallization Kinetics

DSC analysis revealed that PLA did not exhibit crystallization behavior for higher cooling rates due to the formation of amorphous morphology during cooling process. This situation was also obtained for A-POSS, T-POSS and O-POSS filled PLA nanocomposites. On the other hand, crystallization onset temperature was higher for these nanocomposites compared to pure PLA upon cooling. Moreover, PLA/G-POSS nanocomposites tended to crystallize in cooling even for higher cooling rates.

Avrami kinetic model was studied to determine the nucleation type, growth rate and morphology of the crystallite formed for PLA and PLA/POSS nanocomposites. Avrami exponent, n, values found in the case of PLA and PLA/POSS nanocomposites were in the range of 2-3 and they should possibly be related with two-dimensional growth. It was obtained from Avrami curves that crystallization rate constant (Z_c) of PLA/POSS nanocomposites was larger than that of pure PLA, and PLA/G-POSS10 nanocomposite had the largest Z_c value at a given cooling rate. The results showed that all the POSS types used in this study increased the crystallization rate of PLA. Moreover, G-POSS had the most excellent nucleating agent for the crystallization of PLA according to the same content of POSS nucleating agents.

Liu et al. [4] proposed a different kinetic equation by combining the Ozawa and Avrami equations. As the degree of crystallinity was related to the cooling rate α and the crystallization time t (or temperature T), the relation between α and t could be defined for a given degree of crystallinity [5]. The values of $F(T)$ is the required cooling rate to reach a certain degree of crystallinity for the PLA and its selected nanocomposites is listed in Table 1. As can be seen from the Table 1 that, the highest $F(T)$ values were obtained for neat PLA at a given degree of crystallinity. High $F(T)$ values indicate higher cooling rates to reach the certain value of crystallinity in the unit time. Therefore, it can be concluded that crystallization rate of PLA/POSS nanocomposites except for 1 and 3 wt% filled O-POSS ones was faster compared to pure PLA.

TABLE 1. Calculated combined Avrami and Ozawa parameters for PLA and selected PLA/POSS nanocomposites

Sample	X(t) (%)	*F(T)*
PLA	10	3.59
	30	7.56
	50	13.49
	70	33.31
	90	53.07
PLA/A-POSS7	10	1.53
	30	2.00
	50	2.33
	70	2.67
	90	3.14
PLA/T-POSS10	10	1.48
	30	2.06
	50	2.49
	70	2.89
	90	3.34
PLA/G-POSS10	10	1.64
	30	2.17
	50	2.38
	70	2.54
	90	2.71
PLA/O-POSS7	10	10.51
	30	8.95
	50	9.01
	70	9.69
	90	9.97

Interfacial Interactions by means of Thermodynamic Adhesion

Work of adhesion (W_{AD}) between contacting phases can be calculated by utilizing the contact angles and surface free energy components. It was obtained that G-POSS has a proton rich structure unlike the other POSS types. Thus, W_{AD}^{TOT} value between G-POSS and PLA was obtained higher than that of others due to the presence of acidic epoxy groups, which can interact with electron donor groups of PLA (Table 2). It was found that the interfacial interaction is better between G-POSS/PLA compared to other ones, resulting in adsorption and immobilization of PLA chains on the surface of G-POSS. Consequently, this situation affected PLA crystallization positively in the presence of G-POSS particles.

In order to investigate the presence of a correlation between crystallization behavior and interfacial interactions, Avrami rate constant were plotted versus work of adhesion for PLA/POSS. The effect of interfacial adhesion on the crystallization rate constant of 3 wt% POSS filled PLA nanocomposites are shown in Figure 1. One should note that values were obtained for 2 °C/min cooling rate from the Avrami analysis. A linear correlation between the crystallization rate and the interfacial adhesion was obtained as expected (Figure 1). The highest values for both work of adhesion and crystallization rate constant were obtained from the PLA/G-POSS nanocomposites.

TABLE 2. Thermodynamic work of adhesion of nanocomposites according to POSS types

Matrix - Particle	Dispersive Component W_{AD}^{LW} (mJ/m^2)	Acid-Base Componenrt W_{AD}^{AB} (mJ/m^2)	Work of Adhesion (W_{AD}^{TOT}) (mJ/m^2)
PLA - A-POSS	71.2	2.1	73.4
PLA - T-POSS	74.2	1.6	75.9
PLA - G-POSS	77.2	7.4	84.6
PLA - O-POSS	74.7	0.9	75.6

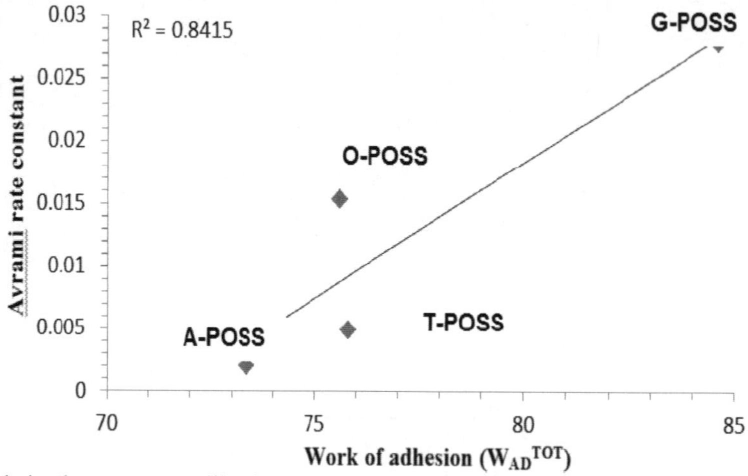

FIGURE 1. Correlation between crystallization rate constant and work of adhesion for PLA/POSS nanocomposites

CONCLUSIONS

It can be concluded that POSS particles increased the crystallization rate of plasticized PLA by acting as a nucleating agent. The crystal growth of PLA was generally in 2-dimensional in the presence of nucleating agents. Moreover, it was obtained that there is a strong correlation between interfacial strength and nucleation activity.

ACKNOWLEDGEMENTS

The authors thank to the Scientific and Technological Research Council of Turkey (TUBITAK) for supporting this study with contract grant number of 111M514.

REFERENCES

1. B. Nuria et al., *Polym. Degrad. Stab.*, **98**, 651-658 (2013).
2. A. Zasadzinska-Grzabka et al., *J. Therm. Anal. Calorim.*, **DOI 10.1007/s10973-016-5311-3** (2016).
3. H. Li and M. A. Huneault, *Polymer* **48**, 6855-6866 (2007).
4. T. Liu et al., *Polym. Eng. Sci.*, **37**, 568-575 (1997).
5. K. Seong Hun et al., *Polymer* **44**, 5625-5634 (2003).

Organic Modification of MMT Clay with Polyaniline: Effect on Electrical and Morphological Properties

Camila P. Presendo[a], Wesley A. Saade[a], Wagner M. Pachekoski[b], Carla Dalmolin[a*] and Daniela Becker[a]

[a]Center for Technological Sciences – State University of Santa Catarina - UDESC
Rua Paulo Malschitzki, 200, Joinville – SC, Brazil
[b]Federal University of Santa Catarina – UFSC
Rua Dr. João Colin, 2700, Joinville – SC, Brazil

carla.dalmolin@udesc.br

Abstract. MMT Clay was modified by replacement of Na^+ with anilinium ion (An^+) followed by chemical oxidation with ammonium persulfate in order to produce polyaniline in a rigid structure. This MMT-Pani composite was tested as conducting filler in polystyrene films. PS/MMT composites were processed in a torque rheometer and hot pressed. MMT-Pani filler and PS/MMT-Pani nanocomposites were characterized by XDR and FT-IR, showing Pani intercalation inside the clay platelets and improved interactions of the conducting polymers with clay. The morphology of the nanocomposites were analyzed by TEM and showed improvement in clay exfoliation when higher quantities of MMT-Pani filler were added. AC and DC electrical conductivity of fillers and nanocomposites were analyzed by two probe methodology and electrochemical impedance spectroscopy (EIS), respectively. Pani alone presented a high electrical conductivity, whereas the presence of the insulanting clay affects all electrical conducting mechanism.

Keywords: nanocomposites, conducting polymer, MMT Clay, electrical properties.
PACS: 62.23.Pq

INTRODUCTION

Polyaniline (Pani) is the most known conducting polymer due to its excellent thermal and environment stability, combined with high electrical conductivity. However, as common in all conducting polymers, Pani exhibits poor mechanical properties, and much work has been done in order to find new processing strategies.

Some of these limitations can be overcome by the intercalation polymerization with nanosized filler, such as clay particles[1]. In the intercalation polymerization, the inorganic host serves as a template within which a guest molecule of nanometer dimension is assembled. The constrained environment is expected to lead to a high degree of polymer ordering and to useful properties such as high anisotropy of electric conductivity [2] and, mostly important, incorporation of Pani into clay platelets may provide characteristics that cannot be attained from pristine conducting polymer such as processability [3]. Among inorganic hosts, montmorilonite (MMT) clays are the most prevalent and interesting due to the nature, abundance, low cost and attractive features such as a large surface area and ion–exchange properties [1-3].

Besides the electrical conducting properties expected by fillers composed with MMT clay modified with Pani (MMT-PAni), delaminated clay layers resulting into exfoliated structure when processing into a polymeric matrix, and have also attracted interest due to enhanced interaction of polymer–clay [4]. Exfoliated nanocomposites are more interesting because it increases the surface area of contact between clay layers and the polymeric matrix.

In present work, MMT clay was modified with aniline in order to produce Pani by intercalation polymerization. The resulting MMT/Pani clay was then used as conducting filler for a polystyrene (PS) matrix, in order to investigate how this new filler acts in the morphological (exfoliation) and electrical properties of the resulting nanocomposite.

EXPERIMENTAL

Montmorillonite clay, named as MMT-Na^+, were provided by Southern Clay Co.; Polystyrene N1841 was provided by the company INNOVA; and Aniline (VETEC) was distilled under vacuum prior to use. All other chemicals were reagent grade and used as received.

For MMT-Pani modification, *in situ* intercalation polymerization of aniline was performed in acid media [2]. Five grams of MMT-Na^+ was added into 250 mL deionized water and stirred for 2 h at 80 °C, then 1 mL of aniline and 3 mL of HCl (conc.) were added. After stirring for 6 h at 80 °C the mixture was washed with a large amount of de-ionized water and cooling down to room temperature. A solution of 0.05 M ammonium persulfate ($(NH_4)_2S_2O_8$) in 1.0 M HCl was added and the polymerization lasted for more than 10 h. The dark green precipitate was isolated by filtration, washed with deionized water thoroughly and dried under vacuum.

PS/MMT (without clay modification) and PS/MMT-Pani nanocomposites samples were produced with different filler contents, according Table 1. Both MMT or MMT-Pani and PS were mixed by melt intercalation in a Haake torque rheometer with a 50 cm^3 mixing chamber and standard rotors, operated at 120°C and 50 rpm for 10 min. The nanocomposites were compressed and molded using a hydraulic press at 120°C for 5 minutes by applying a pressure of 16MPa and cooled with water to room temperature.

TABLE 1. Nomenclature and composition of the samples.

Samples	% MMT	% MMT-Pani
PS	0	0
PS/MMT1	1	0
PS/MMT3	3	0
PS/MMT5	5	0
PS/MMT10	10	0
PS/MMT-Pani1	0	1
PS/MMT-Pani3	0	3
PS/MMT-Pani5	0	5
PS/MMT-Pani10	0	10

Characterization Techniques

X-ray diffraction (XRD) patterns were obtained on a Rigaku D/MAX/2400 X-ray diffractometer; FT-IR spectra were recorded on a Perkin–Elmer 2000 spectrometer using KBr pellets.

The morphology of MMT-Na and MMT-Pani nanocomposites was studied by by field emission electron microscopy (FEG). The images were obtained by a Jeol JSM 6701F FEG. All specimens were coated with gold prior to imaging. TEM photos of nanocomposites with PS and MMT fillers at different composition were taken using JEOL JEM-2100 transmission electron microscope at 200 kV acceleration voltage. Ultra-thin sections of about 50nm were cut with a diamond knife at ambient temperature, using an RCM Power Tome X Ultramicrotome.

Room temperature DC conductivity measurements were performed using a two-probe method. AC conductivity was also measured by Electrochemical Impedance Spectroscopy using an AUTOLAB PGSTAT20 potentiostat / galvanostat with a frequency analyzer (FRA). The blocking electrode method was used, where a thin film was inserted between two stainless steel electrodes at open circuit potential. A DC perturbation of 10 mV (r.m.s.) was applied in the frequency range of 100 Hz to 100 kHz. Powder samples were compressed in form of pellet at 8 T load. The specific conductance (conductivity) was calculated by normalizing the conductance with the contact area of the electrodes ($S = 0.50$ cm^2) and sample thickness ($l = 0.55$ mm).

RESULTS AND DISCUSSION

Characterization of MMT-Pani Nanocomposite

Figure 1a shows XRD patterns obtained by natural and Pani modified MMT. The expansion of clay platelets due to replacement of Na^+ ions by anilinium (An^+) and subsequent polymerization of Pani were evidenced by the shift of peaks referred to the first diffracted plane, d001. The interlayer distance of the anhydrous silicate framework is 12.4 Å, and an expansion to 15.6 Å in the MMT-Pani was observed. This expansion is in agreement with the intercalation of single chains with extended chain conformation rather than coiled conformation of polyaniline in the galleries of the host, and it was already observed for other authors [2]. Besides, sharp and narrow peaks show a crystalline structure, indicating that clay do not suffer exfoliation. Figure 1b shows the FT-IR spectra of MMT-Pani: 910, 870 and 790 cm^{-1} are characteristic vibrations of clay, 1629, 1470, and 1300 cm^{-1} are the characteristic vibrations of emeraldine salt, corroborating the evidence of Pani formation [5].

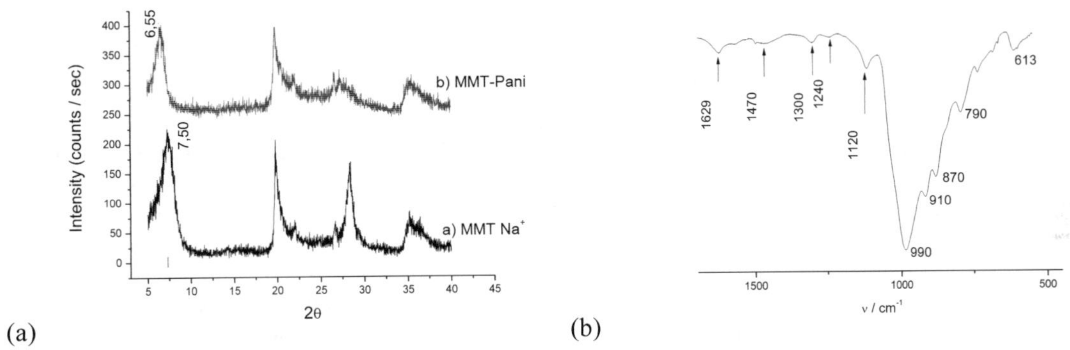

FIGURE 1. (a) XDR patterns of natural MMT and modified clay, MMT-Pani. (b) FT-IR spectra of MMT-Pani

Morphology Analysis

Figure 2 shows SEM photomicrographs of MMT-Pani nanofillers. Globular and high surface area typical of polyaniline are seen on the clay external surface.

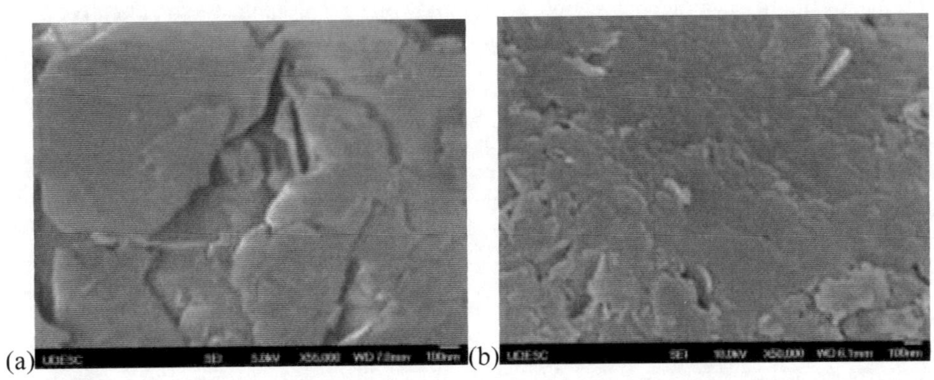

FIGURE 2. SEM photomicrographs of (a) natural MMT, and (b) modified clay, MMT-Pani.

Figure 3 shows the morphology of the nanocomposites PS/MMT10 and PS/MMT-Pani10 obtained by TEM. It is clearly possible to visualize the polymer around the lamellae of clays. In the PS/MMT10 sample, MMT clays are compacted; while in the PS/MMT-Pani10, mostly of the clays were intercaled. This can be explained by the fact that MMT-Pani clays are already organically modified, which facilitates interaction with the polymer matrix.

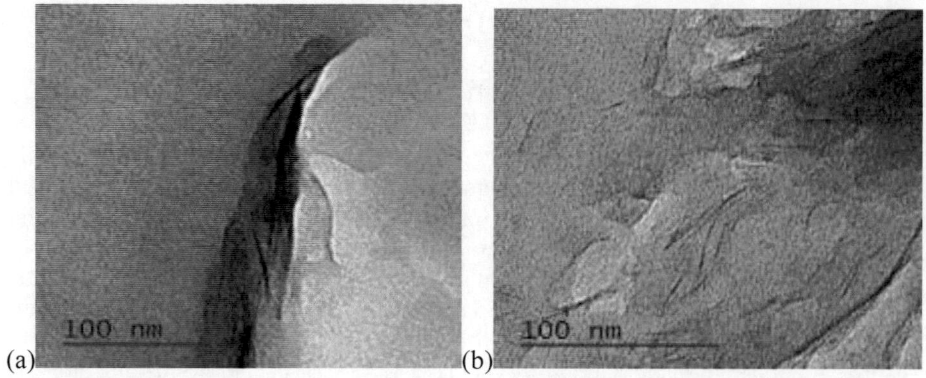

FIGURE 3. TEM photomicrographs of (a) PS/MMT10, and (b) PS/MMT-Pani10 composites.

Conductivity of the Nanocomposites

In order to compare conductivity values of MMT-Pani and pure Pani, DC conductivity was measured by a two-probe method. A DC potential scan was applied between 0 and 1.0 mV and the expected linear response in current was recorded. From the linear regression of these curves, conductivity of Pani was measured as 1.0×10^{-3} $S.cm^{-1}$ and of MMT-Pani as 3.0×10^{-6} $S.cm^{-1}$. Lower conductivity values are explained by the presence of an insulate material (e.g. MMT clay).

The electrical behavior of MMT-Pani fillers were acessed by Electrochemical Impedance Spectroscopy (EIS) and compared with the EIS spectra of Pani. Figure 4 shows both results, where Pani behaves was a pure resistor at lower frequencies, namely, showing constant impedance values independent with the frequency with phase angle equal of zero. In this case, the conductivity can be calculated by the inverse of resistivity, calculated by the relation between impedance and sample dimensions, according Equation 1. The resulting in a value for Pani was 11 $\Omega.m$ or 0,091 $S.m^{-1}$.

$$\sigma = \frac{1}{\rho}, Z = \rho \frac{\ell}{S} \tag{1}$$

On the other hand, the MMT-Pani filler presented a diffusive behavior, namely, the impedance decreases gradually with frequency increase and phase angle different from zero. This is an indicative that the electrical current flows in the Pani chains restrained into de clay galleries, and the nanofiller acts as a mixed electronic-ionic conductor, where electrons flow through Pani chains while a polarization of ionic charges occurs in the clay structure, in the way as a transmission line [6].

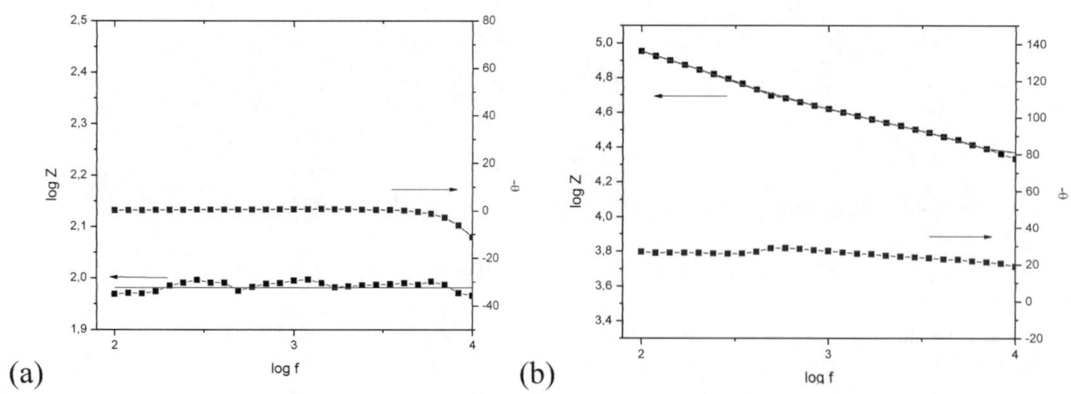

FIGURE 4. EIS spectra of (a) Pani, and (b) MMT-Pani.

Figure 5 shows EIS spectra, both Nyquist (5a) and Bode (5b) plots for PS/MMT-Pani nanocomposites with increased quantity of nanofillers. The electric behavior of insulating polymers such as the PS matrix, filled with conducting particles depends on many factors as composition and agglomeration of particles. Both figures show that samples with low nanofiller contents behave as capacitors (i.e. higher impedance at low frequencies and phase angle near 90°). This means these nanocomposites polarize charges due the applied potential, acting as an insulating. The exception is the nanocomposite with higher MMT-Pani content (PS/MMT-Pani10), that behaves as a mixed ionic-electronic conductor, with a slow infinite-diffusion, often modeled as a transmission line [6]. This may indicate that in the samples with better exfoliation, the insulating matrix contributes to charge polarization, but Pani content can act as an electron carrier inside the polymeric film.

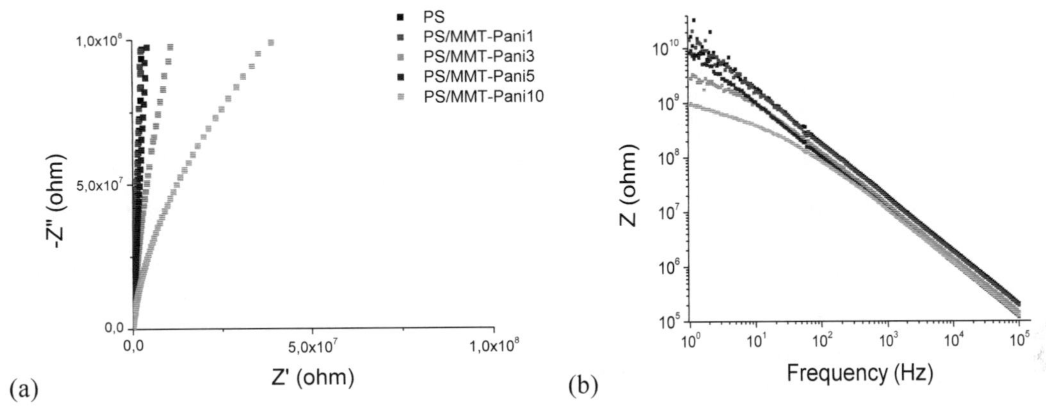

FIGURE 5. EIS spectra (a)Nyquist plot, and (b) Bode plot of PS/MMT-Pani nanocomposites.

RESULTS AND DISCUSSION

PS films with MMT clay modified by polyaniline produced nanocomposites (PS/MMT-Pani) that showed improvement in clay exfoliation when higher quantities of MMT-Pani filler were added. This changed the electric behavior of PS from a capacitive polarization to a diffusive charge transport.

ACKNOWLEDGMENTS

The authors acknowledge the financial support by UDESC – PROEVEN, FAPESC and CNPq.

REFERENCES

1. S. Pande; M. D. Bedre; R. Bhat, R. Deshpande; A. Venkataraman. *Nanoscience and Nanotechnology*, **2**, 90 (2012).
2. Q. Wu; Z. Xue; Z. Qi; F. Wang. *Polymer*, **41**, 2029 (2000).
3. A. F. Baldissera; J. F. Souza; C. A. Ferreira. *Synthetic Metals*, **183**, 69 (2013).
4. N. Srivastava; Y. Singh; R. A. Singh. *Bull. Mater. Sci*, **34**, 635 (2011).
5. J. Stejskal; P. Kratochvil. *Synthetic Metals*, **61**, 225 (1993).
6. I. D. Raistrick; D. R. Franceschetti; J. R. Macdonald, "Theory," in *Impedance Spectroscopy Theory Experiment and Applications,* edited by E. Barsoukov and J. R. Macdonald, New Jersey: John Wiley & Sons, 2005, pp. 27-128.

Rheology and Physical Characterisation of Graphene Nanoplatelet/ Poly (butylene adipate-co-terephthalate) Nanocomposites

Sima Kashi, Rahul K. Gupta, Nhol Kao, Sati N. Bhattacharya*

Rheology and Materials Processing Centre (RMPC), School of Engineering, RMIT University, Melbourne, Victoria 3000, Australia

Presenting author's email:s3406251@student.rmit.edu.au
Reference author's email: sati.bhattacharya@rmit.edu.au

Abstract. Biodegradable poly (butylene adipate-co-terephthalate) (PBAT)/ graphene nanoplatelet (GNP) nanocomposites were prepared and characterised. It was observed that GNP incorporation enhanced the thermal stability, electrical conductivity and Young's modulus of the matrix markedly. Melt flow behaviours of the prepared nanocomposites were investigated via dynamic oscillatory measurements. Viscoelastic properties exhibited significant enhancement with increasing GNP loading. The effect of GNPs was more pronounced in the storage modulus which exhibited weaker frequency dependency in the low frequency region. Rheological percolation threshold of GNPs in PBAT was determined to be in vicinity of 9 wt% at 140 °C. Interestingly, percolation was found to be temperature sensitive, occurring at lower GNP concentrations as the temperature was increased further from the system's melting point.

Keywords: poly (butylene adipate-co-terephthalate); graphene nanoplatelets; rheology; temperature effect
PACS: 81 and 83

INTRODUCTION

Public awareness of the impact of persistent plastic wastes on the environment has generated tremendous attention towards biodegradable polymers as alternatives to the commercial polymers in a variety of applications and products. Poly (butylene adipate-co-terephthalate) (PBAT) is a completely biodegradable aliphatic/aromatic copolyester derived from petroleum. It is synthesized by esterification of 1,4 butanediol with aromatic dicarboxylic acid followed by polycondensation with succinic acid [1]. This polymer exhibits high elasticity, wear and fracture resistance [2] as well as adhesion and compatibility with many other natural polymers [3]. It owes its biodegradability to the butylene adipate group and its stability and mechanical properties to the terephthalate group [2].

In spite of its excellent properties such as very high elongation and good thermal stability, PBAT has been often used as a second phase in polymer blends and nanocomposites due to low mechanical strength. Some researchers, however, have performed studies on PBAT as the sole polymeric matrix for preparation of new materials. It has been demonstrated that incorporation of nano-sized fillers into PBAT can overcome its shortcomings such as low strength, conferring multifunctional enabling properties like enhanced mechanical, thermal, magnetic and electrical properties [4-8]. Mohanty and Nayak [5] studied PBAT/ organically modified nanoclays nanocomposites. Someya et al. [7] focused on the relationship between the morphology and tensile properties of PBAT/ Montmorillonite nanocomposites. Wu [6] studied the chemical and mechanical properties of PBAT/ multi-walled carbon nanotubes composites. Feng et al.[4] investigated the crystallization and creep of PBAT/ graphite nanosheets composites. Mittal et al.[8] studied PBAT/ graphene (up to 5wt%).

Graphene nanoplatelets (GNPs) are graphitic nanoparticles with layered structure which are composed of stacked 2D graphene sheets bonded together with weak Van der Waals forces. Excellent physical properties combined with relatively lower production costs make GNP a promising carbon based nanofiller with potential to replace high-priced carbon nanotubes in many applications [9].

EXPERIMENTAL

Materials

GNPs (grade M) were obtained from XG Sciences (USA) which exhibit an average thickness of 6–8 nm, surface area of 120–150 m^2/g and density of 2.2 g/cm^3. PBAT was Ecoflex F Blend C1200 (BASF, Germany) with density of 1.25-1.27 g/cm^3 and melting range of 110-120 °C.

Nanocomposites preparation

PBAT/GNP nanocomposites were prepared in an internal mixer (Haake Rheomix OS R600) with roller rotors operated at 60 rpm for a mixing time of 10 min at 140 °C. Nanocomposites were prepared with six different GNP concentrations (0, 3, 6, 9, 12, 15 wt%) and were denoted as PB0, PB3,.., PB15 where the number in front of PB indicates the GNP concentration in the nanocomposites. The PBAT/GNP samples were then compression moulded at 140 °C with a compression force of 80 kN for 5 min.

Characterisation

Thermal gravimetric analysis (TGA) tests were conducted using a Perkin Elmer TGA (STA 6000). Samples were first stabilised at 50 °C for 2 minutes and then heated up to 900 °C under nitrogen at a heating rate of 10 °C /min. A FEI Quanta 200 scanning electron microscope with an accelerating voltage of 10 kV was used for morphological studies. AC conductivity was determined from electrical permittivity values of the samples, which were extracted from their scattering parameters collected by using a vector network analyser (Wiltron 37269A) with waveguide setup over X-band frequency range (8.2-12.4 GHz). Instron 4467 Universal testing machine was used to measure the tensile properties according to ASTM D638M with a crosshead speed rate of 5 mm/min.

Dynamic rheological measurements were performed by using a strain-controlled Advanced Rheometrics Expansion System (ARES) rheometer (TA Instruments) with parallel-plate fixture of 25 mm diameter. Frequency sweeps were conducted on the samples within their linear region at different temperatures. Fresh samples were used for measurements at each temperature for all the composition.

RESULTS AND DISCUSSIONS

Thermal stability

Thermal degradation of PBAT and PBAT/GNP nanocomposites under nitrogen atmosphere takes place in a single step process. The onset thermal degradation of pristine PBAT is detected to be about 338 °C (1% weight loss) and the degradation is completed around 426 °C. It can be seen from Figure 1 that addition of GNPs up to 6 wt% does not alter the thermal stability of the matrix considerably. However, as the GNP concentration is raised above 6 wt% to 9 wt% and higher, nanocomposites exhibit enhanced thermal stability. The temperatures at which 5 and 50% of samples' weight loss occur were found to increase by 5 °C and 7 °C respectively with increasing GNP loading from 0 to 15 wt%. This delay in thermal degradation of PBAT could be attributed to the shielding effect conferred by the flake-like shape of GNPs (Figure 2) which hinders the diffusion of volatile decomposition products within the nanocomposites [10].

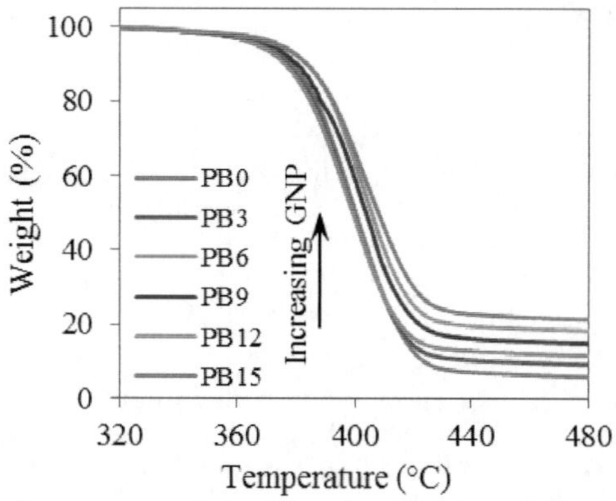

Figure 1. TGA curves of PBAT/GNP nanocomposites.

Figure 2. SEM image of GNPs.

Electrical and mechanical properties

The AC conductivity of PBAT/GNP nanocomposites was calculated at mid X-band frequency (10 GHz) and their Young's moduli were normalised against the modulus of unfilled PBAT. Results are summarised in Table a**Error! Reference source not found.**. Addition of 15 wt% GNPs increased the modulus of pure PBAT from 96 MPa to 460 MPa which is about 5 times higher than that of unfilled PBAT. Conductivity of the matrix was also improved markedly with GNP embedding, reaching a value of 3 S/m at 15 wt% GNPs. This enhancement is due to the formation of electrically conducive pathways of the platelets within the matrix as the GNP content of the nanocomposites increases.

Table a). AC conductivity and Normalised Young's Modulus of PBAT/GNP nanocomposites.

GNPs wt%	0	3	6	9	12	15
Normalised Young's modulus	1	1.5	2.5	2.9	3.7	4.8
AC Conductivity at 10 GHz (S/m)	0.11	0.26	0.46	1.13	1.77	3.02

Rheological properties

Variations of the storage (G') and loss (G") moduli of PBAT with GNP loading at 140 °C are depicted in Figure 3. Loss modulus monotonically increases with increasing GNPs over the entire frequency range. On the other hand, while down to 1 rad/s, the increase in the storage modulus with addition of GNP is uniform, at lower frequencies, a sudden increase of about one decade is observed in G' as the GNP content is raised from 9 to 12 wt%. Loss tangent demonstrates how the elastic and viscous properties of PBAT are changing relative to each other with addition of GNPs. As seen Figure 3c, tan δ of PB3-PB9 have values comparable to that of pure PBAT at all frequencies. However, nanocomposites with GNP content of 12 and 15 wt% demonstrate a much smaller energy dampening ability at frequencies below 1 rad/s.

A comparative illustration of G' and G" of PB9, PB12 and PB15 are demonstrated in Figure 3d. Please note that the curves are shifted along the horizontal axis. It can be seen that G' of PB9 decreases faster than its G" at all frequencies (PB0-PB6 have similar behaviour) which results in the continuous increasing trend of loss tangent (tan $\delta = G''/G'$) with decreasing frequency (Figure 3c). On the other hand, for PB12 and PB15 it is found that at high frequencies, G' decreases faster than G" leading to an increase in tan δ. But, at lower frequencies the terminal relaxation is dampened by the emergence of the slower solid-like response which decreases the slope of G' while G" continues to decrease with roughly the same slope, resulting in the decreasing trend of tan δ over the low frequency region . Therefore, a maximum occurs in tan δ at the frequency where the slopes of G' and G" are roughly equal as shown in Figure 3d.

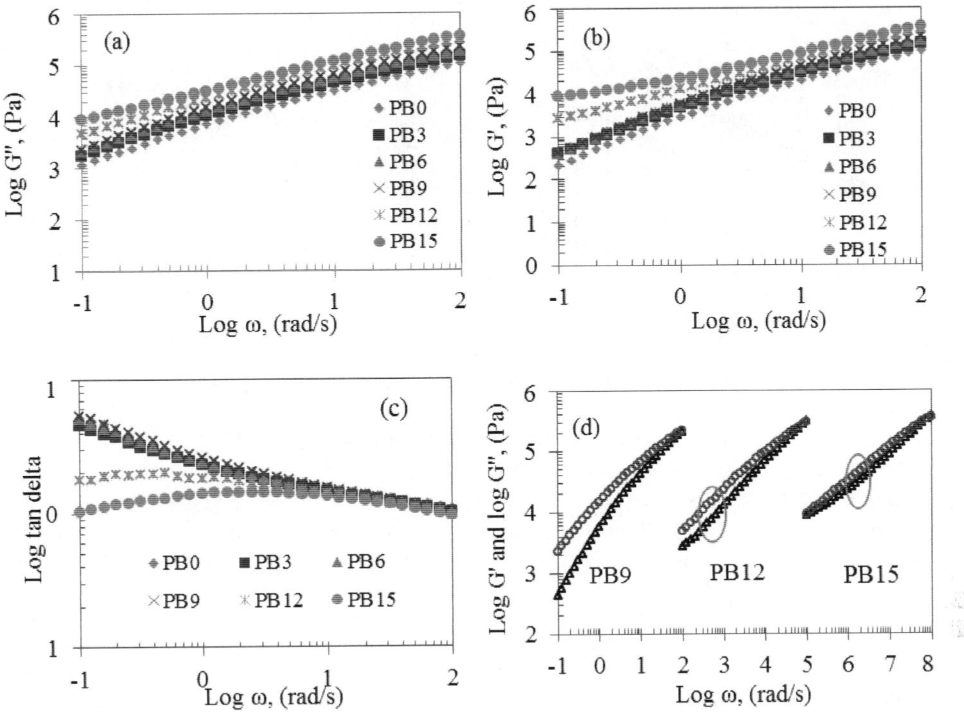

Figure 3. (a) Loss modulus, (b) storage modulus, (c) loss tangent and (d) comparative variation of loss (o) and storage (Δ) moduli of PBAT/GNP nanocomposites at 140 °C.

Figure 4 shows the loss and storage moduli of pure PBAT and PBAT/GNP nanocomposites at 220 °C. Three interesting observatiosn can be made by comparying G' and G" of the nanocomposites at 220 °C with their counterparts at 140 °C. Firstly, the GNP concentration at which the low frequency plateau in G' appears (liquid-solid transition) reduces from 12 wt% to 6 wt% as the temperature increases from 140 to 220 °C, respectively. Detection of a plateau in the storage modulus is an indication of existance of a percolating network throught the matrix which causes the solid-like behaviour of the system. The filler concentration at which the liquid- solid transition occurs is called percolation threshold. Therefore, it can be concluded that the GNP percolation threshold concentration in PBAT decreases with increasing temperature. The second observation is that the effect of GNP addition on the dynamic moduli is more significant at higher temperature; incorporation of 15 wt% GNPs increases G' of the matrix by 4.5 decades at 220 °C while this value is only 1.5 decades at 140 °C. Similarly, G" of PBAT increases by 2 and 1 decades for the same GNP loading at 220 °C and 140 °C, respectively. Lastly, the frequency dependey of the dynamic moduli changes with temperature; G' and G" of PB0-PB6 exhibit stronger variations with frequency as the temperature is raised. On the other hand, frequency dependendy of the moduli of PB9-PB15 especially G' becomes weaker at higher temperature.

Variation of GNP percolation concentration in PBAT with temperature could be attributed to the changes in the interactions (polymer-filler, filler-filler and polymer-polymer) present in the nanocomposite. The decreasing trend of percolation threshold with increasing temperature suggests that a combined percolating network of polymer-filler is responsible for the liquid-solid transition in the viscoelastic behaviour of PBAT/GNP nanocomposites rather than a network formed by only the nanoplatelets, similar to what has been observed for carbon nanotube nanocomposites [11]. It seems that the adsorption of PBAT chains to the surface of GNPs becomes stronger at elevated temperatures, leading to higher the effective GNP volume fraction in the system. These adsorbed chains may also function as bridges between the neighbouring particles[12].

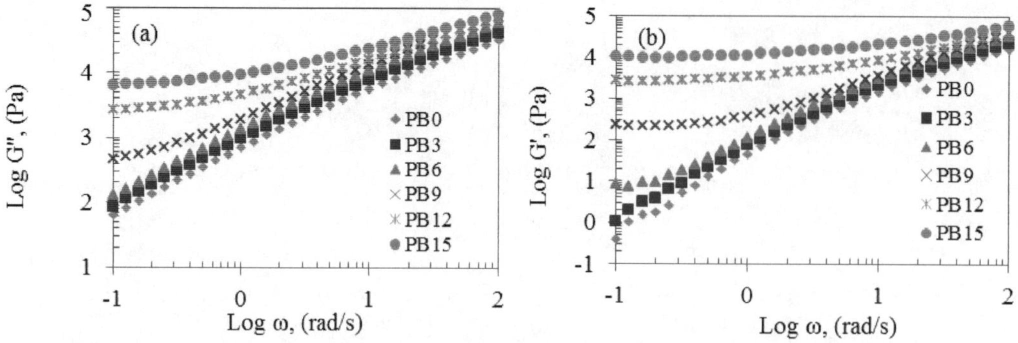

Figure 4. (a) Loss and (b) storage moduli of PBAT/GNP nanocomposites versus frequency at 220 °C.

CONCLUSIONS

Biodegradable PBAT/GNP nanocomposites were prepared with up to 15 wt% GNPs. Physical properties of the matrix were found to enhance with GNP incorporation; nanocomposites with GNP content of above 6 wt% showed improved thermal stability. Furthermore, electrical conductivity and Young's modulus of the matrix increased continuously with increasing GNP loading.

Viscoelastic properties of PBAT were also enhanced markedly with addition of GNPs particularly storage modulus. Increasing the temperature of rheological measurements revealed that the effect of GNPs on the dynamic moduli of the matrix was more pronounced at elevated temperatures. Moreover, the GNP percolation threshold in PBAT was found to decrease with increasing temperature. This temperature dependency of percolation threshold was attributed to the formation of a combined PBAT-GNP network within the nanocomposites.

REFERENCES

1. Hong, S. Y.; Ko, S. W.; Choi, H. J.; Lee, J. H., Journal of Macromolecular Science, Part B 51, 125 2012.
2. Shahlari, M.; Lee, S., Polymer Engineering & Science 52, 1420 2012.
3. Yeh, J.-T.; Tsou, C.-H.; Huang, C.-Y.; Chen, K.-N.; Wu, C.-S.; Chai, W.-L., Journal of Applied Polymer Science, 2009.
4. Feng, S.; Wu, D.; Liu, H.; Chen, C.; Liu, J.; Yao, Z.; Xu, J.; Zhang, M., Thermochimica Acta 587, 72 2014.
5. Mohanty, S.; Nayak, S. K., Polymer Composites, 1194 2009.
6. Wu, C.-S., Carbon 47, 3091 2009.
7. Someya, Y.; Sugahara, Y.; Shibata, M., Journal of Applied Polymer Science 95, 386 2005.
8. Mittal, V.; Chaudhry, A. U.; Luckachan, G. E., Materials Chemistry and Physics 147, 319 2014.
9. Li, B.; Zhong, W.-H., Journal of Materials Science 46, 5595 2011.
10. Murariu, M.; Dechief, A. L.; Bonnaud, L.; Paint, Y.; Gallos, A.; Fontaine, G.; Bourbigot, S.; Dubois, P., Polymer Degradation and Stability 95, 889 2010.
11. Pötschke, P.; Abdel-Goad, M.; Alig, I.; Dudkin, S.; Lellinger, D., Polymer 45, 8863 2004.
12. Kelarakis, A.; Giannelis, E. P., Polymer 52, 2221 2011.

Synergistic Effect of Compatibilizer in Flame-Retardant Polypropylene Compounds

Baltus Cornelius Bonse[a*], Rafael Galdino Bringel[b] and Ana Clelia B. Ferreira[b]

[a]Centro Universitário da FEI – Av. Humberto de A. C. Branco 3972, São Bernardo do Campo, 09850-901, SP – Brazil
[b]Faculdade SENAI de Tecnologia Ambiental - Av. José Odorizzi, 1555, São Bernardo do Campo, 09861-000, SP - Brazil

prebbonse@fei.edu.br

Abstract. Polypropylene (PP) is one of the most consumed thermoplastics and can achieve wide-ranging properties according to polymerization conditions and by means of additives. For instance, its low fire resistance can be improved with the use of flame-retardants. However, this additive may adversely affect mechanical properties. Within this context different flame-retardant PP copolymer formulations, with and without compatibilizer, have been manufactured. Compounding was performed by means of a co-rotating twin-screw extruder, after which specimens were injection molded and subjected to physical, mechanical and thermal testing, as well as scanning electron microscopy analysis, to identify modifications caused by the flame retardant and compatibilizer. The flame retardant showed to decrease tensile, flexural and impact strength, and increased melt flow rate. In contrast, incorporation of the compatibilizer showed to regain the mechanical properties close to the original values, interfering significantly less with MFR, and promoted minimal change in degradation temperature

Keywords: High Impact Polypropylene. Flame Retardant. Compatibilizer.
PACS: 81.05.Lg, 81.40.-z

INTRODUCTION

Polypropylene, one of the most consumed thermoplastics, is commercially produced mainly in the form of PP homopolymer, PP random copolymer and high impact PP (HIPP). The latter consists of approximately 40 wt% ethylene-propylene rubber (EPR) intimately dispersed in a PP homopolymer matrix. Applications include automotive, electrical appliances parts, industrial containers and furniture [1]. Polyolefins are inherently flammable and therefore require flame-retardants to improve their low fire resistance [2]. Flame-retardants can be classified in halogenated (brominated and chlorinated), phosphorus, nitrogen and inorganic (e.g. alumina trihydrate, magnesium oxides or antimony trioxide, which is used in synergy with bromine, phosphorus or nitrogen flame retardants) [3]. However, these additives may impair mechanical properties, especially when used in relatively large amounts [4]. The scope of this investigation was therefore to recover some of the mechanical properties o HIPP, reduced due to flame retardant incorporation (tetrabromobisphenol A bis (2,3-dibromopropyl ether)/Sb_2O_3), by using a suitable compatibilizer without impairing the acquired fire resistance.

EXPERIMENTAL

Materials

All materials used in this study were kindly donated by Aditive Plásticos Ltd (São Paulo, Brazil). The HIPP used was El-Pro P440J from SCG Chemicals, an impact copolymer polypropylene resin suitable for injection molding, with UL-94 HB flammability rating. HIPP flame retardancy was afforded by a T1420® concentrate, a 3:1 combination of tetrabromobisphenol A bis (2,3-dibromopropyl ether) and antimony trioxide (Sb_2O_3). As compatibilizer a maleic anhydride grafted PP Masterfil Adibond® M9584 was used. UV stabilizer SL-L4248/01® and antioxidant M9037C® were also used. Contents of the aforementioned materials in HIPP formulations are shown in table 1, where all values are percentage by weight (wt%).

TABLE (1). Composition and designation of HIPP compounds.

Designation	HIPP	Flame Retardant (T1420®)	Compatibilizer (Adibond®)	UV stabilizer (SL-L4248/01®)	Antioxidant (M9037C®)
Neat HIPP	100	-	-	-	-
HIPP FR	81.7	15.3	-	2	1
HIPP FR **2**C	79.7	15.3	**2**	2	1
HIPP FR **4**C	77.7	15.3	**4**	2	1

Preparation and Characterization of HIPP Compounds

HIPP compounds (3 kg each) were prepared by first tumble mixing HIPP with the additives at the proportions shown in table 1, and then extruding in a Theysohn 20 mm co-rotating twin-screw extruder, L/D=40, with 5 heating zones at temperatures ranging from 170 to 200 °C. The extrudates were subsequently injection molded into tensile, bending, flammability and impact test specimens in a Battenfeld HB 200/45 machine. Processing temperatures ranged from 180 to 210°C, as the flame retardant starts degrading at temperatures exceeding 250 °C.

Melt flow rates (MFR) of neat HIPP and compounds were determined in an MI-1 melt indexer from DSM at 230 °C with a load of 2.16 kg. Fractured surfaces of impact test specimens were gold-coated and analyzed in a CamScan CS 3200 LV scanning electron microscope (SEM), under high vacuum, at 20 kV. Tensile and flexural (3-point bend) testing were performed in an Instron 5567 universal testing machine, according to ASTM D638 and D790, at crosshead rates of 50 and 1.3 mm/min respectively, using at least 5 specimens. Izod impact testing was performed in an XJC 25 D tester from NZ Philpolymer, in accordance with ASTM D256, using a 4J pendulum on at least 10 specimens. Differential scanning calorimetry (DSC) and thermogravimetric analysis (TGA) were performed in TA Instruments testers, Q2000 and Q500, respectively, under constant nitrogen gas flow. DSC at a heating rate of 10 °C/min from 0 °C a 190 °C and TGA at 20 °C/min up to 650 °C. Flammability testing was performed on 3.2 mm thick specimens, in accordance with UL-94 20mm Vertical Burning Test, which measures the self-extinguishing time of a vertically suspended specimen, with its lower extremity 30 mm above a cotton pad. A blue 20 mm high flame is brought into contact with the specimen for 10 s, after which the burner is removed. The material is rated V-0 if the flame extinguishes within 10 s after removal of the burner; V-1 if the flame extinguishes after 30 s after removal; and V-2 if the flame extinguishes after 30 s after removal of the burner and cotton is ignited by dripping flaming particles.

RESULTS AND DISCUSSION

Figure 1 shows results of tensile, flexural and impact tests of neat HIPP and HIPP containing flame-retardant without compatibilizer.

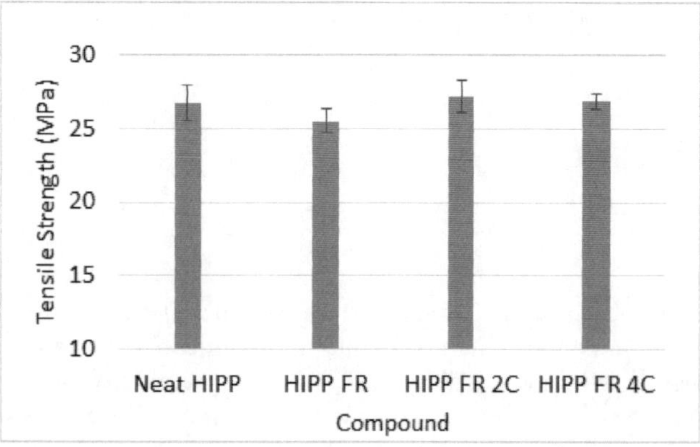

(a)

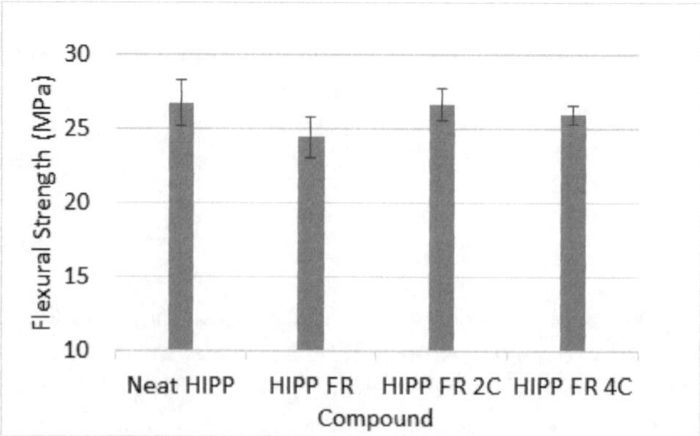

(b)

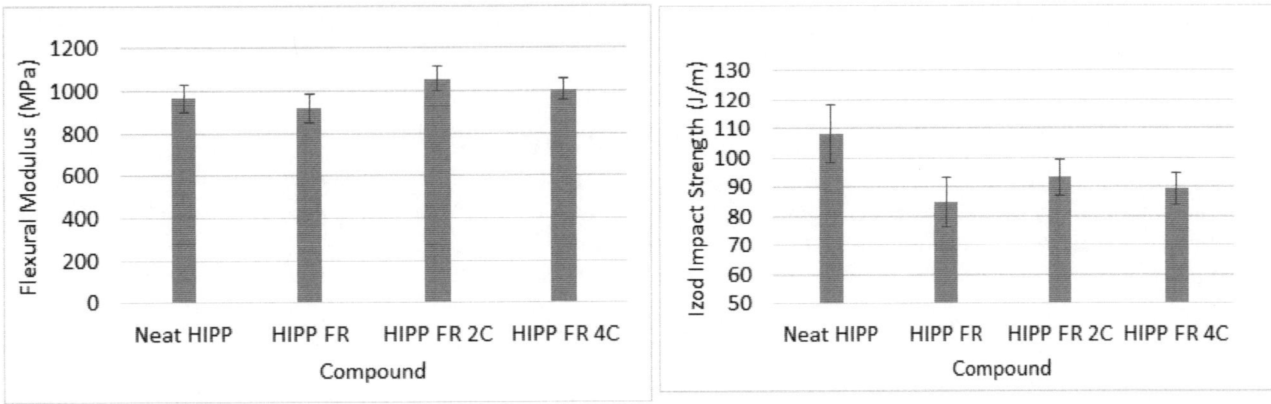

(c) (d)

FIGURE 1. Mechanical properties of HIPP and its compounds containing flame-retardant, with and without compatibilizer: (a) tensile strength; (b) flexural strength; (c) flexural modulus; (d) Izod impact strength

The column graphs in Figure 1 clearly shows that when flame retardant and stabilizers are incorporated into HIPP, all investigated mechanical properties decrease. When compatibilizer is added, losses in tensile strength, flexural strength and flexural modulus are recovered. In the case of impact strength, not all loss is recovered. Increasing compatibilizer content from 2 to 4wt% almost did not affect or slightly decreased recovery.

Figure 2 show MFR and flammability results of neat HIPP and HIPP containing flame-retardant without compatibilizer (HIPP FR), with 2wt% (HIPP FR 2C) and 4 wt% compatibilizer (HIPP FR 4C).

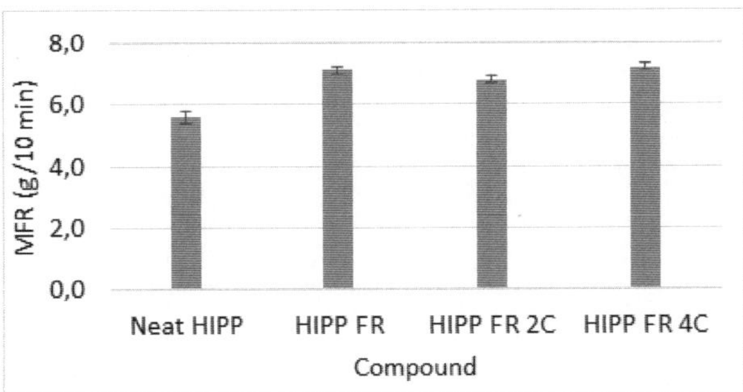

Compound	Rating
Neat HIPP	-
HIPP FR	V0
HIPP FR 2C	V0
HIPP FR 4C	V0

FIGURE 2. MFR and flammability rating of neat HIPP and HIPP containing flame-retardant with and without compatibilizer

Flame retardant and additives are seen to increase MFR, likely due to a lubricating effect, as one would expect a decrease brought about by the inorganic fillers. Despite the high MFR (150 g/10 min), compatibilizer showed little interference with the flame retarded HIPP. With regard to the flammability tests, neat HIPP did not self-extinguish after removal of the flame, burning was sustained and the cotton pad was ignited by dripping flame particles. Therefore, neat HIPP was nonratable by the UL-94 standard. On the other hand, incorporation of flame retardant was successful in rendering HIPP a V0 rating. This rating was maintained after addition of compatibilizer.

CONCLUSIONS

Fire resistance of high impact polypropylene was increased to V0 rating with a 3:1 mixture of tetrabromobisphenol A bis (2,3-dibromopropyl ether) and antimony trioxide (Sb_2O_3). Deterioration in mechanical properties, such as tensile and flexural strength and modulus, due to flame-retardant incorporation, could be successfully recovered by using PP-g-MA compatibilizer. In addition compatibilizer did not affect acquired fire resistance. The mechanical property most affected by flame-retardant was impact strength, with an approximately 20% drop in relation to neat HIPP. Little difference was seen between 2 wt% and 4 wt% compatibilizer, regarding property recovery, with the former indicating improved performance, especially in impact strength, suggesting existence of an optimum content below 4wt%. TGA evidenced two distinct degradation phases in the flame retarded HIPP compounds

ACKNOWLEDGMENTS

The authors acknowledge Aditive Plásticos Ltd. for the donation of the materials used in this study, as well as Plásticos Novacor for permitting to use some of their laboratories.

REFERENCES

1. C. Maier and T. Calafut, *Polypropylene The Definitive User's Guide and Databook,* Norwich, NY:, Plastics Design Library, 1998, pp. 19-25.
1. K. Szpilska, K.Czaja, K. Kudla, *Polimery* **60**, 671-750 (2015). http://dx.doi.org/10.14314/polimery.2015.673
3. H. Mikael S.H. Eldbjorg, H. Dorte, S. Torkjel, P. Stefan, W. Frank, Emerging "new" Brominated flame retardants in flame retarded, Current State of Knowledge and Monitoring requirements, (TA-2462/2008)
4. S. Cavusoglu, E. Akyol, M. Oner, "The Effect of Sb2O3/DBDPE Ratio on the Flame Retardancy and Physico-Mechanical Properties of EPDM Elastomer" International Conference on Chemical and Biochemical Engineering (ICCBE), Paris, 2015

ISOTHERMAL CRYSTALLIZATION BEHAVIOR AND MICROSTRUCTURE OF POLY(BUTYLENE SUCCINATE-CO-ADIPATE)/MODIFIED LAYERED DOUBLE HYDROXIDES NANOCOMPOSITES

G. S. Tsai, Y. A. Chen, E. C. Chen, T. M. Wu

Department of Materials Science and Engineering, National Chung Hsing University, Taichung, Taiwan--tmwu@dragon.nchu.edu.tw

A biodegradable poly[(butylene succinate)-co-adipate] (PBSA) synthesized by polycondensation reaction is solution-mixed with organically-modified layered double hydroxide (m-LDH) to fabricate the PBSA/m-LDH nanocomposite. The crystallization behavior and microstructure of PBSA/m-LDH nanocomposite are studied by differential scanning calorimetry (DSC), wide-angle X-ray diffraction (WAXD), transmission electron microscopy (TEM), and small-angle X-ray scattering (SAXS). Both WAXD and TEM results demonstrate that most of the layered materials are exfoliated and randomly distributed in the PBSA matrix. With the addition of 3 wt % m-LDH into PBSA, the half-time for crystallization of PBSA/m-LDH nanocomposite decreased by 47–76% for isothermal crystallization at 48–56 °C compared with that of neat PBSA. The microstructure of PBSA/m-LDH nanocomposites identified by SAXS shows higher crystal layer thickness is achieved with the presence of 3 wt % m-LDH.

Introduction

Biodegradable polymers, such as poly(butylene succinate) (PBS), poly[(butylene succinate)-co-adipate] (PBSA) and poly(L-lactic acid) (PLLA), have been attracting much research interest from academic and industrial perspectives [1-2]. PBSA, a semi-crystalline aliphatic copolyester with random comonomer sequence, is synthesized by polycondensation of 1,4-butanediol in the presence of succinic acids (SA) and adipic acids (AA) [2]. PBSA has been widely used in life and biomedical applications, including the manure and waste compost, agricultural packing, and food industry. Due to more flexible macromolecular chains, slower crystallization rate, and lower crystallinity of PBSA, practical application has been limited [3].

Layered double hydroxides (LDHs), the synthetic analogs of naturally occurring mineral hydrotalcites, have high charge density and strong interlayer electrostatic interactions between hydroxide layers, which makes the delamination of LDH within polymers more challenging [3]. One of the resolutions is to modify the surfaces of LDH platelets with organo-modifiers not only to expand the interlayer space but also to change the hydrophilic characteristic of LDH into hydrophobic ones [4]. In this study, we have used biocompatible oleic acid and sorbitol to change the surface of LDH (m-LDH) and the m-LDH is then served as the dispersed phase to prepare the PLLA/m-LDH nanocomposites by solution intercalation of PBSA into the swellable layered hosts. The microstructure and crystallization behavior of PBSA/m-LDH nanocomposites were examined by X-ray diffraction, transmission electron microscopy and differential scanning calorimeter.

Experimental

*Fabrication of the PBSA/*m-LDH *nanocomposites*

The nitrate intercalated magnesium/aluminum layered double hydroxides (LDH) were synthesized using the coprecipitation from an aqueous solution of $Al(NO_3)_3 \cdot 9H_2O$ and $Mg(NO_3)_2 \cdot 6H_2O$ dissolved in a deionized water to prepare a total metal cationic concentration of 1.0 M solution, in which the molar ratio of Mg/Al was 2. The as-prepared LDHs were then modified using oleic acid and sorbitol. The PBSA/m-LDH nanocomposites with different m-LDH loadings were fabricated by solution mixing and melt blending process.

Characterization of PBSA/m-LDH nanocomposites

The FTIR spectroscopy was used to characterize the structure of PBSA/m-LDH nanocomposites. The field-emission scanning electron microscopy (FESEM) and high-resolution transmission electron microscopy (HRTEM) were used to characterize the morphology of PBSA/m-LDH nanocomposites.

Results and Discussion

PBSA copolymer was synthesized by a two-stage polycondensation reaction. The first step belonged to transesterification reaction. The reaction temperature at 190 °C was carried out at normal pressure with stoichiometric excess of 1,4-butanediol in the presence of a transition metal complex. Then the reaction was performed by heating the mixture at 220 °C for 8h under vacuum to remove the excess 1,4-butanediol and hence produce chain extension. Figure 1 presents the

^1H–NMR spectrum of PBSA copolymer. The ^1H–NMR spectrum of the PBSA copolymer is shown: –CH_2–, from the 1,4- butanediol unit and the adipoyl unit at δ=1.72; $COCH_2$–, from the adipoyl unit at δ= 2.33; $COCH_2$–, from the succinyl unit at δ=2.63; $OOCH_2$–, from the 1,4-butanediol unit at δ=4.13. The observed ^1H–NMR spectrum is consistent with the chemical structure of PBSA [2]. The molar composition of PBSA copolymer was measured as the area ratio of δ=2.63 and δ= 2.33 peaks. The theoretical values of SA and AA are 75 and 25 mol%, respectively. In fact, the SA residues incorporated into the PBSA copolymer are in an amount about 2 mol% less than expected from the reaction mixture compositions, which is in the range of experimental error. This result indicates that the composition of the PBSA copolymer is in agreement with those expected from the composition of the feed.

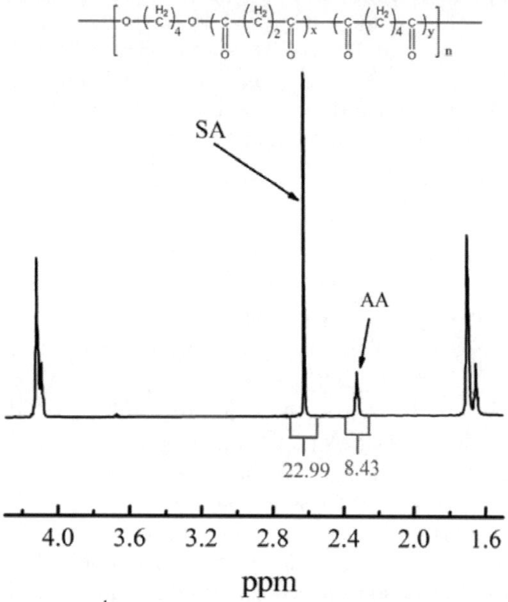

Figure 1: ^1H–NMR spectrum of the PBSA copolyester ([SA]/[AA]=75/25).

Figures 2a and 2f show the WAXD patterns of LDH and m-LDH. The interlayer spacing d_{003} of LDH and m-LDH determined using the strong diffraction peak at 2θ = 10.6 and 1.4° are 8.3 and 58.8 Å, respectively. It is clear that the interlayer spacing of LDH is significantly enlarged with oleic acid and sorbitol through the ionic exchange [5]. As shown in Figure 2b, the diffraction peaks of PBSA at 2θ =19.4°, 21.5°, and 22.5° are corresponding to (020), (021), and (110) crystallographic planes of PBSA, respectively. This data indicates that the crystal structure of PBSA is the same with that of PBS. For the samples of nanocomposites, they are also observed the diffraction peaks at 2θ = 19.4°, 21.5°, and 22.5°, which are attributed to the crystalline structure of PBSA. No diffraction peak of m-LDH is present in the nanocomposite samples. Therefore, the complete disappearance of the basal plane of LDH may be due to the no regular periodicity of the stacked lamellar m-LDH structure, leading to the formation of exfoliated and disordered nanostructure within the PBSA matrix. Consequently, the m-LDH platelets are randomly dispersed throughout the PBSA matrix.

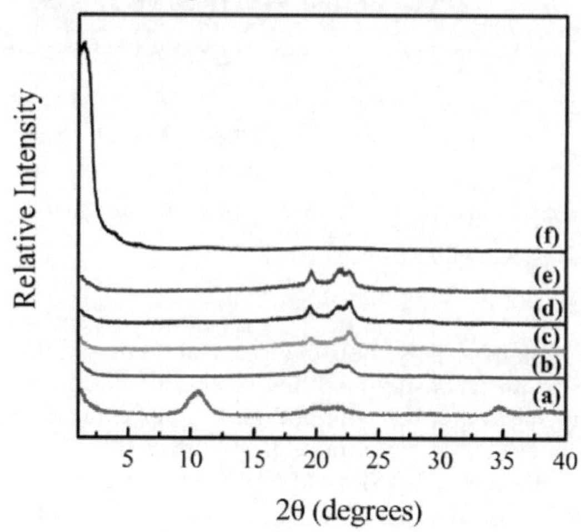

Figure 2: WAXD patterns of (a) LDH, (b) PBSA, (c) 1 wt % PBSA/m-LDH, (d) 3 wt % PBSA/m-LDH, (e) 5 wt % PBSA/m-LDH nanocomposites, and (f) m-LDH.

In order to elucidate the effects of crystallization temperature (T_c) on crystallization behavior of PBSA and PBSA/m-LDH nanocomposites, isothermal crystallization of PBSA matrix with different m-LDH contents were investigated. Isothermal crystallization kinetics was analyzed by Avrami equation. The Avrami plots of PBSA and PBSA/m-LDH nanocomposites show all curves are almost parallel to each other, indicating that the crystallization mechanism of the PBSA and PBSA/m-LDH nanocomposites at different T_{cs} remains the same.

To understand the microstructure of PBSA and PBSA/m-LDH nanocomposites, the SAXS technique is applied to investigate the change of morphology with the addition of m-LDH. The Lorentz-corrected SAXS profiles of PBSA and PBSA/m-LDH nanocomposites isothermally crystallized at 52 °C are presented in Figure 7. These results indicated that the incorporation of m-LDH can enhance the regular chain arrangement of PBSA. By adding more m-LDH into PBSA up to 5 wt%, the L_P and l_c slightly decreases as the content of m-LDH increases. These results reveal that the oleic acid and sorbitol chains of m-LDH inhibit the PBSA crystalline chain packing, which caused the decrease of the crystal layer thickness and the increase of the amorphous layer thickness.

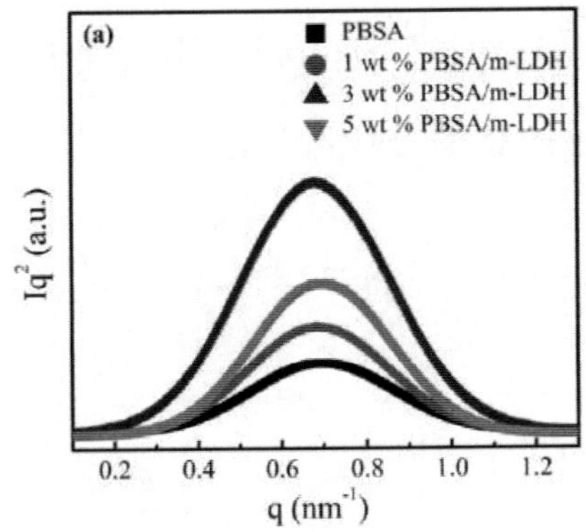

Figure 3: Lorentz-corrected SAXS intensity profiles for PBSA/m-LDH nanocomposites during isothermal crystallization at at 52 °C.

Conclusions

A biodegradable poly[(butylene succinate)-co-adipate] (PBSA) synthesized by polycondensation reaction is solution-mixed with organically-modified layered double hydroxide (m-LDH) to fabricate the PBSA/m-LDH nanocomposite. The crystallization behavior and microstructure of PBSA/m-LDH nanocomposite are studied by differential scanning calorimetry (DSC), wide-angle X-ray diffraction (WAXD), transmission electron microscopy (TEM), and small-angle X-ray scattering (SAXS). Both WAXD and TEM results demonstrate that most of the layered materials are exfoliated and randomly distributed in the PBSA matrix. With the addition of 3 wt % m-LDH into PBSA, the half-time for crystallization of PBSA/m-LDH nanocomposite decreased by 47–76% for isothermal crystallization at 48–56 °C compared with that of neat PBSA. The microstructure of PBSA/m-LDH nanocomposites identified by SAXS shows higher crystal layer thickness is achieved with the presence of 3 wt % m-LDH.

Acknowledgements

The financial support provided by National Science Council through the project MOST 104-2212-E-005-089-MY2 is greatly appreciated.

References

1. Bandyopadhyay J, Sinha Ray S (2010) The quantitative analysis of nano-clay dispersion in polymer nanocomposites by small angle X-ray scattering combined with electron microscopy. Polymer 51: 1437-1449.
2. Nikolic MS, Djonlagic J (2001) Synthesis and characterization of biodegradable poly(butylene succinate-co-butylene adipate)s. Polym. Degrad. Stab. 74: 263-270.
3. Bandyopadhyay J, Al-Thabaiti SA, Ray SS, Basahel SN, Mokhtar M (2014) Unique Cold-Crystallization Behavior and Kinetics of Biodegradable Poly[(butylene succinate)-co adipate] Nanocomposites: A High Speed Differential Scanning Calorimetry Study. Macromol Mater Eng 299: 939-952.
4. Chen G, Yoon J-S (2005) Nanocomposites of poly[(butylene succinate)-co-(butylene adipate)] (PBSA) and twice-functionalized organoclay. Polym Int 54: 939-945.
5. Chen YA, Hang YT, Wu TM (2015) Polymorphism and spherulite morphology of poly(1,4-butylene adipate)/organically-modified layered double hydroxide nanocomposites. J Appl Polym Sci 132.

Influence of Graphene Nanoplatelets on the Properties of Plasticized Poly (lactic acid) Nanocomposites

Josiani Aparecida da Silva[a], Wagner Mauricio Pachekoski[b], Daniela Becker[a*],

[a]*Universidade do estado de Santa Catarina -UDESC*
[b]*Universidade Federal de Santa Catarina - UFSC*

daniela.becker@udesc.br

Abstract. Poly (lactic acid) (PLA) is biodegradable and compostable polymer with great potential for substitution of non-biodegradable polymers, particularly those that are applied in products with quick shelf live. PLA fulfils the packaging industry's requirements for most of the rigid objects but the polymer needs to be plasticized to be used as soft films. In this paper, the aim was evaluate the effect of plasticizer and graphene dispersion in PLA properties. The plasticizers used were polyethylene glycol (PEG) and dibutyl sebacate (DBS). Plasticizers have been used as diluents in graphene nanoplatelets pre-dispersion via sonication and then this mixture was added in melted PLA. Thermal properties were evaluated by differential scanning calorimetry and thermogravimetric analysis and water vapor permeability test were carried out by ASTM method E96. The DBS addition decreased glass transition temperature (Tg) and PLA water vapor permeability in approximately 50%. It is also observed PEG addition decreased thermal stability and accelerates PLA water degradation. The graphene nanoplatelets addition improves water vapor barrier properties and increases the crystallinity degree.

Keywords: Poly (lactic acid), graphene nanoplatelets, plasticizer.
PACS: 82.35.Np.

INTRODUCTION

Currently, most food packaging material are from non-renewable sources, but the environmental concerns and sustainability has promoted the emergence of other measures. The use of bio-based materials is an alternative for reduce the environmental impact at packaging industry [1]. In this context the poly(lactic acid) is a polymer in highlight, due the fact that is a biopolymer, biodegradable and derived from 100% renewable resources such as sugar and corn. Additionally, compared to other biodegradable polymers, PLA presents good processability, excellent mechanical properties and great transparency [2]. However, to substitute petrochemical-derived synthetic polymers in packaging material application, PLA needs to improve some drawbacks that restrict its effective use such as its high brittleness that limits its use in flexible films, its low thermomechanical resistance and its low gas barrier properties to O2, CO2 or H2O which are susceptible to react or damage the food [3].

Many efforts have been directed towards overcoming these limitations, such as the addition of nanoparticles to formation tortuous path and/or increase the crystallinity of the polymer since the crystallites are impermeable and can be able to arrest small molecules; and plasticizers that reduces glass transition temperature and improves the toughness [4,5]. In this paper, the aim was evaluate the effect of plasticizer and graphene dispersion in PLA thermal properties and water vapor permeance. The plasticizers used were polyethylene glycol (PEG) and dibutyl sebacate (DBS). Plasticizers have been used as diluents in graphene nanoplatelets pre-dispersion via sonication and then this mixture was added in melted PLA.

EXPERIMENTAL

Materials

The materials used in this work were PLA Ingeo 3251D manufactured by Nature Works®, in composition with two different plasticizers: polyethylene glycol (PEG, 400g/mol) and dibutyl sebacate (DBS) supplied by Sigma Aldrich. The graphene nanoplatelets were purchased from Strem Chemicals Company with thicknesses of approximately 6-8 nm and wide up to 25 μm with surface area about 150 m²/g approximately 6-8 nm and wide 25 micrometres, with surface area up to 120 to 150 m² /g.

Nanocomposites Preparation

The melt mixing process was performed in a Haake torque rheometer with a 50 cm^3 mixing chamber and standard rotors, operated at 175°C and 100 rpm for 6 minutes. Before the mixture process, PLA was dried for 10h at 60 °C in air-circulating oven. Nanocomposites with plasticizers were performed the graphene pre-dispersion in the plasticizer. These pre-dispersion was performed using a tip sonicator (Sonics, model VCX 750). The energy added to the system was 500 J/g. Table 1 shows the composition of each PLA nanocomposite with their corresponding nomenclature.

Table 1 Formulations of the mixtures calculated in phr (parts per hundred resin)

Sample	PLA(g)	Graphene(g)	DBS(g)	PEG(g)
PLA	40.0	-	-	-
PLA+0.25G	40.0	0.1	-	-
PLA+0.5G	40.0	0.2	-	-
PLA+DBS	40.0	-	8.0	-
PLA+DBS+0.25G	40.0	0.1	8.0	-
PLA+DBS+0.5G	40.0	0.2	8.0	-
PLA+PEG	40.0	-	-	8.0
PLA+PEG+0.25G	40.0	0.1	-	8.0
PLA+PEG+0.5G	40.0	0.2	-	8.0

The samples were compressed and molded using a hydraulic press equipped with heating and cooling system at 180±5°C and 2 kgf/cm² for 15 seconds, and cooled with water to room temperature for 2 min. Aluminum plates with 0.5 mm thickness were used as mold.

Characterization techniques

The melting and crystallization temperature and crystallinity degree were obtained by differential scanning calorimetry (DSC), in a TA Instruments Q100 V9.4 device. The samples were heated from 25°C to 20°C at a rate of 10°C/mim, held at that temperature for 2min, then air cooled to -10°C (cooling rate of 80°C/mim) before second heating scan from -10°C to 200°C at 10°C/mim (under nitrogen atmosphere). All the thermal parameters were determined from second heating scans The crystallinity degree (Xc) was determined as show in Garlotta [6]. The weight loss was obtained by thermogravimetric analysis (TGA), in a Netzsch STA 449 C device with a heating rate of 10°C/min from 25°C to 600°C (under nitrogen atmosphere). Water vapor permeability test were carried out by ASTM method E96.

RESULTS AND DISCUSSION

Figure 1 presents the thermogravimetric curves for all PLA systems and the onset temperature for all stage and peak temperature for the main stage are summarized in Table 1. In general, the graphene addition do not altered significantly thermal degradation of PLA, which starts around 300°C. However, plasticizer addition led to the formation of different degradation stage. In DBS blends, degradation stage begins on 170°C, temperature near the plasticizer flash point (178°C). PEG and PLA mixture leads to a decrease in PLA initial decomposition temperature. Its addition also altered the PLA degradation mechanism with formation of a second degradation stage on the end of process. Besides, the graphene addition in low content (0.25%) decreased significantly the onset temperatures.

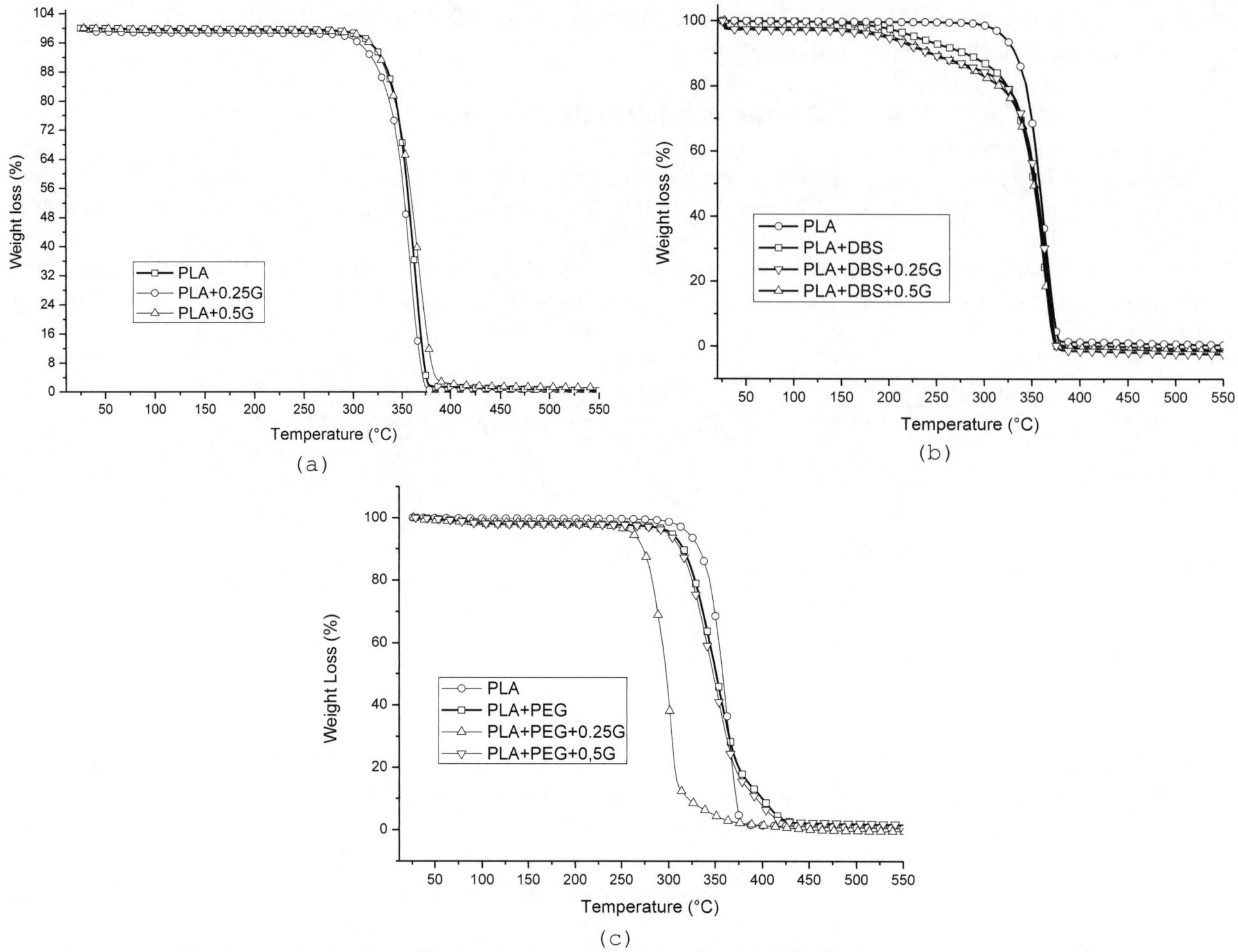

Figure 1 – Thermogravimetric curves for (a) non-plasticizer PLA; (b) PLA with DBS and; (c) PLA with PEG nanocomposites

Table 1 Data obtained from TG analysis for PLA nanocomposites

Samples	T_{onset1} (°C)	T_{onset2} (°C)	T_{peak} (°C)
PLA	-	322	363
PLA+0.25G	-	324	357
PLA+0.5G	-	316	371
PLA+DBS	166	318	366
PLA+DBS+0.25G	170	320	365
PLA+DBS+0.5G	170	320	362
Samples	**T_{onset1} (°C)**	**T_{peak} (°C)**	**T_{onset2} (°C)**
PLA	322	363	-
PLA+0.25G	324	357	-
PLA+0.5G	316	371	-
PLA+PEG	299	351	381
PLA+PEG+0.25G	264	291	304
PLA+PEG+0.5G	294	346	379

Table 2 presents the thermal properties, measured by DSC, for neat PLA, plasticized PLA and PLA nanocomposites, including crystallinity, glass transition temperatures (Tg), crystallization and melting temperature (T_c and Tm). Analyzing the results is observed no modification in PLA thermal transitions with inclusion of graphene nanosheet. It was also observed the plasticizer influence on PLA transitions, reducing its crystallization temperature and increases the crystallinity degree. This behavior is typical of plasticized thermoplastics, where plasticizers may promotes crystallinity due to enhanced chain mobility [7]. PLA+PEG+0.25G presents two peaks for crystallization temperature. This behavior can be attributed as a result of lamellar rearrangement during crystallization of PLA with different crystalline structures [7,8]. The addition of the plasticizer reduces PLA glass transition temperature (Tg). This behavior is justified in the energy reduction for molecular motion which increases free volume and molecular mobility [9]. Some mixtures with PEG (PLA+PEG and PLA+PEG+0.5G) present two glass transition temperatures. These results indicate that occurs phase separation, due to the segments of the PLA and PEGs were partially miscible for PEGs with moderate molecular weights [10].

Table 2 Thermal properties measured by DSC for neat PLA, plasticized PLA and PLA nanocomposites

Samples	Tg^1	Tg^2	Tc^1	Tc^2	Tm	Crystalinity degree (%)
PLA	59	-	96	-	168	22
PLA+0.25G	57	-	96	-	168	25
PLA+0.5G	58	-	96	-	168	19
PLA+DBS	28	-	66	-	162	46
PLA+DBS+0.25G	30	-	69	-	163	48
PLA+DBS+0.5G	30	-	63	-	163	50
PLA+PEG	22	63	85	-	162	55
PLA+PEG+0.25G	28	-	66	84	162	56
PLA+PEG+0.5G	21	63	84	-	161	59

Figure 2 shows water vapor permeance for PLA mixture. In general, graphene addition lead to a decrease in vapor water permeance, this behavior can be explained by the barrier formation by inclusion of platelets that alters the diffusion path of penetrant molecules [11]. The high standard deviation for PEG mixtures are because PEG up the water vapor pressure suffered aging and presented cracks and fissures in the film, damaging the permeation resistance.

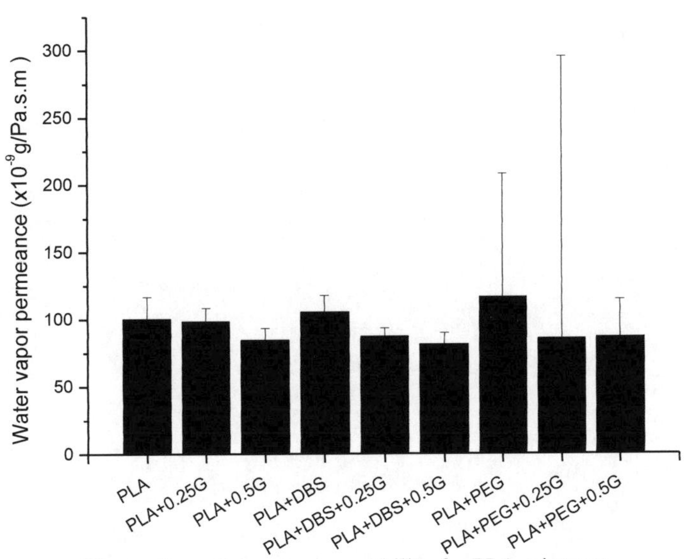

Figure 2 - water vapor permeability for PLA mixtures

CONCLUSIONS

According to the analysis carried out it is concluded the graphene addition do not altered significantly thermal behavior of neat PLA. However, plasticizer addition led to modification on thermal properties. Plasticizes addition led to the formation of different degradation stage. It was also observed the plasticizer influence on PLA transitions, reducing its crystallization temperature and increases the crystallinity degree and reduces PLA glass transition temperature. The DBS addition decreased glass transition temperature (Tg) and PLA water vapor permeability in approximately 50%. It is also observed PEG addition decreased thermal stability and accelerates PLA water degradation. The graphene nanoplatelets addition improves water vapor barrier properties and increases the crystallinity degree of plasticizer PLA.

ACKNOWLEDGMENTS

The authors acknowledge the technical support of EMBRAPA-LNN.

REFERENCES

1 – BIO-based Materials for MAP, *Bio Plastics Magazine*, **08**, 02 (2013).
2 – H.-D. Huang et al., *Journal of Membrane Science*, **464**, 110-118 (2014).
3 – H. Li and M.A. Huneault, *Polymer*, **48**, 6855–66 (2007).
4 – K.S. Anderson et al. *Journal of Polymer Science, Part B: Polymer Physics*, **49**, 1051–83 (2011).
6 – D. Garlotta, *Journal of Polymers and the Environment*, **9**, 63-84 (2001).
7 – O. Martin and L. Avérous, *Polymer*, **42**, 6209-6219 (2001).
8 – L. Cartier et al. *Polymer*, 41, 8909 (2000).
9 – I. Pillin et al. *Polymer*, **47**, 4676-4682 (2006).
10 –Y. Yu et al, *Journal of applied polymer science*, 41808-41817 (2015).
11 – D.R. Paul and L.M. Robeson,*Polymer*, **49**,3187–3204 (2008).

Characterization of Nanocomposites for Hydrogen Storage

Amanda D. Oliveira[a*], Cesar A. G. Beatrice[b], Fabio R. Passador[c], Luiz A. Pessan[b]

[a] *Federal University of Pelotas, Centre of Technology Development, Pelotas – RS,*
[b]*Brazil Federal University of São Carlos, Department of Materials Engineering, São Carlos – SP, Brazil*
[c] *Federal University of São Paulo, Institute of Science and Technology, São José dos Campos – SP, Brazil*

[*] *Corresponding author's email amandaoliveira82@gmail.com*
Presenting author's email: pessan@ufscar.br

Abstract. The use of hydrogen as an energy carrier suitable to replace gasoline and other fossil fuels has been widely discussed as a way to sustainably fuel our civilization. However, hydrogen storage is a major barrier in the establishment of infrastructure for hydrogen technology. The incorporation of nanoparticles to the polymer matrix may be an alternative to obtain materials with promising properties for hydrogen storage. In this work, polyetherimide-based Nanocomposites were prepared using carbon nanotubes doped with sodium alanate (NaAlH4) as filler. The sodium alanate content was fixed at 30 wt% and were studied three carbon nanotubes concentrations: 5, 10 and 20 wt%. The nanocomposites were characterized by differential scanning calorimetry (DSC), thermogravimetric analysis (TGA) and Hydrogen sorption measurements.

Keywords: hydrogen, energy, nanoparticles, storage.
PACS: 81; 82.

INTRODUÇÃO

Hydrogen fuel promises a future of renewable and clean energy. Hydrogen is the most abundant element on the earth and its oxidation product water is environmentally friendly. However, hydrogen is flammable, particularly explosive, highly diffusive, and its volume density is only one three thousandth of that of gasoline. The main challenges of hydrogen economy include hydrogen production, hydrogen transportation, hydrogen storage and hydrogen application. The largest obstacle for the spread use of hydrogen technology today is neither the production nor how to utilize the hydrogen, but rather effective and safe means of storage. Therefore, the ultimate goal related to hydrogen storage is to develop technologies with high energy density, high efficiency and safety [1, 2].

Several materials including carbon nanotubes, metalorganic frameworks, metal hydrides, graphite and activated carbon and metal/carbon nanostructures show promise as potential materials for hydrogen storage [2, 3]. Large specific surface area materials, such as metalorganic zeolites and carbon nanotubes, can produce hydrogen physisorption, but the storage capacity is very low (about 2 wt%) and generally the storage properties totally disappear at temperatures over 100 K, and can be increased up to about 195 K by pressurizing at 100 bar [2].

The addition of filler materials to the matrix material can be used to tailor an ideal nanocomposite material for hydrogen storage. In this work, nanocomposites with different filler materials at different concentrations were prepared and its effects on the thermal behavior and hydrogen storage capacity were evaluated.

EXPERIMENTAL

Materials

PEI (Ultem™ 1000) was purchased from Sabic Innovative Plastics, presents melt flow index of 9g/10min (6.6Kg/337°C) and density of 1.28 g/cm^3. For the modification of the polymer was used a sulfonating agent prepared from N-Methyl-2-pyrrolidinone (NMP) (was distilled before be used), acetic anhydride and sulfuric acid, all commercial grade, in the ratio 20:2:1 (by volume) to form acetyl sulfate reagent.

The MWCNTs used for this study were provided by the Nanostructured and Amorphous Materials Inc., contain 95 wt% purity, diameter less than 8 nm and length of 10 – 30 μm. Sodium alanate ($NaAlH_4$) was purchased from

Sigma Aldrich. Titanium dioxide (TiO_2) was used as a dopant for $NaAlH_4$ particles and was synthetized in our laboratories [4].

Sulfonation of PEI

Firstly, it was prepared the sulfonating agent. A predetermined amount of acetic anhydride was dissolved in NMP and the solution cooled to a temperature of approximately 0ºC. The solution was magnetic stirred and sulfuric acid was added slowly in the solution, the reaction was conducted for 30 min.

The sulfonation of PEI was carried under nitrogen atmosphere. Six grams of PEI was dissolved in 40 ml of NMP at 80ºC, under constant stirring. Then sulfonating agent was added to the PEI solution gradually and the reaction continued for 1 h. The reaction product was precipitated by slow dripping of ethyl alcohol. The final product was washed with ethyl alcohol and filtered on a funnel of Buchner. The sulfonated polyetherimide obtained was dried in a vacuum oven at 80ºC for 3 days.

Preparation of the PEIS/MWCNT/$NaALH_4$ nanocomposites

Since the sodium alanate is highly reactive with the reaction medium used for the modification of polyetherimide, firstly nanocomposites of PEIS/MWCNT were prepared *in situ* with subsequent solubilization and mixture with sodium alanate. The amount of 0.25 g of PEIS/MWCNT was dissolved in 5 ml of NMP and then mixed with $NaAlH_4$ and 2 mol % of TiO_2, at room temperature. This mixture was magnetically stirred for 30 min, precipitated with acetone and sonicated for 10 min. The mixture was then filtered and dried in a vacuum oven at 80º for 2 days.

For the preparation of the nanocomposites it was used three different MWCNT compositions (5, 10 and 20 wt%) and the sodium alanate content was fixed at 30 wt%.

Composites were also prepared from PEIS/NaAlH4+2 mol %. These composites were prepared using the previously described conditions.

Characterization of the materials

The nanocomposites were characterized employing differential scanning calorimetry (DSC), thermogravimetric analysis (TGA) and Hydrogen sorption measurements. Differential scanning calorimetry (DSC) was done on a Q2000 equipment, from TA Instruments and thermograms were obtained from room temperature up to 250 °C under nitrogen flow and heating rate of 10 °C/min. Thermogravimetric analysis (TGA) were carried on a Q50 equipment, also from TA Instruments, under nitrogen atmosphere and heating rate of 20°C/min from room temperature up to 800 °C.

Hydrogen Sorption measurements

Kinetic measurements of hydrogen absorption were performed in a volumetric device (Sievert´s Apparatus) built and designed in the Laboratory of Hydrogen in Metals of the Materials Engineering Department of the Federal University of São Carlos. The absorption measurement was performed at 32 bar of hydrogen and at fixed temperature of 120 °C. The amount of material loaded for hydrogen sorption measurements corresponds to approximately 100 mg.

RESULTS AND DISCUSSION

Figure 1 shows the DSC curves for the PEIS, PEIS/$NaAlH_4$ (70/30%) and PEIS/MWCNT/$NaAlH_4$ nanocomposites.

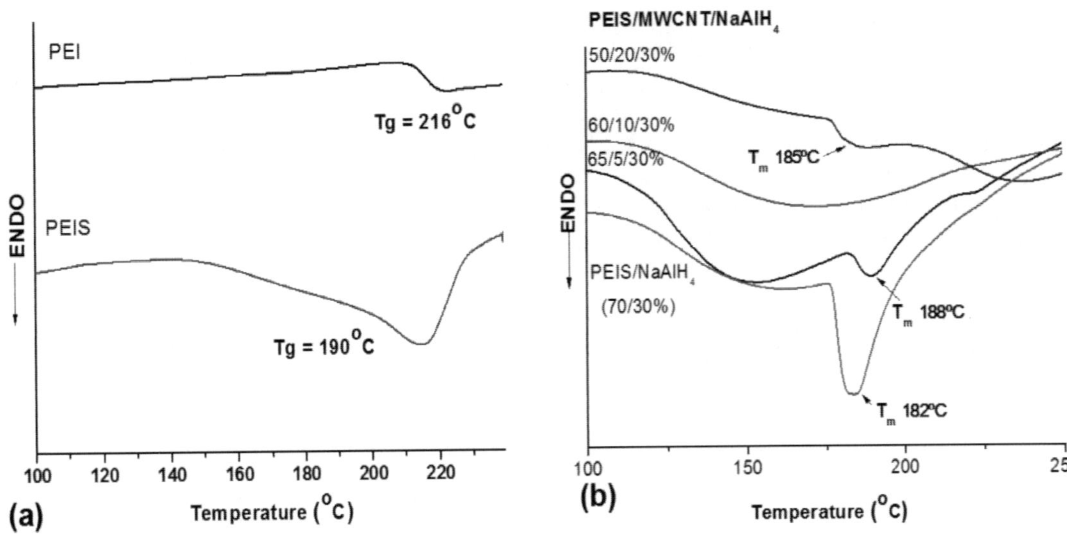

FIGURE 1. DSC curves for: (a) PEI and PEIS; (b) PEIS/MWCNT/$NaAlH_4$.

Neat PEI shows a glass transition temperature (Tg) of approximately 216 °C and after its sulfonation a decrease in Tg to 190 °C is observed, probably due to the introduction of sulfonic groups which increases the free volume in the polymer structure. In the presence of sodium alanate another endothermic peak can be observed, which is correlated to the melting of $NaAlH_4$. The addition of carbon nanotubes (PEIS/MWCNT/$NaAlH_4$) promoted a reduction in the intensity of the melting peak of the sodium alanate and increasing the amount of nanotube an enlargement of this peak was observed. For the composition PEIS/10MWCNT/30$NaAlH_4$ the melting peak is absent. It is also observed in Figure 1 that the nanocomposite with 10 wt% of carbon nanotube has lower glass transition temperature than the PEIS at approximately 177°C.

Figure 2 shows the thermal stability of the PEIS/$NaAlH_4$ (70/30%) and PEIS/MWCNT/$NaAlH_4$ nanocomposites.

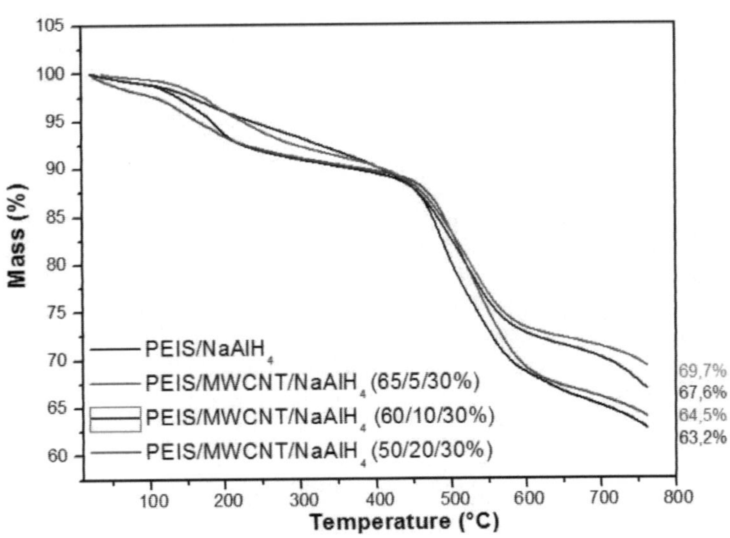

FIGURE 2. TGA curves for PEIS/MWCNT/$NaAlH_4$.

The degradation for all samples occurred in two stages. The first stage corresponds to the degradation of the sulfonic chains and the second stage, around of 440 °C, correspond to the degradation of the PEI chain. The first stage also receives the contribution of the mass loss of $NaAlH_4$ due to decomposition and release of gaseous molecules after melting at temperatures close to 200 °C. It is also observed in Figure 2 that the residues at the end of the analysis increases when increasing the MWCNT content, since the carbon nanotubes remains thermally stable up to higher temperatures.

Figure 3 shows the hydrogen sorption measurements of TiO_2-doped PEIS/$NaAlH_4$ composite and PEIS/20MWCNT/30$NaAlH_4$. Hydrogen sorption of the mixtures was performed at 120°C and lower pressures of 32 bar. For both samples it can be observed that hydrogen sorption increases with the increase of hydriding time from 0.5h to 10h. It can also be observed in Figure 3 that the TiO_2-doped PEIS/$NaAlH_4$ composites and PEIS/20MWCNT/30$NaAlH_4$ can absorb 1.1 and 1.2 wt% of hydrogen in 12 h, respectively. As the hydriding time decrease to 6 h, the hydriding capacity reduces to 0.5 for the TiO_2-doped PEIS/$NaAlH_4$ and to 0.9 for the PEIS/20MWCNT/30$NaAlH_4$. These results show that the materials developed in this work can be used for hydrogen storage as compared to results from the literature.

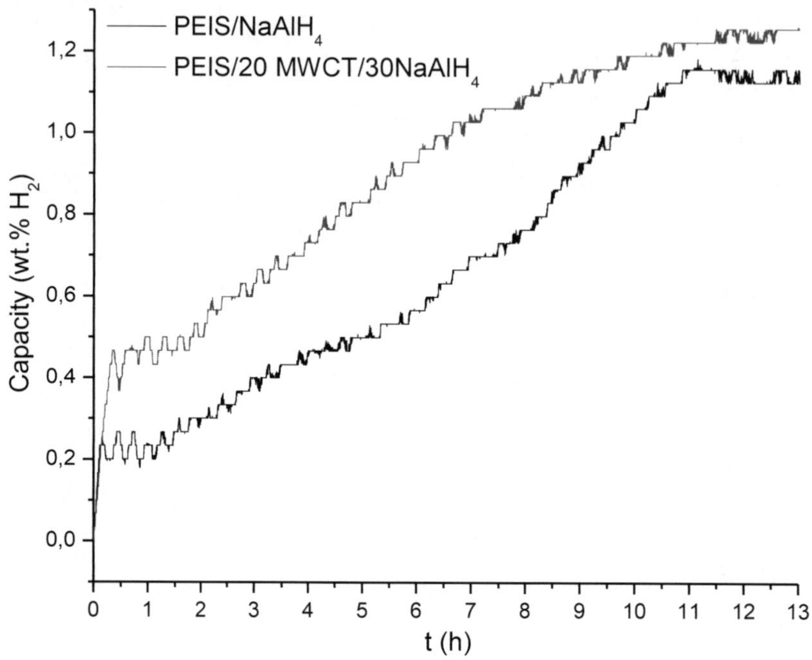

FIGURE 3. Hydrogen sorption for PEIS/$NaAlH_4$ composite with 2 mol % of TiO_2 and PEIS/20MWCNT/30$NaAlH_4$.

CONCLUSION

The hydriding and thermal properties of sulfonated polyetherimide nanocomposite material were investigated. Carbon nanotubes and sodium alanate added to the polymer matrix reduced the glass transition temperature of polymer. Hydrogen sorption tests were carried and it was shown that TiO_2-doped PEIS/$NaAlH_4$ can store 1.1 wt% hydrogen at 120°C. The addition of carbon nanotubes to the composite PEIS/$NaAlH_4$ has small effect on the hydrogen sorption capabilities of the material, though the increase in porosity of the material causes a slight increase in its hydrogen sorption capabilities.

ACKNOWLEDGMENTS

The authors would like to FAPESP (São Paulo Research Foundation) for the financial support through project 2013/23586-0. The authors thank Prof. Dr. Daniel Leiva and Prof. Dr. Tomaz Ishikawa (Federal University of São Carlos, Department of Materials Engineering, Laboratory of Hydrogen in Metallic Materials, Brazil) for their help with the hydrogen sorption measurements.

REFERENCES

1. M. P. Brown and K. Austin, *The New Physique*, Publisher City: Publisher Name, 2005, pp. 25-30.
2. M. P. Brown and K. Austin, *Appl. Phys. Letters* **85**, 2503-2504 (2004).
3. R. T. Wang, "Title of Chapter," in *Classic Physiques,* edited by R. B. Hamil, Publisher City: Publisher Name, 1999, pp. 212-213.
4. C. D. Smith and E. F. Jones, "Load-Cycling in Cubic Press" in *Shock Compression of Condensed Matter-2001*, edited by M. D. Furnish et al., AIP Conference Proceedings 620, American Institute of Physics, Melville, NY, 2002, pp. 651-654.

Polymer Nanocomposites for Hydrogen Storage

Cesar A. G. Beatrice[a], Amanda D. Oliveira[b], Fabio R. Passador[c], Luiz A. Pessan[a*]

[a] *Federal University of São Carlos, Department of Materials Engineering, São Carlos – SP, Brazil*
[b] *Federal University of Pelotas, Centre of Technology Development, Pelotas – RS, Brazil*
[c] *Federal University of São Paulo, Institute of Science and Technology, São José dos Campos – SP, Brazil*

[*] *Corresponding author's email: pessan@ufscar.br*

Abstract. Hydrogen is considered to be a clean, economical and safe renewable energy source that would be ideal to replace fossil fuels, because it is light, highly abundant and its oxidation product (water) is environmentally benign. However, hydrogen is easy to burn (the chemical energy per mass of hydrogen is at least three times larger than that of other chemical fuels), which has the risk of fire and explosion. The problems of transportation and storage restrict the application of hydrogen energy, which has become a key factor in the development and utilization of hydrogen energy. This gas adsorbs at solid surfaces depending on the applied pressure and temperature. For storage purposes in mobile applications, the adsorption of hydrogen has been studied mainly on carbon species, but light and reasonably cheap materials of high surface area should prove to be attractive as well. Porous material is a very promising hydrogen storage material, which stores the gas in the form of molecules at low temperatures and compresses hydrogen into the holes effectively. The purpose of this work was to develop a hybrid porous materials consisting of sulfonated polyetherimide matrix with aluminum nanoparticles and faujasite type zeolite. Dilute solutions were first prepared under stirring at room temperature and the solutions were dried under vacuum. The hybrids were analyzed by differential scanning calorimetry (DSC), thermogravimetric analysis (TGA), transmission electron microscopy (TEM) and hydrogen sorption measurements. The addition of aluminum decreased the glass transition temperature of the hybrids when compared to the sulfonated polymer and the TEM images showed that simply physically mixture occurred between polymer and metallic nanoparticles. Hydrogen sorption tests showed an increase in the amount of hydrogen in the presence of zeolite.

Keywords: hydrogen storage, nanocomposites, polyetherimide, aluminum, zeolite.
PACS: 81; 82.

INTRODUCTION

In an age of mobile technology energy storage has become an extremely important issue as the energy sources such as fossil fuels and natural gas are not infinitely available and even if they were, their combustion or end products are not environment friendly. Exploring sustainable clean energy sources and environmental technology to satisfy the world's growing demand are emerging as one of the foremost challenges for this century [1-2]. One of the alternatives among renewable energy resources is to use hydrogen gas (H_2) as a new energy carrier, once it is clean, highly abundant and non-toxic. Major attracting property of hydrogen is its natural compatibility with fuel cells. The higher efficiency of hydrogen (60%) compared to gasoline (22%) or diesel (45%) improves the efficiency for future energy use. However, a main obstacle impeding anticipated future hydrogen economy is the lack of safe, efficient and economical on-board hydrogen storage [3].

The popularity of porous polymers as new high surface area materials for applications in gas storage and separation is increasing very rapidly owing to their ease of preparation, low cost, chemical tenability and thermal stability [4]. On the other hand, after more than 20 years of research, no material yet possesses adsorbing properties required for applications. All porous structures that exist today are far from being optimal when considered as hydrogen sorbents for mobile applications [5]. However, in most of the cases reported, the hydrogen storage capability relies mostly on the intrinsic porosity of the polymers in the solid state and only very few examples of polymeric materials which have specific interactions with hydrogen have been reported [6]. It has also been shown that introduction of different chemical functions such as metal cations, imine, sulfonate, π-conjugated moieties and heteroatoms into the structure of porous materials could indeed increase significantly the hydrogen uptake [7].

Metallic particles and hydrides of some metals and non-metals can also form a class of materials for hydrogen storage characterized by their ability to store larger amounts of H_2 through the formation of chemical bonds. The incorporation of these nanoparticles into polymers or nanoporous structures can be an alternative to obtain materials with excellent hydrogen storage properties. The metal hydrides chosen for those applications usually provide low reactivity (high safety), lower pressure storage, reasonable gravimetric storage capacity, short storage time (which makes them highly efficient as an on-board storage vehicle but still work needs to be done in order to improve its kinetics), energy density, cycle life, operating temperature and long storage time [4]. Moreira et al. [8] developed composites of polianiline ($NaAlH_4$) and sodium alanate using titanium dioxide as doping agent. The polyaniline presented hydrogen sorption values around 1.0 wt.% after 13 hours of hydrogen sorption. The addition of sodium alanate increased this value in 40 %. Oliveira et al. [9] studied porous nanocomposites of sulfonated polyetherimide with carbon nanotubes and sodium alanate. Both nanoparticles were poorly dispersed in the polymer matrix and the nanocomposites could store 1.1 wt.% of hydrogen at 120ºC.

In this work composites of sulfonated polyetherimide with aluminum nanoparticles and different amounts of faujasite-type zeolite (used as catalyst) were produced. The influence of both nanoparticles on thermal, structural and hydrogen sorption behavior of these materials was evaluated.

EXPERIMENTAL

Materials

A polyetherimide (PEI), grade Ultem™ 1000, was supplied by Sabic Innovative Plastics. It has a density of 1.28 g/cm^3 and MFI of 9.0 g/10 min (6,6 kg/337 °C). The chemical modification of the polymer, in order to produce the sulfonated polyetherimide (PEIS), was done using an acetyl sulfate made of N-Methyl-2-pyrrolidinone (NMP), acetic anhydrid and sulfuric acid (volume ratio of 20:2:1). Aluminum nanoparticles (Al) were purchased from Nanostructured and Amorphous Materials Inc., with 99% of purity, density of 2.699 g/cm^3 (20 °C) and average particle size of 80 nm. A faujasite-type zeolite (ZEO), grade CBV780 was supplied by Zeolyst International and has a SiO_2/Al_2O_3 ratio of 80, surface area of 780 m^2/g and average particle size of 2 μm.

Sulfonation of Polyetherimide

A small round bottom flask equipped with a magnetic stirrer was charged with 6 g of PEI dissolved in 40 ml of NMP at 80 ºC. The acetyl sulfate was added to this solution gradually within 1 h and after that the reaction continued for another hour. At the end, the reaction product was precipitated with ethanol, washed three times with the same solvent, filtered and dried under vacuum at 80 ºC.

Preparation of PEIS/ZEO/Al Nanocomposites

The nanocomposites had to be prepared in two steps because the aluminum powder is highly reactive with the chemical modification environment of the PEI: (1) *in-situ* nanocomposites of PEIS and ZEO were prepared and then (2) dissolved in NMP and mixed with aluminum powder. The concentration of the solutions during the preparation of all nanocomposites was 0.7 g / ml. The content of aluminum was fixed at 30 wt.% in all samples and the zeolite contents were 2.5, 5 and 10 wt.%. Samples without aluminum (only with PEIS and ZEO) were also prepared.

Thermal Characterization

Differential scanning calorimetry (DSC) was done on a Q2000 equipment, from TA Instruments and thermograms were obtained from room temperature up to 250 °C under nitrogen flow and heating rate of 10 °C/min. Thermogravimetric analysis (TGA) were carried on a Q50 equipment, also from TA Instruments, under nitrogen atmosphere and heating rate of 20 °C/min from room temperature up to 800 °C.

Structural Characterization

Morphological analysis was carried out on a transmission electron microscope (TEM), Magelan, model 400L, with an operating voltage of 25kV.

Hydrogen Sorption Measurement

Kinetic measurements of hydrogen absorption (PCT) were collected in a volumetric device (Sievert's apparatus) built and designed in the Laboratory of Hydrogen in Metallic Materials (LHM) of the Federal University of São

Carlos, Brazil. Sorption measurements were performed at fixed temperature of 120 °C and hydrogen pressure of 32 bar for the following samples: PEIS/Al (70/30 wt.%) and PEIS/ZEO/Al (60/10/30 and 65/5/30 wt.%), using 100 mg of each one. Temperature and pressure used in hydrogen sorption was based on literature review [1-2,8].

RESULTS AND DISCUSSION

Figure 1 shows the DSC curves for polyetherimide before and after sulfonating procedure and for the nanocomposites with aluminum and zeolite.

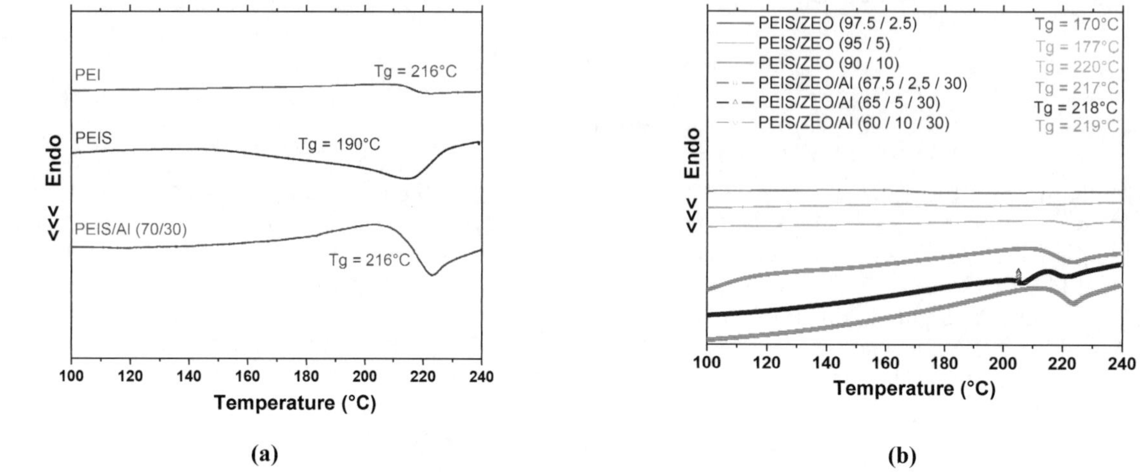

FIGURE 1. DSC curves for: **(a)** PEI, PEIS and PEIS/Al and **(b)** PEIS/ZEO nanocomposites.

From the results above, it is observed that the PEI has a glass transition temperature (Tg) of approximately 216 °C and the sulfonating reaction promoted a decrease to 190 °C, probably due to the introduction of sulfonic groups which increased the free volume between polymeric chains. In the presence of aluminum (PEIS/Al), this transition temperature returned to its initial value prior to the sulfonating procedure. Thus, it can be concluded that the presence of aluminum nanoparticles reduced the mobility of macromolecules [10]. On the other hand, the presence of 2.5 and 5 wt.% of zeolite had an opposite effect; there was a decrease in the glass transition temperature of the polymer. Thus, small amounts of zeolite (less than 10 wt.%) can increase the mobility of the sulfonated polymeric chains, but 10 wt.% had the same effect of aluminum nanoparticles. In all samples with aluminum and zeolite (PEIS/ZEO/Al), only a slightly increase in glass transition temperature of the sulfonated polymer was observed. However, since this variation in Tg is not significant in those cases, it can be stated that the addition of a small amount of zeolite (up to 10 wt.%t), with 30 wt.% of aluminum, does not change the thermal behavior of PEIS.

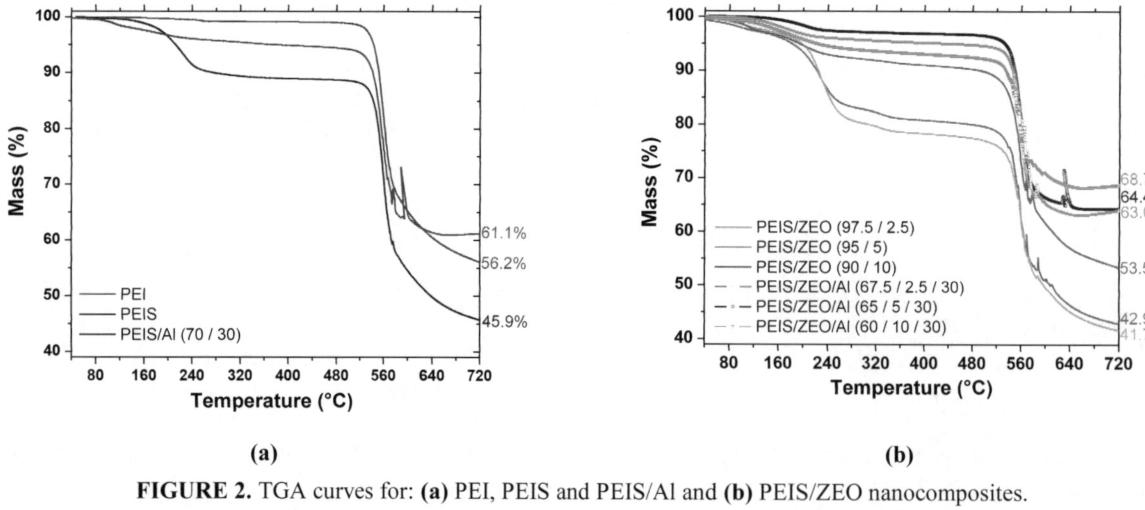

FIGURE 2. TGA curves for: **(a)** PEI, PEIS and PEIS/Al and **(b)** PEIS/ZEO nanocomposites.

In Figure 2 the thermal stability of the polymer matrix and the nanocomposites with aluminum and zeolite can be observed. The degradation of all materials occurred in two stages: (1) the degradation of the sulfonic chains and (2) around 440 °C, the degradation of the PEI main chains. The first stage also contains a small loss of weight due to the presence of zeolite, which contains moisture within its pores and channels. Therefore, decomposition and release of gas molecules begins after the 100 °C [10]. It can be also observed in Figure 2 that the residues at the end of analysis increases with the zeolite content, since its structure is thermally stable up to higher temperatures. The presence of aluminum made the thermal behavior of the nanocomposites more stable at high temperatures, but did not change significantly their initial temperatures of decomposition.

Figure 3 shows the morphology of the PEIS/ZEO/Al nanocomposites obtained by TEM analysis. The aluminum nanoparticles are spheres with large size distribution and average diameter between 30 and 500 nm. The presence of zeolite was difficult to observe due to its wide distribution of particle size and small amount added to the mixtures. It is also possible to observe that the aluminum may be interacting with the zeolite, but it tends to form many small and medium-sized agglomerates within the polymer matrix [10]. The nanocomposite with 10 wt.% of zeolite has a larger number of agglomerates, but with smaller size, when compared to the sample with 5 wt.%. These agglomerates may be contributing negatively to hydrogen sorption of the samples, due to a smaller contact area with the gas during sorption conditions. Thus, the average size, the dispersion and the amount of agglomerates in each sample are important parameters to be determined once they will contribute to the behavior of hydrogen sorption.

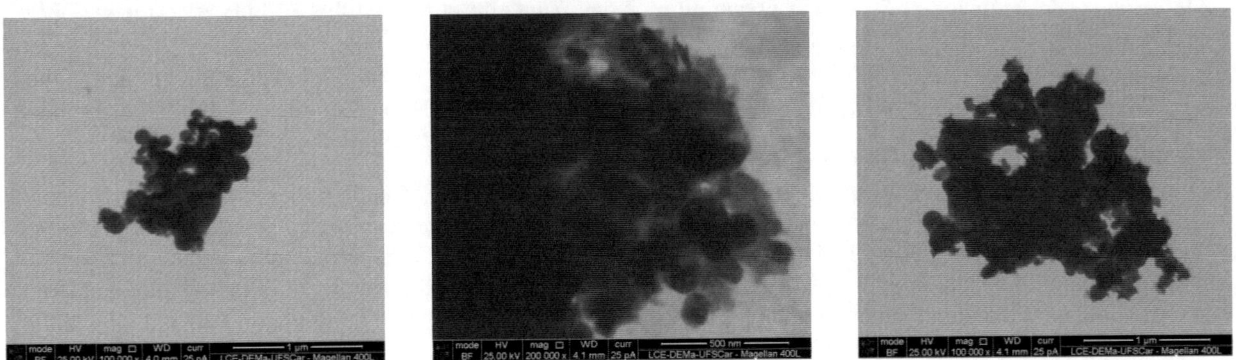

(a) (b) (c)

FIGURE 3. TEM analysis for PEIS/ZEO/Al nanocomposites with: **(a)** 2.5, **(b)** 5 and **(b)** 10 wt.% of zeolite.

Figure 4 shows the PCT curves for the samples analysed after 12 hours measurements.

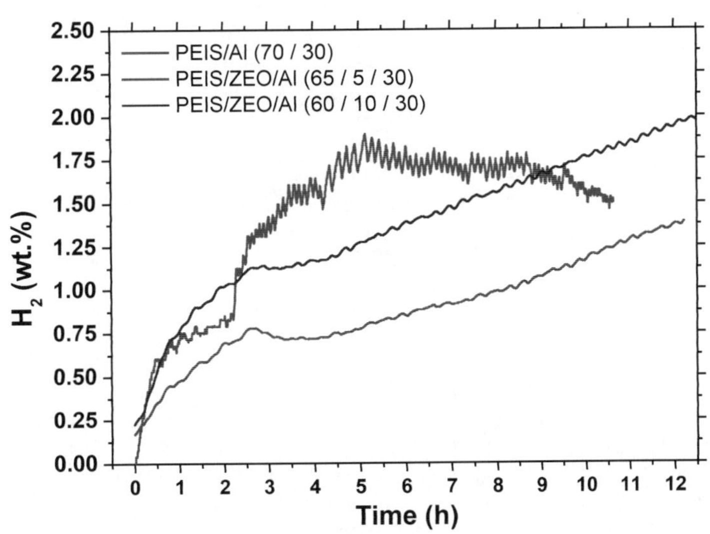

FIGURE 4. PCT curves for PEIS/Al nanocomposites.

For two of the nanocomposites analyzed (the one with 10 wt.% of zeolite and the one only with aluminum) it can be observed that hydrogen sorption quickly increased in the first three hours until reach a maximum value. In the next few minutes the absorption slightly decreased and then, after 4 hours of experiment, it started to increase again with a slower rate. After 12 hours, the sample with 10 wt.% of zeolite could absorb 2 wt.% of hydrogen; near 50 % more than the sample without this porous silicate [10]. On the other hand, it can be observed that the presence of a smaller amount of zeolite (5 wt.%) made the hydrogen sorption much more instable. After 5 hours of measurement this sample absorbed almost 2 wt.% of hydrogen and then the amount of gas started to decrease and reached 1.5 wt.% at the end of the analysis. The presence of larger agglomerates of aluminum and zeolite in this sample, compared to the nanocomposite with 10 wt.% of zeolite, had probably caused this unexpected behavior. These results have evidenced that those hybrid materials can be used for hydrogen storage, but the dispersion of the metallic and inorganic particles is still a problem to be solved in the future development.

CONCLUSION

Polymer nanocomposites were prepared from the mixture of sulfonated polyetherimide with aluminum and zeolite nanoparticles. These hybrid materials were obtained in order to produce new materials to reach greater hydrogen storage capacity for mobile applications.

Zeolite and aluminum added during the preparation of the nanocomposites had few influence on the sulfonated polymer structure. However, they did not change its thermal properties, such as glass transition temperature and the onset temperature of degradation. By TEM analysis it was observed that the zeolite and aluminum nanoparticles were located inside the sulfonated polyetherimide and also on its porous surface, but they were not well dispersed and distributed. The amount of zeolite modified the average size and size distribution of the agglomerates.

Hydrogen sorption tests were carried out and it was observed a significant increase in the amount of hydrogen in the nanocomposite with 10 wt.% of zeolite relative to the one without zeolite. The larger agglomerates in the sample with aluminum and 5 wt.% of zeolite made the hydrogen sorption much more instable. Dehydrogenation tests of these materials will be performed in the future, as soon as better dispersion of the nanoparticles will be reached.

Obtaining polymeric materials for hydrogen storage is important for the scientific and technological development and the polymer matrix and the nanofillers chosen in this work have a great potential to obtain new hybrid materials with ability to store larger amounts of hydrogen.

ACKNOWLEDGMENTS

This work was supported by CNPq (Brazilian Counsel of Technological and Scientific Development, 159187/2014-1) and FAPESP (São Paulo Research Foundation, 2013/23586-0). The authors thank Prof. Dr. Daniel Leiva and Prof. Dr. Tomaz Ishikawa (Federal University of São Carlos, Department of Materials Engineering, Laboratory of Hydrogen in Metallic Materials, Brazil) for their help with the hydrogen sorption measurements.

REFERENCES

1. S. Niaz, T. Manzoor and A. H. Pandith, *Renewable and Sustainable Energy Reviews* (2015), pp. 457-469.
2. Y. Jia, C. Sun, S. Shen, J. Zou, S. S. Mao and X. Yao, *Renewable and Sustainable Energy Reviews* (2015), pp. 289-303.
3. S. S. Muir and X. D. Yao, *International Journal of of Hydrogen Energy* (2011), pp. 5983-5997.
4. E. G. Thibault, D. Cossement, R. G. Nicolas, N. Masoumifard, P. Bénard, F. Kleitz, R. Chahine and J. F. Morin, *Microporous and Mesoporous Materials* (2014), pp. 182-189.
5. R. Roszak, L. Firlej, S. Roszak, P. Pfeifer and B. Kuchta, *Colloids and Surfaces A: Physicochemical Engineering Aspects* (2016), pp. 69-76.
6. Y. C. Zao, D. Zhou, Q. Chen, X. J. Zhang, N. Bian, A. D. Qi and B. H. Han, *Macromolecules* (2011), pp. 6382-6388.
7. A. Li, R. F. Lu, Y. Wang, K. L. Han and W. Q. Deng, *Angewandte Chemie International Edition* (2010), pp. 3330-3333.
8. B.R. Moreira, "Development of Polyaniline / TiO_2 and Polyaniline / $NaAlH_4$ Nanocomposites for Hydrogen Storage," M.Sc. thesis, Universidade Federal de São Carlos, 2013.
9. A. D. Oliveira, C. A. G. Beatrice, F. R. Passador and L. A. Pessan, "Polyetherimide-based Nanocomposite Materials for Hydrogen Storage" in *Polymer Processing Society Conference 2015*, edited by C. Holzer, PPS2015 Conference Proceedings Nan-270, Polymer Processing Society, Graz, Austria, 2016.
10. C. A. G. Beatrice, A. D. Oliveira, F. R. Passador and L. A. Pessan, "Sulfonated Poly(ether imide) / Aluminum Nanocomposites for Hydrogen Storage" in *Polymer Processing Society Conference 2015*, edited by C. Holzer, PPS2015 Conference Proceedings Nan-243, Polymer Processing Society, Graz, Austria, 2016.

Influence Of Matrix Crystallinity On Electrical Percolation Of Multiwalled Carbon Nanotubes In Polypropylene

Marén Gültner, Liane Häußler, Petra Pötschke*

Leibniz-Institut für Polymerforschung Dresden e.V., Hohe Straße 6, 01069 Dresden, Germany

poe@ipfdd.de

Abstract. The effect of matrix crystallinity on the electrical percolation threshold of melt mixed polymer composites containing 0.5 -5.0 wt% of multiwalled carbon nanotubes (MWCNTs) was investigated on the example of polypropylene (PP). Two different polypropylenes having the same low molecular weight and melt viscosity but different crystallinitiy, namely PP amorphous (PPa, 17 % crystallinity) and PP isotactic (PPi, 52 % crystallinity), were compared. In addition an industrial grade PP (49% crystallinity) was applied. The electrical percolation was significantly lower in case of PPi (0.5 wt%) than for PPa (1 wt%) and the commercial PP type HP500N (between 1 and 2 wt%). The comparison of PPa and PPi indicates that a selective localization of CNTs in either the amorphous or the crystalline PP phase can be assumed. The electrical volume resistivity of PPa at 3 wt% loading is about 1E3 Ohm·cm which equals to 1.7 wt% CNT addition in PPi. Using the measured crystallinities, calculations of local CNT contents in the amorphous phases support the assumption of selective localization within this PP phase.

Keywords: conductive polymer composite, polypropylene, carbon nanotube, melt mixing, crystallization
PACS: 82.35.Np, 82.35.Lr, 72.80.Tm, 64.70.dg, 83.80.-k

INTRODUCTION

In polymer composites containing carbon nanotubes (CNTs) the influence the matrix crystallinity on electrical percolation and conductivity is under intensive and controversial discussion. In addition, nanotubes may influence the nucleation in a semi-crystalline matrix and by that influence the matrix crystallinity within the composite. Whereas some authors describe the formation of crystalline layers around the nanotubes, which hinders electrical percolation, others report lower percolation threshold in partially crystalline polymers due to the effect of selective localization of fillers inside the amorphous phase. However, a serious study on that topic is difficult, as many other factors besides matrix crystallinity influence the electrical percolation threshold, which have to set constant to illuminate the effect of crystallinity. This paper should contribute to this discussion focusing on polypropylene as matrix material, a polymer with low polarity which is accepted to show relatively low compatibility towards carbon nanotubes.

In an intensive study Jeon et al. [1] used isotactic PP and different ethylene-propylene copolymers (EPC) to evaluate the influence of PP crystallinity on SWCNT dispersion and electrical conductivity of the composites, whereby the samples were solution mixed and then compression moulded. Thereby, the content of HiPCo-SWCNTs was varied between 0.15 and 1 wt%. At the lowest content of SWCNTs investigated in this study, volume resistivity values in the range 1E9 till 1E7 Ohm·cm were achieved, whereby the matrix with the highest crystallinity (PP, 70%) resulted in the lowest resistivity values. The electrical percolation threshold varied from 0.25 wt% for EPC with 10 % crystallinity towards a value of 0.07 wt% for EPC with 45 and 70 % crystallinity, obtained by fitting the conductivity data using the percolation scaling law. At 1 wt% loading, the sample using PP with 70 % crystallinity showed a resistivity value of ca. 1E3 Ohm cm. In this study, PP and the different copolymers may have slightly different surface energy values and thus different interfacial tensions to SWCNTs. In addition, during compression moulding in the melt, different matrix viscosities may apply which may result in different conditions for secondary nanotube agglomeration, a process which also influences electrical percolation.

Alig et al. found a decrease in the conductivity during the crystallization process of a PP-2 wt% MWCNT composite after melt mixing within the extruder and explained this with the reduction of the amorphous phase on expense of the crystalline phase and/or by crystalline regions in the contact region between tubes [2].

In a study on LDPE and HDPE filled with iron particles Li et al. describe that during crystallization fillers are squeezed out from the forming compact crystallites and stay located within the amorphous phase, where they enrich. Thus, increased crystallinity reduces electrical percolation threshold [3].

Zhang et al. used a beta nucleation agent in PP/MWCNT composites to tune the nanotube network structure [4]. With increasing crystallinity a more homogeneous and compact nanotube network was formed resulting in increased conductivity. However, it was found that if a critical size of crystallites is reached, the nanotubes can be isolated and conductivity decreases.

Another interesting discussion, introduced by Jeon et al. [5] in HDPE/SWCNT composites, describes an improved nanotube dispersion due to nucleation of PE crystallites at the surface of the nanotubes by fast cooling.

This work was aimed to contribute to the discussion on the influence of matrix viscosity on nanotube percolation by using melt mixing of two PP materials with comparable low molecular weight but different crystallinity to produce composites with MWCNTs. On these composites, compared to one based on a commercial, higher viscous PP, a direct comparison of the effect on crystallinity on electrical percolation should be done.

EXPERIMENTS AND CHARACTERIZATIONS

Two well defined products from ALDRICH, named as "amorphous" (PPa) and "isotactic" PP (PPi) were used. The ALDRICH PPs had according to the provider comparable molecular weights of 14,000 and 12,000 g/mol. For comparison, an industrial PP (HP500N, Basell) was applied which shows much higher molecular weight and melt viscosity. Multiwalled CNTs Baytubes® of the type C70P (Bayer MaterialScience) were added in concentrations between 0.5 and 5.0 wt%.

A conical twin-screw microcompounder (DSM Xplore 15) with a capacity of 15 ccm was used for melt mixing. Following an optimization of melt mixing conditions towards best dispersion by varying screw speed and mixing time for Moplen HP500N, a screw speed of 250 rpm and a mixing time of 5 min at 180 °C were selected. Afterwards, the composites were pressed at 180 °C under a pressure of 50 kN into circular plates with a diameter of 60 mm and a thickness of 0.3 mm using a press PW 40 EH (Paul Otto Weber GmbH, Germany). Rectangular samples were manually cut from pressed circular plates and used for the electrical resistivity measurements.

Melt rheology investigations were performed using the ARES oscillatory rheometer (Rheometric Scientific Inc., Piscataway, USA) in nitrogen atmosphere using the plate-plate geometry with ca. 1 mm gap applying 1% strain. Frequency sweeps were performed from 100 to 0.05 rad/s with increasing and decreasing frequency, whereby the second sweep was used for interpretation.

DSC investigation was carried out using a DSC Q 2000 from TA Instruments in the temperature range of -80 °C to 210 °C at a scan rate of 10 K/min in nitrogen atmosphere. A heating-cooling-heating cycle was performed, whereby the matrix crystallinity was calculated from the first heating run using a value of ΔH =207 J/g for 100 % crystalline PP [6].

The filler dispersion was analyzed by optical microscopy (OM, Olympus BH2 transmission light microscope equipped with DP 71 digital camera, booth Olympus Germany) on thin cuts with a thickness of 10 μm. Strands of the composites were cut perpendicular to the extrusion direction by a Leica RM 2265 Microtome (Leica Microsystems) at room temperature.

For conductivity characterization, a 2-electrode test fixture combined with a Keithley electrometer E6517A was used for high resistance samples (above 10E7 Ohm), whereas a 4-electrode test fixture combined with Keithley multimeter DMM 2000 for samples with lower resistance. The reported electrical conductivities are the averaged values of 4 measurements for each sample. Error bars are within the point sizes and not shown.

RESULTS AND DISCUSSION

The DSC investigations (Fig. 1 and Table 1) revealed, that the "amorphous" PPa is not completely amorphous but has a low crystallinity of 17 %. In comparison, the "isotactic" PPi shows a crystallinity of 52 %. The industrial PP type has a comparable crystallinity of 49 %. Interestingly, after the melt mixing the composites based on PPi and PPa with 1 wt% nanotube loading exhibit lower crystallinities than their corresponding polymer matrices, whereas for the industrial PP an unchanged value (compared to the melt processed material) could be observed. In all cases, the onset and maximum crystallization temperatures increased by about 5 - 6 K.

The melt rheological investigations (Figure 2) illustrate that both low molecular PPs, namely PPa and PPi, have comparable melt viscosities of about 0.7 Pa·s at the processing temperature of 180 °C. They show nearly no dependence of melt viscosity on measurement frequency. In contrast the industrial PP has much higher melt viscosity in the range above 1000 Pa·s and indicates shear thinning behavior by values which are decreasing with frequency (ca. 700 Pa·s at 100 rad/s). However, the condition to study PP materials with the same melt viscosity, resulting in similar shear stresses during melt mixing, but different polymer crystallinities, is fulfilled for the two Aldrich types.

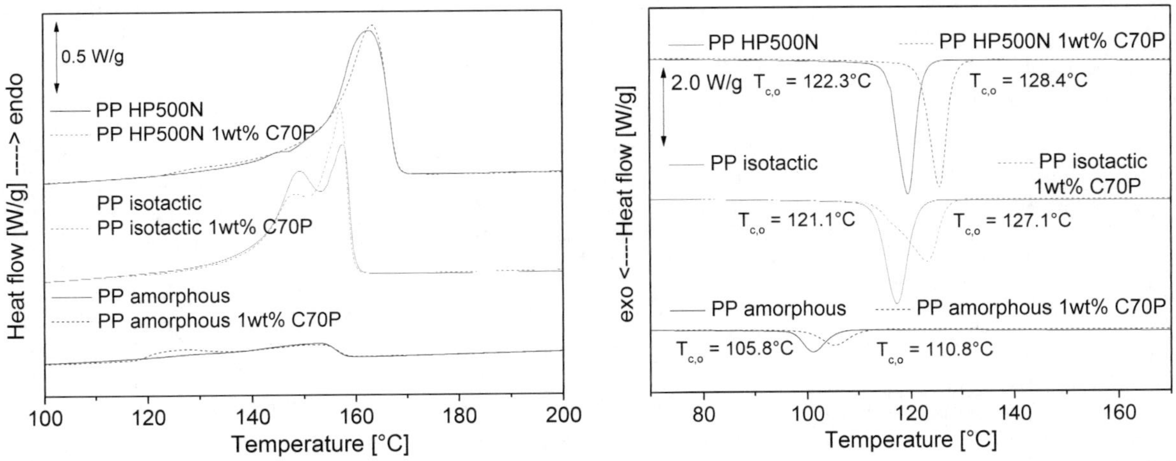

FIGURE 1. DSC curves of 1st heating (left) and crystallization (right) of different polypropylenes and composites containing 1 wt% Baytubes® C70P

TABLE 1. Measured values of enthalpy ΔH and crystallinity α (first heat) for the different PP types as received and after melt processing (180 °C,15 min, 250 rpm) and composites with 1 wt% Baytubes®C70P (ΔH^0 =207 J/g [8])

	PP Moplen HP500N		PP Sigma Aldrich isotactic		PP Sigma Aldrich amorphous	
State of processing	**ΔH [J/g_{PP}]**	**α [%]**	**ΔH [J/g_{PP}]**	**α [%]**	**ΔH [J/g_{PP}]**	**α [%]**
Granules	100.7	49	108.2	52	34.9	17
Extruded Strand DSM	109.7	53	103.0	50	32.1	16
PP C70P 1 wt% DSM	107.9	53	81.3	40	26.5	13

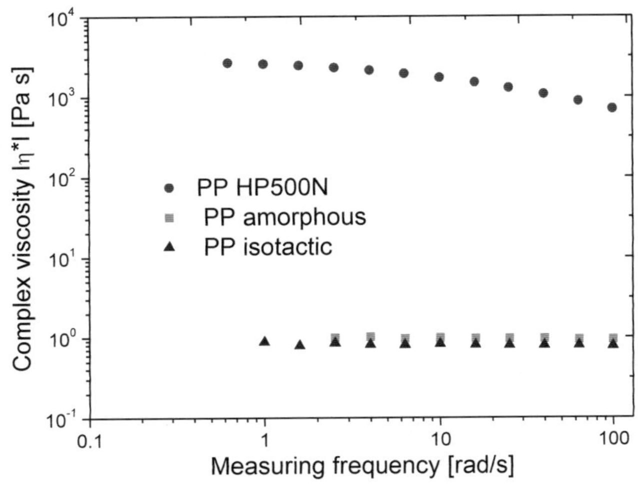

FIGURE 2. Complex melt viscosity |η*| vs. frequency at the processing temperature of 180 °C for the PP materials

The dispersion of the nanotubes was assessed using transmission light microscopy and is shown in Figure 3 for composites containing 1 wt% C70P. Among the samples, the dispersion is best in the industrial high molecular weight PP based composites. In this case, higher shear stresses act during the melt mixing within the microcompounder which favor the dispersion steps of rupture and erosion [2]. Thus, only some and small remaining nanotube agglomerates can be seen. The preparation of thin cuts was more difficult for the composites based on lower molecular ALDRICH types of PP. The cuts appear less homogeneous and larger agglomerates can be seen. This is due the much lower viscosity which facilitates the dispersion step of matrix infiltration into the primary agglomerates but reduces shear stresses which are responsible to individualize these agglomerates into separated nanotubes by rupture and erosion processes. However, there are no significant differences between PPa and PPi so that these composites appear to be comparable also from the point of nanotube dispersion.

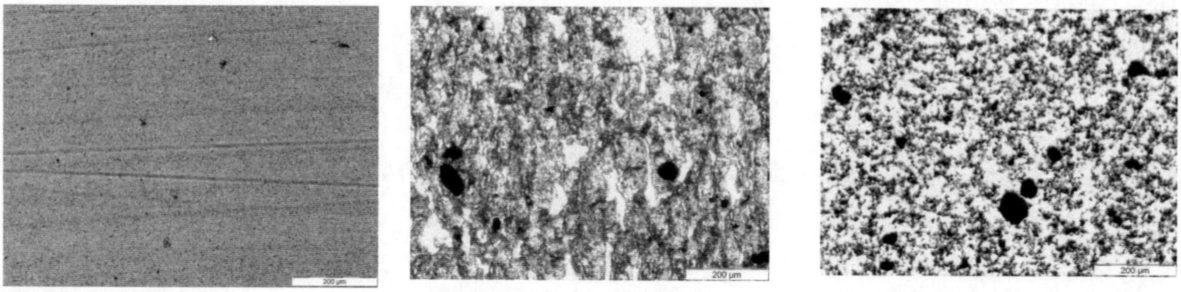

PP Moplen HP500N **PP Sigma Aldrich isotactic** **PP Sigma Aldrich amorphous**

FIGURE 3. Dispersion of MWCNTs in PP based composites with 1 wt% Baytubes® C70P loading as assessed by transmission light microscopy

When now comparing the percolation curves of the nanotubes in the different PPs, as shown in Fig. 4, it is clearly seen that the low molecular isotactic PP results in the lowest percolation threshold which is below 0.5 wt%. The amorphous PP shows percolation of the nanotubes at around 1 wt% and in the industrial PP the nanotubes percolate between 1 and 2 wt%. Even at higher loadings PPi as matrix results in lower values of the electrical volume resistivity than in the other matrices. The highest percolation concentration despite best dispersion in case of the industrial PP may be explained by the possibly more severe shortening of the nanotubes due to the higher matrix viscosity. Shortening of CNTs was shown to be a crucial factor during melt mixing which depends on matrix viscosity and seems to be a main factor in influencing percolation concentration [7].

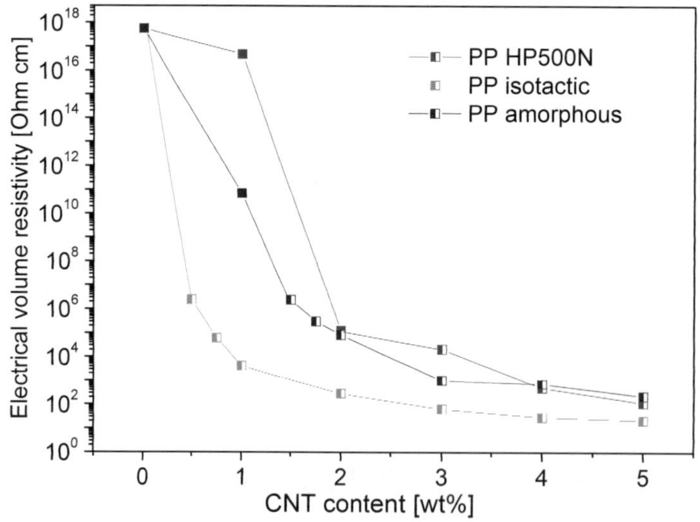

FIGURE 4. Electrical volume resistivity vs. content of Baytubes® C70P nanotubes for different PP matrices

The different percolation concentrations in PPi and PPa may be explained by selective localization of the nanotubes in one of the PP phases - either within the crystalline or the amorphous. The electrical volume resistivity of the composite based on PPa with 17 % crystallinity is about 1 E3 Ohm·cm at 3 wt% CNT loading. The same resistivity is achieved in the system based on PPi with 52 % crystallinity at about 1.7 wt% nanotube loading. If one assumes that in the PPa all CNTs (3 wt%) are localized within the 83 % amorphous phase, then there is a CNT density of 3/83= 0.036. In the isotactic PP the amorphous phase content is 48 %. Assuming the same CNT density a hypothetical CNT content of ca. 1.7 wt% can be calculated which coincidences with the experimental finding for PPi as matrix. Similar calculations can be done when regarding other electrical volume resistivities within the percolation range, e.g. 1 E5 Ohm·cm or 2 E6 Ohm·cm, for which even slightly lower experimental CNT contents for PPi as compared to such calculated ones can be found. This strongly supports the idea that nanotubes localizes in PP within the amorphous phase and are pushed out from the crystalline phase during its crystallization. Thus, PP materials with a lower amorphous fraction have at the same overall CNT content a higher local CNT density within the amorphous phase. As this phase is continuous around the crystallites, percolation occurs at lower overall loadings. These consideration are approximations and neglect the finding of reduced PP crystallinity (1st heat) after nanotube incorporation and the possibility that some of the nanotubes may be covered by crystalline layers, e.g. epitaxial growth on their surfaces, which additionally could insulate them from each other or enhance contact resistances within their percolating networks.

SUMMARY AND CONCLUSIONS

The results on electrical percolation of melt mixed composites based on two PP matrices with different crystallinity, but comparable molecular weight and melt viscosity, indicate that the occurence of crystallinity reduces the electrical percolation concentration of MWCNT. PP with higher crystallinity of 52 % resulted in about half the percolation treshold of PP with 17 % crystallinity. This supports the assumption that in PP, a polymer with generally accepted low compatibility with nanotubes, a localization within the amorphous phase takes place which results in a locally higher nanotube concentration in that continuous phase. Thus, to achieve low electrical percolation threshold, the use of PPs with higher crystallinity seems to be favorable. This is in accordance with findings of Abbasi et al. who proposed that the crystalline structure within PP leads to lower percolation threshold and higher conductivity, when comparing to amorphous polycarbonate composites [8].

ACKNOWLEDGMENTS

We thank Bayer MaterialScience (now Covestro AG) for providing the Baytube® C70P and financial support of this work.

REFERENCES

1 K. Jeon K, S. Warnock, C. Ruiz-Orta, A. Kismarahardja, J. Brooks and R.G. Alamo, *Journal of Polymer Science Part B-Polymer Physics* **48**(19), 2084-2096 (2010).
2 I. Alig, D. Lellinger, S.M. Dudkin and P. Pötschke, *Polymer* **48**(4),1020-1029 (2007).
3 Y.J. Li, M. Xu, J.Q. Feng, X.L. Cao, Y.F. Yu and Z.M. Dang, *Journal of Applied Polymer Science* **106**(5), 3359-3365 (2007).
4 L. Zhang, B. Hu and C.Z. Li *Acta Polymerica Sinica* **12**,1374-1381 (2011).
5 K. Jeon, L. Lumata, T. Tokumoto, E. Steven, J. Brooks and R.G. Alamo, *Polymer* **48**(16),4751-4764 (2007).
6 G.W. Ehrenstein, G. Riedel and P. Trawiel . Praxis der Thermischen Analyse von Kunststoffen. München: Carl Hanser Verlag, 1998.
7 R. Socher, B. Krause, M. T. Müller, R. Boldt and P. Pötschke, *Polymer* **53,** 495-504 (2012).
8 S. Abbasi, A. Derdouri and P. J. Carreau, Effect of crystalline structure on morphology and properties of MWCNT nanocomposites, Proceedings of the 67th Annual Technical Conference of the Society of Plastics Engineers (Antec 2009), 2009-06.

Largely improved toughness of immiscible blend achieved by adding chemically modified graphene oxide

Huang Wen-bin [a], Zhang Ting-ting [a], Mao Han-jun [a], Zhang Nan [a], Huang Ting [a], Yang Jing-hui [a*], Wang Yong [a]

[a]*Southwest Jiaotong University Key Laboratory of Advanced Technologies of Materials (Ministry of Education), School of Materials Science and Engineering, Southwest Jiaotong Universtiy, Erhuan Road, North I, No 111, Chengdu, Sichuan 610031, China*

Email: yangjinghui_84@163.com

Abstract. In this work, chemically modified graphene oxide (m-GO) was successfully prepared by emulsion polymerization of styrene, leading to grafting polystyrene (PS) onto graphene oxide (GO). Different contents of chemically modified graphene oxide (m-GO) were incorporated into immiscible polyamide 6/acrylonitrile-butadiene-styrene (PA6/ABS) blend. SEM pictures revealed the interface localization of m-GO in PA6/ABS. It was suggested that the m-GO located at interface of immiscible blends PA6/ABS and enhanced the interfacial adhesion via the dual interactions with two components, in which one was the electronic stacking interaction between grafted PS on m-GO and ABS while the other one was the interaction between polar groups on m-GO and PA6. Finally, a great improvement on ductility of polymer blends was achieved, in which the elongation at break increased by nearly 400% and the impact strength increased by nearly 200% with only 0.5 wt% m-GO.,

Keywords: chemically modified graphene oxide, interfacial adhesion, toughening.
PACS: 61.41. +e

INTRODUCTION

Immiscible blends of PA6 and ABS are of commercial significance in industrial products. However, inherent immiscibility between ABS and PA6 limits their application in developing new materials, especially interfacial adhesion is still a weak point. As we mentioned above, graphene and its derivatives are good candidates for improving interactions between two polymers due to the "amphiphilic" effects [1, 2] originated from possible interaction between chemically modified graphene and polymer pairs. In this word, polystyrene (PS) grafted reduced graphene oxide is chemically designed to strengthen interfacial adhesion of the immiscible blends PA6/ABS. It is expected that in the blend composites, PS chains on the edge is proposed to strengthen the interaction between chemically modified graphene oxide(m-GO) and ABS; and PA6 could form hydrogen bonding with oxidized groups on the m-GO. It is demonstrated that upon only 0.5 wt% m-GO was incorporated into PA6/ABS, the elongation at break as well as impact strength shows great improvement. The underlying toughening mechanisms are carefully discussed in a viewpoint of interfacial adhesion induced by chemically modified reduced graphene oxide.

FARBIRCATION

Fabrication of chemically modified graphene oxide

About 500 mg GO were dispersed in 200 ml deionized water in a 500 ml round-bottom flask,0.5 g SDS and 30 ml styrene monomer were added into the flask, followed by 30 min of ultrasonic irradiation. Afterwards, 0.5 g $K_2S_2O_8$ was added into the mixture and the reaction mixture was refluxed at 80 °C for 5 h under a nitrogen atmosphere for polymerization. Then, 10 ml hydrazine hydrate was added into the PS-g-GO dispersion and the reaction mixture was again refluxed at 100 °C for 2 h. It should be mentioned that in order to remove the dissolved oxygen, the solution was deoxygenated with oxygen-free nitrogen for 30 min by bubbling nitrogen gas. During this reaction, PS-g-GO was partially reduced by hydrazine to PS grafted reduced graphene oxide (rGO), which in this paper is noted as m-GO. The mixture was cooled to room temperature, and washed with *N, N*-dimethylformamide (DMF) and deionized water to remove the excess surfactants and residual monomer from the reaction mixture. This procedure was repeated more than three times. The purified product was kept in the DMF solution for the next step.

Fabrication of PA6/ABS/m-GO Blend composites

The PA6/ABS/m-GO blend composites were fabricated through a two-step procedure: solution blending of ABS with m-GO, and subsequent melt compounding of PA6 , ABS and the ABS/ m-GO master batch. In the first step, ABS/m-GO master batch with 5 wt% m-GO was prepared via solution blending in DMF, and then the mixture was dried in vacuum oven at 60 °C until the weight of resultant black solid kept constant. In the second stage, the obtained master batch, PA6 and ABS were melt-blended on a twin-screw extruder SHJ-20 (Nanjing Ruiya, China) at a screw speed of 150 rpm, and a temperature profile of 170, 190, 210, 230, 250, 250, 245°C from the feeding zone to die. The extrudate was water-cooled and chopped into pellets, and then dried in a vacuum oven for 24 h at 50 °C. These blends were then injection-molded to obtain the dumbbell specimens and the rectangular specimens using the Haake minilab injection-molding machine (HAAKE MiniJet, Germany) at the melt temperature of 240 °C with the injection pressure of 700 bar. In the blend composites, the weight ratio between PA6 and ABS was maintained at 50:50, and the content of m-GO was varied from 0 to 1 wt%. The sample notation was defined as P5A5Gx, where x represented the content of m-GO in the blend composites. For example, P5A5G0.5 represented that m-GO content was 0.5 wt%. For comparison, the binary PA6/ABS (50/50) blend, which was named as P5A5, was also prepared through the completely same processing procedures as was done for the blend composites

RESULTS AND DISCUSSION

Characterization of chemically modified graphene oxide

The TEM images (FIGURE 1a) of the m-GO distinctly reveal that synthesized polystyrene with a width of several couples of nanometers are attached along the edges of the stacked nanosheets with thickness of several nanometers. The similar morphology can be found with a similar ellipsoidal structure in PS grafted graphene [3]. FTIR spectra analysis was performed to confirm the chemical components of the m-GO. FIGURE 1b illustrates the FTIR spectra of the rGO, and m-GO. As for the m-GO, the newly emerged peaks located at 1580,1490,1446,750,530 cm^{-1} correspond to absorptions of the benzene ring of PS segments, while the peak at 2922 cm^{-1} arises from the attachment of additional methylene groups. It is obviously seen from the FTIR spectra that the PS chains are attached on the rGO. The underlying mechanism for functionalization is derived from the in-situ emulsion polymerization , in which the surfactant micelles were firstly adsorbed onto the edge of GO nanosheets which were full of oxygen-containing groups; Styrene monomers were attached on the hydrophobic end of surfactant, and then initiator was applied to initiate the emulsion polymerization [4].

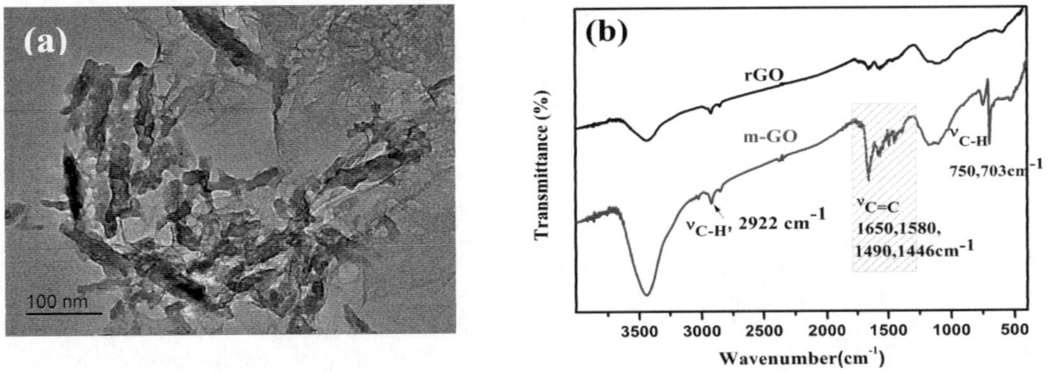

FIGURE 1. (a) TEM photo of m-GO ; (b) FTIR spectra of rGO and m-GO.

Mechanical property of ternary composites

FIGURE 2a shows the representative strain-stress behaviors of the various blended composites. It can be found that all the samples exhibit similar deformation behavior. Elongation at break of the blend without m-GO is only 11%; while it is obvious that incorporation of m-GO has improved the ductility. When 0.3 and 0.5 wt% m-GO were incorporated into blends, the elongation at break increased to 25% and 42% respectively indicating that these blends have a greatly improved ductility. However, a decrease of elongation at break can be observed possibly due to the aggregation of 1 wt% m-GO. On the other hand, seen from FIGURE 2b, it is found that with increasing the concentration of m-GO, the impact strength increases from 9.0 kJ/m^2 to 21 kJ/m^2 when 0.5 wt% m-GO are incorporated into polymer blends. Similar to the elongation at break, when 1 wt% m-GO are incorporated into polymer blends, aggregates of m-GO cause the decrease of impact strength. The increase on impact strength reveals the presence of m-GO in immiscible blend PA6/ABS is also benefit for the energy adsorption during impact strength, and a further study concerning about effects of m-GO on the crack growth during the impact testing are needed in a future work.

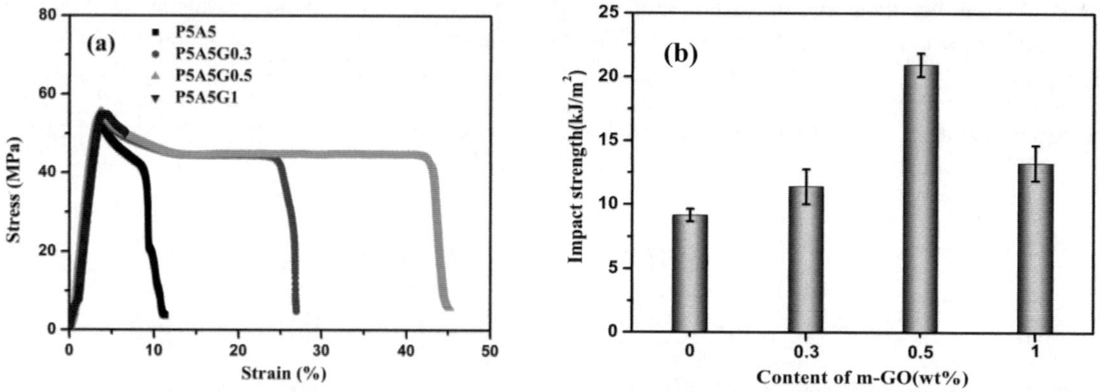

FIGURE 2. (a) Typical strain-stress curves of samples and (b) variation of impact strength of PA6/ABS/PS-GO blend composites versus PS-GO content.

Toughening Mechanism

In order to further explore the toughness mechanism of m-GO in PA6/ABS, the tensile-fractured surfaces along the tensile direction were observed by SEM, showing that finer morphology without severe detaching is observed in PA6/ABS blend with PS-GO,whereas dispersed ABS particles have been pulled out in PA6/ABS blend from the matrix. This obervation indicates stronger interfacial interaction achieved by addition of m-GO. Further the interfaces of the immiscible PA6/ABS blends with and without m-GO have been carefully observed as shown in FIGURE 3, the interfaces for the blends with m-GO (FIGURE 3b) are coarser compared with that of the blank PA6/ABS blend (FIGURE 3a) and some folds of sheets-like structure can be found as shown by arrows. On the other hand, it should be noted that the m-GO with two dimensional planes exhibits high interfacial stability, allowing it to more readily cover the interface [5].

Based on the observations, the toughening mechanism of m-GO in the immiscible PA6/ABS blend can be further understood as indicated below. With the introduction of m-GO, these nanosheets functionalized PS chains can adsorb some PS section of ABS on their basal planes due to the π- π interaction between PS section in ABS and PS chains grafted on GO[6]. On the other hand, the unremoved groups containing oxygen on GO basal planes have strong interaction with PA6 which is originated from the hydrogen bonding between amide group of PA6 and the hydroxyl on rGO [7]. On the other hand, as we know, the fillers in immiscible blends would like to migrate to component with lower viscosity or better interaction with filler. In this case, m-GO are inclined to migrate from ABS/m-GO masterbatch to the PA6 due to the lower viscosity of PA6 and the hydrogen bonding between m-GO and PA6. As we know, due to the π- π interaction between PS section in ABS and PS chains grafted on GO, partial ABS are adsorbed on m-GO in ABS/m-GO masterbatch. When ABS/m-GO blended with PA6, m-GO would like to migrate from ABS to PA6 with lower viscosity. In this case, partial ABS adsorbed on m-GO could solve or diffuse into PA6 domain as accompanied with the migration of m-GO from ABS to PA6; and the phase coarsening may be allowed. It is worthy to be noted that partial m-GO may migrate to the interface and weave across the interface, realizing the formation of entanglements and reducing unfavorable contacts between PA6 and ABS. In this case, the interfacial adhesion was intensified, leading to the great improvement of ductility. On the other hand, in a viewpoint of intrinsic structure of graphene, the wrinkles on m-GO can enhance the interfacial adhesion between polymer and m-GO; and the deflection of crack induced by graphene may also play a positive role on toughening which needs a further discussion in future needs work.

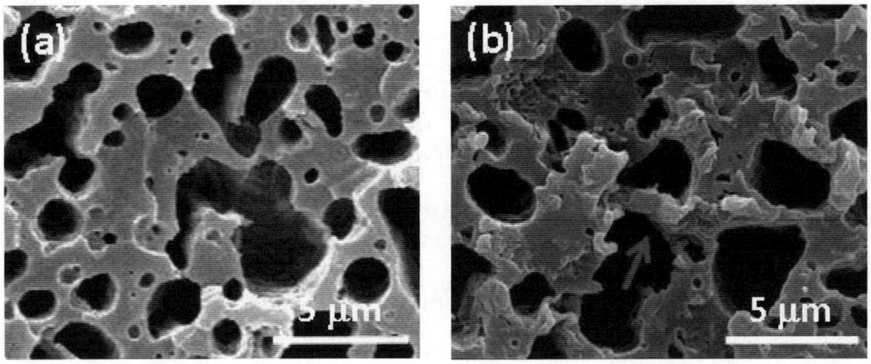

FIGURE 3. Interface of (a) PA6/ABS blend and (b) PA6/ABS/PS-GO blend composite with 0.5 wt% PS-GO.

ACKNOWLEDGMENTS

Authors express their sincere thanks to the National Natural Science Foundation of China (51473137, 51203129 and 50973090) for supporting this work.

REFERENCES

1. H. He, J. Klinowski and M. Forster, Chem. Phys. Lett, 1998, 287, 53.
2. S. Stankovich, R. Piner, X. Chen, N. Wu, S. Nguyen and R. Ruoff, J. Mater. Chem, 2006, 16, 155.
3. Y. T. Liu, J. M. Yang, X.M. Xie and X. Y. Ye, *Mater. Chem. Phys*, 2011, **130**, 794.
4. H. Hu, X. Wang, J. Wang, L. Wan, F. Liu, H. Zheng and C. Xu, *Chem. Phys. Lett*, 2010, **484**, 247.
5. A. Göldel, A. Marmur, G. R. Kasaliwal, P. Pötschke and G. Heinrich, *Macromolecules*, 2011, **44**,6094.
6. *M. Yang, V. Koutsos and M. Zaiser, J. Am Chem. B, 2005, 109, 10009.*
7. X. Zhang, X. Fan, H. Li, and C. Yan, J. Mater. Chem, 2012,22, 24081.

Influence of Plasticizer and Dispersion Methods on the Properties of Poly (Vinyl Chloride) and Clay Nanocomposites

Daniela Becker[a]*, Samara Bohn[a], Palova Santos Balzer[b]

[a] *Center for Technological Sciences- CCT – State University of Santa Catarina - UDESC, Brazil*
[b] *Center University SOCIESC - UNISOCIESC, Brazil*
**daniela.becker@udesc.br*

Abstract. Poly (vinyl chloride) (PVC) is one of the most employed polymers in the world and it is interesting to change some of its properties to broaden its usage. In this paper we aim to evaluate the effect of the plasticizer type and the mixture method of clay dispersion in the PVC matrix. It was employed three plasticizers: dyoctil adipate (DOA), dioctyl phthalate (DOP) and a green plasticizer (poly- ε-caprolactone – PCL). Besides, it was also evaluated the effect of clay pre-dispersion in plasticizer using two mixture methods: sonication and mechanical mixture. It was characterized the morphology by transmition electronic microscopy and mechanical properties by stress-strain test. It was observed a heterogenic dispersion of the clay in PVC matrix and different plasticization regions. However, the pre-dispersion decreased clay agglomerates size and intercalated/expholiated regions were also observed. The clay addition increased 5% and 10% on Young´s modulus for PVC/DOP with clay dispersion using sonication methods and PVC/PCL with clay dispersion using mechanic mixture, respectively. The DOA addition leads to a decrease in Young´s modulus. According to these results, it can be reasoned that clay pre-dispersion in plasticizer may facilitate intercalation but does not eliminate all of clay agglomerates, and the type of plasticizer influences on mechanical properties of plasticized PVC/clay.

Keywords: Poly (Vinyl Chloride), Clay, Plasticizer, Mixture Methods, Dispersion.
PACS: 82.35.Np

INTRODUCTION

Poly (vinyl chloride) (PVC) is one of the most consumed thermoplastics in the world, being a versatile polymer due to the possibility of adding various additives in its formulation, providing variety in its mechanical and physical properties, allowing rigid or flexible characteristic [1]. Flexible PVC are made by incorporating plasticizers into the inherently rigid PVC resin, and among the various groups, the most common plasticizers used in PVC are adipates and phthalates. The adipates are used when it is desired to maintain flexibility at low temperatures, and the most common of them is DOA - dioctyl adipate. The most commonly used phthalate plasticizers is dioctyl phthalate (DOP) due to its low cost, thermal stability, low volatility, heat resistance and low water solubility [3,4]. Currently, it has been looking for alternative to the DOP plasticizer, one strong candidate is polycaprolactone - PCL, which is a polymer with low toxicity [5, 6].

The PVC nanocomposites with clay have gained attention of many researchers [7-16], and they observed an improvement in mechanical, thermal and plasticizers migration using organophilic clay. The preparation of PVC/clay nanocomposites differs in the literature. Some studies, as Zheng e Gilbert [12,13], Silva et al.[10], Alena et al. [11], Wan et al. [7] prepared by meting mixing without clay pre-dispersion in plasticizer. However, Peprnicek et al. [9] and Rodolfo Jr. [14] prepared the nanocomposites by melting method with mechanical pre-dispersion of clay in plasticizer using mechanical mixing. They observed an improvement in the clay dispersion in PVC matrix. Besides mechanical mixing to perform pre-dispersion, it is also possible use sonication to improve dispersion as observed by Sene [15]. The type of plasticizer used in the production of nanocomposites is also very important as observed for Rodolfo Jr [14], Marcilla; Garcia; Quesada Garcia [16] and Peprnicek et al. [9]. Different plasticizers are used, but none of the authors compared the results obtained for different plasticizer, most of them choose only a plasticizer and varies the amount of clay used in the preparation of nanocomposites.

This study was performed with the purpose of obtaining nanocomposites of plasticized PVC matrix with three different plasticizers (DOA, DOP and PCL) and organophilic clay, using different mixing routes of clay in plasticizer, in order to compare the effect of the pre-dispersion method in different plasticizers.

EXPERIMENTAL

Materials

Materials used in this work were poly (vinyl chloride) - PVC - Norvic SP1000 supplied by Braskem, in composition with three different plasticizers: dioctyl phthalate – DOP (ELEKEIROZ-DOP), dioctyl adipate - DOA (ELEKEIROZ-DOA) or polycaprolactone – PCL (PERSTORP - PCL 1000). It is also used as lubricant Drapex 6.8 and stabilizer Naftosafe CZ-4211 supplied by Chemson. The Cloisite® 30B montmorillonite clay (clay modified with quaternary ammonium salts MMT-30B) was supplied by the company Southern Clay. The MMT-30B is an organoclay, with MT2EtOH (methyl, tallow, bis-2-hydroxyethyl, quaternary ammonium) providing its organophilic nature and d_{001} of 18.5Å. Table 1 lists the materials and the amount of each material employed in the preparation of nanocomposites.

TABLE 1. Nanocomposites formulations.

Material	*PHR*	*%w/w*
PVC	100	70.1
Plasticizer	40	28.0
stabilizer	2.5	1.8
lubricant	0.2	0.1
clay	1.4	1.0

Nanocomposites Preparation

Initially, for some samples were performed the clay pre-dispersion in the plasticizer when necessary. These pre-dispersion was performed using sonication or mechanical stirring. In sonication methods, we used VCX 750 Sonics power equipment and added 292J / g in plasticizer and clay mixture. In mechanical methods, the pre-dispersion was performance by magnetic stirring, IKA ® C-MAG HS7, during 15 min and stirring rate of 25% of the equipment total power. Table 2 shows the composition of each PVC nanocomposite with their corresponding nomenclature.

TABLE 2. Composition and nomenclature of PVC/plasticizer/clay samples.

DOP	*DOA*	*PCL*	*MMT 30-B*	*Pre-dispersion methods*	*Samples nomenclature*
X	-	-	-	Without	PVC/DOP
X	-	-	X	Without	PVC/DOP/W
X	-	-	X	Sonication	PVC/DOP/Son
X	-	-	X	Mechanical	PVC/DOP/Mec
-	X	-	-	Without	PVC/DOA
-	X	-	X	Without	PVC/DOA/W
-	X	-	X	Sonication	PVC/DOA/Son
-	X	-	X	Mechanical	PVC/DOA/Mec
-	-	X	-	Without	PVC/PCL
-	-	X	X	Without	PVC/PCL/W
-	-	X	X	Sonication	PVC/PCL/Son
-	-	X	X	Mechanical	PVC/PCL/Mec

After the pre-mixture, all compounds went through the same procedure, and added to a mixer, as follows: all solid components (PVC and stabilizer) were placed in the mixing chamber and allowed to mix until the temperature reached 60°C, then liquid components (plasticizer with or without clay and lubricant) were added and mixing was continued until the temperature reached 120°C. Liquids were absorbed by the solid, and finally a free-flowing powder was produced. Nanocomposites without pre-dispersion, the clay were added with solids components and nanocomposites with pre-dispersion clay were added with plasticizer. After these procedures, the material still in the form of powder was melt mixed on a single screw extruder, CIOLA L/D 25:01. The mixing was carried out at 12 rpm screw speed and temperatures in the heating zones ranged from 145 160°C. The nanocomposites were compressed and molded into a disk of 2mm thickness and 50 mm diameter using a hydraulic press at 180°C for 5 minutes by applying a force of 5T and cooled with water to room temperature.

Characterization Techniques

Ultrathin sections of 30nm were cut from the compression molded disks with a diamond knife at -85°C, using an RCM Power Tome X Ultramicrotome. Sections were collected on the surface of a water-dimethylsulfoxide (60/40 v/v) bath cooled at -60°C. A JEOL JEM-2100 transmission electron microscope was used for the images.
Tensile tests, according to ASTM D 882-02 standards, were carried out in an Oswaldo Filizola - AME-5kN universal testing machine model. Crosshead displacement rate of 50 mm min^{-1} was used. A longitudinal strain gauge was used to obtain the Young modulus.

RESULTS AND DISCUSSION

The Figure 1 presents transmission electronic microscopy images of nanocomposites without pre-dipersion and sonicated for all plasticizers. For all samples was observed a second phase formation (the lighter regions on images), which can be related to heterogeneous plasticizing in PVC matrix. Silva et al. [10] observed more than one glass transition temperature of plasticized PVC samples. According to the authors this fact is related to the heterogeneity in the distribution of the plasticizer into the matrix and/or PVC particles are not completely gelified. It is also noted for samples with DOP and PCL plasticizer with sonication pre-dispersion (Figure 1(b) and (f), respectively) the presence of clay in intercalated/exphloliated state and without pre-dispersion (Figure 1(a) and (e), respectively) some agglomerates. These results indicate the clay pre-dispersion in these plasticizers change the dispersion state in PVC matrix. However, the pre-dispersion in DOA does not significantly alter the state of clay dispersion.

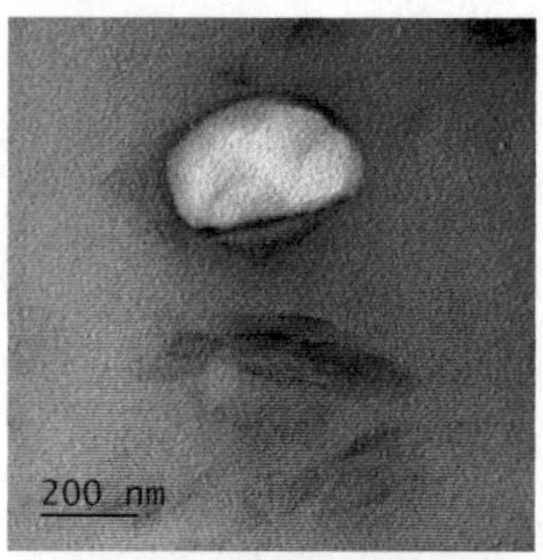

(a)

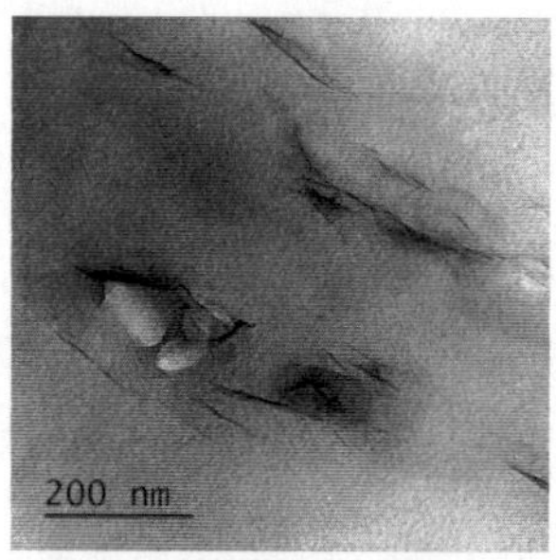

(b)

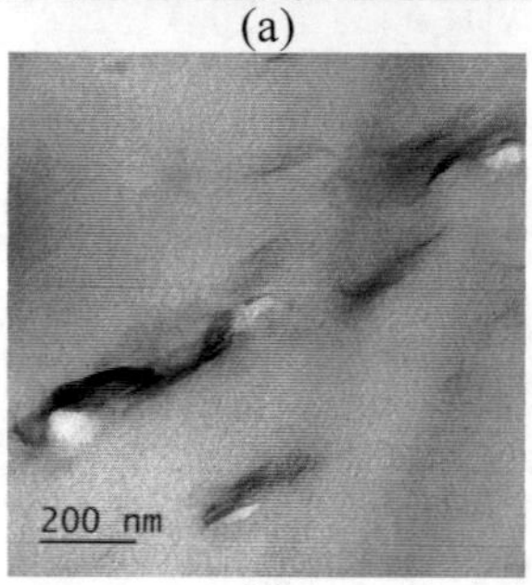

(c)

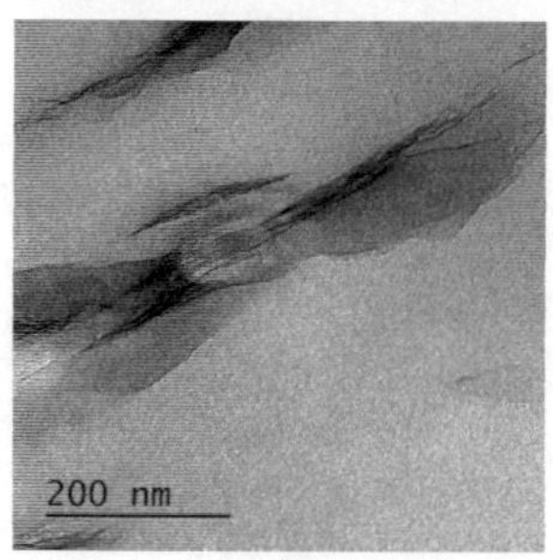

(d)

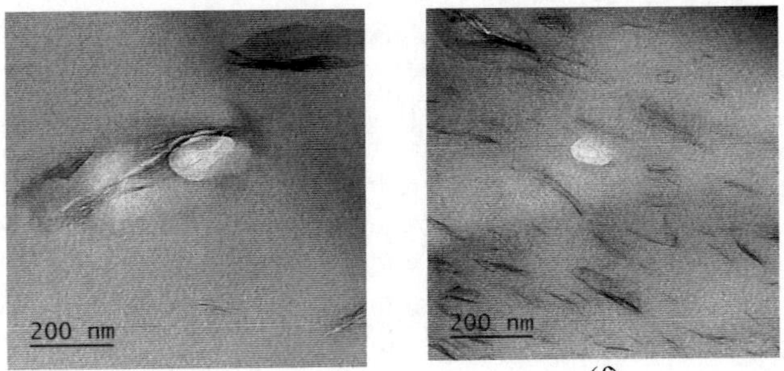

(e) (f)

FIGURE 1. TEM micrographs of (a) PVC/DOP/W; (b) PVC/DOP/Son; (c) PVC/DOA/W; (d) PVC/DOA/Son; (e) PVC/PCL/W; (f) PVC/PCL/Son.

Nanocomposites with DOP prepared by pre-dispersion with sonication present an increase of 10% in Young´s modulus compare to PVC/DOP and mechanical mixture present a decrease of 5% in Young´s modulus compare to PVC/DOP (Figure 2 (a)). A similar behavior was observed in samples with PCL. However, samples with DOA, as plasticizer, present a decrease in all nanocomposites samples. These results could be related with clay dispersion state in nanocomposites as observed in MET micrographs. Others studies also observed that nanocomposites with intercalated or exholiated state present an increase in Young´s modulus and presence of agglomerates decrease Young´s modulus [1, 17, 18]. Regarding the tensile strength, it is observed that only nanocomposites with DOP and clay pre-dispersion present an increase in this property and an increase of 5% to PVC/DOP/Son samples and 10% to PVC/DOP/Mec samples. All nanocomposites with PCL and DOA present a decrease in tensile strength. This behavior can be attributed to some clay agglomerates and / or heterogeneous plasticizing regions. All samples present a reduction in elongation at break, this behavior could be related to brittle characteristic of clay.

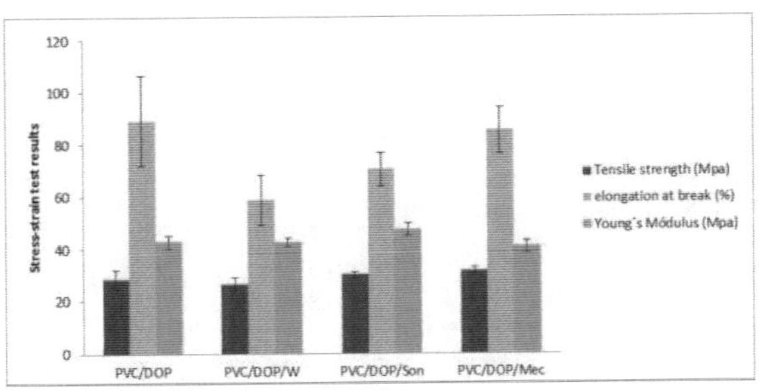

(a)

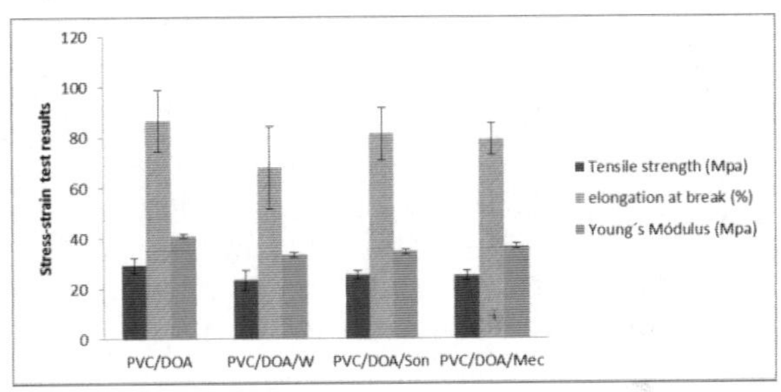

(b)

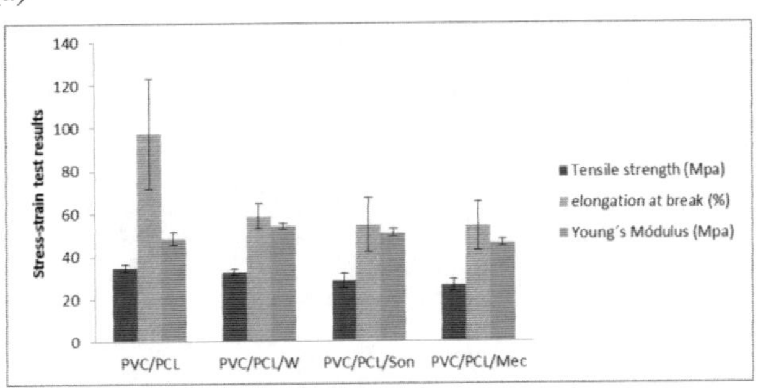

(c)

FIGURE 2. Stress-strain test results for PVC nanocomposites with (a) DOP; (b) DOA; (c) PCL.

CONCLUSIONS

According to the analysis carried out it is concluded that the clay addition on the PVC polymer matrix has different results when we compare its use in plasticized PVC with DOP, DOA and PCL. For DOP and PCL plasticizer, it was found that the pre-dispersion method by sonication produces better results as TEM analysis and mechanical behavior. For PCL and DOP, the method of sonication allowed some intercalation and exfoliation of clay structures, and a reduction in the amount of clay agglomerates. As for the DOA it was found that the clay pre-dispersion in the plasticizer, regardless of the method, has no influence on the final properties of the nanocomposites.

ACKNOWLEDGMENTS

The authors acknowledge the technical support of UNIPOL and PERSTORP for supplied the PCL.

REFERENCES

1. L.Madaleno, et al., Morphology, *Composites Science and Technology*, **70**, 804 – 814 (2011).
2. A. Rodolfo Jr, et al., "Tecnologia do PVC" , 2ed., , São Paulo: ProEditoes, 2006.
3. M.A. Souza, et al., *Polímeros: Ciência e Tecnologia*, **16**, 257-262, (2006).
4. VETTA. Linhas e Produtos, 2014. Disponível em: < http://www.vettaquimica.com.br /linhas-produtos/plastificantes/vettaplast-8-dop-dioctil-ftalato >. Acesso em: Março 2014.
5. E. Madaleno, et al., *Polímeros: Ciência e Tecnologia*, **19**,263-270, (2009).
6. G. Shi, et al., *Polymer Degradation and Stability*, **96**, 1639-1647, (2011).
7. C. Wan, et al., *Journal of Applied Polymer Science*, **89**, 2184 – 2191, (2003).
8. C. Wan, et al., *Polymer Testing*, 453-461, (2003).
9. T. Pepmicek, et al., *Polymer Degradations and Stability*, **91**, 3322 – 3329, (2006).
10. R.P. Silva, et al., *Polímeros: Ciência e Tecnologia*, **20**, 46 – 50, (2010).
11. K. Alena, et al., *Polymer Composites*, 1-7, (2013).
12. X. Zheng, M. Gilbert, *Journal of Vinyl & Additive Technology*, 77 -84, (2011).
13. X. Zheng, M. Gilbert, *Journal of Vinyl & Additive Technology*, 231 -238,(2011).
14. A. Rodolfo Jr, *Polímeros: Ciência e Tecnologia*, **19**, 1-9, (2009).
15. T. S. Sene, "Epoxy matrix nanocomposites with dispersion of carbon nanotubes and clays". Dissertation, Santa Catarina State University, 2012
16. A. Marcilla, et al., *Polymer Testing*, **27**, 221-233, (2008).
17. N.G. Shimpi, S. Mishra, *Journal of Composite Materials*, **45**, 2447 – 2453, (2011).
18. N.G. Shimpi, S. Mishra, *Journal of Applied Polymer Science*, **119**, 148 – 154, (2011).

Thermal History Effects on Moisture Absorption of Fiber-Reinforced Polymer Composites

Youssef K. HAMIDI [a,b,*], Levent AKTAS [b] and M. Cengiz ALTAN [b]

[a] *Ecole Nationale Supérieure des Mines de Rabat, BP 753, Agdal, Rabat, Morocco.*
[b] *School of Aerospace and Mechanical Engineering, University of Oklahoma, Norman, OK 73019, USA*

**e-mail: hamidi@ou.edu*

Fiber-reinforced polymer composites may offer numerous attractive features such as low cost, high specific performance, and ease of production. However, there are concerns about the overall durability of these materials, especially for sustained performance under severe and changing environmental conditions. There also is a general lack of data on the effects of life-cycle thermal history and temporal changes on moisture absorption dynamics in polymer composites. This work investigates the effects of previous life-cycle thermal history on moisture absorption in resin transfer molded glass/epoxy composites. Disk-shaped parts were fabricated using EPON 815C resin and EPICURE 3282 curing agent. Reinforcement was provided by six layers of randomly-oriented planar glass fiber preforms, yielding approximately 32% fiber volume fraction. Samples cut from the molded disks were initially immersed in water at room temperature for 48 hours. The samples were then divided into three groups and subjected to 0ºC, -25ºC, or liquid nitrogen for another 48 hours to impose various levels of life-cycle damage. A set of specimens was subjected to the same absorption cycle at room temperature in order to characterize the baseline behavior of the composite samples. Afterwards, all samples went through a desorption cycle to remove the absorbed moisture. The specimens were then immersed in water at room temperature for 18 months. Their masses were measured at periodic intervals to quantify the amount of water absorbed. Moisture intake during the 18-month period was found to increase considerably in composite samples subjected to harsher environmental/thermal conditions. In addition, short beam shear strength and stiffness reductions after freezing and after moisture absorption were measured. SBSS and stiffness dropped considerably for all composite parts after thermal conditioning. Property drop was observed to be much more significant for the samples that were exposed to lower freezing temperatures before moisture absorption. As much as 28 and 17% reduction in SBSS and stiffness, respectively, was observed for samples subjected to liquid nitrogen. After saturation with moisture, additional property drops were observed in all composite samples.

Keywords: fiber-reinforced polymer composites, moisture absorption, freezing, thermal history.
PACS: 81.05.Qk.

INTRODUCTION

The advantages of fiber-reinforced polymer composites include their light weight coupled with high stiffness and strength. This combination is the basis of their usefulness in aircraft, automobiles, and other structures used in the transportation industry. Other desirable properties include superior corrosion and fatigue resistance compared to metals. Hence, there has been a significant shift toward use of structural composites in the last three decades [1-2].

In recent years, advances in composite manufacturing techniques have permitted manufacturers to move beyond labor-intensive, relatively expensive conventional processes such as wet lay-up. Newer versatile methods such as resin transfer molding (RTM) and vacuum-assisted resin transfer molding (VARTM) have allowed large-scale, high-volume production of cost-effective, near net-shape structural composite components. However, these materials are highly susceptible to environmental elements such as humidity and thermal cycles, to which they are exposed during service. This susceptibility affects both their mechanical integrity and durability. For example, the mechanical properties of fiber-reinforced polymer composites are known to depend significantly on service temperature [2-9]. At high temperatures, the matrix softens severely, thus lowering its ability to transfer loads to the reinforcement and yielding a composite with degraded properties [3-5]. On the other hand, when temperatures are too low, the matrix becomes stiffer and brittle. Moreover, mismatches between the resin and fiber thermal coefficient cause internal stresses that adversely affect the overall mechanical integrity of the composite [6-9].

In addition to classical molecular diffusion, moisture is absorbed and stored in fiber-reinforced composites by two basic mechanisms: either by bonding with polymer chains of the matrix, or by congregating within the existing voids, cracks, and fiber/matrix interphases [9-15]. Moisture absorption behavior of a composite often varies depending on a number of factors, including fiber volume fraction, temperature, and relative humidity of the environment, as well as void content, hydrostatic pressure, and stress state of the composite [10-16].

Moisture absorption into a composite significantly reduces its mechanical performance [10-22]. Moisture is known to affect the composite's interlaminar shear strength (ILSS) [2-10-18], flexural strength and stiffness [2,20], ultimate tensile strength and modulus [2-8,20], energy absorption performance [2], and glass transition temperature

[20], as well as viscoelastic [20-21] and fatigue properties [2,22]. For instance, Gautier et al. [17] reported ILSS loss from 8 to 74% for glass/fiber polyester composites after saturation in water in temperatures ranging from 30 to 100°C. The authors attributed the reduction in ILSS to interfacial debonding induced by differential swelling and by osmosis cracking at the interphase. Gellert and Turley [18] reported a 15, 21, and 25% reduction in flexural strength for glass/polyester, glass/venylester, and glass/phenolic composites, respectively, when immersed in seawater until saturation. The authors also reported that ILSS fell by 12 to 21%. Similarly, our research group [5, 10-15] investigated the effects of several parameters on moisture absorption behavior of E-glass/epoxy RTM composites as well as the moisture effects on mechanical performance. A significant performance reduction with moisture content was observed. For example, the presence of 3.4% wt. moisture was observed to cause a 46.2% reduction in ultimate tensile strength and a 38.5% drop in short beam shear strength (SBSS) of composite samples with 14% fiber content conditioned at 100°C [5]. At 45°C, moisture intake was found to decrease almost linearly with increasing fiber volume fraction, varying between 0 and 24.5% [10-14]. Low fatigue loading cycles were also found to significantly affect moisture absorption of fiber-reinforced composites [15].

Although the effects of moisture absorption on mechanical performance of composites are well documented, little is known about the effects of thermal history on temporal changes of moisture absorption dynamics. In addition, composites used in moisture absorption studies in literature are often fabricated for that purpose and do not have any prior thermal history. During its life cycle, however, a polymer composite typically undergoes many different series of cycles: moisture absorption/desorption cycles, loading cycles, freeze/thaw cycles, and high/low temperature cycles. Understanding how these elements affect the properties of a composite is an essential component of structural design and life-cycle analysis. To the best of the authors' knowledge, there are few studies on the effects of thermal history on the mechanical properties of polymer composites, especially at low temperatures [6-9]. In this study, the effects of thermal history on moisture absorption dynamics are investigated for E-glass/epoxy resin transfer molded composites.

EXPERIMENTAL STUDIES

Molding Procedure and Sample Preparation

The composite disks used in this study are manufactured using a custom-made molding setup, depicted in Fig. 1a. This experimental setup is comprised of a hydraulic press, two reservoirs for resin and the curing agent, a static mixer and a disk-shaped mold cavity. During mold filling, the hydraulic press forces the EPON 815C resin and Epicure 3282 curing agent (*Shell Chemicals*) out of the reservoirs, through an inline static mixer, and into the inlet gate at the center of the cavity. The mold cavity is formed by placing a 3.2mm thick spacer plate with a 152.4mm diameter circle cut out at its center, between two aluminum mold walls. Six layers of randomly-oriented, chopped-strand, E-glass fiber mats with 0.459 kg/m^2 areal density are placed in the mold cavity prior to each filling. After mold filling is complete, curing takes place at room temperature for two days until the disks become rigid enough to be de-molded. This molding procedure is described in detail in earlier publications [10-15]. For this study, three composite disks are fabricated with approximately 32% fiber volume fraction.

As depicted in Fig. 1b, a total of five 114.3mm long composite pieces are cut from each fabricated disk. Then, each of the five pieces is cut into three 38.1mm-long parts as shown in Fig. 2. Thus, a total of 15 samples are obtained per disk. After discarding samples corresponding to disk centers due to the inlet gate surface defects, fourteen 38.1mm-long, 12.7mm-wide specimens are obtained from each molded disk. Initially, 32 samples cut from three molded disks are immersed in water at room temperature for 48 hours. Three different 8-specimen sets are subjected to 0ºC, -25ºC, and liquid nitrogen, respectively, for another 48 hours after the water immersion. A fourth 8-specimen set is left at room temperature to characterize the baseline behavior of these composites. A fifth 10-specimen set, containing at least three samples from each disk, is tested before any conditioning to determine the baseline mechanical properties.

Short Beam Shear Testing

Three point bending tests are performed on all 42 samples cut from the three molded disks in order to evaluate their short beam shear strength (SBSS). The test involves supporting a specimen at two ends and loading it from the center. For this test, a three point bending fixture (MTS series #642), with two adjustable supporting rollers and a loading roller, is mounted on the MTS machine. The maximum force measured until failure is determined and short beam shear strength is evaluated according to the ASTM D 2344/D2344M-00 standard.

Ten samples, including at least three from each disk, are tested to evaluate their short beam shear strength prior to any moisture or thermal conditioning. In addition, four samples from each 8-specimen set are tested after thermal conditioning and desorption. Four specimens from the fourth 8-specimen baseline set are tested without any thermal conditioning. The remaining four samples of all sets undergo moisture absorption. Overall, 16 samples are tested after thermal conditioning, and the remaining 16 are tested after the samples reach moisture saturation.

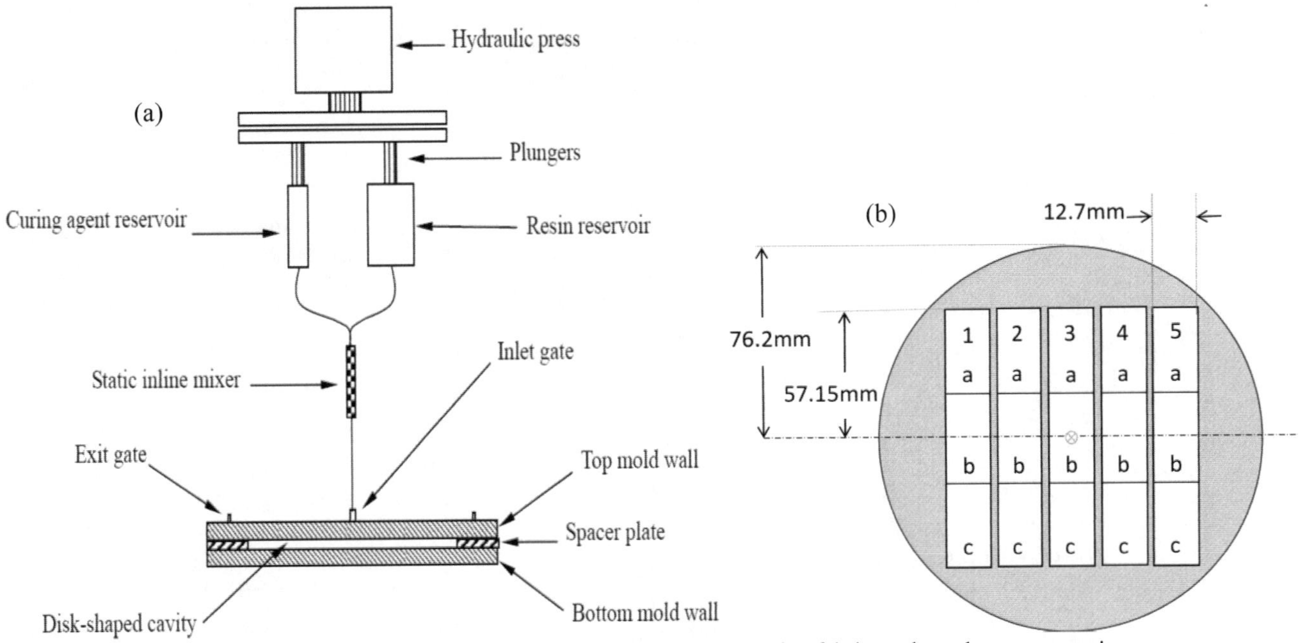

FIGURE 1. Experimental setup: **(a)** Molding apparatus used to fabricate the polymer composite disks, and **(b)** Spatial position of specimens cut from the molded composite disks.

Moisture Conditioning

All 32 specimens are first immersed in distilled water at room temperature for 48 hours. First, samples are immersed in different plastic bags filled with distilled water. Then, these plastic bags are placed in the water bath kept at 25°C. After 48 hours of immersion, three 8-specimen sets undergo freezing at 0, -25 and -195°C, respectively, for another 48 hours. After this thermal conditioning, all samples undergo a full desorption cycle. Next, four of each 8-specimen sets are tested, and the remaining untested 16 specimens are immersed again in distilled water at room temperature. Moisture conditioning is continued for 18 months until moisture saturation is reached. Samples are weighed periodically and percent mass gain for each set is calculated.

Moisture absorption experiments are conducted as specified by the ASTM D 5229/D 5229M standard. Samples are dried in a vacuum oven at 65°C for 7 days before any conditioning or testing. The water bath temperature is constantly monitored with a built-in thermocouple. Samples are weighed by an electronic balance with 0.0001g precision.

RESULTS AND DISCUSSION

Effects of Thermal History on Moisture Absorption

The methodology adopted by the *American Society of Testing and Materials* is the most common way of determining the one-dimensional, Fickian diffusion coefficient in fiber-reinforced polymer composites. For a sample with thickness h, the total percent mass gain, $M(t)$, at a given time t, can be written as:

$$M(t) = \frac{4M_m}{h}\sqrt{\frac{D_z}{\pi}}\sqrt{t}, \tag{1}$$

where M_m is the maximum percent moisture absorption and D_z is the through-the-thickness diffusivity.

Equation (1) depicts a linear mass increase with the square root of time without reaching a final equilibrium level, since the solution is obtained for a semi-infinite domain. At the initial stages of moisture absorption, the amount of absorption through the edges of the sample is negligible and mass gain can well be described by Eq. (1). Consequently, finding the diffusion coefficient can be achieved by equating the initial slope of the experimentally obtained curve and that of Eq. (1). Therefore, the closed-form solution for the diffusion coefficient is given by:

$$D_z = \pi\left(\frac{h}{4M_m}\right)^2\left(\frac{M_2 - M_1}{\sqrt{t_2} - \sqrt{t_1}}\right)^2. \tag{2}$$

The moisture absorption behavior of the thermally-conditioned samples is depicted in Fig. 2. The experiments are stopped after the weights of the samples are stabilized, which corresponds to approximately 18 months of moisture conditioning.

The maximum percent mass gain and the initial slope of each sample can be obtained from the experimental curves plotted in Fig. 2. Sample diffusivities calculated using Eq. (2) are presented in Table 1. Diffusivities and maximum mass gain obtained for these samples agree well with other results obtained in earlier investigations for similar composites [15].

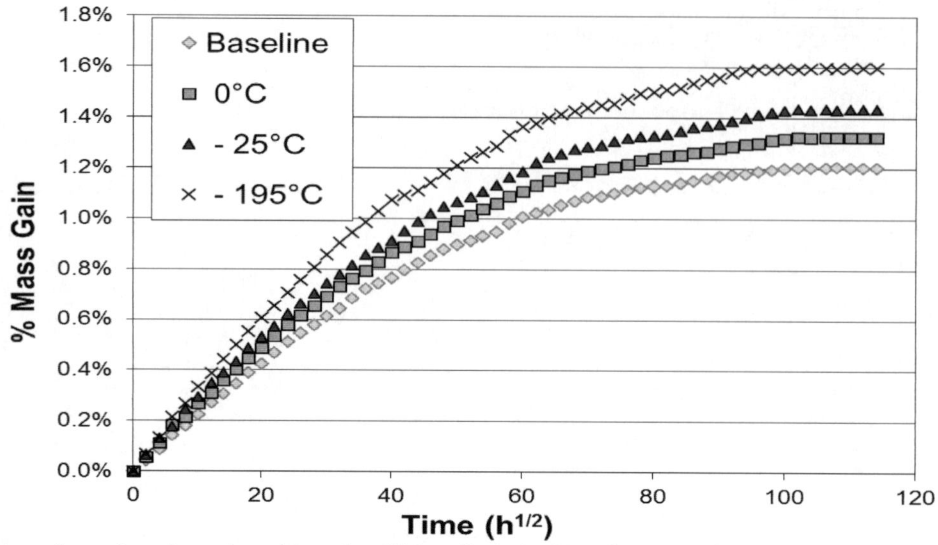

FIGURE 2. Moisture absorption of samples subjected to 25 (baseline), 0, -25 and -195°C after moisture absorption for 48 hours.

As presented in Table 1, both the maximum percent mass gain and diffusivity are observed to increase with thermal conditioning. M_m is observed to rise by 8, 30 and 53% from 1.21 to 1.60% after 48 hours at 0, -25 and -195°C, respectively. Similarly, diffusivity is observed to increase by 24% from 5.04×10^{-10} to 6.24×10^{-10} m^2/h after thermal conditioning at -195°C. Hence, thermal history is observed to significantly increase both the rate and the maximum intake of moisture absorption. Internal stresses due to thermal coefficient mismatches between the matrix and fibers have been reported to cause damage, such as cracks and fiber/matrix debonding, at the micro-structural level [6-9]. In addition, undergoing freezing after moisture absorption is expected to introduce additional damage as the free water expands. Consequently, the damaged areas may become additional sites for moisture accumulation after desorption and thus enable further moisture absorption.

TABLE 1. Effects of thermal history on maximum percent pass gain and diffusivity.

Thermal History	M_m	Diffusivity ($D \times 10^{10}$ m^2/h)
Baseline (25°C)	1.21%	5.04
0°C	1.33%	5.55
-25°C	1.44%	5.80
Liquide Nitrogen (-195°C)	1.60%	6.24

Effects of Thermal History and Moisture on Mechanical Properties

The short beam shear strength and stiffness values of samples subjected to thermal conditioning before and after moisture absorption are compared to the baseline samples, as shown in Figs. 3a and 3b, respectively. The initial cycle of moisture absorption/desorption is not observed to affect the flexural properties of the baseline composite. However, thermal conditioning at or below freezing temperatures is found to degrade significantly both their SBSS and stiffness. Almost 10, 16 and 28% reductions in SBSS are found for samples subjected to thermal conditioning at 0, -25 and -195°C, respectively. A somewhat lower rate of degradation is observed for stiffness with nearly 6, 10 and 17% reductions. This effect is believed to originate from water expansion at sub-freezing temperatures, as well as internal stresses caused by disparities in thermal coefficients of the epoxy and the glass fibers. Another possible cause might occur at the molecular level with plasticizing changes to the moisture-bounded polymer at such low temperatures.

After moisture saturation (i.e., ~ 18-month immersion at room temperature), further reductions are observed in the composite samples. Again, the baseline specimen registered very limited reductions with 2 and 0.3% reductions in SBSS and stiffness, respectively. In contrast, thermally conditioned samples had additional drops in their properties, even reaching a 34% reduction for the liquid nitrogen case. An extra 3% reduction is measured for SBSS for samples treated at 0°C. Similarly, 6 and 7% SBSS additional drops are registered at -25 and -195°C, respectively. In contrast, less than 2% is observed for stiffness for all studied cases. As discussed in the introduction, these property drops are anticipated, as moisture presence in the matrix and the matrix-fiber interphase is known to degrade mechanical properties of composites, mainly by weakening the fiber/matrix interphase.

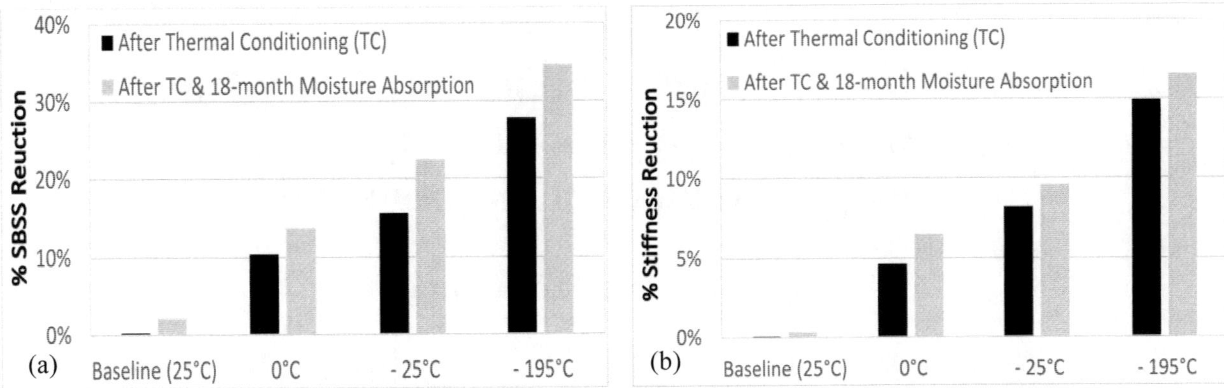

FIGURE 3. Percentage reduction of mechanical properties of composite samples after thermal conditioning (i.e., 48-hours of moisture immersion followed by 48-hour thermal exposure and desorption) and after moisture absorption for 18 months: **(a)** % reduction in SBSS, and **(b)** % reduction in stiffness.

CONCLUDING REMARKS

This work investigates the effects of prior exposure to lower temperatures on moisture absorption in resin transfer molded E-glass/epoxy composites. Thermal history was found to significantly increase both the rate and the maximum intake of moisture absorption. Maximum moisture intake increased by 8, 30 and 53% from 1.21 to 1.60% after 48 hours at 0, -25, and -195°C, respectively. Similarly, diffusivity was found to increase by 24% from 5.04×10^{-10} to 6.24×10^{-10} m^2/h after thermal conditioning at -195°C. These higher amounts of moisture are believed to accumulate inside microstructural damages created by low temperature exposure. Exposure to low temperatures after as little as 48 hours of moisture intake caused significant reductions in both SBSS and stiffness. In addition, lower temperature exposure was found to cause more damage to the composite sample. While the baseline samples showed a 0.2% property loss, as much as 28 and 17% reductions in SBSS and stiffness, respectively, were observed for samples subjected to thermal conditioning at -195°C. Immersing composite samples for another 18 months further reduced their mechanical properties. In contrast, the baseline specimen showed a very limited 2% property reduction. Additional 3, 6 and 7% reductions in SBSS were found for samples subjected to 0, -25 and -195°C temperatures. The higher SBSS reduction observed at lower temperatures correlates well with the higher percentage of moisture absorbed by the composite samples. These results are believed to be applicable to similar fiber-reinforced composites susceptible to thermal cycles and moisture absorption during their life cycle.

REFERENCES

1. Y. K. Hamidi, M. C. Altan and B. P. Grady, "Polymer Composites." In: *Encyclopedia of Chemical Processing,* edited by S. Lee, New York: Taylor & Francis, 2005, pp. 2313-2322.
2. J. Wang, H. G. Rao, R. Liang and W. Liu, J. Reinf. Plast. & Compos. 35, 179–211 (2016).
3. V. M. Karbhari, J. W. Chin, D. Hunston, B. Benmokrane, T. Juska, R. Morgan, J. J. Lesko, U. Sorathia and D. Reynaud, J. Comp. Constr. 7, 238-247 (2003).
4. J. Jedidi, F. Jacquemin and A. Vautrin, *Compos. Struct.* 68, 429-437 (2005).
5. L. Aktas, Y. K Hamidi and M. C. Altan, *IMECE Conference Proceedings,* Art. # 39223, American Society of Mechanical Engineers, New Orleans, Louisiana, 2002.
6. C. Nordin; Z. J. Ma and D. Penumadu, *J. Compos. Constr.* 14, 224-230 (2010).
7. P. K. Dutta and D, Hua, *Composites B* 27, 371-379 (1996).
8. R. D. Adams and M. M. Sigh, *Composites A* 32, 797-814 (2001).
9. V. M. Karbhari, J. Rivera and J. Zhang, *J. App. Polym. Sci.* 86, 2255-2260 (2002).
10. L. Aktas, Y. K Hamidi and M. C. Altan, *J. Mat. Proc. & Manuf. Sci.* 10, 239-254 (2002).
11. L. Aktas, Y. K Hamidi and M. C. Altan, *J. Mat. Proc. & Manuf. Sci.* 10, 255-267 (2002).
12. H. J. Barraza, L. Aktas, Y. K Hamidi, J. Long Jr., E. A. O'Rear and M. C. Altan, *J. Adh. Sci. & Tech.* 17, 217-242 (2003).
13. L. Aktas, Y. K Hamidi and M. C. Altan, *J. Engr. Mat. & Tech.* 126, 427-435 (2004).
14. L. Aktas, Y. K Hamidi and M. C. Altan, *IMECE Conference Proceedings,* Art. # 80385, American Society of Mechanical Engineers, Orlando, Florida, 2005.
15. Y. K Hamidi, L. Aktas and M. C. Altan, *International PPS Conference Proceedings*, Polymer Processing Society, Art. # 563, Marrakech, Morocco, 2011.
16. D. A. Bond, *J. Compos. Mat.* 39, 2113-2141 (2005)
17. L. Gautier, B. Mortaigne and V. Bellenger, *Comp. Sci. & Tech.* 59, 2329-2337 (1999).
18. E. P. Gellert and D. M. Turley, *Composites A* 30, 1259-1265 (1999).
19. S. Pavlidou and C.D. Papaspyrides, *Composites A* 34, 1117–1124(2003).
20. N. Guermazi, A. B. Tarjem, I. Ksouri and H. F. Ayedi, Compos. B 85, 294-304 (2016).
21. J. H. S. Almeida, S. D. B. Souza, E. C. Botelho and S. C. Amico, J. Mat. Sci. 51, 4697–4708 (2016).
22. Y. Hu, A. W. Lang, X. Li and S. R. Nutt, *Polymer Degradation and Stability* 110, 464-472(2014).

Crystallization kinetics of poly (1, 4-butylene adipate)/organically-modified layered zinc phenylphosphonate nanocomposites

Yi-An Chen and Tzong-Ming Wu

Department of Materials Science and Engineering, National Chung Hsing University, 250 Kuo Kuang Road, Taichung 402, Taiwan

Abstract. The organically-modified layered zinc phenylphosphonate (m-PPZn) were prepared by a facile chemical intercalation method. Poly(1,4-butylene adipate) (PBA) belonged to biodegradable polymers and was friendly to environment. Thus, the solution-casting method was used to obtain the new PBA/m-PPZn nanocomposites. Crystallization kinetics and morphology of PBA and 0.5wt% PBA/m-PPZn nanocomposites were studied using differential scanning calorimetry (DSC), polarizing optical microscopy (POM), transmission electron microscopy (TEM), and wide-angle X-ray diffraction (WAXD). Both WAXD and TEM results illustrated that most of the layered materials of partial delamination are randomly dispersed in the PBA matrix. The crystallization time increased with increase in crystallization temperature for neat PBA and 0.5wt% PBA/m-PPZn nanocomposites. When compared to neat PBA, the $t_{1/2}$ of the fabricated 0.5 wt% PBA/m-PPZn nanocomposites demonstrated significant decreased under the same isothermal crystallization conditions. This indicates that the incorporation of the m-PPZn content into PBA significantly enhances the crystallization rate of PBA. With the presence of m-PPZn, the nuclei number of PBA increases and the spherulite size significantly reduces. Therefore, the addition of m-PPZn to PBA induces heterogeneous nucleation.

Keywords: Poly(1,4-butylene adipate), layered zinc phenylphosphonate, crystallization kinetics, nucleation agent

Characterization of the PBA/m-PPZn nanocomposites

As shown in Figure 1d, the WAXD pattern of PPZn reveals three strong diffraction peaks at $2\theta = 6.31°$, 12.48°, and 18.68°, which are the orders of a repeat distance on the (010), (020), and (030) planes, respectively. The diffraction peaks are significantly shifted to lower angles at $2\theta = 2.92°$, 5.91°, and 8.91° because of the intercalation of oleylamine into the interlayer spacing (Figure 1a). According to the Bragg's equation, the interlayer spacing d_{010} in PPZn and m-PPZn was determined at 14.5 Å and 30.3 Å, respectively. These results demonstrate that oleylamine was successfully intercalated into the interlayer spacing of PPZn by a facile chemical intercalation method [1].

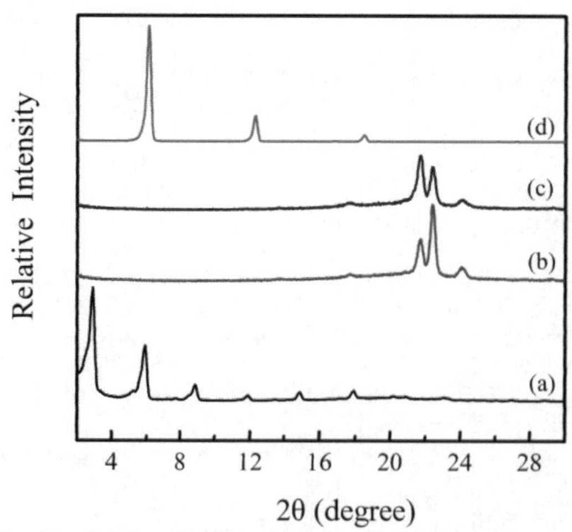

FIGURE 1. .WAXD patterns for (a)m-PPZn, (b) PBA, (c) 0.5 wt% PBA/m-PPZn nanocomposites, and (d) PPZn.

Figure 1b and 1c show the WAXD data of neat PBA and the 0.5 wt% PBA/m-PPZn nanocomposites. The diffraction peaks at 2θ = 21.69°, 22.41° and 24.08° can be observed, and they are attributed to the α-form crystal of PBA [2, 3]. The diffraction peaks of m-PPZn disappear in the 0.5 wt% PBA/m-PPZn nanocomposites. Due to the low quantity of m-PPZn, we only observed that the complete disappearance of the m-PPZn diffraction peaks. Thus, the gallery height of intercalated layers is large enough and the layer correlation is not detected by the X-ray diffractometer. This result suggests that the molecular chains of PBA have been successfully inserted in the m-PPZn galleries, thereby leading to a partial delamination of the m-PPZn with tactoid reduction.

Furthermore, the morphology of 0.5 wt% m-PPZn in the PBA matrix is observed by the TEM analysis, as shown in Figure 2a. Figure 2b shows high magnification TEM images of this sample. It is clear that the m-PPZn layers follow the morphology of the partially delaminated PBA matrix and retain little of their stacking order. Accordingly, the m-PPZn layers' partial delamination morphology in the PBA matrix is in good agreement with the WAXD data.

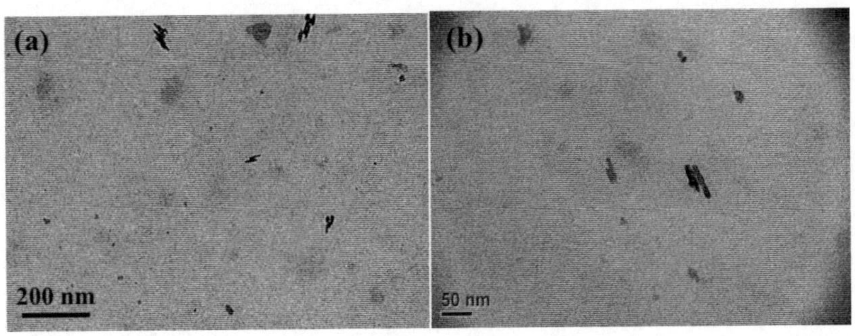

FIGURE 2. (a) TEM micrographs of 0.5 wt% PBA/m-PPZn nanocomposites; (b) the high magnification TEM images of 0.5 wt% PBA/m-PPZn samples.

Isothermally crystallization kinetics of the PBA/m-PPZn nanocomposites

According to the previously literature, we knew that pure α-form crystals of PBA develop at crystallization temperatures above 31 °C and the α-form is thermodynamically stable. Thus, the crystallization temperatures were chosen in the range of 36 to 42 °C. Isothermal crystallization kinetics was often analyzed through means of Avrami equation [4, 5]. The equation is written in the following form:

$$1 - X_t = \exp(-kt^n) \qquad (1)$$

where X_t is relative degree of crystallinity at crystallization time t, k is the crystallization rate constant, and n is the Avrami exponent. In order to convert conveniently with the operation, Eq. (1) can be transformed into its natural logarithm form as given in Eq. (2).

$$\ln[-\ln(1 - X_t)] = n \ln t + \ln k \qquad (2)$$

The crystallization half-time ($t_{1/2}$) is the time at which the extent of crystallization is 50 %, which is defined in Eq. (3).

$$t_{1/2} = \left(\frac{\ln 2}{k}\right)^{1/n} \qquad (3)$$

Figure 3 presents the Avrami plots for neat PBA and 0.5 wt% PBA/m-PPZn nanocomposites crystallized at 42 °C as an example. For comparison, the obtained Avrami parameters are summarized in Table 1 for neat PBA and 0.5 wt% PBA/m-PPZn nanocomposites. From Table 1, it can be seen that n-values are between 2.75 and 3.30 for neat PBA. In general, the n-values close to 3 are attributed to an athermal nucleation process followed by a three-dimensional crystal growth and homogeneous nucleation mechanism. The n-values of 0.5 wt% PBA/m-PPZn nanocomposites are in the range of 2.70–3.10, which are close to those of PBA. Therefore, these results suggest that the introduction of m-PPZn into the PBA matrix does not change the crystallization mechanism of PBA.

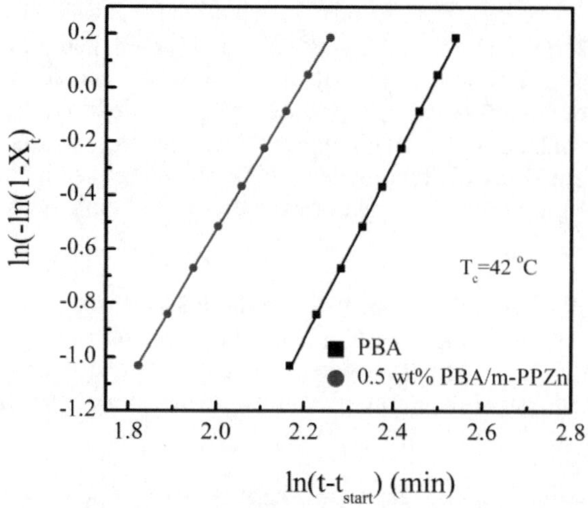

FIGURE 3. Avrami plots of neat neat PBA and the 0.5 wt% PBA/m-PPZn nanocomposites.

Table 1. Kinetic parameters of neat PBA and 0.5 wt% PBA/m-PPZn nanocomposites isothermally melt crystallized at T_c =36–42 °C.

Sample	T_c (°C)	*n*	*k* (min^{-n})	$t_{1/2}$ (min)
	36	2.75	2.74x10^{-1}	1.88
PBA	38	2.96	1.76x10^{-2}	3.46
	40	2.75	4.48x10^{-3}	6.26
	42	3.30	2.77x10^{-4}	10.73
	36	3.10	2.32x10^{-1}	1.42
0.5 wt%	38	3.04	4.50x10^{-2}	2.46
PBA/m-PPZn	40	2.70	1.24x10^{-2}	4.15
	42	2.80	2.17x10^{-3}	7.83

In addition, it is not appropriate to compare the overall crystallization rate from the *k*-values directly because the unit of *k* is min^{-n} and n is not constant in this study for all the samples at different T_{cs} values. Thus, the $t_{1/2}$ is employed to discuss the crystallization kinetics of neat PBA and 0.5 wt% PBA/m-PPZn nanocomposites. As shown in Table 1, it is apparent that the value of $t_{1/2}$ increases with increasing crystallization temperature, suggesting that the isothermal crystallization rate decreases with the increasing T_c, due to the low super-cooling at higher T_c. The $t_{1/2}$ value for 0.5 wt% PBA/m-PPZn nanocomposites is lower than that for PBA. For example, the $t_{1/2}$ values of PBA drastically decrease from 10.73 min to 7.83 min in nanocomposites with 0.5 wt% m-PPZn contents when crystallized at 42 °C. These results signify that m-PPZn accelerates the crystallization of PBA in the nanocomposites.

Spherulitic morphology of the PBA/m-PPZn nanocomposites

POM characterization was carried out in order to study the influence of m-PPZn on the morphology development and size of spherulites of PBA. As shown in Figure 4, the spherulitic morphology of neat PBA and 0.5 wt% PBA/m-PPZn nanocomposites isothermally crystallized at 42 °C. It is clear that the size of PBA spherulites becomes smaller in the presence of m-PPZn. The number of PBA spherulites is greater in the nanocomposites than in neat PBA. This implies that the incorporation of the 0.5 wt% m-PPZn contents into PBA could induce heterogeneous nucleation and change the spherulitic morphology of PBA. According to the POM data, the size and number of PBA spherulites in the nanocomposite are improved due to the nucleation agent effect of m-PPZn, which is consistent with the Avrami results in the previous section.

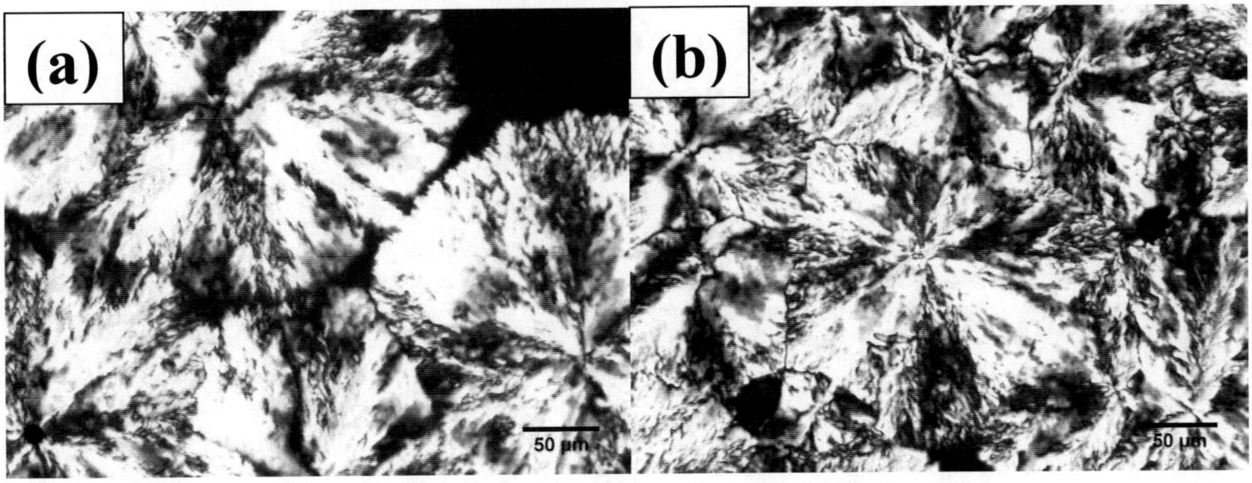

FIGURE 4. Optical micrographs of spherulitic morphology of (a) neat PBA and (b) 0.5 wt% PBA/m-PPZn nanocomposites after complete crystallization at 42 °C. (scale: 50μm)

REFERENCES

1. Y.A. Chen, E.C. Chen and T.M. Wu, *J. Mater. Sci.* 50, 7770–7778 (2015).
2. W. Kai, B. Zhu, Y. He, and Y. Inoue, *J. Polym. Sci. Part B: Polym. Phys.* 43, 2340-2351 (2005).
3. T. Dong, W. Kai, and Y. Inoue, *Macromolecules* 40, 8285-8290 (2007).
4. M. Avrami, *J. Chem. Phys.* 8, 212–224 (1940).
5. M. Avrami, *J. Chem. Phys.* 9, 177–184 (1941).

Synthesis of graphene nanoribbon/reduced graphene oxide/polypyrrole nanocomposites as electrode for supercapacitor

Ching-Yuan Wang and Tzong-Ming Wu

Department of Materials Science and Engineering. National Chung Hsing University, 250 Kuo Kuang Road, Taichung 402, Taiwan

Abstract. A high performance electrode material for a supercapacitor has been developed based on graphene nanoribbon (GNR), reduced graphene oxide (rGO) and polypyrrole (PPy) nanocomposites (GNR/rGO/PPy) synthesized via in-situ chemical polymerization. The morphology and molecular structure of the nanocomposites are characterized by using field-emission scanning electron microscopy (FESEM), transmission electron microscopy (TEM), and Fourier transform infrared spectroscopy (FTIR). Electrochemical characterization of the electrode materials have been carried out by using cyclic voltammetry, galvanostatic charge/discharge and electrochemical impedance spectroscopy. GNR/rGO/PPy nanocomposite reveals that the highest specific capacitance of 591 F g^{-1} at a scan rate of 5 mV s^{-1}, which is better than these of PPy and GNR/PPy nanocopmposite. Furthermore, GNR/rGO/PPy nanocomposite shows a high rate capacity and good cycle stability after 1000 cycles. The excellent electrochemical performance demonstrates that the GNR/rGO/PPy nanocomposite has a great potential in energy storage devices.

Keywords: polypyrrole, graphene nanorribon, reduced graphene oxide, supercapacitor
PACS: 82.45.Yz

Morphological analyses of GNR/rGO/PPy nanocomposites

Figure 1 shows the TEM and SEM images of PPy, GNR, rGO, GNR/PPy and GNR/rGO/PPy. As shown in Figure 1a and 1b, pure PPy particles exhibit a globular structure, forming agglomerates. It could be found that the GNR shown in Figure 1c and 1d were successfully fabricated via the chemical method with sulfuric acid and potassium permanganate. The average diameter of GNR was obtained about 100 nm. Figure 1e and 1f show the particular layer-like structure of the rGO with the wrinkled form. From Figure 1g and 1h, it was found that PPy was tightly coated on the surface of GNR. By comparing the structure of GNR and GNR/PPy, it was found that the diameter was increased from 100 nm to about 350 nm. The morphology of GNR/rGO/PPy nanocomposites shown in Figure 1i and j could be seen that both PPy nanoparticles and GNR were uniformly distributed on the surface and around the edge of the rGO. In addition, some PPy nanoparticles were coated on the surface of GNR and may be acted to connect the rGO and GNR. This structure could play important role for the electric capacity of GNR/rGO/PPy composites [1].

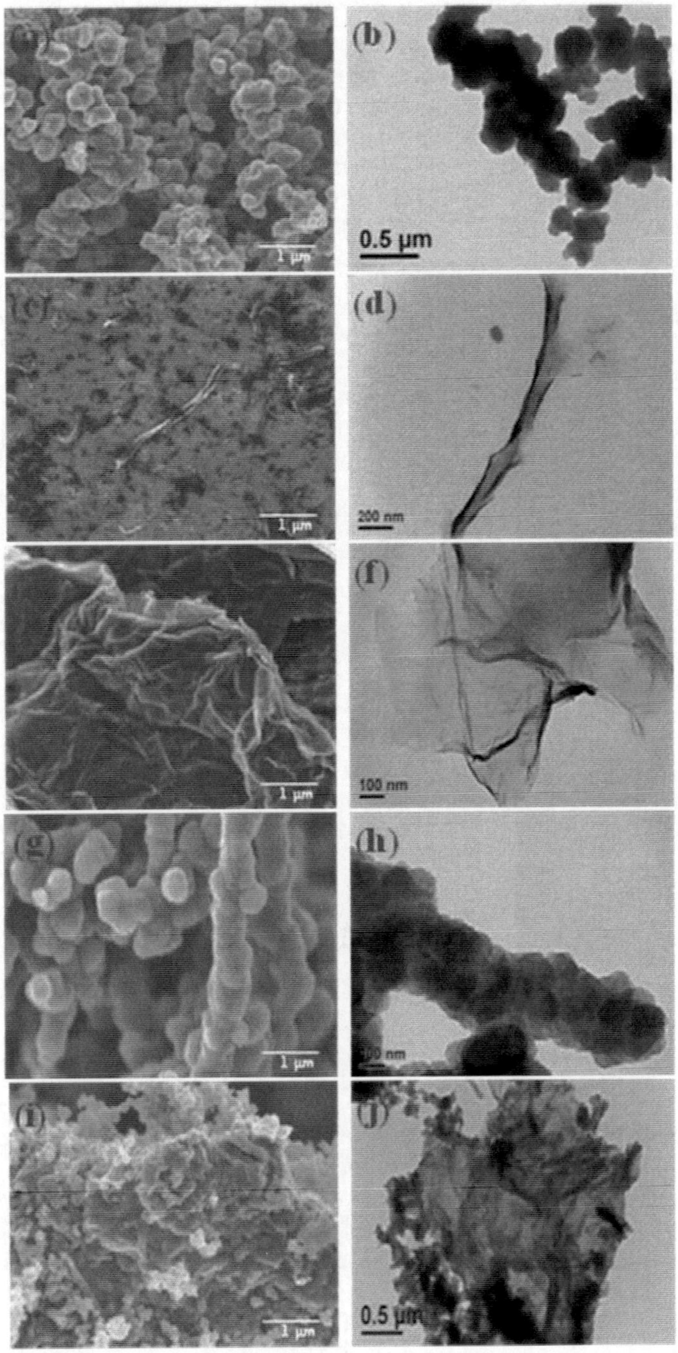

FIGURE 1. SEM image of (a) PPy, (c) GNR, (e) rGO, (g) GNR/PPy, (i) GNR/rGO/PPy and TEM image of (b) PPy, (d) GNR, (f) rGO, (h) GNR/PPy, (j) GNR/rGO/PPy

FTIR analyses of GNR/rGO/PPy nanocomposites

Figure 2 shows the FT-IR spectra of the PPy, GNR/PPy and GNR/rGO/PPy nanocomposites. All samples showed the same absorbance peak at 3440 cm^{-1}, which is associated with -OH stretching vibration. In the spectra of PPy, the peak at 1552, 1460, 1310, 1183 and 1038 cm^{-1} were attributed to C-C stretching vibration, C-N bend vibration, C-O stretching vibration, breathing vibration of the pyrrole ring and C-H deformation, respectively. The peak at 912 cm^{-1} was assigned to the C-H out-of-plane vibration of the pyrrole ring. For the GNR/PPy nanocomposites, the peaks are similar to the pure PPy, indicating that PPy was successfully coated on the GNR surface via the in-situ chemical polymerization. In the case of GNR/rGO/PPy, the absorption peaks at 1518, 1226, 1125 and 994 cm^{-1} were assigned to C-H stretching vibration, C-N stretching vibration, C-O stretching vibration and C-H out-of-plane vibration of the pyrrole ring, respectively. Compared with PPy and GNR/rGO/PPy, the peak attributed to the stretching vibration of the pyrrole rings for the GNR/rGO/PPy nanocomposites was shifted to a low frequency region, which was due to π-π interaction between PPy, rGO and GNR [2]. These results indicated that the GNR/rGO/PPy nanocomposites were successfully prepared using in-situ chemical polymerization.

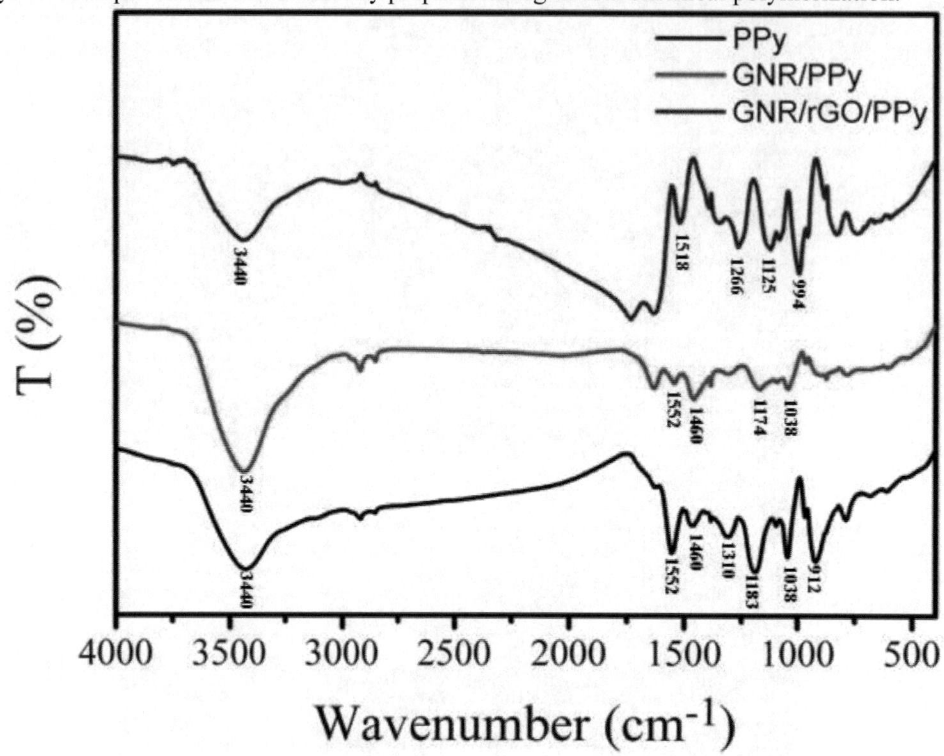

FIGURE 2. FT-IR spectra of (a) PPy, (b) GNR/PPy (c) GNR/rGO/PPy

Electrochemical characterization of GNR/rGO/PPy nanocomposites

Electrochemical test were carried out using CV, GCD and EIS techniques to measure the electrochemical behavior of the prepared samples in three-electrode system. Figure 3a showed the CV curves of PPy, GNR/PPy and GNR/rGO/PPy nanocomposites. The specific capacitance of the electrode could be calculated from the CV curve using the following equation [3, 4]:

$$C = \frac{\int I(V)dV}{2m\upsilon\Delta V} \tag{1}$$

where C was the specific capacitance (F g^{-1}), I was the response current density (A cm^{-2}), υ was the potential scan rate (V s^{-1}), V was the electric potential (V) and m was the mass of the materials on the electrode, respectively. As seen in Figure 3a, the CV curves of GNR/PPy and GNR/rGO/PPy were close to rectangular shape, implying that they have an ideal capacitive characteristics with ion response [5]. However, the CV curve of PPy was not rectangular, indicating resistance-like electrochemical behavior. Similar phenomena were also reported in a previous study [6]. It was found that the CV integral area of GNR/rGO/PPy was larger than those of PPy and GNR/PPy, which indicate that the GNR/rGO/PPy nanocomposites have the highest specific capacitance. The calculated specific capacitance was summarized in Table 1. Figure 3b depicts the charge-discharge curves at the current density of 0.5 A g^{-1} and it was clearly observed that the curves of all samples exhibit a pseudocapacitance performance as the linear shapes are deformed.

Figure 3c shows the Nyquit impedance plots of PPy, GNR/PPy and GNR/rGO/PPy nanocomposites. The semicircles were not observed in the high frequency region, which indicated the lower charge-transfer resistance. In additional, the Nyquit impedance plot of GNR/rGO/PPy has the vertical lines close to 90° at low frequency region, indicating the low diffusion resistance of the electrolyte ions in the electrode structure, which shows the ideal capacitive behaviors [7].

The stability and reversibility of an electrode material are key factors for the application of supercapacitor electrodes. Figure 3d shows the cycling stability of the PPy, GNR/PPy and GNR/rGO/PPy at a scan rate of 100 mV s^{-1} after 1000 cycles. It was demonstrated that the GNR/rGO/PPy shows retention of 82.3 % after cyclic voltammetry for 1000 cycles while PPy only exhibits 62.1 % of the initial capacitance. PPy usually sustains a poor long-term stability during cycling because of the shrinking and swelling of PPy polymer chain leading to degradation. The addition of carbon materials (GNR, rGO) into PPy could sufficiently prevent the damage of the PPy backbone and effectively improved the cycle stability.

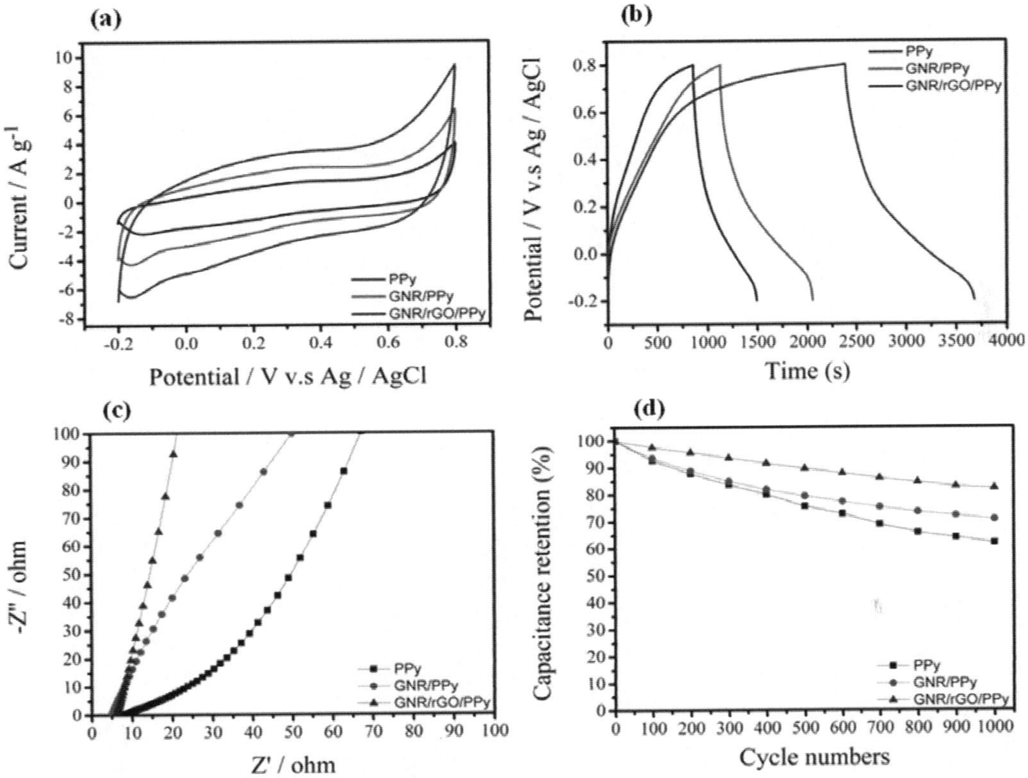

FIGURE 3. (a) CV curve of PPy, GNR/PPy and GNR/rGO/PPy at 10 mV s^{-1} scan rate with 1.0 M H_2SO_4 as the electrolyte. (b) Charge-discharge curve of PPy, GNR/PPy and GNR/rGO/PPy at 0.5 A g^{-1} within the potential window -0.2 to 0.8. (c) Nyquit impedance polts of the PPy, GNR/PPy and GNR/rGO/PPy. (d) Cycling stability test of the PPy, GNR/PPy, and GNR/rGO/PPy electrode as a function of cycle numbers measured at100 mV s^{-1}.

TABLE (1). Specific capacitance (F/g) and Capacitance retention (%) of PPy, GNR/PPy, and GNR/rGO/PPy

Sample	Specific capacitance (F/g)	Capacitance retention (%)
PPy	138.1	62.1
GNR/PPy	367.5	70.8
GNR/rGO/PPy	516.8	82.3

REFERENCES

1. B. Wang, J. Qiu, H. Feng, and E. Sakai, *Electrochim Acta*, **151**, 230-239 (2015).
2. B. Liang, Z. Qin, J. Zhao, Y. Zhang, Z. Zhoua, and Y. Lu, *Journal of Materials Chemistry A*, **2**, 2129-2135 (2014).
3. C. J. Hung, P. Lin, and T. Y. Tseng, *Journal of Power Sources*, **259**, 145-153 (2014).

4. S. Deng, D. Sun, C. H. Wu, H. Wang, J. B. Liu, Y. X. Sun, and H. Yan, *Electrochimica Acta*, **111**, 707-712 (2013).
5. W. Wu, L. Yang, S. Chen, Y. Shao, L. Jing, G. Zhao and H. Wei, *RSC Adv.*, **5**, 91645-91653 (2015).
6. L. Yang, Z. Shi , and W. Yang, *Electrochim Acta*, **153**, 76-82 (2015).
7. J. K. Gan, Y. S. Lim, A. Pandikumar, N. M. Huang and H. N. Lim, *RSC Adv.*, **5**, 12692-12699 (2015)

Comparison analysis of dielectric properties of LLDPE based compounds filled by halloysite nanotubes and montmorillonite

P. Kadlec[a], M. Čermák[a] and R. Polanský[b*]

[a]Department of Technologies and Measurement/ [b]Regional Innovation Centre for Electrical Engineering
Faculty of Electrical Engineering, University of West Bohemia in Pilsen
Univerzitní 8, Pilsen 306 14, Czech Republic

kadlecp6@ket.zcu.cz, rpolansk@ket.zcu.cz*

Abstract. The motivation of this research is to provide an analysis of dielectric properties of two different phyllosilicates intercalated into the polyolefine based composites. For this investigation, halloysite nanotubes and montmorillonite modified with quaternary ammonium salt have been used as a nanofiller in polymer composites with perspective application in a cable industry. Investigation of dielectric properties of composites with two different loading level of nanofiller and comparison with dielectric properties of pure polymer matrix have been made. Experiment was focused on the measurement of dielectric properties in the DC, as well as, in the AC field. The behavior of composites in the DC field is described primarily by values of volume resistivity and polarization indexes. Evaluation of composites behavior in AC field is based on temperature and frequency dependencies of dielectric constant and loss factor. These dependencies are obtained by broadband dielectric spectroscopy. Results show that halloysite nanotubes exhibit different dielectric polarization phenomena compared to the modified montmorillonite. In particular, dielectric losses of composites with halloysite nanotubes are lower at high and expected working temperatures than dielectric losses of the compound with modified montmorillonite. Relaxation and even interfacial polarization were detected based on measurements. The influence of chemical composition and structure of different nanofillers on the dielectric properties is described.

Keywords: Broadband dielectric spectroscopy, dielectric absorption, dielectric constant, halloysite nanotubes, linear low-density polyethylene, loss factor, montmorillonite, nanofillers, polarizations, volume resistivity.
PACS: 77.84.Jd, 77.84.Lf, 61.46.Df, 61.46.Np, 81.40.Tv, 77.22.-d.

INTRODUCTION

Composites based on a polyolefin matrix with intercalated nanoparticles of phyllosilicates become more common in the material research, especially in the field of insulation systems of electrical installation and power lines. The Montmorillonite (MMT) and the Halloysite nanotubes (HNT) can be mentioned as the types of phyllosilicates. Especially, the montmorillonite is currently most common and used phyllosilicate as a nanofiller in this field [1-3]. The application of Halloysite nanotubes is significantly lower, despite of a similar advantages as MMT performing in the polymer nanocomposites. Montmorillonite, As well as, Halloysite nanotubes are known as a perspective nanofiller thanks to their ability to retard fire and minimize smoke generation [4, 5]. During development of a new material in this field is necessary to observe the behavior of material in the electric field in the wide range of temperature. The behavior in the electric field may be significantly different for various types of phyllosilicates. The different behavior of phyllosilicates in an electric field is based on variances in chemical composition, but also on the diversity of structure (e.g., different parameters of bonds between oxygen and hydrogen in the structure of phyllosilicates and the incorporation of water molecules). Dielectric properties of particles of phyllosilicates, which are dispersed (aggregated) in a polymer matrix, clearly affect the electrical conductivity and the formation of polarization phenomena in composites. If the polymer matrix is non-polar, such as polyethylene, similar polarization processes, which are detected in separate phyllosilicates, occur in composites (only attenuated). The incorporation of nanofiller in polyolefine, specifically in polyethylene, is generally associated with an increase in dielectric losses. These losses can increase significantly in AC field due to the achievement of polarization peak at a specific operating frequency and operating temperature.

EXPERIMENTAL

Materials and Preparation of Samples

First used nanofiller is a natural material with trade name Dragonite HP containing more than 95 % of HNT (HNT) [6]. The second nanofiller Cloisite 20A is a montmorillonite modified with a quaternary ammonium salt (MMT). Modification of montmorillonite was used to ensure a better dispersion of particles (to minimize the formation of agglomerates). Halloysite nanotubes exhibit less tendency to agglomeration due to a different surface tension and can be used without surface modification. Structural basis of mentioned nanofillers is aluminosilicate layer, generally consisted of interconnected Si-O tetrahedrons and Al-O octahedrons [7]. Montmorillonite is the phyllosilicate whose basic formations represent aluminosilicate layers as laminated flat plates. The nanotubular structure of HNT is formed as a helically arranged aluminosilicate layer. The helical structure is formed by natural causes, mainly due to a high pressure and temperature during the formation of material in the earth's crust. Between the inner and outer sides of the base layer (in the interlayer of a nanotube) is allocated molecular and weakly bonded water [7]. Chemical and physical properties of HNT (including dielectric properties [8]) are significantly changed when the HNT is dehydrated. Noticeable change in properties due to the temperature induced dehydration exhibits also in MMT. The base polymer, which was used as a matrix of composites, is a linear low-density polyethylene (LLDPE) with a density 0.919 ± 0.003 g·cm^3. Together with the nanofiller a very small amount of additive mixture (compatibilizers and antioxidant) was added during manufacturing of the composites. The first phase of production of samples includes blending of the compound of melted polyethylene, nanofiller and the additive mixture in kneader and plastograph at 190 °C. The multiple pressing was carried out in the second phase. The temperature reached to 200 °C and the maximum pressure was 100 kN. The first sample is the based polymer treated only with a mixture of compatibilizers and antioxidant (hereinafter referred to the LLDPE). Further samples containing 3 wt% or 7 wt% of HNT or MMT. All samples were prepared as foils with a thickness (0.25 ± 0.01) mm.

Measuring Techniques

The course of dielectric absorption at ambient temperature were measured for determining of dielectric properties of tested materials in the **DC field**. During measurement, a charging current was measured by the laboratory device consisting of resistivity test fixture Keithley 8009 and electrometer Keithley 6517A. The resistivity test fixture represents the cell with shielded three-electrode system. The circuit intended for measuring volume resistivity (ρ_V) was used. This circuit allows measuring the current that flows through the sample which is placed between the circular electrodes. Applied measuring voltage is 1000 V and the current which flows through the sample was recorded every second for a period of 100 minutes. The value of the current through the sample is considered as steady at the end of the measured time period (calculated with the average value of the current in the last minute of the measurement). Volume resistivity is calculated from values of steady current, thickness of the sample and the constant of the electrode system. Furthermore, polarization indexes (PI), which represent the ratio of insulation resistances or a directly ratio of charging currents at the beginning and the end of the assessed time interval, were calculated. The first polarization index (PI_1) is the ratio of charging current in the fifteenth and the sixtieth second. Second polarization index (PI_{10}) is the ratio of charging currents which are recorded at the end of the first minute and at the end of the tenth minute after applying the voltage. Evaluated charging (total) current is considered as the sum of absorption (polarization) and conduction (leakage) current. Influence of capacitive current is not mentioned because it disappears sooner than the first evaluated value of current is recorded. More information about implementation of the measurement in the DC field are presented in the Standards [9, 10].

The determination of dielectric properties in the **AC field** was carried out by broadband dielectric spectroscopy (BDS) [11]. Measurements were performed using a High-performance frequency analyzer Novocontrol Alpha-A with active sample cell ZGS. Samples of materials were placed in the electrode system between the parallel circular electrodes with a diameter of 30 mm. The output of the measurements are values of the real part (dielectric constant) and the imaginary part (loss factor) of complex relative permittivity. The values of dielectric constant (ε') and loss factor (ε'') are dependent on the frequency (f) of the applied voltage and on the temperature (T). The results are therefore presented via frequency and temperature dependencies of ε' and ε''. The dielectric constant can be regarded as characteristic of the capacity of the measured material. The loss factor is associated with dielectric loss. The amplitude of measuring AC voltage was set to a value of 1 V_{rms}. All measurements were performed in the frequency range from 5.0E-1 Hz to 1.0E6 Hz and the temperature range from -150 °C to 110 °C. Data was recorded with a temperature step of 5 °C. Based on previously performed measurements, the usage of BDS in reheating mode (two cycles of heating from -150 °C to 110 °C in one measurement) for evaluation of material behavior in an electric field seems to be very useful. Differences in the characteristics, which are recorded for the first and the second heating, may allow distinguishing whether the recorded processes are reversible or irreversible after the exposure to maximal used temperatures.

RESULTS AND DISCUSSION

Results of measurement of dielectric properties in the **DC field** are summarized in Table (1). The values of ρ_V calculated from the steady state value of the charging current, which is ideally equal to the conduction current, decrease slightly with increasing of the nanofiller mass fraction in composites. Values of ρ_V differ in one order of magnitude. The slightly greater decrease of ρ_V, compared to the value for LLDPE, is observed for the composites containing MMT. Polarization indexes are the lowest for pure LLDPE, which corresponds to the assumption of minimal changes in the value of absorption current in a certain time interval for this non-polar material. The increase of polarization indexes after nanofiller adding corresponds to a sharper decline of absorption current. This decline is probably caused by dissipation of bigger quantity of electrical charge which is accumulated on surface of filler particles and on larger inhomogeneities in the measured time interval. Creation of more inhomogeneities corresponds to a higher amount of nanofiller in the polymer matrix. The difference of values of absorption current at the initial time interval is also related to the disappearance of polarization phenomena that occur after adding of nanofiller. PI_1 grows only marginally (decrease of absorption current caused by rapid polarization), but the increase of PI_{10} is very significant (decrease of absorption current caused by the accumulation of charge in composites). The increase of polarization indexes is again more evident after filling with MMT than after filling with HNT.

TABLE (1). Summary of the important parameters from the testing of materials in DC field.

Samples	ρ_V **(Ω·cm^{-1})**	PI_1 **(-)**	PI_{10} **(-)**
LLDPE (matrix)	2.70E18	2.41	1.96
LLDPE + 3 wt% HNT	2.35E18	3.09	4.53
LLDPE + 7 wt% HNT	1.92E18	3.11	6.35
LLDPE + 3 wt% MMT	2.14E18	3.83	7.37
LLDPE + 7 wt% MMT	1.32E18	3.86	9.55

Measurements results of the first heating for all tested materials in **AC field** show Figures 1 and 2 except results for pure LLDPE. Values of ε'' for LLDPE are very low (near the limit of device resolution) due to the non-polar character of this material and, therefore, measured values are not conclusive. For this reason, own results for LLDPE are not shown and the results of similar measurement can be found in the literature [3]. HNT and MMT are very polar materials, therefore, prepared composites also exhibit polar character after filling a very small amount of nanofiller. Figure 1 shows measured values of ε' and ε'' of composites containing a different level of HNT filling. Values of ε' and ε'' grow with increasing level of filling in whole used frequency and temperature range. The trend of decreasing of ε' was recorded when the temperature rose and this trend is determined by the typical behavior for LLDPE [3]. This trend is deviated from the results of pure LLDPE [3] by a slight increase of ε' in the region where the polarization processes, which begin to occur in composite after filling with HNT, are manifested. Figure 1b give information about two peaks of relaxation polarizations and also the formation of interfacial polarization (MWS [3]) can be observed in the region of higher temperatures and low frequencies. The increase of ε'' at the high temperatures and low frequencies is caused primarily by thermally induced electrical conductivity. Figure 2 shows the measurement results for the first heating of composites with MMT. The trend of the ε' at low temperatures is similar to the pure LLDPE, but at a certain temperature (e.g., T = 20 °C for f = 50 Hz) is recorded sharp increase of ε'. A significant difference in the behavior of composites with HNT and MMT is very obvious from the characteristics of ε' and ε''. This difference is most likely caused by the different polarizations which correspond to diverse chemical composition related to the modification of MMT. The reason for the different behavior is also the different structure of used nanofillers (the different character of bonds in the lamellar or helical structure). Values of ε'' become higher with increasing of MMT mass fraction in the whole used frequency or temperature range. Usually, evident signs of electric conductivity at high temperatures and low frequencies are in composites with MMT largely overlapped by presence of polarization. This polarization appears in composites with MMT generally at higher temperatures than in the composites filled with HNT.

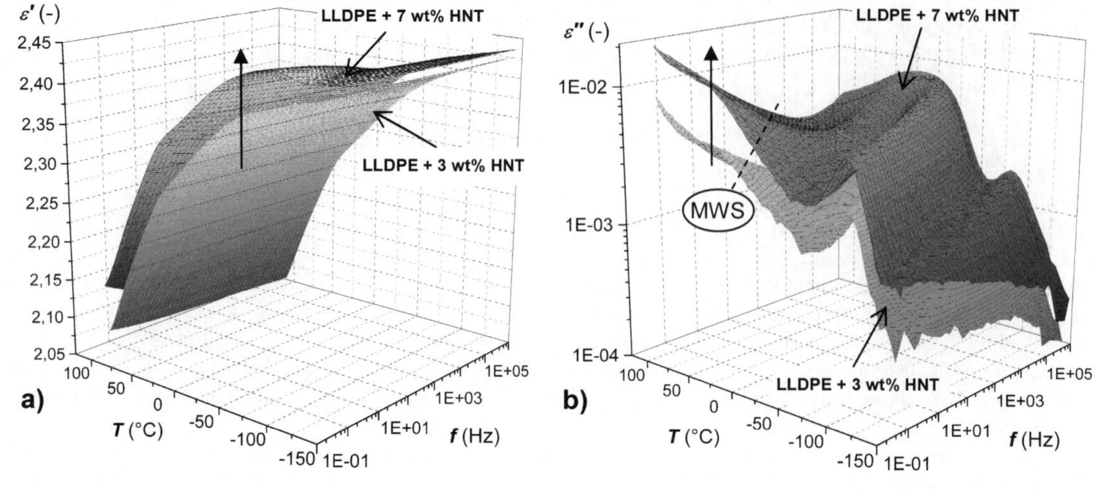

FIGURE 1. 3D temperature-frequency dependencies of a) ε' and b) ε'' for LLDPE with 3 wt% and 7 wt% of HNT.

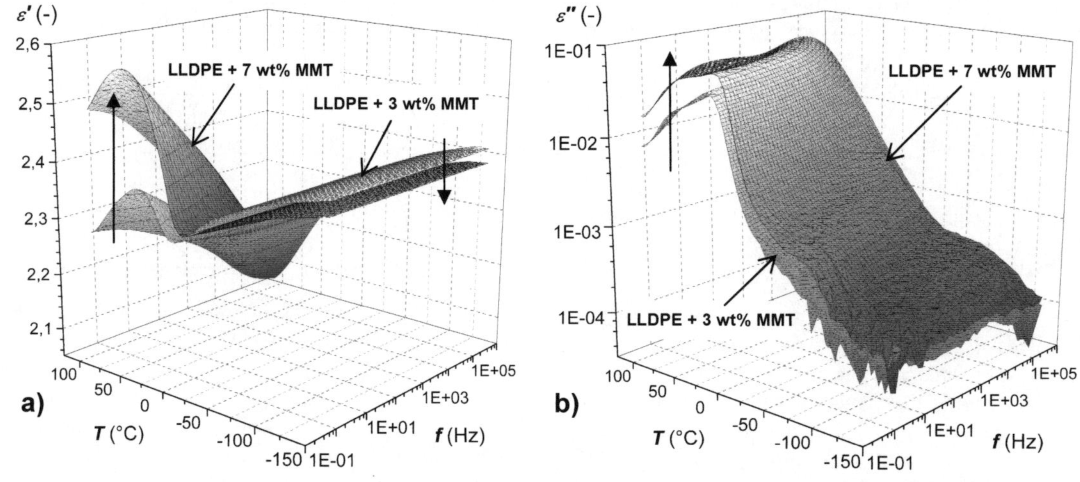

FIGURE 2. 3D temperature-frequency dependencies of a) ε' and b) ε'' for LLDPE with 3 wt% and 7 wt% of MMT.

Comparison of measured data for the first heating and second heating of composites with HNT gives information about change in degree of crystallinity of LLDPE, slight water evaporation and evident disappearance of visible interfacial polarization. Figure 3a shows a change in effects of interfacial polarization in the composite with HNT due to the reheating. It may indicate a slight change in the composite structure at a temperature near the melting point of LLDPE. Additionally, a slight increase of ε' (from 2 % to 2.5 %) at low and medium temperatures appeared. The minimal differences, that manifest as a slight rise in ε' (up to 2 %) and as an attenuation of polarization in the central part of the temperature-frequency dependence, are detected for composites with MMT (not shown in figure).

Figure 3b shows the different values of ε'' at the common power network frequency of 50 Hz and a temperature of 25 °C for composites with different nanofiller. The value of loss factor under the mentioned conditions (f = 50 Hz, T = 25 °C) is ε'' = 4.14E-3 for composite with 7 wt% of HNT and the value for composite with 7 wt% of MMT is ε'' = 2.24E-2. If the temperature rises to 50 °C, the loss factor for composite with the same level of HNT filling reaches the value of ε'' = 4.58E-3 and value for composite with MMT is ε'' = 5.97E-2. Consequently, the usage of HNT as nanofiller in electrical industry (cables for low-frequency power distribution) is more advantageous than the use of MMT in order to minimization of dielectric losses. The value of ε'' for composite with HNT is lower than ε'' for composite with MMT even with a lower filling level. However, the difference between values of ε'' for composites with different nanofiller is reduced according to the decreasing of nanofiller amount.

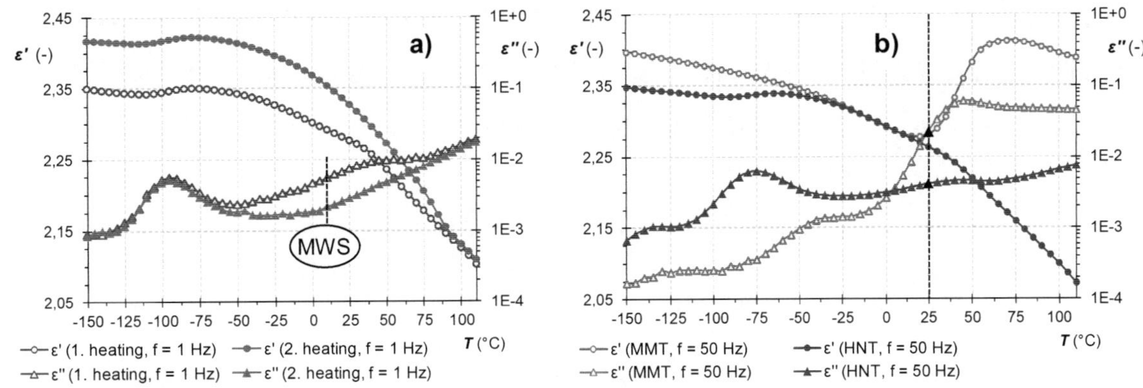

FIGURE 3. Detail a) of minimization of interfacial (MWS) polarization in HNT due to reheating and b) of different values of loss factor for composites with MMT and HNT under common operating conditions.

CONCLUSION

Results of measurements in the DC field indicate that the increasing of nanofiller amount causes a decrease of ρ_V of the composite, whereas the greater decrease is observed after the filling with MMT. The rise of polarization indexes relating to filling level demonstrates the increase in maximum absorption currents in the initial phase of measurement and their subsequent sharper drop. This trend is caused by the increase in manifestations of polarization phenomena and amount of surface charge of filler particles. Measuring of BDS characteristics in the AC field prove a considerable difference in a dielectric behavior of composites with HNT and MMT as nanofiller. This difference is related to the modification of MMT and the different structure of the filler particles. Composites with HNT exhibit lower values of loss factor in the region of common operating temperatures and frequencies for power distribution systems and it can be important for the selection of the material of insulation system for this application.

ACKNOWLEDGMENTS

This research has been supported by the Ministry of Education, Youth and Sports of the Czech Republic under the RICE – New Technologies and Concepts for Smart Industrial Systems, project No. LO1607, and by the Student Grant Agency of the University of West Bohemia, grant No. SGS-2015-020 "Technology and Material Systems in Electrical Engineering".

REFERENCES

1. V. Tomer, G. Polizos, C. A. Randall and E. Manias, "Polyethylene nanocomposite dielectrics: Implications of nanofiller orientation on high field properties and energy storage" in *Journal of Applied Physics*, 2011, vol. 109.
2. X. Li, M. Xu, K. Zhang, D. Xie, X. Cao and X. Liu, "Influence of organic intercalants on the morphology and dielectric properties of XLPE/montmorillonite nanocomposite dielectrics" in *IEEE T. Dielect. El. In.*, 2014, vol. 21, pp. 1705-1717.
3. B. Zazoum, E. David and A. D. Ngô, "LDPE/HDPE/Clay Nanocomposites: Effects of Compatibilizer on the Structure and Dielectric Response" in *Journal of Nanotechnology*. 2013, pp. 1-10.
4. Z. Jia, Y. Luo, B. Guo, B. Yang, M. Du and D. Jia, "Reinforcing and Flame-Retardant Effect of Halloysite Nanotubes on LLDPE" in *Polymer-Plastics Technology and Engineering*, 2009, vol. 48, pp. 607-613.
5. A. B. Morgan and Ch. A. Wilkie, *Non-halogenated flame retardant handbook*. Hoboken, New Jersey: Wiley, 2014.
6. F. J. García, S. García Rodríguez, A. Kalytta and A. Reller, "Study of Natural Halloysite from the Dragon Mine, Utah (USA)" in *Zeitschrift für anorganische und allgemeine Chemie*, 2009, vol. 635, pp. 790-795.
7. P. Yuan, D. Tan and F. Annabi-Bergaya, "Properties and applications of halloysite nanotubes: recent research advances and future prospects" in *Applied Clay Science*, 2015, vol. 112-113, pp. 75-93.
8. M. Adamczyk, M. Rok, A. Wolny and K. Orzechowski, "Dielectric properties of halloysite and halloysite-formamide intercalate" in *Journal of Applied Physics*, 2014, vol. 115.
9. Electric Machinery Committee of the IEEE Power Engineering Society, *IEEE recommended practice for testing insulation resistance of rotating machinery, IEEE Std 43-2000(R2006)*. New York, USA: IEEE, 2006.
10. International Standard IEC 62631 (Part 3-3), *Dielectric and resistive properties of solid insulating materials*.
11. F. Kremer, F and A. Schönhals, *Broadband dielectric spectroscopy*, New York: Springer, 2003.

Investigation of Polypyrrole/Molybdenum trioxide/Graphene nanoribbon nanocomposite with high electrochemical capacity performance for electrode material of supercapacitor

F. H. Hsu and T. M. Wu

Department of Materials Science and Engineering, National Chung Hsing University, 250 Kuo Kuang Rd., Taichung, Taiwan

Abstract. A ternary nanocomposite of polypyrrole/molybdenum trioxide/graphene nanoribbon (PPy/MoO_3/GNR) was successfully synthesized by using *in situ* oxidation polymerization method. The Fourier transform infrared (FTIR) spectra shows the metal oxide of MoO_3 is successfully combined with GNR. The morphology is observed by using field emission scanning electron microscope (FESEM) and transmission electron microscopy (TEM). The electrochemical analysis, which is investigated by using electrochemical analysis instrument, shows the highest specific capacitance of 844 F g^{-1} and good cycle stability after 1000 cycles for PPy/MoO_3/GNR ternary nanocomposite. These results demonstrate that the combination of each component can efficiently enhance the electrochemical capacity performance.
Keywords: Polypyrrole, graphene nanoribbon, molybdenum trioxide, electrochemical capacity
PACS: **82.45.Yz**

Properties and morphology of PPy/MoO_3/GNR nanocomposite

The pure PPy curve is shown in Figure 1a. The peaks at 1550 cm^{-1} and 1455 cm^{-1} are attributed to the C-C and C-N stretching vibration in the pyrrole ring. Others, the absorption peak at 905 cm^{-1} is assigned to C-H out-of-plane vibrations, respectively [1]. In the Figure 1b, three significant peaks at 995 cm^{-1}, 863 cm^{-1} and 554 cm^{-1} for MoO_3/GNR composite, which are assigned to Mo=O, Mo-O and Mo-O-Mo, respectively [2], indicating the existence of MoO_3 in GNR matrix. In the Figure 1c, the spectra of PPy/MoO_3/GNR nanocomposite show the absorption peaks are similar to pure PPy. Additionally, it can clearly observe that the absorption peaks of PPy at 1550 cm^{-1} and 905 cm^{-1} slightly shift to about 1540 cm^{-1} and 885 cm^{-1} in these nanocomposites after combination of each component. This result is attributed to the interaction between PPy and MoO_3/GNR.

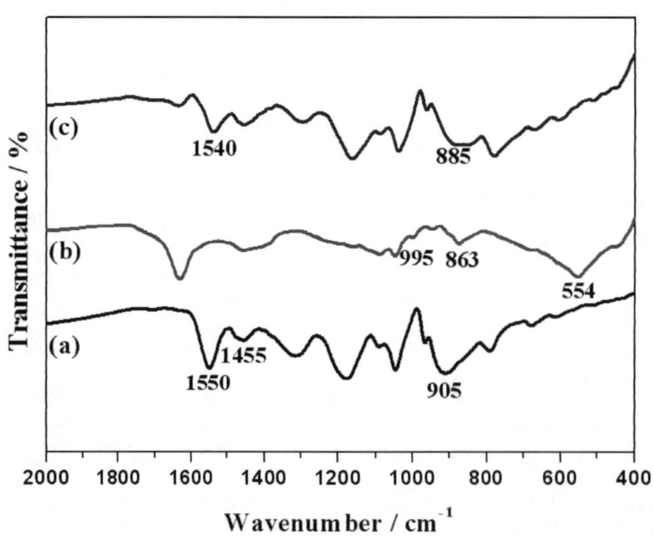

FIGURE 1. The FT-IR spectrum of (a) Pure PPy, (b) MoO_3/GNR, and (c) PPy/MoO_3/GNR nanocomposite

The TEM images of pristine MWCNT and GNR are shown in Figure 2a and 2b. The diameters of MWCNT are about 50 nm with a smooth edge. After chemical modification, the tube-like structure of MWCNT was completely unzipped because the oxidant attacks the MWCNT axial structure to form GNR with a diameter about 100 nm. This result can prove that the GNR is successfully prepared by using chemical oxidation method.

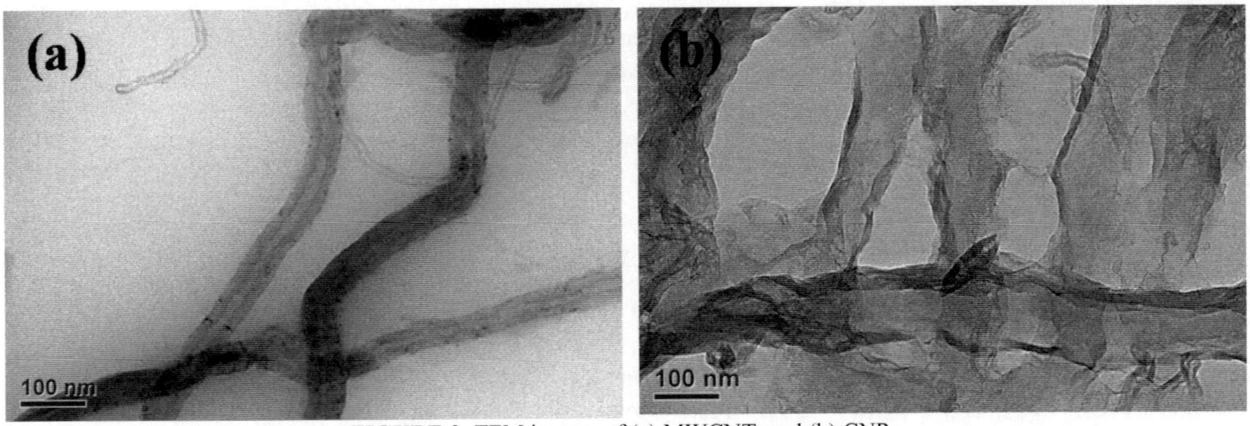

FIGURE 2. TEM images of (a) MWCNT, and (b) GNR

In the Figure 3a, the FESEM and TEM images of pure PPy show the aggregation nanoparticles with a diameter about 150 nm. The PPy/MoO_3/GNR ternary nanocomposite is shown in Figure 3b. One-dimensional morphology of PPy/MoO_3/GNR ternary nanocomposite can be observed. From the TEM images of nanocomposite, the PPy was coated on the surface of GNR and then increased the diameter of GNR.

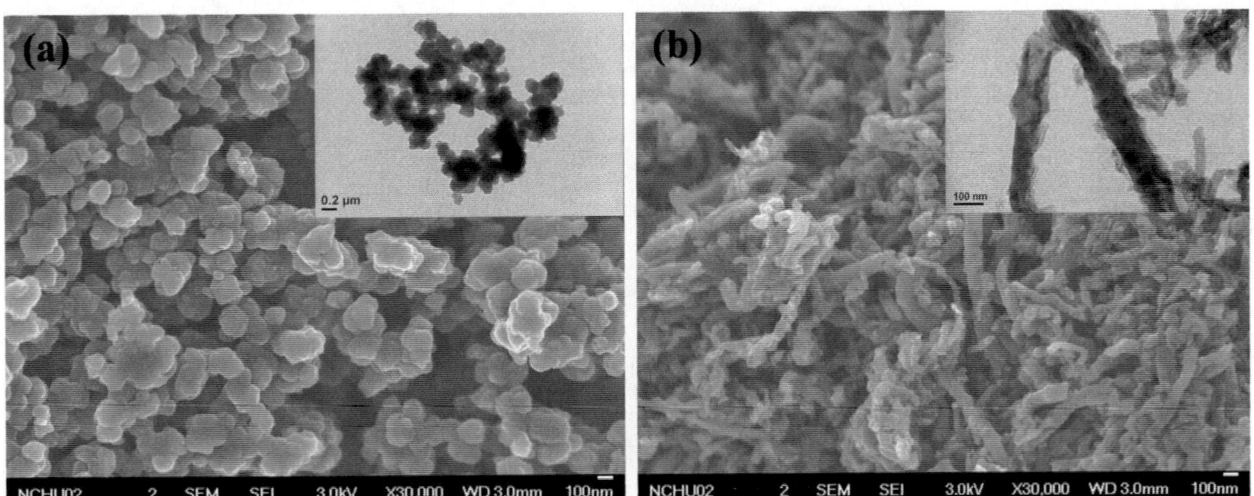

FIGURE 3. FE-SEM and TEM images of (a) Pure PPy, and (b) PPy/MoO_3/GNR nanocomposite

Electrochemical properties of PPy/MoO_3/GNR nanocomposite

The electrochemical properties of the active electrodes were assembled into a three-electrode cell system by cyclic voltammetry (CV) test. The CV test results of pure PPy and PPy/MoO_3/GNR nanocomposite at scan rate of 5 mV/s were shown in Figure 4a. It could be seen that the current density and the area of CV curve for the PPy/MoO_3/GNR nanocomposite was significantly larger than that of pure PPy, indicating that the electrochemical performance and electric capacitance of PPy/MoO_3/GNR nanocomposites were higher than pure PPy. The specific capacitance (C_{sp}) of the pure PPy and PPy/MoO_3/GNR nanocomposites were calculated from CV curves by using the following equation [3]:

$$C_{sp} = \frac{1}{2mvV}\int I\,dV$$

where C_{sp} is the specific capacitance (F/g), ν is the potential scan rate (mV/s), *m* is the mass of active sample on the glassy carbon electrode, $\int$ I dV term is the area of CV curve, and V is the scan potential range.

The calculated specific capacitance is shown in Table 1. It can clearly observe that the PPy/GNR nanocomposite has a specific capacitance about 844 F/g at scan rate of 5 mV/s, which much higher than pure PPy (243 F/g). The improvement of specific capacitance might be attributed to the synergistic effect between PPy, MoO_3 and GNR [4].

Galvanostatic charge/discharge is also a proof for the electrochemical performance and specific capacitance property, which was measured in the potential range of -0.2 V to 0.8 V at a current density of 5 A g^{-1}. In the Figure 4b, it can be clearly seen that the nanocomposite of PPy/MoO_3/GNR shows the longer charge/discharge time than pure PPy. This result also shows that the combination of EDLC and pseudocapacitor could increase the electrochemical performance and specific capacitance by synergistic effect.

The cycle life of pure PPy and PPy/MoO_3/GNR nanocomposite were shown in Figure 4c. It could be seen that the PPy/MoO_3/GNR nanocomposite exhibits excellent cycle stability and decreases only 14 % of specific capacitance after repeating 1000 cycles, which is lower than the same process of pure PPy (40%). The improvement of cyclic life might be attributed to GNR has good electrochemical stability. This result shows the synergistic and interaction between PPy, MoO_3 and GNR would be increased the cyclic stability of nanocomposite [4].

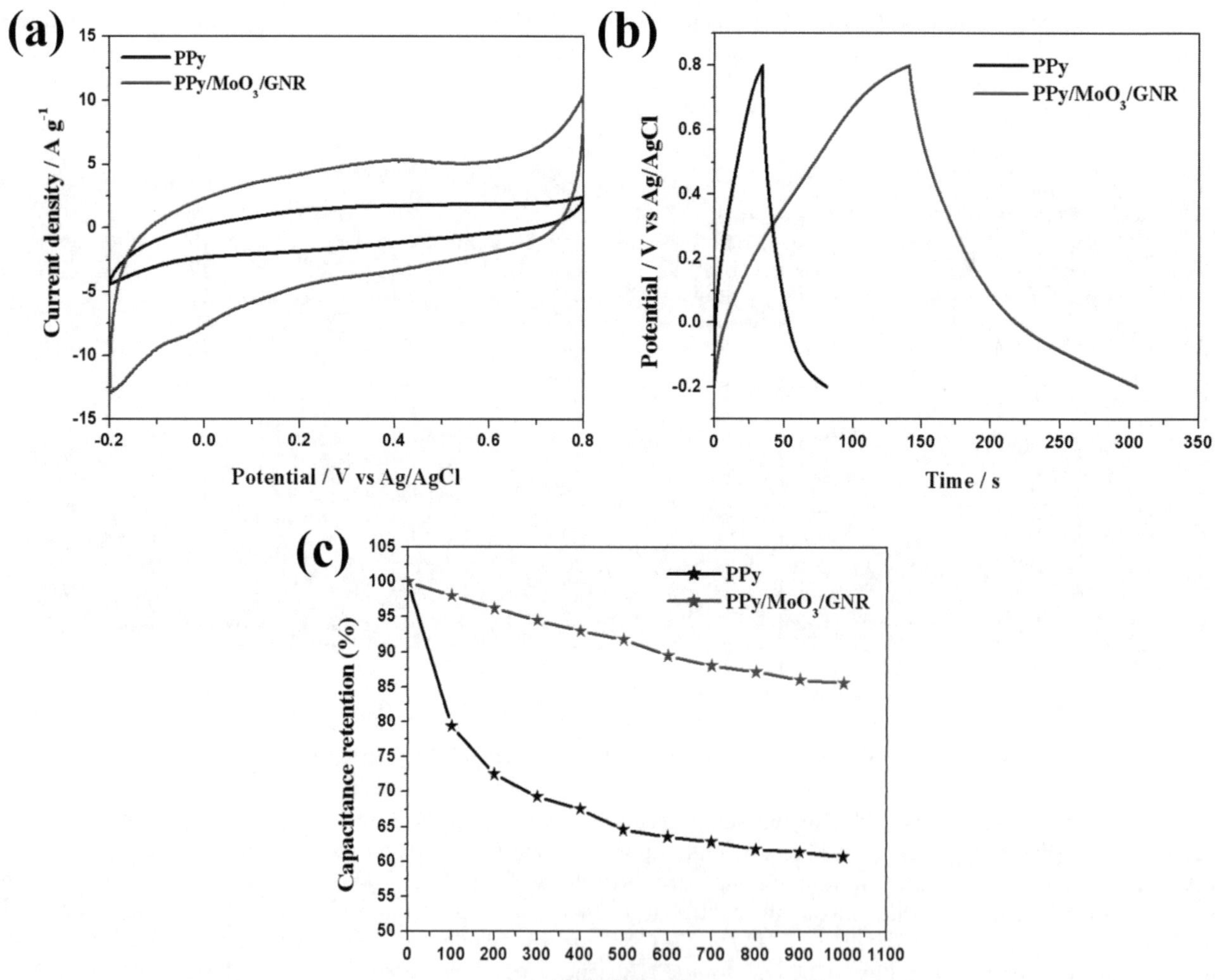

FIGURE 4. Electrochemical measurements of (a) CV test at a scan rate 5 mV s^{-1}, (b) Galvanostatic charge/discharge test at a current density 5 A g^{-1}, and (c) cyclic stability of each component

TABLE 1. Specific capacitance of pure PPy and $PPy/MoO_3/GNR$ nanocomposite at a scan rate 5 mV s^{-1}

	Scan rate (5 mV s^{-1})
Pure PPy	243 F g^{-1}
$PPy/MoO_3/GNR$ nanocomposite	844 F g^{-1}

REFERENCES

1. X. Zhang, X. Z. Zeng, M. Yang, Y. X. Qi, *ACS Appl Mater Interfaces* **6** (2014), 1125-1130.
2. Ch.V. Subba Reddy, E. H. Walker Jr., W. Chen, S. I. Mho, *J Power Sources* **183** (2008), 330-333.
3. A. Davies, P. Audette, B. Farrow, F. Hassan, Z. W. Chen, J. Y. Choi, A. P. Yu, *J Phys Chem C* **115** (2011), 17612-17620.
4. C. Xu, J. Sun, L. Gao, *J. Mater. Chem.*, **21** (2011), 11253-11258.

Oxidization of Carbon Nanofiber and Its Effect on Tensile Properties of PA11 Matrix Composite Reinforced with Carbon Nanofiber

K. Enomoto[a*] and Y. Takahashi[b]

[a] *Department of materials science and engineering, Meijo University (1-501, Shiogamaguchi, Tempaku-ku, Nagoya, Aichi, Japan)*
[b] *Division of materials science and engineering, Graduate school of Meijo University (1-501, Shiogamaguchi, Tempaku-ku, Nagoya, Aichi, Japan)*

Presenting and Reference author's email: enomotok@meijo-u.ac.jp

Abstract. In order to improve interfacial adhesion between the thermoplastic matrix and the carbon nanofibers (CNFs), two-step oxidization by using mixed acid and hydrogen peroxide was performed on the CNFs. Moreover, polyamide 11 (PA11) based composites reinforced with the oxidized CNFs were fabricated, and tensile properties of the composite were evaluated. The effect of 2-step oxidization on the tensile properties was discussed. From the results of the X-ray diffraction and XPS, the functional groups including oxygen were introduced to the surface of CNFs by 2-step oxidation process. However, the oxidization by using strong mixed acid destroyed the structure of CNFs excessively. On the other hand, when the mild oxidization by using a hydrogen peroxide was carried out after strong oxidization in a short time, the structure of CNFs was not destroyed and was able to give functional groups including oxygen. As the result, tensile property of the composites reinforced with mild oxidized CNF improved in comparison with the composites reinforced with strong oxidized CNF. It is suggested that the mild oxidization in the second oxidation process was effective to improve the tensile properties of the PA11 matrix composites.

Keywords: carbon nanofiber, polyamide 11, composite, oxidation, tensile strength
PACS: 81.05.Qk

1. INTRODUCTION

Carbon nanofibers (CNFs) comprise of many overlapping tubular graphite layers, and have diameters in the range of 50 nm to 200 nm; therefore, they can be classified as ultra-multiwalled carbon nanotubes having relatively large diameters. CNFs have outstanding properties such as high modulus, high strength, and high electrical and thermal conductivities along their axial direction due to the high crystallinity of the sp^2 carbon network. Therefore, CNFs are expected to be one of the most promising materials for improving various properties of polymer matrix composites. Further, many studies focus on the possibility of using CNFs for reinforcing the composites[1-4]. However, in order to realize high performance composites, strong interfacial adhesion between the CNFs and the matrix polymer is imperative.

Polyamide 11 (PA11) is one of the bio-based polymers derived from castor oil. PA11 has been used for fuel tubes because of its flexibility, however, it has gathered attention again from a viewpoint of carbon neutral in recent years. Improvement of mechanical properties of PA11 is indispensable to expand the application fields of PA11, and then various studies and investigation were performed such as compounding with fibers and blending with other polymers.

In our previous study[5], we attempted to improve the interfacial adhesion between the CNFs and the PA11 matrix by the oxidization of the CNFs by using an excess of mixed acid. As results, the carboxyl groups were introduced to the surface of CNFs by oxidization process. In the case of adding oxidized-CNF, the increase rate of the maximum tensile stress was improved in comparison with the case of adding CNF. The elongation at break was also improved in comparison with the case of adding CNF. However, the long oxidization process by using a mixed acid destroyed the structure of CNFs excessively, and then sufficient reinforcement effect was not obtained.

In this study, the conditions of the oxidation process was reconsidered, we attempted the mild oxidation by using hydrogen peroxide after the short oxidation process by using a mixed acid. After the oxidation process, PA11 based composites reinforced with the oxidized CNFs were fabricated, and tensile properties of the composite were evaluated. The effect of the 2-step oxidization on the tensile properties was discussed.

2. 2-STEP OXIDIZATION OF CNF

2.1 Materials and procedure

CNFs employed in this study were vapor-grown carbon fibers, which were produced by Showa Denko, K.K. fabricated by thermal chemical vapor deposition and graphitized at 2800 °C in the production process. The typical diameter of the CNFs is about 100 nm with a nominal length of approximately 10 μm to 20 μm.

The procedures of the 2-step oxidization are as follows. In the first step, 5 grams of CNFs were heated with an mixed strong acid (60 ml of nitric acid and 180 ml of sulfuric acid) up to 110 °C and stirred for 0.5 h. After that, oxidized CNFs (o-CNFs) were separated from the mixed acid by using membrane micro-filtration (pore size: 0.5 μm). The separated o-CNFs were washed with distilled water and dried in a vacuum oven for 24 hours, and then finely milled. In the second step, 3 grams of milled o-CNFs were stirred with a hydrogen peroxide (350 ml) and sonication for 10 minutes. After that, the solution including o-CNF was heated up to 80 °C and stirred for several hours. Then, CNFs oxidized by hydrogen peroxide (oh-CNFs) were separated from the acid solution by using membrane micro-filtration (pore size: 0.5 μm). The separated oh-CNFs were washed with distilled water and dried in a vacuum oven for 24 hours, and then also finely milled.

2.2 Characterization of oxidized CNFs

Wide-angle X-ray diffraction (WAXD) was carried out in order to consider the effect of oxidation on structural changes of CNFs. Figure 1 shows changes in diffraction patterns of oxidized CNFs by the differences in oxidation time in the second step. In the case of CNF before oxidation, one strong peak was observed at 2θ=26° due to the graphene sheets. In the case of o-CNF-6h, a small peak was appeared at $2\theta = 24°$. This peak is due to a disturbance or defects of the layered structure of graphene sheets. Therefore, when the oxidation time becomes longer, it is concerned that oxidation process, not only introduces some defects on the surface of CNFs but also destroys the internal layered structure of CNFs. However, in the case of oh-CNFs, although the peak at 2θ=26° became weaker and broader, the peak at 2θ=24° was not appeared. Therefore, this results indicates that 2-step oxidation process with a hydrogen peroxide inhibited the excessive destruction of the layer structure of CNF, and introduced some defects on the surface of CNFs.

Raman spectroscopy was carried out in order to consider the effect of oxidation on crystallinity changes of CNFs. Figure 2 shows changes of I_G/I_D ratio of oxidized CNFs as a function of oxidization time. In the case of CNF, the value of I_G/I_D ratio was large because of the high crystallinity of the sp^2 carbon network by graphitization in the production process. After the first step oxidation, the value of I_G/I_D ratio dramatically decreased because of strong

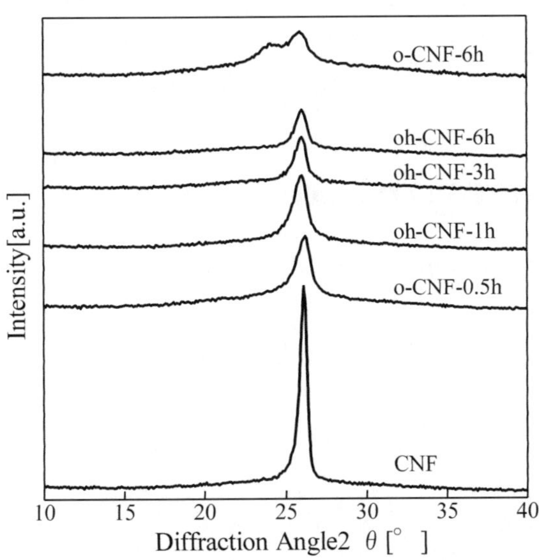

FIGURE 1. X-ray diffraction patterns of CNFs oxidized by using hydrogen peroxide for several hours.

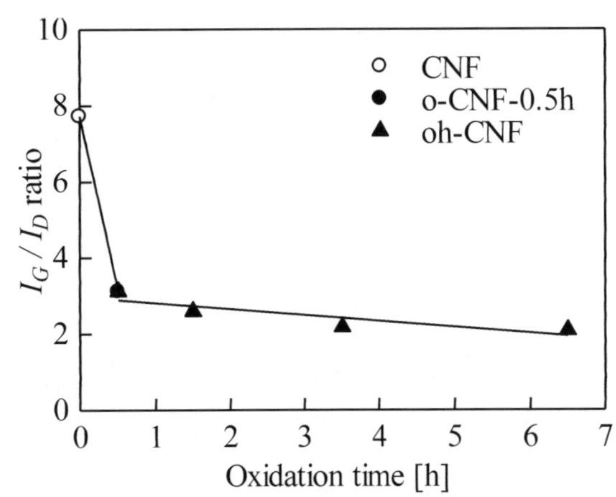

FIGURE 2. Change of I_G/I_D ratio of oxidized CNFs as a function of oxidation time.

oxidation. However, in the mild oxidation process with a hydrogen peroxide, the value of I_G/I_D ratio slightly decreased. This is because that the oxidizability of hydrogen peroxide was weak, so the surface of CNF was destroyed moderately, and some defects were introduced.

In order to observe the changes of the CNF surface, TEM (transmission electron microscope) observation was carried out. Figure 3 shows TEM images of CNF before oxidation, o-CNF-0.5h, oh-CNF-6h and o-CNF-6h. From Fig.3(a), the CNF before oxidation has no defects and has layer structure of fine graphene sheets. However, in the first oxidation process for 0.5 h, although the fibrous form of CNF was maintained, the surface layers were partially destroyed by strong acid. After that the second step of oxidation by a hydrogen peroxide was performed for 6 h, development of the delamination of the surface layer from the minute irregularity formed at the first oxidation process was observed (Fig.3(c)). In our previous study, oxidation for 6 h by a strong mixed acid, CNF was destroyed intensely (Fig.3(d)). Therefore, this result indicate that it is possible to introduce defects to only the surface of CNFs while restraining damages of CNF by 2-step oxidation with a hydrogen peroxide.

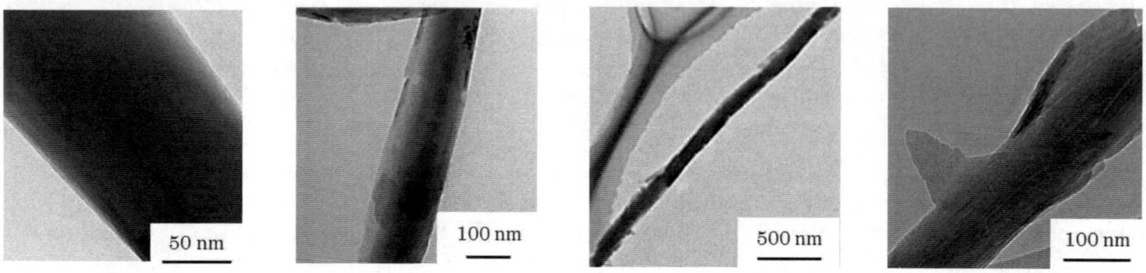

(a) CNF before oxidation (b) o-CNF-0.5h (c) oh-CNF-6h (d) o-CNF-6h

FIGURE 3. TEM images of CNF before oxidation, o-CNF-0.5h, oh-CNF-6h and o-CNF-6h.

In order to characterize the functional groups formed on the defects generated by the oxidization process, X-ray photoelectron spectroscopy (XPS) were carried out. From the results of wide-range measurements, only the peaks derived from C1s and O1s were appeared, so only functional groups including oxygen were introduced by mild oxidation like the case of conventional strong oxidation. Figure 4 shows the changes of the atomic percent of oxygen as a function of the total oxidation time. The atomic percent of oxygen was calculated from the peak area ratio (O_{1S}/C_{1S}) in the spectra of the XPS wide scan measurements[6]. The atomic percent of oxygen increased dramatically by strong oxidation process for 0.5 h in the first step, and then The atomic percent of oxygen slightly increased by mild oxidation for 1 h in the second oxidation process. However, in the case of mild oxidation longer than 1 h, the atomic percent of oxygen decreased. This is because that the delamination of the surface layer introduced functional groups including oxygen was developed from the minute irregularities formed at the first oxidation process as it was indicated Fig. 3(b) and (c) [7]. These results indicate that the mild oxidation in the second oxidation process reduces the crystallinity of CNFs and the amount of functional groups including oxygen, and then introduces the defects and the minute irregularities to only the surface layer of CNFs.

3. EVATUATION OF TENSILE PROPERTIES

3.1 Fabrication of specimens

Polyamide11 is one of the bio-based polymers derived from castor oil. An injection-molding-grade PA11 (Rilsan BMNO, supplied by Arkema) was selected as a matrix polymer. The PA11 matrix composites were obtained via kneading by using a single-screw extruder. The weight fraction of the filler content in the composites was 2.5 %. Tensile-test specimens were fabricated by using injection molding. The dimensions of the specimens were 75 mm in length; 10 mm in width, and 2mm in thickness, 25 mm in length and 5 mm in width at the parallel section.

3.2 Tensile tests

Tensile tests were performed in order to evaluate the fundamental mechanical properties of the CNF/PA11 composites. Tensile tests were carried out in accordance with JIS K 7161 at room temperature in air by using

AUTOGRAPH AG-50kN from Shimadzu Co. Test conditions were as follows; tensile speed was 5 mm/s, initial chuck clearance was 48 mm. Nominal stress – nominal strain curves were obtained from the tests, and the maximum tensile stress before necking was calculated.

Figure 5 shows the maximum tensile stress as a function of the total oxidation time. In the case of the strong oxidation, the maximum tensile stress increased to a peak value at 0.5 h of oxidation time, after that it decreased with the increase of oxidation time. This is because that the layer structure of CNF was destroyed intensely by the strong oxidation as shown in Fig.3(d). On the other hand, in the case of the mild oxidation by using a hydrogen peroxide in the second oxidation process, the maximum tensile stress was almost constant with the increase of the oxidation time. This result suggests that interfacial adhesion between oxidized CNF and the matrix was improved instead of the reduction of the crystallinity and the functional groups including oxygen. This is because that the anchor effect was developed at the region of the minute irregularities formed the mild oxidation in the second oxidation process.

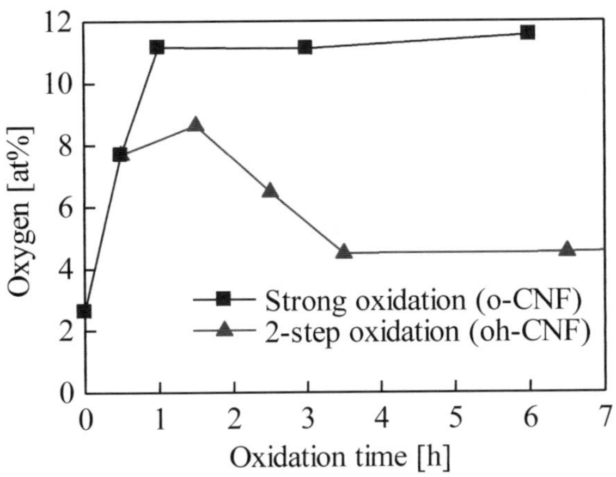

FIGURE 4. Changes of the atomic percent of oxygen calculated from the result of XPS wide scan as a function of the total oxidation time.

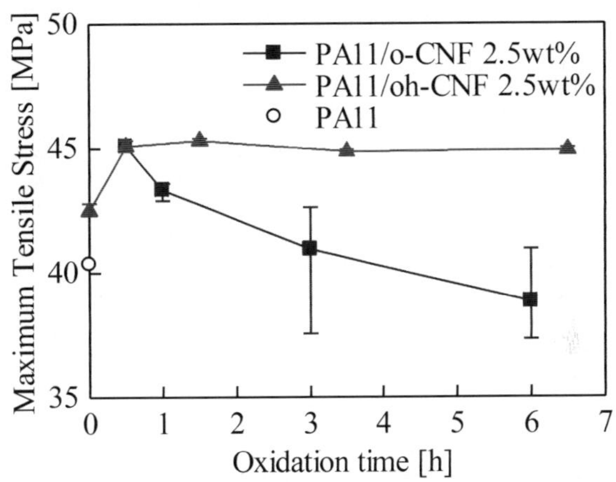

FIGURE 5. Changes of the maximum tensile stress as a function of the total oxidation time.

4. SUMMARY

In this paper, in order to improve tensile properties of polyamide 11 reinforced with carbon nanofiber, the 2-step oxidation of CNF was attempted. From the results of the XRD, Raman spectroscopy, and XPS analyses, the functional groups including oxygen were introduced to only the surface of CNFs by mild oxidization by using a hydrogen peroxide after the strong oxidation by using a mixed acid. The maximum tensile stress of the composites reinforced with mild oxidized CNF improved in comparison with the composites reinforced with strong oxidized CNF. It is suggested that the mild oxidization in the second oxidation process was effective to improve the tensile properties of the PA11 matrix composites.

REFERENCES

1. G. G. Tibbetts and J. J. McHugh, J. Mater. Res. **14**, pp.2871-2880 (1999).
2. O. S. Carneiro, J. A. Covas, C. A. Bernardo, G. Caldeira, F. W. J. van Hattum, J. M. Ting, R. L. Alig, M. L. Lake, Compos. Sci. Technol. **58**, pp.401-407 (1998).
3. R. J. Kuriger, M. K. Alam, D. P. Anderson, R. L. Jacobsen, Compos. A **33**, pp.53-62 (2002).
4. F. H. Gojny, M. H. G. Wichmann, U. Köpke, B. Fiedler, K. Schulte, Compos. Sci. and Technol. **64**, pp.2363-2371 (2004).
5. Y. Takahashi, Y. Suzumoto, K. Enomoto, Conference Paper of the Polymer Processing Society Europe/Africa Regional Conference 2015 Graz, Accepted.
6. N. Tsujioka, Z.Maekawa, H. Hamada, M. Hojo, Journal of the society of Material Science **46**, pp.163-169 (2009), in Japanese.
7. G. Yamamoto, K. Shirasu, Y. Nozaka, M. Omori, T. Takagi, T. Hashida, Transaction of the Japan Society of Mechanical Engineers Series A 78, pp.843-854 (2012), in Japanese.

Preparation and Characterization of Polyethylene Nanocomposites with Clay and Silver Nanoparticles

Washington Luiz Oliani [a], Helio Fernando Rodrigues Ferreto [a], Luiz Gustavo Hiroki Komatsu [a], Ademar Benevolo Lugao [a], Vijaya Kumar Rangari [b], Nilton Lincopan [c,d], Duclerc Fernandes Parra [a]

[a] *Nuclear and Energy Research Institute, IPEN-CNEN/SP, Av. Prof. Lineu Prestes, 2242 – Cidade Universitária - CEP 05508-000 São Paulo – SP - Brazil*
washoliani@usp.br
[b] *Center for Advanced Materials Science and Engineering Tuskegee University, AL - USA.*
[c] *Department of Microbiology, Institute of Biomedical Sciences, University of Sao Paulo, Brazil*
[d] *Department of Clinical Analysis, School of Pharmacy, University of Sao Paulo, SP, Brazil*

Abstract. Polyethylene is widely used for packing applications. Different grades of polyethylene, LLDPE and LDPE play a major role in the film blowing industry. The use of nanoclay in polymer matrices (non-polar) requires the use of a compatibilizer agent, ethylene graft maleic anhydride copolymer (PE-g-MA), which acts as a bridge for nanoclay-matrix-interaction. In our study, blown films of LLDPE, LDPE, and their blends were produced using a twin screw extruder. The combination of PEs as matrix, montmorillonite (MMT) and silver nanoparticles (AgNPs) intended to improve mechanical and thermal properties. Nanocomposites films with 1 wt% of montmorillonite and 0.5 to 1.0 wt% silver nanoparticles were prepared by blow extrusion and evaluated. Nanoclay intercalation/exfoliation degree was investigated by X-ray diffraction spectroscopy (XRD) and scanning electron microscopy (SEM). Mechanical and thermal properties were evaluated. The results showed optimum dispersion at intercalation/exfoliation levels for PEs:MMT:AgNPs, slight increasing in mechanical and especially improvement in thermal properties.
Keywords: polyethylene, montmorillonite, film blowing, silver nanoparticles.

INTRODUCTION

Polyethylene is a general purpose thermoplastic resin with good processing and mechanical properties as well as relative low price [1].The rheology of linear low density polyethylene (LLDPE), low density polyethylene (LDPE) and blends its relevance to the film blowing process. The poor performance of LLDPE in the blowing phase of film making is thought to be related to the low elongational viscosity of LLDPE mostly due to the linear structure of its macromolecules. In contrast, LDPE performs well in such processes due to its high elongational viscosity related to the presence of long chain branched macromolecules in its structure. [2-4].

Thermal processing such as melt blending, extrusion, and injection molding has been applied for incorporating the antimicrobials into polymers, but it should be considered heat stability of active component and chemical compatibility of polymer matrix and antimicrobials, in order to evenly distribute antimicrobial substance [5].

In recent work [6] the low-density polyethylene (LDPE) nanocomposite films containing different concentrations of silver (Ag) nanoparticles (NPs) (0.5 and 1% of polymer weight, w/w) were manufactured via extrusion. The results indicated that LDPE nanocomposite films containing AgNPs could potentially be used as antimicrobial packaging for food applications.

The ultrasound radiation was used to Ag nanoparticles chemical reduction for addition in polyethylene fibers. The effect of reducing reagent, power of ultrasound radiation, reaction time and temperature in growth of the nanometric Ag were studied. Results show a decrease in the particles size as increasing power of ultrasound irradiation [7].

In other work [8], montmorillonite (Mt) was first ion-exchanged with silver ions (Ag+Mt) then organically modified. Active carrier Mt-layers were then compounded into low density polyethylene (LDPE) polymer using a high shear force co-rotating twin screw extruder. Antibacterial activity of the Ag + OMt–LDPE nanocomposite films was then investigated against powerful gram negative *Escherichia coli* (*E. coli*) bacteria. It seems to have strong synergetic effect between silver ions and OMt nanoparticles as evidenced by the bacterial inhibition properties of thin blow molded Ag + OMt–LDPE nanocomposite samples using liquid surface bacterial testing at a much lower active component concentration.

The interested research [9] describes the synthesis and antimicrobial activity of silver–clay nanohybrid structure that was processed to exhibit a combination of accelerated and diffusion-controlled antimicrobial activity, with long term impact. The antimicrobial activity was assessed in terms of interaction with *Escherichia coli*, where the constituents of the nanohybrid structure play a synergistic role. Clay provides a physically stable surface for nucleation of silver nanoparticles. Additionally, the parallel- stacked layered structure of clay facilitates diffusion-controlled antimicrobial activity of in-situ precipitated silver.

An important research work which corroborates the idea of this work was published by [10], where the silver incorporation in MMT plays an important role due to the possibilities of creating an antibacterial material with slow release of silver species in order to obtain a prolonged antibacterial activity. However, additional studies must be done for understanding the chemical nature of those silver species, along with its biological activity and the mechanisms of silver releases over time. Even though the antibacterial activity has been reported for this material, its potential for application depends on how to overcome the aforementioned issues and how the antibacterial activity could be effectively transferred to complex matrices like polymers.

In the present research work, LDPE/LLDPE has been employed as matrix for films nanocomposites. The compounds have been altered by the addition montmorillonite (MMT) and silver nanoparticle and with the use of ultrasound. Such materials can be very valuable in films for biocide activity applications.

MATERIALS AND METHODS

Materials

LDPE with a melt flow index – MFI (190/2.16) of 0.27 dg min^{-1} and density: 0.922 g mL^{-1}; LLDPE with a melt flow index – MFI (190/2.16) of 0.80 dg min^{-1} and density: 0.920 g mL^{-1}, Braskem (Brazil) was provided in the form of pellets. The LLDPE pellets were placed in a nylon container with nitrogen. The irradiation process of the plastic container was carried out at room temperature and at a dose rate of 5 kGy h^{-1}, using a multipurpose ^{60}Co gamma irradiator. The dose used was 20 kGy monitored by a Harwell Red Perspex 4034 dosimeter. The AgNPs were purchased from Sigma Aldrich, and PVP (average molecular weight = 1,300,000 g mol^{-1}), which acted as a surfactant for the AgNPs, was Plasdone. Silver nanoparticles (AgNPs) were purchased from Sigma Aldrich. The commercial nanoclay Cloisite 20 was provided by BYK Additives Company and antioxidant (BASF Irganox B225ED) was added in small quantities, 2.0 wt.%, to prevent the polyethylene from oxidizing and thermally cross-linking at elevated temperatures. Three different formulations containing the LDPE/LLDPE were prepared and are represented in Tab. 1.

TABLE 1. Composition of constituents of polyethylene nanocomposites (wt.%).

Samples	Matrix	Dose/kGy	Irganox	Cloisite	AgNPs
1) PE 1	LDPE/LLDPE	20	2	-	-
2) PE 5	LDPE/LLDPE	20	2	1	0.5
3) PE 9	LDPE/LLDPE	20	2	1	1.0

The PVP solution was prepared by adding 2 g PVP in 150 mL of absolute alcohol at 60 ° C for 20 minutes under stirring until complete dissolution. The PVP solution was added 0.5 to 1.0 wt.% of AgNPs. Then we used the ultrasound Unique with power 65W for 30 minutes to dispersing and deagglomeration the silver nanoparticles.

Preparation of the Nanocomposites

The LDPE/LLDPE (90/10) pellet were mixed with Irganox B 215 ED in a rotary mixer and maintained under this condition for 24 hours. Then the mixture was processed with the addition of clay (MMT 1% by weight) and silver nanoparticles (AgNPs 0.5 and 1% by weight) in a twin-screw extruder Haake co-rotating, model Rheomex PTW 16/25, Figure 1, with the following processing conditions: the temperature profile (feed to die) was 145-170 °C, with a speed of 100 rpm. After processed, the nanocomposites were granulated in a granulator Primotécnica W-702-3. The PE/MMT-AgNPs films were produced in planar sheet extruder and the material was placed directly into the hopper of the extruder with a temperature profile (feed to die) of 150-175 ºC, screw speed of 50 rpm and torque of 33-45 Nm. The films were produced with a thickness of ~ 0.05 mm.

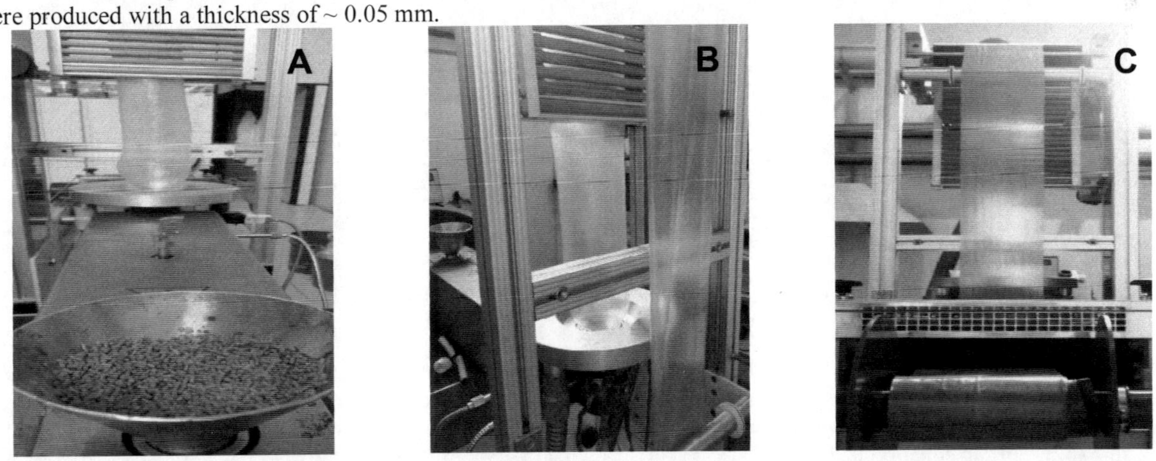

FIGURE 1. (A) Haake film blowing unit used to prepare the blown films, (B) Stable bubble formation of polyethylene nanocomposite, and (C) The blown film take-off transports the extruded film in an upward direction. The film is guided up to the nip rolls where the tube is flattened to create a 'lay-flat' tube of film.

CHARACTERIZATION

X-ray diffraction spectroscopy (XRD)

X-rays diffraction (XRD) measurements were carried out in the reflection mode on a Rigaku diffractometer Mini Flex II (Tokyo, Japan) operated at 30 kV voltage and a current of 15 mA with CuKα radiation (λ = 1.541841 Å).

Scanning electron microscopy and Energy dispersive spectroscopy

Specimens were examined with a Hitachi TM 3000, coupled with a Bruker Quantax 70 for the collection of EDS data. SEM coupled with backscattered electron detector (BSE) and energy dispersive X-ray spectroscopy (EDS). Sample sections for the EDS analysis were taken at 15 keV, and the acquisition period was 120 s.

Mechanical Test

The tensile tests were carried out on Zwick/Roell Z2.5 materials testing system equipped with TestXpert data acquisition system at 2.5 kN load. Thin films were cut according to the ASTM D882 standard in rectangular strip of 70 mm × 10 mm. Film thickness were ranged between 10-50μm and was measured by the optical microscope. Samples at about ten replicates were tested for each film in separation grips of 50 mm. The cross-head speed was set at 0.01 mm s^{-1}. The tensile strength, modulus and toughness were determined by the TextXpert software.

Differential scanning calorimetry

Thermal properties of specimens were analyzed using a differential scanning calorimeter DSC 822, Mettler Toledo. The thermal behavior of films was obtained by: (1) heating from 25 to 280 °C at a heating rate of 10 °C min^{-1} under nitrogen atmosphere; (2) holding for 5 min at 280 °C, and (3) cooling to 25 °C and reheating to 280 °C at 10 °C min^{-1}. The crystallinity was calculated according to the equation (1):

$$X_C = P\,x\,\frac{\Delta H_f\,x\,100}{\Delta H_0} \qquad (1)$$

Where: ΔH_f is melting enthalpy of the sample, ΔH_0 is melting enthalpy of the 100% crystalline PE which is assumed to be 280 kJ kg^{-1} [11, 12], and P was the fraction content of PP in the sample.

RESULTS AND DISCUSSION

X-ray diffraction spectroscopy (XRD)

The X-ray powder diffraction profiles PE samples are shown in Fig.2.

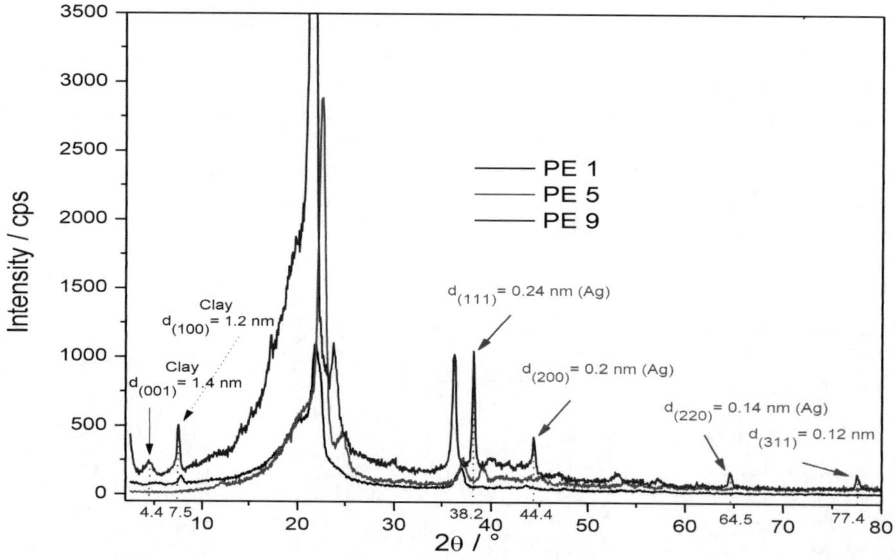

FIGURE 2. Typical XRD patterns from polyethylene: LDPE/LLDPE (PE1), LDPE/LLDPE/MMT/AgNPs 0.5 wt%(PE5), and LDPE/LLDPE/MMT/AgNPs 1wt% (PE9).

The finite layer expansion associated with the polyethylene intercalation results in the appearance of a new basal reflection corresponding to larger gallery space[13].The sample PE9 showed intercalated and flocculated morphology. Considering the diffraction peaks related to the (001) reflexions, PE 9 film showed a basal spacing $d_{(001)}$= 1.4 nm for clay according to the Bragg equation ($2d\sin\theta = n\lambda$, λ=1.541841Å), and crystallite size of 20 nm.

The XRD profile presented in Fig.2, film PE9, exhibit characteristic peaks at 2θ of 38.2, 44.4, 64.5, and 77.4 corresponding to scattering from the (111), (200), (220), and (311) crystallographic planes. These diffraction peaks firmly represent a face centered cubic structure of crystalline AgNPs [14,15].

Disappearance of the nanoclay interlayer diffraction peak (d_{001}) indicates possible exfoliation of the nanoclay platelets and broadening of the peak is considered to be the result of partial exfoliation[16]. The results indicate the formation of microstructures predominantly exfoliated in film PE5.

Scanning electron microscopy and Energy dispersive spectroscopy

The images of Fig. 3 and 4 are related to the PE film with AgNPs and silicon.

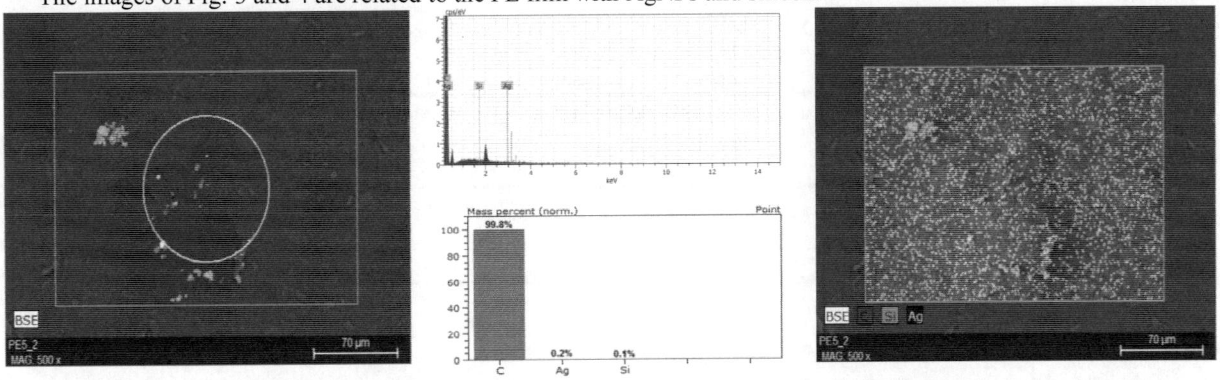

FIGURE 3. Elemental maps, magnification 500X from SEM-EDX of (A) The surface of the PE5 nanocomposite film showing the distribution of silicon (green) and silver (blue) atoms.

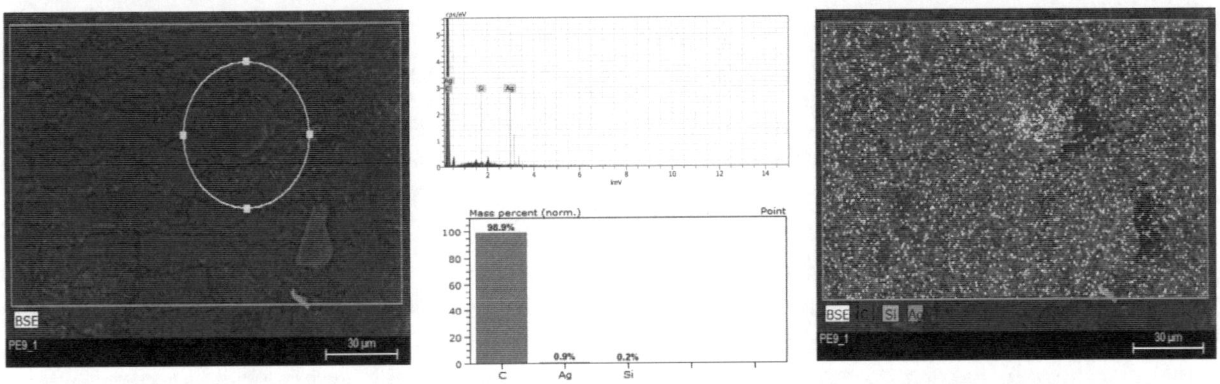

FIGURE 4. Elemental maps, magnification 1000X from SEM-EDX of (B) The surface of the PE9 nanocomposite film showing the distribution of silicon (blue) and silver (green) atoms.

The degree of dispersion of nanofillers plays an important role in influencing the properties of the resulting nanocomposites and energy dispersive X-ray spectroscopy (EDX), combined with scanning electron microscopy, was employed to analyse the elemental composition on the virgin surface, with a PE matrix, comprising carbon, the presence of silver and silicon (Fig. 3 and 4). The evidence of the elemental analysis suggests that the nanofillers are well dispersed into the bulk of PE matrix without large agglomerates on the whole. The blending process was thorough and effective.

Mechanical Test

A representative stress–strain-curve of nanocomposite films of the LDPE/LLDPE is illustrated in Fig.5.

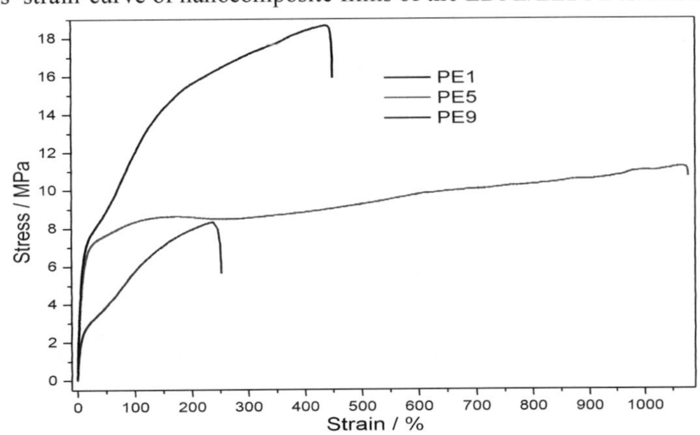

FIGURE 5. Stress-strain-curves of the polyethylene samples.

The introduction of 0.5 wt.% of AgNPs increases the elongation at break, but further addition decreased. The reduction in the tensile strenght can be explained by the increased volume of PE that process a stress concentration with

AgNPs content. Introducing AgNPs at a content 0.5 wt.% increases the elongation to break but the addition 1wt.% decreased.

Differential scanning calorimetry

The DSC results are shown in Tab. 2.

TABLE 2. DSC parameters for various polyethylene PE1, PE5 and PE9.

Samples	Melting peak temperature, T_{m1} / °C (±0.1%)	Crystallization peak temperature, T_C /°C (±0.1%)	Melting peak temperature, T_{m2} /°C (±0.1%)	Degree of crystallinity, X_C /% (±0.5%)
PE 1	113.1	97.9	112.1	32.0
PE 5	115.2	98.0	112.1	33.8
PE 9	111.8	98.0	111.9	34.1

The incorporation of the MMT-AgNPs increased the degree of crystallinity (%) of PE1 from 32.0% to 34.1%. This fact suggest that the AgNPs impart a high efficiency to the heterogeneous nucleation of PE, even at a very low AgNPs content (0.5 wt.%). A research work which corroborates with the idea of this work was published by Benchacine et al. [17], the silver incorporation in MMT plays an important role in the crystallinity degree X_C of PE that increases when Ag–MMT is added

CONCLUSIONS

X-ray diffraction spectroscopy the sample PE9 showed intercalated morphology. The results with MMT indicated the formation of microstructures predominantly exfoliated in film PE5. The film PE5 showed increase of elongation at break above 1000%, which suggests the application in packaging. In SEM/EDS microscopy the morphology of the silver/clay particles with good distribution and some clusters points it were observed.

ACKNOWLEDGMENTS

The authors acknowledge financial support for this work from CAPES-Process: CSF-PVE's - 88887.115684/2016-00, Centre of Science and Technology of Materials – CCTM/IPEN, for microscopy analysis (SEM), the technicians Mr. Eleosmar Gasparin and Nelson R. Bueno, for technical support and multipurpose gamma irradiation facility at the CTR/IPEN.

REFERENCES

1. J. Z. Liang, L. C. Tan, K. J. Wang, F. J. Li, S. D. Zhang, *Polym. Test.*, **41**, 133-139 (2015).
2. P. Micic and S. N. Bhattacharya, *Polym. Int.*, **49**, 1580-1589 (2000).
3. P. Micic, S. N. Bhattacharya, G. Field, *Int. Polym. Proces.*, 12, 2, 110-115 (1997).
4. J. L. White, H. Yamane, *Pure Appl. Chem.*, **59**, 193-216 (1987).
5. J. Maryam, R. A. Rahman, N. A. Ibrahim, L. C. Abdullah, C. P. Tan, *Food Bioprocess Technol.*, **5**, 719–728 (2012).
6. S. A. Hasim, M. C. C. Romero, M. A. Morris, E. Cummins, J. P. Kerry, *Food Packaging and Shelf Life,* **4**, 26–35 (2015).
7. A. R. Abbasi, H. Kalantary, M. Yousefi, A. Ramazani, A. Morsali, *Ultrasonics Sonochemistry*, **19**, 853–857 (2012).
8. L. A. Savas, M. Hancer, *Appl. Clay Sci.*, **108**, 40–44 (2015).
9. B. Girase, D. Depan, J. S. Shah, W. Xu, R. D. K. Misra, *Mater. Sci. Eng. C.* **31**, 1759–1766 (2011).
10. L. F. Giraldo, P. Camilo, T. Kyu, *Current Opinion Chem. Eng.,* **11**, 7–13 (2016).
11. J. E. Mark, "Crystallization Kinetics of Polymer". *Physical Properties of Polymers Handbook*, Springer, EUAi, 2007, 636.
12. Brandrup, J., Immergut, E. H., Grulke, E. A., eds., *Polymer Handbook*, New York, NY, Wiley-Interscience, 4th, v/13, 1999.
13. S. S. Ray, M. Okamoto, *Prog. Polym. Sci.*, **28**, 1539–641 (2003).
14. R. Kakkar, E. D. Sherly, K. Madgula, D. K. Devi, B. Sreedhar, *J. Appl. Polym. Sci.*, 126, E154-E161 (2012).
15. A. Swami, A. Kumar, P. R. Selvakannan, S. Mandal, R. Pasricha, M. Sastry, *Chem. Mater.*, **15**, 17 (2003).
16. M. Ataeefard, S. Moradian, *Appl. Surf. Sci.*, **257**, 2320-2326 (2011).
17. F. Benhacine, A. S. Hadj-Hamou, A. Habi, *Polym. Bull.* DOI 10.1007/s00289-015-1543-9 (2016).

Effect Of Additives On MWCNT Dispersion And Electrical Percolation In Polyamide 12 Composites

Robert Socher, Beate Krause and Petra Pötschke*

Leibniz-Institut für Polymerforschung Dresden e.V. (IPF), Hohe Straße 6, 01069 Dresden, Germany

poe@ipfdd.de

Abstract. The aim of this study was to decrease the electrical percolation threshold of multiwalled carbon nanotubes (MWCNTs) in a polyamide 12 matrix by the use of additives. Different kinds of additive were selected which either interact with the π-system of the MWCNTs (imidazolium based ionic liquid (IL) and perylene-3,4,9,10-tetracarboxylic dianhydride (PTCDA)) or improve the MWCNT wettability (cyclic butylene terephthalate, CBT). The composites were melt mixed using a DACA microcompounder. The electrical percolation threshold for PA12/MWCNT without additives, measured on compression molded plates, was found between 2.00 and 2.25 wt%. With all used additives a significant reduction of the electrical percolation threshold could be achieved. Whereas the addition of IL and CBT resulted in MWCNT percolation at around 1 wt%, a slightly higher percolation threshold between 1.0 and 1.5 wt% was found for PTCDA as additive. Interestingly, the electrical resistivity at higher loadings was decreased by nearly two decades when using CBT and one decade after application of PTCDA, whereas IL did not contribute to lower values in this range. In all cases macrodispersion as assessed by light microscopy was not improved and even worse as compared to non-modified composites. In summary, the results illustrates that these kinds of additives are able to improve the performance of PA12 based MWCNT nanocomposites.

Keywords: carbon nanotubes, polyamide, additives, ionic liquid, electrical properties
PACS: 72.80.Tm, 82.35.Np, 82.35.Lr

INTRODUCTION

For the preparation of electrical conductive thermoplastic composites by incorporation of carbon nanotubes (CNTs) melt compounding is a suitable and scalable method. During the process of dispersion of as produced primary CNT agglomerates into the matrix, the wetting and infiltration of the matrix polymer chains into the agglomerate structures are important initial steps, followed by the dispersion steps involving rupture and erosion of agglomerates [1]. Low viscous additives can help to facilitate the different step, e.g. by reducing the interfacial tension between CNTs and matrix to promote wetting and by facilitating the infiltration step. Better infiltrated nanotube agglomerates have lower resistance against the rupture and erosion steps thus helping to improve dispersion of nanotubes. Also additives which interact with the π-system of the CNTs like ionic liquids seem to be promising to enhance dispersion and stabilize individualized CNTs [2,3].

Composites of polyamide 12 (PA12) containing CNTs are developed for potential application areas like antistatic fuel pipes and parts in fuel pumps for the use in the automotive industry. Due to economic reasons a low electrical percolation threshold is desired. The values so far achieved in literature (e.g. [4]) are still relatively high and it is the aim of this study to reduce this threshold and the values of electrical volume resistivity by adding suitable additives.

EXPERIMENTAL PART

As matrix a low viscous polyamide 12 (Vestamid®) having acid group excess (details on this material see [4]) and multiwalled carbon nanotubes (MWCNTs) of the type Baytubes® C150P (Bayer MaterialScience, Germany) were used for melt compounding. PA12 was dried for >6 hours at 70°C in vacuum, CNTs for at least 1 hour at 120°C. Four different kinds of additives were selected: perylene-3,4,9,10-tetracarboxylic dianhydride (PTCDA), cyclic butylene terephthalate (CBT® 100, Cyclics Corporation), 1-butyl-3-methylimidazolium tetrafluoroborat (IL1), and 1-methyl-3-octylimidazolium tetrafluorborate (IL2) (Sigma Aldrich).

The composites were melt mixed using a conical twin-screw microcompounder (DACA) at 210°C, a rotation speed of 250 and a mixing time of 5 min. MWCNTs and additive were premixed in the required composition and added in charges with the polymer granules to the running compounder. In case of IL a mortar was used to mix it with the nanotubes, whereby a gel like structure was achieved. For electrical measurements the composites were compression molded at 220°C for 1 min using a hot press (PW40EH) to circular plates with 30 mm diameter and 0.5 mm thickness.

Electrical measurements were done on the compression molded plates. For composites with a volume resistivity $>10^7$ Ohm cm a Keithley test fixture 8009 (open symbols) and for higher conductive samples a 4-point test fixture with gold electrodes combined with a Keithley electrometer E6517A (closed symbols) were used, whereby for the latter strips with about 3 mm width were cut from the plates. Results show mean values of at least 4 measurements.

The study of CNT macrodispersion was done on thin sections (5 μm) which were cut at room temperature using a Leica microtom RM2265 and a diamond knife. The light microscopy investigations were performed using a microscope Olympus BH2 and a camera Olympus DP71. The quantification was done by calculation of agglomerate area ratio A_A (%) by division of area of agglomerates A through area of complete image A_0 [5].

RESULTS

PA12/MWCNT Composites with Ionic Liquid

The electrical percolation of MWCNT in the used PA12 type was found to occur between 2.00 and 2.25 wt% [4]. Ionic liquids were previously reported to improve both the dispersion and electrical conductivity of CNTs for melt mixed polymer composites, e.g. in PMMA [6], rubber [7] and polyamide [8] and therefore appeared to be promising for this study. Test experiments using IL directly with PA12 without the addition of nanotubes indicated only very low influence on electrical resistivity. The volume resistivity decreased towards about 1E14 Ohm·cm at 5 wt% and 2E12 Ohm·cm at 10 wt% loading.

Figure 1 shows the percolation curve of PA12/MWCNT together with the electrical volume resistivity of composites prepared using different ratios of IL and CNTs. With the addition of IL the electrical percolation threshold decreases significantly towards values of around 1 wt%. Thereby, for a CNT:IL ratio of 1:1 no significant differences between the use of IL1 and IL2 were found. When varying the CNT:IL ratio between 1:1, 1:2, and 1:4 it was found that the values of electrical resistivity for the PA12 composites containing CNT:IL ratio of 1:1 and 1:2 were nearly similar. At the higher CNT:IL ratio of 1:4, slightly reduced resistivity values could be observed. It may be assumed that a further increase of IL content would even decrease the electrical percolation threshold of the PA12 composites. However, a higher content of the ionic liquids leads due to the immiscibility of PA12 and ILs to a worse material quality, which is revealed in inhomogeneities and bad appearance of the sample surface of the PA12 composites. Furthermore, IL is liquid at room temperature and if applied in great excess it tends to diffuse on the surface of the composite plates. Therefore, only PA12 composites filled with maximum 5 wt% IL were prepared. At higher CNT contents the values of electrical resistivity of PA12/MWCNT composites with or without IL were nearly similar with only slightly lower values for the IL treated nanotubes.

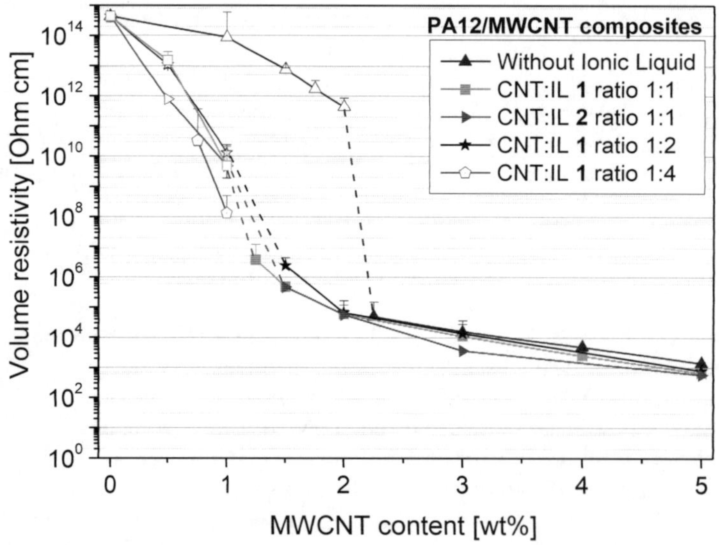

FIGURE 1. Electrical volume resistivity of PA12 composites containing different CNT:IL ratios.

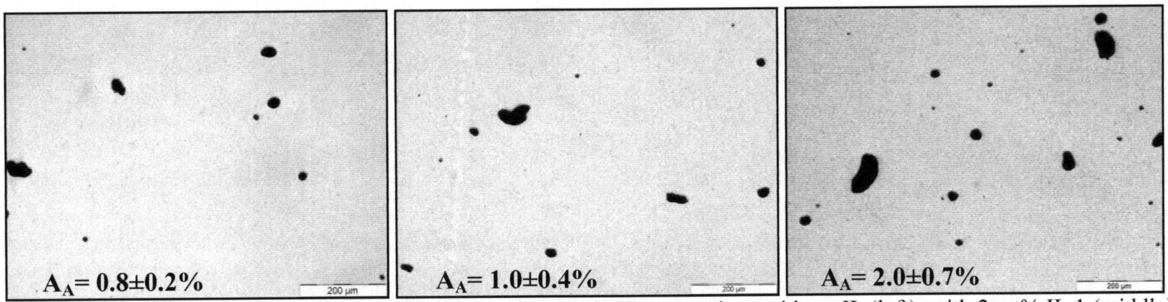

FIGURE 2. Light microscopy images of PA12/2 wt% MWCNT composites: without IL (left), with 2 wt% IL 1 (middle), and with 2 wt% IL2 (right)

As reason for the reduced percolation threshold when using IL treated MWCNTs improved nanotube dispersion may be assumed. However, as illustrated in Figure 2 the CNT macrodispersion of the nanotubes within the PA12 composites seems to be not the main reason for improved electrical properties. The use of IL1 does not change the macrodispersion significantly indicated by nearly the same value for the agglomerate area ratio of PA12/MWCNT and PA12/MWCNT/IL1 in the light microscopy images. With IL2 even more remaining CNT agglomerates and a higher value of A_A compared to the one of PA12/MWCNT was found. However, one has to consider that possibly better infiltrated primary nanotube agglomerates may appear to be larger in the thin cuts. In addition, it may be possible that the IL treatment reduces nanotube breakage, which was shown to occur during melt mixing [5]. This may be due to viscosity reducing effects of IL on the PA12 matrix or a lubrication effect due to IL covering about nanotubes. Less pronounced shortening resulting in longer nanotubes lowers percolation thresholds [5].

PA12/MWCNT Composites with PTCDA

As another additive PTCDA was selected, which is an organic dye molecule and an organic semiconductor with dark red color. Due to its perylene units it is also able to interact with the π-system of the CNTs. Figure 3 illustrates that also PTCDA was suitable to decrease the electrical percolation threshold of PA12/MWCNT composites significantly up to 1-1.5 wt%. Interestingly, at higher MWCNT contents the resistivity values of PA12/WMCNT/PCTDA composites were one decade lower compared to the PA12/MWCNT composites. The result indicates that the presence of PTCDA changes the structure of the conductive network. The study of the CNT macrodispersion using light microscopy again shows that there is no significantly improved macrodispersion due to the PTCDA addition (Figure 4 left). The agglomerate area ratio is only slightly lower as compared to PA12/MWCNT without additive (Figure 2 left).

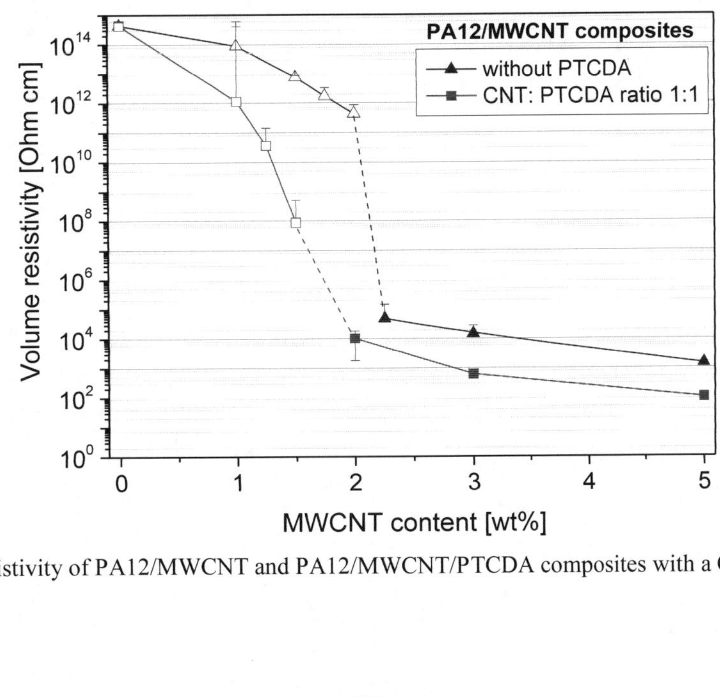

FIGURE 3. Volume resistivity of PA12/MWCNT and PA12/MWCNT/PTCDA composites with a CNT:PTCDA ratio of 1:1.

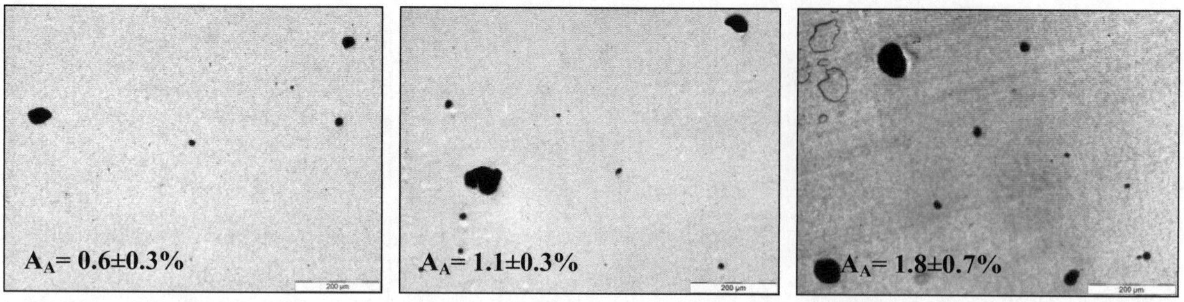

FIGURE 4. Light microscopy images of PA12/2 wt% MWCNT composites: with 2 wt% PTCDA (left), 2 wt% CBT (middle), and 5 wt% CBT (right) (micrograph of the corresponding composite without additive see Fig. 2)

PA12/MWCNT Composites with Cyclic Butylene Terephthalate (CBT)

The third additive was a low viscous CBT which can help to facilitate the CNT wetting and to reduce the interfacial tension between CNTs and the PA12 matrix. This additive is recommended by the supplier as an excellent base resin for masterbatch concentrates due to its broad compatibility with a wide range of thermoplastics and fillers. In addition to excellent filler wetting, the low melt viscosity enables higher loadings and easier processability. CBT was added with a fixed amount of 5 wt% CBT in PA12/MWCNT composites having various CNT contents.

It was found that the CBT incorporation also decreases the electrical percolation threshold from 2.00 to 2.25 wt% to around 1 wt% (Fig. 5). Already for the PA12/0.5 wt% MWCNT composite with 5 wt% CBT a decrease of volume resistivity by nearly 2 decades could be observed. Interestingly, at higher MWCNT contents the values of PA12/WMCNT/CBT composites are nearly two decades lower compared to the PA12/MWCNT composites without additives and reaches values of about 50 Ohm·cm at 5 wt% MWCNT loading. When adding only 2 wt% CBT at 2 wt% MWCNT loading the same resistivity value as for 5 wt% CBT addition was found. Possibly, the addition of 2 wt% CBT would be sufficient for the improvement of electrical properties of PA12 composites.

The investigation of CNT macrodispersion for PA12/2 wt% MWCNT composites showed a slightly higher value of the agglomerate area ratio if 2 wt% CBT was added and a significant higher value if 5 wt% CBT were incorporated (Fig. 4, middle and right). Here, the same reasons as discussed before for IL addition may apply.

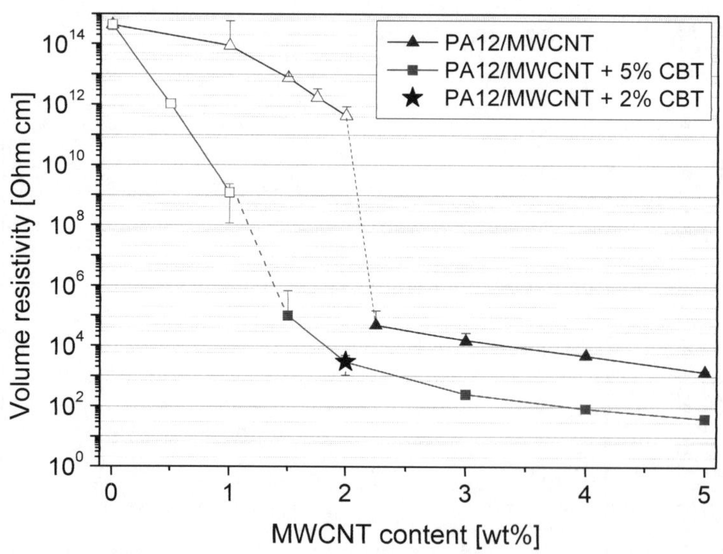

FIGURE 5. Volume resistivity of PA12/MWCNT composites without and with a constant CBT content of 2 or 5 wt%.

SUMMARY AND CONCLUSION

The study presented the use of different additives to improve the electrical properties of PA12/MWCNT composites. The addition of the ionic liquids 1-butyl-3-methylimidazolium tetrafluoroborate (IL1) and 1-methyl-3-octylimidazolium tetrafluorborate (IL2) as well as perylene-3,4,9,10-tetracarboxylic dianhydride (PTCDA) and cyclic butylene terephthalate (CBT) leads to a significant reduction of electrical percolation threshold to about half the concentration of composites without additives. Percolation thresholds for compression molded composites of about 1 wt% could be achieved.

Furthermore, an additional decrease of values of electrical volume resistivity at higher nanotube contents was found when using the additives PTCDA and CBT.

Interestingly, in contrast to expectations, the macrodispersion of CNTs was not improved and in some cases even worsened. Thus, the effects may arise from changed nanotube network structures in the nanoscale and/or better retained nanotube length due to the viscosity effects of the low molecular additives. However, to clarify these reasons further investigations are needed.

ACKNOWLEDGMENTS

We thank the German Federal Ministry of Education and Research (BMBF) for financial support within the Innovation Alliance Inno.CNT, project CarboDis, BMBF 03X0042M. In addition, we thank Evonik Degussa GmbH (now Evonik Industries AG) for providing the PA12 material.

REFERENCES

1 I. Alig, P. Pötschke, D. Lellinger, T. Skipa, S. Pegel, G. R. Kasaliwal and T. Villmow, *Polymer* **53** (1), 4-28 (2012).
2 T. Fukushima, A. Kosaka, Y. Ishimura, T. Yamamoto, T. Takigawa, N. Ishii and T. Aida, *Science* **300** (5628), 2072-2074 (2003).
3 T. Fukushima and T. Aida, *Chemistry – A European Journal* **13** (18), 5048-5058 (2007).
4 R. Socher, B. Krause, R. Boldt, S. Hermasch, R. Wursche and P. Pötschke, *Composites Science and Technology* 7**1** (3), 306-314 (2011).
5 R. Socher, B. Krause, M. T. Müller, R. Boldt and P. Pötschke, *Polymer* **53** (2), 495-504 (2012).
6 L. Zhao, Y. Li, X. Cao, J. You and W. Dong, *Nanotechnology* **23**, 255702 (2012)
7 K. Subramaniam, A. Das, K. W. Stöckelhuber and G. Heinrich, *Rubber Chemistry and Technology* **86**, 367-400 (2013)
8 R. Socher, PA12-MWCNT-Nanokomposite: Wege zur effektiven MWCNT-Dispergierung und zu niedrigen elektrischen Perkolationsschwellen, phd Thesis Technische Universität Dresden (2012), Verlag Dr. Huth, ISBN 9783843909013.

Effect of Processing Sequence on the Mechanical and Tribological Properties of Ternary Nanocomposites (VGCF-X/PA6/HDPE)

Yu Osada[a], Yosuke Nishitani[b*] and Takeshi Kitano[c]

[a] *Department of Mechanical Engineering, Graduate school of Engineering, Kogakuin University, 2665-1 Nakano, Hachioji, Tokyo 1920015, Japan*
[b] *Department of Mechanical Engineering, Faculty of Engineering, Kogakuin University, 2665-1 Nakano, Hachioji, Tokyo 1920015, Japan*
[c] *Polymer Centre, Faculty of Technology, Tomas Bata University in Zlin, Czech Republic, T. G. M. 275, Zlin, 762 72 Czech Republic*

Reference author's email: at13152@ns.kogakuin.ac.jp

Abstract. The aim of this study is to develop the new tribomaterials for mechanical sliding parts with sufficient balance of mechanical and tribological properties. The effect of processing sequence on the mechanical and tribological properties of ternary nanocomposites (VGCF-X/PA6/HDPE) was investigated: the polymer blends of polyamide 6 (PA6) and high density polyethylene (HDPE) filled with vapor grown carbon fiber (VGCF-X) which is one of the carbon nanofiber and has 15nm diameter and 3μm length. Five different processing sequences, (1) VGCF-X, PA6 and HDPE were mixed simultaneously (Process A), (2) Re-mixing (Second compounding) of the materials prepared by Process A (Process AR), (3) VGCF-X was mixed with PA6 (VGCF-X/PA6 composites) and then these composites were blended with HDPE (Process B), (4) HDPE was blended with PA6 (PA6/HDPE blends) and then these blends were mixed with VGCF-X (Process C), (5) VGCF-X were mixed with HDPE (VGCF-X/HDPE composites) and then these composites were blended with PA6 (Process D) were attempted for preparing of the ternaery nanocomposites (VGCF-X/PA6/HDPE). These ternary polymer nanocomposites were melt mixed by a twin screw extruder and injection-molded. Their mechanical properties such as tensile, Izod impact and Durometer hardness, and tribological properties were evaluated. Tribological properties were measured by a ring on plate type sliding wear tester under dry condition. The mechanical and tribological properties changed in according with the processing sequence of ternary nanocomposites (VGCF-X/PA6/HDPE). These may be attributed to the change of the internal structure formation of these ternary nanocomposites, which is dispersibility and localization of VGCF-X in these ternary nanocomposites by different processing sequences.

Keywords: Processing sequence, Ternary nanocomposites, VGCF-X, Mechanical properties, Tribological properties
PACS: 81.05.Qk, 81.40.Pq, 83.50.Xa

INTRODUCTION

There has been a great interest on engineering materials such as structural materials and tribomaterials for mechanical sliding parts based on nanosized filler filled polymeric nanocomposites in recent years[1)-5)]. Especially, the new engineering materials based on carbon nanofiber(CNF) filled polymer nanocomposites are strongly required. However, little study has been done to develop tribomaterials suitable for micro size devices with sufficient balances between mechanical and tribological properties. In previous studies, we carried out the studies on the mechanical and tribological properties of ternary blends of engineering plastics such as polyamide (PA) and polybutylene terephthalate (PBT), which are blends of various grafted copolymers with various functionalized groups, other thermoplastics of both of these with PA or PBT)[6)]. Especially, we conducted the studies on the effect of processing sequences on the mechanical and tribological properties of vapor grown carbon nanofiber (VGCF), which is one of CNF, filled polymer blends of polybutylene terephthalate (PBT) and thermoplastic elastomer (TPE) (VGCF/PBT/TPE)[6)], VGCF filled polymer blends of PA6 and maleic anhydride styrene-ethylene/butylene-styrene copolymer (SEBS-g-MA) (VGCF/PA6/SEBS-g-MA)[7)], and VGCF-X, which is one of the CNF and its diameter is smaller than VGCF, reinforced PA6 and SEBS composites (VGCF-X/PA6/SEBS)[8), 9)]. It was found that the influence of the addition of CNF and TPE on the mechanical and tribological properties of PA differs for each property. In particular, tribological properties such as frictional coefficient and specific wear rate of these ternary nanocomposites are influenced by the processing sequence. This may be attributed to the change of internal structure formation, which is a dispersibility of SEBS particle and VGCF-X in these ternary nanocomposites. Therefore, in order to develop new tribomaterials with further higher performance based on carbon nanofiber filled multi-component nanocomposites, it is necessary to control the morphologies, which are the internal structure formation such as phase structure and filler dispersion of these ternary nanocomposites. This is because the internal structure formation correlates closely with the mechanical and tribological properties of these ternary nanocomposites. It is well known that the morphologies of multi-component nanocomposites are strongly influenced by the processing sequences at melt mixing by twin extruder[7)-9)]. However, there is not enough knowledge of reliable relations between the internal structure formation and the various properties such as mechanical and tribological properties of these ternary nanocomposites according to the type of blending materials and so on. The aim of this study is to develop new tribomaterials for mechanical with sufficient balance of mechanical and tribological properties of ternary polymer nanocomposites based on VGCF-X filled the polymer blends of PA6 and high density polyethylene (VGCF-X/PA6/HDPE). Especially, we investigated the effect of processing sequence on the mechanical and tribological properties of these ternary nanocomposites (VGCF-X/PA6/HDPE).

EXPERIMENTAL

The materials used in this study were ternary nanocomposites of polyamide 6 (PA6, 1013B, Ube industries Ltd., Japan), VGCF-X (Vapor Grown Carbon Fiber, Showa Denko K.K., Japan, d=15nm, L=3μm), which is one of CNF, and high density polyethylene (HDPE, Novatec HD HJ560, Japan Polyethylene Co., Japan, MFR = 7g/10min). The composition of PA6 and HDPE was fixed as 80/20 by weight fraction, and three kinds of VGCF-X volume fraction were selected as 0, 1 and 5vol.%. Prior to mixing, VGCF-X, PA6 and HDPE were dried in a vacuum oven at 80°C for 12h. Five different processing sequences: (1) VGCF-X, PA6 and HDPE were mixed simultaneously (Process A), (2) Re-mixing (Second compounding) of the materials prepared by Process A (Process AR), (3) VGCF-X was mixed with PA6 (VGCF-X/PA6 composites) and then these nanocomposites were blended with HDPE (Process B), (4) HDPE was blended with PA6 (PA6/HDPE blends) and then these blends were mixed with VGCF-X (Process C), and (5) VGCF-X were mixed with HDPE (VGCF-X/HDPE composites) and then these nanocomposites were blended with PA6 (Process D) were attempted for preparing of the ternary nanocomposites. Fig. 1 shows the schematic diagram five different processing sequences of VGCF-X/PA6/HDPE ternary nanocomposites. These materials were melt mixed at 85rpm and 240°C on a twin screw extruder (TEX-30, Japan Steel Works, Ltd., Japan). After mixing, the extruded strands of these ternary nanocomposites were cut by a pelletizer and were dried again at 80°C for 12h in vacuum oven. Various shaped samples for mechanical and tribological properties testing were injection molded (NS20-A, Nissei Plastic Industrial, Japan). Their mechanical properties such as tensile, Izod impact and Durometer hardness and tribological properties were evaluated for these ternary nanocomposites. Tensile test was carried out with dog-bone samples (12mm x 60mm x 3mm, length of parallel part 20mm) on a Toyo Seiki testing machine V-10, and performed in accordance to the JIS K 7161, and at the cross-head speed of 50 mm/min. Impact characteristics were determined by Izod impact test. Izod impact tests were conducted using coupon specimen (12.7mm x 64mm x 3mm) with a notch depth of 2.5mm on a Toyo Seiki impact machine DG-IB at room temperature, in accordance to JIS K 7110. Hardness characteristics were determined by Durometer hardness (type D). Durometer hardness tests were carried out using coupon specimen above mentioned on Toyo Seiki digital hardness testing machine in accordance to JIS K 7215. Tribological properties such as frictional coefficient μ and specific wear rate V_s were measured by a ring-on-plate type sliding wear tester (EFM-3-EN, Orientech Co., Japan) under dry condition. A carbon steel (S45C) ring with the surface finished by No.240 and No.800 polishing paper was used as a counterpart. Tribological properties were evaluated by sliding speed of 0.5m/sec, sliding distance of 3000m and normal load of 50N. The surface of samples fractured cryogenically in liquid nitrogen and their polished surface were observed using scanning electron microscope (EDX-WETSEM, JSM-6360LA, JEOL Ltd., Japan) for understanding the internal structure formation of ternary nanocomposites.

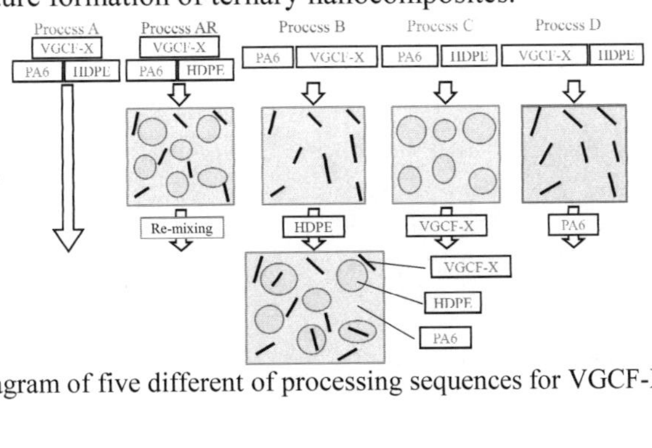

FIGURE1. Schematic diagram of five different of processing sequences for VGCF-X/PA6/HDPE composites.

RESULTS AND DISCUSSION

First, the effect of processing sequences on the mechanical properties of ternary nanocomposites (VGCF-X/PA6/HDPE) is discussed. Fig. 2 shows the influence of volume fraction of VGCF-X V_f on the tensile properties of these ternary nanocomposites prepared by five different processing sequences: tensile strength σ_t (Fig. 2(a)), tensile modulus E_t (Fig. 2(b)) and elongation at break ε_t (Fig. 2(c)), respectively. The influence of V_f on the σ_t has a complex behavior according to the processing sequence. σ_t of the nanocomposites by Process A and C has a minimum peak at V_f = 1vol.%, and then increases with increasing V_f. In contrast, σ_t of the nanocomposites by Process B shows a contrary tendency. On the other hand, σ_t of the composites of Process AR and D increases with increasing V_f. Next, the influence of V_f on E_t of the composites by Process A, AR, C and D has the same tendency as σ_t. However, that of Process B shows a different tendency and has a minimum peak at V_f=1vol.%. In particular, the improvement of tensile properties such as σ_t and E_t is related to the number of mixing VGCF-X, and the twice mixing methods of VGCF-X, which are the re-mixing of process AR, B and D except for V_f=5vol.% of Process B, are more effective than uni-methods such as Process A and C. These tendencies may be attributed to the change of internal structure formation of ternary nanocomposites (VGCF-X/PA6/HDPE) by different processing sequences, which is a dispersibility of VGCF-X in the composites. On the other hand, σ_t of the nanocomposites by all processing sequences decreases with increasing V_f and the type of processing sequences seems remarkably not to affect to σ_t.

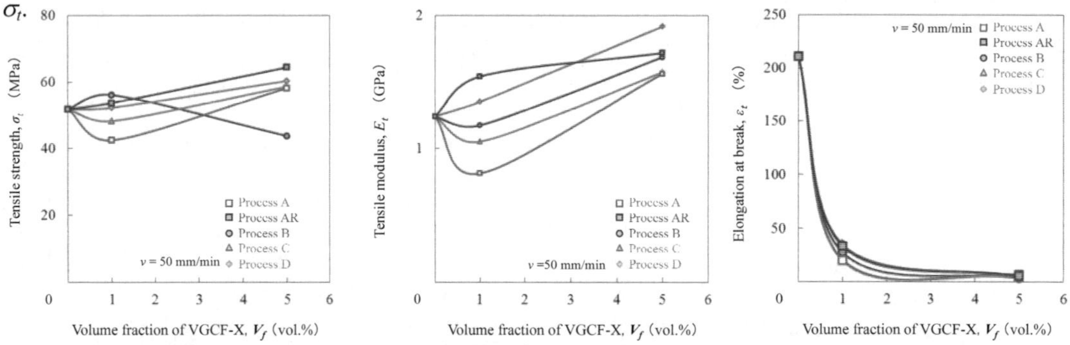

(a) Tensile strength (b) Tensile modulus (c) Elongation at break

FIGURE2. Influence of volume fraction of VGCF-X on the various tensile properties of VGCF-X/PA6/HDPE composites.

Second, the effects of processing sequences on the Izod impact strength and Durometer hardness of ternary nanocomposites (VGCF-X/PA6/HDPE) prepared by five different processing sequences are discussed. Fig. 3(a) shows the notched izod impact strength a_{iN} of various processing sequences of VGCF-X/PA6/HDPE ternary nanocomposites. a_{iN} of all type of processing sequences of VGCF-X/PA6/HDPE composites decreases with increasing V_f and the effect of processing sequences on a_{iN} seems not to influence. Fig. 3(b) shows the relationship between durometer hardness *HDD* and V_f of various processing sequences of VGCF-X/PA6/HDPE ternary nanocomposites. *HDD* of the composites by Process AR and D slightly increases with increasing V_f. On the other hand, *HDD* of the composites by other process has a maximum peak at V_f=1vol.%. From these results, it is found that the mechanical properties of ternary nanocomposites (VGCF-X/PA6/HDPE) change in accordance with the type of processing sequences. This may be attributed to the change of the internal structure formation of these ternary nanocomposites[8), 9)].

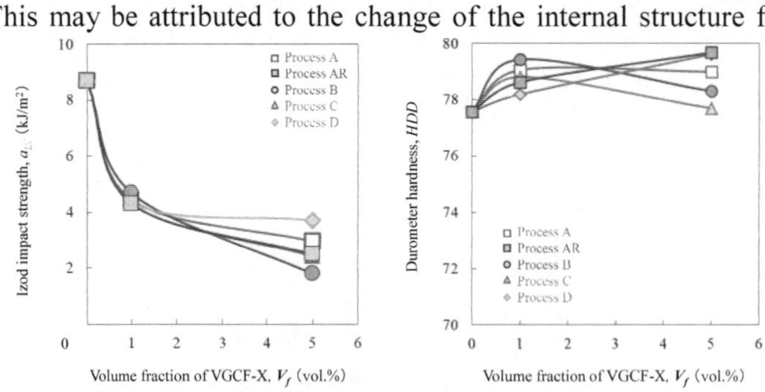

(a) Izod impact strength (b)Durometer hardness

FIGURE3. Influence of volume fraction of VGCF-X on the Izod impact strength and the Durometer hardness of VGCF-X/PA6/HDPE composites.

Third, the tribological properties under dry conditions, which are most important for the development of engineering materials such as tribomaterials, are discussed. Fig. 4(a) shows the relationship between frictional coefficient μ and V_f of various processing sequences of VGCF-X/PA6/HDPE ternary nanocomposites measured by sliding wear testing at constant normal load and sliding speed. μ of all VGCF-X/PA6/HDPE composites is lower than that of PA6/HDPE blends and decreases with increasing V_f. Especially, in V_f=5vol.% of VGCF-X, the improvement of μ for these ternary nanocomposites decrease in the following order: Process AR > Process D > Process C > Process B > Process A. On the other hand, Fig. 4(b) shows the relationship between specific wear rate V_s and V_f of various processing sequences of these ternary nanocomposites obtained by the same testing method for μ in Fig. 4(a). V_s of all processing sequences has a maximum value at 1vol.%, and then decreases with increasing V_f although the effect of processing sequences on V_s of these ternary nanocomposites has a complex behavior. In other words, the tribological properties show the different behavior according to the type of processing sequences and the fiber contents.

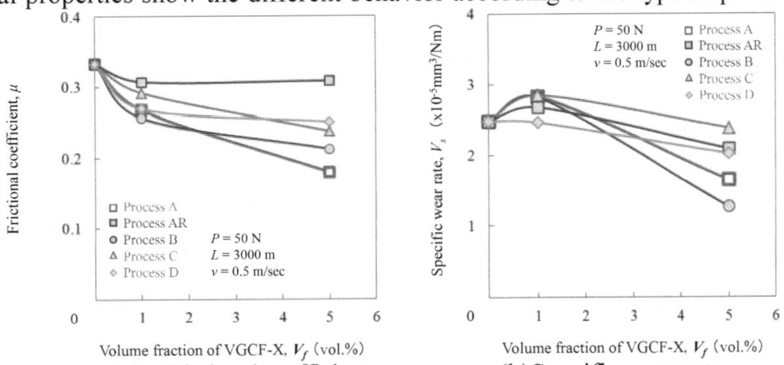

(a) Frictional coefficient (b)Specific wear rate

FIGURE4. Influence of volume fraction of VGCF-X on the tribological properties of VGCF-X/PA6/HDPE composites.

For better understanding the relationship between processing sequences and various properties such as mechanical and tribological properties of VGCF-X/PA6/HDPE ternary nanocomposites, we observed the internal structure formation, which is a dispersibility of HDPE particle and VGCF-X in these ternary nanocomposites. Fig. 5 shows the SEM photographs of cryogenically fractured surfaces of various VGCF-X/PA6/HDPE ternary nanocomposites (V_f=1vol.%) and their polished surface at the same magnification rate (x20,000) and Fig. 6 shows the schematic diagram of morphology of processing sequences for VGCF-X/PA6/HDPE composites from the observation in Fig. 5. Each ternary nanocomposites indicates the typical separate spherical phases (dispersed HDPE particles) in continuous matrix domains (PA6) and there are VGCF-X in both PA6 and HDPE domains. In particular, the size of dispersed HDPE particles changes according to the processing sequences. Furthermore, the dispersibility of VGCF-X in these ternary nanocomposites also changes in accordance with ones. The composites by type A (Fig. 6(a), Process A and C) have some agglomerations of VGCF-X in continuous matrix domains (PA6) although there are no obvious agglomerations of VGCF-X in the composites by type B (Fig. 6 (b), Process AR, B and D). This phenomenon indicates high correlation between the internal structure formation of the agglomerations of VGCF-X in composites and the decrease of mechanical properties such as tensile strength and tensile modulus above mentioned. However, the tribological properties are not related to the agglomerations of VGCF-X in the composites though these tendencies are different from VGCF-X/PA6/SEBS ternary nanocomposites in the previous work[8), 9)]. This is explained in terms of the difference with the localization of VGCF-X in the ternary nanocomposites. The discrepancy is due to the various parameters of internal structure formation in both systems, which are viscosity, interfacial energy, wetting and so on[3), 4)].

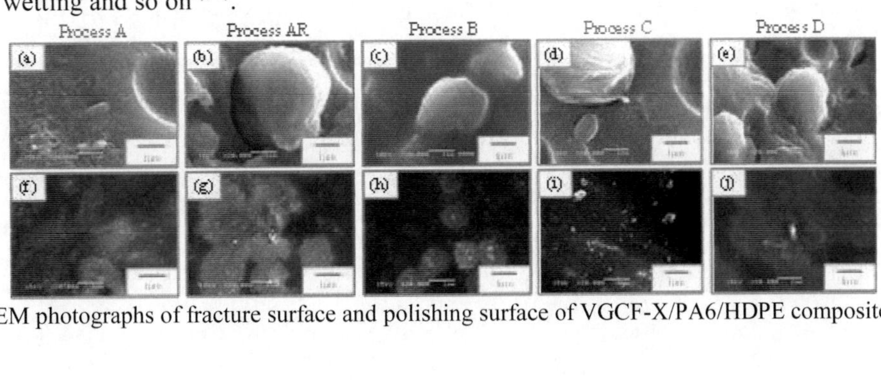

FIGURE5. SEM photographs of fracture surface and polishing surface of VGCF-X/PA6/HDPE composites (V_f=1vol.%).

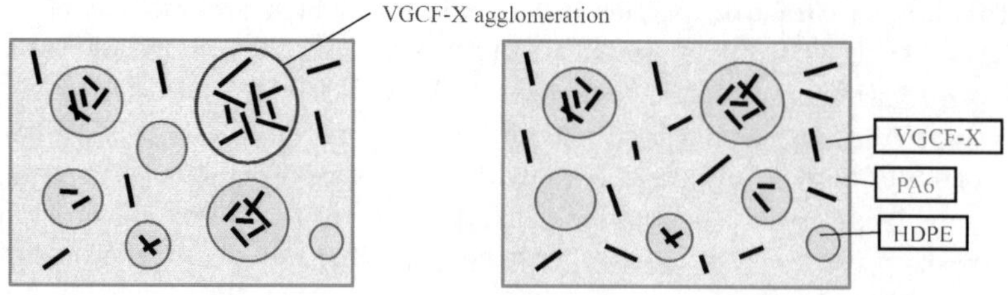

(a) Type A (Process A, C) (b) Type B (Process AR, B, D)

FIGURE6. Schematic diagram of morphology of processing sequences for VGCF-X/PA6/HDPE composites.

CONCLUSION

The aim of this study is to develop new tribomaterials for mechanical with sufficient balance of mechanical and tribological properties of ternary polymer nanocomposites based on VGCF-X filled the polymer blends of PA6 and high density polyethylene (VGCF-X/PA6/HDPE). Especially, we investigated the effect of processing sequence on the mechanical and tribological properties of these ternary nanocomposites (VGCF-X/PA6/HDPE). It was found that the mechanical and tribological properties change in according with the processing sequence. These results may be attributed to the change of internal structure formation, which is dispersibility and localization of VGCF-X in ternary nanocomposites (VGCF-X/PA6/HDPE) by different processing sequences.

ACKNOWLEDGMENTS

We would like to thank the FMS (Functional Microstructured Surfaces Research Center, Japan) of Kogakuin University, the Ogasawara Foundation for the Promotion of Science & Engineering for funding this study and partial support by national budget of Czech Republic within the framework of the Centre of Polymer Systems project (reg.number:CZ.1.05/2.1.00/03.0111). The authors would thank Showa Denko for donating the VGCF-X, Asahi Kasei Chemical for donating the HDPE and Ube industries for donating the PA6 for this study.

REFERENCES

1. F. Klaus and L. Zaiping, *Wear*, **190**, 139-144(1995)
2. W. Philipp, A. Volker, J. Romy, J. Olaf, S. Jan, S. Milo and W. Alan, *Wear*, **257**, 1006-1014(2004)
3. R. Wiwattananukul, Y. Hachiya, T. Endo, S. Nobukawa and M. Yamaguchi, *composites*, **B78**, 409-414(2015)
4. S. Sathyanarayana, M. Wegrzyn, G. Olowojoba, A. Benedito, E. Giménez, C. Hübner and F. Henning, *eXPRESS Polymer Letters*, **7**, 621-635(2013)
5. M. Yamaguchi, *Polymer Processing Society PPS31*, Jeju, Korea, KLG4-04(2015)
6. J. Tsurubuchi, T. Naito, Y. Nishitani, T. Gotoh and T. Kitano, *International Tribology Conference ITC*, Hiroshima, P05-03, (2011)
7. Y. Takenaka, S. Sano, Y. Nishitani and T. Kitano, *The proceedings of Kobunshi Kakou Gijyutsu Touronkai*, Nagoya, 24-27 (2014) (in Japanese)
8. Y. Osada, Y. Nishitani and T. Kitano, *The proceedings of Seikei-Kakou 2015*, Tokyo, F-213 (2015) (in Japanese)
9. Y. Osada, Y. Nishitani and T. Kitano, *Polymer Processing Society PPS31*, Jeju, Korea, S11-286 (2015)

Characterization of organically modified layered zinc phenylphosphonate/ Poly(butylene succinate/adipate) nanocomposites

Dong-Lin Kuo and Tzong-Ming Wu

Department of Materials Science and Engineering. National Chung Hsing University, 250 Kuo KUuang Road, Taichung 402, Taiwan

Abstract. In this study, the organically-modified layered zinc phenylphosphonate (m-PPZn) prepared using coprecipitation method were sucessfully synthesized to intercalate hexadecylamine into the interlayer spacing of PPZn. The interlayer spacing of m-PPZn determined by wide-angle X-ray diffraction (WAXD) was increased from 14.5 Å for PPZn to 27.9 Å for m-PPZn. These results suggest that the hexadecylamine were successfully intercalated into the interlayer spacing of PPZn. The poly(butylene succinate/adipate)(PBSA)/m-PPZn nanocomposites were then prepared by melt-mixing. The structure, morphology, and crystallization behavior of the PBSA/m-PPZn nanocomposites were characterized using WAXD, transmission electron microscopy (TEM), and differential scanning calorimetry (DSC). Both WAXD and TEM results demonstrate that most of the partial-delamination m-PPZn was randomly dispersed in the PBSA matrix. With the addition of 5wt% m-PPZn, the crystallization temperature (T_c) of the PBSA/m-PPZn nanocomposites was increased about 3.84 °C compared with that of neat PBSA using non-isothermal crystallization method. It has been shown that m-PPZn containes an effective nucleating effect for PBSA.

Keywords: Poly(butylene succinate/adipate) 、layered zinc phenylphosphonate 、nanocomposites 、crystalline behaviors

Structure and morphology of PBSA/m-PPZn nanocomposites

As shown Figure 1a, X-ray diffraction peaks of PPZn are at 2θ = 6.08°, 12.3° and 18.5°, which are the orders of a repeat distance on the (010), (020), and (030) planes, respectively [1]. For m-PPZn (Figure 1b), the diffraction peaks are shifted to lower angles at 2θ = 3.16°, 6.37° and 9.57° due to the increase of interlayered distances of PPZn with the incorporation of hexadecylamine. The interlayer spacing determined by Bragg's equation was increased from 14.5 Å for PPZn to 27.9 Å for m-PPZn. These results suggest that the hexadecylamine were successfully intercalated into the interlayer spacing of PPZn. Figure 1c shows XRD pattern of neat PBSA. The diffraction peaks from (020), (021), and (110) planes are observed at 2θ =19.4°, 21.5°and 22.5°, respectively [2]. The XRD pattern of the nanocomposites is shown in Figure 1d. It is clearly observed that the m-PPZn layers are intercalated into the PBSA matrix. To support the XRD results, the morphologies of PBSA/m-PPZn nanocomposites are analyzed by TEM.

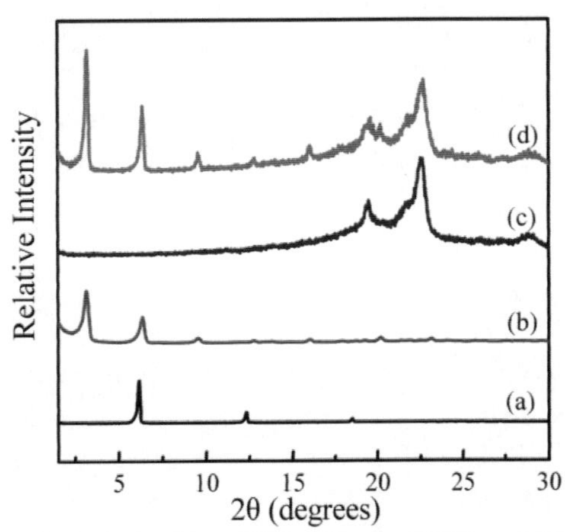

FIGURE 1. XRD pattern of (a) PPZn, (b) m-PPZn, (c) neat PBSA and (d) 5 wt% PBSA/m-PPZn.

The TEM analysis gives direct information about the intercalation of PPZn in the PBSA matrix. For the 5 wt% PBSA/PPZn nanocomposite (Figure 2a), the PPZn were aggregated and unevenly distributed within the PBSA matrix. It seems that the PPZn layers tend toward large agglomerated tactoids. The result could be ascribed to the narrow spacing between the interlayers and/or the strong interlayer forces [3]. Figure 2b shows the TEM image of 5 wt% PBSA/m-PPZn nanocomposite. Although the m-PPZn was inhomogeneously dispersed in PBSA, this data exhibits that the stacks of multiple layers of m-PPZn is distributed in the PBSA matrix with an intercalated morphology. The result indicates that the partial delamination of PPZn layers is distributed in the PBAS matrix.

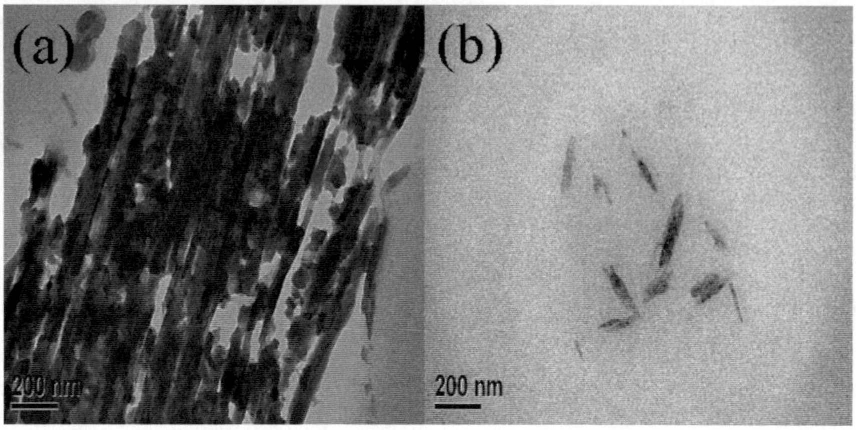

FIGURE 2. TEM images of (a) 5 wt% PBSA/PPZn and (b) 5 wt% PBSA/m-PPZn.

Isothermal crystallization kinetics of nanocomposites

To understand the effect of m-PPZn on the crystallization of PSBA, the isothermal crystalline was perferred using DSC. Figure 3a shows the relative degree of crystallinity, X_t, as a function of time t, for neat PBSA and nanocomposite. The isothermal crystallization kinetics for pure PBSA and its nanocomposites were analyzed using Avrami equation [4, 5]. The Avrami equation is written as:

$$1 - X_t = \exp(-kt^n) \quad (1)$$

where X_t is the relative crystallinity, k is the overall isothermal crystallization rate constant connected with both nucleation and crystal growth contributions, and n is the Avrami exponent on the nature of nucleation and growth geometry of crystals. The linear form of Eq. 1 can be expressed by taking a natural logarithmic transformation:

$$\ln[-\ln(1 - X_t)] = \ln k + n\ln(t) \quad (2)$$

The half-life time ($t_{0.5}$) of crystallization, defined as the time required to half completion of the final crystallinity, can be obtained by Eq. (3)

$$t_{0.5} = (\frac{\ln 2}{k})^{1/n} \qquad (3)$$

Consequently, the Avrami plots of PBSA and 5wt% PBSA/m-PPZn composites are shown in Figure 3b at 68 °C as an example, and the results are listed in Table. 1.

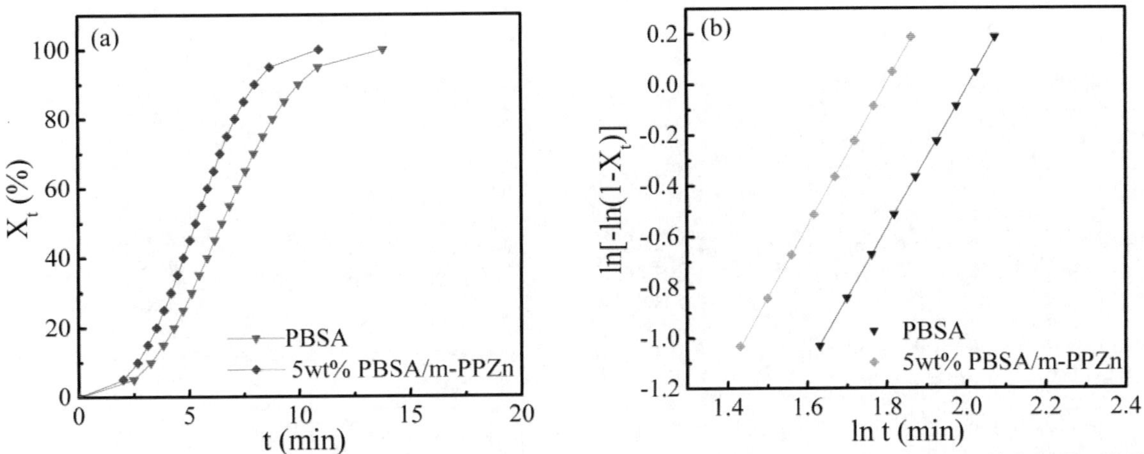

FIGURE 3. (a) Equivalent time-dependent relative degree of crystallinity X_t, (b) Avrami plots of $\ln[-\ln(1 - X_t)]$ versus ln t for the neat PBSA and its nanocomposites at 68 °C.

It is apparent that the value of $t_{0.5}$ increases with increasing T_c but $t_{0.5}$ decreases with increasing loading of PPZn. It is clear that the overall crystallization rate decreases with increasing Tc for PBSA and its nanocomposites. In addition, the overall crystallization rate in the nanocomposites is faster than neat PBSA at a given T_c, indicating that the isothermal melt crystallization is a nucleation control process. However, the value n for neat PBSA and its nanocomposites are between 2.52 to 3.25, suggesting that the crystallization mechanism of PBSA does not change after nanocomposites preparation. The result indicates that PBSA may correspond to three-dimensional sphere growth [6]. Furthermore, the k values of nanocomposites are strongly dependent on the m-PPZn contents. When crystallized at the same T_c, the k values of nanocomposites increase with increasing m-PPZn contents.

TABLE 1. Kinetic parameters based on the Avrami equation.

Sample	Tc(°C)	n	k(min^{-n})	$t_{0.5}$(min)
PBSA	62	2.82	5.90 x 10^{-2}	2.39
	64	2.95	1.82 x 10^{-2}	3.43
	66	3.09	4.81 x 10^{-3}	5.00
	68	2.74	4.10 x 10^{-2}	6.51
	70	2.52	2.88 x 10^{-3}	8.80
5wt% PBSA /m-PPZn	62	2.75	1.07 x 10^{-1}	1.98
	64	3.07	4.11 x 10^{-2}	2.51
	66	3.25	6.83 x 10^{-3}	4.15
	68	2.81	6.35 x 10^{-3}	5.30
	70	2.77	2.10 x 10^{-3}	8.13

Crystallization behavior of PBSA/m-PPZn nanocomposites

The spherulitic morphology of neat PBSA and its nanocomposites was observed with POM. According to Figure. 4, the size of PBSA spherulites are smaller, and the number of PBSA spherulites increases when the m-PPZn loading increases. It is distinct that m-PPZn provides more nucleating sites for PBSA to crystallize, leading to the limited growth of spherulites. The study of spherulitic morphology represented that the nucleation density of PBSA spherulites in the nanocomposites has been improved.

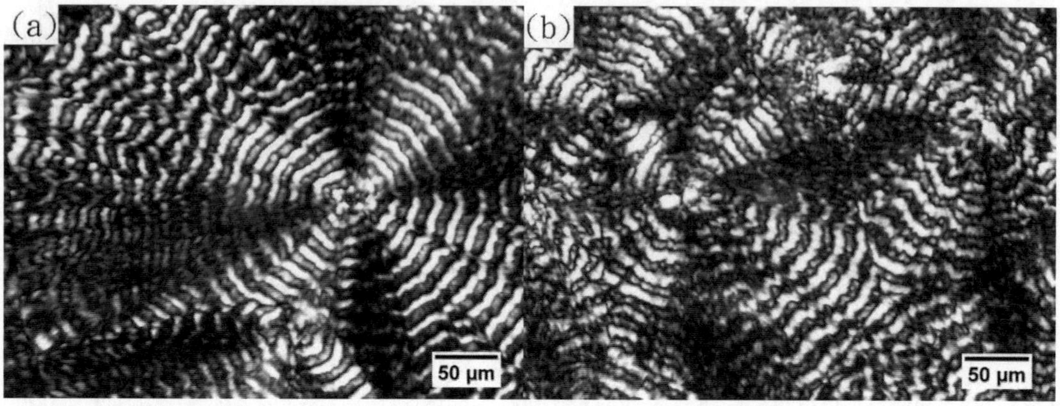

FIGURE 4. Polarized optical micrographs from melt at 68 °C (a) neat PBSA and (b) 5wt% PBSA/m-PPZn (scale: 50 um) after complete crystallization.

REFERENCES

1. K. J. Scott, Y. Zhang, Ren-Chain Wang, and A. Clearfield, *Chem. Mater.*, 1995, 1095–1102
2. B. D. Ahn, S. H. Kim, Y. H. Kim, J. S. Yang, *J. Appl. Polym. Sci.,*2001, 28, 2808–2826
3. P. Pan , Z. Liang, A. Cao, Y. Inoue, *Applied Materials &interfaces*, 2009, 1, 402–411
4. M. Avrami J. *Chem. Phys.* 1940, 8, 212–224.
5. M. Avrami J. *Chem. Phys.* 1941, 9, 177–184.
6. B. Wunderlich, *Macromol. Physic*, 1976, 2, 115–347

Experimental and Theoretical Study on Starve Fed Single Screw Extrusion of Polymer Blends

K.J. Wilczyński (reference author), A. Nastaj, A. Lewandowski, and K. Wilczyński (presenting author)

, *Warsaw University of Technology, Institute of Manufacturing Technologies, Polymer Processing Department 02-524 Warsaw, Narbutta 85, Poland*

Presenting author's email: k.wilczynski@wip.pw.edu.pl
Reference author's email: wilczynski_k@wp.pl

Abstract. An extensive experimental study has been performed for starve fed single screw extrusion of polymer blends. Several materials have been used in the study: LDPE/PS, PP/PS, PP/PMMA, and HDPE/PS blends. "Screw Pulling-out Technique" has been used to study the melting mechanism and flow in the machine. CSM melting mechanism (contiguous solids melting) has been observed for flood fed extrusion with an obvious presence of minor component inclusions. This mechanism has not been observed for starve fed extrusion. A new melting mechanism for starve fed extrusion of polymer blends has been proposed. This includes conductive melting in a partly filled region, and dispersed melting in a fully filled region, and takes into account the solid and molten phases of major and minor components. A forward/backward scheme of computation has been applied for global modeling of the extrusion process of polymer blends. A modified global SSEM_SF computer model has been applied for simulation of experimental runs.

Keywords: Extrusion, Polyblends, Modeling.
PACS: 47.85.md, 81.20. Hy.

INTRODUCTION

Single screw extruders are usually flood fed. Twin screw extruders are generally starve fed and for these machines mixing and melting actions are improved. These have led to the concept of starve feeding for single screw extruders. In starve fed extruders, the polymer is metered into the machine with a feeder. The extrusion output is determined by the feeder and independent on the screw speed. The first turns of the screw are not fully filled with a polymer, and pressure is not developed here. When the screw gets fully filled, the pressure is developed to push the polymer through the die.

Studies on starve fed single screw extrusion are limited. At the beginning, they were focused on experimental evaluation of mixing and melting capacities [1-6], with little attention to modeling of the process [3]. Recently, Wilczyński et. al [7,8] first proposed a model for melting in starve fed single screw extruders. Two stages of melting were distinguished. In the partially filled region of the screw, the polymer granules collect at the active flight and are mainly melted by conduction. In the fully filled region, the unmolten polymer particles are suspended in the previously molten polymer, and melting progresses through heat dissipation, mainly. Using this model the authors developed a global model for conventional starve fed single screw extrusion process [9] and later for non-conventional screw configurations [10,11].

At present, due to an extensive development of new advanced polymeric materials, such as polymer blends, polymer composites, highly filled compounds etc., the extruder design must allow the processing of much more complex polymeric systems than it was previously. Melting of such systems is much more complicated than melting of single polymers.

Our concern is with melting of polymer blends in a starve fed single screw extrusion. It is known that melting of polyblend components influences blending process and it is especially complex when the viscosity and melting temperature of the components are substantially different. Studying and modeling of polyblend morphology development in the single screw extrusion under starve fed conditions is in evident contrast with single screw machines operating under flood fed conditions [12,13]. However, up to the present, the only melting of the major polymer is considered for flood fed single screw extruders since a melting model considering the presence of two materials is not currently available. The melting and flow behavior of polymer blends in a starve fed single screw extruder has not been discussed in the literature.

EXPERIMENTAL

Extensive experimental studies have been performed for starve fed single screw extrusion of polymer blends using three different screw configurations: conventional three-sectional screw of compression ratio CR=8/3, mixing screw equipped with distributive toothed disc element, and mixing screw equipped with dispersive Maddock element. Several materials were used in the study: LDPE/PS, PP/PS, PP/PMMA, and HDPE/PS blends of composition of 85%/15%. "Screw Pulling-out Technique" was used to study the melting mechanism and flow in the machine. Pressure profiles along the screw were measured.

CSM melting mechanism (contiguous solids melting) has been observed for flood fed extrusion with an obvious presence of minor component inclusions. This mechanism has not been observed for starve fed extrusion. A new melting mechanism for starve fed extrusion of polymer blends has been proposed. This includes conductive melting in a partly filled region, and dispersed melting in a fully filled region, and takes into account the solid and molten phases of major and minor components (Fig.1).

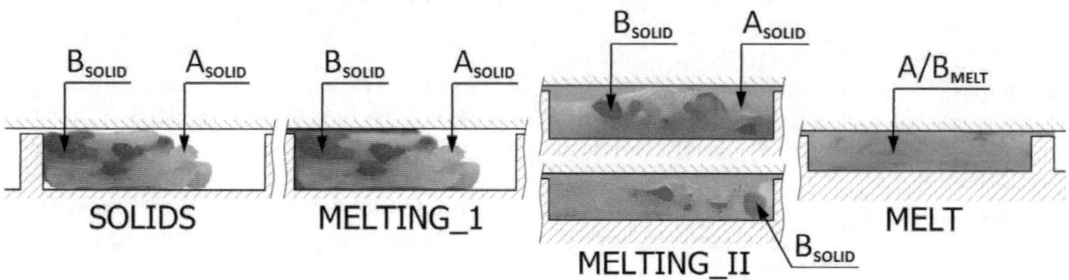

FIGURE 1. Melting mechanism for starve fed extrusion of polymer blends: Solids – solids conveying, Melting_I – conductive melting in a partially filled region, Melting_II – dispersed melting in a fully filled region, Melt – polyblend melt flow, A – major component, B – minor component, A/B – polyblend [14].

SIMULATION & EXPERIMENTATION

Simulations and experimentations have been performed for starve fed extrusion of LDPE/PS blend. The polymers used in the study were a low density polyethylene LDPE (Purell 3020D, Basel Orlen) and a polystyrene PS (Polystyrol 158K, BASF). The low density polyethylene has density 0.927 g/cm^3, melt flow index MFI = 0.3 g/10min (2.16 kg, 190 °C), and a melting point of about 114 °C. The polystyrene has density 1.04 g/cm^3, melt flow index MFI = 3 g/10min (5 kg, 200 °C), and a glass transition temperature of about 109 °C.

The viscous flow properties were determined using a capillary rheometer, at four different temperatures (140 °C, 160 °C, 180 °C, and 200 °C), and Klein equation have been applied to model viscosity curves

$$\ln\eta = A_0 + A_1\ln\dot{\gamma} + A_{11}\ln^2\dot{\gamma} + A_{12}T\ln\dot{\gamma} + A_2T + A_{22}T^2 \qquad (1)$$

where η is the viscosity, $\dot{\gamma}$ is the shear rate, T is the temperature.

Other material parameters of the polyblend were calculated using the parameters of components (i.e. polyethylene and polystyrene) and the particular composition of the melt with simply-structured mixing rules. Data for the components were taken from handbooks.

Since any melting model considering the presence of two materials for single screw extrusion of polymer blends (starve fed or flood fed) is currently not available we have modified our software SSEM_SF [9-11] to simulate polyblend extrusion. This software is based on the locally lumped parameter model of the single screw extrusion process. In the model, the screw channel is divided into elementary volumes (increments), where the input data, e.g. temperature and pressure, are the output data from the previous increment. The output data from the increment are the input data for the next increment. Within each increment, the local parameters are assumed to be constant. Mixing rules have been applied in each elementary volume to compute the local thermo-mechanical properties of the polyblend. A forward/backward scheme of computation has been applied for global modeling of the process. The important point here is to define the position of the transition partially/fully filled channel. The location of the transition point is not known a priori, and has to be searched iteratively involving process modeling. The condition for fill factor FF=1 is checked at each iteration of the simulation.

An example of simulation is shown in Figs.2-5 (Maddock screw). The results are presented in the dimensionless form, including pressure p, temperature T, solid fraction SF which is a measure of melting rate, and fill factor FF.

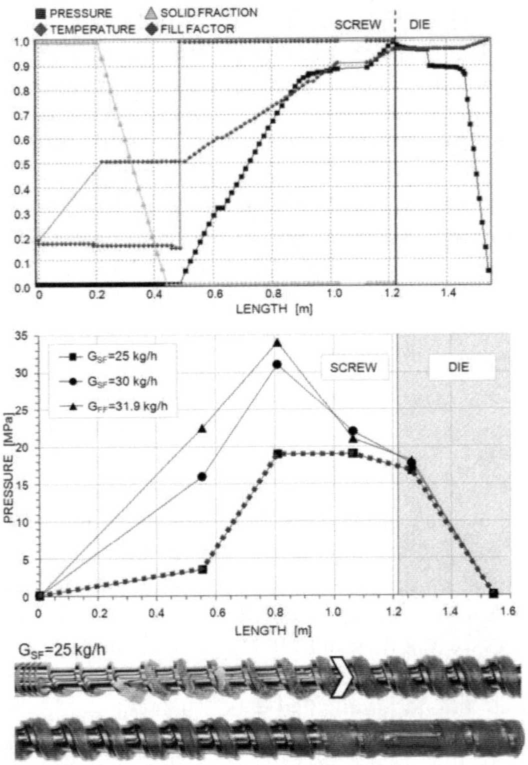

FIGURE 2. Simulation and experimentation for starve fed extrusion of LDPE/PS polyblend, N=80 rpm, G_{SF} = 25 kg/h.

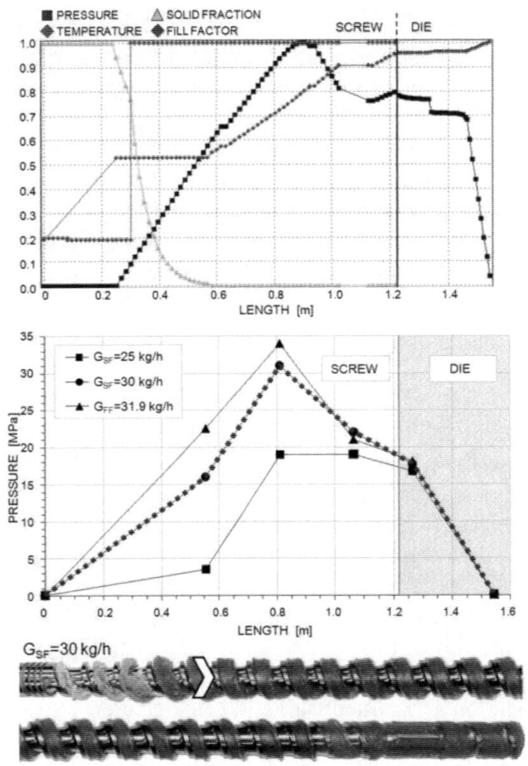

FIGURE 3. Simulation and experimentation for starve fed extrusion of LDPE/PS polyblend, N=80 rpm, G_{SF} = 30 kg/h.

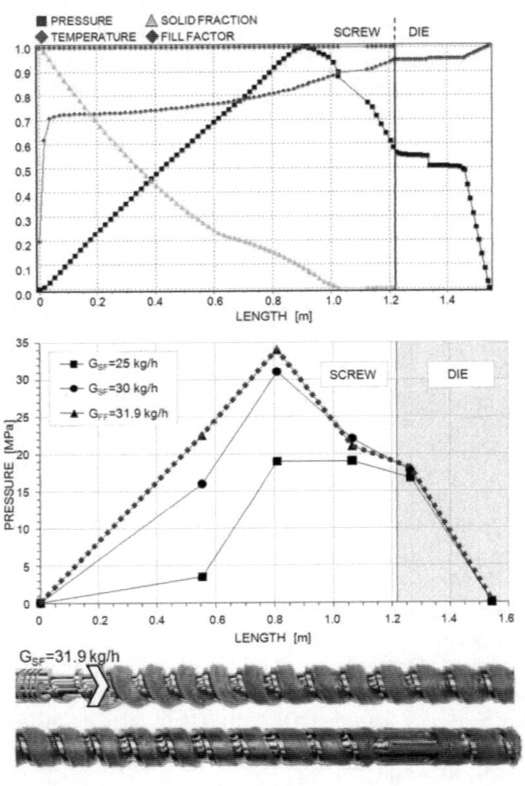

FIGURE 4. Simulation and experimentation for flood fed extrusion of LDPE/PS polyblend, N=80 rpm, G_{FF} = 31,9 kg/h

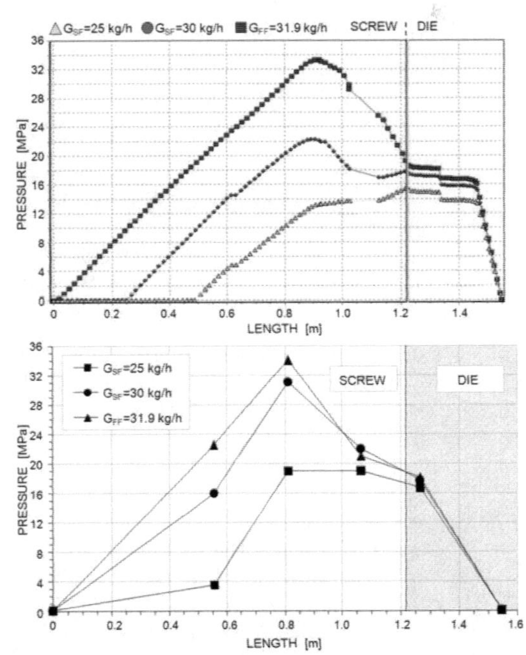

FIGURE 5. Pressure profile: simulation and experimental for starve/flood fed extrusion of LDPE/PS polyblend.

It is seen the pressure is higher for flood fed extrusion than for starve fed extrusion. When flow rate increases, the pressure increases, and the fully filled region also increases. The pressure drops to zero in the partially filled region (fill factor FF<1). At metered flow rate G_{SF} = 25 kg/h, the material is melted according to the first mechanism only, and at metered flow rate G_{SF} = 30 kg/h both mechanisms take place, and the second stage of melting is clearly seen. The filling of the screw was validated by comparing the fully filled regions. The beginning of the fully filled region is indicated by an arrow. This position corresponds to the fill factor FF=1 at the simulation chart. It is seen that filling of the screw channel and the length of the fully filled region increase with an increase of the flow rate, and these are generally well predicted. The melting predictions are worse to compare with screw filling predictions. However, faster melting for starve fed extrusion is clearly observed comparing with flood fed extrusion. The pressure is developed in the fully filled region only, and the pressure profile is well predicted.

CONCLUSIONS

Experimental and theoretical studies of melting mechanism of polymer blends in a starve fed single screw extrusion have been carried out for polymers of low viscosity and high viscosity. CSM type of melting mechanism (contiguous solids melting) was observed for flood fed extrusion of polyblends, but it was not observed for starve fed extrusion. The two-stage melting mechanism with conductive melting in the partially filled region of the screw channel and dispersive melting in the fully filled region was clearly seen for all the polyblends under study. Substantial differences between melting of neat polymers and polymer blends have been observed. In the case of polyblend extrusion, in the partially filled region a mixture of two solid polymers is melted by conduction mainly, and in the fully filled region a dispersion of solid particles of one or two polymers (minor or major component) in a molten matrix may be observed. A computer model SSEM_ST has been modified to simulate the flow of LDPE/PS polyblend in a flood fed/starve fed single screw extruder. The model has been validated experimentally.

Modeling of melting of polymer blends in a starve fed single screw extrusion requires a different approach comparing with melting of single polymers. However, conductive melting in the partly filled region can be modeled similarly considering the mixture of two solids (major and minor components) with equivalent thermo-mechanical properties of these. Dispersed melting in the fully filled region in the case of one solids (minor component) dispersed in the molten matrix of major component (or in the molten matrix of the polyblend) can be modeled as for single polymers, i.e. considering the melting of individual particles and applying an energy balance at the particle surface. However, there is a substantial difference in the case of two solids (minor and major components) dispersed in the molten matrix of polyblend. In this case modeling requires considering multi-phase flow. In addition, in a starve fed single screw extrusion, the location of the transition partially/fully filled screw is not known a priori (it is known in the co-rotating extrusion), and it has to be searched iteratively by checking the fully filled position along the screw at each iteration of process simulation. These are not trivial issues, and will be discussed in the next paper.

ACKNOWLEDGMENTS

The authors would like to acknowledge support from National Science Center, Poland (DEC-2012/07/B/ST8/03327)

REFERENCES

1. L. Lopez-Latorre and J.M. McKelvey, *Adv. Polym. Technol.*, **3**, 355-364 (1984).
2. D.P. Isherwood, R.N. Pieris, and J. Kassatly, *Trans. ASME,* **106**, 132 (1984).
3. S.R. Strand, M.A. Spalding, and S.K. Hyun, *SPE ANTEC,* **38**, 2537-2540 (1992).
4. R.E. Jerman, *SPE ANTEC*, **41**, 279 (1995).
5. M. Gale, *Adv. Polym. Technol.*, **16**, 251-262 (1997).
6. M.R. Thompson, G. Donoian, and J.P. Christiano, *Polym. Eng. Sci.*, **40**, 2014-2016 (2000).
7. K. Wilczyński, A. Lewandowski, and K.J. Wilczyński, *Polym. Eng. Sci.*, **52**, 1258-1270 (2012).
8. K. Wilczyński, A. Nastaj, and K.J.Wilczyński, *Int. Polym. Process.*, **28**, 34-42 (2013).
9. K.J. Wilczyński, A.Nastaj, A. Lewandowski, and K.Wilczyński, *Polym Eng Sci*, **54**, 2362-2374 (2014).
10. K.J. Wilczyński, A.Nastaj, A. Lewandowski, and K.Wilczyński, *Adv. Polym. Technol.,* DOI 10.1002/adv.21570 (2015).
11. K.J. Wilczyński, A. Nastaj, A. Lewandowski, and K.Wilczyński, *Int. Polym. Process.*, **31**, 82-91 (2016).
12. K. Wilczyński, A. Tyszkiewicz, and Z. Szymaniak, *J. Mater. Proc. Technol.*, **109**, 320-323 (2001).
13. N. Dominques, A. Gaspar-Cunha, and J.A. Covas, *Polym. Eng. Sci.*, **50**, 2195-2204 (2010).
14. K.J. Wilczyński, A. Lewandowski, and K.Wilczyński, *Polym. Eng. Sci.* (submitted for publication).

Synthesis and Tribological Investigation of PVD Hard Coatings deposited via Hybrid dcMS/HPPMS for Application in Plastics Processing

K. Bobzin, T. Brögelmann, N.C. Kruppe, M. Naderi*

Surface Engineering Institute, RWTH Aachen University
Kackertstr. 15, 52072 Aachen, Germany

naderi@iot.rwth-aachen.de

Abstract.
In the plastics processing industry extrusion process is a widely used manufacturing process. One of the most important requirements for an extruder is the highest possible mass flow rate, low wear on the extruder screw, a corrosion resistance towards plastics and the provision of a high melt quality throughout the lifetime of an extruder. In order to influence the interactions between the plastic melt and extruder surface, the extrusion tools can be modified using innovative coating technology such as physical vapor deposition (PVD). In this regard, direct current magnetron sputtering (dcMS) and high power pulse magnetron sputtering (HPPMS) hybrid technology offers many advantages, e.g. high ion flux, dense coating morphology, smooth surfaces and uniform coating thickness distribution on complex geometry components. Within the scope of this paper, ternary and quaternary chromium based nitride (Cr,Al)N and oxynitride (Cr,Al)ON PVD hard coatings, were used as protective coatings due to their mechanical, chemical and tribological properties to reduce wear and adhesion forces between extrusion tools and plastic. These hard coatings were deposited by means of hybrid dcMS/HPPMS technology in an industrial scale coating unit under oxygen gas flow variation ($F(O_2) = 25$ sccm and $F(O_2) = 50$ sccm) on plasma nitrided steel 34CrAlNi7-10 (1.8550). The influence of the oxygen content on the coatings morphology and its chemical composition as well as the tribological and the mechanical properties were characterized. An industrial plastic Makrolon® ET3113, PC, was considered for tribological investigations. Adhesion behavior of molten PC (corresponding the processing temperature) on PVD hard coatings was analyzed by high temperature contact angle measurements. Wear tests were performed by using pin-on-disc-tribometer measurements at room temperature T = 23 °C and higher temperature T = 150 °C against solid PC. A reduction in the adhesion tendency and friction coefficients through the coating in comparison to the uncoated substrate was observed.

Keywords: Extrusion, extruder screw, plastics industry, polycarbonate, PVD, hybrid dcMS/HPPMS, oxynitride, (Cr,Al)N, (Cr,Al)ON, wear, tribology.
PACS: 81.15.Cd

INTRODUCTION

Considering the global increase of requirement for plastics products, the plastics processing industry must be able to respond flexibly to the demands of the market. Therefore, an economical improvement of the extrusion process is required. During the extrusion process, interactions occur between the plastic melt and the surfaces of the extrusion die. Through selective modification of the surface properties it is possible to influence the interaction between the plastics melt and the mold surface. A reduction in the adhesion between the plastics melts and the mold surface results in decreasing the pressure loss and a narrow residence time spectrum of the molten plastics. In order to reduce wear on the molding tools, high alloy steels are used, which are expensive and complicated to process. In order to simultaneously achieving a high stability and cost effectiveness of molding tools protective coatings such as PVD (physical vapour deposition) are developed through the use of vacuum coating processes [1-5]. PVD hard coatings are intended to reduce the adhesion between the molten plastics and molding tool surfaces as well as their susceptibility to corrosion and wear. In this regard, direct current magnetron sputtering (dcMS) and high power pulse magnetron sputtering (HPPMS) hybrid technology offers many advantages, e.g. high ion flux, dense coating morphology, smooth surfaces and uniform coating thickness distribution on complex geometry components [6-7]. In industrial applications chromium based nitride PVD hard coatings are already being used as protective coatings on mold tools. Binary CrN was one of the first chromium containing hard coatings developed by PVD technology. In addition to binary CrN and ternary chromium based nitrides (Cr,Al)N, the quaternary oxynitride (Cr,Al)ON can be deposited by using PVD [8-11]. These hard coatings can be used as protective coatings in many applications due to their mechanical, chemical and tribological properties in order to reduce wear, corrosion and adhesion forces between mold insert and plastic. (Cr,Al)ON coatings belong to a class of materials with a combination of properties, such as high hardness and chemical inertness, which are of interest for numerous industrial applications. Moreover, they are suitable for use on machine elements due to their hardness and corrosion resistance in aqueous media [12-13].

Within the scope of this paper, chromium based nitride (Cr,Al)N and oxynitride (Cr,Al)ON hard coatings were deposited by means of hybrid dcMS/HPPMS. The influence of the oxygen content on the coating and system properties was characterized. Tribological investigations as an application-oriented model tests were carried out. This includes high temperature contact angle measurements in order to analyze the wetting behavior of nitride and oxynitride hard coatings against Makrolon® ET3113, PC, Covestro Deutschland AG. Moreover, pin-on-disc-tribometer measurements were carried out for investigation of the friction behavior of nitride and oxynitride hard coatings towards PC at two different temperatures such as room temperature T = 23 °C and at half of the plastics processing temperature of PC T = 150 °C.

EXPERIMENTAL

Coating deposition

Chromium based nitride (Cr,Al)N and oxynitride $[(Cr,Al)ON]_{F(O_2)}$ were deposited by means of hybrid dcMS/HPPMS in an industrial scale coating unit, CC800/9 Custom, CemeCon AG, Würselen, Germany, under oxygen gas flow variation, $F(O_2)$ = 25 sccm and $F(O_2)$ = 50 sccm, on plasma nitrided steel 34CrAlNi7-10 (1.8550). The substrates had a roughness of Ra = 0.02 µm. As schematically illustrated in FIGURE 1 (a) the coating chamber is equipped with two HPPMS cathodes and four dcMS cathodes. In all, three different coatings (FIGURE 1 (b)) were prepared by process parameters, which can be taken from [14] and FIGURE 1 (c).

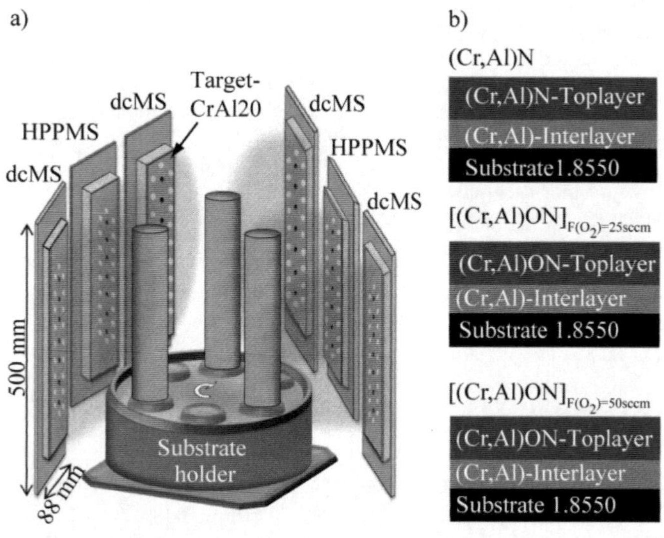

c)

Parameters	Unit	Value
Substrate temperature T_s	°C	420
Pressure p	mPa	620
Ar-flow F(Ar)	sccm	200
N_2-flow $F(N_2)$	sccm	Pressure controlled
O_2-flow $F(O_2)$	sccm	0/25/50
Heating power P_H	kW	4
Substrate Bias Voltage U_{Bias}	V	-100
Average cathode power HPPMS P_{HPPMS}	kW	5
Average cathode power dcMS P_{dcMS}	kW	3
Pulse frequency f	Hz	500
Pulse time t_{on}	µs	40

FIGURE 1. Configuration of industrial coating unit CC800/9 Custom (a), deposited nitride and oxynitride coatings (b) and process parameters (c)

Characterisation of deposited nitride and oxynitride hard coatings

The influence of the oxygen content on the properties of the deposited nitride and oxynitride hard coatings was studied comprehensively. The mechanical properties of the coatings such as the universal hardness (HU) and indentation modulus (E_{IT}) were investigated by means of a nanoindenter XP, MTS Systems Corporation, Oak Ridge, Tennessee, USA. Analysis of the indentation curves was carried out following the method proposed by Oliver and Pharr [15]. Analysis of the surface morphology and the topography of nitride and oxynitride hard coatings were performed by means of scanning electron microscope (SEM) ZEISS DSM 982 Gemini, Jena, Germany at Laboratory RWTH (GfE), micrographs of fractured cross section were taken using a secondary electron (SE) detector. By using a confocal laser scanning microscope (CLSM) Keyence VK X210, Tokio, Japan, the arithmetic average surface roughness Ra and the mean roughness depth Rz was determined according to ISO 4287. Tribological investigations were carried out between coatings and Makrolon® ET3113, PC, Covestro Deutschland

AG, which can be considered as an application-oriented model tests. This includes the investigation of wetting behavior of molten PC (corresponding the processing temperature T_p) on PVD hard coatings by means of high temperature contact angle measurements according to DIN EN 828:1998-01via device DSA 10, Kruess GmbH, Hamburg, Germany (FIGURE 2(a)). The friction and wear behavior of investigated coatings were performed according to DIN 50324 using a pin-on-disc-tribometer (PoD, CSM Instruments, Peseux, Switzerland, at room temperature T = 23 °C and T = 150 °C against solid PC-pin (Ø = 6 mm) as counterpart. After PoD measurements the wear track depth was determined by means of CLSM, which was used to calculate the wear rate k. To calculate the wear rate, the wear volume, W_v, was evaluated and divided by the normal load, F, and the distance, s.

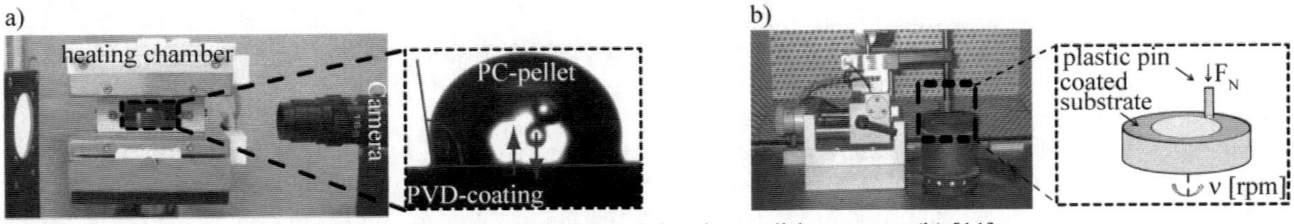

FIGURE 2. Principle of high temperature contact angle (a) and the pin-on-disk wear test (b) [11]

RESULTS AND DISCUSSION

FIGURE 3 (a) illustrates the cross section fractures of SEM micrographs for coating morphology and topography analysis in dependence of oxygen content as well as the roughness and the thickness of the coatings. It can be seen that the coatings exhibit a columnar morphology and compact coating surface. The total coating thickness was kept constant at (d = 2.6±0.1) µm. From the results of the surface roughness can be observed that, the coatings exhibit significant differences concerning the Rz values with an increasing oxygen gas flow. In contrast, no significant difference can be determined for the roughness values of Ra. Moreover, it can be noticed that with increasing oxygen flow, universal hardness increases from HU = (24.1±2.8) GPa to HU = (26.1±3.3) GPa, although the indentation modulus decreases from E_{IT} = (355±26.3) GPa to E_{IT} = (345±35) GPa with increasing oxygen flux in the coating process.

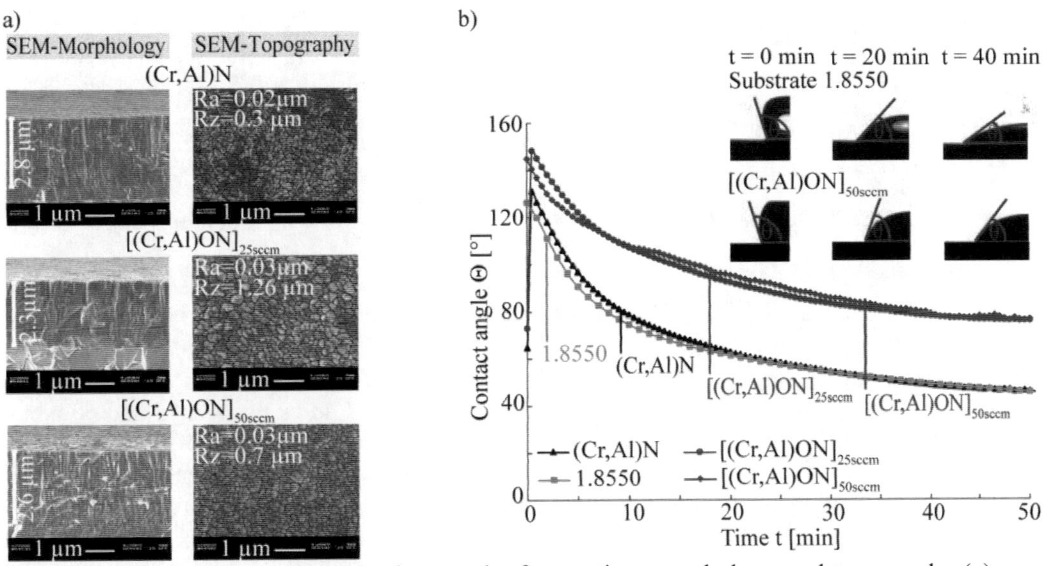

FIGURE 3. Cross sections of SEM micrographs for coating morphology and topography (a), contact angles and wetting behavior of molten of PC against uncoated comparative 1.8550 and nitride and oxynitride coatings (b).

Time dependent development of high temperature contact angle measurements were considered to obtain adhesion behavior of molten PC on the coatings. Uncoated substrate 1.8550 was considered as reference. FIGURE 3 (b) reveals that a lower adhesion affinity, i.e. a higher contact angle, of PC on coated surfaces with oxynitride $[(Cr,Al)ON]_{25sccm}$ and $[(Cr,Al)ON]_{50sccm}$ coatings in comparison to uncoated substrate 1.8550 and nitride (Cr,Al)N coating is achieved. This can be also determined from the wetting behavior of molten PC on nitride and oxynitride hard coatings.

Tribological investigations were performed using a pin-on-disc-tribometer (PoD) to observe the friction behavior of nitride and oxynitride hard coatings and uncoated substrate 1.8550 as a reference against counterparts of solid Makrolon ET3113, PC, at two different temperatures, T = 23 °C and T = 150 °C, which corresponds to the half of the processing temperature in the plastics processing. Moreover, the investigations of the wear track, which have been performed via CLSM to obtain wear rate after PoD measurements at these two different temperatures as well as average coefficient of friction CoF after PoD measurements, can be taken from FIGURE 4 (a) and (b). From FIGURE 4 (a) it can be concluded, that at T = 23 °C the coefficient of friction, CoF, of all nitride and oxynitride hard coatings with different oxygen contents is in a range between μ ~ 0.5-0.55 significantly lower compared to the uncoated substrate 1.8550. Obviously, (Cr,Al)N exhibits a high wear rate k (T = 23 °C) = $0.03 \cdot 10^{-14}$ [$m^3N^{-1}m^{-1}$] in comparison to $[(Cr,Al)ON]_{50sccm}$ with k (T = 23 °C) = $0.01 \cdot 10^{-14}$ [$m^3N^{-1}m^{-1}$]. However, it should be noted that at room temperature the CoF is for nitride and oxynitride hard coatings in a range between μ ~ 0.61-0.56, which is lower than for the uncoated substrate with μ ~ 0.78. The same tendency can be observed for wear rate at high temperature, T = 150 °C, in FIGURE 4 (b).
By comparing the results of the high temperature contact angle measurements and the high temperature pin-on-disc-tribometer tests for nitride and oxynitride hard coatings against PC, a correlation can be determined. It could be observed that $[(Cr,Al)ON]_{25sccm}$ and $[(Cr,Al)ON]_{50sccm}$ with the highest contact angles in contact with Plexiglas PC (FIGURE 3 (b)) show the lowest coefficient of friction in tribological PoD testing for PC (FIGURE 4 (a) and (b)).

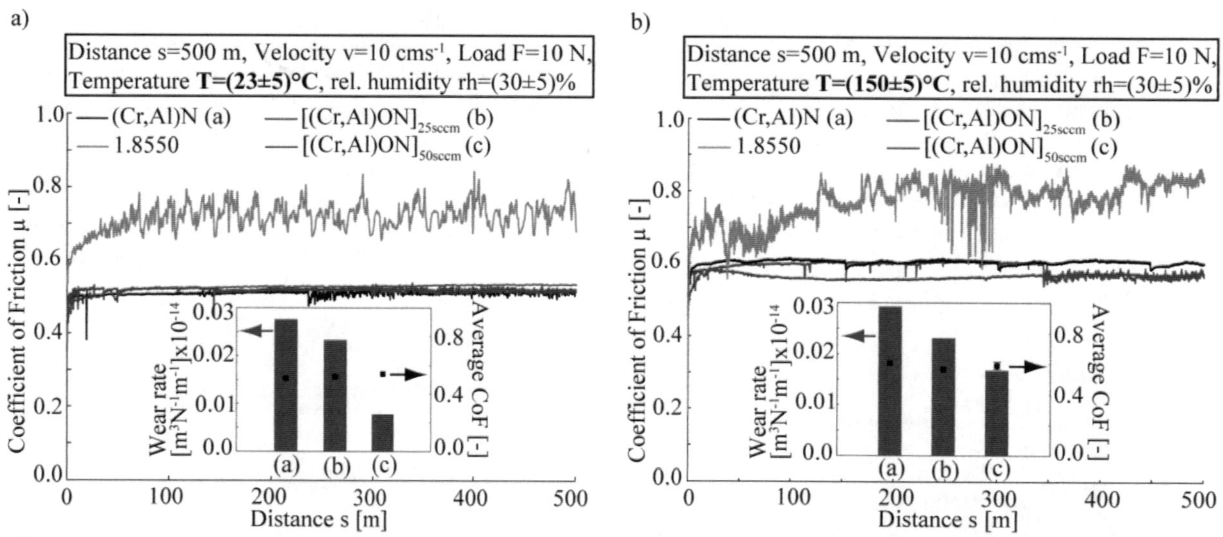

FIGURE 4. Wear rate, average CoF and PoD measurements of nitride and oxynitride hard coatings at 23 °C (a) and 150 °C (b)

Conclusion

In the scope of this work the ternary and quaternary chromium based nitride coatings (Cr,Al)N and the oxynitride coatings $[(Cr,Al)ON]_{F(O2)}$ were deposited at two different oxygen gas flows $F(O_2)$ = 25 and 50 sccm by means of hybrid process dcMS/HPPMS in an industrial scale coating unit. The investigation of coating properties showed that an increasing oxygen gas flow leads to an increase of coating hardness. The investigation of tribological properties revealed that by high temperature contact angle measurements a lower adhesion of polycarbonate Makrolon ET3113 on coated surfaces could be determined in comparison to uncoated substrate. The results are implying that the oxynitride coatings exhibiting the highest contact angles in contact with PC and, therefore, the lower wetting behavior shows the lowest coefficient of friction in PoD-tribometer measurements. Lower adhesion, e.g. higher contact angle, in high temperature contact angle leads to lower coefficient of friction and, therefore, might contribute to a decreasing of the release forces in plastic processing. This is in agreement with the investigations on wear rate of nitride and oxynitride hard coatings. A relation between wear rate and average coefficients of friction was observed for PC. The lowest wear rate and average CoF was measured for PC in contact with coating system $[(Cr,Al)ON]_{50sccm}$ at T = 23 °C and T = 150 °C. The results indicate a high potential of the investigated oxynitride coatings to be used as protective coatings against abrasive wear during processing of polycarbonate.

ACKNOWLEDGMENTS

The research project IGF 497 ZN of "Forschungsvereinigung Kunststoffverarbeitung e.V.", which this work is based upon, has been supported financially within the scope of "Industrielle Gemeinschaftsforschung" (IGF) by "Bundesministerium für Wirtschaft und Technologie" (BMWi) via „Arbeitsgemeinschaft industrieller Forschungsvereinigungen „Otto von Guericke e.V."" (AiF). The authors would like to express their thanks to the BMWi and the AiF for their financial support.

REFERENCES

1. K. Bobzin, N. Bagcivan, N. Goebbels, K. Yilmaz, B.R. Hoehn, K. Michaelis, M. Hochmann, Surface and Coatings Technology **204** (2009), S. 1097-1101
2. N. Bagcivan, K. Bobzin, G. Grundmeier, M. Wiesing, O. Ozcan, C. Kunze, R.H. Brugnara, *Thin Solid Films* **549** (2013), S. 192-198
3. N. Bagcivan, K. Bobzin, S. Theiß, Thin Solid Films **528** (2013), S. 180-186
4. K. Bobzin, E. Lugscheider, R. Nickel, N. bagcivan, A. Krämer, *Wear* **263** (2007), S. 1274-1280
5. E. Lugscheider, K. Bobzin, K. Lackner, *Surface and Coatings Technology* 174-175 (2003), S. 681-686
6. G. Grecczynski, J. Lu, J. Jensen, I. Petrov, E. E. Greene, Journal of vacuum science and technology **A** (2012) 30
7. K. Bobzin, N. Bagcivan, M. Ewering, R.H. Brugnara, *Surface and Coatings Technology* **205** (2011), S. 8-9
8. H.C. Barshilia, N. Selvakumar, B. Deepthi, K.S. Rajam, *Surface and Coatings Technology* **201** (2006) 2193–2201
9. H.C. Barshilia, B. Deepthi, K. S. Rajam, K.P. Bhatti, S. Chaudhary, *Journal of Vacuum Science and Technology* A **27** (2009) 29
10. Z. Changwei, H. Lingling, W. Jia, X. Shuwen, *Solar Energy Materials and Solar Cells* **137** (2015) 243-252
11. N. Bagcivan, K. Bobzin, T. Brögelmann, C. Kalscheuer, *Surface and Coatings Technology* **260** (2014) 347-361
12. L.v. Yanhong, L. Xiaohong, Li Hongxuan, Z. Huidi, C. Jianmin, *Surface and Coatings Technology* **206** (2012) 3961-3969
13. A. Khatibi, J. Sjölen, G. Greczynski, J. Jensen, P. Eklund, L. Hultman, Acta Materialia **60** (2012) 6494–6507
14. K. Bobzin, T. Brögelmann, N.C. Kruppe, submitted to *Surface and Coatings Technology (2016)*
15. W.C. Oliver, G.M. Pharr, Journal of Materials Research Society 1 (2004) 3-20

Development of a method for a realistically and reproducible contamination of polymer melt filters

Volker Schöppner, Peter Meilwes

Kunststofftechnik Paderborn, University of Paderborn, Warburger Str. 100, 33098 Paderborn, Germany
volker.schoeppner@ktp.upb.de; peter.meilwes@ktp.upb.de

Abstract. In the field of polymer processing, extrusion is one of the most common processing methods. Not only in the processing of recycled materials, also when using virgin polymers there can be contaminations during the storage or processing of the material. This may adversely affect the melt quality and by this also the quality of the final extrusion product. Examples of possible contaminations are metal particles which are caused by wear and tear of the extruder, or degradation products of the processed material itself.

As a result of increasing demands on the quality of extrusion products, especially in the field of fibre and film extrusion, the filtration of polymer melts is widely used and state of the art. To remove unwanted debris out of the melt, different metal filter media is used. Some examples of typical filter media are different kinds of wire mesh, filters made of nonwoven metal fibres or sintered metal powder discs. By using this filter media, it is possible to remove foreign particles like solid particles, as well as soft components, so called "gels" from the melt. This ensures a high quality extrusion result.

In order to compare the filtering effect of different filter media and to assess the contamination and selective filtration, it is necessary to develop a possibility for reproducible rapid contamination of filter materials. In this work this should be elaborated as a part of the PET processing. For this, a real filter contamination is analysed. After this, a definition of an adequate substitute dirt is effected. The focus is both on solid particles, as well as on gel contamination. By this, it is possible to generate realistic and reproducible filter soiling so that pressure rise curves can be generated in order to compare the behaviour of different filter media during the extrusion process.

Keywords: Polymer Melt Filtration, Extrusion
PACS: 47.56.+r; 81.20HY; 83.80Sg

INTRODUCTION

As a result of increasing quality requirements for extrusion products, especially in the field of fiber and film extrusion, filtration of polymer melts is widely used today. An example of the need of a melt filtration is the fiber extrusion of polyethylene terephthalate (PET). Even the smallest contaminations of the melt in the micron range can lead to a weakening of the fiber strand [1].

In order to compare the filtering effect of different filter media and to assess the contamination and selective filtration it is necessary to develop a possibility for reproducible rapid contamination of filter materials. In industrial extrusion processes the filter lifetime lasts several days to weeks depending on the degree of contamination. However, in a laboratory investigation, this contamination should ideally can be done in a few minutes or hours.

During extrusion processes there are multiple possibilities of contaminating the polymer material. On one hand the contamination can be brought in by the raw material itself. During storage or transportation of the Material there are lots of possibilities of contaminations. Also the process itself can contaminate the polymer melt. Examples of possible contaminations are metal particles which are caused by wear and tear of the extruder, or degradation products of the processed material itself. Degradation processes in the extrusion process can cause the formation of so called gel particles in the melt. These are areas of similar material but different viscosity. In extrusion products, for example films, these gels can result in so called fish-eyes or discolorations of the material. Basically there are two possibilities of contaminations. On one hand the contamination can consist of inorganic particles like metal abrasion. This contamination leads to an accretion of particles in the filtration media. Also in addition, there may be a structure of a filter cake on the filtration media. On the other hand the contamination can consist of organic particles, so called gels. These can consist of the processed polymer itself or different polymers with a different viscosity. These "soft" contaminations lead to an effective clogging of the filter media [1, 3].

In relation to the suitability of various filter media, the two types of contaminations behave fundamentally different. In wire mesh or powder filters gel particles are easily pressed through the structure of the filter or are cut in small pieces depending on the prevailing pressure. This causes a contamination of the melt behind the filter medium, but can also be used consciously for the comminution of gels [1, 2, 4].

INVESTIGATION OF BACK-FLOW-MASS FROM INDUSTRIAL FILTER SYSTEMS

To get an idea of the realistic contamination in industrial extrusion processes, Back-Flow-Mass out of a commercial polyethylene terephthalate (PET) processing line was tested for including contaminations. First of all micrographs were made by the Back-Flow-Mass. In addition to this, the PET was dissolved with glycol to expose the solid particles in the material. During solving the PET in glycol a gelatinous mass remained, which was only soluble in acetone. This mass is probably caused by cross-linking processes in the material. This indicates, that a significant part of the pollution actually occurs to gels in the polymer material.

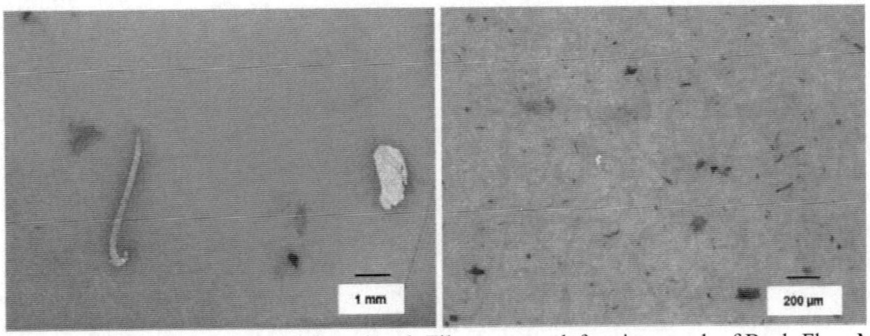

Figure 1. Contamination of PET out of an industrial used melt Filter system; left: micrograph of Back-Flow-Mass; right: residues of dissolved Back-Flow-Mass

The evaluation of the micrographs showed a wide range of solid contaminations in the polymer melt. This is shown in figure 1. These Particle size distribution is shown in figure 2. Besides small particles with a size of a few microns big metal chips with a length of several millimeters can be seen. Generally there are various inclusions in the Back-Flow-Mass samples.

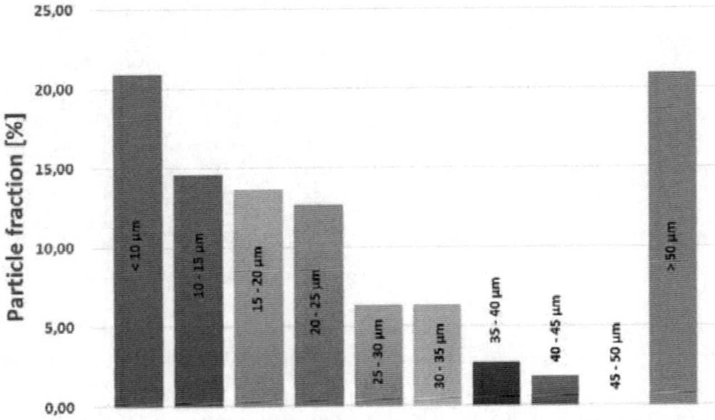

Figure 2. Particle size distribution of solid particles out of a back-flow-mass from industrial PET processing

The particle sizes which are bigger than 50 microns and smaller than ten microns were combined in the analysis because the studies for melt filters were designed for filter dimensions of 20 microns. In order to achieve a defined pollution, a dirt constellation was developed, which takes into account the gel particles as well as the solid part of the dirt. Due to the high range of possible contaminants, the particle size distribution of the test soil must be adapted to the expected size of the tested filter. In the present studies, filters are examined with a filter fineness of 20 microns, so that the dirt particles must also be adapted on this relevant area.

PRACTICAL INVESTIGATIONS FOR THE DEVELOPMENT OF THE ARTIFICIAL CONTAMINATION

The results of the investigation of Back-Flow-Mass should be used to develop a test contamination for a reproducible and fast filter contamination. For this suitable substitutes for a realistically contamination behaviour had to be found. In addition, the test soil has to be evaluated in terms of reproducibility and realistic contamination behavior.

Determination of suitable substitutes for a realistically an reproducible contamination

The challenge was to find suitable substitutes, which on the one hand ensure a realistic picture of a gel and solid particle contamination, but on the other hand simultaneously perform no undesirable effect on the extrusion process itself, such as excessive wear, aggressive behaviour towards steel or greater viscosity changes. Ideally, there should be a material that is already used in extrusion technology. In addition to that, it must be ensured, that the selected material is available in various particle sizes and can be obtained in a very fine fractionated form. Also, it should be available in sufficient quantities. Because of this, a test dust of calcium carbonate (Escal 30) was selected. This dust is commercially available in different particle size ranges in a very fine fractionated size distribution. Also calcium carbonate is often used as filling additive in the extrusion technology, therefore no complications in relation on the system technology has to be expected. Additionally, the fine particle size distribution of the test dusts allows a very defined contamination of the polymer melt.

The next challenge was a suitable substitute for the artificial contamination of the melt with gel particles. The formation of gels can basically be attributed to four potential mechanisms [1,3,4]:

- cross linking processes
- material degradation / condensation
- different melting temperatures
- changes in crystallinity

To study the reproducible production of gel particles in a polymer melt, it must first be tested which mechanism causes the formation of the gel particles in case of PET processing. As already mentioned, from the processing of PET it is known that a high fraction of polymer dust in the granulate leads to an increase of gel particles in the melt. This has also been demonstrated in film extrusion tests (Figure 3).

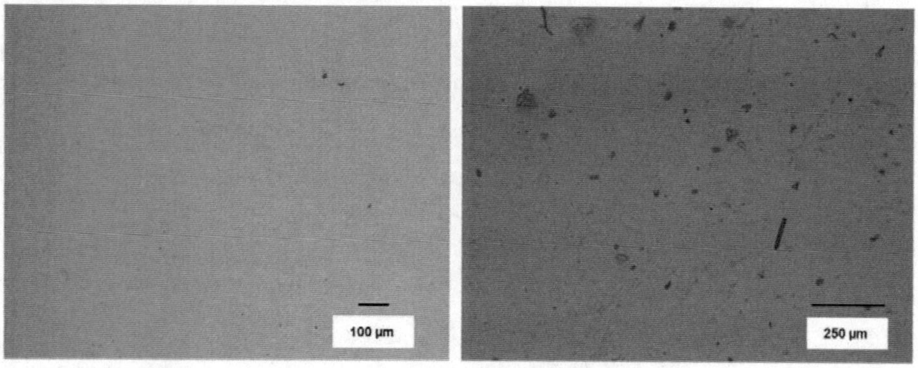

Figure 3. Formation of gel particles; left: uncontaminated PET film; right: PET contaminated with a high fraction of PET dust

For this reason, the production of gels by PET dust was considered in more detail. For this purpose, various studies have been done in order to analyze the above mentioned potential formation mechanisms for the example of gel particles of PET dust.

PET-dust samples from drying processes were analyzed by thermal analysis (DSC), from which conclusions on the crystallinity and the melting temperature may be possible. Neither the crystallinity, nor the melt temperature show significant differences to the virgin material, so that in the case of PET, it is not expected that these mechanisms cause the formation of gel particles.

For checking a possible material degradation measurements of intrinsic viscosity have been done. These investigations can give information on the changes of the molecular chain length in the tested polymer compared to the virgin material. Here also no significant changes could be observed.

For investigations three of four potential mechanisms for the formation of gels in PET can be excluded. It is likely that the formation of gels in PET extrusion processes can be attributed to cross linking processes in the material which leads to a local increase of the viscosity of the polymer melt.

With the use of PET dust from material drying processes, it cannot be guaranteed, that there are constant and reproducible dust properties. Because of this, a similar behaving alternative had to be found. As a result of the investigations of PET dust, a substitute has to be selected, which crosslinks in the process and forms a reproducible gel contamination. For this crosslinking polyethylene powder has been selected (ICORENE 1505 ©). This is commercially available and allows a defined contamination of the melt which is very close to the contamination with gels from polymer dust contamination (Figure 4).

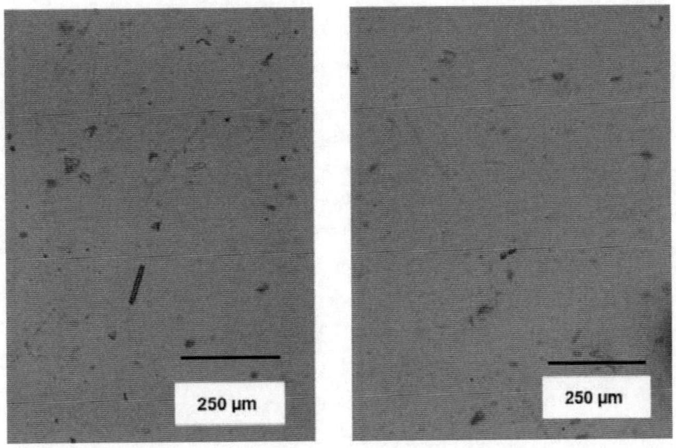

Figure 4. Imitation of a gel contamination by crosslinking polyethylene; left: contamination of PET with PET dust; right: contamination of PET with crosslinking polyethylene powder

The developed test contamination consisting of calcium carbonate and crosslinked polyethylene powder were investigated in extrusion tests on wire mesh filters on its soiling behavior. Characteristic behaviors for the pollution species appeared (Figure 5).
The investigations were done with wire mesh filters with a filter fineness of 25 microns. Polyethylene terephthalate with a 1% contamination with the correspondingly contamination type has been used. While Calcium carbonate showed an exponential increase of pressure drop from a certain degree of extrusion time, the contamination with "soft" dirt runs to a pressure limit at which no further increase of pressure drop can be detected. This can be attributed to a pressing of the gel particles through the filter medium at a certain pressure value. The combination of solid particles and "soft" contaminants is expected to a hybrid of the clogging behavior. For the test contamination, a mixture of 30% calcium carbonate and 70% cross-linking PE-powder was set as the basis of previous investigations.

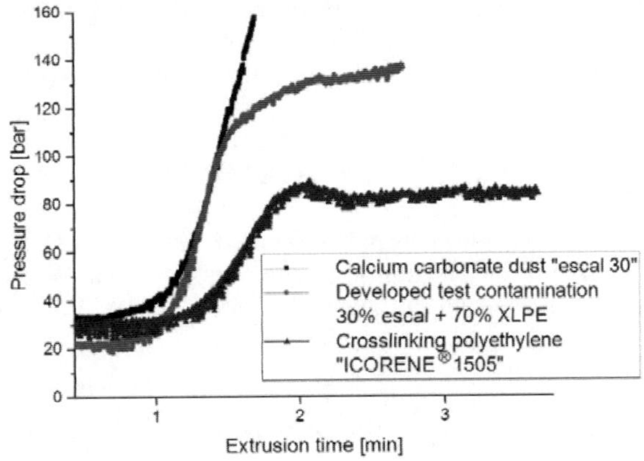

Figure 5. Soiling behaviour of filter screens processed with different types of contamination

Extrusion Trials to validate the reproducibility of filter contamination

The developed test contamination had to be evaluated in terms of reproducibility and realistic soiling behaviour. For this, multiple extrusion tests have been performed with different filter sizes. First, extrusion tests were carried out on small filter screens with a diameter of 30 mm. Screens of wire mesh filters with a filter fineness of 25 microns and 36 microns were applied. The pressure loss through the filter medium during extrusion time can be seen in Figure 6.

It becomes clear, that the dirt generated sufficiently reproducible results regarding the filter contamination. Especially in the filters with a filter fineness in the size of the contaminating particles (25 microns), the pressure loss reveals a very reproducible behaviour.

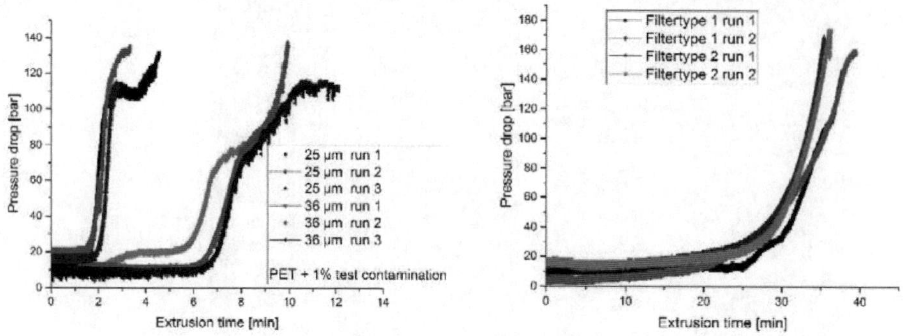

Figure 6. Reproducibility of filter contamination; left: 30 mm filter screens with wire mesh; right: 12 inch Leaf Disc Filters with fleece

Also, in order to validate the processing properties of the test soil in industrial relevant equipment, investigations were carried out on a large-surface melt filter with filter discs. In the filter housing three filter discs with a diameter of 12 inches are mounted. Two runs were carried out with one filter type to validate the contamination behaviour. The used discs where fleece filter discs with a filter fineness of 20 microns. PET was processed with a contamination of 1 % of the developed test contamination. The flow rate was constant at 50 kg / h. The results of the investigations show a reproducible increase in pressure during the pollution time (Figure 6). Both disc types seem to clog defined using the newly developed test soil. This allows a general investigation of the contamination behaviour of polymer melt filters.

CONCLUSION AND OUTLOOK

In this paper a new possibility of fast and reproducible contamination of polymer melt filters was developed. Using the newly generated contamination, it is possible to examine the contamination behaviour of polymer melt filters in the extrusion process in more detail. This is intended to work on a model in which the contamination behaviour of filter units can be calculated as a function of the degree of contamination of the polymer melt, the dimension of the pollution and the process parameters.

ACKNOWLEDGEMENTS

The research project (KF2363833SL3) described here has been sponsored as part of the "Zentrale Innovationsprogramm Mittelstand (ZIM)" by the German Bundesministerium für Wirtschaft und Energie (BMWi). The research Project is conducted in cooperation with "SCHWING Fluidtechnik GmbH".

ZIM
Zentrales Innovationsprogramm Mittelstand

REFERENCES

1. VDI Verlag, *Filtrieren von Kunststoffschmelze,* Düsseldorf: VDI-Verlag 1981, pp. 1-6, 57-58, 129-133, 141-146
2. FKUR GmbH, *Schmelzefilter und Siebwechslersysteme*, Willich: F-Kur Forschungsinstitut Kunststoff und Recycling GmbH, 1996
3. J. Scheirs; *Compositional and Failure Analysis of Polymers – A Practical Approach*, Chichester: John Wiley & Sons Ltd, 2000, pp. 270-273, 275-278, 288-297
4. D. Gneuß; *Filtrieren von Kunststoffschmelzen – 2. Aktualisierte Auflage*, Hiddenhausen: Feist & Lehbrink, 2007, pp. 3-11, 15-16, 61-69

Turn-Key Extrusion Lines For Highly-Filled Compounds: Compounding Of $CaCO_3$ Formulations

Richard Kretz, Michael Thummert

Leistritz France Extrusion
12B Rue Charles Guillon
F01250 Ceyzériat
France

Leistritz Extrusionstechnik GmbH
Markgrafenstraße 36 – 39
90459 Nuremberg
Germany

rkretz@leistritz.com
mthummert@leistritz.com

Abstract: Turn-Key Extrusion Lines For Highly-Filled Compounds: Compounding Of $CaCO_3$ Formulations

Keywords: Extruder, Compounding, Fillers, Twin Screw Extrusion
PACS: 81, 89

COMPOUNDING OF $CACO_3$ FORMULATIONS

Setting up extrusion lines for highly-filled compounds is a very challenging task since the raw materials – and in particular $CaCo_3$ – oftentimes have differing quality aspects and material characteristics. The various raw materials, setup of the line components and parameters of the extrusion process itself open up a wide field of options - not just in terms of improving the final product, but in the worst case also in terms of making things worse. Thus the extrusion process is affected by many factors that need to be taken into account for achieving optimum product quality and a stable process of the extrusion line.

Raw materials vary considerably as regards characteristics like e.g. hardness, particle shape or density, also influencing their particular flow characteristics. These special characteristics in turn influence process technology as well as the final product.

FIGURE 1. Example of materials' varying hardness characteristics

The optimum final product can only be obtained if the extrusion line layout complies exactly with the requirements of the raw materials. All stations involved require tailor-made accuracy, like e.g. material handling, platform design and handling of finished goods. Flexibility and safe operation are other key factors for the extruder line layout.

Leistritz built an extrusion plant for the production of highly filled compounds with up to 85% $CaCo_3$. Since the Leistritz ZSE MAXX extruders have a specific volume/torque ratio (OD/ID = 1.66 and 15.0 Nm/cm³) this plant not only can realize maximum throughputs, but also facilitates a very energy efficient production. In this way the plant is a good example of state-of-the-art technology. The concept also stands out when it comes to flexibility. With slight adjustments it can run various applications with e.g. talcum, titanium dioxide or barium sulphate. In the field of flame retardants, materials such as aluminium hydroxide or magnesium hydroxide can be processed.

Since those compounds are not only rather cost-efficient but prove to have very good mechanical properties, they have experienced a real boom lately – especially e.g. in the automotive and packaging industries or for white goods. But their production is very challenging for compounders. The transport of the material from the big bag stations/silos to the extruder plays a crucial role: This has to be done with minimum dust granting enough settling time and minimum air content in the material.

The art of producing highly filled compounds lies definitively in the optimum distribution of the material streams. Proven expertise in process technology is needed to incorporate a large quantity of filler homogenously into a polymer matrix. Particularly, one needs to control the air streams that are brought in with the material feeding. Another challenge is material moisture which can complicate the process. Therefore, processing unit, screw geometry as well as feeding have to be designed according to the job.

Special attention in this process was paid to the stable feeding of the dosing units which is done by means of pneumatic suction conveyors. The collective feeding system used here is designed in a very energy efficient way and thus works with optimum conveying speed. If necessary, the polymer can be combined from various feeding points. Each component is sucked in according to its formula, accurately weighed and homogenously mixed. Colours and additives are also fed into the process directly above the extruder.

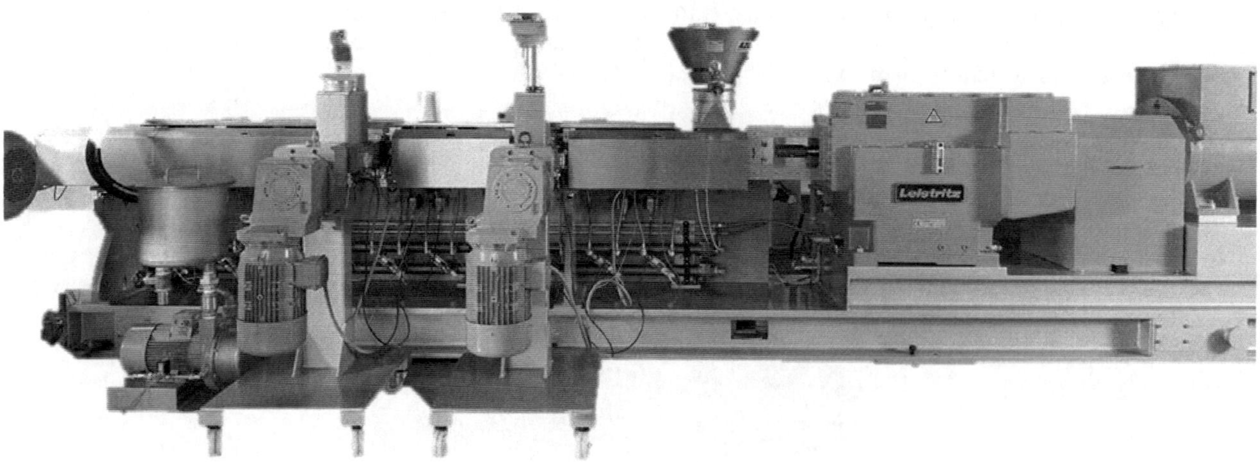

FIGURE 2. Example of an extruder layout with two side feeders

In summary several important parameters have to be taken into account to ensure long-term and constant production. With many years of experience in compounding highly-filled compounds, the paper offers a wide repository of good clues and ideas how to optimize extrusion lines and what aspects have to be taken into consideration to achieve an optimum layout of the extrusion line for such demanding compounding tasks.

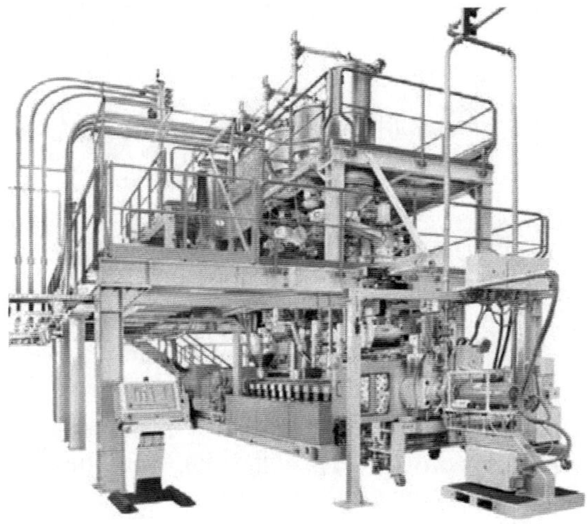

FIGURE 3. Example of a high-tech compounding line

REFERENCES

1. Sven Wolf, Leistritz Extrusionstechnik GmbH @ Technologietag Leistritz 2014

OPPORTUNITIES AND CHALLENGES OF PROFILE EXTRUSION DIES PRODUCED BY ADDITIVE MANUFACTURING PROCESSES

Christian Hopmann[a], Nafi Yesildag[a], Christian Windeck[a], Sebastian Bremen[b], Konrad Wissenbach[b], Simon Merkt[b]*

[a] *Institute of Plastics Processing (IKV) at RWTH Aachen University, Germany*
[b] *Fraunhofer Institute for Laser Technology (ILT) at RWTH Aachen University, Germany*

**Corresponding author: nafi.yesildag@ikv.rwth-aachen.de*

Abstract: The design and manufacture of profile extrusion dies is characterised by costly running-in trials. Significant cost and time savings can be achieved by replacing the experimental running-in trials by virtual ones. A simulative optimisation, however, often leads to complex, free-formed flow channels. A feasible manufacture of such dies is only possible with additive manufacturing processes such as the Selective Laser Melting (SLM). Against this background, the manufacture of profile extrusion dies by SLM is investigated.
A major challenge is to ensure a specific surface quality of the extruded plastics profiles. The roughness of SLM surfaces does not meet the high demands that are placed on the surface quality of extrusion dies. Therefore, in case of the SLM die a concept for the surface finishing of the flow channel is required, which can be applied to arbitrarily shaped geometries. For this purpose, plastics profiles are extruded both with a conventionally and an additively manufactured die. In case of the SLM die only the die land of the flow channel was reworked by polishing. The comparison of PP profile surfaces shows that the SLM die with polished die land leads to the same surface quality of the extruded profile as the conventional die ($R_a \approx 1$ µm).
Another important task in the design of profile dies by SLM is the optimisation of the die topology. The efficiency of the SLM process largely depends on the volume of the part being produced. To ensure the highest possible efficiency, it is necessary to adapt the die geometry to its mechanical loads and minimise its mass. For this purpose, the internal pressure in the die was numerically calculated and used for a first optimisation of the die topology. The optimisation, however, leads to a free-formed outer die wall so that the die cannot be tempered with heating tapes anymore. This problem is solved by using the high potential of SLM for functional integration and integrating contour adapted tempering channels into the extrusion die.

Introduction

The most common production process for plastics profiles is the profile extrusion as it combines high throughputs with a high level of automation. Here, the design and manufacture of the profile extrusion die is a key element for the product quality and process efficiency. To date, extensive running-in trials and reworking steps are required for the design and manufacture of the die and its flow channel [1]. Due to the high fixed costs resulting from these iterations, profile extrusion is only profitable for high lot sizes. A promising approach to minimise the necessary amount of design and manufacture iterations consists in the usage of numerical flow simulations in the design process before the die manufacture [2].

However, a numerical optimisation of the flow channel often leads to complex, free-formed channel geometries [3]. In this case, conventional processes for the manufacture of the die, such as milling or wire eroding [1], reach their limit and lead to high manufacture costs. Depending on the complexity of the free-formed geometries, it may even be that a conventional manufacture of the die is not possible at all. Remedy can be provided by additive manufacturing (AM) process, in which the product, in this case the profile die, is manufactured without a shape-giving die or mould by layer-wise application of material [4].
The main advantage of an additive manufacture of profile extrusion dies is the possibility to economically manufacture arbitrarily shaped flow channels, especially ones with free-formed shapes.

Furthermore, AM enables the integration of additional functional elements, such as geometrically adapted tempering channels, into the die without additional process steps. Another advantage of AM of the die is the possibility to implement a monolithic die design. Currently, profile extrusion dies usually possess a slice design, i.e. they consist of several, screwed together slices in which the respective flow channel is incorporated [1]. This design is necessary in order to produce and rework complex flow channel geometries with conventional manufacture processes. In contrast to this, with AM the die can be built as a single part even in case of a free-formed flow channel. In this way, the costs for the die manufacture can be decreased and the die quality can be increased as there are less parting lines in the die where melt leakage may occur.
A promising AM process for the profile extrusion die is Selective Laser Melting (SLM). This process enables the production of metal parts with series identical mechanical properties without the need for part specific tooling or downstream sintering processes [5]. Against this background, the use of profile extrusion dies manufactured by SLM is being investigated as part of the Cluster of Excellence "Integrated Production Technologies for High-Wage Countries" at RWTH Aachen University. In this article, investigations about two main challenges in the additive die manufacturing are presented: The surface quality of the plastics profiles extruded with additively manufactured dies and the topology optimisation of the die.

Principle of the Selective Laser Melting

The SLM is an AM technology developed by the Fraunhofer Institute for Laser Technology, which enables the layer by layer manufacture of both metal and ceramic components according to a 3D-CAD model. With SLM individual parts with nearly unlimited complexity can be produced without need of part specific tooling. Hereby, the process can be divided into the process steps illustrated in Figure 1: In the first step, the geometrical data is prepared. The 3D-CAD model is cut into layers (slicing) and the scanning vectors of the laser beam are defined. In the next step, a defined thin layer of the powder material is deposited on the substrate of the building platform. Subsequently, the geometric data of the CAD-layer-model is transmitted to the powder bed with the help of a high-power laser beam. The laser beam scans the layer regions in which solid material is to be contained, creating a solid layer of the piece to be produced. After lowering the building platform by the thickness of a single layer, the described process steps are repeated until the part is finished.
The main advantage of SLM in comparison to other AM processes for metal parts is the series identical mechanical properties of the parts produced. This is due to the fact, that the metal powder is completely melted by the laser beam, which enables a density of approximately 100 % in the whole part [5].

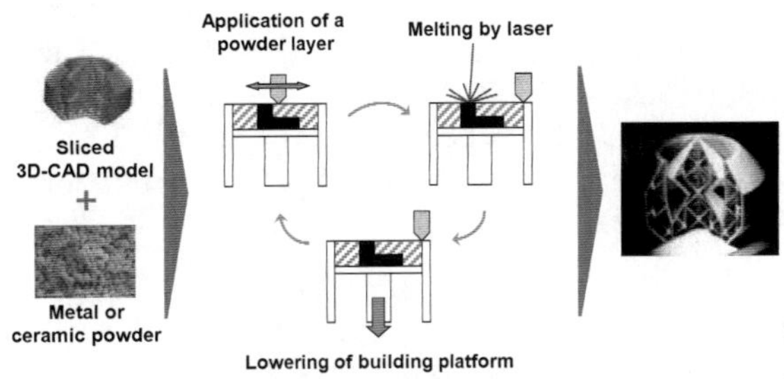

Figure 1: Principle of the SLM process

Investigation of the surface quality

A major challenge in the usage of profile extrusion dies manufactured by SLM is the surface quality of the die. Due to the high optical and haptic requirements placed on plastics profiles, usually the complete flow channel of the profile extrusion die is polished after die manufacture. A conventional requirement is hereby to achieve a roughness R_Z of maximum 1 μm in the flow channel [6]. By contrast, the roughness of parts manufactured by commercial SLM machines out of the steel 1.2709 have a value of at least 20 μm [7]. Accordingly, the flow channel surface requires reworking in case of SLM manufactured dies in order to produce plastics profiles with smooth surfaces. However, a polishing of the whole die is to be avoided, since the SLM process is especially suited to cases with free-formed flow channels.
That is why a concept for surface finishing is required, which both ensures a sufficient quality of the extruded plastics profiles and is applicable to arbitrarily shaped flow channels. A possible solution for this problem is to rework only the die land of the flow channel. The die land is the flow channel area with the biggest influence on the dimensional accuracy and the surface quality of the extruded profile. At the same time, the die land possesses a relatively simple geometry compared to

the remaining flow channel and is therefore the flow channel part easiest to rework. In the investigations presented in the next section, which are described in more detail in [8], the effectiveness of this surface finishing concept is analysed.

For this, a die for the manufacture of an L-shape profile was manufactured both conventionally by milling and additively by SLM (Figure 2). In case of the conventional die the whole flow channel is polished whereas in case of the SLM die only the die land is polished. Extrusion trials are conducted with both dies on a laboratory extrusion line consisting of an extruder, a vacuum-assisted calibrator in a water bath and a take-off unit. The extruder used is a laboratory scale extruder 19/10DW by Brabender GmbH & Co. KG, Duisburg, Germany, with a diameter of 19 mm and a length of 25D.

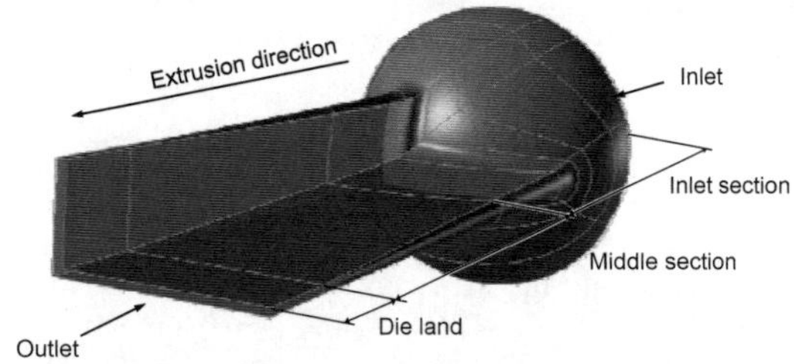

Figure 2: Flow channel for the production of an L-shape profile [8]

In the extrusion trials the process parameters mass throughput (1.4 kg/h, 1.7 kg/h, 2 kg/h) and die temperature (225 °C, 230 °C, 235 °C) are varied. The material used in the trials is Polypropylene (PP) Moplen HP400H of the company Basell Polyolefine GmbH, Frankfurt am Main, Germany. After the extrusion trials, the surface roughness R_a of the extruded profiles are measured with a laser confocal microscope type VK-X200 of the company Keyence Deutschland GmbH, Neu-Isenburg, Germany. For each test point, the roughness R_a of several samples is measured and subsequently averaged in order to determine the mean roughness.

Figure 3 shows the measured roughness of the profiles obtained with the two different dies. First of all, it can be seen that the varied process parameters have a slight influence on the roughness in both cases. Hereby, an increase of the throughput leads to a rougher profile surface since the higher internal die pressure leads to a stronger replication of the rough flow channel surface. An increase of the melt temperature on the other hand improves the surface quality. The reason for this is the decreasing melt velocity. This leads on the one hand to a stronger replication of the protrusions in the flow channel by the melt. On the other hand, a higher mass temperature also means a lower viscosity at the inlet of the calibrator so that the smoothing effect of the calibration is also improved. In this case, the latter effect dominates so that the roughness decreases with increasing temperature.

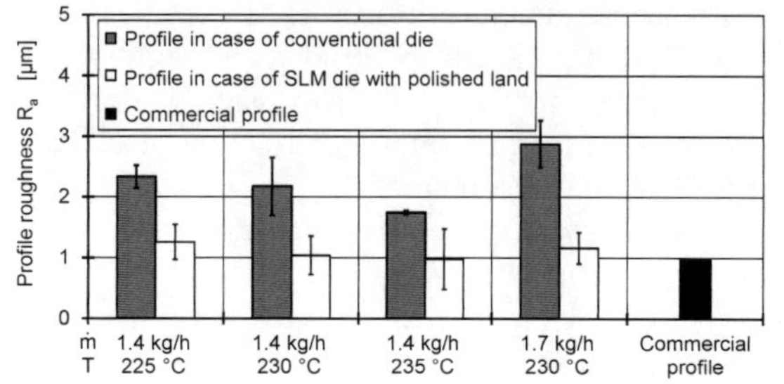

Figure 3: Average roughness of the profiles manufactured with conventionally and additively manufactured dies [8]

When comparing the conventional die and the SLM die, it can be seen that the SLM die with polished die land leads to profiles with a surface quality comparable to that of the ones from the conventional die. In this case, the profiles extruded with the SLM die even have a smoother surface than the ones extruded with the conventional die. This observation, however, cannot be generalised: Roughness measurements on the investigated dies show, that the die land of the SLM die was polished better (R_Z = 7 µm) than the flow channel of the conventional die (R_Z = 11 µm). Hereby, the average roughness in the flow channel of the conventional die is approx. 48 % higher than in the die land of the SLM die. The smoother profile surfaces in case of the SLM die are entirely due to this difference in the surface finishing of the dies.

In order to evaluate the suitability of the profiles extruded with the SLM die for industrial applications, the roughness of a commercial profile was determined as a reference (Figure 3). The comparison of the commercial profile and the profiles from the SLM die shows that the roughness values of the profiles are nearly identical. This confirms the assumption that the surface quality of extruded plastics profiles is mainly influenced by the surface finish of the die land and less by the rest of the flow channel. In this case, profiles with sufficient surface quality for industrial applications can be produced with a die manufactured by SLM by simply polishing the die land of the extrusion die. However, in case of plastics other than PP the rough flow channel of the SLM die could lead to depositions in the flow channel. This aspect will be investigated in future works.

Topology optimisation of the die

Another important topic which has to be considered in the manufacture of profile extrusion dies by SLM is the optimisation of the die topology. The costs of the SLM process are mainly dictated by the powder costs and the build-up time. Both factors increase with increasing volume of the die to be produced. Conversely, the highest possible efficiency of the SLM process is given when the die volume is minimised by adapting the die topology to the mechanical loads on the die.

The die for the production of an L-shape profile, which was used in the previous chapter, was not topology optimised. Instead, it possesses a massive design which is usual for profile extrusion dies (Figure 4). In order to improve the efficiency of the die manufacture by SLM the topology of the die in Figure 4 is to be optimised. The requirement for this is that the mechanical loads on the die, i.e. the internal pressure in the flow channel and the bolt forces, are determined in a first step.

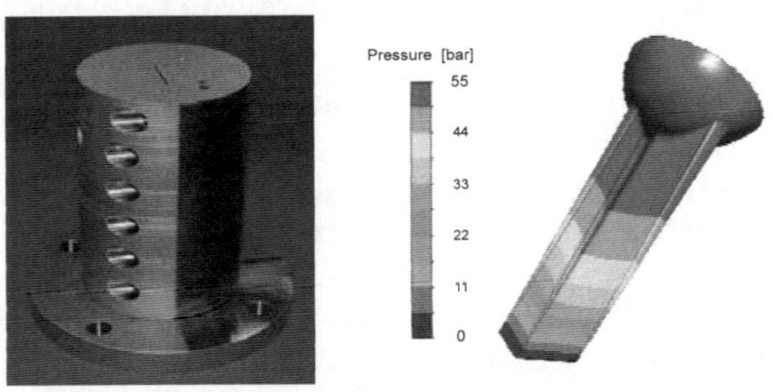

Figure 4: SLM-die without topology optimisation and simulated pressure drop

The bolt forces equal the pretension forces of the used ISO metric bolts taken from [9]. In order to determine the internal pressure in the flow channel, a flow simulation with the commercial Finite-Element software Polyflow of the Ansys Germany GmbH, Darmstadt, Germany, is used. The simulation is conducted for a throughput of 10 kg/h of the material Moplen HP400H at a temperature of 210 °C. The shear thinning behaviour of the plastics is modelled with the Carreau model, whereby the corresponding Carreau parameters were determined by high-pressure capillary rheometery (A = 10'934 Pas, B = 0.43, C = 0.69 at 200 °C).

The simulation results show that the maximum pressure in the flow channel is 55 bar (Figure 4). For a safe design, this pressure is increased in the topology optimisation by a safety factor of 1.5 for the running-in process. Furthermore, it is assumed that this pressure is present along the entire flow channel, as the largest pressure drop is at the die outlet.

Now that the mechanical loads on the die are determined, the next step is the actual topology optimisation. The optimisation is conducted with the commercial Finite-Element software Abaqus of the Dassault Systémes Deutschland GmbH, Düsseldorf, Germany. In the first attempt for the topology optimisation no manufacturing and handling restrictions are considered in order to allow the maximum possible die volume reduction. This leads to the new die topology in Figure 5.

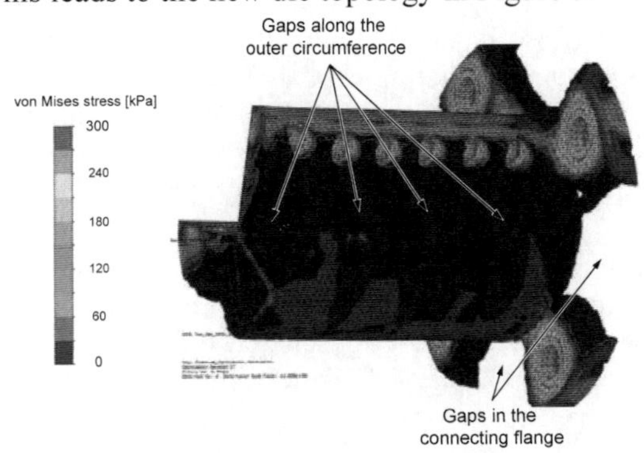

Figure 5: Optimised die topology without consideration of manufacturing restrictions

The topology optimisation has reduced the volume of the die by approximately 50 %, thereby significantly improving the efficiency of the SLM manufacture of the die. What is noticeable, however, is that the die now possesses several free-formed gaps in the area of the connecting flange and along the outer circumference. Because of this an adaptation of the die tempering is necessary. Profile extrusion dies are usually tempered with electrical heating tapes which are spanned around the cylindrical or rectangular outer surface of the die. This method of tempering is not applicable in case of the optimised die due to its irregular surface. A possible solution for this problem is to integrate tempering channels into the die in order to realise a fluid tempering as is used in rubber extrusion dies [1].

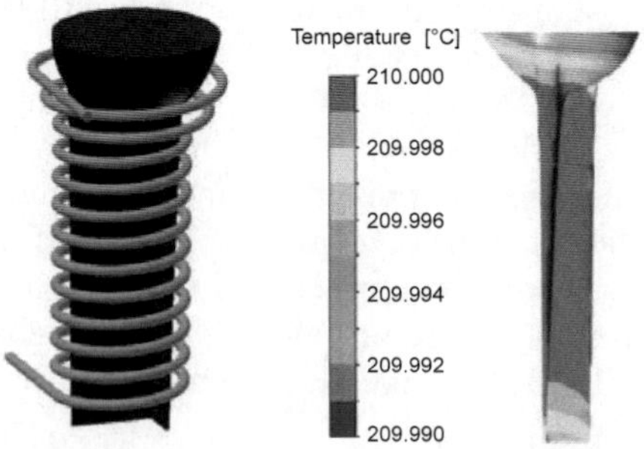

Figure 6: Spiral shaped tempering channel and simulated temperature distribution at the flow channel wall

Here, the high potential of SLM for functional integration can be used in order to integrate contour adapted tempering channels into the die during its manufacture without further processing steps. In case of the die for the L-shape profile a single spiral shaped tempering channel could be placed around the melt flow channel (Figure 6). In this way a homogeneous tempering of the flow channel wall can be achieved as shown by a thermal simulation with Polyflow on the right side of Figure 6. In an upcoming next step, the topology optimisation has to be repeated under consideration of both the designed tempering channel and geometrical restrictions with regards to an ergonomic handling of the die during its assembly, operation and disassembly.

Conclusion

A major challenge in the SLM manufacturing of profile extrusion dies is to ensure a sufficient surface quality of the extruded plastics profiles. Here, a concept for the surface finishing of the flow channel is required, which can be applied to arbitrarily shaped geometries. The presented investigations show that it is possible to achieve a sufficient profile surface quality ($R_a \approx 1$ µm) with a SLM die by only polishing its die land.

Another important aspect in the AM of the die is the optimisation of its topology. The topology optimisation, however, makes it necessary to also adapt the die tempering. A possible solution is the integration of a fluid tempering into the die. The corresponding tempering channels can be incorporated into the die during its SLM manufacture without further processing steps.

Acknowledgments

The depicted research has been funded by the Deutsche Forschungsgemeinschaft (DFG) as part of the program Cluster of Excellence “Integrative Production Technology for High-Wage Countries”. We would like to extend our thanks to the DFG.

Furthermore, we would like to thank the Basell Polyolefine GmbH, Frankfurt am Main, Germany, for the provided test materials and the Döllken Kunststoffverarbeitung GmbH, Gladbeck, Germany, for the reworking of the SLM die.

References

1. Michaeli W (2003) Extrusion Dies for Plastics and Rubber. Carl-Hanser Verlag, Munich, Vienna
2. Brecher C (2012) Integrative Production Technology for High-Wage Countries. Springer, Berlin
3. Szarvasy I, Sienz J, Pitman J, Hinton E (2000) Computer Aided Optimisation of Profile Extrusion Dies. *Int Polym Process* 1: 28-39, doi: 10.3139/217.1577
4. Gibson I, Rosen D, Stucker B (2015) Additive Manufacturing Technologies. Springer, Berlin
5. Schleifenbaum JH (2012) Verfahren und Maschine zur individualisierten Produktion mit High Power Selective Laser Melting. Ph.D. Thesis, RWTH Aachen University
6. Herne KV (2002) Instandhaltung von Extrusionswerkzeugen. *Plastverarbeiter* 53 (2): 78-79
7. Meiners W (2011) Selective Laser Melting – Additive Manufacturing for series production of the future? Intermat 2011, Luxembourg, 01.02.2011.
8. Hopmann C, Yesildag N, Bremen S, Wissenbach K, Merkt S (2015) Surface quality of profile extrusion dies manufactured by Selective Laser Melting. *RTeJournal - Forum für Rapid Technologie*, Vol. 2015
9. Gomeringer UFR, Kilgus MHR, Oesterle FNS, Stephan HPA (2011) Tabellenbuch Metall, Verlag Europa Lehrmittel, Haan-Gruiten

Groove Geometry and Bulk Material Properties Impacting on Pressure Build-up in Single Screw Extruders

Matthias Düngen[a*], Michael Müller[a] and Michael Koch[a]

[a]*Department of Plastics Technologies, Technische Universität Ilmenau, Germany*

matthias.duengen@tu-ilmenau.de

Abstract. Single screw extruders with grooved feed sections effect a forced conveying of solid particles leading to high specific throughputs. Requirements for such a forced conveying were discussed in literature, and models to describe them were presented. Actual visual observations of the processes inside the feed zone are difficult due to high pressures and thus are scarce to find. Therefore this study deals with a simplified experimental setup based on an unwound flat screw channel implementing a kinematic reversal. It allows the visual observation of solid movements and measurement of the force to move the movable wall of the setup. It features variable screw channel heights, groove geometries and different bulk materials (LLDPE, HDPE, PS). This setup allows insight into the actual formation of the particles in the groove and the interaction of the particles in the channel with those in the groove. The experiments show particles agglomerating into wedges above the groove for certain configurations. These wedges support the movement of the particles towards the end of the channel. The experimental setup allows the material to be compressed towards the end of the channel which leads to a pressure build up in the channel. For each combination of channel height, groove geometry and material type a maximum pressure level exists which can be achieved until the wedge collapses. Relative to material kind, dimension and shape, varying conditions are observed. Accordingly, particular material characteristics are identified to reflect the difference in solid conveying. In this study a modeling approach based on the Hertzian theory for contact stresses is presented to calculate this maximum pressure level for different configurations of bulk materials, channel height and groove geometry.

Keywords: Extrusion, Single screw
PACS: 81.20.Hy

1 INTRODUCTION

Single screw extruders (SSE) with a grooved feed section effect in forced conveying of solid particles leading to high specific throughputs. MOL and DARNELL [1] presented a general model for the throughput of bulk material in single screw extruders considering forces acting on the solids plug in the screw channel. MENGES et. al. [2] adapted this model for SSE with an axially grooved feeding section by adjusting the coefficient of friction between bulk material and barrel wall. This model also considers the groove geometry and the coefficient of friction between the bulk material itself. This approach was further refined by PEIFFER [3], POTENTE [4] and KOCH [5]. SCHÖPPNER [6] was able to model the throughput in the grooves. He also distinguished between different conveying mechanisms based on the relation of particle diameter, groove geometry and screw channel height. PAN, JIA ET. AL [7, 8] concluded that the shear strength of the bulk material has a strong influence on the conveying mechanisms. Experiments by SIKORA [9] with a variable taper angle of the groove show nonlinear dependency of the groove taper angle on the throughput. Actual visual observations of the processes inside the feeding zone are difficult due to high pressures and thus are scarce to find.

Further insight in solids conveying in SSE is offered by discrete particle simulation [10-14]. However, the pressure build-up in a grooved feed zone could not be simulated correctly [15]. This is much likely due to the challenging simulation of plastic deformations and shearing of particles. The characteristics of thermoplastic bulk material were investigated by HWANG [16] by measuring the changing compressibility at different temperatures. LÄNGAUER et al. [17] showed the influences of the pellet shape and the viscoelasticity on the compressibility. These complex material properties are currently not fully included in numerical or analytical models.

Therefore further experimental methods are of interest.

2 EXPERIMENTAL

This study deals with a simplified experimental setup based on an unwound flat screw channel implementing a kinematic reversal. It allows the visual observation of the moving solid particles and the measurement of the force to move the movable wall of the setup in z-direction (figure 1). It features variable screw channel heights, groove geometries and different bulk materials (LLDPE, HDPE, PS). This setup allows insight into the actual formation of the particles in the groove and the interaction of the particles in the channel with those in the groove.

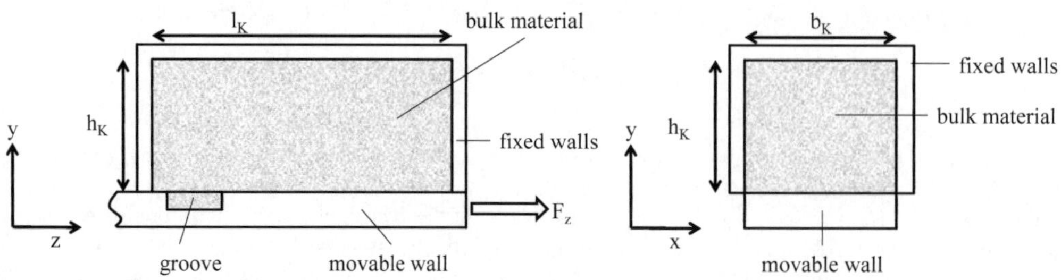

FIGURE 1. Experimental setup comprising an unwound screw channel with an exchangeable groove

The length of the channel in z-direction is 300 mm, which allows a maximum movement of 250 mm of the movable wall in z-direction. The channel width b_K (x-direction) is 165 mm and the possible channel height h_K ranges from 0 to 35 mm. The groove extends only in the x-direction. The grooves are characterized by their height h_G and width b_G. The groove variations used in this study are (h_G x b_G in mm): 2x10, 2x20, 4x10, 4x20.

During the experiment the channel is filled with pellets. The pellets start to follow the motion of the movable wall. Since they cannot leave the channel they get compressed or slide over the movable wall, however, not as much as it would occur if the movable platen were moved off the channel direction under the angle comparable to a screw flight.

The movable wall was driven by a tensile testing machine. The channel walls are made of Aluminum for easy manufacturing and assembly. High pressures inside the channel wall occur, when the bulk material is compressed. This can lead to elastic deformations of the channel walls. Hence a maximum force of 800 N was chosen for F_Z to ensure reproducibility.

Three bulk materials are used in this study: PS, LLDPE and HDPE (table 1). They show different combinations of mechanical and tribological properties.

TABLE (1). bulk material material properties

	Young's modulus *E* [MPa]	**internal coefficient of friction bulk material/bulk material μ_i [-]**	**particle diameter *d* [mm]**	**bulk density ρ_b [g/cm³]**	**density ρ [g/cm³]**
Polystyrol 168 N (PS)	3300	0,43	3,26	0,54	1,04
Dowlex 2045 G (LLDPE)	235	0,45	3,96	0,58	0,92
Dowlex 2377 (HDPE)	645	0,35	4,02	0,62	0,94

3 RESULTS & DISCUSSION

3.1 Results

One side of the channel has a window of PMMA allowing the observation of the pellet movement and its formation. The different formations distinguished by SCHÖPPNER [6] and their experimental counterparts are given in figure 2. SCHÖPPNER also considers powders (case 2b) which are not part of this study.

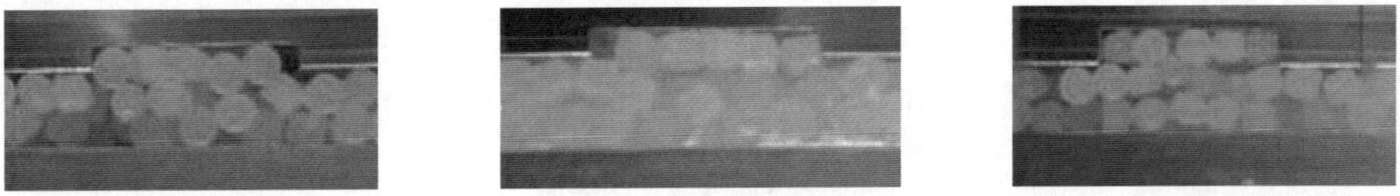

Case 1a: $h_K<2d$, $h_G<d$
Interlocking

Case 1b: $h_K>2d$, $h_G<d$
Interlocking & Friction Locking

Case 2a: h_K, $h_G>d$
Friction Locking

FIGURE 2. Different pellet formation cases according to [6], with channel height h_K, groove height h_G and pellet diameter d

The directly measured parameter in the experiments is the force F_Z to pull the movable channel wall in z-direction (figure 1). This value is used to calculate the channel pressure p_K

$$p_K = \frac{F_Z}{A_W} = \frac{F_Z}{h_K * b_K * R} \tag{1}$$

with the dimensionless density of the bulk material $R = \rho_b/\rho$. For some combinations of material and channel parameters it took considerably more time (i.e. way) to reach a similar channel pressure level. For a SSE this means that the wedges might not be stable fast enough to support the solid conveying process effectively. In order to take this into account the channel pressure gradient $\Delta p_K/\Delta s$ is used further.

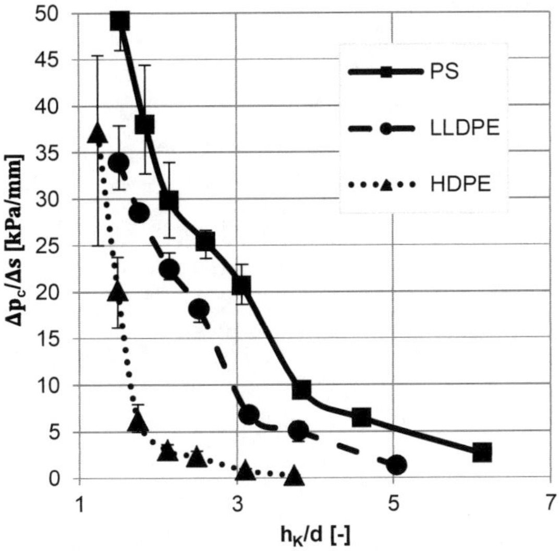

FIGURE 3. Channel pressure gradients for different channel heights and $h_G = 4$ mm $b_G = 20$ mm

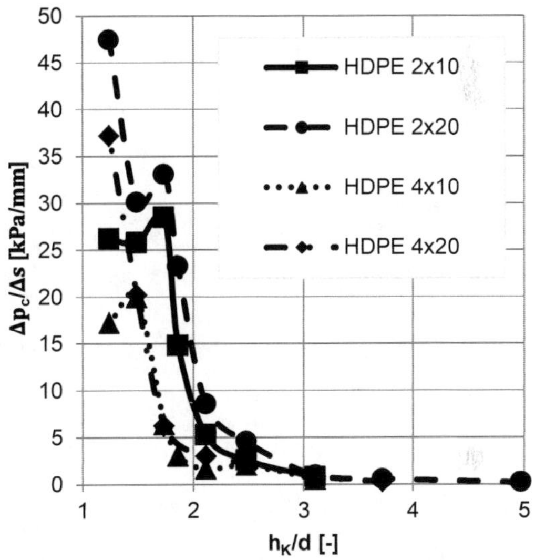

FIGURE 4. Channel pressure gradients for HDPE with different channel groove geometries h_G [mm] x b_G [mm]

Figure 3 shows the results for different channel heights. The curves have a typical L-shape which corresponds to the change from friction locking with a shear interface (low channel pressure gradient) to interlocking. This can be clearly observed through the window in the channel wall. In the case of interlocking the pellets agglomerate to wedges which support the movement of the pellets in front of them.

The highest channel pressure gradients are achieved by PS which has the highest stiffness (the least thermoelastic plasticity) and coefficient of friction. The lowest values account for HDPE which has the lowest coefficient of friction but a much higher stiffness than LLDPE.

The influence of the groove geometry can be seen in figure 4. The highest channel pressure for HDPE is achieved for the 2x20 groove which provides the pellet formation of case 1a & b in figure 2. The same formations with a small groove result in lower values for the 2x10 groove. The 4x10 and 4x20 groove show the pellet formation of case 2a in figure 2. In this case interlocking between pellets is less likely. The changeover from friction locking to interlocking occurs for signifcant lower h_K/d-ratios and leads to lower channel pressure gradients compared to deeper grooves.

3.2 Analytical Model

The analytical model for the possible pressure build-up considers the limiting factors for interlocking:

I. There is enough free volume for pellets to slide off of each other
   a. when the deformation of two adjacent pellets is sufficient
   b. when the deformation of the whole agglomerated pellet wedge is sufficient

II. When the forces are sufficient to shear off a pellet

The relevant symbols are given in figure 5. For this analytical model the pellets are treated as spheres.

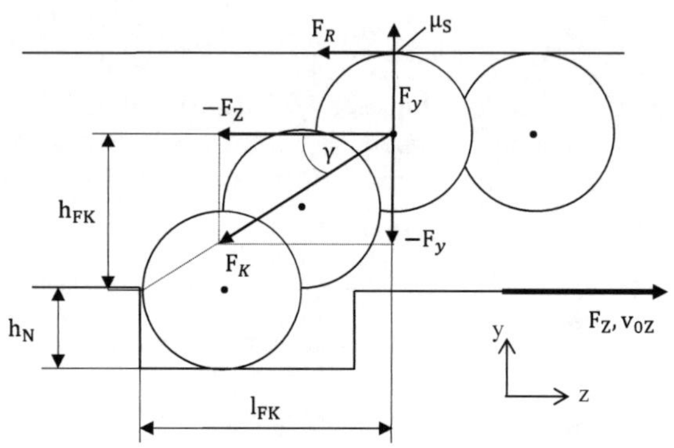

FIGURE 5. Relevant dimensions and forces of a cross-section of the channel

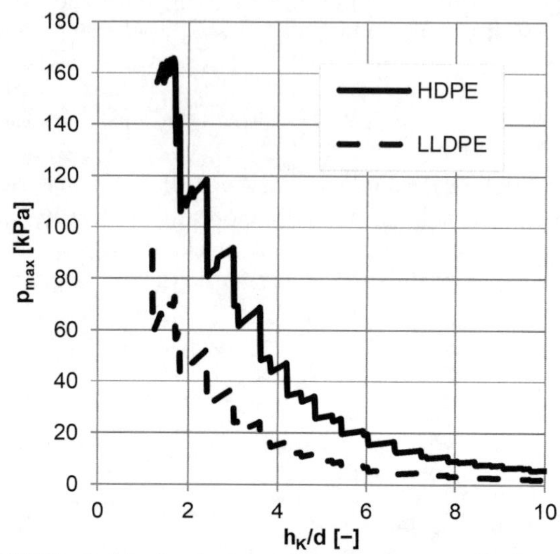

FIGURE 6. Maximal channel pressures according to the mathematical model

The maximum pressure in the interlocking case is supported by the pellets in the groove and given as

$$p_{vmax} = \frac{F_K^*(\cos(\gamma) - \mu\sin(\gamma))}{h_K b_K {}^{\rho}/_{\rho_S}} \tag{2}$$

with the force F_K^* as the minimal value of the cases Ia, Ib and II:

$$F_K^* = \min(F_{KHa}, F_{KHb}, F_{KS}) \tag{3}$$

The angle of the agglomerated pellet wedge is

$$\gamma = \arctan\left(\frac{l_{FK}}{h_{FK} - (n_K - 1) * \frac{a_K}{4}}\right) \tag{4}$$

with $n_K = \left\lfloor\frac{h_K - d}{a_K * 2}\right\rfloor + 1$ and $a_K = \sqrt{2} * \mathrm{d}$. Case II can be calculated as

$$F_{KS} = \frac{\tau_{smax}}{A_s} = \frac{4\tau_{smax}}{\pi(dh_S - h_S{}^2)} \tag{5}$$

with $h_S = \frac{d}{2} - \frac{a}{4}$; in case I: $a = \sqrt{2}d$; in case II: 2: $a = d$. τ_{smax} is the shear strength.
The forces resulting for the required deformations ε in case I can be calculated using the Hertzian contact model:

$$F_{KH} = \sqrt[3]{\frac{\varepsilon}{n_d} \cdot \frac{E\sqrt{\frac{4}{d}}}{1{,}5(1-\nu^2)}} \quad (6)$$

ν is the Poisson ratio and n_d is given as $n_d = \left[\frac{h_G - d/2}{h_{FK}}\right] + 1$.

The results of this model in figure 6 show a non-continuous function which results from eq. (3). The different failure cases dominate the function for short intervals. Especially when the channel height is changed to a level where another pellet fits into the wedge in the y-direction another step occurs in the graph. The calculated values for PS are much higher and would have exceeded the chart. This is due to the influence of the stiffness in eq. (6). Qualitatively the results agree with the curves in figure 3 and 4. Especially the non-continuous nature of the experimental results can be confirmed. Quantitatively the results for HDPE and LLDPE fail to match the experimental values by up to 41 %. This might be due to the assumption of a perfect spherical shape in the mathematical model.

4 CONCLUSION

A novel experimental setup is presented which allows the direct observation of pellet flow and formations in channels and measurement of the resulting pressure build-up. The new key figure *channel pressure gradient* gives information about how fast and up to which pressure level a stable locking mechanism is created. The experiments confirm pellet formation cases known from literature. Due to the choice of pellet materials the influence of the stiffness and the coefficient of friction could be shown. While it is possible to model the influence of the stiffness in a rather simple mathematical model with good qualitative agreement, the influence of the pellet shape has also to be taken into consideration.

5 ACKNOWLEDGEMENTS

The results presented were obtained in the context of the research project 03WKBY03C sponsored by the Bundesministerium für Forschung und Bildung (BMBF). We thank the BMBF for their support. Materials were provided by Basell Polyolefine GmbH and Dow Europe GmbH.

REFERENCES

1. W. H. Darnell, E. A. J. Mol, *SPE journal*, 1956, April; S. 20–29.
2. G. Menges, R. Hegele, G. R. Langecker, *Plastverarbeiter*, **23**, 455-460 (1972).
3. H. Peiffer, "Zum Förderproblem in der genuteten Einzugszone von Einscheckenextrudern", PhD-Thesis, RWTH Aachen, 1981
4. H. Potente, *Kunststoffe*, **75**, 439-441 (1985).
5. M. Koch, "Berechnung und Auslegung von Nutbuchsenextrudern", PhD-Thesis, Universität Paderborn, 1987.
6. V. Schöppner, "Simulation der Plastifiziereinheit von Einschneckenextrudern", PhD-Thesis, Universität Paderborn, 1994
7. M.-Y. Jia et al., *Intern. Polym. Proc.*, **28**, 267-273 (2013).
8. L. Pan et al., *expr. Polym. Letters*, **6**, 543-560 (2012).
9. J. W. Sikora, *Polym. Eng. Sci.*, **41**, 1636-1643 (2001).
10. J.-S. Leßmann, V. Schöppner, "Validation of Discrete Element Simulations in the Field of Solids Conveying in Single-Screw Extruders", PPS-30 Proceedings (2014).
11. J.-S. Leßmann, R. Weddige, V. Schöppner, A. Porsch, *Intern. Poly. Proc.*, **27**, 469-477 (2012).
12. P. A. Moysey, K. L. Cloet, M. R. Thompson, *Intern. Polym. Proc.*, **23**, 301-311 (2008).
13. O. P. Michelangelli, A. Gaspar-Cunha, J. Covas, "Effect of the Coefficient of Friction on the Granular Transport in a Single Screw", PPS-29 Proceedings (2013).
14. V. Schöppner, R. Weddige, "Improving the Feeding Zone of Single Screw Extruders at High Rotation Speed by Using the Discrete Element Method" ANTEC 2010 Conference Proceedings (2010).
15. O. Celik, C. Bonten, "Dreidimensionale Simulation der genuteten Einzugszone eines Einschneckenextruder" 24. Stuttgarter Kunststoffkolloquium, Stuttgart (2015).
16. C.-G. Hwang, J. M. McKelvey, *Adv. Polym. Tech.*, **9**, 227-251 (1989).
17. M. Längauer et al., *J. Appl. Poly. Sci.*, **132**, 42197 (2015).

First insight into the High Speed Extrusion of Polycarbonate

V. Schöppner [a], M. Pohl

[a] *Kunststofftechnik Paderborn, Universität Paderborn – volker.schoeppner@ktp.upb.de; max.pohl@ktp.upb.de*

Abstract. Increasing the economical efficiency of single screw extrusion is an important factor in the area of research and development in extrusion technology. Enlarging processing machines is an approach which raises the mass flow rate but at the same time increases costs for maintenance or batch changes. This is why the focus of this essay lays on an increasing speed at a constant diameter, in order to improve the output. Investigations with polyolefins show that in terms of high-speed extrusion (from 2m/s on) particular attention has to be paid on the partially poor feed behavior and high processing temperatures. Based on the results of previous studies, the investigation of amorphous thermoplastics (polycarbonate) in combination with high-speed extruders is promoted at the KTP institute. Due to the different molecular structure from amorphous to partially crystalline (polyolefin) it can be assumed that the state of knowledge in areas of feed behavior, temperature development or mixing homogeneity can be transferred only partially. The reason for the lack of melting is the enthalpy of amorphous polymers. This essay introduces the first observations of temperature homogeneity and pressure development partially at high-speed extrusion with polycarbonate. These observations are based on the evaluation of various speed ranges on an extruder (D = 30 mm) with vacuum degassing and torque direct drive. The measurements occur with a temperature sensor in the area of the screw tip and with several pressure sensors along the axial screw length. This arrangement allows making a statement about temperature homogeneity, pressure development and the melting process. Based on these results, approaches to process optimization are introduced.

Keywords: Single Screw Extrusion, High Speed Extrusion, Polycarbonate
PACS: 89.20.Bb, 89.20.Kk, 89.90.+n, 81.20.Hy

INTRODUCTION

The throughput of a single screw extruder can be increased by raising the screw speed. It is interesting to reach circumferential speeds over 2 m/s, because this method generates the same throughput as on bigger machines [1]. This proceedings lead to use a direct power engine to turn the screw, due to the fact that the energy input is transferred from torque to screw speed. Also, the specific energy turnover can be reduce by using smaller screw sizes [2].In the past grooved feed zones and powerful drives were often used to enlarge the output, however, the boundaries for maximizing throughput are hereby largely maxed out [3]. Besides the classic extrusion concepts, there are special machine types like S-truder (separation of melt and solid material) or Helibar (barrier screw with grooved barrel), which also use higher screw speeds [3,4]. But in this thesis a smooth barrel extruder (ø 30 mm) is investigated. Different scientific studies have been made with semi-crystalline materials like PP or PE [1-8]. It was found that factors like melt temperature are important and a special process control is needed. Many of this materials had a strong structural viscosity which leads to easier handling at higher screw speeds because of low shear stress. For this reason, this paper discusses amorphous thermoplastics with a pronounced Newtonian flow behavior like Polycarbonate. It might be possible to transfer some of the results to this deviating behavior or for example totally new screw geometries. Moreover a different process understanding is needed. To receive a basic knowledge, various screws speeds from 250 rpm till 1000 rpm will be analyzed. For all these operation points the axial temperature and pressure will be recorded, as well as the mold flow rate.

HIGH SPEED EXTRUSION OF POLYCARBONATE

High speed extrusion offers benefits of low maintenance costs due to the small machines as well as the possibility of using an energy-efficient direct drive. Investigations have been basically carried out focused in the field of semi-crystalline materials (especially polyolefins) by [1-8]. The results verify that high speed extrusion processes are often limited by the mechanical and the thermal stress influencing the melt quality [2,4]. In addition to that, dissipative shear heating has to be considered especially for using polycarbonate at high screw speeds [8]. For the consideration of polycarbonate in high speed extrusion, these effects provide the limiting factor. Therefore, the screw geometry must be designed specifically for polycarbonates. Thereby in previous studies a high speed polycarbonate degassing screw (ø 30 mm) was designed. Here, no mixing elements have been used due to their high shear rates which can lead to a massive increase in temperature [7]. The aim is to increase the throughput with a consistent melt quality. The investigations have been carry out using a polycarbonate with an MVR of 6 cm3/10 min (300 °C; 1.2 kg). This type is characterized by its high strength, impact strength and stiffness. Due to its transparency caused by the amorphous structure, polycarbonate is often used as a glass alternative. The viscosity behavior can approximately be described by the Newtonian equation, which processing requires a special process control at high speeds. Figure 1 shows the shear rate against the viscosity of a common polycarbonate type.

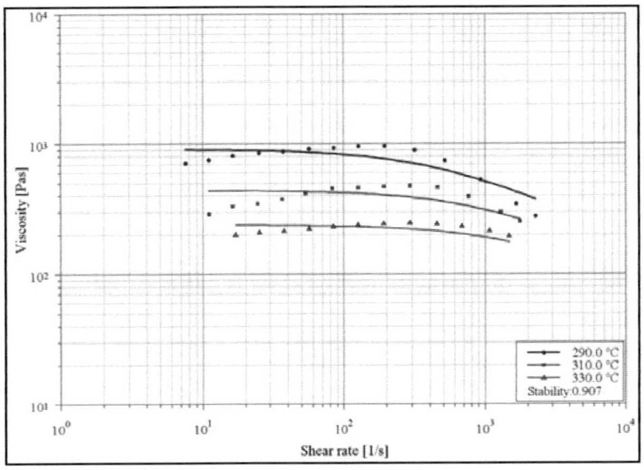

FIGURE 1. Rheology – Viscosity and Shear rate of Polycarbonate with an MVR of 6 cm^3/10min (300 °C; 1.2kg)

As shown in figure 1 a, linear range approaches which is on the one hand an issue for the aspired high speed investigations because the shear rates caused by high circumferential speeds lead specifically to an increase in heating caused by dissipation. On the other hand, further investigations [9] show that high shear rates lead to a devastation of the polymer chains and to a decrease of the viscosity. As consequence of the presented investigations, it is absolutely necessary to analyze the damage that takes place in the material structure at screw speeds of more than 1 m/s. Firstly, MVR measurements are carried out in which a relation can be found to viscosity and material degradation. Additionally, further studies are planned in which parameters such as the Yellowness index (ASTM E313-05) and the measurement of the intrinsic viscosity (Staudinger index) will give a deeper understanding. Direct thermal damage to the material occurs in the absence of oxygen at 340 °C [10]. However, high speed extrusion is determined by a very low residence time [2] in which the thermal influence impacts for a very short period and in this manner the thermal energy does not damage the extruded polycarbonate.

EXPERIMENTAL INVESTIGATIONS

Experimental investigations are carried out with a high speed extruder from ESDE GmbH. The machine has a diameter of 30 mm and a drive 500 Nm torque. In figure 2 the basic structure of the barrel is shown including the positions of the measuring technique for pressure and temperature.

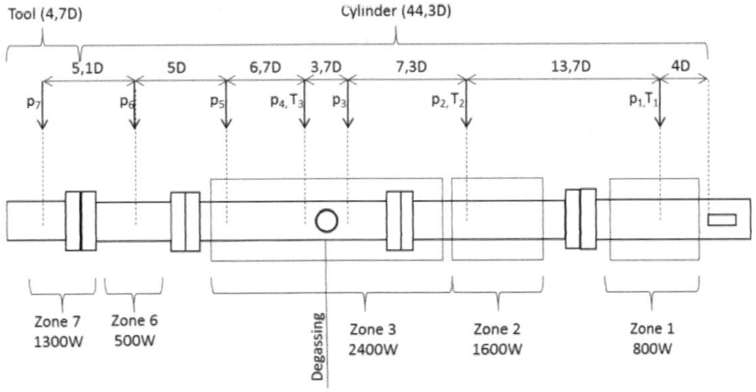

FIGURE 2. Measurement setup – Pressure and temperature profile

Beginning with the hopper outlet, the barrel has a total length of 44.3 L/D. Because undried polycarbonate is used in the investigations, a degassing dome in combination with a vacuum pump is installed at a length of 27.7 L/D. Temperature sensors (Type J) are used to measure the temperature of the melt at the barrel. Furthermore, temperatures are measured at the screw tip manually. The axial pressure profile is recorded in the feeding zone with an optical pressure sensor followed by usual membranes transducers in zones in which melt dominates. In order to keep the investigations close to industry standards, a counter pressure has been set up via an adjustable nozzle tool in typical ranges for polycarbonate (film extrusion of PC: 35- 80 bar).

The screw geometry (40 L/D) is based on a standard degassing screw that is made up of a two 3-zone-screw concept (two times of feed, compression and metering section). The decompression zone is hedged with cylindrical shearing section to prevent the partial filled zone from solid material and a pressure higher than zero. Without this action, the degassing system could be damaged.

In figure 3, an overview of the used screw geometry is shown. Done to a confidentiality agreement, more detailed geometric data cannot been shown.

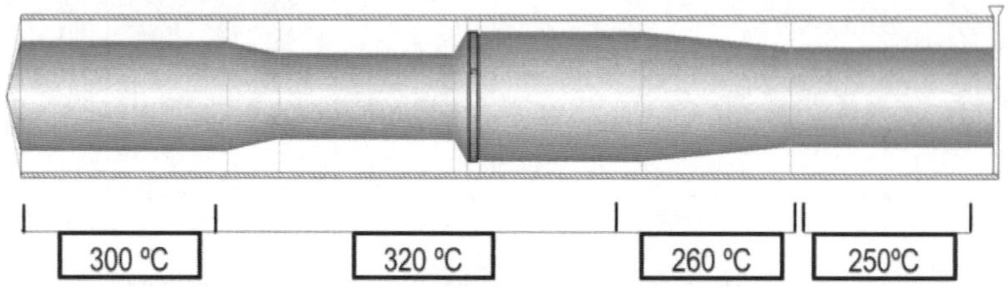

FIGURE 3. Sketch of the screw design and barrel temperature profile

In order to prevent flooding, a special "hump" temperature profile from [11-12] was used for the barrel heating. This originally used setting to optimize throughputs of barrier screws was applied to create a high temperature in the locking ring area to ensure that the material can pass the ring through the degassing zone without plugging. The aim of this investigations is to achieve throughput rates above 180 kg/h. Calculations of the simulation software REX recommend a screw speed of 1000 rpm to reach the throughput rate on a ø30mm machine. In addition to that, the maximum speed is approached in the range of 250 rpm and 750 rpm. Table a) provides an overview of the measured parameters. It can be said that the results are in good agreement with the simulations and the desired throughput rates were almost reached.

The objective of this examination series was to achieve the throughput rate of about 180 kg/h. With the simulation software REX a screw speed of 1000 rpm has been calculated to reach this goal. Therefore, the speeds of 250, 750, 1000 rpm have been analyzed. The following table indicates the speed, torque, throughput and the temperature at the screw tip.

TABLE a). Summary measurement results – Screw speed, torque, throughput and melt temperature

Operation point	Screw speed [Rpm]	Torque [Nm]	Throughput [kg/h]	Melt temperature [°C]
#1	250	370	48	330
#2	750	355	126	343
#3	1000	360	170	357

Table a) shows that with an increasing speed, the torque slightly decreases. This is due to the increasing melt temperatures and the shear stress which effects the viscosity. With the highest screw speed the melt temperature rises above 340 °C which is a known limit from the literature [9]. Figure 4 (b) shows that the limit is reached at screw speeds of 750 rpm. At this speed, all temperatures have been measured above the set barrel temperature values for the whole screw length (fig. 4 (b)). Responsible for this is the dissipative heat input due to the Newtonian behavior of the polymer that increases with the screw speeds.

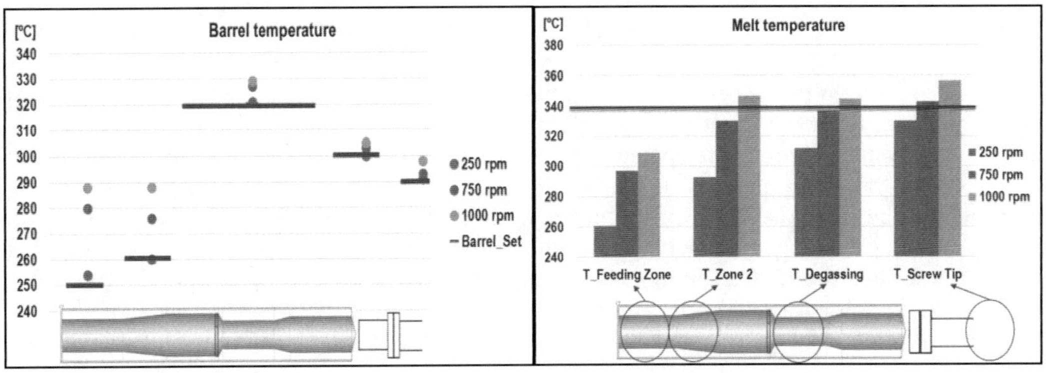

FIGURE 4. (a)Barrel temperature – Is and set values, (b) Measured melt temperature

Because of this process reaction, for each operating point, samples have been gathered, formed and cooled down to room temperature with a special designed removal unit. This unit allows a transfer of the melt directly into a round shape in which the melt can be pressed and cooled in form. These samples were pre-dried (4 hours at 120 °C) and measured with the MVR method according to ISO 1133 (300 °; 1.2kg). Figure 5 provides an overview of the measured values.

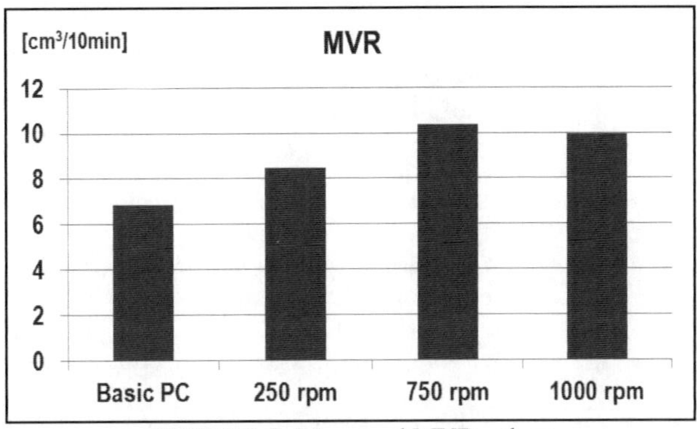

FIGURE 5. Measured MVR values

The results show an increasing damage of the molecular chains with an increasing screw speed. However, the highest value, is at 750 rpm after the MVR decreases and the viscosity increases again. Moreover, not only the shear stress and temperature impact but also the residence time is another important parameter. The problem of an increasing temperature development could thus be compensated by a very short residence time. For this reason, even higher screw speeds have been investigated in this study seems to be possible. Furthermore, the pressure profiles are analyzed. The profiles show that the maximum pressure is located in the area of the locking ring which met the expectations because the material was piled up. With an increasing screw speed the maximum pressure drops, because the melting process is more progressed due to the higher shear stress (dissipative warming) which lowers the viscosity of the material.

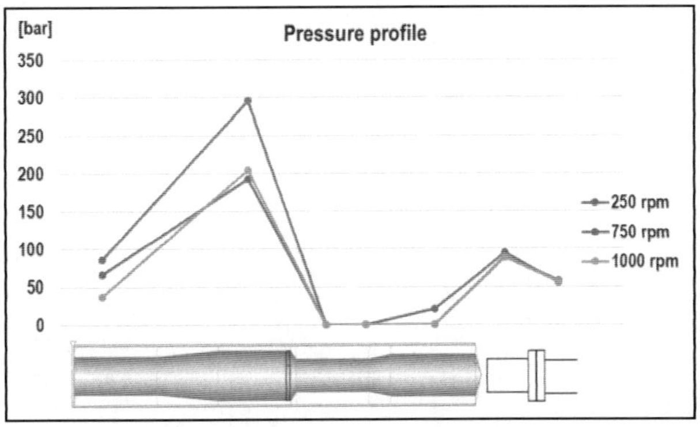

FIGURE 6. Axial pressure profile

SUMMARY AND OUTLOOK

These studies presented in this paper, have shown that an increasing of the screw speed in combination with the processed polycarbonate leads to the anticipated difficulties of damaged polymer due to high temperatures. With a throughput of 170 kg/h, the processing temperature reaches the critical value of 340 °C. However, at the same time the viscosity increases at this point that automatically leads to a lower chain damage. At this point, further investigations need to be carry out. Moreover, studies are planned in which the screw speed will be raised above 1000 rpm and in addition to that, the melting behavior and the Yellowness index is to be measured. It is also necessary to define critical limits for the discussed values because a precise definition of "bad" melt quality is absolutely necessary for analyzing high speed extrusion processes with polycarbonate. The knowledge gained here will then be used for optimizing the screw geometry and to realize even higher throughput rates with a firmer melt quality.

ACKNOWLEDGMENTS

We like to thank the Deutsche Forschungsgemeinschaft (DFG) which has made these studies possible by financing the SCHO 551/19-1 project.

REFERENCES

1. T. Pohl, Entwicklung schnelldrehender Einschneckensysteme für die Kunststoffverarbeitung auf Basis theoretischer Grundlagenuntersuchungen, Dissertation, University of Paderborn, 2003.
1. H. Hörmann, *Theoretische und experimentelle Betrachtung schnelllaufender Einschneckenextruder*, Dissertation, University of Paderborn, 2014.
2. M. Großmann, *Leistungs- und Effizienzsteigerung in der Einschneckenextrusion durch alternative Plastifiziertechnik*, Dissertation, University of Duisburg, 2011.
3. J. Wortberg and G. Karrenberg, *Alternative Plastifizierkonzept für die High-Speed-Extrusion – Der High-Speed-S-Truder mit rotierender Schneckenhülse*, Journal of Plastics Technology 11, Munich, 2015.
4. E. Grünschloß, *HELIBAR® - AN UNIVERSAL PLASTICATING SYSTEM FOR SINGLE-SCREW-EXTRUDERS AND INJECTION-MOULDING-MACHINES*, Stuttgarter Kunststoff-Kolloquium, Institut für Kunststofftechnologie der University of Stuttgart, 2013.
5. H. Rahal, *Alternative Methoden zur Feststoffförderung und Plastifizerung in der Extrusion*, Dissertation, University of Duisburg, 2007.
6. P. Gorczyca, *Analyse und Optimierung von Einschneckenextrudern mit schnelldrehenden Schnecken*, Dissertation, University of Duisburg, 2011.
7. C. Bonten and O. Kast, *Untersuchung eines schnelllaufenden Einschneckenextruders mit wendelgenuteter Plastifizierzone*, Journal of Plastics Technology 11, Munich, 2015.
8. K. Abbas, *Degradation of PC during recycling in a Capillary Rheometer*, PES 7/1980, 1980.
9. B. Dolezel, *Die Beständigkeit von Kunststoffen und Gummi*, CHV, 1978.
10. C. Rauwendaal, *Time to learn about Dynamic Optimization of Extruder Barrel Temperatures*, Plastics Technology, 2008.
11. T. Womer, *Why - and how – to Baseline Your Extruder*, Plastics Technology, 2011.

Study of the Rheological Behavior of Polypropylene/Polyethylene Extruded Mixture Using an Instrumented Die

Ghinwa El hajj sleiman, Isabelle Petit, Nadine Allanic, Sofiane Belhabib, Yannick Madec, Julien Launay, Rémi Deterre

Laboratoire GEPEA, UMR CNRS 6144, Université-IUT de Nantes, 2 avenue Pr. Jean Rouxel, BP 539, 44475 Carquefou Cedex

ghinwa.el-hajj-sleiman@univ-nantes.fr

Abstract. During the recycling of polypropylene (PP), some pollutants are likely to be present in the polymer. Most of these pollutants are polyethylene (PE) which tends to form a heterogeneous blend with PP due to their incompatible nature. In the present study, PP and PE were extruded with different volume fractions of PE. The pressure drop and the temperature variation within the flow in a cylindrical die were studied. A finite element model was used in order to predict the rheological behaviors of the studied polymer mixture ratios. The results obtained in this work show that the rheological response of the blend is very sensitive to the viscosity of the minor polymer when it is lower than the viscosity of the major polymer.

Keywords: recycling, material processing, viscous dissipation, PP/PE blend.
PACS: 83

INTRODUCTION

The growing volume of plastics contained in municipal waste and the demand for solutions for the social and environmental problems deriving therefrom has triggered the industries to work more and more on recycling.

Regardless the type of the recycled plastic, the final product will always contain impurities and pollutants that will affect the mechanical and physical properties of the processed parts. Knowing that polypropylene (PP) and polyethylene (PE) represent over 45% of the world consumption of plastic[1], the study of the recycled matter behavior derived from these two polymers is therefore of a great interest. Indeed, during the recycling of PP, pollutants are likely to be present. PE is the major pollutant, which tends to form a heterogeneous mixture with PP due to the incompatible nature of these two polymers[2]. For that, the determination of the composition of the plastic waste streams is therefore important, because it determines the appropriate management procedure for the end-of-life material and it is also a critical parameter for the future performance of the recycled products in second market applications.

Several types of sorting are widely used in many plastic facilities during recycling. These systems use Fourrier-transform near-infrared (FT-NIR) spectroscopy for polymer type analysis, optical color recognition camera systems to sort the streams into clear and colored fractions, X-ray detection, sink/float separation in water, air elutriation for removing low-density films from denser ground particles[3], to name a few.

Another method for detecting the composition of recycled polymers is the DSC[4] (Differential Scanning Calorimetry). This method uses the glass transition temperature T_g, the melting temperature T_m and the enthalpy of fusion ΔH for the identification and quantification of various polymers in post-consumer waste plastics and post processing recycled polymers.

Despite the various kinds of polymer mixture identification, all are considered to be offline measurements. The aim of this study is to detect the presence of the impurities as well as their concentration fluctuations in recycled thermoplastics during their processing, perform hence an online measurement. For that, a cylindrical die (developed in a previous study[5]) was mounted at the exit of an extruder. The die is equipped with pressure sensors that allow the measurement of pressure drop variations. Polluted polymers, prepared by mixing different percentage of PE and PP, were studied.

Multiphase Flow Models

The products resulting from the melt processing of polymer blends can be very different due to both the materials involved and the processing conditions. Morphology development and de-mixing are the two primary contributors to this difference. Morphology is the relative size, shape and location of each distinguishable phase present in the flow.

Blend morphology can be affected by the way in which the two components flow during forming. Based on the degree of phase separation, there are two main types of multiphase flows[6]. The first one is stratified multiphase flow, in which two polymers form continuous phases separated from each other by continuous boundaries. The second is dispersed multiphase flow, in which one polymer exists as a discrete phase dispersed in the polymer forming the continuous phase (cf. figure 1).

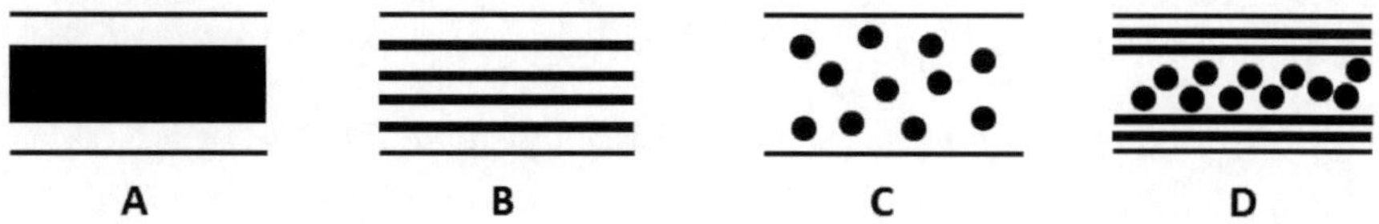

FIGURE 1. Possible morphological structures for an immiscible blend (A) core-sheath (B) multi-layers (C) dispersed phase (D) mix of dispersed phase and multi-layers[7]

Theoretical studies of the rheological properties of Newtonian fluid blends started with Lees[7] who developed various simple models of two phase flow. The simplest of these models involves the viscosity of parallel lamellae oriented in the direction of flow. The dependency of the mean shear rate in a volume element on the volume fraction leads to a monotonic variation of the mixture viscosity versus the volume fraction:

$$\frac{1}{\eta} = \frac{\phi}{\eta_{PE}} + \frac{1-\phi}{\eta_{PP}} \tag{1}$$

Where η_{PP} and η_{PE} are the viscosities of the pure polymers and ϕ is the volume fraction of PE in the mixture.

Another study was held by Taylor[8], on dispersed multiphase flow, who modified Einstein's equation of viscosity of dilute suspensions to include internal circulations in the drop due to viscous shearing in the matrix. Extending Taylor's model developed for dilute suspension to our case, with PE as the dispersed phase and PP as the matrix, we obtained the following expression for the mean viscosity:

$$\eta = \eta_{PP}\left(1 + 2.5\phi\left(\frac{\eta_{PE}+\frac{2}{5}\eta_{PP}}{\eta_{PE}+\eta_{PP}}\right)\right) \tag{2}$$

Min et al.[9] applied Taylor mixing law on dilute mixtures and Lees mixing law on intermediate compositions in an attempt to realize viscosity-composition plots. They show acceptable results for several polymer blends. It is obvious that this mean viscosity description neglects the deformation of the dispersed particles and presumes them to be spheres. This assumption is more or less justified by the wide variation between viscosities.

In this study, we extended to pseudo-plastic polymers by considering local equivalent viscosities the Newtonian fluids approximations mentioned above. For that, we consider the dependency of the viscosities of the pure polymers on shear rate and temperature by the Cross-WLF rheological law:

$$\eta_i = \frac{\eta_0}{1+\left(\frac{\eta_0}{\tau_S}\dot{\gamma}\right)^{1-n}} \qquad (i = PP, PE) \tag{3}$$

where:

$$\eta_0 = D_1 \exp\left[\frac{-A_1(T-T^*)}{A_2+(T-T^*)}\right] \tag{4}$$

and A_1, A_2, T^* D_1, τ_S and n are constant parameters identified for each polymer.

MATERIALS AND METHODS

Materials

The raw materials that were studied in this work are the polypropylene Total Petrochemicals PPC9642 and the polyethylene Dowlex 2042E. Table 1 shows the properties of these two polymers at 190°C. The heat capacity of the polymers was determined by the DSC from Mettler Toledo. The MFI (Melt Flow Index; Norms: NF T 51-016, ISO 1133) device was used to determine the density of the polymers. As for the polymers thermal conductivity, the values were taken from previous studies[10]. The same properties were calculated for the blends by applying the standard law of mixing.

TABLE (1). Properties of Materials

Polymer	Specific Heat Cp $J.kg^{-1}.K^{-1}$	Thermal Conductivity λ $W.m^{-1}.K^{-1}$	Density ρ $kg.m^{-3}$	Viscosity(eq. 3-4)					
				A_1	A_2	D_1	T*	τ_s	n
PPC 9642	2721	0.17	905	19.59	51.6	$1.041*10^{10}$	263.15	$1.91*10^{5}$	0.26
PE 2042E	2362	0.3	795	9.25	51.6	$1.657*10^{7}$	263.15	$4.86*10^{6}$	0.3

The viscosities of the polymers (PE and PP) and their blends, with 10% of one in the other, have been characterized by a rheometric nozzle mounted on an injection molding machine at 190°C. Figure 2 presents the results of the rheometric characterization. These experimental values of the viscosities were smoothened by a Cross WLF law. The parameters of the viscosities of the pure polymers are presented in table 1.

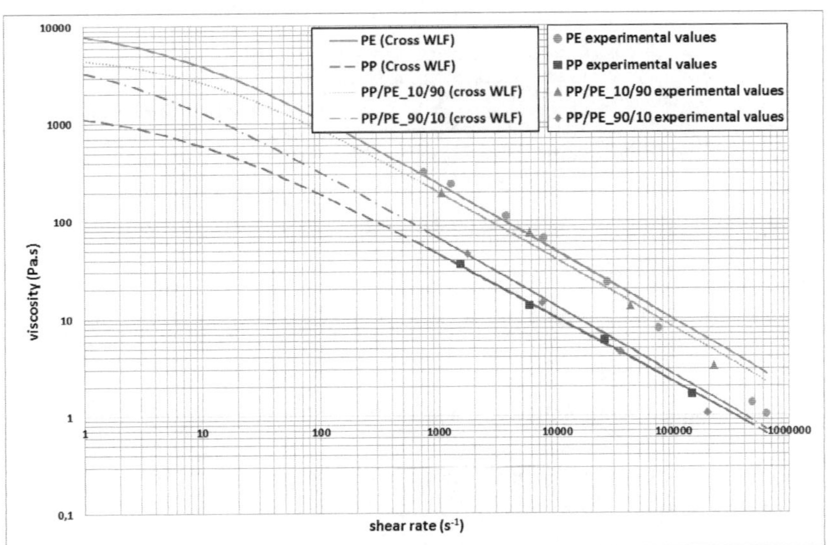

FIGURE 2. The experimental values of the viscosities of polymers with their identified Cross WLF viscosities

Experimental Setup

Extrusions were carried out using an extruder (SCAMIA manufacturer, France). At the exit of the extruder, a cylindrical die was mounted as shown in figure 3. The die was equipped with two pressure transducers, one placed at the entrance of the die and the other at its exit. The device was heated at 190°C by helical coil heating elements that surround the die. Temperatures were measured at its wall and used to control the power delivered to the heating coils.

In addition to the pure polymers, mixtures of PP/PE with various volume fractions ϕ = [0.1, 0.3, 0.4, 0.5, 0.7, 0.9] were extruded at a temperature of 190°C and at a rotational speed of 40 rpm.

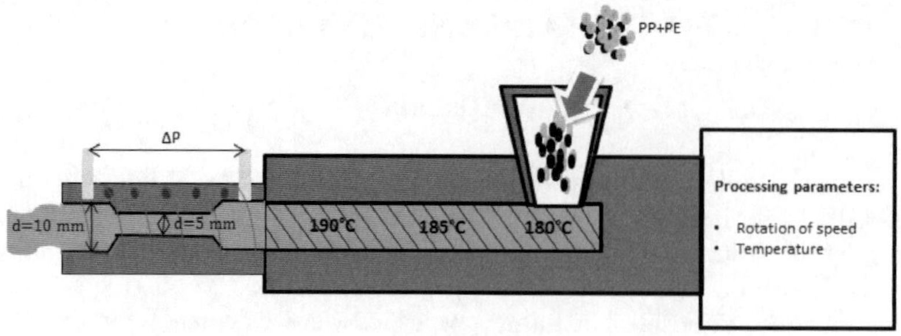

FIGURE 3. Schematic representation of the experimental setup

Numerical Modeling

The extrusion is a process where the flow can be considered to be laminar and as a first approach, incompressible. In this work, a non-slip condition is imposed. The studied geometry, a convergent-divergent channel, is shown schematically in figure 4.
The rheological and thermal behavior of the thermoplastics is described by Navier-Stokes and thermal equations in the flow:

$$\rho(\vec{V}.\vec{\nabla})\vec{V} = -\vec{\nabla}p + \rho\vec{g} + \nabla\left[\eta.\left(\overline{\overline{\nabla\vec{V}}} + {}^t\overline{\overline{\nabla\vec{V}}}\right)\right] \tag{5}$$

$$\rho C_p(\vec{V}\vec{\nabla})T = \nabla(\nabla\vec{T}) + \eta\dot{\gamma}^2 \tag{6}$$

The problem is solved by Comsol Multi-Physics© in axisymmetric coordinates frame.

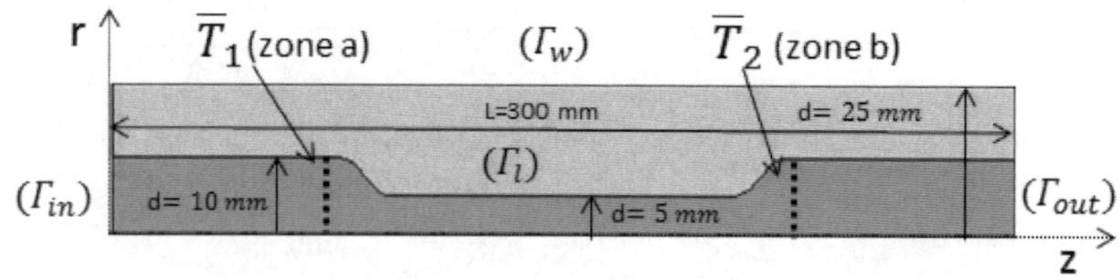

FIGURE 4. Key dimensions of the studied geometry in the axisymmetric coordinates frame

The boundary conditions are described by the equations below:

$$\Gamma_{in}: \vec{V} = (0, 0, w(r)); \quad T(r, z = 0) = T_{in} = 190°C \tag{7}$$

$$\Gamma_{out}: p = 0; \quad \frac{\partial\vec{V}}{\partial\vec{n}} = \vec{0}; \quad \frac{\partial\vec{T}}{\partial\vec{n}} = \vec{0} \tag{8}$$

$$\Gamma_l: \vec{V} = \vec{0}; \quad T(r = R, z) = T_{out} = 190°C \tag{9}$$

$$\Gamma_w: T_w = 190°C \tag{10}$$

Where $\vec{V}$ is the velocity, p is the pressure, $\vec{g}$ is the gravity constant, Γ_{in} is the entrance section, Γ_{out} is the exit section, Γ_l is the lateral surface and Γ_w is the outer surface of the die (figure 4).

RESULTS AND DISCUSSIONS

The measured pressure drop measured depicted in figure 5 shows that the presence of PE in PP ($\Delta P_{\phi=0} - \Delta P_{\phi=0.1} = 13\, bars$) is more difficult to detect than that of PP in PE ($P_{\phi=1} - \Delta P_{\phi=0.9} = 33\, bars$). Numerical simulations run with Cross WLF viscosity law show the same behavior, concerning the pressure drops, as that of the experimental results.

In the flow described previously (figure 4), there are reserved places dedicated to temperature measurement(figure 4, zone a, zone b) by a TMC (Temperature Measuring Cell) [5]. Inserting these intrusive sensors disturbs the flow therefore exploiting these measures require a numerical correction. Before making these delicate measurements, simulations were conducted to assess the relevance of a thermal measurement to the detection of low percentages of PP or PE. Table 2 shows that the difference in the temperatures calculated is of the order of 0.2°C (10% of PE) to 0.4°C (10% of PP). The low range of variation of temperature is related to the low flow rate in extrusion and the large diameter of the channel.

TABLE (2). Temperature difference due to the viscous dissipation

%PE	0	10	90	100
$\Delta T\ (\overline{T}_2 - \overline{T}_1)$	0,29°C	0,49°C	1,39°C	1,77°C

Due to the difficulty in predicting the behavior of the flowing blends of PP and PE, complementary simulations were performed. A Finite Element Model was conducted for every volume fraction by introducing the viscosity law behavior determined from different mixing laws described by Taylor and Lees. Despite the differences in the pressure drops between the experimental and simulated results, the model gives us an early idea on the repartition of the polymers in the blends. The obtained results show that as the volume fraction of PE is less than 0.75, Taylor model is more convenient although it doesn't perfectly predict the pressure drops. This means that the flow configuration of PE is more likely to be dispersed in the flow. As the volume fraction of PE crosses 0.75, the behavior of the mixture probably changes from being dispersed into stratified.

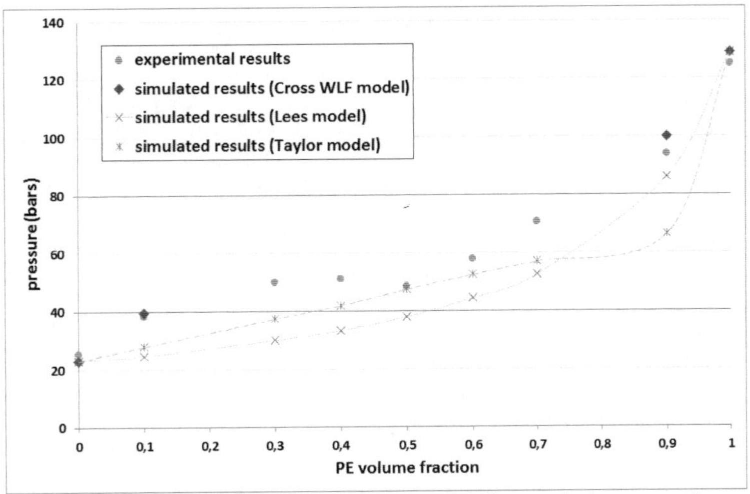

FIGURE 5. Comparison between experimental and numerical pressure drop

CONCLUSION

The obtained results presented here have shown that online measurement for the detection of pollutants in recycled thermoplastics is a promising technique. Finite element simulation is a powerful tool that can be used to design such measurement systems. However, simulating polymer blends remains a difficult challenge due to the lack of knowledge of the behavior of immiscible mixtures. Further investigations are in progress concerning the multiphase mixture flow modeling using level set approach.

REFERENCES

1. *PlasticsEurope - Plastiques 2010 : faits et chiffres*. (2009).
2. Teh, J. W., Rudin, A. & Keung, J. C. A review of polyethylene–polypropylene blends and their compatibilization. *Adv. Polym. Technol.* **13,** 1–23 (1994).
3. New technique automates sorting of plastics for recycling.
4. Identification and quantification of polymers in waste plastics using differential scanning calorimetry.
5. Launay, J., Allanic, N., Mousseau, P., Deterre, R. & Madec, Y. Intrusive measurement of polymer flow temperature. *Polym. Eng. Sci.* **54,** 2806–2814 (2014).
6. HAN, C. D. in *Multiphase Flow in Polymer Processing* 79–80 (Academic Press, 1981).
7. Askar, K. Polymer blend de-mixing and morphology development during tube flow. (University of Toronto, 1999).
8. Taylor, G. I. The Formation of Emulsions in Definable Fields of Flow. *Proc. R. Soc. Lond. Math. Phys. Eng. Sci.* **146,** 501–523 (1934).
9. Min, K., White, J. L. & Fellers, J. F. Development of phase morphology in incompatible polymer blends during mixing and its variation in extrusion. *Polym. Eng. Sci.* **24,** 1327–1336 (1984).
10. Karkri, M. Transferts de chaleur dans un écoulement stationnaire de polymère fondu dans une filière d'extrusion. (Université de Nantes, 2004).

Wasted Tire Ground Rubber/ Wood Flour Optimized Ratio Toward the High Mechanical Properties in Polypropylene Composite

Ratiporn Thongthanom[a] and Hathaikarn Manuspiya[a,*]

[a]*The Petroleum and Petrochemical College, Chulalongkorn University, Soi Chula 12, Phyathai Road, Pathumwan, Bangkok 10330, Thailand*

r.thongthanom@hotmail.com
**hathaikarn.m@chula.ac.th*

Abstract. Tire rubber has become a significant problems among the environmental concern because of its crosslink network and non-biodegradable character. Through this problem, the application of wasted tire rubber as filler in wood plastic composite (WPC) was considered for a production of toughness improvement products. Ground Tire Rubber (GTR) or crumb rubber is the waste tire rubber in a small size and can be used as additive in rubber and plastic compounds to gain the impact properties. In this study, Polypropylene was used as a matrix filled with fillers (WF:GTR) in 70:30 and 60:40. The fillers were prepared by vary the weight ratio of wood flour (WF) to GTR in 1:3 1:1 and 3:1. The result showed optimized mechanical properties in 70:30. Young's modulus and tensile strength was increased when WF increased but it still lower than neat PP may be due to the lower interfacial adhesion. Impact strength was improved in 1:3 WF to GTR so this ratio was selected to improve the interfacial adhesion by using PP-g-MA as a compatibilizer. After PP-g-MA was added, SEM results showed better interfacial adhesion and higher tensile strength after PP-g-MA was added.

Keywords: Polypropylene; Ground Tire Rubber; Wood Flour; Composite; Wood Plastic Composite
PACS: 81.05.Qk

INTRODUCTION

This research is targeted to develop a green pallet for a plastic company that aims to reduce the use of virgin plastic resin. Pallet is a flat structure essential for various industrial in order to handling or transporting the products. There are several types of pallet material for example wooden pallet, paper pallet, metal pallet, plastic pallet and composite pallet. Composite pallet has become an attractive choice of pallet material because it is combine the advantages and properties of plastic and wood together. In addition, stiffness and impact resistance are required simultaneously in structural applications but only natural fiber as fillers give only stiffness. In order to solve this problem, elastomer or impact modifier was introduced to achieve this limitation [1].

Rubber tire is one type of polymeric material which has three dimensional crosslinked network structures that need a long time for natural degradation and make rubber tire become a non-biodegradable polymer cause a serious environmental problem [2]. The amount of rubber tire production and consumption is large in Thailand. As a result, the amount of waste tire accumulation has increased every year especially from tire rubber. To solve tire accumulation problem, recycle technology and reuse of waste tire has been introduced. Crumb rubber or ground tire rubber (GTR) is a product from recycle waste tire rubber through the processes and machineries such as pulverizor, granulator and hammer milled. It can use as additive in rubber manufactures and others in order to reduce cost, modify some properties of product and reduce environmental problems by reuse. GTR can use as additive in polymer matrix in order to improve impact strength [3].

In this study, polypropylene is selected to polymer matrix, ground tire rubber (GTR) is chosen as a fillers to improve impact strength properties and wood flour is chosen as fillers to improve mechanical properties to prepare composite materials use in pallet application. The objectives of this work are prepare composite materials for pallet application and optimize the generic between ground tire rubber and wood flour as well as the ratio between plastic resin and fillers and also study the structure-property relationship of the composite to achieve the pallet.

EXPERIMENTAL

Materials

Polypropylene (HPP1102K) was purchased from IRPC Public Company Limited. Polypropylene-grafted-Maleic Anhydride (PP-g-MA) (EXXELOR PO 1020) with MFI 117 g/10 min and 0.72% grafting obtained from Exxon Mobile Chemical. Ground Tire Rubber in 40 mesh size obtained from Q2Thailand company and Wood flour obtained from Union link Co., ltd.

Composite Preparation

Wood flour and Ground Tire Rubber were first dried at 100 °C for 24 h. Each of the pure PP, composite compound and composite compound with PP-g-MA were built in the same conditions after physically mixing the components. Composite ingredients were pre-mixed together in each batch by a drum-type laboratory blender based on the weight percentage followed by Table (1) and were incorporated simultaneously in intermeshing /co-rotating twin screw extruder. This extruder had ten temperature zones with L/D ratio was 32 and D = 25 mm. The barrel temperatures were: 180/180/220/220/220/220/220/220/215/215 °C. The screw speed was 50 rpm. The extrudates were cooled in water bath and crushed into particles using a pelletizer with a screen of 5 mm. The granulated materials were dried at 80 °C for 24 h to eliminate residual humidity and then injection molded into impact and tensile specimens. Injection molding was carried out in a temperature profile of Injection molded to standard tensile specimen at 182/178/175/168 °C zone with screw speed at 235 rpm and injection pressure: 95 bar also use cooling time 26 s. All obtained materials were kept at 25 °C with 25% humidity for 24 h before testing.

Characterization

Tensile tests were determined by Universal Testing Machine (Instron 4206) on type-I specimens according to ASTM D-638 at crosshead speed of 5 mm/min. Izod-notched impact test was measured according to ASTM D-256 using impact tester at 2.7J energy level. The dynamic mechanical properties were studied using a Dynamic Mechanical Analyzer (GABO Eplexor) in tension mode using rectangular sample (10 x 40 x 3 mm^3). The test was done in a temperature range of -100°C to 100°C at heating rate of 2°C/min at frequency of 1 Hz. Studying of composites microstructure, dispersion of fillers and quality of adhesion between two phases were done by a Hitachi (S4800) Field Emission Scanning Electron Microscope (FESEM) at 15 kV. The samples were first fractured in liquid nitrogen and the surfaces were coated with a gold alloy.

TABLE (1). Composition of the composites

	Content (%wt)					Content (%wt)			
Nomenclature	**PP**	**WF (W)**	**GTR (R)**	**PP-g-MA (g) (phr)**	**Nomenclature**	**PP**	**WF (W)**	**GTR (G)**	**PP-g-MA (g) (phr)**
xPP	100	-	-	-	70:30(1W:3R)_1g	70	10	20	1
70:30(1W:3R)	70	10	20	-	70:30(1W:3R)_3g	70	10	20	3
70:30(1W:1R)	70	15	15	-	70:30(1W:3R)_5g	70	10	20	5
70:30(3W:1R)	70	20	10	-	70:30(3W:1R)_1g	70	20	10	1
60:40(1W:3R)	60	10	30	-	70:30(3W:1R)_3g	70	20	10	3
60:40(1W:1R)	60	20	20	-	70:30(3W:1R)_5g	70	20	10	5
60:40(3W:1R)	60	30	10	-					

RESULT AND DISCUSSION

Polypropylene/ Wood flour/ Ground Tire Rubber Composite

Morphological properties

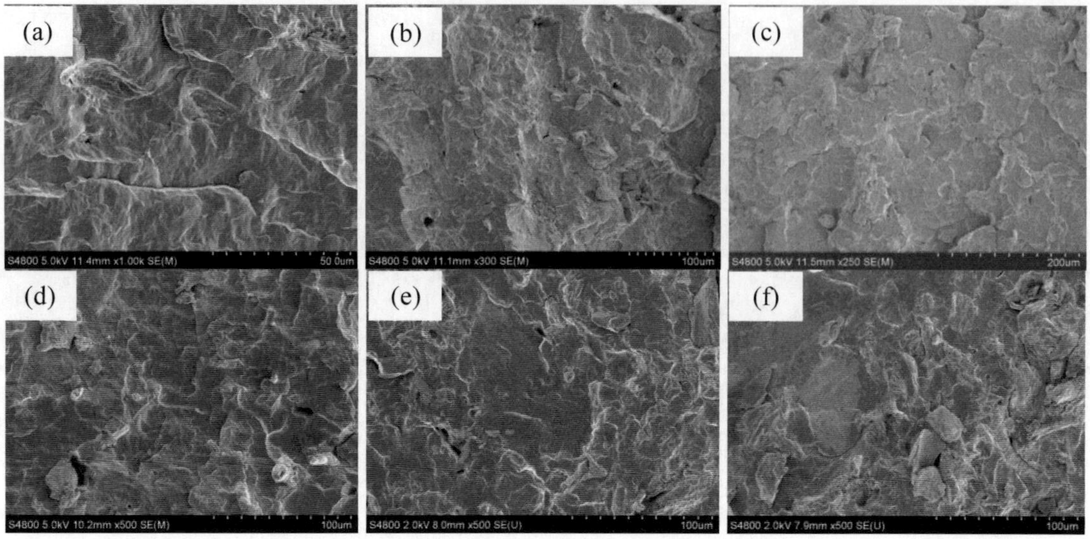

FIGURE 1. FESEM micrographs of PP composites in 70: 30 (a) 1W:3R (b) 1W:1R (c) 3W:1R and 60:40 (d) 1W:3R (e) 1W:1R (f) 3W:1R

Figure 1 shows SEM images taken from the fracture surfaces of sample after breaking them in liquid N_2. When compared at 70:30 (a, b, c) with 60:40 PP:Fillers (d, e, f), it can be observed that for 40% of fillers showed large amount of fillers at surface with worse dispersion, less wettability and bad interfacial adhesion created hole and fillers pulled out. As the increase of weight percentage of wood flour, from Figure 1(a)-1(c), the material become more rigid due to large amount of wood flour was added and there was a change from ductile in 1W:3R to brittle surface behavior in 3W:1R [4]. In addition, the fracture surfaces of composite occurred without plastic deformation due to presence of rigid particle of wood flour [5].

Mechanical Properties

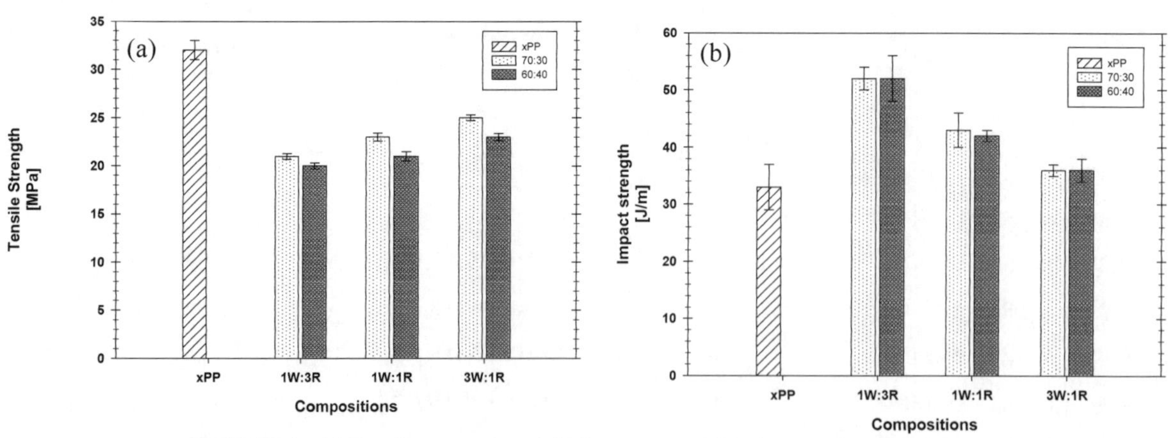

FIGURE 2. (a) Tensile strength and (b) Impact strength of composite materials.

Figures 2(a) and (b) presented the mechanical properties of PP-based composites filled with WF and GTR in different ratios compared to neat PP. As can be seen, tensile strength of composites are increased when increased

WF content but it still showed lower in all ratios when compared to neat PP. This behavior ascribed to poor interfacial adhesion between hydrophilic fillers and hydrophobic polymer [6]. Impact strength was increased since incorporation of ground tire rubber due to its rubber plasticize properties of rubber phase as shown in 1W:3R. These results are corresponded to the expected since ground tire rubber can apply for impact modification. In the composition of 1W:3R has the impact strength approximately to 52 J/m that increased about 57% from neat PP.

Effect of PP-g-MA compatibilizer on properties of Polypropylene/ Wood flour/ Ground Tire Rubber composite

Morphological properties

From the previous results, the composition of 70:30 of PP to Fillers was chosen to further study due to the processibility and overall mechanical properties and the ratio of 1:3 and 3:1 wood flour to ground tire rubber were chosen to study the effect of compatibilizer to improve the properties of composites. In this study, PP-g-MA compatibilizer was added to the matrix for increase the compatibility of wood flour and ground tire rubber with PP matrix. PP-g-MA is maleic anhydride functionalized with homo-polypropylene. It has been primarily designed to improve the polarity to polypropylene matrix.

The morphological changes of composite filled with PP-g-MA compatibilizers are shown in Figure 3. These figures are used to investigate the quality of interfacial adhesion between WF, GTR and PP-g-MA.

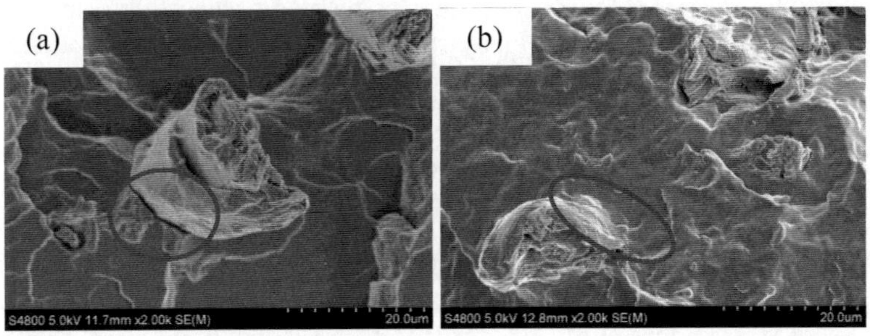

FIGURE 3. FESEM micrographs of PP/WF/GTR composite in 70:30 with 1:3 WF to GTR a) without PP-g-MA b) with PP-g-MA 1phr c) 3phr and d) 5phr.

As can be seen from Figure 3(b), wettability of polymer matrix was changed according to the incorporated fillers. There is stronger interfacial adhesion between wood flour and PP matrix. The gaps and void between fillers and polymer interfaces disappeared and it improved wettability of fillers when PP-g-MA was added showed in Figure 5 (b). This observations suggested that the fillers are well embedded in the matrix resulted in better adhesion and a compatibility was improved.

Mechanical properties

The effect of incorporation of PP-g-MA compatibilizer into PP/WF/GTR composites on mechanical properties is plotted in Figure 4. PP-g-MA greatly affected the tensile strength of composites as can be seen in Figure 4(a). The composites with PP-g-MA compatibilizer exhibited higher tensile strength compared with composites without PP-g-MA. Tensile strength of composites increased with increasing in PP-g-MA content but it has no significant change between 3 and 5 phr of PP-g-MA. Those results can be explained in terms of improved filler wettability by the polymeric resin with consequently improved interfacial adhesion and good stress transfer at the interfaces which confirmed by the FESEM analysis presented earlier

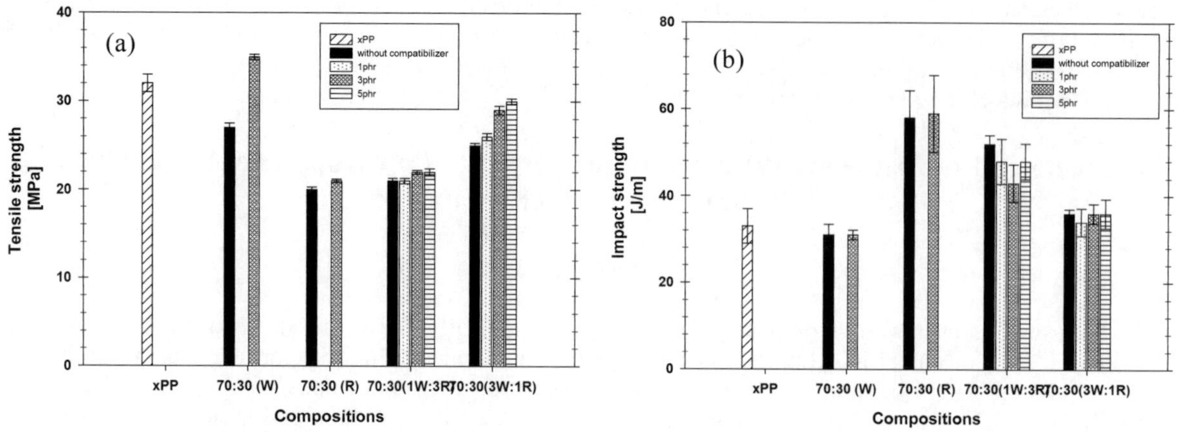

FIGURE 4. Mechanical Properties of PP/WF/GTR composite in 70:30 with PP-g-MA compatibilizer (a) Tensile strength (b) Impact strength.

The notched Izod impact energy of composites is plotted in Figure 4(b). The main idea for addition of ground tire rubber to composites is to increase ductility. The results showed successfully improved of impact strength in PP/WF/GTR at 1:3 WF to GTR but when considered the effect of PP-g-MA, the addition of PP-g-MA has no significant effect on impact strength.

CONCLUSIONS

In this research, the appropriate ratio of WF and GTR effect on mechanical properties of PP was investigated. The addition of WF into PP leads to increase in tensile strength due to presence of rigidity fillers but is still has lower than neat PP may be due to poor adhesion at the interface and poor dispersion of fillers according to the results from FESEM. However, the addition of GTR leads to increase in elongation at yield and impact strength due to elastic nature of soft rubber phase. Considering the effect of adding wood flour and ground tire rubber on the mechanical properties of PP composites, it is possible to conclude that 30% of fillers give higher in overall mechanical properties than 40%. The addition of PP-g-MA as a compatibilizer aims to improve the interfacial adhesion between WF, GTR fillers in PP matrix and improve the dispersion of filler phase. From the FESEM micrographs showed more wettability of fillers by PP matrix and well embedded of filler into the matrix indicated that the interfacial adhesion was improved lead to higher tensile strength but has no significant change in impact strength.

ACKNOWLEDGMENTS

The authors acknowledge to The Petroleum and Petrochemical college Chulalongkorn University for the research instrument supported. I also acknowledge Unionlink Co.,ltd for supplying the raw materials in this study

REFERENCES

1. A. Sudár, C. Burgstaller, K. Renner, J. Móczó, B. Pukánszky, *Composites Science and Technology* **103**, 106-112 (2014).
2. O. Bijaisoradat, N. Kongkatigumjorn, K. Wattanavit, "Developing a Sustainable Waste Tire Management Strategy for Thailand", Unpublished Bechelor Degree's Project Report, Chulalongkorn University, 2013.
3. F.A. Ghasemi, G. Payganeh, M. Rahmani, M.R. Kalaee, *Digest Journal of Nanomaterials and Biostructures* **7**, 1859-1868 (2012).
4. A.J. Nuñez, P.C. Strum, J.M. Kenny, M.I. Aranguren, N.E. Marcovich, M.M. Reboredo, *Journal of Applied Polymer Science* **88**, 1420-1428 (2003).
5. E. Pérez, L. Famá, S.G. Pardo, M.J. Abad, C. Bernal, *Composites Part B: Engineering* **43**, 2795-2800 (2012).
6. A.R. Kakroodi and D. Rodrigue, *Polymer Composites* **35**, 2280-2289 (2014).

Electrification of Polymers during Capillary Extrusion

A. Perolo*, A. Castiglioni and D. Ferri

Claudio Buonerba Research Centre, Versalis S.p.A., Via G. Taliercio, 14, 46100 Mantova (MN) – Italy

andrea.perolo@versalis.eni.com

Abstract. In this work the phenomenon of electrification of polymer melts during extrusion has been investigated. In order to measure the electrostatic charge Q a custom made measurement apparatus has been built consisting of a Faraday cup similar to those used in the literature connected to an electrometer and to a digital acquisition system. Different commercial polystyrene grades with different molecular weights M_w were examined. Each one has been extruded at different flow rates, temperatures and capillary geometries. The results have been expressed and discussed in terms of electric charge per unit volume Q_L (nC/cm^3) of the extrudate. For the different grades investigated a strong effect on Q_L of both temperature and die geometry has been highlighted.

Keywords: Polymer melt, electrification, electric double layer, Faraday cup, capillary rheometry.
PACS: 73.30.+y , 81.20.Hy, 83., 83.80.Sg, 83.85.Cg.

INTRODUCTION

The electrification of simple hydrocarbon liquids during pipe flow is a well-known phenomenon and it has many influences on industrial applications. For these reasons there is a huge literature on this topic that gives a satisfying description of charge accumulation mechanisms. Electrification of polymers during extrusion is on the contrary less known and studied. In this case non-Newtonian flow behavior and sometimes unsteady flow conditions undeniably complicate the phenomenon comprehension [1-3]. Electrification measurements on jets extruded from a capillary at different pressures constitute an alternative way to demonstrate the existence of macroscopic slippage at the interface polymer/die wall. In particular a sharp increase of charge accumulation was observed during spurt instability [4].

The fact that during capillary extrusion most polymer melts exhibit a significant amount of charging can be ascribed to the formation of a double-layer at the solid/liquid interface (based on the Stern model) [1].

The total charge density Q_L (defined as the total charge per unit volume of extruded material) convected by the flow at the outlet of the die is determined by the charge profile $\rho(r,l)$, for a given radial distance r from the die center and die length l, and by the velocity profile v(r) of the fluid flowing inside the die. The analysis could be simplified by supposing that the polymer melt flow during capillary extrusion is laminar and that the double-layer is thin compared to the capillary diameter. In long capillaries ($l=\infty$), where the full equilibrium double-layer can be established, the total charge density at the outlet of the die only depends on the radial position r and can be calculated with the following equation [1-3]:

$$Q_L(\infty) = \frac{2}{R^2 v_{mean}} \int_0^R \rho(r,\infty) v(r) r dr \tag{1}$$

Where R is the capillary radius, $\rho(r,\infty)$ is the charge density at a radial distance r, v_{mean} is the mean flow velocity inside the capillary and v(r) is the velocity profile.

The shear-thinning behavior at high shear rates of non-Newtonian liquids like polymer melts can be approximated by the Ostwald-de Waele power law:

$$\tau = K\dot{\gamma}^n \tag{2}$$

Where τ is the shear stress and $\dot{\gamma}$ is the shear rate, K is the flow consistency index and n is the flow behavior index.

The resulting velocity profile is a plug flow-type, instead of a parabolic one as expected for Newtonian liquids, which can be calculated from this equation:

$$v(r)=\frac{R^{(n+1)/n}}{(2K)^{1/n}}\left(\frac{\partial P}{\partial l}\right)^{1/n}\frac{n}{n+1}\left[\left(\frac{r}{R}\right)^{(n+1)/n}-1\right] \tag{3}$$

Where $\partial P/\partial l$ is the pressure gradient along the capillary length.

The calculations to evaluate the charge profile $\rho(r,\infty)$ can be found elsewhere [1]. The development of the electrical double-layer depends on many physical properties like the diffusion coefficient of the charge species D, the electrical conductivity σ and permittivity ε of the melt. Each one of these properties play a role in determining the charge density profile formation on the capillary wall which may be expressed as follow:

$$\frac{d^2\rho}{dr^2}-\frac{B}{\sigma_0^2+B\rho^2}\rho\left(\frac{d\rho}{dr}\right)^2+\frac{1}{r}\frac{d\rho}{dr}-\frac{\rho}{\varepsilon}\frac{\sqrt{\sigma_0^2+B\rho^2}}{D}=0 \tag{4}$$

Where B is a constant whose value depends on temperature, D and on the sign and the amount of the ionic charges. No analytical solution exists for equation (4) which can be solved only numerically [1].

EXPERIMENTAL PART AND RESULTS DISCUSSION

Materials and Measurement Method

In this work different polystyrenes with different molecular weight distributions were examined as reported in table 1 (grades supported by Versalis S.p.A.). Experiments were conducted on a capillary rheometer (CEAST model RHEOLOGIC 2500) equipped with tungsten carbide round hole capillaries with different geometries (L/D=10/1, 20/1 and 30/1). Extrusion temperatures were set at 180, 190 e 200 °C.

In the rheometer the polymer melt flows through a capillary die under defined piston speeds and the melt pressure is measured in the barrel above the die entrance. The apparent shear stress on the wall is then calculated from the pressure gradient between the atmospheric pressure and barrel pressure ΔP_L.

TABLE 1. Molecular weights and polydispersity index values of the two polystyrene samples used in this study together with typical electrical conductivity and permittivity standard values [2,3] (with $\varepsilon=\varepsilon_0\cdot\varepsilon_r$).

Material	M_w [g/mol]	M_w/M_n	σ(200°C) [S/m]	ε(200°C) [C²/N·m²]
PS-A	190,000	2.15	~$3\cdot10^{-11}$	~$2{,}5\cdot10^{-11}$
PS-B	130,000	2.05		

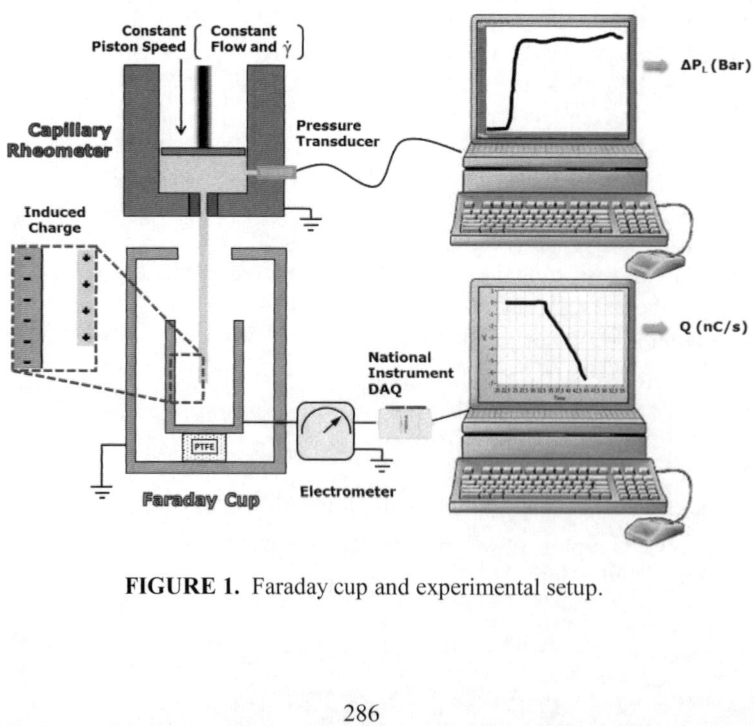

FIGURE 1. Faraday cup and experimental setup.

To measure the charge density the extrudate samples were collected into a Faraday cup. The inner cylinder was connected by a shielded BNC wire to an electrometer (Keithley, model 610C) and the external cylinder as well as the tungsten carbide capillary were earthed (Figure 1). The charge measured by the electrometer is acquired by an USB-6210 from National Instruments and the voltage signal is analyzed using a specific customized LabView™ software. Thus the accumulated charge Q (nC) can be plotted vs. time. Using the ratio between the slope of electrification curves dQ/dt (nC/s) and the applied volumetric flow rate $\dot{V}$ (cm^3/s) it is possible to calculate the charge per unit volume Q_L (nC/cm^3).

Results and discussion

In figure 2 the flow curves obtained at different temperatures for the two polystyrene samples are reported. It is evident that, in the shear rate range investigated, a power law can well describe the flow behavior of these melts.

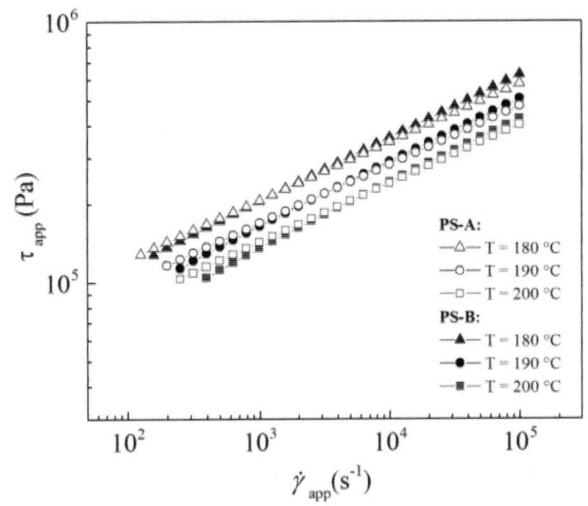

FIGURE 2. Flow curves obtained at different temperatures for PS-A (open symbols) and PS-B (solid symbols) samples.

The graph in figure 3 on the left shows the electrification curves obtained for capillary extrusion of PS-A at 190°C with different exit velocities v_{out} (mm/s), these velocities are calculated from the ratio between $\dot{V}$ and the capillary section (v_{out} is chosen supposing it is comparable to v_{mean} inside the capillary). As can be noticed dQ/dt increases with increasing exit velocity. The graph in figure 3 on the right also reports the temperature effect at the same v_{out}.

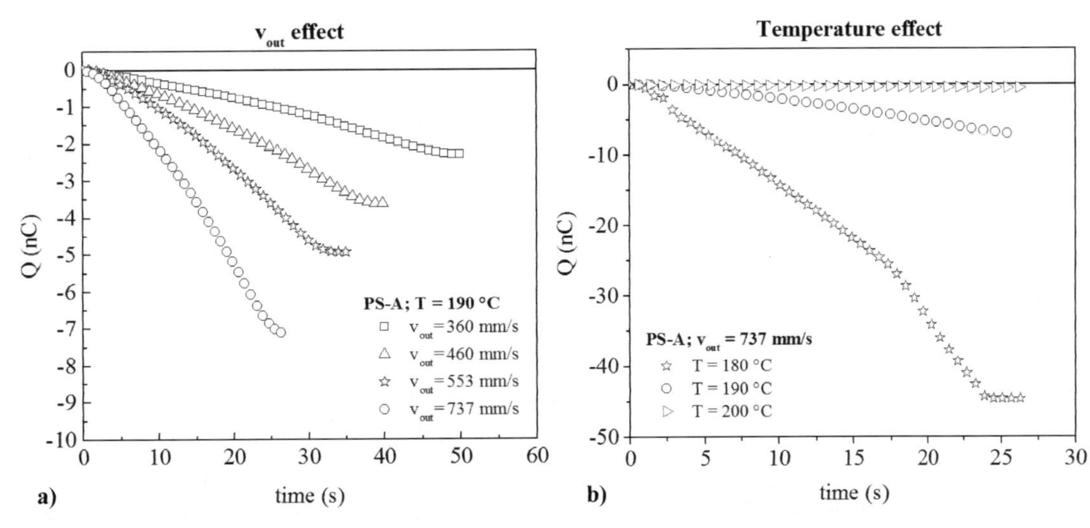

FIGURE 3. Electrification curves (Q vs t) obtained with a round hole capillary (L/D=30/1) for PS-A as a function of v_{out} at 190°C (a) and for different extrusion temperatures at v_{out}=737 mm/s (b).

To compare results from tests conducted with different extrusion conditions is necessary to consider the charge per unit volume Q_L [3]. In figure 4 Q_L values as a function of the inverse of the exit velocity $1/v_{out}$ (s/m) are reported for PS-A and PS-B extruded at different temperatures. The data show a strong temperature dependence, the charge per unit volume increasing with decreasing extrusion temperature. The temperature influence is better highlighted in the left hand graph in figure 4 for PS-A where a significant effect of v_{out} on Q_L is also evident. For high temperatures and low v_{out} values both samples exhibit charge sign inversion.

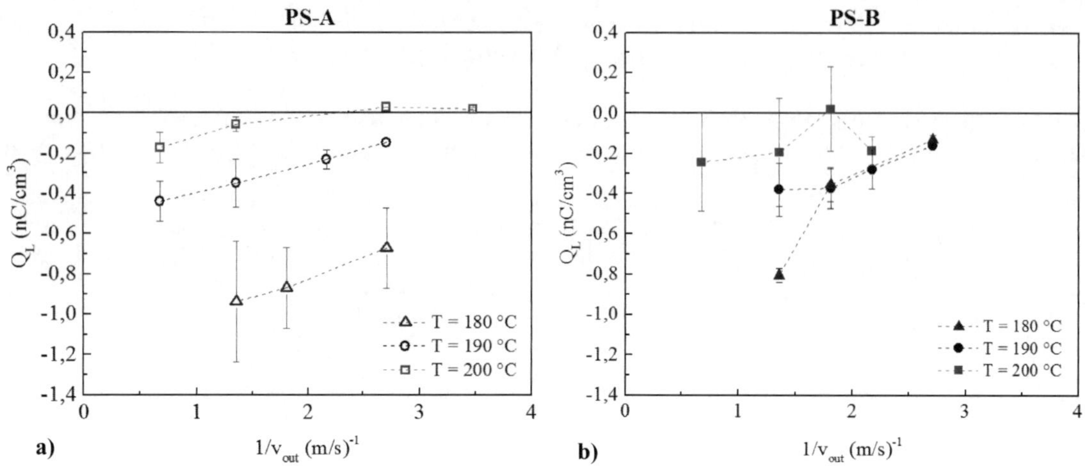

FIGURE 4. Plots of Q_L against $1/v_{out}$ for PS-A (a) and PS-B (b) showing the effect of the extrusion temperatures. Values are obtained using a round hole capillary (L/D=30/1). The dashed lines are only guides for the eyes.

Figure 5 highlights a Q_L dependence on the capillary length L for different extrusion temperatures and velocities. For PS-A samples a charge sign inversion is visible while for PS-B this effect is less evident. In both cases Q_L goes through a maximum.

The Q_L dependence on L may be due to the incomplete double-layer formation as a result of shear stress differences and/or different velocity profiles.

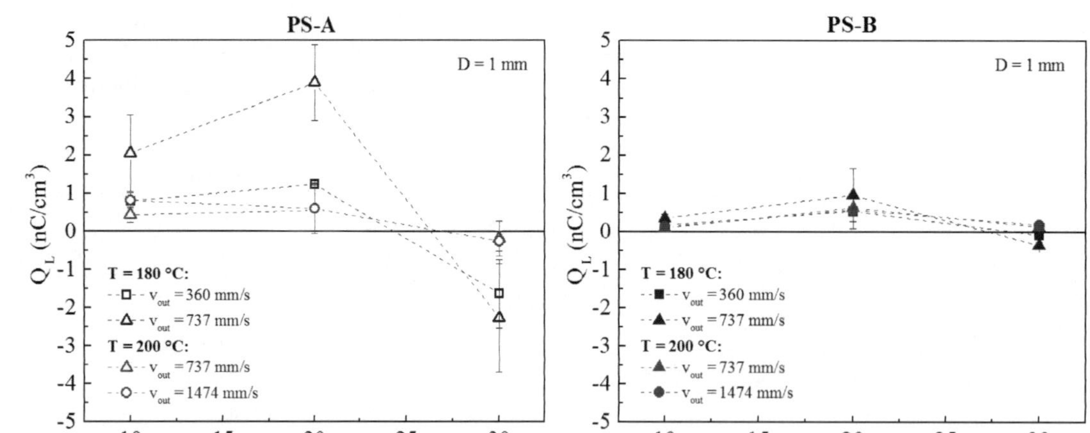

FIGURE 5. Capillary length dependence of Q_L (D=1 mm) for different extrusion temperatures and velocities (PS-A and PS-B reported on the left (a) and right (b) respectively, dashed lines as guides for the eyes).

In figure 6 Q_L values obtained during extrusions at 190°C and v_{out}=360 mm/s for the two polystyrenes are reported together with a PS-B/carbon black concentrated master batch (CB content=50 wt.%). The master shows a much greater charge accumulation than that for the two polystyrenes. Although at room temperature the resistivity of this compound is only of ~10^6 Ωcm the resistivity of the melt significantly increases. During extrusion induction and/or triboelectrification phenomena could be responsible for a charging of the conductive particles. The charge accumulated is then preserved because these particles are enclosed in non-conductive polymer matrix.

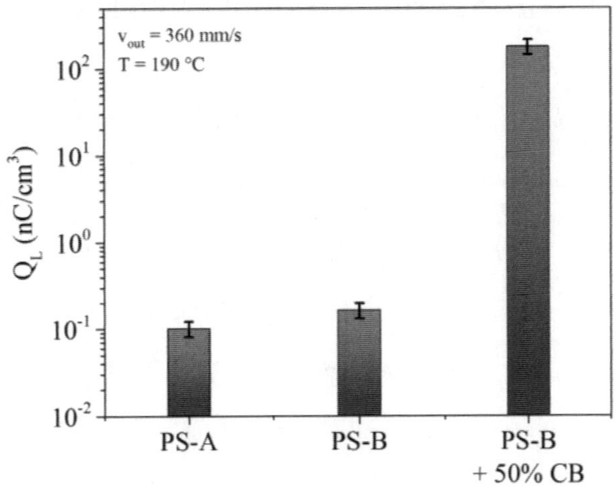

FIGURE 6. Charge per unit volume Q_L obtained for PS-A, PS-B and for a PS-B/CB master batch (CB=50 wt.%) during extrusion at 190°C and v_{out}=360 mm/s.

CONCLUSIONS

In this work an instrumental apparatus to measure the electrification of molten polymers during capillary extrusion has been realized it was demonstrated that charging takes place when commercial polystyrene grades are extruded. The preliminary experimental tests presented in this work suggest that the amount and sign of the charge both depend on the molecular weight of the polystyrene at study, on the capillary geometry and on the extrusion velocity and temperature. In particular it has been documented that Q_L tends to increases as T decreases and as v_{out} increases. It has been also observed that the magnitude of the extruded charge increases when polystyrene is compounded with carbon black.

ACKNOWLEDGMENTS

The authors would like to acknowledge all researchers and technicians of the Versalis S.p.A. from Claudio Buonerba Research Centre of Mantova, in particular Castellani Leonardo and Fregni Doriano for their support for the realization of this research.

REFERENCES

1. F. Flores, D. Graebling, A. Allal, C. Guerret-Piécourt, *Journal of Physics D: Applied Physics* **40**, 2911-2919 (2007)
2. D. M.Taylor, *Journal of Physics D: Applied Physics 7*, 394-402 (1974)
3. D. M.Taylor, T. J. Lewis, T. P. T. Williams, *Journal of Physics D: Applied Physics 7*, 1756-1772 (1974)
4. G. V. Vinogradov et al., *J. Polym. Sci. Part A-2: Polymer Physics **10***, 1061–1084 (1972).

Shell-mimicking structure of polylactide fabricated through layer-assembly extrusion: toward high strength and toughness

Longfei Yi (presenting author)[a], Jiabin Shen[a], Shaoyun Guo[a*]

Polymer Research Institute of Sichuan University, State Key Laboratory of Polymer Materials Engineering, Chengdu, Sichuan, P. R. China 610065

Presenting author's email: yilongfei1987@qq.com; Reference author's email: nic7702@scu.edu.cn.

From the perspective of balancing performances, achieving high strength and toughness in a polymer material has caught great attention in past decades. In this work, shell-mimicking polylactide (PLA) was fabricated through a layer-assembly extrusion which combined an assembly of layer-multiplying elements (LMEs) with an extruder. By applying nine LMEs, multilayer-assembled shish-kebab morphology could be observed which was ascribed to iterative stretching and laminating effects occurring in the layer-multiplying process. When the number of LMEs was increased from zero to nine, substantial increments of 60% and 200% in tensile strength and elongation at break were achieved and a transition from brittle to ductile damage was observed. Besides, accompanied with the evolution of crystalline morphology, the surficial roughness became smaller when experiencing a scratch process. The critical normal load, a significant parameter for evaluating the scratch-resisting ability, was distinctly enlarged from 35.4 to 65.2N indicating that the scratch resistance of PLA got enhanced. High orientation of molecules was considered to be capable of resisting the mechanical deformation, while the interfacial shearing occurring between layers contributed to restraining crack propagation. Thus, it is believed that the shell-like multilayer structure developed in the layer-assembly extrusion shows a great potential in fabrication of superior PLA products with high mechanical strength and toughness.

Keywords: **polylactide (PLA), mechanical property, multilayer structure, shish-kebab**

1. Introduction

Due to the serious environmental pollution caused by petrochemical-derived polymers, there is an urgent need for the development of biodegradable polymers with unusual mechanical properties to meet the growing demand of the sustainable society. As the front runner of such biodegradable polymers, Polylactide (PLA) shows a great potential as an environmentally friendly alternatives in tissue engineering, biomedical, and packaging applications. However, PLA still fails to offer adequate toughness and ductility, which has become the paramount bottleneck of usage in structural materials and biomedical fields for PLA. Thus there is a clear task requiring the optimization of mechanical property of PLA. The past decades have witnessed extensive developments in modifying PLA to offer adequate ductility, the traditional approaches toward modifying PLA are normally characterized by formulating and associating with flexible biopolymers and plasticizers. However, it appears from the existing literature that performances of these common blend and composite systems are out of balance. This can be exemplified by the largely sacrificed strength and stiffness after plasticization process.

As a relatively new but rapidly rising approach, bionics signifies a promising strategy to build hierarchical constructions, by virtue of mimicking biologically derived materials which carry a special set of properties. Biomimetism offers, in principle, the ability to make materials with generated structures having true three-dimensional order and multiple contributions to the bulk properties. The most famous example is probably the nacreous part of shells, which has been long perceived as a rigid biological model for structural composites thanks to its renowned combinations of strength, stiffness and toughness.

In this work, we attempt to develop shell-like PLA by virtue of layer-assembly extrusion process, which combined an assembly of layer-multiplying elements (LMEs) with an extruder. Due to the intensive shear flow and iterative laminating effects, formation pf highly oriented and alternatively packed shish-kebab like morphology was observed through SEM, the influence of morphology evolution on mechanical property as well as scratch resistance was researched, and the mechanism behind the phenomenon was further discussed.

2. Materials and methods

2.1 Materials

Commercially available polylactide (PLA, trade name 4032D, 1.2−1.6% D-lactide content) was purchased from NatureWorks LLC, U.S.A. The weight-average molecular weight (Mw) and polydispersity index are 2.11×10^5 g/mol and 1.74, respectively.

2.2 preparation of shell-like PLA

Prior to melt processing, PLA was dried under vacuum for 12 h at 60 ℃. Subsequently, PLA was simultaneously extruded from different extruders and combined as two-layer melt in the coextrusion block, then flowed through an assembly of LMEs. In our experiment, samples prepared through n LMEs were named as PLA-n.

3. Results and Discussion

3.1 Morphology analysis

The construction of shell-like PLA was confirmed via SEM. As illustrated in Fig. 1, unlike the isotropic spherulites for PLA-0, by applying 9 LMEs, multilayer-assembled shish-kebab morphology could be observed along the flow direction, which was ascribed to iterative stretching and laminating effects occurring in the layer-multiplying process. These highly oriented shishs was grouped into an ordered high-strength phase like the well-organized multiscale ceramic phase existing in the shell, while kebabs decorating the shish are considered as a tenacious interphase like the organic components closely connecting the ceramic phase.

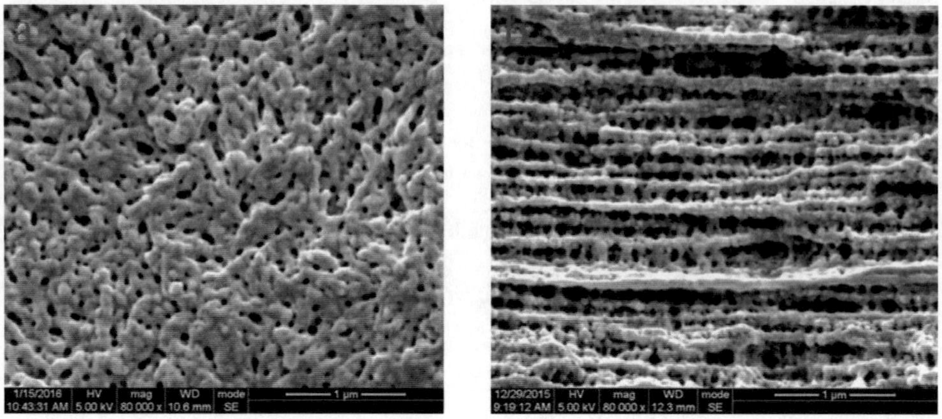

Fig.1 The SEM images of (a) PLA-0; (b) PLA-9

3.2 Mechanical property

Mechanical test indicates the simultaneous enhancement of strength and toughness following the formation of shell-like structural PLA. Unllike the brittle fracture for PLA1, the appearance of stress whitening and shearing band clearly indicates an ductile fracture behavior for PLA5 and PLA9. By increasing the LMEs number from one to nine, substantial increments of 60% and 200% in tensile strength and elongation at break were achieved and a transition from brittle to ductile damage was observed. Thus, it is believed that our layer assembly extrusion provides a successful example of preparing semi-crystalline polymer with super toughness by tailoring crystal superstructure without sacrificing strength.

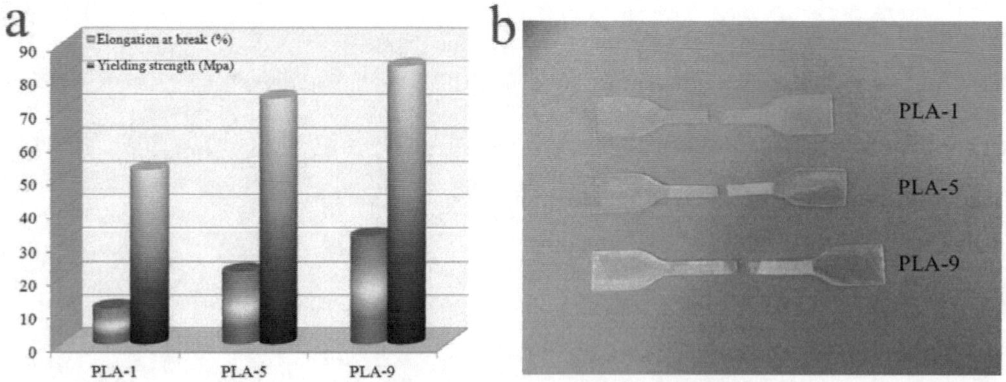

Fig 2. Mechanical performance of structured PLA: (a) detailed mechanical results regarding yielding strength and elongation at break; (b) digital images of tested tensile bars

3.2 Scratch resistance

The scratch experiments were carried following the ASTM D7027-05 standard by employing a linearly increasing normal load. With the evolution of crystalline morphology from isotropic spherulites to highly oriented shish-kebab like crystals, the surficial roughness became smaller when experiencing a scratch process. The critical normal load, a significant parameter for evaluating the scratch-resisting ability, was distinctly enlarged from 35.4 to 65.2N indicating the greatly enhanced scratch resistance of PLA. Such great enhancement is mainly ascribed to the formation of shell-like bionic structure, which is prone to absorb much energy and restrain crack propagation. That is to say, growing cracks are arrested in the fibrillar shish-kebabs and catastrophic failure is postponed in the tenacious interface. Thus, it is believed that the shell-like multilayer structure developed in the layer-assembly extrusion shows a great potential in fabrication of superior PLA products with high mechanical strength and toughness.

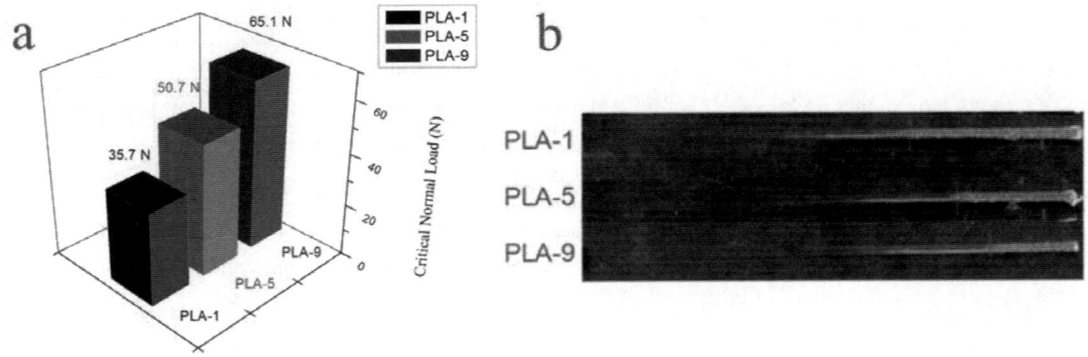

Fig 3. Scratch results of PLA sample: (a) detailed critical normal load for material damage; (b) scanned images of scratch grooves

Transformation of Homocrystallite/Stereocomplex Crystallite of PLLA/PDLA Blended Filament during the Heating Process

Huan-Chieh Ko, Tien-Wei Shyr*

Department of Fiber and Composite Materials, Feng Chia University, Taichung 40724, Taiwan

twshyr@fcu.edu.tw

Abstract. The transformation from homocrystallite to stereocomplex crystallite of PLLA/PDLA (LD) filament with the weight ratio in 50:50 was studied using wide angle X-ray diffraction (WAXD) and differential scanning calorimeter (DSC). DSC results showed that the endotherm of both of homocrystallite and stereocomplex crystallite can be observed in the LD filament. Real-time WAXD was used to observe the transformation from homocrystallite to stereocomplex crystallite of the LD filament during the heating process. The results showed that the homocrystallite was recrystallized at the temperature after T_g first and the stereocomplex crystallite formed at heating process about from 105 to 181 °C. It indicates that the molecular chains were rearranged to form stereocomplex crystallite during the melting of the homocrystallite.

Keywords: Polylactide filament, homocrystallite, stereocomplex crystallite, crystalline transformation
PACS: 65.40.Ba

INTRODUCTION

Polylactide (PLA) is aliphatic polyester with excellent biocompatibility and biodegradability. The decomposition products of PLA were carbon dioxide and water, which will not cause any environmental pollution [1-4]. The monomer, Lactic acid exists in two enantiomeric forms, L -lactic acid and D -lactic acid, results in a wide range of properties from completely amorphous to semicrystalline PLA, depending on the ratio and distribution of enantiomeric units along polymer chains [5].

Tsuji et al. [6] investigated the crystallization behavior of poly L-lactic/poly D-lactic (PLLA/PDLA) blends with different ratios. The stereocomplex crystallite was observed using differential scanning calorimeter and wide angle X-ray diffraction. The melting temperature of stereocomplex crystallite is higher than that of PLLA homocrystallite. The presence of stereocomplex crystallite increases the total crystallization rate of PLLA/PDLA blend. Fujita et al. [7] reported the homocrystallite of PLLA/PDLA blend exhibits the transition to stereocomplex crystallite during lamellar thickening process. This phenomenon has not been fully understood yet. In this study, the transformation from homocrystallite to stereocomplex crystallite of PLLA/PDLA filament with the weight ratio in 50:50 was studied using WAXD and DSC.

EXPERIMENTAL

PLLA (Synterra® PLLA 1510, Mw =180000, R.V. =2) and PDLA (Synterra® PDLA 1010, Mw =110000, R.V. =1.69) were supplied in chip forms by Synbra Technology BV. Melting, crystallization, and glass transition temperatures of PLLA and PDLA were 171, 112, 57 °C, and 172, 108, 60 °C, respectively, measured by DSC. The LD filament with the weight ratio in 50:50 was prepared by a two-step melt spinning process. PLLA and PDLA were melt-extruded to form chips (Step 1), and then heated above the melting temperature to produce filaments (Step 2).

The crystalline structure analysis was performed using an X-ray instrument (Bruker D8 SSS) with a Cu-Kα radiation (40 kV and 100 mA). The scanning angle (2θ) covered a range between 5° and 30° at a rate of 2° min^{-1}. The crystallization and melting behaviors were observed using a Du-Pont DSC Q10 instrument with a TA2000 thermal analysis system. The weight of the sample was in the range of 4–6 mg. The flow rate of the purge gas N_2 was approximately 50 cm^3 min^{-1}. The heating rate was 10 °C min^{-1}. The crystal evolution measurement was carried out at the Beamline BL17A1 (wavelength, λ=1.333 Å) of Nation Synchrotron Radiation Research Center (NSRRC), Hsin-Chu, Taiwan. The diffraction patterns were recorded using a Mar345 imaging plate area detector (Fuji Bas 2500 IP, Bicrondetector). The sample was set in the sample holder and heated from room temperature to around its melting points at a heating rate of 10 °C min^{-1}.

RESULTS AND DISCUSSION

Figure 1 showed WAXD curve of the LD filament. It showed only a broad curve for the amorphous phase. This could mean that there was no homocrystallite and stereocomplex crystallite in the LD filament. Figure 2 exhibited DSC heating curve of LD (first heating). Two endothermic peaks were found in the LD filament. The lower melting endothermic peak located in the range from 153.8 to 181.8 °C was contributed by the melting homocrystallite and the higher melting endothermic peak located in the range from 191.1 to 254.0 °C was attributed to the melting stereocomplex crystallite. The obvious recrystallization was found in the heating curve of LD filament. According to WAXD, it is supposed that there was no enough time to form crystallite during spinning process and the molecular chains obtained enough energy to fold into lattice in a crystallite in the subsequent heating process. Interestingly, WAXD analysis showed that no crystallite in the LD filament. DSC result showed that the endothermic peak at heating processes of both homocrystallite and stereocomplex crystallite can be observed, but only one obvious exothermic peak occurred during the heating process.

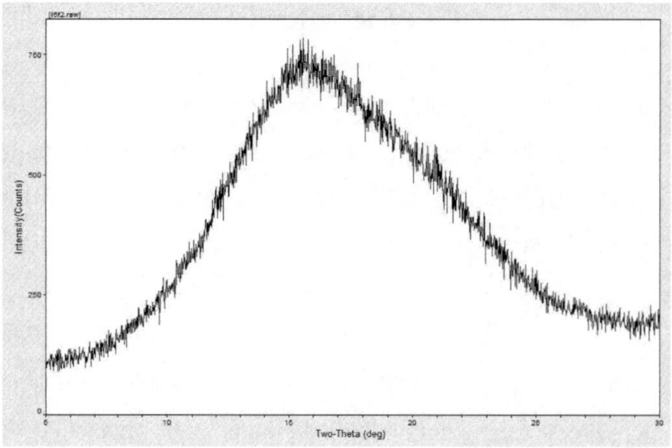

FIGURE 1. WAXD Profiles of the LD filament.

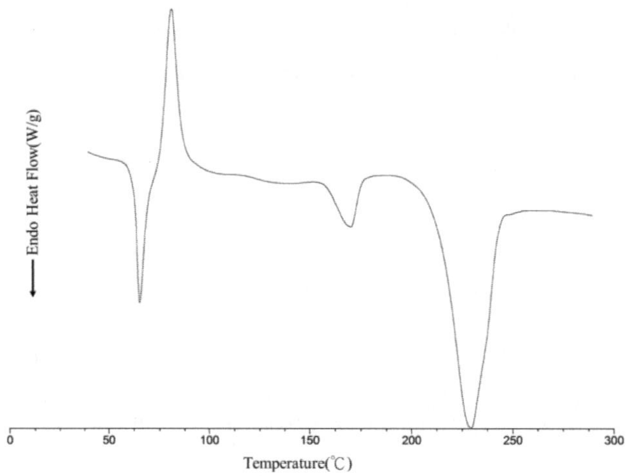

FIGURE 2. DSC Profiles of the LD filament (first heating).

The transformation from homocrystallite to stereocomplex crystallite of the LD filament during the heating process was further analyzed using a real-time WAXD (Figure 3). The WAXD intensity curve at room temperature (RT) did not find any diffraction peak, indicating that the sample was in the amorphous state. At temperature from 70 to 105 °C, (110/200) and (203) reflections related to the homocrystallite appears and the diffraction intensity of the peaks increases, suggesting that the content of homocrystallite increased. Meanwhile, a new diffraction peak related to (110) reflection of stereocomplex crystallite emerged at around 105 °C. As temperature further increased, diffraction peak assigned to the (300/030) and (220) reflections related to the stereocomplex crystallite appeared and the diffraction intensity of the peaks related to the stereocomplex crystallite increased. In the meantime, the diffraction intensity related to the homocrystallite decreased. When the temperature reached 181 °C, the diffraction peaks of homocrystallite disappeared. It means that the molecular chains were rearranged to form stereocomplex crystallite during the melting of the homocrystallite.

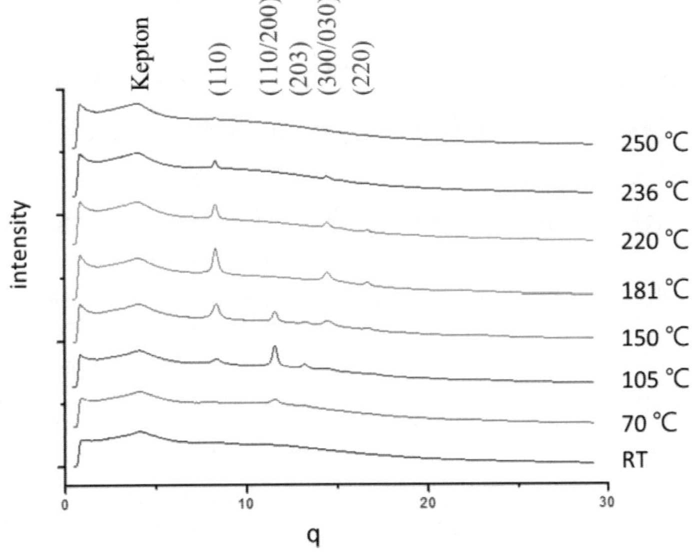

FIGURE 3. Real-time WAXD pattern of the LD filament during the heating process.

CONCLUSION

The WAXD analysis showed that there were no homocrystallite and stereocomplex crystallite in the LD filament. DSC results showed that the endotherm of both of homocrystallite and stereocomplex crystallite can be observed in the LD filament. The Real-time WAXD results showed that the homocrystallite was recrystallized at heating process after the temperature of T_g first and the stereocomplex crystallite formed about from 105 to 181 °C. It means that the molecular chains were rearranged to form stereocomplex crystallite during the melting of the homocrystallite.

ACKNOWLEDGMENTS

This work was supported by the Ministry of Science and Technology, Taiwan, grant number MOST 103-2221-E-035-96-MY3,and the National Synchrotron Radiation Research Center, Taiwan.

REFERENCES

1. Y.-C. Chien, C. Liang, S.-H. Yang, *Atmos. Environ.***45**, 123-127 (2011).
2. N. Ljungberg, B. Wesslén, *Biomacromolecules* **6**, 1789-1796 (2005).
3. B. Li, F.-X. Dong, X.-L. Wang, J. Yang, D.-Y. Wang, Y.-Z. Wang, *Eur. Polym. J.* **45**, 2996-3003 (2009).
4. M. Murariu, A. D. S. Ferreira, M. Alexandre, P. Dubois, *Polym. Adv. Technol.* **19**, 636-646 (2008).
5. S. Saeidlou, M. A. Huneault, H. Li, P Sammut, C. B. Park, *Polymer* **53**, 5816-5824 (2012).
6. H. Tsuji, H. Takai, S. K. Saha, *Polymer* **47**, 3826-3837 (2006).
7. M. Fujita, T. Sawayanagi, H. Abe, T. Tanaka, T. Iwata, K. Ito, T. Fujisawa, M. Maeda, *Macromolecules* **41**, 2852-2858 (2008).

Effects of Reactive Polymer as Modifier on Impact Strength and Hydrolytic Stability of PC/ABS Blend

Kohhei Nishino[a], Yuichi Shindo[a], Tetsuo Takayama[b], and Hiroshi Ito[b]

[a]*Denka Company Limited, 6 Goi-minamikaigan, Ichihara, Chiba, Japan*
[b]*Graduate School of Science and Engineering, Yamagata University, 4-3-16 Jonan, Yonezawa, Yamagata, Japan*

Abstract. PC/ABS blends have been widely used for specific applications such as in automotive interior & exterior parts and office automation equipment parts. In this study, the effects of a reactive polymer as a modifier on the properties of PC/ABS blends were investigated. The reaction between PC and maleic anhydride group cannot be expected usually because the end hydroxyl group of PC is capped with end-capping agents such as t-butyl phenol to improve the properties of PC. However, we have found that a reactive polymer which has maleic anhydride group reacted with the end hydroxyl group of PC which was hydrolyzed with metal salts. From the results, the impact strength of PC/ABS was improved with the reactive polymer.

Keywords: PC/ABS, Reactive compatibilizer, Polymer blend, Polymer alloy.
PACS: 81.05.Lg

EXPERIMENTAL

Commercial grade of Polycarbonate (PC) resin with a melt flow rate (MFR) of 10 g/10 min at 300 °C and 11.8 N was used. Two ABS resins by emulsion polymerization were used, which were manufactured by Denka Co. Ltd. with same conditions except for coagulation method. Styrene-Acrylonitrile (SAN) copolymer was used with an acrylonitrile content of 24 wt% and a weight-averaged molecular weight (Mw) of 107,000 g/mol. RP-1, that is our newly developed copolymer with styrene, methyl methacrylate, and maleic anhydride, was used as a reactive copolymer for PC/ABS. RP-1 had a maleic anhydride content of 18 wt% and a Mw of 161,000 g/mol. PC/ABS blends in Table 1 were extruded using a twin screw extruder with 35mm screw diameter with a cylinder temperature of 260 °C, a screw rotation speed of 200 rpm, and a discharge rate of 30 kg/h.

RESULTS AND DISCUSSION

Table 1 presents the composition and the properties of PC/ABS, comparing the existence of magnesium salts in ABS and the addition of RP-1 as a reactive copolymer. Deterioration of PC caused by metal salts is well known. Charpy impact strength of Sample-B which included Mg salts derived from ABS-2 was lower than that of Sample-A. Additionally, the increase in MFR was observed. This result suggests that PC was hydrolyzed with Mg salts in ABS-2, and the low molecular weight component of PC was generated consequently.

On the other hand, the addition of RP-1 to PC/ABS with Mg salts inhibited the hydrolysis of PC because the increase in MFR was not observed. Furthermore, Sample-C indicated higher impact strength than Sample-A. These results suggest that maleic anhydride group in RP-1 reacted with the end hydroxyl group of PC which was hydrolyzed with Mg salts derived from ABS-2, and the compatibility between PC and ABS could be improved.

TABLE 1. The properties of PC/ABS compatibilized with RP-1

Sample Name	A	B	C	D
PC, wt%	50	50	50	50
ABS-1, wt% without Mg salts	15	-	-	15
ABS-2, wt% with Mg salts	-	15	15	15
SAN, wt%	35	35	25	25
RP-1, wt%	-	-	10	10
MFR, g/10	12.7	28.3	6.2	8.0
Charpy Impact Strength kJ/m^2	83	42	99	92
Vicat Softening Temperature 50N, °C	114	114	121	119

Synthesis of poly(ethylene terephthalate-co-naphthalene) to Achievement the Hot Fill Properties

Abbasi Atiyeh[a], Sheikh Nejad Moghadam Abdollah[b], Amir Aslani Saba Reza[a], Keshavarzian Ali[c*]

[a] *National Petrochemical Company-Research & Technology Co. (NPC-RT), Pazhoohesh Blvd., 15th Km Karaj Highway, Tehran, Iran.*
[b] *Shahid Toundgoyan Petrochemical Complex (STPC), Mahshar, Iran.*
[c] *Iran polymer & Petrochemical Institute (IPPI), Pazhoohesh Blvd., 15th Km Karaj Highway, Tehran, Iran.*

Atiyeh_abbasi@yahoo.com
Ali_keshavarz@aut.ac.ir

Abstract. Polyethylene terephthalate (PET) is the one of the most important thermoplastic materials with wide field of applications. The production of PET items such as fibers, films and bottles for food and household packaging involves many processing steps. These steps result in changes of morphological and structural characteristics. Increase or decrease in the crystallinity caused changes in the physical properties of the end product. Shrinkage, dimension stability, stiffness, dye-ability, mechanical behavior and gas barrier properties of the final products are related to the degree of crystallinity and the glass transition temperature of PET. Chemical modification via copolymerization is one of the most attractive techniques in order to differentiate the properties of PET. Hot filling procedure used for the production of shelf-stable juice involve pouring the heat-treated juice, without significant cooling, directly into the package. Usually, the temperature for hot filling of juice in PET is 80 to 88°C. In this research, synthesis of poly(ethylene terephthalate-co-naphthalene) (PETN) copolymer was studied in presence of different amounts of 2,6- naphthalene dicarboxylic acid (NDA) to Achievement the Hot Fill Properties. The microstructure of the synthesized polymer was studied with Fourier Transform Infrared (FTIR) and Nuclear Magnetic Resonance (NMR). Differential scanning calorimetry (DSC) was used to study the thermal properties. The DSC results have shown that the modification of PET with low levels of naphthalene comonomer (3-8 mol%) increases the Tg. This property has a significant effect on the thermal stability potential of many hot applications to resist higher temperatures without shrinkage. Based on the FTIR and NMR results, the synthesized copolymer was Poly(ethylene terephthalate-co-2,6- naphthalene dicarboxylic acid). Results showed that PETN was suitable for use in the manufacture of hot fill containers.

Keywords: Polyethylene terephthalate, Crystallinity, Characterization, Thermal Stability, Hot Fill properties.
PACS: 83.80.Tc

INTRODUCTION

Poly(ethylene terephthalate) (PET) has become a major synthetic polymer during the past forty years. Significant commercial markets have been developed for its application in textile and industrial fibers, films, and foamed articles, containers for carbonated beverages, water and other liquids, and thermoformed applications [1]. The modification of PET with low levels of naphthalate comonomer increases the Tg and enables optimally oriented articles (films, fibers, containers, etc.) to resist higher temperatures without shrinkage. Heat setting under tension may be applied to further extend thermal stability. In addition, when retention of optical transparency is required, such copolymers crystallize less readily than PET, and may readily be quenched from the melt to the transparent, amorphous state. Thus, thicker, transparent sections are possible. Optimal strength and gaseous-barrier properties on orientation, however, require higher stretch ratios and/or higher molecular weights, for strengths equivalent to PET. The beneficial absorption of UV light is still present in copolyesters of PET modified with low levels (3–7 mol%) of naphthalate. This property may be utilized to protect foodstuffs in naphthalate-modified PET containers and films [2].

PET/PEN blends offer advantages over PET in their mechanical, thermal and barrier properties. These blends are immiscible, but may undergo interchange reactions, which create a copolymer that acts as a compatibilizer. A certain level of copolymerization is necessary to prevent phase separation in these blends. As the interaction of light with phase boundaries retards optical clarity, miscibility is important to achieve desirable properties in PET/PEN blends. Three types of interchange reactions can occur in polyesters having chains terminated by hydroxyl or carboxyl groups. These reactions include alcoholysis, acidolysis, and ester exchange that are referred to as transesterification in this study. Rate determining parameters are temperature, time, catalyst, and mixing processes. At the start of transesterification, larger blocks are formed as in block copolymers. At higher extents of transesterification, sequence lengths of PET and PEN decrease, leading to random copolymers [3].

PEN has a glass transition temperature of 122 °C, which is about 40 °C higher than that of PET, and a melting temperature of 267 °C against 254 °C for PET. Poly(ethylene naphthalate) (PEN) also exhibits an increased tensile strength, as well as elasticity modulus and elongation at break. Its gas barrier properties in oxygen and CO2 are about 70–80% increase compared to PET. Finally, PEN exhibits better UV resistance. However, there are some drawbacks for the use of PEN, which limit its use in bottling. High melt viscosity of PEN results in difficulties in extrusion. Furthermore, increased solid-state polymerization times are needed for high molecular weight PEN. So, at moment, the main problem is its higher price. PEN is about three times more expensive than PET, due to the high cost of 2,6- naphthalene dicarboxylic acid. Reactive blending, as well as thermal and mechanical properties of PET/PEN blends, was studied last years. Only recently some works on PETN copolymers were published [4].

In this research, synthesis of poly(ethylene terephthalate-co-naphthalene) (PETN) copolymer was studied in presence of different amounts of 2,6- naphthalene dicarboxylic acid (NDA) to Achievement the Hot Fill Properties. The microstructure of the synthesized polymer was studied with Fourier Transform Infrared (FTIR) and Nuclear Magnetic Resonance (NMR). Differential scanning calorimetry (DSC) was used to study the thermal properties.

EXPERIMENTAL

Terephthalic acid (TPA) and ethylene glycol (EG) were supplied by Shahid Toundgoyan Petrochemical Complex (STPC, Mahshar, Iran). Antimony oxide and dichloro acetic acid (DCA as solvent in intrinsic viscosity measurement) were purchased from Merck Co. Also NDA was purchased from Aldrich Co. All materials were used as they were supplied. A homemade laboratory scale set up was used for PET synthesis. This set up includes three distinct sections: reaction unit, cooling section, and instrumentation and digital control system. An internally written Visual Basic code was used for all data acquisition and control tasks. Fourier Transform Infrared Spectroscopy (FTIR), NEXUS 670 made by Nikolet and Hydrogen Nuclear Magnetic Resonance (HNMR), Brucker Avance 400, were used to study microstructure of polymer. The thermal behaviour of the polyesters was examined by DSC with a Mettler–Toledo (Columbus, OH) 822e calibrated with indium.

For the synthesis of copolymer: In the beginning, required amount of TPA, EG, NDA and catalyst were mixed in the stainless reactor for 45 min at 90-95°C to have a uniform paste with saturated EG with TPA.

PET and PET-co-PEN were synthesized in two steps reaction. In the first Step, reaction was conducted at 235-260°C, 4.5 atm and 60 rpm of mixing rate. These conditions were determined based on a series of experiments to reach the highest value of intrinsic viscosity (IV). Water, as esterification by product, was gathered and measured as progress of reaction. This step was conducted until water was not produced, normally between 3 and 3.5 hours. Subsequently, mixture was stirred at 260°C under atmospheric pressure for one hour to complete esterification and remove of water residue. In the second step, vacuum was gradually applied over 15 min in order to reach 1 mbar pressure. At the same time, temperature was increased to 280-285°C. EG was removed in this step as polycondensation by-product. This step longs 3-3.5 hr. With starting of esterification, Produced water was gathered and weighted every 15 min.

RESULTS AND DISCUSSION

Table 1 shows comparison of performance of various samples under same reaction conditions. Intrinsic viscosity was measured using DCA in a 1C Ubbelohde viscometer at 25±0.1°C.

TABLE(1). IV and other results of PET and PETN samples

Component	Sample 1	Sample 2	Sample 3	Sample 4
NDA mole%	0	3	5	8
IV(dL/gr)	0.64	0.48	0.44	0.43
DEG(Wt%)	1.48	1.23	1.29	1.31
Tg (° C)	75.23	80.69	82.15	78.22

Moreover, with increase in amount of NDA percent, we can show decrease in IV of samples. Figure 1 shows a typical FTIR spectrum of synthesized PETN.

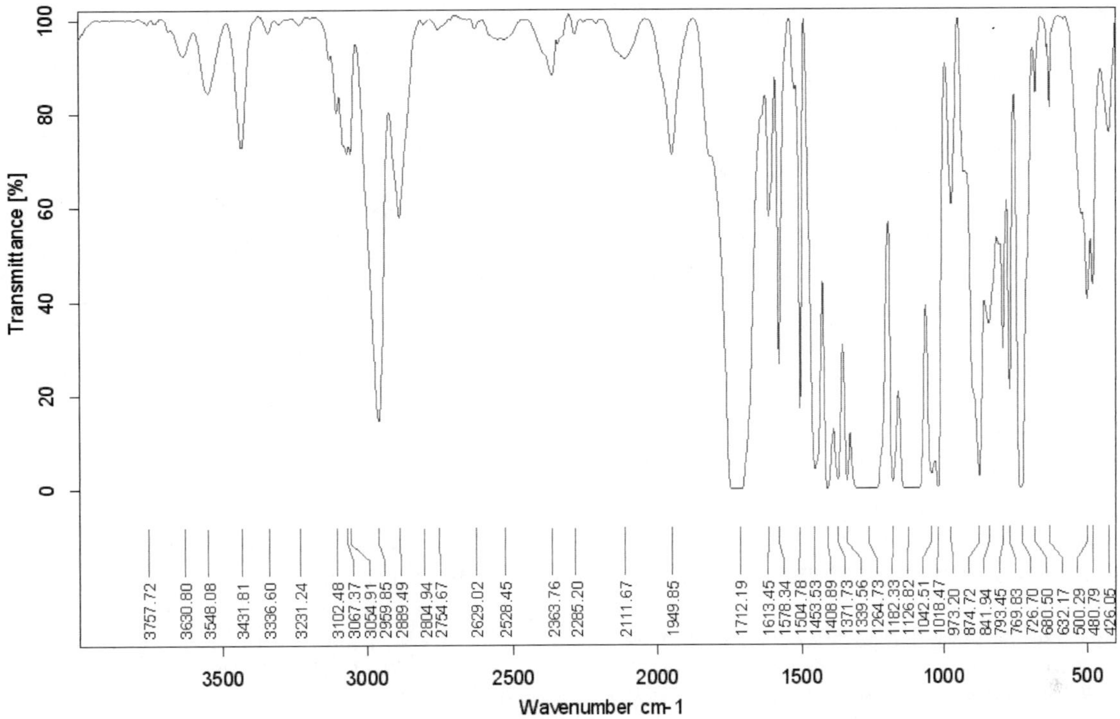

FIGURE 1. FTIR spectrum of the PETN (8 mol% NDA) product by homemade setup

The peak around 1020-1280 cm^{-1} corresponds to C-O bond in ester group. The peak at 1712 cm^{-1} represent the C=O in ester group. Also the peak at 1578 cm^{-1} was assigned to terephthalate group. There are a lot of peaks, while the most characteristic are those in the region of 1720, 1600–1500, 1120, 1613 cm^{-1} for C=O stretching vibration of the ester carbonyl group, stretching vibration of backbone of benzene, C-O stretching vibration and C=C of the backbone of naphthalene, respectively. Hence, based on the FTIR result, the synthesized polymer is PETN.

Figure 2 shows ^{1}HNMR spectrum of PETN synthesized with 5 mol% of NDA.
According to the ^{1}HNMR spectrums, the contents of NDA comonomer in the produces PETN were about 3-8 mol%.

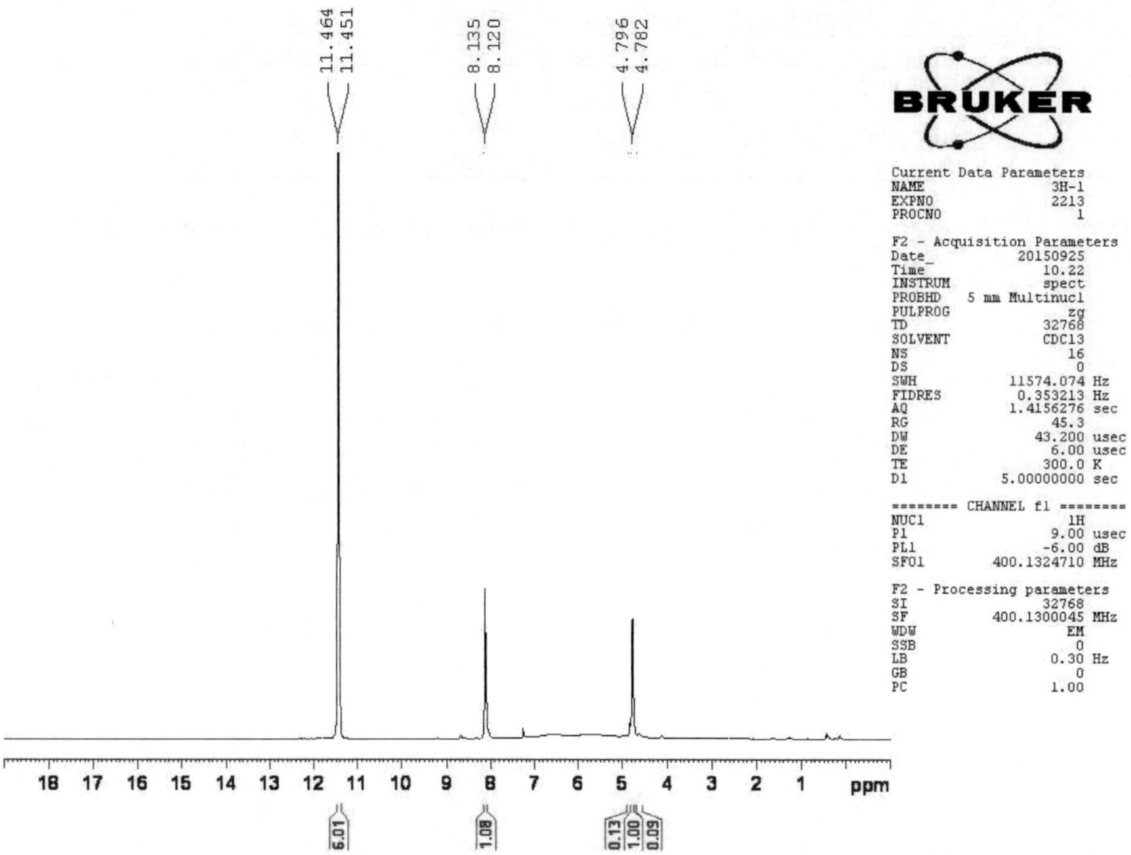

FIGURE 2. ^{1}HNMR spectrum of the PETN (5 mol% NDA)

The thermal behaviour of the copolyesters was studied with DSC. Figure 3 shows the heating and cooling curves for the determination of the crystallization temperature (Tc), Tg, Tm, and heat of fusion.

Table 2 shows the results obtained from the DSC traces for all of samples.

TABLE(2). Thermal behaviour of the samples based on DSC

parameter	Tg(°C)	Tm(°C)	Tc(°C)	Crystallinity(%)
Sample 1	75.23	250.88	181.67	22.82
Sample 2	80.69	251.18	202.95	32.64
Sample 3	82.15	247.50	202.89	29.92
Sample 4	78.22	232.55	202.65	29.32

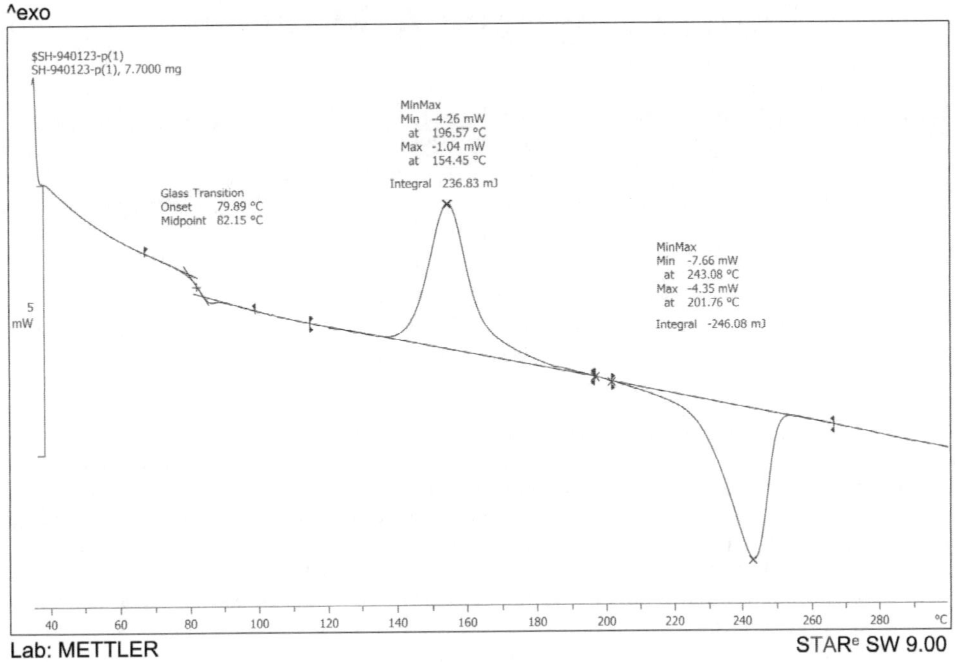

FIGURE 3. DSC trace of PETN product by 5 mol% of NDA

Results have shown that the modification of PET with low levels of naphthalate comonomer, increases the Tg, Tc and crystallinity percent. This property has a significant effect on the thermal stability potential of many hot applications to resist higher temperatures without shrinkage.

CONCLUSION

In this research, synthesis of PETN copolymer was studied in presence of different amounts of 2,6- naphthalene dicarboxylic acid to Achievement the Hot Fill Properties. The DSC results have shown that the modification of PET with low levels of naphthalene comonomer (3-8 mol%) increases the Tg. Based on the FTIR and NMR results, the synthesized copolymer was Poly(ethylene terephthalate-co-2,6- naphthalene dicarboxylic acid). Results showed that PETN was suitable for use in the manufacture of hot fill containers.

REFERENCES

1. S. Sepehri, M. Rafizadeh and F. Afshar-Taromi, *Journal of Applied Polymer Science* **113**, 3520-3532 (2009).
2. K. Pang, R. Kotek and A. Tonelli, *Prog Polym Sci* **31**, 1009-1037 (2006).
3. K. Gunes, A. Isayey, X. Li and C. Wesdemiotis, *Polymer* **51**, 1071-1081 (2010).
4. G. Karayannidis, N. Papachristos, D. Bikiaris and G. Papageorgiou, *Polymer* **44**, 7801-7808 (2003).

Morphology-Rheology Relationship in Thermoplastic Starch/ Polylactic Acid Blends

Mohsen Esmaeili[a], Gholamreza Pircheraghi[a*] and Reza Bagheri[a]

[a] *Polymeric Materials Research Group, Department of Materials Science and Engineering, Sharif University of Technology - Tehran - Iran*

mohsen_esmaeili@mehr.sharif.ir
pircheraghi@sharif.edu
rezabagh@sharif.edu

Abstract. In this work, we have prepared thermoplastic starch (TPS) / polylactic acid (PLA) blends, and investigated the rheological properties as well as morphology of the blends. Thermoplastic starch, prepared using a combination of sorbitol and glycerol plasticizers, was melt blended with PLA in the concentration range of 20 to 80 wt.% in a laboratory twin screw extruder. Compatibilized blends were also obtained with substitution of 20 wt.% of PLA by maleic anhydride grafted PLA (PLA-g-MA). Scanning electron microscopy (SEM) images revealed co-continuous morphology in blends containing 40 and 60 wt.% PLA, while in blends with 20 and 80 wt.% of PLA, matrix-dispersed morphology was observed. Incorporation of compatibilizer resulted in formation of finer morphology in the entire blends. Moreover, in the presence of the compatibilizer, morphology of the blend containing 60 wt.% turned to matrix-dispersed. It was observed that the rheological properties of compatibilized blends were lower than those of their non-compatibilized counterparts. This observation was attributed to the chain scission of PLA during maleic anhydride grafting reaction as evidenced by lower complex viscosity of PLA-g-MA compared to the neat PLA. In addition, it was found that in the samples with co-continuous morphology, rheological properties are strongly dominated by the TPS phase. Moreover, in the blend samples containing 80 wt.% PLA, the rheological properties are negatively deviated from mixing rule; probably due to migration of glycerol and sorbitol plasticizers into the PLA matrix.

Keywords: Thermoplastic Starch, Polylactic Acid, Morphology, Rheology.
PACS: 83.80.Tc

SAMPLE PREPARATION

The mixture of 58 wt.% corn starch, 28 wt.% sorbitol and 14 wt.% glycerol were extruded for TPS granules preparation by a laboratory co-rotating twin extruder, SHJ-20 Nanjing Giant, with L/D ratio of 20 and temperature profile of 120°C-130°C-130°C-135°C from hoper to die. In order to achieve enough time and shear rate for mixing starch and plasticizers, the screw speed was set on 100 rpm. The prepared TPS granules were melt blended with proper amount of PLA for preparation TPS/PLA blends. For preparation of compatibilized blend samples, 20 wt.% of PLA in TPS/PLA blends was replaced by PLA-g-MA which previously prepared by melt grafting of maleic anhydride to PLA chains in the presence of dicumyl proxide (DCP) initiator. Blend samples were prepared in the same extruder with temperature profile of 135°C-170°C-180°C-140°C from hoper to die and screw speed of 50 rpm. The composition of samples is presented in Table (1).

TABLE (1). Material's abbreviations and corresponding sample compositions

Sample's abbreviation	PLA content (wt.%)	PLA-g-MA content (wt.%)	TPS content (wt.%)
TPS	-	-	100
20PLA	20	-	80
40PLA	40	-	60
60PLA	60	-	40
80PLA	80	-	20
100PLA	100	-	-
20PLA-g	16	4	80
40PLA-g	32	8	60
60PLA-g	48	12	40
80PLA-g	64	16	20
100PLA-g	80	20	-

MORPHOLOGICAL OBSERVATIONS

The blend morphology was observed by scanning electron microscopy (SEM). Blend samples were quenched in liquid nitrogen and fractured for obtaining smooth surface which subsequently etched with hydrochloric acid (HCl, 6N) for 3 hours to dissolve TPS phase selectively. While, 20PLA and 20PLA-g samples were not etched by HCl and TPS phase was distinguished by swelling under the electron beams.. The SEM micrographs of 20PLA and 20PLA-g samples are shown in Figure 1. For the non-compatibilized blends, 20PLA sample, the morphology is very coarse as reported by Martin and Averous [1,2], while compatibilization caused smaller PLA domains dispersed in TPS matrix. On the other hand, the TPS phase is dispersed in PLA matrix in the 80PLA and 80PLA-g samples and compatibilization plays the same role, i.e. in the 80PLA-g sample the fine morphology consisting of small domain of the TPS phase was observed. As it was expected, for the samples containing 40 and 60 wt.% PLA, the co-continuous morphology was observed. Interestingly, in the compatibilized sample containing 60 wt.% PLA, the co-continuous morphology turned into the matrix-dispersed morphology. This behavior can be attributed to the better interaction of grafted maleic anhydride group in PLA-g-MA chains with the neighboring TPS chains.

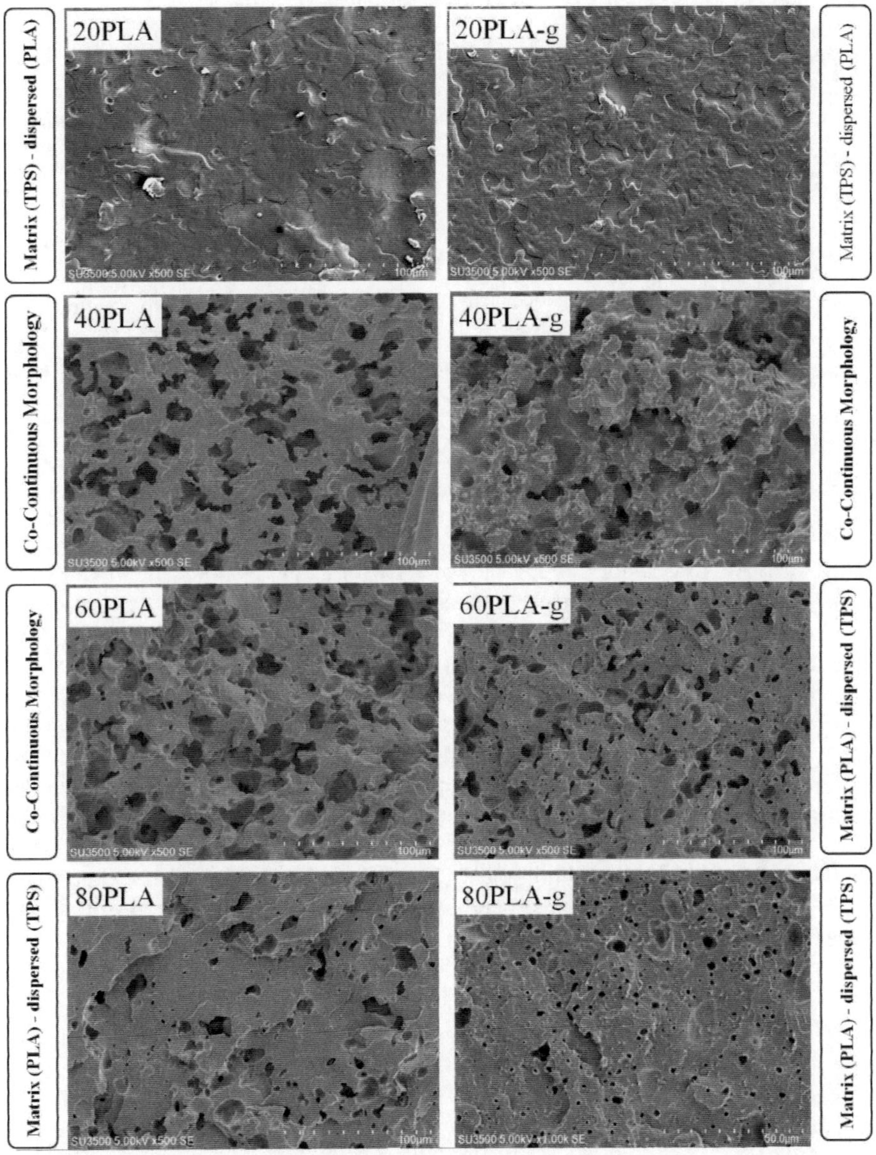

FIGURE 1. SEM micrographs for compatibilized and non-compatibilized TPS/PLA in 500x magnification (The PLA phase in 20PLA and 20PLA-g has been shown by red arrows).

RHEOLOGICAL PROPERTIES

Rheological properties of TPS/PLA blends were measured using a parallel plate rheometer in oscillatory shear flow at 180°C in a frequency range from 100 to 0.1 Hz. The complex viscosity of various blends is presented in Figure 2. As shown in this figure, the complex viscosity of compatibilized blend samples is decreased mainly because of chain scission of PLA during maleic anhydride grafting reaction as also evidenced by the lower viscosity of 100PLA-g sample compared to 100PLA sample. The similar behavior has been reported by Huneault and Li [3]. In the 20PLA and 20PLA-g samples, complex viscosity is controlled by TPS matrix due to formation of three-dimensional network of rigid starch domains [4, 5]. On the other hand, PLA matrix dominated the viscosity behavior of the 80PLA and 80PLA-g samples. Moreover, the decreasing of complex viscosity in the 80PLA and 80PLA-g samples compared to the neat PLA sample might be attributed to diffusion of plasticizers from TPS phase into PLA matrix leading to lubricating the PLA chains motions [6]. The branched or highly entangled structure of starch chains caused the non-Newtonian behavior in TPS, 20PLA, 20PLA-g, 40PLA and 40PLA-g samples, in which the rheological properties are strongly controlled by TPS, the dominant phase. Comparing the complex viscosity of 60PLA and 60PLA-g samples, one can conclude that the interaction between two phases has been increased due to addition of PLA-g-MA chains working as a compatibilizer through chemical reaction with favored group in starch chains [3].

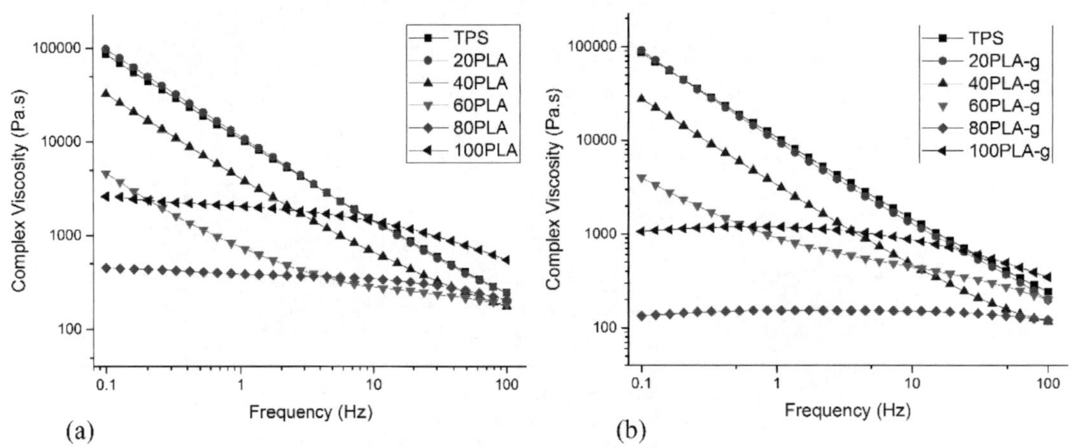

FIGURE 2. Complex viscosity of PLA/TPS blend with different ratios in compatibilized and non-compatibilized states at 180°C.

The storage modulus of compatibilized and non-Compatibilized TPS/PLA blend has been shown in Figure 3. The decreased storage modulus of the 100PLA-g sample compared to the 100PLA sample is mainly due to the simultaneously occurred chain scission reactions during grafting the maleic anhydride to the PLA chains. Moreover, considering the potential of plasticizer migration to the PLA matrix, the storage modulus of the 80PLA and 80PLA-g samples was measured lower than their PLA matrix. On the other hand, the constant storage modulus is observed in the TPS, 20PLA, 20PLA-g, 40PLA and 40PLA-g samples indicating the formation of the 3D elastic network due to dominant role of TPS chains in controlling the rheological properties of these samples [7]. The rheological behavior of different samples is in line with their morphological observation. Further, based on rheological measurements, it was observed that in the blend samples with co-continuous morphology the dominant phase in controlling the viscoelastic properties is the TPS phase.

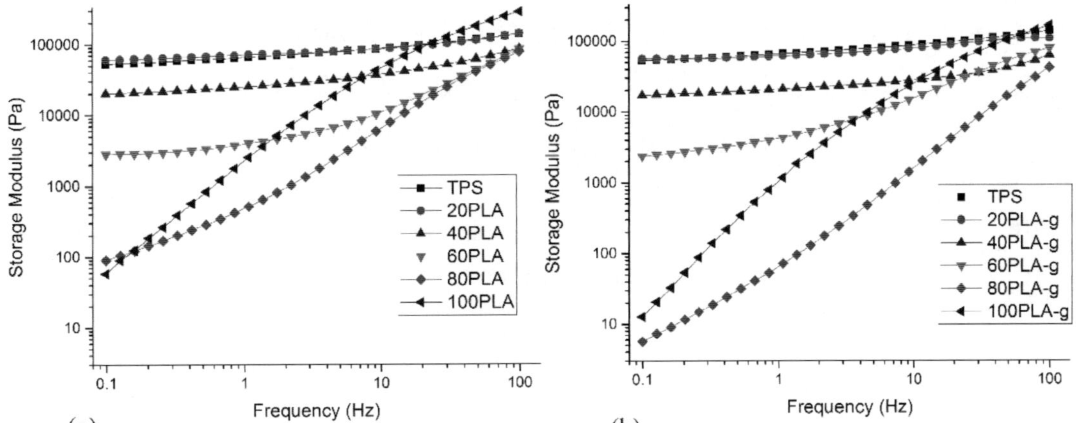

(a) (b)

FIGURE 3. Storage modulus of PLA/TPS blend with different ratios in compatibilized and non-compatibilized states at 180°C.

CONCLUSION

In this study, thermoplastic starch (TPS)/ polylactic acid (PLA) blends were prepared using a laboratory twin screw extruder. The morphology of blend samples was studied by means of scanning electron microscopy. While, the melt rheological properties were measured in linear viscoelastic region using a parallel plate rheometer in oscillatory shear flow field. Based on SEM images, different morphologies including matrix-disperse, co-continuous and phase inversion were observed in blend samples composed of various ratio of TPS/PLA. While, in the compatibilized blend samples, containing PLA-g-MA, the finer morphology with smaller dispersed phase was observed. Also, the co-continuous morphology of the sample containing 60wt% PLA was turned into the matrix-dispersed morphology mainly due the better interaction between phases. The PLA chain scission was confirmed based on the decreasing of the complex viscosity and storage modulus of PLA-g-MA compared to the neat PLA sample. Rheological measurements of blend samples showed the dominant role of TPS phase in linear melt viscoelastic behavior of blend samples with TPS matrix or co-continuous morphologies. In other words, in those samples the storage modulus is independent of frequency due to formation of an elastic 3D network composed of un-plasticized TPS domains. Also, the negative deviation behavior observed in the rheological properties of blend samples with PLA matrix was attributed to the migration of plasticizers from TPS dispersed phase into the PLA matrix.

REFERENCES

1. O. Martin and L. Averous, *Poly (lactic acid): plasticization and properties of biodegradable multiphase systems* Polymer, 2001, pp. 6209-6219.
2. Ning Wang, Jiugao Yu and Xiaofei Ma, *Preparation and characterization of thermoplastic starch/PLA blends by one-step reactive extrusion*, Polymer International, 2007, pp. 1440-1447.
3. Michel A. Huneault, and Hongbo Li, *Morphology and properties of compatibilized polylactide/thermoplastic starch blends*, Polymer, 2007, pp. 270-280.
4. Mirta I Aranguren, Mora Elsi, V. DeGroot Jr. Jon and W. Macosko Christopher, *Effect of reinforcing fillers on the rheology of polymer melts*, Journal of Rheology, 1992, pp. 1165-1182.
5. F. Romani, R. Corrieri, V. Braga and F. Ciardelli, *Monitoring the chemical crosslinking of propylene polymers through rheology*, Polymer, 2002, pp. 1115-1131.
6. Hongbo Li, and Michel A. Huneault, *Comparison of sorbitol and glycerol as plasticizers for thermoplastic starch in TPS/PLA blends*, Journal of Applied Polymer Science, 201, pp. 2439-2448.
7. Zuo Yingfeng, Gu Jiyou, Cao Jun, Wei Shuangying, Tan Haiyan and Zhang Yanhua, *Effect of starch/polylactic acid ratio on the interdependence of two-phase and the properties of composites*, Journal of Wuhan University of Technology-Mater. Sci., 2015, pp. 1108-1114.

Inducing Conductivity in Immiscible PS/PP Blends by a Percolated Polyaniline/PA Filler Selectively Localised by Specific Interactions

Avanish Bharati[a], Prakhyat Hejmady[b], Ruth Cardinaels[a, b], Jin Won Seo[c] and Paula Moldenaers[a*]

[a]*Soft Matter Rheology and Technology, Department of Chemical Engineering, KU Leuven, Celestijnenlaan 200F, box 2424, 3001 Leuven, Belgium*

[b]*Polymer Technology, Department of Mechanical Engineering, TU Eindhoven, P.O. Box 513, 5600 MB Eindhoven, The Netherlands*

[c]*Surface and Interface Engineered Materials, Department of Materials Engineering, KU Leuven, Kasteelpark Arenberg 44, box 2450, 3001 Leuven, Belgium*

Avanish.Bharati@cit.kuleuven.be
Paula.Moldenaers@cit.kuleuven.be

Abstract. We describe an approach to develop conducting immiscible blends of polystyrene (PS) and polypropylene (PP) with a percolated polyaniline/polyamide (PANI/PA) filler in the presence of a SEBS-g-MA compatibilizer. The underlying principle of the work is the ability of the compatibilizer to concurrently stabilize the bi-phasic morphology to develop co-continuous blends and to improve the dispersion of the filler by specific interactions. From the latter perspective, we have chosen a percolated composite of PANI in PA as the filler and prepared its master batches with the compatibilizer expecting that specific interactions between the amino end groups of PA and the MA anhydride groups in SEBS-g-MA would result in a better dispersion of PANI. The effect of master batch preparation method using either melt mixing or solution casting on the conductivity was studied, and the latter technique was found to be superior. As a result, the concomitant effect of morphology stabilization with improved PANI dispersion led to an improved connectivity of PANI in the phase of the blend where it selectively localizes (PS) for 40/60 PS/PP blends. To further improve the connectivity of PANI, we studied the effect of different sequence additions of the solution casted master batch to the blend components. Melt mixing the solution casted master batch with PS followed by PP resulted in double percolated PANI selectively localized in the continuous PS phase owing to their specific interaction in 50/50 PS/PP blend. This resulted in an increase of the dc conductivity by 6 decades in comparison to the neat blend, already at low PANI concentrations (6 wt%). The connectivity of the blend was affirmed by the power law behavior of the viscoelastic response, more in particular the storage modulus G', of the percolated blends. Furthermore, the effect of the compatibilized blend morphology on the dc conductivity and viscoelastic response was investigated by varying the blend composition. In summary, the proper choices of using a master batch preparation technique by solution casting and subsequently melt mixing the master batch with the polar PS phase followed by PP led to the formation of a double percolated network of filler in the continuous PS phase at low PANI concentrations.

Keywords: Immiscible blends, Compatibilization, Electrical percolation, Morphology, Specific interactions
PACS: 83.80.Tc; 83.60.Bc; 72.80.Le; 47.57.Ng; 83.85.Vb; 72.80.Le; 72.80.Tm

INTRODUCTION

Following the rising demand of flexible conducting polymeric materials in EMI shielding, electronic industries and robotics, immiscible polymeric blends with conducting nanoparticles have generated tremendous interest. This popularity stems from the ability to tune the electrical properties of immiscible blends by tailoring their morphology combined with selective nanoparticle localization, thereby facilitating the formation of a percolated particle network. Another class of conducting materials that finds widespread use in batteries, display devices and sensors are intrinsically electrically conducting polymers, which combine the chemical and mechanical properties of polymers with the electronic properties of metals.[1] However, immiscible blends with electrically conducting polymers have far less been investigated. This is due to the fact that conducting polymers have limited thermal stability and difficult processability, and therefore its seems a good alternative to use them as fillers in polymer blends similar to the use of conducting particles.[1]

In our work, we have developed conducting immiscible PS/PP blends containing a conducting polymer (PANI) in order to retain the electrical properties of PANI at a low concentration in the polymer blend. The approach adopted was to employ a SEBS-g-MA compatibilizer to concurrently generate and stabilize a co-continuous morphology and to improve the dispersion of PANI by specific interactions. To achieve the latter, we have chosen a percolated composite of PANI in PA as the filler and have prepared its master batches by melt mixing or solution casting with the compatibilizer. Our conjecture of an improved dispersion of the filler is based on possible reactive compatibilization of the amino end groups of PA and the anhydride group of –MA.[2] Furthermore, a study of the effect of sequence addition of the suitable master batch to the blend components was used to tune the connectivity of the filler based on the specific interactions between the filler and the blend components. Thus, we have exemplified a novel method for tailoring a percolated network of a conducting polymer in an immiscible blend.

EXPERIMENTS

Master batches (MB) of compatibilizer (SEBS-g-MA) (20 wt%) and filler (F) (80 wt%), the latter being a percolated composite of PANI (30 wt%) in PA were prepared by either melt mixing or solution casting. Melt mixing of the MB was performed at 150 °C and 60 rpm for 3 min under inert nitrogen with simultaneous addition of the components. The processing temperature was below the melting point of the filler and PANI was thermally stable during the mixing. To prepare the MB by solution casting, toluene was used, which is a non-solvent for PANI/PA whereas SEBS-g-MA dissolves readily in toluene. A mixture of SEBS-g-MA and PANI/PA in toluene was first stirred for 1 hr at room temperature (RT) and subsequently bath sonicated for 1 hr at RT, after which drying at 80 °C for 8 hr resulted in solution casted films of the MB.

As the state of dispersion and specific localization of the filler are contingent on the specific interactions of the filler with the compatibilizer and the blend components, we systematically performed various sequence additions of both types of master batches to the blend components. All subsequent melt mixing was performed at 150 °C and 60 rpm. The viscosity ratio of the PS/PP blend at 150 °C was 0.16. In the first sequence, 40/60 (wt%/wt%) PS/PP blends were melt mixed for 3 min, after which MB (5 wt%) was added and further allowed to mix with the immiscible blend for 3 min. For the second sequence, PS (40wt%) and MB were melt mixed for 3 min and afterwards PP (60 wt%) was added to the blend and mixing proceeded for 3 min. As the polar PS phase has more favorable interactions with the MB as compared to PP, the study aimed to develop a percolation of PANI in the PS phase,[3] due to which the possible sequence of mixing the MB with PP was not pursued. In addition to the 40/60 PS/PP blends, blends with the solution casted master batch were prepared for various blend compositions in which the PS content was varied between 30 and 70 wt%. All prepared samples were compression molded at 150 °C before characterization of the blend conductivity, linear viscoelastic moduli and morphology.

RESULTS AND DISCUSSION

Effect of Sequence Addition: Melt Mixing versus Solution Casted Master Batches

In this section, a comparison of the conductivity exhibited by 40/60 PS/PP blends containing master batches prepared by either melt mixing or solution casting is presented. The MB has been added to the blend in different sequences, which allows choosing the most optimal master batch preparation technique as well as sequencing addition. The conductivity spectra of the neat compatibilized blend, the PANI/PA filler (F) and the blends prepared by adding various MB to a premixed PS/PP blend are shown in Figure 1a. The neat blend has a nearly insulating character owing to the non-conducting components. Upon addition of the melt mixed MB to the blend, an increase by a decade of the frequency independent conductivity (dc conductivity) is attained, which further increases by one more decade for the blend with solution casted MB. The low frequency conductivity is a signature of the connectivity of the filler. As the dc conductivity of the blend is significantly lower than that of the filler, the dispersion and/or distribution of the MB within the blend is clearly far from optimal, thus suggesting the need for a different sequence addition. The conductivity spectra of blends in which different MB were first premixed with PS followed by addition of PP are shown in Figure 1b and are compared with that of the neat blend and the filler. Upon addition of the solution casted MB, the dc conductivity is about 2 decades higher as compared to that of blends containing the melt mixed MB. This could be corroborated by the presence of elongated and partially connected PS domains laden with filler, as shown in the STEM micrograph of Figure 1c. From Figure 1, it is imperative to emphasize the near insulating character of the blends with the melt mixed MB and the fact that the blend conductivity remains unaffected upon changing the blending sequence. In addition, the dc conductivity of the blend with solution casted MB is consistently higher than that of blends with the melt mixed MB. Also, for blends with solution casted MB, the second sequence (Figure 1b) results in a slightly improved conductivity as compared to the first sequence (Figure 1a). The latter can be attributed to an improved state of dispersion of the filler owing to van der Waals interactions between the phenyl group of PS and the PANI.[3] In summary, blends with solution casted MB exhibit an enhanced dc conductivity as compared to blends with the melt mixed MB and the addition sequence of the MB facilitating its specific interaction with the blend component PS further yields an improvement in the conductivity. We have thus used solution casted master batches for subsequent blend preparation and the effect of the sequence addition of the MB will be investigated further.

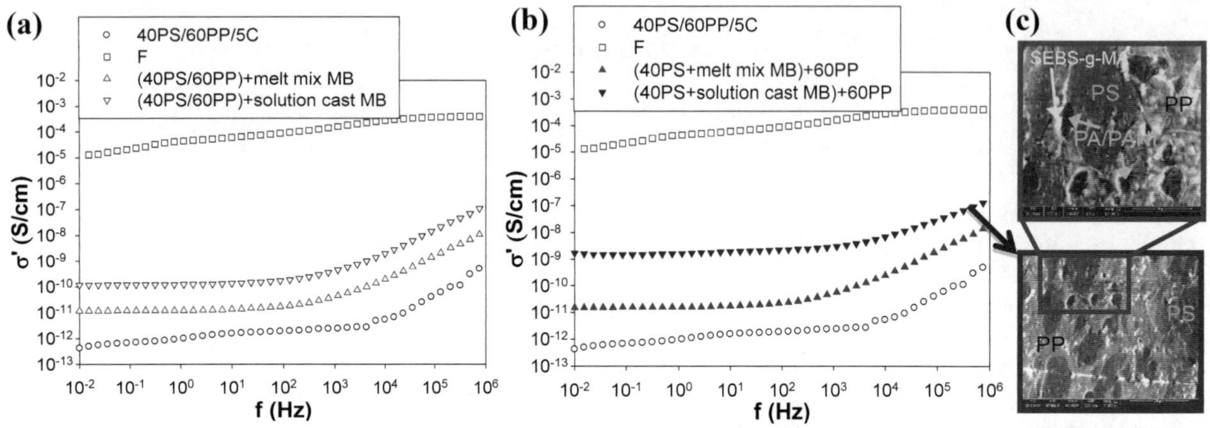

FIGURE 1. Comparison of conductivity spectra of neat blend with compatibilizer (C), filler (F) and blends with different sequence additions (a, b) of melt mixed and solution casted master batches (MB). (c) STEM image of blend with solution mixed MB using the sequence of (b).

Effect of Sequence Addition for Solution Casted Master Batch: Consequence of Blend Composition

To further elucidate the effect of the sequence addition of the MB to the blend constituents, the study has been extended to blends with equal composition (50/50 PS/PP blend) to possibly augment the relative changes in the dc conductivity. Hence, the conductivity spectra for blends prepared with various sequences of solution casted master batches have been shown in Figure 2a for 40/60 and 50/50 PS/PP blends, in which PS is the continuous phase in the former and PP in the latter (from STEM image). Upon comparing the conductivity of the various blends with that of the neat blend and the MB, it is apparent that the sequence of adding the MB to PS followed by addition of PP is optimal for both the blend compositions. In addition, it is imperative to emphasize that the difference in dc conductivity between the different sequences is significantly higher for 50/50 blends (over 3 decades) as compared to that of 40/60 blends (one decade) prepared under identical conditions. The former also exhibits a frequency independent conductivity identical to the dc conductivity of the solution casted MB. This is a manifestation of an electrical percolation of PANI with an effective concentration of 6 wt% (20 wt% filler containing 30 wt% PANI in PA) owing to the formation of a double percolated network of the filler in the continuous PS phase. The blend morphology will be subsequently studied by the storage modulus G' of the blends. In summary, a percolation of PANI at 6 wt% in an immiscible PS/PP blend was achieved with suitable choice of master batch preparation technique and the addition sequence of the MB to the blend components.

To gain further insight in the morphology of the blends whose conductivity is provided in Figure 2a, the viscoelastic response of the blends is shown in Figure 2b. The blend components display a typical terminal behavior at low frequencies ($G' \approx \omega^2$). For 40/60 blends prepared with different addition sequences, a pronounced elastic contribution in G' is observed in comparison to neat blend primarily owing to the increased interface upon compatibilization. However, there is no remarkable effect of the sequence addition on the elastic response. This is in agreement with the mitigated effect of the sequence addition on the conductivity owing to the presence of disconnected PS domains laden with the filler. On accessing the viscoelastic response (G') of the 50/50 blend with MB prepared by addition to the premixed PS/PP blend, a power law behavior is displayed with $G' \approx \omega^{0.32}$. This weak frequency dependence of G' is indicative of connectivity of the blend components. Conversely, the MB does not percolate in Figure 2a in the continuous PS phase. This could be attributed to a lack of specific interactions of the MB and the blend components, which might lead to agglomeration of the filler. Our argument can further be verified by studying the viscoelastic response of the 50/50 blend exhibiting electrical percolation (Figure 2a) in which a nearly frequency independent response of G' is observed ($G' \approx \omega^{0.17}$) for blends with solution casted MB extruded first with PS followed by addition of PP. This is ascribed to the occurrence of PP and PS domains with different length scales and thus disparate relaxation times, characteristic of a co-continuous morphology of the blend.[4] In summary, a continuity of the phase in which the filler specifically localizes (PS) combined with an improved dispersion of the filler owing to its favorable interactions with PS and the compatibilizer unequivocally led to a double percolated network of the filler in Figure 2a. Hence, the sequence of melt mixing solution casted MB and PS followed by mixing with PP will be subsequently investigated for other blend compositions.

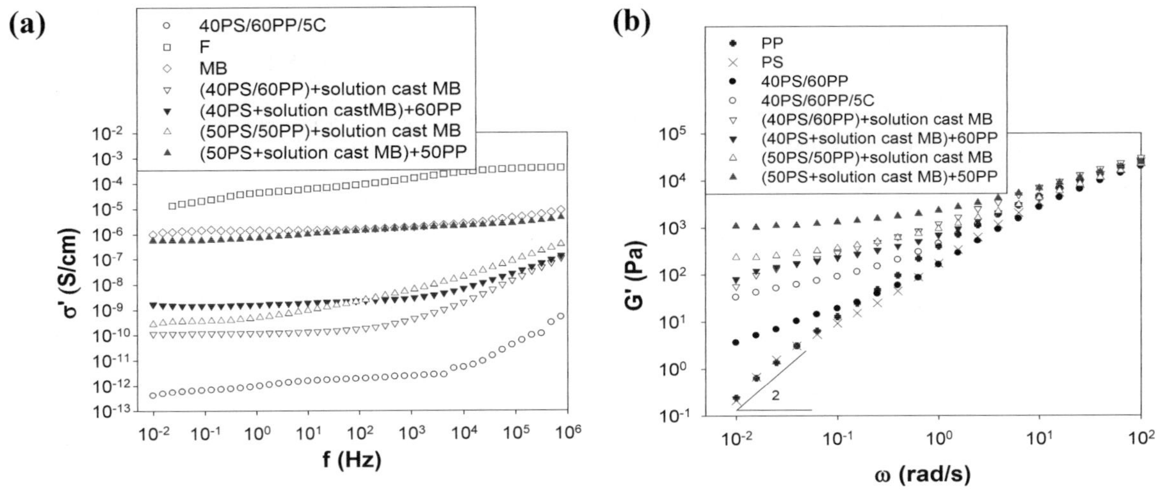

FIGURE 2. Comparison of (a) conductivity spectra and (b) linear viscoelastic response (storage moduli G') for 40/60 and 50/50 PS/PP blends with different sequence additions of solution casted master batches (MB). For comparison in (a) the conductivity spectra of the neat blend, filler and the MB are shown and in (b) G' of the blend components is provided.

Effect of Blend Composition: Blends with Solution Caster Master Batch and Optimal Sequence Addition

In order to obtain the optimal blend properties, it is highly interesting to investigate the effect of blend composition, resulting in different morphologies on the connectivity of the filler for blends with solution casted MB melt mixed with PS followed by addition of PP. Hence, the dc conductivity obtained from the conductivity spectra of blends for various compositions (% PS) is shown in Figure 3a and compared with the dc conductivity of the filler (F), MB and blend components. As expected blends with nearly equal compositions exhibit a pronounced increase in the conductivity owing to the double percolation of the filler in the continuous PS phase. To probe the connectivity of the filler, the low frequency storage moduli (at 0.02 rad/s) for various compositions (% PS) are shown in Figure 3b. A pronounced increase in the G' of 50/50 blends is due to formation of a bi-continuous network of the filler in the continuous PS phase. The fact that the 60/40 blend exhibits electrical percolation in Figure 3a and an attenuated G' in Figure 3b suggests that we are not probing the percolated network of the filler but the co-continuous morphology of the blend. In addition, an increase in G' for the 30 wt% PS blend could possibly be ascribed to the presence of elongated and perhaps connected domains of PS, as 20 wt% filler selectively localizes in the PS. Hence, a one to one correspondence of dc conductivity and elastic response is not obtained as we are probing the percolated PANI network in 50/50 blends and the blend morphology in 60/40 blends. In summary, we have developed conducting blends by formation of double percolated networks of PANI (6 wt%) for several blend morphologies with solution casted MB melt mixed first with the blend component having specific interactions.

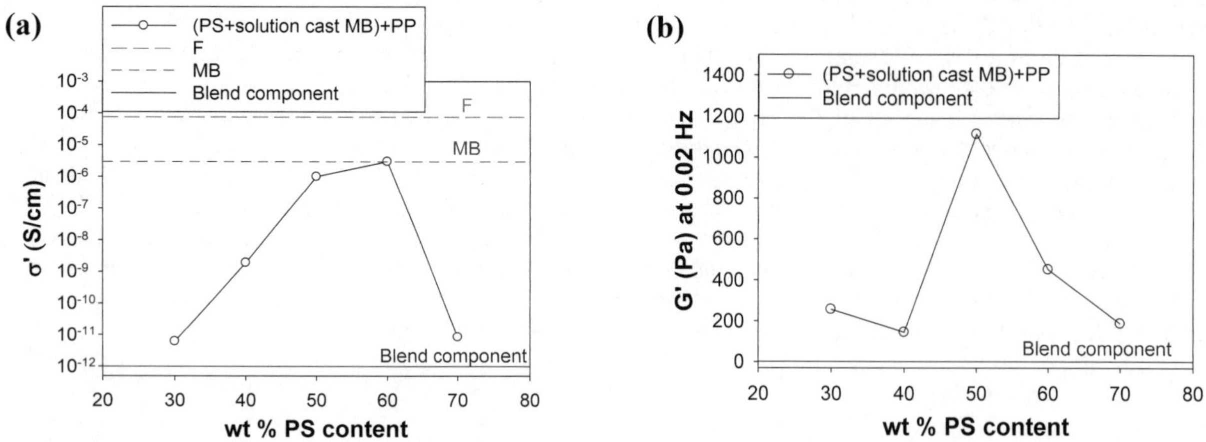

FIGURE 3. Comparison of (a) dc conductivity and (b) linear viscoelastic response (storage modulus G') at 0.02 rad/s for blends with varying PS/PP concentrations with filler (F) and solution casted master batch (MB) (a) and blend component (a and b).

CONCLUSION

In our study, we have successfully demonstrated how master batch preparation techniques consisting of melt mixing or solution casting can be used to tune the dispersion of a conducting polymer, percolated PANI in PA. This was achieved by interactions of the amino end groups of the PA of the filler and the MA anhydride group of the compatibilizer. Furthermore, the sequence of addition of the solution casted master batch to the PS phase helped to specifically localize the filler in the PS phase due to their specific interactions. In summary, the compatibilizer concurrently led to a stable continuous network of the PS phase with an improved PANI dispersion, thereby facilitating the formation of a double percolated network of PANI in the continuous PS phase. As a result a synergistic increase in the conductivity of the percolated blend by 6 decades as compared to that of the neat blend was achieved. The connectivity of the blend was further verified by the power law behavior of the dynamic viscoelastic moduli. Using suitable choices of solution casted master batch and the sequence addition of the master batch to the PS phase, we investigated the effect of various compatibilized blend morphologies on the connectivity and the viscoelastic properties of the blends in order to reduce the effective percolation threshold of PANI in the blend. In conclusion, a suitable choice of master batch preparation technique combined with the addition of the master batch to the blend component having specific interactions can be used to tune a percolated network of a conducting polymer at low concentrations for blends with various morphologies. To further elucidate the contributions of the specific interactions between the MA anhydride group of the SEBS-g-MA and the amino groups of PA, FTIR and TGA analysis of the MB as well as study of blends with SEBS as a non-functionalized compatibilizer or pure PANI as a non-reactive filler will be performed.

ACKNOWLEDGMENTS

This research has been partially funded by GOA/09/002. Ruth Cardinaels is indebted to the Research Foundation Flanders (FWO) for a Post-doctoral Fellowship at KU Leuven.

REFERENCES

1. J. Anand, S. Palaniappan, D. N. Sathyanarayana, *Progress in Polymer Science*, **23**(6), 993-1018, (1998).
2. K. Dedecker, G. Groeninckx, *Macromolecules*, **32**(8), 2472-2479, (1999).
3. M. Zilberman, A. Siegmann, M. Narkis, *Journal of Macromolecular Science, Part B*, **39**(3), 333-347, (2000).
4. S. Steinmann, W. Gronski, C. Friedrich, *Rheologica Acta*, **41**(1-2), 77-86 (2002).

Development of high-performance thermoplastic PPS/PES blends as matrices in fiber reinforced composites

Tim Krooß (presenting author, reference author), Martin Gurka and Ulf Breuer

Institute for Composite Materials IVW GmbH, Erwin-Schrödinger-Straße 58, D-67663 Kaiserslautern

tim.krooß@ivw.uni.kl.de

Abstract. The share of high performance thermoplastic composites within airframe structures is relatively small, although these materials offer the potential of very short cycle time processing. Only polyether ether ketone (PEEK) and polyphenylene suflide (PPS) are used for primary structure applications, however PEEK has a high price and PPS cannot deliver the whole range of properties. This investigation focuses on the development of thermoplastic polymer blends on the basis of PPS and polyether sulfone (PES) as substitutes for high-cost PEEK and neat PPS in fiber reinforced composites. It is targeted to deepen the understanding of PPS/PES blends processing and resulting properties. First studies, utilizing injection molding processing for preparation of the blends, revealed a strong dependency of the resulting property profile on processing conditions and the relative polymer portions in the blend [1]. In this study we focused on the relation of the rheological behavior of the materials and resulting morphology of PPS/PES blend films, made via a film casting/extrusion process. The blend film extrusion process was defined on the basis of rheological analysis concerning the single polymers temperature/viscosity and shearing/viscosity behavior. Analytical effort was primary focused on characterization of the crystallization kinetics and morphology in thin films. Solid plates, made by stacking of thin films and a subsequent hot pressing process, were characterized regarding the processing and composition influences on the mechanical (strength and ductility) and thermo-mechanical (storage modulus at high temperature) properties. It was found that the blend composition has noticeable influence on the degree of crystallization of the semi-crystalline PPS blend component and the ductility and stiffness of the blends are affected by this. It is further shown that thermomechanical properties could successfully be transferred from films into solid plates. This work demonstrates the high potential of tailored PPS/PES blends for film extrusion and stacking processes.

Keywords: PPS blends, film extrusion, morphology, temperature resistance, matrix polymers

PACS: 81.05.Lg, 81.05.Qk, 81.20.Hy, 81.70.Bt, 81.70.Pg, 83.80.Tc

INTRODUCTION

Thermoset resin systems are dominating composite applications of load carrying primary airframe structures of modern civil aircraft (B787, A350XWB). Thermoset carbon-fiber prepregs are typically processed by means of automated tape laying and subsequently curing in an autoclave at a temperature of approx. 180 °C for several hours. The share of high performance thermoplastic composites within airframe structures, however, is still relatively small, although these materials offer the potential of very short cycle time processing [2].

In this investigation the further development of blends made from polyphenylene suflide (PPS) and polyether sulfone (PES) as potential substitutes for neat polyether ether ketone (PEEK) and PPS in carbon fiber reinforced composites is targeted. Sample manufacturing was realized via film extrusion and a subsequent film stacking process without carbon fibers initially. The current findings concentrate on the neat blend properties as a potential matrix material.

METHOD & EXPERIMENTAL

Selection of polymer grades was related to fundamentals of polymer blending and the theory of droplet deformation and breakup [3]. It was mainly focused on the determination of the lowest viscosity ratio of the polymer grades being considered [1, 3, 4]. Film extrusion was examined with the polymer grades *Fortron PPS 0320 P0*, kindly supplied by *Celanese*, and *Ultrason PESU E1010* provided by *BASF Germany*. As a reference for thermo-mechanical analysis injection molded PEEK was chosen *(*type *Victrex 450 G)*. Manufacturing of PPS/PES blend films was performed on a co-rotating *Berstorff Twin-Screw Extruder*. The extrusion parameters were related to the rheological findings of selected polymer grades. There was no preliminary blending of the components. Polymer mass temperature averaged out at approximately 330°C to 340°C. Low rotation speed of 20 rpm was chosen. The hot melt in front of the extruders die was cast on a chill-roll with a constant surface temperature of 80 °C then cooled and spooled on a roll. Film thickness was regulated to 150 - 200 μm. All compositions were processed with and without a modified PES-OH grade, acting as compatibilizer. An overview of polymer properties and film compositions discussed in this paper is given in table (1).

TABLE (1). Polymer properties of PPS and PES (left) and resulting PPS/PES blend compositions manufactured in film extrusion process (right). PES-OH is a OH-modified PES grade, acting as compatibilizer.

	PPS	PES	PEEK	Blend	PPS (wt.-%)	PES (wt.-%)	PES-OH (wt.-%)
Tg [°C]	90	220	140	100/0/0	100	0	0
Tm [°C]	280	-	340	85/15/0	85	15	0
Young´s Modulus [GPa]	4200	2700	3700	85/12/3	85	12	3
Yield Stress [MPa]	90	90	100	50/50/0	50	50	0
Yield Strain [%]	8	6.7	40	50/45/5	50	45	5

Via stacking and hot pressing the films were consolidated to solid plates of thickness 3-4 mm at 300 °C with applied cooling/heating rate of 15 K/min. Morphology investigation was realized via Scanning Electron Microscopy (machine type: *Zeiss Supra 40*). Extruded film samples were exposed to liquid nitrogen to receive brittle fractured surfaces of the specimens. By means of Differential Scanning Calorimetry (machine type: *Mettler-Toledo*) the degree of crystallization was determined for thin films before and after a crystallization step as well as for solid plate samples. In two heating sequences from room temperature up to 300 °C the melting enthalpies of the PPS phase were detected and related to its weight fractions in the blends. For the calculation of crystallinity of the semi-crystalline PPS a melting enthalpy for 100 % crystalline polymer of 80 J/mg was used referring to [5]. Additionally, the effects of a thermal annealing at 204 °C for 2 h on crystallinity of thin film samples was examined related to [6]. A qualitative and quantitative comparison of mechanical properties of films and plates was envisaged by tensile testing of rectangular samples cut out from thin films and dog-bone shaped specimens, machined from plates. The thermo-mechanical behavior of blend compositions was investigated in Dynamic Mechanical Thermal Analysis (machine type: *Gabo*) on both thin films and consolidated plates at a heating rate of 2.5 K/min in the temperature range from 45 °C up to 240 °C. DMTA measurements were force regulated for thin films with a static force of 2 N and a dynamic force of 1 N, and strain regulated for solid plates with 1.4% static and 0.1% dynamic strain portion.

RESULTS AND DISCUSSION

SEM observation of extruded PPS/PES films was used to control morphology formation of the blend partners. Exemplary figure 1 shows the cryo-fractured cross-sections of blend films with 15 wt.-% and 50 wt.-% PES.

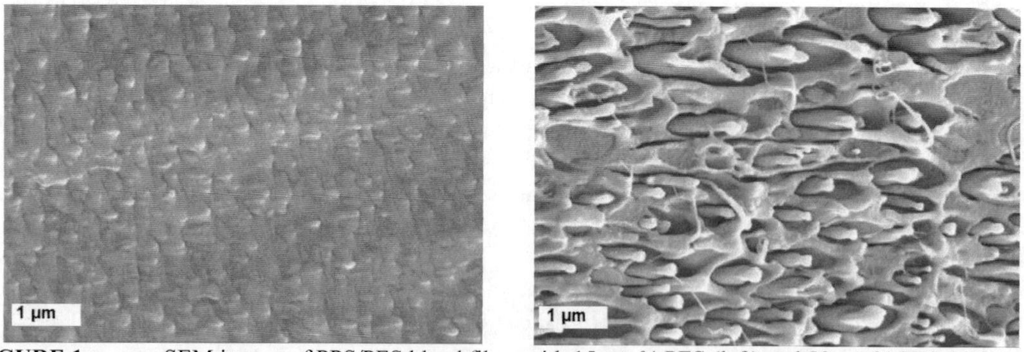

FIGURE 1. SEM images of PPS/PES blend films with 15 wt. % PES (left) and 50 wt. % PES (right).

Low PES amount lead to spherical droplets of PES in a PPS matrix with slight orientation in processing direction of the film extrusion line (left to right). By increasing the amount of PES to 50 wt.-% the morphology changes into a quasi-co-continuous matrix with embedded fibrils or fibrillary network of both components with significant orientation in the processing direction. The presence of the compatibilizing PES-OH component did not result in different morphology characteristics.

The crystallinity of single films in the as extruded state as well as of stacked and consolidated films was determined after annealing and after hot pressing respectively by Differential Scanning Calorimetry (figure 2).

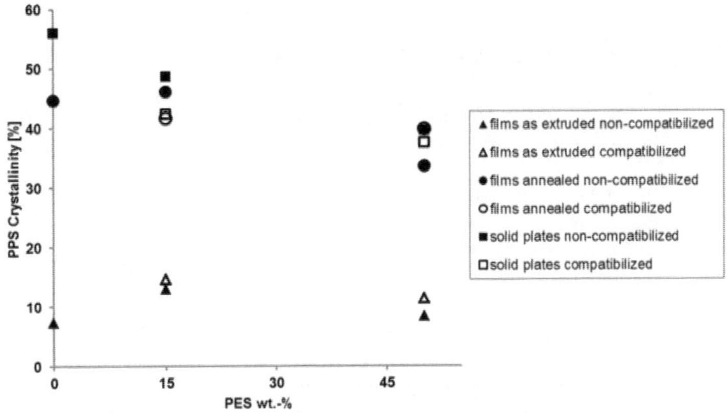

FIGURE 2. Crystallinity of extruded films and stacked and consolidated films as a function of PES weight fraction.

During the film extrusion process the hot polymer melt was cast directly from the film die onto a chill roll with a constant surface temperature of approx. 80 °C. The high temperature difference between extrusion die and chill-roll resulted in a fast cooling of the polymer film, thus retarding crystallization of the PPS component. DSC measurements underline the low degree of crystallization in the blend films, which additionally decreases by trend with a higher amount of amorphous PES fraction. Additional thermal annealing of the films at 204 °C for 2 h resulted in a crystallinity of the PPS component between 34 % and 46 %, following the same trend with increasing amount of PES. In comparison to the single film samples, the crystallinity in consolidated plates, cooled at 15 K/min, showed only a slightly higher value. Neat PPS exhibited highest degree of crystallinity of approx. 56 %. No significant influence of compatibilizing PES-OH fraction was determined (open symbols).

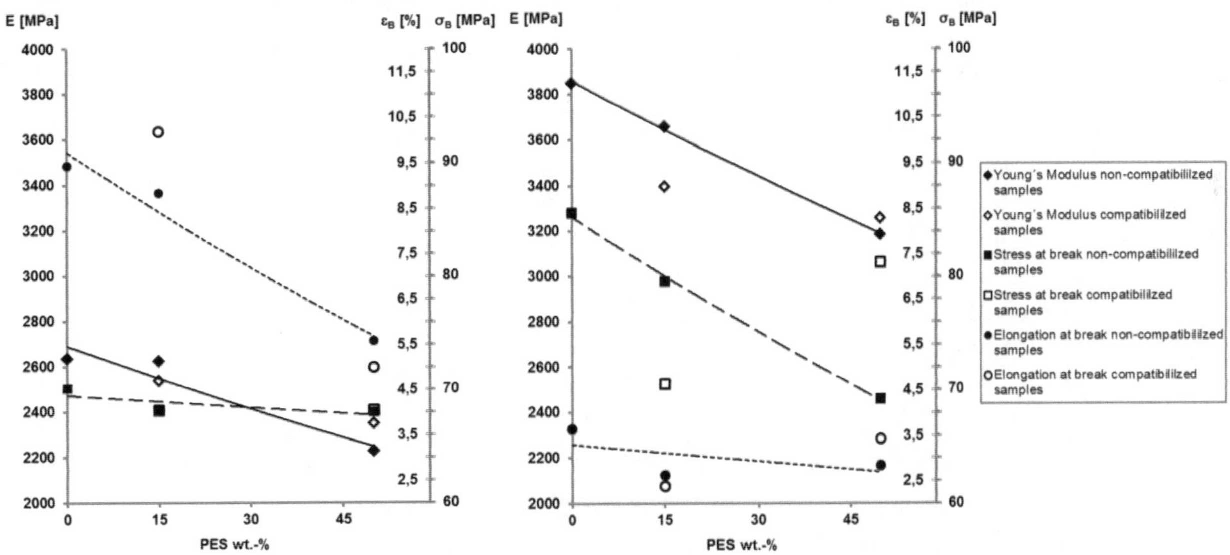

FIGURE 3. Averaged tensile properties (5 samples tested) of crystallized PPS/PES blend films (left) and solid plates (right) as a function of PES amount.

Figure 3 shows the mechanical properties - Young´s modulus, elongation at break and tensile strength for the annealed (re-crystallized) film and solid plate samples depending on PES content of the blend. In Table (2) property standard deviation of tensile tested films and solid plates are listed. Most significant the elongation, strength and modulus of crystallized films as well as strength and modulus of solid plates exhibit a clear dependency of the relative amount of PES in the blend. Films with low PES amount exhibit higher modulus and slightly higher tensile strength than compositions with high amounts. The highest modulus for films was determined at 2.6 GPa, which was almost the same for the specimen out of neat PPS and the blends with 15 wt.-% PES. Tensile strength of all films was nearly constant at 70 MPa in average. A clear drop of elongation at break was observed for samples with higher PES amounts. In general PPS films show the highest deviation of all mechanical properties (table (2)). For the blends the standard deviation of Young´s modulus is positively influenced by the addition of PES-OH.

TABLE (2): Standard deviation of tensile properties of films and solid plates.

	Thin films			Solid plates		
Blend	**E [MPa]**	**σ_B [MPa]**	**ε_B [%]**	**E [MPa]**	**σ_B [MPa]**	**ε_B [%]**
100/0/0	534,8	13,9	3,3	44,3	2,5	1,1
50/50/0	270,9	7,5	0,3	59,6	16,8	1,1
50/45/5	122,7	7,1	0,5	30,3	5,2	0,6
85/15/0	233,7	10,1	1,7	49,9	5,6	0,3
85/12/3	184,4	8,8	2,0	186,2	7,2	0,3

The solid plates, processed out of stacked films, do not show this tendency. Strength and modulus of the plates increased significantly compared to the films. Elongation instead was lower but slightly increased from approx. 2.3 % up to 3.4 % with growing PES fraction supported by compatibilizer. However, neat PPS even shows a maximum strain of 3.6 % in average. Modulus and strength decreased with higher PES amounts. The only exception was the compatibilized blend with 50 wt.-% PES exhibiting an averaged strength of 81 MPa which is almost equal to neat PPS samples. The standard deviations of solid plate properties do not show similar tendencies as tested films. The addition of PES-OH has positive influence on deviation of modulus, strength and elongation at break for the 50/45/5 composition compared to the non-compatibilized 50/50/0 blend.

Regarding the themo-mechanical behavior a comparable dependence of the composition can be observed. The dynamic mechanical thermal analysis shows the storage moduli as a function of temperature for PPS/PES blend films (figure 4) and plates (figure 5) in comparison with neat PPS and PEEK samples.

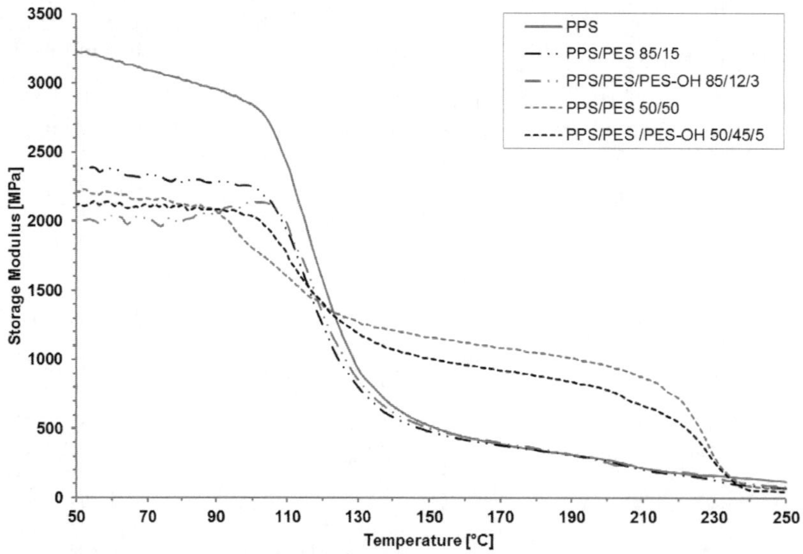

FIGURE 4. Storage Modulus of thin PPS/PES films compared to PPS.

Most prominent observation regarding the DMTA film testing is the difference in storage modulus at the beginning of the measurement. Neat PPS exhibits an initial storage modulus of approx. 3.2 GPa compared to significantly lower values for the blends ranging from 2.0 GPa to 2.4 GPa. Possible explanation for the differences in storage modulus of the blends and PPS can be local inhomogeneity of the PES dispersion or crystallinity differences of the PPS fraction. This is verified by the deviation of films modulus. Above temperatures of 110 °C the PES phase dominates the storage modulus, while both compositions with low PES fraction do not show improvement of storage modulus compared to PPS. The plateau of storage modulus for the 50 wt.-% compositions is significantly higher than PPS and other blends, at least by the factor of 2.

The consolidation of thin films into solid plates at 300 °C and heating/cooling rate of 15 K/min positively affected the thermos-mechanical performance of the blends (figure 5). Firstly, the blend samples again show a significant plateau between 110 °C and 210 °C approximately. Compared to neat PPS the composition with

15 wt.-% PES does not exhibit an advantage in this temperature range, however, no significant loss of storage modulus is observed, neither at high temperatures nor in the range of 50 °C up to 110 °C. Regarding the compositions with high PES fractions (50 wt.-%) the thermal stability is increased at temperatures above the PPS glass transition temperature (approx. 80 °C). No significant influence of PES-OH addition was observed. When being compared to PEEK the initial storage modulus of low PES amount blends is comparable. A reduction of the initial value is observed for higher PES fractions. However, the improvement for these blends at temperatures above 175 °C compared to PEEK and PPS again is significant.

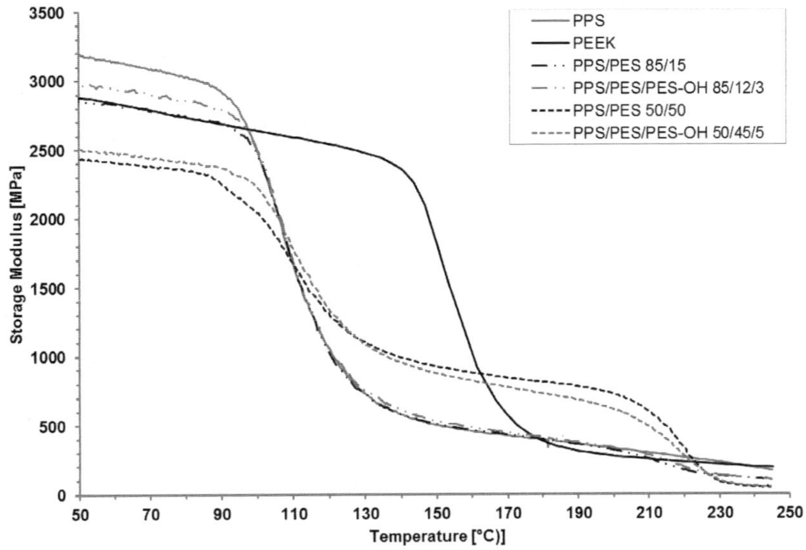

FIGURE 5. Storage Modulus of solid PPS/PES plates compared to PEEK and PPS.

CONCLUSIONS

Manufacturing of PPS/PES blends via film extrusion and film stacking and blend characterization was investigated in this study. Morphology and crystallization was found to be clearly influenced by PES fractions and processing conditions, also affecting mechanical properties. Crystallinity of PPS fraction decreases with growing PES amount. Significant loss of elongation ability was observed for film samples with raising PES fraction; however stress at break remained nearly constant. The mechanical properties of solid plate samples were found to be positively influenced only by high PES amounts in combination with compatibilizing PES-OH fraction. Regarding thermo-mechanical properties the influence of 15 wt.-% PES on the thermal stability at high temperatures did not result in improvement compared to neat PPS or PEEK, neither for thin films nor for solid plates. The influence of 50 wt.-% PES was more significant and thermal stability clearly higher at temperatures above 110 °C when compared to PPS and above 175 °C when compared to PEEK. The thermal analysis evidences the successful transfer of temperature resistance from extruded films into solid plates with similar composition-related material behavior. In future works it will be targeted to investigate the influence of blend properties on carbon fiber reinforced composite properties, such as fiber/matrix adhesion, fracture toughness and damage tolerance in order to develop alternative thermoplastic matrix materials for high performance applications.

REFERENCES

1. T. Krooß, M. Gurka, V. Dück, and U. Breuer, "Development of cost-effective thermoplastic composites for advanced airframe structures" in *ICCM20 Conference Proceedings*, Copenhagen, 2015.
2. A. Gillot, "From carbon fibre to carbon-fibre-reinforced thermoplastics", *JEC Composites Magazine No. 71*, (March 2012).
3. Grace, H. P., "Dispersion phenomena in high viscosity immiscible fluid systems andapplication of static mixers as dispersion devices in such systems" in *Chemical Engineering Communications, Vol. **14***, 225–277 (1982)
4. B. D. Favis, "Factors Influencing the Morphology of Immiscible Polymer Blends in Melt Processing" in *Polymer Blends Volume 1: Formulation* edited by D. R. Paul and C. B. Bucknall, New York: John Wiley & Sons, Inc 2000, pp. 509-513.
5. G. P. Desio and L. Rebenfeld, *Journal of Applied Polymer Science, Vol.* **44**, 1989-2001 (1992).
6. D. G. Brady, *Journal Of Applied Polymer Science Vol.* ***20,*** 2541-2551 (1976).

Enhancement of Impact Strength at Ambient and Low Temperature for Polycarbonate-based Polymer Alloy without Reduction of Flexural Modulus

Tomotaka Saito[a*], Yan Hao[a], Nobuhiro Sumiya[a] and Tomokazu Moriyama[a]

[a]*Research Laboratories, Toyota Boshoku Corporation, 1-1, Toyoda-cho, Kariya, Aichi, Japan*

tomotaka.saito@toyota-boshoku.com

Abstract. For developing novel impact strength enhancement method without reduction of flexural modulus, polycarbonate (PC)-based polymer alloy was prepared with adding 0.1 to 20 wt% of low density polyethylene (LDPE). From SEM micrograph, the alloy contains particles and pores in the matrix. The particles are made of LDPE, and the pores are formed by the immiscibility and the cooling shrinkage difference between PC and LDPE. The prepared polymer alloy containing 1 wt% of LDPE increases 20% and 40% of the impact strength at room and low temperature, respectively. And flexural modulus of the alloy (2.4 GPa) is almost same as PC (2.2 GPa).

Keywords: polycarbonate, polymer alloy, impact strength
PACS: 81.05.Lg

INTRODUCTION

The impact strength is important property for polymer materials not only at ambient but also at low temperature. Research and trial to improve the property has been conducted in several decades. In 1973, Allen et al reported the impact strength of polycarbonate (PC) with variation of temperature [1]. The result shows PC change to brittle at about 0 °C. Yee et al conducted research for PC/polyethylene (PE) alloy, and reported adding 4% of PE improves the impact strength at ambient temperature [2, 3]. Zhou and Osby prepared PC/ultra-high molecular weight polydimethylsiloxane (PDMS) blends [4]. They presented PC/PDMS blends are ductile at -40 °C even though PC is brittle at the temperature. They mentioned the changing to ductile mechanism relates to the PDMS particle size but they didn't present data of the relation. To the best of our knowledge, impact strength enhancement for PC/LDPE alloy at low temperature has not investigated.

In generally, flexural modulus and impact strength are considered as trade-off, therefore satisfying both high modulus and impact strength are still difficult. Adding the rubber, it is general method to enhance impact strength, decrease flexural modulus of the material [5, 6]. Recently, Moriyama reported impact strength enhancement without reduction of flexural and tensile modulus for polyamide 11 (PA11)/poly(butylene succinate) (PBS) alloy [7]. Charpy impact strength of PA11/PBS alloy is 3times higher than neat PA11. We investigated applicability of the impact strength enhancement method for PC/LDPE [8] and the impact strength at ambient temperature was enhanced. In this paper, we report impact strength enhancement at low temperature and flexural modulus of PC/LDPE alloy.

EXPERIMENTAL

Polycarbonate (PC) used in this study was Jupilon E-2000 (Mitsubishi Engineering Plastics Corp.) and low-density polyethylene (LDPE) was purchased from Aldrich. These pellets were dried under vacuum for one night before compound. The pellets were mixed 99.9/0.1 to 80/20 in weight ratio of PC/LDPE. The mixed pellets and neat of them were melt compounded using internal mixer (Laboplastomill V4C150, Toyo Seiki Seisakusyo, Ltd.) with 75 rpm at 240°C (PC and alloys) or 130°C (for neat LDPE). The melt compounded mixtures were formed into rectangular bars by using laboratory hot-press. Charpy impact test was conducted according to ISO 179-1 at room temperature for 23°C and -45°C samples. Flexural modulus was measured by using Universal Tester (Shimadzu, Autograph, AG-250kNG). Morphologies of the alloys were observed at platinum coated cross-sectional surface using scanning electron microscope (JEOL JSM-6510LV).

TABLE 1. Physical properties of polycarbonate and low-density polyethylene.

	Glass transition and melting temperature (°C)	Shrinkage ratio (%)	Charpy impact strength (kJ/m^2)	
			23°C	-45°C
PC	151.2 (T_g)	0.68	65.0	8.7
LDPE	110.5 (T_m)	1.45	55.8	2.3

RESULTS AND DISCUSSION

Representative SEM micrograph of PC/LDPE alloy is shown in **FIGURE 1**. Pores and particles are seen in the micrographs. The particles are made of LDPE and some particles have cavities. It is considered the cavities were formed when freeze-fracturing in liquid nitrogen reported as cavitational process on polymer/ rubber blends by several papers [9, 10].

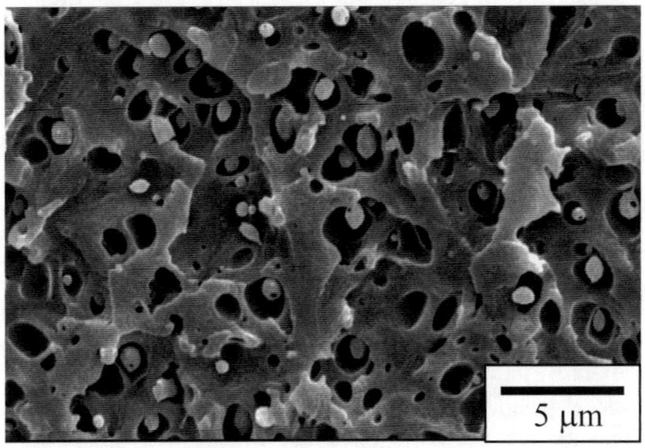

FIGURE 1. SEM micrograph of PC/LDPE alloy. 4 wt% of LDPE is containing the polymer alloy..

Impact stress propagation to LDPE particles is needed for causing the cavitational process. Therefore, interface between matrix PC and LDPE particles was contacted before the freeze-fracturing. From SEM micrographs of PC/LDPE alloys, average particle diameter and number density of the particle were estimated. The diameter and the number density were increased with increasing LDPE content (**FIGURE 2**).

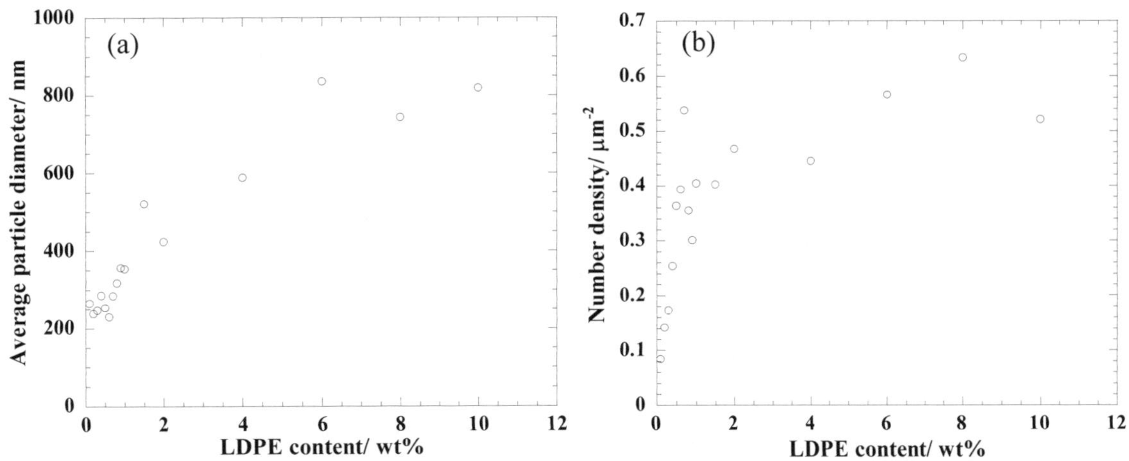

FIGURE 2. Plots of (a) average particle diameter and (b) number density versus LDPE content.

FIGURE 3 shows Charpy impact strength of PC/LDPE alloys at 23°C and -45°C with a factor of LDPE content. At 23°C, Charpy impact strength of PC/LDPE alloy has maximum value at 0.7 wt% of LDPE and 27% higher than neat PC. On the other hand, at -45°C, maximum value is given at 10 wt% of LDPE content. Moriyama et al explained impact strength enhancement was achieved by dilational stress field at around the particle [8].

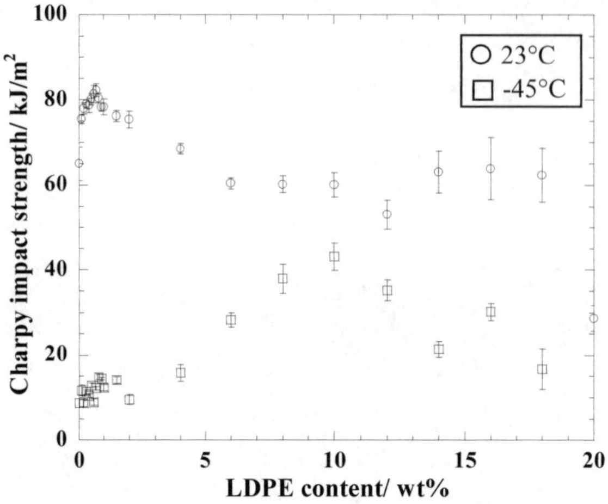

FIGURE 3. Charpy impact strength of PC/LDPE alloy. The impact strength was measured for 23°C and -45°C samples.

FIGURE 4 is schematic image of the dilatational stress field. At the field, volume of the matrix is increased and matrix polymer chains are extended by shrinkage of LDPE particles. Volume fraction of dilatation stress field is considered to increase with increasing the particle size and number density of the particle. The impact strength decreasing of PC/LDPE alloy over 0.8 wt% of LDPE content is caused by the increasing the volume fraction of the LDPE particle, because of the relatively lower impact strength of LDPE (**TABLE 1**). At -45°C, impact strength is enhanced with increasing LDPE content. Below about 0°C, matrix PC shows ductile-brittle transition and the impact strength declines [1]. As Hashima et al reported, local segment motion of matrix polymer chains in dilatational stress field is enhanced at low temperature [11]. Therefore, volume fraction increment of dilatational stress field is considered to enhance the impact strength at low temperature.

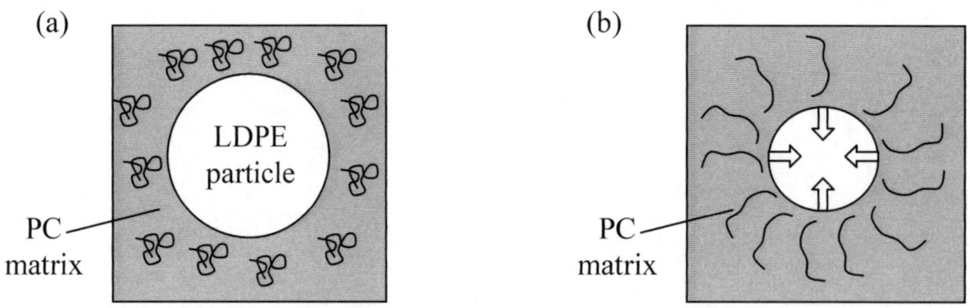

FIGURE 4. Schematic images of (a) before and (b) after shrinkage of LDPE particle. Random coils of matrix polymer chains are extended by particle shrinkage.

Flexural modulus of the PC/LDPE alloy is shown in **FIGURE 5**. From **FIGURE 5**, the flexural modulus was enhanced by adding small amount of LDPE. Orientation of the matrix polymer chains would contribute to flexural modulus enhancement.

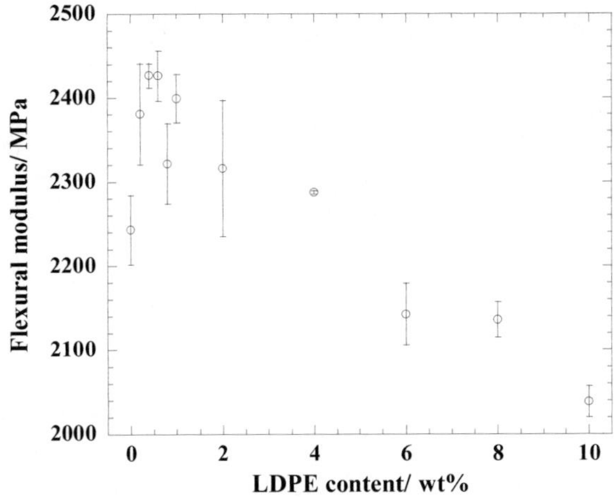

FIGURE 5. Flexural modulus of PC/LDPE alloy.

CONCLUSION

PC/LDPE alloys contain 0.1 to 20 wt% of LDPE were prepared. Charpy impact strengths of the alloys at 23°C and -45°C are enhanced up to 27% and about 5 times compared with neat PC, respectively. The impact strength enhancement is considered to be achieved by dilatational stress field at around LDPE particles. Flexural modulus of PC/LDPE alloy also enhanced. Flexural modulus would be enhanced by matrix polymer chain orientation in the dilatational stress field.

REFERENCES

1. G.Allen et al., *J Mater Sci*, **8**, 1449-1452 (1973)
2. A. F. Yee, *J Mater Sci*, **12**, 757-576 (1977)
3. H. J. Sue et al., *Polymer*, **33**, 4868-4871 (1992)
4. W. Zhou and J. Osby, *Polymer*, **51**, 1990-1999 (2010)
5. Y. Ruksakulpiwat et al., *Compos Part B-Eng*, **40**, 619-622 (2009)
6. Y. C. Ahn and D.R. Paul, *Polymer*, **47**, 2830-2838 (2006)
7. T. Moriyama et al., *Polym J*, **48**, 221-224 (2016)
8. T. Saito et al., *Proceeding of Polymer Processing Conference 2015*
9. C. B. Bucknall et al, *J Mater Sci*, **29**, 3377-2282 (1994)
10. D. Dompas and G. Groeninckx, *Polymer*, **35**, 4743-4749 (1994)
11. K. Hashima et al., *Polymer*, **51**, 3934-3939 (2010)

Production of PBT/PC Particle Systems by Wet Grinding

Jochen Schmidt[a*], Stefan Romeis[a] and Wolfgang Peukert[a]

[a]*Friedrich-Alexander-Universität Erlangen-Nürnberg, Institute of Particle Technology, Cauerstr. 4, D-91058 Erlangen, Germany*

jochen.schmidt@fau.de

Abstract. Comminution of plastic or visco-elastic materials is known to be energy-intensive and costly. Typically impact mills, cutting mills and jet mills are applied for comminution and the feed material frequently is pre-cooled using liquid nitrogen or solid carbon dioxide. Therefore, especially the production of fine polymer particles by the aforementioned dry grinding methods is challenging. We demonstrate that wet grinding of polymers is a feasible approach for production of fine polymer particles that e.g. may be used as a starting material in a process chain for production of spherical polymer particles that can be used in in additive manufacturing. Wet comminution of polymers will be discussed for polybutylene terephthalate (PBT) and polycarbonate (PC). The products are thoroughly characterized by means of vibrational spectroscopy, X-ray diffraction and dynamic scanning calorimetry. The dependency of the grinding result on fundamental process parameters like stirrer tip speed, grinding media size, process time, process temperature and system composition will be discussed. In any case high load conditions, i.e. high stress energies are needed to initiate breakage of polymer materials. Optimization of the process with respect to process time and energy consumption by appropriate selection of process temperature and solvent will be addressed. The selection of appropriate solvents allows for remarkable reduction of mass-specific comminution energy needed.

Keywords: wet comminution, stirred media mill, polybutylene terephthalate, polycarbonate
PACS: 81.05.Lg; 82.35.Lr; 81.20.Wk; 81.15.-z

INTRODUCTION

In recent years an increasing demand for fine polymer powders that may be used for additive manufacturing processes such as selective laser beam melting of polymers (LBM) has been recognized. According to the state of the art, commercially available LBM powders like e.g. polyamide (PA) are obtained by sophisticated precipitation methods that cannot be transferred readily to other types of polymers. As a more general method to produce fine polymer powders, frequently dry grinding has been applied [1-7]. Typically rotor impact mills or cutting mills are used and the polymer feed material is pre-cooled e.g. with liquid nitrogen or solid carbon dioxide. Thus, the comminution of materials showing viscous or plastic deformation behaviour is costly and energy-intensive. Moreover, the minimum product particle sizes that are obtained by dry grinding frequently are in the range of several 10 micrometers. In this contribution wet grinding [8] is discussed as an alternative and versatile approach to produce fine micron-sized polymer particles. Dependencies of process parameters are outlined and the applicability of the wet grinding process to polybutylene terephthalate (PBT), polycarbonate (PC), as well as PBT-PC mixtures is demonstrated. The products are characterized by means of vibrational spectroscopy, X-ray diffraction and dynamic scanning calorimetry. No remarkable structural changes occur during size reduction. The wet grinding process may be employed as the first step in a process chain [9] that allows for production of innovative LBM powders.

METHODS

Particle Sizing

Particle size distributions were determined by laser diffraction particle sizing using a Mastersizer 2000 / Hydro 2000S (Malvern). The product suspensions have been diluted as appropriate prior to measurement. Small amounts of sodium dodecyl sulphate solution (SDS, 98% (Merck)) have been added to assure dispersion stability. Prior to measurements ultrasonication of the dispersions has been performed.

Differential Scanning Calorimetry

DSC characterization was done at a heating rate of 20 K min^{-1} using a DSC8000 (Perkin Elmer).

X-Ray Diffraction

Structural analysis of the polymers by means of powder X-ray diffraction (XRD) was performed using a Bruker AXS D8 Advance diffractometer. The device was equipped with a VANTEC-1 detector and a Ni filter. Cu K_α radiation (λ = 154.06 pm) was used. The X-ray diffractograms were recorded in Bragg-Brentano geometry in the range $20° \leq 2\theta \leq 80°$. The step width was 0.014° and the accumulation time was 1 s for each step.

Infrared Spectroscopy

Infrared spectra were recorded in transmission (KBR method) in the spectral range from 4800 cm^{-1} to 400 cm^{-1} at of 2 cm^{-1} resolution using a FT-IR spectrometer FTS 3100 (Varian). Spectra have been normalized to their respective maximum absorbance at around 1715 cm^{-1} ($\nu_{C=O}$). Baseline corrections were done manually.

RESULTS

Wet Grinding of PBT

Wet comminution of PBT (Ultradur, BASF) was performed in a custom-made agitator bead mill equipped with a torque meter which allows for quantification of the mass specific energy E_m consumed to obtain a certain product particle size $x_{50,3}$. The feed material for wet grinding of all polymers was obtained by grinding of the respective polymer granules in a rotor-impact mill using a 0.5 mm sieve ring. Figure 1 depicts the dependency of the product particle size on the mass specific energy for different stressing conditions. In the grinding experiments a feed mass concentration of 12.5% PBT was applied, the process was done at ambient temperature and the solvent was ethanol and –for SE = 5.4 mJ- also butanol. The variation in stressing conditions is expressed in terms of stress energy SE. SE equals the maximum kinetic energy of a grinding bead [10]. Variation of the stressing conditions was accomplished by changing the stirrer tip speed v_{tip} (7.1 m s^{-1} vs. 10.6 m s^{-1}) and the diameter d_{GM} (1.25 mm, 2.0 mm, 3.0 mm) of the yttrium-stabilized zirconia grinding media of a density of 6050 kg m^{-3}. Obviously, fracture of the stressed feed material under these conditions only occurs of stress intensities of more than 2 mJ are chosen. In the case of d_{GM} = 1.25 mm (c.f. red symbols in Figure 1) virtually no fracture has been observed even for the maximum stirrer tip speed 10.6 m s^{-1} of this type of mill. If grinding media of 2.0 mm or 3.0 mm size are used and ethanol was the solvent no remarkable differences in the course of $x_{50,3}$ over the mass specific energy was observed for $E_m > 10$ MJ/kg. For constant stressing conditions of SE = 5.4 mJ the effect of the solvent on the mass specific energy needed to obtain a certain product particle size was assessed. Obviously using ethanol as compared to the more viscous butanol is beneficial with respect to energy consumption: if an energy input of 30 MJ/kg applies, in the ethanol system product particles of about 10 µm size are obtained whereas in the butanol system larger particles of about 17 to 18 microns are obtained. The behavior qualitatively may be understood in terms of grinding media dampening [8]: the energy transferred to a stressed particle is less for higher viscous systems because a higher portion of energy is dissipated by viscous forces prior to grinding bead collisions. In general low viscous solvents are advantageous and in certain cases [8] also the reduction of the process temperature (embrittlement of the polymer) may improve the grinding kinetics and reduce the energy needed to initiate fracture. Compared with typical mass specific energies needed for polymer grinding and typical minimum product particle size obtainable by dry grinding [1] one can conclude that the wet grinding process allows for smaller product particles (about one order of magnitude) at reasonable energy consumption. The wet grinding process especially is superior to dry grinding, if fine products ($x_{50,3} < 25$ µm) are desired. For brittle polymers product particles with $x_{50,3} < 5$ µm readily may be obtained.

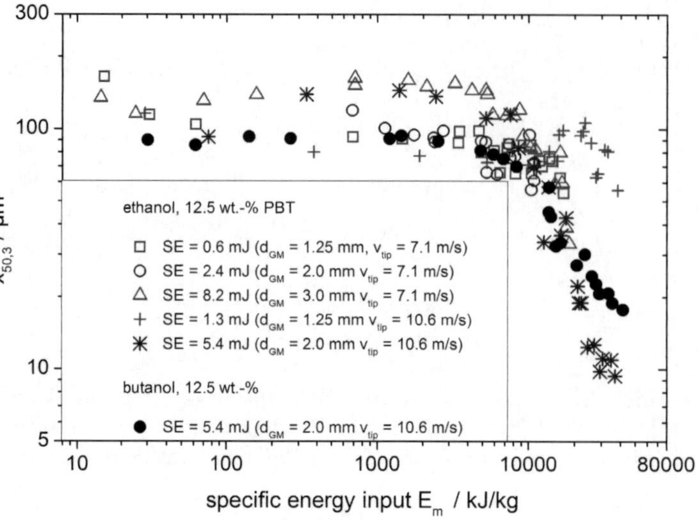

FIGURE 1. Evolution of product particle size $x_{50,3}$ over process time during wet comminution of PBT in ethanol and butanol at ambient temperature (for stressing conditions see figure legend).

Wet Grinding of PBT-PC mixtures

The wet grinding process also has been applied to PC (Makrolon, Bayer) and a mixture (1:1 (wt./wt.)) of PBT and PC. Figure 2 summarizes the evolution of the particles sizes $x_{90,3}$, $x_{50,3}$ and $x_{10,3}$ of stressed PBT and PC and $x_{50,3}$ for the PBT-PC mixture, respectively, over process time for wet comminution in ethanol at 20 °C for a feed concentration of 18.5 wt.-%. A stress energy SE of 1.9 mJ was applied. PC (filled symbols) shows a narrower product particle size distribution (c.f. smaller span = $(x_{90,3} - x_{10,3}) / x_{50,3}$) as compared to PBT (open symbols) for the same SE and process time which reflects differences in mechanical properties and fracture behavior. The grinding kinetics of PBT-PC mixtures, i.e. the evolution of product particle size over process time, as exemplified for a 1:1 mixture has been found to be a simple (volume-weighted) superposition of the grinding kinetics of PBT and PC at the respective mass concentration: no interaction between the polymers occurs during grinding and the fracture mechanism are not influenced by the presence of the respective other polymer. Also no blend formation / transesterification has been observed in wet grinding even for long process times. Similar grinding kinetics like for the PBT-PC mixture have been observed for a PBT-PC blend system of the same PBT : PC mass ratio.

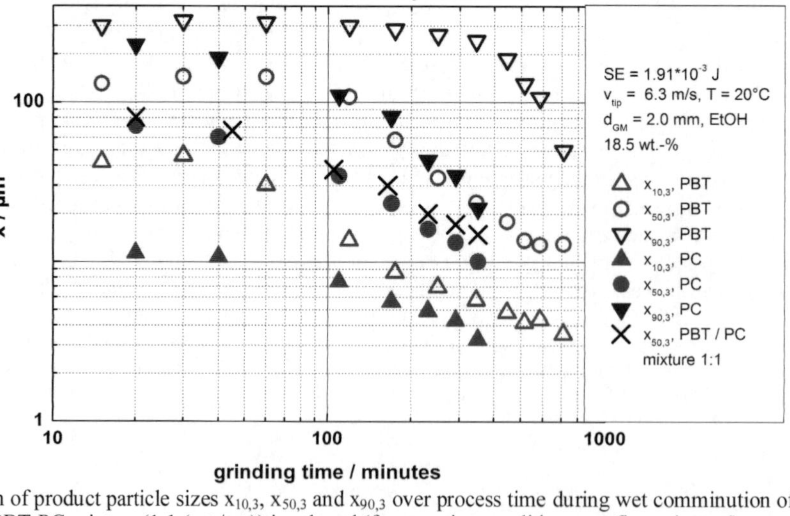

FIGURE 2. Evolution of product particle sizes $x_{10,3}$, $x_{50,3}$ and $x_{90,3}$ over process time during wet comminution of PBT, PC and a PBT-PC mixture (1:1 (wt./wt.)) in ethanol (for stressing conditions see figure legend).

Structural Characterization of Grinding Products

The grinding products amongst others have been characterized with dynamic scanning calorimetry, X-ray diffraction and infrared spectroscopy. With respect to production of powders for an application in LBM [9], it isimportant that no significant changes of crystallinity occur during powder production. Table 1 summarizes the melting temperatures of PBT feed material and product particles obtained after different process times (for stressing conditions see legend of Figure 2) and the crystallinity of the respective samples. The degree of crystallinity was calculated from the melting enthalpies of the respective powders, taking the melting enthalpy of fully crystalline α-PBT of 145.5 J/g [11] into account. No significant change in crystallinity over process time is observed. The determined crystallinities in the range of 34.3 to 36.3 wt.-% of α-PBT, are typical for commercial PBT grades. Therefore, we can conclude that no significant amorphization of the semi-crystalline polymer occurs during stressing under the chosen conditions.

TABLE (1). Melting temperature and degree of crystallinity (α-PBT) estimated from the heat of fusion of PBT powders obtained by wet comminution for different process times.

process time / minutes	melting temperature / °C	degree of crystallinity / wt.-%
0 (feed material)	227.3	35.0
60	226.4	34.3
180	227.9	34.9
240	228.4	36.3
345	228.4	36.1

Also X-ray diffraction and infrared spectroscopy prove that no pronounced structural changes occur: The XRD diffractograms (Figure 3) of the feed material and wet ground PBT are almost the same. The patterns are typical for α-PBT [12, 13], no significant changes due to grinding occur. Infrared spectra of polymer powders obtained for different process times are depicted in Figure 4. For all samples spectral features being typical for PBT [14], i.e. the C=O vibrational band at around 1715 cm^{-1}, the CH2 bending mode at 1503 cm^{-1}, the symmetric and asymmetric C-O-C stretching modes at 1206 cm^{-1} and 1103 cm^{-1}, and modes assigned to the aromatic ring at 1018 cm^{-1}, 873 cm^{-1} and 727 cm^{-1} are found. The spectra are almost identical, no pronounced changes in molecular structure occur during grinding of PBT under the conditions studied.

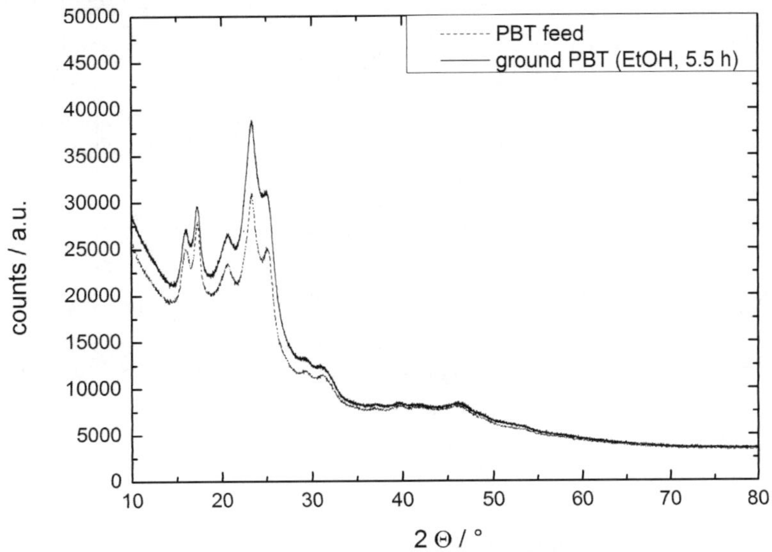

FIGURE 3. X-ray diffractograms of PBT feed material and PBT particles obtained by wet comminution in ethanol.

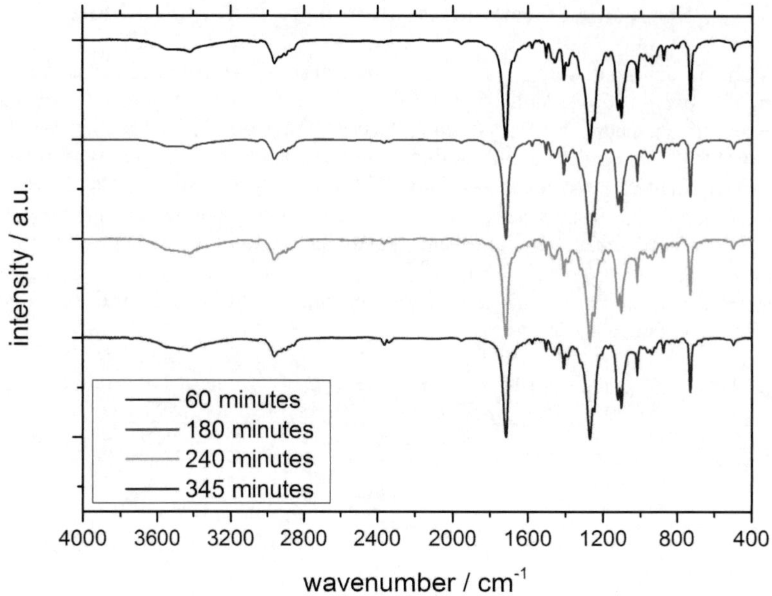

FIGURE 4. IR spectra of PBT particles obtained by wet comminution in ethanol in dependency on process time.

CONCLUSIONS

A wet grinding method [8] was successfully applied to produce PBT, PC as well as PBT-PC mixed systems with particle sizes $x_{50,3}$ below 20 microns. Preservation of the crystallinity and structure of the stressed solids during stressing was proven by dynamic scanning calorimetry, infrared spectroscopy and X-ray diffraction. No significant amorphization takes place during wet grinding which makes the process applicable to successfully produce PBT and PBT-PC based particle systems for additive manufacturing employing a previously proposed process chain [9].

ACKNOWLEDGMENTS

This study has been supported by Deutsche Forschungsgemeinschaft (DFG) in the framework of the collaborative research center CRC 814 "Additive Manufacturing" (project A1). Financial support is gratefully acknowledged.

REFERENCES

1. M. Wilczek, J. Bertling, D. Hintemann, *Int. J. Min. Proc.* **74** S425-S434 (2004).
2. K. Schönert, *DECHEMA-Monographien* **79** 67-88 (1976).
3. J. Bertling, C. Eloo, Verdichtetes Kohlendioxid als Prozessadditiv zur Herstellung polymerer und mikronisierter Nanokomposite. Fraunhofer-Institut für Umwelt-, Sicherheits- und Energietechnik -UMSICHT-, Oberhausen, technical report (2009).
4. W. Peukert, L. Vogel, *Chem. Eng. Techn.* **24** 945-950 (2001).
5. A. Weber, U. Teipel, H. Nirschl, *Chem. Eng. Technol.* **29** 642-648 (2006).
6. M. Liang, C. Lu, Y. Huang, C. Zhang, *J. Appl. Polymer Sci.* **106** 3895-3902 (2007).
7. S. Molina-Boisseau, N. Le Bolay, *Powder Technol.* **105** 321–327 (1999).
8. J. Schmidt, M. Plata, S. Tröger, W. Peukert, *Powder Technol.* **228** 84–90 (2012).
9. J. Schmidt, M. Sachs, C. Blümel, B. Winzer, F. Toni, K.-E. Wirth, W. Peukert, *Powder Technol.* **261** 78–86 (2014).
10. A. Kwade,. *Powder Technol.* **105** 382–388 (1999).
11. F. J. Vallejo, J. I. Eguiazàbal, J. Nazàbal, *J. Appl. Polym. Sci.* **80** 885–892 (2001).
12. C. A. Boye, J. R. Overton, *Bulletin of the Amercian Physical Society* **19**, 352 (1974).
13. R. Jakeways, I. M. Ward, M. A. Wilding, I. H. Hall, I. J. Desborough, M. G. Pass, *J. Polym. Sci. Polym. Phys.* **13** 799–81 (1975).
14. N. Tomar, S. N. Maiti, *J. Appl. Polym. Sci.* **113** 1657–1663 (2009).

Enhancement of Impact Strength of PA11 by Dispersing PBS Particles without Reduction of Flexural Modulus

Tomokazu Moriyama[a*], Nobuhiro Sumiya[a], Tomotaka Saito[a] and Yan Hao[a]

[a] *Research Laboratories, Toyota Boshoku Corporation, 1-1, Toyoda-cho, Kariya, Aichi, Japan*

tomokazu.moriyama@toyota-boshoku.com

Abstract. 10-30 wt% poly(butylene succinate) (PBS) dispersing into polyamide 11 (PA11) without adding a compatibilizer was improved the impact strength and maintained the flexural modulus comparison with PA11. The impact strength was controlled by the difference in shrinkage ratios between PA11 and PBS, and was three times that of neat PA11 (an increase from 13 to 40 kJ/m^2). The PA11/PBS alloy had a sea-island structure, with a PA11 sea and PBS islands. Gaps of several tens of nanometers were observed at the PA11-PBS interfaces, and the sea and islands were connected by string-shaped bridge phases 10 nm in diameter. These bridge phases suggest the formation of dilational stress fields in the PA11 because of tensile stress due to PBS particle shrinkage during the cooling of the melt compounded PA11/PBS alloy. The mechanism of impact strength enhancement is discussed in terms of the formation of these dilational stress fields.

Keywords: bio-plastic, impact strength, polymer alloy, thermal shrinkage.
PACS: 81.05.Lg

INTRODUCTION

Bio-plastics are widely used in dishes, packaging containers, automotive parts, and electronic products [1]. To expand bio-plastic usage, improvements in impact strength, heat resistance, and flame resistance are required. Polyamide 11 (PA11) has a relatively high impact strength for a bio-plastic, and is made from inedible plant oil extracted from castor seeds. However, even for PA11, the impact strength (8 - 15 kJ/m^2) is less than one-fifth that of polycarbonate (70 – 80 kJ/m^2).

As a method of improving the impact strength of plastics, alloying a matrix polymer with a compatibilizer and soft materials such as rubber has been studied [2-5]. In contrast, elastic modulus reduction due to the rubber component is pointed out [6]. Satisfying both impact strength and elastic modulus is required. However, effective method has not been developed.

Recently, Moriyama *et al.* reported a novel method to enhance the impact strength of PA11 while maintaining the elastic and the flexural modulus by adding PBS [7]. The enhancement is considered to utilize dilational stress field at around added PBS. In this paper, we discuss from the perspective of controlling the shrinkage ratio of the PA11/PBS alloy component.

MATERIALS AND METHODS

A PA11/PBS alloy was prepared by melt mixing. **TABLE 1** shows the thermal properties of PA11 and PBS used in this work. The PA11 was purchased from Arkema Japan and the PBS was supplied by Mitsubishi Chemical. The PA11 and PBS pellets were dried in a vacuum oven for 6 h at 80 °C to eliminate water. The dried pellets were melt mixed (barrel temperature: 210 °C, screw rotation speed: 75 r.p.m., mixing time: 4 min) using an internal mixer (Toyoseiki, Tokyo, Japan; Labo Plsastomill model 4C-150) at a predetermined mixing ratio. The resulting PA11/PBS compounds were formed into rectangular bars using a laboratory hot press (210 °C, 20 MPa).

The Charpy impact strength of V-notched specimens (depth of 2mm, apex radius of 0.25mm) was measured using an impact tester (Yasuda Seiki Seisakusho; No. 258-PC) based on ISO 179-1. The flexural modulus was measured using an Autograph (Shimadzu; AG-250kNG) with a crosshead speed of 1mm per minute. The sample morphology was investigated using a scanning electron microscope (JEOL; JSM-6510LV) with an acceleration voltage of 5 kV. The specimens were fractured in liquid nitrogen and coated with platinum.

TABLE 1. Thermal properties of PA11 and PBS. T_c and T_m are crystallization and melting temperature, respectively.

Polymer	T_c (°C)	T_m (°C)	Shrinkage ratio (%)	Supplier	Product code
PA11	168	187	0.81	Arkema	RilsanB BESN O TL
PBS	90	114	1.31	Mitsubishi Chemical	Gs Pla FZ91PD

RESULTS AND DISCUSSION

The Charpy impact strengths of several PA11/PBS alloys are shown in **FIGURE 1**. The impact strengths of alloys with 90/10 to 70/30 PA11/PBS weight ratios were higher than that of PA11. The impact strength of the 80/20 alloy was the maximum observed, with a value (32.8 kJ/m^2) approximately 2.5 times that of PA11 (12.9 kJ/m^2). The flexural modulus of several PA11/PBS alloys are shown in **FIGURE 2**. The flexural modulus of 90/10 to 70/30 alloys were slightly higher than that of PA11. The impact strength of PA11 was significantly improved by the addition of an appropriate amount of PBS without any compatibilizer or soft material, and no deterioration of the flexural modulus occurred. In contrast, adding more than 30 wt% of PBS decreased both the impact strength and the flexural modulus.

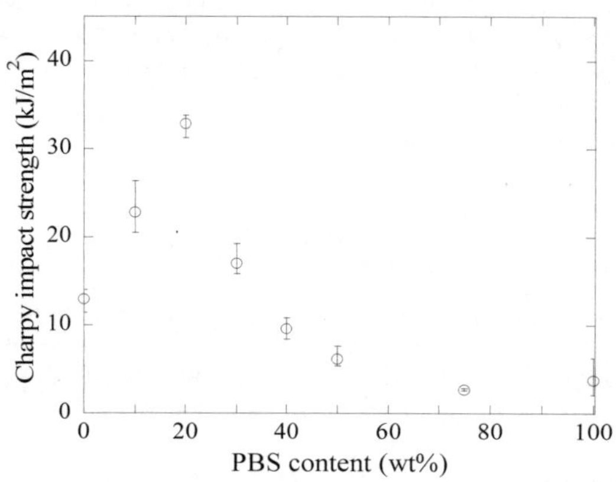

FIGURE 1. Charpy impact strengths of several PA11/PBS alloys.

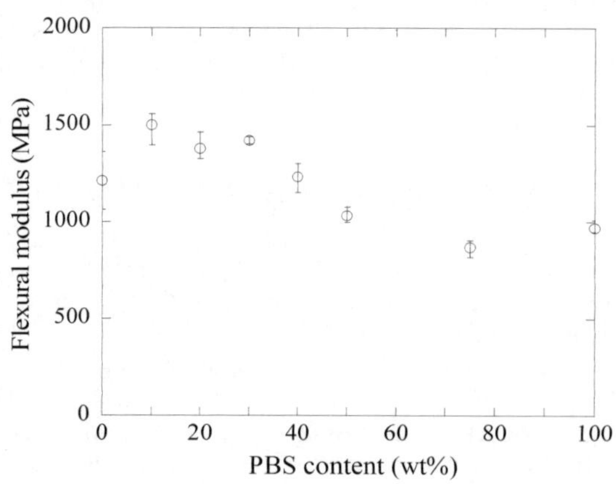

FIGURE 2. Flexural modulus of several PA11/PBS alloys.

Scanning electron micrograph of 80/20 alloy is shown in **FIGURE 3 (a)**. A sea-island structure with a PA11 sea and PBS islands with a particle size of less than 1 μm was observed, and gaps were formed at the interface between the PA11 and the PBS. In contrast, the 25/75 alloy contained PA11 islands in a PBS matrix, no gaps were observed at the interface **(FIGURE 3 (b))**. **FIGURE 4** shows a magnified image of the interface between the PA11 matrix and a PBS particle in the 80/20 alloy, which had the highest impact strength. String-shaped bridge phases approximately 10 nm in diameter were observed in the gap, and connected the PA11 and the PBS. In the alloys with high impact strengths, 1 μm or smaller PBS particles were dispersed in the PA11 matrix, and bridge phases were observed at the interface between the PA11 and the PBS. The impact strength enhancement was probably related to the bridge phase formation.

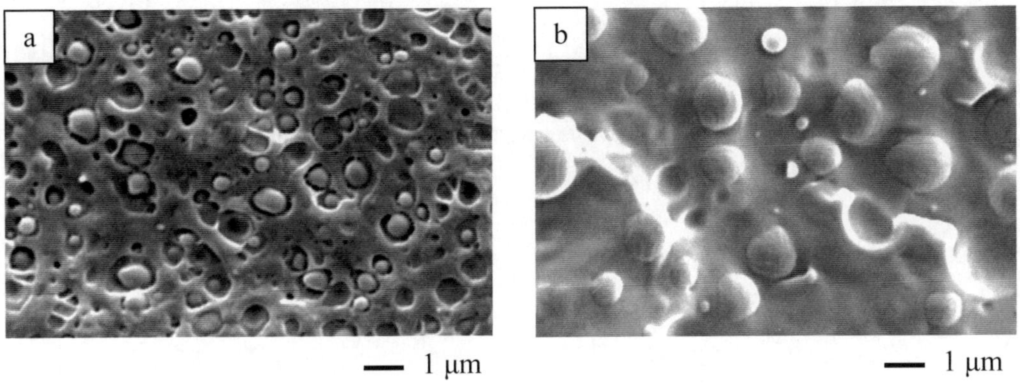

FIGURE 3. Scanning electron micrographs of the fractured surfaces of PA11/PBS alloys: (a) 80/20, (b) 25/75 in weight ratio.

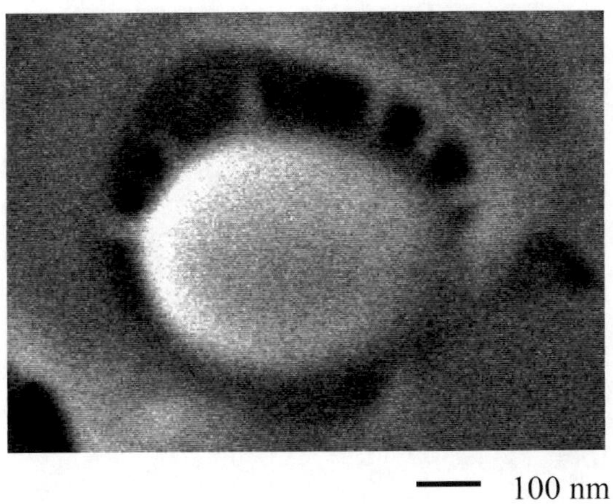

FIGURE 4. Magnified Scanning electron micrograph of interface between the PA11 matrix and a PBS particle. The weight ratio is 80/20.

The relationship between bridge phase formation and impact strength enhancement can be explained as follows. During the cooling process from the molten PA11 and PBS compound, PA11 begins to crystallize around 168°C, and the amorphous region is still in a rubbery state. Next, PBS begins to crystallize and shrink around 90°C. Because the shrinkage ratio of PBS is larger than that of PA11, the PA11 matrix is pulled by the PBS particles. The tensile stress leads to form a dilational stress field in the PA11 around the PA11-PBS interfaces. Hashima *et al.* reported that this dilational stress would lead to a loose segment-segment aggregation that may be favorable for local segmental motion, and making the plastic more ductile [8]. The impact strength improvement of PA11 can be attributed to the dilational stress field formed by the shrinkage ratio difference between PA11 and PBS, and the bridge phases are evidence that the PBS particles pulling the PA11 matrix.

This explanation also leads to the proposition that a larger shrinkage ratio difference between PA11 and PBS forms a larger dilational stress field, resulting in a higher impact strength, and vice versa. Since PA11 and PBS are crystalline polymers, the shrinkage ratio for the polymers is controllable by the cooling rate near the crystallization temperature. To control the shrinkage ratio difference between PA11 and PBS, two cooling conditions were attempted in the preparation of the PA11/PBS alloys. **FIGURE 5** shows temperature variation of the original rapid cooling condition (blue line) and conditions for increasing (red line) and decreasing (green line) shrinkage ratio difference, respectively. Shrinkage ratio difference between the PA11 (shrinkage ratio, S_{PA11}, of 0.78%) and the PBS (shrinkage ratio, S_{PBS}, of 1.78%), $\Delta S = S_{PBS} - S_{PA11}$, of red line cooling condition was 1.0%. Shrinkage ratio difference of green line cooling condition was 0.26% (S_{PA11} = 1.01%, S_{PBS} = 1.27%). These shrinkage ratio differences were larger and smaller than the original difference (ΔS = 0.5%, blue line in **FIGURE 5**). The relationship between the impact strength and the shrinkage ratio difference is shown in **FIGURE 6**. The impact strength was improved by increasing the difference in shrinkage ratio, and the impact strength was approximately 3 times that of neat PA11 for ΔS = 1.0%. That is, the impact strength is controlled by the shrinkage ratio difference.

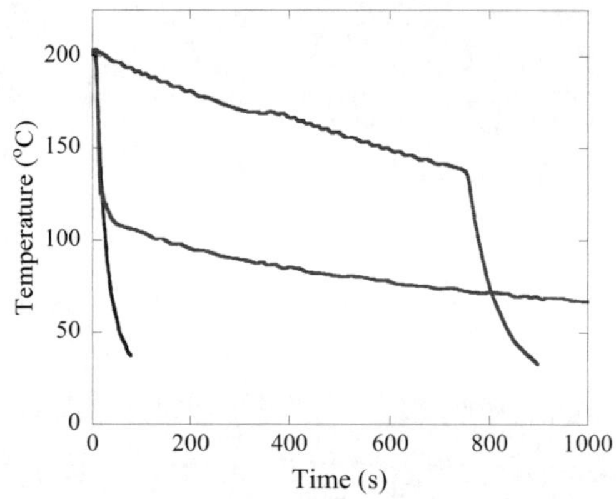

FIGURE 5. Temperature variation during the cooling from melt to solid state.

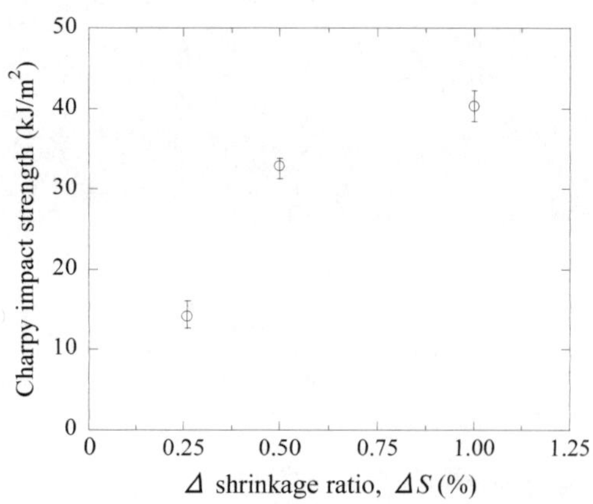

FIGURE 6. Influence of Δ shrinkage ratios, ΔS, on charpy impact strength. The weight ratio of PA11/PBS was 80/20 in all samples.

CONCLUSION

The impact strength of PA11 was successfully improved while maintaining the flexural modulus by dispersing of PBS particles without adding any compatibilizer. The morphology of the PA11/PBS alloy had a sea-island structure, and bridge phases were formed between the PA11 matrix and the PBS particles. The presence of this bridge phase suggests formation of a dilational stress field that enhanced the impact strength of the PA11. Since the dilational stress field in the PA11 matrix is formed by tensile stress due to shrinkage of the PBS, the impact strength improved with increasing shrinkage ratio difference between the PA11 and the PBS. An impact strength of 40.3 kJ/m^2, about 3 times that of neat PA11, was obtained for PA11/PBS with an 80/20 weight ratio. The conditions required for this method to form a dilational stress field are both a shrinkage ratio difference and a melting point difference between the matrix and the dispersed polymer alloy particles.

REFERENCES

1. M. M. Reddy, et al., *Prog. Polym. Sci.*, **38**, 1653-1689 (2013).
2. H. Chen, et al., *J. Appl. Polym. Sci.*, **77**, 928-933 (2000).
3. M. Schneider, et al., *J. Mater. Sci.*, **32**, 5191-5204 (1997).
4. K. Cho, et al., *Polymer*, **39**, 3073-3081 (1998).
5. S. Wu, *Polymer,* **26**, 1855-1863 (1985).
6. A. Arostegui, et al., *Polymer*, **42**, 9565-9574 (2001).
7. T. Moriyama, et al., *Polym. J.*, **48**, 221-224 (2016).
8. K. Hashima, et al., *Polymer*, **51**, 3934-3939 (2010).

High performance millimeter wave absorbers derived from uneven distribution of conducting and magnetic engineered nanoparticles in biphasic polymeric blends

Shital Patangrao Pawar, Mounika Gandi, Suryasarathi Bose*

Department of Materials Engineering, Indian Institute of Science, Bangalore-560012
Presenting author's email: spp.pawar@gmail.com
Reference author's email: sbose@materials.iisc.ernet.in

Abstract. High performance millimeter wave absorbers were designed by the uneven distribution of electromagnetic (EM) active nano-inclusions like conducting multiwall carbon nanotubes (MWNTs) and magnetic ultra-small ferrite (Fe_3O_4) particles (of 2-3 nm) nucleated on the surface of reduced graphene oxide sheets (rGO-Fe_3O_4) in immiscible blends of polycarbonate (PC) and poly(styrene -co-acrylonitrile (SAN). This further resulted as *micro-absorbers* leading to exceptional millimeter wave absorption (95.4%). The attenuation of millimeter wave was achieved by enhanced bulk electrical conductivity of 5 S/cm along with electrically and magnetically lossy nano-inclusions through large interfacial scattering. The prepared blends containing MWNTs and engineered rGO-Fe_3O_4 manifested in total shielding effectiveness (SE_T) of -50.7 dB suggesting negligible transmission of millimeter wave radiation. By considering all EM parameters, it is revealed that the unique strategy of uneven distribution and engineered nano-inclusion results in extraordinary synergy in millimeter wave absorption.

Keywords: MWNTs, reduced graphene oxide sheets, ferrite nanoparticles, micro-absorbers
PACS: 81

INTRODUCTION

Recently, the use of electronic devices and wireless equipment has led to an unavoidable EM pollution resulting in significant impact on civil and military sectors. Since, the EM radiation emitted by electronic devises can interfere with nearby equipment and leads to device malfunctioning. Therefore, shielding of this radiation becomes an indispensable task. Owing to its tailor-made electrical and EM properties polymeric nanocomposites containing EM active nano-inclusions have gained great attention as EMI shielding material [1]. Since absorption is a well appreciated mechanism of millimeter wave attenuation, various studies have been focused to develop microwave absorbers [2-4]. Henceforth, the focus of this study has been concentrated on minimization of the reflection of high frequency electromagnetic wave from material surface and to facilitate the absorption in broad band frequency range.

In this study, we have made an attempt to design light weight composites based on PC/SAN blends as EMI shielding materials with enhanced millimeter wave absorption. The high aspect ratio MWNTs was heterogeneously dispersed in one of the phases of blends to establish conducting pathway which play key roles in millimeter wave attenuation. In this study, ultra-small magnetite ferrite (Fe_3O_4) nanoparticles were nucleated *insitu* on the surface of graphene oxide sheets (rGO-Fe_3O_4) and heterogeneously dispersed in the blends. Furthermore, the synergistic effect of MWNTs and engineered nano-inclusions on millimeter wave attenuation/absorption was systematically assessed in the PC/SAN blends

EXPERMENTAL SECTION

Materials

Polycarbonate was obtained from Sabic. Poly(styrene-co-acrylonitrile) (M_W =165,000 g/mol) and iron oxide (Fe_3O_4) particles were procured from Sigma Aldrich. MWNTs (Avg. diameter = 9.5 nm and Avg. length = 1.5 μm) was obtained from Nanocyl, Belgium. Iron chloride ($FeCl_3$), ascorbic acid and sodium hydrogen bicarbonate ($NaHCO_3$) were procured from Sigma Aldrich. Analytical grades of solvents were procured from commercial sources.

Synthesis of Ferrite Nanoparticles Nucleated on Graphene Sheets (rGO-Fe_3O_4)

As reported earlier [5], the graphene oxide (GO) sheets were prepared using modified Hummer's method. Chen et al. [6] achieved ultra-small iron oxide nanoparticles on the surface of GO. Similar approach was adopted here with few modifications to decorate magnetite nanoparticles on the surface of GO. Thus obtained nanoparticles were annealed under hydrogen atmosphere and referred to as rGO-Fe_3O_4.

Blend Preparation

PC/SAN (60/40) blends were prepared by melt extrusion process at 260 °C and counter rotating speed of 60 rpm using micro compounder (Haake Minilab-Π) for 20 min. The polymers were dried at 80 °C for 12 h under vacuum prior to melt mixing.

Characterization

Perkin Elmer FT-IR was employed to assess the functionalization of graphene sheets. The X-ray diffraction (XRD, PANalytical) was employed to assess structural properties of nanoparticles. Transmission electron micrographs of nanoparticles were recorded using a JEOL transmission electron microscope (TEM). The phase morphology of blends was assessed using A ULTRA 55 scanning electron microscope (SEM). Thermal properties of nanoparticles was studied using NETZSCH STA 409 thermo gravimetric analyser. The room temperature electrical conductivity was studied using a broadband dielectric spectrometer (Novocontrol). The saturation magnetization of nanoparticles was assessed using vibrating sample magnetometer (VSM). Vector network analyzer (VNA) MS4642A from Anritsu was employed to evaluate the scattering parameters.

RESULTS AND DISCUSSION

Synthesis and Characterization of the Nanoparticles

Figure 1a shows the TEM images of rGO-Fe_3O_4 and uniformly decorated ultra-small Fe_3O_4 nanoparticles on the surface of graphene sheet is clearly indicated by arrows. The high magnification TEM micrograph shows discrete Fe_3O_4 nanoparticles present on the graphene sheet (see inset of figure 1a). The mean size of the magnetite nanoparticles was ca. 2-5 nm. Figure 1b depicts the XRD spectra of rGO-Fe_3O_4. The XRD spectra supports formation crystalline Fe_3O_4 nanoparticles and all diffraction peaks closely resembles the JCPDS card no. 01-086-1360. Additional peak present at 25.6° and 44.7° represents the reduced graphene oxide. Figure 1c shows the TGA for rGO-Fe_3O_4.The residual weight of 49% obtained corresponds to Fe_3O_4 nanoparticles present on the surface of graphene sheets. Figure 1d depicts magnetic hysteresis loop for synthesized rGO-Fe_3O_4 nanoparticles. The synthesized rGO-Fe_3O_4 showed a saturation magnetization (M_s) of 23.5 emu/gm along with large coercivity of 346 Oe.

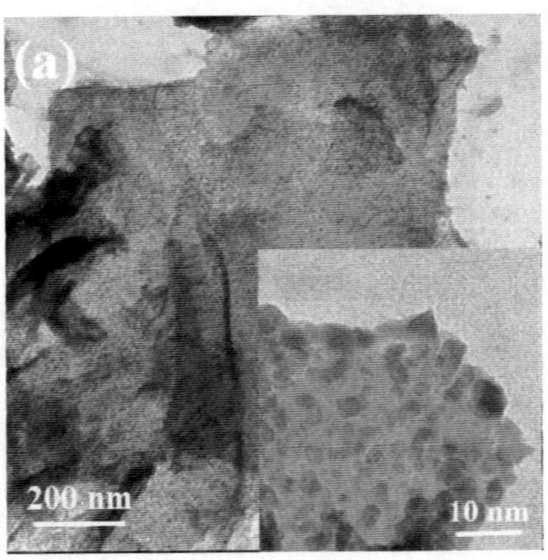

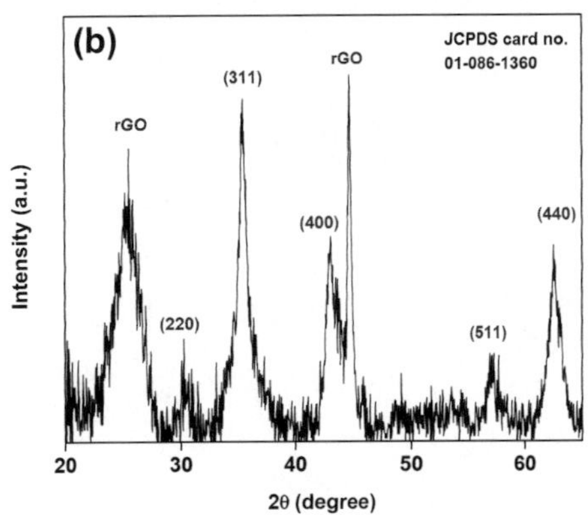

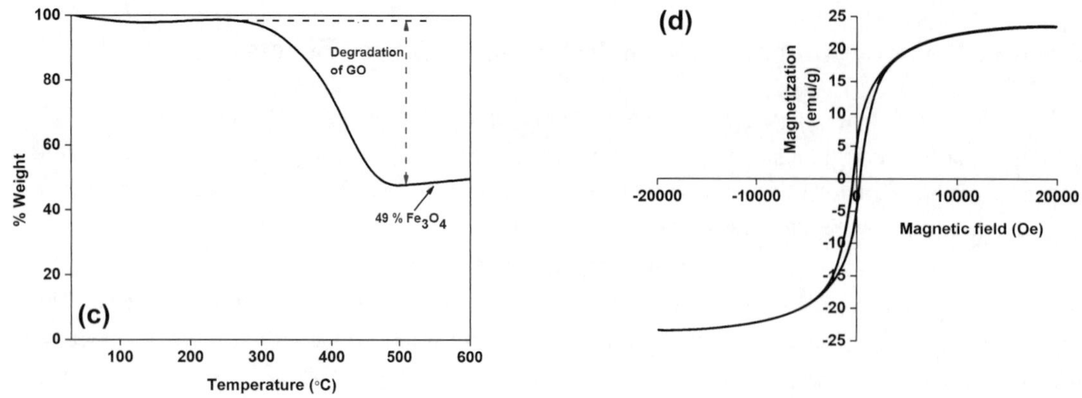

FIGURE 1. (a) TEM micrographs, (b) XRD, (c) TGA and (d) magnetization hysteresis loop for rGO-Fe_3O_4 nanoparticles.

Uneven Distribution of 'Micro-Absorbers' and Electrical Conductivity

Figure 2a depicts SEM micrographs of blends containing 10 wt% rGO-Fe_3O_4 along with MWNTs. The selective localization of MWNTs and rGO-Fe_3O_4 was realized mostly in PC phase of the blends. Herein, the GO sheets were used to tune the localization of Fe_3O_4 particles preferentially in the PC phase. This approach was quite successful to drive Fe_3O_4 particles to a given phase which otherwise might re-distribute during blending. This further facilitates the dense connecting network of high aspect ratio conducting MWNTs aided by rGO-Fe_3O_4.

Figure 2b depicts room temperature electrical conductivity for PC composites and blends. The PC composites containing MWNTs depicted dc electrical conductivity of $2.8*10^{-6}$ S/cm, whereas, the PC/SAN blends resulted in significantly higher conductivity over PC composites at a similar concentration of MWNTs loading. Interestingly, the blends containing MWNTs and rGO-Fe_3O_4 manifested further increase in electrical conductivity. This significant enhancement in conductivity ascribed to the synergistic network of high aspect ratio MWNTs aided by conducting rGO-Fe_3O_4.

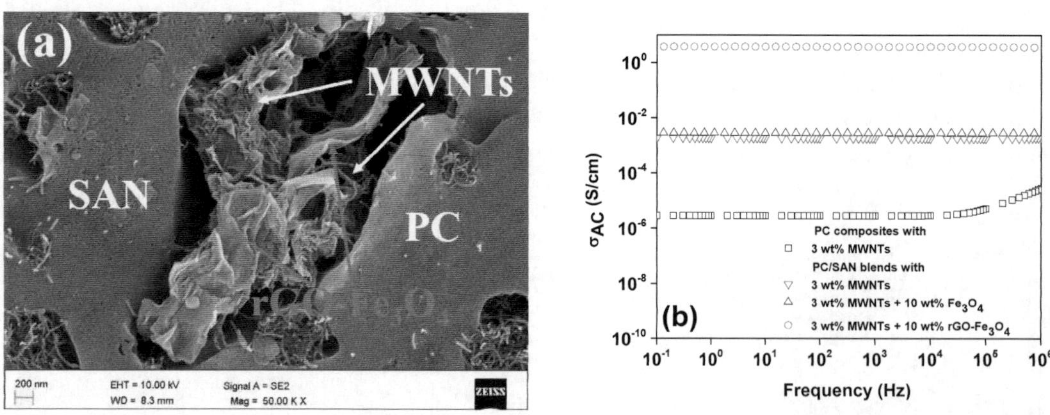

FIGURE 2. (a) SEM image of PC/SAN (60/40) blends with MWNTs and rGO-Fe_3O_4 nanoparticles and (b) Frequency dependent ac electrical conductivity.

Attenuation of Microwave Radiation: Absorption through "Micro-Absorbers"

Figure 3a depicts SE_T over broad band frequency range. The PC composites manifested in SE_T of -17.3 dB at 18 GHz. Interestingly, the 60/40 blends showed strikingly enhanced SE_T of -25.1 dB at a given concentration of MWNTs (at 18 GHz). Furthermore, addition of 10 wt% Fe_3O_4 particles along with MWNTs manifested further increase in SE_T of -37 dB. Among all the nanocomposites, blends consisting of MWNTs and rGO-Fe_3O_4 manifested highest SE_T of -50.7 dB suggesting more than 99.99% attenuation of the incoming millimeter wave radiation. This exceptional enhancement in SE_T corresponds to the synergetic effect of enhanced electrical conductivity facilitated through dense connecting network of MWNTs and rGO-Fe_3O_4. Figure 3b shows SE_A at 18 GHz frequency. The blends containing various nano-inclusion manifested higher SE_A over composites. This is mainly ascribed to inter-phase and inter-particles multiple-scattering within the shield material. Moreover, synergistic absorption of both electrical and magnetic field was realized in the blends containing conducting (i.e. MWNTs) and magnetic (i.e. Fe3O4) nano-inclusions. As realized from SEM study, the EM active nano-inclusions are preferentially resided in PC phase of the blends manifesting heterogeneous dielectric media and large interfaces for scattering of millimeter wave. This uneven distribution of nano-inclusions and large dielectric and magnetic losses resulted in extraordinary millimeter wave attenuation mostly through absorption.

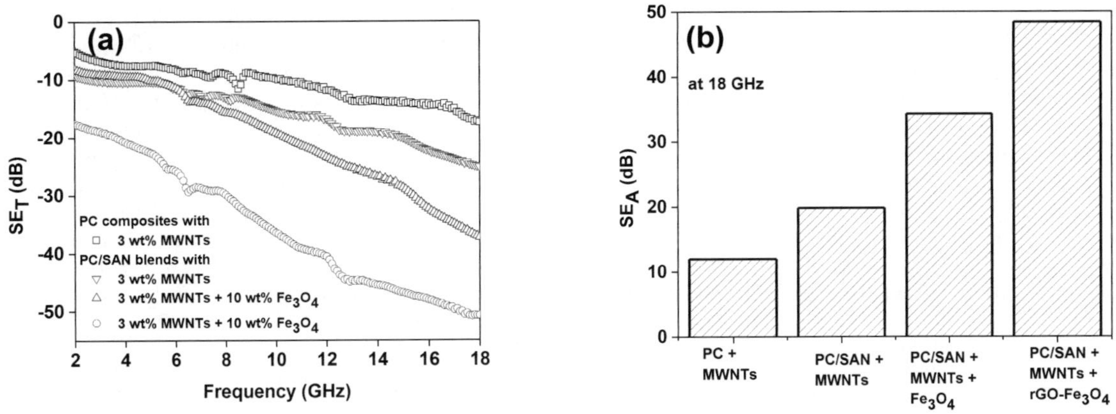

FIGURE 3. (a) SE_T over broad band frequency range and (b) SE_A for PC composites and various blends.

CONCLUSIONS

Taken together, an exceptional millimeter wave absorber was developed with uneven distribution of EM active lossy nano-inclusions like MWNTs, rGO and magnetic Fe_3O_4 in immiscible polymeric blends. This unique strategy paved way for exceptional absorption (95.4%) of millimeter wave radiation. The blends containing rGO-Fe_3O_4 along with MWNTs exhibited a high SE_T of -50.7 dB, suggesting more than 99.99% blocking of the incident millimeter wave radiation.

ACKNOWLEDGMENTS

The authors would like to acknowledge the Department of Science and Technology (DST), India, for providing the financial assistance.

REFERENCES

1. S. P. Pawar, S. Biswas, G. P. Kar and S. Bose, *Polymer* **84**, 398-419 (2016).
2. A. Ohlan, K. Singh, A. Chandra and S. Dhawan, *Appl. Phys. Lett.* **93**, 053114-053111 (2008).
3. S. P. Pawar and S. Bose, *ChemNanoMat* **1**, 603-614 (2015).
4. M.-S. Cao, W.-L. Song, Z.-L. Hou, B. Wen and J. Yuan, *Carbon* **48**, 788-796 (2010).
5. S. Liu, T. H. Zeng, M. Hofmann, E. Burcombe, J. Wei, R. Jiang, J. Kong and Y. Chen, *ACS nano* **5**, 6971-6980 (2011).
6. Y. Chen, B. Song, L. Lu and J. Xue, *Nanoscale* **5**, 6797-6803 (2013).

Combined effects of ultrafine full-vulcanized powdered rubber and β nucleating agent on mechanical properties and heat distortion temperature of isotactic polypropylene

Yanling Zhu, Qiang Fu*

College of Polymer Science and Engineering, State Key Laboratory of Polymer Materials Engineering, Sichuan University. Chengdu 610065, China.

Author's email:sdzylsc@163.com
Reference author's email: qiangfu@scu.edu.cn

Abstract. In this work, ultrafine full-vulcanized powdered rubber (EA-UFPR) and β nucleating agent were incorporated into isotactic polypropylene (iPP) in order to investigate the combined effects on its mechanical properties and its heat distortion temperature (HDT). A large improvement of Izod impact strength verified as a synergistic effect had been obtained by introducing a small amount of EA-UFPR ($\leq$ 2 wt %) into β-nucleated iPP. Decreased crystallization kinetics was found with increasing the content of EA-UFPR. More interestingly, the transformations of crystalline morphologies from β-spherulites to bundle-like β-crystals were detected due to a possible suppressed self-assembly of nucleating agent after adding small amounts of EA-UFPR. On one hand, the largely improved Izod impact strength can be attributed to the synergistic effects between EA-UFPR particles and β-crystalline morphology on promoting the plastic deformation of matrix. On the other hand, the unique interconnected structures formed by the bundle-like crystals and dispersed EA-UFPR particles were proposed for the enhancement in toughness and HDT. Our work provides an effective way to improve the toughness and heat distortion temperature of iPP by adding small amounts of EA-UFPR ($\leq$ 2 wt %), which are crucial for the industrial application of iPP.

Keywords: mechanical properties, heat distortion temperature, crystallization behaviour, crystalline morphology
PACS: 62.20.-x Mechanical properties of solids

INTRODUCTION

Polypropylene is one kind of commonly used semicrystalline plastics which was processed easily, lightweight, inexpensive, and comprehensive mechanical performance.[1,2] However, the dominating crystal of isotactic polypropylene processed under traditional conditions is thermodynamic stable α-crystal which is undesirable for final impact toughness especially low temperature toughness and low heat distortion temperature.[3] Poor toughness especially low temperature toughness of products always significantly restricts the further application in industrial application. The research on achieving the polypropylene with balanceable toughness and higher heat distortion temperature has always attracted attention from scholars for decades.[4-7]

It is well known that the generation of β-crystal with loose-arranged lamella and tight connection between spherulites can improve the impact strength in polypropylene, but there was a weakness that the formation of β - crystal decreases the tensile strength and modulus.[4,8] Ultrafine powdered fully vulcanized rubber (UFPR), a new kind of highly cross-linked nano-particle rubbers by irradiating rubber latex and spray drying. Many works have proved that blending with UFPR is an effective way to balance the toughness and tensile strength of polypropylene.

Introducing β-nucleating agent or UFPR alone into iPP can improve the toughness and heat distortion temperature. However, saturation points of the improvement emerge with the increase of the concentrations. In this work, a small amount of UFPR and β-nucleating agent was incorporated into iPP simultaneously to further improve the toughness and heat distortion temperature of iPP. On one hand, exploring the combined effects of UFPR and β-nucleating agents (β-NA) on toughness and heat distortion temperature of iPP, and on the other hand, investigating the influence of UFPR on the crystallization kinetics and crystalline morphology of β nucleated isotactic polypropylene (β-iPP).

EXPERIMENTAL SECTION

A commercially available isotactic polypropylene, purchased from Dushanzi Petroleum Chemical Co. China, with a melt flow rate (MFR) of 3 g/10 min (230 °C, 21.6 N) was used as the basal resin. Ultrafine Full-Vulcanized Powdered Rubber abbreviated to EA-UFPR (Narpow VP-362) with an average single particle size of about 30 nm, was kindly supplied by Beijing BHY chemical Industry New Technology Corp. The rare earth β-nucleating agent (β-NA), marked as WBG-II, was provided by Guangdong Winner Functional Materials Co. (Foshan, China).

RESULTS AND DISCUSSION

Effects of EA-UFPR concentrations on mechanical property of β-iPP

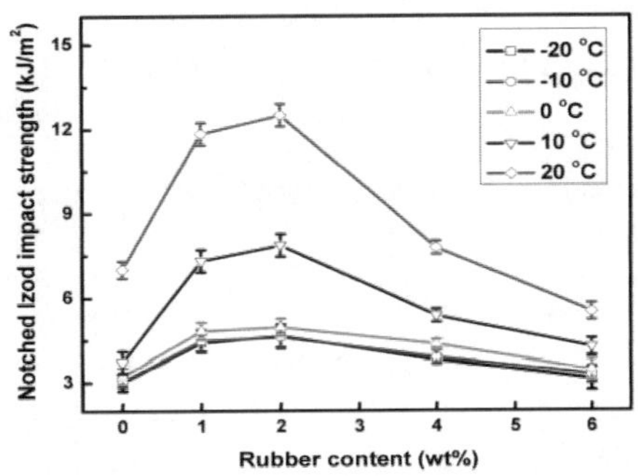

FIGURE 1. Concentration dependence of notched Izod impact strength tested at different temperatures

In order to explore the effects of EA-UFPR on toughness of β-nucleated iPP (Ternary compounds), the notched Izod impact strength were measured at different temperatures as shown in **FIGURE 1**. For pure β-iPP, the notched Izod impact strength remains unchanged at the temperature under 0 °C, while, an obvious improvement of the value occurs at 20 °C, indicating that an enhancement in impact strength of β-nucleated iPP can only be achieved above the glass transition temperature of iPP ($T_{g,iPP}$≈0 °C). Similar results have been extensively reported in the previous literatures. A slight increase in notched Izod impact strength is obtained from 3.0 kJ/m^2 to 4.5 kJ/m^2 at -20 °C after adding 1 wt % EA-UFPR into β-nucleated iPP, while an significant increase in impact strength is achieved (from 6 kJ/m^2 to 13 kJ/m^2) at 20 °C, and similar results are observed in I0.05G2R. However, with the further increasing of EA-UFPR content in β-iPP, the impact strength decreases. All in all, positive interaction between EA-UFPR and WBG can be achieved for systems with lower content of EA-UFPR, while inter-hindered effect generates results in a drop of impact strength for samples with excessive EA-UFPR. Therefore, adding a small amount of EA-UFPR (less than 2 wt %) into β-iPP is effective for the improvement of toughness, especially when the testing temperature is above the glass transition temperature of iPP.

Dependences of heat distortion temperature on EA-UFPR

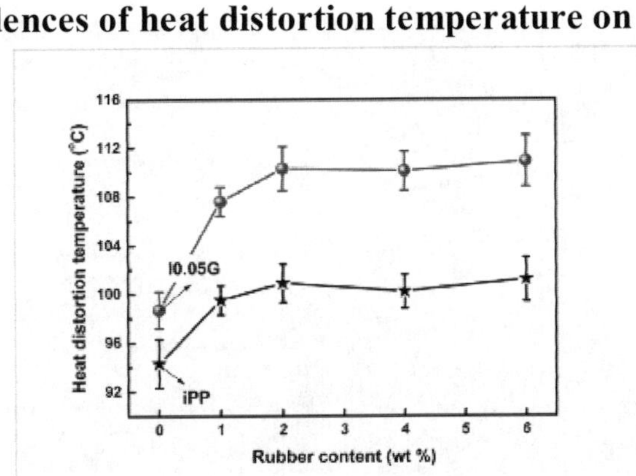

FIGURE 2. Effects of EA-UFPR contents and β nucleating agent on heat distortion temperature (HDT).

Heat distortion temperature (HDT) is a measurement of the upper boundary for the dimensional stability of a material under a specific load and temperature, and a critical parameter for product design and it is widely used in automotive applications. Hence, the HDT of samples was measured in our work as shown in **FIGURE 2**. Obvious enhancements of HDT were obtained after adding small amounts of EA-UFPR for both α-iPP and β- iPP. However, there seems to be a saturated enhancement in HDT at the point with 2 wt % EA-UFPR, while no more obvious increase of HDT is observed above this value. Clearly, the HDT of β-iPP is higher compared with that of pure iPP, which is well accordance with previous studies. For example, the improvement of HDT for α-iPP is only 4 °C (from 94 °C to 98 °C) after adding1 wt % of EA-UFPR, while it comes to 10 ℃ (from 98 °C to 108 °C) for β-nucleated iPP (I0.05G1R). In this case, a synergistic effect on HDT is observed between EA-UFPR and β-NA. Several factors including flexural modulus, the degree of crystallinity, Tg, thermal history and the fillers has been studied for influencing the HDT of polymeric materials. The reasons for the increments in HDT of our study will be discussed below.

Dependence of crystallization behavior on EA-UFPR

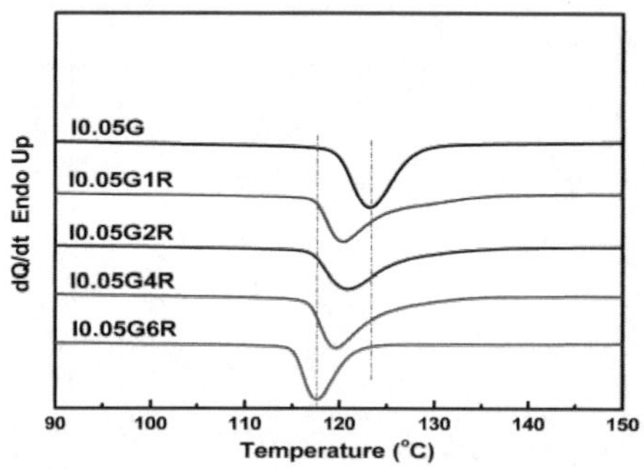

FIGRUE 3. DSC thermograms of crystallization curves for β-iPP containing various EA-UFPR concentrations.

To investigate the effects of EA-UFPR concentration on non-isothermal crystallization behavior of β-iPP, the designed programs were carried out on DSC, as shown in **FIGURE 3**. The peaks of crystallization curves decrease with increase of EA-UFPR concentration indicating the decreased crystallization kinetics after adding EA-UFPR. It is proposed that the decrease in crystallization kinetic after adding EA-UFPR can be attributed to two aspects: (1) the slowing down of the assembly of β-nuclei to generate big nucleation spots because of the interaction between EA-UFPR particles and the nuclei of β-nucleation, and (2) the obstruction effect of rubber particles on the diffusion of iPP chain.

Evolutions of crystalline morphology during isothermal crystallization

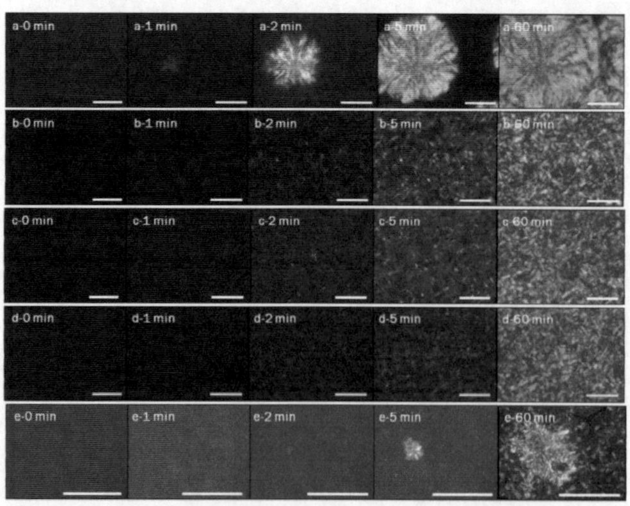

FIGURE 4. POM photographs showing the evolution of crystalline morphology at 138 °C (a) I0.05G, (b) I0.05G1R, (c) I0.05G2R, (d) I0.05G4R, (e) I0.05G6R. Scale bars represent 100 μm.

To explore the effect of EA-UFPR on crystalline morphology of β-nucleated iPP, the evolutions in crystalline morphologies during isothermal process were detected by polarized optical microscopy (POM), as shown in FIGURE 4. Typical dominating β-spherulites are obtained of β-nucleated iPP (I0.05G) in Fig. 9 (a). However, the nucleus density increase dramatically after adding 1 wt % EA-UFPR I0.05G1R (**FIGURE 4** (b-1 min)). Besides, the size of the crystal bundles of β-iPP with EA-UFPR is much smaller than that of I0.05G. It is demonstrated that the adding of EA-UFPR refines the size of the crystalline superstructures because of the increment in nucleus density, and promote the transformation of crystalline morphology from spherulitic to needle-like type. For I0.05G6R, isolated β-spherulites surrounding with large numbers of α-crystals was obtained probably involved with hindering effect of EA-UFPR on the diffusion of iPP chains.

CONCLUSIONS

In our work, synergetic effects between β-NA and EA-UFPR on notched Izod impact strength, HDT, crystallization behavior and crystalline morphology of iPP are investigated in detail. Notched Izod impact strength significantly enhances after adding a small amount of EA-UFPR into β-iPP, and an obvious improvement of HDT with adding small amounts of EA-UFPR is also observed. The crystallization kinetic of β-iPP decrease and reducing size of crystal bundles is obtained after adding EA-UFPR. It is tentatively proposed that the improvement in toughness could be attributed to the combined effects of EA-UFPR and β-NA on plastic deformation due to the largely reduced superstructures size and possible formation of interconnected structure composed of the bundle-like crystals and dispersed EA-UFPR particles.

ACKNOWLEDGMENTS

This work was supported by the National Natural Science Foundation of China (51421061 and 51210005) and the Special Funds for Major State Basic Research Projects of China (2011CB606006).

REFERENCES

1. Y. H. Chen, Y. M. Mao, Z. M. Li and B. S. Hsiao, *Macromolecules* **43**, 6760-6771 (2010).
2. H. Huo, S. Jiang, L. An and J. Feng, *Macromolecules* **37**, 2478-2483 (2004).
3. L. Wang and M. B. Yang, *RSC Advances* **4**, 25135-25147 (2014).
4. F. Luo, C. Xu, K. Wang, H. Deng, F. Chen and Q. Fu, *Polymer* **53**, 1783-1790 (2013).
5. J. Varga, A. Menyhard, *Macromolecules* **40**, 2422-2431 (2007),.
6. F. Luo, C. Geng, K. Wang, H. Deng, F. Chen, Q. Fu and B. Na, *Macromolecules* **42**, 9325-9331 (2009).
7. Y. Zhu, F. Luo, H. Bai, K. Wang, H. Deng, F. Chen, Q.Zhang and Q. Fu, *J. Appl. Polym. Sci.* **129**, 3613-3622 (2013).
8. A. M. Díez-Pascual and M. Naffakh, *ACS Appl. Mater. Inter.* **5**, 9691-9700 (2013).

Poly(lactic acid)/poly(vinyl alcohol)-chitosan Obtained by Melting Processing

Rafael Grande*, Antonio J. F. Carvalho

Departamento de Engenharia de Materiais, Escola de Engenharia de São Carlos, Universidade de São Paulo, 13566-590 São Carlos, SP, Brasil

rafaelgrandee@gmail.com

Abstract Chitosan is a very promising material in functional blends due to its antimicrobial activity and biocompatibility properties. However, blending chitosan in a polymer blend by conventional means of melt processing is still very limited since chitosan cannot be melted under normal conditions. In a recent work we described a novel technique to process chitosan/PLA blends by melt processing in which a miscible blend of chitosan and poly(vinyl alcohol) produced by solution blending was dispersed in a poly(lactic acid) matrix by melt processing. In this work we introduce new processing technique to produce the chitosana/PVA such as spray drying and freezing drying and the evaluation of the processing conditions influence on the morphology and thermal stability of the ensuing blends. The produced blends showed a high thermal stability and more homogeneous morphology in the both, freeze drying and spray drying.

Keywords: Chitosan; Poly(lactic acid); poly(vinyl alcohol); melt blending; freeze-drying; spray drying.
PACS: 81

INTRODUCTION

Chitosan is a natural polymer obtained from chitin, a major constituent of the exoskeleton of insects and crustaceans. Thus, chitosan is completely biodegradable and presents microbial and biocompatibility [1-4]. Although Chitosan is widely studied, there are limitations in its processing since chitosan cannot be melted under normal conditions, making their application in pure unlikely. One possibility for processing this material is employed as a functional component in a polymeric blend.

The poly(lactic acid) (PLA) is a biodegradable synthetic polymer whose monomer is obtained from renewable sources and presents interesting physical properties that can replace many synthetic polymers such as polystyrene. The development of polymer blends using these materials can unite the physical properties of PLA with the special characteristics of chitosan. However, chitosan and PLA are incompatible either by melting or by solution [5-7]. Furthermore, chitosan has a very high melting temperature, in general decomposing before this is achieved. This characteristic is related to strong intermolecular hydrogen bonds of chitosan, which makes it impossible to process by conventional methods, where the material is processed in the molten state. Recently, we developed an innovate method methodology in two steps to melt process chitosan [8]. This method consist in first, blend chitosan by solution with poly(vinyl alcohol) (PVA) forming a compatible blend [8] and glycerol is added to solution as a plasticizer to improve the blend processability. PVA act as a compatibilizer between the chitosan and PLA. Then a PVA-Chitosan solution blend is incorporated with PLA and melt processed forming a ternary melting blend of PLA/PVA-chitosan. The biggest challenge working with this new methodology is the presence of residual solvent that may cause degradation of the materials. The degradation occurs in both steps of the proposed methodology; during the drying step of solution blends of PVA-chitosan as the melt processing step whose degradations is catalyzed by residual solvent of chitosan, the acetic acid [8]. In this work, tests were performed to obtain blends of PVA / chitosan in two different methods, spray drying and freeze-drying. The main goal is an improvement in quality of the PVA/Chitosan and avoid the degradation by eliminating the residual solvent more efficiently.

EXPERIMENTAL SECTION

MATERIALS

Chitosan used in this work was characterized with approx. 89,5% deacetylation, and average molecular mass M_v = 59.800 g mol^{-1}

PVA was purchased from Sigma–Aldrich, with a low molecular weight M_w 9000–10,000 g mol−1 and 80% degree of hydrolysis. The additive Glycerol and acetic acid solvent were reagent grade and used as received.

BLENDING PREPARATION

For the blends prepared by the traditional method, first, chitosan and PVA were blended by mixing a 1% solution of PVA in water with a 1% solution of chitosan in 1% aqueous acetic acid. Glycerol was added to this mixture in 10%w/w as a plasticizer. It is important to highlight that glycerol is a common plasticizer PVA, Chitosan. The resulting compositions of solution blends of PVA-Chitosan are showed in table 1.

TABLE (1). Composition of PVA-Chitosan solution blends

PVA-Chitosan blend	PVA % w/w	Chitosan % w/w	Glycerin % w/w
Binary 1	82	9	9
Binary 2	70	23	9
Binary 3	45.5	45.5	9

The resulting solution was dried in a forced air oven at 38°C for 72h and then dried in a vacuum oven for 24h at 50 °C. Then PVA-chitosan solution blend was cryogenically ground. The resulting powder was again dried under vacuum for 24h at 50°C to remove the remaining moisture.

For spray-dring technique, PVA-chitosan solutions were dried in a Mini Spray Dryer B-290 Buchi.at 125 °C hot air flow and 300 L/h of Nitrogen as atomization gas

For the freeze-drying PVA/chitosan solutions were frozen at -90° C for 24h then freeze-dryed in a Liotop model LP 510 equipment with has an initial vaccum of 0,18mmHg. The freeze-drying cicle took 52 hours.

RESULTS AND DISCUSSION

SEM

Figure 4 shows blends PVA-Chitosan obtained by alternative methods there tested; traditional drying in vacuum oven, spray-drying and freeze-drying.

It can be seen in Figure 4 a) that PVA-chitosan solution blend ground presents a light brown color. It becomes darker as the drying process is conducted is for longer periods, probably caused by degradation catalyzed by acid acetic reside. The particles obtained with grinding shows irregular shapes sizes with particles range of 15-250 mm mm. The degradation of the components may cause complications for the melt processing as also decrease in the quality of the blend and especially reduction in mechanical/optical properties. The first alternative examined to the traditional solution blending method was spray-drying technique. Spray drying is a method that produces a dry powder from a liquid spray, drying it with hot gas. Due the small drop size of the spray, the drying occurs faster and requires a lower temperature than traditional method. Figure 4 b shows PVA-Chitosan binary 1 dyed by spray-drying technique. It can be seen that the resulting material is significantly whiter than the figure 4 a). In addition, the powder obtained by this method showed smaller and more regular particles in comparison with figure 4a. Particles had spherical shape with diameter range of 1 – 10 μm as it can be observed in SEM image in figure 4b.

FIGURE 1. PVA-Chitosan blend; a) vacuum oven ground; b) spray dried; c) freeze dried.

The second method examined was freeze-drying blending. Freeze-drying is a relatively new dewatering process that consists of freezing material, then apply high vacuum to allow the sublimation and extraction of the solvent, resulting in a light and highly porous material. In this case, PVA/Chitosan freeze-dryed resulted in a slightly yellowish material, characteristic color of chitosan as showed in figure 4 c. The morphology obtained in PVA-Chitosan freeze-dryed is particularly interesting showing a layered structure with high porosity as evident in the SEM image in figure 4 c.

CONCLUSIONS

PVA-chitosan blends had a significant improvement as thermal stability, degradation and homogeneity by using different processing methods. The spray dryer technique resulted in a fine powder with small particle size and uniform distribution when compared to a milled blend solution. The freeze-drying technique also resulted in a much lighter material than the traditional method. The resulting morphology, porous and in layers, makes this interesting for biological applications, as cells scaffold, once both, PVA and chitosan, are biodegradable and biocompatible. Obtaining blends evidently lighter by both techniques, is a strong indication that the degradation was reduced, expressively improving the quality of blends.

ACKNOWLEDGMENTS

R. G. Thanks FAPESP for a Postdoctoral grant (2015/07744-0). The authors thank FAPESP and CNPq for financial support. Thank Liobras company for the support with freeze-drying materials and use of their facilities.

REFERENCES

[1] KUMAR, M.N.V. Reactive & Functional Polymers, v. 46, p. 1-27, 2000.

[2] NAM, C. W.; KIM, Y. H.; KO, S. W. Journal of Applied Polymer Science, v. 82, p. 1620-1629, 2001.

[3] BERGER, J.; REIST, M.; MAYER, J.M.; FELT, O.; PEPPAS, N.A.; GURNY, European Journal of Pharmaceutics and Biopharmaceutics, v. 57, p. 19-34, 2004.

[4] EUGENE K.; LEE, Y. Biomaterials, v. 24, p. 2339-2349, 2003.

[5] SÉBASTIEN, F.; STÉPHANE, G.; COPINET, A.; COMA, V., Carbohydrate Polymers, v. 65, p. 185-193, 2006.

[6] GRANDE, R., CARVALHO, A. J. F. Biomacromolecules, v. 12, p. 907–914, 2011.

[7] DRUMRIGHT, R.E.; GRUBER, P.R.; HENTON, D. E. Advanced Materials, v. 12, p. 1841-1846, 2000.

[8] GRANDE, R.; PESSAN, L. A.; CARVALHO, A. J. F. Industrial Crops and Products, v. 72, p. 159-165, 2015.

The Study of Rheology and Morphology of High Density Poly Ethylene (HDPE)/ Ethylene,1- Octene Copolymer (EOC)/ Nano Silica Nanocomposites

Karimi Samira[a], Javadi Azizeh[a*] and Nazokdast Hossein[a]

[a] Department of Polymer and color Engineering, AmirKabir University, Tehran, Iran

*Email: ajavadi@aut.ac.ir

Abstract. *In this research, HDPE/EOC blends containing 20, 40 and 80 Wt. % of EOC, also their Nanocompositses with 5 Wt. % Nano Silica were prepared via melt mixing using a brabender. The rheology and Morphology of samples were studied using Rheomechanical analysis and Atomic Force Microscopy (AFM). The results showed that HDPE/EOC blends are incompatible and show two phases morphology in all studied compositions. AFM micrographs showed that in blends containing less amount of EOC (20 Wt. %), the morphology is matrix-dispersed and this result was confirmed by palierne's model, too. The HDPE/EOC (20/80) blend showed a different morphology and HDPE again remained as a matrix. In HDPE/EOC/Nanosilica samples, it was found that the Nanosilica particles are located in EOC phase. In the HDPE/EOC (20/80) containing 5 Wt.% Nanosilica, the elasticity of EOC will be get closer to HDPE, Due to the presence of nanosilica in EOC phase. So EOC phase can be replaced in the matrix and nanosilica forms a network in the matrix that causes an increase in storage modulus at low frequencies.*

Keywords: *EOC, Nanosilica, Nanocomposites blends, HDPE.*
PACS: *83*

INTRODUCTION

Polyethylene (PE) and their copolymers form a part of commodity plastics, which cover a large part in the present days plastics consumption[1]. It has been demonstrated that the copolymerization of ethylene with α-olefins (such as 1-pentene, 1-hexene, 1-octene, etc.) using single-site metallocene catalysts leads to the formation of homogeneous ethylene/α-olefin copolymers with uniform distribution of comonomers and a narrow molar mass distribution, which have a dramatic effect on their mechanical properties[1-4]. These features make these new copolymers a suitable choice for fundamental studies on structure-property relationships of ethylene copolymers blended with other semicrystalline polymers. Their blends with conventional polyolefins such as poly(propylene) and PEs produce new materials easily process able and having tailored properties for specific purposes[3, 5].

It should be expected that the properties of HDPE/EOC blends with not very high content of comonomer; mostly depends on crystallinity, morphology and degree of dispersion. On the contrary, it has been confirmed that polymer miscibility in the melt state considerably effects on the final properties[3].

Several researchers have studied the miscibility and phase behavior of PEs blends. In the melt state, the average branch content of LLDPE plays an important role in determining its miscibility with other polyethylenes. More branches induce immiscibility [1, 3, 4, 6-8].

In the present Study we investigated morphology and rheology of HDPE/ EOC blends and effect of nano silica addition on morphology and rheology of these blends.

EXPERIMENTAL

HDPE (CRP100 HM) with MFI of 0.2 g/10 min and EOC (LC670) with MFI of 5 g/10 min, melting point of 50°C (Contained 35 Wt. % Octene) used as recieved; Nanosilica (SiO_2), Aerosil® r816, with a specific surface of 190 m^2/g, was supplied by Evonik Company. All blends and composites were prepared via melt mixing using an internal mixer in 180°C and 60 rpm for 10 min. The weight ratios of blend components are shown in Table 1.

The morphology of blends was investigated by AMBIOS Technology atomic force microscopy (AFM) in intermittent contact mode at room temperature.

Rheological characterization was carried out on a Rheological MC200 SN253628 rheometer using 25 mm parallel plate fixtures, with a gap of 1 mm at 190°C. The rheometer was operated in the dynamic oscillatory mode in the linear viscoelasticity region.

TABLE 1). compounding formulation

Sample code	HDPE(Wt.%)	EOC(Wt.%)	Nanosilica(Wt.%)
P80E20	80	20	0
P60E40	60	40	0
P20E80	20	80	0
P80E20N5	76	19	5
P20E80N5	19	76	5

RESULTS AND DISCUSSION

Morphology of blends

The results showed that the HDPE/EOC blends are incompatible in all studied compositions. This finding is in agreement with previous research that showed HDPE are incompatible with the same EOC of our research (containing 35Wt.% 1-octene)[1]. The AFM images of HDPE/EOC blends with different compositions are shown in figure 1a-c. Dark and bright areas are represented the soft EOC and stiff HDPE phase, respectively. In blends containing 20 Wt. % EOC, The matrix-dispersed morphology is observed. By increasing EOC content up to 40 Wt. %, the morphology is changed to co-continuous, figure 1-b. The HDPE/EOC (20/80) blend shows a different morphology and HDPE again remaine as a matrix. This result is attributed to the difference in elasticity of the polymers, which is observed by the other researchers in PS/PVME blend too[9].

Rheology of blends

As it is shown in figure 2, storage modulus- frequency curves of blends with different compositions are embedded between pure component curves. The same results are reported according to previous studies [5].

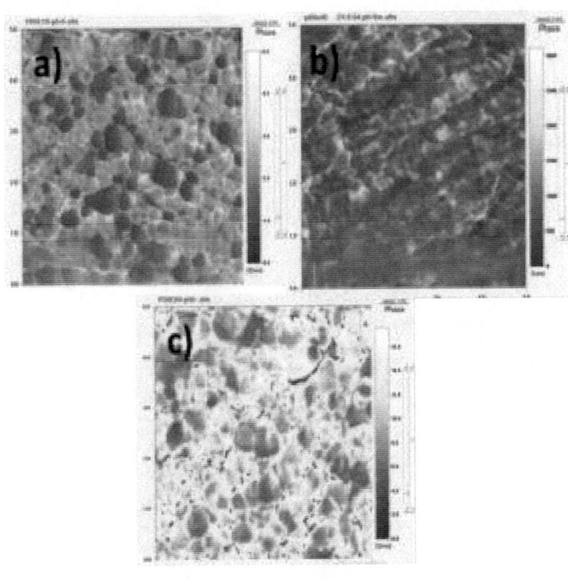

***FIGURE 1.** AFM images: a)P80E20 b) P60E40 c) P20E80*

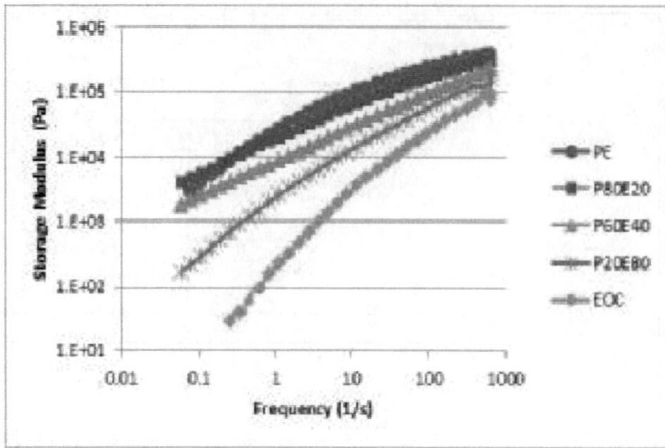

***FIGURE 2.** The variation of storage modulus of HDPE/EOC blends with frequency.*

The palierne's model is used to compare the experimental results too. Figures 3 represents the comparison between experimental data and results obtaining from palierne's model for P80E20 and P20E80 blends.

As it is shown, the results of palierne's model for P80E20 confirm the experimental finding for storage modulus and matrix-dispersed morphology, figure 3.a. In other hand, the results of P20E80 blend have deviated from palierne's model. It could be concluded that in this blend, although the major phase is EOC, this phase does not act as a matrix. This result are confirmed by AFM images too, figure 1.c.

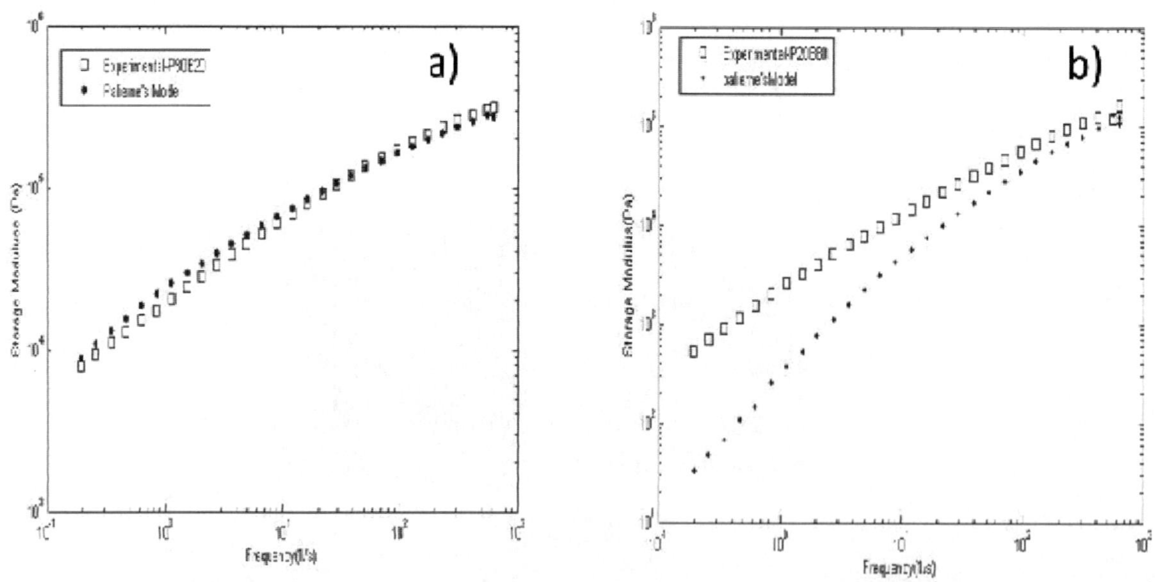

***FIGURE 3.** Comparison between storage modulus of a) P80E20 blend and b) P20E80 blend with palierne's model.*

Nano composite characterization

The variation of storage modulus with frequency for P80E20 and P80E20N5; also P20E80 and P20E80N5 are shown in figure 5 and 6 respectively. As it is seen in figure 5, there is not a big difference between these two curves. It could be concluded that nanosilica particles haven't been located in HDPE, which is dominant phase in this blend. So adding nanoparticles does not have a major effect on storage modulus of blend. On the other hand, in the P20E80N5, it seems that the nanosilica particles are located in EOC phase, and that is why the storage modulus of EOC and HDPE have been closer to each other and the EOC phase can be replaced as a matrix. Also nanosilica could form a network in the EOC matrix which causes an increase in storage modulus at low frequencies, figure 6.

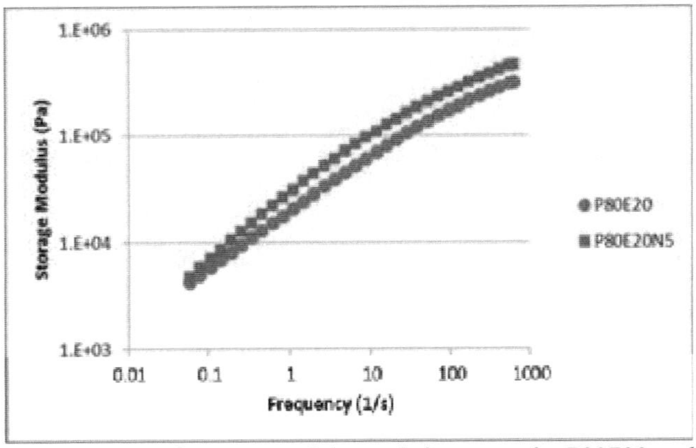

FIGURE 4. *The variation of storage modulus with frequency for P80E20 and P80E20N5.*

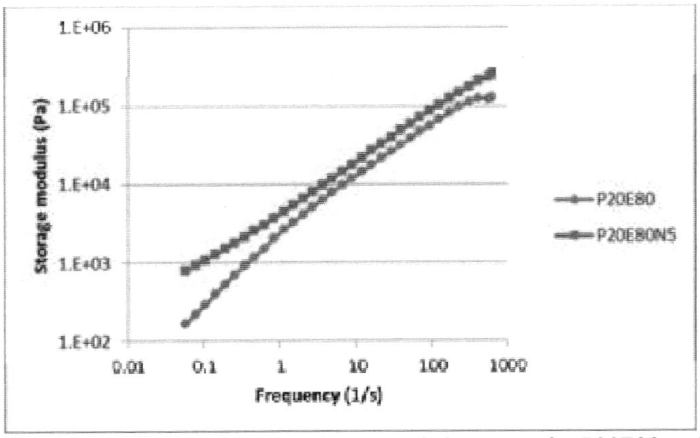

FIGURE 5. *The variation of storage modulus with frequency for P20E80 and P20E80N5.*

CONCLUSION

In this study, the blends of high density polyethylene (HDPE) and ethylene 1-octene copolymer (EOC) containing 5 Wt. % of nanosilica was prepared by melt mixing using an internal mixer. Rheological and AFM methods were applied to investigate miscibility and microstructure change of the blends. The results indicated that the blends of HDPE/EOC were immiscible in all compositions. The HDPE/ EOC (80/20) blend showed matrix-dispersed morphology and the rheological results matched well with Palierene's model. The HDPE/EOC (20/80) blend showed a different morphology remaining HDPE as a matrix again. This results were attributed to the difference in elasticity of the components. It was also found that by addition of nanosilica to HDPE/ EOC (20/80) blend, due to the approach of elasticity of EOC and HDPE, the EOC phase could be replaced as a matrix phase.

REFERENCES

1. *R. Adhikari, R. Godehardt, W. Lebek, and G. H. Michler, Journal of applied polymer science, vol. 103, pp. 1887-1893, 2007*
2. *M. Guimarães, F. Coutinho ,M. Rocha, M. Farah, and R. Bretas, Polymer testing, vol. 22, pp. 843-847, 2003*
3. *D. Mileva, H. J. Radusch, and C. Betchev, Macromolecular Materials and Engineering, vol. 292, pp. 319-328, 2007*
4. *M. J. O. Guimarães, F. Coutinho, M. C. Rocha, and M. E. F. Garcia, Journal of applied polymer science, vol. 81, pp. 1991-1995, 2001*
5. *M. J. O. Guimarães, F. Coutinho, M. C. Rocha, M. Farah, and R. E. Bretas, Journal of applied polymer science, vol. 86, pp. 2240-2246, 2002*
6. *M. Hill, P. Barham, and J. Van Ruiten, Polymer, vol. 34, pp. 2975-2980, 1993*
7. *H. S. Lee and M. M. Denn, Polymer Engineering & Science, vol. 40, pp. 1132-114, 2000.*
8. *A. Tabtiang, B. Parchana, R. A. Venables, and T. Inoue, Journal of Polymer Science Part B: Polymer Physics, vol. 39, pp. 380-389, 2001.*
9. *J. K. Yeganeh, F. Goharpey, and R. Foudazi, RSC Advances, vol. 2, pp. 8116-8127, 2012.*

The Effect of Melt/Gas Mixture Compressibilty on the Cell-Nucleation Behaviors in High-Pressure Foam Injection Molding

Vahid Shaayegan[a], Guilong Wang[a] and Chul B. Park[a*]

[a] *Microcellular Plastics Manufacturing Laboratory, Department of Mechanical and Industrial Engineering, University of Toronto, 5 King's College Road, Toronto, Ontario, Canada M5S 3G8*

Presenting author's email: park@mie.utoronto.ca
Reference author's email: park@mie.utoronto.ca

Abstract. We investigated the formation of gas bubbles in foam injection molding (FIM) using an in-situ visualization technique. In this regard, an innovative visualization mold, equipped with cavity pressure transducers, was designed and manufactured. We used a polystyrene/CO_2 system to carry out high-pressure FIM experiments. To perform high-pressure FIM experiments, we removed all the gate-nucleated cells using a melt-packing pressure. A higher gate-nucleated cell density increased the melt-packing pressure required to remove all the gate-nucleated cells. Therefore, the required melt-packing pressure increased as the injection speed, the gate resistance, and/or the blowing-agent content increased. The shrinkage-induced cells governed the final cellular structure in high-pressure FIM after complete dissolution of the gate-nucleated cells, whereas the gate-nucleated cells would govern the cell structure in case of incomplete dissolution during packing. The two-phase polymer/gas mixtures with un-dissolved gate-nucleated cells could not result in a higher cell density because of their higher compressibility. In contrast, the one-phase polymer/gas mixtures, formed by removing the gate-nucleated cells, could obtain a higher pressure-drop rate because of its lower compressibility (equivalently higher bulk modulus), and thereby a higher cell density.

Keywords: High-pressure foam injection molding, Cell nucleation mechanism, Polystyrene, Visualization

INTRODUCTION

The low-pressure conventional foam injection molding (FIM) offers such advantages as producing lightweight parts, and manufacturing large products with dimensional stability at fast cycles [1-4]. While a large void fraction, up to 35%, is easily obtainable, nevertheless, this technology suffers from considerable non-uniformity in the cellular structure. Formation of large cells due to cell coalescence, presence of elongated cells sheared during mold filling, and nucleation of cells under variable pressure-drop rates deteriorate this structural non-uniformity [4,5]. In low-pressure FIM, a short-shot of the gas-charged polymer melt, i.e. a shot size which is less than the size of the mold cavity, is injected and the expansion of the nucleated cells fills out the mold cavity completely. Here, the pressure-drop required for the cell nucleation is obtained over the gate. In other words, the high pressure is removed from the *one-phase* melt/gas mixture once it passes the gate and enters the mold cavity. Hence, a large number of cells nucleate at the gate (i.e. gate-nucleated cells) and travel in injection direction towards the end of the mold cavity. These cells get sheared and coalesce during mold filling. In addition, the cavity pressure increases as the mold filling proceeds, reducing the pressure-drop rate for further nucleation [4].

However, a uniform microcellular structure can result in improved mechanical properties [5-9], and improved electrical and charge storage capability [10,11]. One effective strategy to minimize the structural heterogeneity in FIM is to *decouple* the mold filling and the cell nucleation stages. In the light of this strategy, high-pressure FIM technology can be employed. This paper explains the mechanism of cell nucleation in high-pressure FIM, and discusses the effect of processing parameters on cell nucleation mechanism in this technology.

EXPERIMENTAL AND METHODOLOGY

In-situ mold visualization technique was employed to investigate the governing cell nucleation mechanisms in FIM. In this regard, an innovative visualization mold was used whose details can be found in [12]. Briefly, a prism was incorporated into the stationary plate of the mold to reflect mold images to a camera (Figure 1a). The FIM experiments were carried out using a 50-ton Arburg ALLROUNDER 270/320C injection molding machine, equipped with MuCell technology. A 135×111×3.2 mm rectangular cavity was fed with a fan gate. Different locations of the mold cavity, and so the foamed part, are named as A, B, and C (with respect to their distance from the gate) and are shown in Figure 1b. Image recording was done by using a CV M10 camera from JAI with a Navitar magnifying lens. Polystyrene, PS675 (MFI 7 gr/10min) from Americas Styrenics was used as the resin, and carbon dioxide (CO_2) from Linde Gas Canada was used as the physical blowing agent. The processing parameter for FIM experiments are summarized in Table1.

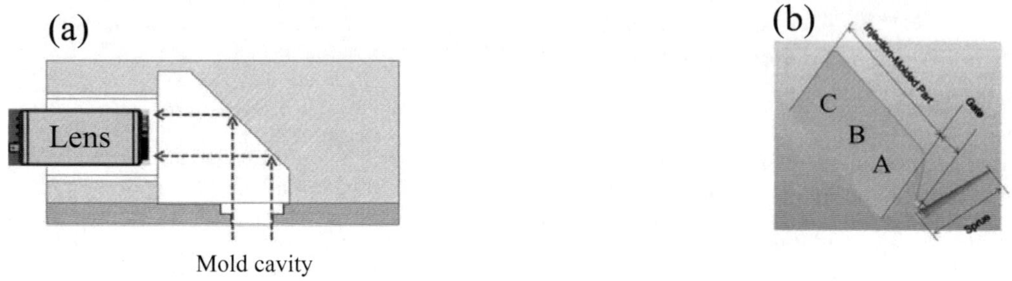

FIGURE 1. (a) Schematic of the prism-insert; (b) Injection molded sample locations (A: near the gate, B: middle cavity, and C: cavity end).

TABLE (1). Processing conditions for FIM experiments

Parameter	**CO_2**	**Injection speed**	**Gate thickness**	P_{pack}	t_{pack}	T_{melt}	T_{mold}
Value	2 wt%	80 cm^3/s	1.7 and 2.78 mm	2.5-20 MPa	1 s	230 °C	30 °C

RESULTS AND DISCUSSION

Unlike the low-pressure FIM, a full-shot is used to perform high-pressure FIM. Therefore, the cavity pressure quickly increases and results in a poor and slow cell nucleation at the gate. Once the mold filling is completed, new cells may nucleate during the cooling and melt shrinkage stage (i.e. shrinkage-induced cells)[13]. Figure 2a shows the visualization snapshots from Location B (see Figure 1b) of a full-shot FIM experiments using PS 675-2wt% CO_2 mixture at an injection speed of 80 cm^3/s. A packing pressure of 2.5 MPa was used in this trial. As demonstrated in successive visualization snapshots, the cells nucleated at the gate and during the filling withstood the cavity pressure. In other words, the cavity pressure was not enough to re-dissolve them back into the melt. These gate-nucleated cells are elongated and sheared. Also, they consume most of the available gas in the melt-gas mixture, and their expansion compensates for the further melt shrinkage. In this scenario, the prevailing cell nucleation mechanism is the nucleation due to the pressure-drop obtained at the gate. By increasing the melt-packing pressure from 2.5 to 8 MPa, the cavity pressure increased. Therefore, more nucleated cells (during filling) were dissolved back into the melt. As shown in Figure 2b, most of the nucleated cells were removed from the system and a few sustained the cavity pressure (see snapshot at t = 6.6 s in Figure 2b). As the time elapsed, new cells were nucleated during cooling, as observed in snapshot at t = 12 s and later. The pressure-drop for the nucleation of the new cells was provided by the melt shrinkage, and the overall cellular structure was comprised of both: (i) cells nucleated at the gate during the filling (fate-nucleated cells), and (ii) new cells nucleated during the melt shrinkage (shrinkage-induced cells). By increasing the packing pressure up to 16 MPa, all nucleated cells during the filling were dissolved back into the melt. As seen in Figure 2c, no cell was observed in visualization snapshots for a considerable time span, after which new (shrinkage-induced) cells nucleated during the melt shrinkage. Hence, under the processing condition of Figure 2c, the governing cell nucleation mechanism is the nucleation due to the pressure-drop obtained during the melt shrinkage. As the nucleation occurred after the mold filling was completed, i.e. no melt flow, the cells were all circular and there was not any evidence of elongated/sheared cells.

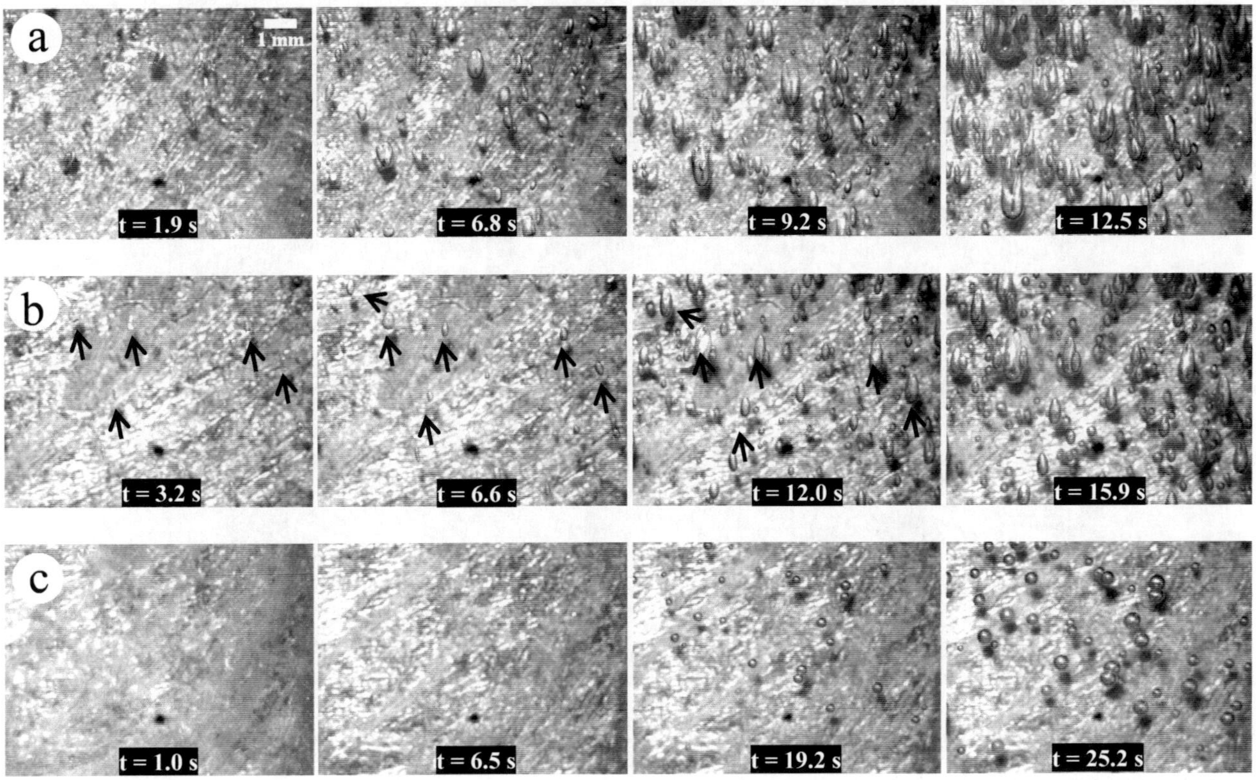

FIGURE 2. Cell morphology in high-pressure FIM at different melt-packing pressures (Gate thickness: 1.7 mm) a) packing pressure = 2.5 MPa; b) packing pressure: 8 MPa; c) packing pressure 16 MPa (PS675-2 wt% CO_2, injection speed: 80 cm^3/s; packing time: 1 s; $T_{melt} = 230°C$, Location B; arrows show typical un-dissolved cells).

Figure 3 shows visualization snapshots from high-pressure FIM experiments, carried out at the same processing parameters used in Figure 2, but using a gate with a higher resistance (i.e. a gate thickness of 2.78 mm). The visualization results revealed that a larger packing pressure, about 20 MPa, was required to completely remove cells which nucleated at the gate during the filling. The higher resistance of the gate resulted in a faster pressure-drop rate during the mold filling. Park et al. demonstrated that a faster pressure-drop yields a higher cell density [14]. Hence, a larger number of cells were nucleated using the high-resistance gate and consequently a higher packing pressure was needed to re-dissolve nucleated cells back into the melt, before the nucleation mechanism entirely switches to the nucleation due to the melt shrinkage. Similarly, an increase in the injection speed and the percentage of the blowing agent increases the required packing pressure to remove gate-nucleated cells.

In high-pressure FIM, the presence of un-dissolved cells increases the compressibility of the melt/gas mixture. Thus, the achievable pressure-drop rate reduces which results in a lower cell density during melt shrinkage. A comparison between the two nucleation mechanisms, i.e. the nucleation at the gate and the nucleation during melt shrinkage, reveals that a higher cell density is obtained when the nucleation occurs at the gate. This indicates that the pressure-drop rate at the gate is much higher compared with that obtained during shrinkage. In order to promote the cell nucleation and increase the void fraction, a high-pressure FIM can be followed by a mold-opening step. In this case, the cells nucleated at the gate are firstly removed by using a proper melt-packing pressure. Then, the mold opens and the required pressure-drop rate is attainable by the mold-opening speed. Using this strategy, uniform high-cell density foam with high void fraction is obtainable.

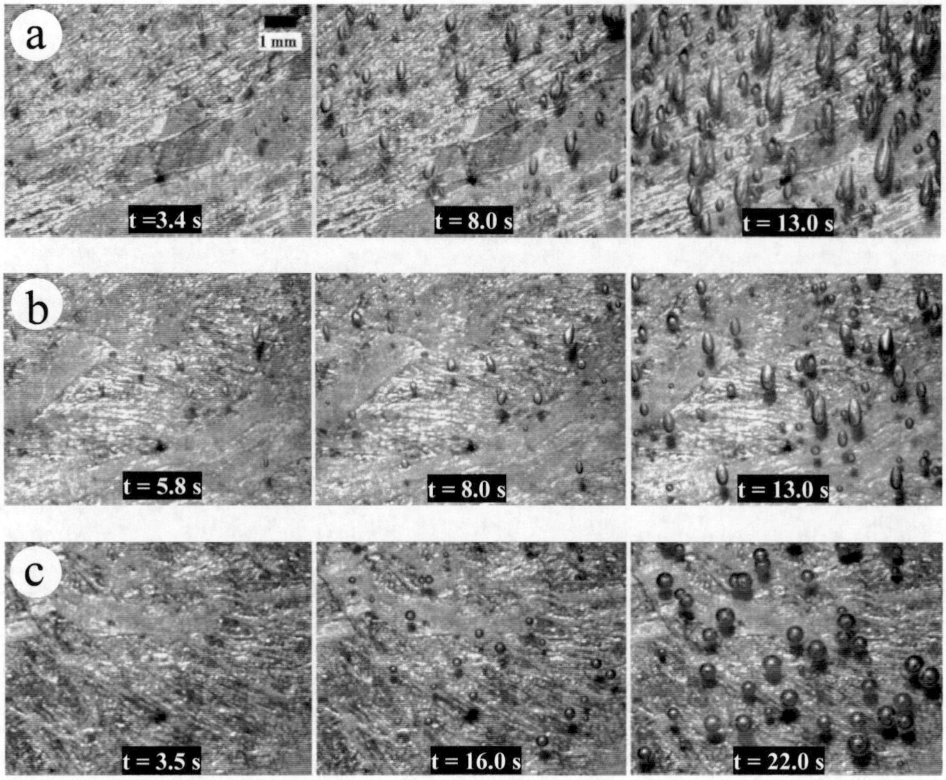

FIGURE 3. Cell morphology in high-pressure FIM at different packing pressures (Gate thickness: 2.78 mm) a) packing pressure: 2.5 MPa; b) packing pressure: 8 MPa; c) packing pressure 20 MPa (PS 675-2 wt% CO_2; injection speed 80 cm^3/s; packing time: 1 s; $T_{melt} = 230°C$; Location B)

CONCLUSION

The mechanism of cell nucleation in a high-pressure foam injection molding was investigated. It was concluded that the governing nucleation mechanism is highly dependent on the cavity pressure. If the cavity pressure is low, the cells nucleated at the gate dominate the structure. The cell nucleation mechanism in this scenario is the nucleation due to the pressure-drop obtained at the gate. If the cavity pressure is high enough to re-dissolve the cells which nucleated during the filling, then the nucleation of new cells occurs during the cooling stage and the melt shrinkage. In other words, the nucleation mechanism switches from “nucleation at the gate” to the “nucleation during melt shrinkage”. If the cavity pressure is insufficient, a melt-packing pressure can be used to remove the nucleated cells at the gate. The required packing pressure increases by parameters such as injection speed, blowing agent percentage, and the gate resistance which promote the cell nucleation at the gate.

ACKNOWLEDGMENTS

This research was funded by the Natural Science and Engineering Research Council of Canada. The authors are grateful to the Consortium of Cellular and Micro-Cellular Plastics (CCMCP) for their support. The resin used in this work was provided by Americas Styrenics, TX, United States.

REFERENCES

1. P.R. Hornsby, *Mater. Eng.* **3**, 443-455 (1982).
2. A. A. Ahmadi, and P.R. Hornsby,. *Plast. Rubber Process. Appl.* **5**, 35-49 (1985).
3. K. Hikita, *JSAE Rev.* **23**, 239-244 (2002).
4. J.W.S. Lee, J. Wang, J.D. Yoon, and C.B. Park, *Ind. Eng. Chem. Res.* **47**, 9457-9464 (2008).
5. M. Yuan, L.S. Turng, S. Gong, D. Caulfield, C. Hunt, and R. Spindler, *Polym. Eng. Sci.*, **44**, 673-686 (2004).
6. S. Wong, J.W.S. Lee, H.E. Naguib, and C.B. Park, *Macromol. Mater. Eng.*,**293**, 605-613 (2008).
7. J.W.S. Lee, C.B. Park, and S.G. Kim, *J. Cell. Plast.* **43,** 297-312 (2007).
8. J.W.S. Lee and C.B. Park, *Macromol. Mater. Eng.* **291**, 1233-1244 (2006).
9. S. Pilla, A. Kramschuster, S. Gong, A. Chandra, and L.S. Turng, *Int. Polym. Proc.*, 22, 418-428 (2007).
10. A. Ameli, M. Nofar, C.B. Park, P. Pötschke, and G. Rizvi, *Carbon*, **71**, 206–217 (2014).
11. A. Ameli, S. Wang, Y. Kazemi, C.B. Park, and P. Pötschke, *Nano Energy* **15**, 54–65 (2015).
12. V. Shaayegan, L.H. Mark, A. Tabatabaei, and C.B. Park, *eXPRESS Polym. Lett.,* **10**, 462-469 (2016).
13. V. Shaayega, G. Wang, C.B. Park, European Polymer Journal, 76, 2-13 (2016).
14. C.B. Park, D.F. Baldwin, and N.P. Suh, *Polym. Eng. Sci.* **35** , 432-440 (1995).

A New Foam Injection Molding Technology - Use Non-Supercritical Fluids as Physical Blowing Agents-

Long Wang[a], Atsushi Yusa[b], Satoshi Yamamoto[b], Hideto Goto[b], Hiromasa Uezono[c], Fumiya Asaoka[a], Shota Ishihara[a], Yuta Hikima[a] and Masahiro Ohshima[a]*

[a] *Dept. of Chem. Eng., Kyoto University, Katsura, Kyoto 615-8510, Japan*
[b] *Technology Development Department, Hitachi Maxell, Ltd., Ibaraki, Osaka, 567-8567, Japan*
[c] *Injection Molding Machine Division, Japan Steel Works Ltd., Hiroshima 736-8602, Japan*

wangl.kevin@cheme.kyoto-u.ac.jp
oshima@cheme.kyoto-u.ac.jp

Abstract. A new foam injection molding technology was developed to produce microcellular foams without using supercritical fluid pump units. In this technology, physical blowing agents (PBA), such as nitrogen (N_2), carbon dioxide (CO_2), and air, do not need to be brought to their supercritical fluid (SCF) state. PBAs can be delivered directly from their gas cylinders into the molten polymer through an injector valve and/or a venting vessel, which can be controlled by a specially designed screw configuration and operation sequence. The excess gas, which is the residual PBA existing as a gas in the molten polymer, is discharged from the molten polymer through the venting vessel. Alternatively, additional PBA is introduced through the venting vessel when the polymer does not get saturated with PBA at the positon beneath the venting vessel. The amount of gas delivered into the molten polymer is controlled by the gas dosing time of the injector valve, the secondary pressure of the reducing valve of the PBA gas cylinder, and the outlet (back) pressure of the venting vessel. Microcellular polypropylene foams were prepared using the developed foam injection molding technology with either 5-6 MPa CO_2, 5-8 MPa N_2, or 7MPa air of room temperature as the physical foaming agents. Five-fold expansion foams with their average cell size of less than 50 μm were prepared with any of those PBAs using the developed technology. The developed technology dispels arguments for the necessity to pressurize N_2 or CO_2 to the SCF state to produce microcellular foaming.

Keywords: Microcellular Foam, Foam Injection Molding, Supercritical Fluid
PACS: Material synthesis and processing methods of, 81.20.-n

INTRODUCTION

Microcellular foaming technology has been developed since the 1990s to reduce material usage with little or no deterioration in the mechanical properties [1-4]. The definition of microcellular foams was given as cells less than 100 μm in size with a number of cells per unit volume greater than 10^8/cm3 in density [3]. The first commercial microcellular foaming technologies were MuCell® extrusion and injection molding processes [3,4]. They were developed based on MIT research [3] and have experienced great commercial success. Especially, the field of applications for microcellular foam injection molding has expanded in the automotive, computer and home appliance industries. These processes use supercritical fluid CO_2 (sc CO_2) or N_2 (sc N_2) as physical blowing agents. In the original and most commonly used microcellular foaming processes, the gas was firstly pressurized to supercritical pressure levels using a supercritical fluid (SCF) pump unit, and then it was introduced through an injector valve into molten polymer in a barrel to dissolve the gas into the polymer prior to extrusion or injection.

After the invention of MuCell® technology, different microcellular foaming technologies such as Optifoam [4] and ProFoam [4, 5] were successively developed. However, those technologies still need a system to pump CO_2 or N_2 up from its cylindrical pressure to a supercritical pressure level to use the PBA in a supercritical state. This supercritical fluid pumping system increases the machine and operation costs. Although it has been proven experimentally that CO_2 and N_2 solubility follows Henry's law, their diffusivity of these gases follows Fick's law at processing temperatures, and microcellular foams can be produced with low concentrations of CO_2 and N_2 when the equilibrium pressures are lower than their critical pressures [6, 7], there is still a groundless belief that supercritical CO_2 or N_2 can provide special effects in the production of microcellular foams and that the usage of supercritical CO_2 or N_2 is key to microcellular foaming technologies [8].

In this study, a new foam injection molding technology was developed to simplify machine and design screw configurations for introducing CO_2, N_2 or Air into the barrel of an injection machine without pressurizing the PBA to the supercritical state level. The most distinctive feature of the foam injection machine is the design of the screw and its sequence of operation allowing the PBA to be delivered directly into the molten polymer from their gas containers (cylinders). The excess gas, which is the residual PBA existing as a gas in the molten polymer in the barrel of the injection molding machine, can be discharged from the molten polymer through a venting vessel. Alternatively, additional PBA is introduced through the venting vessel when polymer is not saturated with PBA at the position beneath the venting hole. The pressurized venting vessel plays a role of controlling and stabilizing the concentration of the PBA in the molten polymer. The developed microcellular foaming technology was tested on polypropylene (PP) foams using CO_2, N_2 or Air as PBAs to demonstrate that the performance of the foam-ability is equivalent, or in some aspects superior, to conventional microcellular foaming technologies.

DESIGN & OPERATION OF FOAM INJECTION MOLDING

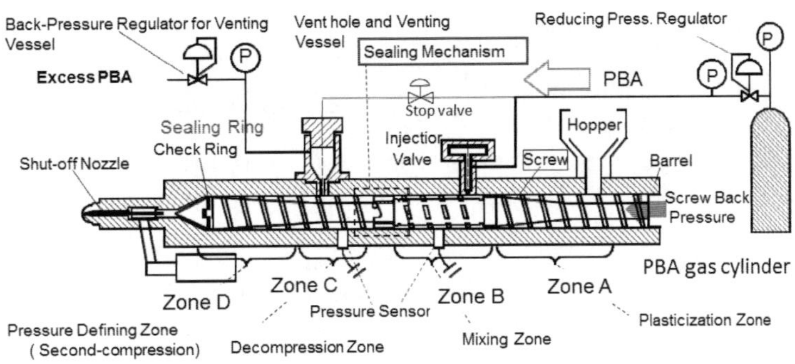

FIGURE 1. T Schematic diagram of the developed microcellular foam injection molding machine.

Schematics Diagram of the Molding Machine

Figure 1 shows a schematic diagram of the developed foam injection-molding machine. There is no supercritical fluid pumping unit, as is required in conventional microcellular foam injection molding machines. In the developed machine, the physical blowing agents (PBAs), such as N_2, CO_2, and Air are directly supplied from their gas cylinders and delivered into the molten polymer through an injector valve and a venting hole. The cylinder pressures are normally 12 MPa for N_2 and Air and 5-6 MPa for CO_2 at room temperature. The pressure of the supplied PBA gas is controlled by adjusting a reducing valve on the cylinder, which is kept at a specified value in the range of 8-12 MPa for N_2 and Air, and 5-6 MPa for CO_2. The vent hole with a venting vessel is provided in the middle of injection molding machine. Through the venting hole, the excess PBA gas, which is the residual PBA existing as a gas in the molten polymer, can be discharged from the molten polymer to the atmosphere or additional PBA gas can be introduced to the molten polymer when the concentration of PBA in the molten polymer does not reach a saturated condition. By tuning the pressure in the venting vessel, the concentration of PBA dissolved in the molten polymer can be adjusted. The pressure of the venting vessel is manipulated by a back pressure regulator at the end of the discharge line from the venting vessel. In this study, it was set to a designated value in the range of 2-8 MPa for N_2, 7 MPa for Air and 2-6 MPa for CO_2.

Screw Configuration and Sealing Mechanism

The screw has four configuration zones as illustrated in Fig. 1. With this configuration, PBA delivery and venting can be achieved. Zone A is a plasticization zone, in which the polymer pellets supplied by the hopper are plasticized and compressed. In Zone B, the physical blowing agent, either CO_2 or N_2, is injected through an injector valve and dissolved into the molten polymer under high pressure and shear rate. The amount of gas injected into the molten polymer is controlled by manipulating the time period that the injector valve is open and the secondary pressure of the supplied PBA, i.e., the pressure set point of the reducing valve on the cylinder. While the injector valve is open, the polymer pressure at the location beneath the injector valve should be lower than the pressure of the supplied PBA so that the PBA gas can be introduced via the pressure difference. This pressure difference can be adjusted by the back pressure of the screw and by the change in direction of the screw rotation. Zone C is called the decompression zone, where the depth of the screw channel is designed to be deeper to decrease the pressure of the molten polymer to a certain level and release the excess PBA into the atmosphere or introduce the additional PBA into the molten polymer. The pressure level is tuned by the pressure of the venting vessel, and it should be set to a level slightly higher than the equilibrium pressure of the dissolved PBA concentration to prevent the polymer from foaming in this zone. The polymer with the dissolved PBA is again compressed in Zone D (the pressure defining zone), and a single phase of molten polymer and PBA is established.

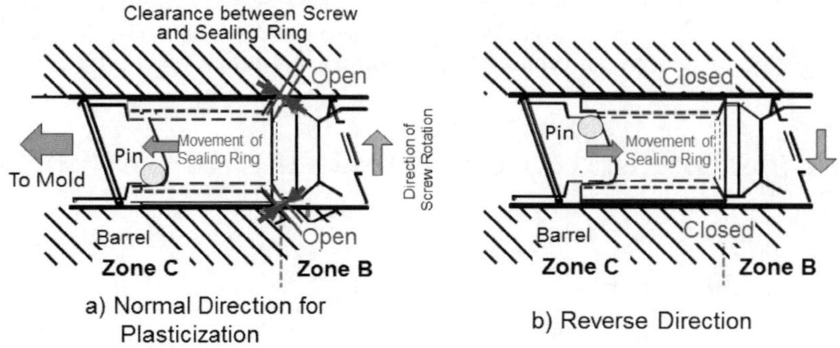

FIGURE 2. Sealing ring movement and sealing operation associated with the screw movement.

Figures 2(a) and 2(b) show an enlarged picture of the sealing point of the screw. As shown in Fig. 2, a sealing ring is placed at the interface between Zones B and C to prevent molten polymer from back-flowing from Zone C to Zone B and the delivered gas from running through Zone B to Zone C without being dissolved into the polymer. When the injector valve is closed, the sealing ring is moved forward to create a clearance-space between the screw, and the screw ring and is opened by rotating the screw in the normal direction. Then, the molten polymer is conveyed forward through the clearance-space from Zone B to Zone C (Fig. 2(a)). When the injector valve is open to introduce the PBA from its gas cylinder, the sealing ring is moved backward to seal Zone B from Zone C by rotating the screw in the reverse direction (Fig. 2(b)). By rotating the screw at low rpm in the reverse direction, the polymer pressure in Zone B is temporarily reduced. Furthermore, because the molten polymer is not conveyed from Zone B to Zone C by the sealing ring, the polymer is starved in Zone C, and an open space is created for excess PBA to diffuse from Zone C or for additional PBA to introduce through the venting hole and vessel.

Figure 3 shows the schematic procedure of the injection molding process of the developed machine as described below:

- **Step 1**. After the polymer is injected into the mold cavity, plasticization begins. The screw is rotated in the normal direction for metering. The molten polymer is moved forward and accumulates in front of the screw (Fig. 3(a)).
- **Step 2**. In the middle of the metering or accumulation, the screw stops rotating for a short period of time and then rotates in the reverse direction. Then, the sealing ring moves backward to seal Zone C from Zone B. The injector valve is opened for a given period of time, and the PBA is delivered to the molten polymer in Zone B (Fig. 3(b)).
- **Step 3**. The reverse screw rotation stops for a short period of time and starts rotating in the normal direction to complete

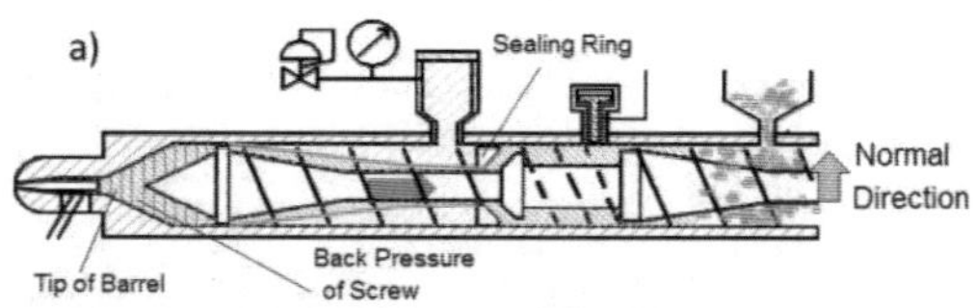

Step 1 The metering and plasticizing process.

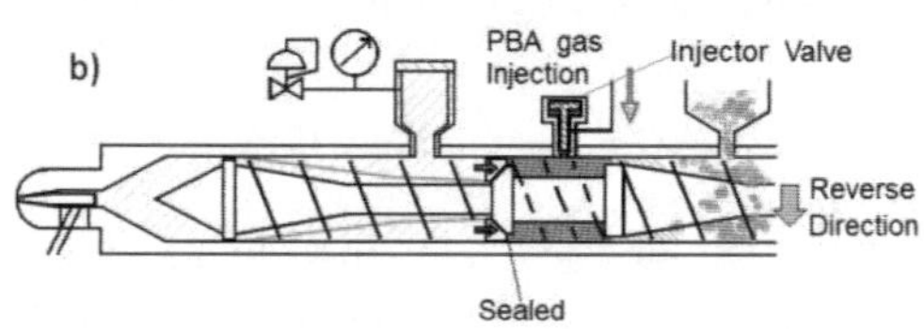

Step 2 PBA gas injection process during reverse screw rotation

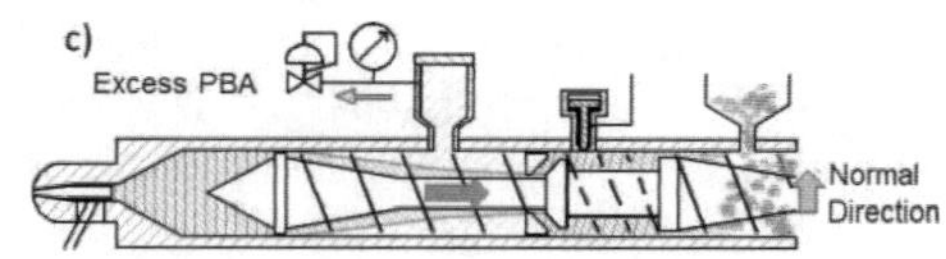

Step 3 Continuation of the plasticization and metering processes.

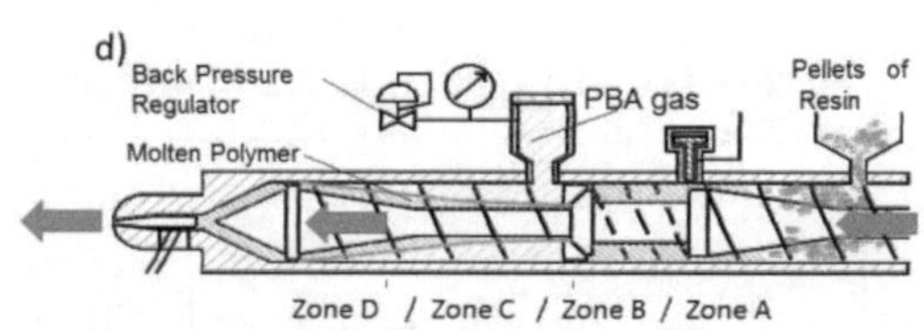

Step 4 Processes for injecting and holding the molten polymer in the mold.

FIGURE 3. Operation sequence of the developed injection molding machine.

the metering and accumulation of the polymer. In Zone C, where the polymer pressure is controlled by the back pressure of the venting vessel, polymer starvation occurs, and excess PBA is diffused out or additional PBA is introduced from the venting hole. During plasticization, using the back pressure of the screw, the pressure of the polymer accumulated in front of the screw is controlled to a constant level higher than the pressure of the venting vessel to prevent the accumulated polymer from foaming in front of the screw. The delivered PBA is mixed with the molten polymer and dissolved into the polymer under high pressure and shear (Fig. 3(c)).

- **Step 4**. After a specified amount of the molten polymer has accumulated, the polymer containing the dissolved PBA is injected into the mold cavity at a given injection rate and injection pressure by opening the shut-off nozzle. After the molten polymer fills the cavities of the mold, the dwell pressure is controlled. The position of the change from speed control to pressure control is set at a point at which the screw position reaches a given constant value. Then, the process returns to Step 1.

When conducting the core back operation (mold opening operation), a portion of the mold is moved at the end of the process to hold the dwell pressure and to quickly increase the mold cavity volume and decrease the polymer pressure in the mold cavity. The sudden pressure drop initiates foaming, enhances the bubble nucleation and achieves the fine cell structure of the foams.

EXPERIMENTS

Materials

The polymer used in this study was a high tacticity isotactic polypropylene, i-PP (F133A, Prime Polymer Co., Ltd.). The melt flow rate was 3 g/10 min (230°C). Its molecular weight, M_w, was 379,000 and its tacticity was 97%. A crystal nucleating agent, 1,3:2,4 bis-O-(4-methylbenzyliden)-D-sorbitol gelling agent (Gel-all MD, New Japan Chemical co., Ltd), was used as a bubble nucleating agent, as in a previous study [9,10]. A nucleating agent was used to enhance the crystallization during the cooling process of the injection molding and to ensure that the crystals could create a structurally inhomogeneity. Nitrogen (N_2) with 0.999 purity, carbon dioxide (CO_2) with 0.99 purity and artificial Air with 80 % N_2 and 20 % O_2, were independently used as a physical blowing agent (Showa Denko Gas Products Co., Ltd.). The pressures of the gas cylinders were 12 MPa for N_2, 6 MPa for CO_2 and 14 MPa for Air.

Molding Condition

The foam injection molding was conducted without using supercritical fluid pumps. The operating condition of the molding machine is summarized in Table 1.

Table 1. Operating conditions of the foam injection molding machine.

Manipulated Variable	Set point values	Manipulated Variable	Set point values
Pressure of gas cylinder [MPa] CO_2 / N_2 /Air	6 / 12/ 12	Pressure of venting vessel [MPa] CO_2 / N_2 /Air	5 / 5, 7/ 5, 7
Secondary reducing press [MPa] CO_2 / N_2 /Air	6 / 8 / 7	Screw back pressure [MPa]	11
Screw rotation speed [rpm] 1) Metering period 2) Gas dosing period	 100 10 / 5	Screw rotating period [s] 1) 1st metering period 2) Gas dosing period 3) 2nd metering period	 2 0.9 19
Injection speed [mm/s]	200	Injector valve opening period [s] CO_2 / N_2 /Air	0-0.7/ 0-0.5/0-0.1
Dwell pressure [MPa]	30	Injection pressure [MPa]	160
Barrel temperature [°C] CO_2 & N_2 /Air: 200, 200, 200, 200, 210, 230, 220, 190, 60	190,190,190,190,195, 195,195, 195, 60 for Air	Mold temperature [°C]	40
Dwell time [s]	5.6 -7.8	Coreback rate [mm/s]	20

RESULTS AND DISCUSSION

Figures 4, 5 and 6 show SEM images taken from perpendicular and parallel views to the core-back direction of the foam. The foams in Fig. 4 were prepared at three different expansion ratios, 2, 3 and 5, using CO_2 as the PBA. The foams in Fig. 5 were prepared at the three different expansion ratios using N_2. The foams in Fig. 6 were also prepared at the three different expansion ratios using Air. In the case of air, the delivery of air through the injector valve is stopped by setting the valve opening period to zero. The air was provided through the venting hole by setting the vent vessel pressure at 7 MPa. This was needed to prevent the oxidization of polymer. The longer mixing time of air with polymer causes the decomposition of polymer.

As can be seen in Figures, it is clear that the microcellular foams were successfully prepared with either CO_2, N_2 or Air. In any of the CO_2, the N_2 and the Air foaming examples, at the upper left corner of each image, the foaming temperature is indicated. With the increase of expansion ratio, cell size observed from perpendicular view to coreback direction, the cell size is slightly increased and the aspect ratio of the cell is increased.

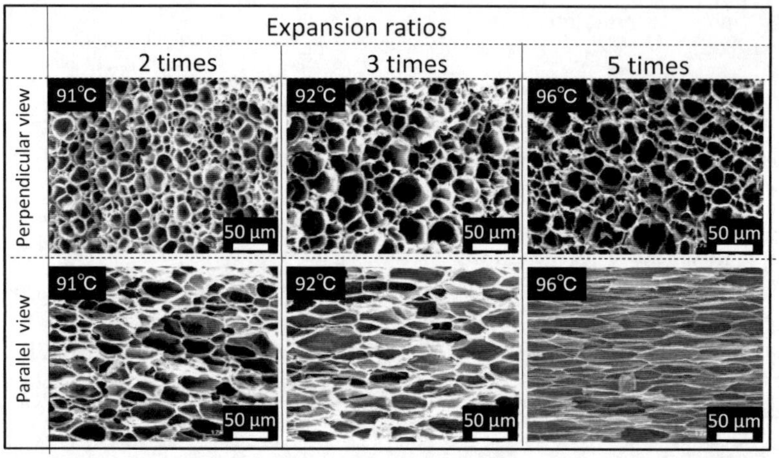

FIGURE 4. SEM micrographs of iPP/MD (0.5%) foams with CO_2 at different expansion ratios. The polymer temperature at which the coreback operation was conducted is indicated in the upper left corner of each micrograph.

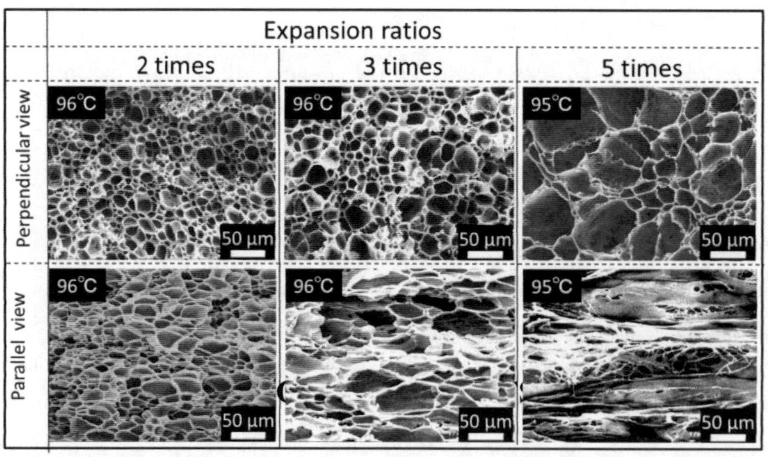

FIGURE 5 SEM micrographs of iPP/MD (0.5%) foams with N_2 at different expansion ratios. The polymer temperature at which the coreback operation was conducted is indicated in the upper left corner of each micrograph.

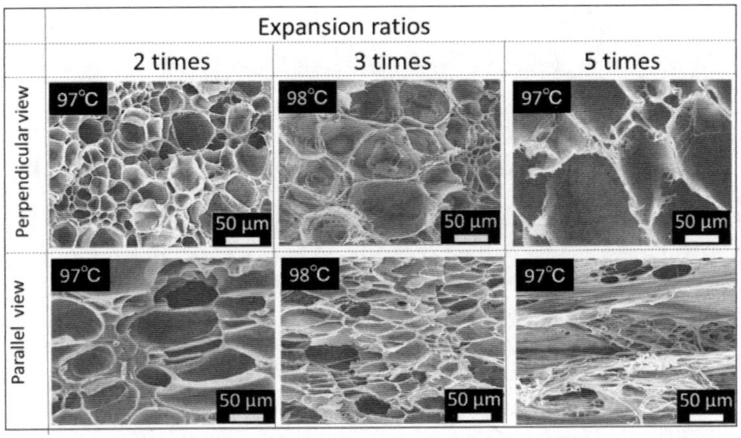

FIGURE 6 SEM micrographs of iPP/MD (0.5%) foams with 7Mpa Air at different expansion ratios. The polymer temperature at which the coreback operation was conducted is indicated in the upper left corner of each micrograph.

CONCLUSION

A new foam injection molding technology was developed. One of the features of this technology is that the foam injection molding machine does not need a high pressure pump to bring the PBA, i.e., N_2, CO_2 or Air to a supercritical state. When the PBA could not be dissolved in the molten polymer, the excess PBA gas is released from the injection machine. Alternatively, when the PBA concentration does not reach a saturation point, the additional PBA is dissolved in the polymer at the position beneath the venting hole. The pressurized venting vessel and hole play a key role of controlling and stabilizing the PBA concentration in the molten polmer. With the developed technology,microcellular foams were stably produced without using a supercritical fluid.

ACKNOWLEDGMENT

The authors in academia are thankful for financial support from Grants-in-Aid for Scientific Research (B) 26289289 and Advanced Low Carbon Technology Research and Development Program (ALCA).

REFERENCES

1. S.-T. Lee, Foam Extrusion Principles and Practice, ed. S.-T. Lee, Technomic Publishing Company, Lancaster, Chapter 1,1-4 (2000)
2. J. Xu, Microcellular Injection Molding, John Wiley &Sons, New Jersey, Chapter 1, pp.1-11, 2010
3. N.P. Suh, Innovation in Polymer Processing, ed. J.F. Stevenson, Hanser Publication, Cincinnati, Chapter 3, 93-149 (1996).
4. M. Berry, Applied Plastic Engineering Handbook, ed. M. Kutz, Elsevier, Chapter 14, 215 (2011)
5 S. Habibi-Naini, Foam Injection molding the flexible way, Sulzer Technical Review 2, 10, (2004)
6. Y. Sato, K. Fujiwara, T. Takikawa, Sumarno, S. Takishima, H. Masuoka, Fluid Phase Equilibria,162, 1 261 (1999),
7. Y. Sato, T. Takikawa, S. Takishima, H. Masuoka, J. Supercritical Fluids,19, 187, (2001),
8. C.L.Higginbotham, J. Lyons, J. Kennedy, Advances in polymer processing, ed. S. Thomas and Y. Weimin, CRC Press, Florida, Chapter 13.5 390, (2009)
9. R. Miyamoto, S. Yasuhara, H. Shikuma, M.Ohshima, 54, 9, 2075 (2014)
10. T. Kobayashi, T. Hashimoto, Development of Self-Assembling Nucleators for Highly Transparent Semi-Crystalline Polypropylene, Bull. Chem. Soc, Jpn, 78, 218 (2005)

Hysteresis Measurements as a Tool to Monitor the Non-Linear Mechanical Deformation Behavior of Bead Foams Under Fatique Loading

Jan-Hendrik Keller[1] and Volker Altstaedt[1,a)]

[1] *University of Bayreuth, Universitaetsstr. 30, 95447 Bayreuth, Germany*

[a)] Corresponding author: altstaedt@uni-bayreuth.de

Abstract. Today, technical components, such as crash absorbers in the automotive industry or running shoes for sports produtcs, which are made by bead foams, are very often exposed to complex and large scale cyclic loading conditions. In order to characterize the time and temperature dependent mechanical behavior during fatigue loading, the principles of the mid-curve and hysteresis evaluation are best suited. These recently further-developed tools can provide instantaneous feedbacks about the state of fatigue the specimen or the component is in. In addition using these methods, complementary information about the prevailing deformation mechanisms of the sample can be derived (e.g. linear, non-linear, failure) at any time during the fatigue test. Within this paper, the results of low density EPS and EPP bead foams (30 g/l) under several conditions are discussed, where non-linear mechanical behavior could be seen even at low deformation amplitudes.

INTRODUCTION

Several expanded beads are welded together to form a complex shaped part. This is one of the main advantages of polymer bead foams [1]. By now, different kind of polymers are being used commerically. Parts made of expandable polystyrene (EPS) for example, can be found as blocks for the insulation of buildings and houses. More importantly, it can be used for the transportation of thermal sensitive goods (e. g. frozen food) . Lately, parts made of EPS and also expanded polypropylen (EPP) can be found in helmets or in other devices for shock absorbing purposes. In addition, EPP and EPS are being used in automobile parts, e. g. as bumpers or as sun visor blades [2]. Unlike in quasi-static loading cases, the mechanical loading situation while driving the car can be quite complex. So for a good life time prediction of these parts, other testing methods are needed. This new tool needs to be capable of processing the collected measurement data for both linear and non-linear material behavior. This is important as most foamed polymers show non-linear visco-elastic behavior upon certain (and sometimes very small) loading conditions.

For low level deformation, the dynamic mechanical analyses (DMA) is well established in polymer science. Despite of the good results generated by this technique, there is always one major drawback: If the material behavior is not linear anymore, which is true for foams even at low loading amplitudes, the calculated results from the DMA experiment will not be exact anymore. The mateial behavior is either under- or over-estimated. Alternative methods are needed for describing the non-linear behavior of polymers under cyclic loading more precicesly.

The tool of the midcurve, which has its foundation on the evaluation of the hysteresis loops, is best suited for describing the linear and non-linear material behavior of polymer solids. The foundation of this principle was lately adopted for polymer melts within the large amplitude oscillatory shear (LAOS) testing [3]. For foams, it is possible for the first time to show the deformation state the material is in, by systematically varying the test conditions (frequency, temperature, loading amplitude, pre-force, etc.). These states could be either pure linear or non-linear visco-elastic. The midcurve tool and the exact evaluation of the hysteresis loops can give deeper understanding, what happens within the material while cyclic experiments are being conducted.

The construction of the midcurve and the math behind it, can be found in the publications of Benjamin Lazan

[4], who did empirical investigations on both metals and polymers. Several our publications, where the midcurve was used, show the importance of this tool (see e. g. [5] and [6]).

MATERIALS AND METHODS

Bead Foams

Commercially available expandable polystyrene (EPS) for mainly insulation purposes was bought from a local supplier in form of 20 mm thick sheets. The specified density is 30 g/l. The specimens were cut into cubes of 20 mm edge length using a circular saw. The faces were sanded by hand.

Expanded polypropylene (EPP) with a density of 30 g/l was kindly supplied by New Materials Bayreuth GmbH, Bayreuth, Germany. The specimens wer also cut using a circular saw to cubes of 20 mm edge length and sanded afterwards. No further treatments were not done. The density of these materials were not changed during the tests.

Density Measurements

The density of the specimens was determined using a density scale (AG254, Mettler Toledo, Greifensee, Switzerland), which uses the archimedic principle. From each material, at least ten different sample were measured and the mean was calculated.

Compression Testing

Prior to the dynamic testing, simple compression test were done on an universal testing machine (Z7050, Zwick Roell, Ulm, Germany). The velocity of the crossbar was set to a constant speed of 100 mm/min. The displacement was determined using the crossbar, while the force was measured using a load cell (500 N, Zwick Roell).

Hysteresis Measurements

A sevo-hydraulic testing machine (PSA 010, Instron Schenck, Darmstadt, Germany) was used in compression mode for detecting the long-term mechanical properties. The displacement was taken from the position sensor of the piston, while the force was taken by a load cell (1kN, Type S2, HBN, Germany). Additionally, the hysteresis loops were obtained with this testing device using a sophisticated software. The evaluation of these loops shows the linear or non-linear stage of the sample while testing.

RESULTS

Morphology

The results show, that the density is in good accordance with the datasheets provided by the suppliers. The results are listed in Table 1. The deviation can be explained by difficult measurement of the exact sample geometry.

More importantly, the results of the compression testing are also listed in the Table 1. The compressive strength at 10 % is displayed as well as standard deviation. Although both materials show the same density, the quasi-static behavior is different. EPS has a 50 % higher strength value at 10 % deformation and is hence stiffer.

TABLE 1. Results of the compression test of the tested bead foams. The density is in good accordance with the data sheets. The compressive strength is taken at the 10 % compression niveau.

Material	Observered density g/l	Given density g/l	Compressive strength kPa at 10%
EPS 30	27	27-29	150 ± 12
EPP 30	29	29-31	100 ± 11

Hysteresis Measurements

When a polymeric material is subjected to cyclic loading, the stress strain behavior can give deeper understanding of the deformation state of one material. This is only possible, if the curves for force and for displacement are made visible for each cylce individually. Figure 1 shows the hysteresis loops for both EPS 30 and EPP 30. The load level was set according to the compressive strength as the upper limit and the lower limit was set to 0.1 times the compressive strength. So for EPS 30 the load was 15-150 kPa and 10-100 kPa for EPP 30. These experiments were conducted in stress controlled mode, so the given force values were used as control input. The frequency was set to 2 Hz.

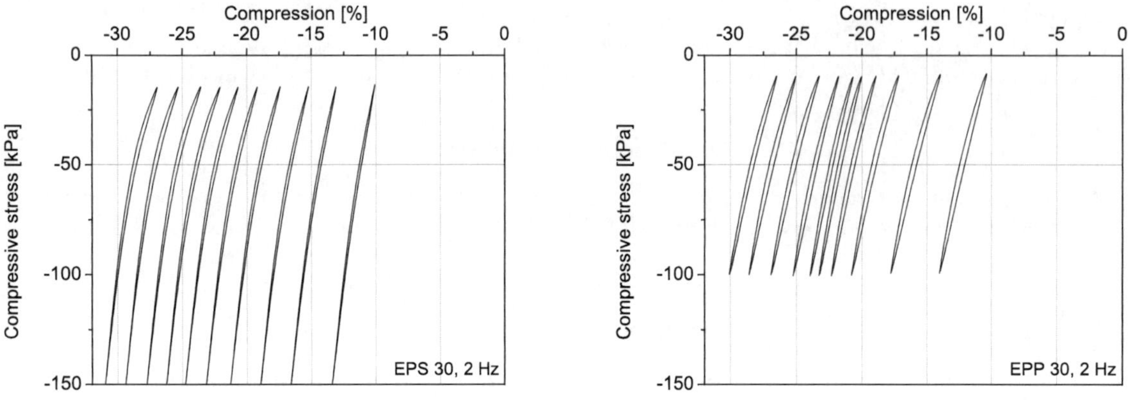

FIGURE 1. Graphical representation of the deformation behavior of EPS 30 (left) and EPP 30 (right). The load level was set to the critical value. The loop on the very right represents the behavior after 10.000 cylces, and each another loops after 10.000 cylces more.

To show changes in the hysteresis loops, different loops for EPS 30 and EPP 30 are shown in Fig. 1. Please note that the origin is in the top right corner, as the material was compressed. Therefore, the loop on the right hand side of each diagramm represents the mechanical behavior at very much of the beginning of the long term loading (10.000 cycle). The following loops are taken at every 10.000 cycles. The position relative to each other is different due to different creep rates per cycle. For EPP 30 however, the distance between the loops are more irregular. This can be explained by the fact, that EPP is very sensitive to changes in ambient temperatures. The loops shifts to the left while loading. The foam specimen gets more and more compressed over time. This cyclic behavior can be interpreted as creep. Please note, the quasi-static creep takes place in much longer periods of time.

In Fig.1 on the left hand side, the loop on the very right shows the mechanical behavior of EPS 30 after 10.000 cycles. This loop represents already non-linear visco-elastic behavior. Because the turn points of the loops are quite sharp, some friction effects occur inside the material. The loop is also not 100 % symmetric, so there are already different non-linear effects starting. These effectx can also be seen when smaller amplitudes are used for testing (not shown here). As the experiments goes on, the shape of the loops change in more curved ones. The non-linear effects inside the material get stronger.

What else can be seen from the loops in Figure 1 (left) is, that the width of the loops is increasing, when going from right to left. This is due to increasing damping. The damping is a number and is calculated over the surface area of the hysteresis loop. The damping of the material describes the amount of energy, which has to be put in extra to keep the system oscillating. The higher the damping, the more heat needs to be dissipated, though.

In Fig. 1 on the right hand side, EPP 30 shows a little bit more linear viso-elastic behavior. The loops are quite symmetric. However, the width of the loop is larger, which indicates also a larger damping. The loops do not change that dramatically over time and do not get such curved.

To further characterize the cyclic creep of these materials, the midstrain is plotted over the number of cycles in Fig. 2. The midstrain represents a value, which is halve the deformation level in one cycle. The stress levels were adopted so that the border of creeping is made visible. It seems, that there is a force limit, up to which the foams only deformes slighty under cyclic loading. In Fig. 2, the solid lines represent the mid strain over the cycles when the force value does not exceed the critical niveau. The dashed line however, represents the mid strain if the critical level is exceeded. Furhter testing showed, that not the mid stress is responsible for the amount of creep rate, but it is a combination of amplitude, upper force limit, and mid stress. Furhter studies will be published elsewhere. However, the limit, up to which the foams do not creep only little over time was identified to roughly $0.75 * \sigma_X$. X stands for the deformation point, where the stress strain curve from quasi-static experiments deviated from the beginning linear slope. In numbers, for EPS 30 the critial niveau is $112.5kPa$ and for EPP 30 it is $75kPa$.

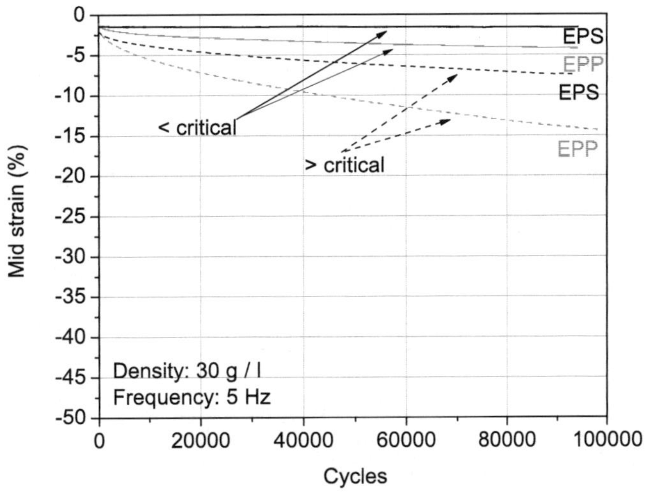

FIGURE 2. Influence of critical load level on the dynamic creep of EPS (black) and EPP (orange). The solid lines represent the mid strain with undercritical loading conditions, whereas the dashed lines repretent the mid strain with critical load levels.

With the critical number in mind, Fig. 3 on the left hand side clearly shows high changes in the mid strain over the numbers of cycles. In this special experiment, the mid stress was set to 100 kPa constant and the amplitude of loading was changed from low (20 kPa) to high (80 kPa). As expected, the 100 kPa ± 20 kPa load level (black) leads to only low dynamic creep, as this load level overshoots the critical niveau little. If the material is loaded with higher amplitudes (100 kPa ± 50 kPa), the dynamic creep increases dramatically. However, a limit arose during testing. The load level of 100 kPa ± 80 kPa induced an increase in temperature inside the sample, which was more than 5 °C. This test was stopped at only a few numbers of cycles. Nevertheless, one can cleary see the trend of this high load level. As this material was tested under ambient temperature conditions, the polymer chains were in complete energy elastic state. This leads to very high creep rates at high deformation levels.

Due to testing under ambient conditions, the specimens made of EPP 30 were tested above T_G, which means the polymer chains were partly in entropy elastic mode. This means, they can deform elastically up to a cerain point and recover more easily to the original state. This could affect the dynamic creep rate and make it less sensitive to changes in the loading amplitudes. Figure 3 on the right hand side shows this experiment for EPP 30. The load levels are adopted accordingly to the critical niveau, so one could compare those to EPS 30.

First of all, the mid strain curves lie much closer to each other. Secondly the heat induction due to loading was not as severe as seen for EPS 30. There is a correlation of mid strain and amplitude visible. However, it is not that promient as seen for EPS. EPP 30 seems to be very sensitive to changes in the ambient temperature of less than 1 °C. The wavy curve of the most bottom curve (load level 66 kPa ± 50 kPa) could be explainted by a slight change in the ambient temperatures.

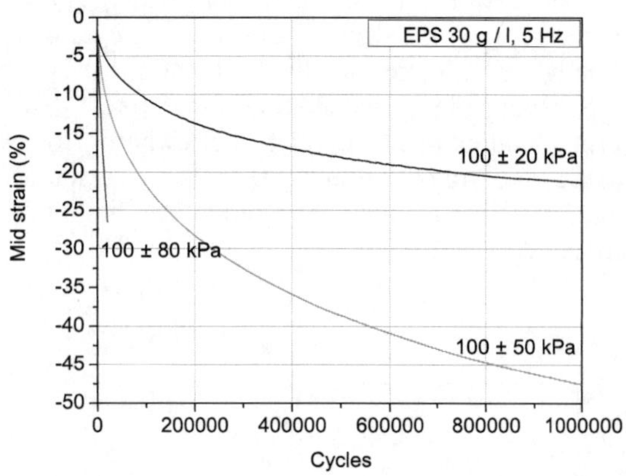

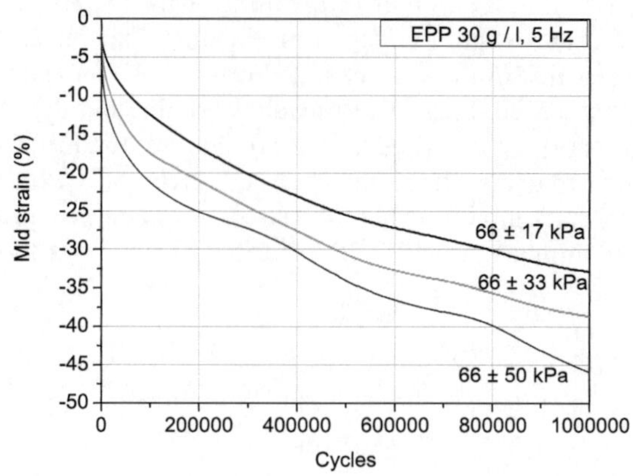

FIGURE 3. The load level for both materials is set to 66 % of σ_{10}. The amplitude is set to 20-25 % of this value (black), 50 % (orange), and 75-80 % (blue), respectively. EPS 30 (left) and EPP 30 (right) are loaded for 1 mio. times at the given load levels at 5 Hz. Testing EPS 30 with 80 % loading amplitude (blue curve) was stopped due to a temperature increase of more than 5 °C.

CONCLUSION

The dynamic creep of low dense bead foams was investigated using the midcurve and hysteresis methods. It has been shown, that EPS 30 is strongly affected by the loading amplitude and shows high dynamic creep rates. This could be attributed to the energy state EPS is in. Also this could explain the high plastic deformation in quasi-static experiments. For EPP 30, the load amplitude only has little influence in the creep rates. For both materials it is valid, that if the critical stress is reached, the materials react with very high creep rates.

ACKNOWLEDGMENTS

The collaboration of the members of the department of polymer engineering at the university of Bayreuth is appreciated. The material supply by New Materials Bayreuth GmbH is also highly appreciated.

REFERENCES

[1] D. Raps, N. Hossieny, C. B. Park, and V. Altstaedt, Polymer **56**, 5–19 (2014).
[2] C. Trassl and V. Altstaedt, Kunststoffe **2**, 73–76 (2014).
[3] K. Hyun, M. Wilhelm, C. O. Klein, K. W. Cho, J. G. Nam, K. H. Ahn, S. J. Lee, R. H. Ewoldt, and G. H. McKinley, Progress in Polymer Science **36**, 1697–1753 (2011).
[4] B. L. Lazan, *Damping Of Materials And Members In Structural Mechanics* (Pergamon Press, 1968).
[5] M. El Fray and V. Altstaedt, Polymer **44**, 4643–4650 (2003).
[6] J. E. Puskas, M. El Fray, M. Tomkins, L. Dos Santos, F. Fischer, and V. Altstaedt, Polymer **50**, 245–249 (2009).

Physical Foam Extrusion of Polyolefin/Cycloolefin Copolymer Blends

Matthias Walluch[a], Anna Uray[a], Clemens Holzer[b], Günter R. Langecker[b]

[a] *Polymer Competence Center Leoben GmbH., Roseggerstr. 12, 8700 Leoben, Austria*

[b] *Montanuniversitaet Leoben, Department of Polymer Engineering and Science, Chair of Polymer Processing, Otto Gloeckl-Strasse 2, 8700 Leoben, Austria*

matthias.walluch@pccl.at

Abstract. There are many different ways to influence the properties of extruded polymer foams. On the one hand the processing conditions, like the temperature and pressure profile in the extrusion line or the die geometry have a major influence on the resulting foam morphology. On the other hand a lot of opportunities for an improvement of the cell structure are provided by different changes in the used material formulation: The type of the matrix polymer and its rheological properties, the amount and type of the nucleating agent as well as the foaming agent determine the achievable foam properties. Often polymer blends are used to improve the foaming performance of some polymers. For example polypropylene (PP) is frequently blended with low density polyethylene (LDPE), because of its higher melt strength, to improve the foamability. In this study the foaming behavior of PP blends was investigated. But the target of the used blend partner was not to improve the melt strength of the PP, but the nucleation rate of the formulation. Several studies show that the combination of polyolefins and cycloolefin copolymers (COC) leads to blends with an unusual phase morphology and special properties, which may be useful for the foaming behavior for example as an improvement of the cell nucleation. Due to this interesting behavior, in this study two different types of neat PP, one homopolymer and one copolymer, and different PP/COC blends were foamed with nitrogen as a blowing agent and talcum as a nucleating agent. The influence of the COC on the foam morphology was examined. A positive influence of COC on the foam morphology of physically foamed polypropylene could be shown.

Keywords: foam extrusion, polypropylene, cycloolefin copolymer, blends.
PACS: 81.05.Lg, 83.80 TC

INTRODUCTION

The number of applications and products where foamed polymers are used is increasing continuously. Thus the requirements in terms of mechanical properties, thermal resistance, and cell structure may differ a lot. A common way to fulfill such high quality demands in addition to a good foaming performance of the matrix is to blend different polymers. Polypropylene (PP) offers a lot of interesting properties like low material costs, a good thermal and chemical resistance as well as food-safety [1]. Foamed polypropylene also offers good mechanical properties [2] why foamed PP is often used in packaging applications. But polypropylene has also disadvantages for foaming applications. The molecular structure and the rheological properties of conventional PP types lead to an instable cell growing and cell coalescence [3]. An improvement of the foaming behavior can be achieved by using high-molecular-strength PP (HMS-PP), but these types are much more expansive then conventional PP types. Another approach followed by some companies is to use PP/LDPE blends to improve the rheological behavior and the melt strength.

The investigations in this work follow a different idea. Several studies show that blends of polyolefins and cycloolefin copolymers (COC, amorphous copolymers of ethane and norbornene) have very special properties and morphologies. It was shown that COC forms fibrous phases in different polyolefines and that the shape of the COC fibers is influenced by the amount of COC in the blend. Šlouf [4], Kolařik [5, 6] and Vavková et al. [7, 8] did plenty of work on the phase morphologies and the mechanical properties of PP/COC and HDPE/COC blends. Fambri et. al [9] studied the rheological properties of such polymer blends.

The approach of this study is that this special inhomogeneous phase morphology of such polymer blends may have a positive influence on the nucleation rate in the physical foam extrusion process. Therefore blends of two different types of polypropylene, one homopolymer and one copolymer were foamed without COC, and also with 10 and with 20 wt.-% of COC. As a nucleating agent talcum was used. The blowing agent, nitrogen, was added in two different dosages. It could be shown that the foam morphology is influenced very strongly by changes in the formulations of the polymer matrix. By the addition of COC to both PP types an improvement of the cell structure was achieved. Even for investigated formulations where no talcum was used the cell sizes could be reduced significantly.

USED MATERIALS

In this study two different types of polypropylene (see table 1) and blends of them with one type of cycloolefin copolymer were physically foamed in an extrusion process. The used nucleating agent was talcum Mistrcocell M90 by Imerys Talc with a D50 of 6.4 µm. As a blowing agent 0.05 wt.-% and 0.1 wt.-% of nitrogen (99.8% pure, Linde Gas, Austria) was used.

TABLE (1). Overview of the used polymers and their basic properties.

	PP 1	PP 2	COC
Manufacturer	Borealis	Unipetrol	Topas Advanced Polymers
Grade	BB412E	Mosten GB005	8007S-04
Type	Copolymer	Homopolymer	Copolymer
MFR (230°C, 2,16 kg)	1,3 g/10min	5 g/10min	29 g/10min

To realize a homogeneous distribution of the talcum in the different formulations nucleating agent masterbatches of all three polymers were produced. These compounds contain 40 wt.-% of the talcum and were produced on a twin screw extruder (D = 25 mm, L/D = 40) by Werner & Pfleiderer (Germany).
All the investigated material formulations are shown in table 2.

TABLE (2). Overview of the composition of foamed material formulations in wt.-%.

PP 1	PP 2	COC	Talcum
100	0	0	0
100	0	0	2
100	0	0	4
90	0	10	0
90	0	10	2
90	0	10	4
80	0	20	0
80	0	20	2
80	0	20	4
0	100	0	0
0	100	0	2
0	100	0	4
0	90	10	0
0	90	10	2
0	90	10	4
0	80	20	0
0	80	20	2
0	80	20	4
0	0	100	0
0	0	100	2
0	0	100	4

All shown formulations were foamed by adding an amount of 0.05 and 0.1 wt.-% of nitrogen.

FOAM EXTRUSION

Foam extrusion was performed on a grooved single screw extruder (Rosendahl, Austria). This extrusion line is equipped with a special screw for physical foaming of polymers. It has a diameter of 45 mm and a length of 32D, including a 4D long shear element and a mixing element with a length of 8D. To improve the homogeneity of the melt an additional static mixer was mounted before the die. For all experiments a round die with a diameter of 2.5 mm was used.

A positive displacement pump (Teledyne Isco syringe pump 260D) was used to inject the blowing agent into the extruder above the shear element. Due to the changing pressure profiles in the extrusion line by using different material formulations the needed volume of nitrogen was calculated for each material formulation. Some of the most important processing parameters are shown in table 3.

TABLE 3. Settings for the foam extrusion of all samples: Temperature profile and screw speed

Temperatures in °C			Scew speed in rpm
Feeding zone	**Gas injection**	**Die exit**	
30	200	190	10

For the foaming process the polymers and the nucleating agent were dry blended and carefully mixed. Then the mixtures were fed into the barrel through the hopper. A sufficient time was given to stabilize the process before the injection of the nitrogen was started. When a steady state was reached again the samples were collected carefully. It is important not to deform or stretch the extruded samples because this would affect the foam morphology.

The last three formulations shown in table 2, those without polypropylene, could not be foamed because the pressure at the point of the gas injection could not be kept constant enough to guarantee a correct amount of blowing agent in the foam.

FOAM CHARACTERIZATION

To analyze the cell morphology of the different foamed formulations, the cross sectional area of the extruded strands was examined by using the optical microscope Alicona InfiniteFocus G4 (Alicona, Austria). To determine the mean cell diameter (in µm) and the cell density N (in number of cells per cm²) the provided Alicona image analysis software was used. The cell densities were calculated using equation 1 [10], were N_b is the number of cells per area (A in cm²), $\rho_{polymer}$ is the density of the unfoamed matrix polymer or blend (in kg/m³), and ρ_{foam} is the density of the foamed samples. (in kg/m³).

$$N = \left(\frac{N_b}{A}\right)^{\frac{3}{2}} \frac{\rho_{polymer}}{\rho_{foam}} \quad (1)$$

RESULTS AND DISCUSSION

The comparison of the cell sizes of the foamed material formulations without talcum as a nucleating agent shows, as expected, a weak foaming behavior. That means that there is a very small number of very large bubbles. But as it is shown in figure 1 and figure 2, the cell size could be reduced significantly for both polypropylene types by adding COC. For PP 1 the achieved cell size reduction is, depending on the amount of COC and gas in the formulation, between 10 and 30 % compared to the neat polypropylene. By adding COC to PP 2 the average cell size was approximately halved for all investigated samples. The difference between 10 wt.-% and 20 wt.-% COC and 0.05 wt.-% and 0.1 wt.-% is less considerable than for the other Polypropylene.

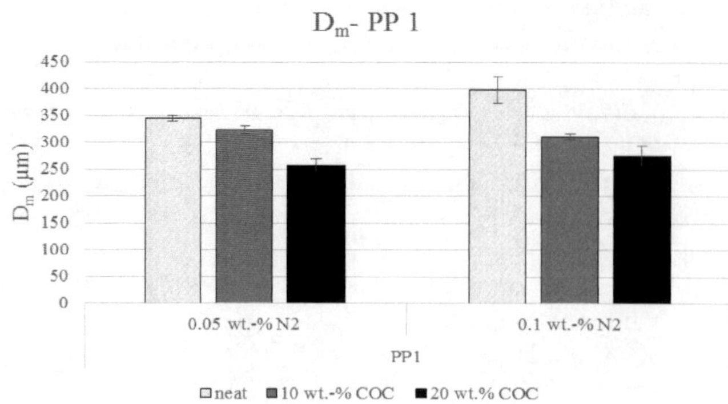

FIGURE 1. Cell sizes of the foamed PP 1 and its blends with COC without any talcum as nucleating agent using 0.05 and 0.1 wt.-% N_2 as physical blowing agent

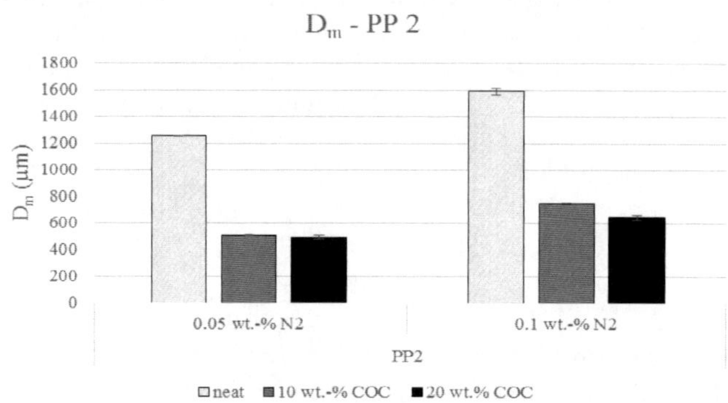

FIGURE 2. Cell sizes of the foamed PP 2 and its blends with COC without any talcum as nucleating agent using 0.05 and 0.1 wt.-% N_2 as physical blowing agent

When talcum is used as a nucleating agent the average cell size can be reduced significantly. This expected effect could be observed for all samples. But figure 3 and figure 4 show that the effect can even be enhanced by blending the polypropylenes with COC. The effect of the cycloolefin copolymer is more oblivious and significant for PP 1, this can be explained by the higher impact of the nucleating agent on the second polypropylene. Nevertheless a slight improvement of the average cell sizes can also be observed here.

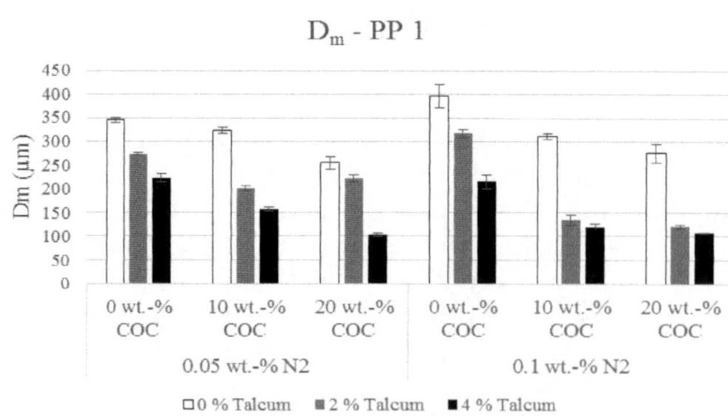

FIGURE 3. Comparison of the average cell sizes of the foamed PP 1 and its blends with COC using 0, 10 and 20 wt.-% of talcum as nucleating agent foamed with a gas amount of 0.05 and 0.1 wt.-% N_2.

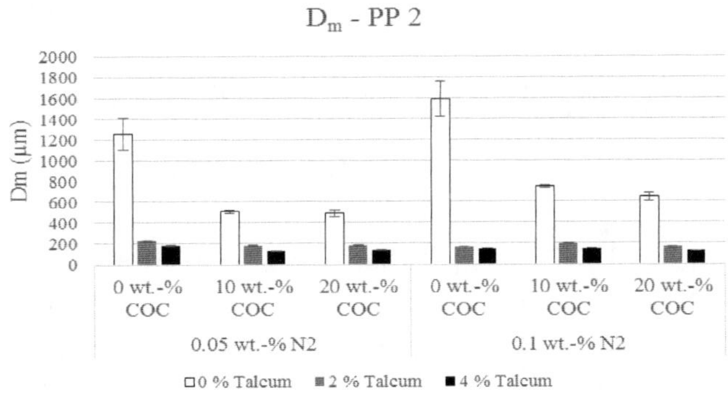

FIGURE 4. Comparison of the average cell sizes of the foamed PP 2 and its blends with COC using 0, 10 and 20 wt.-% of talcum as nucleating agent foamed with a gas amount of 0.05 and 0.1 wt.-% N_2.

CONCLUSION AND OUTLOOK

In this work it could be shown that the foam morphology of physically foamed polypropylene can be improved by blending it with COC. The cell size and the cell densities of the blends were positively influenced by adding 10 wt.-% COC and could be further improved when 20 wt.-% COC were added to the formulation. The most significant effects of the COC could be achieved for PP 1. The smallest average cell sizes were reached with 20 wt.-% COC and 4 wt.-% talcum. The effect of the COC on the foam morphology of the samples were no talcum was used was also quite significant. Particularly the blends with PP 2 show an impressive cell size reduction.

The low melt flow rate and the low elasticity of the COC suggest that this improvement of the cell morphology can only be explained by an improvement of the nucleating rate. The better performance of the blends where no nucleation agent was used may also lead to the conclusion that the COC provides additional spots where a cell nucleation may happen.

The next steps to get a better insight into the mechanisms that are responsible for the improved foaming behavior will be rheological investigations of the polymers and the polymer blends, as well as the investigation of the phase structures of the blends.

ACKNOWLEDGMENTS

The research work was performed at the Polymer Competence Center Leoben GmbH (PCCL, Austria) within the framework of the COMET-program of the Federal Ministry for Transport, Innovation and Technology and Federal Ministry for Economy, Family and Youth with contributions by the University of Leoben. The PCCL is funded by the Austrian Government and the State Governments of Styria, Lower Austria and Upper Austria.

REFERENCES

1. P. Elsner, P. Eyerer and T. Hirth, *Kunststoffe Eigenschaften und Anwendungen, 8th ed. Springer-Verlag, Heidelberg, Germany, pp. 213 – 253.*
2. R. Buix, P. Viot and J.L. Lataillade, *Inter J Impact Eng, 36 (2009), pp. 329-342.*
3. M.B. Bradley and E.M. Phillips, *Novel Foamable Polypropylene Polymers, SPE Antec Tech. Papers, 36: pp. 717-720 (1990).*
4. M. Šlouf, J. Kolařik and L. Fambri: Phase Morphology of PP/COC Blends, *Journal of Applied Polymer Science, Vol. 91, pp. 253-259 (2004).*
5. J. Kolařik, Z. Kruliš, M. Šlouf and L. Fambri, *High-Density Polythylene/ Cycloolefin Copoplymer Belnds. Part 1: Phase Structire, Dynamic Mechanical, Tensile, and Impact Properties, Polymer Engineering and Science – 2005.*
6. J. Kolařik, A. Pegoretti, L. Fambri and A. Penati, *High-Density Polythylene/ Cycloolefin Copoplymer Belnds, Part 2: Nonlinear Tensile Creep, Polymer Engineering and Science – 2006.*
7. T. Vacková, M. Šlouf, M. Nevoralová, L. Kaprálková, *Processing-Improved Properties and Morphology of PP/COC Blends, Journal of Applied Polymer Science, Vol. 122, pp. 1168-1175 (2011).*
8. T. Vacková, M. Šlouf, M. Nevoralová, L. Kaprálková, *HDPE/CPC Blends with Fibrous morphology and their properties, European Polyer Journal, Vol. 48, pp. 2031-2039 (2012).*
9. L. Fambri, J. Kolařik, E. Pasqualini, A. Penati and A. Pegoretti, *Rheological Study on Polypropylene/Cycloolefin Colpolymer Blends, Marcromol. Symp. Vol. 263, 114-120 (2008).*
10. M. Mihai, M. Huneault, and B. Favis, *Polymer Engineering and Science, 50, 3 (2010).*

Welding Quality in Polymer Bead Foams: an *in situ* SEM Study

Julia Gensel[a], Christin Pawelski[b] and Volker Altstädt [a,b,*]

[a] *Neue Materialien Bayreuth GmbH, D-95448 Bayreuth*
[b] *University of Bayreuth, Polymer Engineering, D-95440 Bayreuth, Germany*
Julia.Gensel@nmbgmbh.de
Volker.Altstaedt@uni-bayreuth.de

Abstract. Polymer bead foams are lightweight materials consisting of fused microcellular beads. Their multiscale structure makes bead foams, on the one hand unique in terms of a free choice of shape combined with extreme low density, excellent thermal insulating properties, and good energy absorption properties [1]. On the other hand, fusion of the beads plays a crucial role in determining the performance and thus the application range of bead foams. Still, the question how processing affects the structure and the mechanical strength of the fused regions lacks a clear answer today. Conventional tensile tests and fracture surface analysis give some general information on the fusion strength. However, the proper understanding of the bead fusion requires local weld structure analysis and the fracture path monitoring on the nano- and microscales. In this contribution, we present an *in situ* study of the interbead failure behaviour by means of tensile experiments in the Scanning Electron Microscope. This experimental study allows for direct observation of both, the cell and bead deformation, and more importantly the evolution of the fracture in space and time. Combined with Atomic Force Microscopy measurements on the weld-zone, we discuss the influence of moulding conditions on the interfacial structure and bonding quality in polypropylene bead foams. It was found that low moulding temperatures and short steaming time leads to a pull-out of polymer chains at the beads interface during the tensile test. In contrast, with increasing time and temperature, polymer chains at the interface were solidly anchored resulting in higher mechanical strength of the weld and thus, elastic stretching of the foam cells and beads and finally cohesive matrix break down.

Keywords: bead foams, expanded polypropylene, *in-situ* mechanical analysis, fusion quality, interdiffusion, Scanning Electron Microscopy
PACS: 68.35.Ct ; 68.35.Fx; 68.35.Gy; 68.37.Hk; 68.37.Ps; 81.40.Lm; 81.40.Np; 81.70.Bt

INTRODUCTION

The properties of polymer foams mostly depend on the type of matrix polymer, the density, and the cell morphology. Bead foams show a multi-scale structure where mesoscopic beads, which consist of microscopic cells within a polymer matrix material, are fused into macroscopic foam parts [1]. Due to this assembly out of single mesoscopic beads, bead foams have a special hierarchical morphology. Thus, in contrast to other polymer foams, an additional major parameter controlling the performance of the bead foams has to be taken in account: the interbead bonding. Playing a crucial role, a good fusion of beads is necessary in order to achieve favorable mechanical properties.

The fusion behaviour in amorphous polymer bead foams (expanded polystyrene, EPS) has been extensively studied [2, 3], and it is generally agreed that molecular interdiffusion and entanglements at the interbead contacts during the moulding process are the predominant mechanisms.

For expanded polypropylene (EPP), which is a semi-crystalline polymer, the fusion behaviour is, in addition to interdiffusion, affected by crystallization. Although significant progress in the direction of fusion mechanism in EPP has been made [4, 5], however, it is still unclear, how the micro- and nanostructure can be (i) controlled by processing conditions and (ii) related to the fusion strength in EPP bead foams.

In this study, the fusion quality in EPP bead foams is investigated, by means of *in situ* tensile experiments in the scanning electron microscope (SEM). The local fracture evolution is monitored and the influence of moulding conditions on the interbead bonding is discussed. In addition, for the first time, the morphology of the fused zone is studied by atomic force microscopy (AFM) and the interbead failure behaviour is correlated to its morphological features. The aim of the study is to understand the relations between the moulding conditions, the resulting interbead structure and the corresponding fusion strength, which is greatly important for choosing suitable process parameters and would help improving both the efficiency of the process and the performance of moulded parts.

EXPERIMANTAL SECTION

Materials and Processing

Expanded polypropylene beads Neopolen P 9225 with an average bulk density of 25 g/L were supplied by BASF SE.

Commercially available steam-chest moulding equipment Transtec72/52PP from Teubert Maschinenbau GmbH was used in this study. The dimensions of the mold cavity were 285×195×22 mm. The moulding was carried out at various steam pressures (absolute pressure) 2 bar < p < 4 bar, *i.e.* ~120°C < T < ~145°C, and different total steaming time 3 s < t < 30 s including cross-steaming and autoclave steaming. The silo pressure (2.1 bar), the dynamic pressure (1.7 bar) and the cooling temperature (80°C) were kept constant for all samples.

Characterization Methods

SEM measurements were obtained on a Zeiss EVO MA 15 instrument operating at 5 keV. Samples were sputtered with a ca. 10 nm thin gold layer.

AFM images were taken employing a Dimension™ 3100 AFM equipped with a NanoScope® V controller (both Bruker AXS Inc., USA) operating in TappingMode™ using silicon cantilevers from Bruker with a typical spring constant of ~26 N/m and a typical resonance frequency of 300 kHz (OTESPA-R3). Prior to characterization, samples were treated with oxygen plasma for 10 min at 0.2 mbar using a Flecto 10 from Plasma Technology.

Differential scanning calorimetry (DSC) measurements were conducted under nitrogen atmosphere at a scan rate of 10 K/min with a Mettler Toledo DSC. About 10 mg of the EPP bead sample were sealed in aluminum standard 50 mL crucibles and heated from 25°C up to 200°C.

Tensile tests were conducted according to ISO 1798 on a Zwick Z020 universal testing machine at a constant crosshead speed of 25 mm/min with an initial load of 20 N.

For the *in situ* tensile experiments, tensile/compression module from Kammrath & Weiss GmbH with a load cell of 5 kN was installed inside the SEM chamber. The load speed was 10 μm/s (= 0.6 mm/min). Samples were water-jet cut (dimensions 9×4×15 mm) and sputtered with a ca. 10 nm thin gold layer.

Optical microscopy measurements were performed on ZEISS Axioskop 2 MAT using a λ plate.

RESULTS AND DISCUSSION

Fusion Strength in EPP

Generally, for welding or sintering processes, efficiency and fusion quality strongly depend on the material properties and the selected temperature. In order to find proper moulding condition for EPP, we performed DSC measurements on the beads (Fig. 1a), which show a typical double-peak melting behaviour [1, 4-6]. During the steam-chest moulding, crystals associated with the low melting peak contribute to the fusion, while the unmolten high melting peak crystals are supposed to maintain the cellular structure of individual beads [1]. Therefore, we selected the relevant temperature range between 120-145 °C. According to both theoretical and experimental observations (data not shown) this range corresponds to ~ 2-4 bar steam pressure.

Fusion strength can be obtained through macroscale mechanical testing on moulded bead foams. Therefore, the tensile strength of specimens moulded at selected steam pressures (*i.e.* temperatures) for various durations has been evaluated. The tensile mechanical characteristics of the moulded EPP parts as a function of steaming pressure and time are summarized in Fig. 1b-c. The tensile strength increases with increasing pressure, which corresponds to an increase in temperature. In agreement with the relevant literature [4, 5] our findings indicate significant increase of tensile strength with increasing pressure/temperature and reaches asymptotically a plateau value (Fig. 1b). Figure 1c illustrates the evolution of tensile strength as a function of steaming time for two different temperatures. There is no pronounced increase of tensile strength with increasing time at lower temperature (2 bar /~120°C). However, with increasing temperature (3.5 bar / ~140°C), tensile strength exhibits an increasing trend with moulding time. Figure 1d-e shows the corresponding fracture surfaces of tensile samples indicating an interbead fracture behaviour after moulding at 2 bar (Fig. 1d) and intrabead/cellular fracture after moulding at 3.5 bar (Fig.1c). According to DSC data in Fig. 1a, such transition in the failure mechanism has been suggested to be the consequence of hindered interdiffusion due to polymer chains locked within non-melted crystals (T_1~120°C) *versus* time-dependent interdiffusion and cocrystallization effects upon cooling (T_2~140°C).

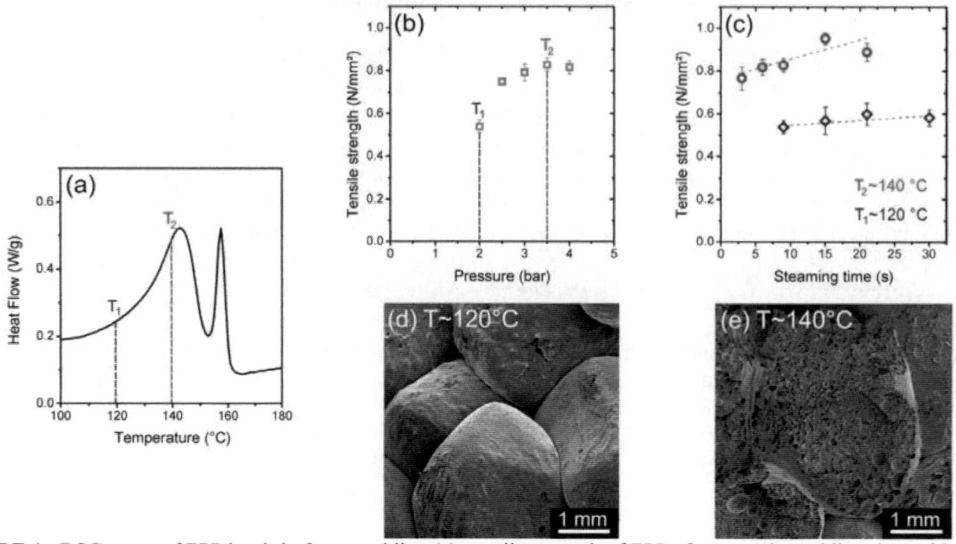

FIGURE 1. DSC curve of EPP beads before moulding (a), tensile strength of EPP after a total moulding time of 9 s as a function of steam pressure (b), tensile strength of EPP moulded at 2 bar (T ~120°C, ◊) and 3.5 bar (T~140°C, ○) for different steaming times (c) and SEM micrographs of fracture surfaces of EPP moulded at T ~120°C (d) and T~140°C (e) at a total steaming time of 9s.

Interface Morphology on the Nanoscopic Length Scale

A detailed knowledge of the morphology on the nanoscale is necessary to understand the effect of crystallization at the interface of the beads (Fig. 2a). For this purpose, AFM, being an important method for characterization of the nanostructure of semicrystalline polymers, was used to study the relative fractions of the amorphous and crystalline phases, as well as the shape and size of crystals directly on the fused zone.

Figure 2b,c shows two selected examples of AFM images on plasma-etched samples produced under low temperature (~120°C, Figure 2b) and high temperature (~140°C, Figure 2c). Due to the selective etching in semicrystalline polymers, amorphous phase is removed much faster than the crystalline. Therefore, the bright regions in the AFM images correspond to the crystalline and the dark to the amorphous phase.

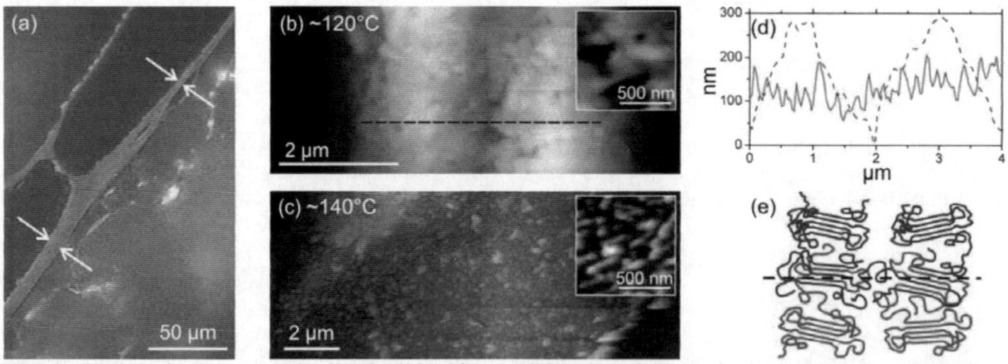

FIGURE 2. Optical microscopy image of the bead interface (a). The arrows indicate the fused skins of two individual beads. The AFM height images (z-scale: 0 – 1000nm) on the fused skins of the EPP moulded at ~120°C (b) and ~140°C (c). The insets are the corresponding 1×1 µm height images (z-scale: 0 – 200nm) with the same magnification. Cross-sectional height profiles (d) through the lines in (b, dashed) and (c, continuous). Schematic illustration of a proposed interface structure involving re- and cocrystallization in the interface (e).

The interface between two beads is still visible after moulding at ~120°C (see cross-sectional height profile in Fig. 2d, dashed line). In contrast, the structure of the interface in Fig. 2c (~140°C) is very homogeneous containing nodular/cylindrical domains with a thickness of ca. 50 nm and length of 50 – 300 nm which are embedded in an amorphous matrix without superstructure of its own. The morphological features observed in Fig. 2c are similar to the well-known mesophase structure of PP which results from the fast cooling of the welded polymer [7]. As shown in Fig. 2d (continuous line), the former interface is not identifiable. According to these observations, we assume that at ~140°C, the increase in temperature is high enough to involve the melting of the thermodynamically metastable crystal domains and subsequent interdiffusion. Upon cooling, fusion strength develops through re- and cocrystallization at the interface (schematically depicted in Fig. 2e) leading to the full healing of the interface.

Our findings are consistent with DSC simulation results published by Park and coworkers [4]. They reported on melting of crystals that were associated with the low-temperature melting peak (see Fig. 1a) during the heating process at temperatures of 135-145 °C and formation of new crystals during the cooling.

In situ Interbead Failure Monitoring

Conventional mechanical testing and fracture surface analysis of moulded bead foams give some general information on the fusion strength. However, processes occurring at the interfaces of fused beads cannot be fully understood based solely on macroscale testing. Therefore, local fracture path monitoring during the deformation must be carried out for the proper understanding of the fusion mechanism. In recent years, new *in situ* techniques such as digital image correlation, SEM, AFM or x-ray computed tomography *etc.* have enabled the exact characterization of microstructure-dependent stress-strain behaviour [8]. *In situ* SEM allows not only analizing the evolution of the fracture with deformation, but also for a direct observation of both, the cell and bead deformation with good time and spatial resolution. Therefore, we performed an *in situ* study of the interbead fracture behaviour and failure mechanism by means of tensile experiments in the SEM.

Selected SEM micrographs of the *in situ* tensile test of EPP moulded at ~120°C and ~140°C bar are shown in Fig. 3 and 4, respectively. The question being addressed is to determine how the fracture behaviour of an interface across which there was initially no recrystallization (T~120°C, Fig. 2b and Fig. 3) compares with that of re- and cocrystallized interface (T~140°C, Fig. 2c and Fig. 4).

Note, that in both cases, crack initiation occurs at the micro-voids (marked with arrows in Fig. 3 and Fig. 4). The subsequent crack propagation and the consequent failure mechanism, however, highly differ between the processing conditions.

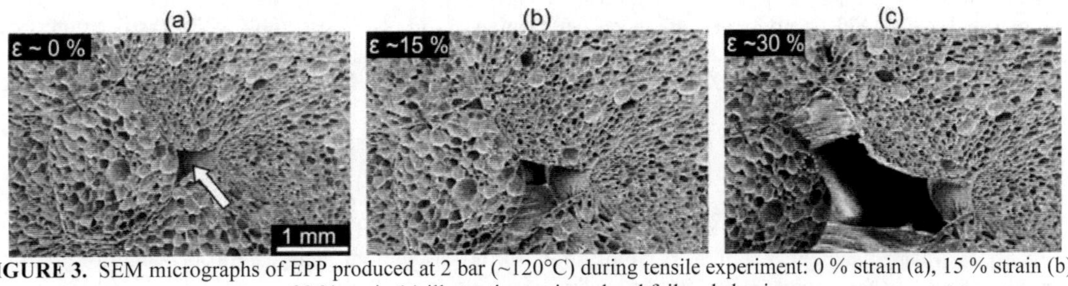

FIGURE 3. SEM micrographs of EPP produced at 2 bar (~120°C) during tensile experiment: 0 % strain (a), 15 % strain (b) and 30 % strain (c) illustrating an inter-bead failure behaviour.

After the moulding at ~120°C, the failure behaviour is inter-bead (Fig.3). No significant cells deformation was observed during the tensile experiment (see for example the dotted marking in Fig. 3). Since there is no crystal melting at this temperature, there is also limited interdiffusion across the interface during the moulding. This narrows the interpenetration depth and likely leads to pull-out of chain ends [9], consequently resulting in a weak interface and, therefore, in poor mechanical properties of the EPP part.

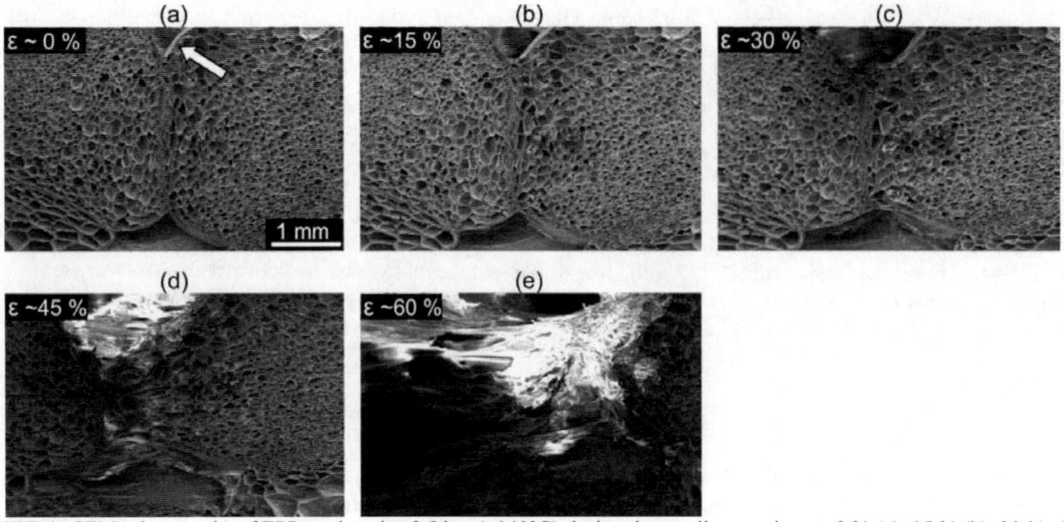

FIGURE 4. SEM micrographs of EPP produced at 3.5 bar (~140°C) during the tensile experiment: 0 % (a), 15 % (b), 30 %, 45 % (d) and 60 % strain (d).

By contrast, sample produced at high temperature (~140°C), shows the intra-bead fracture behaviour (Fig. 4). The cells near the micro-voids exhibit pronounced deformation until ~30% strain. Exemplarily, we have marked two cells in Fig. 4. With further increasing strain, the stress acting in cracks draw the interface into craze-like fibrils and gradual fracture through individual cells, which can be clearly seen in Fig. 4c-d. This behaviour results from the interface healing accompanying by re- and cocrystallization in the interface, which strength the former interface and leads to intra-bead failure behaviour [9].

CONCLUSIONS

In summary, the moulding conditions strongly influence the fusion strength in EPP. The dominant role of re- and cocrystallization in the fused zone on the fusion strength of EPP was examined in relation to direct observation of interface morphology by AFM. Further, experimental verification of the fusion strength in EPP has been presented through *in situ* SEM tensile tests. The findings provided here show evidence on the significant role of crystallization phenomena on the fusion strength of EPP, which change the nature of the interfacial fracture from inter to intra.

ACKNOWLEDGMENTS

This work was supported by Federal Ministry for Economic Affairs and Energy. The authors thank P. Schreier and T. Neumeyer (Neue Materialien Bayreuth GmbH) for fruitful discussions. They are also thankful to L. Klug (University of Bayreuth) for conducting the tensile experiments.

REFERENCES

1. D. Raps, N. Hossieny, C. B. Park and V. Altstädt, *Polymer* **56**, 5-19 (2015).
2. P. R. Stupak, W. O. Frye and J. A. Donovan, *J Cellular Plastics* **27**, 484-505 (1991).
3. J. Rossacci and S. Shivkumar, *J Mat Sci* **38**, 201-206 (2003).
4. W. Zhai, Y.-W. Kim and C. B. Park, *Ind Eng Chem Res* **49**, 9822-9829 (2010).
5. W. Zhai, Y.-W. Kim, D. W. Jung and C. B. Park, *Ind Eng Chem Res* **50**, 5523-5531 (2011).
6. M. Nofar, Y. Guo and C. B. Park, *Ind Eng Chem Res* **52**, 2297-2303 (2013).
7. R. Androsch, M. L. Di Lorenzo, C. Schick and B. Wunderlich, *Polymer* **51**, 4639-4662 (2010).
8. M. Legros, D.S. Gianola and C. Motz, *MRS Bulletin* **35**, 354-360 (2010).
9. R. P. Wool, B.-L. Yuan and O. J. McGarel, *Pol Eng Sci* **29**, 1340-1367 (1989).

In-situ Observation of One-shot Chemical Foaming Process by X-ray Transmission Imaging

Junichiro Tateishi[a], Norihiko Taniguchi[a], Tsuyoshi Nishiwaki[a]
Sathish K. Sukumaran[b], Masataka Sugimoto[b]

[a] *Institute of Sport Science, ASICS, Hyogo 651-2271, Japan*
[b] *Graduate School of Science and Engineering, Yamagata University, Yamagata 992-8510, Japan*

junichiro.tateishi@asics.com

Abstract. Visual observation experiments of the one-shot chemical foaming process were performed by the X-ray transmission imaging method. Specimen compound was composed by the metallocene linear -low -density poly-ethylene (LLDPE) containing the chemical foaming agent (CFA), the CFA activator and the cross-linker. Foaming was carried out in the sealed cavity with the X-ray detecting system and the pressure sensor. The visualized image changed from the homogeneous melt state to the white spot pattern. It was shown that the low electron density structure (i.e. bubble) was formed by the generated gas resulting from CFA thermal decomposition.

Keywords: chemical foaming, X-ray imaging, cell structure, polymer foam
PACS: 83.50.Uv

INTRODUCTION

Chemically cross-linked poly-olefin foams by the one-shot chemical foaming process have been widely used for shock attenuation parts of shoe soles. A typical one-shot chemical foaming process involves several steps. First, polymer compounds including the chemical foaming agent (CFA), the CFA activator and the polymer cross-linker are filled into the mold at a certain temperature. Second, cross-linking reaction of the polymer and thermal decomposition reaction of the CFA proceed simultaneously, while suppressing expansion of compounds by the mold clamping pressure. Finally, as the mold clamping pressure is released, the compounds immediately expand, and are cooled. Then cross-linked polymer foams are obtained. Concerning the one-shot chemical foaming process, the forming mechanism of the cell structure has not been revealed yet. In previous studies, foaming behaviors on one-shot chemical foaming process have been investigated by considering molding conditions and viscoelasticity of compounds [1, 2]. And we reported that thermal decomposition behaviors of the CFA during the one-shot chemical foaming process by the constructed in-situ measuring system [3]. In this study, visual observation of the one-shot chemical foaming process was performed by an X-ray transmission imaging method to investigate the cell forming mechanism on the one-shot chemical foaming process.

EXPERIMENTAL

In-situ observation experiments were performed at SPring-8 BL24XU (Hyogo Prefecture Beam Line, Proposal No. 2014A3329, 2015A3267 and 2015B3267). The visual observation system consists of a transmitted X-ray detector, and a cavity equipped with a pressure sensor and a temperature control unit. The cavity constructed by the stainless steel is cuboid shape with 5mm length, 5mm width and 1.5mm thick, and it has a pressure sensor (KISTLER, 6190CA). A part of the cavity is constructed by the glass to ensure X-ray transmittance. The transmitted X-ray is detected by the CMOS camera (HAMAMATSU PHOTONICS K.K., Orca-Flash4.0, 2048 × 2048pix, view size is 1.3 × 1.3mm) at 0.5s intervals. The X-ray energy is adjusted to 15keV. The cavity temperature was controlled at 433K. The specimen compound is based on a commercial metallocene linear -low -density poly-ethylene (LLDPE, MFR = 3.8g / 10min). Dicumyl peroxide (DCP, 1 min half-life temperature = 448.2K), azodicarbonamide (ADCA, median size = 6μm), and zinc oxide (ZnO) were used as the cross-linker, the CFA and the CFA activator, respectively. The specimen was blended by using open two roll mills at 393K on following weight ratio; LLDPE/DCP/ADCA/ZnO = 100/1.2/10/5. The specimen compound did not leak from the cavity during observation experiments.

RESULTS AND DISCUSSION

First, it was verified to reproduce the one-shot chemical foaming process in this measurement system. Fig. 1 shows the cavity pressure *P* profile at 433K. Arrows denote the timing corresponding to images of Fig. 2. *P* showed s-shaped behavior; after the induction phase, *P* increased rapidly, then, reached the saturated value. This *P* profile resulted from the CFA thermal decomposition behavior as described previous studies [3, 4]. The obtained polymer foam after process observation experiment had a closed cell structure. We confirmed that this one-shot chemical foaming process has reproducibility in our constructed observation system of this study.

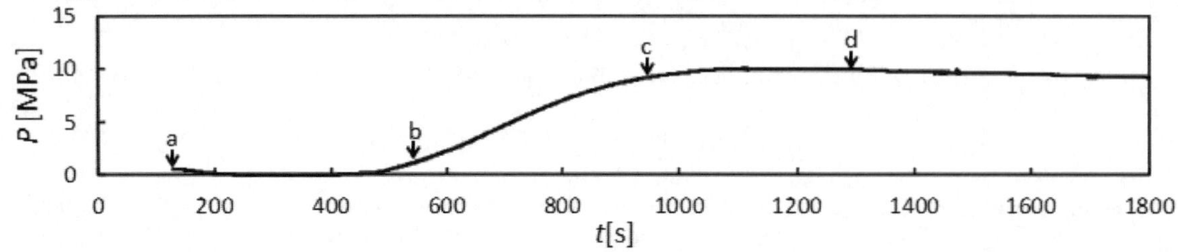

FIGURE 1 Cavity pressure profile of LLDPE compound at 433K. Arrows denote the timing corresponding to images of Fig. 2.

A series of cavity internal image at 433K is shown in Fig. 2. Fig. 2(a) at the induction phase shows a homogeneous image. The CFA particle was not able to be confirmed clearly in this observation system. Then, in Fig. 2(b), white spots with a diameter of tens microns appeared from lower right of the image. The appearance of the white spot pattern indicated that the low electron density structure was formed at this stage in the cavity. Considering *P* profile in Fig. 1, CFA thermal decomposition started at around 530s. We have concluded that the low electron density structure resulted from the generated gas by CFA thermal decomposition; the appearance of the white spots indicated the forming an initial cell structure in the cavity at this stage. AS can be seen in Fig. 1, from this stage the pressure steeply increased toward the pressure saturate phase. In Fig. 2(c), the number of white spots increased, and appeared all over the observation area despite of high pressure. Finally, in Fig. 2(d), there was little change in the white spot morphology with that of Fig. 2(c). The cell collapse and coalescence couldn't be observed in this observation level and we consider that the formed initial cell structure was maintained during foaming process in the high pressure cavity.

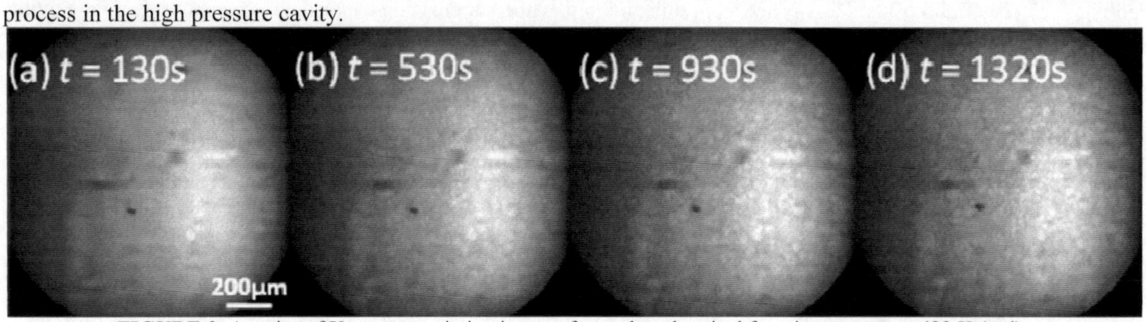

FIGURE 2 A series of X-ray transmission image of one-shot chemical foaming process at 433 K (a-d).

CONCLUSIONS

The one-shot chemical foaming process was investigated by the in-situ X-ray visual observation method. We showed that cells generated by the gas of CFA thermal decomposition were formed in the early low pressure stage and number of the cells increased in the pressuring stage in the cavity, while there was little change in the cell size.

REFERENCES

1. C. J. Benning, *Journal of Cellular Plastics*, 1967, 3, 62
2. R. Gendren, C.Vachon, *Journal of Cellular Plastics*, 2003, 39, 71
3. J.Tateishi et al, Conference paper of ANTEC2015, Orlando, 2015
4. J.Tateishi et al, Conference paper of FOAMS2015, Kyoto, 2015

Reducing the Thermal Conductivity of Polymeric Foams Made from Polystyrene/Expanded Graphite

Minh-Phuong Tran[1, b)], Pengjian Gong[1], Christophe Detrembleur[2], Jean-Michel Thomassin[2], Piyapong Buahom[1], Mehdi Saniei[1], Samuel Kenig[3] and Chul Park[1, a)]

[1] *Microcellular Plastic Manufacturing Laboratory - Mechanical and Industrial Department, University of Toronto 5 King College Road, Toronto, ON M5S 3G8, Canada.*
[2] *Chemistry Department, University of Liege. Allée de la Chimie, B6a, Sart Tilman, 4000 Liège, Belgium.*
[3] Engineering Faculty, Shenkar College. Anne Frank Street 12, Ramat Gan, Israel
[a)] Corresponding author: park@mie.utoronto.ca
[b)] mptran@mie.utoronto.ca

Abstract. Radiative thermal conductivity has long been recognized as an important heat transfer mechanism. In the literature, a heat transfer of 20% to 40% through low density thermal insulation materials has been reported. In addition to its cost effectiveness and easy processability, expanded graphite was proved to absorb thermal radiation more efficiently than carbon nanotubes with the same weight (or volume) content. Using the Rosseland approximation, we quantitatively undertook a first-time study of how expanded graphite significantly reduces the radiative thermal conductivity. We found that particle-added polystyrene foams with a 0.018%vol of expanded graphite nanoplatelets and a 25-fold volume expansion ratio could block 92% of the overall thermal radiative thermal conductivity. Radiation with short wavelengths in the infrared range of 2.5 – 10μm was totally blocked. When we used the Glicksman model for thermal conduction together with the Rosseland approximation (applicable to foam >17 volume expansion), the calculated values of the total thermal conductivity were all in good agreement with the experimental data, and the tolerance was less than 5% (<2 mW/m.K).

INTRODUCTION

Micro/nano cellular foams have a great potential for use in many areas of application such as home construction [1], thermal insulation [2-4], acoustic absorption [5], electromagnetic interference shielding (EMI shielding) materials [6, 7, 8], and charge storage[9]. Closed-cell polymeric foams are known to have the lowest thermal conductivity of any conventional non-vacuum insulation foams. However, the major drawback of low density bulk polymeric foams is the large amount of thermal radiation (20% – 40%) present in their total thermal conductivity [4]. Expanded graphite has great potential for industrial applications due to its low cost, easy processability, non-toxicity, excellent mechanical properties and using as infrared attenuated agents when it was evenly dispersed within host polymer matrix. The improvement in reducing the thermal conductivity of PS/EG foams is illustrated quantitatively using an analytical model taking Knudsen's effect into account. This describes the thermal conduction of gas molecules in a cavity based on the gas kinetics theory, and their thermal conduction through a cellular structure, which is based on the Rosseland approximation theory of thermal radiation. This analytical model qualifies the contribution of each term of the total thermal conductivity. The calculated results based on the noted models are compared to the experimental data. This study is a continuing part of our previous work on heat transfer through polymeric foams [10].

EXPERIMENTS

Polystyrene (PS Styrolution – BASF, 1.04 g/cm^3 density), with a glassy temperature of 101.7°C, a melt flow index of (200°C/5kg) MFI 10g/10min., and 0.18 W/m.K thermal conductivity was received from the supplier. Expanded graphite (EG1-micro) was supplied by Asbury Carbon. Expanded graphite (EG2-nano) was supplied by XG-Science. These nanoplatelets were composed of several sheets of graphene with an overall thickness ranging from 1 nm to 15 nm. The carbon nanotubes (average outer diameter: 10 nm, average length: 1.5 μm, density = 1.44 g/cm^3) were purchased from Nanocyl®, Belgium. Trans-1-Chloro-3,3,3-trifluoropropene (Solstice® 1233zd(E) denoted as *S-propene* gas) was purchased from Honeywell USA, with a boiling point of 19°C, a liquid density of 1.27 g/cm^3 and a vapor thermal conductivity of 10.2 mW/m.K at 20°C. Carbon dioxide was purchased from Linde Gas Canada with a purity of 99.8%, 501298 cylinder 48L.

The neat PS, PS/EG, and PS/CNT composite were placed into a high-pressure autoclave for saturation by the physical co-blowing agent. The autoclave was then heated to 50°C – 70°C so that the *pentane* or the *S-propene* phase became gas and was miscible with the CO_2 phase. This system was kept for 3 – 6 days at a pressure of 5.86 – 13.78 MPa. Due to the strong plasticization effect of the CO_2 – *pentane* or the CO_2 – *S-propene*, foaming was induced by a rapid depressurization. Then, all of the foamed samples were quickly placed into 100°C boiling water for 2.5 minutes for further cell expansion.

A transient plane source (TPS) Hot Disk Thermal Constants Analyser (Therm Test Inc., TPS 2500, Sweden) was used to measure the total thermal conductivity of both the solid bulk and the foamed samples. Scanning Electron Microscopy (SEM) was used to observe the foam morphology. The samples were frozen in liquid nitrogen to create cryogenic fractural surfaces and were then coated with gold for 180s before SEM (JEOL JSM – 606) observation. Image-J processing software was used to calculate the cell density. The samples' bulk densities before and after foaming were measured using the water displacement method based on the ASTM D792 – 00. Fourier Transform Infrared Spectroscopy (FTIR) (Perkin Elmer Spectrum One) was used to measure the spectral transmittance of radiation through the foams. The spectral transmittance was collected by averaging 8 scans in a spectral range of 4000 – 400 cm^{-1}. To calculate the radiation, we measured at least 6 plate samples of different thicknesses.

RESULTS AND DISCUSSION

All composite foams should have a volume expansion of at least greater than 17–fold when using a co-blowing agent of CO_2 – *pentane* (or *S-propene*) in order to apply the Glicksman's model to the calculation of their solid thermal conductivities (λ_{solid}) was used to foam the samples. The foaming system was fixed at 850 – 2000 psi (5.86 – 13.8 MPa) and saturated for 3 – 6 days at 50°C – 70°C to ensure the co-blowing agents are miscible [11]. Due to the co-blowing agent's strong plasticization effect, embryonic nuclei had already occurred when a rapid depressurization was applied at 50°C – 70°C. The nucleated bubbles were further expanded by being heated in water steam (100°C). This expansion was caused by the slow diffusion of the pentane (or *S-propene*) out of the specimen.

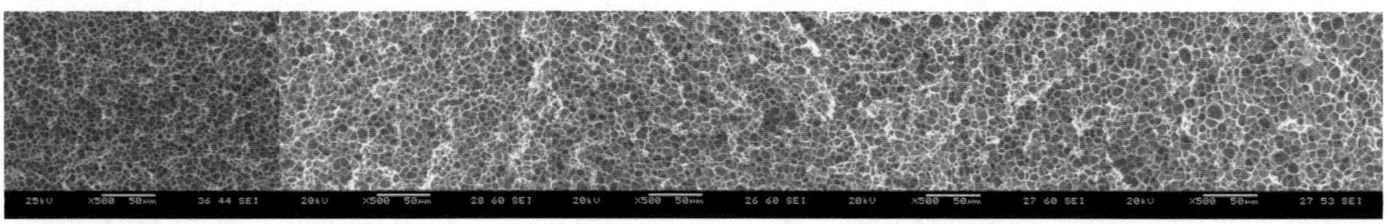
(a) (b) (c) (d) (e)

FIGURE 1. SEM micrographs of neat PS foam and PS/EG1-micro composite foams with different EG1-micro content; Neat PS (a); PS/0.07 vol%EG1-micro (b); PS/0.17 vol%EG1-micro (c); PS/0.34 vol%EG1-micro (d); PS/ 0.72 vol%EG1-micro (e); magnifications of 500 (Initial-vol%).

Figure 1, for example, shows the foam morphology of the neat PS and the PS/EG1-micro composite foams, which had been prepared by soaking in the CO_2 and 25 mL pentane mixture, at 2000 psi, 50°C for 4 days. The SEM micrographs demonstrated a good uniformity, a narrow distribution of cell size, and a penta/hepta polygonal cell shape. With increased EG1-micro content, the foam morphology still maintained its uniformity. A slight increase of 2 – 4 μm was observed when the EG1-micro reached 0.72 vol%. All those samples showing a vol% of 0.07, 0.17, 0.34, 0.72 and 1.7 represent the vol% of the *initial* composite solid. The foaming process is a volume phenomenon; therefore, a dilution of vol% EG filler, which is reversely proportional to its volume expansion, was noticed after foaming. Figure 2 summarizes the cell size and cell density (Figure 2(a)), and the expansion ratio, and the strut fraction (Figure 2(b)). The EG1-micro particles were well dispersed as nucleating agents in the PS matrix. Thus, the cell morphology of the PS/EG1-micro foams was similar to that of the neat PS foam (that is, cell size 4.7 – 7.4 µm). The expansion ratio increased from 17 to 20 when the EG1-micro content was increased from 0 (neat PS) to 0.72 vol%, which had benefited from the larger cell size, as previously noted. The cell density of these foams was around 2×10^{11} cell/cm^3.

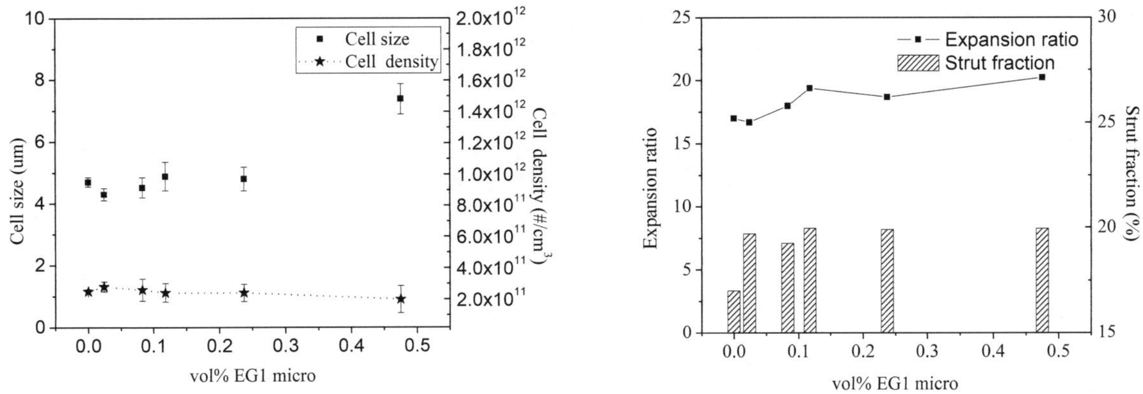

(a) (b)

FIGURE 2. (a) Cell size and cell density, (b) expansion ratio and strut fraction of neat PS and PS/EG1-micro foams.

Thermal Properties of PS/EG Composite Foams

The thermal conductivity of graphite varies from 25 – 450 W/m.K. But a much higher thermal conductivity for graphene (2000 – 6000 W/m.K) and carbon nanotubes (2900 – 6600 W/m.K) has been reported [12-14]. Thus, the thermal conductivity of a PS/EG1-micro composite is lower than a PS/CNT composite at the same weight percent. For polymeric solid conduction, it must be kept in mind that using EG1-micro within PS matrix for thermal insulation is better than using CNTs. The high thermal conductivity of bulk solid of other carbonaceous fillers is a hindrance when it is contained in the polymer composites used to produce thermal insulated foams.

According to the heat transfer theory of thermal conductivity of the solid in polymeric foams, their solid thermal conduction is increased in proportion to the composite's bulk solid thermal conductivity and is inverse to the volume expansion. The larger volume expansion of PS/EG2-nano (25-fold) exhibited the lowest solid thermal conductivity, and the value varied from 3.6 to 4.9 mW/m.K.

The heat transfer via the gas conduction (λ_{gas}) is based on the volume expansion and cell size of the PS/EG composite foams. This assumes that the air has completely filled the cells (25.7 mW/m.K) and is independent of either the EG or CNT content presented within the PS matrix. In our study, the volume expansion varied from 17-fold for the neat PS foam to 20-fold for the PS/EG1-micro foams (and 25-fold for the PS/EG2-nano foams). And the foams' cell sizes all ranged from 4.7 – 7.4 µm. Due to their similar foam morphology (4.7 – 7.4 µm cell size and 17 – 25 volume expansion), the PS/EG1-micro foams (containing EG1-micro of 0.012 – 0.024 vol%) had nearly the same gas conductivities (~ 23.5 mW/m.K) as the PS/CNTs foams.

The IR transmittance, which was recorded by FTIR spectrometry, through the neat PS or the PS/EG foams can be regarded as radiation energy; the higher the transmittance is, the higher the radiative thermal conductivity is. Figure 3 (a) shows the spectral transmittance of the PS foam at sample thicknesses of 0.13, 0.42, 0.96, 1.75, and 1.80 mm, respectively. As the graph shows, the IR transmittance decreased with increased sample thickness. A thicker sample means that there are more cell wall numbers through which radiation reflects and/or is absorbed before it reaches the FTIR detector. The EG2-nano was more efficient than the EG1-micro in blocking radiation. This was evident when the EG2-nano fillers exhibited significant attenuations of their IR transmittance to nearly zero at a wavelength of 2.5 – 10 µm. Figure 3 clarifies the EG1-micro and the EG2-nano's efficiency in blocking radiation.

Knowledge of the Rosseland extinction coefficient ($K_{e,R}$) made it possible to quantitatively calculate the thermal radiation. To be able to compare their thermal radiation, all of the neat PS, the PS/EG1-micro, and the PS/EG2-nano chosen foams needed to have closely shared the same foam morphology (cell size, cell density, volume expansion).

This then made it possible to evaluate the blocking radiation efficiency of the expanded graphite's fillers. The EG gained more by using an IR absorber than the CNTs did.

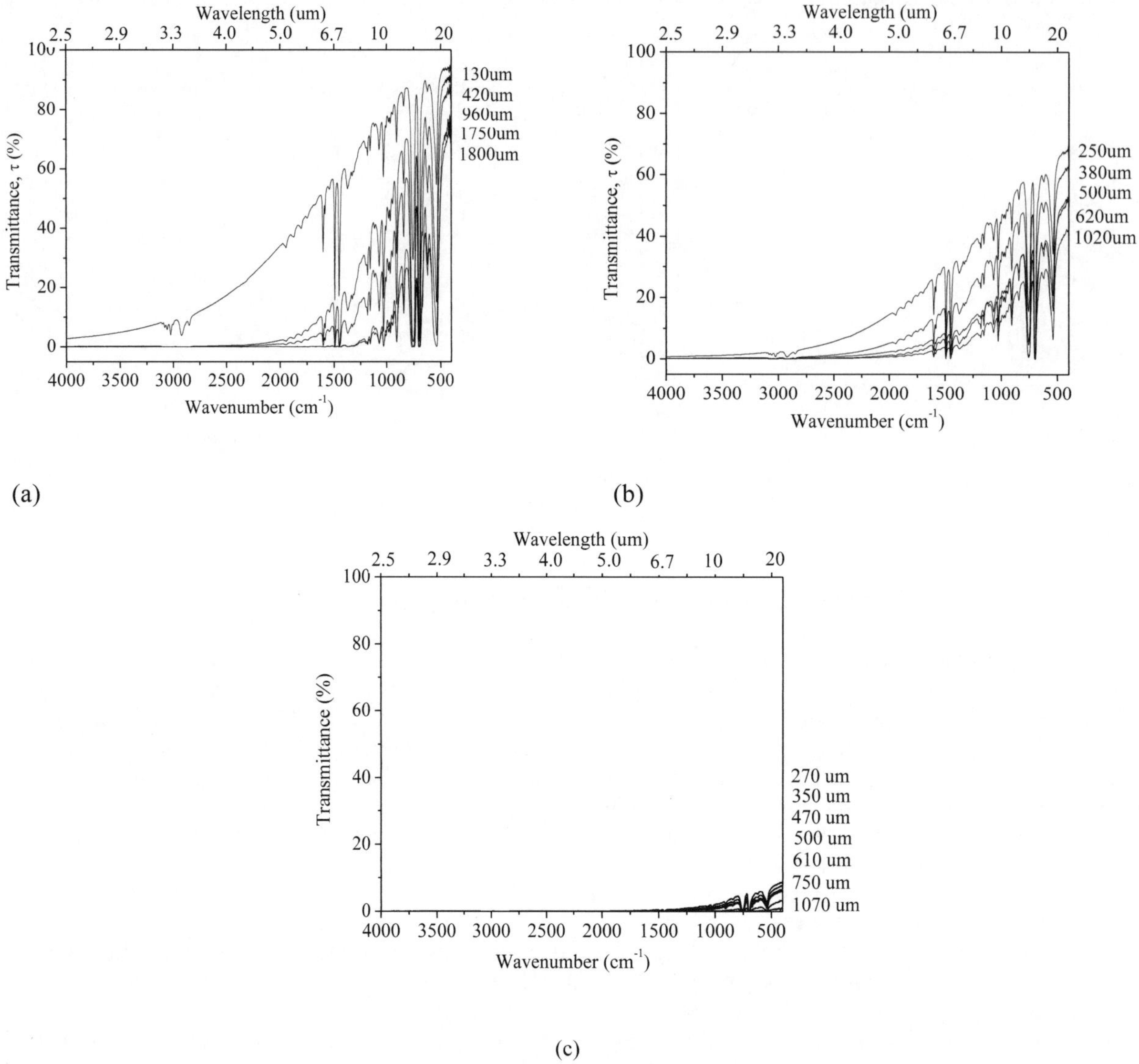

FIGURE 3. FTIR spectral transmittance of (a) neat PS foams of 0.13, 0.42, 0.96, 1.75 and 1.8 mm; (b) PS/EG1-micro 0.023 vol% foams of 0.25, 0.38, 0.50, 0.62 and 1.02 mm; (c) PS/EG2-nano 0.018 vol% of 0.27, 0.35, 0.47, 0.5, 0.61, 0.75, and 1.07 mm (with varying foam thicknesses).

To demonstrate the IR blocking radiation efficiency, the radiation absorption percentage was plotted against the vol% fillers, as Figure 4 (a), 4 (b) shows. The percentage was boosted from 0% to 80% in the range of 0 – 0.05 vol%. The absorbed radiation became saturated and presented as a plateau at a vol% concentration in the foams that exceeded 0.05 vol%. A higher radiation absorption can also occur when more filler content is added. However, a much higher amount of filler content was required to achieve a higher radiation absorption since the PS/EG composite foams had reached their plateau.

Figure 5 shows each thermal conduction mode that contributes to the total thermal conductivity. The graph's *calculated thermal conductivity* represents the sum of three components: (1) gas thermal conduction (λ_{gas}), (2) solid thermal conduction (λ_{solid}) and (3) radiative thermal conduction (λ_{rad}). But the two terms of (1) thermal λ_{gas} and (2) thermal λ_{solid} were theoretically calculated from Glicksman model, where foam morphology parameters were applied. The term of (3) λ_{rad} was calculated using the Roseland approximation and was based on the FTIR experimental data.

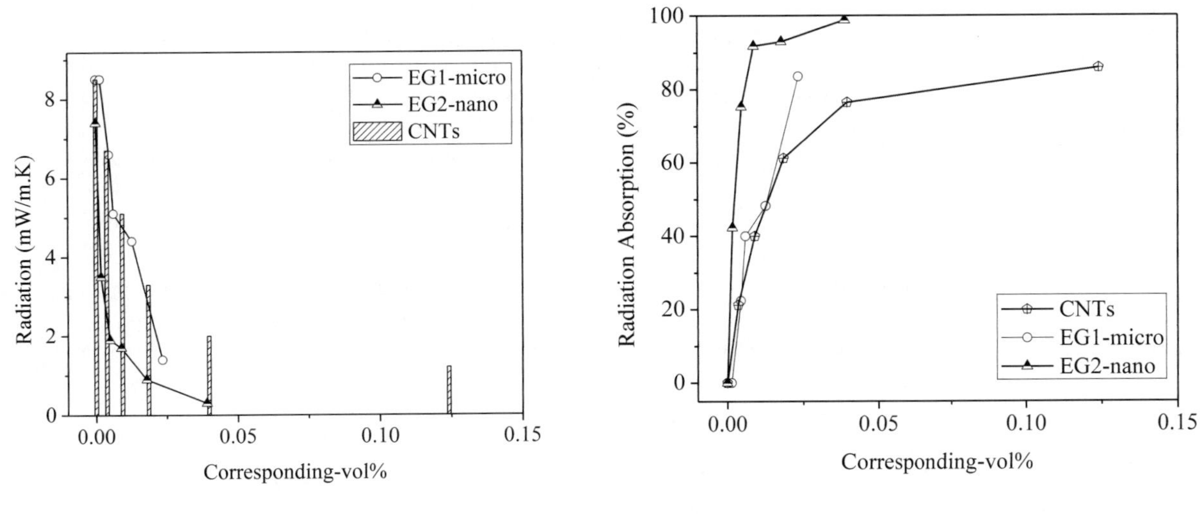

(a) (b)

FIGURE 4. (a) Thermal radiation, (b) radiation absorption of PS/CNTs, PS/EG1-micro and PS/EG2-nano.

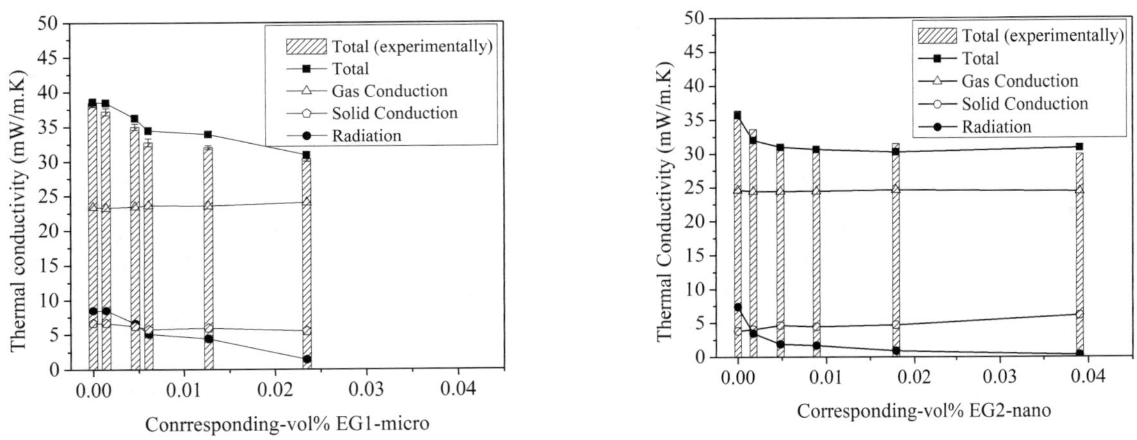

(a) (b)

FIGURE 5. Contribution of each heat transfer term in mW/m.K in total thermal conductivity and the comparison of calculated versus experimental thermal conductivity of (a) PS/EG1-micro, (b) PS/EG2-nano.

The *experimental thermal conductivity* in the graph represents data taken from the experimental measurements. A direct comparison between the *calculated* and *experimental* results of the PS/EG composite foams was made. The EG1-micro's effect as an IR absorber was more evident in the PS/EG1-micro 0.023 vol% sample, where the radiation decreased from 8.5 mW/m.K (neat PS) to 1.4 mW/m.K. This outcome led to a lower calculated total thermal conductivity in the PS/EG1-micro 0.23 vol% of 30.7 mW/m.K compared with the neat PS foam at 38.5 mW/m.K. Both of these foams had the same closed-cell and penta-hepta polygonal cell structure, and they also had nearly the same cell sizes. An excellent match was found between the calculated and the experimental total thermal conductivity with the neat PS and the PS/EG1-micro 0.23 vol%. Similarly, the PS/EG2-nano composite foams exhibited a lower total thermal conductivity than the neat PS foams. This was due to its higher efficiency in blocking radiation. As a result, the PS/EG2-nano foam achieved a total thermal conductivity of ~ 30.7 mW/m.K with a lower content filler (0.01 vol% EG2 instead of 0.023 vol% EG1-micro). A higher amount of EG2-nano was not helpful in further decreasing the total thermal conductivity while the plateau had been reached by exceeding 0.01 vol%. Overall, the calculated values of all the samples were in very good agreement with the experimental data when the Glicksman and Rosseland models were used (that is, when applicable to foam with a > 17 volume expansion), and the error rate was less than 5% (<2 mW/m.K). The Rosseland model does not take the foam morphology (that is, spherical, honeycomb or polygonal cell shapes) into account. However, this model can be used for evidence and can provide guidelines for decreasing radiation when carbonaceous fillers such as CNTs and EG are increased.

ACKNOWLEDGMENTS

The authors are grateful to the Ontario and National Sciences and Engineering Research Council of Canada (NSERC) and to the Korea Institute of Energy Technology Evaluation and Planning (KETEP) for their financial support through the Korea Institute of Construction Technology (KICT).

REFERENCES

1. M. Nofar and C. B. Park, Progress in Polymer Science **39** (10), 1721-1741 (2014).
2. P. Ferkl, R. Pokorny, M. Bobak and J. Kosek, Chem. Eng. Sci. **97**, 50-58 (2013).
3. M. C. Basso, X. Li, V. Fierro, A. Pizzi, S. Giovando and A. Celzard, Adv. Mater. Lett. **2** (6), 378-382 (2011).
4. O. A. Almanza, M. A. Rodriguez-Perez and J. A. De Saja, J. Polym. Sci., Part B: Polym. Phys. **38** (7), 993-1004 (2000).
5. M. Nofar, A. Ameli and C. B. Park, Materials & Design **83**, 413-421 (2015).
6. M.-P. Tran, J.-M. Thomassin, M. Alexandre, C. Jerome, I. Huynen and C. Detrembleur, Macromolecular Chemistry and Physics **216** (12), 1302-1312 (2015).
7. M.-P. Tran, C. Detrembleur, M. Alexandre, C. Jerome and J.-M. Thomassin, Polymer **54** (13), 3261-3270 (2013).
8. A. Ameli, M. Nofar, C. B. Park, P. Potschke and G. Rizvi, Carbon **71**, 206-217 (2014).
9. A. Ameli, S. Wang, Y. Kazemi, C. B. Park and P. Pötschke, Nano Energy **15**, 54-65 (2015).
10. P. Gong, P. Buahom, M.-P. Tran, M. Saniei, C. B. Park and P. Pötschke, Carbon **93**, 819-829 (2015).
11. G. J. Besserer and D. B. Robinson, Journal of Chemical & Engineering Data **20** (1), 93-96 (1975).
12. J. H. Seol, I. Jo, A. L. Moore, L. Lindsay, Z. H. Aitken, M. T. Pettes, X. Li, Z. Yao, R. Huang, D. Broido, N. Mingo, R. S. Ruoff and L. Shi, Science **328** (5975), 213-216 (2010).
13. S. Ghosh, I. Calizo, D. Teweldebrhan, E. P. Pokatilov, D. L. Nika, A. A. Balandin, W. Bao, F. Miao and C. N. Lau, Applied Physics Letters **92** (15), 151911 (2008).
14. S. Berber, Y.-K. Kwon and D. Tománek, Physical Review Letters **84** (20), 4613-4616 (2000).

Foaming Behavior of Novatein and Blends with Polyethylene Compatibilised by Maleic Anhydride

Chanelle Gavin[a], Mark C. Lay[a*] and Casparus J.R. Verbeek[a]

[a]*University of Waikato, Knighton Road, Hamilton, New Zealand*

chanellegavin@gmail.com
mclay@waikato.ac.nz , *jverbeek@waikato.ac.nz*

Abstract. Novatein® thermoplastic protein (NTP), a bloodmeal based thermoplastic, was successfully foamed without blowing agent or blending with other polymers using a BOY-35A injection moulder in free expansion mode. Previously, only soy protein has been successfully continuously foamed and zein and gluten batch foamed without blending or rheological modification. The greatest expansion ratio for NTP was 4.4 at 165°C. Blending NTP with compatibilised LLDPE and LDPE and adding blowing agent reduced expansion ratios. The foams exhibited a fibrous nature, with cell structures similar to those reported in literature for extrusion foamed thermoplastic soy protein isolate. SEM pictures suggested a possible role of either sodium sulphite, sodium dodecyl sulphate or sodium sulphate as a nucleating agent as particulates were found on the cell surfaces. Triethylene glycol and urea were thought to contribute to NTP foaming, along with water present in NTP, with two or more acting as the blowing agent.

Keywords: Free Expansion Foaming, Single Screw, Protein Bio-polymer, Polyethylene, Compatibilised Blend
PACS: 82.35.Pq, 83.80.TC, 83.80.Iz, 81.20.Hy

INTRODUCTION

The waste disposal problems of using non-biodegradable foamed polymeric materials in short-term/limited-life applications such as packaging has resulted in increased use of biodegradable, bio-derived and protein based thermoplastics. One potential biodegradable, protein based thermoplastic is Novatein® Thermoplastic Protein (NTP), developed from bloodmeal by the University of Waikato and Aduro Biopolymers, New Zealand. Bloodmeal is plasticized by adding urea, sodium sulphite, sodium dodecyl sulphate, water and triethylene glycol, and extruded to produce a consolidated thermoplastic with similar mechanical properties to linear low density polyethylene.

Only a small number of thermoplastic proteins have been successfully foamed using extrusion. Examples include soy protein isolate which was successfully foamed into sheets at 150-160°C [1], blends of pearl millet flour and zein (30:70) foamed at 105°C [2], and blends of soy protein concentrate and polylactic acid (PLA) at 150-160°C [3]. In all cases, an additional chemical blowing agent was employed to promote foaming. In comparison, injection foaming is usually conducted into a mould and either a short shot or core pull is used to enable the required expansion. Injection foaming has been applied to a wide range of biopolymers including PLA, PHBV, PBAT and starch [4], however it has only been applied to one other protein thermoplastic (soy) [5].

Previously, the extrusion foaming ability of NTP blended with compatibilised linear low density polyethylene (LLDPE) was investigated using the internal moisture content and sodium bicarbonate as the blowing agent [6]. Two twin screw extruders were used with 10 mm diameter dies. Expansion ratios of < 2 were observed and foam morphology consisted of large irregular closed cells up to 2 mm in size. A foaming grade LDPE was also tried with NTP. While the material was expected to foam better when compared to using LLDPE, because of its different melt flow index and strain hardening behavior, no difference in foaming ability of the blends was observed at 120°C. The low expansion ratios were attributed to the significant losses of blowing agent through the feed hopper and to gases escaping through the unconsolidated material at the die.

To prevent the loss of gases and improve foaming, this work explored the use of a single screw BOY 35A injection moulder in free expansion mode for extrusion foaming. The foaming ability of NTP, LDPE, LLDPE and 50:50 blends was explored over a temperature range of 150-165°C. The materials, blends and resulting foams were characterized by examining cell morphology and expansion ratios.

MATERIALS AND METHODS

Pre-extruded NTP (PNTP) was prepared in a Labtech bench scale high speed mixer at 1400 RPM by blending bloodmeal with a denaturing solution and triethylene glycol as a plasticizer. The denaturating solution was prepared by dissolving sodium sulphite (SS), sodium dodecyl sulphate (SDS) and urea in water by heating and agitation (**TABLE 1**). This solution was added to the bloodmeal in the high speed mixer in two stages with two minutes of mixing between additions. The plasticizer component was then added in the same manner. The PNTP was then stored overnight at 4 °C in a zip lock bag until extrusion.

TABLE 1: Pre-extruded NTP composition

	Reagent	Source	Grade	Weight (g)
Protein Source	Bloodmeal (< 710 μm)	Wallace Corporation Limited, NZ	Agricultural	1500
Denaturing Solution	Water	Produced on-site	Distilled	600
	Sodium Sulphite	Merck, Germany	Analytical	45
	Sodium Do-decyl Sulphate	Merck, Germany	Synthesis	45
	Urea	Ballance Agri-Nutrients, NZ	Agricultural	150
Plasticiser	Triethylene Glycol	Merck, Germany	Industrial	300

Consolidated NTP was produced by extruding PNTP using a Labtech Scientific Twin Screw Extruder across a temperature profile from feed to die of 70,100,100,110,120°C. This extruder has an L/D ratio of 40, 26 mm diameter screws and was operated at a torque of 55-60%. The extruded material (ENTP) was then granulated to a size of less than 4 mm using a tri-blade granulator from Castin Machinery (NZ).

The ENTP and the polyethylene phase (either LLDPE or LDPE) was then extruded with polyethylene grafted maleic anhydride (PE-g-MAH) compatibiliser (10 % on a weight basis). Cotene 3901 (Elastochem, NZ) with a melt flow index (MFI) of 4.0 g/10 min and annealed density of 0.905 g/cm³ was used for LLDPE. LDPE used was Lotrene FD0274 (MFI of 2.4 g/10 min) commonly used for foamed polyethylene products. This material was supplied by Interplas Agencies Limited via Nulon Limited, Auckland, New Zealand. PE-g-MAH was supplied by Sigma Aldrich (CAS 9006-26-2).

Following the second extrusion the material was granulated to less than 4 mm for foaming in the BOY 35A injection moulder. The single screw has a diameter of 24 mm and an L/D ratio of 22. The machine was operated by withholding the die from the mould to enable free expansion of the material.

The foaming behavior of NTP, LDPE, a 50:10:40 blend of NTP: PE-g-MAH:LDPE blend, LLDPE and a 50:10:40 LLDPE blend was investigated in this study with respect to temperature and sodium bicarbonate content (**TABLE 2**). Temperature profiles of 100, 125, X, 135 and 125 °C from feed to die were investigated for foaming where temperature X was set to either 150, 155, 160 or 165 °C

TABLE 2: Summary of extrusion foaming blends, compatibiliser and blowing agent content.

Trial	Material	PE-g-MAH Content (wt %)	Sodium Bicarbonate (pph)
1	NTP	0	0
2	NTP	0	2
3	NTP	0	4
4	NTP: PE-g-MAH: LDPE 50:10:40	10	0
5	NTP: PE-g-MAH: LDPE 50:10:40	10	2
6	NTP: PE-g-MAH: LDPE 50:10:40	10	4
7	LDPE	0	0
8	LDPE	0	2
9	LDPE	0	4
10	NTP: PE-g-MAH: LLDPE 50:10:40	10	0
11	NTP: PE-g-MAH: LLDPE 50:10:40	10	2
12	NTP: PE-g-MAH: LLDPE 50:10:40	10	4
13	LLDPE	0	0
14	LLDPE	0	2
15	LLDPE	0	4

The density and expansion ratio of the material was measured after foaming. Density was determined by measuring a sample block with vernier calipers and weighing the material. The sample mass was then divided by sample volume to obtain sample density (ρ_f) in g/cm³. The expansion ratio (ER) was determined relative to the unfoamed material density (ρ_0). The expected unfoamed density of the given compositions was determined using a weighted average of the density of the individual components. The density for the compatibiliser was assumed to be equal to that of the LLDPE (0.905 g/cm³) and a density of 1.2 g/cm³ was used for NTP.

Cell morphology was observed at low magnification with a Wild M38 optical microscope and at high magnification using a Hitachi S-4700 scanning electron microscope. Samples were first dried at 70°C over 48 hours, then sputter coated with platinum before imaging using an acceleration voltage of 20 kV.

RESULTS AND DISCUSSION

NTP and NTP/PE blends were successfully foamed using extrusion foaming using the BOY35A injection moulder. It was generally observed that expansion ratio increased as the melt temperature increased, with the exception of PE (which was very hard and difficult to cut for analysis), with a processing temperature of 165°C giving the greatest expansion ratio for NTP (4.4). At greater temperatures the material was too fluid to retain the gases and was spitting from the die.

While 165°C is close to the protein degradation temperature for NTP, processing at just 10-15°C lower (150°C) resulted in little to no expansion. LDPE and LLDPE did not foam without a blowing agent, and blends of NTP with PE showed lower expansion ratios than 100% NTP. NTP blended with LDPE at 165°C had an expansion ratio of 3 while NTP blended with LLDPE had an expansion ratio of 2.5. Adding 2 pph blowing agent resulted in a finer cell structure but lower expansion ratio for NTP of 3.91, and 4 pph blowing agent gave an expansion ratio of 2.53 and the material looked highly degraded. Adding blowing to the blends also reduced expansion ratios from to 3 to 2.35 for LDPE blends at 4 pph while the expansion ratio did not change for LLDPE and remained at 2.5. Adding blowing agent to LDPE and LLDPE did result in some foaming (**FIGURE 2**), but expansion ratios were unable to be obtained, because the samples were not sufficiently foamed to be able to be cut and measured.

Based on the above observations, NTP can be foamed (**FIGURE 1A**) using the internal moisture content and possibly urea and triethylene glycol, which are likely to flash off with the water. NTP does not need to be blended or have a blowing agent added. Blending NTP with PE required a second extrusion step which reduced the amount of water present for foaming by about 6%. In addition, the amount of water present was lower because the amount of NTP was lower in the blends overall. Adding the blowing agent resulted smaller cells due to increased bubble nucleation, but resulted in lower expansion ratios.

The resulting NTP foam appeared fibrous on a macro scale and the cell morphology was not completely uniform (**FIGURE 1B**). In 100% NTP foams, distinct regions of large and small cells were observed. Thin walls were associated with highly foamed areas with small cells while the larger cells are often surrounded by a region of unfoamed material. The addition of 2 pph blowing agent modified the morphology such that a smaller more uniform cell structure was observed. However, the addition of blowing agent did not result in greater expansion. SEM images of NTP with 2 pph sodium bicarbonate revealed a series of cells which were more defined in their cell boundary, but larger than those small cells observed in NTP alone. The material directly surrounding these cells is clearly devoid of smaller cells and many of the cells are elliptical in shape, most likely due to the direction of material flow during foaming. However at higher magnifications very thin cell walls are visible suggesting that the overall expansion of this material can be improved further.

For pure LDPE, as sodium bicarbonate content increased the number of visible cells also increased, as expected. Their size however remains very similar. A similar trend was observed for LLDPE with blowing agent although there appears to be more cell coalescence occurring, with localized regions of cells with very thin walls for one or two regions of the cell perimeter with the rest surrounded by unfoamed material. Overall, a more cellular structure was observed for LDPE as opposed to LLDPE.

The NTP:PE blends also developed cells which increased in number and decreased in size with increased blowing agent. The cellular structure of these blends is chaotic with a wide distribution of cell sizes in both cases. However, a more defined cellular structure is observed for NTP:LDPE blends as opposed to NTP:LLDPE blends and once again the cells have thinner wall structures. SEM images at higher magnification showed some very thin wall structures which could almost be described as open cell. However the cells appear to be open only on the surface with no fibrous crosslinked structure below (complete walls visible). This is most likely due to cell rupture which occurred when a vacuum was pulled on the sample for SEM analysis.

SEM analysis at high magnification also revealed a possible nucleating agent in embedded in the walls of the cells (**FIGURE 1C**). Elemental analysis of these particles indicated sodium, sulphur and oxygen was present. This is most likely residual sodium sulphite or sodium dodecyl sulphate which did not dissolve fully during mixing. Alternatively sodium sulphate is formed when sodium sulphite oxidizes and sodium sulphate peaks have previously been noted in powder x-ray diffraction and Fourier transform infra-red analysis of NTP.

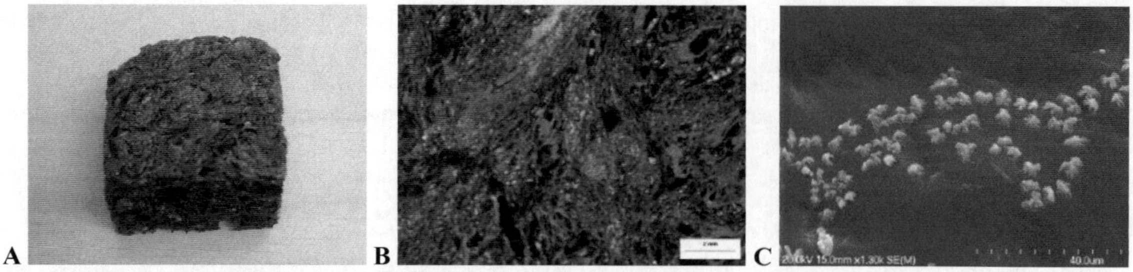

FIGURE 1: **A.** Block of NTP foam **B.** Cell structure by optical microscopy x6.4, **C.** SEM image of particles present in cell wall on which elemental identification was conducted.

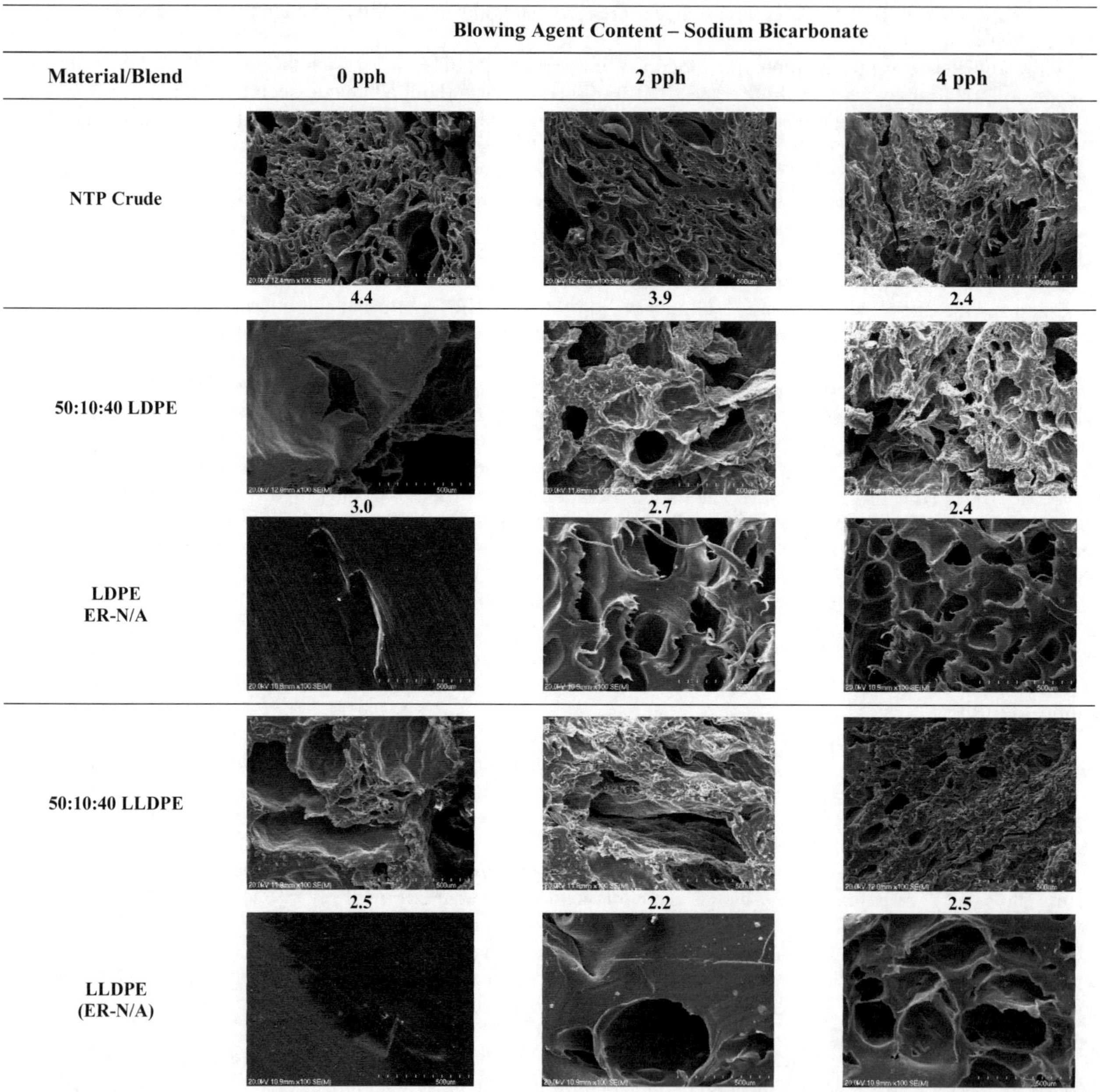

FIGURE 2: Comparison of observed cellular morphology for foams produced at 165°C. Images are at x150 magnification. Expansion Ratios (ER) are shown below images

An ammonia smell was noticed when foaming NTP and NTP containing blends. Like sodium bicarbonate urea is capable of thermally decomposing to form gaseous reaction products. For sodium bicarbonate water and CO_2 are direct products [7], and although urea can thermally decomposes to ammonia, it is more likely to undergo a hydrolysis reaction due to the presence of water and heat and will produce NH_3 and CO_2 [8]. Likewise, both sodium bicarbonate and urea could potentially react with exposed R groups on the protein chain (carboxylic) to evolve the same gaseous products [8], explaining the ammonia smell observed.

The greatest expansion ratio was observed for 100% NTP, this sample contained no added sodium bicarbonate as a blowing agent and therefore would have resulted in the lowest CO_2 evolution by any mechanism. As sodium bicarbonate content increased in any blend containing NTP the expansion ratio decreased, suggesting that an increased ability to generate CO_2 was not beneficial and may be linked to the ratio of gaseous NH_3 to CO_2 where low CO_2 is favored (if these species are not just a by-product of this method). CO_2 is known to be an excellent blowing agent for PLA in both batch and continuous systems [4], while batch foaming of thermoplastic zein and gelatin determined that N_2 or high N_2/low CO_2 mixtures were preferable for these two proteins [9]. The rapid reduction in pressure at the die may also result in a flashing of water and TEG (dictated by vapour –liquid equilibrium).

Future work will determine which component of the system is the primary blowing agent in the case of NTP and NTP – bicarbonate blends. This will be determined by examining the amount of urea lost from the system, the moisture of the resulting products and the loss of TEG through phase–equilibria and changes in mechanical properties of these foams. The effect of other processing variables such as injection pressure, injection speed, cycle time, shot size and back pressure will also be investigated to further improve the expansion ratio of NTP foams.

CONCLUSIONS

NTP was successfully foamed without blowing agent or blending using a BOY-35A injection moulder, one of the few proteins for which foaming has been successful without blending or rheological modification. The greatest expansion ratio for NTP was 4.4 at 165°C. Blending with LLDPE and LDPE and adding blowing agent reduced expansion ratios. Foam morphology appeared fibrous with a mixture large and small cells. SEM pictures suggest a possible role of sodium sulphite, sodium dodecyl sulphate or sodium sulphate as a nucleating agent. Future work will investigate the role of internal water, urea, and TEG in NTP foaming.

ACKNOWLEDGEMENTS

The authors would like to acknowledge the contributions of the Biopolymer Network Ltd (BPN) and Aduro Biopolymers (Novatein® commercialization), and the funding received from the New Zealand Ministry of Business, Innovation and Employment (BPLY1302 contract). Also Dr. Jim Bier for practical help with the injection moulder.

REFERENCES

1. Mungara, P., J. Zhang, and J. Jane, *Extrusion processing of soy protein-based foam.* Polymer Preprints, 1998. **39**(2): p. 148-149.
2. Filli, M., et al., *Development and Characterization of Extruded Biodegradable Foams based on Zein and Pearl Millet Flour.* Annual Transactions of the Nordic Rheology Society, 2011. **19**.
3. Liu, B., L. Jiang, and J. Zhang, *Extrusion Foaming of Poly (lactic acid)/Soy Protein Concentrate Blends.* Macromolecular Materials and Engineering, 2011. **296**(9): p. 835-842.
4. Jeon, B., et al., *Microcellular foam processing of biodegradable polymers—review.* International Journal of Precision Engineering and Manufacturing, 2013. **14**(4): p. 679-690.
5. Jane, J.L. and S. Wang, *Soy Protein-Based Thermoplastic Composition for Preparing Molded Articles*, U.P. Office, Editor. 1996, Iowa State University Research Foundation, Inc" Ames, Iowa: US.
6. Gavin, C., M.C. Lay, and C.J. Verbeek. *Extrusion foaming of protein-based thermoplastic and polyethylene blends*. in *Proceedings of PPS-31: The 31st International Conference of the Polymer Processing Society–Conference Papers*. 2016. AIP Publishing.
7. Janković, B., *Kinetic analysis of isothermal decomposition process of sodium bicarbonate using the weibull probability function-estimation of density distribution functions of the apparent activation energies.* Metallurgical and Materials Transactions B: Process Metallurgy and Materials Processing Science, 2009. **40**(5): p. 712-726.
8. Brooks, B., et al., *Apparatus for quantitatively converting urea to ammonia on demand.* 2004.
9. Sessa, D.J., et al., *Viscosity control of zein processing with sodium dodecyl sulfate.* Industrial Crops and Products, 2006. **23**(1): p. 15-22.

Preparation and Properties of Low Density Polyethylene-Activated Carbon Composite Foams

Darunee Aussawasathien[a*], Kotchaporn Jariyakun[b], Thongchai Pomrawan[b], Kittipong Hrimchum[a], Rungsima Yeetsorn[b], and Walaiporn Prissanaroon-Ouajai[b]

[a]*Plastics Technology Lab, Polymer Research Unit, National Metal and Materials Technology Center, 114 Thailand Science Park, Phahonyothin Road, Khlong Nueng, Khlong Luang, Pathum Thani 12120, Thailand*
[b]*Department of Industrial Chemistry, King Mongkut's University of Technology North Bangkok, 1518 Pracharat 1 Road, Wongsawang, Bangsue, Bangkok 10800, Thailand*

daruneea@mtec.or.th

Abstract. Low density polyethylene (LDPE) containing activated carbon (AC) was foamed with azodicarbonamide (ADC), an exothermic chemical foaming agent, through an extrusion process. The effects of ADC and AC contents on the cellular structure, density, void fraction, and mechanical properties of LDPE-AC composite foams were investigated. The density of composite foams decreased but the void fraction increased as the content of ADC increased due to higher gas liberation. Moreover, the large cell size was also observed in composite foams with high amount of ADC. At low AC dosages (5-10 wt%), the density of composite foams decreased with increasing the void fraction compared to the composite foam without AC because more foaming nucleation centers was induced by adding AC in the composite foam leading to more gas bubble generation. However, the addition of AC up to 15-20 wt% decreased the cell formation and the average cell size tended to decrease with increasing AC content. This attributed to AC particles that assisted to generate more foaming nucleation sites in the polymer matrix, resulting in high viscosity, which enhanced the resistance against cell growth and produced cells with smaller dimensions. Maximum reduction of density by 30% with the void fraction of 30% was achieved when ADC and AC were applied at 7 wt% and 10 wt%, respectively. Larger content of ADC and AC led to composite foams with lower tensile properties as well as impact strength, because of the formation of thinner cell walls, AC agglomeration, and the reduction of polymer matrix.

Keywords: Low Density Polyethylene; Activated Carbon; Azodicarbonamide; Composite Foam; Extrusion Process
PACS: 81.05.Qk

INTRODUCTION

The mixture of functional fillers and polymer foaming technologies has created a new class of materials known as polymer composite foams. Those materials exhibit enhanced thermomechanical properties and specific performances as a result of the multifunctional role played by fillers in such systems. Fillers act as nucleating agents for bubble formation promoting cell morphology improvement which simultaneously lead to an overall increase in macroscopic properties [1, 2]. A wide variety of polymeric matrices such as polypropylene, polylactic acid, polystyrene, polyurethane, polycarbonate, poly(methyl methacrylate), and polyethylene, mixed with several types of filler particles such as nanoclays (mainly montmorillonite), carbon nanotubes, carbon nanofibers or silica particles [3-8] have been recently investigated. To our best knowledge, there is no research work investigating the LDPE-AC composite foam prepared by extrusion method. Although LDPE does not have adequate crystallinity to warrant a fast solidification to sustain thin cell walls, its branching nature makes it unique in maintaining adequate stretching strength to allow continuous extrusion of low density foam sheet. Many chemical foaming agents have been used in commercial polyolefin foams. Among them, ADC is the most widespread use. This is due to its high gas yield and the possibility of matching its decomposition temperature to the polymer processing temperature through the use of suitable activators, for example, zinc oxide. The gas released by ADC consists mostly of nitrogen and carbon monoxide. AC is a highly porous material, which is extensively used to remove a variety of pollutants such as dyes, heavy metals, pesticides and gases, due to its large specific surface area and high adsorption capacity [9]. Therefore, LDPE-AC composite foams may have a potential to be used as packaging to deaccelerate the ripeness and protect agro-products from impact damages during packaging and transportation.

Our present works aim to develop the polymer composite foams based on LDPE and AC through sheet extrusion method. The effects of ADC and AC on cell morphology, density, void fraction, and mechanical properties were investigated.

EXPERIMENTAL PROCEDURE

Materials

Low density polyethylene (LDPE, LD2426K) was supplied from PTT Polymer Marketing Company Limited, Thailand. Its melt flow index was 4 g (10 min)$^{-1}$ at 190°C and 2.16 kg and its density was 0.924 g/cm^3 at room temperature. Activated carbon (AC, CGC-11A 200C) was purchased from Gigantic Carbon Co., Ltd., Thailand. It has surface area and particle size of 1000-1100 m^2 g^{-1} (N_2 BET method) and 74 µm respectively. Azodicarbonamide (ADC, Sunfoam SB00P) with particle sizes range from 8 to 11 µm was purchased from Lautan Luas (Thailand) Co., Ltd., Thailand. Zinc oxide (ZnO, 99% purity, industrial grade, Thai-Lysaght Co., Ltd., Thailand) and Annox20 (Optimal Tech Co., Ltd., Thailand) were utilized as ADC activator and primary antioxidant respectively.

Compounding and Fabrication

The composition of LDPE-AC composite foams is presented in Table (1). The mixture of all materials were premixed in a blending machine with the speed of 500 rpm for 10 min and then compounded by melt mixing process in a twin screw extruder (Labtech, Thailand). The screw diameter, screw length, and screw L/D ratio were 20 mm, 64 cm, and 32 respectively. The set-up temperature of mixing screws was in the range of 130-165°C. The rotation speeds of feeding screws and mixing screws were 12 and 140 rpm respectively. The extrudate was cooled in a water bath and subsequently granulated by pelletizer. Composite foam sheets were prepared by a single screw extruder (HAAKE Polylab OS, L/D ratio of 25) at 220°C with the screw speed of 240 rpm. The composite foam sheet was passed through the rollers with the rolling speed of 35 rpm at 30°C.

TABLE (1). The composition of LDPE-AC composite foams.

Sample	ADC (wt%)	AC (wt%)	ZnO (wt%)	Anox20 (wt%)
0ADC-5AC	0.0	5.0	1.0	0.2
1ADC-5AC	1.0	5.0	1.0	0.2
3ADC-5AC	3.0	5.0	1.0	0.2
5ADC-5AC	5.0	5.0	1.0	0.2
7ADC-5AC	7.0	5.0	1.0	0.2
7ADC-0AC	7.0	0.0	1.0	0.2
7ADC-5AC	7.0	5.0	1.0	0.2
7ADC-10AC	7.0	10.0	1.0	0.2
7ADC-15AC	7.0	15.0	1.0	0.2
7ADC-20AC	7.0	20.0	1.0	0.2

Characterization

Apparent density measurements of foams and solids with the size of 1 cm × 1 cm were performed by helium gas displacement technique using the Automatic gas (Ultrapyc 1200e) at 17.0 psig and 30°C. The void fraction (V_f) of foams is determined by Eq. (1) from the densities of the neat LDPE sheet (ρ_p) and of the composite foam (ρ_f).

$$V_f = 100 \times \left[1 - \frac{\rho_f}{\rho_p} \right] \tag{1}$$

Cellular structure of samples was characterized by scanning electron microscope (SEM SU5000). Samples were freeze-fractured in liquid N_2 and the fractured surface was made conductive by sputtering deposition of a thin layer of platinum. The dispersion degree of AC particles in composite foams was determined as well.

The tensile properties were determined according to ASTM D638 using a universal testing machine (Instron 55R4502), equipped with a 10 kN capacity load cell and with a crosshead speed of 50 mm min^{-1}. The specimen dimensions were 60 mm × 6 mm × 1 mm. The impact strength was measured by using the impact tester (ITR2000) following the ASTM D6110. The sheet sample with a size of 5 cm × 5 cm × 0.1 cm was placed on the sample holder and a ball was dropped on the sample with the speed of 3.46 m s^{-1}. The reported results were the averaged value from 10 samples measurements.

RESULTS AND DISCUSSION

SEM Images

FIGURE 1 and 2 show SEM micrographs of composite foams with different ADC and AC dosages respectively. All examples had closed cells and the cell size and cell size distribution were influenced by different amounts of ADC and AC. The cell size increased as the ADC concentration was increased because higher ADC concentration means more gas production caused by the thermal degradation of the foaming agent. The largest cell size was observed in composite foams containing higher amounts of the ADC. In FIGURE 2, the addition of AC up to 20 wt% decreased the cell formation and the average cell size tended to decrease with increasing AC content. The minimum cell size belongs to the sample containing 20 wt% of AC and 7 wt% of ADC. This attributed to the AC particles that assisted to generate more foaming nucleation sites in the polymer matrix, resulting in high viscosity which enhanced the resistance against cell growth and produced cells with smaller dimensions. The AC agglomeration was also observed, especially at high AC level.

Density and Void Fraction

LDPE-AC composite foams showed the reduction of density (0.9104-0.6649 g cm^{-3}) with increasing ADC content (0-7 wt%) because amount of liberated gas was proportional to the ADC content. It is commonly believed that high amount of released gas generated many cells in the composite foam, resulting in high degree of void fraction (22.95-27.70%). The density of composite foams decreased from 0.6686 to 0.6411 g cm^{-3} whereas the void fraction increased (27.30-30.29%) as the AC dosage increased from 0 to 10 wt% since more foaming nucleation centers was induced by adding AC in the composite foam leading to more gas bubble creation and further cell growth [10]. However, the density of composite foams tended to increase to 0.6551 and 0.7230 g cm^{-3} with the decrease of void fraction from 28.77 to 21.39% at AC contents of 15 and 20 wt% respectively, corresponding to the SEM results.

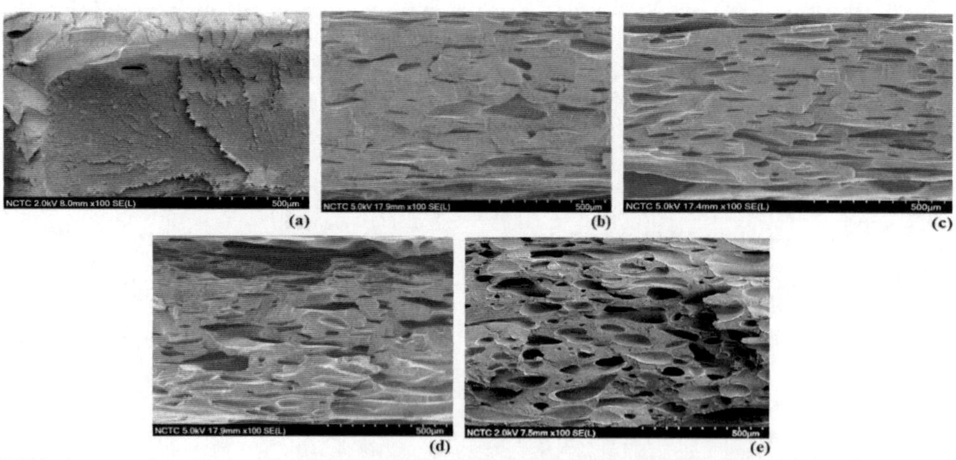

FIGURE 1. SEM photographs of composite foams with various ADC contents at 5.0 wt% of AC loading: (a) 0 wt%, (b) 1 wt%, (c) 3 wt%, (d) 5 wt%, and (e) 7 wt%.

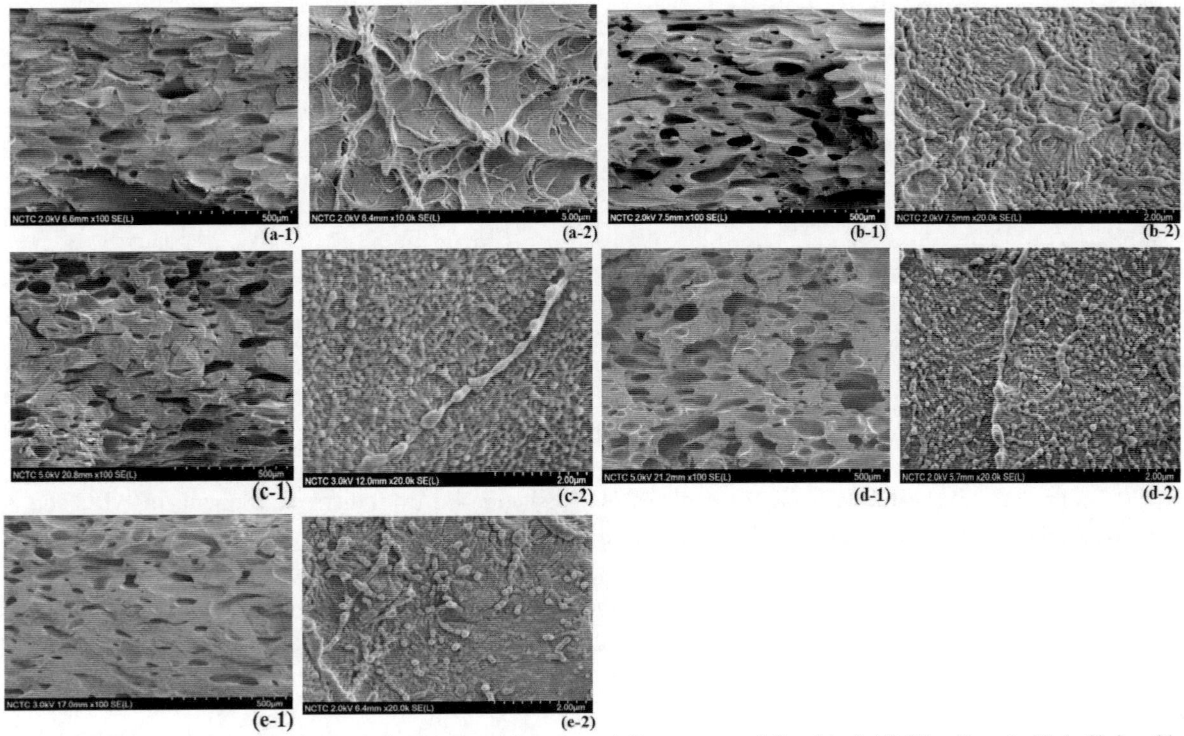

FIGURE 2. SEM photographs of composite foams with various AC contents at 7.0 wt% of ADC loading: (a-1)-(a-2) 0 wt%, (b-1)-(b-2) 5 wt%, (c-1)-(c-2) 10 wt%, (d-1)-(d-2) 15 wt%, and (e-1)-(e-2) 20 wt%.

Tensile Strength and Elongation at Break

The tensile strength at yield and elongation at break in machine and transvers directions of composite foams containing different amounts of ADC and AC are respectively presented in FIGURE 3 (a) and (b) and FIGURE 4 (a) and (b). The tensile properties of composite foams decreased compared with LDPE due to the high stiffness of AC materials and the cellular structure of composite foams. In addition, the proper connection between the polymer and AC did not generate in the interface due to the bubble formation, preventing a suitable transferring of stresses. The tensile strength and elongation at break of composite foams gradually decreased with an increase in ADC and much lower than those of LDPE. By adding AC in composite foams, the tensile strength and elongation at break slightly decreased but much lower than those of LDPE. With the increase in AC up to 20 wt%, a significant reduction in strength and elasticity modulus occurred probably due to the formation of AC agglomeration, the presence of voids and low LDPE content. These agglomerates indeed act as defects in a composite, which can induce an untimely failure, when the interfacial adhesion to the surrounding matrix is poor [11].

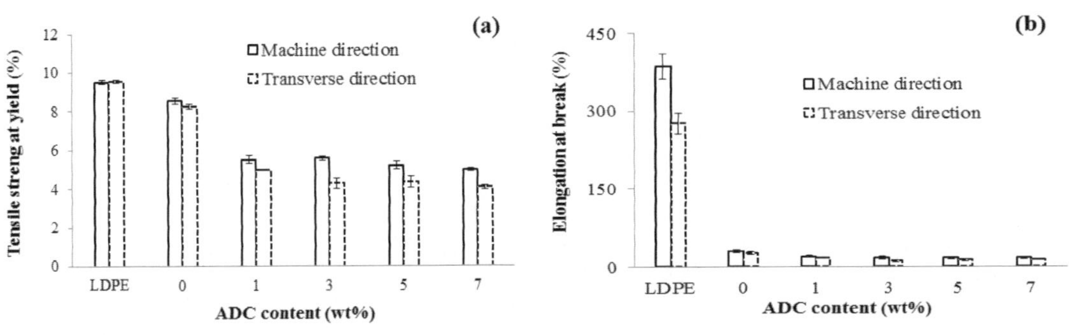

FIGURE 3. Tensile properties of LDPE and composite foams with various ADC contents at 5.0 wt% of AC loading: (a) Tensile strength at yield and (b) Elongation at break.

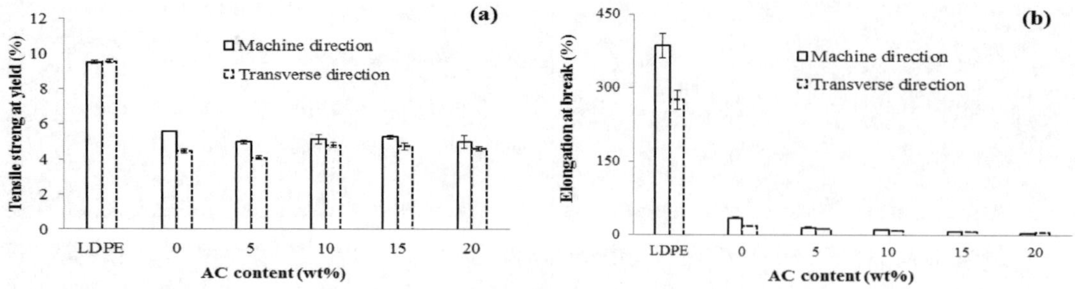

FIGURE 4. Tensile properties of LDPE and composite foams with various AC contents at 7.0 wt% of ADC loading: (a) Tensile strength at yield and (b) Elongation at break.

Impact Strength

FIGURE 5 (a) and (b) shows the impact strength of composites made from different amounts of ADC and AC respectively. Impact strength decreased with the increase of ADC content, significantly lower than LDPE. ADC produces nitrogen and carbon monoxide leading to bubble growth and prevents proper bonding between the polymer and AC, resulting in the development and propagation of the crack. This is also in congruency with research done by Matuana and Mengeloglu [12]. At high AC content, e.g. 20 wt%, the impact strength more reduced which related to the presence of AC agglomerations in the polymer matrix which serves as the stress concentration points.

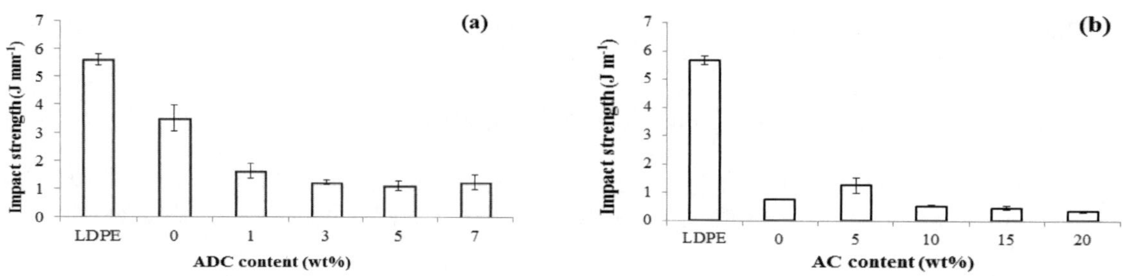

FIGURE 5. Impact strength of LDPE and composite foams with (a) various ADC contents at 5.0 wt% of AC loading and (b) different AC contents at 7.0 wt% of ADC loading.

ACKNOWLEDGMENTS

The authors would like to acknowledge National Metal and Materials Technology Center (MTEC) under National Science and Technology Development Agency (NSTDA), Thailand for the financial support and sample preparation and characterization.

REFERENCES

1. C. C. Ibeh and S. W. Lee, *J. Cell. Plast.* **44**, 498-515 (2008).
2. L. J. Lee, C. Zen, X. Cao, X. Han, J. Shen and G. Xu, *Compos. Sci. Technol.* **65**, 2344-2636 (2005).
3. W. Zhai, T. Kuboki, L. Wang, C. B. Park, E. K. Lee and H. E. Naguib, *Ind. Eng. Chem. Res.* **49**, 9834-9845 (2010).
4. S. M. Seraji, M. K. R. Aghjeh, M. Davari, M. S. Hosseini and S. H. Khelgati, *Polym. Compos.* **32**, 1095-1105 (2011).
5. J. Shen, C. Zeng and L. J. Lee, *Polymer* **46**, 5218-5224 (2005).
6. M. C. Saha, Md. E. Kabir and S. Jeelani, *Mat. Sci. Eng. A-Struct.* **478**, 213-222 (2008).
7. L. Chen, L. S. Schadler and R. Ozisik, *Polymer* **52**, 2899-2909 (2011).
8. W. Zhai, J. Yu, L. Wu, W. Ma and J. He, *Polymer* **47**, 7580-7859 (2006).
9. G. Crini, *Bioresour. Technol.* **97**, 1061-1085 (2006).
10. Z. Zakaria, Z. M. Ariff and C. S. Sipaut, *J. Vinyl. Addit. Technol.* **1**, 641-647 (2009).
11. A. Ashori and A. Nourbakhsh, *J. Compos. Mater.* **43**, 1869-1875 (2009).
12. F. Mengeloglu and L. M. Matuana, *J. Vinyl. Addit. Technol.* **9**, 26-31 (2003).

Laser Transmission Welding of Foamed Thermoplastic Injection Moulded Parts

Christian Hopmann, Suveni Kreimeier* and Maximilian Schöngart

Institute of Plastics Processing (IKV) at RWTH Aachen University
Seffenter Weg 201, 52074 Aachen, Germany

**suveni.kreimeier@ikv.rwth-aachen.de*
maximilian.schoengart@ikv.rwth-aachen.de

Abstract. Laser transmission welding is an innovative way to join plastics parts which have been produced in the foam injection moulding process. However, until now there are no experiences regarding the laser transmission welding of foamed plastics parts. For the use of the laser transmission welding process on a larger scale it is important to acquire knowledge about how far the structure of the foam influences the welding process, the bond strength as well as the structure of the joint. This knowledge allows an application-specific adjustment of the injection moulding process and the joining process to guarantee a constant quality of the parts. In the first step of a research project the dependence of the foam structure on the injection moulding parameters has been determined. In this way it is possible to produce parts which have a defined thickness of the surface layer as well as a specified average cell size. Therefore the integral foam parts are analyzed with light microscopy. It is shown that the injection flow rate has a major influence on the cell structure and the injection time has a significant influence on the thickness of the surface layer. Furthermore, the opening velocity while the mould breathing has an influence on the cell size. In addition to these investigations the transparent parts have been analysed regarding their transmission properties to show that the cell structure influences the optical path of the laser beam and therefore the intensity distribution of the laser beam in the weld zone. In further research the welding suitability of foamed parts was investigated. For this purpose the energy input into the material is varied to determine process windows and to find suitable process parameters for the welding process. In addition it is necessary to research the effect of the welding pressure since a high pressure can expel the melt from the joining area and destroy the cell structure.

Keywords: Laser transmission welding, Foamed parts, Injection moulding

PACS: 81.20.Vj

INTRODUCTION

In the Mid 1960s the first economically relevant processes were developed for foaming thermoplastics [1]. The first mechanical blowing process with a physical foaming agent presented had the aim to minimize sink marks on large-scale products [2]. Meanwhile, in addition to material specific advantages such as high dimensional stability at low weight, the foaming of plastics provides good damping properties and a high economic potential [3]. Due to material savings and cycle time reductions, production cost can be saved.

Foamed plastics are used as housing, technical components, trim parts for white goods and as automotive-parts [1, 4]. The trend is towards more complex, functional and module-integrated parts that cannot only be manufactured by expensive special tools, but in one process step by injection moulding. For an efficient production of these complex components an additional process step in form of a joining process is essential. One of these processes is the laser transmission welding. So far, there is no published experience regarding the laser welding of foamed parts. In order to apply this joining process for foamed parts, knowledge about the influence of the foam structure on the welding process and the weld seam must be obtained.

PRODUCTION OF FOAMED PARTS USING THE MUCELL®-PROCESS

The most common process variant of the foam injection moulding of thermoplastics is the patented MuCell®-Process by Trexel, Inc., Wilmington, Massachusetts/USA [2, 5, 6]. During the MuCell®-Process, a physical foaming agent, generally CO_2 or Nitrogen in the supercritical state is supplied to the melt in the plasticising unit. In contrast to the foam injection moulding with chemical foaming agents, the foaming agent system does not have to match to the plastics material. The plasticising unit is a screw with downstream gas mixer. The gas is supplied via injectors in the front of the screw, which is separated with a non-return valve. A dynamic mixer homogenizes the gas-melt-mixture so that an equally distributed foam structure is generated across the flow path. The process is mainly characterized by the formation of a fine-celled foam structure having an average pore size of 5 to 50 μm [2].

LASER TRANSMISSION WELDING

Laser transmission welding is a technology that has been improved towards serial production during the last years. The special advantages of this welding process are the precise, contactless heat input and the small heat affected zone (HAZ). Therefore, laser welding is applied in areas of application like electrical engineering and medical technology. It is a single-stage process, in which the heating and the joining of the plastics parts take place simultaneously. One of the joining parts must have a high transmittance in the range of the laser wavelength, the other part must exhibit a high absorbance. Before welding, both parts are placed in the desired final position and the joining pressure is applied [7, 8]. The transparent joining component transmits the laser beam without any considerable heating. In the second component, the laser beam is absorbed in a thin layer at the surface, whereby the laser energy is transformed into heat and the plastics is melted. The principle is outlined in FIGURE 1.

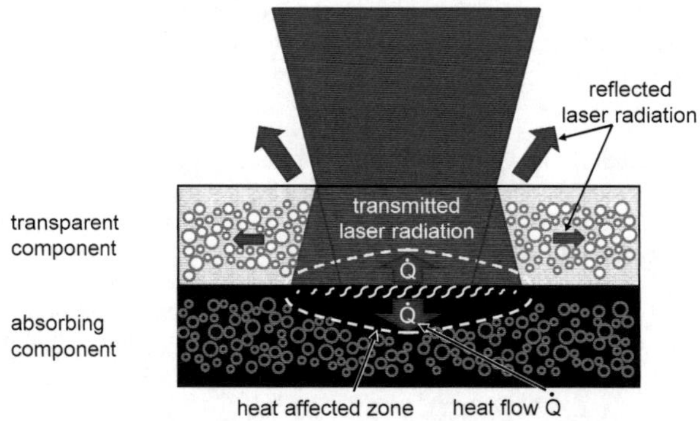

FIGURE 1. Principle of the radiation fluxes in the laser transmission welding of foamed thermoplastics

CONDUCTED INVESTIGATIONS

For the production of the transparent component an unfilled Polypropylen (PP) of the type Moplen EP 3307 was used. The material of the absorbing component is a 15 wt-% talcum-filled PP Hostacom EYC 136 N. Both materials were provided by Basell Polyelefine GmbH, Wesseling/Germany. The materials are used in automotive applications.

For the production of the samples, plates with varying moulding parameters were manufactured in the foaming injection moulding process on an injection moulding machine of the type Arburg Allrounder 520 A 1500-400-ALLDRIVE by Arburg GmbH+Co KG, Loßburg/Germany, using a plate mould with the dimensions 165 x 165 mm^2. From these plates, samples with the dimensions 30 x 80 mm^2 are manufactured. The aim was to create four different foam structures of the transparent material, which differ in cell structure and edge layer thickness (ELT). All the lower parts have the same foam structure.

The term cell structure refers to the size and number of foam cells. One cell structure may consist of many small cells, another one can consist of a few large cells. In order to influence the cell structure, the opening-speed of the breathing mould and the mould temperature were varied. Many small cells are produced with the combination of fast cooling and a fast opening stroke, as the cell growth is inhibited.

The expression "edge layer thickness" describes the dimension of the compact edge layer of the part. To affect it, the mould temperature was varied by 10 °C up or down. A higher mould temperature should lead to a less pronounced edge layer, as the edge layer solidified slower and more cells are formed in the edge layer. This results in the following experimental plan for the production of the four foam structures (TABLE 1).

TABLE 1: Experimental plan for the production of the samples (unfilled PP)

Foam structure [-]	1	2	3	4
Melt temperature T_M [°C]	260			
Mould temperature T_W [°C]	55	65	35	45
Opening speed v [mm/s]	2		5	

A part of the laser beam is reflected or absorbed by the transparent joining partner. To determine the share of losses, the transmittance and reflectance of the laser-transparent sample were measured with the spectrometer Lambda 1050 by PerkinElmer Inc., Waltham, Massachusetts/USA.

During the experiments simultaneous welding was performed, where a rectangular focused laser beam (1.5 x 27 mm^2) irradiates the material. In the course of the experiments the laser power P_L along with the irradiation time t_L were varied as welding parameters.

Six parts were welded for each parameter combination. The two laser systems LDF 400-90 ($P_{L,max}$ = 90 W) and LDM 400-40 ($P_{L,max}$ = 400 W) are high-power diode lasers produced by Laserline GmbH, Mülheim-Kärlich/Germany. Both devices emit laser radiation at a wavelength of λ = 940 nm.

For the evaluation of the results of the welding experiments short-time tensile tests are used to determine the bond strength. Furthermore microscopic analyses are carried out to determine the cell structure and heat affected zon.

RESULTS AND DISCUSSION

In order to describe and analyse the cell structure of foamed plastics several characteristics can be used, which allow a qualitative comparison. In [9] and [10] the cell density is introduced as a normed quality feature. This can be determined from a prepared foam sample. Therefore, all cells n are determined in a defined area A_0. The quotient is the area cell density n_F and is obtained by Eq. 1:

$$n_F = \frac{N_0}{A_0} \quad \left[\frac{1}{m^2}\right] \tag{1}$$

Assuming a uniform cell distribution within the foam the cell density results in space n_R for [9] to:

$$n_R = n_F^{\frac{3}{2}} \quad \left[\frac{1}{m^3}\right] \tag{2}$$

The edge layer thickness is measured by using the image analysis tool ImageJ of the National Institutes of Health, Bethesda, Maryland/USA. The results of the measurements of the cell density and the edge layer thickness for the four different combinations of injection moulding parameters are shown in FIGURE 2.

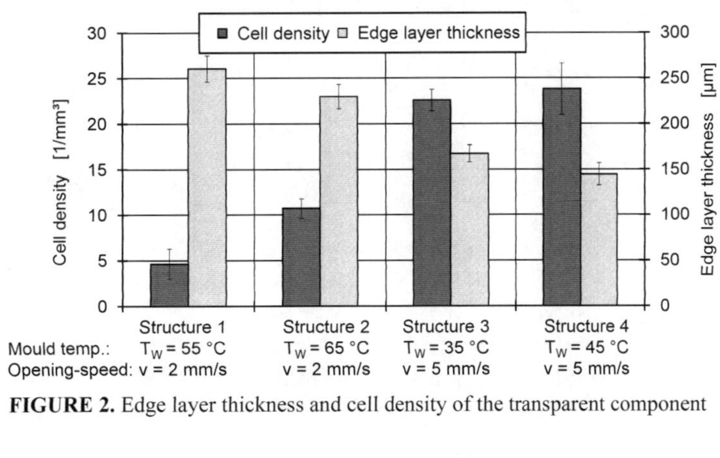

FIGURE 2. Edge layer thickness and cell density of the transparent component

The cell density of structure 1 and structure 2 are low and high for structure 3 and 4. The edge layer thickness is decreasing from structure 1 to structure 4. Basically it can be stated that a higher mould temperature at the same opening-speed leads to a lower edge layer thickness. At the same time a higher opening speed at the same mould temperature leads to a higher cell density.

For an efficient laser welding process a very high transmittance of the transparent component is required. When the transparent component is a foamed part, the distribution of the laser beam is affected. The cells represent additional phase transitions in the optical path of the laser, which influence the energy input in the joining zone negatively. In addition, the strength of the joining partner is weakened by the cellular structure, which also affected the weld strength negatively. The measurement of the transmittance and the reflectance by spectroscopy is executed for all four foam structures as well as the compact sample, which serves as a reference value. The results are shown in FIGURE 3.

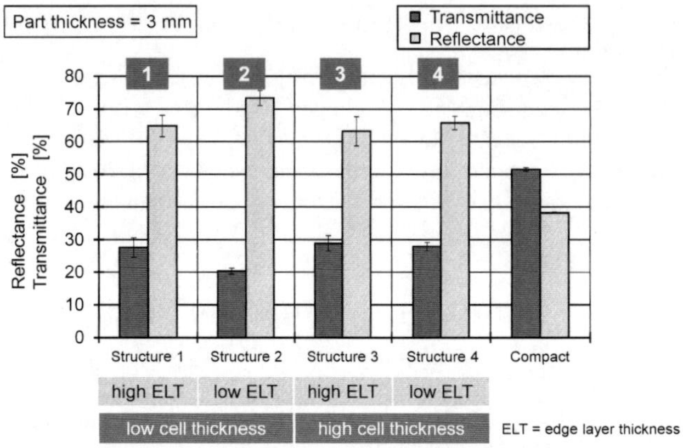

FIGURE 3. Transmittance and reflectance of the foam structures and the compact part

It can be seen that the foamed samples have a significantly lower transmittance than the non-foamed, compact reference-sample. While the compact material transmits 50 % of the incident radiation, the foamed samples only transmit 20 to 30 %. In particular, foam structure 2 has a low transmittance of about 20 %. The reduced transmittance is emerged from a higher reflectance. In contrast to the compact material with a reflectance of about 38 %, the foamed samples have a reflectance-value is about 60 %. This is attributed to a strong increase in internal reflection. The foam structure includes an additional phase transitions between polymer and locked gas.

The weld strength was determined in accelerated tensile tests (FIGURE 4). During the execution of the tensile test, there were two types of failure of the samples observed.

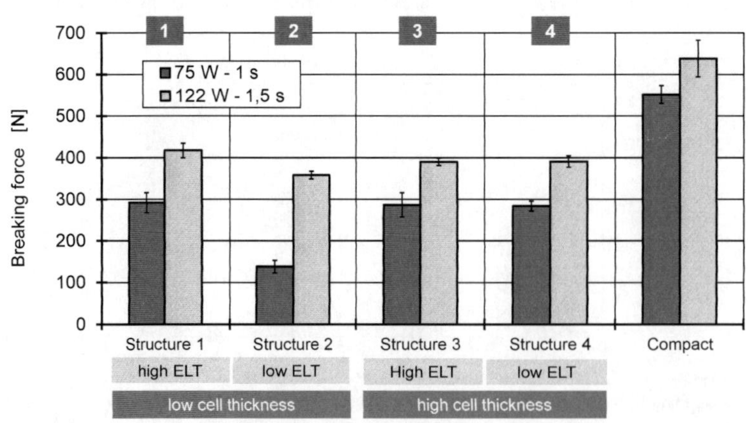

FIGURE 4. Breaking forces of the joints in dependence of the welding parameters

If a weld has a low strength (75 W, 1 s) there is an adhesive failure of the weld. An increasing energy input into the joining zone increases the weld strength and there is a cohesive failure of the foam structure. Due to the simultaneous tensile and shear stress in accelerated tensile tests the foam structure rips open at the weld. This leads to a crack growth through the cell structure and the samples fail. The breaking forces of the compounds of the foamed samples are considerably below the weld strengths of the compact materials. Although, the process point with a laser power of 122 W and an irradiation time of 1.5 s reaches higher breaking forces. The compounds of foam structure 2 reach the lowest breaking forces. In particular, at a laser power of 75 W only a low weld strength is achieved. This results from the lower transmittance of foam structure 2. In how far the cell structure is responsible for this effect must be considered in further investigations.

For the microscopy analysis, a very low sample thickness is needed. This can be achieved with thin-sections of the samples. For this purpose, the samples are placed in resin and cured overnight. The target thickness is the average size of the cells, so that only one layer of the cells is visible in the micrographs and they do not overlap. The capture of the images is performed with a light microscope in combination with a polarizing filter to illustrate structural changes.

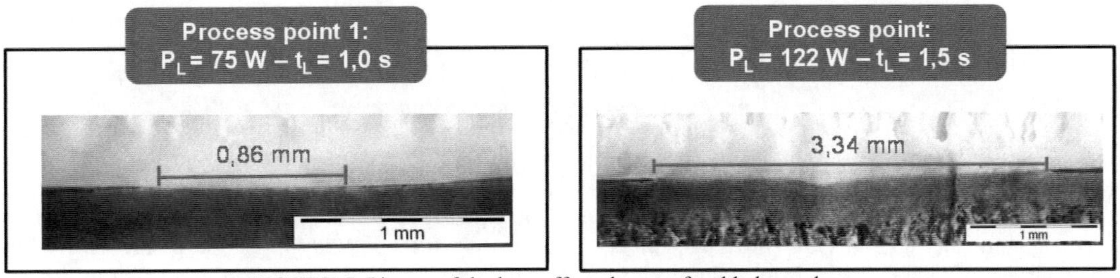

FIGURE 5. Picture of the heat affected zone of welded samples

FIGURE 5 shows a microscopic image of the foam structure 2, which has been welded with different laser powers and irradiation times. With the process point 2 (P_L = 122 W; t_L = 1.0 s), most of the heat affected zone is located in the compact edge layer. Some of the cells, which were located close to the compact edge layer, have been destroyed during the welding process and have been filled with mould. Overall it can be noted, that the identification of the heat affected zone via thin-section microscopy is very challenging.

CONCLUSIONS

The joining of foamed injection moulded parts can be realised by laser transmission welding. A combination of laser power and irradiation time can be found for each foam structure in the simultaneous variant in order to produce high-quality joints. The cell structure leads to a higher internal reflection and a greatly reduced transmittance of the transparent joining partner. This results in a higher amount of energy required during the joining process as in compact components.

REFERENCES

[1] POTENTE, H.; LIMPER, A.: *Schäumen von Thermoplasten mit zellularer Struktur*. Kunststofftechnik Paderborn, Universität Paderborn, Abschlussbericht zum AiF-Vorhaben Nr. 13.595 N, 2005

[2] ALTSTÄDT, V.; MANTEY, A.: *Thermoplast-Schaumspritzgießen*. München, Wien: Carl Hanser Verlag, 2011

[3] SASTRE, L. F.: *Influence oft the Foam Morphology on the Mechanical Properties of Structural Polymer Foams.* RWTH Aachen, Dissertation, 2010 - ISBN: 3-86130-980-7

[4] SCHLOTT, S.: Mit Innenraumwerkstoffen den Sinnen schmeicheln. *ATZ - Automobiltechnische Zeitschrift* 113 (2011) 11, S. 884-887

[5] SCHRÖDER, U.: Pole Position – Mit MuCell® spritzgießen. *Kunststoffe* 93(2003) 10, S. 30-33

[6] WURNITSCH, C.; WÖRNDLE, R.; SPIEGL, B.; STEINBICHLER, G.; EGGER, P.: Thermoplastische Träume durch Schäume: Wirtschaftliches Spritzgießen für individuelle Bauteileigenschaften. *Kunststoffe* 91 (2001) 5, S. 64-67

[7] ACHERJEE, B.; KUAR, A.; MITRA, S.; MISRA, D.: Effect of carbon black on temperature field and weld profile during laser transmission welding of polymers: A FEM study. *Optics & Laser Technology* 44 (2011) 3, S. 514-521

[8] DEVRIENT, M.; DA, X.; FRICK, T.; SCHMIDT, M.: Experimental and simulative investigation of laser transmission welding under consideration of scattering. *Physics Procedia* 39 (2012) 7, S. 117-127

[9] RAMESH, N.S.; RASMUSSEN, D. H.; CAMPBELL, G.A.: The Heterogeneous Nucleation of Microcellular Foams Assisted by the Survival of Microvoids in Polymers Containing Low Glass Transition Particles. Part I: Matemathical Modeling and Numcerical Simulation. *Polymer Engineering and Science* 34 (1994) 22, S. 1685-1697

[10] PETERS, R.: *Schaumstrukturanalyse mit digitalen Bildverarbeitungsmethoden.* RWTH Aachen, Dissertation, 2003

Natural Unidirectional Sheet Processes for Fibre Reinforced Bioplastics

Ahmed-Amine Ouali [a], Roman Rinberg [a], Wolfgang Nendel [a], Lothar Kroll [a], Carolin Siegel [b], Beate Buchelt [b], André Wagenführ [b], Katja Trommler [c], Katja Schreiter [c], Rico John [c] and Stefan Spange [c]

[a] *Technische Universität Chemnitz, Institut für Strukturleichtbau und Kunststoffverarbeitung. Reichenhainer Str.31-33, 09126 Chemnitz, Germany*

[b] *Technische Universität Dresden, Institut für Holz- und Papiertechnik, Marschner Str. 32, 01307 Dresden, Germany*

[c] *Technische Universität Chemnitz, Polymerchemie, Straße der Nationen 62, 09111 Chemnitz, Germany*

ahmed-amine.ouali@mb.tu-chemnitz.de

Abstract. Technical Natural Fibre semi-finished products are mainly used in lightweight automotive components as random fiber mats which are often combined with petrochemical matrix resins for large series production process. However, random fiber arrangement and limited fiber length restrain the complete utilization of the material-inherent strength reserves of natural fibers. With high-levels of petroleum-based plastics (about 20-50 wt -%) in the structure, these applications have a clear potential in terms of eco-friendliness and sustainability. The development of new fully bio-based composites and efficient manufacturing methods suitable for series processing is the purpose of the current sub-project C4 "Flexible textile / plastics processes with renewable raw materials" in the framework of the Excellence Cluster MERGE EXC 1075, funded by DFG (Deutsche Forschungsgemeinschaft). A high-efficiency in terms of mass-production, reproducibility and flexibility requires the performance of successive steps in the manufacture of semi-finished and final products. The appropriate process line comprises the steps of the preparation of the thermoplastic matrix sheets (such as from Bio-polyethylene or Bio-polyamide) and a specific chemical modification of the reinforcing materials (wood veneer and unidirectional flax fibers) on the concept of a discontinuous or continuous production of unidirectional reinforced Natural Fibre Composites (NFC) prepregs. Moreover, the shaping of hybrid lightweight structures from NFC prepregs makes possible the incorporation of these innovative technologies into new concepts and potential applications in several fields such as automotive industry or sport equipment. The mechanical properties profile of the flax - natural unidirectional sheet for fibre reinforced bioplastics has been here studied in detail with press-engineered samples and has confirmed their high application potential as an alternative to glass fiber-reinforced composite sheets.

Keywords: Biocomposite, bio-based polymer, thermoplastic polymer, natural fibre, endless reinforcement, interface matrix/reinforcement, continuous process, automotive application, sport equipment

PACS: 81.05.Qk

INTRODUCTION

In recent years, the concerns about environment and pollution are constantly increasing. The multitude of investigations on composites, whose aim is mainly weight reduction, demonstrates a great tendency to acquire more environmentally friendly technologies [1-4]. In the field of lightweight structure technologies, the optimization of the structure weight and of the processing implementation is the key of saving raw materials and energy consumption. Funded by the DFG (Deutsche Forschungsgemeinschaft) and based at the TUC (Technische Universität Chemnitz) in Germany, the cluster of excellence MERGE EXC 1075 has as objective to merge many skills in order to develop new technologies for multifunctional lightweight structures. MERGE focuses on large-scale energy-efficient processes for products, made of metal and plastic-textile technologies, and micro-systems. Among the 6 MERGE interactive research domains (IRD), the IRD C "the textile plastic based technologies", through the sub-project C4, gives the opportunity to deepen this subject by using renewable raw materials. The realization of new biocomposites, being able to compete with traditional composites such as glass reinforced composites [3,5-7], is divided into several steps. The selection of raw renewable materials and the investigation on improvement of the adhesion between the matrix and the reinforcement belong to the first part of this project. The second one consists in improving and implementing manufacture processes suitable for series production. After investigating on mechanical properties, the ultimate step is the incorporation of the innovative technologies into applications in diverse domains such as automotive industry or sport equipment.

MATERIALS

The goal to process bio-based sustainable composites leads us to define the selection criteria of the raw renewable materials. As regards the matrix materials, the development of thermoplastic semi-finished products represents the most appropriate way in terms of process flexibility and recyclability. The reshaping and the remolding capabilities, the high impact strength and the blending with other thermoplastics make the use of thermoplastic polymers meaningful. For the ecological point of view, the PA11 Rilsan® (here, called Bio-PA), from the company Arkema, and the HDPE I'M Green™ (here, called Bio-PE), from the company Braskem, are produced from, respectively, castor oil feedstock and ethanol of sugarcane. Both of them are bio-based, thermoplastic and non-degradable. Bio-PA demonstrates an interesting thermal stability and high mechanical properties suitable for high-efficient materials. Bio-PE has a relative low cost for a bio-based polymer and keeps interesting characteristics. For this study, Bio-PA and Bio-PE have been selected as polymer matrix. The production of polymer films (100 μm thick) is performed through an extruder followed by a calender system at the TUC (Figure 1 a & b). As regards the reinforcement material, the selection is headed towards the unidirectional continuous endless flax fibres (300g/m² - Bcomp Ltd.) (Figure 1 c) and towards the wood veneer (spruce or beech - 0,3-0,6 mm thick) (Figure 1 d). Both have different interesting intrinsic characteristics and can widen the scope of applications by using them separately or together.

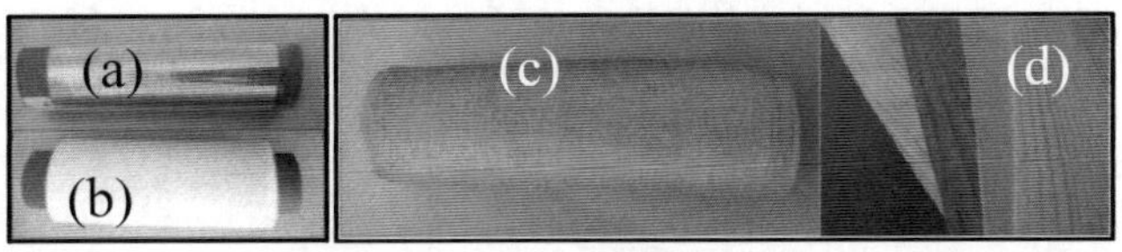

FIGURE 1. Polymer films made of Bio-PA (a) and Bio-PE (b), UD flax fibers (c), several wood veneers (d)

PROCESSING

Regarding the processing methods, many steps are actually required to obtain a lightweight structure and to develop new high-efficient composites with raw renewable resources. The reproducibility of processes is an important parameter; by being continuous in terms of production, the process displays an energy- and cost-efficient aspect. The first one is an installation from Coatema® permitting us to coat and to embed continuously the reinforcement surface with an adhesive agent in order to improve the adhesion between both hydrophilic reinforcement and hydrophobic polymers, such as Bio-PE. This improvement requires preliminary investigations on the adequate adhesive agent to insert. Afterwards, a continuous production of pre-impregnated fibres, called UD-prepregs – a one-layer composite sheet defined as semi-finished product whose reinforcement is unidirectional (UD) – can be also carried out, thanks to the Cepreg machine (from Cetex® institute present at the TUC). The reinforcement is firstly pre-dried and then impregnated between two melted polymer films by means of heated cylindrical presses. The prepregs (Figure 2 a) is processed without blocked air, which could be present by usual film staking process. With the pre-impregnated fibres, the processing of a multi-layer biocomposite with many different configurations (different layer number or fiber orientations) is carried out by a compression molding (Figure 2 b). The shaping and the design of lightweight structure are then possible by back-injection molding (Figure 2 c & d).

FIGURE 2. Roll of flax-BioPA prepregs (a), multi-layered flax-BioPA sheet (b) and a back-injected molded component with flax-BioPA (c) & with wood veneer-BioPE (d)

INVESTIGATIONS AND RESULTS

For the biocomposite characterization, investigations such as tensile, shear and izod impact tests, to determine the mechanical properties, and microscopic pictures, in order to evaluate the fiber-matrix compatibility, were carried out. Among the obtained results, the figure 3 shows the values of elastic modulus E and the tensile strength Rm (according to DIN ISO 527-5) for different kinds of biocomposite.

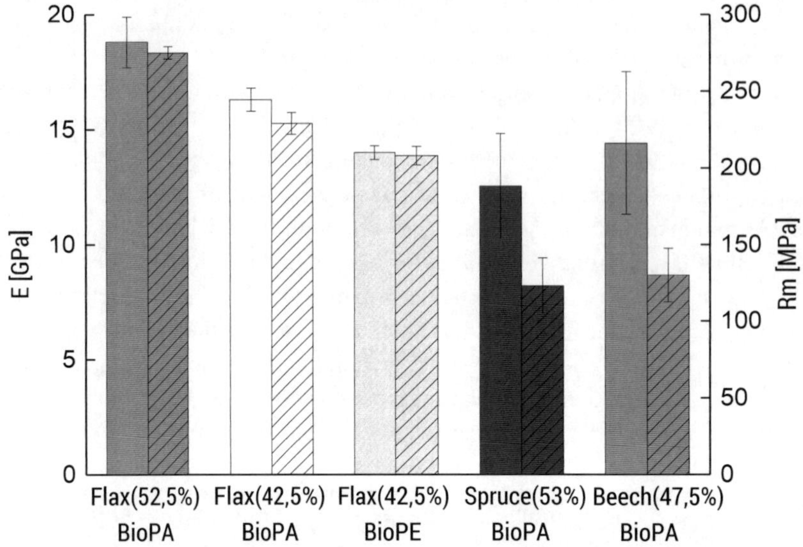

FIGURE 3. E-modulus [GPa] and tensile strength [MPa] of different possible combinations with the selected raw renewable materials

For those biocomposites, the mechanical properties depend on the reinforcement type (flax/wood veneer), the matrix nature (BioPA/BioPE) and the volume fiber content. The range of the values are respectively between [14-20] GPa for E and [120-280] MPa for Rm. By considering the low natural reinforcement density (Flax: 1,4; Spruce: 0,365; Beech: 0,59), the specific properties (mechanical properties / density) of the biocomposites become really interesting and comparable to the conventional composites (glass fiber : 2,55 as density).

APPLICATION

As regards the applications, some prototypes are produced in order to show the potential of those biocomposites (Figure 4). For instance, a veneer layer composite as a core material, a saddle with integrated sensors and a snowboard [5-6] with a core made of Bio-PA and veneer and outer layers made of Bio-PA and unidirectional flax fibres. By the support of KraussMaffei Technologies GmbH, a series production of lightweight load-bearing element made of raw renewable materials has been performed. With the Fiber-Form injection molding machine, a multi-layered biocomposite made of flax fibres and bio-PA 11 is pre-heated in IR-field and, then, placed on the mold with a gripper system. Finally, a third bio-based polymer (here the bio-based PA 10-10 Vestamid® from the company Evonik Industries) is back injected and shapes the biocomposite. These tests, near to series production, have permitted us to define the advantages, such as the material robustness or the good surface quality and some restrictions to be improved for a future technology transfer. In the near future, this bio-based technology will be introduced as components in a car concept related to the MERGE project "Chemnitz Car Concept". For example, a center console with integrated sensors or a reinforced exterior structure at the door.

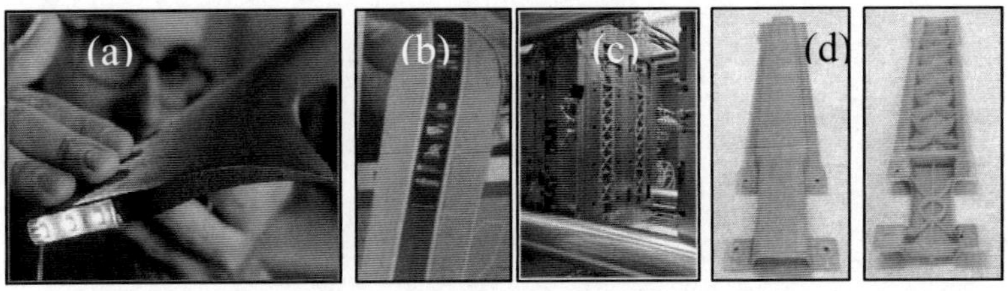

FIGURE 4. Veneer saddle with integrated sensors (a), snowboard with a veneer core and flax outer layers (b) and lightweight load-bearing elements processed in KraussMaffei Technologies GmbH (c) & (d)

CONCLUSION AND OUTLOOK

This work, related to the cluster of excellence MERGE, has as objective to produce composite from raw renewable materials by a resource-, cost- and energy-efficient way at a large-scale for several applications, especially in automotive industry. Composed of unidirectional flax fibres or wood veneer as reinforcement and Bio-PA or Bio-PE films as matrix, biocomposites are processed according to successive methods which are implemented and developed. The continuous coating of reinforcement surface for the improvement of the adhesion at the interface, the continuous production of UD-prepregs, enabling to have a roll-formed semi-finished product without blocked air, the optimization of final products processes, such as compression molding and back-injection molding, take an important part of work in this project. Thanks to the weight reduction by keeping interesting mechanical properties and to the continuous aspect for the manufacture processes, these technologies with raw renewable materials could be able to compete with the technologies of well-established composite, in terms of resource and energy efficiency. In the near future, more investigations will be conducted in order to validate the new potential of this type of biocomposite. The optimization of manufacture processes, the determination of other mechanical properties, the characterization at the interface, the comparison with conventional composites and the diversification for new applications represent the main improvement lines of this project.

ACKNOWLEDGMENTS

Acknowledgments to the cluster of excellence MERGE EXC 1075, funded by DFG (Deutsche Forschungsgemeinschaft)

REFERENCES

1. L. Kroll et al., "Biokunstoffen den Weg in den Markt erleichtern", *Plastverarbeiter,* 3(2015), pp. 66-68
2. F. Tautenhain et al., *Biopolymers in multicomponent injection-moulding application*, Euro BioMAT 2015: European Symposium and Exhibition on Biomaterials ad Related Areas, 2015
3. M. Reinhard et al., "PLA-Viscose-Composite with Continuous Fibre Reinforcement for Structural Application", In *Procedia Materials Science*, 2(2013), pp.137-143
4. K. Oksman et al., "Natural fibres as reinforcement in polylactic acid (PLA) composites", *Composites Science and Technology*, 63(9), 2003, p. 1317-1324.
5. J. Kaufmann and C. Siegel, *Verwendung von neuartigen biobasierten Halbzeugen im Snowboard*, Forum Sonderschau Faserverbund-Werkstoffe auf den Weg in die Serie INTEC 2015, Leipzig, 2015
6. J. Kaufmann, "New Materials for sports equipment made of anisotropic fiber-reinforced plastics with stiffness related coupling effect", *Procedia Engineering*, 112 (2015), pp 140-145
7. A.-A. Ouali et al., "New biocomposites for lightweight structures and their processes", In *Edtmaier, C., Requena, G.* (Ed): 20th Symposium on Composites. Materials Science Forum, 825-826 (2015), Part 2, pp. 1055-1062

Development of Thermally Conductive Polymer Matrix Composite Foams via Supercritical CO_2 Foaming

Yanting Guo[a], Marco Chu[b] and Siu Ning (Sunny) Leung[a*]

[a]*Department of Mechanical Engineering, Multifunctional Materials/Micro-and-Nanostructuring Laboratory, Lassonde School of Enhineering, York University, Toronto M3J 1P3, ON, Canada*
[b]*Division of Engineering Science, University of Toronto, Toronto M5S 1A4, ON, Canada*

Presenting and Reference author's email: sunny.leung@lassonade.yorku.ca

Abstract. In this study, thermally conductive polymer matrix composite (PMC) foams aiming for high-performance heat management applications have been developed by physical foaming process using supercritical CO_2 as the blowing agent. The PMC foams consist of ultra-high-molecular-weight polyethylene (UHMWPE) as the polymer matrix and sub-micron-scale hexagonal boron nitride (hBN) as the thermally conductive filler. Parametric studies have been conducted to elucidate the effects of processing conditions (e.g., saturation pressure, saturation temperature and cooling) on the foam morphology (e.g., cell population density, cell size, and foam expansion) and hence the effective thermal conductivity (k_{eff}) of the UHMWPE/hBN composite foams. From the scanning electron microscope (SEM) micrographs, it is evident that foaming assisted filler localization and filler alignment along cell walls via cell expansion. The highest k_{eff} of PMC foams with 9.2 vol.% hBN reach 0.97 W/(m·K), which represented 20% increase over their solid counterparts. This is consistent with our earlier study of PMC foams using Expancel® microspheres as the foaming agent, and verifies that polymer foams can be turned into thermally conductive by foaming-assisted filler networking. In short, the findings of this work provide guidelines for developing thermally conductive PMC foams for high-performance heat management applications with low hBN loading via physical foaming process.

Keywords: polymer composites; hexagonal boron nitride; thermally conductive; foams; morphology
PACS: 81

INTRODUCTION

Due to the unceasing progress towards innovative materials and manufacturing technologies, as well as the unique properties of polymers (e.g., light weight, low cost, excellent electrical insulating property, good processablity), thermally conductive polymer composites have been extensively used as packaging materials for heat dissipation in micro-electronics industry. Thermally conductive polymer matrix composites (PMC) are commonly produced by embedding the thermally conductive fillers, such as metal[1], ceramic[2], and carbon-based particles[3], into polymer matrices. The heat dissipation ability of PMC hinges on their effective thermal conductivity (k_{eff}). A large number of studies have been conducted to design and fabricate thermally conductive PMC, and most of them promoted the k_{eff} of PMC by loading high contents of thermally conductive fillers. Kume *et al.*[2] reported significant increases in polyimide composites' k_{eff} up to 9.3 W/(m·K) by heavily loading polyimide with 49 vol.% of aluminum nitride granules and 21 vol.% of hexagonal boron nitride (hBN) powders. Such high filler loading is detrimental to the applications of PMC in different thermal management systems since it dramatically compromises the advantages of polymers, such as low cost, light weight, and good processability. Therefore, it is imperative to explore new strategies and directions to develop thermally conductive PMC with minimal filler loadings.

The k_{eff} of PMC relies on the development of thermally conductive filler network in polymer matrices. In light of this, researchers have attempted to optimize the filler network within PMC by various micro-and-nanostructuring techniques, such as sonication[4], ultra-drawing[5], electric field-induced filler alignment[6], use of hybrid fillers[7], foaming[8]. Using Expancel® microspheres as the foaming agent, our recent study[8] reported thermally conductive PMC foams with k_{eff} higher than their solid counterparts for the first time. This study demonstrated that foaming-assisted filler networking is a feasible processing strategy to enhance PMC's k_{eff}, especially at a low filler loading. To investigate the potential for extending the concept of foaming-assisted filler networking to fabricate thermally conductive foams through the physical foaming process (i.e., normally using the inert gases as the blowing agents), this paper uses PMC consists of ultra-high-molecular-weight polyethylene (UHMWPE) and submicron-scale hBN platelets as a case example to elucidate the processing-structure-property relationships of PMC foams. UHMWPE with a molecular weight of 2 – 6 million is composed of extremely long chains. Due to its high molecular weight,

UHMWPE is known to possess a number of important properties, including superior notched impact strength, excellent wear resistance and stress-cracking resistance, outstanding energy absorption capacity, high heat deflection temperature and wide service temperature range. Therefore, UHMWPE has been widely used in a large variety sectors, as a backbone polymer for aqueous mediums[9], crystal nucleating agent for polymeric foams[10], a polymer matrix for electrically conductive composites[11]. Development of thermally conductive UHMWPE-based PMC could help to expand PMC's market to some high-performance thermal management systems that require tougher materials, such as military and aerospace sectors. In this study, the effects of processing conditions on foam morphology of UHMWPE-hBN composite foams, as well as the influences of foam morphology on the k_{eff} of composite foams were examined. The findings of this work would inspire researchers with new directions to fabricate thermally conductive PMCs for next generation thermal management systems with excellent mechanical properties and wear resistance.

EXPERIMENTAL

Materials

Commercially available UHMWPE in powder form (Celanese, GUR 4150) was selected as the polymer matrix material. The hBN platelets with submicron-scaled thickness (Momentive Performance Materials, PolarTherm, AC6041) was used as the thermally conductive filler in the PMC. The physical blowing agent used in this study was carbon dioxide (CO_2) of 99.8% purity (Linde Gas Inc.). The physical properties of UHMWPE and hBN are summarized in Tables 1 and 2, respectively.

TABLE 1. Physical properties of UHMWPE

Property	Value	Unit
Density	930	kg/m^3
Melting Temperature	125 – 138	°C
Thermal Conductivity	0.41	W/(m·K)
Powder Size	200 – 300	μm

TABLE 2. Physical properties of hBN

Property	Value	Unit
Density	2280	kg/m^3
Shape	Platelets	°C
Lateral Size	~6	μm
Thickness	~0.1 – 0.5	μm
In-Plane Thermal Conductivity	>300	W/(m·K)
Out-of-Plane Thermal Conductivity	~3	W/(m·K)

Preparation of UHMWPE-hBN Composites

Based on the findings of our previous study[9], the effect of foaming-assisted filler networking on improving PMC's k_{eff} was more significant when the filler loading was relatively low. As a result, for the UHMWPE-hBN composite samples prepared for subsequent physical foaming, the hBN content was maintained at 9.2 vol.%. Initially, calculated amount of UHMWPE powders and hBN platelets were weighed and dry-blended at room temperature. The mixtures were then loaded into a circular disc mold of 115 mm in diameter and 500 μm in thicknesses and transferred to a compression molding machine (Carver Inc., 4386 CH) preset at 210 °C. The sample was equilibrated at 210 °C under desired pressure for 20 minutes to completely melt and sinter the UHMWPE powders. The compression-molded samples were moved to a cooling module with flowing water channels for solidification. In addition, the unfoamed composite samples with a hBN content of 5 vol.% and 15 vol.%, respectively, were prepared using the same procedures aforementioned for comparison purpose.

Physical Foaming of UHMWPE-hBN Composites Using Supercritical CO_2

Composite samples were cut into rectangular shapes of 2.5 inch by 1 inch and then physically foamed by supercritical CO_2. First, the samples were placed into a high-pressure and high-temperature chamber and saturated with CO_2 at designated pressure and temperature for 30 minutes. After that, the high-pressure CO_2 was released from the chamber by rapidly opening the outlet valve. This sharp pressure drop of CO_2 inside the chamber led to thermodynamic instability of the saturated samples and caused gaseous cells to nucleate and grow within the samples. Finally, the foamed samples were cooled down to stabilize the foam morphology. In order to investigate the effect of cooling on the morphology and k_{eff} of the composite foams, the results of foam samples with immediate cooling after gas depressurization and delayed cooling of 75 seconds were compared. The processing conditions used in the foaming experiments are summarized in Table 3.

TABLE 1. Processing parameters of foaming experiments

Parameters	Value	Unit
Saturation Pressure	1200, 2000	psi
Saturation Temperature	125 – 160	°C
Saturation Time	30	minute

Sample Characterization

The morphologies of composite foams were examined by scanning electron microscopy (FEI Company, Quanta 3D FEG). The cross-sections of composite foams were exposed by cryo-fracturing the samples under liquid nitrogen, and sputter-coated with gold (Denton Vacuum, Desk V Sputter Coater). By analyzing the micrographs, the cell size and cell population density were estimated. The cell population density with respect to unfoamed volume (N_0) was determined by the following equation:

$$N_0 = \left(\frac{nM^2}{A}\right)^{3/2} \times \phi \tag{1}$$

where n is the number of cells in the micrograph, A and M are the area and magnification factor of the micrograph, respectively. ϕ is the volume expansion ratio evaluated using the water-displacement technique in accordance to ASTM D792-13.

The k_{eff} of unfoamed and foamed UHMWPE-hBN composite samples were measured by a thermal conductivity analyzer (C-Therm Technologies Ltd., Tci Thermal Conductivity Analyzer) based on the modified transient plane source technique. For each composition and condition, the k_{eff} of three samples were measured to ensure the reliability and repeatability of the measurement.

RESULTS AND DISCUSSION

Effects of Foaming Parameters on the Morphology of Composite Foams

Parametric studies were conducted on the critical parameters of foaming process, including saturation pressure, saturation temperature and cooling after depressurization. Figure 1 plots the volume expansion ratio of the composite foams fabricated under various temperature, pressure and cooling conditions. For both saturation pressures (i.e., 1200 and 2000 psi) with immediate cooling, the PMC foams showed a mound-shaped volume expansion ratio-to-saturation temperature relationship, and the highest volume expansion ratio occurred at 135°C saturation temperature. When the saturation temperature was 150°C or higher, the volume expansion ratio of PMC foams was almost unchanged. Figure 2 shows the morphologies of UHMWPE-hBN composite foams fabricated at 135°C and 150°C under 1200 psi and different cooling conditions. It can be observed that the foam morphology deteriorated due to the cell collapse as saturation temperature increased and led to a decrease in PMC foam's volume expansion ratio. When the saturation pressure and temperature were maintained at 1200 psi and 150°C, respectively, as a result of the cell opening and cell coalescence, the composite foams fabricated by delayed cooling after gas depressurization had a few relatively large cells.

To investigate the influence of foaming conditions on the morphology of the PMC foams, the cell density of cell size of the PMC foams were examined and plotted in Figure 3. It can be seen that PMC foam's cell population density-to-saturation temperature relationship is consistent with their volume expansion ratio. However, regardless of the saturation temperature, the cell density of PMC foams underwent immediate cooling after depressurization was within the same order of magnitude while PMC foams fabricated by delayed cooling had a significant decrease in cell density as saturation temperature increased. Surprisingly, the cell size of PMC foams fabricated under various foaming and cooling conditions was no larger than 15 μm. The only exception was for the case of saturation at 150°C with delayed cooling, where a few large cells (approximately 100 μm) were achieved.

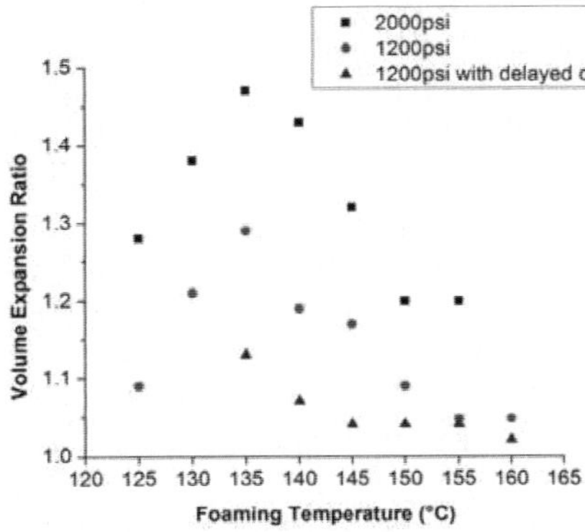

FIGURE 1. Effects of saturation temperature and pressure on the volume expansion ratio of UHMWPE-hBN composite foams

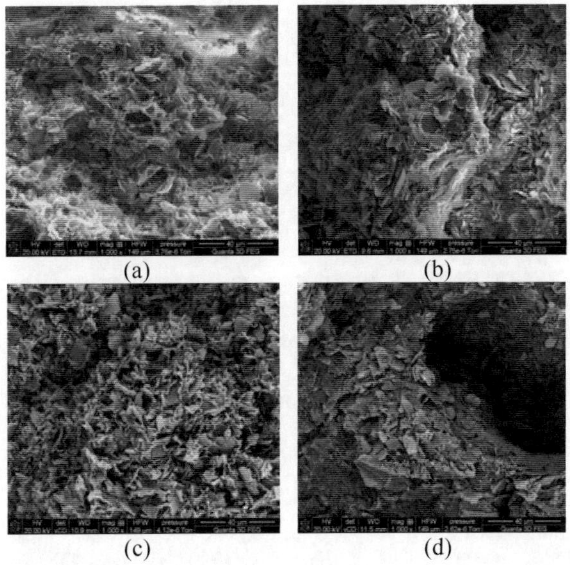

FIGURE 2. Micrographs of composite foams fabricated at 1200 psi and different cooling conditions: (a) 1200psi-135°C; and (b) 1200psi-150°C; (c) 1200psi-135°C-delayed cooling; (d) 1200psi-150°C-delayed cooling

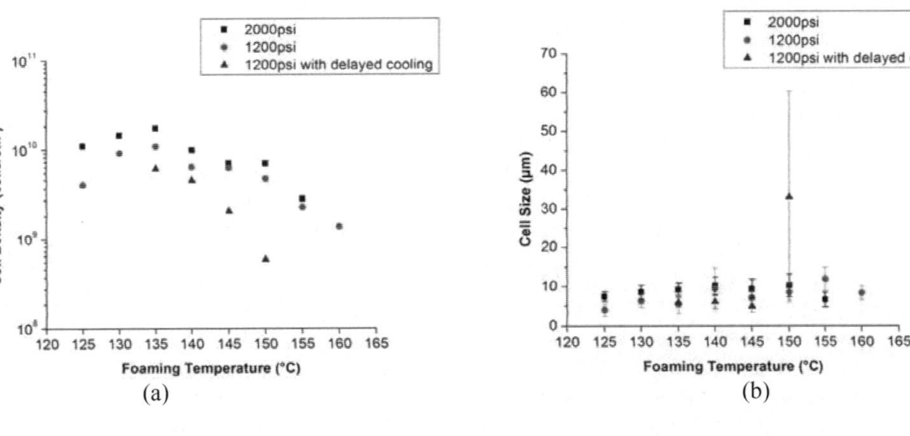

FIGURE 3. UHMWPE-hBN foams fabricated under different conditions: (a) cell population density; (b) cell size

Effects of Foam Morphology on the k_{eff} of UHMWPE-hBN Composite Foams

The k_{eff} of the composite foams fabricated under various foaming conditions are plotted in Figure 4. Experimental results showed that composite foam's k_{eff} decreased monotonically as their volume expansion ratio increased while the saturation temperature was fixed. The UHMWPE-hBN composite foams fabricated at 2000 psi and 1200 psi had highest k_{eff} with approximately 20% and 5% volume expansion, respectively. It is notable that the composite foams fabricated at 1200 psi with delayed cooling showed higher or comparable k_{eff} over the unfoamed composites with the same hBN content (i.e., 9.2 vol.%). The maximal increase in PMC foam's k_{eff} was achieved at 150°C, which was 20% higher than that of the unfoamed composites with the same filler loading. In addition, this highest k_{eff} was close to that of the unfoamed composites loaded with 15 vol.% filler. It is worth noting that the actual filler content of the PMC foams is lower than that of unfoamed composites with the same filler content. For instance, the actual hBN content of the PMC foams fabricated at 150°C with delayed cooling was 8.8 vol.%.

The SEM micrograph in Figure 2 shows that the hBN platelets are partially oriented along the cell walls due to the cell expansion. However, the connection of hBN platelets is not satisfactory when the cell size within composite matrix is relatively small (15 μm). In contrast, when the PMC foams have a few large cells as showed in Figure 2(d), the alignment and connection of hBN platelets along the walls of large cells are significantly improved. As a result, the networking of thermally conductive hBN platelets is promoted, and thereby leading to higher k_{eff} of PMC foams over their solid counterparts.

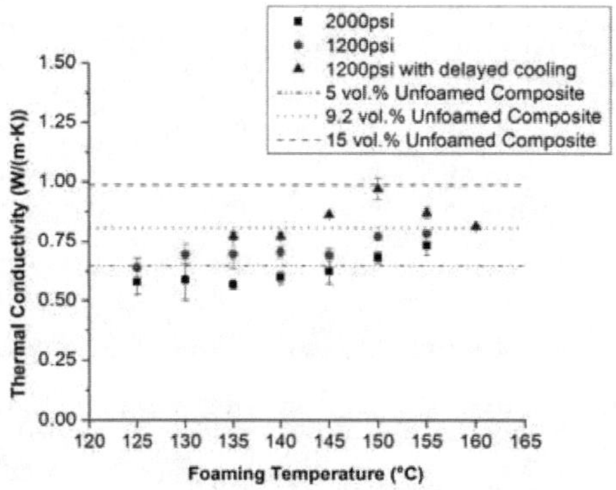

FIGURE 4. The k_{eff} of UHMWPE-hBN composite foams fabricated at different foaming and cooling conditions

CONCLUTIONS

This paper reports the fabrication of thermally conductive PMC foams composed of UHMWPE and hBN by physical foaming process. The k_{eff} of PMC foams loaded with 9.2vol.% hBN reached as high as 0.97 W/(m·K), which represented 20% increase over their solid counterparts. The results of this study is inspiring for the development of PMC with low filler loading in high-performance heat management applications.

REFERENCES

1. Y. P. Mamunya, V.V. Davydenko, P. Pissis and E. V. Lebedev, *Eur. Polym. J.* **38**, 1887-1897 (2002).
2. S. Kume, I. Yamada, K. Watari, I. Harada and K. Mitsuishi, *J. Am. Ceram. Soc.* **92**, S153-S156 (2009).
3. Z. Han and A. Fina, *Prog. Polym. Sci.* **36**, 914-944 (2011).
4. A. Yu, P. Ramesh, X. Sun, E. Bekyarova, M. E. Itkis and R. C. Haddon, *Adv. Mater.* **20**, 4740-4744 (2008).
5. S. N. Leung, M. O. Khan, H. E. Naguib and F. Dawson, *Appl. Phys. Lett.* **104**, 081904 (2014).
6. C. A. Martin, J. K. W. Sandler, A. H. Windle, M. K. Schwarz, W. Bauhofer, K. Schulte and M. S. P. Shaffer, *Polymer* **46**, 877-886 (2005).
7. S. N. Leung, M. O. Khan, E. Chan, H. E. Naguib, F. Dawson, V. Adinkrah and L. Lakatos-Hayward, *J. Appl. Polym. Sci.* **127**, 3293-3301 (2013).
8. H. Ding, Y. Guo and S. N. Leung, *J. Appl. Polym. Sci.* **133**, 42910 (2016).
9. B. Aydinli and T. Tincer, *J. Appl. Polym. Sci.* **127**, 1077-1083 (2013).
10. C. Xin, Y. He, Q. Li, Y. Huang, B. Yan and X. Wang, *J. Appl. Polym. Sci.* **119**, 1275-1286 (2011).
11. A. Mierczynska, J. Friedrich, H. -E. Maneck, G. Boiteux and J. K. Jeszka, *Cent. Eur. J. Chem.* **2**, 363-370 (2004).

Effects of saturation temperature/pressure on melting behavior and cell structure of expanded polyethylene bead

Peng Guo*, Yaohui Xu, Mingfu Lu and Shijun Zhang

Beijing Research Institute of Chemical Industry, Sinopec, Beijing 100013, P. R. China

guopeng.bjhy@sinopec.com

Abstract. The expanded polyethylene (EPE) beads with microcellular structure were prepared by an autoclave-based batch foaming process using non-supercritical CO_2 as the foaming agent. Herein, smaller PE micropellets (~1mm) were obtained from linear low density polyethylene (LLDPE) copolymerized with 1-hexene and 1-butene resin by underwater micro-pelletizer system. The results from differential scanning calorimeter indicated that the double melting peaks of beads moved to higher temperature with increasing saturation temperature and pressure. The kind of comonomer exerts noticeable influence on the double melting behaviors. The relationship between cell morphology and saturation temperature/pressure of EPE beads are discussed preliminarily by scanning electron microscope.

Keywords: expanded polyethylene; melting behavior; cell structure
PACS: 83.80.Iz

INTRODUCTION

Polyethylene (PE) are homopolymer of ethylene or copolymer of ethylene with copolymerized α-olefin. Since Pechmann casually discovered polyethylene in 1898, it has been arousing a considerable interest worldwide for its great performance such as toughness, flexibility and relative transparency, and used in various applications. Considering increasing industry requirements, it is estimated that the demand and capacity will attain 94 M and 105 M tons until 2015, respectively [1]. Among various PE, linear low density polyethylene (LLDPE) is a totally linear polymer with significant numbers of short branches, commonly made by copolymerization of ethylene with short-chain α-olefin including 1-butene, 1-hexene and 1-octene. Among them, ethylene-1-hexene copolymer is regarded as better cost performance due to more excellent properties than ethylene-1-butene and lower monomer price compared to ethylene-1-octene copolymer [2].

The unique structure of ethylene-1-hexene copolymer is potentially beneficial to preparation of expanded polyethylene including gas-loading, cell stability, and free from cell coalescence. Till now, as a kind of important commercial foaming product, expanded polyethylene (EPE) beads have been focused on and investigated by industry and research group [3]. However, few reports about effects of foaming parameters on the melting behavior and morphology of EPE obtained with autoclave processing. Thus, we will introduce a kind of preparation method for LLDPE copolymerized with 1-hexene (name after H-LLDPE) with longer short branched chain and higher melt strength. Then, we use this H-LLDPE and common commercial LLDPE (utilized 1-butene as monomer, named after B-LLDPE) as basic resin to prepare the EPE beads with autoclave-based batch process. Herein, the EPEs prepared from H-LLDPE and B-LLDPE are named after H-EPE and B-EPE, respectively. Furthermore, we will investigate the effects of saturation temperature and pressure on melting behavior and cell structure of H-EPE and B-EPE simply.

Experimental

Materials

Two kinds of PE were selected to study in this work. H-LLDPE is a kind of LLDPE copolymerized with 1-hexene and B-LLDPE is a kind of LLDPE copolymerized with 1-butene. All of them were obtained from Beijing Research Institute of Chemical Industry, SINOPEC. The crystallinity and melting temperatures of PE and EPE were measured by using differential scanning calorimetry (DSC, Perkin Elmer). DSC test procedure applied the three processing steps of heating/cooling/heating at the same rate of 10 °C/min by using a DSC-7 apparatus (Perkin Elmer) calibrated with indium. Carbon dioxide (CO_2, purity: 99.99% w/w) was purchased from Longhui Air Products Co., Beijing and used as received. The section images of EPE beads are characterized by scanning electron microscope (SEM FEI SL-30, USA).

Preparation of EPE beads

An extruder and an underwater micro-pelletizer system (BKG Labline 100, Germany) are used to produce the PE micropellets with ~0.6mm in diameter [4]. Herein, H-LLDPE and B-LLDPE are utilized as base resins in the preparation of micropellets, and talcum powder is used as cell-nucleating agents. Furthermore, other functional additives such as antistatic agent and anti-oxidants are mixed in the system at the same time. As-prepared micropellets were sieved by vibrating screen to remove the undersized and oversized pellets.

Following the preparation of the PE micropellets, the next step is the autoclave foaming process. The pellets are mixed with suspension stabilizers and water in a pressure vessel. CO_2 is introduced to the chamber to infiltrate the PE micropellets floating in the water for about a half hour at a certain temperature. This temperature is around melt point of PE according to previous report [4]. After the saturation of gas in the pellets, the pellets are released from the pressure vessel rapidly so that they can expand due to the thermodynamic instability. The EPE beads are washed by deioned water due to a number of suspension stabilizers on the surface of PE beads. Then foam bead are dried at 70 °C for 6 hours in the oven. Finally, the dried beads are sent into an autoclave, and stayed until the internal cell pressure reaches the atmospheric pressure for other tests.

Results and Discussion

Effect of saturation temperature/pressure on melting behavior

Figure 1 exhibits the melting point (T_m) of H-LLDPE and B-LLDPE. It is noteworthy that H-LLDPE exhibits lower melting temperature compared with that of B-LLDPE. This melting point depression could be attributed to increasing mole fraction of the short chain branch comonomer in metallocene catalyzed polyethylene [5]. The low temperature peak of H-LLDPE is ascribed to the long branching structure due to the addition of 1-hexene.

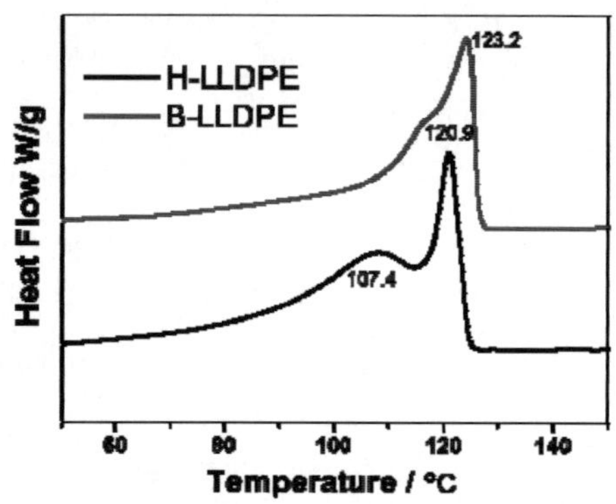

Figure 1. DSC thermograms of H-LLDPE and B-LLDPE at the first heating process.

First of all, we investigate the effects of saturation temperature on the melting behavior as shown in Figure 2. All the DSC curves exhibit two obvious melting peaks: a sharp high melting peak (where temperature noted as T_{mhigh}) and a broad low melting peak (where temperature noted as T_{mlow}). Figure 2 exhibited that the T_{mhigh} was 10~16 °C higher than melt point (T_m), and T_{mlow} was 8~11 °C lower than T_m. And the total crystallinity of H-EPE and B-EPE were higher than PE micropellets (H-LLDPE and B-LLDPE). Herein, in order to figure out the crystalline degree, the enthalpy of crystallization per gram of 100% crystalline PE is considered as 288 J/g [6]. With the increment of temperatures, the crystallinity of two PEs increased. It can be ascribed to the annealing of the molecules at higher saturation temperatures that increased the crystal growth with more perfection and close-packed structure. In short, considering the energy conservation and better sintering of EPE beads during steam-chest molding process (commercial method to obtain EPE products), lower melting temperature and wide range of low temperature peaks is in favor. Thus, when saturation pressure is constant, lower saturation temperature is suggested for these two kinds of EPE beads.

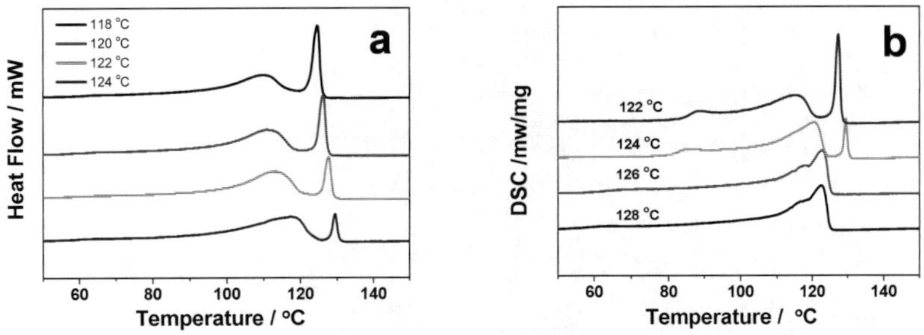

Figure 2. DSC curves of EPE at the same saturation pressure: a) H-EPE; b) B-EPE

From Figure 3, we investigated the effect of various saturation pressures on double crystal melting peak at 121 (around T_m of H-EPE) and 123 °C (around T_m of B-EPE) while the saturation time was fixed as 30 min. By increasing the saturation pressure from about 3 to 6 MPa, the onset temperature of crystallization peak could be depressed significantly by the compressed CO_2, especially at high pressures [7]. Thus, the presence of compressed CO_2 might provide us an opportunity to conduct the melt foaming process at much lower temperatures while keeping PE from crystallizing. Thus, increasing saturation pressure equaled to decreasing saturation temperature. So the foaming temperature could decrease with enlarged saturation pressure. PE micropellets with higher melt strength would keep the formed bubbles from bursting during foaming to produce PE foams with high expansion ratio. And the plasticization of CO_2 dissovled in melt during the saturation time contributes to melt the rest of unmelted crystals. Therefore, fewer unmelted crystals reformed to construct higher perfect crystals and thicker lamellas with a higher melting peak. Hence, the amount of crystals formed during cooling as the low temperature melting peak increased.

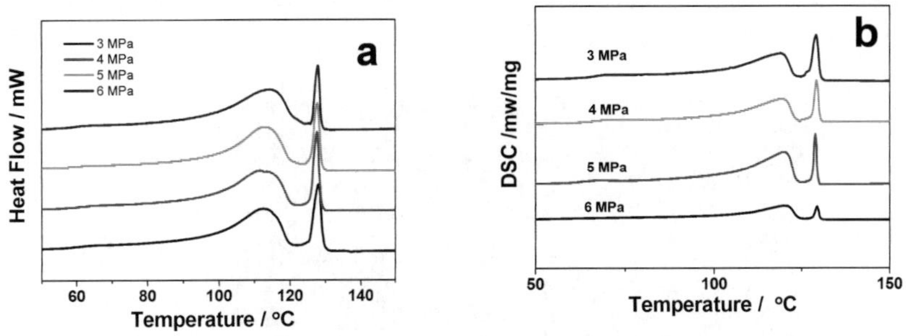

Figure 3. DSC curves of EPE at the same saturation temperature: a) H-EPE; b) B-EPE

Furthermore, as displayed in Figure 3, at around 6 MPa saturation pressure, the major portion of the crystals formed during cooling as the low melting peak. The increasing saturation pressure was useful to promote more crystals melting, thus the intensity of high temperature melting peaks decreased in the meantime. The results represented that the plasticization of CO_2 could compensate for the decreased saturation temperature.

Effect of saturation temperature on cell structure

In order to further investigate the effects of foaming process on the preparation and properties of PE foam, we prepared EPE beads by auto-clave process with H-LLDPE and B-LLDPE (7042). The saturation temperature exerts significant influence on the cell nucleation during PE foaming. Moreover, the intrinsic molecular structure of LLDPE also determined the cellular structure of as-obtained EPE beads. Thus, we can investigate the effects of saturation temperature on the preparation and morphology of EPE beads with comparing the cellular morphology of two kinds of EPE.

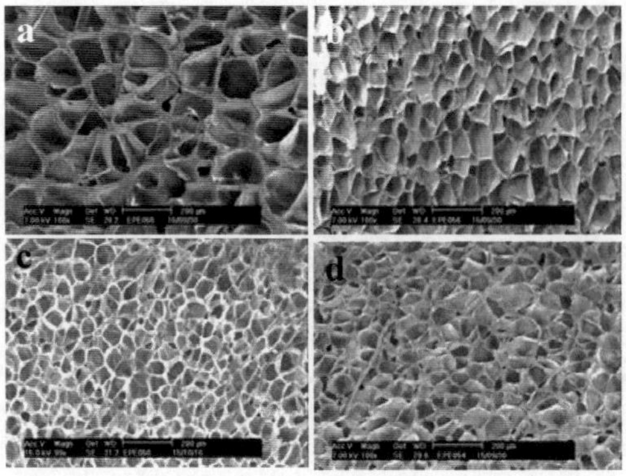

FIGURE 4. The cellular morphology of the H-EPE at 5MPa (a)116°C; (b)118°C; (c)120°C; (d)122°C.

The effects of saturation temperature on the properties of H-EPE at the same saturation pressure (5MPa) are studied in Figure 4. The cell diameter first decreased and then enlarged. Inversely, cell density first enhanced and then reduced. It is explained that with increase of temperature cell nucleation enhanced and then melt strength decreased at higher temperature. When saturation temperature is low, H-LLDPE micropellet has a higher surface tension and melt strength. Herein the energy barrier is too high so that cell nucleation could not generate in a large scale (Figure 4a). With increase of temperature, both surface tension and melt strength decreased. The small and intensively distributed cells are obtained because CO_2 molecular in each cell also decreased (Figure 4b). When saturation temperature attains 120 °C (melt point), most large and well-distributed cells are obtained as seen in Figure 4c. It might be explained that most of crystalline has been melt so that more CO_2 molecule could dissolve in the most domain consisting of PE chains. With further increase of saturation temperature, the decreased melt strength make much more CO_2 get into pellets leading to formation larger cells and thinner cell wall when saturation temperature attains 122 °C (Figure 4d).

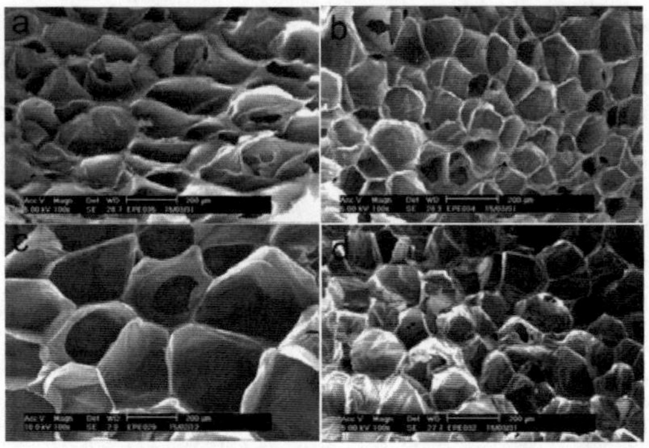

FIGURE 5. The cellular morphology of B-EPE at 5MPa (a)120°C; (b)122°C; (c)124°C; (d)126°C.

In the meantime, we also investigate the effects of saturation temperature on the cell structure for B-EPE in Fig. 5. We can find that the cell distribution of EPE from LLDPE is not as well as those of H-EPE. It might be ascribed that diffusion of carbon dioxide in the B-LLDPE micropellets is worse than that in the H-LLDPE. Lower melt strength is harmful for the preservation of gas in the bubble. Through comparing section photos of two kinds of EPE obtained at higher temperature, we found that obvious cell coalescence and rupture for B-EPE. In addition, B-EPE exhibits larger cell size compared with that of H-EPE. It is suggested that oversized cells do not facilitate the improvement of mechanical properties. Thus, molecular structure is significant factors for the growth and formation of cells in the EPE. It is suggested that melt strength of H-LLDPE is not sensitive to temperature.

Conclusions

This paper introduced the preparation of expanded polyethylene (EPE) beads from two kinds of LLDPE by an autoclave-based batch foaming process using non-supercritical CO_2 as the foaming agent. This H-LLDPE exhibits a longer chain branching structure and a narrow molecular weight distribution. H-LLDPE indicates higher melt strength than that of traditional B-LLDPE (7042). Through preparation of EPE, the effects of foaming temperature on the morphology and melting behavior of EPE are studied preliminarily. Through investigating the processing technologies of EPE, we find that for obtaining the EPE bead with higher cell density and smaller diameter (microcellular structure - potentially better mechanical properties), the moderate saturation pressure (about 5 MPa) and temperature (around melt point) is referred. Through investigating two kinds of EPE, the results indicates that melt strength and molecular structure is key factor for preparation of EPE beads. The initial results can contribute fundamental data for developing the industrial technique in the field of LLDPE and EPE.

ACKNOWLEDGMENTS

The authors want to thank Sinopec for permission to publish this original research article. This work was supported by SINOPEC (No. 214093, No. G6001-14-ZS-0464).

REFERENCES

1. P. S. Chum and K. W. Swogger, *Prog. Polym. Sci.* **33**, 797–819 (2008).
2. F.M. Mirabella Jr. and E. A. Ford, *J. Polym. Sci. B* **25**, 777–790 (1987).
3. A. H. Behravesh, C. B. Park, and E. K. Lee, *Polym. Eng. Sci.* **50**, 1161–1167 (2010).
4. P. Guo, Y. Xu, M. Lu, and S. Zhang, *Ind. Eng. Chem. Res.* **54**, 217–225 (2015).
5. R. G. Alamo, B. D. Viers, and L Mandelkern, Macromolecules 26, 5740-5747 (1993).
6. J. M. Barrales-Rienda and J.M.G. Fatou, Polymer, 13, 407-411 (1972).
7. C. B. Park, D. F. Baldwin and N. P. Suh, Polym. Eng. Sci. 35: 432-440 (1995).

Depressurization and Warpage in Glassy PMDI Foams

Kevin N. Long[a1], Christine C. Roberts[a], Lisa A. Mondy[a], Rekha R. Rao[a]

[a]*Engineering Sciences Center, Sandia National Laboratories, Albuquerque NM USA*
knlong@sandia.gov, ccrober@sandia.gov, lamondy@sandia.gov, rrrao@sandia.gov

Abstract. We investigate how manufacturing conditions result in the warpage of moderate density PMDI polyurethane foam (12-40 lb/ft^3). Previous efforts hypothesized that the manufacturing stresses and viscoelasticity of the matrix were the main drivers of dimensional change in glassy foamed parts. In this work, we now examine the effects of over-pressured CO_2 in the foam microstructure that originates from the simultaneous foaming and curing reactions, the role of density variation, and the associated model characterization. We present experimental measurements of specific components made with realistic manufacturing conditions and their long-term warpage behavior. After the initial manufacturing stresses, gas depressurization appears to be an important mode of deformation, giving the correct time scale of warpage compared to experimental data. Depressurization can be straightforwardly incorporated into our finite element cradle-to-grave framework for chemically blown foams, and we discuss varying levels of model complexity and their associated predicted responses.

Keywords: Coupled Physics Modeling, Bubble Depressurization, Shape Stability, and Mechanical Behavior
PACS: 44, 46, 47, 83

INTRODUCTION

Dimensional changes over time can be problematic for PMDI polyurethane foamed components with geometric tolerance requirements. We are working to develop an understanding of the connection between manufacturing conditions and dimensional stability, to one day mitigate it. Several rigid and flexible foams have been examined in the literature in the context of aging and dimensional change [1-4], with a variety of proposed mechanisms: depressurization of excess blowing agent, residual stress and viscoelasticity of the matrix, additional (thermoset) curing, and humidity induced swelling. We are interested in rigid PMDI foams for encapsulation and structural components. These exhibit post-manufacture warpage over times scales of at least a year [1]. Polystyrene foams have also seen dimensional changes over similar time scales [2].

The PMDI foams of interest are glassy at room temperature (T_g >100°C); thus we anticipate dimensional change due to additional matrix curing to be negligible. Moreover, our experiments maintain a desiccated environment to remove moisture effects. We are left with two viable mechanisms: viscoelasticity and bubble depressurization of the blowing agent. Outside the scope of this paper, we are studying the former, although the time scale of relaxation appears to be long compared to one year. Here, we focus on the depressurization mechanism. We note that effective CO_2 diffusion in polyurethane foams, for the thicknesses of our parts, produce time scales similar to our experimental measurements. We conjecture that spatial variations in density and CO_2 bubble pressurization result in spatially dependent warpage [2]. Our approach differs from the cited literature in two ways: first, we develop a simple micromechanics model that connects density and gauge pressure to the (local) macroscale warpage, and extend this model to a 3D constitutive equation useful for finite element simulation. Second, we control the manufacture and then measure the dimensional change over the course of 200 days of a foam geometry with both thick and thin regions. We compare theory and measurements where possible

A SIMPLE MICROMECHANICS MODEL FOR DEPRESSURIZATION IN FOAMS

We seek a simple means of estimating the extent of warpage possible due to gas depressurization as a function of excess CO_2 pressure, foam density, and matrix phase mechanical properties. The model is based on the idea that excess pressure in the foam bubbles diminishes as the gas phase pressure equilibrates with the surrounding ambient

[1] Sandia National Laboratories is a multi-program laboratory managed and operated by Sandia Corporation, a wholly owned subsidiary of Lockheed Martin Corporation, for the U.S. Department of Energy's National Nuclear Security Administration under contract DE-AC04-94AL85000. Document Report Number SAND2016-3234 C.

pressure over time. The time scale is set by CO_2 diffusion out of the gas bubbles though the solid matrix and into the surrounding environment. (Note, counter diffusion of air into the bubbles occurs but does not lead to over-pressurization.) Under suitable conditions, this phenomenon is a gas diffusion problem at the component length scale with an effective diffusion coefficient. However, our focus in this work is on the associated mechanical consequences of gas leaving the component. We ignore the transport time scale in this section and estimate the net depressurization warpage subject to three simplifying assumptions: uniform density, uniform excess gas pressure, and statistically isotropic foam microstructure at the macroscale. Later, we will relieve the first two assumptions by recasting this theory in 3D constitutive form. Under these simplifying assumptions, the net depressurization strains are volumetric and the same everywhere in the component. To compute this volume strain response, we idealize the foam microstructure with a spherical representative volume (RV) illustrated in Figure 1(a), and we compute the net volume change between the pressurized and depressurized configurations.

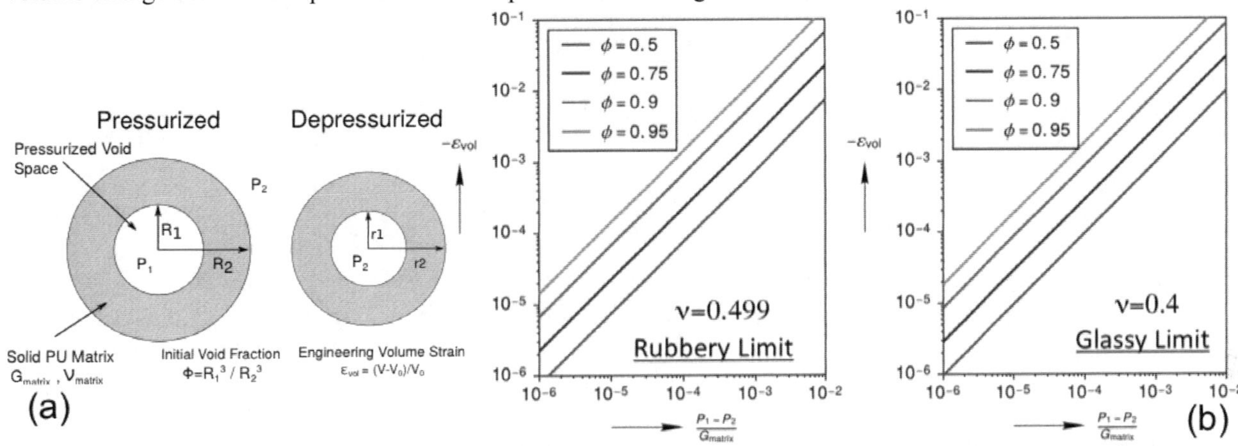

FIGURE 1. (a) Spherical representative volume of the average foam microstructure. The RV illustrates the key model parameters: void volume fraction, matrix phase properties, and initial gauge pressure. (b) Map of the net volume shrinkage due to depressurization for typical rubbery and glassy foams associated with the RV in (a) and its underlying assumptions.

Our depressurization problem is a re-interpretation of the pressurization of an elastic shell that is standard in the literature [5]. The shell deforms in a spherical symmetric manner with a radial displacement field characterize by:

$$u = A_1 + \frac{A_2}{R^3}, \qquad A_1 = \frac{P_2R_2^3 - P_1R_1^3}{3K_{matrix}\left(R_1^3 - R_2^3\right)}, \qquad A_2 = \frac{R_1^3R_2^3\left(P_2 - P_1\right)}{4G_{matrix}\left(R_1^3 - R_2^3\right)}. \tag{1}$$

Here, K_{matrix} and G_{matrix} are the bulk and shear moduli of the matrix phase. Evaluating the radial displacement field at R_2 and asserting that the deformation is spherically symmetric, one can compute the volume strain associated with full depressurization of this RV:

$$\lambda_R = 1 + \frac{u|_{R_2}}{R_2} = 1 + \left(\frac{4G_{matrix}\left(P_1\phi - P_2\right) - 3\phi\left(P_1 - P_2\right)}{12G_{matrix}K_{matrix}\left(1-\phi\right)}\right), \qquad \varepsilon_{vol} = \log\left(\lambda_R^3\right) \tag{2}$$

We reduce the depressurization response to three dimensionless variables: the ratio of the initial gauge pressure to the shear modulus $(P_1 - P_2)/G_{matrix}$, void volume fraction $\phi = R_1^3 / R_2^3$, and matrix phase Poisson Ratio ν. A map of the net volume strain magnitude due to warpage is provided in Figure 1(b) for the bounding cases of typical rubbery and glassy foams. From the map in Figure 1(b), we make several observations. First, in the small strain limit of the RV, the extent of warpage is directly proportional to the dimensionless gauge pressure for the different foam densities considered. Second, foam density has a significant impact. Low-density foams may warp orders of magnitude more than higher density foams for the same dimensionless gauge pressure. Finally, the effect of the matrix Poisson's ratio is secondary relative to the gauge pressure.

EXTENDING THE MODEL TO A 3D CONSTITUTIVE RELATIONSHIP

We relieve two of the three overly simple assumptions of the previous section by extending the analytic representation of foam depressurization to a 3D macroscale constitutive relationship suitable for finite element analysis. The present formulation still assumes isotropic behavior at the macroscale. Consider the motion of a foam body that carries material points within it from a reference state, which we will take to be the undeformed state just after manufacture of the component, to the current configuration of the body at a later time. We assume that the gradient with respect to the referential coordinates of the current position (the deformation gradient) may be multiplicatively split into volumetric and isochoric pieces, and we compute the symmetric rate of deformation following standard methodology [5].

$$\mathbf{F} = \mathbf{F}^{iso}\mathbf{F}^{vol}, \qquad \mathbf{D} = \mathbf{D}^{iso} + \mathbf{D}^{vol} \tag{3}$$

Following the previous section, RV depressurization determines only a volume strain response. We therefore further decompose the volumetric portion of the deformation gradient and rate of deformation into thermal, mechanical (stress producing), and depressurization contributions:

$$\mathbf{F}^{vol} = \lambda_\theta \lambda_M \lambda_d \mathbf{1}, \qquad \mathbf{D}^{vol} = \left(\frac{\dot{\lambda}_\theta}{\lambda_\theta} + \frac{\dot{\lambda}_M}{\lambda_M} + \frac{\dot{\lambda}_d}{\lambda_d} \right) \mathbf{1} = \mathbf{D}_\theta^{vol} + \mathbf{D}_M^{vol} + \mathbf{D}_d^{vol}. \tag{4}$$

The key ingredient is the depressurization rate of deformation. Using the RV from the previous section, we compute the derivative of the radial displacement with respect to the gauge pressure, which itself is a function of time t:

$$\dot{\lambda}_d = 1 + \frac{1}{R_2} \frac{d\left(u|_{R_2}\right)}{dP_1} \frac{d(P_1 - P_2)}{dt}, \qquad \frac{d\left(u|_{R_2}\right)}{dP_1} = \frac{3\phi\left(4G_{matrix} + K_{matrix}\right)R_2}{4G_{matrix}K_{matrix}(1-\phi)}. \tag{5}$$

The chain rule in equation (5) relies on the assumption that P_2 is constant, which we assign to zero. We recognize that a different choice may be more appropriate. Next, consider an isotropic thermal model with an empirical linear thermal expansion coefficient, α_L , which now allows the calculation of the mechanical rate of deformation:

$$\mathbf{D}_\theta^{vol} = \alpha_L \dot{\theta} \mathbf{1}, \qquad \mathbf{D}_m^{vol} = \mathbf{D} - \mathbf{D}_d^{vol} - \mathbf{D}_\theta^{vol}. \tag{6}$$

We assume the foam behaves macroscopically as an isotropic elastic material with elastic constants derived from average properties of the Weaire-Phelan microstructure rather than from the RV in Figure 1(a) [6]:

$$\dot{\sigma} = 2G_{WF}\, \text{dev}\mathbf{D}_M + K_{WF}\left(\text{tr}\mathbf{D}_M\right)\mathbf{1}, \qquad G_{WF} = 0.113(1-\phi)E, \qquad K_{WF} = 0.434(1-\phi)E \tag{7}$$

Here, E is the matrix phase Young's modulus and $\dot{\sigma}$ is the time derivative of the macroscale Cauchy stress, which is not objective, and so we use the Green-Naghdi stress rate in conjunction with equation (7) to satisfy objectivity of the momentum balance (as is standard for a hypoelastic model). We note that the Weaire-Phelan microstructure produces mildly anisotropic elastic constants. Here, we have averaged the [111] and [100] directions. Moreover, these constants apply only to the low relative density limit $(1-\phi) \sim 0$. Since both thermal expansion and depressurization produce volumetric deformation in this model, all other motion is stress-producing in equation (7). The constitutive model inputs are: $\{G_{matrix},\ K_{matrix},\ \alpha_L,\ f_0,\ (P_1 - P_2)\}$. Other elastic constants can be derived from the matrix shear and bulk moduli. The last two quantities vary spatially, and the last varies in time. While the first three may be uniform for the foam, the last two must be initialized and provided to the constitutive model everywhere in the foam component as a function of time if one seeks to capture the time and spatial dependence of the gas depressurization and warpage process. Although future plans are to predict these spatial dependencies using a model for foam blowing and cure, here we discuss initialization based on a model experiment in the next section.

MEASUREMENTS OF WARPAGE IN A TEST GEOMETRY OVER 200 DAYS

Experimental Conditions

A carefully controlled experiment was devised to provide long-term observations of foam shrinkage. Experimental conditions aimed to create a foam part with both thin and thick sections: a staple shaped geometry was chosen as illustrated in Figure 2(b). The mold is primarily aluminum with an acrylic faceplate that allows visualization of the flow path of the material. Foam is injected with the mold in a "U" orientation at its base so that it reacts, expands, and rises symmetrically into the two points. Vents are located at the point tips to allow air to escape as it is displaced. Six calibrated thermocouples (Omega, K-type) and two pressure transducers (Omega, T900) are placed at various locations in the mold flush with the wall (Figure 2(a)) so that the temperature and pressure of the foam is recorded as it reacts exothermically to fill the mold.

A polyurethane foam kit (BKC 44306 PMDI-10, obtained from the National Security Center, Kansas City, Mo) with a free-rise theoretical density of 10 lb/ft^3 is used. This formulation initially consists of separate resin and curative liquids that are mixed and injected into a pre-heated mold that is temperature controlled in a 40°C oven throughout the experiment. Enough material is injected to slightly over-pack the mold to 13 lb/ft^3. The flow path of the foam rising into the mold is videoed through a window in the oven door. Once the foam has finished expanding, the acrylic faceplate is replaced with an aluminum plate and the sample is cured at 120°C for 4 hours per standard procedure. Specimens are then removed from the mold, cooled to room temperature, bolted to a fixture to stand upright, and stored in a dry environment for long periods of time. Periodically the specimen dimensions are measured using a Zeiss Prism Coordinate Measurement Machine (CMM) using a 100 mN probe force. This instrument has a calibrated linear accuracy of 5 μm for the 100 mm length of the base of the geometry. Traces along both the female and male portions of the specimen were made at five separate thicknesses. Additionally, an X-ray CT scan was made of the specimens (North Star Imaging) in order to measure their interior density and identify voids.

Results

Triplicate specimens were foamed in the mold with good reproducibility of the temperature and pressure profiles. As expected, the base of the staple, which is reached by foam first and is the thickest part of the geometry, reaches the highest temperatures (90°C peak temperature). A slight temperature rise is also observed at the foam tips (50°C). Gauge pressures were observed up to 14 psig at the mold walls; pressures and temperatures are predicted to be higher internally to the sample. X-ray CT images (Figure 2(c)) identify large bubbles that are primarily located in the two upright arms. A dense skin reaching over 20 pcf is also observed at the mold walls as in the front image. Coalescence and formation of large bubbles has been observed previously in regions of high shear in other experiments. A greater packing fraction for this foam would reduce these defects. Finite element simulations were performed on this geometry from the room temperature state onward using the X-ray CT density. We used the peak measured pressure, ~14 psig, from Figure 2(a), and applied it uniformly to the part though likely the internal initial gauge pressures are much higher and are functions of density. Typical glassy polymer matrix elastic constants were used ($G \sim 1$ GPa, $\nu \sim 0.4$). We simulated a uniform decay of the gauge pressure and fixed the collections of nodes corresponding to the bolts at the specimen center from the experiment. Similar measurements of dimension change at the waist and 80% along the staple tip were extracted. We found that warpage was an order of magnitude or more lower than experiments, likely a result of too simple an initial gauge pressure. We found that predicted warpage was very sensitive to that initial pressure. Qualitatively, the simulations are useful in that they show large regions tend to warp uniformly, but in thin regions, warpage varies spatially. The final warped shapes of the mid section and front are shown in Figure 2(d).

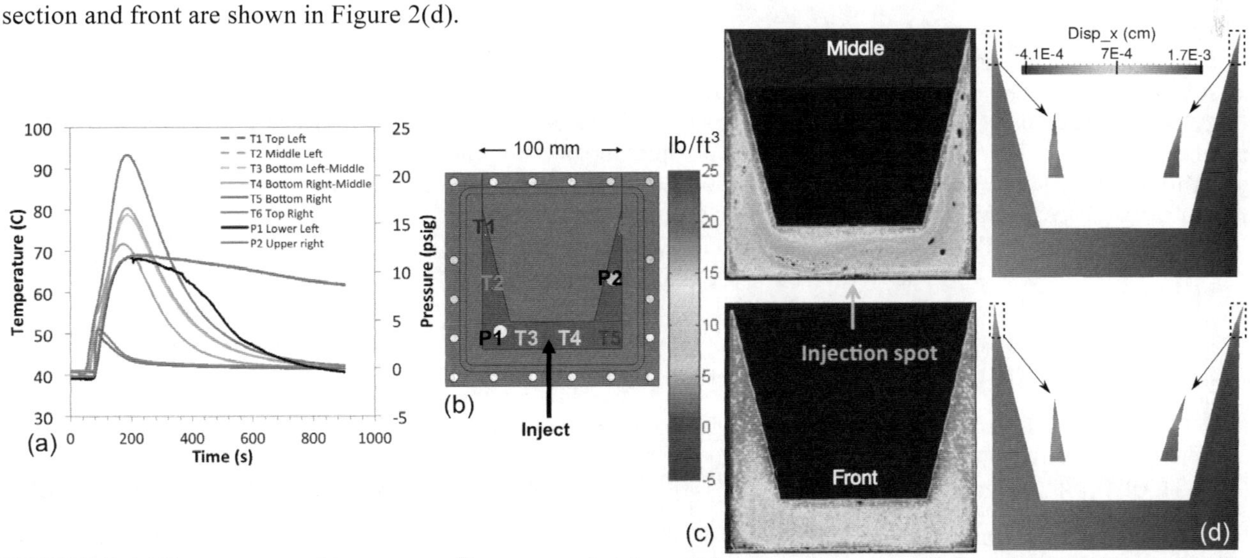

FIGURE 2. (a) Temperature and pressure profiles measured at the wall from locations in the staple mold geometry (a) showing locations of thermocouples (T), pressure transducers (P), and the injection port. Time is measured from the mixing time of the two components. (c) False color X-ray CT image of the density of a specimen halfway through the middle of the sample (into the

page), showing many large bubbles in the arms and a high density skin near the walls. (d) Corresponding FEA showing the fully depressurized warped state along the middle and front slices. Note the distinct warpage at the tips between the two slices.

The dimensional change of a foamed specimen is measured at both the base and 80% of the distance to the tips. The dimensional change at each location is displayed in Figure 3 with respect to the aluminum mold width at the cure temperature of 120°C. At the base, a large initial shrinkage of about 0.6% is observed at short times that is attributable to the thermal contraction of the sample. Then, over approximately 14 days, a slow contraction is observed that adds an additional 0.1% shrinkage. This contraction diminishes over time until it is approximately equal to the error in the measurement. The width of the staple at the arms of the specimen follow an unpredictable, unsteady movement. This movement could be due to the anomalous large bubbles contained within the arm or some other unknown factor.

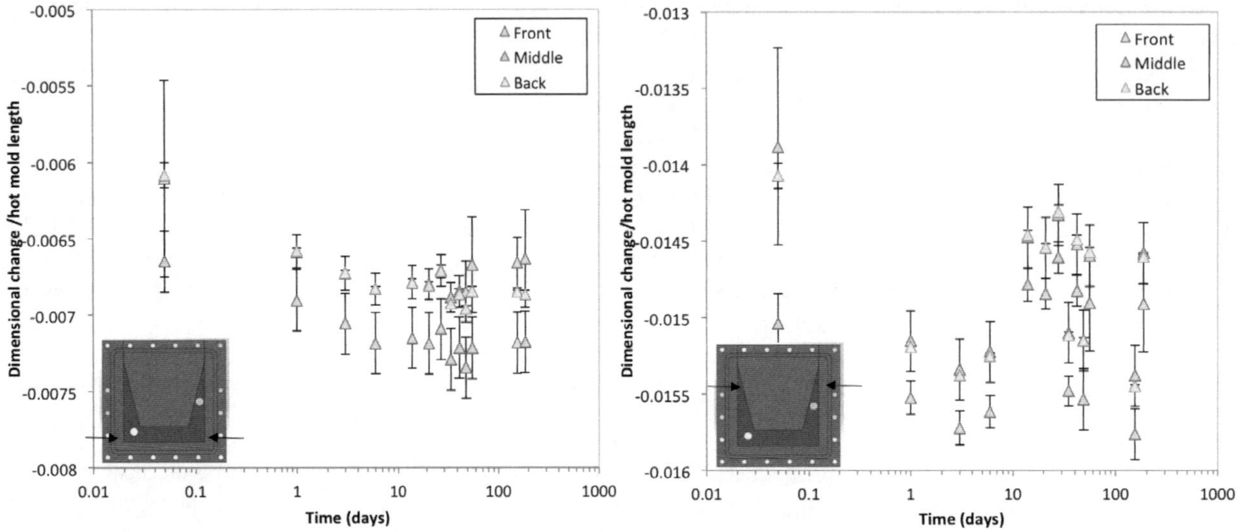

FIGURE 3. Staple width dimensional change normalized by the predicted mold width at the curing temperature (120 oC) for (left) the base of the staple and (right) 80% up the staple arms. Front, middle, and back refer to the location of the trace in the depth of the mold (into the page as pictured)

CONCLUSIONS

We have measured the dimensional change of a chemically blown PMDI 13 lb/ft^3 foam over the course of 200 days, and these measurements show that, despite large defects within the foam from processing, overall shrinkage is more uniform in thick regions than in thin regions. Accompanying FEA simulations using a constitutive model based on a simple micromechanics model of foam depressurization corroborates this finding. We note that under uniform density/gauge pressure conditions, the model is analytic, and it shows a direct trade offs between gauge pressure, foam porosity, and volume change due to depressurization.

REFERENCES

1. R. D. Gilbertson, B. M. Patterson, and Z. Smith. Accelerated aging of BKC44306-10 rigid polyurethane foam: FT-IR spectroscopy, dimensional analysis, and micro computed tomography. *LA-UR-14-20007*, Los Alamos National Laboratory, January 2014.
2. T. Fen-Chong, E. Herve, and A. Zaoui. Micromechanical modeling of intracellular pressure-induced viscoelastic shrinkage of foams: application to expanded polystyrene. *European Journal of Mechanics, A/Solids*, 18, 1999.
3. M. Rampf, O. Speck, T. Speck, and R. H. Luchsinger. Structural and mechanical properties of flexible polyurethane foams cured under pressure. *Journal of Cellular Plastics*, 48:53–69, 2011.
4. C. T. Yang, K. L. Lee, and S. T. Lee. Dimensional stability of LDPE foams: Modeling and experiments. *Journal of Cellular Plastics*, 38:113– 128, 2002
5. I. S. Sokolnikoff. *Mathematical Theory of Elasticity*. McGraw-Hill. 2nd Ed. 1956.
6. J. F. Sadoc and N. Rivier. *Foams and Emulsions*. Nato Science Series E: Applied Sciences Vol. 354. Springer, 1999.

Preparation and Characterization of Conductive Foams based on PBS, Carbon Nanofibers and Expanded Graphite Nanocomposites

Livia Cafiero[a], Maria Oliviero[a], Giovanni Landi[b], Luigi Sorrentino[a,*], Andrea Sorrentino[a,*], Heinz C. Neitzert[b], Salvatore Iannace[c]

[a]Istituto per Polimeri, Compositi e Biomateriali (IPCB-CNR), Consiglio Nazionale delle Ricerche
Piazzale E. Fermi 1, 80055 Portici (NA) - Italy
[b]Dipartimento di Ingegneria Industriale, Università di Salerno
Via Giovanni Paolo II 132, 84084 Fisciano (SA), Italy
[c]Istituto per gli Studi sulle Macromolecole (ISMAC-CNR), Consiglio Nazionale delle Ricerche
Via Corti 12, Milano 20133, Italy

[*]luigi.sorrentino@cnr.it

Abstract. Recently, conductive polymeric foams have aroused considerable research interest owing to their attractive properties associated with conventional polymers and unique electronic properties of metals or semiconductors. Large surface area, lower density and higher specific properties make them promising candidates for broad applications in energy conversion and storage, sensors, actuators, and biomedical devices. This work reports on the preparation and characterization of novel conductive polymeric foams based on a biodegradable polymer (Polybutylene succinate, PBS) and carbon nanoparticles (carbon nanofibres and expanded graphite). Foaming has been performed on PBS/CNF and PBS/CNF/EG nanocomposites using a batch process by using supercritical CO_2 as blowing agent. The control of foaming parameters allowed to prepare foams with tailored morphologies, and cellular structures with macro to micro sized cells were obtained. An in deep discussion about the general design rules, advantages, and also the actual limitations of such novel conductive polymeric foams are provided. Results demonstrate their potential applications as active electrode materials for next-generation biodegradable energy storage.

Keywords: Biodegradable Polymers; Foaming; Carbon Nanoparticles; Electrical Properties
PACS: 81.05.Rm, 81.05.Lg, 72.80.Tm

INTRODUCTION

The electrical conductivity of carbon based nanocomposites has been extensively investigated in the literature [1, 2]. The electric conduction in such composites can be explained by considering the percolation theory. Percolation theory deals with the effect of varying, in a random system, the number of interconnections present. The conduction in nanocomposites has been explained by considering that conductive paths, causing the material to convert from an insulator to a conductor, are formed in the composite when the filler concentration increases over a threshold value. The percolation theory describes the dependence of the conductivity σ on the filler concentration by a scaling law of the form $\sigma = \sigma_0 (X-X_c)$ t, where X is the actual filler content, X_c is the percolation threshold and t a factor depending on the system dimensionality [3-7]. However, a complete understanding of the relations among electrical properties, physical and chemical characteristics of the composite components (especially those pertaining to the interface between matrix and conducting inclusions) and the topological structures is still to be achieved. Therefore, additional efforts aimed at providing further information on the dependencies among electrical characteristics and the above mentioned parameters seems valuable.

Zhang et al.[1] studied the effect of the processing conditions on the morphology of PBS foams. Authors used ammonium bicarbonate as foaming agent and found that low density foam can only be obtained by adding talc as nucleating agent. Since conventional PBS grades do not show good foaming properties, efforts have been made to improve the elongational viscosity of the polymer to improve its foamability. Son et al.[9] obtained PBS foams using supercritical CO_2 as blowing agent. Authors used a reactive compounding technique to increase the molecular weight of PBS and its melt strength. Similarly, Lim et al.[10] used two types of polyisocyanate as branching agents into PBS matrix. Large cells were obtained, with average cell size not lower than 500 micron.

PBS in combination with PLA and MMT [11] and in blends with biodegradable gelatin [12] was foamed by using the extrusion process and CO_2 as blowing agent. The dispersion of nanometric fillers increased the mechanical and impact strengths of the cellular structure. In order to improve the mechanical behavior, nanocomposite matrices were investigated. PBS/clay systems [13] exhibited improved tensile strength and elongation at break (best at 2%wt), while the cell size was reduced down to 300 micron. PBS/carbon nanofiber [14] nanocomposites exhibited a different behaviour. The tensile strength was improved but elongation at break was significantly reduced with the increase of the filler content. Cell size of nanocomposite samples was reduced down to 100 microns, with blowing ratios comparable to the neat systems.

The aim of this work is to shed some light on the dielectric behaviour of biodegradable polymer/carbon nanoparticles after foaming.

EXPERIMENTAL

Materials

Carbon nanofibers (CNF) were purchased by Sigma-Aldrich (Italy). Expanded graphite (EG) was supplied by GrafTech International (TG-741 and its availability exclusively from GrafTech International Holdings, Inc.). PBS Bionolle 1903 produced by Showa High Polymers (Japan) was supplied by Toyo Plastics Co. Osaka, Japan. CO_2 (purity higher than 99.99%) was purchased from Air Liquide (Italy) and used as physical blowing agent for gas foaming experiments.

Nanocomposite and Foam Preparation

CNF, EG and PBS were melt-mixed as received by using an internal mixer (Rheomix 600 Haake, Germany) controlled by a measuring drive unit (Haake Polylab). Different compositions, reported in Table 1), were prepared at 120°C, 80 rpm for 6 min. A compression moulding machine was used to prepare specimens for characterizations and foaming experiments. Polymer slabs were moulded at 120°C under a pressure of 30 bar in a hot plate press (P300 Collin, Germany). Processing times were limited to 8-10 min in order to prevent any thermal degradation of polymers [36].

Foaming experiments were carried out by using a batch process by expanding disc-shaped samples having size of 16 mm in diameter and 1 mm in thickness. Samples were put into confining moulds made of polyamide to produce cylindrical samples suitable for dielectric measurements. In each experiment, samples were placed inside a high-pressure vessel that was firstly evacuated for about 5 min to minimize the presence of oxygen and humidity. After that step, the vessel was heated up to the sorption temperature (T_s = 122°C) and immediately filled with CO_2 up to the sorption pressure (P_s = 90 bar). Sorption lasted 24 hours to reach the solubilization limit. The vessel was then cooled to the desired foaming temperature (T_f = 95°C) and pressure was released to room pressure in a time interval of 55 seconds to allow sample foaming. Finally, the samples were immediately removed from the vessel and allowed to cool at room conditions in order to stabilize the cellular structure.

Sample Characterization

Electrical measurements of the PBS composites have been measured in the voltage range ± 100 V using a Keithley Picoammeter/Voltage source model "487" in both coplanar and non-coplanar configuration (Insets in Fig. 1 and 2). In coplanar configuration, measurements have been performed in two-point contact geometry. The film has been electrically contacted by using a silver paste to ensure a good ohmic contact with an average area (A) of 0.1 cm^2. All the measurements were performed at room temperature by using disk-shaped specimens of 2 mm thick and 15 mm in diameter. The foam density has been evaluated as the mean measured by using geometric parameters.

RESULTS AND DISCUSSION

Foams of PBS/EG and PBS/CNF were successfully prepared and medium densities were obtained. Considering that the density of the neat matrix is ρ = 1.26 g/cm^3 and ρ = 1.27 g/cm^3 for all PBS nanocomposites, the expansion ratio of PBS nanocomposites was comparable to that of the neat PBS and ranged between 3.0 and 4.2 (Table 1). The addition of expanded graphite to CNF did not hinder the foaming process, and in two cases allowed a significant increase of the expansion ratio. Foams morphologies are shown in Fig. 1. Large cells were developed in the used foaming conditions but a qualitative analysis shows that the addition of EG slightly reduced the mean cell size.

TABLE 1. Density of PBS foams for both mould heights.

Sample	Density (g/cm^3)	Expansion Ratio
Neat PBS	0.34	3.7
PBS+0.1 wt%EG	0.36	3.6
PBS+0.5 wt%CNF	0.36	3.5
PBS+1.0 wt%CNF	0.37	3.5
PBS+3.0 wt%CNF	0.37	3.4
PBS+0.5 wt%CNF+0.1 wt%EG	0.32	3.9
PBS+1.0 wt%CNF+0.1 wt%EG	0.43	3.0
PBS+3.0 wt%CNF+0.1 wt%EG	0.30	4.2

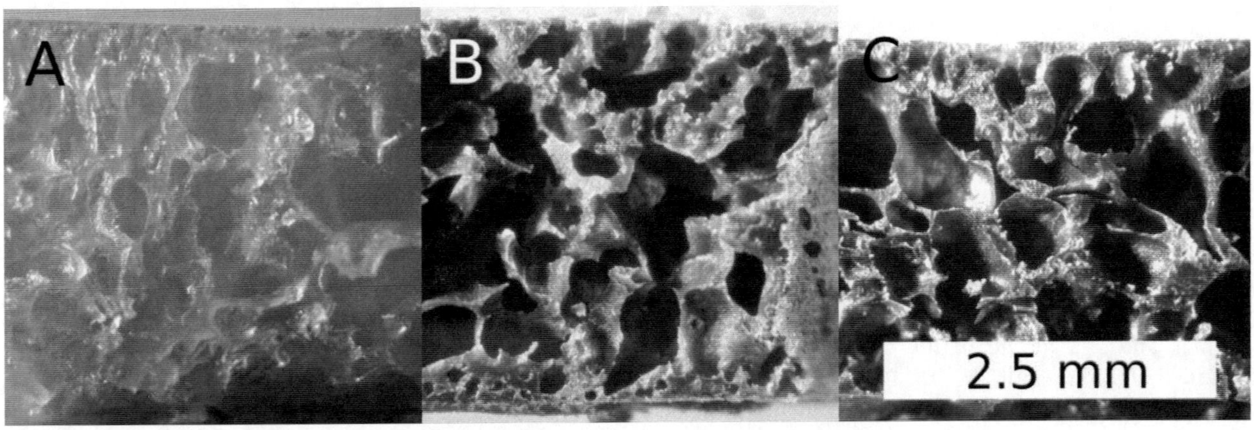

Figure 1. Foam morphologies from selected foams: A) Neat PBS, B) PBS+0.5 wt% CNF, C) PBS+0.5 wt% CNF+0.1 wt% EG

The coplanar conductivity (σ_c) versus the nanoparticle concentration for both series of PBS nanocomposite foams at room temperature is shown in Fig. 2. Both types of PBS nanocomposites exhibit the typical abrupt increase of the conductivity from the neat PBS value ($2.6 \cdot 10^{-15}$ S) to a very high one ($7 \cdot 10^{-5}$ S), in case of 3.0 wt % CNF content. This means that the percolation threshold for conductivity lies within 0.5 wt% and 3.0 wt% of CNF content. It is worth noting that the measured values of nanocomposite conductivities never reach the values obtained for dense films of CNFs (about 10^3 S/m), which is generally ascribed to the role of tunnelling resistance in the electrical conductivity of nanocomposites [15].

In particular, the surface percolation threshold for PBS/CNF nanocomposite foams can be estimated as 0.9 wt%. The conductivity of samples obtained by adding 0.1wt% of EG shows a broader transition. In that case, the estimated percolation threshold occurs at CNF content higher than 1.7 wt%. Such a phenomenon, already reported in the scientific literature for other nanocomposite systems, can be justified as a break-down of the conductive network induced by the presence of expanded graphite particles.

The non-coplanar conductivity (σ_n) shows a completely different behaviour with respect to the coplanar conductivity (Fig. 3). Although the conductivity of the nanocomposite foams increased with respect to the neat polymeric foam, all samples show almost constant non-coplanar conductivity by varying the CNF content. The addition of 0.1 wt% of EG induces an increase of about one order of magnitude in the electrical conductivity for all compositions. The higher difference between coplanar and non-coplanar conductivity has been detected in samples with higher CNF concentration (3.0 wt%), for which a nine order of magnitude gap has been observed.

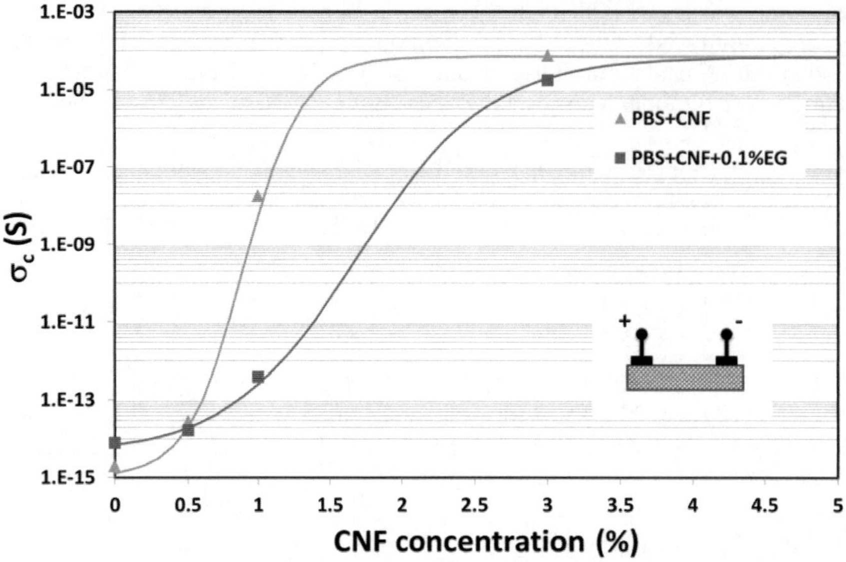

Figure 2. Coplanar conductivity versus CNF wt% concentration at room temperature for the nanocomposite foams.

Since the intrinsic properties of the matrix is constant in the two orthogonal measuring directions and the test conditions (geometric gap and electric contact) are comparable in both systems, such behaviour has to be ascribed to the presence of cellular structure. In particular, the spatial distribution of the nanofibres in the cell walls and struts is changed by the peculiarity of the foaming process, which induces a stretching of the polymer melt during the cell growth. Such stretching tends to align the nanoparticles in turn increasing the percolation threshold. The contact resistance between carbon nanofibers reduces in the cells walls and struts with respect to what happens in the bulk polymer and a lowering of the conduction capability results. This behaviour is mitigated by the presence of expanded graphite, which helps in increasing the conductive interconnections and hence in promoting the transport of electric charges through the cellular structure.

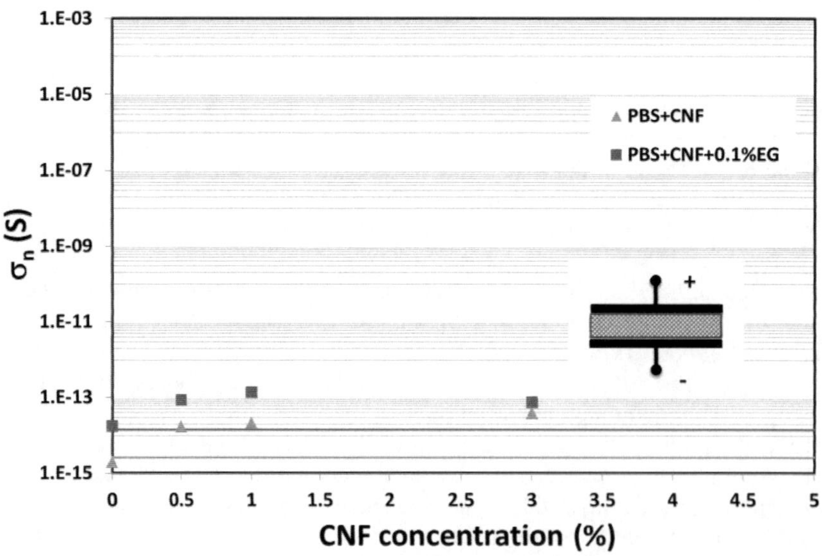

Figure 3. Non-coplanar conductivity versus CNF wt% concentration at room temperature. Solid lines represent the conductivity of the PBS and PBS+0.1 wt% EG matrices.

CONCLUSIONS

Conductive nanocomposites based on a biodegradable polymer (PBS) and carbon based nanoparticles (carbon nanofibres and expanded graphite) has been produced and successfully foamed. Medium density foams and expansion ratios higher than 3 have been obtained. The control of foaming parameters allowed to prepare foams with adequate morphologies and a consistent cellular structure.

Dielectric measurements demonstrate that significant differences between coplanar and non-coplanar conductivities are present in foams. This is due to the modification of the spatial arrangement of the nanofibres in the cell walls and struts with respect to the bulk polymer. In fact the foaming process induces a stretching of the polymer melt during the cell growth that tends to align the nanoparticles and to increase the percolation threshold. This behaviour is mitigated by the presence of expanded graphite, which helps in increasing the conductive interconnections and hence in promoting the transport of electric charges through the cellular structure.

ACKNOWLEDGEMENTS

The authors wish to acknowledge the BIOPOLIS project (Grant number: PON03PE_00107_1). Fabio Docimo (IPCB-CNR) is kindly acknowledged for the technical support.

REFERENCES

[1] Biagio De Vivo, Patrizia Lamberti, Vincenzo Tucci, Liberata Guadagno, Luigi Vertuccio, Vittoria Vittoria, Andrea Sorrentino. Comparison of the Physical Properties of Epoxy-Based Composites Filled with Different Types of Carbon Nanotubes for Aeronautic Applications. Advances in Polymer Technology, Vol. 31, No. 3, 205–218 (2012)

[2] Giuliana Gorrasi, Roberta Di Lieto, Giovanni Patimo, Salvatore De Pasquale, Andrea Sorrentino. Structure-property relationships on uniaxially oriented carbon nanotube/polyethylene composites. Polymer 52 (2011) 1124-1132. 10.1016/j.polymer.2011.01.008

[3] Kim YJ, Shin TS, Choi HD, Kwon JH, Chung YC, Yoon HG. Electrical conductivity of chemically modified multiwalled carbon nanotube/epoxy composites. Carbon 2005;43(1):23–30.

[4] Du F, Fischer JE, Winey KI. Effect of nanotube alignment on percolation conductivity in carbon nanotube/polymer composites. Phys. Rev. B 2005;72(12):121404-1, 4.

[5] Guadagno L, Naddeo C, Vittoria V, Sorrentino A, Vertuccio L, Raimondo M, et al. Cure behavior and physical properties of epoxy resin-filled with multiwalled carbon nanotubes. J Nanosci Nanotechnol 2010;10(4):2686–93.

[6] Munson-Mcgee SH. Estimation of the critical concentration in an anisotropic percolation network. Phys Rev B 1991;43(4):3331-3336.

[7] Oliviero Maria, Sorrentino Luigi, Cafiero Livia, Galzerano Barbara, Sorrentino Andrea, Iannace Salvatore. Foaming behavior of bio-based blends based on Thermoplastic Gelatin and Poly(butylene succinate). Journal of Applied Polymer Science, Volume 132, Numero 48 (2015).

[8] Y. Zhang, B. Lu, F. Lv, W. Guo, J. Ji, P. K. Chu and C. J. Zhang, Appl. Polym. Sci. 126, 756 (2012).

[9] J. M. Son, K. B. Song, B. W. Kang and K. H. Lee, Polymer (Korea) 36, 34 (2011).

[10] S. K. Lim, S. I. Jang, S. I. Lee, K. H. Lee and I. J. Chin, Macromol. Res, 16 (3), 218-223 (2008).

[11] P. Ma, X. Wang, B. Liu, Y. Li, S. Chen, Y. Zhang and G. Xu, Journal of Cellular Plastics 48, 191 (2012).

[12] M. Oliviero, L. Sorrentino, L. Cafiero, B. Galzerano, A. Sorrentino and S. Iannace, Journal of Applied Polymer Science 132, 48 (2015).

[13] S. K. Lim, J. J. Lee, S. G. Jang, S, I. Lee, K. H. Lee, H. J. Choi and I. J. Chin, Polym. Eng. Sci. 51, 1316–1324 (2011).

[14] S. K. Lim, S. I. Lee, S. G. Jang, K. H. Lee, H. J. Choi and I. J. Chin, Journal of Macromolecular Science, Part B: Physics 50, 100–110 (2011).

[15] Li C, Thostenson ET, Chou TW. Dominant role of tunneling resistance in the electrical conductivity of carbon nanotubebased composites. Appl Phys Lett 2007;91(22):223114-1, 3.

Effect of Foaming Conditions on PLA/Starch/Nano Calcium Carbonate Nanocomposite Foams Structure

A. Javadi[a*], *E. Vaziri Nasab*[a] *and H. Garmabi*[a]

[a]*Department of Polymer Engineering and Color Technology, Amirkabir University of Technology, Tehran, Iran*

*Email: ajavadi@aut.ac.ir

Abstract. *In this research, the influence of foaming conditions on poly (lactic acid) (PLA)/starch/nano calcium carbonate (NCC) nanocomposites containing epoxidized soybean oil (ESO) foams is investigated. Hybrid nanocomposites were prepared using an internal mixer at 180 °C for 12 min. Nanocomposite of PLA/starch/ESO/NCC (75/12.5/12.5/4) was saturated in an autoclave at room temperature using a pressure of 100 bar for 8 hours. Then the samples were transferred to silicone oil bath and they were foamed at 75, 90 and 105 °C for 10, 20 and 40 s respectively. The result showed that the increasing in time and temperature of foaming process, leads to void fraction enhancement. Moreover, it was found that the average cell size is increased with increasing the time and temperature too. SEM micrographs showed that in the samples which were foamed at 75 °C for 10 and 20 s, the small cell size foams were formed mostly on the surface of the samples. It could be because of the low foaming temperature and insufficient time for heat transfer from surface to the bulk of the nanocomposites. So the complete foam structure was not formed in the bulk of the sample. The maximum void fraction occurred at 105 °C, but in this condition, because of decreasing the modulus of the samples, the cells began to coalescence, and the average cell size is increased. It seems that the best foaming conditions are 90 °C and 40 s, on the basis of maximum cell densities and minimum cell size. It is evidenced that the presence of nano calcium carbonate leads to increase the cell nucleation and forming the small cell size and uniform cell size distribution.*

Keywords: *Nanocomposite, Poly (lactic acid), Nano calcium carbonate, Foam*
PACS: *82.35.Np*

INTRODUCTION

Nowadays, there has been growing interest on biodegradable materials especially poly (lactic acid) and its composites and nanocomposites. PLA foams are one of the best candidates for substitution of petroleum based foams such as polystyrene and polyethylene. Due to its competitive material and processing costs, and comparable mechanical properties, this environmentally friendly biopolymer is considered as a promising replacement for polystyrene (PS), especially for PS foam products in such commodity applications as packaging [1]. Mihai et al. [2], showed that foam of PLA has relatively uniform cell size distribution with cell size between 50-100 μm. They also showed that the addition of 33 wt. % of thermoplastic starch (TPS) to PLA, dramatically changed the foam morphology. In the absence of modified PLA, the foam had a bimodal morphological structure with finer cells around 30mm and larger ones around 150–200mm. For the foam with 50 wt. % of PLA-g-MA in the PLA phase, the cells were more uniform and finer, at around 40mm, than in the pure PLA foams. All foams had a high proportion of open cells.

NCC (nano calcium carbonate) is one of the common nano-fillers which has two constructive effects on the foam properties. First, NCC, as the nucleation agent, improves the nano-structure. Second, it prevents the cell coalescence at higher temperatures by increasing the melt strength. Huang et al. [3], showed that adding of NCC leads to an improvement in the foaming properties of polypropylene. NCC acted as a non-homogeneous nucleating agent and the cell nucleation happened at the boundary of the nano-filler and the matrix. Consequently, the cell density increases severely and the average cell size decreases. On the other hand, they also showed that the presence of 20 wt. % of the nano-filler causes the aggregation of the particles, so the resulted foam has non-uniform cell structure.

Ji et al. [4], showed that an increase in saturation pressure significantly affected the cell morphologies of PLA foams, especially for cell size distribution. Increasing the pressure leads to decrease cell size and non-uniform cell size distribution. Increasing the foaming temperature resulted in a significant increase of the cell size. Matuana et al. [5], investigated the influence of foaming time and temperature on PLA foam. Increasing the foaming temperature resulted in a significant increase of the cell size. Higher temperature accelerates the rate of gas diffusion, which favors bubble growth. Shorter foaming time appeared to be ideal to achieve the highest void fraction and volume expansion ratio when the samples were foamed at 150°C. Longer foaming time tended to decrease the void fraction and expansion ratio, probably due to the accelerated gas loss and cell coalescence phenomenon.

This research investigated the influence of foaming condition and the amount of nano calcium carbonate (NCC) on PLA/Starch/NCC with epoxidized soybean oil hybrid nanocomposite foam.

EXPERIMENTAL

Poly (lactic acid) used in this study, PLA 2003D, was supplied by NatureWorks. The food grade of native corn starch was purchased from Glucosan Company. The nano-sized precipitated calcium carbonate with an average particle size of 80 nm was supplied by Omya GmbH (Austria). The chemical pure grade of ESO was purchased from Faci Company. The foaming agent was Nitrogen. Starch, NCC and PLA were dried at 70°C for 18 hr., prior to mixing. Hybrid nanocomposites of PLA/starch/NCC with epoxidized soybean oil were prepared using an internal mixer in 180°C for 12 min under Nitrogen atmosphere, then they were compression molded at 190°C and 50 bar pressure. Samples of PLA/starch/ESO/NCC (75/12.5/12.5/4) were saturated at room temperature and 10 MPa pressure for 8 hr. using an autoclave. Finally the samples were transferred to silicone oil bath and were foamed at 75, 90 and 105°C for 10, 20 and 40 s.

RESULTS AND DISCUSSION

Table 1 indicates the effect of foaming time and temperatures on the void fraction, volume expansion ratio and average cell size of the foam samples. As it is seen, in all temperatures, in short times of foaming, the samples will be cold, prior to complement of the foaming process. So there is no enough time for the cells to grow significantly. This leads to the formation of foams with small average cell size and low void fraction. In other hand, increasing the time of foaming, causes the reduction in modulus of samples, which leads to decrease in cell wall thickness and increase the diffusion rate of gas among the cells and as a result, the average cell size will be increased. This phenomenon is seen in other researches too [5].

Also it was found that the average cell size of samples which are foamed in 75°C, for 10 and 20 s are very small and the foam structure is almost formed on the surface of the samples. As it is shown in the table 1, there is no chance for the same sample to form complete foam, even using longer foaming time too. It seems that the high modulus of nanocomposite in 75°C, does not let the perfect gas diffusion to the sample and the reduction of cell growth is occurred.

One of the main impressive effects on the foam density and resulted cell void fraction of a foam sample is the competition between cell nucleation and growth. This competition is directly influenced by the process temperature. As it is obvious in table 1, the average cell size increases with increasing the temperature and usually the cell structure becomes non-homogeneous too. Actually, like the longer process time, higher temperature leads to reduction in polymer modulus too. As a result, by using higher time and temperature in foaming process, the cell growth will dominant to nucleation. This cell growth may be continued and causes adjacent cells coalescence, which leads to increasing the cell size. As it is shown in table 1, the maximum void fraction is occurred at 105 °C, but in this condition, because of decreasing the modulus of the samples, the cells began to coalescence, and the average cell size is increased.

It can be concluded that one of the best foaming conditions for this nanocomposite is 90 °C for 40 s. In this condition, the samples form complete foam and there is no evidence of the cell coalescence. The prepared foam in this condition has high void fraction and low average cell size too.

Table 1- *The effect of foam processing conditions on resulted foam specifications*

Run	**Temperature (°c)**	**Time (s)**	**Volume expansion ratio**	**Void fraction (%)**	**Average cell size (μm)**
1	75	10	1.20	16	11
2	75	20	1.22	18	12
3	75	40	1.43	30	14
4	90	10	1.74	42	12
5	90	20	1.87	46	19
6	90	40	2.31	56	24
7	105	10	2.19	54	16
8	105	20	2.72	63	22
9	105	40	2.85	65	32

In order to indicate the effect of NCC on the resulted foam of the hybrid nanocomposite, the foams of PLA/starch/ESO (75/12.5/12.5) and PLA/Starch/ESO/NCC (75/12.5/12.5/4) were produced and compared with each other. The intended samples were saturated at 10 MPa pressure and foamed in the silicon oil at 90 °C for 40 s.

As it can be seen in Fig. 1, the average cell size decreases with the addition of 4 wt. % of NCC. It was found that the average cell size of PLA/starch/ESO (75/12.5/12.5) foams is 97 μm, but for the same composite containing 4 wt. % of NCC, the average cell size is 24 μm, which shows 75% reduction. Actually, using a nano-filler in a foam process, leads to a drastic decrease in the average cell size, a significant increase in the cell density, and homogeneous cell structure which are due to the nucleation effect of the nano-filler. The average cell size is controlled by the cell growth and cell coalescence which are significantly influenced by the storage modulus of the samples. By addition a nano filler to a polymer matrix, the storage modulus will be increased and the cell coalescence is prevented. So the cell growth will be limited and the average cell size is decreased significantly. On the other hand, the nucleation effect of NCC particles, leads to increase the crystallinity of samples during the foaming process and nucleating more cells. So the cell density will be increased and the average cell size will be decreased. These results are in agreement with that reported of polypropylene foam with nano calcium carbonate as a nucleation agent [3].

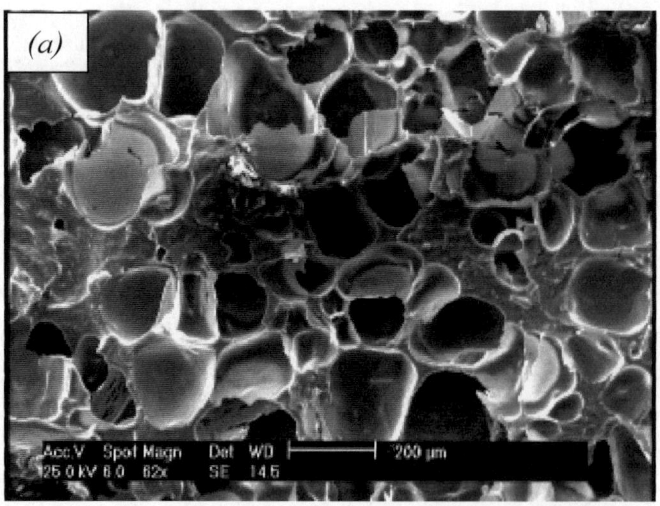

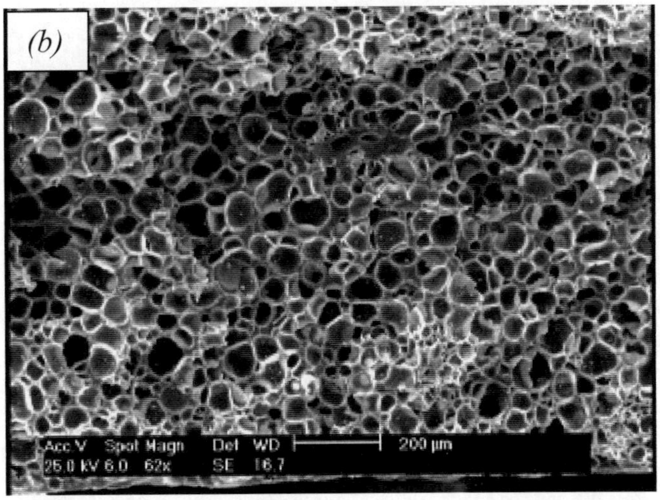

Figure 1 *SEM micrographs of PLA foamed at 90°C for 40 s: a) PLA/starch/ESO (75/12.5/12.5), b) PLA/starch/ESO/NCC (75/12.5/12.5/4).*

CONCLUSION

The cell size and void fraction will be increased with increase in time and temperature of foaming process. The maximum void fraction and volume expansion ratio are achieved using 105 °C and 40 s. But in this condition, the reduction in modulus, leads to cell coalescence, so the average cell size is increased significantly. One of the best foaming condition is using 90 °C and 40 s. Addition of 4 wt. % of NCC to the PLA/starch/ESO (75/12.5/12.5) sample, leads to 75% decrease in the average cell size and drastic increase in the cell density. It means that NCC is an appropriate nucleation agent for this hybrid nanocomposite. Consequently, the presence of NCC causes cell size reduction and significantly increase the cell density of resulted foams.

ACKNOWLEDGEMENTS

The authors are grateful to the valuable help from the supervisors of Elastomer and Plastic Workshop of Department of Polymer Engineering and Color Technology of Amirkabir University of Technology.

REFERENCES

1. Nofar, M. and C.B. Park, Poly (lactic acid) foaming. Progress in Polymer Science, 2014. 39(10): p. 1721-1741.
2. Mihai, M., et al., Extrusion Foaming of Semi-Crystalline PLA and PLA/Thermoplastic Starch Blends. Macromolecular bioscience, 2007. 7(7): p. 907-920.
3. Huang, H.X. and J.K. Wang, Improving polypropylene microcellular foaming through blending and the addition of nano-calcium carbonate. Journal of Applied Polymer Science, 2007. 106(1): p. 505-513.
4. Ji, G., et al., Tensile properties of microcellular poly (lactic acid) foams blown by compressed CO2. Journal of Cellular Plastics, 2013. 49(2): p. 101-117.
5. Matuana, L., Solid state microcellular foamed poly (lactic acid): morphology and property characterization. Bioresource technology, 2008. 99(9): p. 3643-3650.

Mechanical and thermal properties of carbon foams with respect to cell morphology of the precursor foam

Seung A Song[a], and Seong Su Kim[a*]

[a]*Department of Organic Materials and Fiber Engineering, Chonbuk National University, 567 Baekje-daero, Deokjin-gu, Jeonju-si, Jeollabuk-do, Republic of Korea*

ssa@jbnu.ac.kr
sskim@jbnu.ac.kr

Abstract. Carbon foams have been widely used in the field of carbon materials because of their potential application. To obtain the suitable properties at the application fields, it is important to control cell morphology of precursor foam. In this work, phenolic-based carbon foams with controllable pore structure was prepared by foaming of phenolic resin with different types of particles and their weight fraction. Graphene and multi-walled carbon nanotubes (MWNTs) with controlled aspect ratio were used to alter the foam morphology in the phenolic foams. Pitch particles were added to improve the low carbon yield of the phenolic precursor. The phenolic precursors were carbonized at 800, 1000, and 1200°C respectively to investigate the effect of carbonization temperature on the cell morphology. Thermal and electrical properties of the carbon foams with respect to carbonization conditions were estimated by using thermal conductivity test and electric resistance measurement.

Keywords: Polymer foam, carbon foam, Carbon nanotube, Graphene, Thermal properties,.
PACS: *88.20.rr*

INTRODUCTION

Carbon foams are attractive carbon materials due to their advantages such as light weight, high strength, temperature tolerance and high thermal and electrical conductivity [1-2]. These advantages of the carbon foams have potential application in industrial field including the energy absorption material, high temperature thermal insulation, electrodes for energy storage and filters [2]. Generally, carbon foams were fabricated by carbonization of polymeric foams which are thermoset plastics having the high thermal stability. Phenolic resin has excellent flame resistance, high self-ignition temperature of 480°C and high thermal stability. In addition, the phenolic foam does not generate toxic gases during carbonization to fabricate the carbon foam. This has led to a broad usage of the phenolic foam at the application of carbon foam [3-4].

The carbon foam derived from phenolic resin can be used for high strength thermal insulation foam by controlling the various condition such as content of reinforcement, carbonization temperature, heating rate, cell density and morphology of the phenolic precursors [3-5]. Li et al. studied the influence of short carbon fiber on the structure and properties of the carbon foam [3]. Luo et al. improved the mechanical properties and densification of the carbon foam by adding various fibers [4]. Chen et al. adjusted the strength of the carbon foam by controlling heat pretreatment and anisotropic domain size [5]. Many studies have focused on the carbonization condition and the content of reinforcement to improve the mechanical properties of carbon foams. However, the cell morphology of precursor before carbonization is the most important factor affecting mechanical and thermal properties of the carbon foam.

Therefore, in our previous research, the optimal foaming point and weight fraction of the particles of the phenolic foam were suggested to improve the mechanical and thermal properties by controlling the cell morphology. From the results, 1.0 wt% of multiwall carbon nanotube (MWCNT) and 1.0 wt% of graphene show the high porosity and low thermal conductivity due to their high cell density [6]. In this study, the carbon foams were fabricated via carbonization of the phenolic foams reinforced with MWCNT and grapheme under various temperature conditions. Pitch particles were added to improve the low carbon yield of the phenolic precursor. To estimate the insulating performance of the carbon foams, compressive strength and thermal conductivity were measured by compressive test and laser flash method, respectively.

MATERIALS AND EXPERIMETNAL PROCEDURES

Materials

Resole-type phenolic resin (OG-5000, Kangnam Chemical Co., Korea) was mixed with an cure catalyst (PTSA 65%, Kangnam Chemical Co., Korea) in a mixing ratio by weight of 9:1. The phenolic resin and cure catalyst were kept at 20°C to control the viscosity change before mixing process. The mixture was stirred at 500 rpm for 1 min using the impeller. Multi-wall carbon nanotube (724769, Sigma-Aldrich Co. LLC., USA) and graphene (C500, XG Sciences Inc., USA) of 1.0 wt% were added into the phenolic resin, respectively.

Preparation procedures of precursor phenolic foam

Based on our previous research, 1wt% of MWCNT and graphene was added in the phenolic resin, respectively.

The resole and accelerator with and without the particles were mixed using an impeller at 500 rpm for 3 min. Aging of the mixture under constant room temperature was performed to control the initial degree of cure before microwave foaming, and then microwave using an effective intensity per unit mass of 12 kW/kg at 2.4 GHz was applied to the mixture to fabricate the phenolic foams.

Carbonization of the phenolic foam

The phenolic foams were carbonized at the high temperature furnace under 800, 1000, 1200˚C in an argon atmosphere at a heating rate of 10 ˚C/min and held at each temperature for 2 h. The carbonized samples were named from precursor foam and carbonization temperature as shown in Table 1.

CHARACTERIZATION

Porosity and cell density

The porosity of the phenolic foams with respect to the test variables was calculated using the following equation (1):

$$\text{porosity}=\frac{\rho_s-\rho_f}{\rho_s}\times 100\% \qquad (1)$$

where ρ_s and ρ_f are solid density and foam density, respectively.

The cell density which is number of cells per unit volume was calculated by equation (2).

$$N(\text{cell/cm}^3)=\left(\frac{nM^2}{A}\right)^{3/2}\times\left(\frac{\rho_s}{\rho_f}\right) \qquad (2)$$

where, N is number of cell per unit volume (cell density), n is number of cell in a 2-dimension SEM image, M is magnification of the SEM images, A is area of the SEM images (cm^2), ρ_s and ρ_f are solid and foams density, respectively [23].

Compressive strength

To evaluate the load-carrying capacity of the prepared phenolic foams, a uniaxial compression test was conducted to measure the compressive strength. The compressive strength of the each sample with respect to the test variables was measured according to ASTM C 365 using a computer-controlled material testing system (4206, INSTRON, USA). The cross head speed during the test was 0.5 mm/min. To remove the water content, the specimens were dried at 60°C for 5 h before the test. After the drying process, the specimens were cut to the regular hexahedron-shape of 25 mm in sides and 25 mm in length using a cutting machine.

Thermal conductivity

Thermal conductivity of the carbon foam was measured at the room temperature by using the laser flash method (Laser PIT-M2, ULVAC. Inc., USA). To measure the thermal conductivity of carbon foam, the specimens of cylindrical shape with diameter of 5 cm were cut to the thickness of 2 cm using the cutting machine.

RESULTS AND DISCUSSION

FIGURE 1 shows the morphologies of the carbon foam with and without particles carbonized under various temperature. The average cell size of carbon foams was decreased by increasing the carbonization temperature as shown in FIGURE 1. Especially, the carbon foams carbonized at the 1000°C show the highest decrease rate of cell size. This means that volume shrinkage rate of the carbon foam are the largest at the 1000°C.

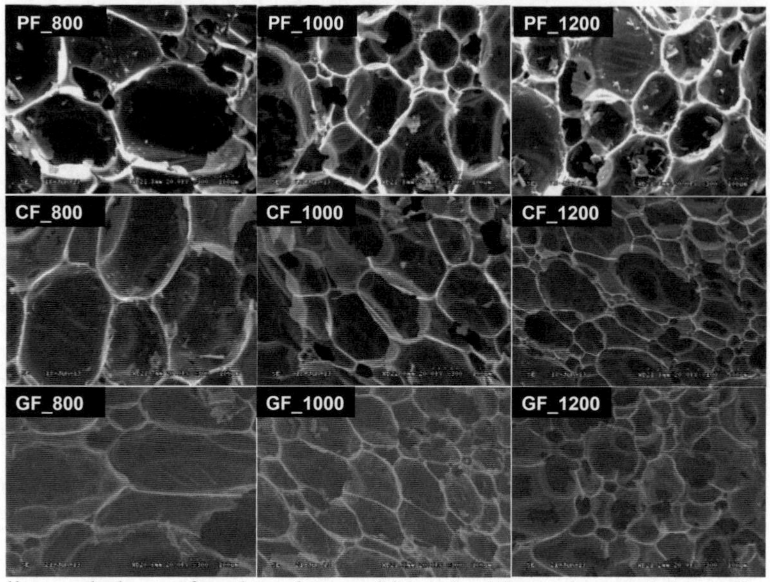

FIGURE 1. The cell morphology of carbon foam with and without the particles carbonized under various temperature.

FIGURE 2 shows the compressive strength and specific compressive strength (strength/foam density) of the carbon foams with respect to carbonization temperature. The carbon foams reinforced with the particles presented the high compressive strength compared to that of the PF because the precursor foams reinforced with the particles have high cell density due to high viscosity before foaming process. The PF and GF specimens have high specific compressive strength when they were carbonized at 800°C. However, in the case of CF specimens, the specific compressive strength increased as the carbonization temperature increases.

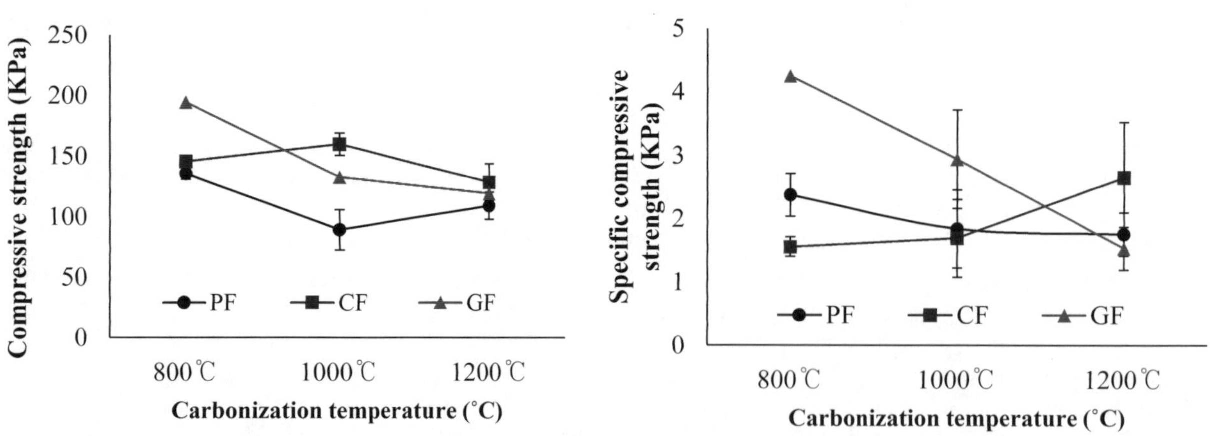

FIGURE 2. (a) Compressive strength and (b) specific compressive strength of the carbon foams with respect to carbonization temperature.

The thermal conductivity of porous structures is affected by various factors such as the porosity, cell density, reinforcement, and thermal conductivity of raw material. Thermal conductivity of the carbon foams was similar or slightly increased compared with precursor foam (0.05 W/m·K) until the carbonization temperature reached to 1000°C due to their increased carbon yield. However, carbonization temperature increased up to 1200°C, the thermal conductivity was decreased by high cell density and low porosity of the carbon foams.

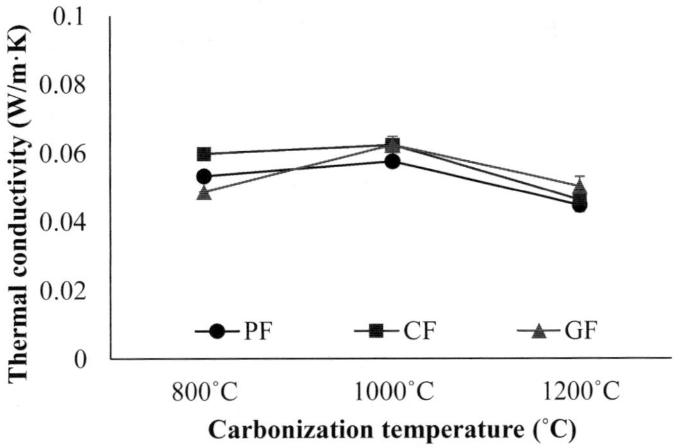

FIGURE 3. Thermal conductivity of the carbon foams with respect to carbonization temperature.

CONCLUSION

In this work, the carbon foams were fabricated by carbonizing the phenolic foam reinforced with and without particle with respect to carbonization temperature. The effect of the particles and the carbonization temperature on the mechanical and thermal properties of the carbon foams was explained. Based on the experiments, the following results were summarized:

(1) The PF and CF at 800°C showed the highest specific compressive strength and GP shows the highest specific compressive strength at 1200°C.

(2) The carbonization temperature increased up to 1200°C, the thermal conductivity was decreased by high cell density and low porosity of the carbon foams.

(3) Based on the specific compressive strength and thermal conductivity, the proper carbonization temperatures of PF, GF, and CF are 800°C, 800°C, and 1200°C, respectively.

From this study, it was concluded that the carbonization under the proper carbonization temperature with respect to precursor foam type are promising ways to obtain the optimum mechanical and thermal properties.

ACKNOWLEDGMENTS

This research was supported by Basic Science Research Program through the National Research Foundation of Korea (NRF) funded by the Ministry of Science, ICT and Future Planning (2011-0010156); it was also supported in part by the Korea Institute for Advancement of Technology (A000600052); and it was also supported by Kolon Industry.

REFERENCES

1. S. Lei, Q. Guo, J. Shi and L. Liu, Carbon, 2644-2673 (2010).
2. C. Chen, E. B. Kennel, A. H. Stiller, P. G. Stansberry and J. W. Zondle, Carbon 44, 153-1543 (2006).
3. W. Q. Li, H. B. Zhang, X. Xiong, F. Xiao, Materials science and engineering A, 7274-7278 (2010).
4. X. Wang, R. Luo, Y. Ni, R. Zhang, S. Wang, Materials letters, 25-27 (2009).
5. C. Chen, E. B. Kennel, A. H. Stiller, Carbon, 1535-1543 (2006).
6. S. A. Song, Y. S. Chung, S. S. Kim, Composites science and technology, 82-93 (2014).

Effect of processing conditions on the cell morphology distribution in foamed injection molded PLA samples

Valentina Volpe[a], Martina De Filitto[a], Vera Klofacova[b], Felice De Santis[a], Roberto Pantani[a*]

[a]*Department of Industrial Engineering. University of Salerno Via Giovanni Paolo II, 132, 84084 Fisciano (SA), Italy*
[b]*Centre of Polymer Systems, University Institute, Tomas Bata University in Zlin, Trida Tomase Bati 5678, 760 01 Zlin, Czech Republic*

rpantani@unisa.it

Abstract. Foam injection molding uses physical blowing agents under high pressure and temperature to produce structural foams having a cellular core and a compact solid skin. This technology is particularly interesting for biodegradable polymers, which often present a very narrow processing window, with the suitable processing temperatures close to the degradation conditions. The addition of a supercritical gas can lead to the reduction of both the viscosity and the glass transition temperature of the polymer melt, which therefore can be injection molded adopting lower temperatures and pressures. In this work, the effect of different processing parameters on foam morphology of Poly(lactic) Acid, PLA, was studied. In particular, two commercial grades of PLA having different rheological properties were adopted to obtain foamed parts by injection molding process with nitrogen as physical blowing agent. For both PLA grades, the effect of mold temperature on the crystallinity and the resulting cell morphology was assessed.

Keywords: Poly(Lactic) Acid, foam injection molding, morphology, crystallinity.
PACS: 83.50.Uv

INTRODUCTION

A significant attention has been recently devoted on environmentally friendly biopolymers, such as Poly(lactic) acid (PLA), as alternative to the petroleum-based polymers. Poly(lactic) acid is a biodegradable thermoplastic currently used in a lot of different applications due to its properties, such as high strength, high modulus and transparency. However, its applications are limited in many manufacturing processes because of the low melt strength, the narrow processing window and the slow crystallization rate [1-7]. Supercritical nitrogen utilized in the foam injection molding process reduces the melt viscosity due to its plasticization effect, allowing the processability of the PLA at decreased pressure and temperatures [8, 9]. Strong interconnection exists between the foaming and the crystallinity. Several authors [10, 11] studied the effects of dissolved gas on the crystallization kinetics of the PLA. It was observed that the final crystallinity of a foam injection molded part is strongly enhanced by presence of dissolved gas and by the strain-induced crystallization during foam expansion [12, 13]. Another important parameter that can influence the crystallinity of the PLA is the percentage of the D-lactide: on increasing the D-lactide percentage, the crystallization kinetics becomes slower and the maximum attainable crystalline content reduces [14]. In this work, the effect of different processing parameters on crystallinity and foam morphology of two commercial grades of PLA having different rheological properties were studied.

MATERIALS AND METHODS

Two different commercial grades of poly(lactic) acid (PLA) produced by NatureWorks were adopted in this work: PLA 4032D and PLA 3251D. The main features of these two PLA grades are reported in Table 1. Before any test or processing, all the materials were dried for 24 h under vacuum at a temperature of 60 °C, because of dramatic effects of water during processing and storage [15-16].

A rheological characterization of the two PLA grades was carried out by a rotational rheometer in dynamic mode in parallel plates configuration at the injection temperatures (200 °C for PLA 3251D and 220 °C for PLA 4032D). A differential scanning calorimeter DSC 822e from Mettler Toledo was used for determination and measurement of the thermal behavior of PLA. The calibration of the temperature was done with the extrapolated onset temperature of the phase transition of indium. The samples, neat or obtained from foam injection molded specimens, with a weight of less than 10 mg were put into an aluminum pan and hermetically sealed. All the experiments were conducted in nitrogen, flow rate 50 ml min^{-1}, in order to prevent oxidative degradation at high temperature. The crystallization kinetics was investigated in isothermal conditions starting from the melt after holding the sample at 200°C for 5 min in order to erase the effect of previous thermomechanical history [17-19].

TABLE (1). Main features of the PLA grades.

	PLA 4032D	**PLA 3251D**
D-lactide content	2.0 %	1.4 %
Specific Gravity (ASTM D792)	1.24	1.24
Melt Flow Rate (210 °C - ASTM D1238)	7 g/10min	80 g/10min
M_w	$210 \cdot 10^3$ g/mol	$90 \cdot 10^3$ g/mol
M_w/M_n	1.7	1.6

The physical blowing agent utilized for the foaming process is nitrogen. A 70 ton Negri-Bossi injection molding machine with screw diameter of 25 mm and L/D=22, modified to carry out the microcellular injection molding process, was adopted. The sample geometry is shown in Fig. 1. The particular geometry of the sample allows to observe the properties variations in all three characteristic dimensions. In this work, we will refer to the top of the sample as "Crown" and to the lower part as "Brim". The injection point is located in the center of the bottom of the brim.

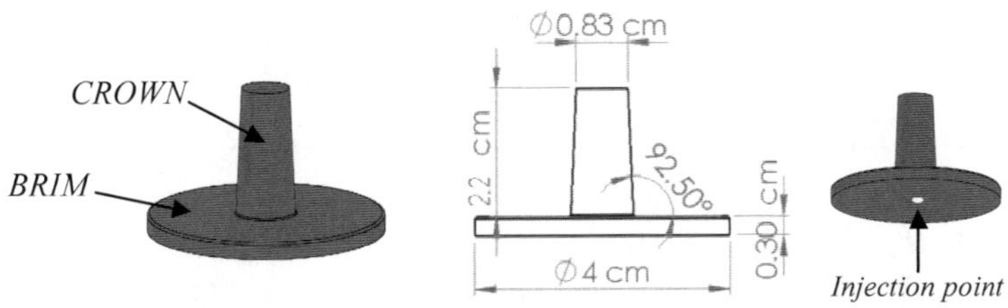

FIGURE 1. Sample geometry.

The experimental parameters were chosen in order to obtain the optimal conditions of molding of the foamed parts. In particular, the set temperature profile of the PLA 4032D was 20 °C higher than that of the PLA 3251D, due to the very different rheological properties of the two PLA grades. The experimental condition adopted for both PLA grades are reported in Table 2.

TABLE (2). Experimental condition.

	PLA 4032D	**PLA 3251D**
Temperature profile [°C]	175 - 170 - 165 - 160	195 - 190 - 185 - 180
Nozzle temperature [°C]	200	220
Mold temperature [°C]	25 - 40	25 - 40
Filling pressure (hydraulic system) [bar]	100	100
Screw rotation speed [rpm]	150	150
Back pressure (hydraulic system) [bar]	2	2
Injection speed [%]	30	30
Gas injected [g]	0.24	0.24
Shot size [cm^3]	8.8	5.9

RESULTS

Fig. 2 shows the comparison between the results of rheological measurements obtained for both PLA grades, in terms of a master curve built at the injection temperatures (200 °C for PLA 3251D and 220 °C for PLA 4032D). The values of the thermal shift factor α T are reported in Table 3. As clear from Fig. 2, in spite of the different temperature, PLA 3251D, namely the grade with the higher melt flow rate, has a viscosity significantly lower than that of the PLA 4032D. Density measurements were carried out on the molded samples at 25 °C. Subsequently, the samples were sectioned by dividing the crown from the brim in order to measure the difference in density between the two parts of the sample. Fig. 3 shows the density measurements made on the foamed samples of PLA 3251D (a) and PLA 4032D (b). For both, the graph shows the density of the whole sample, the density of the brim and the density of the crown. It is possible to observe that the density of the foamed samples in always less than 1 g/ccm, while the density of the unfoamed PLA is 1.24 g/ccm.

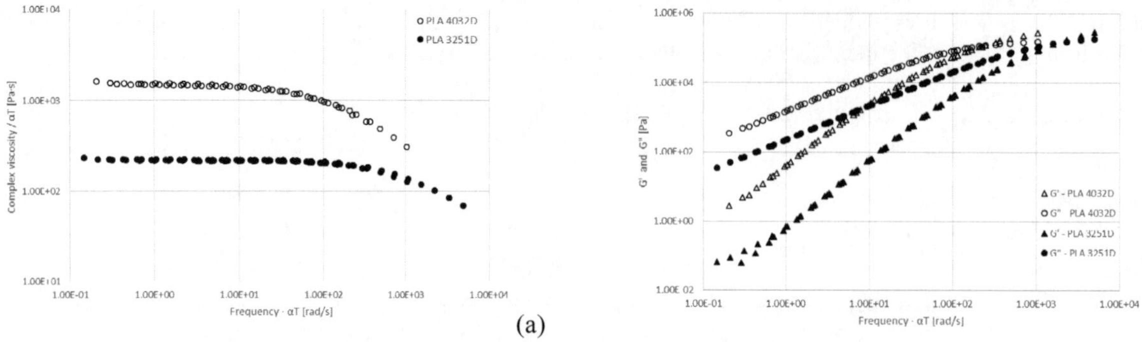

(a) (b)

FIGURE 2. Rheological measurements on PLA 4032D - at 220°C - and PLA 3251D – at 200°C.

TABLE (3). Thermal shift factors αT.

Temperature	αT for PLA 4032D	αT for PLA 3251D
140 °C	-	34
160 °C	7	7.1
180 °C	2.46	2
200 °C	1.41	1
220 °C	1	-

The density of the PLA 4032D samples are lower than those of the PLA 3251D. Furthermore, for both PLA grades at mold temperature of 25 °C the density of the crown is higher than the density of the brim. At higher mold temperature (40 °C) this trend is reversed. Therefore, for both PLA grades, at high mold temperature the crown is more foamed than the brim.

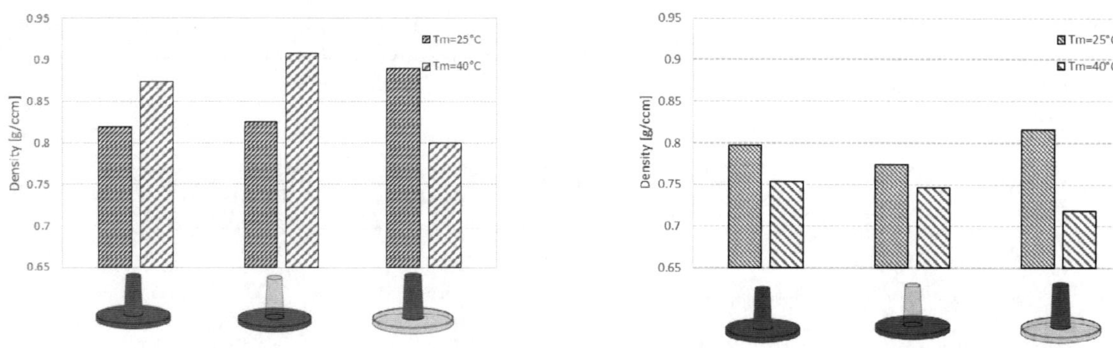

(a) PLA 3251D (b) PLA 4032D

FIGURE 3. Density of PLA 3251D and PLA 4032D foamed samples (whole sample, brim and crown) obtained with mold temperatures of 25 °C and 40 °C.

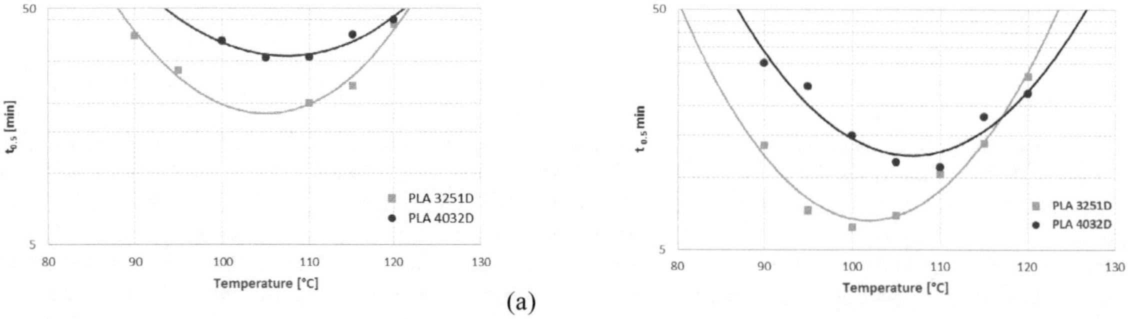

(a) (b)

FIGURE 4. Crystallization half-time $t_{0.5}$ at different temperatures for pure materials (a) and processed by foam injection molding materials (b). Lines are drawn to guide the eye.

Fig. 4 reports the isothermal crystallization half times evaluated by DSC analysis at different temperatures. The DSC analyses were conducted on the two PLA grades, both in form of pellet and after processing by foam injection molding. As expected, the crystallization half-times of the foam injection molded materials are lower than those of the pure material. In both cases, the PLA 3251D presents faster crystallization kinetics. The different morphology between the crown and the brim can be explained by analyzing the filling and cooling stages. The process simulations carried out by Moldflow show that the crown is filled before the brim (Fig. 5). The lower thickness of the brim ensures the faster cooling with respect to the crown, which thus remains warm for a longer time. This lower cooling rate can obviously induce a different morphology.

The molded samples were analyzed by tomography by means of a SkyScan1174 compact micro-CT which uses computer-processed X-rays to produce tomographic images or slices of a specific part. In Fig. 6 it is possible to observe the tomographic images of the PLA 3251D and PLA 4032D samples obtained with mold temperature of 25°C (left column) and 40 °C (right column).

(a) (b) (c)

FIGURE 5. Cross section of a sample of PLA 3251D during the filling and the cooling phase, as simulated by Moldflow. Time from the entrance into the cavity: (a) 0.54 s; (b) 1.26 s; (c) 7.61 s.

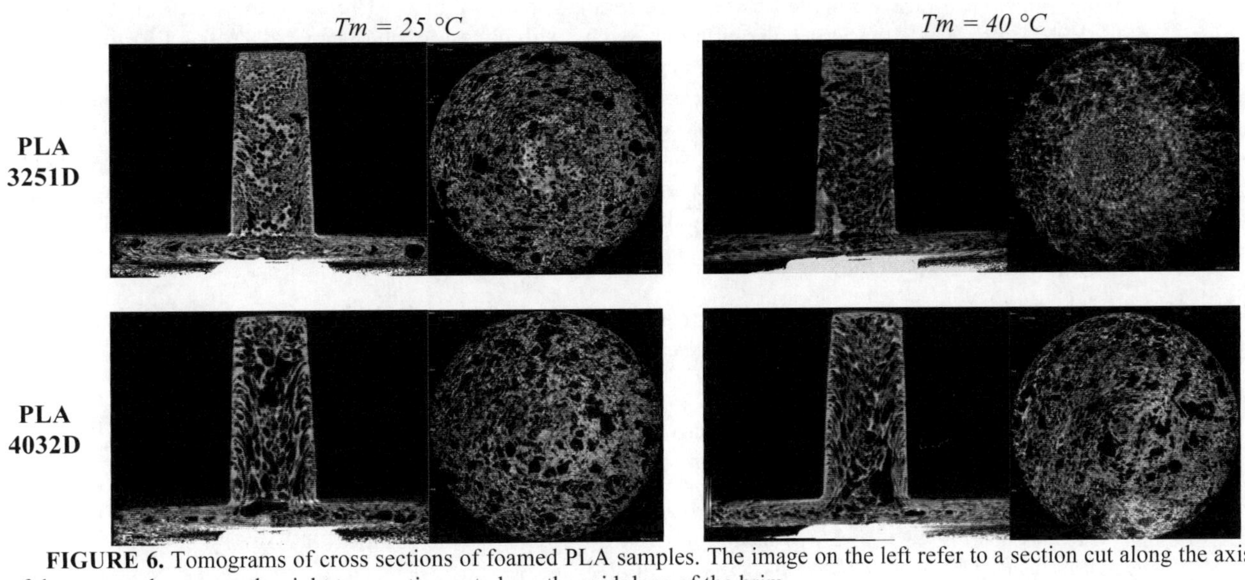

FIGURE 6. Tomograms of cross sections of foamed PLA samples. The image on the left refer to a section cut along the axis of the crown, the one on the right to a section cut along the midplane of the brim.

The images confirm the trend obtained from the density measurements, namely the percentage of voids is much higher for 4032D than for 3251D. As a further information, the cross section taken along the axis of the crown (on the left of each image) shows a different shape of the cells into the crown for the two PLA grades. In particular, the samples of PLA 4032D present cells elongated in the direction of the brim. The samples of PLA 3251D, vice versa, present cells with a circular shape. The crystallinity content, obtained by using DSC analysis, was assessed in two different positions of the samples: in the middle section of the crown and at half-width of the brim, in both cases far from the wall of the sample (Fig. 7). It was observed that the foamed parts are slightly less crystalline that the unfoamed one. For both PLA grades, in the foamed parts obtained at mold temperature 25 °C the brim is more crystalline that the crown. By increasing the mold temperature at 40 °C, the crystallinity of the crown does not vary, while the crystallinity of the brim decreases. In fact, the large thickness of the crown leads to long cooling times that allow the crystallization. The thinner thickness of the brim with respect to the crown does not allow the crystallization. However, the shear flow established into the brim probably determines an orientation that promotes the crystallization. By increasing the mold temperature, the effect on the crown remain almost the same, while the crystallinity of the brim decreases probably because of the lower degree of orientation due to the higher temperature. In all cases, the crystallinity degree of the PLA 3251D is larger than that of the PLA 4032D, consistently with the crystallization kinetics of the materials.

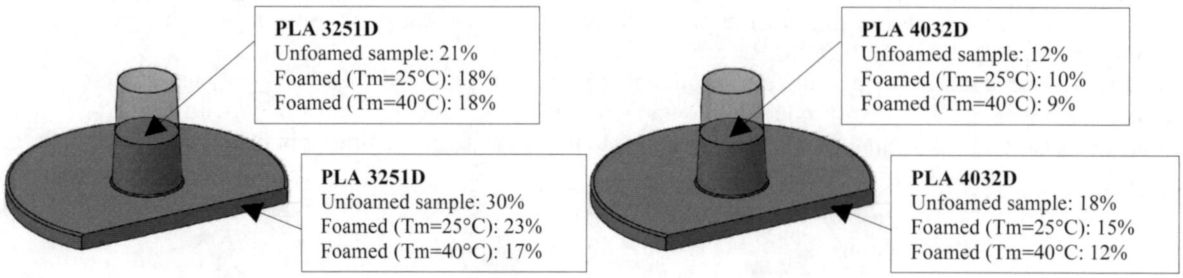

FIGURE 7. Crystallinity degree obtained by DSC analysis.

CONCLUSIONS

In this work, two commercial grades of poly(lactic) acid, PLA 4032D and PLA 3251D, having different rheological properties and D-lactide percentage, were adopted to obtain foamed parts by injection molding process with nitrogen as physical blowing agent. For both PLA grades, the effect of mold temperature on the resulting cell morphology was assessed. The foamed PLA 4032D samples have a lower density with respect to the samples of PLA3251D, but cells with not well defined shape, elongated in the direction of the brim. For both PLA grades, at higher mold temperatures the crown resulted to be more foamed than the brim. The large thickness of the crown induces long cooling times that allow crystallization. The smaller thickness of the brim with respect to the crown should induce a lower crystallinity degree. However, the shear flow established into the brim determines an orientation that can promote crystallization. By increasing the mold temperature, the crystallinity inside the crown does not change significantly, whereas the crystallinity of the brim decreases, probably because of a lower orientation. In all cases, the crystallinity degree of the PLA 3251D is higher than that of the PLA 4032D, consistently with the crystallization kinetics of the materials.

REFERENCES

1. A. Ameli, M. Nofar, D. Jahani, G. Rizvi and C.B. Park, *Chemical Engineering Journal* **262**, 78-87 (2015).
2. M. Nofar, A. Tabatabaei, A. Ameli and C.B. Park, *Polymer* **54**, 6471-6478 (2013).
3. Rasal, R.M., A.V. Janorkar and D.E. Hirt, *Progress in Polymer Science* **35**, 338-356 (2010).
4. G. Gorrasi and R. Pantani, *Polymer Degradation and Stability* **98**, 1006-1014. (2013).
5. R. Pantani, F. De Santis, A. Sorrentino, F. De Maio and G. Titomanlio *Polymer Degradation and Stability* **95**, 1148-1159 (2010).
6. F. De Santis, R. Pantani and G. Titomanlio, *Thermochimica Acta* **522**, 128-134 (2011).
7. F. De Santis and R. Pantani, *Journal of Polymer Research* **22**, 242 (2015). Doi: 10.1007/s10965-015-0885-1
8. W.D. Ding, R.K.M. Chu, L.H. Mark and C.B. Park, *European Polymer Journal* **71**, 231-247 (2015).
9. M. Nofar, A. Tabatabaei, C.B. Park, *Polymer* **54**, 2382-2391 (2013).
10. V. Speranza, A. De Meo, R. Pantani, *Polymer Degradation and Stability* **100**, 37-41 (2014).
11. J. Wang, W. Zhu, H. Zhang, C.B. Park, *Chemical Engineering Science* **75**, 390-399 (2012).
12. R. Pantani, V. Volpe, G. Titomanlio, *Journal of Materials Processing Technology* **214**, 3098-3107 (2014).
13. V. Volpe and R. Pantani, *Journal of Applied Polymer Science* **132**, (2015).
14. M. Mihai, M.A. Huneault, and B.D. Favis, *Journal of Applied Polymer Science* **113**, 2920-2932 (2009).
15. R. Pantani, F. De Santis, F. Auriemma, C. De Rosa, R. Di Girolamo, *Polymer* (2016). Doi: 10.1016/j.polymer.2016.07.008
16. R. Pantani, F. De Santis, *AIP Conference Proceedings* **1695** (2015). Doi: 10.1063/1.4937344
17. F. De Santis, R. Pantani, G. Titomanlio *Thermochimica Acta* **522**, 128-134 (2011).
18. F. De Santis, G. Gorrasi, R. Pantani *Polymer Testing* **44**, (2015). Doi: 10.1016/j.polymertesting.2015.03.015
19. G. Gorrasi, R. Pantani, M. Murariu, P. Dubois *Macromolecular Materials and Engineering* **299**, (2014). Doi: 10.1002/mame.201200424

Artificial Aging and Failure Modes of TPU-Metal Hybrid Composites

J. A. Puentes, A. Leuteritz, I. Kuehnert*,

Processing Department, Leibniz Institut fuer Polymerforschung Dresden e. V., Dresden, Germany

**Corresponding author: Ines Kuehnert, kuehnert@ipfdd.de*

Keywords: overmolded TPU, polymer-metal composites, artificial aging

Thermoplastic elastomers are being continuously used in industrial and commercial applications due to its versatility in terms of a wide range of mechanical properties and also because they can be easily processed. One of the main uses of these materials are as bumping elements, also as decorative parts with a soft-touch feel, or in high performance parts due to its toughness and abrasion resistance as well as chemical and oil-solvent resistance in the automotive or aeronautical industries. The soft-elastic polymer can also improve impact resistance and balance thermal deformation of two components with dissimilar thermal expansion coefficients. However, adhesion of polymers onto metal substrates is a task which has not been completely solved, especially in terms of long term applications and/or in humid or corrosive environments. The idea behind this research is to investigate the aging and failure modes of a polymer-metal composite. Samples were prepared by applying an adhesive coating to a metal substrate with two different surface treatments, phosphated and plain steel, which were subsequently cured in an oven for 150°C. Then, the coated plate was overmolded with a TPE and subjected to an artificial aging process at different temperature and humidity atmospheres. Evaluation of the progressive aging and final rupture of the component is done mainly by microscopy, thermal treatment, spectrometry and mechanical tests.

Mathematical Modeling and Visual Observation of a Chemical Batch Foaming and Crosslinking of Polymer System

Tomohiro Shigeta, Shota Ishihara and Masahiro Ohshima*

Department of Chemical Engineering, Kyoto University, Nishikyo-ku, Kyoto, 615-8510 Japan

t.shigeta@cheme.kyoto-u.ac.jp
oshima@cheme.kyoto-u.ac.jp

Abstract. This study was conducted for understanding the mechanism of a chemical foaming process for thermoplastic polymer, where the chemical crosslinking reaction occurs simultaneously with foaming. Using a temperature-controlled view-cell with an optical microscope, the visual observation experiments were conducted to monitor the bubble nucleation and growth behaviors of foaming polymers with crosslinking agents. The experiments elucidated that the effects of processing temperature (heating rate), the reaction kinetics of chemical blowing agent (CBA) decomposition, and the change in polymer viscosity on the nucleation time and the growth rate of CBA-blown bubbles were significant. The visual observation also showed that the gas generated from decomposing CBA and the bubbles were produced at the site of CBA. Based on the observations, a set of mathematical models of bubble nucleation and growth in chemical foaming process was developed. A kinetic reaction model and a temperature-dependent viscosity model were incorporated with the foaming models. The reaction kinetics were measured by using thermogravimetric analysis (TGA) and the change in polymer viscosity during the crosslinking reaction was also measured by a rheometer in order to determine the values of the model parameters. Several numerical simulations were performed with the developed models and compared with the imaging data of the visual observation experiments. The developed models were not perfect in term of the sensitivity of reaction kinetic parameters but could simulate some aspects of chemical foaming behavior of plastics.

Keywords: chemical foaming, modeling, kinetic reaction, crosslinking

PACS: 82. 20. Wt Computational modeling; simulation

INTRODUCTION

Polymer foams have superior properties, such as lightweight, high impact resistance and low thermal conductivity, which are suitable and useful in many industries. The preparation method of polymer foams can be roughly divided into two, i.e., physical foaming and chemical foaming. In the physical foaming processes, a physical blowing agent (PBA) such as hydrocarbons, nitrogen or carbon dioxide is used to create the pores in polymer. PBA is dissolved into polymer at a given temperature and pressure and the foaming is induced as a phase separation phenomenon by reducing the pressure or increasing the temperature. In the chemical foaming processes, the pores are created in polymer with the gas produced by chemical reaction or temperature-decomposition of chemical blowing agents (CBA) such as oxydibenzenesulfonyl hydrazide (OBSH) and azodicarbonamide (ADCA). The temperature-decomposition type CBA generates gaseous nitrogen (N_2) or carbon dioxide (CO_2) in course of heating. When rubber or polyethylene are chemically foamed, for example, the crosslinking reaction is conducted simultaneously with the foaming so as to increase polymer viscosity and stabilize bubbles.

In both chemical and physical foaming processes, controlling cellular structure is important so as to keep the superior properties and foam quality high. It is well-known that solubility, diffusivity of gas in polymer and rheological properties of polymer significantly affect the cellular structure. But it is still difficult to understand the foaming processes completely because the effects of these physical and rheological properties on cell morphology are changed by the processing temperature and pressure and these processing conditions are dependent on the processing devices and machines, such as compression molding, extrusion and injection molding machines. There have been many researches on modeling and simulation on foaming and the relationships among polymer properties, processing condition and processing machines. In the early stage of the modeling researches, a growth of single gas bubble in polymer was focused and the classical nucleation theory was incorporated with the bubble growth model for describing physical foaming behaviors [1-3]. The researches were then extended to describe the physical foaming behaviors in foam extrusions and foam injection molding processes [4-7]. Several visual observation experiments of foaming have been also conducted to advance the modeling and increase the simulation accuracy [8-10]. Recently, the attention of modeling was paid to microcellular foaming and nanocellular foaming where CO_2 and N_2 are used as a physical blowing agent [11,12].

As for the modeling and simulation of chemical foaming processes, there are a limited number of researches. Wang and Ogale [13] introduced a simple bubble growth model to describe the structural foam molding process. Xu et al. [14] proposed a diffusion-induced bubble growth model. Their model can describe the bubble growth and collapse phenomena in CBA foaming under atmospheric pressure but did not discuss the sites of bubble nucleation.

In this study, visual observation of CBA-based foaming process was conducted using a hot stage with optical microscope imaging system and the mechanism of the chemical foaming process was analyzed and modelled. The location of bubble nucleation sites and the starting behavior of bubble growth at the CBA site were primarily focused in the analysis and modeling.

CHEMICAL FOAMING MODELS

1.1 Chemical foaming process to be modelled

The chemical foaming process investigated in this study was a press forming process, which is often used for foaming tires or shoe soles. In the process, CBA and a cross-linking agent are mixed with a polymer and homogenized in a batch mixer and a roller. The polymer containing the CBA and the cross-linker is loaded into a mold. Then, the polymer is press-formed or shape-formed by heating. During the heating stage, the cross-linking agent is reacted with the polymer, the viscosity is increased to control the bubble growth, and then CBA is decomposed to produce a gas for expanding the polymer. There are two types of the press-forming process for polymer foaming, single and dual step foaming. In the single step foaming process, a press platen is open rapidly to expand the cavity volume and reduce the polymer pressure dramatically. The pressure drop leads to foam expansion. In the dual-step foaming process, the sheets of mixture of polymer, CBA, and crosslinking agent are prepared as a product from the first stage. The sheets are transferred to a mold and then fully-expanded by steam or hot air heating in the mold. One of the major differences of these chemical foaming processes from CBA-based foam extrusion or injection processes is that CBA is mixed with polymer before foaming process and no foaming is conducted in the presence of shear flow.

1.2 Bubble nucleation site

Based on the visual observation, we assumed that the bubbles are originated from the chemical blowing agent. The CBA are distributed in polymer as solid particles and the particle has a size distribution. Without using the classical bubble nucleation rate model, the number of bubble was given by a function of the number density of chemical blowing agent of in polymer. Thus, the number density of CBA was derived from the image of visual observation for non-foamed mixture of polymer, CBA and cross-linker.

Assuming that the particle shape of CBA is spherical, the initial particle size is described by the radius, R_{CBA}. To describe the distribution, a probability density function, $p(R_{\mathrm{CBA}})$, is given by Eq. (1). The function could be determined from the visual observation of the non-foamed mixture as shown in Fig.1.

$$\int_0^{\infty} p(R_{\mathrm{CBA,}})\, dR_{\mathrm{CBA}} = 1 \tag{1}$$

Then, the total number of CBA particles, N_{CBA}, can be calculated from the density of CBA, ρ_{CBA}, and weight of total amount of CBA, W_{CBA}, mixed with an unit weight of polymer by Eq.(2)

$$N_{CBA} \int_0^{\infty} p(R_{\mathrm{CBA,0}}) \frac{4\pi}{3} R_{\mathrm{CAB,0}}{}^3 dR_{\mathrm{CBA,0}} = \frac{W_{CBA}}{\rho_{CBA}} \tag{2}$$

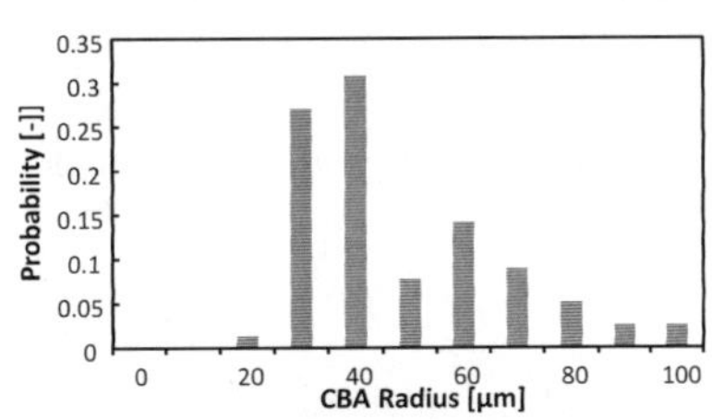

Figure 1. A probability density function of CBA particles size distribution

Assuming that the total number of CBA particles or domains is given by N_{CBA}, the volume of the polymer that is affectively expanded by a single CBA particle or domain is termed as influence volume V_{influ}. It is calculated by dividing the total volume of polymer $V_{total\_polymer}$ by N_{CBA}.

$$V_{influ} = \frac{V_{total\_polymer}}{N_{CBA}} = \frac{M_{polymer}}{\rho_{polymer}} \frac{1}{N_{CBA}} \quad (3)$$

where $M_{polymer}$ is the weight average molecular weight of polymer and $\rho_{polymer}$ is the density of polymer at the foaming temperature and pressure.

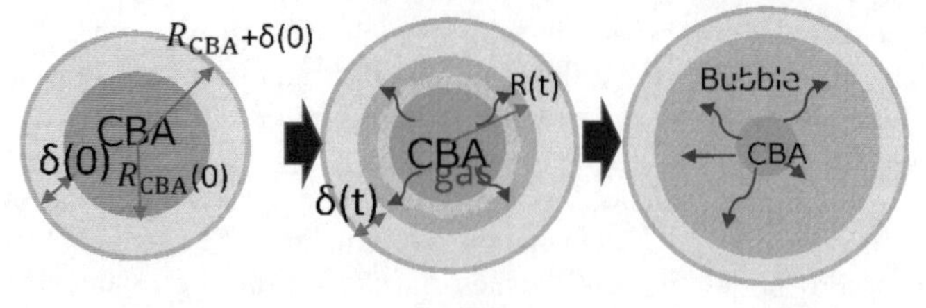

Figure 2 Bubble growth behavior around a shrinking CBA

Here it is assumed that each CBA particle is surrounded by the polymer with the volume of influence region, V_{influ}, as shown in Fig. 2. The bubble is initiated around the CBA particle, the radius of the bubble, $R_{\mathrm{bub}}(t)$, is increasing from $R_{CBA}(0)$ and the thickness of influence volume of polymer, δ(t), is thinning.

$$\frac{4\pi}{3}\{(R_{\mathrm{bub}}(t) + \delta(t))^3 - R_{\mathrm{bub}}(t)^3\} = V_{influ}, \qquad R(0) = R_{CBA}(0), \qquad (4)$$

As the bubble is expanding, the thickness of influence region, δ, is thinning and the overall volume of an expanding bubble, $\frac{4\pi}{3}(R(t) + \delta(t))^3$, is increasing as shown in Fig. 2.

1.3 Bubble Growth Model

The bubble growth behavior is modeled with the consideration of the equations of motion, mass balance over a bubble and the rate of gas generation from the CBA decomposition. With the assumption given below, the mathematical models of bubble growth were developed.

1) The initial radius of the bubble is equal to the radius of CBA particle; $R_{\mathrm{bub,0}} = R_{\mathrm{CBA,0}}$
2) The shapes of CBA as well as bubble are spherical and symmetric throughout the bubble-growth process.
3) The initial pressure ($P_{\mathrm{bub,0}}$) inside bubble is equal to pressure of polymer, which is equal to the polymer system pressure (P_{sys}) in the visual observation experiment.
4) The inertia is negligible.
5) The gas inside the bubble follows the ideal gas law in the bubble growth process.
6) The gas is generated from the decomposed CBA. It is directly used to expand the bubbles. In other words, the gas does not diffuse out or dissolve into polymer.
7) The viscosity of polymer is not constant and is changed depending upon the crosslinking degree of polymer.

1.3.1 Momentum balance

The bubble growth is governed by the conservation of momentum in the radial direction,

$$\frac{dR_{\mathrm{bub}}}{dt} = \frac{R_{\mathrm{bub}}}{4\eta}\left(P_{\mathrm{bub}} - P_{\mathrm{sys}} - \frac{2\gamma}{R_{\mathrm{bub}}}\right) \qquad (3)$$

In the simulation, the bubble cannot start expanding, while $P_{\mathrm{bub}} - 2\gamma/R_{\mathrm{bub}} \le P_{\mathrm{sys}}$ is hold,

1.3.2 Mass balance over a bubble

Assuming the CBA is encapsulated in a bubble, the CBA is decomposed to generate gas. The gas is used to expand the bubble. The bubble volume is needed to subtract $4\pi R_{\mathrm{CBA}}^3/3$ from $4\pi R_{\mathrm{bub}}^3/3$.

$$\frac{4\pi}{3}\frac{d}{dt}\left(\frac{P(R_{\mathrm{bub}}^3 - R_{\mathrm{CBA}}^3)}{R_g T_{\mathrm{sys}}}\right) = \frac{dn_{\mathrm{gen}}}{dt} \qquad (4)$$

The term of the right hand side describes the rate of gas generation from the CBA particle.
When the shrinkage of CBA is taken into account, Eq. (4) is rearranged to

$$\frac{dP_{\mathrm{bub}}}{dt} = \frac{3R_g T_{\mathrm{sys}}}{4\pi(R_{\mathrm{bub}}^3 - R_{\mathrm{CBA}}^3)}\frac{dn_{\mathrm{gen}}}{dt} - \frac{3P_{\mathrm{bub}}}{(R_{\mathrm{bub}}^3 - R_{\mathrm{CBA}}^3)}\left(R_{\mathrm{bub}}^2\frac{dR_{\mathrm{bub}}}{dt} - R_{\mathrm{CBA}}^2\frac{dR_{\mathrm{CBA}}}{dt}\right) + \frac{P_{\mathrm{bub}}}{T_{\mathrm{sys}}}\frac{dT_{\mathrm{sys}}}{dt} \qquad (5)$$

To solve equation (5) the models for giving dR_{CBA}/dt and dn_{gen}/dt are needed.

1.2.3 Rate of generation of blowing gas from the decomposed CBA, dn_{gen}/dt

In order to calculate $dR_{\mathrm{CBA}}/dt$and dn_{gen}/dt, the equation for kinetic analysis of solid-state decomposition of CBA has to be formulated. Assuming that the Arrhenius equation can be applied to express the decomposition of CBA, we use the following equation to describe the relationship between the degree of conversion and temperature. The degree of conversion can be given from the TGA data as a function of temperature.

$$\frac{dn_{gen}}{dt} = n_0\frac{d\alpha}{dt} = n_0A\exp\left(-\frac{E_a}{R_gT}\right)f(\alpha) \tag{6}$$

$f(\alpha)$ is the differential form of the kinetic model, and α is conversion defined as the ratio $(m_0 - m_t)/(m_0 - m_f)$ where m_0, m_t, and m_f refer to the initial, actual, and residual mass of the sample, respectively.

1.2.4 Rate of CBA particle shrinkage from decomposition

Assuming the CBA is a spherical shape particle, the equation for calculating the radius of the decomposed CBA particle is given by

$$\frac{dR_{CBA}}{dt} = -\frac{(R_{CBA,0})^3}{3R_{CBA}^2}\frac{d\alpha}{dt}(1-\delta) \tag{7}$$

EXPERIMENT

2.1 Materials

The base polymer used for the visual observation of batch foaming experiments were ethylene vinyl acetate copolymer resins (Eve Flex EV150, DuPont). The compositions of the samples were listed in Table 1. The samples were provided from Hitachi-Chemicals. The reason why EVA and AIBN was chosen is the high transparency. To clearly observe the bubble growth behavior, a certain degree of transparency was required for the specimen. Thus, each sample was prepared to be a sheet 0.55-0.75 mm in thickness for visual observation.

TABLE 1. Composition of sample

Sample	Resin	CBA	Crosslinking	Crosslinking	Foam
	Eve Flex EV150	AIBN	Agent	Assistant	Stabilizer
No.1	100.0	13.0	1.0	1.0	1.0
No.2	100.0	-	1.0	1.0	1.0

2.2 Foaming Experiment

A high-pressure view-cell has two sapphire windows on both walls. A microscope (Keyence VQ-Z50) with high-speed digital camera (Detect HAS-L1) is attached to one of the windows. The heaters and temperature controller are equipped to heat up the view-cell. The foaming experiments were conducted in the following way:

1) At first, a rectangular-shaped sample (size 5×5 mm^2) was placed on the base sapphire glass window of the view cell and sandwiched by the other sapphire glass.
2) Then, the view cell was heated from 298.2 K (25 °C) to 423.2 K (150 °C) at the heating rate around 10 K·min^{-1}. The temperature profile is recorded through a thermocouple equipped to the view-cell.
3) A high-speed digital camera with a microscope, which can shoot fifty pictures every second, was installed to the view cell as. The operating magnification of the microscope is 75~200. The bubble growth phenomenon was recorded in computer and the recording was started when the first bubble was observed.

2.3 TGA measurement

The decomposition rate of CBA (AIBN) was measured by TGA (DTG-60H, Shimadzu). The measurements were carried out at three isothermal conditions and at four different heating rates, 8, 10, 15, and 20 K/min with 1 mg sample under nitrogen purge at the rate of 50 cm^3/min.

2.4 Rheological measurement

The viscoelasticity of the samples without containing any chemical blowing agents was measured by using a rheometer (ARES, Rheometric Scientific, TA Instrument, USA) equipped with parallel plate geometry. The complex viscosity (η^*) was used as the viscosity (η) of Eq. (4) in the computer simulation.

RESULT & DISCUSSION

3.1 Experimental results

3.1.1 TGA Measurement

Figure 3 shows TGA data measured at isothermal conditions as well as at different heating rates. Several function of kinetic model, $f(\alpha)$, have been reported in corresponding to the kind of CBA. From Figure 3 (left), $f(\alpha)$ is considered to be 1 for describing a linear relationship between α and time. The parameters, E_a and A, were determined by the method proposed by Jankovic et al. [15,16].

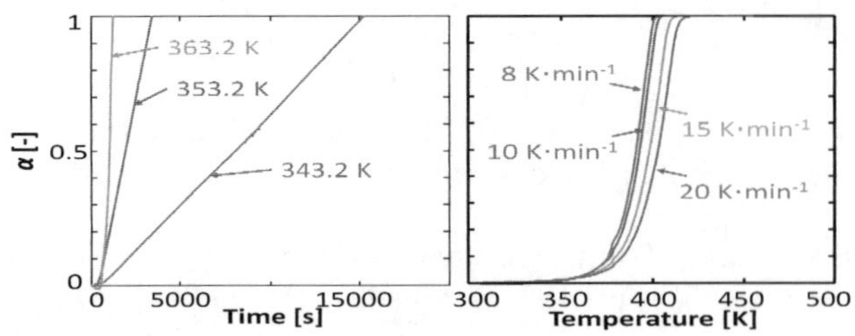

Figure 3. TGA results (left : isothermal condition, right : different heating rate)

3.1.2 Micrographs of Chemical Foaming

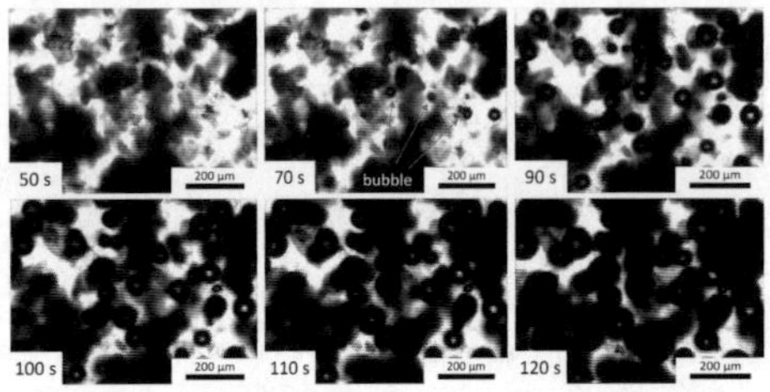

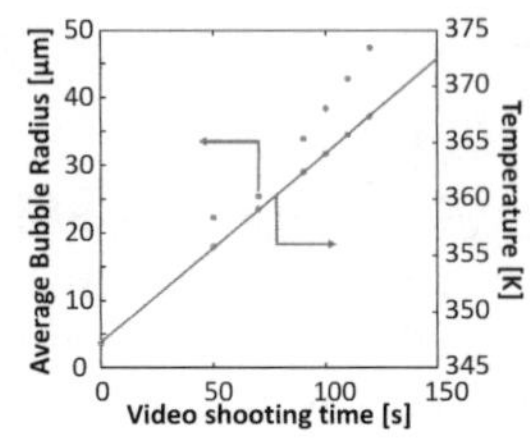

Figure 4. Series of snapshots of visual observation experiments of sample No.1
Figure 5. Average bubble size and temperature vs. time when sample No.1 was foamed

Figure 4 shows a series of micrographs obtained through visual observation experiments of the batch chemical foaming of sample No.1, which could show a dynamic behavior of bubbles' birth and growth. Due to the reflection of the light coming from the opposite side of the cell, bubbles were observed as black dots. The averaged radius of the bubbles were calculated from several snapshots of the movie and plotted against temperature and time. The bubbles appeared at temperature around 355K (82°C). The foaming temperature coincides with the CBA decomposition temperature observed in Figure 5.

3.1.3 Rheological Measurement

Figure 6 presents the temperature dependency of absolute value of the complex viscosity |η*| of sample No.2, measured with the frequency of 0.63 rad· s^{-1}. The viscosity was decreased with the increase of temperature up to around 390K. Because of the crosslinking reaction, the viscosity started increasing when the temperature exceeded 390 K (117 °C). Using these viscosity data directly in Eq. (4) as a function of temperature, the simulation calculation was conducted. However, since the visual observation could cover the bubble growth behavior in early stage, where the temperature was around 350 ~ 380 K, the effect of cross linking on the bubble growth behavior could not be observed clearly for the sample used in this study.

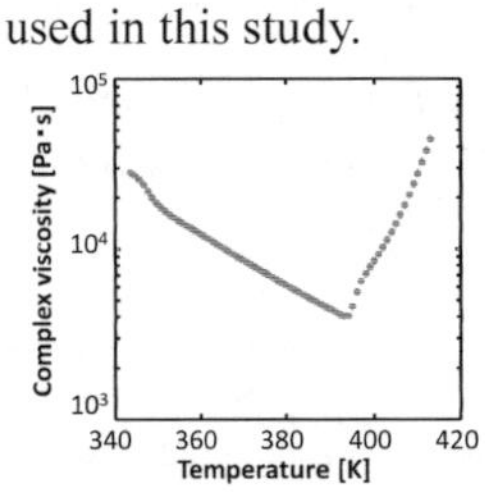

Figure 6. Temperature Dependency of complex viscosity (@0.63 rad· s^{-1})

3.2 Computer simulation

Because of the nonlinearity of the governing equations, an analytical solution was not available. Hence, a numerical simulation algorithm, which integrates the explicit finite difference scheme and the 4th order Runge–Kutta-Gill method, was employed to solve equations.

Figure 7 shows the simulation results. As observed in Figure 7, the bubble growth rate of simulation (red line) is slightly slower than experimentally observed one. However, the onset of foaming was simulated at lower temperature than the experimentally observed temperature. By tuning the value of kinetic parameters, E_a and A of Eq. (6) so as to fit the simulated bubble growth rate to the experimental data, the better agreement of simulation (green line) with experimental data was obtain as illustrated in Figure 7. This indicates the kinetic parameters are very sensitive and the identification of the parameter values are key factors of simulation.

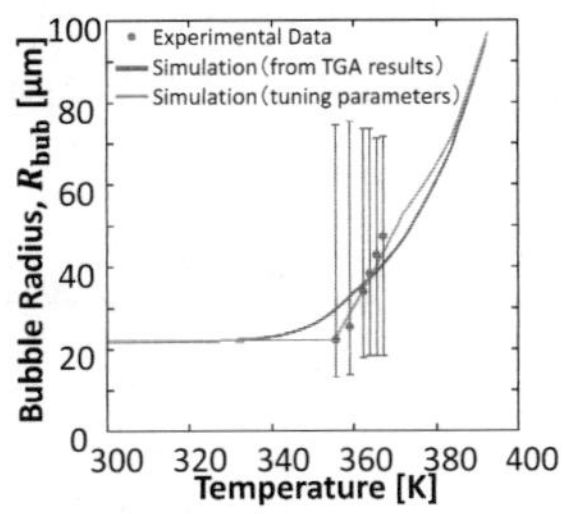

Figure 7. Experimental and Simulation Result

CONCLUSION

The growth behaviors of CBA-blown bubbler were observed using a temperature-controlled view-cell with an optical microscope digital imaging tool. The mathematical models were developed and applied to the chemical foaming system. The estimation of the rate of CBA decomposition was very much critical for describing the foaming behavior of CBA-blown process. Having the precise estimation of the kinetic reaction parameters, the developed model could estimate the foaming behaviors.

NOMENCLATURE

A = pre-exponential factor, s^{-1}; E_a = activation energy, $J \cdot mol^{-1}$; n_0 = initial amount of blowing agent produced from CBA, mol; N_{CBA} = total number of CBA, kg-sample^{-1}; n_{gen} = amount of blowing gas, mol; P_{bub} = bubble pressure, Pa; P_{sys} = mechanical pressure in the case of foaming, Pa; r = the axis of the radius direction; R_{bub} = bubble radius, m; R_{CBA} = radius of CBA, m; R_g = gas constant, $J \cdot mol^{-1} \cdot K^{-1}$; t = time, s; T_{sys} = temperature, K; V = particle volume of CBA, m^3; W_{CBA} = weight of CBA added to the sample or unit volume of the sample, kg-CBA · (kg-sample)$^{-1}$; α = conversion coefficient showing how much ratio of CBA is decomposed ($=(m_0 - m_t)/(m_0 - m_f)$); m_0 = the initial mass of the sample, kg; m_t = the actual mass of the sample, kg; m_f = the residual mass of the sample, kg; γ = surface tension, $N \cdot m^{-1}$; δ = shrinking factor; η = viscosity of polymer, $Pa \cdot s$; ρ_{CBA} = density of CBA, $kg \cdot m^{-3}$

ACKNOWLEDGMENTS

Thanks are due to Hitachi-Chemicals for providing sample for this study.

REFERENCES

1. J. R. Street, L. F. Arthur, L.P. Reiss, *Ind. Eng. Chem. Fundam.*, 10,54-64 (1971)
2. J. S. Colton and N. P. Suh, *Polym. Eng. Sci.*, 27,7,500-503 (1987)
3. M. A. Shafi, K. Joshi, R. W. Flumerfelt, *Chem. Eng. Sci.*, 52,4,635-644 (1997)
4. C. D. Han, C. A. Villamizar, *Polym. Eng. Sci.*, 18(9), 687-698, 699-710 (1978)
5. S-T. Lee, edited, Foam Extrusion, *Technomic Publishing*, (2000)
6. A. Osorio, L-S. Turng, *Polym. Eng. Sci.* 44 (12), 2274-2287 (2004)
7. L.Y. Chang, Y-J. Chang, C.H. Hsu, H. Tanaka, M. Ohshima, *FOAMS 2015* (2015)
8. K. Taki, T. Nakayama, T. Yatsuzuka, M. Ohshima, *J. Cell. Plast.*, 39(2), 155-169 (2003)
9. Q. Guo, J. Wang, C. B. Park, M. Ohshima, *Ind. Eng. Chem. Res.* 45(18)6153-6161 (2006)
10. T. Ishikawa, M. Ohshima, *Polym. Eng. Sci.*, 51(8) 1617-1625 (2011).
11. S. Costeux, I. Khan, S. P. Bunker, H. Jeon, *J. Cell. Plast.* 51(2) 197-221(2015)
12. S. Costeux, H. Jeon, S. Bunker, I. Khan, *FOAMS2015* (2015)
13. S. F. Wang and A. A. Ogale, *Polym.-Plast. Technol. Eng.*, **29**(4), 355-369 (1990)
14. Donglai Xu et al., *Journal of Cellular Plastics*, **41**, 519-538 (2005)
15. B. Janković et al., *Journal of Physics and Chemistry of Solids*, **69**, 1923-1933 (2008)
16. B. Janković, *Metallurgical and Materials Transactions B*, **40B**, 712-726 (2009)

Properties and Structure Analysis of Polymer Compositions Obtained in Cellular Injection Molding Process

Aneta Tor-Świątek[a],

[a]*Lublin University of Technology, Department of Polymer Processing*
36 Nadbytrzycka Str., 20-618 Lublin, Poland

a.tor@pollub.pl

Abstract. Currently produced polymer products, because of physical, mechanical and functional properties are made from solid thermoplastic materials by conventional injection molding. In order to make shapes diphase structure i.e. plastic-pores conventional injection molding must be replaced with the cellular injection molding. This occur through the change of process conditions as temperature of particular plasticizing system zones and temperature of the mold and owing to inserting into plastic blowing agent (porophor). Porophor increase its volume repeatedly and then ensue change of the product structure from solid into cellular along the whole or partial (solid topcoat and a cellular core) section.

Keywords: Injection molding, cellular plastic, polypropylene, low density poliethylene
PACS: 81

INTRODUCTION

Process of modification of processing and functional properties of polymer materials is widely used in processing. Currently, progress in that area is connected with new method of plastic modification, often involving the change of the structure from homogenous and solid to cellular. By of cellular injection molding method is obtained products with lower density, which also caused changes in selected properties of the product specially processed and functional [1, 2].

Cellular injection molding allow obtained parts with solid top coat and cellular core. The most important feature of this parts are [3, 4]:

- low thermal conductivity,
- low density,
- the thickness of the part is in the range of 6,6 – 15 mm,
- there is no collapses in the Surface (during cooling, the sample surface is pressed to injection mold walls) ,
- higher stiffness according to the samples with the same mass but obtained from solid material,
- there is no oriented structure (are free of internal stress, and thus do not warp and have uniform shrinkage in all directions resulting in an increased resistance to the formation of stress cracks).

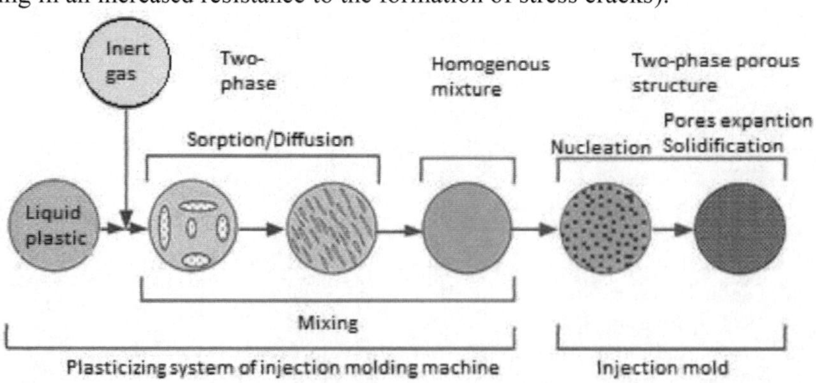

FIGURE 1. Scheme of formation of cellular structure in injection molding process

Cellular injection molding process can be divided into the methods [4]:

- high pressure,
- low pressure,
- with gas line.

High pressure injection lies in the fact that the cavity is completely filled with plasticized plastic, then when the skin produce, a cooled mold is partly opened. By increasing the volume of the cavity it is possible expansion of the gas and creating a porous structure. For larger volume slot can also be used forms with moving cores, which are withdrawn after filling plastic cavity. If the change of the degree of porosity is necessary the compacts should be appropriately chosen time and path length of mold opening or displacement of movable inserts [5, 6].

The low pressure injection foaming process for injecting a cavity portion of material (with a volume smaller than the volume of the cavity) dispersed in the gas. The inserting of gas into the material occurs in a plasticizing system or originated from the decomposition of the blowing agent. In this method, the cellular injection the cavity is filled with 60 - 80% by volume. Low pressure cellular injection is carried out at a high speed and a low injection pressure (approx. 50% lower than the pressure used during the solid plastic injection). Injection molding cycle time obtained by using a large injection speed is reduced, while using a small injection pressure makes it possible to carry out the whole process with less clamping force of the mold and simplified construction of the opening and closing of the mold. Emitted gas in the mold cavity, at low pressure, contributes to the formation of air bubbles in the material. These bubbles in the growth effect of increasing the volume of the material and an increase in pressure sufficient to continue the flow of plasticized plastic into the mold walls and the complete filling of the mold. The amount of plastic injected into the slot depends on the porosity of the final product. The clamping function, fulfill the expanding gas located in the mold. The resulting cellular structure of the material contributes to a reduction in thermal conductivity, and this leads to longer cooling time of the molding. With this method, the finished products obtained with small residual stress (low pressure in the mold cavity and low cooling rate facilitates the process of stress relaxation in the retaining molding) [7, 8, 9].

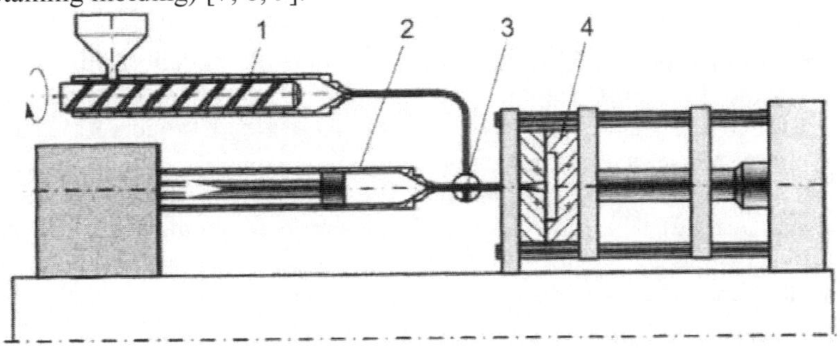

FIGURE 2. Scheme of technological line of low pressure injection molding of cellular plastic: 1 – plasticizing system, 2 – system of injection, 3 – shut-off valve, 4 – injection mold [1]

The methods of injection with the gas line can be divided into different varieties of its implementation [1, 2, 5]:
- with the flow of material from the main cavity to the secondary,
- with using a moving cores,
- with bringing gas through a nozzle injection,
- with flow return material to the plasticizing cylinder of injection machine,
- with bringing gas directly into the cavity,
- with the gas flow through the cavity across the cooling phase of the molding.

In the process of injecting the protective gas is obtained hollow moldings, and the process begins with the injection of portion of the plasticized plastic specified into the cavity (usually a smaller amount than required to completely fill the slot). The next step is injecting a gas into the interior of the plasticized plastic which pushes it along the axis of the cavity and in the transverse direction towards the walls of the cavity. By the pressurized gas material is pressed against the walls of the latter, giving compact of definitive shape, and contacting the plasticized plastic with cold mold causes the rapid solidification. After completion of the cooling phase followed by reduction of the molding pressure of the gas and it is discharged from the molds just before the mold opening [1].

EXPERIMENTAL

The cellular injection process was carried out using a screw injection molding machine CS - 88/63 Czech production. It is an injection molding machine designed for processing thermoplastics.

FIGURE 2. View of injection mold with samples of cellular PP

The device is adapted for use in automatic, semi-automatic and manual control activities (each phase of the cycle can be implemented at different times). This is the injection molding tool with the system closed horizontally using a hydraulic mechanism.

Materials

To cellular injection molding process the plastic was used:

a) Low Density Polyethylene PE-LD with the trade name Malen E FABS 23 D022. Producer Basell Orlen Polyolefins Sp. z. o. o.
b) Polypropylene PP with trade name Moplen EP440G. Producer LyondellBasell.

The both plastics was modified with the blowing agent in the form of microspheres, with the trade name Expancel® 951 MB 120. Producer AkzoNobel (Sweden). Microspheres Expancel 951 MB 120 is not a hazardous substance or mixture. During its use, may form flammable or explosive vapor-air mixture. It contains no substance considered to be persistent, bioaccumulative or toxic. It is a blowing agent, which contains 65% microspheres in the carrier EVA (ethylene vinyl acetate). The main characteristic features of microspheres used was shown in table1.

TABLE (1) Characteristic of microspheres (producer data)

Characteristic feature	Value
Particle size	28 – 38 μm
Amount of solid	65 ± 1 %
Gravity	1000 kg/m^3
Start of decomposition	138 – 148°C
Temperatures of processing	180 - 200°C
Dosage during injection process	1,5 – 9%

Process Conditions

Injection foaming process was realized with the developed processing conditions, taking into account the temperature of the PVC processing and decomposition temperature inside miceospheres. They are as follows:

- Temperature in cylinder heating zones: I - 180°C, II – 190°C, III – 200°C
- Temperature of injection die: 210°C
- Injection pressure in hydraulic system od the machine: 9 MPa
- Clamping pressure in hydraulic system of machine: 6 MPa
- Content of microspheres: 0%; 0,5%; 1%; 2%

The values of process factors changed during injection molding process was shown in table 2.

TABLE (2) Values of the process conditions

Sample Series.	Injection Time [s]	Cooling Time [s]
A	3	7
B	7	20
C	10	7
D	10	20

RESULTS

In the frame of the studies the results of density and tensile strength of solid and cellular samples was obtained. Results are shown on Figures 3÷6.

The density (Fig. 3) of the modified LDPE decreased with the increasing of the content of microspheres. The maximum change is 3%. The changed process factors also have the influence on obtained density results, but this change is not significant. The same situation occurs in results of PP density (Fig. 4). According to the content of microspheres the density is reduced by 6%. I both cases the chenge of process factors, injection and cooling time, have a main influence during processing LDPE and PP with content of microspheres more than 1% mas. Increasing injection time increase the density destrease.

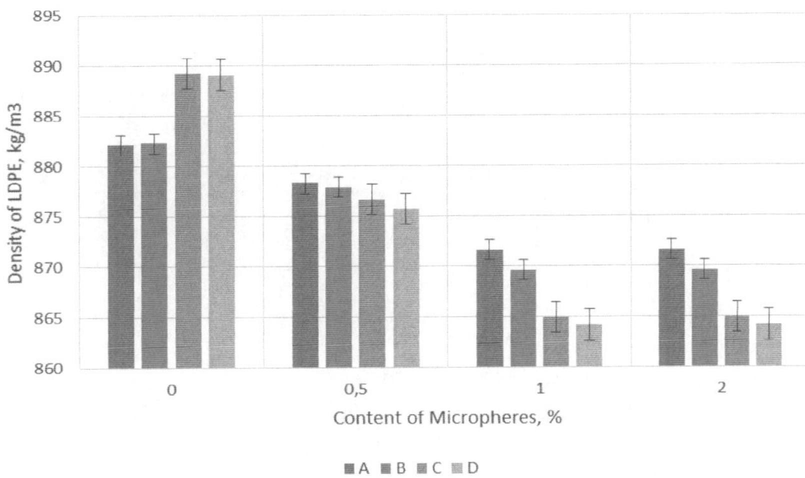

FIGURE 3. Dependence of content of the microspheres on the density of cellular LDPE for changed process conditions

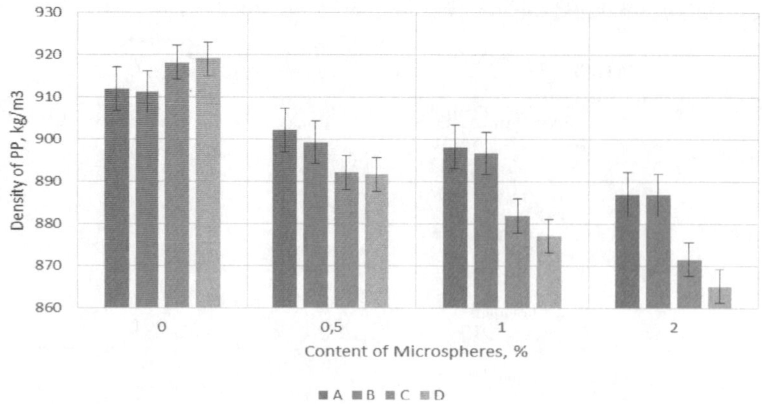

FIGURE 4. Dependence of content of the microspheres on the density of cellular PP for changed process conditions

According to the mechanical properties the tensile strength and elongationa at breake was measured. The result shown that both the content of microspheres and changed process conditions have the influence on studied property, but this influence do not coused significant changes in the material. The tensile strength of LDPE (Fig. 5) stay at the same level and the average value is 10.6 MPa. In case of obtained tensile strength of PP (Fig. 6) the value decreased about 5MPa with the increase the content of microspheres.

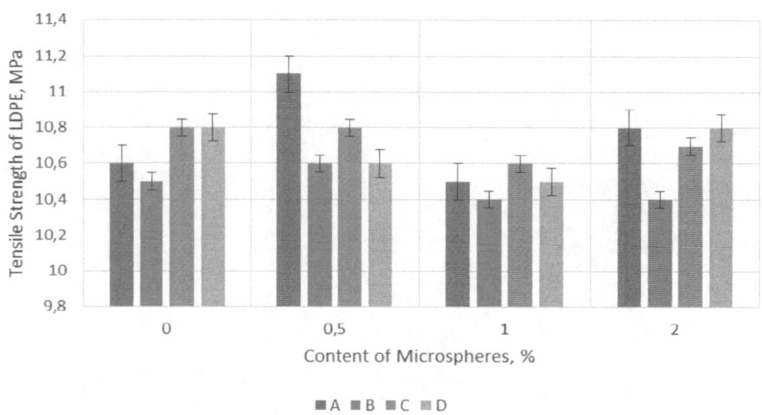

FIGURE 5. Dependence of content of the microspheres on the tensile strength of cellular LDPE for changed process conditions

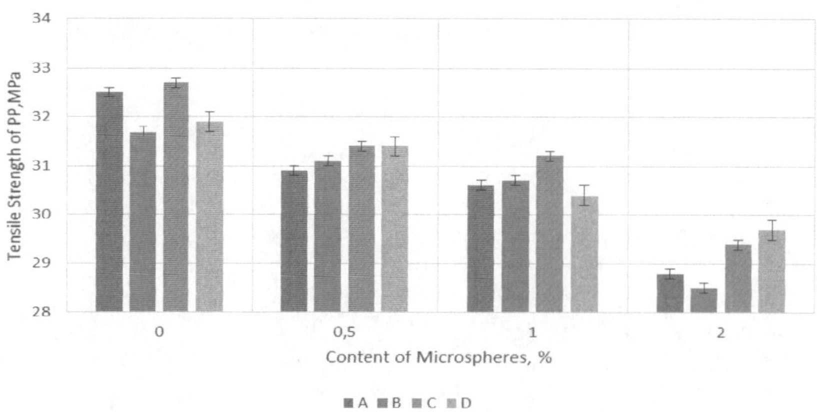

FIGURE 6. Dependence of content of the microspheres on the tensile strength of cellular LDPE for changed process conditions

In the frame of the studies the research of the materials structure was carried. The morphology of the sampes are shown on Figures 7.

The image analysis results enabled determining the distribution of micropores in the cross section of the samples. Cellular injection process resulted the cellular molded products with closed micropores. The growth of micropores is significantly lower when the microspheres are added at a quantity 0.5%, respectively. However, the amount of micropores increased, mainly in the core of the sample.

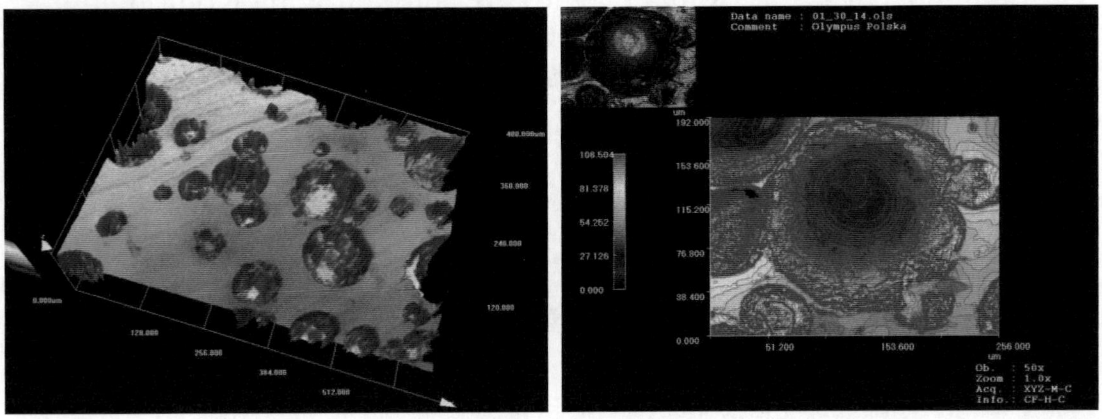

FIGURE 7. View of cellular LDPE structure with visible microspheres

SUMMARY

A dynamic growth of the polymer processing sector, new directions in polymer applications and higher product requirements have lead to the development of innovative technologies and enhancement of auxiliaries used in polymer processing. Cellular injection molding processes are fast developing methods for processing thermoplastics.

Presented results of research on changes in physical and mechanical properties and in the structure of obtained cellular molded parts are reflected in the literature. However, the changes in the properties and structure indicate the need for a broader analysis of the impact of the measures conducted on the individual parameters of the process, taking into account otherwise proportioning of the microspheres more than 2%mas. and process conditions.

REFERENCES

1. E. Bociaga, Special methods of injection molding, WNT, Warszawa, 2008, pp. 101-121.
2. E. Bociaga, P. Palutkiewicz, *Polimery*, **57**, 38-48 (2012).
3. T. Garbacz, *Cellular Polymers,* **33,** 71-79 (2014).
4. T. Garbacz, *Polimery*, **58**, 295-303 (2013).
5. T. Jachowicz et al., Materials Science, **50**, 296-302 (2014).
6. A. Tor – Świątek et al., *Materials Science,* **49,** 812-818 (2014).
7. Tor – Świątek A., *Eksploatacja i Niezawodnosc-Reliability and Maintenance,* **15,** 225-229 (2013).
8. Hu Guanghong and Wang Yue, Microcellular Foam Injection Molding Process, Some Critical Issues for
9. Injection Molding, Dr. Jian Wang (Ed.), INTECH, Shanghai, 2012, pp. 175-202.

Reactive Extrusion Of Poly (lactic acid): Effect Of Coagent Type on the Rheological and Thermal Properties

Praphulla Tiwary, Karolina Dawidziuk and Marianna Kontopoulou*

Department of Chemical Engineering, 19 Division St, Queen's University, Kingston, Ontario Canada, K7L 3N6

**Kontopm@Queensu.ca*

Abstract. Poly (lactic acid) suffers from poor melt strength and slow crystallization kinetics, hindering its application in various polymer processing operations, such as injection molding, film casting etc. The objective of this work was to implement peroxide mediated reactive extrusion in the presence of different coagents: triallyl trimesate (TAM), trimethylol propanetrimethacrylate (TMPTMA), trimethylol propanetriacrylate (TMPTA), and pentaerythritol triacrylate (PETA). The type and amount of coagents and peroxide was varied to prepare formulations with improved melt strength and better thermal properties. The chain branching and thermal properties of the reactively modified formulations were found to be dependent on the type and amount of coagent with TAM and PETA being the most efficient in improving both the rheological and thermal properties of PLA.

Keywords: reactive extrusion, poly (lactic acid), coagent modification, branching.

PACS: 81.05.Lg

INTRODUCTION

Poly (lactic acid), or polylactide (PLA) is one of the major commercially available bioderived, biodegradable thermoplastic polyesters [1]. PLA suffers from low melt strength and very slow crystallization rates. This restricts its processability under common polymer processing operations (injection molding, film casting, etc.) and has limited its applications to commodity food packaging. PLA is also used in biomedical applications where biocompatibility and biodegradability are desired [2]. Various PLA grades have been approved by the FDA (Food and Drug Administration – U.S.) and EU-Plastic Regulations (Europe) for food and drug packaging applications [3].

PLA is currently synthesized by ring opening polymerization of lactide, a cyclic dimer of dehydrated lactic acid produced by fermentation [4-6]. Reactive extrusion can be further implemented to introduce branching. Recently it has been shown that long chain branching can be produced through free-radical mediated dicumyl peroxide (DCP)-initiated grafting of triallyl trimesate (TAM) coagent in the melt state [7,8]. In addition to improving the melt strength of PLA, this approach also resulted in enhanced crystallization kinetics. The final product demonstrated significant improvements in strain hardening, and increased crystallinity and crystallization rates, indicating that this is a promising approach to enhance the processing characteristics and properties of PLA [7].

The objective of this work is to investigate the effect of different coagent types on the thermal and rheological properties of PLA.

MATERIALS AND METHODS

The coagents used in this study were pentaerythritol triacrylate (PETA), trimethylolpropane triacrylate (TMPTA) trimethylolpropane trimethylacrylate (TMPTMA), and triallyl trimesate (TAM). PETA, TMPTA, TMPTMA, acetone, tetrahydrofuran (THF), ethanol, and chloroform were obtained from Sigma-Aldrich®. TAM was obtained from Monomer-Polymer and Dajac Labs®. All the coagents and solvents were used as received. To prepare reactively modified PLA formulations, PLA 3251 D from NatureWorks® LLC, which is an FDA approved grade, was ground into powder, mixed with appropriate ratios of DCP and coagent then coated with an acetone solution to yield a batch size of 30 g. Samples were dried in a vacuum oven at 60 °C for 12 h to remove residual solvent and moisture. This procedure yielded various compositions containing 0.1 – 3.0 wt.% coagent and 0.3 – 0.6 wt.% DCP. The various formulations were designated based on the amount of coagent and DCP (e.g. 1% PETA 0.3 DCP denotes PLA reacted with 1 wt.% coagent, in this case PETA, and 0.3 wt.% DCP). Neat PLA was processed under the same conditions outlined above, to provide a suitable basis for comparison. The formulations, containing between 0.1 – 3.0 wt.% coagent and 0.3 – 0.6 wt.% DCP were charged to a DSM micro-compounder, equipped with twin-co-rotating screws, at 180 °C at 100 rpm for 6 min. After compounding, the strands were chopped into pellets for further characterization. Compression molded discs, 25 mm in diameter and 2 mm thick, were prepared using a Carver hydraulic press at 180°C. The linear viscoelastic (LVE) properties were measured in the oscillatory mode using a stress controlled rheometer (Visco Tech by Rheologica). Frequency sweeps were conducted at 180 °C using 20 mm parallel plates. The average of three measurements are reported throughout this work. The obtained data was fitted to the Cross model.

$$\eta(\omega) = \frac{\eta_o}{1 + |\lambda\omega|^{1-n}} \quad (1)$$

where η is the shear viscosity, η_o is the zero shear viscosity, λ is the relaxation time, n is a constant related to the shear thinning behaviour, and ω is the frequency in rad/s. DSC experiments were conducted according to ASTM D3418 using a DSC Q 100 by TA Instruments. Samples weighing 8-10 mg were sealed in aluminum hermetic pans and scanned between 0 and 200 °C at a heating rate of 5 °C.min^{-1}. After the first heating, each sample was held isothermally at 200 °C for 3 min. before cooling at the rate of 5 °C.min^{-1}, to determine the crystallization onset and peak. The percentage crystallinity of the polymer, X_c, was estimated using equation (2).

$$X_c = \frac{\Delta H_m - \Delta H_{cc}}{\Delta H_{100}} x100 \quad (2)$$

where ΔH_m is the apparent fusion enthalpy, ΔH_{cc} is the exothermic enthalpy that is absorbed by crystals formed during the heating scan and ΔH_{100} is the theoretical fusion enthalpy of a 100% crystalline polymer, which is 93 J.g^{-1} for PLA [9].

RESULTS AND DISCUSSION

The thermal properties of PLA were significantly altered upon addition of coagents (Table 1). Reactive modification of PLA with DCP and PETA coagent resulted in formulations that exhibited sharp crystallization peak, which was absent in neat PLA (Figure 1a). This observation is consistent with the nucleation effect reported previously in PLA modified with TAM [7]. It is also evident that the concentration of PETA and DCP had an effect on the shape of crystallization peak. Furthermore, the cold crystallization peak occurring in the DSC endotherms at around 95°C in neat PLA disappeared (Figure 1b).

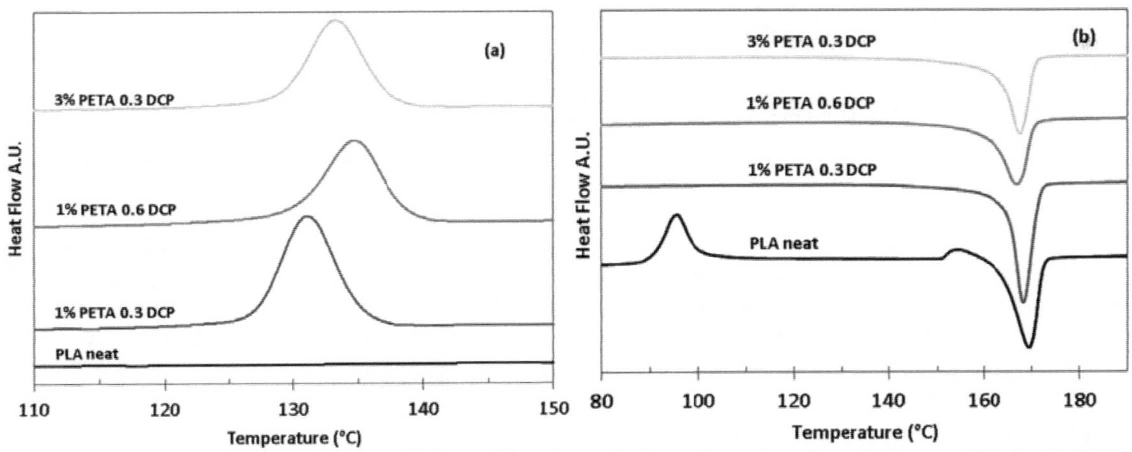

FIGURE 1. (a) Cooling exotherm (b) Second heating endotherm for various formulations modified with PETA in comparison to neat PLA.

Similar property changes were noted when PLA was reacted with the rest of the coagents (Table 1). At similar concentrations of coagent, PETA showed a maximum shift in the crystallization peak followed by TMPTA and TMPTMA (Figure 2). Figure 2 also shows that TAM is significantly more efficient, as it results in the most dramatic changes at a concentration of only 0.1% TAM 0.3 DCP. Depending on the concentration of coagent and peroxide used in the reactive extrusion, the broadness of the melting peaks varied. 1% TMPTA 0.3 DCP and 1% TMPTMA 0.3 DCP formulations showed a double peak in the melting DSC curve (Table 1).

TABLE 1. Thermal properties of PLA and reactively modified counterparts. T_c – crystallization peak, T_m- melting peak.

Material	T_c (°C)	T_m (°C)	Crystallinity (%)
PLA neat	N/A	169	[a]20
0.3 DCP	N/A	163	37
0.1% TAM 0.3 DCP	134	169	43
1% PETA 0.3 DCP	131	168	52
1% PETA 0.6 DCP	135	167	48
3% PETA 0.3 DCP	133	167	47
1% TMPTA 0.3 DCP	124	[b]167, 171	53
1% TMPTA 0.6 DCP	134	168	44
3% TMPTA 0.3 DCP	134	167	44
1% TMPTMA 0.3 DCP	111	[b]164, 169	47
1% TMPTMA 0.6 DCP	127	166	51
3% TMPTMA 0.3 DCP	135	167	45

a) cold crystallization peak present in 2nd heating curve. b) double peak in melting region

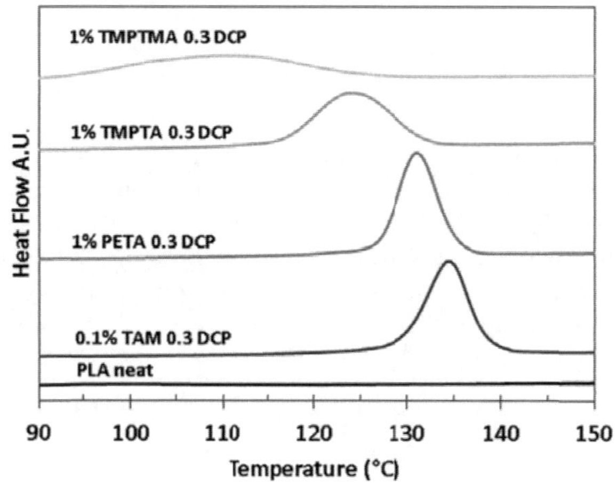

FIGURE 2. Cooling exotherm of various coagents at similar composition of 1% coagent and 0.3 DCP compared with respect to 0.1% TAM 0.3 DCP.

The concentration of the coagent also had an effect on the rheological properties of the samples. Increasing the concentration of PETA or DCP changed the dynamic oscillatory response of the samples with disappearance of the Newtonion plateau and pronounced shear thinning (Figure 3a). The Van-Gurp plot (Figure 3b) shows that the samples with 3% PETA 0.3% DCP and 1% PETA 0.6% DCP deviate from the terminal value of 90°. This hints the presence of branched structures in these formulations. Furthermore, the deviation from slope of 2 (Figure 3c) shows enhanced relaxation time in the aforementioned formulations owing to the presence of branching. All the samples showed negligible amount of gels showing absence of any crosslinking reactions.

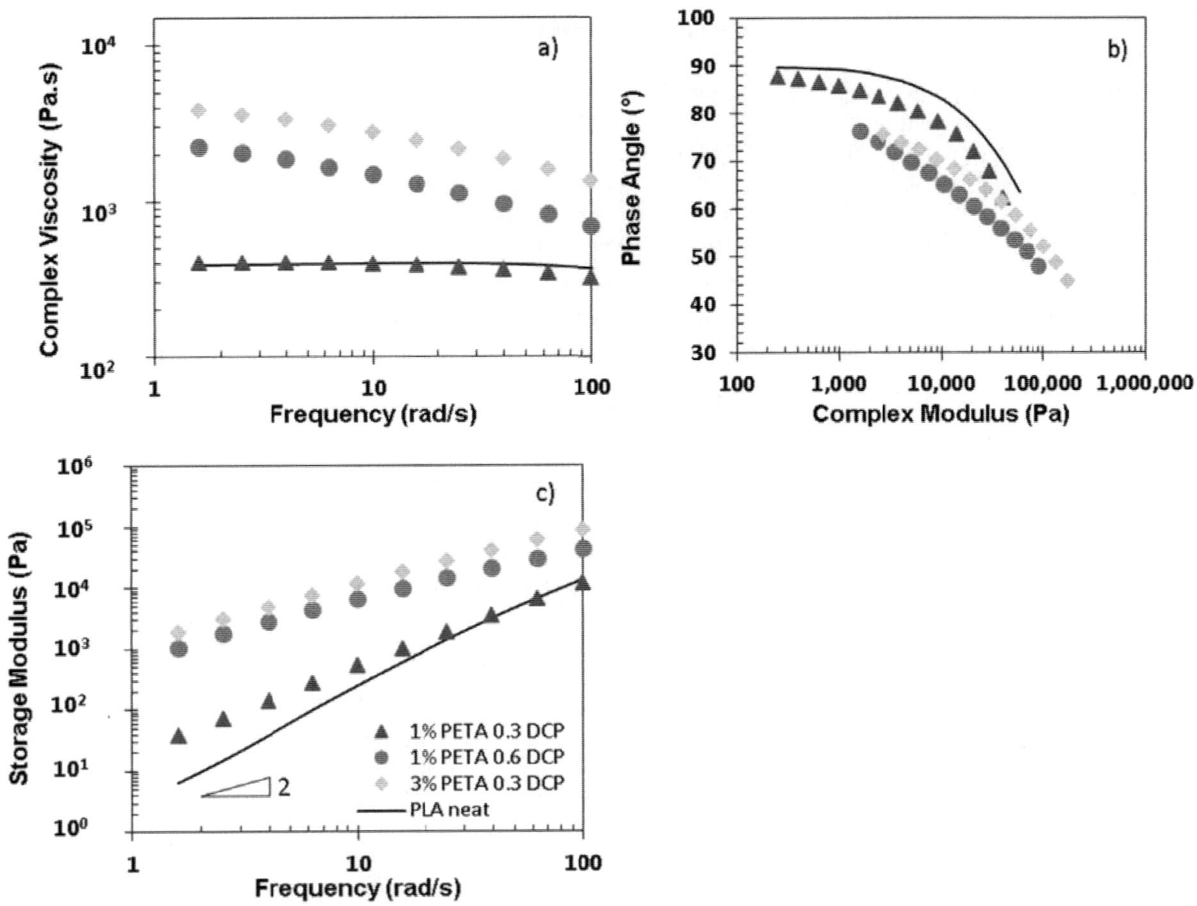

FIGURE 3. (a) Complex viscosity as a function of frequency, (b) phase degree as a function of complex modulus, and (c) storage modulus as a function of frequency of PETA coagent at 180°C.

The linear viscoelastic properties are sensitive to the structural change of material and have been used to verify the existence of branching in linear PLA. Overall, coagent modification resulted in higher zero shear viscosities, more shear thinning and longer relaxation times when compared to neat PLA. The 3% PETA 0.3 DCP formulation showed the highest zero shear viscosity and longest relaxation time at similar compositions (Table 2). It should be noted that once again the TAM coagent appears to be the most effective, with only 0.1 wt% showing a substantial increase in viscosity.

TABLE 2. Cross model fits at 180°C obtained by fitting of the dynamic oscillatory data. ηo is the zero shear viscosity, λ the relaxation time at zero shear, and n is a constant related to shear thinning behavior

Material	η_o (Pa s)	λ (s)	n
PLA neat	410	0.0006	0.25
0.1% TAM 0.3 DCP	2,162	0.06	0.51
3% PETA 0.3 DCP	5,322	0.085	0.49
3% TMPTA 0.3 DCP	5,000	1.58	0.61
3% TMPTMA 0.3 DCP	760	0.0034	0.30

CONCLUSION

This research showed that the amount and type of coagent plays a critical role in modifying the rheological properties and inducing a nucleation effect in reactively modified PLA. Of all the coagents studied in the present work, PETA and TAM are the most efficient in fulfilling the objectives. Though all the reactively modified coagents showed presence of crystallization peak which was absent in neat PLA.

ACKNOWLEDGMENTS

The authors will like to thank NSERC (National Science and Engineering Research Council of Canada) / Collaborative Research and Development (CRD) program and E.I. DuPont Canada for the financial assistance.

REFERENCES

1. S. Park, T. Kim, M. Kim, S. Lee and S. Lim, *Biotechnology Advances*, **30**, pp. 1196 (2012).
2. R. Platel, L. Hodgson and C. Williams, *Polymer Review*, **48**, pp. 11-63 (2008).
3. NatureWorks, "Ingeo Biopolymer 3251D Technical Data Sheet".
4. R. Drumright, R. Gruber and D. Henton, *Adv Mater*, **12**, pp. 1841-1846 (2000).
5. A. Sodergard, M. Stolt, *Prog Polym Sci*, **27**, pp. 1123-63 (2002).
6. E. Vink, K. Rabago, D. Glassner, B. Springs, *Macromol Biosci*, **4**, pp. 551-64 (2004).
7. M. Nerkar, J.A. Ramsay, B. A. Ramsay and M. Kontopoulou, *Macromol Mat Eng*, **299**, pp. 1419-1424 (2014).
8. M. Nerkar, J. A. Ramsay, B. A. Ramsay, A.A.Vasileiou and M. Kontopoulou, *Polymer*, **64**, pp. 51-61 (2015).
9. J. Pan, Y. Wang, S. Qin, B. Zhang and Y. Luo, *J Biomed Mater Res B Appl Biomater*,**74**, pp. 476-80 (2005).

Innovative Phosphate-Free Alternatives for Silicate Resins Applied in Sewer Rehabilitation

Mark Kopietz, Sergiy Grishchuk and Bernd Wetzel

Institut für Verbundwerkstoffe GmbH, Erwin-Schrödinger-Straße 58, D-67663 Kaiserslautern

Mark.Kopietz@ivw.uni-kl.de

Abstract. Innovative organic-inorganic polyurea/polysilicate (denoted as 2P) hybrid resins were made from water glass (WG) dispersed in a mixture of polymeric methylene diphenyl diisocyanate (PMDI) containing epoxidized linseed oil (ELO) in different concentrations acting as emulsifier, plasticizer and phase-transfer-agent at the same time. ELO, made from renewable resources, replaces unfavorable chemically non-bonded organic phosphates used in common reference resins. The ratio of PMDI/WG maintained constantly at 2:1 parts by volume, where different concentrations of ELO were added (5, 10, 15, 20 and 25 vol%). The resulting 2P hybrid resins were tested by different methods. For the evaluation of the processability of new resins the initial viscosity and its progress during reaction time at room temperature (RT) was determined and monitored by Brookfield viscometry. The applicable pot life was set to reach 15 Pa·s. The progress of post-curing during 14 days at RT was investigated using Shore D hardness measurements. Fracture mechanical, three-point bending and compression tests were carried out to compare the hardened materials. In this work it is shown that ELO accelerates the curing and post-curing reactions and improves mechanical properties at concentrations with 5-15 vol%. Slight reduction of fracture toughness was observed for all concentrations of ELO except 10 vol%.

Keywords: 2P hybrid resins, phosphate-free silicate resins, polyisocyanate/water glass, polyurea, polysilicate, renewable resources, composites for trenchless sewer rehabilitation

PACS: 81.05.Qk, Reinforced polymers and polymer-based composites

INTRODUCTION

Polyurea/polysilica hybrid polymers and composites made from highly reactive polyisocyanates and water glass (WG, an aqueous solution of alkali metal silicate: $m\text{Me}_2\text{O} \cdot n\text{SiO}_2$) has gained big acceptance in repair of pipes (trenchless sewer rehabilitation) as well as in strengthening of walls and rocks (mining industry, tunnels) in the past years. The research interest for such organic-inorganic hybrid materials is mostly driven by improvements in mechanical properties and fire resistance [1].

Dispersing water component (in this case water glass, WG) in polyisocyanate yields in highly reactive water-in-oil emulsions (W/O-emulsions) to form carbon dioxide and primary amines as intermediates (eq. (1)), reacting afterwards with remaining isocyanate groups to form polyurea (eq. (2)). At the same time, WG serves as a cheap precursor for the in situ sol-gel formation of the silica particles. The silification reaction induced by reaction-generated CO_2 (eq. (1)) is represented by eq. (3) [2]. In addition to better stability of W/O-emulsion, depression of foaming due to chemical bonding of CO_2 is an additional benefit of WG compared to pure water. Note that functionality of an isocyanate higher than two results in a highly crosslinked polyurea-based thermoset.

$$\text{R{-}N{=}C{=}O} + \text{H}_2\text{O} \longrightarrow \text{R{-}NH}_2 + \text{CO}_2\uparrow \quad (1)$$

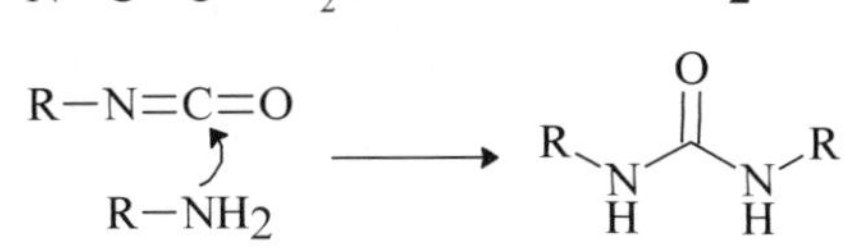

(2)

$$m\text{Me}_2\text{O} \cdot n\text{SiO}_2 \cdot x\text{H2O} + m\text{CO}_2 \rightarrow n\text{SiO}_2 \cdot (x-y)\text{H}_2\text{O} + m\text{Me}_2\text{CO}_3 + y(\text{H}_2\text{O}) \quad (3)$$

Nevertheless, without suitable emulsifiers it is hardly possible to control the stability and reactivity of W/O-emulsion. Therefore, commercially available indirect produced polyurea/polysilica resins contain liquid organic phosphates which are also regulating the rapid curing reaction and act as phase-transfer-agents for the reaction components [3]. Because slow diffusion and/or extraction of liquid phosphates into the application media, the properties of cured resins can change with time. Furthermore, it could be a problematic issue for several application in the future due to the regulations of chemicals and their safe use decided by the European Union (EC 1907/2006, REACH - Registration, Evaluation, Authorisation and Restriction of Chemicals) [4]. Recently, the high potential of melamine-formaldehyde (MF) [5], epoxy [6] and vinyl ester [1] resins as reactive emulsifiers and hybridization components for 2P resins was reported before. However, all investigated resins were petroleum based. Accordingly, our strategy was to replace chemically non-bonded organic phosphates with reactive emulsifiers made from renewable resources – in this case epoxidized linseed oil (ELO). Linseed oil consists of triglyceride esterified with organic fatty acids in different amounts (oleic acid, linoleic acid and linolenic acid). To involve reactive epoxy groups it is possible to introduce them at unsatisfied positions traditionally via Prileschajew reaction with peroxy acids [7] or via chemo-enzymatically reaction with lipase [8], respectively. With a renewable carbon content of 100 % ELO increases the bio-content of the matrix and decreases greenhouse effect via reduction of CO_2 emission.

To avoid diffusion of ELO after curing its chemical bonding to the matrix components is necessary. Epoxy groups are able to be incorporated into the structure of polyurea network via several reactions. The intermediate amine (eq. (1)) is able to undergo a nucleophilic addition to the electrophilic carbon atom in the highly reactive epoxy ring with subsequent ring opening and formation of secondary hydroxyl as shown in equation (5).

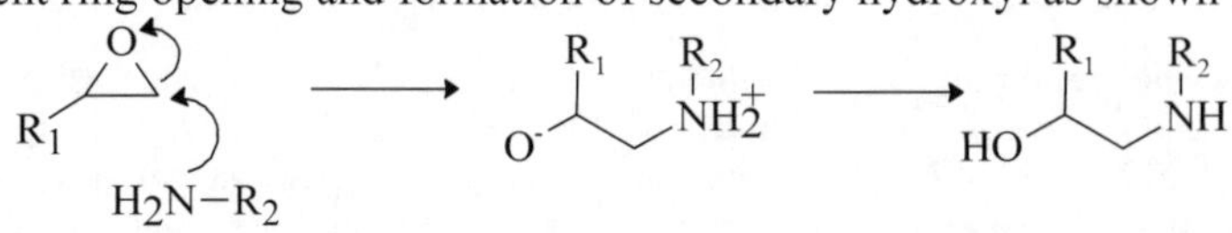

(5)

In the next step the formed hydroxyl group is able to co-react in a nucleophilic addition at the electrophilic carbon atom in the isocyanate group resulting in additional crosslinks via urethane network hybridization (eq. (6)).

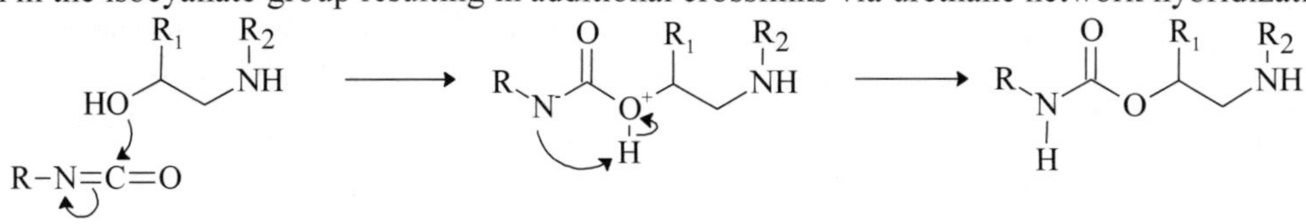

(6)

In another reaction the epoxy group is able to directly react with an isocyanate group to form five-ring oxazolidone linkage (eq. (7)).

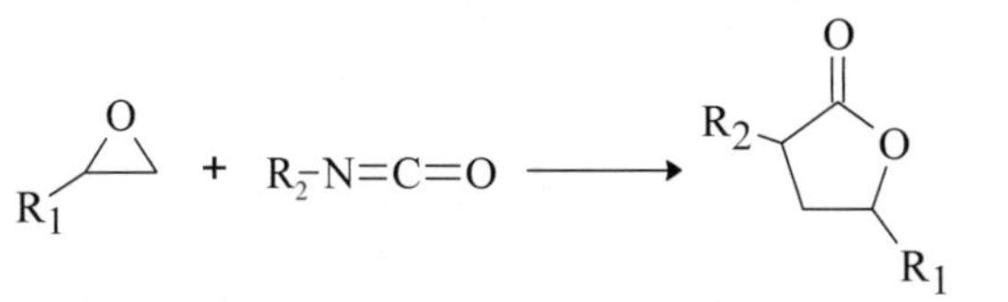

(7)

So the aim of this contribution was to check whether bio-based ELO can overtake the role of an 'active emulsifier' in 2P formulations and to determine whether the incorporation of ELO is accompanied with improvement with respect to the mechanical and fracture mechanical properties of the corresponding hybrid resins.

EXPERIMENTAL

Materials and Formulations

As polyisocyanate source 4,4'-diphenylmethane diisocyanate in a mixture with higher functionalized prepolymers (PMDI) was used. PMDI Ongronat® 2500 with NCO content 30-32 wt%, viscosity 520-680 mPa·s and density of 1.24 g·cm^{-3} was purchased from Borsodchem Zrt. (Kazincbarcika, Hungary). Water glass (WG) PI-Liner 3P Komp. A, a ~50 wt% aqueous sodium silicate solution with a Na_2O/SiO_2 ratio of 1 : >1.6 – <2.6, viscosity 600 mPa·s and density 1.55 g·cm^{-3}, was purchased from Polinvent Ltd. (Gyál, Hungary). As emulsifier and plasticizer ELO (Lankroflex™ L) was purchased from Akcros Chemicals Ltd. (Manchester, United Kingdom) with 100 % renewable carbon content, epoxy equivalent weight 182 g/eq., viscosity 1300 mPa·s and density 1.034 g·cm^{-3}. All chemicals were used as received.

The resins were prepared at room temperature (RT) and under air atmosphere. First of all, PMDI and the required amount of ELO were added into a plastic beaker (V = 800 ml) and homogenized with a turbine stirrer for 1 min at 250 rpm (revolutions per minute). To avoid entrapped air the resulting mixtures were degassed and afterwards WG was added and dispersed for 1 min at 500 rpm. Before casting the mixture into steel-molds it was tested by Brookfield viscometer to collect viscosity data (see below). The samples were cured at RT for 14 days. The present matrix formulations were denoted as 2P (from polyisocyanate/polysilicate).

The PMDI/WG ratio was constantly at 2:1 parts by volume, the additive content of emulsifier ELO increased (5, 10, 15, 20 and 25 vol%, respectively).

Characterization

After mixing all components the viscosity of the reactive liquid formulation was determined with a Brookfield rotational viscometer HBDVII+ (Lorch, Germany) at 10 rpm with spindle number 2. The test was stopped by reaching either 60 minutes or 15 Pa·s.

The post-curing of the materials was determined by Shore D hardness for duration of 14 days after mixing. Therefore a device from Zwick/Roell (Ulm, Germany) with a 30° cone was used.

To determine (fracture) mechanical properties a universal testing machine of Zwick/Roell Type 1474 (Ulm, Germany) was used. The fracture toughness (K_{Ic}) was measured according to ISO 13586-1 standard with $v = 1$ mm·min^{-1}. Compact tension (CT) specimens with dimensions of 35 x 35 x 4 mm³ were prepared and notched by saw. Then sharp cracks were initiated by a razor blade with a special tapping device. The critical stress intensity factor K_{Ic} was calculated by eq. (8) and (9), where B [mm] represents the specimen thickness, W [mm] the distance from load introduction line to the specimen end and a [mm] the distance from load introduction line to crack end.

$$\mathrm{f}(\alpha) = \frac{(2+\alpha)\,(0.886 + 4.64\,\alpha - 13.32\ \alpha^2 + 14.72\,\alpha^3 - 5.6\,\alpha^4)}{(1-\alpha)^{\frac{3}{2}}},\quad \alpha = \frac{a}{W} \tag{8}$$

$$K_{Ic} = \frac{\mathrm{f}(\alpha)\left(F_{max} \cdot 10^{\frac{3}{2}}\right)}{B \cdot \sqrt{W}} \tag{9}$$

Flexural properties were determined by three point bending tests according to DIN EN ISO 178. Rectangular specimens (60 x 10 x 3 mm³) were tested with $v = 2$ mm·min^{-1} until fracture.

Compression tests were carried out according to DIN EN ISO 604. Cubical specimens (15 x 15 x 15 mm³) were compressed with $v = 1$ mm·min^{-1}.

Flexural and compression modulus, strength and strain at maximum load, respectively, were measured. Flexural and compression moduli were estimated from the linear elastic zone of the stress-strain curves after Hooke's law (eq. (10)).

$$E = \frac{\sigma}{\varepsilon} \tag{10}$$

RESULTS AND DISCUSSION

After mixing all components the W/O-emulsion starts to react chemically what is expressed in an increasing viscosity. The data are significant for the application as glass fiber reinforced plastic in sewer rehabilitation. The influence of the different ELO concentrations on the development of Brookfield viscosity during curing for first 60 min and Shore D hardness during post-curing of related systems for 14 days at RT is shown in FIGURE 1.

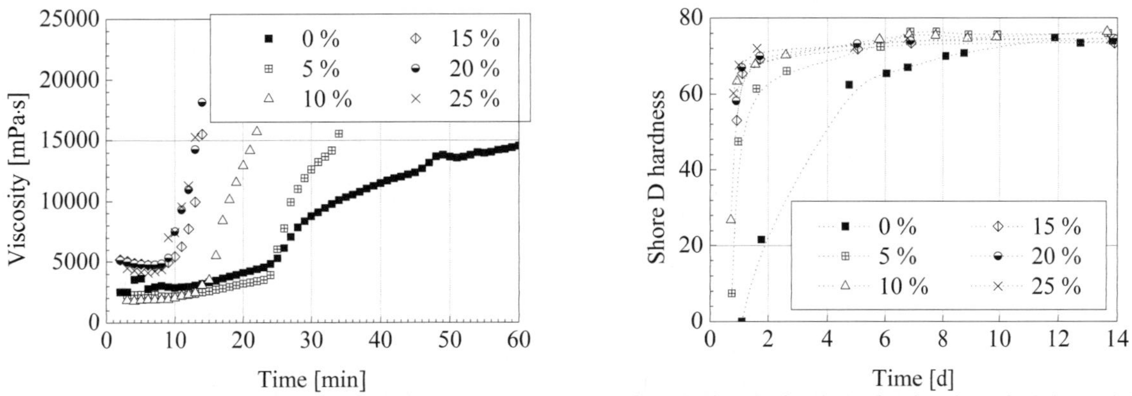

FIGURE 1. Brookfield viscosity for duration of 60 minutes (left) and Shore D hardness for duration of 14 days (right) of 2P hybrid resins with 0, 5, 10, 15, 20 and 25 vol% ELO

Brookfield measurements show initial viscosities of 2-3 Pa·s for 0-10 vol% of ELO whereas higher content increases the initial viscosity to more than 5 Pa·s caused by both the higher viscosity of ELO and higher reaction speed of components in its presence. An increasing ELO concentration shifts the pot life of the 2P hybrid resins, as defined to reach 15 Pa·s, to obviously shorter times. Note that processing related requirement of application in sewer rehabilitation is 10-30 min to reach viscosity of 15 Pa·s. Equally, the post-curing process, which is related to time for reaching final performance of 2P resins not later than after 14 days of curing, was determined by Shore D hardness. It is clearly accelerated with increasing ELO content. After two days all emulsified 2P hybrid resins are reaching hardness values >60 Shore D. This allows the repaired sewer to be put into operation just shortly after renovation, whereas, the non-emulsified material remains very soft (~25 Shore D) after 2 days of curing. This also indicates that ELO is an effective phase-transfer-agent for main components of 2P resins. The final hardness of ELO-containing systems is reached after 7-10 days of post-curing at RT whereby emulsifier ELO is expected to be chemically bonded to the matrix network as described above and switches his function to be a plasticizer. This change is expressed in mechanical properties of the 2P hybrid resins. The flexural and compression properties (FIGURE 2), respectively, as well as the fracture toughness in terms of critical stress intensity factor K_{Ic} (FIGURE 3), are presented below.

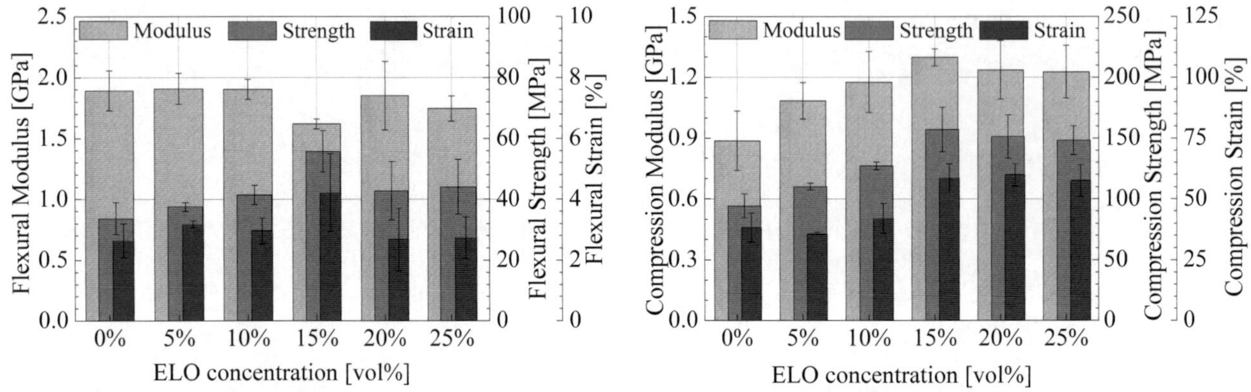

FIGURE 2. Flexural (left) and compression properties (right) of 2P hybrid resins with 0, 5, 10, 15, 20 and 25 vol% ELO

Positive influence of increasing ELO content on flexural and compression properties of 2P hybrid resins is obvious. Improvement of both flexural strength and flexural strain values at maximum load is observed, whereas, the flexural modulus remains nearly constant (except for system with 15 vol% ELO). It is also noteworthy that by exceeding 15 vol% ELO the flexural strength and strain decreases compared to those for 15 vol% ELO. On the other hand, all parameters of compression properties (i.e. modulus, strength and strain) are increasing with raised ELO content up to 15 vol% and remaining in similar range by further increase of ELO content up to 25 vol%. In conclusion, the best mechanical performance in terms of both flexural and compression properties of 2P hybrid resins was reached using 15 vol% ELO.

To obtain perceptions of fracture toughness the propagation of a deliberately introduced sharp crack in the loaded material was measured under increasing load. As a characteristic value for fracture toughness the critical stress intensity factor K_{Ic} was calculated using eq. (8) and (9).

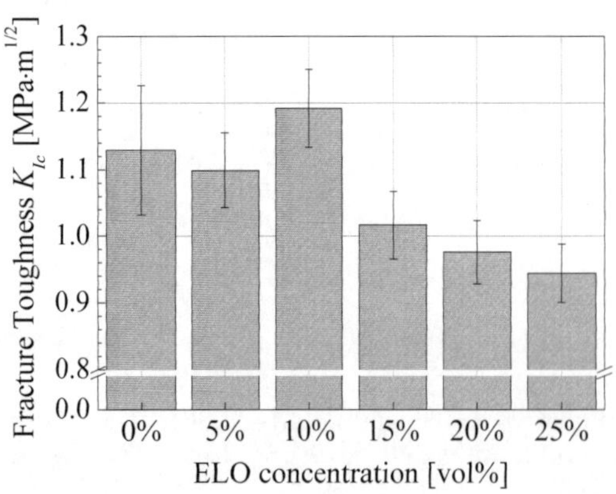

FIGURE 3. Critical Stress Intensity Factor K_{Ic} for 2P hybrid resins with 0, 5, 10, 15, 20 and 25 vol% ELO

With a raising content of ELO there is primary a comparatively increase of fracture toughness detectable until 10 vol% ELO, however, remaining similar to that of non-emulsified system. Unfortunately, higher amounts of ELO resulted in tendentially decrease of the fracture toughness. Nevertheless, in terms of fracture toughness, all developed systems are well acceptable for potential application in repair of pipes, strengthening of walls, protective coatings, etc.

CONCLUSION

In this work the influence of different concentrations of epoxidized linseed oil (ELO) on 2P hybrid resins, consisting of polyisocyanate (PMDI) and water glass (WG), is demonstrated. ELO, as co-reactive emulsifier made of renewable resources with 100 % renewable carbon content, confirmed its high potential to replace chemically non-bonded organic phosphates in common indirect produced polyurea/polysilica resins established in sewer rehabilitation sector.

In respect of application requirements the viscosity during first 60 minutes and also the hardness (Shore D) were investigated. Furthermore the changes of mechanical and fracture mechanical properties, i.e. flexural and compression properties as well as fracture toughness were examined.

It could be shown that the use of ELO as emulsifier has positive impact on curing and post-curing of the 2P W/O-emulsions at RT. The more ELO is applied, the faster the viscosity increases and desired pot-life is reached earlier according to application requirements. Similarly, development of Shore D hardness is influenced positively towards faster hardening of the 2P hybrid resins. Flexural and compression properties were improved by incorporation of ELO. The best mechanical performance was achieved by application of 15 vol%. Fracture toughness was marginally influenced with 5 vol% ELO, tendentially increased with 10 vol% ELO, but decreased above 15 vol% ELO. The respective values are collected in TABLE 1.

TABLE 1. Summary of processing and mechanical properties for 2P hybrid resins with 0, 5, 10, 15, 20 and 25 vol% ELO

ELO	[vol%]	0	5	10	15	20	25
$t_{\eta > 15\,Pa\cdot s}$	[min]	>60	34	22	14	13	12
$t_{Shore\,D > 60}$	[d]	4.3	1.1	1.0	1.0	0.9	0.8
E_F	[MPa]	1892 ± 163	1908 ± 127	1903 ± 83	1619±41	1852±279	1446 ± 166
σ_{Fmax}	[MPa]	33.7 ± 5.3	37.5 ± 1.5	41.4 ± 3.2	55.7±6.9	42.7±9.6	44.1 ± 9.0
ε_{Fmax}	[%]	2.64 ± 0.56	3.18 ± 0.12	2.96 ± 0.43	4.21 ± 1.28	2.69 ± 1.03	2.74 ± 0.68
E_C	[MPa]	887 ± 146	1084 ± 90	1177 ± 151	1298±42	1236± 144	1226 ± 130
σ_{Cmax}	[MPa]	94.3 ± 9.5	110.1 ± 2.8	127.2 ± 3.1	157.1 ± 18.4	151.2 ± 18.0	148.3 ± 11.7
ε_{Cmax}	[%]	38.2 ± 5.8	35.6 ± 0.7	42.1 ± 6.0	58.5±5.8	59.9±4.6	57.5 ± 6.5
K_{Ic}	[MPa·m$^{1/2}$]	1.13 ± 0.10	1.10 ± 0.06	1.19 ± 0.06	1.02±0.05	0.98±0.05	0.95 ± 0.04

ACKNOWLEDGMENTS

The authors acknowledge the EUREKA/COST-Office for permission of EUREKA-Project !Σ 8353 "COMPONAT" and the German Federal Ministry for Economic Affairs and Energy for funding and possibility to realize the project "COMPONAT" in frame of the program "Zentrales Innovationsprogramm Mittelstand", (ZIM, Project 2088328SA3). Furthermore, authors acknowledge the project partners Fluvius GmbH (Düsseldorf, Germany) and Polinvent Ltd. (Gyál, Hungary) for fruitful collaboration.

REFERENCES

1. Karger-Kocsis, J.; Erdélyi, S.; Nagy, G. *J. Appl. Polym. Sci.* **2007**, *103*, 853–859.
2. Grishchuk, S.; Castellà, N.; Karger-Kocsis, J. *European Polymer Journal* **2007**, *43*, 1245–1257.
3. Nagy *et al.*: US Patent No: 5,622,999: "Polysilicic acid/polyisocyanate basic materials, binding materials and foams and process for preparing same." Priority date: 3 June, 1991 [HU]. (DE: P 41 21 153.7).
4. EC 1907/2006: REACH, **2006**.
5. Castellà, N.; Grishchuk, S.; Karger-Kocsis, J.; Schehl, M. *J. Appl. Polym. Sci.* **2011**, *119*, 7–14.
6. Castella, N.; Grishchuk, S.; Karger-Kocsis, J.; Unik, M. *J Mater Sci* **2010**, *45*, 1734–1743.
7. Prileschajew, N. *Ber. Dtsch. Chem. Ges.* **1909**, *42*, 4811–4815.
8. Björkling, F.; Frykman, H.; Godtfredsen, S.E.; Kirk, O. *Tetrahedron* **1992**, *48*, 4587–4592.

Functionalisation of Wood Particles, Cellulose and Hemicellulose using Reactive Extrusion

Marc Gaugler, Alankar A. Vaidya and Dawn A. Smith

New Zealand Forest Research Institute – Scion; 49 Sala Street, Rotorua 3010, New Zealand

marc.gaugler@scionresearch.com

Abstract. Twin-screw reactive extrusion was used to functionalise sander dust (SD), cellulose and hemicellulose with cyclic anhydrides. These materials were processed without any polymeric carrier, unlike other published work in this area. Succinic, dodecenyl succinic and maleic anhydride were used as reactants and under optimum conditions, esters between these anhydride and hydroxyl groups of the biomaterials were formed. The influence of extrusion conditions and catalyst on the reaction yield was assessed by acid titration and spectroscopy. The highest degree of esterification was obtained for the succinilation of cellulose. The order of reactivity for succinic anhydrides was cellulose > SD > hemicellulose, while for maleic anhydride the order was SD > hemicellulose > cellulose. The yield was independent of the extrusion temperatures assessed in formulations with catalyst.

Keywords: Reactive Extrusion, Wood, Cellulose, Hemicellulose, Anhydride, Succinilation
PACS: 81.05.Lg

INTRODUCTION

Using extruders as green and economic continuous chemical reactors has gained increased popularity over the last decades. Reactive extrusion is already used for various chemical reactions such as bulk and graft polymerization, inter-chain copolymer formation, coupling or branching reaction or functionalization (Wang et al. 1997). It can also be used as an environmentally benign route for surface modifications of particles (Gibril et al. 2013) due to the absence of solvents. Recently, extrusion has been used in bio-refinery application to pre-treat biomass and improving the enzymatic hydrolysis of cellulose (Senturk-Ozer et al. 2011). It has also been used to prepare starch succinates (Wang et al. 1997), succinylated cellulose in the presence of ionic liquid (Gibril et al. 2013) and succinylated lignin (Bridson et al. 2013). Biocomposites have future commercial application that would unlock the potential of currently underutilized renewable biomass. A strong bond between the composite matrix and biomass, e.g. wood, is required to achieve good properties. Using unmodified woody or other lignocellulosic biomass can cause poor adhesion with the matrix. Therefore, compatibiliser or coupling agents are often used to improve the filler/matrix interphase. However, an alternative route is the modification of the filler particle (Ichazo et al. 2001). The chemical modification of biomass using acid anhydrides with and without catalyst has been reported by Marcovich et. al. (1996). Using cyclic anhydrides has the advantage that no acids are generated as a reaction product. However, such reactions are commonly done in organic solvents. They are often toxic and corrosive which prevents such modification approaches from being established on a large scale. In light of this, green and industrial scalable solvent-free modification processes of biomass are sought (Carlborn and Matuana 2006). The present work uses reactive extrusion for the modification of lignocellulosic material based on cyclic anhydrides to develop a green route to modify bio-based fillers.

MATERIALS

Analytical grade succinic anhydride (SA), maleic anhydride (MA), dodecenyl succinic anhydride (DDSA), 1-methylimidazole (MI), acetone, ethanol, toluene were purchased from Aldrich (Milwaukee, USA) and used as received. The sander dust (SD) was sourced from a plywood mill (Tokoroa, New Zealand), sieved to 250-500 μm and oven dried at 105 °C for 18 h before extrusion. The cellulose was prepared from SD following a reported chlorite delignification method (Donaldson et al. 2014) and air-dried at 27 °C for 48 h prior to extrusion.

The hemicellulose was isolated from the SD by softening it using steam and extracting a hemicellulose rich liquid using compression. This liquid phase was analysed by ion chromatograph using a Dionex ICS 3000 instrument and contained 96.1 g/L total sugars in which the galactan:glucan:mannan ratio was 0.6:1:2.6. Negligible amounts (< 1 g/L) of arabino-xylan were found. The liquid was freeze-dried for 24 h prior to reactive extrusion.

METHODS

The reactive extrusion was done on a 19 mm co-rotating twin-screw extruder by O.M.C, Italy (L:D ratio 30:1). The extruder barrel was set to 150 to 220°C depending on the reagents. MI was used as a catalyst at 3% w/w. The screw speed was 300 rpm resulting in an estimated mean residence time of about 90 seconds. The reagents were dry blended at room temperature in a high shear mixer for 1 min. at 2500 rpm and fed into the extruder using a volumetric feeder. Only the second half of the extruded samples was retained for analysis to ensure that any residue from previous samples was purged from the barrel. 2 g of the extruded samples were transferred to a soxhlet and extracted with a toluene:acetone:ethanol (4:1:1 v/v/v) mixture for 18 h to purify and remove residual unreacted anhydride and catalyst. The purified samples were vacuum dried at 60°C for 24 h prior to further characterisation. The degree of substitution (DS) and acid value of the extruded samples were assessed by suspending 0.200 g sample in 10 mL of 0.1 M sodium hydroxide (NaOH). The excess NaOH was titrated back to pH 7 with a 0.025 M hydrochloric acid (HCl) using a Metrohm Titrando 888 auto-titrator. The degree of substitution was calculated using 162 g/mol is the molar mass of an anhydro-C6 sugar unit, 100 g/ mol as the net increase in the mass of an anhydro-C6 sugar for each succinyl moiety (this value was replaced by 98 g/ mol for maleated product and 226 g/ mol for DDSA substituted product) based on a published methods (Liu et al. 2007).
The solid state nuclear resonance spectroscopy (CP/MAS 13C NMR) spectra were recorded using a Bruker 200DRX spectrometer at the frequency of 5 KHz with 4 mm Bruker SB magic-angle spinning probe. CP/MAS experiments were conducted using an acquisition time of 30 ms, pulse delay of 1.5 s and proton 90° pulse time of 1.4 μs.

RESULTS AND DISCUSSION

Sander dust (SD) was selected for this study as a cheap, sustainable and renewable potential biomass and filler for polymer composites. While wood can be regarded as a viscoelastic material it obviously does not have the melt flow properties that plastics exhibit (Ranta-Maunus, 1975). In this work the sander dust particles were conveyed in a twin-screw extruder without the presence of any carrier polymer. The intense mixing during extrusion ensured that all reagents were dispersed evenly and that there was good heat transfer from the barrel to the sander dust. The high shear energy input, especially in the kneading sections, is likely to reduce the particle size. However, for non-fibrous filler, e.g. sander dust, this was deemed of minor importance. Our objective was not to optimise the screw design but to assess the feasibility of reactive extrusion as a green and solvent-free way to chemically modify the surface of wood, cellulose particles and hemicellulose. SA was selected because it is a cyclic anhydride, has been reported to increase the photo-stability of biomass and is readily available on large scale. DDSA was used to increase the hydrophobicity of the biomass and its similarity with commonly used coupling agents like MAPP (Safou-Tchiama et al. 2007). MA can introduce vinyl functionality in the modified biomass which can be used for subsequent reactions.

Effect of succinic anhydride level

The acid value of the reactive extruded SA modified SD (150 °C extrusion temperature) increased with SA addition up to 30 % with and without catalyst as shown in Figure 1. The increasing acid value indicates a higher degree of modification of the biomass. Further SA addition to 50 % did not affect the acid value of un-catalysed samples significantly. This levelling off effect above 30% SA may result from most reactive groups being derivatised already or insufficient residence time (90 s) in the reactive extrusion process. However, in the presence of MI catalyst, the acid value increased from 30 to 50 % SA. It is known that MI activates hydroxyl groups which can react faster and further at higher degree of modifications, thereby ensuring a high degree of modification and acid number can be achieved (Liu et al. 2008).

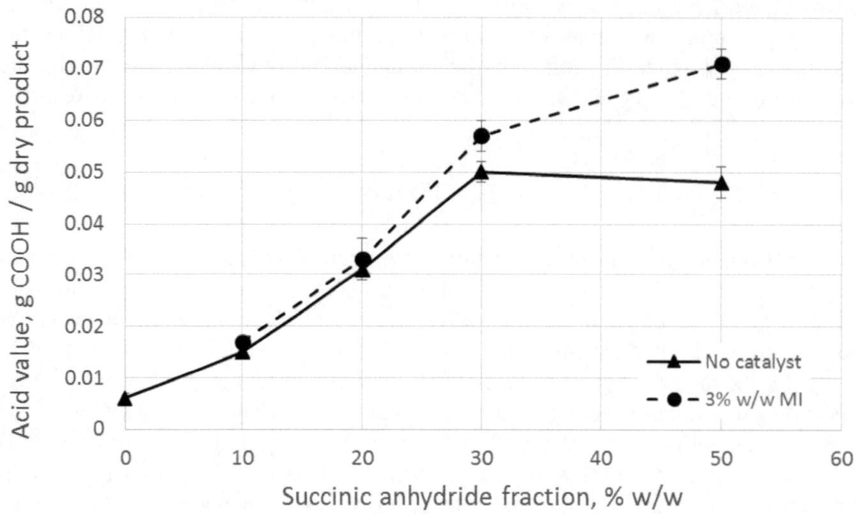

FIGURE 1. Acid value of SA-modified SD extruded with and without MI catalyst.

Effect of extrusion temperature

In order to evaluate the effect of the extrusion temperature on the reactive extrusion efficiency, 30 % anhydride in SD was used. The acid values increased from 150 to 190 °C extrusion temperature for the uncatalysed system (Figure 2). At temperatures above 190 °C the acid value did not increase significantly anymore. Thus, increasing the reaction temperature from 150 to 190 °C, enhanced the mobility and diffusion of the anhydride molecules into the lignocellulosic structure of the wood particles which concomitantly increased the acid value and degree of modification (Liu et al. 2008). Interestingly, in the presence of catalyst, the acid value of the succinylated product decreased compared to non-catalysed reactions at temperatures above 205 °C. Based on these results, an extrusion temperature of 190 °C was used for MA-based reactions. DDSA has a low boiling point, therefore was extruded at 150 °C. The maleated samples had a lower acid value than the comparable SA products. The acid value of MA and DDSA treated SD did not differ (0.037 g/g) for un-catalysed reactions. However when the catalyst was used, this acid value reduced to 0.033 and 0.007 g of acid groups/g of the product for MA and DDSA, respectively. The lower reactivity of DDSA with the SD could be due to steric hindrance caused by the alkyl chains of DDSA which lowers its access to wood cell walls. This is supported by a similar study based on maleated polyethylene (Carlborn and Matuana 2006).

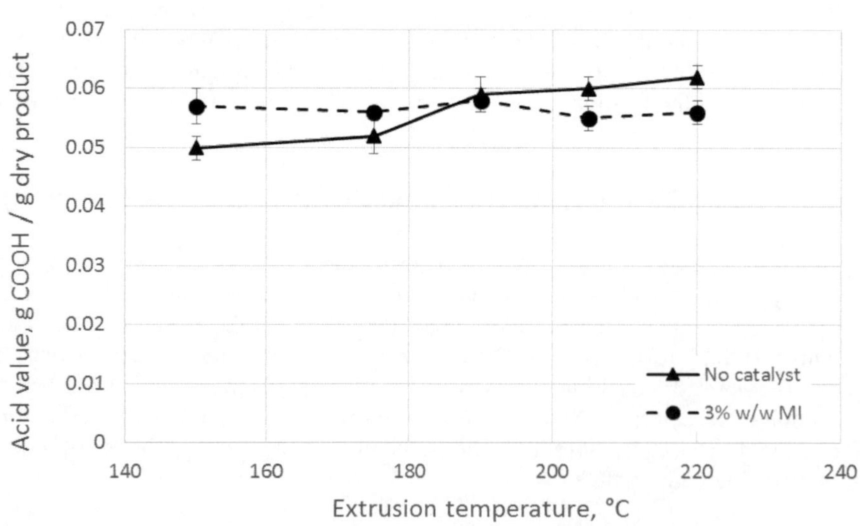

FIGURE 2. The effect of the reactive extrusion temperature on the acid value of the modified sander dust

NMR is the most effective method for the characterization of lignocellulosic materials and is capable of providing detailed information directly on the solid samples (Liu et al. 2011). The increase of the peak at 173–175 ppm indicates that new carbonyl groups have been introduced to the sander dust during succinilation which corroborates with the acid value measurement discussed above. This carbonyl peak was shifted to 167.0 ppm for MA-based reactions (Figure 3). The decrease of the less hindered C6-OH peak of the cellulose chains at 62–63 ppm suggests that these hydroxyl groups of the carbohydrate complex in the wood have been esterified. Safou-Tchiama et al. (2007) reported that changes of the 74–75 ppm peak associated with the C2, C3 and C5 of polysaccharides which indicates reactions of anhydrides with the polysaccharides in the wood cell walls.

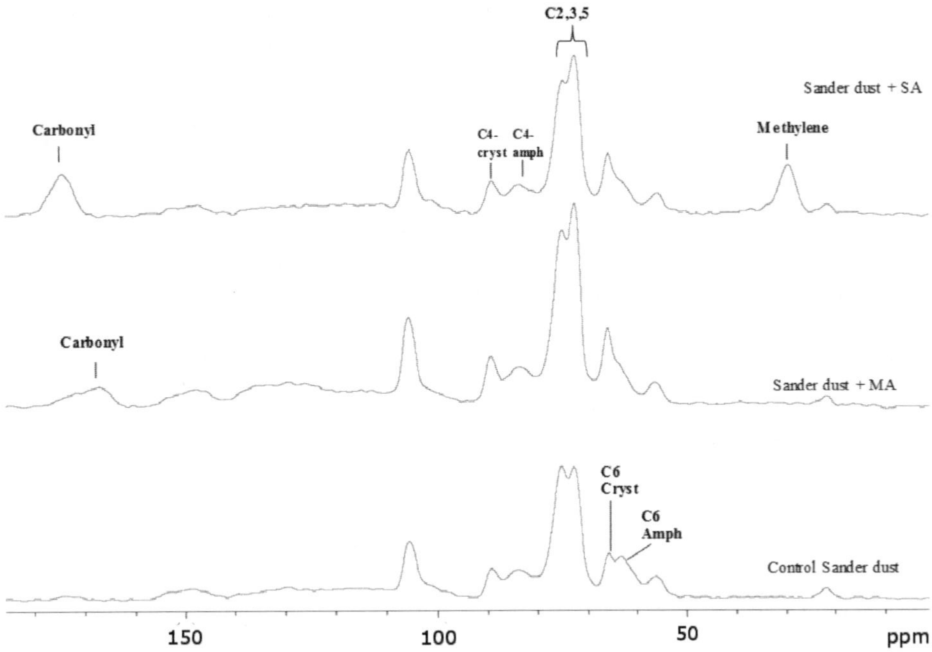

FIGURE 3. 13C NMR spectra of control, MA- and SA-modified sander dust samples

Reactive extrusion of cellulose and hemicellulose

The reactivity of the anhydrides on cellulose and hemicellulose, two components of wood, were also assessed. The degree of substitution could be calculated for these substrates due to their more defined chemistry compared to SD. In the presence of the catalyst, the cellulose succinylated at 190 °C had a higher degree of substitution (DS), but lower acid value compared to the un-catalysed sample (Table 1). This suggests a higher rate of di-ester linkages due to the presence of the catalyst during reactive extrusion. A significantly higher acid value was obtained for the succinylation of cellulose compared to SD and hemicellulose. In catalysed MA and DDSA reactions, higher acid values and DS were measured compared to un-catalysed samples.

Hemicellulose undergoes a phase transition from solid to liquid at about 150 °C. Therefore, the reactive extrusions were performed at 150 °C to maintain it in solid state. The catalyst had no effect on the succinylation and maleation of hemicellulose as shown by the acid value and DS results (Table 1). Compared to cellulose, the maleation of hemicellulose was more favoured in reactive extrusion. When DDSA was used as anhydride, the catalyst increased the modification efficiency. However, the acid value and degree of substitution were small compared to the succinylated and maleated hemicellulose. The hemicellulose isolated from SD is less reactive during extrusion with cyclic anhydrides compared to cellulose as shown by the results summarised in Table 1.

13C NMR indicated that the C-4 crystalline carbon in cellulose has undergone modifications during reactive extrusion with MA and SA. Reduction in the C-6 and C-2,3 carbon peaks indicate that the esterification during reactive extrusion predominantly occurs at these sites. Generally, changes in 13C NMR spectra of cellulose and hemicellulose at 29 and 175 ppm also confirm the esterification.

TABLE 1. Acid values and degree of substitution (standard deviation) for the reactive extruded cellulose and hemicellulose

Substrate	Reactant	Catalyst, % w/w	Temperature, °C	Acid value, g/g	DS, a.u.
Cellulose	SA		190	0.122 (0.009)	0.350 (0.010)
Cellulose	SA	0.3	190	0.106 (0.008)	0.380 (0.005)
Cellulose	MA		190	0.015 (0.007)	0.100 (0.015)
Cellulose	MA	0.3	190	0.018 (0.000)	0.130 (0.003)
Cellulose	DDSA		150	0.010 (0.005)	0.040 (0.003)
Cellulose	DDSA	0.3	150	0.014 (0.003)	0.046 (0.000)
Hemicellulose	SA		150	0.043 (0.002)	0.290 (0.005)
Hemicellulose	SA	0.3	150	0.042 (0.003)	0.290 (0.001)
Hemicellulose	MA		150	0.034 (0.010)	0.270 (0.002)
Hemicellulose	MA	0.3	150	0.030 (0.003)	0.210 (0.001)
Hemicellulose	DDSA		150	0.004 (0.008)	0.022 (0.000)
Hemicellulose	DDSA	0.3	150	0.008 (0.002)	0.056 (0.000)

CONCLUSIONS

A solvent-free, green and industrially-scalable route to prepare functionalized wood particles (sander dust), cellulose and hemicellulose was developed. An addition level of 30 % w/w and a reactive extrusion temperature of 190 °C achieved the highest degree of modification using succinic, maleic and dodecenyl succinic anhydride in sander dust based on the acid values. The highest degree of succinilation was achieved for cellulose followed by sander dust and hemicellulose, while for maleic anhydride the order was sander dust > hemicellulose > cellulose. A mixture of mono- and di-esters is expected based on acid values and degree of substitution.

ACKNOWLEDGEMENT

This research was funded by the New Zealand Ministry of Business, Innovation and Employment under contract C04X1205. We wish to thank Sylke Campion for conducting titrimetric analysis and Evamaria Gaugler for setting up the solid state NMR experiments.

REFERENCES

1. Wang L, Shogren RL, Willett JL. 1997 Preparation of Starch Succinates by Reactive Extrusion. Starch/Stiirke 49:116-120.
2. Gibril ME, Huan L, haiFeng L, Xin Da L, Yue Z, Han K, Muhuo Y. 2013 Reactive extrusion process for the preparation of a high concentration solution of cellulose in ionic liquid for in situ chemical modification. RSC Advances 3:1021-1024.
3. Donaldson LA, Newman RH, Vaidya A, 2014. Nanoscale interactions of polyethylene glycol with thermo-mechanically pre-treated Pinus radiata biofuel substrate. Biotechnol Bioeng 111:719–725.
4. Senturk-Ozer S, Gevgilili H, Kalyon DM. 2011. Biomass pretreatment strategies via control of rheological behavior of biomass suspensions and reactive twin screw extrusion processing. Bioresource Technology 102: 9068–9075.
5. Bridson JH, van de Pas DJ, Fernyhough A. 2013 Succinylation of Three Different Lignins by Reactive Extrusion. J. Appl. Polym. Sci. 128: 4355–4360.
6. Ichazo, M.N., Albano, C., Gonzales, J., Perera, R., Candal, M.V., 2001.Polypropylene/wood flour composites: treatment and properties. Comp. Struct. 54, 207–214.
7. Marcovich, N.E., Reboredo, M.M., Aranguren, M.I. 1996 Sawdust modification: Maleic anhydride chemical treatment. Holz als Roh- und Werkstoff 54(3): 189-193.
8. Carlborn K, Matuana LM. 2006 Functionalization of Wood Particles through a Reactive Extrusion Process. J Appl Polym Sci 101:3131–3142.
9. Safou-Tchiama R, de J´eso B, Akagah AG, S`ebe G, P´etraud M. 2007 A preliminary survey of the interfacial bonding of some tropical hardwoods towards succinic anhydride and 2-octen-1-yl succinic anhydride molecules: Impact of lignin and carbohydrate polymers structure on the chemical reactivity. Industrial Crops and Products 26:173–184.
10. Liu CF, Sun RC, Zhang AP, Ren JL, Wang XA, Qin MH, Chaod ZN, Luod W. 2007 Homogeneous modification of sugarcane bagasse cellulose with succinic anhydride using a ionic liquid as reaction medium. Carbohydrate Research 342: 919–926.
11. Ranta-Maunus, A. 1975. The viscoelasticity of wood at varying moisture content. Wood Science and Technology, 9, 189–205.

Cellulose-Xyloglucan Composite Film Processing Using Ionic Liquids as Co-Solvents

Amine Bendaoud[a], Jean-Eudes Maigret[a], Eric Leroy[b], Bernard Cathala[a] and Denis Lourdin[a*]

[a]UR1268 Biopolyméres Interactions Assemblages, INRA, F-44300 Nantes, France
[b]Laboratoire de Génie des Procédés-Environnement-Agroalimentaire, GEPEA, F-44606- France
amine.bendaoud@nantes.inra.fr
**denis.lourdin@nantes.inra.fr*

Abstract. The aim of the present work is to develop a high-performance composite using biomimetic approaches that will exploit the interactions between two biopolymers, cellulose and xyloglucan, and these interactions play an important role but still not fully understood in the biomechanical properties of primary plant cell walls [1]. In this study, the co-solubility of cellulose and xyloglucan in an ionic liquid (1-ethyl-3-methylimidazolium acetate (EMIMAC)) is used to prepare regenerated films by subsequent precipitation with a non-solvent. Films of pure cellulose, pure xyloglucan and various cellulose/xylogucan blends were obtained and characterized by Fourier transform-infrared spectroscopy (FTIR), X-ray diffraction (XRD) and hygro-mechanical analysis. The biopolymer blends show an increase of the storage modulus comparatively to the regenerated pure biopolymers.

Keywords: cellulose, xyloglucan, ionic liquids, polymers blends, biomimetic approaches, hydro-mechanical properties.

PACS: 83.85, 87.15, 87.14, 83.85, 82.35, 83.80, 87.15

1.Introduction

The use of natural polymers to design new environment-friendly materials is a promising solution to overcome the environmental concerns and decrease of petroleum resources [2]. Cellulose is one of the most abundant polymers, and it has a number of interesting characteristics, such as renewability, low price and biodegradability [3]. However, cellulose cannot be melt processed [4]. This drawback is due to the native crystalline structure involving a compact network of intra- and intermolecular hydrogen bonds.

Consequently, solvents based process are used to transform cellulose. Since the fact that cellulose is difficult to dissolve, a number of adequate solvents have been developed [5]. Ionic liquids are currently a promising solvents, for the processing of cellulose by dissolution and regeneration [6]. However, the regenerated cellulose materials present some drawbacks, like low thermal stability, high sensitivity to relative humidity and poor mechanical properties [7]. To address these issues, a number of authors have used the solubilization step to blend cellulose with other polymers [8].

In the present study, we combine cellulose with xyloglucan. This hemicelluloses biopolymer is acknowledged to be an important constituent of primary plant cell wall. Our objective is to obtain a bioinspired material by taking advantage of the interactions occuring between cellulose and xyloglucan [9, 10].

The developed blend films are characterized by FTIR, XRD and hygo-mechanical analysis in order to study the effect of xyloglucan addition on the properties of the cellulosic materials and increase our knowledge on interactions between plant biopolymers.

2.Materials and Methods

Microcrystalline cellulose (Avicel PH-101, 50 µm) was purchased form Sigma Aldrich. Non fucosylated xyloglucan (Tamarind, 225 Kd (MAALS), purity 95%) was purchased from Megazyme. An ionic liquid, EMIMAC, (purity 98%), was purchased from Solvionic.

2.1. Dissolution and Regeneration Process

The cellulose, xyloglucan and EMIMAc were first dried (80°C-24h) in a vacuum oven. These components were used to prepare biopolymer(s) solutions (~1%wt in EMIMAc) of pure cellulose and blends with the following weight cellulose/xyloglucan ratio: 0.5, 0.75, 1, 1.25, 1.5, 2 and 3. The biopolymer powders were dispersed in EMIMAc under stirring at 100 °C until homogenous solutions were obtained. After this solubilisation step, the different solutions were casted in glass petri dishes. These dishes were placed in a water saturated box for 24h in order to obtain films. These films were then immersed in a distillated water bath during 3 days in order to remove the ionic liquid. Finally, the films were dried at 40°C during 12h. Homogenous and transparent films were successfully obtained for all initial solutions using this method.

2.2. FTIR Spectroscopy

Fourier transform-infrared spectroscopy studies of the initial powders and the regenerated films were carried out by using ATR Perkin FTIR spectrophotometer. The spectra's were recorded from 4000 cm^{-1} to 400 cm^{-1}.

2.3. X-Ray Diffraction Analysis

The X-ray diffraction measurements were provided by using an XRD analyzer (Brucker D8). Wide-angle X-ray intensities were recorded from 3° to 40° in 2θ with a scanning rate of 2°/min. An approximation of the degree of crystallinity of the different samples was calculated as the ratio of crystalline area to the total area under the diffraction peaks.

2.4. Dynamical Mechanical Analysis (DMA) with Controlled Humidity (RH)

The dynamical mechanical analysis is conducted using a Metravib DMA 50N connected to a humidity controller. The thicknesses of films (average of three measurements) were measured using a micrometer. The sample films (20 m x 5 mm x 0.02 mm) were loaded in the testing chamber at 1 Hz frequency, 0.5 N preload force, 0.01% strain and the storage modulus is recorded with the increase of relative humidity from 10% to 70%.

3. Results and Discussion

The regeneration process gives homogenous, thin (~20 μm) and transparent films. No visual differences were observed between the different samples.
In order to have informations about the absence of ionic liquid and presence of xyloglucan in the blend films, infrared spectroscopy measurements are carried out.

Figure 1 shows the spectra of cellulose, xyloglucan, EMIMAc and an example of a blend films, here, Cellulose(CE)-Xyloglucan(XG)-Ratio(R)=1.
Comparatively to native cellulose, XG shows a characteristic peak at 940 cm^{-1} related to the presence of side chains [11,12]. This characteristic peak could be used in the blend films spectras in order to show the presence of xyloglucan.
For the blend films (CE-XG-R1), we can observe the presence of a peak around 940 cm^{-1}, which is linked to the presence of xyloglucan. This result confirms the effectiveness of the process and suggest that interactions between cellulose and xyloglucan occur, indeed xyloglucan remains in the films, despite the fact that xyloglucan is soluble in water, used here as a solvent to remove the ionic liquid.

X-ray diffraction analysis is used to determine the crystallinity of the different samples. The values of the crystallinity related to the different samples are given in the table 1.
Native cellulose shows a typical diffraction spectra of cellulose I. The solubilisation and the regeneration process of cellulose in ionic liquid leads to the transformation of cellulose I to cellulose II [13]. The blend films show an amorphous halo with a small intensity peak at around 20°.

Furthermore, the degree of crystallinity of cellulose films are greatly reduced comparatively to native cellulose, respectively lower than 10% and as high as 50% (table 1). For blend films, the value of crystallinity is low and close to that obtained for the regenerated cellulose film.

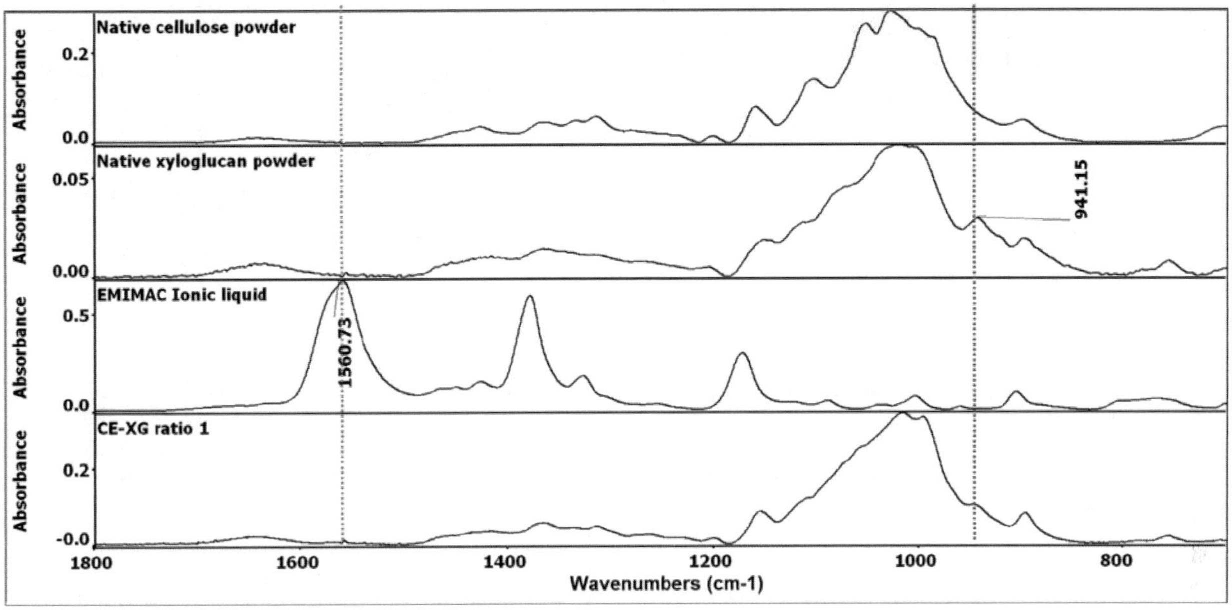

FIGURE 1. FTIR spectra of native cellulose and xyloglucan, EMIMAc and regenerated blend film

TABLE (1). The value of crystallinity of the different studied samples

Samples	Crystallinity (%)
Cellulose powder	50
Cellulose Film	<10
CE-XG-R2 Film	<10
CE-XG-R1 Film	<10
CE-XG-R0.5 Film	<10
CE-Xg-R0.7 Film	<10
CE-XG-R1.5 Film	<10

In order to analyze the hygro-mechanical properties of the different films, a dynamic mechanical analysis with controlled relative humidity is used.
Figure 2 shows the evolution of the storage modulus as a function of the relative humidity for regenerated cellulose, xyloglucan and an example of a blend films (Here, CE-XG-R1.5).

For all formulations, the storage modulus decreases with the increase of relative humidity. This phenomenon is related to the plasticizing role of water, which reduces the polymer polymer interactions. Moreover, we note that the blend cellulose-xyloglucan formulation has a higher storage modulus than the modulus of pure cellulose or xyloglucan films over all the relative humidity range. For example, at 10% RH, CE-XG-R1.5 film shows a storage modulus of 4.8 GPa, while pure cellulose and xyloglucan films, gives respectively 4.3 GPa and 3.6 GPa.

This synergy between the two biopolymers could be ascribed to the interactions between cellulose and xyloglucan. Nevertheless, it is striking that theses interactions take place between the two (mainly) amorphous polymers in the present materials. As an opposition, interactions between xyloglucan and cellulose microfibrils crystalline surface are known to occur in plants [9]. Consequently, the new model material produced open possibilities for a better understanding of the interactions between the two biopolymers.

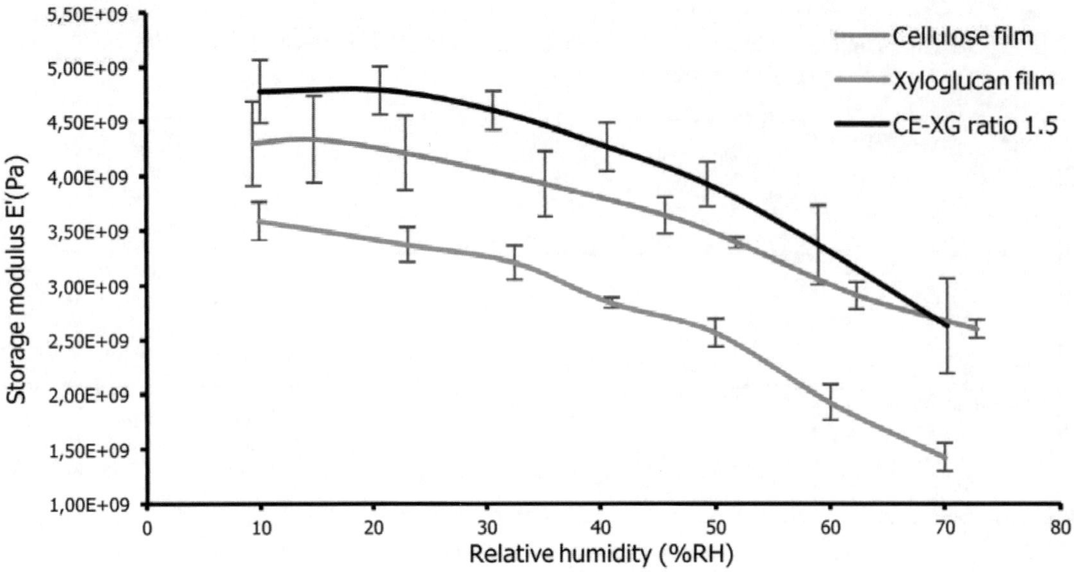

FIGURE 2. Evolution of storage modulus (E') with the increase of relative humidity (%RH) for different regenerated films.

4. Conclusion

In this present work, we successfully obtained thin films of quasi amorphous CE-XG blends. Anionic liquid is able to dissolve both biopolymers (EMIMAC) was used as a processing aid. The obtained CE-XG films are transparents and homogeneous. The FTIR analysis shows that the regenerated films are free of ionic liquid. Moreover, the presence of XG in the blend films can be observed, even after 3 days washing in water. This indicates strong interactions with cellulose, since XG itself is water soluble.

In addition, the blend films show better mechanical properties than cellulose pure films. These new materials thus potentially open new question about XG/CE interactions that have to be questioned in relation the role of XG in plant cell walls. They could be used in applications where there would have interest in modulating the mechanical properties by changing the relative humidity.

ACKNOWLEDGMENTS

We are grateful for the financial support from the MATIERES (Materials and Interfaces for Environment and Resources Exploitation) project funded by the "Région Pays de la Loire" council. We are grateful for the financial support from the project MATIERES. We also thank Sylvie DURAND and Bruno PONTOIRE for their technical support on FTIR and XRD analysis, respectively.

REFERENCES

[1] Y.B. Park and D.J. Cosgrove. *Plant Cell. Physiol.* ***56,*** 180-194 (2015).
[2] R.J. Moon, A. Martini, J. Nairn, J. Simonsen and J. Youngblood. *Chem. Soc. Rev.* **40**, 3941-3994 (2011).
[3] D. Klemm, B. Heublein, H.P Fink and A. Bohn. *Angew. Chem. Int. Ed. Engl.* **44**, 3358-3393 (2005).
[4] B. Medronho, A. Romano, M.G. Miguel, L. Stigsson and B. Lindman. *Cellulose.* **19**, 581-587 (2012).
[5] B. Medronho and B. Lindman. *Curr. Opin. Colloid Interface Sci.* **19**, 32-40 (2014).
[6] H. Wang, G. Gurau and R.D Rogers. *Chem. Soc. Rev.* **41**, 1519-1537 (2012).
[7] Y. Cao, H. Li, Y. Zhang, J. Zhang and J. He. *J. Appl. Polym. Sci.* **116**, 547-554 (2010).
[8] N. Hameed and Q. Guo. *Cellulose.* **17**,803-813 (2010).
[9] K. Mazeau and L. Charlier. *Cellulose.* **19**, 337-349 (2012).
[10] F.L. Hatton, M. Ruda, M. Lansalot, F. D'Agosto, E. Malmstrom and A. Carlmark. *Biomacromol.* (2016).
[11] M. Kacurakova, P. Capek, V. Sasinkova, N. Wellner and A. Ebringerova. *Carbohydr. Polym.* **43**, 195-203 (2000).
[12] M. Kacurakova, A. Ebringerova, J.Hirsch and Z. Hormadkova. *J. Sci. Food. Agric.* **66**, 423-427 (1994).
[13] M. Adsul, S.K. Soni and V. Bansal. *Biomacromol.* **13**, 2890-2895 (2012).

Effects of Epoxy-POSS Nanoparticles on the Properties of PLA/TPU Blends

Nilay TUCCAR KILIC, Buse Nur CAN, Mehmet KODAL, Guralp OZKOC*

Department of Chemical Engineering, Kocaeli University, 41380, Kocaeli/Turkey

nilaytuccar@yahoo.com.tr
**guralp.ozkoc@kocaeli.edu.tr*

Abstract: In this study, the influence of epoxy functionalized polyhedral oligomeric silsesquioxanes (G-POSS) loading level on the mechanical, thermal and rheological properties of poly(lactic acid) (PLA)/thermoplastic polyurethane (TPU) blends were investigated. The blends were prepared in an Xplore Instruments MC15 twin-screw microcompounder at 200°C barrel temperature and 100 rpm screw speed. Mechanical properties were characterized by tensile and impact tests, thermal properties were determined via differential scanning calorimeter (DSC) and thermogravimetric analysis (TGA). Rheological properties were examined by oscillatory rheometer. A decreasing trend in tensile strength and thermal degradation temperature were obtained for PLA/TPU blends at high G-POSS loading levels. G-POSS loading level did not exhibit significant effect on glass transition, crystallization and melting temperatures of the blends.

Keywords: polylactide, thermoplastic polyurethane, blends, POSS.
PACS: 61.25.hk

INTRODUCTION

Petrochemical based non-degradable polymers have negative affects to the ecological system. Biodegradable polymer usage is a solution to solve this problem [1]. Poly(lactic acid) (PLA) is a linear, semi crystalline, aliphatic, biodegradable polyester that can be produced from lactic acid by the fermentation of renewable sources such as whey, corn, potato, or molasses [2]. PLA has a relatively high strength and high modulus but its brittleness and low toughness confines its application. Blending PLA with flexible polymers is a practical and economical way to toughen PLA products. Elastomeric thermoplastic polyurethanes (TPUs) can be used to toughen PLA; since the soft segment of a TPU is mainly polyester, therefore; it is expected to have a good compatibility with PLA. Nanofillers can enhance the mechanical properties of toughened systems [3].

Polyhedral oligomeric silsesquioxanes (POSS) are organic/inorganic hybrid materials used as nanofillers with possibly reinforcement of polymers. The incorporation of POSS with reactive groups into a polymer matrix creates a possible nanoscale reinforcement, in which POSS molecules may link to polymer chains covalently [4].

In the current study, it was aimed to investigate the influence of epoxy functionalized POSS (G-POSS) loading level on mechanical, thermal, and rheological properties of PLA-TPU blends. The reason of choosing epoxy functionality is its potential of making covalent bonding with –COOH functional groups of PLA and –CNO functional groups of TPU.

EXPERIMENTAL

Materials

Poly(lactic acid) (PLI005, MFI: 10-30 g/10 min at 190°C/2.16 kg) was purchased from NaturePlast Company, France. Caprolactone diol based commercial thermoplastic polyurethane (TPU) (75 Shore A hardness) was obtained from Lubrizol, USA. Glycidylisobutyl-POSS (G-POSS) was purchased from Hybrid Plastics Company, USA. G-POSS can be dissolved in THF, chloroform and hexane as PLA and TPU does, which points out a possible compatibility. PLA and TPU are in pellet form; G-POSS is in crystalline white powder form.

Sample Preparation

PLA/TPU/G-POSS blends having different POSS contents (0.25, 0.5, 1 and 3 wt%) were prepared via melt blending in a twin-screw microcompounder (15 mL Microcompounder, Xplore Instruments, The Netherlands) at a screw speed of 100 rpm. PLA/TPU ratio was kept constant as 90/10 by weight. Before compounding, PLA and TPU were dried in a vacuum oven at 65 °C for 24 h. Prior to processing, predefined amount of PLA, TPU and POSS were dry-mixed. The barrel temperature of microcompounder was set to 200 °C. The mixing time was kept constant at 3 min. In order to prevent thermo-oxidative degradation of PLA, the barrel was continuously purged with Argon. The melts were directly transferred with preheated transfer cylinder to DSM Xplore 12 mL laboratory injection-molding machine to obtain standard test bars according to ISO527/5a. The melt and mold temperatures were set 200°C and 30°C, respectively. The samples were abbreviated with respect to POSS loading level, for instance G-POSS 0.25 represents the composite including 0.25 wt % glycidylisobutyl-POSS.

Characterization

A universal tester (Instron Universal Testing Machine) was used to analyze tensile behavior of the blends. The load cell used was 5 kN and the crosshead speed for the measurements was fixed at 10 mm/min. Measurements were performed, at room temperature, on dog-bone shaped tensile bars having dimensions according to ISO 527 standard. Notched Izod impact tests were performed based on ISO 180 standard with an Instron CEAST 9050 model impact testing machine. Differential scanning calorimeter (DSC) analyses were performed using Mettler Toledo DSC1 Star System in nitrogen atmosphere. The analysis was conducted from -90°C with a heating rate of 10°C/min to 200°C with a heating rate of 10°C/min. Thermogravimetric Analysis (TGA) were performed Mettler Toledo TGA1 Star system. Samples were heated from room temperature to 600°C in nitrogen atmosphere. Rheological tests were performed on Anton Paar Modular Compact Rheometer MCR 102 with plate geometry of 25 mm in diameter. Frequency sweep measurements were carried out at 190°C, 0.1-600 rad/s angular frequency, 1% shear strain and 1mm gap.

RESULTS

Izod impact strength, energy absorbed prior to failure, and elongation at break of PLA/TPU and PLA/TPU/G-POSS blends were higher than that of neat PLA because of the higher intrinsic flexibility of TPU. The impact strength of PLA/TPU blends first lowered by the addition of G-POSS; then started to increase by further addition of G-POSS. In addition, increasing G-POSS content demonstrated a decreasing trend in tensile strength in the presence of G-POSS.

TABLE 1. Mechanical properties of PLA, TPU, PLA/TPU, PLA/TPU/G-POSS blends.

Material	Izod Impact Strength (kJ/m^2)	Tensile Strength (MPa)	Energy absorbed in tensile test prior to failure (J)	Elongation at break (%)
Neat PLA	2.4 ±0.3	55.4 ±3.8	1.1 ±0.1	9.1 ±1.6
Neat TPU	27.7 ±1.0	37.5 ±2.8	49.2 ± 3.8	857.0 ±20.2
PLA/TPU	4.5 ±0.3	52.8 ±1.7	3.0 ±0.1	22.9 ±4.7
PLA/TPU/G-POSS0.25	4.0 ±0.3	49.9 ±0.7	2.1 ±0.3	23.4 ±7.5
PLA/TPU/G-POSS0.5	4.1 ±0.2	48.8 ±0.8	3.6 ±0.5	27.5 ±8.6
PLA/TPU/G-POSS3	4.4 ±0.2	45.6 ±0.8	3.7 ±0.4	28.9 ±3.3

DSC thermograms (Figure 1) showed that there was no significant effect of GPOSS addition on glass transition, cold crystallization and melting temperatures of PLA/TPU/G-POSS blends. Thermal decomposition temperatures of PLA/TPU and PLA/TPU/G-POSS blends were higher than that of neat PLA (Figure 2). TPU addition to PLA provided better thermal stability (Table 2).

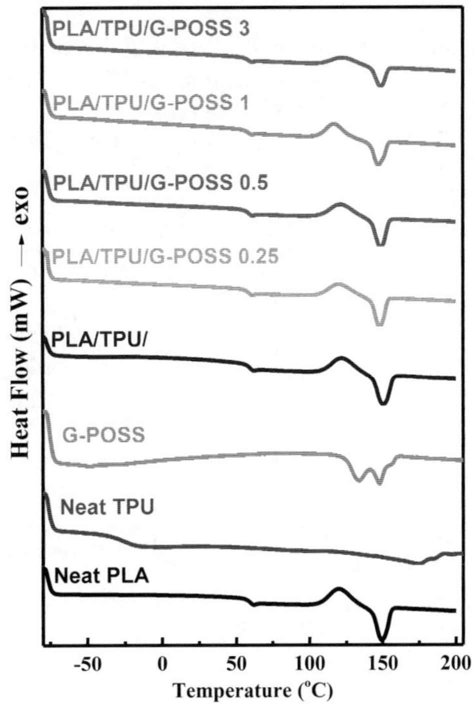

FIGURE 1. DSC thermograms of PLA, TPU, PLA/TPU, PLA/TPU/G-POSS blends.

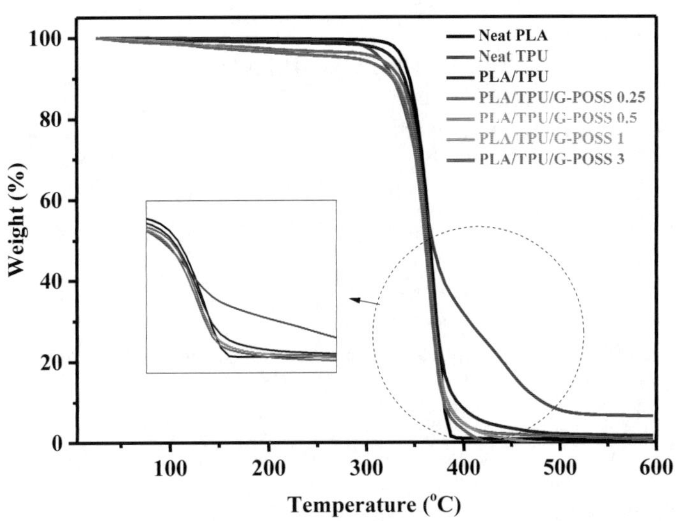

FIGURE 2. TGA thermograms of PLA, TPU, PLA/TPU, PLA/TPU/G-POSS blends.

TABLE 2. TGA results of PLA, PLA/TPU, PLA/TPU/G-POSS blends.

Material	T_{d10} (°C)	T_{d90} (°C)	$T_{decomposition}$ (°C)
Neat PLA	345.7	377.4	347.1
PLA/TPU	340.1	392.5	364.3
PLA/TPU/G-POSS0.25	333.8	383.2	362.6
PLA/TPU/G-POSS0.5	328.4	382.6	361.3
PLA/TPU/G-POSS1	329.6	381.9	361.4
PLA/TPU/G-POSS3	324.4	377.2	362.4

*T_{d10}, T_{d90} represent the decomposition temperature at which 10%, 90% weight loss occurs, respectively.

Rheological measurements demonstrated that both PLA/TPU and PLA/TPU/G-POSS blends had higher viscosities in comparison to neat PLA (Fig 3). The maximum viscosity at 190°C was obtained for neat TPU. The increasing content of G-POSS enhanced the melt viscosity of PLA/TPU. This can be attributed to a possible interaction of GPOSS with PLA or TPU via physical or chemical bonding which restricts the flow of polymer chains.

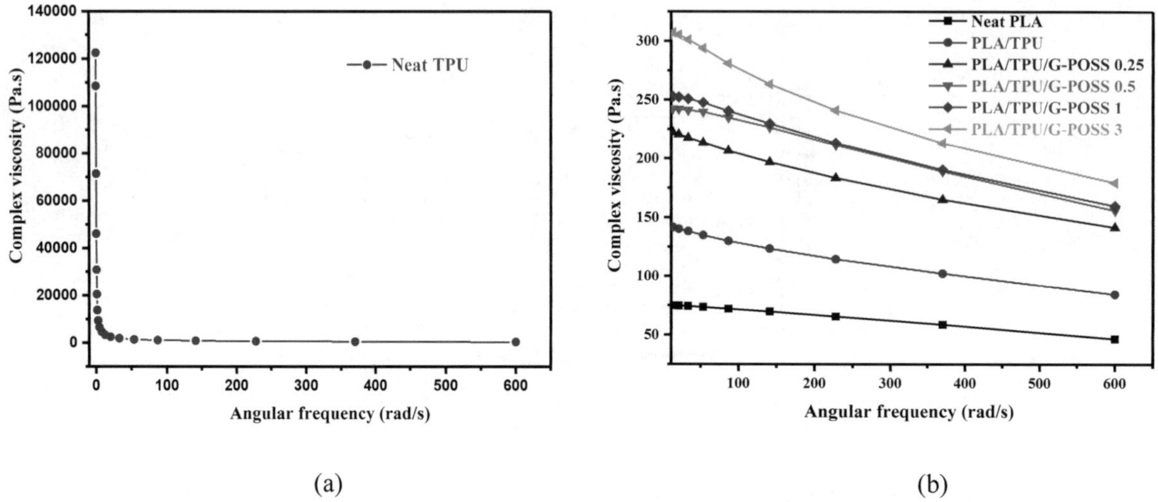

FIGURE 3. Angular frequency vs. viscosity plot of (a). TPU, (b). PLA, TPU, PLA/TPU, PLA/TPU/G-POSS blends.

CONCLUSIONS

Mechanical, thermal, rheological properties of PLA were improved by melt blending with TPU. The incorporation of 0.25% GPOSS to the PLA/TPU blend decreased impact strength with respect to neat PLA/TPU, but further addition of GPOSS improved the impact strength. The tensile strength was decreased with GPOSS addition. The rheological properties were also affected by the addition of GPOSS to PLA/TPU system.

ACKNOWLEDGEMENTS

Authors thank to the Scientific and Technological Research Council of Turkey (TUBITAK) for financially supporting this study (Grant number: 115M576).

REFERENCES

1. S. Jia, Z. Wang, Y. Zhu, L. Chen and L. Fu, *RCS Adv.***5** ,98915-98924 (2015).
2. B. Imre and B. Puzansky, *European Polymer Journal* **49**, 1215-1233 (2013).
3. S. K. Dogan, E. A. Reyes, S. Rastogi, G. Ozkoc, *J. Appl. Polym. Sci.***40251**, 1-10 (2014).
4. E. Ayandele, B. Sarkar, P. Alexandridis, *Nanomaterials* **2**, 444-475 (2012).

Modeling Approach for the Degradation Mechanisms of PLA Resulting from Processing

Matthias Düngen[a*], Stefan Keller[a] and Michael Koch[a]

[a]*Department of Plastics Technologies, Technische Universität Ilmenau, Germany*

matthias.duengen@tu-ilmenau.de

Abstract. Polylactides (PLA) are processed in single or multiple screw extruders where thermal and mechanical stresses occur. These stresses can lead to melt degradation resulting into molecular weight loss. The susceptibility of PLA to degradation during processing is well known. A study of PLA degradation on extruders relative to varying key impacting factors such as temperature, shear rate or residence time is complicated due to the interdependency of these factors and varying screw geometry, melt temperature, pressure and melt velocity gradients along the screw channel(s). Therefore experiments with a rotational rheometer were performed. This experiment allows a differentiated examination of shear and thermally induced degradation. Considerable effects as a result of varying shear rate, temperature and residence time on melt degradation were observed. Mathematical models describe non-linear degradation mechanisms found in rheometer experiments. An approach to transfer the rheometer results to a single screw plasticizing extruder (SSE) is presented. It shows a good qualitative agreement. Screw design and processing conditions are found to have a significant impact on degradation of processed PLA. During processing in a SSE a wide distribution of processing histories for different particles in the screw channel occur due to longitudinal and transversal mixing of the melt. Thus a distribution of residence times, thermal and shear stresses have to be taken into account when using models derived from rheometer experiments. Results of this model are presented.

Keywords: PLA, Degradation, Extrusion, Single screw
PACS: 81.20.Hy

1 INTRODUCTION

Biodegradable polymers such as Polylactide (PLA) are used in a wide range of applications and availability due to installed production capacity is steadily increasing [1]. However, processing of PLA is different from conventional plastics like Polypropylene (PP) or Polystyrene (PS) in particular relative to melt degradation due to e.g. shear or thermal exposure.

Degradation of thermoplastic materials during processing in extrusion or injection molding can have several causes such as thermal degradation, thermo-oxidative degradation, in case of polycondensates hydrolysis, or mechanical i.e. shear induced degradation [2]. These different mechanisms of degradation are often difficult to distinguish in actual production processes. Thus, the total degradation has usually been identified as resulting from processing in general [3-7]. Degradation of PLA-types due to processing leads to molecular weight reductions of up to 88% [8].

For PLA, the degradation caused by processing in single screw extrusion [5], twin screw extrusion [7] and injection molding [3, 4, 6, 8] processes have been studied. A relatively low degradation for twin screw extruded material compared to extreme degradations for material that was injection molded several times indicates a significant influence of the different processing conditions – particularly shear. Mathematical models describing this degradation during processing by single screw extruders have been presented for PP [9, 10] and PS [11] by POTENTE, SCHÖPPNER et al. These models include the results of basic and indispensible processing conditions in a screw process and a wide range of its parameters and resulting molecular weight reduction for specific materials. The quantitative impact of different degradation mechanisms is not distinguished by these models.

The approach of this study is to distinguish the effect of thermal and shear induced degradation for PLA.

2 EXPERIMENTAL

The PLA used was Ingeo 3052D. Moisture content for all experiments was 0,1 wt-% in order to achieve a reproducible and relatively small influence of the hydrolysis reaction. The fundamental approach of this study is to control temperature and shear in the experiments to a defined level in order to analyze their impact on material degradation. Therefore, a special setup in a rotational rheometer was chosen. The applicability of these results was studied to consider a relation to a plastification unit.

2.1 Rotational Rheometer

A rotational rheometer can provide viscosity data as a function of shear rate by performing a shear rate-sweep. This procedure was used in this study to measure the Newtonian viscosity η_0. The Mark-Houwink-equation describes a relation for the mean molecular weight and the Newtonian viscosity for polymers [12]. Thus a decreasing Newtonian viscosity describes a decreasing mean molecular weight. Using this relation makes the rotational rheometer a powerful tool to study the degradation in terms of a change of molecular weight. Two different types of experiments were conducted with the rotational rheometer: one without a continuous rotational motion so that no significant shear induced degradation could occur. In this experiment the degradation was a result of the thermal exposure and thus called thermal degradation. It is expressed by the measured change in the Newtonian viscosity $\eta_{thermal}$. The other type of experiment featured a continuous rotational motion at a predefined shear rate. The thermal influence was always present. By neglecting any interdependencies between thermal and shear induced degradation the shear influence can be defined as

$$\Delta\eta_{shear}\,[\%] = \Delta\eta_{total}\,[\%] - \Delta\eta_{thermal}\,[\%] \quad (1)$$

The parameter range of the experiments was 1, 2, 5, 10, 15, 20 minutes of residence time at temperatures of 180, 190 and 200 °C and shear rates of 0 (thermal exposure with different residence time), 10, 30, 50, 100, 500 and 1000 1/s. A cone-plate configuration according to ISO 3219 was chosen for the rheometer setup. An even shorter residence time would have been of interest since the residence time in plasticating extruders in standard machinery is typically shorter than several minutes, however, given the setup thorough melting and sufficient thermal homogeneity of the melt could not be ensured for shorter residence times. Higher shear rates are relevant for injection molding but did not deliver reproducible results.

2.2 Single screw extrusion

A single screw extruder with a screw diameter of 45 mm was used with two different screw designs as given in tables 1 and 2. For a detailed parameter explanation and discussion of screw designs and their parameters [13] is recommended.

TABLE (1). Parameters for screw "S1"

	Unit	Zone 1	Zone 2	Zone 3
L/D	[-]	8	8	7
h	[mm]	7,2	7,2-2,8	2,8
e	[mm]	4,5	4,5	4,5
t/D	[-]	1	1	1
r_1; r_2	[mm]	2,5; 3	2,5; 3	2,5; 3

TABLE (2). Parameters for Screw "S3 WNT"

	Unit	Zone 1	Zone 2	Zone 3	pitched Maddock mixer (4 channel pairs)
L/D	[-]	8	5	8	2
h	[mm]	7,5	7,5-2,6	2,6	5 (barrier gap: 0,4)
e	[mm]	4,5	4,5	4,5	(barrier gap: 3,8)
t/D	[-]	1	1	1	5,4
r_1; r_2	[mm]	2,5; 3	2,5; 3	2,5; 3	-

Screw "S1" is a conventional three zone design with feeding, compression and metering zone. Screw "S3 WNT" had a similar setup, however, with a pitched Maddock mixer added and different screw channel height in the metering zone, thus both impacting significantly on shear effects. The parameter range of the screw speed was 20, 40, 60, 80 rpm which results in circumferential speeds of 0,05; 0,09; 0,14 and 0,19 m/s. A higher screw speed did not yield acceptable melt quality and process stability. The heating or cooling of the individual barrel heater zones was chosen to produce two different melt temperatures: 135 °C and 150 °C. A melt temperature sensor behind the screw tip measured the temperature of the melt in the middle of the screw channel.

3 RESULTS & DISCUSSION

3.1 Results rotational rheometer

Figure 1 shows an increasing degradation for increasing residence times and temperature levels. According to the van-'t-Hoff rule for chemical kinetics the exponential character of the curves is a typical pattern for a temperature driven chemical reaction.

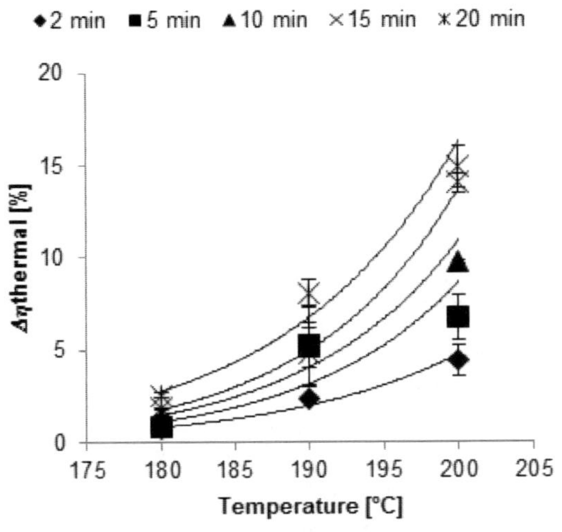

FIGURE 1. Influence of thermal degradation for different temperature levels and residence times

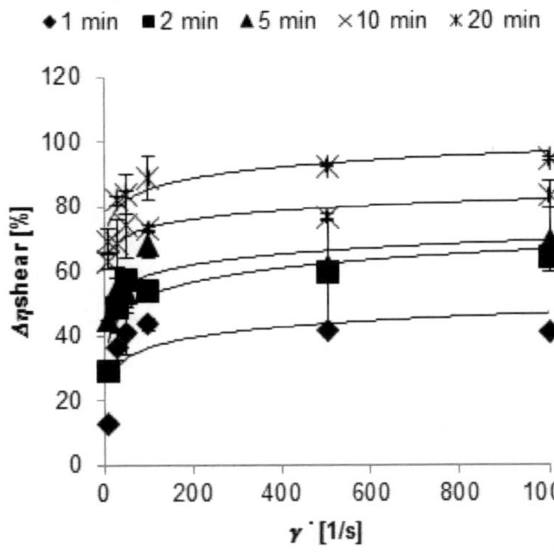

FIGURE 2. Influence of shear induced degradation for different shear rates and residence times

Figure 2 demonstrates an increasing degradation for increasing shear rates and increasing residence times. The shear induced degradation reaches significantly higher levels than the solely thermally induced degradation. The character of this curve is totally different from the thermal degradation indicating no chemical reaction. The slope of the curves is decreasing which correlates with the shear thinning behavior of PLA melt. For high shear rates the melt viscosity is lower and thus shear stresses which can cause molecular chain brake-up do not increase proportionally with the shear rate.

These results show dependencies which are transformed into mathematical models. As boundary conditions, the function have to yield no change in degradation for a residence time of 0 as well as for a shear rate of 0 or a temperature level which does not differ from the reference temperature level of 175 °C at 0 residence time, respectively. Thus basic functions were chosen to describe thermal and shear induced degradation separately.

$$\Delta\eta_{\mathrm{thermal}} = \mathrm{A} \cdot \mathrm{t} \cdot \mathrm{e}^{\mathrm{B} \cdot \mathrm{T}} \tag{2}$$

with a time coefficient for thermally induced degradation *A* [1/s], temperature coefficient for thermally induced degradation *B* [1/°C], residence time *t* and temperature *T*.

$$\Delta\eta_{shear} = (\ln(C \cdot \dot{\gamma})^{D}) \cdot (E \cdot t^{0,2}) \tag{3}$$

with shear rate coefficient for shear induced degradation *C* [s], exponent for shear induced degradation *D* [-], residence time coefficient for shear induced degradation *E* [1/s], shear rate $\dot{\gamma}$ and residence time *t*.

Their coefficients were fitted to the experimental results using a Downhill-Simplex-method (Nelder-Mead-method) resulting in

$$\Delta\eta_{thermal} = 7{,}16 \cdot 10^{-8} \cdot t \cdot e^{0{,}082 \cdot T} \tag{4}$$

$$\Delta\eta_{shear} = (\ln(0{,}14 \cdot \dot{\gamma})^{0{,}14}) \cdot (40{,}15 \cdot t^{0{,}2}) \tag{5}$$

3.2 Results plasticating extruder

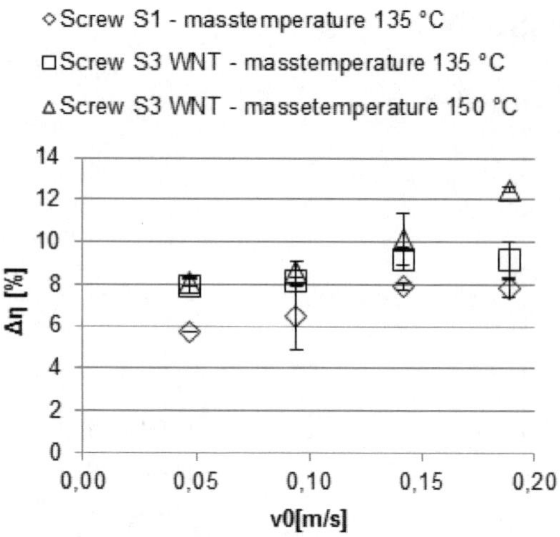

FIGURE 3. Degradation results for different screw designs

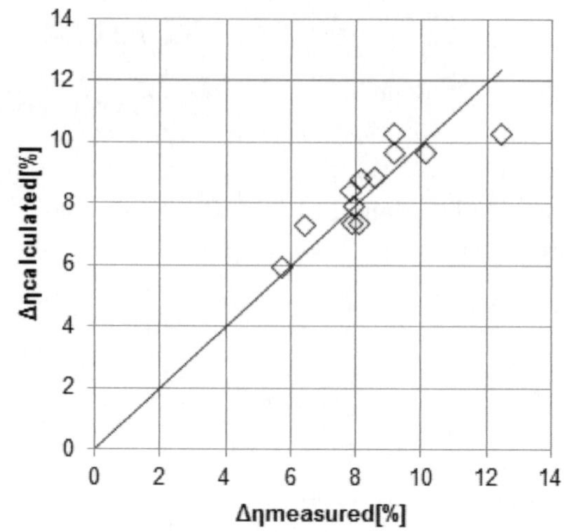

FIGURE 4. Comparison of measured and calculated (eq. 9) Newtonian viscosity change for extrusion experiments

In alteration of the ideal conditions on the rotational rheometer a plasticating extruder does not allow the choice of independently set influence factors as the processing conditions result in complex effects. The residence time depends inversely proportional on the screw speed while the shear rate depends directly proportional. The results in fig. 3 show that these effects do not cancel each other out. The degradation is increased for higher screw speeds (i.e. higher shear rates) even though the residence time is shortened. The results also show higher degradation for higher melt temperatures. For the screw with higher shear influence, higher degradation was expected and measured. Thus the results qualitatively comply with the rotational rheometer results.

The level of quantitative agreement can be calculated using the developed equations 4 and 5. Therefore, the variables residence time, shear rate and melt temperature have to be known.

As melt temperature the temperature of the melt behind the screw tip was measured by a temperature sensor. This temperature does not represent a mean temperature of the total material in the screw channel. Since the material in the screw channel – in particular in the melting zone – partly consists of solid material the average melt temperature of the material is most likely overpredicted by using the melt temperature from the chosen sensor position.

In a first assumption, the mean residence time is calculated using

$$\bar{t} = \frac{m}{\dot{m}} \tag{6}$$

with the measured throughput $\dot{m}$ and the mass in the screw channel m. It is understood that the actual residence time is in fact following a distribution curve for multiple flow paths according to the ration of drag to pressure flow and power law exponent of the polymer.

The shear rate is calculated for each zone with the simplified relationship

$$\dot{\gamma}_i = \frac{v_0}{h_i} \tag{7}$$

which is used to calculate an average shear rate. Further work will have to reflect the variability of residence time and shear rate in order to describe process stability and spread of degradation effects. Therefore, deviations in calculated results are to be expected.

The calculated degradation for the plasticating extruder experiments is 6-7 times higher than the measured degradation. The calculation was executed with the equations derived from the rheometer experiment (eq. 4 and 5). Possible causes for this deviation are:

- temperature profiles differ between the rheometer gap and the screw channel,

- the plasticating extruder features a residence time distribution [14] while the rheometer shows a constant residence time and uniform distribution,
- longitudinal mixing of the extruder allows minimum degradation for a portion of the material that passes the flow channel at the minimum residence time t_1,
- the melting process in the extrusion process is dominated by shearing effects while in the rheometer set-up, heating is accomplished by heat conduction only,
- temperature, shear rate and residence time can change considerably in every section or subsection of the screw, i.e. in particular in the pitched Maddock mixing element.

According to the findings in the extruder experiments, the coefficients of the shear degradation model were adapted to result in the quantitative degradation with the simplified approach of eq. 6 and 7

$$\Delta\eta_{shear} = (ln(5{,}49 \cdot \dot{\gamma})^{2,2}) \cdot (0{,}26 \cdot t^{0,2}) \tag{8}$$

Using this equation (8), very good agreement was obtained comparing measured and calculated results as shown in fig. 4 while confirming the correctness of the assumptions made with the base functions set in eq.2 and 3.

4 CONCLUSION

Rotational rheometer experiments show considerable influence of shearing on the degradation of PLA compared to solely thermal degradation. Mathematical models can describe this behavior accurately. While these insights can be confirmed qualitatively for plasticating extrusion experiments the quantitative analysis shows the importance of several differences compared to the experimental setups. Especially, the distribution of residence time and the complex velocity respectively shear rate and temperature fields in the screw channel have to be taken into further consideration in future work.

The study shows also the significant influence of specific screw designs and the chosen operating condition on the degradation of processed PLA, representing the effects of different shear rates and melting parameters.

5 ACKNOWLEDGEMENTS

Financial support by the Thuringian Ministry for Economy, Employment and Technology with co-funding by the EFRE-fond of the European Union (grant number No. 331/2007, TAB project number 2010 FE 9051) is gratefully acknowledged.

REFERENCES

1. M. Carus, W. Baltus, D. Carrez, H. Kaeb, J. Ravenstijn and S. Zepnik, “Market study and Database on Bio-based Polymers in the World”, Huerth, 2013 .
2. Schmiederer, D., “Schonende Spritzgießverarbeitung von Thermoplasten“, PhD-Thesis, Universität Erlangen-Nürnberg, 2008
3. I. Pillin, N. Montrelay, A. Bourmaud and Y. Grohens, Polym. Degrad. Stab., 93, 321 (2008).
4. F. Carrasco, P. Pagès, J. Gámez-Pérez, O.O. Santana and M. L. Maspoch, Polym. Degrad. Stab., 95, 116 (2010).
5. C. Hopmann, and S. Schippers, “Recycling of polylactide for packaging applications”, Proceedings of ANTEC 2013, Cincinnati (2013).
6. A. Jaskiewicz, A. K. Bledzki and A. Meljon, *Polym. Degrad. Stab.*, **101**, 65 (2014).
7. Q. Meng, M.-C. Heuzey and P. J. Carreau, *Polym. Degrad. Stab.*, **97**, 2010 (2012).
8. S. Gogolewski, M. Jovanovic and S. M. Perre, *Polym. Degrad. Stab.*, **40**, 313 (1993).
9. T. C. Pohl and H. Potente, *J. Plast. Technol.*, **2**, 3 (2006).
10. S. Littek, M. Schneider, K. Huber and V. Schöppner, *J. Plast. Technol.*, **8**, 415 (2012).
11. V. Schöppner and S. Littek, “Measurement and Calculation of the Material Degradation of Polystyrene and Modeling of the Degradation by using REX”, Proceedings of PPS 30, Cleveland (2014).
12. G. Rafler, J. Dahlmann and K. Wiener, *Acta Polymerica*, **41**, 328 (1990).
13. J. L. White and H. Potente, „Screw Extrusion“, Hanser, München, 2003.
14. M. Koch, „Berechnung und Auslegung von Nutbuchsenextrudern“, PhD-Thesis, Universität Paderborn, 1987.

Compatibilizing Effects of Itaconic Anhydride on Blends between Poly(lactic acid) and Novatein® Thermoplastic Protein

Anuradha Walallavita, Casparus J.R. Verbeek, Mark Lay

University of Waikato, Hamilton 3240, New Zealand

asw15@students.waikato.ac.nz, jverbeek@waikato.ac.nz, mclay@waikato.ac.nz

Abstract. Novatein®, a protein-based thermoplastic, was blended with poly(lactic acid) (PLA) using reactive extrusion to produce a cheap and degradable material with improved mechanical properties over neat Novatein. Various compositions of Novatein/PLA were prepared with and without a compatibilizer, PLA grafted with itaconic anhydride (PLA-g-IA).

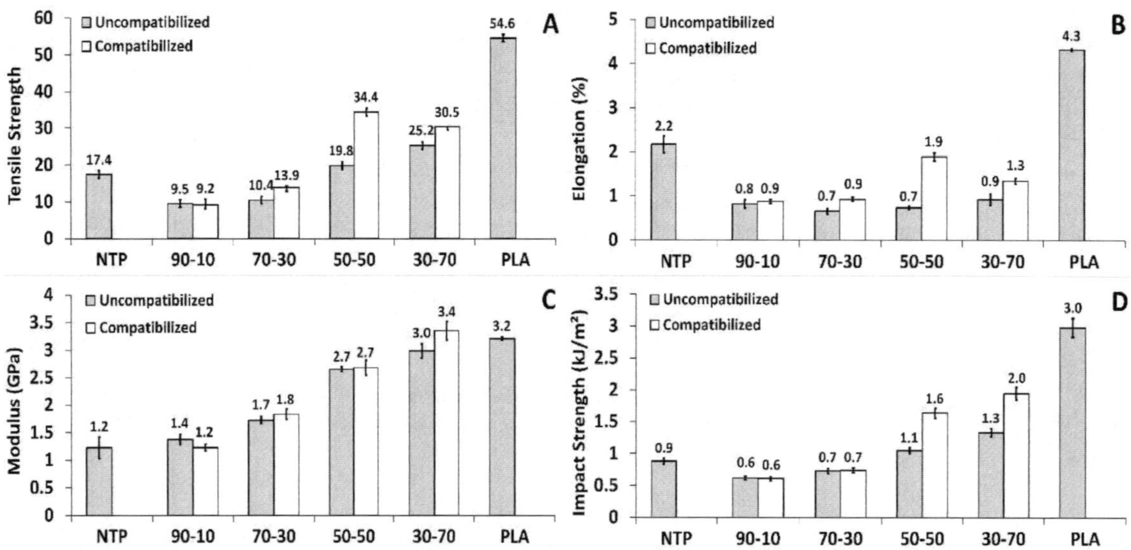

FIGURE 1. Tensile strength (A), percent elongation (B), secant modulus (C) and impact strength (D) of Novatein/PLA blends with and without compatibilizer.

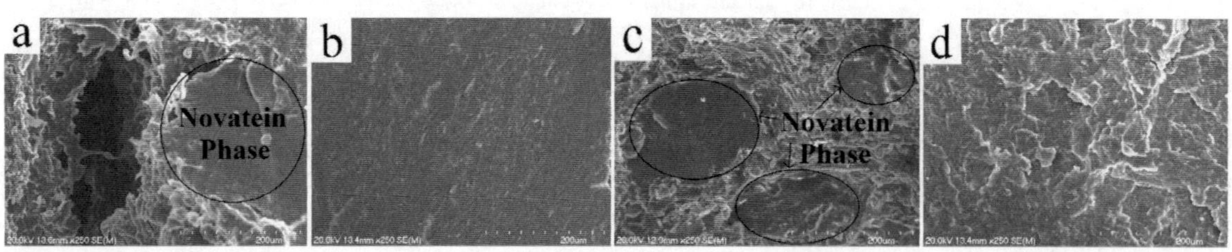

FIGURE 2. SEM images of tensile tested fracture surfaces at magnification x 250 a) 90-10 b) 90-10 IA/DCP c) 50-50 d) 50-50 IA/DCP.

Novatein had a tensile strength of 17.4 MPa while a blend of Novatein with PLA-g-IA at 50 wt. % loading increased the tensile strength to 34.4 MPa (Figure 1.A). All Novatein and PLA-g-IA blends showed an improvement in elongation at break i.e. from 0.74 to 1.88 at 50-50 composition (Figure 1.B). Impact strength of 50-50 IA/DCP was higher`than 30-70 blend without compatibilizer (Figure 1.D). Scanning electron microscopy revealed that adding compatibilizer reduced the presence of cracks and voids in fracture surfaces, resulting in a more homogenous microstructure (Figure 2.a and b). In the 50/50 blend, the Novatein phase became finer and more evenly dispersed within the PLA matrix (Figure 2.c and d). Blending Novatein with PLA can widen its use in many agricultural and packaging applications, but require itaconic anhydride as compatibilizer.

Keywords: Novatein® Thermoplastic Protein, biodegradable polymers, blends, polylactic acid, itaconic anhydride.

Preparation and Characterization of Modified-TiO_2/PLA Nanocomposite Films with Enhanced Microstructure, Photo-degradability and Antibacterial Properties

Kamand Rabiei Tabriz, Ali Asghar Katbab

Amirkabir University of Technology, Department of Polymer Engineering and Color Technology, 15875-4413, Tehran, Iran

kamandrabiee@yahoo.com

katbab@aut.ac.ir

Abstract. Today bio-nanocomposites of biopolymers are from great interest from the point of view of environmentally friendly concerns (1). Among all types of PLA nanocomposites, the ones including oxides of semi-conductive materials such as TiO_2 have attracted the attention of many research centers and industries due to their photocatalytic and antibacterial properties (2). Although there seems to be a great interest about producting the nanocomposite of PLA and TiO_2 nanoparticles, but aggregation of TiO_2 nanoparticles in an organic matrix like PLA, makes it difficult to reach a homogenous film with good properties (3). Therefore, using a method in order to modify TiO_2 nanoparticles surface before introducing into the PLA matrix, seems to be necessary. In this study first, PLA chains were grafted onto TiO_2 nanoparticles surface via reactive melt mixing using a novel approach in presence of $SnCl_2$ as catalyst. In this step, PLA grafted chains were characterized on the TiO_2 nanoparticles using FTIR analysis. Then, PLA-grafted-TiO_2/PLA nanocomposite films with different amounts of modified TiO_2 nanoparticles, were produced using a twin screw extruder. The microstructure of nanocomposite films was characterized using SEM technique, and this way, it was observed that the dispersion state of TiO_2 nanoparticles has been greatly improved after modifying TiO_2 surface. Finally, the PLA-grafted-TiO_2/PLA nanocomposite films were put under UV irradiation, in the range of 360-420 nm, for 2 months and the weight loss of the films was followed during this period. The nanocomposite films containing modified TiO_2 nanoparticles, showed higher amount of weight loss in comparison with the film containing bare TiO_2 nanoparticles. The PLA-grafted-TiO_2/PLA nanocomposite films were also used to investigate the antibacterial properties against both gram negative and gram positive bacteria, under UV irradiation for 2 hours, using a combined method of JIS Z2801:2000 and ASTM E2180-07. The nanocomposite films showed high antibacterial properties against E-coli gram negative bacteria, although this amount was lower in case of St.aureous gram positive bacteria, due to the natural structural differences between gram negative and gram positive bacteria's cellular membrane.

Keywords: PLA/TiO_2 nanocomposite, photodegradability, food packaging, antibacterial films, surface grafting.
PACS: 82.35.NP

INTRODUCTION

For a long time polymers have supplied most of common packaging materials because they present several desired features like softness, lightness and transparency (1). However, increased use of synthetic packaging films has caused a serious ecological problem due to their total non-biodegradability (1). Although their complete replacement with eco-friendly packaging films is just impossible to achieve, at least for specific applications like food packaging the use of bioplastics should be the future (1). Until now petrochemical-based plastics such as polyethylene terephthalate (PET), polyvinylchloride (PVC), polyethylene (PE), polypropylene (PP), polystyrene (PS) and polyamide (PA) have been increasingly used as packaging materials because of their large availability at relatively low cost and due to their good mechanical performance such as tensile and tear strength, good barrier to oxygen, carbon dioxide, anhydride and aroma compound, heat sealability, and so on (1). But nowadays their use has to be restricted because they are not totally recyclable and/or biodegradable, so they pose serious ecological problems (1). The growing environmental awareness imposes to packaging films and process both user-friendly and eco-friendly attributes (1). Among all biocompatible polymers, poly(L-lactic acid) (PLA) is a highly versatile material and is made from 100% renewable resources like corn, sugar beets, wheat and other starch-rich products (2). It has many potential uses, including many applications in the textile and medical industries as well as in the packaging industry (2). Therefore, as for the PLA, a number of basic studies concerning thermal properties, crystallization, and applied studies concerning improvement of thermal and mechanical properties and applications for tissue engineering and so on, has been reported (2). Although biodegradation of PLA is advanced in the ground where moisture and bacteria exist, 2-3 months is needed generally speaking for the decomposition, and biodegradation does not advance in air (4). As a solution, by adding photodegradability to a biodegradative polymer, degradability can be efficiently promoted under any condition (4). Among all the available photodegradation agents, titanium dioxide (TiO_2) nanoparticles have been investigated in recent years because of its photocatalytic effects that decompose various polymers and organic chemicals. The photocatalytic reaction is well known to active oxygen species, e.g. O_2^-, HO_2 radicals, HO radicals, from H_2O or O_2 by oxidative or reductive reactions under UV exposure (3). These active radicals lead to a degradation reaction by attacking polymer chains and accelerating chain cleavage (3). Although the use of TiO_2 nanoparticles in polymeric matrices lead to better degradation properties, there still remain unsolved problems related to uniform dispersion of the photocatalyst in PLA matrix (3). Since there is numerous amount of hydroxyl groups on TiO_2 nanoparticles surface, there is a severe tendency for them to aggregate in hydrophobe matrices such as PLA. This severe aggregation of TiO_2 nanoparticles can significantly reduce the efficiency of photodegradation by the decrease of interfacial areas between TiO_2 and PLA chains (3). The dispersibility of TiO_2 nanoparticles in PLA matrix can be improved via modifying particles surface by grafting PLA chains onto them (3). There is a number studies in which PLA chains have been grafted onto TiO_2 nanoparticles surface using in situ polymerization method; the lactic acid monomers were polymerized from the hydroxyl groups existing on the surface of TiO_2 nanoparticles, in solution state using THF and chloroform as TiO_2 and PLA solvents respectively (5). Though these studies achieved useful results this way, practically it is not an applicable way to use solution methods in industrial scale. In the present work, we applied a new method in order to modify TiO_2 nanoparticles surface using melt processing in presence of stannous chloride as catalyst. PLA chains were grafted onto nanoparticles surface through reactive melt mixing by an internal mixer, using carboxylic acid terminals existing at the end of PLA chains, esteric bonds present in PLA backbone and TiO_2 hydroxyl groups, as the active sites needed for grafting reaction. The surface grafted PLA chains were characterized by FTIR. After modifying TiO_2 nanoparticles surface, the PLA-g-TiO_2/PLA nanocomposite films were fabricated by a twin screw extruder and used to study the photodegradation properties.

The other concern related to food packaging industries is to maintain the quality and to extend the shelf-life of food products during storage by eliminate unfavorable factors such as spoilage microorganisms and bacteria (6). Antimicrobial food-packaging systems are polymeric materials in which a certain amount of additives with pronounced antimicrobial properties are embedded into the packaging and act with the aim mentioned above (6). One of the most interesting antimicrobial agents which have been extensively used during recent decades is TiO_2 photocatalyst (6). By producing very active radicals under UVA irradiation, TiO_2 is able to attack chemicals in bacteria cell wall structure and destroy the bacteria (6). In this study, the fabricated PLA-g-TiO_2/PLA nanocomposite films were introduced to gram negative E-coli and gram positive St.Aureous bacteria under UV irradiation to study antbacterial property of the films.

EXPERIMENTAL

Preparation of PLA-g-TiO₂/PLA Nanocomposite Films

In this study, PLA chains have been grafted onto TiO2 nanoparticles surface via reactive melt mixing using an internal mixer, in presence of $SnCl_2$ as catalyst. Furthermore, modified-TiO_2 nanoparticles used to produce PLA-g-TiO_2/PLA nanocomposite films via melt mixing using a twin-screw extruder. The extruder was equipped with a volumetric feeder and a slit die to reach nanocomposite films. The extruder had six controlled temperature zones which were set to range from 175 next to the feeding section to 195 ºC in die adaptor. The screw speed was maintained at the speed of 30. Before extrusion, PLA pellets and modified-TiO_2 nanocomposite chips were dried in vacuum at 65 ºC for 10 h. The neat PLA was subjected to the mixing treatment so as to have the same thermal history as the blends. For comparison, 1wt.% of bare TiO_2 nanoparticles were blended with PLA using the same technique and labeled as PLA/neat-TiO_2-1. After preparing the nanocomposite films, all the mentioned samples were cut into an equivalent surface and have been stored for photo-degradation test.

Photocatalytic Degradation Test

After preparation of PLA-g-TiO_2/PLA nanocomposite films mentioned in previous step, samples were cut into an equivalent surface and were exposed to UV irradiation in range of 360-420 nm for 2 months, using a cabin equipped with two 8W black-light UV tube. The photodegradation extent of the films can be evaluated directly by their weight loss, since the degradation process of PLA chains follows with mass decrease due to release of a mol of CO_2 gas per hydrolysis of one mol ester bond. The weight decrease trend of nanocomposite films were followed every 4 days during 2 months. Furthermore, scanning electron microscopy technique used to show the change of films surface morphology after exposure to UV for 2 months.

RESULT and DISCUTION

Characterization of Surface Modified TiO_2 Nanoparticles

The FTIR spectra of PLA and TiO_2 nanoparticles after grafting with different amount of stannous chloride are shown in fig. 1.In this figure, the spectroscopic band which is observed around 3500 cm^{-1} is attributed to the stretching vibration of hydroxyl group (Ti--OH) of the TiO_2 nanoparticles. A broad absorption band between 500 and 1000 cm^{-1} is ascribed to the vibration absorption of the Ti—O—Ti linkages in TiO_2 nanoparticles. The peak which is observed around 1645 cm^{-1} is attributed to adsorbed water. As it can be seen in fig . 2, after surface grafting, a new absorption band appears at 1745 cm^{-1} belonging to the carbonyl group (C=O) of PLA, which could be consider as an evidence that PLA chains have been successfully grafted onto TiO_2 surface via reactive melt mixing process. Since the free PLA chains have been completely washed away by excessive chloroform during the isolation process, the peak at 1745 cm^{-1} cannot arise from the PLA physically adsorbed on the surface of TiO_2 nanoparticles. Furthermore, new bands appeared at 1400 cm^{-1} and 1555 cm^{-1} due to the bidentate coordination between Ti atoms and the carboxylic groups of PLA , and peaks at 2870 cm^{-1}, 2998 cm^{-1} and 2945 cm^{-1} are related to the symmetric and asymmetric stretching of C---H_2 and C---H_3 side groups of PLA chains and C---H bond of PLA backbone, respectively. Finally, the new peaks between 1360 cm^{-1} and 1380 cm^{-1} were caused by the bending vibration of C---H group in PLA backbone.

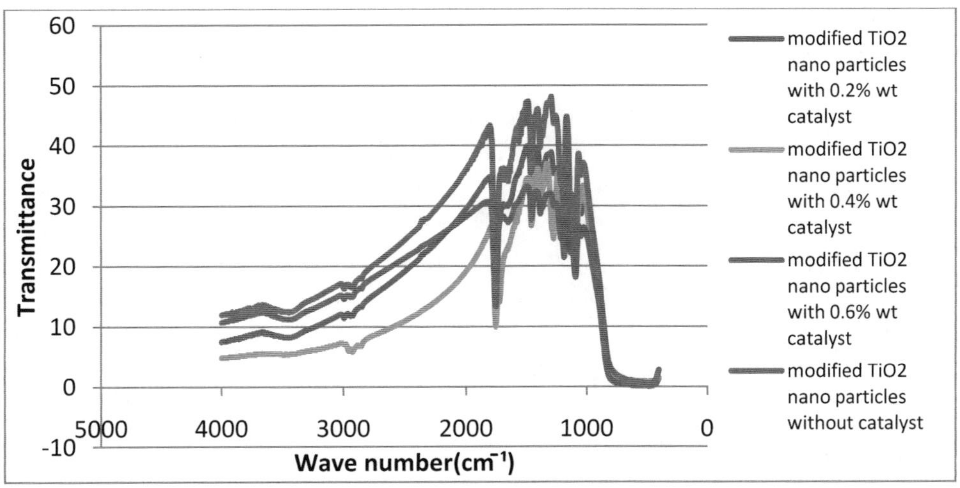

FIGURE 1. FTIR analysis of surface-modified TiO_2 nanoparticles in presence of different amount of $SnCl_2$ as catalyst

Characterization of PLA-g-TiO_2/PLA Nanocompoite Films

The SEM micrograph of fracture surface of PLA-g-TiO_2/PLA nanocomposite films produced at 30 rpm, have been shown in fig.2. As it's been represented in fig.2a, the nonmodified-TiO_2 nanoparticles formed large aggregates with the average size in micron scale (about 1300 nm). It could be seen clearly from fig.2b that after surface modification the average size of TiO_2 nanoparticles diminished from 1300 nm to 650 nm. In fact after surface modification, the dispersion and distribution state of TiO_2 nanoparticles, have been greatly improved.

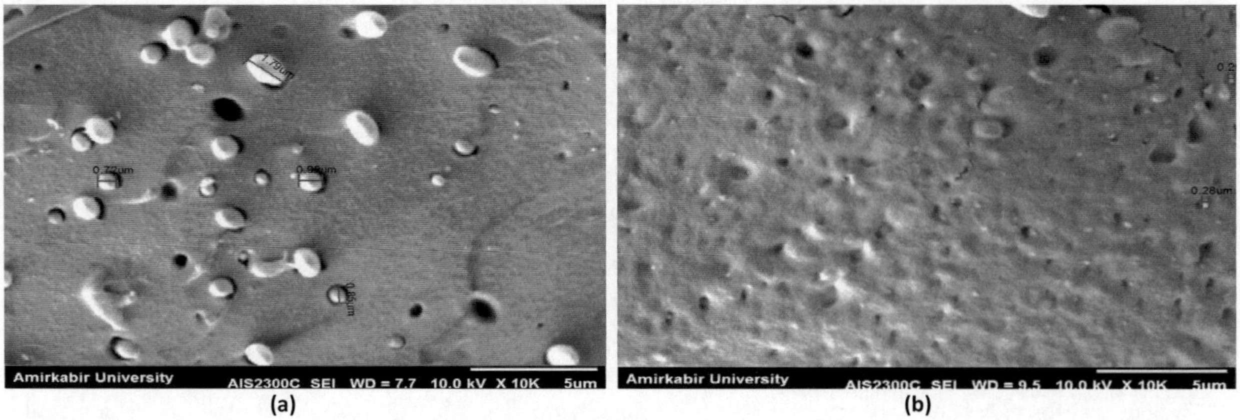

FIGURE 2. SEM micrograph of (a) fracture surface of PLA/TiO_2 composite film and (b) fracture surface of PLA-grafted-TiO_2/PLA nanocomposite film

Photo-Degradability Studies

Fig.3 demonstrates the photo-induced weight loss curves of the pure PLA film, bare-TiO_2/PLA composite film and PLA-g-TiO_2/PLA nanocomposite film. The overall weight loss of the pure PLA film and the composite film including bare-TiO_2 is respectively about 0.5% and 10.5% after 2 months irradiation, which shows that PLA itself has no photocatalytic activity. The overall weight loss of the PLA-g-TiO_2/PLA nanocomposite film which reached about 35% after 2 months irradiation, is the highest among these three kinds of films. These results prove that surface modification of TiO_2 nanoparticles has promoted the photocatalytic degradation of nano-TiO_2-containing PLA films.

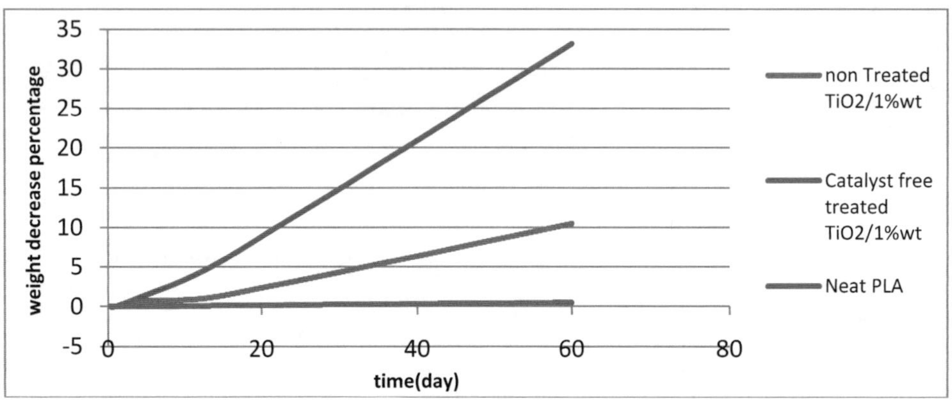

FIGURE 3. weight loss trend of films under exposure to UV for 2 months

Antibacterial Studies

The antibacterial behaviour of the nanocomposite films was tested via colony counting method. Fig.4 shows the antibacterial efficiencies of the different nanocomposite films surfaces against gram negative E. coli and gram positive St. aureous bacteria during 2 hours under UV irradiation. In order to compare the results, similar test was operated on a thin film of pure PLA produced using solvent casting method. In case of reference sample, the amount of colonies decreased due to UV irradiation; the bacteria membrane is vulnerable against UVB and UVC, whereas the maximum peak of UV lamps irradiation range was at UVA. As it's been shown in fig.4, the amount of E. coli colonies have been intensively decreased after 2 hr irradiation. In case of St. aureous bacteria, the decrease of colonies was less than the case of E. coli, which is contributed to the structural differences between their cellular walls; the presence of a thick peptidoglycan layer between the inner and outer layers in cellular membrane of gram positive bacteria, makes it stronger than gram negative ones against external threats.

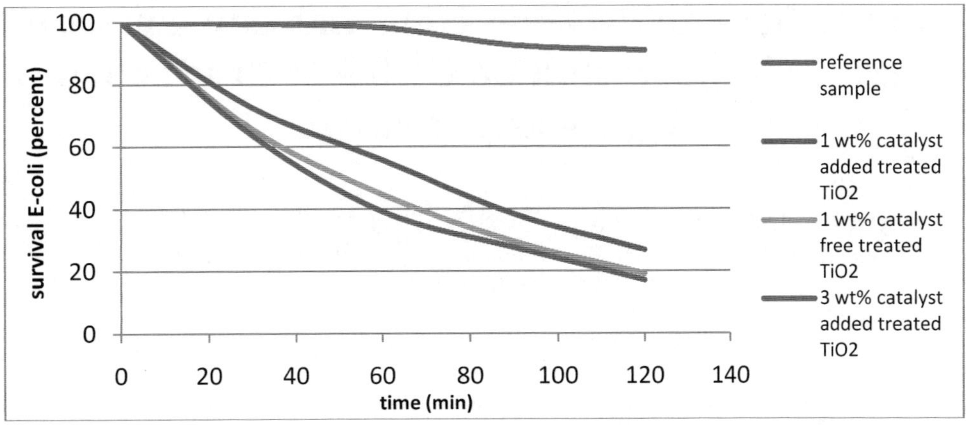

(A)

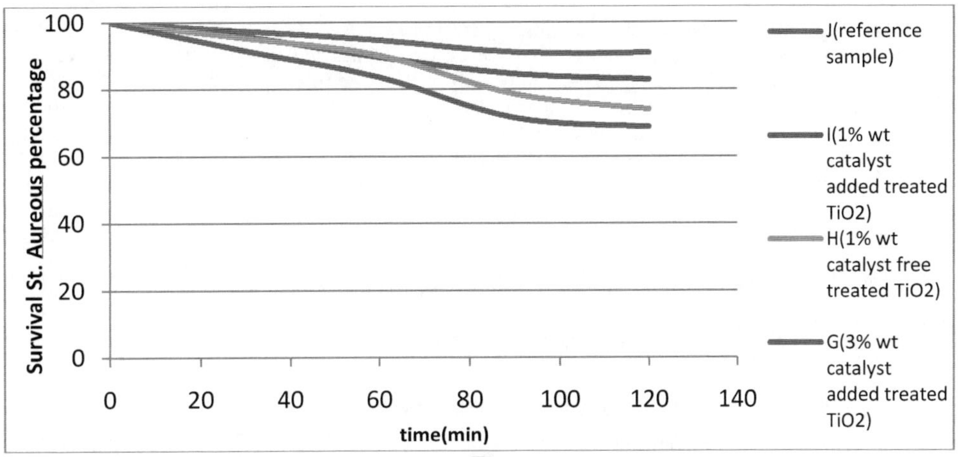

(B)

FIGURE 4. amount of survival (A) gram negative E-coli bacteria and (B) gram positive St.Aureous bacteria, in presence of PLA-grafted-TiO_2/PLA nanocomposite film after 2 hr under UV irradiation

ACKNOWLEDGMENTS

This research was carried out under cooperation of the research and technical centers of the Polymer Engineering and Color Tech. Department of Amir Kabir University of Technology, and Biology Department of Shahid Beheshti University, Tehran, Iran.

REFERENCES

1. V. Siracusa, P. Rocculi, "*Biodegradable Polymers for Food Packaging*", Catania, Italy, Trends in Food Science & Technology, 2008, pp. 634-643
2. J.M. Raquez, Y. Habibi, "*Polylactide (PLA)-based nanocomposites*", Mons, Belgium, Progress in Polymer Science, 2013, pp. 1504-1542.
3. N. Nakayama,T. Hayashi, "*Preparation and characterization of poly(L-lactic acid)/TiO2 nanoparticle nanocomposite films with high transparency and efficient photodegradability*", Japan, Polymer Degradation and Stability, 2007, pp. 1255-1264.
4. J. Rhim, H. Park, *"Bio-nanocomposites for food packaging applications"*, Republic of Korea, Progress in Polymer Science, 2013.
5. X. Lu, *"Nanocomposites of Poly(L-lictide) and surface-grafted TiO_2 nanoparticles: Synthesis and Characterization"*, People's Republic of China, European Polymer Journal, 2008, pp. 2476-2481.
6. H. Azeredo, *"Nanocomposites for food packaging applications"*, Fortaleza, Food Research International, 2009, pp. 1240-1253.

An Investigation between High and Low Pressure Processes for Nanocrystalline Cellulose Production from Agro-waste Biomass

[a]Md. Sakinul Islam, [a]Nhol Kao*, [a]Sati N. Bhattacharya and [a]Rahul Gupta
[a]School of Engineering, Department of Chemical Engineering, RMIT University, Melbourne, VIC 3001, Australia

Presenting & Reference Author's Email: nhol.kao@rmit.edu.au

Nanocrystalline cellulose (NCC) has widely been used for miscellaneous applications due to its superior proprieties over other nanomaterials. The major problems associated with the production of NCC are long reaction time, low production rate and inefficient process. The mass production of NCC within short time remains a great challenge currently. The main objective of this study is to overcome the present problems by investigating the performance of a high pressure process (HPP) and a low pressure process (LPP) for NCC production from raw rice husk biomass (R-RHB) by chemical extraction. The general steps of chemical extraction are delignification, bleaching and hydrolysis. Delignification of R-RHB for HPP and LPP was conducted at 80 °C using a stainless steel reactor and a jacketed glass reactor, respectively. During the delignification in the HPP, 5 bar pressure was applied; while atmospheric condition was used for the LPP. Delignification of HPP and LPP was run for 12 hours using aqueous solution of NaOH (4M) for the separation of cellulose from R-RHB by the dissolution of lignin and other amorphous impurities. Delignified samples were collected after delignification reaction, then washed, bleached and finally hydrolyzed in order to convert cellulose to NCC. NaOCl (20%) and H_2SO_4 (4M) solutions were used for bleaching and hydrolysis processes, respectively. The resultant NCC products of HPP and LPP were analyzed by FTIR, XRD, SEM, EDX, AFM and TEM. The chemical composition of R-RHB and NCC was estimated by TAPPI methods to observe the product purity. High purity NCC has been produced in HPP with crystallinity approximately 55 % and production rate of 62% whereas LPP 51%. It was found that, while both processes produced NCC of diameter 10-20 nm and length 1-2 μm, the HPP was faster and more efficient than the LPP to produce quality NCC from R-RHB.

Key Words: Nanocrystalline Cellulose, High Pressure Process, Low Pressure Process, Rice Husk Biomass

INTRODUCTION

Nanocrystalline cellulose (NCC) has gained significant attention of researchers and scientists all over the world due to its miscellaneous applications. There is plenty of research that has been reported on the production of NCC from lignocellulogic biomass by chemical extraction process, however, all at a lab scale basis. From the last two decades, researchers have been focusing on the use of lignocellulosic biomass for various value-added products such as bio-oil, bio-fuel, bioethanol and biomolecule, etc. due to its low cost and abundant supply all over the world. Agricultural waste biomass such as coconut husk [4], cassava bagasse [5], banana fibre [6], mulberry bark [7, 8], soybean pods [9], wheat straw, and soy hulls [1], and cornstalks [10] have been used in the production of NCC. The hierarchical structure of natural fibres has also been shown that it is a most valuable resource to produce nano-materials [1-3]. Various terminologies have been used in the open literature to describe these cellulose nanoparticles, which include cellulose nanocrystals, nanowhiskers, nanocrystalline cellulose, crystalline nanocellulose and nanofibrilated cellulose [11] - all of which are produced from lignocellulosic biomass. The inefficiency of the extraction process, low production rate and prolong reaction time has hindered the mass production of NCC to date. There is still no benchmark method for NCC production. In addition NCC has only been produced in small scale by physical, mechanical and microbial processes [12, 13]. For this reason NCC is not readily available commercially. Therefore it is very important to resolve this problem. In order to develop an efficient process some key factors need to be considered; these include process parameters, experimental design, material cost and the selection of biomass. The high cellulose containing biomass, such as rice husk is a good choice for mass production of NCC. R-RHB is an agro waste material which is available in abundant all over the world and it contains high cellulose content [15]. Obviously, R-RHB would be cost effective for mass production of NCC. On the contrary the optimization of delignification is a key factor for reducing process cost and reaction time. Delignification can also impact the quality of NCC produced as well. Therefore, both high pressure process (HPP) and low pressure process (LPP) were carried out for the delignification stage in order to optimize NCC production from R-RHB.

METHODOLOGY

Materials: R-RHB provided by Downes Rice Hulls Pty. Ltd. Australia, was used as a main raw material for the production of NCC. Sodium hydroxide, sulphuric acid and sodium hypochlorite were used for delignification, bleaching and hydrolysis, respectively. Other chemicals such as acetic acid, sodium acetate, phosphoric acid, ammonium oxalate, benzene, ethanol and potassium hydroxide were used for the determination of chemical constituents of R-RHB by TAPPI methods. All chemicals are in reagent grade purchased from Science Supply Australia, manufactured by Sigma Aldrich.

Production of NCC: Production of NCC by chemical extraction processes involved the three generic steps such as delignification, bleaching and hydrolysis. In this study the delignification was basically carried out under HPP and LPP conditions to scale-up the NCC production, where R-RHB powder of particle size 75-700 μm was used to produce delignified rice husk pulp (D-RHP). The D-RHP was then separated from black liquor and washed using deionized distilled water (DDW). The detailed information about HPP and LPP for delignification experiments are given in the Table (1). A typical view of HPP and LPP are shown in the following Figure (1).

TABLE (1). Experimental conditions of delignification

Process	Reactor	C_{NaOH}	V_{NaOH}	Temperature	Duration	R-RHB	RPM	Pressure
HPP	SSR	4.0 M	80.0 ml	80 °C	12 hr	10.0 g	--	5.0 bar
LPP	JGR	4.0 M	2000.0 ml	80 °C	12 hr	250.0 g	400	1.0 atm

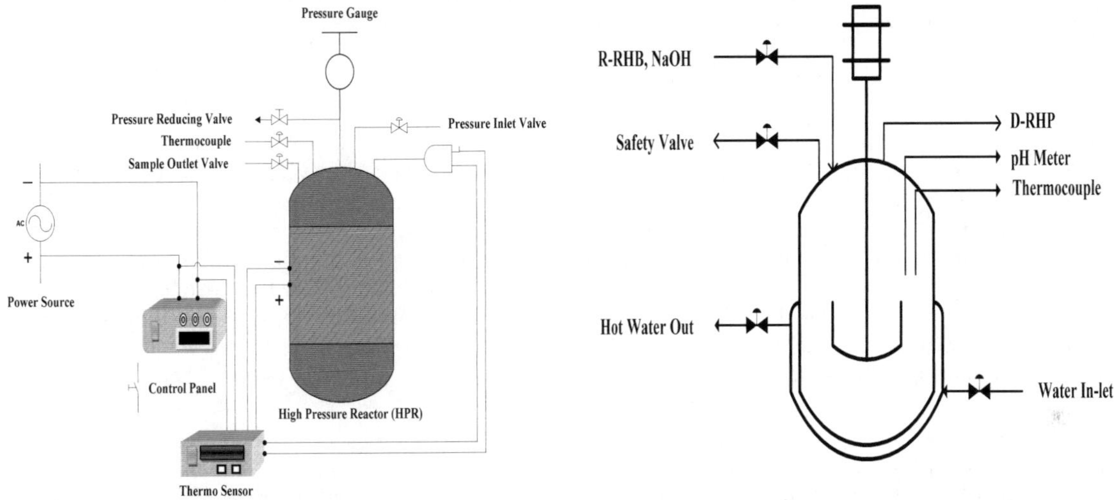

FIGURE (1). Delignification processes (HPP and LPP) of raw rice husk biomass (R-RHB)

For bleaching, approximately 10 g of wet D-RHP sample from both processes (HPP & LPP) was added to 200 mL sodium hypochlorite solution (20%) in a 500 mL reaction vessel separately. Both reactions were performed at 60 °C with constant stirring of 800 rpm for 60 minutes. The operating conditions of the bleaching process are given in the Table (2) for HPP and LPP samples. In this process, the extracted cellulose is converted into white pulp (or paste) due to removal of remaining lignin and hemicelluloses. The bleached rice husk pulp is coded as B-RHP. At the end of this process, B-RHP was washed using DDW and neutralized by acetate buffer for further modifications. This final product from bleaching is B-RHP which is referred to as cellulose pulp (CP). Finally, CP was used for NCC production by acid hydrolysis.

TABLE (2). Experimental conditions of bleaching process

Sample ID	Reactor	C_{NaOCl}	V_{NaOCl}	Temperature	Duration	D-RHP	RPM	Pressure
$B\text{-}RHP_{HPP}$	Vessel	4.0 M	200.0 ml	60 °C	1 hr	10.0 g	800	1.0 atm
$B\text{-}RHP_{LPP}$	Vessel	4.0 M	200.0 ml	60 °C	1 hr	10.0 g	800	1.0 atm

Approximately 10 g of wet B-RHP sample was mixed with 4M sulphuric acid in order to degrade defibrillated cellulose to NCC suspension. The hydrolysis was performed separately for both samples ($B\text{-}RHP_{HPP}$ and $B\text{-}RHP_{LPP}$) in a 500 mL reaction vessel for sixty minutes at 60 °C with continuous stirring of 800 rpm. The detailed experimental conditions of acid hydrolysis are given in the Table (3). Later on, NCC suspension was diluted ten-fold with DDW and neutralized with buffer solution. Finally the raw NCC suspension was separated and dialyzed for 30 minutes against ultra-pure water followed by the method as reported by Lu and Hsieh [17]. The resultant NCC suspension was then transferred into a plastic container with close tighten stopper and preserved in the refrigerator under 5 °C for further analysis. The yield of NCC production (NCC_Y) was estimated using Equation (1), where $W_{R\text{-}RHB}$ is the weight of R-RHB and W_{NCC} is the weight of NCC. The yield of NCC in HPP and LPP were found to be 62 % and 51%, respectively on dry basis.

$$NCC_Y\ (\%) = \frac{(W_{NCC})}{W_{R-RHB}} \times 100 \ \ldots\ldots\ldots\ldots\ldots\ldots\ldots\ldots\ldots\ldots\ldots\ldots\ldots\ldots\ldots\ldots\ldots\ldots\ldots\ldots\ldots\ldots\ldots\ldots \ (1)$$

TABLE (3). Experimental conditions of acid hydrolysis

Sample ID	Reactor	C_{H2SO4}	V_{H2SO4}	Temperature	Duration	B-RHP	RPM	Pressure
NCC_{HPP}	Vessel	4.0 M	200.0 ml	60 °C	1 hr	10.0 g	800	1.0 atm
NCC_{LPP}	Vessel	4.0 M	200.0 ml	60 °C	1 hr	10.0 g	800	1.0 atm

CHARACTERIZATION

The chemical composition of R-RHB, D-RHP, B-RHP and NCC were determined in accordance to ASTM and TAPPI (Technical Association of Pulp and Paper Industry) standard methods [18, 19]. In order to investigate the chemical change of various treated samples FT-IR spectra were measured using a Spectrum 100 FT-IR spectrometer (PerkinElmer, USA) at ambient conditions. The spectra were collected in the transmittance mode from an accumulation of 128 scans at a 4 cm^{-1} resolution over 4000–400 cm^{-1} range. The crystallinity index (CrI) of R-RHB and NCC was measured by Bruker AXS D8 ADVANCE Wide Angle X-Ray Diffraction instrument using Equation (2), which was proposed by Segal and his co-workers [15, 16].

$$CrI\ (\%) = \frac{(I_{002} - I_{am})}{I_{002}} \times 100 \ \ldots\ldots\ldots\ldots\ldots\ldots\ldots\ldots\ldots\ldots\ldots\ldots\ldots\ldots\ldots\ldots\ldots\ldots\ldots\ldots\ldots\ldots\ldots \ (2)$$

Where, I_{002} is the maximum intensity of the (002) lattice diffraction peak and I_{am} is the intensity scattered by the amorphous part of the sample. The diffraction peak for plane (002) is located at a diffraction angle around 2θ = 22 ˚C and the intensity scattered by the amorphous part is measured as the lowest intensity at a diffraction angle around 2θ = 18 ˚C. The microstructures and surface morphology of NCC was examined by a scanning electron microscope (Philips XL 30, FEI/Philips, USA). The samples were mounted on aluminum stubs with conductive carbon tape and sputtered with gold under vacuum at 20 mA for 2 min. To measure the particle dimensions of NCC under high resolution, JEOL 1010 Transmission Electron Microscope (TEM) was used which operated at a 100 kV accelerating voltage. In addition atomic force microscopy (AFM) was also used for morphological analysis. AFM imaging of NCC was performed using the D3100 Atomic Force Microscope.

RESULTS AND DISCUSSION

Table (4) shows the chemical composition of R-RHB, D-RHP and NCC at HPP and LPP experimental conditions. It is clear that the chemical modifications of R-RHB significantly increase the cellulose content by removing amorphous components (hemicelluloses, lignin, silica as well as small amount of pectin and waxes that are grouped together as others). HPP produces higher cellulose content (93.68 wt%) than LPP (75.23 wt%). This confirms that during the delignification process substantial breakdown of lignocellulogic structure took place in HPP - resulting in improved defibrillation. The partial hydrolysis of hemicelluloses and depolymerization of lignin unit occurred during the delignification process as well, which gave rise to sugars and phenolic compounds that are soluble in water as reported in the literature [6]. From this observation it can be concluded that the HPP process can properly cook R-RHB and break the intermolecular bond between fiber and impurities. As a result, the HPP produced greater quantity of cellulose content through delignification than LPP – leading to better conversion rate of NCC production.

TABLE (4): Chemical composition of R-RHB, D-RHP, and NCC sample

Components	R-RHB	$D\text{-}RHP_{HPP}$	$D\text{-}RHP_{LPP}$	NCC_{HPP}	NCC_{LPP}
Cellulose (wt%)	36.87	61.13	52.25	93.68	75.23
Hemicellulose (wt%)	26.43	9.29	11.37	--	5.51
Lignin (wt%)	19.25	6.79	8.28	--	3.62
Silica (wt%)	13.06	3.74	6.73	--	2.35
Others (%)	4.39	19.05	21.37	6.32	13.29

The physical properties such as colour, fibre diameter and volume also changed in every stages of chemical modification as illustrated in Figure (2). Figure 2(a) is the black solution (BS) from LPP process, while Figure 2(b) is from HPP. From these two images it is clearly apparent that delignification in the HPP is well cooked, hence it is darker. Figures 2(c) and (d) are for dry NCC produced from the LPP and HPP, respectively. NCC_{LPP} is basically light yellow due to impurities such as lignin and chromophoric compounds. NCC_{HPP} is lighter colour due to separation of impurities from R-RHB. This indicated that the LPP cannot remove these compounds within short time, while the HPP can do this more effectively. The removal of non-cellulosic constituents from R-RHB by chemical treatment increases the crystallinity of CP and NCC.

(a) BS-LPP **(b) BS-HPP** **(c) NCC_{LPP}** **(d) NCC_{HPP}**

FIGURE (2). Optical images of black solution and NCC from LPP and HPP

The FT-IR spectra of R-RHB and D-RHP for HPP and LPP are shown in Figure (3). The absorption bands at 800-950 cm^{-1} region were observed for all samples due to the presence of C-O and C-H stretching vibration. The spectrum of D-RHP for both processes clearly shows the presence of the characteristic band of the C=C group in the region of 1600–1700 cm^{-1} and a C–O stretching band at the region of 1300–1000 cm^{-1}. In this spectrum the absorption band at the region near 1700 cm^{-1} may be due to the carboxyl group of acetyl ester in cellulose and the carboxyl aldehyde in lignin [17]. The intensity of this peak in R-RHB was not observed, which may be due to impurities or very poor intensity. The broad absorption band with peaks located from 3300 to 3400 cm^{-1} is due to stretching of –OH groups and other one near 2900 cm^{-1} is related to the C–H stretching vibrations.

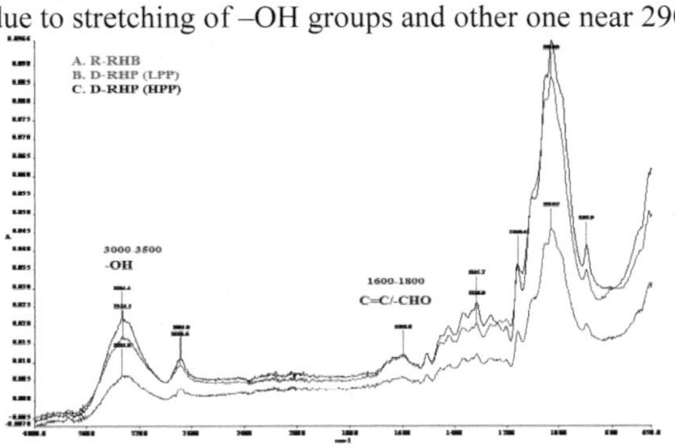

FIGURE (3). FT-IR spectra of D-RHP and R-RHB

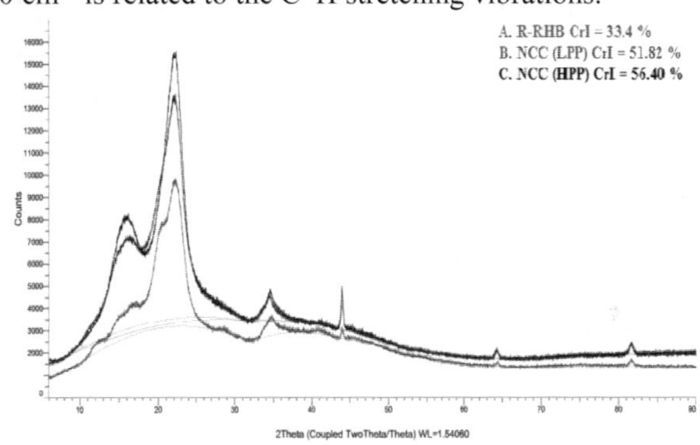

FIGURE (4). XRD analysis of NCC and R-RHB

XRD studies of R-RHB and NCC were conducted in order to investigate the crystalline behavior of the resultant products from HPP and LPP. It is clear that NCC suspension shows high crystalline structure as compared to R-RHB in Figure (4). The crystallinity index (CrI) of NCC_{HPP} and NCC_{LPP} are 56.40% and 51.82 %, respectively. Both values are greater than R-RHB. From this evaluation it can be concluded that the HPP significantly removed impurities from R-RHB and increased the crystallinity. The morphological investigation of NCC as produced by the LPP and HPP was conducted by AFM, SEM (with EDX mode) and TEM. The surface images of both NCC samples are shown in Figure (5). The AFM images in Figures 5(a1) and 5(a2) are for LPP and HPP, respectively. The average particle height was observed from this analysis to be approximately 20 nm. The smallest particles are observed in the HPP but both are in nano-scale. The SEM images of NCC are shown in the Figures 5(b1) and 5(b2) where the smallest diameter is observed in HPP. In the SEM image, the longer fibers are also observed in 5(b2) other than 5(b1). This is may be due to RPM system of LPP. LPP was conducted in JGR which was designed with RPM system, therefore during delignification particle collided among themselves as well as with the rotating blades. SEM has been simultaneously performed with EDX for elementary analysis. Trace amount of S, O, Cl and N were detected due to chemical modification. R-RHB contains 13% SiO_2 which can be removed by chemical treatment and EDX was used to quantify Si in NCC. NCC still contains 1-3% Si which needs to be separate by treatment. On the other hand the HPP has no RPM system therefore fibers are longer but narrow due to well delignification. The TEM images of HPP and LPP are shown in the Figures 5(c1) and 5(c2) and a similar trend was observed as well. The HPP produced smaller particle dimension than LPP. In the LPP the fiber agglomeration was clearly apparent whereas in the HPP fibers were clearly isolated from each other. The average length of nano-fibers in TEM image as produced in HPP is approximately 1-2 μm and diameter is 10-50 nm. On the contrary, the LPP produced NCC of similar length but a bit large diameter due to impurities. The morphological results as presented in this section are quite similar with many published works [4, 5, 7, 16, 19].

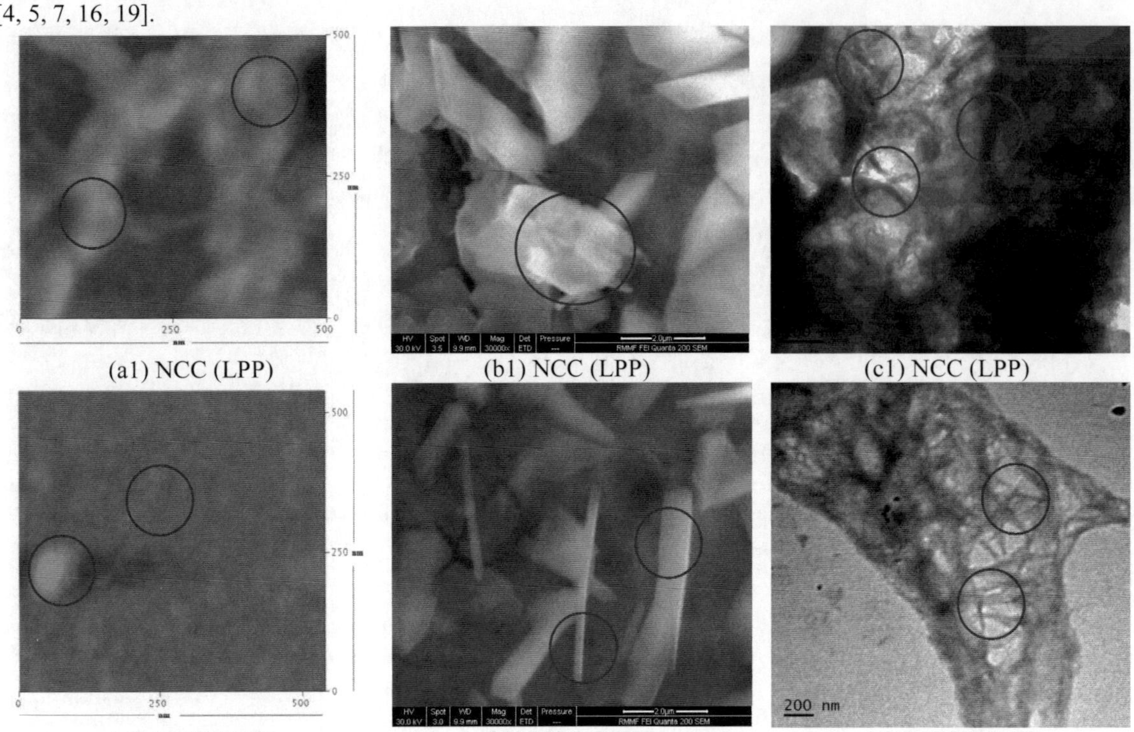

(a1) NCC (LPP) (b1) NCC (LPP) (c1) NCC (LPP)

(a2) NCC (HPP) (b2) NCC (HPP) (c2) NCC (HPP)

FIGURE (5). Surface images of NCC from AFM (a1 & a2), SEM (b1 & b2) and TEM (c1 & c2)

CONCLUSIONS AND RECOMMENDATION

The production of NCC using the HPP is the new approach used in this study. Both the HPP and LPP have been used to successfully produce NCC from R-RHB with 93 % and 75% purity, respectively. The maximum yield was found to be 62 % from the HPP and 51% from the LPP. The average particle dimension of NCC in this production, as observed from morphological analysis, is about 10-20 nm diameter and 1-2 μm in length.

ACKNOWLEDGEMENT

The research work reported in this paper was funded by the RMIT University, Melbourne, VIC 3001, Australia. The authors would like to acknowledge all technical staff of chemical engineering department and RMFF at RMIT for their support during lab work.

REFERENCES

1. A. Alemdar and M. Sain. *Bioresour. Technol.* 99, pp. 1664-1671, 2008a.
2. M. A. S. A. Samir et al., *Biomacromolecules* 6, pp. 612-626, 2005.
3. Y. X. Wang et al., *Macromol. Biosci.* 6, pp. 524-531, 2006
4. M. F. Rosa et al., *Carbohydr. Polym.* 81, 83-92, 2010.
5. D. Pasquini et al., *Ind. Crop Prod.* 32, pp. 486-490, 2010.
6. R. Zuluaga et al., *Cellulose,* 14, pp. 585-592, 2007
7. R. Li et al., *Carbohydr. Polym*. 76, pp. 94-99, 2009b.
8. Q. Li et al., *J. Polym. Sci. Polym. Phys.* 47, pp. 1069-1077, 2009a.
9. B. Wang and M. Sain. *Compos. Sci. Technol.* 67, 2521-2527, 2007.
10. N. Reddy and Y. Yang. *Polymer*, 46, pp. 5494-5500, 2005
11. G. Siqueira et al., *Cellulose* 18, pp. 57-65, 2010.
12. F. Lopez et al., *Chem. Eng. J.* 166, pp. 22–29, 2011
13. Jiang and Hsieh. *Carbohydr. Polym*. 95, pp. 32– 40, 2013.
14. N. Bhammathidas and P. K. Mehta. CANMET-Chennai, India, 199, pp. 379-391, 2004
15. M. G. Jackson. *Anim. Feed. Sci. Technol.* 2, pp. 105-130, 1977
16. A. P. Mario et al., *Carbohydr. Polym.* 62, pp. 164-169, 2005
17. P. Lu et al., Carbohydra. Polym. 87, pp. 564– 573, 2012.
18. E. Abraham et al., *Carbohydr. Polym.* 86, 1468– 1475, 2011.
19. N. Johar et al., *Ind. Crops Prod.* 37,pp. 93-99, 2012.

Influence of Process Route on the Mechanical Properties of Polymer Based Intravaginal Drug Delivery Systems

K. Eggenreich[a], S. Schrank[a,b], G. Koscher[a*], K. Nickisch[c], E. Roblegg[b] and J. Khinast[a,d]

[a]Research Center Pharmaceutical Engineering GmbH, Graz, Austria
[b]University of Graz, Institute of Pharmaceutical Sciences, Graz, Austria
[c]Evestra Inc. San Antonio, Tx, USA
[d]Graz University of Technology, Institute for Process and Particle Engineering, Graz, Austria

Presenting author: karin.eggenreich@rcpe.at
Reference author: gerold.koscher@rcpe.at

Abstract. Intravaginal rings (IVRs) are designed for controlled release of one or more active pharmaceutical ingredients (APIs) to the vaginal tract, and finally into systemic circulation over extended time periods [1]. The concept is built on the principle of using polymers as the carrier, in which the APIs show a certain permeability, yielding specific release rates [2, 3]. Technologically, IVRs are classified into matrix and reservoir types. In matrix systems, the API is homogeneously dispersed throughout the polymer. A route to produce matrix IVRs is hot melt extrusion to obtain drug-loaded polymer intermediates, which are then injection-molded to yield the ring shape. In a reservoir system, the polymer core contains the homogeneously dispersed API and is surrounded by a drug-free skin. The production route of reservoir systems includes co-extrusion of the layered structure (drug-loaded core, drug-free skin), followed by a joining process to obtain the ring. The IVR's mechanical characteristics are critical for its application. Other than performing side-by-side comparison tests with existing products which have demonstrated high user acceptability, there is no model to directly determine the ideal range for mechanical IVR acceptability. This is due to the wide range of vaginal shapes and sizes among the female user population. Under normal physiological conditions, the vaginal tract is a low-friction environment due to the presence of vaginal fluid and cervicovaginal mucus. The IVR ring is required to be flexible. Prior to and during insertion, the IVR has to be easily compressed, followed by fast recovery, to remain in the vaginal cavity without causing damage. Yet, the junction deriving from the ring closure needs to be strong enough to prevent ring opening. An IVR's mechanical properties can be changed via its formulation. In this study, the impact of process parameters, polymers used (polymer crystallinity) and their content on the mechanical properties of EVA-based rings was investigated.

Keywords: intravaginal drug delivery systems, polymer, matrix system, reservoir system, ethylene vinylacetate, mechanical properties
PACS: 81 http://www.aip.org/publishing/pacs/pacs-2010-regular-edition

INTRODUCTION

Until recently, intravaginal drug delivery systems have received comparatively little attention. However, the use of intravaginal elastomeric ring-shaped devices, to deliver therapeutic molecules, is not a revolutionary application strategy. Initial reports about the vaginal administration route date back to the late 1910s, for the vaginal administration of morphine [4]. Already four decades ago [5-8], at the dawn of modern drug delivery technology [9], a contraceptive intravaginal ring (IVR) was one of the first controlled release systems. In the mid-1990s, an elastomeric vaginal device, based on ethylene vinylacetate (EVA), was studied for the prevention of a sexually transmitted disease (herpes simplex). The most innovative field of application is in the area of HIV prophylaxis, with ongoing preclinical studies underway. IVRs are designed to continuously release one or more active pharmaceutical ingredients (APIs), in a controlled way, to the vaginal tract over extended time periods [10]. The polymer is used as a carrier, in which the API(s) show(s) certain permeability, yielding specific release rates [3, 11] and API permeation in the vaginal epithelium [12]. The main advantages of the vaginal rings are cost-effectiveness, patient compliance, and therapeutic efficacy due to the systems' ability and versatility in providing long-term and continuous release of drug at constant, pre-determined rates. In contrast to subdermally implantable systems, which need surgical operation for placement, these rings can be placed in the vagina by woman herself. Today, there are five commercially available IVRs for contraception (Progering®, Fertiring®, Nuvaring®) and hormone replacement therapy (Femring®, Estring®).

IVRs are classified into matrix and reservoir systems. The majority of marketed IVRs are matrix systems based on silicone. Only one reservoir IVR exists, comprising two types of EVA co-polymer. In matrix systems, the API is homogeneously dispersed throughout the polymer. The drug release rate occurs via drug diffusion through the polymer. The release rate, following Higuchi's law (proportional to the square root of time), hence, decreases with time. In contrast, reservoir systems provide more constant release rates. The polymer core contains the homogeneously dispersed API and is surrounded by an, initially, drug-free polymer skin, controlling the release rate, which is governed by drug diffusion through the skin and remains rather constant (i.e., zero order release). One route to produce matrix IVRs is hot-melt extrusion to obtain drug-loaded polymer intermediates, which are subsequently injection-molded to yield ring-shaped devices. Manufacturing steps of reservoir systems include co-extrusion followed by a shaping process or co-injection molding.

The IVR's mechanical behavior is a critical product characteristic for its acceptability. In order to support its application, the IVR needs to be flexible. Prior to and during insertion, the IVR has to be compressed, followed by a fast recovery to remain in the vaginal cavity without causing any damage in the vaginal epithelium. Too little rigidity may lead to expulsion during daily activities. Due to the lack of quantitative models, side-by-side performance tests using existing products with high user acceptability are applied, which directly determine the ideal mechanical range. In addition, the joint, derived from manufacturing, needs certain strength to avoid ring failure. The present study addresses the impact of process parameters, polymer type (polymer crystallinity) and polymer content on mechanical properties of EVA based matrix IVRs.

EXPERIMENTAL

Materials

The EVAs investigated in this study are listed under their trade names in Table 1 and were purchased at Celanese. A cellulose based hydrophilic polymer (HP) was manufactured by Ashland. API 1, a progestin, and API 2, an estrogen, were both produced by Shenzhen Nexconn Pharmatechs Ltd. A marketed EVA-based reference product (NuvaRing®) was purchased from local pharmacies in Graz (Austria) and from a Hungarian wholesaler.

TABLE 1. List of investigated EVA types – VA content and melt flow index (MFI)

EVA Type	VA Content [%]	MFI [g/10 min.]
Ateva 1075	9	8.0
Ateva 1807	18	0.7
Ateva 2604	26	4.0
Ateva 2810	28	6.0
Ateva 2820	28	25.0

Production of Matrix based IVRs

Different formulations were investigated containing two contraceptive, in water practically insoluble, active ingredients, i.e., API 1, a progestin, and API 2, an estrogen. The final drug load was < 0.35% for API 1 and < 0.1% for API 2 per IVR. All formulations were processed on a co-rotating twin screw extruder (Coperion) with 2 intermeshing screws, 18 mm in diameter (D_i/D_a: 1.55), and a processing length of 40D. The screw configuration was specially designed for processing the selected formulations. Drug-loaded pellets were produced in a 2-step process: In a first step, the powder blend was prepared by pre-mixing the hydrophilic polymer (HP) and the two APIs in a high shear blender (Stephan Machinery), which was subsequently transferred using a KT20 loss in weight feeder (K-Tron) for further extrusion. The extruder was operated at a screw speed of 200 rpm and a throughput of 1.0 kg/h. The resulting melt temperature was around 145 °C. The produced strand was transported on a conveyer belt (Geppert) and then cut into drug-loaded, hydrophilic polymer based pellets of ~ 1 mm using a Primo 60E strand pelletizer (Automatik Plastics Machinery). In a second extrusion step, the drug-loaded pellets were further processed

at a screw speed of 200 rpm and a total throughput of 1.0 kg/h with pure EVA pellets at different ratios in a range of 5-15% to yield intermediates with the final formulation. Dosing was performed via split feeding: A KT20 loss in weight feeder (K-Tron) was used for feeding the pure EVA to the extruder, the drug loaded HP-pellets were dosed using the DDW-MD0-MT-1 HYD feeder (Brabender). The melt temperature was 105 °C. Again, the resulting strand was transported on a conveyer belt and then cut into pellets of ~ 1.5 mm using a Primo 60E strand pelletizer (Automatik Plastics Machinery).
The drug-loaded pellets were finally injection-molded to yield ring-shaped delivery systems. Injection molding (IM) was performed on an e-motion 170/55, a small production-scale injection molding machine (Engel), equipped with a 25 mm single screw. The barrel was composed of 3 zones. A 4-cavity hot runner mold was used to shape the desired ring geometry. Hence, in a single shot, four IVRs of 54 mm in outer diameter and 4 mm in cross-sectional diameter (Fig. 1) were produced simultaneously. The selected formulations were processed within a hot channel temperature range of 160- 210 °C.

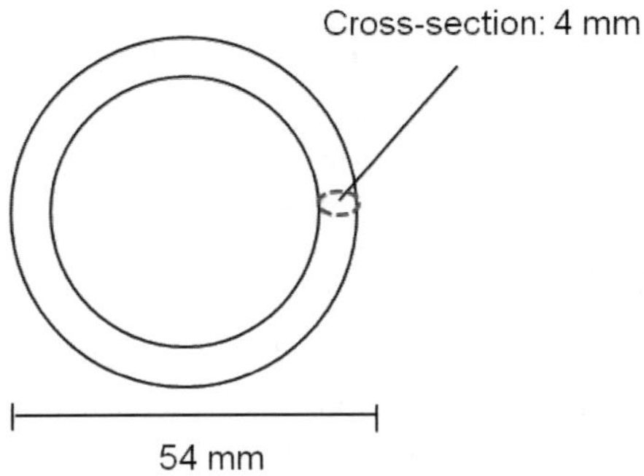

FIGURE 1. EVA based intravaginal ring – product size

Mechanical Testing of IVRs

The manufactured IVRs were evaluated for their mechanical properties using a Z005 tester (Zwick/Roell), studying the compression and relaxation forces and the IVRs' tensile strength. For compression and relaxation, the ring was placed vertically in the system loaded with a 2.5 kN cell. During compression, the probe moved downward to exert a compressive force. During retraction, the probe moved back upward to allow the retractile forces of the IVR to bring it back to its starting position. After descending at a rate of 2 mm/s to compress the IVR by 50% of its initial diameter (i.e., 27 mm), the probe ascended back at a rate of 2 mm/s to its original position. Over the entire cycle, both force and displacement data were collected. A minimum of 6 IVRs were tested for each formulation.
For tensile strength testing, the rings were placed in the system loaded with a 50 kN cell with the IVR oriented vertically and stretched at a rate of 1 mm/s until failure was observed, at which point the net extension and the net load were recorded. To quantify the tensile strength, the maximum tensile stress was calculated using equation 1.

$$\sigma_{\max} = \frac{F}{r^2 \times \pi \times 2} \tag{1}$$

A minimum of 6 IVRs were analyzed for each formulation. The marketed product was used as reference product for both, compression and relaxation and tensile strength measurements. A minimum of 6 reference products was tested.

RESULTS AND DISCUSSION

The mechanical properties of an IVR are crucial as it has to be i.) rigid enough to be retained in the vaginal cavity and to avoid expulsion during daily activities, but ii.) soft enough to be inserted and not to damage the vaginal tissue. Three of the marketed IVR devices approved by the FDA (Food and Drug Administration) are fabricated from silicone, only one IVR (NuvaRing®) is manufactured using EVA. These silicone IVRs require a larger cross-sectional diameter (i.e., 7.6-9.0 mm) compared with an EVA IVR, in order to achieve the necessary mechanical strength due to its soft backbone. Clinical trials have revealed high expulsion rates leading to discontinuation of the IVR usage as an obstacle in IVR acceptability, emphasizing the importance of having the appropriate mechanical properties for the IVR acceptability. Given that NuvaRing's mechanical properties successfully allow the IVR to stay in the vagina without tissue damage [13, 14] and showed high patient tolerance and compliance during acceptability studies [14, 15, 16], this IVR was used as a reference ring for dimensions (Figure 1) and mechanical performance in this study. Depending upon the formulation, delivery systems with specific mechanical parameters may be tailored. In this study, the effect of the polymer vinyl acetate (VA) content and hydrophilic polymer (HP) content on the mechanical properties of the IVR was investigated. Further experiments studied the impact of the processing parameters on both, tensile strength and compression and relaxation. Here, the hot channel, representing the temperature of the melt, showed a rather strong effect. Table 2 provides an overview of the resulting tensile strength and compression and relaxation of selected formulations.

TABLE 2. Mechanical properties of selected formulations and reference product

IVR	Compression and relaxation (N)	STD (N)	Tensile strength (N/mm^2)	STD (N)
NuvaRing®	3.0	0.28	6.5	0.59
Ateva 2604	2.9	0.06	6.9	0.12
90% Ateva 2604+10% HP/APIs	4.8	0.19	5.9	0.22
93% Ateva 2604+ 7% HP/APIs	3.4	0.33	5.6	0.49

To investigate the impact of VA content on elasticity, a range of EVA types within 9-28% VA was tested in this study. The VA content of a polymer determines its crystallinity: A higher VA content provides reduced crystallinity, yielding increased drug permeability in the EVA polymer and increased elasticity. Produced IVRs containing EVA types in the range of 9-18% VA were excluded from mechanical testing. Independent upon the selected hydrophilic polymer content and IM processing temperatures, decreasing the VA content resulted in reduced elasticity. For a VA content of 18% or less, elasticity was significantly decreased and was no longer acceptable due to its high rigidity. Compression and relaxation did not markedly differ between IVRs based on EVA with 26 and 28% VA content, respectively. The measured force to compress an EVA-based ring with 26% VA to 50% of its original diameter was 2.9 N (Table 2) and hence, identical to the reference product (i.e., 3.0 N). The tensile strength was 6.9 N/mm^2 for the EVA ring. Consequently, elasticity of the pure EVA ring was again close to the reference product (6.5 N/mm^2) and within the reference product's specified tensile strength limit of ≥ 3 N/mm^2. A significant impact of the hydrophilic polymer content in the matrix and the hot channel temperature on the tensile strength was observed. Temperatures above 180 °C yielded a change in the inner structure of the IVRs' matrix system. IVRs containing more than 10% HP/APIs showed a very poor tensile strength. Here, independent upon the hot channel temperature, the rings were broken easily, even during manual testing, and hence, were not suitable and excluded from further studies. The hot channel temperature showed a significant effect on the tensile strength of formulations with 5-10% HP/APIs. The optimized hot channel temperature was observed above 180 °C, where IVRs with the specified tensile strength were produced for both, 7 and 10% HP/APIs, without affecting the drug degradation. Regarding elasticity, IVRs containing 10% HP/APIs required ~1.5 times more mechanical force (i.e., 4.8 N) to be compressed than the commercially available ring, whereas for rings with 7% HP/APIs in the system, the measured force was 3.4 N and hence, close to the reference product. No significant difference to rings with reduced HP/APIs content (i.e., 5%) was detected for both, tensile strength and compression and relaxation (data not shown). Additional IM parameter such as mold temperature, screw speed and injection speed were further optimized, but showed a less pronounced effect.

CONCLUSION

This presented study shows that the mechanical properties of an IVR can significantly be affected via the formulation and the hot channel temperature during IM. A marketed IVR was used as reference product to determine both, tensile strength and elasticity (measured as compression and relaxation) to allow for the production of an IVR that is shown to be retained in the vaginal cavity without causing any tissue damage. The VA content in the EVA strongly impacted the compression and relaxation behavior of the IVR. An increase in VA content, leading to reduced crystallinity, was beneficial. Here, VA contents of 18% or below did not yield IVRs with acceptable elasticity. Nevertheless, the VA content did not markedly impact the IVR's tensile strength. Independent upon the process temperature, hydrophilic polymer contents below 10% in the final ring were identified to be ideal for the production of IVRs which show acceptable tensile strength. In addition to the hydrophilic polymer concentration, the applied hot channel temperature showed the most significant impact on the tensile strength of the manufactured IVRs. Below hot channel of 190 °C, the process yielded IVRs with a low tensile strength. Here, produced IVRs were broken easily. On the other hand, the hydrophilic polymer had a negative impact on compression and relaxation: Increasing contents resulted in increased forces to compress the ring, reflecting increased stiffness. This was improved by reducing the hydrophilic polymer content to 7% in the final ring. Summarizing, the presented findings demonstrate that matrix IVRs based on EVA and the hydrophilic polymer can be produced, which show mechanical properties similar to a highly accepted marketed reservoir reference product via HME and injection molding as route of manufacturing.

ACKNOWLEDGMENTS

The authors want to acknowledge the project partner ENGEL AUSTRIA GmbH.

REFERENCES

1. A.D. Woolfson, R.K. Malcolm, and R. Gallagher, Drug Delivery by the Intravaginal Route. 2000. 17(5): p. 47.
2. J.A.H.van Laarhoven, M.A.B. Kruft, and H. Vromans, In vitro release properties of etonogestrel and ethinyl estradiol from a contraceptive vaginal ring. Int J Pharm, 2002. 232(1–2): pp. 163-173.
3. K. Malcolm, D. Woolfson, J. Russell, P. Tallon, L. McAuley, and D. Craig, Influence of silicone elastomer solubility and diffusivity on the in vitro release of drugs from intravaginal rings. J Contr Release, 2003. 90(2): pp. 217-225.
4. D. Macht, On the absorption of drugs and poisons through the vagina. J Pharmacol Exp Ther, 1918. 10: pp. 509-522.
5. P. F. Kiser, T. J. Johnson, and J.T. Clark, State of the Art Intravaginal Ring Technology for Topical Prophylaxis of HIV Infection. AIDS Rev, 2012. 14: pp. 62-77.
6. D. Mishell, M. Talas, A. Parlow, and D. Moyer, Contraception by means of a silastic vaginal ring impregnated with medroxyprogesterone acetate. Am J Obstet Gynecol, 1970. 107: pp. 100-107.
7. A.D. Woolfson, R.K. Malcolm, and R. Gallagher, Drug Delivery by the Intravaginal Route. Crit Rev Ther Drug Carrier Syst, 2000. 17(5): pp. 509-555.
8. R. Malcolm. The intravaginal ring. Drug Pharm Sci, 2003. 126: pp. 775-790.
9. V. Brache, and A. Faundes, Contraceptive vaginal rings: a review. Contraception, 2010. 82: pp. 418-427.
10. A.D. Woolfson, R.K. Malcolm, and R. Gallagher, Drug Delivery by the Intravaginal Route. 2000. 17(5): p. 47.
11. A.H.van Laarhoven, M.A.B. Kruft, and H. Vromans, In vitro release properties of etonogestrel and ethinyl estradiol from a contraceptive vaginal ring. Int J Pharm, 2002. 232(1–2): pp. 163-173.
12. D.P. Benziger, and J. Edelson, Absorption from the Vagina. Drug Metabolism Reviews, 1983. 14(2): pp. 137-168.
13. H. Fang, I. Dicker, R. Dalterio, Y.F. Gong, T. Wang, Z. Yin, Y. Ueda, J. Matiskella, J. Kadow, P. Clapham, J. Robinson, R. Colonno, and P.F. Lin, Envelope conformational changes induced by human immune deficiency virus type 1 attachment inhibitors prevent CD4 binding and downstream entry events. J Virol, 2006. 80(8): pp. 4017-4025.
14. A. Novak, C. de la Loge, and E.A. van der Meulen, The combined contraceptive vaginal ring, NuvaRing: An international study on user acceptability. Contraception, 2000. 67(3): pp. 187-194.
15. F.J. Roumen, D. Apter, T.M. Mulders, and T.O. Dieben, Efficacy, tolerability and acceptability of a novel contraceptive vaginal ring releasing etonogestrel and ethinyl oestradiol. Hum Repord, 2001. 16(3): pp. 469-475.
16. M. Kaur, Delivery of Entry Inhibitors using Intravaginal Rings for Preventing the Sexual Transmission of HIV, Thesis The University of Utah, 2009.

Electrospun Core-Shell Nanofibers based on Thermoplastic Urethane Elastomer and Gelatin: Evolution of Cell Behavior

Mahsa Jafaie Menarbabzari [a], Ali Asghar Katbab[a]

[a] *Department of Polymer Engineering, Amir Kabir University of Technology, Hafez Ave, Tehran, Iran*

mahsa.jafaie@gmail.com
katbab@aut.ac.ir

Abstract. Electrospun scaffolds based on hydrophobic and biodegradable polymers exhibit good mechanical properties but are not suitable for cell attachment. In contrast, scaffolds with natural polymers show inherent tendency for cell attachment, but they are mechanically brittle. Thus core-shell nanostructured scaffolds are expected to be the best candidate. In the present work, attempts have been made to design and fabricate nonwoven mat based on nanofibers with core-shell structure comprising thermoplastic urethane elastomer (TPU) and gelatin as core and shell, respectively. For this purpose, electrospun process was employed with a core-shell type needle, and the two polymers were processed in the form of solution in several solvent. We've tried to use many solvents such as: DMF, DMAC, DCM, THF, HFIP and their mixture for TPU and TFA, TFE, AA, FA, DMF, HFIP and their mixture for Gt, but just HFIP as a solvent for two solvents, lead to produce core- shell structure. Effects of solution and processing parameters upon the developed nanofibers and corresponding mat micro morphology were investigated. Scanning electron microscopy (SEM), Transmission electron microscopy (TEM) and Attenuated total reflectance Fourier transform infrared spectroscopy (ATR-FTIR) evidenced the formation of the gelatin layer on the TPU fiber as the core. The bulk mechanical properties of the obtained mat was evaluated and correlated with the diameter and orientation of the core-shell nanofibers. The surface of the prepared mat exhibited a good hydrophilic behavior with excellent surface softness in the presence of an aqueous media, indicating the potential for the preparation of scaffolds with desired bulk mechanical strength as well as biocompatibility for mussel and tendon regeneration.

Keywords: Core- Shell Nanofibers, Thermoplastic Urethane elastomer, Gelatin, HFIP.
PACS: 87.85.J-

INTRODUCTION

Within the last decade, electrospinning has been used to fabricate fibrous membranes for various biomedical applications [1]. The nanofibrous structure of the electrospun membranes essentially resembles the topography of extracellular matrix (ECM) and could promote the adhesion and proliferation of various cells [2]. The electrospun membranes based on biodegradable hydrophobic polymers, such as polyurethane display good mechanical properties [3], weak for cell attachment. In contrast, natural-occurring bio macromolecules, such as gelatin (Gt), could be electrospun into the fibrous membranes with inherent cell-binding domains. However, these membranes are mechanically brittle [4]. The combination of the two types of polymers to form Core-shell fiberous scaffolds is an effective strategy to circumvent the above limitations [5]. Core- shell membranes combine the advantages of biodegradable hydrophobic polymers and natural-occurring biomacromolecules, thereby having both high mechanical strength and excellent bio-specific properties. Several biodegradable polymer pairs have been electrospun into hybrid fibrous membranes [5, 6, and 7]. In the present work, biodegradables thermoplastic polyurethane (TPU) and gelatin (Gt) have been employed to prepare electrospun nanofiberous membrane with core-shell micro structure. The TPU phase as the core imparted good mechanical strength and the gelatin as the skin could enhance the cell compatibility of the fibers.

EXPERIMENTAL

Gelatin (bovine skin, type B) was purchased from Sigma-Aldrich. Polyurethane was supplied by Bayer Material Science, DMF, DMAC, DCM, THF, TFA, TFE, AA, FA, DMF and 1,1,1,3,3,3-Hexafluoro-2-propanol was purchased from Merck. In this paper, we just explain about solutions prepared by HFIP. Gelatin and TPU nanofibers of various solvent concentration were fabricated using the electrospinning method. Briefly, the Gt powder and the TPU granules were mixed at different percentages as 8-10 wt% and 2-5 wt% in HFIP, respectively. First, TPU and Gt solutions were electrospun separately; then optimum solutions selected for co-axial electrospinning by observing SEM images and electrospin ability.

The morphologies and diameters of the nanofibers were determined with the use of SEM (AIS2100, Seron Technology INC, Korea). Samples were coated with gold using a sputter-coater. Image analysis software (Image pro. Plus. 6) and OriginPro. 9.1 were used to analyze the diameter and the distribution of fiber diameters, respectively.

The core–shell structure of the prepared nanofibers was characterized by transmission electron microscopy (TEM, EM 208, Philips, Czech). For TEM observation, fibers were directly deposited onto a TEM copper mesh.

The chemical structures of the TPU and core-shell nanofibers scaffolds were characterized by ATR-FTIR spectroscopy (NEXUS 670, Thermo- Nicolet, USA). The transmittance of each sample was recorded at room tempreture with resolution of 4 cm−1 between 4000 and 400 cm−1.

To evaluate the hydrophilic/hydrophobic properties of scaffolds, we examined the contact angle test. For measuring the contact angle of a gelatin, TPU core- shell nanofiber scaffold, samples were cut into a rectangular shape (15 mm × 15 mm). Static contact angle (KRUSS G10, Sessile drop, Germany) was used. The contact angle was measured at room temperature. Briefly, the rectangular-shaped samples were placed on the stage, and distilled water was dropped on the samples from a motorized syringe.

RESULTS AND DISCUSSION

In our electrospinning experiments, continuous nanofibers were electrospun from all the solutions mentioned above. But in this paper we just explain about solutions prepared with HFIP. 2-5% wt. TPU solutions and 8-10% wt. Gt solutions were prepared. All of them were electospun and after observing SEM images, Solutions with 2% wt. of TPU and 10% wt. of gelatin were selected for co-axial electrospinning.

Morphology of the Coaxial Electrospun Nanofibers

The nanofibers morphology was examind by using SEM images. Figure 1 shows the SEM image and associated diameter histogram of the core- shell nanofibers. Defect-free, beadless, and geometrically uniform nanofibers could be prepared whit formation of the coaxial structure. As we can observe in Fig. 1, oriented nanofibers could be create under high rotation rate rotator and it occurred due to elastic properties of TPU. Oriented nanofibers could promote cell orientation that it is favorable especially for align tissue like tendon and mussels. The diameters of the nanofibers were calculated by image analysis software. These were found to be in the range from 1930±160 nm. It follows the normal distribution curve diameter.

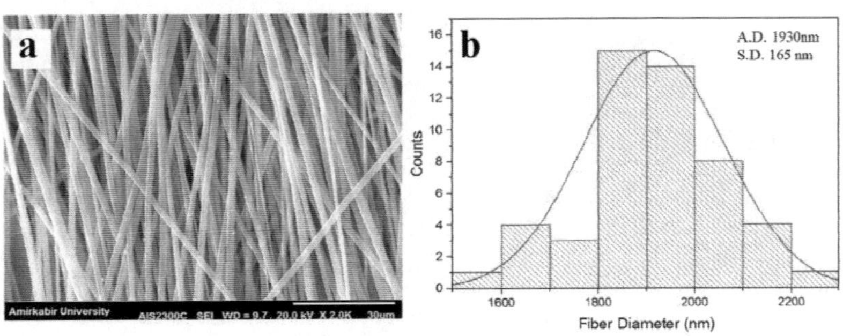

FIGURE 1. (a) SEM micrograph, (b) diameter histogram of TPU- Gelatin core- shell nanofibers.

Transmition Electron Microscopy

The detailed morphology of core- shell nanofibers is shown in a TEM micrograph in figure 2. The contrast which is created by electron beam diffraction represents the distinctive phases in the nanofibers.

FIGURE 2. TEM micrograph of core- shell structured TPU- Gelatin nanofibers.

Attenuated Total Reflectance Fourier-Transform Infrared Spectroscopy (ATR-FTIR)

In order to investigate the surface characterization of core-shell nanofibers, ATR-FTIR analysis has been performed on the sample. Infrared spectra for the TPU-related peaks were detected for the scaffolds containing TPU. These included both the 1110 cm−1 soft segment ether peak and the 1080 cm−1 hard segment ether peak.

As shown in figure 3, the absorbance intensity at 1646 (amide I) and 1280 cm^{-1} (N-H Stretching vibration in collagen) shows that gelatin is present on the surface of nanofibers.

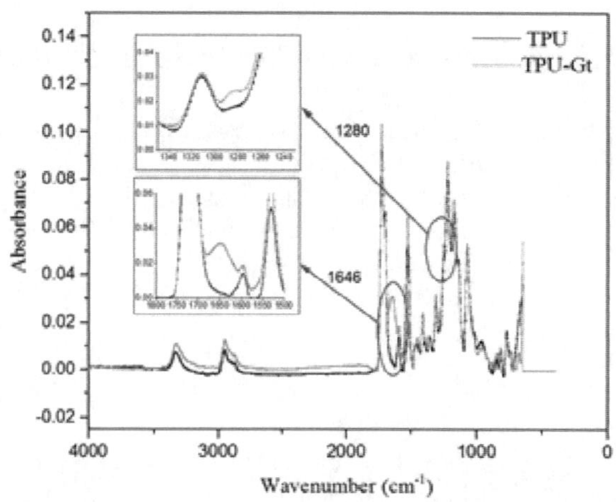

FIGURE 3. ATR- FTIR spectra of TPU and TPU- Gelatin core- shell nanofibers.

Contact Angle

Cells adhesion and spreading on the surface of scaffolds are governed by the extent of surface hydrophobicity [8]. Therefore, the wettability of the electrospun nanofiber scaffolds was measured by the water contact angle test. Contact angle test for 3 types of samples including core- shell fibers, Gelatin and TPU fibers were studied that shown in figure 4.

The contact angle of the pure TPU nanofibers scaffold was 110◦. The high contact angle indicates that the water droplet does not spread on the substrate due to the hydrophobicity of TPU. On the gelatin nanofibers scaffold, water drop exhibited good spreading onto the scaffold surface, indicating that gelatin has a much higher wettability than that of TPU, owing to the inhibition of multiple hydrophilic functional groups [24].

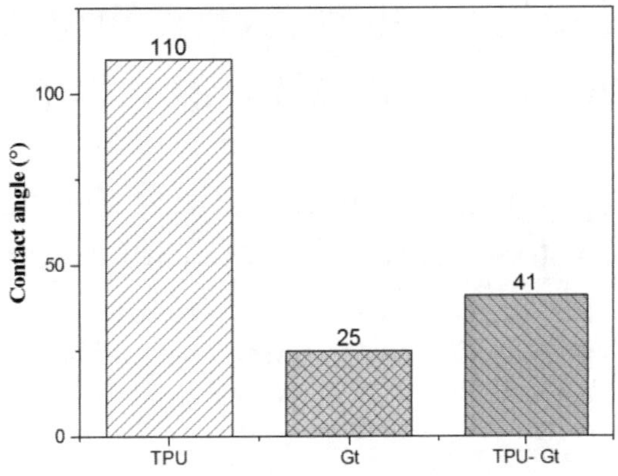

FIGURE 4. Water drop contact angle of gelatin, TPU and core- shell membrane.

CONCLUTION

Generally, natural and synthetic polymers have complementary properties. Unlike synthetic biomaterials, natural polymers possess high cytotropism due to good hydrophilicity and an abundance of surface cell-recognition sites but have poor mechanical properties. Core- shell electrospinning technique showed to be potential for the fabrication of porous scaffolds with good mechanical properties together, with high cell compatibility for the application of tissue regeneration.

ACKNOWLEDGMENTS

This research was carried out under cooperation of the research and technical centers of the Polymer Engineering and Color Department of Amir Kabir University of Technology,Tehran, Iran.

REFERENCES

1. Liang D., Hsiao B. S., Chu B., Adv. Drug Deliv. Rev., 59, 1392, 2007.
2. Pham Q. P., Sharma U., Mikos A. G., Tissue Eng., 12,1197, 2006.
3. Guan J. Fujimoto K.L., Sacks M.S., Biomaterials 26, 3961–3971, 2005.
4. Schiffman J. D., Schauer C. L., Polym. Rev., 48, 317, 2008.
5. Choi J. S., Lee S. J., Christ G. J., Atala A. Yoo J. J., Biomaterials, 29, 2899, 2008.
6. Hong Y., Ye S. H., Nieponice A., Soletti, L., Vorp D. A., Wagner W. R., Biomaterials, 30, 2457, 2009.
7. Li M., Mondrinos M. J., Chen X., Gandhi M. R., Ko F. K., Lelkes P. I. J. Biomed. Mater. Res. Part A, 79, 963, 2006.
8. M.P. Prabhakaran, J. R. Venugopal, T. T. Chyan, , Tissue Engineering Part A, 1787-1797, Volume: 14, 2008.
9. S.E. Kim, et al., Biomedical materials (Bristol, England), Volume 19, Issue 3, p.044106, 2009.

Opportunities and issues in extruding biodegradable polyester blends for packaging

Maria-Beatrice Coltelli [a,b], Patrizia Cinelli[b,c], Seyed Mohammad Kazem Fehri[a], Norma Mallegni[a], Vito Gigante[a], Laura Aliotta[a], Andrea Lazzeri[a,b]*

[a] *University of Pisa, Department of Civil and Industrial Engineering, Largo Lucio Lazzarino 1, 56122, Pisa, Italy*
[b] *National Interuniversity Consortium of Materials Science and Technology, Via Giusti 9, 50121, Florence, Italy*
[c] *Institute for the Chemical and Physical Processes, National Research Council of Italy, Via Moruzzi 1, 56124, Pisa, Italy*

Presenting author's email mb.coltelli@ing.unipi.it
Reference author's email andrea.lazzeri@unipi.it

Abstract. In the present paper some general strategies are discussed for achieving in poly(lactic acid) (PLA) based compounds the right balance of stiffness and impact resistance for properties modulation in both flexible and rigid items. The use of additives, eventually in combination to achieve synergistic effects, or the blending with other thermoplastic polymers gives the opportunity of tuning and controlling both processability and final properties of PLA based blends, composites and compounds. Other biodegradable polyesters, such as poly(butylene succinate) (PBS), up to now not yet developed on a large industrial scale, can offer opportunities for processing natural polymers coming from food or agro-waste, such as proteins, much affected by thermal degradation during processing. The advantage of this approach is the possibility of adopting traditional extruders for processing such blends, although the modification of the process and the development of new technologies is often necessary.

Keywords: **biodegradable polyesters, poly(lactic acid), poly(butylene adipate-co-terephthalate), poly(butylene succinate)**
PACS: 83.50Xa; 83.80.Tc; 89.20.Bb

INTRODUCTION

Poly(lactic acid) (PLA) is one of the most promising and ready to market material for applications in packaging [1]. Since it is rigid and brittle, a right balancing of stiffness and impact resistance is necessary for its properties modulation in both flexible and rigid materials. The use of additives [1-3], also in combination [4-8], the blending with other thermoplastic polymers [9-12], or the addition of natural fillers [13-14] to obtain specific composites gives the opportunity of controlling both processability and final properties of PLA based blends and compounds.

Other biodegradable polyesters, such as poly(butylene succinate) (PBS), up to now not yet developed on a large industrial scale, can be extruded with natural polymers coming from food or agro-waste, such as proteins, much affected by thermal degradation during processing [15-17]. The advantage of this approach is the possibility of using traditional extruders for processing such blends, although the modification of the process and the development of new processing methods and technologies is often necessary. In the latter case, although this topic is still at an early stage of development, it seems evident that it is necessary the control of melt viscosity and compatibility to develop biodegradable blends with controlled and reliable properties.

MATERIALS AND METHODS

Several types of PLA granules of Nature Works, poly(butylene adipate-*co*-terephthalate) (PBAT) granules from BASF and PBS granules from Mitsubishi were dried at 60°C for one night in a ventilated oven before extrusion.

Biodegradable polyesters granules were produced by using a COMAC extruder mod. EBC 25HT/44D High Torque (D/d ratio ≈ 1.65, Screw number of lobes: 2, Numbers of Heating zone: 11, Output: up to 50 kg/h, Italy). All the blends were prepared in a single extrusion step in order to both allow a more sustainable processing and limit polymers chain scission.

Plasticizers were added by using a calibrated peristaltic pump and solid additives were added by using the different feeders of the extruder. The extruder is equipped with one main feeder for polymer introduction as pellets, one precision feeder for adding powder additives in small amount (1-5% wt) and two feeders for introducing fillers or additives (fibers, powders) in high amount (10-30% wt).

Preliminary tests to set up extrusion conditions were performed in a smaller twin screw MiniLab II Haake™ Rheomex CTW 5 conical twin-screw extruder (Thermo Scientific Haake GmbH, Karlsruhe, Germany). After extrusion, the molten materials were transferred through a preheated cylinder to the Haake™ MiniJet II mini injection moulder (Thermo Scientific), to obtain Haake type-III specimens that were used for measurements and analysis.

Tensile tests were performed at room temperature, at a crosshead speed of 10 mm/min, by means of an Instron 4302 universal testing machine equipped with a 10 kN load cell.

RESULTS AND DISCUSSION

In extrusion of polyesters the preliminary drying is particularly important to avoid chain scission due to reaction of water with ester groups during extrusion. An important parameter is the temperature profile that should grant the suitable viscosity in the melt for an optimized transfer of the shear stresses of screws to the molten compound in order to grant a homogeneous dispersion of additives in the melt. Moreover, it is necessary to avoid excessive high processing temperatures not to induce thermal degradation, but on the other hand it is also necessary to grant the correct fluidity of the molten material for allowing a good dispersion and a performing extrusion. Furthermore, it is necessary to achieve a good stability and reproducibility in the strand production. In fact, at the die exit the consistency of the polymer melt should be adequate for rapid cooling and efficient cutting of the extruded strand to obtain granules. The correct balance of mass flow, screw rate, temperature profile, cooling methodology for the strand and cutting rate should be determined in order to achieve the full control of the process.

Extrusion of PLA based compounds

The correct addition to PLA of different additives can allow using it as main component in several compounds suitable for flexible or rigid packaging formulation. In particular, a summary about the effect of composition of PLA based materials on the final properties and applications is reported in table 1.

TABLE (1): different additives for extruding PLA based granules suitable for different applications in the packaging sector

Additives	Properties	Applications	Processing technologies	References
Plasticizers	Decreased Modulus and increased strain at break	Flexible packaging	Flat die extrusion, blow extrusion	[4-8]
Nucleant agents	Increased crystallinity and elastic Modulus	Rigid packaging	Injection moulding	[5-7]
Chain-extenders	Increased melt strength	Flexible packaging	Flat die extrusion, blow extrusion	[8]
Polymers	Increased impact properties, increased strain at break	Rigid packaging	Flat die extrusion (followed by thermoforming), injection moulding	[9-12]
Fillers	Increased Elastic Modulus and tensile strength	Rigid packaging	Flat die extrusion (followed by thermoforming), injection moulding	[13-14]

Plasticizers can be added to PLA to decrease its glass-transition temperature (T_g), which results in a lower stress at yield and a higher elongation at break at room temperature, with these conditions being necessary to improve the flexibility of films and sheets. Acetyl Tributyl Citrate [3-4], triacetine [11] and oligoethers [8], oligo lactic esters [6] and oligo adipic esters [7] resulted much efficient. Interestingly, if the plasticization was carried out in the presence of PBAT as polymeric additive, a preferential migration of the Acetyl Tributyl Citrate in the PBAT phase was observed [4] suggesting the necessity of selecting a plasticizer on the basis of its preferential affinity with the PLA matrix.

In the industrial production usually PLA is melt processed and then rapidly cooled below its glass transition temperature. As the crystallization of PLA is slow, final products (such as injection moulded or blow moulded containers) are essentially amorphous. The increase of temperature above PLA glass transition during the further processing steps (e.g. in the packaging of hot products) or during the use of the material, can enable cold crystallization, resulting in dimension instability and deformation of the items. In addition, an increase in brittleness and change in optical properties can be observed as consequence of crystallinity increase. Thus the use of proper nucleating agents, allowing increasing the crystallization rate of PLA during the rapid cooling is particularly interesting on a technological point of view. In PLA nucleating agents are reported to much affect the crystallinity content because they promote the formation of a high number of crystallization nuclei, thus consequently influencing also crystal growth and transparency [18-19]. Nucleated PLA thus results stabilized and its optical and mechanical properties are not modified as a consequence of heating in the temperature range of its glass transition. The addition of nucleating agents can be important also for stabilizing the extruded strand thus making easier the cutting operation to obtain granules, as well as the regulation of the temperature of cooling media and cutting rate.

In applications such as flat die extrusion or blow extrusion it is important the fine tuning of melt viscosity and melt strength. In these cases the use of chain extenders, much used in the field of polyester processing [20], can be fundamental. In some cases the use of plasticizers with reactive groups can allow better modulating PLA based compounds viscosity, limiting also the migration of the plasticizer from the final film thanks to its partial grafting to PLA [8]. The use of these additives can also affect the elasticity of the melt that much influence the shape of the extruded strand at the extruder die.

The blending of PLA with other polyesters has been much investigated in the past and interesting mechanical properties were observed for PLA/PBAT blends. In fact, these blends showed improved elongation at break and improved impact properties in a large composition range with respect to pure PLA. Several techniques of compatibilization have been investigated in the last decade [2, 10] with the aim of improving the properties of these blends granting also transparency [8] of final blends.

In order to increase elastic modulus in PLA based material the addition of cellulosic fibers to produce PLA based composites is a good strategy [13]. The increased Modulus can allow reducing the thickness of items, thus saving material and reducing the weight of packaging. This can be an environmental advantage considering the waste management issues related to packaging [14]. The addition of fibers during extrusion much influences the melt viscosity. Moreover, the geometry of the fibers, influencing final properties, is affected by the shear stresses of the extrusion.

In many applications peculiar combinations of properties are necessary. It is hence fundamental to investigate about the synergistic behavior of different additives onto final properties. Shi et al. [5] and Coltelli et al. [4] investigated the effect of plasticizer/nucleating agent and plasticizer/polymer respectively. Statistical techniques such as Design of Experiment can be also applied, as in the case of Fehri et al. [6]. These authors determined the range of composition that allowed maximizing the crystallization rate and crystallinity during injection moulding of PLA compounds containing both oligo lactic esters as plasticizer and different nucleating agents.

Extrusion of polymers with reduced thermal stability

Several natural materials that represent a waste of food industry can be used for producing packages. Proteins coming from whey are a waste of the cheese industry and have been investigated in the last years as barrier polymers for the production of multilayer sheets or films with improved recyclability. The use of a protein as a structural material for producing a package is difficult because of its low degradation temperature and low compatibility of its rheological and mechanical properties with traditional processing technologies. Hence the accessibility to traditional technologies such as extrusion for these materials can be a key point for boosting their application. It is possible to plasticize proteins by using mixtures of glycerol and water. By this way it is possible to obtain stable extruded strands and granules. However, the properties of such granule do not allow the preparation of films for flexible packaging. The blending with a biodegradable polyester can be thus fundamental. PBS, having a melting peak at 110°C and thus being compatible with natural polymers like whey showing degradation at temperature higher than 140°C, was blended with plasticized whey and the obtained blends resulted stable and with tunable properties. The full process for producing biodegradable films consisted in a first extrusion for whey plasticization, in a second extrusion for PBS blending and a final extrusion for addition of other additives, such as potato pulp filler [21]. Preliminary trials in a laboratory Minilab extruder and successive mechanical characterization of the obtained material allowed predicting that the granules could be produced in only one extrusion step. In fact in Table 2 it can be observed that the Torque value and tensile properties were similar.

Hence the extrusion was carried out contemporaneously plasticizing the protein, blending with PBS in the presence of additives and the final granules used for producing films by flat die extrusion [22].

TABLE (2). Torque, Elastic Modulus (E), stress at break (σ_b) and elongation at break (ε_b) of the plasticized whey/PBS blend prepared in two and one step in a Minilab laboratory extruder

Blend	Torque (N*cm)	E (GPa)	σ_b (MPa)	ε_b (%)
Plast. whey/PBS two step	57 ± 4	0,25 ± 0,02	16 ± 2	41 ± 4
Plast. whey/PBS one step	50 ± 1	0,22 ± 0,02	11,1 ± 0,5	26 ± 7

CONCLUSIONS

The extrusion of biodegradable polyesters is very important for developing new packaging products. The extrusion of PLA compounds, containing one or more additives necessary to modify its properties was addressed and performing methods developed.

The effect of different additives was discussed for achieving in poly(lactic acid) (PLA) based compounds the right balance of stiffness and impact resistance and for controlling processability and also final properties in both flexible and rigid items. Interestingly the combination of different additives can be an opportunity, although the deep understanding of the influence of composition onto processability and final properties requires systematic studies. The use of a laboratory extruder and the further characterization of the obtained materials was fundamental to select the best conditions for the extrusion and further processing of PLA based compounds.

Biodegradable polyesters with a lower melting temperature with respect to PLA, such as poly(butylene succinate) (PBS), can be extruded with natural polymers coming from food or agro-waste, such as proteins, much affected by thermal degradation during processing. The advantage of this approach is the possibility of using traditional extruders for processing such blends. However, as proteins can result in cross-linking and degradation during processing, the melt viscosity of the process must be thoroughly controlled. For boosting the development of such blends, as the PBS is not yet produced on a large scale and its cost is still high compared to other biodegradable polyesters such as PLA, the development of methods of rapid extrusion can be useful and successful. The use of a laboratory extruder and the further characterization of the obtained materials, thus keeping into consideration their properties, was fundamental to select the most convenient method for extruding the blends in an industrial extruder.

ACKNOWLEDGMENTS

The authors want to thank the European Community that granted the work described in this paper thanks to the funding of several concluded projects in the FP7 program such as FORBIOPLAST (GA n° 212239), DIBBIOPACK (GA n° 280676), N-CHITOPACK (GA n° 315233) and BIOBOARD (GA n°315313).

REFERENCES

1. R. A. Auras, S. P. Singh and J.Singh, *J. Packag Technol Sci*, **18**, 207 (2005)
2. M. B. Coltelli, S. Bronco and C. Chinea, *Polym. Degr. Stab.* **95,** 332-341 (2010)
3. F. Signori, M.B. Coltelli, S. Bronco and F. Ciardelli. *Polym. Degr. Stab.* **94**, 74-82 (2009)
4. M.B. Coltelli, I. Della Maggiore, M. Bertoldo, S. Bronco, F. Signori and F. Ciardelli. *J. Appl. Polym. Sci.* **110(2)**, 1250-62 (2008)
5. X.Shi, G. Zhang, T. V. Phuong and A. Lazzeri, *Molecules*, **20**, 1579–1593 (2015)
6. M. K. Fehri, C. Mugoni, P. Cinelli, I. Anguillesi, M. B. Coltelli, S. Fiori, M. Montorsi and A. Lazzeri, *Expr. Polym. Lett.* **10 (4),** 274–288 (2016)
7. M. K. Fehri, P. Cinelli, M. B. Coltelli, I. Anguillesi and A. Lazzeri, *Int. Jour. Chem. Engin. Appl.*, **7(2)**, 85-88 (2016)
8. A. Lazzeri, V. T. Phuong and P. Cinelli, WO2013164743 A1 (7 nov 2013)
9. S. Farsetti, B. Cioni and A. Lazzeri, *Macromol. Symp.* **301**, 82–89 (2011)
10. M.B.Coltelli, C. Toncelli, S. Bronco and F. Ciardelli, *Polym. Degr. Stab.* **96,** 982-990 (2011)
11. V. T. Phuong, M. B. Coltelli, P. Cinelli, M. Cifelli, S. Verstichel and A. Lazzeri, *Polymer* **55,** 4498-4513(2014)

12. P. Cinelli, A. Lazzeri, M. Penco, T.V. Phuong, WO2012025907 A1 (1 mar 2012)
13. A. Lazzeri and V.T. Phuong, *Compos. Sci. Technol.*, **93**, 106–113 (2014)
14. M. B. Coltelli, V. Gigante, S. Salvadori, P. Cinelli, I. Anguillesi and A. Lazzeri, *Eurofillers and Polymer Blends 2015 Conference Proceedings*, C8-9.4
15. M.M. Reddy, A.K. Mohanty, M. Misra, *Macromol. Mater. Engin.*, **297**, 455–463 (2012)
16. P. Cinelli, M.Schmid, E. Bugnicourt, J. Wildner, A. Bazzichi, I. Anguillesi and A. Lazzeri, *Polym. Deg. Stab.* **108**, 151-157 (2014)
17. M. B. Coltelli, F. Wild, E. Bugnicourt, P. Cinelli, M. Lindner, M. Schmid, V. Weckel, K. Müller, P. Rodriguez, A. Staebler, L. Rodríguez-Turienzo and A. Lazzeri, *Coatings* **6(1)**, 1 (2016),; doi:10.3390/coatings6010001M.
18. V. Nagarajan, K. Zhang, M. Misra, A. K. Mohanty , *ACS Applied Material Interfaces*, **7**, 11203–11214 (2015).
19. L. Wang, Y.N.Wang, Z-G. Huang, Y-X. Weng, *Materials and Design*, **66**, 7–15 (2015).
20. P. Raffa, M. B. Coltelli, S. Savi, S. Bianchi, V Castelvetro, *Reactive and Functional Polymers*, **72(1)**, 50–60 (2012)
21. M. Schmid, C. Herbst, K. Müller, A. Stäbler, D. Schlemmer, M. B. Coltelli and Andrea Lazzeri, *Polymer-Plastics Technology and Engineering* **55(5),** 510-517, 2016
22. E. Bugnicourt, Publishable Final Summary of Bioboard Project, http://bioboard.eu/files/2016/02/Publishable-Summary.pdf

Combination of sonication and hydrothermal for depolymerisation of cellulose in ionic liquid pressurized with CO_2 and N_2

Sumarno Sumarno[a], Mayangsari Novi Eka[b] and Trisanti Prida Novarita[a]

[a]ITS, Institut Teknologi Sepuluh Nopember - East Java - Indonesia,
[b]PPNS, Politeknik Perkapalan Negeri Surabaya - East Java - Indonesia

onramus@chem-eng.its.ac.id

Abstract. Cellulose is a polysaccharide compound having a crystal structure so difficult to dissolve in water. Degradation of cellulose to obtain useful substances (glucose, glyceraldehyde, and others) can be done by acid hydrolysis, ionic liquid, enzymatic / fermentation, and hydrothermal processes. In this study the degradation of cellulose is done with a combination of sonication pretreatment and followed by cohydrothermal processing in ionic liquids (NaCl and oxalic acid). Hydrothermal conditioin was conducted under CO_2 and N_2 gas at 200 bar. Temperature processing was 125 ºC and 200 °C for various degradation time from 30 -180 minutes. After the degradation process the sample was separated as liquid and solid part, where the liquid is analyzed by the dinitrosalicylic acid method (DNS method) using a UV-Vis spectrometer to determine the reducing sugar. The solids were analyzed by X-Ray Diffraction (XRD) and Scanning Electron Microscopy (SEM) to understand the structure change. The results show that the glucose concentration increase with time of hidrothermal proessing, and have a maximum at 120 minutes, and 90 minutes for 200 ºC. The crystallinity and morphology of cellulose shown by XRD results and SEM. Product substances resulted in the degradation process was showed by LC-MS. Keywords: Polysaccharide, sonication, hydrothermal, ionic liquid, CO_2, N_2.

Keywords: Polysaccharide, sonication, hydrothermal, ionic liquid, CO_2, N_2.
PACS: 82.35.Pq

INTRODUCTION

Biomass is renewable natural resources that are abundant, and can be converted into energy, chemicals, food and other raw materials [1, 2]. In general, this material contains 35-50% cellulose, 25-35% hemicellulose, and lignin 10-25% [3] with the major components in the form of cellulose. Cellulose is an organic compound polysaccharide difficult to dissolve in water and has a complex structure and strong for their hydrogen bonding inter- and intra-molecular [3,4].

The insolubility of cellulose in solvent (water) is a problem in the processing and for effective utilization. The benign method that has been developed to degrade cellulose is hydrothermal method. Sasaki et al (2000) conducted a cellulose hydrolysis in subcritical and supercritical conditions, wherein the supercritical water conditions, rate of decomposition into glucose and cellobiose is greater than subcritical water. Additionally, Stein et al (2010) also tried to degradation of cellulose using NaCl and oxalic acid in the system closed reactor (T = 125ºC and P = 70 bar) to obtain product in the form of soluble oligomers higher than simply using an acid, while for the product glucose of 2.80 to 3.90 g/L [5].

Based on previous research publications, degradation of cellulose into low molecular weight material need to be investigated further. The reaction of oxalic acid , sodium chloride salt, and carbonic acid resulting from the dissolution of carbon dioxide, effect on cutting glyosidic bond. In this study, carbon dioxide and nitrogen are applied for suppression the water and ionic liquid as degradation agent for cellulose. We use nitrogen as a relatively inert gas in this system. Both gases have different solubility, interaction with the ionic liquid, and the effect on the degradation reaction. We studied the effect of the degradation time of the reducing sugar and cellulose morphological changes.

METHODOLOGY

Cellulose material (Avicell Microcrystalline PH 102, Asahi Kasei Chemical Corp), carbondioxde and nitrogen (PT Aneka Gas, East Java, Indonesia), and 3,5-Dinitrosalicylic acid (DNS) from Sigma Aldrich were used in this work. Ionic solution made by mixing oxalic acid (0.1 M) and NaCl so that the concentration of NaCl in the solution to 20 % by weight. Cellulose and the solution are mixed to get cellulose concentration of 20 g L-1. 3 ml of mixture was inserted into the hydrothermal reactor. The nitrogen or carbon dioxide flows up to a pressure of 200 bar. The processing were conducted under 125ºC and 200 ºC, with degradation time for 30-180 minutes. The reactor was then decompressed into atmospheric. The product was separated as liquid and solids for analysis. The solid part was analyzed by Scanning Electron Microscopy (SEM) and X-Ray Diffraction (XRD), while for liquid analyzed by DNS (Dinitrosalicylic Acid) using a UV-Vis spectrophotometer. The experimental apparatus used is shown in Figure 1.

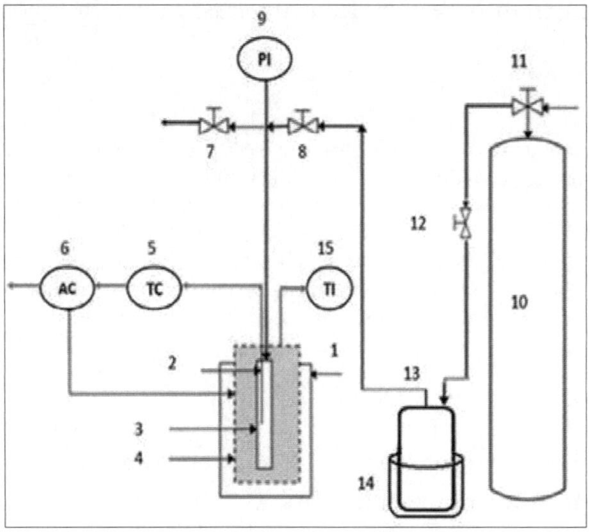

1. Insulator
2. Thermocouple
3. Reactor
4. Band heater
5. Temperature controller
6. AC current source
7. Valve into the air
8. Valve towards the reactor
9. Pressure gauge
10. Gas storage
11. Valve
12. Valve head Gas Supply
13. Gas Supply
14. Water vessel
15. Temperature Indicator

FIGURE 1. The Experimental apparatus used for hydrothermal depolymerisation of cellulose.

RESULTS AND DISCUSSION

The N_2 and CO_2 gas as pressurization media for depolymerisation gave different results as shown below.

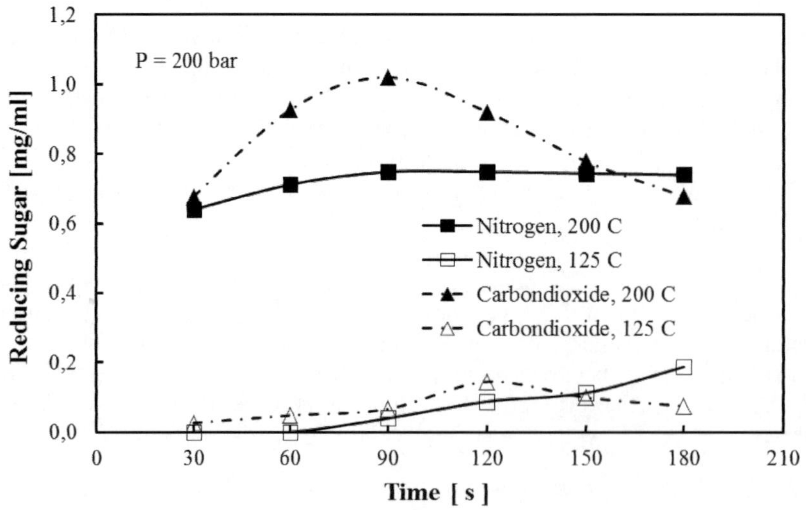

FIGURE 2. The effect of operating condition to reducing sugar, for N_2 and CO_2 as pressurization gas.

The ability of the NaCl ions to degrade the cellulose has been reported by Feng and Chen (2008) [6], and oxalic acid by Rempel and Whiters (2008) [7]. The use of ionic liquid together with CO_2 or N_2 gas and its effect on reducing the production of sugar has not been widely reported. At the same operating conditions CO_2 able to degrade cellulose better than the N_2. The presence of CO_2 in solution to change the solution becomes more acidic due to the formation of carbonic acid. Moreover, the operating conditions (200 bar, 125 °C and 200 °C) are far above the critical point of the gases.
Ionic component being capable of cutting glycosidic bonds rose on the use of carbon dioxide while the nitrogen gas is inert and only for pressurization only. The higher solubility of CO_2 and the formation of carbonic acid contribute to break the bond glyosidic significantly, moreover at higher temperature of 200 ºC. The carbonic acid breaks down into ions are then shared with the ions of oxalic acid and NaCl salt reacts with cellulose.

Figure 3 is the image of SEM of cellulose where (a) is the image of SEM of pure cellulose, (b) is processed under 200 bar and 200 °C, for degradation times of 60 minutes, (c) and (d) are the cellulose processed under 200 bar, for degradation times of 180 minutes and processing temperature is 125 °C and 200 °C respectively. Figure 3 (a) shows that the morphology of the pure cellulose surface is more smooth and intact. After the processing time, the higher the hydrothermal temperature, cellulose surface structure becomes coarser and the size is become smaller. The changes in the solid surface becomes rough means there is a loss of some parts of the surface layer (peeling) solids dissolved into solution [8]. These evidences show that the particle size shrank that might be caused by the surface reaction (conversion of the cellulose in to reducing sugar) that preceded by the release or dissolution of cellulose molecules from the particle. Micrographs showed that the longer the reaction time (c and d) and the higher the temperature (b and d) the particle size is become smaller. When the temperature increase lead to an increase of the reactivity of the water and ionics, or the ionized substance increase.

(a) (b) (c) (d)

FIGURE 3. The morphology change of cellulose particle after degradation in NaCl and Oxalic acid with pressurization under 200 bar of CO_2, where (a) pure cellulose; pressurization at, (b) 200 °C with degradation time is for 60 minutes, (c) 125 °C and (d) 200 °C with degradation time is for 180 minutes.

Processing of cellulose in ionic liquid under high pressure of gas CO_2 and N_2 lead to change in crystallinity. Figure 4 shows that the CO_2 cause the decrease in crystallinity and change from Type I become Type II. Processing with N_2 result in the crystalline change, and the crystallinity was lower than as processed with CO_2. The intensity of crystalline position in 2θ is shift as shown in 27, 32 and 45, as shown in Figure 5. Decreasing in crystallinity is bigger when processed with N_2.

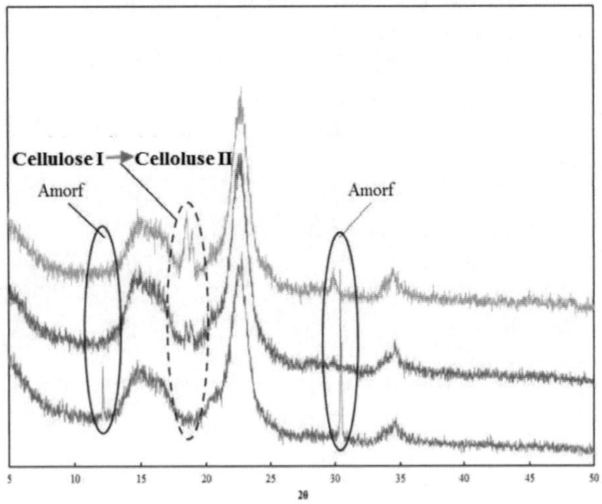

(a)

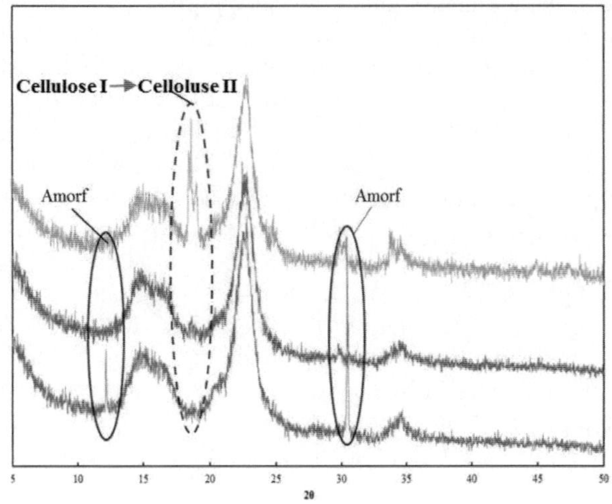

(b)

FIGURE 4. The XRD result of effects of hydrothermal processing under CO_2 pressurization under 200 bar with degradation time is for, (a) 60 minutes and (b) 180 minutes; (— Pure cellulose, — 125°C, — 200°C)

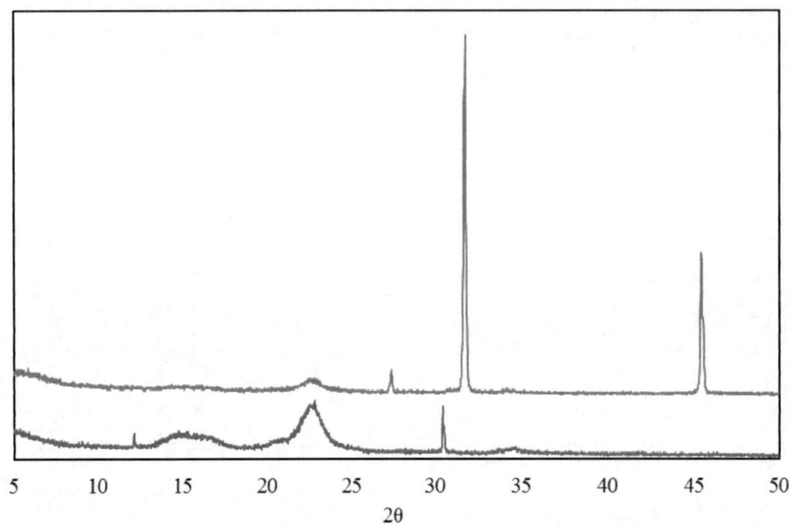

FIGURE 5. The XRD result of effects of hydrothermal processing under N_2 pressurization under 200 bar with degradation time is for 180 minutes; (— Pure cellulose, — 200°C)

CONCLUSSION

The cellulose degradation by hydrothermal method can produce reducing sugar. The pressurization with CO_2 result in reducing sugar which is higher than N_2. The crystallinity and type are change from Type I be II after hydrothermal processing with CO_2. The particle size of cellulose decrease by degradation time cause by reaction with ionic liquid. The N_2 gas cause the decrease crystallinity is higher than CO_2.

ACKNOWLEDGMENTS

The authors are thankful for the financial support of The Ministry of Research, Technology and Higher Education (RISTEKDIKTI) through ***Penelitian Strategis Nasional 2016.***

REFERENCES

1. Minowa T., Zhen F., Ogi T., *Cellulose Decomposition in Hot Compressed Water with Alkali or Nickel Catalyst*, The Journal of Supercritical Fluids, 1998, pp. 253-259.
2. Sasaki M., Fang Z., Fukushima Y., Adschiri T., Arai K., *Dissolution and Hydrolysis of Cellulose in Subcritical and Supercritical Water*, Ind. Eng. Chem, 2000, pp.2883-2890.
3. Zhou C. H., Xi Xia, Lin C. X., Tong D. S., Beltramini J., *Catalytic Conversion of Lignocellulosic Biomass to Fine Chemical and Fuels*, Chem Soc Rev, 2011, pp. 5588-5616.
4. Yu Yun, Hongwei Wu, *Significant Differences in The Hydrolysis Behaviour of Amorphous and Crystalline Portions within Microcrystlline Cellulose in Hot-Compressed Water*, Ind. Eng. Chem, 2010, pp. 3902-3909.
5. Stein T., Grande P., Sibilla F., Commandeur U., Fischer R., Leitner W., Maria P. D.. 2010. "*Salt-Assisted Organic-Acid-Catalyzed Depolymerization of Cellulose*". Green Chemistry. 12, 1844-1849.
6. Feng L. And Chen Z., *Research Progress on Dissolution and Functional Modification of Cellulose in Ionic Liquid*, The Journal of Mol Liq., pp. 1-5.
7. Rempel B. P. and Withers S. G., *Covalent Inhibitor of Glycosidases and Their Application in Biochemistry and Biology*, Glycobiology, 2008, pp. 570-586.
8. Hsu W., Lin M., Hsu J., *Dissolution of Solid Particle in Liquids: A Shrinking Core Model*, International Journal of Chemical and Biological Engineering, 2:4.

The Chitosan Degradation by Ozonation Process to Produce Glucosamine and Oligomers Chitosan

Emma Savitri[a,c], and Edy Purwanto[b*]

[a,b,c]*Chemical Engineering Department, University of Surabaya*
Raya Kalirungkut, Tenggilis, Surabaya,East Java, Indonesia 61290

savitri_ma@staff.ubaya.ac.id

Abstract. Chitosan is a linier copolymer (1→4)–linked 2-acetamido-2-deoxy-β-D-glucan (GlcNAc) and 2-amino-2-deoxy-β-D-glucan (GlcN) units in varying compositions. It is considered to be non-toxic, reactive and abundant biodegradable material. Because of the advantages of chitosan, it has received much attention for many diverse applications in biomedicine, pharmaceutics, cosmetic, biomaterial, agriculture, food processing, and waste water treatment. Chitosan has high molecular weight and also strong intra and intermolecular hydrogen bonding. Ozonation becomes alternatives to degrade chitosan into chitosan oligomers and glucosamine. In this study, chitosan was treated by ozonation process with various composition and temperature process. After ozonation, chitosan was characterized by viscosimetry and HPLC to determine molecular wight and also the dissolved product. Low molecular weight of chitosan was also characterized by FTIR to analyzed wether there was changing in side group of product.

Keywords: Ozonation, degradation, chitosan, glucosamine
PACS: 87.15.rp

INTRODUCTION

Chitosan is biopolymer generated by deacetylation of chitin. Many applications has been arisen from this polymer such as food [1], pharmaceutics [2], biomaterials [3,4], medicine and cosmetics [5]. Eventhough chitosan has much attentions in many applications, but its properties such as high molecular weight and low solubility on neutral solution has limited in its usages. Researches conducted several years found the low molecular weight of chitosan and chitosan oligomers is potential substances which possessing antimicrobial activity,immunostimulating and antigenic effect [6-8].

Chemical hydrolysis [9-11] and enzymatic treatments [12] were generally used to produce low-molecular-weight chitosan (LMWC) and chitosan oligomers (COS). The important problems related with producing low-molecular-weight chitosan and its degradation are the harmful of reagent used for degradation and the difficulty to separate the product from the degradation system. Ozonation might provide an alternative method to degradade chitosan and produce a valuable product. Ozon is a highly oxidative reagent and it can be easily removed from a reaction medium.

In the present work, the study of oxidative degradation of chitosan by ozone was conducted. The study concerns on the molecular weight and chemical changes in chitosan and the degradation product of chitosan after ozonation.

MATERIALS AND METHOD

The medical-grade chitosan purchased from Biotech Surindo (Cirebon, Indonesia) was used directly without any treatment. It has degree of deacetylation (DD) and the viscosity-average molecular weight (Mv) of chitosan are 81,81 % and 2,700 kDa, respectively. The acetic acid used was pure analysis grade purchased from Merck, Germany. High-performance liquid chromatography (HPLC) grade standard glucosamine, N-acetyl glucosamine, chitobiose, and N,N-diacetyl glucosamine were obtained from Sigma Aldrich, Pte Ltd., Singapore.

The ozonation of chitosan in the form of solution in acetic acid was carried out in 1.2 L thermostated bubble reactor at 20 – 50 °C. The mixture of ozon – oxygen (5.8 % mol) generated by ozone generator was flown into reaction system (chitosan diluted on acetic acid solution). The flow of gas out of the reactor was trapped by the KI trap storage to convert ozon into oxygen. The reaction was conducted for 150 mins and sampled every 30 mins.

The low-molecular-weight chitosan was recovered by adding an alkali solution, washed, filtered and freeze dried before being subjected to Ubbelohde viscosimeter, and Fourier transform infrared spectroscopy to characterize the viscosity-average molecular weight and chemical changes in terms of functional group. Dried chitosan were solved in 1 % acidic acid solution and the kinematic viscosity of solutions was measued for different concentrations. Fourier transform infrared spectroscopy (FTIR 8400S Shimadzu, Japan) was done using the KBr disc technique in the range of 4000 – 400 cm-1. All of the spectra were recorded at room temperature with a resolution of 4 cm-1 for 45 scans The filtrate from the samples subsequent to neutralization by the alkali solution was analyzed using a high-performance liquid chromatography equipped with a refractive index detector to analyze the dissolved product.

The schematic diagram of the ozonation process was drawn as shown Fig.1

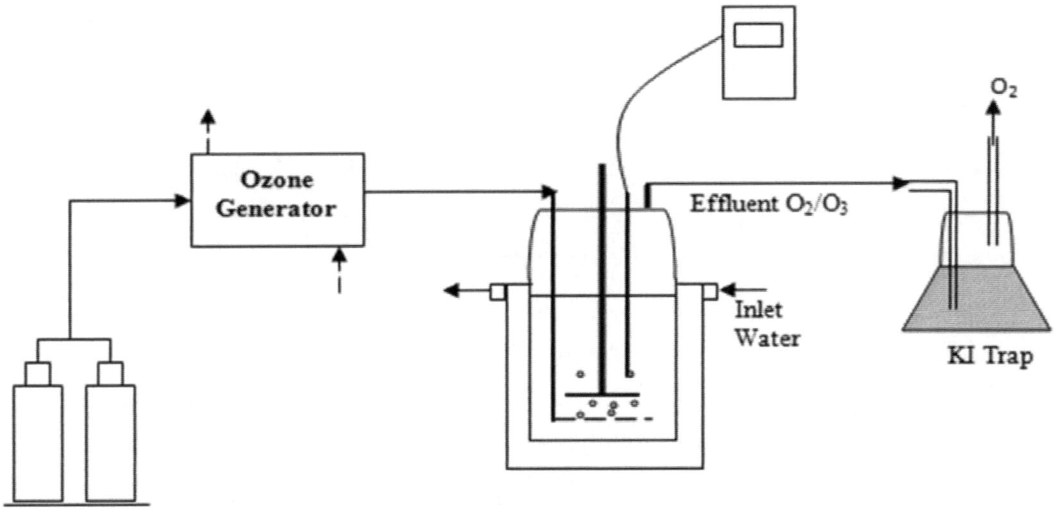

FIGURE 1. Schematic diagram of the Ozonation of chitosan experimental setup

RESULTS AND DISCUSSION

The ozonation of chitosan was conducted to degrade chitosan into low molecular weight of chitosan, oligomers chitosan and also glucosamine. The ozonation was carried out on various chitosan concentration (1-3 % w/v), ozonation temperatures (20 – 50 °C) for 150 mins ozonation process. The effects of the process was observed through the changes in the viscosity average molecular weight, chemical structure and also the product generated.

The viscosity - average molecular weight of the chitosan samples at various concentrations of chitosan and temperatures were determined (Fig. 2). Fig. 2 shows that the viscosity-average molecular weight of the chitosan in all the samples gradually decreased as the ozonation time and chitosan concentration increased. Fig 2 shows that the viscosity- average molecular mass of the chitosan decreases appreciably in proportion to ozonation time. The longer the ozonation time, the more ozone reacts with chitosan molecules by attacking β-D glycosidic bonds beetwen units in macromolecules of chitosan. According to the study [13] the oxidative reaction is initiated by interaction of ozone with chitosan molecule through its electrophilic attacking on C(1)-H bond with the formation of labile hydroxides, destruction of which leads to degradation of chitosan. While the contacting time between ozone and chitosan molecule is longer, the number of β-D glycosidic bonds destructed is also higher and the molecular mass is getting lower. Decreasing of chitosan molecular mass can be observed by decreasing in viscosity of chitosan reaction system. The effect of chitosan concentration on the molecular weight was also studied. It should be noted that decreasing chitosan concentration will lowering the viscosity-average molecular weight. It is caused by increasing number of chitosan molecule in the reaction system. Increasing chitosan molecules in the system will cause chitosan viscosity increase significantly. High viscosity of system inhibited the reaction between ozone and chitosan molecules. In this case, the mass transfer process of ozone to chitosan molecule become slowly and it controlled the reaction process. Because of it, the viscosity-average moleculare weight change was small. Moreover, because the number of chitosan molecules increased, the possibilities of contacting between ozone and chitosan molecules occurred became lower. From the study, the lowest chitosan molecular mass was achieved at 1 % w/v for 150 mins.

The molecular mass reached as high as 16,000 Da. The comparable molecular mass of chitosan can be reach by 1 % w/v and 2 w/v for 120 mins and the limiting molecular weight reached below 60,000 Da.

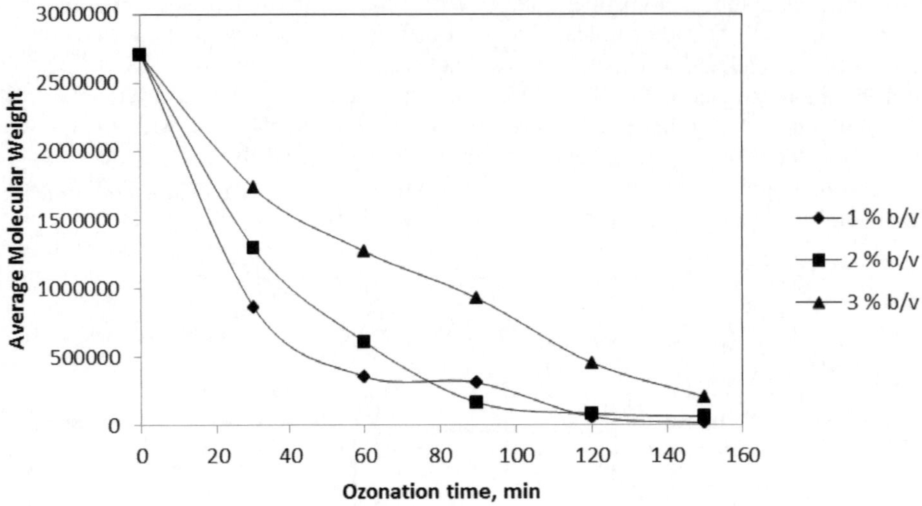

FIGURE 2. The effect of ozonation time and chitosan concentration on the viscosity-average molecular weight of chitosan after ozonation process

It can also be seen in Fig. 3 that the chitosan molecular weight profiles at various temperatures (20 °C - 50 °C) are similar. The increase in ozonation temperature leads to more rapid decrease in the chitosan degradation and accelerates the process. This indicated that increasing the ozonation temperature can lead to breaking the chitosan chain and significantly decreasing its molecular weight significantly. At higher temperatures, the solution is more dilute; this increases the ozone mass transfer into chitosan molecules hence the degradation increases. Therefore, increasing the ozonation temperature during the treatment will enhance the depolymerization effect.

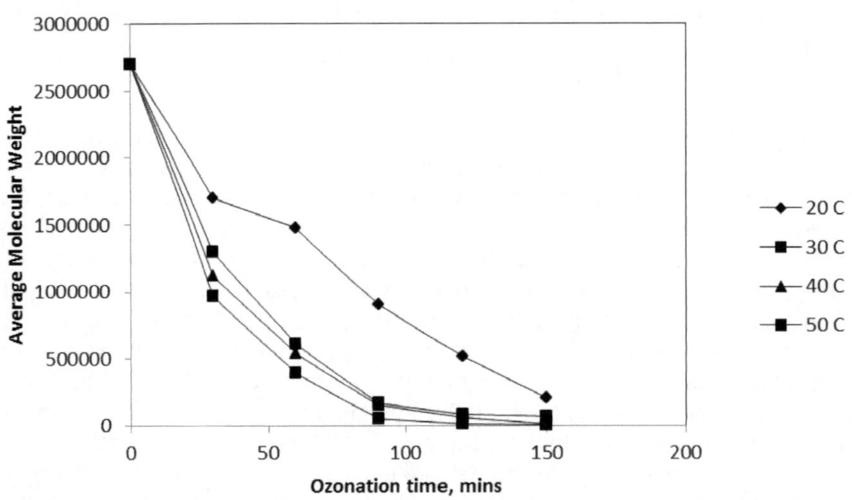

FIGURE 3. The effect of ozonation time and ozonation temperature on the viscosity-average molecular weight of chitosan after ozonation process

According to liquid chromatography analysis the dissolved products as the results of chitosan degradation was glucosamine (RT on 8.2 s) and also substance which have retention time (RT on 9.3 s). On the initial stage, the dominated product was glucosamine, but on the prolong ozonation time the unknown product was gradually increase and reached comparable amount with glucosamine. Based on this results, it showed that the oxidative degradation by ozonation tended to rupture the glucosamine unit selectively not random scission. According to the product profiles, we predicted that the unknown substance was product produced by the further degradation of glucosamine.[14]

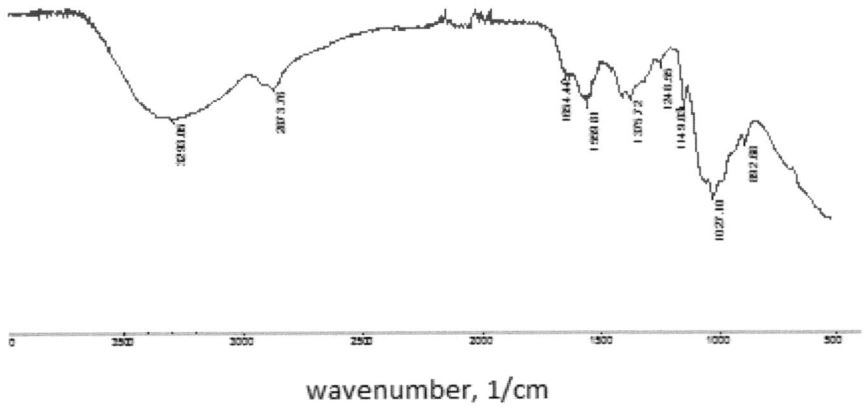

FIGURE 4. FTIR spectra of degraded chitosan via for 60 mins at temperatures of 30 °C and chitosan concentration on 2 % w/v

The FT-IR spectra of the degraded chitosan (low-molecular weight of chitosan) after ozonation process is displayed in fig. 4 and were recorded to investigate any structural changes during the ozonation process. The IR spectra of chitosan showed peaks from the saccharide structure due to C-O-C stretching at 1000 – 1200 cm-1.[15] The band at approximately 1424 cm-1 was attributed to –CH2- bending and to the orientation of the primary hydroxyl group in the polysaccharides. The absorption band at 1596 cm-1 corresponded to the amino deformation mode, and the absorption bands at 1650 and 1324 cm-1 were due to the amide I C=O bending and to amide III. The band at 1324 cm-1 corresponded to the GlcNAc residues. The peaks at 3340 – 3570 and 3230 - 3310 cm-1 were due to the $-NH_2$ bond stretching peak overlapping with the valence vibration peak of the H-bonded O-H, indicating intermolecular and intramolecular hydrogen bonding of the chitosan molecules. It can also be observed that no band is observed between 1650 and 1900 cm-1, which allows us to conclude that groups such as the carboxylic or aldehyde groups do not exist in the chitosan residue. Moreover, the C–H binding vibration of β-pyranose is manifested in the peak at 900 cm-1.

CONCLUSIONS

We conducted experiments involving the chitosan degradation through ozonation and evaluated the molecular weight, and the product upon ozonation. The oxidative degradation via ozonation process reduced the molecular weight with prolonged ozonation time, decreasing chitosan concentration and increasing ozonation temperature. The ozonation process tends to produce glucosamine and further degraded product of glucosamine. The FTIR spectra of low molecular weight of chitosan produced after ozonation still has similar spectra of its parent chitosan.

ACKNOWLEDGEMENTS

The authors are thankful for the financial support of the Directorate General of Higher Education of Indonesia through Hibah Fundamental 2016. The authors also want to express our gratitude to Marcella Oktaviani Biantari and Boby Lianto Gunawan for doing laboratory experiments as part of this project.

REFERENCES

1. F. Shahidi, J.K.V. Arachchi, Y.J.Jeon, *Trends Food Sci. Technol.* **10**, 37-51 (1999)
2. M. George, T.E. Abraham, *J. Controlled Release*. 114, 1-14 (2006)
3. E. Khor, L.Y.Lim, *Biomaterials,* **24,** 2339 – 2349 (2003)
4. J.K.F. Suh, H.W.T. Matthew, *Biomaterials,* **21,** 2589 – 2598 (2000)
5. M.Rinaudo, *Prog. Polym Sci.* **31**, 603 – 632 (2006)
6. J. Rhoades, G. Gibson, K. Formentin, M. Beer, R. Rastall, *Carbohydr Polym*, **64**, 57 – 59 (2006)

7. J.A. Elias, R.J. Homer, Q. Hamid , C.G. Lee. J Allergy Clin Immunol 116, 497 – 500 (2005)
8. K. Suzuki, T. Mikati, J. Okawa, *Carbohydr. Res*, **151**, 403 – 408 (1986)
9. A. Eibu, H. Grasdalen, K.M. Vårum. *Carbohydr Res*. **342**, 1055 – 1062 (2007)
10. K.M. Vårum, M.H. OttØ, O. SmidsrØd O, *Carbohydr Polym*. **46**, 89 – 98, (2001)
11. G.G. Allan, M. Peyron *Carbohydr Res,* **277**, 257 – 272 (1995)
12. S.K. Kim, N. Rajapakse, *Carbohydr Poly.* **62**, 357 – 368 *(*2005)
13. V.A. Demin, N.N. Kabal'nova, G.I. Osipova, V.V.Shereshovetz, *Russ. J. Appl. Chem.* **66,** 2562 (1993)
14. E. Savitri, S.R. Juliastuti, A. Handaratri, Sumarno, A.Roesyadi, *Polym Degrad. Stab*., **110**, 344-352 (2004)
15. V.K. Mourya, N.N. Inamdar, Y.M. Choudhari, *Nat. Polym*, **53**, 583 - 612 (2011)

Mechanical and durability properties of soft PVC plasticized by cardanol derivatives

Antonio Greco[a], Francesca Ferrari[a], Alfonso Maffezzoli[a]

[a] *Department of Engineering for Innovation - University of Salento, Via per Monteroni, 73100 Lecce, Italy*

antonio.greco@unisalento.it

Abstract. This work is aimed to study the plasticizing effectiveness of different cardanol derivatives, characterized by different yields of epoxidation, for the production of soft PVC. The properties of the cardanol derived plasticizers were compared to the properties of commercial plasticizers, both phthalates and natural derived. Mechanical properties of the developed plasticizers were shown to be comparable to those of the phthalate, and in many cases better than those of bio-based commercial plasticizers. The efficiency of the plasticizers, were significantly improved by increasing the yield of epoxidation. Also, mechanical properties performed after ageing showed the excellent stability of the properties of the cardanol derived plasticizer characterized by the higher yield of epoxidation. Evaluation of the property retention index after ageing indicated that such plasticizers showed an improved stability of properties compared to other commercial plasticizer. The results obtained highlight the relevance of an high conversion of the double bonds into epoxies in order to produce high quality plasticizers.

Keywords: plasticizer; renewable resources; diffusion; property retention
PACS: 82.35.Lr; 82.35.Pq

INTRODUCTION

In recent years, conventional oil derived phthalate plasticizers are being replaced by bio-plasticizers in many industrial applications, particularly in soft PVC industry. Despite this, bio-plasticizers are still considered too expensive in many soft PVC applications, which makes the substitution of phthalates still very difficult.

On the other hand, cardanol is a natural resource, characterized by low cost and low toxicological impact, obtained by vacuum distillation of ''cashew nut shell liquid'' (CNSL) [1]. The potential of cardanol derivatives as PVC plasticizers was recently demonstrated. In particular, it was shown that acetylation of the cardanol hydroxyl group improves the compatibility with PVC. Therefore cardanol acetate (CA), can be used as secondary plasticizer, in addition to conventional plasticizers, for the production of soft PVC [2]. Nevertheless, the migration of CA is very high compared to the migration of conventional plasticizers [3]. Further epoxidation of cardanol acetate leads to the production of epoxidated cardanol acetate (ECA), a plasticizer characterized by a very good compatibility with PVC [3]. Compared to other bio-plasticizer, the cardanol derived plasticizers do not require the use of primary resources, and therefore do not contribute to the subtraction of resources to food chain.

The aim of this work is the optimization of the properties of cardanol derived plasticizers by optimization of the procedure used for epoxidation. To this purpose, the properties of different plasticizers, characterized by different yields of epoxidation, were compared to the properties of commercial plasticizers, either phthalates or natural derived. In particular, a relevant part of the work is dedicated to the correlation between the functional properties of the plasticizer and the extent of epoxidation attained.

MATERIALS AND METHODS

The PVC used is Vestolit B7021 (Huls, Germany) in powder form. Cardanol derived plasticizers (CDP) were provided by Serichim (Torviscosa, Udine, Italy). All CDP were subjected to two reaction steps, esterification and epoxidation, with the aim of producing epoxidated cardanol acetate (ECA). All the plasticizers are characterized by a complete acetylation, and varying yields of epoxidation, as reported in Table (1).

TABLE (1). Chemical properties of the CDP and mechanical properties of the soft PVC obtained.

Plasticizer	Yield of epoxidation	Shore A hardness

CDP27	27	63.8±1.8
CDP54	54	63.1±2.1
CDP81	81	62.0±1.4
DEHP		63.3±2.6
ESBO		74.7±2.1
SNS		66.6±1.4

Epoxidated soybean oil (ESBO) and Grindtsed soft-n-safe by Danisco, mainly constituted by fully acetylated glycerol monoester of 12-hydroxystearic acid (SNS) were provided by Kommi srl, Lequile, Lecce, Italy, while bis-(2-ethylhexyl) phthalate (DEHP) was given by Huls. Finally, PVC stabilizer (Reagens CL/123) and costabilizer (Reaflex EP/6) were kindly supplied by Reagens (Italy).

Soft PVC samples at 70 phr plasticizer were obtained by mixing PVC and plasticizers for 10 min at 180 °C in a Haake Rheomix 600\610 mixer, with a rotor speed of 60 rpm. Thermal stabilizers and costabilziers, Reagens (5% weight of PVC) and Reaflex (3% weight of PVC) where added in the mixing chamber. After mixing, the materials were pre-heated in a forced convection oven up to 180 °C, and then compression molded in a Campana hot press with a pressure of 20 tons, using cold plates. Samples 0.4 mm thick were obtained for tensile, tear and durability test. Also, 6 mm thick plates were produced for hardness tests.

The hardness of soft PVC was measured according to Shore A procedure on a Gibitre Instruments Digital Manual Hardness Check.

Ageing tests were performed according to ASTM D2288, by keeping the discs of soft PVC in an oven at 105 °C for 15 days. Loss of weight of the samples was periodically measured in order to evaluate the rate of migration/evaporation of the plasticizer.

Tensile and tear tests were performed on soft PVC samples, before and after ageing, using a Lloyd LR5K dynamometer. Samples obtained by compression molding were subjected to tensile tests according to ASTM D638, using a crosshead speed of 50 mm/min. Tear tests were performed according to ASTM D1938, using a crosshead speed of 250 mm/min.

RESULTS AND DISCUSSION

Hardness tests results, reported in Table (1), clearly show that the plasticizing effectiveness of the different CDP is comparable to that of DEHP, and much better than that of ESBO and SNS. This is due to the comparable molecular weight of CDP and DEHP, which is much lower than that of DEHP and SNS, imparting a lower hardness and lower glass transition temperature to the soft PVC.

Weight loss M_t of the soft PVC samples during ageing tests was fitted by a Fickian diffusion, coupled with a constant surface concentration model [4]:

$$M_t = 2\frac{M_0}{L}\sqrt{\frac{Dt}{\pi}} \tag{1}$$

Where M_0 is the initial mass of the plasticizer, L is the thickness of the sample, D is the diffusivity of plasticizer in soft PVC and t is the time.

The value of diffusivity estimated for the different plasticizers is reported in Figure 1. For the CDP, the diffusivity decreases with increasing yield of epoxidation. For the CDP characterized by the higher yield of epoxidation, the diffusivity is lower than that of DEHP, and comparable to that of ESBO and SNS. Despite the lower plasticizing effectiveness, SNS and ESBO show a reduced diffusivity compared to DEHP, attributed to their higher molecular weight (500 g/mole for SNS and about 950-1000 g/mole for ESBO compared to 390 g/mole of DEHP). In contrast, CDP plasticizer shows a reduced diffusivity, despite its lower molecular weight (about 360 g/mole), which is a further indication of its excellent compatibility with PVC.

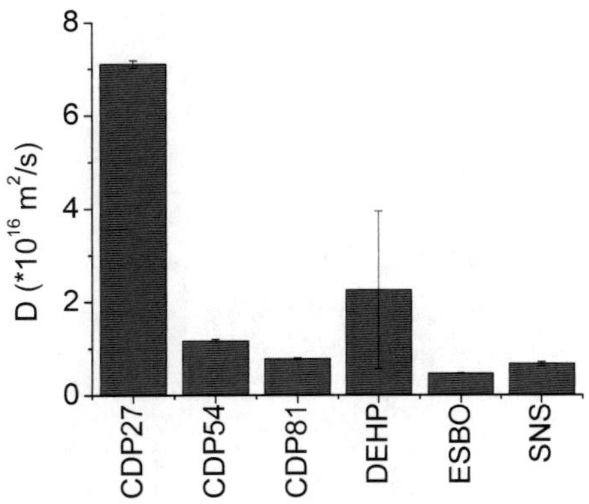

FIGURE 1. Diffusivity of CDP and commercial plasticizer estimated by ageing tests

The values of tensile modulus before and after ageing are compared for the different soft PVC samples in Figure 2. Before ageing, all the CDP impart to soft PVC a modulus comparable to that of PVC obtained by commercial plasticizers. In Figure 2 it is clearly observed that ageing involves an increase of the modulus, which is a direct consequence of the reduction of plasticizer content. In addition, the plasticizer characterized by the lower yield of epoxidation, and therefore higher migration during ageing tests, shows the more significant increase of the modulus. The CDP characterized by the higher yield of epoxidation, and lower diffusivity during ageing tests, shows an excellent stability of the properties, even better than that of the other commercial plasticizers.

A comparison of the tensile strength before and after ageing is reported in Figure 3. The tensile strength of soft PVC samples plasticized by CDP is comparable to that of soft PVC prepared with commercial plasticizers. As in the case of the modulus, ageing involves an increase of the tensile strength, due to the reduction of plasticizer content. For the CDP characterized by the lower yield of epoxidation, a significant increase of the tensile strength is a consequence of the high plasticizer loss. In contrast, the CDP characterized by the higher yield of epoxidation show a negligible increase of the strength. For such plasticizer, the retention of the tensile strength is better than that of the commercial plasticizers.

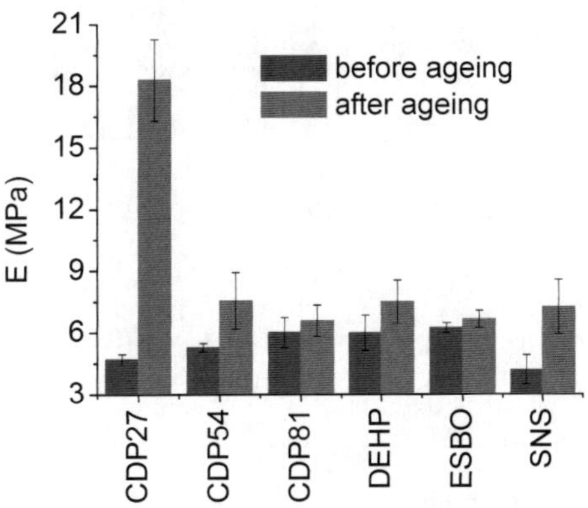

FIGURE 2. Tensile modulus before and after ageing for soft PVC samples

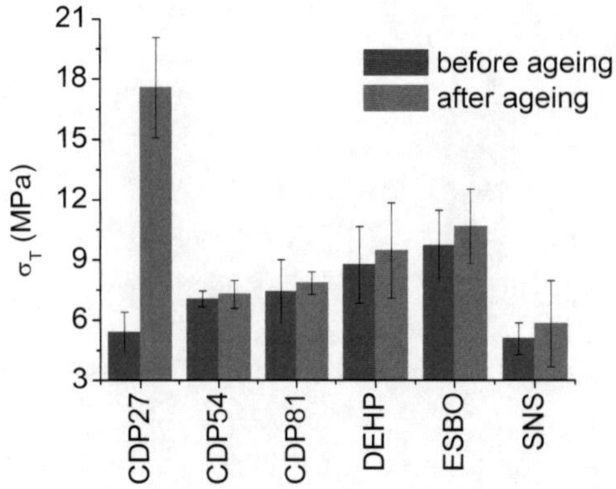

FIGURE 3. Tensile strength before and after ageing for soft PVC samples

The values of tear strength before and after ageing are compared for the different samples in Figure 4. Before ageing, all the CDP impart to soft PVC a tear strength which is comparable to that of PVC obtained by commercial plasticizers. Figure 4 shows that ageing involves an increase of the tear strength. As previously discussed for the modulus, the increased tear strength is a direct consequence of the reduction of plasticizer content. As in the case of the tensile modulus and strength, the retention of the tear strength is directly correlated to the yield of epoxidation: an higher yield of epoxidation involves a better retention of the tear strength. The plasticizer with the higher yield of epoxidation shows a better retention than the commercial plasticizers.

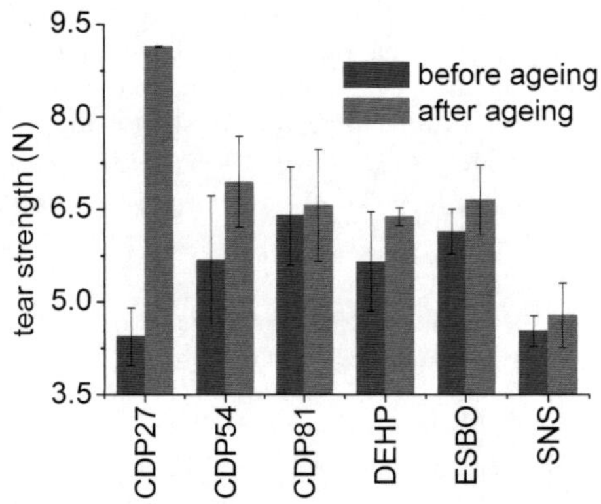

FIGURE 4. Tear strength before and after ageing for soft PVC samples

Evaluation of the mechanical properties was performed before and after ageing. The stability of the plasticizers was estimated though the property retention index (PRI), as defined by ASTM D5870:

$$PRI = \frac{P_{i,t}}{P_{i,0}} \tag{2}$$

Being $P_{i,t}$ the i-th property evaluated after ageing at time t (about 2 weeks), and $P_{i,0}$ the initial i-th property. As defined by eq. (2), the PRI in absence of any change of a property is 1, and it can take values higher or lower than 1 as the stability of the plasticizer decreases, since a generic property after ageing can be higher or lower than the initial value. In order to obtain a direct correlation between the stability of a plasticizer and the PRI, a modified expression was used for the PRI:

$$PRI_i = \frac{max(P_{i,t}; P_{i,0})}{min(P_{i,t}; P_{i,0})} \tag{3}$$

Which is always >1, and increases with decreasing stability of the plasticizer. The overall retention index (ORI) of the different plasticizers was obtained by considering the product between three different property retention indexes, relative to tensile modulus, tensile strength and tear strength. The results reported in Figure 5 show that the ORI decreases with increasing yield of epoxidation and decreasing diffusivity. Moreover, the ORI of the CDP with the higher yield is the lower one, among all tested plasticizers, which indicates that CDP81 is the plasticizer characterized by the better stability of properties, not only among other cardanol derived plasticizers, but also among commercial plasticizers.

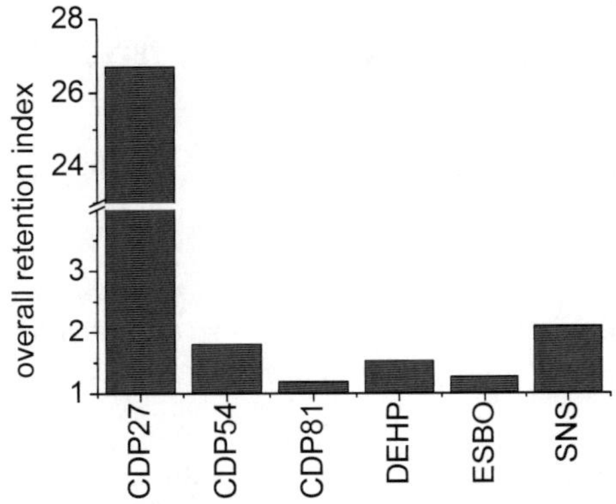

FIGURE 5. overall retention index for the tested plasticizers

CONCLUSIONS

In this work, the properties of cardanol derived plasticizers were correlated to the conditions of epoxidation, and compared to the properties of commercial plasticizers, both phthalates and bio- plasticizers.

The mechanical properties of the soft PVC obtained by the cardanol derived plasticizer were, particularly in the case of the higher yields of epoxidation, comparable to those of soft PVC obtained with commercial plasticizers.

Ageing tests showed that higher yield of epoxidation causes a reduction of the plasticizer leaching, due to a decrease of the diffusion coefficient of the plasticizer. The plasticizers characterized by the higher yield of epoxidation show a diffusion coefficient even lower than that of DEHP, and comparable to that of ESBO and SNS. In the case of the commercial bio-based plasticizers, the reduction of diffusivity compared to DEHP is attributed to the higher molecular weight. The very low diffusivity of the higher yield cardanol derived plasticizer, obtained despite the relatively low molecular weight, is a clear indication of the excellent compatibility with PVC.

Mechanical tests performed after ageing allowed the correlation of the property retention index to the weight loss of plasticizer. The property retention indices decrease with decreasing diffusivity, for cardanol derived and commercial plasticizers. On the other hand, the property retention indices decrease with increasing yield of epoxidation. Evaluation of an overall retention index, obtained by combination of individual property retention indices, allowed to highlight the excellent stability of the cardanol derived plasticizer characterized by the higher yield. For such material, in facts, the lower value of the ORI was obtained among all the other tested plasticizers, also including commercial ones.

REFERENCES

1. K. Sathiyalekshmi, S. Gopalakrishnan S, Evaluation of performance of polyurethanes based on hydroxy alkylated cardanol formaldehyde resins under aging conditions *Adv. Polym. Tech.* **23**, 91-102 (2004)
2. A. Greco, D. Brunetti, G. Renna, G. Mele, A. Maffezzoli, Plasticizer for poly(vinyl chloride) from cardanol as a renewable resource material *Polym. Degrad. Stab.* **95**, 2169-2174 (2010)
3. E. Calò, A. Greco, A. Maffezzoli, Effects of diffusion of a naturally-derived plasticizer from soft PVC *Polym. Degrad. Stab.* **96**, 784-789 (2011)
4. J. Crank, The mathematics of diffusion. Oxford: Clarendon Press, 1975

Mechanical and rheological properties of different plasticized poly(lactic acid) grades filled with talc

J. Seesorn[a], M. Seadan[b], Y.K. Wee[c] and S. Suttiruengwong[a*]

[a]*Department of Materials Science and Engineering, Faculty of Engineering and Industrial Technology, Silpakorn University, Sanamchandra Palace Campus, Nakhon Pathom 73000, Thailand*
[b]*Department of Physics, Faculty of Science, Silpakorn University, Nakhon Pathom 73000, Thailand*
[c]*Roquette Singapore Pte Ltd - 298 Tiong Barhu Road Central Plaza - Singapore*
Presenting author's email : seesorn_j@silpakorn.edu
Reference author's email : suttiruengwong_s@su.ac.th

Abstract. Polylactide (PLA) is a biodegradable aliphatic polyester with a good transparency and high tensile strength. However, its brittleness and cost become major drawbacks. Balancing between the mechanical properties and cost is a key to making PLA more competitive to existing commercial plastics. The aim of this research was to study the effect of plasticizer content and filler on the mechanical properties of poly(lactic acid). The different poly(lactic acid) grades, PLA 4043D, PLA 2003D and PLA 3052D, were selected and melt-mixed with 2 to 14 wt% of Polysorb®ID46 as a plasticizer via an internal mixer. The melt flow index increased with the amounts of plasticizes added. The tensile results demonstrated that upon increasing concentrations of plasticizers, it resulted in a decreased Young's modulus and tensile strength whereas elongation at break increased drastically when adding of the plasticizer above 12 wt% (~330%) for all PLA grades. PLA 4043D and PLA 2003D were showed slightly decreased Young's modulus when increasing plasticizer content, whereas elongation at break was still high. Tanδ peaks of plasticized PLA were shifted to the lower temperature with increasing plasticizer concentrations. 3-15 phr of Talc was then added to 12 wt% and 14 wt% of the plasticizer in PLA. Dynamic mechanical properties of Talc filled-plasticized PLA samples were recorded. The results showed that when Talc contents were increased for both plasticized PLA ratios, the storage modulus and the complex viscosity tended to increase. The effect on mechanical properties was less pronounced for 12 wt% plasticized PLA compared to 14 wt% plasticized PLA because it showed the slow decreasing rate of Young's modulus and tensile strength whereas the elongation at break was still very high compared with neat PLA. For PLA 4043D, the addition of even higher amounts of Talc could increase Young's modulus and tensile strength while maintaining the elongation property.

Keywords: polylactide(PLA); plasticizer; filler; bioplastics
PACS: 82.35.Pq, 87.15.rp

1 INTRODUCTION

Polylactide (PLA) is a biodegradable aliphatic polyester with a good transparency and high tensile strength. It has been recently used in a wide spectrum of applications; from single-used items to durable products. However, its brittleness and cost become the major drawback. Balancing between the mechanical properties and cost with being biodegradable are important factors for PLA to be more competitive to existing commercial plastics. To improve its brittleness, many methods such as copolymerization, blending with other flexible polymers and plasticization have been extensively applied. A number of plasticizers have been reported for PLA. These included glycerol [1], sorbitol [1], lactic acid oligomers [2], citrate esters [3], and poly(ethylene glycol) [4].

C.Courgneau. et al. [5] studied Acetyl tributyl citrate (ATBC) and poly(ethylene glycol) 300 (PEG) as plasticizers for PLA. The results showed the decrease in T_g due to ATBC addition. When modeled with Fox's law, it showed the absence of phase separation in spite of adding up to 17 wt% of plasticizer. This confirmed the miscibility of PLA and ATBC. The contents of equal or higher than 13 wt% of ATBC yielded a substantial improvement of the elongation at break, becoming higher than 300%. The choice of plasticizers employed should also take into account its biodegradability, non-toxicity, low migration rates and good miscibility and solubility with PLA matrix by producing a homogeneous blend and most preferable from renewable resources. In the recent years, Roquette Company has introduced biobased plasticizers Polysorb®, which is a composition of isosorbide diesters made from renewable raw materials. It was first recommended for plasticizing polyvinyl chloride. Adding the plasticizer alone could to some extent improve the brittleness whereas the cost of compound is higher. The addition of fillers is one of the simple ways to reduce the cost of the compound without any expense of desired properties. As reported in the literature, T.F. Cipriano. et al. [6] studied the rheological and morphological properties of PLA and Talc composites.

The results showed that the addition of the Talc to the PLA matrix decreased system viscosity, thus highlighting the filler's lubricant effect. As Talc content increased, system's viscosity also increased, providing material with higher elastic behavior. F. Yu. et al. [7] studied the effects of Talc on the Mechanical Properties of PLA. The results of tensile test showed that the elongation at break increased with Talc content up to 18.1 wt% with obvious changes in fracture behavior, changing from brittle to ductile, when Talc content increased to 24.3%. The addition of both PEG and nano-Talc in PLA was reported by X. Shi. et al. [8]. The results showed that the addition of Talc increased Young's modulus. PEG had the plasticization effect of increasing the break deformation while sharply decreasing the stiffness and the strength of PLA. The synergistic effect of nanofillers and plasticizer achieved the balance between the stiffness and the toughness of PLA.

Amounts these reported, there have not yet been any investigations for comparing different grades of PLA. The addition of both Polysorb® and Talc has not yet been reported elsewhere. Therefore, this work was aimed to study the effect of Polysorb® contents with and without Talc on different grades of polylactic acid. The mechanical and rheological properties of Talc-filled plasticized PLA was investigated.

2 EXPERIMENTAL

2.1 Materials

PLA NatureWorks® Ingeo™ 4043D, 2003D and 3052D having MFI 6, 6 and 14 g/10 min (210 °C, 2.16 kg) from specification data sheet of NatureWorks, respectively were purchased from NatureWorks LLC, USA. Liquid Polysorb® ID46 (isosorbide diesters, ID) was kindly supplied by Roquette Singapore Pte Ltd. Talc with particle size <45 microns was purchased from Haichen Minchem Co., Ltd, Thailand.

2.2 Melt-mixed of plasticized PLA and plasticized PLA filled with Talc

PLA and Talc were dried at 60 °C for 6 hours and 80 °C for 4 hours, respectively prior to use. PLA/Polysorb® and PLA/ Polysorb®/Talc prepared by melt-mixed in an internal mixer with screw speed 60 rpm, at 190 °C for 15 min. The contents of Polysorb® were 2, 4, 6, 8, 10, 12 and 14 wt% of PLA, which were then referred as PID2, PID4, PID6, PID8, PID10, PID12 and PID14, respectively. Talc-filled plasticized PLA was prepared with 3, 6, 9, 12 and 15 phr of Talc in 12wt% and 14 wt% plasticized PLA. These were referred to PID12+3T, PID12+6T, PID12+9T, PID12+12T, PID12+15T.

2.3 Characterization and testing

Melt flow index (MFI) was measured 190 °C and 2.16 kg according to ASTM D1238.

The tensile specimens were prepared by compression molding at 190 °C. Tensile testing was carried out by universal testing machine (Instron, Model 5969, USA) equipped with a 5 kN load cell at a crosshead speed of 10 mm/min, according to ASTM D638 (type V).

Dynamic mechanical analysis of plasticized PLA and plasticized PLA filled with Talc was performed by using a Modular Compact Rheometer (MCR302), Anton Paar GmbH, Plasticized PLA samples were analyzed from 30-90 °C temperature range at a heating rate of 3 °C/min and 1 Hz frequency. Rectangular samples of size 40 x 10 x 1.16 mm^3 in temperature scan mode. Plasticized PLA filled with Talc were analyzed from 0.1-1000 Hz frequency range at 190 °C and using a gap size 1 mm in frequency sweep test.

3 RESULTS AND DISCUSSION

3.1 Plasticized PLA

3.1.1 Mechanical properties

It should be noted that three PLA grades investigated in this work differ in terms of D-isomer contents and molecular weights. PLA 4043D, 2003D and 3052D have comparable D-isomer contents. The molecular weight of PLA 4043D is similar to PLA 2003D whereas the lowest molecular weight is PLA 3052D, as reported elsewhere [8, 9]. These differences may influence the plasticization. The effect of plasticizer contents on mechanical properties of three investigated PLA grades was represented in Figure 1. Young's modulus, tensile strength and elongation at break of all neat PLA grades were around 1,600 MPa, 100 MPa and 8%, respectively. It was found that upon increasing concentrations of plasticizers (Polysorb®ID46), it resulted in the decrease in Young's modulus and tensile strength whereas the elongation at break increased when compared to neat PLA. It was also observed that the mechanical behavior of plasticized PLA abruptly changed from brittle to ductile when adding 12 wt% (PID12) and 14 wt% (PID14) of plasticizers. This was due to the fact that the plasticizer reduced the intermolecular forces and caused the increased mobility of polymer chains, thus enhancing the flexibility of the plasticized PLA. M. Baiardo, et al. [10] investigated PLA/PEG 1.5K and PLA/ATBC systems and found that when increasing plasticizer content, the tensile modulus and tensile strength decreased, while the elongation at break increased. Inspection of the elongation at break values showed that a very significant change of elongation (5% to 194% of PEG 1.5K and 7% to 218% of ATBC) occurred at 12.5 wt% plasticizer content, depending on the types of plasticizer employed. In this work, for all PLA grades, the elongation at break was dramatically improved above 330% when the contents of the plasticizer were above 12%wt. However, Young's modulus and tensile strength were slightly decreased when compared to neat PLA, as shown in Figure 1.

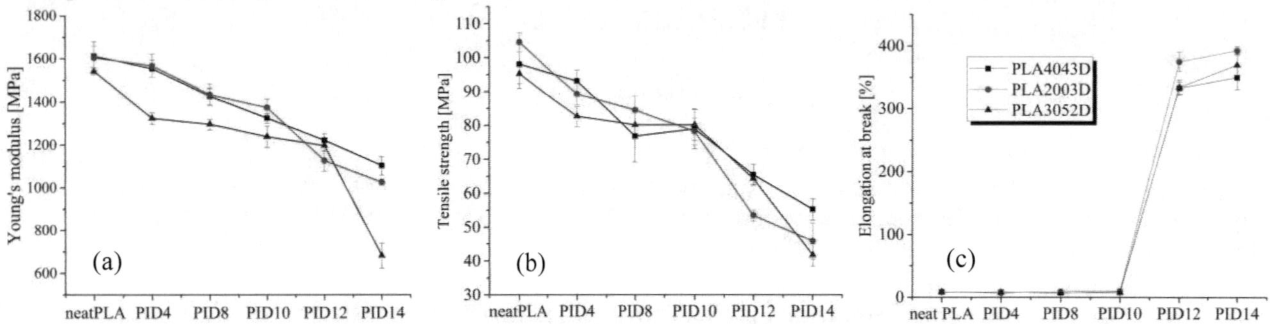

FIGURE 1. Young's modulus (a), tensile strength (b) and Elongation at break (c) of neat PLA and plasticized PLA for PLA 4043D, 2003D and 3052D.

3.1.2 Melt Flow Index (MFI)

Figure 2a illustrated the melt flow index (MFI) of neat PLA; PLA 3052D, PLA 2003D and PLA 4043D, and plasticized PLA 4043D. MFI values tended to increase with the increase of plasticizer contents. This result could be observed for all PLA grades (data not shown). The addition of plasticizers increased the polymer chain mobility which implied a reduction in viscosity, hence increasing MFI of plasticized PLA 4043D.

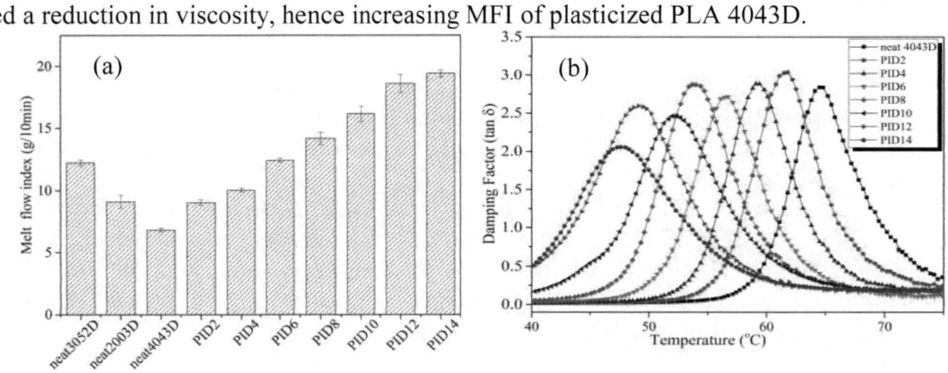

FIGURE 2. (a) Melt Flow Index (MFI) of neat PLA 3052D, 2003D and 4043D, and plasticized PLA 4043D and (b) damping factor (tan δ) with temperature of the neat PLA and plasticized PLA 4043D.

3.1.3 Dynamic Mechanical properties

The damping peaks occurred in the region of T_g is associated with the movement of small groups and chains of molecules within the polymer structure. Figure 2b also showed that damping factor (tanδ) of the samples as a function of temperature. T_g could be determined from the tanδ peaks. Thus, T_g decreased with increasing plasticizer content added. Tanδ peaks of plasticized PLA 4043D were shifted from 65 °C to 47 °C with increasing plasticizer concentrations. The width of the tanδ peaks also became broader than that of the neat PLA with increasing plasticizer contents, suggesting that there were higher molecular relaxations in plasticized PLA compared to that of neat PLA. In addition, the width of the tanδ peaks was indicative of the increased volume of the interface with further addition of plasticizer. Similar results were also reported by V.S.Giita Silverajah. et al. [4]. These results are in good agreement with the elongation at break from mechanical properties and melt flow index.

3.2 Plasticized PLA filled with Talc

3.2.1 Mechanical properties

Figure 3 showed the mechanical properties of all PLA grades plasticized with 12 wt% and 14 wt% and filled with and without Talc. For Young's modulus and tensile strength of Talc-filled plasticized were presented Figure 3a and 3b, respectively. It was found that Young's modulus and tensile strength tended to increase with increasing Talc loading for all PLA grades. The elongation at break as shown in Figure 3c, it resulted that changed depended on PLA grade. Based on this finding, it could be concluded that different PLA grades also exhibited different balances of these properties. In the case of PLA 4043D showed increased Young's modulus and tensile strength when increasing Talc content while elongation at break was high value. For the mechanical properties of Talc-filled plasticized PLA 2003D. Young's modulus and tensile strength were similar to that of PLA 4043D, but the elongation at break dropped when adding up to15 phr of Talc compared to plasticized PLA. The effect of Talc on the elongation at break of Talc-filled plasticized PLA 3052D was most pronounced. It was drastically dropped after adding 6 phr of Talc (dropped by 85.17% and 78.34% from 12%wt and 14%wt plasticized PLA, respectively). The results suggested that Talc acted as the reinforcement fillers while the plasticizer was effective at enhancing the PLA chain mobility. The balance of the stiffness and toughness of PLA was achieved by the combination of both Talc and plasticizer [8]. However, this balance was lost only after adding small amount of Talc for PLA 3052D. This effect was less pronounced for PLA4043D and 2003D, probably due to their higher molecular weights.

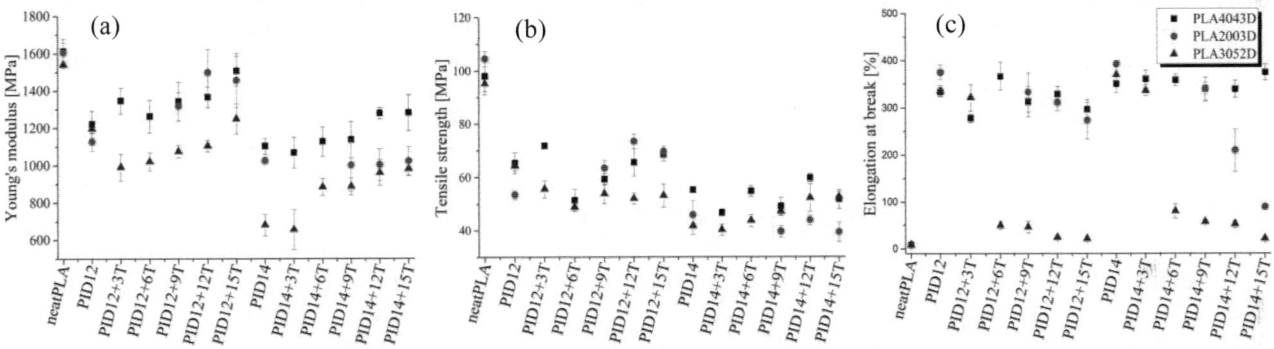

Figure 3. Young's modulus (a), tensile strength (b) and Elongation at break (c) of neat PLA, plasticized PLA and Talc-filled plasticized PLA for PLA 4043D, 2003D and 3052D.

3.2.2 Dynamic Mechanical properties

Figure 4 showed the complex viscosity, storage modulus and loss modulus on the investigated frequencies of plasticized PLA and plasticized PLA filled with Talc compared with those of different neat PLA grades. Neat PLA exhibited the complex viscosity value around 2,500 Pa·s. The complex viscosity (Figure 4a) of all samples exhibited the Newtonian behavior at angular frequencies lower than 7 rad/s. After such values, the complex viscosity decreased showing shear thinning behavior. Upon increasing amount of plasticizer it resulted in the decreased complex viscosity. This was the effect of more disentanglement of PLA chain due to increased segmental mobility. On the other hand, the addition of the higher amount of Talc could cause an increase in the complex viscosity, where this such an increase generally depended on the concentration, particle size, particle size distribution and shape of the fillers. The presence of Talc particles could disrupt the polymer flow and hinder the mobility of polymer chain segment [11]. The corresponding storage and loss moduli were shown Figure 4b and 4c, respectively. The moduli of neat PLA was higher than that of the other samples in the all frequency range and showed more shear thinning behavior. As expected, the moduli of plasticized PLA decreased with increasing plasticizer loading at all frequency. Whereas the increasing Talc loading resulted in the increased moduli. It might be due to the fact that the use of plasticizer reduced the intermolecular force and increased the mobility of polymer chain, at the same time Talc hindered the mobility of polymer chain. All samples exhibited the rheological of typical terminal behavior of polymer melt as characterized by lower the storage modulus compared to the loss modulus. In addition, the storage modulus and loss modulus showed an increased difference with decreasing frequency.

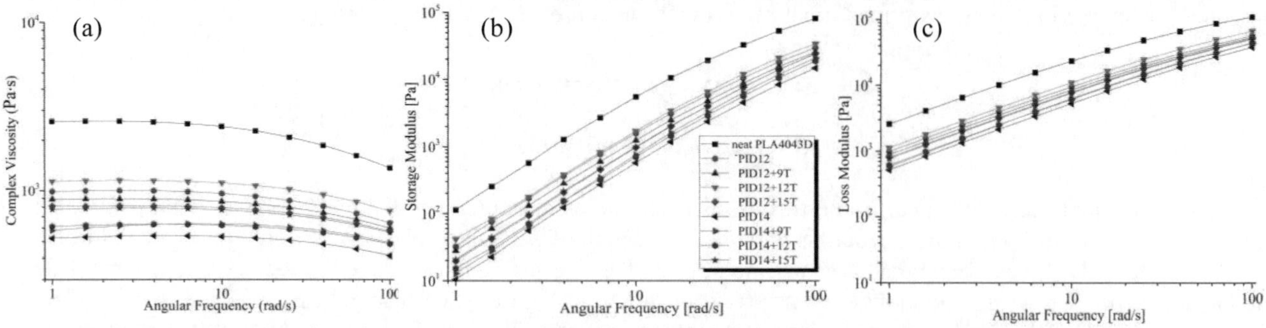

FIGURE 4. Complex viscosity (a), storage modulus (G') (b) and loss modulus (G") of neat PLA, plasticized PLA and plasticized PLA filled with Talc for PLA 4043D.

4 CONCLUSION

The tensile results demonstrated that upon increasing concentrations of plasticizers, it resulted in a decreased Young's modulus and tensile strength whereas the elongation at break increased drastically when adding of the plasticizer above 12 wt% (~330%) for all PLA grades. PLA 4043D and PLA 2003D were showed slightly decreased Young's modulus when increasing plasticizer content, whereas elongation at break was still high. Tanδ peaks of plasticized PLA were shifted to the lower temperature with increasing plasticizer concentrations. Dynamic mechanical and mechanical properties of Talc filled plasticized PLA samples were recorded. The results showed that when increasing Talc contents in plasticized PLA of both ratios, the complex viscosity, storage modulus and loss modulus tended to increase. The effect on mechanical properties was less pronounced for 12 wt% plasticized PLA, compared to 14 wt% plasticized PLA because it showed the slow decreasing rate of Young's modulus and tensile strength whereas the elongation at break was still very high compared with neat PLA. For PLA 4043D, the addition of even higher amounts of Talc could increase Young's modulus and tensile strength while maintaining the elongation property.

ACKNOWLEDGEMENTS

The authors would like to gratefully thank Department of Materials Science and Engineering, Faculty of Engineering and Industrial Technology, Silpakorn University, Thailand for financial supports and Roquette Singapore Pte Ltd for kindly supplying Polysorb®.

REFERENCES

1. H.Li, M. A., and Huneault, *Comparison Sorbitol and Glycerol as Plasticizers for Thermoplastic Starch in TPS/PLA Blends.* Appl. Phys, 2010. **119**: pp. 2439–2448.
2. Burgos, et al., *Characterization and ageing study of poly(lactic acid) films plasticized with oligomeric lactic acid.* Polymer Degradation and Stability, 2013. **98**(2): pp. 651-658.
3. V. DAVE´, et al., *Citrate Esters as Plasticizers for Poly (lactic acid).* Appl. Phys, 1997. **66**: pp. 1507–1513.
4. V. S. Giita Silverajah, Nor Azowa Ibrahim, and N. Zainuddin, *Mechanical, Thermal and Morphological Properties of Poly(lactic acid)/Epoxidized Palm Olein Blend.* Molecules, 2012. **17**: pp. 11729-11747.
5. C.Courgneau, et al., *Analysis of the Structure-Properties Relationships of Different Multiphase Systems Based on Plasticized Poly(Lactic Acid).* polymer environment, 2011. **19**: pp. 362–371.
6. T. F. Cipriano and A.L.N.d. Silva, *Thermal, Rheological and Morphological Properties of Poly (Lactic Acid) (PLA) and Talc Composites.* Polímeros, 2014. **24**: pp. 276-282.
7. F. Yu, et al., *Effects of Talc on the Mechanical and Thermal Properties of Polylactide.* Appl. Phys, 2011. **125**: pp. 99-109.
8. X. Shi, et al., *Synergistic Effects of Nucleating Agents and Plasticizers on the Crystallization Behavior of Poly(lactic acid).* Molecules, 2015. **20**: pp. 1579-1593.
9. B. Mallet, Khalid Lamnawar, and A. Maazouz, *Improvement of Blown Film Extrusion of Poly(Lactic Acid): Structure–Processing–Properties Relationships.* Polymer engineering and science, 2013: pp. 1-18.
10. M. Baiardo, et al., *Mechanical Properties of Plasticized Poly(L-lactic acid).* Appl. Phys, 2003. **90**: pp. 1731–1738.
11. D. SHUMIGIN, et al., *Rheological and Mechanical Properties of Poly(lactic) Acid/Cellulose and LDPE/Cellulose Composites.* Materials science, 2011. **17**(1): pp. 32-37.

Biodegradable polyesters reinforced with eucalyptus fiber: effect of reactive agents

P. Nanthananon[a], M. Seadan[b], S. Pivsa-Art[c], H. Hiroyuki[d] and S. Suttiruengwong[a*]

[a]Department of Materials Science and Engineering, Faculty of Engineering and Industrial Technology, Silpakorn University, Sanamchandra Palace Campus, Nakhon Pathom 73000, Thailand
[b]Department of Physics, Faculty of Science, Silpakorn University, Nakhon Pathom 73000, Thailand
[c]Department of Material and Metallurgical Engineering, Faculty of Engineering, Rajamangala University of Technology Thanyaburi, Klong 6, Thanyaburi, Pathumthani 12110, Thailand
[d]Kyoto Institute of Technology, Matsugasaki, Sakyo-ku, Kyoto City 606-8585, Japan

p.nanthananon@hotmail.com
suttiruengwong_s@su.ac.th

Abstract. The aim of this work was to investigate the effect of reactive agents on the rheological behaviors, morphology and mechanical properties of two biodegradable polyesters; poly(lactic acid) (PLA) and poly(butylene succinate) (PBS) reinforced with eucalyptus fiber. Two types of reactive agents, chain extender (Joncryl®) and peroxide (Perkadox® 14), were selected. Biocomposites consisting of a constant fiber loading of 10 wt% with and without reactive agents were prepared by one-step melt-mixing methods via an internal mixer with the rotational speed of 60 rpm and the temperature of 150 and 190 °C for PBS and PLA, respectively. The biocomposite samples were then compression-molded for tensile testing. Rheological and morphological properties of the biocomposites were also investigated by means of dynamical mechanical analysis (DMA) and scanning electron micrograph (SEM). The results revealed the presence of Joncryl® in PLA biocomposites caused the increase of mixing torque, complex viscosity, storage modulus and mechanical properties when compared to the addition of peroxide. The increase of complex viscosity of PLA biocomposites added Joncryl® was due to the increase in the population of the higher molecular weights as confirmed by GPC analysis. In the case of PBS biocomposites, it showed a dramatic increase in the complex viscosity and storage modulus with decreasing crossover frequencies when adding peroxide rather than adding Joncryl®. The SEM micrographs demonstrated that the compatibility between the polymer matrix and fiber was improved when adding a small amount of Joncryl® and peroxide for both polymers, except for 0.5 phr peroxide loading, leading to an improvement of their modulus and tensile strength.

Keywords: Biocomposite, Biodegradable polymer, Natural fiber, Reactive agent
PACS: 81.05.Lg, 81.05.Qk, 82.35.Gh, 82.35.Pq, 83.80.Ab, 83.85.Cg

INTRODUCTION

Biocomposites based on biopolymers and natural fibers have caught much attention in the last few decades due to the combination of a good mechanical performance as well as less environmental impact with being fully biodegradable and more sustainable in terms of renewable resources. Many commercially available biodegradable polyesters have readily been produced on an industrial scale with a wide range of grades and applications. These include polylactic acid (PLA), polybutylene succinate (PBS), poly(caprolactone) (PCL) and poly(butylene adipate-co-terephthalate) (PBAT). Depending on the desired purposes, the properties of these classes of polymers could be effectively modified through polymer blending or composites. As far as the biocomposites are concerned, the major challenge is the improvement of an interfacial adhesion between natural fiber and biopolymer matrix. The better interfacial adhesion would result in an improvement in mechanical and rheological properties of composites [1]. In order to enhance the natural fiber-matrix adhesion; chemical treatments, *in-situ* reactive blending, and coupling agents have been considered as solutions with the aim of changing the surface tension and polarity through the modification of fiber surface, and thus facilitating coupling with the matrix [2]. The treatments can be achieved usually based on the use of reactive groups such as maleic anhydride (MA), silanes, isocyanate, and peroxide [2]. However, some treatments are robust and time-consuming. The interface properties of fiber and matrix could be improved by peroxide treatment as it initiated free radicals reacted with the hydrogen group of the cellulose fiber and matrix [2]. Alternatively, the presence of a chain extender, reactive multi-epoxy groups like Joncryl®, was used as a chain extension/branching agent in PLA/PBAT blends, and it showed an improvement of their thermal stability as reported by Al-Itry et al. [3]. The epoxide groups of Joncryl® could theoretically react with both hydroxyl and carboxyl end groups of the polyesters [3]. Meng et al. [4] reported that the addition of 1 wt% Joncryl® in PLA nanocomposites with 2 wt% organo-modified clay improved the PLA thermal stability by a chain extension reaction with PLA chain end groups as well and the possible reaction of Joncryl® epoxy groups with the hydroxyl groups in the organo-modifier clay. To the best of our knowledge, there have been no reports of using Joncryl® and peroxide as the *in-situ* compatibilizer in natural fiber to produce biocomposite. Thus, this work was aimed to propose the simple way for the reactive blending of the eucalyptus fiber and biodegradable polyesters (PBS or PLA) using two types of reactive agents; Joncryl® and Perkadox by one-step melt-mixing methods. The influence of reactive agents on rheology, morphology and mechanical properties of biocomposites were investigated.

EXPERIMENTAL

Materials and Biocomposites Preparation

Poly(butylene succinate) (PBS) (1001MD blown film grade) was purchased from Showa High Polymer Co., Ltd., Japan. Poly(lactic acid) (PLA) (3052D injection grade) was purchased from NatureWorks® LLC. (USA). The eucalyptus pulps (500-600 μm in length and 4-10 μm in diameter) were kindly provided by SCG Packaging PLC., Thailand. Peroxide (Perkadox® 14-40B-pd) (containing 40% with $CaCO_3$ and SiO_2) was purchased from AkzoNobel. Joncryl® (ADR-4368F), an epoxy-based chain extender, was obtained from BASF The Chemical Co., Thailand. Paper pulps were dried at 80 °C and ground to obtain eucalyptus fibers using a high-speed grinder. Prior to melt mixing, PLA and PBS were dried at 60 °C in a hot air oven for 6 hr and eucalyptus fibers were dried again at 80 °C overnight. The biocomposites without reactive agents were first prepared by adding 10 wt% of eucalyptus fibers into PLA or PBS (referred as PLA/FB10 or PBS/FB10 respectively) using an internal mixer (Chareon tut CO. LTD., model MX 105-D40L50) with the rotational speed of 60 rpm and the temperature of 190 °C and 150 °C for PLA and PBS, respectively. The concentrations of 0.1, 0.5, 1.0 phr of Joncryl® and 0.05, 0.1, 0.5 phr of peroxide were used for the PLA/FB10 and PBS/FB10 biocomposites. Samples for the rheological, morphological and mechanical analysis were prepared by compression molding at 150°C for PBS and 190 °C for PLA respectively.

Measurement and Testing

Rheological behavior measurements of PLA and PBS biocomposites were performed using an ANTON PAAR (MCR302) equipped with 25 mm parallel plate flow geometry. The over a frequency range of 0.1–100 rad/s (from high to low frequency) and 0.1% strain amplitude were carried out at 190 °C and using a gap size of 1 mm. The molecular weights of the polymer after compounding was determined using gel permeation chromatography technique (GPC) (Waters 2414) and the GPC columns were eluted using THF with a flow rate of 1.0 mL/min at 40°C and calibrated with polystyrene standards. Morphological observations were taken using a scanning electron microscope (SEM, TM3030 Hitachi, Japan) at an accelerating voltage of 15 kV. The samples were fractured in liquid nitrogen and then coated with a thin layer of palladium prior to observations. For mechanical testing, dog-bone-shaped specimens of 70 mm in total length, type V based on standard ASTM D638 were tested using an Instron tensile machine, model 5969 with a load cell of 5 kN, to measure Young's modulus, tensile strength, and elongation at break. All tests were carried out under ambient conditions using a cross-head speed of 1 mm/min.

RESULTS AND DISCUSSION

Rheological and Molecular Weight Analysis

The plot of mixing torque values against time of PBS and PLA biocomposites with and without Joncryl® and peroxide were shown in Fig. 1. It was observed that the final torque values of both PBS and PLA reinforced with fiber 10 wt% showed the 15% increase when compared with their neat ones as seen from Fig. 1(a) and (c), respectively. In addition, with the addition 0.1, 0.5 and 1.0 phr Joncryl®, it showed an increase of final torque 17%, 27% and 22%, respectively for PBS composites and 24%, 37% and 64% for PLA composites when compared with their neat. PBS biocomposites with 0.05, 0.1 and 0.5 phr peroxide (Fig. 1(b)) showed the increase in the torque values after loading all ingredients into an internal mixer, indicating some degree of chemical reaction. The torque curve slightly decreased to the final torque values of 23%, 159% and 223% with increasing the peroxide contents, but was still higher than the neat PBS. In the case of PLA biocomposites with 0.05, 0.1 and 0.5 phr peroxide, from

Fig. 1(d), the final torque values increases to 12%, 9.5%, and 57% respectively, when compared with neat PLA. It was generally known that an increase of torque value was related to an increase of molten polymer viscosity, which was caused by an increase in molecular weight. On the other hand, it was also reported that torque value increased with the presence of various bonding agents in composites (fiber-polymer adhesion enhancing) [1]. The complex viscosity, storage modulus and the plot of storage, loss modulus (G', G") crossover frequencies of biocomposites having different reactive agents were shown in Fig. 2 (PBS) and Fig. 3 (PLA). The complex viscosity of neat PBS and PLA gradually decreased with increasing angular frequency, which is the typical shear thinning behavior. The incorporation of fiber and reactive agents gave rise to the higher viscosity for the whole frequency range and enhanced the shear thinning behavior comparing to neat polymers, therefore, shifted the Newtonian plateau to the lower angular frequency. It was well established that the addition of fiber into the polymer matrix increased the molten viscosity caused resistance against the flow. At an angular frequency of 0.1 rad/s in Fig. 2(a), the complex viscosity of PBS biocomposites with and without Joncryl® and 0.05 phr peroxide were about 4-fold but with the higher amount of peroxide exhibited 16-fold greater than neat PBS. The addition of the higher amount of peroxide in PBS biocomposite might cause more polymer-polymer interactions and increased resistance to flow, which was consistent with mixing torque values discussed earlier. In Fig. 2(b), the storage modulus values of PBS

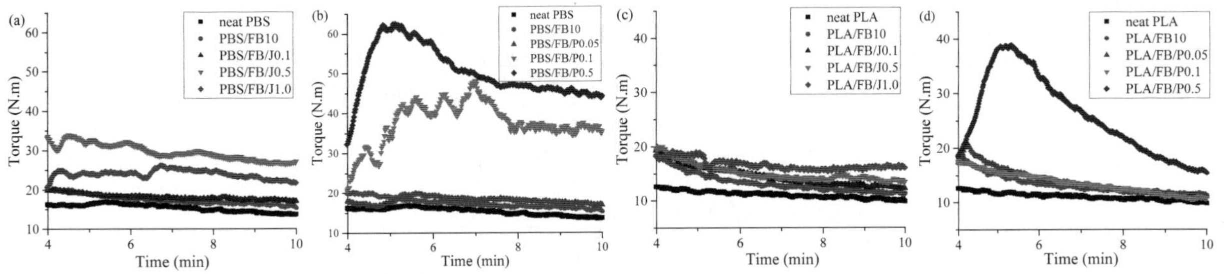

FIGURE 1. Mixing torque versus time after loading all ingredients together and continuous mixing until ten minutes of PBS and PLA composites with various content of Joncryl® (a), (c) and peroxide (b), (d), respectively.

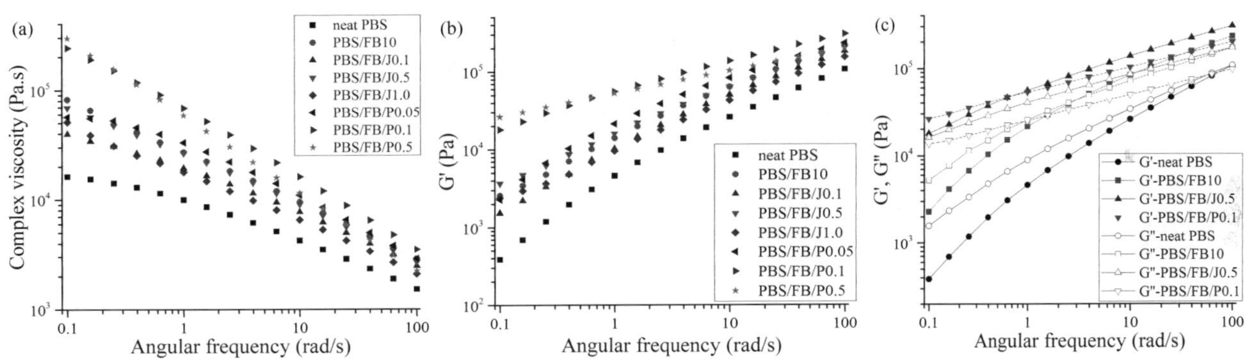

FIGURE 2. Complex viscosity (a), storage modulus (G') (b) and G'/G" cross-over frequencies of PBS biocomposites having different reactive agents and their contents (c), and having various content of peroxide (d) as a function of frequency.

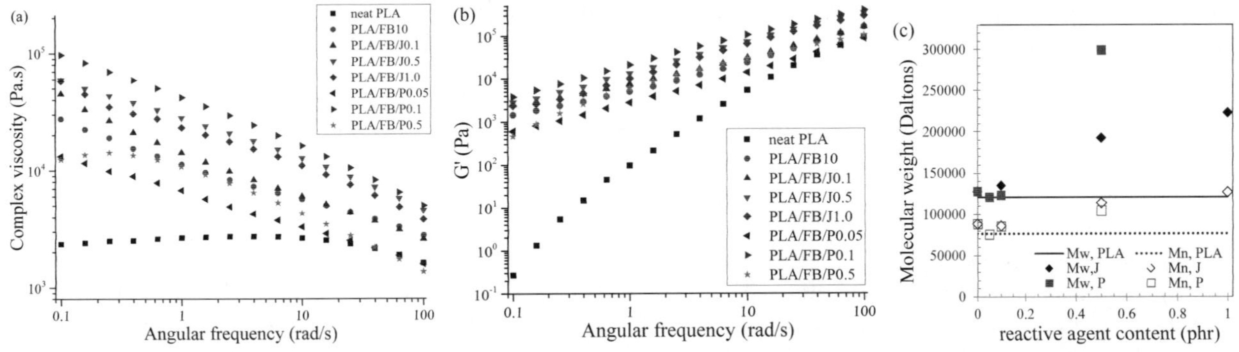

FIGURE 3. Complex viscosity (a) and storage modulus (G') (b) as a frequency dependence of PLA biocomposites having different reactive agents and their contents, and molecular weights of extracted PLA after compounding (c).

biocomposites showed a similar tendency to the complex viscosity. In Fig. 2(b), 3(b), both neat PBS and PLA showed the characteristic terminal liquid-like behaviors in the low-frequency region (G' α ω^2). However, the G' of the PBS and PLA biocomposites displayed a deviation from the liquid-like to solid-like behavior because of the slow relaxation process as noted in the decrease in the slope of G' at low frequencies. For the peroxide-PBS biocomposites system in Fig. 2(c), the G', G'' crossover frequency (ω_c) shifted towards lower frequency with the increased peroxide contents, suggesting the slower relaxation processes (higher relaxation time; $\tau_R = 1/\omega_c$) and enhance of melt elasticity. Therefore, it was apparently seen that the terminal region of PBS added 0.5 phr peroxide dramatically shifted to low frequencies and the visually widened plateau modulus referred as entanglements between polymer chains and M_W increased [5]. This implied that adding high peroxide content could result in crosslinking or branching of PBS. The G', G'' crossover frequency of PBS biocomposites with and without Joncryl® was insignificantly changed, similarly to PLA biocomposites with reactive agents (data not shown). From Fig. 3 (a-b), the complex viscosity and G' values increased with the Joncryl® contents increase and higher than adding peroxide (0.05 and 0.5 phr), compared to PLA biocomposites. This might be due to Joncryl® could epoxy groups reach chain extension reaction with PLA chain end groups as well and the possible reaction with the hydroxyl groups in the fiber [4]. This was confirmed by GPC results in Fig. 3(c), the molecular weight increased with the Joncryl® contents increased. Although adding 0.1 phr peroxide to the PLA biocomposites showed the highest complex viscosity and G' values, the only high content of peroxide added showed the high molecular weight of PLA. This might be caused by a good fiber-PLA adhesion and/or long-chain branches of PLA occurred when 0.1 phr peroxide was added. As reported by Henton et al. [1], the small amounts of peroxide resulted in long-chain branches of PLA via free radical.

Morphological Observations

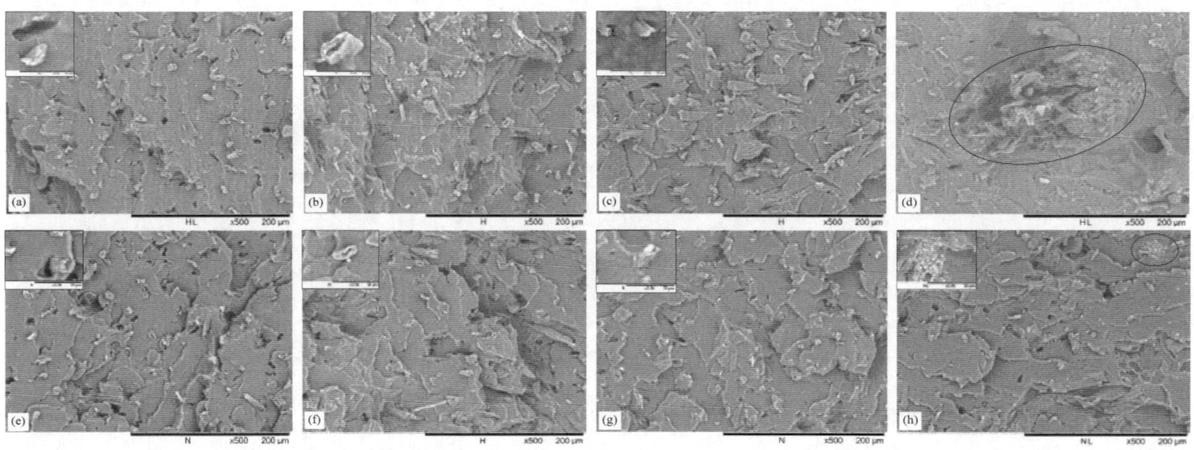

FIGURE 4. SEM micrographs of PBS biocomposites with 10 wt% fiber (a), having Joncryl® 0.5 phr (b), peroxide 0.1 phr (c) and 0.5 phr (d), and PLA biocomposites with 10 wt% fiber with (a), Joncryl® 0.5 phr (f), peroxide 0.1 phr (g) and 0.5 phr (h).

Fig. 4 depicted the SEM micrographs of the PBS and PLA biocomposites with and without reactive agents. When Joncryl® and peroxide were added, less voids and cavities were observed except for 0.5 phr peroxide in PBS biocomposites showing some aggregation of fibers. The decrease in fibers dispersion might be due to a high amount of peroxide affected to PBS matrix reaction, which led to a great increase in torque and complex viscosity values. Whereas, adding 0.5 phr peroxide to PLA biocomposites displayed an aggregation of peroxide. This indicated that small amount of Joncryl® and peroxide might improve the interfacial adhesion between fiber and matrix.

Mechanical Properties

Mechanical properties of PBS and PLA biocomposites with and without reactive agents were shown in Fig. 5. Fig. 5(a) presented an increase of modulus of both PBS and PLA biocomposites, about 1.5-fold and 1.2-fold, respectively, higher than for their neat. It could be seen that more increased modulus when Joncryl® and peroxide were added, except for 0.5 phr peroxide loading. In Fig. 5(b), the tensile strength values of PBS biocomposites only showed an increase of 2% and 4% compared to neat PBS when adding 0.1 phr peroxide and 0.5 phr Joncryl® respectively. From Fig. 5(c), it displayed 6% increase in tensile strength for PLA with 10 wt% fibers compared to neat PLA. The significant increase in tensile strength of 16-19% was obtained upon adding Joncryl® while the presence of the small amount of peroxide caused the increase in tensile strength of 13% compared to neat PLA. A general increase in the modulus and strength was observed in the Joncryl®/PLA system due to the increased molecular weights [6] and also the good fiber-polymer adhesion. The elongation at break of PBS biocomposites decreased, in particular with the addition of 0.5 phr peroxide whiles the elongation at break of PLA biocomposites was maintained. The decrease in mechanical properties when adding 0.5 phr peroxide of both biocomposites might be due to a high-stress concentration from the fiber aggregation (in PBS) and peroxide (in PLA) as observed by SEM micrographs.

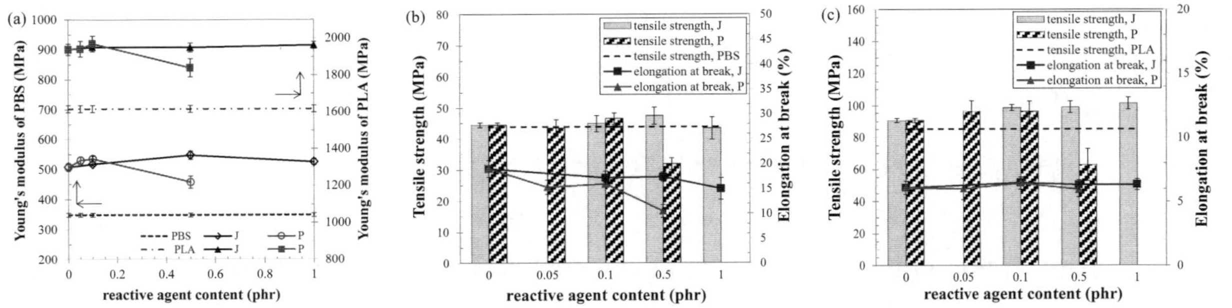

FIGURE 5. Effects of fiber with and without reactive agents on Young's modulus of PBS and PLA biocomposites (a), tensile strength and elongation at break of PBS biocomposites (b) and PLA biocomposites (c).

CONCLUSION

The presence of Joncryl® in PLA biocomposites caused the increase of molecular weights, mixing torque, complex viscosity, storage modulus and mechanical properties when compared to the addition of peroxide. The increase of complex viscosity of PLA biocomposites added Joncryl® was due to the increase in the population of the higher molecular weights as confirmed by GPC analysis. In the case of PBS biocomposites, it showed a dramatic increase in the mixing torque, complex viscosity and storage modulus with decreasing in crossover frequencies with the increase of peroxide content rather than adding Joncryl®. Peroxide was extremely reactive with PBS than with PLA at the same concentrations. The SEM micrographs demonstrated that the compatibility between the polymer matrix and fiber was improved when adding a small amount of Joncryl® and peroxide for both polymers, leading to an improvement of their modulus and tensile strength. However, it revealed some aggregated fibers when adding the high contents of peroxide up to 0.5 phr, which resulted in a decrease in modulus and tensile strength.

ACKNOWLEDGMENTS

The authors would like to gratefully thank the department of Materials Science and Engineering, Faculty of Engineering and Industrial Technology, Silpakorn University, Thailand for financial supports. We are aslo indebted to SCG Packaging PLC., Thailand for paper pulps supply. National Research Council of Thailand was gratefully acknowledged for financial supports under Bioplastics framework.

REFERENCES

1. A.K. Mohanty, M. Misra, and L.T. Drzal, *Natural Fibers, Biopolymers, and Biocomposites*. 2005: CRC Press.
2. M.M. Kabir, et al., *Chemical treatments on plant-based natural fibre reinforced polymer composites: An overview.* Composites Part B: Engineering, 2012. **43**(7): pp. 2883-2892.
3. R. Al-Itry, K. Lamnawar, and A. Maazouz, *Improvement of thermal stability, rheological and mechanical properties of PLA, PBAT and their blends by reactive extrusion with functionalized epoxy.* Polymer Degradation and Stability, 2012. **97**(10): pp. 1898-1914.
4. Q. Meng, M.-C. Heuzey, and P.J. Carreau, *Control of thermal degradation of polylactide/clay nanocomposites during melt processing by chain extension reaction.* Polymer Degradation and Stability, 2012. **97**(10): pp. 2010-2020.
5. C. Liu, et al., *Evaluation of different methods for the determination of the plateau modulus and the entanglement molecular weight.* Polymer, 2006. **47**(13): pp. 4461-4479.
6. N. Najafi, M.C. Heuzey, and P.J. Carreau, *Polylactide (PLA)-clay nanocomposites prepared by melt compounding in the presence of a chain extender.* Composites Science and Technology, 2012. **72**(5): pp. 608-615.

PROPERTIES AND PROCESSING OF PLA/ETHER-AMIDE COPOLYMER BLENDS

THIBAULT JAUZEIN[A], MICHEL A. HUNEAULT[B], MARIE-CLAUDE HEUZEY[A]

A) Chemical Engineering Department, École Polytechnique de Montréal, Montréal, Canada
B) Chemical and Biotechnological Engineering Department, Université de Sherbrooke, Sherbrooke, Canada

Biobased polymers are widely studied as substitutes for petroleum-based materials. Among these polymers, polylactide (PLA) has gained commercial acceptance in thermoformed and blow molded packaging applications because of its transparency, high tensile modulus and strength, and its cost competitiveness. However, the low impact strength of PLA and its poor temperature resistance have prevented its widespread use in some industrial applications.

Many attempts were made to increase the impact strength of PLA using impact modifiers. Brittle matrices can be toughened by mixing with polyolefins, polyolefin based copolymers, poly-ε-caprolactone and ether-amide copolymers. At low content, the latter seem to be the most interesting. Important enhancements were reported in impact strength of PLA using such additives, while the mechanical properties retained relatively high values. Toughening semi-crystalline matrices has also been shown to be more effective than for their amorphous counterparts.

The present study focuses on the improvement of impact properties and particularly on the interaction between crystallinity development and mechanical properties of PLA-based blends. The PLA was toughened by the addition of a random linear ether-amide copolymer (Pebax 3533™) that acted as an impact modifier that we will refer to as EAC. A random copolymer of ethylene, methyl-methacrylate and glycidyl-methacrylate (Lotader AX8900™) was also used to reactively compatibilize the ether-amide copolymer with the PLA matrix. The rheological, mechanical and thermal properties of quenched and annealed blends and their morphology were investigated. The crystallinity of the samples was increased by annealing. The annealing time was varied to investigate the evolution of the mechanical and thermal properties of PLA and its blends versus crystallinity.

In terms of melt rheology, the blends were investigated in oscillatory shear to determine if the impact modifiers could increase the elasticity of the material. The blends exhibited higher viscosity and storage modulus at low frequency than the neat resins (PLA and EAC). The formation of a network could be the reason of this improvement, since the compatibilizer and EAC can chemically react with PLA. The extended chains will have more entanglements, which will lead to higher viscosity at low frequency. As shown on Figure 1, a micron-size dispersion of the impact modifier was achieved in the presence of the reactive compatibilizer. Besides, the compatibilized blend exhibited better adhesion between the additives and the matrix. Tensile properties and the notched Izod impact resistance were measured using standard testing protocols. No change in fracture mechanism was observed between the compatibilized and non-compatibilized blends. The blends exhibited relatively high tensile properties with a Young's modulus higher than 2 GPa. Except for their elongation at break, the tensile properties of the blends remained constant at different crystallinities. However, large increase in impact strength was observed for the blends with 10% of additives. The annealed blends showed even greater improvement. Annealed samples showed the highest impact strength with values attaining 68 kJ/m².

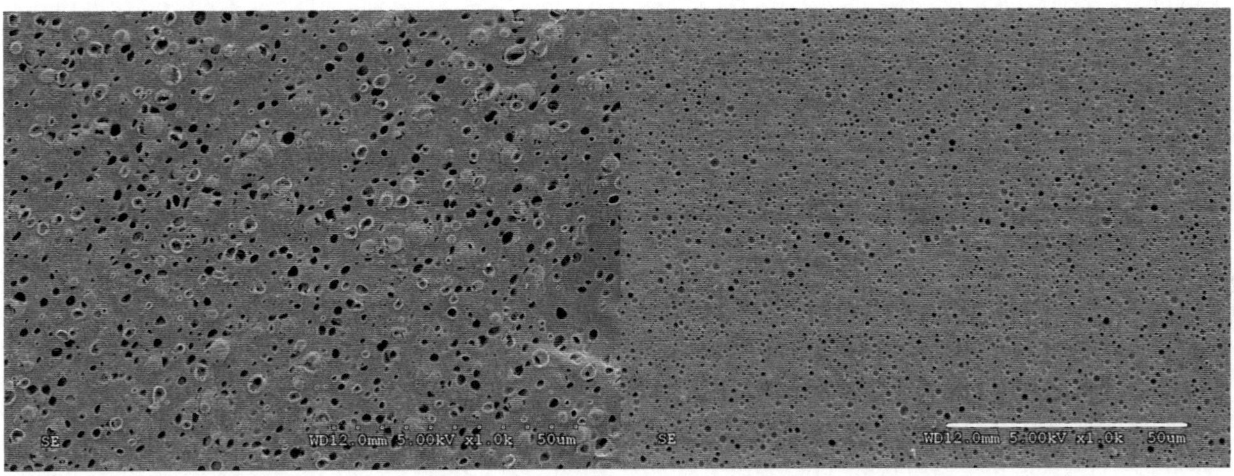

Figure 1 : SEM images of microtomed (left) non-compatibilized blend and (right) compatibilized blend etched with formic acid for 6h. Shown on the bottom right of the figure a 50 μm scale bar.

Sol-Gel Transition and Scaling Laws for κ-Carrageenan in Aqueous Solution

Sijun Liu, Lin Li*

School of Mechanical and Aerospace Engineering, Nanyang Technological University,
50 Nanyang Avenue, Singapore 639798

Presenting author: sjliu@ntu.edu.sg
Reference author: mlli@ntu.edu.sg

Abstract. Rheological properties and scaling laws for κ-carrageenan in aqueous solution were studied by rheology. The critical gel concentration, c_g, obtained using the Winter-Chambon criterion, was found to be proportional to temperature as expressed by $c_g \sim T^{0.85}$ where T is the temperature in °C. At the gel point, the critical relaxation exponent n obtained was a constant (n = 0.62) and independent of temperature. The critical gel strength S_g increased with increasing c_g. A constant gel strength S_g/c_g was obtained by normalizing S_g with c_g to eliminate the effect of temperature, showing a unique character of κ-carrageenan in aqueous solution during gelation. The molecular structure of the junctions at the gel point was analyzed using the modified Eldridge-Ferry model, which supported the similarity of the fractal structure in the κ-carrageenan gels.

Keywords: κ-carrageenan, viscoelasticity, scaling law, rheology, gelation
PACS: 83.60

SOL-GEL TRANSITION

κ-Carrageenan is able to form a strong gel at a high concentration. Figure 1 displays the dependence of storage modulus G' and loss modules G" on the angular frequency ω for aqueous solutions of various κ-carrageenan concentrations at a fixed temperature of 20 °C. A frequency dependence of G' and G" was observed for the low κ-carrageenan concentrations such as 0.75, 0.83 and 0.92 wt%. The solution of the lowest κ-carrageenan concentration (0.75 wt%) shows a liquid-like terminal behavior at low frequencies, $G' \sim \omega^2$ and $G'' \sim \omega^1$ (at $\omega \to 0$). When κ-carrageenan concentration increases to 1 wt%, both G' and G" are higher than those at 0.92 wt% in the whole frequency region and the slope of G' tends to 0, indicating the formation of a gel network from the solution. With further increasing κ-carrageenan concentration, G' and G" become independent of frequency, showing the characteristic of a solid-like gel. The large increase in slope from the G' curve at 0.92 wt% to that at 1 wt% at low frequencies implies that the sol-gel transition of κ-carrageenan should take place at a κ-carrageenan concentration between 0.92 wt% and 1 wt%.

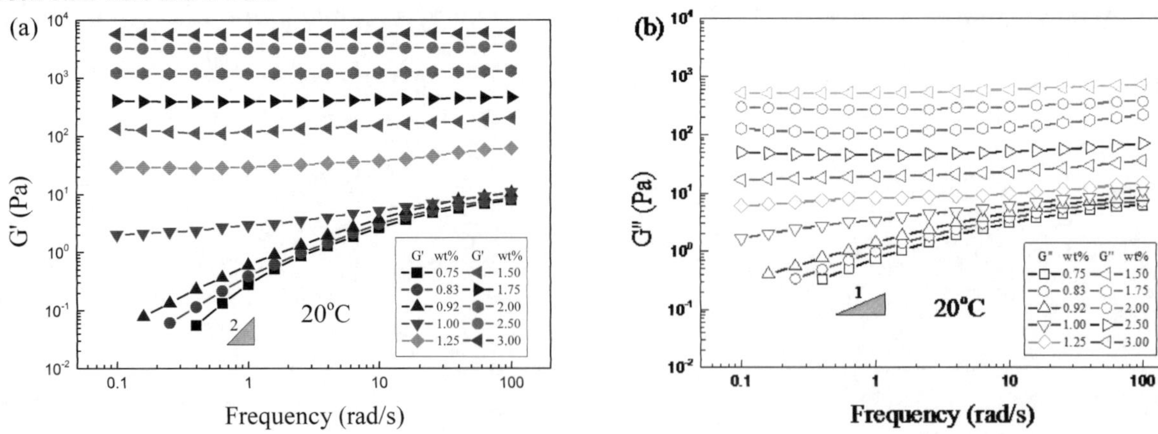

FIGURE 1. Dependence of (a) storage modulus G' and (b) loss modulus G" on angular frequency ω for the solutions of various κ-carrageenan concentrations at a fixed temperature 20 °C and a shear strain amplitude 1 %.

From Figure 1, it is apparent that the sol-gel transition takes place between 0.92 wt% and 1 wt% κ-carrageenan at 20 °C. To determine the exact critical gel concentration, we follow a procedure defined by the Winter-Chambon criterion.[1] This method was first found experimentally in chemical gels with a scaling law of $G'(\omega) = G''(\omega) \sim \omega^{1/2}$ at the gel point, and then generalized to satisfy $G'(\omega) \sim G''(\omega) \sim \omega^{n}$ $(0 < n < 1)$ for other gel systems under a sufficient low frequency, ω. This implies that the phase angle becomes frequency independent at the gel point. Many researchers have shown that this method is reliable and valid on determination of the gel point for various polymeric gels with different gelation mechanisms, such as poly(vinyl chloride)/bis(2-ethylhexyl) phthalate (PVC/DOP) gel by micro-crystalline,[2] poly(ethylene oxide)-poly(propylene oxide)-poly(ethylene oxide) (PEO-PPO-PEO) hydrogel by the arrangement of micelles,[3] and dibenzylidene sorbitol gel by self-assembly.[4] However, it has not been reported whether the gel point can be determined by this method for a κ-carrageenan gel formed through the aggregation of double helices. Figures 2(a) shows the application of such a criterion to κ-carrageenan solution within the sol-gel transition range at 20 °C. The gel points were determined through the multi-frequency plots of loss tangent versus κ-carrageenan concentration. All tangent delta curves pass through the common point at a certain κ-carrageenan concentration 0.97 wt%, which is defined as the critical gel point, c_g. In this way, we obtained the critical gel concentrations of 0.55, 1.25 and 1.87 wt% of κ-carrageenan at 10, 30 and 40 °C, respectively. Through a linear fitting to the data, we obtained an approximate power law for the temperature dependence of the critical gel concentration as shown in Figure 2(b).

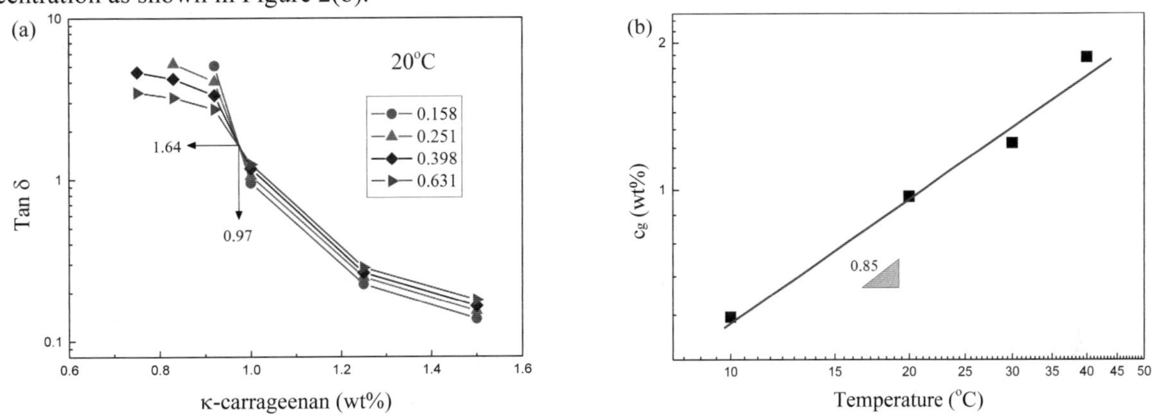

FIGURE 2. (a) Dependence of loss tangent, tan δ, on κ-carrageenan concentration at 20 °C, and (b) Change of critical gel concentration c_g with temperature.

Scaling Laws

As is well-known, a polymeric gel is a three-dimensional network formed from flexible chains through ionic interaction, hydrogen bonding, van der Waals forces, hydrophobic association, self-assembly, or cross-linking by crystalline segments. The critical change from a liquid state to a solid state is known as the sol-gel transition at which the zero shear viscosity and relaxation time of the system diverge. The well-used experimental and theoretical methods for description of the gelation are based on the scaling laws. To examine and explain rheological properties in the vicinity of the gel point, the following rheological scaling law has been widely used[5]

$$G''(\omega)/G'(\omega) = \tan\delta = \tan(n\pi/2) \quad (1)$$

which is based on the power law behavior of the shear relaxation modulus G(t) at the gel point

$$G(t) = S_g t^{-n} \quad (2)$$

where n is named as the critical relaxation exponent, which determines the stress relaxation rate at the gel point. S_g is denoted as the critical gel strength. A similar expression can also be applied for G' and G" at the critical gel point

$$G'(\omega) = G''(\omega)/\tan(n\pi/2) = S_g\omega^{n}\Gamma(1-n)\cos(n\pi/2) \quad (3)$$

Here Γ(1-n) is the Gamma function. By knowing n, one can calculate S_g from G'(ω) or G"(ω) at the critical gel point using equation 3.

Using equation 2, we obtained n values at different temperatures as shown in Figure 3(a). It is apparent that n is independent of temperature. The universal value of n means that the sol-gel transition takes place at the same viscoelastic distance from a completely viscous liquid (n=1) or from a completely elastic solid (n=0). From Figure 2, we know that c_g increases with increasing temperature. Therefore, it can be inferred that more junctions were formed during gelation at a higher temperature in order to keep the same viscoelastic distance at the gel point. On the other hand, the physical nature of a gelling system at the gel point can also be described by the critical gel strength S_g defined by equation 3. We found that S_g increases as temperature increases, and a linear relationship between S_g and temperature was obtained. This result indicates that S_g is related to the total mass of κ-carrageenan during gelation. For a constant n, increasing the cross-link density of a chemical gel usually leads to an increase in the gel strength. The increase of S_g with increasing temperature may be considered to be due to the increase of the junction density at the gel point. Meanwhile, it is surprising to find that the slope 0.85 in the relationship of c_g with temperature is very close to the slope 0.87 in the relationship of S_g with temperature. That is, S_g is only related to c_g in the system and thus the gel strength per unit of mass of κ-carrageenan should be the same. Here, we normalized S_g by the corresponding c_g to obtain the result in Figure 3 (b). As expected, S_g/c_g is almost constant and independent of temperature. Hence, it can be concluded that the increase in c_g with increasing temperature just induces the increase in the junction density but it has a weak influence on the junction size. This result also suggests that the critical κ-carrageenan gels have the similarity in their fractal structures.

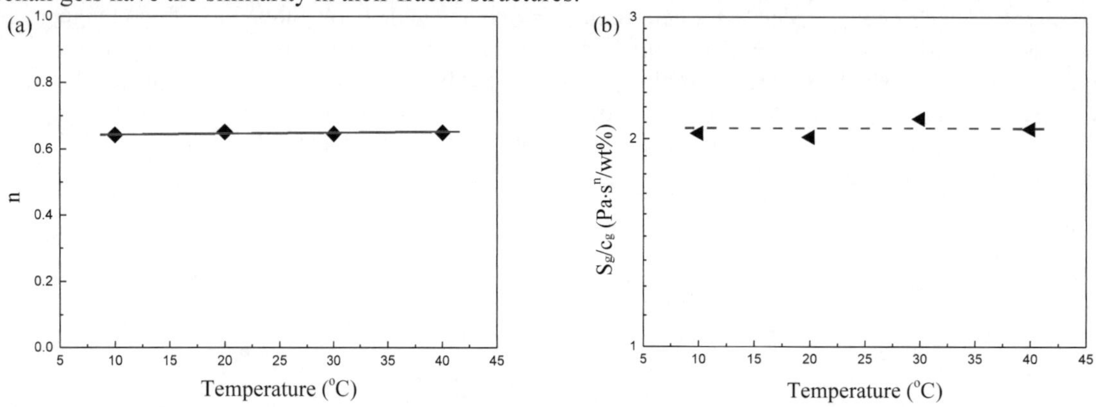

FIGURE 3. (a) Dependence of the critical gel exponent n on temperature, and (b) Relationships of the normalized critical gel strength S_g/c_g with temperature.

Junction Structure

It is of interest to know the junction structure at the gel point because the junction structure may be linked with the gelation mechanism and also has a relationship of viscoelastic properties. Experimentally, it is quite technically difficult to measure the junction structure because of the complexities in gel networks. A theoretical model based on the junction multiplicity has been proposed to analyze the junction structure in thermoreversible gels. A modified Eldridge-Ferry model has been developed and used to describe a multiple junction structure that is composed of s chains and ζ monomers (repeat units) of each chain,[6] where s is called the junction multiplicity and ζ is the junction length, as follows:

$$\ln c_g = \zeta \Delta h/(k_B T) + \ln M/(s-1) + \text{constant} \qquad (4a)$$

Here Δh is the enthalpy of bonding per mole of the repeat units taking part in the junction, k_B is the Boltzmann constant, T is the absolute temperature, and M is the molecular weight of the given gelling polymer. Although a real junction structure may not be as simple as that described by this model, the relation derived from the model is still of value as a measure for estimation of the junction structure, especially for the case where the junction structure is invariable with concentration, temperature, and molecular weight. In the present study, because the molecular weight of κ-carrageenan is constant, it is hard to evaluate s by the dependence of c_g on M at a given temperature. The modified Eldridge-Ferry model can be rewritten

$$\ln c_g = \zeta\, \Delta h/k_B T + \text{constant} \qquad (4b)$$

Now, c_g is just a function of temperature. Using equation 4b, the temperature dependence of a gel with a constant molecular weight M allows us to calculate the junction length ζ. Using the values of c_g in Figure 2(b), c_g was plotted against $10^3/T$ in Figure 4. The solid straight line was given by a linear fitting to the data and then we directly obtained the slope of -3.48. According to equation 4b, the slope is related with ζ by $\zeta = |\text{slope}| 10^3 R/\Delta h$, where R is the gas constant [$10^3 R = 8.31$ kJ/(mol K)] and Δh has a unit of kJ/mol. To calculate ζ, one needs to know

Δh. In this work, we use micro-differential scanning calorimetry to determine the heat of gel melting. The results show that ζ is equivalent to 17.4, which means that the junction structure comprises 17.4 repeat units or monomers of each polymer chain.

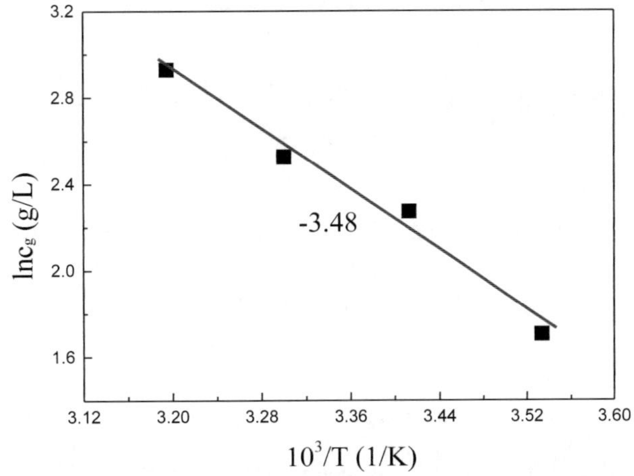

FIGURE 4. lnc_g against $10^3/T$ for the κ-carrageenan gels.

On the basis of our experimental results and the modified Eldridge-Ferry model, a schematic mechanism is proposed in Figure 5 for κ-carrageenan gelation with increasing c_g. At a high temperature, κ-carrageenan molecules exist as random coils in aqueous solution. As temperature decreases, κ-carrageenan molecules at low concentrations are still present as coils, and the solutions show a Newtonian fluid behavior. For the aqueous solutions of κ-carrageenan with high concentrations, however, temperature induced the conformational transition of coils to helices, followed by the aggregation of double helices. The aggregation of double helices leads to the formation of small domains containing a limited number of double helices, which then become physical cross-links of a three-dimensional gel network. With increasing temperature, more κ-carrageenan molecules are required to form a denser gel network while the junction size is independent of c_g. Based on the modified Eldridge-Ferry model, the junction structure calculated contains 17.4 repeat units in each polymer chain.

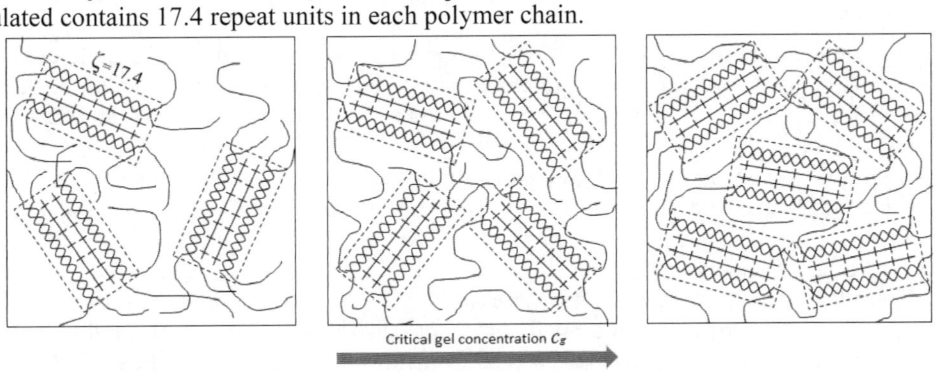

FIGURE 5. A schematic diagram describing the sol-gel transition mechanism of κ-carrageenan in aqueous solution with increasing the critical gel concentration

ACKNOWLEDGMENTS

This work was supported by the Academic Research Fund Tier 1 (RG100/13) from the Ministry of Education, Singapore.

REFERENCES

1. M. Mours, HH. Winter, Macromolecules **29**, 7221-7229 (1996).
2. L. Li, Y. Aoki, Macromolecules **30**, 7835-7841 (1997).
3. SJ. Liu, L. Li, ACS Applied Materials & Interfaces **7**, 2688-2697 (2015).
4. SJ. Liu, W. Yu, CX. Zhou, Soft Matter **9**, 864-874 (2013).
5. HH. Winter, F. Chambon, Journal of Rheology **30**, 367-382 (1986).
6. F. Tanaka, K. Nishinari, Macromolecules **29**, 3625-3628 (1996).

Processing of Cassava Pulp to Glucose and Oligosaccharides by Sonication Process

Cucuk Evi Lusiani [a], Eva Oktavia Ningrum [a], Prida Novarita Trisanti [a], Sumarno Sumarno [a*]

[a] *ITS, Institut Teknologi Sepuluh Nopember - East Java - Indonesia*

cucukevilusiani@gmail.com
onramus@chem-eng.its.ac.id

Abstract. One of tapioca industry solid waste known as cassava pulp having a high amount of biopolymers i.e. 65.5% starch, and 8.1% cellulose. These biopolymers could be processed by sonication, using H and OH radicals to attack glycosidic bond of starch and cellulose, to obtain glucose and oligosaccharides. The aim of this research is to study the effect of reaction time and temperature of sonication process on the glucose and oligosaccharides product, and cassava pulp morphology. Initially, suspension of cassava pulp in water (1/20 w/v) was heated continued by sonication process for desired time. The sonication were done from 0 to 90 minutes at temperature 40 and 50°C. The result indicated that the H and OH radicals could break polysaccharide chains during sonication process so that the concentrations of reducing sugars increased. Moreover, the starch granules can be forced out from fibrous matrix of treated cassava pulp as shown by its morphology.

Keywords: Cassava pulp, sonication, glucose, oligosaccharides.
PACS: 82.35.Pq

INTRODUCTION

Tapioca industry produces about 10-15% solid waste known as 'cassava pulp [7]. The compositions of cassava pulp are 65.5% starch, 8.1% cellulose, 2.8% hemicellulose, and 2.2% lignin (dry basis) [1]. Starch and cellulose from cassava pulp, non-food carbohydrate sources, could be converted to be chemicals and other products such as glucose and oligosaccharides.

In laboratory scale, the degradation methods of cassava pulp have been developed using acid [8] and enzymatic method to convert it to be glucose and oligosaccharides. Therefore in case of acid process, *5-hydroxymethylfurfural* is formed (further degradation) and produce acid waste. Moreover, enzymatic process require a long reaction time, difficulty of maintenance and high operational cost [9].

To obtain a higher yield of reducing sugars, it is necessary to develop the degradation process with pretreatment in order to facilitate the breaking of glycosidic bonds of polysaccharides. Recently, the cassava pulp degradations have been done by enzymatic process with pretreatment sonication [7], and combination of sonication and acid process [6]. Sriroth, et al. (2000) investigated about cassava pulp processing to improve biomass utilization by sonication and enzymatic process [7]. Sonication process was employed to disrupt cassava pulp physically. This study reported that starch granules could be forced out from fibrous matrix completely after 40 s of sonication. Simanjuntak, et al. (2014) studied about producing reducing sugars from cassava pulp by simultaneous sonication and acid hydrolysis [6]. This study was carried out under 40 kHz ultrasound irradiation, and the pH of solution was varied from 1 up to 4. The result indicated that the highest reducing sugars concentration was obtained at pH 2, 90 min of hydrolysis time, and 80°C.

Chemical and physical structure of polysaccharides such as crystallinity degree, viscosity, solubility and pH of solution are changed by sonication process[5]. Ultrasound waves (> 20 kHz) from sonication process could force out starch granules from fibrous matrix of cassava pulp[7], cut intermoleculer bond of starch[5], and improve solubility cassava pulp in water.

From previous studies, degradation of casssava pulp was done by sonication process as pretreatment and adding chemicals to produce reducing sugars. In this work, the cassava pulp in water is degraded by using sonication process. The effect of reaction time and temperature of sonication process was studied by determination the concentration of reducing sugars. The morphological change of cassava pulp was also observed by *Scanning Electron Microscopy* (SEM).

Materials and Methods

A. Materials

Cassava pulp was obtained from home industry located in Kediri, East Java, Indonesia. Dried cassava pulp was ground and sieved through a 40 mesh screen. The DNS method for estimating the concentration of reducing sugars in a sample was originally invented by G. Miller [3] in 1959 using chemicals such as D-Glucose as standard solution, 3,5-dinitrosalicylic acid (DNS) reagent, Sodium Hydroxide, and Potassium Sodium Tartrate. These chemicals were obtained from Sigma Aldrich Corporation.

B. Sonication process

Thirty grams of 40 mesh cassava pulp were mixed in 600 mL of distilled water (1:20 w/v). The suspension of cassava pulp in water was heated while stirred (250 rpm). Probe of sonicator was installed into the reactor after the temperature of system was reached. Cassava pulp samples were then sonicated under continuous stirring (250 rpm) for ~90 minutes at 40 and 50°C. Sonicator is ultrasonic processor VCX 500 Sonics and Materials Inc., USA (500 W, 20 kHz, 50 % Amplitude) completed with Titanium Alloy probe transducer. The sonication process is showed in **FIGURE 1**.

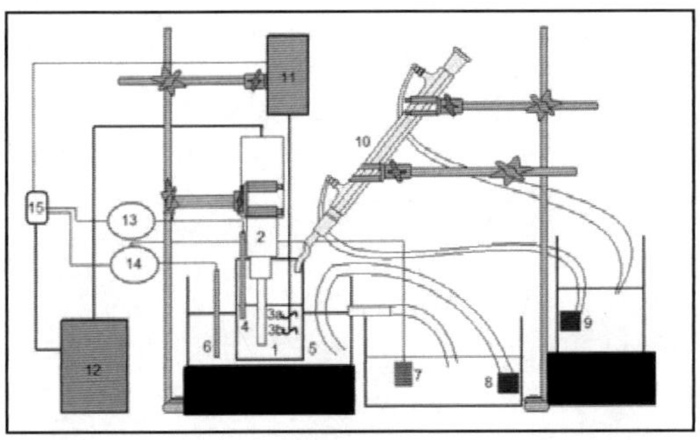

FIGURE 1. Schematic Diagram of Sonication Process

1. Ultrasonic reactor
2. Ultrasonic probe
3. Impeller
4. Thermocouple (reactor)
5. Waterbath
6. Thermocouple (waterbath)
7. Heater
8. Pump (water bath)
9. Pump (condenser)
10. Reflux condenser
11. Stirring Motor
12. Ultrasonic Generator
13. Controller Temperature
14. Controller Temperature
15. Power

C. Product Analysis

Solid product of sonicated cassava pulp was separated from liquid product by centrifuge. The solid sample was freeze-dried and analyzed by *Scanning Electron Microscopy* (SEM) to observe the morphology change. The liquid product from sonication process was analyzed by spectrophotometer UV-Vis using DNS method to measure reducing sugars concentration.

Results and Discussion

Cassava pulp was sonicated for various sonication time from 0 up to 90 minutes at 40 and 50°C. The effect of sonication time on the concentration of reducing sugars is shown in **FIGURE 2**.

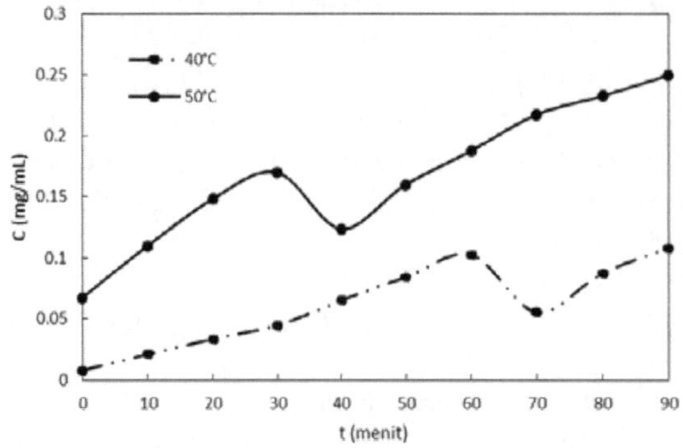

FIGURE 2. Effect of Sonication Time (t) on The Concentration of Reducing Sugars (C)

The reducing sugars concentration increase with sonication time for 0 to 90 minutes at 40 and 50°C. However, the reducing sugars decrease at 70 minutes (40°C) and 40 minutes (50°C). It was caused starch granules that were outside of fiber matrix began further degraded. The concentrations of reducing sugars increase again at 80 minutes (40°C) and 50 minutes (50°C). This can be explained due to the starch granules that were trapped in the fiber matrix were forced out because of fiber matrix swelling then the glycosidic bond of starch were attacked by ultrasound waves and degraded it to be reducing sugars. The degradation of starch by sonication is shown in **FIGURE 3**.

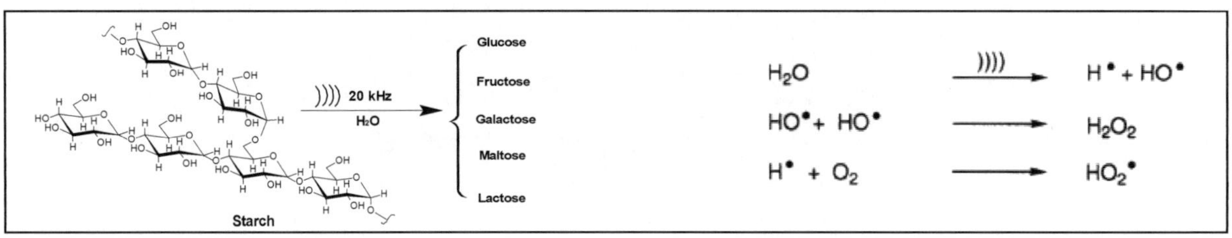

(a) (b)

FIGURE 3. Reaction Mechanisms of (a) Starch by Sonication Process and (b) Water in Sonication Process (Kardos and Luche, 2001)

The solid sample of sonicated cassava pulp was analyzed by SEM to observe morphology change before and after sonication process. It is shown in **FIGURE 4**.

(a) (b)

FIGURE 4. The Morphology of (a) Native Cassava Pulp and (b) Sonicated Cassava Pulp at 40˚C for 60 minutes

Morphological analysis of cassava pulp shows that the structure of native cassava pulp is intact. Starch granules partially trap in the fibrous matrix of cassava pulp and some of it is still outside of fiber matrix. After sonication process, the fibrous matrix of cassava pulp was swollen and the starch granules inside of fiber matrix could be forced out from the matrix. The illustration of this process is shown in **FIGURE 5**.

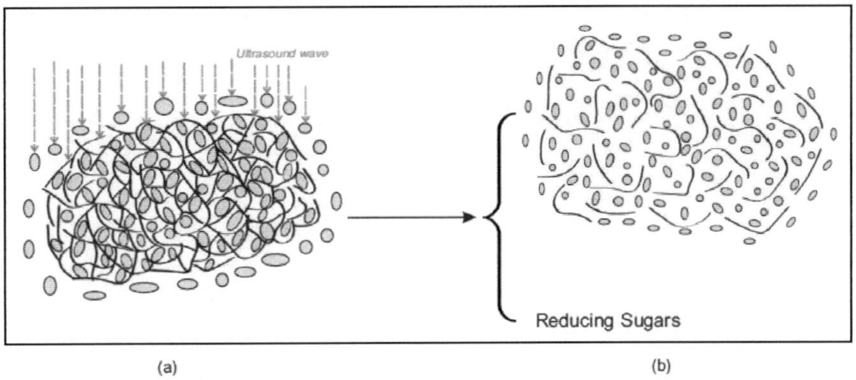

FIGURE 5. The Effect of Ultrasound Waves on (a) Native Cassava Pulp

There are three possibilities phenomena that might occur during sonication process on cassava pulp. First, the breaking of glycosidic bond of starch granules in the outside of fiber matrix to be reducing sugars. Second, the intensity of ultrasound lead to numerous bubble jets produced around the fibrous matrix, and cause swelling the matrix. Third, forcing out of the starch granules from fiber matrix. Ultrasound can degrade the starch since the granules forced out from fiber matrix become easier broken to be reducing sugars.

Conclusions

The concentration of reducing sugars increase with sonication time because of the breaking of glycosidic bond of starch granules. It is proved by morphological of cassava pulp after sonication process. The starch granules outside of fiber matrix were absent in the sonicated cassava pulp due to this starch degraded to be reducing sugars. Moreover, the starch granules inside of fiber matrix were forced out because of fiber matrix swelling then the glycosidic bond of starch were attacked by ultrasound waves and degraded it to be reducing sugars.

ACKNOWLEDGMENT

The authors gratefully acknowledge to Indonesian Scholarship Institution, LPDP (Lembaga Pengelola Dana Pendidikan) for financial support in this research.

REFERENCES

1. Djuma'ali et al., *Cassava Pulp as A Biofuel Feedstock of An Enzymatic Hydrolysis Process*. Makara Teknologi, 2011; pp. 183-192.
2. E. Hermiati et al., *Improvement of Microwave-assisted Hydrolysis of Cassava Pulp and Tapioca Flour by Addition of Activated Carbon*, *Carbohydrate Polymers*, 2012, pp. 939-942.
3. G. Miller, *Use of Dinitrosalicylic Acid Reagent for Determination of Reducing Sugar*, *Analytical Chemistry*, 1959, pp. 426-428.
4. R. Hoover, *Acid-Treated Starches*, *Food Reviews International*, 2000, pp. 369-392.
5. N. Kardos, and J. L. Luche, *Sonochemistry of Carbohydrate Compounds*, *Carbohydrate Research,* 2001, pp. 115-131.
6. Z. Luo, Fu, X, He, X, Luo, F, Gao, Q and Yu, S. Effect of Ultrasonic Treatment on The Physicochemical Properties of Maize Starches Differing in Amylose Content. *Starch/Starke*, 2008, pp. 646-653.
7. W. Simanjuntak, H. Satria, and N. Utami, *Production of Reducing Sugar from Cassava Solid Waste by Simultaneous Ultrasonication and Acid Hydrolysis*, *Indo. J. Chem*, 2014, pp. 233-238.
8. K. Sriroth et al., *Processing of Cassava Waste for Improved Biomass Utilization*, *Bioresource Technology*, 2000, pp. 63-69.

9. A. L. Woiciechowski et al., *Acid and Enzymatic Hydrolysis to Recover Reducing Sugars from Cassava Bagasse: an Economic Study*, *Brazilian Archivesof Biology and Technology*, 2002, pp. 393-400.
10. Y. Yu et al., *Some Recent Advances in Hydrolysis of Biomass in Hot-Compressed Water and Its Comparisons with Other Hydrolysis Methods*, *Energy & Fuels*, 2007, pp. 46-60.

Production of hydrogel wound dressing with neomycin using different gamma radiation doses

Angélica Tamião Zafalon [*], Vinícius Juvino dos Santos, Duclerc Fernandes Parra and Ademar Benevolo Lugão

Nuclear and Energy Research Institute, IPEN-CNEN/SP, Av. Prof. Lineu Prestes, 2242, Cidade Universitária, CEP 05508-000, São Paulo – SP,

angelicatamiao@gmail.com

Abstract. Hydrogels are synthetic systems that are highly biocompatibility with soft tissues, this materials can be processed using gamma radiation. The crosslinked-network materials have been extensively used for wound dressing due to swelling capacity in presence of water and drug release system. The aim of the presente study was to prepare a wound dressing containing nanosilver (NpAg) in composition of poly (N-vinyl-2-pyrrolidone) (PVP), poly (ethylene glycol), agar using gamma irradiation to corsslink and sterilization. The influence of irradiation process dose were investigated. The samples were irradiated at dose of 15, 20 and 25 kGy. The gel fraction and maximum swelling were performed using physicochemical methods. The neomycin drug release was determined by liquid chromatography – mass spectrometry.

.

Keywords: Hydrogel, Poly (N-vinyl-2-pyrrolidone), Neomycin
PACS: 81

INTRODUCTION

Hydrogels are three-dimensional networks formed from polymers and copolymers crosslinked to form insoluble polymeric matrices. Hydrogels are usually soft, elastic, with ability to retain great quantities of water without losing their integrity. They are used in various biomedical applications, as vehicles for drug delivery of anti-inflammatory, antibiotic and growth factor, and tissue engineering 1-6.
Several processes including ionizing radiation process using ^{60}Co gamma source can perform the synthesis of hydrogels. Currently, the hydrogels are used in dressings as a transdermal drug delivery, they provide a moist environment for accelerated the wound healing, allowing the drug release gradually and steadily, avoiding the disadvantages of systemic use or successive applications 7,8.

Hydrogel Poly (N-vinyl-2-pyrrolidone) (PVP) is formed from synthetic polymer with highly polarity, amphoteric characteristics, electrical properties, complexing ability, stability, processability and high transparency. These characteristics are in highly demanded in the pharmaceutical industry for the manufacture of polymers surfactants, emollients, solubilizers, thickeners and wound dressing [9-16]. Researches have emphasized the use of PVP in nanocomposites, such with nanosilver, and as antibiotic carriers for wound therapeutic membrane [7,17]. Drug delivery hydrogels are of interest for the development of biomaterials [18,19]

AgNP are employed for the treatment of topical infections due high surface area, ability to adhere and penetrate into the membranes of bacteria, fungi and viruses by altering biochemical processes and leading to cell lysis. Neomycin is an aminoglycoside antibiotic of broad activity spectrum, produced by *Streptomyces fradiae* and consists of two main components, neomycin B and C. It has high polarity, solubility in water, low volatility and capability for topical use in treatment and prophylaxis of skin infections, including traumatic wounds and burns [20,21].

The development of new techniques for the treatment of wounds without colaterial effects makes the hydrogels promising materials for biomedical application

The aim of the present study was to prepare a wound dressing containing nanosilver (NpAg) and neomycin in composition of poly (N-vinyl-2-pyrrolidone) (PVP), poly (ethylene glycol), agar using gamma irradiation to crosslink and sterilization. at dose of 15, 20 and 25 kGy.

Material and Methods

Materials sample preparation and irradiation

Poly(vinyl pirrolidona) (PVP) provided by Exôdo Científica, Agar provided by Oxoid, Neomycin and NpAg provided by Sigma Aldrich. The formulations prepared for the crosslinking radiation process were obtained by dissolving PVP (6% w/v) in water using a hot plate with magnetic stirrer and temperature between 80 and 85 °C. Neomycin (0.5 %) and NpAg (22 ppm) were dissolved in water. Stirring up to total solution the mixture procedure was performed at 50 °C. The solutions were placed in dishes and irradiated with ^{60}Co gamma radiation at 15, 20 and 25 kGy dose.

Determination of gel fraction

Extraction of the sol fraction was carried out by Soxhlet apparatus for 6 h using water with solvent. Then dried at 50°C to a constant weight. The gel fraction was determinated gravimetrically by using the following equation.

$$G= (Wg/ Wo) \times 100\% \quad (1)$$

Where G is the gel fraction (%), Wg the weight of sample after extration and Wo, before.

Swelling behavior

Synthesized hydrogels were dried to constant weight. After they were immersed in distilled water and weighed each hour for 10 hours and after 24, 48 and 72 hours. The water absorption was calculated as follows:

$$S= (Ws\text{-}Wo/ Wo) \times 100\% \quad (2)$$

Drug release and LC–MS/MS analysis

Hydrogel samples were placed in a beaker and immersed in 100 mL distilled water and incubated at 37 °C under agitation of 80 rpm. 2 mL aliquots were retained at times 2, 5, 8, 10 and 15 min, 1; 1,5; 2; 4; 8; 24 and 48 hours, and were replaced by 2 mL of distilled water at aliquot. Samples were analyzed by LC – MS/MS.

The HPLC equipment consisted of Agilent Technologies 1290 Infinity. The chromatographic separation was performed on Aquasil C18 column, 100 x 2.1mm, 5µ. The following mobile phase gradient was applied with a mixture of acetonitrile/formic acid. The MS–MS system used was AB Sciex 3200 Q Trap equipped with an electrospray interface (ESI). MS data were recorded in the full scan mode from 250 to 620 m/z.

Results and Discussion

Hydrogel synthesis

The crosslinking induced by ionizing radiation was shown to be a suitable method for the synthesis of hydrogels, inducing the formation of a three-dimensional structure and insoluble PVP. The hydrogels presented softness, elasticity and suitable transparency. Radiation technique has advantages as easy to control the process, relatively low cost process and the concomitant sterilization of the membrane [9, 10].

Determination of gel fraction

After irradiation process, gel fraction were determined. In Fig 1 illustrated the gel fraction (%) hydrogels obtained at different doses. It is observed that the gel fraction increases with increasing dose, it is known that irradiation dose increases the crosslink density of the polymer The gel fraction (%) at 25 kGy dose was the highest, around 98%.

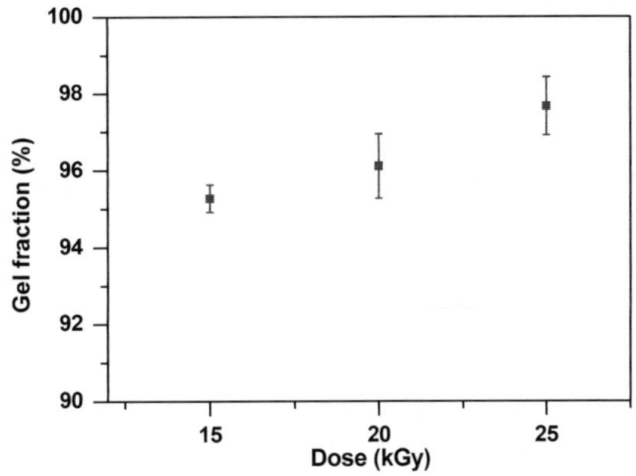

FIGURE 1: Gel fraction (%) of hydrogel irradiated at 15, 20 and 25 kGy doses.

Swelling

The swelling test aimed at determining the mass increase of hydrogel membranes by water absorption to reach an equilibrium. This is known as equilibrium swelling equilibrium rate. The synthesized hydrogels were immersed in distilled water and weighed at intervals. The result of swelling versus time is illustrated in "Fig. 2".

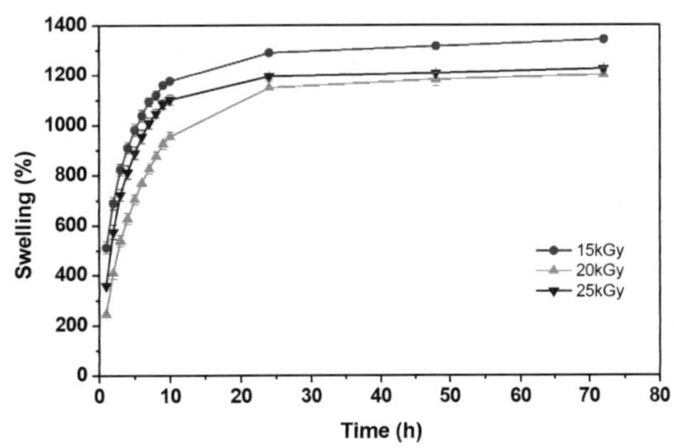

FIGURE 2: Swelling kinetics of hydrogels irradiated at different doses.

The maximum swelling of the hydrogles dressing versus time were reached at 15 kGy dose. The swelling decreases with increasing dose. The gamma irradiation increases crosslink density polymer and difficult the penetration of water by crosslinking polymer, and thus the maximum swelling decreases. It can be seen that all hydrogels reaches the equilibrium swelling after 24 hour of soaking, independent of dose.

Drug release determination

It was observed a rapid release of neomycin and the release maximun were reached in 8 hours of experiment and irradiated membranes at 20 and 25 kGy, remaining stable up to 48 hours. The two membranes achieved the same release profile as concentration of neomycin released by the two curves are within the limits of error bar.

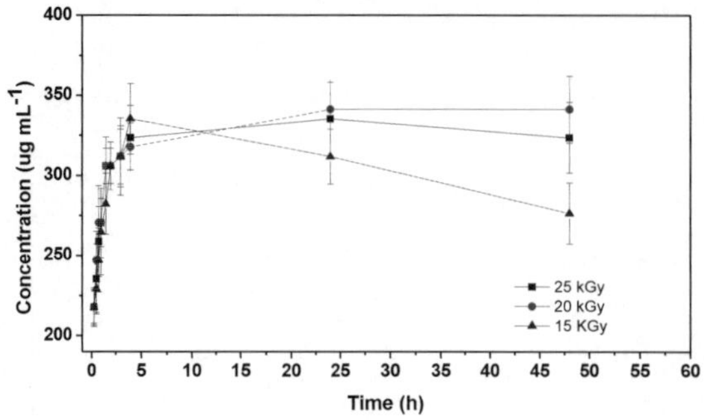

FIGURE 3: Kinect of neomycin release in hydrogels obtained by gamma irradiation at 15, 20 and 25 kGy.

The evaluation of the release membrane neomycin is related to the swelling of the matrix. When water penetrates into the hydrogel occurs a relaxation of dimensional structure and drug is released from the system. The irradiated membrane at 15 kGy showed not be suitable for drug delivery system, because not kept stable concentration of the drug.

CONCLUSION

The hydrogel membranes obtained by gamma radiation are very studied due to potencial medical applications. Hydrogels of PVP/Neomycin/AgNp formulated and characterized. The hydrogel showed transparency, elasticity, a good swelling. The maximum swelling and gel fraction were obtained when irradiated with doses of 15 and 25 kGy, respectively. The dose of ionizing radiation influences the crosslinking of the polymer and the higher the dose to crosslink and gel fraction and less swelling. Neomycin was released from the different membrane quickly during the first hours, and reached equilibrium after 8h. The dose of 15 kGy not showed adequate to obtain wound dressings because does not keep the release constant neomycin, on the other hand, dose of 20 and 25 kGy indicate suitable for controlled drug release.

ACKNOWLEDGMENTS

FAPESP and CAPES for financial support.

REFERENCES

1. N.A. Peppas, T. Huang, M. Torres-Lugo, J.H, *Annual Revs. Biomedical Engineering* **2,** 9-29 (2000).
2. N.A. Peppas, P. Bures, W. Leobandurg and H, Ichikawa, *Eur J Pharm Biopharm.* **50**, 27-46 (2000).
3. E.S. Dragan, *Chemical Engineering Journal* **243**, 572-590 (2014).
4. J.M Rosiak, P. Vlanski, *Radiat. Phys. Chem.*, 55 139-151 (1999)
5. S. Kadlubowski, A. Henke, P. Ulanski and J.M.Rosiak, *Radiat. Phys. Chem.* **79** 261-266 (2010).
6. G. Tae, Y. Kim, W.I. Choi, M. Kim and P.S. Stayton, *Biomacromolecules*, **8**, 1979-1986 (2007).
7. D. Singh and R. Singh, *Radiat. Phys. Chem.* **81,** 1781-1785 (2012).
8. J.M Rosiak, I. Janik, S. Kadlubowski, M. Kozicki, P. Stasica, P. Ulanki, *Nucl. Instrum. Methods, Phys. Rev*. 208, 325-330 (2003).
9. D. J.T. Hill, A. K. Whiyyaker, *Radiat. Phys. Chem*. 80, 213-218 (2011).
10. M. T. Razzak, D. Darwis; *Radiat. Phys. Chem*. 62, 107-113 (2001).

11. O. Wictterle, D. Lim, *Nature* **185,** 117-118 (1960).
12 B. Singh, R. Bala, *Int J Biol Macromol.* **65,** 524-533 (2014).
13 E. Skorupska, A. Jeziorna, S. Kazmierski and M.J. Potrzebowski, *J. Solid State* **58** 2-16 (2014).
14. K. Halake, M. Birajdar, B.S. Kim, H. Bae, C. Lee, Y.J. Kim, S. Kim, S. Ahs and J. Lee, *J. Ind. Eng. Chem.* **20** 3913-3918 (2014).
15. J.C.O. Villanova, L.O. Rodrigo and S.C. Armando, *Polímeros,* 20 51-64 (2010).
16. Z. Ajji, I. Othman and J.M. Rosiak, *Nucl. Instr. and Meth in Phys.* **229,** 375 – 380 (2005).
17. R. Gottlieb, T. Schimidt and R.T. Arndt, *Nucl. Instr. and Meth. in Phys.* **236**
236-371 (2005).
18. A.J. Huh and Y.J. Kwon, *J. Control. Release* **156,** 128-145 (2011).
19. B.S.F. Bazzaz, B. Khamene and M.J. Behabadi, Cont. Lens *Anterior Eye.* **37,** 149-152 (2014).
20. K. Plungpongpan, K. Koyanukkul, K.; A. Kaewvilai, N. Nootsuwan, P. Kewsuwan and A. Laobuthee, *Energy Procedia* **34,** 775-781 (2013).
21. K.C. Ashara, J.S. Paun, M.N. Soniwala and J.R. Chavada, Asian Pac. J. Trop. Dis., **4** 27-32 (2014).

Tailoring The Oxygen Gas Barrier Property In Biodegradable Poly(propylene carbonate) Composites

Yucai Shen*, Tingwei Wang

College of Material Science and Engineering, Nanjing Tech University, Nanjing, 210009, China

ycshen@njtech.edu.cn

Abstract. Poly(propylene carbonate) (PPC) was melt mixed with poly(lactic acid) (PLA) and poly(butylenes adipate terephthalate) (PBAT) in a laboratory internal mixer. The mechanical properties, thermal properties and oxygen barrier properties of the composites were investigated by tensile test, differential scanning calorimetry (DSC) and oxygen permeation measurement. The results showed that the tensile strength of the composites decreases with decreasing of PPC concentration, while the elongation of the composites increases with decreasing of PPC concentration. The oxygen barrier properties of the PPC composites decrease with decreasing of PPC concentration, hence it could be possible to tailor these properties.

Keywords: Biodegradable polymers; Blends; oxygen barrier
PACS: 81.05.QK

INTRODUCTION

The mass emission of CO2 produced by various industries is considered as the main reason to potentially cause the global warming. Meanwhile, the utilization of CO2 as resource has also attracted more researchers' interests in the recent years, and the copolymerization of CO2 with epoxide to form biodegradable poly(propylene carbonate) (PPC) is one of the exciting ways.[1,2] PPC has many particular characteristics. It is amorphous, and has good oxygen barrier property but poor thermal property. On the other hand, to produce PPC-based films, it is also very important to improve its melt and static strength. To expand the application of PPC, great efforts have been put to improve the mechanical and thermal properties, including chemically modifying the end group, blending it with other polymers or introducing inorganic fillers.[3-5] Among these methods, blending with other biodegradable polymers is commercially advantageous as the full biodegradable properties can be kept and the oxygen barrier and thermal properties are readily manipulated by the concentration of polymers. Poly(lactic acid)/poly(butylenes adipate terephthalate) (PLA/PBAT) composites or composite films have already been studied intensively, but those materials have relatively low oxygen barrier properties which limit their applications in some areas, such as food packaging. In this study, PPC was melt mixed in a laboratory internal mixer with PLA and PBAT with different concentrations as shown in Table 1. The ratio of PLA and PBAT was kept constant at 4:6. All samples for measurement were compression moulded.

TABLE 1. Sample formulations

Sample	PPC (%)	PLA (%)	PBAT (%)
1	100	0	0
2	60	16	24
3	40	24	36
4	20	32	48
5	0	40	60

RESULTS

Tensile properties of the samples are shown in FIGURE 1. The pure PPC have a tensile strength of about 20 MPa, with a quite high elongation. The tensile strength of the composites decreases with decreasing of PPC concentration, while the elongation of the composites increases with decreasing of PPC concentration.

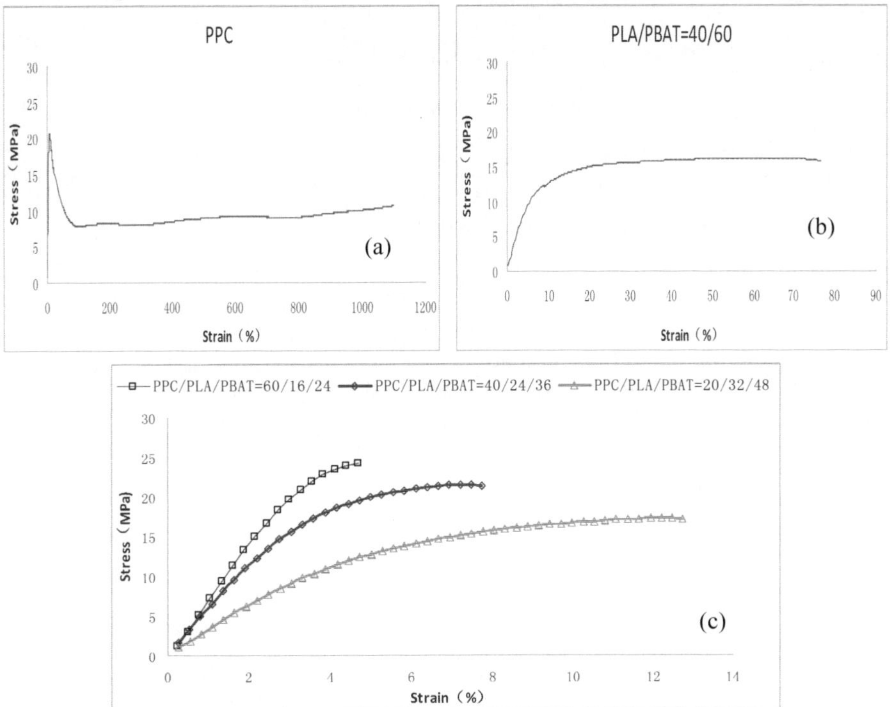

FIGURE 1. Tensile properties of the samples

The oxygen barrier properties of the samples are shown in FIGURE 2. The oxygen barrier properties of the composites decrease with decreasing of PPC concentration.

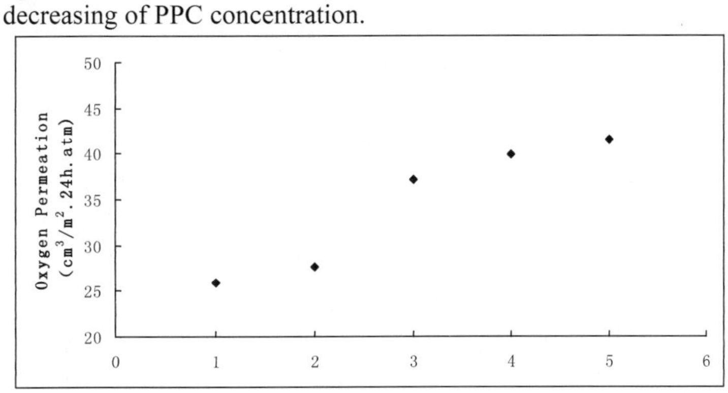

FIGURE 2. Oxygen barrier properties of the samples

The DSC results from the first heating are shown in TABLE 2. Materials with low PPC concentration (20%) or without PPC have only one melting peak, while materials with high PPC concentration (40%, 60%) have two melting peaks.

TABLE 2. Sample formulations

Sample	**Tm1** (℃)	**Tm2** (℃)	**ΔH** (J/g)
1	/	/	/
2	141.1	149.0	0.9
3	146.3	151.7	4.4
4	149.4	/	6.0
5	148.7	/	1.6

ACKNOWLEDGMENTS

This work was supported by the Priority Academic Program Development of Jiangsu Higher Education Institutions (PAPD).

REFERENCES

1. S. Inoue, H. Koinuma and T. Tsuruta, *Makro. Chem.* **130**, 210-220 (1969).
2. R. L. Paddock and S. T. Nguyen, *Macromolecules* **238**, 6251-6253 (2005).
3. X. F. Ma and N. Wang, *Carbohyd. Polym.* **71**, 229-234 (2008).
4. M. Yao, F. Mai and H. Deng, *J. Appl. Polym. Sci.* **120**, 3565-3573 (2011).
5. D. X. Wang, J. Yu and J. Zhang, *Comp. Sci. Tech.* **85**, 83-89 (2013).

Bio-based Thermoplastic Elastomer Containing Polylactide and Natural Rubber via Admicellar Polymerization

Phornkamol Tiewtragool and Rathanawan Magaraphan*

The Petroleum and Petrochemical College, Chulalongkorn University,
Soi Chula 12, Phyathai Road, Pathumwan, Bangkok 10330, Thailand

phornkamol.t@gmail.com
**rathanawan.k @chula.ac.th*

Abstract. Recently, the development of materials for reducing the environmental problems is an important issue. So to reduce the environmental impact from waste, bio-based materials should be used to produce thermoplastic elastomer materials. In this study, polylactide and natural rubber latex (NRL) were chosen as raw materials for a new thermoplastic elastomer via admicellar polymerization. By using constant concentration of NRL substrate of 0.05 g/ml, the lactide monomer content (200, 500, 700, and 1000 mM) was varied to find the suitable concentration of lactide monomer for bio-thermoplastic elastomer applications. Fourier transform infrared spectra of the admicellar synthesized natural rubbers showed the characteristic peak of polylactide. The core-shell structure of polylactide coated onto natural rubber latex particles was revealed by transmission electron microscope (TEM). The bio-thermoplastic elastomers were characterized by dynamic mechanical analysis (DMA) for the glass transition temperature (T_g), behavior of materials and by melt flow index tester for measuring of the ease of flow of the molten bio-thermoplastic elastomers.

Keywords: Thermoplastic Elastomer, Admicellar Polymerization, Natural Rubber, Polylactide
PACS: 82.35.Pq

INTRODUCTION

Nowadays, biodegradable materials from renewable resources have been of interest because they can be alternatives to petroleum based materials. Moreover, the development of materials for reducing the environmental problems is an important issue now. Polylactide (PLA) is one of the biodegradable and biocompatible materials produced from renewable resources [1]. Nevertheless, brittleness, high stiffness, and thermal properties are the limitations for using this biodegradable polymer [2]. In order to overcome these problems, blending or copolymerization of PLA with other biodegradable polymers have been widely used.

Thermoplastic elastomers (TPEs) are new raw materials that also play important role in the industry because they combine the good properties of thermoplastics and elastomers [3]. TPE materials have ability to return to original length when stress is released (rubber-like) but is still easy to process and recycle (plastic-like). To decrease environmental impact from waste, biodegradable materials should be used to produce the thermoplastic elastomer, so called “bio-thermoplastic elastomer”.

In this study, a novel bio-thermoplastic elastomer is prepared via admicellar polymerization of lactide on natural rubber latex particles. This product is special in that its structure contains-the soft and elastic natural rubber and the rigid PLA which is polymerized from L-lactide (L-LA) via ring opening mechanism. Admicellar polymerization is the technique to coat polymer thin film onto the substrate surface by using bilayer of surfactants which act as a reaction template for polymerization [4]. This technique provides a good adhesion at interface of substrate and the synthesized polymer. The goal of this work is to study the effect of PLA content on the production of bio-thermoplastic elastomer. Although it is not recommended to carry out the ring opening polymerization in aqueous media [5], we report the most progress achieved that shows the information to potentially obtain bio-thermoplastic elastomer.

EXPERIMENTAL

Materials

Natural rubber latex (NRL) (60% dried rubber content) was purchased from Rubber Research Institute of Thailand. Lactide monomer (LA) with 99.5% purity was purchased from Shenzhen Bright China Industrial Co., Ltd. Ethylene glycol 99.8% and tin(II) 2-ethylhexanoate ($Sn(Oct)_2$) 95% were purchased from Sigma-Aldrich. The surfactant cetyltrimethylammonium bromide (CTAB) was purchased from Merck Millipor.

Preparation

NR latex of 5 %w/v dried rubber content was dispersed in distilled water at pH8. Then, CTAB was added into diluted NR latex until the concentration of CTAB was 2800 μM. The mixture was stirred at 500 rpm for 12 hr and then ethylene glycol (1.5 mol% of monomer) and $Sn(Oct)_2$ (2 mol% of monomer) were added into the mixture as initiator and catalyst respectively. The mixture was heated up to 90°C then LA monomer (varying concentrations 200, 500, 700, and 1000 mM) was starve-fed into the mixture and the reaction was continued until NR was coagulated. The rubber part was washed with distilled water to remove outer layer of CTAB and dried in vacuum oven at 60 °C for 3 days. Finally, the admicelled PLA-NR (NR-ad-PLA) was obtained.

Characterization

The Fourier transform infrared spectra were recorded with resolution of 4 cm^{-1} using Nicolet, Nexus 670 operating in ATR (Attenuated Total Reflectance) mode with a diamond crystal having a refractive index = 2.4, at incident angle θ = 45°. The number-average molecular weights (M_n), weight-average molecular weights (M_w), and molecular weight distribution (MWD) values were determined at 40°C by gel permeable chromatography (GPC), SHIMADZU GPC-RID10A using styragel®HR5E THF columns (WAT044228) with THF as a mobile phase at flow rate 1 ml/min. The samples were prepared by soxhlet extraction with THF and PLA solutions were injected with a volume 20μL. The core-shell morphologies of samples were investigated by transmission electron microscopy, JEOL, JEM-2100, operated at an acceleration voltage of 100 kV. The samples were deposited on copper grid and dried overnight before testing. The dynamic mechanical properties were studied by a Dynamic Mechanical Analyzer (GABO Eplexor 100N) in tension mode using rectangular sample (10 x 40 x 1 mm^3). The test was done in a temperature range of -100 °C to 100 °C at heating rate of 2 °C/min and frequency of 1 Hz. Melt flow index of samples was measured by Dynisco melt flow indexer. The test followed ASTM D1238 at temperature 190 °C using 10 kg weight.

RESULTS AND DISCUSSION

Attenuated Total Reflectance (ATR-FTIR)

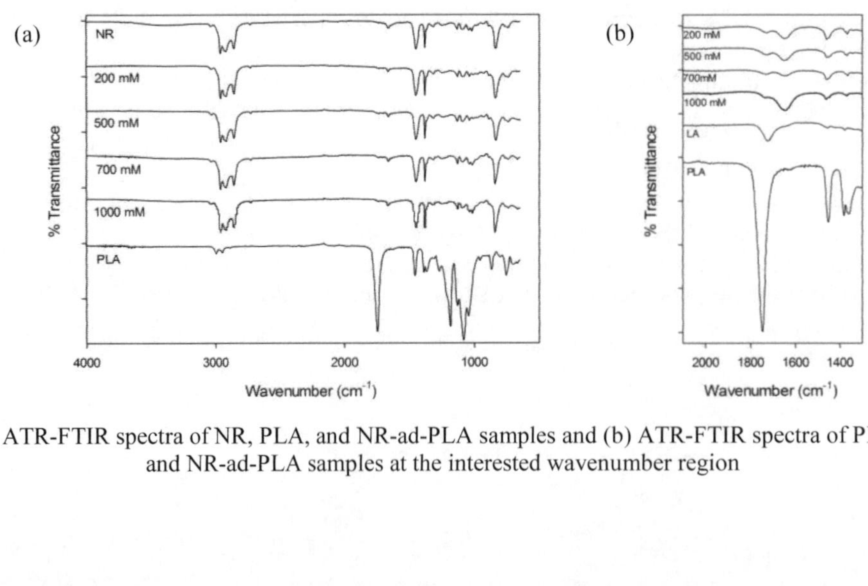

FIGURE 1. (a) ATR-FTIR spectra of NR, PLA, and NR-ad-PLA samples and (b) ATR-FTIR spectra of PLA, LA monomer, and NR-ad-PLA samples at the interested wavenumber region

Figure 1 shows the characteristic peaks of NR observed at 2964, 2929, and 2856 cm^{-1} for C-H stretching. The absorption band at 1454 cm^{-1} is C–H deformation of $–CH_2–$. Moreover, the bands at 1377 and 841 cm^{-1} were corresponded to C-H deformation of $–CH_3$ and C–H deformation of cis –C=C–H, respectively [6]. The characteristic peak of PLA was found at 1734 cm^{-1} assigned to C=O stretching. It is possible to observe a band shift related to the C=O stretch in the monomer in 1727 cm^{-1} to 1757 cm^{-1} in the polymer [7]. So, the peak at 1734 cm^{-1} found in NR-ad-PLA samples refers to low molecular weight PLA. The characteristic peak of PLA was slightly shifted to higher frequency with increasing concentration of lactide monomer. These absorption spectra confirmed that NR-ad-PLA containing both parts of NR and PLA.

Molecular Weight Measurement

M_n, M_w, and MWD values of the extracted PLA obtained from GPC analysis are 374-1122, 940-1583, and 1.21-2.51 as shown in Table 1, respectively. In this study, conversional percentages were so low about 1.02-1.83% (corresponding to amount of PLA produced 0.60-1.14 wt% in NR-ad-PLA) due to the system was in hot aqueous media so most LA was susceptible to transform to lactic acid, noticed by lowering of pH. Moreover, the hydrolysis of PLA is dependent on the ester and water concentrations [8]. Therefore, the high molecular weight of PLA is hardly achieved. The percentage of PLA in NR-ad-PLA likely increased with increasing monomer concentration. So, synthesized PLA by admicellar polymerization technique gave very low yield and low molecular weight PLA. In term of composition, these NR-ad-PLA products were still rubber-like materials.

TABLE 1. Conversion of LA and molecular weight of PLA

Samples	*Corresponded LA content* $^{\$}$ *(wt%)*	*% Conversion**	*wt% PLA in NR-ad-PLA***	M_n	M_w	*MWD*
200 mM NR-ad-PLA	36.57	1.83	0.67	971	1179	1.21
500 mM NR-ad-PLA	59.04	1.02	0.60	374	940	2.51
700 mM NR-ad-PLA	66.86	1.39	0.75	1122	1583	1.41
1000 mM NR-ad-PLA	74.24	1.54	1.14	791	1329	1.68

$^{\$}$ $\text{weight\%} = \dfrac{\text{weight of LA}}{\text{total weight of LA and NR}} \times 100$

* $\text{\%conversion} = \dfrac{\text{weight of dried NR-ad-PLA} - \text{weight of dried NR}}{\text{weight of monomer}} \times 100$

** $\text{wt\% PLA in NR-ad-PLA} = \dfrac{\text{weight of dried NR-ad-PLA} - \text{weight of dried NR}}{\text{weight of dried NR+PLA}} \times 100$

TEM Images

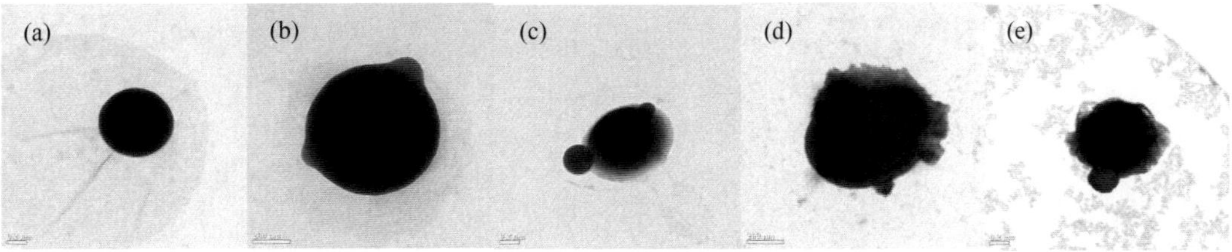

FIGURE 2. The TEM micrographs of (a) NR, (b) NR-ad-PLA 200 mM, (c) NR-ad-PLA 500 mM, (d) NR-ad-PLA 700 mM and (e) NR-ad-PLA 1000 mM

The NR particles had smooth surface and round shape as shown in Figure 2 (a). Figures 2 (b)-2(e) show TEM micrographs of NR-ad-PLA particles where the NR particle (the dark circle) was surrounded with the gray layer of PLA. The NR-ad-PLA particles presented with a thin rough surface showing non-uniform coverage of PLA shell onto the surface of NR particles implying non-uniform distribution of LA monomer in the bilayer of surfactant during the polymerization. As increasing monomer content, the morphology gets more distorted from its typically spherical structure and shows an increase in content of shell [9].

Dynamic Mechanical Analysis

Dynamic mechanical properties can be expressed in terms of storage modulus (E'), loss modulus (E"), and damping factor (tan δ) term. In dynamic mechanical analysis, tan δ is commonly used for predicting the miscibility of polymer systems. Tan δ versus temperature curve is used to verify glass transition temperature (T_g). T_g of pristine NR occurs at -54.10 °C and the T_g of PLA is approximately 60 °C [10]. In Figure 3, all NR-ad-PLA samples show only one T_g value in range about that of T_g of pristine NR. This phenomenon suggested that the NR-ad-PLA samples produced by admicellar polymerization provided good miscibility of two components and behave like rubber [9]. In glassy state, the NR-ad-PLA samples show higher E′ than the pristine NR. This can confirm the presence of PLA which is a rigid polymer. The motion of the NR-ad-PLA samples starts at lower temperature with higher loss modulus than that of NR. Moreover, beyond the glass transition region, the NR-ad-PLA samples possess lower storage modulus (about 1 MPa at room temperature) than NR. The high tan δ in NR-ad-PLA samples as compared to that of NR at high temperature region indicates high chain mobility. These behaviors support that all NR-ad-PLA samples contain short PLA molecules to contribute flow ability better than NR. In this work, the sample using 500 mM LA shows the highest E′ and rather low flow ability while the one using 200 mM LA is the lowest E′. The ones using 1000 mM and 700 mM LA have intermediate E′. This is attributed to the broad MWD in the sample using 500 mM LA (the high MW molecules give high modulus and low flow ability) and the PLA content which is the highest (1.14 wt%) for the sample using 1000 mM LA. The lower content of PLA results in lower E′ and flow ability.

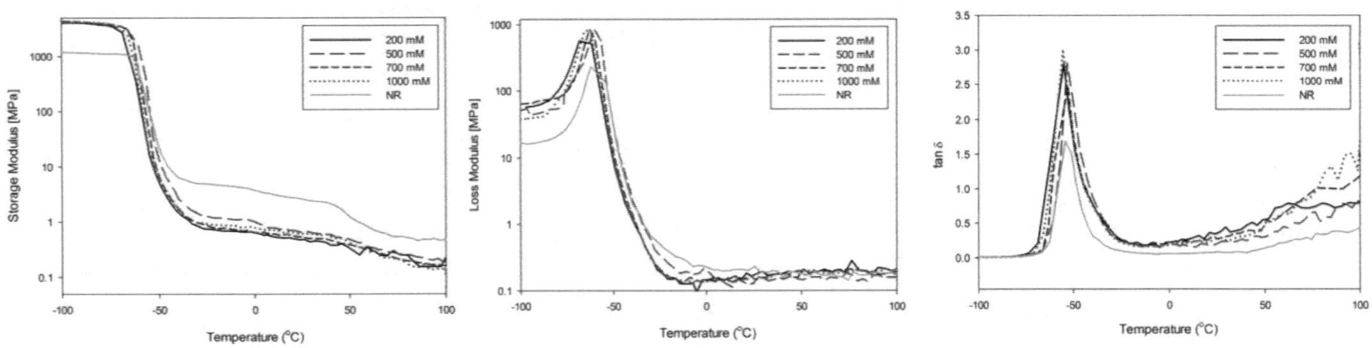

FIGURE 3. Dynamic mechanical properties of NR and NR-ad-PLA samples

Melt Flow Index

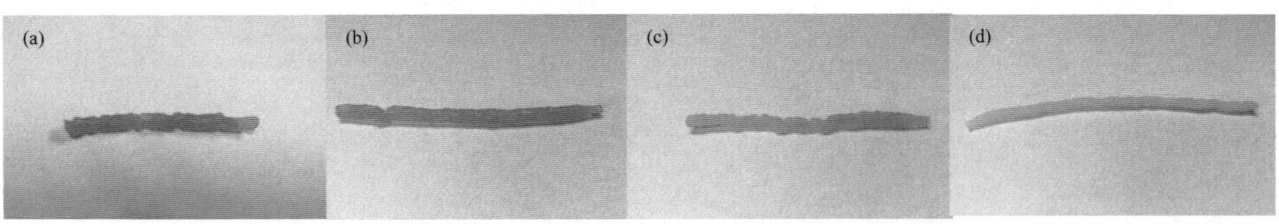

FIGURE 4. NR-ad-PLA samples after passing through melt flow tester (a) 200 mM, (b) 500 mM, (c) 700 mM, and (d) 1000 mM

TABLE 2. The melt flow index values of NR-ad-PLA samples

Samples	*MFI*
NR	0.25±0.01
200 mM NR-ad-PLA	0.32±0.02
500 mM NR-ad-PLA	0.58±0.10
700 mM NR-ad-PLA	0.69±0.05
1000 mM NR-ad-PLA	2.26±0.15

Processability as melt when raised temperature is one of essential characteristics that thermoplastic elastomer must have [3]. Figure 4 shows extrudates of NR-ad-PLA after passing through the die by load 10 kg. Sample using 200 mM LA contains the smallest amount of PLA and thus it performs as rubber with melt fracture. Melt fracture on polymeric extrudate infers that the polymeric material cannot process as thermoplastic [11]. Increasing content of PLA makes the NR-ad-PLA samples extrude with smoother surface and less swell resembling the better behavior of thermoplastic flow. Table 2 shows the MFI values of NR and NR-ad-PLA samples. The MFI increases with the increasing of content of PLA. From this test, the NR-ad-PLA samples prepared via admicellar polymerization tend to perform as bio-thermoplastic elastomer. If the LA conversion to PLA was enhanced, the product would flow better and behave more like thermoplastic elastomer. Next work, the direct polycondensation [12] may be used to converse lactic acid in the aqueous phase to PLA such that more yield of PLA could be obtained.

CONCLUSIONS

The admicellar polymerization of polylactide on natural rubber particles was confirmed by ATR-FTIR and GPC results. The obtained PLA content was very low about 0.60-1.14 wt%. The molecular weight of PLA from this technique was low (M_w 1257±269 and MWD 1.21-2.51) because the loss of LA to lactic acid as well as the hydrolysis of PLA by water can be occurred. TEM image revealed the morphology of core-shell was obtained. Although the amount of PLA produced as shell was so small, it tended to increase with the increase of monomer concentration. The good miscibility of NR-ad-PLA was confirmed by DMA result as shown by the single T_g in range of pristine NR (-54.10 °C). An increase in storage modulus in glassy region and high loss tangent or chain mobility at high temperature revealed the presence of PLA. Furthermore, the melt flow testing showed NR-ad-PLA samples can melt and flow (with less swell) like thermoplastic with a little increase in PLA content. These results suggest the promising success in preparation of bio-thermoplastic elastomer with only small content of PLA plastic via admicellar polymerization.

ACKNOWLEDGMENTS

The authors acknowledge Ratchadaphiseksomphot Endowment Fund, Chulalongkorn University (Bangkok) for STAR no. GSTAR 58-003-23-002 and Petroleum and Petrochemical College, Chulalongkorn University for the full scholarship.

REFERENCES

1. R. M. Rasal, A. V. Janorkar, and D. E. Hirt, *Progress in Polymer Science* **35**, 338-356 (2010).
2. S. L. Yang, Z. H. Wu, W. Yang, and M. B. Yang, *Polymer Testing* **27**, 957-963 (2008).
3. J. E. Mark, and B. Erman, *Science and Technology of Rubber*, (1994).
4. P. M. Karlsson, N. B. Esbjörnsson, and K. Holmberg, *Journal of Colloid and Interface Science* **337**, 364-368 (2009).
5. A. C. Albertsson, and I. K. Varma, *Biomacromolecules* **4**, 1466-1486 (2003).
6. H. M. Nor, and J. R. Ebdon, *Polymer* **41**, 2359-2365 (2000).
7. A. J. R. Lasprilla, G. A. R. Martinez, and B. Hoss, *Chem. Eng* **24**, 985-990 (2011).
8. G. L. Siparsky, K. J. Voorhees, and F. Miao, *Journal of Environmental Polymer Degradation* **6**, 31-41 (1998).
9. A. Pongpilaipruet, and R. Magaraphan, *Materials Chemistry and Physics* **160**, 194-204 (2015).
10. C. Migliaresi, D. Cohn, A. De Lollis, and L. Fambri, *Journal of Applied Polymer Science* **43**, 83-95 (1991).
11. T. Bremner, A. Rudin, and D. G. Cook, *Journal of Applied Polymer Science* **41**, 1617-1627 (1990).
12. F. Achmad, K. Yamane, S. Quan, and T. Kokugan, *Chemical Engineering Journal* **151**, 342-350 (2009).

Welding of Bio-Based Plastics for Applications in the Field of Injection Moulding

Christian Hopmann, Martin Facklam* and Maximilian Schöngart

Institute of Plastics Processing (IKV) at RWTH Aachen University
Seffenter Weg 201, 52074 Aachen, Germany

**martin.facklam@ikv.rwth-aachen.de*
maximilian.schoengart@ikv.rwth-aachen.de

Abstract. Due to intensive research in the field of material development considerable progress in terms of processing and performance characteristics could be achieved during the last years. Therefore, bio-based plastics are increasingly being used for the production of durable injection moulded components. Several applications can be found in the automotive sector, in the field of consumer electronics or the sports industry. For an economic production, joining technology often plays a key role to realise modular designed components. Detachable connections, which are frequently used in plastics processing (e. g. snap hooks or screw connections), often reach their limits with respect to reliability and tightness of the joint. In order to meet the increasing requirements on the joint, plastics joining processes can allow suitable component assembly. Bio-based plastics may differ from conventional plastics regarding their processing properties, such as the melting and solidification behaviour. For these reasons, investigations were carried out to gain extensive process knowledge about the welding of bio-based plastics.
To perform a scientific examination of the influence of the moulding process on the mechanical weld seam quality, cellulose and lactic acid based plastics were selected and injection moulded with varying parameters like melt temperature, mould temperature and injection flow rate. In subsequent welding experiments using ultrasonic welding, materially bonded connections have been produced and analysed with regard to mechanical and morphological properties. The investigations show that all of the selected bio-based plastics can be welded. Depending on the material and the welding method the choice of the moulding parameters can have different influences on the weld seam strength. For example, a higher mould temperature during injection moulding can lead to higher weld seam strengths for ultrasonic welded test specimens of lactic acid-based plastics.

Keywords: bio-based plastics, infrared welding, ultrasonic welding
PACS: 81.20.Vj

INTRODUCTION

For many years research and development activities have been performed in the field of bioplastics. With the aim of new approaches for the recycling of plastic waste, the focus has been on the biodegradability of the material. First, material development focused on the production of short-life products, such as packagings or disposable cutlery. However, due to limited material properties, no significant market for bioplastics was able to develop in the first few years. During the last years fluctuating prices of fossil-based raw materials, an increased climate and environmental awareness as well as political regulations have made a successful market entry possible [1-2]. Particularly in the packaging industry bio-based plastics have been established for film production. Meanwhile, in the field of bioplastics the focus is on the development of materials for the production of technical products. Series applications of injection moulded components can be found in the automotive-, medical-, electrical-, sports- and stationery industries. For any plastic manufacturers injection mouldable bioplastics became an integral part of the product portfolio. This is also reflected in the number of processors and the new products made of bioplastics. In order to provide a clear distinction from the conventional plastics, a definition has increasingly established itself. A bioplastic is a polymer material which satisfies at least one of the following properties [1]:

- material, which is based on bio-based (renewable) raw materials
- and/or is biodegradable.

As a renewable raw material base, e. g. sugar, starch or cellulose can be used for the production of bio-based plastics. Figure 1 shows typical injection mouldable bioplastics, which can be distinguished based on raw material base and degradability [1].

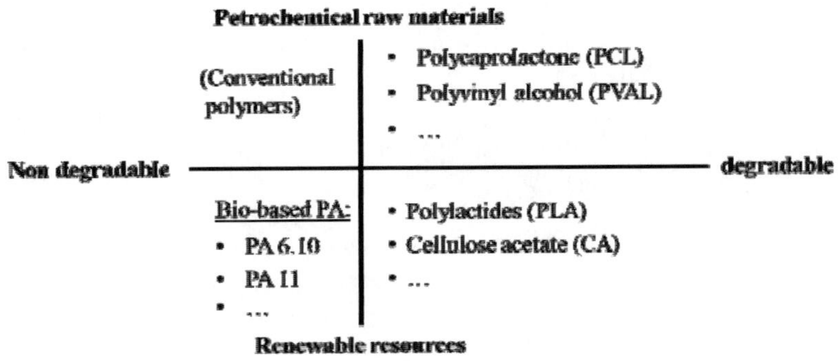

FIGURE 1. Classification of injection mouldable biopolymers in terms of raw material base and degradability [1]

With respect to the processing and use properties, an optimisation of materials plays a key role in the production of long-life products made from bio-based plastics. Due to improved material properties, such as higher temperature resistance compared to unmodified bioplastics, the variety of possible application spectra increase. On the other hand, component geometries are becoming more complex. Often, a modular design is required, which can usually be realised with a joining process. However, in comparison with conventional plastics, material modifications lead to a different processing performance of bio-based plastics, such as the melting and solidification behavior.

The welding of injection moulded joining parts made of two types of bio-based plastics (PLA and CA) was investigated. The aim was to analyse in systematic tests, how rheological and thermal effects of the injection moulding process affect the joining part characteristics and thus also the welding process and the weld seam quality.

Ultrasonic Welding of Plastics

Compared to detachable joints, welded joints are often superior regarding reliability and tightness. For joining plastics, the ultrasonic welding is established for many years in industry especially due to its short process times. A mechanical longitudinal oscillation in the ultrasonic frequency range is coupled in the area of the joint part surface, leading to a vibrational heating of the plastics. The heating mechanism of ultrasonic welding for thermoplastics is mainly based on viscoelastic dissipation under cyclic loading [3-6]. At the same time a joining force has to be applied. In order to achieve a targeted rapid melting of the plastics in the joining area, an energy director is frequently provided on the contact surfaces of one of the joining partners. It serves to focus the oscillation and thereby the energy. During heating, high temperature gradients occur. This leads to a rapid melting of the plastics. The heat affected zone is locally limited to the joining region [6].

Experimental

For the experiments two bio-based plastics were selected and provided by FKuR Kunststoff GmbH, Willich, Germany. Biograde® C 9550 is a compound based on cellulose acetate (CA) and includes a high proportion of mineral fillers. Analysis by differential scanning calorimetry (DSC) showed that the material does not have the typical melting and solidification behavior of semi-crystalline thermoplastics. No defined melting and recrystallisation peak could be detected. The second material is a lactic acid-based polylactide (PLA) called Bio-Flex® V150701. This material is more resistant to hydrolysis and has a higher heat deflection temperature compared to unmodified PLA. For processing, mould temperatures between 90 - 110 °C are recommended by the manufacturer in order to induce crystallisation processes to increase the performance characteristics of the material. Both materials are bio- or semi-bio-based and biodegradable. To perform welding tests, DVS test specimens were produced by injection moulding (see fig. 2).

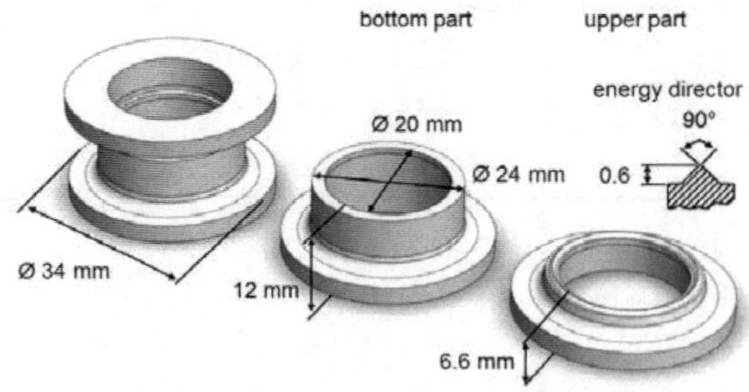

FIGURE 2. DVS test specimen [10]

The production of the test specimens was carried out on a fully electronic IntElect 100-340 injection moulding machine of Sumitomo (SHI) Demag GmbH, Schwaig, Germany. As seen in table 1, injection flow rate (v_{in}), melt temperature (T_m) and mould temperature (T_w) have been varied. By varying these parameters, the largest influence on the formation of microstructures was expected. The selection of the injection moulding parameters is based on the processing guidelines of the manufacturer.

TABLE 1. Varied moulding parameters for test specimens production

	PLA			CA		
v_{in} **[mm/s]**	30	65	100	30	65	100
T_m **[°C]**	190	205	220	190	205	220
T_w **[°C]**	30	50	75	90	100	110

To perform the welding tests, a HS DIALOG digital control 4000 ultrasonic welding machine by Herrmann Ultraschalltechnik GmbH & Co. KG, Karlsbad, Germany, was used. The system-specific welding frequency is 20 kHz. In order to obtain reproducible weld seam strengths, the ultrasonic welding machine has been operated in the mode "weld travel distance". The melt down distance has been measured from the Reference Point Numeric (RPN), which corresponds to a trigger point. The welding process starts after a predetermined trigger force of 150 N is applied to the joining parts. The ultrasound generator switches off after a defined weld travel distance s_f is reached. To ensure a complete melting of the energy director a welding length of 0.7 mm has been chosen. The holding time was set to 2 s.

In a first step, optimum welding parameters for the ultrasonic welding process were determined in preliminary tests. The used test specimens were prepared using the central set of parameters (see table 1). The welding amplitude (25, 30 and 35 µm) and weld force (200, 300 and 400 N) have been varied in three steps. The holding force during the cooling phase corresponds to the set welding force. As an evaluation criterion for determining the welding parameters, the maximum weld strength is selected in combination with the smallest dispersion of measured values, which are determined in short-time tensile tests. A universal testing machine of type Z10 Zwick Roell GmbH & Co. KG, Ulm, Germany, has been used. The test speed was 10 mm/min. Figure 3 exemplarily shows the evaluation for welded specimens of CA.

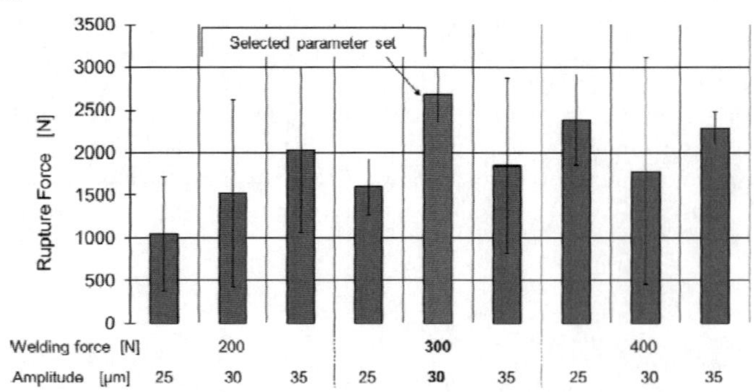

FIGURE 3. Selection of welding parameters based on the determined weld seam strengths of CA

For the cellulose-based material, the maximum rupture force of the weld seams is approximately 2670 N. The set of parameters are a welding force of 300 N and an amplitude of 30 µm (see fig. 3). Similar to the procedure for determining the optimum welding parameters for CA, the parameter set was determined for welding of PLA. For the welding tests, the highest rupture force has been achieved by using a welding force of 300 N and an amplitude of 25 µm. All further test specimens, which were produced with varying moulding parameters, were welded with the optimal welding parameters (see table 1). Hereinafter, the test evaluation of welding tests is presented. These tests were carried out to determine the effect of the influence of the moulding process on the weld seam quality.

Figure 4 shows the results of the mechanical weld seam characterisation depending on the injection molding parameters.

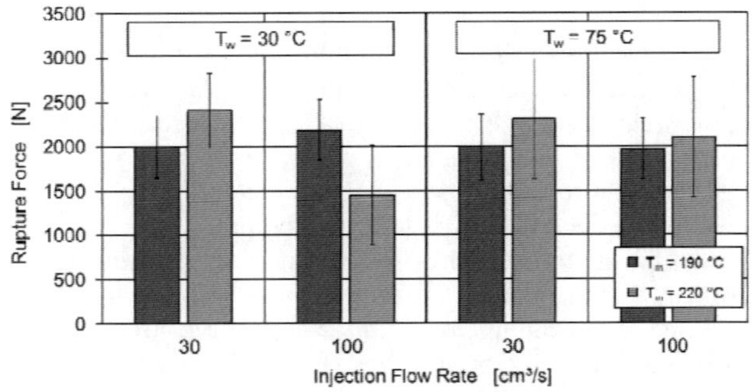

FIGURE 4. Influence of injection moulding parameters on the weld seam strengths of CA

Independent from the variation of mould temperature and injection flow rate, for a melt temperature of 190 °C no significant influences on the weld seam strength appear. Only for a mould temperature of 30 °C and a melt temperature of 190 °C the variation of the injection flow rate from 30 to 100 cm³/s tends to result on higher weld seam strengths.

Generally, for a mould temperature of 30 °C the biggest influences are caused by a variation of melt temperature and injection flow rate. Due to the higher mold temperature of 75 °C, moulding effects on the weld seam strength can be almost compensated. With a mould temperature of 30 °C and an injection flow rate of 100 cm³/s, a mass temperature change from 190 to 220 °C leads to a reduction of the rupture force of about 730 N. Based on the observations described, it is apparent that the temperature gradient in the contact zone of the melt and cavity wall is a crucial factor for influencing the mechanical weld quality.

A high temperature gradient causes rapid cooling of the melt and thus affects the microstructure formation. The effect of the high temperature gradient is enhanced by high shear rates, which occur in the filling process due to rapid injection. Therefore, the greatest influence on the weld seam strength results for test specimens, which were injection molded with a melt temperature of 220 °C and a low mold temperature of 30 °C. In the case of increasing the injection speed from 30 to 100 cm³/s, a decrease in the average rupture force of about 950 N occurred. Additional distinct orientations by a stretching of the melt in the region of the joining surfaces may cause a reduction in the mechanical weld seam quality. Finally it should be mentioned that a high mass temperature tends to reduce the reproducibility, which is reflected in a higher dispersion of the measuring values.

In the diagram below, the results of the welding tests with test specimens of PLA are documented (see fig. 5).

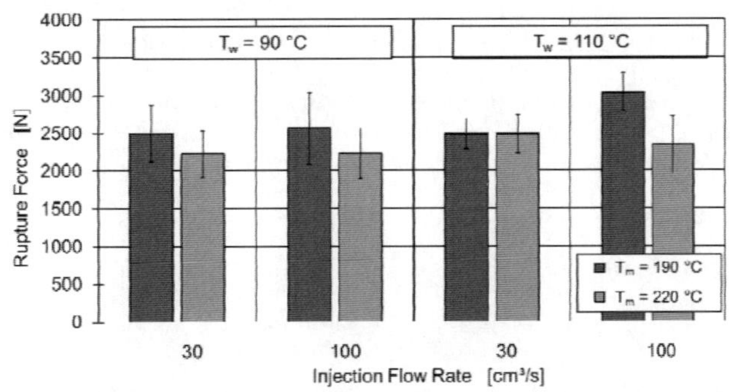

FIGURE 5. Influence of injection moulding parameters on the weld seam strengths of PLA

The evaluation of the short-time tensile tests shows that the average weld seam strength is in a narrow range between about 2200 and 2500 N. In this case, no significant influences of a variation of injection moulding parameters can be observed. For the test point with a high mould temperature (110 °C), injection rate (100 cm^3/s) and a melt temperature of 190 °C, the average measured rupture force is about 3040 N. The material requires mould temperatures in the range of 90 and 110 °C in order to specifically induce crystallisation processes. An explanation for this is that the high mould temperature in combination with the high shear influences due to the high injection flow rate cause an increased thermally and shear-induced crystallisation, which positively affects the formation of the joining zone during welding.

CONCLUSIONS

It has been shown that bio-based plastics based on cellulose and lactic acid could be welded by ultrasonic vibration. The test evaluations show that the injection moulding process can have a significant influence on the weld seam strength. Among others, the welding process is influenced by the microstructure formation in the joining zone. Thus, a reduction of the weld seam strength results for CA, when during the injection moulding a high temperature gradient due to a high melting temperature and a low mold temperature in the joining zone occurs. This effect is enhanced by a strong shear stress of the material during injection molding.

The PLA which has been used for the tests requires a high mould temperature in order to induce crystallisation processes to increase the performance characteristics of the material. For this material the weld seam strength can be increased by a high mould temperature in combination with the high shear influences due to the high injection flow rate cause, which seems to lead to an increased thermally and shear-induced crystallisation. In further investigations, the formation of microstructures in the welding zone will be closer investigated.

ACKNOWLEDGMENTS

The authors would like to thank FKuR Kunststoff GmbH, Willich, Germany who supported the research project with test materials, Herrmann Ultraschalltechnik GmbH & Co. KG, Karlsbad, Germany, for technical support with welding machines and Zwick Roell GmbH & Co. KG, Ulm, Germany, for technical support in tensile testing.

REFERENCES

1. ENDRES, H. J.; SIEBERT-RATHS, A.: *Engineering Biopolymers Markets, Manufacturing, Properties an Apllications*. Munich: Carl Hanser Verlag, 2011
2. COLES, R.; KIRWAN, M.: *Food and Beverage Packaging Technology*. Oxford, UK: Wiley-Blackwell, 2011
3. SURESH, K.S.; ROOPA RANI, M.; PRAKASAN, K.; RUDRAMOORTHY, R.: Modeling of temperature distribution in ultrasonic welding of thermoplastics for various joint designs. *Journal of Materials Processing Technology* 186 (2007) 1, p. 138 - 146
4. MENGES, G.; POTENTE, H.: Studies on the Weldability of Thermoplastic Materials by Ultrasound. *Welding in the World* 41 (1997) 9, p. 47 - 55
5. NONHOF, C. J.; LUITEN, G. A.: Estimates for Process Conditions During the Ultrasonic Welding of Thermoplastics, *Polymer Engineering and Science 36* (1996) 9, p. 1177 - 1183
6. GUTNIK, V.G.; GORBACH, N. V.; DASHKOV, A.V.: Some Characteristics of Ultrasonic Welding of Polymers. *Fibre Chemistry* 34 (2002) 6, p. 426 - 432
7. HEIL, M.*: Heizstrahlerschweißen thermoplastischer Kunststoffe. Dissertation*. Universität-Gesamthochschule Paderborn, 1994
8. FUHRICH, R.: *Infrarotschweißen von Kunststoffen mit thermischen Stahlungsemittern*. Dissertation. Technische Universität Chemnitz, 2013
9. HOFMANN, A.: *Hybrides Laserstrahlschweißen von Kunststoffen*. Dissertation. Universität Erlangen-Nürnberg, 2006
10. N.N.: *DVS-Technical Code 2216-11 – DVS test specimens with an energy director for the ultrasonic welding of thermoplastics*. Düsseldorf: Verlag für Schweißen und verwandte Verfahren, 2008

Assessment of structural changes and rheological behavior of potato and pea starches under extrusion conditions.

LOGIE N.[a*], DELLA VALLE G.[b], DESCAMPS N.[c], SOULESTIN J.[d], COLLINET M.[a]

[a] *IFMAS SAS, Parc scientifique Haute Borne, 60 avenue Halley, 59650 Villeneuve d'Ascq, France*
[b] *INRA-BIA, rue de la Geraudière, 44000 Nantes, France*
[c] *Roquette Frères, rue de la Haute Loge, 62136 Lestrem, France*
[d] *Ecole des Mines, 941 rue Charles Bourseul, 59500 Douai, France*

nicolas.logie@ifmas.eu

Abstract. The effect of extrusion conditions on starch structural changes was investigate in order to improve the understanding of starch melting and its rheological behavior.

Keywords: Starch, extrusion, rheology, destructuration, granules, melting.
PACS: 81

INTRODUCTION

During extrusion, in presence of plasticizers, starch can be converted into a homogenous molten state. As shown in **FIGURE 1**, process conditions may lead to different structural changes, describing the destructuration of starch, such as fragmentation of granules, crystals melting, molecular degradation (depolymerization). All of them influence the final properties of the material. Therefore, a good understanding of starch melting and its rheological behavior is necessary to determine optimal processing conditions and to better control the quality of the products.

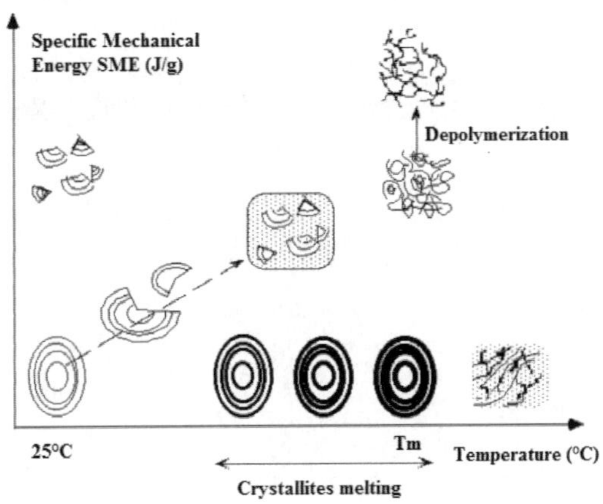

FIGURE 1. Schematic representation of starch melting under extrusion conditions [ref. 1].

Viscosity dependency on starch transformation suggests an analogy with a suspension of solid particles in a fluid. In this purpose, it may be of interest to test a model where, during melting, the decreasing fraction of solid fillers (ϕ) is defined by residual starch fragments embedded in a molten starch matrix [ref.1]. Then, the dependency of consistency (Ks) can be described by a classical sphere suspension model developed by Krieger and Dougherty [ref.2] and generalized by Quemada [ref.3] (Eq. 1).

$$Ks = K\left(1 - \frac{\phi}{\phi m}\right)^{\alpha} \tag{1}$$

Where ϕm is the maximum volume fraction of a packing of spheres (≈0.74) and $\alpha = 2.5$, K is the consistency of molten starch matrix. Thereby, residual starch fragments (ϕ) could be assimilated to a starch destructuration level which enables to evaluate the consistency.

EXPERIMENTS

In this work, two starches of different botanical origin (pea, 33% amylose, and potato, 20% amylose) were processed on a twin-screw extruder under various specific mechanical energy inputs (SME) and temperature (T). They were processed in presence of water (25, 30 and 35% on total wet basis) in order to monitor the melting temperature of starch Tm, determined beforehand by DSC. Processing temperature must be low enough to limit water evaporation and starch expansion at the die. Starch destructuration was investigated at different structural scales: molecular (chromatography), crystal (X-ray diffraction, DSC), granular (microscopy and paste rheology). Finally, the viscosity of partly molten starch was determined on a pre-shearing rheometer, Rheoplast®, which allows simulating extrusion under controlled conditions.

Materials

Smooth pea and potato starches were supplied by Roquette Frères (Lestrem, France). They were conditioned to desired moisture content (25, 30 and 35%) by progressive addition of water during mixing, and storage for 24h. The initial moisture content, respectively 13.0% and 16.5% for smooth pea and potato starch, was determined using Karl-Fisher titration. Smooth pea starch showed typical C-type X-ray diffraction, and B-type for potato starch.

Extruder and extrusion procedures

Thermomechanical treatment was performed by a twin screw co-rotating extruder Thermo-scientific (HAAKE Polysoft OS RheoDrive7). The diameter of the screw was 16mm with a length to diameter (L/D) ratio of 40. The screw configuration was shown in **FIGURE 2** and detailed in **TABLE 1**.

FIGURE 2. Schematic of screw configuration from twin screw extruder.

TABLE (1). Detailed description of screw profil.

Position	Length (mm)	Description
1	64	Transport (pitch 32 mm)
2	496	Transport (pitch 24 mm)
3	24	90° kneading block
4	8	Reverse (pitch 32 mm)
5	16	Transport (pitch 32 mm)
6	24	Transport (pitch 32 mm)

The extruder was divided into ten separately controlled zones: 9 barrel zones and the die zone. Conical die was used with a single opening of 3 mm diameter and 20 mm length. Hydrated starch was fed into the extruder using twin-screw feeder (Brabender) and output rate was controlled every two minutes by sample weighing. Thermocouple and pressure transducer were inserted into the die in order to measure pressure and material temperature, torque was given by the extruder software. Sample collecting starts after 4 minutes of system stabilization (torque, temperature and output rate). It does not last longer than 6 minutes to avoid heterogeneity due to self heating. The extrudates were cooled at room air for one hour and store at controlled humidity (59%) and temperature (20°C). Then, part of samples was converted into powder by cryogenic grinding.

Extrusion campaign was conducted and 35 samples were elaborated with different mechanical energy values (300-1400 kWh/t), by monitoring screw speed and output rate, formulation (25-35% water) and temperature process (80-95°C). Detailed extrusion conditions were given in **TABLE 2**.

TABLE (2). Formulation (moisture content) and process temperature of extruded starch samples.

	Moisture content (%)	Process temperature (°C)	Melting temperature (°C)	Tm-T (°C)
Potato starch	30	95	95	0
	30	80	95	15
	25	90	105	15
Pea starch	30	95	120	25
	35	95	110	15

Macromolecular degradation

Asymmetrical flow field-flow fractionation (A4F chromatography) gave us information concerning molecular degradation. Chromatograms of extruded samples exhibit several features making them different from native. Firstly, there is a sharp decrease of amylopectine peak, this peak was shifted towards lower elution time. Secondly, in most cases, amylose peak is much higher than native. Molecular mass loss compared to the native was evaluated between 20 and 80% for potato starch, and between 40 and 70% for pea starch. As shown in **FIGURE 3** molecular mass loss seems to be governed by SME.

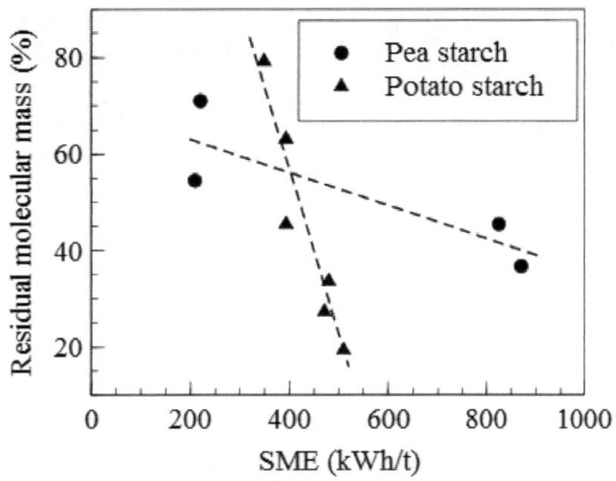

FIGURE 3. Residual molecular mass (% relative to native starch) versus SME.

Granular structure and cristallinity

Gelatinization enthalpy in excess of water was also determined by differential scanning calorimetry Starch samples of 10-15mg mass were heated from 25°C to 95°C at a rate of 5°C.min^{-1}. Medium pressure caps were used to avoid water disturbance. In comparison with native starch enthalpy, residual gelatinization percentage describes a level of destructuration of each sample. The results demonstrated that potato starch has no residual enthalpy unlike pea starch (between 0 and 30%), which confirms the influence of botanical origin. The decrease of pea starch residual enthalpy seems to be governed by material temperature as shown in **FIGURE 4**.

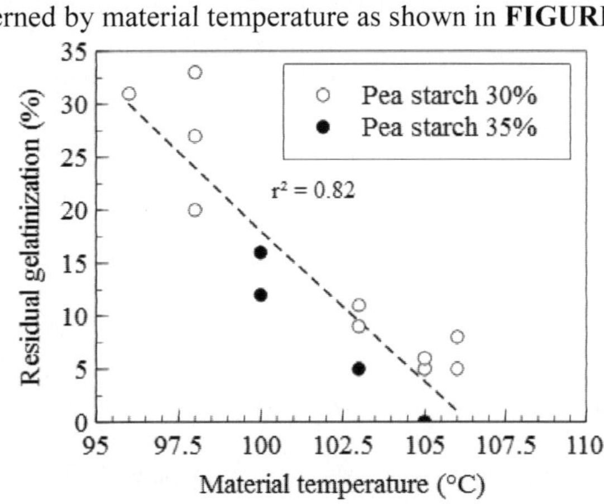

FIGURE 4. The effect of material temperature on residual enthalpy for pea starch samples at different moisture content.

X-ray diffraction measurements were performed to evaluate the crystallinity loss occur during extrusion. The crystallinity ratio between extruded sample and native was determined according to the procedure describe by Wakelin [ref. 4]. The residual crystallinity was evaluated between 0 and 15% for pea starch samples, 0% for potato starch samples except for extrusion at 80°C (0-10% residual crystallinity).

Residual starch granules assessment

The purpose of this work was to bring out the influencing factor between thermal (melting) and mechanical contribution (fragmentation). Light microscopy was performed with a Leica DM2500M, hydrated samples were cut at 10μm thick with a LEICA RM2165 microtome. Sections were observed under polarized light to detect loss of birefringence. Image analysis was used to determine size distribution diagram of uncooked granules or fragments.

Preliminary results show us botanical origin influence: pea starch contains generally more granules than potato starch as shown in **FIGURE 5**. Moreover, large differences among samples regarding extrusion variables, especially SME, were observed (**FIGURE 6**).

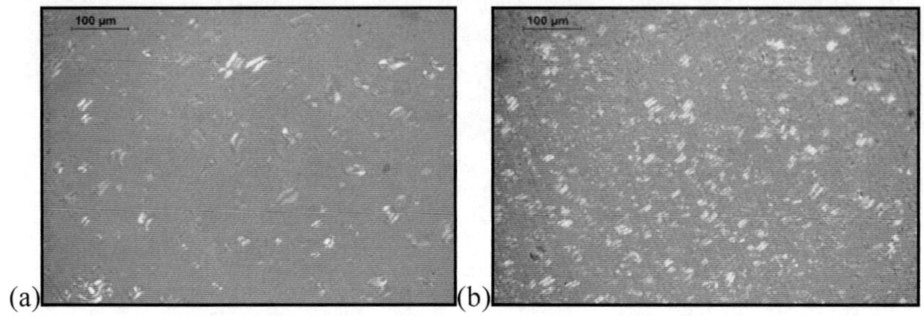

FIGURE 5. Microscopic observation under polarized light of potato (a) and pea (b) starch samples section transformed with the same process conditions.

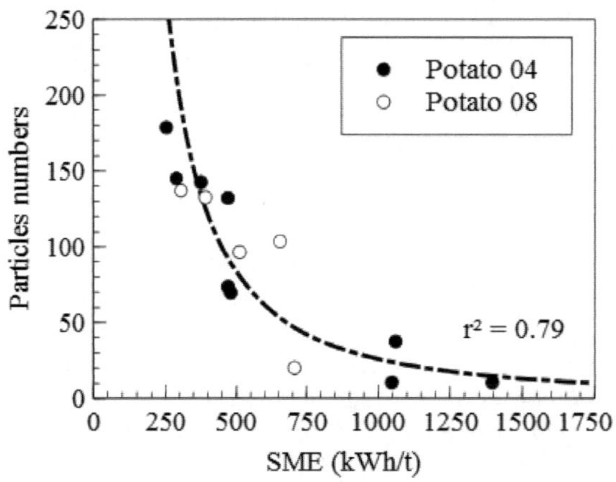

FIGURE 6. The effect of SME on residual particle number for potato starch samples (potato 04: 30% of water, 95°C; potato 08: 25% of water, 90°C).

Starch paste viscosity

Rheology in excess water (starch paste) was performed on Physica MCR 301 rotational rheometer equipped with vane geometry turning at 160rpm. Starch suspension (5% mass) in water was mixed during 24h prior measurement. Samples were heating from 20°C to 90°C at a rate of 4°C.min^{-1}, include isothermal step at 40°C during 5 minutes to correctly homogenize the sample. Next to 90°C isothermal step during 5 minutes, sample was cooled from 90°C to 20°C.

Paste rheology brings us additional information about gelatinization (during heating) and gel formation (during cooling) that could be correlated with molecular degradation. In agreement with differential scanning calorimetry, potato samples do not present significant viscosity peak, that would reflect gelatinization. Then, gel viscosity of potato starch seems to be governed by thermomechanical treatment (SME). This result is consistent with the effect of SME on molecular mass degradation evaluated by chromatography.

Molten starch rheology (Rhéoplast)

The viscosity of partly molten starch was determined in order to assess relationships between starch's structural changes and viscosity. Moistened starches (30%) were processed on a pre-shearing capillary rheometer (Rhéoplast®) fully described by Vergnes [ref. 5; ref. 6].

In order to emulate extrusion conditions, starch was sheared at different rates (100 and 200 rpm) and different temperatures (80 and 95°C for potato starch, 95 and 115°C for pea starch). Flow curve establishment demonstrates the thermomechanical treatment effect on viscosity as shown in **TABLE 3**. Materials followed a typical power law equation (2):

$$\eta = K\dot{\gamma}^{m-1} \tag{2}$$

TABLE (3). Viscosity parameters (m and K) determination for different process conditions.

	Process conditions			Viscosity parameters	
	Temperature (°C)	**Rotation speed (rpm)**	**SME ($J.g^{-1}$)**	**m**	**K ($Pa.s^m$)**
Potato starch	80	100	75	0.24	111 810
	95	100	65	0.22	95 190
	95	200	190	0.27	56 897
Pea starch	95	100	55	0.15	157 253
	95	200	220	0.20	101 173
	115	100	80	0.18	69 915

CONCLUSION

When compared with extrusion conditions, these results first show that the gap between process and melting temperatures (Tm-T) is an important variable to assess changes, and that fragmentation may be induced by shear, whereas depolymerization may occur before complete melting.

Moreover, thermomechanical treatment is a critical factor for viscosity estimation. By combining flow curves with experimental results of starch destructuration, it enables to ascertain melting mechanism and test the suspension model.

REFERENCES

1. C. Barron, G. Della Valle, P. Colonna, B. Vergnes, *J. Food Sci.*, 67, 1426-1437 (2002).
2. I. M. Krieger, T. J. Dougherty, *Trans. Soc. Rheol.*, 3, 137-152 (1959).
3. D. Quedama, *Eur. Phys. J.*, AP 1, 119-127 (1998).
4. J. H. Wakelin, J. Appl. Phys., 30, 11, 1654-1662 (1959).
5. B. Vergnes, J. P. Villemaire, P. Colonna, J. Tayeb, *J. Cereal Sci.*, 5, 189-202 (1987).
6. B. Vergnes, J. P. Villemaire, *Rheol. Acta*, 26, 570-576 (1987).

The Effects of Gamma Irradiation on the Mechanical and Thermal Properties of Poly(lactic acid)

Mehmet Kodal[a], AbdulmounemElchekwis[a], GüralpÖzkoç[a*]

[a]*Department of Chemical Engineering, Kocaeli University, 41380, Kocaeli/Turkey*

mehmet.kodal@kocaeli.edu.tr
guralp.ozkoc@kocaeli.edu.tr

Abstract. Poly (lactic acid) (PLA) is attractive as a substitute for classical polymers due to its biodegradability and sufficient mechanical properties. In the present work, various blend ratios of poly (lactic acid) (PLA) and triallylcyanurate (TAC) were prepared by melt compounding in an Xplore microcompounder and molded by Xplore model microinjection molding device. TAC was used as a coagent. After compounding process, PLA/TAC composites were exposed to a gamma irradiation at various adsorbed doses ranging from 0 to 100 kGy. The mechanical properties of PLA and PLA-TAC were performed by tensile testing. The chemical structure of PLA and TAC was visualized by Fourier Transform Infrared (FTIR) spectroscopy. The thermal properties of the composites were investigated by Differential Scanning Calorimetry (DSC). Results showed that neat PLA was strongly degraded by gamma irradiation while PLA-TAC composites are less affected because gamma irradiation promoted the TAC layer distribution within the PLA matrix. The elongation at break of composites decreased with the increasing TAC content. However, E-modulus has improved about 15% compared to pure PLA. DSC results indicated that a sharp decrease in degree of crystallinity (X_c%) of PLA/TAC composites was obtained for higher gamma irradiation doses. FTIR analysis showed that gamma irradiation of PLA lead to oxidation reactions of ester groups inducing the formation of hydroxyl groups which can be present in hydro peroxides or alcohols.

Keywords: PLA, crosslinking, gamma irradiation
PACS: 82.35.Pq

INTRODUCTION

Gamma irradiation is the most widely-used cold sterilization technique because of its high penetration power [1]. It is well known that ionizing radiation produces physical and chemical changes in materials. On the other hand, most synthetic polymers used for application such as food packaging, biomedical, and agricultural have often short service life, so they end up mostly in landfills. Petroleum-derived plastics have caused many environmental problems associated with their disposal. However, the growing accumulation of petro-polymers in the environment is a global environmental issue and has spurred interest in replacing these with biodegradable polymers from renewable resources [2, 3]. Among the group of biodegradable polymers, polysaccharide, polyesters and their derivatives, poly(ε-caprolactone) and polylactic acid (PLA) have been studied extensively [4]. PLA has the potential for an attractive combination of mechanical properties, high degree of transparency, biocompatibility and cost. Therefore, PLA are considered as ideal candidates to replace non-biodegradable polymeric materials for several applications [5]. Moreover, PLA can be processed in equipment commonly used for synthetic polymers. There are published studies reporting on the thermal stability of this polymer, which has been shown to undergo random chain scission [6]. Moreover, Nugroho et al. has demonstrated that the isothermal degradation of PLA is a random chain scission in which two kinetically independent units take part in the degradation [7]. Several papers reported that γ-irradiation of polymers generally produces simultaneous scission and cross-linking of the polymeric chains [8-11]. The latter causes the formation of a macroscopic network, which reduces gas permeability and improves mechanical properties, whereas the former decreases mechanical properties, on the other hand; increases the biodegradation rate. In this work, blends of PLA/TAC have been subjected to different doses, from 0 to 100 kGy. The molecular characteristics, mechanical, thermal, structural and morphological properties were studied as a function of irradiation does and the mechanism of radiation induced degradation is discussed.

EXPERIMENTAL

Materials and Sample Preparation

Poly(lactic acid) (PLI005, MFI: 10-30 g/10 min at 190°C/2.16 kg) was purchased from NaturePlast Company, France. PLA/TAC with two different concentrations (3 and 7 wt %) was prepared via melt blending in a twin-screw microcompounder (15 ml Microcompounder, Xplore Instruments, The Netherlands) at a screw speed of 100 rpm. PLA was dried in an oven at 50 °C for 12 h under vacuum. The barrel temperature of microcompounder was set to 180 °C. The mixing time was kept constant at 3 min. Prior to processing, predefined amount of TAC and PLA were dry-mixed. At the end of this period, the extrudate was taken by changing the position of the valve to guide the polymer through the die. Extruded composites were injection molded on an Xplore 10 mL injection molding machine at 180 °C and holding pressure of 10 bar.

Pure PLA and PLA/TAC composites were irradiated in a self-contained gamma research irradiator, Gammacell 220 Excel (GC-220E) of MDS Nordion (4.9 kCi). Target doses of 0, 25, 50, and 100 kGy were performed at 3.86 kGy/h dose rate by a 60Co-γ source, under ambient conditions (25 °C)

Characterization

The irradiated samples (6×1 cm^2) were evaluated by using a universal testing machine (Instron 3345 model). The crosshead speed was set at 10 mm/min. A minimum of five specimens were used to determine the average of mechanical parameters (tensile strength, elongation at break, and tensile modulus).

The infrared spectra (FTIR) were obtained at room temperature using Perkin Elmer spectrum 100 FTIR. The samples were analyzed in the range of 4000–600 cm^{-1}, with 4 cm^{-1} of resolution, and 100 scans.

The differential scanning calorimetry (DSC) of irradiated samples was performed on a Mettler Toledo DSC1 star system under nitrogen atmosphere. A sample of 5 mg was heated from 25 to 200 °C with a heating rate of 10 °C/min. Glass transition temperature (T_g), enthalpies of melting (ΔH_m) and cold crystallization enthlapies (ΔH_c), and melting temperature of PLA and PLA/TAC composites (T_m) were determined via DSC thermograms. The percent crystallinity ($\%X_C$) and the nucleation rate of the samples were calculated according to Eq. 1:

$$X_c\% = (\Delta H_m - \Delta H_c) / (\Delta H_m^* ø) * 100 \qquad (1)$$

where ΔH_m is melting enthalpy (J/g), ΔH_c is cold crystallization enthalpy (J/g), ΔH_m^* is heat of fusion for PLA (93 J/g) [12] and ϕ is the fraction of the PLA in the composite,

RESULT AND DISCUSSION

Mechanical Properties

Irradiation usually has notable effects in the mechanical properties of polymers. These effects depend on whether crosslinking or degradation takes place due to irradiation. Figure 1-3 shows the elongation at break (%), E-Modulus (MPa) and maximum load (N) of neat PLA and irradiated PLA/TAC composites, respectively. Irradiation of PLA samples shows a significant decrease in elongation at break (%) and maximum load values for neat PLA, but with slightly decrease in PLA/TAC composites. E-Modulus of PLA increased significantly with gamma irradiation. The results of the mechanical properties indicated that chain-scission could be occurring in the amorphous regions of PLA, therefore; reducing the PLA's molecular weight and severely diminishing the intermolecular forces. Incorporation TAC to the PLA decreased in the chain scission of PLA after irradiation. Irradiation caused the long polymer chains and cross-linked structures to be break down into shorter chains and smaller structures, thereby; decreasing the crosslinking density. Short chains have better mobility and little restriction from the rigid crosslinking networks.

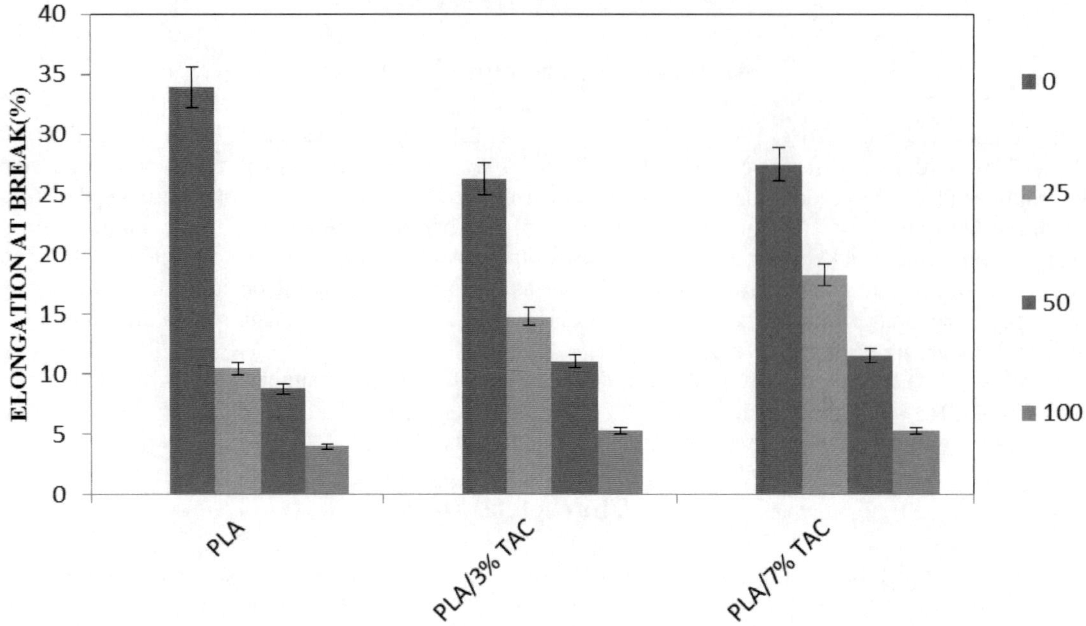

FIGURE 1. Elongation at break values of PLA and PLA/TAC composites at diffirent irradiation doses

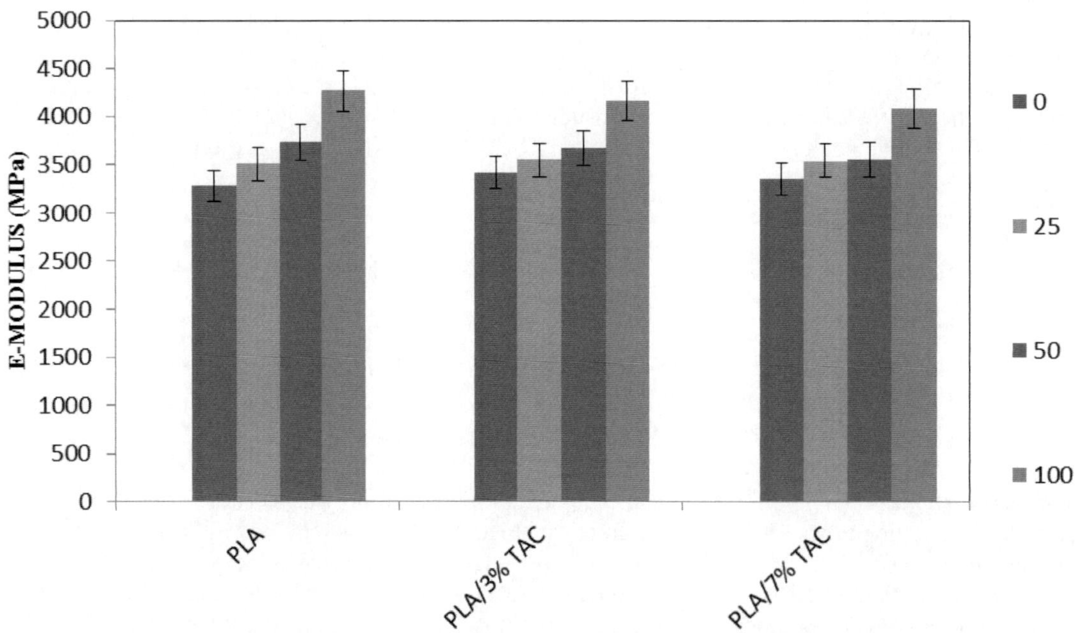

FIGURE 2. E-Modulus values of PLA and PLA/TAC composites at diffirent irradiation doses

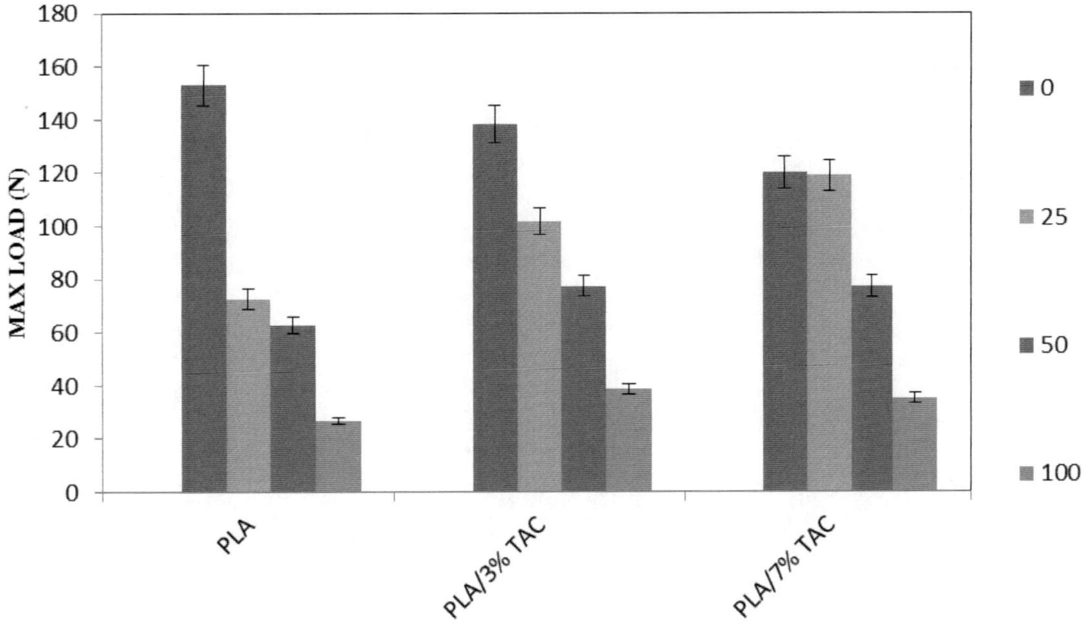

FIGURE 3. Maximum load values of PLA and PLA/TAC composites at diffirent irradiation doses

FT-IR Spectra Analysis

The changes in the chemical structure induced by γ-irradiation (dose of 0, 25, 50, and 100 kGy) exposure of neat PLA, PLA/TAC composites were analyzed by FT-IR is shown. The region 4000–2700 cm^{-1} in the spectra was observed particular bands located at 3506 cm^{-1}, that corresponds to –OH which correspond to the hydroperoxide groups probably resulting from the thermal degradation of the polymer composites during melt processing It was attributed to the formation of hydroperoxide derivatives which degrade in compounds containing a carboxylic acid end groups. Three bands situated at 2995, 2945 and 2875 cm^{-1} were attributed to the saturated CH_3 stretching bands and symmetric/asymmetric deformation bands of $–CH_3$ of PLA, respectively. After γ-irradiation exposure, the bands became stronger in intensity as the absorbed doses increased. It was obtained that a strong absorbance at 1756 cm^{-1} confirmed the presence of the ester via the carbonyl stretch (C=O). Significant changes in the chemical structure have occurred in irradiated samples, especially at 100 kGy. This indicated that γ-irradiation of the PLA/TAC leads to oxidation reactions of ester groups resulting in the formation of hydroxyl groups contained in hydroperoxides, carboxylic acids and alcohols. The amount of each species depended on the radiation dose, especially for the concentration of carboxylic groups, which were predominant when the dose was equal to 100 kGy. Similar conclusions were obtained for neat PLA.

Thermal Analysis

Before γ-irradiation, DSC thermograms of neat PLA display one single melting peak at approximately 153 °C. The melting peaks shifted to lower temperatures and became broader when the irradiation dose increased as can be seen in Figure 4. According to the literature [13], it could be the result of the polymer chains re-organization during γ-irradiation and also the occurrence of different crystal populations with different lamellar thickness. Moreover, the remaining PLA chains could have high mobility and reorganization capacity themselves lead to slightly shift to lower values in the crystallinity.

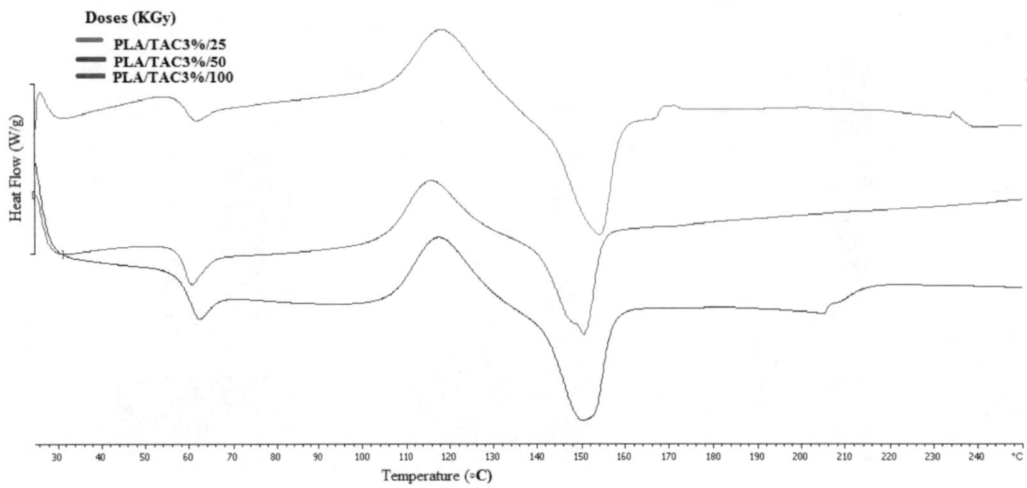

FIGURE 4. DSC thermograms of PLA/TAC composites with different doses of irradiation

CONCLUSIONS

It can be concluded that the effects of γ-irradiation resulted in chain scission and oxidative degradation in the amorphous regions. The irradiated samples turned to yellowish (at 50 kGy) in the fragmented sample. The mechanical properties were greatly dependent on the dose in the samples. The oxidation of PLA chains was considerably intensified at high doses, generating molecular species with higher mobility and induced the formation of crosslinking in the PLA matrix. These results clearly demonstrated that, strong modifications in the chemical structure, especially in 100 kGy of absorbed dose in PLA, and enhancement of these properties were observed in the presence of TAC.

REFERENCES

1. G. Johnsy et al., *Radiat. Phys. Chem.*, **76**, 1205–1212 (2007).
2. H. S. Kim et al., *J. Therm. Anal. Calorim.*, **81**, 299-306 (2005).
3. S. K. Mohanty et al., *J. Polym. Environ.*, **10**, 19-26 (2002).
4. J. Lunt, *Polym. Degrad. Stab.*, **57**, 145-152 (1998).
5. E. Hermida et al., *Macromol. Symp.*, **263**, 102-113 (2008).
6. L. Zaidi et al. *Polym. Degrad. Stab.*, **98**, 348-355 (2013).
7. P. Nugroho et al., *Polym. Degrad. Stab.*, **72**, 337-343 (2001).
8. E. C. L. Cardoso et al., *Radiat. Phys. Chem.*, **94**, 249-252 (2014).
9. N. García-Huete et al., Radiat. Phys. Chem., **102**, 108-116 (2014).
10. H.M. Said, *J. Radiat Res. Appl. Sci.*, **6**, 11-20 (2013).
11. A. Buttafava et al., *Polym. Degrad. Stab.*, **89**, 133-139 (2005).
12. C. D. Migliaresi et al., *J Appl Poly Sci.*, 43, 83-95 (1991).
13. P. N. Thanki et al., *Polymer*, **42**, 535-538 (2001).

Electrocatalytic behavior of chitosan/polyaniline nanocomposite prepared via different acidic aqueous solutions for biosensing

Ching-Feng Li and Tzong-Ming Wu

Department of Materials science and enginnering, National Chung Hsing University, No.250, Kuo kuang Rd., south dist., Taichung 402, Taiwan

chingfengli82@gmail.com

Abstract. A simple sensor based on glassy carbon electrode (GCE) modified by chitosan/polyaniline nanocomposite has been studied. The casted nanocomposite film on GCE was prepared from formic acid, acetic acid and citric acid aqueous solutions. The morphology and electrocatalytic activity of chitosan/PANI nanocomposite were characterized by scanning electron microscope (SEM), cyclic voltammetry (CV) and differential pulse voltammetry (DPV), respectively. CV results revealed that the polyaniline in the chitosan/PANI film exhibited electrocatalytic activity to redox reactions of $Fe(CN)_6^{-3/-4}$ used as probe. Among the three acids, formic acid exhibited the most appropriate polyaniline dispersion in chitosan film, which resulted in the highest electrocatalytic activity. Therefore, chitosan/PANI nanocomposite casted on GCE via formic acid, as working electrode, was used to detect dopamine (DA). The biosensor provides a linear DPV response in the range of 0.25mM to 1.75mM for DA detection.

Keywords: chitosan, polyaniline, biosensor, dopamine

Morphology of Chitosan/PANI nanocomposite

Chitosan can be dissolved in low pH values (approximately < 4).[1] In this research, purified chitosan was add into formic acid, acetic acid and citric acid, respectively. The morphology and electrocatalytic activity was observed. Since the dispersibility of the nanoparticles in the chitosan film would affect its electrocatalytic activity to redox reactions of the analytes, the SEM image of the composites were investigated. As shown in figure 1, the chitosan particles are both evenly dispered in formic acid and acetic acid, but the dispersion of PANI nanoparticles in formic acid was greater than that in acetic acid. The particle size of PANI, as shown in Figure 1(c) and (d), reveals smaller size in formic acid, which indicates a better electron transfer environment due to low surface area. The effect of surface morphology on their electrocatalytic activity and the detection ability for the analyte (DA) was investigated and presented in the following section.

Cyclic voltammetry analyses of Chitosan/PANI via different acid aqueous solutions modified GCE

Figure 2 shows the CV curves record from $Fe(CN)_6^{3-/4-}$ aqueous solutions with a pH of 7.4 at bare GCE, chitosan/PANI prepared via formic acid, acetic acid and citric acid modified GCE. As shown in Figure 2(b), the pure chitosan film modified GCE exhibited a decreased anodic peak for oxidation of $Fe(CN)_6^{3-/4-}$ as a probe compared with the bare GCE because chitosan functions as an insulator, which hinders the electron transfer. Compared with the bare GCE and chitosan modified GCEs, the chitosan/PANI modified GCE exhibited a relatively high anodic peak, which indicates that PANI is electrocatalyzed by the oxidation of $Fe(CN)_6^{3-/4-}$. This electrocatalysis can be associated with the conductivity of PANI and its particle size, which increase the surface area of the electrode.

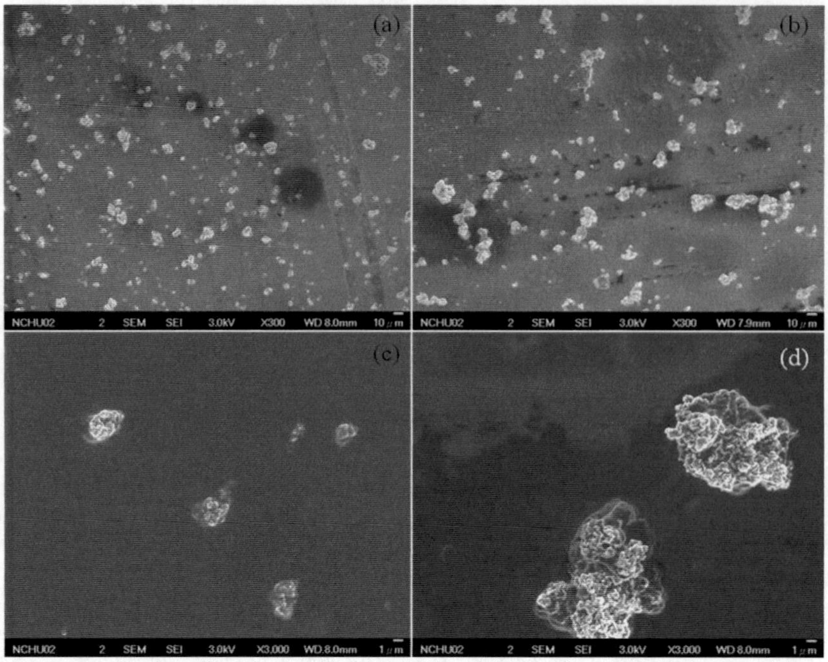

FIGURE 1. SEM image of polyaniline in (a) formic acid and (b)acetic acid. High magnificant of SEM image of polyaniline in both solution are shown in figure 1(c) and 1(d)

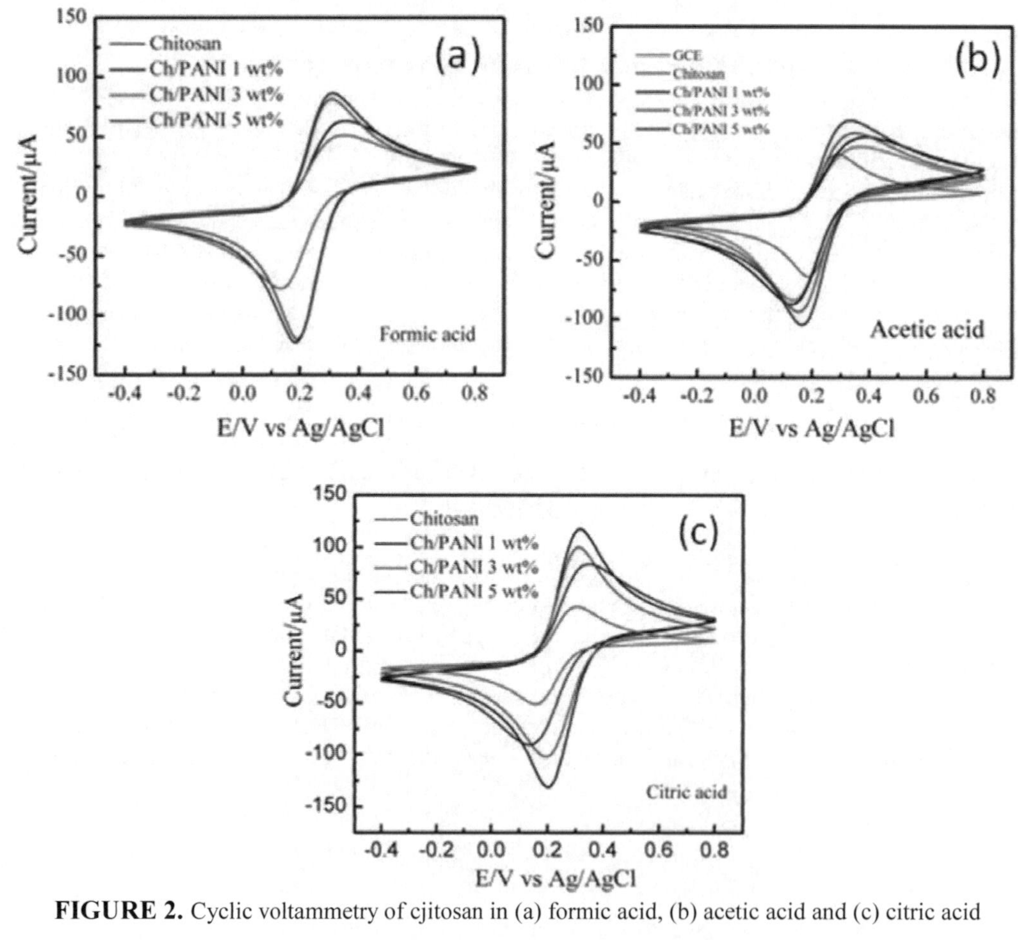

FIGURE 2. Cyclic voltammetry of cjitosan in (a) formic acid, (b) acetic acid and (c) citric acid

The effect of chitosan/PANI nanocomposite film prepared via different acid was also shown in table (1). The potentials of peak which prepared by both formic acid and acetic acid are increased when the proportion of PANI increased. The peak potential of formic acid is about 87.5V with the proportion of PANI is 5 wt%, higher than 70.25V prepared via acetic acid. Both acid can bind strongly with chitosan. The difference was caused by the dispersion and particle size of PANI. Another sample prepared via formic acid shows a different way. That is because of the -COOH and -OH group of citric acid would form a hydrogen bond with the $-NH_3$ group of chitosan, which led to aggregate between citric acid and chitosan that cause a difficult electron transfer.[2]

TABLE 1. Potentials (vs. Ag/AgCl) of anodic peaks (E_{pa}, in V) and currents of anodic peaks (I_{pa}, in μA), obtained from C-V curves recorded using chitosan/PANI-modified GCE electrodes in 5mM $Fe(CN)_6^{3-/4-}$ aqueous solution containing 0.1M PBS.

Sample Code	Acetic acid		Formic acid		Citric acid	
	E_{pa}	I_{pa}	E_{pa}	I_{pa}	E_{pa}	I_{pa}
Chitosan	0.288	41	0.340	51.75	0.298	42.48
Ch/PANI 1 wt%	0.377	25.50	0.350	63.5	0.314	119.12
Ch/PANI 3 wt%	0.343	59.25	0.307	82.25	0.309	100.70
Ch/PANI 5 wt%	0.326	70.25	0.309	87.50	0.344	84.72

* E_{pa} and I_{pa} of GCE in 5mM $Fe(CN)_6^{3-/4-}$ aqueous solution containing 0.1 M PBS is 0.361 V and 48μA, respectively.

Differential pulse voltammetry analysis of CS/PANI film cast from formic acid aqueous solution

In order to prove the sensitivity of this research, differential pulse voltammetry (DPV) was used for monitoring of DA. [3-5] The DPV signals obtained for different concentrations of DA as well as the plotted calibration curve are shown in figure 3(b). Figure 3(a) shows the peak current was increased as the DA concentration increased. The linear relationship is obtained over DA concentrations ranging from 0.5 to 1.75 mM.

The linear regression equations is:

$I_p = 0.3166 + 3.4928\ C_{DA}$ (mM)

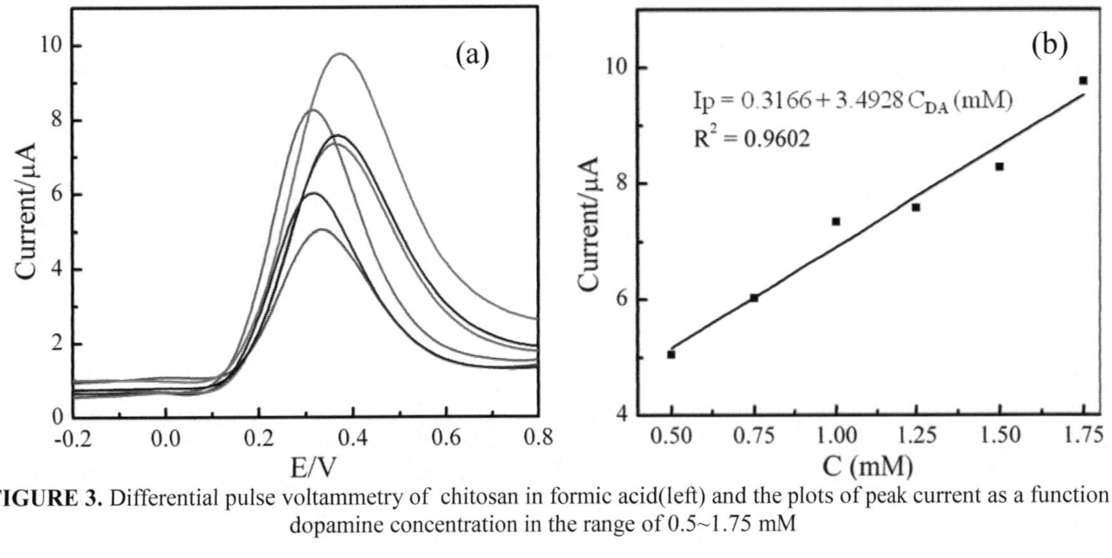

FIGURE 3. Differential pulse voltammetry of chitosan in formic acid(left) and the plots of peak current as a function of dopamine concentration in the range of 0.5~1.75 mM

References

1. Y.T. Shieh, Y.C. Tsai, andY.K. Twu, *Int. J. Electrochem.* 8, 831-845 (2013)
2. Y.T. Shieh, and H.F. Jiang, *J. Electroanal. Chem.* 736, 132-138 (2015)
3. B. Rezaei, M.K. Boroujeni, and A.A. Ensafi, *Biosens. Bioelectron.* 66, 490-496 (2015)
4. S. Dehdashtian, M. B. Gholivand, M. Shamsipur, S. Kariminia, *Mater. Sci. Eng. C* 58, 53-59 (2016)
5. X. Cao, L. Luo, Y. Ding,X. Zou, R. Bian, *Sensor. Actuat B-Chem.* 129, 941-946 (2008)

Improvement Of Mechanical Properties Due To Increased Fibre Matrix Adhesion In Flax Fibre Reinforced Polypropylene

Malfatti Andrea, Steuernagel Leif and Meiners Dieter

Institute for Polymer Materials and Plastics Engineering, Clausthal University of Technology
Agricolastraße 6, 38678 Clausthal-Zellerfeld, Germany

Andrea.Malfatti@tu-clausthal.de

Abstract. The hydrophilic nature of natural fibres and the hydrophobic behaviour of standard polymers such as polypropylene and other polyolefins are the major cause for the lack of adhesion between fibre and matrix in natural fibre reinforced plastics (NFRP's). To overcome this incompatibility of fibre and matrix, different coupling agents can be incorporated into the compound. Our investigations revealed the bonding capability of three dendritic polyesters and one thermoplastic polyurethane in a flax fibre/PP compound. The treatment with different amounts of coupling agent led to an increase in mechanical properties such as bending strength, tensile strength and most of all CHARPY impact strength. In addition, it was possible to make a correlation between IR spectra and the chemical and physical interactions between fibre, matrix and adhesive. Therefore a relationship between intermolecular forces and mechanical properties could be concluded.

Keywords: Natural fibres, adhesives, Charpy impact strength, tensile strength, bending strength, IR spectroscopy
PACS: 81.05.Lg, 81.05.Qk, 68.35.bm, 88.20.rp

INTRODUCTION

NFRP's have become an intensively researched subcategory of fibre reinforced plastics lately. Their low carbon footprint, the rising awareness of global warming and the trend to environmental friendly products increases the interest of industry, costumers and researchers in these so called green composites [1]. In spite of these opportunities challenges arise e.g. fluctuating mechanical properties and quality of the fibres, recycling of non-biodegradable matrixes and last but not least adhesion between fibre and matrix, in case of hydrophobic matrixes.

To improve the adhesion of hydrophilic fibre (e.g. flax) and hydrophobic matrix (e.g. PP) different fibre treatments are known [1-3]. In case of thermoplastic matrixes it is possible to incorporate adhesives like maleic anhydride grafted polypropylene (MAPP) into the matrix. It consists of a polypropylene main chain with irregularly grafted maleic anhydride groups as functional groups. The maleic anhydride groups react with the **OH**-groups of the natural fibre and form covalent bonds through a ring opening addition reaction. The hydrophobic polypropylene backbone interacts with the matrix polymer via secondary valence forces.

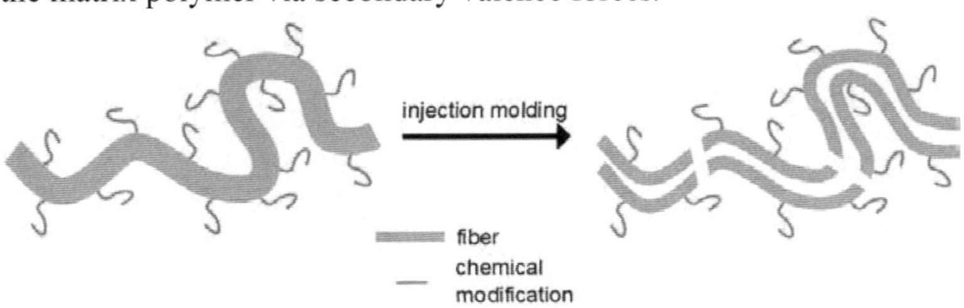

FIGURE 1. Nozzle effect on natural fibres

A chemical treatment of the fibre or a treatment of the matrix during compounding both lead to one major problem: The modification of fibre matrix adhesion is only active on the original interface of fibre and matrix. Previous results at the Institute for Polymer Materials and Plastics Engineering [4] show that natural fibres fan out during injection moulding (fig. 1). The new formed surface areas are not modified and thus are not compatible with the polymeric matrix. Thereby the expected increase of the mechanical properties due to the reduction of fibre diameter does not result in a measurable improvement because of a lack of fibre matrix adhesion. The aim of current research is to find adhesives that allow a chemical or physical interaction with the fibre surface after the nozzle of the injection moulding machine where the fibres fan out.

EXPERIMENTAL WORK

To reach the described aim four different materials where chosen to be evaluated: three dendritic polyesters and one polyurethane which is commonly used as a hot melt adhesive. The materials where provided by COVESTRO (former Bayer Material Science) and PERSTORP. The polyurethane (below PUR) has a density of 1.2 g/cm³ and a mainly linear, highly crystalline structure. The dendritic polyesters (below DP1, DP2 and DP3) vary in their molecular weight and the number of terminal **OH**-Groups. An overview of the number of end groups, the molecular weight, number of hydroxyl groups on the outside of the molecule which are able to react or interact with other molecules and the ratio of these groups per weight is given in table (1).

TABLE (1). Structural properties of the dendritic polyesters DP1, DP2 and DP3

Property	DP1	DP2	DP3
Theoretical molecular weight [g/mol]	1749.8	3607.6	7323.3
Number of end groups (**OH**)	16	32	64
Number of outer end groups (**OH**)	10	20	54
Outer end groups per weight [mol/g]	0.0057	0.0055	0.0073

The possible adhesives were used to process samples with 1 wt-%, 2 wt-%, 4 wt-% and 8 wt-% of these materials in combination with 30 wt-% flax short fibres manufactured by SACHSENLEINEN GMBH with a density of 1.4 g/cm³, having a medium diameter of 68 µm and a semi-crystalline polypropylene H734-52RNA by DOW CHEMICAL with a density of 0.9 g/cm³ and a MFI of 52 g/10min at 230 °C and 2.16 kg. Half of the prepared samples also contained 3 wt-% MAPP to investigate interactions between MAPP and the possible adhesives. The MAPP SCONA TPPP 8012 FA was supplied by KOMETRA, has an MFR of 50 g/10min - 100 g/10min at 190 °C and 2.16 kg and a MA content ≥ 0.9%.

Testing samples were prepared by compounding the mixtures in a laboratory kneader PolyLab OS with the mixing unit Rheomix OS by THERMO HAAKE at 180 °C and 50 rpm, milling the mixtures in a cutting mill by DREHER and injection moulding samples (Type 1BB, according to DIN EN ISO 527-1) with an injection moulding machine ARBURG Allrounder 220S. The temperature of the melt was 190 °C, the nozzle was heated to 200 °C. Before further testing all samples were stored for 24 h at 23 °C and 50 % r.h.. For mechanical testing at least five samples were tested.

IR-spectroscopy was performed with an Infrared Vacuum Spectrometer and an ATR-Unit Platinum ATR Diamond F Vacuum from BRUKER and a BRUCKER Alpha T-IR. Tensile testing was carried out with a universal testing machine by ZWICK ROELL according to DIN EN ISO 527-4. The free clamping length was 22 mm, the testing velocity was 2 mm/min. Creep behaviour was investigated using a dynamic mechanical analysis DMA Q800 by TA INSTRUMENTS. The sample was constantly loaded at 40°C for 500 min. Charpy impact strength was measured with a pendulum machine Zwick 5113 according to DIN EN ISO 179 with a 1 J hammer and an effective span of 20 mm.

RESULTS

IR-Spectroscopy

To investigate possible reactions of the raw materials different combinations were mixed and melted using a hot stage. The spectrum of DP1 mixed with MAPP is shown in figure 2 including the spectra of the raw materials. Since the dendritic polyesters all possess **OH**-groups, a reaction similar to the reaction of MAPP and the hydroxyl-groups of the flax fibre surface may take place. Evidence that this reaction happens is the loss of intensity of the **OH**-peak at about 3300 cm^{-1}. The hydroxyl band is still visible in the spectrum of the mixed sample but it is clearly weakened compared to the pure DP1 spectrum. The characteristic peak of the O=C-O-C=O-function of the MAPP remains with a lessened intensity in the mixed spectrum, although the reaction should result in a ring opening of the MA-group. From these observations it is possible to infer that a reaction of DP1 and MAPP takes place. Due to a not quantitative mixed sample not all functional groups are able to react. The spectra of the other dendritic polyesters DP2 und DP3 and their mixtures with MAPP look analogue to the given spectra in figure 2.

For PUR and MAPP a similar reaction of the MA-group and the NH-group is theoretically conceivable. The IR-spectra show no evidence for such a reaction. If an interaction between those two ingredients takes place it has to be a physical interaction due to a lack of detectable covalent bonds.

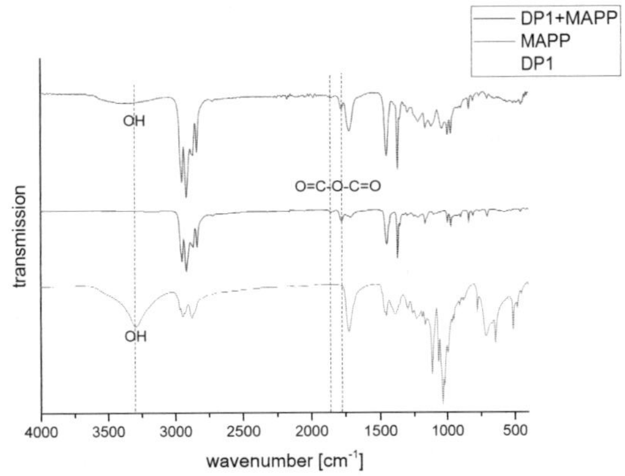

FIGURE 2. IR-spectra of DP1+MAPP, MAPP and DP1

To prepare samples that show the interactions on the surface of the fibre without peaks of PP the originally compounded mixtures dissolved in decalin at 160°C for 2 h, filtered and dried. This procedure is normally used to determine the fibre volume content of composites because it dissolves all not bonded matrix molecules.

The spectra of the so treated fibres all show different wavenumbers for the carbonyl band in the region from 1734 cm^{-1} to 1722 cm^{-1}. The C=O-group can be found in the chemical structure of the dendritic polyesters, the polyurethane as well as the fibres. Therefore the chemical shift can be correlated with interaction of these components. Regarding that, it is clear that the fibre matrix adhesion originates only physical interactions like hydrogen bonding. The strength of the secondary valence forces can be linked to the value of the chemical shift compared to an untreated flax fibre. The data in table (2) show that a combination of 1 wt-% DP1 and 3 wt-% MAPP leads to the highest shift and thereby to the highest amount of interactions. Shifts for DP2 and DP3 have similar values, while the PUR samples show a lower shift and interaction. Noteworthy is the fact that the samples with MAPP have higher shifts than the samples without MAPP with the same total amount of adhesive.

TABLE (2). Chemical shift of the carbonyl band for differently treated fibres

Property	**No treatment**	**1 wt-% DP1 + 3 wt-% MAPP**	**4 wt-% DP1**	**1 wt-% PUR + 3 wt-% MAPP**	**4 wt-% PUR**
C=O-band [cm^{-1}]	1734	1722	1724	1729	1730
Chemical shift compared to flax fibre [cm^{-1}]	0	12	10	5	4

Tensile Testing

Figure 3 shows the YOUNG's modulus and the tensile strength of the prepared composite samples. As a reference a flax fibre composite with 30 wt-% fibre content and a composite with 3 wt-% MAPP are included in the graphs.

The use of the DPs tends to result in a higher YOUNG's modulus but also decreases the tensile strength. Although there is no significant relation between YOUNG's modulus and the mass fraction of DP it is clear to see that both DP with and without MAPP increase the modulus. When looking at the tensile strength only the combination of 1 wt-% DP and 3 wt-% MAPP achieves an improvement compared to the sample without DP with MAPP. In addition, the samples with DP and MAPP show an apparent correlation between the tensile strength and the chemical structure of the polyesters. The small DP1 with least terminal groups also has the lowest strength, followed by the bigger and more polar DP2 and the biggest and most polar DP3 which offers the highest strength. Since this behaviour is more distinct in combination with MAPP, an interaction between MAPP and the DPs can be concluded.

For the polyurethane a small increase in the YOUNG's modulus can be noted. Without MAPP the decrease of the tensile strength with growing amount of PUR is visible. With MAPP, the level of the tensile strength is nearly constant which leads to the conclusion that the combination of PUR and MAPP is able to improve the fibre matrix interactions. Still the measured mechanical properties are lower than those of the DP composites.

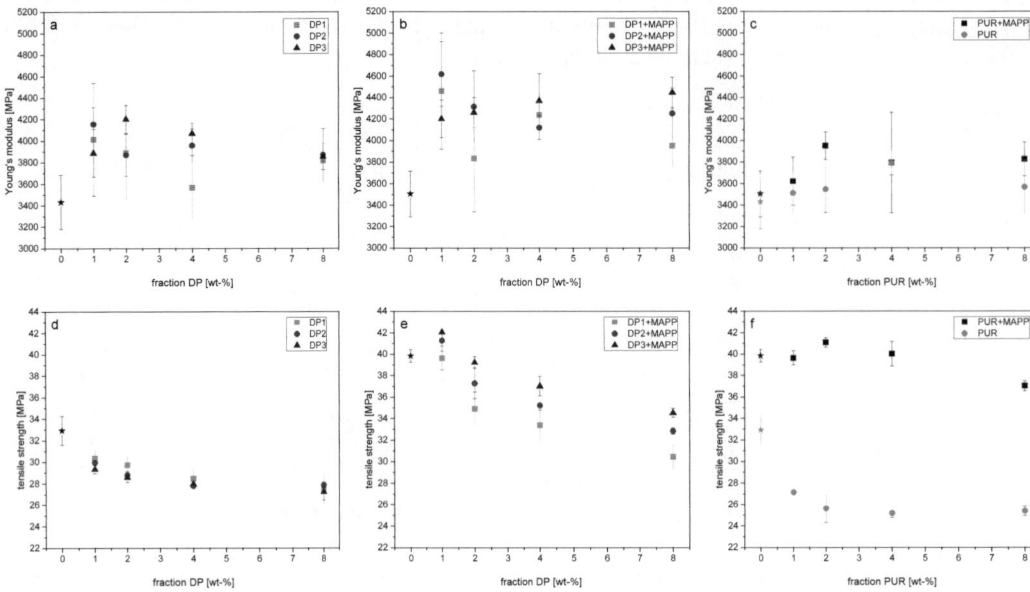

FIGURE 3. YOUNG's modulus and tensile strength of the dendritic polyester composites without (a and d) and with MAPP (b and e) and of the polyurethane composites (c and f)

Creep Behaviour

The creep compliance over time of the investigated adhesives is depicted in figure 4. All of the DPs soften the material and reduce the long-time mechanical stability. In combination with MAPP only the mixture with 1 wt-% DP1 reached the creep compliance of a reinforced sample, which is a reduction in comparison with the sample without DP1. For the PUR composites without MAPP, an increase of the creep compliance can be noted. The samples with MAPP show a better creep behaviour than the ones without PUR or with DPs.

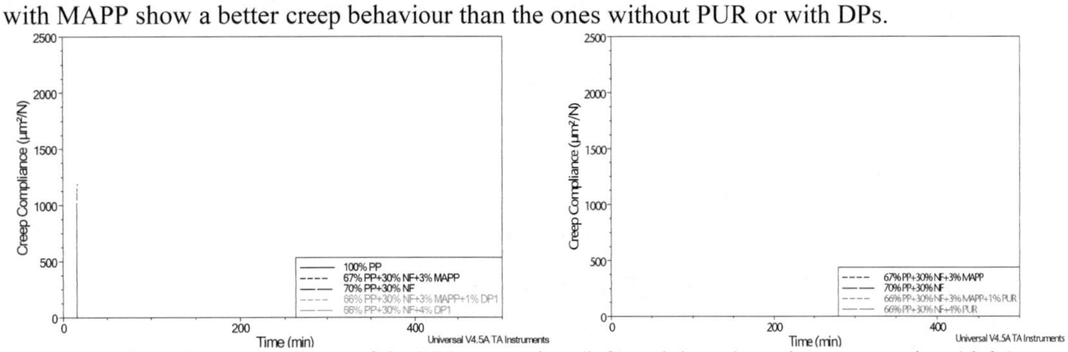

FIGURE 4. Creep compliance of the DP1 composites (left) and the polyurethane composites (right)

CHARPY Impact Strength

The DPs have a major influence on the CHARPY impact strength which can be seen in figure 5. Even without MAPP one percent of the DPs can increase the impact strength by a factor of approximately three. With 3 wt-% MAPP this rise is advanced to a factor up to four. The softening effect observed during tensile testing and DMA now has a positive effect on the distribution of the impact energy through the sample. Still a too high content of the polyesters weakens the composite so that the optimum of the impact strength is marked at 1 wt-%.

In contrast to the DP samples the mixtures with PUR show a lower CHARPY impact strength that is nearly constant in the examined weight fraction range. The addition of PUR leads to an embrittlement of the natural fibre composite. As observed before the samples with MAPP lead to better mechanical properties than those without.

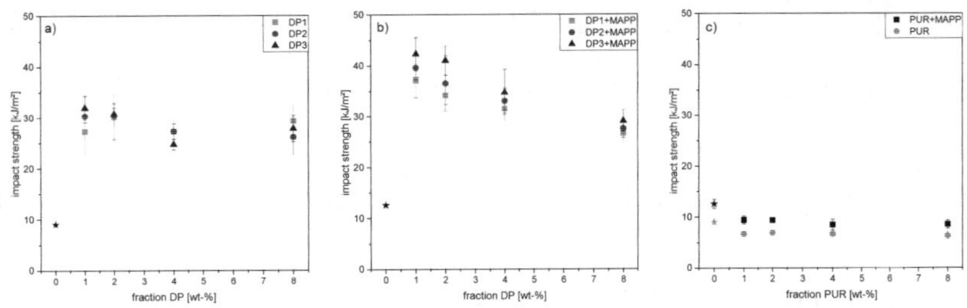

FIGURE 5. CHARPY impact strength of the composites with the DPs (a and b) and PUR (c)

CONCLUSION

The presented results show, that the use of the DPs and the PUR as adhesives for flax fibre reinforced polypropylene is possible with certain restrictions. Figure 6 shows a schematic representation of possible interactions between the different components.

For the DPs, a small increase of the YOUNG's modulus and the tensile strength was observed. Creep resistance drops while the impact strength increases highly. The best results were reached in combination with MAPP. Based on the IR-spectroscopy results the increase in mechanical properties can be correlated with a reaction of MAPP with the DPs resulting in a partly covalent/partly physical fiber matrix adhesion. A too high amount of the DPs leads to accumulations of the relatively small molecules that in turn cause a softening effect which supports the impact strength but also creeping and a loss of the tensile strength.

For the PUR, no covalent interactions with other components were detected. Interactions with the fibre can only occur at the urethane-group. The rest of the molecule is a nonpolar linear chain that is better compatible with the PP matrix than the polar DPs. This is the main reason why both increase and decrease of mechanical properties compared to the reference material is less strongly pronounced. Only under bending load the PUR composites surpass the DP samples which may be correlated with a higher stiffness.

Both types of adhesives should be used in combination with and not as a replacement of MAPP and should be selected in dependence of the future application.

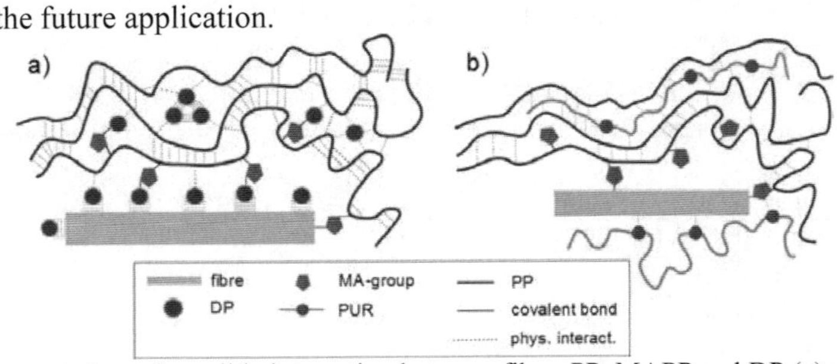

FIGURE 6. Schematic representation of possible interaction between fibre, PP, MAPP and DP (a) and PUR (b) respectively

ACKNOWLEDGMENTS

Special thanks go to Mrs. Dipl.-Ing. Gabriella-Paula Brunotte (Institute for Polymer Materials and Plastics Engineering), Mrs. Karin Bode (Institute for Inorganic and Analytical Chemistry) and the employees of the Institute for Organic Chemistry (all TU Clausthal) for the assistance with the performed measurements. We also like to thank Dr.-Ing. habil. Bernd Weidenfeller for providing the kneader for compounding.

REFERENCES

1. F.P. La Mantia, M. Morreale, *Composites: Part A* **42**, 579-588 (2011).
2. A. El-Sabbagh, *Composites: Part B* **57**, 126-135 (2014).
3. X. Li, L.G. Tabil, S. Panigrahi, *J. Polym. Environ.* **15**, 25-33 (2007).
4. L. Steuernagel, "Natural fibre Reinforced Plastics – Effect of Injection Molding and Primer on Mechanical Properties" presentation in narotec, Erfurt 2014

Effect of Sorbitol/Glycerol Ratio on Hydrogen Bond Formation and Crystallization of Co-Plasticized Starch

Mohsen Esmaeili[a], Gholamreza Pircheraghi[a] and Reza Bagheri[a*]

[a]*Department of Materials Science and Engineering, Sharif University of Technology - Tehran - Iran,*

mohsen_esmaeeli@mehr.sharif.ir
rezabagh@sharif.edu

Abstract. Thermoplastic starch (TPS) samples have been prepared in the present study using glycerol and sorbitol as co-plasticizers. Two different amounts of plasticizer, i.e. 36 and 42 wt.%, and sorbitol/glycerol ratios of 3:0, 2:1, 1:2 and 0:3 were incorporated for plasticizing granular corn starch using a laboratory co-rotating twin screw extruder. The hydrogen bond formation between plasticizers and starch chains was investigated by Fourier Transform Infrared Spectroscopy (FT-IR). X-ray diffraction (XRD) was also employed to study the crystalline structure of TPS over a period of two months to investigate the possible retrogradation process of starch. FT-IR results illustrated that the strength of starch-plasticizer hydrogen bond enhances by increasing the total amount of plasticizer and sorbitol/glycerol ratio. XRD spectra revealed A-type crystals of native starch are turned to anhydrous Va-type crystals after gelatinization. However, retrogradation and formation of hydrated Vh-type crystals may occur in the samples over storage time. Based on XRD results, glycerol plasticized starch samples are extremely prone to retrogradation as evidenced by formation of Vh-type characteristic peak. While, in the co-plasticized starch samples the mentioned peak is weakened and totally disappeared if only sorbitol is used for plasticization. In other words, the retrogradation can be effectively suppressed in the samples plasticized with higher amount of sorbitol. This is attributed to the strong hydrogen bonding between the starch chains and the sorbitol molecules.

Keywords: Thermoplastic Starch, Sorbitol, Glycerol, Retrogradation.
PACS: 82.35.Pq

SAMPLE PREPARATION

Starch flakes were dried at 80°C for 12 hours before the extrusion process. Dried starch was mixed by proper amounts of sorbitol and glycerol and prepared for melt blending. The composition of samples is presented in table (1).

TABLE (1). Material's abbreviations and corresponding sample compositions

Sample's abbreviation	Total plasticizer (wt. %)	Sorbitol:glycerol (wt./wt.)
TPS363600	36	3:0
TPS360036	36	0:3
TPS424200	42	3:0
TPS422814	42	2:1
TPS421428	42	1:2
TPS420042	42	0:3

A co-rotating twin extruder, SHJ-20 Nanjing Giant, with L/D ratio of 20 and temperature profile of 120°C-130°C-130°C-135°C from hoper to die, was used for melt compounding of different samples. In order to achieve enough time and shear rate for mixing starch and plasticizers, the screw speed was set on 100 rpm.

HYDROGEN BOND FORMATION

The inferared spectra of the native starch and the extruded plasticized samples were measured with ABB Bomen MB-100 spectrum scanner by mixing the fine powder with KBr and pressing. The transparent slices tested by transmission method. The analysis of FT-IR spectra of samples provides the identification of hydrogen bonds[1]. According to harmonic oscillator model, the decreasing of force constant (ΔF) could be calculate by eq. (1):

$$\Delta F = f_p - f_{np} = \frac{\mu(\upsilon_p^2 - \upsilon_{np}^2)}{4\pi^2} \quad (1)$$

Where μ=m1m2/(m1+m2) related to the declined mass of oscillator, υ was the frequency of oscillator and f was the force constant. The subscripts p and np corresponded plasticized and non-plasticized oscillator, respectively. υ_{np} related to wavenumber of native starch and μ was constant for the specific two oscillator, thus variations of ΔF directly corresponded to υ_p which is wavenumber of plasticized starch. So, the lower the wavenumber means the stronger interaction[2].

The results of FT-IR spectrums of native starch and plasticized samples are illustrated in Figure 1 for evaluation of hydrogen bonds between starch and plasticizers and comparing effect of sorbitol with glycerol in plasticization. There were three characteristic peaks between 900cm-1 and 1200cm-1: the characteristic peaks 1159cm-1 and 1084cm-1 corresponded to stretching C–O of C–O–H group (OH group of starch which participate in the hydrogen bonding), 1016cm-1 was characteristic peak of stretching C–O of C–O–C group that corresponded in starch with C–O–C group in the anhydroglucose ring[3].

As illustrated in Figure 1 both peaks of stretching C–O of C–O–H shifted to lower wavenumbers that means more stable and stronger hydrogen bond formation. One can see that, in the samples plasticized only with sorbitol, the H-bonding characteristic peaks are more shifted to lower wave numbers compared to counterpart samples plasticized with glycerol. This result indicates the formation of stronger hydrogen bonds between sorbiotol-starch compared to glycerol-starch, as schematically illustrated in Figure 1.

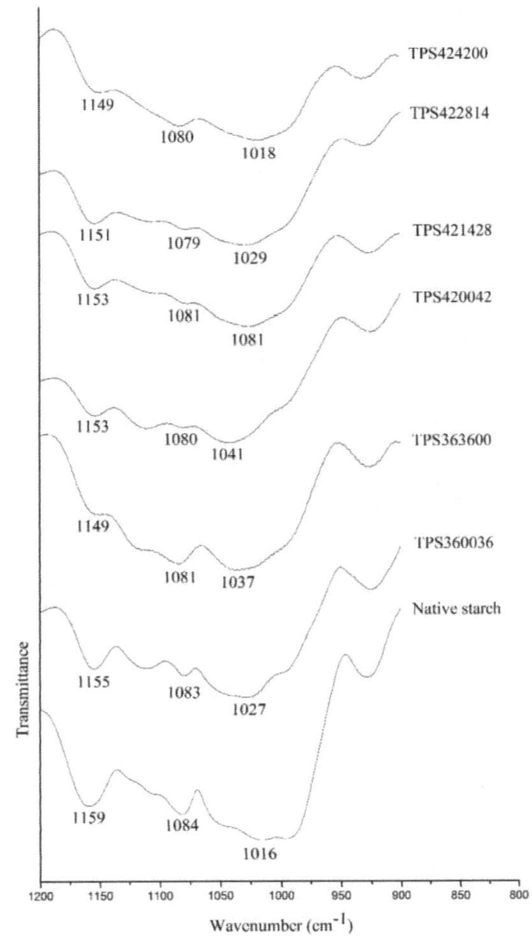

FIGURE 1. FT-IR spectrums of native starch, TPS363600, TPS360036, TPS420042, TPS421428, TPS422814 and TPS424200.

CRYSTALLIZATION AND RETROGRADATION

After two months of storage time, the XRD pattern of the extruded TPS pellets and native starch were recorded with STOE Stadi p diffractometer (Germany) over a scattering angle (2θ) from 5° to 45° at 1°/min. For evaluation of retrogradation, the X-Ray diffractometry was performed on different plasticized samples as well as native starch and the XRD results obtained after 2 month storage, are shown in Figure 2.

Compared to native starch the plasticized samples showed different peaks. Furthermore, crystal types and crystallinity were different in GTPS and STPS samples. It is reported that [4] the crystal type of native starch is double helix A-type which can be turned into Va-type crystal after plasticization. Va-type crystal is non-stable crystal in starch and can be transformed to Vh-type crystal on storage.

GTPS is extremely prone to retrogradation and reorganization to Vh-type crystal which consist of amylose and glycerol chains. While, as shown in figure 2, Vh-type crystals were not formed in STPS and Va-type crystals stayed stable after 2 months, indicating the more resistance of sorbitol plasticized starch against retrogradation. This phenomenon can be attributed to relatively long molecular length and more hydrogen bonding formed between sorbitol molecules and starch chains which caused prevention of Vh-type reforming. The XRD results are in agreement with FTIR and DSC results.

The retrogradation is heavily dependent on hydrogen bond formation between starch molecules and plasticizers and as it presented in FT-IR results; STPS shows stronger hydrogen bonds. Moreover, Va-type crystal contain lesser water molecules rather than Vh-type. The more glycerol percentage, the more Vh-type crystal and the lesser Tg temperature and more flexibility as indicated in DSC curves and tensile results. In the other words, XRD results demonstrated that presence of Va-type crystal by adding sorbitol can decrease the hydrophilicity of TPS and keep TPS resistant against recrystallization.

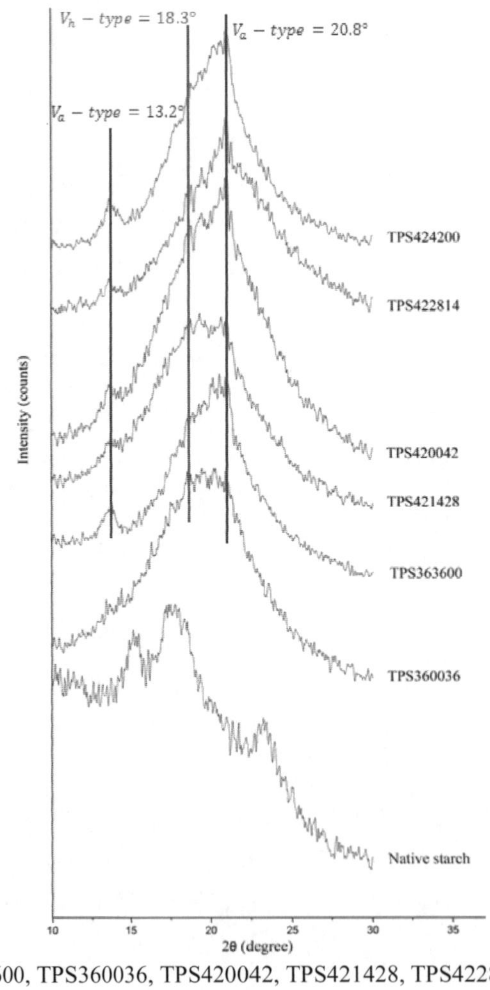

FIGURE 2. XRD curves of TPS363600, TPS360036, TPS420042, TPS421428, TPS422814 and TPS424200 after 2 months.

REFERENCES

1. K. Aoi, A. Takasu, M. Tsuchiya and M. Okada, *New chitin-based polymer hybrids, 3. Miscibility of chitin-graft-poly (2-ethyl-2-oxazoline) with poly (vinyl alcohol)*, Macromolecular Chemistry and Physics, 1998, pp. 2805-2811.
2. A. Pawlak and M. Mucha, *Thermogravimetric and FTIR studies of chitosan blends*, Thermochimica acta, 2003, pp. 153-166.
3. J. M. Fang, P. A. Fowler, J. Tomkinson, and C. A. S. Hill, *The preparation and characterisation of a series of chemically modified potato starches*, Carbohydrate Polymers, 2002, pp. 245-252.
4. JJG Van Soest and JFG Vliegenthart, *Crystallinity in starch plastics: consequences for material properties*, Trends in biotechnology, 1997, pp. 208-213.

PBAT/Thermoplastic Starch Blends: ''Effects of Oxidized Starch and Compatibilizer Content''

Buse Nur CAN, Guralp OZKOC*

Department of Chemical Engineering, Kocaeli University, 41380, Kocaeli Turkey

busenurcan@gmail.com
guralp.ozkoc@kocaeli.edu.tr

Abstract. In this study, the biodegradable blends of thermoplastic starch/poly(butylene adipate-co-terephtalate) (TPS/PBAT) and oxidized thermoplastic starch/poly(butylene adipate-co-terephthalate) (OTPS/PBAT) were melt compounded in the presence of styrene-acrylic multi-epoxidized polymeric compatibilizer. The need for OTPS is to arrange the flow properties of starch phase and to balance the hydrophilicity. The compatibilizer is used to enhance the interfacial properties and the phase dimensions. The TPS/PBAT, OTPS/PBAT and the compatibilizer ratio were selected as material parameters. The TPS/PBAT and OTPS/PBAT blends with and without compatibilizer were characterized by tensile, impact tests; differential scanning calorimeter (DSC), thermogravimetric analysis (TGA) and Fourier transform infrared spectroscopy (FTIR). The rheological behavior was investigated by means of oscillatory rotational rheometer. The morphology investigated by means of scanning electron microscopy (SEM). It was founded that the multi-epoxide compatibilizer improved the mechanical and rheological properties of the blends.

Keywords: Thermoplastic starch, poly(butlylene adipate-co-terephthalate), compatibilization, blending.
PACS: 61.25.hk

INTRODUCTION

The usage of plastics parallel to the modernization of life has increased in daily life in the last decade. This increase of using plastics causes solid waste problems. In that case, in order to prevent the growth of environmental pollution, many researches have been done on biodegradable polymers. There is a strong market need to develop biodegradable plastic packaging products, which can be safely and effectively disposed in soil or in compost, but retaining the satisfied material properties. Because of the demand for renewable resources, interest in polymers produced using non-renewable sources such as poly(butlylene adipate-co-terephthalate) (PBAT) has increased.

PBAT, an aromatic-aliphatic copolyester, is a typical example of biopolymer obtained from petrochemical resources, which is non-renewable [1]. The physical and mechanical properties of PBAT are close to those of polyethylene. However, the application of PBAT is limited due to its high cost. Starch, an example of biopolymer from renewable resources, has been considered as filler with a high potential to be blended with synthetic biodegradable polymers to reduce their cost [2]. Thermoplastic starch (TPS) is generally utilized in the polymer blends. In TPS, the starch granules are plasticized using plasticizers such as water or glycerol under heating. On the other hand, the hydrophilicity difference between starch and PBAT makes these blend components incompatible and adversely affect the resulting properties [1]. In order to overcome the problem of incompatibility of starch and synthetic polymer blends, the reactive compatibilizers can be used [3]. Moreover, the oxidation of starch also decreases its hydrophilicity and can help in compatibilization of TPS/PBAT blends.

The aim of this study is to compare the properties of TPS and OTPS in blends and to investigate the effect of compatibilizer on the properties of TPS/PBAT and OTPS/PBAT blends.

EXPERIMENTAL

Materials and Methods

The biodegradable poly(butylene adipate-co-terephthalate) (PBAT) was kindly provided by SASA Polyester Industry Company (Adana, Turkey). The native and oxidized starches were obtained from SUNAR Starch Company (Adana, Turkey). The compatibilizer (Joncryl ADR) was kindly supplied by BASF in Germany.

The native - oxidized starches and PBAT were dried in an owen at 80°C and at 100°C, respectively prior to processing. TPS/PBAT and OTPS/PBAT blends were prepared in a laboratory twin-screw microcompounder (15ml Microcompounder, Xplore Instruments, The Netherlands) having different TPS/PBAT-OTPS/PBAT and the styrene-acrylic multi-epoxy functional compatibilizer (Joncry) loading level by using one-step melt-compounding process at a screw speed of 100 rpm for 2 min. The TPS/PBAT-OTPS/PBAT ratio (0.25, 50, 75, 100 wt%) and the compatibilizer ratio (0-5 wt%) were selected as a material parameters. The standard dog bone and Izod samples, were obtained via Xplore micro-injection molding device for mechanical and thermal analysis.

Characterization

The tensile properties were determined using an Instron Universal Testing machine following the procedure ISO 527. The crosshead speed of 10 mm/min was used. The Izod impact strength was measured with an Instron/Ceast 9050 test instrument according to ISO 180.

Contact angles were measured by using an Attension Theta Lite model contact angle machine. The distillate water was used as a probe liquid in order to obtain the hydrophilicity of TPS, OTPS and PBAT.

Differential scanning calorimetry analysis (DSC) were run using a Mettler Toledo DSC-1 Star Model instrument at the temperature range of -60°C to 200°C with the heating rate of 10 °C/min under nitrogen flow.

Thermogravimetric analysis (TGA) was performed using a Mettler Toledo TGA1 instrument to determine the thermal stability of the blends under nitrogen atmosphere. Test samples were heated from room temperature to 800 °C at a heating rate of 20°C/min.

RESULTS AND DISCUSSIONS

Tensile strength, elongation at break and Young modulus values of blends improved as compatibilizer content increased regardless of TPS or OTPS content in the blends. It was obtained that incorporation of 5% compatibilizer to the blends improved the tensile strength and elongation at break as a result of better interfacial adhesion in OTPS blend. The compatibilization was seen to be effective on the impact strength of blends in a positive way. This could be associated to the finer morphology in the presence of compatibilizer.

TABLE 1. Mechanical properties of the blends

Blend Composition	Compatibilizer Ratio (%)	Notched Charpy Impact Strength (kJ/m^2)	Tensile strength (MPa)	Tensile Modulus (MPa)	Elongation at Break (%)
50/50 PBAT/OTPS	0	5.4	0.5	30.6	156
	2	3.5	6.2	39.7	33.5
	3	4.0	6.5	37.7	46.7
	5	4.1	6.1	41.3	47.4
50/50 PBAT/TPS	0	2.5	6.5	74.7	17.7
	2	4.3	5.0	47.2	41.1
	3	4.4	7.1	41.9	31.5
	5	4.0	7.4	40.1	32.8

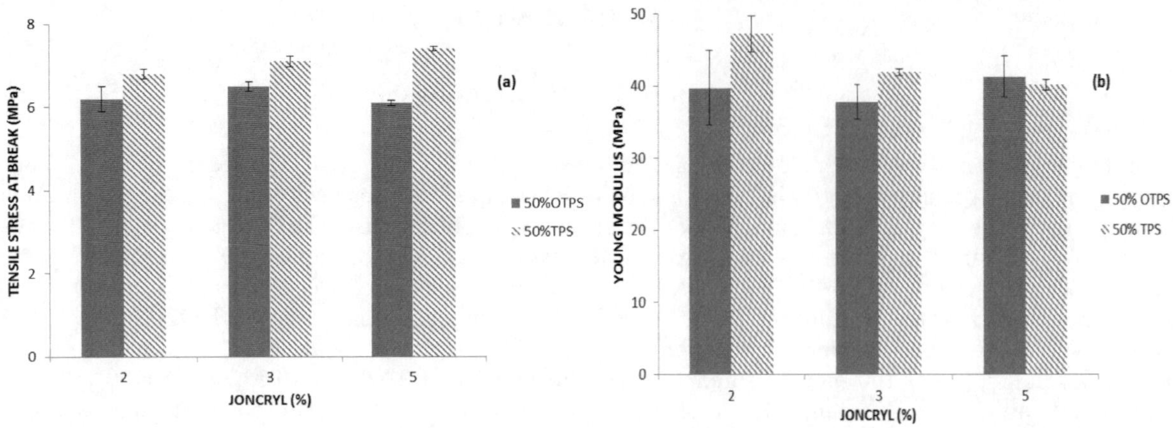

FIGURE 1. Mechanical properties of 50% TPS/PBAT and OTPS/PBAT blends including 2, 3, 5 wt% compatibilizer (a) tensile strees at break and (b) Young modulus

It was obtained from contact angle results that the hydrophilicity of neat OTPS was higher than TPS, meaning that it was more compatible with PBAT, which has a hydrophobic nature (Figure 2).

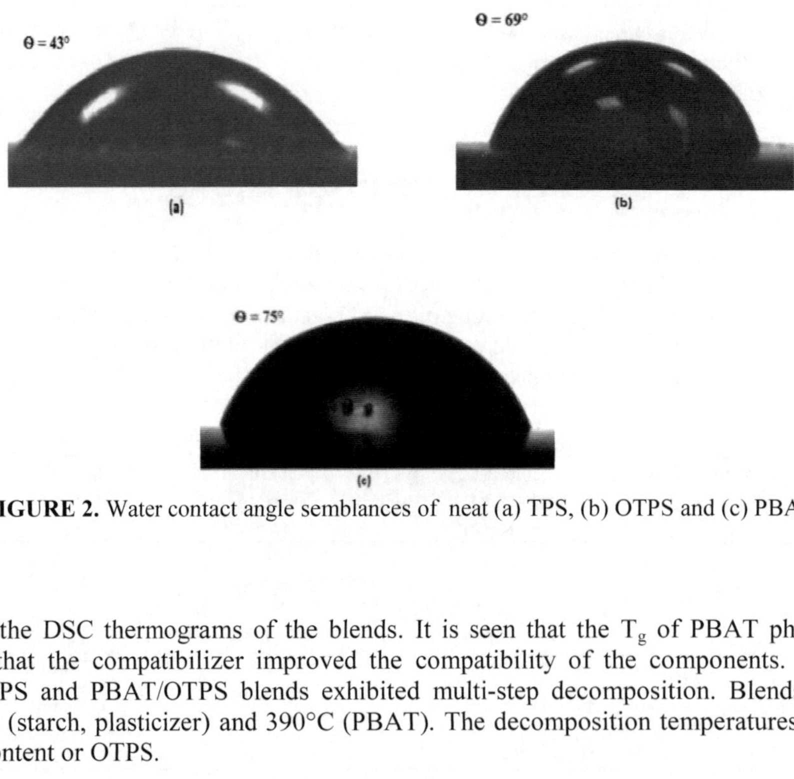

FIGURE 2. Water contact angle semblances of neat (a) TPS, (b) OTPS and (c) PBAT.

Figure 3-a shows the DSC thermograms of the blends. It is seen that the T_g of PBAT phase shifted towards starch. This implies that the compatibilizer improved the compatibility of the components. TGA thermograms showed that PBAT/TPS and PBAT/OTPS blends exhibited multi-step decomposition. Blends pyrolyzed around 120°C (water), 270°C (starch, plasticizer) and 390°C (PBAT). The decomposition temperatures did not vary much with compatibilizer content or OTPS.

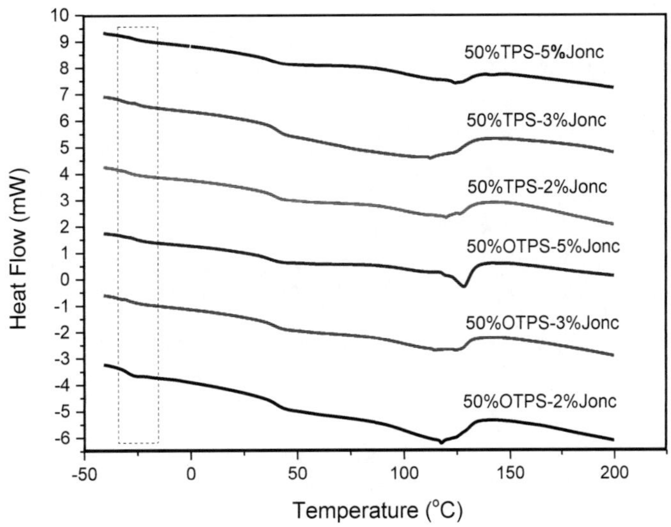

(a)

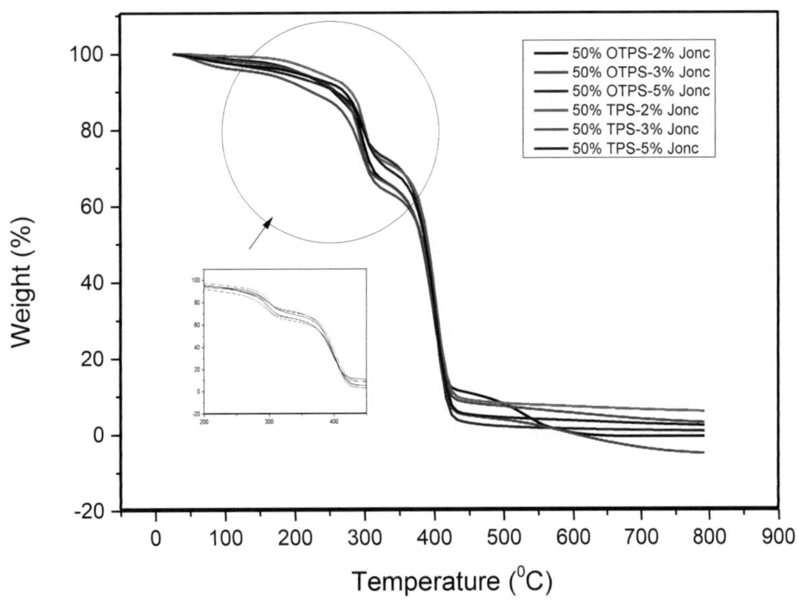

(b)

FIGURE 3. Thermal properties of 50% TPS/PBAT and OTPS/PBAT blends having 2, 3, 5 wt% compatibilizer (a) DSC thermogram (b) TGA curves

CONCLUSIONS

In general, it can be concluded that PBAT/OTPS blends were successively compatibilized by melt compounding. The properties of this blend system are seen to be highly depended on compatibilizer ratio. It was observed that the multi-epoxide compatibilizer improved the mechanical, thermal and rheological properties which is not shown in this abstract.

REFERENCES

1. D. Wei, H. Wang, H. Xiao, A. Zheng and Y. Yang, *Carbohydrate Polymer* **123,** 275 (2015).
2. Y. Zhang, C. Rempel, Q. Liu, *Critical Reviews in Food Science of Nutrition* **54:10**.1353-1370 (201,4).
3. A.R. Rahmat, W.A.W.A. Rahman, L.T. Sin, A.A. Yusuf, Material Science and Engineering C **29**, 2370-2377 (2009)

Preparation of PLLA/PDLA Blends with High Stereocomplex Crystals Content through an Efficient Thermal Treatment Method

Zheng-Chi Zhang[a], Gan-Ji Zhong[a], Zhong-Ming Li[a]

[a]*College of Polymer Science and Engineering, State Key Laboratory of Polymer Materials of Engineering, Sichuan University, Chengdu 610065, China*

Introduction

Polylactides (poly D-lactide and poly L-lactide), produced by abundant renewable naturally occurring corns, sugars or beets, are biodegradable and compostable. It has been used in many applications because of their biocompatibility and biodegradability. However, in comparison with other commercial thermoplastics derived from petroleum, it exhibits a relatively low heat resistance because of its low crystallization rate, which restricts its application in various fields. Fortunately, since the 1990s and the pioneer works of Ikada et al., poly(L-lactide) (PLLA) and its enantiomeric opposite, poly(D-lactide) (PDLA), have been known to form specific stereocomplex crystal (SC) upon mixing in solution or in bulk.[1] These SCs melting at a temperature of about 50 °C higher than the melting point of PLA homochiral crystals (HC) provide a promising solution to high thermal resistant PLA products.[1,2] Nevertheless, it is not an easy task to obtain samples with high SC content especially in melting blened molecular weight PLLA/PDLA blends due to the microscopic phase separation into a PDLA-rich and a PLLA-rich phase.[3,4]

Under these circumstances, several researchers are trying to develop an efficient method to prepare well-stereocomplex materials from high-molecular weight PLLA and PDLA through melting blend. In the present work, thermal treatment conditions were carefully determined by using DSC tests to obtain complete SC samples. The optimal thermal treatment temperature seemed to be higher than the melting temperature of HC but lower than that of SC. Employing time-resolved wide-angle X-ray diffraction (WAXD), we traced the crystallization process during thermal treatment under the optimal treatment conditions and proposed the template effect of residual SCs, which is an efficient and promising method to be applied into conventional processed samples.

Experiments

Materials. PLLA (trade mark L130 with $M_w = 17.3 \times 10^4$ g/mol, $M_n = 8.9 \times$

10^4 g/mol and PDI=1.95 according to GPC tests) and PDLA (trade mark D1010 with M_w=8.5 × 10^4 g/mol, M_n=4.7 × 10^4 g/mol and PDI=1.82 according to GPC tests), a semicrystalline grade PLLA containing around 0.2% D-LA or L-LA, respectively, was supplied by Corbion Company (Netherland).

Preparation of PLLA/PDLA injection-molded parts. To avoid the degradation due to hydrolysis and the formation of voids during processing, PLLA and PDLA pellets were dried at 80 °C under vacuum overnight before extrusion. The PLLA/PDLA blends were melt mixed by employing a twin-screw extruder to produce pellets for subsequent injection molding. The screw speed of the twin-screw extruder was kept constant at 240 rpm, and the processing temperature profile was 190-230°C from hopper to die. The blended pellets were injection molded into dumbbell bars by utilizing a commercial injection molding machine. The temperature profile for injection molding was 210, 240, 250, 250, and 230°C from hopper to nozzle, respectively. The mold temperature and packing pressure was 120 °C and 60 MPa, respectively.

Characterization of SC crystallization during thermal treatment.

DSC measurements. DSC experiments were carried out by utilizing DSC Q2000 (TA Instrument, US) to determine optimal thermal treatment temperature and treatment time. The endothermic and exothermic information of samples cut from injection-molded bars were collected during the following procedure: (i) treating at different temperatures for 3min or at 210 °C for different durations with a foregoing heating process (30 °C/min) and subsequent cooling process (-10 °C/min); (ii) heating at a rate of 10 °C/min to 260 °C to determine the final crystalline information (crystal forms and crystallinities).

In situ WAXD measurements. Synchrotron wide-angle X-ray diffraction (WAXD) experiments were carried out at BL16B (λ=0.124 nm), Shanghai Synchrotron Radiation Facility (SSRF) in order to investigate the crystallization behavior during heat treatment of samples cut from injection-molded PLLA/PDLA bars. A MAR CCD detector (MAR-USA) with a resolution of 1024 ×1024 pixels (pixel size =80μm) was used to acquire data. A modified Linkam THMS-600 stage was employed to precisely control thermal history of the samples. The detail experimental procedures were set as follows: (i) heating at a rate of 30 °C/min from room temperature to 210 °C; (ii) holding the temperature at 210 °C for 10 min; (iii) cooling at a rate of 10 oC/min down to 60 oC. WAXD patterns were collected at 10s/frame during the whole thermal treatment procedures.

Results and Discussion

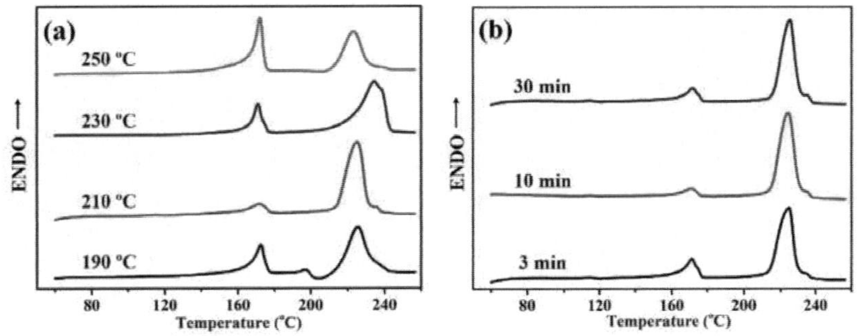

Figure 1. DSC heating curves of samples after thermal treated at (a) different temperatures (190, 210, 230, 250 °C) for 3min and (b) different durations (3, 10, 30 min) under 210 °C.

In order to achieve the desirable results, we determined the proper thermal treatment conditions, i.e. temperature and time, carefully. As shown in Figure 1a, two endothermic peaks are observed during the melting process of samples after thermal treatment at different temperatures, which are ascribed to the melting of HC and SC, respectively. No cold crystallization peaks of HC are noticed in all heating curves. An obvious exothermic peak is observed in samples after thermally treated at 190 °C locating just before the melting of SC, indicating the cold crystallization of SC. As shown in Figure 1a, with the increase of thermal treatment temperature from 190 to 210 °C or 230 °C, the area of SC's melting peak increases accompanied by the decrease of that of HC's melting peak, indicating the enhance of SC content and depress of HC content. Further increase the treatment temperature to 250 °C, an opposite tendency is observed, i.e. SC content decreases and HC content increases. Taking the degradation of PLA, energy costs during heating and dimensional stability of treated samples into consideration, the thermal treatment of around 210 °C with adequate duration is highly desirable in our research, which is much close to the temperature for PLA normal processing (around 200°C).

When it comes to thermal treatment durations, as shown in Figure 1b, the original HC dominated samples (not shown for brief) are turned into SC dominated one after a simple thermal treatment within a short time, which is of great significance to prepare PLA samples with high SC content and heat resistance through commercial processing methods. This phenomenon indicates the high efficiency of residual SCs in inducing the formation of new SCs during thermal treatment process from another aspect, i.e. the saturation value is achieved after only 10 min thermal treatment.

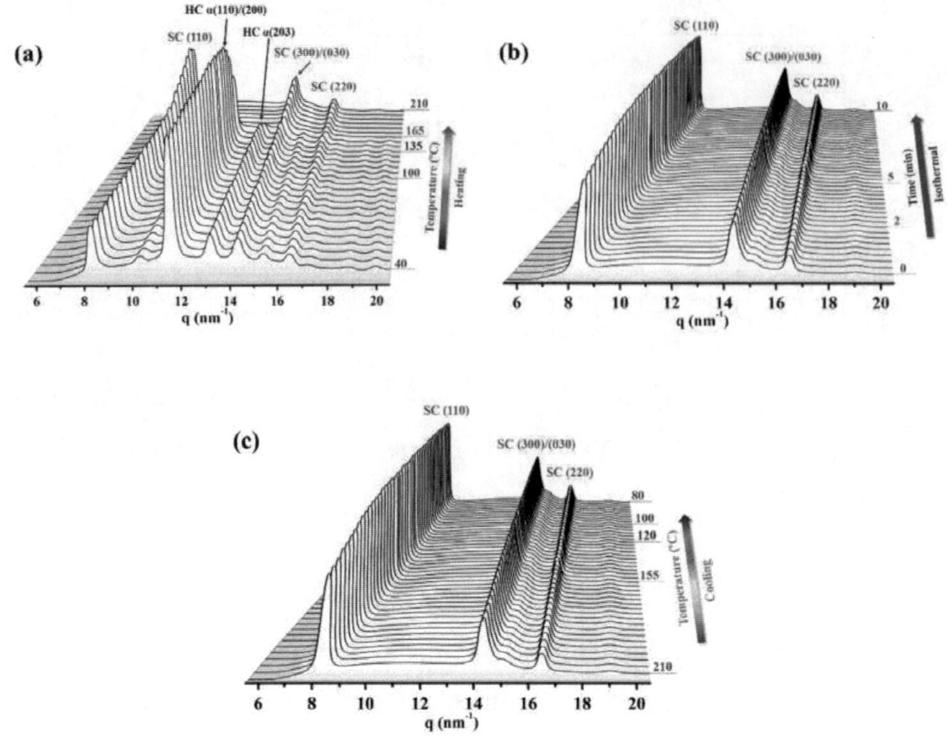

Figure 2. (a) WAXD profiles of the samples at different temperatures during heating process of thermal treatment, (b) WAXD profiles of the samples at different durations during isothermal process of thermal treatment and (c) WAXD profiles of the samples at different temperatures during cooling process of thermal treatment.

To further investigate the crystallization process of SC during thermal treatment, in-situ WAXD was employed to trace the crystallization process. The WAXD profiles obtained from circularly integrated intensities of 2D-WAXD patterns (data not shown here) during thermal treatment are plotted in Figure 2. The WAXD profiles at 40 °C in Figure 2a shows five main scattering peaks corresponding to the crystal planes (110)/(200) and (203) of HC α crystal form[5] located at 11.6 and 13.3 nm^{-1} and the crystal planes (110), (300)/(030) and (220) of SC crystals[6] located at 8.2, 14.4 and 16.7 nm^{-1}. It is easy to conclude that the sample before thermal treatment is HC dominated judged by the height of diffraction peaks. With the increase of temperature, the intensities of HC scattering peaks decrease gradually indicating melting process of HC. Meanwhile, the intensities of SC scattering peaks are gradually enhanced when the temperature is further promoted, which is resulted from the recrystallization of SC. Following the end of heating process, the samples are held at 210 °C for 10 min, i.e. the optimal thermal treatment condition suggested by DSC results. The WAXD profiles of samples during isothermal process of thermal treatment is presented in Figure 2b. No diffraction peaks of HC are observed during the whole isothermal process and the intensities of SC scattering peaks increase gradually with time elapsing. During the following cooling process, similar results are observed as shown in Figure 2c. The intensities of SC scattering peaks are gradually promoted and finally reach a plateau. In addition, even when the temperature reaches the crystallization temperature of HC, i.e. around 120 °C, no HC is observed during further cooling process. It is worthy to note that the absence of HC's diffraction peaks doesn't mean the absence of HC but that the low HC content (could be traced by DSC as mentioned above) is out of the scale of WAXD technique. In the end, the HC dominated PLLA/PDLA blends are transferred into exclusively stereocomplex ones after a simple thermal treatment including heating, isothermal and cooling.

Conclusions

To obtain complete SC samples through a simple thermal treatment process, the optimal thermal treatment conditions were determined by using DSC in the present work. The optimal thermal treatment temperature is 210 °C and the optimal thermal treatment time is 10 min according to our DSC results. The crystallization process was investigated by in-situ WAXD characterization during thermal treatment process under the optimal treatment conditions, i.e. at 210 °C for 10 min. The template effect, new formed SCs are induced by the surface of residual SCs during thermal treatment, was proposed in terms of in-situ WAXD results.

Acknowledgement

The authors gratefully thank the financial support from the National Natural Science Foundation of China (Grants 51120135002 and 51473101).

References

(1) Ikada, Y.; Jamshidi, K.; Tsuji, H.; et al. *Macromolecules* **1987**, 20, 904-906.

(2) Tsuji, H. *Macromol. Biosci.* **2005**, 5, 569-597.

(3) Tsuji, H.; Horii, F.; Hyon, S. H.; et al. *Macromolecules* **1991**, 24, 2719-2724.

(4) Tsuji, H.; Yamamoto, S. *Macromol. Mater. Eng.* **2011**, 296, 583-589.

(5) Pan, P. J.; Kai, W. H.; Bo Zhu, B.; et al. *Macromolecules* **2007**, 40, 6898-6905.

(6) Sawai, D.; Tsugane, Y.; Tamada, M.; et al. *J. Polym. Sci. Polym. Phys.* **2007**, 45, 2632-2639.

Influence of Initial Fiber Length on the Mechanical and Tribological Properties of Hemp Fiber Reinforced Plants-Derived Polyamide 1010 Biomass Composites

Yosuke Nishitani[a*], Jun Mukaida[b], Toshiyuki Yamanaka[c], Tetsuto Kajiyama[d] and Takeshi Kitano[e]

[a]*Department of Mechanical Engineering, Faculty of Engineering, Kogakuin University, 2665-1 Nakano, Hachioji, Tokyo 1920015, Japan*
[b]*Department of Mechanical Engineering, Graduate School of Engineering, Kogakuin University, 2665-1 Nakano, Hachioji, Tokyo 1920015, Japan*
[c] *Tokyo Metropolitan Industrial Technology Research Institute, 2-4-10, Aomi, koto, Tokyo 135-0064, Japan*
[d] *Jonan Branch, Tokyo Metropolitan Industrial Technology Research Institute, 1-20-20, Minamikamata, Ota, Tokyo 144-0035, Japan*
[e] *Polymer Centre, Faculty of Technology, Tomas Bata University in Zlin, T. G. M. 275, Zlin, 762 72 Czech Republic*

Presenting and Reference author's email: at13152@ns.kogakuin.ac.jp

Abstract. The aim of this study is to investigate the influence of initial fiber length on the mechanical and tribological properties of hemp fiber filled plants-derived polyamide 1010 biomass composites in order to develop the new engineering materials such as structural materials and tribomaterials based on all plants-derived materials. Plants-derived polyamide 1010 (PA1010) was made from plants-derived castor oil. Hemp fiber (HF) was previously cut into length of 5mm, 10mm, 20mm and 50mm, and was surface-treated by two types of treatment: alkali treatment by NaOH solution and surface treatment by ureidosilane coupling agent (HF-S). The volume fraction of HF-S in the composites was fixed with 20vol.%. HF-S/PA1010 biomass composites were extruded by a twin screw extruder and injection-molded. Their mechanical properties such as tensile, bending, and Izod impact test, and tribological properties by ring-on-plate type sliding wear testing were evaluated. It was found that the mechanical properties of HF-S/PA1010 biomass composites change with the initial fiber length of HF. The critical fiber lengths of various mechanical properties of these biomass composites decrease as follows: strength > modulus > elongation at break > Izod impact strength. Tribological properties such as frictional coefficient and specific wear rate also change with the initial fiber length. The frictional coefficient of these biomass composites slightly decrease with increasing the fiber length. On the other hand, the specific wear rate of these biomass composites dramatically improved with increasing the fiber length, and has a minimum peak at 20mm. It follows from these results that it may be possible to develop the new higher performance engineering materials by selecting the suitable fiber length.

Keywords: Biomass polymer composites, Plants-derived Polyamide, Natural fiber, Tribology, Mechanical properties
PACS: 81.05.Qk, 81.40.Pq, 81.70.Bt, 83.80.Mc

INTRODUCTION

Recently, the interest for biomass polymers, which are made from carbon neutral raw materials, and their biomass composites has grown because of increasing environmental problems such as uncertainty of the petroleum supply and global warming [1, 2]. However, their low performance compared to traditional petroleum polymers has limited their applications, and supply of raw materials is unstable since many of biopolymers are made from edible biomass like as corn and starch. In order to solve these problems, the new engineering materials based on all inedible plants-derived materials are strongly required. The investigation of short natural fibers, which are banana, flax, hemp, kenaf, ramie, sisal fiber and so on, reinforced biopolymer composites have attracted more and more in recent decades [3-7]. These natural fibers have some ecological advantages over inorganic fibers such as carbon and glass fibers since they are renewable and can be incinerated. However, there are only but few published studies on the fabrication of these natural fiber reinforced biomass polymer composites based on the inedible plants-derived polymers and their various properties [8, 9]. In previous studies, we conducted the studies on the rheological, mechanical and tribological properties of hemp fiber (HF), which is one of the natural fiber, reinforced polyamide 1010 (PA1010) biomass composites [9-17]. PA1010 was made from sebacic acid and decamethylenediamine, which are obtained from plants-derived caster oil [18, 19]. Since castor oil is not used for food, there is no competition with human food consumption. It was found that HF and has a good reinforcing effect on the rheological, mechanical and tribological properties of the compoistes. And the surface treatment by the alkali treatment (mercerization) on HF and the addition of third component such as elastomer and other polymer significantly improved these properties. However, in order to develop new biomass composites with further higher performance using all plants-derived polymer composites, it is necessary to investigate systematically the effect of various factors such as the type of surface treatment, surface treatment method, the addition of third component, fiber content and so on, on the fabrication of these biomass composites and their properties. Especially, a significant factor to design these biomass composites is the lack of experimental results on the composites properties as a function of fiber length. The aim of this study is to investigate experimentally the influence of the initial fiber length on the mechanical and tribological properties of hemp fiber reinforced plants-derived polyamide 1010 biomass composites.

EXPERIMENTAL

The materials used in this study were hemp fiber reinforced plants-derived polyamide 1010 biomass composites (HF/PA1010). Polyamide 1010 (PA1010, Vestamid Terra DS16, Daicel-Evonik Ltd., Japan) was used as the matrix polymer. PA1010 was made from sebacic acid and decamethylenediamine, which are obtained from plant-derived caster oil. Hemp fiber (HF, ϕ50-100μm, Hemp Levo Inc., Japan) was used as the reinforcement fiber. The volume fraction of fiber V_f was fixed with 20vol.%. Hemp fiber was previously cut into length of 5mm, 10mm, 20mm and 50mm, and were surface-treated by two types of treatment: (a) alkali treatment by NaOH solution and (b) surface treatment by ureidosilane coupling agent (HF-S). Alkali treatment by NaOH was employed as follows: a 5% solution of alkali agents was taken in a stainless beaker. The chopped hemp fibers were added into the beaker and stirred well. This was kept at room temperature for 4h. The fibers were then washed thoroughly with water to remove the excess of alkali agents sticking to the fibers. The alkali treated fibers were dried in air for 12h and in a vacuum oven at 80°C for 5h. The surface treatment of hemp fibers with the concentration of 1wt.% ureidosilane coupling agent (3-ureidopropyl-trimethoxysilane, A-1160, Momentive Performance Materials Inc., USA) was carried out in 0.5wt.% of acetic acid aqueous solution in which the pH of the solution was adjusted to 3.5 and stirred continuously for 15 min. Then, the fibers were immersed in the solution for 60 min. The surface-treated hemp fibers (HF-S) were removed from the solution and kept in air for 12h and in a vacuum oven at 80°C for 5h. All the components which were dried for 12h at 80°C in vacuum oven were dry blended in the small plastic bottle, subsequently melt mixed at 85 rpm and 220°C on a twin screw extruder (TEX-30, Japan Steel Works, Ltd., Japan). After mixing, the extruded strands of various HF/PA1010 composites were cut by a pelletizer, and were dried again at 80°C for 12h in vacuum oven. Various shaped samples for mechanical and tribological properties testing were injection molded (NS20-A, Nissei Plastic Industrial, Japan). The molding conditions were as follows: cylinder temperatures of 220°C, mold (cavity) temperature of 30°C and the injection rate of 13cm^3/s. Mechanical such as tensile bending and Izod impact testing, and tribological properties were evaluated. Tensile testing was carried out with dog-bone samples on a Toyo Seiki testing machine V-10, in accordance to the JIS K 7161, at the cross-head speed of 50 mm/min. Three point bend testing was carried out with coupon specimens on the V-10, in accordance to JIS K 7171, at the cross-head speed of 2mm/min. Impact characteristics were determined by Izod impact test. Izod impact tests were conducted using coupon specimen with a notch depth of 2.5mm on a Toyo Seiki impact machine DG-IB at room temperature, in accordance to JIS K 7110. Tribological properties were measured by a ring on plate type sliding wear tester (EFM-3-EN, Orientech Co., Japan). The tests were conducted at room temperature under dry condition in accordance with JIS K 7218a (in which test is a type of full contact between the test specimen and metal counterpiece) at the constant normal load P of 50N), sliding speed v of 0.2m/sec) and sliding distance L of 600m. A carbon steel (S45C) ring with the surface finished by No.240 and No.800 polishing paper was used as a counterpiece. Transfer film on the counterface, worn surface and wear debris were observed by scanning electron microscope (SEM, VE-8800S, Keyence Co. and JSM-6360LA, JEOL Ltd., Japan).

RESULTS AND DISCUSSION

First, the influence of initial fiber length on the tensile properties of hemp fiber reinforced plants-derived polyamide 1010 biomass composites (HF/PA1010 composites) is discussed. Fig. 1 shows the influence of initial fiber length on the mechanical properties of HF/PA1010 biomass composites. In Fig. 1(a), the tensile strength σ_t and modulus E_t are plotted against the initial fiber length l of these biomass composites. σ_t increases with l up to l=20mm, and then increases gradually. On the other hand, E_t increase monotonously with increasing with l. Elongation at break ε_t and Izod impact strength a_{iN} are plotted against l in Fig. 1(b). ε_t shows the different tendency as σ_t and E_t, and has a maximum peak at l=20mm. a_{iN} also shows the same tendency as ε_t, however a_{iN} has a maximum peak at l=10mm and then decreases abruptly. From these results, the mechanical properties of HF/PA1010 biomass composites is found to increase with increasing initial fiber length. The influence of the initial fiber length on the mechanical properties of these biomass composites differs for each mechanical property and their critical fiber lengths exist in each one. Thus, the mechanical properties of these biomass composites change with the initial fiber length of HF. In particular, the critical fiber lengths of various mechanical properties of these biomass composites decrease as follows: strength > modulus > elongation at break > Izod impact strength. These tendencies agree with the results of wood pulp reinforced polypropylene composites [20] and glass fiber reinforced polypropylene composites [21]. It is necessary to further clarify these improving mechanisms to carry out detailed observations of the internal structure of these biomass composites such as fiber length in composites, its distribution, its dispersion and the interfacial adhesion between fiber and polymer matrix.

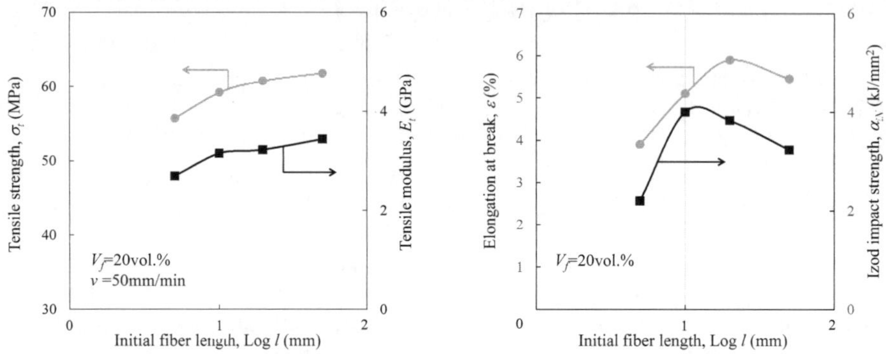

(a) Tensile strength and modulus (b) Elongation at break and Izod impact strength

FIGURE 1 Influence of initial fiber length on the tensile strength and modulus of PA1010 biomass composites.

Second, the tribological properties under dry conditions, which are most important for the development of tribomaterials for mechanical sliding parts, are discussed below. The influence of initial fiber length l on the tribological properties such as the frictional coefficient μ and the specific wear rate V_s of HF/PA1010 biomass composites by constant load and sliding speed testing using a ring on plate type sliding wear tester is shown in Fig. 2. The frictional coefficient μ of these biomass composites slightly decrease with increasing the fiber length l. On the other hand, the specific wear rate V_s of these biomass composites dramatically improved with increasing the fiber length l, and has a minimum peak at 20mm. These results may be attributed to the change in the mode of wear mechanism by the initial fiber length caused by the good lubrication of hemp fiber and the good interaction and interphase adhesion between fiber and matrix polymer. Because the tribological properties of polymer composites are strongly influenced by their ability to form the wear debris, a transfer film on the counterface and the worn surface state of the materials, it is essential to observe these factors for understanding the mechanism of tribological properties [22-24]. Fig. 3 presents SEM photographs of the wear debris of various HF/PA1010 biomass composites. The shape and size of wear debris of HF/PA1010 biomass composites change with the initial fiber length l. First, the shape and size of wear debris of l=5mm filled composites (Fig. 3 (a)) is some filamentary, flaky and granular

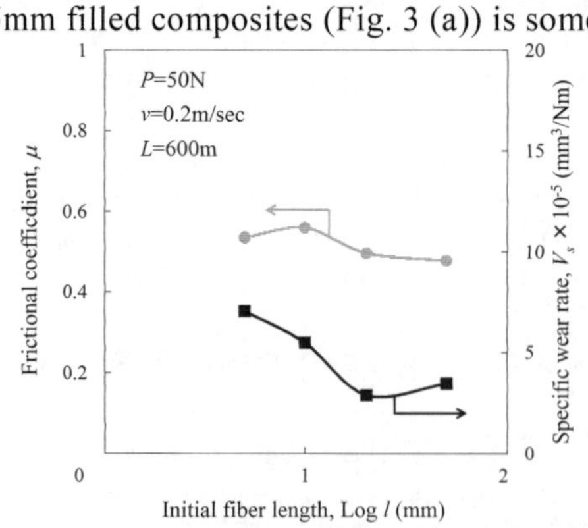

FIGURE2. Influence of initial fiber length on the frictional coefficient and specific wear rate of PA1010 biomass composites.

particles. Second, that of l=10mm filled composites (Fig. 3(b)) is many flaky particles. Third, that of l=20mm (Fig. 3(c)) and l=50mm filled composites (Fig. 3 (d)) is big flaky ones. Thus, the longer the initial fiber length is, the bigger and the flakier the wear debris become. Next, Fig.4 shows SEM photographs of the counterface (S45C) after sliding wear testing against various HF/PA1010 biomass composites. The morphology after sliding wear test of l=5mm filled composites (Fig.4 (a)) shows thin transfer films and small wear debris. On the other hand, that of l=10mm (Fig. 4 (b)), 20mm (Fig.4 (c)) and 50mm filled composites (Fig.4 (d)) shows thin transfer films surface overall, and the thickness of the transfer films on the counterface gradually increases with increasing the initial fiber length. Fig. 5 presents the SEM photographs of the worn surface of various HF/PA1010 biomass composites. The worn surface of l=5mm filled composites (Fig. 5 (a)) is rough surface, which probably associated with the removal of the hemp fibers. That of l=10mm filled composites (Fig. 5 (b)) is rougher than that of l=5mm ones. To the contrary, that of l=20mm (Fig. 5(c)) and l=50mm ones (Fig. 5 (d)) is smooth although there are smaller frictional marks. These tendencies mentioned above are thought to effect of the initial fiber length on the tribological properties of HF/PA1010 biomass composites.

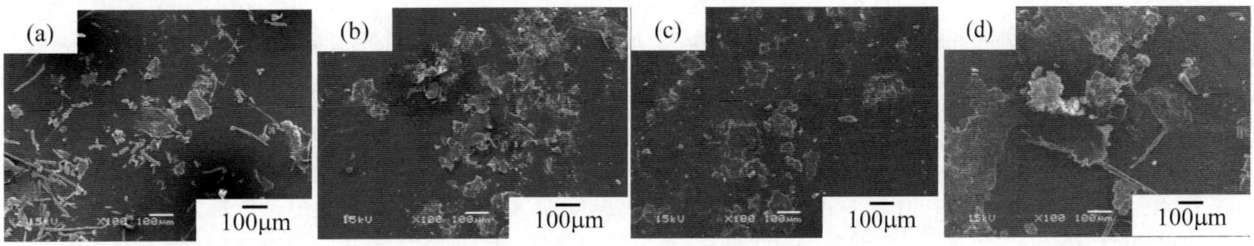

(a) l = 5mm (b) l = 10mm (c) l = 20mm (d) l = 50mm

FIGURE3 Image of SEM photographs of wear debris of various PA1010 biomass composites

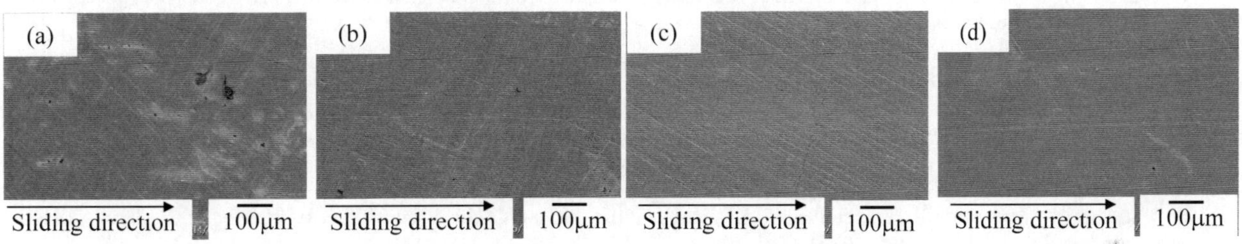

(a) l = 5mm (b) l = 10mm (c) l = 20mm (d) l = 50mm

FIGURE4. Image of SEM photograph of counterface of various PA1010 biomass composites

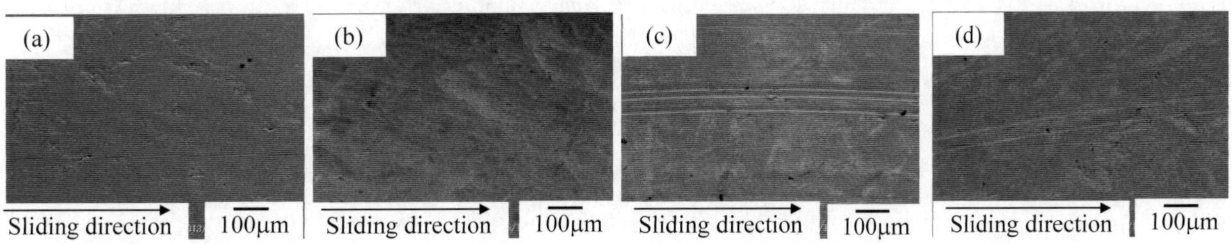

(a) l = 5mm (b) l = 10mm (c) l = 20mm (d) l = 50mm

FIGURE5 Image of SEM photograph of worn surface of various PA1010 biomass composites.

CONCLUSION

The aim of this study is to investigate the influence of initial fiber length on the mechanical properties of hemp fiber filled plants-derived polyamide 1010 biomass composites in order to develop the new engineering materials such as structural materials and tribomaterials based on all plants-derived materials. It was found that the mechanical properties of HF-S/PA1010 biomass composites change with the initial fiber length of HF. The critical fiber lengths of various mechanical properties of these biomass composites decrease as follows: strength > modulus > elongation at break > Izod impact strength. Tribological properties such as frictional coefficient and specific wear rate also change with the initial fiber length. The frictional coefficient of these biomass composites slightly decrease with increasing the fiber length. On the other hand, the specific wear rate of these biomass composites dramatically improved with increasing the fiber length, and has a minimum peak at 20mm. It follows from these results that it may be possible to develop the new higher performance engineering materials by selecting the suitable fiber length.

ACKNOWLEDGMENTS

This work was supported by JSPS KAKENHI Grant Number 16K06750 and 25420735. We would like to thank the Functional Microstructured Surfaces Reserch Center (FMS, MEXT, Japan) of Kogakuin University, and the Project Rresearch of Research Institute for Science and Technology of Kogakuin University for funding this study, and partial support by national budget of Czech Republic within the framework of the Centre of Polymer Systems project (reg. number: CZ. 1.05/2.1.00/ 03.0111).

REFERENCES

1. A. C. Wibowo, A. K. Mohanty, M. Misra and L. T. Drzai, *Ind. Eng. Chem. Res.* **43**, 4883-4888 (2004).
2. L. Petersson and K. Oksman, *Compos. Sci. Technol.* **66**, 2187-2196 (2006).
3. N. Chand and M. Fashim, *Tribology of natural fiber polymer composites*, Cambridge: Woodhead Publishing Limited, 2008, pp.1-58
4. L. Yu., K. Dean and L. Li, *Prog. Polym. Sci.*, **31**, 576-602 (2006)
5. X. Li, L. G. Tabil and S. Panigrashi, *J. Polym. Environ.*, **15**, 25-33 (2007)
6. Y. Xie, C. A. S. Hill, Z. Xiao, H. Militz and C. Mai, *Composites: Part A*, **41**, 806-819 (2010)
7. O. Faruk, A. K. Bledzki, H.-P. Fink and M. Sain, *Prog. Polym. Sci.*, **37**, 1552-1596 (2012)
8. S. Kuciel, P. Kuzniar and A. Liber-knec, *Polimery*, **57**, 627-634 (2012)
9. Y. Nishitani, M. Hasumi and T. Kitano, *AIP Conference Proceedings*, **1664**, 060007 (2015)
10. M. Hasumi, Y. Nishitani and T. Kitano, The proceedings of Seikei-kako symposia'11, Akita, 491-492 (2011) (in Japanese).
11. M. Hasumi, Y. Nishitani and T. Kitano, The proceedings of Tribology Conference Autumn'12, Muroran, 211-212 (2012) (in Japanese).
12. M. Hasumi, Y. Nishitani and T. Kitano, The proceedings of the Polymer Processing Society 28th Annual Meeting (PPS-28), Pattaya, P-07-324 (2012).
13. J. Mukaida, Y. Nishitani, and T. Kitano, *AIP Conference Proceedings*, **1664**, 060008 (2015)
14. Y. Nishitani, J. Mukaida, T. Kajiyama, T. Yamanaka and T. Kitano, The Proceedings of the 31st International Conference of the Polymer Processing Society (PPS-31), Jeju, 1116-1120 (2015).
15. Y. Nishitani, J. Mukaida, T. Yamanaka, T. Kajiyama and T. Kitano, The Proceedings of Regional Conference of the Polymer Processing Society (PPS-2015), Graz, Nat-81 (2015).
16. J. Mukaida, Y. Nishitani, T. Yamanaka, T. Kajiyama and T. Kitano, The Proceedings of Regional Conference of the Polymer Processing Society (PPS-2015), Graz, Nat-104 (2015).
17. Y. Nishitani, J. Mukaida, T. Kajiyama, T. Yamanaka and T. Kitano, *J. Mater. Test. Res. Assoc. Japan*, **61**, 3-11 (2016) (in Japanese).
18. H. Mutlu and M. A. R. Meier, *European J. Liquid Sci. Technol.*, **112**, 10-30 (2010)
19. T. Shimizu, *JETI*, **59**, 71-83 (2011) (in Japanese).
20. A. Thumm and A. R. Dickson, *Composites: Part A*, **46**, 45-52 (2013)
21. J. L. Thomason, *Composites: Part A*, **33**, 1641-1652 (2002)
22. S. Bahadur, *Wear*, **245**, 92-99 (2000)
23. S. Yu, Z. Yu and Y.-W. Mai, *Tribology International*, **40**, 855-862 (2007)
24. Y. Nishitani, Y. Yamada, C. Ishii, I. Sekiguchi and T. Kitano, *Polym. Eng. Sci.*, **50**, 100-112 (2010)

Preparation and Characterization of Polysaccharide Hydrogels Based on Modified Starch

Jongshin Park (presenting author, reference author)[a*], Jungmin Lee[a]

[a]*Department of Biosystems & Biomaterial Science Engineering, Seoul National University, 1 Gwanak-ro, Gwanak-gu, Seoul, Korea*

jongshin@snu.ac.kr

Abstract. Hydrogels are hydrophilic, three-dimensional, and expandable matrices that are produced through chemical and/or physical crosslinking of certain polymers. In some cases, polysaccharides such as carboxymethyl cellulose and starch were used in hydrogel preparation. In this study, chemical modification of starch was carried out in order to improve the functionality of compound. Epichlorohydrin or citric acid were used for crosslinking, respectively. Finally, the polysaccharide hydrogels were prepared by chemical reaction of carboxymethyl cellulose and starch with crosslinker. The chemical structures of modified starch and polysaccharide hydrogels were established using FT-IR spectroscopy. The swelling behavior of hydrogels was studied.

Keywords: polysaccharide, starch aldehyde, carboxymethyl cellulose, hydrogel.
PACS: 82.35.Pq

INTRODUCTION

Hydrogels are hydrophilic, three-dimensional, and expandable matrices that are produced through chemical and/or physical crosslinking of certain polymers [1]. These crosslinked polymers have excellent characteristics, such as swelling, mechanical, permeation, surface and optical properties. Because of such unique properties, hydrogels are widely used in many fields, such as agricultural, disposable diapers, and pharmaceuticals.

In some cases, polysaccharide-based hydrogels have been prepared from a single polysaccharide such as carboxymethyl cellulose (CMC) and starch. CMC is an anionic water-soluble natural polymer derivative, which is widely used in detergents, oil exploration, and in the food, paper, and textile industries because of its viscosity-increasing properties. Recently, CMC has been used in preparation of hydrogels as a biomaterial due to their nontoxicity, degradability, and good biological compatibility [2].

In addition, superabsorbent hydrogel based on starch has been prepared by free radical graft copolymerization of vinyl monomers onto polysaccharides [3]. Starch is a renewable and degradable polymer that can be obtained from various botanical sources [4]. It is composed of an amylose, a linear polymer of α-(1-4) glucose units, and an amylopectin, a linear polymer of α-(1-4) glucose units with periodic branches of α-(1-6) linkages [5]. Starch is widely used in different fields such as food, environmental-friendly plastics, and medicine due to its low cost, biodegradability, and renewability. Chemical modification of starch relies on the reactivity of hydroxyl groups and introduction of new functional groups into polysaccharide chains [6]. In particular, oxidative modification of starch is carried out in order to improve the functionality of compound. The hydroxyl groups, primarily at C-2, C-3, and C-6 positions, are transformed to aldehyde and carboxylic acid by oxidation [7].

In this study, the polysaccharide hydrogels were prepared by chemical reaction of CMC and starch (native or modified) with crosslinker. The starch aldehyde was formed by oxidation. Epichlorohydrin (ECH) or citric acid were used for crosslinking. The structure of product was established using FT-IR spectroscopy. The swelling behavior of hydrogel was studied.

EXPERIMENTAL

Materials

Potato starch and ECH were purchased from Samchun Pure Chemical Co., Ltd., Korea. CMC was purchased from from Sigma Aldrich, Inc, USA. Citric acid was purchased from Showa Chemicals, Inc, Japan. Sodium periodate was used as oxidant of starch and purchased from Samchun Pure Chemical Co., Ltd., Korea. Hydroxylamine hydrochloride and methyl orange were used for titration and purchased from Samchun Pure Chemical Co., Ltd., Korea. Sodium hydroxide was purchased from Duksan Pure Chemical Co., LTD., Korea. Ethanol and methanol were used for precipitation and purchased from Samchun Pure Chemical Co., Ltd., Korea.

Preparation of Starch Aldehydes

First, sodium periodate was dissolved completely in 400 ml of distilled water at 20 °C, and then 40 g of starch were added. The content of sodium periodate was 10g, 20g, and 30g, respectively. This mixture was charged into a three-necked flask fitted with an impeller. Under continuous stirring, the vessel was heated to 35 °C. The pH was adjusted to 7. After 12 hours, the suspension was formed and poured into ethanol. The precipitated product was dried for 24 hours at 50 °C.

Preparation of Hydrogels

Distilled water was employed for the hydrogel preparation. Polysaccharides were dissolved in 50 mL distilled water in a round bottom flask. The ingredient of polysaccharides was shown in table 1. The flask was immersed in a thermostated water bath with stirring. The temperature of flask was maintained to 70 °C for 2 hours. After complete dissolution of polysaccharides, a definite amount of crosslinker was added into the mixture. After 2 hours, the obtained suspension was poured into 200 mL of methanol. Then, the product was filtered and dried in an oven at 50°C.

TABLE 1. The ingredient compositions that were used in preparation of hydrogel.

Sample Code	Content of Polysaccharides		Content of Crosslinker
1	Native starch 2g	CMC 2g	ECH 1.14 g
2	Starch aldehyde($NaIO_4$ 10g) 2g	CMC 2g	ECH 1.14 g
3	Starch aldehyde($NaIO_4$ 20g) 2g	CMC 2g	ECH 1.14 g
4	Starch aldehyde($NaIO_4$ 30g) 2g	CMC 2g	ECH 1.14 g
5	Native starch 2g	CMC 2g	ECH 2.28g
6	Starch aldehyde($NaIO_4$ 10g) 2g	CMC 2g	ECH 2.28g
7	Starch aldehyde($NaIO_4$ 20g) 2g	CMC 2g	ECH 2.28g
8	Starch aldehyde($NaIO_4$ 30g) 2g	CMC 2g	ECH 2.28g
9	Native starch 2g	CMC 2g	Citric acid 2.37g
10	Starch aldehyde($NaIO_4$ 10g) 2g	CMC 2g	Citric acid 2.37g
11	Starch aldehyde($NaIO_4$ 20g) 2g	CMC 2g	Citric acid 2.37g
12	Starch aldehyde($NaIO_4$ 30g) 2g	CMC 2g	Citric acid 2.37g
13	Native starch 2g	CMC 2g	Citric acid 4.74g
14	Starch aldehyde($NaIO_4$ 10g) 2g	CMC 2g	Citric acid 4.74g
15	Starch aldehyde($NaIO_4$ 20g) 2g	CMC 2g	Citric acid 4.74g
16	Starch aldehyde($NaIO_4$ 30g) 2g	CMC 2g	Citric acid 4.74g

Characterization

FT-IR spectrometer (Nicolet iS5, Thermo Scientific, USA) was used to confirm the chemical structure of starch aldehyde and various hydrogels. Before analysis, the samples were dried in an auto-desiccator (SANPLY DRY KEEPER, SANPLATEC Corp., Japan) for 24 hours. The samples were analyzed with a 4 cm^{-1} resolution and 32 scans between 650 and 4000 cm^{-1}.

The aldehyde content of modified starch was determined according to a method described in the literature [8]. First, 17.5 g of hydroxylamine hydrochloride was dissolved in 150 ml of distilled water, and then 0.05% methyl orange solution was added. This solution was diluted to a volume of 1 liter and the pH adjusted to 4.0. Then, 0.1 g of dry sample was dissolved in 25 ml of hydroxylamine hydrochloride-methyl orange solution. The mixture was allowed to stand for 2 hours and titrated with sodium hydroxide solution until the red-to-yellow end point was achieved.

The swelling ratios of hydrogels were calculated by the following equation.

$$\text{Swelling ratio} = (W_S - W_D) / W_D \qquad (1)$$

Where W_S is the weight of swollen hydrogels and W_D is the initial weight of dried hydrogels.

RESULTS

Starch Aldehydes

Synthesis of starch aldehyde was verified by FT-IR spectroscopic studies. The resulting FT-IR spectra were compared with the spectrum of native starch. As seen in figure 1 (b), the FT-IR spectra showed the characteristic IR bands at ~1735 cm^{-1}, which confirm the oxidation of starch. The intensity of –OH peaks at ~3300 cm^{-1} was decreased as the content of sodium periodate was increased.

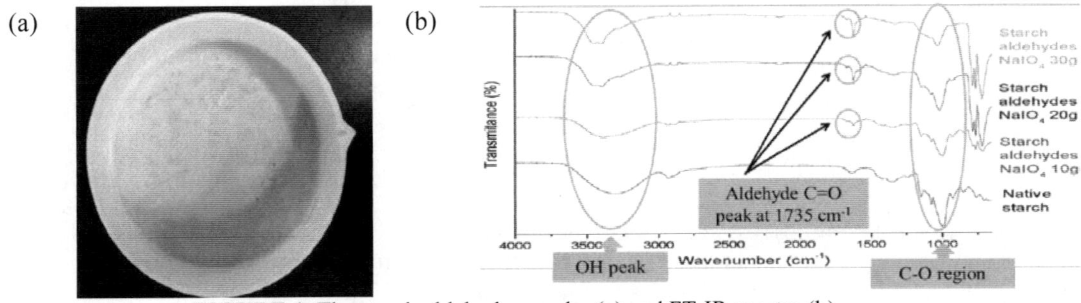

FIGURE 1. The starch aldehyde powder (a) and FT-IR spectra (b).

The number of aldehyde substitution (DS) was estimated from the amount of sodium hydroxide used in titration. The titration mechanism and calculation method are as below (figure 2). Depending on the amount of sodium periodate, each DS was 0.47, 1.46, and 1.52.

$$R\text{-}(CHO) + NH_2OH \cdot HCl \rightarrow R\text{-}(CH{=}N\text{-}OH) + H_2O + HCl$$

aldehydes — Hydroxylamine hydrochloride — Oxime

$$HCl + NaOH \rightarrow NaCl + H_2O$$ Neutralization of acid

$$DS = \frac{mol\ of\ (\text{-CHO})}{mol\ of\ AGU} = \frac{mol\ of\ Na^{+}}{mol\ of\ AGU} = \frac{M_{Na^{+}}(mol/L) \times V_{Na^{+}}(mL) \times \dfrac{1L}{1000mL}}{W_{sample}(g) \times \dfrac{1AGU}{162g/mol}}$$

FIGURE 2. The titration mechanism and calculation of DS.

Polysaccharide Hydrogels

In case of ECH crosslinking, carboxylate and hydroxyl groups of polysaccharides attack to either epoxide or CH_2Cl groups of ECH to form ester linkage [9]. As a result, as seen in sample 1 and 3 of figure 3, the peaks at 1738 and 1027 cm^{-1} showed stretching of C=O and C-O bonds of the ester linkage, which confirm chemical crosslinking of hydrogel. In case of citric acid crosslinking, sample 9 and 11 showed the intense C=O peaks at 1738 cm^{-1}. It is related with ester linkage due to anhydride formation and crosslinking between polysaccharide and citric acid [10]. The broad band at 3200-3400 cm^{-1} was due to stretching of hydroxyl groups of polysaccharides.

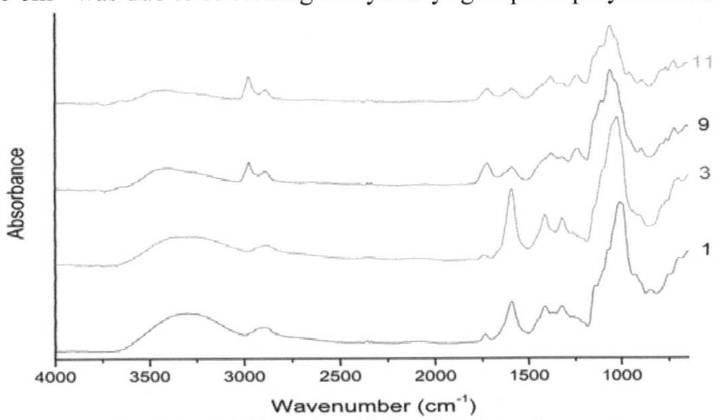

FIGURE 3. The FT-IR spectra of various hydrogels. (number: sample code)

As shown in table 2, the maximum value was about 32. In case of ECH crosslinking, major factor to determine the swelling ratio of hydrogel was crosslinker concentration. Sample 1~ 4 were dissolved, but sample 5~8 showed that swelling ratio is inversely proportional to starch DS. When citric acid was used as a crosslinker, it was hard to find the relation between content of crosslinker and swelling ratio. Further studies will be needed to demonstrate relation between swelling ratio and other materials such as CMC, starch, and citric acid.

TABLE 2. The equilibrium swelling ratios of various hydrogels.

Sample Code	Swelling Ratio	Sample Code	Swelling Ratio	Sample Code	Swelling Ratio	Sample Code	Swelling Ratio
1	dissolution	5	dissolution	9	11.81	13	6.41
2	dissolution	6	32.39	10	8.41	14	11
3	dissolution	7	23.93	11	6.53	15	30.63
4	dissolution	8	15.5	12	5.13	16	14.33

CONCLUSION

The starch aldehyde formed by oxidation was confirmed by FT-IR spectra, which showed the presence of IR bands at ~1735 cm^{-1}. The maximum number of aldehyde substitution analyzed by titration was 1.52. In addition, the chemical crosslinking by ECH or citric acid and the preparation of hydrogels were successful. The maximum swelling ratio of starch aldehyde based-hydrogels was about 32.

ACKNOWLEDGMENTS

This research was supported by the Fundamental R&D Program for Technology of Materials & Components (No. 10050526) funded by the Ministry of Trade, Industry and Energy, Republic of Korea, "Development of High Stiffness and Thermal Resistant Materials for High Performance Automotive Lighting System" funded by KOPRA Co., Ltd., and the Research Institute for Agriculture and Life Sciences at Seoul National University.

REFERENCES

1. W. E. Hennink and C. F. van Nostrum, *Adv. Drug. Deliv. Rev* **54**, 13-36 (2002).
2. P. Liu, M. Zhai, L. Li, J. Peng, and J. Wu, *Rad. Phys. Chem* **63**, 525-528 (2002).
3. D. H. Kim, S. K. Na, J. S. Park, K. J. Yoon, and D. W. lhm, *Eur. Polym. J* **38**, 1199-1204 (2002).
4. S. D. Zhang, X. L. Wang, Y. R. Zhang, K. K. Yang, and Y. Z. Wang, *J. Polym. Res* **17**, 439-446 (2010).
5. A. Serrero, S. Trombotto, P. Cassagnau, Y. Bayon, P. Gravagna, S. Montanari, and L. David, *Biomacromolecules* **11**, 1534-1543 (2010).
6. D. M. Zdanowicz, B. Schmidt, and T. Spychaj, *Pol. J. Chem. Technol* **12**, 14-17 (2010).
7. Y. R. Zhang, X. L. Wang, G. M. Zhao, and Y. Z. Wang, *Carbohydr. Polym* **87**, 2554-2562 (2012).
8. H. Zhao and N. D. Heindel, *Pharm. Res* **8**, 400-402 (1991).
9. H. Hosseinzadeh, *Middle-East J. Sci. Res* **12**, 1521-1527 (2012).
10. C. Demitri, R. D. Sole, F. Scalera, A. Sannino, G. Vasapollo, A. Maffezzoli, L. Ambrosio, and L. Nicolais, *J. Appl. Polym. Sci* **110**, 2453-2460 (2008).

A new process for the preparation of composites and nanocomposites of cellulose pulp and cellulose nanofibers for rigid panels fabrication

Emanoele Maria Santos Chiromito[a], Eliane Trovatti[a,b] and Antonio José Felix Carvalho[a]

[a]Departamento de Engenharia de Materiais, Escola de Engenharia de São Carlos, Universidade de São Paulo, 13566-590 São Carlos, SP, Brasil
[b] Instituto de Química de Saõ Carlos, Universidade de Saõ Paulo, PB 780, 13560-970, São Carlos, SP, Brasil

emanoelechiromito@usp.br
elianetrov@yahoo.com.br
toni@sc.usp.br

Abstract. Composites and nanocomposites of acrylic matrix filled with cellulose pulp fibers (CF) and cellulose nanofibers (NFC) intended for the preparation of panels with high content of cellulose 50 %wt were prepared and characterized. The problem of low compatibility of the hydrophilic cellulose with hydrophobic matrices normally leads to non homogenous materials. In order to overcome this effect, a new process was proposed, where CF or NFC suspensions were mixed with an aqueous acrylic resin emulsion and the composite was obtained by dewatering assisted with vacuum. The process led to the formation of homogeneous composites that were molded by hot pressing. The composites were characterized by SEM, DMA, water absorption and mechanical tests. SEM showed the good dispersion and interaction of the CF and NFC with the matrix, with no pull out for both composite types. DMA and tensile tests confirmed the dispersion and good adhesion of CF and NFC with the matrix, by an extraordinary reinforcing effect, specially for the composites based on NFC.

Keywords: Composites, nanocomposites, cellulose, nanofibers, acrylic matrix.
PACS: 81.05.Qk

INTRODUCTION

The development of sustainable and renewable natural polymer based composites has been gaining attention last decades because the need of reducing the oil based materials. Also, the use of natural fibers such as cellulose fibers has been increasing the technological innovation in terms of increasing the number of new materials with new properties and applications [1]. Regarding to the properties of cellulose use in composite manufacture, they are considered good reinforcement because of its low cost, low density, biodegradability and renewability [2,3].

Cellulose is the essential component of all plant-fibers, it is a natural polymer of d-anhydroglucopyranose units joined together by b-1,4-glycosidic bonds. In general, mechanical properties of cellulose fibers are very good, but particularly, it depends on the source, mainly because each type of cellulose has its own cell geometry and, the geometrical conditions determine the mechanical properties. The term "cellulose pulp" (CF) refers to the purified pulp obtained from pulping process where most of the lignin is extracted. Their dimensions are about 20 um in diameter and 1-4 mm length. Nanofibers of cellulose (NFC) are typically the elemental fibrous component of cellulose with less than 100 nm in diameter and lengths up to several micrometers [4]. When used as filler, the high aspect ratio of cellulose fiber or nanofiber enhances the mechanical behavior of the composites provided by the stress transfer from the matrix to the reinforcing fibers.[5] Besides all the advantages, cellulose displays hydrophilic character and is low compatible with hydrophobic polymers. The low chemical compatibility results in low adhesion between the polymeric matrix and the fibers, representing a challenge in preparation of reinforced composites [6,7,8]. The major problem is found, however, when pulp or cellulose nanofibers are dehydrated because they tend to agglomerate in an irreversible process and become almost impossible to redisperse within the matrix. Several chemical modification and preparation methods have been studied to open up possibilities of combining cellulose with hydrophobic polymers, however, many times with no success.

Here, we propose the development of an innovative method for the preparation of cellulose composites or nanocomposites with polymethylmetacrylate emulsion (PMMA). The formation of the composite was carried out in aqueous medium, the pristine environment of both cellulose suspension and PMMA emulsion generating very homogeneous composites. The material was dried and hot pressed for the preparation of the plates. Regarding to the composition of the materials, very large amounts of cellulose were used, 50 %wt. Thus, these non-conventional cellulose based composites presented emergent properties with high potential of application in the field of engineering, cars and aeronautica.

EXPERIMENTAL SECTION

Materials

Methyl Metacrylate (Sigma), ammonium persulfate (Sigma), sodium dodecyl benzenesulfonate powder (Sigma), aluminum potassium sulfate dodecahydrate (Sigma). Methyl Metacrylate was deprotected before use with 1.0 M NaOH solution and water till neutralization. All reagents, solvents and other chemicals were purchased from Sigma-Aldrich and were used as received. Dried cellulose pulp from Kraft process and cellulose nanofibers were supplied by Suzano Paper and Cellulose SA.

Composite Preparation

The cellulose or nanofibers of cellulose suspension (10%wt) was mixed to the PMMA emulsion (50%wt) and precipitated by addition of 1% Potassium aluminium double salt ($KAl(SO_4)_2 \cdot 12H_2O$) solution, washed with distilled and oven dried at 70ºC. Then the samples were hot pressed at 10 MPa for 2 minutes at 160ºC.

Composite Characterization

DMA measurements were carried out with a Perkin Elmer 8000 DMA equipment working in the tension mode. The tests were performed from -20 to 160 ºC in 5ºC.min-1, at 1Hz, the specimens were 3 x 10 x 40 mm. SEM images were acquired on fractured zones and at the surfaces of carbon coated samples, using a FEI (XL50, Philips) Scanning Electron Microscope. The water absorption was carried out by immersion in a becker cup with 100mL of water, and carefully dried with paper before weight. Tensile tests were carried out in Instron 5969 Universal Material Testing Machine at room temperature, equipped with a 5kN load cell, following ASTM D 638.

RESULTS AND DISCUSSION

Scanning Electron Microscopy

Scanning Electron Microscopy (SEM) micrographs of the fracture of PMMA- NFC (a) and PMMA-CF (b) panels with 50% of filler are shown in Figure 1. PMMA-NFC images clearly showed the nanofibers well dispersed and completely covered by the matrix, indicating the strong interfacial adhesion between NFC and PMMA. PMMA-CF are well distributed within the matrix, however, with low interfacial adhesion than PMMA-NFC.

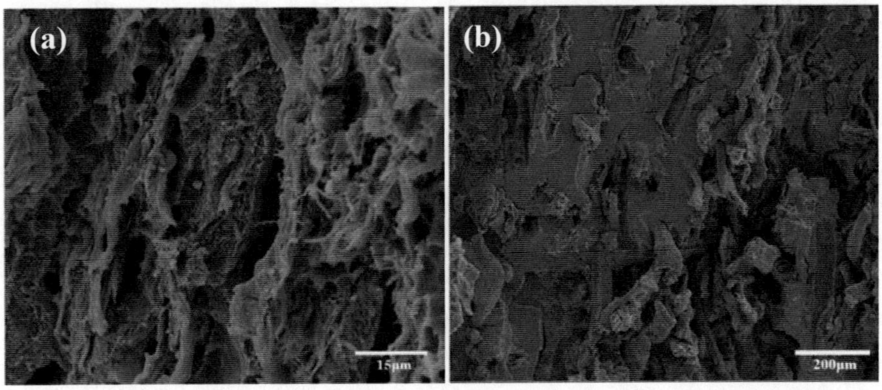

FIGURE 1. SEM micrographs on the fractured surface of (a) NFC-based nanocomposite with 50% of NCF and (b) CF-based composite with 50% of CF.

The dynamic-mechanical properties of PMMA, NFC, CF and their corresponding nanocomposite and composite, are showed in Figure 2. PMMA displayed the typical behavior of thermoplastic polymers. In the glassy state the tensile storage modulus E´ was constant and dramatically dropped when reached the T_g, at about 100ºC. The presence of cellulose fibers and nanofibers in the acrylic matrix increased the storage modulus of both panels above T_g. The increase in storage modulus of the composites can be explained by the strong interaction among the fibers/nanofibers driven by hydrogen bonds that keeps them in contact with each other inside the acrylic matrix, forming in a bi-phasic system. The adhesion of the filler to the matrix also contributed to the increase in dynamic-mechanical properties[9].

DMA Measurements and Mechanical Tests

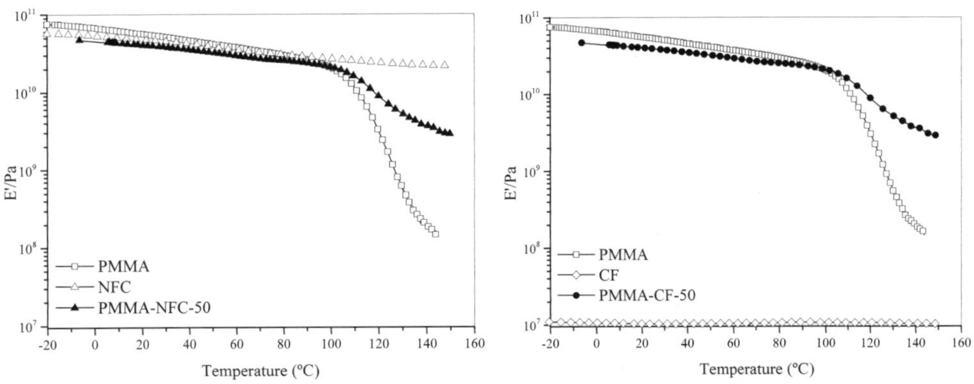

FIGURE 2. Temperature dependence of the storage modulus of the PMMA polymer, NFC, CF and (a) NFC-based nanocomposite with 50% of NCF and (b) CF-based composite with 50% of CF.

The results of tensile tests of PMMA, CF, NFc and their based composites are showed as stress-strain curves in Figure 3. The Young´s Modulus of PMMA matrix was about 1,9 GPa, PMMA-CF and PMMA-NFC nanocomposites were about 2,7 and 3,8 GPa respectively. These values were 42 and 100 % higher than the matrix PMMA. As expected, the presence of NFC decreased the elongation at break of the composites, when compared with the matrix. These effects can be attributed to the presence of cellulose (fibers or nanofibers), which promoted an increment in the materials stiffness. The tensile strength of the composites were lower than the tensile strength of PMMA. This effect can be explained by the intrinsic feature of the composite, in which the amount of fiber and the matrix were equivalent (50 %wt) and could behave quite different of the conventional composites with low amounts of fibrous filler.

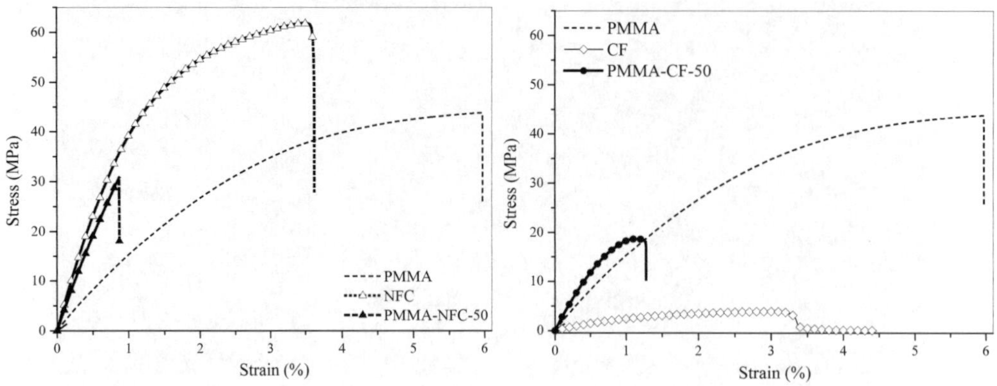

FIGURE 3. Stress-strain curves of PMMA, NFC, CF and their corresponding (a) NFC-based nanocomposite with 50% of NCF and (b) CF-based composite with 50% of CF.

Water Absorption

The water absorption results of PMMA, NFC, CF and the composites are showed in Figure 4. All the materials, except CF absorbed water up to 10 h and then stabilized. As expected, PMMA absorbed the lower amount of wate, while CF and NFc, absorbed the higher amounts of water. PMMA-NFC and PMMA-CF composites exhibited a water absorption of about 12 and 33 wt%, respectively, indicating the hydrophilic cellulose fibers/nanofibers provided an important contribution to the increase of the hydrophilic character of the composites, however, lower if considered its high concentration within the matrix. The lower water absorption can be explained by the coating of the fibers/nanofiber by PMMA, protecting them of hydration, even when completely immersed in water.

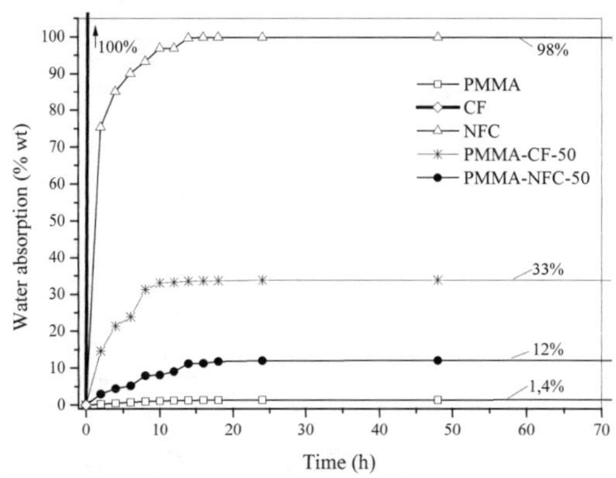

FIGURE 4. Water absorption curves of PMMA polymer, NFC, CF and the corresponding NFC-based nanocomposite and CF-based composite, during 72 hours.

CONCLUSION

The homogeneous dispersion of fibers and nanofibers showed that the method of producing composites and nanocomposites with cellulose fibers was very efficient. The mechanical properties showed that nanofiber is more efficient as reinforcement than micro fibers. The features of the composites of PMMA and NFC or CF are promising in the field of development of thick panels for construction, cars and aeronautics areas. The mechanical properties were satisfactory and end materials presents very homogeneous and attractive aspect, besides its exceptional properties

ACKNOWLEDGMENTS

E. C. thanks CAPES for a MsC. Scholarship. E.T. thanks FAPESP for a Postdoctoral grant (2012/05184-0). The authors thank FAPESP and CNPq for financial support.

REFERENCES

1. Ragauskas AJ, Williams CK, Davison BH, Britovsek G, Cairney J, Eckert CA, et al.The path forward for biofuels and biomaterials. Science 2006;311(5760):484–9
2. Bismarck A, Mishra S, Lampke T. Plant fibers as reinforcement for Green composites. In: Mohanty AK, Misra M, Drzal LT, editors. Natural fibers,biopolymers and biocomposites. Boca Raton: CRC Press; 2005. p. 37–108
3. Oksman K. High quality flax fibre composites manufactured by the resin transfer moulding process. J Reinf Plast Compos 2001;20(7):621–7.
4. Gindl W, Jeronimidis G. J Mater Sci 2004;39:3245–7.
5. Eichhorn, S. J.; Dufresne, A.; Aranguren, M.; Marcovich, N. E.; Capadona, J. R.; Rowan, S. J.; Weder, C.; Thielemans, W.; Roman, M.; Renneckar, S.; Gindl, W.; Veigel, S.; Keckes, J.; Yano, H.; Abe, K.; Nogi, M.; Nakagaito, A. N.; Mangalam, A.; Simonsen, J.; Benight, A. S.; Bismarck, A.; Berglund, L. A.; Peijs, T. Review: Current international research into cellulose nanofibres and nanocomposites. J. Mater. Sci. 2010, 45 (1), 1−33.
6. Andersons, J., Sparnins, E., & Joffe, R. (2006). Stiffness and strength of flax fiber/polymer matrix composites. Polymer Composites, 27(2), 221–229.
7. Trovatti E, Oliveira L, Freire CSR, Silvestre AJD, Neto CP, Pinto JJCC, et al. Novel bacterial cellulose-acrylic resin nanocomposites. Compos Sci Technol. 2010;70(7):1148-1153.
8. Erbas Kiziltas, E., Kiziltas, A., Bollin, S. C., & Gardner, D. J. (2015). Preparation and characterization of transparent PMMA–cellulose-based nanocomposites. Carbohydrate Polymers, 127, 381–389.
7. Siró I, Plackett D. Microfibrillated cellulose and new nanocomposite materials: a review. Cellulose. 2010;.
8. Ahola S, Salmi J, Johansson LS, Laine J, O¨ sterberg M (2008b) Model films from native cellulose nanofibrils. Preparation, swelling, and surface interactions. Biomacromolecules 9:1273–1282.
9. V. Favier, G. R. Canova, J. Y. CavaillC, H. Chanzy, A. Dufresne, and C. Gauthier, Polyrn. Adu. Technot., 6,351(1995).

Preparation of Natural Rubber (NR) /Polycaprolactone (PCL) Bio-Based Shape Memory Polymer Blends

Lai Sun-Mou*, Liu Jung-Liang, Wang Chun-Chieh, Liao Wan-Lin, He Yu-Ting, Huang Yu-Wei, Yeh Hi-Ju, Luo De-Ting, and Chang Chih-Ning

Dept. of Chemical and Materials Engineering, National I-Lan University, I-Lan 260, Taiwan, ROC

smlai@niu.edu.tw

Abstract. PCL (polycaprolactone) is a biodegradable material exhibiting biodegradability and low processing temperature characteristic. NR (natural rubber) is a bio-based natural material without relying on petroleum resources. Shape memory NR/PCL bio-based blends, where NR served as the reversible phase and PCL served as the switching phase, were prepared using a melt blending process. This work aims to further understand the various properties of NR/PCL eco-shape memory blends and foams. Only very limited NR/PCL blends were evaluated in the literature, but rare on their shape memory properties and fracture behaviors. In particular, the effect of crosslinking degree will affect the shape fixing and recovery behaviors in the different phases of the foam blends. The blend of NR and PCL could form thermoplastic elastomeric bio-based blends for processing flexibility with elastomeric characteristics. The effect of crosslinking types including silane and sulfur on the thermal characterizations besides the one-way and two-way shape memory behaviors will be discussed. NR appeared to exhibit two-way shape memory behavior, but the addition of PCL increased the actuation. This work paves the way for new development on the bio-based shape memory polymer blends.

Keywords: Shape memory polymer, polycaprolactone; natural rubber; blend
PACS: 81.05.Lg; 88.20.-j

INTRODUCTION

Shape memory materials are considered to be one type of smart materials which basically adopts the biomimic concept from the nature. Among those shape memory materials, shape memory polymers (SMP) offering high flexibility, low density, and large deformation in comparison with shape memory alloy are on the blink of finding new applications due to the increased interest of novel smart materials. While SMPs subjects to external stimuli, such as temperature, pH, electromagnetic field, etc., they could respond instantaneously and restore their original shapes from temporary shapes [1]. Up to date, most widely investigated SMP systems are triggered by thermal stimulus. Several reviews have reported the recent progress and potential applications of SMP, including biomedical, textile, energy, etc. [1-3].

To program a one-way shape memory effect, the sample was first pre-deformed at various temperatures, and then its temporary shape could be fixed at room temperature after an immediate cooling from the switching temperature. Following the subsequent heating to above the switching temperature (Ts), the sample could energetically regain its original shape via the increased configurational entropy effect. In addition to one-way reversible shape memory polymers, two-way shape memory polymers could change their shape through cooling-induced elongation and melting-induced contraction, instead of melting-induced contraction as seen in one-way shape memory polymers. This additional feature endows SMP with new applications. Several recent works have devoted to design two-way shape memory polymers [4, 5].

To further expand the flexibility in designing novel SMPs, a polymer blending technology, as an environmentally friendly process, has been adopted in several works recently [6-9]. Owing to the environmental concern, the development of bio-based SMPs has been received much attention recently. Those bio-based polymers include biodegradable or biorenewable polymers, which offered ecological advantages toward sustainable development. Our recent works have prepared the first natural rubber/polycaprolactone and PLA/TPU bio-based SMP blends [7, 8]. In this work, we have tried to employ the NR/PCL bio-based blends with different crosslinking types to evaluate their one-way and two-way shape memory performance. Very rare NR-based shape memory polymers were developed until recently, such as NR/PCL [8], and NR/trans-isoprenre rubber blends [10]. The effect of crosslinking types including silane and sulfur on the thermal characterizations besides the one-way and two-way shape memory behaviors will be discussed. This work paves the way for new development on the bio-based shape memory polymer blends.

EXPERIMENTAL

Materials

The materials used in this study were NR and polycaprolactone (PCL). NR was supplied by the Binh Long Rubber Company (SVR-3L, Vietnam). Polycaprolactone (PCL), with a melt index of about 3 g/10 min., was supplied by Solvay Chemicals (Capa™ 6800). Triethoxyvinylsilane (VTEOS, Acros), dicumyl peroxide (DCP, Nippon Oil & Fats Co. Ltd) and dibutyltin dilaurate (DBTL, Tokyo Chemical) were used for grafting VTEOS onto NR. Stearic acid (SA), Zinc oxide (ZnO), 2,2-Dibenzothiazole disulfide (DM), Sulfur (S), Alkylphenol disulfide (TB710) were all purchased from Arkema, France and used as crosslinking reagents.

Sample Preparation

PCL was dried in a vacuum oven at 50 °C for 24 hr prior to processing. NR-g-VTEOS was prepared by mixing NR with 0.2 phr (parts per hundred resins) of peroxide, 0.03 phr of dibutyltin dilaurate, and 1.7 phr of triethoxyvinylsilane at 150 °C under 50 rpm for 5 min. The mixing of NR-g-VTEOS with PCL was melt-blended at 70 °C under 50 rpm for 15 min to form NR-g-VTEOS/PCL blends at the compositions of 80/20 and 60/40. The NR-g-VTEOS/PCL blends were hot-pressed at 150 °C to obtain a ca. 1 mm thick specimen under a molding pressure of 10 MPa for 10 min. The samples were then cured in a water bath at 50 °C for 12 hrs. For the preparation of sulfur-cured samples, the premixed NR containing 1.2 phr of stearic acid, 5 phr of zinc oxide, 1.1 phr of 2,2-dibenzothiazole disulfide, 0.6 phr of sulfur, and 0.6 phr of alkylphenol disulfide was conducted at 90 °C under 50 rpm for 10 min. The premixed NR was melt-blended with PCL at 70 °C under 50 rpm for 15 min to form S-NR /PCL blends at the compositions of 80/20 and 60/40. The S-NR/PCL blends were cured to obtain thin sheets at 160 °C for 20 min. For brief comparison on two-way shape memory process, peroxide cured samples were compared.

Characterization

Scanning electron microscopy (SEM) analysis was performed on the surface morphology of the cryogenically fractured samples with a scanning electron microscope (Hitachi TM3000). The samples were first sputter-coated with a thin gold layer to enhance the conductivity before observation. The operating voltage was 15 kV. Tensile measurements were conducted based on ASTM-D638 standard for test specimens with a gauge length of 25 mm using an Instron Machine model 4469 at a crosshead speed of 200 mm/min. Tensile strength, elongation, and Young's modulus were recorded. Two types of shape memory processes were conducted. For the one-way shape memory behavior, the shape memory test was carried out using an Instron Machine model 4469 according to the following steps, (1) Heating the sample to 65 °C and holding for 5 min, (2) Extending the sample to 100% strain (ε_m) at 65 °C , (3) Cooling down the sample under load to room temperature (about 25 °C) to fix the temporary shape, (4) Unloading the sample to measure the unloaded strain ε_u, (5) Heating the sample to 65 °C again to regain the shape and holding for 5 min to measure the permanent strain ε_p. The shape memory performance was evaluated by determining shape fixation (R_f) and shape recovery ratios (R_r). The equations for R_f and R_r are listed as follows [11],

$$R_f = \varepsilon_u/\varepsilon_m \tag{1}$$

$$R_r = (\varepsilon_m - \varepsilon_p)/\varepsilon_m \tag{2}$$

For the two-way shape memory behaviors, the shape memory test was conducted using a DMA (Perkin Elmer, Pyris Diamond). The gauge length between clamps was at 10 mm. The specimen subjected to cooling-induced elongation and melting-induced contraction across crystallization and melting regions under the fixed load were evaluated according to the following steps, (1) Heating from 40 °C to the highest deformation temperature at 70 °C under non-zero loading, (2) Deforming at various loads, and recording the initial length value at high temperature ($l_{intial\ high}$), (3) Cooling to 15 °C, keeping at 15 °C for 10 min,, recording the length value at low temperature (l_{low}), (4) Finally, heating again to 70 °C, keeping at 70 °C for 10 min, and recording the final length value at high temperature

($l_{final\ high}$). The effectiveness of the two-way shape memory response was expressed in terms of the actuation magnitude, R_{act} (σ) and strain recovery magnitude, R_{rec} (σ) at a given stress (σ) [12].

$$R_{act}(\sigma) = \varepsilon_{low}(\sigma) - \varepsilon_{high}(\sigma) = (l_{low}(\sigma) - l_{intial\ high}(\sigma))/l_0 \quad (3)$$

$$R_{rec}(\sigma) = (l_{low} - l_{final\ high})/(l_{low} - l_{intial\ high}) \quad (4)$$

RESULTS AND DISCUSSION

Mechanical Properties

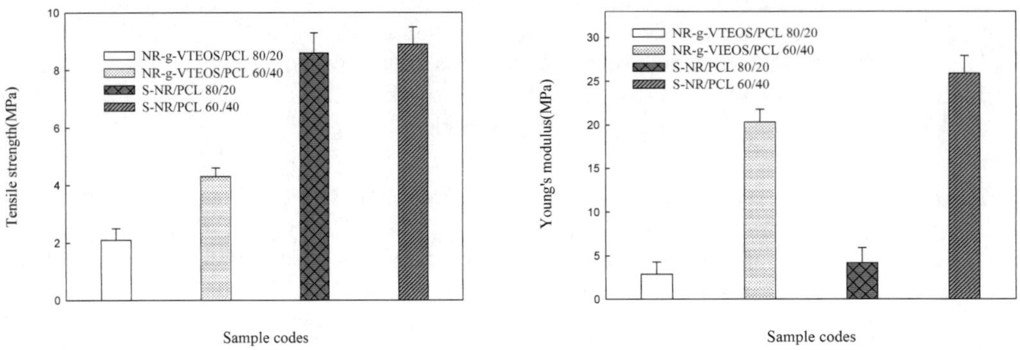

(a) (b)

FIGURE 1. Tensile properties of NR-g-VTEOS/PCL and S-NR/PCL blends, (a) tensile strength, (b) Young's modulus

Mechanical properties of the prepared NR/PCL blends at various curing types and blend compositions were investigated using a tensile test. The results generally suggested that the sulfur-cured blends conferred higher mechanical properties than silane-cured blends. Fig. 1(a) shows that tensile strength of S-NR/PCL blends was higher than that of NR-g-VTEOS/PCL blends, attributed to an improved compatibility and/or higher crosslinking degree for the sulfur-cured blends. Fig. 1(b) shows the results of Young's modulus. Further, the value was higher for S-NR/PCL blends than that of NR-g-VTEOS/PCL blends, as seen in tensile strength.

Shape Memory Behaviors

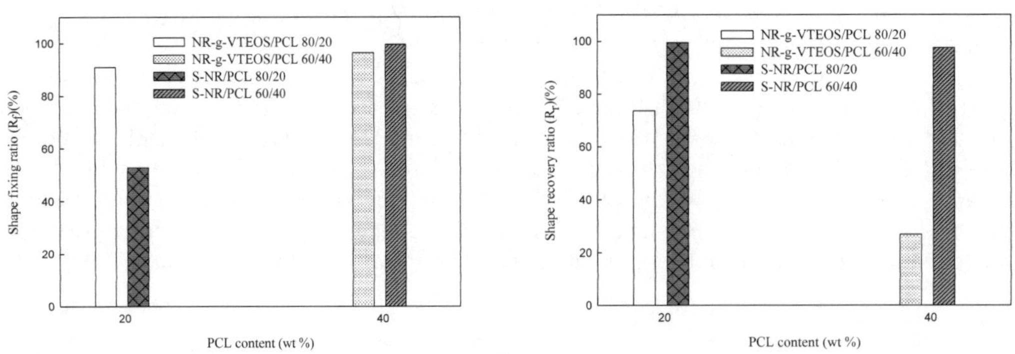

(a) (b)

FIGURE 2. One-way shape memory effect of NR/PCL blends (a) shape fixing ratio, (b) shape recovery ratio

In order to evaluate the effects of crosslink types and PCL content on the one-way shape memory performance of prepared blends, shape memory effects were evaluated by determining R_f and R_r. Fig. 2 shows the shape fixing ratio and the shape recovery ratio for each case. The results indicated that as the PCL content increased, the shape fixing ratio increased up to 99.6±0.7% for the sulfur-cured blend (60/40) due to the fixing capability of PCL component to limit the flow of molecular chain under deformation. In addition, the shape fixing ratios of NR-g-VTEOS/PCL blends were all above 90%, but the lowest shape fixing ratio was found for sulfur-cured blend (80/20). Further, the pendant silanol groups forming hydrogen bonding with polycaprolactone might also enervate this fixing effect as well. To further compare the shape recovery ratios of NR-g-VTEOS/PCL and S-NR/PCL blends, the results indicated a decreased shape recovery ratio with increasing PCL content for silane-cured case, but not for sulfur-cured case. Apparently, the sulfur-cured blends with effective polysulfidic crosslinks outperformed the silane-cured blends with siloxane crosslinks. The pendant silanol groups forming hydrogen bonding with polycaprolactone might limit the recovery effect as well when the temperature was above the recovery temperature.

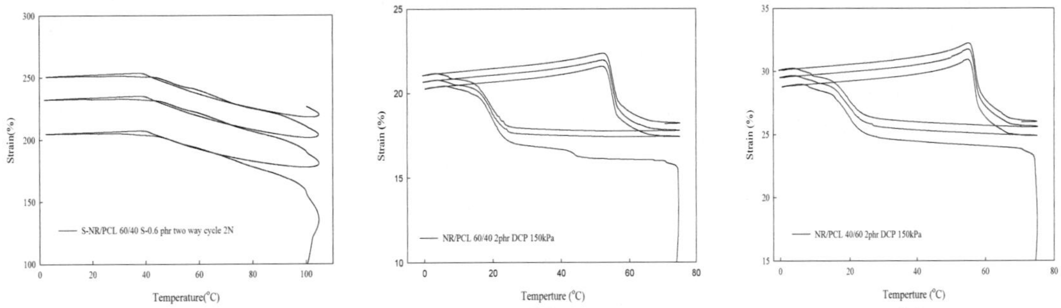

FIGURE 3. Two-way shape memory effect of NR/PCL blends, (a) S-NR/PCL (60/40), (b) P-NR/PCL (60/40), (c) P-NR/PCL (40/60)

Two-way shape memory performance of representative blends for three cycles were evaluated in Figure 3. NR appeared to exhibit two-way shape memory behavior, but the addition of PCL increased the actuation. Although sulfur and silane-cured cases showed two-way behavior with a measureable actuation magnitude, R_{act} (σ), yet the strain recovery magnitude, R_{rec} (σ) at a given stress (σ) was not high enough, even for high NR content. Thus, peroxide cured samples (Fig. 3(b) and (c)) were employed to improve this recovery behavior, mainly attributed to higher gel content for these peroxide cured blends.

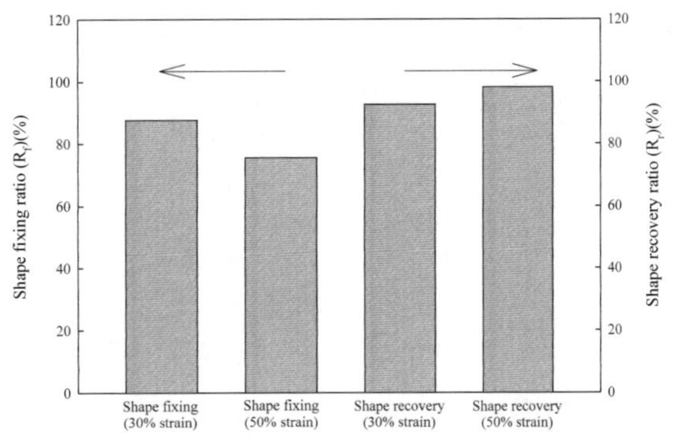

FIGURE 4. Shape memory effect of P-NR/PCL foam blends (40/60) under two compressive strains

To demonstrate the additional merit of peroxide-cured blends, some preliminary foam samples were prepared to evaluate the shape memory performance of foam blends as well. The selected 2 phr peroxide-cured sample NR/PCL (40/60) was foamed using azodicarbonamide (AC) foaming agent, as NR/PCL (60/40) sample was not successfully foamed in the same dosage of AC. The open cell foam with the average cell size of ca. 700 µm dominated in this formulation. However, some areas with the smaller and larger cell size were also observed. Fig. 4 shows the shape fixing ratio and the shape recovery ratio for samples subjected to compressive strains at 30% and 50%, respectively. The high shape fixing and high shape recovery ratio of prepared foam at 30% strain demonstrated the merit of versatile NR/PCL biobased systems. The sulfur and silane-cured foamed blends are under investigation and will be published in the conference as well. This work paves the way for new development on the bio-based shape memory polymer blends.

CONCLUSION

Shape memory NR/PCL bio-based blends were prepared using a melt blending process. The results generally suggested that the sulfur-cured blends conferred higher mechanical properties than silane-cured blends, attributing to an improved compatibility and/or higher crosslinking degree for the sulfur-cured blends. As for the shape memory properties, the results indicated that as the PCL content increased, the shape fixing ratio increased up to 99.6±0.7% for the sulfur-cured blend (60/40). Apparently, the sulfur-cured blends with effective polysulfidic crosslinks outperformed the silane-cured blends with siloxane crosslinks. A preliminary study on two-way shape memory behavior of peroxide-cured samples and the shape memory process under compressive-mode for their foam blends was conducted. The blend of NR and PCL could form thermoplastic elastomeric bio-based blends for processing flexibility with elastomeric characteristics. This work paves the way for new development on the bio-based shape memory polymer blends.

ACKNOWLEDGMENTS

Partial grant support from the Taiwan (ROC) government MOST 104-2221-E-197-031- is acknowledged.

REFERENCES

1. D. Ratna and J. Karger-Kocsis, *J. Mater. Sci.* **43**, 254-269 (2008).
2. Q. Meng and J. Hu, *Composites Part A* **40**, 1661-1672 (2009).
3. T. Pretsch, *Polymers* **2**, 120-158 (2010).
4. S. Chen, S. Yang, Z. Li, S. Xu, H. Yuan, S. Chen and Z. Ge, *Polym. Compos.* **36**, 439-444 (2015).
5. L. Ma, J. Zhao, X. Wang, M. Chen, Y. Liang, Z. Wang, Z. Yu and R. C. Hedden, *Polymer* **56**, 490-497 (2015).
6. O. B. Arnoult and P. T. Mather, ''Crystallization of shape memory polymer blends,'' Annual Technical Conference 1, Society of Plastics Engineers, Cincinnati, Ohio, 2007, pp. 839-844.
7. S. M. Lai and Y. C. Lan, *J. Polym. Res.* **20**, 140 (2013).
8. W. C. Chen, S. M. Lai, M. Y. Chang and Z. C. Liao, *J. Macromol. Sci., Phys.* **53**, 645-661 (2014).
9. I. Kolesov, O. Dolynchuk, S. Borreck and H. J. Radusch, *Polym. Adv. Technol.* **25**, 1315-1322 (2014).
10. G. Y. Lin, S. M. Liu and F. C. Dong, *Appl. Mech. Mater.* **467**, 146-151 (2014).
11. W. Zhang, L. Chen, and Y. Zhang, *Polymer*, **50**, 1311-1315 (2009)
12. S. Pandini, S. Passera, M. Messori, K. Pademi, M. Toselli, A. Gianoncelli, E. Bontempi, and T. Riccò, *Polymer* **53**, 1915-1924 (2012)

Investigation of the Knife-Over-Roll Coating Process for Plastisols Coating Applications

Y.Abdesselam[1,2], Y.Demay[1,3], R.Castellani[1], J.F.Agassant[1], R.Peres[2], D.Gourdin[2]

[1] MINES ParisTech, PSL - Research University, CEMEF - Center for Material Forming, CNRS UMR 7635, 06904 Sophia Antipolis, France
[2] Tarkett GDL, Z.A Salzbaach, L-9559 Wiltz, Luxembourg
[3] Université de Nice Sophia Antipolis, Laboratoire J.A. Dieudonné, UMR CNRS 6621, 06108 Nice, France

Abstract: The rheology of different plastisol formulations has been investigated. The coexistence of smooth PVC particles and rough calcium carbonate aggregates in a Newtonian plasticizer makes this suspension a more complex system than those encountered in the literature. A numerical model has been developed to model the flow of this suspension in a knife-over-roll coating process. It allows predicting the deposited thickness, the pressure field between roll and knife and the separating force.

Key words: plastisol, rheology, coating device, model

INTRODUCTION

Plastisol formulations, composed of PVC particles of various sizes, calcium carbonate filler and several additives in a plasticizer, are widely used for flooring applications. The most commonly used coating process consists in shearing this plastisol formulation at room temperature between a substrate (a glass-mat for example) disposed on a rotating cylinder and a fixed knife of varying geometry. The gap between the substrate and the knife is adjusted in order to deliver the prescribed coating thickness. The coated plastisol is then introduced in an oven for gelation.

The coating ability of a plastisol formulation depends on its rheological behaviour. The complex composition of plastisol formulations, the interaction between PVC resins and plasticizer, the PVC and filler particles sizes, make plastisols a unique kind of suspensions. Many authors have addressed how spherical monodisperse particles behave and influence the flowing properties of a suspension, from dilute to more concentrated systems ([1]–[4]). A rise in viscosity is expected, as particles are added, which diverges when a maximal volume fraction is reached, i.e. packed state prevents particles to flow freely. Additionally, at low shear rates, a shear-thinning behaviour is observed as the sheared suspension microstructure changes with the flow: "ordered" layers of particles are formed when Brownian effects are no more dominant, lowering the viscosity when increasing the shear rate. Marcilla et al. [5] showed on plastisol formulations the existence of a Newtonian plateau at very low shear rates, followed by a shear thinning behaviour and a second Newtonian plateau at intermediate shear rates (around 100 s^{-1}). They pointed out that the polydispersity of PVC resin particles has an influence on the shear thinning behaviour. Indeed, it is well known that adding particles of different sizes to a suspension greatly affects its rheology [6]. First, maximum volume fraction is modified, and the drastic increase in viscosity observed for the monodisperse case is postponed to higher particle concentrations [1]. Using broad particle size distributions has the same effect on rheology [7]. Size distribution impacts also the behaviour of particles suspensions at high shear rates. When volume fractions are high and distributions are narrow, dilatancy and shear-thickening effects are observed ([8]–[15]). Particle clusters that block the flow are formed, leading to instabilities. Viscosity dramatically increases, as well as normal forces, especially in confined geometries. Nakajima et al. [16] measured such dilatancy peak on plastisols at intermediate shear rates. They also mentioned the presence of a second shear thinning behaviour at higher shear rates. The same kind of behaviour was observed by Barroso et al. [17] at low to intermediate shear rates. They used a capillary rheometer to investigate the rheological behaviour up to 5000 s^{-1}. Both authors showed that the viscosity was decreasing when introducing large PVC particles. This is, again, in accordance with the literature of suspensions; adding particles of different sizes usually lessens or prevents dilatancy effects ([18], [19]). However the behaviour of the clusters of particles when the shear rate continues to increase is still open to debate: the high shear rate shear thinning behaviour may be related to the clusters disintegration induced by the high stresses or by a change of the aggregates shape [12]. The capacity of the particles to deform when subjected to high stresses is indeed another parameter that will affect the flow ([1), [20]) and can be expected for plastisols as PVC particles are dispersed in a good solvent.

In this paper the shear rheology of typical plastisol formulations has been measured on a wide range of shear rates using a Couette Rheometer and a capillary rheometer. The phthalate plasticizer (DINP diisononyl phthalate) has a Newtonian viscosity of 0.05 Pa.s. A bi-modal PVC resin (characteristic aggregates diameters 1 and 10 μm) has been used with a volume fraction varying between 45% and the maximum solid volume fraction.

Then three calcium carbonate fillers of different characteristic sizes were substituted in a constant proportion (47.6%) to the PVC particles: filler D10 has a median diameter d_{50} (i.e. 50% of the particles are under this size) of 9 µm, filler D15 has a d_{50}=20 µm and D40 has the highest d_{50} of 30 µm. The shape of the PVC particles is spherical as calcium carbonate particles are much more distorted.

A numerical model has been developed to account for these peculiar behaviours. It allows predicting the deposited thickness, the pressure distribution, the force exerted on the knife, as a function of the processing parameters.

RHEOLOGY OF PLASTISOL FORMULATIONS

Rheology measurements protocol: Plastisol formulations were prepared by adding PVC resins and $CaCO_3$ filler to the plasticizer (DINP). The composition was then mixed at speed 1500 RPM with a Dispermat blender for three minutes. Due to the plastisol sensitivity to temperature, this preparation protocol was fixed to enable a good dispersion and avoid a premature gelling due to temperature rise. Two types of rheometer were used at room temperature: Anton Paar MCR 301 equipped with a Couette (concentric cylinders) geometry. Shear rates range from 0.1 to 4000 s^{-1}. Depending on the formulation, the measurements performed at the highest shear rates are not relevant because they correspond to the saturation of the Couette torque measurement. Gottfert Rheo-Tester capillary rheometer with a 0.93 mm capillary die, 3 length-to-diameter ratios (L/D: 4, 8 and 16) and a 10 MPa transducer. With this type of machine shear rates range from 200 to 20000 s^{-1}.

Typical Rheology curve: Couette and capillary measurements are consistent (figure 1a), providing a continuous rheology curve in a wide shear rate range [0.1-10^4 s^{-1}].

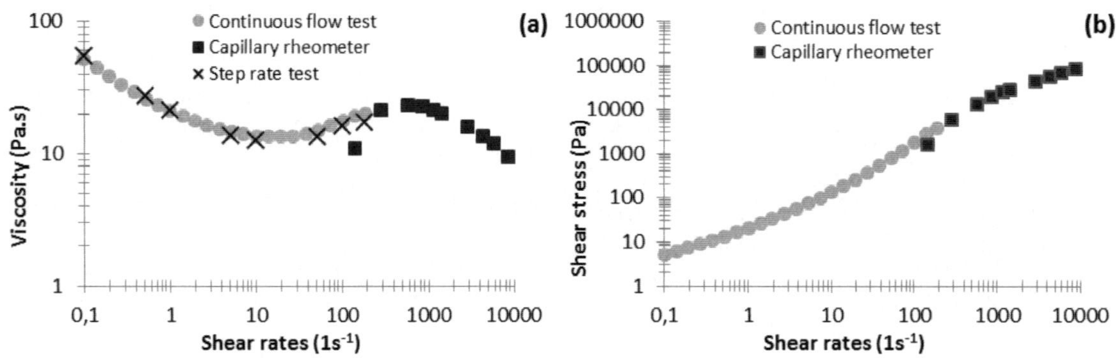

FIGURE 1: Bi-modal resin **(a)** Comparison between Couette flow tests and Capillary tests (63% volume fraction, no filler); **(b)** Shear stress as a function of the shear rate

These results are consistent with Nakajima et al. [16] and Barroso et al. [17] experiments: A first shear-thinning zone at low shear rates [10^{-1} -10 s^{-1}] is followed by a Newtonian plateau [10 – 40 s^{-1}], a shear-thickening peak [40- 600 s^{-1}] and then a second shear-thinning zone [> 600 s^{-1}]. The Newtonian plateau at low shear rate mentioned by Marcilla *et al.* [5] has not been observed.

Figure 2a shows the influence of PVC resin volume fraction: a quasi-Newtonian behaviour is observed at low solid volume fraction. For high solid volume fraction, a slight shear thinning behaviour is observed at low shear rates, followed by a Newtonian plateau at intermediate shear rates, a slight viscosity increase at higher shear rates and finally a viscosity decrease above 500 s^{-1}.

Figure 2b shows the influence of substituting a fraction of the PVC particles by calcium carbonate particles. Substituting the finest $CaCO_3$ particles (9 µm) to the PVC particles results in a sharp increase of the viscosity at low shear rates, but introducing the coarser particles (20 µm and 30 µm) leads on the contrary to a viscosity decrease in the whole shear rate range, as more important as the size of the filler particles increases. The dilatancy peak disappears for all the formulations and this is consistent with the literature on suspensions [18-19]. At high shear rates, and especially in the second shear thinning zone, the viscosity of the formulations with filler content is less than the viscosity of the pure PVC formulation (a factor 3 for the formulation with the coarser $CaCO_3$ particles).

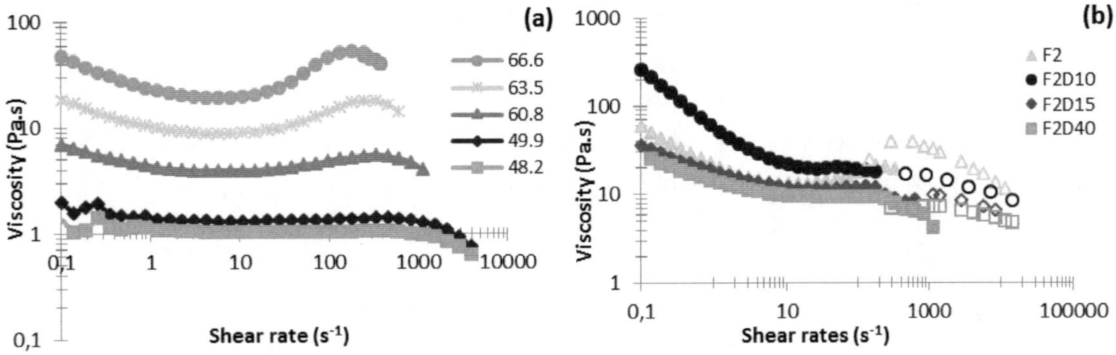

FIGURE 2: Bi-modal resin: **(a)** Couette flow tests at increasing volume fraction; **(b)** Couette (plain symbols) and capillary experiments (empty symbols) for the different bi-modal formulations (neat and with fillers of different sizes; total volume fraction 64%)

MODELING OF THE COATING PROCESS

The knife over roll process is described on figure 3: The roll velocity may be as high as 20m/min. A very narrow gap is adjusted between knife and cylinder (between 100 μm and 200 μm) in order to achieve the target plastisol deposit thickness (figure 3a). A typical industrial knife geometry is presented figure 3b. The flooring gelation is achieved at high temperature (140-200°C) in an oven downstream the knife-over-roll deposit device.

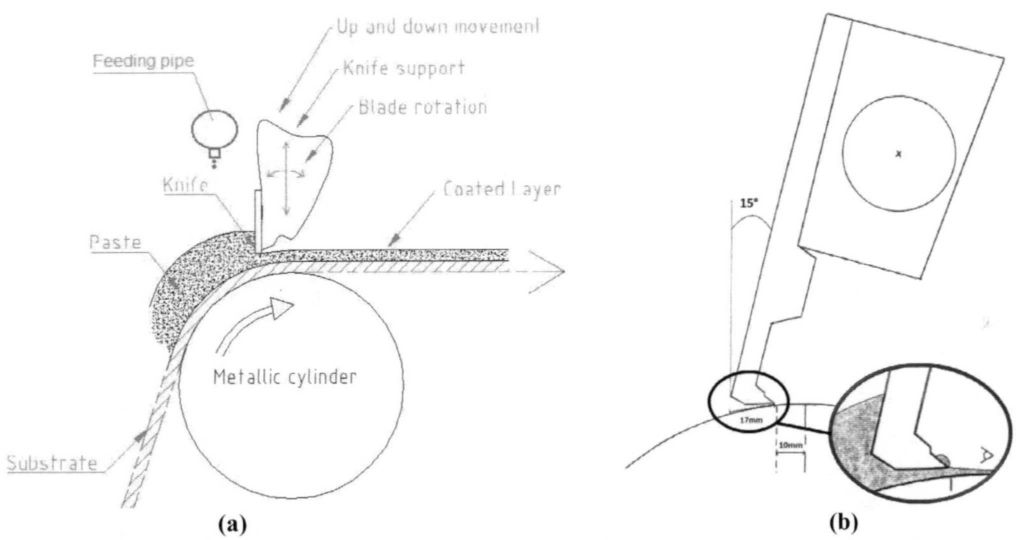

FIGURE 3: **(a)** Knife over-roll process, **(b)** Typical knife geometry

In the following x is the flow direction and y the normal direction to the cylinder surface.As the gap $h(x)$ between the roll and the knife is small compared to the length L of the knife, the lubrication approximations may by applied between the rotating cylinder (y=0) and the knife ($y=h(x)$). It means that the shear rate is mainly represented by $\frac{\partial u}{\partial y}$ and then the corresponding shear stress writes (u is the velocity component in the flow direction):

$$\tau(\frac{\partial u}{\partial y}) = \frac{\partial u}{\partial y}\eta(\frac{\partial u}{\partial y}) \quad (1)$$

The dynamic equilibrium writes:

$$\frac{\partial \tau}{\partial y} = \frac{dp}{dx} \quad (2)$$

which can be integrated as (τ_0 denotes the shear stress on the cylinder):

$$\tau(\frac{\partial u}{\partial y}) = y\frac{dp}{dx} + \tau_0 \qquad (3)$$

As the shear stress is a monotonous function of the shear rate $\frac{\partial u}{\partial y}$ it is thus possible to deduce the velocity gradient by inverting numerically the function represented on figure 1b:

$$\frac{\partial u}{\partial y} = \tau^{-1}\left(y\frac{\partial p}{\partial x} + \tau_0\right) \qquad (4)$$

The velocity profile u, the unknown constant τ_0 as well as the flow rate $Q = \int_0^{h(x)} u(x,y)dy$ are then determined using the boundary conditions: $u(x,0) = V$ and $u(x,h(x)) = 0$

The flow rate conservation $\frac{dQ}{dx} = 0$ and the boundary conditions: $p(0) = p(L) = 0$ allow to deduce the pressure field p.

Figure 4 shows the velocity field distribution (figure 4a) as well as the shear rate along the roll and the knife for the neat bi-modal plastisol formulation (figure 4b). An important recirculating flow is pointed out in the first knife convergent zone. Shear rates are varying in a wide range between the inlet zone and the exit (between 0 and $6000 s^{-1}$) .This points out that it is necessary to master the plastisol behaviour in the whole shear rate range.

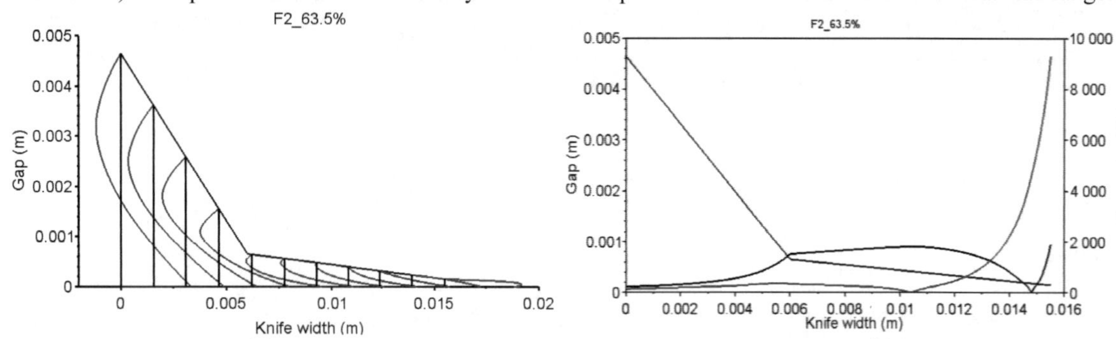

(a) (b)

FIGURE 4: velocity field **(a)** and shear rate distribution **(b)** between the knife and the roll. (red) at the knife surface; (black) at the roll surface (roll velocity: 20 m/min; gap between the roll and the knife: 0.15 mm; final angle between the knife and the cylinder: 15°)

When coating the different formulations which rheology has been presented figure 2b, one observes (table 1) that the deposited thickness varies slightly (for the same final gap). On the other hand, the maximum pressure and consequently the separating force exerted on the knife depend markedly on the formulation.

Formulation	neat bi-modal	bi-modal/D10 filler	bi-modal/D40 filler
Deposited thickness	0.133 mm	0.129 mm	0.124 mm
Maximum pressure	0.66 MPa	0.35 MPa	0.22 MPa
Separating force	2781 N	1516 N	917 N

TABLE 1: Coating characteristic figures as a function of the plastisol formulation; same processing parameters as in figure 6

CONCLUSIONS

The shear viscosity of plastisol formulations is very different than the viscosity of other thermoplastic polymers that can be used for coating. Plastisol formulations behave as suspensions and some phenomena may be explained in the light of the abundant literature on suspensions: influence of particle size and particle size distribution, jamming effects at high shear rates. Nevertheless the interactions between the PVC particles and the plasticizer even at ambient temperature, the coexistence of smooth spherical PVC particles and rough fractal calcium carbonate particles makes this suspension a more complex system than those encountered in the literature and in other industrial applications. The plastisol flow between the knife and the roll induces a wide range of shear rates and, for processing purposes, one needs to master the whole viscosity curve.

A bi-modal resin without filler induces jamming effects in the high shear rate range encountered in the downstream zone of the process and this could block the plastisol flow. Substituting some PVC particles by small diameter calcium carbonate particles supress the jamming effect and even decreases the viscosity at very high shear rates, but it increases the viscosity at the low shear rates encountered in the inlet zone of the knife which could prevent the plastisol to enter in the gap between the knife and the roll. Introducing coarse calcium carbonate particules in a bimodal PVC resin will simultaneously limit the viscosity at low shear rates and suppress the dilatancy peak at high shear rates.

References

[1] D. B. Genovese, "Shear rheology of hard-sphere, dispersed, and aggregated suspensions, and filler-matrix composites", *Adv. Colloid Interface Sci.*,**171–172**, 1–16,(2012).

[2] I. M. Krieger and T. J. Dougherty, "A mechanism for non-Newtonian flow in suspensions of rigid spheres," *Trans. Soc. Rheol.* **3**, 137–152 (1959).

[3] H. J. H. Brouwers, "Viscosity of a concentrated suspension of rigid monosized particles," *Phys. Rev. E*, **81**, (2010).

[4] D. Quemada, "Rheology of concentrated disperse systems II. A model for non-newtonian shear viscosity in steady flows," *Rheol. Acta*, **17**, 632–642 (1978).

[5] A. Marcilla, J. C. Garcia, and M. Beltran, "Study of the flow properties and the ageing process in PVC plastisols from commercial PVC resins," *Eur. Polym. J.*, **33**, 753–759 (1997).

[6] R. J. Farris, "Prediction of the viscosity of multimodal suspensions from unimodal viscosity data," *Trans. Soc. Rheol.,* **12**, 281–301 (1968).

[7] P. F. Luckham and M. A. Ukeje, "Effect of particle size distribution on the rheology of dispersed systems," *J. Colloid Interface Sci.*, **220**, 347–356 (1999).

[8] O. Hess and S. Hess, "Nonlinear fluid behavior: from shear thinning to shear thickening," *Phys. A*, **207**, 517–540 (1994).

[9] B. J. Maranzano and N. J. Wagner, "The effects of interparticle interactions and particle size on reversible shear thickening: Hard-sphere colloidal dispersions," *J. Rheol.*, **45**, 1205- (2001).

[10] Y. S. Lee and N. J. Wagner, "Dynamic properties of shear thickening colloidal suspensions," *Rheol. Acta*, v **42**, 199–208 (2003).

[11] M. E. Cates, M. D. Haw, and C. B. Holmes, "Dilatancy, jamming, and the physics of granulation," *J. Phys. Condens. Matter*, **17**, S2517 (2005).

[12] A. Fall, N. Huang, F. Bertrand, G. Ovarlez, and D. Bonn, "Shear Thickening of Cornstarch Suspensions as a Reentrant Jamming Transition", *Phys. Rev. Lett.*, vol. 100, 2008.

[13] N. J. Wagner and J. F. Brady, "Shear thickening in colloidal dispersions," *Phys. Today*, **62**, 27–32 (2009).

[15] E. Brown, N. A. Forman, C. S. Orellana, H. Zhang, B. W. Maynor, D. E. Betts, J. M. DeSimone, and H. M. Jaeger, "Generality of shear thickening in dense suspensions," *Nat. Mater.*,(2010).

[15]E. Brown and H. M. Jaeger, "Shear thickening in concentrated suspensions: phenomenology, mechanisms and relations to jamming," *Rep. Prog. Phys.*, **77**, 046602, (2014).

[16] N. Nakajima and E. R. Harrell, "Rheology of poly(vinyl chloride) plastisol for superhigh shear-rate processing. I," *J. Appl. Polym. Sci.*, **115**, 3605–3609 (2010).

[17]E. G. Barroso, F. M. Duarte, M. Couto, and J. M. Maia, "High strain rate rheological characterization of low viscosity fluids," *Polym. Test.*, **29**, 419–424 (2010)

[18] C. E. Chaffey, "Mechanisms and equations for shear thinning and thickening in dispersions," *Colloid Polym. Sci.*, **255**, 691–698 (1977).

[19] W. H. Boersma, J. Laven, and H. N. Stein, "Shear thickening (dilatancy) in concentrated dispersions," *AIChE J.*, **36**, 321–332 (1990).

[20] C. Pozrikidis, "Orientation statistics and effective viscosity of suspensions of elongated particles in simple shear flow," *Eur. J. Mech. - BFluids*, **24**, 125–136 (2005). [29]

Effect of Crystallization on PEEK Bearing Components

Richard Pieter Schaake[a], Jiying Fan[a], Jeremy Joseph Weigand[b], Jason Louis Williams[b], Alicyn Marie Rhoades [b]

[a] *SKF Engineering & Research Centre, PO Box 2530, 3430 DT, Nieuwegein, The Netherlands*
[b] *ISchool of Engineering, Penn State Behrend, 5101 Jordan Road, Burke Center 242, Erie, PA 16563*

richard.schaake@skf.com

Abstract. Rolling element bearings are widely used machine components. The understanding of their performance has improved tremendously over the past century and simulations taking into account the dynamic behavior of individual bearing components are common practice. Such simulations allow product developers in the bearing industry to further optimize these components through better understanding of the dynamic loads the components are exposed to. However, to do this requires detailed knowledge of the actual material properties. Where components were originally developed based on datasheet properties and safety margins, it has now become necessary to understand the effect of process on the performance. In this presentation it is illustrated how fiber orientation, porosity and crystallization kinetics add up to explain knock-down factors used in bearing component analysis. The importance of modeling crystallization at high cooling rates is demonstrated.

Keywords: Polyetherketone, Injection-based, Crystallisation, Simulation

PACS: 81,83

INTRODUCTION

FIGURE 1. Rolling element bearing, the rollers are guided by a cage. PEEK is one of the main cage materials.

With its high temperature and chemical resistance, PEEK is a commonly used material for rolling element bearing cages. The function of these cages is to guide the rolling elements; a high clearance between rollers and polymer will lead to excessive shock loads. However, to achieve high accuracy of external dimensions, shrink holes may occur that weaken the properties. Since cages have a pocketed shape, weld lines are unavoidable; the performance of these weldlines will also be dependent on processing conditions.

APPROACH

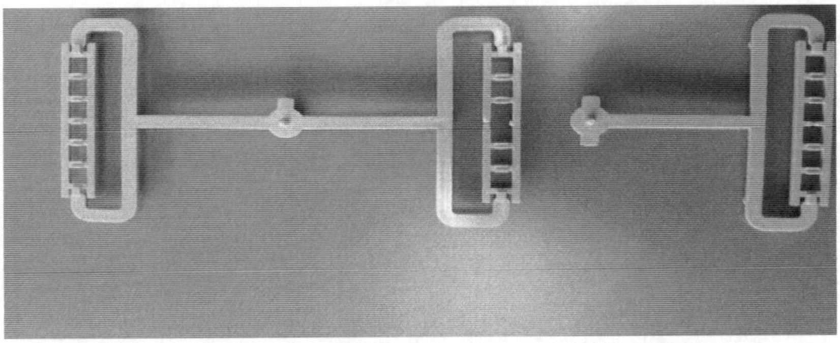

FIGURE 2. Ladders used for studying molding phenomena, as illustrated by these short shots obtained using different configurations. Up-stream effects can be studied by changing the runner lay-out and gating using different mold inserts.

To reduce the mold complexity for generic studies, the cage geometry has been simplified to a straight ladder-like shape. Like the cage, these ladders will have to contain the local geometric features of the cage that lead to local differences in structure of the part. In a two cavity configuration with a central weldline configuration, the stiffness of the ladders is in agreement with the observed porosity, whereas the strength of the weldline is higher in the central weldline that forms first.

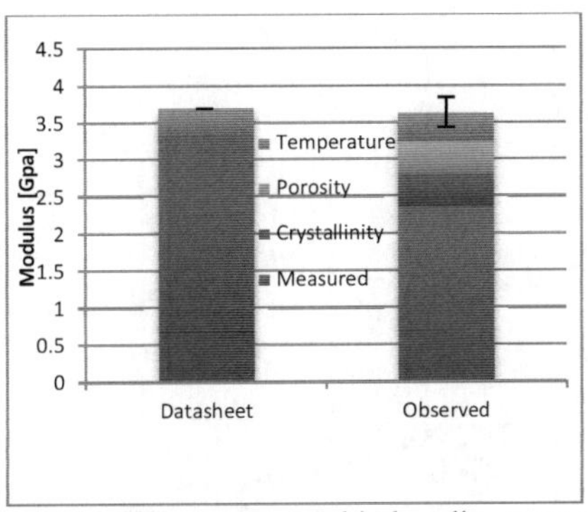

FIGURE 3. Datasheet stiffness compared to stiffness measured in bending tests performed on injection molded ladders.

The properties of molded ladders can be considerably lower than those reported in the material datasheet, see Figure 3. Incomplete crystallization and porosity both account for half of the loss of stiffness compared to the datasheet value.

RESULTS

We measured the quiescent crystallization kinetics of PEEK 150G, using flash DSC and fitted the resulting data to the Nakamura crystallization model as implemented in Moldex3D. This model considers two effects: the induction time for crystallization [1] and the crystal growth as described by the Nakamura model [2]. The induction time can be correlated to an induction temperature at constant cooling rates. This induction temperature was determined with 5°C accuracy from the measurements. Using the results of a preliminary fit of individual cooling rate responses to the Nakamura model a more accurate estimate of Ti(q) could be obtained. After obtaining fits to the individual curves, the best fit to all curves was determined. There are still some differences between the measurement and the response in Moldex3D, see Figure 4.

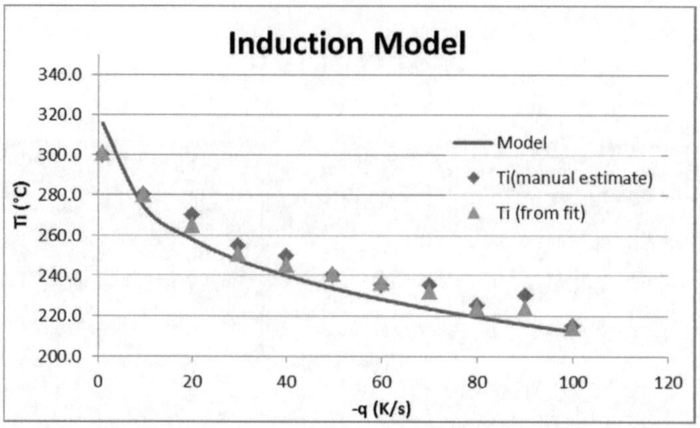

FIGURE 4. Fit of the induction temperature Ti(q)for constant cooling rate to the model used my Moldex3D. Blue marks are the 5°C accurate estimates; the green marks were determined by determining the induction temperature from fitting each cooling rate to the Nakamura model.

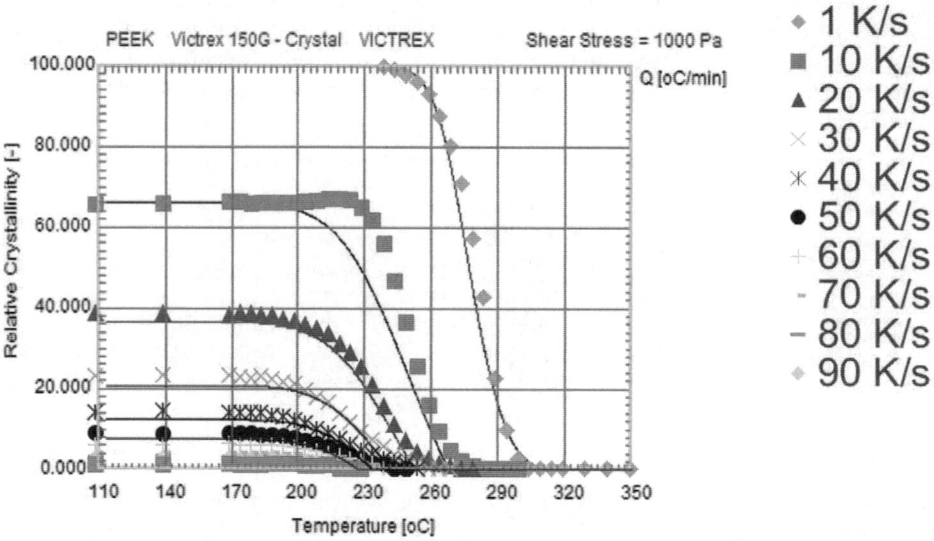

FIGURE 5. Comparison of the FDSC measurements to the values calculated in Moldex3D for different cooling rates.

The crystallization kinetics will affect both the formation of weldlines and the porosity. When comparing the volumetric shrinkage of Victrex PEEK 150G ladders in Moldex3D with and without using a crystallization model, the difference is clear, see Figure 5. A homogeneous volumetric shrinkage distribution correlating to temperature gradients is observed when using the database properties, shrinkage gradients are observed, particularly at intersections.

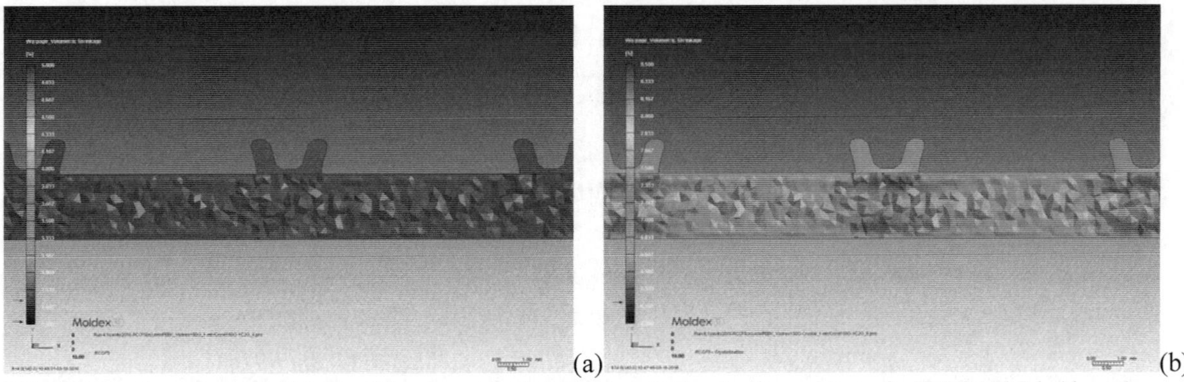

FIGURE 6. Cross section of Moldex3D volumetric shrinkage prediction for the molding of a PEEK 150G ladder using a single cavity configuration with two gates using the datasheet properties (a) and datasheet properties extended with a crystallization model (b)

SUMMARY

We have successfully performed high cooling rate crystallization studies of PEEK using flash DSC to provide input parameters for crystallization calculations in Moldex3D molding simulations. The result is a less homogeneous shrinkage prediction. The next steps are to experimentally verify the relative crystallinity and shrinkage predicted on injection molded samples, as well as to characterize the effect of flow on the crystallization behavior and the properties of the molded parts.

REFERENCES

1. Y.K. Godowsky and G.L. Slonimsky; *J. Polym. Sci.* **12**, pp. 1053-1080 (1974)
2. K. Nakamura, T. Watanabe, K. Katayama and R. Amano, J. Appl. Polym. Sci., **16**, pp.1077-1091 (1972)

Modeling and Optimization of Melt Filtration Systems in Polymer Recycling

Sophie Pachner[a*], Michael Aigner[b] and Jürgen Miethlinger[a]

[a]*Johannes Kepler University Linz, Institute of Polymer Extrusion and Compounding Altenberger Str. 69, 4040 Linz, Austria*

[b]*Erema Engineering Recycling Maschinen und Anlagen Ges.m.b.H Unterfeldstr. 3, 4052 Ansfelden, Austria*

sophie.pachner@jku.at

Abstract. This paper addresses the performance improvement of a double-cavity piston screen changer used for processing slightly contaminated in-house recycling material. Rheological optimization of the flow geometry based on analytical calculations and numerical CFD simulations in order to increase its self-cleaning capacity, decrease the residence time of the melt in the filter and reduce pressure drops, thus enabling gentle processing of a variety of thermoplastic polymers. Various analytical calculations of the overall pressure drop were performed and experimentally verified for a range of in-house recycling materials, throughputs and filter screen types.

Keywords: Extrusion, Polymer Recycling, Melt Filtration, Screen Changer
PACS: 81.20.Hy, 83.50.Uv, 83.50.Ha, 83.80.Sg, 81.05.Lg

INTRODUCTION

Melt filtration plays an important role in polymer recycling, as it ensures that the melt stream is suitably clean and free of contaminants such as paper, metals, wood, foreign plastics, glass, dust particles, and unmelted or degraded polymer formations. Depending on process, throughput rate, contamination and extruder size, a variety of screen changer systems such as slide-plate screen changers, piston screen changers, rotary disc filters and ribbon-type filters are available. [1] Furthermore, various types of metallic filter medium exist, for instance, woven screens, sintered powder, super plates and random fibers, which differ in terms of permeability and ability to hold contaminant and capture gels. [2]

In polymer recycling, double-piston screen changers are used because they allow continuous operation and have a high cleaning capacity. When highly contaminated recycling materials are processed, screens will foul rapidly, so backflush screen changers are used to increase operation time.
Processing applications such as in-house recycling, fiber, blown-film, cast film and pipe manufacturing deal with slightly contaminated materials. These minor contaminants (e.g. dust) can be removed by using a continuously operating direct flow melt filter with two screen cavities. Service life is sufficient even without backflush function of the melt filter, and flow paths can be shortened, which results in gentler polymer processing.

OPTIMIZATION AND MODELING

We improved the performance of a direct-flow double-cavity piston screen changer by optimization based on analytical calculations and numerical CFD simulations. In order to increase the self-cleaning capacity of the melt filter, high wall-shear rates $\dot{\gamma}_W$ are recommended. Short residence times t_{res} and small pressure drops $\Delta p'$ ensure gentle processing.
Designing a melt filter with optimized flow geometry means balancing conflicting requirements: While achieving small pressure drops calls for increased diameters of the flow channels, short residence times and high wall-shear speeds for efficient self-cleaning require diameters to be reduced. So for optimal design of the flow geometry therefore an agreement of all mentioned requirements needs to be found.

Analytical Calculation of Pressure Drop and Wall Shear Rate

For the layout of a filtration process, the initial pressure drop of the polymer melt and the screening performance of the filter must be determined. To this end, the flow geometry is subdivided into cylindrical segments with radius R. Pressure drop Δp_{kav} and wall shear rate $\dot{\gamma}_W$ can easily be determined using the representative viscosity method according to Giesekus [3]:

$$\Delta p_{kav} = \frac{\dot{V} \cdot \eta_{rep}}{k_{res}} \quad \text{and} \quad \dot{\gamma}_{rep} = \frac{4 \cdot \dot{V}}{\pi \cdot R^3} \cdot 0{,}815 \tag{1}$$

$$\dot{\gamma}_W = \frac{4 \cdot \dot{V}}{\pi \cdot R^3}, \tag{2}$$

with volume flow rate $\dot{V}$, resistance factor k_{res} and representative viscosity η_{rep} which is determined using the representative shear rate $\dot{\gamma}_{rep}$, see Equation (1).

Several models exist for calculating the initial pressure drop of a woven screen. According to Brackett-Rozinskiy and Edle et.al. [4, 5], the pressure drop Δp_{screen} of a polymer melt flow through porous media can be determined using Darcy's Law:

$$\Delta p_{screen} = \frac{v \cdot \eta \cdot \delta}{k} \quad \text{with} \quad k = \frac{{d_p}^2 \cdot \varepsilon^3}{150 \cdot (1-\varepsilon)^2}, \tag{3}$$

where k is the relative permeability, v the velocity, and η the fluid viscosity. The relative permeability parameter k is modeled using a Blake-Kozeny relationship according to Bird et.al. [6] and depends on the average pore size d_p and the current filter porosity ε, see Equation (3).

Further, Carley et.al. [7] presented a model for estimating the initial pressure drop of a woven screen:

$$\Delta p_{screen} = F_S \cdot \eta \cdot \dot{\gamma}^{1-n} \left(\frac{4 \cdot \dot{m}}{{D_S}^2 \cdot \pi \cdot \rho} \right)^n, \tag{4}$$

where F_S is the screen resistance factor, η the viscosity, $\dot{\gamma}$ the shear rate, $\dot{m}$ the mass flow rate, D_S the screen diameter, ρ the melt density and n the power-law exponent.

The model by Müller and Piesche [8] is based on similarity laws in the form of dimensionless numbers:

$$\Delta p_{screen} = Eu \cdot \rho \cdot {v_0}^2 \quad \text{with} \quad Eu = \frac{14 \cdot \delta \cdot \phi}{\varepsilon^{5/2} \cdot d \cdot \left(1 + \frac{W}{d}\right) \cdot Re}. \tag{5}$$

Here, the pressure drop is determined based on the Euler Number Eu, the melt density ρ and the undisturbed velocity v_0. The dimensionless Euler number Eu for square woven screens is a function of wire diameter d, screen thickness δ, specific screen surface ϕ, mesh size W, porosity ε and Reynolds number Re.

Flow Geometry Optimization by CFD Simulation

As previously mentioned, we ran CFD simulations using ANSYS Polyflow 16.2© to optimize the flow geometry. In total, 4 different geometry models of a direct-flow double-cavity piston melt filter with a screen diameter of 82 mm were analyzed. Figure 1 shows a schematic diagram of the flow geometry.

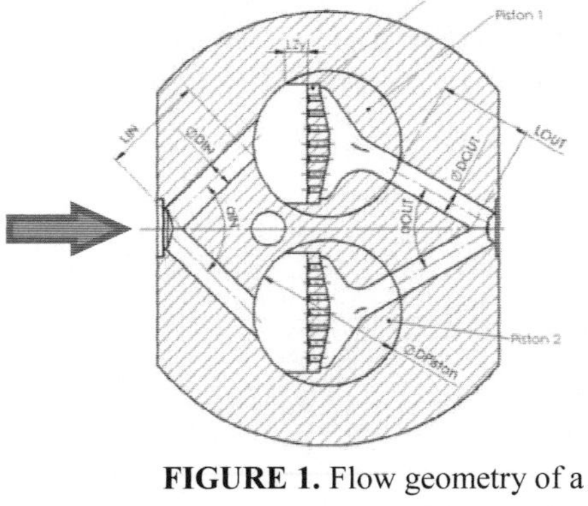

Inflow

D_{IN}	inflow diameter
L_{IN}	length of cylindrical inflow
α_{IN}	angle of inflow separation
L_{Zyl}	cylindrical length in front of the screen pack

Outflow

D_{OUT}	outflow diameter
L_{OUT}	length of cylindrical outflow
α_{OUT}	angle of outflow separation

FIGURE 1. Flow geometry of a direct-flow double-cavity piston melt filter.

CFD Simulations were carried out for a PP-H polypropylene homopolymer intended for thermoforming packaging. Generalized Newtonian material behavior and isothermal flow conditions were assumed. The optimization was performed for minimum and maximum throughputs of 30 kg/h and 120 kg/h, respectively. In total, three optimization steps were carried out and analyzed.
The first optimization step consisted of reductions in in- and outflow diameters and an increase in cylindrical length in front of the breaker plate. In addition, the angle of inflow separation was decreased, which resulted in a shorter length of the cylindrical inflow. These modifications to the flow geometry led to increased wall shear rates in the in- and outflow sections at an even higher total pressure drop.
In the second optimization step, the in- and outflow diameters were reduced to a lesser extent than in step 1, and the inflow angle was further decreased. This resulted, again, in increased wall shear rates in the in- and outflow sections, but at even smaller pressure gradients than in optimization step 1.
The final optimization step used the same in- and outflow diameters and the same inflow angle as in step 2, but the volume in front of the screen pack was reduced. These changes in flow geometry caused increased wall shear rates that are sufficiently high for self-cleaning. Additionally, the residence time and the overall pressure drop were reduced according to step 2 to enable gentle processing even of sensitive polymers such as EVOH, PET and PVC.

EXPERIMENTAL

To verify the analytical calculation of the direct flow melt filter with a screen diameter of 82 mm, we conducted experiments at an Erema recycling plant (Intarema 1108T and Intarema 1108TE) using various in-house recycling materials, filtration fines and mass flow rates. The experimental set-up is summarized in Tables 1 and 2.

TABLE (1). In-house recycling materials for experimental verification:

		LDPE	**HDPE**	**CPP**	**BOPP**	**BOPET**
MFI	g/10min	0,6	0,6	9,4	7	-
		(190°C/2,16kg)	(190°C/5kg)	(230°C/2,16kg)	(230°C/2,16kg)	
melting point	°C	131	135	170	175	258

TABLE (2). Filtration parameters of square woven (*SW*) and twilled dutch (*TD*) metal screens:

Screen		**14**	**16**	**80**	**24/110**	**50/250**	**200/600**
type		SW	SW	SW	TD	TD	TD
fines	Mesh	14	16	80	24/110	50/250	200/600
	µm	1140	1000	200	150	60	32
porosity	%	73	69	68	63	65	55

RESULTS

Flow Geometry Optimization by CFD Simulation

The numerical und analytical calculations yielded an optimal design that takes into account wall shear rate, pressure gradient and residence time. Several simulation results for the velocity distribution, at maximum throughput, of base version 1 and final version 4 are shown in Figure 2. Figure 3 illustrates the cumulative residence time distribution at maximum and minimum throughputs, and Figure 4 presents a comparison of several simulation results for wall shear rate and overall pressure drop at maximum and minimum throughputs.

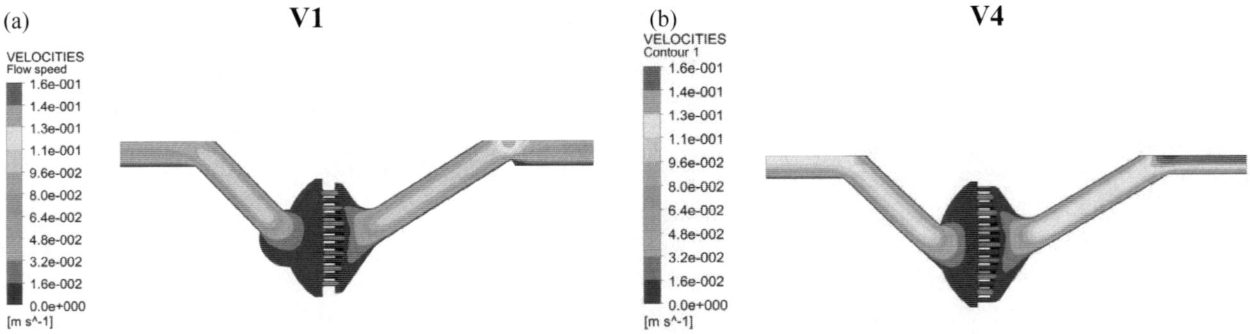

FIGURE 2. Velocity distribution for base version V1 (a) and final version V4 (b); for a max. throughput of 120 kg/h

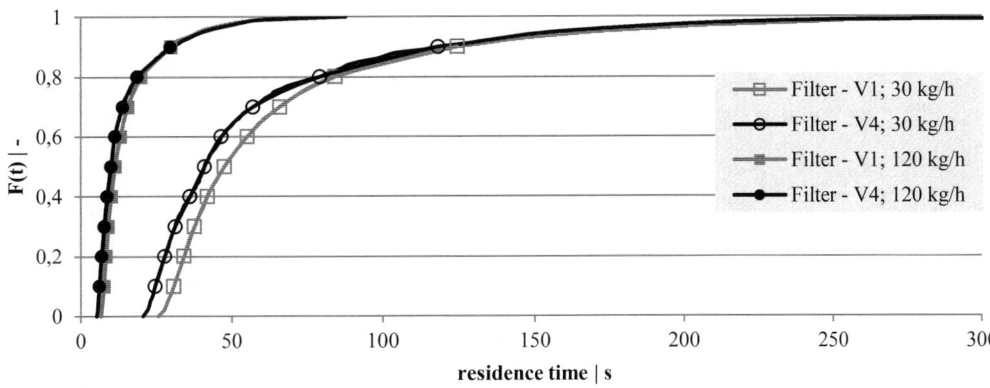

FIGURE 3. Cumulative residence time distribution for base version V1 and final version V4

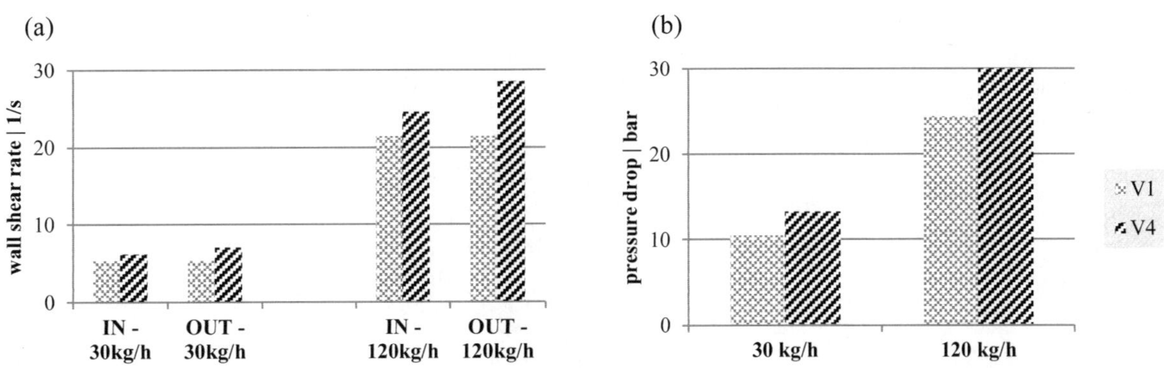

FIGURE 4. Overview of simulation results at min. and max. throughputs: wall shear rate at in- and outflow sections (a); overall pressure drop (b)

The optimization of the flow geometry yielded uniform inflow of the screen pack, and thus optimal utilization of the filter surface was achieved; see Figure 2. Hence, for a given pressure drop, the average residence time could be reduced by 13% and the wall-shear rate increased by 15%.

Experimental Verification of the Analytical Calculation

Figure 5 shows a comparison of the analytically determined pressure drop versus the experimentally investigated data for different throughputs, screen types and materials according to Tables 1 and 2. The analytical calculation was carried out under isothermal conditions because of low Griffith numbers. The results of the analytical calculation for operation without screens are in good agreement with the experimental data, and errors of less than 10% were achieved; see Figure 5a. In the analysis of the pressure loss during operation with woven screens, larger differences between calculation and experiment can be seen (cf. Figure 5b).

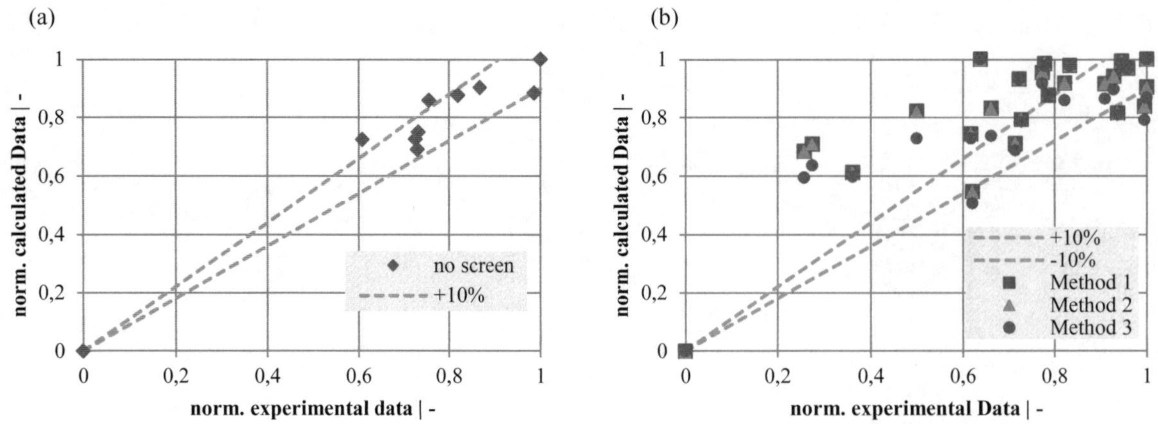

FIGURE 5. Comparison of analytically calculated and experimentally determined pressure drops in normalized presentation; data without screen (a) and data for different screen types (b); Method 1 according to Müller and Piesche [8]; Method 2 according to Carley and Smith [7]; Method 3 according to Bracket-Rozinsky [4]

CONCLUSION

We improved the performance of a double-cavity piston screen changer used for processing slightly contaminated in-house recycling materials. On the basis of analytical calculations and numerical CFD simulations, we optimized the geometry of the flow channels and breaker plates in order to increase self-cleaning capacity, decrease residence time of the melt in the filter and reduce pressure drops. This allows gentle processing of various thermoplastic polymers. The overall pressure drop was calculated analytically and verified experimentally. While the data calculated for operation without screens is in good agreement with the experimentally determined data, larger differences between calculation and experiment were found for operation with woven screens. The analytical calculations did not consider geometry-dependent parameters such as velocity and temperature distribution in front of the screens, which influence the viscosity and result in a deviation of the analytical from the experimental results. Further experiments will therefore be carried out to develop a novel mathematical model based on heuristic optimization algorithms for calculating the pressure drop of polymers flowing through woven screens. This model will consider changes in viscosity caused by flow geometry, velocity and temperature distribution.

ACKNOWLEDGMENTS

Financial support by Erema Engineering Recycling Maschinen und Anlagen Ges.m.b.H is gratefully acknowledged.

REFERENCES

1. H. F. Giles, E. M. Mount, and J. R. Wagner, *Extrusion: The Definitive Processing Guide and Handbook*: Elsevier Science, 2004.
2. C. Rauwendaal, *Polymer Extrusion*: Carl Hanser Verlag GmbH & Company KG, 2014.
3. H. Giesekus and G. Langer, "Die Bestimmung der wahren Fließkurven nicht-newtonscher Flüssigkeiten und plastischer Stoffe mit der Methode der repräsentativen Viskosität," (de), *Rheol Acta*, vol. 16, no. 1, pp. 1–22.
4. N. Brackett-Rozinsky, S. Mondal, K. R. Fowler, and E. W. Jenkins, "Analysis of Model Parameters for a Polymer Filtration Simulator," *Modelling and Simulation in Engineering*, vol. 2011, no. 2, pp. 1–11, 2011.
5. D. Edle and C. Gooding, "Prediction of Pressure Drop for the Flow of Polymer Melts through Sintered Metal Filters," *Industrial and Engineering Chemisty Process Design and Development*, vol. 1985, no. 24, pp. 8–12, 1985.
6. R. B. Bird, W. E. Stewart, and E. N. Lightfoot, *Transport phenomena,* 2nd ed. New York: J. Wiley, 2007.
7. J. F. Carley and W. C. Smith, "Design and Operation of Screen Packs," *Polym. Eng. Sci,* vol. 18, no. 5, pp. 408–415, 1978.
8. M. Müller and M. Piesche, "Ähnlichkeitsgesetze zur Beschreibung des Anfangsdruckverlustes metallischer Drahtgewebe bei der Filtration nicht-Newtonscher Fluide," *F&S Filtrieren und Separieren*, vol. 27, no. 5, pp. 284–291, 2013.

Enhanced Shrinkage Prediction Considering Cooling Rate Effects on PVT

Franco S. Costa, Peter S. Cook and Zhihao Zuo

Autodesk Inc. 259-261 Colchester Rd. KILSYTH, Victoria, Australia

Franco.Costa@autodesk.com

Abstract. The prediction of pressure decay, density and shrinkage of injection molded parts is based on the Pressure-Volume-Temperature (PVT) equation of state. The PVT of materials is typically characterized at equilibrium or near-equilibrium conditions which are far-removed from the rapid cooling rates occurring during conventional injection molding. In this study, a cooling rate dependent PVT model for amorphous polymers is implemented as a user-PVT function interfacing to the Autodesk Moldflow Insight software. Predicted shrinkage values are compared to measured data for two amorphous materials at different packing pressures. Comparisons are made of the shrinkage magnitude and the trend of shrinkage with respect to packing pressure.

Keywords: Injection Molding, Simulation, Software.
PACS: 81.05.Lg, 89.20.Kk, 82.35.Lr, 83.80.Sg

INTRODUCTION

A key outcome sought from injection molding simulation software is the prediction of final part shape after molding. The deviation of the final part shape from the as-designed mold geometry is comprised of both in-plane shrinkage and out-of-plane warp deflections. The prediction of the final part shape is usually accomplished by a finite-element structural analysis of the part geometry with the load condition being the initial local strains occurring due to polymer contraction during cooling. However, in molding practice, this contraction tendency is compensated by the pressurized packing of additional polymer melt into the molding cavity before solidification. Therefore, a warp & shrink calculation must be preceded by a compressible flow simulation of the mold filling and packing stages of the injection molding process. This compressible flow simulation incorporates the counteracting effects of cooling and packing into a volumetric shrinkage result which depends on an equation of state to describe the response of the polymer's specific volume to both pressure and temperature changes. The equation of state, normally referred to as a Pressure-Volume-Temperature (PVT) model, is typically based on measurements performed at near-equilibrium conditions. In contrast, actual cooling rates when hot polymer melt is injected into a relatively cooler mold are as high as 1,000 K/s and typically in the order of 100 K/s.

The present work demonstrates the use of a PVT model for amorphous polymers which depends on the rate of cooling in addition to Temperature and Pressure. This model is used for the prediction of injection molding flow behavior and shrinkage by integration into the commercial Autodesk Moldflow Insight 2017 software package via user-coded PVT routines, and applied to a 3D model of the part and runner geometry. Predictions of shrinkage obtained with both a standard (equilibrium) PVT model and a cooling rate dependent PVT model are compared with measured shrinkage values of samples molded at a range of packing pressures, for two different amorphous materials.

PVT MODELS

Two-Domain Tait PVT Model

The standard PVT model used in most thermoplastic injection molding simulation software is the two-domain Tait Model [1, 2]. In the case of amorphous polymers, the Tait equation for specific volume, V, as a function of temperature, T (in K), and pressure, P (in Pa), is:

$$V(T,P) = V_0(T)\left[1 - C\ln\left(1 + \frac{P}{B(T)}\right)\right] \tag{1}$$

$$V_0(T) = b_1 + b_2(T - b_5) \tag{2}$$

$$B(T) = b_3 e^{-b_4(T-b_5)} \tag{3}$$

$V_0(T)$ is the specific volume at zero gauge pressure and C is a constant, 0.0894. As the polymer cools, it passes through a glass transition at a temperature $T_t(P)$. The transition temperature at zero pressure is considered a material constant, b_5. The transition temperature increases linearly with pressure according to the material constant b_6:

$$T_t(P) = b_5 + b_6 P \tag{4}$$

Although the pressure and temperature dependence of the specific volume of amorphous polymers is said to follow Equations (2) & (3) in both the melt and solid phases, the material constants (b_1, b_2, b_3 & b_4) are assumed to be different for temperatures above $T_t(P)$ (melt phase) to those for temperatures below $T_t(P)$ (solid phase) and are therefore denoted as b_{1m}, b_{2m}, b_{3m} & b_{4m} and b_{1s}, b_{2s}, b_{3s} & b_{4s} respectively.

PVT Model with Cooling Rate Effect

The Tait equations are strictly only applicable for polymers at very low cooling rates, under equilibrium conditions. Numerous authors (for example [3, 4]) discuss how at higher cooling rates the glass transition in amorphous polymers is shifted to a higher temperature and the molecular disorder retained in the solid state is greater. The higher disorder also increases the specific volume relative to that of a slowly-cooled polymer. There is no evidence to suggest that the cooling rate has any effect on specific volume above the shifted transition temperature.

The increase in the glass transition temperature depends on the cooling rate, q, occurring at the time of solidification. It is quite well-fitted to a logarithmic form [4, 5, 6] so that the transition temperature at a non-zero cooling rate can be approximated from equation (4) as:

$$T_t(P,q) = b_5 + b_6 P + \theta \log\left(\frac{q}{q_0}\right) \tag{5}$$

Here θ is a material constant, typically around 3K, and q_0 is a cooling rate sufficiently low that the polymer has time to maintain an equilibrium state. A second approximation suggested by measurement data is that the slopes of the PVT curve above and below the transition temperature (i.e. the derivatives with respect to temperature) are unaffected by cooling rate. With these two assumptions, the Tait PVT equations can be modified to allow for the cooling rate at transition following an approach similar to the one taken by Chang et al [6]. The resulting equations are identical to Equations (1), (2) & (3) with the exception that for the solid phase below $T_t(P)$ the specific volume at zero pressure is given by:

$$V_0(T,q) = b_{1s} + b_{2s}(T - b_5) + (b_{2m} - b_{2s})\theta \log(\frac{q}{q_0}) \tag{6}$$

The temperature and pressure partial derivatives as given in Cook *et al.* [7] are also used because the simulation of cavity filling and packing incorporates material compressibility effects.

SOLVER API FRAMEWORK

The Autodesk Moldflow Solver Application Programming Interface (API) allows user coded functions to be dynamically linked at run-time into the pre-compiled 3D Flow and Warp solvers. The 3D Flow solver is a non-isothermal, compressible flow simulation of the cavity filling and packing which uses three-dimensional tetrahedral elements to represent the part geometry and can use either tetrahedral or one-dimensional beam elements to represent the runner geometry. The compressible Navier-Stokes equations for low Reynolds number flow are coupled with a Generalized Newtonian viscosity model.

The Solver API framework, first proposed by Costa *et al.* [8] for the use of user-coded viscosity functions, is an open environment in which users and researchers can implement customized material models and outputs to meet their individual needs. In the current work, we have coded user PVT functions for the cooling rate-dependent PVT model described in the previous section and performed Flow and Warp analyses of some actual molding cases.

MOLDING CASES

The simulations were compared to molding data obtained in the Autodesk Moldflow Plastics Laboratory for a rectangular cavity with uniform thickness of 2.0 mm, length of 200 mm and width of 40 mm fed from a fan gate at one end of the part (see Figure 1). The mold is instrumented with cavity pressure transducers as indicated and also a pressure transducer at the top of the sprue to record the injection pressure. The cavity surface has an inscribed grid of fine gauge lines as reference points for part dimension measurement. The flow direction shrinkages between intersection points in these gauge lines were measured with an OGP Smartscope Flash 400 system.

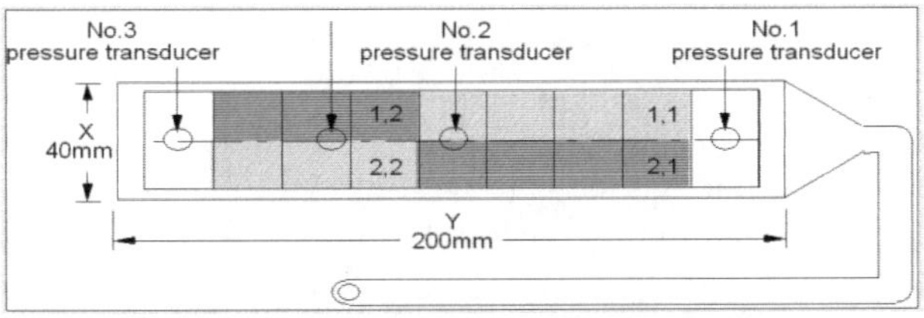

FIGURE 1. Molding geometry for shrinkage study, showing shrinkage measurement zones, pressure and temperature sensors.

Two unfilled amorphous polymers were selected for the molding trials, one being a polycarbonate (Infino EH-1050 from Cheil Industries Incorporated) and the other a Styrene Acrylonitrile (SAN) (Luran 368 R from Styrolution). The standard PVT model constants for this material are shown in Table 1. The two additional constants, θ and q_0, required for the cooling rate dependent PVT model, are also listed as well as the molding process conditions. In all cases the packing time was 15 seconds.

Table 1. PVT model parameters and molding condition details.

Material	PC (Infino EH-1050)	SAN (Luran 368 R)
b_5	408.15 K	373.53 K
b_6	3.955 x 10^{-7} K/Pa	2.846 x 10^{-7} K/Pa
b_{1m}	0.0008749 m^3/kg	0.0009479 m^3/kg
b_{2m}	5.973 x 10^{-7} m^3/kg-K	5.839 x 10^{-7} m^3/kg-K
b_{3m}	1.75099 x 10^8 Pa	1.8201 x 10^8 Pa
b_{4m}	0.004477 1/K	0.003749 1/K
b_{1s}	0.008749 m^3/kg	0.0009479 m^3/kg
b_{2s}	2.466 x 10^{-7} m^3/kg-K	1.862 x 10^{-7} m^3/kg-K
b_{3s}	2.60136 x 10^8 Pa	2.76006 x 10^8 Pa
b_{4s}	0.002954 1/K	0.002802 1/K
θ	2.86 K	2.86 K
q_0	1 x 10^{-3} K/s	1 x 10^{-3} K/s
Melt Temperature	309°C	244°C
Mold Temperature	99°C	53°C
Injection Time	1.0 s	0.9 s
Packing Pressure	52, 67 & 84 MPa	38, 55, 73 & 91 MPa

The influence of cooling rate at transition on the specific volume predictions from the cooling rate dependent PVT model are shown for these two materials in Figure 2.

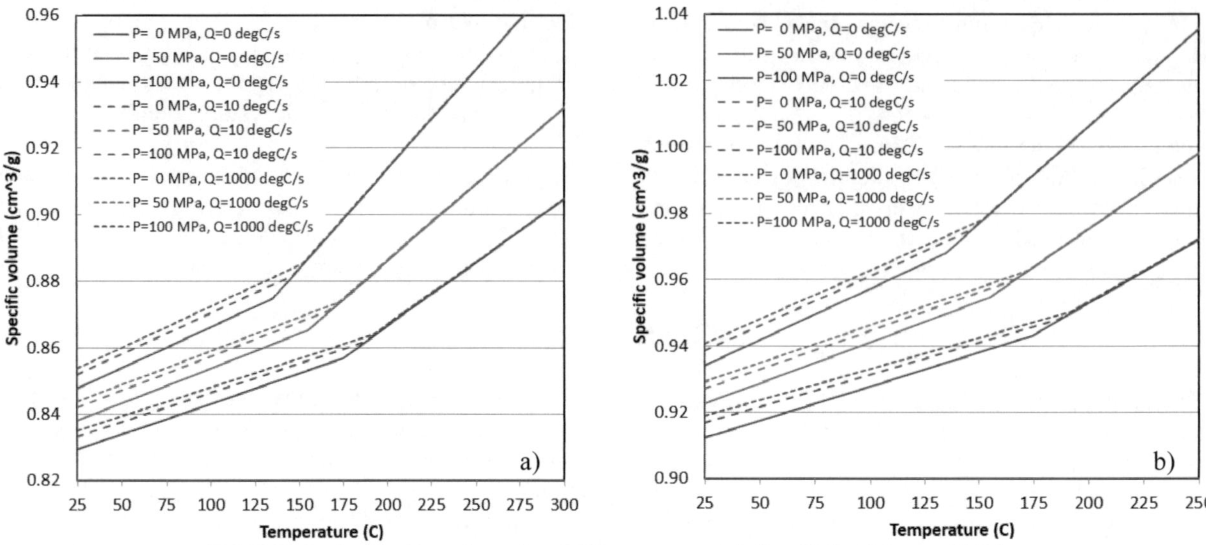

FIGURE 2. Cooling Rate dependent PVT response. a) Infino (PC), b) Luran (SAN)

RESULTS & DISCUSSION

The local cooling rate at transition was recorded by the coded user-routines during the 3D flow solution. Figure 3 displays this result on a cutting plane through the part and runner geometry for the Infino material with 67 MPa packing pressure. Also displayed is the profile of predicted volumetric shrinkage through the part thickness at a point near the center of the part. The volumetric shrinkage profiles from the standard PVT model and the cooling rate dependent PVT model are compared and show two distinct regions. In the skin layer, the two PVT models yield almost identical volumetric shrinkage predictions because the skin layer solidifies while pressurized. In contrast, the core region, which solidified after the cavity pressure reached zero, shows an increased volumetric shrinkage for the cooling rate dependent PVT model.

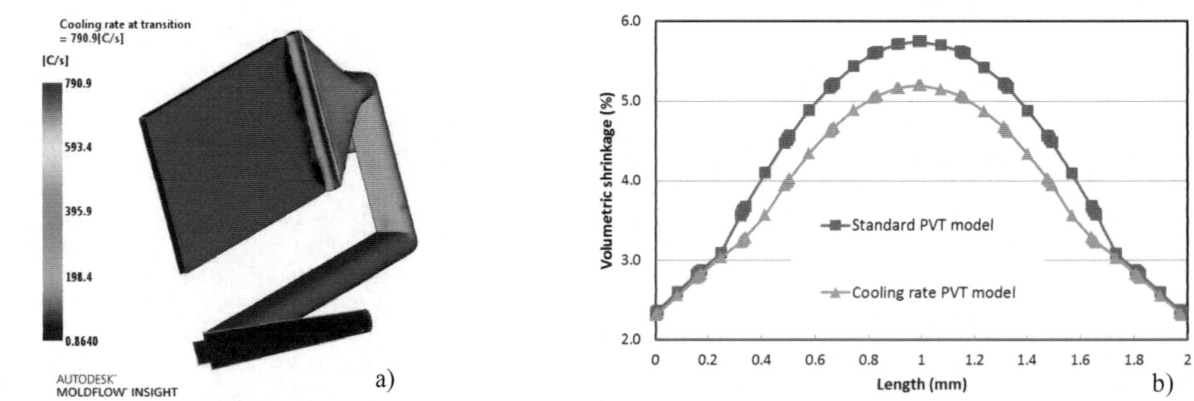

FIGURE 3. a) Cooling rate at transition, b) Volumetric shrinkage profile through thickness.

The cooling rate and volumetric shrinkage results from the mold filling and packing analysis were then used as inputs to the PVT model employed during the 3D Warp calculation. The linear shrinkages predicted from this analysis are compared in Figure 4 to the measured shrinkages for both the Infino (PC) and Luran (SAN) materials over a range of packing pressures. It is apparent that for both materials, the use of the cooling rate dependent PVT model has the effect of consistently lowering the shrinkage prediction. In the case of the Infino polycarbonate material, where the standard PVT model somewhat over-predicted the linear shrinkage, the use of the cooling rate dependent PVT model appears to improve the prediction. However, in the case of the Luran SAN material, where some of the shrinkage predictions with the standard PVT model already under-estimated the shrinkage, and the sensitivity with respect to packing pressure was not correct, the use of the cooling rate dependent PVT model did not improve the prediction accuracy. It appears that other modelling factors may have a more significant impact on solution errors.

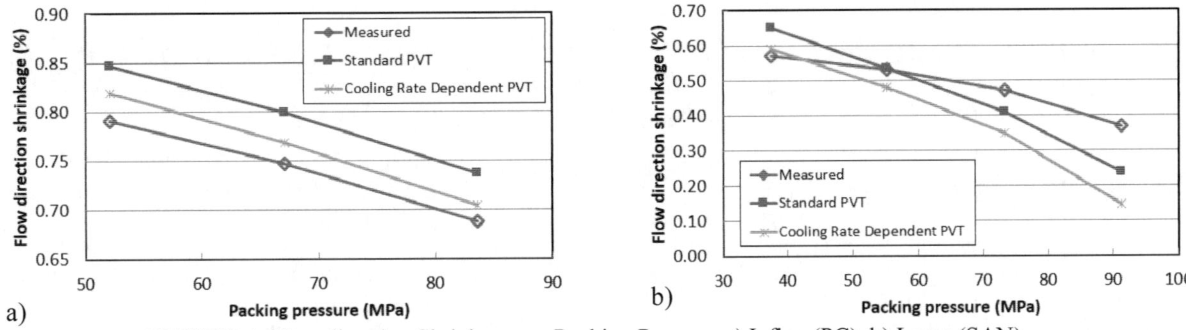

FIGURE 4. Flow direction Shrinkage vs. Packing Pressure a) Infino (PC), b) Luran (SAN)

CONCLUSION

A cooling rate dependent PVT model was successfully integrated into an injection molding 3D filling, packing and warp simulation for two amorphous materials, but was not found to offer a clear and consistent improvement for the shrinkage prediction accuracy. While the cooling rate dependence of PVT does appear to have a role to play in shrinkage prediction for amorphous materials, other modeling factors are likely to be stronger determinants of accuracy.

The use of a Solver API in Autodesk Moldflow Insight was found to be a useful tool to allow the rapid integration of non-standard material models into a 3D injection molding simulation package.

ACKNOWLEDGMENTS

The authors wish to thank A. Prof. József Kovács from Budapest University of Technology & Economics for suggesting the cooling rate dependent PVT model.

REFERENCES

1. D. W. van Krevelen, *Properties of polymers: their estimation and correlation with chemical structure*, 2nd ed. Amsterdam, Elsevier, 1976.
2. P. Zoller and Y. A. Fakhreddine, *Thermochimica Acta*, **238**, 397-415 (1994).
3. G. Menges and P. Thienel, *Poly. Eng. Sci.*, **17**, 758-763 (1977).
4. A. J. Kovacs, J. J. Aklonis, J. M. Hutchinson and A. R. Ramos, *Poly. Eng. Sci.*, **17**, 1097-1162 (1977).
5. J. S. Yu and D. M. Kalyon, *Poly Eng. Sci.*, **31**, 153-160 (1991).
6. R. Y. Chang, C. H. Chen and K. S. Su, *Poly Eng. Sci.*, **36**, 1789-1795 (1996).
7. P. S. Cook, Z. Zuo and F. S. Costa, *SPE-ANTEC Tech. Papers*, **62**, (2016).
8. F. S. Costa, P. S. Cook and D. Pickett, *SPE-ANTEC Tech. Papers*, **61**, 1688-1693 (2015).

Erosion in Extruder Flows: Analytical and Numerical Study

Miron Kaufman (presenting, reference author)* and Petru S. Fodor

Department of Physics, Cleveland State University, 2121 Euclid Avenue, Cleveland, OH 44115, USA

m.kaufman@csuohio.edu

Abstract. We consider the erosion of particles (e.g. carbonblack agglomerates) advected by the polymeric flow in a single screw extruder. We assume a particle to be made of primary fragments bound together. In the erosion process a primary fragment breaks out of a given particle. Particles disperse because of the shear stresses imparted by the fluid. The time evolution of the numbers of particles of different sizes is described by the Bateman coupled differential equations developed a century ago to model radioactivity. Using the particle size distribution we compute an entropic fragmentation index which varies from 0 for a monodisperse system to 1 for an extreme poly-disperse system. The time dependence of the index exhibits a maximum at some intermediate time as the system starts monodisperse (large size particle) and evolves through a poly-disperse regime at intermediate times to a monodisperse (small size particle) at late times..

Keywords: Single-Screw Extruder; Particle Tracing; Particle Fragmentation; Erosion.
PACS: 47.15.Rq; 47.85.md; 47.55.Kf.

INTRODUCTION

We extend and adapt to polymer processes a model that we proposed [1] to describe erosion in microchannel flows. A model of erosion and rupture of agglomerates carried by melt in an extruder was analyzed with Monte-Carlo simulation [2]. The particles are advected by an incompressible flow and erode because of the shear stresses imparted by the fluid. A particle with a larger surface area is further assumed to be more likely to erode than a particle with a smaller surface area. The extruder is modelled using Tadmor's unwound channel model [3]. We consider the creeping flow in an infinitely long extruder driven by the pressure gradient and drag along the extruder axis. We demonstrate that under those circumstances there is an analytic solution to the fragmentation equations, based on the solution of the Bateman equations [4] of nuclear physics. This quantification of fragmentation of particles advected by a fluid in an extruder is important for the design and optimization of this polymer technology. To evaluate the level of fragmentation we use the Shannon-information entropy [5] which was applied [6] also to quantify distributive and component mixing.

FRAGMENTATION MODEL

We model the particles advected by the extruder flow as being made of s primary fragments bound together. An erosion event is the breakage of a particle of size s into a primary fragment and a particle of size s-1. We define an index m to label a certain size: $s(m) = M - m + 1$, where M is the size of the largest particle, and $m = 1, 2, \ldots, M$. The particles are advected by the laminar flow and erode due to the flow shearing. The process of erosion is assumed to be a Poisson random process, in which any aggregate particle erodes during a time interval with a probability that is proportional with the time interval. The rate of change of the mean number of particles of type m is equal to the difference between the mean number of particles of type m-1 and type m, respectively, that erode per unit time. The numbers of aggregate particles of type m, each such particle containing $M - m + 1$ primary particles, satisfy a set of coupled differential equations:

$$\frac{dN_{m,\vec{r}(t)}(t)}{dt} = -\lambda_{m,\vec{r}(t)}(t)N_{m,\vec{r}(t)}(t) + \lambda_{m-1,\vec{r}(t)}(t)N_{m-1,\vec{r}(t)}(t) \qquad (1)$$

where $\lambda_{0,\vec{r}(t)}(t) = \lambda_{M,\vec{r}(t)}(t) = 0$ and $m = 1, 2, \ldots M$. In Eq. (1) λ stands for the erosion probability rate.

Particles disperse because of the shear stresses imparted by the fluid. We assume the erosion probability rate λ is proportional to the largest component of the shear stress tensor σ and inversely proportional to the cohesive stress.

We further assume that a particle with a larger surface area has a higher chance to erode than a particle with a smaller surface area. Particles of the smallest size, i.e. $m = 1$, do not decay. Hence we postulate that for a particle of size s, the erosion probability rate λ is proportional to $(s - 1)^{2/3}$. Thus the erosion rate of a particle of type m is:

$$\lambda_{m,\vec{r}(t)} = \frac{(s(m)-1)^{2/3} \max(|\sigma_{x,y}|,|\sigma_{y,z}|,|\sigma_{z,x}|)}{\tau\sigma_0} \tag{2}$$

where τ is a characteristic time and σ_0 is the cohesive stress of the particle.

We compute the shear stress tensor by solving the Navier-Stokes and continuity equations for incompressible Newtonian fluids, in the limit of zero Reynolds number and assuming that steady state has been achieved. The channel is a parallelepiped with a rectangular cross-section. Since the axial length is much longer than the cross-sectional size, we assume an infinitely long rectangular channel. It follows that the three components of the velocity depend only on transversal coordinates: **V**(X,Y). Furthermore since the walls are fixed or move in z direction there is no drag to induce fluid circulation in the xy plane: $V_x = V_y = 0$. The solution for the velocity along z direction is given in Ref. 3:

$$\begin{aligned} \frac{V_z}{V_B} &= \frac{4}{\pi}\sum_{n=1}^{\infty}\frac{\sinh((2n+1)\pi y/w)}{(2n+1)\sinh((2n+1)\pi/w)}\sin\left((2n+1)\pi x/w\right)+ \\ &+\frac{H^2}{2\eta V_B}\frac{\partial p}{\partial z}[y^2-y+\frac{8}{\pi^3}\sum_{n=1}^{\infty}\frac{\cosh\left((2n+1)\pi(x-w/2)\right)}{(2n+1)^3\cosh\left((2n+1)\pi w/2\right)}\sin\left((2n+1)\pi y\right)] \end{aligned} \tag{3}$$

where V_B is the wall velocity, $x = X/H$, $y = Y/H$ and $w = W/H$. The shear stress tensor components, needed to compute the erosion rates from Eq. (2), are:

$$\begin{aligned} \sigma_{x,z} &= \sigma_{z,x} = \eta\frac{\partial V_z}{\partial X} \\ \sigma_{y,z} &= \sigma_{z,y} = \eta\frac{\partial V_z}{\partial Y} \\ \sigma_{x,y} &= \sigma_{y,x} = 0 \end{aligned} \tag{4}$$

Depending on the flow characteristics the erosion probability rates λ may depend on time. However in the current work, where we study the creeping flow driven by a pressure gradient and drag along the axis in a long extruder, the shear stress does not change along the flow line and thus the probability rates λ are time constants. A flow line is specified by giving the transversal coordinates x and y. Equations (1) are then identical to the Bateman equations [4] used in nuclear physics to model radioactive series. Under the initial conditions:

$$N_{m,x,y}(0) = 0 \text{ where m = 2, 3,..., M} \tag{5}$$

the solutions to the set of differential equations (1) is:

$$N_{n,x,y}(t) = N_{1,x,y}(0)\sum_{i=1}^{n} c(n,i,x,y)e^{-\lambda_{i,x,y}t} \text{ for n = 1, 2,...,M} \tag{6}$$

where:

$$c(m,j,x,y) = 1 \text{ if m = j = 1} \tag{7}$$

$$c(m,j,x,y)=\frac{\prod_{i=1}^{m-1}\lambda_{(i,x,y)}}{\prod_{i=1,i\neq j}^{m}(\lambda_{(i,x,y)}-\lambda_{(j,x,y)})} \quad \text{if } 1\leq j\leq m \tag{8}$$

Using Eqs. (5) – (8), one gets a set of nontrivial identities that have been used [7] to derive sum rules relating light scattering intensities from the different coexisting phases close to a multicritical point in liquid mixtures.

To quantify the fragmentation process we define a measure based on the Shannon entropy:

$$S_{x,y}(t)=-\sum_{m=1}^{M}\frac{N_{m,x,y}(t)}{N_{x,y}}\ln\left(\frac{N_{m,x,y}(t)}{N_{x,y}}\right) \tag{9}$$

where:

$$N_{x,y}=\sum_{m=1}^{M}N_{m,x,y}(t) \tag{10}$$

S as defined in Eq. (9) is a positive quantity with a maximum possible value of *ln(M)* achieved when all populations are equal to each other. We define the fragmentation index *S/ln(M)* which is zero when one population fraction is 1 and the others are zero, i.e. monodisperse system, and it is 1 when all populations are equal in size; i.e. extremely poly-disperse system.

RESULTS

There are two dimensionless numbers associated with this problem, The first number $\mathcal{N}_1$ compares the drag time H/V_B to the fragmentation time $1/\lambda \sim \tau\sigma_0 H/(\eta V_B)$. It is equal to:

$$\mathcal{N}_1=\frac{\eta}{\tau\sigma_0} \tag{11}$$

When $\mathcal{N}_1 << 1$ the fragmentation process is weak as the agglomerate particles are resilient. If $\mathcal{N}_1 >> 1$ the agglomerates are very brittle. The numerical results below are obtained for the intermediate regime where $\mathcal{N}_1 = 1$.

The second number $\mathcal{N}_2$ compares the effect of the pressure gradient on the flow speed to that of the wall drag:

$$\mathcal{N}_2=-\frac{H^2}{2\eta V_B}\frac{\partial p}{\partial z} \tag{12}$$

We model the fragmentation process for the following three choices: (i) pure pressure driven flow: $\mathcal{N}_2 = \infty$; (ii) pure drag flow: $\mathcal{N}_2 = 0$; and (iii) mixed pressure and drag flow $\mathcal{N}_2 = 1$. All results of this section are presented using dimensionless quantities. Distances are expressed in units of the height of the rectangular cross-section size *H* and time is expressed in units of H/V_B. We consider an aspect ratio $w = W/H = 3$ and the size of the largest particle to be $M = 21$.

In Figure 1 we show the time evolution of the number of particles of types 1, 5, 10, 15, and 21, respectively for the flow line defined by $x = 0.2$, $y = 0.3$. $N_{1,x,y}$ drops monotonically from its initial value and $N_{5,x,y}$ builds up from zero (initial value) goes through a peak at $t = 0.2$. Similar dependence is exhibited by the other population numbers with peaks occurring at later times. The knowledge of the time evolution in the number of particles of different sizes is used to characterize the size dispersion of the system and its time evolution based on the entropic fragmentation index given by equation (9).

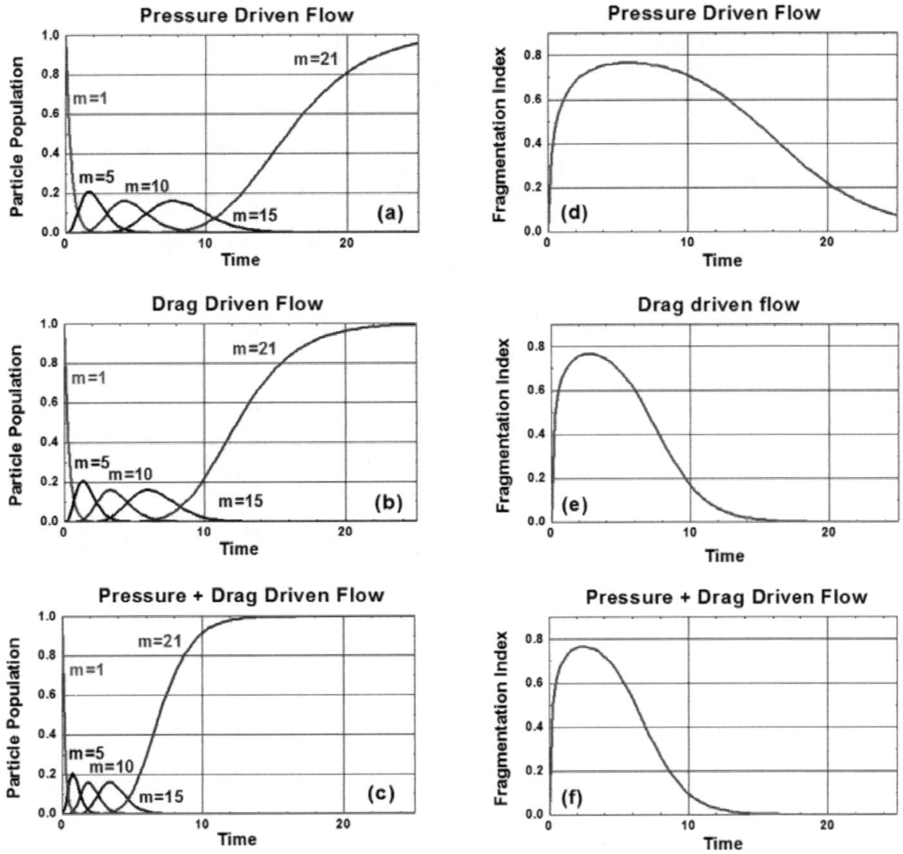

FIGURE 1. (a), (b), and (c) Time evolution of the population fractions of the different type of driven flows considered; (d), (e), and (f) The corresponding entropic fragmentation index vs. time.

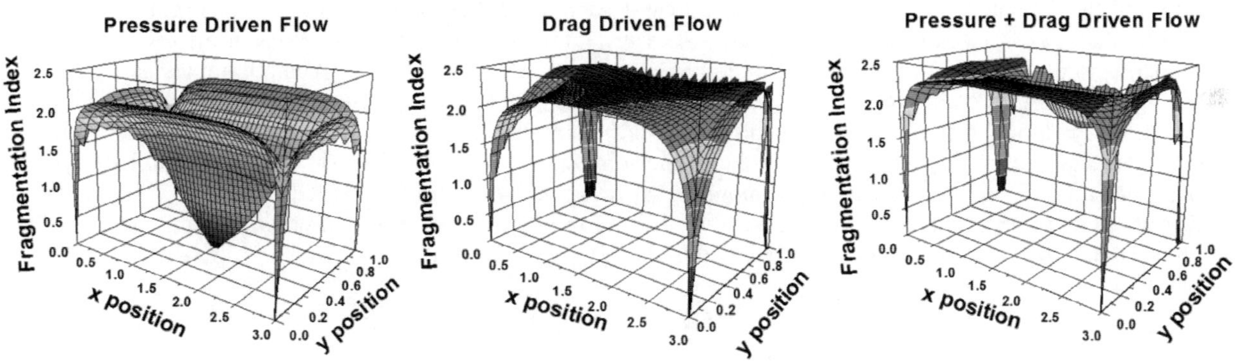

FIGURE 2. Position dependence of the fragmentation index at $t = 1$.

As observed in Figure 1 the index $S/ln(M)$ varies from 0 for a monodisperse system to 1 for an extremely polydisperse system. The time dependence of the index exhibits a maximum at some intermediate time as the system starts monodisperse (large size particles) and evolves through a polydisperse regime at intermediate times to a monodisperse (but with small size particles) at late times. Note that the maximum value of the fragmentation index is less than unity. This is due to the fact that not all size particles are present.

Since the fragmentation entropic index depends on the flow line, we show in Figure 2 the dependence of the entropy ($0 < S < ln21$) on x and y at $t = 1$. We assume for initial condition a uniform distribution in the xy plane of particles of largest size ($M = 21$) only. Note the system is monodisperse at the corners and in the center of the cross-section. This result is explained by the fact that at these points the shear stress is zero.

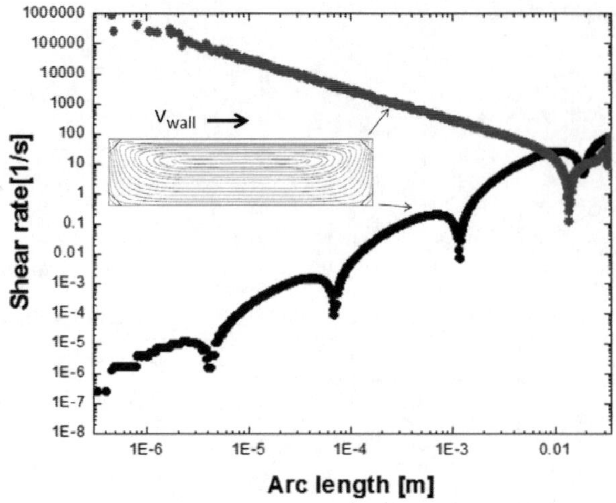

FIGURE 3. Shear rates away from the singular points represented by the extruder corners. The top corner is defined by a moving and a stationary wall, while for the bottom corner both walls are stationary.

SUMMARY

We have presented a model for fragmentation of particles advected by laminar flows in extruders. We have also introduced an index of fragmentation based on the information entropy that is useful for various optimization tasks. For instance we can determine the minimum time needed to reach a desired level of fragmentation, based on the time evolution of this fragmentation index.

Due to certain simplifying assumptions, such as zero Reynolds number, we were able to solve the model analytically by employing the Bateman equations of nuclear physics. It is desirable to improve the model by relaxing some of those assumptions. In particular in an extruder modeled using the Tadmor's unwound channel model [3] the wall of the cavity moves diagonally and thus drag forces are present in the cross section of the system. This leads to the formation of complicated flow structures such as Moffatt eddies [8], associated with singularities imposed by the boundary conditions at the corners. As seen in Fig. 3, numerical simulations of these flows indicate that the shear rates for these systems are quite complex and will affect both the fragmentation rates as well as the distribution of fragments within the volume of the extruder. Given the complexity of these systems a combination of numerical and analytical methods is warranted for tackling the problem [9, 10]. Under this approach, an analytical solution, as described above can be used to obtain the longitudinal velocity field (in the limit of zero Reynolds number). This allows us to devote all the computational resources solely for a detailed numerical solution of the transversal velocity field. The combined 3D solution for the flow field can then be used with the above described model to characterize the fragmentation process within the extruder type system.

REFERENCES

1. M. Kaufman, *Nanoscale and Microscale Thermophysical Engineering* **11**, 129–136 (2007).
2. N. Domingues, M. Camesasca, M. Kaufman, I. Manas-Zloczower, A. Gaspar-Cunha, J. A. Covas, *International Polymer Processing* **XXV**, 251-257 (2010).
3. Z. Tadmor and C.G. Gogos, "Chapter 6 – Pressurization and Pumping," in *Principles of Polymer Processing, 2nd edition*, Wiley-Interscience, Hoboken, New Jersey, 2006.
4. H. Bateman, *Proc. Cambridge Philos. Soc.* **15**, 423-427 (1910).
5. C. E. Shannon and W. Weaver, in *The Mathematical Theory of Communication,* University of Illinois Press, Urbana, Illinois, 1949.
6. K. Alemaskin, I. Manas-Zloczower, and M. Kaufman, *Polymer Engineering and Science* **45**, 1011-1020 (2005).
7. M. Kaufman and M. Ma, *Phys. Rev. A* **35**, 2369-2372 (1987).
8. H.K. Moffatt, *J. Fluid Mech.* **18**, 1 – 18 (1963).
9. M. Kaufman and P.S. Fodor, *Physica A* **389**, 2951-2955 (2010).
10. P. S. Fodor, M Kaufman, *AIP Conf. Proc.* **1664**, 050010 (2015).

A semi-mechanistic model for polyurethane cure thermokinetics : Application to a natural rubber based thermoset

Thi Kieu Nhung Tran[a], Gaël Colomines[a], Arnaud Nourry[b], Jean-François Pilard[b], Rémi Deterre[a] and Eric Leroy (presenting author)(reference author)[c]

[a]*LUNAM Université, IUT de Nantes, CNRS, GEPEA, UMR 6144, OPERP, 2 Avenue du professeur Jean ROUXEL, BP 539, 44475 Carquefou, France.*
[b]*LUNAM Université, Université du Maine, IMMM, UMR CNRS 6283, Institut des Molécules et Matériaux du Mans, Avenue Olivier Messiaen, 72085 Le Mans Cedex 9, France.*
[c]*LUNAM Université, CNRS, GEPEA, UMR 6144, CRTT, 37, Boulevard de l'Université, 44606 St Nazaire Cedex, France*

eric.leroy@univ-nantes.fr

Abstract. The future of thermosetting polymers relies both on the formulation of biobased systems and on the energetic optimization of the molding processes. This second task more than ever relies on accurate thermokinetic modelling of the cure reactions involved. Nevertheless, bridging that gap between mechanistic knowledge and the production of simple kinetic equations that can be used in process simulation softwares remains a difficult issue. In the present work, we revisit the modelling of polyurethane cure kinetics taking the case of a partially biobased system : Isoconversional analysis was applied to non-isothermal bulk cure kinetics of uncatalyzed polyurethane (PU) resins based on telechelic hydoxyl natural rubber (HTNR) and poly[(phenyl isocyanate)-co-formaldehyde] (P-MDI). The dependency of apparent activation energy on conversion led to the identification of two different kinetic models, each involving two activation energies: a phenomenological model based on the Kamal-Sourour equation with non integer reaction exponents; and a new semi-mechanistic model. It assumes a main second order reaction between alcohols and isocyanates forming urethane linkages, and a secondary reversible reaction forming allophanates that is known to become significant above 120°C. Such a model based on mechanistic considerations gives more accurate simulation of experimental data, using less kinetic parameters and allows explaining the variations of reactivity of the HTNR polyols with its average molecular weight.

Keywords: bio-based polyurethane, natural rubber, calorimetry, kinetics, modelling
PACS: Thermosetting polymers, 83.80.Jx

INTRODUCTION

Polyurethane (PU) is one of the most versatile polymers used on diverse applications. By adjusting the key design features of PU at microscale (crosslinking density, hard segment content), at mesoscale (phase morphology) or at macroscale (foam density, cell dimension), a wide range of final products such as thermosetting, thermoplastic elastomers, foams, coatings, adhesives can be obtained[1]. In order to develop "greener" PUs, recent researches have been dedicated to "No Isocyanate"PU (NIPU)[2]. Nevertheless, the PU production from the renewable polyol sources such as vegetable oils[3] is still an active area of research and application. Herein, we took an interest on the design of PU derived from natural rubber. Indeed, hydroxyl telechelic natural rubber (HTNR) could represent a versatile source of soft long chain bio-based polyols, in which the length of rubber chains can be controlled by depolymerisation of natural rubber before telechelic functionalization[4].

In the present paper, we studied the formation of polyurethane networks from such bio-based polyols, from a thermo-kinetic point of view. In order to focus on the influence of the chain length between telechelic hydroxyl groups, two HTNR polyols with different molecular weight (2260 g/mol and 1200 g/mol), were synthetized and reacted with a commercial isocyanate poly[(phenyl isocyanate)-co-formaldehyde] (P-MDI).

EXPERIMENTAL

The structure of the reactants is given on FIGURE 1. Poly[(phenyl isocyanate)-co-formaldehyde] (P-MDI, functionality = 2.7, purchased from Sigma-Aldrich) was used without further purification. Polyols were synthesized from natural rubber (10 CV60, supplied by Hutchinson): HTNR1200 ($\overline{M}_{n,NMR} = 1200 g/mol$, functionality =2, hydroxyl number = 101); HTNR 2260 ($\overline{M}_{n,NMR} = 2260 g/mol$, functionality =2, hydroxyl number = 47). The synthesis of polyols was described elsewhere[5].

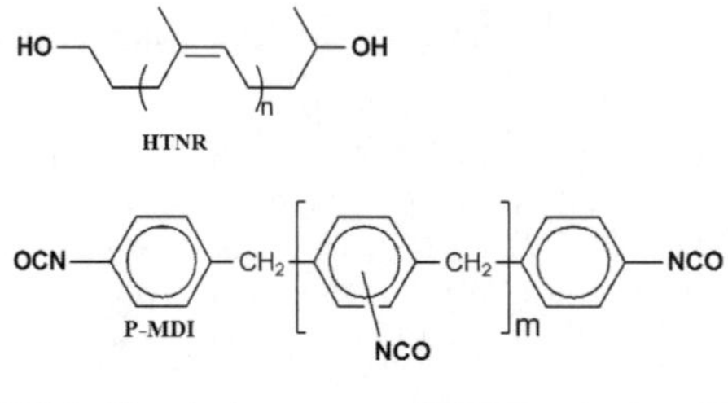

FIGURE 1. Chemical structures of HTNR polyols and P-MDI.

The polymerization of PU (P-MDI/HTNR, NCO/OH ratio of 1) was monitored using the Mettler Toledo DSC1. Non-isothermal scans were performed at 2, 4, 8, 16, 24 K/min heating rates with a temperature range from 0°C to 250°C. The measurements were carried out following the ICTAC Kinetics Committee recommendations for Collecting Thermal Analysis Data for Kinetic Computations[6]. In order to perform the kinetic modelling of the chosen PU system, we followed the ICTAC Kinetics Committee recommendations for performing kinetic computations on thermal analysis data[7]. The original DSC curves were transformed into the conversion degree (α) and rate of reaction (dα/dt) curves by integration and normalization. For the simulations of the reaction rate (dα/dt) and the conversion degree (α) curves by the different kinetic models we used Runge & Kutta method for calculations and the ordinary least squares method was used for model fitting.

RESULTS AND DISCUSSION

FIGURE 2 shows the non-isothermal thermokinetic data (conversion degree (α) and reaction rate (dα/dt)) extracted from DSC thermograms PU-HTNR1200 (similar results were obtained for PU-HTNR2260, not shown for space reasons) at different heating rates. Modelling was first performed using the Kamal & Sourour equation[8]:

$$\frac{d\alpha}{dt} = (k_1 + k_2\alpha^m)(1-\alpha)^n = (A_1 e^{-Ea1/RT} + A_2 e^{-Ea2/RT}\alpha^m)(1-\alpha)^n \qquad (1)$$

where: n, m are kinetic exponents; k_1, k_2 are rate constants ($k_1 = A_1 e^{(-Ea1/RT)}$; $k_2 = A_2 e^{(-Ea2/RT)}$).

After optimization, the kinetic parameters listed in TABLE 1 are obtained. The corresponding simulations of the cure conversion degree (α), the reaction rate (dα/dt) are plotted on FIGURE 2. The global reaction order (m+n =2) is consistent with literature, since this approximate value of 2 was also reported for other PU systems[9]. Nevertheless, the fact that we obtain non integer values for the reaction exponents (n=1.4, m=0.6, very close to those found by Fernandez d'Arlas et al.[9]) renders the physico-chemical meanings of the Kamal & Sourour model questionable. The use of the Kamal-Sourour equation for the description of PU cure kinetics in literature actually derives from mechanistic assumptions. Initially Baker[10] described the alcohol- isocyanate reaction by a second order reaction (n = 2). The Kamal-Sourour equation (with n=2 and m=1) corresponds to an extension of the Baker model proposed by Sato[11] who introduced an autocatalytic term to the model of Baker in order to take into account the experimental deviations from the second order reaction.

For our PU system, as well as for the more classical system studied by Fernandez d'Arlas and coworkers[9], non-integer values of n and m have to be introduced in the Kamal & Sourour equation in order to fit the model to the experimental data. This suggests that the reaction mechanism is either more complex or different from the simple autocatalytic scheme mechanism proposed by Sato[11].

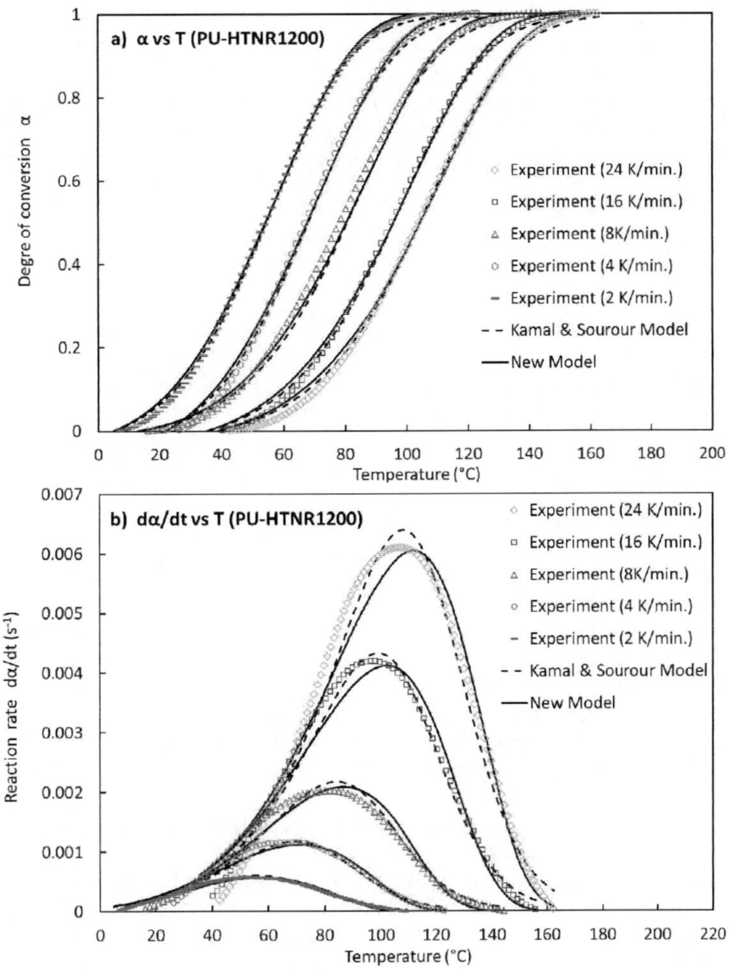

FIGURE 2. Experimental data (symbols) and modelling data (lines) for the cure of PU-HTNR 1200: degree of conversion (a), rate of reaction (b)

TABLE 1. T Kinetic parameters of Kamal & Sourour model for two PU formulations.

Kinetic parameters	PU-HTNR2260	PU-HTNR1200
E_{a1} (kJ/mol)	40	42
E_{a2} (kJ/mol)	69	71
A_1 (s^{-1})	3.32×10^3	7.24×10^3
A_2 (s^{-1})	2.64×10^7	6.45×10^7
n	1.4	1.4
m	0.6	0.6

TABLE 2. Kinetic parameters of the New Semi mechanistic model for the two PU formulations.

Kinetic parameters	PU-HTNR2260	PU-HTNR1200
E_{a1} (kJ/mol)	39	43
E_{a2} (kJ/mol)	56	57
A_1 (s^{-1})	2.73×10^3	1.07×10^4
A_2 (s^{-1})	4.39×10^5	6.92×10^5

Actually, it was shown in 1975 by Anzuino et al.[12] that the experimental deviation of the second order alcool-isocyante reaction can be ascribed to a reversible secondary reaction between urethanes and isocyanates to form allophanates. Such a reaction scheme was successfully used by Krol et al.[13] to model the reaction kinetics of diisocyanates with polyols in solution. The main second order alcool-isocyanate reaction was found to have an activation energy of 40 kJ/mol, whereas the side reaction forming allophanates have an activation energy of 60 kJ/mol. The model used by Krol [13] involves the two following reactions (FIGURE 3):

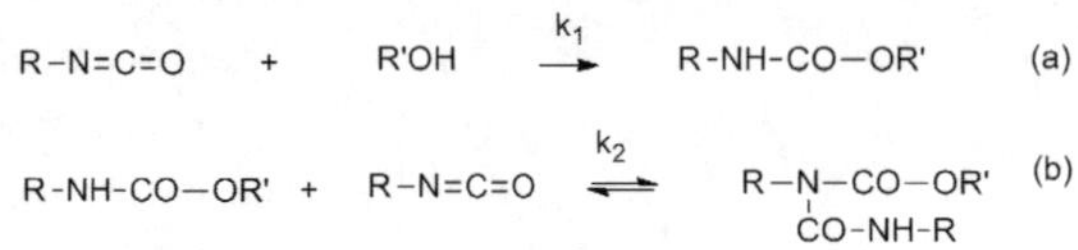

FIGURE 3. Formation of urethane (a) and allophanate (b) functions

Since the allophanate forming reaction is highly reversible (according to Kogon[14] the equilibrium constant is equal to 0.2 at 128°C), we assume as a raw approximation that it does not significantly modify the urethane concentration. We also assume that the isocyanate "reformed" by the reverse reaction of scheme 2(b) is not available for further reactions. These assumptions lead to the following new "semi mechanistic" model:

$$\frac{d\alpha}{dt} = k_1(1-\alpha)^2 + k_2(1-\alpha)\alpha \tag{10}$$

This new Semi Mechanistic model is significantly different from the Kamal-Sourour equation (the second term does not correspond to an autocatalytic mechanism here). After optimization, the kinetic parameters listed in TABLE 2 are obtained. The corresponding simulations of cure conversion degree (α) and reaction rate ($d\alpha/dt$) are plotted on FIGURE 2 (continuous lines). A satisfactory description of all experimental data could be observed for the two PU systems. In addition, the activation energies found are very close to those reported by Krol et al.[13] for the isocyanate and allophanate forming reactions (E_{a1} = 40 kJ/mol and E_{a2} = 60 kJ/mol).

FIGURE 4 shows the calculated values of the ration k_1/k_2 at given temperatures ranging from 0°C to 200°C. It can be seen that for the two PU systems, the reversible allophanate forming reaction becomes significant only above 120°C ($k_1/k_2 = 1$) being in agreement with literature[15]. In addition, if we now compare the rate constants of the two PU systems, in the 50-150°C temperature range, we have the following average values: $[k_{1\,(PU\ HTNR\ 1200)}/k_{1\,(PU\ HTNR\ 2260)}] \approx 1.3$ and $[k_{2\,(PU\ HTNR\ 1200)}/k_{2\,(PU\ HTNR\ 2260)}] \approx 1.5$.

From a qualitative point of view, this increase of the rate constants for lower molecular weight polyols is consistent since the rates of reaction depend on the probability of successful collisions among reactive molecules, which is proportional to their local concentration. By reducing the length of HTNR polyol, the local concentration of reactive groups (OH and NCO) increases since the global NCO/OH ratio was maintained equal to 1. This increase of the concentration of reactive groups in PU-HTNR1200 is likely to result in an increase of the frequency factor A, as observed in TABLE 2.

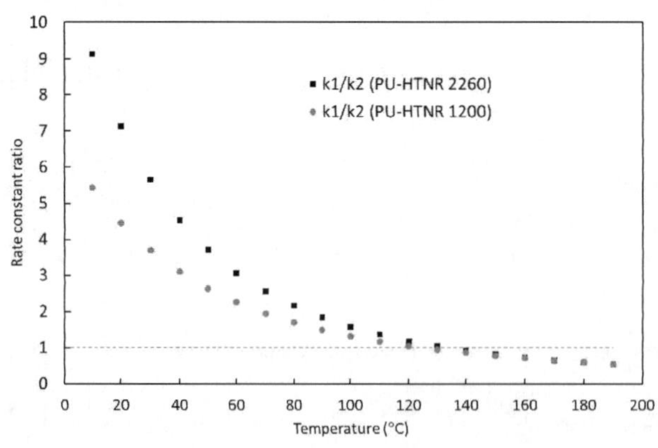

FIGURE 3. Ratio between the rate constants k_1 and k_2 of the new semi mechanistic model

CONCLUSION

Uncatalysed cure kinetics of a renewable resource based-polyurethane synthesized from telechelic hydroxyl natural rubber (HTNR) and poly[(phenyl isocyanate)-co-formaldehyde] has been investigated by non-isothermal DSC. The isoconversional analysis leads to the identification of two possible kinetic models: a phenomenological model based on the Kamal & Sourour equation with non integer reaction exponents, and a new semi-mechanistic model assuming a main second order reaction between alcohols and isocyanates forming polyurethane and a secondary reversible reaction forming allophanates that become significant above 120°C. The semi-mechanistic model allows explaining the variation of reactivity of the HTNR polyol with its average molecular weight. Furthermore, such a model could be used for modeling others commercial PU polymerizations in bulk.

ACKNOWLEDGEMENTS

We are grateful for the financial support from the MATIERES project funded by the "Région Pays de la Loire" council. The authors thank to Mr. Y. Madec for his kind support for DSC operation.

REFERENCES

1. H.-W. Engels, H.-G. Pirkl, R. Albers, R.W. Albach, J. Krause, A. Hoffmann, H. Casselmann, J. Dormish, *Polyurethanes: Versatile Materials and Sustainable Problem Solvers for Today's Challenges*, Angew. Chem. Int. Ed., 52 (2013) 9422-9441.
2. M.S. Kathalewar, P.B. Joshi, A.S. Sabnis, V.C. Malshe, *RSC Advances*, 3 (2013) 4110-4129.
3. C. Zhang, S.A. Madbouly, M.R. Kessler, Biobased, *ACS Appl. Mater. Inter.*, 7 (2015) 1226-1233.
4. F. Sadaka, I. Campistron, A. Laguerre, J.-F. Pilard, , *Polym. Degrad. Stab.* , 97 (2012) 816-828.
5. T.K.N. Tran, J.-F. Pilard, P. Pasetto, , *J. Appl. Polym. Sci.,* 132 (2015).
6. S. Vyazovkin, K. Chrissafis, M.L. Di Lorenzo, N. Koga, M. Pijolat, B. Roduit, N. Sbirrazzuoli, J.J. Suñol, *Thermochim. Acta*, 590 (2014) 1-23.
7. S. Vyazovkin, A.K. Burnham, J.M. Criado, L.A. Pérez-Maqueda, C. Popescu, N. Sbirrazzuoli, *Thermochim. Acta*, 520 (2011) 1-19.
8. M.R. Kamal, S. Sourour, *Polym. Eng. Sci.*, 13 (1973) 59-64.
9. B. Fernandez d'Arlas, L. Rueda, P.M. Stefani, K. de la Caba, I. Mondragon, A. Eceiza, *Thermochim. Acta,* 459 (2007) 94-103.
10. J.W. Baker, J.B. Holdsworth, 135. *J. Chem. Soc.* , (1947) 713-726.
11. M. Sato, *JACS*, 82 (1960) 3893-3897.
12. G. Anzuino, A. Pirro, O. Rossi, L.P. Friz, *J. Polym. Sci., Part A: Polym. Chem.*, 13 (1975) 1657-1666.
13. P. Krol, B. Atamanczuk, J. Pielichowski, *J. Appl. Polym. Sci.* , 46 (1992) 2139-2146.
14. I.C. Kogon, , *J. Org. Chem.*, 24 (1959) 83-86.
15. C. Hepburn, *Polyurethane Elastomers*, Elsevier Applied Science, London, 1991.

Application of the Least Square Finite Element Method in the Interface Capturing of a Two-Phase Flow

Mehdi Mostafaiyan,[1,2] Sven Wiessner,[1,2] Gert Heinrich[1,2]
1 - Leibniz-Institut für Polymerforschung Dresden e. V, D-01069, Dresden, Germany
2 - Institut für Werkstoffwissenschaft, Technische Universität Dresden, D-01069 Dresden, Germany

Abstract

A Least square finite element method is locally formulated to predict the advection of the interface in a two-phase flow. Moreover, an imaginary control volume around each node is introduced, which determines the nodal values of the volume fractions at the onset of the flow and provides a highly satisfactory reconstructed interface at the same time. Also, the effect of the surface tension on the second phase deformation is calculated using the interface tangent, which is estimated by the derivatives of the volume fraction on the Gaussian points in the vicinity of the interface intersections. Finally, some explanatory examples are provided to evaluate our suggested procedure.

1- Introduction

There are many industrial problems which comprise moving boundaries and interfaces, and some examples can be found in [1]. Based on the importance of the two phase flow problems, different methods are developed to capture or track the interface using Eulerian, Lagrangian or mixed approaches [2]. Our concentration in this research is on the volume of fluid method (VOF), which benefits from Eulerian formulation [3] to capture the interface.

2- Governing Equations and the Solution Method

This research is concentrated on a two-dimensional creeping flow; nevertheless, the method can be extended to a three-dimensional case with a more complicated flow field.

2-1- Equations of motion and continuity

The governing equations of the creeping flow of an incompressible fluid are presented in equations (1-a) and (1-b):

$$\nabla^2 \mu \boldsymbol{D} - \nabla P = -\int_S \sigma \kappa \mathrm{n} \delta(\boldsymbol{x} - \boldsymbol{x}_i) ds \qquad \text{1-a}$$

$$\nabla . \boldsymbol{u} = 0 \qquad \text{1-b}$$

where P, $\boldsymbol{D}$ and μ are the pressure, deformation rate tensor and Newtonian viscosity, respectively. The vectors $\boldsymbol{x}$ and $\boldsymbol{u}$ show the location and velocity of any arbitrary point, which are $\boldsymbol{x} = (x, y)$ and $\boldsymbol{u} = (u, v)$, correspondingly; for a two-dimensional flow field in the Cartesian coordinate. The integral on the RHS of equation (1-a) is the contribution of the surface tension [4], in which σ is the surface tension coefficient; also the interface, its curvature and normal are denoted by $\boldsymbol{S}$, κ and n, respectively. δ is a two-dimensional delta function, which is unity on the interface ($\boldsymbol{x} = \boldsymbol{x}_i$) and zero otherwise ($\boldsymbol{x} \neq \boldsymbol{x}_i$). The solution method of the equations of motion (equation 1-a) and continuity (equation 1-b) by the Galerkin finite element method [5] are described in [6, 7], and the surface tension contribution is treated as a source term.

In order to avoid the curvature calculation on the RHS of equation (1-a), Frenet relation is employed [4, 8]; therefore, the surface tension contribution can be expressed as equation (2):

$$\int_{\Omega} \sigma \kappa \mathbf{n} \delta(\boldsymbol{x} - \boldsymbol{x}_i) ds = \frac{\sigma(\mathrm{T}_{s2} - \mathrm{T}_{s1})}{A_{\Omega}} \tag{2}$$

In equation (2), T_{s1} and T_{s2} are the tangent vectors of the interface at its intersections with the boundaries of an active element (Ω) with the area of A_{Ω}, as shown in figure (1). An active element is an element through which the interface passes.

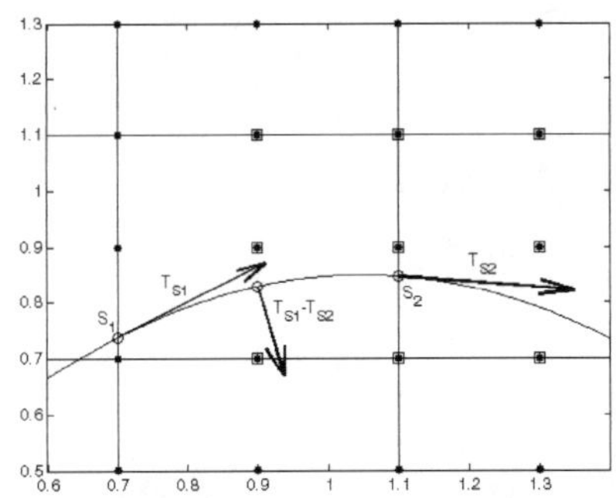

Figure 1: The tangent vectors of the interface (T_{s1} and T_{s2}) at its intersections with the boundaries of an arbitrary active element.

2-2- Equation of Interface Advection

The advection of the interface in a two-phase flow problem (e.g. white and dark fluids); can be described using equation (3) [3]:

$$\frac{\partial F}{\partial t} + \boldsymbol{u}.\nabla F = 0 \tag{3}$$

The value of F is equal to unity or zero, where a control volume is occupied by the dark or white fluid, respectively. Therefore, the interface passes through the elements which contain both of the fluids ($0 < F < 1$). As shown in figure (2), the area portion which is occupied by the dark fluid in each phantom element or the imaginary circle around each node, determines the nodal value of the volume fraction. Also, the volume fraction of a node can be assumed to be zero or unity, when its distance up to the interface is more than the radius of the phantom element (R_e); therefore, the thickness of the interface can be assumed to be $2 \times R_e$.

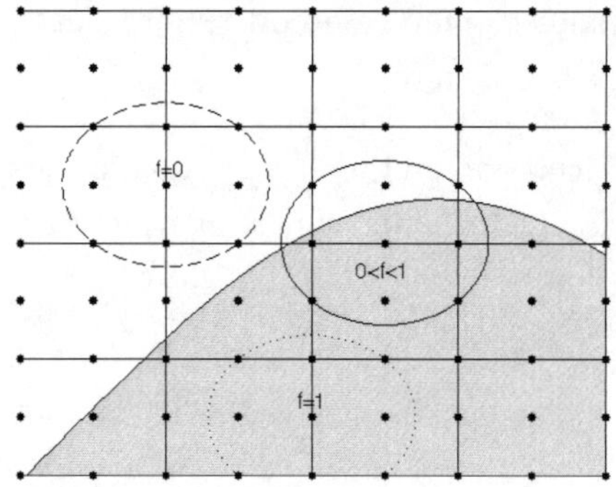

Figure 2: The nodal value of the volume fraction is determined by the dark fluid area in a phantom element

In order to solve the differential equation of the interface advection, a stabilized method shall be employed [2, 9, 10]. In this research we have used least square finite element method [11], as stated in equations (4-a) to (4-e).

$$\frac{\partial\langle\varphi\rangle}{\partial t}\{\tilde{F}\}+u\frac{\partial\langle\varphi\rangle}{\partial x}\{\tilde{F}\}+v\frac{\partial\langle\varphi\rangle}{\partial y}\{\tilde{F}\}=0 \qquad \text{4-a}$$

$$\langle\varphi\rangle\frac{\{\tilde{F}^{n+1}-\tilde{F}^{n}\}}{\Delta t}+u\frac{\partial\langle\varphi\rangle}{\partial x}\{\tilde{F}^{n+1}\}+v\frac{\partial\langle\varphi\rangle}{\partial y}\{\tilde{F}^{n+1}\}=0 \qquad \text{4-b}$$

$$\langle\phi\rangle\{\tilde{F}^{n+1}\}-\langle\varphi\rangle\{\tilde{F}^{n}\}=0 \qquad \text{4-c}$$

$$\langle\phi\rangle=\langle\varphi+u\Delta t\frac{\partial\langle\varphi\rangle}{\partial x}+v\Delta t\frac{\partial\langle\varphi\rangle}{\partial y}\rangle \qquad \text{4-d}$$

$$\int_{\Omega}\{\omega\}\left(\langle\phi\rangle\{\tilde{F}^{n+1}\}-\langle\varphi\rangle\{\tilde{F}^{n}\}\right)d\Omega=0 \qquad \text{4-e}$$

where $\{\tilde{F}\}$ and $\langle\varphi\rangle$ are the vertical and horizontal vectors of the calculated volume fractions and shape function coefficients in each element, respectively. The implicit time discretization of equation (4-a) is shown in equation (4-b), which can be summarized as stated in equation (4-c). The value of ϕ in the latest equation is defined in (4-d). The weighted residual form is also mentioned in equation (4-e), in which ω is the weight function and is equal to ϕ for least square finite element method (LSFEM). In this research, in order to avoid the excessive diffusion of the interface, a local least square finite element method (LLSFEM) is employed, in which equation (4-e) has been solved just for the active elements and elements in their vicinity.

3- Surface tension

To consider the effect of the surface tension on the interface advection and flow field in a two-phase flow, Frenet law is employed, therefore, it would be necessary to calculate the interface tangents at its intersection with element faces. Using the assumption that the nodal volume fractions ($\tilde{F}$) is equal to $1/2$ on the reconstructed interface ($\tilde{\boldsymbol{S}}=\tilde{S}(x,y)$); equation (5-a) can be used:

$$\frac{\partial\tilde{F}}{\partial\tilde{S}}=0 \qquad \text{5-a}$$

$$\frac{\partial\tilde{S}/\partial y}{\partial\tilde{S}/\partial x}=-\frac{\partial\tilde{F}/\partial x}{\partial\tilde{F}/\partial y} \qquad \text{5-b}$$

Using equation (5-b), which is obtained by partial derivation of equation (5-a), the tangent of the interface can be calculated. To avoid the discontinuity of the volume fraction derivative on the element boundaries, this values are calculated on the Gaussian points in each element

4- Explanatory Examples

In this section, the proposed procedure is employed to predict the deformation of an immiscible droplet in a simple shear flow, as shown in figure (3).

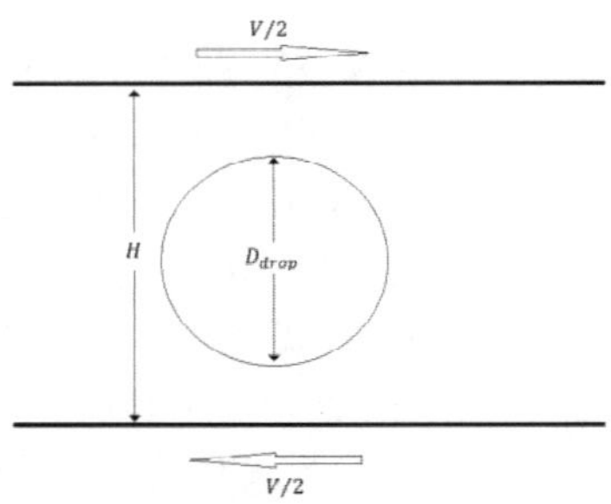

Figure 3 : An immiscible droplet in simple shear flow

In figure (3), the gap between the moving plates in the simple shear flow is denoted by H and D_{drop} is the diameter of the undeformed droplet. The upper and lower plates move with the velocities of $V/2$ and $-V/2$; respectively, and the average shear rate is defined by $\bar{\gamma} = V/H$.

In figure (4) the deformation of two droplets with moderate and big diameters ($D_{drop} = 0.5H$ and $0.9H$) at their equilibrium are illustrated for different Capillary numbers ($Ca = \frac{\mu_1 \bar{\gamma} D_{drop}}{\sigma}$). μ_1 is the viscosity of the matrix, which is equal to the droplet viscosity.

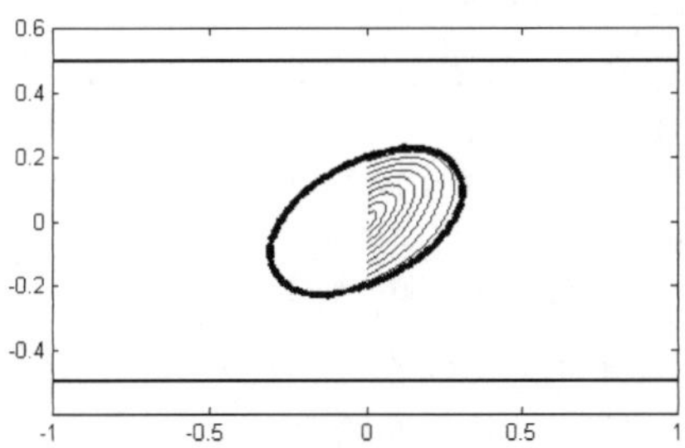

Figure 3-a: Equilibrium state $D_{drop} = 0.5H$ and $Ca = 0.2$

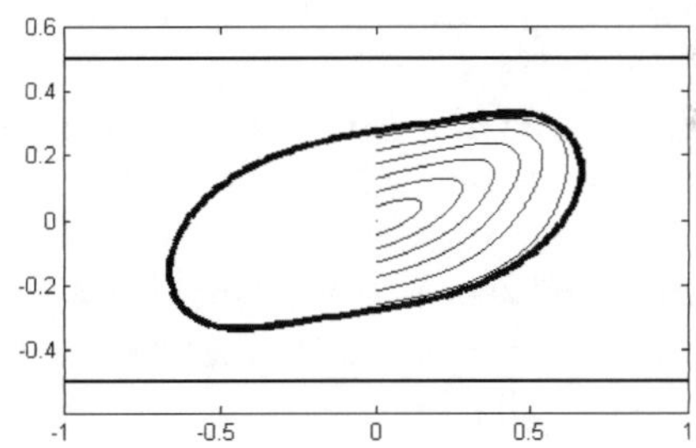

Figure 3-b: Equilibrium state $D_{drop} = 0.9H$ and $Ca = 0.2$

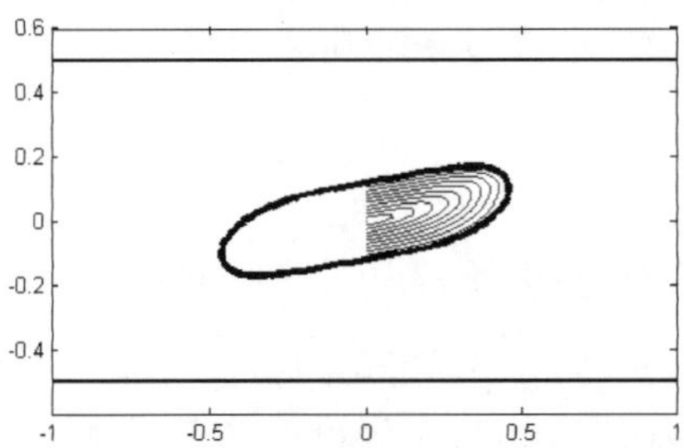

Figure 4-c: Equilibrium state $D_{drop} = 0.5H$ and $Ca = 0.4$

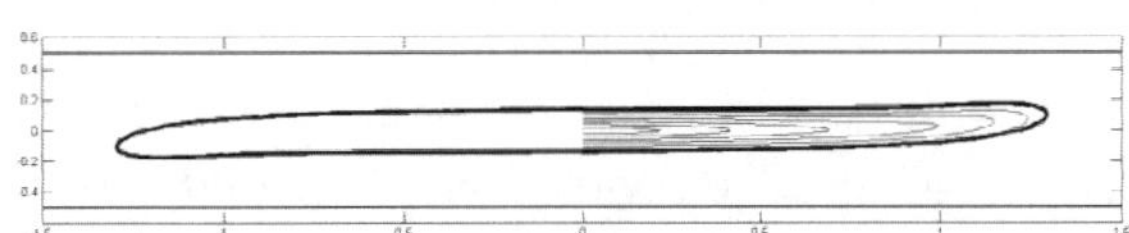

Figure 4-d: Equilibrium state $D_{drop} = 0.9H$ and $Ca = 0.4$

Figures (4-a) to (4-d) show the equilibrium or steady deformation of the droplets for different values of the Capillary numbers. In these figures the interfaces are illustrated by the thick solid lines. As it is expected by increasing the Capillary number, which is accompanied by stronger viscous forces or weaker surface tension forces, a greater deformation is observed. Also, the streamlines, which do not intersect the interface after equilibrium, are depicted by thin solid lines in the right half of the droplet. Similar results for the droplet deformation are also reported in [12, 13].

5- Conclusion

In this research the least square finite element method (LSFEM) has been employed to predict the advection of the interface. In order to constrain the excessive diffusion of the interface, a local least square finite element method has been introduced, in which the advection domain is defined according to the current interface position and the elements in its vicinity. Moreover, a phantom element is introduced to evaluate the nodal values of the volume fraction. Furthermore, the surface tension forces have been calculated by introducing a method for evaluation of the volume fraction derivatives on active element faces. Finally, droplet deformation problem (as an example for two-phase flow field) in the simple shear flow has been studied, which its results are fairly comparable with similar results in the literature. The mentioned method can be extended to three-dimensional cases and for the problem with different viscosity ratios.

References

1. Pilliod, J.E. and E.G. Puckett, *Second-order accurate volume-of-fluid algorithms for tracking material interfaces.* Journal of Computational Physics, 2004. **199**(2): p. 465-502.
2. Tezduyar, T. *Interface-tracking and interface-capturing techniques for computation of moving boundaries and interfaces.* in *Proceedings of the Fifth World Congress on Computational Mechanics, On-line publication: http://wccm. tuwien. ac. at/, Paper-ID*. 2002.
3. Hirt, C.W. and B.D. Nichols, *Volume of fluid (VOF) method for the dynamics of free boundaries.* Journal of computational physics, 1981. **39**(1): p. 201-225.
4. Tryggvason, G., et al., *A front-tracking method for the computations of multiphase flow.* Journal of Computational Physics, 2001. **169**(2): p. 708-759.
5. Reddy, J.N. and D.K. Gartling, *The finite element method in heat transfer and fluid dynamics*. 2010: CRC press.
6. Mostafaiyan, M., S. Wiessner, and G. Heinrich, *The study of the distributive mixing in converging channels: Numerical method and analytical approach.* Polymer Engineering & Science, 2015.
7. Mostafaiyan, M., et al., *A numerical study on deformation of Newtonian droplets through converging cylindrical dies.* e-Polymers, 2013. **13**(1): p. 49-66.
8. Shin, S. and D. Juric, *Modeling three-dimensional multiphase flow using a level contour reconstruction method for front tracking without connectivity.* Journal of Computational Physics, 2002. **180**(2): p. 427-470.
9. Tezduyar, T.E., *Finite element methods for flow problems with moving boundaries and interfaces.* Archives of Computational Methods in Engineering, 2001. **8**(2): p. 83-130.
10. Stein, E., R. de Borst, and T.J. Hughes, *Chapter 17 in Encyclopedia of computational mechanics*. 2004: Wiley.
11. Jiang, B.-n., *The least-squares finite element method: theory and applications in computational fluid dynamics and electromagnetics*. 2013: Springer Science & Business Media.
12. Li, J., Y.Y. Renardy, and M. Renardy, *Numerical simulation of breakup of a viscous drop in simple shear flow through a volume-of-fluid method.* Physics of Fluids (1994-present), 2000. **12**(2): p. 269-282.

13. Janssen, P. and P. Anderson, *Boundary-integral method for drop deformation between parallel plates.* Physics of Fluids (1994-present), 2007. **19**(4): p. 043602.

Cavity Dependence of Core Penetration in Co-Injection Multi-Cavity Molding

Chao-Tsai (CT) Huang[a*], Chih-Chung Hsu[b], Rong-Yeu Chang[b], and Shi-Chang Tseng[c]

a. *Chemical and Materials Engineering, Energy and Opto-Electronic Materials Research Center, Tamkang University, New Taipei City 251, Taiwan*
b. *R and D Division CoreTech System (Moldex3D) Co., Ltd., Hsinchu 302, Taiwan*
c. *National Yunlin University of Science and Technology, Yunlin 640, Taiwan*
**Corresponding author:* cthuang@mail.tku.edu.tw

Abstract. Co-injection molding is commonly used in daily accessories, car parts, and structural-reinforcement product. However, there are too many combinations of designs, materials, core/skin ration, and process condition, how to have suitable control of co-injection is very challenge. Furthermore, co-injection with multi-cavity system which is commonly applied in some forks structure products. However, due to the complicated nature, the inside mechanism for the combination of co-injection and multi-cavity is still not fully understood yet. In this study, we have proposed three kinds of multi-cavity systems to investigate cavity design influence on the core material penetration behavior. In Model 1 with three separated cavity system, the core penetration into individual cavity is strongly dependent on the timing of complete filled of the first cavity. Although flow rate conditions will affect the penetration history during the processing, it doesn't change the final core penetration shape too much. To realize how the flow rate effect is, we have applied slicing technology to discover the core penetration behavior and its history. Clearly, when the flow rate is increased, it is easier for the core material penetrated into thickness direction in the upstream period. It will push more skin material into downstream. It ends up a low core ratio into the cavities and has a more non-uniform skin/core distribution. On the other hand, we have also considered different thickness of connection between cavities in Model 2 (with 3.5 mm thick connector) and Model 3 (with 1.75 mm thick connector). During the early filling age (less than 90% of total volume filled), in Model 3, it is Branch 2 dominant; while in Model 2 it has no preference for Branch 1 and 2. The flow rate conditions have no significant effects for core penetration in this period. However, as more melt keeps flowing into cavities till the end of filling, the core penetration behavior is dramatically different at low flow rate (10 cm^3/s). These results show that in the presence of different thickness of connector, the core penetration history and final shapes are significant different to that of separated cavity system (Model 1). Obviously, cavity design can alter the preference of core penetration from one side to the other. The results can help people for the management of skin/core distribution in co-injection molding.

Keywords: Multi-component molding, co-injection molding, core penetration
PACS: 83

INTRODUCTION

Co-injection molding, one of the multi-component molding methods are commonly used in daily accessories, car parts, and structural-reinforcement product, such as soft in the skin/hard in the core, fresh skin/regrind core, raw skin/reinforcement core [1-3]. The advantages of co-injection are to reduce cost, reuse material, and upgrade production efficiency. Besides, strength of structural reinforcement product is correlated to skin/core material distribution. Therefore the control of internal material distribution is important. The cavity-filling ratio of skin material determines the break-through location. Material viscosity and filling rate affect uniformity of core material distribution. However, there are too many combinations of designs, materials, core/skin ration, and process condition, how to have suitable control of co-injection is very challenge. Furthermore, the multi-cavity co-injection molding has been utilized in many industrial systems. Due to the complicated nature, the inside mechanism of the multi-cavity co-injection is still not fully understood yet [4-8].

In this study, we have proposed three kinds of multi-cavity systems to investigate cavity design influence on the core material penetration behavior. Results show that in the separated cavity system (Model 1), although flow rate can influence core penetration during the early history through the runners, the cavity structure will restrict the core development in Cavity 1. To realize how the cavity effect is, we have applied slicing technology to discover the core penetration behavior and its history. Clearly, when the flow rate is increased, it is easier for the core material penetrated into thickness direction in the upstream period. It will push more skin material into downstream. It ends up a low core ratio into the cavities and has a more non-uniform skin/core distribution. Moreover, when system has a different thickness of connection between cavities in Model 2 (with 3.5 mm thick connector) and Model 3 (with 1.75 mm thick connector), during the early filling age (less than 90% of total volume filled), it is Branch 2 dominant in Model 3; while in Model 2 it has no preference for Branch 1 and 2. The flow rate conditions have no significant effects for core penetration in this period. However, as more melt keeps flowing into cavities till the end of filling, the core penetration behavior is dramatically different at low flow rate (say 10 cm^3/s). These results show that in the presence of different thickness of connector, the core penetration history and final shapes are significant different to that of separated cavity system (Model 1). Obviously, cavity design can alter the preference of core penetration from one side to the other. The results can help people for the management of skin/core distribution in co-injection molding.

THEORY AND NUMERICAL METHOD

In this study, the fluids are considered to be incompressible, Newtonian (for water) or generalized Newtonian (for polymer melt). The governing equations for 3D transient non-isothermal motion are:

$$\frac{\partial \rho}{\partial t} + \nabla \cdot \rho \mathbf{u} = 0 \qquad (1)$$

$$\frac{\partial}{\partial t}(\rho \mathbf{u}) + \nabla \cdot (\rho \mathbf{u}\mathbf{u} + \tau) = -\nabla p + \rho \mathbf{g} \qquad (2)$$

$$\rho C_P\left(\frac{\partial T}{\partial t} + \mathbf{u} \cdot \nabla T\right) = \nabla \cdot (k \nabla T) + \eta \dot{\gamma}^2 \qquad (3)$$

where u is velocity vector, T is temperature, t is time, p is pressure, τ is stress tensor, ρ is density, η is viscosity, k is thermal conductivity, Cp is specific heat and $\dot{\gamma}$ is shear rate. For the polymer melt, the stress tensor can be expressed as:

$$\tau = -\eta\left(\nabla \mathbf{u} + \nabla \mathbf{u}^T\right) \qquad (4)$$

The modified-Cross model with Arrhenius temperature dependence is employed to describe the viscosity of polymer melt:

$$\eta(T, \dot{\gamma}) = \frac{\eta_o(T)}{1 + \left(\eta_o \dot{\gamma} / \tau^*\right)^{1-n}} \qquad (5)$$

$$\eta_o(T) = B Exp\left(\frac{T_b}{T}\right) \qquad (6)$$

where n is the power law index, η_0 is the zero shear viscosity, and τ^* is the parameter that describes the transition region between zero shear rate and the power law region of the viscosity curve.

A volume fraction function f_i is introduced to specify the evolution of the polymer/air front (i=1) and skin/core front (i=2) interfaces. Here, $f_i = 0$ is defined as no-filled region, $f = 1$ as fully-filled region, and finally the interfacial front is located within cells of an f value between 0 and 1. The advancement of f over time is governed by the following transport equation:

$$\frac{\partial f_i}{\partial t} + \nabla \cdot (\mathbf{u} f_i) = 0 \qquad (7)$$

During the polymer melt filling phase, the velocity and temperature are specified at the mold inlet. While the core material is injected, the flow rate setting is specified at the mold inlet. On the mold wall, the non-slip boundary condition is applied, and fixed mold wall temperature is assumed.

MODEL AND INVESTIGATION PROCEDURES

The geometry models and dimensions of the runners and cavities are shown as in Figure 1. It includes (a) Model 1: the separated cavity system, each cavity has 20 mm diameter and 3.5 mm thickness, (b) Model 2: the connected cavity system with 3.5 mm thick connector, (c) Model 3: the connected cavity system with 1.75 mm thick connector. The material applied is GPPS_POLYREXPG-22. The process conditions are including filling time 0.6 s, mold temperature 60°C, melt temperature 220°C, packing time 0.05 s, and so on. Specifically, the core material enters the runner at 72% total volume in Model 1, while it flows into runner at 80% total volume in Model 2 and 3 to avoid the break through.

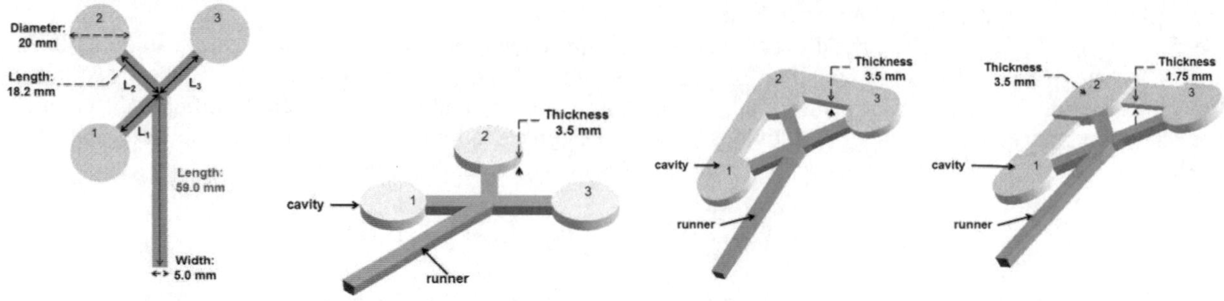

(a) Model 1 (b) Model 2 (c) Model 3

Figure 1. Geometry model and dimensions: (a) Model 1: separate cavity system, (b) Model 2: cavity system with 3.5 mm thick connector, (c) Model 3: cavity system with 1.75 mm thick connector.

RESULTS AND DISCUSSION

Figure 2 (a) shows the core penetration shape as skin/core ratio is 72/28 for Model 1. When the injection flow rate is 10 cm^3/s, the longest core penetration to hit the end of runner happens in Branch 1 (as shown in L1). As the injection flow rate is increased higher enough (for example, 30 cm^3/s or higher), the core penetration is enhanced in Branch 2 to touch the end of runner first. After core material hits the end of the runner, it is continuously to drive melt into the cavity till the end of filling. Clearly, there is very few core material enters the first cavity through Branch 1, as in Figure 2 (b). It is due to the first cavity is filled completely at early age, and then there is no room for further core material to enter. To realize the detailed dynamic feature for the core shape generation,

We further studied the flow rate effect on the core penetration quantitatively as in Figure 3. In Figure 3(a), it shows the direction of slicing the cross section in primary runner. When we increase the flow rate from 10cm^3/s to 50cm^3/s, the core material occupies area has been increased as in Figure 3(b). Moreover, if we switch to the slicing in secondary runner shown in Figure 3(c) and (d), the trend is opposite that the increasing flow rate will reduce the core material occupies area. Clearly, when flow rate is increased, the core material is easier to penetrate into thickness direction during the early filling stage. It is the reason why core material occupied larger space in the primary runner as flow rate increased in Figure 3(b). At the same time, since the total skin/core ratio is fixed, the less core material will go through the secondary runner and cavities as shown in Figure 3(d).

Furthermore, to study the design influence on the core penetration behavior for different types of cavities, Figure 4(a) displays skin and core co-injection molding with 3.5 mm thick connecting between cavities (in Model 2 structure). Here the switch time for core material is switched to 80% of total volume to avoid break through problem. When the first core layer hits the end of runner, Branch 1 and 2 are almost touched at the same time. Besides, there is no significant change from low flow rate (10 cm^3/s) to high flow rate (50 cm^3/s). It is very different as seen in Figure 2(a) for Model 1. If it keeps going to fill more melt till the end of filling, the final core penetration result is in Figure 4(b). It is so surprise that flow rate conditions have little influence in early core development, but it ends up with a very different core shape especially for flow rate at 10 $cm^{3/}$s. It is almost opposite to that of behavior obtained in Model 1 as described earlier in Figure 2(a) and (b). Moreover, Figure 5(a) shows that the core penetration goes through the second branch is the first one to hit the end of runner during the early age, from low flow rate (10 cm^3/s) to high flow rate (50 cm^3/s) in the system with 1.75 mm thick connection between cavities. It is different as we have seen in Fig 2(a) for Model 1, and in Fig 4(a) in Model 2. If it keeps going to fill more melt till the end of filling, the final core penetration result is in Figure 5(b).

Finally, the simulation is validated with the literature result for Model 1 [7]. Figure 6(a) displays the flow front of the skin layer. Clearly, no matter high or low injection flow rate, Branch 1 always reaches first regardless of change in mold temperature and flow rate. Moreover, Figure 6(b) shows the core penetration flow front. For low flow rate (10.2 cm^3/sec) L1 is the longest in runner, and for high injection flow rate (51 cm^3/sec) L2 takes the lead. These results show both simulation and experiments are in a good agreement in trend. Furthermore, the experimental verification for Model 2 and 3 will be performed in the coming future.

CONCLUSIONS

In this study, we have constructed three kinds of multi-cavity systems to investigate cavity design influence on the core material penetration behavior. Results show that in the separated cavity system (Model 1), although flow rate can influence core penetration during the early history through the runners, the separated cavity structure (Model 1) will restrict the core development in Cavity 1. It ends a non-uniform skin/core distribution. The presence of different flow rate does affect the final core shape too much. Moreover, when the system has a different thickness of connection between cavities in Model 2 (with 3.5 mm thick connector) and Model 3 (with 1.75 mm thick connector), during the early filling age (less than 90% of total volume filled), it is Branch 2 dominant in Model 3; while in Model 2 it has no preference for Branch 1 and 2. The flow rate conditions have no significant effects for core penetration in this period. However, as more melt keeps flowing into cavities till the end of filling, the core penetration behavior is dramatically different at low flow rate (say 10 cm3/s). These results show that in the presence of different thickness of connector between cavities, the core penetration history and final shapes are significant different to that of separated cavity system (Model 1). Obviously, cavity design can alter the preference of core penetration from one side to the other. The results can help people for the management of skin/core distribution in co-injection molding.

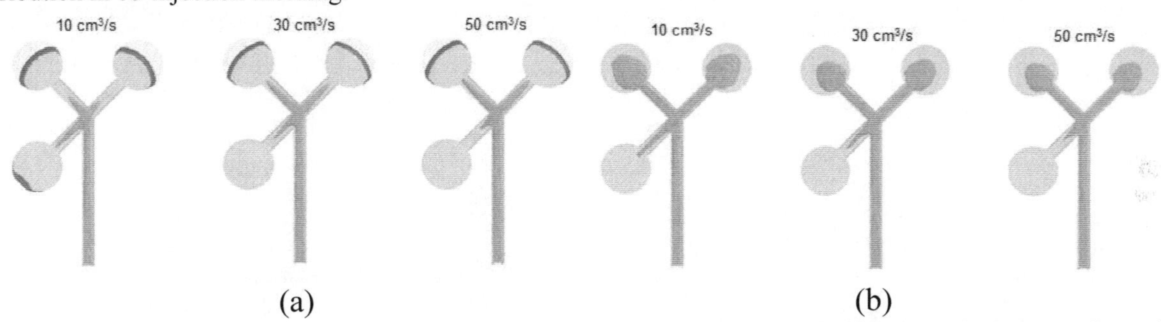

Figure 2. Flow rate effect on core penetration for Model 1: (a) penetration length to hit the end of runner, (b) at the end of filling.

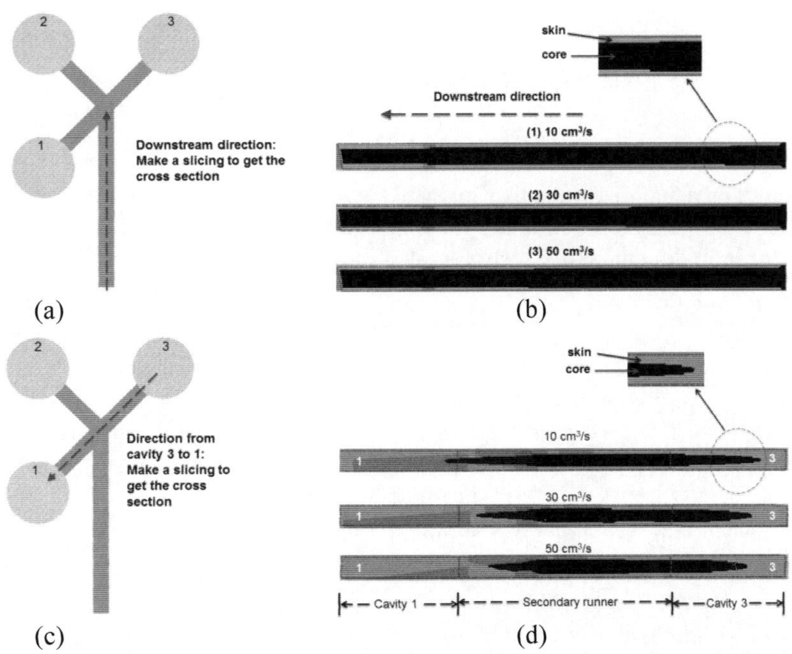

Figure 3. Flow rate effect on the core penetration. (a) the slicing direction in primary runner, (b) the occupied area of the core material in different flow rate in cross section of the primary runner, (c) the slicing direction in secondary runner, (d) the occupied area of the core material in different flow rate in cross section of the primary runner.

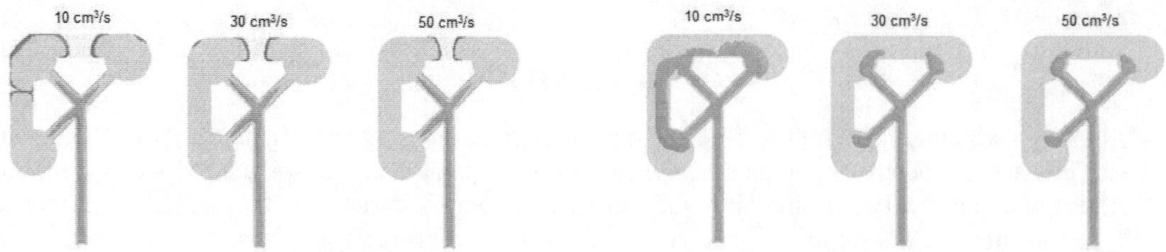

(a) (b)

Figure 4. Flow rate effect on core penetration for Model 2: (a) penetration length to hit the end of runner, (b) at the end of filling.

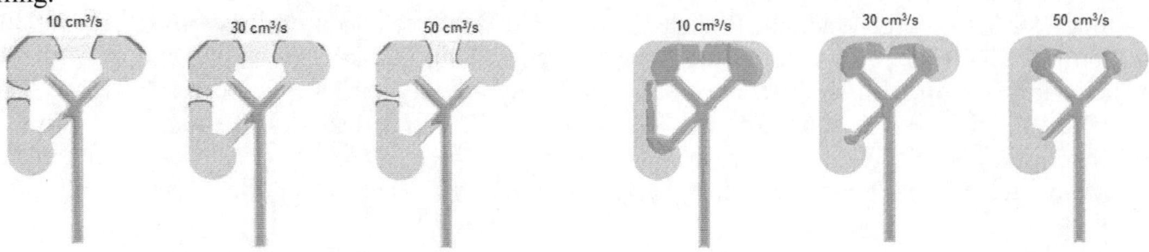

(a) (b)

Figure 5. Flow rate effect on core penetration in Model 3: (a) penetration length to hit the end of runner, (b) at the end of filling.

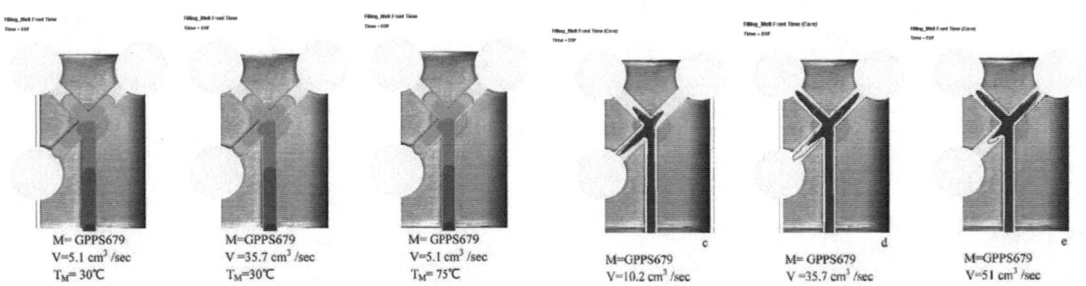

(a) A single layer system (b) A co-injection system

Figure 6. Simulation and experimental validation: (a) displays the skin layer, (b) displays the core layer only.

ACKNOWLEDGEMENTS

The author (Chao-Tsai Huang) would like to thank Ministry of Science and Technology of Taiwan, R.O.C. (Project number: MOST 105- 2218-E-032-002-) for financially supporting for this research. Special thank goes to Mr. Kuan-hao Chen for providing the cross-section plots.

REFERENCES

1. P.J. Garner and D.F. Oxley, British Patent 1,156,217, 1971.
2. V. Goodship and J.C. Love, Multi-Material Injection Molding, 2002.
3. S.-P. Sun, C.-C. Hsu, C.-T. Huang, K,-C. Huang and S.-C. Tseng, SPE-ANTEC Conferences, 2012
4. Jimmy C.Chien, Chao-Tsai Huang, Wen-Hsien Yang, David C.Hsu, SPE ANTEC 2006
5. John P. Beaumont and Jack H. Young, "Mold Filling Imbalances in Geometrically Balanced Runner Systems" Journal of Injection Molding Technology, September 1997, Vol. 1, No. 3
6. John P. Beaumont, “Runner and Gating Design HandBook”.
7. W.M. Yang, H. Yokoi, Polymer Testing, 22, 37, 2003.
8. Jackie Yang, C.-T. Huang, et al, SPE ANTEC2015.

Prediction of Cross-Sectional Shape at Die Exit on Film Casting Process by Viscoelastic Flow Simulation

Shuichi TANOUE[a*], Hideyuki UEMATSU[a] and Yoshiyuki IEMOTO[a]

[a]*Graduate School of Engineering, University of Fukui*
3-9-1 Bunkyo, Fukui 910-8507, JAPAN

tanoue@matse.u-fukui.ac.jp

Abstract. On the film casting process, the following phenomena occur, 1) neck-in, the film width becomes smaller than the die width, 2) Edge bead, the film thickness around the edge becomes larger than that at the film center. One of the requirement points on the film casting process is to produce the film with uniform thickness. In this study, we tried to predict the cross-sectional shape of die exit on the film casting process in order to produce the film with uniform thickness by using a viscoelastic flow simulation. The Phan-Thien Tanner (PTT) model was employed as the constitutive equation. By this simulation, we could get the cross-sectional shape of die exit in order to manufacture the film with uniform thickness. The width of this die exit became large and the thickness around the edge of this die exit became small when the draw ratio increased because the neck-in and edge bead phenomena occurred on the film casting process. The width of the die exit at same draw ratio was almost independent of the fluid characteristics, i.e., the viscosity ratio of uniaxial elongational viscosity to planar elongational one at same elongation rate. However, the flat part region of die thickness at the die exit became increased when this viscosity ratio increased. And the deformation from rectangular cross-section became prominent at high draw ratio.

Keywords: Film casting process, Viscoelastic flow simulation, Finite element method
PACS: Replace this text with PACS numbers; choose from this list: 47.11.Fg, 47.50.-d, 47.85.md, 83.80.Sg

INTRODUCTION

On film casting process, the following phenomena occur, 1) neck-in, the film width becomes smaller than the die width at die exit, 2) Edge bead, the film thickness around the edge becomes larger than that at the film center. These phenomena have the worse influence on the quality of film products. Then, it is necessary to understand these phenomena. However, it is difficult to observe the flow phenomena directly on film casting process. The numerical simulation is useful for understanding a flow state on film casting process. Many research examples were carried out by several researchers. For example, Kajiwara et al. [1] showed that the film width was determined by the ratio of uniaxial to planar elongational viscosity rather than the extension-thickening nature in uniaxial elongation for both viscous and viscoelastic fluids by the quasi-three dimensional viscoelastic or purely-viscous flow simulation. Shiromoto et al. [2] showed that the neck-in normalized by the air gap was in good correlation with the ratio of planar to uniaxial elongational viscosity rather than the strain hardening nature of uniaxial elongational viscosity. Ahmed and Khayat [3] carried out the numerical study about draw resonance by linear stability analysis of film casting process for Newtonian fluid. As mentioned above, the numerical studies of flow phenomena on film casting process have been carried out.

On the contrary, one of the requirement points on the film casting process is to produce the film with uniform thickness. In this study, we tried to predict the cross-sectional shape of die exit on the film casting process in order to produce the film with uniform thickness by using a viscoelastic flow simulation.

CALCULATION MODEL

The steady-state and isothermal flow of incompressible fluid was assumed in the film casting process. The inertia term in the momentum equation was negligible as the focused fluid, a polymer melt, has high viscosity. In addition, the gravity term in the momentum equation could be assumed to be negligible. The Phan-Thien Tanner (PTT) model [4] was employed as a constitutive equation for the viscoelastic fluid. And the Carreau-Yasuda model [5] was also employed as the purely-viscous fluid in order to consider the effect of viscoelasticity on this process. The three kinds of fluids with same frequency sweep testing data of polypropylene (PP) melt [6] were used as the focused fluids (the test fluids). Figure 1 (a) shows the elongational viscosity ratio of the test fluid $((\eta_{EU}/3)/(\eta_{EP}/4)$, η_{EU} is uniaxial elongational viscosity and η_{EP} is planar elongational viscosity) as a function of elongation rate. Figure 1 (b) shows the shear viscosity and first normal stress difference N_1 as a function of shear rate. The Fluid models of Fluid A and Fluid B are the PTT model, and that of Fluid C is the Carreau-Yasuda model. The shear viscosity curves in steady state are almost same each other shown in Fig. 1(b). There is no N_1 curve for Fluid C because Fluid C has no elasticity. And the elongational viscosity ratio of Fluid B is larger than that of Fluid A as shown in Fig. 1 (b). Figure 2 shows the calculation model. The calculation area is assumed to be the region between the die exit and the contacting point on the roll surface (named the Draw point). And the calculation model is constituted in the range of 1/4 symmetry with *xz* plane and *yz* plane as the symmetry planes (Colored part in Fig. 2). In this simulation, the film cross-sectional shape on the Draw point was fixed to the rectangles with fixed film width and uniform thickness. And the cross-sectional shape of the die at the die exit was changed with the calculation algorithm. Then, it is possible to predict the cross-sectional shape of the die exit to produce the film with uniform thickness by this simulation.

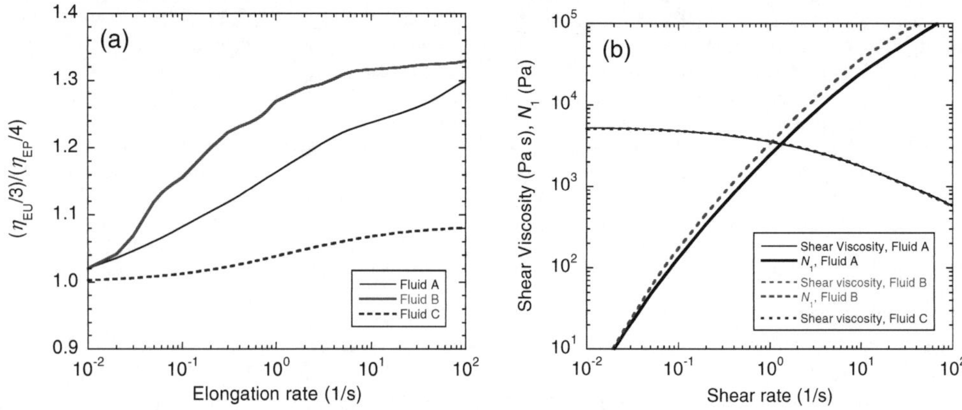

FIGURE 1. Flow characteristics of the test fluids in this study. (a) The ratio of uniaxial elongational viscosity to planar one versus elongation rate, and (b) Shear viscosity and first normal stress difference N_1 versus shear rate.

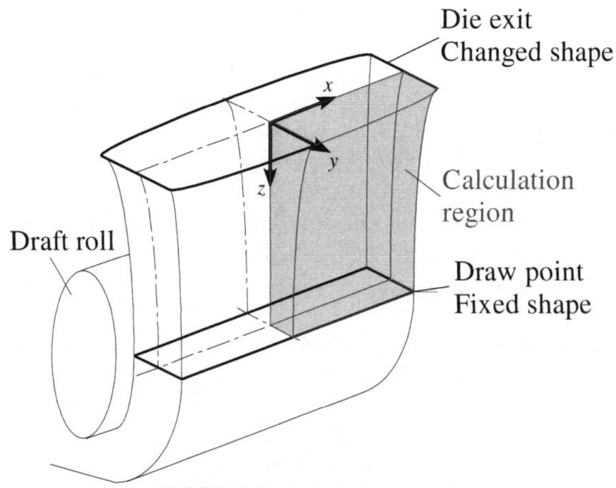

FIGURE 2. Calculation model.

In this study, the following flow problem is created. (1) The initial cross-sectional shape of the die exit before calculation is the rectangle with the width of 2000 mm and the thickness of 1 mm, (2) The reference length H is the half length of the die initial thickness at the die exit (H = 0.5 mm), (3) The air gap length (which is the length between the die exit and the Draw point) is 200 mm, (4) The cross-sectional shape of film at the Draw point is the rectangle with the width of 2000 mm and thickness of $1/Dr$ mm (where Dr is the draw ratio which is defined by the ratio of the velocity of z direction on the roll to the mean velocity of the fluid at the die exit).

RESULTS AND DISCUSSION

Figure 3 shows the whole film shape and the dimensionless velocity $v_z/\langle u \rangle$ ($\langle u \rangle$ is the mean velocity at the die exit) distribution at Dr = 10 for Fluid A. The neck-in phenomenon can be seen in the draw region. The predicted cross-sectional shape of the die at the die exit is as follows, the width of the die becomes larger than film width on the roll (x/H = 2000) and the thicknesses of the die becomes thick at the center and thin around the edge. The edge bead phenomenon can be seen on the actual film casting process. Then, the die thickness at the edge should be thin if the film with uniform thickness is produced. Then, the predicted die shape at the die exit is reasonable.

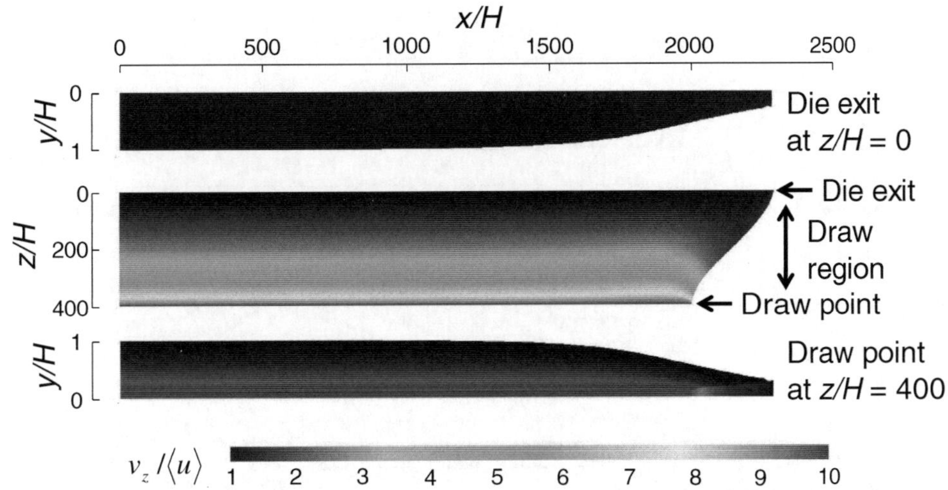

FIGURE 3. The whole film shape and distribution of dimensionless velocity $v_z/\langle u \rangle$ for Fluid A at Dr = 10.

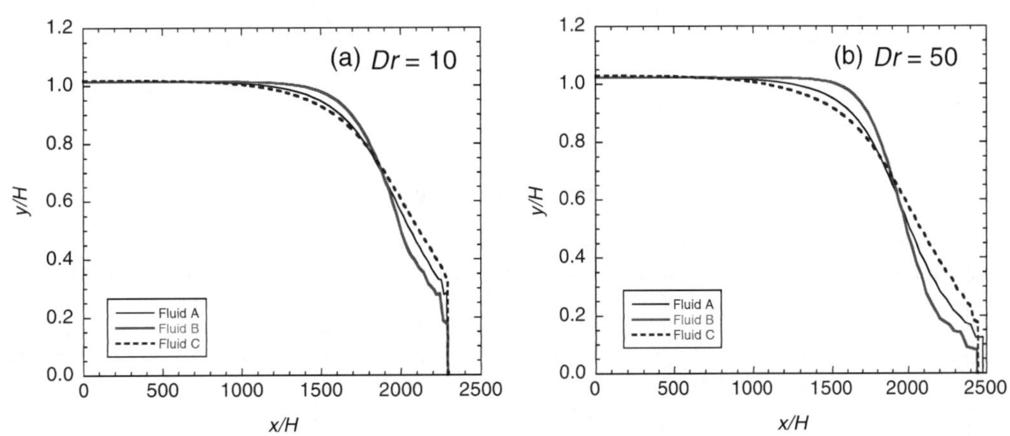

FIGURE 4. The predicted die cross-sectional shape at the die exit for various test fluids.

Figure 4 shows the predicted die cross-sectional shapes at the die exit for various test fluids at Dr = 10 and 50. The die width increased with draw ratio Dr because of neck-in phenomenon. However, the predicted die width at same Dr is almost independent of the kind of test fluids. According to the die thickness at same Dr, the flat part of the die thickness became large and the die thickness around the edge became small when the elongatinoal viscosity ratio is large. The elongational viscosity ratio of Fluid B is larger than that of Fluid A at same elongation rate. Therefore, Fluid B is larger than Fluid A for easiness of doing the planar elongational deformation. This characteristic indicates the predicted result of die exit shape. When the draw ratio Dr was large, the predicted die width became larger than that at Dr = 10 because of the larger neck-in phenomenon. And the predicted die width was also independent of the kind of test fluids, in short, the elongational viscosity ratio of the test fluids. In addition, the flat part in the predicted die thickness became large, and this tendency depended on the elongational viscosity ratio of the test fluids. Figure 5 shows the streamlines in the draw region for various test fluids at $Dr = 50$. The straight streamlines in Fig. 5 show that the planar elongational deformation is caused in the draw region. The ranks of easiness of having a planar elongational deformation were Fluid B, Fluid A and Fluid C. The region with straight streamlines for Fluid B was the largest. This result implies the reason why Fluid B had the largest flat part in the predicted cross-sectional die shape at the die exit. On the actual film casting process, it would be difficult to cause the problem, e.g. polymer melts would not be able to flow around the thin edge in a T-die. Then, one of the future works is the prediction taking into consideration of the flow in the T-die.

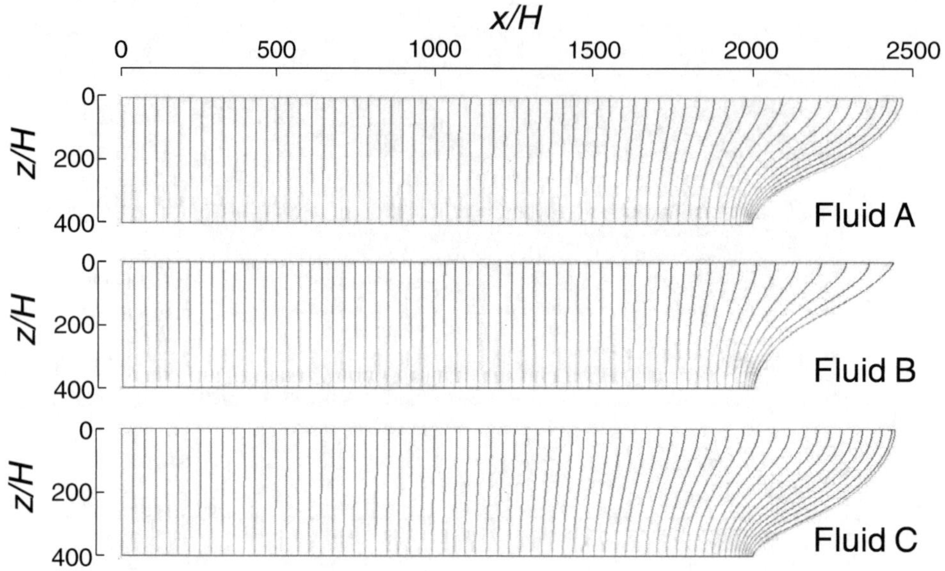

FIGURE 5. The streamlines in the draw region for various test fluids at $Dr = 50$.

CONCLUSIONS

In this study, we tried to predict the cross-sectional shape of the die exit on the film casting process in order to produce the film with uniform thickness by using a viscoelastic flow simulation. By this simulation, we could get the cross-sectional shape of die exit in order to produce the film with uniform thickness. The width of this die exit became large and the thickness around the edge of this die exit became small when the draw ratio increased because the neck-in and edge bead phenomena occurred on the film casting process. The width of the die exit at same draw ratio was almost independent of the fluid characteristics, i.e., the viscosity ratio of uniaxial elongational viscosity to planar elongational one at same elongation rate. However, the flat part region of die thickness became increased when this viscosity ratio increased. And the deformation from rectangular cross-section became prominent at high draw ratio. An experiment is necessary to verify the results obtained in this study.

REFERENCES

1. T. Kajiwara, M. Yamamura and T. Asahina, *Nihon Reoroji Gakkaishi* **34**, 97-103 (2006).
2. S. Shiromoto, Y. Masutani, M. Tsutsubuchi, Y. Togawa and T. Kajiwara, *Rheol. Acta* **49**, 757-767 (2010).
3. Z. U. Ahmed and R. E. Khayat, *J. Fluid Eng.* **134**, 101302-1-11 (2012).
4. N. Phan-Thien and R. I. Tanner, *J. Non-Newton. Fluid Mech* **2**, 353-365 (1977).
5. K. Yasuda, R. C. Armstrong and R. E. Cohen, *Rheol. Acta* **20**, 163-178 (1981).
6. M. J. Solomon, A. S. Almusallam, K. F. Seefeldt, A. Somwangthanaroj and P. Varadan, *Macromolecules* **34**, 1864-1872 (2001).

Experimental Characterisation And Identification Of Elastic-viscoplastic Polymer Properties Used In Hot Embossing Process

Gang Cheng[a], Thierry Barrière[b*], Jean-Claude Gelin[b] and Mohamed Sahli[b]

[a]*Mechanical Engineering Design Department, INSA Lyon, 85 Rue Henri Becquerel, 01100 Bellignat, France*

[b]*FEMTO-ST Institute, Applied Mechanics Department, 24 Rue de l'Epitaphe, 25000 Besançon, France*
gang.cheng@insa-lyon.fr
thierry.barriere@univ-fcomte.fr

Abstract. Micro hot embossing process is widely used for manufacturing of polymeric components with complex surfaces. The physical properties of polymer plate in hot embossing process could be considered as viscoelastic or viscoplastic, depending on materials and temperatures. A large number of hot embossing experimental results have been published, the material modelling and processes simulation to improve the quality of micro replication by hot embossing process are still lacking. This paper consists to characterize the physical behaviours of amorphous polymers used in hot embossing process. The compression tests at the temperature range lightly above T_g have been carried on and an elastic-viscoplastic law has been proposed to describe the polymers' behaviours in the tests. The experimental data fits well by the model with the identified parameters. It shows the further application for the numerical simulation of hot embossing process in the future.

Keywords: hot embossing, polymer, amorphous, viscoplastic, compression, relaxation.
PACS: 60

INTRODUCTION

Hot embossing is considered to be the most promising process in the micro fabrication manufacturing of polymeric component [1]. The embossed material is always heated to a certain temperature, which allows the material flow into the cavities of mould insert. For amorphous thermoplastic polymers, the processing temperature is lightly above their glass transition temperature (T_g), which the massive changes in physical properties of polymer occur during this temperature range, such as activation energy, specific heat, coefficient of thermal expansion. Rheology, creep rate, various mechanical properties, like toughness and strength also exhibit major changes at the T_g [2]. The physical material behaviours of polymer are strongly temperature dependent and time dependent, which combined the elastic, viscoelastic and viscoplastic behaviours. More and more micro-components have been fabricated with thermoplastic polymers by micro hot embossing process during recent years, but the lack of the numerical modelling and simulation restricts the further development of the process [3].

The polymeric material's viscoelastic and viscoplastic properties have been well studied in the literature [4-9]. The term of "elastic" is used when the strain is recoverable and thermodynamically reversible after the deformation [10]. The term of "plastic" is used when the strain is unrecoverable and thermodynamically irreversible, which is assumed to depend only on the magnitude of the local stress or strain. The term of "visco" is used when the strain also changes with time [11]. The polymer's behaviours during the hot embossing process are strongly temperature dependent and time dependent, which combined the elastic, viscoelastic and viscoplastic behaviours. In this paper, some efforts concerning the characterization of amorphous thermoplastic polymers elastic-viscoplastic behaviours in hot embossing temperature range have been effectuated. The uniaxial compression tests have been effectuated in order to characterize the polymers' physical behaviours during the hot embossing process [12]. The uniaxial compression tests have been carried on lightly above T_g, which is correspondent with the hot embossing temperature range for the amorphous thermoplastic polymers. The true stress and true strain of the polymers have been calculated based on the instantaneous deformations of the material. The true stress and true strain curves respect to time have been obtained with two different kinds of amorphous thermoplastic polymers. An elastic-viscoplastic model based on overstress [13-14] has been proposed to describe the amorphous thermoplastic polymers' mechanical behaviour. The materials' parameters in the model have been identified according to the stress strain curves obtained in the compression tests. A proper agreement between the experimental data and the fitting models parameters has been achieved. The elastic-viscoplastic model is possible to be integrated in the finite element software to describe the amorphous thermoplastic polymers' behaviour in numerical simulation of hot embossing process.

UNIAXIAL COMPRESSION TESTS

The uniaxial compression tests were performed by means of a Bose® ElectroForce® test instrument. Two types of amorphous thermoplastic polymer polystyrene (PS) and polymethyl methacrylate (PMMA) samples with the diameter eq. 10 mm and the length eq. 18 mm were used in the uniaxial compression experiments. The WinTest® digital control system was used to control the strain rate or stress rate during the test. The top and bottom surface of the sample were polished to reduce the friction at sample/platen interface. The polymer samples were placed in the middle of two platens in a high temperature furnace. The temperature in the furnace was controlled by a thermocouple. The polymer samples were allowed to anneal at the testing temperature for about half an hour before every compression test. The lower platen was fixed and connected with a load sensor to record the compression force. The load sensor is able to measure the quasi static force up to 300 N with the accuracy of 0.01 N. The upper platen is able to move down towards the lower platen to compress the sample with predefined strain or stress rate. The relative displacement of the upper platen has been recorded by the Bose® high-accuracy displacement sensor. The maximum measuring displacement can reach 6mm with the accuracy level of 10^{-2}-10^{-4} mm.

AN ELASTIC-VISCOPLASTIC MODEL BASED ON OVERSTRESS

The elastic-viscoplastic model described here is derived from the theory of viscoplasticity based on overstress (VBO). The VBO model is originally used to describe the viscoplastic behaviour of the metallic material. Gates et al. [15] have proposed the elastic-viscoplastic constitutive model based on the VBO model to predict the nonlinear, time-dependent behaviour of the fibre reinforced thermoplastic composites. M. Al-Haik with his co-workers has used this elastic-viscoplastic model to analyse the polymeric composites' behaviour observed from several load and temperature levels [13-14]. In this study, the elastic-viscoplastic model has been used to describe the amorphous thermoplastic polymers' mechanical behaviour above their T_g. The total strain ε^{to} is written as a combination of elastic term ε^{el} and plastic term ε^{pl}. The relation between the elastic strain and stress is provided by Hooke's law, and the plastic strain is expressed by a power law:

$$\varepsilon^{el} = \frac{\sigma}{E} \tag{1}$$

$$\varepsilon^{pl} = A\sigma^{n} \tag{2}$$

where σ is the stress and E is the elastic modulus, A and n are the material constants.

For a rate dependent constitutive relation, the total strain rate $\dot{\varepsilon}^{to}$ is divided into elastic and viscoplastic components. The viscoplastic strain rate $\dot{\varepsilon}^{vp}$ is expressed as:

$$\dot{\varepsilon}^{vp} = \dot{\varepsilon}^{pl} + \dot{\varepsilon}^{vp'} \tag{3}$$

Where $\dot{\varepsilon}^{pl}$ is the plastic strain rate presented as:

$$\dot{\varepsilon}^{pl} = \begin{cases} An\sigma^{n-1}\dot{\sigma} & if \quad \dot{\sigma} > 0 \\ 0 & if \quad \dot{\sigma} \le 0 \end{cases} \tag{4}$$

and the second part is viscoplastic strain rate based on overstress $\dot{\varepsilon}^{vp'}$, which could be presented as:

$$\dot{\varepsilon}^{vp'} = \begin{cases} \left[\dfrac{\sigma - g}{K}\right]^{1/m} & if \quad \sigma > g \\ 0 & if \quad \sigma \le g \end{cases} \tag{5}$$

where g is the equilibrium stress, K and m are material constants. $\dot{\varepsilon}^{vp'}$ exsite only when the dynamic stress exceeds the equilibrium stress g. The equilibrium stress g could be found in the quasistatic elastoplastic relation [13]:

$$\varepsilon^{to} = \frac{g}{E} + Ag^{n} \tag{6}$$

There are six parameters in the elastic-viscoplastic model: an elastic modulus E, four material constants A, n, K, m and also the equilibrium stress g. The experimental tests need to be effectuated in order to identify these parameters.

RESULTS AND DISCUSSIONS

The relaxation compression tests of the selected polymers PS and PMMA have been effectuated at T_g + 20 °C in order to investigate the stress response in a relative large deformation range. Several strain levels have been applied on the polymers' samples to observe the evaluation of the stress respect to time. The material constants in the elastic-viscoplastic model have been identified according to the experimental data in the uniaxial compression tests.

Relaxation stress in uniaxial compression test

In the stress relaxation test, the strain remains constant while the stress decreases. Therefore the plastic strain rate is zero, because the stress rate is negative in the stress relaxation tests. The strain remains constant, so the total strain rate is zero. The viscoplastic strain rate in the stress relaxation tests could be expressed as follows:

$$\left[\frac{\sigma - g}{K}\right]^{1/m} = -\frac{\dot{\sigma}}{E} = \dot{\varepsilon}^{vp} \tag{7}$$

The stress rate $\dot{\sigma}$ could be obtained by differentiating the stress-time curve with respect to time. Then the material constant K and m could be identified by the relation between the viscoplastic strain rate $\dot{\varepsilon}^{vp}$ and the stress σ. The equilibrium stress g can be obtained by applying the nonlinear regression methods. The stress relaxation tests can be repeated for different strain levels and the equilibrium stress can be calculated for each test.

The stress relaxation tests were repeated for six different strain levels at T_g + 20 °C. The polymer sample was imposed with a constant true strain rate to the ultimate strain level. And then, the true strain remained constant, the variation of the true stress respect to time was recorded. The relaxation stresses vs. time with different strain levels for the testing polymers are showed in **FIGURE 1**.

Different strain levels have been applied on the polymers samples in order to investigate the stress response of the testing polymers. The relaxation stress have been recorded when the strain level remains constant. According to the values of the relaxation stress at $t = 0$ for each polymer, one can conclude that the true stress increase with the increase of the applied strain level. The true stress of the testing polymers decreases respect to time in the stress relaxation tests. The large decline of the true stress values of PS and PMMA appears at the first period (from 0 s to 40 s). And then the true stress becomes nearly constant. For example, at the strain level $\varepsilon_6 = 0.40$, the relaxation stress values decrease about 92% during the observation time. However, at the same strain level, the relaxation stress decreases about 48% for PS and about 59% for PMMA.

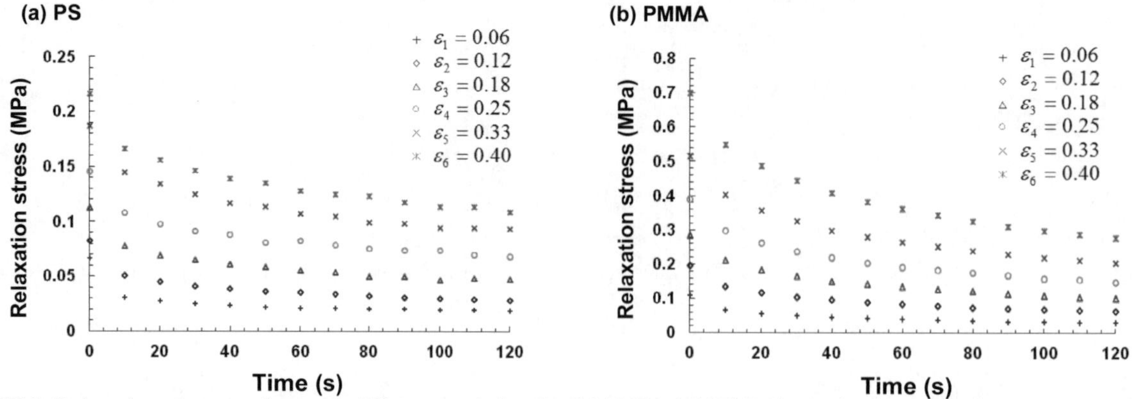

FIGURE 1. Relaxation stress vs. time with different strain levels of (a)PS and (b)PMMA obtained in relaxation tests at T_g+20°C.

Identification of material parameters in elastic-viscoplastic model

The relaxation stress vs. time for the amorphous polymers PS and PMMA has been obtained by the compression stress relaxation tests. The viscoplastic strain rate could be calculated by derivation the relaxation stress, associated with the elastic modulus. The material constants K and m could be identified by modelling the stress vs. viscoplastic strain rate curve. The logarithm of the relaxation stress vs. the logarithm of the viscoplastic strain rate is presented in **FIGURE 2**. The nonlinear relationships between the logarithm of the stress and the logarithm of the viscoplastic strain rate could be observed in the curves obtained in different strain levels. The equilibrium stress could be removed in the total stress in order to generate linear relation between viscoplastic strain rate and the overstress. The curves have been fitted with the linear equations in order to yield the material constant K and m. Note that K and m are independent of the initial applied strain level [13]. K and m have been obtained by fitting the curves in different strain levels and the average value of the parameters are presented in **TABLE 1**.

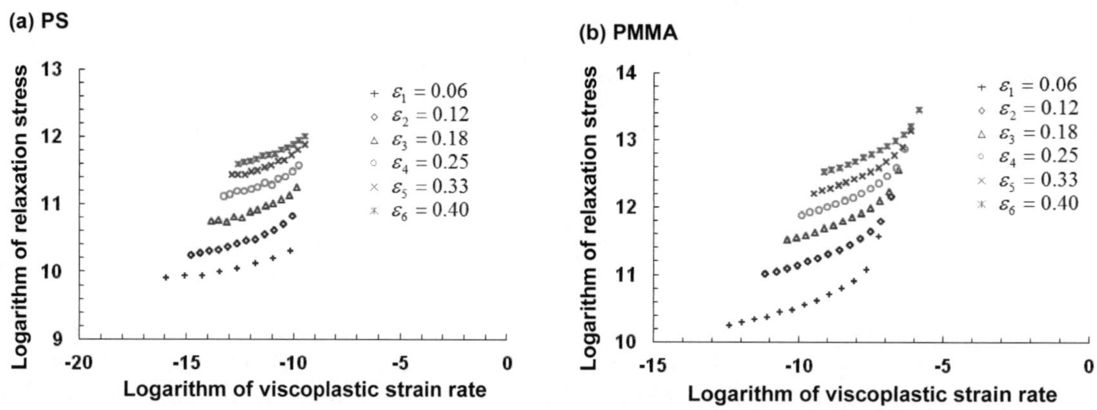

FIGURE 2. Logarithm of the relaxation stress vs. logarithm of the strain rate for different strain levels of (a)PS and (b)PMMA. Stress is in Pa

The material constants A and n can be identified due to the quasitatic elastoplastic relation between stress and strain, which could be written as follows:

$$\ln\left(\varepsilon^{pl}\right) = \ln\left(\varepsilon^{to} - \frac{g}{E}\right) = n\ln(g) + \ln(A) \tag{8}$$

where a linear relationship between the logarithm of the plastic strain and the logarithm of the equilibrium stress is observed. The material constant A and n could be identified by fitting the logarithm of the plastic strain vs. the logarithm of the equilibrium stress curves at different strain levels, which is showed in **FIGURE 3**. The material constants A and n are presented in **TABLE 1**. A good agreement of fitting curve with the experimental data has been observed for PS and PMMA.

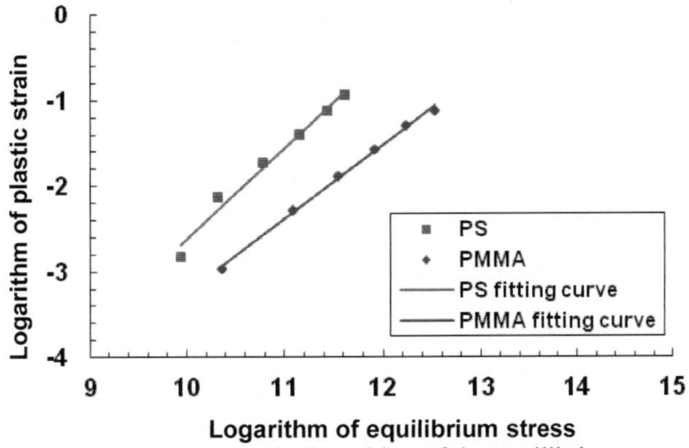

FIGURE 3. The logarithm of the plastic stain vs. the logarithm of the equilibrium stress curves at different strain levels.

TABLE 1. Material constants of the testing polymers PS and PMMA at T_g + 20 °C.

Polymer	E (MPa)	A	n	K	m
PS	23.53	0.7547	4.754×10^{7}	1.0691	1.648×10^{-6}
PMMA	3.16	0.8381	3.338×10^{7}	0.8803	6.629×10^{-6}

CONCLUSIONS

The uniaxial compression tests with the cylindrical PS and PMMA samples have been carried on at the temperature lightly above their T_g, which is corresponding to the hot embossing processing temperature range. The relaxation stress of the amorphous thermoplastic polymers has been investigated by applying six different strain levels. The relaxation stresses vs. time curves have been fitted with exponential equation in order to identify the equilibrium stress and generalize the derivative of the stress. A proposed elastic-viscoplastic model based on overstress has been used to characterize the viscoplastic behaviour of the selected polymers in compression stress relaxation tests. The equilibrium stresses of the selected polymers have been identified for each applying strain level. The evolution of the relaxation stress in function of the viscoplastic strain rate has been obtained, which presents nonlinear relationship between each other. Then the equilibrium stresses have been removed in the total stresses to achieve a linear relationship between the overstress and the viscoplastic strain rate. The material parameters K and m in the model have been identified by fitting these curves in different strain levels. The material constants A and n have been identified by the quasistatic elastoplastic relationship. A proper agreement between the identification and experimental data has been obtained for each selected polymer. It shows the great possibility for the further application of the numerical simulation of hot embossing process.

REFERENCES

1. M. Worgull, "Hot embossing process", in *Hot Embossing: Theory and Technology of Microreplication,* edited by W. Andrew, Burlington: Elsevier, 2009, pp. 137-175.
2. S.H. Kim et al., *Polymer* **48**, 4271-4277 (2007).
3. M. Worgull and M. Heckele, *Microsyst. Technol.* **10**, 432-437 (2004).
4. S.G. Bardenhagen et al., *Mech. Mater.* **25**, 235-253 (1997).
5. G.J. Frank and R.A. Brockman, *Int. J. Solids Struct.* **38**, 5149-5164 (2001).
6. A.D. Drozdov, *Int. J. Solids Struct.* **38**, 8285-8304 (2001).
7. F. Zairi et al., *Eur. J. Mech. A-Solid.* **24**, 169-182 (2005).
8. D.W. Holmes and J.G. Loughran, *Polym. Test.* **27**, 189-203 (2008).
9. G.Z. Voyiadjis et al., *Int. J. Plasticity* **28**, 21-45 (2012).
10. C. Mareau et al., *Int. J. Solids Struct.* **46**, 223-237 (2009).
11. J.-S. Kim and A.H. Muliana, *Int. J. Solids Struct.* **47**, 580-594 (2010).
12. V. Srivastava et al., *Int. J. Plasticity* **26**, 1138-1182 (2010).
13. M.S. Al-Haikk et al., *Compos. Part B-Eng.* **32**, 165-170 (2001).
14. M.S. Al-Haik et al., *Int. J. Plasticity* **20**, 1875-1907 (2004).
15. T.S. Gates and C.T. Sun, *AIAA J.* **29**, 457-463 (1991).

Experimental Study of Polymer Degassing in an Agitator Vessel

Stefan Hirschfeld[a*], Olaf Wünsch[a]

[a]*Institute of Mechanics, Chair of Fluid Mechanics, University of Kassel, Mönchebergstr. 7, 34125 Kassel, Germany*

stefan.hirschfeld@uni-kassel.de

Abstract. The present work deals with the experimental investigation of polymer devolatilization in a partial filled agitator vessel with a blade stirrer focusing on film degassing. The apparatus generate a wiped melt film on the barrel wall and a rotating melt pool at the stirrer blade contributing to mass transfer. A model substance system consisting of high-viscous polydimethylsiloxane as polymer and 1,1,2-trichloro-1,2,2-trifluoroethane as volatile is used. The driving force for the degassing is provided by reducing the partial pressure in the gas phase with a gas flow of nitrogen. The concentration of volatiles in the melt is measured by thermogravimetric analysis. The stripped volatiles were recovered and weighted by use of cold traps. The results are evaluated with models based on the penetration theory and surface renewal theory. Experimental measurements indicate that the gas-side mass transfer resistance cannot be neglected. Possible sources of inaccuracy in the execution of degassing studies and their influence on the results are discussed.

Keywords: polymer degassing, polymer devolatilization, polymeric solution, mass transfer, surface renewal, high viscosity, volatile organic compound, polymer processing, multiphase flow

PACS: 47.55.Ca; 47.85.md; 81.05.Lg; 82.60.Lf; 83.80.Rs; 83.80.Sg; 89.20.Bb; 89.20.Kk

INTRODUCTION

Desorption of volatile organic compounds (VOCs) from molten polymers, commonly known as degassing or devolatilization, is an essential unit operation in polymer processing industry. The controlled removal of these volatiles like solvents, residual monomers, reaction products or additives is desired in order to obtain high quality products and to fulfill compulsory regulations. Due to the special characteristics of polymers, this task usually requires a lot of effort and expensive equipment [1, 2]. Unfortunately, the designing of devolatilizers still demands on many empirical test series and a lot of experience [3]. Therefore, it is preferable to have detailed knowledge about the process to reduce costs. Polymer degassing is a physical separation process that can be performed in non-rotating and rotating devices. Moreover two mechanisms of devolatilization can be distinguished: film degassing and bubble degassing. This study will treat with film degassing in rotating devices that are primarily used for processing of high-viscous polymers, which is notably the case at very low residual concentrations of volatiles.

THEORY

The molar mass transfer rate $\dot{n}_i$ of volatiles (i) from liquid phase (l) to the free surface (*) at steady state is equal to the molar mass transfer rate from the free surface into the gas phase (g). This can be described according to common mass transport models by

$$\dot{n}_i = k_l A\left(c_i - c_i^*\right) = \frac{k_g A}{RT}\left(p_i^* - p_i\right) \tag{1}$$

or more generally by the product of intensity, active surface and driving force [4]. In eq. (1) the intensity is identified with the mass transfer coefficients k_l, k_g the active surface accords to the gas-liquid interface area A and the driving force correlates either to the molar concentration difference Δc_i or to the partial pressure difference Δp_i. Dividing eq. (1) by the fluid volume V leads to an ordinary differential equation

$$\dot{c}_i = \frac{dc_i}{dt} = k_l \frac{A}{V}\left(c_i - c_i^*\right) = \frac{k_g A}{RTV}\left(p_i^* - p_i\right) \tag{2}$$

with the free surface-to-volume ratio A/V as intensive quantity. In polymeric solutions, the gas-liquid equilibrium of the volatile obeys the Flory-Huggins equation

$$\ln a_i = \ln\left(\frac{p_i}{p_i^0}\right) = \ln \phi_i + (1-\phi_i) + \chi(1-\phi_i)^2. \tag{3}$$

The activity a_i and therefore the vapor pressure p_i of the polymeric solution is a function of the volume fraction ϕ_i. In eq. (3) p_i^0 is the vapor pressure of the pure solvent and χ an interaction parameter. For low concentrations this relationship can be simplified to Henry's law

$$c_i = H_i^{cp}\, p_i. \tag{4}$$

Elimination of the experimentally hardly accessible quantities c_i^* / p_i^* in eq. (1/2) with eq. (4) leads to

$$kA = \frac{\dot{n}_i}{c_i - p_i/H_i^{cp}}, \quad \frac{kA}{V} = \frac{\dot{c}_i}{c_i - p_i/H_i^{cp}}, \quad \frac{1}{k} = \frac{1}{k_l} + \frac{H_i^{cp}RT}{k_g}, \tag{5}$$

with the total mass transfer resistance *1/k* as the reciprocal of the total mass transfer coefficient *k* composed of the liquid-side mass transfer resistance and the gas-side mass transfer resistance. This model assumes that there is no surface resistance which means that the gas-liquid equilibrium attain immediately at the interface. In case of $k_l << k_g$ eq. (5) simplifies to $k = k_l$, so that the mass transfer is liquid phase controlled and the gas-side gradient nearly vanishes ($p_i^* \approx p_i$). This is typically the case for polymer degassing since the diffusion coefficient *D* of solvents in polymers is very low.

The classical model to describe the mass transport is the two-film model [5]. Near the interface the mass transfer resistance is assumed as steady-state laminar diffusion layer of thickness δ and with a linear concentration profile, so the mass transfer coefficient *k* becomes

$$k = D/\delta. \tag{6}$$

Particularly for rotating devolatilizers there are models that are more realistic based on the penetration and surface renewal theory from Higbie [6] and Danckwerts [7]. In this case, mass transfer is modeled as an unsteady process for a local fluid element at the free surface *A*. During this exposure time *t*, mass transport will occur due to diffusion. Afterwards the fluid element is assumed to be completely mixed into the liquid bulk phase. Under consideration of Fick' laws and integration follows

$$kA = \frac{\dot{n}_i}{c_i - c_i^e(p_i)} = 2\sqrt{\frac{D}{\pi}}\left(\frac{A_{Film}}{\sqrt{t_{Film}}} + \frac{A_{Pool}}{\sqrt{t_{Pool}}}\right), \quad k = k_l. \tag{7}$$

The indices (*Film, Pool*) indicate the typical regions where degassing in rotating devices will appear. Values for *A* and *t* depend on geometry and filling degree of the devices. The exposure times are proportional to rotational speed.

EXPERIMENTAL

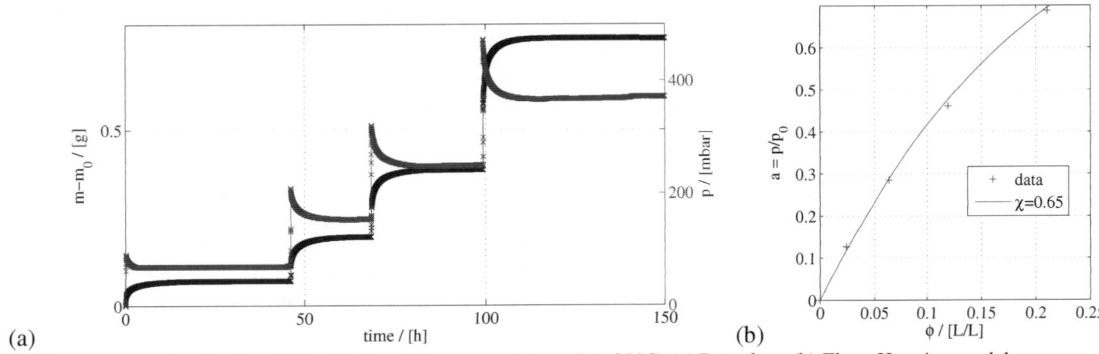

FIGURE 1. Gas-liquid sorption isotherm PDMS-CFC-113 at 30°C: (a) Raw data, (b) Flory-Huggins model

Technical polymer melts are usually processed at temperatures of about 200°C. In this study, all experiments were performed with a suitable inert model substance system that fulfill all essential features of technical systems at ambient temperature. High-viscous polydimethylsiloxane (PDMS) (η_0 = 100 Pas) as polymer and 1,1,2-trichloro-1,2,2-trifluoroethane (CFC-113) as volatile is used. The gas-liquid absorption equilibrium was measured with a magnetic suspension balance at RuboLab GmbH (Ruhr-Universität Bochum, Germany). Fig. 1 show a good agreement of measured data with the Flory-Huggins model (eq. (3)).

Degassing experiments were performed in a partial filled agitator vessel with a blade stirrer. Fig. 2 (a) shows the interior of the used degassing device with view from the agitator shaft during stirring. The apparatus generate a wiped melt film with the height H on the barrel wall and two rotating melt pools at the blade stirrer's flanks.[1] Before starting degassing experiments, the mixture of polymer and volatile was homogenized for at least 8 h.

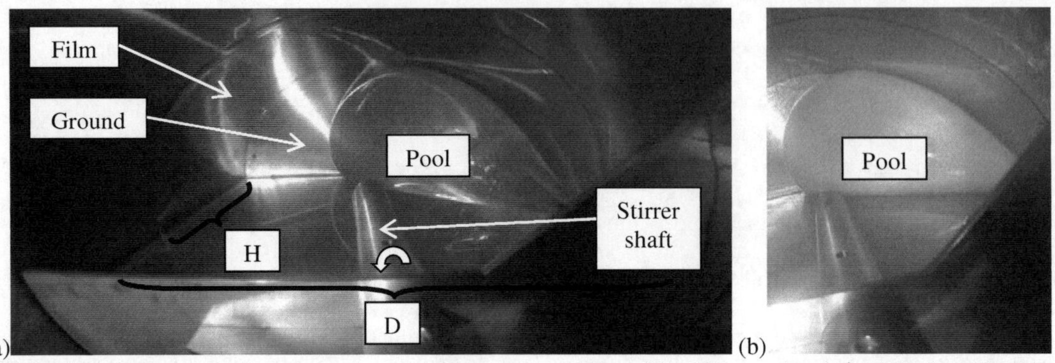

FIGURE 2. (a) Agitator vessel with a blade stirrer, D = 236 mm, H = 170 mm, V ≈ 2.2 L, n = 10 min^{-1}; (b) Pool with bubbles

Fig. 3 shows the process flow chart of the entire experimental plant. The driving force for degassing was provided by reducing the partial pressure in the gas phase with a stripping flow of nitrogen 5.0 from a gas bottle. Two nozzles with an inner diameter of 6 mm serve as gas inlets into the degassing device. Cold traps cooled with liquid nitrogen were used to recover and weigh the stripped volatiles in the loaded gas stream. A degassing test run starts by opening the gas supply. The mass flow rate was regulated manually to a defined value. At several points of time, 10 samples were taken from liquid phase to measure concentration. At the same time, the gas flow was switched with the 3-way valve to the second parallel pipe of 2-staged cold traps to weigh the first pair of cold traps.

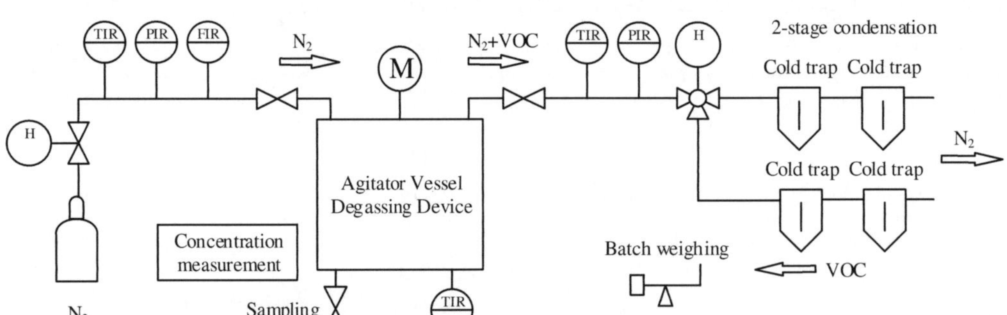

FIGURE 3. Process flow chart of the experimental plant

Temperatures were measured in gas and liquid phase with Pt100 elements. Pressure transmitters of the *CTE8000 series* (*First Sensor AG*) and thermal mass flow meters *F-111AC* (*Bronkhorst GmbH*) provided data of the gas phase to calculate the partial pressure of volatiles. Concentration measurement of volatiles in the melt was performed by thermogravimetric analysis. Samples with a defined amount of 0.5-1.0 g melt were taken into snap ring vials and thermally treated in an oven for a defined time (24 h). Mass fraction subsequently was determined by reweighing.

The mass transfer kA according to eq. (7) is used to assess the devolatilization during each time interval. Necessary conversions of concentration units (mass fraction w_i, molar concentration c_i, volume concentration ϕ_i) for the evaluation were carried out with molar mass M_i and densities $\rho_{i,}\ \rho_{PDMS}$ of the volatile and the polymer:

$$c_i = \bar{\rho}\, w_i / M_i\,, \qquad \bar{\rho} = \frac{\rho_i \rho_{PMDS}}{w_i \rho_{PMDS} + (1 - w_i)\rho_i}\,, \qquad w_i = \frac{\phi_i \rho_i}{\phi_i \rho_i + (1 - \phi_i)\rho_{PDMS}} \tag{8}$$

---

[1] Both degassing regions can be found equally in many devolatilizers like for instance in (double-flighted) screw extruders

Solving of the ODE (cf. eq. (7) / eq. (2)) with constant values for kA [2] and V and the initial condition $c_i(t = 0) = c_{i,0}$ leads to an exponential progression of the bulk concentration c_i:

$$c_i(t) = c_i^e(t) + \left(c_{i,0} - c_i^e(t)\right)\exp\left(-\frac{kA}{V}t\right), \quad c_i^e(t \to \infty) = 0 \tag{9}$$

To calculate kA by eq. (7) it is necessary to know $\dot{n}_i$, c_i and c_i^e. Therefore, the measured thermogravimetric data were fitted to exponential functions and mean values were calculated for $\dot{n}_i$ and c_i by integration of the fits for each time interval. Evaluation is performed is this way to compensate the variation of single concentrations measurements. The free surface concentration c_i^e is calculated with the Flory-Huggins relation (eq. (3)). The required partial pressure p_i of the volatile in the degassing vessel is obtained the with Dalton's law, the ideal gas law, measured data in the gas phase and the calculated value for $\dot{n}_i$.

RESULTS AND DISUSSION

Fig. 4 shows the measured nitrogen stripping mass flow and the pressure in the devolatilizer for three time intervals during a typical degassing test run. Pressure increases in order to reach the desired mass flow of nitrogen. The reason for this is plugging inside the cold traps in consequence of crystallization of volatiles. Higher concentrations of volatiles in the gas phase like in the first time interval leads to faster plugging than lower concentrations in later time intervals. Nevertheless, all test series with 5 L_N/min (under normal conditions) gas flow run to a stable state of about 1800 mbar. Detailed inspections revealed that crystallization occurs at locations of high gas flow acceleration and deflection. This happens typically at the inlet of the inner tube inside the cold finger.

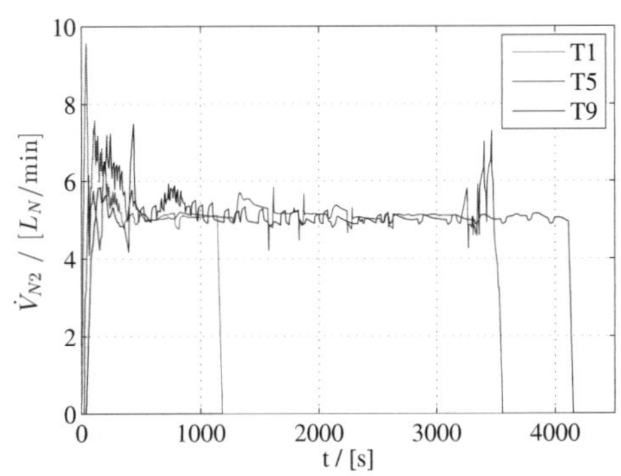

(a)

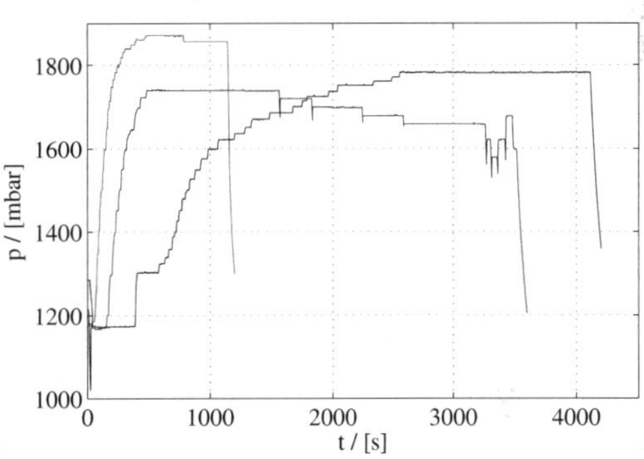

(b)

FIGURE 4. Measured data in gas phase during three time intervals of an exemplary degassing test run (M3); T1...first time interval (0-20 min), T5...fifth interval (160-220 min), T9...last interval (410-480 min): (a) Mass flow of nitrogen, (b) Pressure in the degassing device

At the end of each time interval pressure have to be reduced to take samples for concentration measurement and to regulate the mass flow to the equal value in the clean pair of cold traps in the next time interval. This pressure reduction causes a bubble formation in the polymer melt (cf. fig. 2 (b)), which will collapse at the next pressure increase (cf. fig. 2 (a)). These effects are also observed in pure polydimethylsiloxane without any volatile components and can be explained by solved nitrogen in the polymer. A pressure drop leads to supersaturation of the polymer with nitrogen and only in combination with shear strain, bubble formation occurs. Supersaturation conditions without shear strain are not sufficient to boil a viscous solution [8]. Bubble formation influences the process of film degassing in several ways. On the one hand, degassing will occur inside each bubble although it is uncertain whether this leads to an effective removal of volatiles from the system or the degassed volatiles will be mixed back into the fluid at the next pressure increase. On the other hand, the fluid flow will be affected, which will influence back mixing of the degassed fluid elements at the free surface after exposure time.

Fig. 5 shows a comparison of different evaluated degassing test series with varied parameters. The blue (M1,M2) and green (M3,M4) curves demonstrate the reproducibility of the produced results. The former were performed at continuous degassing. The later were performed with 1 h additional mixing without gas flow after each time interval. During this additional time the polymer solution will still degas a bit in order to reach equilibrium state

[2] This assumption is applicable due to the fact that the free surfaces A_{Film}, A_{Pool} and exposure times t_{Film}, t_{Pool} will not change significantly during degassing and the diffusion coefficient D is expected to be constant over the concentration range

between liquid phase and gas phase concentration. The black curve (M5) represents an experiment with 1 L_N/min gas flow instead of 5 L_N/min. The mass transfer coefficient k is considerable lower than in case of higher gas flow. These results clearly show that the gas-side mass transfer resistance (cf. eq. (5)) cannot be neglected in this experimental setup, which stands in contrast to outcomes of other works [9, 10].

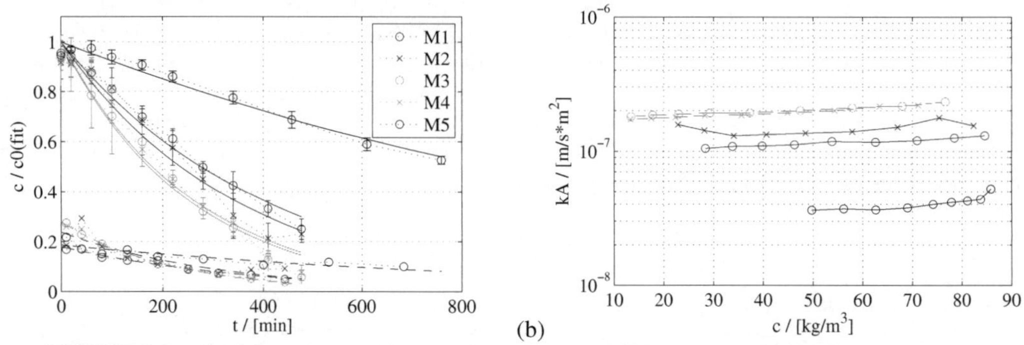

(a) (b)

FIGURE 5. Result of degassing experiments (cf. eq. (7)): $m_0 = 2233$ g, $c_0 = 75...90$ kg/m³, n = 10 rpm
M1-M4: $\dot{V}_{N2} = 5$ L_N/min, M5: $\dot{V}_{N2} = 1$ L_N/min, M3-M4: 1 h additional mixing without degassing after each time interval
(a) normalized concentration progression of c_i and c_i^e (cf. eq. (7)): measured data (marker), exp-fit (solid line)
(b) mean mass transfer kA calculated with eq. (5/7) plotted over mean mass concentration for each time interval

The pool size has been observed during video analysis (cf. fig. 2) of the process. The two pools generated by the blade stirrer were not exactly symmetric for the whole degassing time. The system seems to be very sensitive to the gap size (of about 2 mm) between blade stirrer and barrel wall. However, it is assumed that this behavior does not influence degassing significantly.

CONLUSION AND OUTLOOK

The results presented here show that for an experimental study of polymer degassing a lot of influencing factors have to be considered. That concerns the experimental setup, the test procedure as well as the evaluation of the experiments. A main outcome of this study is that the gas-side mass transfer resistance must be considered in the used experimental setup and could also have an influence on degassing in real technical devices. Video analysis proved to be an important instrument to reveal an undesired bubble formation and an unbalanced pool distribution. Bubble formation occurs as a result of dissolved nitrogen in the polymer and pressure jumps during the experimental procedure because of plugging in the cold traps. In conclusion, all results presented here are plausible and correlate with qualitative expectations.

In future further investigations are planned. In a first step, the experimental procedure will be optimized. Recent studies indicate that cold traps filled with Raschig rings as packing material will drastically reduce the pressure drop of the cold traps and hence will allow to realize degassing experiments with a much higher mass flow of nitrogen. It would be preferable to reach a level at which the gas-side mass transfer resistance can be neglected. Moreover, future investigation could be performed at a constant pressure level to avoid artificial bubble formation. In a second step, parameter studies will be performed to investigate the limits and restrictions of quantitative mass transfer models for film degassing like common models basing on penetration and surface renewal theory.

REFERENCES

1. R. J. Albalak, *Polymer Devolatilization*, New York: Marcel Dekker, 1996.
2. J. A. Biesenberger, *Devolatilization of Polymers*, Munich: Hanser Publishers, 1983.
3. VDI-Gesellschaft Kunststofftechnik, *Aufbereitungstechnik 2006*, Düsseldorf: VDI-Verl., 2006.
4. W. Vauck, H. Müller, *Grundoperationen chemischer Verfahrenstechnik*, Stuttgart: Dt. Verl. für Grundstoffindustrie, 2000.
5. W. K. Lewis, W. G. Whitman, *Ind. Eng. Chem.* **16**, 1215-1220 (1924).
6. R. Higbie, *AIChE J.* **31**, 365-389 (1935).
7. P. V. Danckwerts, *Ind. Eng. Chem.* **43**, 1460-1467 (1951).
8. M. Favelukis, Z. Tadmor, R. Semiat, *AIChE J.* **45**, 691-695 (1999).
9. G. P. Collins, C. D. Denson, G. Astarita, *AIChE J.* **31**, 1288-1296 (1985).
10. I. Gestring, "Entgasen von Polymeren", Ph.D. Thesis, Universität Hannover, 2002.

Investigation of the influence of material and pellet shape on the dissipation in the solids conveying zone of single-screw extruders based on the Discrete Element Method (DEM)

Jan Trippe [a] and Volker Schöppner [a]

[a]*University of Paderborn, Warburger Straße 100, 33098 Paderborn, Germany*

jan.trippe@ktp.upb.de

Abstract. In polymer processing, there are existing currently established models for the conveying and melting behavior of polymer materials. These models record the majority of relevant industry applications well. In the field of conveying solids the models especially relate to Darnell and Mol (1956), who postulate a plug flow in the solid area. The following melting process implies a heat input exclusively from heat conduction at the barrel and from dissipation energy which results from the contact between pellets and screw / barrel. The frictional heat, resulting from relative motion in the solid bed, is being neglected in the models. This is why an overestimation of melting efficiency by the actual models is to be expected. Furthermore, the models do not take the influence of pellet shape and size into consideration, which is why there are expected higher internal frictions from sharp-edged pellets than from round ones with a smooth surface. In terms of cylindrical pellet, alignment and unroll effects play an important role. In order to mathematically describe and analyze complex motions in a solid bed correctly, numerical simulations based on the Discrete Element Method (DEM) have been carried out. In order to do so, a 60 mm solid conveying test stand has been built up in the simulation. The test stand can be fed with pellets of various shapes and sizes. Pellet and material parameters, which have been defined in preliminary investigations and were implemented in the simulation, play a decisive role. Because of the variation of dimensionless geometry and process index, a modelling approach arises, with which now the description of the influence of various pellet sizes and shapes on the dissipation in the solid conveying area is possible for common extruder sizes.

Keywords: solid conveying, discrete element method, dissipation
PACS: 89.20.Bb, 89.20.Kk, 89.90.+n

INTRODUCTION

The task of the solids conveying zone is to provide high output without pulsation and to bring in frictional power via contact between pellets and screw or barrel to initiate the melting process. For the mathematical description of the solid conveying a plug flow in the solid area is assumed where no relative movement occur within the granules. Here, on the one hand the transfer of forces takes place by friction from the screw flight to the first layer of granules. The transfer of forces in the solid block on the other hand is assumed to proceed lossless by friction without relative movements of individual layers of granules. Numerous investigations demonstrate, however, that this simplification does not correspond to reality even at low screw speeds. Furthermore, it is assumed that a block flow explicitly neglects internal dissipation resulting from kinematic motion and internal friction in the granule bed. Here again, material and form-dependent parameters have a significant influence on dissipation. For a more realistic consideration of the flow process in the solids conveying zone, more complex movements have to be considered. By taking into account the movements of the granules many new equations of motion are formed in the modeling of solids movement, whose exact direction in space is unknown. To investigate such effects, the Discrete Element Method (DEM) is currently gaining prominence as a useful tool. This paper will introduce the fundamentals, procedure and conditions of the development of a mathematical model in which the description of the influence of various pellet sizes and shapes on the dissipation in the solid conveying zone is possible for common extruder sizes.

DISCRETE ELEMENT METHOD

The Discrete Element Method was developed by Cundall and Strack and has been applied successfully for many years by Moysey and Thompson in the simulation of extrusion processes. In [1-5] their procedure for modeling the further development of the simulation models is described in detail. In contrast to the Finite Elements Method (FEM), the geometries to be calculated are not linked together by the generated mesh, but are approximated through simple, usually spherical discrete particles, for which positions and velocities are calculated in every iteration step. Because of the simple particle geometry, that allows a diversity of particles to be calculated with moderate calculation times, the process is suitable not only for simulating molecule movements but also for calculating granular systems in geotechnology, process engineering and plastics technology. Calculation of the particle positions and velocities is performed on the basis of a virtual overlapping of two or more particles (Fig. 1 (left)), in which frictional, impulse and repulsion forces are determined. Several contact models exist, including various parameters such as stiffness factors, damping factors and friction coefficients (Fig. 1 (right)). In this study, the DEM software "EDEM" (DEM Solutions Ltd.) is used for the simulations. A Hertz-Mindlin model is used as the contact model because it is available by default in the software and has also shown itself to be effective in other papers [6-8] Next to the selection of suitable material models and the correct input of the model parameters, the simulation quality of the real pellets can, especially with clearly non-spherical pellets, have a major influence on the simulation result because different bulk densities occur depending on the shape of the pallet. Moreover, describing the shape-dependent dissipation strongly depends on the accuracy of the visualized pellets in the simulation. Therefore, one aim is to approximate the real pellets in the best possible way by combining several spheres in the simulation.

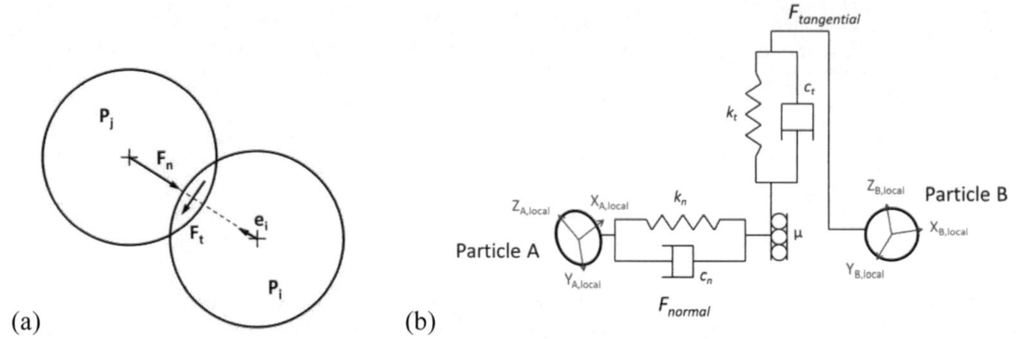

(a) (b)

FIGURE 1. Calculation Basis of Discrete Element Method Simulations
(a) Visualization of two Colliding Particles and the Resulting Contact Forces [9], (b) Contact Force Model [10].

FORM AND MATERIAL PARAMETERS

In order to describe the dissipation in the solids conveying area with a mathematical model, material and granular characteristics have to be determined that are relevant for dissipation. Furthermore, dissipative effects have to be considered for three different areas: The outer dissipation which is caused by contact between granules and the screw or barrel and the inner dissipation which is caused by contacts within the granular bed. For each case the friction and coefficient of restitution between the partners have to be known. Therefore the following section describes the method for determining the material parameters and the definition of granular parameters in which a granule of any shape can be described. The investigations have been conducted using high density polyethylene (HDPE) and polyamide 6 (PA), because both materials vary in form, coefficient of friction and coefficient of restitution. Moreover, both materials can be manufactured on systems with cold cut or hot cut, which means that based on the raw material, cylindrical and spherical forms can be produced. This is an important feature for the studies that are presented in the later chapters. The starting point for the parameterization is the Coefficient of Restitution (COR), which has been determined for specific test cases; to do so, balls of both materials have been manufactured and then have been dropped from a specific height h_0. The resulting bounce height h_1 has been then measured for both plastic-plastic and plastic-steel combinations. The COR is then calculated as

$$e = \sqrt{\frac{h_1}{h_0}} \tag{1}$$

Additionally, friction coefficients have been measured according to DIN ISO 7148 (pin-on-disc method). A surface pressure of 0.7 MPa has been used to create a set-up which is close to the aspired simulations that are assumed with a counter-pressureless conveying behavior. In addition a gliding speed of 0.1, 0.3 and 0.5 m/s has been varied. By calculating with the outer diameter, the gliding speeds therefore correspond to the typical circumferential speed ranges of common extruder sizes. Table a) provides an overview of the material parameters. It can be seen that the results are in good agreement with known values from the literature. The density, the shear modulus and the Poisson´s ratio given here have been taken from the appropriate supplier's data sheet and relevant literature.

TABLE a). Relevant Material Parameter of the Examined Polymers

Material Parameter	HDPE	Literature	PA	Literature
Shear Modulus G [MPa]		0.85		4.1
Internal Friction Coefficient[-]	0.531	0.45	0.696	0.525
External Friction Coefficient[-]	0.247	0.225	0.306	0.415
Internal COR [-]	0.544	0.4	0.694	0.6
External COR [-]	0.628	0.5	0.907	0.8
Density [kg/m³]		956		1300
Poisson´s Ratio [-]		0.425		0.39

For the determination of particle form characteristics different forms were manufactured from the raw materials in a hot cut and cold cut process, measured by a special scanner software and then assigned to geometry characterizing parameters. To analyze the most possible differences in shape and its influence on dissipation, both material types were manufactured in an almost perfectly round shape and a very long ideal cylinder (Fig. 2).

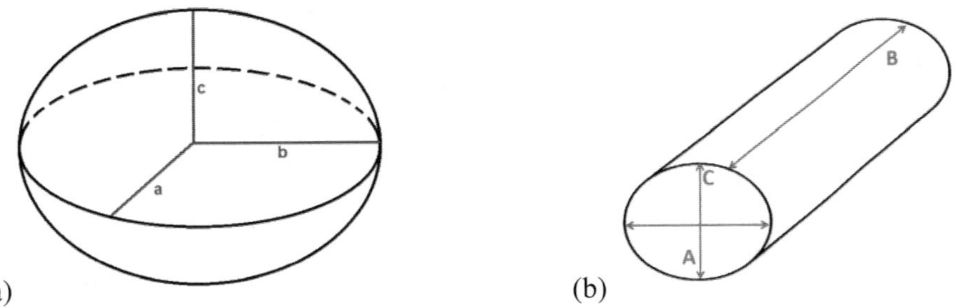

FIGURE 2. Evaluation of particle shapes of (a) an ellipsoidal granule, (b) a cylindrical granule.

According to the recommendation of [11], four geometry parameters have been chosen for the subsequent characterization of the materials; the volume, the sphericity, which describes the ratio of the surface of a sphere with the surface of the comparative form, the L/D ratio and the diameter deviation were determined for 100 particles per material and form. Based on these parametrization, an implementation in the simulation software is possible. Here, the presentation of the cylindrical particles is carried out by overlapping several spheres. The following table b) shows that with this method a good agreement to the real granules is possible.

TABLE b). Imaging Accuracy of the Particles Created in the Simulation

Form Parameter	**HDPE real**		**HDPE sim**		**PA real**		**PA sim**	
Form	**Cyl.**	**Ell.**	**Cyl.**	**Ell.**	**Cyl.**	**Ell.**	**Cyl.**	**Ell.**
	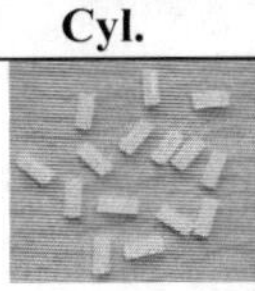		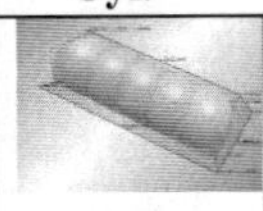	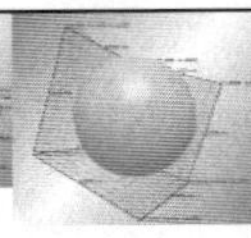	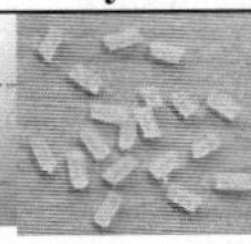	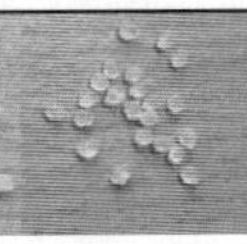	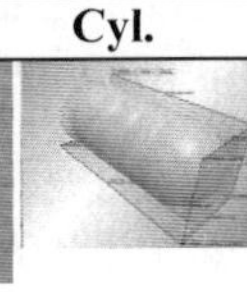	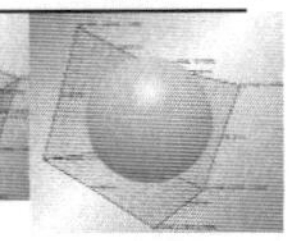
Volume [mm³]	25.48	39.37	23.15	39.36	27.76	13.33	24.43	13.33
Sphericity [-]	0.78	0.99	-	1	0.82	0.99	-	1
L/D Ratio [-]	2.7	1.1	3	1	2.2	1.2	2.4	1
Diameter deviation [-]	0.73	1	1	1	0.85	1	1	1

SIMULATIONS ON DISSIPATION

To model the dissipation in the solids conveying area a dimensional analysis is carried out to determine the parameters affecting the dissipation and to reduce the experimental scale of the time-consuming simulations. Assuming a pressureless solid conveying behavior and that the influence of varying the shear modulus, density and Poisson´s ratio has a rather small effect on dissipation result the dimensionless factors shown in table c).

TABLE c). Varied Ranges of Simulation Parameters

Parameter	Values	Values e.g. for an 30 mm extruder	Dimensions	Values e.g. for an 90 mm extruder	Dimensions
Screw Speed Π_N [-]	8,45E-05…0,00036	150…640	[rpm]	50…213	[rpm]
Channel Depth h/D [-]	0.15…0.2	4.5…6	[mm]	13.5…18	[mm]
Particle Size D_P/D [-]	0.033…0.133	1…4	[mm]	3…12	[mm]
Particle Form L_P/D_P [-]	1…2.8				
Internal COR [-]	0.5…0.86				
External COR [-]	0.6…0.96				
External Friction [-]	0.1…0.4				
Internal Friction [-]	0.3…0.7				

As shown in table c the model will take into account particles varying from spheres with a diameter of D_P to cylinders with a length up to 2.8 L_P/D_P in which D_P is the mean diameter of the cylindrical surface. Effects on dissipation that result from the length or the size of the particles are thus considered sufficiently. Here, especially the differences of rolling effects as well as the orientation of long cylinders in comparison to short ones or sphere pellets matter. The ranges of the internal and external COR and the coefficient of friction include the current values of known polymers. The dimensionless screw speed Π_N made up of the screw speed, the screw diameter, the shear modulus and the density represents material, geometry and process specific parameters in one dimensionless value. Other parameters have been varied in order to cover common geometry sizes as part of the regression model. The model is based on a central composite design in which the simulation effort is further reduced and yet a consideration of square effects between the parameters is obtained. The dissipation as a target parameter is examined as the sum of the internal and external dissipation. The dissipation basically is calculated by the change in enthalpy in which heat is generated by friction.

$$c_P \cdot \Delta T - vdp = dh - vdp = dq + dW_{Diss} \tag{2}$$

In equation 2 c_P is the specific heat capacity of the polymer, T is the temperature, v is the specific volume, p is the acting pressure and W_{Diss} is the dissipation power. If the heat flow q is neglected and under the consideration that the dissipation power is determined by heat generation in the movement processes the dissipation power can be equated with the friction power which is given in equation 3.

$$W_{Diss} = F_R \cdot \Delta v = F_N \cdot \mu \cdot \Delta v \tag{3}$$

In equation 3 Δv is the relative velocity between two contact partners. This means that for the evaluation of a mathematical model, every contact of every particle in the simulation is analyzed within defined time steps in which tangential contact forces and relative viscosities are summarized and then add up for a screw section or the whole system. Differentiating the contacts to the barrel, screw or particles among themselves leads then to a consideration of the inner and outer dissipation. In addition, the shape-dependent flow rates which result from the different bulk densities have to be validated. Since the simulations are not finished yet, reference is made at this point to prospective elaborations.

VALIDATION STUDIES

To validate the dependency of material and granular characteristics on the friction initial experimental investigations have been carried out. For this purpose a screw demonstrator has been developed in which frictional forces and dissipated temperatures can be detected (Fig. 3 (left)).

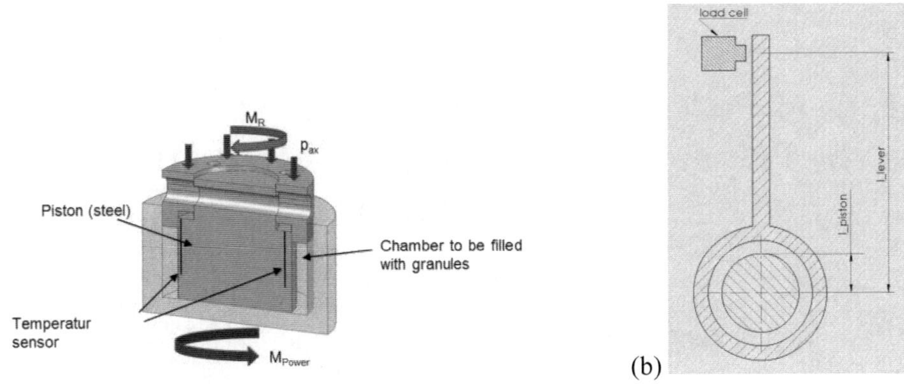

(a) (b)

FIGURE 3. Friction test stand used as screw demonstrator
(a) Cross-section of the test Chamber, (b) Acting lever arm.

The outer cylinder is powered by an electric motor. Here the normal force is generated via the pressure p_{ax} which acts on the piston top surface. Via the granules in the chamber the friction moment M_R is transferred to the piston which can be measured by a load cell by means of a lever arm (transferred tangential force) (Fig. 3 (right)). In addition to that the temperature is measured by a distance from the surface of the piston of 0.5 mm. The data of the materials used for the test are summarized in table a) and b). In figure 4, it can be seen that for the present geometry, the pellet size as a characteristic value has more influence in the transmitted torque then the various L_P/D_P ratios. The corresponding temperatures are correlated with the measured friction forces, which supports the assumption of a shape-size-dependent dissipation. Further studies, however, are mandatory and planned.

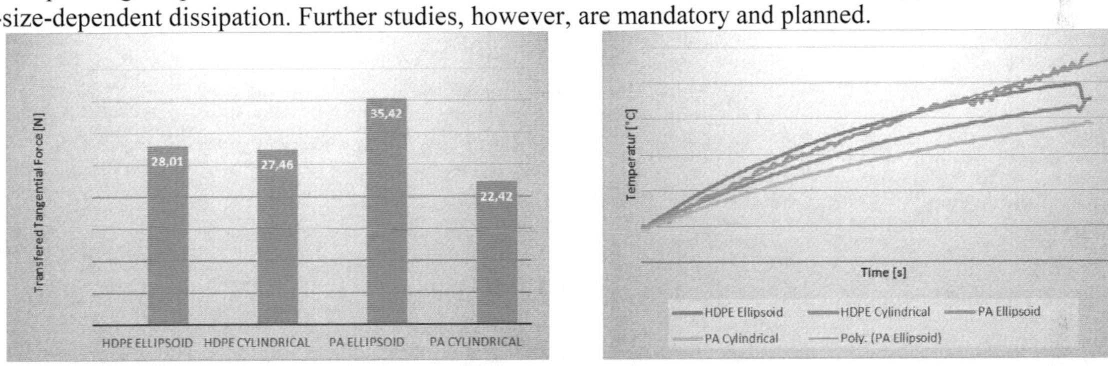

FIGURE 4. Transferred tangential force and measured temperature for HDPE and PA in different forms.

ACKNOWLEDGMENTS

We like to thank the Deutsche Forschungsgemeinschaft (DFG), which has made these studies possible by financing the SCHO 551/24-1 project.

REFERENCES

1. P. A. Moysey and M. R. Thompson, Polym. Eng. Sci 44, 2203-2215 (2004).
2. P. A. Moysey and M. R. Thompson, Powder Techn. 153, 95-107 (2005).
3. P. A. Moysey and M. R. Thompson, Chem. Eng. Sci. 62, 3699-3709 (2007).
4. P. A. Moysey and M. R. Thompson, Polym. Eng. Sci. 48, 62-73 (2008).
5. P. A. Moysey et al., Int. Polym. Proc. 23, 301-311 (2008).
6. J.-S. Lessmann and V. Schoeppner, PPS-30 Conference Proceedings (2014).
7. O. R. Walton and R. L. Braun, Acta Mechanica 63, 73-86 (1986).
8. J.-S. Leßmann et al., Int. Polym. Proc. 27, 469 (2012)
9. DEM-Solutions, Theory Reference Guide (2012).
10. M. Börner, Einführung in die Discrete Elemente Methode, NaWiTec, Institute for Process Engineering (2011).
11. H. Rahal, Alternative Methoden zur Feststoffförderung und Plastifizierung in der Extrusionstechnik, Dissertation, University of Duisburg (2008).

Analysis of Pellet Diameter and Inorganic Filler Influences on the Polymer Melting Behavior in Planetary Roller Extruders

Johannes Rudloff (presenting and reference author)[a], Marieluise Lang[a], Karsten Kretschmer[a], Peter Heidemeyer[a], Martin Bastian[a] and Michael Koch[b]

[a]*SKZ - German Plastics Center, Friedrich-Bergius-Ring 22, 97076 Würzburg/Germany*
[b]*KTI - Department of Plastics Technologies, TU Ilmenau, Gustav-Kirchhoff Straße 5, 98684 Ilmenau/Germany*

j.rudloff@skz.de

Abstract. The planetary roller extruder is well known for its superior mixing performance and its exact temperature control capabilities. Therefore it is widely used in the polymer industry for the compounding of shear- and temperature sensitive materials like PVC or Wood Plastic Composites. In contrast to the considerable amount of scientific work that deals with investigation, modeling or simulation of the process behavior of single and twin screw extruders, there are only few publications about the planetary roller extruder. Due to increased quality requirements and the trend to cost reduction by process optimization, this is increasingly becoming a problem for plant construction and processing companies. In our former work first analysis of the plastics melting process in planetary roller extruders were conducted and models describing these processes were published. However the influences of pellet diameter or inorganic filler were not considered. As a consequence the process models are only valid for a partial amount of the materials processed on planetary roller extruder. This paper presents experimental melting behavior investigations of powder, micro pellets and standard pellets. Furthermore the influence of filler amount is examined with standard pellets. At these experiments the melting initiation and the melting propagation is determined. Based on the experimental results, a theoretical consideration for melting in planetary roller extruders is proposed.

Keywords: Melting, Modeling, Simulation, Planetary roller extruder.
PACS: 44, 47, 89

INTRODUCTION

The increasing requirements for plastic products lead to an ongoing advancement and optimization of production processes. Process simulation is an effective tool for these procedures. Therefore a modeling of all relevant processes in processing machines is required. For extruders an entire view of processing can only be conducted if melting is considered in modeling and simulation. Despite the huge amount of research that exists for single and twin screw extruders, the understanding of melting is still incomplete because of the complexity of this topic. Especially for twin screw extruders, the melting behavior of polymers can be regarded as one of the last unresolved issues [1]. Regarding the plastic melting in planetary roller extruder (PRE), the state of knowledge is even more incomplete. This reduces the competitive position of the PRE and, furthermore, blocks the potential development of new compounds or processes, which require the outstanding temperature control capabilities of this extruder type. To address this topic, first models describing solid conveying, melting initiation and melting propagation were developed in our previous research [2], in which the first two sections of a typical planetary roller extruder used for compounding as shown schematically in FIGURE 1 were observed. In the first section, which is comparable to a feeding section of a single screw extruder, the plastic is fed gravimetrically to the extruder. In this section, the barrel is cooled. The extruder is partially filled and solid conveying occurs. The second section comprises a planetary roller module and a tailback ring. At first, the extruder is partially filled and solid conveying occurs as well. In the heated barrel the temperature of material in contact with metal parts starts to rise, which can lead to melting initiation. The tailback ring leads to a fully filled volume because of back flowing melt at the end of the section. The length of this fully filled volume depends on process conditions, such as throughput and rotation speed, the viscosity and the ring diameter. Because of the contact with back flowing melt, melting initiation takes place at the latest when pellets enter this fully filled volume. When the material reaches section 3, which is where filler is typically added, there should be a melting degree of 100 % for an optimized process.

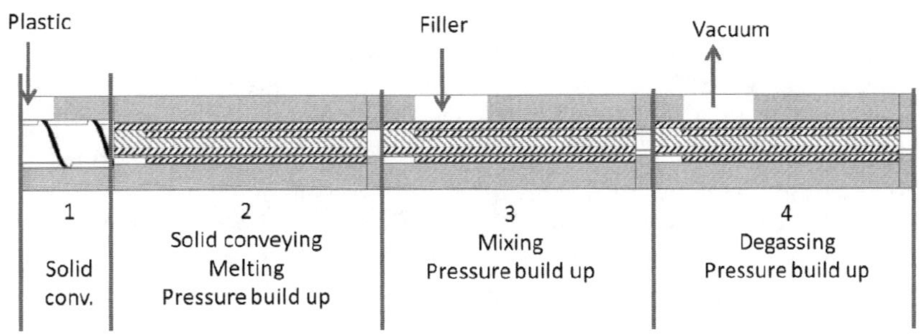

FIGURE 1. Schematic view of a PRE used for compounding

The developed mathematical models for the melting behavior showed a satisfactory agreement between experiment and simulation, but only unfilled pellets with a diameter of approximately 3 mm had been observed. Research conducted for other extruder types i.e. [1, 3, 4 and 5] shows relevant influences of pellet diameter and filler content on the melting behavior. So it can be assumed that the developed process models for planetary roller extruders are only valid for unfilled pellets with a diameter of 3 mm. For micro pellets, powders or compounds, an experimental investigation of the melting behavior and based on that adaption of the models is necessary.

EXPERIMENTAL

In order to analyze the melting behavior experimental investigations with an ENTEX planetary roller extruder TP-WE 70 were conducted. The throughput was set to 18 kg/h, the rotation speed to 75 1/min. The barrel temperature in the feeding section was 23 °C, the barrel temperature in the planetary roller section 200 °C and the screw temperature was 120 °C. PE powders and pellets with different particle size were used. Furthermore, pellets with different filler content were analyzed. Table 1 shows the particle diameter, specific enthalpy, specific heat capacity of solid particles at the feeding temperature (23 °C) and at the melting temperature (133 °C) of the processed materials.

TABLE (1). Materials

Term	Form	Particle diameter	Specific melting enthalpie	Specific heat capacity of solid particles at 23 °C	Specific heat capacity of solid particles at 133 °C
PE 250	Powder	0 – 250 µm	460 J/g	1,89 J/(g*K)	14,37 J/(g*K)
PE 375	Powder	250 – 500 µm	460 J/g	1,89 J/(g*K)	14,37 J/(g*K)
PE 500	Powder	500 – 1000 µm	460 J/g	1,89 J/(g*K)	14,37 J/(g*K)
PE 1500	Pellet	1500 µm	460 J/g	1,89 J/(g*K)	14,37 J/(g*K)
PE 3000	Pellet	3000 µm	460 J/g	1,89 J/(g*K)	14,37 J/(g*K)
PE-C 5	Compound	3000 µm	440 J/g	1,77 J/(g*K)	11,45 J/(g*K)
PE-C 20	Compound	3000 µm	400 J/g	1,75 J/(g*K)	11,81 J/(g*K)
PE-C 35	Compound	3000 µm	360 J/g	1,58 J/(g*K)	10,56 J/(g*K)
PE-C 40	Compound	3000 µm	320 J/g	1,47 J/(g*K)	9,58 J/(g*K)
PE-C 65	Compound	3000 µm	240 J/g	1,29 J/(g*K)	6,53 J/(g*K)

In order to determine the melting degree, a special barrel, which could be opened, was used and a color masterbatch was added for better differentiation of molten and unmolten material. After reaching a steady state operation condition, the machine was stopped abruptly and the barrel was opened so that samples could be taken. These samples were pressed to plates. At these plates the melting degree along the extruder axis could be measured.

RESULTS AND DISCUSSION

In FIGURE 2, the measured melting degrees along the barrel length are shown. The material is feed at 0 mm, the planetary roller section starts at 150 mm and the tailback ring starts at 550 mm.

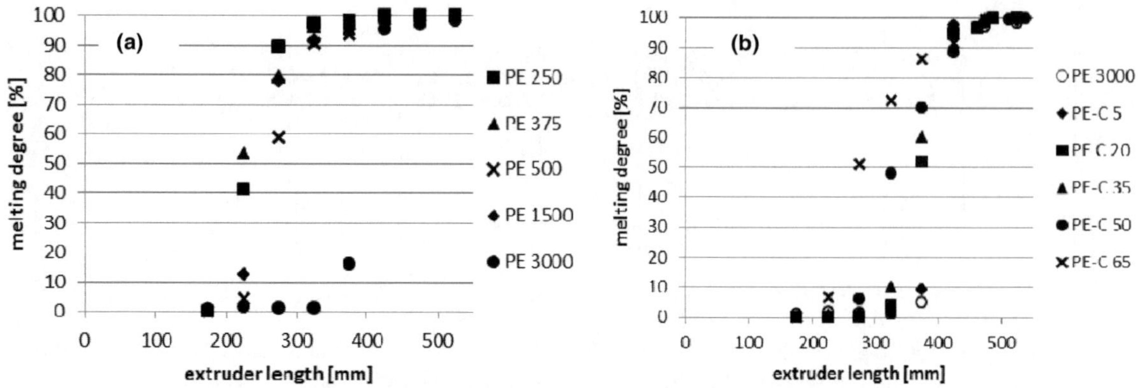

FIGURE 2. Melting degree vs. extruder length as function of particle size (a) and filler content (b)

In the filling section of the PRE, the process behavior of all analyzed material was constant. Only solid conveying and no melting occurred. This behavior is also known for the feeding section of twin screw extruders and is explained by negligible friction due to starvation and the barrel cooling of this section.

The melting initiation takes place in the planetary roller section. Here, material in contact with the heated barrel surface can start to melt. It can be seen that the melting initiation shifts to prior positions with a decreasing particle diameter and increasing filler content. This fits well into the theoretical model for the melting initiation because of the thermal conduction into material that has contact with the barrel surface. This model was developed by Jung for twin screw extruders [5] and was transferred to the PRE in our previous work [2]. In this model, melting can start when the temperature of material in contact with the barrel reaches the softening temperature. The temperature development of the material with barrel contact is described by equation (1)

$$\overline{T}_S(l) = a_B \cdot (T_B - T_0) \cdot \left(1 - e^{-\frac{2 \cdot a_B \cdot l}{\rho_S \cdot c_{p,S} \cdot \overline{v}_S \cdot r_0}}\right) + T_0 \tag{1}$$

where $\overline{T}_S$ is the solid particle temperature, T_B is the selected temperature at the extruder control unit, $\overline{v}_S$ is the average solids velocity, a_B is the heat transfer coefficient between a pellet and the barrel, ρ_S is the density and $c_{p,S}$ is the specific heat capacity of the solid particles. A decrease of the particle diameter r_0 can lead to a faster temperature rise and consequently an earlier melting initiation. Furthermore, a higher filler content leads to a lower $c_{p,S}$ and thus can cause an earlier melting initiation as well.

The melting propagation is influenced by the particle diameter and filler content as well, but the effects are less clear than those for the melting initiation. At first, it has to be considered that at an axial position of approximately 400 mm the partially filled section ends and the fully filled section starts. This leads to a significant increase in the local residence time and thus to an increase in the melting rate over the extruder length. Because of the late melting initiation, the melting of the unfilled 3 mm pellets mainly takes place in the fully filled section. Because of the high residence time, a fast melting rate over the extruder length is found. The melting of the micro pellets, powders and compounds, mainly takes place in the partially filled section. So especially for compounds the melting rate over length is lower because of the lower local residence time. Nevertheless, the melting rate increases with a decrease of particle diameter and an increase of filler content. These effects can be explained by the melting theory introduced in [2] as well, in which the so called disperse melting model published by Melish [3] was transferred to the PRE. In this model, isolated solid particles in surrounding melt are observed. The solid particles are considered as spheres and ongoing melting leads to a reduction of the radius of the spheres. The sphere radius is described by equation 2

$$r_{i+1} = \left\{ r_i^2 - \frac{2 \cdot \lambda_M}{\rho_S \cdot \overline{v}_z \cdot c_{p,M}} \cdot \ln\left[1 + \frac{c_{p,M} \cdot (T_M - T_{Fl})}{\Delta h}\right] \cdot \Delta l \right\}^{\frac{1}{2}} \tag{2}$$

where T_M is the melt temperature, t is the time, λ_M is the thermal conductivity of the melt, ρ_M is the melt density, $c_{p,M}$ is the specific heat capacity of the melt, r_i is the particle radius at a given position, T_{Fl} is the solid melting temperature, $\overline{v}_z$ the average melt velocity and Δh the melting enthalpy. The initial particle radius r_i is a major parameter in equation 2. Thus a decrease in the particle diameter leads directly to an increased melting propagation. The variation of the melting enthalpy h, affected by a variation of the filler content, is covered by equation 2 as well. For powders, there is a phenomenon that is not covered by equation 2. The melting rate decreases significantly at melting degrees above 80 %. The reason for that are powder agglomerates as shown in Figure 3, which can be found in the samples taken out of the barrel. These clusters are formed when not molten powder agglomerates are embedded by melt. Because of the low shear forces, especially in the volume between two planetary spindles, the powder agglomerates are not destroyed during processing. The diameters of the agglomerates found in the samples are between 1 and 5 mm. So they are much larger than isolated powder particles and this leads to a significantly slower melting compared to isolated powder particles.

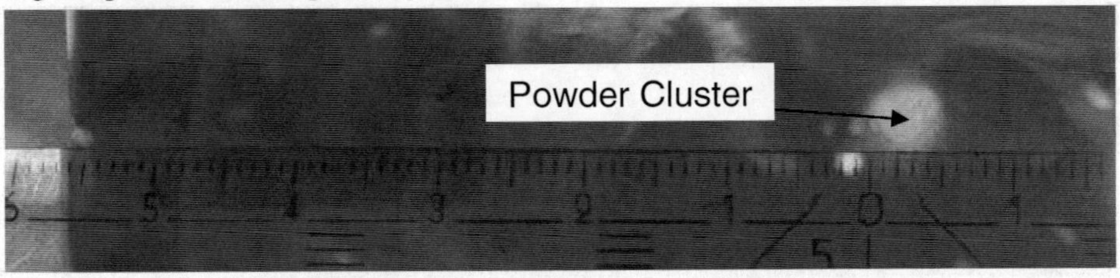

FIGURE 3. Powder agglomerates embedded by melt

Apart from particle radius, heat capacity and melting enthalpy further parameters in equation 1 and 2 are variable as well. So to calculate melting degrees, it was necessary to integrate the functions described above into the planetary roller extruder simulation software SimPla, which was developed by the SKZ [6]. In SimPla an interaction with equations describing the material data, material conveying and melt temperature is possible. Furthermore, the PRE is divided into small axial segments with a length of $l < 1$ mm, so that geometry, material and process parameters can be regarded as constant for the segment. Figure 4 a shows a comparison between measured and calculated melting initiation length. The melting initiation length is calculated with a good agreement to the experimental values for powders and pellets with a particle diameter of 3000 µm. For pellets with a particle diameter of 1500 and for compounds, the calculated melting initiation lengths are overestimated. The reasons for this may be that solid velocities are too high or heat transfer coefficients are too small in the composite model. Furthermore, the neglecting of plastic energy dissipation in the calculation can affect overestimated melting length. In contrast to that, the calculated melting lengths have a good agreement to the measured melting length for pellets and compounds as shown in Figure 4 b. For powders to short melting length are calculated. The error increases with decreasing powder diameter. The reason for that are the powders agglomerates that are not covered by the melting theory. In general equation 2 could be applied for the melting of the powder agglomerates as well, when the agglomerate radius instead of the powder radius is considered. Nevertheless a calculation of an average melting degree is not possible because the amount and diameter of the agglomerates is unknown. To address this problem in a first empiric approach, the initial particle radius used for calculation can be limited to a minimum of 750 µm. This should lead to reasonable melting length for powders.

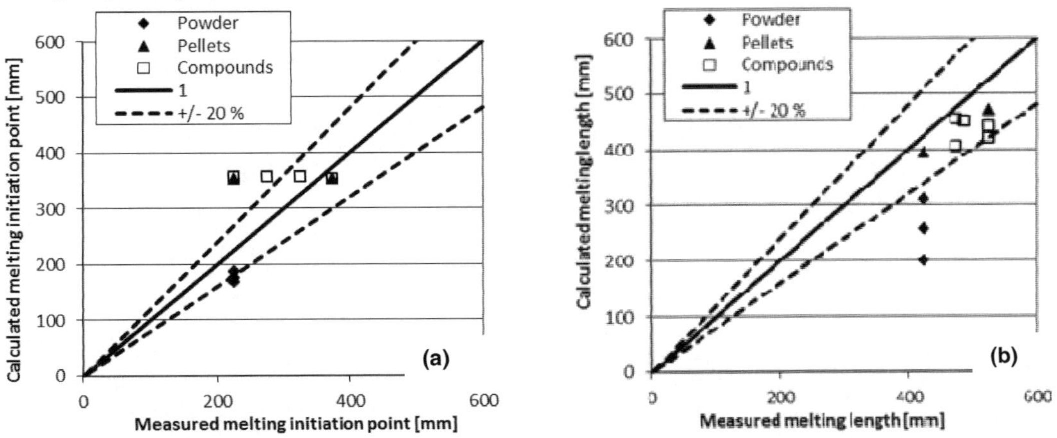

FIGURE 4. Comparison between measured and calculated melting initiation points (a) and melting length (b)

CONCLUSION

The melting behavior of PRE fed with powders, pellets and compounds was analyzed in experimental research work. Based on the experiments the processes were described in theoretical models. The developed models allow a calculation of the melting process in PRE. The comparison between simulated results and experimental data shows a decent agreement for pellets and compounds, considering the simplifications made and the number of assumptions that were used to generate the model. For powders further work is necessary to describe the melting of powder agglomerates that are formed during the melting process. For twin screw extruders, where this problem is known as well, a compact melting model leads to god results for pellets with a diameter under 1500 μm. Alternatively the amount and average diameter of the clusters could be modeled by a coupled mass and force balance. Furthermore, the influence of process parameters like throughput, barrel temperature and rotation speed on the melting behavior of powders and compounds have to be regarded. This will be addressed in our ongoing research about the process behavior of the PRE.

ACKNOWLEDGMENTS

The results presented here were obtained in the research project "BA 1841/7-2", which is supported by the German Research Foundation (DFG). Special thanks go to the DFG. Additional acknowledgements go to ENTEX, BASF SE, SABIC Europe and Lifocolor Farben GmbH & Co. KG for supplying the extruder and the polymer materials.

REFERENCES

1. Z. Tadmor, C.G. Gogos: Principles of Polymer Processing, Hoboken: John Wiley & Sons, 2nd Edition, 2006.
2. J. Rudloff, M. Lang, K. Kretschmer, P. Heidemeyer, M. Bastian, M. Koch "A mathematical model describing the solid conveying and melting behavior of planetary roller extruders", AIP Conference Proceedings 1593, American Institute of Physics, Melville, NY, 2014, pp. 592 – 595.
3. U. Melisch, "Grundlagen zur Simulation des Förder- und Plastifizierprozesses dichtkämmender Gleichdrall-Doppelschneckenextruder", Ph.D. Thesis, Universität Paderborn, 2000.
4. J. Flecke, "Grundlagen zur Simulation des Prozeßverhaltens eines Gleichdrall-Doppelschneckenextruders bei der Aufbereitung mineralisch gefüllter Polymere", Ph.D. Thesis, Universität Paderborn, 2000.
5. H. Jung, "Investigation of melting in modular intermeshing co-rotating twin screw extruders", Ph.D. Thesis, University of Akron, 2004.
6. J., Rudloff; K., Kretschmer; P. Heidemeyer; M., Bastian; M., Koch, "A physical – mathematical model describing the process behaviour of planetary roller extruders", PPS 27 Proceedings, Marrakech, 2011, CD.

AUTHOR INDEX

AUTHOR INDEX

AUTHOR INDEX

AUTHOR INDEX

AUTHOR INDEX

AUTHOR INDEX

AUTHOR INDEX

AUTHOR INDEX

Polymer Processing Society
810 E. 10th St.
Lawrence, KS 66044
USA

ISBN 978-1-5108-4339-4

9 781510 843394

CURRAN ASSOCIATES INC.
proceedings.com

32nd Polymer Processing Society Annual Meeting 2016 (PPS-32)

Lyon, France
25 - 29 July 2016

Volume 2 of 2

32nd Polymer Processing Society Annual Meeting 2016 (PPS-32)

Lyon, France
25 - 29 July 2016

Volume 2 of 2

ISBN: 978-1-5108-4339-4

Printed from e-media with permission by:

Curran Associates, Inc.
57 Morehouse Lane
Red Hook, NY 12571

CURRAN ASSOCIATES INC. proceedings.com

Some format issues inherent in the e-media version may also appear in this print version.

Printed by Curran Associates, Inc. (2017)

Polymer Processing Society
810 E. 10th St.
Lawrence, KS 66044
USA

Phone: +1-800-627-0326
+1-785-865-9403
Fax: +1-785-843-6153

pps@allenpress.com

Additional copies of this publication are available from:

Curran Associates, Inc.
57 Morehouse Lane
Red Hook, NY 12571 USA
Phone: 845-758-0400
Fax: 845-758-2633
Email: curran@proceedings.com
Web: www.proceedings.com

TABLE OF CONTENTS

VOLUME 1

BLOW MOLDING, THERMOFORMING AND ROTOMOLDING

NANOCOMPOSITES AND FILLED POLYMERS

EXTRUSION AND EXTRUSION PROCESSES

POLYMER BLEND AND ALLOYS

POLYMER FOAMS AND LIGHTWEIGHT STRUCTURES

BIOPOLYMERS AND POLYMER FROM RENEWABLE RESOURCES

PROCESS MODELING AND SIMULATION

VOLUME 2

PROCESS MONITORING, CONTROL AND SENSORS

REACTIVE PROCESSING

INJECTION MOLDING AND MOLD

MIXING AND COMPOUNDING

MORPHOLOGY AND STRUCTURE DEVELOPMENT

RECYCLING, PLASTIC SOLUTIONS FOR ECO-CONCEPTION

HIGH PERFORMANCE COMPOSITE PROCESSING

ADDITIVE MANUFACTURING FOR PLASTIC COMPONENTS

MICROTECHNOLOGY AND SMART MATERIALS

BIOBASED PLASTICS AND MATERIALS FOR RENEWABLE ENERGY

Modeling Devolatilization in Single- and Multi-Screw Extruders

Christian Marschik[a*], Bernhard Löw-Baselli[a] and Jürgen Miethlinger[a]

[a]*Institute of Polymer Extrusion and Compounding, Johannes Kepler University Linz, Altenberger Str. 69, 4040 Linz, Austria*

christian.marschik@jku.at

Abstract. Vacuum-vented single- and multi-screw extruders are widely used to remove undesirable residual volatile components from highly viscous polymer melts. To guarantee a smooth separation process and to increase devolatilization efficiency, screw channels in the extraction section of vented extruders are only partially filled. In the governing flow field, volatile components evaporate from the rotating melt pool that develops in front of the active flight flank and from the wiped melt film deposited on the extruder barrel. This paper investigates mass transport phenomena in the rotating melt pool in detail. Based on numerical methods, the flow field in the melt pool is calculated and related to Fick's laws of diffusion. The course of volatile concentration in the melt pool is analyzed taking the circulatory motion of the melt pool into account, and the beneficial impact of surface renewal on devolatilization efficiency is elucidated.

Keywords: single- and multi-screw extruders, devolatilization, venting, surface renewal.
PACS: 47.11.Df, 47.55.Ca, 47.57.Ng, 66.10.C-

INTRODUCTION

Most polymers leave the manufacturing process containing some undesirable residual low-molecular-weight components such as unreacted monomers, solvents, water, and other impurities. To improve the properties of the polymer and to comply with tight environmental regulations, these contaminants are usually removed from the polymeric material in a post-reactor operation commonly referred to as devolatilization. From a scientific point of view, devolatilization is a thermodynamically driven separation process comprising the following steps: (i) transport of volatiles dissolved in the polymeric material to a polymer-vapor interface, (ii) evaporation of the volatiles at the interface, and (iii) removal of the volatiles from the vapor phase [1, 2].

Depending on the viscosity of the polymer solution, a wide variety of commercial devolatizers are used in industry. Employing vented single- and multi-screw extruders in particular provides a competitive solution, as they are able to devolatilize polymers with viscosities ranging across several orders of magnitude. Increasing the devolatilization efficiency in vented extruders requires proper screw design and thus a thorough understanding of the mass transport behavior in the devolatilization section.

In general, mass transport in molten polymeric materials is governed either by molecular diffusion or by a foaming mechanism. At very low volatile component concentrations, however, the viscosity of the polymer melt is relatively high, and thus bubble initiation is inhibited. In this case, the foaming tendency decreases, and the remaining mass transport mechanism is considered to be molecular diffusion [3, 4].

Only a few analytical studies have investigated devolatilization in vented extruders. The first detailed mathematical analysis was presented by Latinen [5]. Other analytical models were developed by Coughlin and Canevari [6], Roberts [7], and Biesenberger [8-10]. Since in all these cases molecular diffusion was assumed to be the rate-determining mechanism of mass transport, diffusion theory was used to describe mass transport phenomena mathematically [11]. According to these studies, devolatilization in vented extruders is governed by two parallel mass transport mechanisms, as indicated in Fig. 1(a): evaporation from the circulating melt pool that develops in front of the active flight flank and from the wiped melt film deposited on the barrel surface. However, these engineering studies considered the melt pool to be stationary and ignored the circulatory motion of the polymeric material accumulated in front of the active flight.

The present work investigated diffusive mass transport in the rotating melt pool in detail. Taking the motion of the melt in the pool into account, we emphasize the significant impact of surface renewal on the devolatilization behavior.

ANALYTICAL MODELING

One-dimensional unsteady-state diffusion in isotropic media is usually described by Fick's second law:

$$\frac{\partial c}{\partial t} = D\frac{\partial^2 c}{\partial x^2}, \tag{1}$$

where c denotes the concentration of diffusing matter and D the diffusion coefficient. The solution of Fick's second law depends heavily on both initial and boundary conditions. In the pioneering research work on devolatilization in vented extruders, however, mainly two specific approaches were applied to describe evaporation from the melt pool mathematically.

The first model considers the melt pool to be a stationary, semi-infinite medium [5, 7-10]. The initial concentration of volatile component in the melt pool at time t=0 is c_0, and the concentration at the polymer-vapor interface for t>0 is maintained at c_e, namely the equilibrium concentration. The basic assumption of the semi-infinite medium model is that the exposure time of a polymer molecule at the free surface of the circulating melt pool is short. Consequently, diffusion time is limited, and the concentration gradient of volatile component in the pool is confined to a very thin layer near the free surface. Applying the semi-infinite medium concept yields the following solution for Eq. (1):

$$c^* = \frac{c(t,x) - c_e}{c_0 - c_e} = \mathrm{erf}\left(\frac{x}{2\sqrt{Dt}}\right), \tag{2}$$

where erf( ) is the Gaussian error function defined in [11] and c^* the fraction of volatile component remaining unevaporated. A characteristic ratio that is frequently used as a measure of diffusion time is the dimensionless Fourier number:

$$\mathrm{Fo} = \frac{Dt}{w_p^2}, \tag{3}$$

where w_p denotes the width of the melt pool, as indicated in Fig. 1(b). It is well known that the applicability of the semi-infinite medium concept is particularly restricted to $\mathrm{Fo} < \mathrm{Fo}_{cr,1} = 0.1$. [12]. As diffusion coefficients of volatile components in polymer melts are generally very low ($D \leq 10^{-8}$ m^2/s) [13, 14], this criterion is usually fulfilled.

However, with increasing exposure time, the penetration depth grows and the concentration change may extend to a measurable extent into the bulk material. As a result, the concentration gradient is no longer confined to a thin layer near the exposed surface, and different boundary conditions must therefore be applied. In this context, a second analytical approach considering the melt pool as a stationary, finite medium was presented [6]. Taking the finite-medium concept into account, the solution of Eq. (1) becomes:

$$c^* = \frac{c(t,x) - c_e}{c_0 - c_e} = \frac{4}{\pi}\sum_{n=0}^{\infty}\frac{(-1)^n}{2n+1}\exp\left(-(2n+1)^2\frac{\pi^2}{4}\frac{Dt}{w_p^2}\right)\cos\left((2n+1)\frac{\pi}{2}\frac{x}{w_p}\right). \tag{4}$$

This infinite series converges rapidly for large values of the Fourier number. If $\mathrm{Fo} > \mathrm{Fo}_{cr,2} = 0.3$, the series can be approximated by using only the first term [12]. However, all earlier analytical studies ignored the motion of the melt accumulated in the pool. In the following section, numerical methods are applied to describe diffusive mass transport in the pool, placing special emphasis on the rotating motion of the polymeric material.

NUMERICAL MODELING

Flow Field Initialization

The mathematical description of the flow pattern in the circulating pool is based on the flat-plate model [15]. Using this reference frame, the helical screw channel is unwound from the screw to yield a straight, rectangular flow channel covered by a flat plate, namely the barrel surface. Further, the kinematic conditions are reversed, that is, the screw is stationary and the barrel surface moves at a constant velocity v_b at an angle φ_b with respect to the down-channel direction. This situation is shown schematically in Fig. 1(b), where h is the channel height, w the channel width, and ξ the degree of filling. As the numerical study focuses exclusively on the analysis of molecular diffusion in the pool, the flight clearance is assumed to be zero and evaporation from the wiped film is ignored.

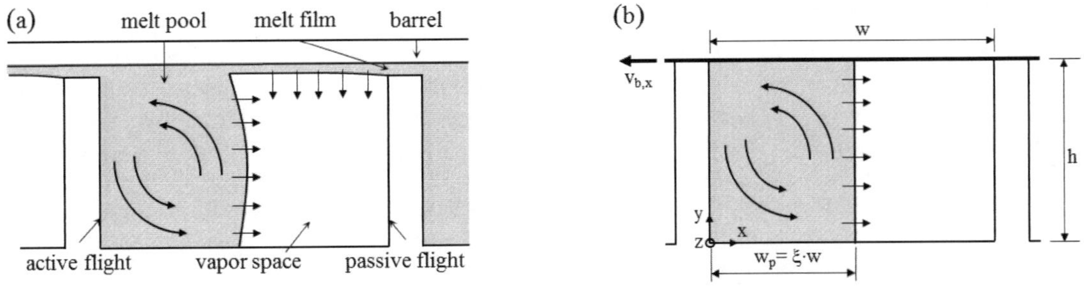

FIGURE 1. Mass transport mechanisms in a partially filled screw channel (a) and basic principles of the flat plat model (b).

Applying the flat-plate model allows the velocity of the barrel surface to be decomposed into a component in the down-channel direction, $v_{b,z}$, and into a component in the cross-channel direction, $v_{b,x}$:

$$v_{b,z} = D_b \pi N \cos(\varphi_b) \qquad \text{and} \qquad v_{b,x} = D_b \pi N \sin(\varphi_b), \tag{5}$$

where D_b denotes the barrel diameter and N the screw speed. Whereas the former drags the melt through the devolatilization section, the latter causes both cross-channel flow and surface renewal effects.

Assuming a constant down-channel velocity $v_{b,z}$, the flow field in the melt pool is expressed in terms of a two-dimensional velocity vector. Owing to the number of boundary conditions available, the corresponding velocity components v_x and v_y are initialized by a sixth-order power series, and the unknown parameters $a_{i,k}$ and $b_{i,k}$ are calculated by taking the boundary conditions in Tab. 1 into account. Furthermore, the fluid is considered to be incompressible, and hence div ($\underline{v}$) = 0. The final velocity vector in dimensionless form is then:

$$\underline{v}(x_1, y_1) = \begin{pmatrix} \sum_{k=0}^{6}\sum_{i=0}^{k} a_{i,k-i} x_1^i y_1^{k-i} \\ \sum_{k=0}^{6}\sum_{i=0}^{k} b_{i,k-i} x_1^i y_1^{k-i} \end{pmatrix} = \begin{pmatrix} -4v_{b,x} x_1^2 \left(3 - 5x_1 + 2x_1^2\right) y_1 \left(-2 + 3y_1\right) \\ 4v_{b,x} x_1 \left(6 - 15x_1 + 8x_1^2\right) y_1^2 \left(-1 + y_1\right) \end{pmatrix}, \tag{6}$$

where $x_1 = x/w_p$ and $y_1 = y/h$.

TABLE (1). Boundary conditions defining the flow field in the melt pool

position	$v_x(x_1, y_1)$	$v_y(x_1, y_1)$
active flight	$v_x(0, y_1) = 0$	$v_y(0, y_1) = 0$
screw core	$v_x(x_1, 0) = 0$	$v_y(x_1, 0) = 0$
free surface	$v_x(1, y_1) = 0$	$\partial v_y(1, y_1)/\partial x_1 = 0$
barrel surface	$v_x(0.5, 1) = v_{b,x}$	$v_y(x_1, 1) = 0$

To describe the amount of residual volatile component in the melt pool mathematically, a characteristic cross-sectional average concentration is defined as:

$$c_{av}^* = \frac{1}{h w_p} \int_0^h \int_0^{w_p} c^*(t, x, y)\, dx\, dy. \tag{7}$$

Diffusion in Moving Fluids

In general, Fick's second law of diffusion in the form of Eq. (1) is valid for stationary systems. In vented screw extruders, however, the polymer melt accumulated in front of the active flight flank rotates according to Eq. (6). Thus, Fick's second law of diffusion needs to be extended to take the motion of the diffusing matter into account:

$$\frac{Dc}{Dt} = \frac{\partial c}{\partial t} + \underline{v} \cdot \underline{\nabla} c = D \Delta c \qquad \text{with} \qquad \underline{v} \cdot \underline{\nabla} c = \underline{v} \cdot \underline{\nabla} c + c \underline{\nabla} \cdot \underline{v} = \underline{\nabla} \cdot (\underline{v} c). \tag{8}$$

Combining these relationships yields the first time-derivative of volatile concentration in the pool:

$$\frac{\partial c}{\partial t} = -\frac{\partial}{\partial x_1}\left(v_x c\right) - \frac{\partial}{\partial x_1}\left(-D\frac{\partial}{\partial x_1}c\right) - \frac{\partial}{\partial y_1}\left(v_y c\right) - \frac{\partial}{\partial y_1}\left(-D\frac{\partial}{\partial y_1}c\right). \tag{9}$$

Eq. (6) and (9) are numerically solved using the finite-volume method. To discretize the analytical relationships in space, methods of second order are applied, and Heun's method is used for time integration. The initial concentration of volatile component in the pool is $c^*(t = 0, x, y) = 1$.

RESULTS AND DISCUSSION

To validate the modeling set-up, numerical results were compared to analytical solutions available for short and long exposure times given in Eq. (2) and (4), respectively. As both analytical models consider the melt pool to be a stationary medium, numerical calculations were carried out assuming that the screw speed is equal to zero. Fig. 2(a) illustrates the concentration distribution of residual volatile component in the melt pool.

Diffusive mass transport occurs exclusively in the region near the exposed surface, where volatiles evaporate into the contiguous vapor phase. With increasing exposure time, the penetration depth grows and the concentration change extends into the center of the bulk. Taking this specific case into account, both the numerical and the analytical study yield similar results, as indicated in Fig. 3(a). If $Fo < Fo_{cr,1}$, the semi-infinite medium model predicts the cross-sectional average concentration in the pool, whereas if $Fo > Fo_{cr,2}$, the finite medium concept reflects the actual mass transport behavior.

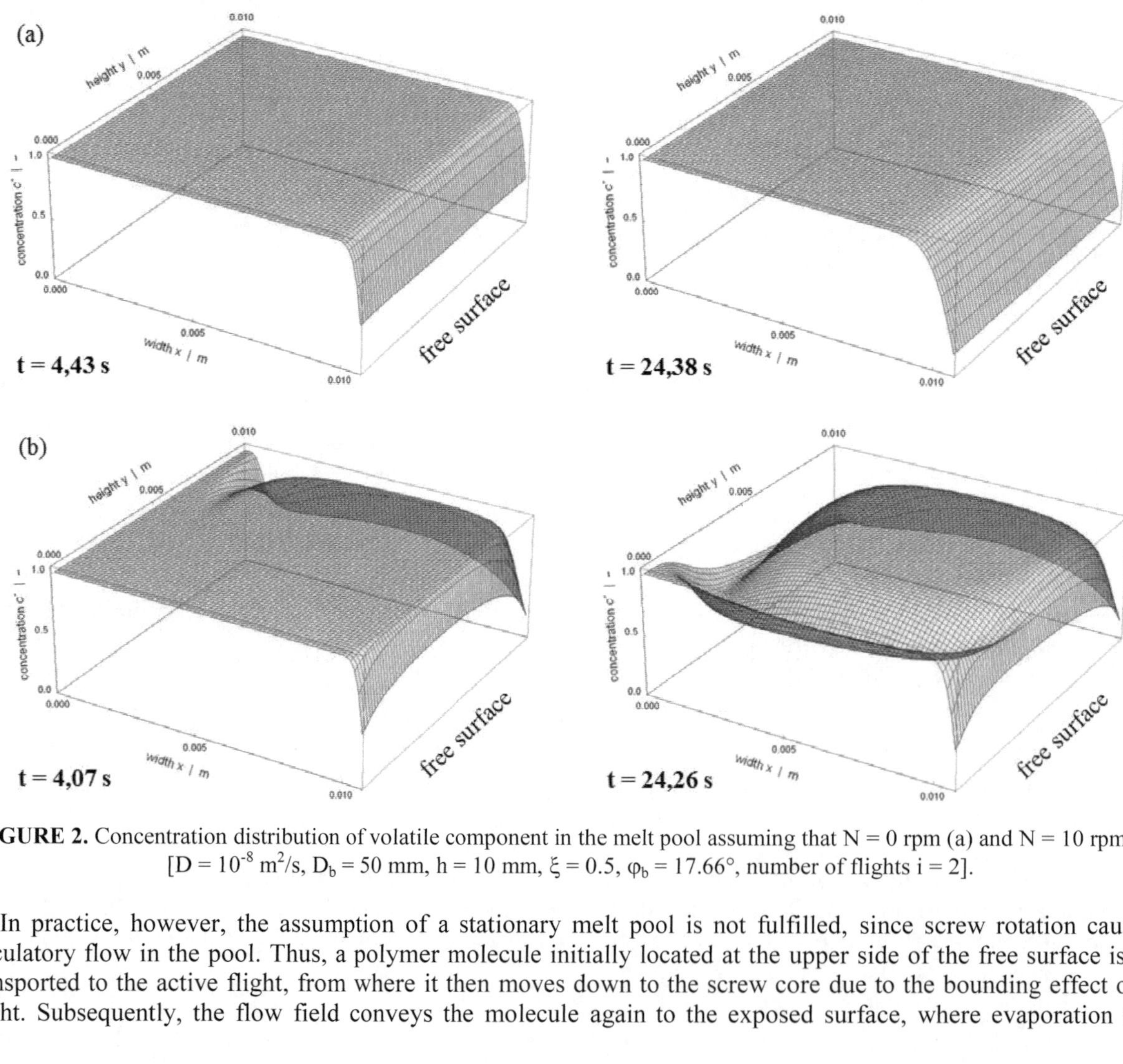

FIGURE 2. Concentration distribution of volatile component in the melt pool assuming that N = 0 rpm (a) and N = 10 rpm (b), [$D = 10^{-8}$ m^2/s, $D_b = 50$ mm, h = 10 mm, $\xi = 0.5$, $\varphi_b = 17.66°$, number of flights i = 2].

In practice, however, the assumption of a stationary melt pool is not fulfilled, since screw rotation causes a circulatory flow in the pool. Thus, a polymer molecule initially located at the upper side of the free surface is first transported to the active flight, from where it then moves down to the screw core due to the bounding effect of the flight. Subsequently, the flow field conveys the molecule again to the exposed surface, where evaporation takes place. Fig. 2(b) plots the concentration distribution of volatile component in the melt pool, assuming a screw speed of 10 rpm.

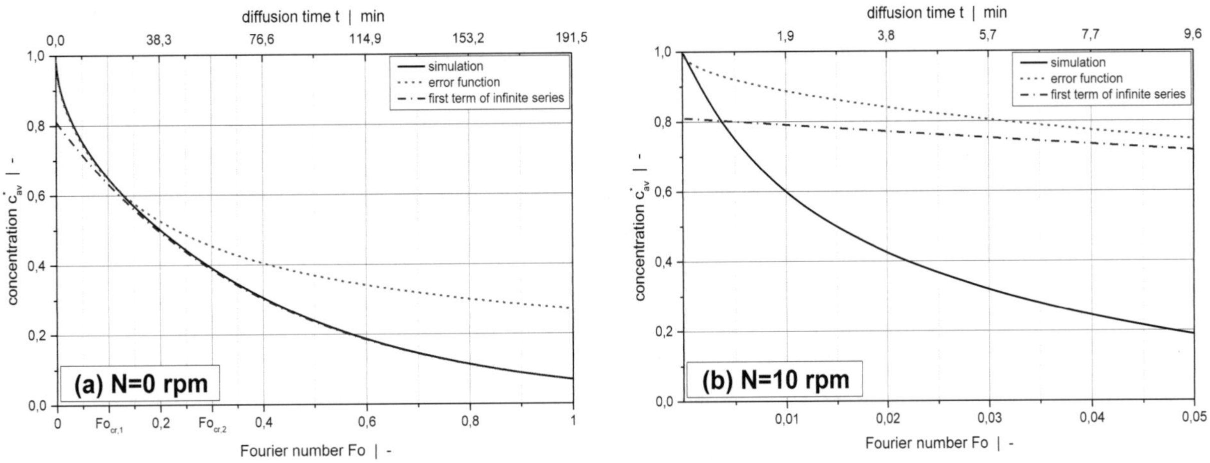

FIGURE 3. Cross-sectional average concentration of volatile component as a function of Fourier number [$D = 10^{-8}$ m²/s, $D_b = 50$ mm, $h = 10$ mm, $\xi = 0.5$, $\varphi_b = 17.66°$, number of flights $i = 2$].

Even at low screw speeds, a significant impact of surface renewal on mass transport behavior is detected, as shown in Fig. 3(b). After a devolatilization time of 1.9 min, nearly 40% of volatile component has evaporated from the pool according to the numerical approach. In contrast, almost 90% of volatile component remain unevaporated when the semi-infinite medium concept is taken into account.

Both cross-channel flow and surface renewal effects must therefore be considered when analyzing molecular diffusion in vented extruders. To extend the knowledge in the field of devolatilization in vented extruders, the numerical tool developed in the course of this research work is currently used to analyze further important issues, especially at higher screw speed values.

ACKNOWLEDGMENTS

This work was supported by the Austrian COMET-K2 program of the Linz Center of Mechatronics (LCM), and was funded by the Austrian federal government and the federal state of Upper Austria.

REFERENCES

1. R. J. Albalak, "An Introduction to Devolatilization," in *Polymer Devolatilization*, edited by R. J. Albalak, New York: Marcel Decker, 1996, pp. 1-12.
2. P. S. Metha, "An Overview of Devolatizers," in *Polymer Devolatilization*, edited by R. J. Albalak, New York: Marcel Decker, 1996, pp. 291-260.
3. A. Tukachinsky, Y. Talmon, and Z. Tadmor, *AIChE J.*, **40**, 670-675, (1994).
4. C. Rauwendaal, *Polymer Extrusion*, Munich: Hanser, 2014, pp. 175-189.
5. G. A. Latinen, *ACS Adv. Chem.*, **34**, 235-246, (1962).
6. R. W. Coughlin and G. P. Canevari, *AIChE J.*, **15**, 560-564, (1969).
7. G. W. Roberts, *AIChE J.*, **16**, 878-882, (1970).
8. J. A. Biesenberger, *Polymer Engineering and Science*, **20**, 1015-1022, (1980).
9. J. A. Biesenberger and G. Kessidis, *Polymer Engineering and Science*, **22**, 832-835, (1982).
10. J. A. Biesenberger and D. H. Sebastian, *Principles of Polymerization Engineering*, New York: John Wiley and Sons Inc., 1983, pp. 573-659.
11. J. Crank, *The Mathematics of Diffusion*, Oxford: Clarendon Press, 1975, pp. 14.
12. R. Marek and K. Nitsche, *Praxis der Wärmeübertragung*, Munich: Hanser Publishers, 2012, pp. 142.
13. Y. Sato, K. Fujiwara, T. Takikawa, Sumarno, S. Takishima, and H. Masuoka, *Fluid Phase Equilibria*, **162**, 261-276 (1999).
14. P. Durrill and R. G. Griskey, *AIChE J.*, **169**, 106-110 (1969).
15. C. Rauwendaal, *Polymer Extrusion*, Munich: Hanser, 2014, pp. 411-418.

Temperature Dependence of the Hydrodynamic Characteristics of the Flow of the Polymer Melt in the Converging Channels

Kuznetsov Alexandr[a], Koshelev Konstantin[b], Tolstykh Mikhail[c],

Naud Jean-Charles[d], Pyshnograi Grigory[a*]

[a]Altai State Pedagogical University - Altai - Russia
[b]Institute for water and environmental problems - Altai – Russia
[c]Polzunov Altai State Technical University - Altai – Russia
[d] Alkante -Bretagne, France
pyshnograi@mail.ru

Abstract. At the presented work the influence of temperature on the dimensions of the vortex flow polymer melt in the region at the inlet of the converging channels investigated. To solve the problem of mathematical modeling of three-dimensional flow of the polymer melt in a converging plane-parallel channel uses a modified Vinogradov and Pokrovskii rheological model generalized to account for the non-monotonic dependence of the viscosity gradient in uniaxial elongation. Getting discrete analog volume control is carried out by a division into physical processes, and numerical implementation is carried out using techniques of parallel computing based on GPU. The temperature dependence of the rheological properties of the polymer melt is Arrhenius form. The numerical experiment shows that the dependence of the size of the vortex of the melt temperature is non-monotonic.

Keywords: rheology, 3D flow, parallel computing technology.
PACS: 83.50.Uv

INTRODUCTION

As rule, mathematical study polymer fluid flows involves two interrelated phases. The first step involves the selection and justification of rheological model. In the second step, it is the algorithmic implementation of the obtained equations and the numerical experiment. Note, that today there are plenty of rheological models of various complexity [1,2,5–12]. All these models describe the main effects observed in viscometric experiments: gradient dependence of shear and elongation viscosity, the first and the second normal stress difference and non-monotonic establishment of time dependence of shear and elongation stresses.

All viscometric flows have a fairly simple structure, since the velocity gradient tensor is known, and all the models show similar accuracy in the description of these flows. Therefore, the issue of adequacy of the rheological model is solved via calculation of two and three-dimensional flows in areas with complex geometry.

In this paper, we apply a modified Vinogradov-Pokrovskii rheological model to solve the problem of mathematical modeling of three-dimensional nonlinear viscoelastic fluid flows in a plane-parallel channel with a sudden contraction. The method of control volume with the division of physical processes is used upon obtaining discrete analogs. The numerical implementation is carried out using GPU-based parallel computing technology CUDA.

THE MATHEMATICAL MODEL

It is well known that linear polymer melts are non-linear viscoelastic media. Nowadays the equations that take into account essential features of polymeric liquids behavior are commonly used to describe their flows [5–9]. The priority in choosing a mathematical model should be given to models which to some extent take into account the structure of polymer molecules. It is quite complicated so the most popular models are those based on the mesoscopic approach. In this case, the behavior of polymer macromolecules is replaced by the behavior of one or more nonlinear relaxation oscillators, and the transition to macroscopic description is carried out via methods of statistical mechanics [5,8,9,12,13]. One such model is a modified Vinogradov and Pokrovskii rheological model [9]

$$\sigma_{ik} = -p\delta_{ik} + 3\frac{\eta_0}{\tau_0}a_{ik}, \tag{1}$$

$$\frac{d}{dt}a_{ik} - \nu_{ij}a_{jk} - \nu_{kj}a_{ji} + \frac{1+(\kappa-\beta)I}{\tau_0}a_{ik} = \frac{2}{3}\gamma_{ik} - \frac{3\beta}{\tau_0}a_{ij}a_{jk}.$$

Where σ_{ik} – stress tensor of the polymer system; a_{ik} – dimensionless extra stress tensor; p – hydrostatic pressure; η_0 – initial value of shear viscosity, τ_0 – initial relaxation time; κ and β – scalar coefficients of anisotropy taking into account shape and size of macromolecular coil; ν_{kj} – velocity gradient tensor; γ_{ik} – symmetrized velocity gradient tensor; I – trace of tensor a_{ik}.

In paper [10] the two- and three-dimensional flows under the constant pressure gradient are considered in channels with a rectangular cross-section. The pressure field gradient has been known. In paper [15] an axisymmetric circular channel was used to model the input flows. However, an attempt to use the numerical method [15] for calculating flat channels failed. For calculating real flows on the basis of this model the following equations of momentum conservation and mass should be added to (1)

$$\rho\left(\frac{\partial}{\partial t}v_i + v_k\frac{\partial}{\partial x_k}v_i\right) = \frac{\partial}{\partial x_k}\sigma_{ik}, \qquad \frac{\partial}{\partial x_k}v_k = 0. \tag{2}$$

Here ρ – the density of the polymer; v_i – i-th component of the velocity vector.

The system of equations (1, 2) is closed with respect to variables a_{ik}, v_i, p. To solve the problem initial and boundary conditions are required. We first discuss the computational domain to establish these conditions.

The computational domain involves two 3D boxes. The first reservoir is channel 14 × 14 (mm) with a square section, the second a slotted channel with a cross section of 14 × 1(mm). Lengths of the parallelepipeds are selected sufficiently large $-80 \le x \le 80$ (mm) to eliminate the influence both of the input into the tank and output from the slotted channel. It is expected that within this area the flow will be three-dimensional, i.e. different directions in different axes. In this case, the main flow is directed along axis *Ox*; whereas along axis *Oy* a strong flow compression will occur. This can cause vortices though along axis *Oz* changes will be immaterial. Therefore, we call axis *Ox* – the flow direction, axis *Oy* – the direction perpendicular to the flow, and axis *Oz* - the direction neutral to the flow.

The main boundary conditions are stick conditions on a solid surface for velocity: $v_i = 0$.

The boundary conditions for the dimensionless stresses are obtained by substituting these conditions in (1) and discarding the corresponding terms.

The boundary conditions $z = 0$ and $y = 0$ are used for terms of symmetry or vanishing of the corresponding partial derivatives. At the entrance to the reservoir at $x = -80$ for the velocity components we use expressions $v_2(-80, y, z) = v_3(-80, y, z) = 0$ and $v_1(-80, y, z) = 9V(z-7)^2(y-7)^2/5488$ (mm/s), where V – volume flow rate.

Flows in a plane-parallel channel with a sudden contraction, are commonly referred to as converging flows or input flows. In order to calculate such flows via model (1, 2) it is necessary to determine the numerical values of the parameters for the rheological model: κ, β, η_0 and τ_0. To do this, we turn to the experimental data on the dependence of steady-state shear viscosity on the shear rate [18] where flows of two samples of polyethylene melts were applied – linear low density polyethylene (LLDPE) and low density polyethylene (LDPE). The model parameters were chosen to give the best agreement between the experimental and theoretical curves. For LLDPE sample we obtained $\eta_0 = 14500(Pa \cdot s)$, $\tau_0 = 0.2(s)$, and for the LDPE $\eta_0 = 18500(Pa \cdot s)$, $\tau_0 = 2(s)$. Anisotropy parameters $\beta = 0.1$ and $\kappa = 0.12$ were the same. Density values were given in [18] and account for $918(kg/m^3)$ for LDPE and $926(kg/m^3)$ for LLDPE.

The numerical method is constructed for relatively easy implementation when creating a computer model using technologies GPU NVIDIA and CUDA. The finite-difference approach is employed to find the solution. For the difference approximation the method of control volume was used that does not cause difficulties in a relatively simple geometry of the domain.

MATHEMATICAL MODELING FOR EXIT FLOWS

To compare the results of calculations we used the experimental data [18] where flows of two samples of polyethylene melts were applied LLDPE and LDPE. The differences between these two examples is that LLDPE shear viscosity changes less than for LDPE, and LLDPE relaxation time is shorter than that of LDPE. We found out that there is an explicit secondary flow for LDPE flows in the corners of the flow channel, which is not observed for LLDPE. These vortices change their shape in the channel cross-sections parallel to the axis, which characterizes the three-dimensional nature of the flow field. Furthermore, it is noted that within the investigated vortex flow there is a helical stream, which is directed to the tank wall. Particular attention is paid to the distribution of velocity along the symmetry axis of the channel. It was found that LDPE maximum velocity is observed directly in the entrance portion of the channel slot. This effect does not appear for LLDPE under the same conditions.

All these effects are found during the numerical experiment. The calculations imply that the increase in speed in the slit of the channel happens due to the three-dimensional nature of the flow field and the stress caused by the increase in the flow of LDPE compared with a sample of LLDPE. The three-dimensional nature of LDPE flow is confirmed by the presence of the neutral component of velocity in the flow direction, which was not observed for LLDPE flow. This additional component of velocity results in increasing the volumetric flow along the centerline at the entrance to the slit channel.

Thus, the velocity, pressure and stress fields were calculated in a steady flow. Herewith circulating areas were found at the entrance of the slit channel. As seen from Figure 1 for LLDPE sample ($\tau_0 = 0.2(s)$), the reverse flow zone is negligible. For LDPE sample ($\tau_0 = 2(s)$) this zone, as seen from Figure 2 is much larger. Note that the value of viscous parameter η_0 in these samples is approximately the same and the difference in their behavior should be attributed to time parameter τ_0.

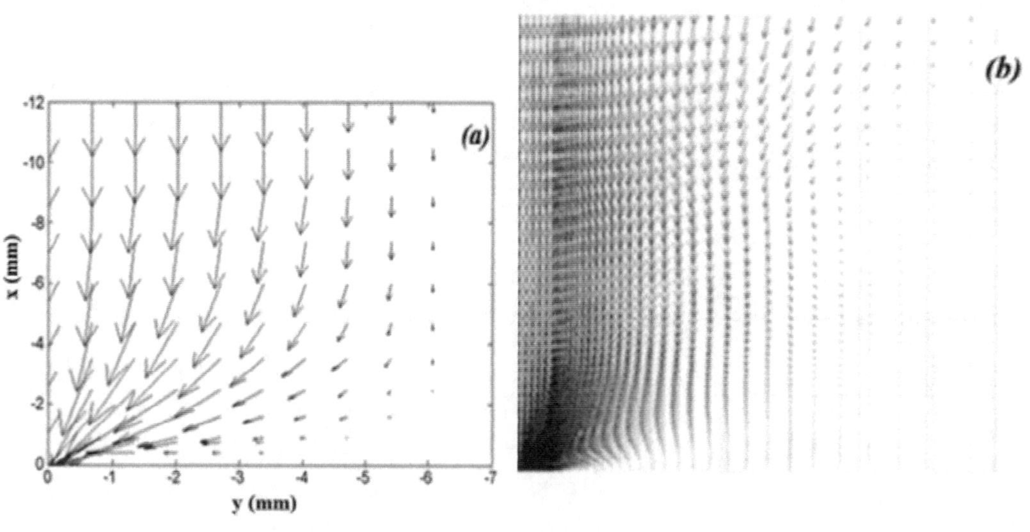

FIGURE 1. Comparison of experimental (a) and theoretical (b) LLDPE velocity vector fields in a slit entrance channel ($V = 0.20[cm^3/s]$).

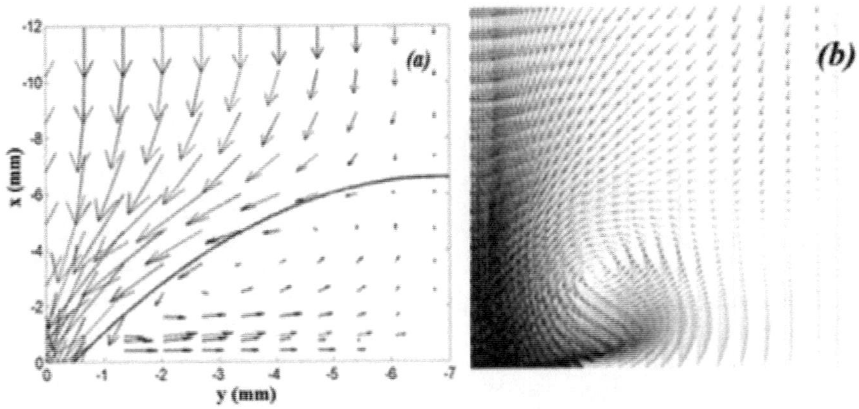

FIGURE 2. Comparison of experimental (a) and theoretical (b) LDPE velocity vector fields in a slit entrance channel ($V = 0.20[cm^3/s]$).

We note that the values of the Reynolds number calculated using the formula ($\mathrm{Re} = \rho v_0 h / \eta_0$) for the considered currents in the slot channel do not exceed 0.02. Such flows are fairly well understood in the classical Newtonian model, and eddy zones in this case are not observed.

Thus, converging flows for some polymer melts may exhibit a substantial three-dimensional picture that is shown by the presence of velocity components in the direction neutral to the flow. This should be taken into account in the design of the experiment, as there are methods that are not able to measure all the components of velocity.

Also one should mention the fact of long-term relaxation of the velocity profile in the gap. Calculations showed that the steady velocity profile is observed at a considerable distance from the entrance slit. This should be considered when conducting measurements in a narrow channel portion.

Previously, it was noted the presence of eddy currents in the corners of the flow channel, the size of which depends strongly on the temperature of the melt and on specific consumption.

One of the essential characteristics of the test flow is the size of the vortex. We calculated the area of the vortex zones when flow section planes $z = const$. Comparing shown in Figure 3 the calculated and experimental vortex sizes depending on the distance from the axis of the channel may be concluded that an increase in the intensity of the vortex flows away from the axis of the channel. This fact can be explained by Weissenberg effect. From this comparison it can be concluded qualitative agreement of the experimental and the payment data. Note that in the calculations carried out for the Newtonian law of behavior ($\tau_0 = 0; \beta_0 = p_0 = 0$) demonstrate the absence of vortex zones, and calculations for viscoelastic Oldroyd-B ($\beta_0 = p_0 = 0$) show low values of the vortex area and the lack of increase in the intensity of the vortex when you remove the cross section on the axis of the channel.

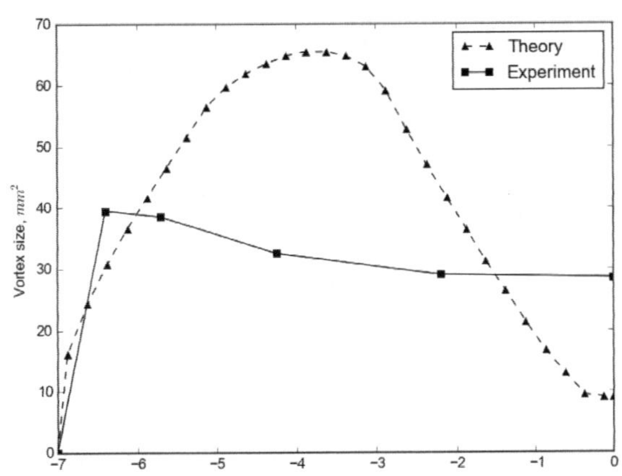

FIGURE 3. The size of the vortex in different sections.

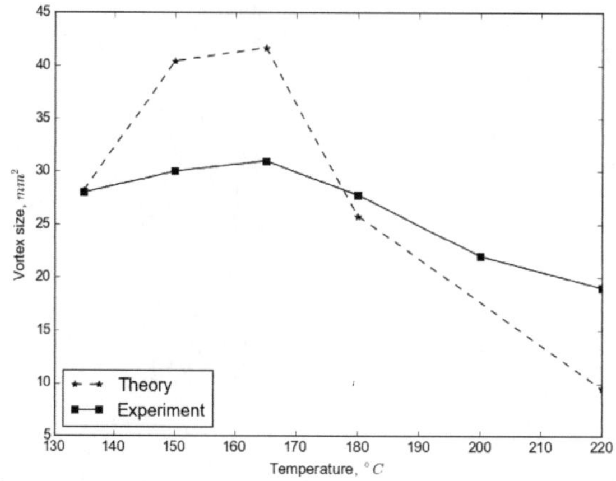

FIGURE 4. The dependence of the vortex size of the melt temperature.

Consider how dependent the size of the vortex zone of the temperature of the polymer melt. On Figure 4 shows calculated values of the vortex size in the axial section at various temperatures. It may be noted that the initially vortex size increases with increasing temperature, but then vortex size begins to decrease. This is not monotonic dependence observed in experiments, the results of which are also shown on Figure 4.

Thus, the comparison of the hydrodynamic characteristics of the polymer melt flows at different temperatures in a convergent channel with a rectangular cross section and it is shown that the sudden narrowing of the canal leading to the appearance of vortex currents, the size of which pass through a maximum with increasing melt temperature, which is detected in the experiments. This is yet another confirmation of the applicability of modified rheological Vinogradov and Pokrovskii model for the flows of polymer melts in areas with complex geometry. Also, the results obtained proofed of the effectiveness of technology CUDA parallel computing for unsteady calculations of three-dimensional flows of nonlinear viscoelastic fluids with rheological behavior of law in differential form.

REFERENCES

1. Wilco M.H. Verbeeten, Gerrit W.M. Peters, Frank P.T. *J. Non-Newtonian Fluid Mech.* **117,** 73-84 (2004).
2. Mitsoulis E., Schwetz M, Mnstedt H. J. *Non-Newtonian Fluid Mech.* **111,** 41-61 (2003).
3. H. Munstedt, M. Schmidt, E. Wassner *J. Rheol.* **44,** 413-427 (2000).
4. Afonso A.M., Oliveira P.J., Pinho F.T., Alves M.A. *J. Fluid Mech.* **677,** 272-304 (2011).
5. McLeish T.C.B., Larson R.G. *J. Rheol.* **42,** 81 (1998).
6. Verbeeten W.M.H., Peters G.W.M., Baaijens F.P.T. *J. Rheol.* **45,** 823 (2001).
7. Leonov A.I., Prokunin A.N. *Chapman and Hall, New York. 1sted* 356-395 (1994).
8. Pyshnograi G.V., Gusev A.S., Pokrovskii V.N. *J. Non-Newtonian Fluid Mech.* **163,** 17-28 (2009).
9. Pokrovskii V.N. *The Mesoscopic Theory of Polymer Dynamics. 2nd Edition. (Berlin: Springer)* 184 (2010).
10. Kuznetsova Yu.L., Skul'skiy O.I., Pyshnogray G.V. *Computational Continuum Mechanics.* **2,** 55-69 (2010) (in Russian).
11. Altukhov Yu.A., Samoylov V.S., Pyshnograi I.G., Pyshnograi G.V. *J. Composite Mechanics and Design.* **18,** 325-332 (2012).
13. Altukhov Yu.A., Pyshnograi G.V., Pokrovskii V.N. *J. Non-Newtonian Fluid Mech.* **121,** 73-86 (2004).
13. Pokrovskii V.N., Altukhov Y.A., Pyshnograi G.V. *J. Non-Newtonian Fluid Mech.* **76,** 153-181 (1998).
14. Altukhov Yu.A., Golovicheva I.E., Pyshnograi G.V. *Fluid Dynamics.* **1,** 3 (2000).
15. Altukhov Yu.A., Pyshnograi G.V. *Journal on Composite Mechanics and Design.* **1,** 16-23 (2001).
16. Koshelev K.B., Pyshnograi G.V., Tolstykh M.Yu. *Fluid Dynamics* **50,** 315-321 (2015).
17. Hertel D., Mnstedt H. *J. Non-Newtonian Fluid Mech.* **152,** 73-81 (2008).
18. Hertel D., Valette R., Mnstedt H. *J. Non-Newtonian Fluid Mech.* **153,** 82-94 (2008).

Polymer Film Production and Rheological Characteristics of Forming Solutions

Pyshnograi Grigory[a*], Naud Jean-Charles[b], Tegubova Julia[c], Kondratieva Olga[c]

[a]*Altai State Pedagogical University - Altai Region – Russia*
[b] *Alkante -Bretagne, France*
[c]*Altai State Technical University - Altai Region - Russia*
[*]*pyshnograi@mail.ru*

Abstract. In the present mathematical model, the process of film formation from polymer solution is described. Rheological parameters of the model were determined by comparing theoretical curves and experimental data measured for simple shear in chitosan solutions. Dimensionless similarity criteria were found for the rheological characteristics of the process. This work considers the effect of these criteria on the dependence of the velocity, concentration and width of the film on the distance from the exit of the die.

Keywords: rheology, polymers, rheological state equation, mass transfer.
PACS: 83.50.Uv

INTRODUCTION

Polymers are widely used in the production of films, fibers, plastics, new constructional and functional materials of various functions. It has led to an increased number of researches in the field of high-molecular-weight compound processing, in particular, in the mathematical simulation of polymer flows and melts [1-6].

Mathematical modeling used in the design and manufacture of products made of polymer materials gives a number of benefits, such as the ability to control the quality of polymer products and to solve a number of production optimization problems. Such problems are based on a mathematical model, which should be fairly simple but reflect all relevant characteristics of the process. The fabrication of polymer films is one of the common processes in polymeric material processing. The process of obtaining films from polymer solution is two-staged and is so complex. This complexity is caused by the fact that an increased polymer concentration results in a sharp increase in the polymer system's viscosity. Thus, while developing mathematical models to describe the processes of polymeric film formation, the characteristics of each forming step should be considered.

RHEOLOGICAL EQUATION OF STATE

The analysis of polymer flows shall lean on a rheological model which establishes a relation between the stress tensor of the polymeric system with the kinematic characteristics and internal thermodynamic variables. This model allows to close the equation set of the polymeric medium dynamics. Such closing on the basis of a microstructural approach has a number of advantages as it allows to trace the dependence between the macro- and micro characteristics of the polymeric system. It is necessary to note that the application of this approach leads to rheological models of various complexity. A review and a comparison with other approaches can be found in works [7,8]. The simplest basic rheological model suggested in these studies consists in modeling of stationary viscometric flows of linear polymer solutions and melts both qualitatively and quantitatively. Therefore, let us considers the model in details. It was designed for the case when macromolecule dynamics was modeled by a non-linear dumbbell to consider the slowest relaxation process only:

$$\sigma_{ik} = -p\delta_{ik} + \eta_1\gamma_{ik} + 3\frac{\eta_0}{\tau_0}a_{ik};$$
$$\frac{d}{dt}a_{ik} - v_{ij}a_{jk} - v_{kj}a_{ji} + \frac{1+(\kappa-\beta)I}{\tau_0}a_{ik} = \frac{2}{3}\gamma_{ik} - 3\frac{\beta}{\tau_0}a_{ij}a_{jk}, \quad (1)$$

where σ_{ik} - is the stress tensor; p - the hydrostatic pressure; η_0 and τ_0 - the initial values of shear viscosity and relaxation time for the polymeric component; η_1 - the residual shear viscosity considering the presence of the solvent; ν_{ik} - the velocity gradients tensor; a_{ik} - the symmetric tensor of anisotropy of the second rank; $I=a_{jj}$ - the first invariant of the anisotropy tensor; $\gamma_{ik}=(\nu_{ik}+\nu_{ki})/2$ - the symmetrical tensor of velocity gradients; κ,β are the phenomenological model parameters considering the size and shape of the molecular ball in the macromolecular dynamic equations [9]. We state that κ,β are independent of molecular weight and concentration. The expression $\kappa=1{,}2\beta$ is the corresponding condition of independence of the asymptotical behavior of stationary shear viscosity of the polymer molecular weight as shown in [10].

These equations differ from those applied in the paper [9] by the introduction of the coefficient η_1. This coefficient allows one to consider the presence of a solvent in the polymeric system. Thus, it is assumed that the solvent contribution can be described by the Newtonian law.

Following expressions for the description of the concentration dependences of the dimensional parameters in rheological model (1) were found in work [11] and :

$$\eta_0(c)=\eta_0(c^*)\left(\frac{c}{c^*}\right)^4,\ \eta_1(c)=\eta_1(c^*)\left(\frac{c}{c^*}\right)^{2,2},\ \tau_0(c)=\tau_0(c^*)\left(\frac{c}{c^*}\right)^2. \quad (2)$$

We can use them, for example, while modeling the process of film formation from a polymer solution. Such a model will be built in a one–dimensional approximation where expressions (2) will allow us to complete the set of differential equations.

MATHEMATICAL MODEL

We apply model (1) to describe the process of film formation from polymer solution. As the velocity gradients are unknown, the equations of conservation of mass and momentum given in the Cartesian coordinates [12-14] should be added to the set of equations (1). In this case, evaporation should be taken into account in the equation of concentration transfer.

Further on, a stationary problem is considered, when there are no time-dependent variables in a fixed coordinate system. Since the film is sufficiently thin, the concentration across the film thickness can be assumed to be constant. The origin is put at the middle of the die, the axis *Ox* is directed along the direction of the film, and variables depending only on x were sought for: $v_x=u(x);c=c(x)$. We consider the stationary problem in a one-dimensional approximation. In this paper we propose to consider the process of film formation from polymer solution in the biaxial elongation mode [15]. It should be noted that, in this case, it is necessary to take into account the additional stress on the axis *Oz*, whose value is determined in accordance with experimental data.

In this case, the rate of change in the width of the film differs from the rate of change in its thickness and we can find the dependence of the film velocity on its width and thickness:

$$w(x)=w(0)\left(\frac{u(x)}{v_0}\right)^{-\frac{1-\alpha}{2}},\ b(x)=b(0)\left(\frac{u(x)}{v_0}\right)^{-\frac{1+\alpha}{2}}. \quad (3)$$

where α is a parameter that characterizes the anisotropy of the flow elongation.

Introduce several dimensionless criteria: the Reynolds number $\mathrm{Re}=\rho u_0 l/\eta_0(c^*)$; the diffusion Peclet number $Pe=u_0 l/\lambda$; the diffusion Nusselt number $Nu=\mu l/\lambda$ (l - is the distance from the die exit to the nip drum); and the Weisenberg number $We=\tau_0(c^*)u_0/l$. From (1), (3) in dimensionless form we obtain the following mathematical model of the process of forming a film from the polymer solution in the one-dimensional approximation:

$$\mathrm{Re}\cdot We\cdot u\frac{du}{dx}=3\frac{d}{dx}\left(\frac{\eta_0}{\tau_0}(a_{11}-a_{33})\right)+\frac{d}{dx}\left(\eta_1\frac{du}{dx}\right),$$

$$u\frac{da_{11}}{dx}-2\frac{du}{dx}a_{11}+\frac{1+(\kappa-\beta)(a_{11}+a_{22}+a_{33})}{We\cdot\tau_0}a_{11}=\frac{2}{3}\frac{du}{dx}-\frac{3\beta a_{11}^2}{We\cdot\tau_0},$$

$$u\frac{da_{22}}{dx}+(1-\alpha)\frac{du}{dx}a_{22}+\frac{1+(\kappa-\beta)(a_{11}+a_{22}+a_{33})}{We\cdot\tau_0}a_{22}=-\frac{1-\alpha}{3}\frac{du}{dx}-\frac{3\beta a_{22}^2}{We\cdot\tau_0}, \quad (4)$$

$$u\frac{da_{33}}{dx}+(1+\alpha)\frac{du}{dx}a_{33}+\frac{1+(\kappa-\beta)(a_{11}+a_{22}+a_{33})}{We\cdot\tau_0}a_{33}=-\frac{1+\alpha}{3}\frac{du}{dx}-\frac{3\beta a_{33}^2}{We\cdot\tau_0},$$

$$Pe\cdot u\frac{dc}{dx}=\frac{d^2c}{dx^2}-2Nu\frac{l}{h}\left(\frac{(1-c)u}{1-c(0)}\right)^{\frac{1+\alpha}{2}}(c-c_0).$$

Here λ is the diffusion coefficient, μ - the evaporation coefficient with the environment, h - the ratio of the sectional area of the film to the perimeter of the cross section ($h=wb/(2w+2b)=b/2$ for thin films $w>>b$), c_0 the maximum solvent concentration in the environment. Instead of equations (2) $\eta_1(c)=0{,}68(c)^{2{,}2}$; $\tau_0(c)=(c)^2$ are used: c - being the dimensionless polymer concentration, referred to $c^*=3\%$ [11]; h/l is dimensionless film thickness, referred to l.

Equations (4) represent a set of five ordinary differential equations, four of which are of the first order and one of the second order. Therefore, it should be supplemented by six boundary conditions. Four of them can be put simply.

For velocity, these conditions read as follows: $u(0)=v_0; u(l)=kv_0.$, where k is the coefficient of film elongation.

The boundary conditions for concentration can be put as follows: $c(0)=c_1$; $c'(l)=0$, where c_l is the concentration of the polymer solution at the outlet of the die. The second condition means that the polymer concentration on the nip drum takes some stationary value.

Additional conditions for stresses can be obtained by calculating the fluid motion in the polymer extruder [16]. They are expressed as: $a_{22}(0)=0$, $a_3(0)=-5/17$.

RESULTS AND DISCUSSION

Based on model (6) it is possible to calculate the dependences of velocity, concentration and stresses on the distance x from the die output. The dependence of the film width on x represents the greatest practical interest of the study. In this paper, the influence of the parameters *Pe*, *Re*, *Nu*, and *We* on these dependences is considered. Ranges of their variation calculated based on the data for the process of forming a film of chitosan dissolved in acetic acid [11].

Dependences of solvent concentration in the film and the film width at various values of *Pe* are shown in Figures 1 and 2. An increase in the *Pe* number leads to a gentle narrowing of the film width and to the gradual reduction of the solvent concentration in the film.

Inverse dependence can be observed with an increase in the diffusion Nusselt number. Despite the fact that the *Re* and *We* numbers are present in the model at variations in the range of 10^{-1} – 10^5 for *Re* and 10^{-8}–10^{-2} for *We*, their influence was insignificant.

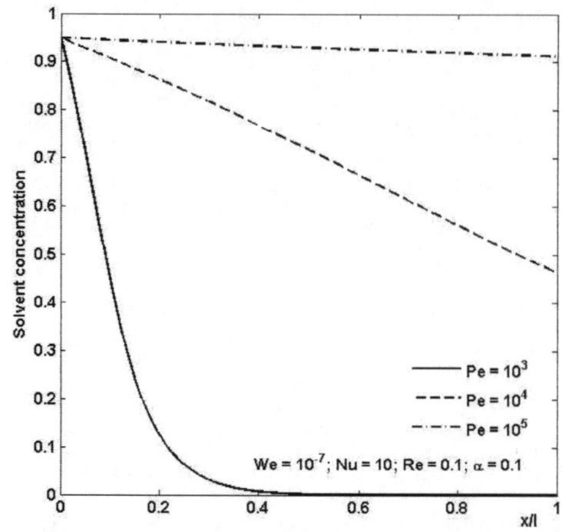

FIGURE 1. Influence of the *Pe* number on the solvent concentration in the film

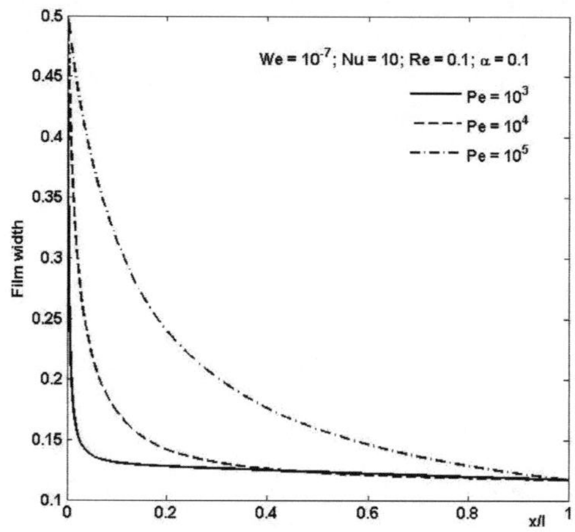

FIGURE 2. Influence of the *Pe* number on the film width

It should be noted that the change in the anisotropy parameter α of flow elongation has the most significant effect on the width of the film (Fig. 3), with greater values of this parameter corresponding to the greater width of the film.

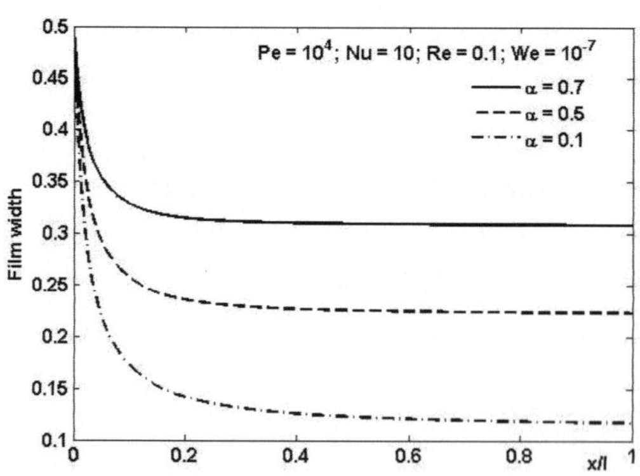

FIGURE 3. Influence of the α on the film width

CONCLUSION

In the presented study the possibility of using the modified Vinogradov and Pokrovskii rheological model for flows of linear polymer solutions based on mass transfer in a variety of deformation modes was tested.

The set of dynamics and mass transfer equations is written in a one-dimensional approximation, when the elongation velocity, the concentration of the solvent, and the non-zero components of the stress tensor are functions of the distance to the die exit, and the rheological parameters of the model are known functions of concentration. The model allows us to predict the final concentration value on the nip drum. Therefore, we hope it will inspire other researchers in this field to carry out further experiments.

A set of ordinary differential equations for the dependence of the width and thickness of the film on its longitudinal velocity for the case of biaxial tension was obtained and solved.

REFERENCES

1. P. Akay and D. Omay *Int. Polym. Proc.* **29**, 287-294 (2014).
2. W. Cheewawuttipong, D. Fuoka, S. Tanoue, H. Uematsu, Y. Iemoto *Int. Polym. Proc.* **29**, 214-220 (2014).
3. C-P. Chu, B-Y Jiang, C. Weng, F-Z. Jiang *Int.'Polym. Proc.* **29**, 245-251 (2014).
4. S. Shiromoto *Int. Polym. Proc.* **29**, 197-206 (2014).
5. G. Zhang, T. Liu, Y. Yi, L. Liu, X. Tan, J. Luo, J. He, H. Ma, A. Lu, Z. Deng, Y. Shu *Int. Polym. Proc.* **29**, 175-183 (2014).
6. C.W. Seay and D.G Baird *Int. Polym. Proc.* **24**, 41-50 (2009).
7. G.V. Pyshnograi, A.S. Gusev, V.N. Pokrovsky *J. Non-Newt. Fluid Mech.* **164**, 17-28 (2009).
8. V.N. Pokrovski, *The mesoscopic theory of polymer dynamics*, 2nd Ed. Springer, Dordrecht-Heidelberg-London-New York, 2010.
9. G.V. Pyshnograi, Yu.A. Altukhov *Polym. Sci.* A **38**, 1185-1193 (1996).
10. S.A. Zinovich, I.E. Golovicheva, G.V. Pyshnograi *J. Appl. Mech. Techn. Phys.* **41**, 154-160 (2000).
11. H.N.A. Al-Joda, G.V. Pyshnograi, A.B. Shipovskaya *BPMS* vol. 9 No. 2, 131-136 (2012).
12. G.V. Pyshnograi, I.V. Tret'yakov, Ju.A. Altukhov *J. Appl. Mech. Techn. Phys.* **53**, 84-90 (2012).
13. G.V. Vinogradov, A.Ya. Malkin, *Rheology of Polymers,* Mir, Moscow, 1980.
14. D.D. Joseph, *Fluid Dynamics of Viscoelastic Liquids,* Springer–Verlag, 2014.
15. A. Gusev, G. Afonin, I. Tretjakov, G. Pyshnogray *Viscoelasticity: Theories, Types and Models,* 197-212 (2011).
16. Yu.A. Altukhov, A.S. Gusev, G.V. Pyshnograi *J. Eng. Phys. Thermophys.,* **3**, 64-67 (2003).

Behaviour of the Thermal Conductivity of Amorphous Polymers

Wilson Nunes dos Santos
Department of Materials Engineering
Federal University of São Carlos
Rodovia Washington Luiz, Km 235,CP 676, São Carlos, CEP 13565-905, São Paulo, Brazil

dwns@ufscar.br

Abstract. The thermal conductivity of four amorphous polymers was determined within a wide range of temperatures, starting at room temperature up to above the glass transition temperature (Tg). In this work, the hot wire technique was adopted for the experimental measurements and the thermal conductivity was directly determined in the temperature interval analyzed. As expected, the experimental results show that Tg exerts a measureable influence on the thermal conductivity of amorphous polymers. For such materials, the dependence of the thermal conductivity with the temperature is quite different from that foreseen by dieletric materials. The thermal conductivity behaviour with temperature displays two distinct regions: in the glassy state before Tg and in the rubbery state after Tg. The influence of this transition temperature on the thermal conductivity behaviour with temperature has been analysed in terms of phonons conduction process, specific heat variations with temperature and modulus of elasticity of the analyzed polymers, in order to explain this apparently anomalous behaviour. This effect is hardly detectable for semi-crystalline polymers, since only the amorphous phase is responsible for this behaviour, and the crystalline phase overrides this influence.

Keywords: Thermal transition temperatures, thermal conductivity, phonons heat conduction, hot wire technique.
PACS:65.60.+a

INTRODUCTION

Thermal conductivity (k) is the property that determines the working temperature levels of a material, determining the temperature gradient inside this material, and so it is an important parameter in problems involving steady state heat transfer. This property assumes a critical role in the performance of materials in high temperature applications. Low thermal conductivity values are required when the purpose is to minimize heat losses. On the other hand, when heat transfer from one site to another is desirable, materials with higher thermal conductivities must be chosen. So, reliable thermal conductivity values are essential for material selection criteria in order to get the best performance of the material in a specific application. However, it is one of the physical quantities of difficult measurement and requires high precision in the determination of the parameters involved in its calculations. Moreover, many factors can affect the thermal conductivity of a material [1-3]: temperature, density, porosity, moisture, degree of crystallinity, orientation of grains, size of molecules, impurities, etc.

The equation relating the thermal conductivity (k, W/m K), thermal diffusivity (α, m^2/s), specific heat at constant pressure (cp, J/kg K) and the bulk density (ρ, kg/m^3) is given by:

$$k = \alpha \rho c_p \qquad (1)$$

Heat conduction in insulating solids (ceramics and polymers) may be considered either as the propagation of anharmonic elastic waves through a continuum or as the interaction between quanta of thermal energy called phonons (lattice thermal conduction) [1]. At high temperatures another mechanism of conduction must be considered: the photon conductivity. However, since this process of conduction is proportional to the fourth power of temperature, it is usually neglected at low temperatures (< 500 ^{0}C). For insulating solids the thermal conductivity is proportional to the c_p and ρ of the material, and to the mean speed (v) and the mean free path (λ) of the phonons,

For amorphous materials, the mean free path may be considered approximately constant, and when temperature increases, thermal conductivity increases in proportion to the specific heat. In this case the thermal conductivity may be represented by an equation of the form [1]:

$$k = cT+d \qquad (2)$$

where c and d are experimental constants to be determined. Although the previous equation is also valid for polymers, since phonons are the main responsible for heat transfer, polymeric materials have some particularities which make the structure of such materials different from that of ceramic structures. The polymer thermal conductivity is also dependent on molecular weight distributions, crystallites orientation, crystallinity degree, macromolecule size, thermo-mechanical history, and other structural features. The glass transition temperature (Tg), exerts a measureable influence on the thermal conductivity of amorphous polymers [2, 3]. This influence is experimentally hardly detectable for semicrystalline polymers. However, for this class of polymers, a singular behaviour of the thermal conductivity is observed within the melting temperature range.

RESULTS AND DISCUSSIONS

As stated before, the hot wire technique was employed in this work. Details of such techniques were given in many occasions elsewhere [4 - 7]. Table 1 displays the polymer samples details along with the data on the glass transition (Tg). Figure 1 shows the experimental results obtained for such polymers studied: polymethylmethacrilate (PMMA), high impact polystyrene (HIPS), crystal polystyrene (PS) and polycarbonate (PC). In this figure the Tg values obtained from DSC analysis are also indicated.

TABLE (1). Polymer samples details with their respective properties.

Material	Resin Supplier	Code	Crystalline melting temperature Tm (°C)	Glass transition temperature Tg (°C)	Crystallinity (%)
PMMA	Unigel	Acrigel	-	112	-
HIPS	BASF	PS495F	-	93	-
PS	BASF	PS145D	-	90	-
PC	Sabic	Lexan grade	-	145	-

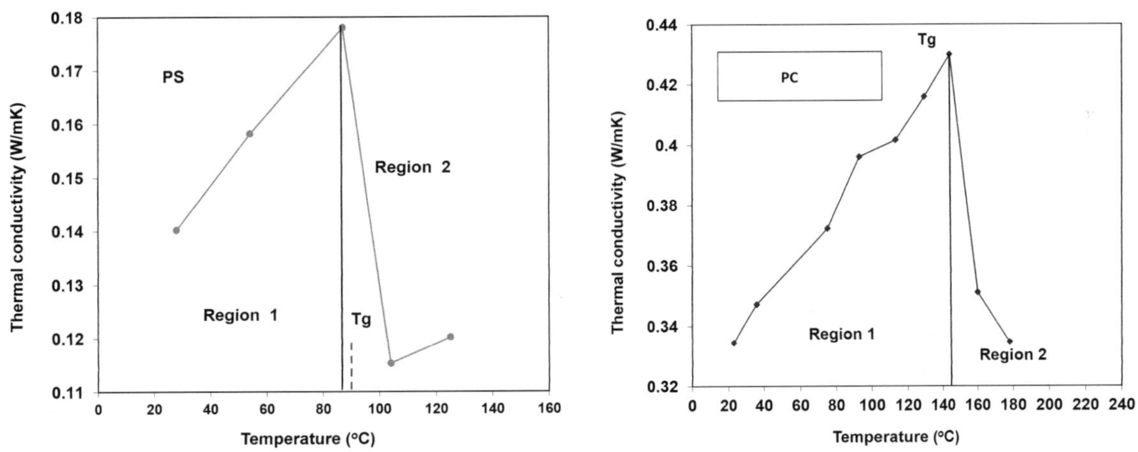

FIGURE 1a. Thermal conductivity as a function of temperature for PS and PC.

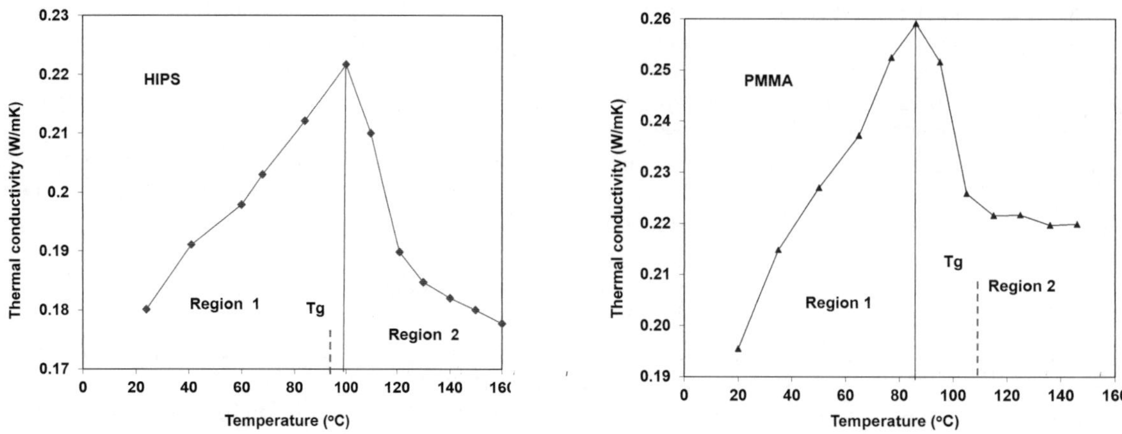

FIGURE 1b. Thermal conductivity as a function of temperature for HIPS and PMMA.

For the amorphous polymers, the dependence of the thermal conductivity with the temperature is quite different from that foreseen by equation (2) for amorphous ceramic materials. The glass transition temperature exerts a detectable influence on the thermal conductivity behaviour with temperature and two distinct regions may be considered: in the glassy state before Tg and in the rubbery state after Tg. In the glassy state region, the thermal conductivity increases with temperature as predicted by equation (2). This is due to the specific heat increase effect, since phonons mean free path is considered practically constant with temperature for amorphous materials. Starting in the rubbery region, an abrupt decrease of the thermal conductivity is observed. As a result, a peak is attained close to the Tg. This sudden drop may be explained in terms of molecular mobility, since in the temperature interval in the neighborhood of Tg the polymer passes from a rigid and brittle glassy state to the soft rubbery state. In the rubbery state, the large-scale mobility of atoms and macromolecules increases and chain segments undergo intensive thermal motion and torsional rotations. This disorder introduced in the material structure contributes to the scattering center phonons and, thereby, decreasing the thermal conductivity. A sudden drop of the thermal diffusivity of amorphous PET in Tg's region also was observed by Chen et al using the flash method [8]. Physically, this phenomenon may be explained as follows.

For amorphous materials, the phonon mean free path is practically independent of the temperature. When the lateral dimensions are much less than the wavelength, an extensional wave is propagated and the phonon mean speed (v) is proportional to the square root of the ratio of the elasticity modulus (E) and the bulk density of the material, i.e.:

$$v \propto \sqrt{\frac{E}{\rho}} \qquad (3)$$

Although the density decreases with temperature rise, the elasticity modulus experiences a sudden and intense drop above Tg, as schematized in Figure 2. This figure displays the sudden drop of E after the Tg for semi-crystalline and amorphous polymers. This sudden drop in E is much more pronounced (2-3 orders of magnitude) in amorphous than in semi-crystalline polymers. The net effect is a decrease in the phonon speed, which leads to a correspondent decrease in the thermal conductivity. After this sudden drop, the thermal conductivity displays a smooth variation with temperature, as shown in Figure 1.

The much higher thermal conductivity values observed for PC in Figure 1, in comparison to those of other amorphous polymers, is probably due to the fact that PC exhibits high molecular stiffness below Tg, with the presence of two p-phenylene groups in its macromolecular skeleton, and thus contributing towards higher phonons conduction process [9]. It is also possible to observe in Figure 1 that the sudden drop in thermal conductivity for PMMA takes place at around 86 °C, which is a temperature considerably lower than that verified for the Tg of this polymer, as determined by the conventional calorimetric technique (112 °C). This fact may be explained having in mind the long dwell time (up to 6h) before each measurement of thermal conductivity in order to get the thermal equilibrium in the hot wire test. After this elapsed time, part of the polymeric mass would reach the rubbery state, even at a temperature below the classical Tg and thus resulting in a decrease of the thermal conductivity. In order to confirm this hypothesis, two DSC runs with the PMMA samples are provided in Figure 3. Initially, the sample (8 mg) remained at 86 °C for a short dwell time of 5 minutes and was then heated up to 140 °C, as shown on Figure 3 (2nd heating run). This curve was then compared to the one obtained in the determination of Tg, from room temperature up to 140 °C (1st heating run). In both cases the heating rate was 10 °C/min.

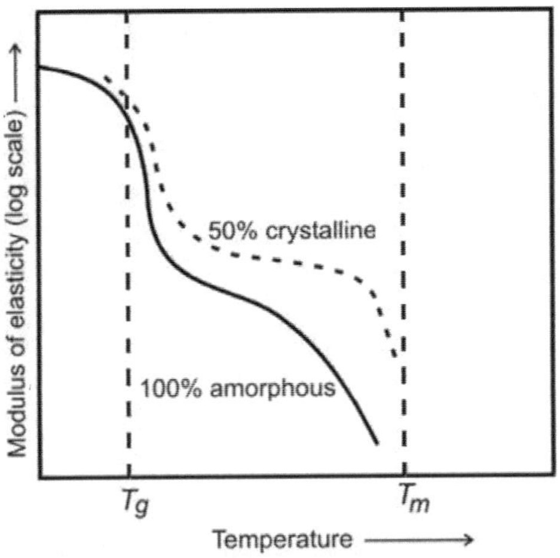

FIGURE 2. Schematic variation of modulus of elasticity as a function of temperature for semicristalline and amorphous polymers.

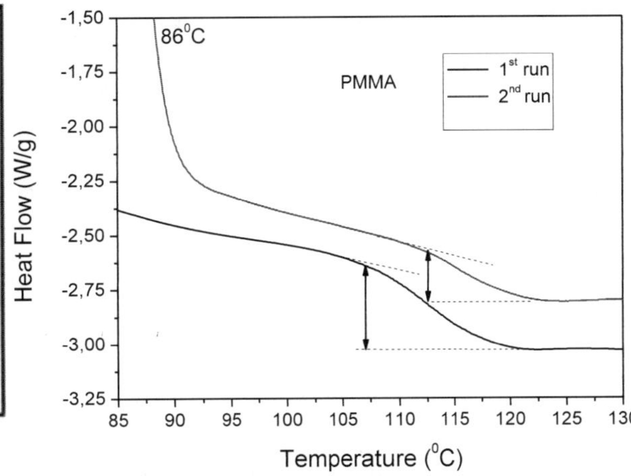

Figure 3. DSC curves for PMMA: heating rate 10 °C/min; 2nd heating after the specimen remained for 5 minutes at 86 °C.

It is evident from Figure 3 that the variation of the base line decreased after PMMA sample dwells at 86 °C for five minutes. This behavior is an indication that part of the polymeric mass had already attained the rubbery state. As a consequence, a drop in its thermal conductivity is observed at this temperature, several degrees before Tg value, as determined by DSC procedure. This fact may be explained having in mind that the conventional Tg is a kinetic (rate-dependent) manifestation of an underlying thermodynamic phenomenon and corresponds to the temperature at which the forces that keep the connected chain segments of a solid polymer are overcome by the thermally induced movement within the experimental time scale. So, the value of such property depends on the heating rate and the time that the material remains at a fixed constant temperature. Figure 3 shows that for PMMA the actual glass transition temperature, in the hot wire thermal conductivity technique, is attained at around 86 °C, and not at 112 °C, as determined by the classical differential scanning calorimetry, as shown in Table 1. For the others amorphous polymers this behavior was not detected. The variation of the base line remains approximately constant even when such materials remained at a dwell time below its classical Tg value, which indicates that this time interval was not enough to lead the polymer to a rubbery state. This difference in the behaviour between amorphous polymers is associated with the type and stiffness of each structure. The effects of number-average molecular weight, chain stiffness and cohesive forces on the value of Tg are different from each other.

ACKNOWLEDGEMENTS

The author acknowledges to FAPESP, grant #2011/02943-4 and 2015/24268-8, São Paulo Research Foundation (FAPESP) for the financial support.

REFERENCES

1. W.D. Kingery and M.C. McQuarrie, *Journal of the American Ceramic Society* **37**, 67-72 (1954).
2. A.B. Bashirov and T.D. Shermergor. *Mekhanika polimerov* **3**, 5538-5542 (1976).
3. P. Dashora and G. Gupta. *Polymer* **2**, 231-236 (1996).
4. A.L. Schieirmacher, Wiedemann, Phys. **34**, 38-42 (1888).
5. E.F.M. Van Der Held and F.G. Van Drunen, Physics **15**, 885-881 (1949).
6. W.N. dos Santos, *Cerâmica* **38,** 86-91 (2002).
7. W.N. dos Santos, *Polymer Testing* **24**, 932-941 (2005).
8. F. C. Chen, Y. M. Poon and C. L. Choy, *Polymer* 18, 129-135 (1977).
9. J. A. Brydson, *Plastics materials*, London: Butterworth-Heinemann Ltd, 1999, pp. 1-920.

Unusual pvT-behavior of Amorphous Polymers under Processing Conditions and Consequences for the Process Simulation

Reinhard Forstner[a*], Rotraud Freytag[a] , Jose Jil-Perez[a] and Claudia Maticka[a]

[a]Franz-Fritsch-Straße 11, 4600, Wels, Austria

reinhard.forstner@tckt.at

Abstract.
The development of high performance products at lowest possible product development-time and costs is the demand of the plastics industry today. A successful use of process simulation to describe the behavior of complex structures and to minimize the technical and economic risks for companies requires application and process-relevant material data. The pvT-behavior of polymers is an essential parameter for process simulations. Especially the dependence of the specific volume of pressure and temperature under process conditions is not mapped to the existing measurement methods.

In this study pvT measurements were performed on selected commercial amorphous polymers (PC, ABS) with a Pirouette pvT device, a combination of a dilatometer and Couette rheometer. The specific volume was determined as a function of temperature (25-300°C) and at pressures ranging between 200-1000 bar. On top of that, the influence of the cooling rate was also investigated by pvT measurements performed at cooling rates of 0.1°C/s, 1°C/s and 100°C/s. The results of the experiments show non expected pvT behavior as an increase of volume was found in the area of the glass transition temperature for all cooling rates and pressure steps applied. The results were confirmed for the low cooling rates with a High Pressure Capillary Rheometer. These results could be an explanation route for several problems reported by industry where huge gaps between simulation results and processing behavior exist.

Keywords: specific volume, transition temperature
PACS: 82.35.Lr - Physical properties of polymers

INTRODUCTION

Polymer melts are highly compressible under pressure. For example - in an injection molding machine the melt can be compressed up to 15% and compression further increases with a higher injection speed. On top of that, the effect of pressure on the specific volume the change of temperature is equally important for the process. The pvT behavior of amorphous polymers differs fundamentally from that of semi-crystalline ones, as shown in Figure 1.

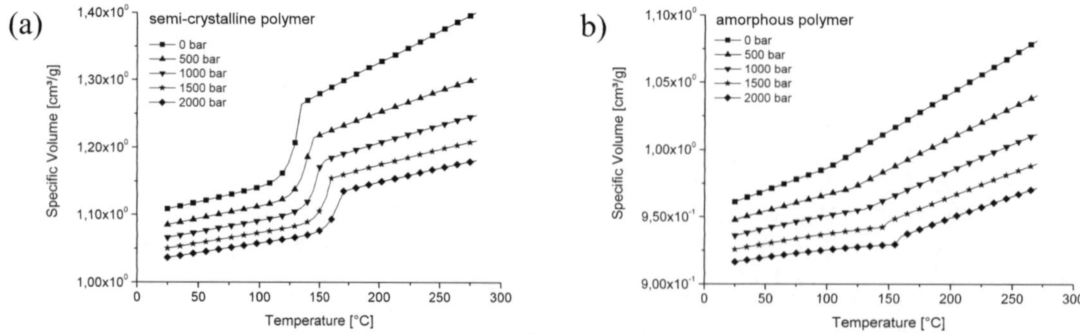

FIGURE 1. pvT-behavior of a semi-crystalline (a) material and an amorphous (b) material

EXPERIMENTAL SETUP AND METHODS

The dilatometer, named Pirouette-pvT-Apparatus (Figure 2), which is a combination of a traditional piston-die-dilatometer and a Couette rheometer, was designed to measure the specific volume of polymers as a function of pressure (up to 1000 bar) and cooling rates (up to 100 °C/s) and shear (Figure 3). So the Pirouette enables measurements of the specific volume under processing conditions. The measurement procedure of the samples performed was starting above the melting temperature for each material with different cooling rates at 3 various pressures. The ring-shaped samples are fabricated with a Babyplast micro injection molding machine (Figure 4).

The pvT data were measured using the Pirouette with ring shaped samples having an outer diameter of 22 mm, a thickness of 0.5 mm, a height of 2.5 mm.

(a) (b)

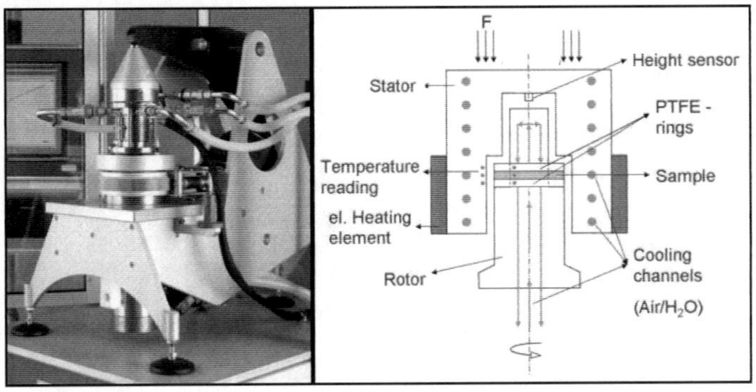

FIGURE 2. measurement chamber of the Pirouette-pvT-Apparatus (a) and cross-section of the measurement cell from the Pirouette (b) [1]

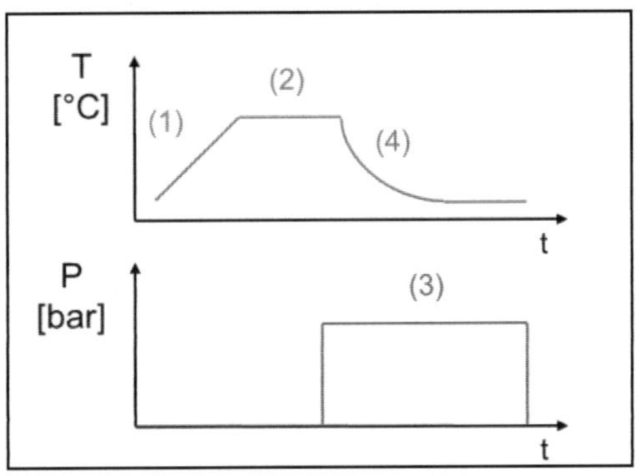

FIGURE 3. test sequence of an experiment under isobaric cooling

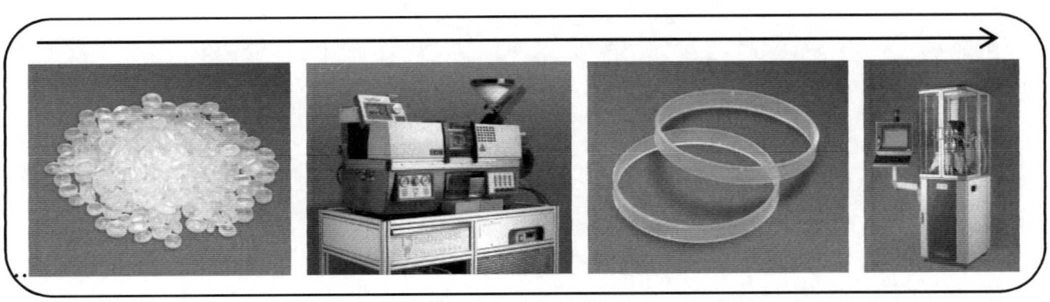

FIGURE 4. sequence of the sample preparation

For comparison reason pvT measurements in High Pressure Capillary Rheometer (HKR), which is a common technique to measure the specific volume as a function of temperature and pressure of polymers, were performed. However, a High Pressure Capillary Rheometer is not equipped to measure polymers under industrial relevant process conditions, as high cooling rates, but is sufficient fair for the low cooling rates.

MATERIALS

In this study selected commercial amorphous polymers such as polycarbonate (PC) and ABS were investigated. The relevant material properties, as glass transition temperature, melting temperature und melt flow rate (MFR) of the tested polymers are given in Table 1.

TABLE (1). Material properties of the amorphous polymers

Polymer	glass transition temperature [°C]	MFR [g/10min]
PC	145	38 (300°C/1.2kg)
ABS	115	15 (220°C/10kg)

EXPERIMENTAL RESULTS

PvT Measurements

The results of the experiments show non expected pvT behavior for both materials as an increase of volume was found in the area of the glass transition temperature for all cooling rates and pressure steps applied. Figure 5 shows the results of experiments performed with PC at low cooling rates of about 0.1°C/s. There is a small shift for the absolute values between the measurement methods, which is not explained in detailed at the moment, but the slopes in the melt are comparable and so in principle one could shift the values in mind and then the curves would collapse. However, it clearly can be seen for both methods that in the area of the glass transition temperature some non-explained effects about non-typical volume behavior can be found. This phenomenon was initially reported by industry as they are aware about PC and ABS grades of which they talk about "zero-shrinkage" materials.

Of course, these results are the first step and much more indication and need more detailed investigations and discussions about the phenomenon found. However there are hints in literature and work topics of other research groups that for example work on processing issues related to these materials, that give reason that this identified "non-shrinkage" behavior could be the initial course for their research area.

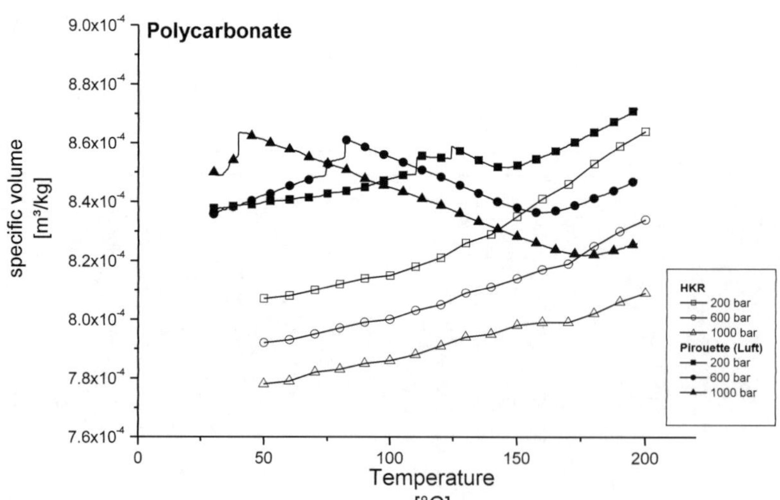

FIGURE 5. pvT measurements of a PC grade at cooling rates of 0.1°C/s and different pressure levels performed in the Pirouette dilatometer and HKR

Comparison of Results with Different pvT Data

In a second step, the fitted pvT data obtained by using the Pirouette and data from the material database were used in Autodesk Moldflow Simulation 2014 to simulate the injection molding process of a special designed demonstrator part [2]. The 2nd domain Tait approach, which is used in the software Autodesk Moldflow was basis to fit the measured Pirouette pvT data. The experimental results of shrinkage and warpage of the produced injection molded parts are compared with calculated simulation results. For both pvT data, the shrinkage prediction is very different. The results of shrinkage calculated with Pirouette pvT data for example of ABS are closer to the experimental shrinkage data.

SUMMARY

This study provided pvT-behavior of two amorphous polymers (PC and ABS) under processing conditions and their implementation in the process simulation. The pvT data were investigated by a Pirouette dilatometer and the influences of cooling rate and pressure on the specific volume were examined. The relevance of the pvT properties in the numeric simulation for injection molding was visualized by a demonstrator part, where deviations between the simulated results and the produced part were evaluated.

REFERENCES

1. R. Forstner, G.W.M. Peters, H.E.H. Meijer, International Polymer Processing XXXIV 2009, 2, 114-121.
2. R. Freytag, J. Perez, R. Forstner, "pvT-behavior of polymers under processing conditions and implementation in the process simulation", 20th Symposium on Composites, Materials Science Forum Vols 825-826 (2015), pp. 677-684

Online Estimation of Permeability in Resin Transfer Molding Using Parallel-Plate Capacitor

Tzu-Heng Chiu[a], Kai-Hong Wang[a], Yuan Yao[a*], Wei Qi[b]

[a] *Department of Chemical Engineering, National Tsing Hua University, Hsinchu, 30013, Taiwan, ROC*
[b] *School of Information and Electric Engineering, Zhejiang University City College, Hangzhou, 310015, PRC*

Presenting author. Email: e6060372@yahoo.com.tw
[*] *Corresponding author. Email: yyao@mx.nthu.edu.tw*

Abstract. Resin transfer molding (RTM) is a popular liquid composite molding technique, which has been widely used for manufacturing polymer composites. In RTM, permeability of fiber reinforcement is a crucial factor affecting the property of resin flow and consequently influencing the final product quality. Hence, it is essential to estimate the permeability of the fiber during the manufacturing process. In past research, the permeability is estimated online using the flow-front information captured by a CCD camera and the local pressure measured by a pressure transducer array. However, the feasibility of such method relies on the visibility of the melt flow. When even the mold is not transparent or the color of the resin melt is similar to that of the fiber, this method is not applicable. To overcome the limitation of the existing method, a parallel-plate capacitor is developed in this research for measuring the flow front, where the capacitance keeps an approximately linear relationship with the position of flow front. The principle of the measurement is as follows. Before mold filling, the medium between the capacitor plates is mainly air. Then it is gradually replaced by resin melt during filling. Therefore, the dielectric constant of the medium is continually varying, resulting in the variation of the capacitor value, i.e., the capacitance. Based on the resin flow-front position, the flow velocity is calculated and substituted into the Darcy's law together with the injection pressure. In doing so, online estimate of permeability is available. The proposed method is illustrated with experiments on an RTM process using glass fiber reinforced plastic (GFRP).

Keywords: resin transfer molding, permeability, parallel-plate capacitor, Darcy's law, glass fiber reinforced plastic.
PACS:

INTRODUCTION

Advances in technology have caused increased demands for materials. Polymer composites have gained attention by virtue of their high strength and low weight, and they are now widely adopted in many fields, such as the wind turbine blade and aerospace industries. Resin transfer molding (RTM) is a common method for manufacturing composites, which often pours resin into a closed mold containing fiber preform with the help of a vacuum pump. To ensure product quality, the permeability of the preform plays an important role in filling the resin in the mold [1, 2]. Thus, accurately calculating the permeability is vital to the quality of the end-product. Some studies have attempted to determine the in-plane permeability of glass fibers [3, 4]. Kim et al. [5] measured the three-dimensional (3D) permeability of fiber preforms by the inverse parameter estimation technique. Hoes et al. [6] presented a two-dimensional (2D) central injection setup, based on the automatic detection of the flow-front position as a function of time, to measure permeability. Wei et al. [7] adopted a method of online measurement of the one-dimensional (1D) local permeability and developed equipment to track local pressure and velocity. It can accommodate variation in local permeability within the preform caused by the irregular arrangement of fibers among different regions. In addition, there are several other studies on permeability measurement [8-10]. What these studies have in common is that they were not carried out in a non-transparent mold. In their experiments, it was necessary to use cameras to detect the resin flow locations. However, most molds used in industries are not transparent.

To estimate the in-plane permeability of fibers, the flow-front of the resin should be recorded. Besides using cameras, various sensor systems have been reported in the literature to track the resin flow-front. Here are some examples. Yenilmez et al. [11] used a grid of 50 dielectric sensors embedded in the mold to monitor RTM. The capacitance of each sensor increases when the resin fills the space between the dielectric sensor plates. Matsuzaki et al. [12] measured the resin flow-front using a multifunctional interdigital electrode array (MIEA) to detect the resin flow. These sensors are usually difficult and expensive to set up and operate. In this study, a parallel-plate capacitor is used to measure the resin flow-front based on the changing capacitance between the plates, and then online calculation based on Darcy's law is conducted for permeability estimation. This design has many advantages: 1) the cost of the parallel-plate capacitor is lower than other instruments and it is easy to operate; 2) the method can measure the flow-front of the resin without a transparent mold; 3) permeability can be estimated online immediately.

EQUIPMENT AND INSTRUMENTATIONS

The mold of the experimental setup is shown in Fig. 1. The mold, with a cavity of 12 cm × 33 cm × 0.3 cm, is made of acrylic. This study integrated the parallel-plate capacitor and other hardware with LabVIEW for recording and immediate estimation. In the experimental procedure, we pasted one copper strip in the upper mold, with the other in the lower mold. In the second step, a circuit board using the RS-232 communication protocol was connected to the copper strips. Finally, the resin was infused into the mold because the air in the mold was pulled out by a pump to prompt resin movement. In the process, the permittivity was changed because the volume of resin in the mold increased as the resin filling progressed, leading to a variation in capacitance. The changing capacitance was captured by the circuit board shown in Fig. 2. The LabVIEW software converted the capacitance signal into displacement according to a calibration curve established in advance. Eventually, by acquiring the information of the displacement, LabVIEW computed the permeability of the fiber with Darcy's law.

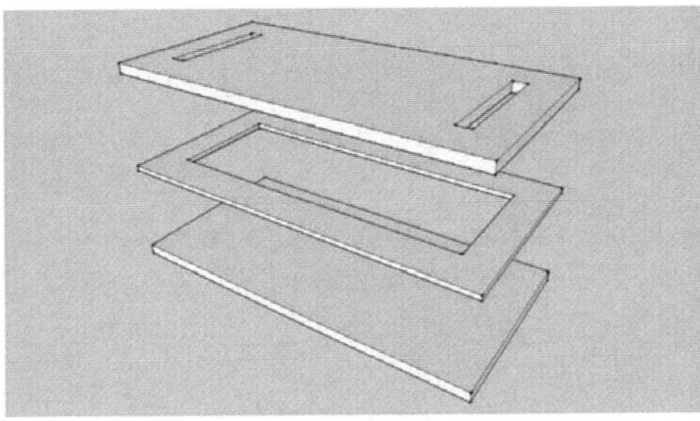

Fig.1. Mold of the experimental setup

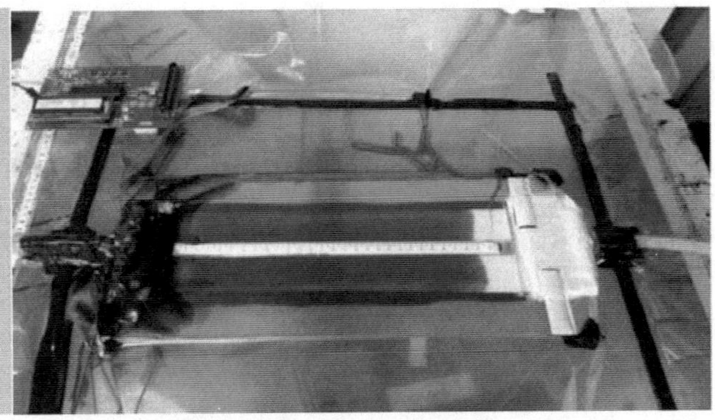

Fig.2. Parallel-plate capacitor and circuit board

ONLINE ESTIMATION OF PERMEABILITY

Parallel-Pate Capacitor

A parallel-plate capacitor utilizes the change of the measured object in physical quantity and converts the change into capacitance, which is composed of two plates parallel to each other, and using the space or dielectric isolating the two thin conductors. Capacitance C is determined by the distance between the plates d, relative permittivity ε, and area A of the two electrode plates:

$$C = \varepsilon \frac{A}{d} \tag{1}$$

The capacitance value is changed by moving the plates up or down (changing d), or shifting the upper plate (changing A). In our study, the upper and lower plates were fixed, and the capacitance value rose linearly when the permittivity was increased due to the resin flowed in the mold between the two plates.

The permittivity ε was determined by a mixed effect of air, resin and fiber. By Lichtenecker's empirical formula [13],

$$log(\varepsilon_{mix}) = V_1\, log(\varepsilon_1) + V_2\, log(\varepsilon_2) \tag{2}$$

where ε_{mix} is the mixed dielectric, and V_i is volume fraction.

In our study, the capacitance was in parallel, so formula (1) could be modified as

$$C = \frac{\varepsilon_{mix1}\, a\, x}{d} + \frac{\varepsilon_{mix2}\, a\, (L-x)}{d} \tag{3}$$

where a is the width of the copper strip, x is the resin flow location, and L is length of the mold (Fig. 3).

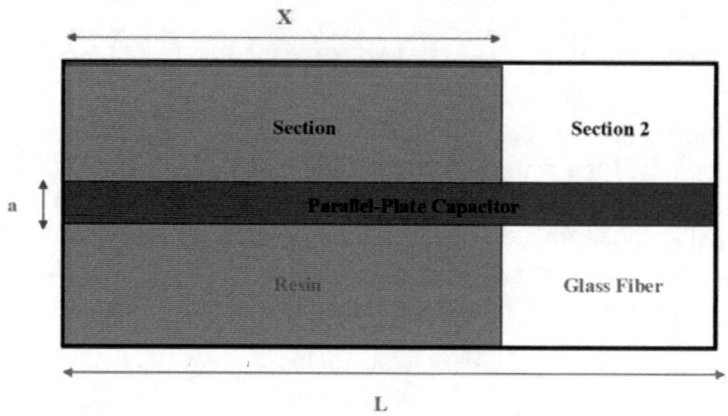

Fig.3 . Mold parameters

Then, substituting (2) into (3), and making the calculation, yields

$$C = \frac{a\varepsilon_g{}^{V_g}\,(\varepsilon_r{}^{1-V_g}-1)}{d}x + \frac{a\,L\,\varepsilon_g{}^{V_g}}{d} \tag{4}$$

where ε_g is the permeability of the glass fiber and V_g is the volume fraction of the glass fiber. The value of the capacitance is derived as a linear function of x.

Darcy's Law

For the RTM process, Darcy's law is widely applied in the research, which is defined as

$$u = -\frac{K}{\mu}\cdot(\nabla P) \tag{5}$$

where u is the Darician velocity, μ is the fluid viscosity, ∇P is the pressure gradient, and K is the overall permeability tensor of the glass fiber. Following some common assumptions, Darcy's law can be simplified as

$$u_x = -\frac{K}{\mu}\cdot\left(\frac{\partial P}{\partial x}\right) \tag{6}$$

The assumptions include: 1) resin flows along a one-dimensional direction, i.e., the x-axis; 2) the z-scale is neglected [14]; 3) the flow is continuous and steady. We also took a constant pressure infusion, i.e., $P_0 = 1$ atm in the injection port and $P = 0$ atm at the vent because the pump exhausted the air in the mold at the fore-end. We made the mathematical calculation by (6). The in-plane permeability of the preform is

$$K = \frac{\mu}{2P_0}\frac{x^2}{t} \tag{7}$$

where L is the resin flow-front of location and t is the infusion time.

Based on (7), we can use the measured parameters of the parallel plate and LabVIEW software to calculate the permeability.

RESULTS AND DISCUSSION

To verify the feasibility of the proposed method, experiments were conducted using a nine-layer fiber preform. In the experiments, the injection pressure was set to be 1.0 atm. In order to find the relation between the capacitance and displacement, a transparent mold was utilized and the flow front was recored by a CCD camera. By plotting the flow front positions versus the corresponding capacitance values and fitting the data with a straight line, a calibration model was obtained as shown in Fig. 4. The slope and the intercept of the line are the $\frac{a\varepsilon_g\,(\varepsilon_r{}^{1-V_g}-1)}{d}$ and $\frac{a\,L\,\varepsilon_g{}^{V_g}}{d}$ terms in equation (4).

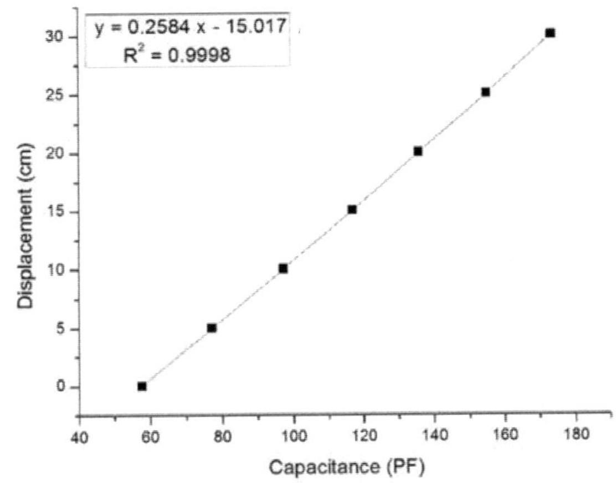

Fig.4. Calibration curve

Then the calibration model was used for estimating the flow front in the subsequent experiments. By substituting the capacitance values into the calibration model (4), the displacement estimates were acquired. As shown in Fig. 5, there was no significant error exhibited in the estimation results.

In the next step, the equivalent permeability between the injection port and the flow front was estimated at each sampling time point using equation (7). The results in Fig._6 showed that the equivalent permeability estimated at the beginning was higher than the average value. This was because the resin flow is not a fully-developed fluid flow when it just entered into the mold. Such entrance effect influenced the estimation results of the permeability. Gradually, the flow became more stable than it was initially. As a result, the permeability can be estimated by considering the resin flow as the steady-state Stokes flow [15, 16]. Hence, the subsequent estimates became consistent.

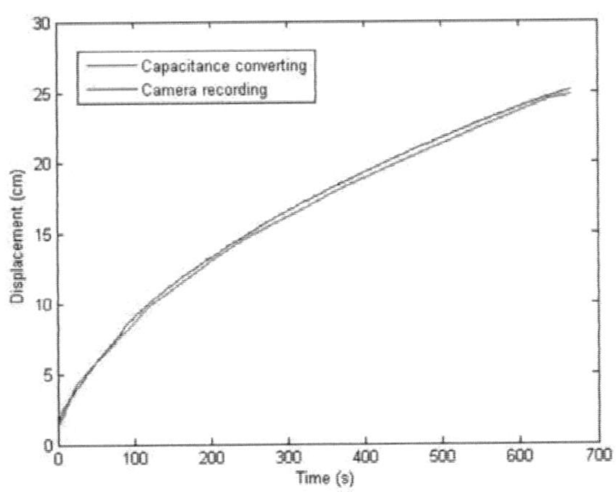

Fig.5. Comparing the results between the camera and capacitor

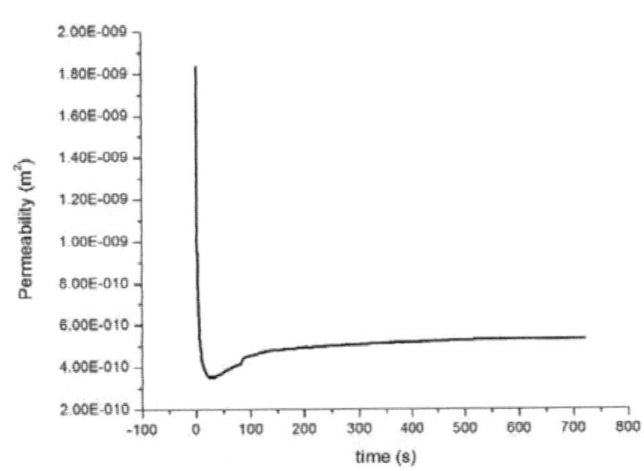

Fig.6. Overall permeability

CONCLUSIONS

In RTM, the permeability of the fiber varies with its geometric composition and is an important parameter that influences the resin flow properties. In this study, we use a parallel-plate capacitor to detect the capacitance values between the top and bottom mold plates, based on which online estimation of permeability can be achieved. A transparent mold was utilized in the experiments. However, it should be noted that it is not necessary to use a transparent mold and a CCD camera to get the calibration line. In fact, the calibration line can be obtained based on the information acquired at the initial and final points, i.e. the capacitance values measured when the flow front reaches the injection port and the vent.

REFERENCES

1. J. Li, X. Fu, C. Zhang, R. Liang and B. Wang, *Journal of Composite Materials*, **43**, 1695 (2009).
2. Y.J. Lee, Y.T. Jhan, C.H. Chung and Y. Hsu, *Engineering*, **3**, 691 (2011).
3. Robitaille F, Gauvin R. *Polymer Composites,* **19**, 198 (1998).
4. Weitzenbock JR, Shenoi RA, Wilson PA. *Polymer Composite*, **20**, 321 (1999).
5. Sun K. Kim, Isaac M. Daniel, *Composites: Part A,* **34**, 421 (2003).
6. Kris Hoes, Daniela Dinescu, Hugo Sol, Marleen Vanheule, Richard S. Parnas, Yiwen Luo, Ignaas Verpoest, *Composite : Part A*, **33**, 959 (2002).
7. Bai-Jian Wei, Yu-Sung Chang, Yuan Yao, Jun Fang, *Polymer Composites*, **37**, 1249 (2016)
8. L. Ding, C. Shih., Z. Liang., C. Zhang and B. Wang, *Composites Part A: Applied Science and Manufacturing,* **34**, 779 (2003).
9. Q. Liu, R.S. Parnas, H.S. Giffard, *Composites: Part A,* **38**, 954 (2007).
10. P. Ferland, D. Guittard and F. Trochu, *Polymer Composites*, **17**, 149 (1996).
11. Bekir Yenilmez, E. Murat Sozer, *Composites: Part A*, **40**, 476 (2009)
12. Ryosuke Matsuzaki, Seiji Kobayashi, Akira Todoroki, Yoshihiro Mizutani, *Composites: Part A*, **42**, 782 (2011)
13. Goncharenko, A. V., V. Z. Lozovski, and E. F. Venger. *Optics Communications,* **19**, 174 (2000)
14. S.G. Advani, E.M. Sozer, *Elsevier Marcel Dekker Inc* (2003).
15. Simon Bickerton, Suresh G. Advani, *Composites Science and Technology* **59**, 2215 (1999)
16. P.Simacek, S.G. Advani, *Composites: Part A*, **80**, 107 (2016).

An Approach to Non-Destructive Testing of Aged Polymers

Elmar Moritzer[a], Matthias Hüttner[a*], Bernd Henning[b], Manuel Webersen[b]

[a] *University of Paderborn, Kunststofftechnik Paderborn (KTP), Paderborn, Germany*
[b] *University of Paderborn, Measurement Engineering Group (EMT), Paderborn, Germany*

*Reference author: *Matthias Hüttner, matthias.huettner@ktp.upb.de*

Abstract. Because of the variably modifiable material qualities, plastics are used in every area of life nowadays. Therefore, the polymers have to meet various requirements to be able to guarantee the suitability for the use of the respective components. Besides, during their operation time the plastics are exposed to steadily changing environmental influences, which lead to the aging of the polymers. Frequent temperature changes, dampness, saline media and mechanical loads are examples of such influences [1]. Strength losses, changes in the material behavior, for example embrittlement, or a degradation of the molecular weight resulted from the aging process, can only be detected after the components' failure. Besides, to be able to prove these properties in aged plastic components, destroying testing methods are necessary, so that the checked components cannot be used anymore and are to be exchanged in any case. [2, 3]

Hence, in the following a measuring procedure will be introduced, with which aging effects in plastics can be detected non-destructively. In addition, polyamide 6 specimens are subjected to artificial hygrothermal aging and their macromolecular changes are made visible by means of ultrasonic testing.

Keywords: AGING, NON-DESTRUCTIVE, ULTRASONIC, HYGROTHERMAL, HYDROLYSIS
PACS: 43.58.+z, 81.40.Cd, 82.35.Lr, 81.05.Qk, *43.35.Zc, 81.70.Cv, 81.70.-q

AGING OF POLYMERS

The aging process of polymers is effected by complex mechanisms. During the lifecycle of thermoplastics, simultaneous physical and chemical aging mechanisms are at work [4]. Physical aging causes a change in the morphological structure of the polymer, e.g. relaxation, which does not influence the chemical constitution of the molecules. In semi-crystalline thermoplastics, for example, post-crystallization may occur. These processes are reversible by re-melting the polymer.

Chemical aging causes changes on the molecular constitution. Molecular chains, for example, may begin to break down under influences from environmental conditions, allowing micro-cracks to propagate starting from the part surface; the resulting notch effect may lead to part failure [5]. One particular type of chemical aging is the hydrolysis of condensation polymerized thermoplastics.

During molecular synthesis of this thermoplastic polymer type, low-weight molecular groups, such as water, are lost as by-products. Therefore, polycondensates display a high affinity to water. They are capable of storing water molecules in their amorphous zones, which leads to a reverse-condensation reaction. This phenomenon causes a splitting of the hydrolysable groups and results in a decrease of the molecular mass. An increase of the temperature accelerates the described decomposition. Other significant influences on the hydrolysis reaction are the degree of crystallinity in semi-crystalline polymers and the amount of reagent available [1, 3].

One of the most well-known polycondensates is polyamide, which belongs to the category of the engineering plastics. Especially polyamide 6, which is used in the following investigations, can absorb a comparatively large amount of water with circa nine mass percent and is thus particularly susceptible to hydrolysis [6].

EXPERIMENTAL INVESTIGATIONS

The experimental investigations are performed in two steps. In step one a standard polyamide 6 (PA6) is object of the aging experiments [7]. For that several values are analyzed on chemical, physical and mechanical level. The non-destructive testing is also performed by using an ultrasonic measurement system [7]. In step two endless glass fiber reinforced thermoplastics (GFRP) are aged in the same way and analyzed by measuring chemical, physical and mechanical values.

The aging process of the specimens is carried out by artificially accelerated hygrothermal aging. For this purpose, a temperature-controlled 60 °C saline bath with a sodium chloride content of 50 grams per liter is used. Samples are taken at selected durations to investigate the effects of hygrothermal aging over time. These intervals are shown in

Table 1. To ensure a commonly defined starting state for each of the following measurements, the aged specimens are brought to stasis in normal atmospheric conditions (23 °C / 50 % humidity).

TABLE 1. Aging stages und sampling intervals

Aging Stages	Time in weeks	Time in hours
1	0	0
2	1	168
3	2	336
4	4	672
5	8	1344
6	12	2016

The measurements carried out in the experiments for the standard PA6 used samples produced from two different semi-finished extruded products which consist out of the same material. The chemical, physical and mechanical properties are characterized using dog-bone specimens (DIN EN ISO 527-2, Type 1B) machined from calandered sheets. The ultrasonic measurements are carried out on hollow cylindrical samples produced from solid extruded rods [7]. The endless glass fiber reinforced thermoplastics consist of three layers of glass fiber fabrics (plain weave, fiber orientation: 0°/90°, thickness per layer: 0.5 mm) and a PA6 matrix. The measurements are carried out with a specimen geometry based on DIN EN ISO 527-4 (Type 3). Figure 1 (left) shows the used specimen geometries for the respective measurements.

To be able to draw additional conclusions about molecular changes in the aged polymer, the viscosity number is determined for each stage of aging according to standard DIN EN ISO 307 (90-percent formic acid, polymer-solvent-concentration: 5 grams per liter). On physical level, the crystallinity is determined by Differential Scanning Calorimetry (DSC) (DIN EN ISO 11357) and the specimens' water content is determined by Karl-Fischer titration after conditioning the material. As a result of these measurements, changes of the mechanical values are inferable. Therefore the mechanical quantities tensile strength und Young's modulus, among others, are recorded based on DIN EN ISO 527.

The non-destructive determination of the standard polyamide samples is carried out via ultrasonic testing. The specimens are clamped between two piezoelectric ultrasonic transducers which acts as a transmitter (at the bottom) and a receiver (at the top). An electric signal is converted into a mechanical ultrasonic wave by the transmitter, which travels through the hollow cylinder and arrives at the receiver. There, the mechanical wave is transferred back into an electric signal. The coupling between the transducers and the specimen is ensured by defined coupling gel layers. With the temporal knowledge of input signal and output signal, the transit time of the longitudinal wave can be determined. Using the length of the given sample the velocity of the longitudinal wave in the sample can be calculated as distance per unit of time. The test setup used for these investigations is shown in Figure 1 (right) and is based on [8].

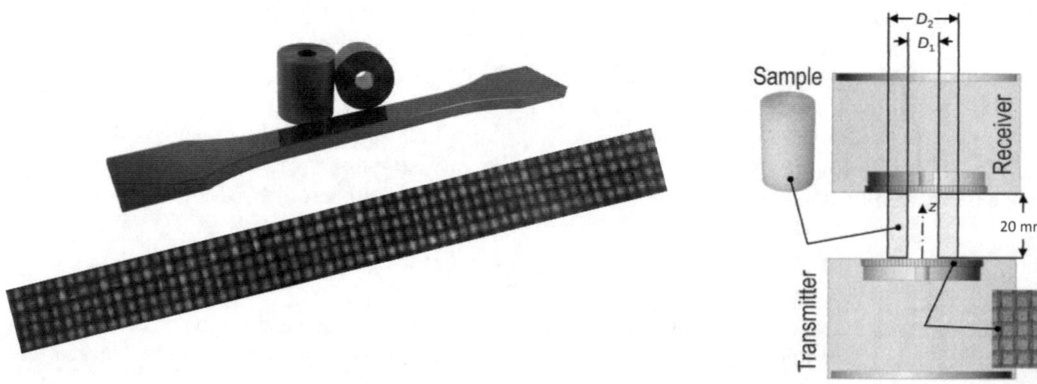

FIGURE 1. Specimen geometries of standard PA6 & endless glass fiber reinforced PA6 (left) and used ultrasonic measurement system (right) with D_1 as inner diameter and D_2 as outer diameter of the hollow cylinder

RESULTS AND DISCUSSION

Before the results of the endless glass fiber reinforced thermoplastic will be presented, the investigation of the the standard PA6 is shown.

Results of the Aged Standard Polyamide 6

Physical aging of the PA6 could not be confirmed based on a comparison of the respective degrees of crystallinity. This comparison is shown in Figure 2 (left). It shows that the crystallinity remains constant over the aging period. This allows to exclude the possibility of post-crystallization in the polymer, meaning that, morphologically speaking, no aging occurs under the given environmental conditions.

The viscosity number, which represents a measure of molecular chain length, decreases further the longer hygrothermal effects are at work, i.e. the longer aging continues. This means that the molecular chains are breaking and getting shorter, which leads to the conclusion that hydrolytic reactions take place. It is notable that a marked plateau can be observed between weeks 4 and 8; during this period, the observed variable does not change measurably. As of week 8, the viscosity begins once again to decrease. Figure 2 (right) shows the normalized viscosity number in relation to the respective original value from the untreated sample.

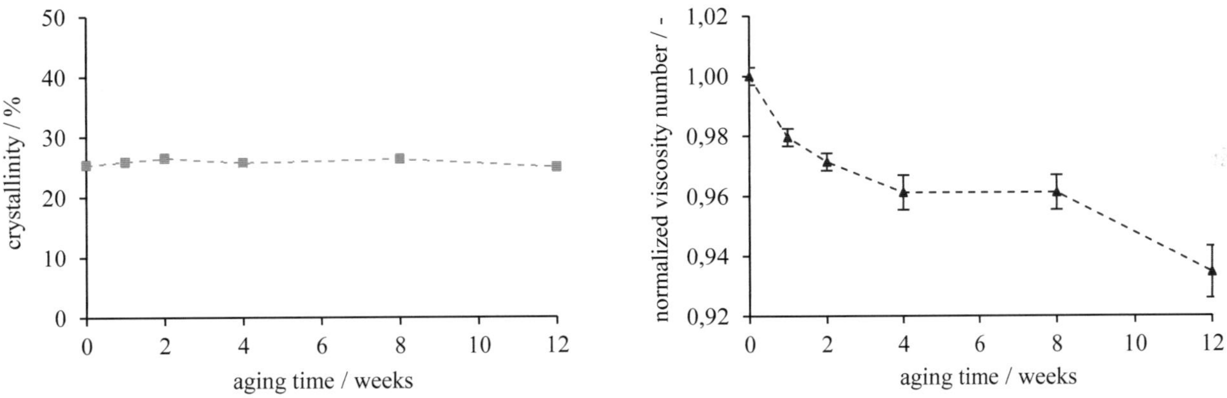

FIGURE 2. Crystallinity (left) and normalized viscosity number (right) over the aging period

The same phenomenon is reflected in the ultrasonic testing results, seen in Figure 3 (left). The normalized curves of the viscosity number and the longitudinal wave velocity, also normalized in relation to the respective starting value, are almost identical. Hence, a causal relation exists between the two values. This relation can be explained by the physical splitting and shortening of the polymer chains, leading to a higher number of chains in total. This fact leads to the conclusion that the aging of the PA6 can be characterized similarly with both or either measurement method.

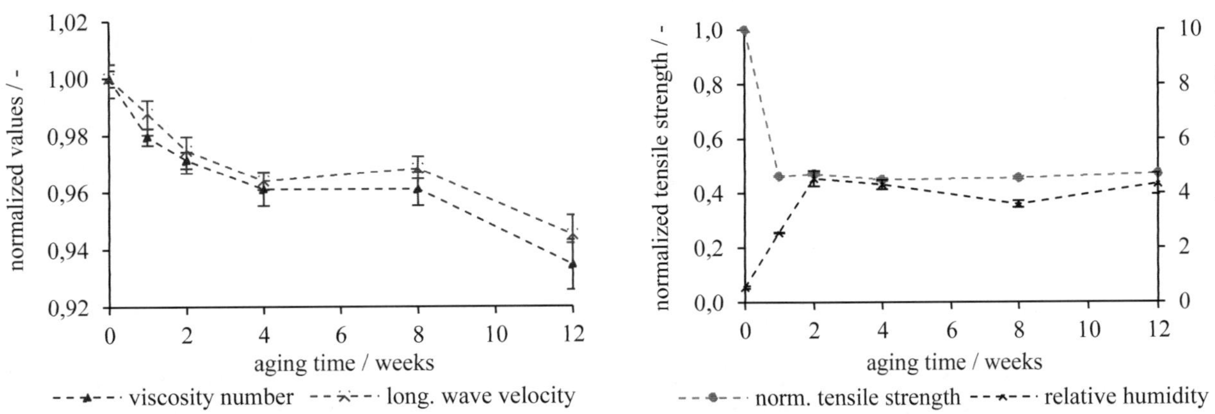

FIGURE 3. Normalized longitudinal wave velocity and normalized viscosity number (left) and normalized tensile strength and relative humidity (right) over aging period

The results of the mechanical investigations show for all recorded values a significant decrease of the respective dimension during the first week of aging time. After this decrease, the values are on a static low level over the whole investigation period. The shown normalized tensile strength in Figure 3 (right) illustrates this trend, which can be explained by the specimens water content, also shown in Figure 3 (right). The water content increases during the first two weeks and is on the same level over the following aging time. After the first week the specimens stored more than two mass percent of water inside, which leads to a softening effect in the polyamide. Consequently the material loses its strength and stiffness as seen in the diagram.

Results of the Aged Endless Glass Fiber Reinforced Polyamide 6

The investigations of the aged fiber reinforced PA6 show equally clear tendencies as the presented standard PA6. The crystallinity is expectably on the same level over the whole aging duration and is accordingly not presented here. The curve of the viscosity number, respective the normalized viscosity number presented in Figure 4 (left), shows the same behavior as the standard PA6. The more significant decrease of the polymer chain length is to be constituted by the PA6 matrix material, which is different from the used standard PA6 one. The hydrolytic reaction progresses faster and leads to an accelerated breakage of the polymer chains. Remarkable is the constant viscosity number between the aging weeks 4 and 8 that was already indicated in the PA6 investigations.

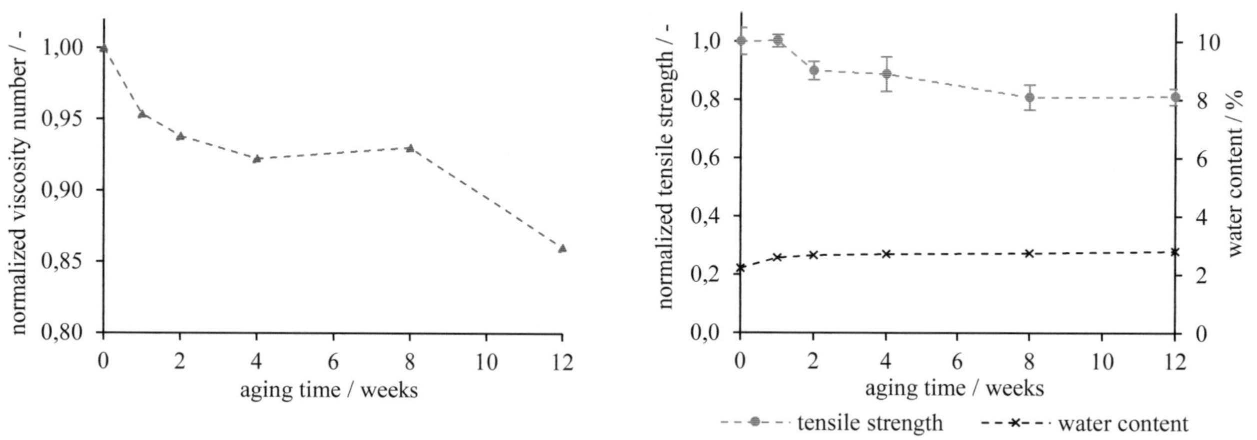

FIGURE 4. Normalized viscosity number (left) and normalized tensile strength in relation to the samples water content (right) over the aging duration of the endless glass fiber reinforced PA6

The results of the mechanical measurements reveal a different material behavior than the unreinforced PA6. The tensile strength decreases until after the first aging week, which cannot be explained by an increase of the specimens' water content. This quantity is on a relatively constant level over the whole aging time. Hence, the humidity does not really have a significant influence on the tensile strength of the endless glass fiber reinforced thermoplastic with a fabric orientation of 0°/90°. This fact is plausible because the prevailing main part of the force transmission is assumed by the fibers orientated in load direction. The decrease of the tensile strength after the first aging weeks has to be investigated particularly. A reason which may explains this phenomenon is that the aging of the polymer leads to a weakening of the laminates inner strength. The molecular degradation may induce a decrease of the fiber matrix interface. Due to that, the resulting forces by rising load cannot be transmitted into the fibers and a breakage at lower loads results. The aging process causes a tensile strength decrease of circa 20 percent over the observed duration. Besides the closer investigations of the mechanical changes during the aging, the non-destructive testing has to be performed. This planned experiments are described in the following prospects.

CONCLUSION AND PROSPECTS

The results of the experiments presented here show that the hygrothermal aging of polymers can be characterized using a variety of methods. The most significant results were obtained within the framework of the chemical and acoustic experiments for a standard PA6. A direct relationship was established between microscopic changes in terms of the viscosity number and changes in the acoustic characteristic velocity of sound propagation. Using these results, a non-destructive testing system can be developed to characterize the degree of aging of polymer materials.

This approach has to be adapted for fiber reinforced thermoplastics. Therefore, a different acoustic measurement system is required. Planned is a testing setup as shown in Figure 5.

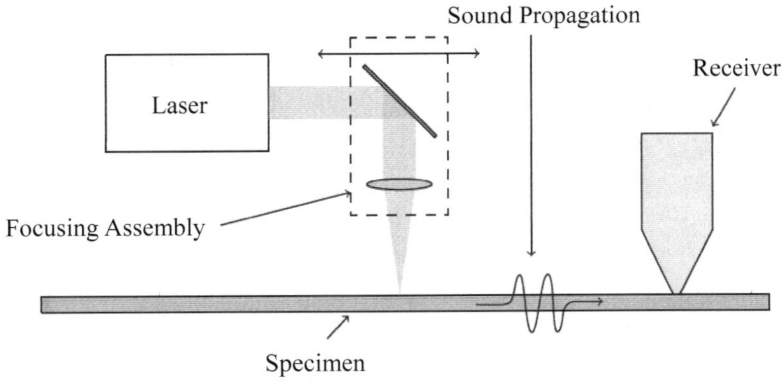

FIGURE 5. Non-destructive measurement system for testing of planar semi-finished products

The principle of function is that a pulsed ultraviolet laser, focused by a specific assembly, sends a high energy ultraviolet laser beam on the polymer specimen for an incremental short time. The energy adsorption leads to an infinitesimal thermal expansion which generates an ultrasonic wave. This ultrasonic wave propagates over the whole sample and is received by a piezoceramic transducer. This receiver converts the ultrasonic wave into an electrical signal, which is processed and can be analyzed. Material changes caused by the aging process can be measured by this setup.

Besides the molecular changes of polymers being measured non-destructively, the mechanical behavior of the investigated materials has to be analyzed by the used testing methods. These values have to be correlated with further acoustic values, so that additional relations may be detected.

Accordingly, the shown experiments have to be evaluated with further materials to validate the results. Therefore investigations with unreinforced and endless glass fiber reinforced polycarbonate are planned. An examination with unreinforced polyethylene terephthalate is currently being carried out.

ACKNOWLEDGMENTS

The results presented here were obtained in the context of the research project MO 685/13-1, sponsored by the Deutsche Forschungsgemeinschaft (DFG). We would like to thank the DFG for their support.

REFERENCES

1. G.W. Ehrenstein, S. Pongratz, *Beständigkeit von Kunststoffen,* Band 1, Carl Hanser Verlag, München, 2007
2. W. Grellmann, S. Seidler, *Kunststoffprüfung,* 2. Auflage, Carl Hanser Verlag, München, 2011
3. G. Menges, E. Haberstroh, W. Michaeli, E. Schmachtenberg, *Werkstoffkunde Kunststoffe,* 6. Auflage, Carl Hanser Verlag, München, 2011
4. DIN 50035, *Begriffe auf dem Gebiet der Alterung von Materialien – Polymere Werkstoffe*, Beuth-Verlag, 2012
5. P. Potthoff, *Alterung und Recycling von HDPE für Flaschentransportkästen,* Kunststoffe, 75 (1985) 8, Carl Hanser Verlag, München, pp. 481-486
6. P. Elsner, *Kunststoffe – Eigenschaften und Anwendungen,* 7. Auflage, Springer Verlag, Berlin Heidelberg, 2008
7. E. Moritzer, M. Hüttner, B. Henning, M. Webersen, *Non-Destructive Characterization of Hygrothermally Aged Polymers*, SPE ANTEC-Conference 2016, (accepted unpublished paper)
8. F. Bause, J. Rautenberg, B. Henning, C. Budde, E. Moritzer, *Utilizing ultrasound material parameter determination for the characterization of different stages of ageing in polyamide 6,* 6th Dresden Airport Seminar, 2013

Compensating Viscosity Fluctuations in Injection Moulding

Dipl.-Ing. M. Theunissen, Prof. Dr.-Ing. Ch. Hopmann and M.Sc. J. Heinisch*

Institute of Plastics Processing (IKV) at RWTH Aachen University, Pontstrasse 49, D-52062 Aachen, Germany

matthias.theunissen@ikv.rwth-aachen.de
zentrale@ikv.rwth-aachen.de

Abstract. A conventional control of the injection moulding process is based on machine values like the position and speed of the plasticising screw. Modern machine technology can reproduce these values with a high precision. However, systematic disturbances like fluctuating environmental conditions or variations in material viscosity still affect the reproducibility of the injection moulding process and the resulting part quality. Consequently adaptive and self-optimising control algorithms have to be applied for a further improvement of product quality. A pvT-optimisation allows a material-tailored online control of the holding pressure. Based on an online determination of the temperature state of the melt in the cavity, the cavity pressure can be closed loop controlled in order to achieve a constant specific volume of the moulded parts. Therefore a pvT-optimisation enables a compensation of thermal disturbances during injection moulding. However, fluctuations of material viscosity have a strong impact on the resulting profiles of the cavity pressure. Therefore under the influence of viscosity disturbances the specific volume and other related quality values cannot be reproduced homogenously using a pvT-optimisation. For the compensation of viscosity fluctuations an online measurement of the current viscosity in the process is necessary. For this purpose an indicator based on the pressure loss within the mould cavity is defined and analysed. On this basis a control of the viscosity is implemented. Within this concept the injection speed is adapted to the current process viscosity. First results of compensating viscosity fluctuations inflicted by the use of recycled material show the feasibility of the concept. In the future this concept has to be validated in combination with the pvT-optimisation. This enables an injection moulding process which is less sensitive to changing boundary conditions.

Keywords: compensation of viscosity fluctuations, cavity pressure control, pvT-optimisation
PACS: 81.05Lg; 82.35Pq; 83.50.Uv; 83.60.Fg; 83.80.Sg

INTRODUCTION

The quality of complex technical injection moulded parts is facing growing requirements regarding tolerances. As the available machine technology allows for a very accurate reproducibility of machine parameters like axis movement, conventional control concepts can only achieve further improvements in part quality with unreasonable efforts [1]. In spite of reproducible machine behaviour, changing boundary conditions to the injection moulding process may cause differences in the pressure and temperature history of moulded plastics material and consequently moulded parts. Adaptive and self-optimising control strategies enable an adaption of the process to the current boundary conditions and hence a further increase in part quality.

As part of the Cluster of Excellence 'Integrative Production Technology for High-Wage Countries' the concept of a model-based self-optimisation (MBSO) has been developed for injection moulding among other production processes (figure 1) [2]. Within this concept, temperature sensors like infrared (IR) sensors in information processing sensor actor system (ISA-system) allow the determination of thermal boundary conditions and the working point of the process. On this basis the optimal course of the cavity pressure can be calculated by a quality model in the model-based optimisation system (MO-system). The desired course is determined by the pvT-behaviour (pressure, specific volume and temperature) of the used polymer, which serves as a grey-box model. By this means, quality variables like the weight or warpage can be controlled regardless of thermal disturbances. The pvT-optimal cavity pressure is set up as a reference for the ISA-system, which realises the desired cavity pressure in a model predictive closed loop control [3]. This pvT-optimisation allows the compensation of fluctuating thermal boundary conditions e. g. resulting from changing environmental conditions or an in time changing cooling system behaviour of the mould [4].

However, differences in the material behaviour can still affect the process. The material viscosity is the most significant material property that affects the quality of processes and parts in plastics processing [5]. Especially in the context of a pvT-optimisation viscosity changes are crucial for the part quality since the pressure loss along the flow path changes along with the viscosity (figure 2). Consequently, different pressure profiles within the part cause different distribution of shrinkage potential for parts with a changed material viscosity. Accordingly, differences in warpage are expected to affect the geometrical dimensions of the part and result in rejects.

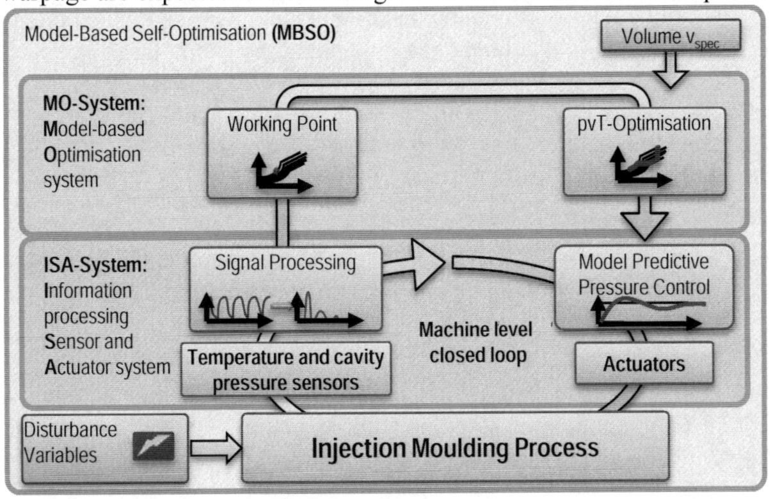

FIGURE 1. Model-Based Self-Optimisation (MBSO) concept for injection moulding

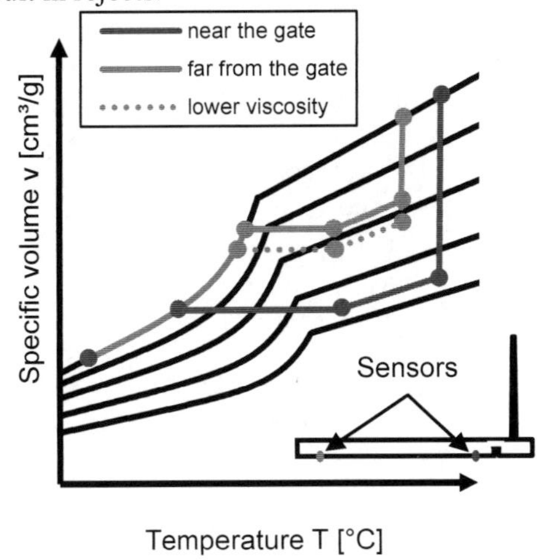

FIGURE 2. Effect of viscosity fluctuations in a pvT-diagram

RELATED WORK

Due to the crucial impact of material viscosity to the part quality serval concepts to evaluate and control the melt condition have been developed [6-10]. Generally, the melt viscosity during injection moulding can be evaluated using machine data, process data from mould sensors or with the use of special inline measuring instruments like rheometer nozzles [6].

Regarding machine data the torque and energy consumption of the plasticising unit during the plasticising phase correlate with the material viscosity [7]. During the injection phase the energy required for injection can serve as an indicator for the viscosity of the plastics melt [8]. The pressure of the screw can either be integrated over time (viscosity index) or the screw position (flow index). However, the closing time of commonly used non-return valves can change with different material viscosities. This can be considered by start and end point of the pressure integral over time. In this way, the so-called filling index allows an adaption of the changeover point and holding pressure to current melt properties and therefore a reduction in process variations [9].

Approaches based on data from mould sensors benefit from not being machine-dependent. An individual pressure sensor in a mould only allows for limited information on material viscosity since the flow behaviour cannot be considered. With the use of a pressure sensor near to the gate and a temperature sensor at the end of the flow path an evaluation of viscosity is possible. Therefore, the time between which the melt front reaches the two sensors is correlated with the shear rate. Additionally, the shear stress is correlated with the pressure value as soon as the melt front reaches the temperature sensor. Consequently, the shear viscosity can be considered as the quotient of shear stress and shear rate. A control of cylinder wall temperature and injection speed can help to reproduce constant process conditions regarding shear rate and stress [10].

In the following investigations an approach using two pressure sensors is chosen to evaluate and control the melt viscosity based on the pressure loss along the flow path. Previous investigations have shown that the injection speed is suitable to manipulate melt viscosity and hence influence the quality variable part weight [11].

EXPERIMENTAL SET-UP

The geometry of the investigated moulded parts is shown in figure 3. The moulded part is a plate specimen with a wall thickness of 2 mm and a cavity volume of 33.5 cm³. In the centre along the flow path two combined pressure-temperature sensors with a diameter of 4 mm are integrated in the mould in a distance of 108 mm (type 6190 CA from Kistler AG, Winterthur, Switzerland). The injection moulding experiments referred to in this paper were carried out on an injection moulding machine of the type 'Allrounder 520A 1500-400' produced by Arburg GmbH & Co. KG, Loßburg, Germany.

For the experiments a polybutylene terephthalate (PBT) was used (type: Ultradur B4520 from BASF SE, Ludwigshafen, Germany). To change the material viscosity recycled material was used. For this purpose, the same plate specimens have been produced and reground. To evaluate the viscosity change the new and the recycled material have been measured with a high pressure capillary rheometer (type: Rheograph 2002 from Göttfert Werkstoff-Prüfmaschinen GmbH , Buchen, Germany) (figure 4). The viscosity of the recycled material was found significantly lower with 98.6 Pa*s in contrast to the viscosity of the new material with 282.0 Pa*s (shear rate: 1,018 1/s, temperature 260 °C). Furthermore, the residual moisture content was assessed with *Karl Fischer* titration. Probably due to the lower initial moisture content because of the drying and processing of the material a week before, the recycled material had a significantly lower moisture content with 0,019 % in contrast to 0,035 % in the new material while processing.

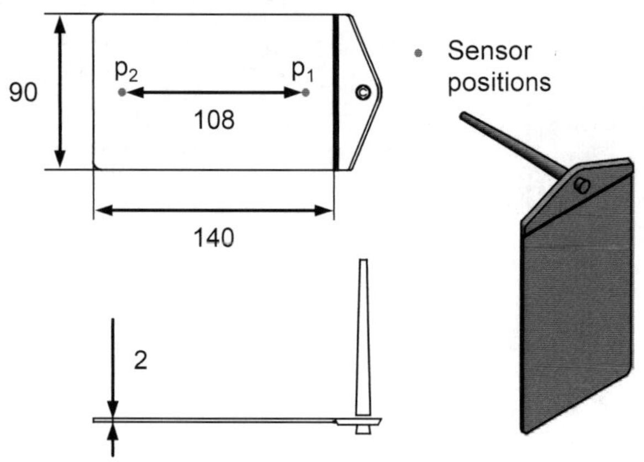

FIGURE 3. Investigated plate specimen with pressure sensor positions

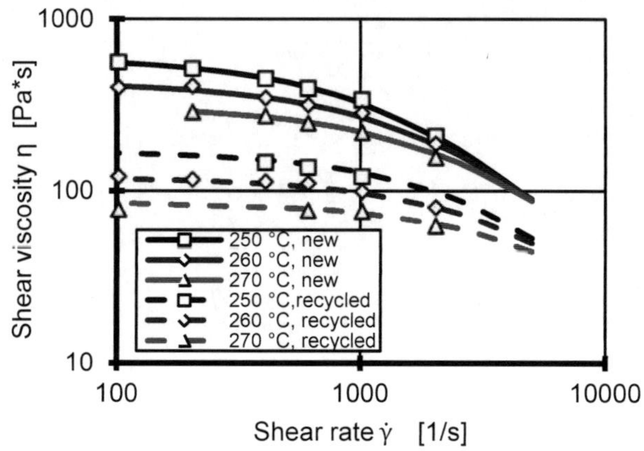

FIGURE 4. Shear viscosity of the new and the recycled PBT according to high-pressure capillary rheometer measurement

To evaluate the viscosity within these experiments the indicator K_{pl} has been defined (equation 1):

$$K_{pl} = \frac{p_1 - p_2}{\dot{V}} \tag{1}$$

The indicator K_{pl} relates the difference between the pressure p_1, which is acquired by the sensor close to the gate and the pressure p_2, which is measured by the second sensor 108 mm in the direction of flow, divided by the volume flow $\dot{V}$. In this case, the volume flow is the set injection speed normalised with the screw cross section. Since the melt flow front has to reach the second sensor before a reasonable viscosity indicator can be obtained, the pressures are measured once an increase in pressure of 5 bar at the second sensor has been detected. The definition of K_{pl} is inspired by the shear viscosity while all constant geometry parameters are neglected. However, the melt temperature and the shear rate are not constant during the cavity filling. Therefore, K_{pl} allows only an assessment of the current viscosity in the cavity and not a quantitative measurement. Whether this is enough for control purposes shall be investigated in the following.

Two series of tests with new and recycled material have been conducted and the moulded part weight has been evaluated. Therefore, in each experiment 1.5 kg new material has been placed in the hopper. After the material is almost finished it has been restocked with 1.5 kg of 100 % recycled material, then again 1.5 kg of 100 % new material and at least a mixture of 50 % new and 50 % recycled material have been restocked. In the first test series the process is controlled as a reference by the machine control. The setting parameters have been selected according to the material manufacture's recommendations. As changeover criteria to the pressure holding phase a cavity pressure of 300 bar has been used. In this case the holding pressure has been set to 250 bar in the plasticising unit.

In the second test series the injection speed has been adapted from cycle to cycle in order to achieve a constant K_{Pl} indicator. The reference value for K_{pl} has been obtained during the process set up with new material and when good part quality has been achieved. The injection speed has been proportionally adapted while the maximum step width has been set to 1 cm^3/s for stability reasons. Besides, the pressure holding phase was controlled by the cavity pressure as it would be for a process using pvT-optimisation.

RESULTS AND DISCUSSION

Figure 5 shows the resulting part weights of the first test series with the conventional machine controlled process. At the beginning, apart from the naturally occurring process variation, the mean value of the part weight is on a constant level. After approx. fifty cycles the part weight rises abruptly. With the holding pressure controlled at the back of the screw the lower material viscosity of the recycled material results in a lower pressure loss along the flow path and a higher pressure level in the cavity. Therefore, the solidification takes place on a lower specific volume and the part weight increases. When new material is processed again, the part weight drops due to the opposite effects. When the 50 % mixture of new and recycled material is processed the weight starts to rise again. However, this time a temporal evolvement is visible. The reason for this is probably the feeding behaviour of the plasticising unit. The round underwater pelletised virgin material is smaller than the sharper reground pellets. As a result, there is some segregation between the two materials and between the material batches successively put in the machine hopper.

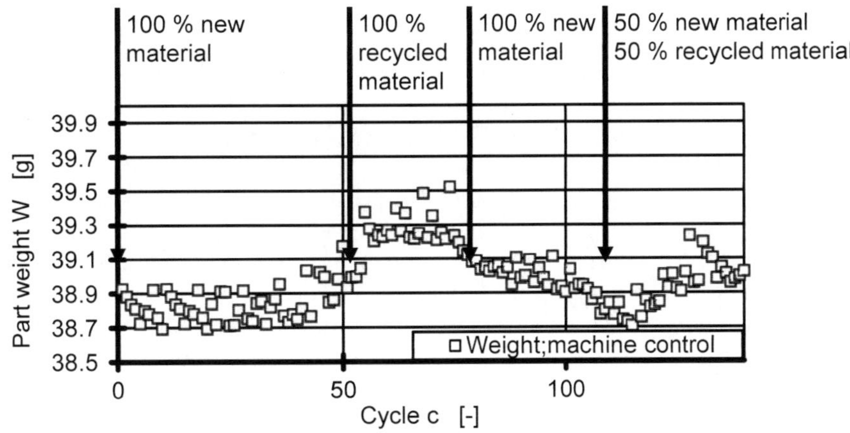

FIGURE 5. Resulting part weight with the conventional machine control

Figure 6 shows the resulting part weights and the injection speed of the second test series with the adapted injection volume flow and the cavity pressure controlled holding phase. Due to the cavity pressure control, the process variations are significantly lower. For the first fifty cycles no significant effects on the part weight have been detected. However, the injection volume flow is first actively lowered and then increased from its initially setting until a constant level is reached after approx. ten cycles. Probably the melt viscosity at the beginning was reduced due to the dwelling time during the start-up process. When the first recycled material is processed, the part weight drops immediately. In this case the effect of the lower viscosity is reversed due to the pressure control in the cavity. The pressure at the first sensor (figure 3) is controlled on a constant level. Therefore, the lower viscosity causes a lower pressure level in the sprue and sprue distributor and in the first 16 mm of the plate. After the first pressure sensor the pressure level is higher. In total, this causes a solidification of the material on a higher specific volume in the pvT-diagram and hence a lower part weight. In the following cycles the injection volume flow is lowered due to the lower viscosity and consequently the lower value of K_{pl}. The slower injection process causes less shear thinning and therefore the viscosity indicator K_{pl} can again reach its reference value which is indicated by a constant injection volume flow in the following cycles. Accordingly the part weight rises until it is back on its initial level.

Once the new material is processed again the viscosity rises abruptly and the injection speed is stepwise increased until it is back on its original level. Contrary to expectations, the part weight drops significantly at this point. Even a short shot is observed. An explanation for that can be the pressure dependent switch over point. With the injection volume flow being on a low level and the abrupt increase in viscosity the flow resistance is so high that the switch over criterion is reached while the cavity is not completely filled. Consequently, the cavity is not completely filled until the pressure holding phase (or in case of the short shot not at all). Due to that, the overall pressure level and the part weights are lower. With the increase of the injection speed the viscosity decreases again until the cavity is completely filled at the switch over point and the part weight is back at its initial level.

The processing of the 50 % mixture of new and recycled material is only detectable by a slight decrease in injection volume flow due to the lower viscosity. An effect on the part weight is not detectable. Whether this is due to the adapted injection volume flow, cannot be determined without a doubt. For this, a reference trial with a cavity pressure controlled holding pressure has to be conducted.

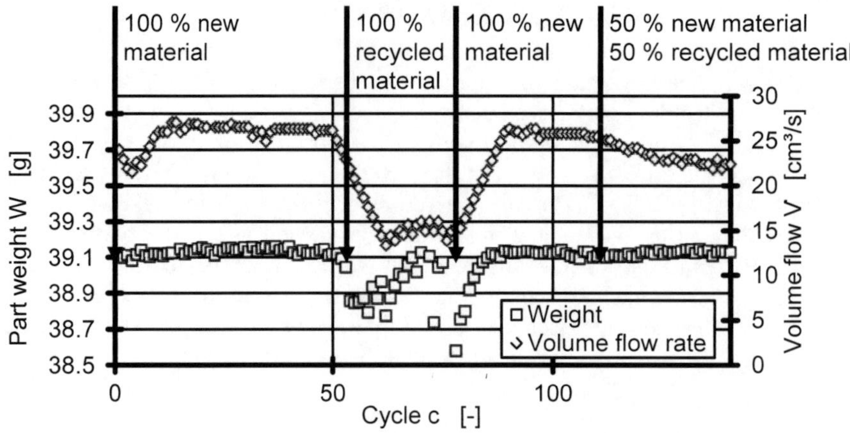

FIGURE 6. Resulting part weight and volume flow with adapted injection speed

CONCLUSION AND OUTLOOK

An evaluation of material viscosity based on the pressure loss of two pressure sensors along the flow path has been used as a reference value to control the part weight of injection moulded plate specimens using the injection volume flow as manipulated value. The effect of viscosity changes inflicted by the use of recycled polybutylene terephthalate on the part weight could be compensated after several cycles.

Further investigation on the effects of the switch over method should be conducted in order to be able to adjust it accordingly. The used proportional adaption method with a constant step wide should be modified to react faster on changing material properties.

ACKNOWLEDGMENTS

The depicted research has been funded by the Deutsche Forschungsgemeinschaft DFG as part of the Cluster of Excellence 'Integrative Production Technology for High-Wage Countries'. We would like to extend our thanks to the DFG. We give thanks to ARBURG GmbH + Co. KG (Loßburg, Germany) and BASF SE (Ludwigshafen, Germany) for supplying the injection moulding machine and plastics material used in the experiments.

REFERENCES

1. J. Wortberg, T. Kamps, and R. Schiffers, *A Comparison of Drive Technologies.* Kunststoffe international, 2002, 102 (12): pp. 8-12.
2. U. Thombansen, et al., *Model-Based Self-Optimization for Manufacturing Systems*, in *International Conference on Concurrent Enterprising*. 2011: Aachen.
3. C. Hopmann, et al., *A self-optimising injection moulding process with model-based control system parameterisation.* International Journal of Computer Integrated Manufacturing, 2015: pp. 1-10.
4. W. Michaeli and A. Schreiber, *Online Control of the Injection Molding Process Based on Process Variables.* Advances in Polymer Technology, 2009, 28(2): pp. 65-76.
5. F. Heinzler, C. Mielicki, and J. Wortberg, *A viscosity model for technical polymers describing chemical and physical aging.* Polymer Engineering & Science, 2015, 55 (7): pp. 1628-1633.
6. C. Gornik, *Viscosity Measurement: Determining Rheological Data Directly at the Machine.* Kunststoffe international, 2005, 95(4): pp. 88-92.
7. F. Heinzler, et al., *Useful Process Data from the Injection Molding Machine.* Kunststoffe international, 2014, 104 (2): pp. 42-46.
8. A. Lampl, K. Gissing, and H. Painsith, *Kontrolle des Formfüllvorganges beim Spritzgießen mit Hilfe der Einspritzarbeit.* Plastverarbeiter, 1983, 34 (10): pp. 1105-1108.
9. R. Schiffers, S. Kruppa, and S. Moser, *The Right Changeover Point for Each Shot.* Kunststoffe international, 2014, 104 (11): pp. 26-30.
10. C. Bader and E. König, *Taking Control.* Kunststoffe international, 2012, 102 (6): pp. 62-67.
11. C. Hopmann, M. Weber, and A. Ressmann, *Effect analysis for compensating viscosity fluctuations by means of a self-optimising injection moulding process.* AIP Conference Proceedings, 2015, 1664 (110004): pp. 1-5.

Swelling and drying process in ionic hydrogels for low and high pH using fluorescence technique

Demet Kaya AKTAŞ

Istanbul Technical University Department of Physics Maslak Istanbul TURKEY

demet@itu.edu.tr

Steady-state fluorescence (SSF) technique was employed for studying swelling and drying of poly(acrylamide-co-acrylic acid) (P(AAm-co-AAc)) composite gels for low and high pH. Disc shaped composite gels were prepared by free-radical crosslinking copolymerization of P(AAm-co-AAc). N, N'- methylenebis (acrylamide) (BIS) and ammonium persulfate (APS) were used as crosslinker and initiator, respectively. Pyranine 4 (4sPy) was introduced as a fluorescence probe. P(AAm-co-AAc) composite gels dried before using for swelling experiments. Fluorescence intensity of 4sPy was monitored during in situ swelling and drying processes of composite gels. It was observed that fluorescence intensity values decreased as swelling is proceeded. Li-Tanaka equation was used to determine the swelling time constants, τ and cooperative diffusion coefficients, D from intensity variations during the swelling processes. It was also observed that fluorescence intensity values increased as drying is proceeded. Gravimetrical and volumetrical experiments were also performed. It was shown that swelling time constants, τ decreased, diffusion coefficients for swelling process, D_F increased and diffusion coefficients for drying process D_G decreased as the pH values are increased.

Keywords: Hydrogels, swelling, drying, fluorescence.
PACS: 60

INTRODUCTION

Generally, hydrogels can be defined as three-dimensional networks of polymers[1,2]. The most favourable property of hydrogels is their ability to swell, when put in contact with a thermodynamically compatible solvent. When a hydrogel is in contact with solvent molecules, at first the solvent molecules attacks the hydrogel surface and then penetrates into the polymeric network. The swelling process of chemically cross-linked gels can be understood by competition between the osmotic pressure and the restraining force[3,4]. The total free energy of a chemical gel consists of bulk and shear energies. In fact, in a swollen gel, the bulk energy can be characterized by the osmotic bulk modulus K, which is defined in terms of the swelling pressure and the volume fraction of a polymer at a given temperature. On the other hand, the shear energy that keeps the gel in shape can be characterized by shear modulus G. Here, shear energy minimizes the nonisotropic deformations in the gel. A typical hydrogel is an example of smart material that means it can change its structure in response to external effects. Volume phase transitions in gels can be controlled by electric field, magnetic field, temperature or other extra external effects. On the other hand, ionic hydrogels made of crosslinked polyelectrolyte display big differences in swelling and drying properties depending on the pH of the environment. Since the pendant acidic and basic groups on polyelectrolytes undergo ionization, the swelling of polyelectrolylate hydrogels is mainly due to electrostatic repulsion among charges present on polymer chain[5]. Many theorical, experimental and macroscopic techniques have been used to investigate volume phase transitions in pH sensitive hydrogels up to now. Fluorescence technique that is of particularly useful for elucidation of detailed structural aspects in moleculer dimension of the polymer gels. This technique is based on the interpretation of the change in anisotropy, emission and/or excitation spectra, emission intensity, and viewing the lifetimes of injected aromatic molecules to monitor the change in their microenvironment [6]. Although most of these studies were focus to use different spectroscopic methods to investigate macroporous surface and morphology; otherwise in situ steady-state fluorescence technique provides a comprehensive tool for understanding of polymer gels, volume phase transitions and also pH effect on volume phase transition in microscopic level. It is obvious that the fluorescence technique may measure the behavior of the microstructure of the gel, here segmental motion of the gel network can be modelling by change in fluorescence intensity of bonded aromatic molecule and therefore swelling and drying process can be monitored at a molecular level. In this study we investigated the effect of pH during swelling and drying processes in poly(acrylamide-co-acrylic acid) P(AAm-co-AAc) ionic composite hydrogel by using fluorescence technique. Here 4sPy was chosen as an aromatic molecule since its emission spectra does not depend on pH. Change in fluorescence intensity of 4sPy was monitored during in situ swelling and drying processes of composite gels. Swelling - drying time constants and diffusion coefficients were determined from modelling with change in fluorescence intensity as a result of solvent effect during volume phase transitions processes. It was shown that swelling time constant, τ decreased, diffusion coefficient for swelling process, D_F increased and diffusion coefficient for drying process, D_G decreased as pH values are increased. The gravimetric measurements for swelling and drying process were also performed to compare results found from fluorescence technique. It is expected that diffusion coefficient, D values measured by using fluorescence technique are at least order of magnitude much larger than the values measured by gravimetric technique, which may present the different behaviors of the gel. Segmental motion of the gel network can be monitored by using fluorescence intensity and monitors the swelling at a molecular level. According to the above argument, one may suggest that chain segments move much faster than the bulk polymeric material during volume phase transition process.

Results and Discussion

Emission and scattered intensity of P(AAm-co-AAc) composite gel were monitored during the swelling process in low and high pH solutions. It can be seen that as the solvent uptake is increased fluorescence intensity, I_{em} decreased and the scattered light intensity, I_{sc} increased. Since the increase in I_{sc} corresponds to increase in turbidity of the swelling gel, then we have defined corrected fluorescence intensity, I as I_{em}/I_{sc}. It was observed that fluorescence intensity values decreased as swelling is proceeded. Plots of water uptake, W found from relation between fluorescence intensity versus swelling time are presented in Figure 1a and b for swelling in pH 4 and pH 10, respectively. It has been suggested [5] that the kinetics of swelling and shrinking of a polymer network and/or gel should obey Li-Tanaka relation,

$$\frac{W_t}{W_f} = 1 - \sum_{n=1}^{\infty} B_n e^{\frac{-t}{\tau_n}} \tag{1}$$

Here W_t and W_f are the solvent uptakes at time t and at infinite equilibrium, respectively.

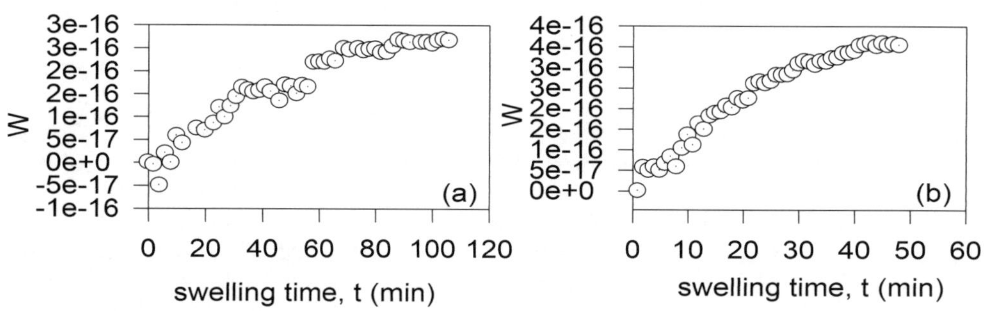

FIGURE 1. Solvent uptake versus swelling time for a) pH=4 and b) pH=10, respectively.

The logarithmic forms of the data in Figure 1 are fitted to the logarithmic form of equation (1) as follows.

$$\ln\left(1 - \frac{W_t}{W_f}\right) = \ln B_1 - \frac{t}{\tau} \tag{2}$$

Using equation (2) linear regressions of curves in Figure 2and b provide us with B_1 and τ values. Taking into account the dependence of B_1 on R, one obtains R values and from α_1-R dependence α_1 values were produced [7]. τ is inversely proportional to the collective cooperative diffusion coefficient D of a gel disk at the surface and given by the relation [5]

$$\tau = \frac{3a^2}{D\alpha_1^2} \tag{3}$$

Then using equation (3), cooperative diffusion coefficients were determined for these disc-shaped composites and found to be around 10^{-8} m^2/s. Here a represent half of the disc thickness in the final infinite equilibrium which can be experimentally determined. Experimentally obtained τ and D values are summarized in Table 1 found from both fluorescence and gravimetric techniques. It was shown that swelling time constants, τ decreased and diffusion coefficients, D increased as the pH values are increased.

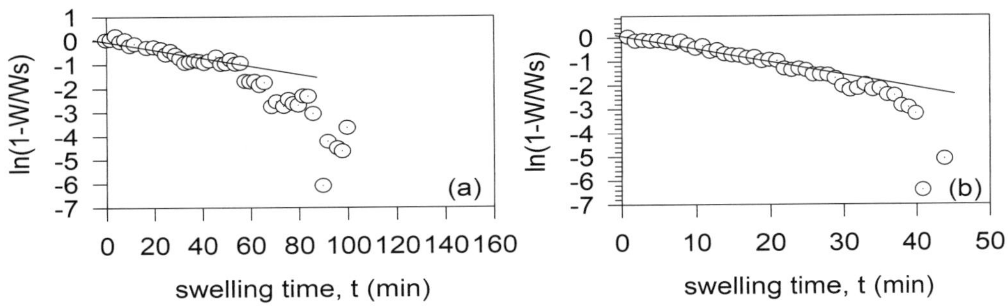

FIGURE 2. The logarithmic forms of the data in Figure 1 are fitted to the logarithmic form of equation (1) for a) pH=4 and b) pH=10.

On the other hand, during drying, it is assumed that two regions exist in the gel, i.e. less concentrated region near the surface separates itself from the high concentrated region by a boundary which moves during drying towards the center of the gel. As the boundary moves, water molecules evaporate from the surface of the gel into air. Then, the behavior of W against t_d can be explained by considering a model in which diffusion occurs in two regions separated by a moving interface [8]. The moving interface can be marked by a discontinuous change in concentration as in the absorption by a liquid of a single component from a mixture of gases or by a discontinuity in the gradient of concentration as in the progressive freezing of a liquid. When the diffusion coefficient is discontinuous at a concentration c; i.e. the diffusion coefficient is zero below c and constant and finite above c; then the total amount, M_t of diffusing substance desorbed from unit area of a plane sheet of thickness a at time t, is given by the following relation

$$\frac{M_t}{M_\infty} = 2\left[\frac{D}{\pi a^2}\right]^{1/2} t^{1/2} \tag{4}$$

where D is a diffusion coefficient at concentration c_1. Here $M_\infty = a\,c_1$ is the equilibrium value of M_t. If one assumes that the diffusion coefficient of polymer segments in the gel is negligible compared to the desorption coefficient, D of water vapor into air, then Equation (4) can be written as follows

$$\frac{W}{W_\infty} = 2\left[\frac{D}{\pi a^2}\right]^{1/2} t_d^{1/2} \tag{5}$$

Here it is assumed that M_t is proportional to the amount of solvent molecules released, W at time, t_d. The plots of W versus $t_d^{1/2}$ for pH=4 and pH=10 are presented in Figure 3a and b, respectively where the fit of the data produced the desorption coefficient, D which are listed in Table 1. It is seen that D values decreased as the pH content is increased.

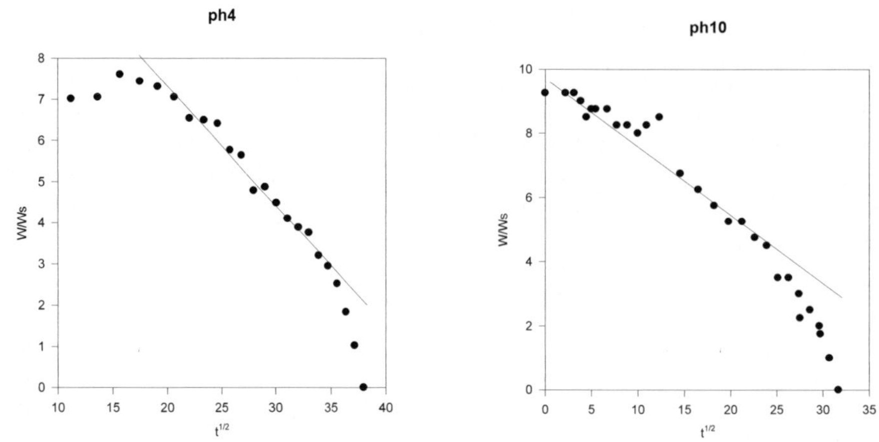

FIGURE 3. The plots of W versus $t_d^{1/2}$ for pH=4 and pH=10. Linear regressions of the data according to Equation 5 produces desorption coefficients.

TABLE (1). Experimentally measured parameters for swelling and drying processes using both fluorescence and gravimetric techniques.

pH	$\tau_{F\text{-swelling}}$ (min)	$\tau_{G\text{-swelling}}$ (min)	$D_{F\text{-swelling}}$ ($x10^{-8}m^2/s$)	$D_{G\text{-swelling}}$ ($x10^{-8}m^2/s$)	$D_{F\text{-drying}}$ ($x10^{-8}m^2/s$)	$D_{G\text{-drying}}$ ($x10^{-9}m^2/s$)
4	60	50	25.5	30.6	12	5.77
10	15	44.11	300	102	5.3	4.18

Here it is seen in Table 1 that D values measured by using fluorescence technique are much larger than the values measured by gravimetric technique, which interested different behaviours of the gel. It is obvious that the fluorescence technique measure the behaviour of the microstructure of the gel i.e segmental motion can be measured by fluorescence because pyranine molecules are bounded to the polymer chains. However, gravimetric measurements may provide us with the behaviour of the macroscopic environment. According to above argument, one may suggest that chain segments move much faster than the bulk polymeric material during swelling process.

REFERENCES

1- T. Tanaka, *Sci. Am.*, **244,** 124-136 (1981).
2- D. Rossi, D. Kajiwara, Y. Osada and A. Yamauchi, *Polymer Gels*, Plenum press New York (1991) .
3-S. Candau, J. Bastide and M. Delsanti, *Adv. Polym. Sci.*, **44,** 27-73 (1982).
4-L. Yong and T. Tanaka, *J. Chem. Phys.,* **92,** 1365-1371 (1990).
5-T. Tanaka, *American Chemical Society.***480** , pp. 1-21 (1992).
6-J.R. Lakowicz, Principles of Fluorencence Spectroscopy, Plenum Press, New York (1983).
7- Y. Li and T.Tanaka, *J. Chem. Phys.*, **92**, 1361371 (1990).
8- J. Cranck, Mathematics of diffusion, Oxford, Clarendon Press (1975).

On the Exploitation of Optical Signal from Raman Spectroscopy for In-Situ Conversion Monitoring of Emulsion Polymerization

Dropsit E.[a,b,e,f*], Chapron D.[a,b], Bourson P.[a,b], Hoppe S.[c,d], Durand A.[e,f]

[a]*LMOPS, EA 4423, Université de Lorraine, 2 rue Edouard Belin, Metz, F-57070 (France)*
[b]*LMOPS, EA 4423, Centrale-Supélec, 2 rue Edouard Belin, Metz, F-57070 (France)*
[c]*LRGP, UMR 7274, Université de Lorraine, 1 rue Grandville, Nancy, F-54001 (France)*
[d]*LRGP, UMR 7274, CNRS, 1 rue Grandville, Nancy, F-54001 (France)*
[e]*LCPM, UMR 7375, Université de Lorraine, 1 rue Grandville, Nancy, F-54001 (France)*
[f]*LCPM, UMR 7375, CNRS, 1 rue Grandville, Nancy, F-54001 (France)*

elise.dropsit@univ-lorraine.fr

Abstract. Raman spectroscopy was used for in-situ monitoring of styrene emulsion polymerization with sodium dodecyl sulfate and potassium persulfate as the surfactant and initiator, respectively. A new method for analyzing Raman data and calculating monomer conversion was proposed on the basis of the shift of peak corresponding to aromatic ring breathing. It was demonstrated that monomer conversion calculated with this method was similar to the one calculated using the area of vinylic bond as well as to off-line gravimetric determination. Moreover, the kinetic intervals were detected. The conversion measured at the transitions was similar to the literature data.

Keywords: Styrene, Emulsion Polymerization, In-situ Monitoring, Raman Spectroscopy, Kinetic intervals.
PACS: 82.70.Kj; 82.80.Gk.

INTRODUCTION

Emulsion polymerization is an industrial process which produces important chemicals like latexes for paints or other coatings. Because of the complexity of the reaction medium and the number of physical and chemical phenomena involved during the reaction there has been also a great deal of academic work about elucidation of mechanism, control of particle morphology in final latex, chemical engineering aspects… The need to monitor several properties of reaction medium during polymerization (particle size distribution, monomer conversion, surface coverage by surfactant) has led to a lot of investigations. Because of the difficulties associated with sampling a heterogeneous reaction medium, in-situ monitoring using spectroscopic techniques has raised interest. In particular, Raman spectroscopy, with its very low sensitivity to water content, has appeared as a very convenient tool for about 25 years[1-3]. As for all spectroscopic techniques, one important question is the exact link between experimental data (peak intensity or position) and quantities of species within reaction medium. Regarding the emulsion polymerization of styrene, the most used method has been to correlate peak ratios to monomer conversion on the basis of signal from vinylic double bond and aromatic ring. Nevertheless, this method has been questioned since it has long been established that even Raman peaks originating from non-reactive groups, like aromatic ring, underwent variations of area during polymerization[4-7]. In that work, we introduced a new method for monitoring styrene conversion based on Ramanshifts instead of ratios of peak areas. Both methods were compared to off-line monitoring based on gravimetry. In addition, main kinetic aspects of emulsion polymerization based on Smith-Evart mechanism were evidenced using both methods.

EXPERIMENTAL SECTION

Materials and Polymerization

All chemicals were purchased from Aldrich® and used as received. Potassium persulfate (KPS) was used as thermal initiator, sodium dodecyl sulfate (SDS) was used as surfactant. Hydroquinone was used as radical scavenger for stopping polymerization when sampling reaction medium. The amount of SDS was varied in the experiments (TABLE 1).

TABLE 1. Recipes used for preparation of polystyrene latexes.

Recipe	Distilled Water (% w/w)	Styrene (% w/w)	SDS / Styrene (g/g)	KPS / Styrene (g/g)
1	90.5	9.5	0.10	0.02
2	90.5	9.5	0.07	0.02
3	90.5	9.5	0.16	0.02

The emulsion polymerizations were performed in a 1 L double-jacketed glass reactor at 60 (+/- 2) °C under a nitrogen atmosphere. Apart from the preemulsification step, the reaction was carried out under magnetic stirring at 600 rpm for avoiding disturbances in optical measurements. The total volume of the aqueous phase was 445 mL.

Raman Spectroscopy

In-situ measurements were performed with the RXN-2 spectrometer (Kaiser Optical System®), with non-contact probes equipped with lenses of 40 mm focal. The spectrometer had a spectral resolution of 1.5 cm^{-1}. The laser excitation wavelength was 785 nm with a power of 400 mW. The selected acquisition time was 10 s and a spectrum was recorded every 15 s. The probe-reactor distance was optimized and set at 22 mm for all experiences.

Off-Line Conversion Measurements

Reaction medium was periodically sampled and polymerization stopped by adding hydroquinone. The monomer conversion was determined gravimetrically using a Metler Toledo® HG 53 infrared balance. The measurements were made at 170°C until the mass of the sample stabilized and the monomer conversion was calculated (EQ 1).

$$X_{Gravimetry} = \frac{(m_{ES} - m_{NVS})}{m_{Styrene,0}} \quad \textbf{(EQ 1)}$$

In EQ 1 m_{ES} stands for the final dry weight of the sample, m_{NVS} is the weight of non-volatile species contained in the sample and $m_{styrene,0}$ is the initial monomer weight (m_{NVS} and $m_{styrene,0}$ were calculated from reactor load).

RESULTS AND DISCUSSION

In Situ Monitoring of Conversion: Description and Comparison of Two Methods

The first method, *Area Method*, was used in most previous studies[8,9]. It is based on the monitoring of a characteristic monomer peak which decreases during the reaction, the peak at 1630 cm^{-1}, assigned to the stretching mode of the vinylic double bond. In order to remove parasitic variations of Raman signal due to other phenomena than the reaction, the peak at 1000 cm^{-1} was used as reference (EQ 2). It was assigned to the trigonal ring breathing mode in styrene as in polystyrene[10] and was assumed to remain unmodified during the polymerization.

$$X_{Area} = \frac{\left(A_{1630}/A_{1000}\right)_0 - \left(A_{1630}/A_{1000}\right)_t}{\left(A_{1630}/A_{1000}\right)_0} \quad \textbf{(EQ 2)}$$

In EQ 2, A_{1630} and A_{1000} are the area of the bands at 1630 cm^{-1} and 1000 cm^{-1} respectively. They were calculated using the iC Raman Software (Kaiser Optical System®) by setting two points for the peak definition and two points to create a linear baseline (TABLE 2).

TABLE 2. Definition of Raman bands.

Peak	Assignment	Band Definition	Baseline Definition
1630 cm^{-1}	C=C monomer stretching mode	1638 cm^{-1} – 1624 cm^{-1}	1640 cm^{-1} – 1620 cm^{-1}
1000 cm^{-1}	Trigonal ring breathing mode	1009 cm^{-1} – 992 cm^{-1}	1010 cm^{-1} – 990 cm^{-1}

The second method, *Shift Method*, was based on the position of the reference band at 1000 cm^{-1} which is common to styrene and polystyrene and shifts during the reaction. As the ring breathing is not the same in a small molecule like monomer and in a macromolecule, the single peak could be deconvoluted into two, separated by around 2 cm^{-1} with areas varying with conversion[4-7]. In an isothermal process, this resulting shift is solely due to the decrease of the styrene peak at 999 cm^{-1} and the increase of the polystyrene one at 1001 cm^{-1}. Consequently we proposed to calculate the conversion of styrene on the basis of the relative shift (EQ 3).

$$X_{Shift} = \frac{(RS_{1000})_t - (RS_{1000})_{Styrene}}{(RS_{1000})_{PS} - (RS_{1000})_{Styrene}} \quad \textbf{(EQ 3)}$$

$(RS_{1000})_t$ is the Ramanshift of the band at the time when the conversion is calculated (t), $(RS_{1000})_{Styrene}$ stands for pure styrene (X = 0) and $(RS_{1000})_{PS}$ for pure polystyrene (X = 1) in an emulsion process.
Thus two peaks of Raman spectra were of particular interest for the Area and Shift Methods (FIGURE 1).

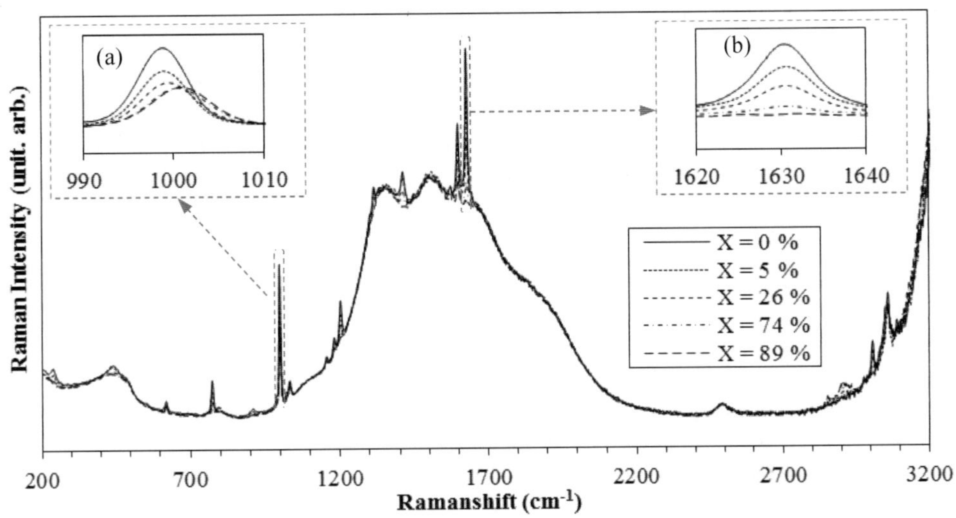

FIGURE 1. Evolution of Raman spectrum during an emulsion polymerization (recipe 1). Insets are zooms onto trigonal ring breathing mode peak (a) and monomer C=C stretching mode peak (b).

Comparison of Area and Shift Methods

The results of in-situ conversion monitoring during an emulsion polymerization (recipe 1) were analyzed using both methods and compared to off-line gravimetric analysis (FIGURE 2).

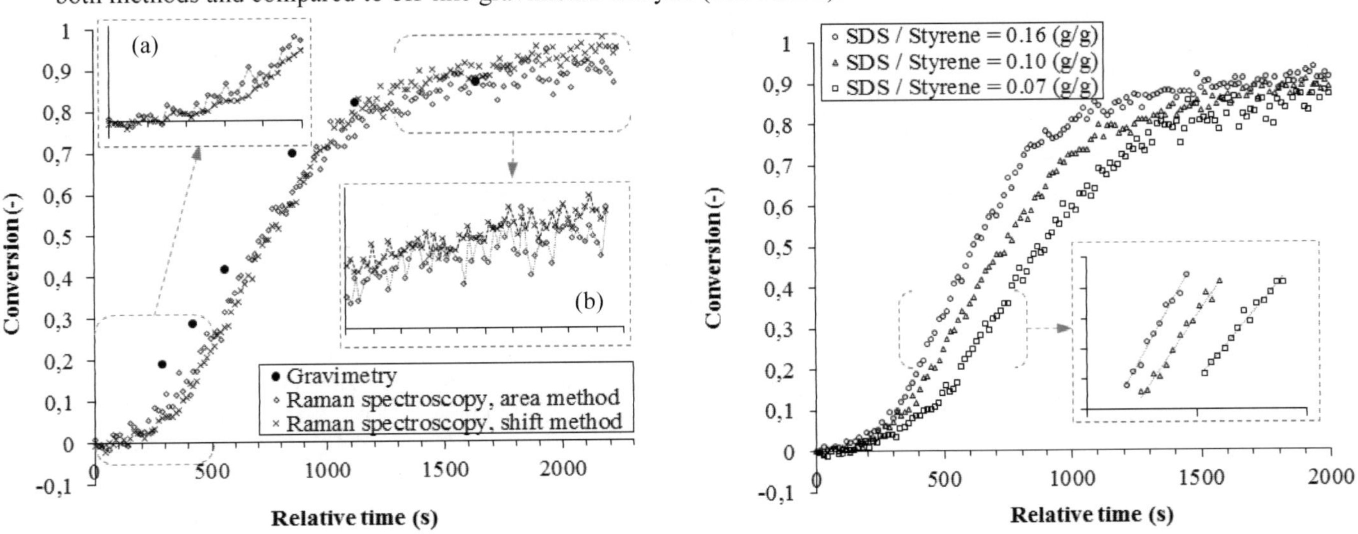

FIGURE 2. Left: Comparison of styrene conversion during an emulsion polymerization (recipe 1) determined by gravimetry and two different Raman methods, with zooms onto the beginning (a) and the end (b) of the reaction. For symbols see graph. Right: Influence of the amount of SDS on the kinetics.

Apart from a time offset, in-situ monomer conversion was in good agreement with off-line results. The time offset was attributed to the sampling time and the efficiency of the hydroquinone to stop the reaction. Both methods led to consistent estimations of styrene conversion. These results demonstrate for the first time that the use of Ramanshift provides reliable estimation of the monomer conversion.
The Shift method gave better results than the Area one especially at the beginning and at the end of the reaction. The smoothing given by monitoring the Raman shift of the peak at 1000 cm^{-1} brings new opportunities to go deeper in the in-situ kinetic studies, especially at the early stage of the reaction.

A Study of Kinetics: Detection of the Three Intervals of Emulsion Polymerization

The emulsion polymerization of styrene stabilized by a surfactant at a concentration above its critical micelle concentration follows the Smith-Ewart kinetics case 2[11]. It is divided into three intervals (I to III), with transitions depending on the amount of surfactant (I - II) and the swelling of polymer particles by monomer (II - III). The three intervals are characterized by different rates of polymerization.

$$R_p = k_p [M]_p \frac{\bar{n}}{N_A} \frac{N_p}{V} \qquad \textbf{(EQ 4)}$$

In EQ 4, k_p, $[M]_p$, $\bar{n}$, N_p, N_A, V stand for the propagation constant ($m^3.mol^{-1}.s^{-1}$), the monomer concentration in the latex particles ($mol.m^{-3}$), the average number of free radicals per particle, the number of particles, the Avogadro number and the volume of the reaction medium (m^3) respectively.
In the nucleation stage (I), R_p increases drastically because of an increasing number of polymer particles. After complete disappearance of surfactant micelles, no more particle can be formed and R_p remains constant, it is the growth of particles stage (II). When the monomer droplets disappear, it is the beginning of the last kinetic stage (III), R_p decreases because of monomer consumption within particles in absence of a gel-effect as in the polymerization of styrene.
By using the two previous methods, these critical conversions at transitions were determined (FIGURE 3).

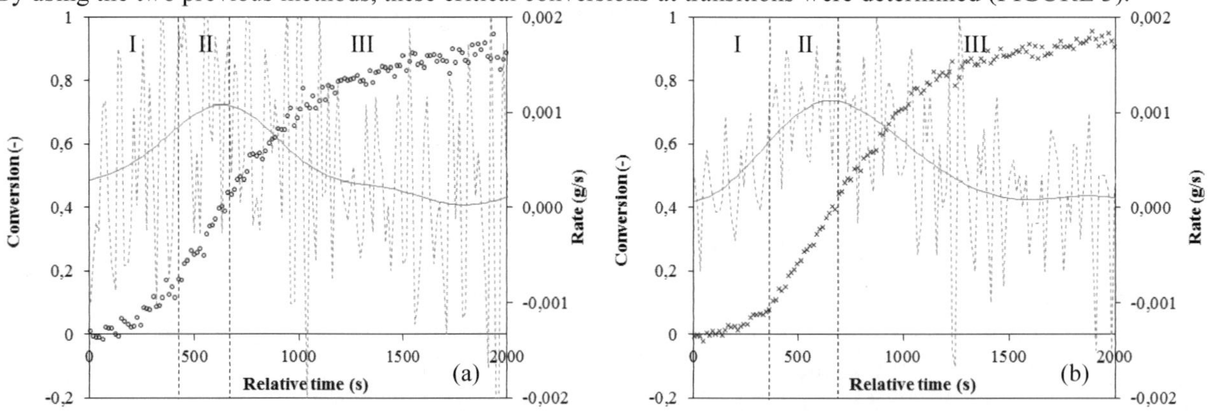

FIGURE 3. Evolution of the monomer conversion and the rate of polymerization given by the Area method (a) and the Shift method (b). The dotted lines stand for the rate raw data.

The I to II transition was determined directly from the conversion data thanks to a linear fit study. For the II-III transition, the conversion data were differentiated and filtered. The conversion corresponded to the end of the plateau and the beginning of the rate decrease. The results were compared to literature data (TABLE 3).

TABLE 3. Critical conversion values obtained by in-situ conversion and rate data studies.

Method	$X_{I\text{-}II}$ (%)	$X_{II\text{-}III}$ (%)
References[12, 13]	5-10	36-42
Area method	17	45
Shift method	6	44
Fitted Area method	9	41
Fitted Shift method	9	37

Because the rate was calculated from conversion, the same effect of the noise appeared: the Shift method gave better data than the Area one. These results were improved by deepening the procedure of signal analysis. In order to reduce the measurement error, a fit was done on each band from raw data by a Gaussian model, with no constraint for the peak at 1000 cm^{-1} and by fixing the width at half maximum height for the peak at 1630 cm^{-1}. Thanks to this additional step, the final results obtained were improved (TABLE 3).

Thanks to the detection of the intervals and to the sensitivity of the Raman spectroscopy even for fast reactions, the influence of the amount of SDS on the kinetics was examined (FIGURE 2). As it is shown on EQ 5, an increase of the amount of SDS involves an increase in the number of particles and thus an increase in R_p.

$$N_p \propto a[SDS]^{0.6}[KPS]^{0.4} \qquad \textbf{(EQ 5)}$$

The comparison of rates of polymerization during interval II could be done by using a linear fit of the conversion data. The relative variation of R_p with SDS concentration is fully consistent with EQ 5 (TABLE 4).

TABLE 4. Influence of SDS on the rate of polymerization during interval II.

Recipe	1	2	3
Rate during interval II (g/s)	$10.7\ 10^{-4}$	$8.6\ 10^{-4}$	$13.3\ 10^{-4}$
Relative rate	0.81	0.65	1.00
Relative rate calculated from EQ 5	0.75	0.61	1.00

CONCLUSION

In-situ monitoring of styrene emulsion polymerization in batch reactor was carried-out using Raman spectroscopy. A new method for calculating monomer conversion was introduced, based on the shift of the band at 1000 cm^{-1} assigned to aromatic ring breathing. On the basis of this new method, kinetics of emulsion polymerization was analyzed and the results were found similar to literature data.

ACKNOWLEDGMENTS

We are grateful to the Région of Lorraine for its financial support to this project.

REFERENCES

1. A. Al-Khanbashi, M. Dhamdhere, M. Hansen, *Appl. Spectrosc. Rev.* **33**, 115-131 (1998).
2. G. E. Fonseca, M. A. Dubé, A. Penlidis, *Macromol. React. Eng.* **3**, 327-373 (2009).
3. G. Ellis, M. Claybourn, S. E. Richards, *Spectrochim. Acta* **46A**, 227-241 (1990).
4. C. Wang, T. J. Vickers, J. B. Schlenoff, C. K. Mann, *Appl. Spectrosc.* **46**, 1729-1731 (1992).
5. M. van den Brink, J.-F. Hansen, P. de Peinder, A. M. van Herk, A. L. German, *J. Appl. Polym. Sci.* **79**, 426-436 (2001).
6. B. Chu, G. Fytas, G. Zalczer, *Macromolecules* **14**, 395-397 (1981).
7. E. Gulari, K. McKeigue, K. Y. S. Ng, *Macromolecules* **17**, 1822-1825 (1984).
8. N. Brun, I. Youssef, M.-C. Chevrel, D. Chapron, C. Schrauwen, S. Hoppe, P. Bourson, A. Durand, *J. Raman Spectroscopy* **44**, 909-915 (2013).
9. T. R. McCaffery and Y. G. Durant, *J. Applied Polymer Science*, **86**, 1507-1515 (2002).
10. W. M. Sears, J. L. Hunt and J. R. Stevens, *J. Chem. Phys.* **75**, 1589-1598 (1981).
11. R. G. Gilbert, “Kinetic and Thermodynamic Precepts” in *Emulsion Polymerization. A Mechanistic Approach*, edited by R. H. Ottewill and R. L. Rowell.
12. C.C. Chern, S.-Y. Lin, T.-J. Hsu, *Polym. J.* **31**, 516-523 (1999).
13. L. V. de la Rosa, E. D. Sudol, M. S. El-Aasser, A. Klein *J. Polym. Sci: Part A: Polym. Chem.* **34**, 461-473 (1996).

Application of Self-Tuning PI Controller to Resin Transfer Molding

Kai-Hong Wang, Yao-Chen Chuang, Yuan Yao *

Department of Chemical Engineering, National Tsing Hua University, Hsinchu, 30013, Taiwan, ROC

Presenting author's email: expipiisminus1@gmail.com
Reference author's email: yyao@mx.nthu.edu.tw

Abstract. Resin transfer molding (RTM) is an advanced manufacturing process for fabricating high-strength and light-weight polymer composites, which injects thermosetting resin into a mold to impregnate the fiber preform. In RTM, effective flow control is necessary to direct the resin flowing in a desired manner and avoid the formation of dry spot and air entrapment. Previous research indicates that a constant flow velocity is beneficial to achieve good product quality. However, the existing flow control strategies for RTM are often based on an impractical assumption that the permeability of the fiber reinforcement is uniform within the mold and known in advance. The traditional proportional-integral-differential (PID) controller is unsuitable either, because of the nonlinear and time-varying characteristics of the system. In this research, a self-tuning proportional-integral (PI) control strategy is proposed for adjusting the flow behavior of the resin inside the mold. To account for the nonlinear and time-varying characteristics, the recursive least squares (RLS) method with an adaptive directional forgetting factor is adopted to identify the input-output relationship of the RTM process under control. Based on the identified models, the ultimate values of the closed-loop control system are computed. Then, the PI parameters are tuned by using the Ziegler-Nichols method. To demonstrate its applicability and effectiveness, the proposed control strategy is applied to an RTM process with a closed mold cavity where the permeability of the fiber reinforcement is uncertain and unknown.

Keywords: Resin transfer molding (RTM), Flow control, self-tuning control strategy, recursive least squares, Ziegler-Nichols method.
PACS: 07.05.Dz, 07.05.Rm, 07.05.Hd.

Introduction

Resin transfer molding (RTM) is a technique for producing high-quality composite materials. Because of the irregularity in fiber arrangement, the resin tends to flow toward the high permeability zones during RTM manufacturing, leading to insufficiently impregnation in low permeability zones. Therefore, it is necessary to conduct flow control to solve this problem.

Flow control for resin transfer molding can be categorized into two types: passive flow control and active flow control. Passive flow control aims to optimize mold design or gate location based on numerical flow simulation before conducting the actual molding process [1-2]. However, passive control cannot handle the unpredictable disturbances occurred during manufacturing. In such cases, active control should be used. Most active flow control approaches are operated by adjusting either of the following two types of manipulated variables: injection pressure and local temperatures in the mold. For temperature adjustment, Johnson et al. [3] developed a movable induction coil to track the flow front, which reduces local resin viscosity by heating the resin, thereby altering the local flow velocity to stabilize the flow front. To monitor and accelerate the resin flow, Matsuzaki et al. [4-5] introduced a multifunctional interdigital electrode array (MIEA) for heating. Considering pressure adjustment, Nielsen et al. [6-8] employed a fuzzy logic and artificial neural network based algorithm to estimate the permeability coefficient. To maintain the flow rate constant, Bender et al. [9] applied to a fuzzy logic PID controller, and Mogavero et al. [10] developed a nonlinear control method.

Most of the flow control methods applied in above literature firstly employ a numerical simulation and then conducting experimental comparisons. However, in actual RTM processes, it is difficult to setup the system parameters due to the uncertainty in fiber permeability and the race tracking phenomenon. This paper aims at establishing an adaptive control process, where the model parameters are updated at each sampling time. Although adaptive control has undergone years of development in the field of injection molding, it is rarely applied to the field of resin transfer molding. This study proposes an architecture for a self-tuning PI controller that can be applied to resin transfer molding. The controller parameters can be adjusted according to time-varying behaviors of the process, thus enhancing the control performance. In the meantime, numerical simulation is not necessary in the proposed control scheme.

Equipment and Instrumentation

The P&ID chart of experimental setup is shown in Fig.1. The National Instruments LabVIEW is employed for hardware integration, flow visualization and realizing the control algorithm. Here, the controller output (i.e. the manipulated variable) is the injection pressure, which is adjusted by a pressure regulator. An air compressor is utilized to deliver the air into the resin bucket. The PID controller of the injection system is connected to the PC by RS485 serial communication through a VISA function board. For the purpose of realizing the vacuum state in the outlet, a vacuum bucket and a vacuum pump are applied. In order to realize flow visualization, a CCD video camera is placed above the transparent mold. On each sampling time point, the process status is captured in an image which is displayed in the LabVIEW program environment. Then, the flow front displacement and velocity can be obtained by applying the image binarization technology. In this study, the process output (i.e. the controlled variable) is the flow front velocity.

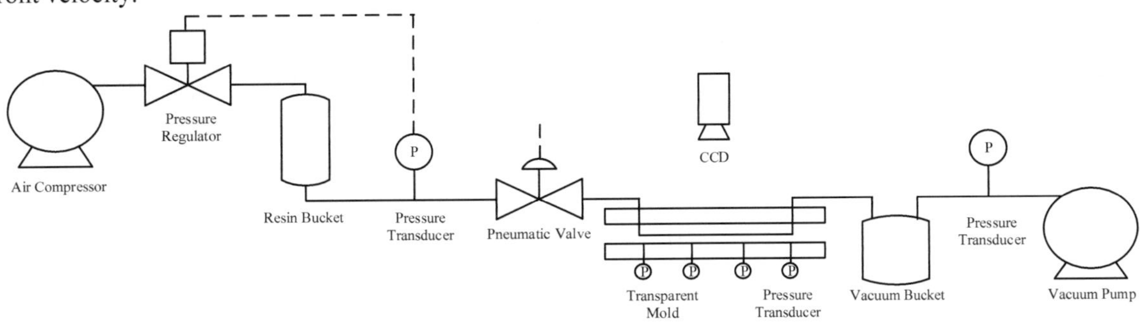

FIGURE 1. P&ID chart of experimental setup

Self-tuning PI controller

The proposed self-tuning PI control algorithm consists of three basic parts: recursive system identification, calculation of critical parameters and design of controller parameters. The structure of the control system is presented in Figure 2.

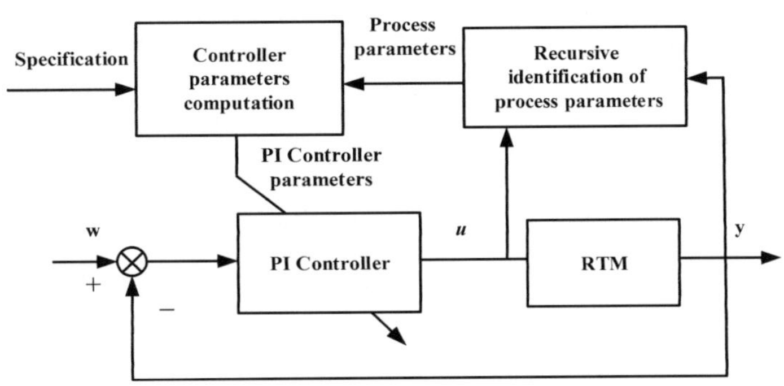

FIGURE 2. Self-tuning PI control system

Given the complexity of resin transfer molding process, it is difficult to describe it with a simple model. This study focused on identifying the relationship between the injection pressure and the flow front velocity. We employed an ARX regression model and the recursive least squares with directional forgetting method [11] to update the model parameters in real time. These changing model parameters describe the time-varying characteristics of the system. The ARX regression model structure is given as follows.

$$A(z^{-1})y(k) = z^{-nd}B(z^{-1})u(k) + \hat{e}(k) \quad (1)$$

where $A(z^{-1}) = 1 + a_1 z^{-1} + ... + a_{na} z^{-na}$ and $B(z^{-1}) = b_1 z^{-1} + b_2 z^{-2} + ... + b_{nb} z^{-nb}$ are shift operator polynomials with orders *na* and *nb*, respectively, *nd* is the order of process delay, *k* standard for the time instant of sampling. $y(k)$ is the system output (i.e. the flow front velocity), $u(k)$ is the system input (i.e. the injection pressure) and $\hat{e}(k)$ is the white noise. Detailed descriptions can be found in [11]. The PI controller is usually designed for applications with a low-order model. Therefore, in this study, a third-order system is considered (i.e., both order *na* and *nb* are chosen as 3 and *nd* is equal to 0).

The ARX model can be rewritten as follows.

$$y(k) = \mathbf{\Phi}^T(k)\hat{\mathbf{\Theta}}(k-1) + \hat{e}(k) \tag{2}$$

where $\hat{\mathbf{\Theta}}(k) = [a_1, ... a_{na}, b_1, ..., b_{nb}]^T$ is the process model parameter vector under examination and $\mathbf{\Phi}^T(k) = [-y(k-1), ..., -y(k-na), u(k-1-nd), ..., u(k-nb-nd)]$ is the regression vector.

The recursive system identification algorithm is given as follows. The process model parameter is updated according to the recursive relation

$$\hat{\mathbf{\Theta}}(k) = \hat{\mathbf{\Theta}}(k-1) + \mathbf{L}(k)\hat{e}(k) \tag{3}$$

where $\mathbf{L}(k) = \dfrac{\mathbf{C}(k-1)\mathbf{\Phi}(k)}{1+\xi(k)}$ is the gain matrix, $\xi(k) = \mathbf{\Phi}^T(k-1)\mathbf{C}(k-1)\mathbf{\Phi}(k)$ is an auxiliary scalar, and $\hat{e}(k) = y(k) - \mathbf{\Phi}^T(k)\hat{\mathbf{\Theta}}(k-1)$ is the prediction error.

The gain matrix is updated by relation

$$\mathbf{C}(k) = \mathbf{C}(k-1) - \frac{\mathbf{C}(k-1)\mathbf{\Phi}(k)\mathbf{\Phi}^T(k)\mathbf{C}(k-1)}{\varepsilon^{-1} + \xi(k)} \tag{4}$$

where $\varepsilon(k) = \varphi(k) - \dfrac{1-\varphi(k)}{\xi(k-1)}$ and $\varphi(k)$ is the adaptive directional forgetting factor. Detailed description of the algorithm can be found in [16].

The digital PI controller is written as follows.

$$u(k) = q_0 e(k) - q_1 e(k-1) + u(k-1) \tag{5}$$

where $e(k)$ is the tracking error, $q_0 = K_p(1 + \dfrac{T_0}{T_I})$, $q_1 = -K_p(1 - \dfrac{T_0}{2T_I})$ are the controller parameters, K_p, T_I and T_0 are the proportional gain, the integral time and the sampling period, respectively.

To set the parameters of the PI controller, Ziegler–Nichols continuous cycling method [12] was adopted. It provided a way to calculate the controller parameters based on the estimation of two critical parameters: the critical proportional gain K_{pu} and the critical period T_u (i.e. $K_p = 0.45K_{pu}$ and $T_I = T_u/1.2$). The transfer function of the system to be controlled can be described as

$$\frac{Y(z)}{U(z)} = \frac{z^{-d}B(z^{-1})}{A(z^{-1})} \tag{6}$$

where *Y(z)* and *U(z)* are the z-transform of the the controller and process outputs, respectively. Suppose the system is controlled by a proportional controller, i.e.

$$\frac{Y(z)}{E(z)} = K_p \tag{7}$$

where *E(z)* = *W(z)*- *Y(z)* is the z-transform of the tracking error and *W(z)* is the z-transform of the reference output. Based on the equations (6) and (7), the closed loop transfer function can be written as

$$\frac{Y(z)}{W(z)} = \frac{z^{-d}K_p B(z^{-1})}{A(z^{-1}) + z^{-d}K_p B(z^{-1})} \tag{8}$$

The denominator stands for the characteristic equation of the closed loop transfer function. For a third-order model,

$$A(z^{-1}) + z^{-d}B(z^{-1}) = z^3 + (a_1 + b_1 K_p)z^2 + (a_2 + b_2 K_p)z + a_3 + b_3 K_p \tag{9}$$

Bobal et al. [11] proposed an identification method for K_{pu} and T_u in their textbook. If the critical poles lie on the real axis, the critical values can be calculated by the equations (10).

$$z^3+(a_1+b_1K_{pu})z^2+(a_2+b_2K_{pu})z+a_3+b_3K_{pu}=(z^2-2\alpha z+1)z(1+e_1z^{-1}) \tag{10}$$

By comparing the coefficients, the following equations can be obtained

$$\begin{aligned} a_1+b_1K_{pu} &= -2\alpha+e_1 \\ a_2+b_2K_{pu} &= 1-2\alpha e_1 \\ a_3+b_3K_{pu} &= e_1 \end{aligned} \tag{11}$$

from which it is derived that

$$K_{pu1,2}=\frac{-r_1\pm\sqrt{d}}{2r_2} \quad \alpha=\frac{a_3-a_1+K_{pu}\left[b_3-b_1\right]}{2} \quad \mathrm{T}_u=\frac{2\pi}{\omega_u} \quad \omega_u=\frac{\arccos\alpha}{T_0} \tag{12}$$

In (12), $r_0=a_3(a_3-a_1)+a_2-1$, $r_1=b_3(2a_3-a_1)+b_2-a_3b_1$, $r_2=b_3(b_3-b_1)$ and $d=r_1^2-4r_0r_2$. The solutions should satisfy the following inequalities: $K_{pu}>0$ and $|e_1|<1$. If the conditions are not met either for K_{pu1} or K_{pu2}, another case is considered, i.e. the critical poles are with the imaginary parts. The critical values can then be calculated by the equation (13)

$$z^3+(a_1+b_1K_{pu})z^2+(a_2+b_2K_{pu})z+a_3+b_3K_{pu}=(z+1)z^2(1+f_1z^{-1}+f_2z^{-2}) \tag{13}$$

By comparing the coefficients for the same power of z, the following equations are derived

$$\begin{aligned} a_1+b_1K_{pu} &= 1+f_1 \\ a_2+b_2K_{pu} &= f_1+f_2 \\ a_3+b_3K_{pu} &= f_2 \end{aligned} \tag{14}$$

while the solutions are

$$K_{pu}=\frac{1-a_1+a_2-a_3}{b_1-b_2+b_3} \quad T_u=2\mathrm{T}_0 \tag{15}$$

After obtain the critical parameters, the proportional gain and the integral time can be calculated by using Ziegler–Nichols continuous cycling method, and the PI controller parameters can be finally deduced via the equation (5).

Results and Discussions

Experiments are conducted to verify the effectiveness of the proposed controller. The results are compared to those based on the traditional PI controller, as shown in Fig. 4 and Fig. 5. A comparison of Fig. 4(a) and Fig. 5(a) clearly shows that the output of the self-tuning PI controller is less oscillatory than that of the traditional PI controller. This is because that the controller parameters of the self-tuning PI controller can adapt to the time-varying proccess behaviors. The flow front velocities displayed in Fig. 4(b) and Fig. 5(b) further confirm the effectiveness of the adopted algorithm. To compare the performances of both controllers quantatively, integral of the absolute value of the error (IAE) and integral of the squared error (ISE) were calculated and shown in Table 1. Obviously, both the IAE and ISE values of the self-tuning PI controller are much lower than those of the traditional PI controller.

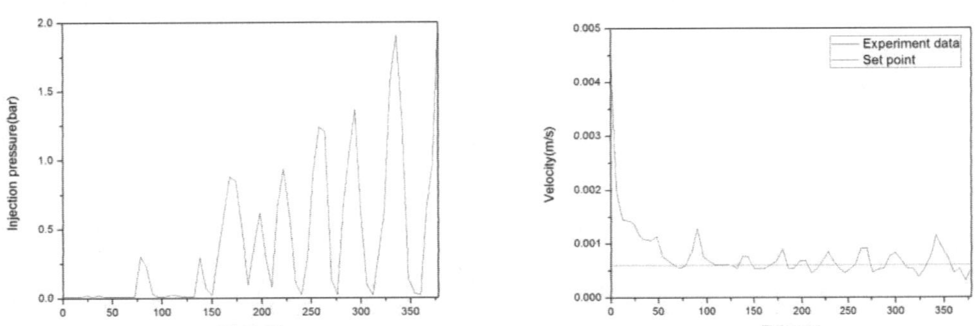

(a) (b)

FIGURE 4. Results of the traditional PI controller

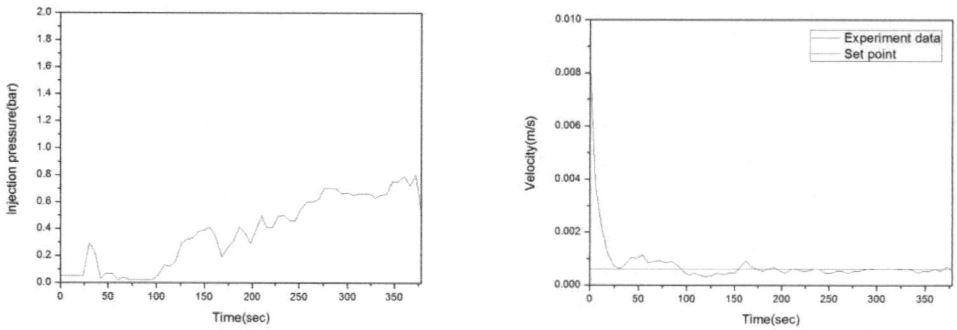

(a) (b)

FIGURE 5. Results of the self-tuning PI controller

TABLE 1 Comparison of the self-tuning PI versus traditional PI responses for IAE and ISE values

Controller	IAE	ISE
Traditional PI controller	0.0413	1.0765×10^{-5}
Self-tuning PI controller	0.0232	3.7588×10^{-6}

Conclusions

In this study, an adaptive control algorithm is adopted to control the flow front velocity in resin transfer molding. The recursive least squares with directional forgetting algorithm is utilized for system identification. Based on the identified process model that is time-varying, the control parameters can be adjusted in an adaptive manner. The adaptive PI control system has a simple structure and significantly outperforms the traditional PI controller. The feasibility of the proposed control strategy is verified by experiments.

REFERENCES

1. Bickerton, S., Stadtfeld, H. C., Steiner, K. V., and Advani, S. G. "Design and application of actively controlled injection schemes for resin-transfer molding." Composites Science and Technology 61, 1625-1637(2001)
2. Hsiao, Kuang-Ting, and Suresh G. Advani. "Flow sensing and control strategies to address race-tracking disturbances in resin transfer molding. Part I: design and algorithm development." Composites Part A: applied science and manufacturing 35, 1149-1159(2004)
3. Johnson, R. J., and Pitchumani, R. "Flow control using localized induction heating in a VARTM process." Composites Science and Technology 67, 669-684(2007)
4. Matsuzaki, R., Kobayashi, S., Todoroki, A., & Mizutani, Y. "Control of resin flow/temperature using multifunctional interdigital electrode array film during a VaRTM process." Composites Part A: Applied Science and Manufacturing 42, 782-793 (2011)
5. Matsuzaki, R., Kobayashi, S., Todoroki, A., & Mizutani, Y. "Flow control by progressive forecasting using numerical simulation during vacuum-assisted resin transfer molding." Composites Part A: Applied Science and Manufacturing 45, 79-87(2013)
6. Nielsen, D. R., & Pitchumani, R. "Intelligent model-based control of preform permeation in liquid composite molding processes, with online optimization." Composites Part A: Applied Science and Manufacturing 32, 1789-1803 (2001)
7. Nielsen, D. R., & Pitchumani, R. "Closed-loop flow control in resin transfer molding using real-time numerical process simulations." Composites Science and Technology 62, 283-298(2002)
8. Nielsen, D. R., & Pitchumani, R. "Control of flow in resin transfer molding with real-time preform permeability estimation." Polymer composites 23, 1087-1110(2002)
9. Bender, D., Schuster, J., & Heider, D. "Flow rate control during vacuum-assisted resin transfer molding (VARTM) processing." Composites Science and Technology 66, 2265-2271(2006)
10. Mogavero, J., Sun, J. Q., & Advani, S. G. "A nonlinear control method for resin transfer molding." Polymer composites 18, 412-417(1997)
11. Bobál, V., Böhm, J., Fessl, J., and Machácek, J. "Digital Self-tuning Controllers: Algorithm." Implementation and Applications. Springer Science & Business Media, 2006.
12. Seborg, D. E., Mellichamp, D. A., Edgar, T. F., & Doyle III, F. J. "Process dynamics and control. " John Wiley & Sons, 2010.

Development of a Novel and Adaptable Injection Unit for an Automated and Quality Controlled Manufacturing of RTM parts for Aerospace Applications

Christian Hopmann[a], Marc Linus Fecher[a], Arne Boettcher[a]*, Kai Fischer[a] and Lionel Winkelmann[b]

[a]*Institute of Plastics Processing (IKV) in Industry and Skilled crafts at RWTH Aachen University (IKV), Seffenter Weg 201, D-52074 Aachen, Germany*
[b]*Airbus Operations GmbH, Airbus-Strasse 2, 21684 Stade*

arne.boettcher@ikv.rwth-aachen.de

Abstract. Fibre reinforced plastics (FRP) are increasingly being implemented in aircraft and automotive applications for structural lightweight components. A decisive problem for the high volume production of structural parts made of FRP with high fibre volume contents of up to 60 % is, that current processing procedures are only partly automated and have a low process robustness. Additionally the procedures are typically not controlled. For a quality controlled qualification of the parts, time consuming post processes are necessary. In order to increase process robustness and reduce time consuming post processes a novel injection unit has been developed by the Institute of Plastics Processing (IKV) at RWTH Aachen University together with Airbus Operations GmbH, Stade, and FILL GES.M.B.H., Gurten, Austria. This injection unit shall enable a fully automated and quality controlled manufacturing of Resin Transfer Moulding (RTM)-parts for aerospace applications. Therefore, different sensors are integrated into the injection unit to fulfill the quality assurance: pressure sensors, thermocouples and dielectrical sensors. This allows for a measurement of all necessary process parameters which cannot all be measured and controlled by common injection units or in-mould sensors. The selected arrangement of the pressure sensors enables the build-up of an online rheometer for a controlled injection of the resin system depending on the current viscosity. Furthermore, an online determination of the parts porosity and the curing degree is envisaged. This paper explains the functionality of the developed injection unit with all integrated measurement systems. Additionally, the online rheometer is presented more in detail and results of the online viscosity measurement for different volume flows are presented.

Keywords: Composites, process control, inline quality assurance
PACS: 61.41.+e; 81,05.Ni; 81.20.Wk; 06.20.F-

INTRODUCTION

In the aerospace industry fibre reinforced plastics (FRP), also referred to as composites, are used for over 50 % of the structural components in current civil aerospace generations of the companies Airbus (A 350) and Boeing (B787) due to their high lightweight potential [1]. The increasing use of composites is driven by operating cost factors and regulations to reduce pollutant emissions. The high lightweight potential of the material helps to reduce fuel consumption and CO_2-emissions. Furthermore, the payload can be increased by lower empty weights of the aircrafts.

A well-established manufacturing process for FRP with high fibre volume contents (FVC) and good mechanical properties in aerospace applications, but also in the automotive industry, is the Resin Transfer Moulding (RTM)-process. Numerous variants of the common RTM-process have been developed due to the high potential of the RTM-process for high component quality and high volume production gained by its good suitability for automation [2, 3, 4]. Still, the robustness of the process has to be increased by process control and quality assurance systems have to be developed in order to further reduce processing steps and scrap costs [5, 6, 7]. Therefore, a fundamental knowledge of the complex interactions of material, process and component parameters for the manufacturing processes is necessary. Especially an integration of a quality assurance system in the manufacturing process would be beneficial in order to reduce post processing steps and quality control effort.

Individual process parameters like pressure, temperature, curing degree and the flow front development can already be detected with state of the art sensor technologies [8, 9, 10]. For example, vents can be closed automatically when reached by the flowing resin. This can help to prevent manufacturing failures which can occur due to effects like race-tracking and the vents can be closed automatically at the end of the process [11]. However, these sensor technologies only allow for a rudimentary process control. Additionally, common approaches for process control and measurement of process parameters require an integration of cost intensive and sensible sensors in each single mould. This results in high costs especially for small to medium series production and productions with a high number of needed moulds. Moreover, available process control technologies and sensor technologies still require post processing steps for a qualification of the parts. This is because an inline determination of quality parameters like the component porosity is not implemented in state of the art processes [12]. These post processing steps are very time consuming, leading to high manufacturing cycle times and thus component costs [13]. In order to overcome these drawbacks a novel injection technology developed at the Institute of Plastics Processing (IKV) in Industry and the Skilled Crafts at RWTH Aachen University is presented in this paper. It enables an automated and quality controlled manufacturing of structural composite components for aerospace applications, by measuring the degree of cure, resin viscosity and resulting component porosity inline.

APPROACH – DEVELOPMENT OF A NOVEL INJECTION UNIT

The approach to enable an inline quality assurance is to combine different measurement systems and sensors in the injection unit for the RTM-process. The integrated measurement systems are determining the resin viscosity, the degree of cure of the resin system and the porosity of the component. The common way to integrate an inline measurement into a RTM-process is the integration of sensors in the RTM-mould. The disadvantages of in-mould sensors are a limited robustness of the sensors, the costs for equipping every single mould with the sensors and markings of the sensors on the manufactured components. In contrast, by integrating the sensors into the injection unit, the measurements are independent from the used RTM-mould, which makes the injection unit adaptable to several moulds in the production environment. Such an injection unit (Intelliject) was developed by IKV together with the partners Airbus Operations GmbH, Stade, and Fill Ges.M.B.H., Gurten, Austria. The injection unit establishes a basis for the automated and quality assured manufacturing of composite parts. This inline measuring injection unit can be connected directly to the RTM-mould without hoses or flexible connections (Figure 1).

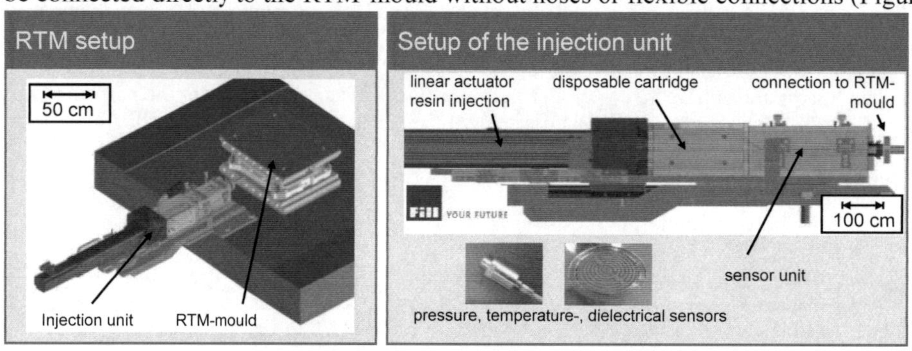

FIGURE 1. Setup of the developed injection unit

In this example, the one component resin system is stored in disposable cartridges to reduce cleaning labour and enable repeatable conditions for the resin system with heating only the needed amount of resin. The concept is easily transferable to two component resin system which are processed with mixing heads. Between the cartridges and the RTM-mould the sensor unit for measurement of different process parameters (pressure, temperature) and quality assurance parameters (curing degree, porosity, viscosity) is installed.

INLINE QUALITY ASSURANCE

The sensor unit mainly consists of a flow channel along which two pressure sensors and one dielectrical (DEA)-sensor are installed. The pressure sensors are piezoresistive sensors of the type 4001 from the company Kistler Instrumente AG, Winterthur, Switzerland, which allow for a measurement of low pressure values (≤ 20 bar) at high temperatures (~180 °C) without drift. Furthermore, the pressure sensors enable a measurement of the temperature directly at the surface to the resin system in the flow channel. This is beneficial, because a direct temperature measurement in the contact with the resin system is possible without additional sensors. Beside measuring the injection pressure, the pressure sensors are used to determine the resin viscosity (η) based on the pressure loss (Δp) along the flow path between both sensors. As DEA-sensor a monotrode sensor of the type 4/3RC from the company

NETZSCH Geraetebau GmbH, Selb, Germany, is used. The DEA-sensor is used to determine the degree of cure of the resin system. Therefore, the sensor measures the ionviscosity of the resin system which can be correlated with the degree of cure. Along the flow channel the sensors are specially aligned to enable the viscosity and degree of cure measurement (Figure 2).

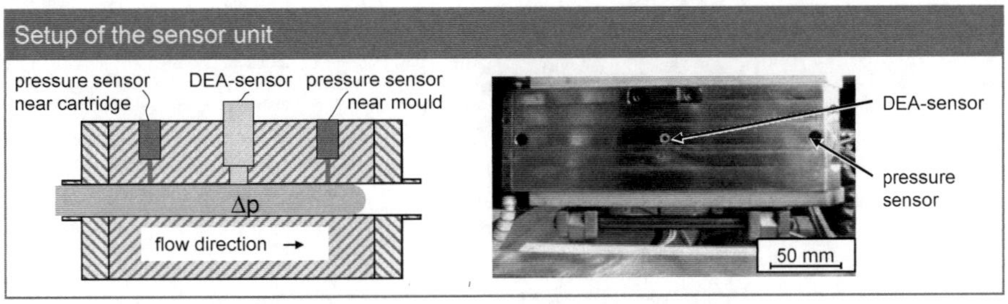

FIGURE 2. Setup of the developed sensor unit

In addition to viscosity and curing degree measurement, the porosity of the manufactured components should be measured inline. The measurement is based on the principles for different compression behaviour of fluids and gases. Due to the design of the injection unit all sealings and connections, it is possible to increase the resin pressure subsequent to the injection process without a change in volume of the whole manufacturing process, like it is possible by using hoses or commonly used o-ring notches. Hoses or commonly used o-ring notches allow for an increasing system volume by expanding the used hoses or compression of the o-rings in the o-ring notches. By using disposable cartridges and a piston based resin injection, the change in volume of the injected resin system can be measured by increasing the injection pressure after the vents are closed. Using a piston based injection system enables the measurement of the change in volume by measuring the movement of the piston. Actual investigations also focus on the definition of a correlation between the compressibility of the injected resin system and the enclosed porosity [14]. Following, the measurement principle and selected results for the inline determination of the resin viscosity are presented.

Inline Viscosity Measurement

The inline measurement of the resin viscosity during the injection is realised by an inline capillary rheometer. Inline capillary rheometers are state of the art in extrusion processes but are not used in state of the art RTM-processes [15]. For the build-up of an inline capillary rheometer, the two pressure sensors are positioned in the flow channel of the sensor unit with a defined distance to measure the pressure loss along the defined flow path. The pressure loss depends on the viscosity of the resin system, the flow velocity of the resin system and the cross section of the flow channel [16, 17]. To set up the inline capillary rheometer for the RTM-process the cross section of the flow channel and the distance between both pressure sensors have to be adapted to enable the measurement of the pressure loss between both sensors. A minimum pressure loss of 0,1 bar is envisaged to guarantee a measurable pressure difference between both pressure sensors. Especially the low viscosities of commonly used RTM resin systems lead to a low pressure loss along a flow path, which requires an analytical design of the flow channel.

The pressure loss in dependency of the geometry of a flow channel can be described by the equation of *Hagen-Poiseuille*, which is valid for laminar currents of newtonian fluids in a circular flow channel [16]. The pressure loss in the flow channel will be measured by two pressure sensors. The pressure sensors have a planar surface which cannot be adapted to circular flow channels. To prevent from undercuts and to enable a sufficient cleaning of the flow channel, the flow channel has a rectangular cross section. The equation of *Hagen-Poiseuille* for rectangular flow channels with a low channel height (h) simplifies to (channel width >> channel height) [17]:

$$\Delta p = \frac{12 \cdot \eta \cdot v \cdot l}{h^2} \tag{1}$$

In this equation (η) represents the fluids viscosity, (v) the flow velocity, (l) the length of the flow channel along which the pressure difference is measured and (h) the flow channel height. Based on equation (2) the pressure drop is calculated for different flow channel lengths and heights to define a suitable flow channel geometry and distance between both pressure sensors. Therefore, calculations were made for 0.1 l/min, 0.2 l/min and 0.3 l/min, channel heights of 1-7 mm, channel lengths of 100-300 mm and a minimum resin viscosity of 30 mPas. The aspired pressure sensors have a diameter of 9 mm. Due to the size of the pressure sensors the channel width was selected to 10 mm for the calculation of the flow velocity in dependency of the volume flow. The results of the calculations are visualized in figure 3.

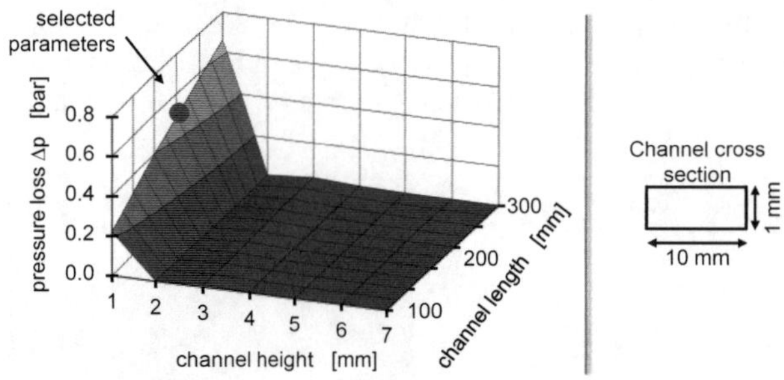

FIGURE 3. Calculated pressure loss depending on the flow channel height and length (volume flow 0.1-0.3 l/min/viscosity 30 mPas)

It becomes obvious, that a flow channel height of 1 mm has to be selected at a flow channel width of 10 mm (figure (3), right) to generate a measurable pressure loss. For the flow channel length a compromise of sensor box dimensions and measurable pressure loss has to be selected. Thus, the flow channel length is selected to 200 mm. As mentioned above the *Hagen-Poiseuille* equations are valid for laminar currents of newtonian fluids. To guarantee a laminar current the Reynolds number has to be less than 2300 [18]. Therefore, the Reynolds numbers for different flow velocities have to be calculated. Table 1 summarizes the expected pressure losses, and Reynolds numbers for the flow channel with a length of 200 mm between the pressure sensors, a cross section of 10 x 1 mm², resin viscosities of 30 mPas and 130 mPas as well as the above mentioned volume flow rates.

TABLE 1. Expected pressure losses and Reynolds numbers

flow rate [l/min]	pressure loss [bar] (30 mPas)	pressure loss [bar] (130 mPas)	Reynolds number [-] (30 mPas)	Reynolds number [-] (130 mPas)
0.1	0.12	0.52	11.12	2.56
0.2	0.24	1.04	22.24	5.13
0.3	0.36	1.56	33.36	7.7

To verify the flow channel design, measurements of the pressure loss along the flow channel with the resin system RIMR135 / RIMH137 of the company Hexion GmbH, Duisburg, Germany, were carried out. The measurements were carried out with the above mentioned volume flow rates in the described sensor box. Exemplary presented are the measurements at a volume flow rate of 0.2 l/min and a temperature of 40 °C (resin and flow channel temperature). Measurements of the flow channel height after the manufacturing of the sensor box have shown, that the flow channel height is 1.15 mm, which has to be taken into account for the calculation of the expected pressure loss. At a flow rate of 0.2 l/min, a flow channel height of 1.15 mm and a resin viscosity of 100-120 mPas (measured in a plate plate rheometer) a pressure loss of 0.53-0.63 bar is expected. The measured pressure values and the depending pressure loss are visualized in figure 4 in comparison to the calculated pressure loss.

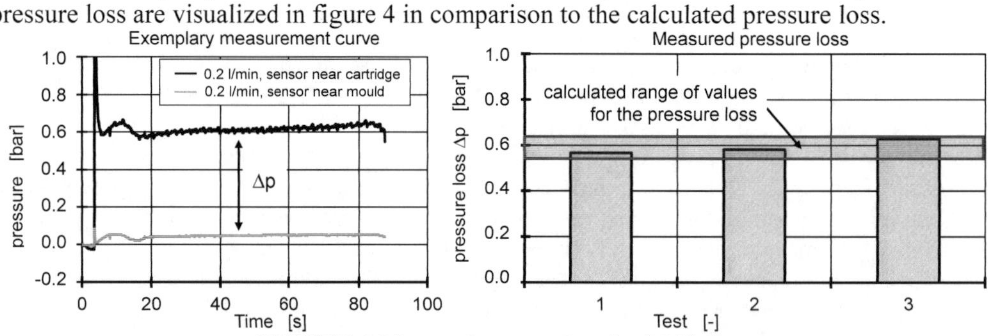

FIGURE 4. Measured pressure loss for 0.2 l/min

The measured values are consistent with the calculated values for the pressure loss. Small deviations of the measured values from the calculated values are possible due to the real resin viscosity, which can vary from the expected viscosity for the calculation. Measurements of the temperature at the outlet of the sensor box show, that a shear heating and a corresponding decreasing of the resin viscosity is only very low. At the outlet a resin temperature of 39.8 °C was measured and the resin system had a temperature before the injection of 39.6 °C. Measurements in a plate-plate rheometer have shown, that the viscosity of the resin system can vary at 40 °C from 100 mPas to 120 mPas. In general the measurements of the viscosity with the build-up inline capillary rheometer are valid and the theoretical model (*Hagen-Poiseuille*) can be chosen for the calculation of the viscosity based on the measured pressure loss.

Following, measurements during the resin injection have to be carried out. The reinforcing fibres will lead to a counter-pressure during the injection due to the preform permeability. It will be investigated which influence the counter-pressure has on the viscosity measurements. Based on these results a viscosity controlled resin injection will be implemented.

CONCLUSION AND OUTLOOK

In this paper a novel approach for the inline measurement of quality assurance parameter of a RTM-component inline subsequent to the injection process was introduced and the specifically developed injection unit and inline measurement was described. Beside the measurement of temperatures and processing pressures, the injection unit enables the measurement of quality assurance parameter like the resin viscosity during the injection. It was shown, that the inline determination of the resin viscosity is possible by measuring the pressure loss along a flow path, as it is common in a capillary rheometer. Measurements with a resin system also showed that the measured pressure loss is consistent with the calculated pressure loss. Variations in the measured pressure loss can be explained by viscosity variations of the resin system, which are also measured in verifying measurements in a plate-plate rheometer. In further investigations the measurement principle has to be validated in RTM-processes. Furthermore, the measurement and analysis of the data has to be automated and the data will be used to control the injection process based on the actual resin viscosity.

ACKNOWLEDGMENTS

The depicted research has been funded by the Deutsche Forschungsgemeinschaft (DFG). We would like to extend our thanks to the DFG and our partners in the project.

REFERENCES

1. Laessig, R.; Eisenhut, M.; Mathias, A.; Schulte, R.T.; Peters, F.; Kühmann, T.; Waldmann, T. and Begemann, W.: Serienproduktion von hochfesten Faserverbundbauteilen. 2012
2. Chaudhari, R, Pick, M., Geiger, O., Schmidt, D., Elsner, P. and Henning, F., "Compression RTM - A new process for manufacturing high volume continuous fiber reinforced composites", Conference Proceedings of the 5th International CFK-Valley Stade Convention, 2011, Stade
3. N.N., Leichtbau macht Reaktionstechnik lebendig. URL: http://www.k-zeitung.de/leichtbau-macht-reaktionstechnik-lebendig/150/1190/88898/1, 20.07.2015
4. Hopmann, C., Fischer, K. and Bastian, R., *CFK-Motorhaube in einem Schuss gefertigt*. MM COMPOSITESWORLD (2014) , pp. 18-21
5. Drechsler, K., "CFK-Technologie im Automobilbau: Was man von anderen Märkten lernen kann". Conference Proceedings of the C.C.e.V. Automotive Symposium. Neckarsulm, 2010
6. Drechsler, K., Stute, U. and Herrmann, A.S.: *Die Grossserie wird greifbar*. Laser Community (2011) 1, pp. 12-13
7. Thoma, B., Weidenmann, K.A. and Henning, F., *Chemical-Stitching, ein vieversprechender Ansatz für die automatisierte Preform-Fertigung*. Zeitschrift Kunststofftechnik 8 (2012) 5, pp. 490-514
8. Abrahan, D. and McIlhagger, R., *Glass fibre epoxy composite cure Monitoring using parallel plate dielectric analysis in comparison with thermal and mechanical testing techniques*. Composites Part A: Applied Science and Manufacturing 29 (1998) 7, pp. 811-819.
9. McIlhagger, A., Brown, D. and Hill, B.: *The development of a dielectric system for the on-line cure monitoring of the resin transfer moulding process*. Composites Part A: Applied Science and Manufacturing 31 (2000) 12, pp. 1373-1381
10. Yenilmez, B. and Sozer, E.M., *A grid of dielectric sensors to monitor mold filling and resin curie in Resin Transfer molding*. Composites Part A: Applied Science and Manufacturing 40 (2009) 4, pp. 476-489
11. Devillard, M., Hsiao, K.T. and Advani, S.G., *Flow sensing and control strategies to address race-tracking disturbances in resin transfer molding. Part II: automation and validation*. Composites Part A: Applied Science and Manufacturing 36 (2005) 11, pp. 1581-1589
12. Kochan, A.: "Untersuchungen zur zerstoerungsfeien Pruefung von CFK-Bauteilen fuer die fertigungsbegleitende Qualitaetssicherung im Automobilbau". Ph.D. Thesis, University Dresden, 2012
13. Ullmann, T., Schmidt, T., Hofmann, S., and Jemmali, R., "In-line quality assurance for the manufacturing of carbon fiber reinforced aircraft structures". Conference Proceedings of the 2nd International Symposium on NDT in Aerospace, 2010.
14. Hopmann, CH, Fecher, M.L., Boettcher, A. and Fischer, K. " Intelligente RTM-Injektionseinheit für endlosfaserverstärkte Luftfahrtanwendungen". Conference Proceedings of the 28. Internationales Kolloquium Kunststofftechnik, Aachen, 2016
15. Shenoy, A.V., *Rheology of Filled Polymer Systems*. Springer Science & Business Media, 2013
16. Grote, K.-H. and Feldhusen, J., Dubbel: Taschenbuch fuer Maschinenbau, Springer-Verlag, Berlin Heidelberg, 2014
17. Michaeli, W., *Extrusionswerkzeuge für Kunststoffe und Kautschuk – Bauarten, Gestaltung und Berechnungsmöglichkeiten*.Carl Hanser Verlag, München
18. Herwig, H., *Stroemungsmechanik – Einfuehrung in die Physik von technischen Stroemungen*. Sprigner-Verlag, 2008

Spatially Resolved Temperature Measurement in Injection Moulding using Ultrasound Tomography

Christian Hopmann, Jens Wipperfürth* and Maximilian Schöngart

Institute of Plastics Processing (IKV) at RWTH Aachen University
Seffenter Weg 201, 52074 Aachen, Germany

**jens.wipperfuerth@ikv.rwth-aachen.de*
maximilian.schoengart@ikv.rwth-aachen.de

Abstract. Current simulations of the injection moulding process insufficiently consider the thermal interactions between melt, solidified material and mould. To improve the prediction accuracy in terms of shrinkage and warpage these interactions cannot be neglected and require a precise observation of the temperature field within the ongoing process. Nowadays temperature measurements at the transition of the polymer melt to the mould or near the surface of the polymer melt are applicable and do not allow either a non-invasive analysis or a determination of the temperature field. In the approach of ultrasound tomography presented here, an ultrasound beam is emitted into the melt and the time-of-flight (TOF) is detected by a set of transducers, which are radially arranged around the melt. Subsequently the measurement is repeated from different directions. Using algebraic reconstruction techniques, a distribution of the ultrasound velocity can be calculated based on the TOF-dataset. With additional information about the polymer from pvT-curves, the distribution of the ultrasound velocity can be converted into a temperature field. To evaluate the applicability of ultrasound tomography, a prototypical measurement device was constructed. The device consists of a hollow cylinder in which eight brass forerun elements are integrated in order to scatter the ultrasound beam. A polymer melt can be placed within the hollow cylinder and ultrasound tomography can be performed. The results show the applicability of commercially available transducers and the operability of the brass forerun elements, since each of the seven transducers was able to detect a signal that was sent out from the one remaining transducer. However, a useful reconstruction of the temperature distribution is, due to the inhomogeneous melt in the prototype, not yet possible. Currently an injection mould is designed which fulfills all prerequisites for ultrasound tomography and overcomes the difficulties from the prototype.

Keywords: Tomography, Ultrasound, pvT, Measurement Device, Injection Moulding
PACS: 81.70.Cv, 82.35.Lr

INTRODUCTION

In-situ temperature measurement is needed to observe manufacturing processes and validate simulation models. This is challenging because manufacturing processes like injection moulding have a highly dynamic temperature field and strong forces are applied on the melt [1]. This is problematic because the increasing accuracy of newly developed simulation models requires high precision measurements for validation. Most techniques use contact measurements to determine the melt temperature. Since these techniques only provide punctual information and influence melt flow, contact measurements appear increasingly unfavourable. In addition, high forces are applied during the manufacturing process and hinder the mounting of sensors. Because of these issues contact measurements are mostly used for the determination of the contact temperature and pressure between mould and melt. These disadvantages have led to an increasing need for non-invasive measurement techniques in science and industrial application, such as pyrometry [2], fluorescence spectroscopy or ultrasound. *Praher et al.* [3] suggested to use ultrasound signals to measure the temperature field in the screw ante-chamber of an injection moulding machine. Their method is similar to the approach we suggest in this work to measure the temperature field in the mould. They presented a prototype to measure the temperature distribution in water. Their results about the ultrasound time of flight measurement to reconstruct the temperature distribution are promising.

In this work we present a similar approach to measure the temperature distribution in a polypropylene (PP) melt.

PRINCIPLES OF ULTRASOUND TIME-OF-FLIGHT TOMOGRAPHY

In ultrasound time of flight (TOF) tomography an ultrasound wave is emitted from an ultrasound transducer. A set of receiving transducers is arranged around the region of interest. The arrival time of an ultrasound wave at the receiving transducers is measured and used for the reconstruction of a sound velocity distribution. Typically tomographic methods require consecutive measurements from different directions. This requires a rotation of either the sample or the sensor-detector-arrangement. In case of ultrasound tomography a rotation is not required when the transducers are able to both emit and detect ultrasound signals.

The arrival time of a detected ultrasound wave can be described under assumption of discrete area elements s by the sum of the time t_i which the ultrasound wave needs to traverse each element. In addition the time t_i can be expressed by the ultrasound velocity c_i divided by the length of the discrete element Δs.

$$t_{TOF} = \sum_i b_i t_i = \sum_i b_i \frac{c_i}{\Delta s} \tag{1}$$

The preliminary factor b_i describes the position with respect to a chosen coordinate system and scales the time the ultrasound wave needs to traverse a specific element. Applying equation 1 to every emitter-receiver-combination and substituting $b_i \Delta s = a_i$ equation 1 can be expressed as the linear set of equations shown in equation 2.

$$\begin{pmatrix} a_{11} & \cdots & a_{1n} \\ \vdots & \ddots & \vdots \\ a_{n1} & \cdots & a_{nn} \end{pmatrix} \begin{pmatrix} 1/c_1 \\ \vdots \\ 1/c_n \end{pmatrix} = \begin{pmatrix} t_1 \\ \vdots \\ t_n \end{pmatrix} \tag{2}$$

Solving the linear set of equations for $\vec{c}$ by inversion yields the spatial distribution of the ultrasound velocity. Since the matrix is highly complex, common inversion techniques cannot be used and iterative algorithms must be applied [4]. The representation of the measured values in matrix form and the solution of the resulting linear set of equations are referred to as Algebraic Reconstruction Techniques (ART) [4, 5]. Since we use a special sensor design and a radial cavity the used algorithms have to be customised, a challenge which has yet to be solved.

To convert the distribution of ultrasound velocities into a temperature distribution one can use the dependencies shown in equation 3 [3]:

$$c(x,y) = \left(\frac{v(x,y)}{\kappa} \right)^{1/2} \tag{3}$$

In equation 3 v is the specific volume and κ is the adiabatic compressibility. Both material properties are temperature and pressure dependent and can be derived from pvT-data [3]. Thus a pressure measurement during injection moulding is required and subsequent each ultrasound velocity can be converted into a temperature.

CHALLENGES OF ULTRASOUND TOMOGRAPHY IN INJECTION MOULDING

The mathematical description requires that each transducer is able to detect an emitted ultrasound wave independent of the location of the emitter. In case of radially arranged transducers, this is fulfilled if each transducer produces an ultrasound wave with a fan shaped propagation profile. Since commercial ultrasound transducers do not exhibit such a wave profile [6], the scattering has to be created artificially. This can be done by using curved interfaces to the melt. Mathematically, the transition of ultrasound waves at curved interfaces can be described by the lens equation [7].

$$\frac{1}{f} = \frac{1}{b} + \frac{c_1 / c_{melt}}{g} \tag{4}$$

In equation 4 f is the focal length, b is the image distance, and g is the distance between the transducer and the angular point of the curvature. The scattering of ultrasound waves thus depends on the ultrasound velocities of an arbitrary media c_1 and the melt c_{melt}. From equation 4 and with respect to the orientation of the curvature it is possible to deviate four different scattering behaviours that are shown in figure 1.

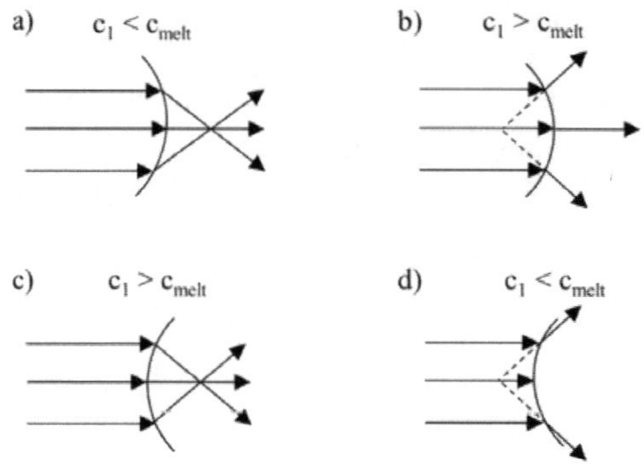

FIGURE 1. The behaviour of ultrasound waves at curved interfaces

Since the ultrasound velocity of a polymer melt is typically very small ($c_{PP} \approx 1600$ m/s) the best way to achieve a favourable scattering behaviour is figure 1 b), which can be realized using materials like brass ($c_{brass} \approx 3500$ m/s) or steel ($c_{steel} \approx 5900$ m/s).

EXPERIMENTAL SETUP

In a previous work the experimental setup has already been introduced [8] and subsequently modified due to the immense scattering effect. The current setup is shown in figure 2. The principle of the two different parts, that are the heating part (I) and the measurement area (II), was adopted from the previous measurement setup. This is still necessary since the commercially available ultrasound transducers are not applicable in the temperature range necessary to obtain a PP melt. The heating part consists of a hollow Cu-cylinder surrounded by a silicon heating band (Horn GmbH, Gottmadingen, Germany). It is used to heat the cavity up to nearly 220 °C to produce a polymer melt. After melting the material it is pushed into the measurement area (II) by moving the two pistons.

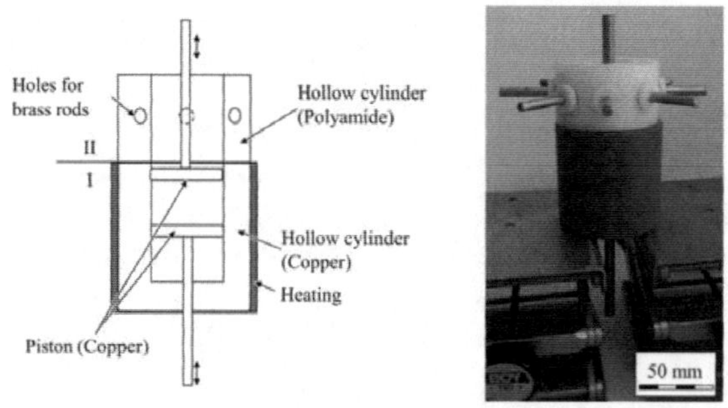

FIGURE 2. Schematic illustration of the ultrasound measurement device (left) and picture of the ultrasound measurement device (right)

The measurement area is illustrated in figure 3 in more detail. It consists of a polyamide 6.6 hollow cylinder with an outer radius of 70 mm, an inner radius of 30 mm and 40 mm in height. Eight brass forerun elements (diameter d = 6 mm) are placed in the hollow cylinder in 45° intervals with respect to the center of the cavity. The parts of the forerun elements which directs to the cavity have a half cylindrical shape in order to achieve the scattering behaviour described in the previous section.

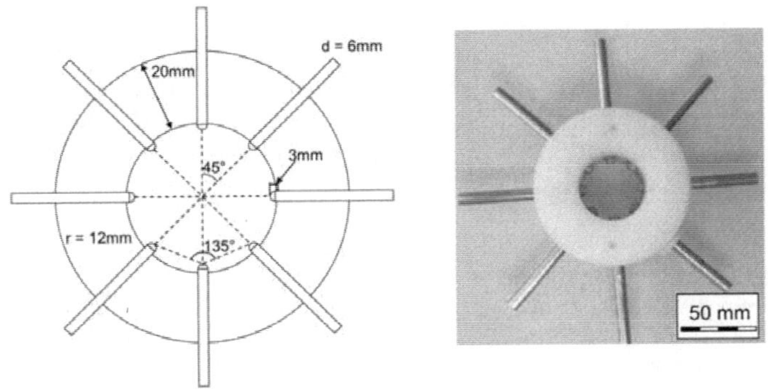

FIGURE 3. Schematic illustration (left) and picture (right) of the measurement area

The ultrasound transducers are attached to the forerun elements. To improve ultrasound transmission the contact area between transducers and the forerun elements is coated with a coupling gel.

To obtain the complete data from all directions each transducer will act as emitter, while the remaining transducers act as detectors. This measurement is repeated until each transducer has acted as emitter. The complete dataset can be used for the reconstruction of the distribution of ultrasound velocities. Signal detection is done with the mobile ultrasound measurement device EPOCH 600 (Olympus Scientific Solutions Americas Inc., MA, USA).

RESULTS

Preliminary Tests with Water

To test the measurement concept preliminary tests were done with water under standard pressure and temperature conditions (p = 1 bar, T = 20 °C). Since water is a homogeneous medium and has approximately the same ultrasound velocity at these conditions as PP does at around 180 °C the results should show the functionality of the forerun elements and give an impression of the behaviour of a homogeneous PP melt. The results of the preliminary tests with water are shown in figure 4. The measured transit times are in good agreement with the theoretically calculated ones. The increasing measurement inaccuracy with decreasing angles to the emitter position is caused by surface and geometry imperfections of the half cylindrical forerun elements. Those imperfections lead to a signal loss because the scattering of the ultrasound wave deviates from an ideal fan shape and make signal identification difficult. The slight difference between the calculated and the measured transit times arises from deviations of the real forerun element positions from those used for calculation.

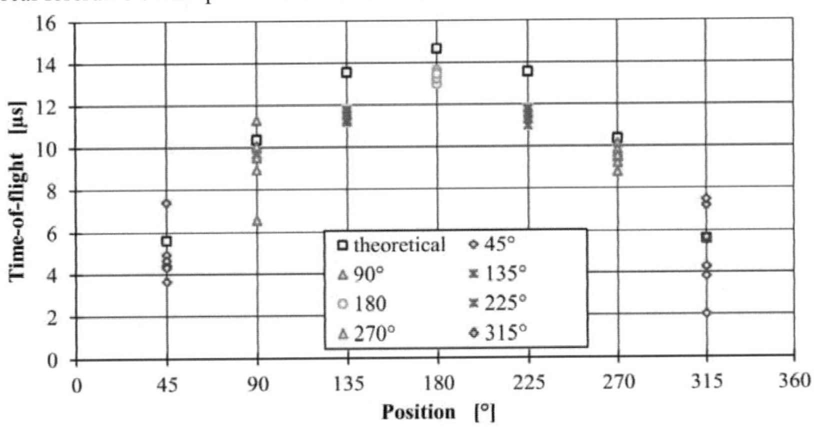

FIGURE 4. Measured and calculated ultrasound transit times in water under standard conditions

Measurements in PP Melt

The results of the measurement in PP melt are shown in table 1. Due to fast cooling after transferring the material into the measurement area not every emitter position could be measured. Furthermore transit times are higher than theoretically calculated. While melting and during solidification air pockets are formed which show a significantly lower ultrasound velocity, elevating transit times in comparison to a homogeneous melt. It could be shown that scattering of the ultrasound wave at the interface is strong enough to detect signals at adjacent transducers. For the usage of ultrasound tomography in injection moulding this is a promising approach.

TABLE 1. Measured transit times (μs) in the PP melt

		Emitter				
		0	**1**	**2**	**3**	**4**
Receiver	**0**	x	33,22	59,15	49,89	45,23
	1	48,12	x	61,53	65,87	54,60
	2	80,84	60,14	x	29,11	79,11
	3	67,86	75,53	63,48	x	25,15
	4	70,22	64,95	59,13	29,18	x
	5	61,46	25,95	62,76	47,45	39,78
	6	71,49	55,60	38,49	56,53	70,50
	7	30,26	44,30	57,25	56,31	46,29

CONCLUSION

The presented work for temperature measurements using ultrasound TOF measurements provides a non-invasive method for determining the temperature distribution in polymer melts. The results show the applicability of commercially available transducers and the operability of brass forerun elements, since each of the seven transducers was able to detect a signal that was sent out from the one remaining transducer. However, a useful reconstruction of the temperature distribution is, due to the inhomogeneous melt in the prototype, not yet possible. Currently an injection mould is designed which fulfills all prerequisites for ultrasound tomography and overcomes the difficulties from the prototype. The injection mould will be equipped with 20 ultrasound transducers to reach a spatial resolution smaller than 1 mm. A special measurement device is needed to do fast signal acquisition.

ACKNOWLEDGMENTS

The presented research is funded by the *Deutsche Forschungsgemeinschaft* (DFG) as part of the Collaborative Research Centre "Precision Melt Engineering" (SFB 1120). We would like to extend our thanks to the DFG.

REFERENCES

1. LAI, G.; RIETVELD, J. X.: Role of Polymer Transparency and Temperature Gradients in the Quantitative Measurement of Process Stream Temperatures During Injection Molding via IR Pyrometry. *Polymer Engineering and Science* 36 (1996) 13, S. 1755-1768
2. BENDADA, K.; COLE, K.; LAMONTAGNE, M.; SIMARD, Y.: A hollow waveguide infrared thermometer for polymer temperature measurement during injection moulding. *Journal of Optics A: Pure and Apllied Optics* 5 (2003) 464-470
3. PRAHER, B.; STRAKA, K.; STEINBICHLER, G.: An ultrasound-based system for temperature distribution measurements in injection moulding: system design, simulations and off-line test measurements in water. *Measurement Science and Technology* 24 (2013) 084004, S. 1-9
4. BUZUG, T.: *Computed Tomography.* Berlin, Heidelberg: Springer-Verlag, 2008
5. KAK, A. C.; SLANEY, M.: *Principles of Computerized Tomographic Imaging.* Philadelphia: Society for Industrial and Applied Mathematics, 2001
6. HENDEE, W. R.; RITENOUR, E. R.: *Medical Imaging Physics.* New York: Wiley-Liss, Inc., 2002
7. KRAUTKRÄMER, J.; KRAUTKRÄMER, H.: *Werkstoffprüfung mit Ultraschall.* Berlin, Göttingen, Heidelberg: Springer-Verlag OHG., 1961
8. HOPMANN, C.; SPEKOWIUS, M.; WIPPERFÜRTH, J.; SCHÖNGART, M.: A concept for Non-Invasive Temperature Measurement During Injection Moulding Precesses. *The 31st International Conference of the Polymer Processing Society.* Jeju, Korea, 2015

Tensile Impact Testing on Polymer Materials Considering the Force-Oscillation Phenomenon

Christian Hopmann, Jan Klein* and Maximilian Schöngart

Institute of Plastics Processing (IKV) at RWTH Aachen University
Seffenter Weg 201, 52074 Aachen, Germany

**jan.klein@ikv.rwth-aachen.de*
maximilian.schoengart@ikv.rwth-aachen.de

Abstract. Thermoplastic materials are extensively used as a light weight replacement for metal alloys, especially in automotive applications. Furthermore polymer materials can be used to enhance the safety of passengers and pedestrians in a car accident. The state of the art design process for plastics parts is based on Computer Aided Engineering (CAE). Using Finite Element Analysis (FEA), highly dynamic systems can be simulated with a very high accuracy. Considering crash applications, the influence of strain rate on the mechanical behaviour is of paramount importance. To determine the effect of strain rate on the mechanical behaviour, tensile impact tests are conducted at different haul-off velocities. With an increase of the haul-off velocity, the force-oscillation phenomenon becomes significant. The force-oscillation phenomenon is an artifact caused by the dynamic load application in highly dynamic tests. It can be monitored as a superimposition of measured force data with oscillations which increase in amplitude to higher impact velocities.

This study looks at the phenomenon of force oscillation in tensile impact testing on viscoelastic materials and its dependency on longitudinal stress wave propagation. Aside a detailed analysis of its origin, a new approach is presented to measure a nearly oscillation free force signal in tensile impact testing on polymer materials, basically independent of the considered haul-off velocity. For this purpose a modification of a standard dumbbell specimen was designed.

Keywords: Force oscillations, Material testing, Impulsive load, Dynamic properties, Data logging

PACS: 81.70.Bt

INTRODUCTION

Due to the extensive use of thermoplastic materials as a lightweight respectively low price alternative to metal alloys in automotive applications, tailored material models where designed to simulate effects such as the strain rate dependency and the lateral contraction behavior of plastics using FEA [1]. To calibrate these models, resilient material data are required as input data. Aside inverse approaches [2], SEĖ-methods [3] present a very effective procedure to determine material data for crash simulations. In general, SEĖ-methods determine true stress data dependent on strain and strain rate which are used to define true stress-strain curves at different constant strain rates [4]. These data can be used to calibrate material models for crash simulations.

The knowledge of the mechanical response of a structure to impact loadings is of particular importance, especially in the development process of automotive applications. Strain rates in a vehicle crash can exceed local magnitudes of about 500 s^{-1} [5]. Inverse procedures as well as SEĖ-methods are not restricted to a certain maximum strain rate, though material data measurement is. High-speed tensile testing is the standard test procedure to determine material data for crash simulations [6]. Servo-hydraulic test setups are able to obtain haul-off velocities of more than 10 m/s which is sufficient to reach local strain rates of more than 500 s^{-1}. However, a resilient force measurement is not possible at high haul-off velocities using standard piezoelectric load cells. The limiting factor to force measurement in tensile impact testing is the force-oscillation phenomenon. The force-oscillation phenomenon is an artifact that can be monitored in tensile impact tests at high haul-off velocities, independent of the considered material. It describes a superimposition of force measurements with an oscillating signal [7]. FIGURE 1 illustrates force-oscillations in tensile impact tests on polyamide 6 for different haul-off velocities.

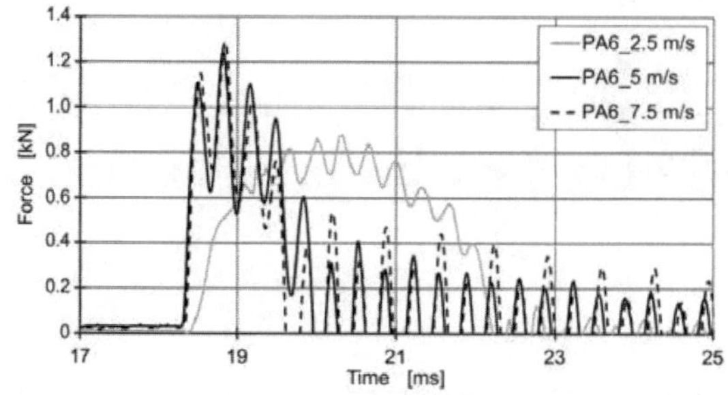

FIGURE 1. Force-oscillation phenomenon in tensile impact testing on polyamide 6

OSCILLATION FREE FORCE MEASUREMENT

The superimposition becomes predominate with an increase of the haul-off velocity. A distinct course of the force characteristic is not given anymore. Since material data measurement for crash simulation depends on force data, the force-oscillation phenomenon presents a parasitic error with a direct impact on FEA.

In addition to the extensive research on the force-oscillation phenomenon, published within the last decades [5, 7], the authors conducted fundamental investigations on its origin. It was found that: (1) It is basically impossible to achieve force equilibrium in tensile impact testing. Due to the inertia of the clamping device attached to the load cell, piezoelectric force measurement in tensile impact testing is always defected. (2) The actual loading of a specimen in tensile impact testing is nearly oscillation free. (3) The force characteristic of a specimen in tensile impact testing can be determined by measuring its deformation. Though, elastic material behavior is required. Therefore the authors designed a step-dumbbell specimen and a tailored evaluation approach based on Digital Image Correlation (DIC) to conduct a nearly oscillation free force measurement, independent of the considered haul-off velocity [8]. The principles are outlined in FIGURE 2.

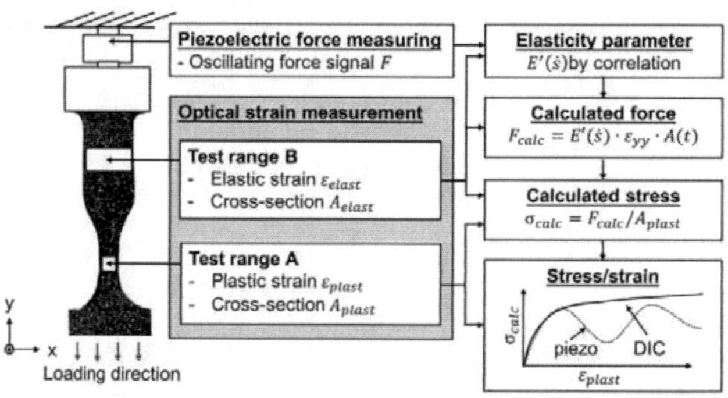

FIGURE 2. Approach for oscillation free force measurement in tensile impact testing

The basic idea of the developed approach is to determine the time dependent strain-field of the elastic area within test range B to determine a force-time curve based on Hook's law, see Eq. 1. The time dependent cross-section *A(t)* of the evaluation area B can be determined using the strain characteristic $\varepsilon_{xx}(t)$ and the initial cross section of the evaluation area A_0, assuming isotropic deformation behavior in width (x-direction) and thickness direction (z-direction), see Eq. 2. Using Eq. 3, anisotropic deformation behavior can be taken into account. However, it requires a setup of two 3D-camera systems to measure three-dimensional strain fields with an acceptable accuracy in tensile impact testing [4].

$$\sigma = F/A = \varepsilon \cdot E \tag{1}$$

$$A(t) = A_0(1 - \varepsilon_{xx}(t))^2 \tag{2}$$

$$A(t) = b_0(1 - \varepsilon_{xx}(t)) \cdot w_0(1 - \varepsilon_{zz}(t)) \tag{3}$$

Due to the strain rate dependency of thermoplastic materials [9], the application of the Young`s modulus in Eq. 1 is restricted to quasi-static tensile tests and therefore not suitable for the developed approach. Hence, a variation of the Young's modulus E named elasticity parameter $E'(\dot{s})$ is introduced, see Eq. 4. $E'(\dot{s})$ is determined for every testing using a regression between the piezoelectric force signal $F(t)_{piezo}$ and the DIC-based force measurement $F(t)_{DIC}$, see Eq. (5). A range in strain from 0.0005 to 0.001 in test range B proved to give a very good approximation for the determination of $E'(\dot{s})$. The time dependent force characteristic $F(t)$ is determined using the strain in loading direction $\varepsilon_{yy}(t)$, see Eq. 6.

$$E'(\dot{s}) = F(t)/(\varepsilon_{yy}(t) \cdot A(t)) \tag{4}$$

$$F(t)_{piezo} = F(t)_{DIC} \tag{5}$$

$$F(t) = E'(\dot{s}) \cdot \varepsilon_{yy} \cdot A(t) \tag{6}$$

SCALE CONCEPT

Due to a shift of the transition to nonlinear material behavior dependent on the considered polymer, the dimensions of the specimen are directly correlated to the material. FIGURE 3 (left) shows the correlation between the strain in test range B and the width of test range A and B for the investigated polyamide 6. The simulations were conducted using LS-DYNA Version 971, by Livermore Software Technology Corporation, Livermore, CA, USA, and the material model MAT_SAMP_1. The stress at failure was taken from preliminary testing. The polymer analysed in the presented study is an unreinforced polyamide 6 B30S by LANXESS GmbH, Leverkusen, Germany.

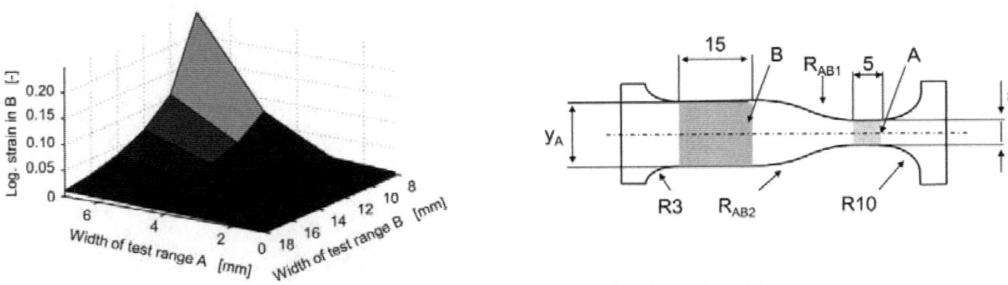

FIGURE 3. Strain distribution and specimen design for polyamide 6

The basic geometry of the specimen is presented in FIGURE 3 (right). Test range A was designed in dependency on a standard dumbbell specimen for tensile impact testing [10] which proved to be a usable geometry for high-speed material data measurement. To adapt test range B to a certain polymer, it is required to determine the maximum strain at which irreversible deformations can be neglected. The critical strain, i. e. the strain at which irreversible processes initiate [11], can be used to determine the approvable strain for test range B. Since the critical strain is determined under static conditions and the transient to plastic material behavior shifts to larger strain with an increasing strain rate, it is a reliable parameter for the specimen design.

The maximum strain for test range B was determined to 0.02. To avoid an extensive iterative process for the specimen design, Eq. (11) is introduced to give an approximation for the required cross-section in test range B. The deviation is given by Eq. (7) – Eq. (10). Aside the critical strain, the Young's modulus, the Poisson's ratio and the maximum force expected in tensile impact testing is required as input data. The maximum force as well as the elastic modulus can be taken from a simple tensile impact test using a standard dumbbell specimen, see FIGURE 1. As presented afterwards, the maximum force isn't superimposed with force-oscillations. When the maximum force is given by F_{max}, the cross-section of test range B A_0 can be determined using the Young's modulus E, the Poisson's ration ν_{xy} and the critical strain ε_{crit}.

$$\sigma_{crit} = E'(\dot{s})\, \varepsilon_{crit} = \frac{F_{max}}{A_{crit}} \tag{7}$$

$$\nu_{xy} = \frac{-\varepsilon_{xx}}{\varepsilon_{yy}} \tag{8}$$

$$\varepsilon_{yy} = \varepsilon_{crit} \tag{9}$$

$$A_{crit} = A_0(1 - |\varepsilon_{xx}|)^2 = A_0(1 - |\nu_{xy}\varepsilon_{crit}|)^2 \tag{10}$$

$$A_0 = y_A \cdot t = \frac{F_{max}}{E\varepsilon_{crit}(1-|\nu_{xy}\varepsilon_{crit}|)^2} \tag{11}$$

With an elasticity parameter of 2900 MPa at 7.5 m/s, a maximum force of a rounded 1400 N, a Poisson's ration of 0.35 and a critical strain of 0.02, Eq. (11) gives a cross-section of 24,48 mm which is rounded to 26 mm, giving a y_A=13 mm at a thickness t=2 mm.

Since the mechanical behavior of thermoplastics is also highly dependent on the stress state [12], it's necessary to analyze the stress state of the evaluation areas. Changing the dimension of test range B influences the distribution of force and therefore the stress state in both evaluation areas. FIGURE 4 (left) presents the stress state for the designed specimen, characterized by the p/q-rate. The p/q-rate describes the stress state using the invariance p (hydrostatic pressure) related to q (v. Mises effective stress), see Eq. 7 and Eq. 8. A value of p/q= 0.33 indicates uniaxial compression, a value of p/q= -0.33 is according to uniaxial tensile stress and a value of p/q= 0 is according to pure shear loading. Values over p/q= 0.33 respectively under p/q= -0.33 indicate that the uniaxial stress state changes into a biaxial loading. The simulations were also conducted using LS-DYNA Version 971 and the material model MAT_SAMP_1. Radius R_{AB1} and R_{AB2} are both designed to 10 mm.

$$p = \frac{1}{3}(\sigma_I + \sigma_{II} + \sigma_{III}) \tag{12}$$

$$q = \sqrt{\frac{1}{2}[(\sigma_I - \sigma_{II})^2 + (\sigma_{II} - \sigma_{III})^2 + (\sigma_{III} - \sigma_I)^2]} \tag{13}$$

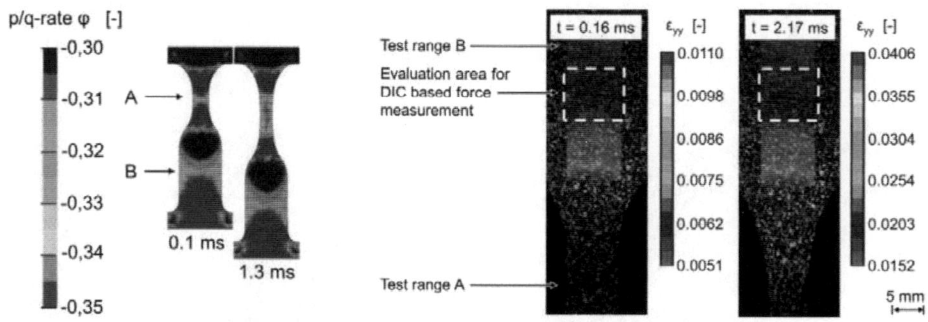

FIGURE 4. Stress state and strain distribution

Test range A shows a slight deviation from perfect uniaxial tensile stress with increasing deformation. This is due to the significant lateral contraction behavior of thermoplastics and therefore unpreventable. In contrast, test range B exhibits an adequate area with a basically constant stress state of about -0.33 until failure. FIGURE 4 (right) shows the strain distribution in test range B at 2.5 m/s. The calculated maximum strain of 0.02 is not exceeded.

Experimental

Tensile impact tests were conducted to analyze the applicability of the developed approach. The mechanical investigations were executed using a Zwick REL 1863 (Zwick GmbH & Co. KG, Ulm, Germany) testing machine with a Kistler 9331B (Kistler Instrumente AG, Winterthur, Switzerland) piezoelectric load cell. The response frequency of the load cell is about 45 kHz. The load signal was amplified using a Kistler 5011 charge amplifier at a sample rate of 100 kHz. High-speed strain measurement was conducted at 100 kHz using a 3D setup consisting of two high-speed cameras SA 1.1 by Americas and Antipodes Photron USA Inc., San Diego, California, USA, and Digital Image Correlation. DIC is a full-field deformation measurement procedure with a very high spatial resolution [4] which is based on gray scale correlation [13]. Strain measurements were executed using ViC-3D by Correlated Solutions Inc., Columbia, South Carolina, USA.

A comparison of piezoelectric and DIC-based force measurement is outlined in FIGURE 5. At 0.5 m/s the general course of both characteristics is basically congruent. Though, the force signal determined using the piezoelectric load cell (broken line) shows slight oscillations and an offset at about 9 mm. This offset is due to the inertia of the clamping device connected to the load cell. At 10 m/s there is only a slight deviation in both force-displacement curves up to 2 mm. Afterwards, the piezoelectric force signal (broken line) shows a significant superimposition with an amplitude of about 500 N. The force signal measured using DIC exhibits a distinct course. The local peak at 3 mm is expected to be caused by the oscillations of the test setup. Hence, it is no parasitic error but an actual loading of the specimen [8].

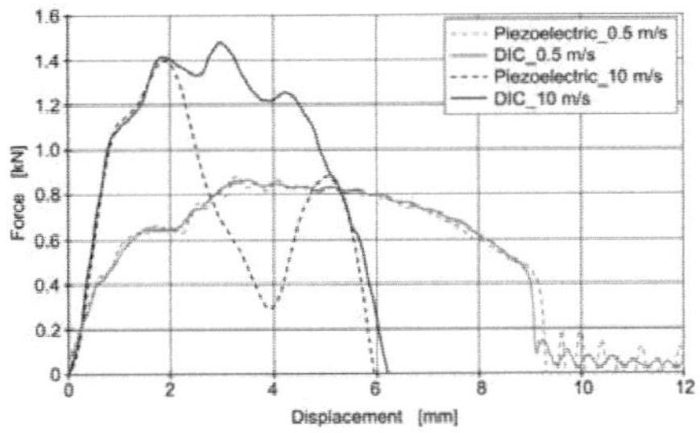

FIGURE 5. Comparison of piezoelectric and DIC-based force measurement

CONCLUSIONS

The force-oscillation phenomenon is a parasitic error in piezoelectric force measurement. A specimen in tensile impact testing is exposed to a nearly oscillation free load. Hence material data measurement for crash simulation is only limited by the applied measuring systems. DIC-based force measurement presents a resilient alternative that requires equipment which is already necessary to characterize material data for crash simulations. The presented scale concept enables an application of the presented specimen geometry to any given polymer.

ACKNOWLEDGMENTS

The depicted research has been funded by the Deutsche Forschungsgemeinschaft (DFG) as part of the program Cluster of Excellence "Integrative Production Technology for High-wage Countries", as part of the research group ICD-C3. We would like to extend our thanks to the DFG. The authors are very grateful to LANXESS GmbH, Leverkusen, Germany who supported the research project with test materials and to Limess Messtechnik & Software GmbH, Krefeld, Germany, for technical support in optical deformation measurement.

REFERENCES

[1] KOLLING, S.; HAUFE, A.; FEUCHT, M.; BOIS, P. A. D.: SAMP-1: A Semi-Analytical Model for the Simulation of Polymers. *4. LS-DYNA Anwenderforum.* Bamberg, 2005

[2] ZHANG, W.; CHO, C.; XIAO, Y.: An effective inverse procedure for identifying viscoplastic material properties of polymer Nafion. *Computational Material Science* (2014) 95, S. 159-165

[3] LAURO, F.; BENNANI, B.; MORIN, D.; EPEE, A. F.: The SEĖ method for determination of behaviour laws for strain rate dependent material: Application to polymer material. *International Journal of Impact Engineering* 37 (2010) 6, S. 715-722

[4] HOPMANN, C.; KLEIN, J.: Determination of strain rate dependent material data for CEA crash simulation of polymers using digital image correlation. *Computational Material Science Part B* 100 (2015) S. 181-190

[5] XIAO, X.: Dynamic tensile testing of plastic materials. *Polymer Testing* 27 (2008) 2, S. 164-178

[6] GRYTTEN, F.; DAIYAN, H.; POLANCO-LORIA, M.; DUMOULIN, S.: Use of digital image correlation to measure large-strain tensile properties of ducile thermoplastics. *Polymer Testing* 28 (2009) S. 653-660

[7] KRAGER-KOCSIS, J.; BENEVOLENSKI, O. I.: Toward understanding the stress oscillation phenomenon in polymers due to tensile impact loading. *Journal of material science* 36 (2001) 14, S. 3365-3371

[8] KLEIN, J.; HOPMANN, C.: On the Origin and Handling of the Force Oscillation Phenomenon in Tensile Impact Testing of Polymer Materials. *DOI 10.1007/s11340-015-0124-z* (2016) S.

[9] ŞERBAN, D. A.; WEBER, G.; MARŞAVINA, L.; SILBERSCHMIDT, V. V.; HUFENBACH, W.: Tensile properties of semi-crystalline thermoplastic polymers: Effects of temperature and strain rates. *Polymer Testing* 32 (2013) 2, S. 413-425

[10] JUNGINGER, M.; WERNER, H.; THAM, R.; THOMA, K.: Schnellzerreißprüfung von Thermoplasten und Bestimmung der mechanischen Werkstoffeigenschaften unter Berücksichtigung lokaler Dehnungen. *Amsler Symposium - World of Dynamic Testing.* Gottmadingen, 2001

[11] ALEKSANOGLU, H.; SCHOLZ, A.; OECHSNER, M.; BERGER, C.; RUDOLPHI, M.; SCHÜTZE, M.; STAMM, W.: Determining a critical strain for APS thermal barrier coatings under service relevant loading conditions. *International Journal of Fatigue* 53 (2013) S. 40-48

[12] HACHOUR, K.; ZAIRI, F.; NAIT-ABDELAZIZ, M.; GLOAGUEN, J. M.; ABERKANE, M.; LEFEBVRE, J. M.: Experiments and modeling of high-crystalline polyethylene yielding under different stress states. *International Journal of Plasticity* 54 (2014) S. 1-18

[13] LAVA, P.; COPPIETERS, S.; WANG, Y.; HOUTTE, P. V.; DEBRUYNE, D.: Assessment of measuring errors in strain fields obtained via DIC on planar sheet metal specimens with a non-perpendicular camera alignment. *Applied Mechanics and Materials* 70 (2011) S. 165-170

New Understanding on the Formation of Phase Structure of Thermoplastic Vulcanizate (TPV) During Dynamic Vulcanization

Ming Tian[1,2], Hanguang Wu[1], Nanying Ning[1,2] , Hongchi Tian[2], Youping Wu[1,2], and Liqun Zhang[1,2]

1. Key Laboratory of Beijing City on Preparation and Processing of Novel Polymer Materials, Beijing University of Chemical Technology, Beijing 100029, China

2. Key Laboratory of carbon fiber and functional polymers, Ministry of Education, Beijing University of Chemical Technology, Beijing 100029, China

In our study, the minimum size of the rubber phase in the EPDM/PP blend at the early stage of dynamic vulcanization was firstly calculated to be in the range of 25 nm and 46 nm by using the critical breakup law of the viscoelastic droplets in matrix. Meanwhile, the real size of rubber phase in Thermoplastic Vulcanizate (TPV) at both the early stage and the final stage of dynamic vulcanization were observed by using Peak Force Tapping Atomic Force Microscopy (PF-AFM). The results indicated that EPDM phase indeed broke up into nano-scale particles at the early stage of dynamic vulcanization, consistent well with the calculated result. More interesting, it was firstly revealed that the micro-meter sized rubber particles commonly observed in TPV were actually the agglomeration of nano-scale rubber particles with a diameter of 40-60 nm. We further revealed a new mechanism for the morphological evolution and microstructure formation of TPV during dynamic vulcanization. Our results indicated that the occurrence of phase inversion of the dynamically vulcanized blend was dominated by the formation and agglomeration of rubber nanoparticles rather than the increased viscoisty ratio of EPDM and PP as previously reported. This study provides guidance to control the microstructure of TPV to prepare high performance TPV products for a wide range of industrial applications.

Introduction

Thermoplastic vulcanizates (TPVs) are a particular group of thermoplastic elastomers produced by dynamic vulcanization, which consists of the selective crosslinking of the elastomer and its fine dispersion throughout the thermoplastic phase under intensive mixing.[1] TPVs combine the excellent resilience of ordinary vulcanized rubbers and the high processability of thermoplastics, and the shape and size of the rubber phase in the matrix has a significant influence on the mechanical properties and processability of TPV. Many studies have reported that the rubber phase dispersed in the matrix consists of spherical particles with a diameter of 0.5-3.0 μm.[2, 3]

Experimental

Preparation of the sample

First, the EPDM rubber and PP were fed into the two-roll mill in the required mass ratio (60:40) at 180 °C. Meanwhile, the antioxidant was added. Subsequently, the homogeneous blend was transfered to another two-roll mill at ambient temperature, 0.7 phr crosslinking agents and 0.15 phr antioxidants were added. The premix obtained was fed into the Haake rheomix (HAAKE Rheomix 600 OS, Thermo Fisher Scientific, America) at 170 °C and a rotor speed of 80 rpm. In order to follow the mechanism by which the final microstructure of EPDM/PP TPV is formed, samples were taken at different mixing times and immediately cooled down in liquid nitrogen to stop the crosslinking reaction and further morphological changes.

Measurement and Characterization

In order to calculate minimum size of the rubber phase in the EPDM/PP blend at the early stage of dynamic vulcanization, we characterized the rheological behavior of EPDM and PP by dynamic rheological measurements in a stress-controlled rotational rheometer (AR-G2, TA Instruments, America) using parallel-plate geometry (radius = 10 mm; gap = 1 mm), and calculated the interfacial tension between EPDM and PP at 170 °C from the polarities (χ) and surface tensions (γ) of EPDM and PP at 170 °C. The volume swell ratios (Q) of the EPDM/PP blends were measured as a method of measuring the crosslink degree of the EPDM phase by extracting the samples in cyclohexane at the ambient temperature. The Payne effects of the EPDM/PP blends was analyzed by Rubber Process Analyzer (RPA 2000, Alpha, America), and the result was used to reflect the distribution of the EPDM rubber network in PP matrix.

The samples morphology was examined by using a new type of peak force tapping atom force microscope (PF-AFM) (Nanoscope IIIa, Bruker corporation, Germany). In order to observe the size and shape of the individual rubber particles, the dynamically vulcanized TPV was dissolved in hot xylene, and the suspension obtained was spin coated on a clean coverslip by using a spin coater (KW-4A; China). The sample obtained was observed by the PF-AFM as well.

Results and Discussion

The breakup of the rubber phase in EPDM/PP TPV occurs when the premix was just fed into the Haake rheomix.[4] Because of its extremely low crosslink density, the deformation and breakup of the rubber phase at this time can be approximated by those of unvulcanized EPDM in EPDM/PP blend in a rotational shear field. For an immiscible polymer blend such as unvulcanized EPDM/PP blend, the minimum size of the EPDM dispersed phase can be obtained from the critical breakup law proposed by Wu:[5]

$$\frac{4}{3}\pi\left(\frac{\alpha}{2}\right)^3 = \frac{4}{3}\pi L B^2 \qquad (1)$$

$$4\left(\dot{\gamma}\eta_m / F(p)\right)\sin(2\phi) \geq \gamma_{12}\left(1/L + 1/B\right) \qquad (2)$$

$$F(p) = \frac{16p+16}{19p+16} \qquad (3)$$

where α is the initial diameter of the spherical droplet, L and B are the major semi axis and short semi axis of the deformed ellipsoid droplet, respectively; $\dot{\gamma}$ is the shear rate, η_m is the viscosity of the matrix, p is the viscosity ratio of droplet to matrix (η_d and η_m), ϕ is the angle between the major axis of the deformed ellipsoid droplet and the axis perpendicular to the flow direction (see Figure 1), and γ_{12} is the interfacial tension between the droplet and the matrix.

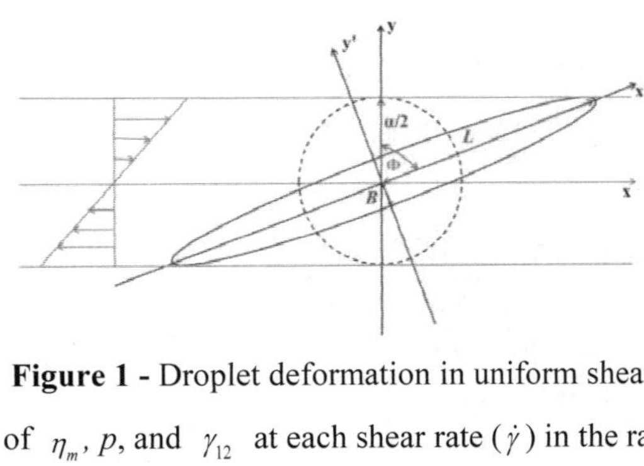

Figure 1 - Droplet deformation in uniform shear

With the measured values of η_m, p, and γ_{12} at each shear rate ($\dot{\gamma}$) in the range of 7.1 s^{-1} to 108 s^{-1}, the minimum diameter (α_{min}) of the dispersed EPDM droplet (α_{min}) is calculated by using equations (1), (2), and (3) and is shown in Table 1. It can be seen that α_{min} ranges from 25~46 nm.

Table 4 - Measured parameter values and calculated α_{min}

$\dot{\gamma}$ (s^{-1})	η_d (Pa. s)	η_m (Pa. s)	$F(p)$	$a/2L$	$sin(2\triangleleft)$	α_{min} (nm)
7	4.4×10^4	3.3×10^4	0.903	0.249	0.482	46
50	2.3×10^4	1.2×10^4	0.890	0.188	0.369	25
108	1.2×10^4	5.2×10^3	0.884	0.170	0.315	35

Figure 2 shows the variation of the torque and temperature as a function of mixing time during dynamic vulcanization of EPDM/PP blend in Haake rheomix. Six samples were selected according to the torque-time curve, and these samples were immediately cooled down in liquid nitrogen for morphology observation by using PF-AFM.

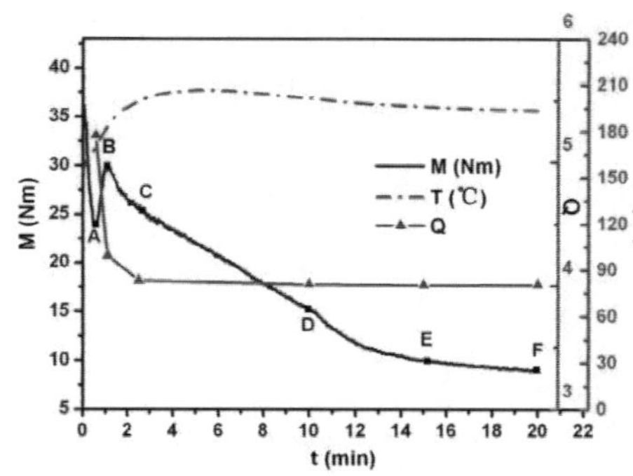

Figure 2 - The variation of the torque (M), the temperature (T) and the swell ratio (Q) of EPDM/PP blend in Haake rheomix

The AFM phase image of EPDM/PP blend taken at point A is shown in Figure 3, where the lighter regions represent PP phase and the darker regions represent the crosslinked EPDM phase. The agglomerate of a great many rubber nanoparticles can be observed from this figure, and the diameters of the rubber nanoparticles are in the range of 25-70 nm (see Figure 3(b)). This indicates that the EPDM phase indeed break up into nanoparticles at the early stage of dynamic vulcanization, in good agreement with the calculated results. Figure 4 shows that the micro-meter sized rubber particles in EPDM/PP TPV taken at the end of dynamic vulcanization are actually the agglomerates of many EPDM nanoparticles with diameters in the range 40-60 nm, and the single rubber nanoparticles are more clearly observed through dissolving TPV in the hot xylene and spin caoting the obtained suspension.

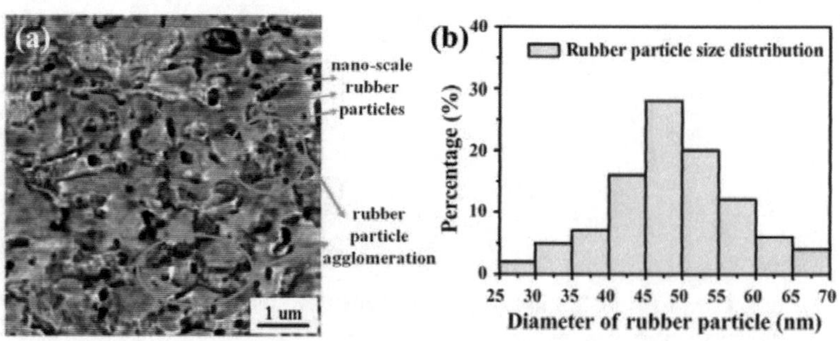

Figure 3 - (a) AFM phase image of EPDM/PP blend taken at point A and (b) size distribution of rubber nanoparticles

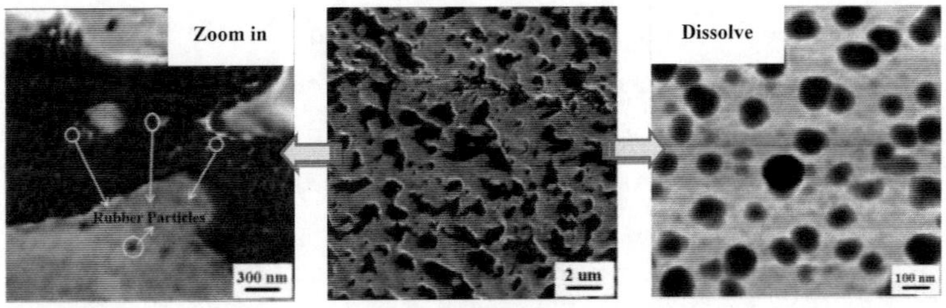

Figure 4 - AFM phase images of dynamically vulcanized EPDM/PP blend (sample F) and the dissolved rubber nanoparticles

The height images of the selected six samples observed by using PF-AFM are shown in Figure 5, in which the darker regions represent PP phase and the lighter regions represent the EPDM rubber phase. A co-continuous structure is observed for sample A and sample B, which is ascribed to the coalescence of some EPDM rubber droplets at these two stages due to the low crosslink degree of the rubber phase. The EPDM continuous phase is transformed into the dispersed phase in continuous PP matrix for sample C, indicating the occurrence of phase inversion, as shown in Figure 5(c). This is ascribed that all the rubber nanodroplets has been transformed into rubber nanoparticles, and then form rubber agglomerates because of the high crosslink degree, facilitating the break up of EPDM continuous phase and the formation of EPMD dispersed phase in the PP matrix. During the further dynamic vulcanization (C Point to F Point), the size of the rubber agglomerates increases significantly, and the PP layers encapsulating the rubber agglomerates are thicker. From the RPA result shown in Figure 6, it can be seen that the rubber network in the EPDM/PP blend is disrupted severely during dynamic vulcanization.

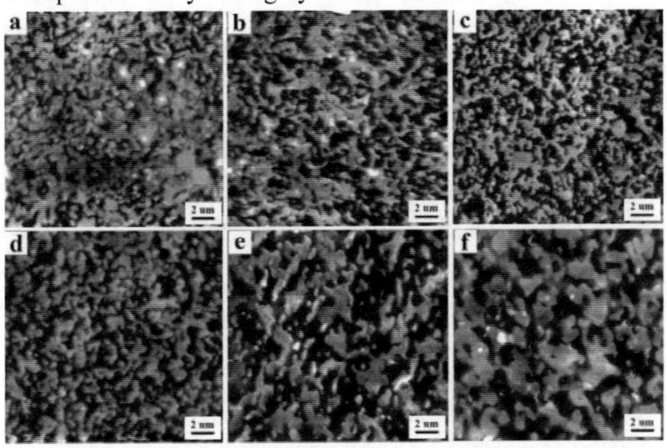

Figure 5 - AFM height images of the PP/EPDM blend taken at different times: (a) A point; (b) B point; (c) C point; (d) D point; (e) E point; (f) F point

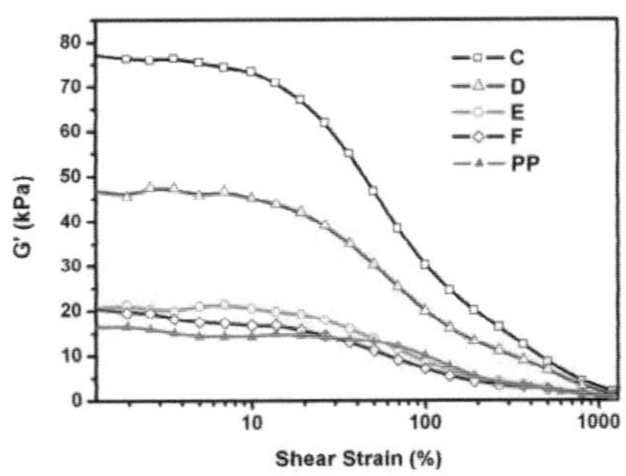

Figure 6 - Storage modulus vs. shear strain for PP and EPDM/PP blends taken at different stages (from C to F point)

Then, the schematic illustration of the morphological evolution of the EPDM/PP TPV during dynamic vulcanization is summarized, which is shown in Figure 7. The rubber phase breaks into nano-scale droplets at the initial stage of dynamic vulcanization. Because of the inhomogenious of crosslink degree at the initial stage of vulcanization, some broken rubber nanodroplets transform into nanoparticles, whereas some other rubber nanodroplets with lower crosslink degree can still coalesce to form rubber aggregates (Figure 7 (b)). The phase inversion occurs when all the broken rubber nanodroplets have transformed into nanoparticles and spontaneously agglomerate with the further increase in crosslink degree (Figure 7 (c)). At the later stage of dynamic vulcanization (Figure 7 (d)), rubber nanoparticles spontaneously agglomerate more densely, larger and regular with the increase in the dynamic vulcanization time, leading to the thickening of the PP layers encapsulating the rubber agglomerates and the significant weakening of the rubber network.

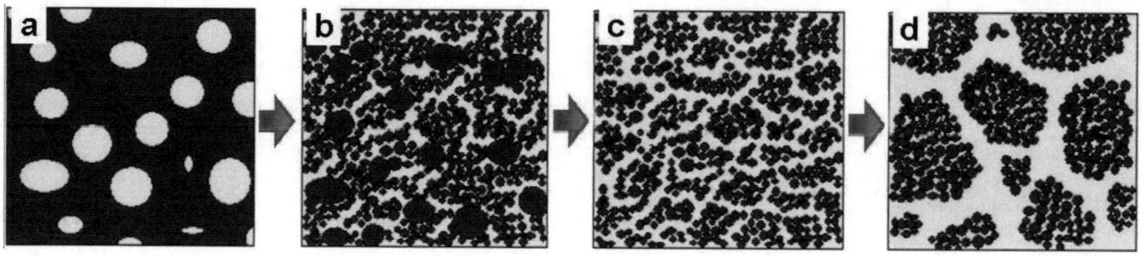

Figure 7 - Schematic illustration of the morphological evolution of TPV during dynamic vulcanization

Conclusions

The micrometer-sized rubber particles in TPV obtained at the end of dynamic vulcanization were actually the agglomerates of rubber nanoparticles. We further revealed a new mechanism for the morphological evolution of TPV during dynamic vulcanization as following. First, the breakup and crosslinking of the rubber phase mainly occured at the early stage of dynamic vulcanization. The phase inversion of the rubber phase and thermoplastic occured when all the rubber nanodroplets transformed into interconnected nanoparticles, which formed a strong rubber network. At the later stage, these rubber nanoparticles agglomerated more densely because of the requirement of thermodynamic stability, leading to the formation of larger and more regular agglomerates and significantly weakened rubber network.

Acknowledgements

We gratefully acknowledge the National Basic Research Program of China (Grant No. 2011CB606003) and the National Natural Science Foundation of China under (Grant No. 51221002) for financial supports.

References

1. C. F. Antunes, A. V. Machado, and M. V. Duin, Eur. Polym. J. 2011, 47, 1447.
2. M. V. Duin, Macromolecule 2006, 233, 11.
3. A. Mousa, Int. J. Polym. Mater. 2005, 54 , 619.
4. F. Goharpey, A. A. Katbab, and H. Nazockdast, J. Appl. Polym. Sci. 2001, 81, 2531.
5. S. Wu, Polym. Eng. Sci. 1987, 27, 335.

InnoREX – European Project Reveals Impact of Microwave and Ultrasound Energy on Polymerization of PLA via Reactive Extrusion

Björn Bergmann[a*], Jan Diemert[a], Jean-Marie Raquez[b], Thomas Hielscher[c], Markus Reichmann[d], Chantal David[e], Satya Dubey[f], and Carolina Losada Fernández[g]

[a]Fraunhofer Institute for Chemical Technology, Joseph-von-Fraunhofer Strasse 7, 76327 Pfinztal, Germany
[b]Center of Innovation and Research in Materials & Polymers, (CIRMAP), Laboratory of Polymer and Composite Materials, University of Mons-Hainaut, Place du Parc 20, B-7000 Mons, Belgium
[c]Hielscher Ultrasonics GmbH, Oderstr. 53, 14513 Teltow, Germany
[d]Muegge GmbH, Hochstrasse 4 – 6, D - 64385 Reichelsheim, Germany
[e]Sciences Computers Consultants, 10 rue du Plateau des Glières, F-42000 Saint-Etienne, France
[f]Advanced Vehicle Engineering Centre , Cranfield University, Cranfield MK43 0AL, Bedfordshire, UK
[g]AIMPLAS · Instituto Tecnológico del Plástico; Valencia Parc Technologic, Calle Gustave Eiffel, 4, Aparto de correos 51, 46980 Paterna Valencia, Espana

Bjoern.Bergmann@ict.fraunhofer.de

Abstract. The InnoREX project, funded by the EU under FP7, was initiated by twelve partners in December 2013. The project aim is the continuous, highly precise, metal-free polymerization of PLA using alternative energies for reactive extrusion. This presentation will discuss the project's innovations regarding the incorporation of microwave and ultrasound into the extrusion process, the resulting process opportunities and the influence of these energy sources on process performance and material characteristics.

Keywords: PLA, Biopolymer, Reactive Processing, Ultrasound, Microwave, Simulation

THE INNOREX PROJECT

Within the InnoREX project a novel reactor concept using alternative energies for the continuous, highly precise, metal-free polymerization of PLA in a continuous extrusion line has been developed.

A pilot line has been assembled which includes the energy sources and ensures the polymerization and purification of the polymer from the residual monomer. Using real-time analytics the product quality and the process stability is ensured, and a scientific interpretation of the obtained data provides insight into the underlying molecular mechanisms. Simulation tools have been developed enabling fast upscaling of the process to industrial scale, which allows further additivation of the material and validation of the packages produced.

Development of a Potentially Ecofriendly Catalyst

As tin-based catalysts may be regarded as problematic for some applications, the InnoREX project aims to replace tin-containing catalysts by organic alternatives. Project partner University of Mons synthesized different types of catalysts. Retaining the optical purity in bulk polymerizations proved to be a particular challenge. Nevertheless, a new class of catalyst has been found, allowing robust conversion of the monomer, optimal optical purity and, most importantly, reproducible results at bulk scale and under extrusion conditions.

Alternative Energy Sources Ultrasound and Microwave within the Processing Length of a Twin Screw Extruder

In a cooperation with ultrasound specialist and project partner Hielscher Ultrasonics, ultrasound was incorporated into the processing length of a twin screw extruder. The specialized sonotrode is integrated directly into the extruder barrel. The use of an adapted screw configuration ensures that a large fraction of the melt is in direct contact with the surface of the sonotrode. As a result, the mechanical vibrations of the ultrasound are effectively transported into the melt, efficiently enhancing the polymerization.

Analytics revealed that the resulting molecular weight of the material produced with and without using ultrasound is equal within the error range. However, the residual monomer content of the resulting material could be drastically decreased by incorporating ultrasound into the melt (see figure 1 (b)). Figure 1 (a) shows the extruder barrel containing the sonotrode (invisible), booster horn, and ultrasound generator.

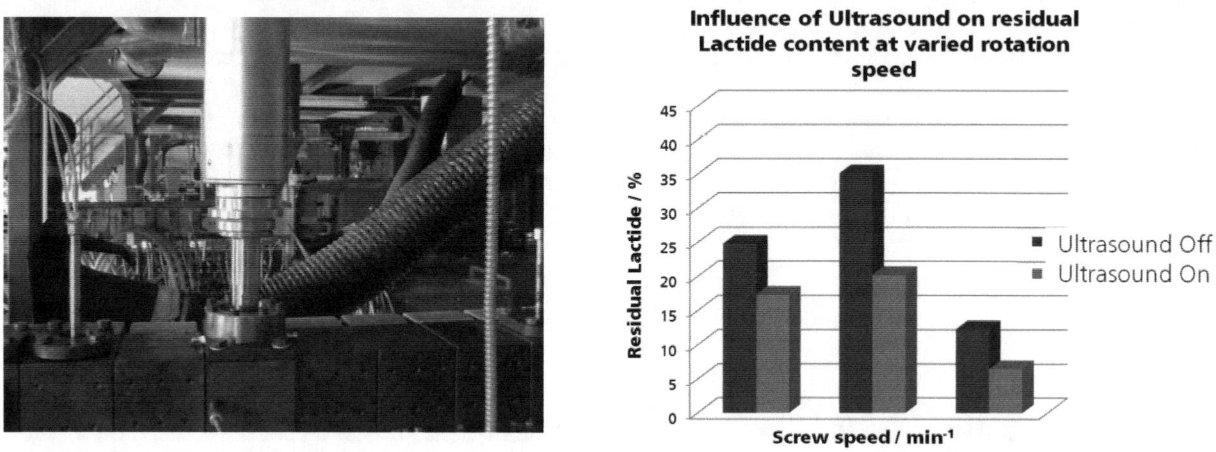

(a) (b)

FIGURE 1. (a) Extruder barrel containing the sonotrode (invisible), booster horn and ultrasound generator (on top)
(b) Residual lactide content of produced material with and without ultrasound treatment

In cooperation with microwave specialist and project partner Muegge, microwaves were incorporated into the processing length of a twin screw extruder. Using an optimized design concept (see figure 2 (a)) and microwave wavelength the microwaves can be introduced into the steel surrounding of a commercial extruder without the need for specialized ceramic machine parts.

Analytics revealed that the resulting molecular weight of the material produced can be positively influenced by microwaves. Especially for energy-saving process settings, the additional incorporation of microwaves can lead to an increase in molecular weight without material degradation (see figure 2 (b)).

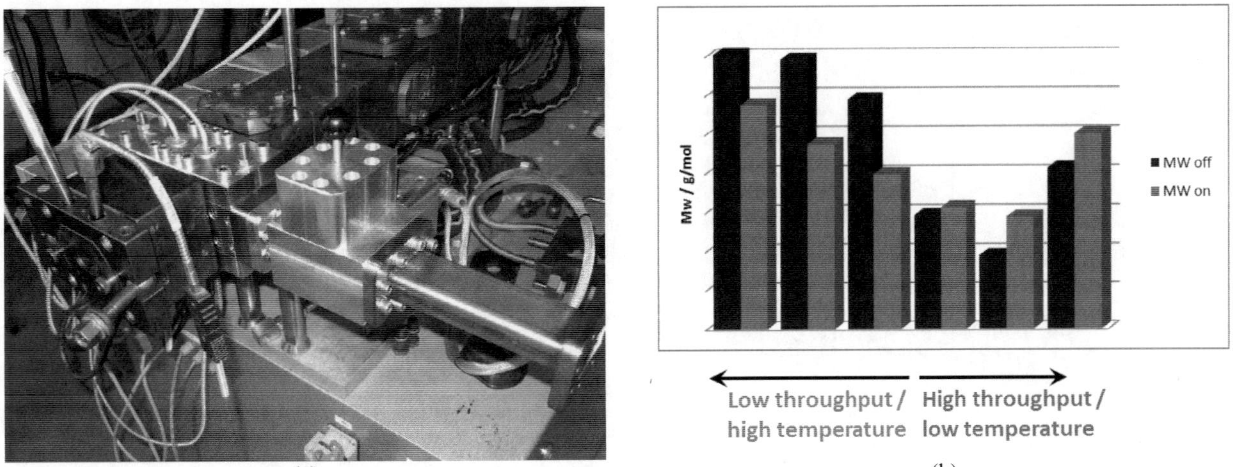

(a) (b)

FIGURE 2. (a) Extruder barrel for microwave incorporation (b) Resulting molecular weights of the materials produced with and without microwave treatment

Online Analytics and Simulation

In cooperation with project partner Gneuss an online viscometer based on the principle of capillary rheometry was designed for use within the processing length of a twin screw extruder (see figure 3 (a)). Using this device the dynamic viscosity of the material could be followed at different points of the process. Combining these viscosity measurements with molecular weights measured offline (see figure 3 (b)) allows correlations to be established to ensure product quality and the success of the reaction, without the need for resource-consuming offline characterization.

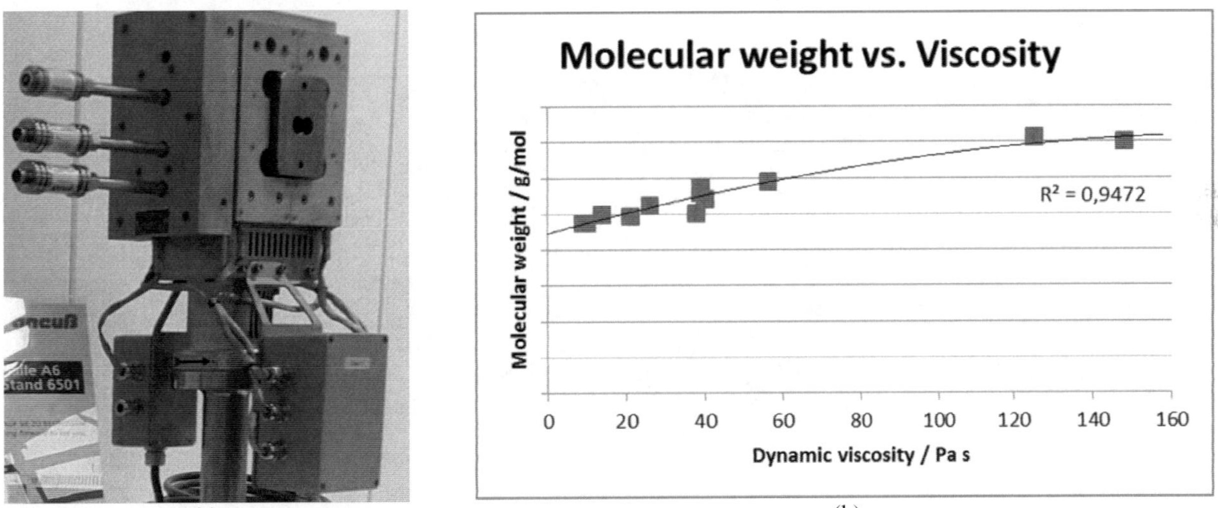

(a) (b)

FIGURE 3. (a) Online viscometer to be mounted in the processing length of a twin screw extruder (b) Correlation of dynamic viscosity values measured online within the process and molecular weights measured offline.

Feeding all the data obtained into molecular simulations allowed the development of predictive tools. Starting with a modelling approach at the University of Cranfield, isothermal curves showing the development of molecular weights and conversion were developed for the reactive extrusion bulk polymerization [1].

These equations were transferred into the commercial Ludovic© simulation software by software developer and project partner SCC. This enhanced the Ludovic© software with new PLA kinetics and new alternative energy inputs, as now the application of microwave energy can be reproduced by the simulation software as well (see figure 4).

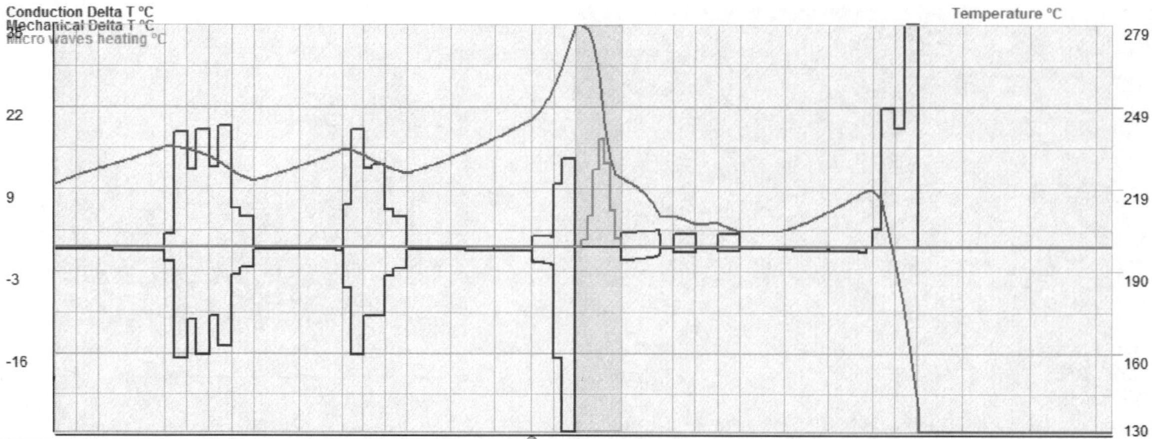

FIGURE 4. Screenshot of simulated output of the Ludovic© software showing energy input of the extruder screw (red curve), response of the extruder cooling (blue curve), energy input of the microwave (green curve), and resulting material temperature (black curve). The zone in which the microwaves were incorporated is marked in yellow.

Material Validation

Most importantly, the newly developed material grades have been further optimized by means of additivation. Developing an optimized additive package, project partner AIMPLAS succeeded in further enhancing the material properties, gaining an increase of 315% in impact resistance with a decrease of the heat deflection temperature of only 9.7% for the InnoREX PLA. Figure 5 (a) shows some sample PLA granules produced before and after additivation.

Project partner Talleres Pohuer further processed the optimized PLA grades by means of injection molding, obtaining food packages as one of the case studies within the InnoREX project (see figure 5 (b)).

In a second case study, project partner AIMPLAS produced sheets from the newly developed PLA grade, which were further processed by means of thermoforming (see figure 5 (c)).

(a) (b) (c)

FIGURE 5. (a) InnoREX PLA granules before and after additivation (b) Injection molded package using new PLA grade (c) Sheet extrusion of new PLA grade for further processing in thermoforming.

CONCLUSION

After 3 ½ years of project work the InnoREX project successfully demonstrated the continuous, highly precise polymerization of PLA using alternative energies for reactive extrusion. For this purpose a new class of potentially eco-friendly catalysts was developed, which allows robust monomer conversion under bulk scale extrusion conditions while maintaining an optimal optical purity.

The alternative energy forms ultrasound and microwave were integrated into the reactive extrusion process, and a high process stability was achieved. Both technologies showed benefits: while the molecular weight could be influenced by the application of microwaves, it appeared that ultrasound had a significant effect on the residual lactide content of the materials produced.

Using online analytics such as online viscometry, product quality could be ensured and scientific understanding of the process could be developed. Translating this understanding into mathematical equations allowed us to further develop the commercial Ludovic© software, which can now handle the reactive extrusion polymerization of PLA with additional microwave energy input.

Finally, the produced materials were further additivated and validated for use in food packaging by means of injection molded and thermoformed packages.

ACKNOWLEDGMENTS

The authors wish to acknowledge the financial support provided by European Commission. This project has received funding from the European Union's Seventh Framework Programme (FP7/2007-2013) under grant agreement no. 309802.

REFERENCES

1. S. Dubey et al., *International Journal of Chemical, Molecular, Nuclear, Materials and Metallurgical Engineering* **9(2)**, 329 - 333. (2015)

An Engineering Approach to Improve the Prediction of Melt Flow Behaviour for Injection Moulded Rubber Parts

A. Amine El Sayed (presenting author)[a], R. Iervolino[b] (reference author)

[a b] *SKF Engineering & Research Centre,*
Kelvinbaan 16, 3439MT, Nieuwegein, The Netherlands

achraf.amine.el.sayed@skf.com
rossana.iervolino@skf.com

Abstract. While injection molding simulation softwares are becoming more and more popular in thermoplastic industries due to their ease of use and advantages, many rubber molding companies have not yet fully benefit from this technology, known as reactive injection molding (RIM). This is mainly due to the complexity of material characterization, the lack of material data and appropriate description of the material behaviour. Specifically, the analysis of the filling stage of the RIM process requires the determination of the cure kinetics and the rheological, PVT, and thermal behavior of the material. Often one of the hurdles is characterizing and modeling the viscosity of rubber compounds reinforced with particulate fillers. This type of compounds usually does not obey to the Cox-Merz rule [1] due to the presence of fillers [2]. In addition, the cure rate dependence of viscosity has to be properly addressed. In this work, we have shown that the destruction of filler network at high shear rates makes the steady state shear viscosity, η, measured by capillary rheometer, lower than the complex viscosity, η*, obtained via RPA. In order to obtain an effective methodology to model compound viscosity, an engineering approach has been developed. It combines material characterization and modeling, numerical simulation, and experimental verification. By applying this methodology, numerical simulation results have shown good prediction of melt front time and sprue pressure for different injection molds and rubber compounds.

Keywords: Rubber, Injection Moulding Simulation, Rheology.
PACS: 83.10.Gr, 83.50.Ax, 83.50.Uv, 83.60.Fg.

METHODOLOGY: EXPERIMENT AND MODELING

Rheological Behaviour

Two filler-reinforced rubber compounds of nitrile- (NBR) and fluoro-rubber (FKM) have been investigated in this study. An Alpha technologies RPA2000 oscillatory rheometer has been used to measure the complex shear viscosity as a function of angular frequency at temperatures in the range of 80-120°C for NBR compound and 70-110 °C for FKM compound, using an oscillating angle of 0.1 deg. The temperatures used for the frequency sweep measurements were chosen in order to prevent curing during testing. The oscillating angle used for testing was chosen in such way that material remained in its linear viscoelastic region. The latter was assessed by measuring the complex viscosity as a function of oscillating angle at a frequency of 1Hz. A capillary rheometer (Göttfert 2002) was also used to measure the steady-state shear viscosity as a function of shear rate at the same temperatures as employed in the dynamic testing.
Concerning the viscosity model, a well-known rheo-kinetic model, called Cross Castro-Macosko, which describes well the dependence of viscosity on shear rate, temperature, and curing conversion, was used. The model is already built into Moldex3D and is based on the following equations:

$$\eta = \frac{\eta_0 \cdot \left(\dfrac{\alpha_g}{\alpha_g - \alpha} \right)^{C_1 + C_2 \alpha}}{1 + \left(\dfrac{\eta_0 \cdot \dot{\gamma}}{\tau^*} \right)^{1-n}}, \quad \eta_0 = A \exp\left(\frac{T_b}{T} \right), \quad T_b = \frac{E_a}{R} \tag{1}$$

where C_1, C_2, A, n, τ^* are material constants, E_a is the activation energy, α is the degree of cure, α_g is the conversion at gel point, γ is the shear rate and η_0 is the viscosity at zero shear rate.

During moulding process of rubber compounds, the material generally starts curing after the filling phase is completed. This is crucial, because rubber compounds are no longer deformable and cannot be shaped or processed once vulcanized. Based on equation (1), this means that defining the shear rate and temperature dependencies of viscosity is generally sufficient to simulate the filling phase. Only in certain cases, for example when there is an effect of shear heating in small cavities, the curing conversion dependency of viscosity must be considered. By modeling curing kinetics, explained in the following paragraph, one can numerically predict the state of cure of molded part at the end of filling phase, therefore assessing whether or not the effect of curing on the viscosity equation must be considered.

Curing Kinetics

RPA2000 oscillatory rheometer was also used to measure the torque response of the compounds during vulcanization. Time sweep measurements were carried out with an oscillating angle of 0.1 deg and a frequency of 1Hz at temperatures in the range of 160-180°C. From these measurements, the degree of cure α could be calculated and plotted as a function of time.

Concerning kinetic rate determination of a chemical reaction, the traditional chemical method is based on a mechanistic approach: the concentrations of reactants or products are monitored throughout the reaction and the rates of change in concentrations are fitted to an equation [3]. However, vulcanization process is quite complex, involving multiple competing reactions with several kinetic equations, thus requiring very complicated mechanistic models [3,5]. For this reason, phenomenological models are more attractive and have proven their value in describing curing kinetics [6]. Phenomenological models use properties, which are related to the reaction, such as the change in viscosity or material torque response. For the vulcanization reaction, a well-known phenomenological kinetic model, called Combined Model, is used to describe the kinetics. It combines the kinetic behaviour of n-th order chemical reaction and that of autocatalytic chemical reaction, as follows:

$$\frac{d\alpha}{dt} = K_a(1-\alpha)^n + K_b\alpha^m(1-\alpha)^n, K_a = A\exp\left(\frac{-E_a}{RT}\right), K_b = B\exp\left(\frac{-E_b}{RT}\right) \tag{2}$$

where K_a, K_b are rate constants following Arrhenius law; E_a, E_b are the activation energies; R is the gas constant; T is the temperature; m, n, A and B are constants.

RESULTS AND DISCUSSION

Viscosity Model as a Function of Temperature and Shear Rate

To determine the viscosity model to be used to predict the filling stage, at least three viscosity curves at different temperatures are needed. Temperatures were chosen in such way that material does not cure during testing. This simplifies Equation (1) substantially, as the effect of cure degree is removed (α=0). First, the validity of the Cox-Merz rule was checked for both NBR and FKM compounds by comparing the steady and the dynamic viscosity in the temperature range of 80-120°C for NBR compound and 70-110°C for FKM compound, as shown in Figure 1. For both compounds, the shear rate dependence of steady state shear viscosity does not correspond to the angular frequency dependence of complex viscosity. As a consequence of the Cox-Merz rule failure and due to the lack of direct measurement of viscosity in the injection press, calibration of the viscosity model via curve fitting becomes problematic. Indeed, due to difference between capillary rheometer and oscillatory rheometer viscosity data, in principle three different fitting curves could be obtained, as depicted in Figure 2 obtained at 90°C for FKM compound. Consequently, direct effect on the simulation results can be expected. To assess which experimental data are to be used for reliable fitting and, in general, to obtain an effective methodology to model compound viscosity, a reverse engineering approach was applied. The sprue pressure was measured at the injection press during the injection moulding of a T-ring made of FKM compound. Results were compared with numerical simulation carried out using the same mould geometry and injection parameters as in the experiments. The comparison when using dynamic viscosity data is shown in Figure 3. By applying this reverse engineering approach, the material data to fit the viscosity model could be assessed. In other words, the accuracy of the viscosity data from RPA and capillary rheometer was evaluated. Specifically, numerical simulation results have shown good prediction of melt front time and sprue pressure for different injection molds when a viscosity model based on complex shear viscosity data was used. These experimental data along with the fitting curves are reported in Figure 4, while the material constants of the Cross-Castro-Macosko model are contained in Table 1. On the other hand, the prediction of sprue pressure was not corresponding to experimental results when a viscosity model based on steady state shear viscosity was employed. This has been observed also in the case of NBR compounds and for other mould geometries.

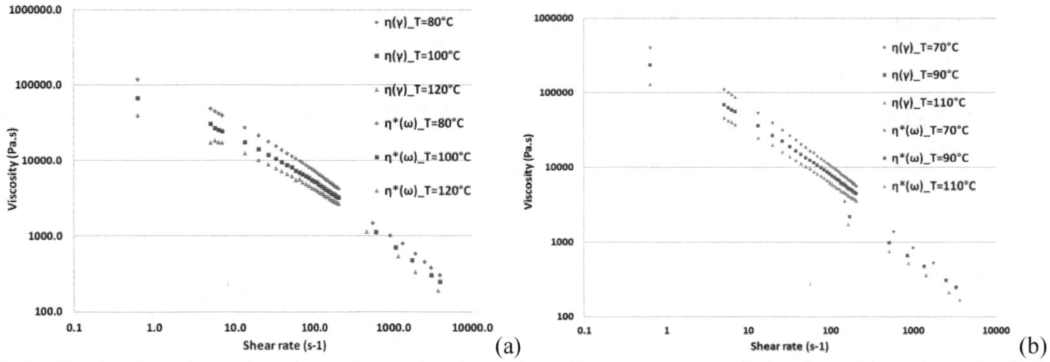

FIGURE 1. Steady state shear viscosity and complex shear viscosity versus respectively shear rate and angular frequency at three different temperatures for filler-reinforced NBR compound (a) and filler- reinforced FKM compound (b).

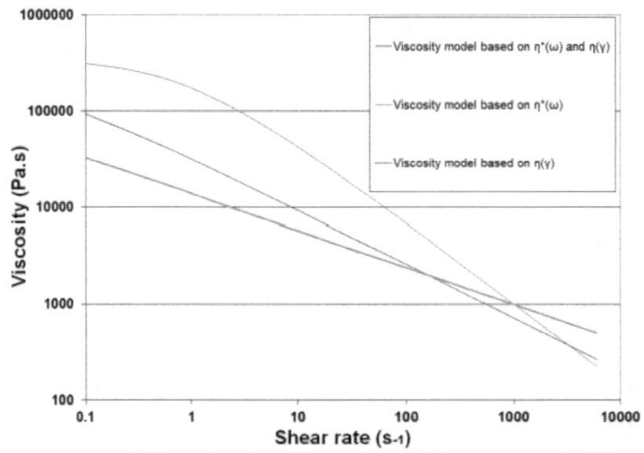

FIGURE 2. Three possible fitting curves could be obtained based on data shown in Figure 1 (b) for FKM compound. Cross *Castro–Macosko* model assuming zero conversion is used for modeling the dependency of viscosity on shear rate and temperature.

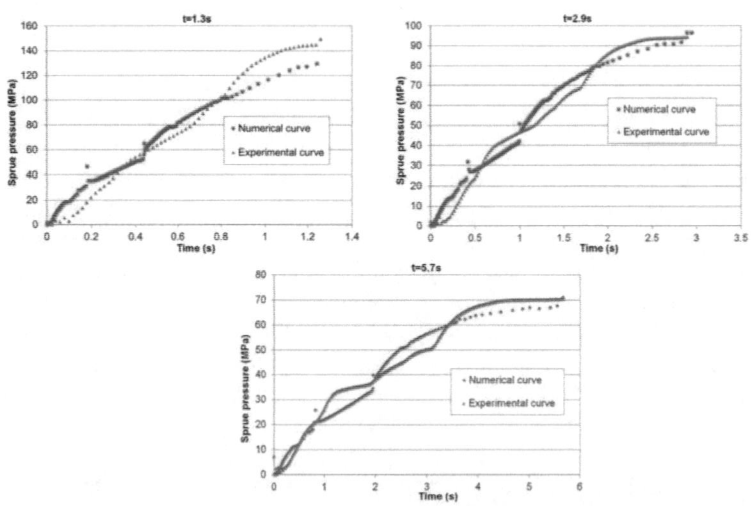

FIGURE 3. Comparison between sprue pressure calculated via numerical simulation and measured experimentally at different filling times for an injection-moulded T-ring made of FKM compound. A viscosity model based on complex shear viscosity has been used for the numerical calculations.

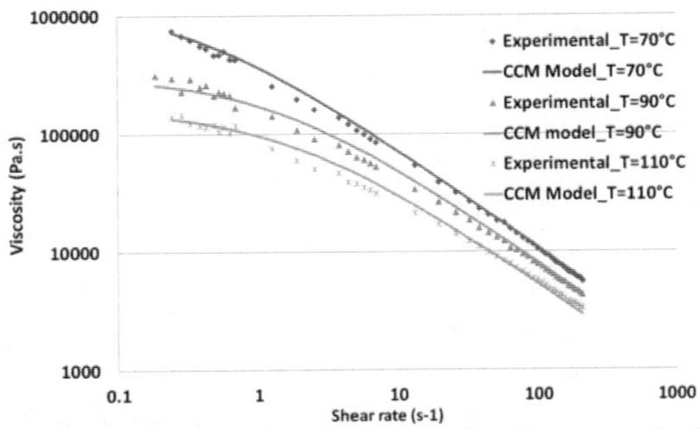

FIGURE 4. Complex shear viscosity measured as a function of angular frequency for FKM compound at three different temperatures. An oscillating angle of 0.1 deg (linear region) has been used. Data are fitted with Cross Castro-Macosko model.

TABLE (1). Material constants of Cross-Castro-Macosko model for FKM compound.

Material constants	
n	0.2
Tau(Pa)	369853.4
Tb(K)	8152.1
A(Pa.s)	6.76205E-05

Modeling Curing kinetics

In order to fit the Combined Model on the curing kinetic data, at least three vulcanization curves measured at different curing temperatures, are required. From each curve, one can plot the time derivative of the cure degree versus α. Figure 5 shows the experimental results obtained in the case of FKM at temperature of 160, 170 and 180°C compared with the fitting model. The material parameters are contained in Table 2.

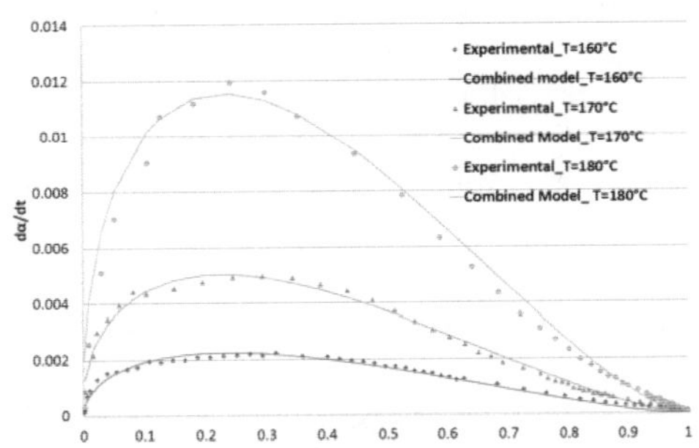

FIGURE 5. Experimental curing kinetic data of FKM compound compared with the Combined model fitting curves. Tests were done with an oscillating angle of 0.1 deg and a frequency of 1Hz.

TABLE (2). Material constants of Combined model for FKM compound.

Constants	
m	0.478
n	1.527
A (1/s)	2.53E+37
B (1/s)	3.3E+13
Ea (J/mol)	349770
Eb (J/mol)	130130.7

Injection Molding Simulation Results

Based on rheological properties, vulcanization kinetics and other thermal properties of the rubber compounds, the degree of cure of moulded parts could be predicted. This prediction could help to optimize the processing parameters and the mould design. For instance, one can check the effect of melt temperature on the state of cure of moulded parts at the end of filling phase (see Figure 6), or the effect of curing time on the state of cure at the end of curing phase (see Figure 7). A comparison of the melt front time obtained via computation with short shots taken out of the mould is reported in Figure 8. Results indicate excellent agreement between numerical predictions and experimental data once more showing that reliable material data inputs are the key for accurate simulations.

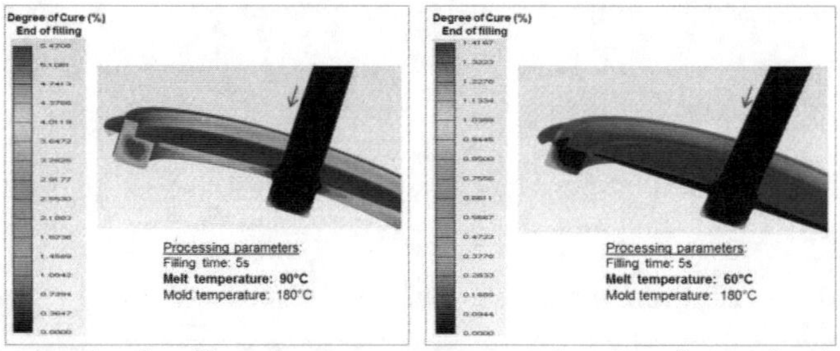

FIGURE 6. Simulation results showing the effect of melt temperature on the state of cure of an injection moulded rubber part at the end of filling phase. The average degree of cure at the end of filling phase increases with an increase of melt temperature. A high average of degree of cure at the end of filling phase would lead to parts with quality issues.

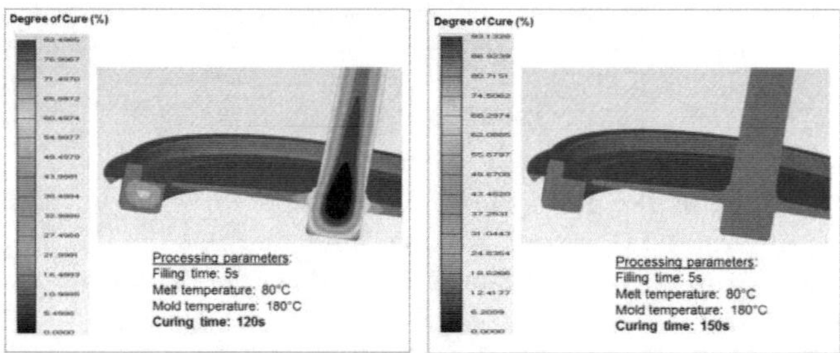

FIGURE 7. Simulation results showing the effect of curing time on the state of cure of an injection moulded rubber part at the end of curing phase. For rubber moulding process, the cycle time is mainly dictated by the curing time. A low curing time would lead to parts with quality issues if the curing is not completed.

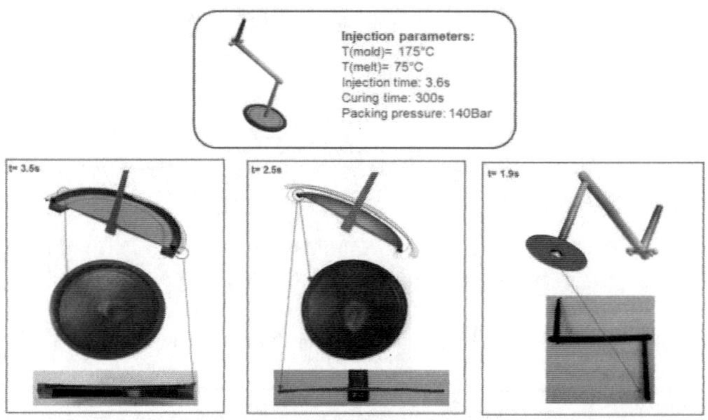

FIGURE 8. Comparison between melt front time obtained experimentally and that calculated via numerical simulation for an injection-moulded T-ring made of NBR compound. A viscosity model based on complex shear viscosity has been used. Injection parameters are indicated in the above panel.

REFERENCES

1. W.P. Cox and E.H. Merz, Journal of Polymer Science, 28, 619 (1958).
2. J. E. Mark, B. Erman and C. M. Roland, The Science and Technology of Rubber (Fourth edition), pp. 322.
3. B. Bilyeu, W. Brostow, and K.P. Menard, Epoxy thermosets and their applications: Kinetic equations and models, Journal of Materials Education Vol. 23 (4-6): 189-204, 2001.
4. Gregory IH, Muhr AH, Imizan AB, Prediction of state of cure throughout rubber components. J Rubber Res, 1-22, 1999.
5. S.L. Simon and J.K. Gillham, J. Appl. Polymer Sci., 53, 709, 1994.
6. J.M. Kenny and A. Trivisano, Polymer Eng. & Sci., 31, 1426, 1991.

Reactive Compounding of Poly(dimethylsiloxane-Urea) Copolymer based Masterbatch Colorants

Hans-Joachim Radusch[a,c*], Matthias Huebner[b], Friedhelm Roeber[b], André Wutzler[c]

[a]University Halle-Wittenberg, Center of Engineering Sciences, Polymer Technology, 06099 Halle, Germany
[b]DAW SE Werk Nerchau, 04668 Grimma OT Nerchau, Germany
[c]Polymer Service GmbH Merseburg, 06217 Merseburg, Germany

hans-joachim.radusch@iw.uni-halle.de

Abstract. Optical and esthetic modification is a basic need for polymer application in all technical and consumer fields. Colorants are used in broad volume to satisfy the demand of costumers for that purpose. Mainly, colorants are included in a carrier polymer, which should help to dose the colorant precisely, to mix it homogeneously, to be compatible with the matrix, and it should not influence the mechanical and rheological properties unfavourably. A colorant masterbatch on the basis of polydimethylsiloxane-polyurea copolymer (PDUC) was proposed, which should warrant these demands in high manner. The masterbatch was generated by a specific twin-screw extruder with liquid feeding unit for the liquid initial components polydimethylsiloxane (PDMS) and diisocyanate, and an optimized segmented mixing screw. As PDMS a pigment containing PDMS prepared in a pre-dispersion process was inserted. The precise adherence of the stoichiometric ratio of the initial components as well as the optimization of residence time of the polymer in the extruder in concern to the reaction time of the components allowed the generation of the PDUC based masterbatch. Different isocyanates and aminopropyl terminated PDMS with different molecular mass were used. The investigation of coloring effectivity and homogeneity was realized by a single screw extruder with flat die and chill roll device. A higher number of thermoplastics was involved into the investigation of the suitability of the generated masterbatches. Beside the nearly universal applicability of the masterbatch, the improvement of processibility - expressed by lowering of the viscosity and reduction of energy consume of the extruder - could be demonstrated. The good dispersion of the pigment particles is demonstrated by optical microscopy. Stress-strain measurements proved that no significant negative influence of the new masterbatch in concentrations < 5 % on the mechanical properties of the investigated thermoplastics could be observed.

Keywords: masterbatches, poly(dimethylsiloxane-urea) copolymer, colorants

TWO-STEP TECHNOLOGY FOR POLY(DIMETHYLSILOXANE-UREA) COPOLYMER BASED MASTERBATCHES

Poly (dimethylsiloxane-urea) copolymers (PDUC) are thermoplastic alternating blockcopolymers, which are synthesized by the polyaddition of α,ω-aminopropyl terminated polydimethylsiloxane (PDMS) and diisocyanate. It was found, that PDUC can be used advantageously as matrix polymer for universal masterbatches [1].

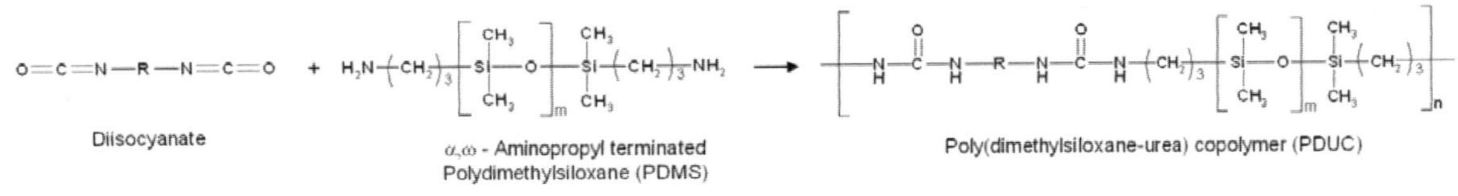

FIGURE 1. Reaction scheme of PDUC formation

Generally, the masterbatch technology can be considered as state of art, concerning the coloring and additivation of polymers. Masterbatches are defined as polymer bonded color and/or additive concentrates with an amount of active agent that is higher than in the final product [2]. The standard method for the production of masterbatches is the compounding of a carrier polymer with a pigment or another particulate solid in a melt mixing device, preferably in a twin-screw extruder. Basically, the main purpose of the compounding step is the dispersion of the solid particles in the polymer matrix, realizing the separated processes of desagglomeration/desaggregation, of wetting and distribution of the particles. A limitation of the twin-screw extrusion technique is the restricted residence time. Due to the time dependence of all dispersion steps, in special cases it is not possible to reach an optimum of dispersion within the masterbatch compound. Thus, an alternative technological approach for the preparation of PDUC-based masterbatches was developed to overcome these limitations [3] (Fig. 2).

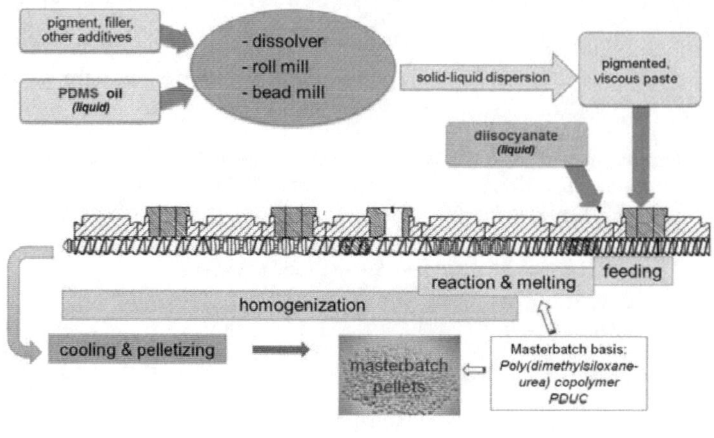

FIGURE 2. Concept of manufacturing of PDUC-based masterbatches

According to this approach, the masterbatch production is subdivided into two processing steps: The *first step* comprises the predispersion of a particulate solid material in the liquid α,ω-aminopropyl terminated polydimethylsiloxane. This preparation step can be realized using different solid-liquid dispersion techniques, such as dissolver technology, bead grinding mills or roller mills.

The dispersion of the pigment or filler in the PDMS-precursor comprises several advantages. In comparison to the common melt compounding technology by means of continuously working extruders, the solid-liquid dispersion techniques are discontinuous but time independent. Hence, the filler dispersion in such devices can be better controlled and adjusted to a desired level. Furthermore, by means of solid-liquid dispersion all the three fundamental dispersing processes *desagglomeration/ desaggregation, particle break-up and distribution* can be realized in the process with high intensity due to the impact and milling mechanisms. In contrast, at the extruder based compounding process mainly distribution takes place initiated by melt deformation, but shear induced dispersion is low. Furthermore, because of the amphiphilic character and the relatively low molecular weight of the α,ω–aminopropyl terminated PDMS no dispersing agent is required in most cases.

The *second step* comprises the reactive manufacturing of the thermoplastic masterbatch, using a modified co-rotating twin-screw extruder. The liquid reactive initial components are fed into the extruder gravimetrically at the molar end group ratio of 1:1. After feeding of the components and beginning of mixing the liquids still in the feeding zone, the reaction starts for the formation of the PDUC in presence of the pigment particles. As the result of the polyaddition reaction a thermoplastic PDUC melt is generated in that the pigment particles are embedded.

The building reaction of the masterbatch carrier polymer, i.e. the PDUC, follows the reaction scheme demonstrated in Fig. 1. The step growth polymerization takes place in the extruder at the temperature of 70°C without an additional catalyst. The metering zone of the extruder is heated up to 170°C for transferring the generated compound (masterbatch) into a gently flowing state. After finalizing the reaction, subsequent homogenization and feeding the die, cooling and pelletizing of the compound follows.

Compared to the conventional twin screw compounding, more degrees of freedom are given for the dispersion step. So, it is possible to achieve an optimal dispersion in the masterbatch even with substances, which are to disperse only very difficult, e.g. organic pigments, additives or nano-fillers.

Compared to the conventional melt compounding, here, the reactive extrusion step is only required for the PDUC polyaddition reaction and not for pigment or filler dispersion anymore. The reactive extrusion under the existing time conditions becomes possible, because the step growth polymerization of diisocyanate and α,ω–aminopropyl terminated polydimethylsiloxane belongs to the fastest building reactions in polymer chemistry. Furthermore, the reaction of these two components results in polymers with rather high molecular weight that is advantageous for the masterbatch properties [4]. Furthermore, strong hydrogen bonds are formed within the urea hard segments [5], thus, thermoplastic PDUC with low number of hard segments but sufficient properties can be synthesized that is the basis for solid masterbatches with practice-friendly handling.

VARIATION OF COPOLYMER COMPOSITION AT PDUC SYNTHESIS

α,ω–aminopropyl terminated polydimethylsiloxanes (PDMS) with different number average molecular weight (Fluid NH15: Mw~1030g/mol, Fluid NH40: Mw~2800g/mol, Fluid NH130: Mw~11970g/mol) obtained from Wacker Chemie AG, Burghausen, Germany were used as the silicon containing components. Different isocyanates have been tested for investigation of the influence of the type of isocyanates on reaction rate and copolymer properties under specific consideration of their use as masterbatch carrier. 1,6-diisocyanatohexane (HMDI), 3-isocyanatomethyl-3,5,5-trimethylcyclohexylisocyanate (IPDI), 4,4'-diisocyanatodicyclohexylmethane (H12MDI), 2,4-toluylendiisocyanat (TDI) purchased from Sigma-Aldrich Chemie GmbH as well as Tetramethylxylene diisocyanate (TMXDI) supplied by Cytec Industries Inc. were used (Fig. 3).

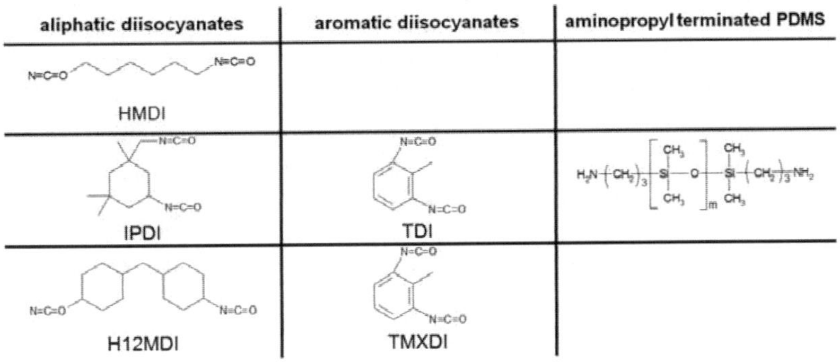

FIGURE 3. Components for PDUC synthesis

Unfilled PDUC were prepared using the components listed in Fig. 3 according to the reactive extrusion process described above without the use of any catalyst. Measurement of reaction kinetics at synthesis of PDUC was carried out by means of the method of adiabatic temperature rise. The components were mixed at the molar end group ratio of 1:1. To ensure good mixing, the α,ω–aminopropyl terminated polydimethylsiloxane was diluted in acetone at the ratio of 1:2. The investigation was carried out at room temperature (25°C). Temperature rise was measured with a thermocouple and a sampling rate of 10s-1. The temperature curves for the different diisocyanates reacting with the α,ω–aminopropyl terminated PDMS are presented in figure 4. The curves demonstrate that the reaction time for all systems is in the range of less than 60 seconds. Nevertheless, the differences in the progress of the curves correlate with the increase of complexity of the diisocyanate structure (Fig. 4a), the reaction becomes slower. Furthermore, the aromatic TDI shows a much higher reaction rate than the analogous aliphatic HMDI. With increasing molecular mass of the fluid the reaction rate is reduced (Fig. 4b,c).

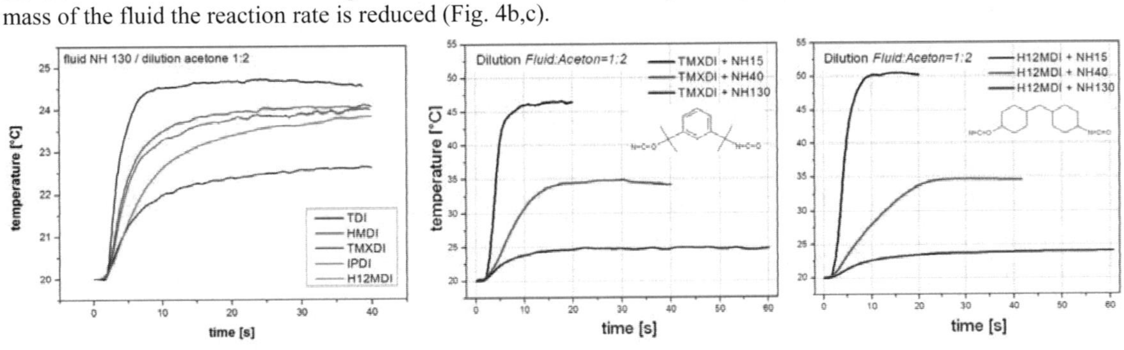

a) b) c)

FIGURE 4. Adiabatic temperature rise curves in dependence on (a)d ifferent diisocyanates and Fluid NH130, different fluids and (b) aromatic TMXDI and (c) aliphatic H12MDI.

In technical scale, the masterbatches were produced by means of the technology described above using a twin-screw extruder Berstorff ZE25/46D. The pigment (titanium dioxide) was predispersed in the fluid by a laboratory dissolver (Dispermat CA40, VMA-Getzmann GmbH), and the isocyanate was injected directly into the extruder barrel. The average residence time of the compound in the twin-screw extruder was determined of about 128 seconds for an output of 10 kg/h or 85 seconds for an output of 15 kg/h, i.e. that the residence time in the extruder is manifold higher than the reaction time for complete generation of the PDUC. Using the Fluid NH130 rather high molecular masses could be realized; e.g. the PDUC from Fluid NH130 and the diisocyanate IPDI or H12MDI had an average molecular mass of 30000 and 25000, respectively.

APPLICATION OF PDUC-BASED MASTERBATCHES FOR COLORING OF THERMOPLASTICS

The produced masterbatches with different components and recipe, respectively, were investigated in concern to their applicability for coloring of various thermoplastics (PE, PP, PA6, PET, PVC). The compatibility of the masterbatches with the polymer was theoretically evaluated by help of the Flory-Huggins equation. While the urea-hard segment shows only a poor compatibility to all of the investigated polymers, the PDMS-soft segment is rather good compatible to PE, PP and PET, but quite incompatible to PVC and PA6.

The PDUC-based masterbatch was applied for coloring the polypropylene using a Brabender laboratory single screw extruder 19D. The neat polymer and respectively 1,2,3,4 and 5 wt.-% of each masterbatch were premixed in the granular state and subsequent processed at 200°C and 80 rpm. The extrusion torque was recorded for each composition. The melt passed out through the die was immediately flattened by a chill-roll to a strap with a thickness of about 400 µm. These straps were used as basis for all further characterisations.

The dispersion of the titanium dioxide in the polypropylene was investigated using an optical microscope Leica RM-DX. 5µm thick slices were prepared using a microtom (Microm, HM 360) and investigated in transmitted-light mode (Fig. 5). A complete dispersion of the pigment particles is visible, also at higher concentrations (Fig. 5c).

a) b) c)

Figure 5. Pigment distribution in polypropylene using PDUC-based masterbatch NH130/TMXDI with (a) 1%, (b) 3 % and (c) 5 % titanium dioxide

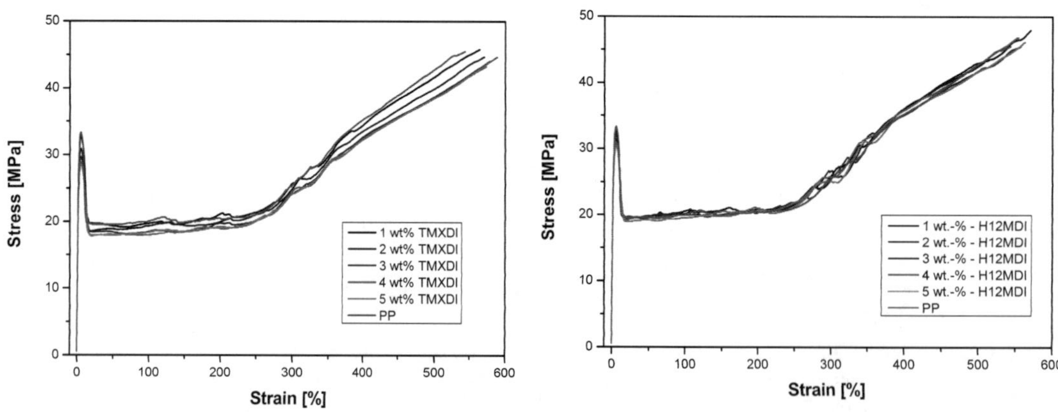

a) b)

Figure 5. Stress-strain behavior of PP-PDUC-based masterbatch compounds with (a) Fluid NH130/TMXDI and (b) Fluid NH130/H12MDI

An important requirement to a good masterbatch is the non-influence on the application properties of the colored polymer. Therefore, the stress-strain behavior of PDUC-based masterbatch colored plastics was investigated in dependence on the content of the PDUC-based masterbatch produced from Fluid NH130 and TMXDI or H12MDI isocyanate (Fig. 6). From the stress-strain curves becomes clear that there is no essential influence of the masterbatch on the mechanical behavior, neither in the case of the aromatic TMXDI (Fig. 6a) nor in the case of the aliphatic H12MDI (Fig. 6b). Both diagrams demonstrate that the modulus, the yield strain, the strain at break as well as the stress at break is nearly the same for all compositions. Slight differences can be found for the yield strength. With an increase of the masterbatch concentration, the yield strength slightly decreases. It is assumed that the PDUC forms separate domains in the amorphous regions of the PP compounds. The domains act as imperfections, and due to these imperfections the stress induced yielding starts at lower forces. This effect becomes more intensive at higher PDUC concentrations.

REFERENCES

1. M. Hübner, F. Röber, F.; D. Chmielewski, H.-J. Radusch, Verfahren zur Herstellung eines Farbmasterbatches, Farbmasterbatch sowie damit eingefärbter Polymerwerkstoff, European Patent EP 2 113 522 A1, 2009.
2. K. Ullmann, Grundlagen zu Masterbatches, in: Farb- und Additivmasterbatches in der Praxis, VM Verlag Köln, 2003.
3. M. Hübner, Multifunctional colorant masterbatches on the basis of thermoplastic polydimethylsiloxane-polyurea copolymer Ph.D. Thesis, Martin Luther University Halle-Wittenberg, Halle 2014.
4. F. Wang, Polydimethylsiloxane Modification of Segmented Thermoplastic Polyurethanes and Polyureas, Ph.D. Thesis, Virginia State University, 1998.
5. E. Yilgor, I. Yilgor, Hydrogen Bonding: a critical parameter in designing silicone copolymers; Polymer 42, 7953-7959, 2001.

Electron Induced Reactive Processing Of Polypropylene/Ethylene Octene Copolymer Blends – A New Method To Produce Toughened Polypropylene

Carsten Zschech[a*], Uwe Gohs[a] and Gert Heinrich[a, b]

[a]*Leibniz-Institut für Polymerforschung Dresden e.V., Hohe Straße 6, 01069 Dresden, Germany*
[b]*Technische Universität Dresden, 01069, Dresden, Germany*

zschech@ipfdd.de

Polypropylene (PP) belongs to one of the most used commodity polymers due to its exceptional properties including good chemical resistance, mechanical properties and processability at low price level. In addition to traditional applications, it is increasingly used as functional material in automotive industries, but its low impact resistance limits this application. Consequently, it is blended with a rubber in order to increase its toughness. PP/ethylene octene copolymer (EOC) blends show higher impact strength in comparison to other PP/rubber blends. The main drawback of this method is a reduction of E-modulus and tensile strength. In addition, the weak interfacial interaction between EOC and PP phases negatively effects the final properties of this blend. Thus, the PP/EOC interface has to be modified. Based on previous works on PP/EOC thermoplastic vulcanizates, the authors firstly tested a continuous pilot plant for electron induced reactive processing (EIReP) to prepare EOC based toughened PP. This novel reactive processing method offers a high efficiency as well as a local and temporal precise generation of radicals. The mechanical, morphological and rheological properties of toughened PP are studied. The tensile strength of toughened PP is at the same level compared to virgin PP due to increased interfacial adhesion between sub-micrometer EOC phase in the PP matrix.

Keywords: toughness, polymer blends, processing parameters, electron induced reactive processing
PACS: 81, 82, 34

INTRODUCTION

The use of polypropylene (PP) in electrical and automotive industries is limited by its low impact resistance [1]. The blending of PP with a rubber is one method to improve its impact properties. In this study, ethylene octene copolymer (EOC) was used for blending. It is available in form of pellets and PP/EOC blends show higher impact strength in comparison to blends of PP/EPDM and PP/NR [2]. Nevertheless, the use of EOC at higher temperatures is limited due to its low melting temperature [3] and the weak interfacial adhesion between EOC phases and PP matrix influences the blend properties [4,5]. These disadvantages can be reduced by crosslinking of EOC as well as PP-EOC graft-links in the interface. The exposure of PP/EOC blends to high energy electrons is one possibility to overcome these drawbacks. High energy electrons are used in a wide spectrum of applications for polymerization and crosslinking as well as degradation and functionalization of polymers [6]. It has to be taken into account, that a modification of PP at room temperature leads to trapped radicals and so called post treatment alteration of properties. The novel method of electron induced reactive processing (EIReP) avoids this drawback and was used to prepare high performance PP/EOC thermoplastic vulcanisates (TPV) by tailored processing conditions of discontinuous EIReP [7, 8]. Based on these studies EOC toughened PP was prepared by continuous EIReP.

EXPERIMENTAL

Polypropylene (PP HG 455FB; M_n = 84800 g/mol, polydispersity = 2.64, MFI (230 °C/2.16 kg) = 27 g/10 min, Borealis AG, Australia) and ethylene octene copolymer (Engage 8100; M_n = 86600 g/mol, polydispersity=1.58, MFI (190°C/2.16 kg) = 2.16 g/10min, octene comonomer content = 23 mol%, Dow Chemical, USA) were used as received.

Figure 1 shows the pilot plant for continuous EIReP. It includes the non-reactive processing by a twin screw extruder (TSE DSE18, Leistritz, Germany), the EIReP in a high viscosity reactor (IPF, Dresden, Germany) coupled to an electron accelerator (ELV-2 BINP, Novosibirsk, Russia) as well as a water-bath and granulator (Collin, Germany).

Based on a previous study [9], EOC toughened PP was prepared by non-reactive blending of PP with 7.5 wt.% of EOC in a twin screw extruder at a temperature of 180 °C and a screw speed of 120 rpm. Afterwards, the molten blend was exposed to dose values of 9 and 18 kGy across its passage through the high viscosity reactor at a temperature of 180 °C and under nitrogen gas in order to reduce oxidative reactions. Finally, the chemically modified molten strand passed the water bath and the granulator in order to form a solid strand and pellets, respectively. In addition, PP/EOC blend was processed without EIReP modification in order to prepare reference materials with the same processing history as well as samples for state of the art electron treatment at 180 °C with dose values of 18 to 100 kGy.

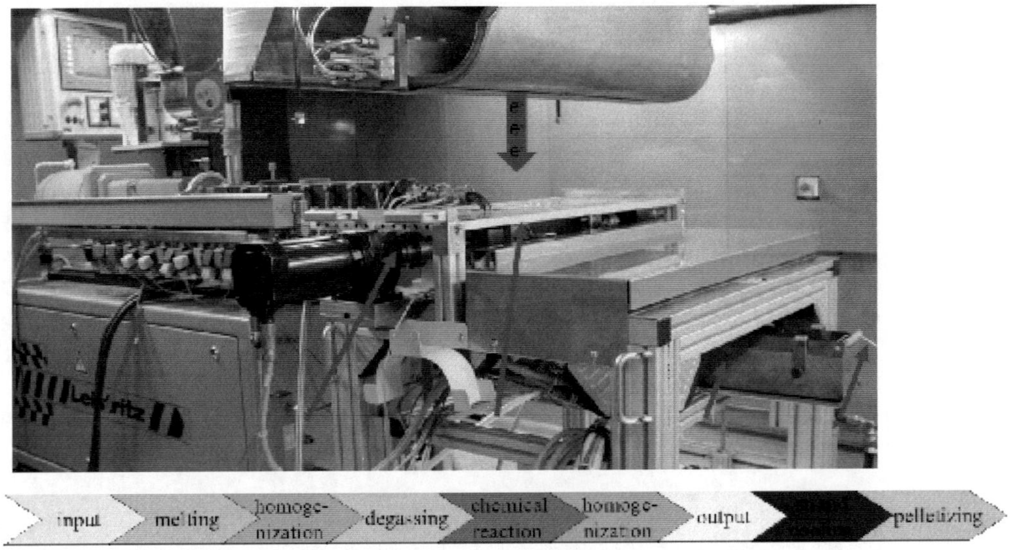

FIGURE 1. Pilot plant for continuous electron induced reactive processing

Scanning electron microscopy (SEM; NEON 40 EsB, Carl Zeiss, Oberkochen, Germany) was used to characterize the morphology of the prepared blends. After cryo-fracturing in liquid nitrogen at -160 °C, an extruded strand of each sample was microtomed to obtain a smooth surface which was sputter coated with 3 nm platinum prior to the SEM analysis.

Standard samples for tensile and impact tests were prepared by injection molding (180 - 210 °C from hopper to nozzle, 100 bars). The tensile tests were performed on a tensile tester (Zwick/Roell,Germany) in accordance to the ISO 527-2 standard at ambient temperature with crosshead speeds of 1 mm/min (measurement of E-modulus) and 50 mm/min (at high strains) using an applied force of 1000 N. For each sample the average value of five measurements was reported. Charpy impact tests were done at room temperature in accordance to ISO 179 standard using an impact tester (Zwick/Roell,Germany). The average values of five measurements were reported.

The gel content of EOC and toughened PP was measured by extraction in boiling xylene (140 °C, 24 h). After extraction, all samples were dried in a vacuum oven and weighed. The gel content (X_g) in percentage was calculated in accordance to $X_g = m_1 * 100 * m_0^{-1}$. In this equation, m_0 and m_1 is the sample mass before and after extraction, respectively.

RESULTS AND DISCUSSION

Using the Charlesby-Pinner equation [10] for the evaluation of experimental data of gel content, the crosslinking behavior of crosslinkable polymers (here: EOC and PP/EOC blend) can be characterized. Below the gel dose long chain branching is dominating and no gel content is detectable by gel extraction. In the case of EOC and PP/EOC blend, the gel dose amounts to about 18 kGy (**Figure 2b**). Consequently, an EB treatment of EOC and PP/EOC blend with a dose below 18 kGy mainly leads to long chain branching.

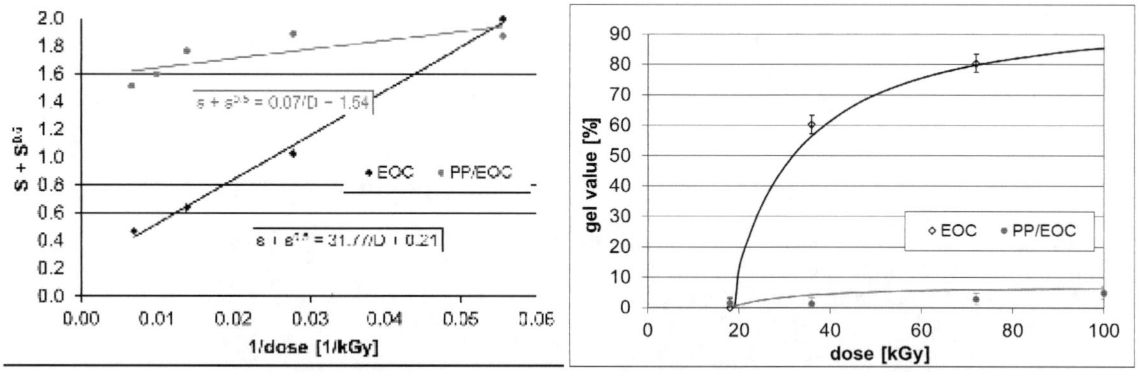

FIGURE 2. Charlesby-Pinner plot (a) and gel value (b) for EOC and toughened PP

Figure 3 represents SEM images of PP/EOC blends at 0, 9 and 18 kGy. In all blends, the EOC minor phase was dispersed uniformly within the PP matrix. An uniform distribution of EOC phases within PP matrix, a lower polydispersity of EOC and submicron size of EOC phases are microstructural characteristics leading to best toughening performance [11, 12]. All these microstructural characteristics were attained. Finally, it was observed that EIReP do not change the size of EOC phase or homogeneity level within the PP/EOC blends.

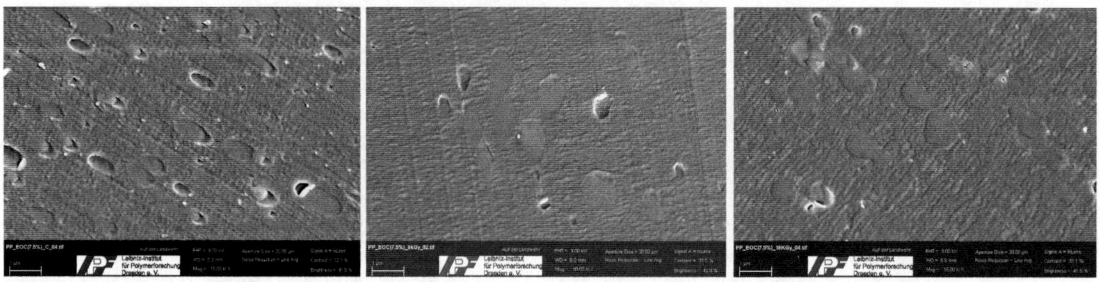

FIGURE 3. SEM micrographs of PP/EOC blends at 0 (a), 9 (b) and 18 kGy (c) PP (scale: 1 μm)

Figure 4 shows the relative values of tensile and impact properties for non-modified and EIReP modified PP/EOC blends with respect to neat PP as reference material. All the samples show a ductile behavior, represent yield behavior and fail in subsequent stress hardening. In addition, all tested mechanical parameters of EIReP modified PP/EOC blends are higher in comparison to non-modified PP/EOC blend.

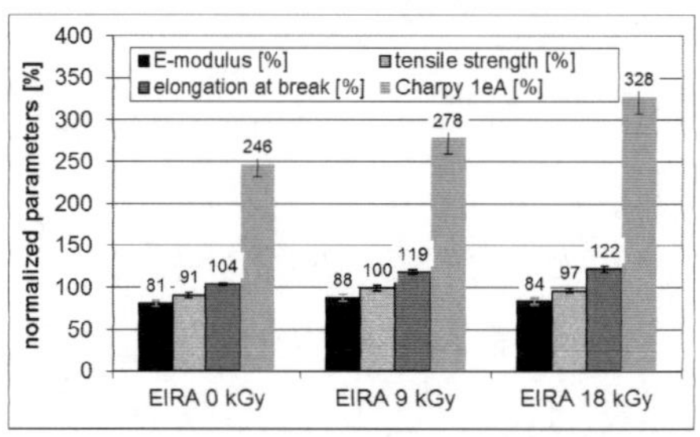

FIGURE 4. mechanical properties of PP/EOC blends at 0, 9 and 18 kGy

At the tested blend ratio, the increase of modulus of EOC phase can be explained by long chain branching of EOC phase. This observation is in accordance with the reports of Nunes [13]. In addition, there is a significant enhancement of impact strength by EIReP, which is not related to a changed morphology. Consequently, another mechanism should lead to this enhancement. A further mechanism leading to enhanced impact strength of EIReP modified PP/EOC blends is a change of interface structure. Due to [14], in EIReP modified PP/EOC blends graft-links between EOC-phases and PP matrix were generated leading to shear yielding and enhanced toughness.

Consequently, the property of PP-EOC interphase was changed with respect to non-modified PP/EOC blend by increased compatibility and miscibility of EOC and PP polymer chains.

CONCLUSION

A pilot plant for continuous EIReP was used for commercial preparation of EOC toughened PP. The processing conditions of continuous EIReP based on previous experiences of discontinuous EIReP. Continuous EIReP does not change the size of EOC phase or homogeneity level within the PP/EOC blends, but leads to increased mechanical properties. The improved impact properties of EIReP modified PP/EOC blends based on graft-links between PP matrix and EOC phases. A micro-mechanism was proposed in [14] in order to explain the differences in impact strength of EIReP modified and non-modified PP/EOC blends.

ACKNOWLEDGMENTS

The EOC toughened PP was developed within the Collaborative Research Centre SFB 639 "Textile-reinforced composite components for function-integrating multi-material design in complex lightweight applications". Consequently, the authors acknowledge the German Science Foundation DFG for funding this work. Finally, the authors would like to thank Jacqueline Hampel and Holger Scheibner from Leibniz-Institut für Polymerforschung Dresden e.V. for gel content and mechanical measurements.

REFERENCES

1. X. Yan, X. Xu, T. Zhu, C. Zhang, N. Song, L. Zhu, *Materials Science and Engineering* **A 476 (1-2)**, 120-125 (2008).
2. C. Grein, M. Gahleitner, K. Bernreitner, *eXPRESS Polymer Letters* **6 (9)**, 688-696 (2012)
3. P. Svoboda, S. Poongavalappil, R. Theravalappil, D. Svobodova, P. Mokrejs, K. Kolomaznik, T. Ougizawa, T. Inoue, *Journal of Applied Polymer Science* **121**, 521-532 (2011).
4. R. R. Babu, N. K. Singha, K. Naskar, *Journal of polymer research* **218 (1)**, 31-39 (2011).
5. P. Harnischfeger, P. Kinzel, B. J. Jungnickel, *Die Angewandte Makromolekulare Chemie* **175 (1)**, 157-168 (1990).
6. R. Clough, *Nuclear Instruments and Methods in Physics Research Section B: Beam Interactions with Materials and Atoms* **185**, 8-15 (2001).
7 R. Rajeshbabu, U. Gohs, K. Naskar, M. Mondal, U. Wagenknecht, G. Heinrich, *Macromolecular Materials and Engineering* **297**, 659-669 (2012).
8. R. Rajeshbabu, U. Gohs, K. Naskar, V. Thakur, U. Wagenknecht, G. Heinrich, *Radiation Physics and Chemistry* **80**, 1398-1405 (2011).
9. C. Zschech, F. Miersch, U. Gohs, G. Heinrich, Journal of Plastics Technology **12 (2)**, 112-134 (2016)
10. A. Charlesby, S. H. Pinner, Proc. Roy. Soc. (London) **A 249**, 367-386 (1959)
11. M. R. Aghjeh, M. Nazari, H. A. Khonakdar, S. H. Jafari, U. Wagenknecht, G. Heinrich, *Materials & Design* **88**, 1277-1286 (2015).
12. T. Omonov, C. Harrats, G. Groeninckx, P. Moldenaers, *Polymer* **48**, 5289-5297 (2007).
13. R. W. Nunes, J. R. Martin, J. F. Johnson, *Polymer Engineering & Science* **22**, 205-212 (1982).
14. M. Razavi-Aghjeh, H. A. Khonakdar, S. H. Jafari, C. Zschech, U. Gohs, G. Heinrich, *RSC Advances* **6**, 24651-24660 (2016)

Micro-Compounding as Reliable Screening of Reactive Extrusion (REX) Processes

Boudewijn Scholtens[a], Ton Loontjens[b] and Güralp Özkoç[c*]

[a] *Boudewijn's Consultancy, 6151 JA Munstergeleen, Netherlands*
[b] *University of Groningen, Nijenborgh4, 9747AG Groningen, Netherlands*
[c*] *Kocaeli University, Izmit/Kocaeli, 41380, Turkey*

bscholtens@home.nl
guralp.ozkoc@kocaeli.edu.tr

Abstract. Recycling of polyesters and polyamides is frequently hampered by inferior mechanical properties due to various possible degradation processes. Chain extension by reactive compounding or extrusion (REX) can become a new way to repair such deficiencies on an industrial scale. However, process optimization of REX processes needs many variables to be tested such as chain extender type, concentration, catalyst and many process conditions. In this paper a new way of reliable and very fast screening of many REX variables is shown: micro-compounding with vertical force detection. This enables to study more options in a shorter time frame. In addition, results of this study led to new insights how to interpret the vertical force during micro-compounding as a direct melt viscosity control under processing conditions.

Keywords: Reactive extrusion, chain extender, micro-compounder, vertical force, melt viscosity, PET, PLA, PA, CBC, PBO, PDI.
PACS: 81.05.Lg, 82.35.Lr, 83.50.Uv, 83.50.Xa, 83.80.Sg, 83.85.Cg, 83.85.Jn.

INTRODUCTION

Recycling of plastics intends to reprocess plastic waste into a new product life, thereby reducing raw material needs, plastic waste streams and plastic waste in nature. However, a main hurdle for this noble endeavour is the usually inferior mechanical properties of the so reprocessed plastics due to many possible degradation processes during their first life cycle and during reprocessing. For end functional plastics such as polyesters or polyamides, chain extension by solid-state postcondensation (SSP) or reactive compounding/extrusion (REX) may restore the original molecular weight/length and, as a consequence, mechanical properties. SSP is well known, both as continuous process (mainly for virgin polycondensates as it is large scale, but inflexible, and cheap), or as batch process (tumble drier, which is more flexible due to its smaller scale, but also more expensive in operation). As alternative for SSP, REX with various chain extenders is studied here. Main potential advantages are flexibility and speed. As chain extension needs many screening tests for optimization of process formulation and conditions, we introduce here a reliable, fast and convenient screening method: micro-compounding with vertical force detection.

EXPERIMENTAL

Materials: Polymers and Chain Extenders

For this study several commercial virgin and recycle grades of PET (polyethylene terephthalate, ex DSM EP and ex REKO, respectively), virgin grades of PA (polyamide, ex DSM EP) and a virgin grade of PLA (poly(lactic acid), ex Natureplast) were selected. PET and PLA was always carefully dried in vacuum at 70 and 60 C, respectively, for 12 h and immediately fed into the MC 15 to minimize hydrolytic degradation during compounding.

The chain extenders were CBC (carbonyl biscaprolactam, ex DSM NBD), that reacts with –OH and $-NH_2$ end groups, 1,3-PBO (phenylene bisoxazoline, ex DSM NBD) [1,2], 1,4-PBOX (phenylene bisoxazine, ex DSM Licensing Center), that react with –COOH end groups, PDI (1,4-phenylene diisocyanate, ex Sigma-Aldrich), that reacts with –OH and –COOH end groups and Joncryl® ADR-4368 (ex BASF, Germany) that reacts with –COOH end groups [3].

Methods: Compounders and Extruders

Some reference experiments were performed on a ZSK 30 parallel twin-screw laboratory extruder.

Most REX experiments were performed with an Xplore MC 15 micro-compounder, which is a vertical, conical twin-screw compounder with a batch volume of 15 ml. This MC 15 is equipped with a proprietary vertical force (VF) measurement, it has an excellent temperature control (7 thermo couples, 6 thermo heaters) and extremely durable and scratch and abrasion resistant barrel and screws with a wide window of operation (up to 450 °C, 0 < pH < 14). In addition, it is easy to quantitatively fill an MC 15 with either a front hopper or a water cooled top hopper, the compound time can varied by switching a simple valve (from recirculation to extrusion) and all processing data are stored in a data acquisition system. Finally, the MC 15 has a small footprint so that it can be used on a lab bench. These specifications make the MC 15 suitable for fast screening of REX conditions, as we shall show below.

PET was typically processed at 280 – 300 C, PA 6 at 240 – 260 °C, PA 66 at 280 °C and PLA at 200 °C.

Characterization

Relative viscosity measurements of PET were performed in m-cresol (25 °C) and of PA in formic acid (25 °C); intrinsic viscosities of PLA were determined in chloroform (30 °C), all using an Ubbelohde capillary viscometer.

The reaction in the REX experiments performed with the Xplore MC 15 was mostly followed via the vertical force, and frequently also by relative or intrinsic viscosity measurements after a certain time interval.

RESULTS AND DISCUSION

First REX and PA experiments were performed on a ZSK 30 parallel twin-screw laboratory extruder. Table 1 displays the effect of an extrusion run and of REX with CBC and with CBC and PBO.

TABLE 1. Relative viscosity (RV) of PET and PA 6, virgin, after extrusion and after REX on a ZSK 30 (3 min at 290 °C and 2 min at 260 °C, respectively) with stoichiometric concentrations of chain extender.

Sample history	PET RV	PA 6 RV
Virgin	1.55	2.6
After extrusion	1.46	2.6
REX with CBC	1.83	3.5
REX with CBC + PBO	1.90	4.0

It is clear that PET, despite its drying, degrades during extrusion (mainly hydrolysis), and REX with CBC can significantly increase the RV, even well above the value of unprocessed PET. Use of a mixture of complementary chain extenders CBC and PBO (although they react with complementary groups they do not react with each other!) can further increase this RV.

For PA 6 no significant change of RV by extrusion is observed and a similar increase by REX with CBC and PBO, although the effect of PBO is larger in PA 6 than in PET. In both polymers CBC is more effective than PBO which is probably due to the unequal concentration of end groups and/or lower reactivity of PBO with -COOH.

For optimization of a specific REX formulation and its processing conditions such experiments in a ZSK 30 are very time and material consuming (it took almost 2 weeks and some kg of polymer to obtain the data in Table 1). Therefore some test experiments were carried out in an Xplore MC 15 with vertical force (VF) detection, see Figure 1 below. These two tests were performed within one hour with only ca. 30 g of material.

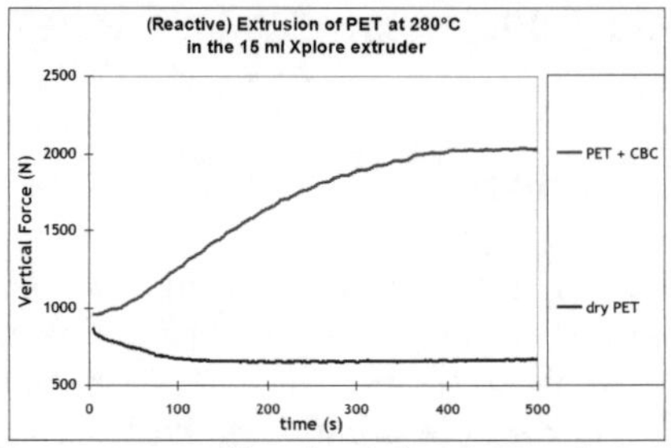

FIGURE 1. Vertical force (N) during compounding in recirculation in MC 15 as function of time for dry PET (blue) and dry PET with CBC (REX).

The decrease in blue VF shows the hydrolytic degradation of PET, the increase in red VF the chain coupling by CBC, which takes about 6 min in the MC 15. These effects were verified by independent RV measurements of the extrudate after 8 min. This effective chain extension of PET by CBC is confirmed in several virgin PET grades (Figure 2) and recycle PET grades (Figure 3). In these graphs the RV of the extrudate after 6 min is shown.

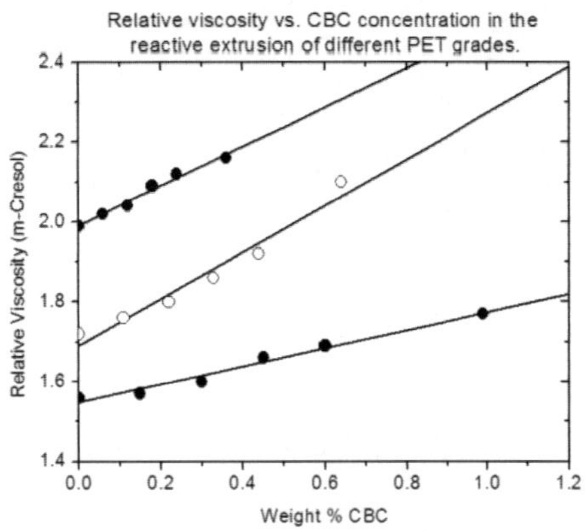

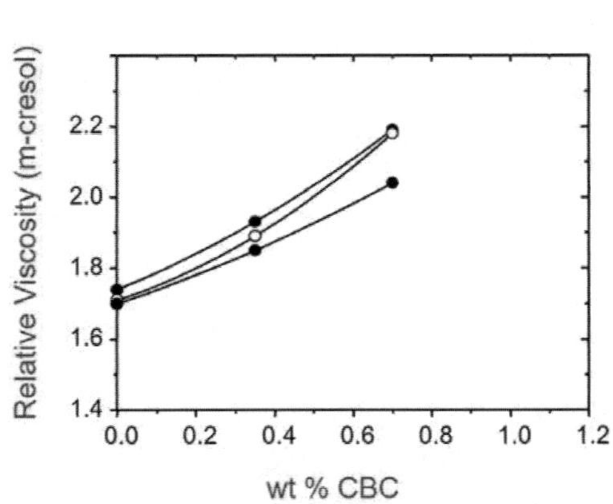

FIGURES 2, 3. Relative Viscosity RV of 3 virgin (2) and recycle (3) PET grades as function of concentration CBC. RV is measured after 6 min REX at 280 °C in MC 15. The highest concentration of each set is the stoichiometric concentration.

Later in our chain extender project, the measurement of RV values of the many PET samples also became too time consuming and costly. Therefore we tried to empirically relate the RV of many PET extrudates with their vertical force measured in the MC 15 just before extrusion, see Figure 4 below.

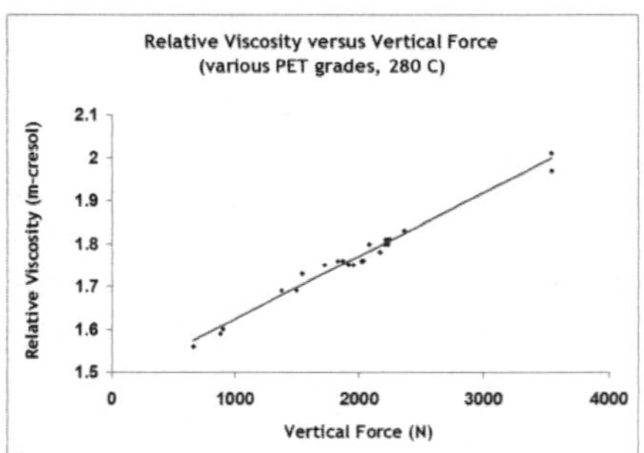

FIGURE 4. Empirical relationship of RV of PET-CBC extrudates VF in the MC 15 (both measured after 6 min REX at 280 °C).

This convincing empirical relationship between RV and VF was used for all subsequent PET-REX experiments, no RV was measured anymore, which saved much time and costs. In addition this finding initiated to model the flow in the micro-compounders and derive approximate, very useful relations for Xplore's proprietary rheological software under guidance of prof. H. Meijer, TU/e, Netherlands. This new software gives the screw torque in the melt, melt viscosity (proportional to VF), shear rate and shear stress under actual extrusion conditions.

The effect of various concentrations CBC in PA 6 and PA 66 is shown below in Figures 5 and 6, respectively.

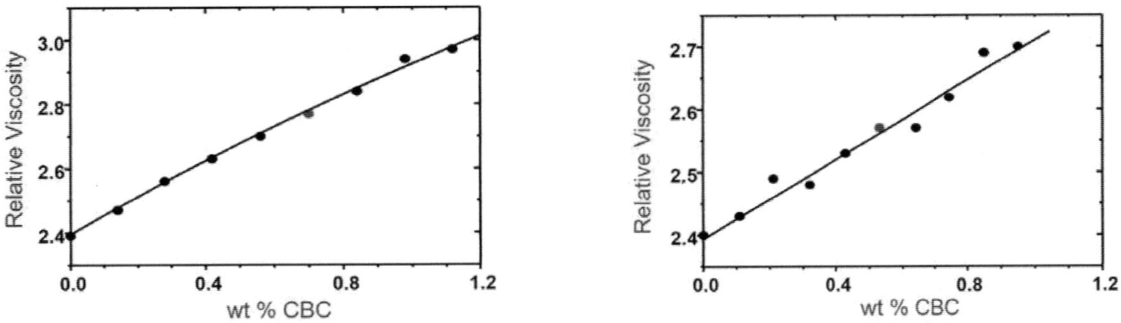

FIGURES 5, 6. RV of PA 6 (5) and PA-66 (6) as function of concentration CBC; RV is measured after 3 min REX at 240 °C and 280 °C, respectively in the MC 15. The red point in each set is the stoichiometric concentration.

Chain extension of PLA with CBC did not work under the conditions tested, probably due to its lower processing temperature. Therefore PDI and Joncryl were applied, using the VF of the MC 15 to monitor the melt viscosity. In these experiments the chain extender was added after 4 min compounding, see Figure 7. This qualitative behavior was confirmed by IV measurements after 1, 2 and 4 min, for 1 and 3 wt % of chain extender, see Figures 8 and 9.

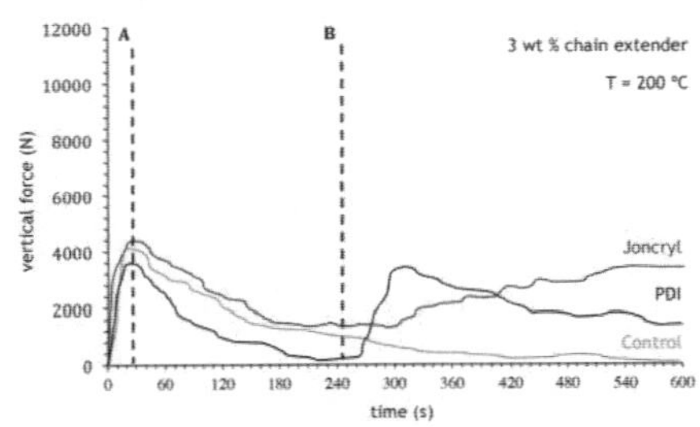

FIGURE 7. VF during compounding in recirculation in MC 15 as function of time for pure PLA (=control), with PDI (black) and Joncryl (blue). Chain extender was added after 240 s to the melt (at B).

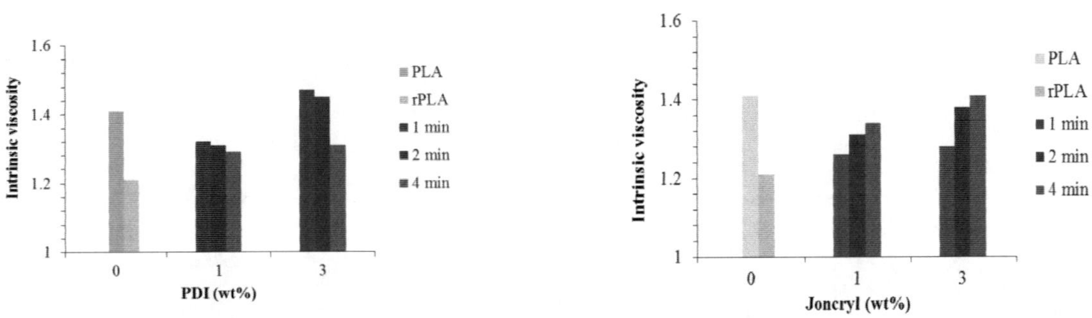

FIGURES 8, 9. IV of PLA with PDI (8) and Joncryl (9) at 2 concentrations after 1, 2 and 4 min REX at 200 °C in the MC 15. rPLA stands for the control PLA after 4 min REX.

These VF and IV data show that PDI gives a fast reaction with PLA with a maximum at ca. 1 min, after which the VF and IV decrease due to degradation. On the other hand, Joncryl is slower (to be expected since isocyanate is more reactive than epoxy), giving a steady increase in VF and IV and masking any degradation.

Our data of CBC in PA and PET show that the reaction in PA is complete in less than 1 min (fast reaction with -NH_2 end groups), whereas in PET it takes longer. In order to make CBC an industrially attractive chain extender for PET this reactivity is too low. Therefore a long list of potential catalysts was tested for the combination of PET and CBC. The detection of chain extension via the VF appeared to be very useful, reliable and fast, see Figures 10 and 11.

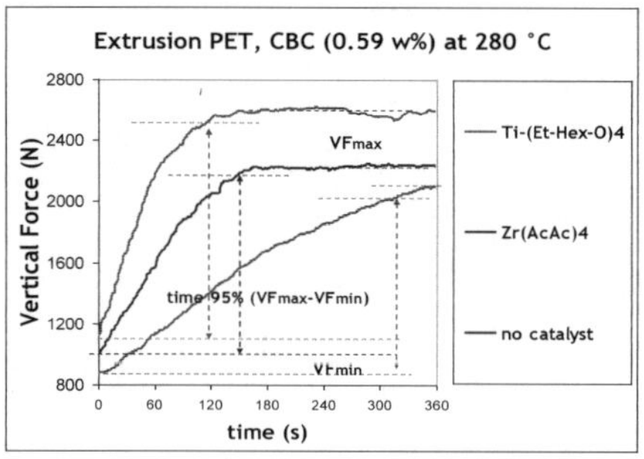

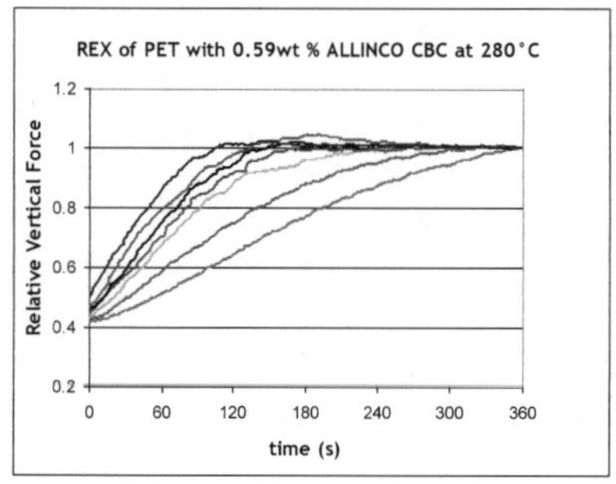

FIGURES 10, 11. VF and relative VF (= (VF at 360 s)/(VF at time t)) as function of REX reaction time.

In Figure 10 the time to reach 95 % of $VF_{max} - VF_{min}$ is used to rank catalyst formulations. In Figure 11 the relative VF shows to be very convenient and fast to discriminate between 7 different formulations, measured and analysed in half a day. As shown by Figure 11, it is easy to do a fast and effective screening of many different catalysts by plotting relative VF-time curves of the many REX formulations and reaction conditions. By this method we could test more than 50 different catalysts in less than 1 week.

CONCLUSIONS

Our data show that the Xplore MC 15 with vertical force detection is a convenient, fast and reliable instrument to do many REX studies in a short time. The data show a consistent empirical relation between vertical force measured just before extrusion and RV of the extrudate. This relation was later confirmed theoretically by modelling the flow in the MC, which led to the development of a rheological software package. CBC is an effective chain extender, fast in PA, slower in PET and possibly not effective in PLA. PLA can be chain extended with fast PDI or slower Joncryl.

ACKNOWLEDGMENTS

Basak Tuna (Kocaeli University, Turkey) is kindly acknowledged for the many careful experiments performed for this study.

REFERENCES

1. B.J.R. Scholtens and J.A. Loontjens, *Polyester 2000, 5th World Congress,* Maack Business Services, November 2000; II/4.
2. B.J.R. Scholtens and J.A. Loontjens, *Polyamide 2001, 2nd World Congress,* Maack Business Services, June 2001; II/4
3. B. Tuna and G. Özkoç, *Effects of diisocyanate and polymeric epoxidized chain extenders on the properties of recycled poly(lactic acid),* Submitted, 2016.

Improving the actuated performance of graphene/dielectric elastomers composites by using some new methods

Nanying Ning [1,2], Qin Ma[2], Suting Liu [2], Xiaoqing Zan [2], Ming Tian [1,2] and Liqun Zhang[1,2]

1. Beijing University of Chemical Technology, State Key Lab of Organic-Inorganic Composites, Beijing 100029, P.R. China;
2. Beijing University of Chemical Technology, College of Materials Science and Engineering, Center of Advanced Elastomer Materials, Beijing 100029, P.R. China

Abstract: Dielectric elastomers (DEs) can give rise to surprisingly large deformations by applying an electric field. Because of the lightweight, large strain, and high energy density etc., DEs find many applications in industry such as artificial muscles and sensors. Among various DEs, graphene based DEs have attracted much attention because of its high dielectric constant (*k*) and unique layered structure. In this study, some new methods were used to improve the electromechanical performance of graphene based DEs. First, we prepared carboxylated nitrile rubber (XNBR) DE with extremely high dielectric constant at quite a low percolation threshold via the directionally distribution of graphene oxide nanosheets (GONS) in XNBR latex and the in-situ thermal reduction of GONS at a low temperature. Second, we prepared GONS-encapsulated carbon nanosphere (GO@CNS) hybrid/XNBR dielectric composite with high *k*, low dielectric loss and large actuated strain at a low electric filed by latex compounding. Third, we prepared poly (dopamine) encapsulated GONS/elastomer composites with low dielectric loss and improved breakdown strength and actuated strain at a low electric field. The improvement in the electromechanical performance of graphene based DEs facilitates the wider application of DEs.

Introduction

Dielectric elastomers (DEs), as a kind of electroactive polymers, are capable of converting electric energy to mechanical energy and working efficiently over a broad frequency range [1-3]. DEs have good performance such as large active strain, high energy density, high electromechanical coupling efficiency, reliability, durability, fast response, and easiness of processing[4, 5]. Thus, DEs find applications in various devices, such as eyeball actuators[6], tactile displays[7], and inchworm robots[8], and have attracted much attention in the past decades [4, 9]. The preparation of DEs with high dielectric constant, low dielectric loss and large actuated strain is big challenges for DEAs. In this study, some new methods were used to improve the electromechanical performance of graphene based DEs.

Results and Discussion

1. We prepared graphene encapsulated rubber latex composites with high dielectric constant, low dielectric loss and low percolation threshold.

First, we prepared carboxylated nitrile rubber (XNBR) DE with extremely high dielectric constant at quite a low percolation threshold via the directionally distribution of graphene oxide nanosheets (GONS) in XNBR latex and the in-situ thermal reduction of GONS at a low temperature. The encapsulation of GONS on XNBR latex particles was mainly realized via the hydrogen bonding interactions between GONS and XNBR during latex mixing. A segregated graphene network was obtained at a low content of thermally reduced graphene (TRG), resulting in a low percolation threshold (0.25 vol.%). The dielectric constant at 100 Hz obviously increased from 23 for pure XNBR to 2211 and 5542 for the composite with 0.5 vol.% and 0.75 vol.% of TRG, respectively. The dielectric loss of the composites retained at a low value (less than 1.5). Meanwhile, the elastic modulus only slightly increased with the presence of 0.1–0.5 vol.% of TRG, keeping the good flexibility of the dielectric composites.

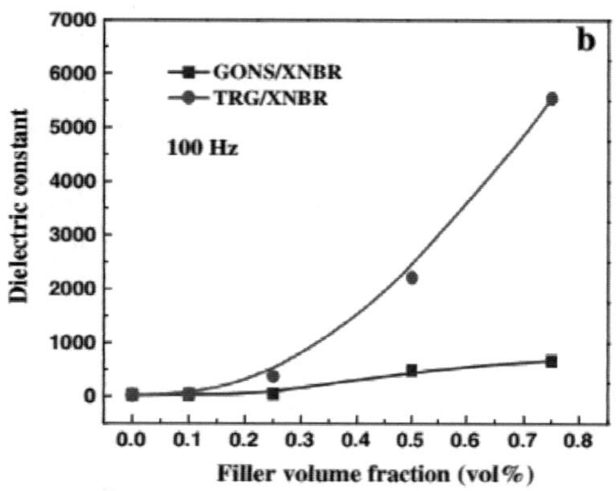

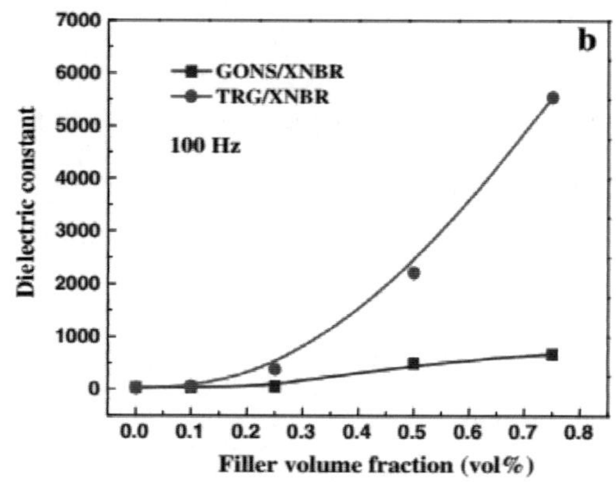

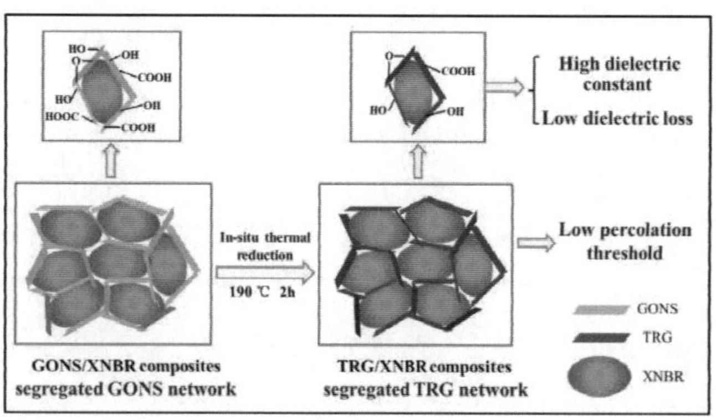

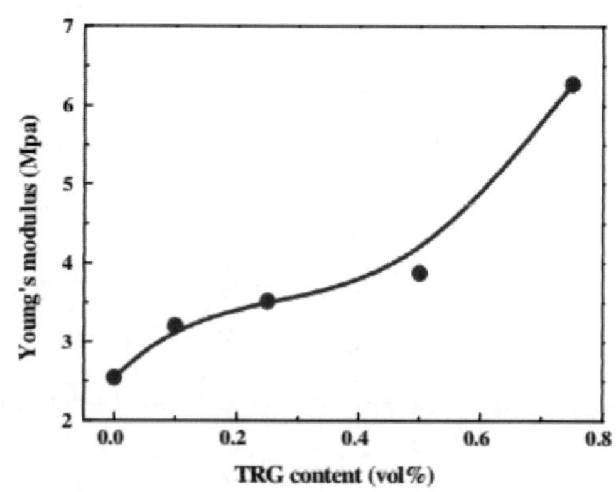

2. We prepared high performance dielectric composites by latex compounding of graphene oxide encapsulated carbon nanosphere hybrids with XNBR

We prepared the graphene oxide nanosheet(GONS)-encapsulated carbon nanosphere (GO@CNS) hybrid/XNBR dielectric composite via p–p interaction and hydrogen bonding interaction by simply mixing the CNS and GO suspension. The (GO@CNS) hybrid/XNBR dielectric composite is high *k*, low dielectric loss and large actuated strain at a low electric filed by latex compounding. After vulcanization and in situ thermal reduction, the thermally reduced GO (RGO)@CNS/XNBR composites were then obtained with the formation of a segregated filler network. The results showed that k at 10^3 Hz obviously increased from 28 for pure XNBR to 400 for the composite with 0.75 vol% of the hybrids while dielectric loss is still low (<0.65 at 10^3 Hz). In addition, the elastic modulus only mildly increased with the addition of 0.75 vol% of the hybrids. More interestingly, the actuated strain at a lower electric field (2 kV mm^{-1}) largely increased from 0.23% for pure XNBR to 3.06% for the composite with 0.75 vol% of RGO@CNS.

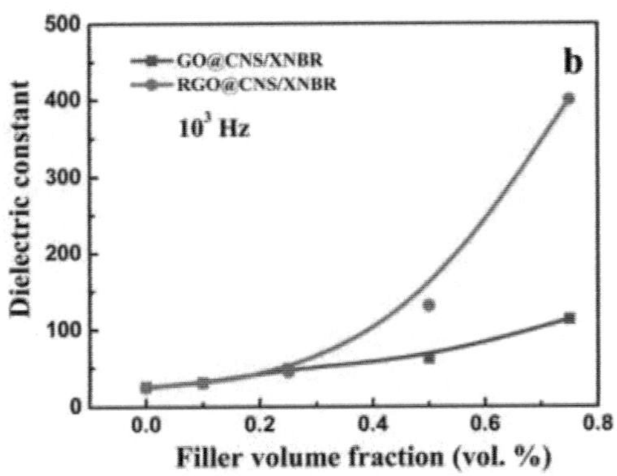

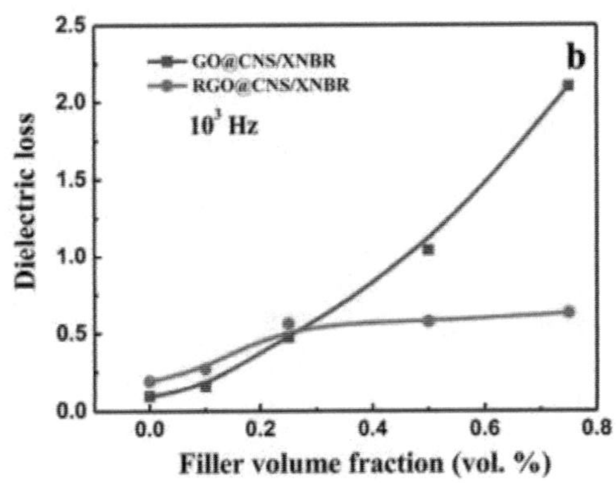

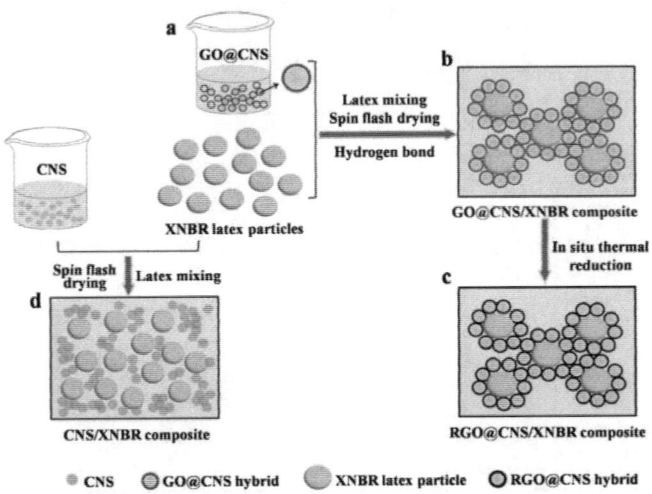

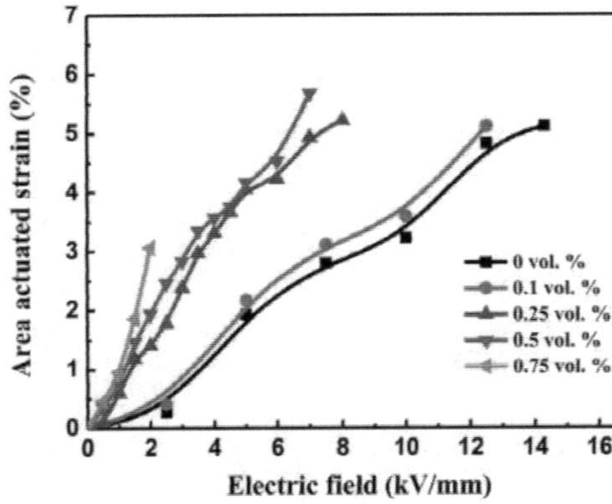

3. We tailored dielectric and actuated properties of elastomer composites by bio-inspired poly (dopamine) encapsulated graphene oxide

Third, we prepared poly (dopamine) (PDA) encapsulated graphene oxide nanosheets (GONS)/elastomer composites with low dielectric loss and improved breakdown strength and actuated strain at a low electric field by using dopamine (DA) to simultaneously coat the (GO) and partially reduce GO by self-polymerization of DA on GO. The PDA coated GO (GO - PDA) was assembled around rubber latex particles by hydrogen bonding interaction between carboxyl groups of carboxylated nitrile rubber (XNBR) and imino groups or phenolic hydroxyl groups of GO - PDA during latex compounding, forming a segregated GO - PDA network at a low percolation threshold. The results showed that the introduction of PDA on GO prevented the restack of GO in the matrix. The dielectric and actuated properties of the composites depend on the thickness of PDA shell. The dielectric loss and the elastic modulus decrease, and the breakdown strength increases with increasing the thickness of PDA shell. The actuated strain at a low electric field (2kV/mm) obviously increases from 0.2% for pure XNBR to 2.3% for GO - PDA/XNBR composite with the PDA thickness of 1.1 nm, much higher than that of other DEs reported in previous studies.

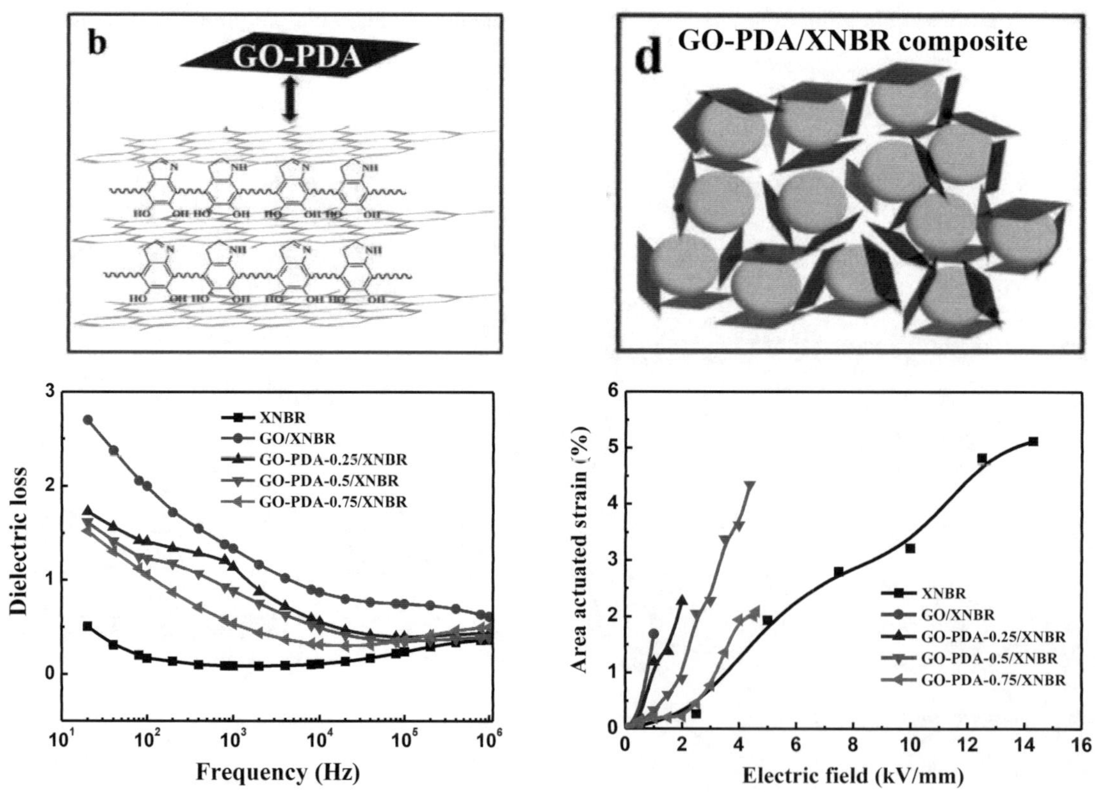

Conclusion:

We have successfully improved the electromechanical performance of graphene based DEs. First, we prepared carboxylated nitrile rubber (XNBR) DE with extremely high dielectric constant at quite a low percolation threshold via the directionally distribution of graphene oxide nanosheets (GONS) in XNBR latex and the in-situ thermal reduction of GONS at a low temperature. Second, we prepared GONS-encapsulated carbon nanosphere (GO@CNS) hybrid/XNBR dielectric composite with high *k*, low dielectric loss and large actuated strain at a low electric filed by latex compounding. Third, we prepared poly (dopamine) encapsulated GONS/elastomer composites with low dielectric loss and improved breakdown strength and actuated strain at a low electric field. The improvement in the electromechanical performance of graphene based DEs facilitates the wider application of DEs.

Acknowledgements

We would like to express our sincere thanks to the National Science Fund for Distinguished Young Scholars of China (Grant No. 51525301), the National Natural Science Foundation of China (Grant No. 51173007, 51221002) for financial support.

References:

1. Rao, Y.; Wong, C. P.. *J. Appl. Polym. Sci.* **2004**, 92, 2228-2231.

2. Song, Y.; Shen, Y.; Liu, H.; Lin, Y.; Li, M.; Nan, C. *J. Mater. Chem.* **2012**, 22, 16491-16498.

3. Stoyanov, H.; Kollosche, M.; Risse, S.; McCarthy, D. N.; Kofod, G.. *Soft Matter* **2011**, 7, 194-202.

4. Shankar, R.; Ghosh, T. K.; Spontak, R. J.. *Soft Matter* **2007**, 3, 1116-1129.

5. Pelrine, R.; Kornbluh, R.; Joseph, J.; Heydt, R.; Pei, Q.; Chiba, S.. *Mat. Sci. Eng. C-Mater.* **2000**, 11, 89-100.

6. Suo, Z.. *Acta Mech. Solida Sin.* **2010**, 23, 549-578.

7. Todd, M. G.; Shi, F. G.. *J. Appl. Phys.* **2003**, 94, 4551-4557.

8. Li, Z.; Fredin, L. A.; Tewari, P.; DiBenedetto, S. A.; Lanagan, M. T.; Ratner, M. A.; Marks, T. J.. *Chem. Mater.* **2010**, 22, 5154-5164.

9. Potts, J. R.; Dreyer, D. R.; Bielawski, C. W.; Ruoff, R. S.. *Polymer* **2011**, 52, 5-25.

Reactive extrusion of PLA/PC blends and composites

Vito Gigante[a], Maria-Beatrice Coltelli[a], Vu than Phuong[b], Patrizia Cinelli[c], Andrea Lazzeri[a]

[a] *University of Pisa, Department of Civil and Industrial Engineering, Largo Lucio Lazzarino 1, 56122, Pisa, Italy*
[b] *Can Tho University, Department of Chemical Engineering, Can Tho, Vietnam 56122, Pisa, Italy*
[c] *Institute for the Chemical and Physical Processes, National Research Council of Italy, Via Moruzzi 1, 56124, Pisa, Italy*
Presenting author's email: vito.gigante@dici.unipi.it
Reference author's email: andrea.lazzeri@unipi.it

Abstract. Poly(lactic acid) (PLA) is one of the most promising renewable polymer to be employed in injection molded parts. However, the heat deflection temperature and the impact properties should be improved. The reactive blending with polycarbonate (PC) in the presence of an interchange reactions catalyst was demonstrated to be a successful strategy. This process was applied to both virgin and recycled-PC/PLA blends and interesting differences were evidenced as a function of different viscosity in the melt, of the polymers. The blends were prepared in a laboratory extruder, tensile and impact properties were measured, and the phase morphology was investigated even by electron microscopy. Cellulose/polymer composites were also prepared by using PLA/PC blends as polymeric matrix. Fibres having different thickness and aspect ratio were selected and used as fillers in extrusion trials. The elastic modulus increased thanks to the reinforcing behavior of the fibres and several models were applied to fit the experimental trends as a function of composition.

Keywords: poly(lactic acid), Poly(carbonate), biocomposites, cellulosic fibers, predictive models
PACS: 81.20.Hy; 83.80.Tc; 82.35.Pq; 81.05.Qk

INTRODUCTION

PLA has been considered, in several scientific articles, as the most promising bio-based polymer to be employed in injection molded parts, but its mechanical properties should be improved.
In fact PLA, as compostable material derived by renewable resources, shows advantages when compared to traditional plastics, like high elastic modulus and transparency. However it presents very important limitations such as high hydrolytic susceptibility, low thermal stability, brittleness and low crystallization rate. Hence there is still considerable request for cost-effective methods to enhance PLA properties [1].
The blending of PLA with polycarbonate (PC), having a high glass transition, above 150°C, can be a successful strategy for limiting its tendency to decrease its resistance above its glass transition of about 60°C.
In the patent literature blends of PLA with PC as well as with some other commercially-available products [2-3] have been already described. PC has been widely used as an engineering plastic as it presents high thermal stability, high tensile strength and elongation at break.
However, PLA/PC blends are immiscible, and adhesion between the two polymers is weak due to high interfacial tension. Hence a proper compatibilization method should be selected.
A new glass transition temperature intermediate between those of pure PLA and PC was evidenced by DMA measurements, employing both a trans-esterification catalyst tetrabutyl ammonium tetraphenyl borate (TBATPB) and triacetin. The novelty of this approach of Phuong et al. [4-5] consisted in the fact that only a very short residence time of 60 seconds was necessary for observing the improvement of compatibilities and final properties of the material thus making the process compatible with traditional methods of industrial extrusion. In fact, triacetin favored the interchange reactions between PLA and PC probably because it can merely diffuse in the melt during extrusion reacting with both components. This deep analysis about reactivity and properties of PLA/PC blends showed that the blending of PLA and PC in the presence of a catalytic system based on Triacetin and TBATPB is a good strategy to improve the properties of PLA.
For improving the ecological content of the material it could be interesting to follow two different strategies: increasing the PLA content in the blend, thus employing a higher content of renewable material, and replacing virgin PC with recycled PC. The use of recycled PC in a second series of trials with PLA content increased up to 60% was tested in terms of tensile and impact properties. Moreover, also a recycled PLA was employed, thus maximizing the ecological content of the developed materials.

In order to develop natural composites, trials in the presence of three different types of cellulose fibers were carried out. Thanks to the analysis of tensile tests results it was also carried out a theoretical work for selecting the best mathematical model (usually used for aligned fiber composites) to predict Elastic Modulus of these biocomposites .

MATERIALS AND METHODS

Materials used for this work are:

- Poly (L-lactic) acid was purchased from NatureWorks LLC having a nominal average molecular weight Mw = 199590 Da (NatureWorks® IngeoTM 2002D Extrusion Grade), density 1,24g/cm^3.
- Bisphenol A Polycarbonate (Mitsubishi Chemical Co. - Japan, grade S3000) had a density of 1,20 g/cm^3
- Corresponding post-industrial materials were also selected. In particular, the thermoforming of PC is a process producing high amount of waste, as the thermoforming companies use sheets and cut from them the thermoformed parts but can not recycle the residual parts in their own facilities. Hence recycling companies collect these wastes, which have a post-industrial and not a post-consumer origin, and grind them in order to allow the reintroduction in an industrial cycle.
- Three kinds of cellulosic fibers (with different aspect ratios) were added to prepare composites with PLA-PC matrix
- Triacetin (TA, also known as glycerin triacetate or 1,2,3-triacetoxypropane, CAS # 102-76-1) was purchased from Aldrich Chemicals.
- Tetrabutylammonium tetraphenylborate (TBATPB, CAS # 15522-59-5), used as an interchange reactions catalyst, were purchased from Aldrich Chemicals.

After drying in a ventilated oven, the two polymers (PLA and PC) were mechanically mixed for about 10 minutes with different ratios of micro-cellulose by means of a high-speed mixer. After this mixing stage, the Triacetin and TBATPB (5% and 0,2% by weight respectively) were added and mixed for an extra 10 minutes using the same equipment. The resulting mixtures were processed with a Haake MiniLab II Micro Compounder conical twin-screw extruder (Thermo Scientific Haake GmbH, Karlsruhe, Germany). The mixing was carried out at a temperature of 230°C with a screw speed of 100 rpm, for a recirculating time of 1 minute. After extrusion, the molten materials were transferred through a preheated cylinder to the Haake MiniJet II mini injection molder (Thermo Scientific Haake GmbH, Karlsruhe, Germany) to obtain Haake type III dog-bone tensile bars used for tensile tests and ISO 179-2000 samples to make impact tests.
Tensile tests were performed at room temperature at a crosshead speed of 10 mm/min by means of an Instron 4302 universal testing machine (Canton MA, USA), equipped with a 10 kN load cell and interfaced with a computer running the Testworks 4.0 software (MTS Systems Corporation, Eden Prairie MN, USA). The morphology of the composites was studied by scanning electron microscopy (SEM) using a JEOL JSM-5600LV (Tokyo, Japan) and by analyzing the fracture surfaces of the samples broken in liquid nitrogen. Prior to SEM analysis all the surfaces were sputtered with gold.
Notched Charpy tests were performed following the ISO 179 standard by using a Ceast 9050 Charpy machine equipped with DAS 8000 Junior to acquire data. The specimens were obtained by using a dedicated standard ISO179 mold in the Haake Minijet injection molder and notched in accordance with ISO 179.

MECHANICAL PROPERTIES OF PLA-PC BLENDS

The correlation between properties and composition of PLA/PC blends was deeply investigated comparing the blends obtained with virgin and recycled polymers and evaluating the effect of the addition of the TBATPB. The investigated compositions of PLA/PC blends were 70/30, 60/40, 40/60 and 70/30. The results related to 40/60 and 60/40 blends were reported in Table 1. In this study virgin PC was replaced with recycled one to increase the ecological content of the blends. Moreover some blends containing also recycled PLA were examined.
On the basis of these results it was possible to deduct that it was possible to increase the content of PLA in the blend from 40% up to 60% by replacing pure PC with recycled PC. This result can be explained by considering that the viscosity in the melt of the recycled PC was lower than the one of virgin PC. For this reason the phase distribution was different in R-PC/PLA blends (having a viscosity ratio closer to 1) with respect to PC/PLA blends. The different phase morphology evolution was demonstrated by comparing several SEM micrographs recorded onto cryofractured surfaces of the examined blends (data not shown).
It was also demonstrated that it is possible to replace PLA with recycled PLA. When these recycled polymers were used interesting properties were obtained also without the catalytic system. Nevertheless the triacetin/TBATPB catalytic system was demonstrated to influence blends properties not only thanks to the compatibilization occurring because of the formation of PLA-PC copolymers, but also because of the modification of the molecular weight of PLA and PC in the blends.

TABLE 1). Tensile properties of PLA/PC blends, PLA/R-PC blends and R-PLA/R-PC blends obtained without and with the catalytic system (CAT) consisting in triacetin and TBATPB. The weight percentage of each polymer is indicated after the name.

blend	Tensile Strength (MPa)	Young's Modulus (GPa)	Elongation at break (%)
PC	57.2 ± 0.8	2.25 ± 0.06	84 ± 4
R-PC	65 ± 1	2.6 ± 0.4	54 ± 6
PC70PLA30	55.3 ± 0.3	2.86 ± 0.07	125 ± 3
R-PC70PLA30	58.2 ± 0.2	2.62 ± 0.08	58 ± 5
PC70PLA30CAT	58.0 ± 0.6	3.01 ± 0.09	110 ± 5
R-PC70PLA30CAT	60,0 ± 0.2	2.76 ± 0.4	68.8 ± 0.6
PC40PLA60	56.1 ± 0.7	2.98 ± 0.03	3.3 ± 0.5
R-PC40PLA60	56 ± 3	2.8 ± 0.2	50 ± 7
PC40PLA60CAT	51.9 ± 2.4	3.3 ± 0.1	2.0 ± 0.1
R-PC40PLA60CAT	64 ± 1	3.3 ± 0.3	75 ± 6
R-PC40R-PLA60	64 ± 3	3.5 ± 0.3	72 ± 6
R-PC40R-PLA60CAT	56 ± 3	3.4 ± 0.3	3.5 ± 1
PLA	60.4 ± 0.3	3.5 ± 0.1	4.1 ± 0.5

Notched Charpy impact properties and density were also measured and reported in Table 2.
The impact properties of PC based blends are better than those of PLA based blends. In the case of R-PC40R-PLA60 blend the Charpy Impact strength is more than doubled with respect to pure virgin PLA, moreover the presence of a significant content of R-PC in the blend grants the maintenance of resistance even in the PLA glass transition temperature region, as demonstrated by DMA measurements (data not shown).

TABLE 2). Charpy impact properties and density of PLA-PC blends

blend	Charpy Impact Strength (MPa)	Density (g/cm^3)
PC	10.9 ± 0.2	1.148 ± 0.006
R-PC	9.4 ± 0.5	1.161 ± 0.012
R-PC70PLA30	9.3 ± 1	1.158 ± 0.005
R-PC70PLA30CAT	5.30 ± 0.18	1.170 ± 0.004
R-PC60-RPLA40	7.5 ± 1.9	1.175 ± 0.003
R-PC40PLA60	5.71 ± 0.48	1.173 ± 0.004
R-PC40PLA60CAT	4.61 ± 0.88	1.177 ± 0.001
R-PC40R-PLA60	6.25 ± 0.82	1.171 ± 0.008
R-PC40R-PLA60CAT	4.64 ± 0.18	1.175 ± 0.002
PLA	2.87 ± 0.16	1.194 ± 0.001

The determination of the density of the blends was quite important to boost the applications of these blends. In fact it is important to evidence that polymeric materials show densities low with respect to other materials such as metals. Hence their use in injection molded part in different types of vehicles to replace metal part can be particularly convenient for reducing fuel/energy consumption.

PREPARATION OF COMPOSITES WITH CELLULOSIC FIBERS AND PREDICTION OF THEIR ELASTIC MODULUS THROUGH MATHEMATICAL MODELS

Lignocellulosic fillers are widely used as reinforcing fillers in thermoplastic composites to provide positive environmental benefits with respect to ultimate disposability. The major advantages of these composite materials are low cost, light weight, high specific strength, and recyclability.
Processing temperature of PC is high, so degradation of the natural fibers can occur during processing. Nevertheless, it usually does not happen because both PLA and PC matrices interact very poorly with natural fibers thanks to their high surface tension.
The addition of fibers lead to an increase in Elastic modulus and tensile Strength and a strong reduction of elongation at break. The use of the catalytic system shown before results in a slight increase in both Elastic Modulus and Tensile Strength for all the compositions with respect to the not catalyzed composites.
In this work different mathematical models have been tested to predict the elastic behavior of these biocomposites with different compositions of misaligned short fibers. The equations are based on the elastic equilibrium for a volume element in a circular cross section fiber, immersed in a matrix cylinder, on which a tensile stress in the fiber direction is applied. The experimental data reported in Figure 1 (first bar on the left indicated by the arrow) for composites obtained by using cellulosic fibers with different aspect ratio are compared with the data calculated by using different predictive

models (Cox, Halpin-Tsai, Clyne, Starink and Kim). The Halpin-Tsai model (represented by the third bar, indicated by the arrow, in fig.1) is the one that provided values closest to the experimental data. The reason can be that the ability of this latter model to calculate both the longitudinal elastic modulus and the transverse one (and then provide empirical equation to monitor the random orientation of the fibers) probably provides a more accurate response with respect to the other equations

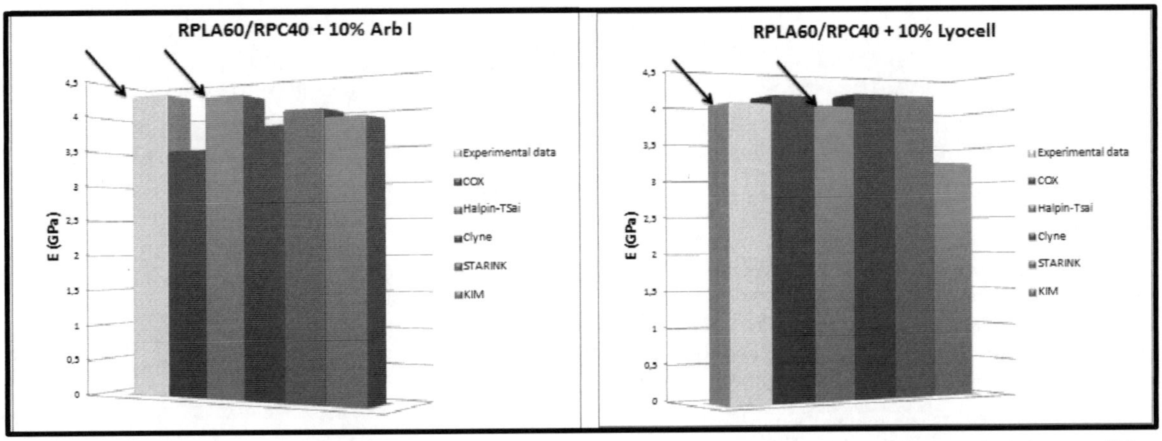

FIGURE 1. Comparison of experimental data about Young's Modulus of R-PLA/R-PC based composites containing two different types of cellulosic fibers with calculated data obtained by using different predictive models.

CONCLUSIONS

In conclusion it is evident that PLA based blends and composites represent a great opportunity for developing materials renewable and with reduced impact onto environment. In particular PLA has the characteristics to be employed in injection molded parts, but its mechanical properties should be improved. For this purpose the blending with PC was demonstrated to be a good strategy for obtaining materials with improved properties in the glass transition region. The current work evidences also that this method can be applied by using recycled PC and recycled PLA. The key issue is represented by the control of phase morphology thanks to consideration about viscosity in the melt of the different polymeric components. It was evidenced that blends containing 60% of PLA showed impact resistance and tensile properties better than pure PLA (in some cases with an increase above of 100%).
Biocomposites represented a good alternative for developing light weight materials. It is possible to obtain materials with Young's Modulus up to 4,3 GPa with a content of 10%wt of Cellulosic fibers added in the recycled PLA- recycled PC matrix. The mathematical method that predicts more accurately the Elastic Modulus of these composites was demonstrated to be the Halpin-Tsai model.

REFERENCES

1. F. Elmaghor, L. Zhang, F. Run, L. Hangquan, *Polymer* **45,** 2004, pp. 6719–6724
2. H. Li, MA. Huneault, *Polymer* **43**, 2007 pp. 6855-6866
3. A. Takehara, T. Onoki, F. Kageyama, *Japan Published Patent Application* 2007, 131795
4. V.T. Phuong, M.B. Coltelli, P. Cinelli, M. Cifelli, S. Verstichel, A. Lazzeri, *Polymer*, **55(12),** 2014, pp. 4498-513
5. P. Cinelli, A. Lazzeri, M. Penco, V.T. Phuong, *WO2012025907 A1 (1 mar 2012)*

The Interplay between Molecular Dynamics and Reaction Kinetics during Curing Reactions

Jürgen E.K. Schawe

Mettler-Toledo GmbH, Analytical, Sonnenbergstrasse 74, CH-8603 Schwerzenbach, Switzerland

juergen.schawe@mt.com

Abstract. For the example of a curing reaction of an epoxy resin the curing dependence of the relaxation times is measured by DMA and temperature modulated DSC and described by the conversion dependence on the glass transition temperature. The relaxation time change slows down during isothermal reaction, if vitrification occurs. The interaction between the relaxation time and the curing process is described by the diffusion controlled function. The prediction of the reaction kinetics agrees well with the experimental results.

Keywords: Thermoset, reaction kinetics, vitrification, relaxation.
PACS:

INTRODUCTION

The network formation during the curing reaction of a thermosetting resin causes a large increase of the characteristic relaxation time and therefore an increase of the glass transition temperature [1,2]. If the reaction time of an isothermal curing reaction is slower than the reaction at the resulting glass transition temperature of the fully cured material, the material vitrifies. This vitrification has an important influence on the reaction kinetics. The reaction turns from a chemically controlled to a diffusion controlled reaction.

In this contribution we study the isothermal curing reaction of an epoxy resin (diglycidylether of biphenol A, DGABA, cross-linked with diaminodiphenyl methane, DDM) by Differential Scanning Calorimetry (DSC), Temperature Modulated DSC (TMDSC) and Dynamic Mechanical Analysis (DMA). The focal point is the change of the molecular dynamics during cure and the description of the influence of the diffusion control on the reaction kinetics.

EXPRIMENTAL

Sample

Commercial diglycidylether of bisphenol A (DGEBA) (Shell Chemical, Epikote 828) and diaminodiphenyl methane (DDM) (Aldrich) were used as a thermosetting system. The substances were poured together in the stoichiometric amounts of 2 mol DGEBA per 1 mol DDM and heated up to 120 °C for 20 s. During this time the sample was stirred to get a homogeneous mixture. The mixture was then rapidly cooled to room temperature and specimens of approximately 3 mg were prepared in hermetically sealed 40 µl aluminum crucibles. The specimens were stored at –30 °C until the measurement. The rest of the sample was stored at -30 °C in a glass container covered with aluminum foil.

Instrumental

The DSC measurements were performed using a METTLER TOLEDO DSC 2 with IntraCooler. The software option Model Free Kinetics (MFK) was used for the evaluation. The TMDSC measurements were performed using the stochastic temperature modulation technique TOPEM [3].

The DMA shear measurements were performed using the METTLER-TOLEDO DMA861. The DMA was preheated and thermally equilibrated at the reaction temperature. The sample holder together with the sample was installed in the preheated instrument. The measurements are performed at frequencies between 0.1 Hz and 750 Hz.

For the measurement of the mechanical relaxation behavior of the fully cured material, isothermal frequency sweeps between 0.01 Hz and 1000 Hz were performed at different temperatures between 100 and 200 °C.

RESULTS AND DISCUSSION

Mechanical Relaxation Behavior

From the frequency sweep measurements, the complex shear compliance $J^*(\omega) = 1/G^*(\omega) = J'(\omega) - i\,J''(\omega)$ was determined. G^* and J^* are the complex shear modulus and compliance, respectively. J' is the storage component, J'' the loss component, ω the angular frequency and i is the imaginary unit. From the maximum of the J'' peak the relaxation time τ is given by $\tau(T) = 1/(2\,\pi\,f_{max})$. The high temperature range is the region of the dynamic glass transition, which follows the Vogel-Fulcher-Tammann (VFT) behavior [4]:

$$\log\left(\frac{\tau}{\tau_0}\right) = \log\left(\frac{B}{T - T_V}\right) \qquad (1)$$

where τ_0 is the limiting relaxation time at high temperatures, B is the curvature parameter and T_V the Vogel-temperature ($\tau_0 = 4\cdot10^{-10}$ s, $B = 1086$ K and $T_V = 390$ K). At lower temperatures the activation curves approach Arrhenius behavior; the material is in the vitrified glassy state.

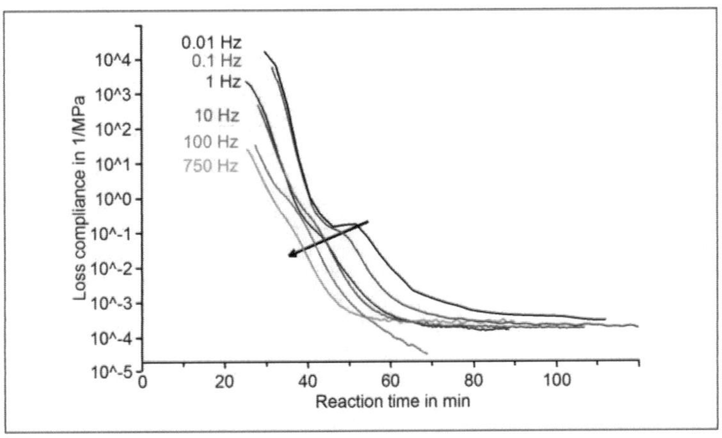

FIGURE 1. Loss compliance as a function of the reaction time for an isothermal cure at 100 °C, measured at different frequencies. The arrow indicates the peak.

To measure the relaxation times during a reaction, isothermal DMA measurements at 100 °C are performed at frequencies between 0.01 Hz and 750 Hz. The external sample preparation causes a small experimental error in the time of the temperature equilibration. From the maximum of the loss compliance curves (Fig. 1), the relaxation time is determined as a function of the reaction time.

Vitrification and Thermal Relaxation during Reaction

The heat capacity and the heat flow due to the isothermal curing reaction are simultaneously measured by stochastically temperature-modulated DSC (TOPEM) [3]. This technique enables the simultaneous determination of the frequency dependent complex heat capacity [3, 5].

Figure 2a shows a set of heat flow and static heat capacity curves measured during isothermal cure at temperatures between 60 and 160 °C. The total specific enthalpy of reaction, Δh_1, is 407 J/g [6]. The degree of conversion during the reaction is calculated from the enthalpy curves. The specific heat capacity curves show the initial increase due to the difference of the heat capacity of the initial mixture and the reaction product. As the reaction progresses, the heat capacity decreases stepwise with vitrification. The vitrification time t_c is the time at the inflection point. From the set of heat capacity curves the glass transition temperature, T_g, versus conversion (Fig. 2b) is determined ($T_g = T_{react}$). The temperature-time-transition (TTT) diagram is derived from the isothermal TOPEM measurements (Fig. 2c). It describes the transition from the liquid reactive system into the glassy state.

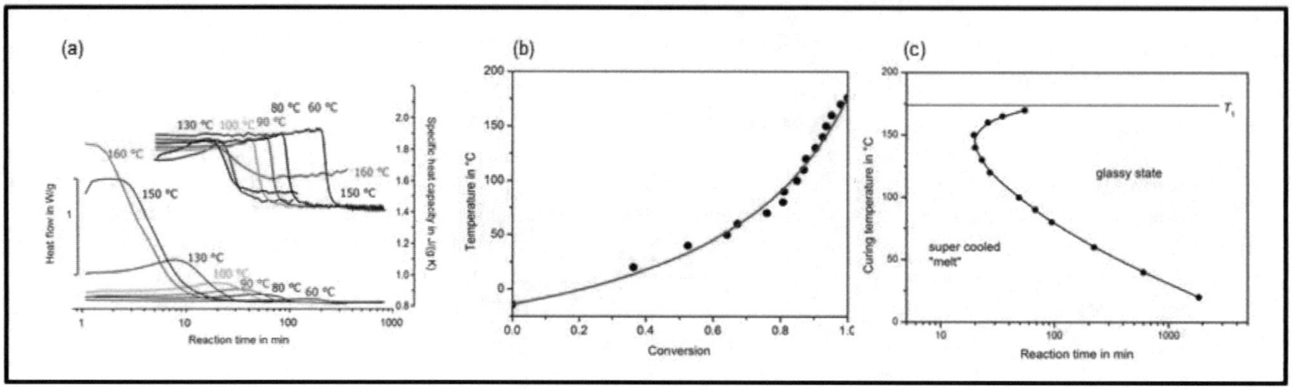

FIGURE 2. a) Heat flow and quasi-static specific heat capacity versus the reaction time; b) Glass transition temperature versus conversion; c) TTT-diagram.

The conversion dependence of T_g (Fig. 2b) is fitted by the DiBenedetto equation [7]:

$$T_g(\alpha) = \frac{\lambda\,\alpha\,(T_1 - T_0)}{1-\alpha(1-\lambda)} + T_0 \quad (2)$$

The fit-parameters are T_g of the uncured mixture, $T_0 = -14$ °C; T_g of the fully cured thermoset, $T_1 = 176$ °C; and the interaction parameter, $\lambda = 0.3$.

From the isothermal TOPEM measurement at 100 °C the curves of the specific heat capacities at several frequencies, f, are determined simultaneously. The thermal relaxation time is $\tau_{th} = 1/(2\,\pi f)$. In Fig. 3 the mechanical and thermal relaxation times are plotted versus the reaction time. The solid curve is given by a combination of eq. (1) and (2):

$$\tau(\alpha) = \tau_0 \exp\left(\frac{B}{T_{react} - T_g(\alpha) + C_2}\right) \quad (3)$$

If we take for C_2 the universal WLF parameter $C_2 = 51.6$ K [8], eq. (3) has no independent parameter. This equation describes well the curing time dependency of the relaxation time just by the conversion dependency of T_g.

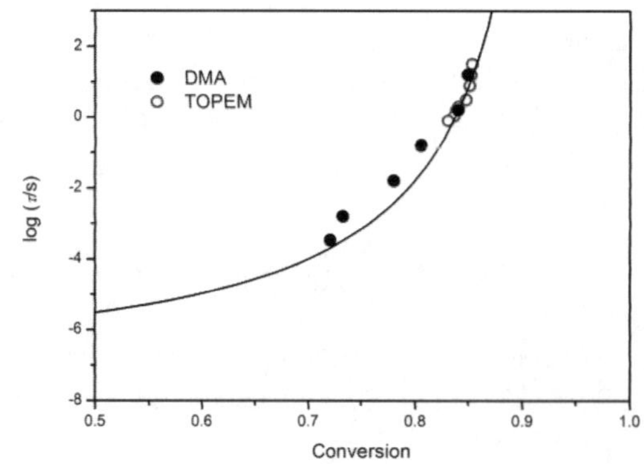

FIGURE 3. Relaxation time versus reaction time for an isothermal reaction at 100 °C.

Vitrification and Reaction Kinetics

The kinetics of a chemically controlled reaction is usually described by

$$\left[\frac{d\alpha}{dt}\right]_{\text{chem}} = f(\alpha)\,k_0 \exp\left(-\frac{E_a}{RT}\right) \qquad (4)$$

$f(\alpha)$ is the conversion function describing the reaction model, E_a denotes the activation energy, k_0 the pre-exponential factor of the rate constant and R the gas constant. With DSC measurements only the mechanism of the reaction cannot be investigated in detail. An alternative is the use of the iso-conversion methods using model free kinetics (MFK) [9].

The reaction kinetics is strongly influenced by the vitrification process because of the limited molecular transport in the glassy state. This influence can be described by introduction of a diffusion control function f_d [6]:

$$\frac{\mathrm{d}\alpha}{\mathrm{d}t} = \left[\frac{\mathrm{d}\alpha}{\mathrm{d}t}\right]_{\text{chem}} f_{\mathrm{d}}(\alpha) \qquad (5)$$

For a chemically controlled reaction f_d is unity. In the case of a system with completely frozen, long-scale mobility the diffusion control function tends towards zero.

According to eq. (3) the course of τ during a reaction can be described using the change of the glass transition. We therefore use the conversion dependence of T_g also for the diffusion control function [6]:

$$f_d(\alpha, T_{\text{react}}) = 1 - \left(1 + \frac{1}{2}\left(\frac{T_{\text{react}} + \Delta T - T_{\mathrm{g}}(\alpha)}{\Delta T}\right)^3\right)^{-1} \qquad (6)$$

where $\Delta T \approx 10$ K represents the width of the thermal glass transition. E. (6) describes the influence of vitrification during reaction on the reaction kinetics by the curing dependence of the glass transition temperature. The course of reaction can be predicted from the kinetics of the chemically controlled reaction and the diffusion control function, f_d. A set of isothermal predictions in Fig. 4 shows a good agreement with the measured curves.

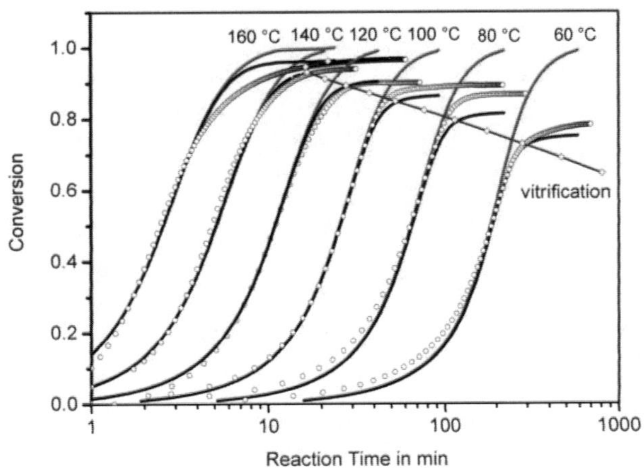

FIGURE 4. Conversion as a function of the reaction time for isothermal reactions at different temperatures. (black lines: measured curves, red lines: predictions of the MFK according to eq. (4), dotted curve: prediction using the diffusion control function (eq. (6)), blue curve: prediction of the vitrification process (according to TTT-diagram)).

REFERENCES

1. J.K. Gilham, *Polym. Eng. Sci.***26**, 1429-1433 (1986).
2. S. Montserrat, *J. Appl. Polym. Sci.***44**, 545-554 (1992).
3. J.E.K. Schawe, T. Hütter, C. Heitz, I. Alig, D. Lellinger, *Thermochim. Acta* **446**, 147-155 (2006).
4. E. Donth, *The Glass Transition: Relaxation Dynamics in Liquids and Disordered Materials*, Berlin: Springer, 2001, pp. 26-64.
5. I. Fraga, S. Montserrat, J. Hutchinson, *J. Thermal. Anal. Calorim.* **91**, 687-695 (2008).
6. J.E.K. Schawe, *Thermochim. Acta* **388**, 299-312 (2002).
7. A.T. DiBenedetto, *J. Polym. Sci.,Part B: Polym. Phys.***25**, 1949-1969 (1987).
8. M.L. Williams, R.F. Landel, J.D. Ferry, *J. Am. Chem. Soc.* **77**, 3701-3707 (1955).
9. S. Vyazovkin, A.K. Burnham, J.M. Criado, L.A. Pérez-Maqueda, C. Popescu, N. Sbirrazzuoli, *Thermochim. Acta* **520**, 1-19 (2011).

Production of CFRP Components with Clear Surface Layer in the PU Spray Impregnation Process

Ch. Hopmann, Ch. Karatzias, R. Wagner, A. Böttcher* and Kai Fischer

Institute of Plastics Processing (IKV) at RWTH Aachen University
Seffenter Weg 201, 52074 Aachen, Germany

arne.boettcher@ikv.rwth-aachen.de

Abstract. The high volume capable production of continuous fibre reinforced plastic components using carbon fibres (CRFP) for visible areas with high fibre volume contents still presents a challenge. One of the most important disadvantage is the low economic efficiency due to the required work subsequent to part manufacturing (grinding, painting). To overcome this disadvantage, a process for a high volume capable production of visible components with polyurethane (PU) matrix systems has been developed at the Institute of Plastics Processing (IKV) at RWTH Aachen University. Two approaches based on the manufacturing processes PU impregnation and wet pressing are pursued. Both approaches rely on the generation of a clear PU surface layer (coating) on the visible component surface. As process variants, a one-step pressing process with an application of the coating directly on the non-cured PU-impregnated preform has been developed, as well as a process with an application of the coating on the final-cured CFRP component. This paper describes results of the process integrated application of an aliphatic PU coating system in two different developed process variants (one-step and two-step) in the wet pressing process. The focus of the investigations was the generation of a low surface waviness of PU based CFRP components to increase the surface quality. The results of investigations regarding the surface quality show that the coating material can compensate the characteristic waviness of the neat CFRP component.

Keywords: carbon fibre reinforced plastics, PU impregnation, surface quality, wet pressing
PACS: 81.05

INTRODUCTION

Continuous fibre reinforced plastics (FRP) with high fibre volume content (FVC) $\geq$ 45 % exhibit a high lightweight potential. For their use in visible areas with highlighted fibre structures, e.g. in automotive components such as roof structures [1] or interior parts [2], the realisation of a high surface quality is of high importance [3-5]. One of the main surface defect in the production of FRP components is the shrinkage of the resin system coupled with the different thermal expansion coefficients of fibres and cured resin system, which results in the formation of surface waviness from local accumulations of resin e.g. at the crossover of two rovings in a woven textile preform [6-8]. Therefore, for the use in visible areas, the surface waviness has to be compensated, e.g. by additional post-production processes like grinding and polishing [3, 11], by process optimisation of the FRP production [7-10] or by coating processes integrated in the mould (e. g. In-Mould-Coating [6], Surface RTM [12]). However, in most cases a painting of the part with lacquer is still necessary to achieve very high surface qualities what lead to high additional cost in the FRP part production for visible areas. A possibility to avoid additional steps in the post-production is the use of an aliphatic lightfast elastomer based polyurethane (PU) system as coating material which has excellent properties for UV stability and scratch resistance. Therefore, it can be used for FRP components with visible fibre structure in interior areas without additional painting processes. By this, the investigations regarding the production of carbon fibre reinforced plastics (CFRP) with the aim to compensate the characteristic surface waviness were conducted by using the aliphatic PU system as coating material. For the matrix system an aromatic PU system was used which was applied on the fibres in an open PU impregnation process. The part production subsequently takes place in a wet pressing process. The combination of fibre impregnation with PU and further processing in a wet pressing process is based on technology developed at IKV within previous studies with monolithic components without surface optimization [13].

This paper focusses on the process integrated application of the aliphatic PU coating system in two different developed process variants (one-step and two-step) in the wet pressing process. The aim of the investigations described in this paper is to increase the surface quality for PU based CFRP components by compensating the surface waviness. For that, CFRP components with the coating material have been produced in both variants and investigated using micrographs for analysing the impregnation quality and the coating thickness and a surface measuring system for analysing the surface waviness. The parts with coating material will be compared to parts without coating.

EXPERIMENTAL

Materials

All investigations described in this paper were conducted with a compact aromatic PU system (puropreg®185 IT / puronate® 900) as matrix for the CFRP component and a compact aliphatic PU system (puroclear® 3098 IT / puronate® 960/1) as coating material, both provided by Ruehl Puromer GmbH, Friedrichsdorf, Germany.

The preforms used in the CFRP components have a symmetric layup with four carbon woven fabrics (SIGRATEX® C W245, 2x2 twill weave, 245 g/m², 3K - 200 tex, thread count per cm 6x6) provided by SGL Technologies GmbH, Meitingen, Germany. The fabrics have been fixed using a thermoplastic binder fleece PA 1300 (6 g/m²), Keuchel Associates Inc., Cuyahoga Falls, USA.

Process Technology

The process sequence of the combined PU impregnation and wet-pressing process for the production of CFRP components in a dimension of 150 x 300 mm² with the additional coating material is shown in figure 1. For both process variants (one-step and two-step) the dry carbon fibre preform is placed on a perforated supporting structure where a vacuum can be applied under the preform (figure 1 left). The aromatic PU matrix system is then dispensed on the top of the preform. Due to the applied vacuum the PU matrix system impregnates the preform in thickness direction. The vacuum is active until a homogeneous impregnation of the preform is completed. Based on impregnation studies the impregnation time for the presented components has been set up to 80 sec to achieve a FVC of 50 Vol.-% for a component thickness of 1.2 mm. After the impregnation process, the impregnated preform is placed in a wet pressing mould.

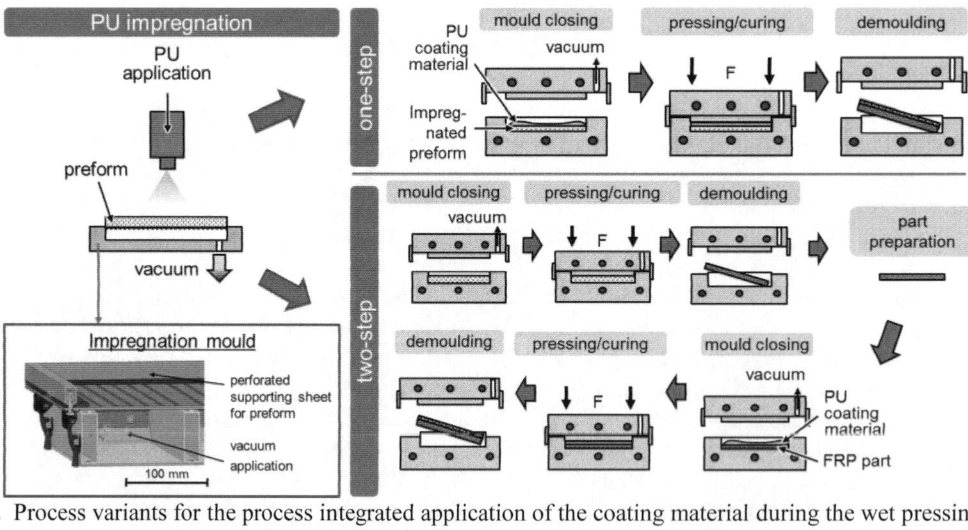

FIGURE 1. Process variants for the process integrated application of the coating material during the wet pressing process.

As a two-step process variation (figure 1 lower right), the CFRP component is first cured for 20 min within a pressure of 25 bars in the closed mould. Afterwards the mould is opened and the surface of the component is refined by grinding of the external release agent (Frekote® 700-NC, Henkel AG & Co. KGaA, Düsseldorf, Germany) as this will affect the later bonding between the CFRP component and the coating material. Then the coating system is applied on the refined surface and the component is again cured for 20 min before demoulding the component. As a one-step process variation (figure 1 upper right) the PU coating material is dispensed directly on the wet impregnated preform. The mould is then closed and both PU systems are parallely cured completely for 20 min, as it is recommended in the data sheet. This variation has the advantage that no external release agent can accumulate between the surface coating and the reinforced component. Based on the curing temperature recommendation of the PU materials, for both process variations the upper half of the mould has a temperature of 75 °C and the lower half of the mould has a temperature of 65 °C. The reference components have been manufactured without a coating material. Table 1 summarizes the investigated process variants and used parameters. For each process variation two parts have been produced with the followed parameters.

TABLE (1). Investigated process variants and used parameters

Process variation	Impreg-nation time [sec]	Mould cavity pressure [bars]	Mould temperature (upper/lower mould side) [°C]	Curing time [min]	Weight of applied PU coating system [g]	Fibre volume content (FVC), absolute [%]
One step process	80	25	75/65	20	39	35
Two step process	80	25*	75/65*	20*	14	38
Reference	80	25	75/65	20	0	48

*Both curing steps

Methods

For the analysis of the CFRP components different methods have been used. For a descriptive illustration of the coating effect, photographs of the components surface are made with a defined light reflection under reproducible conditions using a fluorescent tube. For the analysis of the binding between coating layer and component as well as for measuring the resulting thickness of the coating layer micrographs are prepared, too.

For the analysis of the surface waviness of the components, a scanning system of the type MicroGlider® by FRT, Fries Research & Technology GmbH, Bergisch Gladbach, Germany, is used. This contact-free surface scanner is able to measure the shape, the roughness and the topography of a surface using the white-light interferometry as optical measuring principle. For the measurement of the transparent PU coating material, there is so far no experience about the accuracy of the measurement results using the Micro Glider® regarding the surface waviness of the components. Accordingly, the results of the optical measurements are currently validated at IKV using contact measurement techniques in additional investigations.

However, the proceeding for the determination of the surface waviness is shown in figure 2. By this, the topography along one profile line can be measured by the MicroGlider®. Based on the generated profile the determination of the roughness as well as the determination of the wave length and the wave depth (W_t) can be done. As a characteristic value for the waviness the average of the wave depths W_z is chosen.

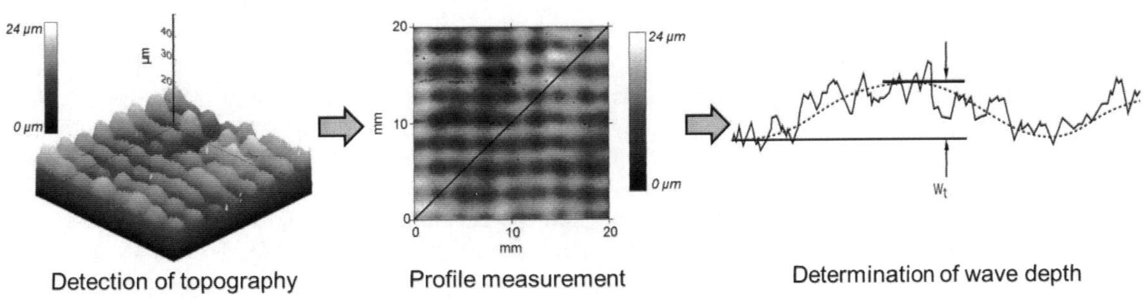

FIGURE 2. Proceeding for determination of the surface waviness of a fiber reinforced component.

RESULTS AND DISCUSSION

Figure 3 shows the photographs and the micrographs of the reference process without a coating and the two investigated process variations with a coating. The micrographs are exemplary for the process variants and have been analysed in different sections of each part.

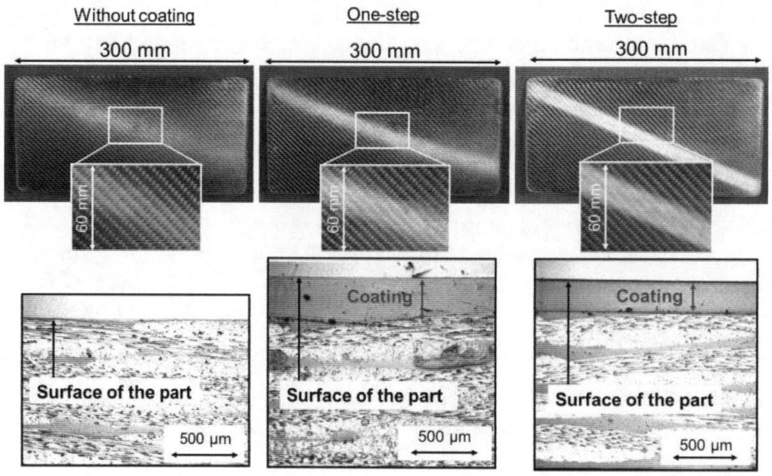

FIGURE 3. Photographs and micrographs of CFRP components without a coating (reference, left), produced in a one step process (middle) and produced in a two-step process (right).

In the photographs the sharpness of the light reflection shows the effect of the coating and the difference between the coating qualities for both process variants. On the photograph for the component without coating material the reflection of the light has no clear shape, especially at the edge. This diffused reflection results from a high surface waviness which can be already visually detected by the eye. Considering the photographs of the components with coating material the reflection is sharper for the one-step process and even sharper for the two-step process. Taking into account that a definition of a high surface quality can be also a subjective evaluation of the viewer, the two-step process shows the best results with regard to a better surface quality.

For the micrographs shown also in figure 3 no significant differences can be determined for the impregnation quality. Within all three process variants the components show a fully impregnated structure. For the process variants with the coating application the coating material is clearly visible in the cross section and the thickness can be determined. Therefore, table 2 shows the resulting thickness of the PU coating material for both process variations and associates them with the applied amount.

TABLE (2). Amount and resulting thickness of the applied PU coating system for both process variations.

Process variation	Weight of applied PU coating system	Thickness of the coating layer
One step process	39 g	0.35 mm
Two step process	14 g	0.27 mm

Analysing the interface between the coating material layer and the CFRP component, the CFRP component using the two-step process variant is already completely cured and has a planer surface before the coating material is applied. So, the dispersion is more homogeneous compared to an application on the impregnated preform (one-step process variant) and a reproducible layer thickness of the coating material can be achieved. Accordingly, a smaller amount of PU coating material can be used for the two-step process variant. For the one step process there is no straight interface between the coating material and the CFRP component. So, the given layer thickness is an average value. In return, the non-straight interface with different coating thicknesses using the one-step process variant can lead to a better adhesion between coating and component. Figure 4 shows the average weave depth W_z for two components per process variant as a characteristic value for the surface waviness. Using the optical measuring system, the presented values clearly show the effect of the coating material in comparison to the reference component as well as the difference between the two process variations. Therefore, the surface waviness compared to components without coating (2.32 µm) can be reduced in the one-step process variant up to 28 % (1.67 µm) and in the two-step process variant up to 78 % (0.71 µm). This coincides with the recognition of the analysis of the photographs in figure 3. It is notable that within the two-step process variant a higher reduction of the surface waviness can be achieved compared to the one-step process variant even if the coating thickness layer is lower. This results from the thermal shrinkage of the aromatic PU matrix material. Within the one-step process variant the shrinkage of the PU matrix material still affect the coating material, as within the two-step process the PU matrix is final cured and cannot affect the curing processing of the coating material. Therefore, a complete compensation of the surface waviness with the one-step process variant has not been reached yet.

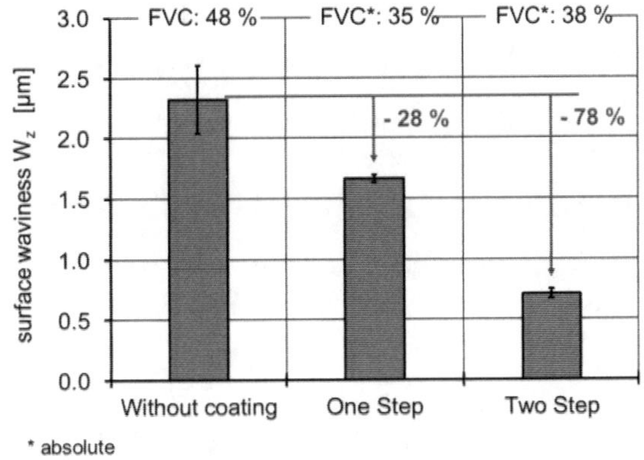

FIGURE 4. Surface waviness for the two process variations and a reference component without coating measured with the MicroGlider®.

CONCLUSION AND OUTLOOK

The presented results show that the wet pressing process with a separate preceding impregnation of the dry fibre preform enables to produce CFRP components with higher surface qualities by applying a PU coating material during the production process. The results regarding the surface waviness measured with an optical measuring system will be validated using contact measuring techniques. In further investigations, the one-step process will be optimised for the realisation of better surface qualities. The produced components will be also investigated regarding mechanical properties and UV stability. After the investigation on planar components, the results will be transferred to a three dimensional structure.

ACKNOWLEDGMENTS

The depicted research has been funded by the *Deutsche Forschungsgemeinschaft (DFG)*. We would like to extend our thanks to the DFG. The project is also supported by the partner companies Mubea Carbo Tech GmbH, (Salzburg, Austria) and BMW Group AG (München, Germany). We would also like to extend our thanks to the SGL Technologies GmbH (Meitingen, Germany) and the Ruehl Puromer GmbH (Friedrichsdorf, Germany) for supporting material.

REFERENCES

1. N.N., "Aston Martin | Vanquish | Design", URL: http://www.astonmartin.com/en/cars/the-new-vanquish/vanquish-design, 25.11.2015.
2. G. Fuchslocher, "Carbon-Splitterschutz von Eissmann und Porsche", URL: http://www.automobil-produktion.de/technik-produktion/fahrzeugtechnik/carbon-splitterschutz-von-eissmann-und-porsche-278.html, 25.11.2015.
3. H. Ferkel, H. Kurz, J.-W. Becke, "Faserverbundwerkstoffe als wirtschaftliche Option für die Großserie". CCeV Automotive Forum, Neckarsulm, Germany, 2010.
4. M. Schultz, M. Thienel, "Leicht und Modular – Ein CfK Türsystem für den intelligenten Materialmix", Conference Carbon Composites, Augsburg, Germany, 2012.
5. W. Reuter, "Hochleistungs-Faser-Kunststoff-Verbunde mit Class-A-Oberflächenqualität für den Einsatz in der Fahrzeugaußenhaut", Ph.D. Thesis, University Kaiserslautern, 2001.
6. W. Rogers, C. Hoppins, Z. Gombos and J. Summerscales, *Journal of Cleaner Production* **70**, 282-291 (2014).
7. M.L. Herring, J.I. Mardel and B.L. Fox, *Journal of Materials Processing Technology* **210**, 926-940 (2010).
8. L. P. Canal, M. Benavente, M. Hausmann, and V. Michaud, *Composites: Part A* **78**, 264–273 (2015).
9. Ch. Hopmann, W. Michaeli, R. Bastian, *GAK : Gummi, Fasern, Kunststoffe* **65,** 45-53 (2012).
10. Ch. Hopmann, R. Bastian, R., "Optimierung der Oberflächenqualität von Außenhautbauteilen aus kohlenstofffaserverstärkten Kunststoffen", Conference Carbon Composites, Bindlach-Bayreuth, Germany, 2013
11. T. Hoppe, „Bearbeitungs- und Lackierkonzept für CFK-Automobilbauteile im Sichtbereich". CCeV Automotive Forum, Ingolstadt, Germany, 2011.
12. J. Renkl, S. Schmidhuber and E. Fries, *Kunststoffe international* **03,** 52-55 (2014).
13. Ch. Hopmann, Ch. and C. Karatzias, "New compression molding technology for high volume production of continuous fiber reinforced structural composites", 28th Annual Meeting of the Polyer Processing Society, Pattaya, Thailand, 2012.

Low-Field RheoNMR: Newly Developed Combination of Rheology and Time Domain-NMR to Correlate Mechanical Properties with Molecular Dynamics in Polymer Melts

Volker Räntzsch[a], Mürüvvet Begüm Özen[a], Karl-Friedrich Ratzsch[a,b], Gisela Guthausen[c] and Manfred Wilhelm[a*]

[a]Karlsruhe Institute of Technology (KIT), ITCP, Karlsruhe, Germany
[b]University of Freiburg, Freiburg Materials Research Center (FMF), Freiburg, Germany
[c]Karlsruhe Institute of Technology (KIT), Pro^{2}NMR, MVM, IBG, Karlsruhe, Germany
Manfred.Wilhelm@kit.edu

Abstract. NMR is a useful technique for characterization of the microscopic dynamics and structure of polymeric material, while rheology provides access to the macroscopic mechanical properties of polymer melts. To achieve greater insight into the interplay of these two domains, especially with regard to the effects of shear, it is desirable to combine these two methods in one setup. Here we present a low-field RheoNMR set-up based on a portable TD-NMR unit (ν_L = 30 MHz) that was integrated into a commercial high-end strain-controlled rheometer. This unique combination can simultaneously conduct a full rheological shear characterization (G', G'', $|\eta^*|$, FT-Rheology: $I_{3/1}$, Q_0) while monitoring molecular dynamics *in-situ* via ^{1}H TD-NMR. To display the possibilities of this new technique, studies on the crystallization of isotactic polypropylene as a function of the crystallization temperature are presented.

Keywords: RheoNMR, hyphenated methods, low-field NMR, polymer crystallization, polypropylene.
PACS: 83.85.Fg

INTRODUCTION

By combining rheometry with other characterization techniques, measurements of molecular properties such as structure, orientation or dynamics can be performed simultaneously during rheological experiments on soft matter. Different hyphenated set-ups based on high-end rheometers have been designed in the past, e.g. RheoSAXS, RheoIR or RheoOptics where these have mainly focused on the microscopic structure [1, 2]. The combination of rheometry with Time Domain (TD)-NMR measures the molecular dynamics instead, which permits for studies into how dynamic molecular processes are translated into macroscopic flow behavior. Rheometry generates shear deformations on length scales of millimeters and time scales of seconds, whereas TD-NMR provides information on length and time scales that are several orders of magnitude smaller.

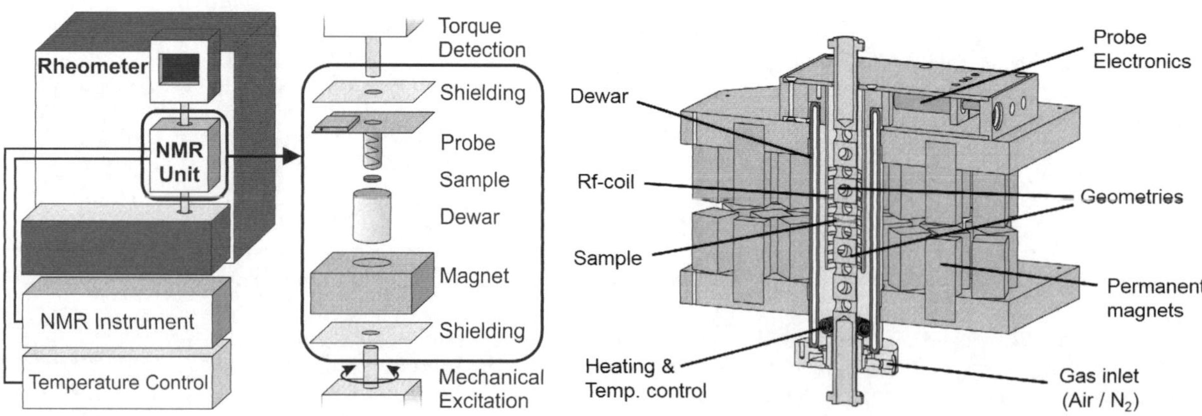

FIGURE 1. Low-Field RheoNMR: Portable TD-NMR unit (30 MHz) implemented in a strain-controlled rheometer.

Technical Details

The newly developed low-field RheoNMR setup is based on a commercial high-end strain-controlled rheometer (TA/Rheometrics ARES) which can perform a large variety of rheological tests including small and large amplitude oscillatory shear (SAOS, LAOS, FT-rheology) [3-5], steady shear and complex shear experiments. The TD-NMR unit consists of a Halbach array of 92 permanent magnets that permits its facile integration into a rheometer due to its light weight, small dimensions and low stray field (Fig. 1) [6,7]. A silvered dewar was employed to achieve good thermal insulation between the sample space and permanent magnets, which can be controlled from -15 – +210 °C using a N_2-based convective heating system. ^{1}H NMR experiments require non-conducting and proton-free geometries which were made from a ceramic (13 mm cone-plate and parallel-plate). A broadband NMR probe (tank circuit: 33 x 16 mm solenoid coil and two capacitors) exhibits a dead time of 12 μs and pulse lengths of 3.5 μs (90°) and 6.2 μs (180°), respectively. Pulsing, data acquisition and temperature control are carried out by a Bruker *the minispec* electronic unit and a Bruker *Variable Temperature* unit. Such a portable TD-NMR unit was also used for other purposes, e.g. in combination with an extruder to monitor flow instabilities and polymer crystallization simultaneously by a pressure-sensitive detection system and ^{1}H TD-NMR [8].

Applications

Low-field RheoNMR can be employed to study different types of soft matter including a quantitative measurement of the composition of crystallizing polymers and multiphase systems during the application of linear and non-linear shear deformations (e.g. shear-induced crystallization). In general, processes during which microscopic molecular dynamics as well as macroscopic mechanical properties change significantly are well suited for studies with LF-RheoNMR. In the case of polymer crystallization, the build-up of crystalline domains in a polymer melt has a major effect on the flow behavior, which can be treated as a special case of a suspension of rigid, potentially interacting particles in a viscoelastic fluid. Studies from Winter *et al.* [9] showed a gelation-like transition during polymer crystallization from the melt that has not yet been fully understood. Due to the extremely high temperature dependence of polymer crystallization, the correlation of data from separate methods is practically impossible [10]. With low-field RheoNMR, the polymer crystallinity and mechanical properties are determined simultaneously allowing for a substantially better correlation of the data.

RESULTS AND DISCUSSION

Isothermal crystallization of isotactic polypropylene (PP Isplen PP070G2M, M_w = 294 kg/mol, *PDI* = 5, Repsol, Spain) from the melt was investigated to identify relations between the flow properties and evolving crystallinity during solidification. A multi-echo-based TD-NMR sequence (Mixed Magic Sandwich Echo/CPMG) [11] was employed to monitor molecular dynamics which undergo a transition from a mobile liquid-like to a rigid solid-like behavior during crystallization. Based on the change in molecular dynamics, the polymer crystallinity was obtained using an indirect data analysis method. A bi-exponential was fitted to the CPMG data and the back-extrapolated intensities were compared with the signal of the super-cooled melt at $t + \Delta t = 0$ (Δt = time needed for achieving a stable and homogeneous crystallization temperature):

$$X_c(t) = 1 - \frac{I_{CPMG}(t)}{I_{CPMG}(t + \Delta t = 0)} \tag{1}$$

As the CPMG multi echo sequence is selective for relatively mobile protons, the polymer crystallinity is determined indirectly by the loss of CPMG intensity during solidification. The mass crystallinity was converted into volume crystallinity using the densities of the fully amorphous and fully crystalline material (ρ_c = 0.95 g/ml) [12]:

$$\phi_c(t) = \frac{\rho_a}{\dfrac{\rho_c}{X_c(t)} + (\rho_a - \rho_c)} \tag{2}$$

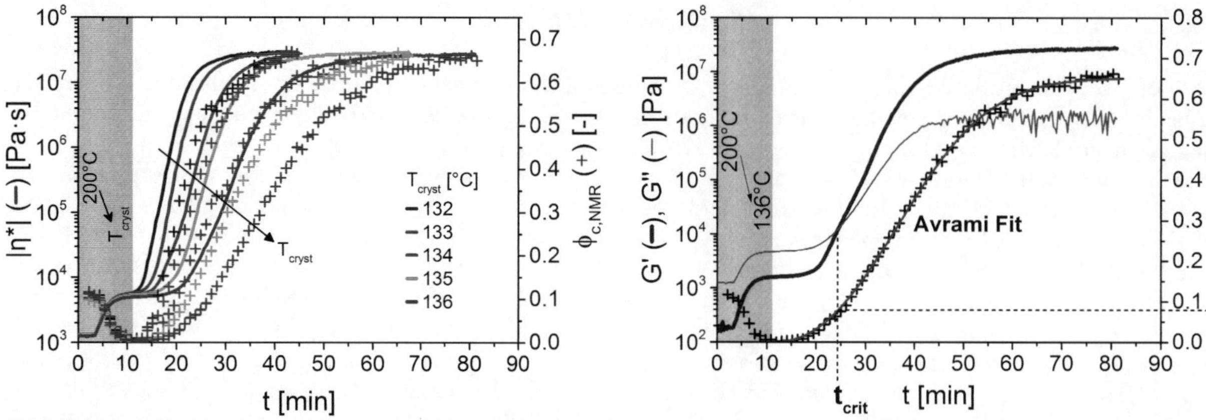

FIGURE 2. Left: Simultaneously acquired RheoNMR data (absolute complex viscosity $|\eta^*|$ and crystallinity $\phi_{c,\mathrm{NMR}}$) during the isothermal crystallization of isotactic polypropylene (i-PP) from the melt at different crystallization temperatures. The rheological test conditions were $\gamma_0 = 2$ %, $\omega_0 = 1$ rad/s using a cone-plate geometry and the TA ARES auto strain mode. The polymer crystallinity was obtained using an indirect data analysis method based on a Mixed MSE-CPMG NMR pulse sequence (eq. 1) [11]. Right: For every crystallinity curve an Avrami fit was performed to determine the kinetic parameters and to facilitate the correlation of rheological parameters with polymer crystallinity from TD-NMR. The critical crystallinities $\phi_{c,\mathrm{crit}}$ at $G' = G''$ were determined to compare the influence of crystallization temperature on the initial flow behavior during crystallization.

In Fig. 2, experimental results are depicted for the crystallization of i-PP at various crystallization temperatures. The sample was kept at 200 °C for 15 min (3 min shown) to erase the thermal history and then rapidly cooled to the respective crystallization temperature which resulted in an initial increase in the viscosity. At $t = 12$ min all viscosity curves reached a plateau value of ~5 x 10^3 Pa·s and the crystallinity curves showed a minimum which indicated a stable and homogeneous temperature in the sample (Δt = 12 min). The crystallization set in and the viscosity / crystallinity increased at a rate that was dependent on the crystallization temperature (higher temperature – slower crystallization). All crystallinity curves were fitted using an Avrami model approach to obtain the kinetic parameters and facilitate the correlation of the rheological data with the crystallinity:

$$\phi_c(t) = 1 - \exp\left\{-\left[K(t - \Delta t)\right]^n\right\} \quad \text{(Avrami model)} \tag{3}$$

An Avrami fit for T_{cryst} = 136 °C is shown on the right of Fig. 2. The model describes the entire curve very well which is an indication that there was no significant secondary crystallization and allows for a comparison of the kinetic parameters K and n. The rate parameter K decreased with increasing crystallization temperature from 0.09 to 0.03 min^{-1}, whereas the exponent n increased moderately from 2 to 2.5 (left side of Fig. 3).

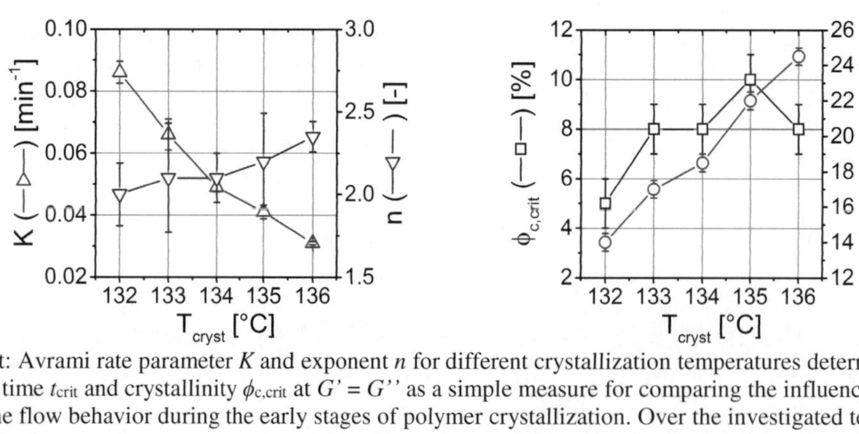

FIGURE 3. Left: Avrami rate parameter K and exponent n for different crystallization temperatures determined by RheoNMR. Right: Critical time t_{crit} and crystallinity $\phi_{c,\mathrm{crit}}$ at $G' = G''$ as a simple measure for comparing the influence of crystallization temperature on the flow behavior during the early stages of polymer crystallization. Over the investigated temperature interval, a typical crystallization temperature – growth rate dependence is seen: higher temperature leads to slower crystallization. The critical crystallinity $\phi_{c,\mathrm{crit}}$ stays rather constant which is probably due to a similar number density and size distribution of spherulites for the different temperatures investigated.

As a second method to compare the crystallinities, values for $\phi_{c,crit}$ at $G' = G''$ were evaluated, which are straightforward to compare as long as the same rheological test conditions (γ_0, ω_0) were applied. On the right of Fig. 3, the time t_{crit} exhibits a monotonic increase with crystallization temperature which follows the crystallization temperature – growth rate dependence as observed for the Avrami rate K. The critical crystallinity $\phi_{c,crit}$ shows no clear temperature-dependent behavior, which is possibly due to the small temperature interval investigated. As seen in optical microscopy, low crystallization temperatures lead to numerous, but small, spherulites which should have a larger impact on the flow behavior during the early stages of crystallization relative to the situation where there are few, but large, spherulites at high crystallization temperatures.

A closer look at the influence of the crystallization temperature on the flow behavior during crystallization can be made by plotting a rheological parameter, e.g. the absolute complex viscosity $|\eta^*|$, as a function of the crystallinity $\phi_{c,NMR}$, thereby removing the effect of temperature on the time dependence of polymer crystallization (Fig. 4). For $0.1 < \phi_{c,NMR} < 0.2$, a systematically higher viscosity is seen at lower crystallization temperatures before the curves collapse onto one line with the exception of the curve for the highest crystallization temperature. In the latter case, an imperfect filling of the geometry might have led to lower measured viscosities relative to the other experiments. For $\phi_{c,NMR} > 0.2$, a very strong increase in $|\eta^*|$ is seen, which is qualitatively similar to other systems such as suspensions, gel-formers or filled polymer systems and will be further investigated using different polymers, additives and crystallization conditions.

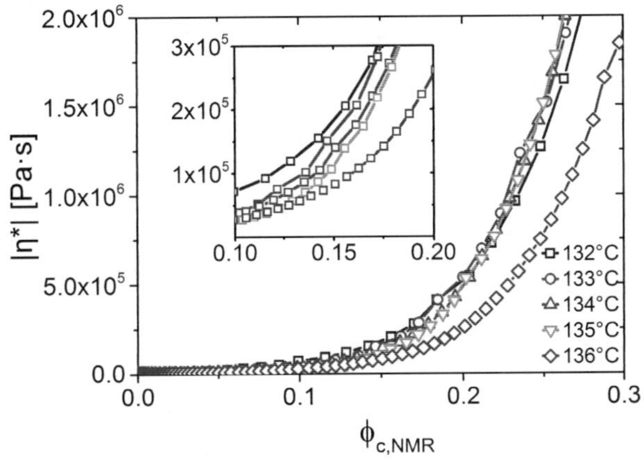

FIGURE 4. Absolute complex viscosity $|\eta^*|$ vs. crystallinity $\phi_{c,NMR}$. By focusing on the crystallinity dependence of the flow behavior, a close examination can be made on the influence of number density and spherulite size as set by the chosen crystallization temperature on the resulting viscosity. This removes the effect of temperature on the time dependence of polymer crystallization. A very strong increase in $|\eta^*|$ is seen for low values of $\phi_{c,NMR}$, i.e. at an early stage of crystallization. This gel-like behavior is qualitatively similar, but quantitatively stronger, to what is observed for e.g. suspensions ("jamming").

CONCLUSION

Low-field RheoNMR based on a portable TD-NMR unit and a commercial high-end strain-controlled rheometer is a novel characterization method that can simultaneously make a full rheological shear characterization while monitoring molecular dynamics *in-situ* via ^{1}H TD-NMR. To explore the possibilities of this new technique, studies on the crystallization of isotactic polypropylene were presented. The build-up of crystallinity was measured using a multi-echo-based TD-NMR sequence in combination with an indirect data analysis method. An Avrami analysis was performed to determine kinetic parameters (K, n) which follow the trends commonly seen for crystallizing polymers (e.g. slower crystallization for higher crystallization temperature). The crystallinity $\phi_{c,crit}$ at $G' = G''$ was independent of the crystallization temperature interval investigated. By correlating the absolute complex viscosity $|\eta^*|$ with the crystallinity $\phi_{c,NMR}$, a detailed investigation was made with respect to the influence of crystallization temperature on the flow behavior. In the intermediate stage ($0.1 < \phi_{c,NMR} < 0.2$), the viscosity was found to be higher for lower crystallization temperatures, indicating a potential effect of the nucleation density / spherulite size. By further studying these phenomena using different polymers with inherently different spherulite sizes, additives and crystallization conditions, a better picture of the flow behavior during polymer crystallization will be achieved. Low-field RheoNMR could furthermore be employed to study the crystallization of low molecular weight substances, gelation or curing reactions and to investigate how the application of shear can affect the nature of these processes.

ACKNOWLEDGMENTS

The project was funded by Deutsche Forschungsgemeinschaft (DFG, SPP1369-B10, Pro²NMR) and by Bundesministerium für Bildung und Forschung (BMBF, 'FUNgraphen'). The authors thank Dr. J. Kübel for proofreading the manuscript and Prof. A. Müller (University of the Basque Country, Donostia-San Sebastián, Spain) for providing the samples and fruitful discussions.

REFERENCES

1. T. Meins, K. Hyun, N. Dingenouts, M. Fotouhi Ardakani, B. Struth and M. Wilhelm, *Macromolecules* **45**, 455-472 (2012).
2. M. Boulet-Audet, F. Vollrath and C. Holland, *Phys. Chem. Chem. Phys.* **13**, 3979-3984 (2011).
3. M. Wilhelm, *Macromol. Mater. Eng.* **287**, 83-105 (2002).
4. K. Hyun and M. Wilhelm, *Macromolecules* **42**, 411-422 (2009).
5. M. Cziep, M. Abbasi, M. Heck, L. Arens and M. Wilhelm, *Macromolecules* (2016) submitted.
6. S. Kahle, W. Nussbaum, M. Hehn, H. P. Raich, M. Wilhelm and P. Blümler, *Kaut. Gummi Kunstst.* **61**, 92-94 (2008).
7. V. Räntzsch, M. Wilhelm and G. Guthausen, *Magn. Reson. Chem.* (2015). doi:10.1002/mrc.4219
8. K. F. Ratzsch, R. Kádár, I. F. Naue and M. Wilhelm, *Macromol. Mater. Eng.* **298**, 1124-1132 (2013).
9. N. Pogodina and H. H. Winter, *Macromolecules* **31**, 8164-8172 (1998).
10. G. Lamberti, G. W. M. Peters and G. Titomanlio, *Int. Polym. Proc.* **22**, 303-310 (2007).
11. A. Maus, C. Hertlein, and K. Saalwächter, *Macromol. Chem. Phys.* **207**, 1150-1158 (2006).
12. D. W. van Krevelen and K. te Nijenhuis, *Properties of Polymers*, 4. Ed., Amsterdam: Elsevier, 2009, p. 86.

Elongational Rheology of Glass-fiber and Natural-fiber-reinforced Polypropylenes with a Novel Online Rheometer

Hans-Jürgen Luger [a,c*], Thomas Köpplmayr[a,b], Lukas Sobczak[c], Andreas Haider[c] and Jürgen Miethlinger [a]

[a] *Institute of Polymer Extrusion and Compounding,*
Johannes Kepler University Linz, Altenberger Str. 69, 4040 Linz, Austria
[b] *Profactor GmbH, Im Stadtgut A2, 4407 Steyr-Gleink, Austria*
[c] *Kompetenzzentrum Holz GmbH, Altenberger Str. 69, 4040, Linz, Austria*

hans-juergen.luger@jku.at

Abstract. The aim of the present work was to study the influence of fiber content and fiber length distribution on the shear and elongational viscosity of polypropylene compounds measured with a novel online rheometer. The recently developed device consists of two slit sections and a hyperbolic contraction part in between, which allows for continuous monitoring of both shear and elongational viscosity during an extrusion process. Due to the favorable design of the hyperbolic contraction, pressure transducers can be incorporated directly into the flow channel, which prevents material accumulation in pressure holes. A polypropylene homopolymer intended for thermoformed packaging applications was compounded with different fiber content of either glass or cellulose fibers. While the measured shear viscosity only changed slightly, the elongational viscosity was found to be sensitive to the fiber content. Furthermore, the variation of fiber length distribution in a glass-fiber reinforced polypropylene random copolymer, intended for pipe applications, was detected by the elongational viscosity. Therefore the developed elongational rheometer is recommended as online process rheometer for quality control as well as for R&D. These results were confirmed by other methods such as Cogswell analysis of High Pressure Capillary Rheometer (HPCR) data or measurement on a wedge-shaped slit die. In addition, fiber length distributions of selected compounds were determined to study the fiber shortening during processing before the rheological measurements.

Keywords: Elongational rheometry, hyperbolic slit die, glass fiber, cellulose fiber, compounding
PACS: 83.80.-k; 83.80.Ab; 83.80.Mc; 83.50.Ax; 83.85.Vb

INTRODUCTION

Over the last decades fiber reinforcement of thermoplastics and thermosets has becomes increasingly important, since the addition of fibers leads to improved mechanical and thermal material properties. In recent years many efforts have been made to replace the commonly used glass fibers with fibers from renewable resources, such as wood or cellulose fibers. On the one hand, the processing must ensure sufficient dispersion of the fibers, i.e. break-up of fiber agglomerates, and on the other hand should involve minimal fiber breakage and shortening. Several studies [1,2,3] have shown that the average fiber length usually decreases in processing.

In contrast to geometrically isotropic fillers, the rheology of fiber-filled polymer melts becomes quite complex due to fiber-fiber, wall-fiber and polymer-fiber interaction. Köpplmayr et. al. [4] studied the influence of pre-shearing on the shear viscosity measured in a parallel plate rheometer. Crowson and Folkes [5] investigated the effect of fiber concentration, fiber length and temperature on shear viscosity and die swell. Both fiber length and volume fraction hardly affected viscosity at high shear rates, but were found to have a considerable effect at low shear rates, which was attributed to fiber clustering.

While shear viscosity is defined as the resistance to flow due to a force which is perpendicular to the nodal of the plane on which the force acts, elongational viscosity is the resistance to flow due to a force which is parallel to the nodal of the plane on which the force acts. For the measurement of the elongational properties several types of rheometers are commercially available, such as Sentmanat Extensional Rheometer (SER) [6], RME (Rheometric Scientific) [7] or Rheotens instrument (Göttfert) [8]. In addition, analysis of the entrance flow and entrance pressure drop in a high pressure capillary rheometer (HCPR) can be used. As an alternative to the rheometers mentioned above, Haake X-Die [9] was developed to measure shear and elongational viscosity with one die. The advantage of this method is that the elongational viscosity can be calculated from data obtained with standard measurement equipment directly in the production process. Köpplmayr et al. [4] and Aigner et. al. [1] showed that measurements of elongational viscosity are preferable when determining the influence of the aspect ratio of glass fibers in polypropylene matrix. Recently, we reported on a novel rheometer [10], which allows for shear viscosity measurements in two slit sections and elongational viscosity measurements in a hyperbolic section at constant elongation rates. Moreover, due to the favorable design of the hyperbolic contraction, pressure transducers can be incorporated directly into the flow channel, which prevents material accumulation in pressure holes.

EXPERIMENTAL

Materials

A Borealis HB600TF polypropylene homopolymer (MFR 2.0 g/10 min, 230°C/2.16 kg) was used as a base polymer and compounds with different weight portions of either glass or cellulose fibers were produced:

- PP-GF10 and PP-GF20: 10 and 20 wt% of glass fibers (diameter 13 µm, max. fiber length: 4.5 mm),
- PP-NF06 and PP-NF13: 6 and 13 wt% of cellulose fibers (diameter 10.5 µm, av. fiber length: 0.39 mm).

The compounds were prepared by means of a Brabender DSE20 laboratory twin-screw extruder with a length of 42 times the diameter and two vacuum venting units in combination with underwater pelletizing.

In addition, two compounds with a Borealis RA130E polypropylene random copolymer (MFR 0.3 g/10 min, 230°C/2.16 kg) containing both short (max. length: 1.5 mm) and long (max. length: 4.5 mm) glass fibers were prepared by means of a Coperion ZSK 70 industrial twin-screw extruder with a length of 42 times the diameter and three venting units (two atmospheric and one under vacuum) in combination with underwater pelletizing at a throughput of 900 kg/h and screw speed of 400 rpm. The weight proportion of glass fibers was 24% in each compound:

- PP-GF24 70:30: 70% long and 30% short fibers,
- PP-GF24 20:80: 20% long and 80% short fibers.

Shear rheometry

The shear viscosity was measured using a Thermo Haake Rheomex system, which consists of a 19 mm single screw extruder (L/D 33) equipped with a 2.4 cm³ melt pump, a by-pass valve, and a slit die of defined gap size of 0.8 mm. The extruder provides the polymer melt and the melt pump guarantees a defined melt flow for measuring the viscosity curve based on the pressure drop along the die. In addition, a high pressure capillary rheometer Rheograph 25 (Göttfert Werkstoff-Prüfmaschinen GmbH) was used, which also allows for evaluation of the elongational viscosity according to Cogswell [11].

Elongational rheometry

For our experimental studies we used both the wedge-shaped slit die (X-Die) [9] and the novel hyperbolic die [10] in combination with four pressure transducers and one temperature sensor. The elongational dies used in this study were mounted on the Thermo Haake Rheomex system and consist of two slit sections (pure shear) and a converging section in between (elongation and shear). In contrast to the wedge-shaped slit die, the hyperbolic elongational slit die provides a uniform elongation rate (Eq. (1)) along the contraction part.

$$\dot{\varepsilon} = \frac{\dot{V}}{L} \cdot \left(\frac{1}{H_2 \cdot W_2} - \frac{1}{H_1 \cdot W_1} \right) \tag{1}$$

Shear viscosities were determined by the pressure drops in the two slit sections and power law parameters were calculated pointwise. All measurements were conducted at a constant processing temperature of 240°C. The pressure drop within the converging section $\Delta p_{\text{measured}}$ consists of three contributions, the pressure drop Δp_{shear} due to viscous shear flow, the pressure drop due to viscous elongational flow $\Delta p_{\text{elong.}}$ and the pressure drop dedicated to the elasticity of the melt $\Delta p_{\text{elast.}}$:

$$\Delta p_{\text{measured}} = \Delta p_{\text{shear}} + \underbrace{\Delta p_{\text{elong.}} + \Delta p_{\text{elast.}}}_{\Delta p_{\text{E}}} . \tag{2}$$

The viscous shear part within the converging section with the width *W(z)* and height *H(z)* can be calculated by:

$$\frac{\Delta p_{\text{shear}}}{z} = \left(\frac{2^{m+1} \cdot (m+2) \cdot \dot{V}}{W(z) \cdot H(z)^{m+2} \cdot \Phi} \right)^{1/m}, \tag{3}$$

which is based on the assumption that the flowing material exhibits power-law behavior (fluidity Φ and flow index m). The elongational stress σ is equal to the entrance pressure drop Δp_E and can be obtained by knowing the shear part in the converging section with the assumption that elastic properties are neglected. The elongational viscosity is then given by:

$$\eta_{\text{elong.}} = \frac{\sigma}{\dot{\varepsilon}} = \frac{\Delta p_{\text{E}}}{\dot{\varepsilon}} = \frac{\Delta p_{\text{measured}} - \Delta p_{\text{shear}}}{\dot{\varepsilon}}. \tag{4}$$

Detection of the fiber length distribution

In order to measure the fiber length distribution the fibers were extracted from samples through incineration or dissolving in the case of glass fibers or cellulose fibers, respectively. Afterwards, water-fiber suspensions were prepared and analyzed with the image processing tool FASEP [12], which provides a complete statistical evaluation of the fiber length distribution.

RESULTS AND DISCUSSION

Fiber length distributions

The fiber length was determined by means of FASEP for the compounds shown in Tab. 1 at the given positions. For the evaluation only fibers longer than 50 μm were considered. For the glass fibers in the polypropylene melts, the two processing steps lead to substantial fiber shortening compared to the manufacturer´s specifications. The cellulose fibers were hardly shortened. In order to investigate the effect of the melt pump used to control the volume flow rate in the rheological characterization, samples were taken before and after the melt pump. Those showed only small differences and therefore shortening of cellulose fibers in the melt pump can be neglected. The measurements of the compounds with 24% of fibers with different lengths show distinct differences in the fiber length distributions after compounding.

TABLE (1). Fiber properties of different polypropylene glass and cellulose fiber compounds used in this study.

Sample		**PP-GF20** before melt pump	**PP-NF13** before melt pump	**PP-NF13** after melt pump	**PP-GF24-70:30** compound	**PP-GF24-20:80** compound
weighted av. fiber length	μm	504.8	394.4	378.1	487.8	329.5
av. fiber length	μm	340.5	272.7	262.4	252.7	186.7
max. fiber length	μm	1391.4	1445.0	1154.9	1600.7	1382.5
total number of measured fibers		2527	3650	3619	3185	3196

Shear viscosity

The shear viscosity curves of all investigated materials are shown in Figs. 1a and 1b. The measurements with the slit capillary rheometer showed no difference in the shear viscosity of the PP-HB600TF compounds with different weight proportions of glass or cellulose fibers. Moreover, the PP-RA130E glass fiber compounds could not be distinguished by means of shear rheology with different characterization methods (HCPR, wedge-shaped slit die and hyperbolic slit die) and all methods yield comparable results. Since the non-isotropic fibers are oriented in flow direction, the difference in the fiber content, type and length distribution does not influence the flow behavior in pressure-driven shear flow.

Elongational viscosity

In contrast to shear rheology, elongational measurements provide an efficient method to distinguish and characterize fiber-reinforced polymers (Figs. 2-3). As expected, the elongational viscosity increases with increasing fiber loading (Fig. 2), which was detected by both the wedge-shaped slit die and the hyperbolic slit die. The novel hyperbolic die provides a well-defined, but lower elongation rate for the same volume flow rate compared to the wedge-shaped slit die. Furthermore, the hyperbolic die provides a lower total Hencky strain, which results in a lower entrance pressure drop and, thus, a lower level of the elongational viscosity. Our results show that determining the elongational flow properties of fiber-reinforced polymer melts with the novel hyperbolic die at low and constant elongation rates is beneficial in order to detect differences in fiber loadings.

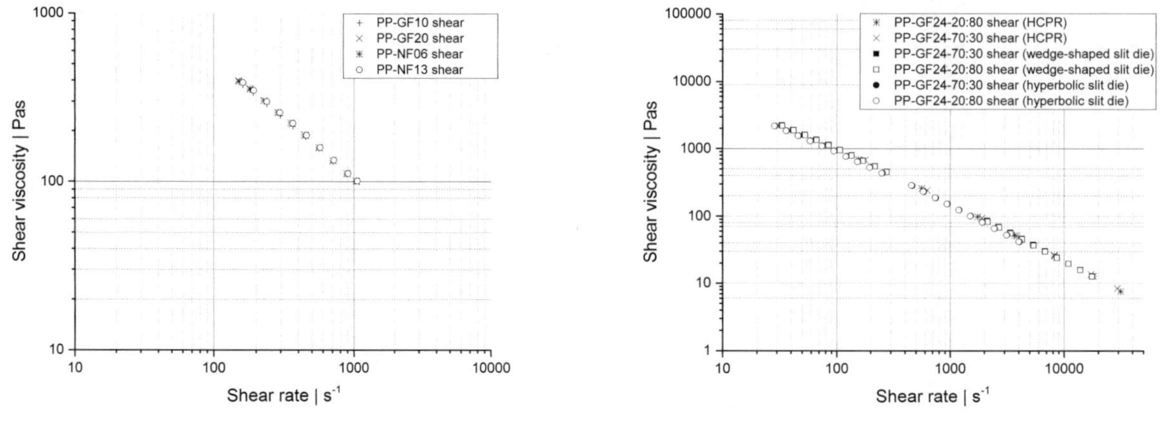

(a) (b)

FIGURE 1. Comparison of the shear viscosity curves at 240°C of (a) PP-HB600TF compounds with different weight portions of short glass or cellulose fibers measured with the capillary slit die and (b) PP-RA130E glass fiber compounds with different fiber length distributions using different characterization methods.

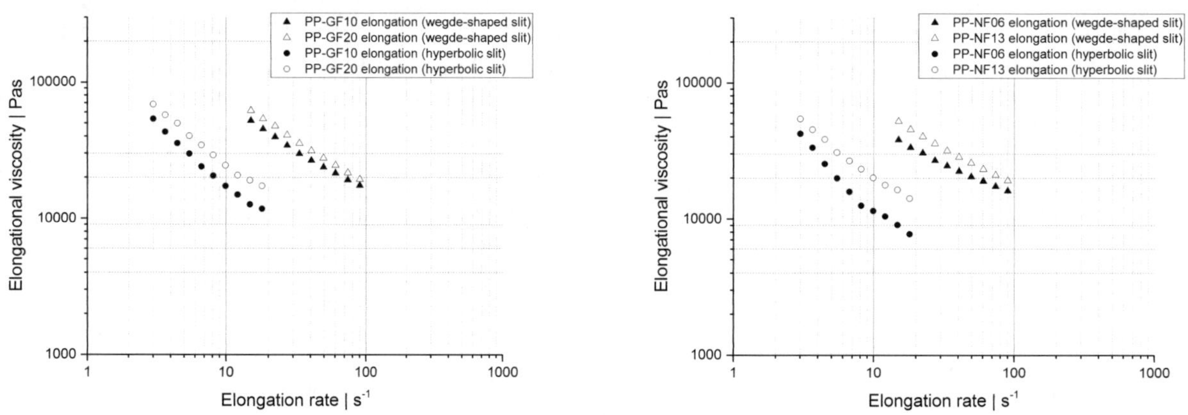

(a) (b)

FIGURE 2. Comparison of the elongational viscosity curves at 240°C of the PP-HB600TF compounds measured with the wedge-shaped slit die and the hyperbolic slit die: (a) shows the influence of the increase in glass fiber content from 10 to 20 wt%, (b) the influence of the increase in cellulose fiber content from 6 to 13 wt%.

Fig. 3 shows that the elongational viscosity is sensitive to the fiber length distribution, even when the same quantity of fibers (24 wt% in total) is used. The gradient of the viscosity curves strongly depends on the elongation rate and, in case of the hyperbolic die, also on the fiber length distribution. This could be due to the formation and destruction of agglomerates at low and high deformation rates, respectively.

CONCLUSION

We used several characterization techniques to measure shear and elongational viscosity of different polypropylene glass and cellulose fiber compounds. The shear viscosity was determined with both an online extrusion slit capillary rheometer and a HCPR, and was found to be insensitive to different fiber loadings and fiber length distributions. We have shown that elongational rheology with online extrusion rheometers (wedge-shaped slit and hyperbolic slit) is as efficient in characterization of fiber-reinforced polypropylenes as the offline method HCPR. The results obtained with the novel hyperbolic slit die are very promising, because a constant elongation rate along the converging flow section can be realized. This makes elongational rheology a valid tool for online quality control in compounding. Future work will focus on characterization of different polymers and compounds with the novel rheometer, as well as long-time monitoring of polymer compound quality.

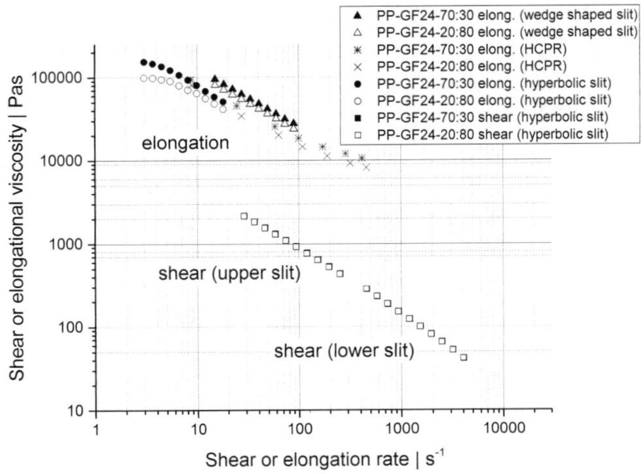

FIGURE 3. Comparison of the elongational and shear viscosity curves at 240°C of the PP-RA130E compounds measured with different characterization methods.

ACKNOWLEDGMENTS

Financial support by the Austrian Research Promotion Agency (FFG) within a national research project ("Polymer-Compounds mit erhöhter Homogenität und Performance durch gezielten Einsatz von Dehnströmungen", project number 843507) is kindly appreciated. Part of this work was performed within the WOOD K-plus project, which is funded Austrian Government and the Provincial Government of Upper Austria within the COMET program.

REFERENCES

1. M. Aigner, D. Salaberger, T. Köpplmayr, B. Heise, S. E. Schausberger, A. Buchsbaum, D. Stifter, and J. Miethlinger, "Determining the orientation, distributions and concentration of glass fibers in polymer matrix using X-Ray computed tomography and optical coherence tomography images," in *Conference on Industrial Computed Tomography - Non destructive Testing, 3D Materials Characterisation und Dimensional Measurement*: Shaker Verlag, 2014, pp. 395–402.
2. B. Fisa, "Mechanical Degradation of Glass Fibers During Compounding with Polypropylene," *Polym. Comp,* vol. 6, pp. 232–241, 1985.
3. V. B. Gupta, R. K. Mittal, P. K. Sharma, G. Mennig, and J. Wolters, "Some Studies on Glass Fiber-Reinforced Polypropylene. Part I: Reduction in Fiber Length During Processing," *Polym. Comp,* vol. 10, pp. 8–15, 1989.
4. T. Köpplmayr, I. Milosavljevic, M. Aigner, R. Hasslacher, B. Plank, D. Salaberger, and J. Miethlinger, "Influence of fiber orientation and length distribution on the rheological characterization of glass-fiber-filled polypropylene," *Polymer Testing*, vol. 32, no. 3, pp. 535–544, 2013.
5. R. J. Crowson and M. J. Folkes, "Rheology of Short Glass Fiber-Reinforced Thermoplastics and its Application to Injection Molding: II. The Effect of Material Parameters," *Polym. Eng. Sci,* vol. 20, pp. 934–940, 1980.
6. M. L. Sentmanat, "Miniature universal testing platform: from extensional melt rheology to solid-state deformation behavior," *Rheol Acta*, vol. 43, no. 6, pp. 657–669, 2004.
7. J. Meissner and J. Hostettler, "A new elongational rheometer for polymer melts and other highly viscoelastic liquids," *Rheola Acta*, vol. 33, no. 1, pp. 1–21, 1994.
8. M. H. Wagner, B. Collignon, and J. Verbeke, "Rheotens-mastercurves and elongational viscosity of polymer melts," *Rheol Acta*, vol. 35, no. 2, pp. 117–126, 1996.
9. A. Ruthard, G. Chaidron, and P. Reinheimer, "New die design for a global and rapid rheological characterization of polymers, in the lab and on-line," in *Proceedings PPS-19*, 2003.
10. T. Köpplmayr, H.-J. Luger, I. Burzic, M. G. Battisti, L. Perko, W. Friesenbichler, and J. Miethlinger, "A novel online rheometer for elongational viscosity measurement of polymer melts," *Polymer Testing*, vol. 50, pp. 208–215, 2016.
11. F. N. Cogswell, "Converging flow of polymer melts in extrusion dies," *Polym. Eng. Sci,* vol. 12, no. 1, pp. 64–73, 1972.
12. M. R. Hartwich, N. Höhn, H. Mayr, K. Sandau, and R. Stengler, "FASEP ultra-automated analysis of fibre length distribution in glass-fibre-reinforced products," in *SPIE Europe Optical Metrology*: SPIE, 2009, p. 738921.

Measuring the Effects of Filler Variations on Compound Viscoelastic Properties as Measured by the Capillary Rheometer and the RPA

John S. Dick[a], Edward Norton[b], and Tianxiang Xue[c]

[a]3030 Gilchrist Road, Akron, OH 44305 USA john.dick@alpha-technologies.com [b]3030 Gilchrist Road, Akron OH 44305 [c]Shanghai Roper Industries Trading Company, LTD, Room 601, 567 Lan Gao Road, Shanghai, 200333 China

Die swell and other viscoelastic characteristics imparted to a rubber compound by different elastomers and reinforcing fillers are very relevant to compound performance in extrusion, calendering, molding, and other factory operations.

In this study, a capillary rheometer was used to measure the die swell and other rheological behavior at different shear rates. Model rubber compounds were prepared and tested. Processability characteristics were also measured for these mixed stocks using the RPA with EDR. Comparisons were made of the performance behavior for these compounds. Statistical analyses were performed on *running die swell* and *relaxed die swell* vs. tan delta and G under a wide range of applied frequencies and strain amplitudes with the Rubber Process Analyzer (RPA). Observations were made and discussed for the *tan delta inversion crossovers* which may prove useful in quantifying state of mix and degree of *in-situ* silanization during the mixing process.

Rubber compounds are viscoelastic materials that are subject to die swell (also called extrudate swell or the *Barus* Effect) and dimensional stability problems when engaging in extruding or calendering in the factory.[i] Viscosity values such as the complex dynamic viscosity (*) from the RPA or the apparent viscosity (η_{APP}) from the capillary rheometer are important for measuring processability and throughput, but they are not the only processing properties that should be measured.[ii] Compound uncured elasticity or nerviness can be equally important. This property relates to such properties as dimensional stability of uncured rubber in extrusion and calendering operations. The rubber compound elasticity can be measured with either the uncured elasticity G´ or tan with the RPA or the running die swell from the capillary rheometer. It is very important to measure both the compound viscosity and the compound elasticity up stream in a rubber factory because rubber processability depends on these viscoelastic properties.[iii,iv]

EXPERIMENTAL

A series of mixed rubber recipes with variations in the filler loading levels of N234, N330, and N660 carbon blacks in addition to variations in non-black fillers which included Hi-Sil 210 precipitated hydrated silica (with and without organosilane coupling agent TESPT), clay, and ground calcium carbonate (whiting), are shown in variations of Table 1 below.

Table 1 - Variations in Subject Filler

Compound	1 phr	2 phr	3 phr	4 phr	5 phr
SBR 1500	100	100	100	100	100
Subject Filler	15	30	45	60	75
Naphthenic Oil	15	15	15	15	15
Stearic Acid	2	2	2	2	2
6PPD	2	2	2	2	2
Zinc Oxide	3	3	3	3	3
Total	137	152	167	182	197

Subject Filler is N234 carbon black, N330 carbon black, N660 carbon black, Hi Sil 210 precipitated, hydrated silica, hard clay, and finely ground limestone (whiting).

Silica was used with and without OFS 6945 Organo Silane (TESPT). When the organosilane was used with the silica, it was used in the following proportions.

Hi Sil 210 (phr)	15	30	45	60	75
OFS-6945 Silane/Carbon Black (50:50)	2.25	4.5	6.75	9	11.25

With these compounds, comparisons were made between the Alpha Technologies RPA 2000® with Extended Dynamic Range (EDR) and the Alpha Technologies ARC 2020® Capillary Rheometer.

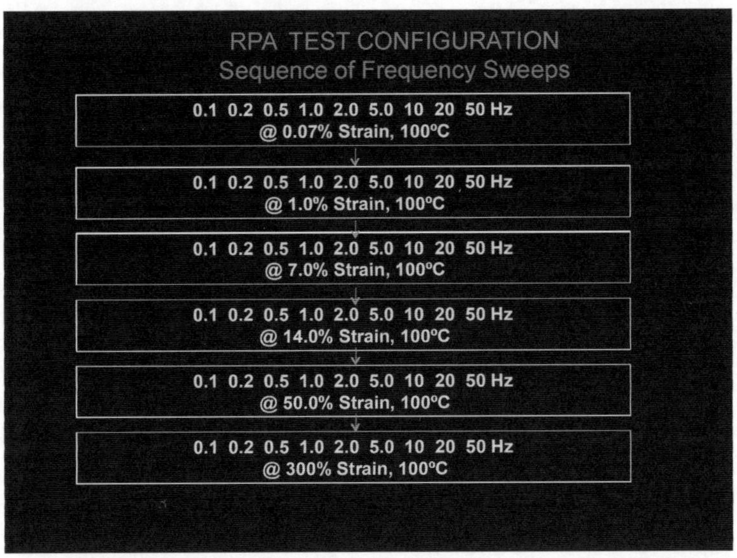

The figure below shows the resulting running die swell values for all fillers loaded at 75 phr in the model SBR test recipe.

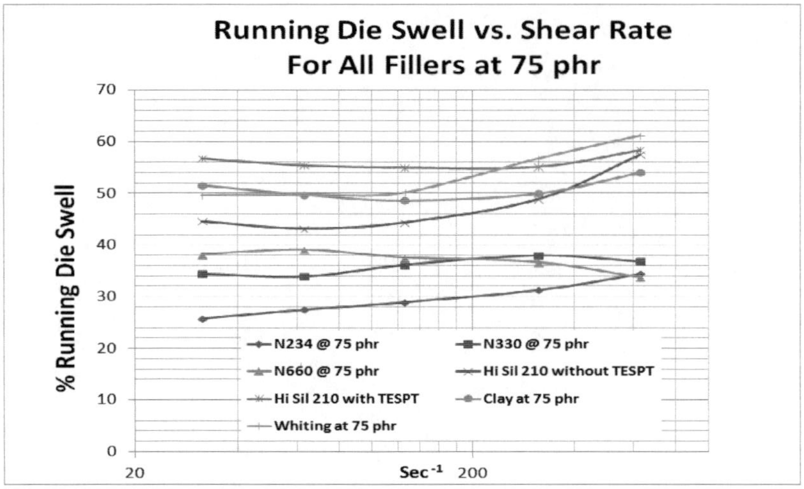

Finer particle size fillers impart less die swell than larger particle size fillers.

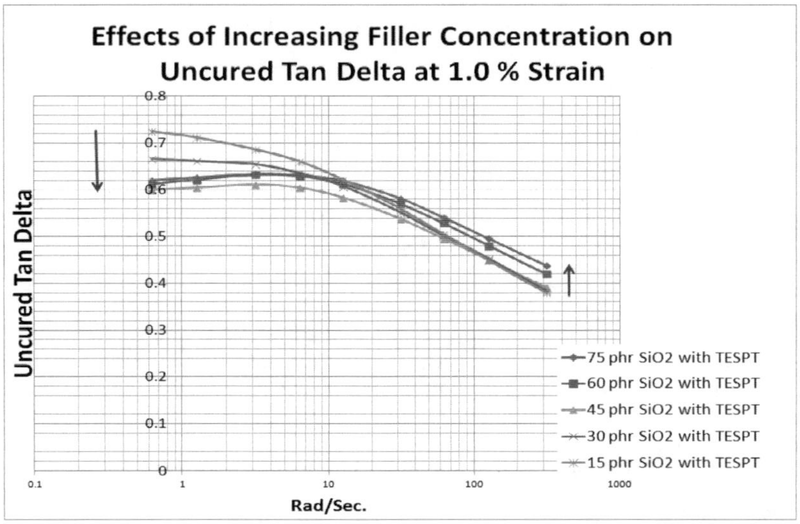

The figure above shows that with the silica with TESPT silanization, the uncured tan values overall are clustered closer together and there is a well defined *uncured tan delta inversion crossover point* at 15 radians / second. This is a signature that the silica filler network is very much destroyed from the silanization reaction with the TESPT. Also this inversion crossover point could be useful in the future for quantifying the degree of silanization that has actually occurred.

The Figure below shows that the RPA uncured tan inversely correlates to the capillary rheometer running die well, if it is on the high frequency side of this inversion crossover point, but actually has a positive correlation with die swell if the tan point is on the low side of this crossover point.

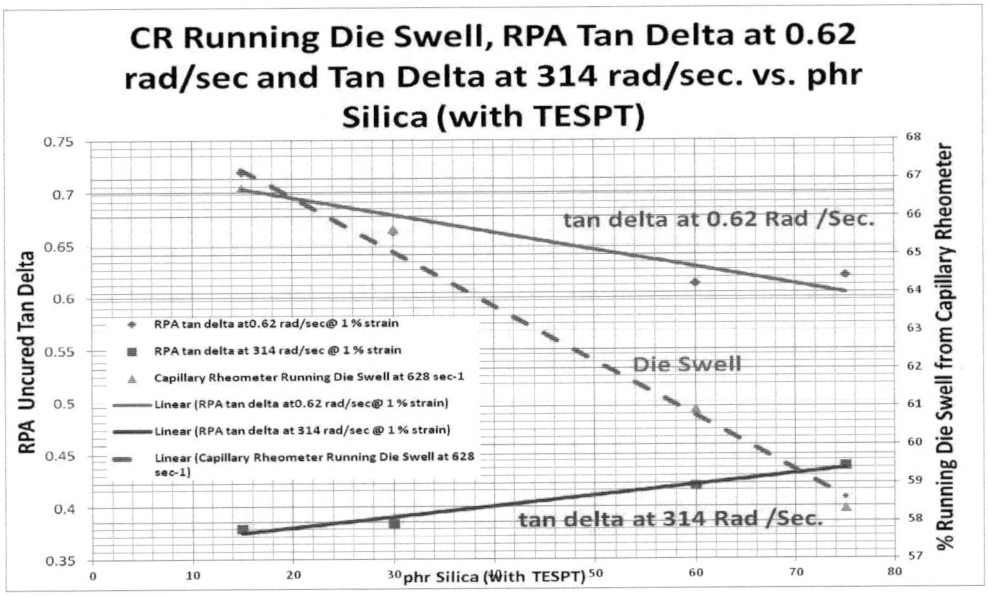

CONCLUSIONS

1. Repeatability is three times better with the RPA to measure rheological properties relating to die swell than with the Capillary Rheometer.
2. RPA rheological properties such as tan and G´ can correlate to capillary rheometer die swell.
3. Capillary rheometer running die swell has an R square correlation of 72 % explained variation with relaxed die swell.
4. The RPA is three times faster and more productive than the capillary rheometer
5. The Alpha Technologies Extended Dynamic Range (EDR) for the RPA 2000 exhibits much better measurement of the Payne Effects for commercial compounds.
6. The ARC 2020 capillary rheometer effectively measured increases in die swell from higher applied shear rates, the use of lower filler concentrations, or the use of larger particle size fillers (instead of smaller particle size fillers).
7. Silanization of precipitated hydrated silica significantly increased the compound s die swell.
8. The specific strain amplitude applied during the performance of an RPA frequency sweep of a filled rubber compound is very important. Increasing the applied strain in a sequence of multiple frequency sweeps can measure an uncured *tan δ inversion.* Also in a single frequency sweep, at a properly set strain, a specific frequency may be recorded where *the tan δ inversion crossover* occurs.
9. The uncured tan inversion crossover may prove useful in quantifying the degree of silanization that has occurred from a factory silica rubber mixing process.
10. Even though increasing strain amplitude and/or frequency applied during RPA testing of filled rubber compounds can give additional useful rheological information, running at ultra low strain with ultra low frequency gives no useful information, just noise.

References

[i] Kejian Wang, Chapter 4, "Die Swell of Complex Polymeric Systems", *Viscoelasticity – From Theory to Biological Applications*, INTECH, 2012
[ii] J. Dick, E. Norton, T Xue, "Effects of Variation in Strain Measurements with the Extended Dynamic Range (EDR) of the RPA on the Accuracy of Shear Thinning Measurements for Rubber Compounds", Paper No. 88 presented at the Rubber Division, ACS Meeting, October 14, 2014
[iii] J. Dick, Overview of Capillary Rheometry, Part 1, Rubber World, January, 2007, p.30
[iv] J. Dick, Overview of Capillary Rheometry, Part 2, Rubber World, February, 2007, p.36

Processability and rheological properties of extensively recycled high density polyethylene (HDPE)

P. Oblak[a], J. Gonzalez-Gutierrez[a], B. Zupančič[a], A. Aulova[a] and I. Emri[ab*]

[a] *University of Ljubljana, Faculty of mechanical engineering, Center for Experimental Mechanics, Pot za Brdom 104, Ljubljana, Slovenia*

[b] *Institute for Sustainable Innovative Technologies, Pot za Brdom 104, Ljubljana, Slovenia*

igor.emri@fs.uni-lj.si

Abstract. In plastics industry it is common practice to mechanically recycle waste material arising from productions. However, mechanical recycling affects material inherent structure and consequently its rheological properties and processability; therefore it needs to be quantified.

Mechanical recycling of high density polyethylene (HDPE) was simulated by one-hundred consecutive extrusions. During the extrusion also processing parameters, namely melt pressure and torque were recorded. After every cycle, portion of material was removed for the purpose of characterization. After the processing, rheological properties of the material were characterized in terms of measuring melt flow index and viscosity. In addition, gel permeation chromatography and solubility tests were performed in order to study structural changes.

The results of rheological measurements, i.e. MFI and complex viscosity, showed unfavourable effect of mechanical recycling on HDPE processability. With increasing number of the reprocessing cycles MFI drops practically to zero after 30th extrusion cycle. Further, viscosity significantly increases between 5th and 20th cycle. Moreover, changes in processability were directly monitored through the melt pressure and extrusion torque. Both values rise in first 30 extrusion cycles, pressure reaches more than 500% in respect to first extrusion. GPC measurements and solubility tests suggest that chain-scission and chain branching occur simultaneously from the start of the reprocessing. Through the first 30 extrusion cycles the molecular structure is being dominated by chain branching, while after 30th extrusion cycle chain scission becomes predominant over the chain branching. After 60th extrusion cycle, the branched molecule chains begin to cross-link.

Keywords: High density polyethylene, mechanical recycling, processing pressure, melt flow index, viscosity, gel permeation chromatography, solubility

PACS: 83

INTRODUCTION

During the second half of the 20th century, plastics became one of the most universally-used multipurpose materials in the global economy. With continuous growth of plastic industry also amount of generated waste is increasing, not only as post-consumer waste but also as waste generated during the production process. Runners, sprues and off-specification products are common in production; they are easy to identify and are of high quality. Instead of being rejected as waste they are granulated and reprocessed; this makes economic sense as it reduces both production waste and utilisation of raw materials. Reprocessing of production waste is particularly important in cases where amount of waste is high in respect to the product. As an extreme case, micro injection moulding can be exposed; in its case the volume of delivery system, i.e. sprue and runner systems, which represent the waste material, can exceed that of the injected part by a factor 100 or more [1]. Another example is extrusion blow-moulding technology where amount of the scrap during production can reach 40%.

Properties of mechanically recycled polymers do not remain the same because of degradation from heat, mechanical stresses and oxidation during their reprocessing. Degradation affects molecule structure and consequently material rheological properties are changed. This has direct effect on the processability and needs to be considered during the following technological processes.

In line with the information mentioned above, present work is focused to investigation of the processability and rheological properties of HDPE when exposed to extensive mechanical recycling, i.e. to one hundred (100) consecutive extrusions cycles. In addition, this investigation supports changes of processability and rheological properties by observed structural changes. Results of complete investigation are presented in [2].

MATERIAL AND METHODS

Material

For the purposes of the investigation, high density polyethylene (HDPE) LANUFENE®HDI-6507UV, produced by Ras Lanuf Oil & Gas Processing Co. was used. The material has density 0.965 g/cm^3 (23°C) and melt flow index 7.5 g/10min (190°C and 2.16kg). Furthermore, producer provides information on suitability of material for the manufacturing of crates, containers, trays and other similar injection moulded objects where good weather-ability and mechanical properties are needed. In addition, the producer gives information that the material contains UV light stabilizers and recommends its processing temperature to be 220 – 270°C.

Methods

Simulation of Mechanical Recycling

The process of mechanical recycling was simulated by extensive extrusion using twin-screw extruder (PTW 16/40 OS, Thermo Scientific). Material was extruded at screw rotation of 150min^{-1} using horizontal rod die, at processing temperature of 240 °C and throughput between 1200 and 1300 g/h. During the processing, melt pressure and extrusion torque were recorded. After material was extruded, it was pelletized and a portion was removed for the purpose of afterwards characterization. The rest of material was submitted to a new extrusion cycle.

Melt flow Index Measurements

Measurements of melt flow index (MFI) were carried out according to ASTM D 1238-04c. All the tests were performed on commercial indexer (D4004DE, Dynisco). The MFI was measured at the temperature of 190°C by applying pressure with a load of 2.16 kg. For the virgin material and its recyclates, MFI was measured in triplicate.

Viscosity measurements

Viscosity measurements were performed in oscillatory mode by varying the frequency of excitation i.e. frequency sweep (Haake MARS, Thermo Scientific). Tests were implemented using a plate-plate measuring geometry at 240°C and frequencies ω increasing from 0.1 Hz (0.628 rad/s) to 100 Hz (628.32 rad/s). All measurements were performed by applying a shear stress τ of 300 Pa, previously determined as high enough to obtain good resolution of results without leaving the linear viscoelastic region of molten HDPE. For every material six repetitions were done. Within the current study, results on viscosity measurements are presented with the magnitude of complex viscosity, $|\eta^*(\omega)÷$, which is related to steady state viscosity, η , by the Cox-Merz rule:

$$|\eta^*(\omega)| \Leftrightarrow \eta(\dot{\gamma}) \quad \text{, when} \quad \omega\left[rad/s\right] \Leftrightarrow \dot{\gamma}\left[s^{-1}\right] . \qquad (1)$$

Gel Permeation Chromatography

Changes in the molecular weight were studied by implementation of high temperature gel permeation chromatography (HT GPC). Samples were introduced into the eluent, a trichlorobenzene (TCB), at concentration 2.0 mg/ml. The analysis was carried out at 150 °C with set of three Olexis columns of total length 900 mm. Injection volume was 200 μl and flow rate 1.0 ml/min. A differential refractometer (DRI detector) was employed and data analysed using Agilent Cirrus software. Presented results, namely molecular weight distribution, MWD, are the result of one injection per material studied.

Solubility tests

To expand the study on changes of molecular structure, solubility test was also implemented to define the insoluble part which confirms presence of cross-linked portion. Solubility tests were performed by extraction in Soxhlet assembly. A portion of 1.5 g of a specimen material was introduced into the Soxhlet thimble and extracted by xylene for 24 hours at 6 to 7 cycles per hour. Mentioned technique provided the information on a weight percentage of the insoluble part of a sample.

Results and discussion

Extrusion torque and pressure

Real time melt pressure and extrusion torque are presented in Fig. 1. Both curves exhibit similar trend; during the first 30 extrusion cycles pressure and torque are linearly rising and start to decrease after the 35th extrusion. Several authors reported that during the reprocessing of HDPE, two competing processes took place, i.e., chain scission and chain branching [3, 4, 5], besides also cross-linking might be present [6, 7]. Continuous rising of extrusion pressure and torque through the first 30 extrusion cycles indicates reduction of a material flowability at a given shear rate which could be a consequence of cross-linking and/or chain branching process.

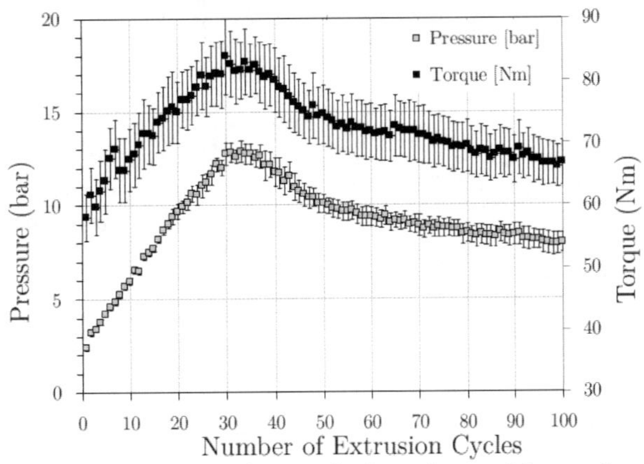

FIGURE 1. Recorded extrusion pressure and torque in dependence of the number of extrusion cycles

Melt Flow Index

Results of our investigation, presented in Fig. 2, show significant drop of MFI during the first extrusions, thereby the maximum drop of 98.8% is observed at the 30th extrusion cycle. After 30th cycle MFI practically does not change.

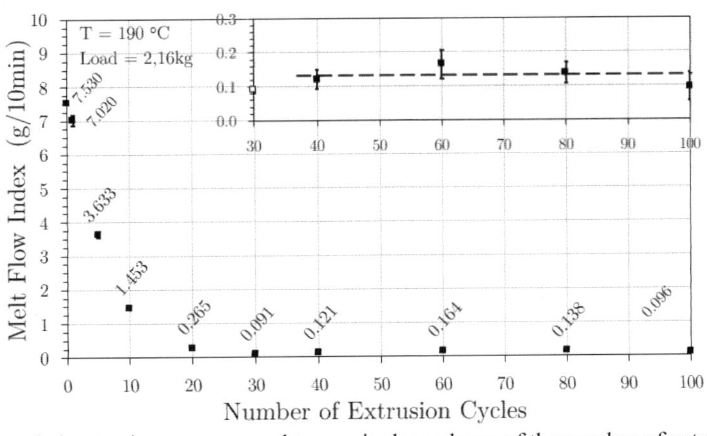

FIGURE 2. Recorded extrusion pressure and torque in dependence of the number of extrusion cycles

Drastic decrease of MFI shows lowering of chain mobility which suggests a branching process present throughout the first 30 reprocessing cycles, similarly as suggested by processing parameters presented previously.

Viscosity

The complex viscosity is presented in Fig.3, left for low frequency (0.628rad/s), and right for high frequency (291.6 rad/s). In both cases, significant leap is evident between 5th and 20th extrusion cycle. Increase is more evident at low frequency where the value at 20th cycle exceeds 600% of the initial one, while under high frequency an increase of 123% was recorded. After the 20th reprocessing cycle, complex viscosity at low frequency constantly fluctuates between 400 and 700%, on the other hand, decreasing trend is recorded at high frequency.

Increase of complex viscosity between 5th and 20th shows lowering the chain mobility which may result from chain branching.

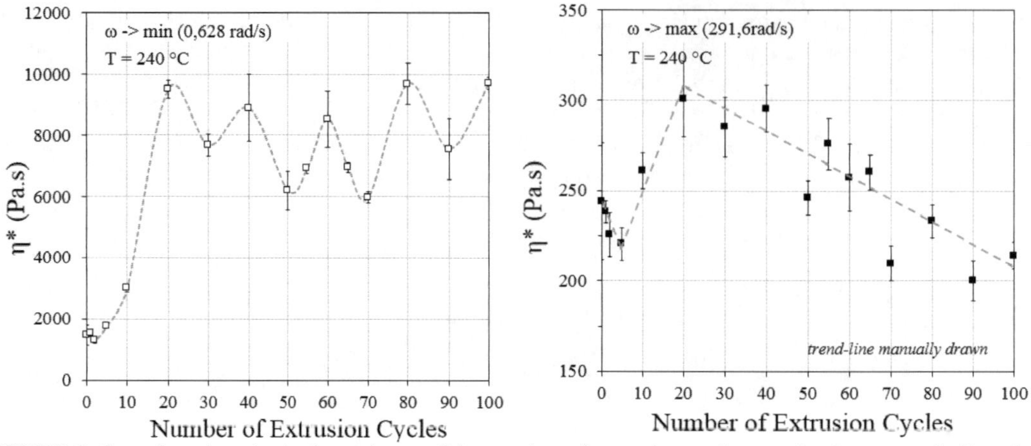

FIGURE 3. Complex viscosity in dependence of the number of extrusion cycles at min. frequency (left) and max. frequency (right)

Molecular weight and structure

Finally, to support suggested changes in the material structure, gel permeation chromatography technique was used. In Fig. 4 normalized molecular weight distribution curves are presented for the virgin material and five extrudates: 20, 30, 50 70 and 100 times extruded material. It can be seen that reprocessing affects the molecular weight distribution in two ways; (i) it slightly broadens with number of extrusion cycles, this is evident from the beginning of the reprocessing and (ii) the peak shifts towards smaller values, this becomes evident after 30th extrusion cycle. Recorded broadening of the molecular weight distribution can be attributed to chain branching and cross-linking process while drifting to smaller values suggest that there are processes which reduce the molecular weight, like chain scission. From the results we may assume that chain branching process, as defined in continuation, took the place from the beginning of the reprocessing, meanwhile chain-scission is simultaneously happening after 30th extrusion. Identical is suggested by the results on other characterization methods previously presented in this report.

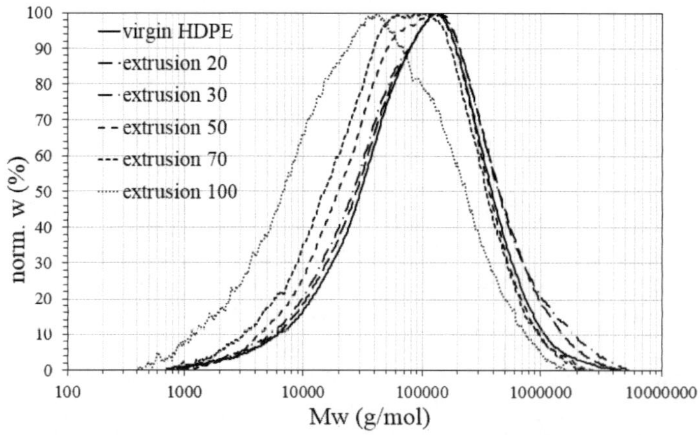

FIGURE 4. Normalized molecular weight distribution for virgin HDPE and its extrudates.

Portions of insoluble parts defined by extraction are presented in the Table 1. It can be seen that insoluble part was detected only after the 60th extrusion cycle, while only for 100-times extruded material the amount of insoluble part becomes significant i.e. 24.5 wt%. Insoluble portions confirm presence of cross-linking mechanism after 60th reprocessing cycle which intensity is highly increased after 100th extrusion.

Results on GPC measurements and solubility test, similarly as other results presented in this report, suggest molecular enlargement to be present during reprocessing, and after the 30th extrusion cycle also chain-scission process. Results on solubility test support the idea that molecular enlargement is occurring through chain branching, and that the cross-linking occurs only after the 60th reprocessing cycle.

TABLE (1). Insoluble portion of the virgin HDPE and its recyclates

Material	Insoluble portion (wt.%)
Virgin HDPE	0
20 times recycled	0
30 times recycled	0
50 times recycled	0
60 times recycled	1.0
70 times recycled	1.3
100 times recycled	24.5

Conclusions

Effect of mechanical recycling on processability and rheological properties of HDPE was investigated. Results on rheological measurements, i.e. MFI and complex viscosity showed unfavourable effect of mechanical recycling on HDPE processability. With number of the reprocessing cycles MFI drops practically to zero after 30th extrusion cycle. Moreover, changes in processability were directly monitored through the melt pressure and extrusion torque. Both values rise in the first 30 extrusion cycles, pressure reaches more than 500% in respect to the first extrusion.

Researchers suggested two competing mechanisms of degradation during reprocessing of HDPE, chain scission and side-chain branching [3, 4, 5]; in addition, some suggested also presence of cross-linking [6, 7]. Results presented in this report, GPC measurements and solubility tests, also suggest that chain-scission and chain branching occur simultaneously from the start of the reprocessing. Through the first 30 extrusion cycles the molecular structure is being dominated by chain branching, while after the 30th extrusion cycle chain scission becomes predominant over the chain branching. After the 60th extrusion cycle, the branched molecule chains begin to cross-link.

It can be concluded that mechanical recycling worsens HDPE processability already through the first 30 reprocessing cycles. We can expect the need for resetting the processing parameters, increased burden on equipment and extrudate swelling when using recycled HDPE, especially if the material was recycled more than 10 times.

ACKNOWLEDGMENTS

This work was conducted with the financial support of the Slovenian Research Agency (ARRS) of program P2-0264. We would like to thank Odelo Slovenija and BASF Company for their help at MFI and GPC measurements.

REFERENCES

1. C. Jeggy, "Micro-injection moulding: From process to modelling," Lueven, UCL Presses Universitaires de Louvain, 2004.
2. P. Oblak, J. Gonzalez-Gutierrez, B. Zupančič, A. Aulova and I. Emri, "Processability and mechanical properties of extensively recycled high density polyethylene," Polymer degradation and stability, vol. 114, pp. 133-145, 2015.
3. T. Kealy, "Rheological Analysis of the Degradation of HDPE During Consecutive Processing Steps and for Different Processing Conditions," Journal of Applied Polymer Science, vol. 112, no. 2, pp. 639-648, 2009.
4. M. K. Loultcheva, "Recycling of high density polyethylene containers," Polymer Degradation and Stability, vol. 57, no. 1, pp. 77-81, 1997.
5. L. A. Pinheiro , "The role of chain scission and chain branching in high density polyethylene during thermo-mechanical degradation," Polymer Degradation and Stability, vol. 86, no. 3, pp. 445-453, 2004.
6. S. Apone , "Effects of thermomechanical treatments on HDPE used for TLC ducts," Polymer Testing, vol. 22, no. 3, pp. 275-280, 2003.
7. C. N. Kartalis , "Mechanical recycling of postused high-density polyethylene crates using the restabilization technique. I. Influence of reprocessing," Journal of Applied Polymer Science, vol. 73, no. 9, pp. 1775-1785, 1999.

Effect of the wave length and the intensity of UV LED on UV-curable LSR

Joonsung Tae[a], Kyunggyu Yim[a] , Jae B. Kwak[b] , Jinsu Gim[a] and Byungohk Rhee[a*]

[a]*Ajou University, 210 east hall, 206 Worldcup-ro, Yeongtong-gu, Suwon, 16499, South Korea*
[b]*Samsung Electronics, Global Technology Center,129 Samsung-ro, Yeongtong-gu,Suwon, 16677, South Korea*

muska8901@gmail.com
rhex@ajou.ac.kr

Abstract. UV-curable LSR can be molded at relatively low temperature condition, which has advantages for an over-molding with inserts of temperature-sensitive materials. As the light sources for the radiation-cure process, halogen or mercury UV lamps have been used conventionally. However, the conventional lamps dissipate a lot of heat and their service life is relatively short. As substitution for the lamps, UV LED lamps provide a longer service life and a lower heat dissipation. In this research, UV LEDs were tested to analyze the rheological behavior of UV-curable LSR during curing process. UV-LEDs of various wave lengths and intensities were tested. The steady shear test was applied to find the starting time of cure and the SAOS was applied to find the ending time of cure. In addition, the hardness change along the curing time was compared with the rheological test result to make sure the reliability of the rheological test.

Rheological Analysis of 3D Printability of Alginate Hydrogel

Huijun Li[a,b], Lin Li[a,b*]

[a]Singapore Center for 3D Printing, 50 Nanyang Avenue, Singapore 639798, Singapore

[b]School of Mechanical and Aerospace Engineering, Nanyang Technological University, 50 Nanyang Avenue, Singapore 639798, Singapore

Presenting author's email: HLI018@e.ntu.edu.sg
Reference author's email: mlli@ntu.edu.sg (L. Li)

Abstract: With advanced 3D bio-printing technologies, fabrication of complex tissues or even organs becomes possible. One of the challenges for successful 3D bio-printing is the selection of a suitable scaffold material with a good printability. Hydrogels have attracted great attention as biomaterials for printing three dimensional (3D) scaffolds for tissues or organs. However the relations between rheological properties of hydrogels and 3D printability have not been extensively studied. In this study, alginate-based hydrogels were prepared and used as a model material for 3D printing of scaffolds using an extrusion-based printer. Rheological studies were performed for the hydrogels with different formulas. We deduced the range of shear rates for the hydrogels that underwent during the 3D printing process. Furthermore, thixotropic properties of the hydrogels were investigated by applying and removing a shear rate. The results will be shown and discussed. This presentation will provide some new understanding of the relation between rheological properties of hydrogels and 3D printability.

Keywords: Rheology; hydrogel; printability; 3D printing

1. INTRODUCTION

The traditional fabrication technologies can fabricate artificial tissues with relatively simple structures. The combined potential of 3D printing technologies and tissue engineering has been an exciting route in developing tissue engineered scaffolds with desired complicated structures [1-3]. Alginate based hydrogels have attracted a great attention as biomaterials for 3D printing due to their soft and wet nature as well as similarities to biological tissues. Many researchers have noted that rheological properties of potential hydrogels for an extrusion based printer are a key factor in controlling the printability of the hydrogels. For example, Jia et al. evaluated the printability of hydrogel by measuring the area of printed dots. They found that the plot of viscosity against dots area can directly reflect the relationship between viscosity and printability of different samples. Furthermore, the optimal range of viscosity was defined for a piston driven printer [4].

In this study, we present a novel approach to analysis of the printability of hydrogels. Alginate-based hydrogels were prepared as a model printing material for an extrusion-based 3D printer. The range of shear rates generated during the printing process was deduced theoretically. This range of shear rates helped us to select a proper shear rate to investigate the thixotropic properties of the hydrogels. Then we investigated the printability of the model material to find whether it is a suitable material for an extrusion based printer.

2. MATERIALS AND METHODS

2.1 Materials and Sample Preparation

Calcium chloride and sodium alginate were purchased from Sigma-Aldrich, Singapore. Aqueous solutions of alginate with various concentrations (2, 4, 6, 8 and 10 wt%) were prepared. Calcium chloride solutions with a fixed $CaCl_2$ content of 25mM/L were added to each alginate solution to obtain alginate/calcium chloride hydrogels. The prepared hydrogels samples were labelled with Algn, where n was the weight fraction of alginate.

2.2. Rheological Evaluation of Hydrogels

The rheological properties of the hydrogels samples were measured by using a rotational rheometer (DHR, TA Instruments, USA). Steady-state flow tests and recovery tests at room temperature were conducted for evaluating the thixotropic properties of samples.

2.3. Fabrication of 3D Structures

An extrusion-based printer, bioplotter (TechnoDigm, Singapore) was used in this study. The 3D constructs were printed with the prepared hydrogels along the preprogrammed tracks with an adjustable speed (15 mm/s). The piston was driven with a controllable speed (0.009 mm/s) to extrude the hydrogel from the syringe onto a transparent glass slide. The height between the nozzle and contact surface was fixed as 0.25mm because the diameter of nozzle was 0.25mm.

3. RESULTS AND DISCUSSION

3.1 The Shear Rate in an Extrusion-based 3D Printing Process

In this study, the range of shear rates in the syringe and in the nozzle is estimated from a given piston speed of the 3D printer. For an extrusion based printer, the hydrogel' flow behavior can be described through balancing of forces within a finite volume. Based on the fluid mechanics [5], consider a laminar and steady flow fluid in a circular pipe of radius, R, as shown in Figure 1.

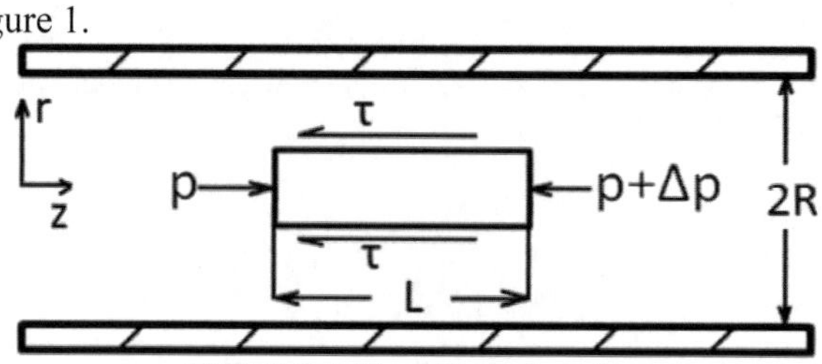

FIGURE.1. Flow through a pipe

where p is the pressure, τ is the shear stress on the surface (at the radius r) of the cylindrical element and L is the length of the element.

The force balance in the z direction on a fluid element situated at a radius r (0 < r <R) can be described as

$$p(\pi r^2) - (p + \Delta p)\pi r^2 = 2\tau\pi rL \tag{1}$$

Based on the power-law model described as

$$\tau = m(\dot{\gamma})^n \tag{2}$$

where n is the flow behavior index , m is the fluid consistency coefficient.
In a simple pipe, the volumetric flow rate (Q) can be expressed as

$$Q = \pi R^2 V = \pi\left(\frac{n}{3n+1}\right)\left(\frac{-\Delta p}{2mL}\right)^{\frac{1}{n}} R^{\left(\frac{3n+1}{n}\right)} \tag{3}$$

where V is the speed of piston. From equations (1)-(3), the shear rate in the pipe can be described as follows

$$\dot{\gamma}^n = r\left[\frac{VR^2}{\left(\frac{n}{3n+1}\right)\left(R^{\frac{3n+1}{n}}\right)}\right]^n \tag{4}$$

As the diameter of the syringe is not the same as that of the nozzle, the shear rate in the syringe and in the nozzle must be different. Assuming that the volume of the hydrogel does not change before and after printing, we get the average speed of hydrogel in the nozzle. Finally, we can deduce the range of shear rate in the nozzle using the average speed of hydrogel in the nozzle and radius of the nozzle.

3.2 Rheological Evaluation

An ideal printable hydrogel should be highly thixotropic, which means that viscosity of the hydrogel becomes low quickly when applying a shear force but the viscosity can recover quickly after the shear force is removed. But before conducting the rheological test we should know the range of shear rate generated during the printing process.

As discussed in the section 3.1, the shear rate can be calculated using equation (4). The procedure of the calculation is: 1) steady-state flow tests were performed on the hydrogel samples with different formula using a rheometer; 2) based on the power-law model and experimental data, the constants m and n for each sample can be obtained through curve fitting (data not shown here). The maximum shear rate in the nozzle for each sample is around 100 s^{-1} under the certain conditions (i.e. the piston speed is 0.009 mm/s, the diameter of the syringe and the diameter of the nozzle are 10 mm and 0.25 mm respectively). The whole test process included three steps: Step 1 simulated the initial state of a hydrogel before printing. Step 2 simulated the condition for a hydrogel under a certain shear rate (i.e. 100 s^{-1}) during the printing process. Step 3 simulated a condition similar to the final state of the hydrogel after printing. Figure 2 shows the recovery behavior in terms of the shear viscosity for the hydrogel. In our study, the viscosity of the hydrogel could recover after a short period of rest. However, the hydrogel could only recover its viscosity by 85% of the initial value after 30s (for Alg10). It is considered that the broken crosslinks caused by shearing need some time to recover and rebuild a gel network. Other tested samples also have the similar thixotropic properties.

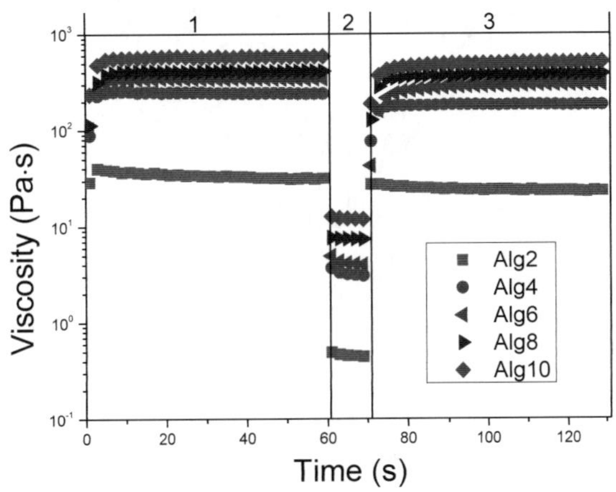

FIGURE.2. Effect of various alginate concentrations on the recovery behavior of alginate hydrogels

3.3 Fabrication of 3D Structures

The printed constructs shown in Figure 3 had 9 layers and these pictures were taken at the initial time of printing. It is apparent that for the construct printed with a higher alginate concentration, the printed filaments were of more uniform width and the shape was more stable. If the filament width was defined as d, it can be found from Figure 3(f) that d decreases with increasing alginate concentration. This is because that the hydrogel with a higher concentration of alginate was stronger and not easy to collapse compared to the hydrogel with a low alginate concentration. Thus a smaller width implies a better printing quality of the scaffold.

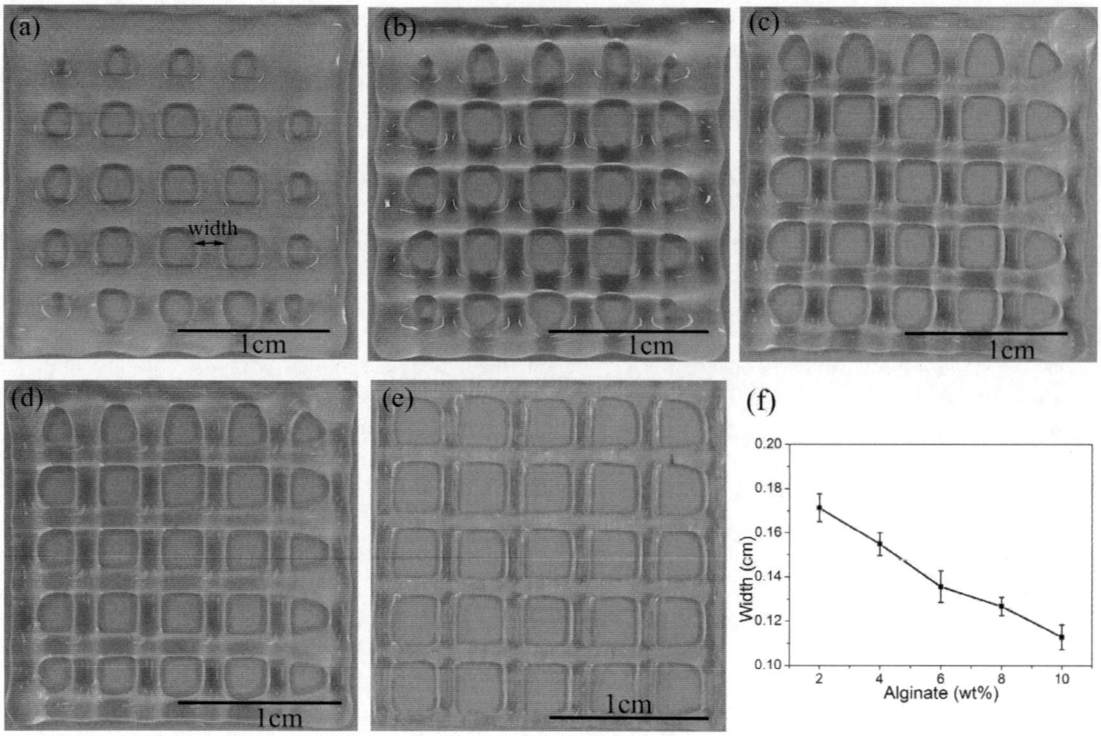

FIGURE.3. The images of printed hydrogel constructs s with various alginate concentrations (a) 2 wt%, (b) 4 wt%, (c) 6 wt%, (d) 8 wt%, (e) 10 wt%,and (f) the effect of various alginate concentrations on filament width.

4. CONCLUSIONS

This study presents a novel approach to analysis of the printability of hydrogels. Firstly, alginate-based hydrogels were prepared as the model material. Then, the range of shear rates generated during the printing process was deduced to help us to study the thixotropic properties and recovery behavior of the hydrogels. Finally, the printability of hydrogels was evaluated by two parameters: thixotropic properties and recovery time. We found that the alginate hydrogel is not an ideal biomaterial for 3D printing with an extrusion-based printer due to its low ability of viscosity recovery.

ACKNOWLEDGMENTS

We thank the support from the Singapore Center for 3D printing.

REFERENCES

1. Ozbolat IT, Yu Y. Bioprinting toward organ fabrication: challenges and future trends. Biomedical Engineering, IEEE Transactions on 2013;60:691-9.
2. Zhang Y, Tse C, Rouholamin D, Smith P. Scaffolds for tissue engineering produced by inkjet printing. centeurjeng 2012;2:325-35.
3. Murphy SV, Atala A. 3D bioprinting of tissues and organs. Nature biotechnology 2014;32:773-85.
4. Jia J, Richards DJ, Pollard S, Tan Y, Rodriguez J, Visconti RP, et al. Engineering alginate as bioink for bioprinting. Acta biomaterialia 2014;10:4323-31.
5. Chhabra RP, Richardson JF. Non-Newtonian flow and applied rheology: engineering applications: Butterworth-Heinemann; 2011.

Strain recovery of polyolefin polymer blends Based on PP/HDPE

Shafiei Sararoudi Saeed [(a)*], Salehi Ashtiani Saeed [(a)]

[(a)] *Islamic Azad University, South Tehran Branch - Tehran - Iran*

S_Shafiei@azad.ac.ir
saeedsalehiashtiani@gmail.com

Abstract. The rheological properties of blends consist of various polyolefin have received much attention in recent years. In present work, strain recovery after shear cessation was studied in immiscible blends composed of Polypropylene/ polyethylene and EPR as compatibilizer. The effect of blend composition (ranging from 20 to 50 wt. % of dispersed phase), and EPR addition (concentration up to 10 wt. % with respect to the dispersed phase) was evaluated and correlated to the evolution of the morphology of the blends. The blends showed three relaxation stages; a first fast relaxation which was attributed to the relaxation of the pure phases, the second one with slower rate, named plateau, which related to droplets relaxation, and the last one with higher relaxation which was interpreted in the present work. On the third step, by increasing the dispersed phase concentration, the relaxation modulus increases, whereas the relaxation rate decreases. The compatiblized blend showed higher relaxation rate at third region that can be related to smaller droplet and narrower droplet size distribution. During the step of preshearing, the droplets elongate and ruptured until a steady state is obtained and the morphology becomes finer and the droplets size smaller. Two important factors affect the third stage, preshear and waiting time. At various preshear rates, the third stage was strongly affected by the magnitude of the shear rate.

Keywords: Rheological behavior, Stress relaxation, Polyolefin blends.

INTRODUCTION

Polymer blends get a considerable attention today as they have improved properties compare to their components. One of the most important group among polymer blends is polyolefin blends as their Varity, low cost and good processing. Most of poly olefin blends are immiscible due to hydrophobic and nonpolar particles. [1-2]

Properties of immiscible polymer blends are strongly affected by phase morphology which can be controlled and improved during processing. Morphology development during processing can be detected by Rheology data such as stress relaxation measurement. Studies on this field have been done to investigate the morphology and its transmutation as the pioneer work of Takahashi that observed the evolution of a single droplet when submitted to step shear of different amplitudes. Later observation of the evolution of two droplets to understand the effect of coalescence was done by other [8-11]. Recently other studies have investigated the effect of compatibilizer on stress relaxation of polymer blends. Yamaha et al. have observed the single droplet of polyisobutylen in poly dimethyl siloxane and investigate its stress relaxation behavior by droplet arrangement angle during recovery stage [8]. Okamoto et al. observed the shape recovery of a PS/PC blend after application of step strain [3]. Iza and Bousmina investigated the relaxation behavior of poly (1-butene) and poly (dimethylsiloxane) blend after the application of various strains [4].By adding the compatibilizer to the blend the situation gets more complicated. There are few studies on stress relaxation measurements for compatibilized systems. Most of the researchers detect three stages for stress relaxation behavior: a first fast relaxation which was attributed to the relaxation of the pure phase, a second one which characterized by the presence of a plateau and a third fast one, but there is no accurate data about the third stage [3-14]. In the present work, we observed the stress relaxation behavior of PE/PP blend with and without EPR as compatibilizer and tried to interpreted third stage trend of relaxation modulus curve.

EXPERIMENTAL

Material and sample preparation

As the main phases, commercial PP (525J, Jam Polypropylene P.C, Iran) and HDPE (5030, Tabriz P.C, Iran) and as compatibilizer Random co-polymer Ethylene Propylene Rubber, EPR (provided by Dutral Co. Grade 043) were used. Various weight concentrations (80/20, 70/30 and 40/60) of PP/HDPE blend were prepared. Concentration of EPR ranging from 0 to 20 wt% with respect to dispersed phase (HDPE) for PP/HDPE was used. The blends were prepared by a Brabender twin-screw extruder; model TSE 20. Compatibilized blends were prepared through two steps. First the compatibilizer was mixed with the miner phase (HDPE) and then blended with matrix (PP). In order to rheological and morphological analysis, samples were obtained by compression molding.

Morphological analysis

Scanning electron microscope (SEM) micrographs by SEM Cambridge S360, England were used for morphological observations. In order to avoid large deformations in the surface to be examined by SEM, the sample were fractured in liquid nitrogen to have brittle fracture. The resulting fractured surfaces were coated with gold prior to observation.

Rheological analysis

Rheometrics Mechanical Spectrometer, RMS (MCR 501 made by Anton Paarl) were used to rheological measurements of PP/HDPE compatibilized and uncompatibilized blends. Relaxation experiments for blends were carried out using disk like parallel plates with 25 mm dimeter and 1mm gap at 200c.

	Interval 1		Interval 2	Interval 3	
Test number	Presheare S^{-1}	Time S	waiting time S	Strain %	Time S
T1	0.4	400	50	20	1200
T2	0.2	400	50	20	1200
T3	0.6	400	50	20	1200
T4	0.4	400	20	20	1200
T5	0.4	400	10	20	1200

Table 1: relaxation experiment 1- Step of preshearing. 2-Step of waiting time 3- Step of relaxation

RESULT AND DISCUSSION

Step strain experiments and morphlogy

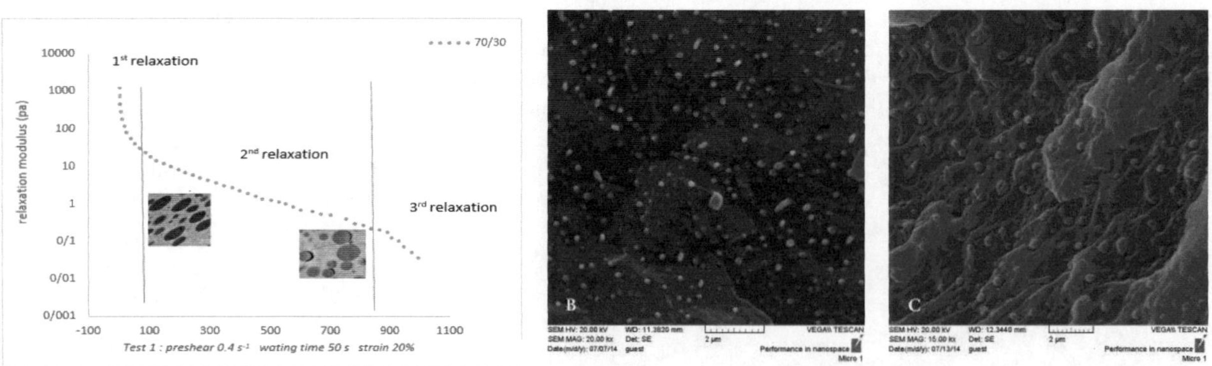

Figure 1 A typical curve of stress-relaxation modulus, Morphology of non-compatibilized PP/HDPE blends: (B) 80/20; (C) 70/30;

Figure 1A shows a typical curve of relaxation modulus for PP/HDPE blend. This curve consists of three stages. The first one is a fast relaxation associated to pure phases; the second one is a plateau that is related to the relaxation of the deformed droplets, and the last one is start with an increase in relaxation modulus with respect to the pure phase. At the beginning of the plateau the droplets are starched to form an ellipsoidal shape and at the end of the plateau they recover to the spherical shape. [3, 4, 14] The third stage is fast relaxation step which investigated in the present work. Figure 1 B and C shows the SEM micrographs of polymer blends 80/20 and 70/30. Increasing of particle size by increasing dispersed phase concentration can be seen.

Influence of blend composition and copolymer addition

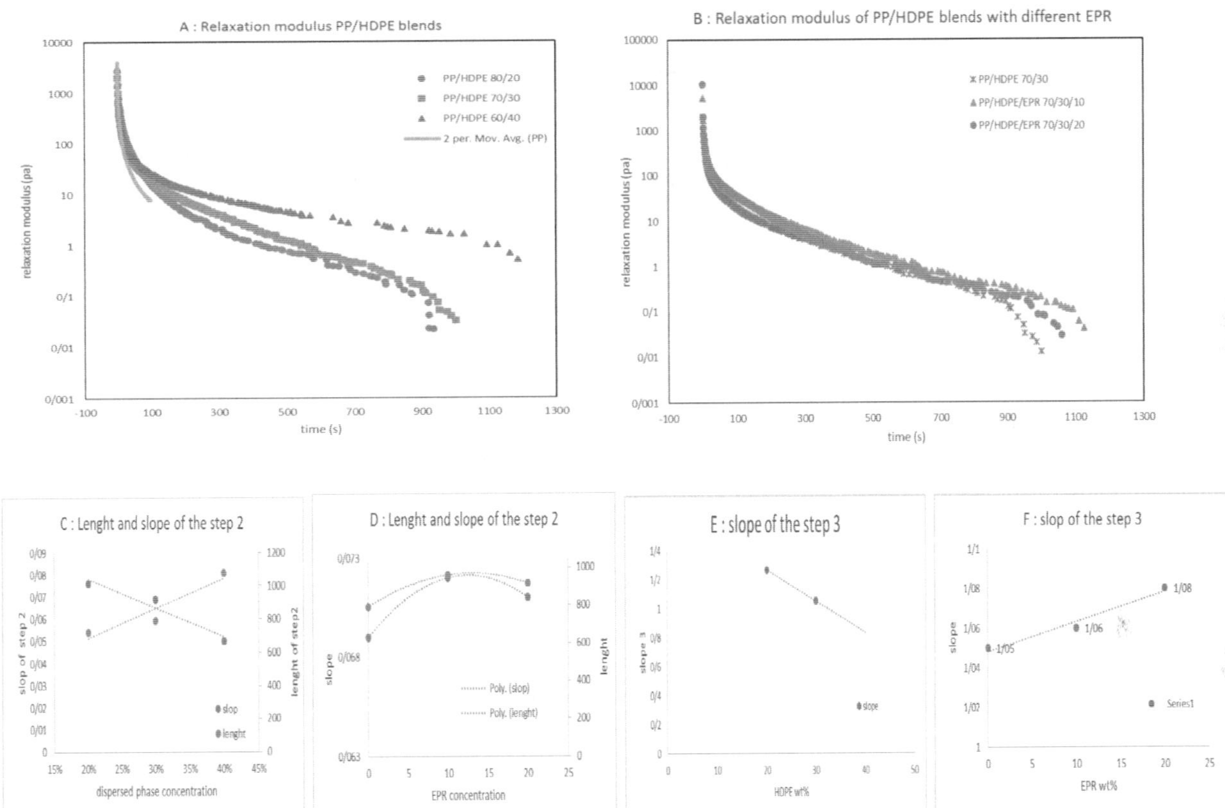

Figure 2: A: Relaxation modulus of PP and PP/HDPE blends with different HDPE wt % .B: relaxation modulus for PP/HDPE (70/30) blends to which EPR copolymer has been added at similar condition .C-F: slope and length of steps Test conditions is chosen according to test 1 (table 2).

Figure 2.A shows a typical relaxation modulus curve for PP/HDPE blend in various HDPE concentrations. The first stage is almost independent of dispersed phase concentration and behaves same as PP as matrix, whereas at the second and third stages, the magnitude of relaxation modulus increases by increasing of HDPE concentration. As was shown in Figure 2.C, the length of the plateau increases with increasing of the dispersed phase concentration, whereas slope of the plateau reduces. It is due to the droplets size increase with increasing of the HDPE amount, therefore relaxation becomes slower, because the bigger droplets stretched more than smaller ones; consequently they need more time for recovery. In addition to, the probability of coalescence of droplets increases above 20% wt. of HDPE, therefore relaxation modulus of the blend contain 40 percent of HDPE is higher than lower concentration blends [8, 13, 14]. It can be seen in Figure 2.E, the slope of the third stage reduces with increasing of the dispersed phase concentration. These results clearly showed that the third stage is dependent on dispersed droplets size.
The relaxation modulus for the blend (30% of HDPE) versus of time for various EPR (as a compatibilizer) concentration was shown in Fig. 2B. It shows that the first step is independent of EPR concentration. While in the

second and third stages, the EPR concentration affects on the magnitude of the relaxation modulus. It can be seen that by increasing of the EPR concentration up to 15 percent, the magnitude of the relaxation modulus in the second stage increases, however at higher percent of EPR, it reduces. In general, adding of EPR causes the relaxation modulus rises compare to uncompatibilized blend as a result of improvement of the interface. However, in concentration higher than 20% , after saturated concentration of EPR in the interface, the excess EPR can form third phase and because of lower modulus of EPR than HDPE, the magnitude of the modulus of the blend decreases.
Length and slope of the second step versus EPR concentration are shown in figure 2.D. By increasing of the EPR concentration, interface is improved and it leads to smaller particles. It could affect on slope of the plateau and diminish it, in other words, relaxation carry out much faster.The length of the plateau for the compatibilized blend is longer than uncompatibilized which is attributed to improved interface region. However, for EPR concentration more than 15%, both slope and length of the second stage reduce, because the copolymer could act as an individual phase [14-18].
Figure 2.F shows the slope of the third step versus EPR concentration. As it was shown, the slope of the third stage slightly increases by addition of the copolymer, which is due to obtain the narrower droplets size distribution and smaller droplets size, which lead to increase recovery rate of the droplets shape.

Influence of preshear rate and waiting time

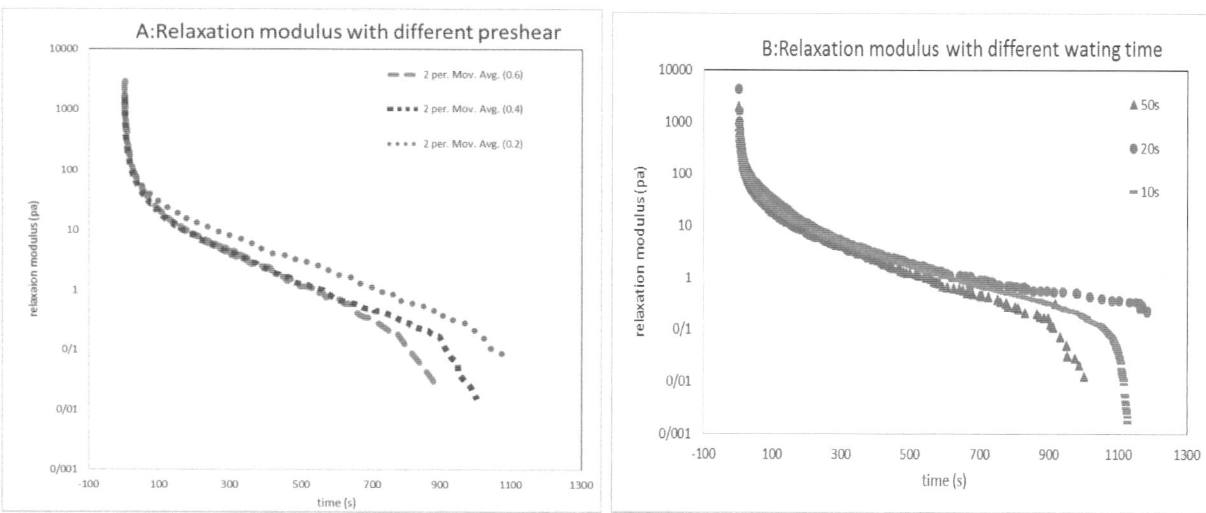

Figure3: A: *Relaxation modulus of PP/HDPE (70/30) blends at different preshear rates (test 1, 2, 3) B: Relaxation modulus of PP/HDPE (70/30) blends with different waiting times (test 1, 4, 5)*

Figure 3A shows a typical relaxation modulus curve for PP/HDPE (70/30) blend at different preshear rates. It was reported that after preshear, droplets turn from spherical shape to ellipsoidal. During the preshear, the droplets elongate and probably rupture until a steady state is obtained, also the coalescence of the droplets would happen and compete with break up. Also after cessation of the shear, during the waiting time (the interval of time between cessation of the preshear and imposing a sudden strain), the droplets break up and become finer. By increasing of the shear rate from 0.2 to 0.6, the morphology become finer and the droplets size smaller. At preshear rates of 0.4 and 0.6, slopes of the plateau are same but higher than it at preshear 0.2, which is attributed to droplets size as mentioned above. The length of the plateau decreases with increasing of the shear rate which is due to smaller droplets extended less under shear rather than bigger ones. However, narrow droplets size distribution at preshear rate of 0.6 leads to shorter relaxation time at the third step [3, 4, 19-21].
Figure 3B shows the stress relaxation curve for PP/HDPE (30/70) blend in various waiting time. As can be seen, =\\\waiting time is not an affecting factor first step, but in the second and third step, when waiting time increases from 10 to 20 seconds, relaxation time increases which is due to residual stresses in droplets after cessation of preshear. Iza and Bousmina observed that before shear, the morphology of the blend is spherical with high

polydispersity in size. During the shear, the spherical particles deform to kind of ellipsoids of revolution oriented at nearly 45° with respect to the direction of the flow. The aspect ratio increases with the deformation and the morphology becomes highly fibrillary. After cessation of shear, a rapid retraction of highly extended elliptical droplets takes place. In the other words, the droplets relax a portion of the residual stresses immediately after shear cessation. The retraction occurs during the short time, though some droplets start undergoing breakup mainly by end pinching. It was obvious that in the short waiting time, the relaxation time increases and then it decreases in longer waiting time. The residual stress relaxes by recovery of droplets which is accompanied with a slower relaxation. Slow relaxation of residual stress causes relaxation time increases at waiting time 10 to 20 seconds, however at waiting time 50 seconds, breakup of droplets increases, therefore droplets become finer and it leads to relax faster[4, 8]. As a result of mentioned discussion, the waiting time has affected much more intensely on third step rather than second step which could conclude that the shape recovery is the main factor which impact on the third step.

CONCLUSIONS

In this work, stress relaxation behavior of PP/HDPE blend compatibilized and uncompatibilized was investigated. The main aim was investigation of the trend of the third stage of relaxation modulus curve and determination of the main important factors which affect on it. Test was started with a preshear rate, then in the waiting time shear was ceased and test was finished with step strain experiment. For all the blends, three steps were observed. First step was fast which was attributed to pure phases. The Second step is characterized by a plateau that has a slow rate and related to relaxation of the deformed droplets, in the beginning of the plateau, the droplets are extended and at the end, they recover to spherical shape. Third step is fast, according to our results it depends on droplets size distribution and shape recovery. The narrower droplets size distribution at higher preshear rate leads to shorter relaxation time at the third step. Also the waiting time has affected much more intensely on third step rather than the second step, which could conclude that the shape recovery is the main factor which impact on third step. Thus it can be concluded that the third stage is related to shape recovery of droplets or retraction droplet. Also the results showed that any factors which make the droplets size smaller cause the relaxation time to decrease.

Acknowledgement

The authors cordially thank Dr. Leila Barangi (National *Petrochemical Company, Petrochemical Research and Technology Co., Tehran, Iran) for* her help in the RMS Calibration for the strain step test and her opinion in the analysis of the data.

References

1. D.Nwabunma and T.Kyu, Polyolefins Blends, Published by John Wiley & Sons, Inc., Hoboken, New Jersey, 2008, pp. 59-64.
2. M.M. Dumoulin, P.J. Carreau. *Polymer Engng* **27**, 1627(1987)
3. Okamoto K, Takahashi M, Yamane H, Kashihara H, WatanabeH, Masuda T. J Rheol 43(4):951–965(1999)
4. Iza M, Bousmina M J Rheol 44(6):1363–1384(2000)
5. Gramespacher H, Meissner J .J Rheol 36(6):1127–1141(1992)
6. Wang J, Velankar S. Rheol Acta 45:297–304 (2006a)
7. Wang J, Velankar S. Rheol Acta 45:741–753 (2006b)
8. Yamane H, T.M., Hayashi R, Okamoto K, Kashihara H, Masuda T. *J Rheol* **42**(3):567–580 (1998)
9. Okamoto K, Iwatsuki S, Ishikawa M, and Takahashi M. Polymer 49(8):2014–2017 (2008)
10. Handge UA, Okamoto K, Münstedt H. Rheol Acta 46:1197–1209 (2007)
11. Hayashi R, Takahashi M, Kajihara T, Yamane H. J Rheol 45(3):627–639(2001b)
12. A.M.C. Souza, N.R.D., *Polymer* 43, 3959-3967(2002)
13. A.M.C. Souza, N.R.D., *Polymer* 43, 1313-1321(2002)
14. Demarquette, M.Y.A.M.C.S.b.S.V.N.R. *Rheol Acta* 48, 527–541(2009)
15. Bousmina, M.I.a.M. *J. Rheol* 44, 1363-1384 (2000)
16. R.-E. Riemann, H.-J.C., C. Friedrich. *Polymer Bulletin* 36, 637-643 (1996)
17. R.-E. Riemann, H.-J.C., and Chr. Friedrich. *Macromolecules* 30, 5476-5484 (1997)
18. U. Jacobs, M.F., J. Winterhalter, and Chr. Friedrich. *J. Rheol* 43, 1495-1509 (1999)
19. Vinckier, I., P. Moldenaers, and J. Mewis, J. Rheol. 40, 613–631 (1996)
20. Vinckier, I., P. Moldenaers, and J. Mewis. J. Rheol. 41, 705–718 (1997a)
21. Vinckier, I., P. Moldenaers, and J. Mewis, Rheol. Acta 36, 513–523 (1997b)

Dynamic Heterogeneity of PMMA/PVDF Blends: Composition Dependence and Role of Interphase

Bo Lu[a], Khalid Lamnawar[a*] and Abderrahim Maazouz[a,b]

[a] *Université de Lyon, CNRS, UMR 5223, Ingénierie des Matériaux Polymères, INSA Lyon, F-69621, Villeurbanne, France*
[b] *Hassan II Academy of Science and Technology, Rabat, Morocco*

lubo.mse@gmail.com (B.L.)
khalid.lamnawar@insa-lyon.fr (K.L.)

Abstract. We demonstrated the dynamic heterogeneity in PMMA/PVDF blends, where its composition dependence and the role of interphase were probed. Firstly, composition dependence of thermorheological complexity of blends in the melt was revealed and found to be governed by the molecular entanglements. Dynamic heterogeneity especially in molten state was assessed from microscopic dynamics to macroscopic properties by combing dielectric relaxation spectroscopy (DRS) and small-amplitude oscillatory shear (SAOS). Interestingly, local heterogeneity was further demonstrated to exist in the melt-state blends with intermediate compositions, and its molecular origin was depicted to be an interphase. We demonstrate herein that this interphase shows some unfavorable interchain entanglement, augmenting the overall thermorheological complexity. Hence and referenced with our recent work (Zhang et al., *J. Rheol.*, **2016**, 60, 1-23), the reduced viscosity and speed-up relaxations in the present blends could be explained. Besides, two apparent glass transition temperatures were resolved in light of segment relaxations in the miscible phase and crystal-amorphous interphase, and further examined in the framework of "self-concentration" concept. The presence of a crystal-amorphous interphase, likely leading to three distinct segmental dynamics, was supposed to contribute to the dynamic heterogeneity in segment relaxations for blends at the solid state. Effects of dynamic heterogeneity on dynamic mechanical properties were also addressed (Lu et al., *Soft Matter*, **2016**, 12, 3252-3264).

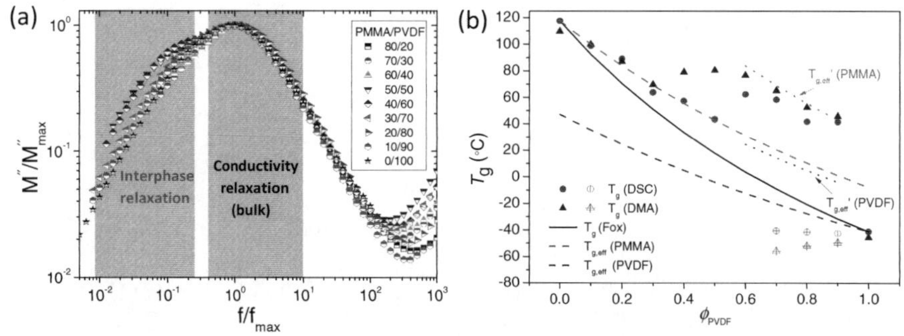

FIGURE 1. Normalized frequency dependence of dielectric loss modulus at a melt temperature of 180 °C; (b) glass transition temperatures as determined by experiments (scatters) and predicted by theories (lines)

Keywords: blends; dynamic heterogeneity; chain entanglement; interphase; properties.

REFERENCES

1. B. Lu, K. Lamnawar, A. Maazouz, and H. Zhang. *Soft Matter*, 2016, **12**, 3252-3264.
2. H. Zhang, K. Lamnawar, A. Maazouz and J. M. Maia, *J. Rheol.*, 2016, **60**, 1-23.
3. H. Zhang, K. Lamnawar and A. Maazouz, *Macromolecules*, 2013, **46**, 276-299.

Rheological Characterization of Macromolecular Colloidal Gels as Simulant of Bronchial Mucus

Olivier Lafforgue (presenting and reference author)[a,b*], Sébastien Poncet[a], Isabelle Seyssiecq[a] and Julien Favier[a]

[a] *Laboratoire M2P2 UMR 7340, Université d'Aix-Marseille, CNRS, Ecole Centrale Marseille*
[b] *SAS Physio-Assist*

olivier.lafforgue@etu.univ-amu.fr

Abstract. Mucus is mainly composed of water (90-95%) and mucins (2-5%), these mucins being high molecular weight macromolecules forming a 3D cross-linked matrix. It makes it a complex non-Newtonian fluid, displaying viscoplasticity, viscoelasticity, shear-thinning and thixotropy. These properties were qualitatively and separately tested by different authors. Due to the difficulties to collect samples and the extreme sensibility of the tested materials, the results are widely variable. A complete and intrinsically consistent characterization remains to be done to develop a reliable rheological model for future numerical simulations of mucus displacements in airways, in the case of pathologies such as cystic fibrosis. For this purpose, samples of mucus simulants were tested using a controlled stress rheometer. The material consists of macromolecular colloidal gels at different concentrations in macromolecules to mimic the variability in mucin production depending on the disease state and environmental factors. The rheological properties at rest were investigated using small amplitude oscillatory shear tests as a function of stress amplitude, frequency and temperature. They revealed that mucus simulant behaves as a gel within a defined linear viscoelastic region and as a viscoelastic liquid above the yield stress zone. To characterize the behavior of mucus in response to in vivo shearing induced by cough or by air flows produced by clearance helping devices, steady state flow tests have been performed. The steady state flow curves for various polymer concentrations are well fitted using a Herschel-Bulkley law. To account for the time dependent behavior of mucus, 3 intervals thixotropy tests were also implemented to monitor structure breakdown and buildup processes. The combination of all these measurements finally designs a reliable procedure accounting for the rheological complexities of mucus, that now needs to be applied to real mucus for validations.

Keywords: Rheology, viscoelasticity, thixotropy, shear-thinning, mucus, gel.
PACS: 83.60.Pq, 83.60.Rs, 83.60.La, 83.60.Bc, 83.80.Hj, 83.85.Cg, 87.19.rh

INTRODUCTION

Respiratory mucus is composed of water (90-95%), mucins (2-5%), lipids (1-2%), salts (1%) and 0.02% of DNA and other molecules such as cells debris [6]. Mucins are high molecular weight macromolecules (glycoproteins) insuring a structural protection function. The entangled and cross-linked network of these branched chains of sugar forms a 3D matrix spanning the gel layer [5]. As a consequence, it is a complex non Newtonian fluid which shows viscoplasticity, shear-thinning, viscoelasticity and thixotropy. The complex association of the latter properties is responsible for the vital clearance function of the mucus layer coating the airways. The clearance role of the mucus layer could be summarized by trapping particles in the gel where enzymes and antibodies start the cleaning and transporting it thanks to beating cilia and air/mucus interaction towards the pharynx, where it is either digested, or expectorated. In disease conditions (e.g. Cystic Fibrosis), the clearance function is affected by modifications of the mucus composition. This makes the rheology of mucus a challenging but crucial topic of study. Previous works on mucus have studied some of these properties but not all at the same time, and often qualitatively [3]. Furthermore, the method for samples collection and the natural variability of this complex biological fluid lead to large differences in the existing literature, in terms of results for mucus rheological characterization [6]. As a consequence, a complete and intrinsically consistent characterization campaign remains to be performed in order to develop a reliable rheological model that can be used for future numerical simulations of mucus displacements in airways, in the case of pathologies such as cystic fibrosis. This constitutes the framework of the present study.

MATERIALS AND METHODS

Two of the main challenges for mucus rheological characterization are the large behavioral variability from one sample to another, and the difficulty to collect it. This implies that a complete and consistent characterization of the same mucus, with several different protocols is not easy to achieve. The present work thus aims at developing a whole protocol to realize a complete characterization of mucus using simulants that can be prepared in large quantities. The simulants were prepared according to the protocol set by Zahm *et al.* [7]. The simulants are made of branched macromolecular chains (branched polysaccharides) from locust bean gum (Viscogum™ Cargill™) and from *Sclerotium* fungus fermentation (Actigum™ Cargill™) in salted water (physiological proportions). The high molecular weight branched macromolecules of sugar, cross-linked by a Sorensen buffer, build a 3D macromolecular network mimicking the mucin matrix patterning the native mucus. To approach the diversity of real mucus, gels with various scleroglucan concentrations from 0.5 to 2wt% were prepared.

The experiments were performed on a stress-controlled rheometer AR 550 (TA Instruments) equipped with a 50 mm 2° steel cone. The temperature was regulated by a Peltier plate. Dehydration was prevented by using a wet steel cover insuring a water saturated atmosphere around the sample.

The viscoelastic properties of the simulant unbroken structure were investigated through a series of dynamical shear experiments (Small Amplitude Oscillatory Shear). The results were interpreted from the evolution of the elastic and viscous moduli and the loss angle (G', G'', δ) in response to the sinusoidal load input.

The stress dependency of the structure was observed via stress amplitude sweeps at constant frequency (1 rad.s^{-1}). A Yield point that delimits the linear viscoelasticity range (LVR) was deduced from the G' departure from linear behavior, this is the stress limit above which the measurements no longer reflect the native structure since a breakdown starts to occur. A flow point was also identified as the moduli crossover, i.e. the point above which the material becomes more viscous than elastic due to a critical structure breakdown.

The frequency dependency (from 0.1 to 100 rad.s^{-1}) of the mucus simulant structure before flow occurs was probed at constant stress amplitude, within the LVR according to the amplitude sweep results.

The temperature dependency of the gel unbroken structure between 10 and 45°C was investigated through a series of up and down temperature ramps at constant frequency and stress amplitude in the LVR (0.2 Pa, 10 rad.s^{-1}).

To characterize the behavior of the mucus layer in response to in vivo-like solicitations such as cough or air flows induced by clearance helping devices, rotational shear tests have been preset to investigate the rheological properties of the simulants under flow conditions. These analyses are not limited to the study of the structure at rest but also account for the breakdown of the gel network. The gel properties were interpreted from the measured shear rate, shear stress or apparent viscosity as a function of the imposed load (shear stress or shear strain rate depending on the preset). The time-dependency of the flow curves was also recorded to characterize the thixotropic behavior.

The plasticity and the non-linear viscoelastic behavior of the materials were investigated throughout controlled shear stress (CSS) flow tests. For every sample, a stress ramp was applied in the largest possible range. Each stress value was applied until a steady flow was considered to be reached. The evolution with increasing shear stress of the shear rate, the viscosity and the strain were recorded to be interpreted in terms of flow behaviors.

Thixotropy was probed by the three interval thixotropy test (3ITT). It consists in a stepwise change of shear rate (CSR) or stress (CSS) to monitor the alteration and recovery of the thixotropic structure: In the first interval, the sample undergoes low shear conditions (in the LVR) to build a reference for the structure at rest. Then, high-shear conditions are imposed to break down the internal structure. During the third interval, the sample is allowed to regenerate under low shear conditions. The state of regeneration can thus be probed at different periods of rest time.

Several of these tests were realized at 20°C and 37°C. No significant differences were noted concerning the influence of the temperature on the results, hence the series of isothermal experiments were performed at 20°C to limit dehydration issues.

RESULTS AND DISCUSSIONS

Small Amplitude Oscillatory Shear: Structure at Rest

The stress sweep tests conducted on simulants gels at different concentrations all showed the same tendencies, however with moduli and yield stresses both increasing according to the polymer concentration. Figure 1 shows the evolution of the dynamic parameters for a 1.5% of scleroglucan chains sample. The plateau region of G' and G'' defines the linear viscoelasticity domain at the preset frequency. In this range, the stress is too small to entail irreversible structural changes of the sample. The structure breakdown only occurs beyond a certain amount of stress: the yield point τ_y. The yield point was defined as the last stress value before deviating of more than 5% of a straight fitting line of the G' plateau.

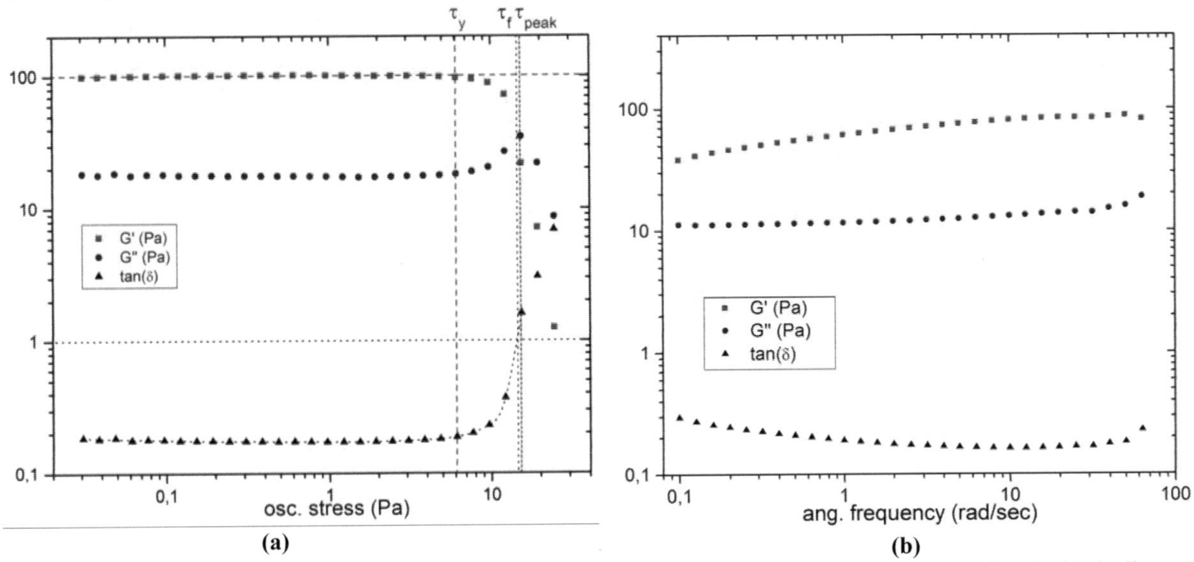

FIGURE 1. SAOS results for a 1.5% sample. (a) Stress sweep 1rd.s^{-1}. The plateau region of the moduli delimits the linear viscoelasticity range. At higher stress values the structure becomes to yield and the moduli cross over (tan(δ)=1). This point characterizes the end of the solid-like state and the beginning of flow (G''>G'). (b) Frequency sweep at 1Pa.

Beyond the yield stress, the moduli are no longer constant due to the breakdown of the gel network. In a first range of values, the stress is large enough to initiate the structure breakdown but too small to allow flowing (G'>G"). The flow occurs beyond the crossover point (G'=G") which ends the yield zone and above which the viscous component of the mucus simulant overcomes the elastic one (G">G'). The structure is then broken and starts to flow.

Above the yield stress, the loss modulus shows an overshoot before decreasing. According to Mezger [4] this is a common phenomenon in gels. The increase of deformation energy inducing the overshoot could be related to a progressive collapse of the network due to structure components with some freedom of motion. The resulting friction would thus be responsible for the high dissipated energy (increase in G'') until the final breakdown occurs at the peak value. This kind of behavior has been classified as Type III ("weak strain overshoot") behavior by Hyun *et al.* [1]. A review of Hyun *et al.* [2] emphasizes the very various possible structural causes of this behavior depending on the class of the soft material and claims the necessity of LAOS study to probe the meaning of this behavior.

The frequency sweep tests showed a gel-like behavior (G'>G'') from 0.1 to 40 rad.s^{-1} for every sample. The moduli both follow a power law under the form $G = K\omega^A$. The power parameter A was found independent of the frequency and equal to 0.1 for the elastic modulus and 0.04 for the loss modulus. The amplitude parameter K increases with increasing polymer concentration.

The samples also showed a gel-like behavior over the whole range of temperature sweep (10-45°C) with a linear decrease of the moduli with increasing temperature. The absolute slope increases with the concentration in scleroglucan chains (0.1 for a 0.75% sample and 0.7 for a 1.50% sample). Beyond 40 rad.s^{-1}, the tests were not reliable probably due to wall slip effects.

Steady-state Shear Flow: Structure in Flow Conditions

Figure 2 plots the flow curves of the mucus simulants under steady-state conditions. Each test were repeated three times with fresh samples to check reproducibility, standard deviations are plotted as error bars. All the samples showed the same profile. First, unsteady flow results are recorded at very low shear. Since the material is multiphase, the shear load could be too small to be homogeneously distributed within the whole volume. It would result in the coexistence of regions of different viscosities leading to self-oscillations. Once steady-state conditions are achieved, the sample remains in a solid-like state (almost constant stress) until a critical yield stress corresponding to the structure collapsing is reached. Beyond this yield point, the measurements show a characteristic shear-thinning behavior with a power law-like profile. The experimental data are well fitted with a Herschel-Bulkley model (Figure 2 (b) and Table 1). Nevertheless, one has to keep in mind that this experiment, as well as the HB model, does not account for the time dependent behaviors. Besides, the yield stress values observed here are not inherent properties of the material since the structural strength of the gel is time dependent as it will be discussed in the next section. The characterization of the steady-state behavior constitutes a first step towards a more complete characterization accounting for time dependent behaviors, it also allows defining a base for the modeling work.

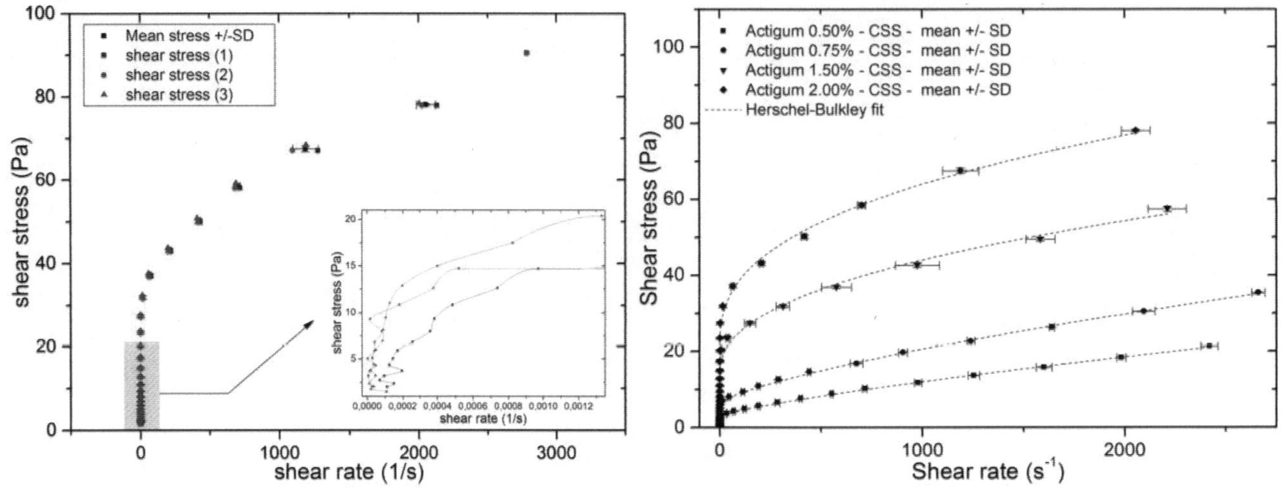

(a) (b)

FIGURE 2. (a) Steady state flow curves of a 2% sample, stress oscillations occurs at very low shear due to inhomogeneous load distribution (inset), Hershel-Bulkley shear-thinning profile when steady-state is achieved. (b) Results for concentrations from 0.5% to 2% of scleroglucan chains, each curve is obtained from the mean of three tests, error bars from the standard deviation.

TABLE 1. Fit results of the Hershel-Bulkley law: $\tau = \tau_0 + K\dot{\gamma}^n$

Concentration (%)	τ_0 (Pa)	K	n
0.50	3	0.04	0.78
0.75	6	0.10	0.72
1.50	14	1.67	0.42
2.00	20	3.29	0.37

Stepwise Change of Shear: Structure Time-dependence

As most colloidal gels, synthetic mucus is prone to time dependent behaviors. Thixotropy was qualitatively observed during hysteresis loop tests and can be properly studied in terms of time dependent destructuration and time dependent regeneration throughout stepwise changes of shear rate. Figure 3 shows the stress evolution with time during a 3ITT of a sample with 2% of scleroglucan macromolecules. The low shear intervals are preset to $0.29s^{-1}$ in order to entail a stress level within the LVR and remain as close as possible to rest conditions. The equilibrium stress value in the first interval constitutes the reference stress level at rest. After the sudden start of flow at $100s^{-1}$, a stress overshoot occurs reaching 4 times the reference value and then progressively decreases toward a new steady level after approximatively 30s. After the brutal cessation of high shear conditions back toward close to rest conditions, the shear stress drops far under the reference value. Then, the sample rapidly recovers 70% of its properties before a progressive and complete recovery of the reference state (see Table 2).

Note that the final steady value of the first low shear interval and the extrapolation of the second interval curve at the same instant provide the stress levels of the unbroken structural state at $0.029s^{-1}$ and $100s^{-1}$. This constitutes the basis for further progressing series of stepwise changes of shear rates leading, among other things, to isostructure flow curves and thixotropic model characterization.

TABLE 2. Progressive recovery of structure properties at rest after cessation of $100s^{-1}$ flow

Recovery (%)	74	90	98	100
Duration (min:sec)	00:05	00:17	02:13	19:00

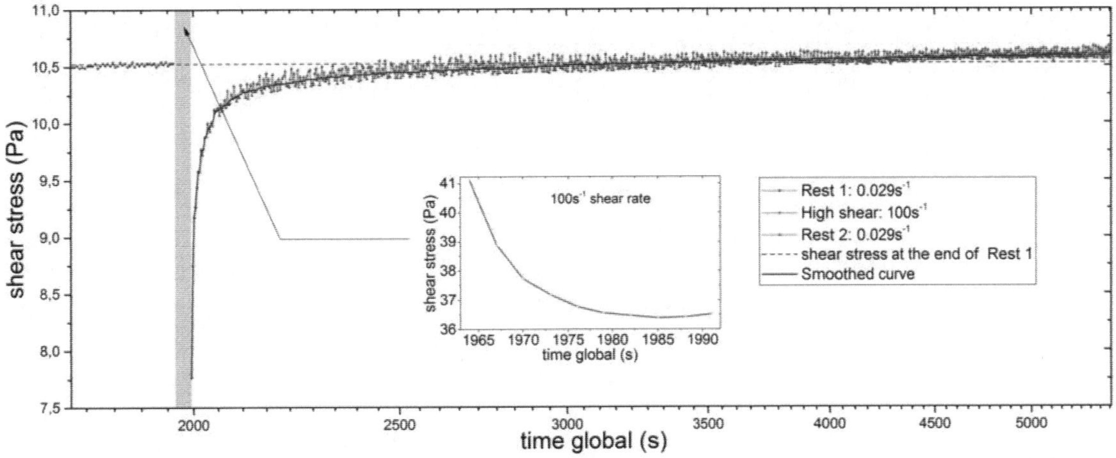

FIGURE 3. Three Interval Thixotropy Test of a 2% sample. The first low shear step gives a reference stress level (structured state). After sudden start of flow at $100 s^{-1}$ a sudden stress overshoot occurs before progressive decrease toward a new steady state (inset). Stress drops at cessation of high rate and increases progressively to reach the reference value.

CONCLUSION

The respiratory mucus plays a crucial role in the protection of vital organs. Its airways clearance function is insured thanks to its complex rheological properties. The difficulty to exhaustively test samples from a same mucus collection first motivated the establishment of a procedure on synthetic mucus samples. The rheological properties at rest were investigated using small amplitude oscillatory shear tests as a function of stress amplitude, frequency and temperature. They revealed that mucus simulant behaves as a gel within a defined linear viscoelastic region and as a viscoelastic liquid above the yield stress zone. To characterize the behavior of mucus in response to in vivo shearing induced by cough or by air flows produced by clearance helping devices, steady state flow tests have also been performed. The steady state flow curves for various polymer concentrations are well fitted using a Herschel-Bulkley law but do not account for time dependent phenomena. To account for the time dependent behavior of mucus, 3 intervals thixotropy tests were also implemented to monitor structure breakdown and buildup processes and further stepwise rate tests are in progress to establish a complete model. The combination of all these measurements finally design a reliable procedure accounting for the rheological complexities of mucus, that now needs to be applied to real mucus for validations.

ACKNOWLEDGMENTS

The authors are indebted to the Physio-Assist company and ANRT for their financial supports through a CIFRE grant and to Laserson France for providing the polymers.

REFERENCES

1. K. Hyun, S.H. Kim, K.H. Ahn, S.J. Lee, "Large amplitude oscillatory shear as a way to classify the complex fluids", *J. Non-Newtonian Fluid Mech.* **107**, 2002, pp. 51-65.
2. K. Hyun, M. Wilhelm, C. O. Klein, K. S. Cho, J. G. Nam, K. H. Ahn, S. J. Lee, R. H. Ewoldt, and G. H. McKinley, "A review of nonlinear oscillatory shear tests: Analysis and application of large amplitude oscillatory shear (laos)". *Progress in Polymer Science* **36**, 2011, pp. 1697-1753.
3. S. K. Lai, Y-Y. Wang, D. Wirtz, and J. Hanes. "Micro- and macrorheology of mucus". *Advanced Drug Delivery Reviews* **61**, 2009, pp. 86-100.
4. T.G. Mezger. *The Rheology Handbook: For Users of Rotational and Oscillatory Rheometers.* European Coatings. Vincentz Network, 2014.
5. D. J. Thornton and J. K. Sheehan. "From mucins to mucus: toward a more coherent understanding of this essential barrier". *Proceedings Of The American Thoracic Society* **1**, 2004, pp. 54-61.
6. P. A. Vasquez and M. G. Forest. "Complex fluids and soft structures in the human body". *Complex Fluids in Biological Systems*, Springer Science, S.E. Spagnolie (ed.), New-York, 2015, pp. 53-110.
7. J.M. Zahm, M. King, C. Duvivier, D. Pierrot, S. Girod, and E. Puchelle. "Role of simulated repetitive coughing in mucus clearance". *European Respiratory Journal* **4**, 1991, pp. 311-315.

Rheological Behavior Modelling for the Copolymer Hydrogenated Poly (isoprene-co-styrene) Solutions

Ioana Stanciu[a]

[a]*University of Bucharest, Faculty of Chemistry, Department of Physical Chemistry, 4-12 Elisabeta Blvd, 030018, Bucharest, Romania*

istanciu75@yahoo.com

Abstract. In this paper I undertook to determine the rheological characteristics of the copolymer hydrogenated poly (isoprene-co-styrene) solutions. I have studied the 3% concentrated solutions for a temperature interval from 40°C and 90°C in 10 degree steps, and for the shear rate from 3 to 1312 s^{-1} in the second domain. The measurements were performed using a Haake VT 550 rotational rheometer with a HV_1 sensor. The rheology of the solutions has been impacted by the shear rate, the temperature and the concentration of the copolymer hydrogenated poly (isoprene-co-styrene). I have found one model that successfully describes the rheological behaviour of the solutions compared to other models that do not consider the copolymer's concentration.

Keywords: Dynamic, Viscosity, Shear Rate, Copolymer

INTRODUCTION

Viscosity is the most important physical feature of the lubricants for it varies with the temperature and the shear rate and determines their performance.

Multi-grade oils contain organic polymer additives to improve the viscosity index. The organic polymers are base oil soluble, have molar mass from 10^4 to 10^6 g/mol and are able to reduce the dynamic viscosity's dependence on the temperature [1-3].

The viscosity index additives are those polymers whose chains may be blocked by twisting them either in between or with the oil's, which alters the speed, the geometry and the molecular movement of the both components and varies along with the temperature [4-6]. The chains get blocked by twisting and therefore cannot move easily at high temperature as compared to no twisting and thus we have a loss in viscosity [7].

The performance of a polymer used as a viscosity index additive are being influenced by the structure, the concentration, the chemical stability, the oxidation thermal resistance, and the shearing stability or strength. For any given polymer, the shearing stability and the effect of viscosity increase or oil thickening varies inversely with the molar mass, while the shearing stability decreases with the increase in the latest, and the higher the mass the higher the viscosity [8]. When choosing a polymer to use as oil viscosity additive, one must keep a balance between its efficiency as a viscosity booster and its shearing stability [9].

The behaviour of an oil film between two surfaces that travel with respect to each other is quantified by the dynamic viscosity, a measure that correlates the shearing tension to the shear rate [9]. To determine the actual efficiency of a given polymer as a viscosity booster, one must know the rheological behaviour of its solutions over the widest range of shear rates.

While the rheological behaviour of certain viscosity boosters has been studied for low and high shear rates (10^4 - 10^6 s^{-1}), it has not been considered for intermediate shear rates. Usually, diluted solutions display a Newtonian behaviour while those semidiluted and concentrated behave like non-Newtonian fluids [10].

As the modification of the temperature-induced viscosity variation has a significant impact on the bearings' load capacity and the oil flowing rate, in many applications we need to know the viscosity-temperature relations [11].

In this paper I undertook to determine the rheological characteristics of the copolymer hydrogenated poly (isoprene-co-styrene). I have studied the 3% concentrated solutions for a temperature interval from 40°C and 90°C in 10 degree steps, and for the shear rate from 3 to 1312 s^{-1} in the second domain. The measurements were performed using a Haake VT 550 rotational rheometer with a HV_1 sensor. The rheology of the solutions has been impacted by the shear rate, the temperature and the concentration of the copolymer hydrogenated poly(isoprene-co-styrene). I have found one model that successfully describe the rheological behaviour of the solutions compared to other models that do not consider the copolymer's concentration.

EXPERIMENTAL

The chemical and physical properties of copolymer hydrogenated poly (isoprene-co-styrene) are: physical state – solid, form – white solid blocks, colour – compressed crumbs, odourless, flashpoint > 150°C, water insoluble, not a hygroscopic matter, stable, density (15°C) – 0.272 g.cm^{-3} and no hazardous decomposition.

The SAE 10W oil's physical and chemical properties are: form – homogeneous, yellowish brown liquid, specific odour, density (15°C) – 0.875-0.910 kg·m^{-3}, kinematic viscosity at 40°C - 90 cSt, kinematic viscosity at 100°C - 8 cSt, dynamic viscosity (20°C) - 65 cP, viscosity index – 70-100, solubility in organic solvents, petroleum, fat, water-insoluble, flash point > 200°C, melting point - (-10) ÷ (-15)°C, amount of coke – 0.03 – 0.5%. The utilized SAE 10W oil is mostly paraffinal and contains 75% saturated hydrocarbons.

The of 3% copolymer hydrogenated poly(isoprene-co-styrene) solutions concentrations have been prepared at the room temperature through continuous mixing for several weeks.

A Haake VT 550 rotational drum viscosity meter has been used to determine the rheological features of the copolymer hydrogenated poly(isoprene-co-styrene) solutions concentrated. The rheological behaviour of the copolymer hydrogenated poly(isoprene-co-styrene) solutions in the SAE 10W oil has been studies on a Haake VT 550 viscosity meter with a HV_1 sensor that provides for shear rates from 3 to 1312 s^{-1} for a temperature range from 40 to 90°C.

RESULTS AND DISCUSSION

The Figure 1 shows variation dynamic viscosity with temperature of 3% copolymer hydrogenated poly(isoprene-co-styrene) solution concentration for the 437.9 s^{-1}, 729 s^{-1} and 1312 s^{-1} shear rates. The solution was studied in the 40-90°C temperature range for the 145.8 s^{-1}, 243 s^{-1}, 437.9 s^{-1}, 729 s^{-1} and 1312 s^{-1} shear rates. As one can see from the curves, the solution's dynamic viscosity decreases with the increase of the temperature and the shear rate.

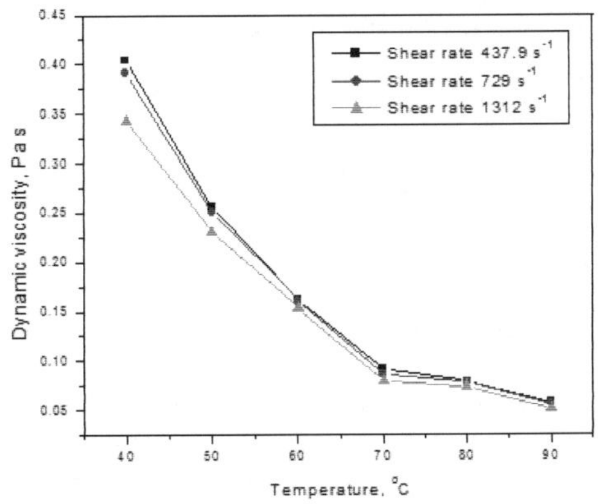

FIGURE 1. Dynamic viscosity temperature variation for a 3% copolymer hydrogenated poly (isoprene-co-styrene) solution concentration for certain shear rates

For a 40°C temperature and a 145.8 s^{-1} shear rate, the 3% copolymer hydrogenated poly (isoprene-co-styrene) solution displays a dynamic viscosity of 0.4278 Pa s. If we increase the shear rate to 243 s^{-1} and keep the temperature constant, the solution's viscosity goes down to 0.4083 Pa.s. When keeping the shear rate constant at 243 s^{-1} and increasing the temperature of the copolymer hydrogenated poly(isoprene-co-styrene) solution, the dynamic viscosity reaches 0.2566 Pa s.

For a shear rate of 437.9 s^{-1} and a temperature range of 40 - 90°C, the dynamic viscosity of the copolymer hydrogenated poly(isoprene-co-styrene) solution is 0.4044 Pa.s for a temperature of 40°C and goes down to 0.0577 Pa. s at 90°C.

For a shear rate of 729 s^{-1} and a temperature range of 40 - 90°C, the dynamic viscosity of the solution reaches 0.3912 Pa.s at 40°C and decreases to 0.0555 Pa.s at 90°C.

For a shear rate of 1312 s^{-1} and a temperature range of 40 - 90°C, the dynamic viscosity of the copolymer hydrogenated poly (isoprene-co-styrene) solution is 0.3436 Pa.s at 40°C and goes by a polynomial curve down to 0.05083 Pa.s at 90°C. As one can see, at the temperatures of 80°C and 90°C the solution's dynamic viscosity displays a insignificant reduction even if we increase the shear rate.

The Table 1 shoes the shear rates for which we determined the dynamic viscosities of the 3% copolymer hydrogenated poly(isoprene-co-styrene) solution in relation with the temperature, the parameters A, B and C obtained by the polynomial fitting of the curve, the correlation coefficients obtained by the polynomial fitting, and the statistically obtained correlation coefficients.

The equation that describes the dynamic viscosity temperature of dependence for all of the solutions studied is (1):

$$\eta = A + Bt + Ct^2 \tag{1}$$

where η – the solution's dynamic viscosity (Pa.s), t – temperature in °C, A, B, C – parameters that depend on the polymer's and solvent's nature, and R^2 – correlation coefficient.

TABLE (1). The shear rate and the parameters A, B and C of described by the equation (1), and the correlation coefficients obtained by the polynomial fitting and statistically

Shear rate, s^{-1}	*A*	*B*	*C*	Correlation coefficients, R^2 equation (1)	Statistical correlation coefficients, R^2
437.9	1.2829	-0.0289	1.7149E-4	0.9955	0.8724
729	1.2360	-0.0277	1.6359E-4	0.9949	0.8763
1312	1.0441	-0.0228	1.3105E-4	0.9944	0.8949

For the 3% copolymer hydrogenated poly(isoprene-co-styrene) solution, the value of parameter A decreases from 1.2829 to 1.0441 for shear rates between 437.9 s^{-1} and 1312 s^{-1}. The parameter B displays very narrow values over the shear rate range in which we studied the solution. The parameter C goes down with the increase in the shear rate and the working temperature. The correlation coefficients determined by the equation (1) are close to one, which shows that the equation accurately describes the rheological behaviour of the solution.

The statistically determined correlation coefficients show every point's deviation from the straight line. As expected, the statistical correlation coefficients are much lower as the solution's dynamic viscosity exponentially decreases with the increase in the solution's temperature.

CONCLUSIONS

In this paper, I have studied the rheological behaviour of a copolymer hydrogenated poly (isoprene-co-styrene) solution of various concentrations for certain shear rates and temperatures. The reduction in the dynamic viscosity has been more visible at lower shear rates than at higher rates. The dynamic viscosity increases with the increase in the temperature and the increase in the concentration of the copolymer hydrogenated poly (isoprene-co-styrene) solution. Nevertheless, the results show that the viscosity of the copolymer solution has been more sensitive to higher temperatures. The equations (1) provide a good correlation between the dynamic viscosity and the shear rate, the concentration and the temperature. The computed parameters can be used to describe the rheology of the solution in accordance with the shear rate, concentration and temperature.

REFERENCES

1. S. Zhong, M. T. Zuber, L. Moresi, M. Gurnis , *Journal of Geophysical Research: Solid Earth (1978–2012)* **105**, 11063-11082 (2000).
2. Jr, T. G. Fox, P. J. Flory, *Journal of the American Chemical Society* **70**, 2384-2395 (1948).
3. A. K. Doolittle, *Journal of Applied Physics* **22**, 1031-1035 (2004).
4. E. M. A. Elbashbeshy, M. A. A. Bazid, *Journal of Physics D: Applied Physics* **33**, 2716 (2000).
5. O. O. Fasina, H. Hallman, M. Craig-Schmidt, C. Clements, *Journal of the American Oil Chemists' Society* **83**, 899-903 (2006).
6. O. O. Okoturo, T. J. VanderNoot, *Journal of Electroanalytical Chemistry* **568**, 167-181 (2004).

7. C. T. Nguyen, F. Desgranges, G. Roy, N. Galanis, T. Mare, S. Boucher, H. Angue Mintsa, *International Journal of Heat and Fluid Flow* **28**, 1492-1506 (2007).
8. C. J. A. Roelands, J. C. Vlugter, H. I. Waterman, *Journal of Basic Engineering* **85**, 601-607 (1963).
9. N. G. Kafoussias, E. W. Williams, *Acta Mechanica* **110**, 123-137 (1995).
10. S. Yasutomi, S. Bair, W. O. Winer, *Journal of tribology*, **106(2)**, 291-302 (1984)
11. K. U. Hess, D. B. Dingwell, E. Rössler, *Chemical geology*, **128(1)**, 155-163 (1996).

Rheological Properties of Graphene/Nylon 6 Nanocomposites Prepared by Masterbatch Melt Mixing

Chang-Mou Wu*, Sh-Shen Cheong, Wen-You Lai, Teng-Hsiang Chang

[a] *Department of Materials Science and Engineering, National Taiwan University of Science and Technology, Taipei, 106 Taiwan,ROC*

E-mail:cmwu@mail.ntust.edu.tw

Abstract. In this study, rheological behaviors of graphene/nylon 6 (GP/PA6) nanocomposites were investigated. Graphene filled nylon 6 nanocomposites were prepared through diluting a GP/PA6 masterbatch in a nylon matrix by melt compounding with a kneader mixer. The viscosity curves display a much greater decrease with frequency and exhibit shear-thinning behavior which was confirmed by the flow behavior index (n) analysis. The viscous activation energy of nanocomposites decreased as the concentration of graphene increased, showing that the composites exhibit less temperature dependence at elevated temperature. It is believe that the GP/PA6 interactions increase with increasing the graphene loadings.

Keywords: Graphene, Nylon 6, Rheological properties, Activation energy, Flow behavior index, Master batch.
PACS: 83

1. INTRODUCTION

The prospective widespread usage of graphene (Gp) in industrial workshops will cause emissions to the environment and result in an increase of human exposure to Gp. Therefore, handling of Gp in plastic parts manufacturing workshops is however often a key issue. These difficulties represent a cause of concern for human health and indicate that strict preventive and protective measures should be taken to limit inhalation exposure to Gp in occupational settings. Using commercial masterbatch in the production of Gp based polymer nanocomposites would be a better choice as it offers adjust free, no health and safety risks in comparison to bulk Gp dispersion process. Other benefits of using masterbatch include elimination of dispersion difficulties, formulation development and easy handling. However, literature survey reveals that, reports on the detailed investigations on the properties of Gp/PA 6 masterbatches are limited when compared to the studies of nanocomposites prepared by direct incorporation of Gp in nylon and this prompted us to take up this work. Therefore, in the present work Gp/PA 6 composites were produced with varying Gp content by diluting Gp/PA 6 masterbatch. The prepared nanocomposites are characterized for their rheological properties under various Gp contents, shear rates and temperatures.

2. MATERIALS AND METHODS

Nylon 6 (N140, Libolon, Taiwan) and graphene (P-ML20, Enerage Inc. Taiwan) were used in this study. The Gp nanoparticles consist of few layers Gp having a lateral dimension of 9-40μm and a thickness of 50 nm. Gp is prepared with oxygen content around 2.5 wt% and specific surface area 20 m2/g.

Four Gp concentrations (0, 3, 5, 10 wt%) of Gp/PA 6 composites were prepared. At first, we prepared 10 wt% Gp/PA 6 masterbatches by twin-screw extruder (Y-REG-02042, Yuh-Dak Machinery CO., LTD) with a screw diameter (D) of 21.7 mm and length to diameter ratio (L/D) of 42:1. The barrel and nozzle temperature were set at 230 ℃ and 240 ℃. The screw speed is 30 rpm. Then, we prepared 3 and 5wt% of Gp/PA 6 composites by using 10 wt% Gp/PA 6 masterbatches blended with pure nylon 6 in kneader mixer (JKM-DK50G, Jian Kwang Machinery).The mixing rotation speed and temperature were fixed at 30 rpm and 230 ℃, respectively.

Rheological specimens were first prepared by using the Polytetrafluoroethene (PTFE) film to wrap the Gp/PA 6 composites and then put it in the hot¬-press at 230℃, 50 kg/cm3. Finally, we can get the samples with average thickness 0.5 mm. All specimens were dried in vacuum oven at 50℃ for 8 hours. Rheological measurements were performed on rotary rheometer (MCR-102, Anton Parr). A gap of 0.2 mm between two parallel-plates was used for all tests and the diameter of the plate was 25 mm. Steady shear sweeps were performed in a stepwise manner from

0.1 to 500 s-1, with the sample being held at each step for 10 s in test. Samples are left to equilibrate for 1 min. prior to measurement. Three temperatures: 240, 245, and 250 ℃ were used.

3. RESULTS AND DISCUSSION

3.1 The viscous behaviors in different shear rates and temperatures

The effect of shear rate on the viscosity of various concentrations of Gp/PA 6 composites at 245 ℃ is depicted in Fig. 3. The viscosities of all concentration composites decrease with increasing shear rate, indicating pseudo plastic flow behavior of the Gp/PA 6 composites. The pseudo plastic nature of polymers arises from the randomly oriented and entangled nature of the polymer chains which, on application of high shear rates become oriented and disentangled resulting in reduction of viscosity. Under low shear rate (<100 s^{-1}) (Fig. 3(a)), the viscosity increased with increasing of Gp addition; the 10 wt% Gp/PA 6 composite showed highest viscosity: 4910 Pa·s. While the viscosity tested at high shear rate (100 s^{-1} to 500 s^{-1}) as shown in Fig. 3(b), decreased with increasing of Gp addition; the pure nylon6 showed highest viscosity: 106 Pa·s. It may be attributed to slip processes localized at interface between Gp and nylon 6 molecules aligned in the direction of shear, chemical degradation and de-entanglement of nylon 6 chains at high shear rate.

In order to quantitate the flow behaviors of Gp/PA 6 composites under different temperature and shear rate, the power-law was applied. A Power-law fluid, or the Ostwald–de Waele relationship, is a type of generalized newtonian fluid for which the shear stress, τ, is given by

$$\tau = K\dot{r}^{n} \qquad (1)$$

or

$$\eta = K\dot{r}^{n-1} \qquad (2)$$

Where η is viscosity, K is the flow consistency index, ṙ is the shear rate or the velocity gradient perpendicular to the plane of shear, and n is the flow behavior index. Power-law behavior is characterized by a power (flow behavior index n) relationship between shear stress and shear rate. Where K reflects the viscous properties of the polymer melt, with higher values representative of more viscous materials, and n is the power-law index giving a measure of the pseudo plasticity, with greater departures from unity showing more pronounced shear-thinning characteristics. The n gives an idea about the nature of flow, i.e., whether it is newtonian or non-newtonian. Most polymers show pseudo plastic behavior with n less than 1. By plotting ln η vs. ln ṙ curve, the power-law index n and K were determined from the slope and the intercept respectively.

The influence of shear rate and Gp concentration on the viscous flow behavior of nylon 6, as shown in Fig. 1, was determined quantitatively by the n and K value and summarized in Table 1. As Gp concentration increased, the n value decreased from 0.99 to 0.51. The results revealed that the flow behavior of the Gp/PA 6 composites changed from newtonian to pseudo plastic flow. As for the flow consistency index, K, the K value (viscosity) increased dramatically at 10 wt%, whereas slight increased from 0-5 wt%. The 10wt% Gp/PA 6 composite showed highest K value (1857 at 240 °C). The results indicated high intermolecular friction resulted from strong interaction between Gp and nylon 6 molecules, which restricted the motion of nylon 6 molecules. It is believed that filler-polymer and filler-filler interactions increase with increasing the filler loadings. The filler-filler interactions play a dominant role in the flow behavior for the composites, resulting in an increase in the viscosity without the newtonian plateau region. The physical network of the Gps caused by the filler-filler interaction can be understood as the so-called rheological percolation. The n values for various Gp/PA 6 composites were not influenced by test temperature between 240-250 °C. The results indicated that the flow behaviour does not influence by the tested temperatures. As temperature increased, the K value decreased, while the trend for the effects of Gp concentration on the K value remained. It is attributed to the chains un-entangled at high temperature, while the temperature was not high enough to destroy the physical interaction between Gp and nylon 6 molecules.

3.2 Effects of temperatures on the viscous flow

The plots of viscosity to shear rate in different temperature for various Gp concentrations composite samples are shown in Fig.2. We can notice that the rise of temperature results in the viscosity decrease. For filled polymers, the viscosity of composite system is determined by how molecules constituting the system interact. Viscosity is a measure of the resistance of flowing movement due to internal friction between polymer chains and fillers. This resistance may be influenced by temperature, and thus the Arrhenius' equation is applied to determine the viscous flow activation energy.

$$\ln\eta = \ln A - \frac{E_a}{RT} \quad (3)$$

Where A is constant, Ea is viscous flow activation energy, R is gas constant, T is absolute temperature. Ea is the energy consumed for breaking up the interactions among the chain segments [7]. Ea is also the sensitivity of viscosity to temperature which means when Ea becomes smaller, viscosity becomes more sensitive to temperature. A plot of lnη v.s reciprocal of temperature for various Gp concentrations composite samples tested at shear rate: $0.1s^{-1}$ was shown in Fig.3. Thus, Ea was determined from the slope of line shown in Fig. 3 and listed in Table 2. Similarly, the Ea values for all samples tested at 1 s^{-1} and 10 s^{-1} shear rates were determined and listed in Table 2. We found that the Ea of Gp/PA 6 composites decreased with increasing of Gp concentration (from 232 to 53 kJ/mol). The results indicated that the viscosity becomes more sensitive to temperature for high loading Gp samples. That means the interaction forces between molecules of Gp and nylon are sensitive to temperature. The interactive forces are easier to break at high temperature. Thus, it is important to be paid more attention on processing of the Gp/PA 6 composite systems to maintain a steady flow behavior.

Besides, according to the time-temperature superposition principle [8], we know that shear rate may also affect the Ea. The flow behavior of the Gp/PA 6 systems at high shear rate can equivalent to low temperature. Thus, the Ea value increased with the shear rate increased, and agrees to our result as shown in Table 2.

The viscosities we derived from the rotary rheometer which help us seek an important rheological property: shear stress though the formula:

$$\tau = \eta\dot{r} \quad (4)$$

The resulting shear stress exerted by the nylon 6 enhances the exfoliation of Gp. It is believed that high levels of shear stress advances to exfoliate the Gp platelets. Based on results in Fig. 2 and Eq (4), plots of shear stress vs shear rate were determined as shown in Fig. 4. In low concentration Gp samples (≤ 5 wt%), the shear stress increased with the shear rate increased. The most suitable processing parameters for best exfoliation stress level was found in the range 25-30 kPa at 100 s-1share rate and 240 ℃. Maximum share stress (14 kPa) was found at 50 s-1share rate and 240 ℃ for 10 wt% Gp/PA 6 samples, which was suggested as the best processing parameters. The lower stress level for high Gp concentration samples was attributed to the shear-thinning flow behavior. The results provide important flow information for guiding process of Gp/PA 6 composites.

4. CONCLUSIONS

In this study, effects of Gp content on the rheological behavior and viscous activation energy of Gp/PA 6 composites were investigated. Twin-screw extruder was used to produce 10 wt% Gp/PA 6 masterbatch. A kneader mixer was used for dilution process to provide another 3 wt% and 5 wt%. The DSC and PLM results showed heterogeneous nucleation induced by Gp and good dispersion of Gp in the nylon 6 matrix. The viscosities of Gp/PA 6 composites increased with increasing Gp concentration. The interaction between nylon and Gp which results in a strong restriction of movement in nylon 6 molecule was revealed. As the concentration of Gp increased, the flow behavior index (n) decreased from 0.99 to 0.51, indicated the flow behavior changed from Newtonian to pusedoplastic flow. The viscosity became more sensitive to temperature for high loading Gp samples. The viscous activation energy of composites decreases as the concentration of Gp increases. The most suitable processing parameters for best exfoliation stress level was found in the range 25-30 kPa at 100 s-1share rate and 240 ℃.

REFERENCES

1. G. Mittal, V. Dhand, K. Y. Rhee, S. J. Park, W. R. Lee, J. Indust. Eng. Chem.
2. A. B. Sulong, M. I. Ramli, S. L. Hau, J. Sahari, N. Muhamad, H. Suherman, Composites B: Eng. **50**, 54-61 (2013)
3. M. Abdel-Goad, Composites B: Eng. **42**, 1044-1047 (2011)
4. K. Prashantha, J. Soulestin, M. F. Lacrampe, P. Krawczak, G. Dupin, M. Claes, Compos. Sci. Technol. **69**, 1756-1763 (2009)
5. K. Q. Xiao, L. C. Zhang, I. Zarudi, Compos. Sci. Technol. **67**, 177-182 (2007)
6. S. Mishra, S. S. Sonawane, N. G. Simpi, Appl. Clay Sci. **46**, 222-225 (2009)
7. A. Wiśniewska, K. Sozański, T. Kalwarczyk, K. Kędra-Królik, C. Pieper, Polymer **55**, 4651-4657 (2014)
8. R. M. Guedes, Polym. Testing **30**, 294-302 (2011)

TABLE (1). Flow behavior index (n) and consistency index (K) of Gp/PA 6 composites at different temperatures

Gp conc. (wt%)	Temperature (℃)					
	240		245		250	
	n	K	n	K	n	K
0	0.99	342	0.98	254	0.99	119
3	0.91	469	0.95	292	0.89	206
5	0.84	682	0.80	452	0.87	432
10	0.51	1857	0.52	1832	0.53	1368

TABLE (2). Viscous flow activation energy of various Gp/PA 6 composites at different shear rates

GP Conc. (wt%)	0.1 (s^{-1})	1 (s^{-1})	10 (s^{-1})
0	**232**	**234**	**242**
3	**163**	**181**	**208**
5	**120**	**109**	**92**
10	**53**	**73**	**73**

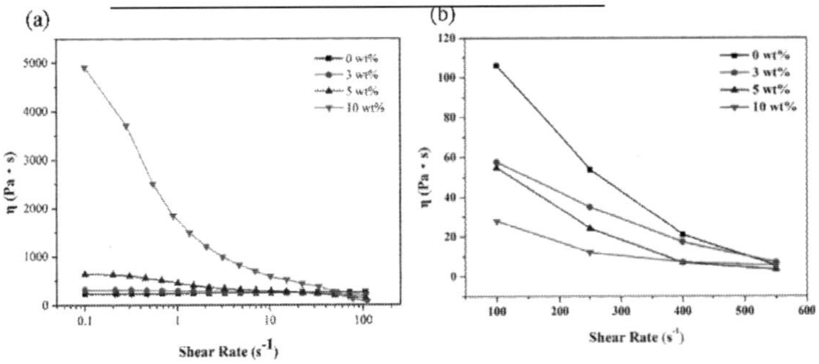

FIGURE 1. The relationship of viscosity and shear rate for GP/PA 6 at (a) low shear rates (b) high shear rates.

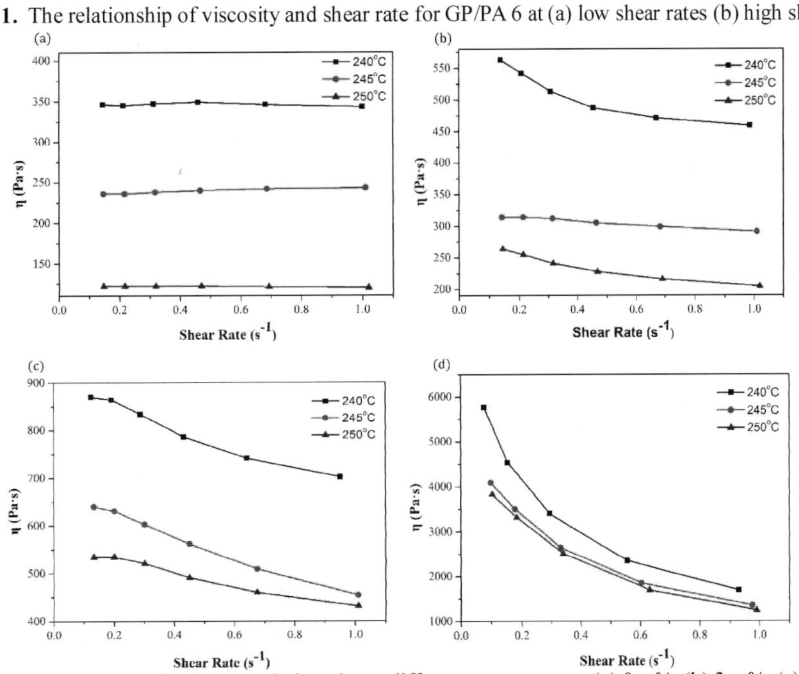

FIGURE 2. The relation between shear rate and viscosity at different temperatures. (a) 0wt%, (b) 3wt%, (c) 5wt%, (d)10wt%

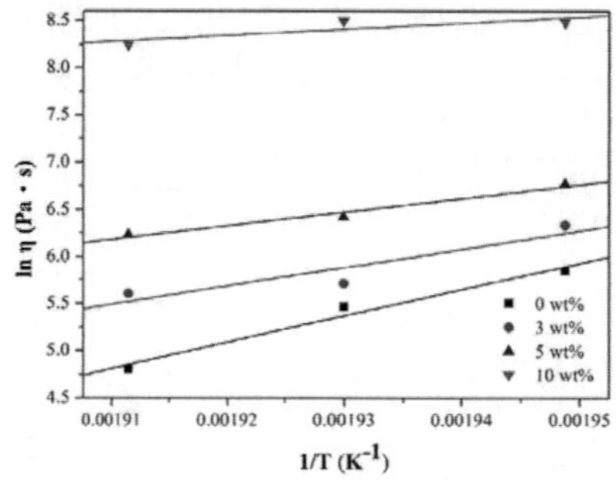

FIGURE 3. The viscosity against the reciprocal absolute temperatures for various GP/PA6 composites at shear rate: 0.1 s^{-1}.

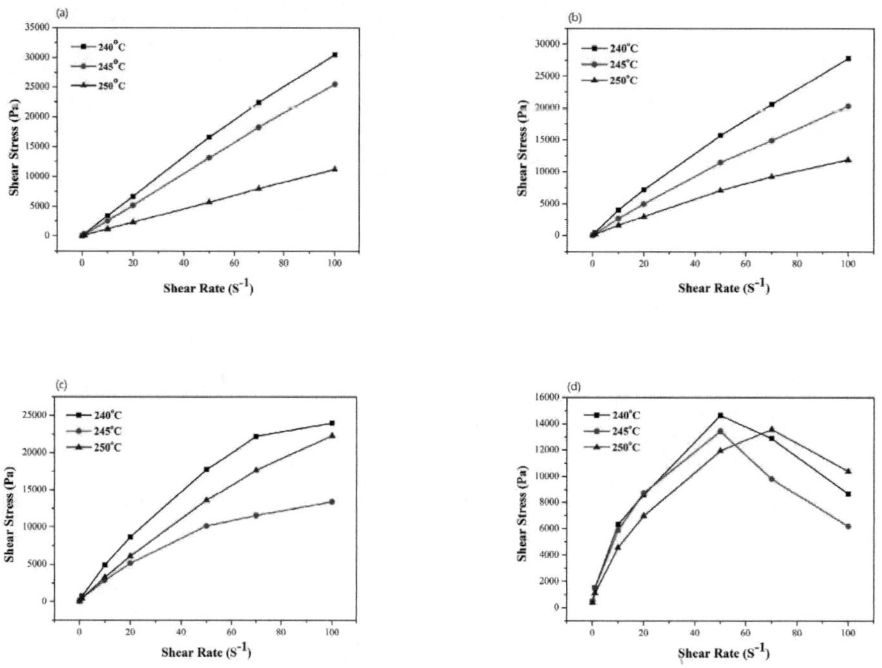

FIGURE 4. Share stress against shear rate under various temperature for various GP/PA6 composites (a) 0 wt% (b) 3 wt% (c) 5 wt% (d) 10 wt%.

The Effect of Lipid Absorption on the Mechanical Properties of Poly(Styrene-Block-Isobutylene-Block-Styrene) for Use in Biomedical Applications

Alexandra Damley-Strnad, Mauro Fittipaldi, Brigitte Morales, Landon R. Grace*

Department of Mechanical & Aerospace Engineering
University of Miami, Coral Gables, FL, 33146, USA

a.damleystrnad@umiami.edu
l.grace@miami.edu

Abstract. The decrease in tensile strength and increase in specimen weight due to lipid diffusion in a biocompatible thermoplastic elastomer was studied and quantified. Mechanical and viscoelastic properties of poly(styrene-isobutylene-styrene) (SIBS) block copolymer are critical to determine feasibility of certain load bearing in vivo applications. Moreover, changes of these properties due to the presence of lipids must be well understood for long-term bio implantation. Dumbbell specimens were thermoformed via injection molding and weights were recorded. Lipid uptake in the body was simulated by specimen immersion in palm and castor oils at 25 °C and 37 °C. After only 96 hours of immersion at body temperature (37 °C), dumbbell weight increased by 6% and 0.3% for palm oil and castor oil, respectively. These values correspond to a reduction in ultimate tensile strength of approximately 30% and 10%, respectively. These results restrict the use of this biocompatible polymer in certain critical components due to the high concentration of lipids in vivo. Based on these significant and rapid reductions in tensile strength in the presence of lipids, it is of vital importance to fully understand the bio-durability and lipid uptake characteristics of SIBS for future design and performance prediction of implantable devices. Further, the results highlight the necessity of improving lipid resistance in order to fully exploit the biocompatibility of SIBS.

Keywords: SIBS, Lipid Absorption, Tensile Strength, Bio-degradation
PACS: 66.30.hk

INTRODUCTION

Poly(Styrene-block-Isobutylene-block-Styrene) (SIBS), a thermoplastic elastomer, has become an increasingly important material in a variety of applications in the medical field due to its high degree of biocompatibility [1-5]. SIBS has been found to interact well with the human body: The Food and Drug Administration approved the material for ultra-long term vascular device applications and it is currently used in the TAXUS® drug-eluting coronary stent [1,4]. Despite favorable results in these applications, some in vivo studies have found SIBS to be susceptible to viscoelastic creep as well as surface cracking [4-9]. Among the more likely causes of these deficiencies is exposure of the material to lipids, which are an integral component of the in vivo environment. As such, careful study of the effects of lipids on the material's mechanical and viscoelastic properties is necessary. Although there is ample research detailing the effects of lipids on other elastomers, there has been no widely-published work that reports the effects of lipids on SIBS. The nature of these types of materials makes it a challenging issue that has yet to be adequately addressed in the open literature. Few of these studies have focused on thermoplastic elastomer block copolymers, though there are various reports of significant degradation in elastomers similar to SIBS after exposure to lipids. Lipid-induced degradation of SIBS is a significant impediment to the use of this extremely biocompatible polymer in vivo, particularly in lipid-rich areas of the body. It is extremely important to fully understand the mechanical properties of the material in order to take advantage of its excellent biocompatibility.

In order to efficiently assess the susceptibility of SIBS to lipid uptake and lipid-induced degradation, a suitable substitute fluid was chosen. Palm oil is an edible vegetable oil that is derived from the fruit of oil palms. This oil has multiple uses ranging from the manufacturing of soaps and lubricants to food and preservatives. Palm oil was chosen to represent in vivo conditions in this experiment as it contains both saturated and unsaturated fats, vitamin E, and beta-carotene, all commonly found in the human body. Palm oil is often used in this capacity. A study on the effect of palm oil on poly(urethane urea) polymer was performed on HFL9-PU1, HFL13-PU2, and HFL16-PU3 polymers. The results show that palm oil significantly decreases the ultimate tensile strength of the material while increasing specimen weight [10]. This significant change in mechanical properties was attributed to the absorption of the lipid, its plasticizing effect, the weakening of virtual cross links (VCL) and reformation and realignment of the molecular structure with VCLs of the material. Based on these results for materials with similar characteristics, it is important to understand how SIBS reacts when exposed to lipids, as there are high concentrations of lipids in certain areas of the body where this material might be implanted. It is of vital importance to fully understand the bio-durability and lipid uptake characteristics of SIBS for future design and performance prediction of implantable devices, as well as quantify tensile strength degradation and the expected onset of these effects.

The purpose of this study is to quantify the effects of palm oil absorption, and by extension in vivo lipids, on the mechanical and viscoelastic properties of the poly(styrene-isobutylene-styrene) (SIBS) block copolymer and to establish a preliminary relationship between lipid absorption and ultimate tensile strength.

EXPERIMENTAL

The SIBS pellets used in this experiment have a composition by weight of 30% styrene and 70% isobutylene. The block copolymer, trade name SIBSTAR™ 103T, was used in the as-received condition from Kaneka Corporation. Pellets were injected using a bench top injection molder that was preheated to 450 °F to ensure adequate fusion and flow into the dumbbell mold, minimizing entrapped air bubbles. Injection speed is fully controllable via a linear actuator and variable speed controller built in to the injection molder. Before injecting the test samples, the injection molder was flushed to minimize specimen variability. The dumbbell mold used to create the samples is in compliance with Type V of ASTM D638 Standard Test Method for Tensile Properties of Plastics. The parameter levels used for this study were as follows: melt temperature of 450 °F, mold temperature of 50 °C, injection rate of 100 mm^3/s, and packing time of 10 seconds. These levels were based on previous experience with this material and allowed a relatively consistent outcome for each specimen in terms of dimensions and strength. All specimens (128) were produced in one sitting to reduce the risk of any accidental variations in injection. Specimens that had visible imperfections or trapped air bubbles were replaced and discarded.

The specimens were numbered and weighed; 28 specimens were immersed in a distilled water control group at 25 °C and 36 in palm oil at 25 °C; 28 specimens were immersed in a distilled water control group at 37 °C and 36 in palm oil at 37 °C. All specimens were immersed between 0 and 550 hours prior to tensile testing. Lipid absorption was quantified via gravimetric analysis. Gravimetric data was obtained by removing the SIBS samples from palm oil immersion, followed by removal of surface oil with oil absorption towels. After most of the surface oil had been removed with the towels, each sample was washed in acetone to remove the remaining residue. After the acetone evaporated, each sample was weighed using a precision analytical balance, and then re-immersed after all samples had been recorded. The water-immersed specimens were dried and then weighed in a similar fashion, using water absorbent towels and brief exposure to ambient air to remove the liquid on the surface via evaporation. Testing of 120 specimens was performed using an Instron® 5966 Tensile Testing Machine at a rate of 500 mm/min. Four specimens were tested for each temperature and fluid immersion condition. All values represent the average of all four specimens tested for each immersion temperature, and all specimen groups are identified by the exposure temperature and fluid type.

RESULTS AND DISCUSSION

Table 1 shows the fluid content (by weight) after 550 hours of full immersion. Based on these results, it is clear that the specimens immersed in palm oil were much more susceptible to fluid uptake. Moreover, the specimens immersed in palm oil at 37 °C absorbed more fluid than equivalent samples exposed at 25 °C. This result is not unexpected, and conforms to the general behavior of fluid uptake in polymers at elevated temperatures.

TABLE 1. Fluid content by weight percentage after 550 hours immersed

Water 25 °C (%)	Palm Oil 25 °C (%)	Water 37 °C (%)	Palm Oil 37 °C (%)
0.1	8.3	0.2	13.8

Figure 1 shows the tensile strength as a function of the number of hours immersed in palm oil at 37 °C and 25 °C. The curve for the specimens at 37 °C is much steeper from the very first measurement, which is an indication of the sensitivity of tensile strength to the increase in temperature.

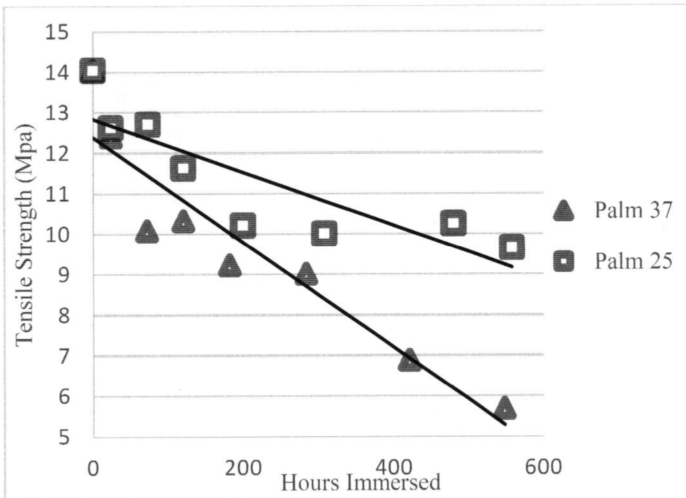

FIGURE 1. Tensile strength vs. immersion time in palm oil at 37 ºC and 25 ºC

Figure 2 shows that the behavior of SIBS in water is quite unlike that in palm oil. Serving as our control group, the water was expected to have minimal effect on the material. Despite the lack of water absorption, the specimens were still somewhat affected by the liquid.

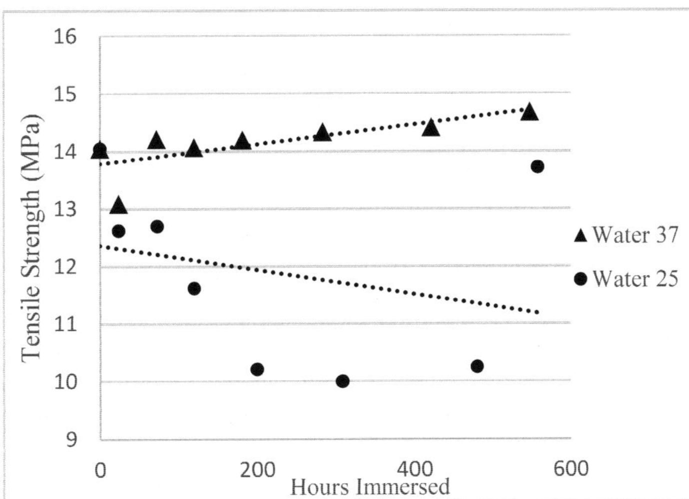

FIGURE 2. Tensile strength vs. immersion time in water at 37 ºC and 25 ºC

The percent weight increase of the specimens as a result of the immersion fluids at their respective temperatures in comparison to the percent decrease in tensile strength after 550 hours is shown in Table 2. A correlation can be seen between the two values, especially at a higher temperature. There is a direct correlation between the immersion temperature and degree of fluid absorption, which is itself directly related to the magnitude of the decrease in ultimate tensile strength. A predictive model based on these experimental results would enable estimations of the performance of SIBS-based implants.

TABLE 2. Percent weight increase and percent decrease in tensile strength of specimens after 550 hours immersion

Specimen Group	Weight Gain (%)	Decrease in Tensile Strength (%)
Water 25	0.1	-2.4
Water 37	0.2	-4.3
Palm 25	8.3	-31.3
Palm 37	13.8	-59.5

Figure 3 shows that the higher the percent weight palm oil content in the specimen, the lower its tensile strength. Despite this trend being slightly more pronounced at 37 °C, specimens at both temperatures exhibit the same behavior.

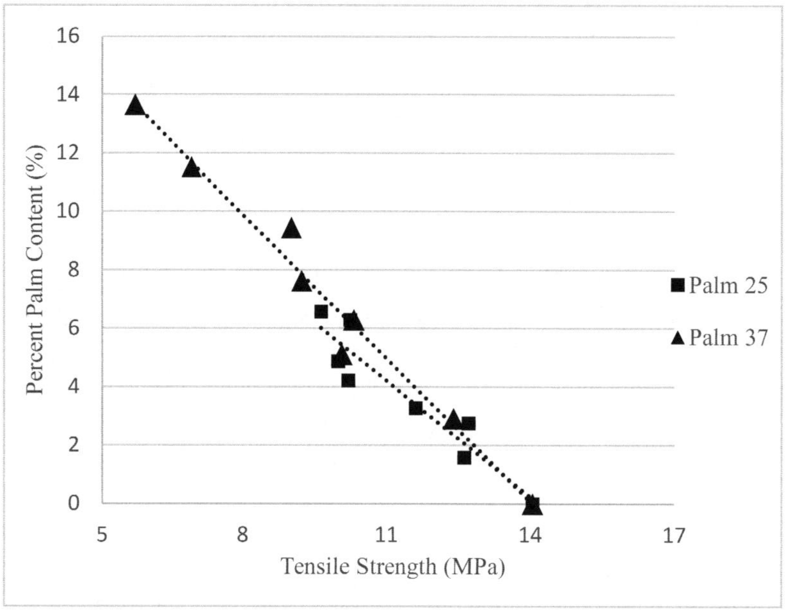

FIGURE 3. Percent palm oil content vs. tensile strength at 37 °C and 25 °C

CONCLUSION

In this study, the effects of immersion in palm oil and water at 25 °C and 37 °C on the ultimate tensile strength of a SIBS copolymer were analyzed and quantified. The total percent weight gain was approximately 8% and 14% for samples immersed in palm oil at 25 °C and 37 °C, respectively, after 550 hours. The percent decrease in tensile strength was about 31% and 60% for those same samples. A direct correlation was found between absorbed fluid and loss of ultimate tensile strength. These preliminary results bolster the feasibility and necessity of developing an accurate model of material property degradation based on both temperature and lipid concentration. Given the potential for the use of SIBS in implantable devices, these results suggest that careful consideration should be given to the long-term performance of these materials in the presence of elevated lipid concentrations.

REFERENCES

1. M. Boden, R. Richard, M.C. Schwarz, S. Kangas, B. Huibregtse, and J.J. Barry, *Journal of materials science. Materials in medicine* **20,** 1553-62 (2009).
2. M. El Fray, P. Prowans, J.E. Puskas, and V. Altstadt, *Biomacromolecules* **7,** 844-50 (2006).

3. S.L. Gallocher, A.F. Aguirre, V. Kasyanov, L. Pinchuk, and R.T. Schoephoerster, *Journal of biomedical materials research. Part B, Applied biomaterials* **79,** 325-34 (2006).
4. L. Pinchuk, G.J. Wilson, J.J. Barry, R.T. Schoephoerster, J.M. Parel, and J.P. Kennedy, *Biomaterials* **29,** 448-60 (2008).
5. F. Strickler, R. Richard, S. McFadden, J. Lindquist, M.C. Schwarz, R. Faust, G.J. Wilson, and M. Boden, *Journal of biomedical materials research. Part A* **92,** 773-82 (2010).
6. K.R. Kamath, J.J. Barry, and K.M. Miller, *Advanced drug delivery reviews* **58,** 412-36 (2006).
7. S. St. Lawrence, D.M. Shinozaki, M. Gerchcovich, U. Myler, J.E. Puskas, and G. Kaszas, *Rubber Chemistry and Technology* **74,** 601-613 (2001).
8. R.F. Storey, B.J. Chisholm, and M.A. Masse, *Polymer* **37,** 2925-2938 (1996).
9. Q. Wang, A.J. McGoron, R. Bianco, Y. Kato, L. Pinchuk, and R.T. Schoephoerster, *The Journal of heart valve disease* **19,** 499-505 (2010).
10. Vinoy Thomas and Muthu Jayabalan, *Biomaterials*, **23-1**, 273-282 (2002).

Novel Silicone Thermoplastic Elastomers with Tailored Permeation Properties

Gisbert Riess[a*], Katrin Berger[b] and Wolfgang Kern[a, b]

[a]*Polymer Chemistry, Department of Polymer Engineering and Science, Montanuniversitaet Leoben -Styria- Austria*
[b]*Polymer Competence Center Leoben - Styria - Austria*

corresponding and presenting author: riess@unileoben.ac.at

Abstract. The paper deals with the synthesis and properties of thermoplastic silicone elastomers containing soft segments of dimethylsiloxane and hard urea segments. In order to determine structure-property relations, regarding to permeation properties, the chemical structure was varied. In this study thermoplastic silicone elastomers, based on polydimethylsiloxane oligomers with amino groups and different diisocyanates, were synthesized. The polymers were characterized by nuclear magnetic resonance (NMR) and infrared (FTIR) spectroscopy. Thin membranes of thermoplastic silicone thermoplastic elastomers were prepared and investigated with respect to the permeation properties of organic solvents. Vials were filled with a mixture of an aromatic (toluene) and an aliphatic hydrocarbon (iso-octane) and were sealed with the polymer films. The weight loss of the vials was measured .The concentration of residual solvent mixture was detected by the relation of the peak area of the two solvents. This was investigated by the GC-MS-technique. A separation of the solvent mixture occurred. The synthesized thermoplastic silicone elastomers are highly permeable for organic solvents and have selective barrier properties. The barrier properties are influenced by chemical structure of the elastomers. Therefore these siloxane polymers can be used as semipermeable membranes for separation of solvent mixtures i.e. in oil field applications.

Keywords: Thermoplastics elastomer, silicone, permeability
PACS: 81.05.Lg

INTRODUCTION

Polysiloxanes such as polydimethylsiloxane (PDMS) combine several unique properties. They exhibit a low glass transition temperature, high thermal and oxidative stability, high UV-resistance, low surface energy and strong hydrophobicity as well as high permeability to many gases [1]. These characteristics are independent over a wide temperature range. Polydimethylsiloxane (PDMS) as low Tg-polymer is also the basis for silicone elastomers. Here, the PDMS chains are chemically crosslinked and a permanent network exists. Therefore no further thermal forming of the final part is possible. If the crosslinks between the chains are reversible nature like thermoplastic elastomers, these elastomers can be processed from the melt [1,2].

In thermoplastic silicone elastomers the soft segment is PDMS and the hard segment consists of moieties with either high melting temperatures or high glass transition temperatures. The type of the hard segment, the relative molecular weight of the soft segment and the ratio of the hard and soft segments tailor the properties of the polymer. The hard segments can be crosslinked via hydrophobic interactions, liquid crystalline interactions, hydrogen bonding or donor/acceptor interactions, but such polymer can be melted and processed.

Our research deals with the synthesis and properties of polymers containing soft segments of dimethylsiloxane and hard urea segments. In order to determine structure-property relations, regarding to permeation properties, the chemical structure was varied

EXPERIMENTAL

In this study thermoplastic silicone elastomers, based on polydimethylsiloxane oligomers with amino groups and different diisocyanates, were synthesized according to figure 1. The details of the synthesis is described in [3].

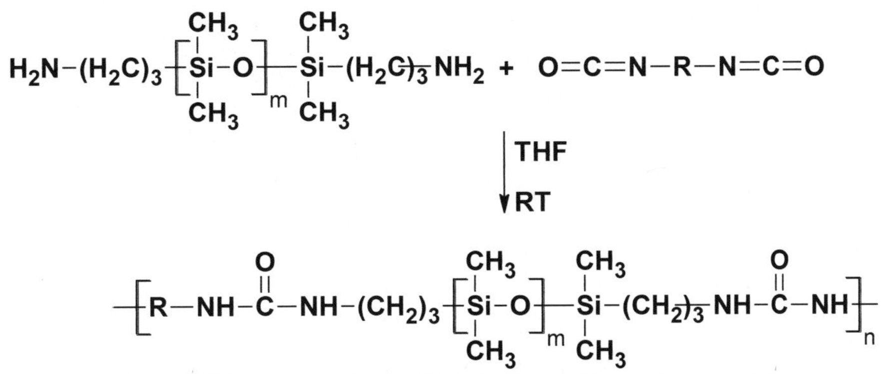

FIGURE 1. Scheme of synthesis of thermoplastic silicone elastomer [3]

The polymerization was carried out with THF as solvent at room temperature. It was found that this solvent leads to high molecular weight, which is in agreement with the synthesis of similar systems described in the literature [4].

The polymers were characterized by nuclear magnetic resonance (NMR) and infrared (FTIR) spectroscopy. The IR-spectra of polymers show characteristic signals, such as the NH-peak at 3320 cm^{-1} and the signals of the urea linkage appear at 1630 cm^{-1} and 1577 cm^{-1}. No peak at 2270 cm^{-1} indicates the absence of isocyanate endgroups.

With ^{1}H-NMR-spectroscopy the polymerization can also be confirmed. The signal at 2.6 ppm, which can be assigned to the methylene group next to the amino group in the telechelic diamino siloxane, is no more visible. Instead a signal at 3.1 ppm appears, which is assigned to the same methylene group linked to the urea group.

The thermoplastic silicone elastomer listed in table 1 were synthesized and characterized in the described manner.

TABLE (1). Synthesized thermoplastic silicone elastomer.

	diamino siloxane (1400 g/mol)	**diamino siloxane (2800 g/mol)**
hexamethylene diisocyanate	HMDI-1400	HMDI-2800
isophorone diisocyanate	IPDI-1400	IPDI-2800
2,4-toluene diisocyanate	TDI-1400	TDI-2800

Thin membranes of thermoplastic silicone elastomer were prepared and investigated with respect to the permeation properties of organic solvents. Vials were filled with a mixture of an aromatic and an aliphatic hydrocarbon (mixture of toluene and iso-octane: 50:50 vol%) and were sealed with the polymer films. All polymer films had same thickness (200 µm).

The weight of the vials was periodically measured to investigate possible barrier properties of the polymers. The weight loss of solvent over a period of 7 days was determined.

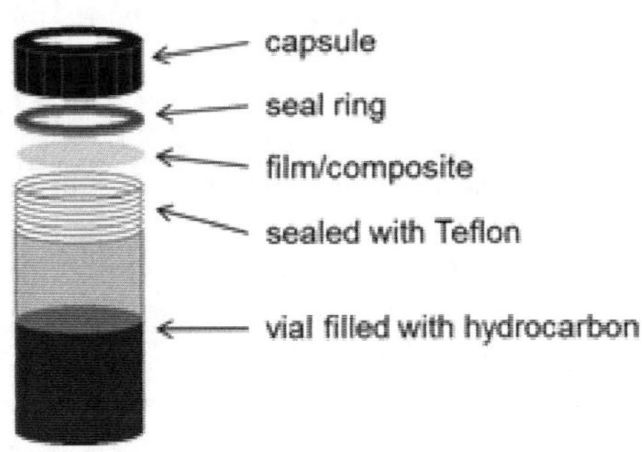

FIGURE 2. Experimental setup for gravimetric measurement

After a certain time the remaining mixture was determined with GC-MS measurements (2 samples per elastomer, duration: 7 d). The GC-MS permeation measurements were performed with a Shimadzu GC-MS-QP2010 Plus equipped with an Autosampler AOC-20i. The GC-MS system is provided with a Optima-5-Accent-0,25 µm fused silica capillary column. The following parameters were used for the determination of the single solvent component:

TABLE (2). Parameters of the GC-MS measurements of the hydrocarbon mixture.

GC		MS	
Injection Temperature:	250°C	Ion Source Temperature	200°C
Injection Mode:	Split	Interface Temperature	200°C
Split Ratio:	400	Solvent Cut time	1,5 min
Start temperature:	50°C/5 min	Acquisition Mode:	SIM (m/z = 57, m/z = 91)
Heating rate:	10°C/min		
End temperature:	150°C/1 min		
Sample volume	1 μL		

RESULTS AND DISCUSSION

The figure 3 shows the weight loss of the hydrocarbon mixture through the elastomer film dependent on diisocyanate component. The influence of the hard segment on the permeation properties was investigated. The siloxane soft segment has a molar mass of 2800 g/mol.

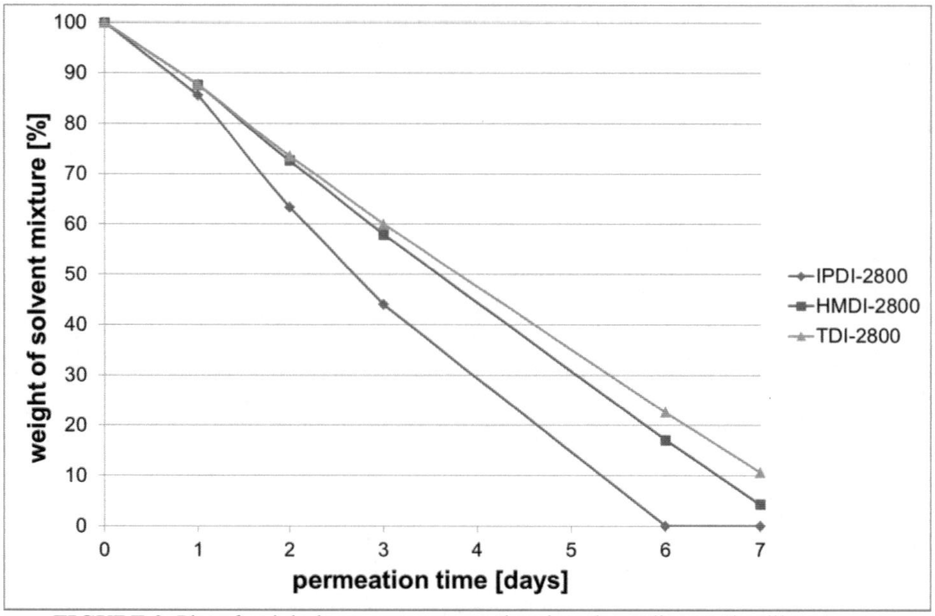

FIGURE 3. Plot of weight loss versus permeation time depending on the diisocyanate

The Polymer based on TDI have the lowest weight loss of the solvent mixture, the weight loss of the polymer with HMDI is slightly higher. The most weight loss can be detected by the polymer based on IPDI. Therefore these vials are empty after 6 days. The weight loss correlates with permeation rate, the permeation rates of the polymers can be ordered in the following manner:

IPDI-2800 > HMDI-2800 > TDI 2800

This can be explained by a denser package of the hard segment in the polymer with TDI (aromatic ring) compared to the polymers with aliphatic isocyanates such as IPDI and HMDI. Higher density leads to less permeation in the hard segment. But overall all polymers are highly permeable to the solvent mixture of iso-octane and toluene.

In the figure 4 the influence of soft segment is shown. Therefore the molar mass of soft segment in the urea, the siloxane unit was varied. One siloxane unit had a molar mass of 2800 g/mol, the other one a molar mass of 1400 g/mol. But also the influence of the hard segment was tested by using of IPDI and HMDI.

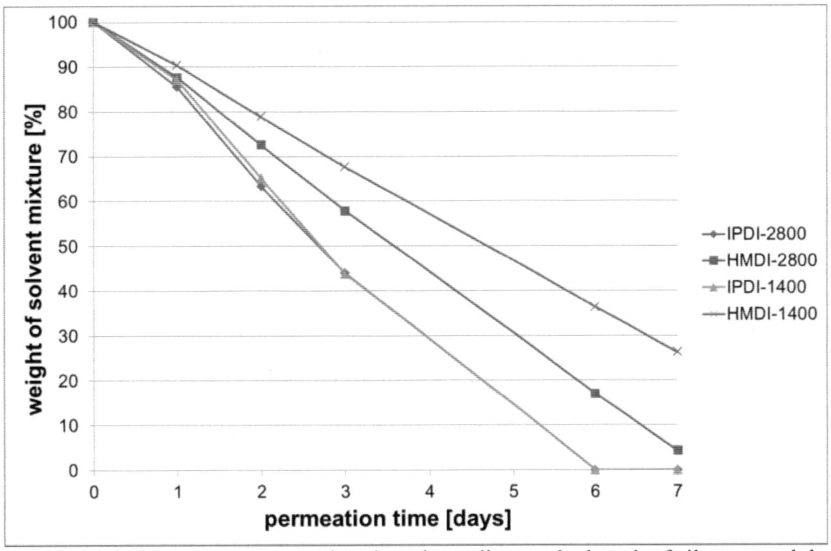

FIGURE 4. Plot of weight loss versus permeation time depending on the length of siloxane and the diisocyanate

Shorter siloxane units lead to lower weight loss and permeation rates in case of HMDI. The concentration of the soft segment in HMDI-1400 is lower compared to HMDI-2800. Additionally the soft segment is more permeable compared the hard segment. This explains the different permeation rates. However the polymer with IPDI does not show the same behavior, because the weight loss of the solvent mixture is equal. Therefore it is important to determine the permeation behavior of single solvents. By using of GC-MS the residual solvent in the vials are analyzed.

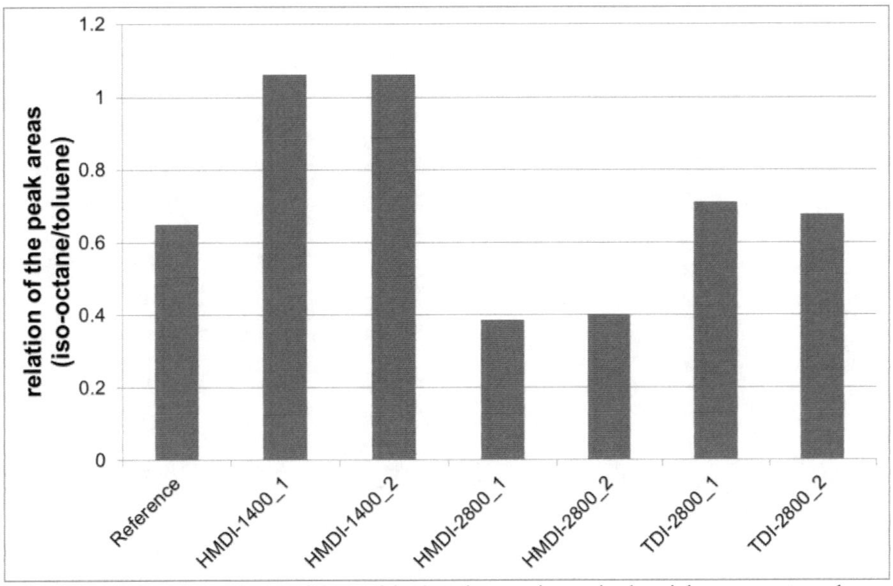

FIGURE 5. GC-MS data of the residuals solvent mixture in the vials per two samples

The concentration of residual solvent mixture was measured by the relation of the peak area of the two solvents. Figure 5 depicts that a separation of the solvent mixture occurred. The HMDI-polymer with short siloxane unit (HDMI-1400) exhibited a higher permeability of aromatic solvent, while the opposite effect took place for the sample with longer siloxane-chains. Toluene diisocyanate with longer siloxane units showed no effect.

Therefore the permeation properties can be tailored with regard to the field of application. These siloxane polymers can be used as semipermeable membranes for separation of solvent mixtures, but the effect cannot be explained in a simple manner at this time.

CONCLUSIONS

In conclusion, we have successfully synthesized novel thermoplastic silicone elastomers with selective barrier properties. These polymers are highly permeable for organic solvents depending on the chemical structure. The permeation properties are important with respect to the application as membrane material in oil field applications. But more research is necessary for the correlation of the permeation behavior with the molecular structure of such main chain urea-siloxane elastomers.

ACKNOWLEDGMENTS

The authors want to thank Natascha Andraschek for supporting of the polymer synthesis.

REFERENCES

1. R.J. Spontak, N.P. Patel, *Current Opinion in Colloid & Interface Science* **5**, 333-340 (2000)
2. B.R. Nair, V.G. Gregoriou, P.T. Hammond, Polymer **41**, 2961-2970 (2000)
3. G. Riess, H.-W. Schmidt, *Chem. Mon.* **137**, 935-941 (2006)
4. I. Yilgör, A. K. Sha´aban, W. P. Steckle, D. Tyagi, G. L. Wilkes, J. E. McGrath, *Polymer* **25**, 1800-1806 (1984)

Characterization of the Temperature Profile during Dosing under Steady State Conditions by Means of a Specially Designed Test Stand

R. C. Kerschbaumer[a*], B. Lechner[a,b], G. Graninger[a] and W. Friesenbichler[b]

[a]*Polymer Competence Center Leoben GmbH, Roseggerstrasse 12, 8700 Leoben, Austria*
[b]*Montanuniversitaet Leoben, Department of Polymer Engineering and Science, Injection Molding of Polymers, Otto Gloeckel-Strasse 2, 8700 Leoben, Austria*

**roman.kerschbaumer@pccl.at*

Abstract. The material temperature distribution significantly influences (1) the reduction of the incubation time during processing and (2) the curing degree homogeneity of the produced part. Thus, it is substantial to know the temperature distribution of the dosed rubber compound prior to injection to be able to ensure on the one hand the processability of the used material and on the other hand to achieve the required part quality. Consequently, a specially designed test stand was constructed and used for the determination of the temperature profile of the dosed material. The test stand consists of a plasticizing cylinder, which is equipped with a pressure sensor, and a throttle valve that offers the ability to adjust the back pressure as well as to measure the temperature of the dosed material. To investigate the temperature distribution it is necessary to perform the measurement under quasi steady state conditions. Therefore, a defined injection molding cycle was repeated four times to check the quasi steady state conditions. In addition, all experiments were performed under processing conditions.

Keywords: rubber compound, temperature profile, injection molding, dosing phase, steady state conditions.
PACS: 81.05.Lg

INTRODUCTION

Rubber compounds essentially consist of the base polymer, fillers, plasticizers and the curing system. It is well known that the rate of the curing reaction is significantly depending on the material temperature. Excess material temperature prematurely starts the crosslinking reaction which negatively affects processability and part quality as a consequence. Furthermore, the distribution of the relative curing degree is one of the dominant structural and quality characteristics in injection molded rubber parts. It influences almost all part properties including tensile strength, hardness, compression set and dynamic properties. Accordingly, the material temperature is widely acknowledged as being a key parameter because of its effects on (1) viscosity, (2) reduction of the incubation time (Scorch-Index), (3) rate of curing reaction, (4) process stability and (5) final product quality.
From a rubber processing viewpoint it is of utmost importance that the temperature of the dosed material in the injection chamber is homogenously distributed and constant over each cycle.

In any case, achieving a constant and homogenous temperature is a very challenging task. Plastification of rubber compounds is achieved in the plasticizing cylinder via heat conduction of the heated or cooled cylinder wall and by shear heating along the length of the screw. On the one hand, the problem is that cycle times in rubber processing are long and the dosing step is a discontinuous process. During process idle times, a significant heat flow occurs from the material to the plasticizing cylinder wall. This leads to a temperature distribution due to the low thermal conductivity of the rubber compound. On the other hand, rubber particles in the area of the screw tip undergo less or no shear heating in comparison to particles in the feeding section, which are sheared along the entire screw length. Consequently, it can be expected that the dosing phase results in an axially and radially pronounced temperature profile.

A number of melt temperature measurement methods can be found in literature, including thermocouple meshes [1-3], infrared sensors [4-6] and ultrasonic measurement [7]. Due to the high viscosity of rubber compounds during the dosing in comparison to thermoplastic materials and the high amount of fillers (e.g. carbon black) some problems would occur with the above mentioned temperature measurement methods. To overcome some of this problems and deficiencies, a fast and reliable temperature measurement system and methodology were developed within this work.

EXPERIMENTAL EQUIPMENT

The apparatus used in this work was a specially designed test stand (Fig. 1a), which consists of a dosing unit of the "first in first out" rubber injection molding machine MTF750/160 edition manufactured by MAPLAN GmbH (A-2630 Ternitz) and a removable throttle valve (Fig. 1b) manufactured by EM Gummitechnik (A-7024 Hirm). This test stand provides means to regulate the (1) temperature of the plasticizing cylinder, (2) screw rotational speed and (3) back pressure as well as the ability (4) to measure the temperature of the dosed material with a specially designed measurement bar (see Fig. 1b).

The throttle valve is equipped with a measurement bar (Fig. 1b) consisting of polyamide 6.6 with 30 % of glass fibers featuring five fast responding and soldered thermo-wires across the diameter. However, this work deals only with the temperature development in the middle of the channel (Fig. 1b, red circle). Furthermore, the plasticizing cylinder is equipped with a pressure sensor (located 32 mm from the tip of the screw) used to verify the conditions during the experiments.

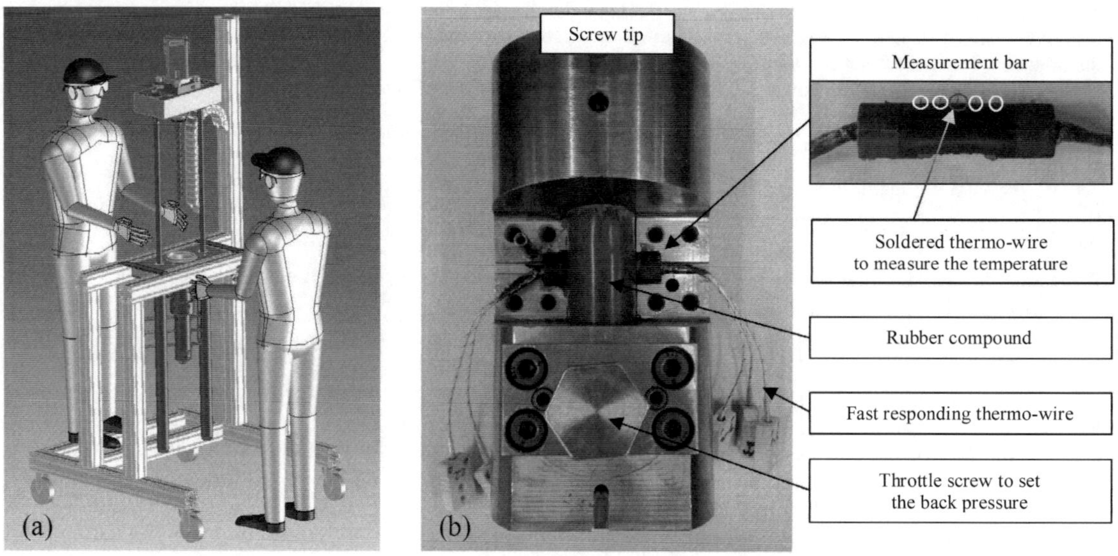

FIGURE 1. Apparatus used for the process-orientated determination of the mass temperature during the dosing phase (a) test stand, (b) throttle valve featured with a fast responding thermo-wire measurement bar.

MATERIAL AND EXPERIMENTAL CONDITIONS

The experiments were carried out using an industrial EPDM rubber compound (hardness of 67 Shore A, unknown quantities of mineral fillers, sulfur based curing system and a density of 1.519±0.004 g·cm^{-3} at 23 °C) under experimental conditions given in Tab. 1. Since the experimental design has to reflect industrial processes as close as possible, the cycle time of an injection molded rectangular shaped rubber part (100 x 100 x 15 mm, Fig. 2a) has been determined and was reproduced with the test stand (Fig. 2b).

TABLE (1). Normalized test stand settings for the determination of the temperature profile of the dosed material (T_{Feed} = Temperature of the feeding section, T_{Plast} = Temperature of the plasticizing cylinder, n = screw rotational speed, p_{bp} = back pressure).

	T_{Feed}	T_{Plast}	n	p_{bp}
Setting 1	1	1.33	high	low
Setting 2	1	1.33	high	high
Setting 3	1	1.33	low	low
Setting 4	1	1.33	low	high

EXPERIMENTAL

In conventional injection molding processes, the back pressure is controlled by the injection molding machine and is defined as the pressure in the injection chamber (Fig. 2c left). With the current setup of the test stand, this is not possible, since the material is dosed into the atmosphere instead of into an injection chamber. Therefore, the back pressure p_{bp} is determined as follows (Fig. 2c right):

(1) Equipping the plasticizing cylinder with a no return valve to determine the pressure p_{nrv} for each setting.

(2) Replacing the no return valve with the throttle valve and setting the required pressure p ($p = p_{nrv}+p_{bp}$).

With this engineering approach, it is possible to correlate the back pressure in the injection chamber with the measured pressure p of the experimental setup.

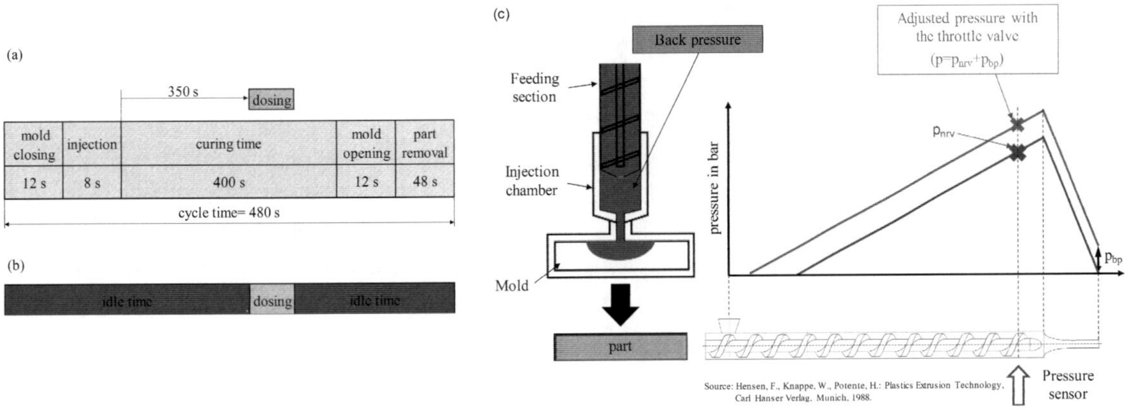

FIGURE 2. (a) Cycle time to produce a rectangular shaped part on the injection molding machine, (b) Reproduced cycle time on the test stand to dose a volume of 150 cm³, (c) Determination of the corresponding back pressure with the test stand.

In a further step, the dosing time for the experiments has to be determined since the test stand is only able to dose material for a defined time instead of a defined volume. The determination of the dosing time is performed as follows:

(1) Dosing material for 10, 20 and 30 s.

(2) Cutting it at the throttle valve, weighing and calculating the corresponding volume with the help of the pvT – diagram.

With this method, the dosing times were calculated for a dosing volume of 150 cm³. Fig. 3a shows the linear regressions for each setting.

To carry out the experimental investigation of the mass temperature distribution under quasi stationary conditions, the response of the pressure and temperature signals of the fourth cycle per setting were analyzed and representatively shown for setting 1 and the measured temperature in the middle of the flow channel. It could be shown that the pressure response is in the same range for cycle 1 to 4 (Fig. 3b).

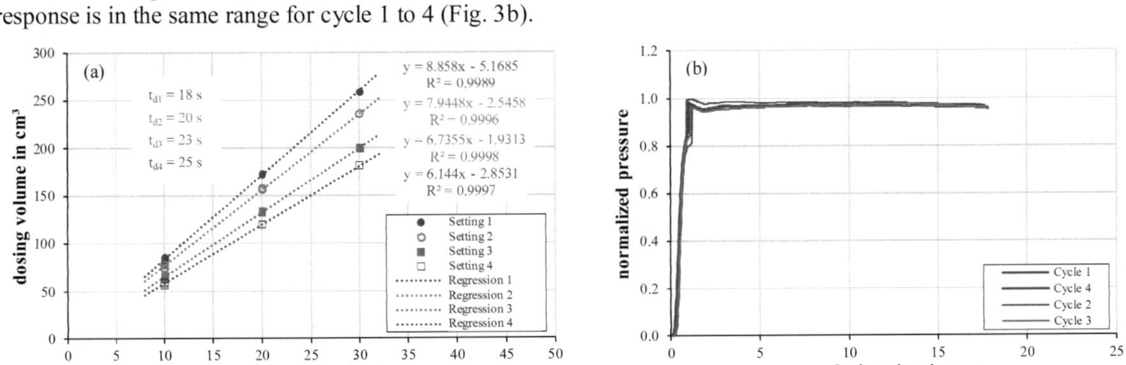

FIGURE 3. (a) Experimental determination of the dosing time, (b) Quasi steady state conditions of the pressure (setting 1).

In contrast, the temperature profile of the dosed material provides information regarding how many cycles are required to achieve quasi steady state conditions. As depicted in Fig. 4a, quasi steady state conditions can be achieved after only three cycles, because cycle 3 and 4 lead to very similar temperature profiles.

In order to determine the reproducibility of the temperature measurements, three experiments were carried out (Fig. 4b) wherein the fourth cycle of every single experiment was used for the comparison. Up to a dosing time of 6 s, the maximum deviation is about 1.8 °C and afterwards about 0.8 °C. Therefore, the achieved reproducibility of the temperature can be considered as remarkably good.

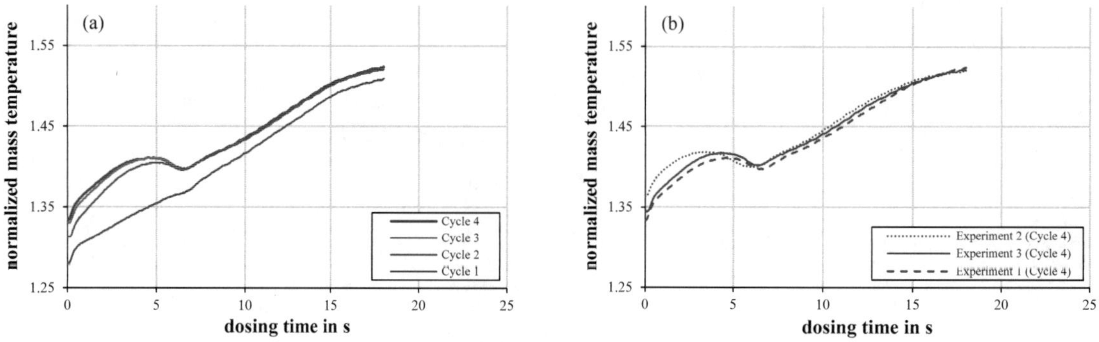

FIGURE 4. Measured mass temperature in the middle of the channel for setting 1 (a) Determination of quasi steady state conditions, (b) Reproducibility for three single experiments.

RESULTS AND DISCUSSION

The axial temperature profile in the middle of the dosed material was measured with an accuracy of less than ±0.9 °C (shaded area in Fig. 5 and 6) and after the local minimum with a deviation of less than ±0.4 °C. Regardless of the used setting, a local maximum and minimum of the mass temperature could be detected (in the range of 4 to 9 s). This results from the measurement setup, because the tip of the screw and the measurement bar are located in a distance of 76 mm from each other. Accordingly, the mass temperature up to the local minimum can be considered as residual temperature of the last cycle and can be neglected for the further investigations. Thereafter, the temperature rises monotonically with increasing dosing time (shear heating in the screw channel) and reaches the maximum temperature at the end of dosing.

At high screw rotational speeds (Fig. 5a), the level of the back pressure (low or high) significantly influences the temperature profile. Furthermore, for low screw rotational speeds (Fig. 5b) no influence of the level of the back pressure could be observed. Accordingly, there should be a lower limit for the screw rotational speed. Above this value, changes in the back pressure strongly influence the resulting temperature profile.

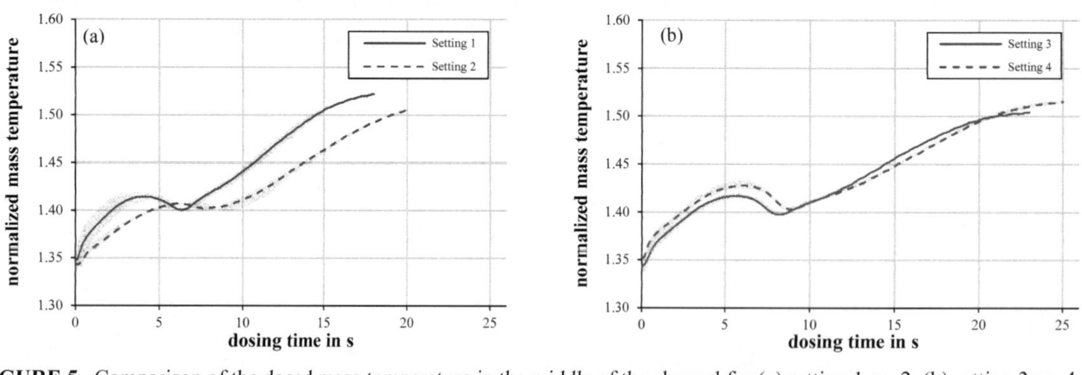

FIGURE 5. Comparison of the dosed mass temperature in the middle of the channel for (a) setting 1 vs. 2, (b) setting 3 vs. 4.

The same effect as described in Fig. 5a, could be observed when the back pressure is at low level and the screw rotational speed is varied from high to low values (Fig. 6a). At high back pressure levels (Fig. 6b) the temperature profile isn't influenced by a change in the screw rotational speed.

The results found can be summarized as follows: (1) a significant temperature rise could be observed for all settings and (2) a strong interaction between the back pressure and screw rotational speed can be expected. These effects will be investigated in more detail in the future via statistical design of experiments (DoE) to enable simple modelling of the occurring effects and interactions. Furthermore, the chosen settings significantly influence the dosing time (differences up to 7 s).

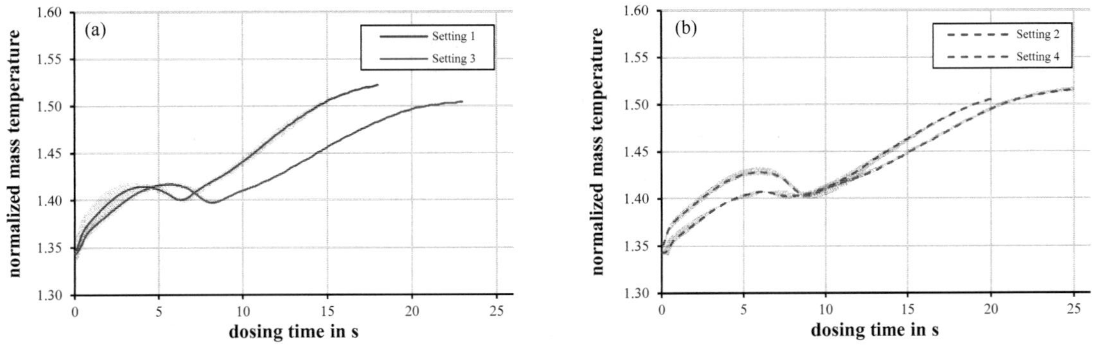

FIGURE 6. Comparison of the mass temperature as a function of the dosing time in the middle of the channel for (a) setting 1 vs. 3, (b) setting 2 vs.4.

CONCLUSION

The dosing phase of a conventional rubber injection molding cycle was accurately reproduced with the developed test stand. Furthermore, the accuracy in heating and cooling of the plasticizing cylinder is remarkably good because quasi steady state conditions could be observed after only three cycles. On the one hand, a significant temperature rise could be observed for all settings and on the other hand a strong interaction between the back pressure and the screw rotational speed has been identified.

ACKNOWLEDGMENTS

The research work of this paper was performed at the Polymer Competence Center Leoben GmbH (PCCL, Austria) within the framework of the COMET-program of the Federal Ministry for Transport, Innovation and Technology and the Federal Ministry of Economy, Family and Youth with contributions by Montanuniversitaet Leoben (Injection Molding of Polymers) and MAPLAN GmbH, Semperit Technische Produkte Gesellschaft m.b.H., Woco Industrietechnik GmbH., Erwin Mach Gummitechnik and CAS-Computerunterstützte Automatisierungs-systeme GmbH & Co. KG.

The PCCL is funded by the Austrian Government and the State Governments of Styria, Lower Austria and Upper Austria.

REFERENCES

1. N. Sombatsompop, M. Panaploy, *Die geometry effects on the temperature profile measurement of flowing PP melt in a twin screw extruder*, Polym. Test. 21, 17–25 (2002).
2. E.C. Brown, A.L. Kelly, P.D. Coates, *Melt temperature field measurement in single screw extrusion using thermocouple meshes*, Rev. Sci. Instru. 75, 4742–4748 (2004).
3. L. Shih-Jung, S. Po-Chang, *Novel three-dimensional in-cavity transient temperature measurements in injection molding and fluid-assisted injection molding*, Polymer Testing 28, 66-74 (2008).
4. G.Y. Lai, J.X. Rietveld, *Role of polymer transparency and temperature gradients in the quantitative measurement of process stream temperature during injection via IR pyrometry*, Polym. Eng. Sci. 36, 1755–1768 (1996).
5. C. Maier, *Infrared temperature measurement of polymer*, Polym. Eng. Sci. 36, 1502–1512 (1996).
6. A. Bendada, K. Cole, M. Lamontagne, Y. Simard, *A hollow waveguide infrared thermometer for polymer temperature measurement during injection moulding*, J. Opt. A: Pure Appl. Opt. 5, 464–470 (2003).
7. E.C. Brown, P. Olley, P.D. Coates, *In line melt temperature measurement during real time ultrasound monitoring of single screw extrusion*, Plast. Rubber Comp. 29, 3–13 (2000).

Two-Component Injection Moulding of Thermoplastics with Thermoset Rubbers: Process Development

G.J. Bex[a], J. De Keyzer[b], F. Desplentere[c], A. Van Bael[a]

[a]*Materials Technology TC, KU Leuven Technology Campus Diepenbeek, Wetenschapspark 27, 3590 Diepenbeek, Belgium*
[b]*Sustainable Chemical Process Technology TC, KU Leuven Technology Campus Diepenbeek, Wetenschapspark 27, 3590 Diepenbeek, Belgium*
[c]*Materials Technology TC, KU Leuven Technology Campus Ostend, Zeedijk 101, 8400 Oostende, Belgium*

Corresponding author: gertjan.bex@kuleuven.be

Abstract. Two-component injection moulding is a manufacturing process for combining polymers with different properties in a single injection moulding process. The process is typically used to combine thermoplastics with another thermoplastic or with a thermoplastic elastomer to create colour differences or hard and soft areas respectively. The present study aims at the development of a two-component injection moulding process for the combination of a thermoset rubber and a thermoplastic. Currently products that consist of those two materials (e.g. wheels, syringes and other products with gaskets) are made by assembling separate components. Implementing the two-component injection moulding technique for these products will result in better interphase properties, savings on rubber and avoiding costs related to the assembly process. A technological challenge is posed by the fact that injection moulding of rubbers and thermoplastics is fundamentally different. The injection moulding of a rubber requires a heated mould (140°C-180°C) whereas thermoplastics need a relatively cold mould (20°C-100°C) for the polymer to solidify. In this study a versatile two-component mould is proposed in which the mould cavities for the rubber and the thermoplastic are thermally separated and equipped with facilities to control the temperature of both cavities individually. The design of the mould also makes it possible to vary the sequence of injection. In this way it is possible to test several processes variations. Preliminary test results will be presented for specific rubber-thermoplastic combinations.

Keywords: Two-component injection moulding, thermoset rubbers, thermoplastics.
PACS: 81.20.Hy

INTRODUCTION

Several applications require a combination of parts with different characteristics like hard/soft, conductive/non-conductive or different colours. This diversity in requirements is only possible by combining different materials. Two-component (2K) injection moulding is a manufacturing process that allows to combine different polymers with different properties in a single injection moulding process. 2K injection moulding is often used to combine multiple thermoplastics with colour differences. In other applications, like in toothbrushes, a combination of a soft rubberlike material and a stiff material is required. In these cases the rubberlike material is a thermoplastic elastomer (TPE).

A large number of products combine thermoset rubbers with thermoplastics, e.g. wheels, syringes and other products with gaskets. Currently these products are made by assembling separate components. Implementing the 2K injection moulding technique for these products should result in better interphase properties and the avoidance of costs related to the assembly process. Furthermore it is expected that a redesign will result in savings on rubber. Such 2K process poses considerable challenges, since injection moulding of thermoplastics and thermoset rubbers requires fundamentally different process parameters. Thermoset rubbers are injected in a heated mould (140°C-200°C) to start the curing process whereas thermoplastics are injected in a relatively cold mould (20°C-100°C) to solidify the product [1,2]. High-end thermoplastics like polyamide (PA) can withstand high temperatures making it possible to insert the thermoplastic part in a completely heated mould and then overmould it with rubber [3, 4]. When this technique is applied with commodity thermoplastics like polyethylene (PE) or acrylonitrile butadiene styrene (ABS), it is expected that the thermoplastic part will deform under the high temperatures and pressures necessary for the injection moulding process of the thermoset rubber. This study aims at the development of a novel 2K injection moulding process that can be applied with commodity thermoplastics. Two possible solutions are proposed. In the first, deformation of the thermoplastic part is prevented by cooling the thermoplastic part during vulcanization of the rubber. This technique requires special moulds where the mould cavities for the rubber and the thermoplastic are thermally separated and equipped with facilities to separately control the temperature of both cavities. In the other solution, the thermoplastic material is injected after vulcanization of the rubber part.

MATERIALS AND METHOD

Material

For this research HDPE (Sabic, grade M80064) is selected as thermoplastic material. HDPE is a commonly used thermoplastic characterized by a rather low maximum service temperature (110°C-130°C) since it easily deforms at high temperatures. The thermoset rubber selected for this research is EPDM (Hercorub, grade 005K), a frequently used material for seals, gaskets, hoses, roofing and cable insulation. The selected EPDM rubber is a sulphur vulcanizing rubber that needs approximately 20 minutes of curing at 160°C.

Method

Injection moulding is conducted on an Engel ES330H/80V/80HL-F 2K injection moulding machine with a vertical rubber injection unit and a horizontal thermoplastic injection unit.

A versatile mould is required to enable both proposed 2K injection solutions. First of all it is designed in such a way that the injection sequence can be reversed. For this purpose, the mould is based on a movable core system. Instead of using a real moving core, a metal plate that can be manually inserted was used (figure 1). Moving the metal plate from one mould cavity to another is required to enable a different injection sequence. Secondly the mould cavities for the rubber and the thermoplastic are thermally separated and equipped with facilities to control the temperature of both cavities separately. In this way, it was possible to cool the thermoplastic during the in-mould vulcanization of the rubber. Thermal separation was achieved by using insulation and air gaps.

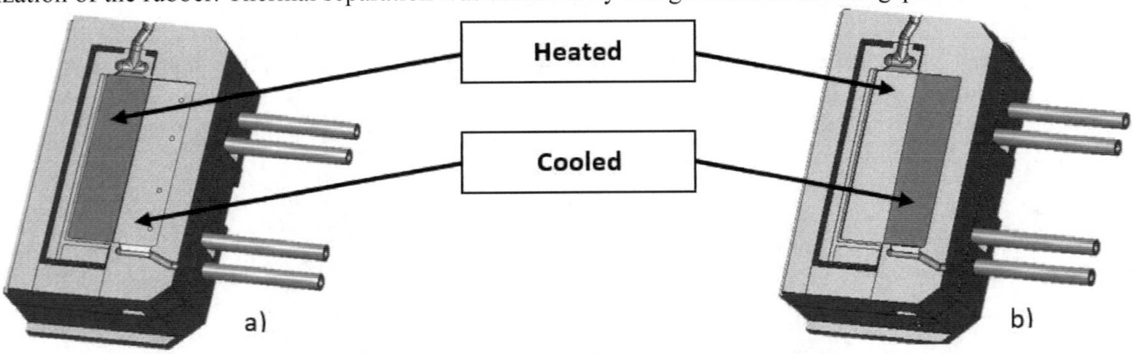

FIGURE 1. Mould system based on a movable core system. The metal plate, indicated in green, is used to change the order of injection: (a) shows the configuration where the thermoplastic is injected first followed by the rubber, (b) shows the configuration where the rubber is injected first. Both mould cavities are thermally separated. The left cavity is heated, the right cavity is cooled.

Pressurized water is used both for heating the mould cavity for the thermoset rubber and for cooling the mould cavity for the thermoplastic part. Design rules found in literature were used for the cooling channels [5] in order to obtain evenly distributed temperatures.

To achieve even more flexibility a combination of two temperature control units and a valve system is used to heat the mould cavity of the rubber (Figure 2). This set-up makes it possible to quickly cool the rubber after vulcanization. The same set-up is used for rapid heat cycling of the mould in injection moulding of thermoplastics to improve mouldability, strength of weld lines and surface quality [6].

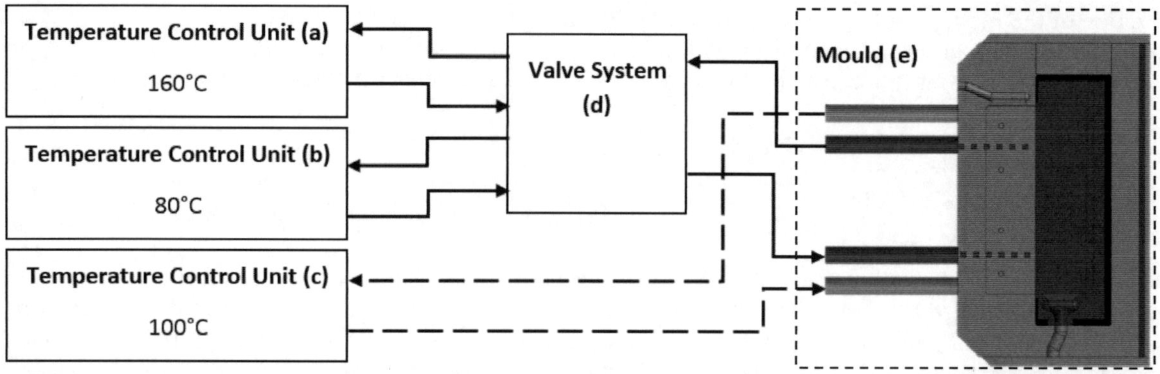

FIGURE 2. Temperature control set-up with possibility for rapid heat cycling on the mould cavity for the rubber. (a) and (b) are respectively temperature control units for the upper and lower temperatures of the mould cavity for the rubber. (d) is the valve system that switches between hot and cold water. (c) is the temperature control unit for the mould cavity for the thermoplastic part. (e) is the mould used in this study.

Temperature control units (a) and (b) determine respectively the upper and lower temperatures of the mould cavity for the rubber. The valve system (d) allows either the hot or the cold water to flow through the mould cavity for the rubber. In this way it is possible to vulcanize the rubber first and then quickly cool it to the temperature of the thermoplastic part. Temperature control unit (c) controls the temperature of the mould cavity for the thermoplastic part.

EXPERIMENTAL

Temperature Distribution Within the Mould

To verify thermal separation between the mould cavity of the thermoset rubber and the thermoplastic, an infrared camera (OPTRIS PI400) is used. Black adhesive tape with an emissivity of 0.95 is used in order to avoid reflection of the metal surface. The temperature control unit for the mould cavity of the rubber was set at 160°C. The temperature control unit for the cavity of the thermoplastic was set at 100°C.

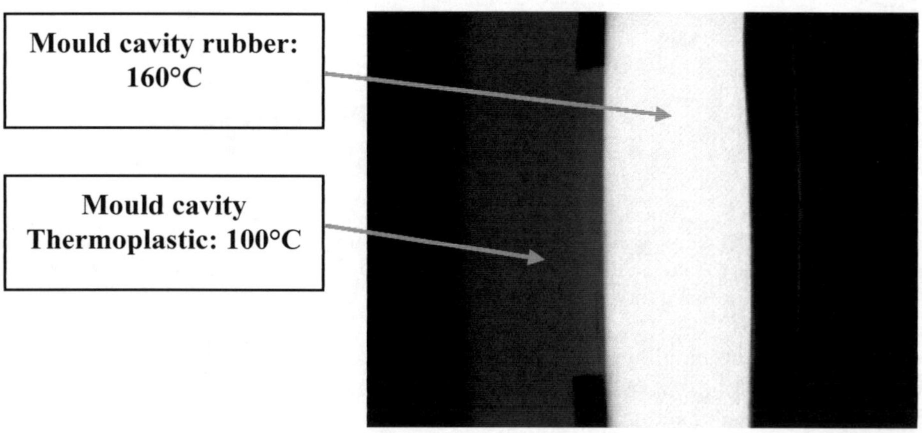

FIGURE 3. Thermal image of both mould cavities during heating and cooling. The left cavity is the colder cavity (100°C) for the thermoplastic part. The right cavity is the hot (160°C) cavity for the thermoset rubber.

At the left side of the thermal image is the cavity for the thermoplastic. It can be seen that the temperature of this cavity is successfully maintained at 100°C. At the right side of the thermal image is the cavity for the rubber. The temperature of this cavity is 160°C. The thermal image shows a uniform temperature distribution within the cavity for the rubber, which allows the rubber to vulcanize evenly and prevents warpage of the rubber part.

Rubber – Thermoplastic injection sequence

In this section the process of first injecting and vulcanizing the rubber is described. The rubber is vulcanized for 20 minutes. After cooling to a temperature of 40°C, the polyethylene is injected. Slow injection speeds (34 cm^3/s) and a low holding pressure (274 bar) were used for the polyethylene in order to minimize the deformation of the rubber part. In figure 4 (a) it can be seen that the rubber part is largely deformed during the injection of the polyethylene. The deformation is caused by the elastic behaviour of the rubber. Due to the deformation of the rubber, it is impossible to accurately control the dimensions of the finished product. Adhesion between the rubber and the polyethylene was found to be very poor. The stress within the rubber, caused by the deformation of the rubber, was large enough to break the adhesion.

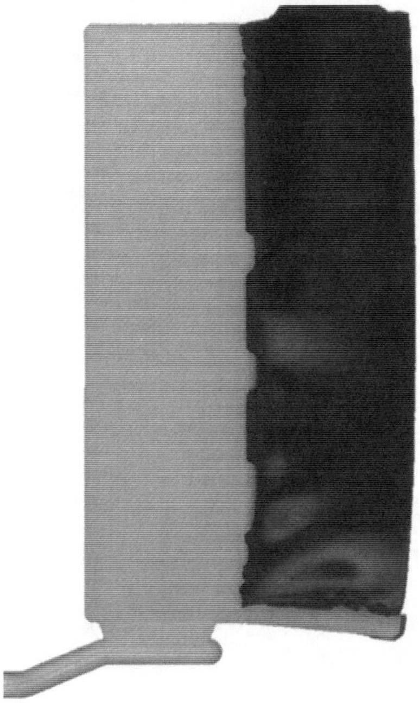

(a)

(b)

FIGURE 4. 2K injection moulded test samples of thermoset EPDM rubber and polyethylene. In (a) the thermoset rubber was injected first. Due to the elastic properties of the rubber, the rubber part is largely deformed. In (b) the injection moulding process started with the injection of the polyethylene.

Thermoplastic – Rubber injection sequence

In a second solution the thermoplastic part is moulded first and kept at a lower temperature during the vulcanisation process of the thermoset rubber. The result is shown in figure 4 (b). It is important to mould the thermoplastic part with minimal shrinkage by selecting optimal process parameters. Thanks to thermal separation between the two cavities, the cavity for the thermoplastic part can be kept at a lower temperature than the cavity for the rubber. When the thermoplastic part is smaller than its the cavity due to shrinkage, the interface between the rubber and the thermoplastic will be in the colder cavity. Because of the lower temperatures, the rubber at the interface will not vulcanize. The next step in the process is the moulding of the rubber part. It is important to keep the temperature of the thermoplastic part at an intermediate temperature during the moulding of the rubber part. The intermediate temperature should not be too high in order to prevent deformation of the thermoplastic part. On the other hand when the temperature is too low, the thermoplastic part will absorb heat at the interface resulting in an incomplete vulcanisation of the rubber at the interface. In this research it was found that for the specific type of polyethylene, 100°C is a good value for the temperature of the thermoplastic part. The rubber is now vulcanized for 20 minutes. After vulcanisation the rubber is quickly cooled down till 100°C using the valve system (d) in figure 2. The last cooling step of the rubber part is necessary for the solidification of the polyethylene at the interface. Without this cooling step the interface is weak and fails when ejected.

CONCLUSION

In this study a versatile mould is developed for 2K injection moulding of thermoset rubbers and thermoplastics. The sequence of injection is controllable by moving a metal plate. The cavities for the rubber and the thermoplastic are thermally separated in order to control the temperatures of both cavities separately. With the aid of a valve system it is possible to quickly cool the rubber after vulcanisation. Several 2K injection moulding variants have been tested. In this research it was found that the deformation of the rubber part is too large when the rubber is injected first. Adhesion between the rubber and the thermoplastic part was very weak for this injection sequence. By first injecting the thermoplastic it was possible to successfully make two-component products out of thermoset rubber and thermoplastics. In this process it was necessary to minimize shrinkage of the thermoplastic part by selecting optimal process parameters, in order to completely vulcanize the rubber at the interface. The thermoplastic part was kept at an intermediate temperature during the vulcanisation process of the rubber. After vulcanisation the rubber was quickly cooled down before ejection in order to solidify the thermoplastic at the interface. In future research this process will be optimized in order to maximize the adhesion and minimize the deformation. In order to further validate the proposed 2K injection moulding process, more combinations of different thermoset rubbers and thermoplastics will be tested.

ACKNOWLEDGMENTS

G.J. Bex acknowledges Research Foundation – Flanders (FWO) for funding his PhD grant Strategic Basic Research. The authors also acknowledge the company Hercorub NV for providing the rubber material.

REFERENCES

1. J.A. Lindsay, *Practical guide to rubber injection moulding*, London: Smithers Rapra, 2012.
2. T. Osswald, L-S Turng and P. Gramann, *Injection Molding Handbook*, Munich: Carl Hanser Verlag, 2008, pp. 65, 73.
3. D. Schneegans, R. Gattringer and Emmerich, *KGK Kautschuk Gummi Kunststoffe* **58**, 16-21 (2005).
4. D. Schneegans, R. Gattringer and R. Bauer, *GAK* **60**, 494-499 (2007).
5. W. Guilong, Z. Guoqun, L. Huiping and G. Yanjin, *Materials and Design* **31**, 3426-3441 (2010).
6. W. Guilong, Z. Guoqun and W. Xiaoxin, *International Journal of Heat and Mass Transfer* **78**, 99-111 (2014).

The Recent Developments in Tire Uniformity

Kayhan Koca[a*], Mehmet Kodal[a,b], Husnu Gerengi[c] and Guralp Ozkoc[b]

[a] *Köseköy Vocational School, Department of Plastic and Tire Technology, Kocaeli University, Kocaeli/Turkey*
[b] *Department of Chemical Engineering, Kocaeli University, Kocaeli/Turkey*
[c] *Department of Mechanical Engineering, Düzce University, Düzce/Turkey*

kkoca@kocaeli.edu.tr

Abstract. Tires are highly engineered structural composites whose performance can be designed to meet vehicle manufacturers' ride, handling and traction criteria, plus the quality and performance expectations of the customer. There are several factors and parameters influencing the quality and safety of a tire. One of them is the "uniformity". Uniformity is the performance of tires, which should be the same on the circumference direction under static and dynamic conditions. It is very important to determine the tire uniformity after curing process. It is carried out by the specially designed tire uniformity machines. In this manuscript, the factors such as radial and lateral force variation, conicity, bulge and depression that affect the tire uniformity are reviewed from the recent literature. The recent advancements were summarized and the future trends were enlightened. In addition, the methods that can be used to improve the tire uniformity were discussed.

Keywords: Tire uniformity, tire quality, tire performance
PACS: 82.35.Lr

INTRODUCTION

A tire consists of several parts. During the production of a tire, these parts are produced in different departments and combined to the first and the second stage production machines. The curing process is the last section of the tire production. After the curing process, produced tires are controlled by the quality control machines called Tire Uniformity Machines.

A tire includes many parts such as carcass, belts, tread, and beads as can be seen in Figure 1. They are overlapped on the drum. If the belts are not exactly positioned in the center of a tire, non-uniformity will occur in the tire. Quality of tires meaning uniformity is very important for automotive sector. To produce a quality and a safety tire, there are lots of factors and parameters take into accounts. One of them is uniformity the performance of tires, which should be the same on the circumference direction under static and dynamic conditions. In this research, effects of radial force variation, lateral force variation, conicity, radial run out and ply steer on the tire uniformity were discussed.

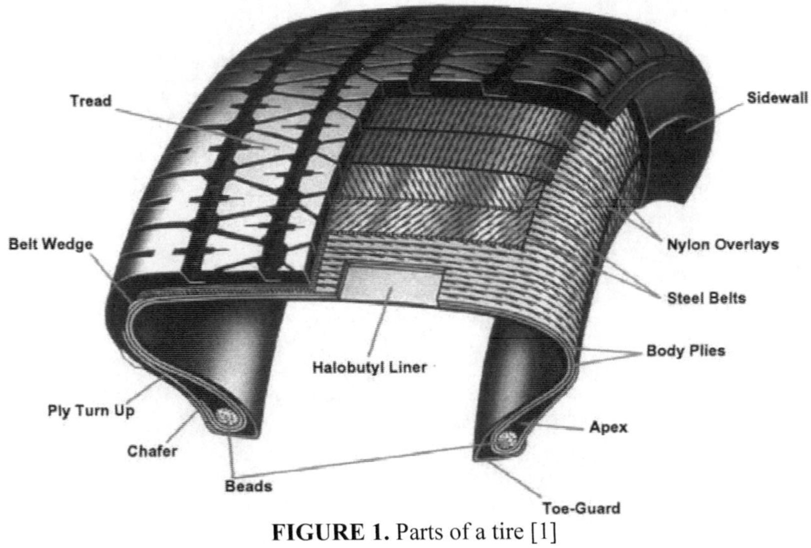

FIGURE 1. Parts of a tire [1]

RESULT AND DISCUSSION

In general, a tire includes rubber compound, steel and textile fabrics and bead wire. The calendaring method uses covering the inner liner, steel belt and ply cord. Sidewall and tread part of the tire are shaped through extruding system. The thickness of inner liner has to be same for every area. Irregular spread material results from unbalanced tires.

The manufacturing of carcass starts with setting the bead wires in first stage building machine as can be seen in Figure 2. Bead wires joint the sides of drum and they must be parallel each of them. Next, inner liner and body plies are wrapped around the drum (Figure 3). This application enables to hold the air pressure. Finally, sidewalls add manually on the drum as can be seen in Figure 4 [2].

In the second stage building machine, there are two big bobbins namely, nylon belt and steel belt. The bobbins are wrapped on the carcass and finally tread is added (Figure 5 and Figure 6) [2.

FIGURE 2. Joint the bead wears [3]

FIGURE 3. Applying of the body plies [3]

FIGURE 4. Adding the sidewalls [3]

FIGURE 5. Cutting and adding Belts [3]

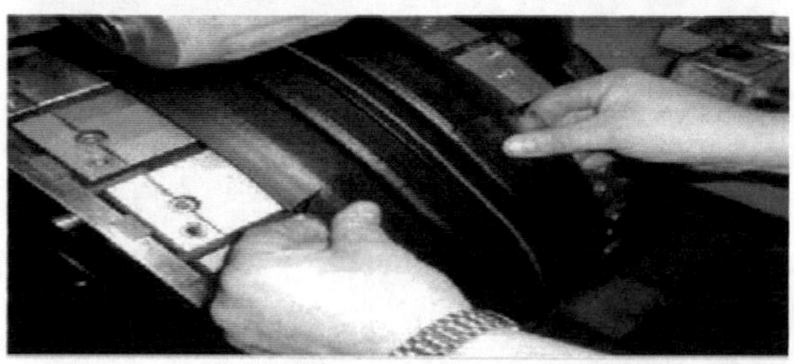

FIGURE 6. Adding the tread [3]

Radial-Lateral Force Variation

Figure 7 and Figure 8 show the forces/axis systems of a tire and modelling of radial forces, respectively.

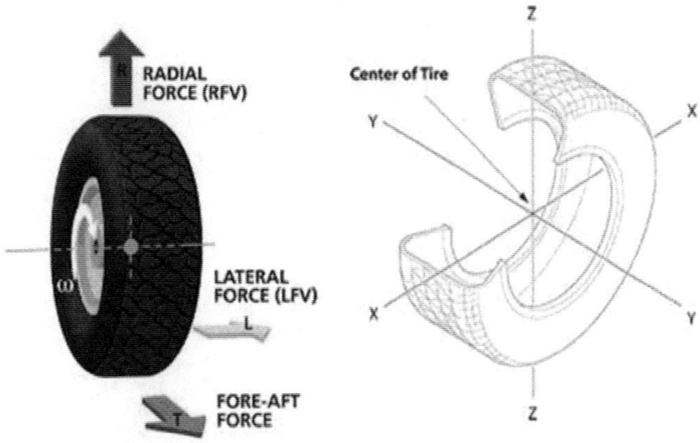

FIGURE 7. Forces and axis systems of a tire [4]

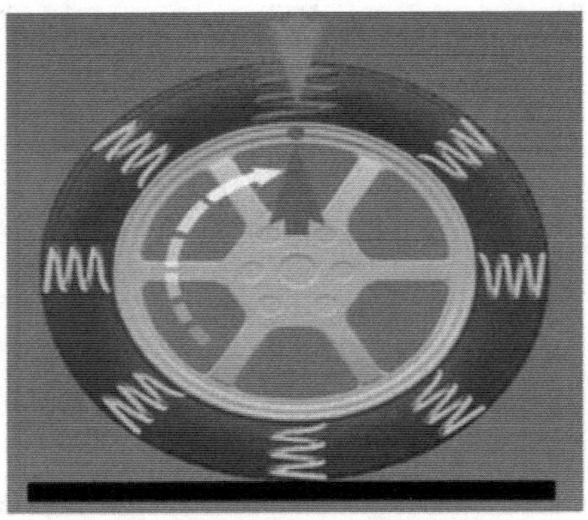

FIGURE 8. Modelling of radial forces

When a tire rolls on the road, it flexes and gives a vibration. Stiffness of the sidewall is not same along the surface of the sidewall and it cannot be rectified through balancing. The tire moves to go up and down, which results from Radial forces. Therefore, side to side forces affects the tire, which is called Lateral Force Variation. Radial and Lateral forces results from as follows:

a- Packet of belts and bead are not in the same axes system,
b- Irregular thickness of semi-finished products such as variation of thickness of sidewall or inner liner or tread etc.,
c- The unstandardized width of semi-finished product arising from the textile cutting machine or worker,
d- Wrong positions of the joints,
e- Over stretching of semi-finished product on the drum,
f- Oval shapes of beads, which have to be in circular shape,
g- Wrong positioned of carcass to the curing press.

CONCLUSIONS

Material, construction and manufacturing parameters affect the tire uniformity. Design and type of manufacturing are very important to produce uniform tire. Every tire manufacturer aims to make a tire with acceptable force variations. In a summary, it can be concluded that;

a- Uniformity size of a tire must be defined properly,

b- The position of the bead wires, inner lines and body plies have to be true in the tire,

c- Every stage production machines must be continuously controlled during processing,

d- Laboratory tests must be reliable for every stage and results are evaluated by experienced technicians and results must be between tolerances,

e- After manufacturing, all of the tires must be controlled by the quality control machines measure uniformity,

f- Storage conditions must be acceptable standards.

REFERENCES

1. B. Rodgers and W. Waddell, *Tire Engineering. The Science and Technology of Rubbers*, Waltham: Elsevier, 2013, pp. 656.
2. Pirelli, *TPM Tire Uniformity*, Kocaeli, Turkish Pirelli Company, 2002, pp. 21-27.
3. J. Suffelers, "Tyre non uniformities and their effect on chassis vibration", Master's Thesis. Technische Universiteit Eindhoven, 2010.
4. Akron Standart (http://www.gsp9700.com/technical/itw.pdf).

THE EFFECT OF ADDITIVES ON HIGH TEMPERATURE WATER RETENTION AND THERMAL STABILITY OF PFSA POLYMER COMPOSITES.

Bhawana Agrawal[a], Monika Willert-Porada[b]

Chair of Materials Processing, Faculty of Engineering Sciences, University of Bayreuth, 95440 Bayreuth, Germany

bhawana.iitd2006@gmail.com
Monika.Willert-Porada@uni-bayreuth.de

Abstract. The high temperature water retention of new composites with PFSA (polyfluorosulfonic acid) polymer matrix is investigated based on structural analysis of such polymers containing a mineral acid and MgF_2-nanoparticles as additives. As compared to the parent polymer the temperature for water removal is increased by almost 50°C to 250°C. The thermal stability of the PFSA polymer is slightly decreased from 360 to 340°C with respect to the onset temperature of de-sulfonation and back-bone de-polymerization. Different PFSA polymer composite membranes with a fixed amount of sulfuric acid as additive and a varying amount of 3 to 15 weight % of MgF_2-nanoparticles coated glass particles were investigated. Structural details of the PFSA evident from FTIR-ATR analysis of the solid membranes can be associated with two major phenomena: (1) dissociation of the proton from the sulfonic acid group in the presence of small amount of water and (2) changes in the conformation and / or the degree of the crystallinity of the PTFE hydrophobic domains composite membranes in the PFSA as a function of the polymer and the weight ratio of the solid additive. From these results evidence is given that sulfuric acid acts as bridging agent to facilitate direct bonding between the sulfonic acid groups of the PFSA polymer and MgF_2-nanoparticles, which are deposited as coating on the surface of micron-size glass particles used for mechanical strengthening of the polymer. The significantly increased water retention capability is attributed to the intimate bonding between the PFSA polymer and the solid inorganic additive.

Keywords: PFSA polymer, composites, nanoparticles, sulfuric acid, water retention, thermal stability
PACS: 88.30.pd

INTRODUCTION

Perfluoro-ionomer proton exchange polymers have superior oxidation and thermal stability as compared to non-fluorinated ionomers but for achievement of good ionic conductivity water molecules have to be incorporated into the hydrophilic part of the polymer, which are lost by evaporation at temperature around 100°C [1]. Numerous attempts have been made to synthesize the Per-Fluoro-Sulfonic-Acid-based polymers, abbreviated as PFSA-ionomers, with improved high temperature ionic conductivity, while preserving a common structural principle of this materials–superior thermo-chemical stability related to a polytetrafluoroethylene backbone, ionic conductivity enabled by acidic groups in the side chain of the polymer and sufficient water retention at elevated temperature. So far, neither changes in the structure of the monomer, like exchange of a longer branched side-chain present in e.g., Nafion® (EW = 1100-1500) by a shorter linear side chain, like e.g., in 3M PFSA (EW = 700-1100) [2, 3] nor change of the acidic group, like e.g. in Flemion® and Aciplex® with shorter side chain and carboxylic acid groups instead of sulfonic acid groups could improve the water retention at elevated temperature [4, 5].

Other strategies applied so far for improvement of water retention in PFSA polymers are based on composite formation. Inorganic additives are mixed with the polymer [4] namely nano-sized hygroscopic oxides, like e.g., SiO_2, TiO_2 ZrO_2, Al_2O_3, as well as several hetero-poly-acids (i.e. phosphor-tungstic acid) [6].

The underlying concept of "internal" supply of water from a more hygroscopic substance, like nano-scaled oxides, to the PFSA polymer upon heating has not been proven so far by thorough investigation of thermal stability of such complex mixtures of different hygroscopic inorganic and organic materials. Only if water is released from the inorganic additive at temperatures not exceeding the onset of PFSA decomposition a synergetic effect on water retention in the polymer could be expected. In order to supply water which is released by an inorganic additive to the acid group of the PFSA polymer an intimate contact between the hydrophilic part of the PFSA-polymer chain and the inorganic additive is required.

As a new approach to solve this problem we have applied a rigorous removal of the solvent from a PFSA-emulsion to enable chemical interaction of the PFSA-hydrophilic side chain with the surface of an inorganic additive. In addition, the size of the inorganic additive and the size of the hydrophilic domains have to be matched, which means operation at a scale of ~10-20 nm. Homogeneous distribution of such small particles has been attempted in the past with sol-gel and other precursor-based techniques [4]. However, agglomeration of the inorganic additive as well as of polymer micelles could not be prevented [7]. We therefore use a micron size template, namely glass particles, which are coated with the inorganic additive MgF_2 [8]. The MgF_2 particles are truly nano-sized, reaching ~10-18 nm primary particle size. Chemical bonding between the additive and the PFSA-side chain sulfonic acid group via hydrogen-bridging bonds is facilitated by utilization of sulfuric acid, H_2SO_4, as liquid co-additive, as shown in Scheme 1. As polymer matrix the short side chain PFSA polymer from 3M is used, with EW of 800. In our work we have analysed the coupled effect of sulfuric acid, H_2SO_4 and MgF_2 inorganic nano-particles as additives for the PFSA polymer, with the aim to understand the potential of improvement of water retention and its influence on thermal decomposition and underlying chemical degradation mechanism of the PFSA polymer.

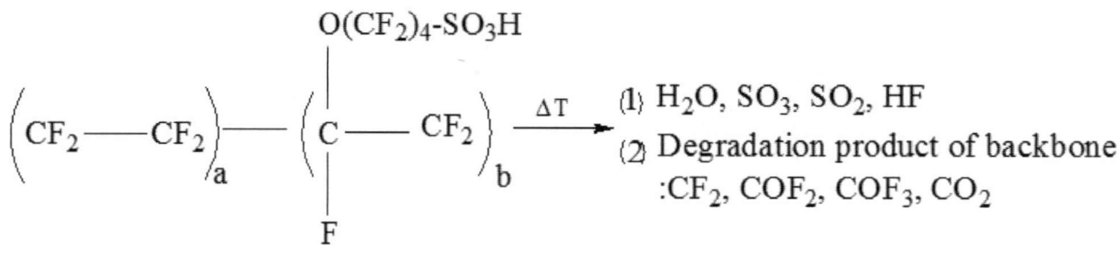

Scheme1. Chemical structure of the 3M polymer and primary decomposition reaction

EXPERIMENTAL

Materials

The 3M PFSA ionomer slurry (Emulsion of 20wt% of Perfluorosulfonic acid polymer with an equivalent weight (EW) of 813 g/mol in n-propanol and water in the ratio of 70:30) was provided by 3M Company, Neuss in Germany. Diethyl monoethyl glycol (DEME), and aqueous hydrofluoric acid (≥40%, HF) were purchased from Sigma-Aldrich. Sulfuric acid (≥95%, H_2SO_4) was purchased from Fischer Chemicals. Commercially available borosilicate (BS) glass flakes (Eckart Luxan E221, ~100-μm size, lateral thickness 1-2 μm and surface area 4 m^2/g) coated with TiO_2 nanoparticles, were used in this study which were further coated by MgF_2-nanoparticles and treated with sulfuric acid for activation. The modified glass flakes called functionalized solid inorganic additives (FSIA).

Preparation of Composite Membranes

For preparation of PFSA-FSIA composite membranes, PFSA slurry is mixed with 96% sulfuric acid and diluted with DEME. The mixture is stirred at 80°C for 20 min and de-aerated; and then, it is ball-milled for 1h, washed subsequently with DEME to remove the excess of H_2SO_4 and to adjust the viscosity for the Doctor Blade casting process. The casting is performed using Kapton® foil as support at a speed of up to 25 mm/min at ambient atmosphere. The composite membrane is further processed by solvent removal at room temperature (RT) followed by annealing at different temperatures (80°C for 60 min, 130°C for 30 min and 140°C for 5 min). After annealing a relaxation step is applied for 20 min at RT followed by removal of the composite from the Kapton® foil by swelling in distilled water (1-3 μS/cm) for 10 min. The composite membrane is finally dried at 80°C for 15 min in an oven.

Instruments

Thermal characterization is performed with thermo-gravimetric analysis (TGA) (Netzsch, Germany and STA 449 C Jupiter) working under Argon flux of 40 ml/min with a heating rate of 10°C/min from 30°C to 650°C. The infrared spectra of the membranes are obtained by using Fourier transform infrared spectroscopy (FT-IR) spectrometer (Bruker Vertex 70) in attenuated total reflectance (ATR) mode with MVP-Pro™ Star single reflection Diamond ATR Accessory.

RESULT AND DISCUSSION

The aim of this work is to reveal the nature of chemical interaction between the FSIA and the PFSA ionomer matrix polymer in presence of water and the mineral acid H_2SO_4. Water, constituent of the PFSA polymer, is crucial for high temperature ionic conductivity. The emphasis is given to the investigation of the different possible chemical interactions of water molecules with PFSA, FSIA as well as the mineral acid.

Water Uptake Analysis by ATR-FTIR

The vibrational spectra of the polymeric composite membranes at different water contents are shown in figures 1 and 2. Formation of H_3O^+/SO_3^-can be detected in an IR-spectra of the solid polymer by the frequency shift of symmetric and anti-symmetric [9] of the S-O vibration at 852/871 cm^{-1} (834 / 883 cm^{-1}). Additional evidence of ion formation comes from a coupled C-O-C / SO_3^- vibrational and C-O-C bending mode, which is detected at 995 and 992 cm^{-1}. The ratio of symmetric C-F stretching at 1130±20 cm^{-1} to asymmetric C-F stretching at 1200±1210 cm^{-1} is changing which indicates that order-disorder conformational changes in the PTFE backbone is occurred. At first, after 8 h of soaking the membrane in water, both intensities of modes are almost equal and after prolonged soaking the membrane in water, the vibration of the CF_2 group shows an increase in intensity as compared to SO_3^- group. As mentioned in literature [10-12], the band at 1057±1 cm^{-1} is not intense in completely dried membrane as compared to 1410 cm^{-1}. But, in case of all H_2SO_4 treated dry polymeric composite membranes, the band at 1410 cm^{-1} is missing and the band at 1057±1 cm^{-1} has high intensity due to the effect of H_2SO_4, which increase as the water content increase. After the formation of the hydronium-sulfonate ion pair with increasing the water content in the membrane, the additional water molecules form a solvation sphere around the sulfonate group and the cation moves away from the sulfonate ion into the bulk of the hydrophilic domains.

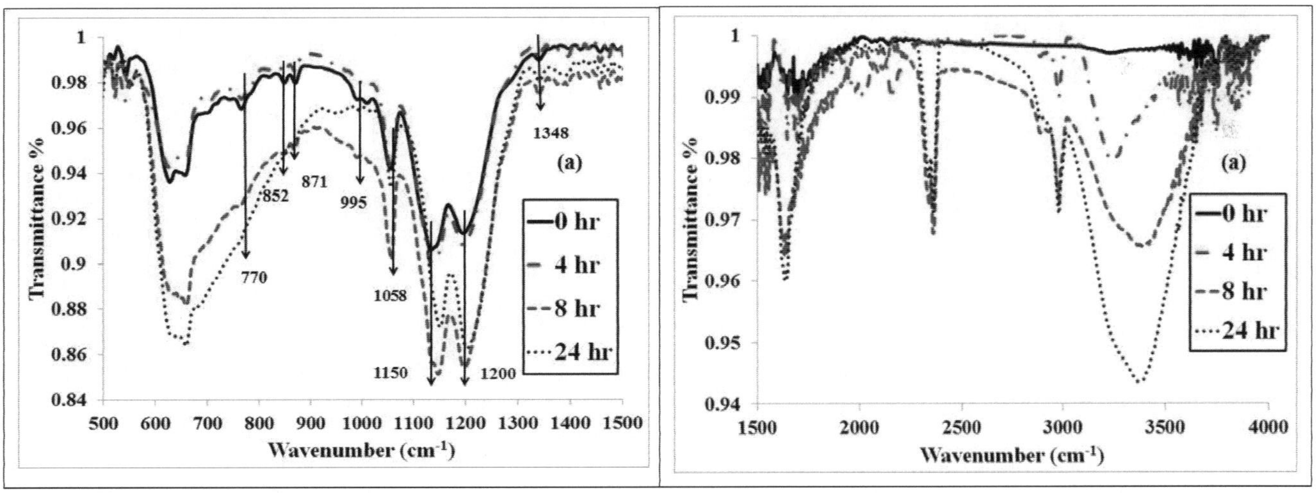

FIGURE 1. FT-IR ATR profiles of the PFSA composite membrane at different water contents (a) D_800_PEM

The H_2O bending frequency, at 1650±20 cm^{-1}, with increasing intensity shows red shift and splitting occurs with increasing the water content, where it is assumed that there is a maximum number of hydrogen bond association occurred. An extremely broad band in the wavenumber region 3000-3700 cm^{-1} is observed and divided in 3 sub-bands similar to liquid water: a faint shoulder near 3600 cm^{-1} in all dry membranes, an intense (the strongest) stretching band at 3400±50 cm^{-1} and another shoulder near 3264±30 cm^{-1}.

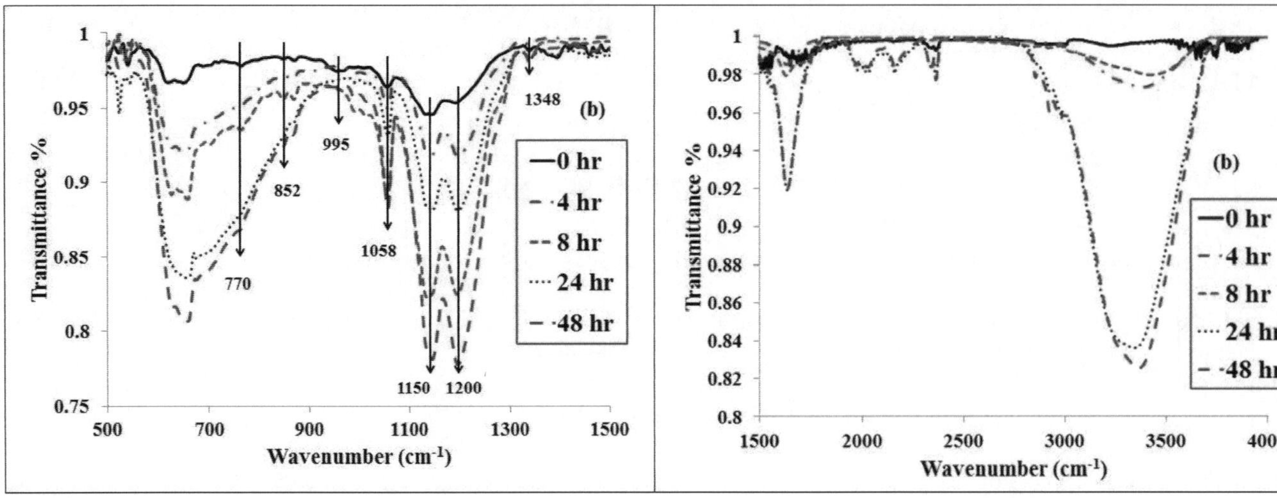

FIGURE 2. FT-IR ATR profiles of the PFSA composite membrane at different water contents (b) D_800_FSIA_3 wt. %

The band at 3400±50 cm^{-1} shows red shift (lower frequency number) that means hydrogen bond strength is increased with increasing the amount of water content and helps to improve the ionic conductivity of the membranes. The maximum red shift is observed in the D_800_FSIA _3 wt. % membrane compare to other membranes. In D_800_PEM treated with H_2SO_4 showed the broadening of the band with increasing the water content that means the water interact with polymeric sulfonic group directly.

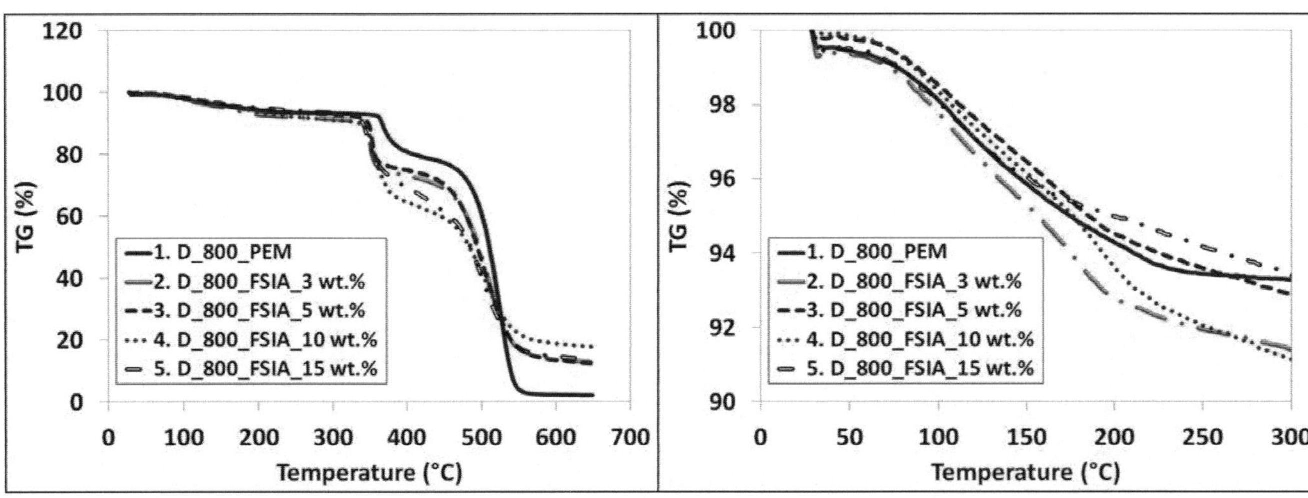

FIGURE 3. TGA profiles of the PFSA polymer composites with different weight % of FSIA.

Thermo-Gravimetric Analysis

The thermal stability of the 3M PFSA polymeric composites is determined with thermo-gravimetric analysis and is shown in figure 3. The thermal stability of the composites is decreased from 360°C to 340°C upon increasing the weight percent of FSIA in comparison to PFSA polymeric membrane because a highly exothermic reaction is occurred by H_2SO_4 which works as a highly oxidizing agent [9, 13-15]. There are four steps mass elimination, the first step in the range 30-130°C, the decomposition is related to removal of the free and bound water molecules and the second step in the range 130-300°C corresponds to the de-sulfonation which attack on the PTFE backbone, while the third step corresponds to the rest of the polyether side chains of the polymer and the removal of the crystalline water bonded with inorganic additives. The fourth step which started from ~370°C to 550°C corresponds to the PTFE backbone pyrolysis quickly follows the third. At temperature above 600°C, the residue is related to the amount of inorganic glass flakes moieties present in the bulk composites.

CONCLUSIONS

This study is based on the coupled effect of mineral acid and FSIA additives on the water retention and thermal stability of the PFSA polymeric composites. Nano coating of MgF_2 particles and sulfuric acid helps to improve the water retention at high temperature because of the strong hydrogen bonding with water molecule and sulfonic groups of the PFSA polymer which acts as bridging agent. As compared to the parent polymer the temperature for water removal is increased by almost 50°C to 250°C.The spectral changes evident in the vibrational spectra of these composites can be associated with two major observations: (1) the dissociation of the acidic proton is most evident by the absence of peak at 1410 cm^{-1} and the band growth at 1059 and 1318 cm^{-1} and (2) The intensity ratio of the band between 1147 and 1240 cm^{-1} is an indication of the change in crystallinity and order-disorder conformation of the PTFE hydrophobic backbone domains.

ACKNOWLEDGMENTS

We thanks to the Bavarian Science Foundation, Germany for financial support of this project (Grant: DOK-133-11). We thank to 3M for providing us the D-800 polymer slurry. We gratefully thanks to Mrs. Ingrid Otto for her great support during analysis.

REFERENCES

1. M. Doyle and G. Rajendran, In Handbook of Fuel Cells: Fundamentals Technology and Applications, edited by W. Vielstich et al, Chichester: John Wiley & Sons, 2003, pp. 351-360.
2. G. A. Giffin, G. M. Haugen, S. J. Hamrock, V. D. Noto, J. Am. Chem. Soc. 135, 822-834 (2013).
3. M. C. Ferrari, J. Catalano, G. M. Baschetti, M. G. De Angelis, G.C. Sarti, Macromolecules 45, 1901-1912 (2012).
4. S. M. J. Zaidi, "Research Trends in Polymer Electrolyte Membranes for PEMFC" in Polymer Membranes for Fuel Cells, edited by S. M. J. Zaidi et al., New York: Springer US, 2009, pp. 7-25.
5. C. J. Heitner-Virguin, Membr. Sci. 120, 1-33 (1996).
6. G. Alberti, M. Casciola, Annu. Rev. Mater. Res. 33, 129-154 (2003).
7. A. Sienkiewicz, "Influence of process parameters on functional properties of slurry derived PFSA membranes", Ph.D. Thesis, University of Bayreuth, Germany, 2012.
8. M. Willert-Porada, H. Wolf, A. Moszczynska, WO 2009/014930 A1 (January 2009).
9. A. Givan, L. A. Larsen, A. Loewenschuss, C. J. Nielsen, J. Mol. Struct. 509, 35 - 47, (1999).
10. W. F. H. Detlef, V. Di Noto, M. Vittadello, J. Phys. Chem. B 113, 632 -639 (2009).
11. M. Laporta, M. Pegoraro, L. Zanderighi, Phys. Chem. Chem. Phys. 1, 4619-4628 (1999).
12. T. A. Ford, M. Falk, Can. J. Chem. 46, 3579-3586 (1968).
13. E. S. Stoyanov, I. V. Stoyanova, C. A. Reed, J. Am. Chem. Soc. 132, 1484-1485 (2010).
14. J. Li, B. Jiang, H. Guo, J. Am. Chem. Soc. 135, 982-985 (2013).
15. M. T. Sucarrat, J. S. Francisco, J. M. Anglada, J. Am. Chem. Soc. 134, 20632 -20644 (2012).

Effect of Blending High and Low Stereo-regularity Components on Crystallization and Melting Behaviors in the Polypropylene Fibers

Kouya Kawai[a], Wataru Takarada[b] and Takeshi Kikutani[b*]

[a]*Department of Organic and Polymeric Materials, Graduate School of Science and Engineering, Tokyo Institute of Technology, 2-12-1, O-okayama, Meguro-ku, Tokyo 152-8552, Japan*
[b]*Department of Materials Science and Engineering, Tokyo Institute of Technology*

kawai.k.ae@m.titech.ac.jp
kikutani.t.aa@m.titech.ac.jp

Abstract. Crystallization and melting behaviors of blend fibers consisting of two types of polypropylene (PP), high stereo-regularity/high molecular weight PP (HPP) and low stereo-regularity/low molecular weight PP (LPP), was investigated. Blend fibers of various HPP/LPP compositions were prepared through the melt spinning process. On-line diameter measurement of the spinning line revealed that the blending of LPP caused the shift of solidification point to downstream. This result indicated the lowering of crystallization temperature by the addition of LPP component. WAXD analysis of as-spun fibers clarified that the crystallinities of HPP/LPP blend fibers of various blend ratios taken-up at 2 km/min or higher matched with the values estimated based on the rule of additivity, whereas the crystallinities of blend fibers taken-up at 1 km/min were lower in comparison with the expected values. On the other hand, in the WAXD analysis during the heating of as-spun fibers, modification of crystalline structure proceeded in the case of HPP single component fibers, whereas the melting of thin crystals mainly consisting of LPP, modification and reorganization of HPP component, and the beginning of the melting of HPP occurred consecutively with the increase of temperature. These results suggested that crystallization of HPP did not occur sufficiently in the spinning process at low take-up velocities because of the addition of LPP component. When the take-up velocity was 2 km/min or higher, development of HPP crystals proceeded sufficiently in the spinning process even though LPP component existing in the system lowered the crystallization temperature.

Keywords: melt spinning, polymer blend, polypropylene, stereo-regularity
PACS: 81.05.Lg, 83.50.Uv, 83.50.Xa

INTRODUCTION

Polypropylene (PP) is one of the most commonly used polymer in the world due to its reasonably good properties and low cost. Recently, PP with extremely low stereo-regularity (LPP) synthesized applying metallocene catalysts was developed and commercialized [1]. The newly developed LPP is being applied either as the resin modifier for improving processability of high stereo-regularity PP (HPP) or as the major component of spun-bonded elastomeric nonwoven fabrics for mask, disposable diaper etc.

We carried out the melt spinning experiment of HPP/LPP blend fibers and confirmed the improvement of spinnability in comparison with the single component spinning of HPP and LPP. Through the DSC analysis of thus prepared fibers, mutual interaction of HPP and LPP components on the melting behavior was investigated. In the heating process of the as-spun blend fibers, it was confirmed that melting of LPP component was hindered by the presence of HPP crystals, while crystallization of HPP component proceeded in the heating process of fibers prepared at low take-up velocities because crystallization of HPP in the spinning process as well as in the conditioning process after the spinning was hindered by the presence of LPP [2]. However, exact mechanisms for the improvement of spinnability has not been clarified yet.

Therefore, in this research, attempts were made to understand the fiber structure development behavior in the melt spinning process of blend fibers. Particular attention was paid for the analyses of crystallization behavior of each component by conducting on-line measurements in the spinning process and structural analysis of the as-spun fibers.

EXPERIMENTAL

Two types of PP, i.e. high stereo-regularity/high molecular weight PP (HPP, MFR = 20 g/10min, T_m = 161.0 °C) and low stereo-regularity/low molecular weight PP (LPP, MFR = 60 g/10min, T_m = 59.6 °C) were used. Single component fibers of HPP, LPP and HPP/LPP blend fibers of various compositions were prepared through the melt spinning process, in which dry-blended polymer pellets were melted at 200 °C and extruded through a single screw extruder, metering pump and a spinning nozzle of 1.0 mm diameter at the throughput rate of 6.0 g/min. Take-up velocity was varied from 1.0 km/min to the attainable highest velocity. The prepared blend fibers with different HPP/LPP compositions were designated as HL*xxyy* with *xx* and *yy* representing the weight percent of HPP and LPP, respectively. In the spinning process, on-line measurement of the fiber diameter was carried out at various distances from the spinneret along the spinning line to analyze the thinning behavior of fibers. Wide-angle X-ray diffraction (WAXD) patterns of the as-spun fibers were obtained using an X-ray generator equipped with a CCD camera. Variation of crystalline structure in the heating process of as-spun fibers was also analyzed through the WAXD measurement.

RESULTS AND DISCUSSION

Spinning Behavior

The attainable highest take-up velocities for LPP, HL1585, HL5050, HL8020 and HPP were 2, 5, 9, 8 and 5 km/min, respectively. This result means that the spinnability of LPP was improved by blending HPP, whereas the spinnability of HPP was also improved by blending LPP. Results of on-line diameter measurement are shown in FIGURE 1. With the increase of LPP composition, solidification point shifted to downstream. This result corresponds to the lowering of crystallization temperature, which may cause the reduction of lamella thickness. Reduction of crystallizability, i.e. hindrance of the crystallization of HPP by the presence of LPP, can be the origin of this behavior. Reduction of viscosity is also attributable for the suppression of orientation-induced crystallization.

In the case of LPP and HL1585, clear solidification behavior was not observed except for HL1585 taken-up at 5 km/min. It can be considered that low crystallizability and low crystallinity are the origins for the elastmeric deformation of the spinning line. This consideration was supported by the fact that LPP fibers stuck each other on the take-up bobbin. Accordingly, we speculated that in the case of LPP-rich blend, improvement of crystallizability by blending HPP probably contributed to the improvement of spinnability, whereas in the case of HPP-rich blend, reduction of solidification speed through the suppression of crystallization by blending LPP contributed to the improvement of spinnability [3].

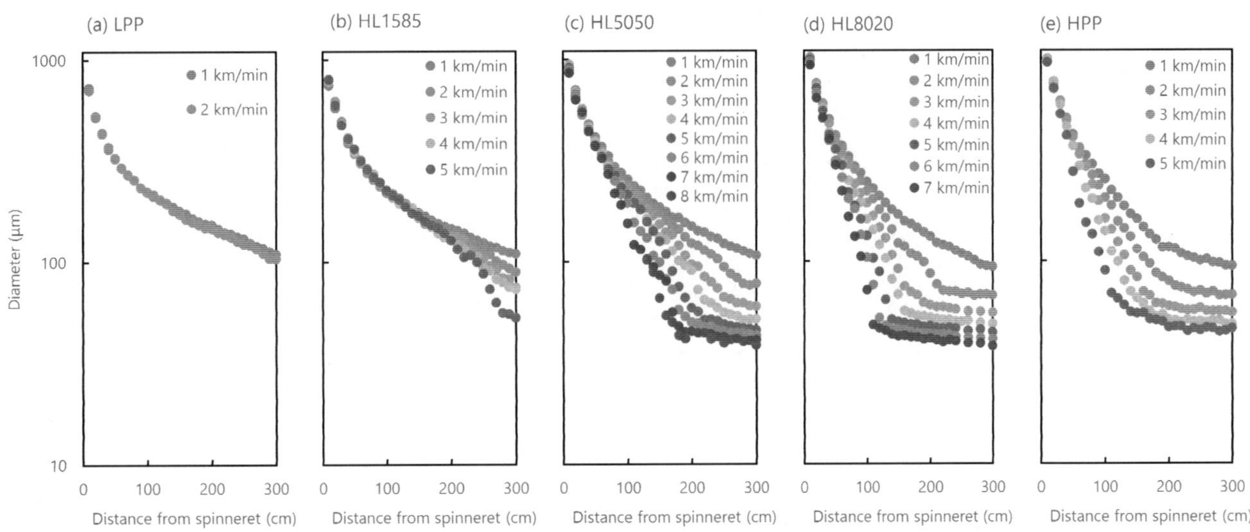

FIGURE 1. Thinning behaviors of fiber diameter along the spinning line of different take-up velocities for LPP (a), HPP/LPP blend of various compositions (b,c, and d) and HPP (e).

WAXD Analysis of As-spun Fibers

WAXD patterns of LPP, HPP/LPP blend and HPP fibers are shown in FIGURE 2. LPP fibers taken-up at 1 km/min showed the crystalline structure with α-form crystal of no orientation. LPP fibers taken-up at 2 km/min, and all other blend fibers and HPP fibers showed the bimodal c- and a*-axis orientation of α-form crystals. When HPP composition and take-up velocity were increased, crystalline orientation increased reflecting the increases of tensile stress applied to the spinning line. The stress increase was caused by the increase of viscosity of polymer melt and the increases of air-friction and inertia forces, respectively.

Crystallinity of as-spun fibers obtained through the analysis of WAXD patterns is shown in FIGURE 3. Crystallinity was estimated firstly by converting the 2-D WAXD data to the diffraction angle versus intensity data through the azimuthal integration with weighing of sin ϕ, where ϕ is the azimuthal angle, and secondly separating the obtained intensity into crystalline and amorphous components. The analyzed crystallinity increased with the increase of HPP composition. It should be noted that the crystallinities of the blend fibers taken-up at 1 km/min were clearly low in comparison with values predicted based on the rule of additivity. This result suggested that the crystallization of HPP component was hindered during the spinning process as well as the conditioning process after spinning because of the presence of LPP [2].

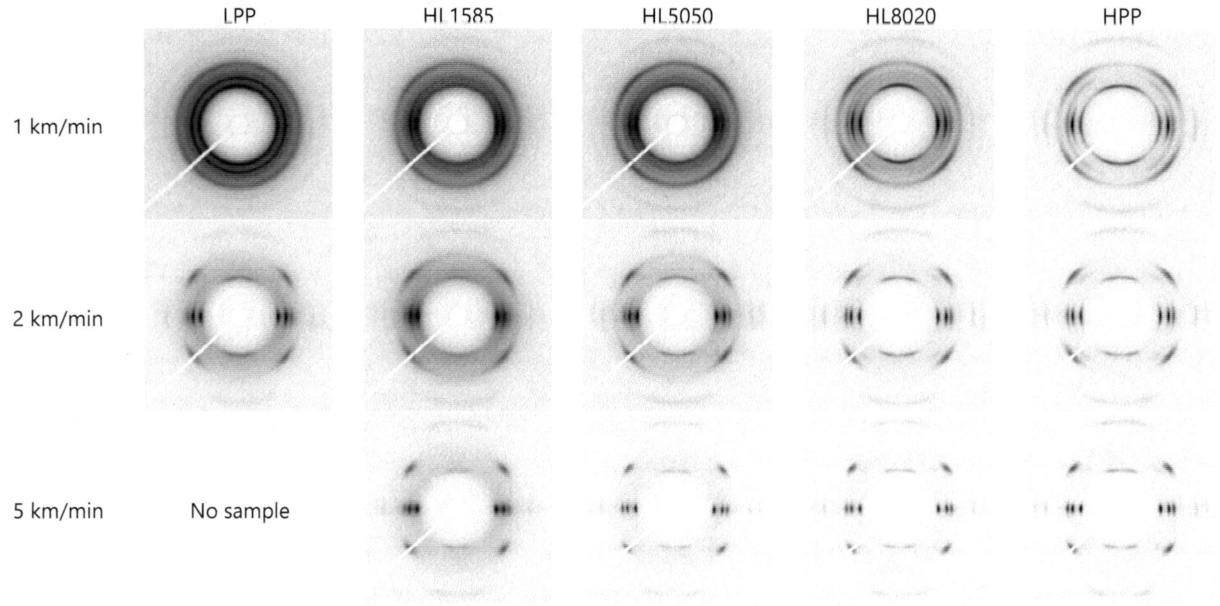

FIGURE 2. WAXD patterns of as-spun LPP, HPP/LPP blend and HPP fibers of selected take-up velocities.

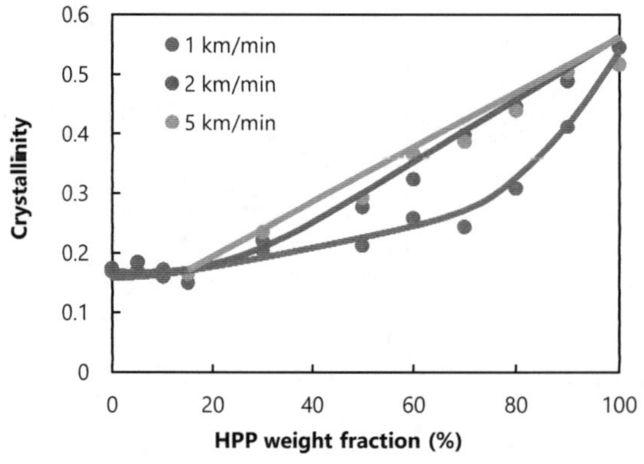

FIGURE 3. Crystallinity of as-spun fibers analyzed from the results of WAXD measurements.

Structural Changes during Heating

WAXD measurement was conducted in the heating process of as-spun fibers at a heating rate of 2 °C/min. Variations of WAXD intensity profiles, which were obtained through the azimuthal integration as described in the previous section, with the increase of temperature for HPP and HL8020 fibers taken-up at 1 km/min are shown in FIGURE 4. In case of HPP, intensities of (110) and (040) reflections did not change upon heating, whereas the intensity of the valley between (110) and (040) reflections decreased. This result indicates the occurrence of crystalline form modification in the heating process. On the other hand, in case of HL8020, intensity of (040) reflection decreased at 90 °C, increased at 150 °C, and finally decreased again at 170 °C. These three steps may correspond to the melting of thin crystals mainly consisting of LPP, modification (reorganization) of HPP component, and the beginning of melting, respectively. It should be noted that this result explains the reason why the crystallinities of blend fibers taken-up at 1 km/min were lower than the expected values. At this take-up velocity, crystallization of HPP component did not proceed sufficiently due to the addition of LPP component. On the other hand, when take-up velocities was 2 km/min or higher, HPP crystals developed sufficiently in the spinning process even though LPP component existing in the system caused the lowering of crystallization temperature.

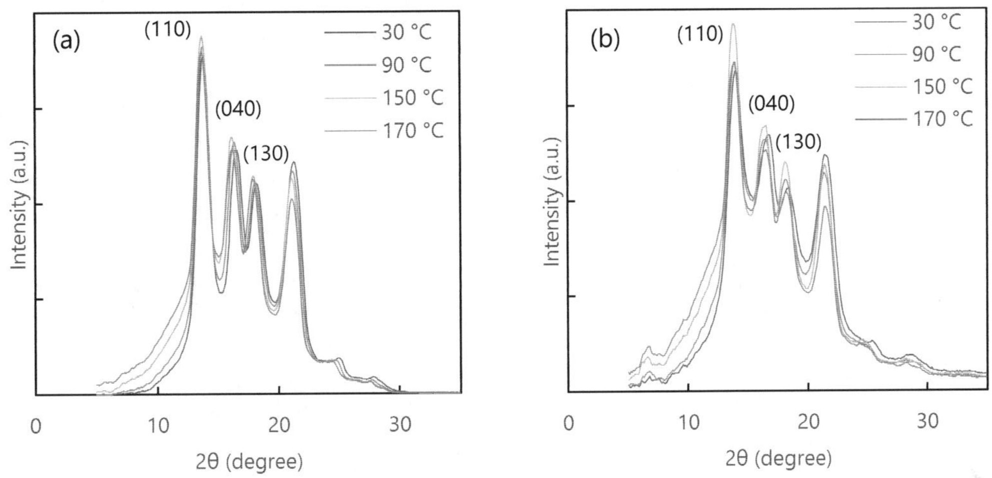

FIGURE 4. Azimuthally integrated WAXD profiles at selected temperatures for (a) HPP 1 km/min and (b) HL8020 1 km/min fibers.

CONCLUSIONS

Investigations on the mechanism of fiber structure development in HPP/LPP blend fibers were carried out. Improvements of spinnability of LPP by the blending of HPP as well as that of HPP by blending of LPP were confirmed. Through the on-line measurement of the thinning behavior of the spinning line and the WAXD measurement of the as-spun fibers, it was confirmed that crystallization of HPP component was hindered by the presence of LPP in the spinning process. At low take-up velocities, HPP crystals could not be fully developed if a certain amount of LPP was blended, however when take-up velocity was increased, crystallization of the HPP component proceeded sufficiently in the spinning process although presence of LPP in the system lowered the crystallization temperature.

REFERENCES

1. Y. Minami, T. Takebe, M. Kanamaru and T. Okamoto, *Polymer Journal* **47**, 227-234 (2015).
2. K. Kawai, Y. Kohri, W. Takarada, T. Takebe, T. Kanai and T. Kikutani, *AIP Conference Proceedings* **1713**, 080002 (2016).
3. Y. Kohri, W. Takarada, H. Ito, T. Takebe, Y. Minami, T. Kanai, T. Kikutani, *Seikei-Kakou* **20**, 831-839 (2008).

A new process to convert proton exchange membranes into high performances and reliable hybrid membranes for fuel cell

J.P. COSAS FERNANDES, L. GONON, V.H. MAREAU AND H. MENDIL-JAKANI.

Univ. Grenoble Alpes, INAC-SPrAM, F-38000 Grenoble, France
CNRS, SPrAM, F-38000 Grenoble, France - CEA, INAC-SPrAM, F-38000 Grenoble, France
E-mail: laurent.gonon@cea.fr

Key Words : Proton exchange membrane, PEMFC, PEMWE, Ionomer, sol-gel

Proton exchange membrane for fuel cell (FC) is a promising environmentally friendly and oil-free technology. One of its key components is the polymer electrolyte membrane acting as separator and proton conductor. Improving both the performances and durability of the membrane is a crucial issue. Different strategies have been explored to improve the mechanical properties of the membrane and to reduce the ageing impact. Despite a lot of progress, their functional properties (conductivity, water sorption, thermomechanical properties, chemical stability…) stay below expectation when operated above 100°C whereas higher operating temperature (150°C). Therefore, there is currently an increasing interest in using reliable alternative membranes that are capable of operating at higher temperatures, low relative humidity to ensure long term performances. Popular candidates are polyaromatic polymers like sPEEK which present good mechanical properties at high temperatures but poor functional properties (proton conductivity and durability in FC). We will demonstrate that the sPEEK performances can be significantly improved using a suitable hydrothermal treatment up to becoming equivalent to that of Nafion the benchmark material. SAXS analyses performed on hydrothermally treated sPEEK show that the increase in molecular mobility induced by the pretreatment (water, temperature) allows a gradual nano-structuration of the membrane and a sharp increase of its proton conductivity. To improve the membrane durability, we have developed an original strategy based on the insertion of a chemical stabilizing network by a sol-gel process. These hybrid membranes will combine excellent proton conductivity at high temperature and low relative humidity and will be stable under hard operating conditions. The process involved should be universal allowing to improve the membrane properties of common ionomer membrane

Using Chemical Synthesis Method to Develop the Polyamide 6 Filament Knitted Fabric with Anti-Mosquito, Anti-UV, and Cooling Functions

C.P. Chin (presenting author)[a], J. Y. Liu [b](reference author), K.B. Cheng [a], J.T. Chang [b], and F.L. Huang [b]

[a]*Department of Fiber and Composite Materials, Feng Chia University, Taichung 407, Taiwan*
[b]*Textile and Material Industry Research Center, Feng Chia University, Taichung 407, Taiwan*

m29615228@gmail.com;
liujenyung@gmail.com

Abstract. The innovation of these fabrics is to use cooling fiber whose raw material is not physical mixed by traditional method, but synthesized directly by one single chemical polymerization process. The functional powder is well-disturbed in the molecular solution while polymerization. The outcome is from the features of shorter processes, save energy and lower cost, etc. The results show that the cooling additive slurry with IR slurry 33wt% and deionized water 67wt% is directly mixed with precursors during polymerization. The best performance of cooling slurry suspension for added polypropylene glycol (PPG) and sodium pyrophosphate (SPP) for dispersant compound ratio is 1 : 2, it has the best average particle size is 345.06nm, and -38.83mV for zeta potential measurement, and the suspension could be conserved more than two days, and also could fabrication of polyamide 6 cooling chips for the relative viscosity is 2.184. Furthermore, we melt-spin the polyamide 6 filament and blend with the composite yarn with the function of anti-mosquitoes and anti-UV fiber. The product is under the test of dynamic thermal image outdoor one hour running that the ΔT is more than 0.9 °C, and static indoor 5 hours running that ΔT is more than 1.0 °C. We also got excellent cooling effect. In order to raise the color fastness to light of this textile, we use the technique of dye select. And we develop the composite and outdoor leisure series products. These products have the cooling feeling, anti-mosquitoes and anti-UV function is over 50+.

Keywords: Polyamide 6, IR slurry, Cool Touchfeeling, Anti-UV

INTRODUCTION

The functional man-made fiber raw materials are usually produced by using the physical mixing functional additives and polymer matrix. However, the polymer matrix would experience multi thermal process during the mixing, and induce consequent thermal cracking. The mixing technique also affects the dispersion of functional additives significantly. If the additives can be added while the chemical polymerization of matrix masterbatch, the above mentioned disadvantages of physical mixing method can be completely avoided. And the production of functional fiber can be shortened in one single process. This would reduce the waste and get high functional performance fabric. It is known that cotton, viscose rayon, nylon, polyester, and Tencel fibers were used as the warp or weft knitted fabric are mainly used in underwear, sport wear, socks, pants, and bra materials etc. Within the scope, the aim of this paper was to fabricate comfortable multifunctional knitted fabrics which have better air permeability, cool touchfeeling property, moisture transfer, and anti-mosquito performance with the raw material produced by chemical synthesis.

In this paper, we use chemical synthesis method to fabricated cooling PA6 chips and cooling PA6 fiber, a new fiber material which contains mineral composite cooling powders (MSCCP) is designed for cooling fabrics. MSCCP were prepared by wet type ball grinding and through spraying dry then make the masterbatch (MB) and added certain amount of MB to spin full draw yarn (FDY) with 40d/24f by melt spinning. Furthermore, the regular draw textured yarn (DTY) with 40d/36f and 40d/34f by Murata nip twister. Finally, the weft knitted fabric was fabricated by circular knitting machine. It is expected that the knitted fabrics exhibit the cooling, anti-UV, reflection index and water absorption properties. The cooling FDY yarns were used as the raw materials to fabricate weft knitted fabrics with the different fiber materials by circular knitted machine with temperature difference, anti-UV, and anti-mosquito function properties. The MSCCP particles, cooling FDY yarns and their inter-knitted fabrics have been fabricated successfully. Furthermore, the objectives of this investigation were to characterize thermal imaging, ultraviolet protection factor, air permeability, and anti-mosquito ability of knitted fabrics with the cooling knitted yarns. Finally, the end applications of the knitted fabrics will also be proposed.

The areal weight, constitutions, knitted structure, density of wale and course yarns per unit length (WPI, CPI) were shown in **TABLE 3**. In order that the fabric codes can be apprehended easily, the codes were prepared by using the numbers of yarn codes which are used in manufacturing of the knitted fabrics. The knitted fabrics were manufactured in single jersey knitted structure (SJ) by Liba circular knitted machine and Tricot knitted structure by Carl Meyer warp knitted machine in Germany. The texture reports of the weft knitted fabrics are same. The weft knitted fabric produced by Liba weft knitting machine was configured to have an areal weight similar among these five fabrics. An interknitting of 40d/24f Nylon 6 FDY with cooling and semidull (43 wt%) and 40d/34f Nylon 6 DTY with bright (43 wt%) and 20d Spandex was used to produce weft knitted fabric 32122A. An interknitting of 50d/72f polyester DTY with semidull (89%) and 20d Lycra (11wt%) was used to produce weft knitted fabric 32301. An interknitting of 40d/36f Nylon 6 DTY with semidull (86%) and 20d Lycra (14wt%) was used to produce weft knitted fabric 32821. An interknitting of 50d/72f polyester DTY with semidull (80%) and 20d Lycra (20wt%) was used to produce warp knitted fabric S615135. Finally, an interknitting of 40'S/1 cotton yarn (94%) and 20d Lycra (6 wt%) was used to produce weft knitted fabric ABVF0191-12.

EXPERIMENTAL PROCEDURES

Before synthesis, the additive needs to be suspended in liquid. Cooling additive slurry with different dispersants are listed in **TABLE 1**. 67g of deionized water and 0.99g of dispersants were mixed and ultrasonicated for 30min, then 33g cooling powders were mixed into solution and ultrasonicated for 30min, cooling additive slurry were obtained. Referring to **TABLE 2**, 300g of ε-caprolactam, 0.83g of terephthalic acid, 5.36g of deionized water and 12g additive slurry were mixed in the 1000ml reaction bottom flask, and heated in nitrogen atmosphere at constant temperature with stirring. The open-ring reaction temperature was maintained at 190 °C for 2h, then the polymerization temperature was increased to 250°C for 8h. The polyamide 6 composite was obtained and water cooled. After drying procedures and crushing, the polymer chips were purified with 110 °C boiling water for 3h, and dried in vacuum at 110°C for 24h , then cooling polyamide 6 chips was obtained.

TABLE 1. The labels of additive slurries with different fractions of dispersants, deionized water and cooling powders.

Sample	Dispersants		Deionized water	Cooling powders
	PPG (g)	SPP (g)	(g)	(g)
IR	---	---	67	33
IRA	0.99	---	67	33
IRB	0.66	0.33	67	33
IRC	0.33	0.66	67	33
IRD	---	0.99	67	33

TABLE 2. The labels of polyamide 6 sample chips synthesized with different cooling slurries.

Sample	CPL (g)	TPA (g)	Deionized Water (g)	IR slurries				
				IR (g)	IRA (g)	IRB (g)	IRC (g)	IRD (g)
PA6	300	0.83	13.4	---	---	---	---	---
IR12	300	0.83	5.36	12.00	---	---	---	---
IRA12	300	0.83	5.36	---	12.00	---	---	---
IRB12	300	0.83	5.36	---	---	12.00	---	---
IRC12	300	0.83	5.36	---	---	---	12.00	---
IRD12	300	0.83	5.36	---	---	---	---	12.00

The areal weight, constitutions, knitted structure, density of wale and course yarns per unit length (WPI, CPI) were shown in **TABLE 3**. In order that the fabric codes can be apprehended easily, the codes were prepared by using the numbers of yarn codes which are used in manufacturing of the knitted fabrics. The knitted fabrics were manufactured in single jersey knitted structure (SJ) by Liba circular knitted machine and Tricot knitted structure by Carl Meyer warp knitted machine in Germany. The texture reports of the weft knitted fabrics are same. The weft knitted fabric produced by Liba weft knitting machine was configured to have an areal weight similar to these five fabrics. An interknitting of 40d/24f Nylon 6 FDY with cooling and semidull (43 wt%) and 40d/34f Nylon 6 DTY with bright (43 wt%) and 20d Spandex was used to produce weft knitted fabric 32122A. An interknitting of 50d/72f polyester DTY with semidull (89%) and 20d Lycra (11wt%) was used to produce weft knitted fabric 32301. An interknitting of 40d/36f Nylon 6 DTY with semidull (86%) and 20d Lycra (14wt%) was used to produce weft knitted fabric 32821. An interknitting of 50d/72f polyester DTY with semidull (80%) and 20d Lycra (20wt%) was used to produce warp knitted fabric S615135. Finally, an interknitting of 40'S/1 cotton yarn (94%) and 20d Lycra (6 wt%) was used to produce weft knitted fabric ABVF0191-12.

TABLE 3. Technical data of the knitted fabrics with different constitutions.

Sample	Description	Constitutions	Weight (g/m^2)	WPI	CPI
32122A SJCBS	SJ	40d/24f SD Cooling PA 6 DTY 43% 40d/34f Bright PA 6 DTY 43% 20D Spandex 14%	126.5	64	124
32301 SJPDL	SJ	50D/72f Polyester DTY 89% 20D Lycra 11%	118.0	55	103
32821 SJNDL	SJ	40d/36f SD PA 6 DTY 86% 20D Lycra 14%	128.5	61	131
S615136 TPDS	Tricot	50d/36f Polyester DTY 80% Spandex 20%	120.03	71	131
VF0191-12 SJCS	SJ	40'S/1 Cotton yarn 94% Spandex 6%	140	39	52

Note: SJ means single Jersey; Tricot: Tricot warp knitted structure; WPI: wales/inch; CPI: Course/inch; SD: Semi Dull; B: Bright

RESULTS AND DISCUSSIONS

The **FIGURE 1.** shows the effects of different dispersant on the cooling powder's particle size in slurries. For all of the samples, IRC slurry's dispersant recipe disperses cooling powders efficiently as compared to IR slurry, IRA slurry, IRB slurry and IRD slurry. We also find the mixed dispersant system disperses cooling powders better than PPG solution and SPP solution alone. the average particle size (D50) is decreased from 517.70nm to 373.73nm and 345.26nm for IRB slurry and IRC slurry respectively. This demonstrates that the mixed dispersant system disperses cooling powders more efficiently as PPG solution and SPP solution alone. Due to volume confinement of PPG molecular segments and ionic charge intensity of SPP.

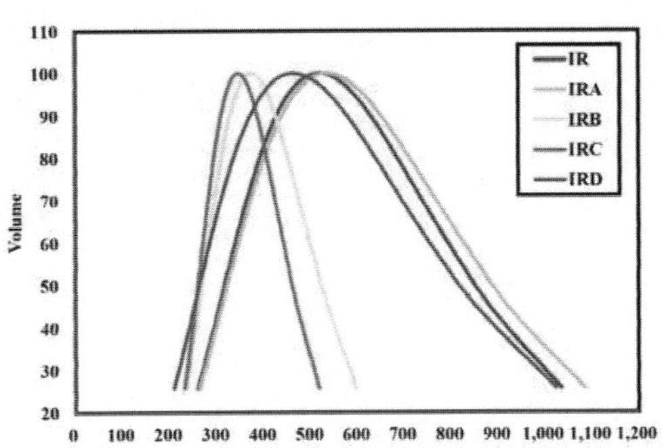

FIGURE 1. The particle size distribution of slurries with different dispersants.

FIGURE 2. present the effect of pH value on zeta potential of cooling powders surface in deionized water with or without dispersants. Pristine cooling powders aggregated seriously **(FIGURE 2.)** in all the pH values because almost the absolute values of the zeta potentials of cooling powders are positive charge. Because of cooling powder's surface is negative charge, it could attract positive charged side of dispersant near to the cooling powders. Whereas addition of dispersants (SPP or PPG+SPP) can make cooling powders are negative charge, as shown in **FIGURE 2.**, zeta potential values of cooling powders are relatively high from pH 7 to pH 12, and at base pH values, zeta potential are efficient with cooling powders. From pH 2 to pH 12, zeta potentials have a tendency to increase, addition of SPP increased, the ionic strength have a tendency to increase.

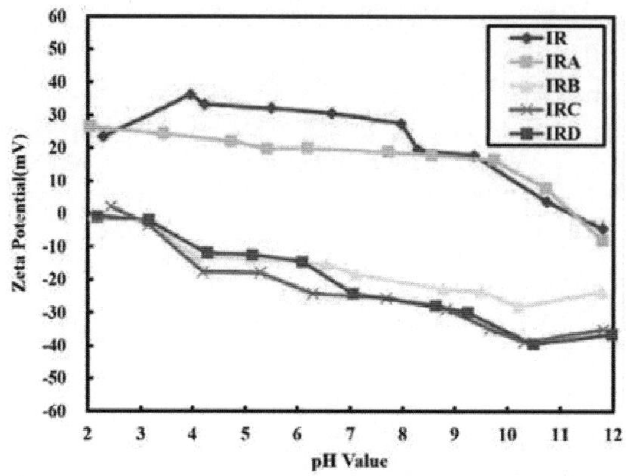

FIGURE 2. The effects of different dispersant recipes on the zeta potential of suspended particles.

FIGURE 3. illustrates the relative viscosity value of cooling polyamide 6 chips. Whereas addition of IR slurry and mixed with CPL, it restricts the polymerization reaction. As shown in **FIGURE 3.**, addition of dispersants (PPG, SPP or PPG+SPP) can make polymerization reaction, the RV is decrease from 0 to 1.473, 2.001, 2.184 and 1.711 respectively for IRA12, IRB12, IRC12 and IRD12. It may be due to the dispersants can adsorption on the cooling powders, avoid to contact the polymer of PA6 chains.

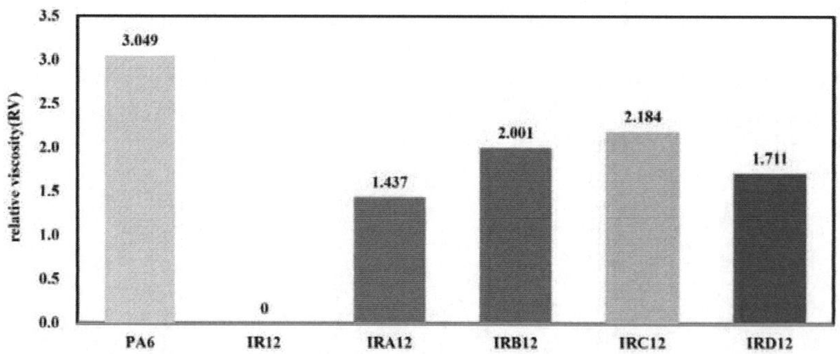

FIGURE 3. The relative viscosity value of cooling polyamide 6 chips.

The temperature difference of the knitted fabrics was shown in **TABLE 4.** All of the knitted fabrics exhibited thermal absorption and diffusivity behaviors. Average increase of temperature difference of fabrics decreases with the content of Nylon 6 cooling draw textured yarn (DTY) in **TABLE 4**. Results indicated that the average temperature increase of SJCBS single jersey knitted fabrics reduces about 30%, which performs a better cool feeling than other fabrics with lower content cooling yarn in weft knitted fabric.

The transmission and shielding rate of ultraviolet of the knitted fabrics was listed in **TABLE 5.** Although the SJCBS knitted fabrics exhibited 43% cooling yarn and have lower thermal conductivity and higher thermal diffusivity, it is benefit for UV-blocking. Results indicated that the transmission and shielding rate of ultraviolet for the SJCBS single jersey knitted fabrics have not significant difference to other four samples. It could be due to all five fabric contain other yarns with titania additives.

TABLE 4. Temperature difference of the knitted fabrics.

Sample	ΔT_1 (T_{10}-T_0)	ΔT_2 (T_{10}-T_{20})	ΔT_3 (T_{20}-T_0)
32122A SJCBS	7.7°C	5.2°C	2.5°C
32301 SJPDL	7.7°C	5.5°C	2.3°C
32821 SJNDL	7.6°C	5.4°C	2.3°C
S615136 TPDS	9.6°C	7.0°C	2.6°C
VF0191-12 SJCS	10.2°C	7.4°C	2.6°C

TABLE 5. Transmission and Shielding Rate of Ultraviolet for the Knitted Fabrics.

Sample	Transmission of ultraviolet (%)		Shielding rate of ultraviolet (%)	
	UVB	UVA	UVB	UVA
32122A SJCBS	2.4	2.9	97.6	97.1
32301 SJPDL	2.7	2.2	97.3	97.8
32821 SJNDL	0.3	0.3	99.7	99.7
S615136 TPDS	0.2	0.2	99.8	99.8
VF0191-12 SJCS	0.1	0.1	99.9	99.9

CONCLUSIONS

This is study develops a chemical synthesis method to synthesize multifunctional polyamide 6 filament for knitted fabric usage. The additives powders are dispersed uniformly in the polyamide 6, and synthesized polyamide 6 chips exhibits enough spinning ability for producing cooling polyamide 6 filament. The functional fiber and regular polyamide 6 fiber materials were used to fabricate the knitted fabrics. The performances of knitted fabrics are compared for better understanding of the factors and testing methods which influences the functions of fabrics and to evaluate their potential applications in functional apparels. The knitted fabrics with single jersey configurations were produced on Liba circular knitting machine and one fabric was produced on a Carl Meyer warp knitting machine. All the fabrics were tested by the thermal imaging, ultraviolet protection factor (UPF) and anti-mosquito index testing methods to assess fabric performance. The results of this experiments indicated that cooling yarn manufacturing method is a viable process for producing the high quality and performance cooling knitted fabrics. To achieve international brand underwear quality, the knitting fabric configuration must incorporate stringent cooling yarn constitutions that will control cooling fiber content, yarn tension, alignment, and configuration stability. More details on the fabricated techniques used for these structures and the mechanical properties are described elsewhere.

ACKNOWLEDGMENTS

The authors appreciate the analysis equipment support from Precision Instrument Support Center and Department of Materials Science and Engineering in Feng Chia University. This study is financial support by Ministry of Science and Technology in Taiwan under MOST 105-2218-E-035-004 project.

REFERENCES

1. C. Hepburn, "Polyurethane Eastomers", Applied Science Publisher, London and New York ,1982, pp.3-6.
2. G. Oertel, "Polyurethane Handbook", Hanser Publisher, Munich Vienna New York, 1985.
3. F.C. Frisch, "Polyurethane Technology", Ed., Bruins, P.F., Interscience Publishers, New York, 1979, pp.1.
4. K. Bergmann, EMS INVENTA-FISCHER, Polyamide 2002 World Congress, Modern Polyamide-6 Production Technology, 2003.
5. M. I. Kohan, Nylon Plastics Handbook, 1995.

Structure and Properties of Melt-spun Fibers of Polyethylene Blended with Cellulose Fibers

Yuta Sekiguchi, Wataru Takarada and Takeshi Kikutani*

Department of Materials Science and Engineering, Tokyo Institute of Technology

Kikutani.t.aa@m.titech.ac.jp

Abstract. Melt spinning of high-density polyethylene (HDPE) blended with cellulose nano-fibers was carried out with the aim of evaluating maximum mechanical properties of the compound. Some types of chemical modification were applied for polyethylene to improve interaction with cellulose fibers. When the fibers prepared at the same take-up velocities were compared, tensile properties of the polyethylene fibers blended with only 1 wt% of cellulose fibers showed significantly higher tensile modulus and strength in comparison with those for the neat polyethylene fibers. Structure analysis of the fibers through the wide-angle X-ray diffraction and birefringence measurement revealed that the improvement of the mechanical properties was caused through the enhancement of molecular orientation of polyethylene rather than the reinforcement by the cellulose fibers. Single fiber pull-out test was also performed to measure the interfacial shear strength between matrix and fiber. The results showed that the interfacial shear strength between HDPE and cellulose fiber was 6.5 MPa, which indicates poor adhesion, while that between modified PE and cellulose fiber increased to 30.6 MPa.

Keywords: Cellulose nano-fiber, Fiber reinforced composites, Polyethylene, Melt-spinning
PACS: 81.07.Bc, 82.35.Gh, 82.35.Lr

INTRODUCTION

Cellulose nanofiber (CeNF) is the crystalline micro-fibril with the dimensions of 3 – 4 nm in diameter and several micrometers in length. One of the promising applications of the CeNF is considered to be the use as the filler for fiber reinforced thermoplastic nano-composites especially in the field of commodity plastics. This is due to the high tensile strength and modulus of the CeNF combined with its low density, biodegradability and renewability. However, application of CeNF reinforced plastics (CeFRP) in the market is still limited because the process for preparing CeNF requires long processing time and high energy consumption. To overcome such drawbacks, simultaneous nano-fibrillation compounding (SFC) process has been proposed recently [1]. In this process, fibrillation of cellulose fibers during the compounding with thermoplastic polymers is promoted through the chemical and physical modifications of the process. In this study, with the aim of investigating the ultimate mechanical properties of the compound, melt spinning of thus prepared CeFRP was performed to obtain fibers with the structure of highly oriented CeNF and matrix polymers. The interfacial shear strength between matrix resin and reinforcing cellulose fiber, which plays an important role for the improvement of mechanical properties of composites, was also evaluated through the single fiber pull-out test to examine the effect of chemical modification.

EXPERIMENTAL

Materials used for the melt spinning are listed in Table 1. Mod PE in this table corresponds to the HDPE blended with around 5 wt% of maleic-anhydride-modified HDPE. Melt spinning of these three types of resin were carried out using a single screw extruder equipped with a metering pump and a single-hole spinneret with the nozzle diameter of 1 mm. Extrusion temperature and mass-throughput rate were set to 200 °C and 5 g/min, respectively. Winding of the extruded fiber was performed at the take-up velocities between 0.5 and 4.0 km/min using

TABLE 1 Composition of resin

	Sample code and content (Mass fraction)		
	Neat PE	CeFRP	CeFRP / Mod PE
HDPE	100	100	100
Pulp	-	1.01	1.01
Mod PE	-	-	5

a high-speed winder placed at 3.3 m below the spinneret.

Crystalline structure of the as-spun fibers was analyzed through the wide-angle X-ray diffraction (WAXD) measurement. Molecular orientation of matrix resin was evaluated through the birefringence measurement using a polarizing microscope. Mechanical properties of as-spun fibers were characterized through the tensile test at a strain rate of 100 %/min.

For the evaluation of interfacial shear strength between matrix resin and cellulose fibers, fibers of HDPE, HDPE modified with maleic anhydride and HDPE containing 0.05 wt% peroxide (Peroxide/PE) were prepared through the melt spinning process. The single-fiber pull-out test sample was prepared by embedding rayon fiber of 11 μm diameter perpendicularly into HDPE fibers, which were placed and melted on a glass-plate [2]. Applying the Greszuzuk model [3], interfacial shear strength τ_{max} was evaluated from the averaged interfacial shear strength τ_{av}, which was calculated from the maximum pull-out force, fiber diameter and embedded length,

RESULTS AND DISCUSSION

Mechanical Properties of As-spun Fibers

In the melt spinning experiment, attainable highest take-up velocities for the neat HDPE, HDPE reinforced with CeNF, and modified HDPE reinforced with CeNF were 4.0, 1.5 and 2.0 km/min, respectively. Addition of only 1 wt% of cellulose fibers significantly deteriorated the spinnability of polyethylene. It should be noted that the melt spinning of CeNF reinforced modified PE containing peroxide was not possible because of high viscosity of the polymer melt.

Stress-strain curves of the as-spun fibers obtained at the take-up velocity 1 km/min using the three types of resin are compared in Fig.1. Fibers from the resins with the reinforcement of CeNF showed significantly higher yield stress and tensile strength in comparison with the fibers of neat HDPE. Variations of tensile modulus, tensile strength and elongation at break with the take-up velocity for these three types of fibers are shown in Fig. 2-(a), (b) and (c), respectively. At low take-up velocities, HDPE fibers reinforced with CeNF generally showed better mechanical properties, i.e. higher tensile modulus and strength and lower elongation at break, in comparison with the fibers of neat HDPE. When take-up velocity was further increased up to 4.0 km/min, mechanical properties of the as-spun fibers of neat HDPE exceeded the highest tensile modulus and strength achieved for the fibers of CeFRP. Effect of the addition of CeNF on the tensile modulus of the fibers was estimated based on the Cox theory [4] with the assumption of perfectly oriented CeNF along the fiber axis. The obtained values are also plotted in Fig.2 (a). Tensile modulus of the as-spun fibers of CeFRP were higher than the predicted values, suggesting that the increase of tensile modulus was not solely due to the reinforcement by the incorporated CeNF.

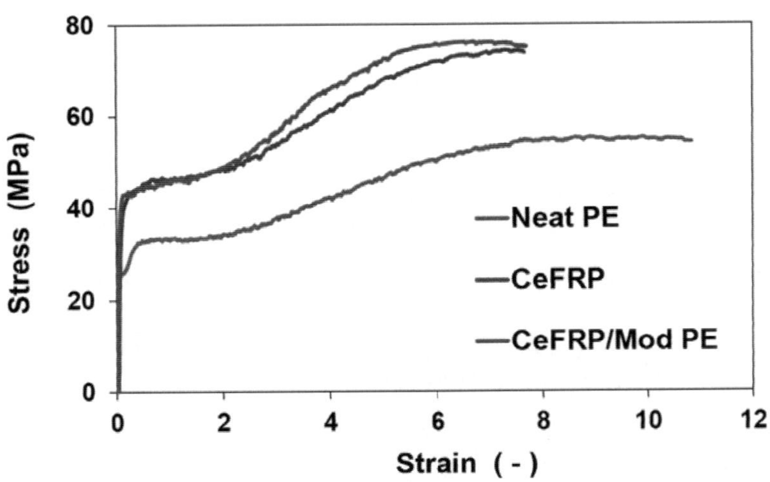

FIGURE 1. Stress-strain curves of the three types of fibers prepared at take-up velocity 1.0 km/min.

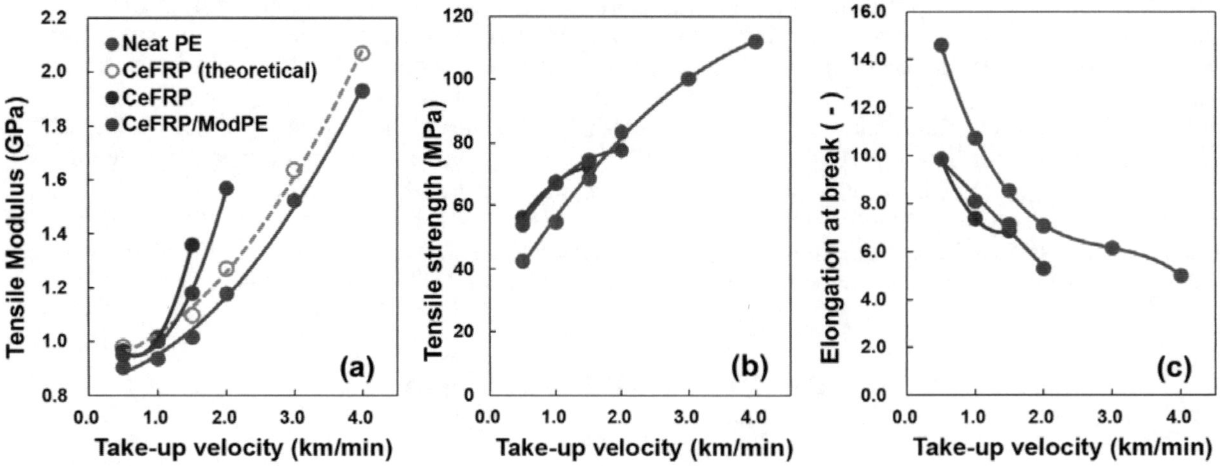

FIGURE 2. Effect of take-up velocity on (a) tensile modulus, (b) tensile strength and (c) elongation at break of the as-spun fibers prepared from three types of resin. Tensile moduli theoretically estimated based on the Cox theory are also plotted.

Structure and Properties of As-spun Fibers

Variation of birefringence with take-up velocity for the three types fibers are shown in Fig. 3. There was a significant increase of birefringence for the fibers with the reinforcement of CeNF. It should be noted that the existence of only 1 wt% CeNF cannot improve the birefringence value to such a high level even when the CeNF are highly oriented along the fiber axis. In other words, increase of birefringence is mainly originated from the improvement of molecular orientation of HDPE. Two-dimensional WAXD patterns of the as-spun fibers obtained at take-up velocity 1.0 km/min are presented in Fig. 3. The inner and outer strong reflections corresponds to the (110) and (200) reflections of PE crystals. It was found that the development of crystalline orientation of HDPE proceeded from lower take-up velocities in case of the as-spun fibers of CeFRP. In addition, existence of a certain amount of HDPE crystals of extremely high orientation along the fiber axis was confirmed from the azimuthal intensity distribution. This result suggested the epitaxial growth of HDPE crystals from the surface of CeNFs, which are considered to be highly aligned along the fiber axis. In summary, it can be concluded that the improvement of the mechanical properties of the HDPE fibers containing the CeNF was caused through the enhancement of molecular orientation of polyethylene rather than the reinforcement by the cellulose fibers.

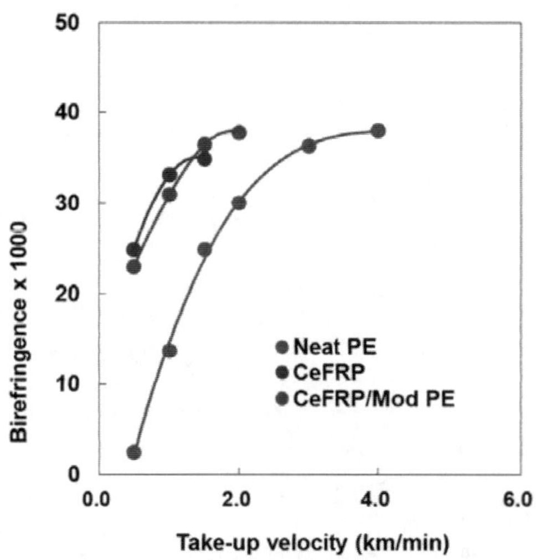

FIGURE 3. Effect of take-up velocity on birefringence of as-spun fibers prepared using the three types of resins.

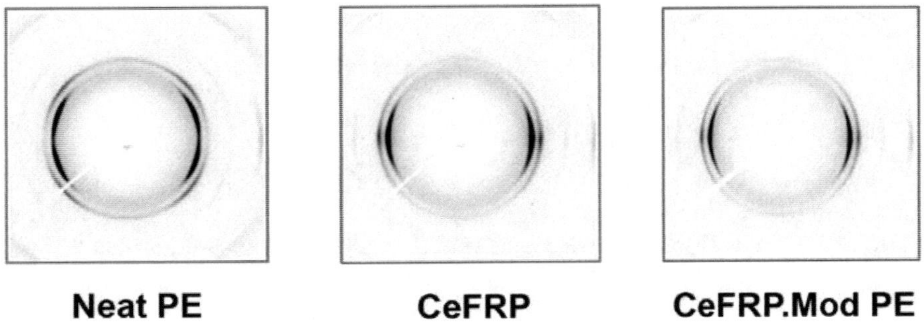

FIGURE 4. WAXD patterns of as-spun fibers of three types of resin prepared at take-up velocity 1.0 km/min.

Interfacial Shear Strength

A typical load versus displacement curve measured during the single-fiber pull-out test is shown in Fig. 5. Photograph of the prepared pull-out test sample is also shown in the figure. Sudden reduction of load at the displacement of around 0.3 mm reflected the debonding of the embedded fiber from the matrix resin followed by the sliding of the cellulose fiber in the matrix resin with the resistance of friction force.

Averaged interfacial shear stress at the moment of debonding, which will be called averaged interfacial shear strength, measured for the samples of various embedded lengths are shown in Fig. 6. It should be noted that the averaged interfacial shear strength slightly decreased with the increase of embedded length. Regarding the effect of the modification of matrix resin, averaged shear strength obviously increased through the modification of HDPE with maleic anhydride, while addition of peroxide did not show significant influence for the averaged shear strength. Through the fitting of theoretical curve based on the Greszuzuk model to the experimental data, maximum interfacial shear stress at the moment of debonding, which is supposed to start at the position where the fiber enters into matrix resin, was estimated. The value corresponds to the interfacial shear strength. It was found that the interfacial shear strength between HDPE and cellulose fiber was improved from 6.5 MPa to 30.6 MPa through the modification of HDPE with maleic anhydride, whereas the result for HDPE containing peroxide was only 9.4 MPa, revealing that the peroxide did not show significant influence for the adhesion of cellulose to PE resin.

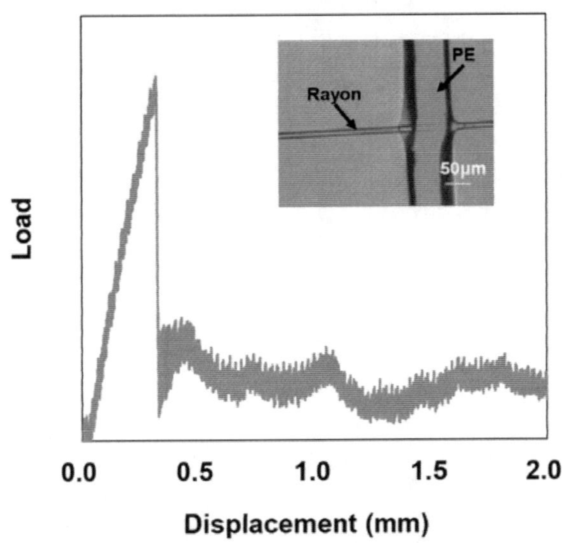

FIGURE 5. Typical load versus displacement curve recorded during the single-fiber pull-out test. Photograph of the pull-put test specimen is inserted in the figure.

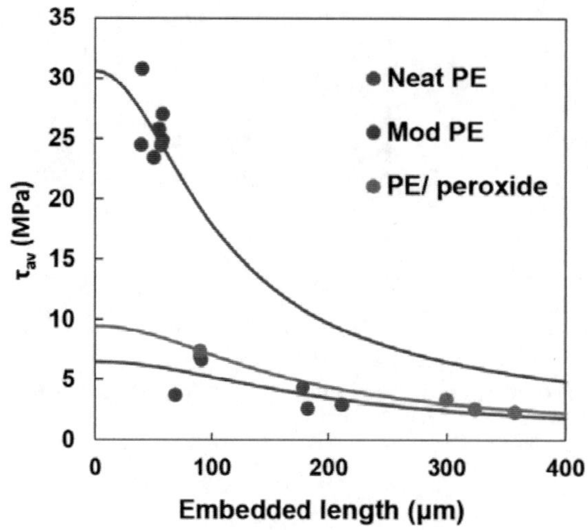

FIGURE 6. Relation between embedded length and averaged interfacial shear strength obtained through the single-fiber pull-out test for the cellulose fiber and three types of resin indicated in the figure. Curves fitted to the data are obtained based on the Greszuzuk model, and the value at the embedded length of zero corresponds to the interfacial shear strength.

REFERENCES

1. Strategic Innovation Promotion Program: http://www8.cao.go.jp/cstp/gaiyo/sip/keikaku/3_zairyou.pdf
2. H. Kobayashi, E. Hayakawa, T. Kikutani, A. Takaku, *Adv.Composite Matr.*, **1** (2), 155-168 (1991).
3. L. B. Greszczuk, *ASTM Special Technical Publication* **452**, 42-58 (1969)
4. H. L. Cox, *British J. appl. Phys.* **3**, 72-79 (1952)

Structural Characterization by X-ray Scattering of PET Film by Near Melting Point Stretching

A. Tanimura[a], K. Horinouchi[a], J. Inagaki[a], K. Funaki[a], T. Shimizu[a], Y. Numata[a] and K. Ito[a*]

[a]Research center, Toyobo Co., Ltd.
2-1-1 Katata, Otsu, Shiga 520-0292, Japan

Presenting author's email: Aya_Tanimura@toyobo.jp
Reference author's email: Katsuya_Ito@toyobo.jp

Abstract. Biaxial oriented Poly (ethylene terephthalate) film is excellent heat resistance and transparency, which is utilized in industrial and packaging application. Heat resistance or other properties are achieved by biaxial orientation method, which is characterized in machine-directional and transverse-directional stretching. The amount of heat, temperature profile, or draw ratio in stretching process improves the film properties, and the stretching temperature is generally set on near glass transition point. In this research, a trial to heighten the biaxial stretching temperature to near melting point was made. PET loses the thickness uniformity during biaxial stretching at near melting point, because tensile stress decreases, which are experimentally apparent. However, this study overcame it by undergoing heat crystallization of uniaxial stretched sheet. The obtained film exhibited characteristic properties compared to usual PET film. The film had high crystallinity, low MD orientation, and exhibited excellent thermal dimensional stability in MD. The amorphous orientation was suggested to be relaxed. Furthermore, *in-situ* characterization of biaxial stretching process revealed that crystallinity was retained during stretching at near melting point, which was not observed during usual stretching. In usual stretching, crystallinity was temporary decreased. Taken together, the stretching at near melting point exhibited notable structural formation and film properties. This study achieved some possibility in developing new biaxial stretching process of PET for the higher functioning film.

Keywords: Poly (ethylene terephthalate), Film, X-ray scattering
PACS: 36.20.-r, 61.05.C-, 61.41.+e.

INTRODUCTION

Essential skill for the improvement of plastic film properties is to control crystal structure or orientation of molecular chain arbitrary. Concerning biaxial stretched film, many researches are seen which are applied to the high functioning film: pursuing of the internal structure to be aimed [1], controlling the birefringence by introducing high crystallinity structure into monomer [2], controlling the thermal shrinkage by optimization of film stress during stretching process [3], and so on. In this research, focusing on biaxial oriented Poly (ethylene terephthalate) film, the new biaxial stretching process was considered in order to make highly functioning homo-PET film. This research is of interest to stretch PET biaxially at high temperature condition near melting point.

This research is notable from the perspective of using only homo PET as a raw material and trying to improve properties only by modifying stretching process. Since no additives are used, the developed film is hazardless, clean, low-cost and recyclable. These characteristics are important for popular use especially in food packaging application.

However, stretching at near melting point has a serious difficulty; tensile stress decreases and stretchability get worse, which are experimentally apparent. PET is normally easy to rupture or unstretched partially when stretched at near melting point. This research overcame it by undergoing heat crystallization of uniaxial stretched sheet. Heat crystallization prevented film from a decrease in tensile stress during stretching and enabled to retain film quality. The obtained film had no unstretched part. Also, the film by new stretching process exhibited excellent thermal dimensional stability.

This study discussed the result of property and structural characterization by X-ray scattering of the new film. Moreover, the structural change during stretching was studied by *in-situ* characterization. The study clarified that there was a specific crystal construction process at near melting point that affects the thermal dimensional stability.

EXPERIMENT

Sample Preparation

Polymer used in this study was Poly (ethylene terephthalate) (PET), having intrinsic viscosities (IV) of 0.62 dl/g (Toyobo Co., Ltd.). The film was prepared in the successive biaxial stretching process showed in Figure 1.

For the beginning PET resin was melted and kneaded at 280℃ in twin screw extruder. The resin was extruded by conventional T-die method and casted at 20℃. The obtained cast sheet was amorphous and its average thickness was 250 μm.

Machine directional (MD) stretching of the cast sheet was performed by using speed difference between rolls. Stretching ratio (MDx) was 3.0. Roll temperature was 70, 80, 90℃ at pre heating roll, 105℃ at fast speed stretching roll, and 40℃ at low speed stretching roll. After stretching, the obtained uniaxial stretched sheet was heated using IR heater for the purpose of heat crystallization. Heating time was 0.15 seconds and the surface temperature of sheet was 100℃.

The heated uniaxial stretched sheet was stretched transverse directionally (TD) in the tenter machine. Stretching ratio (TDx) was 6.0. Temperature was set at 3 conditions as: 1. 90℃ at preheating and 95, 100℃ at stretching 2. 130℃ at preheating and 135, 140℃ at stretching 3. 170℃ at preheating and 175, 180℃ at stretching. After stretching, heat set was performed at 230℃, 10 seconds.

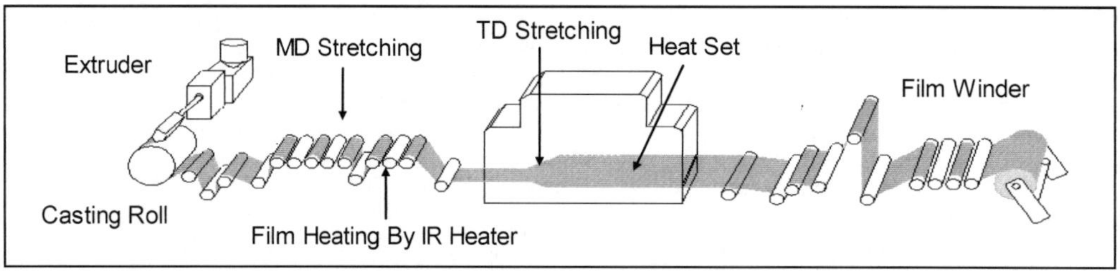

FIGURE 1. Successive biaxial stretching process.

Measurement of Film Properties

Refractive index of MD (Nx) was measured based on ISO489:1999. The measuring device was Abbe Refractometer NAR-4T (Atago Co., Ltd.) and intermediate liquid was diiodomethane.

Bulk crystallinity of film was measured based on ISO1183:1987, using a density gradient column Type A (Shibayama Scientific Co., Ltd.). Crystallinity was calculated by substituting the measured density (D) to the equation (1). For calculation, density of perfect amorphous: d_a=1.335 [4], and density of perfect crystal: d_c=1.501 [5].

$$\frac{d_c[D-d_a]}{D[d_c-d_a]} \tag{1}$$

Crystallite size was measured by wide angle X-ray diffraction (WAXD) method using X-ray diffractometer RINT2500 (Rigaku Co., Ltd.). The output was 40kV-200mA, the target was CuKα, and the filter was Ni. X-ray was exposed to film by the transmission method and (010) diffraction peak was detected. Diffraction angle θ and half-value width B of the peak was substituted to the Scherrer equation (2), and apparent crystallite size ACS of diffraction surface (010) was calculated [6]. For calculation, k was collection constant of 0.9, λ was X-ray wavelength, and b was broadning constant of the machine which was collected by NIST standard Si Power 640b.

$$ACS = \frac{k\lambda}{\cos\theta\sqrt{B^2-b^2}} \tag{2}$$

Thermal shrinkage of MD (MD thermal-shrinkage) was measured based on ISO11501:1995. Test conditions were 150℃ 30 minutes and 200℃ 30 minutes. Sample size was 10mm for TD and 150mm for MD.

Tensile strength of MD (MD tensile stress) and tensile elongation of MD (MD tensile elongation) were measured based on ISO527-1:2012, ISO527-2:2012. Test speed was 100mm/min and sample size was 10mm for TD and 40mm for MD.

in-situ Characterization of Structural Change During TD Stretching

For the purpose of clarification of structural change during TD stretching, in-situ characterization test was performed at the large synchrotron radiation facility SPring-8 (Hyogo, Japan). The originally produced mini stretching machine was placed so that X-ray can be exposed perpendicularly to a sample. The heated uniaxial stretched sheet was set in the machine and stretched at 100℃, 130℃, 180℃ each, irradiating X-ray. Wide angle X-ray diffraction (WAXD) and Small angle X-ray scattering (SAXS) measurements were carried out simultaneously during stretching. A detector was flat panel for WAXD and was CCD image intensifier camera for SAXS.

RESULT AND DISCUSSION

Measured film properties were shown in Figure 2~4.

As TD stretching temperature rose, MD refractive index Nx decreased for 0.2%, from 1.642 to 1.639 (Figure 2(a)). It meant that MD orientation of film stretched at 180℃ was lower than that of usual stretched film. On the other side, crystallinity increased for 11%, from 36.2% to 38.7% (Figure 2(b)), and (010) crystallite size increased for 18%, from 59.7Å to 70.8Å (Figure 2(c)).

MD thermal-shrinkage both decreased remarkably, from 1.9% to 0.7% (150℃) and from 9.7% to 2.7% (200℃) each (Figure 3, 4). In addition, MD tensile strength decreased from 117MPa to 42MPa while MD tensile elongation increased from 53% to 87% (Figure 5, 6). These results proved that film stretched at 180℃ has different property from usual stretched film.

Considering PET triclinic crystallite structure, diffraction surface (010) is known as one of the parallel surface to PET molecular chain [7]. The increase of (010) size corresponds to increase of crystallite size perpendicular to molecular chain. Referred to biaxial oriented PET film, molecular chain have high orientation to TD, and it means that the increase of (010) size corresponds to increase of crystallite size to MD. Namely, PET crystallite size increased to MD as TD stretching temperature increased. Bulk crystallinity is one proof of such change.

However, total MD orientation Nx decreased as TD stretching temperature rose. Nx includes both amorphous and crystal orientation. So it is suggested that amorphous structure changed. Also, MD thermal-shrinkage decreased as TD stretching temperature rose. It is generally known that MD thermal-shrinkage is affected by amorphous orientation and micro-Brownian motion activity under high temperature over grass-transition point [8, 9]. These facts indicated that amorphous orientation of MD was relaxed when stretched at high temperature near melting point, while crystallite size increased. Tensile strength and tensile elongation support this suggestion.

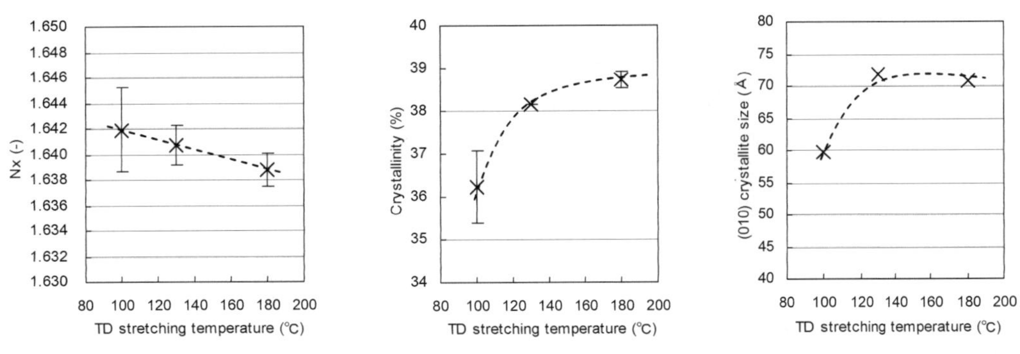

(a) (b) (c)

FIGURE 2. Structural property of obtained film.
(a) Nx, (b) crystallinity, (c) (010) crystallite size.

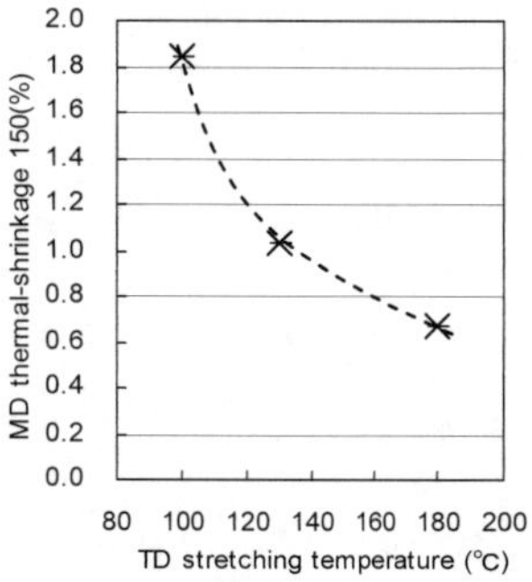

FIGURE 3. MD thermal-shrinkage 150℃30seconds.

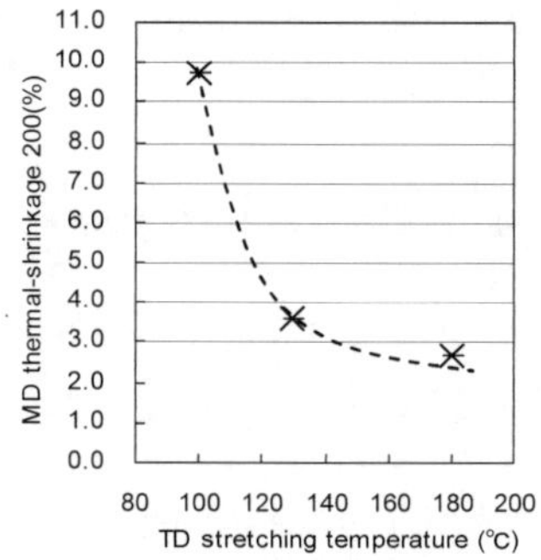

FIGURE 4. MD thermal-shrinkage 200℃30seconds.

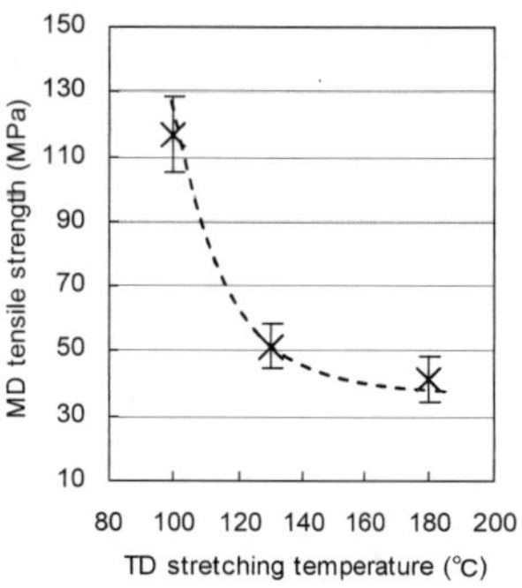

FIGURE 5. MD tensile strength.

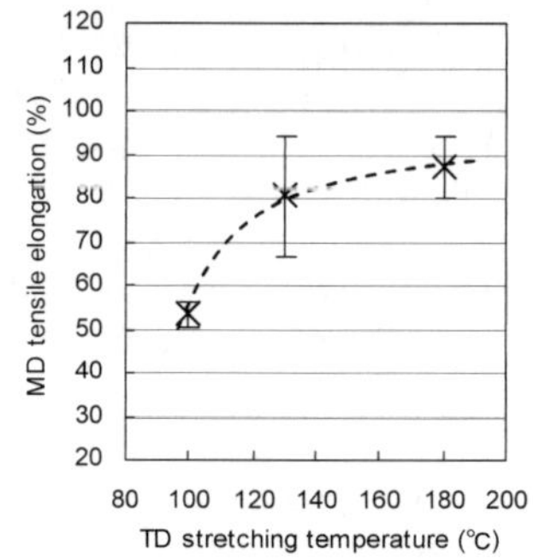

FIGURE 6. MD tensile elongation.

Figure 7 summarizes the results of *in-situ* characterization during stretching. During stretching under 100℃, a long period peak of MD oriented lamellar divided into two peaks at about 50% elongation and formed TD oriented lamellar peak at about 200% elongation, as SAXS view showed. Crystallite diffraction peak was obscure at about 100% elongation and formed TD orientation at about 250% elongation, as WAXD view showed. This suggested that crystallinity and orientation decreased temporary during stretching. After stretching progressed to some extent, TD oriented crystallization occurred. On the other side, during stretching under 180℃, lamellar divide or obscure peak during stretching were not observed, indicating that crystallinity and orientation were sustained during stretching. This results means that TD stretching under high temperature at 180℃ has different structural formation from usual TD stretching, sustaining high crystallinity and orientation. This characteristic stretching affected to film properties: high crystallinity and high (010) crystallite size.

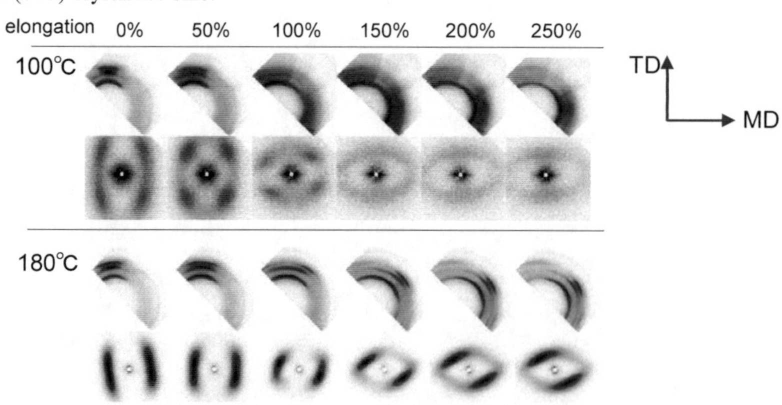

FIGURE 7. WAXD and SAXS views of film during stretching.

CONCLUSION

This high temperature stretching study concluded that: the film stretched at high temperature near melting point has different structure from usual film, with characteristic MD thermal property and MD mechanical property. Such structure was created by characteristic structural formation during TD stretching, retaining high crystallinity.

By TD stretching under 180℃, MD orientation decreased, MD crystallite size and bulk crystallinity increased. Also, MD thermal-shrinkage of 150℃ and 200℃ remarkably decreased. Since refractive index includes both amorphous and crystal orientation, the amorphous orientation of MD is indicated to be relaxed. Taken together, the film stretched at high temperature near melting point has much bigger crystallite and would have lower MD amorphous orientation compared to usual film.

in-situ characterization during stretching revealed the structural formation under 180℃. Both WAXD and SAXS view showed different behavior between 180℃ stretching and 100℃ stretching. No obscure view were seen under 180℃ stretching. The result indicated that crystallinity was highly retained during stretching, while crystalline structure broke temporary during usual TD stretching. These differences resulted in the structural difference.

The achieved film properties were affected by this characteristic structural formation. MD thermal-shrinkage was remarkably low compared to usual stretched film. MD tensile strength was low and MD tensile elongation was high. Low amorphous orientation was suggested to influence these properties. There would be still room for improvement of mechanical property for better film quality, since high tensile strength is in high demand in several applications.

However, what is noticeable is that this research achieved some possibility in developing new stretching process of PET. The developed film has low MD thermal-shrinkage and characteristic structure, which is different from usual PET film. Since no additives are used, this film has possibility to be used to any application which needs thermal dimensional stability: food packaging of retort, magnetic tape base, solar panel protection, circuit board reinforcement, touch panel base, and so on. The establishment of this research enables the film to be applied to these diverse purposes.

REFERENCES

1. H. Ishihara, M. Shibaya and N. Yoshihara, *Nihon Reoroji Gakkaishi* **33(4)**, 161-165 (2005).
2. Y. Ohta, H. Watanabe and M. Kato, *Journal of the Japan Society of Polymer Processing* **27(3)**, 94-97 (2015).
3. Y. J. Kim, G. S. Song and S. M. Kim, U.S. Patent No. 20,110,160,425 (2 March 2011)
4. A. B. Thompson and D. W. Woods, *Nature* **176**, 78-79 (1955).
5. H. Zahn and R. Krzikalla, *Macromol. Chem.* **23**, 31-53 (1957).
6. P. Scherrer, *Nachrichten von der Gesellschaft der Wissenschaften zu Göttingen* **2**, 98-100 (1918).
7. S. Fakirov, E. W. Fischer and G. F. Schimidt, *Macromol. Chem. Phys.* **175**, 2459-2465 (1975).
8. D. Patterson and I. M. Ward, *Transactions of the Faraday Society* **53**, 1516-1526 (1957).
9. Y. Ohya, Y. Yamamoto, Y. Saito and S. Kinoshita, *Seni'i Gakkaishi* **31(11)**, T508-T517 (1975).

Polymer Structure Effects on Crystallization and Properties in Polypropylene Film Casting

Markus Gahleitner[a], Daniela Mileva[a], Dietrich Gloger[a], René Androsch[b], Davide Tranchida[a]

[a]*Borealis Polyolefine GmbH, Innovation Headquarters, St. Peterstr 25, 4021 Linz, Austria*
[b] *Martin-Luther-University Halle-Wittenberg, Institute for Polymeric Materials, Eberhard-Leibnitz-Str. 2, 06217 Merseburg, Germany*

markus.gahleitner@borealisgroup.com

Abstract: One possible way to modify the properties of isotactic polypropylene (iPP) is to use the material's polymorphism which in case of film casting means to balance the content of α-crystalline and mesomorphic phase in the product. A cast film study was performed with several iPP homopolymers of different isotacticity and an ethylene-propylene (EP) random copolymer at two different film thickness levels (50 and 200 μm), varying the cooling rate in solidification by adapting the processing conditions. The results from film characterization can be correlated to findings from a parallel investigation with a combination of DSC and FSC techniques in a cooling rate range of 0,01 to 1.000°C/s

Keywords: polypropylene, processing, crystallization, film, optics, mechanics
PACS: 61.05.C, 64.70.dg, 81.05.Lg

INTRODUCTION

One of the essential reasons for the massive volume growth experienced by isotactic polypropylene (iPP) is the flexibility of this polymer in terms of property design and processability, enabling an extremely wide range of applications. The polymorphism of iPP, i.e. its capability to solidify in so far five documented crystal modifications from α to ε [1-3], and in a mesomorphic state [4] is one key element of this flexibility. In recent years more attention than earlier has been paid to the mesomorphic phase of isotactic polypropylene (iPP), not the least because better calorimetric techniques and instruments had become available, enabling studies at high cooling rate [5,6]. In addition to the scientific interest sparked by the availability of novel differential scanning calorimetry (DSC) equipment enabling cooling rates up to 3.000 K/s (FSC, fast scanning chip calorimetry), the mesomorphic phase also plays an important role in high-speed conversion processes for iPP like cast-film extrusion [7,8], metal or paper coating, fibre spinning [9] and cable insulation. The presence of this phase can be either positive, e.g. for transparency and toughness in case of films, or negative in terms of long-term stability and post-crystallization.

Recent investigations have shown that the reduction of crystallization rate by increasing chain disturbance (stereodefects or comonomers) shifts the growth transitions from the α-crystalline to the mesmorphic phase to lower cooling rates [10,11]. This and the fact that high-temperature treatment of the mesophase will induce a transformation into the α-phase goes together with the common experience that iPP random copolymers with ethylene (EP-RACOs) can be used for achieving highly transparent and more flexible films which, however, may change their optical properties in steam sterilization [12]. As in this previous study the effect of film processing conditions was mostly studied for nucleated materials, we now intended to improve the understanding for standard film materials.

An extensive cast film study was therefore performed with several iPP homopolymers of different isotacticity and an ethylene-propylene (EP) random copolymer at two different film thickness levels (50 and 200 μm). Melt and chill roll temperature were varied systematically together with the processing speed, effectively varying the cooling rate in solidification. The produced films were characterized in order to check the resulting difference of crystallinity, mechanical and optical properties together with the behavior on subsequent annealing and further processing. Parallel to that, the polymers were studied by a combination of DSC and FSC techniques at 0,01 to 1.000°C/s, seeking to use the results for a better understanding of iPP film casting.

EXPERIMENTAL WORK

Cast films of 50 and 200 µm thickness were produced from iPP homopolymers of different isotacticity as well as EP random copolymers, table 1 giving an overview of the key properties of the selected materials. All polymers except the experimental grades HC-EXP1 and HC-EXP2 are commercial products of Borealis, and the two polymer grades used for the 50 µm films were selected from the homo/random series in an earlier study [13], also used by Cavallo *et al.* [10]. A cast film extrusion line equipped with coat hanger die and a conventional chill / pick-up roll assembly was used, choosing, 6 variations of melt temperature, chill roll temperature and line speed / output in a technically relevant range for both series. The produced films were characterized by standard film tests regarding mechanical and optical properties, and by wide-angle X-ray scattering (WAXS) to determine crystallinity. Tensile testing was done in accordance to ISO 527-1 (cross head speed 1 mm/min) while haze, clarity and transparency were determined according to ASTM D1003-00.

The WAXS analysis of crystallinity and polymorphic composition was performed in reflection geometry using a Bruker D8 Discover with GADDS x-ray diffractometer operating with the following settings: x-ray generator: 30 kV and 20 mA; θ1 = 6° & θ2 = 13°; sample-detector distance: 20 cm; beam size (collimator): 500 µm; and duration/scan: 300 seconds. Intensity vs. 2θ curves between 2θ = 10° and 2θ = 32.5° were obtained by integrating the 2-dimensional spectra and the amorphous halo from atactic PP was subtracted. The quantification of intensity vs. 2θ curves were then performed as usual. Parallel to that, a study involving standard differential scanning calorimetry (DSC) and fast scanning calorimetry (FSC) was performed for the materials in order to relate the film properties to an average cooling rate in the process. DSC experiments were performed at rates between 1 and 20 °C/min, collecting heat-flow rate data with a heat-flux DSC 820 from Mettler-Toledo, equipped with the liquid nitrogen accessory for controlled cooling. The sample mass was about 2.5 mg, and for encapsulation the small 20 µl aluminum pans from Mettler Toledo were used. FSC experiments were performed at cooling rates of 2.10^0 and 2.10^3°C/s with a Flash DSC 1 from Mettler-Toledo in conjunction with a Huber intercooler TC45. Prior to cooling, a relaxed melt was obtained by heating to 493 K and isothermal holding for a period of 0.1 s.

TABLE 1. Key parameters of the selected polymer grades and related film thickness.

Grade	MFR 230°C/2.16kg g/10min	Isotacticity (NMR pentads) mol%	C2 (NMR) mol%	Film thickness µm
HD234CF	8	95.6	0	50
RD208CF	8	n.d.	7.3	50
HC-EXP1	3.6	93.5	0	200
HC-EXP2	3.3	96.5	0	200
HC300BF	3.3	97.3	0	200

RESULTS AND DISCUSSION

Figure 1 gives a first impression of chill roll temperature effects on film crystallinity (WAXD) for the 3 iPP homopolymers tested on 200 µm films. The WAXS diffractograms for HC-EXP1 on the left show the gradual transition from predominant α-phase to mesomorphic phase with reducing chill roll temperature for an extruder output of 90 kg/h and a line speed of 12 m/min. Both latter parameters were found to be of secondary importance as shown by the overall crystallinity values in the diagram on the right; their importance however increases at lower chill roll temperatures. In terms of tensile modulus this means a reduction by 23-38% when averaging across the various conditions for the highest and the lowest chill roll temperature. The effect is getting even stronger with reduced isotacticity.

In case of the 50 µm films, the quenching effect is even stronger. Similar effects in terms of mechanical and optical properties were found for the tested homopolymer and random copolymer as shown in Figure 2. The crystallinity effect on tensile modulus is strong but rather comparable for both polymer grades which were selected to have the same MFR and MWD [13], the difference being ~ 200 MPa between both series. Stiffness reduction here is ~ 34% for the iPP and ~ 40% for the iPP-Eth, meaning that the trend of chain disturbance enhancing the quenching is the same as in the 200 µm series above. Effects of the output and line speed on crystallinity are small here as well, but not in case of the film transparency (haze & clarity). At constant chill roll temperature, an increase in output by a factor of 2 results in a haze reduction of ~50%. The reasons for this effect were checked further by surface investigations with atomic force microscopy (AFM). Especially one needs to consider that the total haze depends on a combination of surface roughness and bulk superstructure of the film as shown by Resch *et al.* earlier [14].

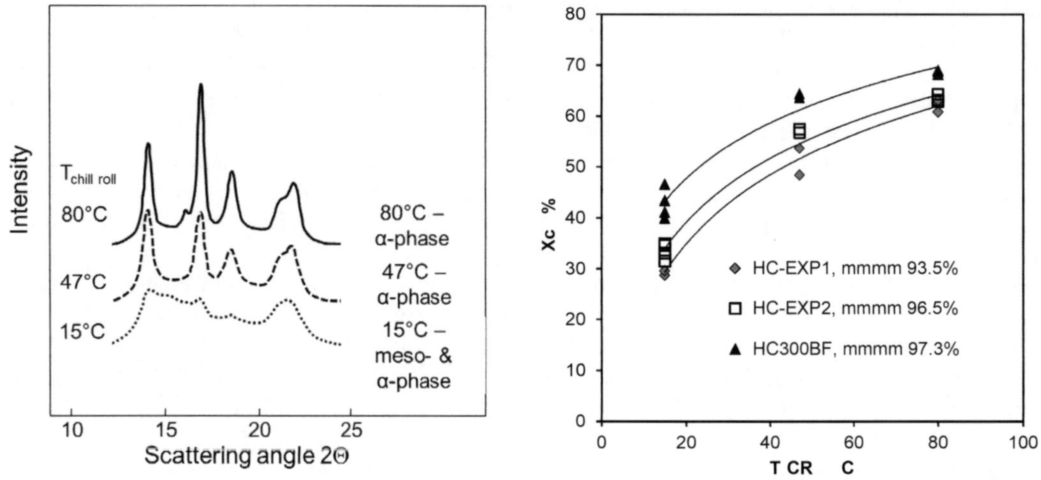

FIGURE 1. Effect of chill roll temperature at constant melt temperature (270°C) on crystallinity of iPP cast films of 200 µm thickness; WAXS diffractograms for HC-EXP1 left and overall crystallinity values right (note: a certain amount of β-phase is found for higher chill roll temperatures, mostly for higher isotacticities).

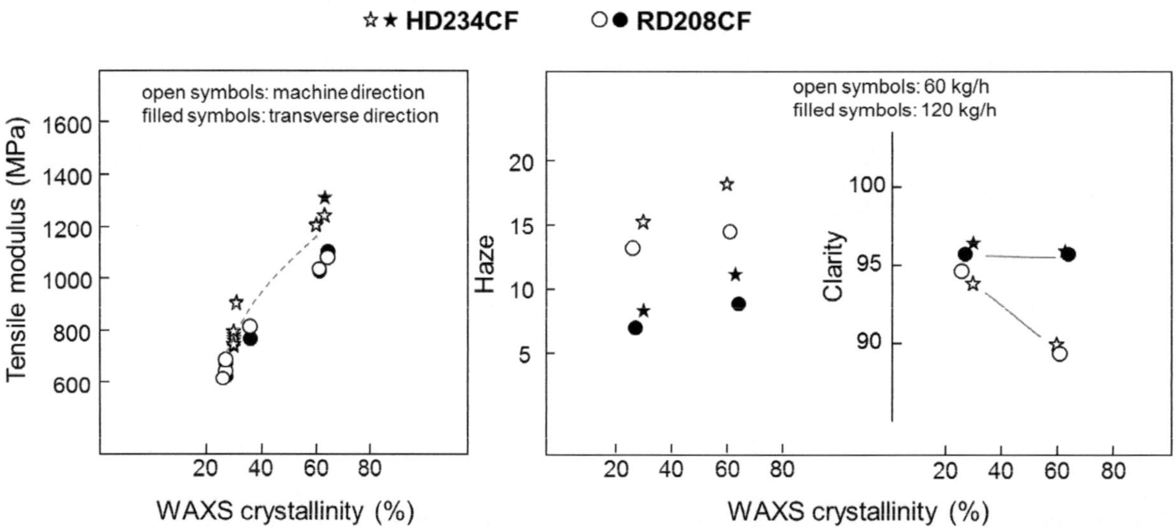

FIGURE 2. Effect of chill roll temperature and extruder output at constant melt temperature (230°C) on crystallinity, mechanics (left) and optics (right) of iPP cast films of 50 µm thickness; stars relate to PP homopolymer (HD234CF) and circles to EP random copolymer (RD208CF).

A difference in the quenching behavior for the two materials from the 50 µm series was also found in the combined DSC/FSC analysis (see Figure 3). The transition from α- to mesomorphic phase is a rather abrupt one occurring ~ 80°C/s for the random copolymer. In contrast to that, the iPP homopolymer shows parallel development of both phases in a certain range with a final transition at ~ 200°C/s. When comparing these results to the figures in Cavallo's CCC (continuous cooling curve) study [10] the agreement is rather good, with the "gradual" α- to meso-transition for HD234CF showing up nicely in Figure 7A of said paper presenting a peak deconvolution analysis of the two phases based on WAXS. When comparing both the FSC and CCC data to film casting results one must consider the effect of flow on the latter, illustrated well by the extrusion speed effects in Figure 2.

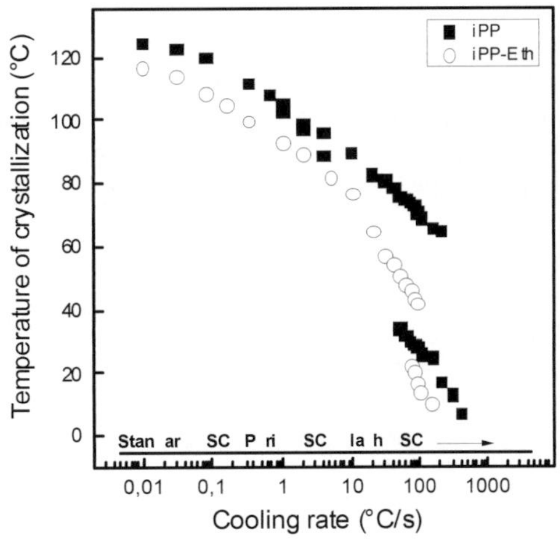

FIGURE 3. Cooling rate effect on crystallization temperature for PP homopolymer (iPP, HD234CF) and EP random copolymer (iPP-Eth, RD208CF); transition from α- to mesomorphic phase at ~ 80°C/s for iPP-Eth and ~ 200°C/s for iPP.

The AFM investigations yielded some interesting results in terms of chill roll temperature effect. In Figure 3 two examples for the surface structures of 50 μm films based on PP homopolymer (HD234CF) are presented, representing the low (15°C) and high chill roll temperature (70°C) versions of the low output rate (60 kg/h). While the higher chill roll temperature (i.e. the lower cooling rate) induces a "spherulitic" surface structure, only rows of small crystalline structures (next to some isolated ones), always arranged in machine (flow) direction, are found at higher cooling rate (lower chill roll temperature). These structures are obviously identical to the ones found in a previous study [14] where they were termed "8-shaped superlamellar crystalline structures". The now available better SFM resolution clearly shows that these are surface imprints of spherulites arranged along row nuclei as postulated e.g. by Lamberti and Brucato [15]. Nodular shaped domains are seen in the vicinity of the ordered rows of crystalline structures, in accordance with the WAXS data showing presence of mesomorphic phase with small inclusions of alpha-crystals of films quenched at 15°C.

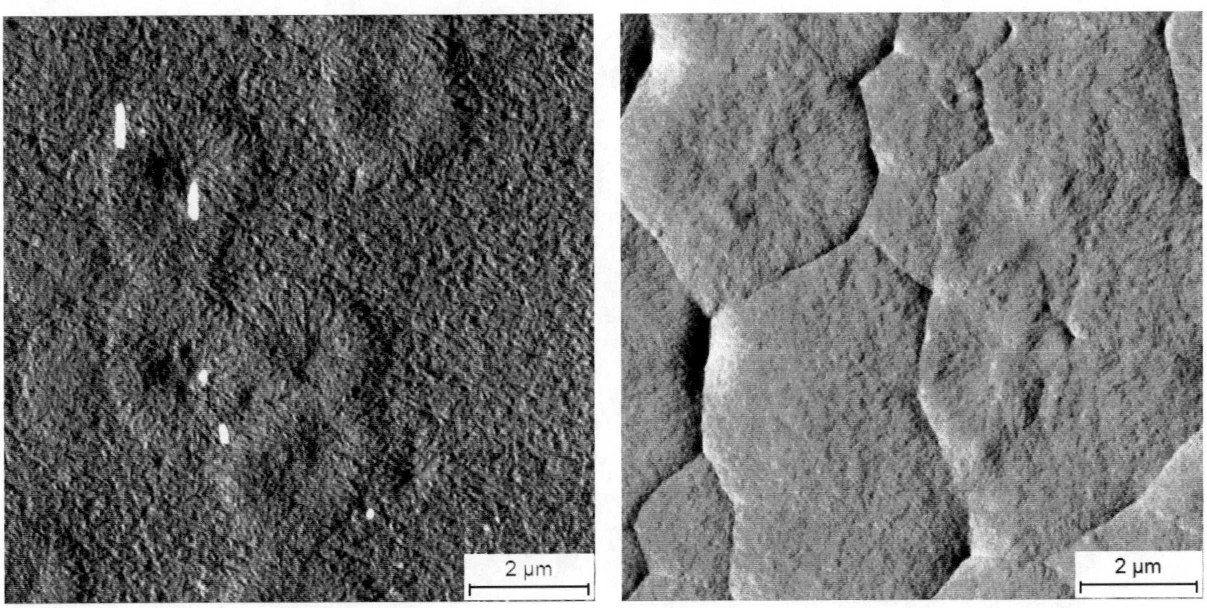

FIGURE 3. Effect of chill roll temperature – left 15°C and right 70°C – at constant extruder output (60 kg/h) and melt temperature (230°C) on surface structure of iPP cast films of 50 μm thickness (HD234CF, machine direction vertical).

CONCLUSIONS

The polymorphism of iPP can be seen as an opportunity to expand the application range of this versatile polymer, but also as a critical factor causing unwanted variations in processing. For understanding the interaction between polymer designs and processing parameters, basic research on iPP crystallization is important. A thorough understanding is required to find suitable combinations of polymer composition, nucleation and process settings. These three factors can and will interact with each other in defining the final application properties. In a new systematic study focussing on iPP cast films we tried to expand the knowledge in this area. In this case a combination of data from standard DSC and fast scanning chip calorimetry (FSC) with results from a laboratory-scale cast-film processing line the possibility to affect the formation of the mesomorphic phase was studied. Modifying iPP crystallinity by copolymerization can clearly be combined with a selection of suitable processing parameters to achieve the desired product performance.

REFERENCES

1. S. Brückner, S.V. Meille, V. Petraccone and B. Pirozzi, *Prog. Polym. Sci.* **16**, 361-404 (1991).
2. C. De Rosa, F. Auriemma, P. Vollaro, L. Resconi, S. Guidotti and I. Camurati, *Macromolecules* **44**, 540-549 (2011).
3. B. Lotz, Macromolecules **47**, 7612-7624 (2014).
2. R. Androsch, M.L. Di Lorenzo, C. Schick, B. Wunderlich, *Polymer* **51**, 4639-4662 (2010).
5. V. Mathot, M. Pyda, T. Pijpers, G. Vanden Poel, E. van de Kerkhof, S. van Herwaarden S, *Thermochim Acta* 45, 36-45 (2011).
6. I. Kolesov, D. Mileva, R. Androsch, C. Schick, *Polymer* **52**, 5156-5165 (2011).
7. G. Lamberti, V. Brucato, V., *J. Appl. Polym. Sci.* **41**, 998-1008 (2003).
8. M. Xu, S. Zhang, J. Liang, H. Quan, J. Liu, H. Shi, D. Gao, J. Liu, *J. Appl. Polym. Sci.* **131**, 41100 (2014).
9. H.P. Nadella, H.M. Henson, J.E. Spruiell, *J. Appl. Polym. Sci.* **21**, 3003-3022 (1997).
10. D. Cavallo, F. Azzurri, R. Floris, G.C. Alfonso, L. Balzano, G.W.M. Peters, *Macromolecules* **43**, 2890-2896 (2010).
11. D. Mileva, R. Androsch, D. Cavallo, G.C. Alfonso, *Eur. Polym. J.* **48**, 1082-1092 (2012).
12. M. Gahleitner, C. Grein, R. Blell, J. Wolfschwenger, T. Koch, E. Ingolic, *eXP. Polym. Lett.* **5**, 788-798 (2011).
13. M. Gahleitner, P. Jääskeläinen, E. Ratajski, C. Paulik, J. Reussner, J. Wolfschwenger, W. Neißl, *J. Appl. Polym. Sci.* **95**, 1073-1081 (2005).
14. K. Resch, G.M. Wallner, C. Teichert, G. Maier, M. Gahleitner, *Polym. Eng. Sci.* **46**, 520-531 (2006).
15. G. Lamberti, V. Brucato, *J. Appl. Polym. Sci.* **41**, 998-1008 (2003).

Visualization of a pillar-shaped fiber bundle in a model needle-punched nonwoven fabric using X-ray micro computed tomography

Tatsuya Ishikawa[1], Kyoung Hou Kim[1] and Yutaka Ohkoshi[1,2*]

[1]*Faculty of Textile Science and Technology, Shinshu University, 3-15-1 Tokida, Ueda, Nagano, 386-8567, Japan*
[2] *Division of Frontier Fibers, Institute for Fiber Engineering, Shinshu University, 3-15-1 Tokida, Ueda, Nagano, 386-8567, Japan*
15st102b@shinshu-u.ac.jp
yokoshi@shinshu-u.ac.jp

Abstract. In the needle-punching process, barbs of a needle catch the fibers and orient them along the thickness direction of fabric. The oriented fibers form a pillar-shaped fiber bundle, which act as a bonding point of fabric. Therefore the structure of pillar-shaped fiber bundle dominates the mechanical properties of needle-punched nonwoven fabric, both of which are largely influenced by needle-punching conditions. However, the three-dimensional (3D) structure of pillar-shaped fiber bundles and its development due to the needle-punching conditions have not been revealed. In this study, we used X-ray micro computed tomography (XCT) and visualized the pillar-shaped fiber bundles by using PET and PE fibers, between which there is a difference in X-ray absorption coefficient. Then, we investigated the 3D structure of pillar-shaped fiber bundles and its development as a function of penetration depth of needle. In 3D images obtained by XCT, we could clearly observe the 3D structure of pillar-shaped PET fiber bundles by erasing the PE bulk fibers, and the pillar structure developed with an increase of penetration depth of needle. Especially, at the penetration depth of 19.0 mm, the pillars protruded from the bottom surface of fabric and the fibers constituting the pillar were interlocked with following pillar. Meanwhile, the number of fibers constituting a pillar largely increased.

Keywords: needle-punching, nonwoven fabric, X-ray micro computed tomography, pillar-shaped fiber bundle.
PACS: 81

INTRODUCTION

In the process of needle-punching, fibers are caught by barbs of a needle and plunged into the web. Through this process, the fibers are oriented along the thickness direction of the fabric and form a pillar-shaped fiber bundle [1], which acts as a bonding point of the fabric. The structure of pillar-shaped fiber bundles such as shape, size and density are largely affected by manufacturing conditions for example the needle gauge, needling density and penetration depth of the needle, and determine the properties of needle-punched nonwoven fabric [2]. In other words, parts of fibers whose both ends are restrained by the pillar-shaped fiber bundles principally bear the tensile force applied to the fabric. Hence the tensile properties of needle-punched nonwoven fabrics rely on the number of restraint fibers decided by the structure of pillar-shaped fiber bundles. Hearle et al. investigated the number of fibers constituting a pillar-shaped fiber bundle by measuring the penetration force of a needle during the needle-punching process [3]. However, the structure of pillar-shaped fiber bundle was not observed in those studies. Furthermore, Hearle et al. [2, 4] directly observed the structure of pillar-shaped fiber bundle using colored fiber as a tracer. They reported the shape, size and number of fibers constituting a pillar-shaped fiber bundle. However, the analysis was destructive and time-consuming. In addition, because they made a two-dimensional (2D) observation, they could not observe the inner structure of the fiber bundle. Thus a three-dimensional (3D) technique of observing the structure of fiber bundle in the needle-punched nonwoven fabric have been desired for a further understanding of the relation of fabric structure to its mechanical properties.

Recently, X-ray micro computed tomography (XCT) has become popular as a method of non-destructive observation for visualizing the internal structure of materials. For example, Shen et al. observed phenolic foam reinforced with glass fiber through XCT [5]. Because the density of glass fiber was more than twice that of phenol resin, they could extract only glass fiber information from the obtained 2D and 3D images, and analyzed the length and orientation angle distribution of glass fibers. XCT is now being used for the analysis of the internal structure of nonwoven fabric. Jeon et al. calculated the fiber orientation distribution, fiber volume fraction and contact level between fibers in a needle-punched nonwoven mat employing XCT and proposed a formula for the prediction of the tensile strength [6]. Furthermore, they studied the changes in the fiber orientation distribution, volume fraction and contact level during the tensile deformation of the needle-punched nonwoven mat [7]. Even though XCT has been applied to analyze the structure of needle-punched nonwoven fabrics, the structure of pillar-shaped fiber bundles, which determine the properties of fabric, has not been investigated directly.

The pixel intensity of an XCT image represents the X-ray absorption coefficient, which is proportional to material density [8]. In this study, the pillar-shaped fiber bundle was visualized in an XCT image using fibers having different density, instead of using colored fibers as described above. For this purpose, needle-punched nonwoven samples produced from polyethylene terephthalate (PET) and polyethylene (PE) webs were used. Moreover, the effects of the penetration depth of the needle on the structure of the pillar-shaped fiber bundle such as the shape, length, and number of fibers of a pillar were investigated. This information would be useful for designing the desired properties of needle-punched nonwoven fabrics.

METHODS

Two kinds of crimped short fibers were used to prepare webs: PET fibers with a diameter of 40 μm and length of 51 mm, and PE fibers with a diameter of 20 μm and length of 35 mm. The two kinds of fibers were separately opened then carded by an opener and a carder. The basis weight of the obtained parallel-laid webs was 40 g/m^2 for the PET web and 160 g/m^2 for the PE web. The obtained PE web was overlaid with a PET web in a parallel manner, then used for needle-punching process.

A needle-punching machine and FPD-1-25 needles were used. The needles are categorized as regular-barb spacing needles and have a barb depth of around 250 μm. Seventy-six needles were planted in two rows on a needle board, where rows and needles in each row were spaced at intervals of 11 and 10 mm, respectively. The web was fed at 2 mm/punch, which corresponds to a needling density of 6.4 punches/cm^2. The penetration depth of the needle was set at 6.4, 12.7, and 19 mm, which corresponds to the first, fourth, and seventh barbs reaching the top surface of the bedplate. The thickness and basis weight were measured for the web and needle-punched nonwoven fabrics. The thickness measurement was made according to ISO 9073-2, and the basis weight was measured by weighing five samples with dimensions of 100 mm × 100 mm.

A sample with dimensions of 20 mm × 20 mm (MD: machine direction × CD: cross direction) was set on a stage of a Skyscan 1272 XCT device, in which X-rays was generated at 30 kV and 200 μA. A series of transmission images with 2452 × 1640 pixels (8 μm/pixel) was obtained by rotating the sample around the normal-direction axis at 0.2 degrees/image. A sequence of 2D tomographic images was reconstructed from the series of transmission images, and 3D images were obtained. Histograms of the pixel intensity for 2D tomographic images were recorded for samples and blank air, with the latter treated as the background.

RESULTS & DISCUSSION

As shown in FIGURE 1, an obtained histogram of pixel intensity of the background was quite different from histograms of the webs. For example, the intensity histogram of the background drops rapidly above a value of 10 while the intensity histograms of the webs have a clear shoulder or plateau. The fiber in the web can thus be distinguished from the background air using a threshold of 10. In addition, a histogram of the PET web was largely different from that of the PE web. For example, the pixel intensity of the PET web extends to 255 while that of PE web extends to less than 120.

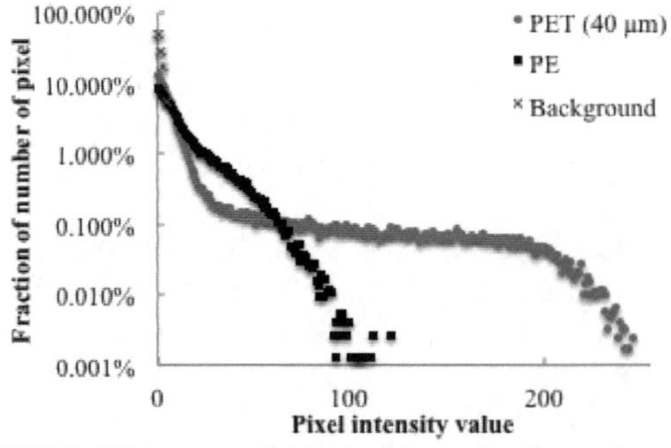

FIGURE 1. Histogram of the pixel intensity for each material.

Using the above-mentioned differences in histograms, only the information of PET fibers was extracted from needle-punched samples. FIGURE 2 shows color-coded 3D images with threshold levels of 10 and 120. It is noteworthy that the structure of pillar-shaped fiber bundles can be analyzed in XCT images using PET fibers as the tracer and PE fibers as the bulk. This is an effective method for analyzing the structure of pillar-shaped fiber bundles without affecting the structure of nonwoven fabric.

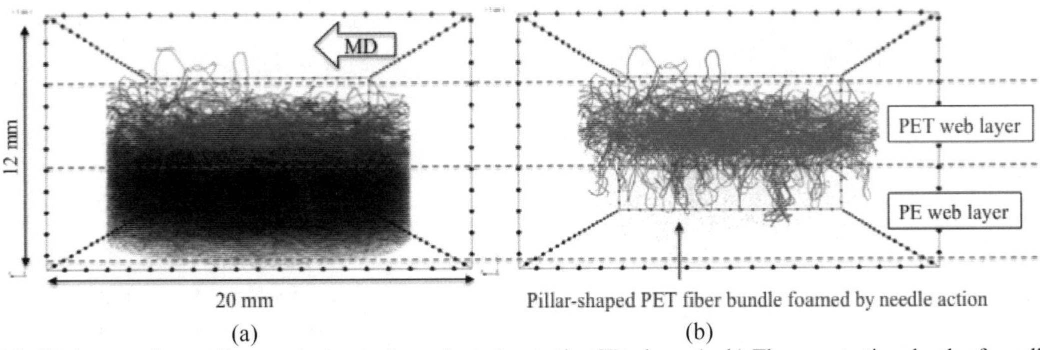

FIGURE 2. 3D image of a needle-punched sample projected onto the CD plane. (a, b) The penetration depth of needle for the sample was 12.7 mm. Pixel intensity is represented by color in the images: (a) black for values of 10-120 and blue for values exceeding 120, (b) transparent for values less than 120 and blue for values exceeding 120.

According to the above-mentioned image processing, the effects of the penetration depth of the needle on the nonwoven fabric structure were investigated. 3D images of needle-punched nonwoven fabrics produced from PET and PE webs are shown in FIGURE 3. In the 3D images, the PET fibers are shown in blue, below which there are the PE fibers shown in transparent (erased). Three images are displayed at the same scale, and the distance between two white points is 1 mm and the bottom surface of the displayed grid matches the bottom surface of the fabric. In order to simplify the visual observation for the structure of a pillar-shaped fiber bundle, a columns of a series of 4-5 consecutive pillars were virtually dissected by selecting a length of 5 mm in the CD plane and 10 mm in MD plane. At a penetration depth of 6.4 mm, a few PET fibers penetrated into the underlying PE layer and formed very shallow loop-shaped pillars. And, both PET and PE layers were bulky. Then, by increasing the penetration depth to 12.7 mm, pillars slightly developed inside the PE layer, and the PE layer was largely compressed. At a penetration depth of 19.0 mm, the pillars largely developed and protruded from the bottom surface of the fabric. Moreover, the PE layer was largely compressed and few PET fibers remained on the top surface of the fabric.

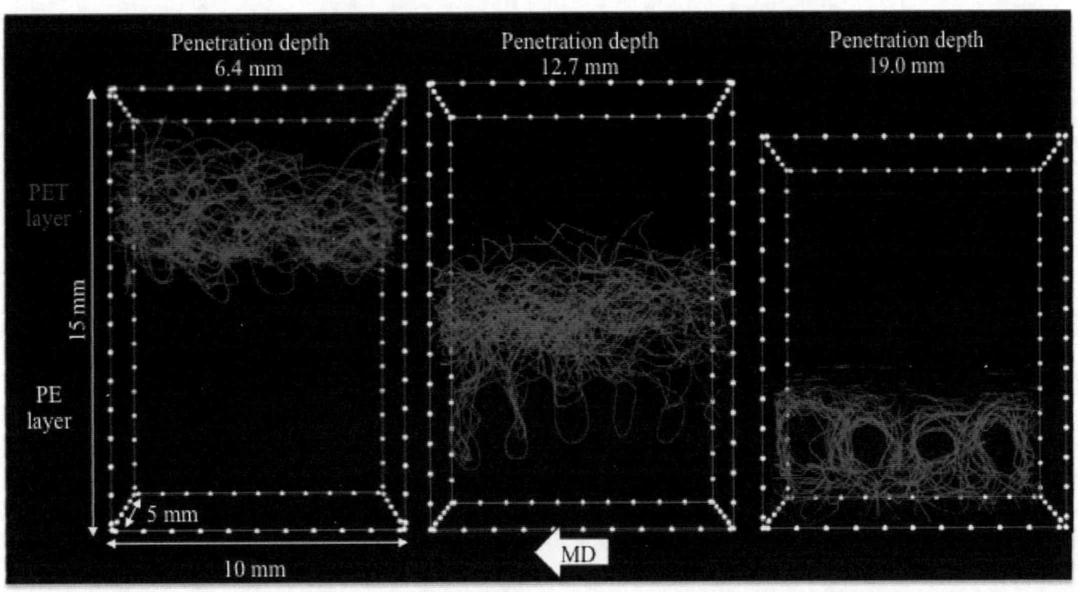

FIGURE 3. 3D images of a needle-punched nonwoven sample produced from 40-μm PET and PE webs. The penetration depth of the needle is noted in the figure: (a) image of a series of pillar-shaped fiber bundles projected onto the CD plane, (b) image of a pillar projected onto the MD plane.

By analyzing the 3D and 2D tomographic XCT images more quantitatively, the thickness of PET and PE layers, and the structure (shape, length, and number of fibers) of the pillar-shape fiber bundle were characterized as a function of the penetration depth of the needle. The results obtained from XCT image analysis and also the basis weight and thickness are summarized in TABLE 1. The thicknesses in an XCT image were defined as the distance between the top and bottom surfaces of each layer. The length of a pillar was defined as the distance from the top of the fabric to the tip of the pillar. The number of fibers was the average number of fibers constituting one pillar at 1 mm below the top surface of the PE layer.

As shown in TABLE 1, total thickness decreased with increasing in penetration depth, but the decreasing phenomena were not the same for the upper (PET) and lower (PE) layers. The thickness of the PE layer largely decreased with the increase in penetration depth from 6.4 to 12.7 mm, while that of the PET layer mainly decreased with the increase in penetration depth from 12.7 to 19.0 mm, at which most of the PET fibers formed pillar-shaped bundles and few remained on the PE layer. With a decrease in thickness, the basis weight of the needle-punched nonwoven samples also decreased, especially at a penetration depth of 19.0 mm. The decrease was due to the drafting of the web during the needle-punching process and spreading of fibers by the needle penetration [1]. The pillar-shaped fiber bundle developed with increasing penetration depth of the needle. At a penetration depth of 6.4 mm, the tip of the loop-shaped pillar consisted of only 2 fibers in average reaching up to 8 mm from the bottom surface of the fabric. By increasing the penetration depth to 12.7 mm, the tip of pillars consisted of 5 fibers in average reaching 2-3 mm from the bottom of the fabric. On the contrary, for the penetration depth of 19.0 mm, the pillars consisted of around 20 fibers and they were protruded from the bottom surface of the fabric, forming a "stitch" at which the fibers constituting the pillar were interlocked with the fibers of the following pillar. Hearle et al. reported the stitch structure, the dependence on the needle-punching conditions, the formation mechanism, and the effects on the properties of nonwoven fabric [1, 9]. In their studies, although there were quantitative differences due to the use of a fiber and needle different from those used in the present study, qualitatively similar behavior, such as an equivalent penetration depth to form the stitch structure, was observed. Furthermore, the development of a pillar structure indicates a similar behavior to that observed with a microscope [2]. For example, the length of a pillar and the number of fibers constituting a pillar increased with depth. Therefore, the results obtained in this study are supposed to be credible. In addition to that, the technique used in this study can be used to analyze the structure of a pillar-shaped fiber bundle more quantitatively than existing techniques; i.e., quantitative determination of the length of a pillar, realistic structure of the stitch, and accurate number of fibers constituting a pillar. This merit of the 3D non-destructive XCT observation allows the more quantitative evaluation of the properties of needle-punched nonwoven fabrics.

TABLE 1. Results obtained from XCT image analysis

Penetration depth (mm)	Basis weight (g/m^2)	Thickness (mm)	Thickness in XCT image (mm)			Structure of pillar-shaped fiber bundle		
			Total	PET layer	PE layer	Shape	Length* (mm)	Number of fibers
Web	200	5.7						
6.4	170	4.9	13.6	3.9	9.7	Loop	5.2	2
12.7	190	5.0	9.3	3.2	6.1	Loop	7.3	5
19.0	160	3.8	4.5	0	4.5	Stich	> 4.5	20

*The distance from the top surface of the fabric to the tip of the pillar.

CONCLUSION

In this study, the pillar-shaped fiber bundle in a needle-punched nonwoven fabric were visualized using X-ray micro computed tomography (XCT) on the basis of the difference in the X-ray absorption coefficient between PET and PE fibers. For a material density ratio less than 1.4 and a diameter of PET fibers of 40 μm, a pillar-shaped bundle of PET fibers was visualized by erasing 20 μm PE fibers in XCT images. Through XCT observations, the development of the pillar-shaped fiber bundle as a function of the penetration depth of the needle was investigated. A remarkable feature of the development of pillar was that the number of fibers constituting a pillar largely increased at a penetration depth of 19.0 mm, at which the pillars protruded from the bottom surface of the fabric and formed a stitch structure. This technique employing XCT is effective for the quantitative analysis of the structure of pillar-shaped fiber bundles without affecting the structure of the nonwoven fabric.

ACKNOWLEDGMENTS

This work was supported by a Grant-in-Aid for the Shinshu University Advanced Leading Graduate Program provided by the Ministry of Education, Culture, Sports, Science and Technology, Japan.

REFERENCES

1. J. W. S. Hearle and M. A. I. Sultan and T. N. Choudhari, *J Text Inst* **59**, 103-116 (1968).
2. J. W. S. Hearle and A. T. Purdy, *Fibre Sci Technol* **5**, 113-128 (1972).
3. J. W. S. Hearle and M. A. I. Sultan, *J Text Inst* **59**, 237-242 (1968).
4. J. W. S. Hearle and A. T. Purdy, *Fibre Sci Technol* **4**, 81-100 (1971).
5. H. Shen and S. Nutt and D. Hull, *Compos Sci Technol* **64**, 2113-2120 (2004).
6. S. Y Jeon and W. R. Yu and M. S. Kim et al., *Fibers Polym* **15**, 1202-1210 (2014).
7. S. Y. Jeon and W. J. Na and Y. O. Choi et al., *Compos Part A Appl Sci Manuf* **63**, 1-9 (2014).
8. Y. Nishikawa and Y. Hatanaka and S. Iizuka et al., *Polymer* **53**, 4287-4292 (2012).
9. J. W. S. Hearle and T. N. Choudhari, *J Text Inst* **60**, 478-496 (1969).

Electrospinning of PVAaq solutions with ZnO nanoparticles and antibacterial properties of the electrospun mats

Masataka Sugimoto[a*], Tongsai Jamnongkan[a, b] Yumiko Takatsuka[a], Tomijiro Hara[a], Sathish K. Sukumaran[a], Kiyohito Koyama[a]

[a]*Graduate School of Science and Engineering, Yamagata University, 4-3-16 Jonan, Yonezawa, Yamagata 992-8510, Japan*
Continue Here
[b]*Present address: Department of Fundamental Science and Physical Education, Faculty of Science at Sriracha, Kasetsart University, Chonburi 20230, THAILAND*

Presenting author's email: sugimoto@yz.yamagata-u.ac.jp
Reference author's email: sugimoto@yz.yamagata-u.ac.jp

Abstract. In order to obtain nano fiber mats which possesses antibacterial properties, we studied the electrospinnability of PVA solutions. Due to less toxiccity of ZnO to humans than silver, they are attractive as antibacterial agents in biomedical applications. First the effect of the nanoparticles on the solution viscosity of PVA solutions, especially we focused on the effect of the particle size and the nanoparticle volume fraction on the viscosity of the suspension. We found that the unusual increase of the solution viscosity of the suspensions as increasing the volume fraction of the nanoparticles and decreasing the size of the particles despite of very low volume fraction of 0.02%. The measured viscosities at the volume fraction of the nanoparticles are significantly higher than those predicted by the Einstein's equation. The increased viscosities affected the electrospinability of PVA/ZnO suspensions. We will also present the influence of the size of ZnO nanoparticles on their antibacterial action against strains of three bacteria: one Gram-negative, Escherichia coli and two Gram-positive, Bacillus subtilis and Staphylococcus aureus.

Keywords: Electrospinning, PVA, nano fiber, antibacterial property, ZnO.
PACS: 81.07.-b

INTRODUCTION

Nanofibers have some significant properties such as high porosity, low density, softness, and variable pore size because of the thin fiber diameter and large surface area to volume ratio. These characteristics make the nanofibers suitable for certain applications, e.g. filtration membranes for water disposal, and air purification systems. Electrospinning process is a relatively simple method to fabricate continuous nanofibers from various polymer liquids. The nonwoven mats and webs are expected to be in biomedical applications, e.g. as drug delivery systems, would dressing systems, and scaffolds in tissue engineering. In terms of biomedical applications poly(vinyl alcohol) (PVA) is attractive due to the biocompatible and biodegradable properties and has been reported in some applications. Another crucial factor is water conservation of PVA owing to the hydrophilicity. We also used ZnO nanoparticle. Among the inorganic particles ZnO particles have desirable properties for biomedical applications, such as robust antibacterial activity and chemical stability. Furthermore nanosized particles can be expected to have stronger antibiotic activity comparing with larger particles due to high surface to volume ratio. Especially, ZnO is essentially nontoxic to humans and environmentally friendly as biomedical application materials. Therefore we came to the idea to fabricate nanofibers made from a suspension including PVA and ZnO nanoparticles. In terms of the effect of the antibacterial activity we focused on the effect of the diameter of the ZnO particles on their antibacterial properties against three bacteria strains by using the hybrid electrospun mats fabricated by the suspensions with various size of the nanoparticles: *Escherichia coli* (Gram-negative), and *Bacillus subtilis* and *Staphylococcus aureus* (both Gram-positive).

First we examined the rheological behavior of the suspensions which may affect the electrospinability of the suspension and spun fiber morphology. Then we will report the results of electrospinning of the suspensions such as the fiber diameter and morphology of electrospun fiber mat. Finally we studied the antibacterial activity of ZnO both in suspensions and electrospun fiber mats for three types of bacteria strains.

EXPERIMENTAL

We used PVA with weight average molecular weight Mw of 140,000, which was hydrolyzed 87-89 %, and ZnO nanoparticles with diameters of 20, 70, and 100 nm. We also used *Escherichia coli* (E. coli) of microorganism library at Yamagata Univerisity. Bacillus subtilis (B. subtilis NBRC 13719) and Staphylococcus aureus (S. aureus NBRC 100910) were purchased from the Biological Resource Center (NBRC), National Institute of Technology and Evaluation, Japan. We prepared a 10 wt% PVA aqueous solution by dissolving PVA in hot water (90 °C) under vigorous stirring. 1 wt% of nanoparticles with a given diameter was added to the suspension. For the rheological analyses the shear-rate dependent viscosity was measured by rotational rheometer (Ares, TA instruments) with Couette geometry at room temperature. The antibacterial activity experiments were examined by a change of bacterial cell number on the nutrient plate with or without ZnO nanoparticles after incubation of the bacterial cells at 37 °C for 24 hr.

RESULS AND DISCUSSION

Figure 1 shows SEM images of electrospun fibers from PVA aqueous solution, PVA/ZnO (100nm) aqueous suspension, and PVA/ZnO (20nm) aqueous suspension. It was possible to produce nonwoven fiber mats under the condition from the PVA solutions with nanoparticles (100, 70, 20 nm). Although the fiber mat morphology seems similar, the diameters of the fibers from the suspensions are larger than that from PVA and increased with increasing the diameter of the particles. When the strength of the electric field is increased from 0.7 to 2.5 kV/cm, the average diameter of the fibers decreased. At a given electric field strength the diameter of the fiber decreased as increasing the diameter of the particles. Figure 2 shows the steady shear viscosity of the PVA solution and the suspensions with different diameters of ZnO. We should note that the viscosity of the suspension indicated much higher viscosity than PVA solution irrespective of very low volume fraction of the particle, and the value was dependent of the particle diameter. PVA/ZnO (20nm) suspension which led to the thickest fiber diameter exhibited highest viscosity. This difference is not trivial, and much higher than that expected from the Einstein's equation for suspension. Although FTIR spectra did not show any chemical reaction between ZnO and PVA, there was preliminary indications that suggested intermolecular hydrogen bonding between them. The difference of the particle size also affects the antibacterial efficacy. We will discuss it in the presentation.

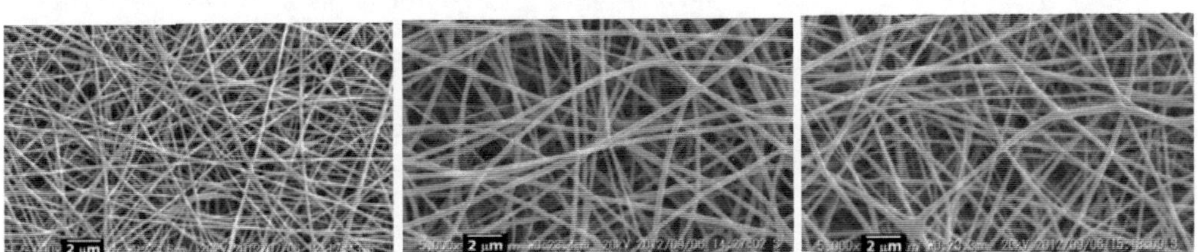

FIGURE 1. SEM images of electrospun fibers from PVA aqueous solution (left) and PVA/ZnO suspensions with diameters of 100 and 20 nm (middle and light).

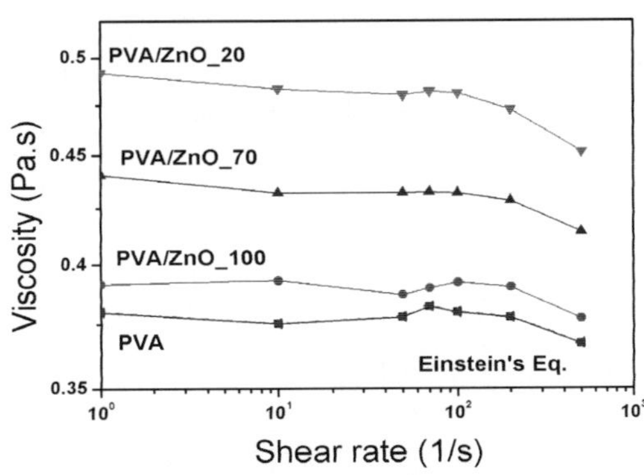

FIGURE 2. Steady shear viscosity of PVA aqueous solution and PVA/ZnO aqueous suspensions with nanoparticles of diameters of 20, 70, and 100 nm at room temperature.

Analysis on Crystallization Behavior of Oriented Poly(ethylene terephthalate) Fibers and Films

Seiji SHIBATA, Wataru TAKARADA* and Takeshi KIKUTANI

Department of Materials Science and Engineering, School of Materials and Chemical Technology, Tokyo Institute of Technology. 2-12-1-S8-32, Ookayama, Meguro-ku, 152-8552 Japan

takarada.w.aa@m.titech.ac.jp

Abstract. It is known that the crystallization rate of polymeric materials increase with its molecular orientation. Conventional differential scanning calorimetry (DSC) measurements and fast scanning chip calorimetry (FSC) measurements were conducted for the quantitative investigation on the crystallization rates of poly (ethylene terephthalate) (PET) fibers and films with various degrees of molecular orientation. The models for crystallization kinetics proposed by Ozawa[1] and Nakamura[2] were combined to calculate the non-isothermal crystallization rate of the PET fibers and PET films. As a result, it was found that highly oriented PET has more than 1000 times higher crystallization rate than isotropic PET. Additionally, from the comparison between crystallization rate of fibers and that of films, it was revealed that uni-axial orientation has more influence on the crystallization rate than planar orientation.

Keywords: poly (ethylene terephthalate), crystallization rate, molecular orientation
PACS: 64.70.dg

INTRODUCTION

Crystalline structure, crystallization mechanism and crystallization rates of polymeric materials have been studied by many researchers for a long time. But most of them are under the isothermal condition without molecular orientation. Crystallization rates of polymeric materials are significantly accelerated by molecular orientation, and that makes it difficult to measure the crystallization behaviors of oriented polymers. However, crystallization usually occurs under flow or stretching during processing. Thus understanding of crystallization behavior and quantitative analysis of crystallization rate of oriented polymer is very important.

Recently, fast scanning chip calorimetry (FSC) has been developed and began to spread rapidly. FSC can heat up and cool down tiny samples at more than 1000 K/s and measure the exothermic heat and endothermic heat from the samples. That enable us to measure very fast phenomenon that could not be detected by conventional differential scanning calorimetry (DSC). In this study, crystallization rate of uni-axially and bi-axially oriented poly (ethylene terephthalate) (PET) were investigated utilizing conventional DSC and FSC measurements.

EXPERIMENTAL

PET fibers with various degree of uni-axial orientation were prepared through the melt spinning process at various take-up velocities. PET films with various degree of planner orientation were prepared through the simultaneous equi-bi-axial stretching of amorphous and isotropic PET films. The maximum take-up velocity in the melt spinning was 6000 m/min and the maximum stretching ratio in the biaxial stretching was 4.00×4.00. Structure these fibers and films were analyzed by wide-angle X-ray diffraction (WAXD) measurements and birefringence measurements. Conventional DSC and FSC measurements were conducted to investigate the crystallization behaviors of oriented fibers and films. Heating rates were 5 – 200 K/min in conventional DSC measurements and 50 – 1000 K/s in FSC measurements.

RESULTS AND DISCUSSION

From WAXD measurement, it was confirmed that PET fibers which were wound at the velocity more than 4,625 m/min had been crystallized in the spinning process. PET film which were stretched at draw ratio higher than 3.0×3.0 had also been crystallized in the stretching process. Results of conventional DSC measurement for stretched films are shown in Fig. 1. Cold crystallization temperature shifted to lower temperature with increment of molecular orientation at the film with stretch ratio 2.0 x 2.0. On the other hand, crystallization behaviors of the film with stretch ratio 4.0 x 4.0 were overlapped with the baseline shift of glass transition, so detailed crystallization analysis was not available for highly oriented samples both in fibers and films.

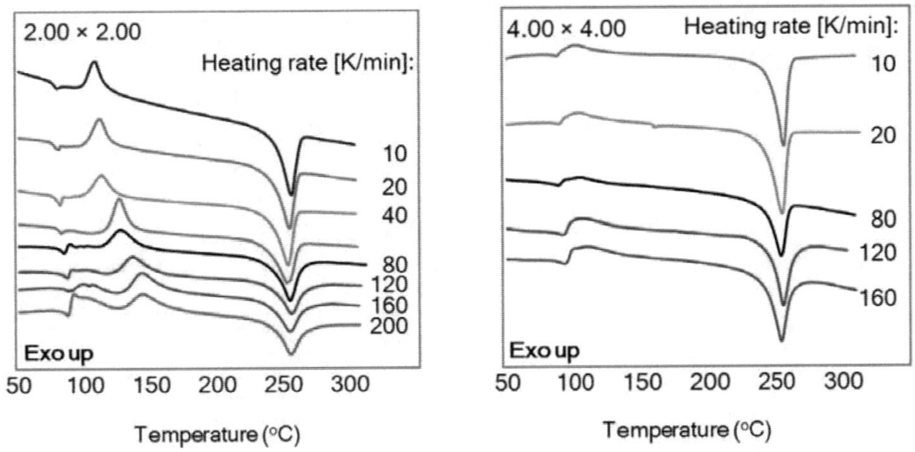

FIGURE 1. DSC thermogram of stretched films (stretch ratio: (a) 2.0 x 2.0, (b) 4.0 x 4.0).

FSC thermogram of the film with stretch ratio 2.5 x 2.5 were shown in Fig. 2. Significant data fluctuation caused by the thermal shrinkage of the oriented fibers and films was observed at around T_g without pre-thermal treatment (Fig. 2 (a)). To avoid this fluctuation, we conduct pre-thermal treatment on oriented samples before first heating. Schematic illustration of measurement method was shown in Fig. 3. During the pre-thermal treatment (maximum temperature was 120 °C), oriented sample could shrink but could not crystallize, so we can obtain cold crystallization cleary in subsequent first heating like Fig 2 (b). In FSC measurement, bending of baseline is another difficulty. We found that the FSC thermogram of amorphous sample could be used as a baseline. To obtain those FSC thermogram, molten sample after first heating was cooled rapidly and heated up again. No crystallization was observed in this second heating because sample lost its orientation in melting. Objective of second run in our method shown in Fig. 3 is taking baseline of first run. Thick lines in Fig. 2 (b) represent first run and thin lines represent second run. Heat of crystallization was calculated from the difference of these two lines.

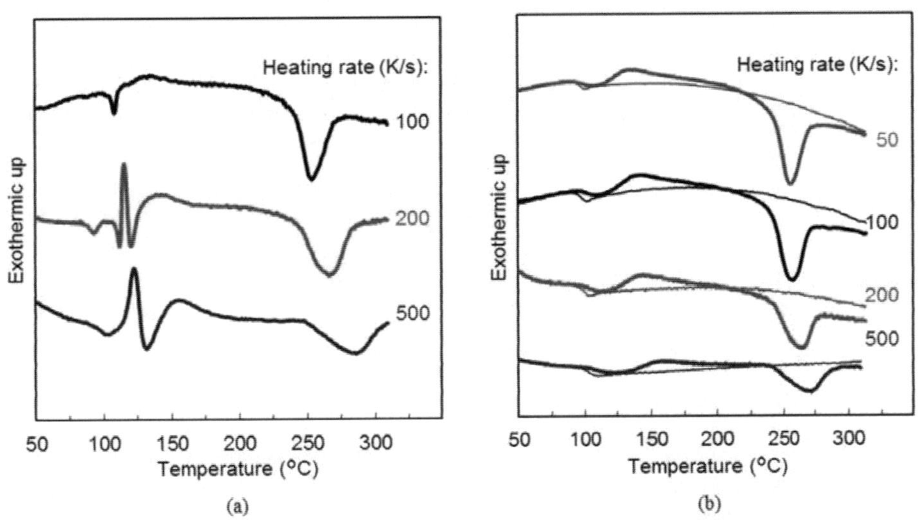

FIGURE 2. (a)FSC thermograms without pre-thermal treatment and (b) those after pre-thermal treatment for PET films stretched to the draw ratio of 2.5×2.5.

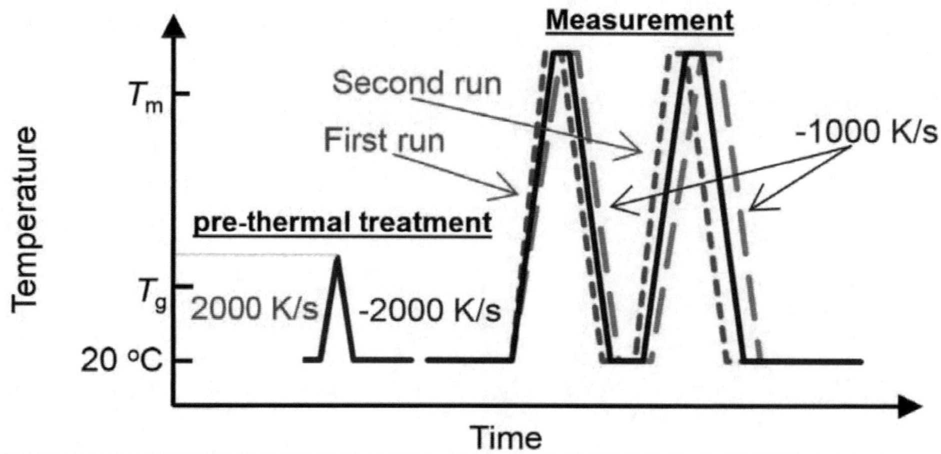

FIGURE 3. Time-temperature profile for FSC measurements including pre-thermal treatment

To calculate the non-isothermal crystallization rates of PET fibers and PET films, we utilized two models for crystallization kinetics proposed by Ozawa[1] and Nakamura[2]. Those two models represent non-isothermal crystallization of polymers in following equations:

$$C(T) = 1 - \exp(-\frac{x(T)}{a^n}) \tag{1}$$

$$C(T) = 1 - \exp\left[-\left\{\int_0^t K(T(\tau))d\tau\right\}^n\right] \tag{2}$$

Here $C(T)$ is relative crystallinity at the specific temperature T, $x(T)$ is the function called heating function[1], a is heating rate, $K(T)$ is non-isothermal crystallization rate function and n is the constant corresponds to Avrami index. Equation (1) can be rearranged as follows:

$$\log[-\ln\{1 - C(T)\}] = \log x(T) - n \log a \tag{3}$$

From equation (3), Avrami index n and $x(T)$ can be calculated from the plot of $\log[-\ln\{1 - C(T)\}]$ against $\log a$ [1]. Table 1 contains Avrami indices of PET fibers and films with various orientation. In this table, condition means take-up velocity for fibers and draw ratio for films. Avrami indices of the fiber with take-up velocity 1,000 m/min and that of un-stretched film were obtained from the conventional DSC measurements and Avrami indices of other samples were obtained from FSC measurements. It was indicated that the samples with higher molecular orientation have lower Avrami index.

By combining equation (1) and (2), the equation for $K(T)$ is described as follows:

$$K(T) = \frac{-\frac{dx(T)}{dT}}{na^{n-1}\frac{\{x(T)\}^{\frac{n-1}{n}}}{a^{n-1}}} = \frac{\frac{dx(T)}{dT}}{n\{x(T)\}^{\frac{n-1}{n}}} \tag{4}$$

From equation (4), non-isothermal crystallization rate functions $K(T)$ were calculated and shown in Fig. 4. It was found that highly oriented PET fibers and films has more than 1,000 times higher crystallization rate than almost isotropic PET, the fiber with take-up velocity 1,000 m/min or un-stretched film. Furthermore, comparing $K(T)$ between fibers and films with similar orientation function f, fiber samples showed higher crystallization rate than film samples. It indicates that uni-axial orientation in fiber samples accelerate crystallization rate more than planar orientation in film samples.

TABLE 1.Avrami index n and orientation function f of PET fibers and PET films

	Fiber	Fiber	Fiber	Film	Film	Film
Condition	1,000 m/min	4,000 m/min	4,500 m/min	un-stretched	2.50 × 2.50	3.00 × 3.00
n	2.1	1.4	1.1	2.1	0.5	0.4
f		0.23	0.34		0.39	0.53

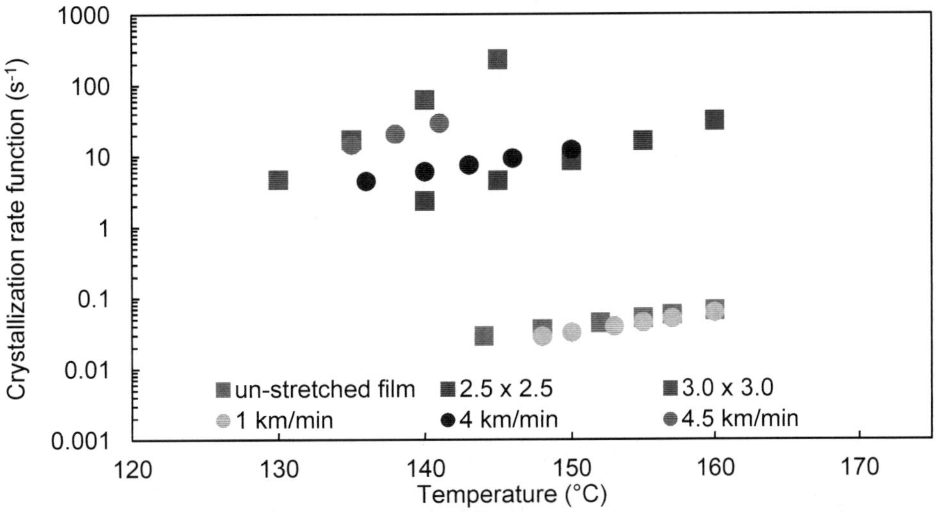

FIGURE 3. Non-isothermal crystallization rate functions for PET fibers and PET films with various degrees of molecular orientation analyzed from the results of DSC and FSC measurements. Square marks indicate crystallization rate function of films, circular marks indicate that of fibers.

CONCLUSION

Data fluctuations were observed at around T_g because of rapid thermal shrinkage in FSC measurements. But by applying pre-thermal treatment, those could be suppressed and thermograms were obtained even for oriented PET samples. Both PET fibers and PET films with high degrees of molecular orientation had more than 1000 times faster crystallization rates than isotropic ones. Furthermore, it was revealed that axial orientation has more influence on the increase of the crystallization rate than planar orientation.

ACKNOWLEDGMENTS

This work was supported by JSPS KAKENHI Grant Number 25288097.

REFERENCES

1. T. Ozawa, *Polymer* **12**, 150-158 (1971)
2. N. Nakamura, T. Watanabe and K. Katayama, *J. Appl. Polym. Sci.*, **16**, 1077-1091 (1972)

Preparation of superhydrophobic/superolephic electrospinning fiber

Fang Qiu [a], Zhimei Wei[a], Xiaojun Wang[b*] and Jie Yang[b*]

[a]*College of Polymer Science and Engineering, Sichuan University, Chengdu, 610065, China.*
[b]*Institute of Materials Science and Technology, Sichuan University, Chengdu, 610064, China.*
493551636@qq.com
ppsf@scu.edu.cn
wangxj@scu.edu.cn

Abstract. Fabrication of materials for water/oil separation is of great importance for industrial application. Herein, superhydrophobic and superolephilic fiber was fabricated by combination of PDA coating on the surface of PES fiber and subsequently chlorosilane modification. The as-prepared membrane exhibit that a statistic contact angle for water is greater than 150° and the oil can be quickly adsorbed, which indicated the property of superhydrophobicity and superolephilicity. The results presented in this work might provide a simple, low-cost and efficient approach for application in the field of water/oil separation.

Keywords: Superhydrophobic, superolephilic, fiber, electrospinning.
PACS: S12-412

INTRODUCTION

With the frequency occurred water pollution caused by chemical leakage and oil spillage, the removal, adsorption and collection of organic contaminant has attracted widely attention [1]. The conventional method of solving this problem include chemical degradation, mechanical extraction, and adsorption[2-4]. Among them, chemical degradation has the disadvantage of produce secondary pollutants, the mechanical extraction is inconvenient to deal with trace pollutants in water. Therefore, consider of the efficiency and economy, absorption is a promising method of removal organic contaminant from oil/water mixture. In recent years, many materials which have opposite wettability for water and oil have attracted great interest for scientific community in the field of oil/water separation[5]. Generally, materials with superhydrophobicity (water contact angle higher than 150° and sliding angle less than 10°) and superoleophilicity(oil contact angle less than 5°) was considered a efficiency adsorbent to dispose of oil/water mixture. Xue et al fabricated superhydrophobic and superoleophilic PET textiles by a simple sol–gel coating usingtetraethoxysilane and 1,1,1,3,3,3-hexamethyl disilazane as precursors[6]. After coating, the fibers were decorated with $Si(CH_3)_3$ functionalized SiO_2 nanoparticles. The modified textile was used to filter the mixture of oil-water, realizing continuous oil–water mixture separation. Liu el at fabricated superhydrophobic and superoleophilic cotton by the combination of SiO_2 nanoparticles on cotton fiber surface and subsequent octadecyltrichlorosilane modification[7]. The as-prepared cotton could be used to selectively absorb various common oils and organic solvents up to above 50 times of its own weight while repelling water completely. Most modification for superhydrophobic and superoleophilic property was focused on SiO_2 particles which was respectively easily fall off from the surface of substrate. Hence, new method of fabrication superhydrophobic surface without easily falling particles is necessary.

Electrospinning is a versatile method of fabrication continuous fibers with diameter ranging from micrometer to nanometer[8]. The nonwoven mats has roughness surface which was beneficial to improve the hydrophobicity of materials. In the paper superhydrophobic and superoleophilic PES fibers was prepared and applied to removal oil from oil-water mixture.

EXPERIMENTAL

Materials. Polyethersulfone (PES, Mw=45,000g/mol) was supplied by BASF chemical Co. Dopamine (99%) was purchased from Aladdin Co. Tris(hydroxymethyl)aminomethane (Tris), Dimethyl sulfoxide (DMSO), HCl (36%) ,

Dodecyltrichlorosilane (DTS) were obtained from Kelong chemical Co. All chemicals were of AR grade and used without further purification unless described. Gasoline was commercially available.

Preparation PES fiber. Solution of PES/DMSO was firstly prepared by adding 2.4 g PES particle in 10 ml DMSO. Then the solution was used to produce fibers. The electrospinning setting was simple and easily conducted. Direct current high–voltage generator (ZGF Chuan Gao electro-tech, China) was used to provide high voltage of 15 KV. A 50ml plastic syringe with a 20 gauge blunt needle was used as the spinneret. Grounded rotating drum covered with aluminum foil was used as the collector and the rotation speed fixed at 0.2 m/s. The flow rate of the solution is 1 ml /h. The superhydrophobic nonwoven mat was obtained as following: the electrospun fiber was immersed in the 0.01 M Tris-buffer solution of dopamine. After 24 h, the PES/PDA fiber was taken out and dried. Then the fiber was immersed in n-hexane of DTS for adequate time. The effects of pH of the buffer solution and concentration of the dopamine on the hydrophobicity of fiber were investigated.

Characterization of method. The scanning electron microscopy (SEM Inspector F, FEI Co. USA) was employed to characterize the morphology of fiber. The pH of solution was measured on pH indicator (Aolilei PHS-3C, China). The statistic contact angel was conducted on the instrument of surface tension (KrÜss K100, Germany).

RESULT AND DISCUSSION

The morphology of fiber was obtained by SEM, as shown in FIGURE 1, and the following features were observed: (1) small hole was formed on the surface of the virgin PES fibers; (2) the holes were disappeared after PDA coating on the surface of the PES fiber; (3) some particles were adhered onto the surface of the ultrafine fibers, which could be ascribe to the adhesion of PDA particles; (4) the surface of PES/PDA/DTS fiber was rough, which was resulted from

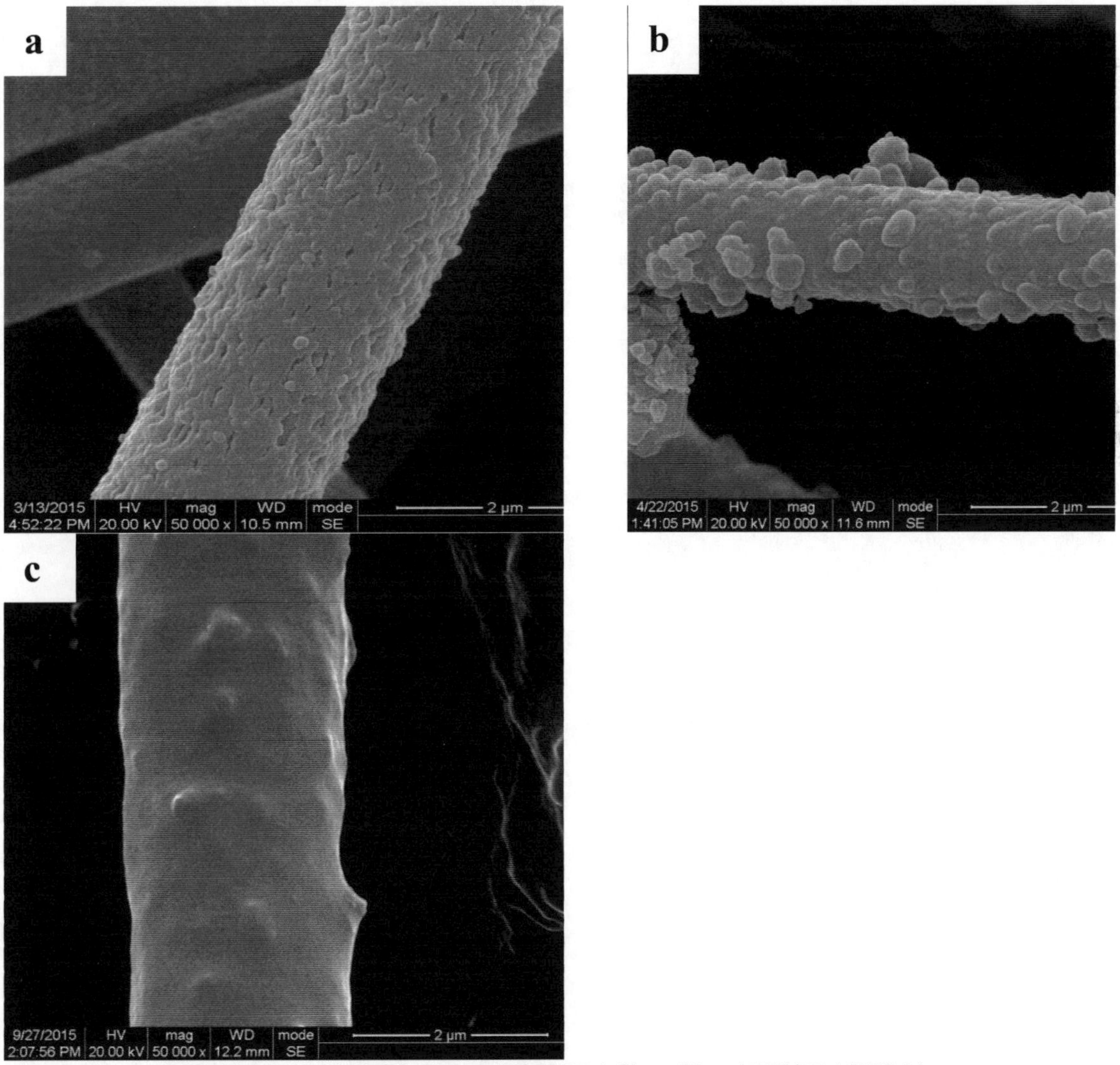

FIGURE 1. SEM of PES fibers (a), PES/PDA fibers (b) and PES/PDA/DTS (c)

some irregular embossment.

The concentration of dopamine and pH of buffer solution have great effect on the coating of PDA on the substrate[9]. Therefore, the two factors were investigated, and the result was shown in FIGURE 2. The alkalinity condition was favor to improve the hydrophobicity of PES/PDA/DTS and the contact angles were over 150° in the condition of the value of pH 8.5 and 10.5. The concentration of dopamine also had influence on the hydrophobicity of PES/PDA/DTS. When the concentration of dopamine was below 2mg/ml, the contact angle of PES/PDA/DTS increased with increasing the concentration of dopamine. But the contact angle of PES/PDA/DTS fiber almost keep immutability when the concentration of dopamine was equal to or greater than 2mg/ml. In the low concentration of dopamine, it was not enough to fully coating the PES fiber so the hydrophobic alkyl couldn't completely decorated the surface of fiber.

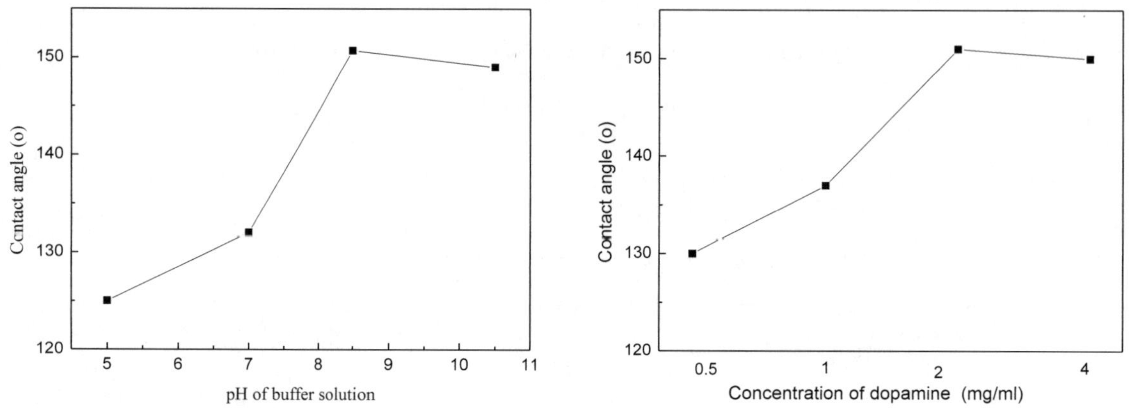

Figure 2. The effect of ph of buffer solution and concentration of dopamine on the contact angle

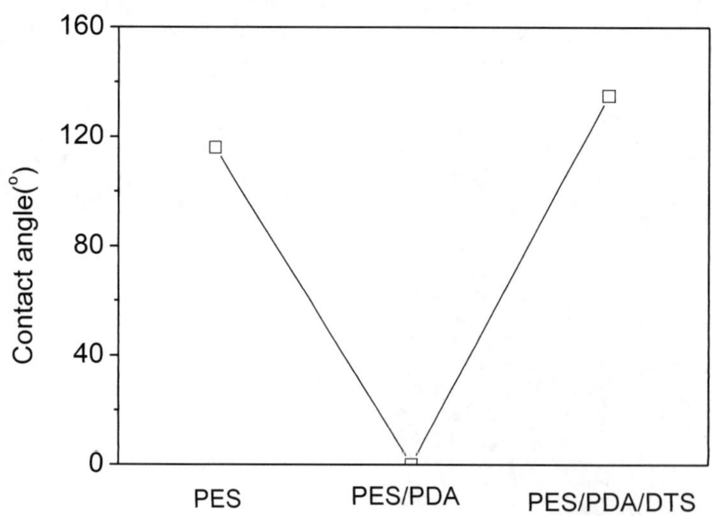

FIGURE 3. The contact angle of fibers before and after modification

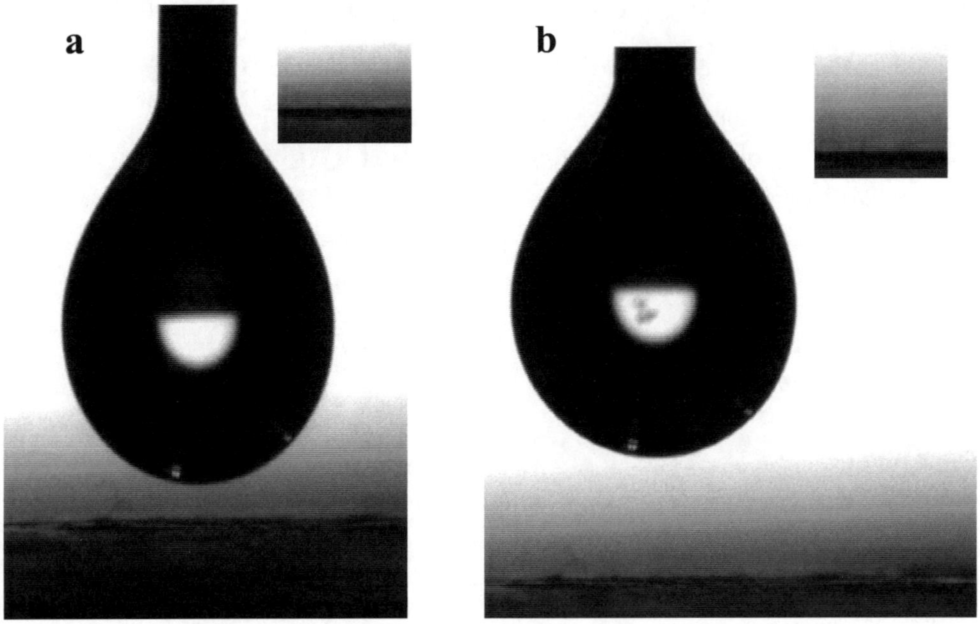

FIGURE 4. Before and after oil droplet dripping on the surface of PES/PDA/DTS fiber gasoline (a) and n-hexane (b)

CONCLUSION

Superhydrophobic and superolephilic fiber membrane were fabricated by combination of PDA coating on the surface of PES fiber membrane and subsequently chlorosilane modification. The pH of buffer solution and concentration of dopamine have great effect on the hydrophobicity of PES/PDA/DTS fiber. Alkalinity was favor to improve the hydrophobicity of PES/PDA/DTS fiber. The as-prepared fiber exhibit that a statistic contact angle for water was greater than 150° and the gasoline and n-hexane can be quickly adsorbed, which indicated the property of superhydrophobicity and superolephilicity.

REFERENCES

1. J. Zhang and S. Seeger, Angewandte Chemie International Edition, 50(29): 6652-6656 (2011).
2. P. Lee and M. A. Rogers, Langmuir, 29(18), 5617–5621 (2013).
3. C. Teas, S. Kalligeros, , F. Zanikos, S. Stournas, E. Lois, and G. Anastopoulos, Desalination, 140(3), 259–264 (2001).
4. J.T. Korhonen, M. Kettunen, R.H.A. Ras, O. Ikkala, ACS Appl. Mater. Interfaces 3, 1813–1816 (2011).
5. A. Steele, I. Bayer, E. Loth, Nano Lett., 9, 501-505 (2009).
6. C. H. Xue, P. T. Ji, P. Zhang, Applied Surface Science, 284: 464-471 (2013).
7. F. Liu, M. Ma, D. Zang, Carbohydrate polymers, 103: 480-487 (2014).
8. W. Ma, Q. Zhang, D. Hua, RSC Advances, 6(16): 12868-12884 (2016).
9. F. Ponzio, P Payamyar, A Schneider, The journal of physical chemistry letters, , 5(19): 3436-3440 (2014).

Effects of molecular weight distribution and stereoregularity on mechanical properties and structure of Isotactic Polypropylene (iPP) fiber

Tatsuma KUNIMITSU[a], Kai TOYODA[a], Toshifumi IKAGA[a], KyoungHou KIM[a], Yutaka OHKOSHI[a,b]*, Katsuhiko KOIKE[c])

[a] *Faculty of Textile Science and Technology, Shinshu University, 3-15-1 Tokida, Ueda, Nagano, 386-8567, Japan*
[b] *Division of Frontier Fibers, Institute for Fiber Engineering, Shinshu University, Japan*
[c] *Prime Polymer Co., Ltd./Mitsui Chemicals Inc., Shiodome City Center, 1-5-2 Higashi-Shimbashi, Minato-ku, Tokyo, 105-0021, Japan*

12f1016a@shinshu-u.ac.jp
yokoshi@shinshu-u.ac.jp

Abstract. Isotactic Polypropylene (iPP) fibers were made by various spinning and drawing conditions with three types of iPP, that is, two different stereoregularity metallocene catalyst iPP and Ziegler-Natta catalyst iPP. As the result, it was revealed that the higher tensile strength and smaller creep strain fiber than the fiber of Ziegler-Natta catalyst iPP could be obtained by using high stereoregularity metallocene catalyst iPP.

Keywords: Isotactic Polypropylene, Fiber Strength, Creep Strain, Molecular Weight Distribution, Stereoregularity
PACS: 81

INTRODUCTION

Isotactic polypropylene (iPP) fiber has the merits of low cost, hydrophobicity and chemical resistance, but also has a weak point of heat resistance than other commodity fibers, such as PET and Nylon. In addition, further improvement in tensile strength and heat resistance are desirable to reduce the weight or expand the use of products.

This study focused on the iPP polymer polymerized by the metallocene catalyst. The commonly used iPP fiber is made by the polymer synthesized by Ziegler-Natta catalyst. By using the metallocene catalyst, iPP having the narrower molecular weight distribution than the iPP synthesized by the Ziegler-Natta catalyst can be obtained [1]. In addition, stereoregularity of iPP is also focused in this study. For the purpose, three iPP polymers, two metallocene catalyst iPP of high and low stereoregularity, and Ziegler-Natta catalyst iPP, were used in this study. The mechanical properties and creep behavior are evaluated for the melt-spun and drawn iPP fibers of these three iPP polymers. Then the influence of polymer, spinning conditions, and drawing conditions to the fiber structure and properties of resultant iPP fibers are discussed in this study.

METHODS

We used three types of iPP polymer in this study, that is, high stereoregularity metallocene catalyst iPP (IPF=98%, Mw/Mn=2.5; H-miPP), low stereoregularity metallocene catalyst iPP (IPF=93, Mw/Mn=2.5; L-miPP), and high stereoregularity Ziegler-Natta catalyst iPP (IPF=98%, Mw/Mn=3.5; ZNiPP). All polymers were provided by Prime polymer Co., Ltd. These three types of iPP polymer were melt-spun through a single hole nozzle of diameter 0.7mm. The melt spinning conditions are shown in TABLE 1, in which the sample name represents the polymers (L, H, ZN), nozzle temperature, mass of flow rate, and melt draw ratio. The as-spun fiber was drawn by using a hot plate drawing system. The maximum heater temperature was selected for each polymer, 140°C for H-miPP and ZNiPP, and 125°C for L-miPP. The fiber feeding speed is 1.0 m/min for all, and taken-up to the maximum draw ratio for each as-spun fiber. For each drawn iPP fiber, tensile test, thermal mechanical analysis (TMA) and wide angle X-ray diffraction (WAXD) were performed. The creep behavior was analyzed by the TMA measurement, with the tensile load and temperature were 50 MPa and 125°C, respectively.

TABLE 1. As-supn fibers.

Sample	Polymer	Nozzle Temperature (°C)	Mass of Flow Ratio (g/min/hole)	Melt Draw Ratio	Spinning Speed (m/min)	Fiber Diameter Ave. (μm)	C.V. (%)
L-220-3.0-14.8	L-miPP					184.7	1.4
H-220-3.0-14.8	H-miPP	220	3.0	14.8	125	182.2	1.9
ZN-220-3.0-14.8	ZNiPP					184.1	1.8

RESULTS & DISCUSSION

According to the results of drawing process for three kinds of as-spun fiber having a different stereoregularity and molecular weight distribution, the maximum drawable temperature was affected by the stereoregularity of as-spun fiber. All as-spun fibers made from H-miPP and ZNiPP could be drawn steadily at 140°C, while that made from L-miPP could not be drawn above 125°C. And the as-spun fibers of H-miPP and ZNiPP could be drawn steady up to 16.0 times, while that of L-miPP could not be drawn over 13.6 times. The difference in the maximum draw ratio should be related to the difference of the maximum drawing temperature. The drawing stress are shown in FIGURE 1. The drawing stress increases with the increase of draw ratio, and takes maximum at the maximum draw ratio for each as-spun fiber. The H-miPP fiber shows the maximum drawing stress of all, while ZNiPP shows the minimum value of the maximum drawing stress. It indicate that the drawing stress should be decided not only by the draw ratio but also by the molecular weight distribution of iPP polymer.

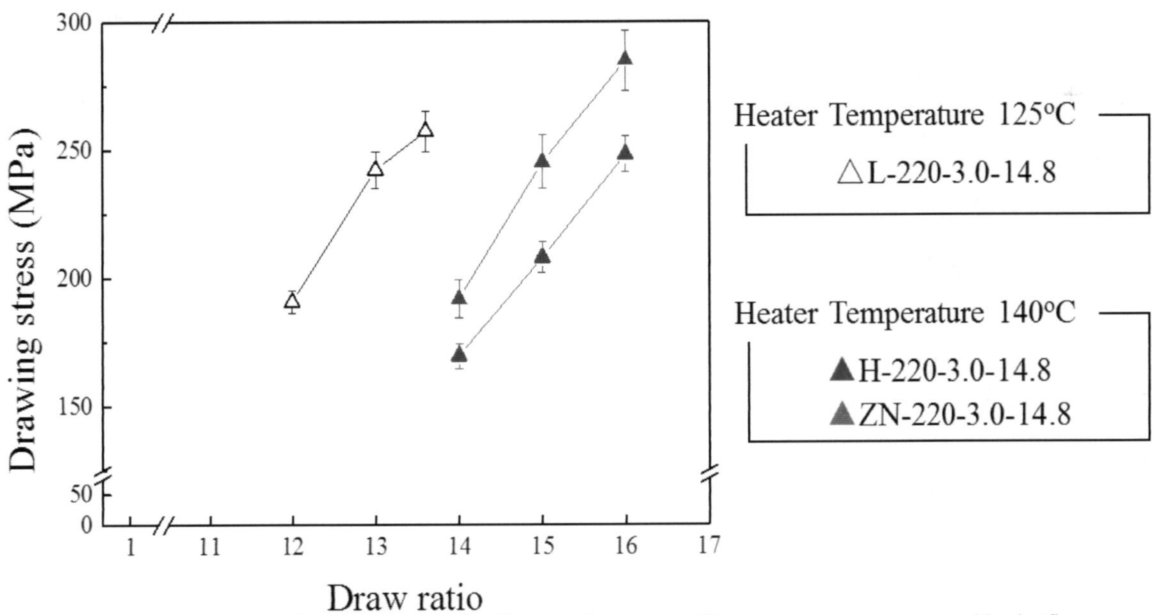

FIGURE 1. Drawing stress plotted against draw ratio. The sample name and heater temperature are noted in the figure.

The tensile properties of drawn fibers as a function of the draw ratio are shown in FIGURE 2. The tensile strength took the maximum at a certain draw ratio for each as-spun fiber. The maximum tensile strength of L-miPP, ZNiPP, and H-miPP fibers were 10.8, 10.8, and 11.4 cN/dtex, respectively. These results suggest that the fiber having higher tensile strength can be obtained by iPP of higher stereoregularity and narrower molecular weight distribution. The initial modulus also increased with the increase of draw ratio. The maximum initial modulus of L-miPP, ZNiPP, and H-miPP fiber were 122.4, 145.9, and 151.5 cN/dtex, respectively. The maximum initial modulus of ZNiPP and H-miPP fiber were predominantly larger than that of L-miPP fiber.

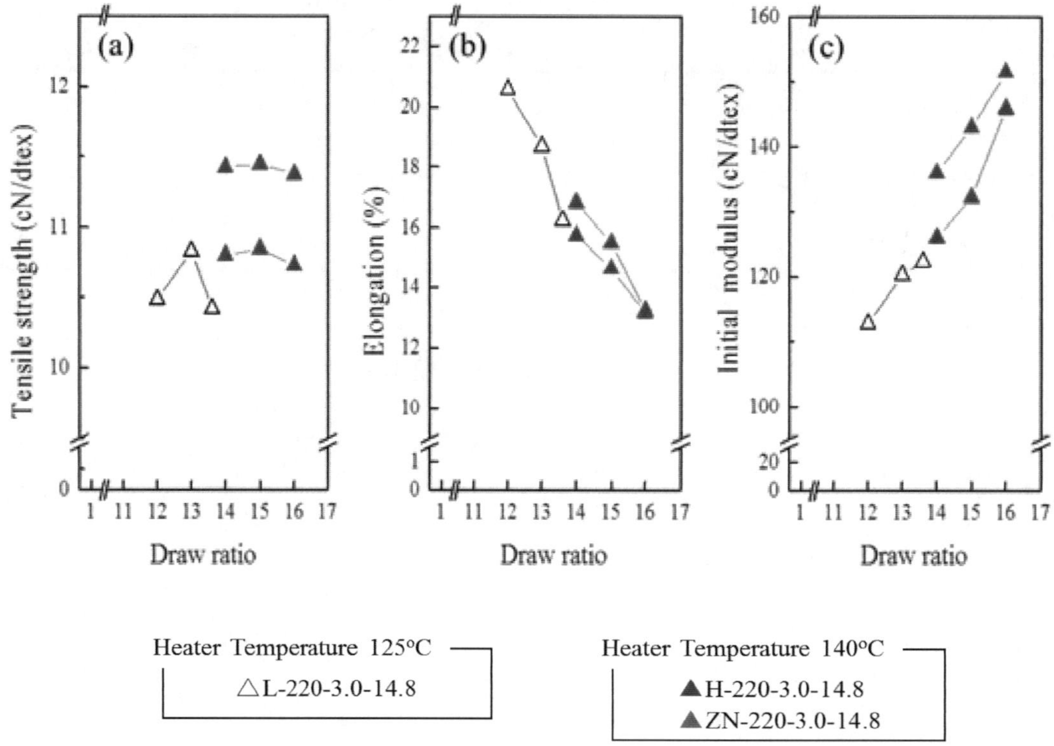

FIGURE 2. Tenshile properties of drawn fibers plotted against the draw ratio;(a) tensile strength, (b) elongation and (c) initial modus. The sample name and heater tempertature are noted in the fighure.

The result of creep test are shown in FIGURE 3. H-miPP and L-miPP fibers showed smaller creep strain than ZNiPP. The ZNiPP drawn fiber was broken around 40 minutes from the start of test, while the H-miPP and L-miPP fibers were not broken until the end.

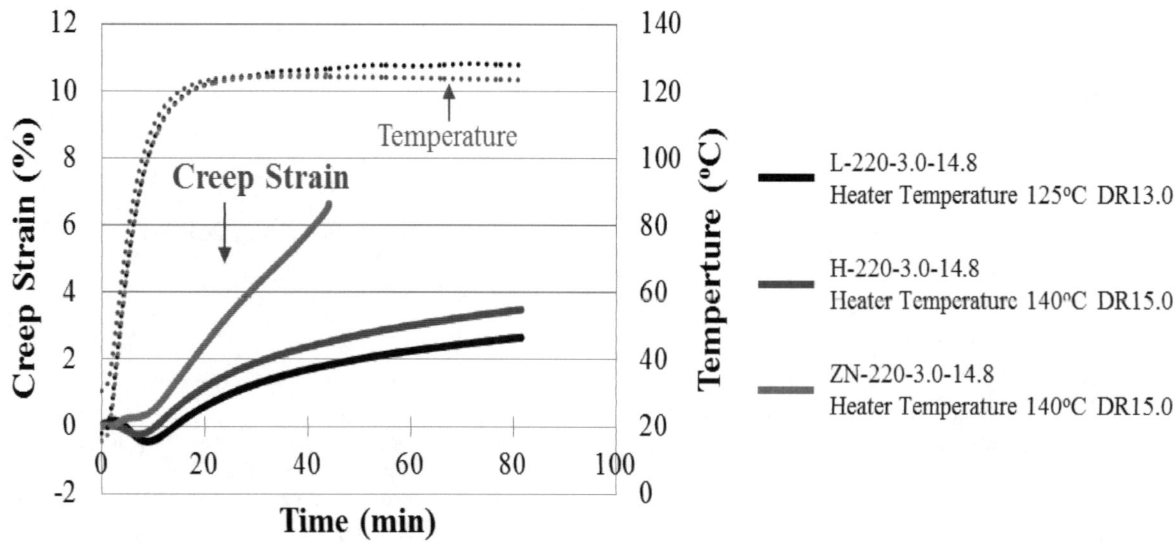

FIGURE 3. Thermal mechanical properties of drawn fibers plotted against the time. The sample name, heater temperature and draw ratio are noted in the figure.

CONCUSION

By using high stereoregularity metallocene catalyst iPP, higher tensile strength and smaller creep strain fiber than the fiber of Ziegler-Natta catalyst iPP could be obtained.

REFERENCES

1. Yuzi F, Higher-Order Structure of Polypropylene, Kubunshi 2005; 54:682-685.

AFM-Raman Co-localization Setup: Advanced Characterization Technique for Polymers

J.P. Cosas Fernandes, V.H. Mareau and L. Gonon*

Univ. Grenoble Alpes, INAC-SyMMES-SPrAM, F-38000 Grenoble, France
CNRS, INAC-SyMMES-SPrAM, F-38000 Grenoble, France
CEA, INAC-SyMMES-SPrAM, F-38000 Grenoble, France

joaopaulo.cosasfernandes@cea.fr
**laurent.gonon@ujf-grenoble.fr*

Abstract. The optimization of the manufacturing process and the investigation of polymer materials during all their life-cycle, up to decomposition mechanisms involved during ageing processes imply an extended knowledge of the physical and chemical properties of the material. To obtain co-localized chemical, morphological and thermomechanical information on the same sample area, we have developed an innovative co-localized AFM-Raman setup and a well suited sample preparation strategy using cryo-ultramicrotomy. The developed sample preparation technique allows perfect cross-sectioning of thin membranes (20-300μm) (or any bulk material) without any epoxy embedding and therefore no chemical contamination. Contrary to surface analysis on both side of the material, the opened surface is characteristic of the material's bulk properties and allows spectroscopic and near-field measurements without sample deformation and topographic artefact. In addition, thin sections can be collected during cutting for complementary Transmission Electron Microscopy (TEM/STEM).

Keywords: AFM, Raman, STEM, cryo-ultramicrotomy, hybrid polymer membrane
PACS: 81.07.-b; 82.47.Gh; 81.07.Pr

INTRODUCTION

The comprehension of the process-structure-properties relationship is mandatory in order to develop high performances materials. The complexity of this understanding demands multiple characterization techniques capable of providing insights about intrinsic chemical and physical properties of the material like, for example, the impact of the process on the dispersion and orientation of charges, the long term behavior of polymeric and composite materials, including the chemical profiles developed in the sample volume during oxidation and the impact on their mechanical properties (crosslinking, chain scissions…). In the literature, characterization of the structure, environment, dynamics, ageing, and mechanical properties of polymeric materials has been obtained using either Raman spectroscopy or Atomic Force Microscopy (AFM) taken as single instruments. Studies of the chemical composition and structure of polymers used in a variety of applications have been done using Raman spectroscopy, like composites for engineering applications, nanocomposites, polymer blends and membranes for electro-dialysis and fuel cell application.[3,4] Nanocomposites and polymer blends have also been studied by AFM, demonstrating its capacity for the identification of the different phases, morphology and the characterization of mechanical properties at the nanoscale.[5,6] All these studies show that these techniques, if co-localized, have a huge potential in the characterization and development of new materials, as they provide qualitative and quantitative complementary information at high resolution.

We have developed an innovative setup of characterization that couples AFM with Raman spectroscopy, making possible successive and complementary analyses of the same sample position providing multiple information at nanometric up to micrometric scale. AFM gives quantitative mechanical properties of the sample at nanoscale, based on the acquisition of about half a million of force-distance curves per image, which provide mechanical information such as Modulus, Adhesion, Deformation and Dissipation, simultaneously acquired with the topographic information of the sample at high resolution. Confocal Raman microspectroscopy is a straightforward complementary technique to access chemical and structural information at micro-scale. Furthermore, position and intensity of Raman bands give qualitative and quantitative information about the composition of the material, such as its chemical nature (the structural units, type and degree of branching, end groups and additives), conformation order (physical arrangement of the polymer chain), crystalline and amorphous phases and orientation (type and degree of polymer chain and side group alignment in anisotropic materials)[1,2] which can be correlated to differences in mechanical properties measured by AFM. Characterization of polymer membranes by AFM and Raman requires going deeper than the exposed surfaces, potentially very interesting but often non-representative of the bulk, especially for multiphase composites, hybrids, copolymers and aged materials. Therefore, the material needs to be opened without generating artefacts, to reveal its inner structure. We have adapted the cryo-ultramicrotomy technique to be able to open very thin membranes (typically 30 µm thick) without any epoxy embedding (which necessitates cutting the membrane very close to its holder, typically 50 to 100 µm). Using a diamond knife and operating in -100°C we are able to properly open cross-sections of polymer membranes and reveal their inner structure, avoiding cryofracture artefacts or chemical contamination by diffusion of epoxy resin. At the same time this sample preparation technique allows to collect ultrathin sections perfectly suited for TEM observations. Therefore, we can access co-localized topographic, mechanical, thermo-mechanical and chemical mappings (AFM-Raman) from the freshly cut surface with morphological, and also elemental and chemical contrast analysis (TEM, STEM, Nano-EDX) from the collected sections.

Experimental

AFM-Raman Setup

The system developed in our lab is composed of a Bruker dimension Icon AFM and a Horiba-Jobin Yvon LabRam HR Raman spectrometer as shown on FIGURE 1 (a). In order to avoid mechanical vibrations that could be transferred from the Raman spectrometer to the AFM, an active piezoelectric table was customized for the AFM making possible the use of both equipments at the same time with no loss of performances. On top of the standard AFM's motorized stage is installed a PI piezoelectric stage (100x100µm) dedicated to the Raman mappings. The AFM motorized stage in addition to its classical use for the AFM is also able to move the sample from the AFM tip to the Raman objective with limited positioning error (less than 3 µm). The Raman spectrometer (Laser He Ne-632,8nm-17mW) is equipped with two confocal microscopes. The one in front of the AFM set-up is dedicated to co-localized AFM-Raman experiments and will be called the co-localized microscope. The second one, which is positioned with direct access in front of the user (standard position for LabRam HR), will be called the regular microscope. It is dedicated for routine experiments, but still equipped with a motorized stage (50x70 mm) for mappings. Both Raman microscopes are equipped with autofocus systems, a piezo one for the co-localized microscope, and a motorized one for the regular microscope. In order to achieve co-localized analysis, two different procedures were developed depending on the intensity of the Raman response of the investigated material. For samples with strong Raman response the AFM motorized stage is simply used to move the sample from the AFM to the co-localized Raman microscope, and reciprocally, making it possible to first optically visualize the sample with the co-localized Raman microscope (better resolution than the AFM optical microscope) to select the area of interest, then, to move the sample below the AFM tip to obtain the AFM images and finally to move back to the co-localized Raman microscope to obtain the Raman mapping (potentially more destructive for the sample: laser heating) as shown in FIGURE 1 (b). This approach is suitable for fast mappings (mapping time below 30 minutes). On this time scale, the thermal drift of the co-localized setup is negligible and do not affect the measurements. The co-localization error induced by the large sample displacement is only of about 3µm and in most of the cases it can be corrected easily by scanning larger surfaces than the co-localized final area. On the other hand, for samples that respond weakly on Raman, like polymers in general, the acquisition time of a Raman mapping (typically 10x10µm) can be up to 10 hours. Therefore any thermal drift on the µm range during Raman mapping can jeopardize the Raman lateral and vertical resolution. Having no other choice than using the regular Raman microscope (no thermal drift for this microscope as both the microscope and its stage are fixed on the Raman, which is not the case for the co-localized Raman microscope) we have developed a shuttle stage that allows transferring the sample from the AFM stage to the regular Raman microscope stage and reciprocally with limited positioning error (less than 3 µm). The sample repositioning is done with the help of two fixed targets on the shuttle stage - FIGURE 1 (c).

Sample Preparation: The cryo-ultramicrotomy strategy

Samples of hybrid membranes about 30µm thick were trimmed in the form of a triangle and installed in a vice (special holder provided by LEICA) as shown in FIGURE 2 below. The top of the triangle was then cut by ultramicrotomy using a Leica EM UC7 ultramicrotome. This approach allows cutting the membrane without epoxy embedding, thus without chemical contamination. Cutting was done at -120°C to avoid any plastic deformation of the material during the process, in order to obtain the best quality for both collected sections (homogeneous thickness) and surfaced membrane's sample (flat and smooth). As the hybrid membranes are very sensitive to humidity, a system was developed to encapsulate the entire holder after cutting, limiting the water condensation on the surface of the material during its warming back to the room temperature. The obtained surfaced membrane's sample allows several different characterizations like Optical Microscopy, Scanning Electron Microscopy, Micro-FTIR, and is perfect for AFM-Raman as well. As the AFM tip will interact mechanically with the surfaced membrane, one has to be careful that the membrane does not protrude more than about 50-100 μm form the vice surface, which requires extra care during cutting to avoid knife contact with the vice. As the process is very stable and enables the control of different parameters like temperature, cutting angle, cutting depth and speed, one can prepare a sample with reduced roughness in which the fracture propagation is not impacted by material heterogeneities (which is not the case when using a simple cryo-fracture technique) and thus is more representative of the inner structure of the material. Especially for membranes, by the means of this technique, one can have access to the entire cross-section of the material, allowing analysis from the outer surfaces to the bulk. Comparisons between both techniques (cryo-fracture/cryo-ultramicrotomy) can be found in the literature.[8,9] Finally, the ultrathin collected sections (thickness about 50 nm) obtained by cryo-ultramicrotomy give access to morphological and chemical contrasts, as well as elemental analysis (EDX) by means of Scanning Transmission Electron Microscopies (STEM) and TEM.

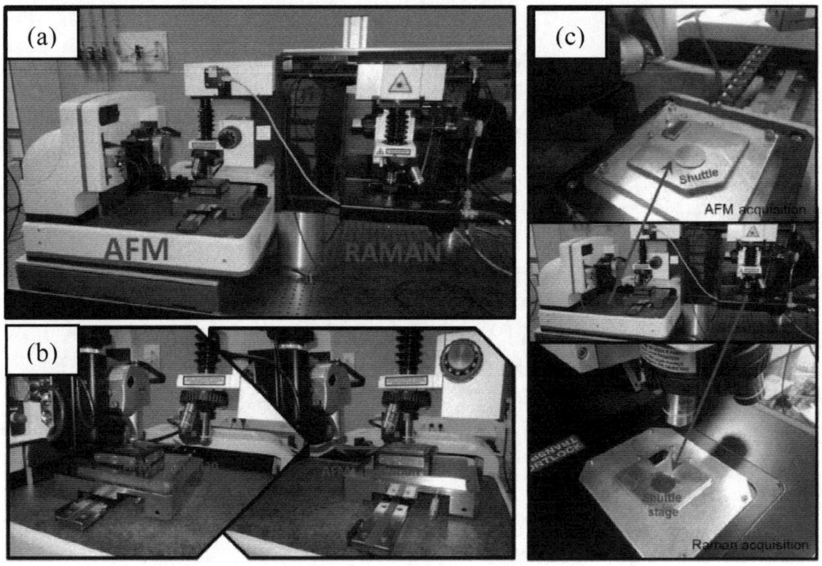

FIGURE 1. AFM-Raman setup and co-localization strategies

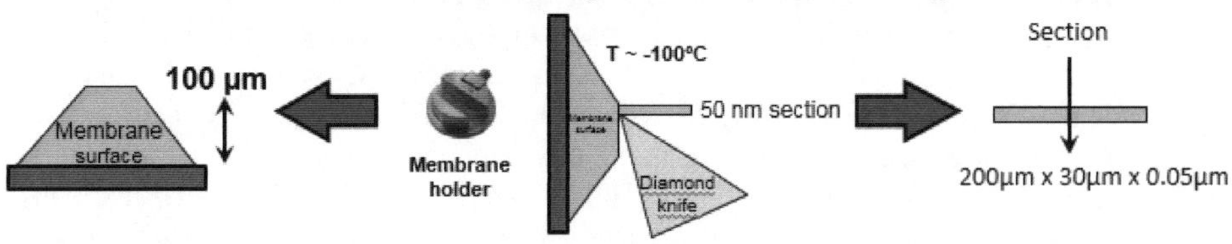

FIGURE 2. Schematic representation of cryo-ultramicrotomy applied to membrane preparation for both: surface analysis on the surfaced sample and STEM/TEM observations through collected sections

Results and Discussions

The interest of the AFM-Raman's co-localization strategy will be exemplified with the analysis of the inner structure of hybrid membranes developed in our laboratory. These membranes are composed of a commercial host membrane of sulfonated Poly(ether ether ketone) (sPEEK) into which a sol-gel (SG) network is grown from (3-Mercaptopropyl)methyldimethoxysilane precursors impregnated into the host membrane. The chemical structure of the materials is shown on FIGURE 3 (a) and (b). A thin (30 μm) hybrid membrane of sPEEK + SG was cryo-ultramicrotomed in order to produce a flat surface of about 200x30μm^2. Thin sections were also collected for complementary STEM analysis. Topography and mechanical images of the hybrid membranes were obtained using PeakForce QNM AFM mode. Images were taken with a resolution of 512x512 pixels, thus each image of the sample corresponds to 512x512 force-distance curves (each being the average of 2 successive acquisition) taken on an area of 10x10 μm² and were analyzed using Nanoscope Analysis version 1.7. A relative calibration method was applied using a dedicated reference samples kit provided by Bruker. The range of spring constant for the cantilevers used in this study varied from 30-40N/m, a good range for the modulus of the two phases that compose our material. FIGURE 4 (a) and (b) show AFM images of the cryo-ultramicrotomed surface of the membrane. On the Young's Modulus mapping image, the microstructure becomes clearly evident highlighting the differences in mechanical properties between sPEEK and SG. sPEEK has a higher Young's Modulus than the SG phase, and appears in light yellow color while the SG appears as brown dark with domains of about 800-1000 nm well dispersed in the membrane, except for the outer surface (right side of the image) that presents a skin-like structure mainly composed of sPEEK (higher modulus).

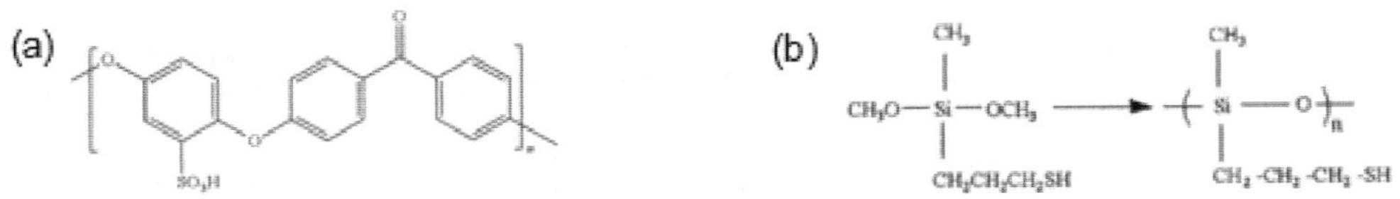

FIGURE 3. (a) sPEEK structure (b) SG structure

The calibration of the tip allows estimating the modulus of each phase which gives respectively about 1,2GPa for the sPEEK and 0,5 GPa for SG. FIGURE 4(c) is a Bright Field (BF) STEM image of a thin section collected from the sample. As sPEEK and SG interact differently with the electron beam due to the presence of heavier Si atoms in the SG, a strong chemical contrast is observed (no need for staining) with the SG phase scattering more electrons and appearing darker than the sPEEK phase on the image.

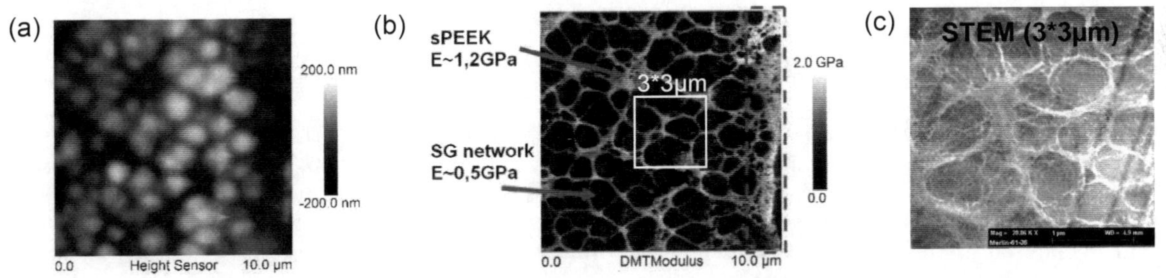

FIGURE 4. (a) AFM topography image of the hybrid membrane (b) AFM complementary modulus image (b) STEM image of the corresponding section

To confirm this pretty good dispersion of the SG phase in the membrane except for the first micrometer near the outer surface, a co-localized Raman mapping (100x long working distance objective) of the same 10x10μm area imaged by AFM (FIGURE 4(a)) was acquired using the shuttle stage procedure. FIGURE 5 (a) shows a typical spectrum of the hybrid membrane with 3 distinct peaks on the range of 2500 to 3200cm^{-1}. The peak centered at ~3080cm^{-1} is assigned to the contribution of the CH groups of the aromatic ring of the sPEEK. The peak centered at ~2900cm^{-1} refers to the aliphatic –CH vibrations of the SG, as well as the peak at ~2580cm^{-1} which is attributed to the –SH functions born by the SG phase structure.

By mapping the relative areas of the peaks between 2785 and 3015cm^{-1} and 3020 to 3120cm^{-1} one can evaluate the dispersion of the SG inside the membrane. FIGURE 5 (b) shows a good correlation between the mechanical properties of the hybrid membrane analyzed by AFM and the Raman chemical signature of sPEEK and SG phases, as evidenced by the lower SG(CH)/sPEEK(CH) peak ratio found on the outer surface of the membrane (right-side), which confirms the existence of a skin-like structure mainly composed of sPEEK observed in the AFM Modulus image of the same area. This depletion is related to the washing step applied during the elaboration of the hybrid membrane to avoid the formation of a SG coating on the membrane surface. Furthermore, except for the outer surface the distribution of the SG phases inside of the membrane is also found to be homogeneous at a micrometric scale, confirming the AFM observations. It's important to highlight that the AFM analysis is based on a surface analysis of the surfaced sample (depth of penetration of the AFM tip of few nm to tens of nm), while the Raman spectra has a deeper volumetric response (depth probed of about 1 to 2 μm) which added to the limited lateral resolution of the confocal Raman microspectroscopy of about 1μm in every direction, explains the better image resolution obtained by AFM. However, this chemical analysis gives precious additional information. The SH functions carried by the SG phase of the membranes were chosen for the protection against oxidation that they provide to the sPEEK. In order to evaluate the distribution of the SH functions in the membrane we have done a mapping of the ratio between the areas of the peak relative to the thiol function and the aliphatic CH of the SG phase, and shown in FIGURE 5 (c). One can see that, in agreement with the previous map, a lower ratio is found at the outer surface of the membrane, due to the SG depletion. However, from the bulk to the outer surface of the membrane, we found a decreasing ratio, which means that an unexpected and unwanted oxidation of the thiol functions took place on the first micrometers of the membrane. This observation was very useful to detect an oxidizing step in the elaboration process which could be corrected to preserve the SH functions and therefore the stabilizing efficiency of the SG phase.

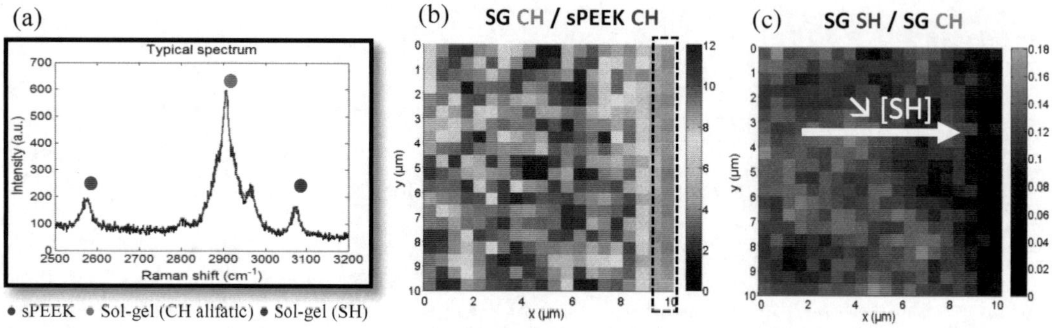

FIGURE 5. (a) Typical Raman spectrum of the hybrid membrane (b) Raman mapping co-localized with AFM data of FIGURE 4(a) showing the distribution of SG in sPEEK (c) same Raman mapping showing the oxidation of the SH moieties in the SG phase

Conclusion

The co-localized AFM-Raman is a powerful technique making possible the acquisition of complementary information in order to: reveal topography and morphology (AFM), identify phases and analyze their distribution (from mechanical properties by AFM or chemical composition by Raman), but also the ageing impact (chemical by Raman and mechanical by AFM) on polymer and composite materials. To acquire quantitative data characteristic of the bulk without any artefacts due to topography (tip interaction for AFM and focus for Raman) and/or sample deformation, a specific sample preparation based on cryo-ultramicrotomy was developed. In addition, this sample preparation technique is well suited for additional characterization techniques like high resolution morphology images provided by Electron Microscopy (STEM, TEM and SEM) but also chemical mapping (EDX). The present methodology could be of interest for a variety of systems, such as nanocomposites, copolymers, polymer alloys and hybrids, where the structure, morphology and physical properties can be related to the chemical composition depending on the processing conditions or before and after ageing, in order to design more efficient materials and optimize processes.

ACKNOWLEDGMENTS

The authors acknowledge the support from Nicolas Mollard and Pierre-Henri Jouneau (LEMMA-INAC-CEA Grenoble) for their collaboration in optimizing the vice used for the co-localization and the technical help with the complementary STEM images, and Ekaterina Pavlenko (SyMMES-INAC-CEA Grenoble) for her help with the treatment of the Raman mappings.

REFERENCES

1. B. H. Stuart, *Vib. Spectrosc.*, vol. 10, 1996, pp. 79–87.
2. U. Schmidt, S. Hild, W. Ibach, and O. Hollricher, *Macromol. Symp.*, vol. 230, no. 1, 2005, pp. 133–143.
3. R. Scipioni et al. *Membranes (Basel).*, vol. 4, no. 1, 2014, pp. 123–42.
4. R. S. Das and Y. K. Agrawal, *Vib. Spectrosc.*, vol. 57, no. 2, 2011, pp. 163–176.
5. B. Foster, *Am. Lab.*, vol. 44, no. 4, 2012, pp. 24–28.
6. D. Johnson and N. Hilal, *Desalination*, vol. 356, 2015, pp. 149–164.
7. D. M. Panaitescu, A. N. Frone, and I. C. Spataru, *Compos. Sci. Technol.*, vol. 74, 2013, pp. 131–138.
8. D. E. Martinez-Tong et al. *Compos. Sci. Technol.*, vol. 104, 2014, pp. 34–39.
9. B. S. Tanem et al. *Polymer (Guildf).*, vol. 44, no. 15, 2003, pp. 4283–4291.

Corn-Bran: Alternative Cellulosic Filler for Polypropylene

Buse Nur CAN, Aylin KUMAN, Cansu GOZUTOK, Guralp OZKOC*

Kocaeli University, Chemical Engineering Department, Kocaeli, 41380, Türkiye

busenurcan@gmail.com
**guralp.ozkoc@kocaeli.edu.tr*

Abstract. In this study, the use-ability of corn bran (CB) as a reinforcing filler in polypropylene (PP) matrix was investigated. The properties of CB/PP composites were compared with wood fiber (WF) filled PP composites. In order to enhance the interfacial adhesion, maleated-PP (PP-g-MAH) (MaPP) was used. The CB/PP and WF/PP composites were compounded in an Xplore laboratory compounder. The CB/PP/MaPP and WF/PP/MaPP composites were characterized by tensile test, differential scanning calorimetry (DSC), thermogravimetric analysis (TGA) and dynamic mechanical analysis (DMA). The morphology investigated by scanning electron microscopy (SEM). It was shown that corn bran can be an alternative candidate to be used as a reinforcing natural filler source for thermoplastic composites.

Keywords: Natural fibers, corn-bran, polypropylene, wood fiber, mechanical properties
PACS: 61.25.hk

INTRODUCTION

Recently, natural fibers from bio-renewable resources have attracted considerable interest from research community owing to their unique intrinsic properties such as biodegradability, availability, environmental friendliness, flexibility, easy processing and very high mechanical properties. Natural fibers have increasingly used as a reinforcement component in polymer matrix in order to add the specific properties to final product [1]. Fiber – reinforced composites have found a large in several areas of technical applications and have been used in many industries such as the aircraft, automobile, electronic and medical industries, due to their superb characteristic, such as low eight, high strength and stiffness [2]. On the other hand, in natural filler composites, weak adhesion may be result because of the poor dispersion and incompatibility between the hydrophilic naturel fillers and the hydrophobic polymer. The interfacial adhesion can be improved by using compatibilizer or coupling agents. Recent study suggests maleated polypropylene as a coupling agent that improves the fiber-matrix bonding significantly [3].

Corn bran is a food by-product made from the tough outer layer of corn. Like the brans derived from other grain crops, it is very high in fiber content, and it can be used in a wide variety of ways. Many commercial food producers use this substance as filler in their foods, and to reduce the calorific value of snack foods. On the other hand, it can be utilized as a cellulosic filler in thermoplastic composites [4].

The aim of this study is to investigate the applicability of corn-bran as an alternative reinforcing filler in polypropylene (PP) matrix and to compared with wood fiber-PP matrix with using maleated polypropylene as a coupling agent.

EXPERIMENTAL

Materials

Corn bran fiber was kindly obtained from Sunar Starch Company (Adana, Turkey). Wood fiber was provided by Kastamonu Entegre A.Ş. Company (Kastamonu, Turkey). Polypropylene used as a polymeric matrix material and maleic anhydrite used as a coupling agent were kindly supplied by EMAS Plastic A.Ş. (Bursa, Turkey).

Sieve Analysis

Corn bran fiber (CB) was separated via a sieve analyser into different particle sizes such as 125 and 500 µm. It was washed with ethanol solution and was dried for 24 hours at 50°C in an oven before processing. Besides, as a reference sample, wood fiber (WF) was separated with pore size of 500 µm. The resin of wood fiber was precipitated with toluene solution and the resin was stripped from wood fiber. Afterwards it was dried for 24 hours at 90°C in an oven.

Compounding and Injection

The CB/PP, CB/MaPP/PP composites were prepared in a laboratory twin-screw microcompounder (15mL Microcompounder, Xplore Instruments, The Netherlands) having 30,50 wt% loading level and with each of particle size of corn bran by one-step melt-compounding process at a screw speed of 100 rpm for 3 min at 200°C. In addition, WF/PP and WF/MaPP composites were prepared with the same process as a reference. The standards dog bone samples were obtained via Xplore model micro-injection moulding device for the mechanical and thermal analysis.

Testing and Characterization

The tensile properties were determined using an Instron Universal Testing machine following the procedure ISO 527. The crosshead speed of 25 mm/min was used.

Differential scanning calorimetry analysis (DSC) were run using a Mettler Toledo DSC-1 Star Model instrument at the temperature range of 25°C to 200°C with the heating rate of 20 °C/min under nitrogen flow.

Thermogravimetric analysis (TGA) was performed using a Mettler Toledo TGA1 instrument to determine the thermal stability of the blends under nitrogen atmosphere. Test samples were heated from the room temperature to 800 °C at a heating rate of 10°C/min.

Dynamic mechanical analysis (DMA) od composites were carried out using a dynamic mechanical analyser (MetraviB, DMA 25/50) at a heating rate of 3°C from 25°C to 150°C, while the testing frequency of 1 Hz.

The morphology of the composites was performed by scanning electron microscopy (SEM). Prior to analysis, the samples were coated via gold to prevent arching.

RESULTS

Table 1 shows the composition of the composites prepared in this study. The thermal degradation temperatures (shown in Table 2) of CB/PP was higher than neat PP. The addition of MaPP into CB/PP matrix provided better thermal stability with a smaller particle size of CB. There was no significant affect of CB or WF addition on melting points of CB/PP composites. However the crystallinity of the composites increased with CB loading level.

TABLE 1. The composition ratios of Corn Bran/Polypropylene Maleic Anhydride in different particular sizes

PARTICLE SIZES (µm)	MATERIAL COMPOSITIONS (wt%)		
	Corn Bran	Polypropylene	PP grafted Maleic Anhydride
	0	100	0
125	30	70	0
	50	50	0
	30	70	0
500	50	50	0
	0	100	2
125	30	70	2
	50	50	2
	30	70	2
500	50	50	2

TABLE 2. . Thermal degradation temperature of Neat PP, CB/PP%, WF/PP% and CB/MA-PP&, WF/MA-PP composites

Composites (wt%)	Temperature (°C)					Char Yield (%)
	T_{10}	T_{1st} derivation*	T_{50}	T_{2nd} derivation*	T_{90}	
Neat PP	470	435	455	-	457.3	2.5
30% CB-125 µm	475	280	375	395.0	276.7	3.3
50% CB-125 µm	670	280	425	433.2	278.2	7.7
30% CB-500 µm	470	365	455	458.2	274.0	7.5
50% CB-500 µm	0	270	440	499.0	277.8	11.8
30%WF	525	340	455	459.6	357.0	8.8
50%WF	418	320	385	495.0	400.0	0.7
30% CB-2% MaPP-125 µm	635	280	420	428.0	277.3	6.4
50% CB-2% MaPP- 500 µm	0	265	455	462.0	278.3	13.3
30% CB-2% MaPP- 500 µm	0	290	450	456.7	279.8	9.6
50% CB-2% MaPP- 500 µm	0	275	455	460.8	277.7	10.4
30% WF-2% MaPP	545	340	452	459.6	357.0	9.1
50% WF-2% MaPP	505	335	430	438.0	356.8	2.0

TABLE 3. The melting points and the percentages of crystallinity of of Neat PP, CB/PP%, WF/PP% and CB/MA-PP&, WF/MA-PP composites

Composites (wt%)	T_{m1}(°C)	T_{m2}(°C)	X_c(%)
Neat PP	164.1	171.6	36.7
30% CB-500 µm	-	168.4	37.9
50% CB-500 µm	-	164.5	42.5
30% WF	-	169.5	43.2
50% WF	166.8	170.8	56.8
30% CB-500 µm-2% MaPP	-	168.7	36.0
50% CB-500 µm-2% MaPP	-	171.4	56.0
30% WF-2% MaPP	-	171.3	46.4
50% WF-2% MaPP	168.4	172.3	52.1

Tensile strength at yield, elongation at break and elastic modulus of CB/PP composite at a lower concentration have better results when it is compared with WF/PP composites. On the other hand, although at a lower concentration, the addition of MaPP in the composites increased the tensile strength at yield, the elongation at break and elastic modulus of CB/PP composites decreased in a presence of MaPP.

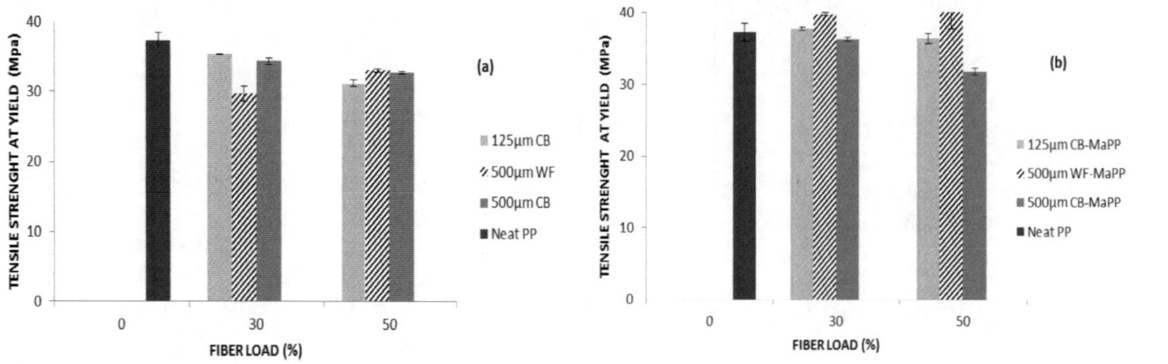

FIGURE 1. Tensile strength at yield of neat PP, (a) 30,50 wt% CB/PP and WF/PP , (b) 30,50 wt% CB /MaPP and WF /MaPP composites

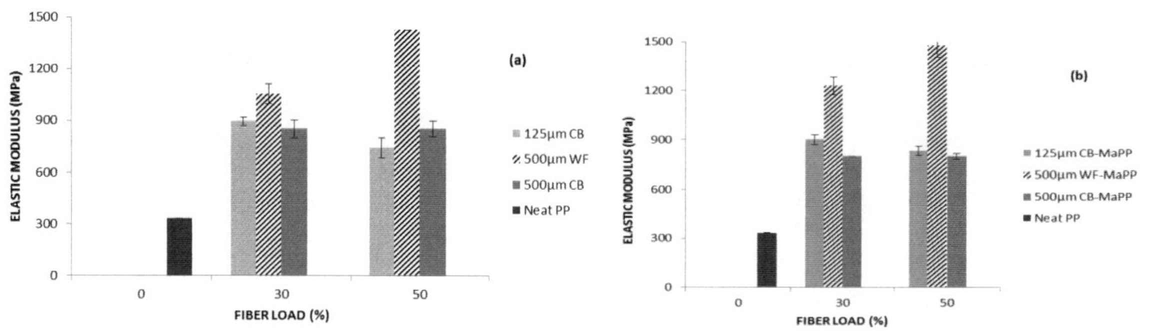

FIGURE 2. Elastic modulus of neat PP, (a) 30,50 wt% CB/PP and WF/PP , (b) 30,50 wt% CB/Ma-PP and WF/Ma-PP composites

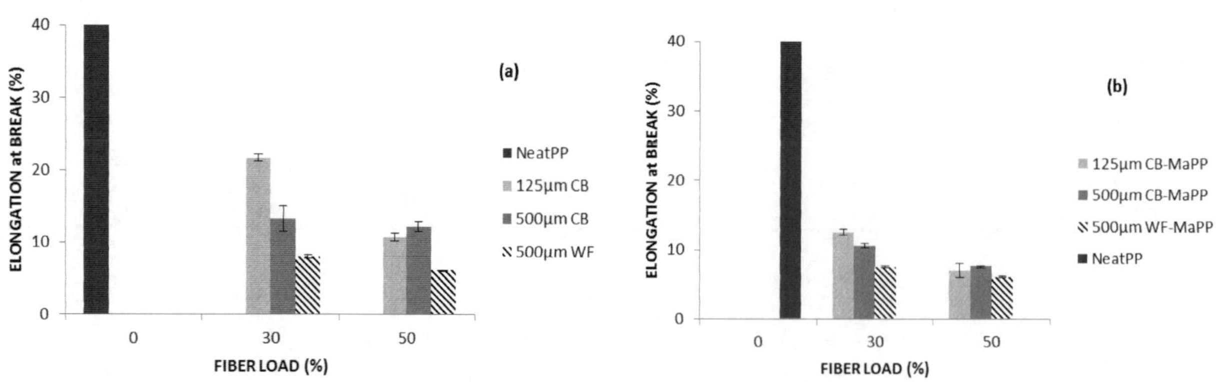

FIGURE 3. Elongation at break of neat PP, (a) 30,50 wt% CB/PP and WF/PP , (b) 30,50 wt% CB/Ma-PP and WF/Ma-PP composites

The DMA results were shown in Figure 4. It is clearly seen that the addition of any filler increased the thermo-mechanical resistance of PP. Maximum modulus was obtained at 50% fiber concentration regardless of filler type. With the addition of MaPP, the modulus of both 30 and 50 % of fillers improved. As the thermo-mechanical properties of CB and WF filled composites were compared, CB was found to be competitive at 30%, whereas at 50%, WF was found to be better than CB.

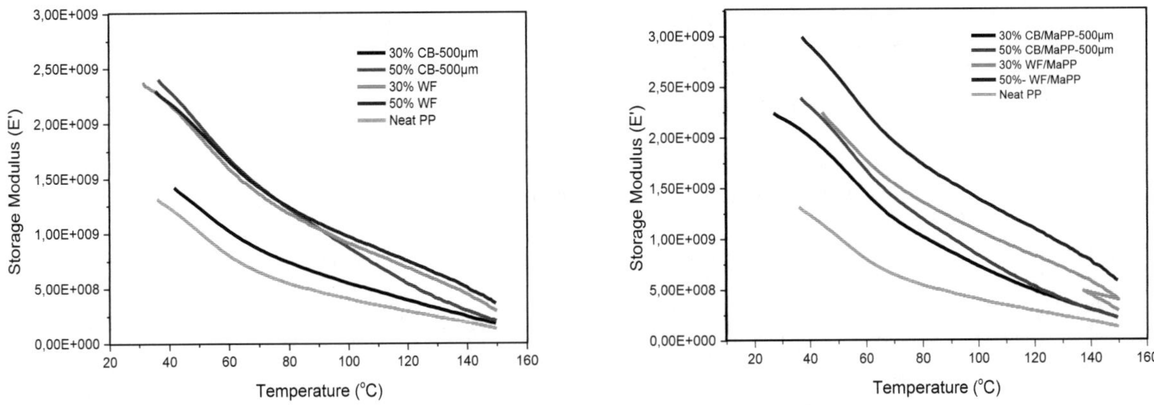

FIGURE 4. Storage modulus of neat PP, (a) 30,50 wt% CB/PP and WF/PP , (b) 30,50 wt% CB/MaPP and WF/MaPP composites

CONCLUSION

It was found that at lower concentration, CB was competitive to WF in terms of mechanical and thermal properties. The response of the fiber in the presence of coupling agent was dominating in the case of WF with respect to CB. It was also shown that CB is an alternative candidate to be used as a reinforcing naturel fiber source for thermoplastic composites.

REFERENCES

1. S.-J. Kim, J.-B. Moon, G.-H. Kim, C.-S. Ha, *Polymer Testing*, **27**, 801-806 (2008).
2. D.H. Mueller, A. Krobjilowski, *Journal of Industrial Textiles*, **33:2**, 111-130 (2003).
3. M.G. Salemane, A.S. Luyt, *Journal of Applied Polymer Science*, **100**, 4173-4180 (2006).
4. M. Marc, "What is corn bran?", Conjecture Corporation (2016)

Interfacial Strength of Twaron Fiber/PCL Composite Materials

Jieng-Chiang Chen[a*], Chang-Mou Wu[b], Yi-An Teng[c] and Jie-Mao Wang[c]

[a*]*Department of Product Design, Vanung University, Chungli, Taoyuan, Taiwan.*
[b]*Department of Materials Science and Engineering, National Taiwan University of Science and Technology, Taipei, Taiwan*
[c]*Department of Fiber and Composite Materials, Feng-Chia University, Taichung, Taiwan*

Presenting author: burnier@mail.vnu.edu.tw
Reference author: burnier@mail.vnu.edu.tw

Abstract. Twaron fibers were treated with H3PO4 solution to rough the surface of Twaron fibers. Three H_3PO_4 solution concentrations (1%, 10% and 30%) and treating time (1 min, 2 min and 3 min) were chosen to rough the fibers. Both treated and untreated Twaron fibers are reinforced polycaprolactone (PCL) matrix. Interfacial shear strength and tensile strength of treated and untreated were carried on a material testing machine (MTS 810)

The results show that the increasing H_3PO_4 solution concentrations and treating time decreased the tensile strength of the treated fibers. However, increasing H3PO4 solution concentrations and treating time increased the interfacial shear strength of the Twaron fibers/PCL materials. The optimal treatment for Twaron fiber with H3PO4 was 35% for 3 minutes. We also found that appropriate coating light-cured HV3 resin on the treated fibers enhanced the interfacial shear strength of Twaron/PCL materials.

Keywords: PCL, Twaron Fiber, Interface, Tension
PACS: 81.05.Qk, 81.70.Bt

INTRODUCTION

A fiber-reinforced composite (FRC) is a high performance composite material made up of three components – the fibers as discontinuous or dispersed phase, the matrix acts as the continuous phase, and the fine interphase region or the interface. The matrix is basically a homogeneous and monolithic material in which a fiber system of a composite is embedded. It is completely continuous. The matrix provides a medium for binding and holding reinforcements together into a solid. It offers protection to the reinforcements from environmental damage, and serves to transfer load.

Twaron fiber is an aromatic polyamide (Figure 1); developed by Teijin co. Aramid has a number of beneficial mechanical properties that make it such an important material in many different fields. Some of these are low density, high strength, good impact resistance, good abrasion resistance, good chemical resistance, good resistance to thermal degradation, and compressive strength similar to E-glass fibres. For these excellent properties of Twaron fiber, the fiber gets much interest both in research and industry. Some applications for aramid fibers are bullet proof vests, protective apparel such as gloves, motorcycle protective clothing, sails for sailboats, belts and hosing for industrial and automotive applications, aircraft body parts, and boat hulls. It is usually used as fiber reinforcement for polymer composites.

Polycaprolactone (PCL) is biodegradeable polyester with a low melting point of around 60 °C and a glass transition temperature of about −60 °C. In this study, we tried to make Twaron fiber reinforced PCL composites. First, we used H3PO4 to make acid treatment for fiber surface and tensile testing. After, fiber combined with PCL and test interfacial strength between fiber and resin. We, also, observed the influence of interfacial strength that used H_3PO_4 with different concentration and different soak time of fiber. Second, we coated light-cured resin HV3 on the surface of the fiber after acid treated. After cured, the fibers are tension on a MST machine to measure interfacial strength between fiber and resin.

EXPERIMENTAL

Materials and apparatus

A standard Twaron yellow filament yarn with 420 dtex in linear density and the PCL resin with 50000g/mole in number average weight were used in the current study. Sodium hydroxide (NaOH), Acetic acid (AA), and Phosphoric acid (H_3PO_4) were supplied by Wako Pure Chemical Industries, Ltd., SHIMAKYU Co. and, SHOWA Co. A Material Test System (MTS 810) was used to measure the tensile strength of fibers and the interfacial strength between the fibers and PCL resin. An Optical microscope was used to detect the surface of Twaron fiber before and after acid treatment.

Acid Treatment and Testing

Prior to acid treatment, the Twaron fibers should be purified to remove oil and impurities. Figure 2 shows the purification process of Twaron fiber before the fiber treated with acid solution. Twaron fibers was soaked in a 10% NaOH solution for 10 minutes with ultrasonic vibration. And then, the fibers were ultrasonically vibrated in a deionized water bath for 5 minutes. Finally, the fibers were dried in an oven at 60 ℃ for 24 hours. Figure 3 indicates the experimental procedure of strength estimation of Twaron fiber reinforced PCL composites while the fibers treated with acetic acid solution and phosphoric acid solution. It contains four steps, preparing acid solution, soaking Twaron fibers in acid solution, drying, and testing. First, three acetic acid solution named AA solutions (100%, 80%, and 60%) and three phosphoric acid solutions named H3PO4 solution (35%, 10%, and 1%) are prepared. Second, the Twaron fibers soaked into these solutions separately for 1min, 2min and 3min. Third, the acid treated Twaron fibers are dried in an oven at 60 ℃ for 24 hours. Finally, measure the tensile strength of Twaron fibers before and after acid treatment on an MTS machine. For measure the Twaron fiber-PCL matrix interfacial bond strength, single-fiber pull-out specimens were prepared. The fiber pull-out was performed at a constant speed of 0.2 μm/s from the matrix using controlled load cell. The pull-out force against displacement was recorded using a computer controlled plotter. The apparent shear strength τ_{IFSS}, was calculated from the maximum load F_{max} and the fiber embedded area in the matrix using the mathematical relation:

$$\tau_{IFSS} = \frac{F_{max}}{\pi dl} \quad (1)$$

Where l is the embedded length and d is the diameter of the fiber.

RESULTS AND DISCUSSION

Tensile Strength

Figure 4 shows that the surface of Twaron fiber became rough after acid treatment. The more acid treatment time, the much rough surface appeared. When the fiber surface damaged, it has low capacity to against tensile force. Figure 5 indicates that the tensile strength of Twaron fiber decreased with increasing the acid treated time and the concentration of acid solution as treated with phosphoric acid solution. For acetic acid treatment (Figure 5a), the effect of treated time on the decreased of tensile strength is quilt low while the solution with low (60%) acetic acid content. For higher acetic acid content, the tensile strength of fiber decreased with increasing treated time. In phosphoric acid (Figure 5b), the tensile strength decreased with increasing treated time in each concentration. This means that phosphate acid has a strong corrosive. The phosphate acid gradually destroys Twaron fibers with increasing soaking time. So, the soaking time and phosphoric acid concentration must be selected carefully.

Interfacial Strength

The results of Twaron fiber-PCL matrix interfacial bond strength are showed in Figure 6. The interfacial strength increased with increasing treated time and acid concentration. In general, the greater the concentration of acid solution the rougher the surface of Twaron fibers can be got. Likewise, the rougher surface of Twaron fibers can be achieved with the longer of treated time. The rougher surface increases the friction between fibers and matrix. So, higher acid concentration and longer treated time lead to higher interfacial strength of Twaron-PCL. This result is completely different from the result of tension strength. So, we must choice an optimum treatment time and acid concentration according to the tensile strength and interfacial bonding strength. Therefore, analysis of variance (ANOVA) was used to estimate the influence of acid treated time and acid concentration on the tensile strength and interfacial strength. From Table 1, for acetic acid treatment, 100% concentration and 3 minutes treatment is a proper experimental parameter. The optimal treated time and acid concentration is 3 minutes and 35% for the Twaron fiber soaked in phosphoric acid solution.

CONCLUSION

The surface of Twaron fibers become roughened after acid solution treatment. The longer the soaking and the higher the concentration of acid solution, the much roughness of the surface of Twaron fiber can be get. Thus, the roughness of fibers improves the interfacial strength between Twaron fibers and PCL matrix. It was found that, the proper soaking time is 3 minutes while the concentration of acetic or phosphoric acid solution is 100% or 35 %, respectively.

ACKNOWLDGMENT

The work was supported by the Ministry of Science and Technology of Taiwan, R. O. C. through the research grant MOST 103-2622-E-238-005-CC3. The authors with to thank all partners involved for the contributions: Shen-Jie Huang, Jie-Mao Wang, Hsiao-Chi Tsai and Ching-Shiun Wu.

REFERENCES

1. T Ramanthan et al., The use of a single-fibre pull-out test to investigate the influence of acidic and basic surface groups on carbon fibres on the adhesion to poly(phenylene sulfide) and matrix-morphology-dependent fracture behavior, Composites Science and Technology, Vol. 61, 12, 2011, pp. 1703-1710.
2. Caixia Jia et al., Effects of Twaron fiber surface treatment by air dielectric barrier discharge plasma on the interfacial adhesion in fiber reinforced composites, Surface and Coatings Technology, Vol. 204, Issues 21–22, 2010, pp. 3668-3675.
3. Xiangyi Zhang et al., Effects of plasma-induced epoxy coatings on surface properties of Twaron fibers and improved adhesion with PPESK resins, Vacuum, Vol. 97, 2013, pp. 1-8.
4. Masoud Fahool, Ali Reza Sabet, Parametric study of energy absorption mechanism in Twaron fabric impregnated with a shear thickening fluid, Inter. J. of Impact Eng., Vol. 90, 2016, pp. 61-7
5. Pieter J. de Lang et al. Controlled interfacial adhesion of Twaron® aramid fibres in composites by the finish formulation, Composites Science and Technology, Vol. 67, Issue 10, 2007, pp. 2027-2035.
6. J.L. Mena-Tun et al., Non-linear characterization of PP/Twaron laminates based on a model of plasticity with damage, Composite Structures, Vol. 93, Issue 11, 2011, pp. 2808-2816.
7. J.P. Mofokeng et al., Effect of layered silicates on the thermal stability of PCL/PLA microfibrillar composites, Polymer Testing, Vol. 50, April 2016, pp. 9-14.
8. Mahdi Forouharshad et al., A low-environmental-impact approach for novel bio-composites based on PLLA/PCL blends and high surface area graphite, European Polymer Journal, Vol. 70, 2015, pp. 28-36.

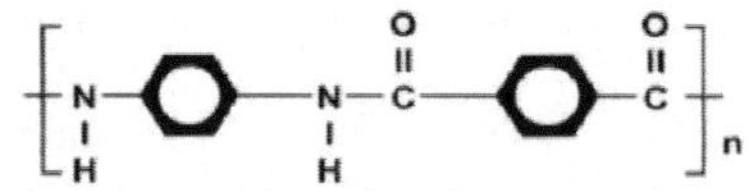

FIGURE 1. Molecular Structure of Twaron Fiber.

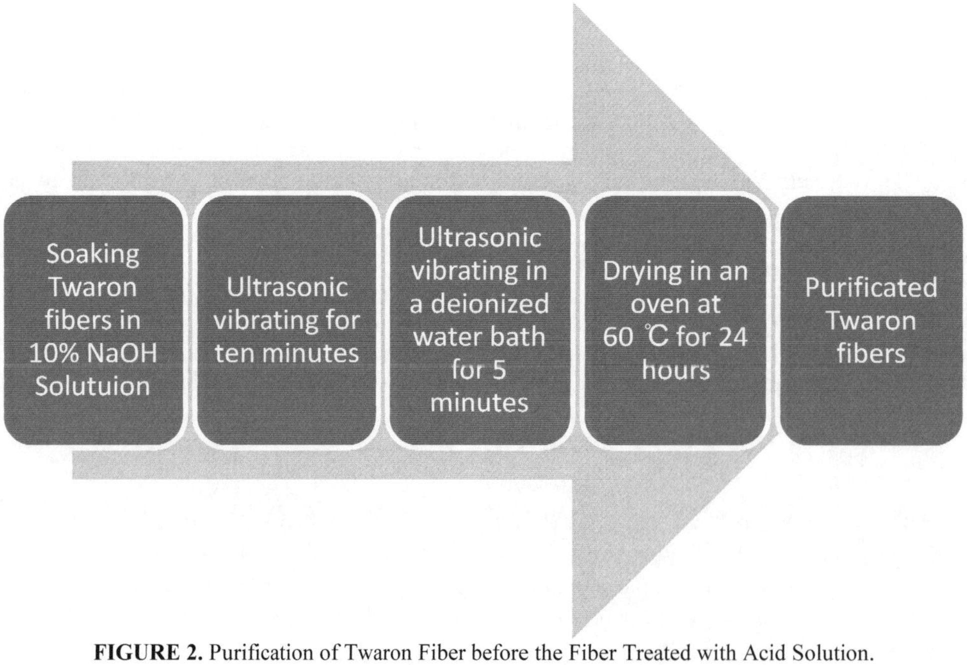

FIGURE 2. Purification of Twaron Fiber before the Fiber Treated with Acid Solution.

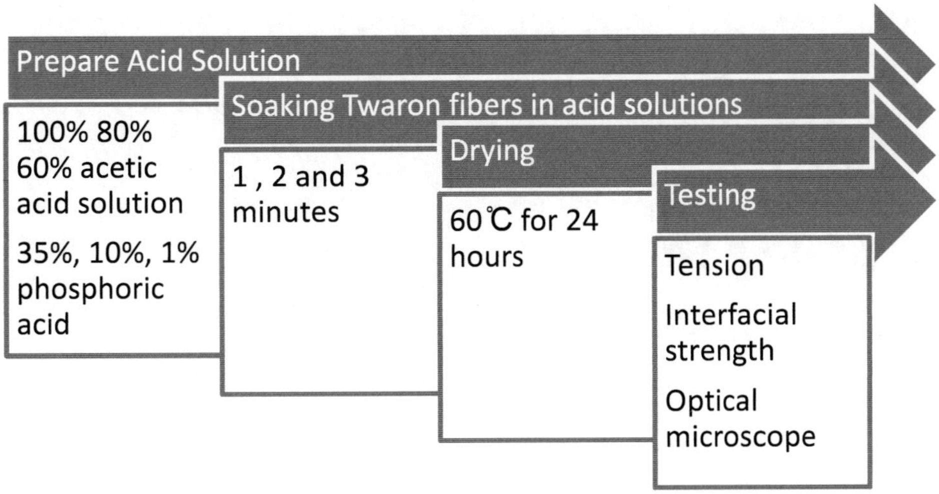

FIGURE 3. Experimental Procedure of Strength Prediction of Twaron Fiber Reinforced PCL Composites while the Fiber Treated with Acetic Acid Solution and Phosphoric Acid Solution.

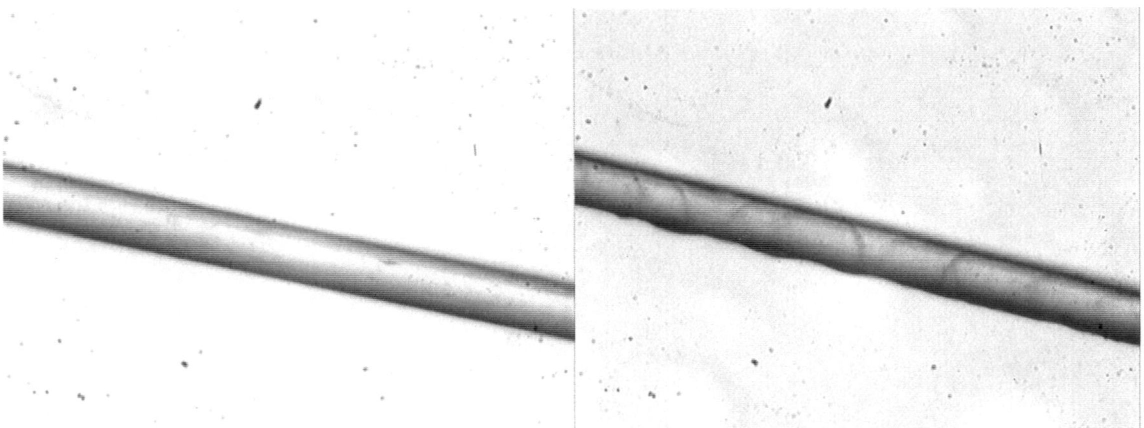

(a) (b)

FIGURE 4. Surface Images of Twaron Fibers before (a) and after (b) Acid Treatment

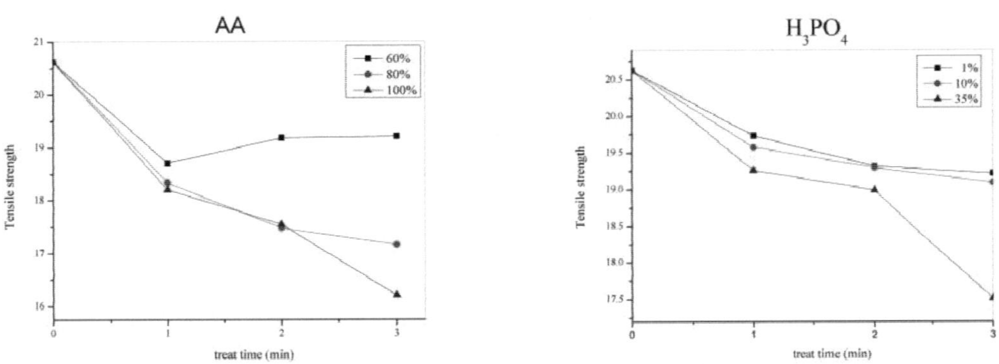

(a) (b)

FIGURE 5. Effects of Acid Solution Concentration and Treated Time on the Tensile Strength of Twaron fibers. (a) Acetic Acid Solution, AA (b) Phosphoric Acid Solution, H_3PO_4

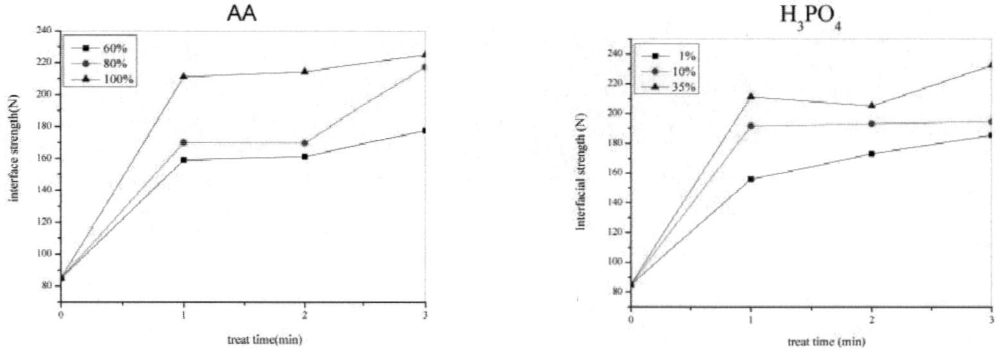

(a) (b)

FIGURE 6. Effects of Acid Solution Concentration and Treated Time on the Interface Strength of TwaronFiber/PCL Composite Materials.
(a) Acetic Acid Solution, AA, (b) Phosphoric Acid Solution, H3PO4

Table 1 Analysis of Variance of Acid Concentration and Treated Time to Tensile Strength and Interfacial Strength under Acetic Acid or Phosphoric Acid Solutions.

			Concentration(C)	**Treated Time(T)**	**C*T**	**Error**
Tensile Strength	Acetic Acid (AA)	SS	5326.178	1300.328	1917.844	6311.404
		DF	2	2	4	40
		MS	2663.089	650.164	479.461	157.785
		F	16.878	4.121	3.0039	-
		P	0.000	0.024	0.028	-
	Phosphoric Acid (H_3PO_4)	SS	1349.724	1402.691	805.248	2220.272
		DF	2	2	4	40
		MS	674.862	701.346	201.312	55.507
		F	12.158	12.635	3.627	-
		P	0.000	0.000	0.013	-
Interfacial Strength	Acetic Acid (AA)	SS	19639.900	6673.422	2421.522	24856.803
		DF	2	2	4	40
		MS	9819.950	3336.711	605.380	621.420
		F	15.802	5.369	0.974	-
		P	0.000	0.009	0.432	-
	Phosphoric Acid (H_3PO_4)	SS	15089.475	2648.081	1658.149	30032.237
		DF	2	2	4	40
		MS	7544.737	1324.041	414.537	750.806
		F	10.049	1.763	0.552	-
		P	0.000	0.185	0.699	-

Influences of Deposition Time of Arc Ion Plated Diamond-liked Carbon Films on the Polyamide 66 Fabric

C.P. Chin (presenting author)[a], K.B. Cheng (reference author)[a,b], C.C. Chang[a], J. Y. Liu[b], J.T. Chang[b], and F. L. Huang[b]

[a] *Department of Fiber and Composite Materials, Feng Chia University, Taichung 407, Taiwan*
[b] *Textile and Material Industry Research Center, Feng Chia University, Taichung 407, Taiwan*

m29615228@gmail.com
kbcheng@fcu.edu.tw

Abstract. An Arc Ion Plating (AIP) system was used to deposit diamond like carbon coatings onto polyamide 66 woven fabrics. Coatings with different thickness were obtained by controlling deposition time. The tensile strength, wear resistance, water contact angle, heat conductive coefficient, and stiffness of different coated fabrics were measured and compared. The results show that, depositing DLC coatings enhances the stiffness, hydrophobic property and heat conductive coefficient of polyamide 66 woven fabrics. The stiffness and thermal conductivity of fabric increased with increasing the coating thickness. The tensile strength of coated sample initially increases with deposition time. The sample of 30 min deposition exhibits a maximum tensile stress of 50.39 N/mm^2. The rubbing resistance of the sample of 10 min deposition is 212.2% higher than uncoated polyamide 66 woven fabrics. The DLC coatings effectively change the hydrophilic polyamide 66 fabric into hydrophobic. The sample of 40 min deposition shows a 17.21% in warp and 45.45% in weft stiffness improvement and better than uncoated fabrics. It also displays a 25.51% higher thermal conductivity and better than uncoated fabrics. However, the accumulated heat during deposition may cause deterioration of polyamide 66 fabrics. The coated polyamide 66 fabrics with optimized tensile strength and rubbing resistance are obtained between 10 to 30 min depositions. These results evidence the feasibility of applying DLC on polyamide 66 woven fabrics and the other polymeric fiber products for several potential applications such as home, Industrial textiles and packet, bag, box, mad, shoes, tent etc.

Keywords: Polyamide 66, diamond-like coated (DLC), arc ion plating (AIP).
PACS: 60

INTRODUCTION

Polyamide is the first commercially synthetic thermoplastic polymer. It exhibits several outstanding properties, such as low density, high mechanical strength and elasticity, better wear and chemical resistance. It is also easily dyed, and can absorb or desorb moisture. These advantages made polyamide become the most applied engineering polymer material. This silky material is similar to natural fiber materials, and suit for blended with other fiber materials to produce high quality composite fibers. Therefore, polyamide has been widely applied on producing fabrics, including cloth, wetsuit, bag, packet, bag, box, mad, shoes, tent, web, rope, strap, braid, convey belt, blanket, parachute, etc. which requires high tensile strength and high wear resistance. Using surface treatments to modify the surface properties of polyamide fiber product not only improves their performance and durability but also enables new functions and potential applications.

Diamond-like carbon (DLC) coatings are known to be amorphous and to mix "diamond" sp^3 with "graphitic" sp^2 bonding, are composed of carbon, hydrocarbon or a small amount of metal [1]. They exhibit well-known physical properties such as mechanical stability, high thermal conductivity, extreme hardness, low friction coefficient, high corrosion resistance, and biocompatibility [2-5]. There are some diamond-like carbon depositions on fabric materials has been reported, such as ion implantation [6,7], arc ion plating [8], plasma-enhanced chemical vapor deposition [9], to enhance the mechanical strength, flexible electric heating, and hydrophobic functions. Considering the low thermal stability of organic fabrics, the deposition on such polymeric materials must be processed at lower deposition temperature. Arc ion plating (AIP) has attracted great attention because of its low equipment cost, high ionization ratio, acceptable deposition rate, strong coating adhesion, and relatively lower deposition temperature. Therefore, in this study, the AIP technique is used to prepare carbon-based coatings on polyamide 66 woven fabrics. The influences of coating conditions on wear resistance, tensile strength, stiffness, thermal conductivity, and hydrophobic properties of coated woven fabric are evaluated.

EXPERIMENTAL PROCEDURES

The 100% polyamide 66 woven fabric is made by 40 denier yarns with a warp density of 124 epi and a weft density of 116 epi. Before depositing the DLC coating, the fabric surface was glow-discharge ion bombardment in C_2H_2 atmosphere of 1.1 Pa for 3 min to remove the residual contaminations and activate the fabric surface. After ion bombardment, the pressure of C_2H_2 atmosphere was lower down to 0.95 Pa, the titanium arc ion source was ignited. During the deposition, the C_2H_2 was decomposed and forming the carbon layer on fabric surface. The depositions were carried out for 10, 20, 30, and 40 min, respectively to obtain different film thickness. The detail of arc ion plating system can be referred to Wang's work [8].

The surface morphologies, tensile strength, stiffness, wear resistance, thermal conductivity, and water contact angle of different samples are compared. Both the tensile strength test and stiffness test were carried out in accordance with ASTM D1059-2001 standard. The wear resistance test is carried out by using a fastness tester. The thermal conductivity of fabric is measured by using a cool touch-feeling tester, KES-F7 Thermo Labo Ⅱ.

RESULTS AND DISCUSSIONS

The surface and cross-sectional morphologies of obtained diamond-like carbon coating on polyamide 66 fabrics are shown in **FIGURE 1**. Since the polyamide 66 fiber in fabrics is twisted in various tensile and angles, it is difficult to determine an exact coating thickness from fabric samples. The coatings exhibit a dense columnar structure. The estimated deposition rate is about 0.08 to 0.1 μm/min. However, it can be found that the coating thickness increases with the deposition time. From **FIGURE 2**, it can be seen that the appearances of fabrics changes from white to black with increasing the deposition time.

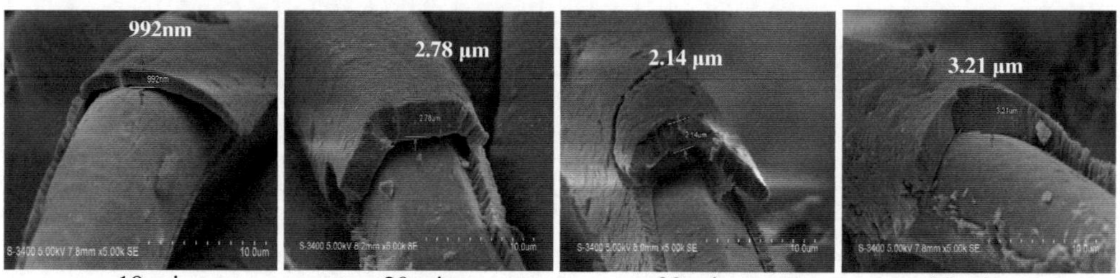

10 min 20 min 30 min 40 min

FIGURE 1. The surface and cross-sectional scanning electron microscope SEM images of samples obtained by 10, 20, 30, 40 min depositions

Uncoated 10 min 20 min 30 min 40 min

FIGURE 2. The appearance of polyamide 66 woven fabric coated with different thickness of diamond-like carbon

The tensile strength of fabrics was tested in accordance with ASTM D1059-2001 standard. The test results were summarized in **TABLE 1**. Only the sample of 30 min shows an enhanced tensile strength, the other coated samples show decreased tensile strength. The samples of 10 min and 20 min show higher tensile strain than the others. The sample of 40 min depositions displays a lowest tensile strength and strain. It could be attributed to the accumulate heat embrittles polyamide fiber. For other samples obtained by shorter deposition time, the thinner coatings with lower hardening effect cause lesser decrease of tensile strength and remain the internal sliding stretchability of fabrics. Therefore, shortening the processing time can suppress the thermal damage on fiber which reduced ultimate strength, and keeps the stretchability of woven fabrics.

TABLE 1. The tensile test results of uncoated and diamond-like carbon coated polyamide 66 fabrics

Deposition Time (min)	Ultimate Stress (N/mm^2)	Ultimate Strain (%)	Improvement Rate (%)
0	46.73	35.92	--
10	40.92	58.80	-12.4
20	43.08	56.64	-7.8
30	50.39	28.52	7.8
40	39.77	25.43	-14.9

The wear test results are shown in **TABLE 2**. Samples of 10 and 20 min deposition exhibit excellent improvement on rubbing resistance, especially the sample of 10 min. Samples with thick coatings show lower wear resistance than uncoated sample. It shows a similar trend as tensile strain. This reveals that adequate thin coating provides hard protection with remaining the stretchability of fabrics is benefit for rubbing resistance. For samples of long deposition time, the polyamide fiber could be embrittled due to accumulated heat. Furthermore, the hard debris chipped from coatings might enhance the abrasive wear on fabrics. It requires further investigation on comparing diamond-like carbon coatings on fabrics which are woven with different density, yarn and texture, to obtain appropriate coating parameters.

TABLE 2. The wear test results of uncoated and diamond-like carbon coated polyamide 66 fabrics

Deposition Time (min)	Rupture Rubbing Cycle	Improvement Rate (%)
0	32	--
10	100	212.0
20	80	150.0
30	30	-6.2
40	20	-37.5

The polyamide 66 is moisture absorbable material. Deposition the hydrophobic diamond-like carbon coatings can changes the moisture absorption behavior of polyamide fabric. It could be used as a water repellent treatment for outdoor fabrics. The measured water contact angle on uncoated sample and diamond-like carbon coated are listed in **TABLE 3**. All coated fabrics exhibit a water contact angle over 125°. It evidences that depositing hydrophobic diamond-like carbon material can make the water absorbable polyamide 66 fabric become hydrophobic.

TABLE 3. The water contact angles on uncoated and diamond-like carbon coated polyamide 66 fabrics

Deposition Time (min)	Water contact angle
0	0
10	128.14
20	125.76
30	127.29
40	125.32

For cloth usage, the fabric usually needs to be soft and comfortable touch feeling. For loading component usage, the fabric needs to be stiff and keep the dimensional stability. The stiffness test results of uncoated and coated sample are shown in **TABLE 4**. The stiffness of fabrics increases with the thickness of diamond-like carbon coatings. The stiffness improvement rate in weft is more significant than that in warp. It could be due to the higher density in warp. The higher warp yarn density provides a higher stiffness in warp direction already, thus contribution from hard coating is not as significant as that in weft direction.

The measured thermal conductivity results of uncoated and coated fabrics are shown in **TABLE 5**. Depositing high thermal conductivity diamond-like carbon material on polyamide 66 fabric does enhance the thermal conductivity of fabrics obviously. The thermal conductivity of coated fabric increases with the coating thickness.

TABLE 4. The stiffness test results of uncoated and diamond-like carbon coated polyamide 66 fabrics

Deposition Time (min)	Improvement Rate in Warp (%)	Improvement Rate in Weft (%)
0	-	--
10	3.27	17.0
20	13.11	35.2
30	14.75	39.8
40	17.21	45.5

TABLE 5. The thermal conductivity of uncoated and diamond-like carbon coated polyamide 66 fabrics

Deposition Time (min)	Average Thermal Conductivity (W/(cm • K))	Improvement Rate (%)
0	1.98	--
10	2.06	4.0
20	2.22	12.1
30	2.24	13.2
40	2.48	25.5

Based on these result, the surface enhancement of polyamide 66 fabric by utilizing arc ion plated diamond-like carbon is feasible. Depositing the hard, thermal conductive and hydrophobic diamond-like carbon is an effective method to modify the polyamide 66 fabric, which may improve the durability of current polyamide 66 products and extend the potential of polyamide 66 material.

CONCLUSIONS

In this study, to evaluate the applicability of diamond-like carbon coatings on artificial fiber and fabrics, carbon coatings with different thickness were successfully deposited onto polyamide 66 woven fabrics by utilizing arc ion plating technique. After comparing the tensile strength, stiffness, wear resistance, thermal conductivity, and water contact angle of coated and uncoated fabrics, several conclusions are summarized.

1. With extending the deposition time, the coating thickness increases with deposition time, and appearances of the fabrics turns from white to dark brown, then black.
2. The diamond-like carbon coating can enhance the tensile strength and stiffness of fabrics, but limits the stretchability of fabrics. The sample of 30 minute deposition exhibits a 7.8% improvement in tensile strength.
3. The sample of 40 minutes deposition exhibits a 17.21% in warp and 45.45% in weft improvement stiffness than uncoated fabric.
4. The wear resistance of sample of 10 minute deposition increases to 212% of uncoated.
5. The thermal conductivity of PA66 woven fabrics increases with the deposition time and thickness of diamond-like carbon coatings.
6. The diamond-like carbon coating turns the polyamide 66 fabrics from water absorbable into hydrophobic. All coated fabrics displays a water contact angles over 120°.
7. The coated polyamide 66 fabrics with optimized tensile strength and wear resistance are obtained between 10 to 30 min depositions. These results reveal the potential of applying diamond-like carbon coating for enhancing

the durability of polymeric fabrics and products. However, the heat accumulated during the long deposition time might damage the polyamide fibers and their strength.

8. Through altering the deposition parameters, DLC-coated polyamide 66 woven fabrics could be used on several potential applications such as home, Industrial textiles and packet, bag, box, mad, shoes, tent etc.

ACKNOWLEDGMENTS

The authors would like to thank Ministry of Science and Technology of Taiwan for financial support under the contract of NSC 101-2632-E-035-002-MY3.

REFERENCES

1. M. Sedlaček, B. Podgornik, J. Vižintin, *Mater. Charact.* **59**, 151 (2008).
2. J. Robertson, *Mater. Sci. Eng. R*, **37**, 129-281 (2002).
3. R. Hauert, *Diam. Relat. Mater.*, **12**, 583-589 (2003).
4. F. Platon, P. Fournier, S. Rouxel, *Wear*, **250**, 227-236 (2001).
5. B. K. Gupta, B. Bhushan, *Wear*, **190**, 110-122 (1995).
6. N. Kitahara, T. Sato, H. Isogawa, Y. Ohgoe, S. Masuko, F. Shizuku, K.K. Hirakuri, *Diam. Relat. Mater.* **19**, 690-694 (2010).
7. K. Baba, R. Hatada, *Thin Solid Films*, **506– 507**, 55 – 58 (2006).
8. J.T. Wang, C.C. Hsu, C.M. Chen, J.L. He, *Thin Solid Films*, **528**, 247–254 (2013).
9. D. Caschera, B. Cortese, A. Mezzi, M. Brucale, G. M. Ingo, G. Gigli, G. Padeletti, *Langmuir*, **29**, 2775–2783 (2013).

Design And Simulation Of Conformal Cooling Channels In SLM Produced Injection Molds: Effects on Process and Product Quality

W.Six[a*], R. Appermont[b], F. Desplentere[a], J. Dekeyzer[b] and A. Van Bael[b]

[a]*KU Leuven Technology Campus Oostende, Zeedijk 101, 8400 Oostende, Belgium*
[b]*KU Leuven Technology Campus Diepenbeek, Wetenschapspark 27, 3590 Diepenbeek, Belgium*

Presenting author: wim.six@kuleuven.be
Reference author: wim.six@kuleuven.be

Abstract. In the injection molding process, the cooling step is the most inefficient and time consuming phase. Moreover, the cooling of the product often leads to defects such as unexpected warp due to unevenly distributed cooling in the mold. These conditions can be attributed to the classical manufacturing techniques for injection molds as drilling, milling and EDM. The major drawback of these techniques is their inability to produces 3D curved cooling channels, preferably parallel to the product wall. In the present study the Selective Laser Melting (SLM) 3D printing technique is used to overcome these limitations. Via numerical simulations and experimental measurements on existing molds, the original situation is characterized. Important parameters as average mold surface temperature, temperature gradients, cycle time, time to reach steady state condition, etc.. are determined In this study mold inserts are developed in order to obtain conformal cooled core and cavity inserts, taking into account constraints of conformal cooling as well as SLM printing such as shape, max diameter, overhangs, …. Thermal and Moldflow simulations are combined to optimize the design of the conformal cooling channels in the core and cavity inserts. Also CFD simulations were used to optimize the flow distribution and pressure drop in the cooling channel itself. With these simulations, improvements on the process (times and temperatures) and on the product will be obtained. With the predictions out of the simulations an economical study was made to determine the profits resulting from the conformal mold.

Keywords: injection molding, conformal cooling, process optimization, SLM printing
PACS: PACS: 83.50, 83.80, 81.20Hy

INTRODUCTION

Producing polymer products by the injection molding process is a well-known technique discussed in many papers and technical magazines. Most interesting for OEM's producing polymer parts is the efficiency of the production. In many cases product design, material and process-settings are optimized but the tempering of the mold is left out. In many cases it can be addressed to a lack of knowledge of cooling and the restrictions of the currently used techniques such as drilling, milling, and EDM to design molds. This results often in undesired warp deformation, long cooling times, etc.… This is often remedied by additional adjustments that increase the cost of the mold and cause a delayed production. However nowadays, new 3D printing techniques as Selective Laser Melting (SLM) make it possible to design 3D curved cooling channels. In that way an optimal cooling geometry can be designed. The study in this paper describes the redesign of an existing mold which produces sorting trays. In this study, restrictions of both techniques, conformal cooling and SLM printing are taken into account.

CONFORMAL COOLING

Conformal cooling is a technique that refers to the cooling channels that follow the shape of the surface of the injection mold thus creating a uniform, conformal surface temperature of the mold cavity. Because the cooling channels are placed close to the cavity surface, it is possible to reach a steady state condition in only one cycle. i.e. $\tau < t_{cycle}$, as illustrated in figure 1.

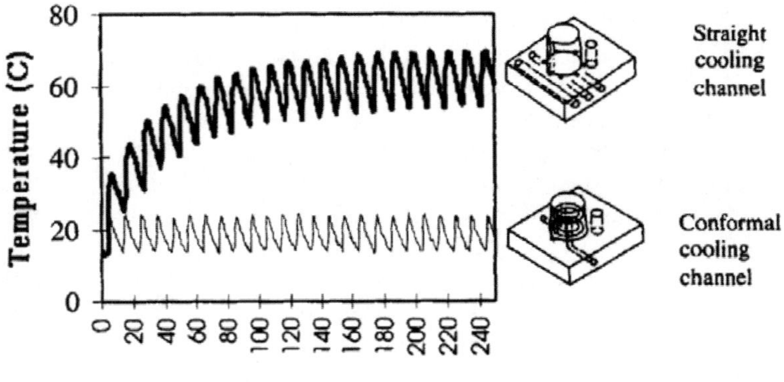

FIGURE 1: Thermal start up behavior of conventional and conformal mold. [1]

Design Rules For Conformal Cooling [2]

To design a conformal cooled mold, literature sets some ground roles, resulting the design of unit cooling cell, based on a geometry with two cooling channels. The most important parameters are the diameter of the channel D, the distance between the channels W and distance form channel to the mold wall l_m. The nomenclature is shown in figure 2.

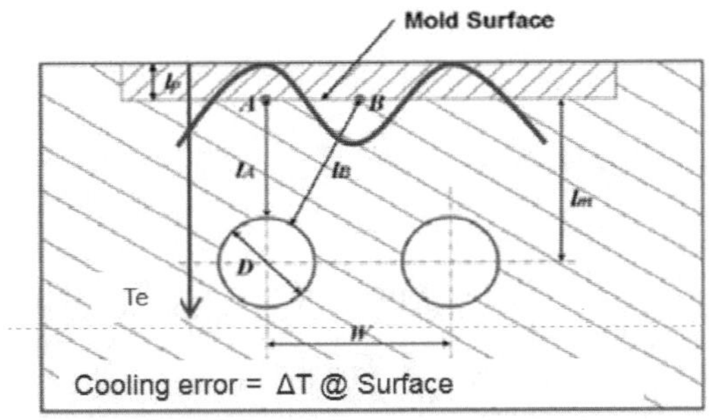

FIGURE 2: simplified cooling cell [2]

Thermal Constraints

In the design guide for injection molding design, the range for l_m is about one to two times the diameter D, i.e. D < l_m < 2D This rule makes it possible to have only a small temperature variation between mold wall and channel wall. These ΔT should be lower than 5°C. i.e. $T_{water} - T_{wall} \leq 5°C$. To obtain a conformal cooled surface also some restrictions are set on the max temperature variations on the mold surface. The temperature deference between point A and point B should be in a small range. i.e. $\overline{T_B} - \overline{T_A} \leq 10°C$ To achieve this restriction the distance between the cooling channels (W) should be between 2 and 3 times the diameter of the cooling channel. i.e 2D < W < 3D

Pressure Drop and Mechanical Stiffness

Next to the thermal constraints, also some other important constraint as pressure drop has to be considered. If a mold consist of various parallel sections within the cooling geometry, an equal pressure drop over the channels is advisable. This can be calculated by:

$$P = \frac{L}{2D}\rho_c V^2 C_f \text{ where } C_f = \frac{0.25}{1.8^2}\left(\log\left(\left(\frac{e}{3.7D}\right)^{1.11} + \frac{6.9}{R_{eD}}\right)\right)^{-2} \text{ for turbulent flow } (R_{eD} > 2500) \quad (1)$$

Also pressure drop should not go over 3bar to avoid overloading the tempering unit. As a last constraint, the mechanical stiffness of the mold should still be high enough. Due to the many cooling channels created in the mold, the resistance of the surface can be affected by the high fatigue loading inherent to the injection moulding process. This can be easily controlled by executing some stiffness simulations. The maximum allowed stress level is dependent on the used mold material.

SELECTIVE LASER MELTING (SLM) [3]

SLM is a 3D metal printing technique, in which the produced part is build up, layer by layer. Within every step a new layer of metal powder is applied. Next, a well-controlled laser beam melts the material on specific locations, dependent on the produced part. This is shown in figure 3(left)

Materials used in this technique are rather limited. In many cases it is limited to variations of steel, titanium, aluminum and stainless steel. The powder particle size is between 10 and 60μm. The speed and the accuracy of the process is dependent on the printer itself and its parameters. The most important parameters are layer thickness and laser intensity. In contrast to the Selective Laser Sintering (SLS) technique, SLM printed products have a relative density of nearly 100% and have the same properties as solid mold blocks. Also the SLM printed parts can be hardened, which is a must for injection molds. Other advantages are the possibility to create free shaped products and cooling channels. However, some constraints need to be taken into account.

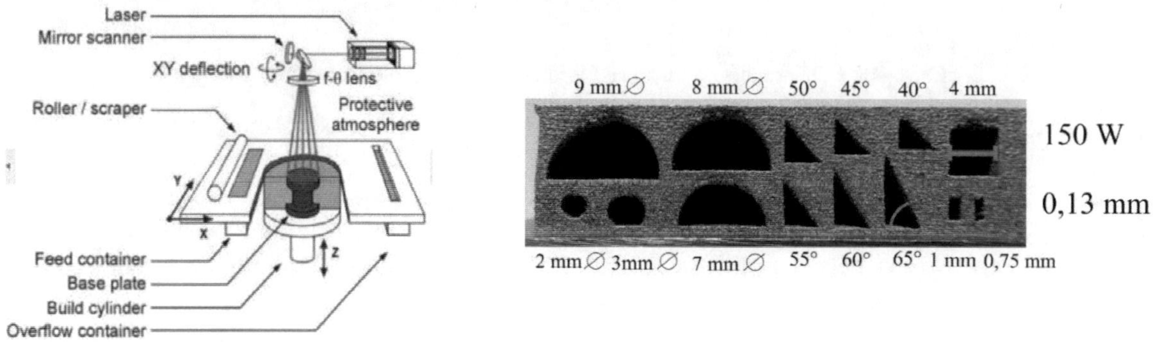

FIGURE 3: left: working principle of SLM [3], right: quality of SLM printed channels

The first constraint is the inability to print large overhangs. In that way it is impossible to print large channels, the maximum allowed channel diameter is 6mm, and large overhangs should be supported with a support angle of 45°. The quality of the openings is shown in figure 3(right). When designing the channels, the possibility to remove the unused powder should be taken into account. Next to this, SLM printing is a very expensive process, this is due to the low printing speed (5-20cm³/h), The design should be made for minimum printed volume.

PRODUCT AND MOLD

The product for the case study is a sorting tray typically produced with PP material and has a wall thickness of 1.5 mm. The original cooling geometry exist out of 2 cooling channels with diameter ø8mm and 2 bubblers in the core with diameter a ø16mm. The conformal designed mold has channels of ø4.5mm. The distance from the surface $l_m = 4.5mm$ and the spacing $W = 9mm$

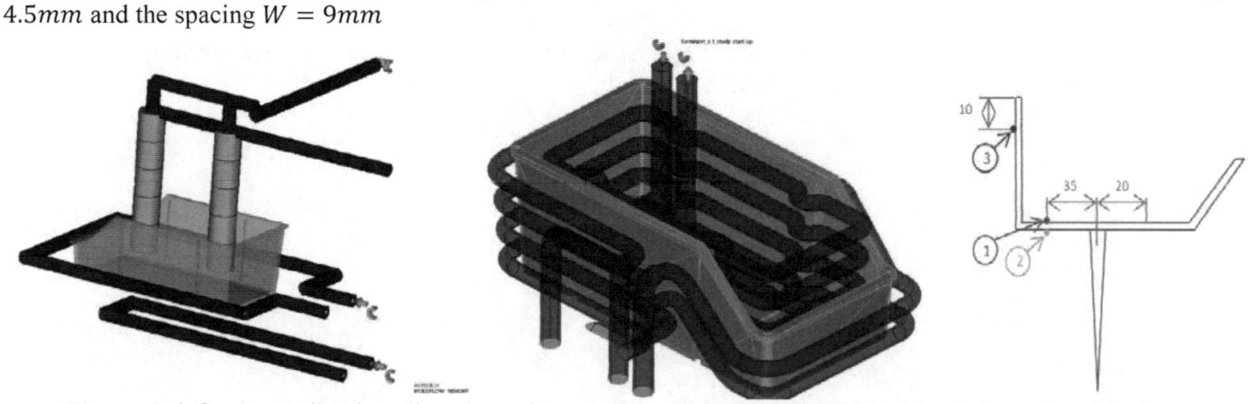

Figure 4: left: conventional cooling channels, center: conformal cooling channels, right: censor locations

NUMERICAL SIMULATIONS

With Autodesk Moldflow Synergy 2016, transient from production start-up simulations were made of both conformal and conventional cooling situation. Also, simulations during the cycle itself were made. the most important results are discussed in the section below.

Start Up Losses

Due to the inefficient cooling of the conventional mold, simulations predict a startup of at least 13 cycles, this is mainly caused by the large distance between the mold surface and the cooling channels. This is shown in figure 5.The simulation software also predicts a large difference in temperature between core and cavity. Al the simulated temperatures are much higher than the desired averaged mold wall temperature of 30°C. In the conformal situation, the startup only takes 1 cycle to get a steady state situation. Furthermore, all average temperatures are the same and very near to the desired mold wall temperature. The sensor locations are shown in figure 4 right

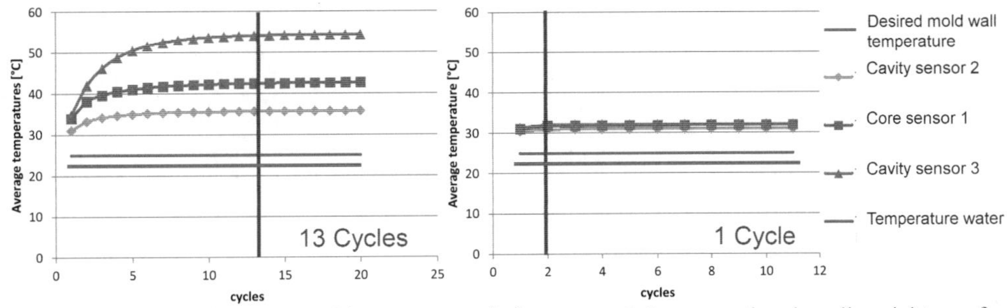

FIGURE 5: prediction of average mold temperature during startup left: conventional cooling right: conformal cooling

Temperature During The Cycle

Also in the cycle itself, it is clear that there is a large temperature difference in core and cavity and even on the surfaces of the cavity itself. In the cycle itself a temperature difference of 20°C is monitored, while in the conformal situation only a variation of 2.87°C is measured on the same nodes. This is shown in figure 6.The lower mold temperature also leads to a shorter cooling time of 4s instead of 9,5s, or a reduction of cooling time by 60% and of total cycle time by 30%. In that way the shorter cooling time increases the production rate form 850Kpcs/y to 1.2MPcs/y

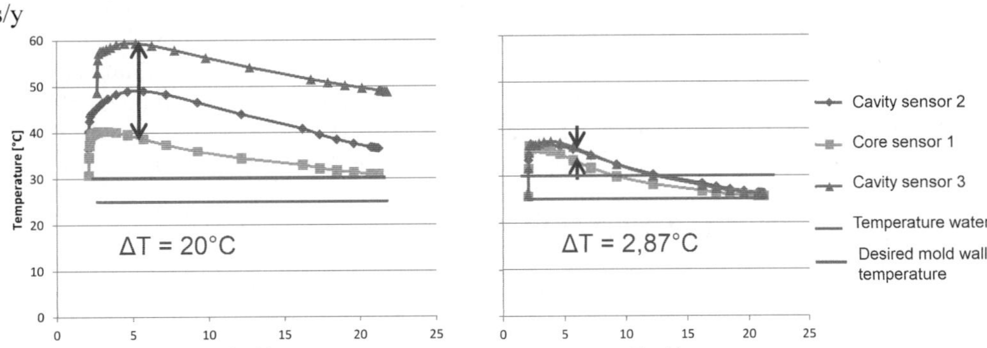

FIGURE 6: predicted temperature variation during the cycle left: conventional cooling right: conformal cooling

Warpage

Due to the inefficient cooling in the core of the conventional mold, undesired warp occurs. The higher temperature in the core causes the box to collapse in the inwards direction, as shown in figure 7. This makes stacking of the trays very difficult. The simulation of the warpage, predicts a inwards movement of 0.8mm for each side of the box. The warp prediction of the conformal mold, shows a large reduction of the warpage., The product even deforms a little bit in the opposite direction, which is beneficial for the stacking.

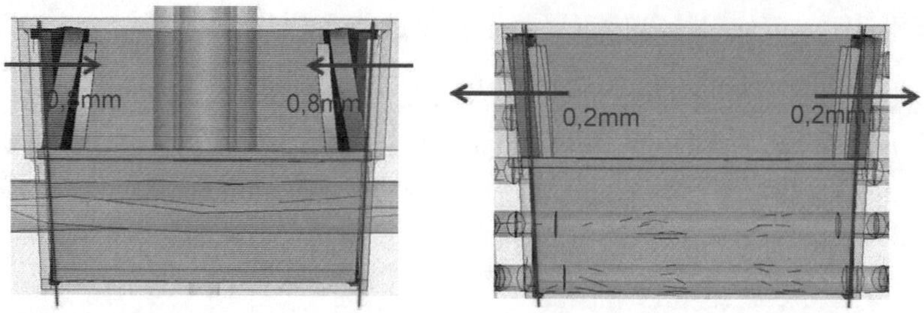

FIGURE 7: predicted warpage, left: conventional cooling, right: conformal Cooling

ECONOMICAL PROFIT

For injection molders, the economical profit is one the most important factor to use conformal, SLM printed molds. Even with a more expensive mold (35K€ as opposed to 25K€ for the conventional mold), conformal cooling with SLM sintering makes an economical viable alternative. This is shown in table 1.

TABLE 1: cost comparison between a one cavity conventional en conformal mold

	conventional	**conformal**
Pcs/year	850000	1225000
	€/pcs	€/pcs
machine	0,1049	0,0728
material and other	0,0208	0,0208
mold	0,0089	0,0062
total	0,1454	0,1106
Benefit /pcs	0,0145	0,049
benefit/year [€]	12325	60025

CONCLUSIONS

Using conformal cooled, SLM printed molds have, as simulations show, many benefits compared to the conventional system. Simulations predict a decrease of 5.5s in cooling time and an equal temperature in cavity and core. This results a lower warpage. Another benefit is the disappearance of start-up losses Also, simulations can have a major benefit while designing the perfect mold. It helps to predict major problems in the existing molds as hotspots, flow problems and the resulting process parameters. In most cases the set ground rules for conformal cooling are sufficient to design a conformal cooled mold. As a last point, it can be proven that this technique has financial benefits

ACKNOWLEDGMENTS

The authors wish to thank the Flemish government for the funding in the IWT TETRA140320 project. The results discussed in the present paper where obtained within this project.

REFERENCES

1. *The Design of Conformal Cooling Channels in Injection Molding Tooling.* **Xu, Xiaorong, Sachs, Emanuel and Allen, Samuel.** 2001. pp. 1265-1279.
2. *Automatic Design of Conformal Cooling Circuits for Rapid Tooling.* **Wang, Yu, et al.** 2011.
3. **Creaghs , Tom.** selectieve laser melting. [Online] national instruments. [Cited: 10 2, 2015.] http://sine.ni.com/cs/app/doc/p/id/cs-13103#prettyPhoto[gallery]/0/.

Experimental Studies and Numerical Simulation of the Foam Injection Molding Process using Gas-counter Pressure

Vahid Shaayegan[a], Chul B. Park[a*], Franco Costa[b], and Sejin Han[c]

[a] *Microcellular Plastics Manufacturing Laboratory, Department of Mechanical and Industrial Engineering, University of Toronto, 5 King's College Road, Toronto, Ontario, Canada M5S 3G8*
[b] *Autodesk, Inc. 259-261 Colchester Rd., Kilsyth, VIC. 3137, Australia*
[c] *Autodesk, Inc. 2353 North Triphammer Road, Ithaca, NY 14850, USA*

Presenting author's email: park@mie.utoronto.ca
Reference author's email: park@mie.utoronto.ca

Abstract. The mechanism of cell nucleation and growth in foam injection molding with gas-counter pressure is investigated through molding experiments and numerical simulation. The molding data includes the visualization of cell nucleation and growth during foam injection molding as well as cavity pressure data. It was found that the application of gas-counter pressure suppresses the cell nucleation during filling. A faster cavity pressure decrease, and consequently a faster nucleation, was detected using gas-counter pressure. A finite element numerical simulation of the foam injection molding process with gas-counter pressure was performed and compared with experimental results. The cell nucleation process was calculated using the fitted nucleation model. Both the predicted cavity pressure profiles and the predicted cell density were in good agreement with the experimental results.

Keywords: Foam Injection Molding, Gas-counter Pressure, Cell Nucleation, Numerical Simulation

INTRODUCTION

Recently, foam injection molding (FIM) has attracted increasing attention in manufacturing industries owing to its ability to produce light-weight parts with high geometrical accuracy. A blowing agent such as carbon dioxide (CO_2) or nitrogen (N_2) is mixed with a polymer melt inside the injection barrel to produce a homogeneous melt/gas mixture before injection. This homogeneous single-phase melt/gas mixture is under a high pressure inside the barrel, well above the solubility pressure of the blowing agent. The pressure on the melt/gas mixtures drops quickly below the solubility pressure of the dissolved gas upon entry into the cavity; consequently some cell nucleation can occur immediately after passing through the gate, leading to non-uniform cell distributions. A uniform microcellular structure is sought to yield improved mechanical properties [1-7], improved electrical and charge storage capability [8,9], as well as improved heat and sound insulation [10-12].

Foam injection molding can be performed in short-shot or full-shot manners. In short-shot injection, also called low-pressure FIM, the mold cavity is only partially filled by the initial injection of melt/gas mixture, and the polymer flow caused by expansion of the nucleated cells fills out the remaining mold cavity. The pressure drop required for cell nucleation and foaming in this scenario is obtainable at the gate. Specifically, the resistance of the gate provides the high pressure required to keep the blowing agent in solution in the melt/gas mixture. Once passing the gate, the pressure drops to the ambient pressure and nucleation occurs. Although the obtainable void fraction is high in low-pressure FIM, the cellular structure is non-uniform within the molded part. In addition, the cells are coalesced and elongated as a result of a simultaneous cell growth and cell travel during injection. Lee et al. proposed strategies to minimize these drawbacks, and improve the uniformity of the cellular structure, in low-pressure FIM by keeping the cavity pressure below the solubility pressure of the blowing agent during injection [7].

In a full-shot injection, also called high-pressure FIM, the entire mold cavity is filled with the melt/gas mixture prior to significant cell growth. The material shrinkage during cooling determines the cell density and the void fraction obtainable. In this full-shot injection approach, the cavity pressure is much higher than in the short-shot injection method, resulting in slower cell nucleation during mold filling. In addition, the presence and expansion of nucleated cells reduces the rate of the obtainable pressure decrease. Shaayegan et al. suggested that a higher pressure drop rate will be obtained during cooling if the cells nucleated during the filling stage are removed from the system, and so suggested the application of a packing pressure sufficiently high to re-dissolve the nucleated cells back into the melt. Consequently, a higher a pressure drop rate is achievable by the reduced melt/gas compressibility [13].

Despite the FIM method, achievement of a uniform micro- or nano-sized cell structure is still very challenging. Non-uniformity in the structure, severe coalescence, and swirl marks are commonplace with conventional processes. In this context, numerous technologies and methods have been developed to improve the cellular structure in FIM [14-22]. Examples of this are the addition of an extra accumulator to separate the mixing and injection stages [14, 15], MuCell technology [16, 17], static mixers [18, 19], modification of the gas-injection nozzle location [20], and application of an autoclave instead of the hopper [21, 22]. However, cell nucleation still occurs during the mold filling stage in all these processes.

In order to improve the uniformity of the cellular structure and the surface quality of the foam products in FIM, a gas-counter pressure (GCP) can be used [23, 24]. In this technology, the mold cavity is charged with an inert gas such as N_2 before injection. It is essential in this process that the cavity pressure remains at pressures higher than the solubility pressure of the blowing agent during filling. The melt/gas mixture is injected into the mold cavity while the GCP is being exhausted from vents. The opposing pressure of the GCP prevents the nucleation of cells during injection, thereby decoupling the filling and nucleation stages.

In this research, an in-situ visualization method has been employed to analyze the mechanism of cell nucleation and foaming in FIM using GCP. A *near* full-shot injection is employed.

EXPERIMENTAL AND METHODOLOGY

A visualization mold was designed and manufactured which enabled us to monitor the entire foaming phenomena occurring inside the mold cavity. A fused-silica prism is incorporated into the stationary plate of the visualization mold to reflect the cavity images to a camera. The visualization mold is also equipped with a number of cavity pressure transducers at different locations of the mold cavity. The visualization mold design is detailed elsewhere [25]. A 50-ton Arburg ALLROUNDER 270/320C injection molding machine was used to perform FIM experiments. The injection molding machine is equipped with MuCell technology in order to inject the blowing agent into the injection barrel. The mold cavity (rectangular shape with nominal dimensions of 135*111*3.2 mm) was fed with a fan gate. A CV M10 camera from JAI equipped with a magnifying lens from Navitar was used to record the visualization movies.

A Polystyrene, MC3650 from Americas Styrenics (with MFI 13.0 gr/10min; 200°C/5 Kg; ASTM D-1238) was blown using CO_2 from Linde Gas Canada. The GCP medium was N_2 from Linde Gas Canada. The GCP was applied by a control module of Caropreso Associates, MA, US. A near full-shot was used to carry out all experiments. The overall visualization line and the principle of the visualization mold and prism action are shown in Figures 1 and 2.

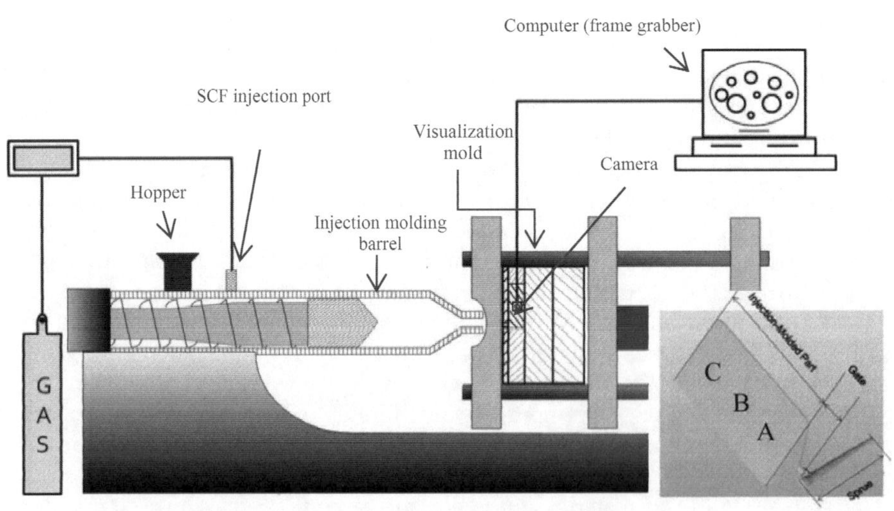

FIGURE 1: Schematic of the FIM visualization line. Different locations of the foamed part are named A to C with respect to the gate location.

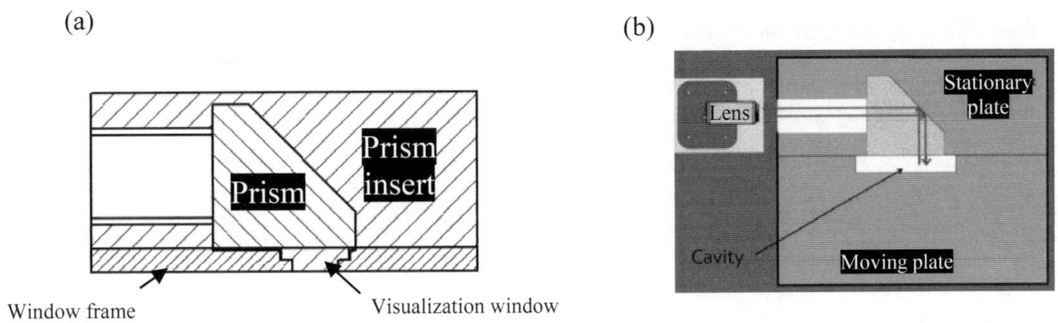

FIGURE 2: (a) Sectional view of the prism insert; (b) Principle of the prism action

To measure the cell density with respect to the unfoamed volume (N'), the volume expansion (VER) was multiplied by the cell density with respect to the foamed volume (N_{foam}) using the following equations: [26]

$$N'(\mathrm{t})= N_{foam}(\mathrm{t})\times VER(\mathrm{t})$$

$$N_{foam}(\mathrm{t})= \left(\frac{N(t)}{A}\right)^{3/2}$$

$$VER(t)= 1+\left[\left(\frac{4}{3}\pi \sum_{1}^{m}\frac{R_i^3(t)}{m}\right) N_{foam}(\mathrm{t})\right]$$

EXPERIMENTAL AND METHODOLOGY

The melt front advancement in FIM experiments of polystyrene (PS) blown with 1% CO_2 at 230 °C is shown in Figure 3a. As the visualization snapshots show, the melt front is associated with a myriad of cells, nucleated at the gate and during the mold filling. This melt front appeared as a foggy front. On the other hand, using GCP resulted in a smooth and clear melt front (Figure 3b), where no nucleated cells were observed. This transparent melt front indicates that the opposing pressure of the GCP effectively suppressed the cell nucleation during cavity filling.

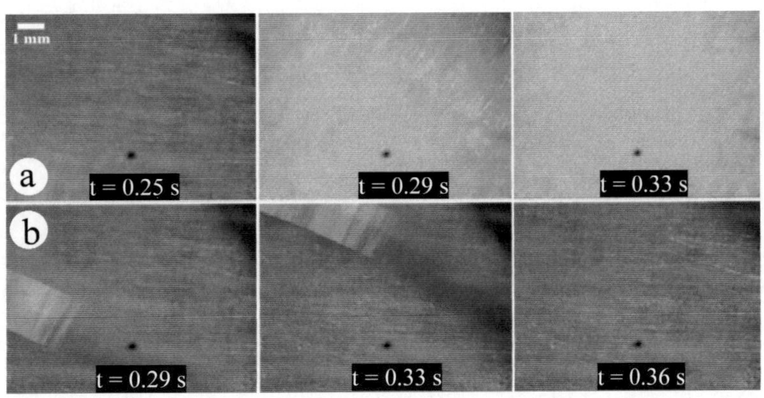

FIGURE 3: Filling in FIM experiments of PS-1%CO_2: a) No GCP; b) GCP.
(Injection speed = 35 cm^3/s at 230 °C; Location A; t = 0 s refers to the moment melt/gas mixture enters the mold cavity).

Figure 4 shows the recorded cavity pressure with and without the application of GCP. By comparing Figure 4a and 4b, it is readily noticeable that the obtained pressure drop rate is much higher by the sudden removal of the GCP in the near full-shot FIM using GCP. While the cavity pressure decays slowly in the near full-shot FIM experiment without GCP (Figure 4a), for the GCP case it remains at a minimum pressure level, dictated by the GCP which is greater than the solubility pressure of the dissolved gas, and dropped rapidly after the removal of the GCP (Figure 4b). Consequently, a faster cell nucleation rate and a higher cell density were achieved with the application of the GCP as shown in Figure 4c.

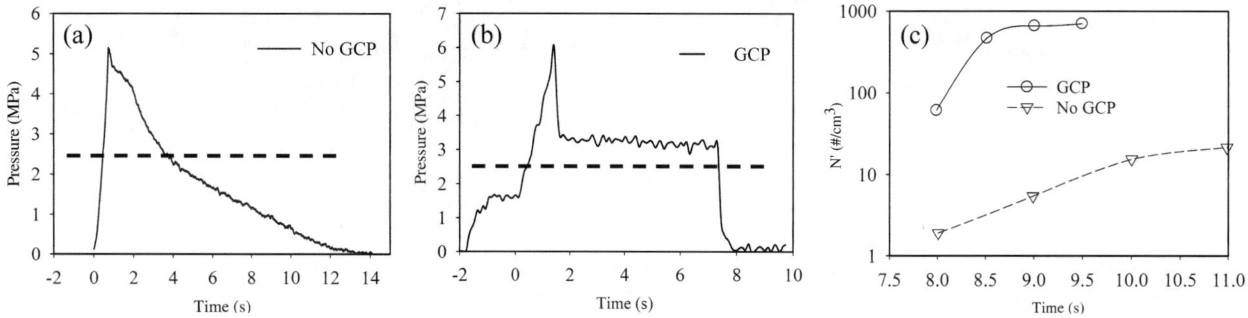

FIGURE 4: (a) Cavity pressure profile in near full-shot FIM without GCP; (b) cavity pressure in near full-shot FIM using GCP; (The solubility pressure of the dissolved gas is shown with dashed lines). (c) Cell density (PS-1%CO_2 foaming system; Injection speed = 50 cm^3/s, T_{melt} = 230 °C, Location A).

Numerical Simulation

A three-dimensional finite element numerical simulation of FIM with GCP has been performed based on the Autodesk Molding Insight software package. The basic simulation method is described in [27] including the fitted nucleation model used to predict the cell nucleation. The simulation used a finite element mesh to represent both the part cavity as well as the feed system. The filling control was switched from flow rate (velocity) control to pressure control near the end of filling. The gas counter pressure was maintained at approximately 3.2 MPa for about 6 seconds. Figure 5a shows a comparison of the pressure traces at Location A obtained from the experiment and the simulation. Location A is in the part cavity and is about 38 mm from the gate. During filling, the pressure values obtained from the simulation were higher than those from the experiment. During post-filling, the values from the simulation and the experiment agreed reasonably well. Figure 5b compares the experimental and calculated cell density at Location A. As shown, the cell nucleation started when the GCP was removed during the post-filling stage. Overall, a reasonable agreement between the experimental and the simulated results was observed.

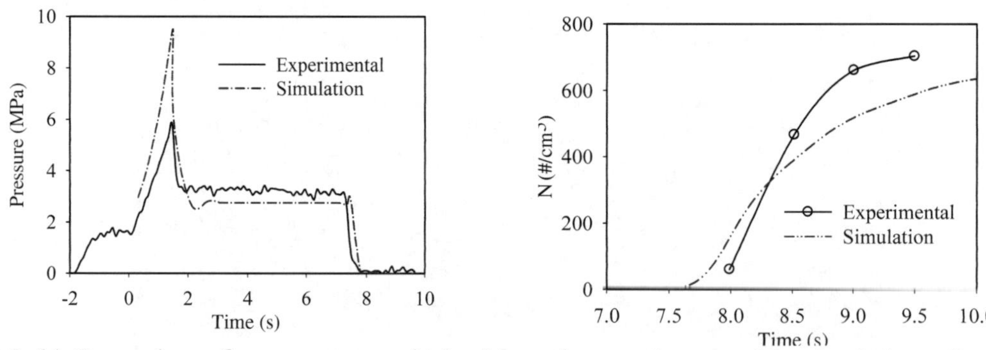

FIGURE 5: (a) Comparison of pressure traces obtained from the experiment and the simulation at Location A; (b) Comparison of cell density traces obtained from the experiment and the simulation at location A.

CONCLUSION

Application of gas-counter pressure in foam injection molding was investigated. It was observed that an effective gas-counter pressure can successfully prevent cell nucleation during the mold filling. The use of GCP decouples the filling and nucleation stages. By sudden removal of the gas-counter pressure, a rapid pressure drop rate is achievable, which results in a higher cell nucleation rate and higher cell density. The simulated results were in good agreement with the experiment, and can be used to design and predict manufacturing processes using gas-counter pressure.

ACKNOWLEDGMENTS

This research was financially supported by the Natural Science and Engineering Research Council of Canada. The authors also thank the Consortium of Cellular and Micro-Cellular Plastics (CCMCP) for their support. The polystyrene was provided by Americas Styrenics, TX, United States.

REFERENCES

1. M. Yuan, L.S. Turng, S. Gong, D. Caulfield, C. Hunt, and R. Spindler, *Polym. Eng. Sci.*, **44**, 673-686 (2004).
2. S. Wong, J.W.S. Lee, H.E. Naguib, and C.B. Park, *Macromol. Mater. Eng.*,**293**, 605-613 (2008).
3. J.W.S. Lee, C.B. Park, and S.G. Kim, *J. Cell. Plast.* **43,** 297-312 (2007).
4. J.W.S. Lee and C.B. Park, *Macromol. Mater. Eng.* **291**, 1233-1244 (2006).
5. S. Pilla, A. Kramschuster, S. Gong, A. Chandra, and L.S. Turng, *Int. Polym. Proc.*, 22, 418-428 (2007).
6. A. Kramschuster, S. Gong, L.S. Turng, T. Li, and T. Li, *J. Biobased Mater. Bioenergy* **1**, 37-45 (2007).
7. J.W.S. Lee, J. Wang, J.D. Yoon, and C.B. Park, *Ind. Eng. Chem. Res.* **47**, 9457-9464 (2008).
8. A. Ameli, M. Nofar, C.B. Park, P. Pötschke, and G. Rizvi, *Carbon*, **71**, 206–217 (2014).
9. A. Ameli, S. Wang, Y. Kazemi, C.B. Park, and P. Pötschke, *Nano Energy* **15**, 54–65 (2015).
10. P. Gong, P. Buahom, M. Tran, M. Saniei, C.B. Park, and P. Pötschke, *Carbon*, **93**, 819–829 (2015).
11. D. Jahani, A. Ameli, P.U. Jung, M.R. Barzegari, C.B. Park , and H. Naguib, *Mater. Des.* **53,** 20–28 (2014).
12. S. Ghaffari Mosanenzadeh, H.E. Naguib, C.B. Park, and N. Atalla, *J. Mater. Sci.*, **50**, 1248-1256 (2015).
13. V. Shaayega, G. Wang, C.B. Park, *European Polymer Journal*, **76**, 2-13 (2016).
14. C.B. Park, X. Xu, Apparatus and Method for Advanced Structural Foam Molding. U.S. Patent Application 11/219,309, filed Sep 2, 2005; Canada Patent Application A2,517,995, filed Sep 2, 2005; International Patent Application PCT-1304865, filed Aug 30, 2006.
15. X. Xu, C.B. Park, J.W.S. Lee, X. Zhu, *J. Appl. Polym. Sci.* **109**, 2855-2861 (2008).
16. J. Xu, D. Pierick, D. Microcellular Foam Processing in Reciprocating- Screw Injection Molding Machines. J. Injection Molding Technol. 2001, 5 , 152.
17. D. Pierick, J. Anderson, S.W. Cha, L. Chen, J. Stevenson, D. Laing, D. Injection Molding of Polymeric Material. U.S. Patent 6,884,823 (2005).
18. S. Habibi-Naini, Foam-Injection Molding the Flexible Way. Tech. Re V . Sulzer 2004 , 86 , 10.
19. S. Habibi-Naini, Mold Technology for Optifoam, Blowing Agents and Foaming Process, Stuttgart, Germany, May 10-11(2005).
20. W. Michaeli, S. Habibi-Naini, Foam Injection Molding: The Search for a Better Method Mod. Plastic Intl . 2004 ,34 (3), 38–39.
21. W. Michaeli, L. Florez, D. Obeloer, and M. Brinkmann, *J. Cell. Plast.* **45,** 321-351 (2009).
22. W. Michaeli, T. Krumpholz, and D. Obeloer. Profoam, A New Foaming Process for Injection Molding, In: Proceedings of the 66th Annual Technical Conference of the Society of Plastics Engineers, Milwaukee, Wisconsin, USA (2008).
23. S.C. Chen, P.S. Hsu, and S.S. Hwang, *J. Appl. Polym. Sci.*, **127**, 4769-4776 (2013).
24. S.C. Chen, W.H. Liao, and R.D. Chien, *Int. Commun. Heat Mass,* **39**, 1125-1131 (2012).
25. V. Shaayegan, L.H. Mark, A. Tabatabaei, and C.B. Park, *eXPRESS Polym. Lett.,* **10**, 462-469 (2016).
26. A. Wong, S. N. Leung, G.Y.G. Li, and C.B. Park, *Ind. Eng. Chem. Res.* **46**, 7107-7116 (2007).
27. S. Han, F. Costa and L. Kishbaugh, *SPE ANTEC Tech. Papers*, March 23-25 (2015) Orlando, FL, USA.

Injection Molding of Micro and Nanostructured Surfaces from Liquid Silicone Rubber

Marisely De Jesus Vega[a], Stephen Johnston[a], Joey Mead[a] and Carol Barry[a*]

[a]*Department of Plastics Engineering Department, University of Massachusetts Lowell One University Ave, Lowell MA, 01854*

**carol_barry@uml.edu*

Abstract. Liquid silicone rubber can be processed very similarly to thermoplastics using injection molding. This material offers several advantages to the micromolding industry such as very low viscosity, biologically inert, high temperature resistance and rapid curing, yet, there is a lack of research on injection molding of microstructures from liquid silicone rubber. This work consists on the design of mold and tooling inserts for the production of micro and nano structured surfaces. PhothＯlithography and E-beam lithography were the methods selected for tooling fabrication. The lack of accuracy between the design and the measured feature sized does not hinder the usability of these tooling in the molding process because tooling was produced with high precision on the feature dimensions.

Keywords: liquid injection molding, liquid silicone rubber, microstructures, nanostructures.
PACS: 81.16.Rf

INTRODUCTION

Cast polydimethylsiloxane (PDMS) is the preferred material for exploratory research and development of new devices with microstructured surfaces [1, 2]; but, it is not practical for manufacturing due to the labor intensive and time consuming nature of the casting process. PDMS' low tensile moduli (usually < 2 MPa) also limits the minimum dimensions that can be molded to the microscale range due to mechanical instabilities that result in collapsing of features in the submicron range [3], or lack of replication fidelity [4]. High-rate manufacturing of microstructures and nanostructures is possible with thermoplastic materials such as polymethylmethacrylate (PMMA), polystyrene (PS), polycarbonate (PC), and cyclic olefin copolymers [2, 4, 5] among others. These materials have been used for the production of devices for high-throughput screening of analytes [2], optical communication and data transmission [6], self-cleaning adhesives mimicking gecko feet [7], and superhydrophobic surfaces [8, 9]. Injection molding has been widely studied for fabrication of these devices and the behavior of these materials is very well understood. Melt viscosity [10-14] and solidification time [10] are the most important factors controlling good replication. Depending on mold design, packing pressure or injection velocity are also important [10, 15]. These materials, however, lack the chemical resistance, flexibility and extended service temperature range of PDMS, which are required for certain applications.

Liquid silicone rubber (LSR), developed at the end of the 1970s [16], with characteristics similar to PDMS, but higher tensile modulus (4-12 MPa) [17] and the processability of thermoplastics materials. Research for devices with microstructured surfaces can greatly benefit from the LSR outstanding properties and fast processing. Novel applications recently developed in research laboratories using PDMS can be produced from LSR to meet the high-rate manufacturing requirements and become commercially significant. More complex 3D-geometries can be produced with injection molding rather than casting, adding countless possibilities for additional developments. LSR parts with microstructures have been replicated for electrode human implants [18], bioelectrodes [19], superhydrophobic surfaces [20], and dielectric electroactive polymer (DEAP) transducers [21]. The performance of LSR when molding microstructures, however, remains unclear with only a few studies focusing on this material [18-20]. Zhang et al. [18] investigated 4-μm-diameter pillars with spacing and aspect ratios of 0.5:1; processing conditions, however, were not discussed. Kaitainen et al. [19] focused on impedance measurements of 80-μm-diameter pillars with an aspect ratio of 1:1 and spacing ratios of 0.8:1 and 4:1; again, processing conditions were not discussed. Hopmann and coworker [20] did report the microinjection molding processing conditions of LSR for 10 and 20-μm-diameter pillars with aspect ratios of 0.25:1 to 1.5:1 and spacing ratios of 1.6:1 and 1.3:1. Although the best replication resulted from high injection velocity, high melt temperature, and low mold temperature, quantification on the replication was not presented. Moreover, in all previous LSR features were in the microscale range and sub-micron scale features have not been investigated.

This work investigated the effects of processing conditions and feature size on LSR micropillars with a consistent aspect ratio of 1:1. A mold was designed and built to hold tooling inserts with the microscale features. Tooling inserts were designed and fabricated with micropillars have diameters of 100 to 0.1 μm and spacing ratios of 1:1. The effect of processing conditions was evaluated during molding and optimized molding parameters were obtained for these features.

EXPERIMENTAL

Mold Design

The B-plate of a mold was designed to work with an A-plate available and suitable for the application required. The A-plate only contained a flat surface to provide a flat part. The B-plate was designed to have a vacuum port, an opening for a thermocouple, and space for holding a mold insert (to hold the tooling insert with microscale features). Drawings of the B-plate design and the mold insert are shown in Figure 1a. The mold insert was designed to hold a 63 mm by 50 mm tooling insert, which is the part with the microstructured surface. This mold design facilitated fast changing of tooling insert for production of parts with different microstructures. The gate location was selected to provide impingement flow of the microstructures and the fan gate provided more uniform flow of the LSR over the microscale features. The B-plate and mold insert were machined from tool steel.

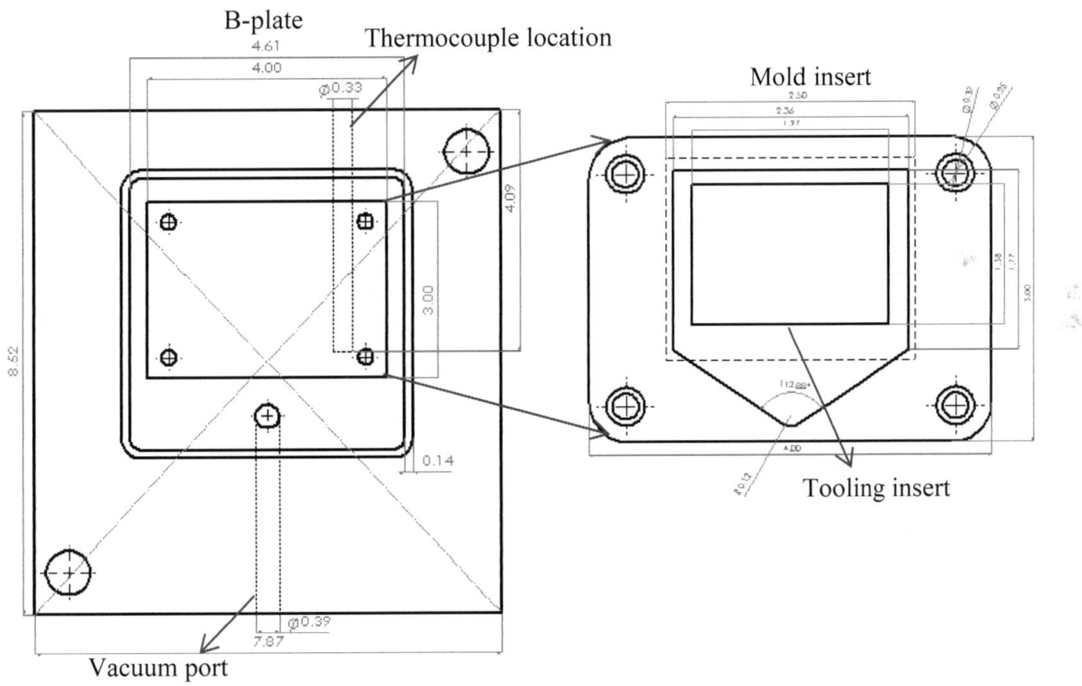

FIGURE 1. Mold design schematics showing the B-plate, its components, and the mold insert.

Tooling Design and Fabrication

The tooling inserts contained the microscale and nanoscale features and were fabricated in silicon wafers using lithographic processes. The designed micro and nanofeatures were an array of holes (to produce pillars) of different dimensions with aspect and spacing ratios of 1:1 (Table 1). Depending of the size of the features, the pillars were fabricated using two different lithographic processes: photolithography for feature sizes of 10 μm and greater and e-beam lithography for feature sizes below 10 μm. For features sizes of 10 μm and above, first, the silicon wafer was spun coated with photoresist (Microposit S1813) and soft baked for 1 min at 115 °C before placing it in a mask writing tool (Heidelberg's Mask Writer Micro PG 1). At this point a master mold was created by softening the designed pattern on the photoresist using the mask writer, removing the softened photoresist in a developer bath (MF-26A Megaposit) for certain period of time, and hard baking the remaining photoresist at 180 °C for 10 minutes. Then, deep reactive-ion etching (DRIE) at cryogenic conditions (-120 °C) was performed to etch the pattern on the silicon wafer using inductively coupled plasma (Oxford Plasma Lab 380). Times for all these steps was optimized depending on the size of the features. Finally, the residual photoresist was removed in a remover bath (Microposit Remover 1165) until visually clean. Further cleaning was performed using reactive ion etching (RIE) (Oxford Plasma Lab 80) for two minutes. Similar steps were followed for E-beam lithography but using polymethylmethacrylate as the photoresist and Raith150 as the mask writing tool.

The dimensions for each tooling were measured by optical profilometry (Wyko NT2000) to obtain the depth of the features and optical microscope (Nikon L200ND) to measure the diameter and spacing between features. Perfluorooctyl-trichlorosilane (PFOTS) was applied as an anti-stiction coating because it had been shown to improve filling and ejection of thermoplastics [22]. An additional tooling was produced (with about 50 μm features as well as Tooling 2) to study the effect of anti-stiction coating, this tooling was not coated. Then, the silicon wafers were cut by cleaving and polished to fit the mold insert.

TABLE 1. Designed Tooling Dimensions

Tooling	Diameter (μm)	Depth (μm)	Spacing (μm)
1	100	100	100
2	50	50	50
3	10	10	10
4	1	1	1
5	0.5	0.5	0.5
6	0.1	0.1	0.1

RESULTS AND DISCUSSION

The machined B-plate is shown in Figure 2a and a sketch of the resultant part is shown in Figure 2b. The white surface in the B-plate shows the opening for the tooling insert. The sprue location selected and the microstructures array location are shown in the part (Figure 2b). A high temperature silicone O-ring was inserted in the groove in order to keep the cavity evacuated while molding.

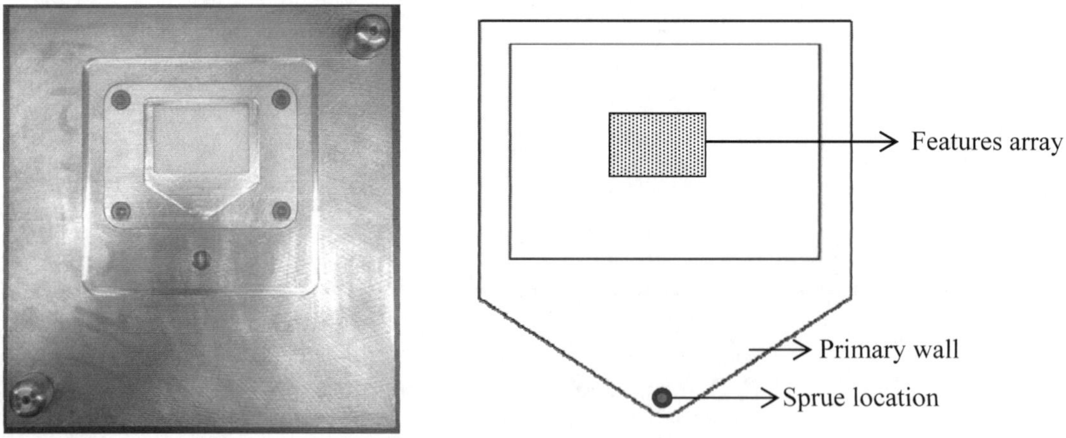

(a) (b)

FIGURE 2. Liquid injection molding (a) fabricated mold and (b) part to be produced.

The measured feature dimensions of the tooling inserts differred from the designed dimensions as shown in Table 2. These differences were a result of variability in the lithographic process, especially in the developing and etching steps. This lack in accuracy and precision on the dimensions of the tooling resulted in aspect ratios and spacing ratios that were not the designed 1:1. Aspect ratios as low as 0.81:1 were obtained for the tooling with a targeted diameter of 100 μm, whereas aspect ratios as high as 1.70 were obtained for the smallest feature size. In contrast, the smallest spacing ratio obtained was for the tooling with a targeted diameter of 50 μm and the greater spacing ratio for the smallest diameter feature as well. In general, the depths were greater than targeted for all samples, except for the features with 100 um of diameter; and diameters were greater than targeted, except for the smallest diameter features tooling. Optical microscope images for tooling produced are presented in Figure 3. Although the dimensions were not as expected, high precision can be seen with great unifomity in the diameter and spacing of the features.

TABLE 3. Designed and Measured Tooling Dimensions

Tooling	Designed Diameter (μm)	Designed Depth (μm)	Designed Spacing (μm)	Measured Diameter (μm)	Measured Depth (μm)	Measured Spacing (μm)	Resultant Aspect Ratio	Resultant Spacing Ratio
1	100	100	100	113.0	91.1	86.7	0.81	0.77
2	50	50	50	64.3	72.9	35.6	1.13	0.55
2-nc	50	50	50	52.0	83.1	48.7	1.60	0.93
3	10	10	10	8.0	13.7	11.9	1.70	1.48

nc = no anti-stiction coating was applied

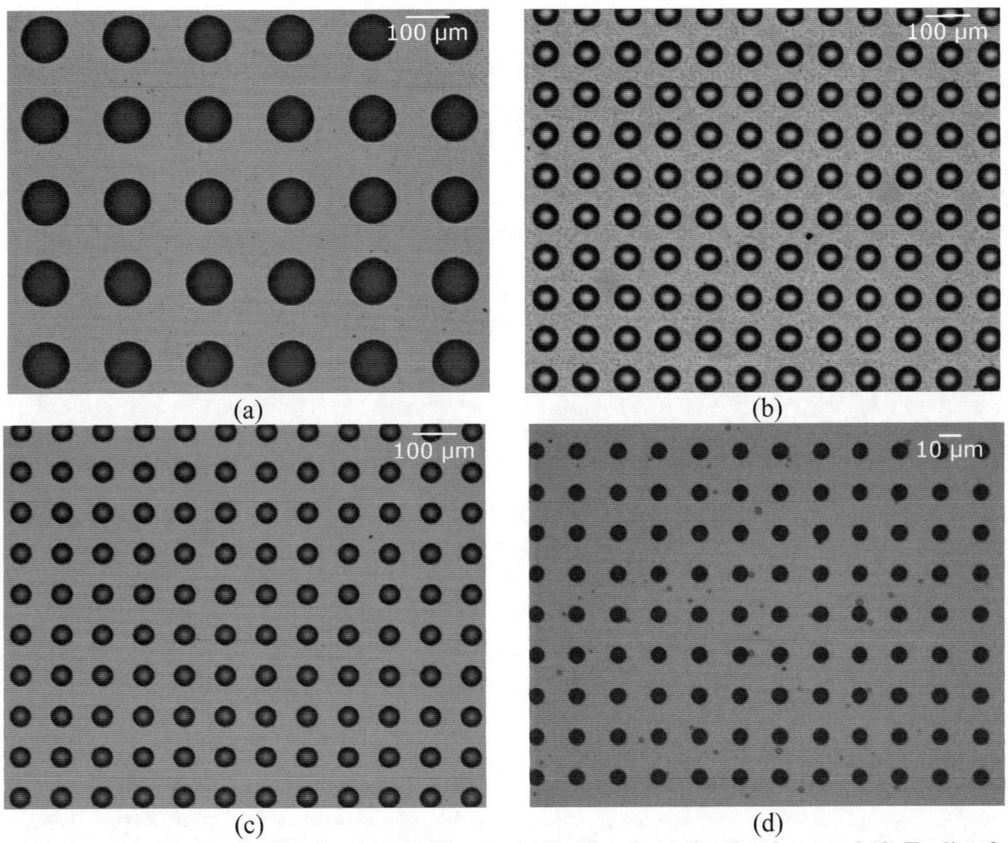

FIGURE 3. Microscopic images of (a) Tooling 1, (b) Tooling 2, (c) Tooling 2-nc, and (d) Tooling 3.

CONCLUSIONS

Although liquid silicone rubber is widely used in the injection molding industry and is able to replicate fine details due to its low viscosity, there is lack of research on replication in the submicron range and on the effect of processing conditions of liquid injection molding on the micro and sub-micron scale range. Mold design was completed and tooling inserts with features as small as 10 µm were produced using photholithography; for smaller feature sizes, however, e-beam lithography was required. E-beam lithography was optimized for the production of feature sizes as small as 0.1 µm. Injection molding of these features using liquid silicone rubber currently is in process.

ACKNOWLEDGMENTS

The authors thank Matthew Bont for his valuable input, Albright Techlnologies for the donation of liquid silicone rubber, and Dave Rondeau and Johannes Moeller for their help with the mold manufacturing.

REFERENCES

1. G.M. Whitesides, *The origins and the future of microfluidics.* Nature, 2006, **442** (7101): pp. 368-373.
2. K.F. Lei, *Chapter 1 Materials and Fabrication Techniques for Nano- and Microfluidic Devices*, in *Microfluidics in Detection Science: Lab-on-a-chip Technologies*. 2015, The Royal Society of Chemistry. pp. 1-28.
3. J.C. Love, B.W. Daniel and M.W. George, *Nanostructures Replicated by Polymer Molding*, in *Dekker Encyclopedia of Nanoscience and Nanotechnology - Six Volume Set (Print Version)*. 2004, CRC Press.
4. C. De Marco, et al., *High-Fidelity Solvent-Resistant Replica Molding of Hydrophobic Polymer Surfaces Produced by Femtosecond Laser Nanofabrication.* Langmuir, 2011. **27**(13): pp. 8391-8395.
5. C.W. Tsao and D.L. DeVoe, *Bonding of thermoplastic polymer microfluidics.* Microfluidics and Nanofluidics, 2008. **6** (1): pp. 1-16.
6. H.Mizuno, et al. *Fabrication method of large-core polymeric optical waveguide through replication process using anisotropically etched silicon.* 2004.
7. A.G. Gillies and R.S. Fearing, *Shear Adhesion Strength of Thermoplastic Gecko-Inspired Synthetic Adhesive Exceeds Material Limits.* Langmuir, 2011. **27** (18): pp. 11278-11281.
8. E. Huovinen, et al., *Mechanically Robust Superhydrophobic Polymer Surfaces Based on Protective Micropillars.* Langmuir, 2014. **30** (5): pp. 1435-1443.
9. E. Lepore and N. Pugno, *Superhydrophobic Polystyrene by Direct Copy of a Lotus Leaf.* BioNanoScience, 2011. **1**(4): pp. 136-143.
10. R. Thiruvenkataswamy, *Controlling Dimensions when Injection Molding Microfluidic Devices.* 2007: University of Massachusetts Lowell.
11. K. Mönkkönen, et al., *Replication of sub-micron features using amorphous thermoplastics.* Polymer Engineering & Science, 2002. **42** (7): pp. 1600-1608.
12. B. Sha, et al., *Investigation of micro-injection moulding: Factors affecting the replication quality.* Journal of Materials Processing Technology, 2007. **183**(2–3): pp. 284-296.
13. C. Srirojpinyo, University of Massachusetts Lowell Thesis 2005.
14. O. Murakami, et al., *Replication of Injection Molded Parts with Micro-Scale Features.* ANTEC, 2006.
15. Y. Kim, et al. *Effect of thermoplastic elastomer properties in molded microscale features.* in *ANTEC.* 2011.
16. J. LeFan and M. Eng, *Liquid Silicone Rubber Injection Molding.* 2011.
17. *Silicone Design Molding.* Albrigth Technologies Inc.
18. Y. Zhang, et al., *Replication of microstructures on three-dimensional geometries by injection moulding of liquid silicone rubber*, in *Proceedings of the 15th International Conference on Metrology and Properties of Engineering Surfaces, ASPE.* 2015, American Society for Precision Engineering.
19. S. Kaitainen, et al., *Liquid silicone rubber (LSR)-based dry bioelectrodes: The effect of surface micropillar structuring and silver coating on contact impedance.* Sensors and Actuators A: Physical, 2014. **206**: pp. 22-29.
20. C. Hopmann, et al., *Injection Molding of Superhydrophobic Liquid Silicone Rubber Surfaces.* Silicon, 2013. **6** (1): pp. 35-43.
21. S. Vudayagiri, L. Yu, and A.L. Skov, *Techniques for hot embossing microstructures on liquid silicone rubbers with fillers.* Journal of Elastomers and Plastics, 2015. **47** (7): pp. 585-597.
22. S.H. Yoon, et al., *Effect of processing parameters, antistiction coatings, and polymer type when injection molding microfeatures.* Polymer Engineering & Science, 2010. **50** (2): pp. 411-419.

Melt Processed and Self-Reinforced Ultrahigh Molecular Weight Polyethylene Used for Total Joint Replacement

Yan-Fei Huang[a], Jia-Zhuang Xu[a,*], Zhong-Ming Li[a,*]

[a] *College of Polymer Science and Engineering, State Key Laboratory of Polymer Materials Engineering, Sichuan University, Chengdu 610065, People's Republic of China.*

Presenting author's email: huangyanfeijfr@gmail.com (Y. F. Huang)
Reference author's email: jiazhuangxu@scu.edu.cn (J. Z. Xu); zmli@scu.edu.cn (Z. M. Li)

Abstract. The longevity of ultrahigh molecular weight polyethylene (UHMWPE) implant faces an ongoing challenge that the number of younger and more active patients needed total joint surgery is growing steadily. It is, therefore, of great importance to enhance the mechanical performance of UHMWPE without sacrificing its wear resistance and fatigue resistance so as to decrease the risk of revision. Besides, melt processing of UHMWPE is a bottleneck due to the extremely high melt viscosity. In this study, we demonstrated that, by addition of 9.8 wt% ultralow molecular weight polyethylene as a flow accelerator, UHMWPE could be easily melt injection molded. During the packing stage of injection molding, a controllable shear flow was exerted, which increased the overall crystallinity and efficiently induced self-reinforced superstructure, i.e., interlocked shish-kebabs and oriented lamellae. As an advantage, mechanical performance of melt-processed UHMWPE (MP-UHMWPE) was significantly enhanced compared to compression-molded UHMWPE (CM-UHMWPE). Especially, yield strength and ultimate tensile strength increased to 46.3 ± 4.4 and 65.5 ± 5.0 MPa, with increments of 128.0% and 57.4%, respectively. Impact toughness climbed from 68.7 ± 1.3 kJ/m2 for CM-UHMWPE up to 90.6 ± 1.6 kJ/m2 for MP-UHMWPE. Simultaneously, wear rate dropped to 7.1 ± 0.3 mg/MC, which was 2.9 mg/MC lower than that of CM-UHMWPE. Our work provides a feasible way to fabricate high-performance UHMWPE by manipulating the crystal structure during melt processing.

Keywords: Ultrahigh molecular weight polyethylene; Melt processing; Self-reinforcement; Mechanical properties.
PACS: 80.87

INTRODUCTION

Ultrahigh molecular weight polyethylene (UHMWPE) has a long successful clinical record as the load-bearing surface material in total joint replacement (TJR) due to its outstanding wear resistance, fatigue resistance as well as recognized biocompatibility.[1] A crucial issue currently remaining unresolved is the low efficiency of fabrication of UHMWPE implants because of its extremely high melt viscosity. Nowadays, the processing of this biomaterial is basically limited to compression molding and ram extrusion, which, however, requires a long molding time at high temperature (e.g., about 200-300 °C for about 3 h) and elevated pressures for the molecular chains to migrate across grain boundaries in order for UHMWPE particles to completely melt and finally consolidate.[2] Our work gives a practical strategy to prepare melt processable UHMWPE by addition of low viscosity polyethylene as a flow accelerator. While melt injection molding, a specific shear flow was applied to induce oriented lamellae and shish-kebab self-reinforcing superstructure[3,4,5].

FORMATION OF SELF-REINFORCED STRUCTURE

Crystalline Morphology

With the incorporation of 9.8 wt% ultralow molecular weight polyethylene acting as a flow accelerator, UHMWPE was successfully melt processed.4 By applying shear flow during injection molding, large quantities of shish-kebabs and orientated lamellae were formed. As shown in Figure 1, numerous interlocked shish-kebab entities are observed in the ULMWPE-rich region in the outer layer (the layer from surface to ca. 2.5 mm along the width direction) of the MP-UHMWPE (see Figure 1a$_1$). Interestingly, in the UHMWPE-rich region, large amounts of oriented lamellae appear in outer layer (Fig. 1a$_2$). As for CM-UHMWPE, only randomly distributed crystalline lamellae are detected throughout the sample, which is not shown here for brevity. Aside from inducing self-reinforced superstructure, the strong flow field also promoted MP-UHMWPE to be well consolidated without any structure defects (Figure. 1b_1), while in the CM-UHMWPE sample (Fig. 4b_2), explicit granule boundaries with some structure defects (voids and cracks) are very visible.

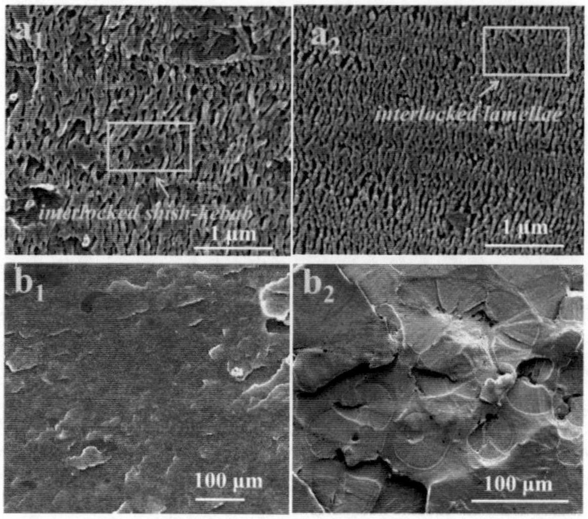

FIGURE 1. SEM images of (a_1) cryo-fracture surface of MP-UHMWPE; (a_2) cryo-fracture surface of CM-UHMWPE; (b_1) out layer of MP-UHMWPE, ULMWPE-rich region; (b_2) out layer of MP-UHMWPE, UHMWPE-rich region. The shear flow direction is parallel.

Crystalline Structure and Molecular Orientation

Figure 2 shows WAXD patterns of MP-UHMWPE and CMUHMWPE, in which two strong diffraction reflections from inner to outer circles represent the (110) and (200) crystal planes of polyethylene, respectively. For MP-UHMWPE, the arc-like diffractions of (110) plane are clearly seen in all the positions (Figure 2a), which is ascribed to the existence of oriented crystals. While for CM-UHMWPE, only isotropic diffraction circles appear (Figure 2b).

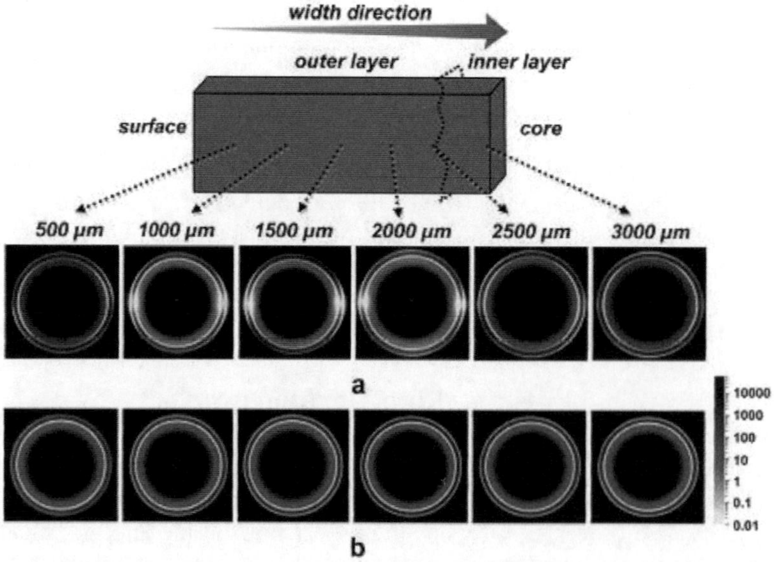

FIGURE 2. WAXD patterns of (a) MP-UHMWPE and (b) CM-UHMWPE along the transverse direction, the shear flow direction is vertical

MECHANICAL AND TRIBOLOGICAL PROPERTIES

With such a unique self-reinforced shish-kebabs and oriented lamellae combined with less structural defects, the mechanical performance of the MP-UHMWPE samples was dramatically improved. Figure 3 shows the mechanical properties of melt-processed UHMWPE (MP-UHMWPE) samples, herein, the results of compression molded UHMWPE (CM-UHMWPE) samples are also presented for comparison. In comparison to CM-UHMWPE, MP-UHMWPE exhibits a dramatical improvement in yield strength, from 20.3 ± 0.5 to 46.3 ± 4.4 MPa (p = 0.0037) with an increment of 128.0 %. The ultimate tensile strength (p = 0.0011) and Young's modulus (p = 0.0002) of MP-UHMWPE also increase remarkably from 41.6 ± 1.4 MPa and 662.7 ± 3.8 MPa for CM-UHMWPE up to 65.5 ± 5.0 MPa and 1248.7 ± 45.3 MPa for MP-UHMWPE, respectively. The impact toughness is also elevated as manifested by impact strength rising from 68.7 ± 1.3 kJ/m^2 for CM-UHMWPE to 90.6 ± 1.6 kJ/m^2 for MP-UHMWPE. Meanwhile, the retention rate of ultimate tensile strength of MP-UHMWPE samples after being fatigue tested is 77.3 ± 7.1 %, which outperforms CM-UHMWPE (53.4 ± 6.0 %) by 44.8 %, demonstrating a superior (p = 0.0001) fatigue resistance to the normal UHMWPE sample. In addition, the elongation at break of MP-UHMWPE (83.9 ± 3.1 %) is reduced compared to CM-UHMWPE, but is still adequate for knee joint implants.

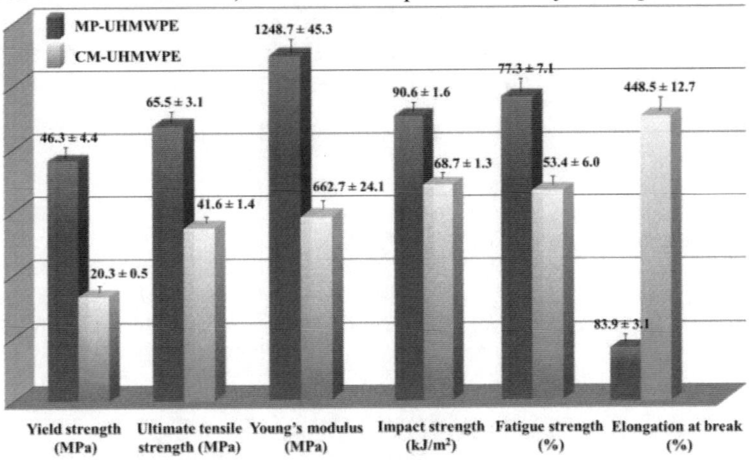

FIGURE 3. The yield strength, ultimate tensile strength, Young's modulus, Izod impact strength, fatigue strength and elongation at break (from left to right) of MP-UHMWPE and CM-UHMWPE.

Figure 4 shows the friction coefficient as a function of distance traveled during wear for MP-UHMWPE and CM-UHMWPE samples, respectively. The friction coefficients remain invariable as the distance exceeds 2800 m. The friction coefficient of MP-UHMWPE (0.15 ± 0.03) is below that of CM-UHMWPE (0.18 ± 0.03, p = 0.0001). Figure 4 also presents the wear rates of MP-UHMWPE and CM-UHMWPE (see the inset). MP-UHMWPE exhibits a notably dropped wear rate (7.1 ± 0.3 mg/MC) in comparison with CM-UHMWPE (10.0 ± 0.4 mg/MC; p = 0.0079).

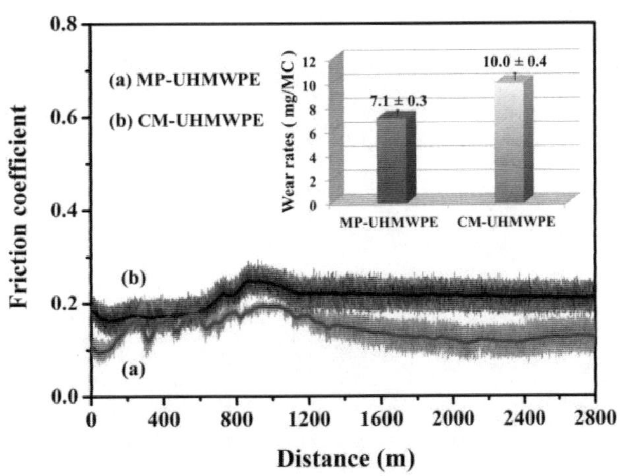

FIGURE 4. Friction coefficient vs. the distance traveled during wear for (a) MP-UHMWPE and (b) CM-UHMWPE samples. The inset presents the wear rate of MP-UHMWPE and CM-UHMWPE.

The low efficiency of fabrication of ultrahigh molecular weight polyethylene (UHMWPE)-based artificial knee joint implants is a bottleneck problem because of its extremely high melt viscosity. We prepared melt processable UHMWPE (MP-UHMWPE) by addition of 9.8 wt% ultralow molecular weight polyethylene (ULMWPE) as a flow accelerator. More importantly, an intense shear flow was applied during injection molding of MP-UHMWPE, which induced the formation of self-reinforcing superstructure, i.e., interlocked shish-kebabs and oriented lamellae. Thus the mechanical and tribological properties were significantly improved. Our work sets up a new method to fabricate high-performance UHMWPE implants by tailoring the superstructure during thermoplastic processing.

ACKNOWLEDGMENTS

The authors gratefully acknowledge the financial support from the Program of National Natural Science of China (Grants 51121001, 51033004, 51273114, 51227801) and Science & Technology Department of Sichuan Province (Grant 2013TD0013). We are heavily indebted to Prof. Benjamin S. Hsiao and Miss Yan-Hui Chen for the help of synchrotron X-ray scattering measurement in the Brookhaven National Laboratory (USA). We also would like to express sincere thanks to the Shanghai Synchrotron Radiation Facility (SSRF, Shanghai, China) for the kind help on WAXD and SAXS measurements.

REFERENCES

1. S. -M. Kurtz, O. -K. Muratoglu, M. Evans, A. -A. Edidin, Biomaterials 20, 1659-1688 (1999).
2. O. -K. Muratoglu, C. R. Bragdon, D. O. O'Connor, M. Jasty, W. H. Harris, R. Gul, et al. Biomaterials 20, 1463-1470 (1999)
3. L. Xu, C. Chen, G. -J. Zhong, J. Lei, J. -Z. Xu, B. -S. Hsiao, and Z. -M. Li, ACS Appl. Mater. Interfaces 4, 1521-1529 (2012).
4. Y. -F. Huang, J. -Z. Xu, J. -Y. Xu, Z. -C. Zhang, B. -S. Hsiao, L. Xu and Z. -M. Li, J. Mater. Chem. B 2, 971- 980 (2014).
5. Y. -F. Huang, J. -Z. Xu, J. -S. Li, B. -X. He, L. Xu and Z. -M. Li, Biomaterials 35, 6687-6697 (2014).

Effects of Fast Mold Temperature Evolution on Micro Features Replication Quality during Injection Molding

S. Liparoti[1,a)], M. Calaon[2], V. Speranza[1], G. Tosello[2], R. Pantani[1], N.H. Hansen[2], and G. Titomanlio[1]

[1]*Department of Industrial Engineering, University of Salerno, Italy*
[2]*Department of Mechanical Engineering, Technical University of Denmark, Lyngby, Denmark*

[a)]corresponding author : sliparoti@unisa.it

The growing demand to manufacture, with high accuracy, functional structures in the micro and sub-micrometer range polymer based microsystem products calls for reliable mass production processes. Being injection molding (IM) the preferential technology employed for polymer mass fabrication and mold temperature one of the most relevant process parameter to enhance polymer replication at the micro meter scale, the present study investigates effects of fast mold temperature evolution on final replication quality of produced injection molded parts. Micro features master geometries were produced by UV lithography and subsequent nickel electroplating. The mold temperature was controlled by a thin heating device (composed by polyimide as insulating layer and polyimide carbon black loaded as electrical conductive layer) able to increase the temperature on mold surface in a few seconds (40°C/s) by Joule effect and let the surface to cool down soon after. This heating device allowed to maintain mold temperature at a constant value for a time that could be equal to the filling time or longer. A fully characterized isotactic polypropylene was used as the polymer material during the injection molding experiments. The experiments revealed that the replication was mostly sensitive to cavity pressure and mold temperature. In particular, an increase of holding pressure and mold temperature enhanced the replication. Also the heating time increased the replication quality.

PAC: 8350Uv

INTRODUCTION

In the last decade a growing interest is focused on microfluidic devices, systems that process or manipulate small (10^{-9} to 10^{-18} litres) amounts of fluids, using channels with dimensions of tens to hundreds of micrometers, thanks to their wide range of possible application in molecular analysis, bio-defence, molecular biology and microelectronics [1]. Despite many advantages presented by microfluidics, the difficulties in mass production have limited their use. Recently, many author proposed to apply UV lithography, nano imprint lithography and e-beam lithography, techniques to produce micro-features on metallic inserts, to injection molding and micro injection molding to obtain high accurate objects for microfluidic devices [2]. Efforts have been devoted to correlate the process parameters of injection molding to replication accuracy [3][4]. One of the most important parameters in controlling replication accuracy seems to be mold temperature [5][6], however, the increase of mold temperature is generally not preferable since it requires additional cycle time. For this reason, many attempts to fast control mold temperature have been reported in the literature. Induction coils [7], radiation [6] or proximity heating [8] have been proposed to increase mold temperature on the cavity surface. However, these technologies suffer of some limitations, as high additional design and tool costs and small temperature increase/decrease velocity that make too long the cycle time. As far as cycle time is concern, a different approach has been proposed in [9] to overcome this issue. The design made of two insulation layers with a resistance layer in between allows for fast and effective heating phase reducing so cycle time compared with other studies [10].

In this paper, a conductive polymeric film, composed by poly(imide-amide) loaded carbon black, have been used to fast control mold surface temperature. This heating device allows to maintain mold temperature at a constant value for a time that could be equal to the filling time or longer, and the reduced thickness allows a fast cooling after heater deactivation. Main focus of the paper is given on the equipment capability of enhancing polymer replication fidelity for produced μm test structures on a nickel shim. Structures widths at different process settings have been quantified. Selected samples have been qualitatively analyzed to study effects of temperature evolution and its effect on molded microstructures morphologies.

EXPERIMENTAL

Molding experiments were carried out using polypropylene (i-PP) grade supplied by Montell (now Basell), with commercial name T30G (non-nucleated, Mw=376 000, Mw/Mn=6.7, tacticity=87.6%). A conductive polymeric film, polyimide loaded carbon black (ENSACO 260G, 30% w), with thickness of 50 μm and 40 Ω/square as sheet resistance [11], is layered between two insulating layers of polyimide and applied on both sides of the cavity, giving rise to the mold surface heating. A steel layer of 0.4 mm thickness protects the heating device from the melt except where the nickel shim is located (see Fig. 1).

The master geometries were originally fabricated using a 365 nm UV mask aligner process with 2 mask layers on a 100 mm silicon wafer. Both layers were defined in AZ5214E photoresist layer and dry etched using a reactive ion etching (RIE), which produced the final structures depth in the silicon substrate. After structure fabrication into the silicon sacrificial layer a titanium/ nickel seed layer with a thickness of 10 nm and 100 nm respectively, was deposited by sputter deposition prior to electroplating of a 320 μm nickel layer. Wire electrical discharge machining was finally employed to cut the nickel shim, later used during the experimental phase, with required dimension to fit the mold design and structure.

A 70-ton Negri-Bossi reciprocating screw, injection molding machine was used for the experiments. The polymer was injected into line-gated rectangular cavities having length L=107 mm, width W=12.7 mm, and thicknesses S=1.5 mm. Fig 1 shows the cavity used for the injection experiments, a sketch of the heater is also reported. The heater has a length of 70 mm. The nickel insert, 19x19 mm^2, is located downstream the gate. Where the nickel insert is not present a steel layer of the same thickness is located. The molding machine and the mold were equipped with five piezoelectric transducers: one in the injection chamber (P0), one just before the gate (P1) and three in the cavity (P2, P3 and P4), located in the non-moving part of the mold (15, 60 and 105 mm downstream from the gate position). Moreover, a temperature sensor was located in the cavity, on the steel layer, in positions P2.

All the experiments were carried out keeping constant the flow rate at 2.8 cm^3/s, the holding time at 8s, the melt temperature at 220°C, and the mold temperature, except on the heated surface, at 28°C. Experiments were carried out supplying two electrical power densities, 7 and 10 W/cm^2, to the heating device, for different times (t_h), 1.3s, 8s and 13s. Table 1 summarizes the experimental conditions (where P=electrical power applied; T_{level}= temperature reached on the steel layer; t_h=time that the heater is activated after the polymer reaches position P2; t_a = time that the heater was activated before the contact with the polymer in P2; P_{hold}=holding pressure).

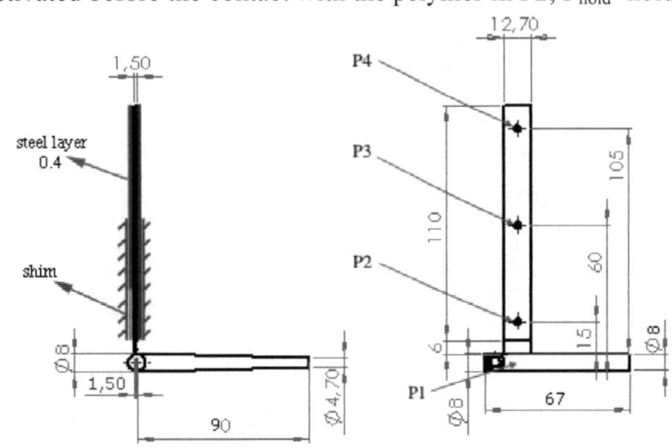

Fig. 1 Geometry of the cavity (P1-P4 indicate the pressure transducers positions) and sketch of the heating device.

Test run	*P (W/cm^2)*	*T_{level} (°C)*	*t_h (s)*	*t_a (s)*	*P_{hold}*
p-passive*l*	0	28	0	0	360
p-passive*h*	0	28	0	0	720
p-1201*h*	7	120	1.3	4	720
p-12013*h*	7	120	13	4	720
p-1501*h*	10	150	1.3	4	720
p-1501*l*	10	150	1.3	4	360
p-1508*h*	10	150	8	4	720
p-15013*h*	10	150	13	4	720
p-15013*l*	10	150	13	4	360

Table 1 Operating conditions adopted for the injection molding experiments.

Fig. 2 shows the pressure evolution at the five transducer positions related to the experiments p-passive*h,* performed without activating the heater (Fig. 2a), and activating the heater for 13s with a power density of 10 W/cm^2 (Fig.2b).

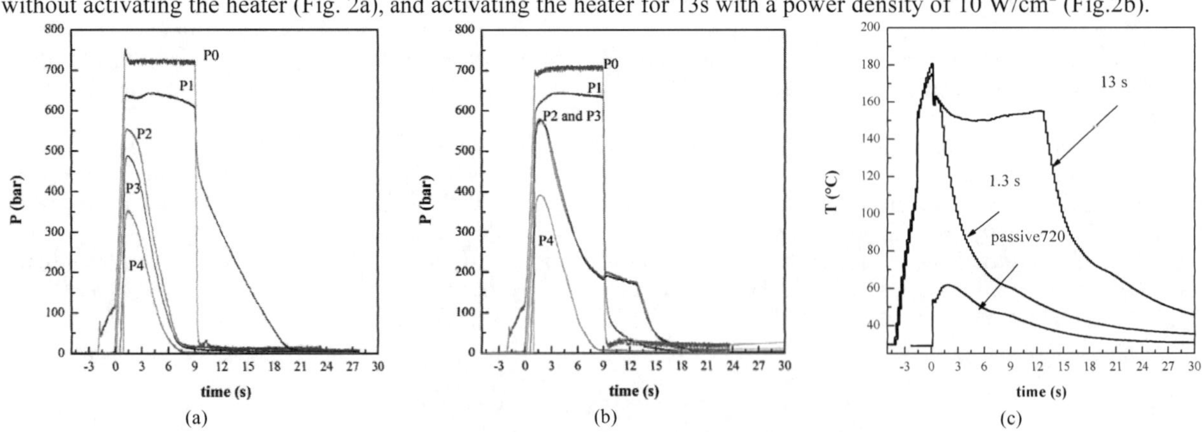

Fig. 2 Pressures in different transducer positions, related to the experiment (a) p-passive*h* and (b) p-15013*h*. (c) Temperature evolutions related to the experiments p-1501*h*, p-15013*h* and for p-passive*h*.

The gate sealing takes place at t=3.5s in both cases. However, when the heater is active pressures in positions P2 and P3 go to zero into two steps, the first one due to the decrease of the average temperature from the melt temperature to the heater surface temperature, 150°C in this case, and the second, that takes place at the heater deactivation, from 150°C to the mold temperature. Also pressure in position P4 goes to zero with a certain delay, about 1.3s, with respect to pressure decay at the same position for the experiment p-passive*h*. Similar pressure evolutions have been obtained supplying a lower electrical power density, 7 W/cm^2, to the heater.

Fig. 2c shows temperature evolutions in position P2 for the three experiments p-passive*h*, p1501*h* and p15013*h*. The heater is activated 4s before the melt reaches position P2 (t=0), during such a time temperature on the surface goes from the mold temperature to 150°C. The temperature increase during the range -1.5s<t<0s is due to the decrease of the contact resistance and the consequent current increase. When the polymer reaches position P2, the temperature reaches the maximum value and soon after it starts to decrease to the selected surface temperature.

RESULTS AND DISCUSSION

To quantify effects of process conditions on the replication fidelity of the molded, micro-crosses, an Atomic Force Microscope (AFM) was employed. Measurements have been carried out in tapping mode with commercial probe tips FIB3D2 from Bruker with nominal spring constants of 42 N m^{-1}, resonance frequencies of 320 kHz and declared tip radius of 10 nm. Fig. 3 shows AFM height images of micro-crosses, with nominal width of 20 μm on the nickel insert and two examples of polymer replicated micro-crosses using p-passive*h* and p15013*h* machine settings, see Table 1.

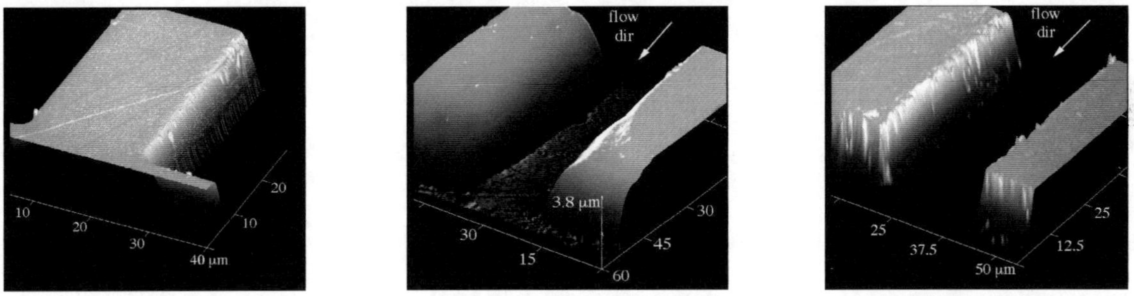

nickel insert | passive720 | 150°C 13s 720bar

Fig. 3 AFM height images of the micro-crosses having a width of 20 μm on the nickel insert and of micro-crosses replicated on the samples p-passive*h* and p-15013*h*. The flow direction is also indicated.

The AFM scans show that mold surface temperature is a very important parameter in achieving good microstructure replication. Indeed, the micro-cross replicated on p-15013*h* sample appears to have a sharper edge than the micro-cross in p-passive*h* sample, enhancing corners replication fidelity.

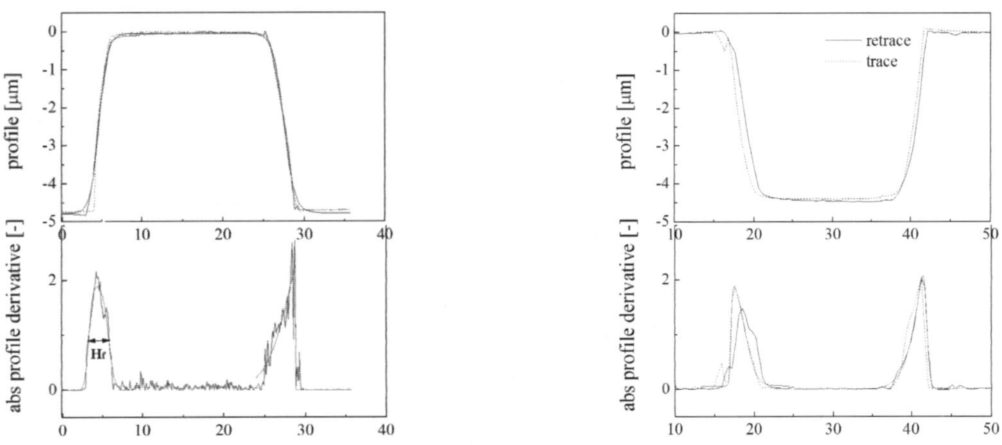

(a) (b)

Fig. 4 Profile of the micro-cross on the shim (a) and on the sample p-15013*h* (b). The absolute value of the profile derivative and their fitting by asymmetric double sigmoidal function are also reported.

To quantify the accuracy of the micro-crosses replication the traces and retraces (at 20 μm from the corner center), obtained by AFM analysis, have been analyzed considering errors due to the interaction between the sample and the tip [12]. The average height of the measured polymer micro-crosses is equal to 5 ± 0.2 μm. Therefore due to the small heights variation among the produced samples a different measurements to establish the replication fidelity as a function of different process parameters has been considered. In particular, the absolute values of the trace derivative have been fitted by an asymmetric double sigmoidal function, and the full width at half maximum of the peak on the fitting curve (H_f) has been chosen as accuracy index. An example of high replication fidelity for a micro feature having a sharp edge is shown if Fig. 4. For this sharp replicated corner (high fidelity between master nickel geometry and polymer replica) a small H_f, close to the value of the shim (2.2±0.8 μm) is measured. Fig. 5 shows the comparison of the H_f obtained fitting the trace/retrace of AFM height data obtained for samples produced in different packing conditions and supplying 7 and 10 W/cm^2 to the heating device. Data have been sorted on the bases of the AFM analyzed position on the micro-crosses as reported in Fig. 5b.

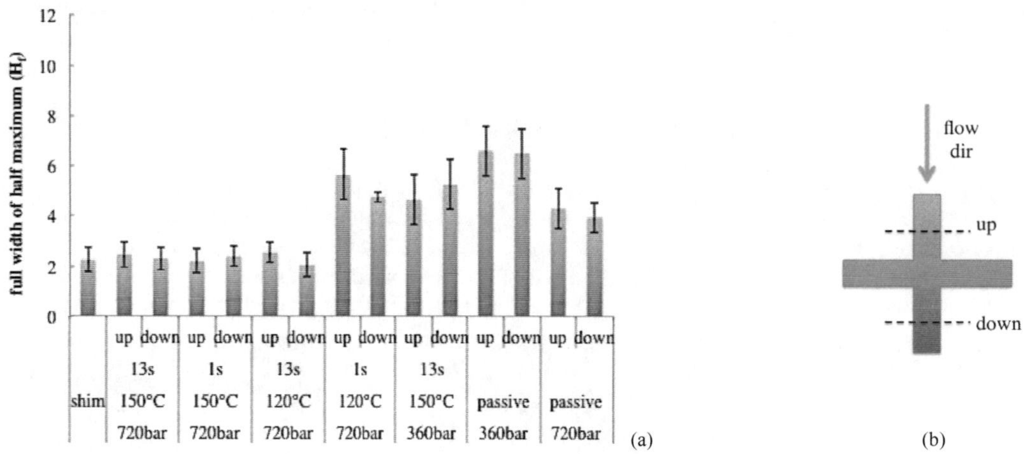

Fig. 5 Comparison of the H_f evaluated for samples produced in different injection molding conditions. (a) Branch parallel to the flow direction. The standard deviation of experimental data is also indicated. (b) Sketch of the analyzed positions on the micro-crosses.

The comparison between p-passive*l* and p-passive*h* suggests that packing pressure has a significant effect on the replicability: the replication is improved by selecting higher packing pressure, however, the H_f evaluated for p-passive*h* sample is still larger than the value evaluated for the nickel insert. Only selecting higher temperatures (120°C and 150C°) the H_f becomes closer to the value of the nickel insert. Fig. 6a shows that H_f is small for all the samples produced selecting 150°C as mold surface temperature at all the considered heating times, whereas, decreasing mold surface temperature to 120°C the replication becomes poor at small heating time (1.3s). When the heater surface temperature is 150°C and the holding pressure is 720 bar the replication is accurate and is not dependent on the heating time. At lower heater surface temperature, 120°C, and the same holding pressure, good replication can be obtained only selecting 13s as heating time, whereas at 1.3s of heating time the replication becomes worst. The reason for this behavior is twice: first, the heater surface temperature is not sufficient to reach good replication, second pressure is not higher enough in the cross position.

The mold surface temperature evolution has a significant effect on the morphology of the molded samples [10,13]. For this reason a qualitative investigation on the molded microstructure morphology has been performed. Slices, cut perpendicular to the flow direction in position P2, where the micro-crosses are also located, have been observed by optical microscope, with crossed analyzer and polarizer. Fig. 6 shows optical micrographs related to the molded samples p-passive*l*, p-passive*h*, p15013*l* and p15013*h*.

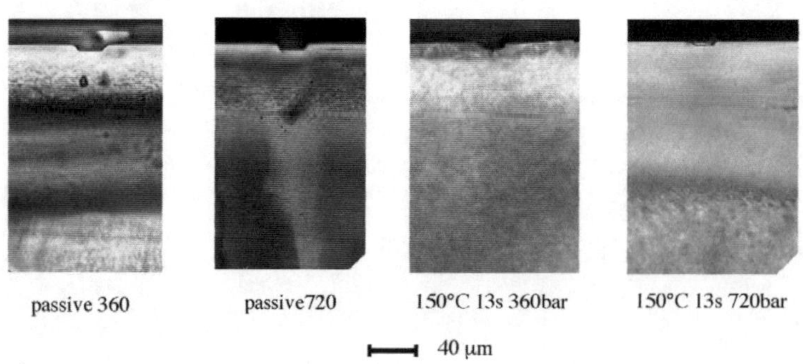

Fig. 6 Optical micrographs of the slices cut from the samples p-passive360, p-passive720, p-15013*l* and p-15013*h*, near the micro-crosses having a width of 20 μm. Flow direction at 45° respect to polarizer direction.

As observed in other works [13–15], the skin layer is drastically reduced when the mold temperature is kept high during the injection molding process; the shear layer, that is also reduced by the use of higher mold temperature, disappears when the lower packing pressure is selected, in this case only spherulitical structures can be observed. These phenomena can be explained taking into account the relaxation of the molecular stretch and orientation that take place at higher temperature [10]. Preliminary conclusions from the qualitative investigation (Fig. 6) indicate no visible effect on the structures morphology.

CONCLUSIONS

In this work, injection molding coupled with fast control of mold surface temperature have been proposed for the mass production of samples with micro-features, since the heating device is able to produce fast temperature evolution giving rise only a reduced cycle time increase. A nickel insert, having microstructures produced by UV lithography, has been located on the mold surface, where the heating device was present.

Packing pressure increase has been found to improve micro-crosses replication. However, additional packing pressure is not sufficient to achieve high accurate replication. Mold surface temperature is the most important parameter in controlling replication of microstructures, since mold surface temperature of 150°C allows to produce samples with microstructures very similar to those located on the master geometry. The presence of micro-crosses on the nickel insert does not seem to influence the morphology of the molded samples.

REFERENCES

1 G.M. Whitesides, *Nature. 442 368–373 (2006)*. http://dx.doi.org/10.1038/nature05058.

2 S. Menotti, H.N. Hansen, G. Bissacco, M. Calaon, P.T. Tang, C. Ravn, *The International Journal of Advanced Manufacturing Technology. 74 907–916 (2014)*.

3 H.-L. Lin, C.-S. Chen, R.-T. Lee, S.-C. Chen, R.-D. Chien, M.-C. Jeng, et al., *Japanese Journal of Applied Physics. 52 44001 (2013)*. http://stacks.iop.org/1347-4065/52/i=4R/a=044001.

4 M. Calaon, G. Tosello, H.N. Hansen, J. Nôregaard, Influence of process parameters on edge replication quality of lab-on-a-chip micro fluidic systems geometries, in: Annual Technical Conference - ANTEC, Conference Proceedings, 2013: pp. 1084–1088. http://www.scopus.com/inward/record.url?eid=2-s2.0-84903544048&partnerID=tZOtx3y1.

5 S. Liparoti, T.M. Hunag, A. Sorrentino, G. Titomanlio, M. Cakmak, *30th International Conference of Polymer Processing Society, Cleveland, Ohio, USA - AIP Conference Proceeding. 1664 (2014)*.

6 S.-C. Chen, Y. Chang, Y.-P. Chang, Y.-C. Chen, C.-Y. Tseng, *International Communications in Heat and Mass Transfer. 36 1030–1035 (2009)*.

7 H.L. Lin, S.C. Chen, M.C. Jeng, P.S. Minh, J.A. Chang, J.R. Hwang, *International Communications in Heat and Mass Transfer. 39 514–522 (2012)*.

8 D. Yao, T.E. Kimerling, B. Kim, *Polymer Engineering & Science. 46 938–945 (2006)*.

9 K.M.B. Jansen, A.A.M. Flaman, *Polymer Engineering & Science. 34 894–897 (1994)*.

10 S. Liparoti, A. Sorrentino, G. Guzman, M. Cakmak, G. Titomanlio, *RSC Advances. 5 36434–36448 (2015)*.

11 S. Liparoti, G. Landi, A. Sorrentino, V. Speranza, M. Cakmak, H.C. Neitzert, *Advanced Electronic Materials. (2016) DOI 10.1002/aelm.201600126*.

12 G. Tosello, A. Antico, H.N. Hansen, F. Marinello, G. Lucchetta, Influence of measuring parameters on the precision of atomic force microscope in industrial applications, in: International Conferences on Multi-Material Micro Manufacture, 4M/International Conferences on Micro Manufacturing, ICOMM, 2009: pp. 401–406.

13 S. Liparoti, A. Sorrentino, G. Titomanlio, *AIChE Journal. (2016)*.

14 S. Liparoti, A. Sorrentino, G. Titomanlio, *Polymer Processing with Resulting Morphology and Properties: Feet in the Present and Eyes in the Future: Proceedings of the GT70 International Conference - Salerno (Italy) 2015. AIP Conference Proceedings. 1695 (2015)*.

15 S. Liparoti, T.M. Hunag, A. Sorrentino, G. Titomanlio, M. Cakmak, *Proceeding of PPS-30: The 30th International Conference of the Polymer Processing Society - Cleveland (Ohio US) 2015. AIP Conference Proceedings. 1664 (2015)*.

16 S. Liparoti, A. Sorrentino, G. Guzman, M. Cakmak, G. Titomanlio, *PROCEEDINGS OF PPS-31: The 31st International Conference of the Polymer Processing Society–AIP Conference Proceeding. 1713 170001 (2016)*.

17 R. Pantani, I. Coccorullo, V. Speranza, G. Titomanlio, *Progress in Polymer Science. 30 1185–1222 (2005)*.

Experimental Investigations on the Fibre Length Degradation in Injection Moulds

Elmar Moritzer[a], Yannick Martin[a*]

[a]*University of Paderborn, Kunststofftechnik Paderborn (KTP), Paderborn, Germany*

yannick.martin@ktp.upb.de

Abstract. The sustainable use of natural resources such as raw materials and fuels, and the carbon footprint of products are becoming increasingly important in many industrial sectors. Due to the substitution of materials such as steel and aluminium, the use of fibre-reinforced thermoplastic materials has increased dramatically in recent years. The properties of injection moulding products are considerably dependent on the resulting fibre length in the components. Increased fibre length results in a significant improvement concerning tensile strength, stiffness and impact strength of the moulded parts. The fibre length reduction occurring in the injection unit is already the subject of current research, while the decrease inside of injection moulds has hardly been investigated. By the use of a special design and arrangement of the cavities, a modification of the runner system, as well as a gate adjustment a reduction in the shortening of the reinforcing fibres in the injection mould can be achieved. Thus, the material potential can be utilized better by these optimizations. In the presented study different injection moulds were examined by determining the length of fibre at different locations along the flow path. From these results, the most important factors influencing the fibre damage can already be identified.

Keywords: reinforced thermoplastics, fibre length degradation, injection moulding
PACS: 81.05.Qk, 81.20.Hy, 82.35.Lr, 83.50.Uv, 83.80.Rs

INTRODUCTION

There will be higher mechanical and thermal requirements for fibre reinforced products, created in the injection moulding process. In particular, the mechanical characteristics such as tensile strength, stiffness, impact strength and heat resistance should be significantly improved by the fibre-matrix composite. To use the full potential of a fibre composite material, it is necessary to achieve good fibre-matrix adhesion and a fibre length that should be as high as possible. [1]

On one hand the fibre-matrix mixture is influenced by high shear stresses during the plastification process. On the other hand the injection operation causes high shear stresses by the flow through the nozzle and the mould geometry. This shear stress leads to fracture and a reduction in length of the reinforcing fibres. The aim of this study is to identify possible influences on the fibre length reduction by experimental investigations of different mould geometries in the injection moulding process. To do so three different injection moulds are selected and various materials a analyzed. [2]

EXPERIMENTAL

System Technology and Process Sequence

For the experimental investigations an injection moulding machine of the type TM 130/750 from manufacture Battenfeld, built in the year of 2005, is used. With a maximum clamping force of 1300 kN, it is one of the smaller machines of this type. The plastification unit is characterized by a screw diameter of 45 mm.

To identify the mould design factors, which influence the fibre length reduction, three different tools are used. These contain special geometrical characteristics regarding the fibre damage. In each case four significant positions are defined within the mould, where samples are taken for measurement of the fibre lengths. Thus, the fibre damage is examined within the injection mould.

First, a spiral flow mould is used, which is equipped with a very long flow path as the characteristic variable. The melt is injected through a sprue gate with a small undercut that ensures controlled demoulding. The flow path cross-section is shaped as a trapezoid having a height of 3 mm and a base width of 10 mm. Due to the trapezoidal configuration, the upper side width is reduced to 7 mm. The flow path has a total length of 1450 mm, the spiral winds around the gate and narrows at the component end (cf. Fig. 1). This mould is used in the following experiments to determine the influence of the flow path length on the fibre length reduction. The measurement points are uniformly distributed across the flow path. The first sample is taken at the gate, the second sample is extracted at 400 mm, the third sample at 800 mm, the fourth sample is taken at a flow path length of 1200 mm (cf. Fig. 1).

FIGURE 1. Fibre length measuring points highlighted on the injection mould "Flow Spiral"

A stacking box is selected as the second mould geometry. It has the dimensions 100x136x48 mm with a wall thickness of 1.5 mm. In contrast to the previous mould in this case a hot runner is used instead of a sprue gate. The thin design is particularly interesting for the investigations because high shear forces occur in the melt over a wide area and a certain length. Therefore a higher interaction between fibre and cavity can be expected. Moreover, the geometric design of the stacking boxes includes several melt deflections. Measurement point 1 is placed at the end of the hot runner. Point 2 is applied to the corner of the base of the box in order to investigate the influence of the low wall thickness without simultaneous deflection. The measuring points 3 and 4 are positioned on the end face of the box, whereby the influences of the deflection can be determined (cf. Fig. 2).

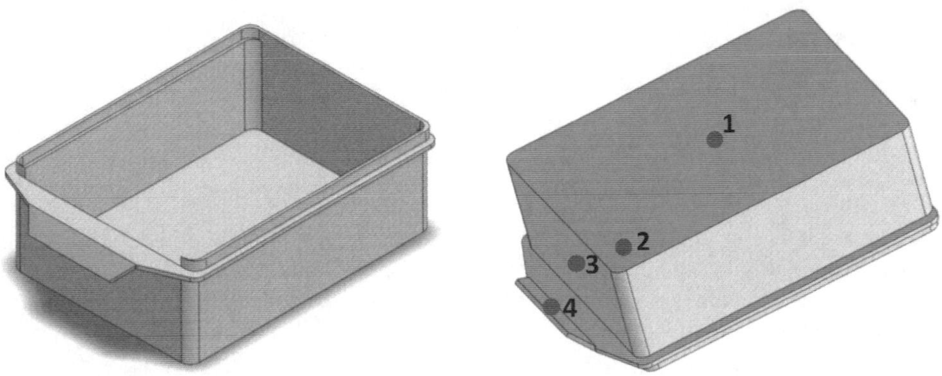

FIGURE 2. Fibre length measuring points highlighted on the injection mould "Stacking Box"

Following the second geometry, the third one is a letter mould (cf. Fig. 3, left). This is characterized by a large runner system and 36 individual cavities for the capital letters of the alphabet and all the digits. Similar to the flow spiral mould also the letter mould has a sprue gate. The individual letters and digits are connected via a balanced runner system. The letters and numbers are linked to the runner by a pinpoint gate. The pinpoint gates have a very small cross-section, which results in high shear forces on the melt during the filling process. Thus these shear forces also have an effect on the fibres. To investigate the influences of the runner system and the pinpoint gate on the fibre length reduction, four measurement points along the flow path are defined (cf. Fig. 3, right).

FIGURE 3. Fibre length measuring points highlighted on the injection mould "Letters"

Components Employed

To determine the influence of different types of materials, various glass fibre reinforced polymers are selected. For this purpose various matrix materials, fibre contents and initial fibre lengths are analyzed in the material. There are six different materials chosen to cover these influential parameters. Table (1) shows the various materials with relevant matrix materials, fibre contents and measured initial fibre lengths.

TABLE (1). Selected fibre-reinforced materials

Material Name	Matrix Material	Fibre Content [%]	Measured Initial Fibre Length [mm]	Measured Melt Cake Fibre Length [mm]
PA6.6 GF15	Polyamid 6.6	15	0.63	0.58
PA6.6 GF30	Polyamid 6.6	30	0.46	0.43
PA6 GF30	Polyamid 6	30	0.49	0.44
PP GF15	Polypropylene	15	1.08	1.00
PP GF30	Polypropylene	30	0.93	0.75
PP LGF30	Polypropylene	30	9.23	3.71

Measurement of Fibre Length

To determine the fibre length distribution in the respective components areas the fibre and the matrix must be separated initially. Pyrolysis is chosen as a separation method, in which the samples are ashed in a muffle oven at 600 °C. The matrix is evaporated leaving only the pure fibres as the ignition residue. [3]

Afterwards the fibres can be evaluated by image analysis. For this, the fibres are first placed in a petri dish with distilled water and digitized with a high-resolution scanner. With a special software, the digitized fibres can finally be evaluated. [4]

Experimental Results

The results of the experimental investigations are shown below. The relative change of length is shown in reference to the measured fibre length in the melt cake, because only the fibre length reduction in the mould is considered.

In Fig. 4, the relative change of length is assigned to the flow path of the flow spiral. In this case significant differences between materials with short glass fibres and the material with long glass fibres are visible. The relative fibre length changes of the five materials with short glass fibres are roughly the same. Up to a flow path length of 800 mm, the fibre length decreases by about 10 %. But at the end of the mould a slight increase in fibre lengths can be seen. However the PP LGF30 has a significantly larger fibre lengths reduction. Thus the fibre length has decreased after 800 mm by about 58 %. At the end of the flow path slightly longer fibres are detectable again.

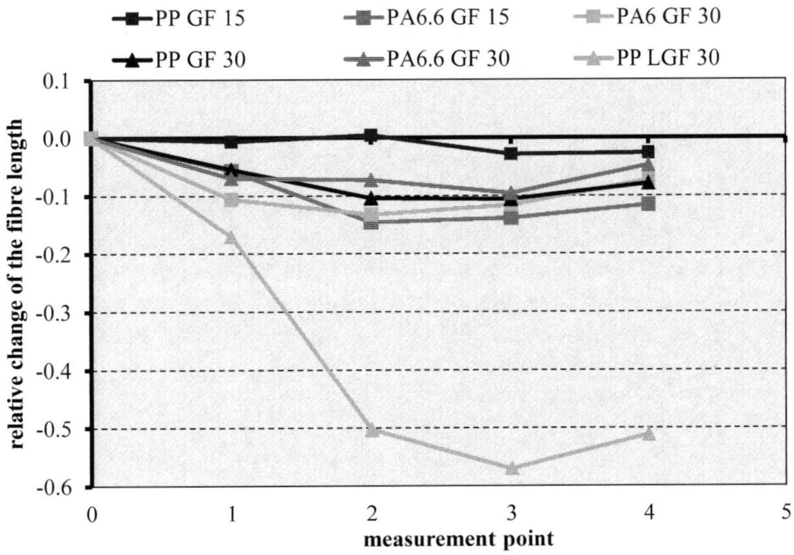

FIGURE 4. Relative change of the fibre lenght in „Flow Spiral" mould

In contrast to the studies of the flow spiral mould, an influence of the matrix material is recognizable at the stacking box mould. In the materials with a matrix of PP the fibre length is reduced by approximately 12 % at the flow path's end. The materials with a matrix of PA have a significantly increased fibre length reduction of up to 25 %. The fibre length reduction of PP LGF 30, relating to the melt cake, is approximately 67 %. In addition, the largest fibre damage is measured in the area of the thin-walled base. This is recognizable independently of the material.

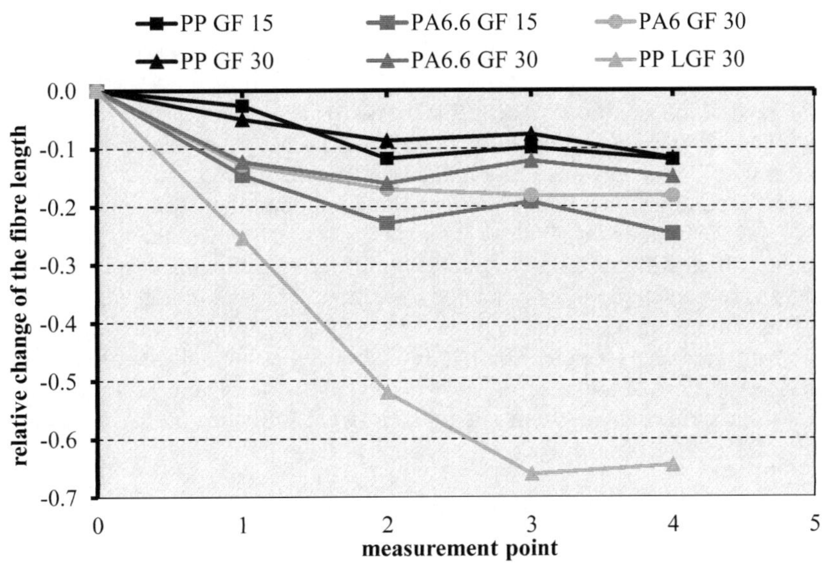

FIGURE 5. Relative change of the fibre lenght in „Stacking Box" mould

Investigations at the letter mould show that the fibre damage of the five materials with short glass fibres are similar to the flow spiral mould regardless of the matrix material. The fibre length is reduced by approximately 15%. The long-fibre material have much larger fibre damage by the flow through the letter tool. First the fibre length is reduced by about 40 % in the runner and finally by a further 25 % through the pin gate. The final fibre length in the letters and numbers is approximately 35 % of the fibre length after plastification.

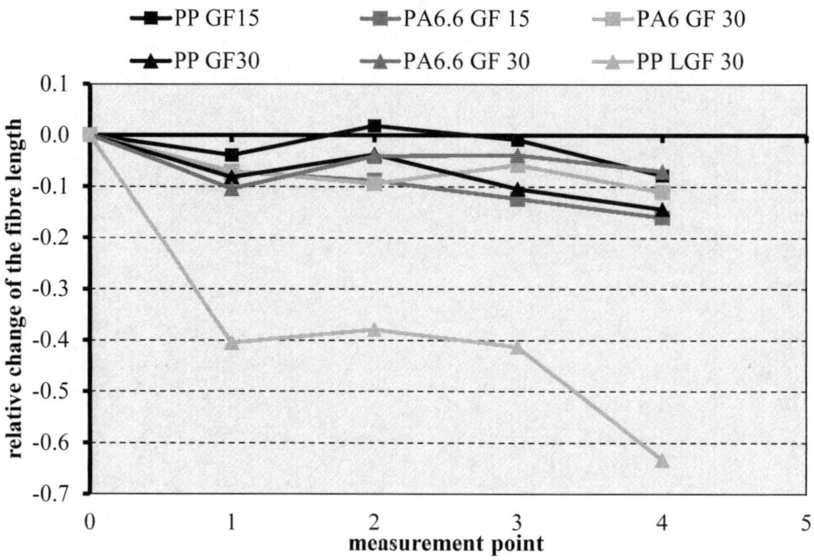

FIGURE 6. Relative change of the fibre lenght in „Letters“ mould

CONCLUSION

In the experimental studies of flow spiral mould, it turned out that the flow length only had a minor influence on the fibre length at a large flow path cross-section. The largest fibre damage occurred at 90° deflections and the associated slight material accumulations from the gate to the flow channel. Along the remaining course of the flow path, no significant change in the fibre lengths was measured. Only at the flow path end minor fibre damage was detected. The damage to the short fibre materials was significantly smaller than the reduction of the long-fibre material. This is caused by the free buckling length of the fibres. This is significantly larger with long fibres, so these are much more likely to break.

In the stacking box mould, the small component thickness was a significant influencing factor. Analogous to the investigations of the flow spiral mould, the long fibre material had a much greater degradation in this case. However, no increase in the fibre length was observed on the flow path end of the mould. In addition differences of the short fibre materials, depending on the matrix materials, have been detected. The fibre damage concerning the materials with a matrix of polyamide was about twice as large as the fibre length reduction of the materials with a matrix of polypropylene. These differences were due to the different structural viscosities of the matrix materials. Because polyamide has a low structure viscosity a wide boundary layer was produced. Therefore, it remained only a small core flow area in the middle part, leading to higher shear forces than in polypropylene.

The experimental investigations of the letter mould have shown that the pin gate of each letter had a strong fibre-damaging effect. This is regardless of the material. This is caused by the extremely high shear forces which occur at this point. The length of the cold runner system had no significant influence on the fibre damage. Due to the large channel cross-section, a wider core flow region was formed. Thereby the shear forces on the fibres were low.

REFERENCES

1. J. L. Thomason, "The influence of fiber length and concentration on the properties of glass fiber reinforced polypropylene", 6. The properties of injection moulded long fibre PP at high fibre content. *Composites Part A: Applied Science and Manufacturing*, Volume 36, Issue 7, July 2005, Pages 995–1003.
2. E. Witten, „AVK – Industrievereinigung Verstärkte Kunststoffe e.V.: „Handbuch Faserverbundkunststoffe/Composites Grundlagen-Verarbeitung-Anwendungen“, 4. Edition, Springer Verlag Frankfurt a.M., 2014
3. T. Pflamm-Jonas, "Auslegung und Dimensionierung von kurzfaserverstärkten Spritzgussbauteilen", *dissertation*, Technische Universität Darmstadt, 2001
4. K. Brast: „Verarbeitung von langfaserverstärkten Thermoplasten im direkten Plastifizier-/Pressverfahren“, *dissertation*, Technische Hochschule Aachen, 2001

Effects of Different Mold Coatings on Polymer Filling Flow in Thin-Wall Injection Molding

M. Sorgato[a*], D. Masato[a] and G. Lucchetta[a]

[a] *Department of Industrial Engineering, University of Padova, via Venezia 1, 35131 Padova, Italy*

marco.sorgato@dii.unipd.it

Abstract. Injection molding of parts having high flow length/thickness ratio is challenging task for both micro and packaging applications. In the filling phase, high-speed injection can cause a raise of cavity pressure, which prevents the complete replication of the mold geometry. Therefore, in this paper, the effect of three mold coatings, such as aluminum oxide, diamond-like carbon and silicon oxide, were investigated. In particular, the filling of a representative micro-part was studied as a function of mold coating, injected polymer and different process parameters, in order to identify the effects of heat conduction and wall slip related to the coatings.

Keywords: Coating, Mold, Micro injection molding.
PACS: 83.50 Uv

INTRODUCTION

Computer, communication, consumer electronics and packaging industries are increasingly demanding lighter and thinner parts due to economics and environmental reasons. In fact, a reduction of the main wall thickness generally leads to a smaller part volume and, therefore, to a lower material cost and impact. A thinner wall also results in a shorter cycle time, decreasing the process environmental and economical burdens [1]. Thin-wall designs are also common in micro-structured components for optical and biomedical applications, such as light guide plates and micro fluidic devices. The very high flow length/thickness ratio of thin-wall parts can cause a raise of cavity pressure in the filling phase, which prevents the complete replication of the mould geometry, especially surface micro features [2]. A reduction of flow resistance can be effectively achieved by increasing the injection speed or the mould temperature. High-speed injection molding permits the melt to fill the cavity before the growth of the solidified skin layer hinders the flowing core. However, it produces a non-uniform pressure distribution in the cavity, which often generates quality problems, such as flow marks, polymer degradation, high residual stress, sink marks and warpage [3]. Alternatively, the flow resistance can be reduced using Rapid Heat Cycle Molding (RHCM) technologies, which maintain the mould cavity surface at a relatively high temperature during the filling phase, counteracting the development of the solidified skin layer [4]. In this way McFarland et al. successfully molded polystyrene cantilever beams having a thickness of 10 μm and an aspect ratio exceeding 170 [5]. As an alternative solution to RHCM, which requires high initial investments for mould making and increases the process cost due to a significantly longer cycle time, Chen et al. proposed the use of cavity surface coatings made of low thermal conductivity materials for improving the quality of ABS parts [6].

EXPERIMENTAL

Mold Design

The filling of a representative micro-part was studied as a function of mould coating. The cavity is an open flow micro channel characterized by a length of 40 mm, a width of 6 mm and two alternative thickness value of 400 and 800 μm (Figure 1). The mould was designed to mount, on its moving half, inserts with different coatings, all characterized by a depth of 400 μm. The thickness of the channel was varied by changing the fixed side from a flat surface to a 400 μm deep channel. The cavity pressure during the injection molding process was directly measured using a Kistler 6182C piezoelectric pressure sensor, positioned at 6 mm from the injection location. The acquired cavity pressure curves were aligned to the beginning of the injection phase, and the steady-state values of the cavity pressure were collected.

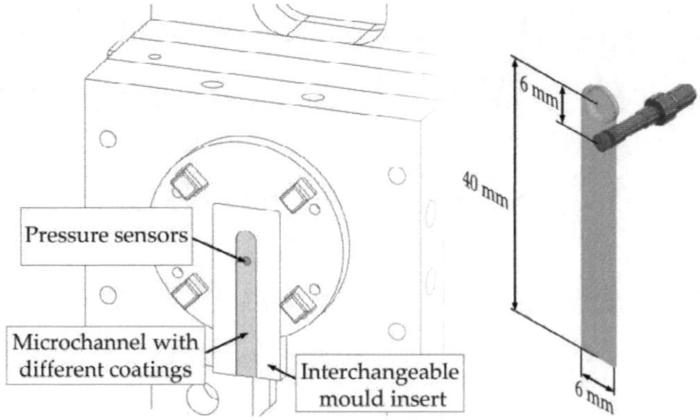

FIGURE 1. Design of the mold cavity and of the part.

Mold Surface Coatings

Three different mould surface coatings technologies were selected according to their tribological and thermal properties: (i) atomic layer deposition (ALD) of Al_2O_3, (ii) diamond like carbon (DLC) and (iii) plasma-enhanced chemical vapor deposition (PECVD) of SiO_x (Table 1). The coated mould inserts were evaluated both in terms of surface quality and channel geometry. Morphology and continuity of the coated surface was observed using scanning electron microscopy (FEI, Quanta SEM 400). The surface roughness of the coated inserts was characterized using a contact stylus roughness tester (Zeiss-TSK Surfcom 1400A), with a cut-off filter of 0.25 mm and a background noise below 7 nm Ra. Results were not showing significant differences in roughness parameters among coated surfaces, with average values of Ra in the range 25-45 nm. For the purposes of the current investigation, no significant differences were observed and therefore it was assumed that surface topography is not influencing the filling flow.

TABLE 1). Main characteristics of the selected mold surface coatings.

Insert	Coating	Coating thickness [μm]
A	-	-
B	ALD	0.02÷0.05
C	DLC	0.5÷5
D	SiO_x	0.5÷5

Materials and Manufacturing System

An amorphous polystyrene (PS, Total PS Crystal 1540) and a semi-crystalline polyethylene terephthalate (PET, Cepsa PET SR08) were considered in the study. The rheological characterization for both the materials was performed by means of rotational (TA Instruments, ARES) and capillary rheometer (Ceast, Rheo 250). Drying cycles were performed for the PET at 180°C for 8 hours, with a dew point of - 45°C. The adsorbed moisture, and its influence on the Newtonian viscosity, was related to the drying time by measuring the residual humidity using a moisture meter (Brabender, Aquatrac). The results indicated that a drying cycle of 8 hours at 180°C produced a residual humidity smaller than 40 ppm. The molding experiments were performed using a micro injection molding machine (Wittmann-Battenfeld, MicroPower 15).

Experimental Design

The flow of the polymer melt was experimentally analyzed by varying the injection speed (V_{inj}) for both polymers, the mould temperature (T_m) only for PS and the melt temperature (T_b) only for PET. The range values for the selected controllable process variables were defined considering the literature, recommendations of the materials supplier and the technological limits of the available experimental setup (Table 2).

The tests were performed using different depth of the coated channels for the two polymers. In particular, a channel depth of 800 μm was adopted for PET, a depth of 400 μm for PS, accordingly with their different rheological properties.

TABLE 2). Processing conditions for the experiments.

Parameter	PS	PET
T_m [°C]	80,120	15
T_b [°C]	240, 230, 210, 200	300, 295, 285, 270 280, 275, 270, 260
V_{inj} [mm/s]	100, 200, 300, 450, 600, 750	

MODELLING

Slit Flow Model

For a straight, rectangular channel having length, L, width, w, and thickness, h, with the assumptions of a fully developed steady state laminar flow with no-slip on the wall, the apparent shear rate and real shear stress in the slit model are given by:

$$\dot{\gamma}_{w(app)} = \frac{6Q}{wh^2} \tag{1}$$

and

$$\tau_{w(real)} = \frac{wh}{2(w+h)}\left(\frac{-\Delta P}{L}\right) \tag{2}$$

where Q is the volumetric flow rate and ΔP the pressure drop. According to the Walter correction [7], the wall shear rate of a non-Newtonian fluid can be determined by:

$$\dot{\gamma}_{w(real)} = \dot{\gamma}_{w(app)}\left(\frac{2}{3}+\frac{1}{3}n\right) \tag{3}$$

where n is the power law index, which can be calculated using the Schümmer approximation:

$$x^* = \left(\frac{3n+1}{4n}\right)^{\frac{n}{n-1}} \tag{4}$$

where x^* is a weak function of n, which is 0.79 for a slit geometry [7].

Numerical Model

The non-Newtonian and non-isothermal filling flow of the polymer melt was modeled using the injection molding simulation software Autodesk Moldflow. The numerical value of the cavity pressure was obtained in a mesh node positioned at the same distance from the injection location as the pressure sensor in the experimental setup. The polymer flow during the injection molding process is inherently transient and includes a free surface moving through narrow cavities, thus a good model for the mould-melt heat transfer coefficient (HTC) is essential to describe the non-continuous temperature distribution at the interface and to predict the cavity pressure.

The numerical model was employed in the study to understand how the cavity pressure is influenced by the injection speed and by the HTC value.

Inverse Analysis

The value of the HTC that forces the numerical results to fit the measured pressure values for each coating was determined using an inverse analysis approach. The differences between numerical and experimental pressure values were minimized using an iterative optimization algorithm. In order to reduce the computational time, artificial neural networks (ANN) were used as a meta-model to locally approximate the simulation results. The artificial neural networks were trained to reproduce the results of the simulation [8], for all the 12 experimental process conditions, using the Levenberg-Marquardt back propagation.

The difference between numerical and experimental results was minimized using an optimization procedure implemented in modeFrontier, based on a MOGA-II genetic optimization algorithm [9]. For each iteration and for the corresponding values of the 12 input factors, the neural networks previously trained approximated the numerically calculated pressure values, in order to compare them to the pressure experimentally measured.

RESULTS AND DISCUSSION

Figures 2 and 3 show the relationship between the steady-state cavity pressure and the injection speed for PET and PS. It is clear that coatings reduce PET flow resistance up to 8% at T_b = 300°C. This effect is still evident but less significant (-5%) at T_b = 280°C. DLC has a lower influence on pressure reduction at both temperatures while Al_2O_3 and SiO_x have a larger effect but it is very sensitive to the melt temperature.

The same coatings have different effects on PS flow resistance at T_m = 80°C. While Al_2O_3 reduces the pressure up to 3% at high injection speeds, DLC and SiO_x increase it up to 6%. At T_m = 120°C the coatings consistently show the same trend but with attenuated effect. In injection molding the wall shear stress increases with shear rate for a non-slip wall-polymer interface. According to Mhetar and Archer, there are three slip regimes: (i) a weak slip regime at low shear stresses when slip exceeds a first critical stress (0.1–0.3 MPa); (ii) a stick-slip regime at intermediate shear stresses and shear stress; (iii) a strong slip regime at high shear stresses [10].

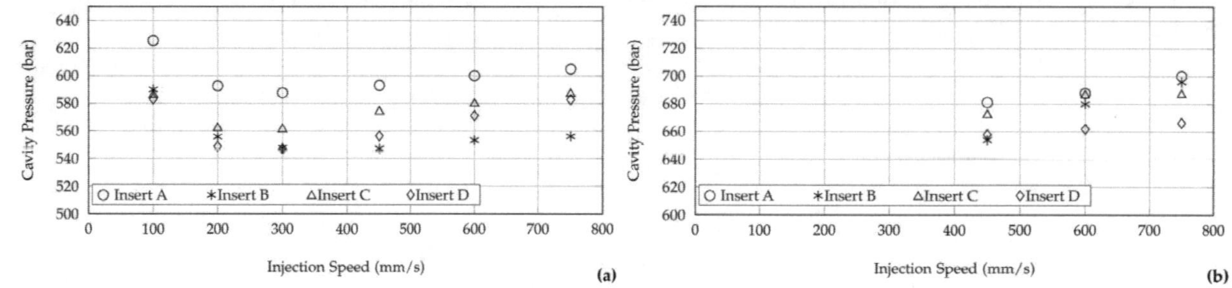

FIGURE 2. Cavity pressure measurements for PET at a) T_b= 300°C and b) T_b= 280°C.

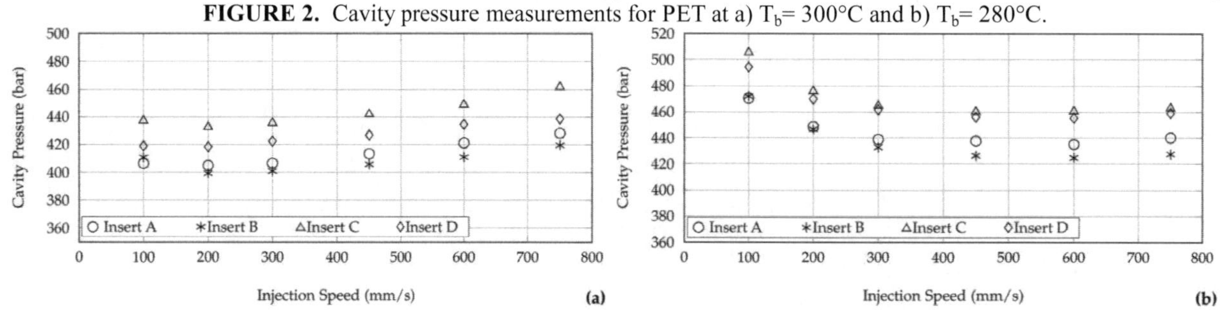

FIGURE 3. Cavity pressure measurements for PS at a) T_m= 120°C and b) T_m= 80°C.

Figure 4 depicts the flow curves of PET melts for the different inserts. For inserts A, C and D the shear stress gradually increases with shear rate with smaller slope and the polymer melt might be in the weak slip regime. For insert B the shear stress oscillates with an increasing shear rate. This indicates that polymer melt might be located in the stick-slip regime.

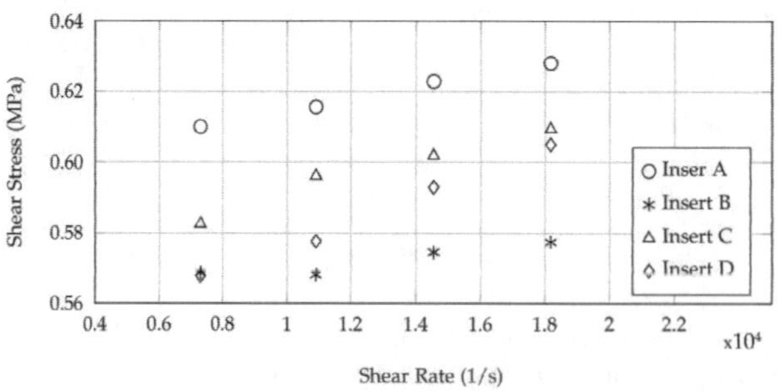

FIGURE 4. Cavity pressure measurements for PS at a) T_m= 120°C and b) T_m= 80°C.

With reference to insert A (uncoated) and C (DLC), Figure 5 shows a comparison between experimental and numerical results, the latter varying only the calibrated HTC value. The numerical simulation fits well the pressure results obtained for the uncoated insert even at high injection speed. However, for insert C a constant value of HTC is clearly incapable of reflecting the complex behavior of the polymer at the mould wall, which is likely to be influenced by wall slip. This calibration strategy allowed to separate the effects of heat conduction from the wall slip related to the low-friction coatings.

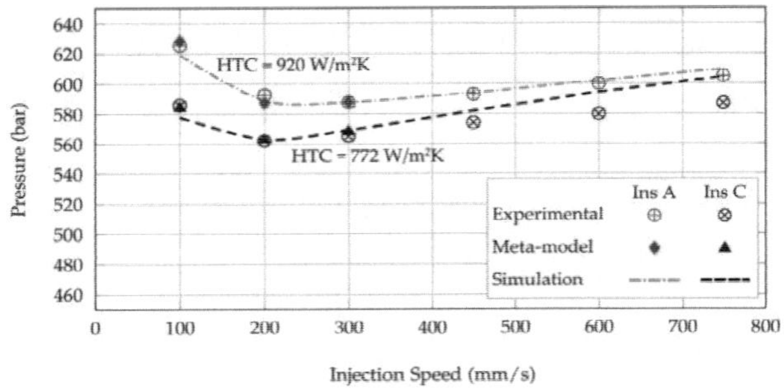

FIGURE 5. Cavity pressure measurements for PS at a) T_m= 120°C and b) T_m= 80°C.

Table 3 reports the values of HTC determined for each coating in relation to their characteristics in terms of thickness and thermal conductivity, as provided by the suppliers. Insert D has the lowest value of HTC, due to the low values of both thermal conductivity and thickness. Even though SiO_x and Al_2O_3 have the same thermal conductivity the coating of insert B is thinner, which results in a higher value of HTC. Eventually, insert C has the highest HTC due to its elevated thermal conductivity, which is still an order of magnitude lower than the mould steel.

TABLE 3). Values of HTC determined for the mold surface coatings.

Insert	**Coating**	**HTC** [W/m²K]	**Thermal Conductivity** [W/mK]	**Coating Thickness** [µm]
A	-	920	25÷30	-
B	Al_2O_3	758	1.5÷3	0.02÷0.05
C	DLC	772	5÷10	0.5÷5
D	SiO_x	736	1÷2	0.5÷5

CONCLUSIONS

This work focused on the effects of aluminum oxide, diamond-like carbon and silicon oxide coatings on the reduction of filling flow resistance in thin-wall injection molding of PET and PS. The flow curves of a slit cavity with thicknesses of 400 and 800 µm were analyzed and modeled. The experimental results indicated that all the coatings could be effectively exploited to reduce the cavity pressure (up to 8%) for PET. For PS only Al_2O_3 caused a reduction of the filling pressure of (up to 3%). The experiments also gave some insight into the influence of the process parameters. In particular, for PET the lower value of melt temperature attenuated the coatings effect. With PS, an increase in mould temperature reduced the influence of the coatings on the filling pressure.

High values of wall shear stress indicated the onset of a slip regime. As wall slip occurs at high shear rates, a numerical model was calibrated using the experimental results obtained at low values of injection speed. In this way a HTC was identified for each coating, separating the effects of heat conduction from the wall slip. Experimental and numerical results clearly indicate that the investigated mould coatings can be effectively exploited to reduce the filling pressure of PET, aiding the complete replication of the mould geometry. In particular, SiO_x is more effective in lowering the HTC at low shear rates while Al_2O_3 promotes the wall slip at the interface at high shear rates.

REFERENCES

1. Lucchetta G, Bariani PF, *CIRP Annals—Manufacturing Technology* **59**(1):33–36, (2010).
2. Lucchetta G, Sorgato M, Carmignato S, Savio E, *CIRP Annals - Manufacturing Technology* **63**(1):521-524, (2014).
3. Schmidt FM, Agassant JF, Bellet M, *Polymer Engineering & Science*, **38**(9):1399-1412, (1998).
4. Yao D, Kim B, *Polymer Engineering & Science*, **42**(12):2471-2481, (2002).
5. McFarland AW, Poggi MA, Bottomley LA, Colton JS, *Nanotechnology*, **15**:1628-1632, (2004).
6. Chen S-C, Chang Y, Chang Y-P, Chen, Y-C, Tseng C-Y, *Int. Com. in Heat and Mass Trans.*, **36**(10):1030-1035, (2009).
7. Laun HM, *Rheologica Acta*, **22**(2):171–185, (1983).
8. Sadeghi BHM, *J. of material Processing Technology,* **103**:411-416, (2000).
9. Quagliarella D, Periaux J, Poloni C, Winter G, *John Wiley and Sons*, (1998).
10. Mhetar V, Archer LA, *Macromolecules*, **31**(24):8607–8616, (1998).

Morphology and Electrical Properties of Injection-Molded PP Carbon-based Nanocomposites

M. Zaccone[a], I. Armentano[b], L. Torre[b], F. Cesano[c], S. Cravanzola[c], D. Scarano[c], A. Frache[d], M. Monti[a]

[a]*Proplast, Strada Comunale Savonesa 9, 15057 Tortona (AL), Italy*
[b]*Università degli Studi di Perugia, Strada di Pentima 4, 05100 Terni, Italy*
[c]*Università degli Studi di Torino, Via Pietro Giuria 7, Torino, Italy*
[d]*Politecnico di Torino, Corso Duca degli Abruzzi 24, Torino, Italy*

marta.zaccone@proplast.it
marco.monti@proplast.it

Abstract. The aim of this work was to investigate the influence of the process condition on the morphology of PP/MWCNT nanocomposites and its effect on the electrical properties, experimentally developed in this study. Electrical and morphological characterization was performed in order to analyze the multilayered morphology skin-core-skin, validated by a theoretical model.

Keywords: nanocomposites; carbon nanotubes; polypropylene; injection molding; skin effect
PACS: 81.05.U-, 81.07.-b, 82.35.Np, 84.37.+q

INTRODUCTION

Carbon-based nanomaterials with different shape and aspect ratio have attracted in recent years considerable attention in several industrial sectors [1, 2], and have been validated as effective fillers for modifying the electrical properties in the polymer matrix, in which they are embedded. Among these fillers, carbon black and expanded graphite are ones of the most used in plastic market for decades, while multi-walled carbon nanotubes (MWCNTs) has been identified in the last years as one of the most promising nanostructures to obtain multifunctional and smart materials [3]. Indeed, because of their high aspect ratio, they have a great tendency to form a conductive network within the polymer matrix, higher than other conductive additives [4].

Many factors can affect the conductivity of the polymer nanocomposites, such as the content and the type of the conducting carbon fillers, the nature of the polymer and finally processing condition [5, 6, 7]. As for the injection molding process, it has been reported that the produced samples show an external layer with a lower electrical conductivity, and this is favored by a low temperature mold [8].

From these considerations, it is clear that the filler dispersion is a key point in determining the conductivity performances of the composite [9, 10, 11, 12, 13].

In this paper, polypropylene (PP) nanocomposites with several content of conductive fillers were developed. Electrical characterization was performed in the two main directions (in-plane and through-thickness directions). The effect of processing condition on the anisotropic electrical percolation threshold and the morphology was deeply investigated and analyzed also by theorizing a conduction model.

EXPERIMENTAL

Materials and Methods

Multi-walled carbon nanotubes (Nanocyl NC7000, hereinafter referred to as MWCNTs), expanded graphite (Timcal Timrex C-Therm001, hereinafter referred to as EG) and two different kinds of carbon black (Timcal Ensaco 250 G and Timcal Ensaco 350 G, hereinafter referred to as CB-C and CB-S respectively) were used as fillers. Polypropylene (PP) Moplen RP 348 R (LyondellBasell) was selected as polymer matrix. It is a random copolymer, with a MFI of 25 g/10min (230°C – 2.16 kg) and a density of 0.9 g/cm^3. Different contents of fillers were added to PP. All fillers were homogeneously mixed with the polymer in a co-rotating twin-screw extruder. The obtained nanocomposites were injection molded to produce square-shaped samples (100 x 140 x 2.5 mm in sizes).

Direct current (DC) electrical characterization was performed in the two main directions, longitudinal to the flux of the material during the mold filling (in-plane) and in the through-thickness direction (through-thickness). Furthermore, a dedicated setup was developed to measure the influence of the z-coordinate in the electrical properties of a cross sectional area of the specimen. Tests were performed at intervals of 250 μm along the whole sample thickness, from the top to the bottom surfaces. The morphology of the samples has been investigated by means of a Zeiss Evo50 SEM.

RESULTS AND DISCUSSION

Electrical Properties and Morphology

Figure 1 shows the electrical resistivity as a function of the conductive filler content for the different produced materials: CB-C, CB-S, EG and MWCNTs. In both graphs, through-thickness direction (filled squares) and in-plane direction (empty squares) are represented. First of all, as it is reported in literature [14, 15], it can be observed that the electrical percolation threshold of EG and CB composites can be found at higher content of filler if compared with MWCNTs. However, the most interesting outcome shown in Figure 1 is related to the comparison of the electrical percolation threshold in the through-thickness direction with the in-plane one. This difference is more evident for CB-C composites, where the electrical percolation threshold is in the range of 14-15wt% for the through-thickness direction and 9-10wt% for the in-plane one, and MWCNT composites, where the percolation threshold is in the range of 3wt% for the through-thickness direction and 1-2wt% for the in-plane one. No significant variation in the percolation threshold was observed for CB-S and EG composites, even though in each nanocomposite at the same content of filler, lower values of resistivity were detected in the in-plane direction measurements.

A deeper investigation of the electrical behavior was performed on MWCNT-based nanocomposites, because among all the tested materials, they have the lowest percolation threshold. For this reason, in order to understand the change in the electrical properties in different directions, a thorough study on the influence of the z-coordinate in electrical properties of the cross sectional area was performed. The 2wt% was selected for this analysis, since it is the content at which the gap between the electrical resistivity in the two directions is highest.

Figure 2 shows the depth in which the test has been performed as a function of the surface electrical resistance. The right and left sides of the graph represent the two external surfaces of the injection molded sample, while the whole x-axis represent its thickness.

From this figure, it can be observed that the electrical resistance is not constant in the sample thickness, being higher moving towards the external regions from the center. From the graph, the thickness of the external less-conductive layers can be roughly estimated to be about 500 μm each, with a resulting thickness of the more-conductive region of about 1.5 mm.

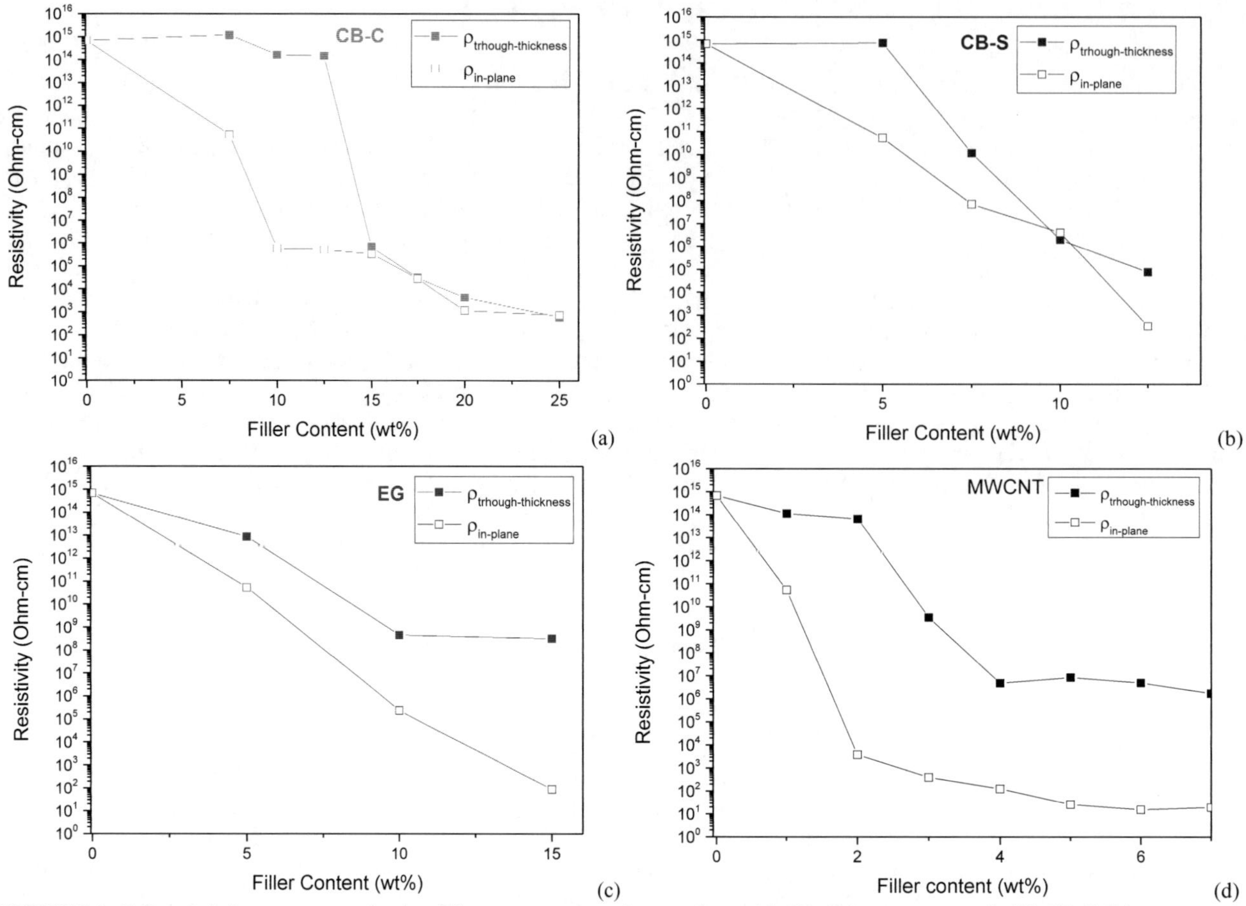

FIGURE 1. DC resistivity versus conductive fillers content in polypropylene: (a) CB-C (orange square), (b) CB-S (blue squares), (c) EG (red squares) and (d) MWCNTs (black squares). In both graphs, through-thickness direction (filled squares) and in-plane direction (empty squares) are represented.

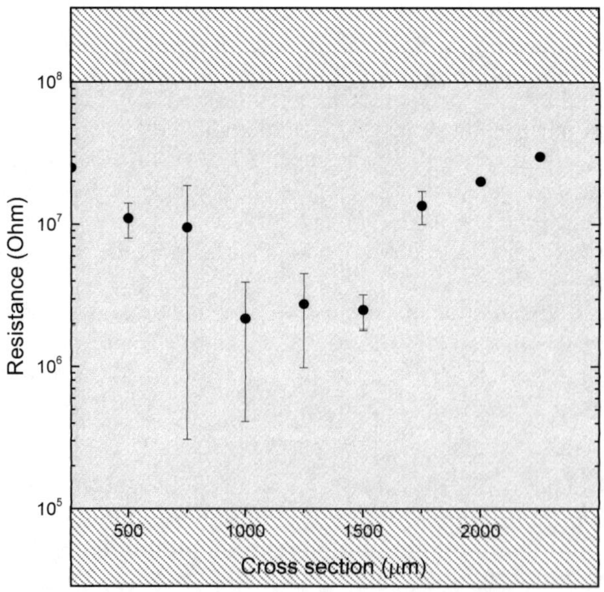

FIGURE 2. Surface resistance versus penetration depth for 2wt% MWCNTs–PP nanocomposite.

In order to explain the obtained results, a morphological characterization was performed by scanning electron microscopy (SEM). Figure 3 reports the micrographs of 2wt%, obtained both in the external layer (a) and in the internal region (b).

The 2wt%-MWCNT nanocomposite morphology is based on internal regions richer of MWCNTs than the external layers. The formation of a conductive network in the internal core leads to a conductive path, which can explain the before discussed electrical properties [16]. Consequently, it is possible to conclude that the skin layers, covering the external regions, are characterized by a low conductivity, while the core region already becomes conductive at low filler content.

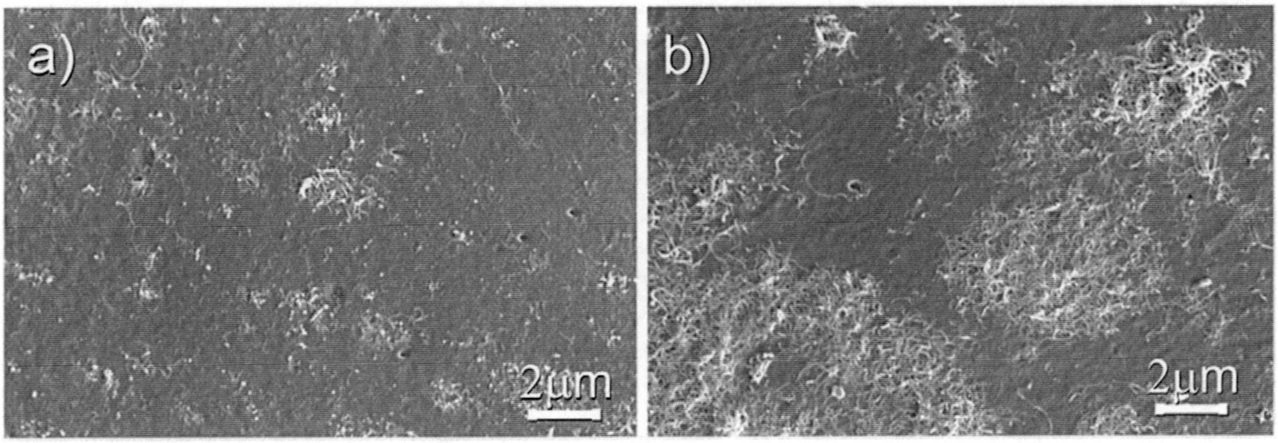

FIGURE 3. SEM images of MWCNT-based PP composites (2wt% MWCNTs) obtained in the external (a) and in the core regions (b) respectively.

Proposed Model for the Electrical Behavior

The results of the electrical characterization (Figures 1 and 2) can be exploited to develop a mathematical model for the through-thickness electrical behavior. The injection-molded component can be modeled as a multilayer system, being the external layers (the skin) less conductive and the internal core more conductive, due to the inhomogeneous distribution of MWCNTs in the two regions. The through-thickness overall electrical resistance can be simulated by a series of resistors corresponding to the four layers in which the cross-sectional area is divided. The in-plane electrical resistance is whilst taken as a parallel of resistors for the same system.

Therefore, the proposed model helps to detect a threshold between the regions with different amount of resistivity. It can be used to identify in the cross-sectional area the skin layers as the less conductive parts and the core region as the more conductive one. The results of the system have been calculated for the 2wt% MWCNT nanocomposite and compared to the resistivity in the through-thickness and in-plane directions, experimentally measured.

The experimentally measured resistivity in the through-thickness direction ($\rho_{through\text{-}thickness}$) is 6.65E+13 Ohm·cm, while the in-plane resistivity ($\rho_{in\text{-}plane}$) is 1.17E+05 Ohm·cm. These values fit in with the results obtained with the proposed theoretical model. In fact, the calculated skin resistivity (ρ_{skin}) is 1.50E+14 Ohm·cm, instead the calculated core resistivity (ρ_{core}) is 1.62E+04 Ohm·cm.

The proposed model allows one to understand the electrical behavior of an injection-molded component, taking into account the skin effect, which is induced by the flow of the molten material inside the mould, during manufacturing.

CONCLUSIONS

The electrical characterization of the produced carbon nanocomposites has shown that the electrical behavior changes when the measurements are performed in different directions. In fact, the electrical percolation threshold is higher for the through-thickness direction than for the in-plane one. This aspect is supported by the morphological characterization, which has highlighted the presence of a percolative conductive network of MWCNTs in the through-thickness area.

Moreover, the electrical resistance in the cross-sectional area increases when moving from the internal area toward the external layers. This can be explained by the skin effect of the external regions, MWCNT-less rich areas.

Indeed, it can be concluded that the injection molding process induces an inhomogeneous distribution of the MWCNTs in the through-thickness direction, which influences the electrical behavior of the injection molded components. This behavior can be predicted and mathematically schematized by the simple phenomenological model proposed in this study.

REFERENCES

1. M.H. Saleh, W.H. Saadeh, U. Sundararaj, *Carbon* **60**, 146-156 (2013).
2. K. Hu, D.D. Kulkarni, I. Choi, V.V. Tsukruk, *Prog Polym Sci.* **39**(11), 1934-1972 (2014).
3. S. Cravanzola, S.M. Jain, F. Cesano, A. Damin, D. Scarano, *RSC Advances* **5**(125), 103255-103264 (2015).
4. Z. Wei, A. Dehghani Sanij, R. Blackburn, *J Mater Sci.* **42**, 3408-3418 (2007).
5. G. Haznedar, S. Cravanzola, M. Zanetti, D. Scarano, A. Zecchina, F. Cesano, *Mater Chem Phys* **143**(1), 47-52 (2013).
6. G. Mittal, V. Dhand, K.Y. Rhee, S-J. Park, W.R. Lee, *J Ind Eng Chem* **21**, 11-25 (2015).
7. N.G. Sahoo, S. Rana, J.W. Cho, L. Li, S.H. Chan, *Prog Polym Sci.* **35**(7), 837-867 (2010).
8. G. Kasaliwal, A. Göldel, P. Pötschke, *J Appl Polym Sci.* **112**(6), 3494-3509 (2009).
9. P-C. Ma, N.A. Siddiqui, G. Marom, J-K. Kim, *Comp Part A: Appl Sci Manufact.* **41**(10), 1345-1367 (2010).
10. S-J. Park, S-Y. Lee, F-L. Jin, "Surface modifications of carbon nanotubes for high-performance polymer composites", in *Handbook of Polymer Nanocomposites Processing, Performance and Application*, edited by K.K. Kar, J.K. Pandey, S. Rana, Heidelberg: Springer, 2015, pp. 13-60.
11. R.M. Mutiso, K.I. Winey, *Prog Polym Sci.* **40**, 63-84 (2015).
12. T. Villmow, P. Pötschke, S. Pegel, L. Häussler, B. Kretzschmar, *Polymer* **49**(16), 3500-3509 (2008).
13. M. Monti, I. Armentano, G. Faiella, V. Antonucci, J.M. Kenny, L. Torre, et al, *Comp Sci Technol.* **96**, 38-46 (2014).
14. S.K.H. Gulrez, M.E.A. Mohsin, H. Shaikh, A. Anis, A.M. Pulose, M.K.Yadav, et al, *Polym Compos* **35**, 900-914 (2014).
15. K. Kalaitzidou, H. Fukushima, L.T. Drzal, *Materials* **3**, 1089-1103 (2010).
16. I. Alig, P. Pötschke, D. Lellinger, T. Skipa, S. Pegel, G.R. Kasaliwal, et al., *Polymer* **53**(1), 4-28 (2012).

Modular Technological Inserts for Injection Molding Machines

D. Schneider, Dr.-Ing. Volker Reichert; A&E Produktionstechnik GmbH

It is easy to mount a mold into an injection molding machine – if all the parts fit perfectly. This means the dimensions of the clamping plates, of the centering hole and centering collar, radii and flat faces of nozzles, sprue bushes and valves all have to make a perfect match. The same applies to the characteristics of the clamping unit and the injection unit. On the one hand they have to fulfill the requirements of the injected part, which specifies the shot weight or the projected area. On the other hand the maximum injection volume of the machine has to be taken into account. [1], [2]

Taking a look into everyday practice makes it very obvious that the assumption of perfectly fitting parts proves to be applicable only to rare occasions. Usually injection molders own a range of general-purpose machines to put them in the position for an economically appropriate yet flexible production.

On the contrary molds usually are designed for one single machine configuration. This leads to the situation, that it is not an uncommon request to use molds in combination with a machine they initially weren't designed to be used with.

This problem usually will be solved in two ways. The first option is to wait for a machine of the needed configuration to have free capacity. The second one is buying another machine of the exact same configuration.

Another way of solving this problem is to add intermediate plates to the machine. This is the ideal solution especially for smaller or middle sized machines with a clamping force up to approximately 4000 kN or 5000 kN making existing machines more flexible in a cost-effective manner by allowing users to install also molds on these machines that were originally not foreseen for them. For larger machines flexibility usually plays a rather minor role as mold changes are rarer, nevertheless they can be equipped with intermediate plates as well when needed. The use of intermediate plates is common in industrial practice when it is required to close gaps between small molds and big machines if intended to use in combination, as it is shown in figure 1.

Grafik: Firma Engel

Fig. 1: Intermediate plates for adjustment of minimum mold installation height (Source: Engel)

In contrast to these distance plates which are solely adjusting installation height the plates developed by the authors contain additional functionality. They act as technological inserts which, integrated as additional modules, can adjust the characteristics of a standard injection molding machine as needed in the process. By this it is possible to enable a general-purpose machine to match special molds or to perform various special injection molding methods. Examples are, to name but a few:

- The integration of one or two custom built hot runners.
- The integration of valves which join different melts to set up a co-injection process or to set up sandwich injection molding creating a core and skin structure.
- To apply the melt with additional energy or to add material during injection or during cycle breaks. Examples are the application of ultrasound, the integration of chemical or physical foaming or the addition of colors, fillers or reinforcing materials and agents.

Hence the modular technological inserts act as technology building blocks which can easily be integrated into existing injection molding machines and processes.

Taking into account the numerous functions and options which the integration of these inserts renders possible a huge application variety evolves which one won't always be fully aware of during decision and development processes. The potential application range those components are able to add to standard injection molding machinery can be captured using a newly developed technology assessment method. The method bundles and focuses technology aspects by the use of modeling techniques to conquer the full potential of the given technological resources.
Similar to the use of modeling techniques and simulation in engineering processes this calculation method aims at the facilitation of complex data to find good results without tremendous effort. Therefore the development process is permanently monitored by the description and rating of its possible decision paths. The technological and economical description and rating steps are realized adapting various methods of the field of business informatics, like for example modeling languages, semantic modeling or business process modeling. [3]

The variety of possible applications for hot runner intermediate plates to draw upon for this process and the implications which have to be taken into account during the process can be explained based on some common intermediate plate applications. These applications involve injection molding machines with different sizes and configurations and requirements starting from the simple adjustment of mechanical interfaces followed by the integration of additional control, sealing and flow enhancement up to special injection molding processes like sandwich injection molding.

In general hot runners as used also in injection molds guide the melt from the machine nozzle to the mold sprue bush or within the mold from the sprue bush to the cavities. Within a mold there usually is a rigid connection between runner and cavity. In intermediate plates, hot runners are movable so that they can absorb the nozzle contact force of the injection molding machine while remaining able to transmit the force of the hot runner nozzle that is applied on the mold sprue bush. This nozzle contact force for the connection between mold sprue bush and hot runner nozzle is generated by springs or a wedge. It is also possible to generate this force by thermal expansion or to allow the latter to act in addition to the spring force. When using an intermediate plate the connection between hot runner nozzle and mold sprue bush must always be designed in a way that ensures a leak-tight melt flow into the mold.

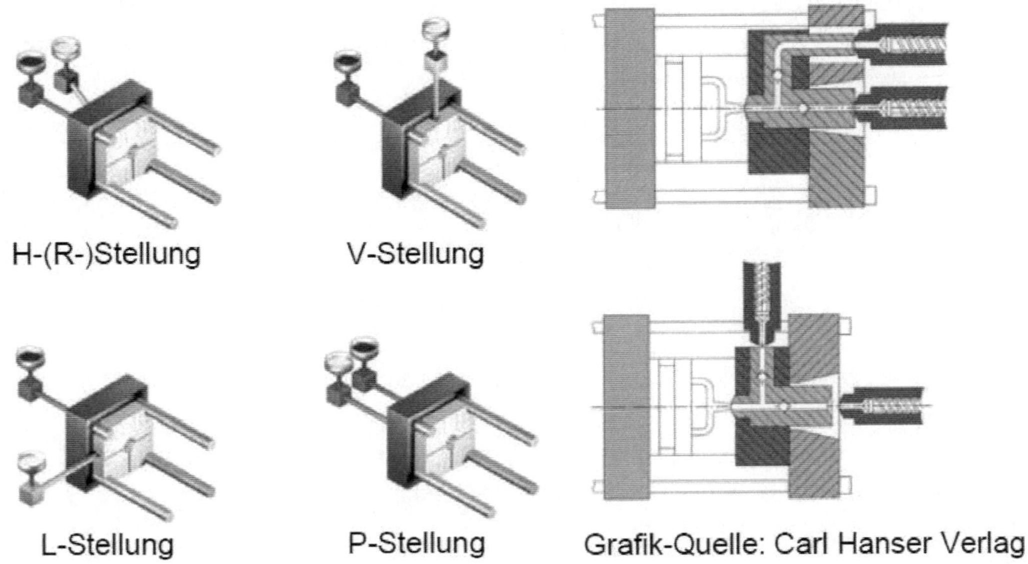

Fig. 2: Variants of sandwich units in intermediate plates for different positions of the injection units of the machine (Source: Carl Hanser Verlag)

The requirements for 1-component (standard) injection moulding to ensure that mold and machine are a perfect match are extended by another aspect that needs to be considered for multi-component injection molding machines: The nozzles on the injection units of the machine have to reach the sprue bushings on the mold, as it is shown in figure 2 and 3. For parallel horizontally located injection units it is also possible that both injection units are offset from the actual machine center line so that two hot runners are required in the intermediate plate.

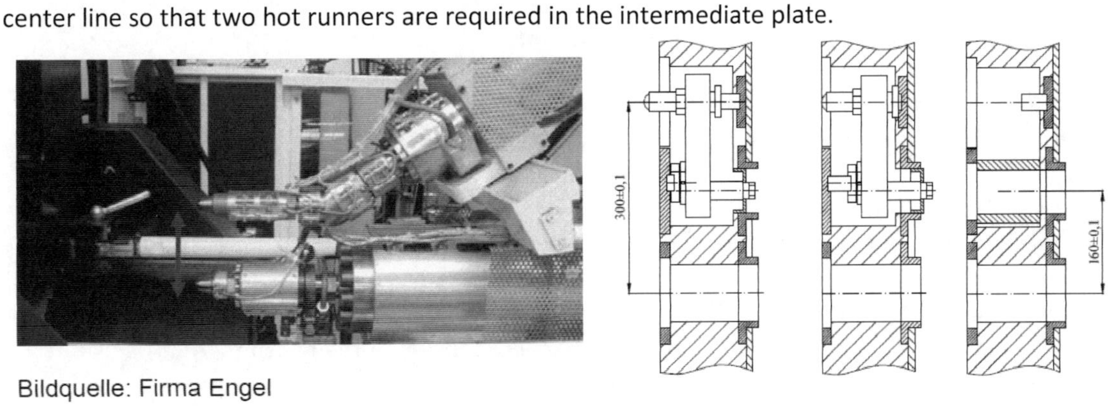

Fig. 3: Application of hot runner intermediate plates for an injection moulding machine with injection units arranged one above the other with a nozzle clearance of 160 mm. As the plastic processor wanted to use also 2-component moulds with a clearance of 300 mm of the sprue bushings arranged one above the other on this machine, this was made possible by an inter-mediate plate with hot runner melt diversion from 160 mm to 300 mm. The dimensions of the hot runner intermediate plate are 1,150 x 1,030 x 136 mm.This intermediate plate has been in use since 2001. (Source: Engel)

One possibility to enhance process reliability and to permanently ensure the precise opening and closing of the melt channel especially when processing melt containing different fillers is given by locking systems of various kinds. They can be integrated as well behind the hot runner sprue bush as behind the hot runner nozzle. Those so-called machine shut-off nozzles or hot runner shut-off nozzles can either be activated by springs, by melt pressure or by hydraulic or pneumatic cylinders according to the available installation space. [4]

Another example of the opportunities given by the use of intermediate plates as modular technological inserts is the implementation of sandwich injection moulding. In the sandwich process, also called co-injection, two different melts are injected successively and/or simultaneously into the cavity via a single nozzle orifice. A skin-core structure forms due to the developing source flow of the polymer melts in the cavity that fills during injection and the relative low thermal conductivity of the plastic melts.
To allow this process on commercial multi-component injection moulding machines melt valves, also called 2-component units, can be integrated in intermediate plates. They bring together two different melt flows to a single nozzle simultaneously or successively.
The authors refrain from a description of the process sequence and special features and refer to their relevant literature. [5], [6]

Typical applications for this technology include [7]:

- The use of regrind/recyclate within the core of the part to decrease material cost, because recycled material can be assumed to be 70 % to 80 % less expensive in comparison with virgin material
- The addition of fillers and reinforcement materials within the core layers while preserving a smooth and ready to paint surface layer
- The foaming of thermoplastic core material by physical or chemical foaming technologies while preserving a solid surface layer without foamed structure.

Also intermediate plates with 2-component sandwich units feature the functions already described for the hot runner intermediate plates:

- Absorption of the nozzle contact force from one or both plasticising and injection units of the machine in the intermediate plate
- Build-up of the nozzle contact force to the mould's sprue bushing by springs, mechanical actuators and/or utilization of the linear thermal expansion of the hot runner in the intermediate plate
- Sealing and control functions of the melt valve of the 2-component sandwich unit cause a clear separation between skin and core material in the moulded part and prevent drops from being formed and melt leakage from entering the mold sprue.

What this all amounts to is that on the one hand a newly developed technology assessment method enables the handling of an overwhelming pool of process possibilities by the systematic allocation and aggregation of already existing or soon to be purchased industrial resources and thus supports a successful strategic targeting. At the same time more than ten years of market experience clearly proof intermediate plates used as modular technological inserts to considerably increase flexibility and application range of injection molding machines while .

[1] Reichert, V.; Becker, H.: Ventile steuern den Fluss. In: Kunstoffe, Nr. 11, 2009, S. 39–42

[2] Reichert, V.; Schneider, D.: Hotrunner intermediate plates for multicomponent injection moulding and hot runner adapter plates for use in standard injection moulding processes. In: Advances in Plastics Technology 2015. Sosnowiec, 2015

[3] Braun, R.; Esswein, W.: Entwicklung einer BPMN-Extension für ressourcen-intensive Prozesse im Maschinenbau. In: *Multikonferenz Wirtschaftsinformatik 2014*. 2014, S. 1574–1586. DOI: http://www.mkwi2014.de/Content/Tagungsband_MKWI2014.pdf

[4] Schüpbach, K.: Maschinenverschlussdüsen System ENG für hohe Prozesskonstanz. In: Unveröffentlichte Firmenschrift, KS Engineering GmbH, Egerkingen 2014

[5] Reichert, V.; Nendel, W.: Einspritzlösung für Sandwich-Spritzgießen. In: Plastverarbeiter, Nr. 12, 2014, S. 28–31

[6] Reichert, V.: Einsatz von 2K-Werkzeugen erweitert. In: K-Zeitung Kunststoff und Kautschuk, Nr.13, 2011, S. 16

[7] Reichert, V.: Leichtbau durch den Einsatz von mehrschichtigen Kunststoffteilen. In: Dresdner Transferbrief, Nr. 3, 2009, S. 19

Induction Heating Simulation and Validation for the Plastic Injection Molding Process

Clinton Kietzmann[a], Guillaume Rigoni [b*], Lu Chen[c] and Stéphane Quilliet[b]

[a]*Autodesk Australia Pty Ltd, Moldflow R&D Center, 259-261 Colchester Road, Kilsyth Vic 3137, Australia*
[b]*RocTool, Savoie Technolac, BP341, Modul R, 73375 Le Bourget du Lac cedex, France*
[c]*Autodesk ACRD, 6F-7F No. 130, Lane 91, E Shan Road, Shangha,i China 200127*

Clinton.Kietzmann@autodesk.com
Guillaume.Rigoni@roctool.com

Abstract. In mold induction heating is proving to be an efficient method of producing high quality low cost injection molded parts. Autodesk Moldflow is the world leader in plastic injection molding simulation products. RocTool® is the foremost technology provider of induction heating equipment for plastic injection and compression molding. Autodesk Moldflow and RocTool have collaborated to develop a three dimensional finite element simulation of the Maxwell equations for the induction heating process in plastic injection molding. This paper describes the key features of this electromagnetic solver together with validation results from tests conducted on existing molds.

In order to simulate induction heating a true three dimensional finite element solution of both the real and imaginary components of the Maxwell equations was obtained, without any simplifications or assumptions. The derivation and the numerical techniques used in implementing this unique solution of the Maxwell equations on the injection mold domain are referenced.

Simulation of induction heating offers the mold designer an insight into the mechanisms of induction heating. Induction heating creates a very hot non-uniform temperature distribution concentrated in the surface skin of the mold body very quickly. Ideally the surface skin layer touching the part is at the material's melt temperature during the filling phase of the injection molding cycle. With most of the heat concentrated in the very thin skin region of the mold touching the part, the mold can be cooled down to ejection temperature very quickly thereby, ensuring a desired short cycle time.

In the paper, induction heating simulation is demonstrated on a real world model. The simulated results are compared to results provided by RocTool that were measured in their test facility. Any discrepancies between the simulated results and the experimental results are discussed in detail.

Keywords: Electrical, Electromagnetic, Injection Molding, Simulation

PACS: 41, 75

INTRODUCTION

Traditional injection molding design aims to maintain the mold at a constant temperature for the entire injection molding cycle. In order to achieve this, coolant is pumped through the mold cooling channels with constant inlet temperatures [1].

New molding technologies described variously as "Rapid Heating and Cooling", "Variotherm" or "Rapid Temperature Cycling" varies the temperature of the mold during the cycle. The aim is to have a mold cavity surface temperature above the glass transition temperature of the melt during the mold filling stage and to reduce the surface temperature to the ejection temperature during the packing and cooling stages of the injection molding cycle to reduce cycle time.

The high mold cavity surface temperature during filling, results in significantly less visible and stronger weld lines in the finished part. Weld lines are inevitable when holes or grills are present in the part or if parts are filled by conventional multi-gated runner systems. Visible weld lines weaken the aesthetic appearance of the part and can significantly weaken the part structurally, especially for fiber filled materials where the fibers do not bridge the weld.

The high mold cavity surface temperature during filling results in high gloss surface finishes of plastic parts avoiding the need for other post molding surface enhancements such as painting. A hotter mold during filling, results in better control of injection pressures. Better control of injection pressures allow molded parts to have reduced wall thicknesses and lower rib ratios avoiding sink marks while providing the molder more choice in relation to gating position, all of which reduce cost.

The benefits of rapid heating and cooling processes are particularly favorable when molding clear components such as lenses and flat screen television housings among other components.

Various rapid heating and cooling technologies use different methods of achieving the same outcome of a hot mold during filling. The most common heating methods considered are saturated steam, electric cartridge and pressurized hot water heating [2]. These methods are widely used in industry and have been reported on before [3]. The disadvantage of these traditional methods is that they have higher cycle times leading to lower productivity. The density of the heat flux is very low resulting in large areas of the mold heating up which have to be cooled again within the same cycle. Steam and hot water are also considered to be hazardous materials in the factory from a health and safety point of view.

Electromagnetic induction heating allows for lower cycle times due to the precise control and positioning of the heat flux density and rate within the mold. Induction heating relies on the cycling of an alternating current through a mold designed specifically for this purpose. The electromagnetic induction heating process, its application to injection molding and the process by which it is simulated is explained.

Autodesk Moldflow, the world leader in plastic injection molding simulation products and RocTool®, the foremost technology provider of induction heating equipment for plastic injection and compression molding have collaborated to develop a three dimensional finite element simulation for the induction heating process in plastic injection molding.

Finally, induction heating simulation results are demonstrated on a model of a part provided by RocTool®, who also provided the relevant material properties and the temperature validation data for this simulation after manufacturing the part on an instrumented mold in the RocTool® test facility.

ELECTROMAGNETIC INDUCTION HEATING

Electrical induction heating is a complex combination of electromagnetic, heat transfer and metallurgical phenomena [4]. The electromagnetic phenomena occur when a rapidly alternating voltage is applied to an induction coil. The alternating voltage results in an alternating source current in the coil. The alternating source current produces a time variable magnetic field within its immediate surroundings of the same frequency as the source current. This magnetic field induces eddy currents in its surrounding electrical conducting metal objects. The induced eddy currents have the same frequency but different direction to the source current. These induced currents produce Joule heating (I^2R) in the surrounding metal objects. The current distribution is never uniform hence the heat source is never uniform; hence the temperature profile is never uniform. Alternating current tends to run at the surface of the body; hence the heat source in induction heating is concentrated close to the surface. This is known as the skin effect [4].

Induction heating is entirely dependent upon the metallurgical properties of the metals on which it is being applied. Metals have electromagnetic properties in the same way as they have thermal and structural properties. The most important electromagnetic properties needed for simulating induction heating are the metal's electrical conductivity and magnetic permeability. The electrical conductivity σ is the materials ability to conduct electrical current. Good electrical conductors have high conductivity or low resistivity like copper. The magnetic permeability μ_r of the material indicates the ability of the material to conduct magnetic flux

Alternating current in a conductor has the maximum current concentrated in the conductor's surface as opposed to direct current that has the maximum current concentration at its center. The current distribution with alternating current is known as the skin effect. The skin effect occurs with the induced current in the surrounding metal objects being magnetized, with no current flow in the center of the object. Approximately 86% of the power is concentrated in the surface layer of the conductor. This surface layer is called the penetration or skin depth δ and can be derived from the Maxwell equations

$$\delta = \frac{1}{\sqrt{\pi f \mu \sigma}} \qquad (1)$$

where f is the frequency in Hertz [Hz]. Typical frequency ranges for induction heating are 5 kHz to 100 kHz. From equation (1) it can be seen that the skin depth depends on the conductor properties and the frequency of the current in the conductor. When simulating the induction heating the skin depth given in equation (1) is a very important parameter when creating finite element meshes.

Electromagnetic field modeling

The electromagnetic phenomena of induction heating are described mathematically by the Maxwell equations. Maxwell's equations in differential form can be written as

$$\nabla \times (\vec{H}) = \vec{J} + \frac{\partial \vec{D}}{\partial t} \quad \text{(from Ampere's law)} \qquad (2)$$

$$\nabla \times (\vec{E}) = -\frac{\partial \vec{B}}{\partial t} \quad \text{(from Faraday's law)} \qquad (3)$$

$$\nabla \cdot (\vec{B}) = 0 \quad \text{(from Gauss's law)} \qquad (4)$$

$$\nabla \cdot (\vec{D}) = \rho^{charge} \quad \text{(from Gauss's law)} \qquad (5)$$

Where $\vec{H}$ is the magnetic field intensity, $\vec{J}$ is the conduction current density, $\vec{D}$ is the electric flux density, $\vec{E}$ the electric field, $\vec{B}$ the magnetic flux density and ρ^{charge} the electric charge density with t being time.

The derivation and reduction of the Maxwell equations that are applicable to induction heating are described in detail in Kietzmann et al ref [5]. The final derived quantity from the Maxwell equations that is used is the Joule heat $\dot{Q}$ in the conductors given by

$$Q = \frac{1}{2\sigma}\left|\vec{J_e}\right|^2 \qquad (6)$$

The Joule heat Q is the volumetric heat source with units (W/m^3) that is induced by the time harmonic induced eddy currents $\vec{J_e}$ in the conductor. The heat transfer phenomena taking place in induction heating is the heat conduction within the conductor and is described by the transient heat conduction equation that is used in all the simulations [6].

$$\rho C_p \frac{\partial T}{\partial t} = k\, \nabla^2 T + Q \qquad (7)$$

Where T is temperature, ρ the density, C_p the specific heat capacity, k the thermal conductivity of the material and Q the Joule heat from equation (6). The reduced Maxwell equations are solved using the finite element method on the same finite element domain as that which equation (7) is solved on with the Joule heat Q being the source term.

Simulation technology

Induction heating simulation is dependent upon the correct specification of the magnetic permeability and electrical resistivity of the mold material along with its thermal conductivity, density and specific heat. The placement of the exact geometry of the induction coils in the mold inserts needs to be modeled precisely. The frequency of the power generator providing the alternating current to the induction coil needs to be specified together with the timer settings. The timer settings refer to the specific time during the injection molding cycle when the alternating current to the induction coil is switched on and when it is switched off. The root mean square (RMS) source current being applied to the induction coil has to be specified too. Induction heating relies on the skin effect; equation (1) which depends upon the material properties in the mold and the frequency of the alternating current in the induction coils. It is advisable to calculate the skin effect thickness on all the bodies and mold inserts in the mold before meshing and then to mesh these components to a matching mesh density. This would ensure the heat generated in the skin thickness would be captured entirely, resulting in a very accurate solution.

Critical areas of the mold are defined as the areas between the induction coils and the part, especially the areas of the mold touching the part. The areas between the induction coils and the mold base that is fastened to the injection molding machine are less critical. It is advisable that all areas of the mold be meshed to these requirements; however this can result in a huge number of elements in the mesh. In order to lower the element count the user can resort to coarser meshes in the less critical areas of the mold.

Validation and verification

In order to validate the induction heating simulation the instrumented mold shown in figure 1 was tested in the laboratory and the temperatures on the surfaces recorded. An identical simulation was performed and the results compared to the actual results. The total cycle time for the manufacture of the part was 48 seconds with a combined injection, packing and cooling time of 35 seconds. After cooling the mold opens and the part gets ejected. The induction coils then turn on with 455 ampere (RMS) source current at a frequency of 48 kHz to heat the mold for 7 seconds. After the heating the mold remains open for a further 4 seconds before closing for a further 2 seconds before the next cycle commences. During the injection molding cycle water at 30 °C is cycled at between 6 and 8 l/min through the cooling channels.

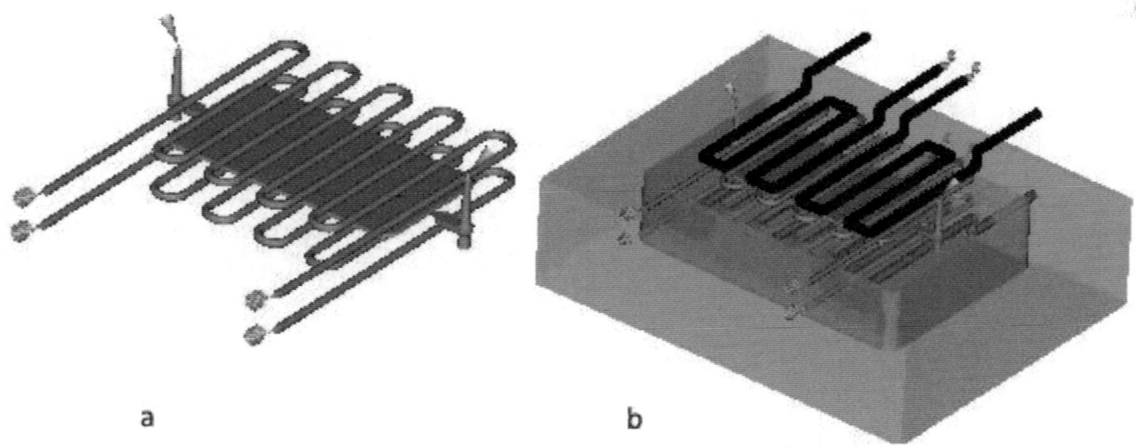

FIGURE 1. Part mold combination used for simulation. (a) Shows the part and the induction coils. (b) Shows the part, induction coils, cooling channels, moving half insert and moving half mold block

Results and discussion

Figure 2 is the plot of a temperature sensor on the mold cavity, part interface for an entire injection molding cycle of 48 seconds. The solid line is the measured result from the experimental mold and the dotted line the equivalent result from the simulation. As can be seen from figure 2 the simulated result matches the experimental result very closely. As no assumptions or simplifications were made the mesh had to be very fine resulting in a slow solver execution time. For this reason the simplification of using the impedance boundary condition, which is based on the surface mesh, will be investigated with the aim of achieving the same accuracy as the current solver shown in figure 2. The current solver is based on the volumetric mesh hence has a slow execution time. To simplify the specification of the current and frequency boundary conditions, provision will be made for the user to directly specify the power generator in the simulation.

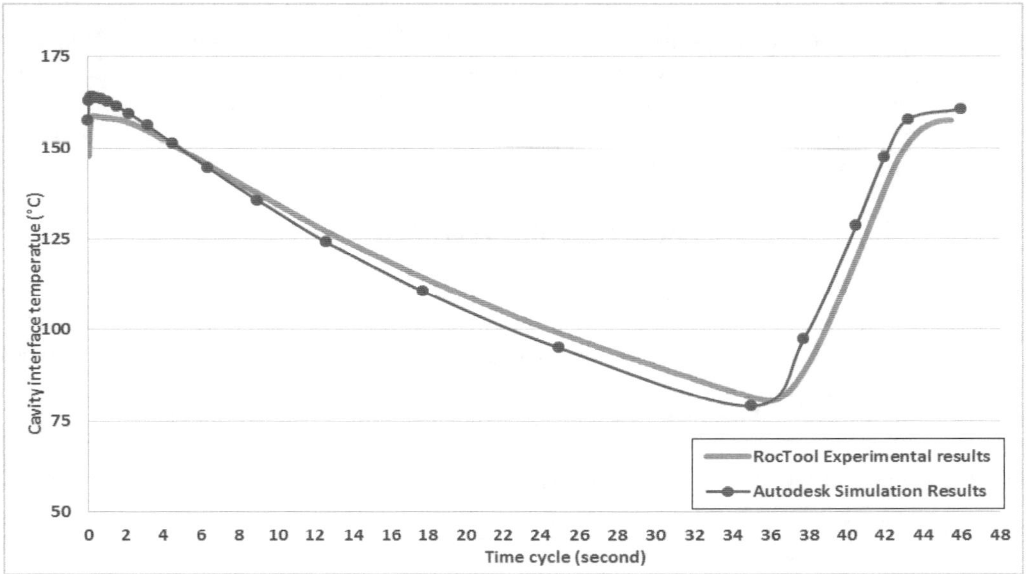

FIGURE 2. Mold cavity surface interface temperature evolution during the injection molding cycle

Conclusion

The simulated results match the experimental results very closely. The excellent results achieved with the current solver can be used as a benchmark for developing the surface mesh impedance boundary condition method which is a simplification of the current volume based solution. This would speed up solution time and simplify the meshing process. Provision will also be made for direct specification of the power generator in the simulation. The current implementation of the induction heating solution provides worthwhile functionality to the existing transient cool solver.

REFERENCES

1. Shoemaker Jay (Editor), (2006), Moldflow *Design Guide. A Resource for Plastics Engineers.* Hanser
2. Kietzmann C, Chen L, ,Lin-Feng H, Costa F, (2012), Rapid Heating and Cooling Simulation for the Injection Molding Process, *SPE ANTEC Proceedings*, Orlando
3. Chen Lu, Kietzmann Clinton, Astbury David, Shao Liang, (2013) Advanced Cooling Simulation Technologies for the Injection Molding Process, *SPE China Proceedings*, Shanghai
4. Rudnee V et al. (2003) Handbook of Induction Heating, Marcel Dekker, Inc
5. Clinton Kietzmann, José Feigenblum, Stéphane Quilliet, Lu Chen, David Astbury, (2015) Induction Heating Simulation for the Plastic Injection molding process, NAFEMS World Congress Proceedings, San Diego
6. Kietzmann C, Chen L, Lin-Feng H, Costa F, Le Goff R, (2011) Transient Mold Cooling Simulation for the Injection Molding Process, *SPE ANTEC Proceedings*, Boston.

Taking Into Account Normal Stresses During Simulation Of Thermoplastic Injection Molding

Alexander Bakharev [a], David Astbury [a], Shishir Ray [a], Franco S. Costa [a] and Russell Speight [a]

[a] *Autodesk, 259-261 Colchester Rd., Kilsyth, Victoria 3137, Australia*

Alex.Bakharev@autodesk.com

Abstract. The article deals with the effect of the first and the second normal stress difference on the predicted flow front and pressure distribution during the filling stage of the thermoplastic injection molding. The normal stresses are predicted using the non-linear Criminale-Ericksen-Filbey model incorporated into a comprehensive 3D simulation software for mold-filling analysis. The model parameters are estimated from the viscosity curves using the Abnormal Cox-Mertz rule. The additional term improves the prediction of the so called ear-flow effect (melt racing on the edge of the cavity) .

Keywords: Finite Elements, Injection Molding, Simulation, Rheology, Normal stress difference, Ear-flow

PACS: 83.50.Uv

INTRODUCTION

Thermoplastic injection molding is the most common manufacturing process for producing plastic parts. Material is fed into a heated barrel, mixed, and forced into a mold cavity where it cools and hardens to the configuration of the cavity. Significant progress has been achieved in three dimensional finite element simulation of plastic filling the mold (mold filling analysis). [1] Typically in commercial simulations polymer melt is considered as generalized Newtonian fluid, there the deviatoric stress tensor is proportional to the deviatoric deformation rate, the scalar coefficient connecting the shear rate and shear stress, known as viscosity, is dependent upon temperature, shear rate invariants, pressure and other factors. [2] However, the generalized Newtonian fluid model does not predict any normal stress differences during simple shear flow whereas real polymers usually exhibit significant normal stress differences.

In this current work, we develop a finite element mold-filling program that allows incorporation of the Criminale-Ericksen-Filbey viscoelastic model that can accurately predict normal stress differences in a wide range of temperatures and shear rates. [3] One of the motivators for this development was potentially improving the prediction of ear-flow, a little understood phenomenon of the more rapid advance of the flow front on the edge of a mold cavity than in the center of the cavity [4].

EAR-FLOW PHENOMENON

Numerous cases have been observed in industrial injection molding practice of amorphous materials which exhibit a race-track or ear-flow effect. This is the more rapid advance of the flow front at the edges of the molding cavity than in the center of the cavity, usually observed at elevated injection speed. This flow-leading effect at the edge cannot be explained by differences in cavity wall section thickness. In the worst cases, this race-tracking leads to air-traps and visual defects in the molded part. A typical flow front propagation demonstrating the ear-flow in a polystyrene material is shown on **FIGURE** 1.

As was originally suggested by experimental observations of Murata *et al.* [5], and confirmed by mold-filling simulations of Costa *et al.* [4], in many cases the ear flow is caused by the higher polymer temperatures in the edge region. The temperature rise is caused by shear heating in the high-shear region (close to the boundary) of the runners and gates. This temperature rise is then convected into the cavity, favoring the cavity edge due to the distribution pattern in the gate. The effect of the preferential convection of the temperature rise from the shear regions of the runners is also similar to the development of flow imbalance in geometrically balanced feed systems explained by Cook *et al.* [6]

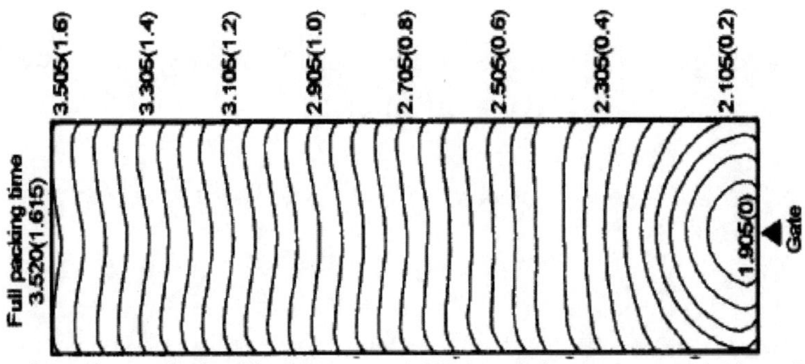

FIGURE 1 Flow front shapes obtained by Murata et al. [5] for a Polystyrene material in a glass insert mold

However in many practical injection molding cases ear flow phenomenon is observed for conditions where very little shear heating is predicted, thus, rising suspicions that there is also another mechanism responsible for the ear flow effect. Since normal stress differences caused by in-plane shear near the cavity edges may push polymer perpendicular to the flow they are a candidate for such a mechanism. Our simulation was used to test this possibility.

MATHEMATICAL MODEL

Filling stages of injection molding process are described by the combination of the momentum equation (1):

$$\frac{\partial}{\partial t}(\rho\vec{v}) = \rho\vec{g} + (\nabla\cdot\underline{\sigma}) - (\nabla\cdot(\rho\vec{v}\vec{v})) \tag{1}$$

mass continuity equation (2):

$$\frac{\partial\rho}{\partial t} + \nabla\cdot(\rho\vec{v}) = 0 \tag{2}$$

and the energy equation (3):

$$\rho c_p\left(\frac{\partial T}{\partial t} + \vec{v}\cdot\nabla T\right) = \beta T\left(\frac{\partial p}{\partial t} + \vec{v}\cdot\nabla p\right) + p\nabla\cdot\vec{v} + \left(\underline{\sigma:\nabla\vec{v}}\right) + \nabla\cdot\left(k\nabla T\right) \tag{3}$$

together with appropriate material equations and boundary conditions [2]. The system of equations (1)-(3) is solved using a specialized finite element method customized for the typical conditions of the injection molding process described in [1]

In order to incorporate the effect of normal stresses we implemented material rheological function connecting stresses (σ) with flow conditions that follows Criminale-Ericksen-Filbey model [3]:

$$\underline{\sigma} = 2\eta(T,\dot{\gamma},p)\underline{D} - \psi_1(T,\dot{\gamma},p)\overset{\nabla}{\underline{D}} + 4\psi_2(T,\dot{\gamma},p)\underline{D}\bullet\underline{D} \tag{4}$$

In equation (4) $\underline{\sigma}$ is the deviatoric stress tensor; $\underline{D}$ is the deformation rate tensor, $\eta(T,\dot{\gamma},p)$ is viscosity dependent upon temperature T , scalar shear rate $\dot{\gamma} = \|\underline{D}\|$ and pressure p; $\overset{\nabla}{\underline{D}}$ denotes the upper convective derivative of the deformation rate tensor:

$$\overset{\nabla}{\underline{D}} = \frac{\partial}{\partial t}\underline{D} + \vec{v}\bullet\nabla\underline{D} - (\nabla\vec{v})^T\bullet\underline{D} - \underline{D}\bullet\nabla\vec{v} \tag{5}$$

In equation (4), $\vec{v}$ is velocity and functions $\psi_1(T,\dot{\gamma},p)$ and $\psi_2(T,\dot{\gamma},p)$ are correspondingly the first and the second normal stress difference functions.

Following established practice for mold-filling simulation we use Cross-WLF model [2] for the viscosity model:

$$\eta(T,\dot{\gamma},p)=\frac{\eta_0(T,p)}{1+\left(\frac{\eta_0(T,p)\dot{\gamma}}{\tau^*}\right)^{1-n}} \tag{6}$$

In equation (6) $\eta_0(T,p)=D_1\exp\left[-\frac{A_1(T-T^*)}{A_2+(T-T^*)}\right]$; T^* is the glass transition temperature, $A_2=A_3+D_3p$; D_1, A_1, A_3, D_3 and n are empirical coefficients. Autodesk Moldflow material library stores quite an extensive data of the Cross-WLF parameters for thousands of material grades.

To evaluate functions $\psi_1(T,\dot{\gamma},p)$ and $\psi_2(T,\dot{\gamma},p)$ when viscosity is known we followed the so called Cox-Mertz Abnormal Rule as described by V. Sharma and G.H. McKinley. [7]

$$\psi_1(T,\dot{\gamma},p)\approx 2\dot{\gamma}^{-1}\eta\left(1-\frac{\eta_c^2}{\eta^2}\right)^{0.5}\left(\frac{\eta_c^2}{\eta^2}\right)^{0.7} \tag{7}$$

In equation (7), $\eta=\eta(T,\dot{\gamma},p)$ is the steady state viscosity and $\eta_c=\frac{\partial}{\partial\dot{\gamma}}\left(\eta(T,\dot{\gamma},p)\dot{\gamma}\right)$ is the tangential or consistency viscosity. As shown in [7] equation (7) allows relatively accurate estimation of the first normal stress difference function if Cross-WLF parameters are known.

Finally, to estimate the second normal stress difference function $\psi_2(T,\dot{\gamma},p)$ we assume that it is proportional to the first normal stress difference function:

$$\psi_2(T,\dot{\gamma},p)=K_\psi\psi_1(T,\dot{\gamma},p) \tag{8}$$

where the coefficient K_ψ takes values between -0.5 and 0 [3].

The set of equations (4)-(8) was integrated into a special build of 3D flow solver of Autodesk Moldflow Insight 2017 software and the resulting program was used for simulation of injection molding processes.

FILLING OF A THIN RECTANGULAR CAVITY

In order to estimate the effect of normal stresses on flow front propagation we use a filling simulation of a simple thin rectangular plaque of 100mm in length, 20mm in width and 2mm in thickness shown on **FIGURE** 2. The plague being filled by the generic polymethyl-methacrylate (PMMA) material. The filling time is 1 second. The cavity is meshed by 4 node tetrahedral elements with at least 10 layers of elements through the thickness. Melt inlet boundary conditions are applied along one of the short edges of the plaques in the style of a film or fan gate.

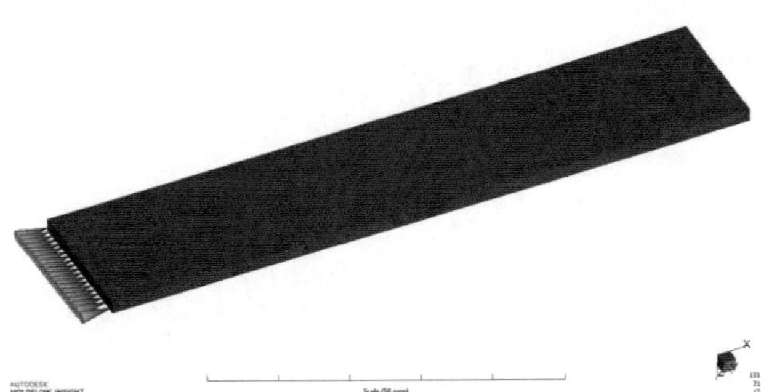

FIGURE 2 Illustrative molding case

Three rheological models were considered:

- No normal stresses
- First normal stress difference estimated from the Cox-Mertz Abnormal Rule but no Second normal stress differences: coefficient K_{ψ} =0
- First normal stress difference estimated from the Cox-Mertz Abnormal Rule and large Second normal stress differences K_{ψ} =-0.5

All other rheological, thermal and mechanical material parameters were taken from the standard material library.

All three cases simulation do not show any significant shear heating in the edge area as was expected because the runners and gates were not modeled. The flow front positions for all three cases are shown in **FIGURE** 3.

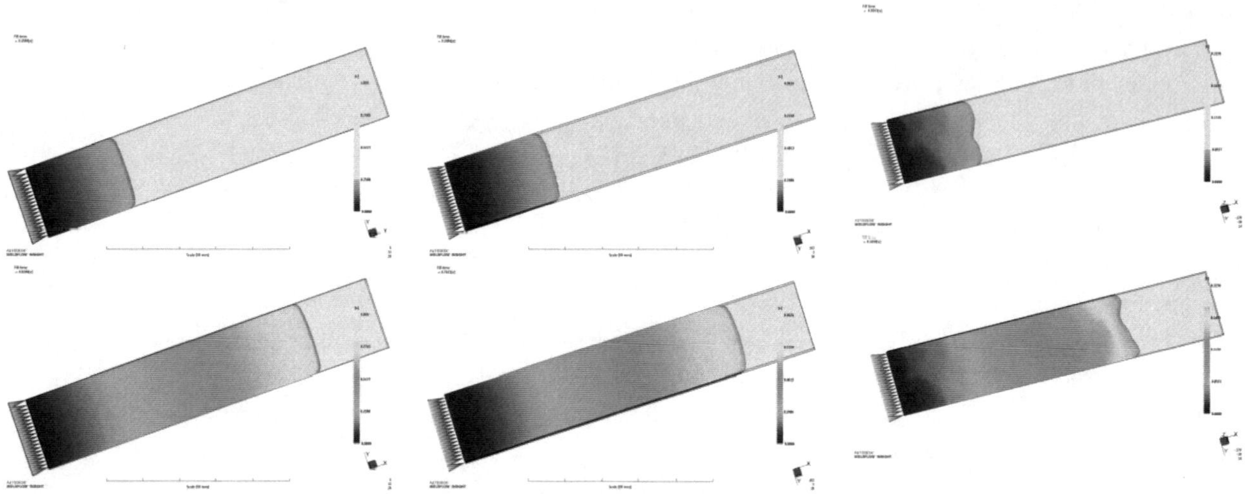

No normal stresses | Only first normal stress differences | First and second normal stress differences

FIGURE 3 Flow front positions for different rheological models

The simulation without normal stresses does not predict any ear flow phenomenon as it was expected because the runners and gates were not modeled.

When only the first normal stress difference is included, the flow front propagation pattern is very similar to the case of no normal stress.

The flow front distribution from the simulation using first and second normal stress differences is shown in the third column on **FIGURE** 3 A significant ear flow effect can be seen. The results of these simulations show that the second normal stress difference can be an important contributing factor to the ear flow phenomenon.

CONCLUSIONS

- An integrated system of 3D mold filling simulation that takes into account nonlinear rheological properties of normal stress differences is presented.
- The second normal stress difference appears to be an important contributing factor to the ear flow phenomenon while the first normal stress difference does not affect much the flow front propagation.

REFERENCES

1. Talwar, K., Costa, F., Rajupalem, V., Antanovski, L and Friedl, C., *Three Dimensional Simulation of Plastic Injection Molding.* SPE ANTEC Proceedings (1998)
2. Kennedy, P. and Zheng, R., *Flow Analysis of Injection Molds 2d edition.* Munich : Carl Hanser Verlag (2013)
3. Macosko, Christopher W. *Rheology: principles, measurements and applications.* New York : Willey-VCH, 1994.
4. Costa, F.S., Yokoi, H., Murata, Y. and Kennedy, P.K., *Numerical Simulation of Ear-flow: The Faster Advance of the Flow Front at the Edge of a Cavity.* Japan : Proc.22nd Annual Conference of the Polymer Processing Society (2006)
5. Murata, Y., Abe, S. and Yokoi, H., *Experimental Analysis of Flow Front Advancing Phenomena.* Seikei-Kakou **15(10)**, pp.706-713 (2003)
6. Cook, P. S., Yu, H. Kietzmann, C. V., Costa, F. S., *Prediction of Flow Imbalance in Geometrically Balanced Feed Systems.* Boston : SPE ANTEC Proceedings, 2005.
7. Sharma, V. and McKinley, G.H., *An Intriguing Empirical Rule for Computing the First Normal Stress Difference from Steady Shear Viscosity Data for Polymer Solutions and Melts.* Rheol. Acta (2012).

Simulation of the Influence of Thermal Properties of the Mould Steel on the Injection Moulding Process

Thomas Lucyshyn[a*], Lara-Vanessa Des Enffans d'Avernas[a], Clemens Holzer[a]

[a]*Chair of Polymer Processing, Department Polymer Engineering and Science, Montanuniversitaet Leoben, Otto Gloeckel-Str. 2, 8700 Leoben, Austria*

clemens.holzer@unileoben.ac.at
thomas.lucyshyn@unileoben.ac.at

Abstract. The thermal properties of the mould steel influence the cooling situation in the injection moulding process. While there are some experimental studies investigating the influence of special mould materials like copper alloys, these studies are limited to few polymers. In the present work, an extensive parameter study with the commercial injection moulding simulation software Autodesk Moldflow Insight was performed to further analyse the influence of the mould steel on the injection moulding process, specifically the cycle time and warpage behaviour. The investigated part was a square box with two thickness variations (1 mm and 3 mm) to see also the influence of the part thickness. A conventional mould steel (1.2343 ESU) was compared with a new steel grade developed by BÖHLER (W620 ISOBLOC), featuring approximately double the thermal conductivity compared to the conventional steel grade. In a systematic parameter study more than 150 simulations were performed with 18 different polymers covering the most common material families including amorphous, semi-crystalline, unfilled as well as glass fibre reinforced polymer grades. The whole mould including the cooling system was modelled with a 3D-mesh and a transient cooling simulation was performed. As expected, the higher thermal conductivity reduced the cycle time – surprising was the range of this effect, though (< 5 % up to 25 % of cycle time reduction), which strongly depends on the used polymers and the related processing conditions, but also the part thickness. For the extent of the cycle time reduction a correlation was found with the ratio of the difference between melt and mould temperature to the difference between ejection and mould temperature. With this correlation it was possible to predict the range of cycle time reduction for other polymers based on those three temperatures. Furthermore, the simulations showed that the higher thermal conductivity of the mould steel resulted in a reduced warpage of the part, but had no critical influence on filling pressure and clamp force.

Keywords: injection moulding, tool steel, thermal conductivity, cycle time, simulation.
PACS: 81.20.Hy; 44.05.+e

INTRODUCTION

Cycle time is a crucial economic factor of the injection moulding process. Several factors affect the cycle time such as the thermal properties of the polymer and the mould material, the geometry of the injection moulded part (specifically the thickness), the layout of the cooling channels or the chosen process settings (especially the melt and mould temperatures). A theoretical description of the dependence of the cooling time on many of these parameters has already been published more than 35 years ago by Zorić and Čatić [1, 2]. With a defined polymer and part geometry the mould material is one potential degree of freedom to reduce cycle time. According to Čatić et al. [3] the thermal properties of the mould affect the cycle time in two ways: On the one hand the thermal conductivity of the mould material influences the contact temperature at the metal/polymer interface and on the other hand the heat transport from the interface into the mould is supported by a higher thermal conductivity of the mould material.
Some experimental work with focus on the influence of the thermal properties of the mould material on cycle time, part quality and energy consumption was presented by Kelly et al. [4]. In their study the authors compared a conventional tool steel to beryllium free copper alloys with rather high copper content from 85 to 96 %. The investigations were performed for a simple plate shaped part with two semi-crystalline materials (high density polyethylene and polybutylterephthalate). In this publication already a difference of the effectivity of the mould material was stated between the two polymers, but not investigated in more detail.

In the presented work a parameter study using a commercial injection moulding simulation software was performed to better understand the influence of the polymer itself on the effectivity of the mould material in terms of cycle time reduction. For that purpose 18 different materials were investigated covering the most common polymer families.

EXPERIMENTAL

For the simulations the commercial software package Autodesk Moldflow Insight 2016 was used. The investigated test part had a box shaped geometry with a bottom area of 100 by 100 mm² and 40 mm height (Figure 1). The thickness was either 1 or 3 mm to see the thickness influence on the results. The gating situation was a hot runner with a short cold gate to the part. The part was meshed with a 3D tetrahedral mesh. The hot runner as well as the cooling channels were modelled with beam elements and the whole mould block was also meshed with a 3D tetrahedral mesh.

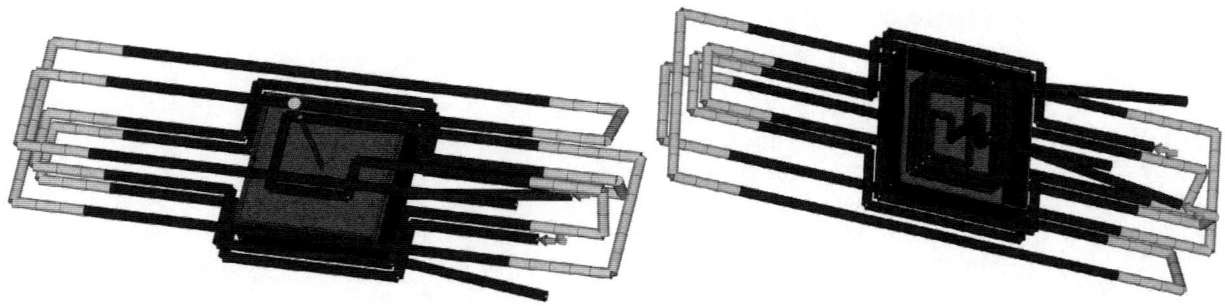

FIGURE 1. Tetrahedral mesh of the box shaped test part (green) with hot runner (red), cooling channels (blue) and connecting hoses (grey). Left: nozzle side view; right: ejector side view.

The whole mould was defined either as tool steel 1.2343 ESU or W620 ISOBLOC. The main difference between the two steel grades is their thermal conductivity with 26 W/(m·K) for 1.2343 ESU and 49 W/(m·K) for W620 ISOBLOC. In order to analyse the influence of the thermal conductivity of the mould material on the cycle time for different polymers, 18 widely used material families were arbitrarily chosen from the Moldflow database (Table 1) and the recommended processing temperatures were used for the simulations. A transient 3D-FEM cooling analysis was performed assuming a stable production process.

TABLE (1). Investigated polymers with filler content (GF30 = 30 wt.% glass fibre), morphology (a = amorphous, sc = semi-crystalline) and recommended processing temperatures (according to Moldflow material data base)

Polymer	Filler	Morphology	Grade name	Manufacturer	Melt temp. (°C)	Mould temp. (°C)	Ejection temp. (°C)
ABS	-	a	Lustran ABS LGA	INEOS ABS	226	44	86
ABS/PC	-	sc	Cycoloy LG8002	SABIC Innov. Pl. US,LLC	277	87	104
PE-HD	-	sc	Marlex 9004 PE	Chevron Philips	234	27	80
PA 6.6	-	sc	Zytel ST801HS NC010	DuPont Perf. Polymers	280	70	220
PA 6.6	GF30	sc	Ultramid A3WG6	BASF	290	85	195
PBT	-	sc	Pocan B 1505	Lanxess	270	100	170
PBT	GF30	sc	Ultradur B 4300 G6 HR	BASF	260	80	110
PC	-	a	Lexan ML7681	SABIC Innov. Pl. US,LLC	285	85	128
PC	GF30	a	Lexan 3413	SABIC Innov. Pl. US,LLC	329	99	120
PET	-	sc	J125	Mitsui Chemicals Co Ltd	270	100	150
PET	GF30	sc	Rynite FR530 NC010	DuPont Perf. Polymers	280	110	195
PMMA	-	a	Plexiglas 8N	Evonik Roehm GmbH	235	70	94
POM	-	sc	Ultraform N2320 003	BASF	200	90	110
PP	-	sc	RJ766MO	Borealis Europe	240	40	111
PP	GF30	sc	Hostacom G3 U01	SF Hoechst	230	50	93
PPS	-	sc	Fortron 6162A7	Celanese	325	147	228
PPS	GF30	sc	Supec G301T	SABIC Innov. Pl. US,LLC	320	150	220
PS	-	a	Polystyrene 804	TOTAL Petrochemicals	230	50	80

For both wall thicknesses (1 and 3 mm) and each of the materials, an individual optimal cycle time was determined for the 1.2343 ESU steel grade in repeated simulations. The found parameters for each polymer were then also used for the W620 ISOBLOC grade. For both wall thickness variations, a hotspot was found in the corner of the box. This region was determining the required cycle time. Therefore, a specific node in this region was picked out for the evaluation of all materials to get comparable results. For the comparison of the cycle times, the part temperature was drawn against process time for this node and the time when the temperature reached the recommended ejection temperature was defined as the shortest possible time for ejection. Adding a mould open time of 5 s resulted in the evaluated cycle time for each individual polymer and wall thickness.

Additionally, the distance of opposite walls on the open side of the box were evaluated in terms of shrinkage and warpage to analyse the influence of the tool steel. Furthermore, the injection pressure and resulting clamp force were also evaluated in order to see a potential negative influence of the higher thermal conductivity of the steel on the processability of the polymers.

RESULTS

As expected, the steel with the higher thermal conductivity resulted in a reduction of cycle time. What was surprising, though, was the range of values from about 3 to almost 25 % for the 1 mm thick box and approximately 6 to 14 % for the 3 mm thick box (Figure 2). The influence of the wall thickness can be explained by the low thermal conductivity of the polymer itself. Generally spoken, the thicker the part the less is the effect of the surrounding steel, as the heat has to be transferred from the centre of the part to the surface, which is mainly determined by the thermal conductivity of the polymer.

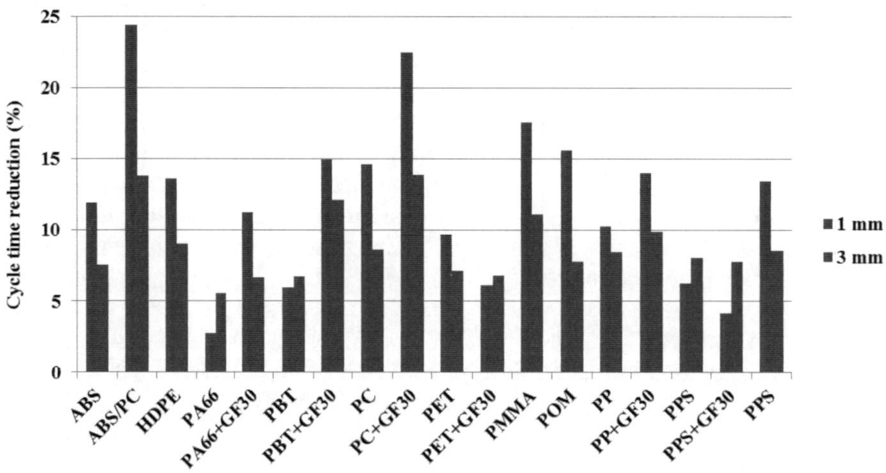

FIGURE 2. Cycle time reduction of W620 ISOBLOC compared to 1.2343 ESU for different polymers and two wall thicknesses.

In an additionally performed design of experiments the melt temperature (T_{melt}), the mould temperature (T_{mould}) and the ejection temperature (T_{eject}) turned out to be the most important factors influencing the effect of the mould steel on cycle time reduction. Therefore, a correlation was searched between the achieved cycle time reduction and a function incorporating these three temperatures. As a function describing the cycle time depending on those three temperatures, well known as "cooling time equation" (confer [1]), already exists, a similar relation was established defining the cycle time reduction as a function of these temperatures according to the following equation (1):

$$\text{Cooling\_time\_reduction} = f\left(\frac{T_{melt} - T_{mold}}{T_{eject} - T_{mold}}\right) \tag{1}$$

After testing some different functions, a simple linear relation between the temperature coefficient defined in equation (1) and the corresponding cooling time reduction turned out to produce the best correlation coefficient R^2. As the cooling behaviour was different for the two wall thicknesses, individual models for both thickness variations were established. In Figure 3 the correlation for both wall thicknesses is shown together with the found correlation function and the corresponding correlation coefficient. With increasing temperature coefficient more cycle time reduction can be achieved. For the thicker part a smaller slope of the function was determined. The correlation coefficient for both thickness variations is satisfying with values higher than 0.8. For an even better correlation, also the thermodynamic properties of the polymer (density, thermal conductivity and specific heat capacity) should be taken into account in the model, but so far no reasonable model was found that further increased the correlation coefficient.

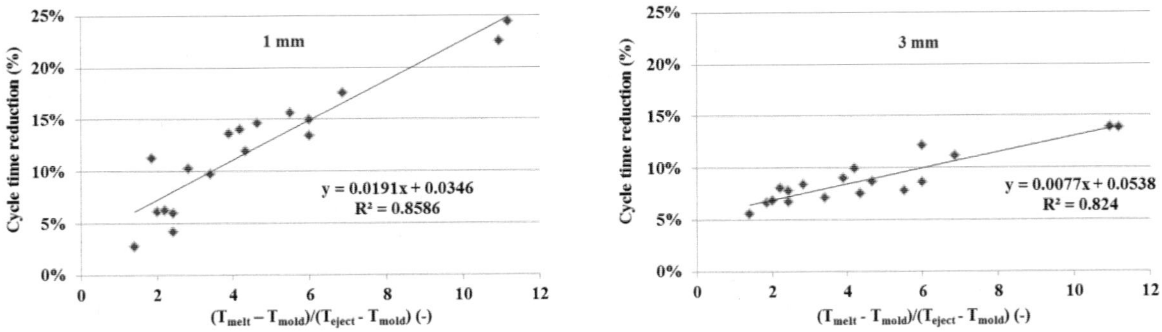

FIGURE 3. Correlation between temperature coefficient and cycle time reduction for two wall thicknesses (left: 1 mm, right: 3 mm).

In order to validate the found correlation model, three additional polymers were randomly chosen from the Moldflow database (Table 2). For these polymers the same simulations were performed and evaluated as already described for the other 18 investigated polymers for both wall thicknesses. Additionally, the expected cycle time reduction was calculated using the three given process temperatures from the Moldflow database and the correlation functions shown in Figure 3.

TABLE (2). Investigated polymers for the validation of the regression model with filler content (GF30 = 30 wt.% glass fibre), morphology (a = amorphous, sc = semi-crystalline) and recommended processing temperatures (according to Moldflow material data base)

Polymer	Filler	Morphology	Grade name	Manufacturer	Melt temp. (°C)	Mould temp. (°C)	Ejection temp. (°C)
PVC	-	a	V-1	Hepworth Building	195	40	68
POM	-	sc	Kepital F10-02	Korea Engineering Plastics Company Ltd.	200	80	135
PP	-	sc	J3021GR	Prime Polymer Co Ltd.	225	40	100

The results of the simulations and the correlation model are compared for both wall thicknesses in figure 4. The model is able to predict the expected cycle time reduction in a satisfying quality, although there are some quantitative differences.

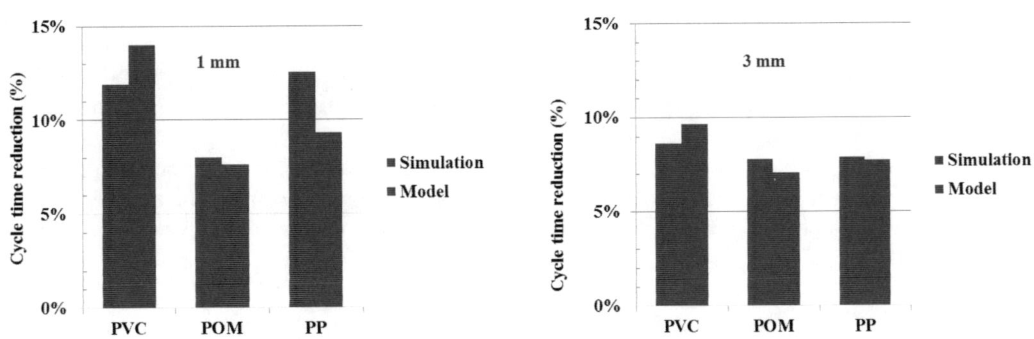

FIGURE 4. Validation of correlation model for three randomly chosen polymers for two wall thicknesses (left: 1 mm, right: 3 mm).

Besides the focus on the cycle time reduction potential of the steel with the higher thermal conductivity, there was also a positive effect expected on the warpage behavior, especially for this box shaped part, where the core tends to overheat. The results shown in figure 5 confirmed this expectation. For each material and both wall thickness variations the warpage (bending inwards of the sidewalls) was reduced when using W620 ISOBLOC.

Additionally, the effect of the higher thermal conductivity of the mould steel on the filling pressure and clamp force was evaluated. Potentially, there could have been a negative effect of the higher cooling efficiency of the steel on the filling behaviour, as an improved cooling of the melt during the filling stage could result in higher viscosities and a thicker frozen layer. In fact, the simulation results proved that this effect can be neglected, as the highest increase of filling pressure as well as clamp force of all variations was smaller than 5 %, which is not problematic for the processability of the parts.

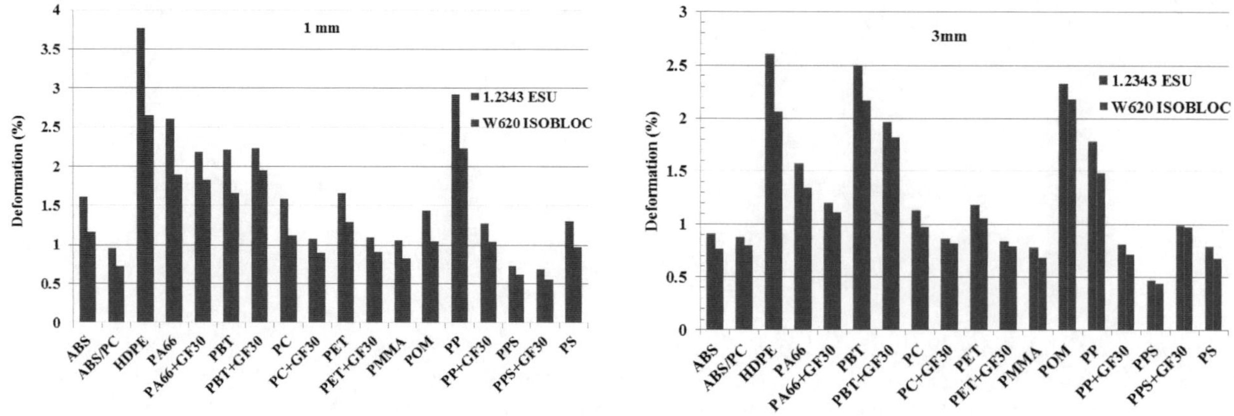

FIGURE 5. Deformation of W620 ISOBLOC compared to 1.2343 ESU for different polymers and two wall thicknesses (left: 1 mm, right: 3 mm).

CONCLUSIONS AND OUTLOOK

In an extensive parameter study using the commercial injection moulding simulation software Autodesk Moldflow Insight the influence of the thermal properties of the mould steel on the cycle time, the warpage, filling pressure and clamp force was investigated for 18 widely used polymer families. The main finding was that the effect of the mould steel on the cycle time reduction varied strongly among the different polymers (in the range of 3 to 24 %), but also an influence of the wall thickness was shown. As the most important influencing factor of the polymer a temperature coefficient was identified, comprising the recommended processing temperatures (melt, mould and ejection). A linear function could be found as correlation model between this temperature coefficient and the cycle time reduction, which was validated using three additional randomly chosen polymers. With this model it was possible to estimate the achievable cycle time reduction without having to perform several simulations, but only by using these three process parameters from the Moldflow database in the found model.

The results showed a positive effect of the higher thermal conductivity of the mould steel on the warpage behaviour, especially for the inward bending of the side walls, which is a result of the overheating of the core. On the other hand, no negative effects on the processability in terms of filling pressure and clamp force were detected.

To summarize the results showed the high potential of using a tool material with higher thermal conductivity, especially for moulds with cores, which may be difficult to cool properly and especially for thin walled parts, but depending on the used polymer.

In order to validate the simulation results, a test mould for the described part has been equipped with mould inserts made from different mould steels. Experimental work is in progress to measure the part temperature after ejection with an infrared-camera for selected polymers to see the real impact of the mould steel on the cooling behaviour.

ACKNOWLEDGMENTS

Thanks go to BÖHLER Edelstahl GmbH & Co KG (Kapfenberg, Austria) for the initiative and the financial support of this study.

REFERENCES

1. J. Zorič, I. Čatić, *ANTEC 1981*, 814-817 (1981).
2. I. Čatić, *Polym. Eng. Sci.* **19/13**, 893-899 (1979).
3. I. Čatić, M. Rujnić-Sokele, M. Šercer, *ANTEC 2005*, 961-965 (2005).
4. A.L. Kelly, L. Mulvaney-Johnson, R. Beechey, P.D. Coates, *Polym. Eng. Sci.* **51/9**, 1837-1847 (2011).

Molecular Orientation and Residual Stresses Impact on Optical Properties of Plastic Lenses

Stefan Schneidmadel, Michael Koch, Xiang Qin

Technische Universität Ilmenau, Plastics Technology Department (KTI)

Stefan.schneidmadel@tu-ilmenau.de

Abstract. The substitution of conventional materials such as glass for optical plastics offers a high potential for the realization of complex optical systems. Large-scale production processes in combination with high-grade optical parts can be realized by using an injection molding process. The processing of these plastics parts influences the optical properties due to the fact that physical effects occur during filling and cooling of the melt. In general, the optical properties can be differentiated in inner properties (refractive index, transmittance) and outer properties (surface, dimensional accuracy). Both influence and change the light path through the optical material. This work aims to analyze the processing of plastic lenses by tracing back the processing parameters to the occurring physical effects. In comparison to experiments and simulations, the difference between calculated and analyzed optical properties is presented in detail. For this purpose, lens-shaped test specimen were processed on an injection molding machine with different processing parameters. In the simulation program MOLDEX 3D, birefringence patterns, warpage and shrinkage were analyzed and compared to the experimental data applying polariscope for residual stresses and an integrating sphere for transmittance grade. Transmittance values of the lenses between 88 and 89, 1% were accomplished for a polycarbonate lense. By taking into account the correlation between processing parameters and optical properties, a forecast of birefringence and transmittance of plastic parts is possible.

Keywords: optical plastics; residual stresses; birefringence; birefingence patterns; transmittance
PACS: 42.70.Jk

OPTICAL PROPERTIES OF PLASTIC LENSES

The replacement of glass material for plastic materials provides new possibilities in light guide, complex part geometry and filtering of special wavelengths. Especially in the context of functionalization, optical plastic material offers a wide-ranging field of application for plastic lenses or light guides. Light consists of high frequent electro-magnetic waves and can be characterized by wavelength λ [1]. For human eyes, the visible range reaches from 400 to 800 nm [2].

$$\lambda = \frac{c}{f}. \tag{1}$$

$$c = \frac{c_0}{n}. \tag{2}$$

Propagation speed c and frequency f are influenced by refractive index n of the substance, which the electromagnetic waves penetrate. The characteristic state of matter of the substance insists of atomic or molecular structure. In correlation with the first law of thermodynamics, the energy of the rays is split into reflection ρ, transmittance τ and absorption α during passing.

$$\rho + \alpha + \tau = 1. \tag{3}$$

Plastics materials insist of molecular chains and can be differentiated in spherulites (crystalline structure) or ravels (amorphous structure). Optical parts need a high transparency and can only be realized by using amorphous structures. The difference between optical plastic materials mainly consists in energy absorption behavior on atomic area and relates to the transformation of the material structure [3]. The dissolution of molecular structures, the shortening of chains and molecular chain orientations effect a different optical or mechanic behavior of the material [4, 5]. On the other side, physical effects during processing of these materials influence the molecular distribution and orientation. The reachable optical properties therefore determine the application of these materials.

PHYSICAL EFFECTS ON OPTICAL PROPERTIES

Optical plastics have in general an amorphous material structure, which means that the material consists of long molecular chains. In dependence with the processing parameters, these chains are strained by thermal and mechanical stresses, which leads to a shortening of the chains, the accumulation in lower stress areas and the orientation towards stress direction [6]. In Figure 1 the general behavior of the molecular chains in dependence with physical and thermal stresses is shown.

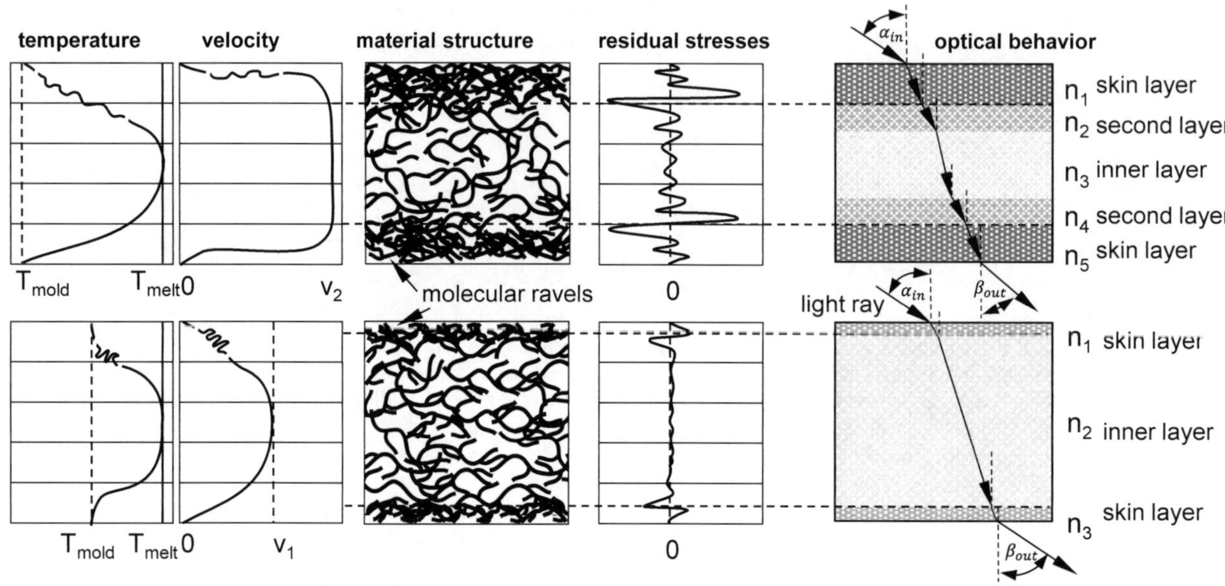

Figure 1: Correlation between physical effects and optical properties

These effects influence the distribution and orientation of the molecular chains within the material, cause a shortening of the chains which reduces the molecular weight and resolves the chain loops of the amorphous material. A difference in the molecular density of the amorphous material causes different refractive indexes and inner optical properties such as light transmittance and birefringence of rays or the polarization of the incoming light [7]. Outer optical properties can be influenced by warpage and shrinkage, which comes into effect during cooling. The molecular chains tend to compensate the inhomogeneous distribution [8]. The compression of molecular chains as a result of high mechanical stresses decreases the mobility of the chains and the difference in molecular distribution cannot be balanced, whereas areas with a lower molecular capacity provide a free movement of the molecular chains. The resulting residual stresses within the material lead to shrinkage and warpage [9].

The challenge of processing optical plastic material consists of a dense interlinkage of these effects which means that correct outer properties such as warpage don't stand in correlation with correct inner properties such as the refractive index. The consideration of the occurent physical effects improves the understanding of influences on the optical properties irrespective of processing parameters.

SIMULATION OF SHEAR STRESSES AND BIREFINGENCE PATTERNS

The simulation of birefingence patterns offers the possibility of forecast the later optical properties of plastics parts. The model basis behind the calculation describes the viscoelastic behavior of molecular chains in the flow [10]. For the calculation of birefingence patterns, viscoelastic properties have to be analyzed regarding the deformation caused by heat loss and the capability of storing elastic energy [11]. This information is also required for the analysis of shrinkage and warpage and is with regard to plastic materials usually known. Occurring birefingence patterns can be traced back to flow-induced and thermally- induced effects which can be evaluated.

PROCESSING OF PLASTIC LENSES

The plastics lenses are processed on an injection molding machine Krauss Maffei 160 SP380 which possesses the clamping force FH = 1600 kN and the screw diameter DScrew = 45 mm. The mold and the dimension of the produced lenses are shown in Figure 2b. The lenses were processed with the amorphous material PC Makrolon 2405.

Influences on Optical Properties

Considering physical effects during processing on an injection molding machine, a process analysis was conducted. Optical properties can be influenced by the material in relation to humidity, molecular weight and impurities. Effects of humidity and impurities can be influenced by an adjusted pre-treatment of the material, whereas the molecular weight determines the refractive index and transmittance through the material.

During processing, the mold influences the surface of the lenses, which mainly impacts the transmittance of the rays. Other process factors such as mold temperature and part geometry influences the distribution and concentration of the molecular chains and different inner optical properties.

The mechanical and thermal strain on the melt can be influenced by screw geometry, nozzle diameter and melt temperature. High resting time or high shear strain leads to a shortening of molecular chains and therefore different optical properties.

Regarding the influence of processing parameters, a differentiation between mechanical strain and temperature effects is necessary. Injection velocity, injection pressure and pressure dwell have influence on mechanical stress on the molecular chain. The resulting forces and torque effectuate a molecular chain distribution and orientation in the plastic part. The temperature profile influences the balance of molecular chain accumulation. This can be traced back on melt temperature, mold temperature and cooling time. All parameters influence the development of a skin layer with different optical properties.

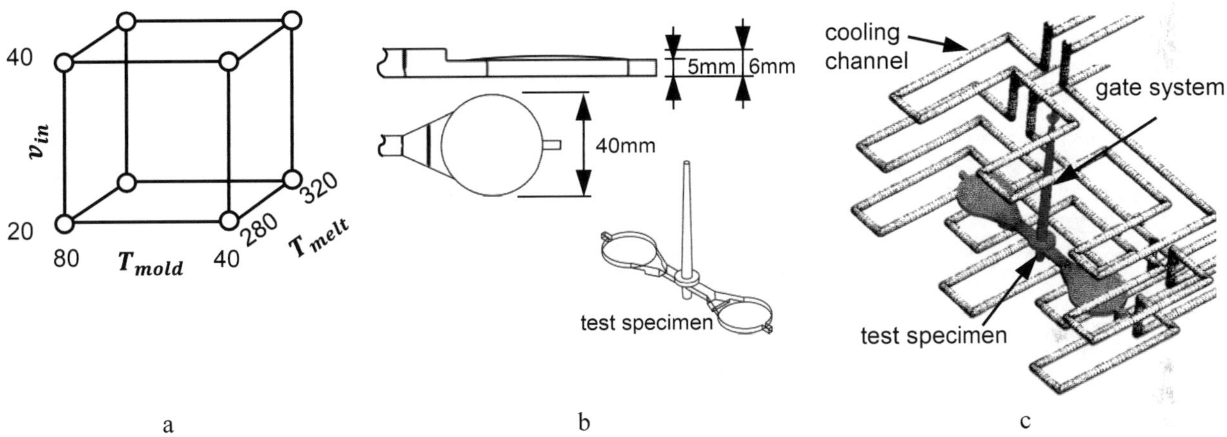

Figure 2: (a) experimental design (b) test specimen (c) simulation model

Experiments using Design-Of-Experiments (DOE)

For the experiments, an experimental design was formulated which allows to correlate the processing parameters to optical properties. The chosen parameters determine the significance of the analysis and are set to the limiting conditions of processing. With regard to the thermal influences, the melt temperature was varied from 280°C to 320°C, the mold temperature was between 40°C and 80°C. Mechanical stresses can be traced back to an injection volume between 20 and 40 cm^3/s. In Figure 2b, the shape of the test specimen is shown. Residual stresses and the transmittance of the lenses were analyzed by using a polariscope and an integrating sphere regarding DIN 5036 [12]. The analysis about residual stresses was based on the number of polarization rings and the area of polarization.

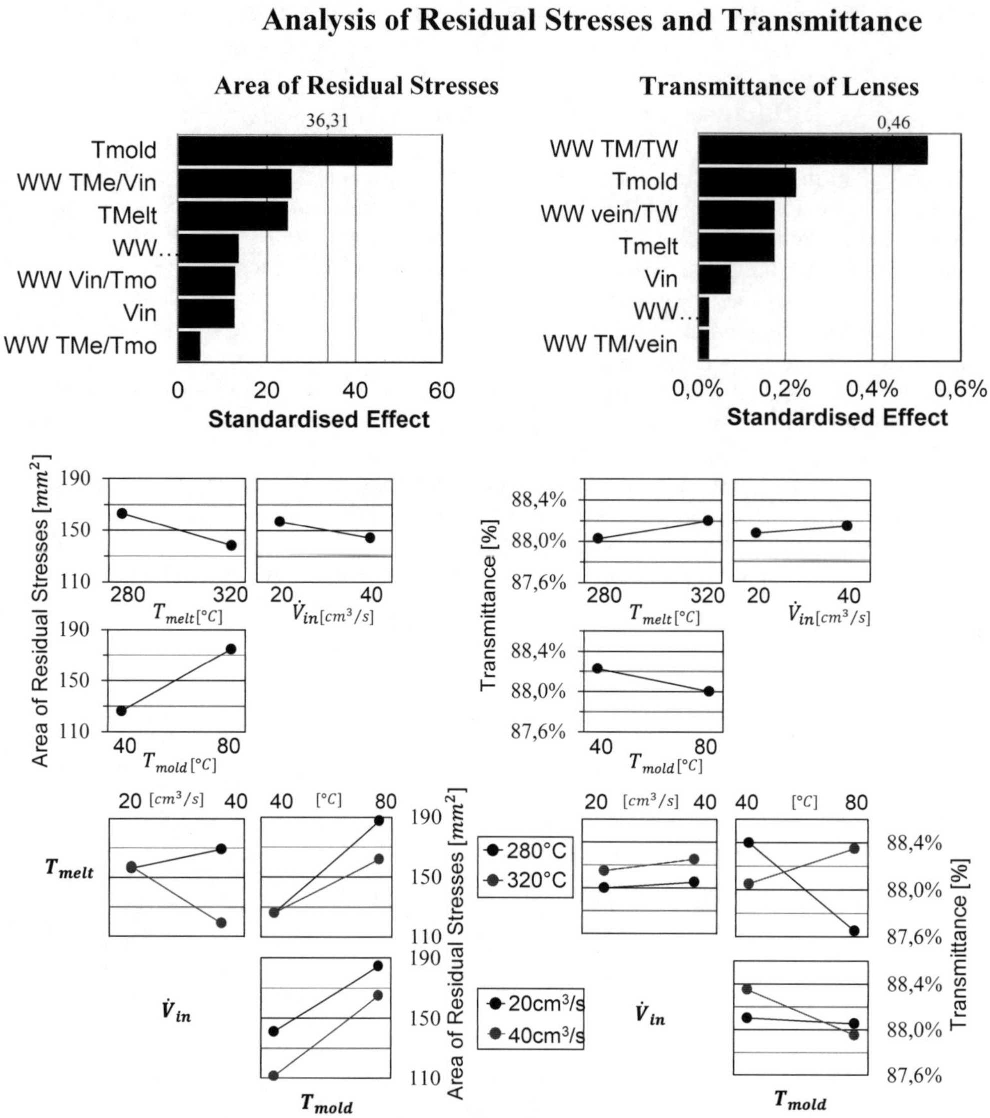

Figure 3: results for area of residual stresses and transmittance

In Figure 3 the main influences on formation of residual stresses and transmittance is shown. Both optical properties mainly depend on temperature of melt and mold. There are also interrelations between mold temperature and melt temperature. Injection volume flow in combination with mold temperature influences the molecular effects due to viscosity. There is a direct relation between residual stresses and the lens transmittance. Increasing melt temperature leads to lower residual stresses and higher transmittance. In contrary a higher mold temperature leads to higher molecular compression due to a later freezing of melt and a lower transmittance. Flow volume is in this context insignificant.

Simulation of Birefingence Patterns

For the simulation of birefingence patterns, the injection molding simulation program MOLDEX 3D was used. The program is endued with a module for illustrating the birefringence patterns, which calculates on the basis of standard viscoelastic material properties. Other part properties such as warpage and shrinkage were used for tracing back the optical properties on physical effects. The model for simulation with cooling channel, gate system and shape of the test specimen is shown in Figure 2c.

RELATIONS BETWEEN CALCULATED AND EXPERIMENTAL DATA

The comparison between the simulated birefingence patterns and the birefringence in the test specimen show a difference. The calculated birefingence patterns have a bigger area compared to what the analysis on the test specimen has shown. This deviation can be related to different viscoelastic material properties which were implemented for calculating the birefingence pattern. Further optical effects depend on additional optical effects on the surface of the test specimen. Although the force of residual stresses is different, the location of residual stresses match with the simulation.

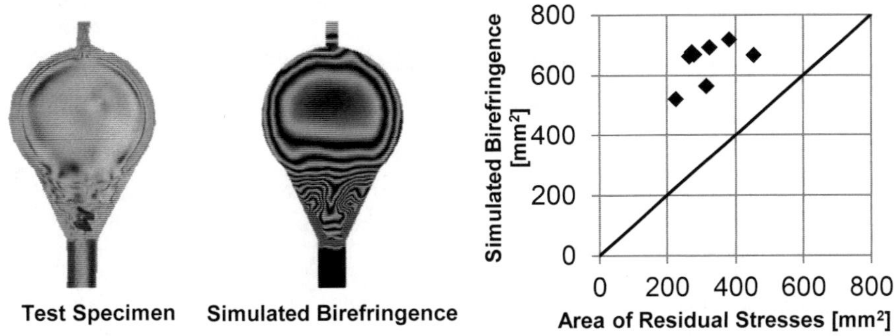

Figure 4: Experimental und simulated birefingence patterns of test specimen

CONCLUSION

In this work, optical properties of plastic parts were analyzed and compared to calculated properties using a flow simulation. The objective was to enable the prediction of the birefingence patterns of lenses as a result of temperature profile and mechanical strain during the injection molding process. The occurring effects on transmittance and residual stresses can be traced back to the physical effects of molecular chain distribution and orientation. The lenses achieve transmittance between 88 and 89,1 %. The comparison between experimental test specimen and simulation data shows a good accordance. It could also be shown, that the existing models have deficiencies in describing the physical effects accurately because of different viscoelastic material properties of the plastic melt. This has to be attributed to the fact that this data is in most cases not available and has to be determined through a material analysis. Considering these factors offer an opportunity to select processing conditions such as injection velocity and the temperature profile to influence optical effects such as birefingence and refractive indexes. Further work aims to analyze different processing conditions applying compression forming and variant thermal cooling of lenses and mold, which influences the molecular chain balance and leads to improved residual stresses.

REFERENCES

1. J. Shamir, "Optical Systems and Processes", Bellingham, NY: SPIE optical engineering press, 1999, pp. 10-14.
2. D. Walker, "Energy, plants and Man", Brighton: Oxygraphics Limited, 1992, p. 27.
3. J. Pospísil, P. P. Klemchuk, "Oxidation Inhibition in organic materials – Vol II", Florida: CRC Press, 1990, p. 143.
4. I.-C. Khoo, J.F. Lam, F. Simoni, "Nonlinear optics and optical physics", New Jersey, World Scientific Publ., 1994 pp. 185
5. K.-S. Lee, "Polymers for Photonics Applications I – Nonlinear Optical and Electroluminescence Polymers", Berlin, Heidelberg: Springer- Verlag, 2002, pp. 165 - 168
6. C. D. Han, "Rheology and Processing of Polymeric Materials – Vol 2 Polymer Processing", New York, Oxford University Press, 2007, p. 368
7. S.R. Holding, E. Meehan, "Molecular Weight Characterisation of Synthetic Polymers", in *Rapra review reports* **7**, edited by R. Dolbey, Shrobshire (UK), Rapra Technology Ltd., 1995, pp. 6-8
8. V.R. Gowariker, N.V. Viswanathan, J. Sreedhar, "polymer science", New Delhi, New Age Int. (P) Ltd., 2005, pp. 93 - 100
9. J. M. Fischer, "Handbook of Molded Part Shrinkage and Warpage", Oxford, Elsevier Inc., 2013, p. 148
10. S.-J. Liu, "Modeling and Simulation of Thermally Induced Stress and Warpage in Injection Molded Thermoplastics" in *Polymer engineering and Science* **36**, 6 , (807 – 818)
11. V.G. Mavrantzas, "Polymer Melt Viscoelasticity: What we can learn from Molecular Simulations", in Rheology Reviews, 1 – 52 (2007)
12: DIN 5036, "Strahlungsphysikalische und lichttechnische Eigenschaften von Materialien – Teil 1: Begriffe Kennzahlen", Deutsches Institut für Normung, Köln, Beuth Verlag, 1978

The Influence of Polymer Viscoelasticity on the Optical Features during an Optical Lens Development

Gwo-Geng Lin[a*], De-Lung Lai[a], Chao-Tsai (CT) Huang[a], Jye Wang[b], and Yuan-Jung Chang[b], and Rong-Yeu Chang[b],

a. Chemical and Materials Engineering Department, Energy and Opto-Electronic Materials Research Center, Tamkang University, Taipei, Taiwan
b. R and D Division, CoreTech System (Moldex3D) Co., Ltd., Hsinchu, Taiwan
**Corresponding author: gglin@mail.tku.edu.tw*

Abstract. In modern life, polymeric optical lens has been utilized in many electronic devices, such as camera, mobile phone, tablet, and so on. Although hundred million lenses are produced annually for those devices, the quality is the key to the main suppliers. Indeed, the optical quality is affected by both the geometric features and the optical birefringence. However, how the material's viscoelasticity influences the optical features is still not fully understood yet. In this study, we have systematically investigated the viscoelastic effects on the optical property of a concave lens made of Zeonex COP 480R material by the traditional injection molding, simulated by use of the White-Metzner model. We assumed that the rheological parameters can be modified artificially based on the White-Metzner model of viscoelasticity, in which either the viscosity or the relaxation time constant was of the Cross-type relation. Results show that if the power-law index (*n*) was changed from 0.1 to 0.6, the total fringed order has no significant difference, but the fringed patterns was varying dominantly in the perpendicular direction, instead of in the flowing direction. This situation was matched with the distribution of the flow-induced residual stresses. Moreover, if the τ^* in the Cross model was varied from 100,000 Pa to 300,000 Pa, the total fringed patterns were apparently changing from being dominated in the flow direction to being in the direction perpendicular to the flow. The influenced width of the fringed patterns was almost linearly increased with the increasing τ^*. To investigate the viscoelastic effect on the optical property, a modified White-Metzner model, in which the relaxation time constant was also described by using a Cross-type relation with the strain rates, was utilized to simulate the injected molding process. It has been found that a large relaxation time constant in the W-M model for the processed material could cause strong birefringence in both the flow and the perpendicular directions to the molded concave lens. The results can be applied as some guideline for either the further material modifications or the new material developments.

Keywords: viscoelasticity, optical birefringence, Cross model, White-Metzner model
PACS: 83

INTRODUCTION

In modern life, polymeric optical lens has been utilized in many electronic devices, such as camera, mobile phone, tablet, and other optical devices. Although hundred million lenses are produced annually for those devices, the quality is the key to the main suppliers. Indeed, the optical quality can be affected by the geometric features and/or the optical birefringence. Usually, the residual birefringence in materials are credited to two mechanisms, *i.e.*, the flow-induced and the thermal-induced birefringence, during the injection molding process based upon the photo-elasticity theory. The former is attributed to molecular-chain orientation caused by shear stresses during the filling and holding stages, and can be relaxed or, otherwise frozen during post-filling processes and after ejection. The latter can be attributed to the internal stresses caused by the spatial non-uniform shrinkage, which is due to the thermo-mechanical history of the processed material at each specific location inside the part during the injection molding process [1-2]. Moreover, according to the previous researches done by Wang et al. [3,4], residual birefringence in COC molded lenses is contributed mostly by the flow-induced effect, because it accounts for 92 % as compared to 8 % for the thermal-induced parts.

Furthermore, some people are trying to develop new materials to have great optical performance to overcome some optical defects, but there is huge gap between the material design to the final application. Indeed, how the material's viscoelasticity influences the optical features is still not fully understood yet. In this study, first we have characterized Zeonex COP 480R material to obtain the $n = 0.4$ and $\tau^* = 217{,}000$ Pa in the Cross model. Then we have investigated the viscoelastic effects using the modified White-Metzner model. Results showed that the total fringed pattern development of the injected optical lens is strongly influenced by the modification of the power law index (shear thinning effect). Also, the τ^* can change the fringed domain. Moreover, using modified White-Metzner model, when material has larger relaxation time, it will cause stronger birefringence effect in both the flow and the perpendicular directions. The details will be addressed as the following sections.

THEORY AND MODELING

In this study, the commercial software Moldex3D® is adopted for virtual injection molding simulation. The polymer melt is assumed to behave with the White-Metzner model. Hence the non-isothermal 3D polymer flow can be mathematically described by the following:

$$\frac{\partial \rho}{\partial t} + \nabla \cdot \rho \mathbf{u} = 0 \tag{1}$$

$$\frac{\partial}{\partial t}(\rho \mathbf{u}) + \nabla \cdot (\rho \mathbf{u}\mathbf{u} - \boldsymbol{\sigma}) = \rho \mathbf{g} \tag{2}$$

$$\boldsymbol{\sigma} = -p\mathbf{I} + \tau \tag{3}$$

$$\rho C_P\left(\frac{\partial T}{\partial t} + \mathbf{u} \cdot \nabla T\right) = \nabla(\mathbf{k}\nabla T) + \frac{1}{2}\tau : (\nabla \mathbf{u} + \nabla \mathbf{u}^T) \tag{4}$$

where **u** is the velocity vector, T is the temperature, t is the time, p is the pressure, σ is the total stress tensor, ρ is the density, τ is the extra-stress tensor, k is the thermal conductivity, and C_p is the specific heat. In present work, τ is obtained by the constitutive equations of White-Metzner and modified White-Metzner models, respectively, as follows:

$$\tau + \lambda \overset{\nabla}{\tau} = \eta(\nabla \mathbf{u} + \nabla \mathbf{u}^T) \qquad \lambda(T,\dot{\gamma}) = \frac{\eta(T,\dot{\gamma})}{G} \qquad \text{(W-M model)} \tag{5}$$

$$\lambda(T,\dot{\gamma}) = \frac{\lambda_0(T)}{1+\left(\frac{\lambda_0(T)}{k}\dot{\gamma}\right)^{1-m}} \qquad \lambda_0(T) = \lambda^* \exp\left(-\frac{A_1(T-T_f)}{A_2+(T-T_f)}\right) \qquad \text{(modified W-M model)} \tag{6}$$

where λ is the relaxation time, and η is the viscosity; λ and η are functions of both shear rate and temperature; G is the storage modulus constant; m is the power law index; k is a fitting parameter. The modified-Cross model with Arrhenius temperature dependence is employed to describe the viscosity of polymer melt:

$$\eta(T,\dot{\gamma}) = \frac{\eta_o(T)}{1+(\eta_o\dot{\gamma}/\tau^*)^{1-n}} \qquad \eta_o(T) = D_1 * Exp\left(\frac{-A_1(T-T_f)}{A_2+(T-T_f)}\right) \tag{7}$$

where n is the power law index, η_o the zero shear viscosity, τ^*(Taus) is the parameter that characterizes the transition region from the Newtonian plateau to the power law region of the viscosity curve.

In our previous study [5], the relationship between the injected part birefringence which is varying as a function of the applied stress and the fringed pattern has been addressed in details. To catch the viscoelastic effects, the geometry and dimension of a concave lens is shown in Figure 1. The key dimensions for the lens are: diameter in 32 mm, thickness in 2 mm, curvature in 75 mm and gate thickness in 0.80 mm. The material used is COP Zeonex 480 R (supplier is Zeon). The operation conditions for injection molding processes includes filling 0.10 s, packing 5 s, cooling 60 s, mold open and close 8.67 s, melt temperature 275°C, mold temperature 124°C.

RESULTS AND DISCUSSION

Figure 2 shows the viscosity measurement data and curve-fitting for the studied material. The material parameters of the Cross model we adopted as our original design with n =0.4, τ^* is 216,890 Pa, and D_1 (zero-shear viscosity at reference temperature, T_f) is 1.129 E+14 Pa*s. Figure 3 (a) displays the total fringed pattern of the original design. Due to the sudden geometrical change around the gate area, there happen dramatic hydrodynamic variations which can be characterized by the flow-induced residual stresses as shown in Figure 3(b). The results we obtained are in a very good agreement with that of literature [5]. Furthermore, we would put emphasis on how the viscoelasticity effects the birefringence pattern. We tried to introduce some artificial parameters based on the Cross model of viscosity, including modification of either the power law index (n) or τ^*, and change of the D_1. Figure 4(a) shows the viscosities with the alteration of power law index from n = 0.1 to 0.6. When the power law index is modified from n = 0.4 (original value) to n = 0.1, the shear thinning effect is enhanced, and to n = 0.6 is lessened. Figure 4(b) described the viscosities against shear rate with real τ^* of 216,890 either increased to 300,000 or decreased to 100,000. Then using these modified material parameters for further injection molding simulation, the influence on the optical property is exhibited in Figure 5 (a) and (b).

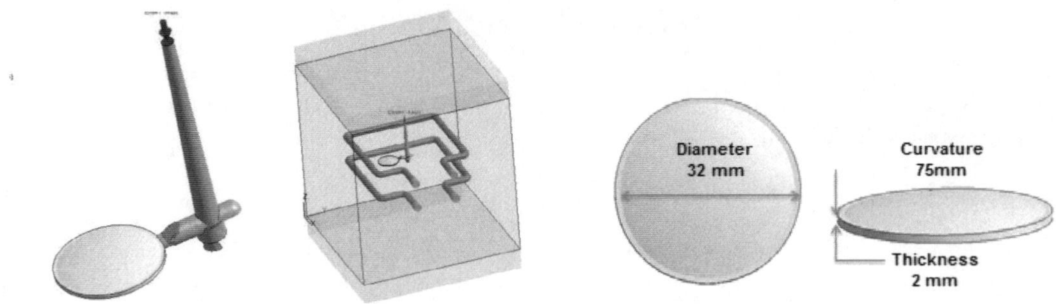

(a) (b)

Figure 1. System construction: (a) the optical model structure and mold layout, (b) the dimension of the optical lens.

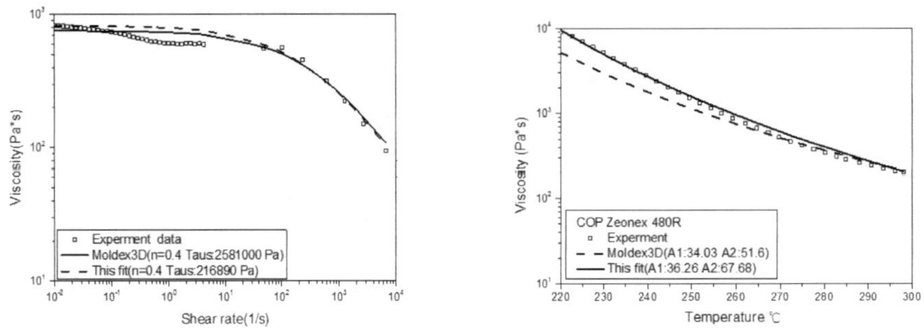

(a) (b)

Figure 2. Data curve-fitting based on experimental viscosity data: (a) viscosity against shear rate, (b) viscosity against temperature.

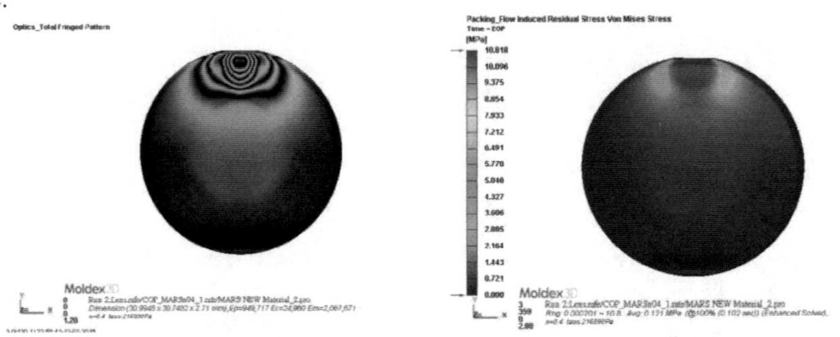

(a) (b)

Figure 3. Fringed pattern of original design: (a) total fringed pattern, (b) Flow-induced residual stress

In Figure 5 (a), when n is increased from 0.1 to 0.6, the total fringed order has no significant difference, but the fringed patterns was varying dominantly in the perpendicular direction, instead of the flow direction (n = 0.1). If pay more attention to the gate area (in Figure 5 (b)), from n=0.4 to n=0.1, the fringed pattern developed area in the flow direction is increased around 30%, but that in the perpendicular direction is decreased by 40%. From n=0.4 to n=0.6, the fringed pattern developed area in the flow direction is not significantly changed, but that in the perpendicular direction is increased by 15%. To investigate what reason to cause this variation, Figure 6(a) for the iso-curves of the melt flow advancement is shown. There is no significant difference among power law indices from 0.1 to 0.6. However, when we move to flow induced residual stress development as seen in Figure 6(b), it is reducing in the flow direction and extending in the perpendicular flow direction. It is exactly similar to what we have found in Figure 5. Furthermore, we also study the influence of the variable τ^*. The effects of τ^* on the optical property is shown in Figure 7. It is found that if τ^* was varied from 100,000 Pa to 300,000 Pa, the total fringed patterns were apparently changing from being dominated in the flow direction to being in the direction perpendicular to the flow. The influenced width of the fringed patterns was almost linearly increased with the increasing τ^*.

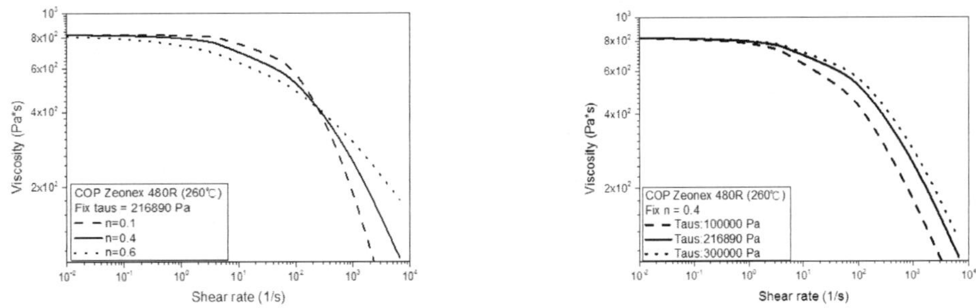

Figure 4. Based on Zeonex COP, (a) the artificial modification of power law index from n=0.1 to 0.6, (b) Tau* effect on the viscosity.

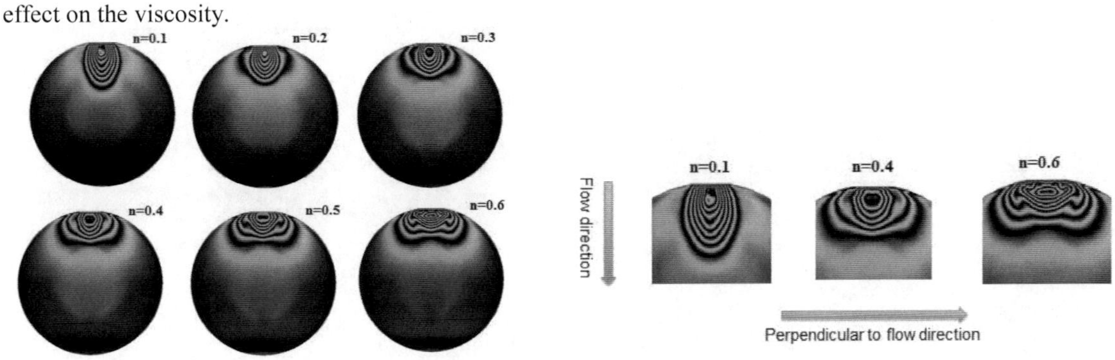

Figure 5 (a). The shear thinning effects on the fringed patterns, (b). From flow direction dominant (n=0.1) to perpendicular direction dominant (n=0.6)

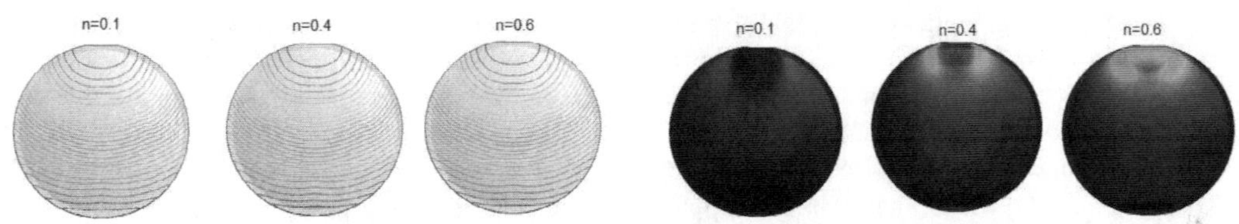

Figure 6. (a) Iso-curves of the melt flow advancement, (b) Flow induced residual stress (0-10.8 Mpa, blue = 0, green = 10.8)

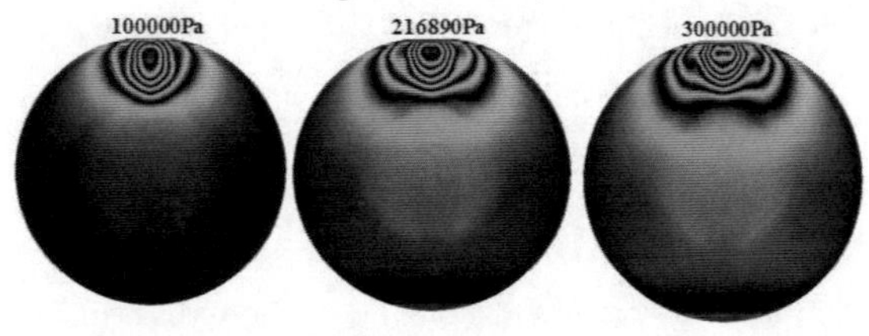

Figure 7. Tau* effect on the fringed pattern.

Moreover, after the manipulation of relaxation measurement and curve-fitting, the material parameters of the Modified White-Metzner model are λ_0=0.0185, m=0.3, k=0.67, A_1=7.95, A_2=153.3, T_0=533.15. It is shown as the solid line in Figure 8(a) and (b). To discover the viscoelastic effects on the optical property, Figure 9 shows the viscoelasticity with the alteration of power law index from m = 0.1 to 0.6. When the power law index is modified from m = 0.3 (original value) to m = 0.1, the shear thinning on the relaxation time is enhanced, and to m = 0.6 is lessened. Then using these modified material parameters for further injection molding simulation, the influence on the optical property is exhibited in Figure 9(a). As m is lower, the relaxation time is shorter, the viscoelastic effect is miner which causes the lighter fringed pattern (as m=0.1). On the other hand, when m=0.6, the longer relaxation time contributes stronger birefringence effect. It dominants in both of the perpendicular and flow directions. Finally, as the k value is varying from 0.1 to 1, when the larger k value (k=1.0) is applied, the slower shear thinning of relaxation time happens. It exhibits stronger relaxation effects to cause significant birefringence as in Figure 9(b).

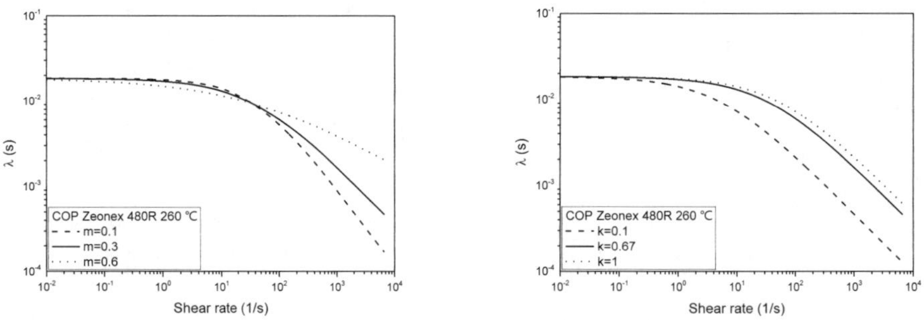

(a) (b)

Figure 8. (a) different power law index for relaxation time against shear rate, (b) different *k* values for relaxation time.

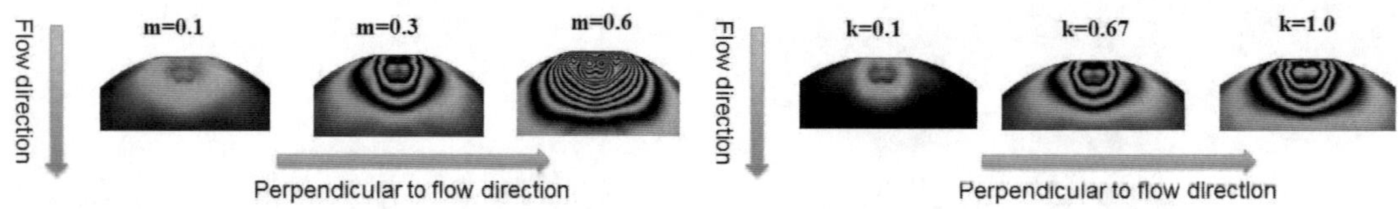

(a) (b)

Figure 9. The shear thinning effects on the fringed patterns, (a) from low relaxation (m=0.1) to high relaxation (m=0.6); (b) from low k (0.1) to high k (1.0)

CONCLUSIONS

In this study, we have investigated the viscoelastic effects on the optical property of a concave lens made of Zeonex COP 480R material by the traditional injection molding, simulated with n = 0.4 and τ^* = 217,000 Pa in the Cross model for the melt viscosity. Results show that if the power-law index (n) was changed from 0.1 to 0.6, the total fringed order has no significant difference, but the fringed patterns was varying dominantly in the perpendicular direction, instead of the flow direction (n =0.1). This situation was matched with the distribution pattern of the flow-induced residual stresses. Moreover, if the τ^* in the Cross model was varied from 100,000 Pa to 300,000 Pa, the total fringed patterns were apparently changing from being dominated in the flow direction to being in the direction perpendicular to the flow. The influenced width of the fringed patterns was almost linearly increased with the increasing τ^*. Furthermore, by using the modified White-Metzner model to investigate the viscoelastic effect on the optical property, it was found that a large relaxation time constant in the W-M model for the processed material could cause strong birefringence in both the flow and the perpendicular directions to the molded concave lens. The results can be applied as some guideline for either the further material modifications or the new material developments.

REFERENCES

1. Yoon, K., An Experimental Study on Precision Injection Molding of Center-Gated Disks, *Korean Journal of Rheology*, **7**, 19-27 (1995).

2. Shyu, G. D., Isayev, A. I., and Lee, H. S., Numerical Simulation of Flow-induced Birefringence in Injection Molded Disk, *Korea-Australia Rheology Journal*, **15**, 159-166 (2003).

3. Wang, P. J. and Lai, H. E., Study of Optical Characteristics for Injection Molded Aspheric Lenses, *Asia Pacific Conference on Optics Manufacture*, (2007).

4. Wang, P. J. and Lai, H. E., Study of Residual Birefringence in Injection Molded Lenses, *Plastics Encounter at SPE ANTEC*, 2494-2498 (2007).

5. Chang, Y. J., Yu, C. K., Chiu, H. S., Wen-Hsien Yang, W. H., Lai, H. E., and Wang, P. J., Simulations and Verifications of True 3D Optical Parts by Injection Molding Process, SPE ANTEC , 253-258 (2009).

Design and Fabrication of Injection Mold Using Gypsum

Chung-Chih Lin[a*], Bean-Yin Lee[a], Jenn-Yih Chen[a] and Guo-How Lee[a]

[a] *Department of Mechanical and Computer-Aided Engineering, National Formosa University, 63208, Hu-Wei Town, Yunlin County, Taiwan*

lincc@nfu.edu.tw

Abstract. Rapid tooling (RT) technology has been recognized as a good approach for the initial trial during the product development process. Lots of RT methods are revealed for making tools with different pilot run requests. Gypsum (calcium sulphate dehydrate) is a well-known mineral obtained from natural resources and burned to calcium sulfate hemihydrates or anhydrite to be used as plaster of Paris in different applications such as buildings, art or ceramics. The inherent property of gypsum is weak and porous so that the mold made from gypsum can only be used for low pressure molding process, i.e., casting. In an attempt to provide a rapid and economical alternative for making mold with low volume production, the potential of gypsum as a mold insert used in injection mold is developed. The design and fabrication process of the mold insert was implemented, starting with the gypsum powder, mixing the powder with water and binder to form plaster. Before the plaster solidified, an additional compression was introduced to enhance its strength. Then the plaster was machined into a mold insert and validated by injection test. The optimal combination of ingredient, 58wt%/37wt%/5wt% of hemihydrate/water/binder, was found to work best. A coin pattern was selected to build the gypsum insert for checking the replication accuracy of the molded part. In addition, the life cycle of the gypsum insert for injection molding was also evaluated in this study

Keywords: rapid tooling, gypsum, injection molding.
PACS: 81.05.Lg

INTRODUCTION

In order to reduce the risk of development process, a small quantity of the prototypes produced by a quicker and cheaper method before mass production is necessary. Rapid prototyping (RP) technology provides a faster and economic approach to validate the design concept in the initial stage of the product development [1]. However, the economic advantage of RP technology is decreased with the increase of the quantity requirement of product. Although the technology is improving and there is increasing use of RP for direct rapid manufacture, fabricating many parts that require high accuracy and good material properties are still difficult [2]. Therefore, an alternative tooling used to produce more components in representative materials as well as still keeping the speed and flexibility of RP technology is highly demanded.

Molds used in injection molding process are typically made from hardened steel to resist the injection pressure for longer tool life consideration. Sometimes the steel molds cannot be justified in case of low quantity requirements Rapid tooling (RT) is typically used to describe a process which either uses a RP model as a pattern to cast a mold quickly or uses the RP process directly to fabricate a tool for a limited quantity of prototype manufacture. In the new product development cycle, there is often need of some intermediate tooling to produce a small quantity of prototypes for pilot run check, design or concept verification. If the products whose quantity requirements are insufficient to manufacture by steel tooling, they may be applied by RT with cost and efficiency considerations. Indirect rapid tooling processes produce molds by forming them from a master pattern. The machining time of this method is better than that of direct rapid tooling. Silicone rubber is a common example of this technique. This process is simple but the mold tends to damage after 10~20 times of produce. The properties of the polyurethane parts are still inferior to injection molded parts as well [2]. Kirksite tooling, a moderate strength zinc-base alloy, is developed primarily as a quick mold manufacture method. Maximum quantities of five to ten hundred pieces are typical prototype runs of Kirksite tooling process [3]. Comparing to conventional steel mold, it provides low-cost tooling because the alloy can be accurately cast, requiring a minimum of finishing. However, it is expensive to set up the production line of this process. For lower quantity requirement, the simplest and most economical methods of producing a tool for the injection molding is cast resin tooling. This method is fast and can be applied to common thermoplastics such as polypropylene and ABS. Due to the low mechanical strength of the molds; complicated products are difficult to be molded by this method. Ma et.al [2] developed a better approach using epoxy composite material to enhance the mechanical property of the cast resin tooling. A suitable composite, 20% alumina powder filled epoxy composite, was found to work best.

Gypsum (calcium sulfate dihydrate) is a well-known mineral that is obtained from natural resources and used as different building materials. Gypsum is burned to calcium sulfate hemihydrates or anhydrite. A reaction to form a solid model by using gypsum is just simply adding water into a mold with the hemihydrates or anhydrite. The energy consumption of the mold making by using gypsum is lower than other methods [2~5]. The objective of this study was to develop an gypsum based material suitable for making injection mold with a range of production quantity, which links the quantity gap between very low volume prototype tooling and full production tooling (like Kirksite tooling). The proposed method can be applied to a wider range of thermoplastic parts with low production quantity. The mechanical property of different ratios of calcium sulfate hemihydrate, water and binder were tested to evaluate suitability for mold material firstly. To check whether the tool making process could create the gypsum insert with acceptable quality, the master pattern of a Taiwanese coin (fifty-dollar) was prepared for mold making. Then the insert made of gypsum was assembled into a mold base for injection molding process. The process of preparation of the gypsum insert for injection mold is illustrated in Fig.1.The geometrical replication of the molded parts was measured for evaluating the dimensional accuracy. The results showed that the application of gypsum can be extended to injection mold successfully.

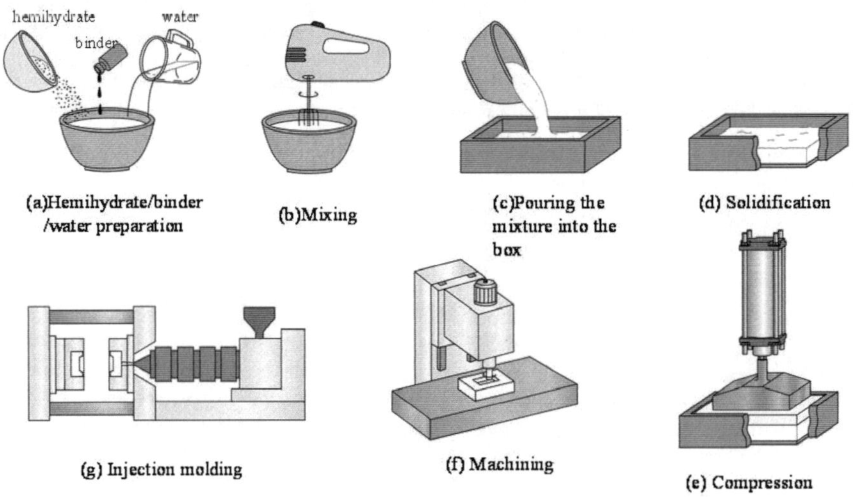

FIGURE 1. the preparation of gypsum inert for injection mold.

EXPERIMENTAL

Material

The main material of gypsum insert is from calcium sulphate hemihydrate (α-type, CK25 from Kuang Pang Gypsum Ltd, Taiwan) [6]. When hemihydrate is mixed with water, a part of it immediately dissolves making the solution saturated with respect to Ca^{2+} and SO_4^{2-} ions. This saturated solution becomes supersaturated with respect to calcium sulphate dihydrate leading to nucleation and crystal growth [4] and then the strength of the gypsum insert is developed. The solidification involves physico-mechanical processes. In fact, the hardened mass is not a compact solid, but a highly porous material with a brittle characteristic. Only low pressure molding process, like vacuum forming or casting can use calcium sulphate dihydrate as the mold material. In order to improve these problems, attempts have been made to determine the mixture ratio of hemihydrate and water with binder to evaluate how effectively they can improve the properties of gypsum. The binder selected in this study was PVA (PolyVinyl Alcohol, W256 from Winroy Industrial Polymers Ltd, Ireland) due to its polarity. PVA is a water-soluble synthetic polymer and is compatible with water during the hydration of hemihydrate. According to the reaction of hydration of hemihydrate [4], the ingredients of hemihydrate/water/binder for making the gypsum insert were listed in Table 1.

TABLE (1). Ingredients of Gypsum Insert

No.	α- Calcium Sulphate Hemihydrate, wt%	Water, wt%	Binder, wt%	Compression Load, kgw (30 min)
1		46	0	
2	54	43	3	2.5, 5.0, 7.5
3		41	5	
4		42	0	
5	58	39	3	2.5, 5.0, 7.5
6		35	5	
7		38	0	
8	62	35	3	2.5, 5.0, 7.5
9		32	5	

Specimen Preparation and Characteristics

A simple test to evaluate the performance of different ingredients as listed in Table 1 on resisting the injection pressure was implemented. In order to have an isotropic compression effect, a cylindrical shape of the specimen was designed and the dimension of the specimen was ϕ22.5mm×35mm in diameter and length, respectively. The specimen geometry and sample were illustrated as Figs. 2(a) and 2(b). A following compression process with different load was introduced before the setting of gypsum specimen for 30 min. Note that the dimension of the specimen was not the same as that of the standard test regulations. However, it provided a comparative test result to determine the optimal ingredient. The test specimens were conditioned at 25°C and 50% relative humidity for 48 hr. before test. The number of the compressive strength test of each ingredient was twelve. The test was measured at room temperature with a speed of 5 mm/min referring to ASTM D695-96 by using the testing machine HongTa (HT-2402, Taiwan).

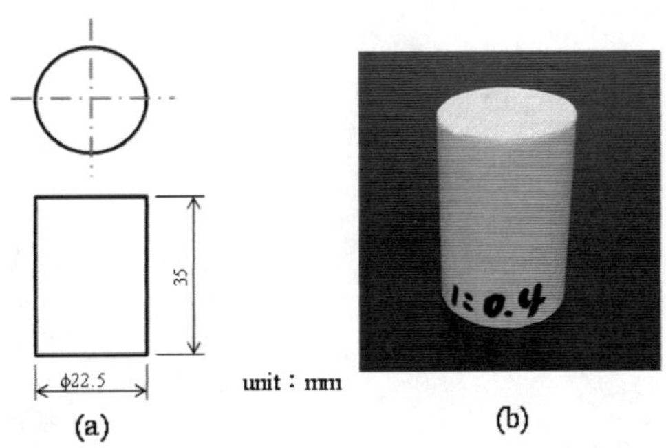

FIGURE 2. (a) the dimensional sketch of the specimen and (b) the photo of the specimen.

Mold Assembly and Molding Process

The gypsum insert whose dimension was 150 mm × 80 mm ×30 mm for length, width and thickness individually was assembly into the test mold as shown in Fig.3(a). The mold contained two cavities for quick change consideration as shown in Fig. 3(b). The steel material for the test mold was P-20 steel. The injection molding machine used in this study was Fanuc Robotshot S-2000i 50A (50 Tons, Japan) and the molding parameters injection process was set the same as conventional mold except cooling time. Since gypsum has poorer thermal conductivity, the corresponding cooling time was likely to be longer. For small production batches, this was unlikely to cause significant time cost.

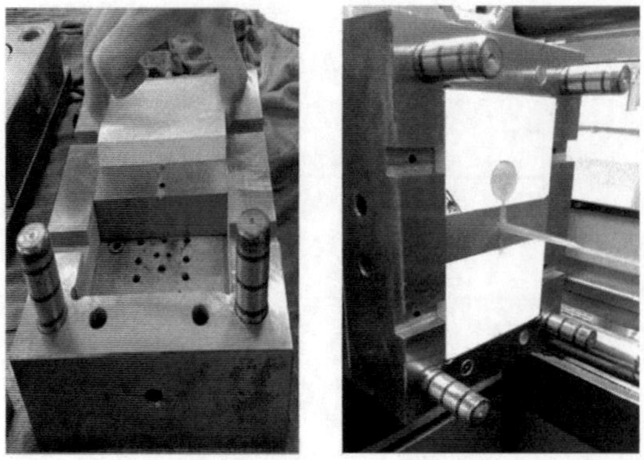

(a) (b)

FIGURE 3. (a) the assembly of gypsum insert into cavity, (b) accomplished injection mold.

RESULTS AND DISCUSSION

Compressive Strength of Gypsum

The compressive strength of gypsum specimens made of different ingredients was shown in Fig. 3(a). The result indicated that the yield compressive strength of the specimen was increase with the increase of binder approximately. It also illustrated that the compressive strength of the specimen was enhanced best as the amount of hemihydrate was 58%. When the hydration occurs in the paste, the plaster solidified and developed its strength. Before the finial setting of the gypsum in solidification process, an additional compression process was introduced to increase the rigidity of the specimen. The increment of the compressive strength of the specimen with the compression process was up to 13% at the compression load of 5 kgw and 7 kgw. It meant that the compression effects of the both loads were the same. The result was shown in Fig. 3(b)

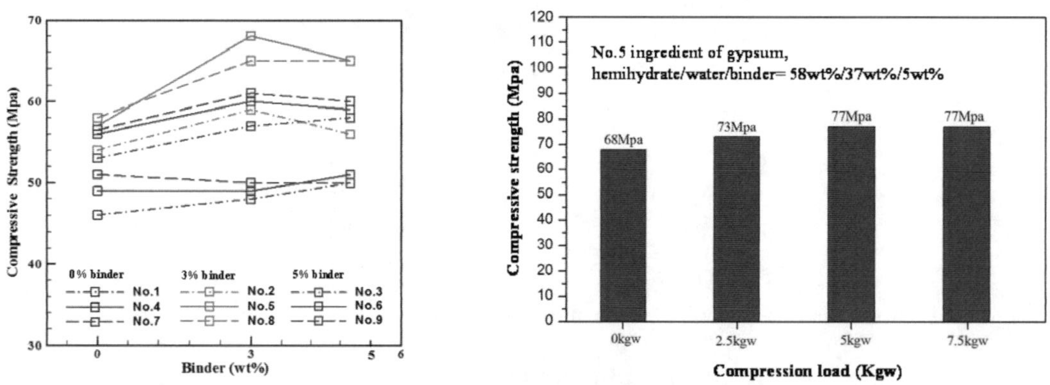

FIGURE 4. (a) the compressive strength of different ingredients and (b)the effect of compression process on the compressive strength of No.5 ingredient of gypsum.

Replication of gypsum insert

A Taiwanese coin (fifty-dollar) shown in Fig.5 (a) was selected as the master pattern for the preparation of the injection mold. The master pattern was put on a plastic plate and four plastic walls were placed and fixed around the master pattern to form a box. According to the first example's result, we selected the optimal ratio of hemihydrate/binder/water for making mold insert. After mixing thoroughly, the mixture was gently poured into the box with the master pattern. A runner system was machined onto the gypsum insert after the gypsum solidification process was completed. As the fabrication of the mold was completed, the resin of PP (Polypropylene, GP6331 from LCY Chemical Corp. Taiwan) was used for injection molding. The main processing parameters can refer to the recommended information of LCY website [7]. The molded part with runner system was shown in Fig.5 (b). The replication of the master pattern, the gypsum insert and the molded part were shown in Figs.5(c) ~ 5(e). Before the gypsum insert cracked, the number of injection run of this case was 256 times.

FIGURE 5. (a) Taiwanese coin, (b) molded part by gypsum mold, (c) detailed view of the coin on red square, (d) detailed view of the gypsum insert, and (e) detailed view of the molded part.

CONCLUSION

Most of gypsum material is applied to mold making only for casting process whose pressure is relative lower than injection molding. In this study, the development and fabrication of gypsum insert of injection mold was presented. It extends the application of gypsum material to high pressure injection molding through the ingredient modification. The compression process introduced into the gypsum insert preparation raised the compressive strength about 13%, so the resistance to brushing effect during injection process was improved. Comparing with other RT methods, the gypsum insert has the advantages of a rapid and low energy consumption approach of mold making. This method proves that gypsum has the potential performance for injection mold utilizations where low batch products are required.

ACKNOWLEDGMENTS

The authors would like to thank for the financial support from National Science Council in Republic of China under the contract number of MOST103-2221-E-150-015-.

REFERENCES

1. S. Rahmati, P. Dickens, *International Journal of Machine Tools & Manufacture.* **45**, 740-747 (2007).
2. S. Ma, I. Gibson, G. Balaji, Q.J. Hu, *Journal of Materials Processing Technology.* **85**, 75-82 (2007).
3. J. Noblea, K. Walczakb, D. Dornfelda, *Procedia CIRP*, **14**, 252-256 (2014).
4. N.B. Singh, B. Middendorf, *Progress in Crystal Growth and Characterization of Materials*, **53**, 57-77 (2007).
5. H. Kima, J. Sunga, F. E. Goodwinb, and T. Altan, *Journal of Materials Processing Technology.* **205**, 459-468 (2008).
6. http://www.kpisi.com/307073316729986216973044637636l.html.
7. http://www.lcygroup.com/.

Morphology and mechanical properties of micro-injection molded polymer products

Shi-Wei WANG[a] , Qian LI [a], Chang-Yu SHEN [a], Guo-Hua HU[b]

[a] *National Center For International Joint Research of Micro-Nano Molding Technology, School of Mechanics & Engineering Science, Zhengzhou University, Zhengzhou, 450001,China.*
[b] *Laboratory of Reactions and Process Engineering (LRGP), CNRS-University of Lorraine, 1 rue Grandville, BP 20451, 54001 Nancy, France.*
Email: qianli@zzu.edu.cn

Abstract. Novel multi layer and stripe morphology were investigated in micro-injection molded isotactic polypropylene gear. The annealing treatment was performed on the gear tooth under different temperature and time to control its morphology evolution. The effects of the mold temperature of micro-injection molding on the microstructures and mechanical properties of isotactic polypropylene gears were studied. The micro-injection molded gears present a skin-core type of morphology. The core layer is much thicker than the core and shear layers. Nanoindentation tests show that the modulus decreases along the flow direction and increases with increasing mold temperature. The highly oriented shear layer has the highest nanoindentation modulus compared with the skin and core layers.

Keywords: Micro-injection molding; Layer structure; Mold temperature; Crystalline structure; Mechanical properties
PACS: 81.05.Lg

INTRODUCTION

Microinjection molding (MIM) is a kind of exactitude and easy control method, as well as many new features compared with normal injection processing[1-2]. Micro processing was wildly concerned because of its micro scale effect, which was apart from normal processing condition. One of the main challenges associated with MIM is the result of the micro size of the product, which is difficult to see clearly without the aid of microscopic techniques. Fundamentally similar to that of conventional injection molding, the morphology of MIM polymer also plays an important part in the shape and arrangement of the final products[3-4].

In this work, a novel morphology of MIM iPP gear was observed by polarized light microscopy, and the annealing was applied to control the morphology evolution, also wide angle X-ray diffraction was used to associate the crystalline structure with novel morphology. PP micro-gears were molded at different mold temperatures. Their morphology was characterized by a polarization optical microscope. The mechanical properties of different layers and/or different positions of the micro-gears were measured by nanoindentation. The relationship between microstructure and mechanical properties provides new insights into MIM polymer parts.

EXPERIMENTAL

Materials and sample preparation

A commercial grade iPP, F401, with a melt flow rate of MFR=2.3g/10min (ASTM D1238, 230^{0}C, and 2.16Kg load) from LanZhou petroleum Chemical Co, Ltd. (PR China), was used in the study. The test sample is a iPP gear, with a thickness of 450 μm, a gear teeth length of 1000 μm, half gear teeth width of 350 μm. It was manufactured by MIM machine (Babyplast 6/10P) at 4MPa and 200 °C. The effect of mold temperature (30, 40, 50, 60, 70, 80°C) on the structure and properties of MIM PP sample was investigated. For annealing experiments, The iPP gears were placed in an oven at temperature ranged from 60 to160 °C for different times (20, 40, 60, 80, 120mins). Four positions with 800μm gaps were selected along flow direction for testing.

Results and discussion

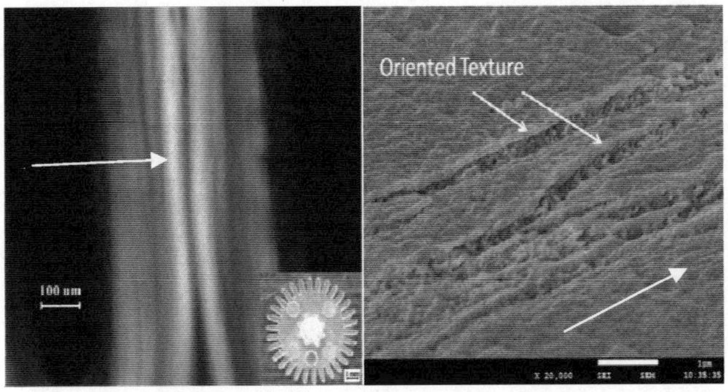

Fig.1. PLM(a) and SEM(b) Morphology of MIM iPP gear tooth without annealing.

Fig. 1(a) shows the polarized morphology of the central part of MIM iPP gear tooth without annealing. The morphology was taken directly from the gear sample placed on the stage using transmission mode. Clear stripes morphology displays on the Fig. 1(a) from central to the both sides, from surface to central, the width of the strip increased, while the brightness of the strip decreased. This novel stripe morphology has not been reported yet. In order to probe the structure in full, surface morphology was investigated besides only measuring the thick samples with PLM. The surface morphology of the etched gear tooth was detected by SEM and a clear oriented texture appeared as shown in Fig 1(b).

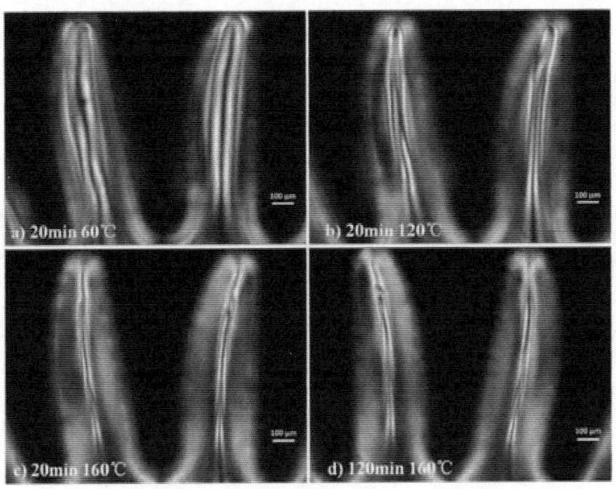
Fig. 2. MIM iPP gear tooth annealing for (a) 20min at 60 °C, (b) 20min at 120 °C, (c) 20min at 160 °C and (d) 120min at 160 °C.

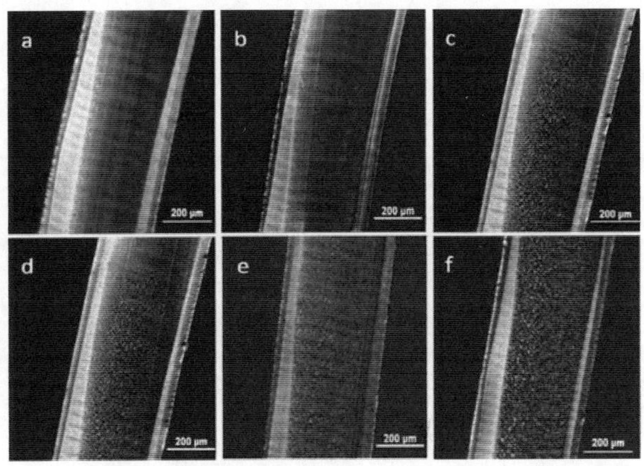

Fig.3. Images observed by the polarized optical microscope of the micro-injection molded isotactic polypropylene gear slices at different mold temperatures (a) 30 (b) 40 (c) 50 (d) 60 (e) 70 and (f) 80°C.

The annealing treatments were performed on the gear tooth under different temperature and different time. Fig. 2(a) shows gear teeth annealed at 60 °C for 20 minutes. Compared with Fig. 1, the width of stripes in Fig. 2(a) increases, also the shape of the stripes shows some disparity. This phenomenon presents more obvious for the gear teeth annealed at 120 °C for 20 minutes as shown in Fig. 2(b). The number of strips increases significantly, especially in the central of the teeth and results in reducing of the dark stripe. With the increase of the annealing temperature, the width of the fuzzy layer increases, as shown in Fig. 2(c). When the annealing temperature rise to 160 °C, the dark strip becomes even narrow and the width of the fuzzy edges enlarges obviously. Extension of the annealing time shows little affection on the morphology of the teeth, as shown in Fig. 2(d), the same temperature while 100 min longer compared with the sample of Fig. 2(c). Fig. 3 shows the morphology of the gears molded at different mold temperatures. The flow direction from the bottom to the top corresponds to the region of the gate and the top of the gear tooth. It is seen that all the samples show a multi-layer structure. It is known that injection molded isotactic polypropylene typically shows three layers, namely the skin layer, shear layer and core layer. For the MIM PP micro-gears, there exists a layer between the shear and skin layers which is rich in small crystals. The formation of this layer may be due to a lower cooling rate and accrued polymer chain mobility by shear induced heat (viscous dissipation) between the shear layer and skin layer. Along the flow direction, the thickness of the shear layer decreases and the crystal size of the core layer increases, indicating that the orientation decreases along the flow direction.

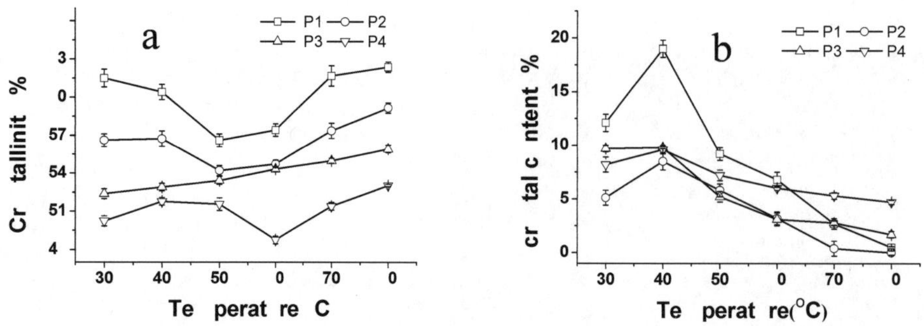

Fig. 4. (a) Crystallinity (b) β-crystal content of different positions of micro-injection molded isotactic polypropylene gears as a function of the mold temperature.

Fig. 4shows the crystallinity and relative content of β crystals for MIM PP micro-gears under different mold temperatures. They are calculated from the WAXD curves [5]. The crystallinity decreases significantly along the flow direction from P1 to P4 as shown in Fig. 4a. This is consistent with the morphology of Fig. 2. After 60°C, a further increase in mold temperature increases the crystallinity.

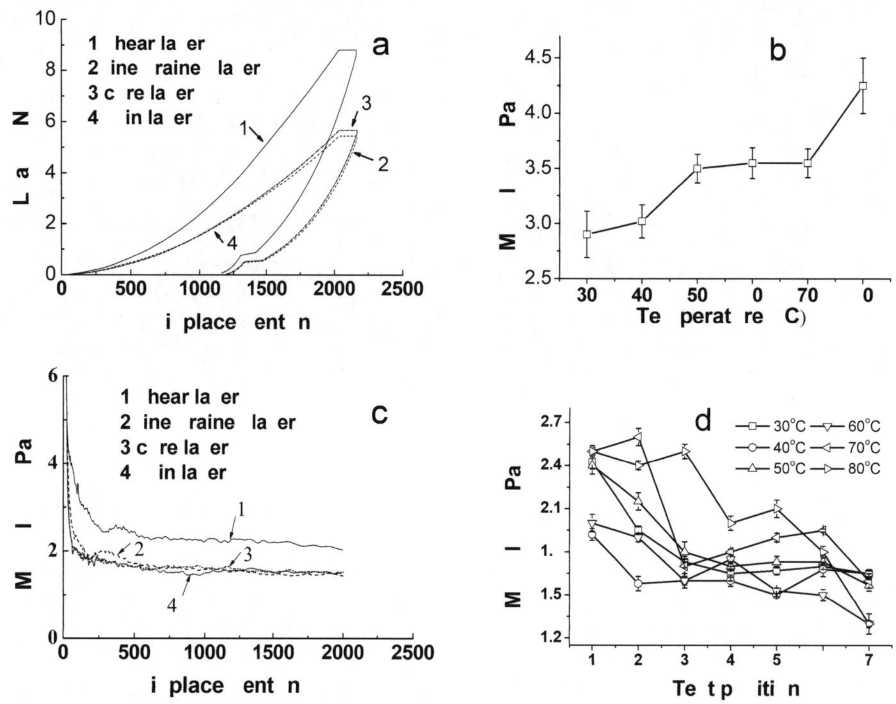

Fig.5. (a) Load-displacement curve (b) Average modulus values of shear layer at four locations vs. mold temperature (c) Modulus-displacement curve of micro-injection molded isotactic polypropylene gears at a mold temperature of 50 ℃; (d)Nanoindentation modulus of the shear layer of micro-injection molded isotactic polypropylene gears at different temperatures.

The nanoindentation test was performed to find the relationship between structure and properties of iPP micro-gears [6-8]. Fig.5a shows the load-displacement curves of four different layers of MIM PP gear molded at a mold temperature of 50℃. The load force of the shear layer is higher than those of the skin and core layers, the latter being very close. On the decompression curve, the residual depth of the shear layer is the smallest. Fig.5b shows that the average modulus of the shear layer increases with increasing mold temperature. This can be attributed to the fact that spherulites are bigger at a higher mold temperature. Fig. 5c shows the modulus-displacement curve of the micro-injection molded isotactic polypropylene gear a mold temperature of 50℃. The shear layer has the highest modulus. The other three layers (fine grain, skin and core layers) exhibit a similar modulus. In order to further study the effect of the mold temperature on the mechanical properties of the shear layer of MIM PP gears, the nanoindentation test was carried out at seven different positions along flow direction. The results are shown in Fig. 5d. The modulus increases with increasing mold temperature, due to larger spherulites. This fact is accordance with the WAXD results.

CONCLUSION

A novel stripe morphology in MIM iPP gear was investigated in this work, and the number and the width of stripes as well as the crystallinity were changed under various annealing temperature and time. Secondary crystallization was rebuilt, molecular chains rearranged and perfect structure further affect the light path of polarized light. Ultimately, the increasing stripes appeared, as well as the decrease of the strip width in the sample. An obvious skin-core structure is formed during micro-injection molding of isotactic polypropylene (PP) gears, high shear and fast cooling. The core layer represents a large portion of the gears, followed by the shear and skin layers. The orientation induced by the wall slip effect decreases especially at the MIM PP gears tooth. The modulus of the shear layer is higher than those of the skin and core layers. Moreover, they are affected by shear to a great extent, compared to those of the skin and core layers.

ACKNOWLEDGMENTS

The authors acknowledge the National Science Fund (No. 11372286), Basic and advanced technology research project of Henan Province (No. 152300410033), International technological cooperation project (2015DFA30550) for their financial support of this project. The Project is sponsored by the Scientific Research Foundation for the Returned Overseas Chinese Scholars, State Education Ministry. .

REFERENCES

1. WS, Guan., HX, Huang, *J. Micromech. Microeng,* **22**, 1-10 (2012)
2. M, Heckele., WK, Schomburg, *J. Micromech. Microeng,* **14**, 1-14 (2004)
3. V, Nagarajan., K, Zhang., M, Manjusri., AK, Mohanty, *ACS Appl. Mater. Interfaces,* **7**, 11203−11214 (2015)
4. W, Michaeli., A, Spennemann., R, Gärtner, *J. Polym. Eng,* **24**, 81-94 (2004)
5. YM, Wang., HH, Zhang., M, Li., et al. *Polym Test,* **41**, 163-171 (2015)
6. AM, Díez-Pascual., MA, Gómez-Fatou., F, Ania., A, Flores. *Prog Mater Sci.* **67**, 1-94 (2015)
7. PM, Wood-Adams., JM, Dealy., AW, DeGroot., OD, Redwine, *Macromolecules,* **33**, 7489-7499 (2000)
8. SW, Wang., W, Yang., MB, Yang, *Polym Test,* **27**, 638-644 (2008)

Measurement of Fiber Orientation Distribution in Injection-molded Composites with High Filler Content

Senji Hamanaka[a], Katsuhisa Yamashita[a], Chisato Nonomura[a], Thanh Binh Nguyen Thi[b], Takayuki Wakano[b], Atsushi Yokoyama[b,*]

[a]Research Center, Toyobo Co., LTD, 2-1-1 Katata, Otsu, Shiga 520-0292, Japan
[b]Kyoto Institute of Technology, Matsugasaki, Sakyo-ku, Kyoto 606-8585, Japan

senji_hamanaka@toyobo.jp
yokoyama@kit.ac.jp

Abstract. Short-fiber-reinforced composites are widely used in a number of industries and applications, including in the transportation industry, and in business machine, durable consumer items, and sporting goods. Properties of fiber-reinforced composite depend on its fiber orientation distribution. Thus, knowing the fiber orientation is of great importance, and a number of researchers have been interested in developing useful and accurate techniques for determining the fiber orientation in injection-molded parts formed from short-fiber composite. However, out-of-plane orientation was preformed manually and difficulties arise when employing the technique in the case of composites with high fiber contents, over 50wt.%. In this research, short-glass fiber-reinforced polyamide 6 specimens produced using two different plate-shaped cavities having three different thicknesses ranging from 2mm to 4mm and with the fiber contents ranging from 30wt.% to 65wt.% are carried out using injection molding. The three-dimensional (3D) fiber orientation observation and measuring are examined by Micro-computed tomography system Y.Cheetah μHD which cooperated with calculation software VG.Studio max 2.2 using a calculation model (ex. Monte-Carlo Model). Herein, detailed 3D fiber orientation distribution from skin to core at different locations in high filler content composite specimens is automatically and fast clarified. This novel measurement methodology may enable us to evaluate the ability of new model theory, the effective of implementation procedure for fiber orientation and property prediction in injection-molded composite.

Keywords: Fiber Orientation, Injection Molding, High Filler Content Composite, Micro-computed tomography.
PACS: 81.20.Hy

INTRODUCTION

Short-fiber-reinforced plastics (SFRP) are widely used in a number of industries and applications. SFRP is more superior in strength, rigidity than other non-reinforced resins, and specific gravity is smaller than metal, therefore, SFRP is used in automotive parts as metal substitute materials for the purpose of weight reduction. Recently, application SFRP with high filler content over 50wt.% is in a tendency to increase for the purpose of giving high functionality. Generally, the mechanical property of SFRP shows anisotropy, because of existence of fibers. Thus, it is great important to understand fiber orientation distribution of injection-molded SFRP, because it influences mechanical properties. Moreover, the fiber orientation distribution and the mechanical property can be predicted by numerical method, which uses the reliable experimental results. It was generally supposed that it can be a very effective supporting for part design phase. Therefore, the fiber orientation distribution of injection-molded SFRP parts has studied by many researchers [1-5], and they developed the numerical prediction of fiber orientation for above purposes. Few publications have been reported the fiber orientation distribution of injection-molded parts which used SFRP with high filler content. However, the quantitative analysis is still lacking.

The ultimate aim of the present studies is to develop numerical prediction of fiber orientation in high filler composites. For this reason, a detailed and automatic method needs to be established on the 3D fiber orientation observation and measurement. In this paper, an effective method was employed to observe and measure directly the 3D fiber orientation distribution in short-glass fiber-reinforced polyamide 6 (GF/PA6) specimens with the fiber contents ranging from 30wt% to 65wt%. These specimens were produced using plate-shaped cavities having three different thicknesses ranging from 2 mm to 4 mm by injection molding. Herein, Micro-computed tomography (μ-CT) system Y.Cheetah μHD was employed, and the reconstruction of X-ray CT images was carried out by

VG.Studio max 2.2 (Volume Graphics GmbH, Heidelberg, Germany). The calculation of fiber orientation in each specimen was carried out by cooperation with the software VG.Studio max 2.2.

EXPERIMENTAL

Injection-molded Specimens

Pellets of GF/PA6 were prepared in twin screw extruder in Toyobo Co., Ltd. The GF/PA6 composites with 30wt.%, 50wt.%, and 65wt.% fiber concentration were fabricated by injecting pellets into injection molding machine (Toshiba: IS-80), and then used in this experiment. The following molding conditions: resin temperature of 290°C, mold temperature of 120°C, filling time of 0.4s, injection speed of 100mm/s, and packing pressure of 40MPa was kept for all sample preparations. Fig.1 (a) shows the shape of the experimental mold cavity which was designed to produce the plate-shaped samples. Planar dimensions of 100mm × 100mm, and cavity thicknesses of 2.0mm, 3.0mm, 4.0mm were designed and manufactured. Gate position was arranged at the edge of cavity as side gate. The 3D fiber orientation distribution of injection-molded parts was examined by μ-CT system Y.Cheetah μHD. Samples for X-ray CT measurement were cut out from GF/PA6 composite samples at A, B, and C positions (see Fig. 1 (a)). During sample preparation, great care was taken to ensure that the flow direction (X direction). The Y direction was determined according to the right-hand rule, and the Z direction is thickness direction. Table 1 shows the detail of experiment conditions of the sample preparation for X-ray CT measurement.

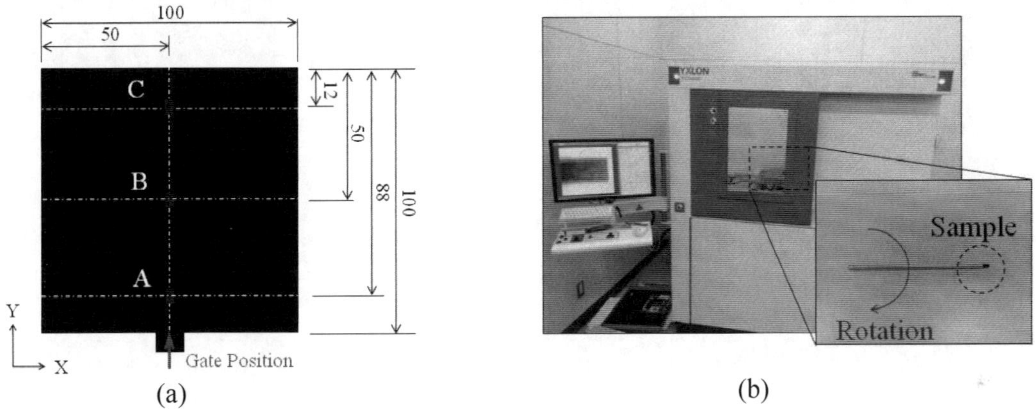

FIGURE 1. Shape of experimental mold cavities with planar size of 100 x 100 [mm^2] and thicknesses of 2.0mm, 3.0mm, 4.0mm and testing positions A, B, and C for X-ray CT measurement (a), μ-CT system and set sample (b).

X-ray CT Imaging and Analysis

μ-CT system YXLON International, Y.Cheetah μHD was employed to observe the fiber orientation on a series of X-ray CT images. Tube voltage was 30kV, and current was 100μA. Measurement range was same with whole size of samples. In this measurement, samples were attached rod which was made from Aluminum. Fig.1 (b) shows μ-CT system and set sample. X-ray absorption occurs because of compton scattering or photoelectric effect. X-rays collide to an electron, and intensity of X-ray becomes smaller than initial statement. Fiber projection images on whole angle were obtained by rotating axis. The reconstruction of fiber slice images on whole angle was carried out by using the VG Studio max 2.2 software (Volume Graphics GmbH, Heidelberg, Germany), and 3D internal structure of samples was obtained. The calculation of fiber orientation in each specimen was carried out by the software which cooperated with a calculation software VG.Studio max 2.2.

RESULT AND DISCUSSION

Fig.2 shows X-ray CT images of cross section and the fiber orientation tensor diagrams in 30wt.% GF/PA6 part with 2mm thickness at the positions A,B, and C. The axis 1, axis 2, and axis 3 are flow direction (MD), transverse direction (TD), and thickness direction (t), respectively. Fiber orientation to MD was relatively strong around surface, and fiber orientation to TD was relatively strong around mid-plane under all conditions. It is considered that the oriented state around surface is occurred by fountain flow near the melt front. The fibers near the melt front is moved outward to the wall by passing through the fountain flow region, and fibers are oriented to flow direction by shear effect between mold and melting resin around surface. On the other hand, around mid-plane, it is considered that fibers are aligned perpendicular to flow direction by stretching flow dominated near the mid-plane. The fiber orientation around mid-plane intensifies as filling advances. These tendency shows good agreement with previous our studying [4, 5].

TABLE 1. The conditions of samples prepared from injection-molded GF/PA6 parts.

Glass Fiber Content	Position	Thickness[mm]
30wt.%	A,B,C, short-shot B	2
50wt.%	A,B,C, short-shot B	2
65wt.%	A,B,C, short shot B	2
30wt.%	B	3, 4
50wt.%	B	3, 4
65wt.%	B	3, 4

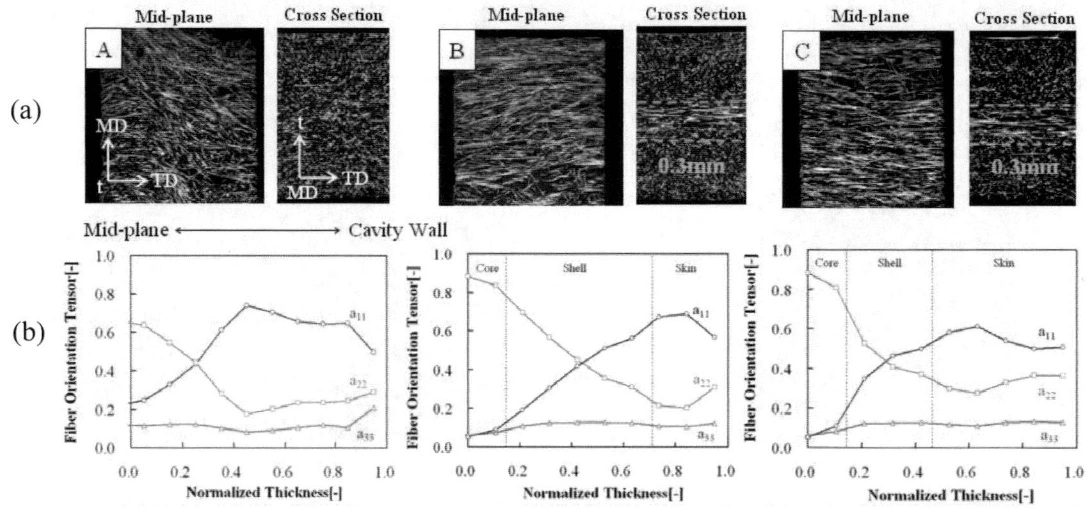

FIGURE 2. X-ray CT images of cross section (a) and the fiber orientation tensor diagrams (b) in 30wt.% GF/PA6 part with 2mm thickness at the positions A,B, and C.

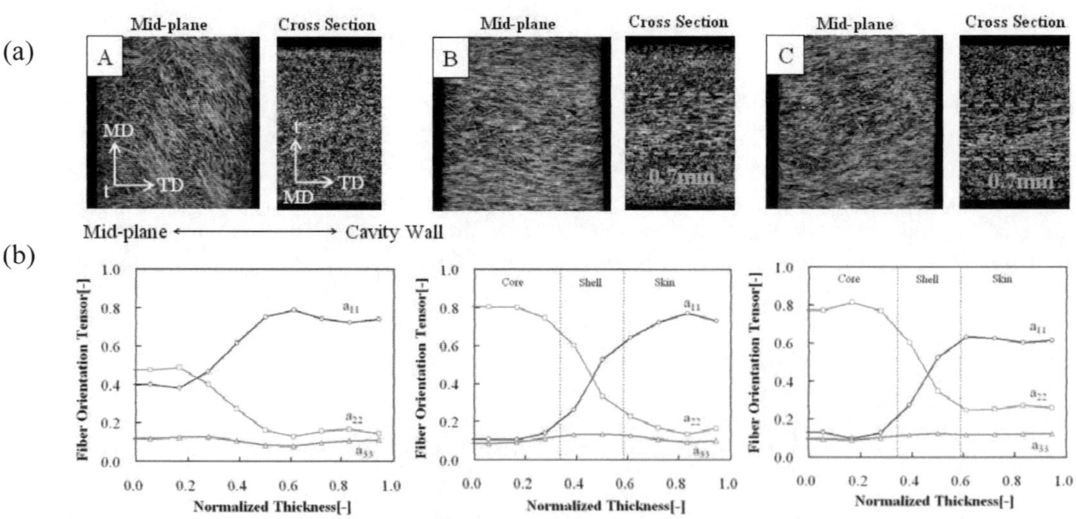

FIGURE 3. X-ray CT images of cross section (a) and the fiber orientation tensor diagrams (b) in 50wt.% GF/PA6 part with 2mm thickness at the positions A,B, and C.

Similarly, Fig.3 shows the results of 50wt.% GF/PA6 part. The general tendency about layer structure in thickness direction and intensification of fiber orientation in mid-plane was similar with GF30wt.%. However, it was confirmed that core layer was thicker than GF30wt.%. In this study, the boundary between core layer and shell layer was defined by directly measurement of cross section of X-ray CT images.

It is considered that the shell layer including cross point of a_{11} and a_{22} means transition zone from core layer to skin layer. This is because the cross point indicate random fiber orientation by fountain flow. Therefore, we assumed that the double distance from the boundary between core layer and shell layer to cross point along thickness direction was transition zone (shell layer). Similarly, Fig.4 shows the results of 60wt.% GF/PA6 part. The general tendency about fiber orientation is similar with GF30wt.% and GF50wt.%. However, it was confirmed that core layer is thicker than GF30wt.% and GF50wt.%.

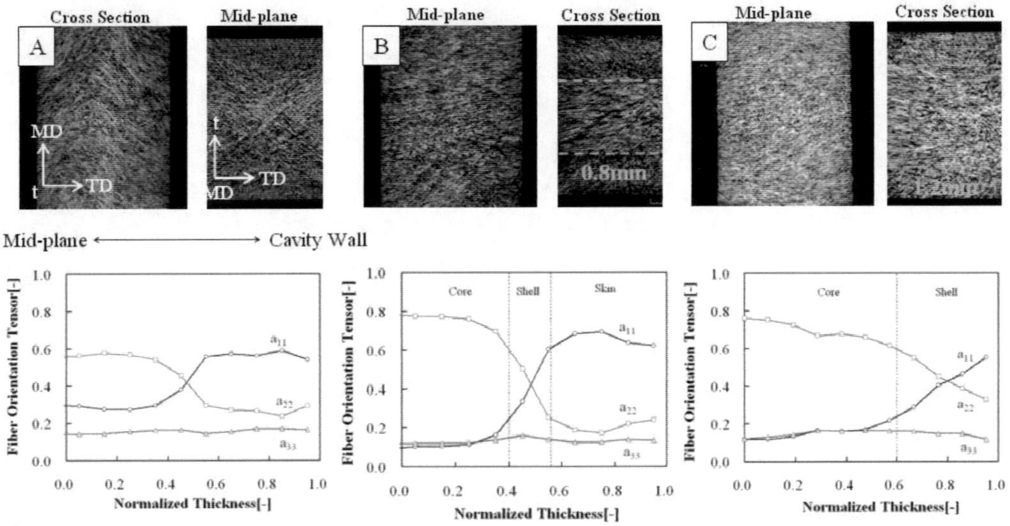

FIGURE 4. X-ray CT images of cross section (a) and the fiber orientation tensor diagrams (b) in 60wt.% GF/PA6 part with 2mm thickness at the positions A,B, and C.

Fig.5 (a) shows X-ray CT images of cross section and observed position measured about short-shot-molded plate with 2mm thickness and fiber content of 30wt.% , 50wt.% and 65wt.%. Fiber orientation distribution in each testing point is shown in Fig.5 (b). Fibers were aligned perpendicular to flow direction strongly across from around mid-plane to surface at all samples with different fiber contents. It is indicated that almost fibers across thickness direction are aligned perpendicular to flow direction by stretching flow dominated near the mid-plane just before occurring fountain flow.

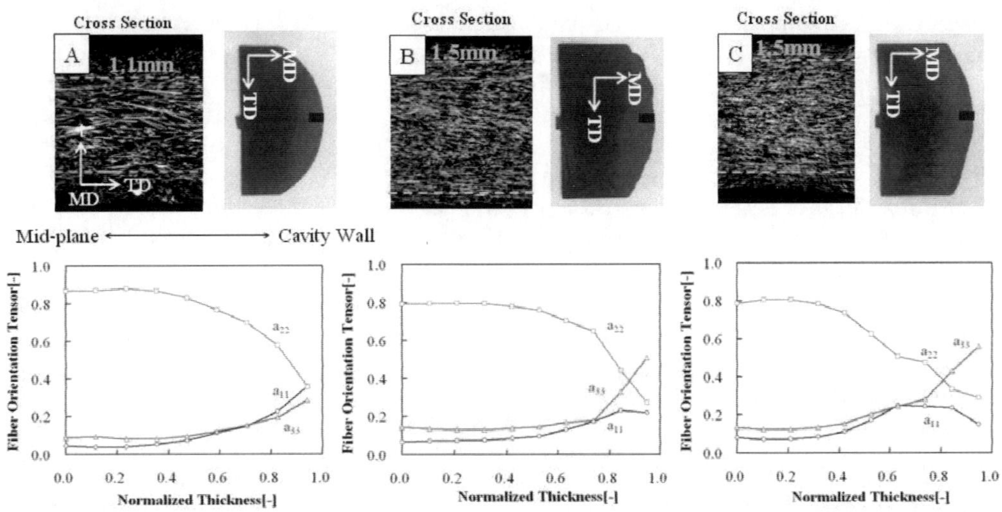

FIGURE 5. X-ray CT images of cross section and observed position measured about short-shot-molded plate with 2mm thickness and fiber content of 30wt.% , 50wt.% and 65wt.% (a); and the fiber orientation distribution in each testing point.

The thickness of core layer was plotted as a function of mold thickness (see Fig.6 (a)) and flow image of GF/PA6 parts is shown with different fiber contents (see Fig.6 (b)). Thickness of cavity wall in all parts is different regarding to the thickness of core layer. The thickness of core layer was 0.3mm, 0.7mm, and1.6mm in the case of cavity thickness of 2mm, 3mm, and 4mm, respectively.

As these results, it is confirmed that the thickness of core layer is larger according to increase of cavity wall. This tendency is also confirmed in the case of high filler content of 50wt.% and 60wt.%. However, thickness of core layer in each cavity wall increased according to higher filler content.

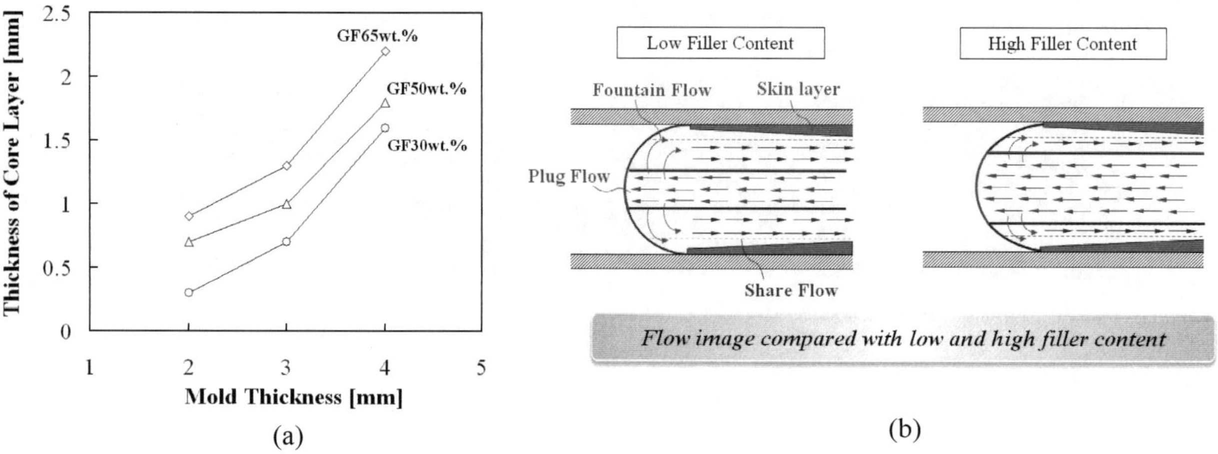

FIGURE 6. The plots of core layer thickness as a function of mold thickness (a) and flow image compared between low and high filler content (b).

Overall, it is considered that the difference of core layer thickness between GF/PA6 parts with different fiber contents was exhibited by the degree of the fiber interaction. In the case of low fiber content like as 30wt.%, it is considered that fountain flow is occurred near center. On the other hand, in the case of high filler contents like as 50wt.% or 65wt.%, it is considered that fountain flow is occurred near mold wall. This is because the behavior of fibers in high filler content isn't free by high interaction degree between fiber and fiber.

CONCLUSION

In conclusion, this present study has demonstrated the feature of fiber orientation distribution about injection-molded composite with high filler content. In injection-molded GF/PA6 part with the fiber contents of 30wt.%, 50wt.% and 65wt.%, the fiber orientation was observed by 3D X-ray computed tomography system Y.Cheetah μHD which cooperated with VG.Studio max 2.2. And the 3D fiber orientation in each specimen was calculated by the software which cooperated with VG.Studio max 2.2. Comparison with experimental results, it is found that the thickness of core layer increases with increasing the filler content, and although fibers are aligned perpendicular to flow direction strongly across from around mid-plane to surface in short shot plate. The present results suggested that in the case of low fiber content like as 30wt.%, it is considered that fountain flow is occurred near center. On the other hand, in the case of high filler contents like as 50wt.% or 65wt.%, it is considered that fountain flow is occurred near mold wall. This is because the behavior of fibers in high filler content isn't free by high degree interaction between fibers. Further studies are needed in order to understand the behavior of fiber in injection-molded parts with high filler content.

REFERENCES

1. A. Ayadi, H. Nouri, S. Guessasma, and F. Roger, *Compos. Struct* **136**, 635–649 (2016).
2. A. Bernasconi, F. Cosmi, P. J. Hine, *Compos. Sci. Tech* **72**, 2002-2008 (2012).
3. R. S. Bay, C. L. Tucker, *Polym. Compos* **13**, 332-341 (1992).
4. T. B. Nguyen Thi, A. Yokoyama, K. Ota, K. Kodama, K. Yamashita, Y. Isogai, K. Furuichi, and C. Nonomura, in The 29th International Conference of the Polymer Processing Society-2013, edited by V. Altstadt, AIP Conference Proceeding 1593, American Institute of Physics, Melville, NY, 2014, pp. 571-577.
5. T. B. Nguyen Thi, A. Yokoyama, S. Hamanaka, K. Yamashita, and C. Nonomura, *J. Mater. Process. Tech* **219**, 1-9 (2015).

Knowledge-Based Approach for an Automatic Rheological Optimization of Injection-Molded Parts

A. Porsch[a], R. Andrae[b], P. Köhler[b] and J. Wortberg[a]

[a]*University of Duisburg-Essen, Institute of Product Engineering, Engineering Design and Plastics Machinery, Duisburg, Germany*
[b]*University of Duisburg-Essen, Institute of Product Engineering, Computer Aided Engineering, Duisburg, Germany*

alexander.porsch@uni-due.de
rene.andrae@uni-due.de
peter.koehler@uni-due.de
johannes.wortberg@uni-due.de

Abstract. The motivation for the presented approach results from the absence of an integrated method of supporting the designing of injection-molded parts. Thereby, the focus lies on the rheological optimization, which is ensured by knowledge transfer from the production engineer to the designer, i. e., the mentioned aim is achieved by creating a knowledge base including the expert knowledge. This method is based on an integrated product simulation containing the interconnection of CAD and CAE-systems. The first step in the process is a filling simulation done by a standard injection molding process simulation software. The gate location is considered as fixed and optimal. Nevertheless, it is still possible that a uniform mold filling is not reached. To determine an uneven mold filling the results of the simulation are exported as scalar values to a post processing software tool. Thus, a knowledge based analysis of the results using user defined parameters based on flow length, pressure distribution, etc., is enabled. To enhance the filling, the original part design is modified by adding flow aids or barriers. Their positioning is calculated by the post processing software tool and this result is returned to the CAD-system. There the preliminary design is done by an analytical approach. By iterations of the mentioned method, the geometry of the flow aids or barriers and thus the mold filling are improved. The presented approach is illustrated exemplarily by an injection-molded part. As a result, the designer is supported methodically designing injection molded parts by implementing and using expert knowledge. So the number of iterations and the development time (costs) are reduced.

Keywords: design for injection-molded parts, knowledge-based engineering, integrated product simulation, process simulation, rheological optimization
PACS: 83.50.Uv - Material processing

INTRODUCTION

Today the optical, haptic or mechanical requirements for injection molded parts increase. Concurrently, the product life cycles decrease and the diversity of variants increases. Therefore, the development process of injection molded parts has to be improved constantly.

To reach these goals the usage of 3D-CAx systems has become state of art during the development process. However, in many areas integrated methods are missing which are supporting a specific design for manufacturing. For example, CAD-software as PTC Creo (PTC, Inc.) or Autodesk (Autodesk, Inc.) have integrated CAE-tools supporting the engineer optimizing the part design. PTC Creo Simulate is capable of simulating the filling of the injection molding process or determining a gating position. However, the interpreting of the results and the modifications of the part design have to be conducted by the engineer based on his knowledge.

Humpa presents in his paper [1] a practicable CAD-based geometry design approach for spiral milling parts. He is using knowledge-based engineering methods such as user defined functions. In combination with hybrid modelling techniques and rule based model design this approach leads to an improved process [1]: "By the use of methods and tools of knowledge-based engineering (KBE) both manufacturing process information and manufacturing knowledge can be integrated into the 3D product model."

An approach for an automatic optimization is presented by te Heesen [2]. Based on three-dimensional numerical flow simulations, a parametric CAD-model of an extrusion die including geometrical and manufacturing restrictions is optimized. A multiple criteria genetic optimization algorithm is coupled with a three-dimensional numerical flow simulation to optimize the die geometry by using defined quality and exclusion criteria calculating new parameter sets to change the geometry [2].

Based on the presented literature, an approach for integrated product simulation (IPS) of injection molded parts has been developed and is presented in the following.

INTEGRATED PRODUCT SIMULATION

The IPS consists mainly of two phases. The first one considers the designing of the part by use of a CAD-software. In the second part rheological and thermal optimization of the part design is conducted by knowledge-based automatic analyzes of injection molding process simulations. If the part design has to be changed, quantitative information will be transferred back to the CAD-software (Figure 1). The described process is repeated iteratively until the part design is optimal.

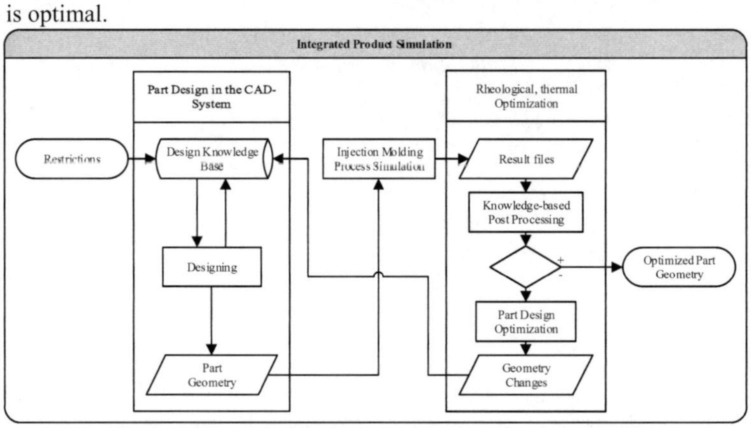

Figure 1. Process of the integrated product simulation (IPS)

There are general designing rules for injection molded parts as draft angles, wall thickness distributions, shape of undercuts, etc., resulting from the process and material selection. This specific knowledge needs to be implemented in the system. It is presented by guidelines, standards and the experience of the user himself. The acquisition is an important precondition for the qualification of the modeling process. The acquired knowledge is filed in a design knowledge base. The implementation is carried out by modern KBE methods. The usage of KBE techniques is application specific. This requires a largely subdivision according to certain criteria, for instance, the integration deep or the desired knowledge representation form [3, 4].

Based on the implemented knowledge, the part is designed with injection molding oriented design features. These features could be created in two different ways. The first is the integration of guidelines and designing rules in given modeling options of the CAD-system. The other one is the development of new injection molding oriented design feature. An example for a rip design feature is given in the following Figure 2. This is implemented in the software PTC Creo 3.0. The height and thickness of the rips is controlled by the wall thickness of the base to improve the part quality regarding sink marks or cooling time. Also injection molding specific values of the draft angle and radius on the top and in the transition to the base of the rips are specified. These guidelines ensure a smooth filling and trouble-free ejection of the part [5].

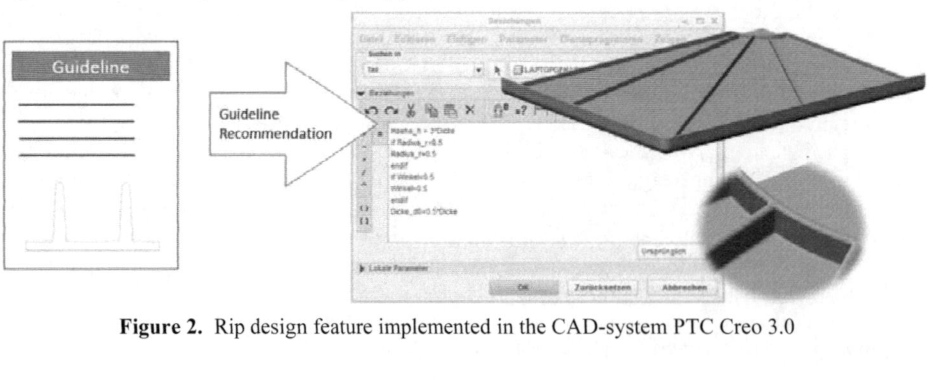

Figure 2. Rip design feature implemented in the CAD-system PTC Creo 3.0

It is also possible to analyze an existing part design. This is carried out similarly to the new design features whereas the knowledge which is used for creating design features implies the knowledge of creating analysis features. For the optimization, both cases assure the useful selection of design variables, in the previously defined boundaries. As an advantage over predefined analyze features as in Autodesk Moldflow Design, the user is able to modify the design knowledge base according to his needs. It is also possible to implement the designing rules and guidelines into other CAD-software.

As result of the first phase of the IPS, a part is designed or optimized by applying the injection molding specific designing rules and guidelines. Without any further analysis (e. g., injection molding process simulations) the part design is considered as optimal. Finally, the part geometry is exported to the CAE-software SIGMASOFT (SIGMA Engineering GmbH) beginning the second phase of the IPS.

For the export of the part geometry a mesh is generated (partially) automated in the CAD-system wherein the mesh strategy is adapted to the SIGMASOFT mesh. This generated h-mesh is created in this example as an Ansys file. In addition to the mesh, the meshing setup parameters are also exported as text file.

As mentioned above, CAD-software are capable of filling simulations and automated gate positioning. Nevertheless, restrictions regarding distance to an adjacent part surface or edge or runner / gate type are not considered. Therefore, a knowledge-based configuration system would be developed in the CAD-System. This is interconnected to the process selection and part geometry. Accordingly, different gate types could be locked out directly. The gate and the gate position would be transferred into the simulation environment together with the part.

RHEOLOGICAL OPTIMIZATION

This paper is focused on rheological optimization and, due to the limited extend of this paper, thermal optimization is not considered.

In general, rheological optimization is used to improve the filling phase of the injection molding cycle. Among other things, this is achieved by:

- reducing the injection time and pressure,
- reducing the number and quality of weld lines,
- and creating an evenly filling pattern [6, 7].

In the following, an approach for a knowledge-based rheological optimization is presented. As a result of this automated post processing, quantitative information for positioning and designing a flow aid (rip) is exported as text file. With this information it is possible to generate automatically an optimized part design in the CAD-software which displays a more evenly filling pattern (Figure 3).

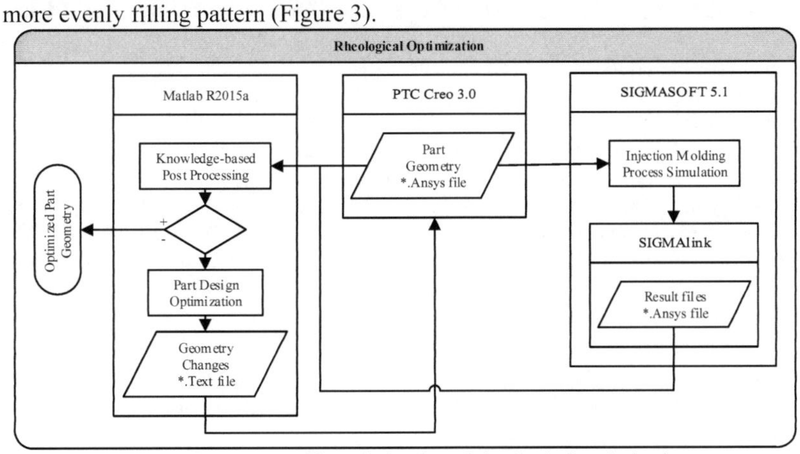

Figure 3. Process of knowledge-based rheological optimization

During the preprocessing the part geometry is imported. Subsequently, the part is meshed and the process is defined. Because the part is to be optimized, the mold temperature is considered as isotherm during the whole filling phase. The flow rate is constant and the used material is a polypropylene of LyondellBassell (Purell_HP371-P). After the filling simulation, the results are exported as scalar values. To do so, the SIGMSSOFT add on SIMGAlink writes the results on the PTC Creo mesh. So for every result and time step one mesh file is exported.

In Figure 4a, the result Material Age in the middle of the example part (constant wall thickness) is shown. This result type illustrates the residence time of the melt in the part. Therefore, the material near the gate has the lowest

Material Age (blue) and the one in the stagnation zones the highest Material Age (red). For the engineer, it is obvious that the filling is unevenly. The challenges are (1) achieving this result interpretation automatically and (2) calculating quantitative values for geometry changes.
Thus, a Matlab R2015a (The Mathworks, Inc.) program has been developed capable of preforming a knowledge-based rheological optimization. The first step of the analysis process is importing the result and mesh files. Then the mesh nodes at boundary of the part, in the following named boundary nodes (BN), are identified.
Here the PTC Creo mesh consists of six layers. The number and the position of the layers are imported from the text file which was mentioned above. For each layer the mean Material Age is calculated and the following analyses are only conducted for the layer with the lowest mean Material Age. This layer is of course in the middle of the part because the melt solidifies at the boundaries of the part. Therefore, there is more stagnation as in the center of the part. So at the end of filling most melt will flow in this layer of the part. Moreover, the maximum values of Material Age are limited. In this way the "flow paths" are identified. Likewise, the analyses are only preformed for the results belonging to the flow paths.
To detect unevenly filling, two conditions are examined. Firstly, the pressure gradients at the BN are calculated. In the case of occurring high gradients before the part is filled to 95 % (Filling Time), the first condition is fulfilled. The pressure loss is the same for every flow path. If the melt reaches one end of the part or the end of a flow path before the end of filling, then there is an isobar pressure distribution and shortly a high pressure gradient on this flow path. Thus, this condition is a hint for unevenly filling.
To verify this proposition, the velocity gradients on the same flow path are analyzed. If there is a high velocity gradient at the same time as the pressure gradient and the mean velocity drops below 1 cm/s, the second condition is fulfilled as well, and therefore the filling is unevenly.

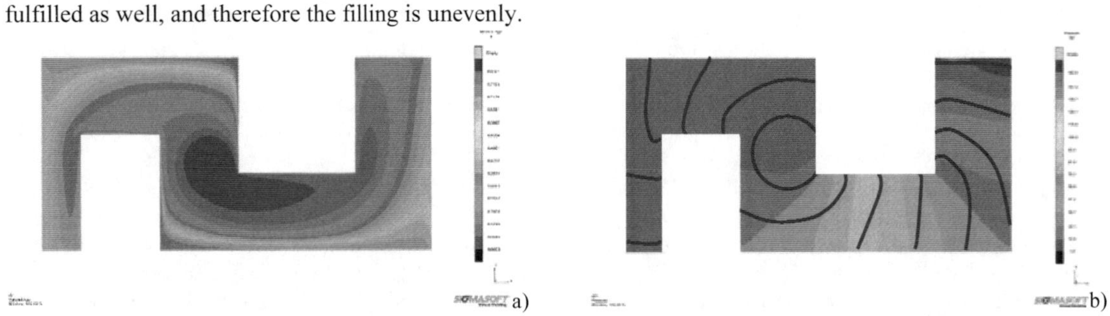

Figure 4. Material Age [s] result at the end of filling (a), Pressure result [bar] at the end of filling (b)

If the gate position is considered as fixed, the only alternative for improving the filling pattern is to add a flow aid or barrier. In this approach a rip is added to the original part geometry. So the pressure losses of both flow paths become nearly the same. Figure 4b shows the pressure distribution at the end of the filling phase and schematically the positions of the melt front. For every melt front result the node where the pressure is lowest is calculated. Finally, the coordinates of the nodes are transferred as text file to the CAD-software.

As shown in Figure 5, some points (orange) are close to an edge and their positions are rather random. Until now, only the wall distance d is checked and modified automatically. The resulting trajectory (grey) on which the new design feature is created and later optimized is created manually. Of course, the flow aid will not be placed on a visible surface.

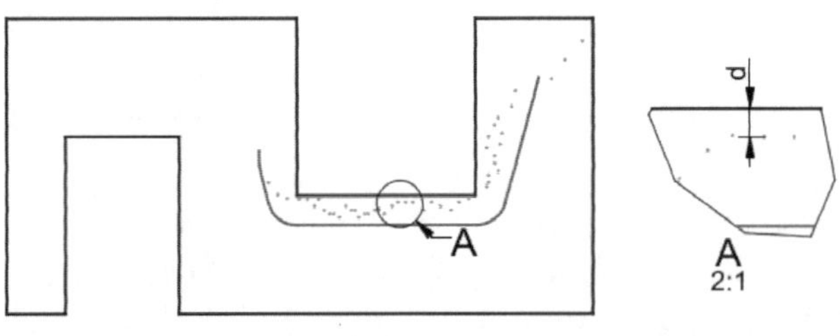

Figure 5. Exported nodes and resulting trajectory

In this example the flow aid is rib having a rectangular cross section. This design has a significant positive influence on the simulation duration and the knowledge-based rheological optimization. The height of the rip is derived from the designing guidelines presented in [5]. It is the half of the part wall thickness and almost constant over the entire length. At the beginning and the end the rib has a chamfer. The adjustment of size of the cross sectional area is realized by changing the width.

Figure 6b shows the result Filling Time of the part geometry of the third iteration. Plainly, the filling is more evenly in comparison to the original part design (Figure 6a). As another positive effect the maximal injection pressure is reduced by about 20 %. In contrast the filling time is increased by 5 % (0.05 sec). Because of the greater part volume, this effect is logical. But by increasing the flow rate, this disadvantage can be leveled. The resulting maximal inlet pressure and shear stress are less than the ones of the original part design.

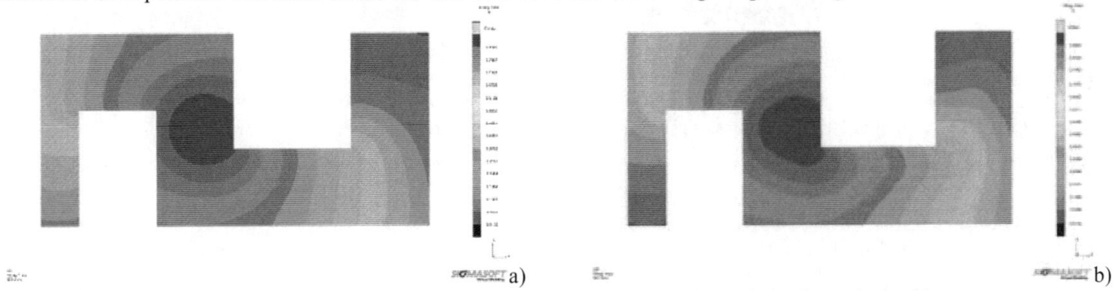

Figure 6. Comparison of the filling pattern of the original part design a) and the optimized part design b)

CONCLUSION

The design knowledge base has been developed and implemented in PTC Creo. Of course, the number of designing and analyze features will be increased step by step. Also the automatic calculation of a regression line and the following validation of the resulting trajectory will be implement into PTC Creo. To use as many model information and existing features, the knowledge base has been implemented by PTC Creo 3.0 Toolkit. Finally, the influence of the knowledge-based designing on the part quality will be evaluated.

The developed analyze program respectively the knowledge based rheological optimization have been successfully tested with different example parts. The resulting iterations differ between two and four. The number of iterations depends mainly on the desired degree of uniformity of the filling. However, these parts had a constant wall thickness and can be considered as 2.5D parts.

Thus, the next step is the verification of the applicability of the presented approach for real three dimensional and more complex parts. Also a method which combines both flow aids and barriers will be developed because by adding a flow aid the material costs per part will always rise. Completing, the determination of the optimal design of the flow aids and barriers will be conducted.

ACKNOWLEDGMENTS

This project is supported by the German National Science Foundation (Deutsche Forschungsgemeinschaft, DFG), project no. WO 302/56-1 and KO 1620/15-1. I would like to thank the SIGMA Engineering GMbH for the support.

REFERENCES

1. M. Humpa and P. Köhler, "Design Approach for Spiral Milling Parts Using Knowledge Based Engineering", 27th International Conference on CADCAM, Robotics of the Future 2012, London, UK
2. O. te Heesen, "Ganzheitlicher Ansatz zur Dimensionierung und Optimierung von Extrusionswerkzeugen am Beispiel von Wendelverteilern", Ph.D. Thesis, University of Duisburg-Essen, 2015.
3. S. Danjou and N. Lupa and P- Köhler, "Approach for Automated Product Modeling Using Knowledge-Based Design Features," in *Computer-Aided Design & Applications*, 2008, No. 5, pp. 622-629.
4. T. Manoharan and M. Humpa and A. Martha and P. Köhler, "Knowledge Integration in CAD-CAM Process Chain" in *Proceedings of CAD'15: International CAD Conference and Exhibition*, London, UK, 2015, pp. 449-454.
5. E. A. Campo, "*The Complete Part Design Handbook – For Injection Molding of Thermoplastics*", Munich, Carl-Hanser-Verlag, 2006.
6. R. A. Malloy, *Plastic Part Design for Injection Molding: An Introduction*, Munich, Carl-Hanser-Verlag, 2011.
7. J. Shoemaker, *Moldflow Design Guide: A Resource for Plastic Engineers*, Munich, Carl-Hanser-Verlag, 2006.

Investigation of Nanoparticles Emitted When Injection Molding Neat and Additive-filled Polypropylene and Polycarbonate

Matthew Theriault[a], Suzanne Yoeuth[a], Jessica Matar[a], John Martin[b], Dhimiter Bello[b], and Carol Barry[a*]

[a]Department of Plastics Engineering and [b]Department of Work Environment
University of Massachusetts Lowell, One University Ave., Lowell, MA, 01854, USA

carol_barry@uml.edu

Abstract. Injection molding of thermoplastic resins is well known to produce volatile organic compounds which can adversely affect operator health, but there is little understanding of particulate emissions. Recent molding of nanocomposites produced no emission of carbon nanotubes, but emission of organic nanoparticulates of unknown composition. This work is an investigation of those particulates with the gaseous phase emissions of airborne nanoparticles collected during injection molding of neat and additive-filled polypropylene and polycarbonate compounds. The starting materials, collected nanoparticles, and molded parts were analyzed using thermogravimetric analysis, thermogravimetric analysis coupled with Fourier-transform infrared spectroscopy, differential scanning calorimetry, and gas chromatography-mass spectrometry. The results were analyzed to determine the composition of the nanoparticulates and compounds lost from the polypropylene and polycarbonate during injection molding. Characterizing these emissions will allow for a better understanding of the precautions and safety equipment needed during melt processing.

Keywords: particle emissions, injection molding, polycarbonate, polypropylene
PACS: 81.20.Hy

INTRODUCTION

Injection molding is a widely used forming process thermoplastic, elastomeric, and thermoset polymers. Additives are typically formulated into the polymer, allowing the feedstock to meet the needs of specific product applications. During the injection molding process, the solid resin (often pellets) is fed to the injection unit and melted or softened using viscous dissipation and some conductive heating. The molten or liquid polymer is forced into a mold cavity and cooled or cured to form parts. The completed parts then are removed from the mold.

The heating of plastics resins during processing, purging, cleaning, and maintenance can expose workers to hazardous emissions. Extensive studies have evaluated the volatile organic compounds (VOCs) created during injection molding. When processing of polypropylene with and without additives, the major expected VOCs are 0.2-19 ppm formaldehyde, 0.01-0.14 ppm acrolein, and acetone[1]. Melt processing of polycarbonate general produces 30-60 ppm monochlorobenzene, 1-7 ppm diphenylcarbonate, 0.2-0.4 ppm bisphenol A, and 0.2-2.0 ppm phenol[1]. For worker safety, the U.S. Occupational Safety and Health Administration (OSHA) and National Institute for Occupational Safety and Health (NIOSH) have permissible exposure limits for VOCs. For example, the permissible expose limit for formaldehyde is 0.75 ppm time-weighted average (TWA), the OSHA exposure limit for acrolein is 0.1 ppm while NIOSH recommends an exposure limit of 75 ppm TWA, acetone has OSHA and NIOSH permissible exposure limits of 250 ppm and 1000 ppm TWA, respectively, monocholorbenzene's exposure limit is 75 ppm, phenol's exposure limit is 5 ppm, and diphenylcarbonate and bisphenol A have no exposure limits[1]. The VOCs generally are controlled by ventilation near the injection molding machines and in the facilities.

Emissions of particulates have been less studied. A recent study[2-3] examining the effects of recycling neat and carbon nanotube-filled polypropylene and polycarbonate nanocomposites showed that no nanotubes were emitted during the 30 cycles of injection molding parts and grinding those parts. This work, however, showed that the injection molding process produced carbon-containing particles of unknown composition. These findings were not unexpected. In a recent study[4] conducted on occupational exposure to nanoparticles at commercial photocopy centers, emissions other than carbon-based particles – i.e., metal oxides – as well as ultrafine particles (< 100 nm as compared to the 5-μm diameter toner particles) were present; these finding indicated the need to addressing indoor air quality problems at various levels.

This work evaluated the emissions of particulates when injection molding neat and additive-containing polypropylene and polycarbonate. The objective was to identify the unknown particulate emissions from the prior study. Particulate emissions were measured during injection molding process. The chemistry of particles collected during injection molding was characterized to determine the identity of these particulate emissions. This work should improve the understanding of the particulate emissions and their possible on worker safety.

EXPERIMENTAL APPROACH

The resins used in this experiment were limited to polypropylene (PP) and polycarbonate (PC), with and without selected additives. The neat PP was Pro-fax PD702 (LyondellBasell), while the neat PC was Makrolon 2405 (Covestro). The neat PC contained a mold release agent. The additive-filled polymers consisted of Pro-fax SG722 PP (LyondellBasell), which contained nucleation, anti-static, and slip agents, as well as Makrolon 2407 PC (Covestro) with mold release and a UV stabilizer.

Thermogravimetric analysis (TGA) was performed on all resins using a TA Instruments' Q50 with ASTM E1131-08(2014). To model the injection molding process, the heated was performed at accelerated rate of 50 °C/min from room temperature to 500 °C and in air atmosphere (purged at 60 mL/min). The test was conducted three times to ensure accuracy of collected data. Differential scanning calorimetry (DSC) was performed in accordance with ASTM D3418-15 using a TA Instruments' Q200. Parameters were set at a temperature range of -100 °C to 500 °C with a heating rate of 10°C/min and completed full heat-cool-heat cycle. To ensure data accuracy and confidence, DSC was repeated three times per resin.

Injection molding was conducted using a 730-kN all-electric machine (Sumitomo's SE 75 DUZ C250) with a 28-mm-diameter screw and a shot size capacity 86cm^3. The CPU housing mold had a single cavity with a volume of 11.5 cm^3. The barrel, nozzle, and mold temperatures are listed in Table 1. Parts PP were molded for three hours and PC parts for two hours using a semi- automatic cycle. The PP resins were used as received, but the PC resins were dried at 121°C for 4 hours in a DRI-AIR Industries desiccant hopper bank system (Dryer model HP4-X 50FM, Hopper model RH60) prior to molding.

TABLE 1. Barrel, Nozzle, and Mold Temperatures used during Injection Molding

Temperature Zones	Unfilled PP (°C)	Unfilled PC (°C)
Barrel Zone 1 (feed)	190	270
Barrel Zone 2	240	275
Barrel Zone 3	240	280
Barrel Zone 4 (nozzle)	240	285
Mold	27	71

For each molding trial, the screw and barrel were cleaned prior to molding to eliminate any potential of cross contamination from previous materials. During three to five consecutive hours of molding, the concentration of particle emissions as a function of particle size diameter (5.6 nm–20 μm) were captured using two real-time instruments: (1) a Fast Mobility Particle Sizer (FMPS, 3091) which measured the particle diameter from 5.6-560 nm and (2) an Aerodynamic Particle Sizer (APS, 3321) that measured particles from 560 nm to 20 μm. Both instruments were factory calibrated and passed a field "zero" calibration test. The particle sizers were hooked up to a laptop computer where the onboard time clock was synchronized and data logging was conducted at one-second intervals. Collection of particles throughout the molding process was conducted from instrument inlets positioned at the nozzle of the injection molding machine as shown in Figure 1. This position was selected because prior work[2-3] showed that the emissions were greatest at the nozzle (compared to the hopper and mold).

The timescale for emissions collection during injection molding is presented in Table 2. Background emissions were was collected for five minutes prior to starting any of the equipment. Emissions were measured during the 30 minutes required for heating of the injection unit, three minutes of initial purging of the injection molding machine, and the molding cycles associated with optimization of the process parameters. This process optimization required 4 minutes for PP, but 86 minutes for PC. Due to molding complications while running the PC, multiple purges were necessary. Using the optimized processing parameters, PP and PC parts were molded and emissions collected for 228 and 125 minutes, respectively. Emissions were also collected during the six minutes required to purge and shut down the machines. Particles collected during molding currently are being analyzed using TGA-FTIR as well as liquid-chromatography mass-spectrometry (LCMS) and gas-chromatography mass-spectrometry (GCMS) at the University of Wisconsin. Analysis of molded parts will not be conducted.

Figure 1: Apparatus of FMPS and APS tubes placement.

TABLE 2. Procedure for Measurement of Particle Emissions

Process	Time for PP (min)	Time for PC (min)
Background collection on FMPS	5	5
Machine turned on	1	1
Heating of injection unit	30	30
Motor turned on	1	1
Hopper filled and purging	3	3*
Processing parameters optimized	4	86
Continuous molding of parts	228	125
Purging and machine shut down	6	6

RESULTS AND DISCUSSION

The TGA was run on all four resins primarily to determine the temperature limits for molding. As shown in Table 3, the DSC results showed that the glass transition temperature, T_g, was similar for the neat and filled polycarbonate. The melting temperature, T_m, and crystallization temperature, T_c, were slightly greater for the filled PP than the neat PP. The area of the melting and crystallization peaks, A, also were different for the neat and filled PP resins. These changes were attributed to the additives in the filled PP which may have caused nucleation of the PP crystallites and increased the mobility of the polymer chains.

TABLE 3. DSC Results for Polypropylene and Polycarbonate

Material	T_g (°)	T_m (°C)	A (J/g)	T_c (°C)	A (J/g)
PP-neat	---	159	73.07	116	84.64
PP-filled	---	163	80.50	123	79.47
PC-neat	135	---	---	---	---
PC-filled	137	---	---	---	---

Table 5 shows the mean, standard deviation (SD), geometric mean (GM), and geometric standard deviation (GSD) of the particle emissions occurred during the processing of PP and PC. The greatest amount of particle emissions for both resins occurred while the barrel was heating up and corresponded to smoke exiting the nozzle as the barrel zones reached their set temperatures. The smoke was probably caused by the anti-seize applied to the threads of the nozzle before it was reinstalled after cleaning the screw and barrel. The lowest amount of particulates emitted for both resins ensued during the purging. During this step, the melt was immediately expelled out into the open, with lower heating causing less oxidative degradation of the PP. The emissions count increased during molding of parts because the resin was soaked in the barrel in between each shot; this hotter resin was more likely to degrade and produce particulate emissions.

TABLE 5. Particle Emissions Present During Molding Cycle

Material	Process	Mean	SD	GM	GSD
PP	Heating	8,495,671	5,838,069	3,567,132	8.95
	Motor On	6,713,833	1,371,839	6,572,923	1.23
	Purge	4,438,611	710,452	4,381,541	1.18
	Injection	53,028	94,829	19,118	3.48
PC	Heating	8,263,961	4,027,289	4,354,502	8.18
	Motor On	5,295,222	2,510,560	4,754,610	1.60
	Purge	1,951,586	4,274,204	32,367	18.3
	Injection	5,548	1,022	5,484	1.15

Figure 2 shows the number of particles collected per cubic centimeter when molding parts from neat PP and neat PC resins with the optimized processing parameters and semiautomatic cycling. During the first 60 minutes of molding PP parts, the particle emissions initially were very high (> 200,000 particles/cm^3), but there was a steady decrease in the amount of particles emitted from the neat PP. After the first 60 minutes of molding parts, the particle emissions were relatively low. This behavior was attributed to the overall length of the molding cycle and the nature of isotactic polypropylene. Since PP parts were molded only seven minutes after filling the hopper, particulates had the potential to separate from the mass of pellets sitting in the hopper. After an hour in the hopper, most of these smaller particles had either settled on surfaces in the room or been removed by the vents. In addition, oxidative degradation of polypropylene produces lower molecular weight products. The particulates may have been these degradation products from the purging. The number of particles emitted during the molding of the PC, however, remained constant at about 6,000 particles/cm^3 during 125 minutes of part molding. This lower level of particle emissions may have been due to the greater stability of PC to oxidation, but the lack of reduction in particle emissions with molding time was probably caused by the multiple purging actions required during parameter optimization. There was one spike about 27 minutes into the molding of the PC that could be due to a combination of multiple factors, but the exact reason is unknown.

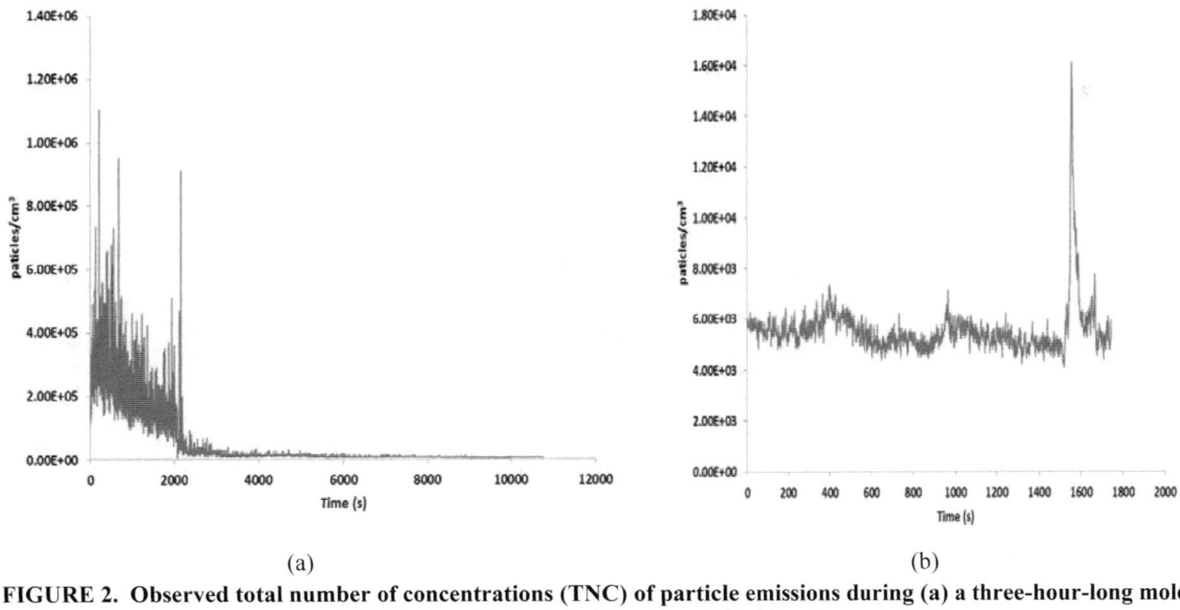

(a) (b)

FIGURE 2. Observed total number of concentrations (TNC) of particle emissions during (a) a three-hour-long molding trial for PP and (b) a two-hour-long molding trial for PC.

The size of the particle collected during molding of PP was much greater than that of PC (Figure 3). The PP showed a small group of particles with an average diameter of about 10 nm and a much larger group of particles with an average diameter of 100 nm. In contrast, the PC resin produced particles with diameters of 10 nm to 70 nm. The larger particulates emitted by PP again suggest lower molecular products from oxidation; (this will be confirmed by chemical analysis of the collected particulates). Although the PC resin had a mold release agent in its formulation, this additive was probably emitted as a VOC.

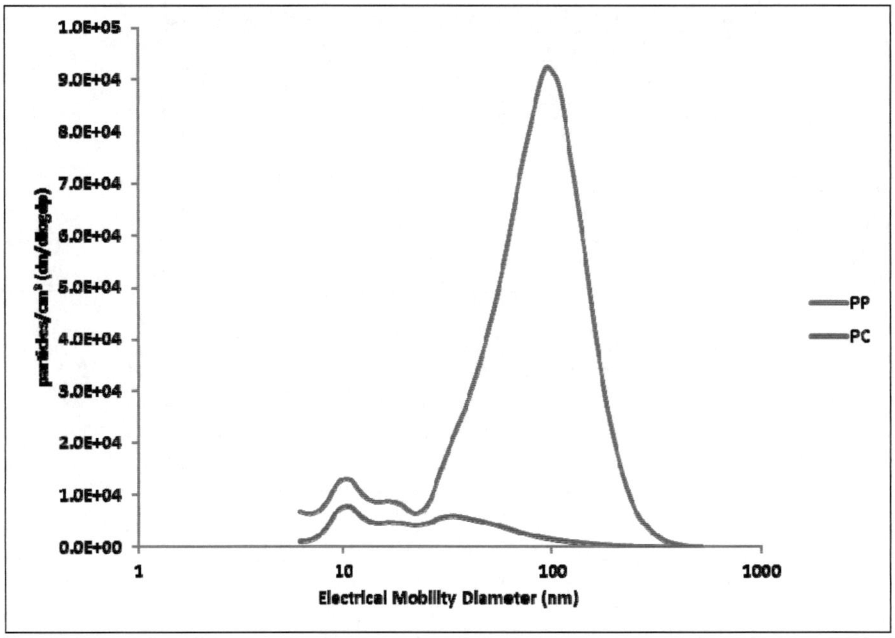

FIGURE 3. Particle size distribution of emissions during the molding of PP and PC.

CONCLUSIONS

During injection molding neat PP exhibited greater particle emissions and emission of particles with larger sizes than a neat PC. This behavior was probably due to the oxidative degradation of PP during purging as well as separation of smaller particles (fines) from pellets after loading of the hopper. Chemical analysis (currently in process) will confirm whether the emitted particles were low molecular weight oxidation products or fines. The particulate emissions from oxidatively-stable PC were much lower, but remained in the air much long after multiple purging events. Overall, injection molding of polymer materials emitted nanoparticulates that must be addressed for the health and safety injection molding machine operators.

ACKNOWLEDGMENTS

The authors thank Sumitomo for their machine donation to the Department of Plastics Engineering and Jean Vangsness for her assistance with the thermal characterization.

REFERENCES

1. R. Dematteo, "Chemical Exposure and Plastics Production" Chemical Exposure and Plastics Production: (n.d.): n. pag. National Network on Environments and Women's Health, Dec. 2011.
2. J. Zhang, A. Panwar, D. Bello, T. Jozokos, J. Isaacs, C. Barry, J. Mead, "The Effects of Recycling on the Properties of Carbon Nanotube-filled Polypropylene Composites and Worker Exposures," *Environmental Science Nano,* 2016, DOI: 10.1039/C5EN00253B
3. P. Boonruksa, D. Bello, J. Zhang, J. Issacs, J. Mead, S. Woskie, "Characterization of Potential Exposures to Nanoparticles and Fibers during Manufacturing and Recycling of Carbon Nanotube Reinforced Polypropylene Composites," *Annals of Occupational Hygiene*, 60(1),40-55, 2016.
4. J. Martin, D. Bello, K. Bunker, M. Shafer, D. Christiani, S. Woskie, P. Demokritou, "Occupational Exposure to Nanoparticles at Commercial Photocopy Centers," *Journal of Hazardous Materials,* 298, 351-360, 2015.

Complex Injection Moulded Components - Bridging the Knowledge Gap

Tino Meyer[a], Paul Sherratt[a], Andy Harland[a], Chris Holmes[b], Tim Lucas[b] and Barry Haworth[a*]

[a]*Loughborough University, Loughborough, LE11 3TU, UK*
[b]*adidas AG, Herzogenaurach, 91074, Germany*

T.Meyer@lboro.ac.uk
B.Haworth@lboro.ac.uk

Abstract. Injection moulding is the predominant manufacturing process enabling the production of precise and consistent polymeric parts at a high volume. The final performance of those parts is critically dependent on their melt flow history and the current approach of testing simplified specimens produced by idealized melt flow conditions to specify new or enhanced materials is therefore not sufficient, since final parts often feature a more complex geometry. The purpose of this research is to highlight this omission by conducting high velocity impact and quasi-static tensile tests on PA-12 specimens obtained from a new concept injection moulding tool. This mould allows controlled modification of the material flow by adding specific mould tool design features which lead to the creation of a weld line, flow hesitation, or combination of both of these irregular flow phenomena and is therefore an improved representation of final injection moulded components. Furthermore, test specimens representing simplified as well as more complex geometries can be obtained from the same moulded samples, guaranteeing identical applied process conditions. The occurring microstructural differences due to the diverse melt flow history are verified using optical microscopy and Differential Scanning Calorimetry.

Keywords: Injection moulding, concept mould/mold, microstructure flow hesitation, weld lines
PACS: 81.70.Cv, 83.50.Uv, 83.80.Ab

ADVANCED MOULD CONCEPT

The process of injection moulding (IM) is one of the most widely employed methods for manufacturing an extremely diverse range of parts, varying in size and degree of complexity, from polymeric materials. The final quality and performance of injection moulded parts is complex and depends on specific material attributes (molar mass, molecular architecture and additive type/ concentration), as well as part and mould design and the specific process parameters required for manufacture. The final mechanical performance of a part is predetermined by its microstructure, morphology and is strongly influenced by the melt flow behavior during injection. Previous investigations dealing with different melt flow modifications were mostly focused on the influence of weld lines per se [1] and the effect of different process parameter [2], [3] or additives/ blends [4] on their mechanical strength. While these publications were often based on the comparison of test samples with and without weld lines produced using different moulds, often requiring different machine settings, the new mould enables the production of samples representing both cases with the same shot of material, ensuring identical settings, important for processing materials which are sensitive to degradation induced by oxidation, hydrolysis, or by excessive residence time. Besides weld lines, the effect of flow hesitation and combinations of both flow phenomena can be investigated by adding or removing corresponding mould features (FIGURE 1). The circular obstacle, dividing the melt flow to create a weld line can be modified in diameter and the side channels causing flow hesitation can be varied in thickness enabling the mould to be customised depending on the final application it is simulating. The side channels can also be used for 'thin to thick' and 'thick to thin' transitional flows. To understand the observed macroscopic behavior and the influence of different flow modifications on the material morphology, optical microscopy and Differential Scanning Calorimetry (DSC) were utilized.

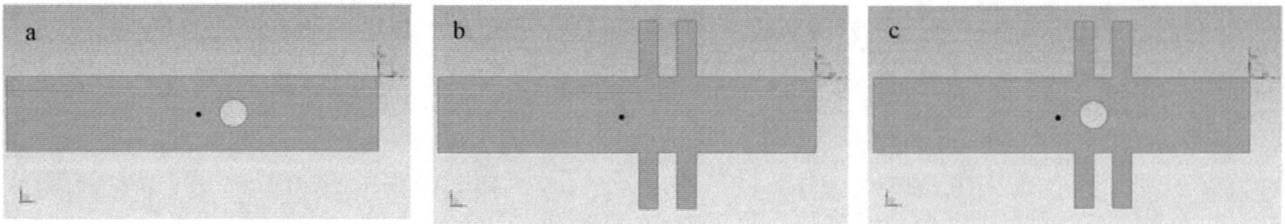

FIGURE 1. CAD representations of possible mould modifications (a) Mod 1 - weld line due to hole obstacle (b) Mod 2 - flow hesitation due to side channels (c) Mod 3 - combination of weld line and hesitation

Besides the central injection gate used within this study, two more injection locations are possible at either end of the mould (all featuring a hot runner system). These can be used to investigate differences in the melt flow path length, as well as for the creation of samples containing additional weld lines when used simultaneous. All investigations were made on Polyamide 12 (PA-12), an engineering plastic with a relatively low crystalline melting point due to its longer aliphatic chain sequences, which also confers relatively low moisture intake and superior chemical and impact resistance compared to other PAs [5]. The influence of different material flow behavior specifically on PAs has been investigated before, dominated by work on PA-6 or PA-66 while less existing literature is available for PA-12. The aim of this research investigation is to highlight the need for a more sophisticated approach when evaluating new or enhanced materials. A solution is presented in the form of a new mould concept. Test results allow predictions with regards to the final behavior of more complex injection moulded parts without being dependent on a time-consuming prototyping phase. The new mould would be even more beneficial for final part predictions when used in combination with CFD software, to obtain simulative data to complement and to predict practical outcomes [6].

Experimental

Material selection: a commercially available PA-12 was selected featuring a high degree of transparency allowed the weld line detection without additional tools and enabled consistent sample preparation. The material was injected at 250°C at an axial injection speed of 20 mm/s into the 60°C mould at three different modifications (FIGURE 1), cooling for 35s. Due to the hygroscopic nature the material was dried at 80°C prior to injection moulding and conditioned prior to the different mechanical tests to ensure reproducible test results. All test specimens were conditioned at 70°C and 62% relative humidity (ISO BS 1110) for 72 hours leading to a weight increase of 0.48%. All tests were performed on 2 mm thick samples with different geometries according to the applied standard.
Quasi-static tensile test data was obtained using an Instron 5569 tensometer operating at a speed of 0.25 mm/min during the first 0.3% strain (detected using a mechanical clip-on extensometer) and 50 mm/min afterwards with an initial clamp separation of 55 mm. ISO BS 527-2 5 Type A samples were tested normal to the flow direction regarding their tensile toughness, which was calculated using the implemented trapezoidal function in MatLab (Mathworks, Natick, USA). Instrumented falling weight impact tests were conducted on an Instron Dynatup 9250 using a 20 mm striker at an impact speed of 4.4 m/s providing an overall impact energy of around 197 J. Circular specimens with 52 mm diameter were tested in accordance with the ISO BS 6603-2 standard.
Optical microscopy was carried out to investigate the microstructure in the weld line regions using a Nikon SMZ 1500 for surface analysis and a Leica DMRX microscope, used in polarised light transmission mode, investigating crystalline microstructures on 10 µm samples prepared using a microtome and Euparal as the mountant. Additional, differences in the sample morphology extracted from the weld line region were investigated and compared to samples from identical locations (all central, 60 mm from injection gate) not containing a weld line by using DSC following the ISO BS 11357 standard on a DSC Q2000 thermal analysis device combined with a RCS 90 cooling system, both from TA instrument Inc. A heat/ cool/ heat procedure was performed on 7.6 mg (± 0.3 mg) samples, heating and cooling at 10°C/ min between 0°C and 270°C with a 5 minute isothermal pause when reaching the limits. All samples were tested with no prior thermal treatment to avoid, respectively minimise, any additional post injection structural change. The degree of crystallinity was calculated by:

$$C = \frac{\Delta H_m}{\Delta H^0{}_m} \times 100\ [\%] \quad (1)$$

where $\Delta H^0{}_m$ = 210J g^{-1} [7] was taken as the heat of fusion for a theoretically 100% crystalline PA-12 and ΔH_m was the measured melting enthalpy for the moulded PA-12 samples.

Results and Discussion

The final properties of an injection moulded part are influenced by the melt flow behavior during the process and the occurrence of weld lines or flow hesitation can significantly change the properties of materials and moulded artefacts. Samples obtained from the new mould concept, representing four different flow modifications were tested and compared regarding their quasi-static and instrumented impact properties. The mechanical testing was followed by material characterisation using optical microscopy and DSC.

Mechanical Testing

Quasi-static tensile data was obtained from the Bluehill software completing the Instron 5569 test system. The detected results, including stress and strain at break and the resulting tensile toughness data are presented in TABLE 1.

TABLE 1. Quasi-static tensile toughness results PA-12: Effect of different material flow modifications

Material Flow Modification	Stress at Break [MPa]	Strain at Break [%]	Tensile Toughness [J]
Mod 1 - Weld Line	29.7 ± 1.1	110 ± 2.6	14.8 ± 0.3
Mod 2 - Hesitation	50.9 ± 5.0	211 ± 25.0	34.7 ± 0.9
Mod 3 - Hesitation + Weld Line	38.1 ± 2.2	136 ± 15.9	18.4 ± 0.7
No Modifications	50.3 ± 3.4	214 ± 11.8	32.8 ± 0.6

The obtained data confirmed a reduction of tensile toughness for PA-12 samples containing a weld line (Mod 1) by up to 55% compared to unmodified and 57% compared to samples produced under the influence of flow hesitation (Mod 2). All weld line samples failed early along the actual weld line caused by the formation of a v-notch [8], consistently (2.7% standard deviation) around 110% strain. This is believed to be caused by an unbalance in spherulite size, different localised crystalline structures (presented in FIGURE 2b) established due to the unique thermal properties along the melt fronts, which probably results in reduced cooling rates and, improved spherulite growth conditions. The addition of a mould design feature (Mod 3) causing flow hesitation reduced the magnitude of this material weakness and necking continued even after the notch formation, enabled due to a supposed more homogeneous morphology within the weld line region. The obtained test data for instrumented impact testing is presented in TABLE 2.

TABLE 2. Impact results PA-12: Effect of different material flow modifications

Material Flow Modification	Peak Force [N]	Total Energy [J]
Mod 1 - Weld Line	3350 ± 10	48.3 ± 0.59
Mod 2 - Hesitation	3560 ± 10	51.2 ± 0.72
Mod 3 - Hesitation + Weld Line	3450 ± 30	46.4 ± 0.74
No Modifications	3360 ± 40	49.5 ± 0.67

The observed differences in impact behavior due to different flow modifications are less obvious compared to quasi-static tensile results. The differences between the highest data for the hesitation samples (Mod 2) and the lowest, for hesitation + weld line (Mod 3) was 4.77 J, a reduction of 9.31% with all samples showing a ductile failure. While flow hesitation increased the tensile toughness for samples featuring weld lines, it reduced the obtained total energy during impact testing. The increased peak force compared to the weld line (Mod 1) and no modification samples was accompanied by the lowest puncture deformation findings. Therefore, the positive influence of hesitation on PA-12 weld line samples is not correlating between quasi-static uniaxial and dynamic multi axial loading scenarios, although the hesitation samples (Mod 2) showed again a superior behavior.

Material Characterization

The occurring mechanical properties of a moulded material sample are predetermined on the microscopic level where certain structures are established during the injection moulding process. The observed v-notch (FIGURE 2a) developing under tension along the weld line, leading to an early breakage during the quasi-static tensile test was investigated by optical microscopy. Under polarized light (FIGURE 2b and c) the weld line area showed an increased spherulite size compared to the surrounding area.

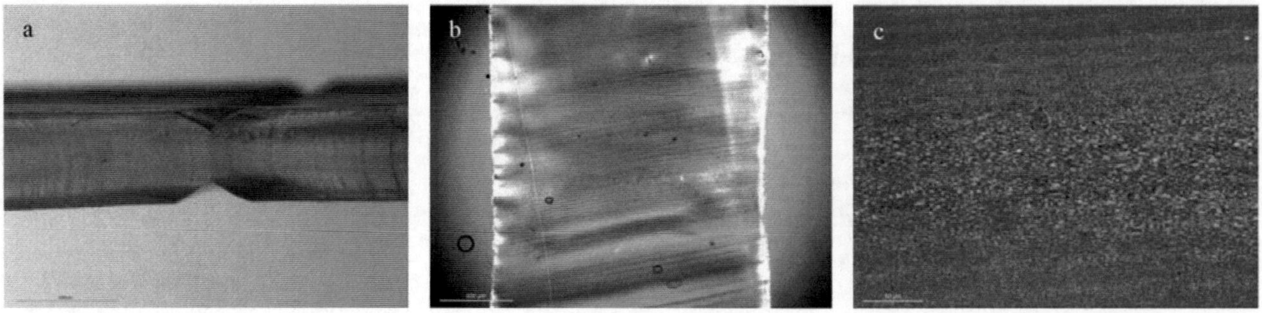

FIGURE 2. Optical microscopy findings weld-line region (a) occurring v-notch quasi-static tensile test (Nikon SMZ 1500) (b) difference in crystalline structure/ spherulite size 5x objective (Leica DMRX) (c) 40x objective (Leica DMRX)

The different crystalline microstructure within the weld line region makes the sample more prone to failure due to the occurring stress concentration under tension. While the developing v-notch on the surface of the samples is the more obvious macroscopic weakness, the final failure is caused by the poorly bonded layer along the weld line characterised by a different crystalline structure. The strength of the weld interface was improved for samples with occurring flow hesitation during the injection moulding process.

Data for the thermal behavior was obtained from DSC analysis and used to interpret the differences in mechanical properties: melt temperature, the detected enthalpy and the calculated degree of crystallinity are each presented in TABLE 4.

TABLE 4. DSC results PA-12: Effect of different material flow modifications and conditioning

Material Flow Modification	T_{Melt} [°C]	Melting Enthalpy [J/g]	Degree of Crystallinity [%]
Mod 1 - Weld Line	178.3 ± 0.4	42.6 ± 0.2	20.3 ± 0.1
Mod 2 - Hesitation	177.8 ± 0.4	52.1 ± 0.7	24.8 ± 0.3
Mod 3 - Hesitation + Weld Line	177.7 ± 0.6	46.3 ± 0.4	22.0 ± 0.2
No Modifications	178.2 ± 0.5	55.0 ± 0.6	26.2 ± 0.3

While the melting temperature is consistently found around 178°C (γ-crystalline form), matching material supplier information, a significant difference in melting enthalpy, respectively degree of crystallinity, was detected. While the samples with superior properties (Mod 2 and no modifications) showed 25% crystallinity, the results decrease for samples containing a weld line, with a higher degree when influenced by hesitation (Mod 3). These differences in morphology due to flow modifications correlate with the distinction of fracture mechanisms observed during mechanical testing, confirming weld lines as a material weakness for the investigated PA-12 material.

Conclusion

The influence of different melt flow modifications during the injection moulding process was highlighted for a commercially available PA-12 material. The mechanical properties for test specimen produced using a new mould differed by up to 57% when tested regarding their quasi-static tensile behavior and 9% with regards to impact properties due to different modifications. This highlights the fact that those investigations are essential to prevent unexpected failure of final parts with a more sophisticated geometry. Melt flow modifications leading to the formation of a weld line exhibited the highest performance decrease, weakening a part significantly compared to unmodified samples. A positive influence of occurring flow hesitation on the resulting weld line strength was observed for the uniaxial tensile properties. However, facing a dynamic multi axial loading scenario, during the impact test, the additional flow hesitation was counterproductive, decreasing the detected values. Structural features were evaluated on the microscopic scale using optical microscopy and DSC. Differences in spherulite size for weld line samples were detected as the reason for reduced tensile properties causing the creation of a notch and a stress concentration to form under tension significantly weakening the sample. A higher degree of crystallinity was detected for samples not containing a weld line, showing superior mechanical properties. Overall it can be concluded that the new tool allows the testing of new or enhanced materials and delivers more realistic results for complex geometries compared to simplified injection moulded test specimen produced by idealised melt flow conditions. More suitable design data are therefore obtained to predict the mechanical behavior of a final part, using practical or computational techniques, by paying sufficient attention to the change in performance due to differences in flow behavior when injecting more sophisticated geometries.

ACKNOWLEDGMENTS

The presented results are part of an industrial funded PhD at Loughborough University (UK). The new mould concept was developed by the LKT Erlangen (Germany) in cooperation with the adidas future engineering team (Germany).

REFERENCES

[1] H. Daiyan, E. Andreassen, F. Grytten, O. V. Lyngstad, T. Luksepp, and H. Osnes, "Low-velocity impact response of injection-moulded polypropylene plates - Part 2: Effects of moulding conditions, striker geometry, clamping, surface texture, weld line and paint," *Polym. Test.*, vol. 29, no. 7, pp. 894–901, 2010.

[2] A. Malguarnera, S. C., & Manisali, "The Effect of Processing Parameters on the Tensile Properties of Weld Lines in Injection Molded Thermoplastics," *Polym. Eng. Sci.*, vol. 21, pp. 586–593, 1981.

[3] R. Selden, "Effect of Processing on Weld Line Strength in Five Thermoplastics," vol. 37, no. 1, 1997.

[4] C. Lu, S. Guo, L. Wen, and J. Wang, "Weld line morphology and strength of polystyrene/polyamide-6/poly(styrene- co-maleic anhydride) blends," *Eur. Polym. J.*, vol. 40, no. 11, pp. 2565–2572, 2004.

[5] I. Y. Phang, T. Liu, A. Mohamed, K. P. Pramoda, L. Chen, L. Shen, S. Y. Chow, C. He, X. Lu, and X. Hu, "Morphology , thermal and mechanical properties of nylon 12 / organoclay nanocomposites prepared by melt compounding," *Polym. Int.*, vol. 464, no. December 2004, pp. 456–464, 2005.

[6] B. Ozcelik, E. Kuram, and M. M. Topal, "Investigation the effects of obstacle geometries and injection molding parameters on weld line strength using experimental and finite element methods in plastic injection molding," *Int. Commun. Heat Mass Transf.*, vol. 39, no. 2, pp. 275–281, Feb. 2012.

[7] R. Puffr and V. Kubanek, *Lactam-Based Polyamides Volume 1*. Boca Raton: CRC Press, 1991.

[8] K. Tomari, H. Hamada, K. Lee, and T. Morii, "The V-Notch at Weld Lines in Polystyrene Injection Moldings," *Polym. Eng. Sci.*, vol. 30, no. 15, 1990.

Study on the Fiber Breakage History through a Barrier Structure Screw in Injection Molding

Chao-Tsai (CT) Huang[a*], Huan-Chang Tseng[b], Rong-Yeu Chang[b], and Sheng-Jye Hwang[c]

a. Chemical and Materials Engineering Department, Energy and Opto-Electronic Materials Research Center, Tamkang University, Taipei, Taiwan

b. R and D Division, CoreTech System (Moldex3D) Co., Ltd., Hsinchu, Taiwan

c. Department of Mechanical Engineering, National Cheng Kung University

**Corresponding author:* cthuang@mail.tku.edu.tw

Abstract. In recent years, using short and long fiber-reinforced thermoplastics (LFRT) to replace some metal components has become one of the most attractive lightweight technologies. However, the microstructures of fiber inside plastic matrix are too complicated to manage and control during the injection molding through the screw, the runner, the gate, and then into the cavity. On the other hand, to enhance the plastification for better melt quality, the barrier type screw is very popular. But how the barrier design will influence the long fiber breakage is still not clear yet. In this study, we have integrated the screw plastification, to injection molding for fiber microstructures investigation. Specifically, we have focused on fiber breakage history through a barrier structure screw in injection molding. Results show that when 13 mm carbon fibers go through the barrier screw, the length of fibers is down to 1.75 mm (the number average) at the exit of screw. It is 10% shorter than that when fibers go through the regular type screw system. Moreover, the fiber breakage phenomena for two types of fibers (glass fiber and carbon fiber) are also investigated. At the exit of the screw, the number average of glass fibers is 2.21 mm. Clearly, the carbon fiber is easier to be broken. Furthermore, the carbon fiber length distribution has higher peak at the exit of the screw. This discovery is consistent with the experimental results of some current literature.

Keywords: long fiber-reinforced thermoplastics (LFRT), fiber breakage, fiber length distribution (FLD), screw plastification, injection molding

PACS: 83

INTRODUCTION

In recent years, lightweight technology has been applied into automotive and many other industries to enhance the fuel efficiency, especially, fiber-reinforced thermoplastics technology. Short and long fiber-reinforced thermoplastics (FRT) are now commercially available materials. FRTs can demonstrate greatly improved mechanical properties of injection molding products in automotive. This advantage is helpful for reducing costs and vehicle weight of FRT manufacturing. The important fiber microstructure variables, including orientation, length, and concentration, determine the degree of enhancement of mechanical properties. Furthermore, it is critical that the microstructure influences final shrinkage/warpage of with anisotropy [1-2]. On the other hand, there has been little attention paid to the material preparation inside the plastification unit. For example, we may have acceptable properties of the final product while poor mixing of color additives in the material makes the appearance unacceptable. In addition, there is potential for even more problems such as excessive shear heating of the melt or break the fiber length [3-6]. Since the microstructures of fiber inside plastic matrix are very complex, it is not easy to visualize. Besides, there are too many combination for material, process conditions, the dynamic features of microstructures (such as fiber length breakage, fiber length distribution, fiber orientation, fiber density) are too complicated to manage and control during the injection molding though the screw, the runner, the gate, and then into the cavity. In this study, we have integrated the screw plastification, to injection molding for fiber microstructures investigation. Specifically, we have focused on fiber breakage history through a barrier structure screw in injection molding. Results show that when 13 mm carbon fibers go through the barrier screw, the length of fibers is down to 1.75 mm (the number average) at the exit of screw. It is 10% shorter than that when fibers go through the regular type screw system. Moreover, the fiber breakage phenomena for two types of fibers (glass fiber and carbon fiber) are also investigated. At the exit of the screw, the number average of glass fibers is 2.21 mm. Clearly, the carbon fiber is easier to be broken. Furthermore, the carbon fiber length distribution (FLD) has higher peak at the exit of the screw. This discovery is consistent with the experimental results of some current literature.

NUMERICAL THEORY AND ASSUMPTION

I. Model for Screw Plastification and Fiber breakage

The model for plastification has been described in detailed in our previous study [5]. Moreover, regarding the model for fiber breakage, a mean fiber length is described by number average L_N and weight average L_W, as follows:

$$L_N = \sum_{i=1}^{N} N_i l_i / \sum_{i=1}^{N} N_i \quad \text{and} \quad L_W = \sum_{i=1}^{N} N_i l_i^2 / \sum_{i=1}^{N} N_i l_i \tag{1}$$

where N_i represent the expected value of the number of fibers of length l_i in a sample taken from a small region. Note that $l_i = i\Delta l$, Δl is the discretized length segment, namely, $\Delta l = L/N$, L is the initial length, N is number of segment. Phelps et al. [6] describe the change rate of the number of fibers with length l_i, as below:

$$\dot{N}_i = -P_i N_i + \sum_{k=1}^{N} R_{ik} N_k \tag{2}$$

$$R_{ik} = N_{norm}(l_i, \frac{l_k}{2}, Sl_k) \; ; \quad R_{ki} = R_{(i-k)i} \; ; \text{ and } \sum_{k=1}^{N} R_{ji} = 2P_i \tag{3}$$

where P_i is a breakage probability for parent fibers of length l_i at time Δt; R_{ik} is the rate of production of child fibers of length l_i by breaking parent fibers of length l_k, thus with the mass conservation law. R_{ik} is given by the normal distribution function, as follows:

$$N_{PDF}(x, \mu, \sigma) = \frac{1}{\sqrt{2\pi\sigma^2}} e^{-\frac{(x-\mu)^2}{2\sigma^2}} \tag{4}$$

where x is variable, μ is mean, and σ is variance; a scaling factor S is used to adjust the profile.

Furthermore, the probability of fiber breakage can then be expressed by the Dinh-Armstrong hydrodynamic compression force F_i and the Euler critical buckling force F_{crit} for a fiber,

$$P_i = C_b \dot{\gamma}[1 - \exp(1 - \frac{F_i}{F_{crit}})] \tag{5}$$

$$F_i = \frac{\varsigma \eta_m l_i^2}{8}(-\mathbf{D} : \mathbf{A}) \quad \text{and} \quad F_{crit} = \frac{\pi^3 E_f d_f^4}{64 l_i^2} \tag{6}$$

where C_b is a scales the breakage rate; ζ is a dimensionless drag coefficient; η_m is matrix viscosity; d_f is fiber diameter; E_f is fiber modulus

INVESTIGATION MODEL AND INFORMATION

I. Regular Screw Design

The regular screw design is shown as in Figure 1(a). For conventional screw, there are three zones, i.e., feeding zone (11D), compression zone (6.5D), and metering zone (4.5D), respectively. Here 11D means the length is 11 times of diameter. Initially, the channel depth of feeding zone is 7.2 mm, and metering zone is 3.0 mm. The screw length to diameter is L/D = 22. The diameter of screw is 36 mm. The compression ratio is about 7.2/3.0.

The injection molding process setting is as following: filling 0.74 sec, packing 3 sec, cooling 10.6 sec, and mold open and close 5 sec. Cycle time is 19.3 sec. Material is SABIC ABS Cycolac CGF20. The barrel is setup with three sections. The temperature setting is 225°C, 235°C, and 245°C, respectively. Screw operation is 100 RPM. The stoke of the screw is 34.91 mm. The original length of the long fiber is 13 mm.

II. Barrier Screw Design

The barrier screw geometry design is similar to that of regular screw is as shown in Figure 1(b). The main difference is the structure of compression zone where it contains primary and secondary channels at the same time. In general, this type of screw can offer better plastification and mixture efficiency. Here we are going to study how it affects the fiber length history along operation.

III. Different Fiber Type Effect

We applied two different types of fibers to study the fiber type effects. The original length of those two types of fibers is 13 mm long, and other key parameters are given as below [9].

- Glass fiber: diameter 15 μm, Young's modulus 50 Gpa

- Carbon fiber: diameter 5 μm, Young's modulus 500 Gpa

To cover full injection molding process, the screw plastification is simulated using Moldex3D R13 ScrewPlus module. Fiber microstructures, including fiber length, fiber orientation, and fiber density distribution are predicted using Moldex3D software.

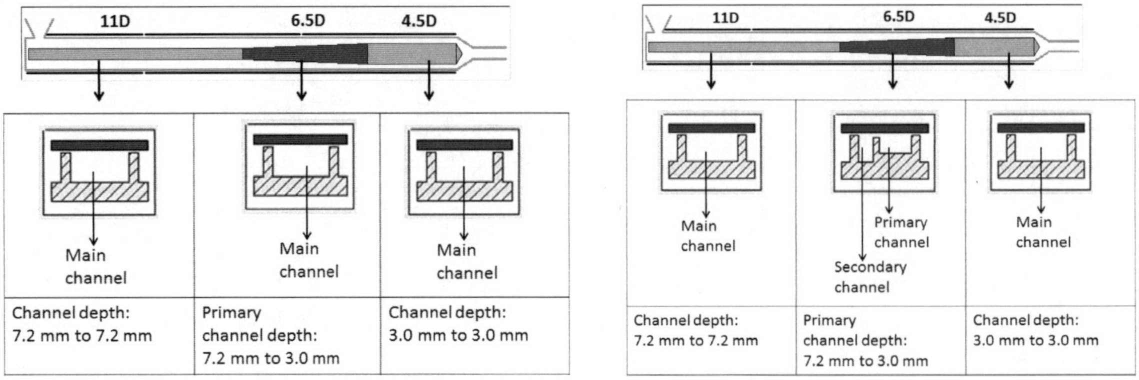

(a) (b)

FIGURE 1. The schematic of a screw design: it contains three sections with compression ratio of 7.2/3.0. (a) a regular screw design, (b) a barrier screw design.

RESULTS AND DISCUSSION

I. Validation of Fiber Breakage Simulation

Firstly, the simulation results are validated with that of Patcharaphun and Opaskornkul [7]. The initial fiber length was 7 mm. Here we performed a comparative numerical analysis with results of fiber degradation under a screw speed at 12 RPM. Due to the nature of statistics of sampling, two length units are used: number-average length (Ln) and weight-average length (Lw). Figure 4 shows fiber degradation in number average length at different screw sections. The predicted fiber length distribution (FLD) is in a good agreement with the experimental FLD result [7].

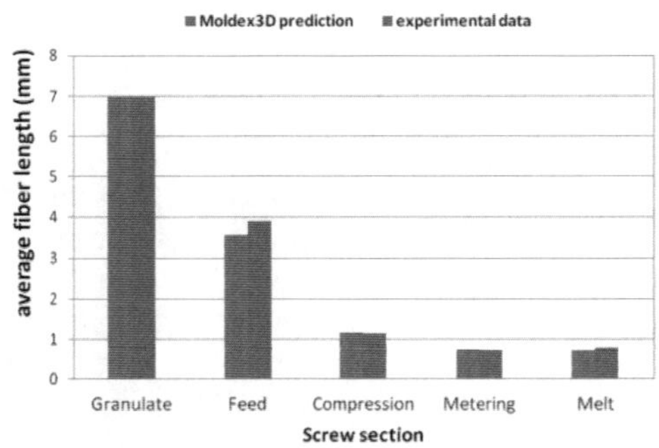

FIGURE 4. Fiber degradation in number-average length with various screw sections.

II. Different Fiber Type Effect

Figure 5 shows the shear rate variation along the screw downstream for the regular and barrier screws. For the regular screw, from feeding zone to metering zone, the highest shear rate happened around the end of compression zone. Then we applied two different types of fibers (glass and carbon fibers) into the screw for plastification. Figure 6(a) shows fiber breakage history for both glass fiber and carbon fiber based on number average. The length of glass fiber is 2.21 mm, while that of carbon fiber is 1.75 mm. Although the strength of carbon fiber is stronger than that of glass fiber, the final length of glass fiber is longer. It displayed that the carbon fiber is easier broken than glass fiber. Similarly, the behavior can be observed based on weight average from Figure 6(b). Moreover, Figure 7 shows FLD at the end of screw. Although the average of fiber length of glass fiber is longer than that of carbon fiber, the carbon fiber exhibits the higher peak for FLD compared to that of glass fiber. It is consistent with the current observation of Chen et. al.[8].

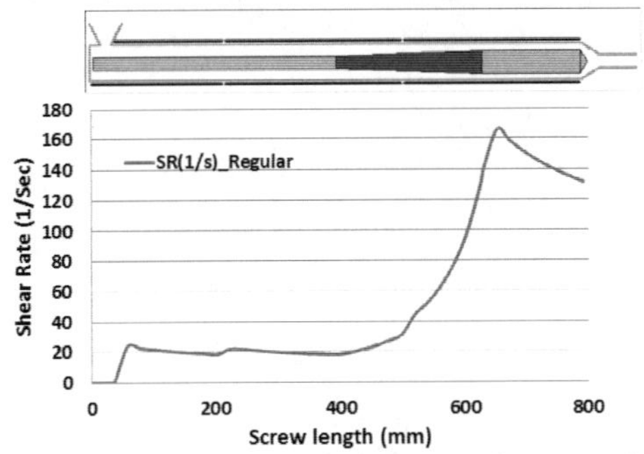

FIGURE 5. Shear rate variation along the screw performance for regular screw.

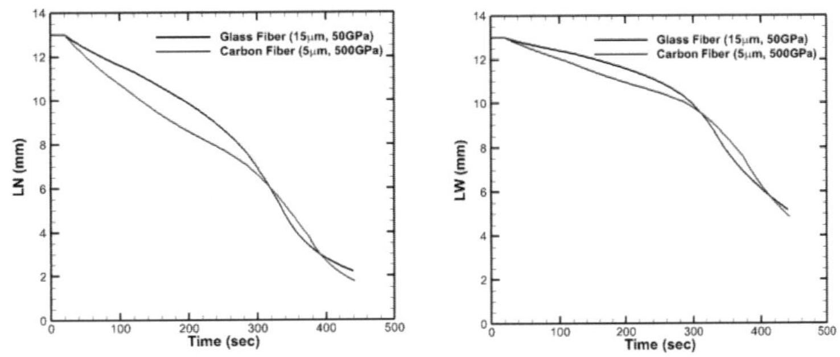

(a) (b)

FIGURE 6. (a) Number average fiber length along the screw direction for different fiber types, (b) weight average fiber length along the screw direction for different fiber types.

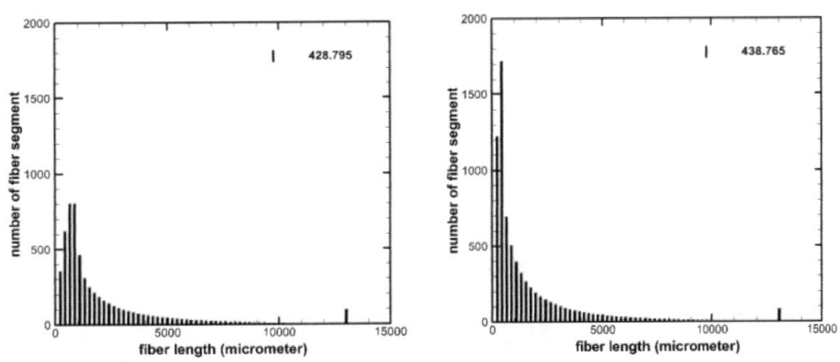

(a) (b)

FIGURE 7. The fiber length distribution for glass fiber at the end of screw, (a) glass fiber, (b) carbon fiber.

III. Barrier Screw Effect on Fiber Breakage

Furthermore, Figure 8 shows the shear rate variation along the screw downstream for regular and barrier screws. In the regular screw design, from feeding zone to metering zone, the highest shear rate happened around the end of compression zone. For the barrier screw, generally, the shear rate variation is similar. However, due to the higher shear force from barrier section, the highest shear rate happened around the similar region but is increased significantly.

CONCLUSIONS

In this study, we have integrated the screw plastification, to injection molding for fiber microstructures investigation. Specifically, we have focused on fiber breakage history through a barrier structure screw in injection molding. Results show that when 13 mm carbon fibers go through the barrier screw, the length of fibers is down to 1.75 mm (the number average) at the exit of screw. It is 10% shorter than that of regular screw system. Moreover, the fiber breakage phenomena for two types of fibers (glass fiber and carbon fiber) are also investigated. At the exit of the screw, the number average of glass fibers is 2.21 mm. Clearly, the carbon fiber is easier to be broken. Furthermore, the carbon fiber length distribution has higher peak at the exit of the screw. This discovery is consistent with the experimental results of some current literature.

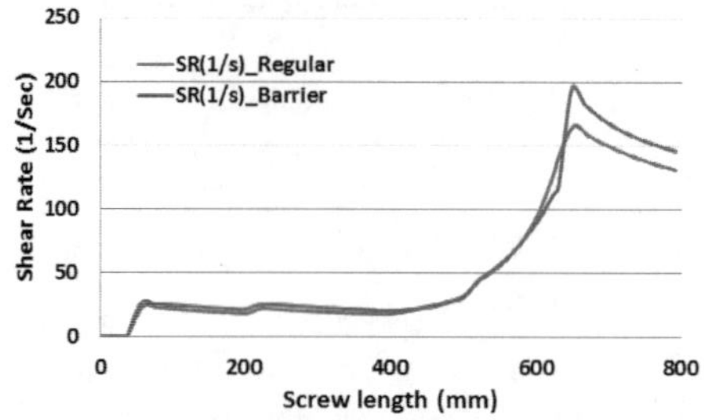

FIGURE 8. Shear rate variation along the screw performance for Screw 1.

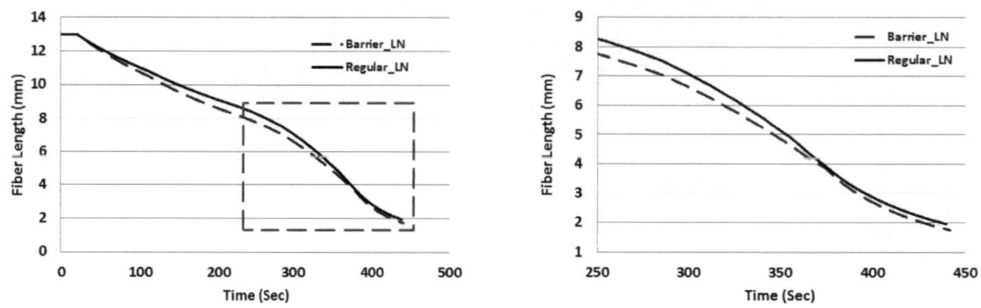

(a) (b)

FIGURE 9. (a) Number average fiber length along the screw direction for different screw design, (b) the enlarge of (a).

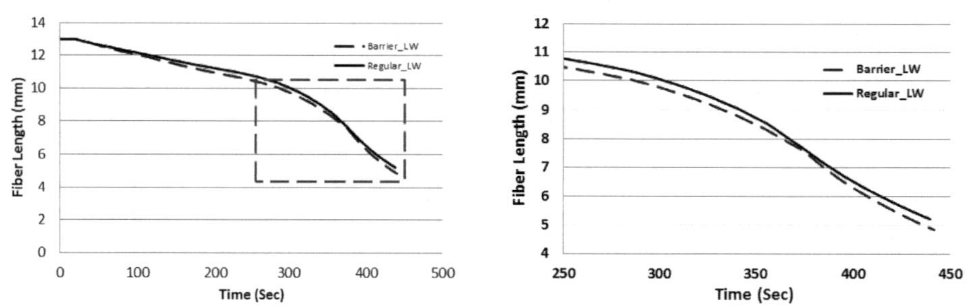

(a) (b)

FIGURE 10. (a) weight average fiber length along the screw direction for different screw design. **(b)** the enlarge of (a).

REFERENCES

1. Advani, S.G., 1994. *Flow and Rheology in Polymer Composites Manufacturing* (Elsevier, New York)
2. Tseng, H.-C., Chang, R.-Y., and Hsu, C.-H., "Phenomenological improvements to predictive models of fiber orientation in concentrated suspensions." J. Rheol. 57 1597 (2013).
3. Vlcek, J., Miller, L., and Huang, C.-T., "Simulation of screws for injection molding", SPE Annual Tech meeting, 2478-2481 (2009).
4. Potente, Helmut; Effen, Norbert; Ujma, A.; Injection molding machines. Optimal use by computer simulation. Kunststoffberater (1996), 41(10), 12-17.
5. Huang, C.-T., Tseng, H.-C., Vlcek, J., and Chang, R.-Y, "Fiber breakage phenomena in long fiber reinforced plastic preparation", IOP Conf. Series: Materials Science and Engineering, 87, No. 012023, pp.1-6.
6. Phelps, J.H., Abd El-Rahman, A.I., Kunc, V., and Charles L. Tucker, I., 2013. Composites Part A: Applied Science and Manufacturing **51** 11-21
7. Patcharaphun, S. and G. Opaskornkul, "Characterization of Fiber Length Distribution in Short andLong-Glass-Fiber Reinforced Polypropylene during Injection Molding Process." Kasetsart J. (Nat. Sci.) 42 392-397 (2008).
8. Chen, H., Cieslinski, M., and Baird, D. G., "Progress in modeling long glass and carbon fiber breaakage during injection molding", AIP Conference Proceeding, 1664, 060004 (2015); doi:10.1063/1.4918422.

Micro Injection Molding: Investigation of Interfacial Fracture Behavior

Matthieu Fischer[a], Ines Kuehnert[a*]

[a]Processing Department, Leibniz-Institut fuer Polymerforschung Dresden e.V., Dresden, Germany
**corresponding author: Ines Kuehnert, kuehnert@ipfdd.de*

In relation to standard injection molding the challenges of testing methods, increasing complexity as well as the process and equipment requirements of micro injection molding will be discussed. The mechanical behavior of micro injection molded specimens was investigated. Therefore, a new tensile test specimen was developed to evaluate systematically the impact of interfaces (e.g. weld lines/ multi material interfaces) on small scale applications. A failure observation of micro injection molded specimen was carried out under a light microscope. The processing effects on spherulitic structure were linked to bonding strength, material properties and injection molding simulation.

Keywords: **micro injection molding, fracture, morphology, interface, weld line, simulation.**

Influence of Processing Conditions and Filler Type on The Crystallinity of PLA Compounds

Gernot A. Pacher[1], Thomas Hutterer[1], Gerald R. Berger[2], Walter Friesenbichler[2]

[1]*Polymer Competence Center Leoben GmbH, Roseggerstraße 12, A-8700 Leoben, Austria*
[2]*Injection Molding of Polymers, Montanuniversität Leoben, Otto Glöckel Straße 2, A-8700 Leoben*

gernot.pacher@pccl.at

Abstract. It is well known that neat PLA is not inclined to crystallization, which results in inferior mechanical properties compared to commodity plastics like polyethylene. To empower PLA for commercial applications, the overall degree of crystallinity has to be increased. In this study, the crystallinity of PLA-compounds was influenced via (a) processing conditions and (b) addition of fillers. Mold temperatures for conventional injection molding on PLA are generally set in the range of 10 °C to 25 °C while the crystallization temperature lies well over 100 °C. In general, this leads to amorphous parts. Due to this fact, tempering is often used as a second manufacturing step in order to generate crystallinity. Another well-known approach to promote crystallization in PLA is the addition of fillers as nuclei. In this work, an in-mold-tempering approach via Rapid-Heat-Cycle-Molding was designed and conducted to enable single-step manufacturing of crystalline PLA parts. To enable crystallization, the mold temperature was elevated above the crystallization temperature before start of injection, held for prolonged times (time for in-mold-tempering) and rapidly cooled down to conventional mold temperatures again. In order to evaluate the effectiveness of the in-mold-tempering approach, different PLA compounds (with varied amounts of silica and talcum) were tested in addition to neat PLA, followed by measurements of the degree of crystallinity via differential scanning calorimetry (DSC).

Keywords: Polylactic Acid, Rapid Heat Cycle Molding, Talc, Layered Silica, Crystallinity
PACS: 81.05.Lg, 81.10.Fq, 83.50.Uv

INTRODUCTION

To induce higher degrees of crystallinity into injection molded PLA than can be obtained via traditional processing methods, a few ways are viable[1,2,3]. In this work, an effort has been made to raise crystallinity by adding fillers acting as nucleating agents as well as manufacturing the specimens using rapid heat cycle molding (RHCM). RHCM is now a well known technique used to conquer various problems occurring during injection molding, such as weak weld lines, visibility of sink marks and low crystallinity.

Furthermore, the chosen filler types, respectively talc and layered silica, are readily available materials, which can be incorporated in the polymer without extensive precautions using twin screw extruders and appropriate scales. To separate the impact of the fillers on crystallinity from RHCM, specimens were also manufactured from neat PLA.

Investigations on the degree of crystallinity were carried out applying DSC analysis to the injection molded specimens as well as to the compounds before injection molding.

MATERIAL PREPARATION AND USED MATERIALS

For this work, Ingeo™ Biopolymer 3251D (by Nature Works LLC, MN55345 Minnetonka, USA) was used as a base material. This type of polymer is recommended for injection molding applications and has a MFR of 80 g/10min (210 °C, 2.16 kg) a crystalline melt temperature of about 155 °C - 170 °C and a glass transition temperature of around 55 °C – 60 °C. It was used for manufacturing the control specimens as well as a base material for compounding the filled material. As the talc filler, Finntalc M03 (Mondo Minerals B.V., NL-1041 AR, The Netherlands) with a Top Cut (d98%) particle size of 5 µm was used. Table (1) lists the materials used for injection molding of the multipurpose specimens in accordance with EN ISO 527-1

TABLE (1). Compounds produced for specimen manufacturing

Compound	Filler Content
PLA-N	0%
PLA-T	5% talc

Compounding was carried out using a co-rotating twin screw extruder (Type ZSE27MAXX-44D, Leistritz Extrusionstechnik Gmbh, 90459 Nürnberg, Germany) with temperatures set to 180° C at 100 rpm. The compounds were pelletized using an underwater pelletizer, quenching it simultaneously. The talc filler for PLA-T was directly weighed in.

EXPERIMENTAL

For the manufacturing of the specimens, an injection molding machine, type HM 110, by Wittmann Kunststoffgeräte GmbH was used. The mold used for the molding process produced two multipurpose specimens compliant with DIN EN ISO 527-1A at once by using two film gates sourced by one injection port. Figure 1 (a) shows the inside of the mold used for specimen manufacturing. Figure 1 (b) depicts the assembly of the heatingpads in the mold. The mold was also equipped with two ceramic heating pads (Ultramic®, Watlow Electric Manufacturing Company, MO 63146, USA) for RHCM situated in the middle of each cavity, while cooling was carried out on both mold halves by water cooling.

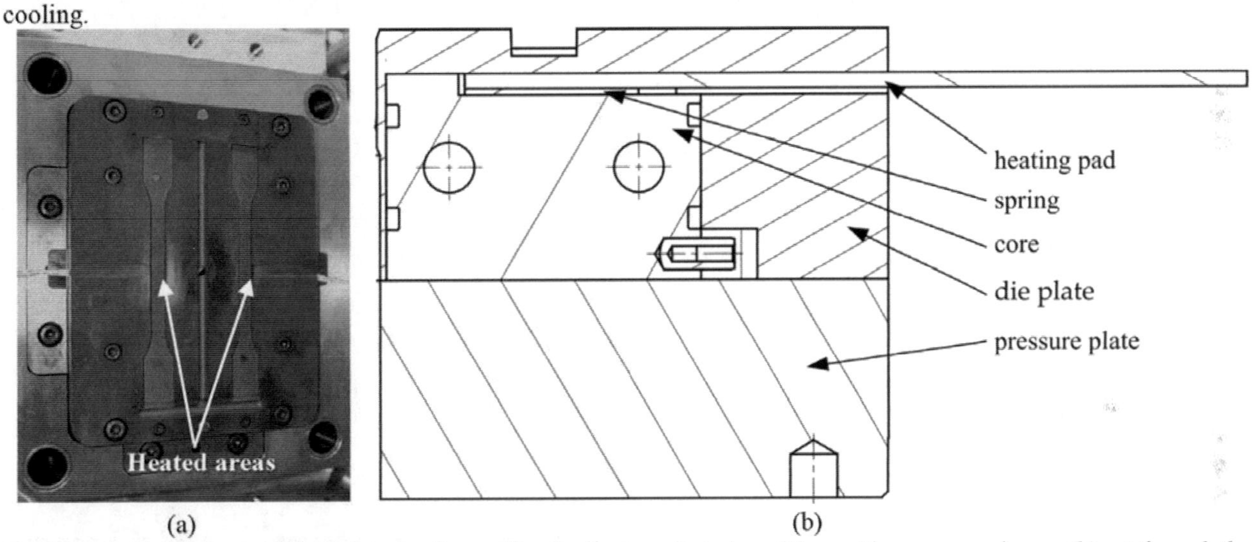

FIGURE 1. (a) Moving mold half showing the cavities for the manufacturing of the multipurpose specimens. (b) cut through the mold at the location of the heating pads.

Injection Molding

During injection molding, a full factorial 3^2 DOE was applied to each material, with annealing time and temperature acting as variable factors. Table 2 lists all DOE factors and their respective settings. Because crystallization rates of neat PLA and nucleated PLA can differ greatly[4], samples of each material, which were annealed for 900s at 110° C were also manufactured. Annealing times and temperatures were chosen according to previous work[5] as well as DSC analysis, where peak crystallization temperatures for the pelletized compounds were determined.

TABLE (2). DOE used for injection molding

DOE Level	Factor 1 Annealing Time [s]	Factor 2 Annealing Temperature [° C]
Low	0	100
Middle	120	110
High	240	120

DSC Analysis

TABLE (3). Measurement program for DSC analysis

Program Step	Program Characteristics
First heating run	25° C to 240°C at 10° C/min
First hold	240°C for 1 min
Cooling run	240°C to 25°C at 10° C/min
Second hold	25°C for 5 min
Second heating run	25°C to 240°C at 10°C/min
Third hold	240°C for 1 min

In order to determine the degree of crystallinity within the specimens achieved by the DOE, DSC analysis was chosen. The experiments were carried out using a DSC 4000 by Perkin Elmer Inc. The measurement program chosen across all types of materials are depicted in Table (3). The following values were then obtained from the DSC curves: glass transition temperature (first heating run, T_g) enthalpy of cold crystallization (first & second heating run, ΔH_{cc1}, $\Delta H_{cc,2}$), melting enthalpy (first & second heating run, ΔH_{m1}, ΔH_{m2}) and crystallization enthalpy during cooling (ΔH_c). Thus, the degree of crystallinity (χ) can be obtained using equation (1)[3], with ΔH_{m0} representing the melting enthalpy of the 100% crystalline material (93 J/g for PLA[6])

$$X = \frac{\Delta H_m - \Delta H_{cc}}{\Delta H_{m0} - \left(1 - \frac{\%wt\ filler}{100}\right)} * 100 \qquad (1)$$

RESULTS AND DISCUSSION

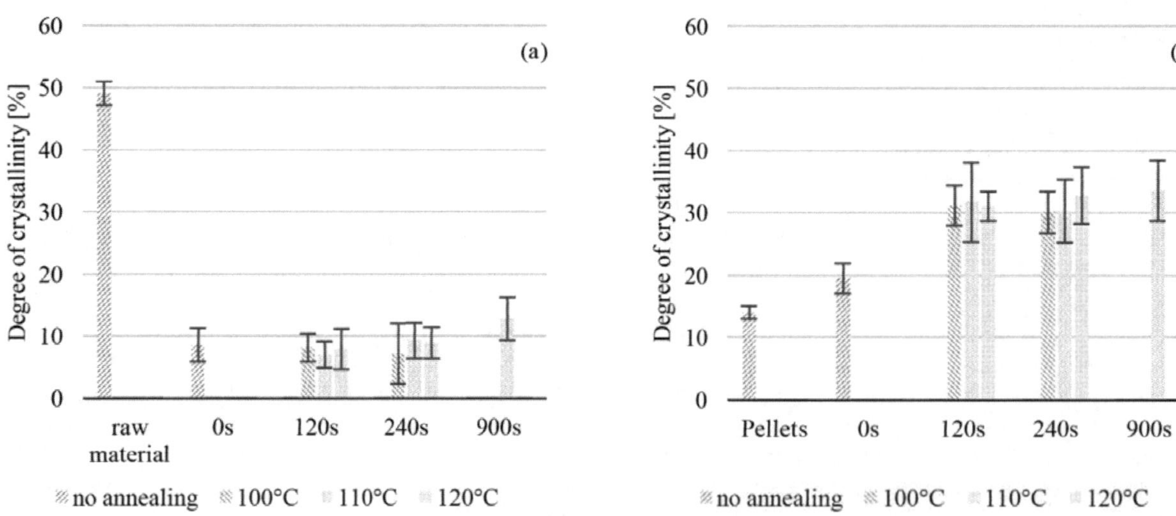

FIGURE 2. Degree of crystallinity of (a) PLA-N and (b) PLA-T, as measured after the first DSC heating run

Figure 2 depicts the arithmetic averages of the degree of crystallinity achieved through injection molding at different factor levels for PLA-T (a) and PLA-N (b). For PLA-T, the degree of crystallinity was also determined for the quenched pellets and for PLA-N, this was done for the pelletized raw material. The values were obtained from the first heating run of the DSC analysis.

For PLA-T, it can be seen that the talc filler raises crystallinity well above the levels that are achievable without filler, even without annealing. Annealing of PLA-T further raises crystallinity, even at low temperatures and short times. However, at annealing times and temperatures greater than 120s at 100° C, no further increase of the degree of crystallinity can be observed.
For PLA-N, a significant effect of this method can only be seen at durations far above 240s.

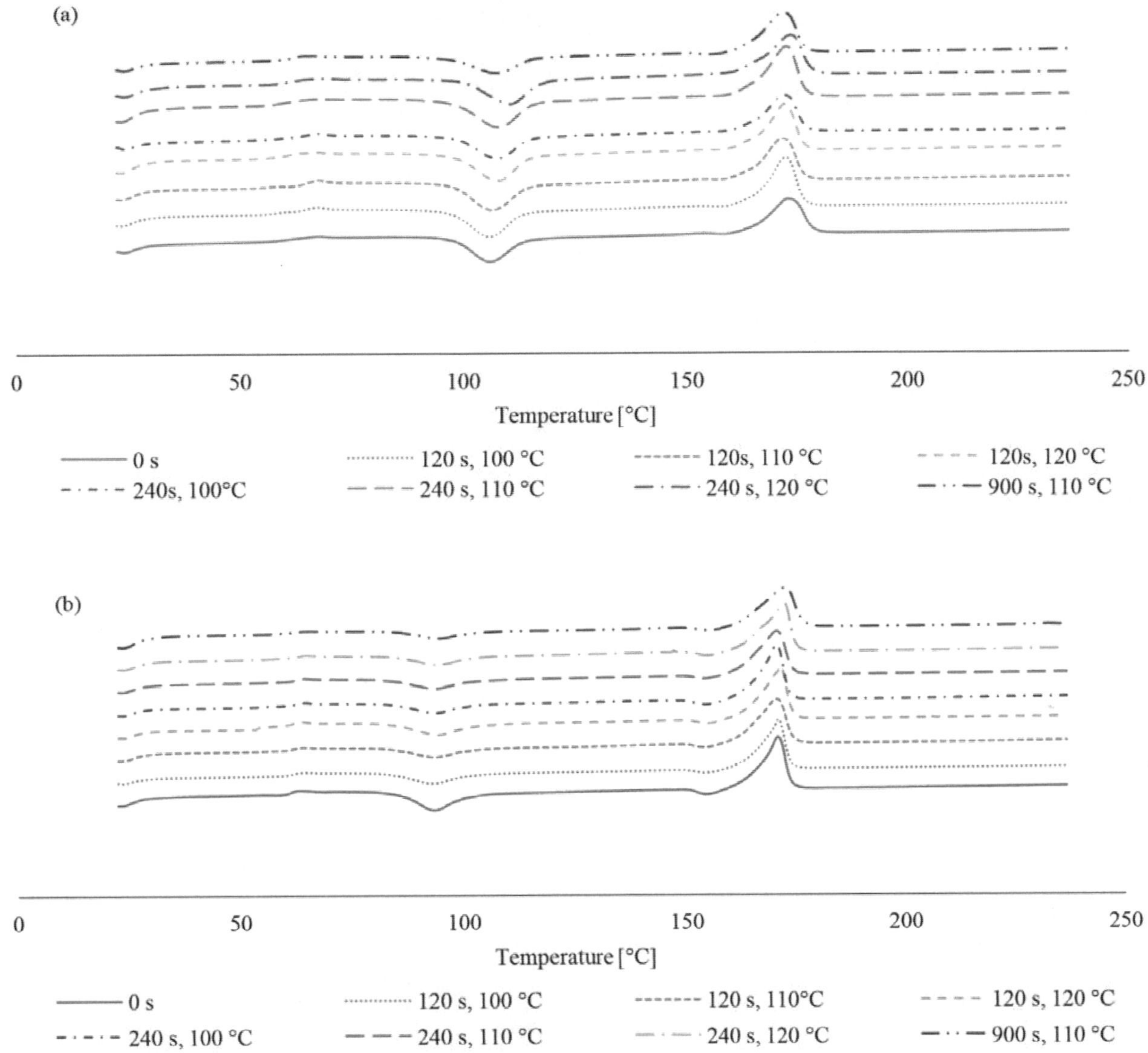

FIGURE 3. First heating run DSC Curves of (a) PLA-N and (b) PLA-T

Figure 3 shows the first heating run DSC data measured at different processing conditions for (a) PLA-N and (b) PLA-T. The curves are not medium values, but rather distinct samples for each DOE setting. All samples show a distinct peak for cold crystallization. Cold crystallization occurs, when crystallization cannot be finished before the polymeric chains freeze[7]. However, it can be seen that this peak is less accentuated for all processing samples of PLA-T. Also, the shaping of those cold crystallization peaks for the annealed PLA-T samples is visibly less than when no annealing occurred. It can also be seen that cold crystallization for PLA-T occurs at much lower temperatures than for PLA-N (94 ± 0,8 °C, respectively 106 ± 1,5 °C). This can be due to a nucleating effect of the talc filler, where crystals can form at lower temperatures. Nucleation can also lead to smaller spherulites, which will be investigated via polarized optical microscopy. At 900 s annealing time, the cold crystallization peak of PLA-N is visible less shaped than at shorter annealing time, however this difference is not as clearly visible for PLA-T.

CONCLUSIONS

It was shown, that while certain processing conditions and techniques (such as RHCM) can influence the crystallization behavior of neat poly(lactic acid) positively, these are not practically applicable. The addition of talc to the polymer can alter the crystallization behavior of PLA in such a way, that the impact of annealing is readily measureable at shorter and more practically relevant time frames.

ACKNOWLEDGEMENTS

The research work of this paper was performed at the Polymer Competence Center Leoben GmbH (PCCL, Austria) within the framework of the COMET-program of the Federal Ministry for Transport, Innovation and Technology and the Federal Ministry of Economy, Family and Youth with contributions by the Institute of Injection Molding of Polymers, Montanuniversitaet Leoben. The PCCL is funded by the Austrian Government and the State Governments of Styria and Upper Austria.

REFERENCES

1. I. Armentano et al., Prog. in Poly. Science **38**, 1720-1747 (2013)
2. H. Bai et al., Polymer **55**, 6924-6934 (2014)
3. D. Battegazzore, Expr. Poly. Letters **5**, 849-858, (2011)
4. Z. Refaa et al., AIP Conference Proceedings **1593**, AIP Publishing, Melville NY, (2014)
5. L. Crema, "A Study of Rapid Mould Temperature Variation Influence on Morphology and Appearance of Injection Moulded Parts", Ph.D. Thesis, Universita degli studi di Padova, 2015
6. J. F. Turner et al., J. of Therm. Analysis and Calorimetry **74**, 257-268 (2003)
7. A. Frick and C. Stern, *DSC Prüfung in der Anwendung*, Munich: Hanser (2006) pp. 64-67.

Advanced Fiber Orientation Prediction for High Filler Content Short-fiber/Thermoplastic Composites

Thanh Binh Nguyen Thi[a], Atsushi Yokoyama[a,*], Senji Hamanaka[b], Katsuhisa Yamashita[b], Chisato Nonomura[b]

[a]*Kyoto Institute of Technology, Matsugasaki, Sakyo-ku, Kyoto 606-8585, Japan*
[b]*Research Center, Toyobo Co., LTD, 2-1-1 Katata, Otsu, Shiga 520-0292, Japan*

nttbinh@kit.ac.jp
yokoyama@kit.ac.jp

Abstract. Properties of fiber-reinforced composite are dominated by the microstructure of the fabricated part rather than the properties of constituent materials. As the microstructure of composite is related to the flow-processing route of the fiber-reinforced suspensions and the geometry of the mold, the microstructure of composite can be tailored in order to achieving high-performance composites by exercising control over the flow processing. Thus, numerical methods are used to model the resin flow, the fiber orientation, and mold design, and they become important challenges during molding process. In our previous research, a theoretical fiber-fiber interaction model with a global fiber interaction coefficient was developed. In this study, the three-dimensional (3D) fiber orientation distribution is predicted by combining our developed fiber interaction model and improved Anisotropic Rotary Diffusion - Retarding Principle Rate (iARD-RPR) model. The fiber orientation calculation started from the gate, and from a random orientation to alignment orientation is investigated at that location as the inlet boundary condition. Moreover, simplified deformation behavior of fountain flow was employed for fiber orientation in the flow front region. In experimental, short-glass fiber-reinforced polyamide 6 specimens produced using plate-shaped cavity having three different thicknesses ranging from 2 mm to 4 mm and with the fiber contents ranging from 30 wt% to 65 wt% are carried out using injection molding. 3D fiber orientation observation and measuring are exactly and automatically examined by Micro-computed tomography system. Comparisons with experimental results, it is showed that this advanced prediction methodology can be effectively used for the 3D fiber orientation prediction in high filler content composite parts.

Keywords: Fiber Orientation, Fiber-Interaction Model, Injection Molding, High Filler Content Composite.

PACS: 81.20.Hy

INTRODUCTION.

The mechanical, functional performances of the short glass fiber thermoplastic composites are significantly dominated by the orientation distribution of reinforcing fibers which produces during injection molding process. Properties of the final part vary throughout the part due to local variations in the orientation of fibers. The reason is that the flow field inside mold is very complex. Thus, flow processing is the key to tailor the orientation microstructure of fiber reinforced composites to suit the desired application with minimum cost and maximum reliability [1, 2]. Mathematical models need to be developed with the emphasis on accurate predicting the final orientation pattern.

The 3D flow at the melt front region, namely fountain flow, carries molten material from the mid-plane to the cavity walls. The molten material is instantly solidified, and then forms a shell structure. Thus, fountain flow plays a prominent role in the orientation of fibers in injection molded geometries. And prediction 3D fiber orientation is a difficult task in which the orientation mechanism, flow field, and solidification near the wall should be considered simultaneously [2, 3].

Recently, Mazahir et. al [2] assessed fiber orientation predictions of Folgar-Tucker (F-T) model and its two modified versions, delayed F-T model and reduced strain closure (RSC) model, in the frontal region of a 30wt.% short glass fiber injection-molded center-gated disk in fountain flow simulation. They found that changes in orientation in the frontal region will affect the fiber orientation along the entire walls, and slow down evolution of orientation in the frontal region using RSC model was not enough to match the even slower evolution observed in experimental data. Moreover, the larger C_I value provided better predictions in the shear-dominated lubrication region but may not be suitable for extension-dominated frontal region. In our previous studies [4, 5], a developed fiber interaction model represented a global coefficient in short fiber suspension as a function of a function of the fiber aspect ratio, volume fraction and general shear rate may help more accurate estimate of the fiber orientation at all regions of the mold cavity. In addition, molten material in high filler content composite becomes more complex, so an advanced prediction of fiber orientation in fountain flow needs to be established.

Consequently, the 3D fiber orientation distribution is predicted by combining our developed fiber interaction model and iARD-RPR model [6]. Simplified deformation behavior of fountain flow [3] was adopted for fiber orientation in the flow front region. In experimental, short-glass fiber-reinforced polyamide 6 (GF/PA6) specimens were produced using plate-shaped cavity having three different thicknesses ranging from 2mm to 4mm and with the fiber contents ranging from 30wt.% to 65wt.% by the injection molding. The 3D fiber orientation observation and measuring are exactly and automatically examined by Micro-computed tomography (μ-CT) system Y.Cheetah μHD. Comparisons with experimental results, it is showed that this advanced prediction methodology can be effectively used for the 3D fiber orientation prediction in high filler content composite parts.

FIBER ORIENTATION THEORY

Jeffery's Hydrodynamic (HD) model and Folgar and Tucker's Isotropic Rotary Diffusion (IRD) model are two representative theoretical models over the past two decades. In general, the standard F-T orientation equation is a combination between HD and IRD model [6, 7], as given by.

$$\dot{A} = \dot{A}_{HD} + \dot{A}_{IRD}(C_I), \dot{A} = \frac{Da_{ij}}{Dt}, \dot{A}_{IRD} = 2C_I\left|\dot{\gamma}\right|(\delta_{ij} - \alpha a_{ij})$$
$$\dot{A}_{HD} = -\frac{1}{2}(\omega_{ik}a_{kj} - a_{ik}\omega_{kj}) + \frac{1}{2}\lambda(\dot{\gamma}_{ik}a_{kj} + a_{ik}\dot{\gamma}_{kj} - 2\dot{\gamma}_{kl}a_{ijkl}). \quad (1)$$

Recently, iARD-RPR model was developed suitable for both short and long fiber filled materials [6], as shown below:

$$\dot{A} = \dot{A} = \dot{A}_{HD} + \dot{A}_{iARD}(C_I, C_M) + \dot{A}_{RPR}(\alpha), \dot{A}_{iARD} = \dot{A}_{IRD}(C_I) + \dot{A}_{SB}(C_I, C_M)$$
$$\dot{A}_{SB} = 2\dot{\gamma}C_IC_M\left\{\left(A - \tilde{L}\right) + 5\left[\left(A.\tilde{L}\right)_{symm} - \left(A_4 : \tilde{L}\right)\right]\right\}, \dot{A}_{RPR}(\alpha) = -R\bullet\dot{\Lambda}_{IOK}\bullet R^t \quad (2)$$
$$\tilde{L} = \frac{L^T.L}{L^T : L}, \left(A.\tilde{L}\right)_{symm} = \frac{1}{2}\left[A.\tilde{L} + \left(A.\tilde{L}\right)^t\right], A = R\bullet\Lambda\bullet R^t, \Lambda = \begin{bmatrix} \lambda_1 & 0 & 0 \\ 0 & \lambda_2 & 0 \\ 0 & 0 & \lambda_3 \end{bmatrix} = R^t\bullet A\bullet R.$$

where C_M describe the fiber–matrix interaction, while parameter α can slow down the rate of fiber orientation development. A new anisotropic term of the steric barrier (SB) is introduced, where the tensor $\tilde{L}$ is a so-called dimensionless flow resistance tensor, depending on the velocity gradient tensor, L. The RPR orientation tensor rate is mathematically described using the similarity transformation of linear algebra. The rotation matrix, R, with its transpose of matrix, R^t, is defined by the orientation tensor's three eigenvector columns; and the diagonal matrix, K, is given by the orientation tensor's three eigenvalues.

Simulation of combined model was carried out with a global C_I [5] using iARD-RPR model.

$$C_I = 0.214\left|\dot{\gamma}\right|\left[1 - e^{-0.224\frac{L}{D}\phi_f}\right] \quad (3)$$

EXPERIMENTAL

The GF/PA6 composites with 30wt.%, 50wt.%, and 65wt.% fiber concentration were fabricated by injecting pellets into injection molding machine (Toshiba: IS-80) under following molding conditions: resin temperature of 290°C, mold temperature of 120°C, filling time of 0.4s, injection speed of 100mm/s, and packing pressure of 40MPa. The experimental mold cavities have their planar dimensions of 100mm × 100mm, and thicknesses of 2.0mm, 3.0mm, 4.0mm, which were designed and manufactured in Toyobo Co., Ltd. The 3D fiber orientation distribution of injection-molded parts at testing position was examined by μ-CT system Y.Cheetah μHD.

μ-CT system YXLON International, Y.Cheetah μHD was employed to observe the fiber orientation on a series of X-ray CT images. The reconstruction of fiber slice images on whole angle was carried out by using the VG Studio max 2.2 software (Volume Graphics GmbH, Heidelberg, Germany), and 3D internal structure of samples was obtained. The calculation of fiber orientation in each specimen was automatically carried out by a calculation software which cooperated with software VG.Studio max 2.2. This method can reduce calculation time compared with the method in our previous research [7, 8].

NUMERICAL

The numerical algorithm for predicting the fiber orientations described earlier was incorporated into the previously simulation program of mold filling for injection molding Moldflow. The injection molding process is simulated using the finite element solver of Moldflow software. In the flow analysis, the non-Newtonian and the Hele-Shaw flow were assumed, and triangle shell element was used. The shear rate distribution was calculated in these assumptions. The fiber orientation distribution was updated based on flow fields at each time step. The effect of fiber orientation upon flow fields is neglected.

The computer model was created has the same dimension as experimental mold cavity and the molding conditions were also simulated as same molding conditions as experimental conditions.

Simplified deformation behavior of fountain flow [3] was adopted for fiber orientation in the flow front region, as shown in Fig. 1. The relationship between z_1 and z_2 for a power-law fluid, and between elementary deformation dz_1 and dz_2, dx_1 and dx_2 can be described as follows:

$$\begin{gathered} z_1\left[1-(z_1/b)^{\frac{n+1}{n}}\right]=z_2\left[1-(z_2/b)^{\frac{n+1}{n}}\right] \\ \left[1-(2+1/n)(z_1/b)^{\frac{n+1}{n}}\right]dz_1=\left[1-(2+1/n)(z_2/b)^{\frac{n+1}{n}}\right]dz_2 \\ \frac{dx_1}{1-(2+1/n)(z_1/b)^{\frac{n+1}{n}}}=\frac{dx_2}{1-(2+1/n)(z_2/b)^{\frac{n+1}{n}}}. \end{gathered} \tag{4}$$

Then change of fiber orientation in fountain flow region can be calculated. Herein, the initial conditions in level 1 (z_{10}) was set up using experimental results.

RESULT AND DISCUSSION

As mentioned in Numerical section, the initial conditions in level 1 was set up using experimental results. Fig.2 shows the core-layer thickness (Tcore) diagrams as a function of mold thickness at center position of all GF/PA6 parts, which was produced with the fiber contents ranging from 30wt.% to 65wt.% using plate-shaped cavities having thicknesses ranging from 2mm to 4mm.

The initial conditions in level 1 (z_{10}) is equal to (Tcore/2) in a half-mold where normalized thickness is from 0 to 1. Using this condition, the fiber orientation in fountain flow region (level 2) can be calculated. Herein, 3D initial condition of fiber orientation is also set based on experimental data at (Tcore/2) layer. This calculation and the numerical algorithm of the combined model was incorporated into simulation program of mold filling. Fiber orientation a_{11} and a_{33} in fountain flow region was calculated.

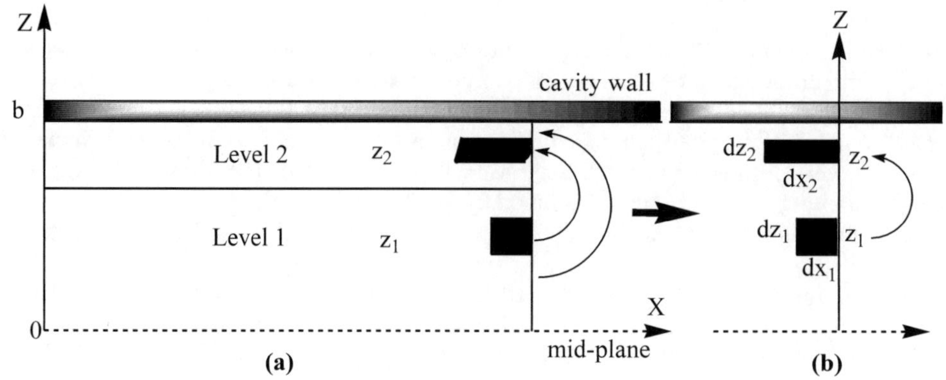

FIGURE 1. Simplified scheme of the fountain flow region (a), an elementary deformation in the fountain flow region (b).

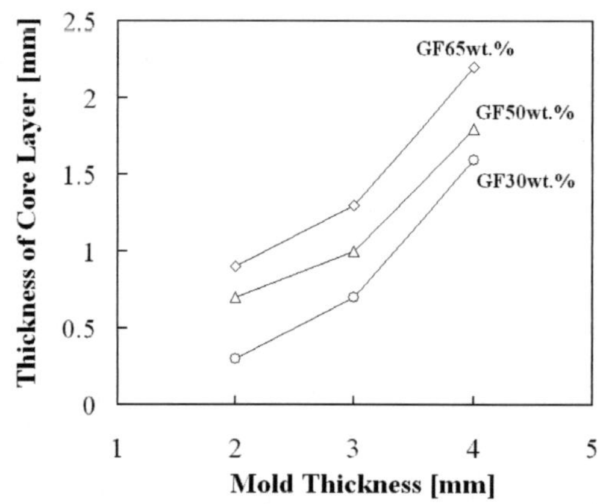

FIGURE 2. The core-layer thickness diagrams as a function of mold thickness at center position of all GF/PA6 parts, which was produced with the fiber contents ranging from 30wt.% to 65wt.% using plate-shaped cavities having thicknesses ranging from 2mm to 4mm.

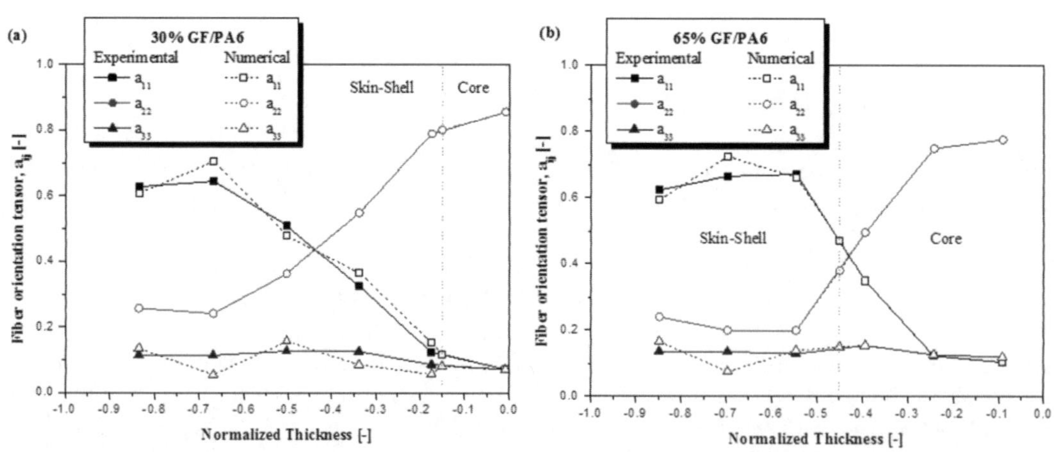

FIGURE 3. Comparion the numerical and experimental results in injection-molded plate 30wt.% (a), and 65wt.% GF/PA6 parts (b) fabricated using 2mm thickness plate-shaped cavity in the skin-shell region.

The fiber orientation calculation started from the gate, and a random orientation is applied at that location as the inlet boundary condition. Then comparing predicted and experimental fiber orientation distributions will lead to know whether numerical method is effective and reliable. Herein, C_M describes the fiber-matrix interaction (use $C_M = 0$), while paramater α can slow down the rate of fiber orientation development ($\alpha = 0.7$) [6]. Numerical algorithm of this combination model for predicting the 3D fiber orientation in injection-molded parts is detailed expanded and then incorporated into the simulation program of mold filling for plastic injection molding Moldflow. As mentioned earlier, there is a possibility to predict more accurately the fiber orientation distribution using combined model.

Fiber orientation a_{11} and a_{33} in fountain flow region was calculated using simplified deformation behavior of fountain flow. Fig.3 compares the numerical and experimental results in injection-molded plate 30 wt% (a), and 65 wt% GF/PA6 parts (b) fabricated using 2mm thickness plate-shaped cavity for C_I value of 0.0112 in the skin-shell region. The results in Fig.3 indicated a reasonable agreement between numerical prediction and experimental validation using combined model and simplified deformation of fountain flow. However, the initial conditions in level 1 (z_{10}) still need to be set by experimental results.

Overall, there is a possibility to predict more accurately the fiber orientation distribution using combined model. And 3D simulation is needed to predict detailed the 3D fiber orientation in injection-molded part.

CONCLUSION

The 3D fiber orientation distribution in skin-shell region is predicted by combining our developed fiber interaction model and iARD-RPR model and simplifying the deformation behavior of fountain flow. In experimental, short-glass fiber-reinforced polyamide 6 (GF/PA6) specimens produced using plate-shaped cavity having three different thicknesses ranging from 2mm to 4mm and with the fiber contents ranging from 30wt.% to 65wt.% are carried out using injection molding. 3D fiber orientation observation and measuring are exactly and automatically examined by Micro-computed tomography system Y.Cheetah μHD. Comparisons with experimental results, it is showed that this advanced prediction methodology can be effectively used for the 3D fiber orientation prediction in high filler content composite parts. However, the initial conditions in level 1 (z_{10}) still need to be set by experimental results.

REFERENCES

1. S.M. Mazahir, G.M. Velez-Garcia, P. Wapperom, and D. Baird, *J. Non-Newto. Fluid* **216**, 31-44 (2015).
2. S. G. Advani, "Molding Short-Fiber Composites, Flow Processing", in *Wiley Encyclopedia of Composites*, Second Edition, edited by L. Nicolais and A. Borzacchiello, John Wiley & Sons, Inc., 2012, pp. 1-12.
3. K-H. Han, and Y-T. Im, *Polym. Compos* **23**, 222-238 (2002).
4. T. B. Nguyen Thi, A. Yokoyama, K. Ota, K. Kodama, K. Yamashita, Y. Isogai, K. Furuichi, and C. Nonomura, "Numerical Approach of the Injection Molding Process of Fiber-reinforced Composite with Considering Fiber Orientation" in *The 29th International Conference of the Polymer Processing Society-2013*, edited by V. Altstadt, AIP Conference Proceeding **1593**, American Institute of Physics, Melville, NY, 2014, pp. 571-577.
5. T. B. Nguyen Thi, A. Yokoyama, S. Hamanaka, K. Yamashita, and C. Nonomura, "Numerical Simulation Of Fiber Interaction In Short-Fiber Injection-Molded Composite Using Different Cavity Geometries" in *31st International Conference of the Polymer Processing Society-2015*, edited by Byungohk Rhee, AIP Conference Proceeding **1713**, American Institute of Physics, Melville, NY, 2016, pp. 040006.
6. P. H. Foss, H. C. Tseng, J. Snawerdt, Y. J. Chang, W. H. Yang, and C. H. Hsu, *Polym. Compos* **35**, 671-680 (2014).
7. T. B. Nguyen Thi, M. Morioka, A. Yokoyama, S. Hamanaka, K. Yamashita, and C. Nonomura, *J. Mater. Process. Tech* **129**, 1-9 (2015).
8. T. B. Nguyen Thi, M. Morioka, A. Yokoyama, S. Hamanaka, K. Yamashita, and C. Nonomura, "Numerical prediction of fiber orientation in injection-molded short-fiber/thermoplastic composite parts with experimental validation" in *30th International Conference of the Polymer Processing Society-2014*, edited by Sadhan C. Jana, AIP Conference Proceeding **1664**, American Institute of Physics, Melville, NY, 2015, pp. 110011.

Effects of Environmental Aging in Polypropylene Obtained by Injection Molding

Rebeca S. Grecco Romano[1], Washington Luiz Oliani[1], Duclerc Fernandes Parra[1], Hélio Fernando Rodrigues Ferreto[1], Ademar Benevolo Lugao[1].

[1] *Nuclear and Energy Research Institute, IPEN-CNEN/SP, Av. Prof. Lineu Prestes, 2242 – Cidade Universitária – CEP 05508 – 000, São Paulo – SP – Brasil*
grecco.rebeca@usp.br

Abstract: UV light, heat, and pollutants can interact with Polypropylene (PP) molecules, mainly with the tertiary carbon producing free radicals which can react with oxygen producing changes in its properties. PP has outstanding chemical and physical properties, and a good processability at very low market price. In addition PP is extensively used for manufacturing various kinds of products, however due to its large scale consumption a lot of waste is generated at the end of their life cycle to the environment with low rate degradation. This study has the objective of comparing the environmental exposure of PP (spheres) neat and PP (spheres) irradiated with 20 kGy. Dumbbell samples were manufactured by injection molding and exposed to the environment during 90 days. Another one set was exposed to gamma irradiation at 20 kGy total dose and exposed at the same conditions. The samples were characterized by Mechanical Testing, Fourier Transform Infrared Spectroscopy (FTIR), Scanning Electron Microscopy (SEM), and Differential Scanning Calorimetry (DSC). The dumbbell samples PP 20 kGy, after environmental aging, showed oxidation with presence of surface cracks more intensely than the PP spheres neat.

Keywords: Polypropylene, Gamma Irradiation, Environmental Aging and Injection Molding

Introduction

Polypropylene (PP) is a polymer which offers a good balance of thermal, chemical and physical properties; plus low cost and processing facilities by injection molding, however it is necessary to deal with their low environmental factors resistance [1]. During exposure to weathering, such as UV, oxygen, pollution, humidity and heat, polypropylene undergoes changes in its main chain that lead to the start of the PP's oxidative degradation and later to the cracking and to lose mechanical properties. These occurrences are associated with the chain scission process [2-4]. The simulation of environmental conditions on the study of polymers degradation is a powerful tool to evaluate and predict the changes expected in real time life. The interaction of environmental factors with the PP influences the degradation and causes breaking in the bond of side groups, forming unsaturated molecules, peroxides, hydroxides and chain scission. Decrease in the elongation at break and tensile strength at break are caused by chain scission under UV exposure. [5-6]. The processing of polymers through ionizing radiation is frequently can be used to modify the molecular structure, mainly by crosslinking and chain scissions mechanism due to free radicals formation as a result of the interaction of gamma rays with polymers. The tertiary carbon present in PP molecules makes it susceptible to oxidation via macro radicals formation P • and consequent oxygen it reaction to a peroxide (represented by POO•) or hydroperoxides which in room temperature may form radicals such as hydroxyl and alkoxyl [7-9]. To explain the effects of gamma irradiation in the presence of oxygen as chain scission and loss of properties, some authors showed in their studies that after irradiation occurs the formation of radicals trapped in the amorphous phase they migrate to the interface between crystalline and amorphous region and react with oxygen and decompose. [10-11]. Previous studies have shown that in the case of PP irradiated at 30 kGy, and exposed to the environment, showed that the tensile strength decreases with time during the 60 days of exposure period, so that the study showed PP exposed to environmental degradation is accelerated by irradiation [12]. Another study conducted by Komatsu et.al showed that polypropylene with clay and irradiated at 12.5 kGy, when exposed to environmental aging for 12 months, showed large cracks and fissures [13]. Regarding the loss of mechanical properties in a study conducted by Oliani et. al samples of HMSPP (High Melt Strength PP) irradiated at 20 kGy were exposed to environmental aging for 120 days, initially presented maximum tensile values at break compared to PP, but after aging, it was found decreased values, and loss of maximum stress at rupture in HMSPP irradiated at 20 kGy [14]. In this work the process of injection molding was chosen because nowadays it is the most important method of polymers processing, and the most widely used in industry. Pieces manufactured by injection molding are present in several sectors like, automotive, medical equipment, and home appliances. The main advantages are minor index of defects induction and high processing velocity [15].

The aim of this work was to study the degradation of polypropylene spheres irradiated to 20 kGy using gamma radiation and compares the effect of environmental aging with PP spheres.

Materials and Methods

Materials

The isotactic polypropylene (iPP) spheres (neat) that were used are supplied by Braskem, suitable for injection with Melt Flow Index of 3.5 dg min^{-1}, density of 0.905 g cm^{-3}, Mw = 300,000 g mol^{-1}.

Methods

PP dumbbell samples (according to ASTM D-638-14 - type I) that were used in this study were manufactured by injection molding process. The PP was irradiated by gamma rays at dose of 20 kGy, in an irradiator with ^{60}Co source, at dose rate of 5 kGy h^{-1}.

Scanning Electron Microscopy

Specimens were coated with gold in a sputter coater prior to examination to avoid charging. The EDAX Philips XL30 SEM was used for collecting secondary electron images from the samples.

Differential Scanning Calorimetry

Assays were performed in equipment Mettler Toledo 822 DSC in a nitrogen atmosphere. The program was used: heating 50-280 °C for 10 °C min^{-1}, keeping at 280 °C for 5 minutes, cooling from 280 to -50 °C at a rate of -5 °C min^{-1} and reheating -50 to 280 °C at a rate of rewarming 10 °C min^{-1} under N_2 (50 mL min^{-1}). The samples were weighed (around 10 mg) and placed in aluminum crucibles, according ASTM D 3418-14.

Mechanical Tests

Tensile strength at break analyses were accomplished according to ASTM D-638-14, by using type I specimen, in Universal Instron machine-model 5567.

Fourier Transform Infrared Spectroscopy

The FTIR was carried out with a Thermo Nicolet 6700 FTIR spectrometer and Smart Orbit ATR accessory, in the wavelength range of 4000 to 400 cm^{-1}. In order to ensure satisfactory contact between the ATR diamond crystal and the sample, FTIR spectra were recorded at various regions on the sample for three times.

Environmental Aging

The samples were exposed in environmental aging device, according to ASTM D 1435-05. Geographical position: Latitude (23° 33° South) Longitude (46° 44° West) and Altitude (750 meters), and North position 45° position. Figure 1 shows the essay structure.

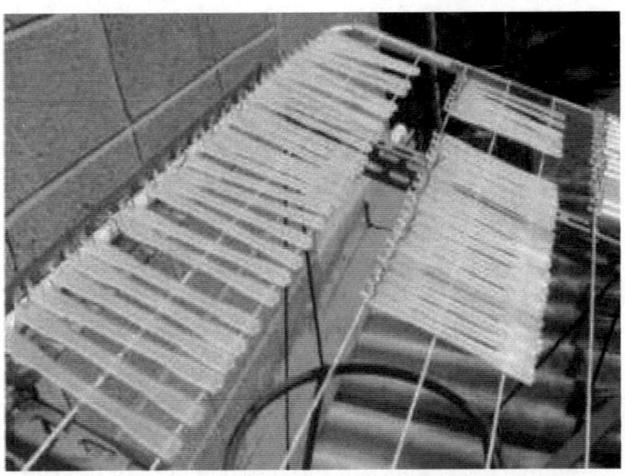

FIGURE 1. Device with dumbbell samples exposed outside for environmental ageing at the polymer processing laboratory IPEN.

Results

Scanning Electron Microscopy

Figure 2 shows the SEM results for the PP and irradiated PP to 20 kGy after environmental aging for 30, 60 and 90 days.

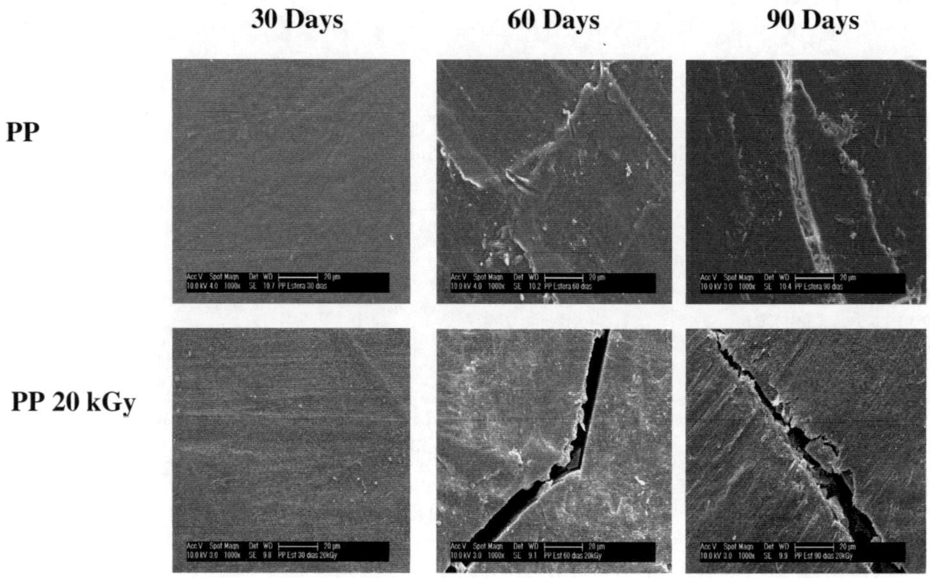

FIGURE 2. Photomicrographs obtained by SEM enlarge 1000x times for PP and PP 20 kGy.

All samples were exposed to same environment conditions such as UV rays, heat, humidity, pollutants and oxygen. When PP is exposed to those factors, changes in its structure occurs, attributed to the phenomenon of chain scission. Visible cracks and fissures can be observed in Figure 2, starting at 60 days of aging for irradiated PP and increases with the aging time, while non-irradiated PP shows visible fissures only at 90 days of exposure to environmental aging. The cracks observed on Figure 2 can be explained by the contraction of oxidized layers on the surface of the samples. The contraction may have been created by the increased crystallinity of the surface layers, consequence of free radicals formation under gamma irradiation and environmental aging with UV rays and atmospheric oxygen. [16-18].

Differential Scanning Calorimetry

The temperatures of second melting (Tm_2) and the degree of crystanillity (X_c) are presented Figure 3 and Table 1.

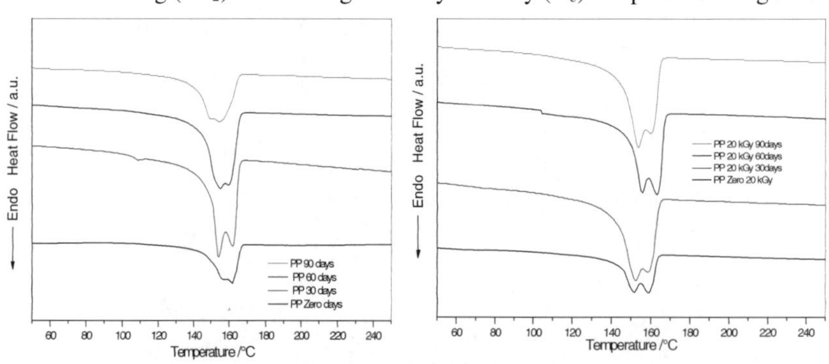

FIGURE 3. DSC curves of samples PP (A) and PP 20 kGy (B) under environmental aging.

TABLE (1) - Melting temperature of samples and the degree of crystanillity.

	T_{m2} / °C (±0.1%)				X_C / % (±0.5%)			
Samples	zero	30 days	60 days	90 days	zero	30 days	60 days	90 days
PP	161.6	154.5	161.4	154.5	41.5	47.7	47.5	44.5
PP 20 kGy	158.9	152.1	161.4	152.6	44.9	48.1	41.4	48.2

The DSC results show that at zero time the melting temperature of the PP 20 kGy is lower than the non-irradiated PP. The displacement the T*m2* at lower temperatures also occurs at the end 90 days for PP and PP 20 kGy of environmental aging. Previous studies showed that this behavior is consequence of oxidative degradation of PP, when is exposed to gamma irradiation, in air atmosphere, and environmental parameters such as UV rays, temperature, pollutants and oxygen which promotes chain scission reactions. Table 1 shows increasing of crystallinity after exposure it can be justified by small crystals formation (phenomenon chemicrystallization), generated by chain scission [19-20].

Mechanical Tests

The mechanical testing is a good method to evaluate the changes in properties and analyze structural modifications from degradation such crosslinking and chain scission. Each result of Tensile Strength at break was an average of eight measurements. Figure 4 show the results of tensile strength at break versus tensile strain at break.

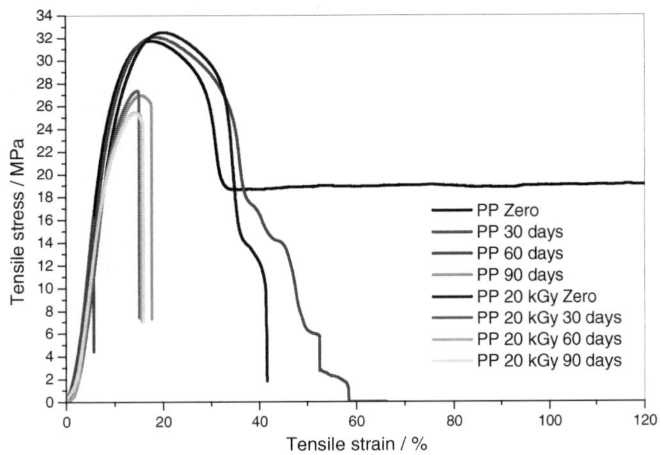

FIGURE 4. Results of tensile strength of samples PP and PP 20 kGy after environmental aging.

It was observed that from 60 days PP, has a low tensile strength at break indicating an influence of weathering at the samples. On the other hand, after 30 days of exposure, occurs a decrease of tensile strength at break in PP 20 kGy. In Figure 4, is shown that after gamma irradiation and environmental aging, the early loss of mechanical properties for PP 20 kGy was higher than PP non-irradiated. Mechanical damage can be justified by chain scission and crosslinking formation caused by combination of gamma irradiation and UV rays [21].

Fourier Transform Infrared Spectroscopy

FTIR spectra of the samples in the environmental aging, Figure 5. FTIR-ATR was used here into identify any functional group that indicating one sign of chain degradation

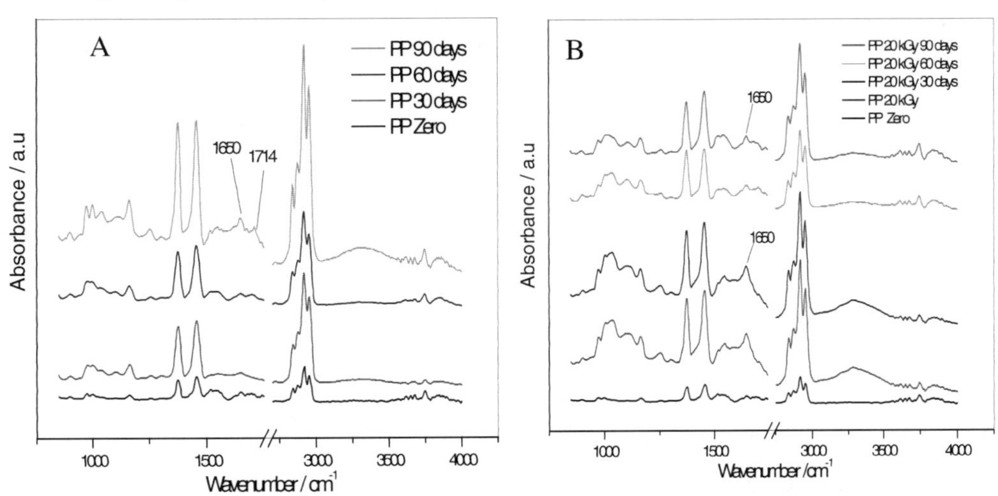

FIGURE 5. Infrared spectra PP (A) and PP 20 kGy (B)

During environmental aging and gamma radiation, PP undergoes oxidative degradation. Carbonyl groups could be found as result of oxidation. Figure 5A shows that PP with 90 days under environmental aging, presented peaks around 1714 cm-1 which correspond to carbonyl groups (C=O) and 1650 cm^{-1} which corresponds to unsatured groups (C=C). In Figure 5B was possible to note that the bands of unsatured groups started after 60 days of environmental aging. In Figure 5B at 90 days of exposure, it was observed that the intensity of unsaturated groups at the same time that the carbonyl groups decreases begin to be evidenced report previously [22].

Conclusion

The PP irradiated at 20 kGy suffered more intense degradation than PP non-irradiated. The mechanism degradation was justified by the displacement of second melting temperature, fissures and cracks on surface, loss mechanical properties and the presence of unsatured and carbonyl groups. All these results showed that gamma irradiation can accelerate the degradation allowing a faster elimination of PP wastes from the environment.

ACKNOWLEDGMENTS

The authors acknowledge CAPES for supported the Project Electronuclear 012/2013, Centre of Science and Technology of Materials – CCTM-IPEN, for microscopy analysis (SEM), the technician Eleosmar Gasparin for technical support and multipurpose gamma irradiation facility at the CTR-IPEN.

REFERENCES

1. J. Karger-Kocsis, *"Polypropylene Polymer Science and Technology"*, R. Ziegler-Natta catalysis and propylene polymerization, edited by, Dordrecht, Kluwer Academic Publishers, 1999, pp.897-919.
2. N. Wanasekara, V. Chalivendra, P. Calvert, *Polym. Degrad. Stab.,* 2011, pp. 1-6.
3. S. Al Malaika, *Inter. Mat. Rev.*, 2003, pp.165-183.
4. Qui, W., Endo, T., Hirostsu, T, *Euro. Polym. J.*, 2005, pp.1979-1984.
5. Singh, B, Sharma, N, *Polym. Degrad. Stab.,* 2008, pp.561- 584.
6. A. Tidjani, *J. Appl. Polym. Sci.*, 1996, pp.2497-2503.
7. Parparita. E, Zaharescu. T, Darie. R.N, Vasile. C, *Indus. Eng. Chem.*, 2015, pp.2404-2413.
8. B. Keene, M. Bourham, V. Viswanath, H. Avci, R. Kotek, *J. Appl. Polym. Sci,* 2014, pp.1-10.
9. Hans Zweifel, "Principles of oxidative degradation," in *"Stabilization of Polymeric Materials,"* edited by Berlin, Springer-Verlag, 1998, pp.1-156.
10. A. B. Lugao, "Estudos da Síntese por Irradiação, da Estrutura e do Mecanismo de Formação de Polipropileno com Alta resistência do Fundido," (PhD thesis, São Paulo University, 2004), pp.95-147
11. A. Rivaton, D. Lalande, J. L. Gardette, *Nucl. Instrum. Meth. Phys. Res.,* 2004, pp.187-200.
12. F. Yoshi, G. Meligi, T. Sasaki, K. Makuiuchi, A. M. Rabie, *Polym. Degrad. Stab.*, 1995, pp.315-321.
13. L. G. H. Komatsu, W. L. Oliani, D. F. Parra, A. B. Lugao, *Rad. Phys. Chem.,* 2016, pp.11-15.
14. W. L. Oliani, D. F. Parra, A. B. Lugao, *Rad. Physics. Chem.,* 2010, pp.383-387.
15. Z. Tadmor, G.Costa, "Molding," in, *Principles of polymer processing* edited by, New Jersey, Wiley 2nd Edition, 2006, pp.753-820.
16. J. Kotek, I. Kelnar, J. Baldrian, M. Raab, *Euro. Polym. J.*, 2004, pp. 2731- 2738.
17. W. L. Oliani, D. F. Parra, A. B. Lugao, *Rad. Phys. Chem.*, 2010, pp.325-328.
18. C. Albano, J. Reyes, M. Ichazo, J. Gonzalez, M. Britob, D. Moronta, M.S Rabello, J.R White, *Polymer*, 1997, pp.6379-6387.
19. G. Meligi, F.Yoshii, T. Sasaki, K. Makuuchi, A.M. Rabie, S. Nishimoto, *Polym. Degrad. Stab.*, 1995, pp.323-327.
20. I. Krupa, A.S Luyt. *Polym. Degrad. Stab.*, 2001, pp.505-508.
21. W. L. Oliani, D. F. Parra, L.F.C.P. Lima A. B. Lugao., Polym. Bull., 2012, pp.2121-2130.
22. C. Deshmane, Q. Yuan, R.S Perkins, R.D.K. Misra, *Mat. Sci. Eng.*, 2007, pp.150-157.

A Study on Manufacturing Thermoplastic CFRP Products Using Injection/compression Molding Process

Euichul Jeong[a,b], Jongsun Kim[b], Kyunghwan Yoon[a], Sangyong Lee[b], Jungeon Son[b], and Sunghee Lee[b*]

[a]*Department of Mechanical Engineering, Dankook University, (152, Jukjeon-ro, Suji-gu Yongin-si, Gyeonggi-do)*
[b]*Mold and Dies Technology Group, Korea Institute of Industrial Technology (397 Seokcheon-ro, Ojeong-gu, Bucheon-si, Gyeonggi-do)*

euicha0817@kitech.re.kr
birdlee@kitech.re.kr

Abstract. In this study, the production of carbon fiber reinforced thermoplastics(CFRTP) using an injection/compression molding process that differed from the conventionally used fabrication methods was investigated. Before the application of composite molding in the injection/compression molding process, a simple compression molding experiment was performed using a hydraulic press to determine the base material impregnation and to obtain basic physical property data for the CFRTP. Based on these results, injection/compression molded products were produced, and an additional insert/over molding process was applied to improve the impregnation rate of the molded products. The results demonstrated that the tensile strength of the molded product prepared using the faster injection/compression process was similar to that of a hydraulic press molded product.

Keywords: CFRP, Thermoplastic, Injection/compression molding
PACS: 81.05.Qk (Reinforced polymer-based composites)

1. INTRODUCTION

Today, with the segmentation and advancement of industries, energy exhaustion and environmental problem have been increasing worldwide. Thus, restrictions on carbon dioxide emissions have been strengthened in developed countries and measures to prevent global warming have been taken, leading to push for the development of eco-friendly and lightweight materials.[(1)] Recently, technological development of composite materials has been achieved using plastics with outstanding processability. In particular, because carbon fiber reinforced composite materials with excellent specific strength and specific stiffness have been highlighted as the new generation of lightweight materials, many studies on fabrication methods and strength evaluation have been conducted.[(2~4)] Based on the type of matrix being used, carbon fiber reinforced composite materials can be categorized as carbon fiber reinforced thermosets(CFRSP) or thermoplastics(CFRTP). Although a CFRTP exhibits excellent productivity compared with CFRSP, impregnation of the matrix into the carbon fibers is difficult because of the high melting temperature and viscosity of the matrix. Accordingly, the fabrication of CFRTP is difficult. To overcome this problem, many studies on the film impregnation process and co-mingled fiber impregnation process have been conducted in an effort to achieve easier impregnation of the thermoplastic matrix into the reinforcement.[(5)] In this paper, the fabrication of a CFRTP using continuous fiber with the application of an injection molding process with excellent productivity was investigated. This process differed from the conventional CFRTP fabrication method.

2. EXPERIMENT

2.1 Simple Compression Molding Experiment Using Hydraulic Press

The most important parameter in CFRTP molding is the impregnation rate of the thermoplastic matrix.[(6)] To determine basic physical property data for the CFRTP and the impregnation rate of a matrix, a simple compression molding experiment using a hydraulic press was performed. Unlike the molding method using a conventional polymer film, the plastic pellet utilized in injection molding was used as the matrix.

The structure of the mold used in press equipment is shown Fig. 1(a), and the molding experiment was performed after lamination using the hand lay-up method with the structure shown in Fig. 1(b). CF3327(plain fabrics) from Hankuk Carbon was used as the reinforcement, and PP(M1850) from LG Chemicals was used as the matrix. The molding result is shown in Fig. 1(c). Scanning electron microscopy(SEM) measurements performed on the cross section of the product were used to determine the impregnation of the matrix into the reinforcement, as illustrated Fig. 1(d). Tensile tests were conducted to determine the tensile strength of the molded products, and the results are presented in Fig. 2.

Two methods used to predict the stiffness and strength using the volume of a composite material are the Voigt and Reuss models, which are described in equation (1):[7]

$$\text{Voigt}: M_{composite} = v_f M_f + v_m M_m, \quad \text{Reuss}: \frac{1}{M_{composite}} = \frac{v_f}{M_f} + \frac{v_m}{M_m} \tag{1}$$

$$v_f = \frac{(A_f - A_p) \cdot t_f}{V_{composite}}$$

$$(M: Physical\ property,\ v: volume\ fraction,\ f: reinforcement,\ m: matrix,$$
$$V: volume,\ A: area,\ p: porosity,\ t: thickness\ of\ layers)$$

The tensile strength of a composite material increases as the volume fraction of the reinforcement increases according to the increased number of carbon fabric layers. The tensile strength of the PP composites increased by 3.0, 5.0, and 6.4 times, respectively, when 2, 3, and 4 carbon fabric layers were inserted compared with the virgin PP resin tensile strength of 28.0 MPa. However, it is difficult to predict the exact behavior of the composite material using the Voigt and Reuss models because the arrangement of the reinforcement and matrix in a particular direction is assumed.

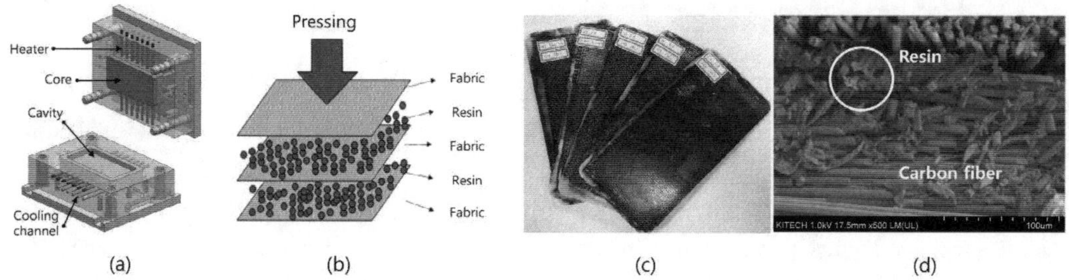

FIGURE 1. (a) Designed mold system, (b) impregnation structure of pellet, (c) formed PP composite specimens, and (d) cross sectional SEM image

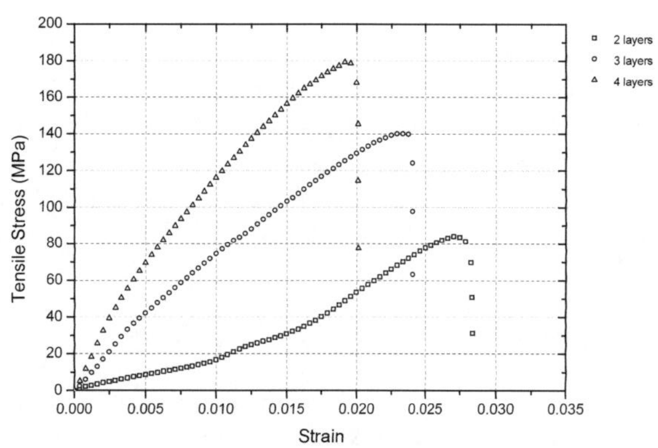

Results			
Number of carbon fabric layers	2	3	4
Thickness of specimen (mm)	2.02	2.12	2.16
Maximum tensile stress (MPa)	84.28	140.3	179.9
Fiber volume fraction	0.256	0.367	0.480

FIGURE 2. Stress-strain curve of PP composites as a function of number of carbon fabric layers

2.2 Molding Experiment Using Injection/compression Molding

When a carbon fiber fabric is used as a reinforcement, impregnation of the matrix occurs through the porosities between the woven carbon fibers. Because this process is similar to the existing micro pattern molding process, the composite material was molded by applying the injection/compression molding process, which is one of micro pattern molding process.[8] The structure of the mold used in the injection/compression molding is shown in Fig. 3(a). An advantage of the injection/compression molding process is that it is easier to impregnate a matrix into the reinforcement compared with the injection molding process during the filling stage because the gap between molds is larger and involves an additional compression process. The composite material molding mold is shown in Fig. 3(b). The product is compressed with advancing core after the filling stage in a slightly open mold with a certain interval. In the composite material molding process, even if including the time until the fabric is inserted into the mold, the product molding was possible within 5 min.

Using the injection/compression molding process, although some resin was impregnated on the surface of the molded product, the carbon fiber fabric was exposed on the surface of the molded product, as observed in Fig. 4 because the impregnation rate was low. This phenomenon is likely caused by the formed solidified resin layer in the porosities between woven carbon fibers.

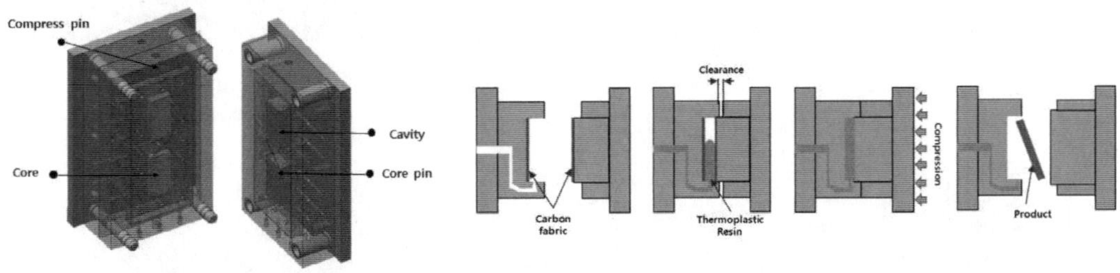

FIGURE 3. (a) Designed mold and (b) schematic of injection/compression molding process

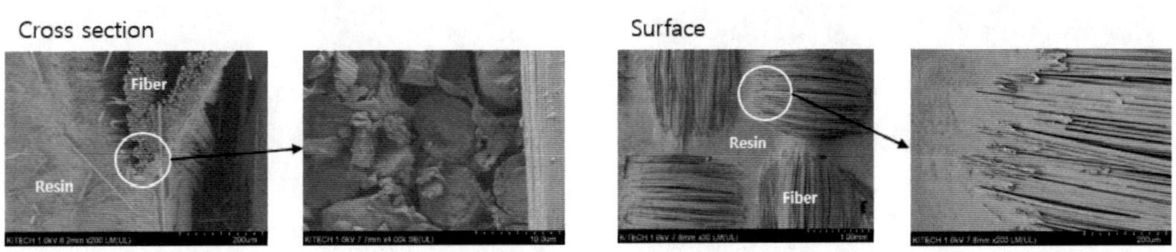

FIGURE 4. SEM images of PP composite for the injection/compression molded specimens

2.3 Improvement of Injection/compression Molded Product

For the molded products prepared using the injection/compression molding process, the insert/over molding process was additionally applied to improve the low impregnation rate on the mold wall. Two molded products were manufactured from one carbon fiber fabric layer using the injection/compression molding process, as illustrated in Fig. 5. The first molded product was inserted into the mold again such that the part with a low impregnation rate could be placed at the center of the mold, and a molded product with two fabric layers was manufactured. SEM images of the surfaces and cross sections of the molded products are presented in Fig. 6, illustrating that the resin filling between two of the first molded products was impregnated between carbon fibers.

The molded products from both the injection/compression process and insert/over molding process are shown in Fig. 7(a). The tensile strengths of the molded products are compared in Fig. 7(b). An improvement in the tensile strength with increasing matrix impregnation rate was confirmed. The tensile strength of the molded product prepared using the insert/over molding method was approximately 79.9 MPa, which is similar to the tensile strength value of the hydraulic press molded product with the insertion of two carbon fiber fabric layers (84.6 MPa).

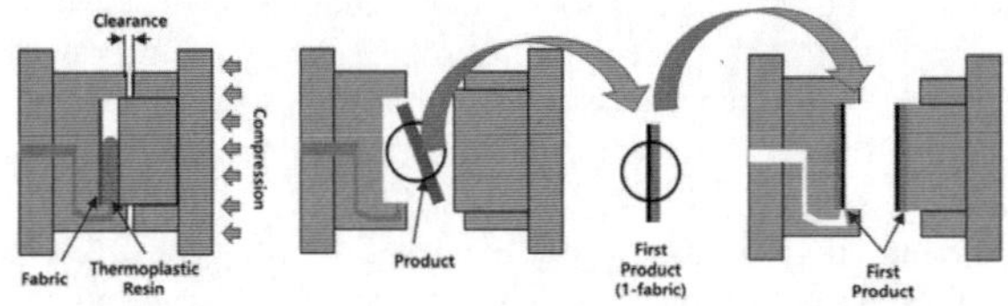

FIGURE 5. Schematic of insert/over molding process

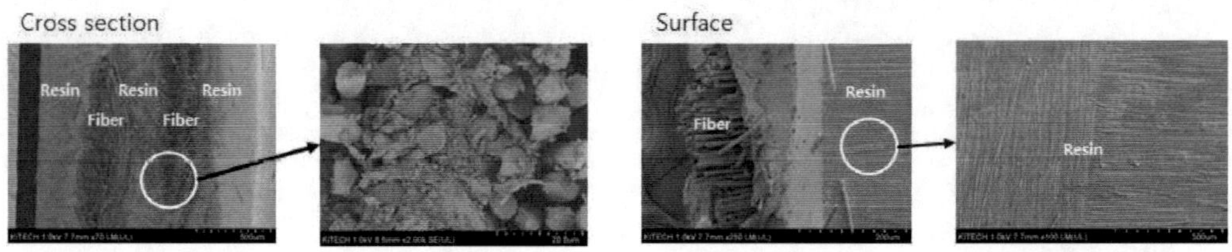

FIGURE 6. SEM images of PP composite for the insert/over molded specimens

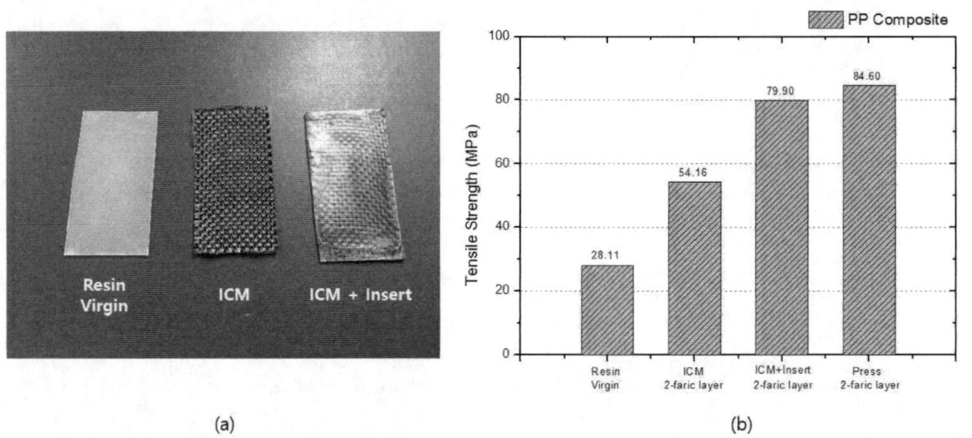

FIGURE 7. (a) Comparison of molded products and (b) tensile strengths of products prepared using different processes

3. CONCLUSIONS

In this study, the fabrication of carbon fiber reinforced thermoplastic composites using hydraulic press molding and injection/compression molding was investigated. The tensile strength of PP composites prepared using hydraulic press molding increased by 3.0-6.4 times depending on the number of carbon fiber fabric layers inserted compared with the base material. For the injection/compression molding process, although PP composites with two fabric layers were produced, a lower tensile strength of 54.17 MPa was obtained compared with the tensile strength of the hydraulic press molded PP composite (84.69 MPa) because of the low impregnation rate of the base material. To improve the impregnation rate, an insert/over molding process was additionally applied. The resulting PP composite exhibited a tensile strength of 79.90 MPa, which is close to that of the hydraulic press molded PP composite (84.69 MPa); in addition, the production time was shortened using the insert/over molding process.

4. ACKNOWLEDGMENTS

This study was conducted with support from the RCOE project and RCMS project of the Korea Institute of Industrial Technology (EO160030, KM150120).

5. REFERENCES

1. G. S. Cole, and A. M. Sherman, "Light weight materials for automotive applications", *Mater. Charact*, 1995, Vol. 35, pp. 3.
2. N. J. Johnston, and P. M. Hergenrother, "High Performance Thermoplastic: A Review of Neat Resin and Composite Properties", *SAMPE symp, Proc,* 32nd Int, 1987, pp. 1400.
3. I. J. Shin, and D. J. Lee, "Formability of Thermoplastic Laminar Composite depending on the Types of Fabric", *Trans. Korean Soc. Mech. Eng. A*, 2003, Vol. 27, No. 8, pp. 1338-2003.
4. J. W. Kim, and D. G. Lee, "Effect of Fiber Orientation and Fiber Contents on the Tensile Strength in Fiber-reinforced Thermoplastic Composites", *J. Korean Soc. Compos. Mater.,* 2007, Vol. 20, No. 5, pp. 13-19.
5. S. T. Kim, "Development Trend for Thermoplastic Polymer Composite", *Int. Polymer Sci. Tech.,* 2013, Vol. 24, No. 1, pp. 25-28.
6. S. H. Lim, "Thermoplastic Resin Composites", *Int. Polymer Sci. Tech.,* 1995, Vol. 6, No. 5, pp. 451-458.
7. W. H. Yoo, J. H. Lim, S. W. Kim, S. H. Kim, and D. S. Hwang, "Prediction of the Equivalent Elastic Properties of Fiber Reinforced Composite materials and Structural Analysis of Composite Satellite Panel", *Proceedings of The Korean Society for Aeronautical and Space Sciences Conference,* 2013, pp. 644-648.
8. S. W. Cho, K. H. Yoon, J. J. Kang, "An Experimental Study on the Shape and Replication Ratio of Micro Patterns of 7 inch LGP using Injection/Compression and RHCM", *Dankook University Master thesis*, 2010, pp. 13-14.

Development of a Highly Segmented Temperature Control in Injection Moulding for Reduced Warpage and Increased Process Stability

M. Theunissen[a], Ch. Hopmann[a], P. Nikoleizig and M. Schmitz[a*]

[a] *Institute of Plastics Processing (IKV) at RWTH Aachen University, Pontstraße 55, D-52062 Aachen, Germany*

[a]*Matthias.Theunissen@ikv.rwth-aachen.de*
[a*]*Mauritius.Schmitz@ikv.rwth-aachen.de*

Abstract. In the plastics processing industry a growing demand for precision moulded parts can be identified, especially regarding avoidance of warpage as well as achieving high process stability. To meet these requirements, an approach based on local material properties within the injection mould during the moulding process is pursued. The goal of the pvT-optimised process is the optimisation of the shrinkage by establishing a constant specific volume of the melt and part during the injection moulding process. This is pursued by controlling the holding pressure according to the in-mould pressure and part temperature on the basis of specific materials pvT-data. Due to the nature of flow dynamics and material characteristics the pressure within the mould is inhomogeneous. Especially in injection and compression phase, pressure varies from ambient pressure at the flow front to a high injection pressure caused by the pressure loss. In packing phase local specific volume varies due to pressure differences and temperature differences. To establish a local control of the specific volume, in addition to the injection pressure the temperature is added as a controlled variable. To achieve an accurate and dynamic temperature control, tempering concepts including fluid-based tempering, CO2-tempering, resistance heating as well as heating by laser radiation and electromagnetic induction are evaluated regarding usability, reaction time and performance. In addition, local temperature and pressure measurements are needed to accurately adapt the temperature in the mould. Therefore, available measurement techniques for injection moulding were evaluated regarding reaction time and efficiency. A mould concept is developed to meet the requirements needed for an inline, online as well as a self-optimising temperature control, based on a model-based measurement correction, predictive temperature control and a system-wide optimisation function.

Keywords: cavity temperature control, pvT optimisation, process stability
PACS: 05, 06, 07, 44, 89

INTRODUCTION

Injection moulding is one of the most important processes in melt-based production of plastic parts [1]. In injection moulding, due to fast cycle times, the mould has to be cooled in order to extract the heat, frequently introduced to the mould by the molten plastics [2]. This is commonly achieved by circulating water or oil through cooling channels. In this process the plastics undergoes high changes in temperature and pressure. Additionally the spatial distribution of temperature and pressure is inhomogeneous within the processing frame. Due to its compressible material behaviour, the specific volume of the plastic material changes drastically for an inhomogeneous distribution of specific volume. The result is a heterogeneous distribution of shrinkage potential, which leads to part warpage. The reduction of part warpage is the most challenging task in injection moulding and the most important requirement for manufacturing of dimensionally stable and precise parts. Common practice is the design of suitable cooling channels according to the local cooling demand. The aim of this technique is to achieve a homogeneous temperature distribution at ejection [2]. With consideration of the spatially and temporal distribution of temperature and pressure, resulting from the pressure loss over the flow path and varying melt temperatures (illustrated in figure 1) within the part, a static cooling channel layout cannot meet the requirement of homogeneous shrinkage and precise part production. Therefore a new approach is presented to handle this most important challenge in injection moulding. The approach is based on the assumption, that a homogeneous distribution of the specific volume within the part over a large process frame leads to a homogeneous shrinkage potential and therefore minimised warpage. The specific volume depends on the local temperature and pressure, which has to be controlled locally. As a first approach, a process control for the in-mould pressure has been developed at IKV to compensate process variations by adapting the holding pressure according to the materials pvT-behaviour [3]. Since it is not possible to apply pressure locally without changing the parts shape or surface, the temperature is taken as correcting variable. Following, the development of a mould concept including a segmented temperature control system and the development of an appropriate control model is pursued. In this mould concept, tempering zones will be arranged at the cavity surface to locally adapt the temperature. Therefore, the most suitable tempering technique regarding dynamics, efficiency, response time and dimensionally scalability has to be found to enable an accurate temperature control. Additionally a suitable sensor technique regarding response time, accuracy, robustness and dimensional size has to be chosen to accurately measure the current process values and adapt the temperature.

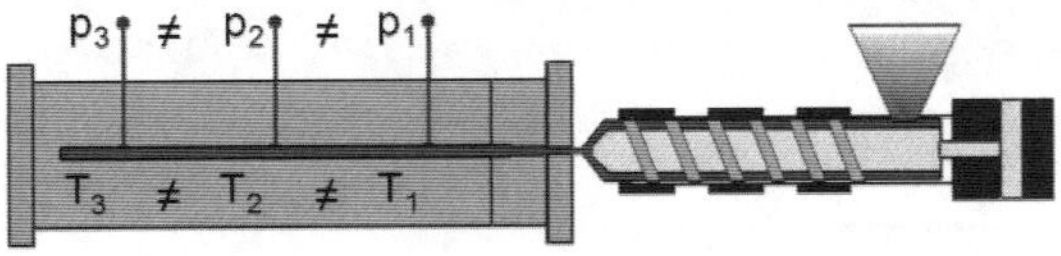

FIGURE 1. Illustration of the locally varying process properties in injection moulding.

PRELIMINARY INVESTIGATIONS

Studies in tempering techniques

With respect to the process and needs of the manufactured part, a set of requirements is defined regarding dimensional scalability, maximum achievable temperatures, heating rates and dynamics. In this approach a setup of at least 3x3 tempering zones per mould half is designed in order to establish a temperature profile of fourth order (including static tempering at all edges) over two axis of the parts surface. An exemplary temperature profile is illustrated in figure 2.

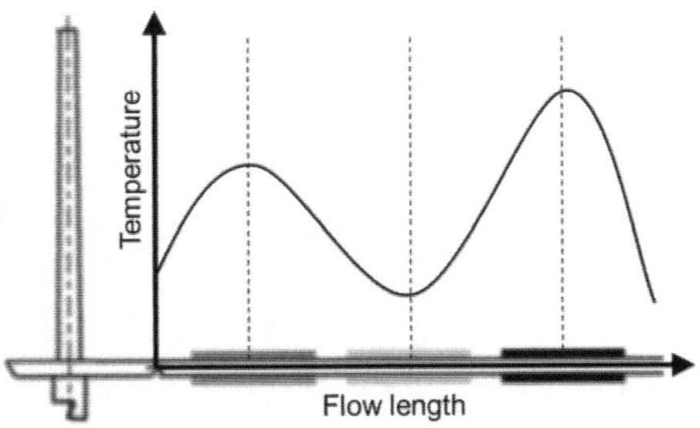

FIGURE 2. Exemplary temperature profile for three tempering zones.

For a part with a parting plane area of 170 mm x 170 mm, resulting by the available machine size at IKV injection moulding laboratory, this consideration leads to a maximum diameter of 50 mm per tempering zone. By use of these dimensions, standardised rapid heating ceramics as well as standardised sensors can be used. Nevertheless, the evaluated tempering techniques should be able to be further scaled down to increase the amount of controllable tempering zones and the resolution of the local temperature control. To effectively be able to influence the local temperature, the local pressure and temperature has to be measured. This results in another geometric restriction for the tempering elements due to the positioning of a combined pressure-temperature sensor.

Known tempering techniques are evaluated including liquid tempering, CO_2 cooling, heating cartridges, rapid heating ceramics, inductive heating and laser radiation [4,5,6]. Based on the geometric restrictions demonstrator elements are produced including the techniques of liquid tempering, rapid heating ceramics, CO_2 cooling and inductive heating (figure 3 (a)).

Additionally a steel plate is manufactured to hold these demonstrator elements and to represent the mould steel. The opposite face of the plate represents the cavity wall and is used to measure the resulting temperature distribution for various load cases via thermal imaging. Minimum, maximum and average temperatures are measured at the cavity surface in a central circular surface with a diameter of 50 mm. Additionally the temperature profile in radial direction is measured, represented by the horizontal line through the center of the test plate. The test setup is illustrated in figure 3 (b).

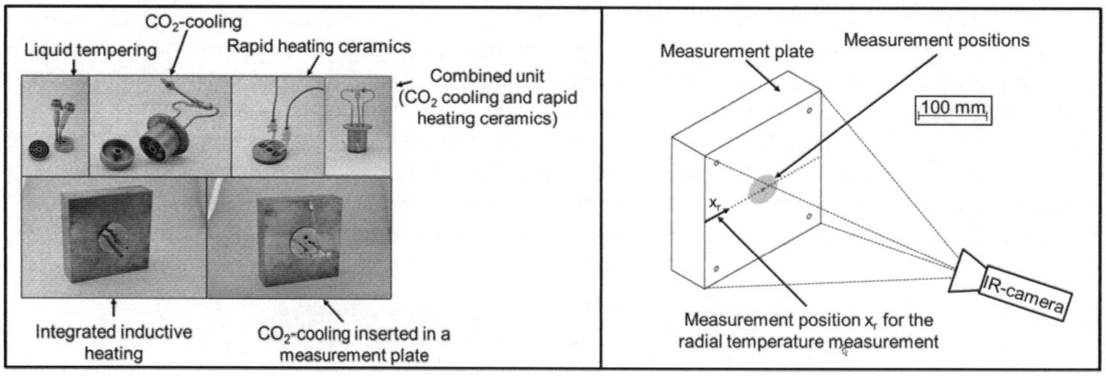

FIGURE 3. Demonstrator elements (a) and Measurement setup (b).

An exemplary comparison of a heating process via liquid tempering and rapid heating ceramics is illustrated in figure 4. Most important aspects are the achievable temperatures, the heating rates over the process time as well as the heating rates over the operating temperature. The comparison of water tempering (150 °C) and rapid heating ceramics (800 W, clocked to 80 % performance) illustrates that rapid heating ceramics are superior with respect to reaction time (figure 1 (a)), highest achievable temperature (figure 1 (a)) and absolute heating rates (figure 1 (b)). Additionally rapid heating ceramics keep high heating rates over a larger operating temperature (figure 1 (c)) and are therefore able to accurately control the temperature at a significant higher level.

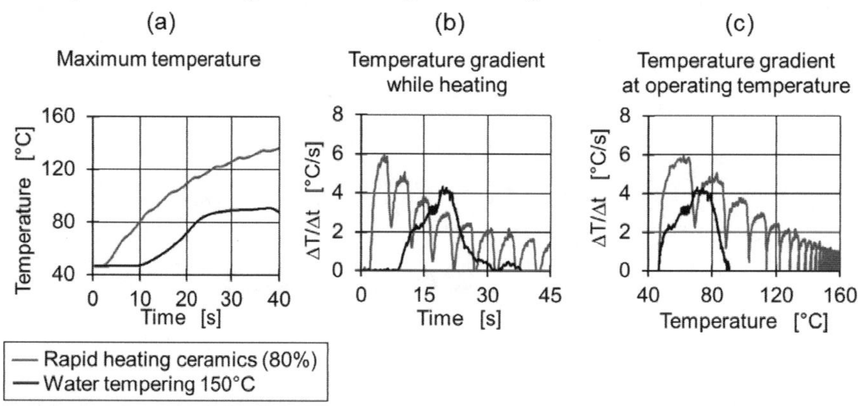

FIGURE 4. Comparison of heating process by liquid tempering and rapid heating ceramics.

These trials were performed for the manufactured different tempering elements and have pointed out, that the most suitable technique for tempering is a combined element, consisting of a rapid heating ceramic and a CO_2 cooling unit. These techniques are the most dynamic, have a low response time and are dimensionally scalable to very small sizes. Additionally ceramics allow a further reduction of the cavity wall thickness, due to their high pressure resistance, which results in an even more dynamic behaviour.

Studies in sensor techniques

As mentioned above, at every tempering zone local specific volume has to be controlled. Based on the local pressure and a prescribed specific volume profile, the local temperature is adapted. Therefore a local measurement of the pressure and temperature is necessary. Since a process control will be created to accurately adapt the local temperature, low response times and a high accuracy is needed for the measurement technique. Regarding pressure measurements, commonly used piezoelectric sensors match these requirements. Regarding temperature measurements, commonly used thermocouples as well as contact temperature sensors show low response times compared to common process control cycle times. This is mainly caused by the thermal diffusion process and the thermal mass of the sensor, which has to be heated. Investigations in different temperature measurement techniques were conducted. Additionally to thermocouple-based measurement techniques, ultrasonic temperature measurement and infrared temperature measurement were analysed. Via ultrasonic sensors, the temperature can be measured between an ultrasonic sender and receiver, because the speed of the ultrasonic waves is dependent on the temperature of the medium [7,8]. By measuring the speed of flight it is possible to calculate the average temperature. This is why this technique only delivers average values, which prevent an accurate temperature measurement over the part thickness. Via infrared temperature sensors, the heat radiation of the material passes through sapphire window and is converted into an electrical signal [9]. By this technique, temperatures in a defined depth can be measured with response times below 20 ms. Illustrated in figure 5 is a comparison between a contact temperature sensor and an infrared temperature sensor. Both sensors have a front diameter of 4 mm and are flush-mounted with the cavity wall.

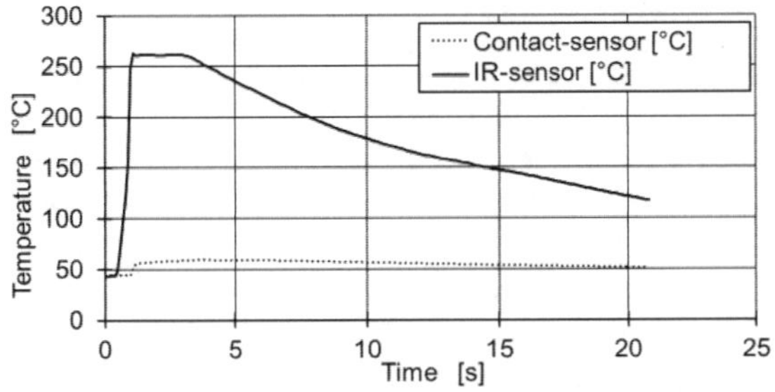

FIGURE 5. Comparison of contact temperature sensor and infrared temperature sensor.

The measurements reveal that the infrared sensor has a shorter response time, which can be seen at the earlier increase of the IR-sensor curve. The IR-sensor additionally is able to measure temperatures inside the melt due to the received heat radiation. Therefore a higher accuracy and a more defined temperature profile can be measured. This sensor type is further used in the developed mould concept.

DEVELOPED MOULD CONCEPT

Based on the derived measurements as well as the restrictions given for the tempering technique and the sensor technique, a mould concept is developed to integrate a segmented temperature control into an injection mould. Illustrated in figure 6 is the overview of the important components developed for this concept. To be able to perform a variety of different examinations, the mould concept relies on a modular approach. Therefore, the mould consists of a carrier mould, including a hot runner system, cooling channels for each edge of an insert, wiring channels for CO_2, power supply and sensor wires as well as general functionalities like the ejector system. The carrier mould enables the use of an exchangeable melt distribution geometry and standardised inserts, providing different cavity layouts as well as different arrangements of tempering elements, based on rapid heating ceramics, CO_2-cooling and IR-Sensors. To be able to insert nine tempering elements into the given space of the mould inserts, a layer principle is used to reduce the complexity of the involved part functionalities to a specific layer. Therefore, the front plate is mainly used to define the geometry of the part, the second layer provides the mounting of rapid heating ceramics and IR-sensors as well as the space for CO_2 evaporation. Between the second and third layer, the complex CO_2 channel layout is positioned, based on a half shell principle to allow an easy manufacturing process. The last layer finally provides the wiring channels for sensor wiring and power supply.

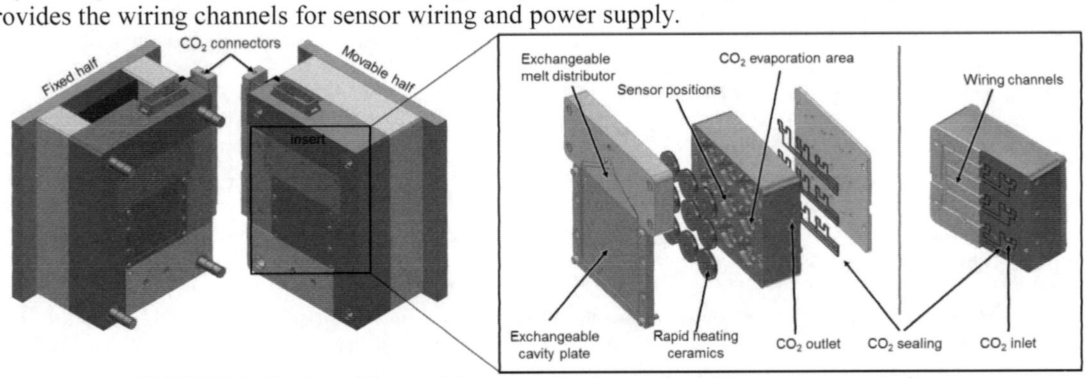

FIGURE 6. Design of the mould concept for the segmented temperature control.

Both halfs of the mould will be equipped with an insert including nine tempering zones, such that both sides of the part can be thermally controlled according to given quality criterion. The current quality criterion prescribes a homogeneous distribution of the specific volume over the whole injection moulding process.

PROCESS CONTROL STRATEGY

To accurately control the temperature of the part, both tempering system and control strategy have to be designed according to the needs of a highly dynamic and short cycle process. Although the design of the mould, especially the tempering system was designed to be most effective and dynamic, thermal processes involve a long response time due to thermal inertia of the material. To establish an online process control, a model predictive control approach will be chosen to overcome the issue of long response times. The approach relies on the assumption, that the thermal state can be controlled most accurately, if the thermal response of the system can be forecasted by an adequate model. A current approach involves a one dimensional finite difference simulation of the temperature distribution within the part, which will be adjusted frequently by measured data. This enables a deeper understanding of the process since an approximation of the current state of the system is known for every point in the process. This simulative approach performed for a prediction of one second for eighteen tempering zones in every control cycle. These calculations currently take less than 4 ms of calculation and open the opportunity for additional model-based refinements to increase the efficiency of the control technique.

CONCLUSION

Thermal processes have a comparably slow reaction time due to the nature of diffusive processes. Therefore investigations are performed regarding tempering techniques, sensor techniques and process control strategies to establish an accurate and dynamic tempering technique designed to locally adjust part properties according to a given target strategy. Only such a holistic approach allows the development of an in-situ temperature control, within the given restrictions of the injection moulding process like high temperature and pressure gradients and short cycle times. Tempering techniques and sensor techniques have been found to develop a suitable mould concept regarding reaction time, dynamics and measurement accuracy. Following investigations will focus on the effectiveness of the current control strategy as well as on further refinements of this highly extendable process control approach.

ACKNOWLEDGMENTS

All presented investigations were conducted in the context of the Collaborative Research Centre SFB1120 "Precision Melt Engineering", subproject B3 "Self-Optimizing Process Control Strategies for a Highly Segmented Temperature Control in Injection Moulding" at RWTH Aachen University and funded by the German Research Foundation (DFG). For the sponsorship and the support we wish to express our sincere gratitude.

REFERENCES

1. N.N., "*Branchenüberblick der Kunststoffbranche*" in *Plastics Europe*, 2006.
2. G. Menges, W. Michaeli and P. Mohren, "*Spritzgießwerkzeuge. Auslegung, Bau, Anwendung*", Munich, Vienna: Carl Hanser Verlag, 2007
3. W. Michaeli and A. Schreiber, "*Online Control of the Injection Moulding Process Based in Process Variables*" in Advances in Polymer Technology, 28, 2006, pp. 65-76.
4. H. Ridder, H.-P. Heim, "*Möglichkeiten und Grenzen variabler Werkzeugtemperierung*" in Kunststoffe 99, 2009, pp. 22-29.
5. M. Koch, „ Rezepte für die Werkzeugtemperierung" in Kunststoffe 92, 202, pp. 28-36.
6. M. Berghoff, „ Perspektiven bei der Temperierung von Problemzonen im Werkzeug", Iserlohner Kunststoff-Technologie GmbH, Iserlohn, 2010.
7. I. Alig, D. Lellinger and K. Wassum, "*Temperaturschwankungen in der Verfahrenseinheit*" in Kunststoffe 93, 2003, pp. 62-65.
8. B. Praher, K.Straka, G.Steinbichler, „ Schmelzetemperatur mit Ultraschall messen" in Kunststoffe 103, 2013, pp. 45-49.
9. FOS Messtechnik GmbH, „ *Perspektiven bei der Temperierung von Problemzonen im Werkzeug*", Technical data sheet, 2008.

Production of plastics parts with partially or fully metallic surfaces using the In-Mould-Metal-Spraying (IMMS)

Prof. Dr.-Ing. Ch. Hopmann[a], Dipl.-Ing. Matthias Theunissen[a,*], M.Sc. Sukran Katmer[a], M.Sc. Philipp Ochotta[a], Prof. Dr.-Ing. Kirsten Bobzin[b], M.Sc. Mehmet Öte[b], Dipl.-Ing. Thomas Frederik Linke[b], M.Sc. Xifang Liao[b]

[a] *Institute of Plastics Processing (IKV) at RWTH Aachen University, Pontstrasse 49, D-52062 Aachen, Germany*

[b] *Surface Engineering Institute (IOT) at RWTH Aachen University, Kackertstrasse 15, D-52072 Aachen, Germany*

matthias.theunissen@ikv.rwth-aachen.de

Abstract. Although various technologies are capable of producing plastics/metal hybrid parts for electrical applications, they all have disadvantages in common due to several manufacturing steps and limitations in productivity or the level of achievable geometrical part complexity. In addition, the material groups are often regarded separately, particularly when it comes to the materials-related forming processes. One approach to overcome these disadvantages, which is currently examined at the Institute of Plastics Processing (IKV) and the Surface Engineering Institute (IOT) at RWTH Aachen University within the scope of the Cluster of Excellence "Integrative Production Technology for High-Wage Countries", is to combine the plastics injection moulding and the thermal spraying of metals to create a new multi-integrated process with regard to the production of plastics/metal hybrid parts.
The In-Mould-Metal-Spraying (IMMS) is intended to enable the integration of metallic areas on plastics parts. In the first step a metallic layer is applied to specific areas of the mould's surface using the thermal spraying, which is immediately followed by the plastics injection. During this second step the metallic layer is transferred onto the plastics part. The plastics part including the transferred partially metallic surface is then removed from the mould.
A strong, permanent connection between the plastics and the metallic layer and an exact outline replication of the metallic layer in the pre-assigned areas are to be realized. To develop and improve the new technology the influence of different plastics types on the transferability of the metal layer onto the plastics part has been investigated systematically. Additional evaluations relating to the component design are to be made with the help of a more complex test part which possesses a structured surface. The results provide valuable information on the favourable process configuration and plastics/metal combination.

Keywords: Coating transfer, multi-component injection moulding, plastics/metal hybrid, thermal spraying
PACS: 81.20.Hy

INTRODUCTION

Plastics parts for electrical and electronic applications are getting more important in everyday life. The current trends of miniaturization and functionality of electronics present new challenges to manufacturing companies particularly in high-wage countries, which have the task to produce highly innovative mass products cost effectively and keep their competitors from emerging markets at bay. Hence an integrative approach, which combines the advantages of different materials and manufacturing processes within one part, is of increasing importance.
Within the scope of this challenges, the so-called In-Mould-Metal-Spraying (IMMS) has been developed, which combines the thermal spraying of metal coatings and the injection moulding into one integrative process. The metal layer is firstly applied onto the cavity insert of an injection mould using the so-called wire arc spraying process (see Figure 1). The assembled insert is then back-moulded with the plastics melt inside the closed mould. During this second step the metal layer is transferred onto the plastics part similarly to the In-Mould-Labelling process. The plastics part with the transferred metal surface is then removed from the mould. Thereby, the integration of thermal spraying and injection moulding provides the possibility to produce established metallised plastics parts more efficiently. Furthermore, electronic components which are conventionally produced out of massive metallic materials can be substituted by new plastics/metal hybrid parts.

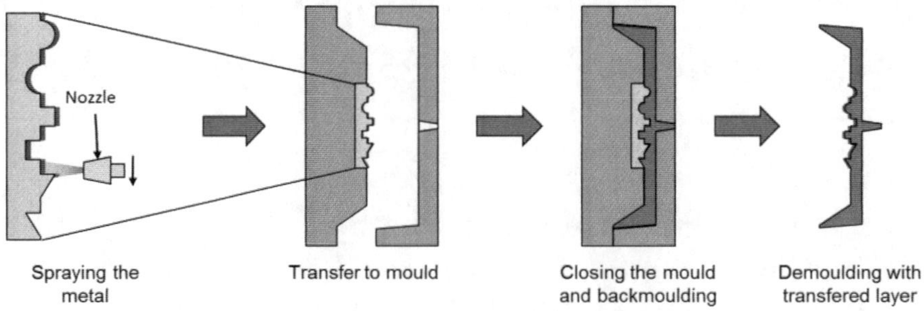

FIGURE 1. Principle of the in-Mould-Metal-Spraying process

In order to form a bond between two different materials such as plastics and metal, different bonding mechanisms are involved. in the case of IMMS, the forming bond cannot be described by a single adhesion model, but through a combination of different effects [1-3].

Basically, the mechanical and the specific adhesion are grouped under the adhesion theory. The mechanical adhesion is based on mechanical interlocking of the two bonding partners. The plastics melt can penetrate e.g. microscopic undercuts, pores or capillaries on the surface of the metal substrate and lock itself after solidification [4]. Specific adhesion as an adhesion phenomenon is based on chemical, physical and thermodynamic laws.

When looking at possible bonding mechanisms regarding the IMMS process, most attention should be paid to mechanical adhesion, the polarisation theory and the adsorption theory. While the influence of mechanical adhesion on the bonding could already be observed for IMMS parts [5, 6]. The influence of other adhesion phenomena is not fully understood yet. The usage of highly polar thermoplastic materials could have a positive influence on the bonding, therefore the wetting ability of the plastics melt on the metal substrate is positively influenced and thus the build-up of intermolecular interactions (hydrogen bonds) increased [7-9].

EXPERIMENTAL SETUP

To investigate the adhesion behaviour between the sprayed metal layer and the plastics part, contact angle measurements (CA measurements) based on DIN 55660-5 were performed with 6 different plastics types. These plastics types are an unfilled Polypropylene (PP) (579S) produced by SABIC Europe B.V., Sittard, Netherlands, a Polycarbonate (PC) (Makrolon 2205) of the Covestro AG, Leverkusen, Germany, an unfilled Polyamide 6 (PA6) (Durethan B30S), a PA6 with 30 % glass fibre content (Durethan BKV30), an unfilled Polybutylene terephthalate (PBT) (Pocan B1305) and a PBT with 25 % white pigments (Pocan B7375) produced by Lanxess AG, Cologne, Germany. Some properties of these materials are given in Table 1.

TABLE (1). Properties of the materials [10-12]

Parameters	**Unit**	**PP**	**PC**	**PA6**	**PA6 GF30**	**PBT**	**PBT WP25**
Melt volume rate, MVR	cm^3/10min	47	40.5	110	18.8	78	78
Yield Stress	MPa	35	65	80	180 (break)	55	55 (break)
Tensile modulus	MPa	1900	2400	3200	9800	2800	3500
Melting / Glass transition temperature, 10°C/min	°C		145	222	222	225	225
Density	kg/m^3	905	1190	1140	1360	1300	1560

The contact angle measurements were carried out using an OCA 20 measuring instrument of the DataPhysics Instruments GmbH, Filderstadt, Germany. In deviation from the DIN 55660-5, granules of the chosen materials were placed onto a steel plate coated with zinc, which was used as substrate material for the measurements. The adjusted measuring procedure is not suitable for the calculation of the surface energies of the bonding partners, but it allow an estimation of adhesion affinity [13]. The specimens were heated up inside a heating chamber until the melting temperature of the plastics was reached. A CCD camera was used for recognizing the shape of the melted granules. Nitrogen gas (N_2) was used to protect the surface of the plastics against oxidation for all plastics types.

The CA measurements were followed by the back-moulding of a sprayed zinc layer with the different plastics types mentioned in Table 1. Pure zinc was used as coating material and sprayed by wire arc spraying. The investigations were carried out using metal sheets as removable mould inserts to produce flat test specimen plates with the measurements 165 x 92 x 3 mm^3. The inserts were pre-treated via glass bead blasting. The surface roughness (Ra) of the coating is 9 – 12 μm, the coating thickness is approximately 165 μm.

The test specimen with the varying geometrical structures is shown in Figure 2. The structures consist of triangular, rectangular and octagonal prisms, spheres, squares and structures with thickness transitions. The structures are symmetric to the sprue gate (point-symmetry). Along the circumference the structure varies in shape, depth and flanging radius. In the figure, embossed and debossed features are identified with numbers and capital letters, respectively.

The used cavity inserts are made of tool steel 1.2312 (AISI P20+S). The pre-treatment of the inserts was done via glass bead blasting. After the application of the zinc layer the cavity inserts were back-moulded with an unfilled polyamide 6 Durethan B30S of the Lanxess Germany GmbH, Cologne.

The deposition and distribution of the zinc coating onto the geometrical structures was investigated via macroscopy.

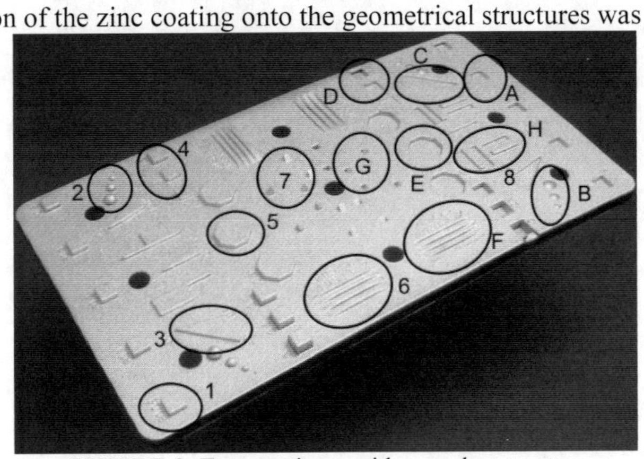

FIGURE 2. Test specimen with complex structure

RESULTS

Influence of the plastics type on the transfer of an arc sprayed zinc coating

According to the results of the contact angle measurements, shown in Figure 3, PP has the smallest contact angle, followed by (in ascending order) PA6, PBT, PBT WP25, PC, and PA6 GF30. It is assumed that a small contact angle between the plastics melt and the coated substrate lead to a high adhesion affinity of both materials and therefore have a high potential of a positive transfer [14].

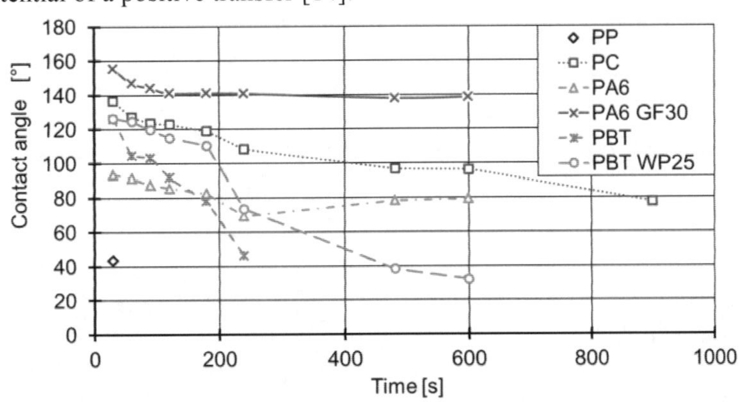

FIGURE 3. Contact angles of different plastics on zinc coated steel substrates (Ra ~10μm)

Figure 4 gives an overview of zinc-coated parts which were back-moulded with the different plastics mentioned above. It is evident that PP (a) and PA6 (d) parts show the highest coating transfer rates. Especially, PA6 (d) parts show a 100% transfer of the coating. In the same figure, PC (c) has the third highest transfer rate. But the adhesion between the coating and the plastics was very low, which can be identified by the wrinkles of the coating.

According to the results of the CA measurements and the made assumption, PBT should have a higher transfer rate, but as shown in the Figure 3, both PBT materials show the lowest transfer rates. The cause for the results could be the release agent and lubricant that were used as additives. The release agent encapsulates the melt like a film and builds up a boundary layer between the melt and the coating, which inhibits the adhesion. Both the PP and the unfilled PA6 do not contain any additives, but show good transfer rates. The parts moulded with PA6 GF30, shown in (b) of Figure 3, show small transfer rates at the border areas of the part. The additives and glass fibre contents could lead to that result. To understand the reason of these results for PBT and reinforced PA6 more experiments with materials without additives such as lubricant and release agent have to be carried out.

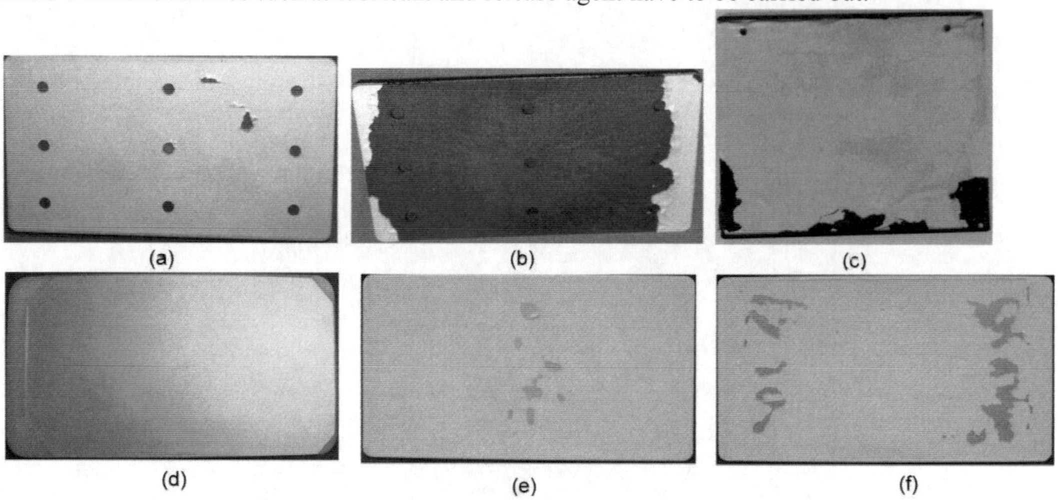

FIGURE 4. Moulded IMMS parts with (a) PP, (b) PA6 GF30, (c) PC, (d) PA6, (e) PBT, (f) PBT WP25

Transferability on plastics components with complex geometrically structured surfaces

Figure 5 shows a hybrid part with the structured coated surface. A transfer of 99 % of the zinc coating could be accomplished. Residuals on the cavity insert could not be observed by visual examination. The experiments reveal the boundaries of the structures that can be coated. The transition regions between the spheres (2; B) and the carrier body are no longer coated. This could be observed for all sizes. The square structures (D) show a delamination of the coating at the side surfaces, which increases when the structures are getting higher or deeper, respectively. A possible explanation relates to the demoulding process. Since the square structures e.g. have a demoulding angle of 3°, the friction between the part and the cavity insert during the demoulding could lead to the damage of the coating. The damage rises, the deeper or higher the structures are.

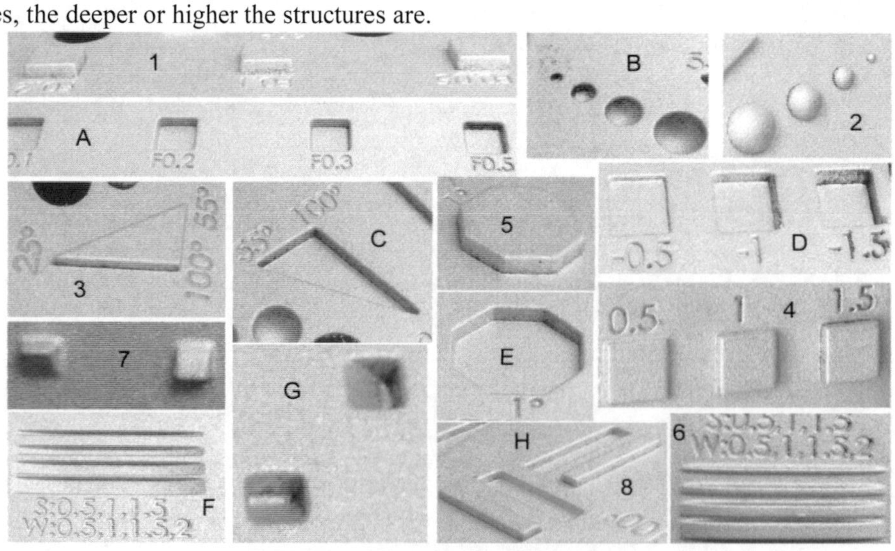

FIGURE 5. View details of the structures

CONCLUSION AND OUTLOOK

The presented results of the contact angle measurement have shown that unfilled semi-crystalline thermoplastics have a promising adhesion affinity to the applied zinc coating during the In-Mould-Metal-Spraying. Nonetheless, the transfer experiments could not confirm these assumptions yet. Further studies regarding the compounding of the plastics types and especially the used additives are necessary.

An almost complete transfer of a zinc coating with complex geometrical structures could be realised. In future studies the cross sections of the structures and the coating distribution will be analysed by means of microscopy and a quantitative detection via image processing software.

ACKNOWLEDGMENTS

The depicted research referring to the In-Mould-Metal-Spraying (IMMS) has been funded by the German Research Foundation (DFG) as part of the program Cluster of Excellence "Integrative Production Technology for High-Wage Countries". We would like to extend our thanks to the DFG.

We would also like to extend our thanks to the Scientific and Technological Research Council of Turkey (TUBITAK).

REFERENCES

1. F. Garbassi, M. Morra and E. Occhiello, *Polymer Surfaces – From Physics to Technology*, Chichester, New York: Wiley-VCH, 1998
2. J. Großmann, "Einfluss von Plasmabehandlungen auf die Haftfestigkeit vakuumtechnisch hergestellter Polymer-Metall-Verbunde", Ph.D. Thesis, Universität Erlangen-Nürnberg, 2009.
3. G. Habenicht, *Kleben – Grundlagen, Technologie, Anwendungen*, Berlin, Heidelberg: Springer-Verlag, 2009.
4. D. Nikolova, „Charakterisierung und Modifizierung der Grenzflächen im Polymer-Metall-Verbund", Ph.D. Thesis, Universität Halle Wittenberg, 2005.
5. K. Bobzin, M. Öte, T.F. Linke, C. Schulz, C. Hopmann, and J. Wunderle, "Integration of Electrical Functionality by Transplantation of Cold Sprayed Electrical Conductive Cu Tracks via Injection Moulding", International Thermal Spray Conference Proceedings, Barcelona, 2014, pp 1-10.
6. C. Hopmann, P. Ochotta, K. Bobzin, M. Öte, T.F. Linke and X. Liao, „Flächige und partielle Metallisierung von Kunststoffbauteilen mit dem In-Mould-Metal-Spraying-Verfahren", edited by C. Hopmann, 28th Internationales Kolloquium Kunststofftechnik Proceedings, Institut für Kunststoffverarbeitung, Aachen, 2016, pp. 721-732.
7. E. Dayss, G. Leps and J. Meinhardt, *Surface and Coatings Technology* **9**, 986-990 (1999).
8. J. Friedrich, M. Pohl and J. Gähde, *Acta Polymerica* **32**, 48-55, (1981).
9. K. Grundke, "Characterization of Polymer Surfaces by Wetting and Electrokinetic Measurements – Contact Angle, Interfacial Tension, Zeta Potential" in *Polymer Surfaces and Interfaces*, edited by M. Stamm, Leipzig: Springer, 2008, pp. 103-136.
10. N. N., Durethan and Pocan Materials Data Sheet, Lanxess AG, Cologne, Germany, 2016
11. N. N., 505P and 579S Materials Data Sheet, SABIC Europe B.V., Sittard, Netherlands, 2016
12. N.N., Makrolon 2205 Materials Data Sheet, Covestro AG, Leverkusen, Germany, 2016
13- G. Maier, DataPhysics Instruments GmbH, Filderstadt, 2015 (private communication)
14. Y. Yuan and T. R. Lee, "Contact Angle and Wetting Properties" in *Surface Science Techniques*, edited by G. Bracco and B. Holst, Leipzig: Springer-Verlag, 2013, pp. 3-34.

The ZSK at 60: A Retrospective of Technology Development with a View to the Future and Growth of Non-Traditional Applications

Paul Andersen

Coperion Corporation, 6 Arrow Road, Suite 202, Ramsey, NJ, 07446

paul.andersen@coperion.com

Abstract. The ZSK co-rotating, fully intermeshing was invented by Rudilf Erdmenger while working for IG Farben in the 1940's and subsequently licensed, developed and commercialized by Werner and Pfleiderer in the 1950's. During the intervening years it has become the primary production unit for compounding of polymer based materials. It also has had a long term presence in processing materials in the polymer, chemical and food industry and more recently in pharmaceuticals. While this equipment might be considered a "mature" technology and should have reached market saturation, it has not experienced a decline in new process development opportunities as might be expected, but rather a significant number of advancements continue to evolve.

This paper will trace the early development of twin-screw devices from units designed for producing ceramic bricks through and beyond the development of the ZSK. The presentation will additionally include highlights some of the more atypical applications such as animal feed, candy, snack food, ethanol, energetics and more.

Keywords: Twin-screw, Co-rotating, Counter-rotating, History

INTRODUCTION

The ZSK co-rotating twin-screw compounding extruder will celebrate 60 years of commercial availability in 2017. It is also celebrating another milestone, approximately 70 years since the development of the self-wiping intermeshing geometry and initial construction of the first laboratory unit [1]. While the Erdmenger ZSK twin-screw geometry is today the dominant system today for processing plastics and other high viscosity materials, it was not developed in a vacuum. The term twin-screw extruder (unlike their single screw cousin) does not define a unique operating principle. There are 3 basic twin-screw concepts, counter-rotating non-intermeshing, counter-rotating intermeshing and co-rotating intermeshing. Along with the Erdmenger ZSK co-rotating rotating design, each of the counter-rotating designs maintains an important place in the processing of plastics today. Interestingly however, none were originally designed for plastics processing. Today, while the majority of volume produced on each of these geometries is related to plastics, there are many unique and challenging non-plastics processes in operation and new ones being developed.

Before looking towards future applications, it is beneficial to take a step back and trace the development of the technology and the various applications which may have spurred its development.

Fundamental Historical Developments for Twin-Screw Extruders

Counter-rotating Intermeshing Geometry

It is interesting to note that the first recorded use of a twin screw extruder type device precedes the Royal single screw patent [2] by well over 50 years. It is reported in "**The Well Taught Brick Maker" by Schaller** (from 7th print run 1874; 1st print run 1828).

This first twin-screw device was not associated with plastics, but rather it was designed to convey a low moisture ceramic mass down a channel. This unit was counter rotating but the screw shafts were only partially intermeshing. The mass was compressed and compacted as the screws pushed it through a restrictive orifice to shape a ceramic rod which was subsequently cut to form bricks.

Even though the screws were only partially intermeshing, this unit is a direct predecessor to the parallel and conical twin-screw devices of today used to extrude PVC and other profiles.

The overall impact of this twin-screw device is as a:

1) Precursor to development of all twin screw devices
2) Initial step in development towards the counter-rotating screw pump
3) First twin-screw profile extruder.

Counter-rotating Non-intermeshing Geometry

In the late 1800's the internal mixer that eventually became know most widely as the Sigma Blade Mixer was patented by Freyburger. Paul Pfleiderer and Herman Werner purchased the patent from Freyburger in 1878 and began production of the device a year later for kneading bread dough.

Based on his experience with the internal batch mixer, Paul Pfleiderer [3] filed for a patent in Germany on July 10, 1881 for a continuous twin-screw kneading device, Figure **1** for dough.

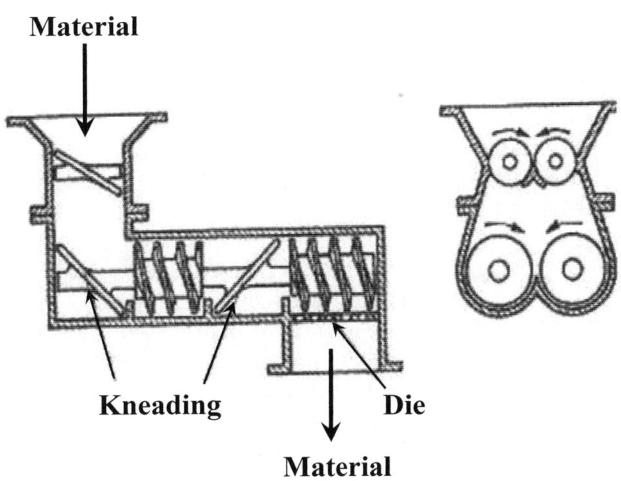

Figure **1**: Schematic of counter-rotating non-intermeshing continuous dough kneader.

This counter-rotating tangential device contained both pumping (conveying) and mixing sections. Just as in the current design for the Continuous Mixer (CM) the material can be dropped as a continuous mass for further processing or as shown in Figure 1, pushed through a die.

The overall impact of this twin-screw device is as a precursor to "Continuous Mixer" technology

Co-rotating Intermeshing Geometry

This last of the three geometries reviewed has had a more significant impact on the compounding of plastics than any other invention in the field. Interestingly, even though not fully developed until the mid of the 20th century, just as with the other two twin-screw geometries, the co-rotating design was also developed for a non-plastic application.

Basic Patent: While some initial concepts were patented in the early 1900's by Wuensche [4] and Easton [5, 6], the co-rotating design used as the basis for essentially all compounding systems marketed today is based on the self-wiping geometry known as the Erdmenger profile. The initial development of the self-wiping profile is described in 1952 patent [7], Figure 2. The objective of the design was for mixing high viscosity liquids already in the fluid state, such as post-polymerization reaction products. It is also interesting to note that the two screws were oriented over/under as opposed to the current standard side by side design.

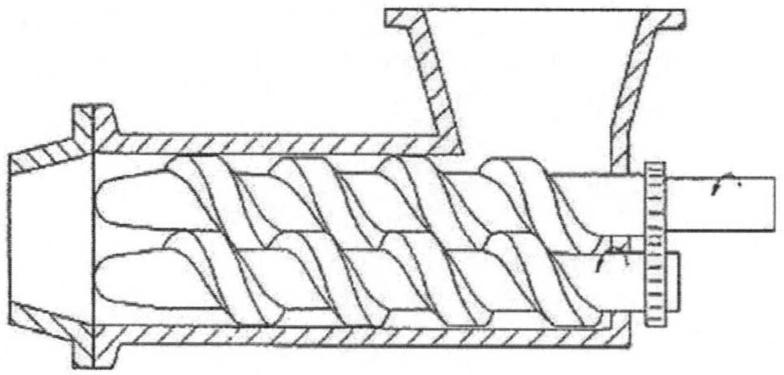

Figure **2**: Schematic of self-wiping profile from basic Meskat and Erdmenger patent.

This patent was the geometry basis for the eventual development and commercialization in 1957 of the ZSK twin-screw extruder as well as the many copies introduced during the intervening 60 plus years. The key feature of the design is the self-wiping characteristic of one screw with respect to the other. This eliminated stagnation and eventual degradation of material as it was transported downstream.

The overall impact of this invention is that it is the basic patent related to the co-rotating twin-screw compounding system predominantly used today in the plastics, food and chemical industry.

Supplemental Basic Patents: The initial co-rotating twin-screw patent provided the self-wiping profile which permitted materials to be mixed and conveyed along the screw channel. However, material distribution and dispersion that occurred during transfer of material from one screw to another was not sufficient for all mixing applications. This initial design also did not provide for the elimination of volatiles. Several additional patents by Edemenger and other colleagues at Bayer provided the key design features that when all assembled in one package are essentially the co-rotating twin-screw used today.

Kneading discs and kneading blocks: The first improvement of the initial patent conveying/mixing screw geometry was the addition of eccentric circular discs, as well as mono-lobe and bi-lobe discs [8,9], Figure 3. The individual discs could be placed on the shaft as single units or as multiple units one following the other. These discs could also be rotated with respect to the unit immediately upstream and downstream. Among the impacts noted was the axial flow in either the up or down channel direction that improved mixing. It is interesting to note that the US patent [9] does not include the bi-lobe geometry. It can be surmised that there may have been some prior art cited during prosecution of the U.S. application.

The next development in screw element design, 3-lobe kneading discs (Figure 4), came about 4 years later [10,11]. According to the patent, the previous eccentric or mono lobe mixing units were sufficient for mixing material already in the liquid phase. However, when trying to process granular material using the mono or bi-lobal discs, the extruder would stall.

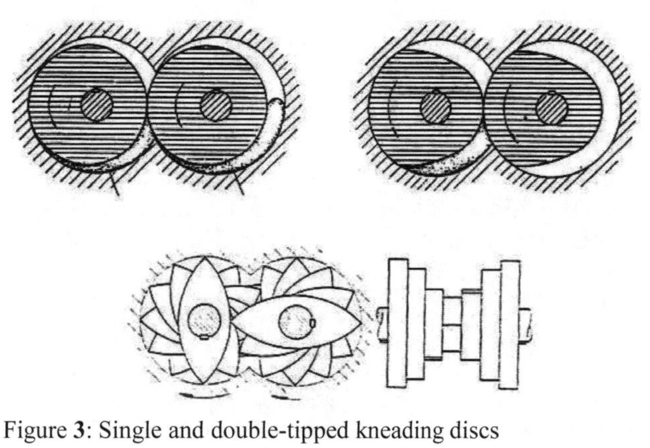

Figure **3**: Single and double-tipped kneading discs

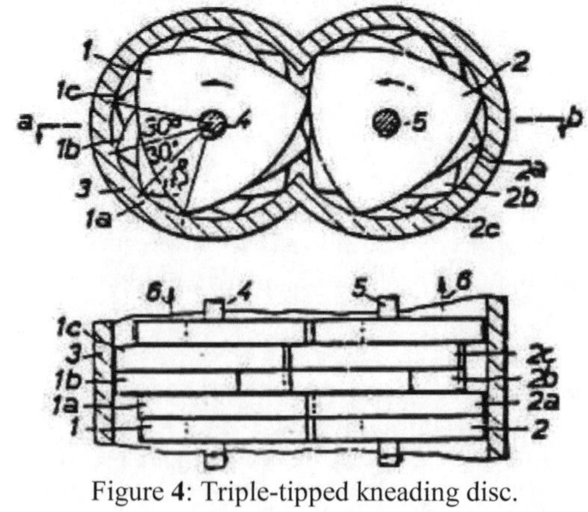

Figure **4**: Triple-tipped kneading disc.

The overall impact of these two patents is two-fold:

1) The mixing discs and sequence of discs defined by these patents form the basis for the mixing functions (both dispersive and axially distributive) provided in the co-rotating twin-screw compounding system cited previously.
2) Additionally, while not specifically patented in the claims, the noted flexibility of sequencing the orientation of the individual discs in the text is the first instance of modular design for the twin-screw extruder.

Modular intermeshing co-rotating twin-screw extruder with conveying and reverse flight elements: In between the patents for the mono / bi-lobe and the tri-lobe mixing and kneading elements, Meskat & Pawlowski were granted a patent [12] for reverse "self-cleaning" elements, Figure 5. As noted in White's book on Twin Screw Extrusion [13], this development was used for the production of spinning solutions and reactive extrusion for the esterification of cellulose.

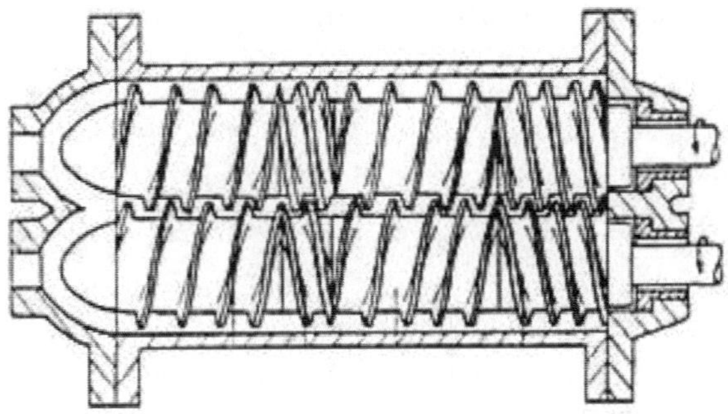

Figure **5**: Reverse self-cleaning elements.

The impact of this invention is several fold:

1) First introduction of reverse elements to provide added mixing
2) Along with the mixing / kneading disc patent of the same years this is one of the first mentions of modular design for screw elements
3) While not explicitly mentioned in the patent, it was later noted that this invention in conjunction with the twin-screw device was used for a reactive extrusion process.

Co-rotating twin-screw "stuffer": While described as a stuffer, this device, Figure 6, was designed as a devolatilization unit [14]. This is not surprising in light of the initial use of the twin-screw as a processor for high viscosity liquid mixing.

The impact of this invention is two-fold:

1) Key component of system to devolitalize "elastic" materials
2) Precursor of twin-screw side de-gassing unit as well as the stuffer for downstream feeding of filler and/or additional polymer

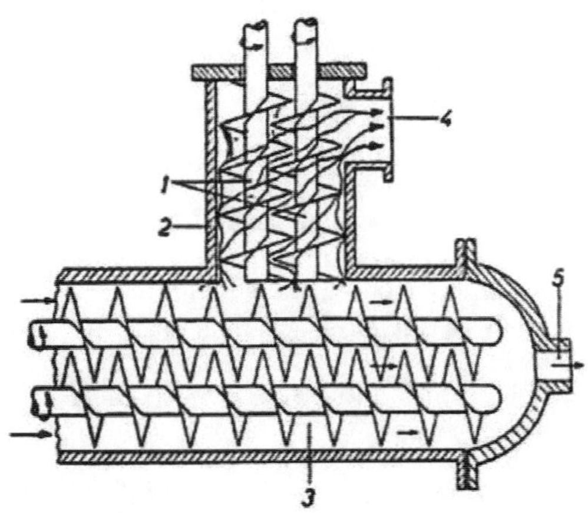

Figure **6**: Twin-screw extruder design for devolatilization.

Modular Co-rotating Twin-screw Extruder: There is no patent with this title. However, Erdmenger [15] was issued U.S. Patent with the benign title of "Worm Arrangement". In the patent he provides recognition that specific sequencing the various elements in relation to the physical form and status of the material will result in a superior mixed material. The following in an excerpt from the patent description:

> As a result of the combined arrangement of the threaded worms and disc elements or kneading discs, there is obtained an improved feeding of substances, a more accurate determination of residence time and a better mixing. It is also possible to process substances that change their state of aggregation in this apparatus during the treatment. Furthermore, costly cutting of grooved screw can be avoided due to the fact that Industrial engineer can make and test their own screw design changes at will with the interchangeable slip-on screw elements on the keyed screw shafts.

Most Recent Developments for Twin-screw Extruders

Since the development of the basics for the co-rotating twin-screw extruder there have been a significant number of incremental improvements to the technology. These include numerous new screw element geometries [16], the two-lobe element profile for increased internal free volume (the initial profile described in the first Erdmenger patent was based on a low free volume 3 lobe geometry), new screw shaft geometries for improved power transmission, and new process applications for the system [17].However a significant step forward was achieved with the identification of the fundamentals of high rpm / high torque compounding technology [18].

High torque, high rpm co-rotating twin-screw compounding technology: The key to the success of this process is the combination of both additional power (torque) transmission capacity as well as increased screw rpm. A system that simply runs at higher rpm will at some juncture impart enough additional energy to the material being processed to cause degradation. However, the higher power transmission capability provides the compounding unit the ability to process at a higher rate per rpm. This in turn reduces the energy input (temperature rise) per kg of product. Therefore the extruder can be run at higher rpm without exceeding material temperature limits.

Feed Enhancement Technology (FET) [19]: In order to fully utilize the higher torque and increased rpm of the latest extruder generations, an improved feed intake system for low bulk density materials was developed. The technology is based on an improved design to a) remove entrained air and b) increase conveying efficiency.

The impact of both of these latest developments resulted in significant increase in productivity (production rates), efficiency and flexibility for the co-rotating twin-screw extruder.

REFERENCES

1. R. Erdmenger, Schneckenmachinen fuer die Hochviskostechnik, Bayer AG (1978)
2. V. Royle and J. Royle, Jr., U.S. Patent 325,363 (September 1, 1885)
3. P. Pfleiderer, German Patent 18,797 (July 10, 1888)
4. A. Wunsche, German Patent 131,392 (1901)
5. R. W. Easton, British Patent 109,663 (1916)
6. R. W. Easton, U.S. Patent 1,468,379 (1923)
7. W. Meskat and R. Erdmenger, German Patent 862,668 (1952)
8. R. Erdmenger, German Patent 813,154 (1951)
9. R. Erdmenger, U.S. Patent 2,670,188 (1954)
10. R. Erdmenger, German Paternt 940,109 (1951)
11. R. Erdmenger, U.S. Patent 2,814,472 (1957)
12. W. Meskat J. Pawlowski, German Patent 949,162 (1956)
13. J. L. White, Twin Screw Extrusion: Technology and Principles (1991)
14. W. Winkelmuller, R. Erdmenger, S. Neidhart, G. Hirschberg, and B. Fortuna, German Patent 915,689 (1954)
15. R. Erdmenger, U.S. Patent 3,122,356 (1964)
16. M. Bierdel, Co-Rotating Twin-screw Extruders: Fundamentals, Technology, and Applications (2007) K. Kohlgrueber & W. Wiedmann Editors
17. P.G. Andersen, Plastics Compounding, D.B. Todd ed., 71-124 (1998)
18. P. Heidemeyer, E. Haring, R. Munz, R. Herter, and U. Burkhardt, U.S. Patent 6,042,260 (2000)
19. P.G. Andersen, M. Hoelzel, T. Stirner, *SPE-ANTEC Tech. Papers*, **57**, (2011)

An Investigation on Dispersion State of Graphene in Polypropylene/Graphite Nanocomposite with Extensional Flow Mixing

Koki Matsumoto[a], Yoji Nakade[a], Keita Sugimoto[a] and Tatsuya Tanaka[a*]

[a]Applied Material Engineering Laboratory, Department of Mechanical Engineering, Faculty of Science and Engineering, Doshisha University, 1-3, Tatara Miyakodani, Kyotanabe-City, Kyoto, 610-0321, Japan

eup1502@mail4.doshisha.ac.jp

Abstract. Graphene nanoplatelets (GNP) have attracted considerable attention because of their excellent mechanical, thermal and electrical properties. Many researchers have tried to exfoliate GNP directly from graphite (Gr) in polymer by shear flow mixing process with twin-screw extruder, internal mixer and so on. By usual shear mixing, however, exfoliation of Gr into GNP is very difficult because of inefficiency process. Therefore, we focused on the extensional flow for exfoliating to GNP from Gr. In this study, the mixing effect of pure extensional flow for exfoliating into GNP was investigated by comparing with shear flow. Primary, PP/Gr masterbatch pellets were extruded by capillary rheometer equipped with orifice die (i.e. extensional flow) and capillary die (i.e. shear flow). Then, mixing effect was discussed by morphological and rheological analysis. As the results, extensional flow can exfoliate into few layer GNP with keeping larger sure face area by comparing shear flow and storage modulus G' was improved as the extensional stress is higher.

Keywords: Nanocomposite, Polymer, Graphene, Extensional Flow, Dispersive Mixing
PACS: 83.50.Jf

INTRODUCTION

Graphene is one of the most attractive nanocarbon materials, because it has excellent mechanical properties and functional properties (i.e. thermal or electrical conductivity and barrier properties). Many companies and researchers have been attempted to manufacture the high quality graphene by chemical process (e.g. Chemical Vapor Deposition (CVD)) or exfoliate to graphene from graphite directly by mechanical process (e.g. ultra-sonication, melt mixing). However, the cost of graphene by chemical process is very expensive, and the quality of the graphene by mechanical process is lower than by chemical process because the graphene is broken into small pieces.

We have been focused on extensional flow to obtain the better dispersion state of nanofillers (i.e. MWCNT [1-3]) with twin-screw extruder. Typically, shear flow is used for compounding the nanofillers and thermoplastic resin by melt mixing process. Experimentally, extensional flow is more efficiently than shear flow in terms of dispersive mixing [4]. The exfoliation to graphene by shear mixing is difficult because the van der Waals force between the layers works strongly. To use this graphene for reinforcement of polymer, graphene should be exfoliated to nanoscale layer with keeping large surface area.

Therefore, we attempt to use the extensional flow for exfoliating graphite to graphene directly in polypropylene (PP) matrix. The purpose of this study is making clear the mixing effect of pure extensional flow by using orifice die which is installed in capillary rheometer. The effect of extensional flow can be evaluated by measured entrance pressure drop ΔP_e while extruding melted composite and entrance pressure drop can be measured directly with orifice die [5]. Furthermore, the mixing effect of extensional flow was compared with that of shear flow in terms of exfoliated (dispersion) state of graphene and particle size (i.e. surface area).

EXPERIMENTAL

Material

PP (J108M, Prime Polymer Co., Ltd.) was used as matrix with MFI 45g/10min at 2.16kg/230°C. Flake Graphite (XD150, Ito Graphite Co., Ltd., Japan) which is 150μm in average particle size was used as filler. The reason of selection this graphite with large surface area is that the morphology observation of graphene is easy. Then, graphite based PP composite pellets (masterbatch) were produced by masterbatch process with twin-screw extruder (ZSK18 MegaLab, Coperion GmbH, Germany). The filler loading was 1wt%. The compounding temperature was 200°C, the screw speed was set at 150 rpm and the through put was 4.5kg/h. To avoid the fracture of graphite, the screw configuration with little mixing effect was used (see Fig.1).

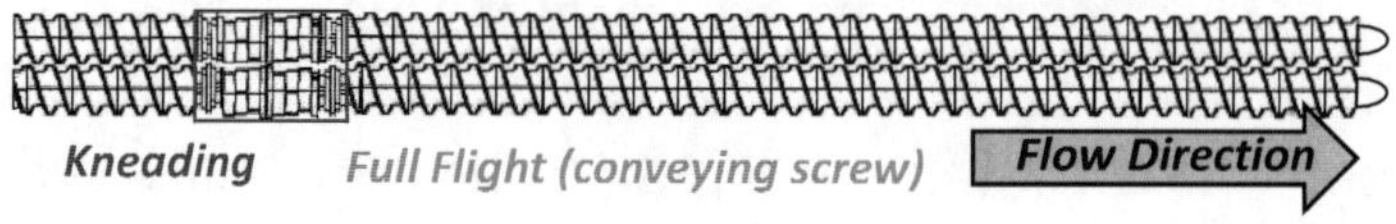

FIGURE 1. Screw configuration of twin-screw extruder with little mixing effect (low shear).

Extrusion for Exfoliation of Graphite into Graphene

Extensional Flow Mixing

The obtained pellets of graphite based PP composite were extruded by capillary rheometer (Rosand RH2200, Bohlin Instruments Ltd.) equipped with orifice die (see Fig.2 (a)). The extruded temperature was set at 200°C. The masterbach was extruded in various piston speeds with different diameter of orifice die to measure the wide range of entrance pressure drop ΔP_e. The entrance pressure drop was measured while extruding the material. The condition of extrusion is shown in Table 1. Then, extensional stress σ was calculated by Cogswell approach:

$$\sigma = \frac{3}{8}(n+1)\cdot \Delta P_e \tag{1}$$

where n is global shear power law index. This index could be obtained from shear viscosity data.

Shear Flow Mixing

At the high shear rate (>10 s^{-1}), the masterbach pellets were extruded by capillary rheometer that capillary die (see Fig.2 (b)) was installed at 200°C. By using capillary die, the materials were applied high shear stress. However, the measured pressure drop ΔP includes entrance pressure drop ΔP_e and shear pressure drop ΔP_{sh}. Therefore, the materials were affected by both extensional and shear stress. The condition of extrusion is shown in Table 1. Shear stress τ was calculated by Hagen-Poiseuille equation with Bagley correction:

$$\tau = \frac{(\Delta P - \Delta P_e)D}{4L} = \frac{\Delta P_{sh} D}{4L} \tag{2}$$

where D is diameter of capillary die and L is length of capillary die.

On the other hand, at the low shear rate (<10 s^{-1}), the rotational rheometer (Gemini II, Bohlin Instruments Ltd.) was used under the steady shear stress condition. The 25mm parallel plate was equipped, the gap height was 1 mm, and the temperature was 200°C. The applied shear stress was 50 Pa, 100Pa, 500Pa and 1000Pa for 30 sec.

TABLE (1). The experimental condition of extensional and shear flow extrusion.

ID	Extensional Flow		(High) Shear Flow	
	Diameter of Orifice die[mm]	Piston speed [mm/min]	Geometry of Capillary Die $D\times L$ [mm]	Piston speed [mm/min]
1	1	20	1 × 16	20
2	1	100	1 × 16	100
3	1	500	1 × 16	500
4	2	20	2 × 16	20
5	2	100	2 × 16	100
6	2	500	2 × 16	500
7	3	20	3 × 16	20
8	3	100	3 × 16	100
9	3	500	3 × 16	500

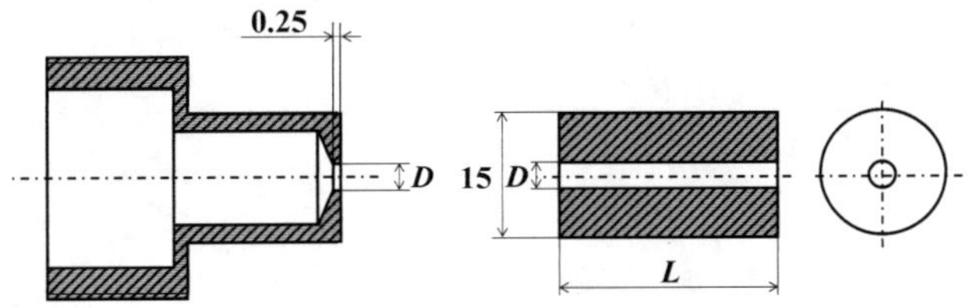

(a) (b)

FIGURE 2. Geometry of extrusion die for capillary rheometer; (a) Orifice die (b) Capillary die.

Morphological and Rheological Analysis

Primary, to investigate the state of graphite particle roughly after extensional and shear extrusion, the particle size above 10μm were measured by laser microscopy (VK-X210, KEYENCE CORP.) for 100 graphite particles. The graphite particle was obtained by burned the composite with Electric Furnace (KDF 300-Plus, DENKEN-HIGH DENTAL Co., Ltd., Japan) at 500°C for 1h. Then, to observe the state of graphene in nanoscale, TEM (JEM-2100F, JEOL Ltd., Japan) investigation was conducted. The specimen for TEM observation was prepared by microtome and the thickness of specimen was about 100nm.

Furthermore, the rheological characterization in oscillation mode was conducted with rotational rheometer equipped with 25mm parallel plate because this method can characterize the nanostructure in nanocomposite more accuracy. The storage modulus G' increases by improving the dispersion of nanofiller [6]. In this study, the storage modulus G' was measured under a strain of 5% at angular frequency 0.5 rad/s for all composite after extrusion.

RESULTS AND DISCUSSION

Pressure Drop Measurement and Stress Calculation

The results of pressure drop measurement and stress calculation are shown in Table 2 for extensional and shear flow mixing. From these results, the entrance pressure drop and shear pressure drop increased as the piston speed is higher and diameter of each extrusion die are smaller. Hence, the tendency of each stress magnitude exhibited the same behavior as that of each pressure drop.

TABLE (2). The results of pressure drop measurement and stress magnitude calculation.

ID	Extensional Flow			Shear Flow		
	Entrance Pressure Drop ΔP_e [kPa]	Stretch Rate[s^{-1}]	Extensional Stress σ [kPa]	Shear Pressure Drop ΔP_e [kPa]	Apparent Shear Rate [s^{-1}]	Shear Stress τ [kPa]
1	253	123	140	2877	600	45
2	768	575	364	5277	3000	82
3	2174	5794	865	8778	15000	137
4	68	11.2	44	387	75	12
5	222	56.3	127	1002	375	31
6	663	267	330	1978	1875	62
7	16	2.86	11	61	22	2.9
8	65	15.9	41	226	111	11
9	301	80.9	167	814	556	38

Morphological Analysis

Primary, the result of particle size measurement for only graphite particle is shown in Fig.3 (a) and observed images of graphite were shown in Fig.3 (b). To evaluate the each mixing effect, the particle size distribution was compared between masterbatch, orifice die with 1mm diameter at 500mm/min piston speed (i.e. maximum extensional stress) and capillary die with 1mm diameter at 500mm/min piston speed (i.e. at maximum extensional stress and shear stress). From this result, the particle size decreased dramatically by high shear stress.

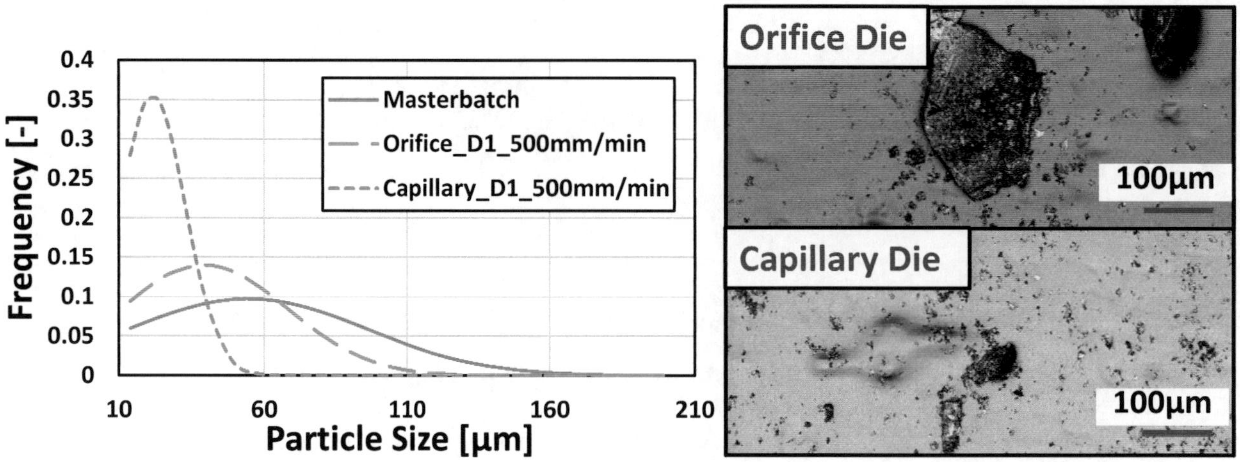

(a) (b)

FIGURE 3. The results of particle size measurement for flake graphite after extrusion; (a) Comparison of particle size distribution (probability density function) (b) Observation images of flake graphite.

Furthermore, observed TEM micrographs were shown in Fig.4 to discuss the mixing effect for each flow in more detail. TEM micrograph of masterbatch pellets was shown in Fig.4 (a). The thick layers which have about 3.5μm particle size can be observed and the layers didn't exfoliate into one layer. Then, TEM micrograph of extruded materials under maximum of extensional stress was shown in Fig.4 (b). From this image, the layers didn't exfoliate into one layer completely yet. However, thinner layers were observed by comparing with thick layers in materbatch. Finally, the TEM image of the extruded materials under maximum of shear stress was shown in Fig.4 (c). The smaller (<500nm) and thinner layers were observed in this image. However, the surface area of graphene layer by shear extrusion was smaller than that of graphene by elongational extrusion. This tendency coincided with the results of particle size measurement for flake graphite.

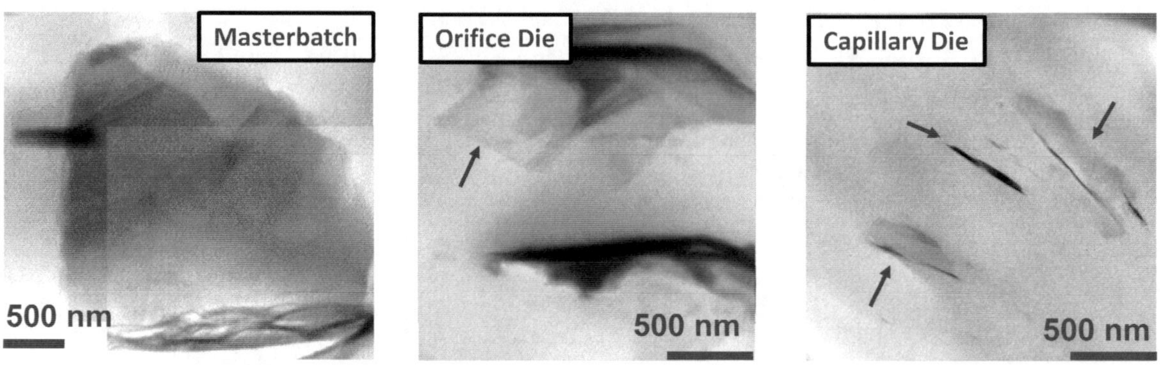

(a) (b) (c)

FIGURE 4. TEM micrographs of the cross-section of the extruded materials; (a) Masterbatch (b) Orifice die (at maximum extensional stress) (c) Capillary die (at maximum shear and extensional stress).

Rheological Analysis

The results of storage modulus G' for each flow were shown in Fig. 5. To separate the mixing effect, equation (3) and (4) were used. From these results, the $\Delta G'$ for extensional flow increased as extensional stress is higher. On the other hand, the $\Delta G'$ for shear flow decreased as shear stress is higher. These results indicates that the $\Delta G'$ was depend on the surface area (i.e. length to thickness ration) of graphene layer and the exfoliate state or dispersion state. Therefore, it is suggested that graphene was destroyed not by extensional flow rather but by high shear flow.

$$\Delta G'_{ExtensionFlowEffect} = G'_{ExtensionExtrusion} - G'_{Masterbatch} \quad (3)$$

$$\Delta G'_{ShearFlowEffect} = G'_{ShearExtrusion} - G'_{Masterbatch} \quad (4)$$

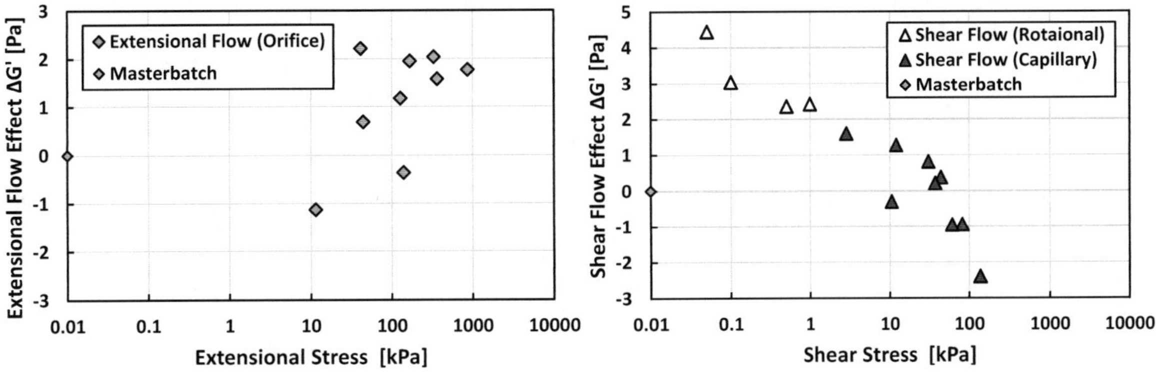

(a) (b)

FIGURE 5. The results of storage modulus G' in rheological analysis; (a) Extensional flow (b) Shear flow.

CONCLUSIONS

This research work investigated the effects of extensional flow mixing for exfoliation the graphite into graphene in PP matrix on morphological and rheological analysis by comparing with shear flow mixing. From morphological analysis, the flake graphite particles and graphene layers were broken by shear flow and the extensional flow can exfoliate the graphite into graphene with keeping large surface area by comparing with shear flow. Furthermore, from rheological analysis, storage modulus G' of extrusion material by extensional flow was improved by 200%. However, storage modulus G' of extrusion material by shear flow was decreased as shear stress is higher. Therefore, it is suggested that extensional flow is more effective for exfoliate graphite or disperse the graphene than shear flow.

ACKNOWLEDGMENTS

The authors acknowledge support from Doshisha University Research and Development Center for Advanced Composite Materials.

REFERENCES

1. K. Matsumoto, N. Kayamori, T. Tanaka and Y. Arao, *Proceedings of PPS-30: The 30th International Conference of the Polymer Proceeding Society-Conference Papers,* **1664**, (02006)1-5 (2015).
2. K. Matsumoto, T. Morita, T. Tanaka and Y. Arao, *ANTEC®2015-Orlando*, 989-993 (2015).
3. K. Matsumoto, Y. Arao and T. Tanaka, *Computational Methods and Experimental Measurements XVII*, **59**, 33-45 (2015).
4. H. P. Grace, *Chem. Eng. Comm.*, **14**, 225-277 (1982).
5. Cogswell, F.N., *Polym. Eng. Sci.* **12**, 64-73 (1972).
6. W. Lertwimolnun, B. Vergnes, *Polymer*, **46**, 10, 3462-3471 (2005).

Thermal Diffusivity Behavior of the High Density Polyethylene with the Addition of Pigmentation Masterbatches

Castro L. Juan Felipe[1] & Medina P. Jorge Alberto (Dr. Ing.)[1]

1- *Grupo de Materiales y Manufactura, CIPP-CIPEM, Departamento de Ingeniería Mecánica, Universidad de los Andes, Bogotá Colombia,*

jmedina@uniandes.edu.co
jf.castro1218@uniandes.edu.co.

Abstract .The thermo-physical properties are useful in designing manufacture processes for polymers, as they supply relevant information about the performance of the material during the different manufacturing stages. The thermal diffusivity (α) is one of the most important of these properties because it contains information about other properties such as the heat capacity (C_p), the density (ρ) and the thermal conductivity (k). Therefore, the analysis of the thermal diffusivity is paramount, especially when fillers are added to the polymer mixture, an example of those are the pigmentation masterbatches. Moreover, several methods and techniques have been developed to characterize the thermal diffusivity, such as the xenon-laser method. Using this method, the aim of this research is to determine the effects in the thermal diffusivity of a high density polyethylene after supplementing it with different kind of pigmentation masterbatches (red, blue and gray) at a variable concentration. The results of the experimentation show the dependence of the diffusivity with temperature changes, which are affected by alterations in the composition and concentration of the added low density polyethylene (matrix of the masterbatch) and the inorganic material (pigment). These alterations are more evident in a domain of high temperatures, near to the melting temperature, in which case the competence between amorphous and crystalline phases is taking place [1]. The experimental data was compared with the results from a theoretical model that predicts the behavior of the thermal diffusivity in a compound polymer [2]. These results are relevant in the manufacture process in polymers because it can explain the difference in the processing times by adding different kind of pigments, that are influenced by the concentration of the masterbatch instead the nature of the inorganic components.

Keywords: High density polyethylene, pigments, masterbatch, thermal diffusivity, laser flash technique.
PACS:

1. INTRODUCTION

The knowledge and control of the nature of the thermo-physical properties of polymeric materials is relevant in the modeling of engineering problems where the mechanical and thermal behavior of materials will be study. One of those is the thermal diffusivity that gives the information of the amount of energy a material can store or transfer [3] defined by the equation (1). The analysis of this property on different scenarios builds up useful information to define conditions of processing and applications for plastic materials.

$$\alpha = \frac{k}{\rho C_p} \quad \textbf{(1)}$$

The thermal diffusivity may be affected by factors like ambient conditions or material morphologies. For example the increase of the ambient temperature causes that the thermal diffusivity of a specific material decreases [4]. In the other hand at a micro-structural level the heat conduction is influenced by the interaction of the phonons (for amorphous materials) or the un-harmonic elastic vibration of networks (for crystalline materials); those are called the mechanisms of heat transfer described by dos Santos et al. Common polymers have semi-crystalline structure therefore the two mechanisms of heat transfer are present in this kind of materials. Therefore the thermal diffusivity would be the result of the competition in the formation or presence of those phases in the material. Recent investigations show the behavior of the thermo-physical properties and some particularities of different polymers (semi-crystalline and amorphous) at variable conditions [1-5].

The addition of pigmentation mastebatch in polymers is very common in industrial processing. Those fillers bring to the polymer physical properties like colors, brightness and absorption or reflection of light. Those materials can affect the mechanisms of heat transfer in the structure of the material, because it can transform the structure of the polymer acting like a nucleating agent [6]. Therefore it is important to make an analysis in the variation of thermal diffusivity and thermo-physical properties with the addition of this kind of fillers to verify the relevance of the concentration in the blended material.

In order to analyze the variations on thermal diffusivity by the temperature, concentration and nature of the pigmentation masterbatch two methods were used in this research. The first one was experimental, using the laser xenon flash technique by the DXF200 from TA Instruments. And the second one was a theoretical approximation made by Mikulenok that can predict the behavior of the thermal diffusivity in composite materials (α_{TP}) using the volume fraction added (φ_i) and the thermal diffusivity (α_i) of each component, this relation is in equation (2).

$$\alpha_{TP} = [\sum_{i=1}^{n}(\frac{\varphi_i}{\sqrt{\alpha_i}})]^{-2} \quad \textbf{(2)}$$

2. EXPERIMENTAL

2.1. Materials

The materials used were HDPE B5429 (Sabic, Saudi Arabia)[7], with melt flow index of 0,3g/10min, and three different kind of pigmentation masterbatches (red B149, blue PL52605167 and gray PL73606512)[8] all of those provided by Clariant. The samples were termed as HDPE (pure high density polyethylene), HDPERX, HDPEBX and HDPEGX (High density polyethylene with Xwt% red, blue and gray pigmentation masterbatch respectively). The different concentrations of masterbatch as well as real pigment (measured with TGA) used are shown in Table 1.

Table 1. Composition of the samples blends of HDPE and masterbatch.

Sample	Matrix polymer	Filler material	Masterbatch wt%	Pigment wt%
HDPE	HDPE	N/A	0	0,0000
HDPER1	HDPE	Red pigmentation masterbatch	1	0,0026
HDPER3			3	0,0077
HDPER6			6	0,0154
HDPER15			15	0,0385
HDPER40			40	0,1025
HDPER75			75	0,1923
HDPER100	LDPE		100	0,2564
HDPEB1	HDPE	Blue pigmentation masterbatch	1	0,11
HDPEB3			3	0,34
HDPEB6			6	0,68
HDPEB15			15	1,70
HDPEB40			40	4,55
HDPEB75			75	8,52
HDPEB100	LDPE		100	11,36
HDPEG1	HDPE	Gray pigmentation masterbatch	1	0,05
HDPEG3			3	0,15
HDPEG6			6	0,31
HDPEG15			15	0,76
HDPEG40			40	2,03
HDPEG75			75	3,81
HDPEG100	LDPE		100	5,09

2.2. Processing

The materials (HDPE and the masterbatch) were mixed in an internal Brabender Plasti-Corder PLE331 mixer at 190°C and 60 rpm, during 20 min. Then the blend materials were pressed by compression molding at 180°C and pressures of 1,5 MPa for two minutes and 8,5 MPa for another two minutes to achieve samples of 12,7 mm (1/2")diameter and thickness of 30 mm.

2.3. Thermogravimetric analysis

TGA measurements were performed in order to characterize the real fraction of inorganic material in the pigmentation masterbatches. Those were made in a TGA NETZSCH STA 409 PC in a nitrogen atmosphere with a heating rate of 10 °C/min from 24°C to 950°C.

2.4. Differential scanning calorimeter

A TA Instruments DSC-Q200 modulated was used to determine the heat capacity of the pure HDPE, the pigmentation masterbatch and the blended material. About 9 mg of each material was analyzed with this technique in a nitrogen atmosphere from -5°C to 160°C at a heating rate of 3°C/min.

2.5. Xenon laser technique

The thermal diffusivity was measured with the DXF200 of TA Instruments. This equipment can provide directly the thermo-physical property shooting a xenon laser into the sample and measuring the temperature on the other side of it. Details of the technique were provided in other researches and the equipment manufacture directly [9]. The main equation used in this measure equation (3) came from the heat transfer equilibrium in a solid wall and with boundary conditions given. The measurements were performed in a nitrogen atmosphere from 25°C to 125°C in intervals of 10°C.

$$\alpha = \frac{1{,}38L^2}{\pi^2 t_{\frac{1}{2}}} \quad \textbf{(3)}$$

Where, L is the simple thickness.

$t_{\frac{1}{2}}$ is the maximum half time that the simple needs to reach the maximum temperature.

3. RESULTS AND DISCUSSION

3.1. Thermogravimetric analysis

Figure 1 shows the results obtained of the concentration of polymer and inorganic constituents of the different pigmentation masterbatches used in the mixing of the samples. It provides the information of the amount of LDPE and pigment on the masterbatch. The variation on the pigment %wt is evident for the HDPER100 0,19%wt, the HDPEG100 5,09%wt and the HDPEB100 11,46%wt.

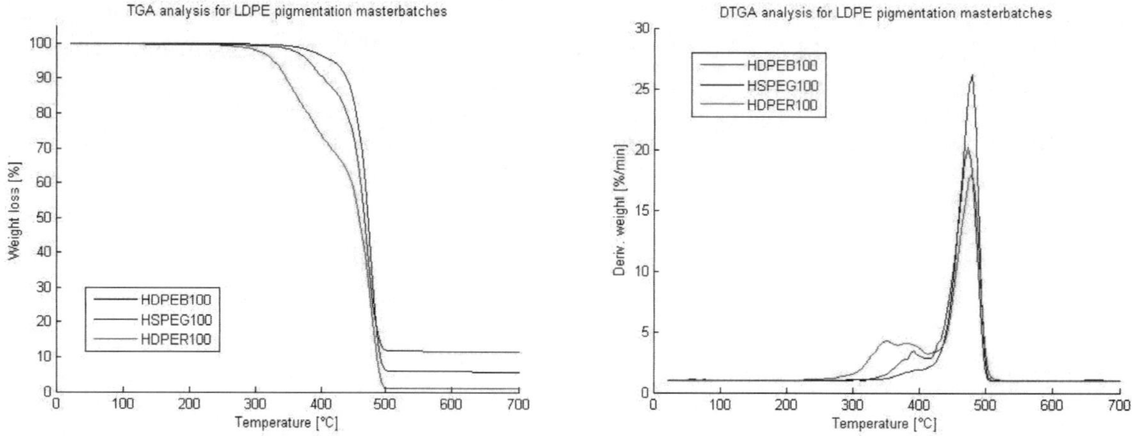

Figure 1. (a) TGA results (b) Derivate of the weight loss for the different masterbatch samples.

3.2. Differential scanning calorimeter

DSC curves obtained for various samples are shown in Figure 2. It provides important information about the changes in the heat capacity of the different mixtures because of the addition of the masterbatch.

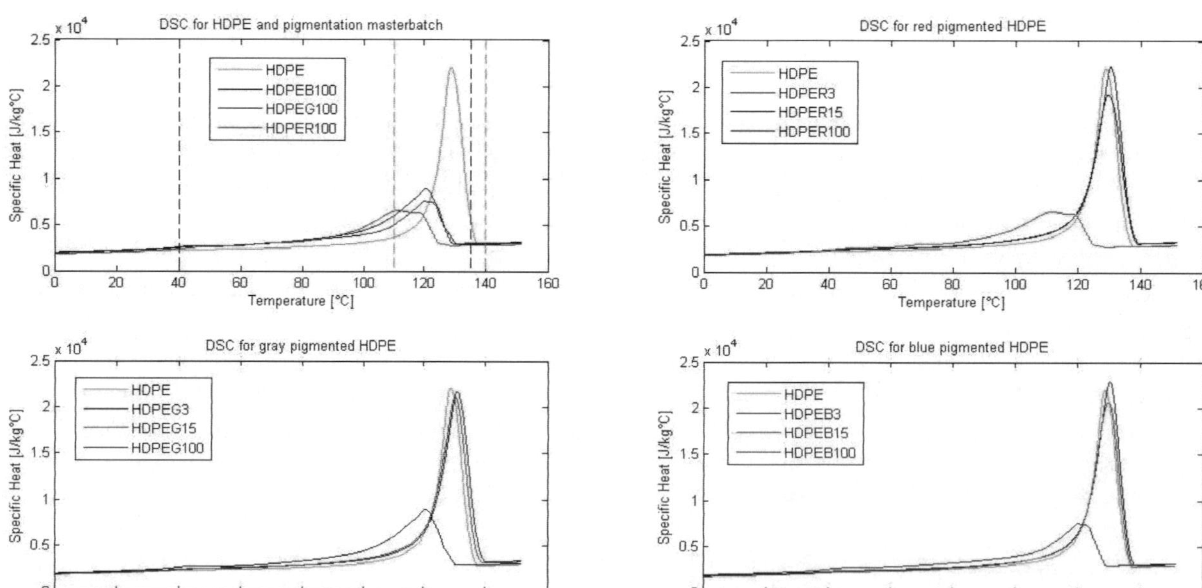

Figure 2. DSC results for different samples of HDPE and pigmentation masterbatch.

3.3. Xenon laser technique

Figure 3 provide the results obtained for the different samples evaluated at different temperatures. For all probes the thermal diffusivity displays a decrease as the temperature rises from room temperature to the beginning of the melting process, as expected from theory [1-4]. This behavior is explained by the change in other properties of the material like the density and the heat capacity as the Figure 2 shows. It can be seen an particular behavior of the thermal diffusivity of the polymers near to the 80°C, where the change in the heat capacity of the materials is more evident as shown in Figure 2.

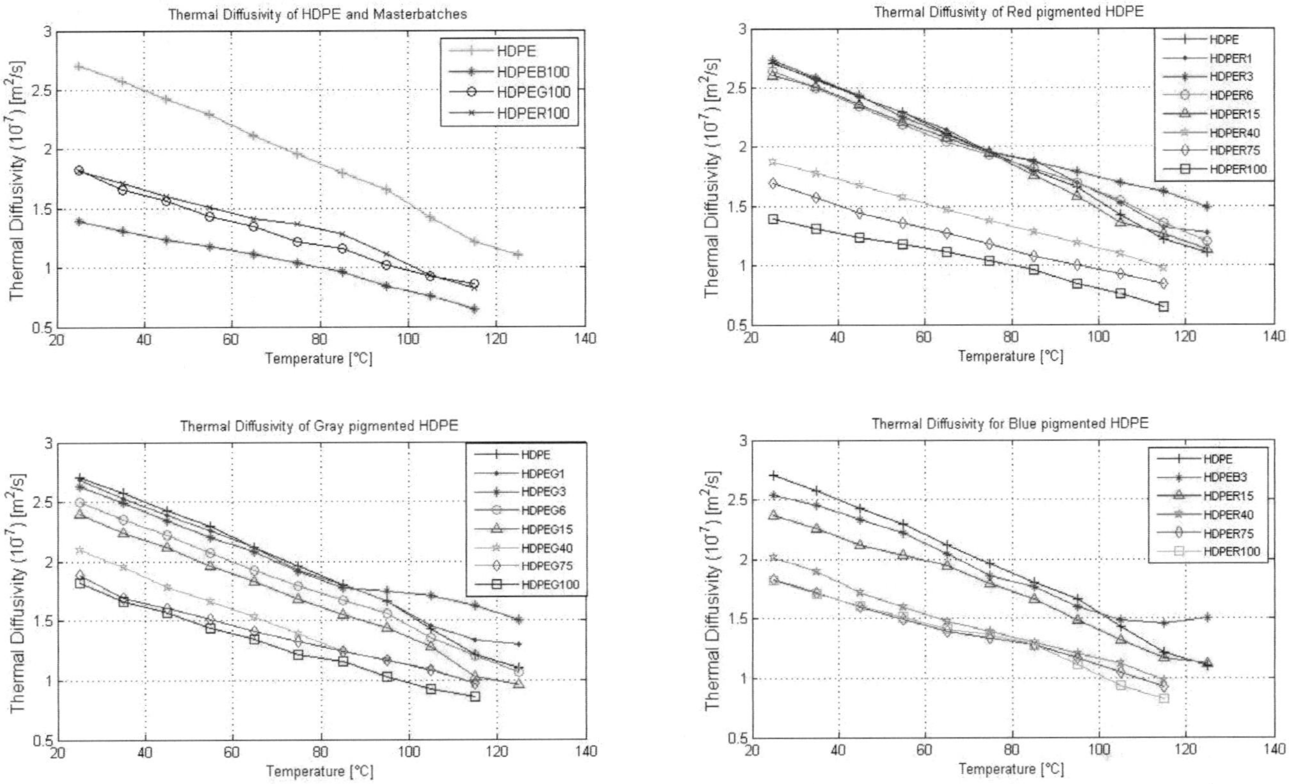

Figure 3. DXF results of thermal diffusivity for the blended samples and blank materials.

The thermal diffusivity presents a smooth decrease with the addition of the masterbatch. This response is because of the fillers that have different thermal properties. Several industry processes attribute those changes to the pigment content, but in fact is mostly affected because the LDPE (the polymer matrix) of the masterbatch which is added at high concentration than the inorganic components. The last described effect can be seen in Figure 3 and 4. In the last figure a decrease because of the addition of LDPE present a similar behavior regardless the kind of aggregate pigment. While the inorganic material addition presents a remarkable effect on the thermal diffusivity of the polymer depend on the kind of pigment. The aforementioned does not an overwhelming change in the thermo-physical property that is most influenced by the LDPE concentration.

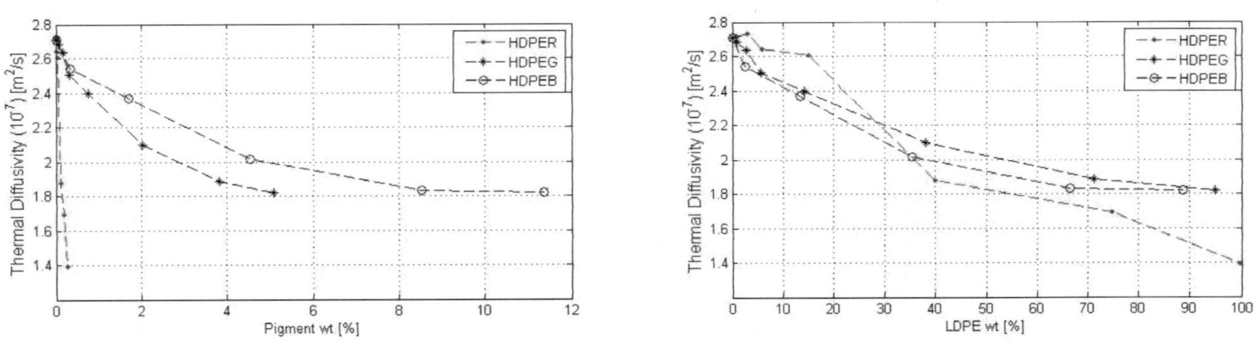

Figure 4. Thermal diffusivity behaviors of blended materials with the variation of LDPE (a) and inorganic material (b) concentration at 25°C.

The addition of LDPE generates changes in the structure formation of the blend material. As it affect the inherent crystallinity of the HDPE. Therefore the dominant mechanism of heat transport in the material may vary with the addition of masterbatch. Then the result thermal diffusivity of the material is the representation of the competition between the amorphous and crystalline phases to govern the heat transport phenomena. In this sense results evident that with the increment of LDPE aggregated the thermal diffusivity of the material will decrease as shown in Figures 3, 4 and 5.

The theoretical model presented in Figure 5 was made using different concentrations of masterbatch and polymer; with their respectively thermal diffusivity measurement. It shows the behavior of the thermal diffusivity of the blend material as a general approximation and trend. Unfortunately it is unavailable to give an accurate value of the thermo-physical property of the mixture, or give the particular behaviors of the material at various concentrations given by the amorphous and crystalline competition of the heat transport mechanisms domains.

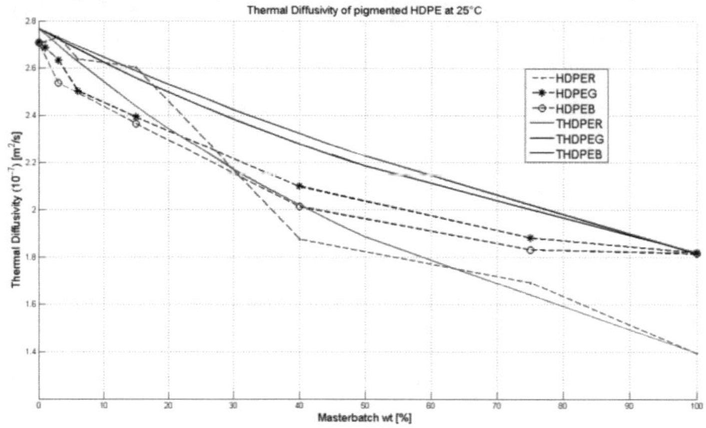

Figure 5. Theoretical model (continuous line) and experimental values of thermal diffusivity for the different mixtures.

4. CONCLUSIONS

The thermal diffusivity of HDPE with the addition of pigmentation masterbatches at variable concentration was determined by the DXF200. The results show a clearly dependence of the thermo-physical properties of a composite polymer with the temperature, these relation is inversely proportional as expected theoretically because of the mechanisms of the diffusion of heat in materials.

It is evident that the nature of the pigmentation masterbatch has an influence in the thermo-physical properties of the polymeric materials. The decrease on the thermal diffusivity because of the addition of the masterbatch is explained by the contribution of the LDPE matrix that induces a higher percentage of amorphous phases to the compound, as shows the results for different masterbatch with variable concentration of inorganic and organic components. In the other hand the increase of the diffusivity at high temperatures (near to fusion) is due to the decrease of the heat capacity of the blended materials at this point compare to the pure HDPE.

Finally the theoretical model for the thermal diffusivity of composites is a good predictor of the behavior of this property. However it is not suitable at low levels of fillers and temperatures near to the fusion of the material. This is because in this stage and concentrations the material would be affected by the competence of the formation of crystalline and amorphous morphologies.

References

1. dos Santos, W. N., de Sousa, J. A., & Gregorio Jr, R."Thermal conductivity behaviour of polymers around glass transition and crystalline melting temperatures". Polymer testing , pp 987-994, 2013.
2. Mikulenok, I. O. "Determining the thermophysical properties of thermoplastic composite materials". International Polymer Science and Technology, pp 9-13, 2012.
3. Y. Cengel, "Heat and Mass Transfer". McGraw Hill, 2007.
4. T. Osswald and G. Menges, *Materials Science of Polymer for Engineers,* Munich: Hanser, 2003.
5. S, Agrawal, *"Effective thermal conductivity of epoxy matrix composites filled with titania powder",* M.Sc. Thesis, NIT Roureka Department of Mechanical Engineering , 2011.
6. M. Oliviera, M. Cramez and R. Crawford, *Effect of pigmentation on the microstructure and properties of rotationally molded polyethylene,* Journal of Material Science, 1998, pp.4869-4877.
7. Clariant International, "The coloration of plastics and rubbers". Muttenz, 2010.
8. Sabic, "Injection HDPE" recover the 25 January 2015, from Sabic: http://www.sabic.com/corporate/en/images/M40060S_tcm12-5908.pdf, 2014.
9. TA Instruments, "Thermal diffusivity by the Flash Method" recover the 21 January 2015, from TA instruments: http://thermophysical.tainstruments.com/PDF/technotes/TPN68%20Thermal%20Diffusivity%20by%20the%20Flash%20Method.pdf

Scale up of Polypropylene/Carbon Nanotubes Nanocomposites: Dispersion Improvement by Means of Simulation Software

L. Roca[1], B. Galindo[2].

[1] AIMPLAS-Plastics Technology Centre, lroca@aimplas.es, Spain.

[2] AIMPLAS-Plastics Technology Centre, bgalindo@aimplas.es, Spain.

In this work Co-rotating twin screw extruder (Co-TSE) simulation software was employed to optimize the dispersion of carbon nanotubes in the polypropylene nanocomposites scale-up from 25 mm diameter (Co-TSE) to 50 mm. To do so different processing conditions and screw configurations were simulated employing the Co- rotating Twin Screw simulation software from SCC Consultants Ludovic©. Materials employed where PP Metocene HM562P from lyondellbassell and CNT NC 7000 from Nanocyl.

Once a good degree of dispersion was obtained running at certain conditions in the 25 mm Co-TSE these conditions where simulated in the 50 mm Co-TSE. In order to assess the degree of dispersion optical microscopy was done following the standard ISO 18553:2002, with the resultant image showed in fig 1. Dispersion was measured by optical microscopy and statistically treated by particle counting software (Leica MW software) were agglomerate lenght and agglomerate density gave the mixing quality index see table 1.

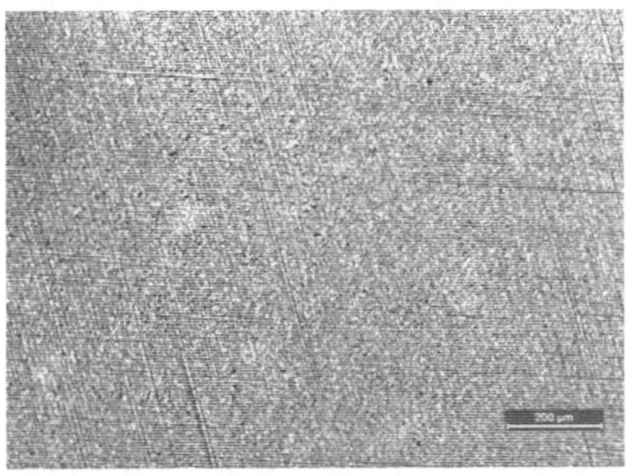

Table 1.Dispersion Quality of 25 mm Co-TSE compounds

Average Length (µm)	Agglomerates Density (%)
10.6	0.4

Figure 1. Quality dispersion of 25 mm Co-TSE PP+1% CNT compound

The procedure followed to make the scale up was to match Specific mechanical Energy (SME) directly linked with the mixing energy and hence with dispersion and residence time. Simulation with Ludovic software aimed to match the same Specific mechanical energy and residence time in both extruders to have the same degree of dispersion, in table 3 were shown the different SME and residence time corresponded to the good dispersion index (figure 1) in the column name AIMPLAS and the other ones to the different screw modifications in the industrial scale Co-TSE. All of them gave very different residence times and SME in the simulation. Processing conditions were shown in table 2

Table 2. Processing conditions.

Screw speed (rpm)	Production rate (kg/h)	Temperature profile (°C)
800	14	190/185/180/175/170

Table 3. SME and residence time comparison simulation

	AIMPLAS	COMPANY screw 1	COMPANY screw 2	COMPANY screw 3	COMPANY screw 4	COMPANY screw 5
Mean residence time (s)	77	30	60	31	27	20
Specific mechanical energy (KWh/t)	715	335	218	365	299	226

From the results of table 3 any of the screws that the company works with can give similar values of RT and SME than AIMPLAS screw profile, then keeping the same processing conditions the screw profile was modified (Screw 6, see Fig 3) giving the results obtained in table 4, where residence time was very similar unlike than SME that was too lower.

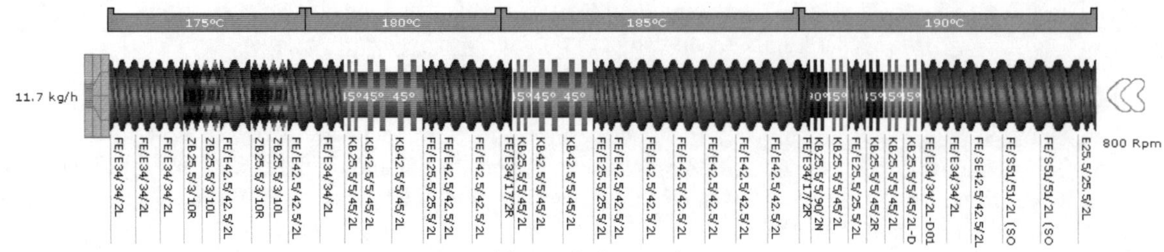

Figure 3. Screw 6

Table 4. SME and residence time comparison simulation

	AIMPLAS	New process parameters
Mean residence time (s)	77	75
Specific mechanical energy (KWh/t)	715	431

This Low SME gave bad dispersion as can be seen in figure 4 and higher values of agglomerate mean size and density, table 5 if compared with the optimum values of table 1.Then was mandatory to increase SME maintaining the RT.

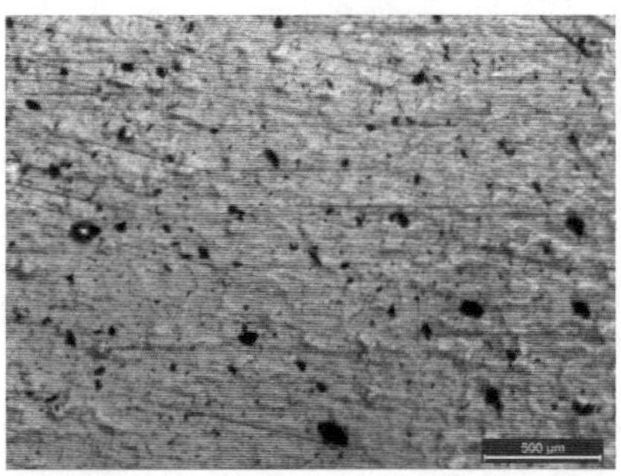

Table 5.Dispersion Quality of 50 mm Co-TSE compounds

Average Length (µm)	Agglomerates Density (%)
23.8	1.9

Figure 4. Quality dispersion of 50 mm Co-TSE PP+1% CNT compound

To improve dispersion new screw elements (reverse ones in red color) where incorporated in order to increase shear to screw 6, screw 7 was shown in figure 5

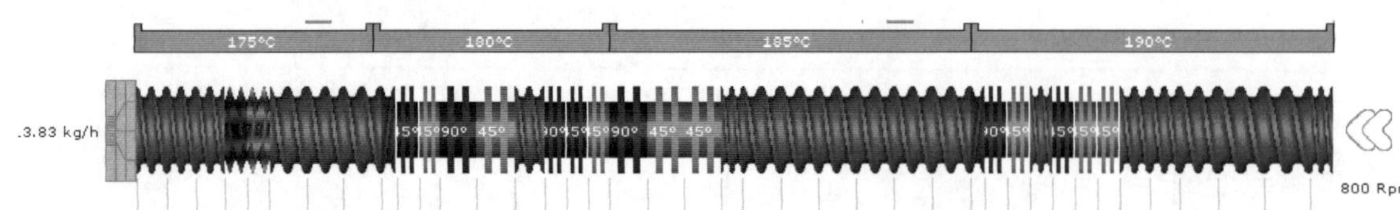

Figure 5. Screw profile 7, with reverse elements

Working with this new screw at the same conditions that where detailed in table 2 the results of SME and RT where very similar to the ones of AIMPLAS lab trials as can be seen in table 6.

Table 6. SME and mean residence time comparison after modifications suggested by simulation

	25 mm Co-TSE	50 mm Co-TSE
Mean residence time (s)	77	66
SME(KWh/t)	715	736

As a result of the SME increasing the dispersion get improved reducing the average length of the agglomerates, table 7 and also its density compared with table 5.

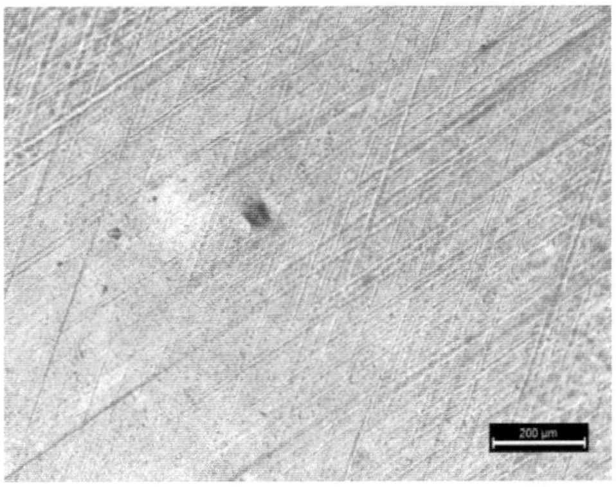

Table 7.Dispersion Quality of 50 mm Co-TSE compounds

Average Length (μm)	Agglomerates Density (%)
14.6	0.11

Figure 6. Quality dispersion of 50 mm Co-TSE PP+1% CNT compound

CONCLUSIONS

Co-TSE Simulation software combined with conventional technics for measure dispersion gave optimum results in the scale up of PP+ 1% of CNT for multifilament melt spinning allowing the obtaining of an antistatic fabric, see figures 7 and 8.
With previous screw configurations due to the high amount of agglomerates the melt spinning die head get blocked not allowing the multifilament production.

Figure 7. Multifilament spinning of PP + 1% CNT

Figure 8. PP + 1% CNT fabric

Compression properties of the inorganic filler-thermoplastic composites

Yukinori Sano[a], Tastuya Tanaka[a] and Midori Betsudan[b]

[a]*Applied Materials Engineering Laboratory, Faculty of Engineering, Doshisha University, 1-3, Tatara Miyakodani, Kyotanabe, Kyoto, 610-0321, Japan – dup0557@mail4.doshisha.ac.jp*
[b]*JOTO TECHNO Co., Ltd., 2-11-1, Shodai - Tajika, Hirakata, Osaka, 573-1132, Japan — betsudan@joto.com*

Abstract. In earthquake-prone country Japan, in order to ensure the earthquake resistance and breathability of housing, the ventilator windows have been provided on the concrete foundation. However, at the time of the earthquake, the foundation for house was collapsed due to stress concentration occurring in the ventilator window. Recent year in Japan, instead for this ventilator window, a packing for house which is used the inorganic filler-thermoplastic composites have been attracted. By separating the foundation for house and the ventilator windows, to prevent the collapse of foundation for house. For the further growing demand for packing, there is a need to clarify the relationship between the inorganic filler-thermoplastic composites and compression strength. However, there have not been studies which focused on this relationship. In this study, we prepared a different particle diameter of the inorganic filler for composites. We compared for dispersion of the filler within the molded specimens which was measured from X-ray images and compressive strength. From the above, we investigated what kind of influence a different particle diameter of the inorganic filler is to compression properties of the inorganic filler-thermoplastic composites.

Keywords: Compression properties, the inorganic filler-thermoplastic composites
PACS: 62.20.-x

INTRODUCTION

It is said that Japan is an earthquake nation. It is impossible to predict a time and place of earthquake occurrence and a magnitude of the earthquake. As result of "January 17, 1995: The Great Hanshin Awaji Earthquake" and "March 11, 2011: The Great East Japan Earthquake", a lot of people have been dead or missing and became historical disasters. There is no telling when an earthquake will occur, so earthquake resistance of housing is paid the most attention now. Especially the Great Hanshin Awaji Earthquake has showed that the earthquake resistance of housing is critically important. Not only to the economic damage of the building itself, it is such as shown in the following was occurred. First, a lot of human damages were caused by house collapse. Second, rescue activities such as evacuation rescue, firefighting activities and restoration was obstructed, since road blockage had caused by debris scattering of collapsed houses. Third, firefighting activities did not proceed smoothly, since the rescue team had to rescue residents trapped in the affected house. [1] In order to ensure the earthquake resistance and breathability of housing, the ventilator windows have been provided on the concrete foundation. However, at the time of the earthquake, the foundation for house was collapsed due to stress concentration occurring in the ventilator window. Recent year in Japan, instead for this ventilator window, a packing for house which is used the inorganic filler-thermoplastic composites have been attracted. By separating the foundation for house and the ventilator windows, to prevent the collapse of foundation for house. For the further growing demand for packing, there is a need to clarify the relationship between the inorganic filler-thermoplastic composites and compression strength. However, there have not been studies which focused on this relationship. Purpose of this study is to clarify the relationship between the inorganic filler-thermoplastic composites and compression strength.

In this study, we prepared a different particle diameter of calcium carbonate (CaCO3) flour using the sieve (38μm, 75μm, 180μm – mesh). The raw material for inorganic filler-thermoplastic composite was Polypropylene (PP) and each of CaCO3 flour. We compared for compressive strength. To investigate the dispersion of CaCO3 flour contained in the compressive specimens, we observed composites internal structure using the X - ray CT. From tomographic image date, dispersion was evaluated using fractal value. We investigated what kind of influence a different particle diameter of CaCO3 is to compression strength of the PP/CaCO3 composite.

Material composition analysis model was created and this model was calculated simulation of compressive test using computer soft. Each analysis model has aspect ratio 0.7, 1.0 and 1.4. We investigated the influence of aspect ratio on compression strength.

MATERIALS

As matrix of inorganic filler-thermoplastic composites, PP (J108M, Prime Polymer Co., Ltd.) was selected. Inorganic filler used in this study was CaCO3 flour (LP-90, OMI MINING CO., Ltd.). This CaCO3 flour was classified by the classification device such as Fig.1. CaCO3 flour was placed the sieve (φ200×60mm/38μm,75μm,180μm – mesh, AS ONE Corporation.). CaCO3 flour was sucked from the bottom of sieve with vacuum cleaner (DYSON V6 TRIGGER, Dyson Limited.). About the specific method of classifying, first, CaCO3 flour was classified with 38μm – mesh sieve. This CaCO3 flour was called for "under 38μm" which could not pass through 38μm – mesh sieve. Second, CaCO3 flour which could pass through 38μm – mesh sieve was classified with 75μm – mesh sieve. This CaCO3 flour was called for "38～75μm" which could not pass through 75μm – mesh sieve. At last, CaCO3 flour which could pass through 75μm – mesh sieve was classified with 180μm – mesh sieve. This CaCO3 flour was called for "75～180μm" which could not pass through 180μm – mesh sieve. Each of CaCO3 flour was observed with optical micro scope (MT9430, MEIJI TECHNO CO., LTD.). From two dimensional image date of each of CaCO3 flour, we measured particle diameter with image processing soft (Sigma Scan Pro). Fig.2 shows a particle diameter distribution curve on a mass basis. From this figure, it understand that CaCO3 flour was able to classify as planned.

To improve the adhesion of the CaCO3 flour, MAPP (Youmex 1010, Sanyo Chemical, Acid number 24.9) was used as a compatibilizing agent.

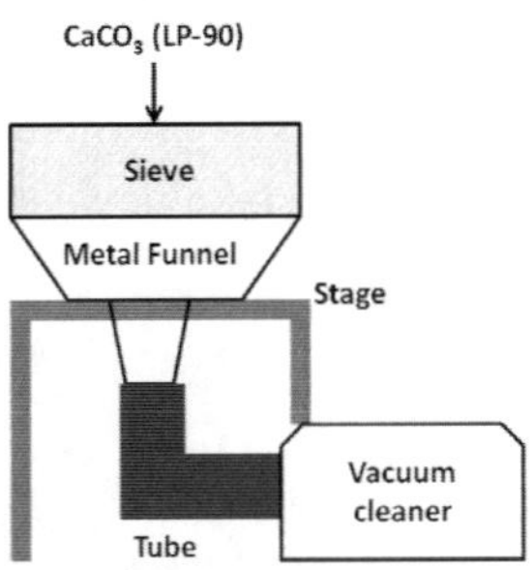

Fig.1. Classification device

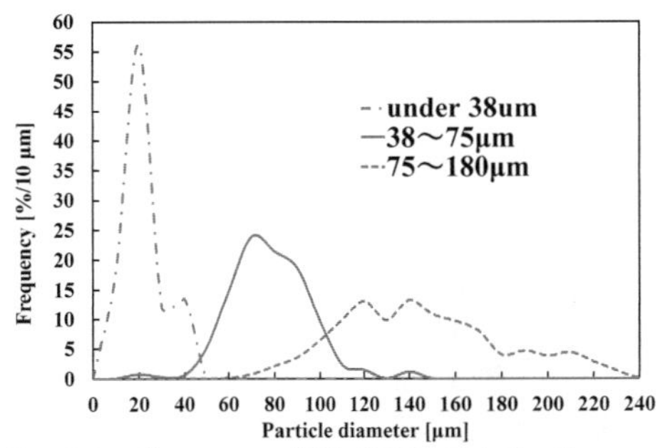

Fig.2. Particle diameter of CaCO3 distribution curve

PREPARATION OF PP/CACO3 COMPOSITE

Preparation of PP 45wt%/CaCO3 55wt% composite and PP 40wt%/CaCO3 55wt%/MAPP 5wt% composite by the twin screw extruder method. A co-rotating twin-screw extruder (ZSK18, Coperion) with a screw diameter of 2 mm and a length-to-diameter ratio of 40 was used for all compounding. The standard compound screw was composed of conveying elements and kneading discs with a screw speed of 200 rpm. The process temperature was set to 230°C from the hopper to the die.

First, the CaCO3 were dried in an oven at 80°C for 24 h. Then, PP, CaCO3 and MAPP were dry mixed thoroughly before they were fed into the twin-screw extruder and compounded under the conditions mentioned above. The CaCO3/PP composites extruders were immediately quenched in water and then cut into strands before being chopped into granules and dried. All specimens were injection molded into cuboid-shaped block using a single screw injection-molding machine (STX2055V, NISSEI PLASTIC INDUSTRIAL CO., LTD.) with a barrel temperature of 230°C. An end-gated mold was used for molding cuboid-shaped samples. The thickness, width and height of the specimens are 10, 10 and 20 mm, respectively. Compressive tests were conducted using these samples.

X - RAY COMPUTED TOMOGRAPHY (X - RAY CT)

To investigate the dispersion of CaCO3 flour contained in PP 45wt%/CaCO3 55wt% composite, we observed composites internal structure using the X - ray CT (SKYSCAN 1172, Bruker Co.). The measurement conditions were following : voltage 80 kV, electric current 100 μA and camera pixel 1.0 μm. Observation samples were cut out from compression specimens and this sample volume was 20×2×2 mm^3. From tomographic image date, dispersion was evaluated using fractal value.

COMPRESSIVE TESTS

The compressive properties of the cuboid-shaped PP 45wt%/CaCO3 55wt% composite and PP 40wt%/CaCO3 55wt%/MAPP 5wt% composite were measured using a universal testing machine (AG-100kNX, Shimadzu) operated at a constant crosshead speed of 10 mm/min. The average compressive strength was calculated by testing five samples. Compression face was $10 \times 10\ mm^2$.

ANALYSIS METHODS

About modelling of mesostructured, three dimensional simplified Representative Volume Element, RVE, composed of soft and hard phase is applied in simulation. RVE was created by multi-scale structure modelling platform, Digimat (e-xstream engineering). PP was modelled as matrix and CaCO3 is modelled as spherical inclusions. Fig.3. shows the modelling of RVE (1×1×1). We prepared four kinds of simplified dual phase RVE model, Figs.3 (a), (b), and (c) which indicate the different mesostructure. Numerical analysis is performed using nonlinear dynamic explicit finite element code, LS-DYNA.

Boundary conditions for computational material testing of dual-phase RVE were following: weight fraction of inclusion 0.55, each mesh node of the inclusions coupled with the mesh node of the matrix. Fig.3 (a), 3 (b) and 3(c), displacement along +x direction is provided at +x face with fixed parallel +y and +z faces. In case of Figs.3 (b) and 3(c), displacement along +z direction is also provided at +z face with fixed parallel +x and +y faces. Piecewise - linear - plasticity behavior for PP and rigid behavior for CaCO3 were assumed with bilinear law and considered so that the result of numerical uniaxial compressive test of Figs.3 (a), (b), and (c).

(a)PP / CaCO3×9 (Aspect ratio 1.0) (b)PP / CaCO3×9 (Aspect ratio 0.7) (c)PP / CaCO3×9 (Aspect ratio 1.4)

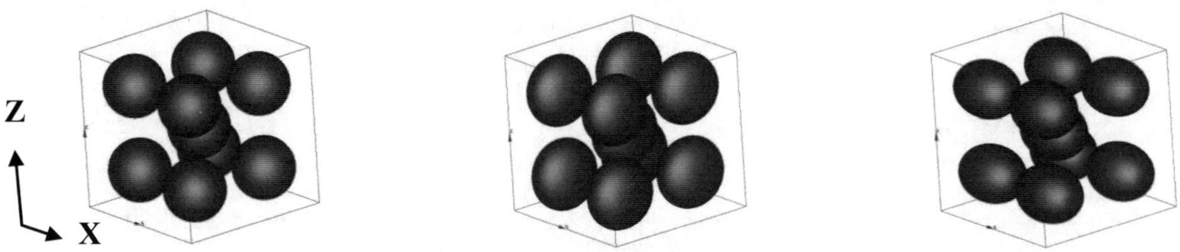

Fig.3. Modelling of RVE (1×1×1)
(a)PP / CaCO3×9 (Aspect ratio 1.0), (b)PP / CaCO3×9 (Aspect ratio 0.7), (c)PP / CaCO3×9 (Aspect ratio 1.4)

EXPERIMENTAL RESULT AND DISCUSSION

Fig.4 shows the particle diameter distribution curve on a mass basis. From this figure, even though CaCO3 flour with under 38μm was molded PP/CaCO3 composites, variation of particle diameter distribution was not seen. CaCO3 with particle diameter of 30～60μm was increased at PP/CaCO3(under 38μm) composites. It is possible that CaCO3 flour with under 38μm was agglutinated. On the other hand, when PP/CaCO3 composites were used CaCO3 flour with 38～75μm and 75～180μm, variation of particle diameter distribution was seen. In case of CaCO3 flour with 75～180μm, PP/CaCO3 composites contained CaCO3 with particle diameter of 20～50μm which was not seen at a simple CaCO3 flour. Since it is not believed that CaCO3 particles were broken by the kneading and molding process, we need to reconsider a way to process X-ray CT images.

Fig.5 shows the result of the dispersion of CaCO3 flour contained in PP 45wt%/CaCO3 55wt% composite. From this figure, the smaller particle diameter, dispersion became better. From Figs.4 and 5, CaCO3 flour with under 38μm was not agglutinated, while maintaining the particle diameter distribution.

Fig.6 shows the result of compressive test. From Fig.6 (a), compressive strength did not differ by particle diameter of CaCO3. From Figs.4 and 5, it had been expected that compressive strength of PP/CaCO3 composites would be improved as particle decrease in diameter. However PP/CaCO3 (under 38μm) composite was slightly smaller than the other CaCO3 flours. As reason for that, adhesiveness on interface between PP and CaCO3 was poor. Even if the particle diameter of CaCO3 is small and good dispersion, when adhesiveness on interface between PP and CaCO3 is poor compressive strength is not able to improve.

From the above, in case of additional 5wt% of MAPP, we investigated the relationship of compressive strength and particle diameter of CaCO3. Fig.6 (b) shows the result of compressive test. From this figure, compressive strength differs by particle diameter of CaCO3. Along with reduction of particle diameter, compressive strength of PP/CaCO3 composites was improved. Comparison of 38～75μm and 75～180μm by compressive strength, even if the smaller CaCO3 flour with 38～75μm was used, compressive strength increased only 1 MPa. On the other hand, when using under 38μm, the compressive strength was increased to 4MPa. Therefore, in case of particle diameter is 75μm or more, it is suspected that compressive strength could not be improved by controlling particle diameter. To obtain compressive strength by controlling particle diameter, it is necessary to at least using CaCO3 flour with under 38μm and improving interface adhesion with PP.

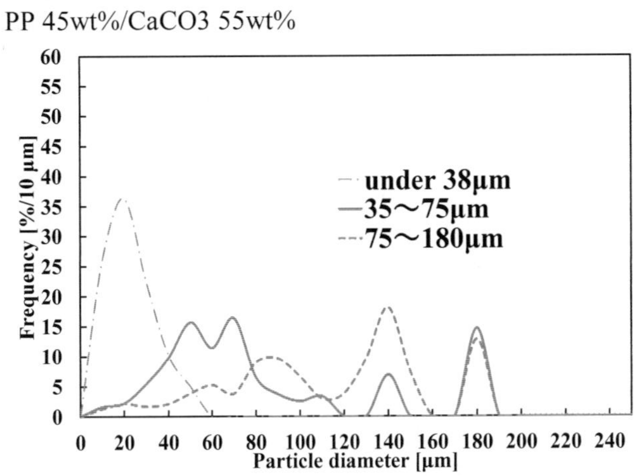

Fig.4. Particle diameter of distribution curve for CaCO3 contained in the compressive specimens

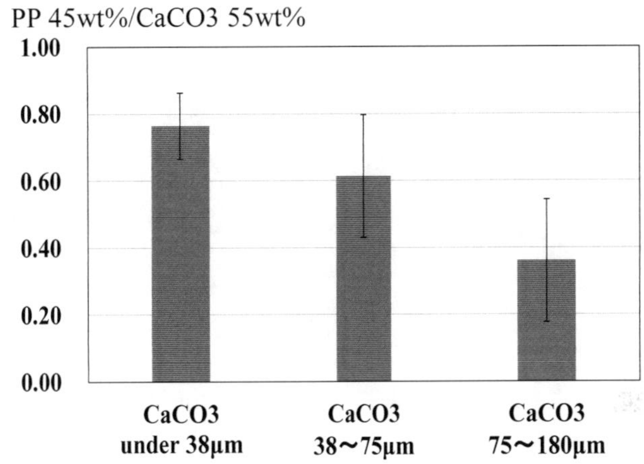

Fig.5. Evaluation of dispersion used fractal value.

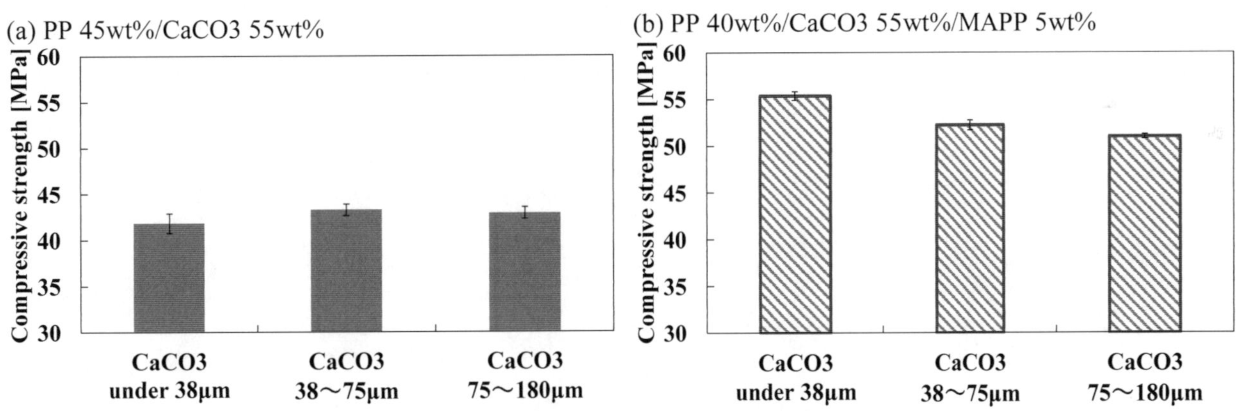

Fig.6. Result of compressive strength
(a) PP 45wt%/CaCO3 55wt%, (b) PP 40wt%/CaCO3 55wt%/MAPP 5wt%

ANALYSIS RESULT AND DISCUSSION

From the image processing of X-ray CT, it was possible to obtain a distribution of the aspect ratio. Fig.7. shows the result. From this figure, at any of CaCO3 flours, it was relatively observed that aspect ratio were 0.6～0.9 and 1.4. Therefore, we considered that the bias of aspect ratio influences compression strength. It was examined the influence of the aspect ratio using the model shown Figs.3 (a), 3 (b) and 3 (c). In all of the models, each mesh node of the inclusions coupled with the mesh node of the matrix.

Fig.8 shows the analysis result of compressive test. From this analysis result, compressive strength differed by aspect ratio. The compressive strength of the model having aspect ratio 1.4 (X-direction) was the highest. However the model having aspect ratio 1.4 (Z-direction) decreased the compressive strength. Even if model has the same high aspect ratio, compressive strength is decreased when displacement direction is different. On the other hand, compared to aspect ratio 1.0, the low aspect ratio was decreased compressive strength. In particular, the compressive strength of the model having aspect ratio 0.7 (X-direction) was the lowest. It was found that compressive strength is influenced by interface adhesion as well as aspect ratio.

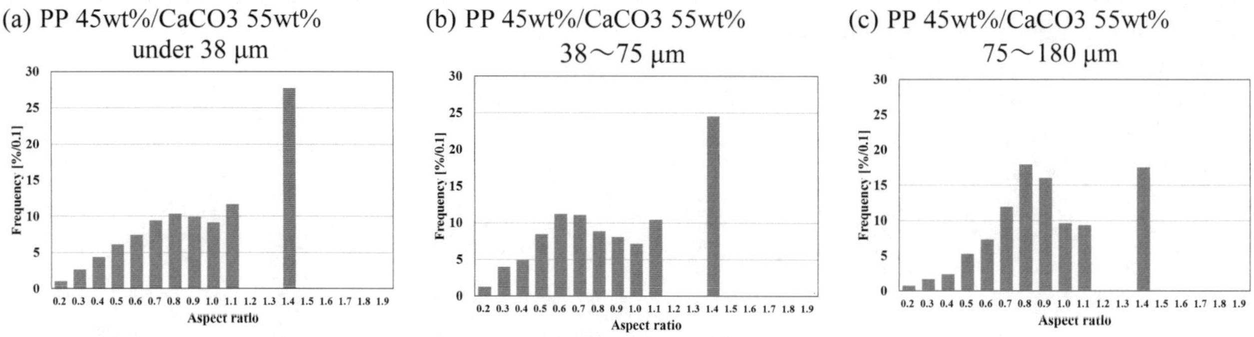

Fig.7. Distribution of the aspect ratio
PP 45wt%/CaCO3 55wt% (a) under 38 μm (b) 38～75 μm (c) 75～180 μm

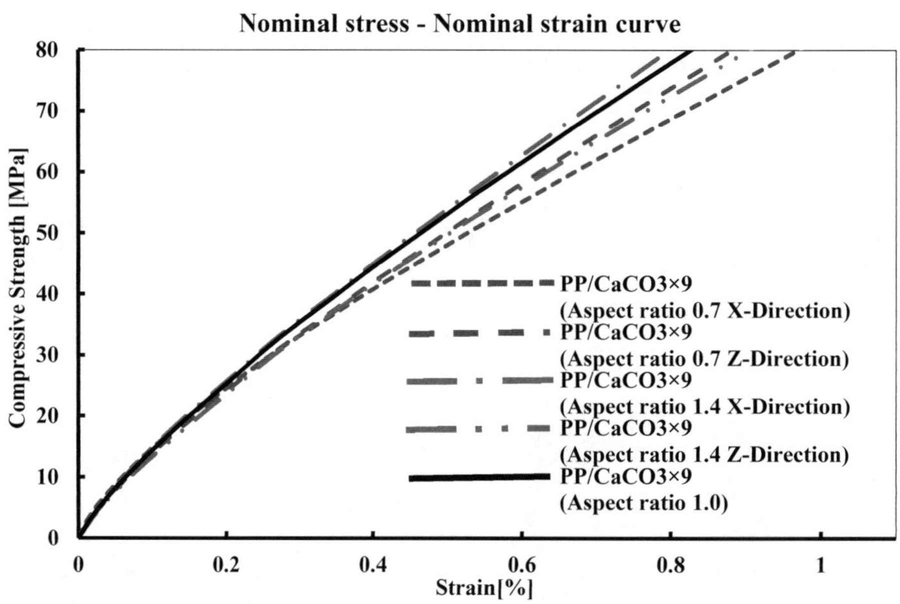

Fig.8. Analysis result of compressive test using the model having difference of aspect ratio

CONCLUSIONS

To obtain compressive strength by controlling particle diameter, it is necessary to at least using CaCO3 flour with under 38μm and improving interface adhesion with a resin. And compressive strength is influenced by interface adhesion as well as aspect ratio.

REFERENCES

1. Masayuki KOHIYAMA, Miho YOSHIMURA and Kimura *MEGURO, INCENTIVES AND DISNCENTIVES TO RETROFITTING Importance of risk communication in programs of earthquake disaster reduction*, J. Environ. Eng., AIJ, NO. 606, 2006, pp. 89-96.

Modelling Mechanical Properties of Glass and Basalt Fiber Filled Thermoplastics

M. Bruchmüller, S. Schneidmadel, J. Geis, M. Koch

Technische Universität Ilmenau, Plastics Technology Department,
Gustav-Kirchhoff-Str. 5, D-98693 Ilmenau – Germany

matthias.bruchmueller@tu-ilmenau.de

Abstract. Filaments and short fibers as fillers are among the most common ingredients for thermoplastics modification in order to impact on the mechanical properties of injection molded parts and maintain cost efficiency at a minimum part weight. In this study, glass and basalt fibers are processed into a thermoplastic matrix to compare achievable mechanical property improvements depending on fiber type, content, length and orientation in various plastic matrix materials.

Keywords: basalt fiber, glass fiber, fiber length, orientation, compound properties, shear lag model
PACS: 81.05.Qk

INTRODUCTION

Mathematical models for compound material properties, like E modulus or tensile strength, are based on the shear lag theory of CHAMIS [1]. The model deals with the mechanical properties of both components and simplifications which can lead to significant deviations compared to experiments.

Additionally, current model values cannot help to improve process conditions to improve compound properties.

EXISTING MODELS FOR MECHANICAL COMPOUND PROPERTIES

Based on the shear-lag theory of endless fiber reinforced plastics by CHAMIS (Formula 1), many modifications where done to improve the precision of injection molded parts.

$$E_C = \phi_f \cdot E_f + (1 - \phi_f) \cdot E_m \tag{1}$$

The first change was made by COX, including fiber dimensions length (Formula 2) and radius (Formula 3) as well as the average distance between fibers [2].

$$\eta_1 = 1 - \frac{\tanh\left(\beta \cdot \frac{l_f}{2}\right)}{\beta \cdot \frac{l_f}{2}} \tag{2}$$

$$\beta = \sqrt{\frac{2 \cdot \pi \cdot G_m}{E_f \cdot A_f \cdot \ln\left(\frac{R}{r}\right)}} \tag{3}$$

Furthermore, KRENCHEL found relations between fiber orientations and mechanical properties at injection molded parts. KRENCHEL created a sum formula to include fiber orientations at their angular offset from the load direction (Formula 4).

$$\eta_o = \sum a_f \cdot cos^4(\varphi_f) \tag{4}$$

The two factors, η_o and η_1, where multiplied with fiber Young's modulus of the shear lag theory formula to consider their influence on compound Young's modulus.

$$E_C = \eta_0 \cdot \eta_1 \cdot \phi_f \cdot E_f + (1 - \phi_f) \cdot E_m \quad (5)$$

This formula represents the calculation model for a compound material made of any type of fibers and matrix materials. Fiber matrix boundary layers are simplified by the assumption of ideal adhesion. [3]

EXPERIMENTS

Materials and General Procedural Method

Glass and basalt fibers were compounded into the matrix materials polypropylene and polyamide 6 using a co-rotating twin screw extruder in a Design of Experiment (DoE). Microscopic and density measurements of granulated material reveal fiber lengths and contents. Correlations between measurements and DoE present regions of high fiber lengths. Process parameters, affecting the highest fiber lengths, were used to produce the basic material for injection molding investigations as well. The compounds were molded into specimen for further measurements.

Measurement Methods

To measure fiber content, calcination methods were used to determine the weight of compounds and sole fibers [Nor98]. After that, fibers were separated and their lengths were measured using a light microscope with a camera. In addition, injection molded tension ties were separated vertical to the tension direction in two equal parts to analyze the fiber orientation and distribution.

Compounding

A co-rotating twin screw extruder with a diameter of 40 mm was used. The screw design is shown in FIGURE 1.

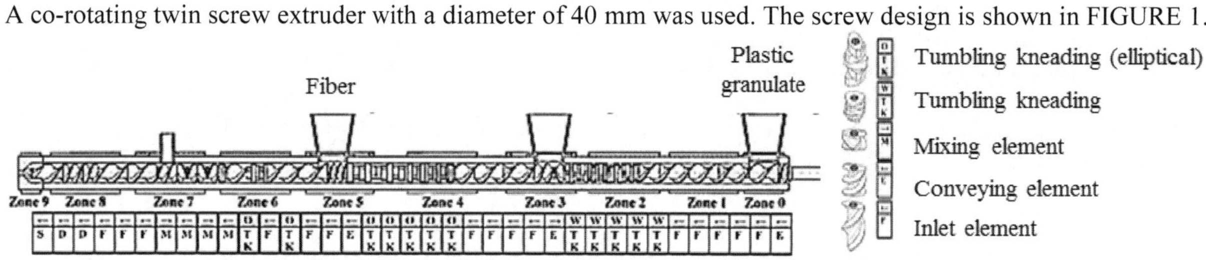

FIGURE 1: Chosen screw design for extruder experiments [4]

The DoE allows drawing conclusions from process parameters and fiber lengths. Thus, rotation, throughput and melt temperatures were documented to find influences of fiber length reduction by experimental design.

The chosen DoE is a combination of a star and a cube and allows drawing conclusions from linear and squared dependencies. Additionally, it shows linear interactions of two or more input values. The experiment was done by using variations of three parameters. Parameters and variations are shown in TABLE 1.

TABLE 1: Varied parameters in a DoE for the compounding process

Compound Parameter	*Variation*
Screw Speed [rpm]	50 – 150
Throughput [g/min]	50 – 250
Melt Temperature [°C]	200 – 260

Shear rate and residence time were identified as primary influences on fiber degradation, distribution and orientation. Thus, rotational screw speed, throughput and melt temperature are the main process parameters to affect them. Shear rate is directly proportional to screw speed. Combined with throughput, fill rates were calculated. The fill rate influences residence time as well. Melt temperatures affect viscosity, melt density and residence times. Ranges of parameters were checked in pilot tests. The throughput was limited by torque rates at 250 g/min and screw speed of 50 rpm. The pilot tests offer a good fiber distribution unaffected by screw speeds, however, high screw speeds shorten fibers intensely. Thus the maximum screw speed was limited to a maximum of 150 rpm. Melt temperature ranges were chosen according to manufacturer's data to avoid decompositions.

Injection Molding

An injection molding machine with a screw diameter of 35 mm, a standardized screw and a clamp force of 80 tons was used. The chosen DoE method equals the configuration of the compounding process of the input values screw speed and melt temperature. It differs in the choice of back pressure and volume flow rate. Because of the chosen method, conclusions could be drawn towards linear and squared dependencies as well as interactions between the values. Fiber length reductions due to injection molding were analyzed with the four parameters by the variation shown in TABLE 2.

TABLE 2: Varied parameters in a DoE for the injection molding process

Molding Parameter	*Variation*
Screw Speed [rpm]	25 – 150
Back Pressure [bar]	5 – 100
Volume Flow Rate [cm³/s]	25 – 96
Melt Temperature [°C]	220 – 260

Main effects of the varied parameters can be linked to shear rate and residence time. The back pressure effects the material compression and the region of the melt bed along the screw, which affects the residence time of molten compounds. Melt temperatures and screw speed dependencies from shear rate and residence time are mentioned in the compounding process description. During the injection process, the flow rates affect both shear rate and injection time. Because of the compounding process results and pretests of injection molding, screw speed variation was changed from 25 rpm up to 150 rpm. Back pressure and volume flow rate are varied up to the maximum ranges of the machine. Melt temperature ranges were chosen according to manufacturer's data, to avoid decompositions.

FIBER LENGTH INFLUENCE DURING PROCESSING

Processing of fibers in plastic materials by twin screw extruders and injection molding machines result in fiber length reduction due to shear stresses. Varied parameters done with DoE result in a significance analysis clearly represented in Pareto charts. Significances of single parameters and their interaction with the influences on fiber lengths are shown in FIGURE 2 for compounding and FIGURE 3 for injection molding. Additionally, the directions of influences are shown.

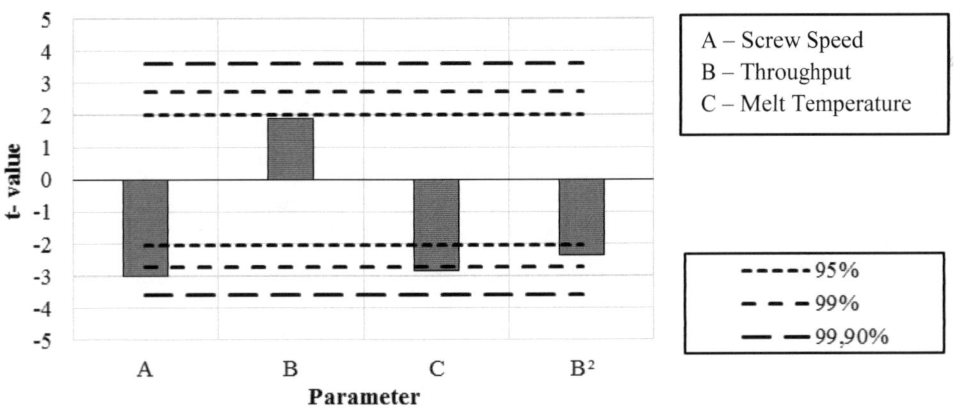

FIGURE 2: Pareto-Chart for classification into significant and indifferent properties during compounding

The results of DoE show that increases in screw speed and melt temperature shorten fibers at a linear dependence. A higher screw speed results in a higher shear rate, which lead to more likely collisions between fibers and machine parts. Throughputs affect fiber lengths at a squared relation. From low throughput values up to the maximum, the increasing filling rate reduces the area to volume ratio. Fiber to machine interactions can be averaged to a bigger volume per spiral. Thus, fiber lengths are preserved at higher throughputs up to a peak. The different spiral dimensions along the screws of a co-rotating extruder combined with an increasing throughput can fill regions of the screw and increase the pressure at and in front of filled elements. This leads to higher shear rates. Thus, further increases of throughput shorten fibers. The knowledge of these dependencies allows correlations between shear rate, residence time and fiber lengths by means of regression models.

$$l_f = (1003 - 6{,}37 \cdot \dot{\gamma}_e - 1{,}38 \cdot \bar{t}_e) \cdot \left(152{,}28 - e^{-0{,}065 \cdot \phi_f}\right) \quad (6)$$

The formula above allows a prediction of fiber length at average shear rate, average residence time and fiber content. Because of the averaged values, actual deviations of the fiber lengths vary up to 20 %.

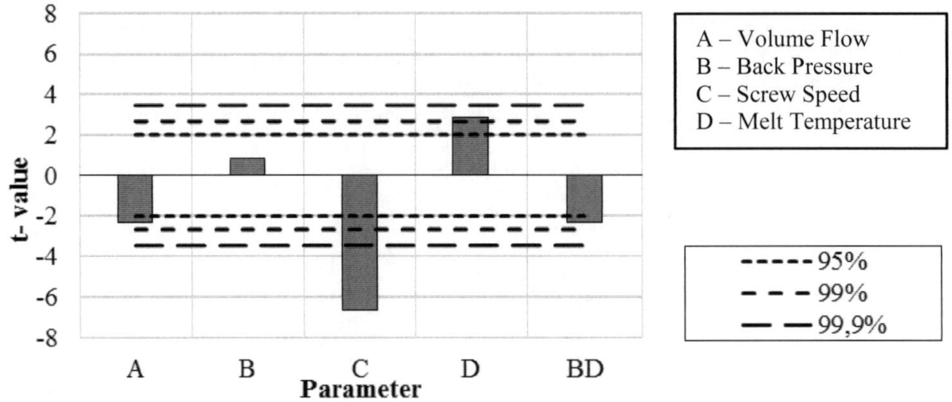

FIGURE 3: Pareto-Chart for classification into significant and indifferent properties during injection molding

Volume flow rates have linear and negative influences on fiber lengths. Low screw speeds and injection velocities create low shear rates, which affect less shortened fibers. Increasing melt temperatures decrease the viscosity of melt in the compound. Well distributed fibers cannot interact with their neighbors and are protected by the melt surrounding them. Thus, increasing melt temperatures preserves the fiber lengths in injection molding processes. Additionally, the melt temperature interacts with the back pressure. Low back pressures, combined with increasing melt temperatures, affect unshorten fibers at a lower compression, while plasticizing granulate into melt. Higher back pressures need to be avoided during plasticizing, which leads to longer plasticizing times. Higher plasticizing times increase residence times at high shear rates. As a result of high melt temperature, melt viscosity decreases and work against the high back pressure cannot be kept. Therefore, high back pressures and melt temperatures shorten fibers.

CORRELATION OF MEASUREMENTS AND MODEL

Using the example of Young's modulus, the tests were compared to the existing models of fiber reinforced plastics. The correlations were proved for the models of COX and KRENCHEL. FIGURE 4 shows the deviation between models and measurements by glass and basalt fiber reinforced polypropylene.

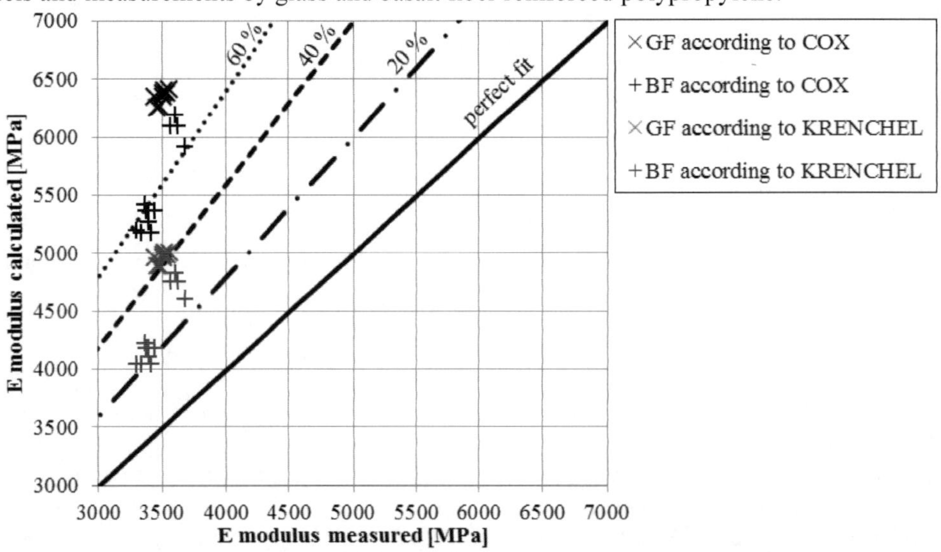

FIGURE 4: Comparison of measured and calculated Young's modulus by COX (black) and KRENCHEL (red)

The COX model overestimates compound Young's modulus by 60 %. The consideration of fiber length and mechanical properties is not significant enough. Thus, KRENCHEL included a fiber orientation factor to adapt the calculation model to the experimental results. In case of fiber reinforced polypropylene, the modification decreases the deviation by 20 to 45 %. In the studied case, this still leads to an enormous difference between calculated and measured values ranging from 550 up to 1500 MPa. Due to the fact that the models consider mechanical properties of both components, the bonding interaction between fibers and matrix material is considered ideal. Considerations of adhesion will improve the validity of the KRENCHEL model to avoid high failure rates.

CONCLUSION

Shear rates and melt residence times of compounding processes can be used to describe compounded fiber lengths. Thus, these parameters allow a prediction of resulting fiber lengths due to design processes of co-rotating twin screws and extruders. A key for improving the existing shear lag model is to include bonding interactions between fibers and plastic using surface energy.

ACKNOWLEDGEMENTS

The results presented are obtained in the context of the research project KF2184742EB4 sponsored by the Bundesministerium für Wirtschaft und Energie (BMWi). We thank the BMWi for their support.

REFERENCES

1. Chamis C.C. Mechanics of composite materials: past, present and future. J. Compos Technol Res ASTM 1989; 11: 3–14
2. H.L. Cox, *The elasticity and strength of paper and other fibrous materials*, Teddington, Middlesex: The National Physical Laboratory, 1951
3. H. Krenchel, *Fibre reinforcement,* Copenhagen: Akademisk Forlag, 1964
4. S. Schneidmadel, *Bedingungen für die Einarbeitung leitfähiger Füllstoffe in Thermoplaste*, Ilmenau: Technische Universität Ilmenau, Fachgebiet Kunststofftechnik, 2012

Processing and properties of PLA-HA nanocomposites: the effect of particle morphology and dispersants

Magdalena M. Tomczynska (presenting author)[a], Michael Ward[a], Gabriel Y. H. Choong[a], Kirsty Walton[a], Davide S. A. De Focatiis[a], David M. Grant[a], Derek J. Irvine[a] and Andrew J. Parsons[a] (reference author)

[a]*Faculty of Engineering, University of Nottingham, Nottingham, NG7 2RD, UK*

Magdalena.Tomczynska@nottingham.ac.uk
Andrew.Parsons@nottingham.ac.uk

Abstract: Biodegradable polylactic acid nanocomposites for orthopaedic implants require optimum particle dispersion and high molecular weight in load bearing applications. Novel coated hydroxyapatite nanoparticles can offer new opportunities for enhanced dispersion during melt compounding. Nanocomposites compounded with a Minilab recirculating twin screw extruder have improved flexural strength when compared to the pure polymer. Dispersion of particles within the matrix was determined from transmission electron micrographs. Chromatography revealed that the polymer matrix of nanocomposites compounded over dry nitrogen had a higher molecular weight than that over air, and hence should exhibit delayed degradation within the body.
Keywords: Polylactic acid, nanocomposites, hydroxyapatite, melt compounding, TEM, GPC, flexural strength.
PACS: 81.05.Qk, 81.20.Hy, 83.50.Xa.

INTRODUCTION

Biodegradable polymeric implants are desirable alternatives to metallic orthopaedic implants such as screws and plates. Advantages include a reduction in stress shielding, radio-opacity and elimination of the requirement for a secondary removal operation, as their biodegradation products are extracted by natural metabolic pathways. An excellent candidate for biomedical applications is high molecular weight polylactic acid (PLA), which is a commercially available biodegradable polyester currently used for maxillofacial restorations, suture anchors and small screws. However, for load bearing conditions PLA requires reinforcement to match bone properties (cancellous bone: compressive strength between 2-20 MPa and elastic modulus between 0.1-2 GPa; cortical bone: compressive strength 100-200 MPa and elastic modulus 7-25 GPa) [1, 2].

Hydroxyapatite (HA), a natural constituent of bone, is favourable bioresorbable filler for PLA. Novel nano-sized hydroxyapatite particles (HANP) produced at Nottingham offer the opportunity to create resorbable nanocomposites with properties closer to those of bone. However, these properties depend on an optimum HA dispersion, which is challenging to achieve. Our previous work [3] introduced PLA-HA nanocomposites that made use of tailored HANP coatings to aid dispersion during melt compounding. This study investigates the effectiveness of these novel tailored dispersants on particle dispersion, melt properties and on macroscopic mechanical properties.

MATERIALS AND METHODS

Resomer LR706S - Poly(L-co-D,L-lactide) (PLDLLA) was supplied by Evonik Industries AG with a molar mass $\underline{M}_w$ of 445.9 kDa [4]. HANP with platelet and rod morphologies were synthesised via a hydrothermal counter-flow process [5 - 7] with dispersant coatings added in-situ [8, 9]. Post-production they were freeze-dried and stored in a dry anaerobic glove box. For dispersion, neat dodecenylsuccinic anhydride (DDSA) was purchased, whilst short-chained PLA with an isosorbide head group (is-PLA) [8, 9] and star shaped PLA 3 kDa (star-PLA) were polymerised via a standard ring-opening route involving lactide, a tin catalyst [10] and an isosorbide initiator. Nanoparticles were functionalised with dispersants with varying efficiency producing coated HANP with 50 wt% is-PLA8, 10 wt % DDSA, 30 wt% HANP is-PLA16, 35 wt% HANP star-PLA and 55 wt% HANP is-PLA24, measured via thermogravimetric measurements (TGA).

Nanocomposites were produced in a twin-screw recirculating extruder HAAKE MiniLab II fitted with co-rotating conical screws. Two pressure transducers in the recirculation channel provided pressure difference measurements used to determine the wall shear stress (τ_w). Materials were compounded at 210 °C with a screw speed of 50 rpm for 15 min in ambient air (Air) or in bottled nitrogen (N_2) at 1 bar and a flow rate of 0.3 l min^{-1} which was in-line dried (N_2D) with a liquid nitrogen moisture trap to remove residual water.

Prior to compounding, PLDLLA was dried at 50 °C under vacuum for a minimum of 8 hours to remove residual moisture, since water causes excessive hydrolysis during melt processing [11]. HANP were additionally dried for compounding in nitrogen: coated platelet HANP at 50 °C under vacuum for 2 hours and neat HANP at 300 °C in conventional oven for a minimum of 8 hours. Dried materials were transferred directly to the MiniLab compounder.

After extrusion, the molecular weights of neat polymer and nanocomposites were analysed using size-exclusion chromatography (SEC) in tetrahydrofuran (THF), calibrated with polystyrene standards. The viscosity average molecular weight (M_v) was used to describe the molecular weight of the nanocomposites:

$$M_{\mathrm{v}} = \left[\frac{\sum M_{\mathrm{i}}^{1+a} N_{\mathrm{i}}}{\sum M_{\mathrm{i}} N_{\mathrm{i}}} \right]^{\frac{1}{a}} \tag{1}$$

where N_i is the number of moles of each polymer species of molar mass M_i and a = 0.736 [12] is the Mark–Houwink exponent for PLA in THF, relating the intrinsic viscosity to molar mass.

TGA tests were performed to confirm the HANP content for all the HANP-PLDLLA extrudates and the neat PLDLLA using a TA Instrument TGA Discovery, with a nitrogen purge gas flowing at 40 ml min^{-1}. Materials were heated up at 5 °C min^{-1} up to 500 °C and burned off isothermally for 20 min at 500 °C. Results were normalised by subtraction of the residual ash mass of neat PLDLLA.

A Transmission Electron Microscope, FEI TECNAI G2 BioTWIN with tungsten filament gun at an accelerating voltage of 100 kV (TEM FEI) was used to investigate the nanocomposite morphology directly after extrusion. Sections ~100 nm thick were cut using a glass knife and supported on copper grids.

For mechanical testing, the extrudates obtained from the MiniLab were compression moulded (CM) at 170 °C for 15 minutes to manufacture isotropic specimens subsequently used to produce minibars. Mechanical tests were carried out using an in-house miniature 3-point bending rig with specimen sizes of 7 x 2 x 0.5 mm. The initially produced air extrudates were injection moulded (IM) directly after compounding, with a HAAKE MiniJet injection moulding machine at 205 °C with 800 bar pressure and a mould at 45 °C. In an attempt to reduce the degree of polymer degradation, the injection moulding step was subsequently removed.

RESULTS AND DISCUSSION

TABLE (1). HANP- PLDLLA nanocomposites and PLDLLA recirculated in MiniLab in Air or N_2D. Summary of: HANP and coating wt% versus nominal loading; viscosity average molar mass M_v; wall shear stress τ_w after 15 minutes recirculation; flexural strength σ measured on moulded bars. ^{a} TGA wt% results; ^{b} wt% of coating in nanocomposite; ^{c} IM; ^{d} CM. Values are listed in ascending order of M_v.

Sample Name	Gas	HANP ^{a} [wt%]	Coating ^{b} [wt%]	M_v [kDa]	τ_w [kPa]	σ [MPa]
(1) 10 wt% HANP	Air	8.75 ±0.03	-	101.6 ±1.8	7.3 ±0.1	n/a
(2) 4.9 wt% HANP is-PLA8	Air	1.72 ±0.07	1.72	117.4 ±0.7	69.9 ±4.7	94.3 ±1.2^{c+d}
(3) 2.8 wt % HANP DDSA	Air	2.37 ±0.04	0.26	121.8 ±1.1	66.7 ±3.8	n/a
(4) 5 wt% HANP	Air	4.42 ±0.04	-	150.8 ±2.6^c	23.6 ±0.2	131.3 ±10.9^{c+d}
(5) 3.6 wt% HANP is-PLA16	Air	2.38 ±0.08	1.02	175.5 ±1.2	49.8 ±0.4	n/a
(6) 2.5 wt% HANP	Air	2.29 ±0.16	-	181.6 ±2.0	44.0 ±0.2	83.6 ±3.0^d
(7) 11.1 wt% HANP is-PLA24	N_2D	4.84 ±0.03	5.92	184.3 ±3.6	42.0 ±0.4	n/a
(8) 5.6 wt% HANP DDSA	N_2D	5.08 ±0.10	0.56	192.7 ±3.2	45.2 ±0.3	n/a
(9) 7.7 wt% HANP star-PLA	N_2D	4.70 ±0.09	2.53	193.3 ±1.2	43.0 ±0.3	n/a
(10) PLDLLA	Air	-	-	193.8 ±4.2	55.0 ±1.6	n/a
(11) 5.6 wt% HANP is-PLA24	Air	2.38 ±0.08	2.91	201.2 ±4.7	44.4 ±0.4	91.3 ±2.4^d
(12) 3.9 wt% HANP star-PLA	Air	2.42 ±0.05	1.30	202.6 ±2.7	46.8 ±0.2	92.8 ±1.6^d
(13) 2.5 wt% HANP rod	Air	2.50 ±0.09	-	206.8 ±2.8	60.8 ±3.4	92.0 ±3.9^d
(14) 1 wt% HANP	Air	0.99 ±0.07	-	216.0 ±4.0	57.7 ±3.1	n/a
(15) 7.2 wt% HANP is-PLA16	N_2D	3.86 ±0.01	1.65	232.8 ±6.0	58.6 ±0.3	n/a
(16) 2.8 wt% HANP DDSA	N_2D	2.60 ±0.09	0.29	271.6 ±5.0	75.6 ±2.4	n/a
(17) 5 wt% HANP	N_2D	4.94 ±0.01	-	274.1 ±7.4	81.4 ±3.5	98.5 ±2.9^d
(18) 3.9 wt% HANP star-PLA	N_2D	2.43 ±0.07	1.31	281.9 ±5.5	76.3 ±2.2	99.6 ±2.8^d
(19) 5.6 wt% HANP is-PLA24	N_2D	2.48 ±0.09	3.03	288.6 ±3.9	79.1 ±4.8	92.7 ±2.6^d
(20) 3.6 wt% HANP is-PLA16	N_2D	1.89 ±0.02	0.81	304.0 ±4.7	88.5 ±4.9	n/a
(21) PLDLLA	N_2D	-	-	309.0 ±4.0	85.0 ±7.1	91.7 ±4.5^d
(22) 2.5 wt% HANP	N_2D	2.40 ±0.08	-	343.7 ±3.7	106.6 ±4.3	96.6 ±3.4^d

Table 1 summarises all measurements on nanocomposites and PLDLLA recirculated in the MiniLab for 15 min in Air or N_2D. Sample names include the nominal total amount of coated or uncoated HANP compounded with PLDLLA. TGA results provide the final values of HANP wt% in the composite and coating wt% is calculated in relation to this, based on TGA tests performed on particles directly after coating. Additionally, τ_w after 15 min of recirculation in the MiniLab is recorded along with the flexural strength (σ) measured on bars produced via IM and CM, or CM alone. M_v of PLDLLA recirculated in N_2D is 37 % higher than in Air and M_v of nanocomposites compounded in N_2D is higher when compared to the nanocomposites in Air with the same nominal loading.

Figure 1 (a) shows a selection of data to illustrate the effect of HANP morphology and dispersant coatings on τ_w during processing in Air. A significant 55 % decrease in τ_w occurs due to PLDLLA degradation during 15 minutes of compounding, which causes reductions in molecular weight. HANP DDSA and HANP rod yield the highest τ_w, which may imply that more work is done in achieving a better dispersion.

Figure 1 (b) illustrates the importance of carefully drying nitrogen when it is used as an inert gas for melt processing of PLDLLA. This prevents excessive decrease of molecular weight caused by the presence of water, which can be a significant contaminant in bottled gas, and in result considerably mitigates reduction in τ_w during compounding. After recirculation in N_2 that has not been dried, τ_w is much lower than for either N_2D (58 % less) or Air (35 % less). Additionally, τ_w for Air decreases at a greater rate than for N_2D, inferring faster degradation causing an eventual larger decrease in molecular weight.

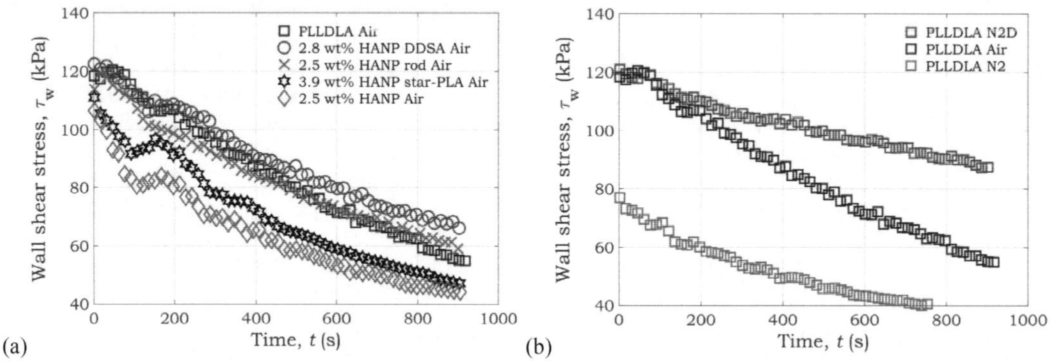

FIGURE 1. (a) Comparison of melt compounding of selected PLDLLA nanocomposites in Air with nominal loading of 2.5 wt% HANP. (b) Influence of moisture on melt compounding. All data filtered with moving average of 5 points.

Figure 2 (a) and (b) show the N_2D measurements of τ_w as a function of compounding time for a selection of nanocomposites from Table 1. The addition of 5 wt% HANP to PLDLLA in comparison to 2.5 wt% HANP leads to a significant decrease in τ_w, e.g. for HANP star-PLA τ_w reduced by 44 % after increasing the amount of HANP, probably indicative of inferior HANP dispersion or a plasticising effect of the particle coating. Additionally, the decreased gradient on Figure 2 (a) when compared to Figure 1 (a) confirms that residual moisture removal from N_2 is essential and drying HANP directly before compounding is beneficial.

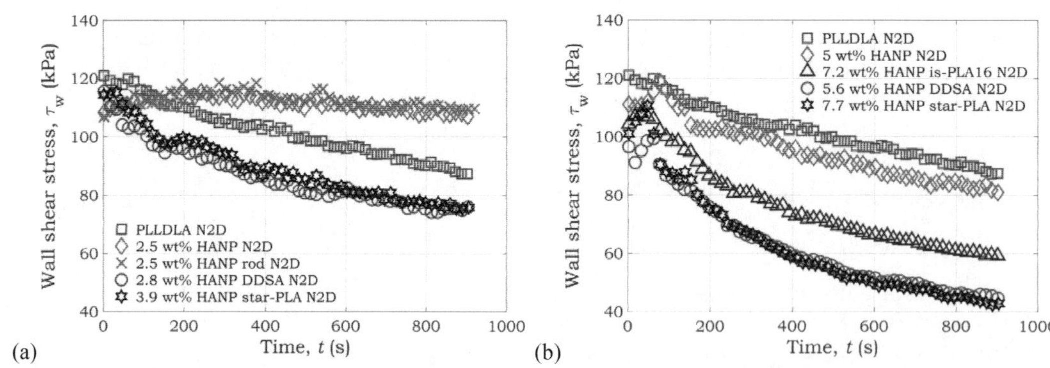

FIGURE 2. Comparison of melt compounding of selected PLDLLA nanocomposites in N_2D: (a) with nominal loading of 2.5 wt% HANP, (b) with nominal loading of 5 wt% HANP. All data filtered with a moving average of 5 points.

TEM images comparing selected nanocomposites from Table 1 compounded in Air and N_2D are shown in Figure 3. A simple method was employed for image analysis, determining the largest open square that can be applied on the image without crossing or encapsulating any particle [13]. This is a useful semi-quantitative approach as it is assumed that the dispersion worsens with increasing free space. Dispersion of (i) 2.5 wt%

HANP looks similar regardless of Air or N_2D compounding. (ii) 2.8 wt% HANP DDSA displayed superior dispersion in Air, which is in agreement with the compounding curve in Figure 1 (a), and (iii) 3.9 wt% HANP star-PLA appears enhanced in N_2D, which is in line with its τ_w in Figure 2 (a) when compared with Figure 1 (a). Applying the aforementioned image analysis method to Figure 3 reveals the greatest dispersion for 2.8 wt% HANP DDSA in Air in comparison to the other nanocomposites.

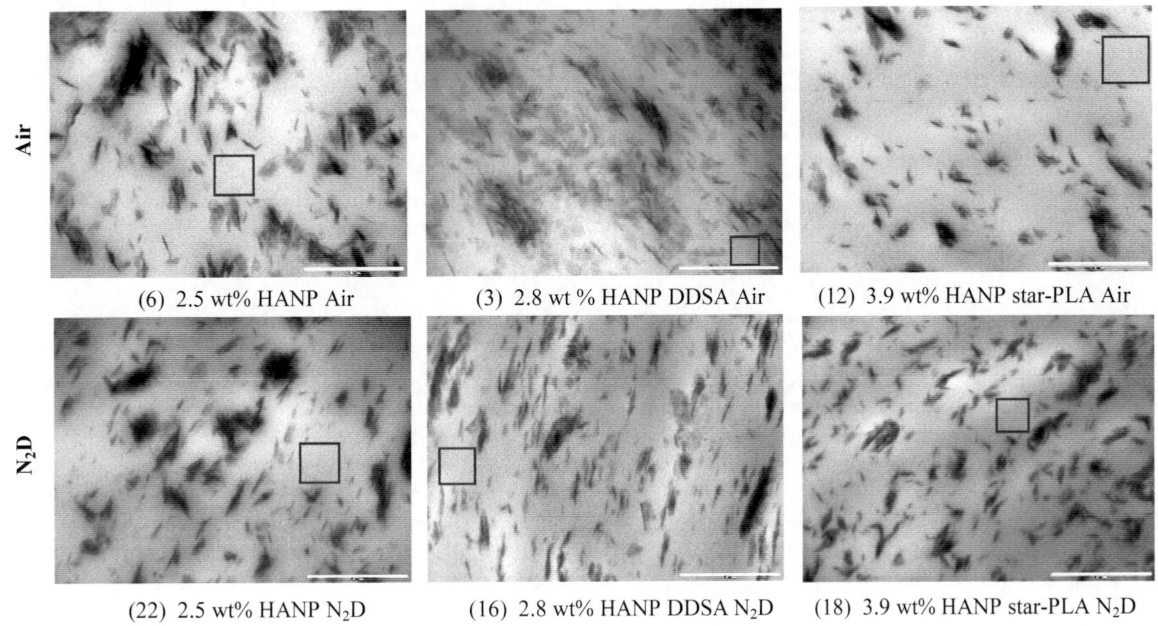

FIGURE 3. TEM images obtained with TEM FEI at x 26,500 magnification, scale bar 1 μm. All nanocomposites with nominal loading of 2.5 wt% HANP (plus coating) compounded in Air: (3), (6), (12) and in N_2D: (16), (18), (22). Numbers correspond to conditions in Table 1.

Figure 4 shows measurements of M_v as a function of τ_w at 15 min for all materials from Table 1, separated by processing gas. The gradient for both linear regressions is similar, which confirms a comparable trend. The close correlation of all the results indicates that τ_w is principally a function of M_v and that the presence of the nanoparticles does not have a significant effect on the polymer viscosity. The only significant exceptions to this appear to be the (2) 4.9 wt% HANP is-PLA8 and (3) 2.8 wt% HANP DDSA specimens. This could suggest improved dispersion in these materials, which will be studied in greater detail in future research.

Materials with the same nominal loading of HANP have higher M_v and τ_w for compounding in N_2D than in Air, demonstrating a clear benefit of N_2D in terms of preserving M_v during processing, e.g. 3.9 wt% HANP star-PLA: (18) in N_2D and (12) in Air. Additionally, decreased M_v and τ_w can be observed in N_2D for coated HANP with 5 wt% nominal loading compared with 2.5 wt%, e.g. HANP is-PLA24: (7) 11.1 wt% and (19) 5.6 wt%.

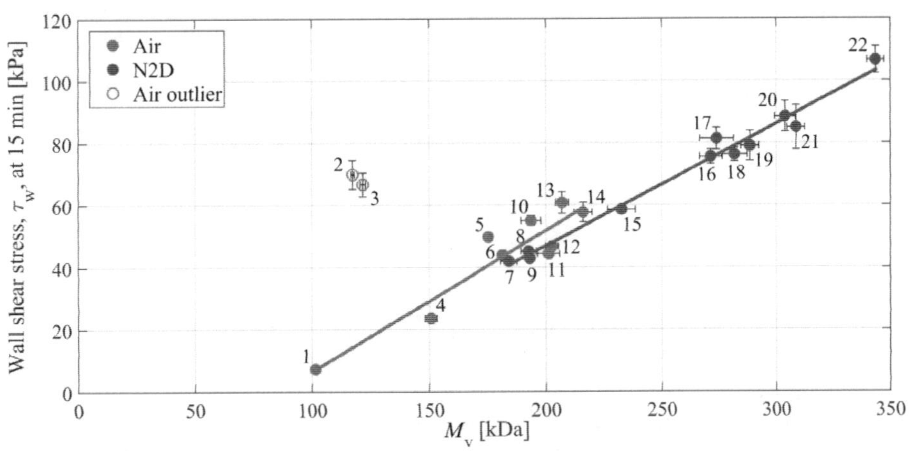

FIGURE 4. M_v for PLDLLA and nanocomposites after 15 min recirculation in the MiniLab extruder in Air and N_2D versus τ_w at 15 min. Linear regression for Air excludes (2) and (3) (4.9 wt% HANP is-PLA8 and 2.8 wt% HANP DDSA respectively).

No substantial improvements in the initial mechanical properties were seen when compounding in N_2D as compared to Air, despite increased M_v. However, increased molecular weight would improve material response during application within the body, where the fall in mechanical strength caused by degradation will be delayed in the manufactured products.

The most significant difference in the mechanical performance was for the 5 wt% HANP, where the Air extruded and IM specimen presented a 33 % increase compared to its N_2D extruded counterpart. All of the other specimens that were tested ranged from 84 MPa to 100 MPa, identifying the significance of the result of the 5 wt% HANP. It is possible that the injection moulding procedure provided additional shear during the manufacture of the specimens, allowing a superior exfoliation to be obtained, enhanced with the presence of the larger quantity of HANP. This effect allows for the improved mechanical properties seen with IM to outweigh the possible loss that would be seen due to the significant loss in molar mass of these specimens. Although the initial properties of this composite are high, the low molecular weight will cause rapid degradation during application. Further work is being undertaken to determine the influence of IM, where the use of a N_2D atmosphere will provide the improved molar mass required for application.

CONCLUSIONS

The effectiveness of novel tailored dispersants, when coated onto HANP, and nanoparticle morphology was investigated during and after compounding over ambient air and dried bottled N_2. Preliminary findings show that drying the N_2 is essential to maintain high molecular weight, which in turn yields high wall shear stress during compounding. The most significant improvement in M_v was observed when the HANP was dried at a high temperature. This is believed to remove any remaining water from the PLDLLA during processing, significantly reducing the effect of the depolymerisation ordinarily taking place at the compounding temperatures. Although the mechanical properties of the nanocomposites are not significantly affected by the moisture content, the more limited drop in molar mass is likely to play a significant role once these materials experience hydrolytic degradation during application within the body.

Preliminary TEM image analysis was carried out with a semi-quantitative approach, and revealed superior dispersion in 2.8 wt% HANP DDSA in Air when compared to other nanocomposites. This will be studied in greater detail in future research and verified with microCT measurements.

ACKNOWLEDGMENTS

This work was supported by grant EP/J017272/1 from the UK Engineering and Physical Sciences Research Council (EPSRC), and by a travel grant - Andrew Carnegie Research Fund from Institute of Materials, Minerals and Mining enabling attendance at the conference. Data created in support of this research is openly available at http://dx.doi.org/10.17639/nott.43.

REFERENCES

1. S. Bose, M. Roy & A. Bandyopadhyay, Trends in Biotechnology, 30, 546-554 (2012).
2. S. I. J. Wilberforce, C. E. Finlayson, S. M. Best & R. E. Cameron, Polymer, 52, 2883-2890 (2011).
3. M. M. Tomczynska, G. Y. H. Choong, D. S. A. De Focatiis, D. M. Grant and A. J. Parsons (in press) "Compounding and rheometry of PLA nanocomposites with coated and uncoated hydroxyapatite nanoplatelets". AIP, PPS2015 Graz Conference Proceedings.
4. G. Y. H. Choong, A. J. Parsons, D. M. Grant & D. S. A. De Focatiis, "Rheological techniques for determining degradation of polylactic acid in bioresorbable medical polymer systems", PPS Conference Proceedings 30, Polymer Processing Society, Cleveland, Ohio, 2014.
5. E. Lester, S. V. Y. Tang, A. Khlobystov, V. L. Rose, L. Buttery & C. J. Roberts, CrystEngComm, 15, 3256-3260 (2013).
6. E. Lester & B. Azzopardi (2005) Counter Current Mixing Device, Patent No. W/O 2005077505. Available at: http://www.google.com/patents/WO2005077505A2 [Accessed 30 March 2016]
7. E. Lester (2010) Hydroxyapatite material and methods of production, Patent No. W/O 2010122354. Available at: http://google.com/patents/WO2010122354A1 [Accessed 30 March 2016].
8. F. Hild, K. Walton, M. Gimeno Fabra, E. Lester & D. Irvine, "Continuous, Single Stage Synthesis of Dispersant Included Nanoparticles", NanoBioTech-Montreux Conference, Montreux, Switzerland, 2014.
9. M. Gimeno-Fabra, F. Hild, P. Dunne, K. Walton, D. Grant, D. J. Irvine & E. Lester, CrystEngComm, 17, 6175-6182 (2015).
10. O. Dechy-Cabaret, B. Martin-Vaca & D. Bourissou, Chem Rev, 104, 6147-76 (2004).
11. L. T. Lim, R. Auras & M. Rubino, Progress in Polymer Science, 33, 820-852 (2008).
12. J. R. Dorgan, J. Janzen, D. M. Knauss, S. B. Hait, B. R. Limoges & M. H. Hutchinson, Journal of Polymer Science Part B: Polymer Physics, 43, 3100-3111 (2005).
13. H. S. Khare & D. L. Burris, Polymer, 51, 719-729 (2010).

A New Sample Preparation Method to Investigate Melting Phenomena within a Partly Filled Melting Zone of Co-Rotating Twin Screw Extruders

Volker Schöppner[a], Kim Jacqueline Scharr[a*] and Tobias Thomas Westhues[a]

[a]*University of Paderborn, Kunststofftechnik Paderborn, Germany*

volker.schoeppner@ktp.upb.de
tobias.westhues@ktp.upb.de
kim.scharr@ktp.upb.de

Abstract. Polymer melting in co-rotating twin-screw processing equipment is an active field of research. Especially melting mechanisms such as plastic energy dissipation due to compression in the kneading blocks or disperse melting in fully and partly filled screw areas have been explored extensively in the recent past. Though melting in these machines has already been described and examined, it is still very complicated to generate experimental quantifiable values for the partial melting of polymer samples and to examine the melting length along the screw. To gain new knowledge and to build up models for the correct calculation within this machine part it is essential to use a quantifying measurement technique. In this paper we present methods to generate experimental quantifiable values for the partial melting of polymer samples from different melting zones of intermeshing co-rotating twin screw extruders. Furthermore, we introduce a method to measure the mean material temperature within solid material with the help of a calorimetric measurement. The advantages are discussed.

Keywords: Melting, twin-screw extruder, temperature measurement, melting degree
PACS: 81.20.Hy

INTRODUCTION

Developed in the 1940s and 1950s, tightly intermeshing co-rotating twin screw extruders have become very important in various industrial sectors. The knowledge of the physical processes in each zone is of fundamental importance for the creation of an appropriate machine design. A particular focus of interest is the melting zone: Here, up to 80 % of the total energy is contributed [1]. The study of melting phenomena in extruders has recently attracted much attention, mainly due to the wide application of extruders in industry. The study of melting not only covers how heat is generated and transferred to polymers, but also how melting is initiated and propagated in the extrusion channel. The first published melting study of an intermeshing co-rotating twin-screw extruder was compiled by Todd in 1992 [5]. The recent investigations are described in the following: Qian and Gogos (2001) [4] concentrate their elaborations on an interparticle creation of frictional heat, irreversible deformation and breakage of solid particles as well as viscous energy dissipation of the resulting melt and the solid-melting-mixture. Thümen (2008) [6] expanded the disperse melting model Melisch 1996 [3] set up. In their explanations of a more extensive, analytical process model, they took the influence of convection on the melting process and the influence of the finite extent of twin-screw channels while taking an optimized internal barrel wall temperature calculation into consideration.

In what follows the technologies applied in the past are mentioned briefly. During Dead-Stop-Tests, the extruder is stopped and cooled down during the running process after which the material on the screws is analyzed. For this purpose, extruders with folding barrels as in [7] or extruders with simplified screw removal are used. The hardened plastic mass can be removed from the screws and examined with e.g. thin sections, in which the melting degree in the different areas has to be determined [8, 2, 6]. In addition, barrel elements with observation windows are used (visualization twin-screw extruder, VTSE) [9, 10, 11], which allows to document and analyze the running process. Furthermore, sample plates can be used between the normal barrel elements. At these locations, the melting degree can be examined. Within the extrusion process, the molten polymeric matrix is colored, e.g. with the addition of pigments, and then experimentally analyzed via thin section or cutting of the samples [12]. All technologies have in common that with them it is impossible to analyze a quantifiable value for the melting degree / the melt within the screw channel within the first two process zones.

EXPERIMENTAL

When an analytical regression is planned, the challenge of a melting examination in twin-screw extruders is to precisely depict the melting degree and the material temperature within the machine with a suitable technology. With conventional measurement techniques, so as the Dead-Stop-Tests to determine the melting degree or the conventional penetration thermometer to measure the particle temperature, it is hardly possible to depict the melting degree in the very first melting zone or to measure the mean temperature of the entire particle. Most of the temperature measurements are only valid when the temperature on the surface is of interest. Therefore, a simple method to precisely determine material temperatures in the solid conveying zone was developed at the University of Paderborn. Furthermore, a technique for determining the melting degree was conceived. In the following the experimental setup is described and some results are presented.

Apparatus

Both measurement techniques have been conceived to perform measurements on a twin-screw extruder ZSK 25 WLE (L/D = 20) from Coperion. The barrel configuration includes five barrels and one final test plate. The element temperatures (°C) were: 20–20–20–100–150–150. The considered parameters were speed (300, 450, 600 rpm) and flow (5, 10, 15 kgh^{-1}). The examined screw configurations are pictured in Figure 1. The feed is taking place in barrel three, barrel three and four and the die plate are run through by the material flow. The first field of investigating the screw configurations (SC 1 to SC 2) allows the view on fully filled melting. The screw configurations SC 3 to SC 6 were chosen in order to quantify the melting degree process with four conveying kneading blocks (length 36 mm, five kneading disks, offset angle 45°) in partly filled melting zones along screw length. The mass flow is examined at the end of every single screw configuration. Samples are taken without any dies at the extruder outlet. Stringing together solutions from SC 3 to SC 6 reveals the melting degree process in configuration SC 6.

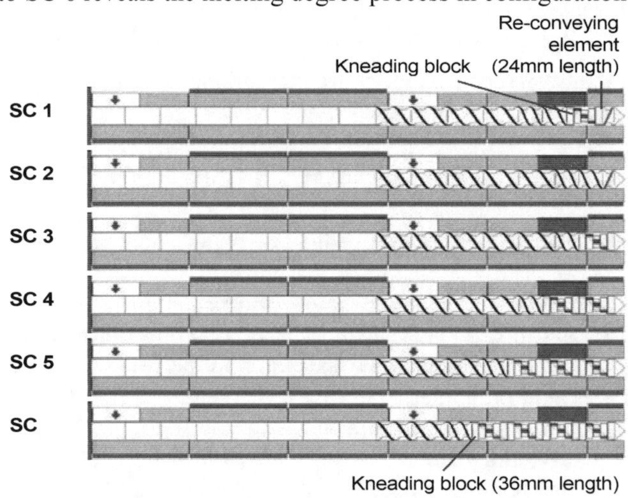

FIGURE 1. Investigated screw configurations.

The investigated materials were polystyrene (PS) and polypropylene (PP). Both materials are colored white with titanium dioxide (PP with 1.3 % by weight; PS with 6 % by weight) and have a lenticular shape. Before the material is incorporated, 2 % carbon black, which sticks on the pellets' surface and colors the melt black, is added to it by weight. The result, a high contrast between solid and melt even in a cooled down sample material stage, is the precondition for a definite classification into different phases.

Sample Preparation

The polymer mass flow is intercepted instantly after it comes out of the screw tip (without using any die) in a square aluminum profile. A precisely fitting, wooden stamp, which is applied after the sample was fed, is compressing the sample material (avoiding of air locks) as it is pressed down with a weight of 17 kg / 0,0652 MPa (Figure 2). In this state the sample material cools down. A rectangular sample is produced and further on described as square pellet. After the material cooled down to room temperature, the square pellet is cut following Figure 3. A perspective view of the square pellet is provided ①, as well as both performed cut variations (cross section ② and der longitudinal section ③). The surfaces described as A and B in Figure 3 ② are relevant in terms of determining the melting degree in the further process.

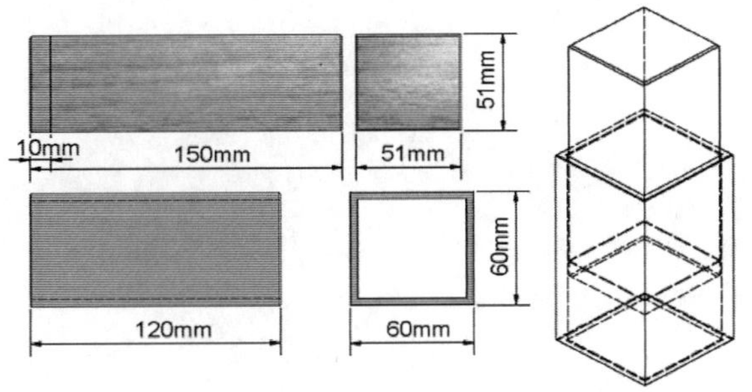

FIGURE 2. Auxiliary device for sample preparation.

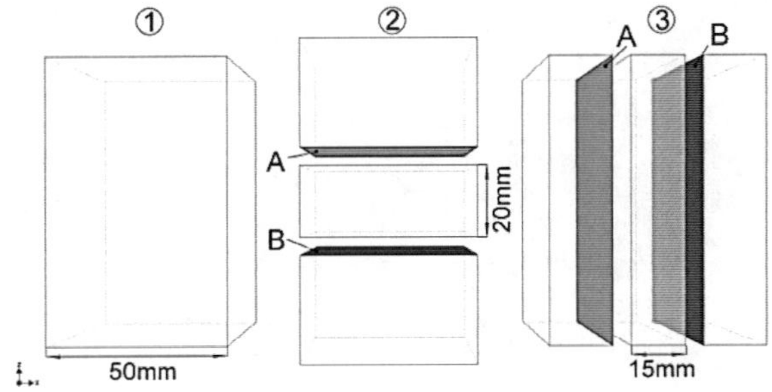

FIGURE 3. Cutting direction of sample.

The grinding process is being prepared depending on the saw cut, with a belt grinder and it is performed manually on two disc sanders with grain sizes 200, 400 and 800 (CAMI grit designation). A flat bet scanner is used to digitalize the grinded intersections. On the left, Figure 4 shows a micro section ready for further evaluations. In order to determine the melting degree for individual samples, the amount of colored melt is being visually determined by analyzing gray and white micro section areas.

Preparing the samples, it is practical to adjust an optical contrast as large as possible between melt and solid or to match colored coordinates to melt and solid, which makes a clear distinction possible. The decision to color the particles will be decisive for an automatable evaluation. At first, the amount of pixels of the micro section's whole surface is defined. The solid content of the sample, which is represented by white areas of the surface and clearly differs from the black colored melt, is identified. The difference of total amount of pixels and solid amount of pixels is the pixel amount that has to be assigned to the molten part.

Figure 4 shows the uncolored sample on the left side, the sample with pixels to be identified as solids in the middle and the whole surface of the micro section on the right side. Hollows might have to be subtracted manually from the total surface amount. The residual heat of the polymer particles does not affect the evaluation of white and black areas because after inserting the sample into the aluminum profile it is guaranteed that no dispersion of the different phases is possible, although naturally, there are temperature balancing processes between different phases.

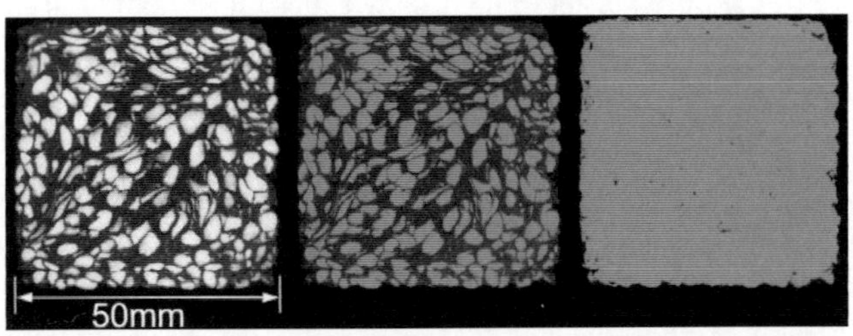

FIGURE 4. Grinded surface (le.); pigmented solid (mi.); pigmented sample surface (ri.).

In order to determine the beginning of the melting process in dependence of the regarding screw configurations and parameters, the occurring mass flow manifestations have to be applied above throughput and speed in grid view. Figure 5 shows micro sections for SC 2 (PS). In rows from left to right the pictures are arranged by throughput (5, 10 and 15 kgh^{-1}) and in the columns top down the pictures are organized by speed (300, 450 and 600 rpm). This alternative way to visually present the data confirms the fact that the melting degree of a sample decreases in case of an increasing relation between throughput and speed.

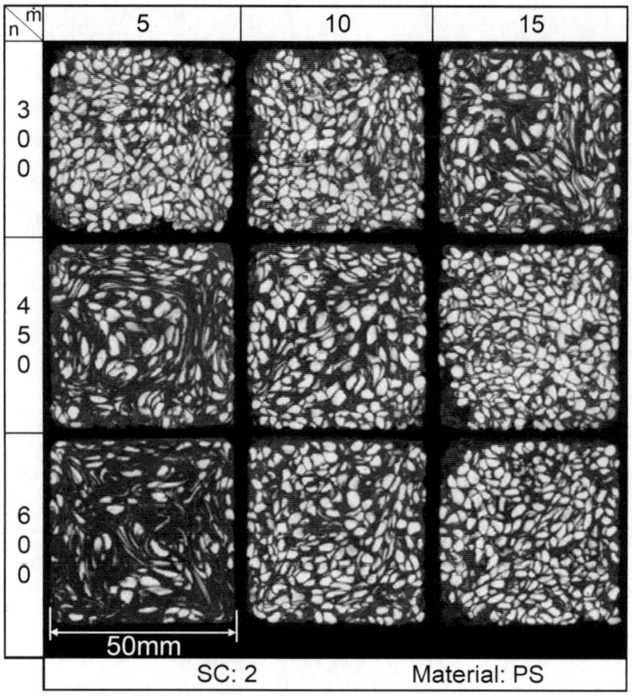

FIGURE 5. Grid view; PS, SC 2.

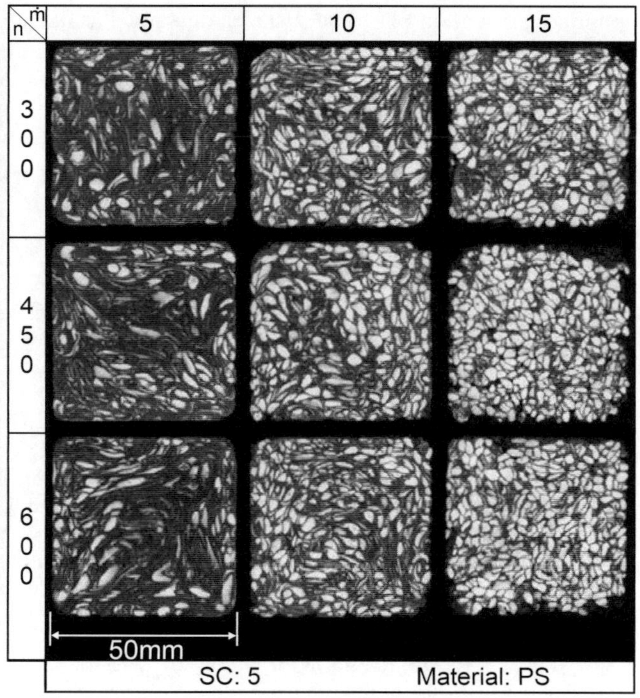

FIGURE 6. Grid view; PS, SC 5.

The advantage of this sample preparation method is its' time-efficient evaluation technique. There are only 60 minutes passing between sampling and result, if the user is trained and software-macros are present.

Calorimetric temperature measurement

This chapter describes how temperature measurements are conducted, using the heat energy determination by isothermal evaporation calorimetry in order to make a qualitative statement about the temperature development of initially molten samples after processing in co-rotating twin screw extruder. Therefore, in accordance with the specific enthalpy-temperature curve of the polymer material, the measured heat content is converted to the mean sample temperature.

For this, the plastic samples are added to a container filled with liquid nitrogen immediately after processing. As the heat content of the plastic sample is removed by the liquid nitrogen, the nitrogen evaporates. Until cooling to the nitrogen's boiling temperature, the plastic sample loses the heat energy already mentioned. This heat energy is directly proportional to the evaporated amount of nitrogen that can be determined using a precise balance system.

Using the nitrogen's specific evaporation enthalpy $h_{N2} = 196.8\ kJ/kg$ and the vaporized quantity of nitrogen $m_{N2,vap,end}$, the heat content difference ΔE between the desired sample's outlet temperature and the boiling point of liquid nitrogen in polymer samples can be determined. m_{Sample} is the weight of the polymer sample. $\Delta h_{T_{Sample}-T_{N2}}$ is the specific enthalpy difference for the sample material between the desired sample temperature and the boiling temperature of nitrogen. The following equation (2) shows, how $\Delta h_{T_{Sample}-T_{N2}}$ can be calculated.

$$m_{N2,vap,end} * h_{N2} = \Delta E = m_{Sample} * \Delta h_{T_{Sample}-T_{N2}} \quad (1)$$

$$\Delta h_{T_{Sample}-T_{N2}} = \frac{m_{N2,vap,end} * h_{N2}}{m_{Sample}} \quad (2)$$

Adding $\Delta h_{T_{Sample}-T_{N2}}$ to the absolute, specific enthalpy of the sample material at boiling temperature leads to the absolute specific enthalpy of the sample material for the desired outlet temperature. Some further small adjustments are required in order to use this method: The specific enthalpy curve for the polymer material, e. g., is not measurable under temperatures of -60 °C – which is why an extrapolation needs to be done. The accuracy of the measurement technique was studied via reference tests. In these experiments, polymer samples of known weight were heated up to a defined temperature within a heating furnace. After removed from the oven they were directly placed into the nitrogen tank. The mean temperature of the samples was calculated via the decrease in mass, according to the calculation above. The figure below shows the quality of the measurement: 16 compared measurements were performed. The y-axis shows the calculated sample temperature, the x-axis shows the real sample temperature. The dashed line marks an absolute deviation of 5 °C from the real temperature.

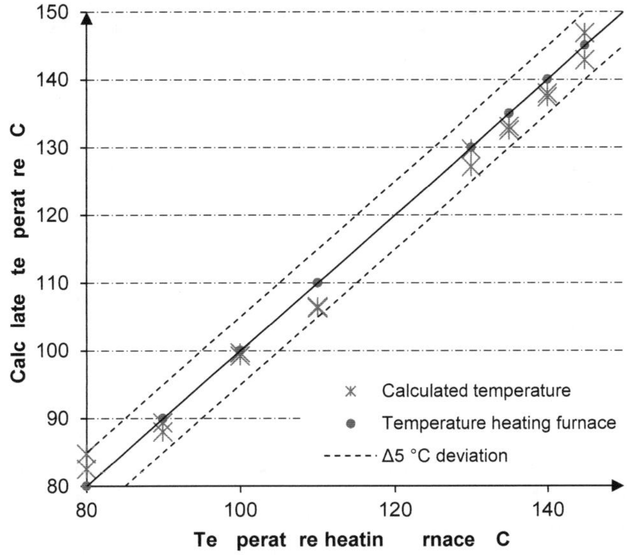

FIGURE 7. Difference between defined and calculated temperature

DISCUSSION AND CONCLUSION

The novel research method considers and resolves the difficulties with regard to polymeric mass melting in the specific zone of a TSE, for what the transition area between solid and melt as well as the melting degree are of high interest. The grid views of sample materials from investigations with different screw configurations allow to determine location and process parameters of initial melting. Comparative observations of melting degrees from screw configurations using up to four conveying kneading blocks may display the melting degree along screw length. The calorimetric temperature measurement technique allows measuring the heat energy of polymer samples between solid temperature and melting temperature. It is a good opportunity for an accurate measurement in processing regions, where the average measurement technique is about to fail.

REFERENCES

1. B. Qian, C.G. Gogos, The Importance of Plastic Energy Dissipation (PED) to the Heating and Melting of Polymer Particulates in Intermeshing Co-Rotating Twin-Screw Extruders, Adv. *in Polymer Tech*., Vol. 19, No. 4, 287-299 (2000).
2. H. Jung, J.L. White, Investigation of Melting Phenomena in Modular Co-rotating Twin Screw Extrusion, *Int. Polymer Processing*, 18, 127–132 (2003).
3. H. Potente, U. Melisch, Theoretical and Experimental Investigations of the Melting of Pellets in Co-Rotating Twin-ScrewExtruders, *Int. Polymer Processing*, 11, 101-108 (1996).
4. B. Qian, C.G. Gogos, Plastic Energy Dissipation (PED) A Major Contributor to Melting of Polymers in Polymer Compounding Equipment, *SPE ANTEC Tech. Papers*, 49, 126-131 (2001).
5. D.B. Todd, Melting of Plastics in Kneading Blocks, *SPE ANTEC Tech. Papers*, 38, 2528-2536 (1992).
6. A. Thümen, Untersuchung und Beschreibung des dispersen Aufschmelzens in Gleichdrall-Doppelschneckenextrudern. *Dissertation, Universität Paderborn, Institut für Kunststofftechnik KTP* (2008).
7. M. Liang, H.E. Huff, F.H. Hsieh, Evaluating Energy Consumption and Efficiency of a Twin-Screw Extruder, *Journal of Food Sience* Vol. 67, Nr. 5, (2002).
8. M. Bastian, Plastifizierung und Morphologie von inkompatiblen Polymerblends bei der Verarbeitung in Gleichdrall-Doppelschneckenextrudern. *Dissertation, Universität Paderborn, Institut für Kunststofftechnik KTP,* (2000).
9. H. Chen, U. Sundararaj, K. Nandakumar, On-line Visualization of PS/PP Melting Mechanisms in a Twin-Screw Extruder. *SPE ANTEC Tech. Papers,* 49, 110-114 (2003).
10. L. Zhu, X. Geng, Experimental Investigation of Polymer Pellets Melting Mechanisms in Co-Rotating Twin-Screw Extrusion. *Advances in Polymer Technology*, Vol. 21, No. 3, 188-200 (2002).
11. G. Spohr, Research of Energy Dissipation in Thermoplastics in High Speed and High Torque Twin-Screw Extruders, *Master-Thesis, Technische Universität Darmstadt* (2012).
12. P.H.M. Elemans, P. Bleimann, J. Blanchard, Evaluation of Melting Performance of a Co-Rotating Twin-Screw Extruder. Society of Plastic Engineers, *Antec Paper* (2002).

Investigating the Impact of Operating Conditions on the Extent of Additive Mixing during Thermoplastic Polymer Extrusion

Dmytro Stratiychuk-Dear[a,b*], Kieran Looney[a], Paul Oliver[a], Stuart Blackburn[b] and Mark Simmons[b]

[a]*DuPont Teijin Films U.K. Ltd., The Wilton Centre, Wilton, Redcar, United Kingdom, TS10 4RF*
[b]*School of Chemical Engineering, University of Birmingham, Edgbaston, Birmingham, United Kingdom, B15 2TT*

Dmytro.Stratiychuk-Dear@gbr.dupont.com

Abstract. Adequate dispersion of the additive in polymer composites is vital to their performance. Dispersion occurs via rupture, erosion and collision induced mechanisms. Organophosphorus salt/PET nanocomposites were manufactured on a co-rotating fully intermeshing pilot scale twin screw extruder to determine the impact of operating conditions on the mixing achieved in order to understand the level of control available over the process. The effect on key dispersion mechanisms is discussed. Dispersion was found to increase with decreasing throughput, increasing screw speed and increasing molecular weight of the polymer. Linear relationships between calculated parameters including specific mechanical energy, degree of fill and specific throughput were also observed. A need for a compromise between achieving the desired level of mixing and maintaining the molecular weight of the polymer is established.

Keywords: Dispersive Mixing, Twin Screw Extrusion, Operating Conditions, Specific Mechanical Energy, Degree of Fill, Specific Throughput, Molecular Weight, Intrinsic Viscosity.
PACS: 81.05.Qk, 81.20.Hy

INTRODUCTION

Polyethylene terephthalate (PET) is a thermoplastic polymer substrate used in a variety of forms; including bi-axially oriented films. Applications of such films range from commodity markets e.g. packaging to specialized uses e.g. electronic displays. For properties of bi-axially oriented PET films and its uses see [1].

Numerous additives are utilized within the polymer in order to modify the bulk and/or the surface properties of the film. Several additive-polymer systems can be formed often with different mixing requirements including miscible mixtures, immiscible blends and arguably the most common polymer composites. Additives such as silica and clay are utilized in composites to generate roughness on the surface of the film, while pigments e.g. TiO_2 are added for color [1]. Good dispersion of the additive is required to achieve the desired functionality within the film.

An overview of mixing mechanisms in polymer blends and composites can be found in [2]. The size reduction of the secondary phase contained within the primary phase occurs via dispersive mixing. In composites three mechanisms exist. Agglomerate *rupture* is a rapid process which occurs when the hydrodynamic stress acting on the agglomerate is larger than cohesive forces holding it together (combination of adhesive, viscous and electrostatic forces) corresponding to a critical stress; see [3]. *Erosion* of agglomerates takes place when small fragments are detached from the overall structure (onion peeling model) as the edges of the agglomerate become infiltrated with polymer; see [4]. Erosion occurs at lower stresses compared to rupture [5] but at a much slower rate; which is proportional to the surface area of the agglomerate. *Collision* induced dispersion results in fragments being generated due to the impact between agglomerates due to the relative velocity between the solid masses [5]. This occurs even at very low shear rates, where the number of fragments is proportional to the shear rate squared and the number of collisions is related to the number of particles per unit volume. In either mode, it is vital that the newly formed fragments are separated and rapidly wetted by the polymer to prevent re-agglomeration.

Polymers are extruded on *single* (SSE) or *twin screw extruders* (TSE). Few studies directly compare the mixing performance of SSE and TSE [3, 6]; with majority focusing on distributive mixing in TSEs e.g. [7, 8]. TSEs offer better distributive mixing performance due to increased melt reorientation in the nip region [3]. Higher stresses generated here and increased areas of extensional flow aid break up of agglomerates [6] delivering improved dispersive mixing performance. The modular design of TSEs enables the extruder to be configured in accordance with the mixing requirements of the product [9]. Since TSEs operate partially filled, in starve fed mode, the output of the machine can be controlled almost independently of the screw speed. This also allows for fillers to be introduced simultaneously with the polymer into the feed pocket without pre-blending or directly to the melt via a side mounted stuffer feeder. Typical configurations of TSEs for compounding are discussed in [10, 11]. In bi-axially oriented film production additives are typically introduced as a compounded masterbatch; pre-dispersed additive uniformly distributed in a polymer carrier.

Extruder screw speed directly affects the shear rate and the shear stress experienced by the material thus influencing dispersion. While, the throughput governs the *specific mechanical energy* (SME) input, both of these control the *degree of fill* (DOF) changing average shear rate and stress experienced by the polymer [12]. It's vital to understand the degree of control these parameters offer over the process/product.

The majority of recent studies examining the effect of operating conditions on dispersion focus on composites of multiwalled carbon nanotubes (MWCNTs) in polycarbonate [13,14], polylactic acid [15] and polycaprolactone [16] substrates. Organoclay/polypropylene composites are addressed in [17,18]. Analysis on PET substrate is limited; with [19] considering the dispersion of standard vs. surface modified $BaSO_4$ particles in nanocomposites produced via in situ polymerization and [20] examining the dispersion of TiO_2 added directly to the polymer melt.

This work focuses on evaluating the impact of operating conditions and melt viscosity on the extent of dispersive mixing achieved in *organophosphorus salt/PET* composites. A comparison between the *particle size distribution* (PSD) of the additive powder and that measured in the polymer composite is utilized to identify the key dispersion mechanisms governing in a given regime.

MATERIALS AND EXPERIMENTAL METHODS

Standard film grade PET (S-PET) produced by DTF was used to manufacture composites containing organophosphorus salt at 5 $\%_{wt}$ loading. A higher molecular weight polymer (H-PET) was used to evaluate the impact of polymer viscosity on the level of dispersion. Intrinsic viscosity (IV) directly relates to number average molecular weight via the Mark - Houwink – Sakurada equation. H-PET has an IV around 30 % higher than S-PET, resulting in melt viscosity 3.5 times higher than S-PET at 280 °C and shear rate of 100 s^{-1}; as calculated from [21].

The additive is a fine grained white powder with particle size below 25 μm, bulk and material density of ~ 180 and 1350 kg m^{-3} respectively. The powder is not hygroscopic and is insoluble in water or common organic solvents.

Berstorff (Germany) ZE 40A co-rotating fully intermeshing twin screw extruder, 28 D in length, was used to manufacture the composites. The TSE is split into 7 zones; each with independent temperature control. Zone 3, 4 and 7 are designated melting, vacuum degassing and mixing while the other zones are used for conveying only. The set up uses a number of kneading blocks and toothed mixing elements with 4.5 D as the cumulative length of these elements. Post extruder, the polymer passes through a gear pump to deliver a steady flow of material, and is extruded though an end fed die onto a casting drum to generate cast film. No filter was used to ensure that any large agglomerates would be seen in the cast film.

Neither material was pre-dried as any moisture contained within is removed in the vacuum degassing zone; thus minimizing hydrolysis. The additive and the polymer were introduced simultaneously to the feed pocket via two separate material streams. Brabender gravimetric twin screw feeder fitted with powder screws was used to dose the additive. S-PET was processed at barrel temperatures set to 275 °C; these were raised to 290 °C for H-PET in order to overcome torque limitations of the machine. In the latter, barrel temperatures of ~ 300 °C were observed in the melting zone, as the barrel cooling capacity was insufficient.

The samples were analyzed at 2.5x magnification under transmitted light using a Leica DMRX microscope fitted with a Leica DFC 425 camera enabling image capture. Open source java script based *ImageJ (FiJi)* software was utilized to perform the image analysis and generate volume weighted PSDs based on the projected area diameter for the additive contained within the polymer. Twelve images per sample were analysed yielding an average particle count of 25,100 (circa 2,100 per image). ISO 13322-1 requires ca. 6,100 particles to specify the mean volume diameter of a distribution at 5 % accuracy and ca. 15,200 particles to specify percentiles of distributions i.e. D_{90} to at 10 % accuracy. These are the confidence intervals used for the reported results.

The additive powder PSD was obtained via a Beckman Coulter LS 13 320 laser diffraction particle size analyser, with the use of Mie scattering optical model. The additive was dispersed in deionized water using a high shear mixer; no dispersive aids were used.

RESULTS

Impact of Extruder Output

The impact of throughput on additive dispersion was studied for S-PET and H-PET polymers; see Figure 1. PSD in the composite was compared to that of the powder (as measured; and provided by the supplier) showing poorer dispersion. The discrepancy between supplier PSD and measured PSD is attributed to the supplier using acetone with dispersive aids for the measurement; unlike DTF. Example microscopy and binary images are in Figure 1b.

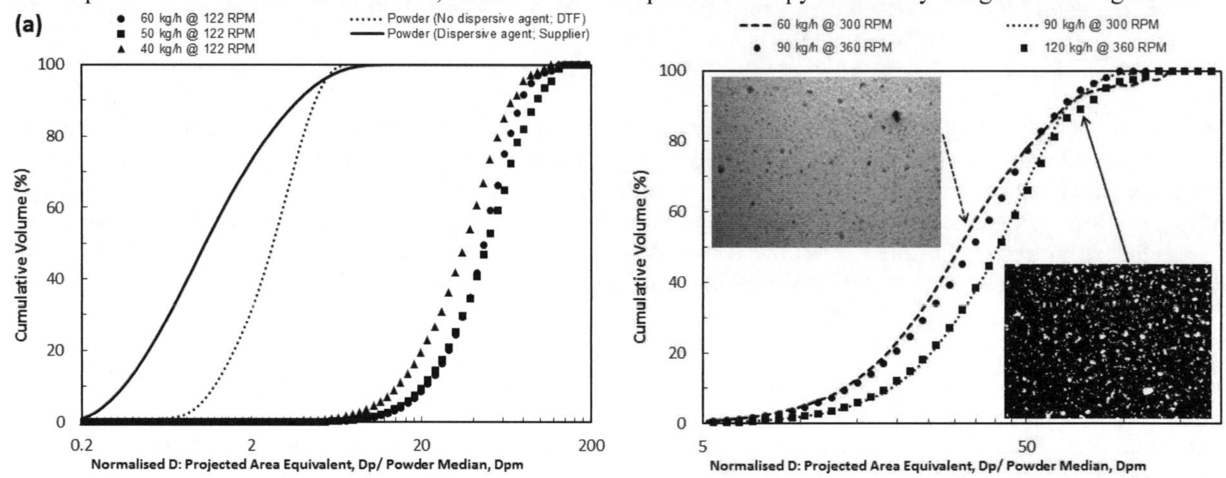

FIGURE 1. Effect of Extruder Output on Dispersion of Organophosphorus Salt in PET: (a) H-PET, (b) S-PET

Figure 1 demonstrates that the level of dispersive mixing achieved can be increased by reducing the extruder output; in agreement with [16, 18, 22]. At the same screw speed (peak shear stress) a reduction in output reduces the average DOF within the screw channel increasing the mean shear stress experienced by the polymer/additive [12] thus increasing dispersion via fast rupture mechanism. A reduction in output also increases the residence time, thus increasing the contribution of the slower erosion mechanism [13].

Whether running at 300 or 360 rpm, a 30 kg h^{-1} reduction in output leads to a similar level of mixing improvement; see Figure 1b. This is attributed to the conditions having similar average DOF; with the parameter reduction being comparable from 43.8 to 35.6 % at 360 rpm, and 42.3 to 32.9 % at 300 rpm. Similar level of mixing is achieved at 60 kg h^{-1} & 300 rpm and at 90 kg h^{-1} & 360 rpm, with comparable IV reduction of 19.5 $\pm$ 1.5 % and 18.2 $\pm$ 2.3 %. Depending on process either condition may be preferred. The discrepancy of the 60 kg h^{-1} leading to slightly better mixing compared to 50 kg h^{-1} sample (in Figure 1a) cannot be fully explained. At the low screw speed the DOF change should have been significant at 61.7 to 48.9 % respectively.

Impact of Screw Speed

It is clear that for both H-PET and S-PET, an increase in screw speed leads to better dispersive mixing with both the mean and the 90th percentile of the distributions decreasing; see Figure 2a.

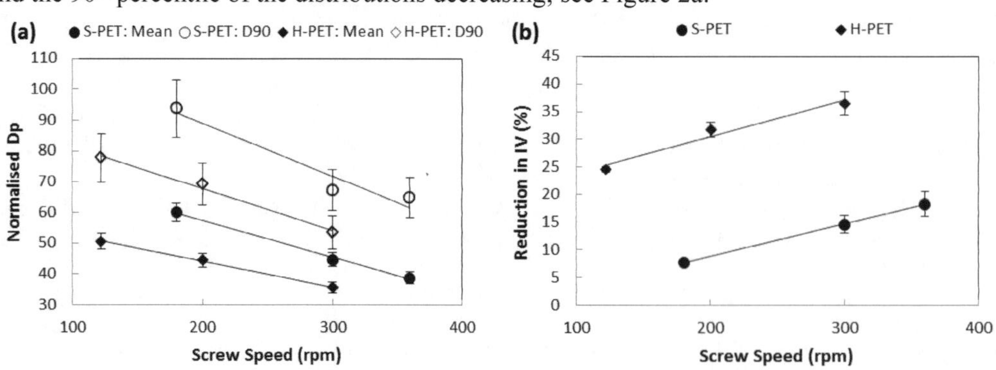

FIGURE 2. Effect of screw speed on (a) normalized particle diameter and (b) polymer intrinsic viscosity

An increase in screw speed raises the peak and the average shear stress at constant DOF, with the average stress increasing even further due to average DOF reduction as a result of constant output, increasing the effectiveness of the rupture mechanism. The potential gain is limited by the increase in melt temperature and the increased molecular weight loss which lead to a reduction in melt viscosity and hence shear stress. Similarly, the reduction in residence time minimizes dispersion via erosion mechanism; although increased shear forces will somewhat counteract this.

Impact of Melt Viscosity (Starting Molecular Weight) and Additive Loading

Superior dispersion was observed in H-PET samples due to the higher melt viscosity enabling transfer of larger stresses to the agglomerates; see Figure 2a. This perhaps is unexpected since H-PET experiences a considerably higher IV reduction during processing, but is the likely result of the additive being introduced to the feed pocket allowing large stresses to be transferred to agglomerates in the melt zone prior to the viscosity reduction. Screw speed has very similar effect on the size of agglomerates for both H-PET and S-PET. This is a likely result of the IV reducing at almost identical rate, irrespective of the starting value.

From the PSDs obtained for S-PET and H-PET at the same operating conditions, it is clear that starting molecular weight has little effect on the lower end of cumulative distribution. One would envisage the lower viscosity leading to improved wetting of the additive resulting in smaller particle size; however a 2.5 times reduction in viscosity at the processing conditions is probably too small to see the effect. The viscosity change is large enough to induce a reduction in the size of the largest agglomerates due to the increased shear stresses transferred to the additive. In [14], smaller viscosity differences (of 1.2 times) had a significant impact on the large agglomerates.

Loadings of 3, 5 and 7 $\%_{wt}$ were investigated at the same operating conditions. The 90th percentile of normalized PSD remained similar at 80.5 ± 8.0, 77.7 ± 7.7 and 87.6 ± 8.7 respectively, as the peak shear stress is the same; assuming any viscosity changes with loading increase are negligible. Any minor differences were attributed to difficulty in distinguishing particles in the higher loaded samples during analysis. The 10th percentile increased with loading from 16.3 ± 1.6 to 21.0 ± 2.1 and 25.7 ± 2.5 respectively, even though increased loading should lead to more fragments via collision mechanism [5]. This is likely due to the increased chance of re-agglomeration during collisions, higher wetting requirements of the increasingly filled composite and lower energy input per agglomerate.

Impact of Calculated Parameters: Degree of Fill and Specific Mechanical Energy Input

Effect of calculated parameters including average DOF, specific throughput and SME input was also considered. Figure 3a shows a linear increase in mixing with a reduction in average DOF. A similar relationship is observed with specific throughput, which is easier to calculate.

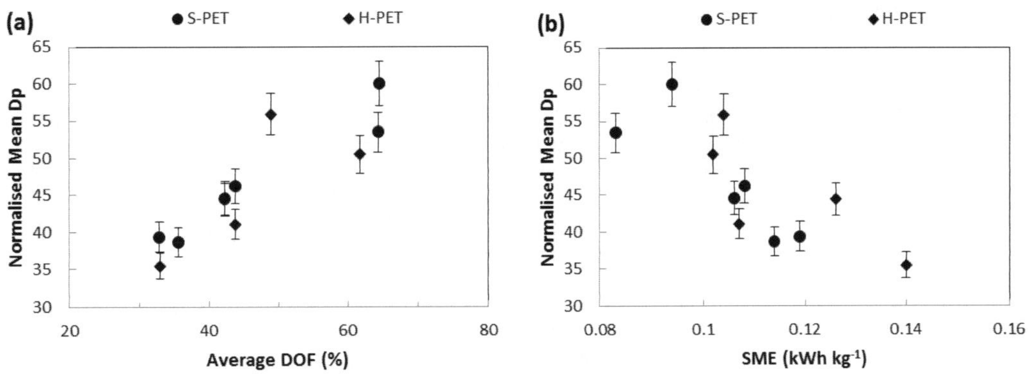

FIGURE 3. Effect of (a) Average Degree of Fill and (b) Specific Mechanical Energy on Dispersive Mixing Achieved

A linear improvement in dispersion is also observed with increasing SME; see Figure 3b. SME is a far more complex parameter which depends on the screw speed, extruder output and torque. Torque itself is dependent on the DOF within the machine (affected by output, screw speed, screw design) and the viscosity of the polymer; which in turn is influenced by the screw speed, barrel temperatures and starting molecular weight. SME increases with decreasing output and increasing screw speed. Similar relationships were observed for the IV (molecular weight) reduction of the polymer. IV was found to reduce with decreasing DOF and increasing SME. This identified a need for a compromise. High SME input and low DOF are required for good dispersive mixing and the opposite is required to preserve the IV of the polymer, which governs the mechanical properties of the finished film. It is these machine independent parameters which can be utilized for effective scale up of the extrusion process for machines with similar screw configurations.

CONCLUSION

A number of trials were performed to evaluate the dispersion of organophosphorus salt in standard and high molecular weight PET utilizing a Berstorff ZE40A co-rotating twin screw extruder. The impact of operating conditions on dispersion was studied to evaluate the level of control available over the process. Although the additive particle size within the composite was significantly larger than that of the powder form, reasons for which were explored but not presented here, key relationships between the level of dispersion and operating conditions were established. Dispersive mixing was found to increase with decreasing throughput (degree of fill), increasing screw speed (increasing shear stress, decreasing degree of fill) and increasing molecular weight (melt viscosity).

Linear relationships between calculated parameters including SME, DOF, specific throughput and the level of dispersive mixing were also observed. It is hoped that these can be utilized for effective scale up of twin screw extrusion processes with the same or similar screw configurations, without having to run experiments to define optimum operating conditions.

ACKNOWLEDGMENTS

The presenting author would like to acknowledge the help, guidance and support received from named authors on this paper as well as all the other personnel at DTF. Particular thanks must go to Sean Shortall for performing the IV analysis, Karl Rakos for help in measuring the additive powder PSDs, Jessica Nix for aid with microscopy analysis and the Semi-Tech operations team at DTF who made the experimental trials possible. The author would also like to express gratitude to DuPont Teijin Films and Engineering and Physical Sciences Research Council (EPSRC) for providing the funding for this work.

REFERENCES

1. W. A. MacDonald, "Polyester Films," in *Encyclopedia of Polymer Science and Technology*, John Wiley & Sons, 2003, pp.30-40.
2. T. Kajiwara and Y. Nakayama, *J. Chem. Eng. Jpn.* **44**, 831-839 (2011).
3. A. Lawal and D. M. Kalyon, *Polym. Eng. Sci.* **35**, 1325-1338 (1995).
4. I. Manas-Zloczower and D. L. Feke, "Dispersive Mixing of Solid Additives," in *Mixing and Compounding of Polymers*, edited by I. Manas-Zloczower, Munich: Henser Publishers, 2009, pp.183-215.
5. O. Seyvet and P. Navard, *J. Appl. Polym. Sci.* **78**, 1130-1133 (1999)
6. I. Manas-Zloczower and H. Cheng, *Macromol. Symp.* **112**, 77-84 (1996).
7. D. B. Todd, *Polym. Eng. Sci.* **15**, 437-443 (1975).
8. D. M. Kalyon and H. N. Sangani, *Polym. Eng. Sci.* **29**, 1018-1026 (1989).
9. D. M. Kalyon, C. Jacob and P. Yaras, *Plast. Rubber Compos. Process. Appl.* **16**, 193-200 (1991).
10. H. F. Giles Jr., J. R. Wagner Jr., E. M. Mount 3rd, *Extrusion: The Definitive Processing Guide and Handbook*, New York: William Andrew Publishing, 2005, pp. 425-433.
11. R. Rudolf, "General Overview of Compoundiong Process: Tasks, Selected Applications, and Process Zones," in *Co-Rotating Twin-Screw Extruders - Fundamentals, Technology, and Applications*, edited by K. Kohlgrüber, Munich: Carl Henser Verlag, 2008, pp.57-89.
12. H. Potente, J. Ansahl and B. Klarholz, *Int. Polym. Proc.* **9**, 11-25 (1994).
13. G. R. Kasaliwal, S. Pegel, A. Göldel, P. Pötschke and G. Heinrich, *Polymer.* **51**, 2708-2720 (2010).
14. G. R. Kasaliwal, A. Göldel, P. Pötschke and G. Heinrich, *Polymer.* **52**, 1027-1036 (2011).
15. T. Villmow, P. Pötschke, S. Pegel, L. Häussler and B. Kretzschmar, *Polymer.* **49**, 3500-3509 (2008).
16. T. Villmow, B. Kretzschmar and P. Pötschke, *Compos. Sci. Technol.* **70**, 2045-2055 (2010).
17. T. Domenech, E. Peuvrel-Disdier, B.Vergnes, "Influence of Twin-Screw Processing Conditions on Structure and Properties of Polypropylene – Organoclay Nanocomposites" in *27th World Congress of the Polymer Processing Society,* Marrakech, Morocco, 2011, pp.1 - 10.
18. T. Domenech, E. Peuvrel-Disdie and B. Vergnes, *Compos. Sci. Technol.* **75**, 7-14 (2013).
19. W. Gao, b. Zhou, X. Ma, Y. Liu, Z. Wang and Y. Zhu, *Colloids. Surf. A. Physicochem. Eng. Asp.* **385**, 181-187 (2011).
20. G. Fortunato, A. Tenniche, L. Gottardo and R. Hufenus, *Eur. Polym. J.* **57**, 75-82 (2014).
21. J. Champion, "Use of Computational Fluid Dynamics to improve the layer thickness control of polyester based multilayered films", Eng.D. Thesis, University of Birmingham, 2015.
22. R. Uphus, O. Skibba, R. H. Schuster and U. Gorl, *KGK.* **53**, 279-289 (2000).

Influence of Compounding Technology on Rheological, Thermal and Mechanical Behavior of Blast Furnace Slag Filled Polystyrene Compounds

Abdelhamid Mostafa[1], Gernot Pacher[1], Thomas Lucyshyn[2], Clemens Holzer[2], Helmut Flachberger[3], Elke Krischey[3], Bertram Fritz[4], Stephan Laske[5]

[1] *Polymer Competence Center Leoben GmbH (PCCL), Roseggerstraße 12, 8700 Leoben, Austria - hamid.mostafa@pccl.at, Gernot.Pacher@pccl.at*
[2] *Chair of Polymer Processing, Montanuniversitaet Leoben, Otto Glöckel-Straße 2, 8700 Leoben, Austria – thomas.lucyshyn@unileoben.ac.at, Clemens.Holzer@unileoben.ac.at,*
[3] *Chair of Mineral Processing, Montanuniversitaet Leoben, Franz-Josef-Straße 18, 8700 Leoben, Austria – Elke.Krischey@unileoben.ac.at, helmut.flachberger@unileoben.ac.at*
[4] *voestalpine Stahl GmbH, voestalpine-Straße 3, 4020 Linz, Austria - Bertram.Fritz@voestalpine.com*
[5] *Research Center Pharmaceutical Engineering GmbH, Inffeldgasse 13, A-8010 Graz, Austria - stephan.laske@tugraz.at*

Abstract - The influence of melt-compounding technique on blast furnace slags (BFS) filled polystyrene (PS) compounds was investigated. BFS are byproducts of iron industry, and are formed during the production of iron via thermo-chemical reduction in blast furnaces. BFS are mineral-structured materials composed of severeal such as silicon oxide (SiO2), calcium oxide (CaO), magnesium oxide (MgO) and alumina (Al2O3) as well as other minor oxides and elements. Such combination of oxides might be of technical advantage if BFS is properly prepared and tailored for use as a functional filler for PS. In addition, BFS is outstandingly inexpensive and require minimal refining costs compared to common mineral fillers used in polymer industry such as calcium carbonate and talc, giving BFS an economic significance. In current study, compounds were produced via melt-compounding approach, where two different processing technologies were used: (1) Laboratory rotor-blade internal mixer (IM) and (2) co-rotating, twin-screw compounding extruder (TSC). It was found that compounding process did not yield a strong influence on the rheological properties, where comparable levels for shear viscosity, storage- and loss moduli were observed for all compounds except for '20G40 TSC' compound. Such deviancy was clear in thermal properties of this particular compound, where slightly lower transition temperature (T_g) as well as higher specific heat capacity (C_p) were reported. For mechanical behavior, comparable stress-strain curves and young's modulus values for both processes were witnessed. Deviant '20G540 TSC' compound showed slightly lower young's modulus compared to corresponding IM.

Keywords: Blast furnace slag, process comparison, melt compounding, polystyrene

Introduction

Considerable efforts have been dedicated to modify functional behavior of polystyrene via incorporation of various oxide and mineral fillers [1–8]. Still, there is an urge to search for novel types of fillers that might further widen the application window of PS. Since common mineral fillers, such as calcium oxide and talc, are naturally derived resources, an abundant byproduct filler could have an advantage. Second, a novel filler might be effective if its composition is rich with oxide elements, which is the case for blast furnace slag (BFS). BFS is an abundant byproduct of iron industry. It is rich-oxide material as it is composed mainly of calcium oxide (CaO), silicon oxide (SiO_2), aluminum oxide (Al_2O_3) and magnesium oxide (MgO) [9] BFS are primarily utilized in cement industry and road building applications. Despite such utilization in industry, massive amounts of BFS are still being landfilled, which is an uneconomical action that imposes serious environmental concern [10].

Accordingly, utilization of BFS as a functional filler into thermoplastics such as PS could introduce some advantages, such as (1) the potential for improving functional property profile of PS, thanks to BFS's rich composition of oxides, (2) providing more utilization routes for BFS to decrease amount of landfilled slags, (3) saving extraction and processing costs associated with common mineral fillers and (4) utilizing BFS's inexpensive raw material cost for further treatment and functionalization steps.

Melt-compounding approach was chosen for PS-BFS systems as it is easy-to-control, time-saving, and easily up-scaling to industry standards. In a previous study, the influence of BFS as a functional filler on rheological, thermal and mechanical properties of PP and PS compounds was thoroughly investigated though mixing BFS with PP or PS in a laboratory rotor-blade internal mixer (IM) [11]. In current study, we compare the influence of IM with that of another melt-compounding equipment TSC, on the rheological thermal and mechanical properties of PS-BFS compounds.

Experimental

Materials

Styrolution PS 116N/L, provided by company Styrolution GmbH, Germany, was utilized as a base polymer with a melt flow rate of 23 cm^3 /10 min;

200 °C/5 kg and density of 1040 kg/m^3. BFS, kindly provided by company voestalpine Stahl GmbH, Austria, was introduced into PS matrix in two types: Granulated (GBS) and air-cooled (ACBS). The elemental composition of BFS is illustrated in Table 1. Milling of GBS and ACBS types into mesh size ranges of 0 µm to 71 µm and 0 µm to 20 µm was performed at Chair of Mineral Processing, Montanuniversitaet Leoben, Austria (LAV). This particular mesh size was decided based on milling equipment and conditions at LAV.

Table 1 – Composition of BFS

Constituent	Composition (%)
CaO	35-42
SiO_2	33-38
Al_2O_3	10-15
MgO	7-12
FeO	< 1.0
MnO	≤1.0
S	1 – 1.5
Cr_2O_3	≤ 0.1

Compound preparation

For IM technique, compounds were prepared via Haake Polylab system 3000P laboratory kneader, Thermo-Fischer Scientific Inc., USA, with mixing mass temperature of 155° C, mixing speed of 60 rpm and residence time of 10 minutes, which was calculated upon starting addition of BFS filler. For TSC technique, ZSK 25 gravimetric-fed, co-rotating, twin-screw extruder (Werner & Pfleiderer, Germany) with a screw diameter of 25 mm and length of 40D. The temperature profile for the TSC barrel was set between 160-210° C. To avoid material degradation due to hydrolysis, PS was always dried for 4 h at 80 °C prior to processing. Condition formulations are shown in Table 2.

Table 2 –Condition formulation code

Code	BFS filler		
	wt.-%	Type	Size (µm)
20A71	20	ACBS	0-71
20A40	20	ACBS	0-40
20G40	20	GBS	0-40

Compression molding

Collin hydraulic vacuum press (Collin 200 PV, Dr. Collin, Ebersberg, Germany) was utilized for compression molding. Compression molding process is divided into four phases where configuration for each phase is shown as (temperature in °C, pressure in bar, time in minutes) as follows: Phase 1 (195, 1, 6), phase 2 (195, 65, 2), phase 3 (195, 100, 1) and phase 4 (30, 100, 10). Produced plates dimensions are 150 x 150 x 4 mm and 150 x 150 x 2 mm.

Rheological investigation

A physica MCR 501 plate-plate type rheometer (Anton Paar Ltd., Austria) was utilized for measurements with a plate diameter of 25 mm and a gap of 1 mm. Plate-like samples of 20 mm diameter and 2 mm thickness were tested under nitrogen atmosphere to prevent degradation. Two testing modes were utilized: (1) dynamic strain sweep to make sure all conditions are within the viscoelastic region of PS, and (2) dynamic frequency sweep to investigate the shear viscosity, storage- and loss moduli. Frequency range of 0.1-500 rad/s was maintained for all tests and at least three replicates per conditions were averaged for statistical significance.

Thermal investigation

Thermal behavior was investigated via a DSC-4000 differential scanning calorimeter (Perkin Elmer, USA). The experiments were conducted using 10 mg sample sealed in an aluminum pan. Heating and cooling rates were fixed at 10 °K/min for all samples. Each sample was subjected to one cooling and two heating cycles.

Mechanical investigation

Zwick/Roell station (Zwick and Co.KG, Ulm, Germany) was utilized for tensile testing experiments according to standard ISO 527-1. A strain rate of 50 mm/min was maintained for determination of stress-strain curves, while a strain rate of 1 mm/min was maintained for estimating young's modulus values. At least three replicate samples for each condition were machined out of 4 mm-thick pressed plates to a standard dog-bone shape and the tensile test was conducted at room temperature. All data were evaluated using testXpert II software.

Results and Discussion

Rheological behavior

Figures 1 and 2 illustrates that both IM and TSC showed a modest influence upon rheological properties. Curves of both processes showed shear-thinning plateaus. Except for the '20G40 TSC' condition, all shear viscosity and moduli curves were almost coinciding. More elaboration on such deviation will be discussed in next section. It was also noticed that almost identical damping factor values, Figure 3, were evident IM and TSC processes, suggesting comparable elasticity for all compounds.

Thermal behavior

Within statistical margin of error, no major difference was witnessed for T_g values of IM and corresponding TSC compounds, except for a slight decrease for the '20G40 TSC' condition as compared to IM twin condition, as shown in Figure 4. Lower T_g implicates more flexible chain mobility and hence lower shear viscosity, as indicated in Figure 1. This is probably due to intensive shear action in TSC, possibly allowing more uniform distribution of slag filler and hence easier polymer chain movement [12, 13]. The slight decrease in T_g in '20G40 TSC' condition was coupled with a clear increase in specific heat capacity as compared to IM, meaning more heat was required for the transition action to occur. It is difficult to explain the reason behind this, since it was specific for this condition. Such increase in the specific heat capacity compared to other compounds could be a function of this particular BFS filler size and/or type as related to the mixing kinetics in TSC. Therefore, more focus on this particular condition is recommended for better understanding.

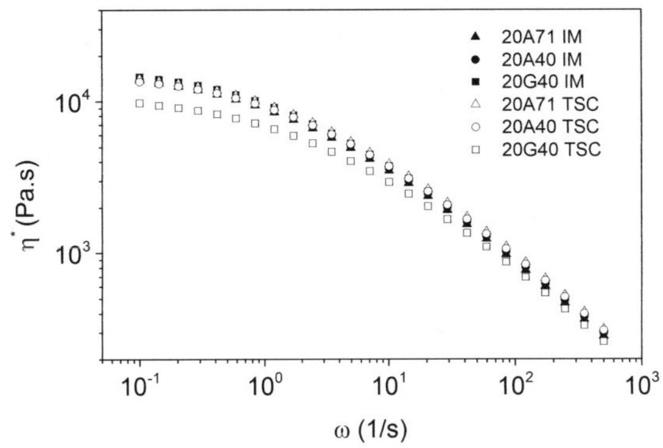

Figure 1 - Shear viscosity curves for different compounds

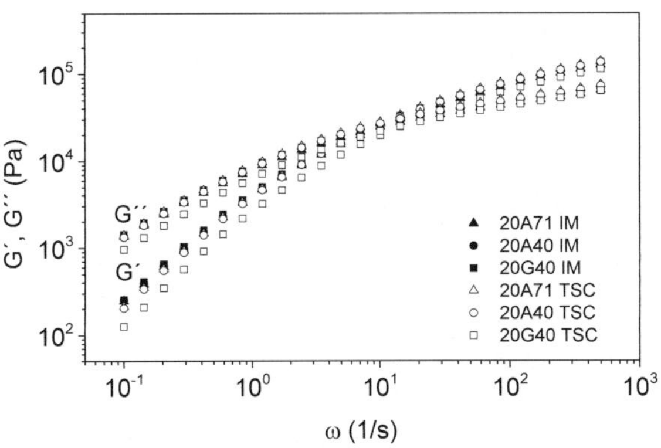

Figure 2 – Storage and loss moduli curves as a function of frequency

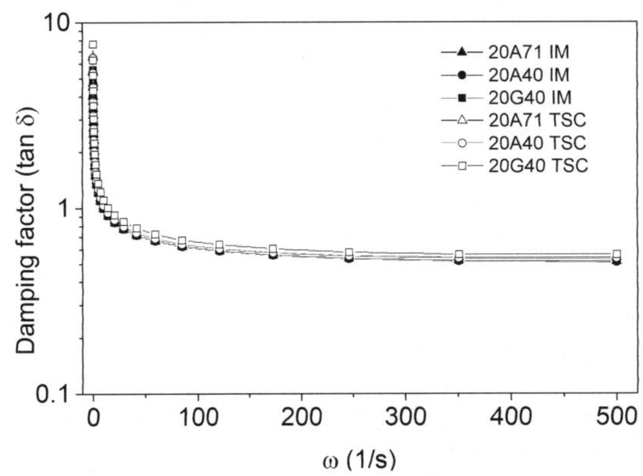

Figure 3 – Damping factor for IM/TSC compounds

Mechanical behavior

Stress-strain curves for respective IM and TSC formulations are shown in Figure 5. It was observed that both IM and TSC-compounds yielded comparable strength and strain behavior. '20G40-TSC' condition showed a slight increase in strength up to 1% strain level.

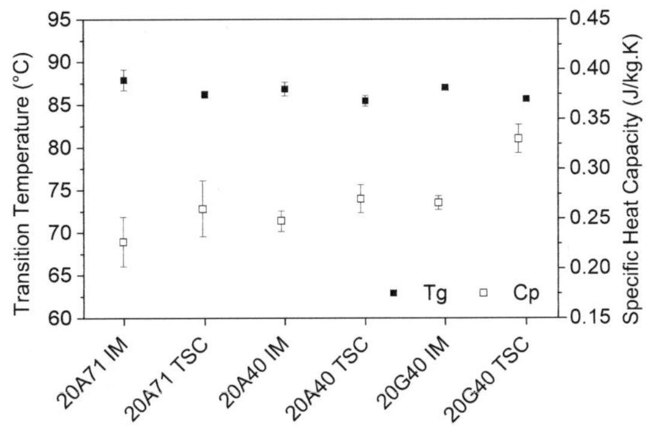

Figure 4 – Transition temperature and specific heat capacity for different compounds

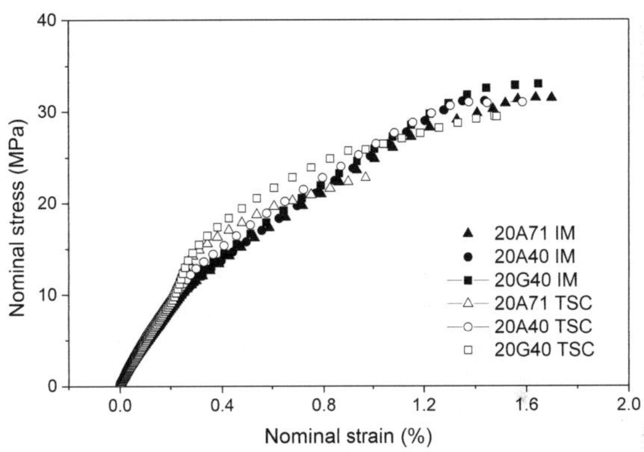

Figure 5 – Stress-strain curves for different compounds

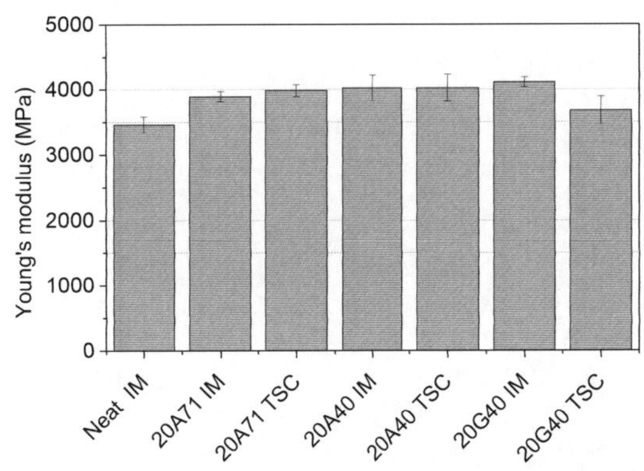

Figure 6 – Young's modulus for different compounds

The mean Young's modulus values, Figure 6, and the statistical error bars show that upscaling from IM to TSC did not show a clear influence upon young's modulus of filled compounds, except for slightly lower value for deviant '20G50 TSC' condition. Following the discussion in rheology and thermal section, this slight decrease in stiffness might be correlated with the aforementioned high specific heat capacity for such condition. Slightly higher Young's modulus level is shown for most BFS-filled conditions as compared with neat PS. Such modest increase is believed to be due to presence of BFS filler as rigid-particle reinforcer [14, 15].

Conclusions

The influence of melt-compounding approach on rheological, thermal and mechanical properties was successfully investigated for PS-BFS compounds. The findings generally showed modest interaction between BFS filler and PS matrix, which was expected due to polymer-filler incompatibility. '20G40 TSC' condition showed deviant behavior, which was reflected in throughout the different findings presented in current study. Therefore, more investigation is recommended of this particular condition. In addition, promoting one or more suitable coupling agents for PS-BFS system is highly advisable. That is, treating BFS would likely lead to more interaction with PS matrix through network building, which can positively affect the compound performance.

Acknowledgements

Current research was achieved at Polymer Competence Center Leoben GmbH (PCCL, Austria) and funded within the framework of the 'Bridge' program of the Austrian Research Promotion Agency (FFG). Invaluable contributions toward the fulfillment of this research were provided by company voestalpine Stahl GmbH, Chair of Polymer processing and Chair of Mineral Processing, Montanuniversitaet Leoben. PCCL is funded by the Austrian Government and the State Governments of Styria and Upper Austria.

References

[1] A. Chandra, L.-S. Turng, S. Gong, D. C. Hall, D. F. Caulfield, and H. Yang, "Study of polystyrene/titanium dioxide nanocomposites via melt compounding for optical applications," *Polym. Compos,* vol. 28, no. 2, pp. 241–250, 2007.

[2] K. J. Kim and J. L. White, "Rheological investigations of suspensions of talc, calcium carbonate, and their mixtures in a polystyrene melt," *Polym. Eng. Sci,* vol. 39, no. 11, pp. 2189–2198, 1999.

[3] S. Siengchin, J. Karger-Kocsis, and R. Thomann, "Alumina-filled polystyrene micro- and nanocomposites prepared by melt mixing with and without latex precompounding: Structure and properties," *J. Appl. Polym. Sci,* vol. 105, no. 5, pp. 2963–2972, 2007.

[4] H. Tanaka and J. L. White, "Experimental investigations of shear and elongational flow properties of polystyrene melts reinforced with calcium carbonate, titanium dioxide, and carbon black," *Polym. Eng. Sci,* vol. 20, no. 14, pp. 949–956, 1980.

[5] S. Tanoue, L. A. Utracki, A. Garcia-Rejon, J. Tatibouët, K. C. Cole, and M. R. Kamal, "Melt compounding of different grades of polystyrene with organoclay. Part 1: Compounding and characterization," *Polym Eng Sci*, vol. 44, no. 6, pp. 1046–1060, 2004.

[6] H. S. Vaziri, I. A. Omaraei, M. Abadyan, M. Mortezaei, and N. Yousefi, "Thermophysical and rheological behavior of polystyrene/silica nanocomposites: Investigation of nanoparticle content," *Materials & Design*, vol. 32, no. 8-9, pp. 4537–4542, 2011.

[7] H. S. Vaziri, I. A. Omaraei, M. Abadyan, M. Mortezaei, and N. Yousefi, "Thermophysical and rheological behavior of polystyrene/silica nanocomposites: Investigation of nanoparticle content," *Materials & Design*, vol. 32, no. 8-9, pp. 4537–4542, 2011.

[8] Z. Wang, G. Li, H. Peng, Z. Zhang, and X. Wang, "Study on novel antibacterial high-impact polystyrene/TiO2 nanocomposites," *J Mater Sci*, vol. 40, no. 24, pp. 6433–6438, 2005.

[9] National.Slag.Association, *Blast Furnace Slag*. Available: http://www.nationalslag.org/blast-furnace-slag (2015, May. 26).

[10] A. A. Francis, "Conversion of blast furnace slag into new glass-ceramic material," *Journal of the European Ceramic Society*, vol. 24, no. 9, pp. 2819–2824, 2004.

[11] A. Mostafa, S. Laske, G. Pacher, C. Holzer, H. Flachberger, E. Krischey, and B. Fritz, "Blast furnace slags as functional fillers on rheological, thermal, and mechanical behavior of thermoplastics," *J. Appl. Polym. Sci,* vol. 133, no. 8, pp. n/a-n/a, 2016.

[12] M. A. Repka, N. Langley, and J. DiNunzio, *Melt extrusion: Materials, technology and drug product design*. New York: Springer, 2013.

[13] K. Kohlgrüber and M. Bierdel, *Co-rotating twin-screw extruders: Fundamentals, technology, and applications*. Munich [Germany]: Carl Hanser Publishers, 2008.

[14] T. P. Selvin, J. Kuruvilla, and T. Sabu, "Mechanical properties of titanium dioxide-filled polystyrene microcomposites," *Materials Letters*, vol. 58, no. 3-4, pp. 281–289, 2004.

[15] S.-Y. Fu, X.-Q. Feng, B. Lauke, and Y.-W. Mai, "Effects of particle size, particle/matrix interface adhesion and particle loading on mechanical properties of particulate–polymer composites," *Composites Part B: Engineering*, vol. 39, no. 6, pp. 933–961, 2008.

PROPERTIES OF POLYAMIDE 6-GRAPHENE-COMPOSITES PRODUCED AND PROCESSED ON INDUSTRIAL SCALE

*Christian Hopmann, Maximilian Adamy, Christian Windeck, Nafi Yesildag**

Institute of Plastics Processing (IKV) at RWTH Aachen University, Germany

**Corresponding author: nafi.yesildag@ikv.rwth-aachen.de*

Abstract: The use of graphene as a filler in thermoplastics has already been investigated extensively. The mechanical properties as well as electrical and thermal conductivity of thermoplastics can be improved due to graphene. However, these studies were carried out in experimental scale, which allows a good dispersion of graphene because of a long residence time during melt mixing and because of the use of almost ideal graphene, which has a high specific surface and low number of layers. In this study the scientific findings are transferred into industrial practice. For that purpose, a co-rotating intermeshing twin screw extruder with limited residence time and limited shear energy input is used to produce graphene based polyamide 6 composites and short carbon fibre reinforced graphene based polyamide 6 composites. These composites are further processed by injection molding to produce specimens in order to determine the mechanical properties. In addition to the scale-up to industrial production, two types of graphene platelets are used, which are commercially available. This investigation reveals that the addition of 1 wt.-% commercially available graphene platelets to polyamide 6 improves Young´s modulus and tensile strength. Compared to pure polyamide 6 an improvement up to 20 % with regard to Young´s modulus and up to 15 % with regard to tensile strength can be achieved. The combination of short carbon fibre with graphene platelets in polyamide 6 enables an increase of Young´s modulus, which is higher than the additive effect.

Introduction

Graphene in its pure and ideal form is the thinnest, stiffest and strongest material known up to now. It is also an excellent conductor for both electricity and heat. Furthermore, graphene has a very high specific surface area, which enables to build strong physical or chemical bonds [1]. These remarkable properties make graphene interesting as functional filler for plastic composites to enhance the mechanical, electrical and thermal properties of various plastic products, especially at low filler contents. Thus, the potential of enhancing the mechanical, electrical and thermal properties of plastics by addition graphene is highlighted in numerous scientific publications [2, 3, 4, 5, 6].

For example, Chatterjee et al. focus on the dispersion of graphene nanoplatelet (GnP). The GnP were physically premixed with a surfactant powder and pestled for 30 min before compounding with PA12 by using a micro extruder. The adding of 0.5 wt.-% GnP leads to an improvement in the yield strength by 32.2 % relative to the yield strength of 36.7 MPa for pure PA12 [7]. Also Zhao et al. observe a large effect on the mechanical properties of polyvinyl alcohol (PVA) at low loadings. The addition of 0.6 wt.-% graphene in PVA improves Young´s modulus by 73 %. In a further study, the addition of only 0.42 vol.-% graphene in polypropylene achieved an increase in yield strength by 75 % and an increase in Young´s Modulus by 74 % [8]. Furthermore, the maximum values regarding to all mechanical strengths and modulus are reached at a graphene content of 1.0 wt.-%.

These three publications are representative in the field of graphene based composites. Even at low filler contents a significant improvement of the mechanical properties is observed. But, the increase of the reinforcing effect of graphene becomes smaller at higher contents (> 2 wt.-%), because graphene agglomerates due to the van der

Waals forces of the nanosheets. This leads to a less increase of the reinforcing effect on the mechanical properties of the composites [8]. In addition these studies perform the preparation of graphene composites on lab scale, which is not scalable to industrial scale. Especially the mixing conditions influence the exfoliation and dispersion of graphene in higher viscosity thermoplastics like PA and this plays a key role in order to exploit the extraordinary properties of graphene in graphene composites. To unlock the potential of graphene for plastic applications, the properties of graphene composites made in industrial scale have to be investigated. Therefore, in this study graphene based composites are produced and further processed at industrial scale. The utilization of industrial scale processes restricts the degree of the dispersion of the graphene due to limited residence time during melt compounding in a twin screw extruder. Thus, the potential of graphene as a filler in composites in industrial practice is evaluted.

Materials

Different compositions of graphene based polyamide 6 (PA6) composites are produced in order to investigate the tensile properties and morphology of the resulting composites. The PA6 type Ultramid B27E01, provided by BASF SE, Ludwigshafen, Germany, is used as matrix material. It has a relative viscosity of 2.7, a density of 1.13 g/cm³ and is a typical compounding grade. Two commercially available graphene types are utilized in this study. The N006-P polar graphene powder, provided by Angstrom Materials, Dayton, USA, is ~10-20 nm thick and has the approximate x-y dimensions of 5.0 µm. The carbon content is around 95.0 % and the oxygen content is less than 4.0 % at a density of $\leq$ 2.20 g/cm³. The specific surface area is $\geq$ 15 m²/g. The second used graphene is provided by XG Science Inc., Lansing, USA. The nanoplatelets Grade M has average thickness of 6-8 nm and specific surface area of 120-150 m²/g. the oxygen content is $\leq$ 1 %. In addition multiwalled carbon nanotubes (MWCNT) are taken as a reference. The used MWCNT type NC7000 have an average diameter of 9.5 nm and an average length of 1.5 µm. The density of the MWCNT is 1.75 g/cm³. They are provided by Nanocyl S.A., Sambreville, Belgium. These three nanofillers are added to the PA 6 with a filler content of 1, 2 and 3 wt.-%. Furthermore, short carbon fibre reinforced (SCF) graphene based PA 6 composites are produced in industrial scale. When SCF and graphene or MWCNT are combined, a graphene or MWCNT content of 1, 2 and 3 wt.-% is added to a constant SCF content of 20 wt.-%. The SCF type HT C493 have a density of 1.82 g/cm³, a fibre diameter of 7 µm and an initial fibre length of 6 mm. They are prepared with 2.7 wt. % sizing suitable for PA6 and are provided by Toho Tenax Europe GmbH, Wuppertal, Germany.

Experimental procedure

The composites are mixed using a co-rotating intermeshing twin screw extruder type ZSK26MC of Coperion GmbH, Stuttgart, Germany, with a screw diameter (D) of 26 mm and a length-to-diameter-ratio of 45 D. The PA6 is fed into the main hopper, while the different graphene types and MWCNT are dosed at the 4th barrel element, whereas the SCF are fed in the 8th barrel element to avoid a strong deterioration of the fibres. All components of the composites are dosed using appropriate gravimetric dosing systems and side feeders. The composite is degassed at the 11th barrel element, before exiting the extruder as a strand. The strands are quenched in a water bath and pelletized. The composites are produced at a throughput of 10 kg/h, a screw rotational speed of 500 min^{-1} and a barrel temperature profile resulting in 240 °C. After melt compounding, the composites are further processed to tensile test specimens by injection molding using an Ergotech 80/420-310 of Demag, Duisburg, Germany. The tensile properties Young's modulus, tensile strength and elongation at break of 10 composite test specimens are determined according to EN ISO 527 using a tensile testing machine type Z150 of Zwick GmbH & Co. KG, Ulm, Germany, at room temperature (23 °C). The notched impact strength of 10 specimens each is evaluated according to EN ISO 179 using a pendulum of Zwick GmbH & Co. KG, Ulm, Germany.

In addition segments of the specimens are removed and further prepared to cross section areas by an ultramicrotome type Leica UCT of Leica Camera AG, Wetzlar, Germany. Afterwards the cross section areas are coated with gold to analyze the dispersion of the graphene, by using a field emission scanning electron microscopy (FESEM) type Zeiss Sigma VP of Carl Zeiss AG, Oberkochen, Germany.

Investigation of tensile properties

The results show, that the production and further processing of graphene based PA 6-composites on industrial scale is possible. The addition of graphene type N006-P and nanoplatelets Grade M into PA6 results in a higher Young´s Modulus in comparison to pure PA6 and to MWCNT reinforced PA6 (Fig. 1). A filler content of 1 wt.-% (N006-P) leads to an improvement of Young´s modulus by 20 % compared to pure PA6. But, an increase of the filler content does not cause in a significant further improvement of Young´s modulus, as the differences measured in the Young´s modulus are within the range of failure distribution. The possible agglomeration of graphene, which is mentioned earlier, could reduce the reinforcing effect of the graphene and therefore Young´s modulus does not further increase with increasing filler content.

The addition of the respective nanofillers causes in a variable standard deviation in tensile strength. A reason could be different degrees of dispersion of the nanofillers in the specimens. The use of MWCNT and graphene type N006-P tends to higher tensile strengths compared with pure PA6, because load is transferred from matrix to the durable filler via high specific surface of the filler. At a filler content of 1 wt.-% the tensile strength is improved from 72.5 MPa (pure PA6) to 83.5 MPa (N006-P) in average, which is an increase of 15 %. The influence of the nanoplatelets Grade M is not clear.

Independent of the used nanofiller, the elongation at break is decreased strongly to ≤ 5 % at a filler content of 1 wt.-% (Fig. 2). The increase of the filler content does not lead to a further significant decrease of elongation at break.

Also the notched impact strength is reduced due to the adding of graphene or MWCNT. But no clear tendency depending on filler content and filler type is visible.

Summarizing, the addition of the used two graphene types and MWCNT results in higher Young´s modulus and tensile strength. The reinforcing effect is lower compared to the abovementioned publications.

This may occur, because the degree of the graphene dispersion is low. In Figure 3 it is visible, that both graphene types exist in form of agglomerates already at a filler content of 1 wt.-%.

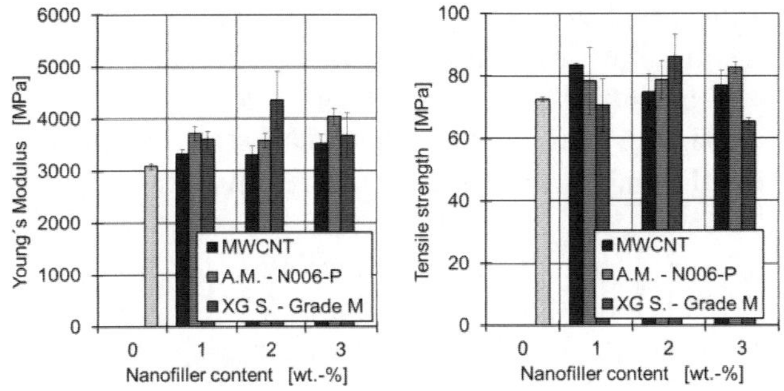

Figure 1: Young´s modulus and tensile strength of graphene/MWCNT based PA6 composites

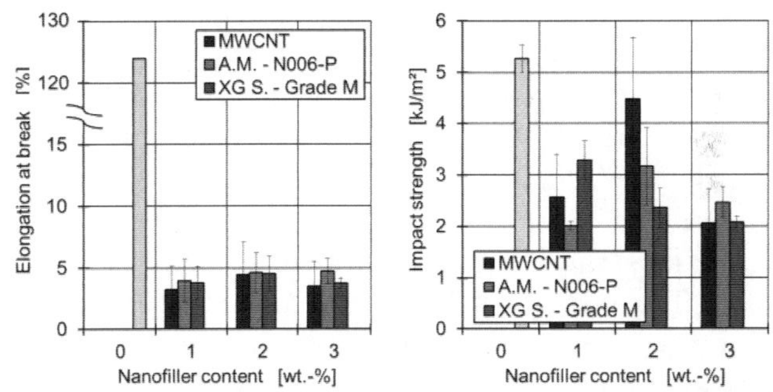

Figure 2: Elongation at break and notched impact strength of graphene/MWCNT based PA6 composites

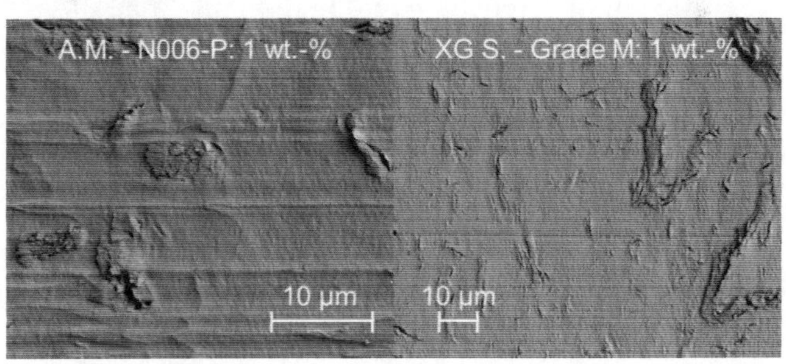

Figure 3: Morphology images of graphene based PA6 composites

Combination of fillers

A combination of fillers in nanometer and micrometer scale should improve the mechanical properties significantly. For this reason SCF reinforced graphene based PA6 composites are produced and the mechanical properties are analyzed. The results show, that Young´s modulus increases strongly from 13,500 MPa (20 wt.-%

SCF) up to 15,200 MPa by adding 1 wt.-% of MWCNT (Fig. 4). The use of graphene type N006-P and nanoplatelets Grade M also enables an improvement of Young´s modulus. The melt viscosity increases due to the addition of fillers in nanometer scale [10]. This leads to a higher carbon fibre orientation in the specimens, which are produced by injection molding. The mechanical properties are analyzed in flow direction of the melt and therefore a higher degree of fibre orientation leads to higher Young´s modulus. Thereby, the reinforcing effect due to the addition of graphene or MWCNT to SCF reinforced PA6 is bigger than the additive effect.
On the other hand tensile strength decreases due to the adding of graphene or MWCNT in comparison to the SCF reinforced PA6 despite of a higher fibre orientation (Fig. 5). Because, the viscosity increase causes in higher filler-matrix-interactions. This leads to a reduction of the short carbon fibre length and therefore to a reduction of tensile strength [10]. Thus no synergy effect is visible.
As before, the addition of graphene or MWCNT to SCF reinforced PA6 causes in a decrease of elongation at break (Fig. 6). But, the use of nanoplatelets Grade M enables a higher elongation at break compared to the use of graphene type N006-P and MWCNT. These results are unexpected, because Fig. 2 shows no significant difference in elongation at break by using different graphene types or MWCNT. Further investigations have to be carried out to settle this relationship.
The notched impact strength of SCF reinforced graphene based PA6 composites decreases by adding graphene or MWCNT. Because of the higher melt viscosity and therefore higher filler-matrix-interactions the carbon fibre length decreases. The notched impact strength also decreases with decreasing fibre length [11]. Therefore, the adding of graphene or MWCNT does not improve the impact strength of SCF reinforced graphene based PA6 composites.

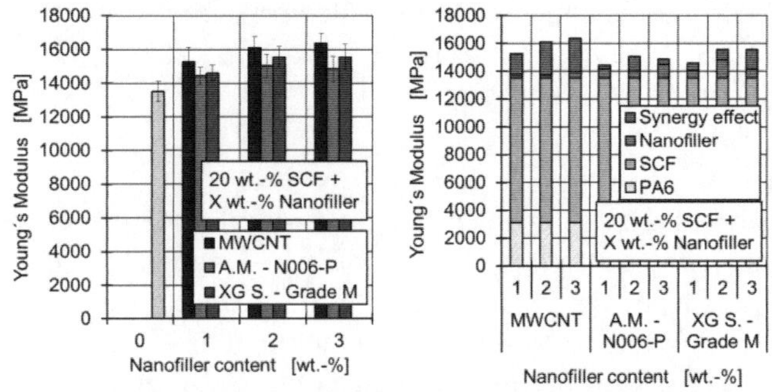

Figure 4: Young´s modulus and synergy effect of SCF reinforced graphene based PA6 composites

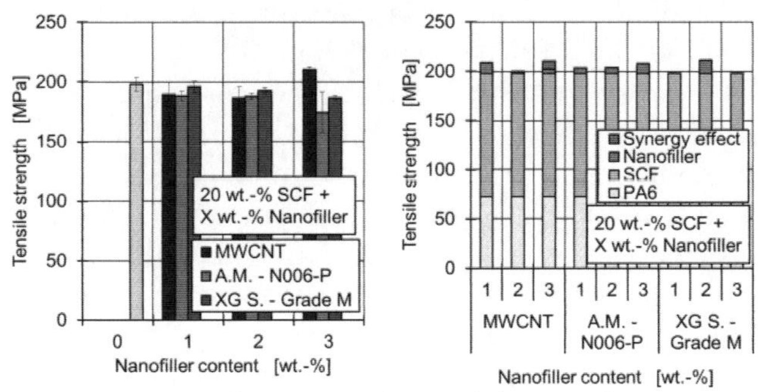

Figure 5: Tensile strength modulus and synergy effect of SCF reinforced graphene based PA6 composites

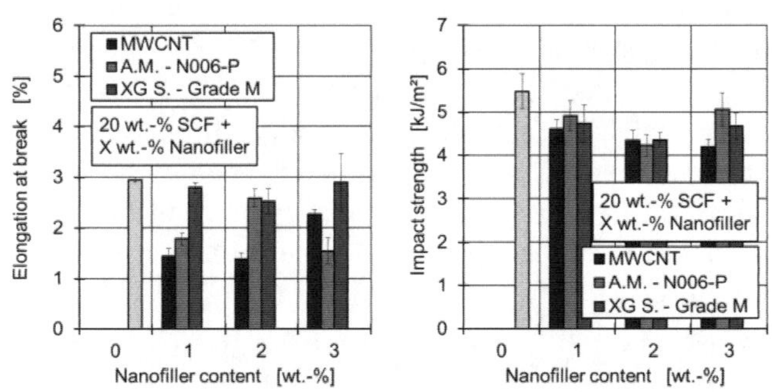

Figure 6: Elongation at break and notched impact strength of SCF reinforced graphene based PA6 composites

Conclusion

This study investigated the mechanical properties of graphene based PA6 composites and SCF reinforced graphene based composites, which are produced using a co-rotating intermeshing twin screw extruder and further processed in injection molding. The results show, that the processing of graphene in industrial scale is possible. The adding of graphene into PA6 leads to an improvement of Young´s modulus and tensile strength. On the other hand graphene based PA6 composites are characterized by brittle material behavior, which is identified by a low elongation at break and low notched impact strength compared to pure PA6. Nevertheless, the reinforcing effect of the used graphene types is less pronounced compared to graphene based composites, which are produced in laboratory scale, because of a low degree of dispersion. Furthermore, the used graphene types were not perfect graphene with high specific surface area and low thickness. Therefore, it is of interest, if the processing of more ideal graphene by using a co-rotating intermeshing twin screw extruder enables a good dispersion of the graphene in thermoplastics and if it is possible to improve the degree of dispersion by using optimized processing parameters during compounding and further processing.

Acknowledgments

The research project (141 EBG) of the Forschungsvereinigung Kunststoffverarbeitung and Schweißen has been sponsored as part of the "Industrielle Gemeinschaftsforschung und -entwicklung (IGF)" by the German Bundesministerium für Wirtschaft und Energie (BMWi) due to an enactment of the German Bundestag through the AiF. We would like to extend our thanks to all organizations mentioned.

Furthermore, we thank Coperion GmbH, Stuttgart, Germany, for providing the machinery and the companies BASF SE, Ludwigshafen, Germany, and Nanocyl S.A., Sambreville, Belgium, for supplying the materials.

References

1. Kim, H.; Abdala, A. A.; Macosko, C. W.: Graphene/Polymer Nanocomposites. Macromolecules, 2010, vol. 43, no. 16, pp. 6515-6530
2. Du, J.; Cheng, H.-M.: The Fabrication, Properties, and Uses of Graphene/Polymer Composites. Macromolecular Chemistry and Physics, 2012, vol. 213, no. 10-11, pp. 1060-1077
3. Potts, J. R.; Dreyer, D. R.; Bielawski, C. W.; Ruoff, R. S.: Graphene-based polymer nanocomposites. Polymer, 2011, vol. 52, no. 1, pp. 5-25
4. Stankovich, S.; Dikin, D. A.; Dommett, G. H. B.; Kohlhaas, K. M.; Zimney, E. J.; Stach, E. A.; R. Piner, D.; Nguyen, S. B. T.; Ruoff. R. S.: Graphene-based composite materials. Nature, 2006, vol. 442, no. 7100, pp. 282-286
5. Young, R. J.; Kinloch, I. A.; Gong, L.; Novoselov, K. S.: The mechanics of graphene nanocomposites: A review. Composites Science and Technology, 2012, vol. 72, no. 12, pp. 1459-1476
6. Verdejo, R.; Bernal, M. M.; Romasanta, L. J.; Lopez-Manchado, M. A.: Graphene filled polymer nanocomposites. Journal of Materials Chemistry, 2011, vol. 21, no. 10, pp. 3301-3310
7. Chatterjee, S.; Nueesch, F. A.; Chu, B. T. T.: Comparing carbon nanotubes and graphene nanoplatelets as reinforcements in polyamide 12 composites. Nanotechnology 2011 , 22
8. Zhao, X.; Zhang, Q.; Chen, D.: Enhanced mechanical properties of graphene-based poly(vinyl alcohol) composites. Macromolecules 2010; 43: 2357–63.
9. Song, P.; Cao, Z.; Cai, Y.; Zhao, L.; Fang, Z.; Fu, S.: Fabrication of exfoliated graphenebased polypropylene nanocomposites with enhanced mechanical and thermal properties. Polymer 2011; 52: 4001–10.
10. Wang M, Wang W, Liu T, Zhang WD. Melt rheological properties of nylon 6/multi-walled carbon nanotube composites. Compos Sci Technol 2008;68(12):2498-502.
11. Wypych, G.: Handbook of Fillers. Toronto: ChemTec Publishing, 2010

Mechanical Properties of Polypropylene/Recycle Polypropylene Blend Prepared By Melt Compounding

Jittiwat Nithikarnjanatharn[a*], Anuchit Khongrit[b] Seksan Hadsadee[a]

Naruedon Jongjaihon[a] and Narongsak Thammachot[a]

[a] *Department of Industrial Engineering, Faculty of Engineering and Architecture, Rajamangala University of Technology Isan, Muang, Nakornratchasima, 30000 THAILAND*
[b] *Indsutrail management Program, Faculty of Industrial technology, NakhonRatchasimaRajabhat University, Muang, Nakornratchasima, 30000 THAILAND*

jittiwat.ni@rmuti.ac.th

Abstract. This research aims to study the mechanical properties of poly propylene and polypropylene (PP) led recovery. by adding reactive coupled substances coupled reactive silane types Vinyl-Triethoxy Silane; VTMS A-171; VTMS A-171) , The reaction began Dicumyl Peroxide; DCP., Substances added stability (Anox PP 18 P and Alkanox 240 were). Preparing a polymer mixture used a twin screw extrusion machines. In order to analyze the proportion of the mix polymer mixture of poly propylene (PP) and copolymer polypropylene grades recycling (r-PP). The ratio of the total mix 5 ratios is 100: 0, 80:20, 50:50, 20:80 and 0: 100 and the addition of four coupled levels 0, 1, 3 and 5 phr. The study showed that the proportion of 80/20 mix PPr-PP can be used instead. Polypropylene pellets again affecting the mechanical properties of poly propylene pellets decrease. The way additives coupled reactive silane the amount of 1 phr will affect the mechanical properties of polymer blends increases. The tensile strength was 29.531 MPa. The Young's modulus was 631.084 MPa. The hardness equal to 65.00 shore D. This microstructure was similar to the PP pellets new.

Keywords: Compounds reactions coupled siren, Polymer mix, Mechanical Properties

Introduction

In the present, Polymer industry was a key significant in the small and large industry. The properties of the polymer are suitable to be applied to multiple forms of production such as containers bottles, tanks, home appliances and auto parts etc [1-2]. The rate of polymer product was extensively and rapidly increasing. The polymer material is the most popular alternative in production and consumption. Polymer has degraded so slowly become a problem for the environment. By the way of waste disposal of these polymers can be done in several ways, such as bringing in soil to landfill to incineration. Recycling and reuse polymer are possible way to manage the plastic waste.

T. Shuichi et al., [3] discussed the effect of the electric irradiation on the properties of Polypropylene (PP)/Vapor-grown carbon fiber (VGCF) composites prepared by melt compounding using a twin-screw extruder. Ethylene-propylene random copolymer was included in PP. Graft copolymer of maleic anhydride of polypropylene (MA-PP) was employed as an additive agent. The Young's modulus of PP and PP/MA-PP increased with the irradiation dose. According to the X-ray diffraction results, the crystallinity of PP is almost independent of electric irradiation. Thermal conductivity of the composites was also independent of the electric irradiation even if the thermal conductivity increased by adding of VGCF.

A. Khongrit and U.[4] Meekum In this studied, Wood Plastic composites based on crosslinked polypropylene and wood flour reinforcement was performed. As the crosslink structure of polymer chain enhances the properties of material such as higher in resistance to heat and outdoor degradation. Accordingly, it will give rise to superior in dimensional stability and durability. The statistical approach by mean of design of experimental(DOE) on the given constituent parameters that effect to the properties of WPC will be the prime interests. From the design of experiment approach to validate the optimal contents of WPC constituents based on crosslinked PP as the matrix, it was found that compounding of Wood Plastic composites at high level(+) of silane, wood flour and talc content produced the composite with generally good mechanical and thermal properties. Sauna cured via silane/water macro crosslinked reaction facilitated the interfacial adhesion between fibre and matrix and hence enhanced the mechanical properties.

A. Khongrit and U. Meekum [5] Wood plastic composite(WPC) based on crosslinked polypropylene(X-PP)/wood flour was explored. The peroxide/silane was used as crosslinking system. The sauna incubation under moisture saturated oven was applied to accelerate the competition of the siloxy/moisture networking reaction. There were three parts of the research work; design of experiment, toughening of WPC and the effect of peroxide, silane and PP copolymer on properties of the WPC, respectively. In this published work, the toughness improvement of the composite was focused. Ultra-high molecular weight polyethylene (UHMWPE) and Ethylene propylene diene terpolymer(EPDM) were employed to improve impact strength via blending with x-PP matrix. Composites were compounded into pellets by co-rotational twin screw extruder and test specimens were prepared by injection molding. Sauna incubation at 105°C for 12 hrs in oven chamber was performed to accelerate the final silane condensation crosslink reaction. MFI, impact strength, flexural properties and heat deflection temperature measurement were conducted. Impact strength, HDT and flexural modulus were improved with increasing UHMWPE content, and the optimal values around 5-10 phr of UHMWPE were achieved. Addition of EPDM elastomer to the matrix blends, reduced flexural strength and modulus but increased impact strength. While incorporation of EPDM into the PP/UHMWPE blends was exhibited much higher impact strength than that of the PP/UHMWPE binary blends. Silane crosslinked through sauna treatment improved the impact strength. HDT were also much risen for the crosslinked composite comparing with the non-crosslinked one.

B. Beltran and C. Mijangos [6] Peroxide initiated vinylsilane grafting of polypropylene in an intensive mixer, and the subsequent water crosslinking process were studied. Different concentrations of vinyl trimethoxysilane and dicumyl peroxide were used. For the hot pressed sheets with concentrations higher than 4 phr of vinyl silane an important increase in the melt viscosity was observed. This increase agrees with the change observed in the mechanical properties, which show a maximum for the water crosslinked samples containing 4 phr of vinyl silane. The modulus increases and silane grafted water crosslinked samples show a more stable thermal behavior than both the silane grafted samples and the unmodified polypropylene.

The study found that most of the above research. With the introduction of the polymer any other kind to improve the properties tested. But rarely seen bringing grain polymer poly propylene (PP) was experiment and test the various features. The researchers have developed the concept of polymer type's poly propylene (PP). The principle is to mix two types of polymers, Poly propylene (PP) with grades recycled poly propylene (r-PP) at various proportions. Then fill Coupled reactive silane compound type Vinyl-Triethoxy Silane; VTMS A-171), the reaction began Dicumyl Peroxide; DCP. Nearby properties to plastic with new tablets.

Experimental

The process of implementation. The way to study the mechanical properties of the polymer mixture between polypropylene (PP) and recycled grade poly propylene (r-PP).

Melt compounding used this study, PP/rPP /Silane composites were prepared in a co-rotating twin screw extruder (TSE). Nine cases of PP/rPP/Silane with the compounding ratios are listed in Table 1, These are prepared in each type of Silane.

TABLE (1). Compounding ratios for various kinds of PP/rPP /Silane composites

Case number of materials	PP	r-PP	Silane
1	100	0	0
2	80	20	0
3	50	50	0
4	20	80	0
5	100	0	0
6	100	0	1,3,5
7	80	20	1,3,5
8	50	50	1,3,5
9	20	80	1,3,5
10	100	0	1,3,5

Compounding was carried out at a uniform barrel temperature of 200 ℃, a screw rotation speed of 150 rpm. Before melt compounding, PC and VGCF were dried in an oven at 80 ℃ for a minimum of 24 h. The feed rate is 1,000 g/h for all extrusion cases.

Mechanical Testing description of the mechanical tests. The tests can be divided into three parts.

1) The tensile strength tests indicate the strength of the polymer when the force slowly. The element analysis tensile strength is the hardness at the Yield Point, the maximum tensile strength, the Young Modulus, and percent Elongation at Break

2) Properties testing hardness. Measure the forces acting on the move. Head pressed down on material .The results of the testing indicate resistance to deformation by use of materials

3) Properties testing Tensile strength. Test results indicate the ability of a material to absorb energy before decisive occurs when a force immediately.

The morphology of the broken surface. Will be analyzed by using Scanning Electron Microscopy; SEM. Morphological analysis of selected polymer blend (PP:r-PP). The amount of reactive silane coupled filler 0, 1 and 3 phr. the ratio (100:0, 20:80, and 0:100). The surface of the specimen to be analyzed was obtained from specimens that tested the impact.

RESULTS AND DISCUSSION

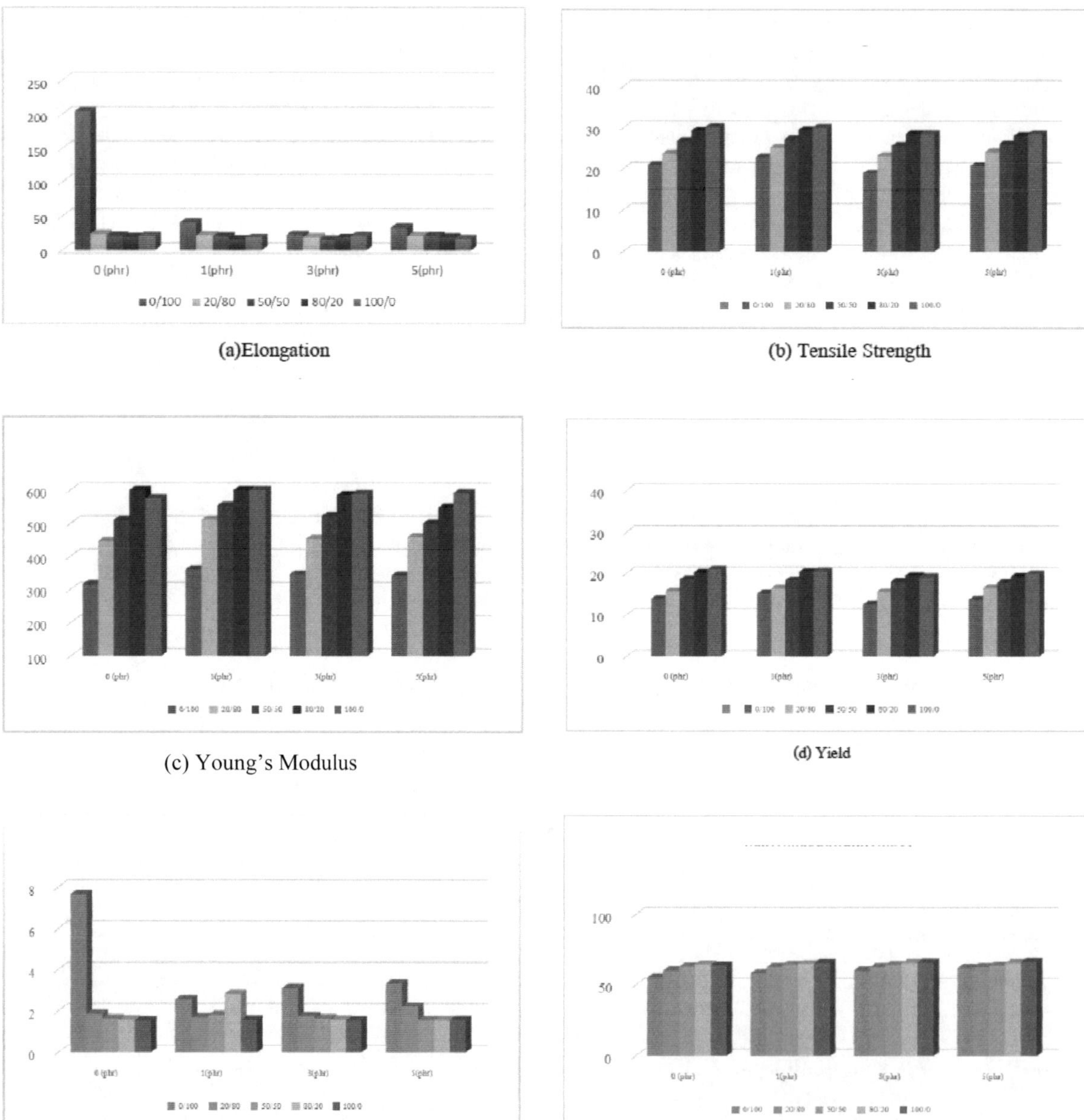

(a)Elongation (b) Tensile Strength

(c) Young's Modulus (d) Yield

(e) Impact (f) Hardness

FIGURE 1. The mechanical tests. The polymer compound (r-PP: PP).

The Tensile test results will show the mechanical properties. Polymer mixture of poly propylene (PP) and copolymer polypropylene grades recycled (PP/r-PP), as shown Figure 1. The mixing ratio between the polypropylene and polypropylene grade recycling (PP/r-PP). When mixing recycled poly propylene pellets (r-PP) in the ratio of 80/20 not resulted the mechanical properties polypropylene pellets decrease. Due to the tensile strength equal 29.333 MPa, the Young's modulus equal 602.566 MPa, the hardness equal 64.73 Shore D. When coupled reactive silane additives in volume 1 phr. Resulted the tensile strength. Young's modulus and hardness of the polymer blend of polypropylene and polypropylene grade recycling (PP/r-PP) increased compared with the polymer blends with no added substances reacted together silane. The addition of substances coupled reactive silane at 1 phr in proportion 80/20. The tensile strength equal 29.531 MPa, Young's modulus equal 631.084 MPa and a hardness equal 65.00 Shore D.

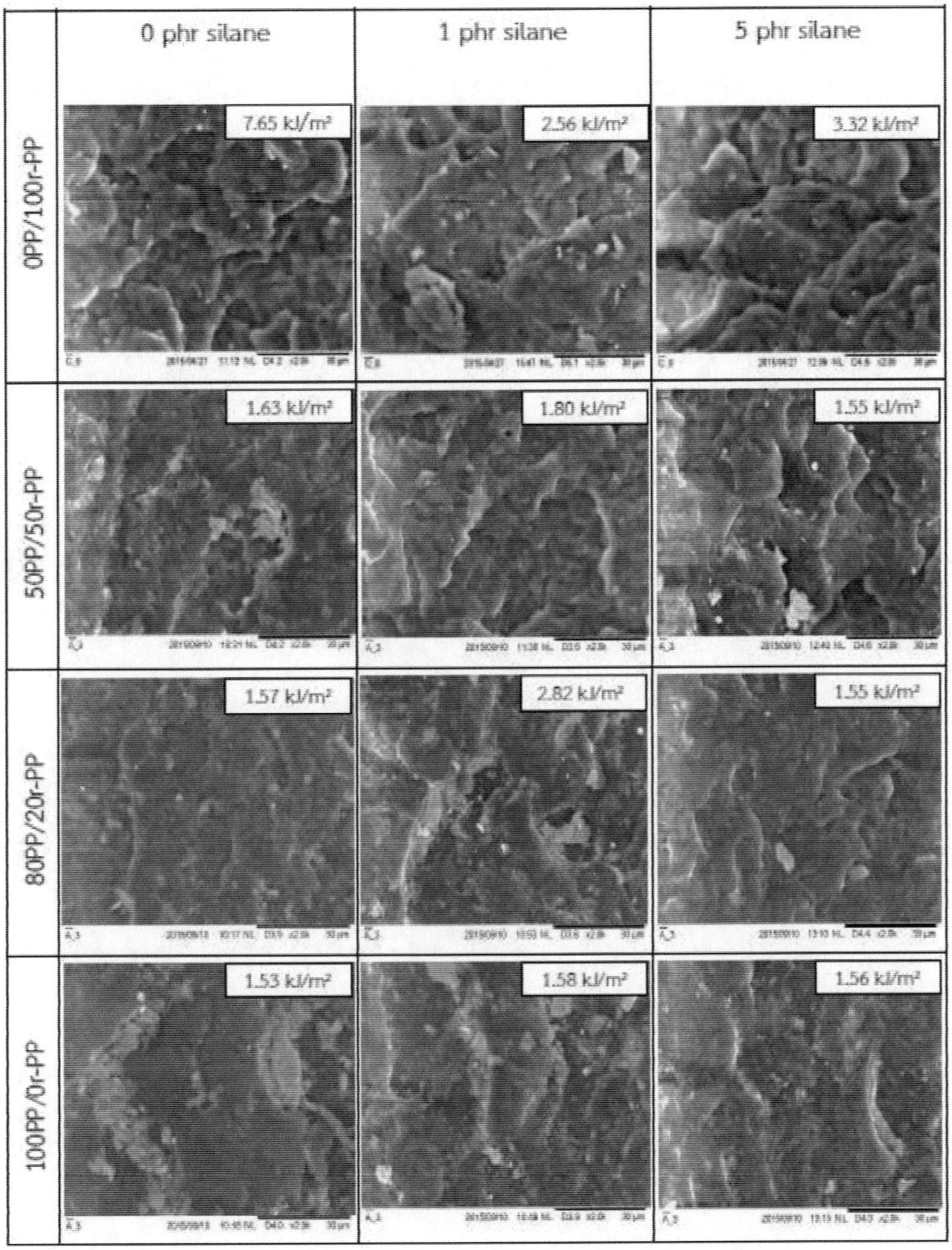

FIGURE 2. SEM photographs of the fracture surface for PP/rPP blend

The Results morphology of the surface the work piece decisive .Polymer mixture of r-PP /: / PP ratio mix. 100: 0,50: 50.0: 100, the amount of reactive silane coupled filler 0, 1 and 3phr as shown in Figure 2. The broken polymer compound (r-PP: PP) substances, when added together reactive silane magnification of 2,000 times. The photos of the broken polymer mixture of polymer blend of poly propylene and poly propylene recycled grades (PP / r-PP). When coupled reactive silane filling ratio of 80/20 in a striped structure of the smooth surface, very similar to the ratio of 100/0, which may result in the mechanical properties in both areas are similar. When coupled reactive silane substances which fills the amount 1 phr as the broken cystic looks like a wave cracks. And it can be seen that -grade poly propylene recycling (r-PP) is distributed in the body poly propylene pellets new (PP) makes polymer mixture is homogeneous.

The Discussing from the study of mechanical properties and morphological analysis of the mix new poly propylene with poly propylene grade recycling. The ratio of 0/100, 80/20, 50/50, 20/80 and 100/0. By adding together amount VTMS 0, 1, 3 and 5 phr. Can be concluded of operations are as follows. The Mixing Ratio of poly propylene and poly propylene recycled grade (PP / r-PP). When mixing recycled poly propylene pellets (r-PP) in the ratio of 80/20 not resulted the mechanical properties poly propylene pellets decrease. And the addition of silane coupled reactions that resulted the amount 1 phr tensile strength,modulus of elasticity and hardness of the polymer mixture poly propylene and poly propylene recycled grade (PP / r-PP) increased compared to polymer blends without replenishment coupled reactive silane.

CONCLUSION

The results found that the ratio of 80/20 (PP/r-PP) could be used as virgin polypropylene which was unaffected toui mechanical propertiesof virgin polypropylene. Yet, adding 1 phr of Silane coupling agents could be affected to the increasing of mechanical properties which had tensile strength as 29.531 MPa, Young's Modulus as 631.084 MPa, and hardness as 65.00 shore Dsimilar to virgin polypropylene. When adding Silane, it was influenced to adhesion and polymer blends.

ACKNOWLEDGMENTS

This research was supported by research grants. From NRCT pass Rajamangala University of Technology Isan. Thank you Korat TRC plastic generous of two polymer types. Proposition preliminary findings and recommendations And the Department of Industrial Engineering, Department of Materials Engineering. Central University, Post Harvest Engineering. Rajamangala University of Technology Isan generous place and accessories in operations research.

REFERENCES

1.B. Erickson et al., "Perspective on Opportunities in Industrial Biotechnology in Renewable Chemicals", Biotechnology Journal 7, 176-185 (2012).
2. S. Mandal, "Porter's Five for CES Analysis of the Indian Plastic Industry", International Journal of Multidisciplinary Research 1, 294-316 (2011).
3.T. Shuichi et al., "Effect of Electric Irradiation on the Mechanical Properties of Polypropylene/Vapor Grown Carbon Fiber Composites Prepared by Melt Compounding", Journal of Textile Engineering 58, 1-7 (2012).
4. K. Anuchit and M. Utai, "Design of experiment : Wood composite based on crosslinked polypropylene", The 18th International Conference on Composite Materials, Korea, 2011.
5. K. Anuchit and M. Utai, "Toughening Of Wood Plastic Composite Based On X-PP", Proceedings of PPS-31: THE 31ST International Conference of the Polymer Processing Society, Korea, 2015
6. B. Beltran and C. Mijangos, "Silane Graftin g and Moisture Crosslinking of Polypropylene", Polymer Engineering and Science 40, 1534-1541(2000)

Enhancing the Conductivity of Carbon Nanotube Filled Blends by Tuning their Phase Separated Morphology with a Random or Block Copolymer

Ruth Cardinaels[a,b], Avanish Bharati[b], Jin Won Seo[c], Michael Wübbenhorst[d] and Paula Moldenaers[b*]

[a]*Polymer Technology, Department of Mechanical Engineering, TU Eindhoven, P.O. Box 513, 5600 MB Eindhoven, The Netherlands*
[b]*Soft Matter Rheology and Technology, Department of Chemical Engineering, KU Leuven, Celestijnenlaan 200f, P.B. 2424, 3001 Leuven, Belgium*
[c]*Surface and Interface Engineered Materials, Department of Materials Engineering, KU Leuven, Kasteelpark Arenberg 44, P.B. 2450, 3001 Leuven, Belgium*
[d]*Soft Matter and Biophysics Section, Department of Physics and Astronomy, KU Leuven, Celestijnenlaan 200D, P.B. 2416, 3001 Leuven, Belgium*

R.M.Cardinaels@tue.nl
Paula.Moldenaers@cit.kuleuven.be

Abstract. Nowadays, there is a large demand for deformable conductive materials in applications such as flexible electronics, EMI shielding or soft robotics. Immiscible blends containing conductive nanoparticles form a versatile class of materials with excellent possibilities to tune electrical functionality by means of tailoring their morphology and particle localization. The conductivity of two-phasic blends consisting of poly–alpha-methyl-styrene-co-acrylonitrile and polymethylmethacrylate (PαMSAN/PMMA) containing multiwall carbon nanotubes (MWNTs) was optimized by tailoring the blend morphology with a polystyrene – polymethylmethacrylate (PS-PMMA) copolymer. The PαMSAN/PMMA blend has a lower critical solution temperature (LCST) and undergoes spinodal decomposition upon heating above the LCST during which the MWNTs selectively localize in the PαMSAN phase. The effects of the molecular weight and architecture (block or random) of the copolymer on blend conductivity, linear viscoelastic moduli and morphology were systematically investigated. The copolymer induces a huge conductivity increase from 10^{-9} S/cm to 10^{-5} S/cm in blends with only 0.5 wt% MWNTs. The increase in conductivity is caused by a morphology refinement and increased degree of cocontinuity with copolymer, leading to double percolated networks. At low molecular weight a block copolymer has a more pronounced effect whereas at high molecular weight the difference between random and block copolymers vanishes. Finally, the dielectric spectra of the blends show an interfacial relaxation peak whose dielectric strength and relaxation time reflect the contact resistance and interfacial capacitance of PαMSAN entrapped between neighbouring MWNTs. From this microcapacitor behavior, an estimate of the gap spacing between the MWNTs was obtained for the different systems.

Keywords: polymer blend, carbon nanotube, copolymer, conductivity, morphology, compatibilization, dielectric relaxation
PACS: 82.35.Np; 83.60.Bc; 83.80.Tc; 83.85.Vb; 72.80.Le; 72.80.Tm; 77.22.Gm

INTRODUCTION

Polymer nanocomposites containing conductive nanoparticles form a viable material class for generating light-weight flexible electronic materials. In recent years, the use of anisotropic nanoparticles such as carbon nanotubes or graphene sheets has allowed to conceive conductivity at relatively low filler fractions, thereby retaining the processability and deformability of polymeric materials [1,2]. A next leap forward in this approach can be realized by blending immiscible polymers and exploiting the selective localization of nanoparticles in one of the phases or at the interphase of the two-phasic blend [3]. In this case, the blend morphology is known to be essential for the final properties. Whereas compatibilization by means of interfacially localized copolymers is a well-known technique to tailor the morphology of blends, its combined use with selectively localized nanoparticles has only received limited interest. Han et al. [4] have shown that addition of a reactive compatibilizer to an immiscible polycarbonate-poly(acrylonitrile-butadiene-styrene) (PC/ABS) blend with carbon nanotubes can improve the dispersion of the latter as well as decrease the droplet size of the ABS inclusions thereby facilitating the formation of a conductive path of carbon nanotubes in the minor phase. In our previous work, we have shown that a random copolymer with a large molecular weight allows to refine the blend morphology and increase the continuity of the components in a phase separating poly–α-methyl-styrene-co-acrylonitrile/polymethylmethacrylate (PαMSAN/PMMA) blend containing carbon nanotubes [5]. In the present work, we extend and refine this approach by investigating the effect of the molecular weight and architecture (block or random) of the copolymer on blend conductivity, linear viscoelastic moduli, morphology and microcapacitor behavior of the carbon nanotubes.

MATERIALS AND METHODS

40/60 PαMSAN/PMMA blends containing 0.5 wt% amine-functionalized multiwall carbon nanotubes (MWNTs) with and without 2 wt% of various types of poly(styrene)-poly(methyl methacrylate) copolymer were prepared in a one-step melt mixing procedure. A 15 cm^3 DSM microcompounder under inert nitrogen atmosphere was used at 200°C with a rotational speed of 60 rpm for 20 min. Since 40/60 PαMSAN/PMMA blends exhibit a lower critical solution temperature of 165 °C [6], melt mixing resulted in phase separated blends. These blends were subsequently compression moulded into disks at 160°C and 50 MPa, resulting in monophasic blends. Finally, phase separation was allowed to proceed under quiescent conditions by annealing the samples at 220 °C. Simultaneously, characterization of the linear viscoelastic moduli (G' and G") and dielectric spectra (ε' and ε" or equivalently σ" and σ') was performed in respectively an AR2000 rotational rheometer and a rheo-dielectric setup consisting of a Novocontrol alpha dielectric analyzer connected to an MCR501 rotational rheometer. The morphology of the phase separated blends was visualized by means of optical microscopy and scanning transmission electron microscopy (STEM) on thin quenched sections.

RESULTS

Electrical Conductivity, Linear Viscoelastic Moduli and Morphology

The electrical conductivity and storage modulus of all blends reached steady state values after less than 4 hours of annealing, demonstrating completion of the phase separation and limited remaining morphology coarsening. The electrical conductivity spectra of the phase separated blends are shown in Figure 1a. The neat blend as well as the blend with 0.5 wt% MWNTs show a limited conductivity of 10^{-9} S/cm, which results from a minor ionic conductivity of the PαMSAN phase. During phase separation, the MWNTs selectively localize in the PαMSAN phase of the blend, as shown in the STEM image in the inset in Figure 1b. The limited electrical conductivity with 0.5 wt% MWNTs implies that either the MWNTs do not form a percolated network in the PαMSAN phase or the latter phase is not continuous, both aspects impeding the formation of a long-range conductive path throughout the blend. When copolymers are added to the blend, the DC conductivity plateau values significantly increase. Figure 1a shows that when the molecular weight of the copolymer is low, only the block copolymer provides a significant enhancement of the conductivity whereas at high molecular weight, both copolymers have a similar effect. The storage modulus of the phase separated blends obtained from small amplitude oscillatory rheology is presented in Figure 1b. It can be seen that all blends show an increased elasticity and decreased slope of G' versus frequency as compared to that of the neat polymers, whose moduli reach the limiting slope of -2 (schematically indicated by the full line in Figure 1b), which is characteristic of terminal behavior. The power law behavior exhibited by the blend moduli is a manifestation of a broad distribution of relaxation times, corresponding to a broad distribution of characteristic lengths from either the domains in the (partially) cocontinuous blends and/or the entangled MWNTs-PαMSAN clusters in the (partially) interconnected MWNTs network. The latter requires a distance between the MWNTs of less than twice the radius of gyration of the polymer chains, thus allowing linkage of MWNTs by means of PαMSAN chains. Figure 1b demonstrates that the increase of the plateau modulus and the decrease of its slope for the different copolymers follow the same trend with copolymer type as the electrical conductivity.

The similar evolution of the electrical conductivity, plateau modulus and limiting slope of G' suggests that the effect of the copolymer is governed by a single phenomenon. Two main causes for the observed trends can be proposed namely an increased dispersion of the MWNTs upon addition of the copolymer, thereby reducing the distance between the MWNTs and increasing their connectivity within the PαMSAN phase or a change in morphology, which increases

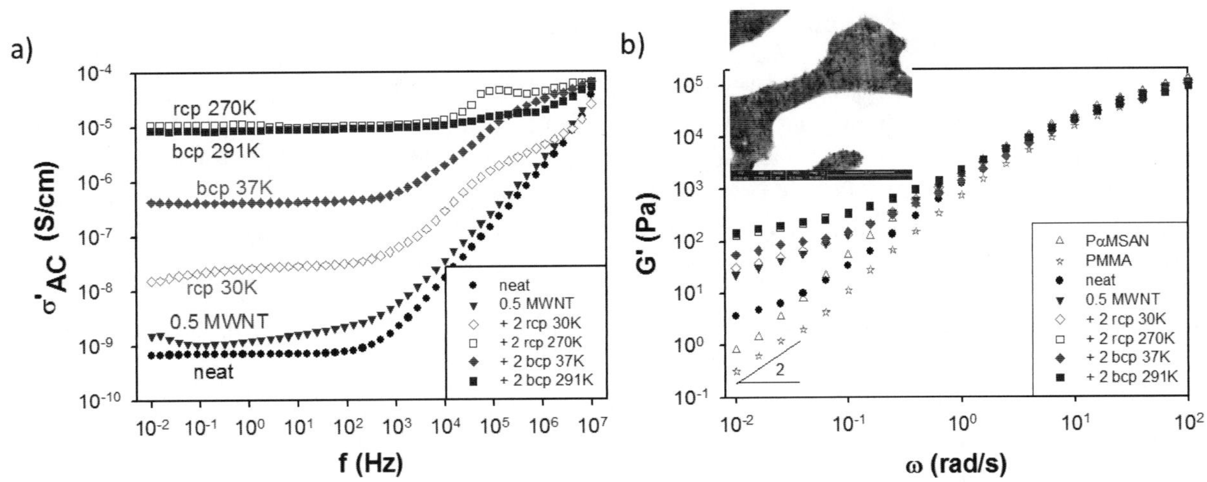

FIGURE 1. Steady state conductivity spectra (a) and elastic modulus (b) at 220°C for phase separated blends with 0.5 wt% MWNTs and 2 wt% copolymer of various architectures and molecular weights. Results for the neat blend and the blend without copolymer are added as a reference. The number in the copolymer name indicates its molecular weight (g/mol). In the inset PαMSAN and PMMA are the dark and bright phases respectively.

the connectivity of the PαMSAN domains. Since the interparticle distance resulting in electrical versus rheological percolation is different, being related to either the required tube-tube distance for electron hopping/tunelling or that for effective bridging of the MWNTs by PαMSAN chains, it is envisaged that a change in morphology is the major cause of the changes in properties after addition of copolymers. PαMSAN/PMMA blends with a 40/60 composition are known to phase separate through spinodal decomposition, initially forming a cocontinuous structure, which undergoes coarsening and breakup during the later stages of phase separation [6]. Figure 2 provides optical images of 40/60 PαMSAN/PMMA blends with 0.5 wt% MWNTs clearly showing the presence of a partially interconnected microstructure in the absence of copolymer (Figure 2a). When using a copolymer with a high molecular weight, a refined morphology with improved interconnectivity is obtained, irrespective of the copolymer type (Figures 2b and 2d). On the other hand, at low molecular weight, the random copolymer is not effective in stabilizing the cocontinuous blend morphology, as shown in Figure 2c, thereby providing limited effect on the blend properties. The copolymers were proven to be interfacially segregated by determination of the interfacial tension of 85/15 PαMSAN/PMMA blends containing copolymer, based on the extra elasticity of the interface of the droplet-matrix structure in these blends [5]. Hence, the effectiveness of the random copolymer appears to be linked to its ability to cross the interface several times and to effectively entangle and anchor the homopolymers. On the other hand, the block copolymer appears to be effective even at low molecular weight. It should be kept in mind though, that in the case of the low molecular weight block copolymer, the length of each block still exceeds the length of each of the blocks in the random copolymer.

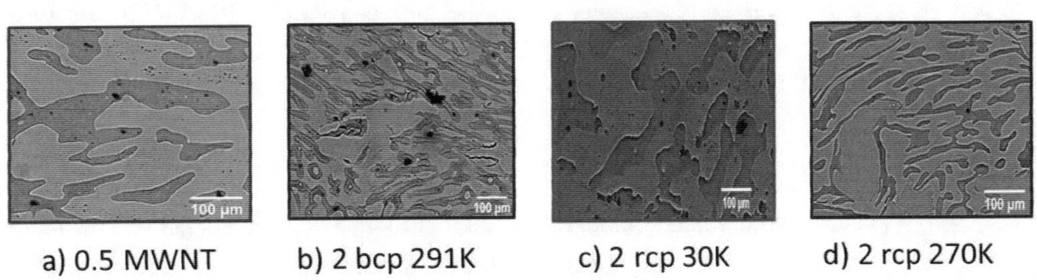

FIGURE 2. Optical images of blends with 0.5 wt% MWNTs and 2 wt% random or block copolymer after completion of phase separation. PαMSAN and PMMA are the dark and bright phases, respectively. The number in the copolymer name indicates its molecular weight (g/mol).

MWNT microcapacitor network and distance between the MWNTs

The dielectric spectra in Figure 1a show a clear relaxation shoulder around 10^3-10^5 Hz. As the temperature at which these dielectric spectra were obtained is far above the components Tg's, this relaxation does not originate from the components but is the consequence of charge trapping at an interface. By comparing its dielectric strength with that of the interfacial relaxation in PαMSAN-PMMA blends without MWNTs [7] as well as that of PαMSAN-MWNT nanocomposites (data not shown), it could be determined that this interfacial relaxation results from the entrapped charges at the interface between the MWNTs and the PαMSAN chains. Figure 3a presents a representative dielectric spectrum for the case of a blend with 0.5 wt% MWNTs and 2 wt% block copolymer of 291 kg/mol. The regions of the dielectric spectrum corresponding to electrode polarization, interfacial polarisation of the MWNT-PαMSAN interface as well as the hopping of electrons between MWNTs are indicated in the graph. Based on the dielectric spectrum, the Nyquist impedance spectrum can be derived, representing the imaginary part of the complex impedance versus the real part of the latter. This way of representing the data amplifies the contribution of the interfacial relaxation, as shown in Figure 3b, in which each polarization occurs as a semi-circular arc. In addition, it allows to model the blend response by means of an equivalent electrical circuit model. In this model, the MWNTs are represented by a parallel coupling of a resistance with a capacitance whereas the electrode polarization as well as the interfacial polarization each correspond to a resistance in parallel with a constant phase element. The impedance of the latter element is represented by $Z^*=1/(Q\omega^{\alpha})$, with α indicating the frequency dependence of the capacitance, resulting from non-ideal capacitor behaviour due to a spatial gradient of properties at the interface. Combining these three elements in series allows to describe the dielectric response of the blend, as shown in Figure 3b, and to determine R_I, Q and α, which characterize the MWNT-PαMSAN interface. The interfacial capacitance C_I can then be obtained as $C_I = Q(\omega_{max})^{(\alpha-1)}$, with ω_{max} the frequency at which the peak of Z_s" occurs.

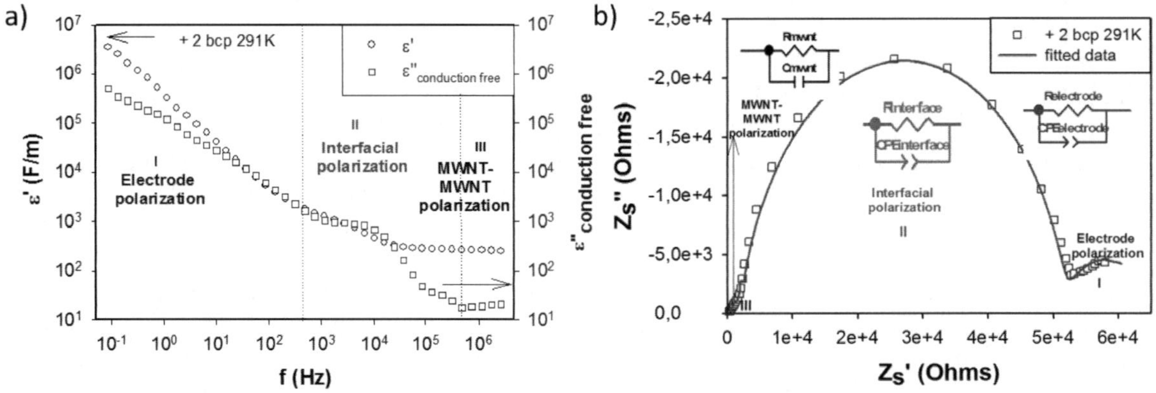

FIGURE 3. a) Dielectric permittivity and "conductivity free" dielectric loss spectrum of blend with 0.5 wt% MWNTs and 2 wt% block copolymer with M_w = 291 kg/mol at 220°C indicating electrode polarization, PαMSAN-MWNT interfacial polarization and MWNT-MWNT polarization, b) corresponding Nyquist impedance spectrum, the full line indicates a fit with an equivalent circuit model consisting of a resistance – capacitance (RC) network in series with two resistance – constant phase element (RQ) networks.

A physical representation of the assembly of MWNTs with entrapped PαMSAN exhibiting interfacial polarization is a microcapacitor with distance δ between the MWNTs. The electron transport between MWNTs in polymer nanocomposites governing their DC conductivity is suggested to be controlled by fluctuation induced tunneling [8], which was verified here by studying the temperature dependence of the conductivity (data not shown). Using the fluctuation induced tunneling model, the capacitance, resistance and relaxation time of the microcapacitor network can be described by [9]:

$$C_I = \varepsilon_0 \varepsilon \frac{A}{\delta}; \quad R_I = \frac{16\pi^2 h}{3e^2} \frac{\delta}{k_0 A} \exp(k_0 \delta) \quad (1)$$

$$1/\tau_I = 1/(R_I C_I) = \frac{3e^2}{16\pi^2 h}\frac{k_0}{\varepsilon_0 \varepsilon}\exp(-k_0\delta) \quad (2)$$

in which ε is the permittivity of the PαMSAN dielectric capacitor material, A is the capacitor area, δ is the thickness of the capacitor, h is Planck's constant, e is the electronic charge, and $k_0 = 2\sqrt{(2m_e V)}/h$, with m_e the mass of an electron and V the potential barrier, which is typically 0.5 eV for multiwall carbon nanotubes [10]. Obviously, the interfacial capacitance C_I and interfacial resistance R_I obtained from the fits of the Nyquist impedance plot are the total values obtained for the network of microcapacitors formed by the MWNT network throughout the blend. Hence, their values reflect not only the properties of each microcapacitor, but also their connectivity. However, the interfacial relaxation time, shown in Eq. (2), only depends on the capacitor thickness δ, or the distance between the MWNTs. Using Eqs. (1) and (2), this distance δ between the nanotubes was found to be approximately 15-20 nm, irrespective of the presence of copolymer whereas the interfacial capacitance increases and the interfacial resistance decreases with the addition of copolymer (data not shown), showing a similar dependence on the copolymer type as the electrical conductivity and linear viscoelastic moduli. Since the determined values of C_I and R_I are that of the complete microcapacitor network, these results confirm the dominant effect of the copolymers on the continuity of the PαMSAN domains in the blends whereas each microcapacitor, represented by its characteristic relaxation time, remains more or less unaltered upon addition of copolymer. The latter conclusion was also confirmed by the fact that the linear viscoelastic moduli of PαMSAN composites with an equivalent concentration of MWNTs as the blends were not affected by the addition of copolymer, suggesting insignificant effects of the copolymer on the MWNTs dispersion.

CONCLUSIONS

In this work, the electrical conductivity and dielectric permittivity of phase separated 40/60 PαMSAN/PMMA blends containing multiwall carbon nanotubes selectively localized in the PαMSAN phase were tailored by addition of polystyrene-polymethylmethacrylate random or block copolymers of various molecular weight. Morphology refinement and increased connectivity of the polymer domains were obtained upon addition of copolymer, with block copolymers already being effective at low molecular weight whereas random copolymers require a higher molecular weight to allow multiple interface crossings and entanglement. The increased connectivity of the PαMSAN domains in the blend results in an increased electrical conductivity as well as increased interfacial capacitance and decreased interfacial resistance of the network of MWNT microcapacitors containing entrapped PαMSAN chains. Hence, compatibilization of polymer blends not only allows to refine and stabilize the blend morphology but also provides a tool to tailor the dielectric response and conductivity of filled polymer blends.

ACKNOWLEDGMENTS

This research has been partially funded by GOA/09/002. Ruth Cardinaels is indebted to the Research Foundation Flanders (FWO) for a Postdoctoral Fellowship at KU Leuven.

REFERENCES

1. H. Kim, A.A. Abdala, C.W. Macosko, *Macromolecules* **43**, 6515-6530 (2010).
2. M. Moniruzzaman, K.I. Winey, *Macromolecules* **39**, 5194-5205 (2006).
3. M. Salzano de Luna, G. Filippone, *European Polymer Journal*, in press (2016).
4. I.S. Han, Y.K. Lee, H.S. Lee, H.G. Yoon, W.N. Kim, *J. Mat. Sci.* **49**, 4522-4529 (2014).
5. A. Bharati, R. Cardinaels, J.W. Seo, M. Wübbenhorst, P. Moldenaers, *Polymer* **79**, 271-282 (2015).
6. H.M. Laun, *Pure & Appl. Chem.* **70**, 1547-1566 (1998).
7. A. Bharati, M. Wübbenhorst, P. Moldenaers, R. Cardinaels, *Macromolecules* **49**, 1464-1478 (2016).
8. K. Kilbride, J. Coleman J. Fraysse, P. Fournet, M. Cadek, A. Drury, S. Hutzler, S. Roth, W. Blau, *J. Appl. Polym. Phys.* **92**, 4024-4030 (2002).
9. J.G. Simmons, *J. Appl. Phys.* **34**, 1793-1803 (1963).
10. J.G. Meier, J.W. Mani, M. Klüppel, *Phys. Rev. B* **75**, 054202 (2007).

Morphology and Structure Development during Injection molding with fast mold temperature evolution

V. Speranza, S. Liparoti, R. Pantani and G. Titomanlio[a)]

Department of Industrial Engineering, University of Salerno – ITALY

[a)] Corresponding author: gtitomanlio@unisa.it

The effect of temperature, pressure and flow on relaxation time, nucleation density, crystallites growth rate, the interrelation among these quantities and the distributions of deformation rate and cooling time during the process determine the morphology distribution in the final injection molded part.

A model linking all these quantities was developed to describe morphology evolution during the process. The effect of flow on nucleation density and growth rate of an iPP was described on the basis of a molecular stretch parameter; the molecular stretch evolution was described by a nonlinear Maxwell model whose relaxation time was determined by temperature, pressure and crystallinity and by the molecular stretch. Following literature experimental indications the transition between spherulitic and fibrillar morphology was related to the local viscous dissipation, which in turn is related to the evolution of rheology during crystallization.

In this work, the predictions of the model are compared with the cross section morphology distributions of samples obtained by injection molding, coupled with fast mold temperature evolution, of an iPP accurately characterized for rheology, quiescent crystallization, effect of flow on nucleation and spherulitic growth rates and spherulitic/fibrillar transition. The model predictions reproduce main characteristics of the experimental results for both pressure evolution and final morphologies distributions

PAC 83.50.Uv

INTRODUCTION

Injection molding process is one of the most widespread processes for mass production of polymeric objects. The properties of molded objects, such as mechanical properties, dimension stability etc, are significantly conditioned by the morphology developed during the process. The morphology distribution is determined by the molecular stretch evolution subsequent to both the filling and packing flows and by its relaxation before solidification [1]. It is well known that crystalline polymers give rise to fibrillar morphology under high molecular stretch and to spherulitic morphology if the molecular stretch is sufficiently small [2]. As a result, final morphology distribution along the thickness direction of semicrystalline injection molded objects is characterized by a sequence of different morphological layers, which from the surface to the midplane are: the skin layer, composed by planar lamellar structures [3] at the surface and a thin oriented nanostructured zone [4]; the shear layer, having a highly oriented fibrillar morphology; the spherulitic core [5].

The molecular stretch distribution at solidification depends on the injection conditions and, therefore, also mold surface temperature [6]. This latter is a key parameter in controlling molecular stretch: if temperature is kept high starting from the filling for a few seconds, the relaxation of the molecular stretch can be calibrated before cooling [1,7]. To that purpose, a thin heating device was allocated just below the mold surface, to obtain a fast evolution of mold temperature [8]. The device has been adopted to prepare iPP samples injected in a cavity whose surface is held at a very high temperature, 150°C, for times ranging from 1 to 20s before cooling down to room temperature.

Moreover, simulations of injection molding process with fast temperature evolution have been performed. Up to now, good results were obtained simulating the injection molding of iPP, in several operating conditions and geometries, in terms of predictions accuracy of the evolution of pressure distribution, molecular stretch distribution along both the flow and the cavity thickness directions, spherulite dimensions and crystalline distributions [2,9,10].

The UNISA code was modified to take into account of a thin heating device held active for a few seconds just below the mold surface. Following previous works [2], the experimental distribution of morphological layers thicknesses is interpreted on basis of the concept that a minimum critical local viscous work is required to obtain shear layer morphologies.

EXPERIMENTAL

The polypropylene (i-PP) grade adopted in this study was supplied by Montell (now Basell), and its commercial name is T30G (non-nucleated, Mw=376 000, Mw/Mn=6.7, tacticity=87.6%); it was extensively characterized as far as rheology and crystallization kinetics [11–14].

Thin heating devices were allocated just below the mold surfaces to increase temperature by Joule effect. The heating devices consists of very thin conductive layer, made of poly(imide-amide) loaded carbon black, electrically insulated from the mold by two polyimide layers; a steel layer (100 micron thickness) was adopted to protect the device from the melt [8,15].

A 70-ton Negri-Bossi reciprocating screw, injection molding machine was adopted for the experiments. The polymer was injected into line gated rectangular cavities having length L=110 mm, width W=12.7 mm, and thicknesses S=1.5 mm. A sketch of the cavity adopted for the injection experiments is reported in Fig. 1.

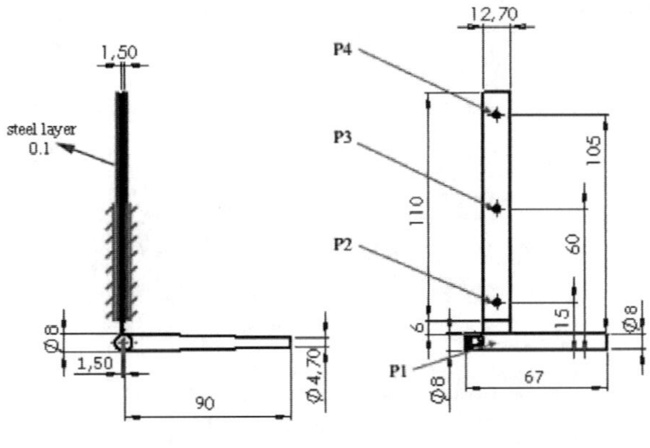

Fig. 1 Geometry of the cavity (P1-P4 indicate the pressure transducers positions) and sketch of the heating device.

Test run	P (W/cm²)	T_{level} (°C)	t_h (s)	t_a (s)	P_{hold}
p-steel	0	28	0	0	720
p-passive	0	28	0	0	720
p-1501	10	150	1.3	2	720
p-1506	10	150	6	2	720
p-15012	10	150	12	2	720
p-15020	10	150	20	2	720

Table 1 Operating conditions adopted for the injection molding experiments.

The molding machine and the mold were equipped with five piezoelectric transducers: one in the injection chamber (P0), one just before the gate (P1) and three in the cavity (P2, P3 and P4), located in the non-moving part of the mold (15, 60 and 105 mm downstream from the gate position). The heating layers were located on both sides of the cavity for a length of 70 mm downstream the cavity entrance and one temperature sensor was located in the cavity, on the steel layer, in pos. P3. The injection molding experiments were carried out adopting 3.2 seconds as injection time (flow rate 2.9 cm^3s^{-1}), a melt temperature of 220°C, a mold temperature of 28°C, 720 bar as holding pressure and 10s as holding time. The other operative conditions are summarized in Table 1 (P=electrical power applied; T_{level}= temperature reached on the steel layer; t_h=time that the heater is activated after the polymer reaches position P2. t_a = time that the heater was activated before the contact with the polymer in P2: P_{hold}=holding pressure).

Experiment were performed applying an electrical power of 10 W/cm^2 on the conductive layers, starting 2s before the melt arrives at pos. P2, and keeping it active for additional times (1.3s, 6s, 12s, 20s) after the first contact of the polymer in pos. P2. The mold wall temperature evolutions of the experiments performed selecting these heating times are reported in Fig. 2a. Temperature evolutions obtained without the heating device (p_steel) and using the de-activated heater (p_passive) are also reported for comparison. In the figure, the instant at which the polymer reaches pos. P2 is set to zero.

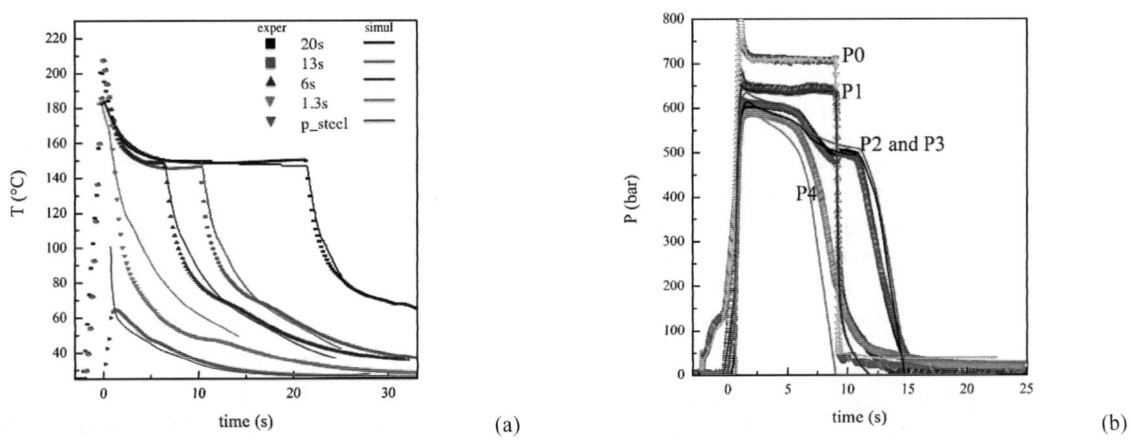

Fig. 2 (a) Simulated and experimental temperature evolutions for the experiments performed applying an electrical power of 10 W/cm^2 for 2s before the melt arrives at pos P2 and for additional heating times. Temperature evolution obtained without the heating device (steel) is also reported for comparison. (b) Simulated and experimental pressure evolutions in the cavity obtained keeping active the heater for 12s after the contact of melt in pos. P2. The model predictions are represented by lines.

For t<0 the temperature increases from 28°C to about 150°C, at t=0, by effect of the contact with the hot polymer, temperature undergoes a further sharp increase, from 150°C to about 210°C. After the contact of the melt in pos. P2 the polymer surface stars to cool down toward a mold surface asymptote temperature (150°C). Obviously, as soon as the heater has been deactivated the temperature decreases very fast to the mold temperature (28°C). Fig. 2a shows that the cooling times at the deactivation are very close to each other, they do not depend on the time the heater has been held active.

Fig. 2b shows pressure evolution obtained for the experiment p_15012, performed holding active the heater for 12 s. Pressures in positions P2 and P3 go to zero into two steps, the first one due to the decrease of the average temperature from the melt temperature to the surface temperature, 150°C in this case, and the second, that takes place at the heater deactivation, from 150°C to the selected mold temperature.

The injected samples were characterized by optical microscopy as far as the distributions of the surface, shear and spherulitic layers. The results are reported in Fig. 3.

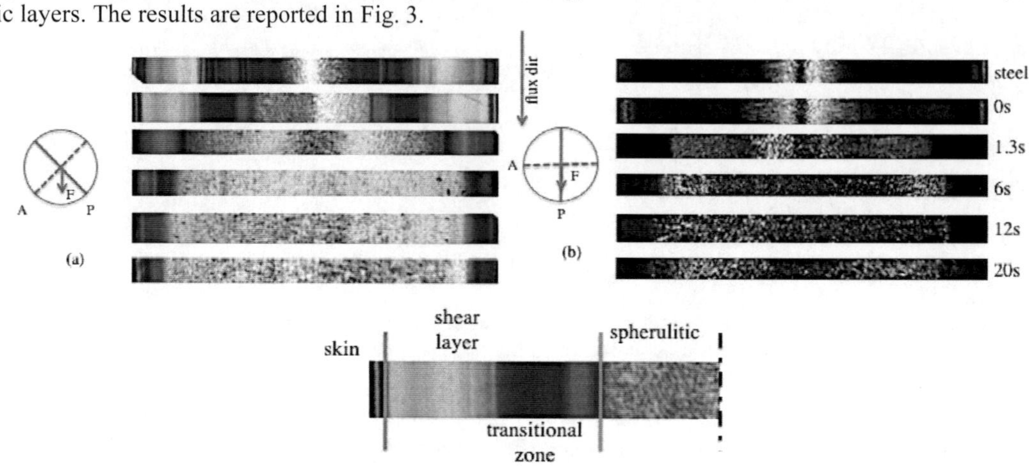

Fig. 3 Optical micrographs of the slices cut from the samples produced supplying 10 W/cm^2 to the heater for different heating times (a) flow direction at 45° respect to polarizer direction; (b) flow direction aligned along polarizer direction.

Surface mold heating determines the morphology distribution inside the molded samples. Fig. 3 shows optical micrographs of sample slices cut parallel to the flow-thickness plane of samples produced selecting the molding conditions reported in Table 1. Samples have morphology distributions typical of an injection molded semi-crystalline polymer, usually denoted as skin-core morphology, characterized by the presence of a series of distinct regions: the skin layer, which includes very thin planar lamellar structures at the surface and a thin oriented nanostructured zone adjacent to it; a highly oriented non-spherulitic zone (the so called 'shear layer'), where molecular orientation and fiber density are both very high; adjacent to the shear layer, towards the symmetry plane of the molding there is the spherulitic layer. Fig. 3 also shows that between shear layer, the brighter part in the optical micrographs, and spherulitic layer there is a transition zone, which appears brown in the optical micrographs performed with the flow direction rotated of 45° with respect to the analyzer [16]. This zone is formed by fibers with a lower density and orientation than those present in the shear layer [1].

PROCESS MODELING BY UNISA CODE

The numerical simulation has been performed by a software developed at University of Salerno described in a previous paper [2]. In the UNISA code the evolution of molecular stretch is described through a simple model for the evolution of the molecular conformation based on an appropriate not linear Maxwell model. Moreover the crystallization of iPP is described considering two crystalline phases (α and mesomorphic) competing for the available amorphous volume. For the evolution of the mesomorphic phase the Nakamura equation is adopted, whereas the Kolmogorov approach is followed for the description of the morphology evolution during spherulitic crystallization of the α phase. The effect of flow on crystallization kinetics is very large for many polymers and specifically for the polymer considered in this work [14]. As reported in [17], it is accounted on the basis of a flow dependent thermodynamic crystallization temperature. In turn, the evolution of the crystallization temperature by effect of flow is related to the evolution of a suitable scalar measure of the molecular stretch [18,19].

To carry out the simulation of the injection tests considered in this work, some software modifications have been implemented. In particular, the possibility of considering the presence of the heating device with a heat generation layer below the cavity surface. Changes of thermal histories due to the heating of cavity mold surface strongly influence the thermo-mechanical history experienced by the polymer and, thus, its morphology, crystallinity and orientation distribution in the final moldings.

Obviously, also the pressure evolution is affected by the surface temperature evolution. The comparison between software predictions for the evolution of pressures in the positions P0-P4 and experimental results is good, an example is shown in Fig. 2b for the experiment p_15012. Good results were obtained in term of surface temperature simulations, as showed in Fig. 2a.

MORPHLOGICAL LAYERS THICKNESS DISTRIBUTION ON THE CROSS SECTION

As far as morphology distribution is concerned, the emphasis in this work is about the distribution of the morphological layers thicknesses on the molding cross section.

A previous work on the morphology distribution of the same injection molded iPP [17] demonstrated that, at 140°C, the polymer solidifies into fibrillar morphology (shear layer) when the viscous dissipation reaches 20 MPa and the molecular stretch is above a minimum value. This criterion was verified by comparison with the distribution of morphological layers in the cross sections of iPP moldings obtained in the same cavity adopted in this work with a surface temperature kept to 150°C for 1.3s after the first contact between polymer and mold [17, 18]. The value of 20MPa for the critical viscous work was confirmed. The distribution on the molding cross section of final values of viscous work and of molecular stretch, Δ, as calculated by the UNISA software, is reported in Fig. 4a for the sample p_passive (see Table 1).

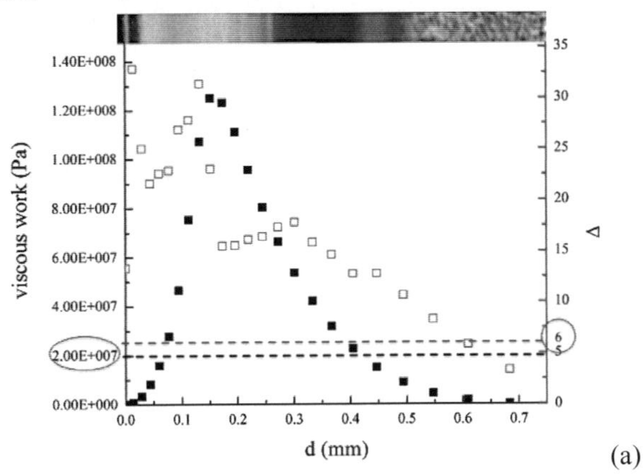

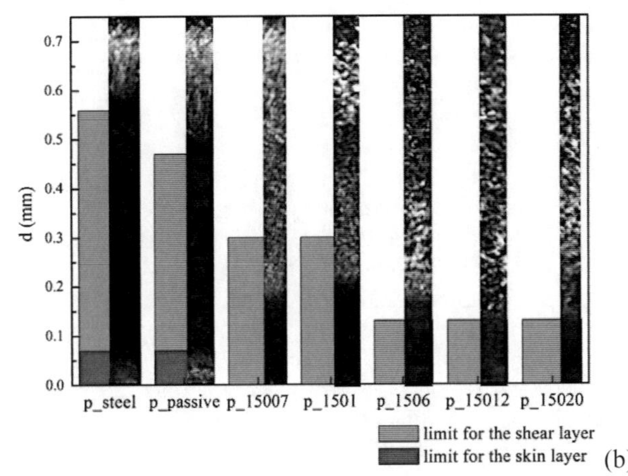

Fig. 4 (a) Viscous work (filled symbols) and molecular stretch, Δ , as calculated by the UNISA software for p_passive sample. The dotted lines indicates the limits for viscous work and Δ which need to be overcome to obtain a fibrillar morphology. (b) Comparison between the shear layer thicknesses of the molded samples, produced as reported in Table 1, and calculated shear layer thickness (red bars indicates the distance from the skin at which skin layer is expected, green bars indicates the distance from the skin at which shear layer is expected).

In this work also the presence of a transitional area was taken into account. This area, that represents the transition from the fully fibrillar (shear) layer toward the spherulitic layer, has a fibrillar character, although the fibers are thicker and their density is lower. The minimum molecular stretch value, Δ_{min}, was correlated to the border between the shear transitional layer and the spherulitic layer. The Δ_{min} adopted in this work is 6; with respect to those reported in [20] this value is obviously smaller because in the previous work the transition zone was not included.

The thicknesses and the positions of the shear layers (including the transitional layers) were identified (for each injection molding condition) from plots similar to that shown in Fig. 4a for p_passive sample. The results are compared in Fig. 4b (where the shear layer thickness is represented by green bars) with the optical micrographs of the cross section morphologies; the comparison appears not very precise but certainly satisfactory.

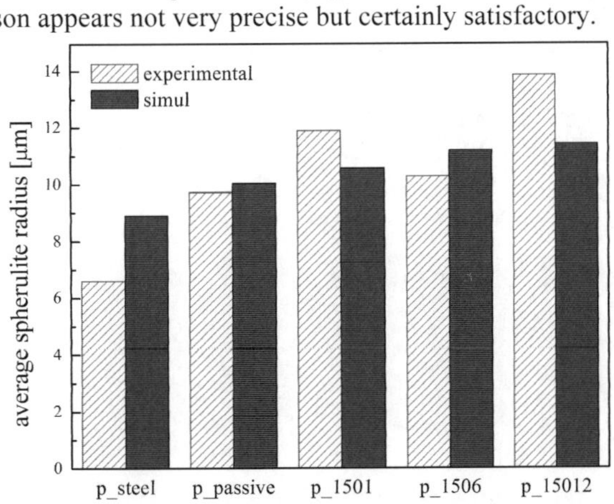

Fig. 5 Experimental and calculated average spherulite dimensions for the molded samples produced as reported in Table 1.

The identification of the shear layer on the cross section allows to identify also the thickness of the other layers: the skin layer (red bars of Fig. 4b) at positions closer to mold surface, and the spherulitic core at the central position.

The predictions of the average spherulitic dimensions are compared in Fig. 5 with the average experimental dimensions inside the spherulitic layers. Also in this case the comparison appears satisfactorily.

CONCLUSIONS

In this work, a software for the simulation of injection molding was adopted to predict the evolution of morphology in molded parts obtained applying a fast surface heating technique. The software was modified to consider the heating system and to simulate the effect of a heating power on the mold surface temperature. The prediction of temperature profiles and of pressure evolution resulted to be good when compared with experimental data. The thickness and the position of the fibrillar-shear layer for the injection molded samples was identified by applying a criterion based on the final viscous work and molecular stretch distributions on the cross section; with suitable choice of transitional value the model correctly captures the dependence of the shear layer thickness in function of the mold conditions assessed by experimental results. The identification of the shear layer on the cross section identifies also the thickness of the other layers: the skin layer on the molded sample surface and the spherulitic core in the center. The average dimensions of spherulites are also correctly predicted.

REFERENCES

1 S. Liparoti, A. Sorrentino, G. Titomanlio, *AIChE Journal. (2016).*

2 R. Pantani, I. Coccorullo, V. Speranza, G. Titomanlio, *Progress in Polymer Science. 30 1185–1222 (2005).*

3 P.G. Lafleur, M.R. Kamal, *Polym.Eng.Sci. 26 92–102 (1986).*

4 S. Liparoti, A. Sorrentino, G. Guzman, M. Cakmak, G. Titomanlio, *RSC Advances. 5 36434–36448 (2015).*

5 I. Coccorullo, R. Pantani, G. Titomanlio, *International Polymer Processing. 20 186–190 (2005).*

6 S. Liparoti, A. Sorrentino, G. Titomanlio, *Polymer Processing with Resulting Morphology and Properties: Feet in the Present and Eyes in the Future: Proceedings of the GT70 International Conference - Salerno (Italy) 2015. AIP Conference Proceedings. 1695 (2015).*

7 G. Wang, G. Zhao, X. Wang, *International Journal of Heat and Mass Transfer. 78 99–111 (2014).*

8 S. Liparoti, T.M. Hunag, A. Sorrentino, G. Titomanlio, M. Cakmak, *30th International Conference of Polymer Processing Society, Cleveland, Ohio, USA - AIP Conference Proceeding. 1664 (2014).*

9 U. Vietri, A. Sorrentino, V. Speranza, R. Pantani, *Polymer Engineering & Science. 51 2542–2551 (2011).*

10 R. Pantani, F. De Santis, V. Speranza, G. Titomanlio, *PROCEEDINGS OF PPS-30: The 30th International Conference of the Polymer Processing Society–Cleveland (OHIO US) 2015 - AIP Conference Proceeding. 1664 50006 (2015).*

11 I. Coccorullo, R. Pantani, G. Titomanlio, *Polymer. 44 307–318 (2002).*

12 R. Pantani, V. Speranza, G. Titomanlio, *Macromolecular Theory and Simulations. 23 300–306 (2014).*

13 H. Janeschitz-Kriegl, E. Ratajski, *Polymer. 46 3856–3870 (2005).*

14 R. Pantani, I. Coccorullo, V. Volpe, G. Titomanlio, *Macromolecules. 43 9030–9038 (2010).*

15 S. Liparoti, G. Landi, A. Sorrentino, V. Speranza, M. Cakmak, H.C. Neitzert, *Advanced Electronic Materials. (2016) DOI 10.1002/aelm.201600126.*

16 K.H. Kim, A.I. Isayev, K. Kwon, *Journal of Applied Polymer Science. 95 502–523 (2005).*

17 R. Pantani, F. De Santis, V. Speranza, G. Titomanlio, *PROCEEDINGS OF PPS-29: The 29th International Conference of the Polymer Processing Society-AIP Conference Proceeding. 1593 636–640 (2014).*

18 R. Pantani, V. Speranza, A. Sorrentino, G. Titomanlio, Molecular orientation and strain in injection moulding of thermoplastics, in: Macromolecular Symposia, 2002: pp. 293–307.

19 R. Pantani, V. Speranza, G. Titomanlio, *International Polymer Processing. (2016).*

20 S. Liparoti, V. Speranza, R. Pantani, G. Titomanlio, Fast temperature evolution on the mold surface: analysis and simulation, in: Proceeding of PPS 2015 Europe-Africa Polymer Processing Society Regional Conference - Graz (Austria), 2015.

21 R. Pantani, A. Sorrentino, V. Speranza, G. Titomanlio, *Rheologica Acta. 43 109–118 (2004).*

Structural Recovery Mechanism of Multi-Walled Carbon Nanotube in a Simple Shear Flow

S.Ghiassinejad[a], B.Ranjbar[a], A.Hosseinpour[a], A.Katbab[a] and H.Nazockdast[a*]

[a] Department of Polymer Engineering and Color Science. Amirkabir University of Technology (Tehran Polytechnic), Tehran, IR Iran

Presenting author's email: sina.ghiassinejad@gmail.com
Reference author's email: nazdast@aut.ac.ir

Abstract. In this work the flow induced orientation and the mechanism governing for structural recovery of multi-walled carbon nanotube filled polypropylene nanocomposites were investigated. A series of melt linear and nonlinear rheological measurements including start-up flow test and time sweep experiments were performed in order to study the structural breakdown and/or nanoparticles orientation and subsequent structural recovery at different temperatures and MWCNT loadings. The results showed that the structural recovery occurs in two stages. The first quick recovery which was independent of temperature could be explained in terms of interparticle van der Waals interaction. This stage which was named as "initial aggregation" and had major role in increasing storage modulus. The second stage of recovery (secondary aggregation) that was much slower, was found to be highly dependent on temperature and therefor was attributed to rotary diffusion of nanoparticles. These result were in good agreement with those calculated using van der Waals and diffusion concepts.

Keywords: MWCNT, Structural recovery, Aggregation, van der Waals interaction, Rotary diffusion.
PACS: 83.60.Bc, 83.80.Hj, 83.85.Cg, 82.35.Np

INTRODUCTION

Multiwall carbon nanotubes (MWCNTs) have attracted remarkable interest from both academia and industry, because of their large aspect ratio, excellent Young's modulus and desired electrical and thermal properties at very low loadings. The flow induced orientation of nanoparticles is a key parameter in determining rheological, thermal, electrical and mechanical properties of nanocomposites. Abbasi et al (1) found that increasing deformation rate in different processing conditions leads to change the degree of nanotube alignment, from random orientation to highly aligned that affects melt viscoelastic behavior of nanocomposites. Reorientation and structural recovery of aligned nanoparticles are subjects that is being studied by many researchers to recognize the actual mechanism. Solomon et al (2) showed the disorientation kinetics is non-Brownian because reconstitution of aligned particles and tactoids trough Brownian motion needs long time. They also noted that structural evolution is attributed to interaction between the clay tactoids. Treece et al (3) showed that the logarithmic dependence of the storage modulus with time is a two-step process during annealing. Khalkhal et al (4) showed that increasing kinetics of structure evolution with temperature is an indication of the importance of Brownian forces in the absence of flow. They found first step of recovery is attributed to van der Waals forces which can be considered independent of temperature. In the present work we used the steady shear transient experiment and time-resolved measurements to study reorientation and structural recovery mechanism of MWCNT nanoparticles. Different test parameters (concentration, shear rate and temperature) were considered to study the origin of recovery after shear induced orientation.

MATERIALS AND METHODS

A commercial HPPP550J polypropylene with melt flow index (MFI /2.16 kg) of 1.3 g/10 min and melting temperature supplied by Jam petrochemical Co was utilized as the matrix. PP-g-MA purchased from Honam petrochemical with melting temperature of was used as compatibilizer in 1:9 ratio to matrix. Multi-walled carbon nanotubes (MWCNTs) with the commercial trade name of Nanocyl NC 7000 was utilized as the dispersed phase.

Nanocomposites were prepared by the direct melt-compounding method using an internal mixer (Brabender Plasticorder W50). Melt compounding was carried out at and rotor speed of 90 rpm for 20 min.

The rheological measurements were carried out using a Paar Physia UDS 200 rheometer. The structural evolution of the nanocomposites was investigated by a series of transient tests. Firstly to investigate the structural recovery behavior of a percolated sample, time test after pre-shearing for 2500 second. Also time tests after per-shearing were performed in different temperatures to investigate temperature effect on structural recovery.To provide more insight into understanding of structural recovery mechanisms, transmission electron microscopy was utilized. Samples were pre-sheared at 0.5 s-1 for 2500 seconds at 190 C. After cessation of shear, time tests were performed at for 180 seconds and the sample was put in cold water immediately in order to freeze orientation. This sample was microtomed in nitrogen atmosphere into ultrathin sections. The recovered and oriented structures of nanocomposites were examined by TEM (FEI, Tecnai 20) with acceleration voltage of 100kV.

RESULT AND DISCUSSION

Figure 1 shows time sweep experiment carried out immediately after pre-shearing for percolated sample. Results show increasing shear rate led to decreasing of the storage modulus which was attributed to the shear induced structural breakdown. As it can be seen increasing shear rate caused intense network breakdown and decrease of aggregation size that led to lower storage modulus. The other important feature of this system which can be seen in figure 1 is two slopes of the storage modulus that could be attributed to a two-stage recovery mechanism. The first stage occurred at short time (180 seconds) and had a significant contribution to the increment of storage modulus. The second stage was much slower (3000 seconds) and had negligible contribution to storage modulus.

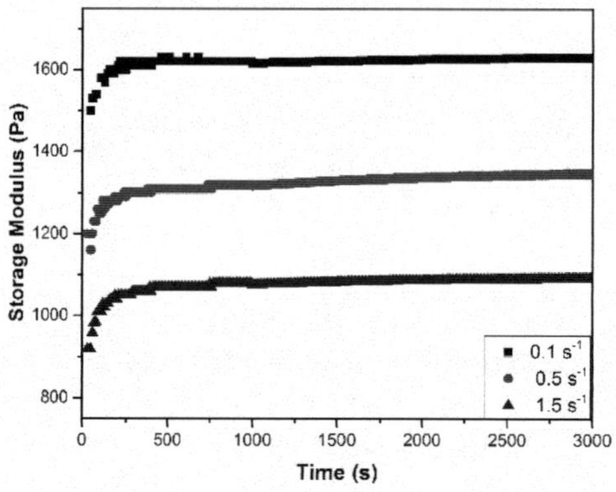

FIGURE 1. Storage modulus recovery of PP containing 2 wt % MWCNT as a function of annealing time after different shear rates for 2500 sec at 190 C.

In order to better understanding the origin of structural recovery mechanisms, a series of experiments were designed. A test that was designed for percolated samples to study temperature dependency of each stage, was time sweep experiments after pre-shearing at different temperatures. Figure 8 shows almost no temperature dependency for the first quick stage (180 sec). Based on figure 2 first recovery stage is originated from a kind of interparticle interaction that had no temperature dependency. This could be explained in terms of van der Waals interactions (4).

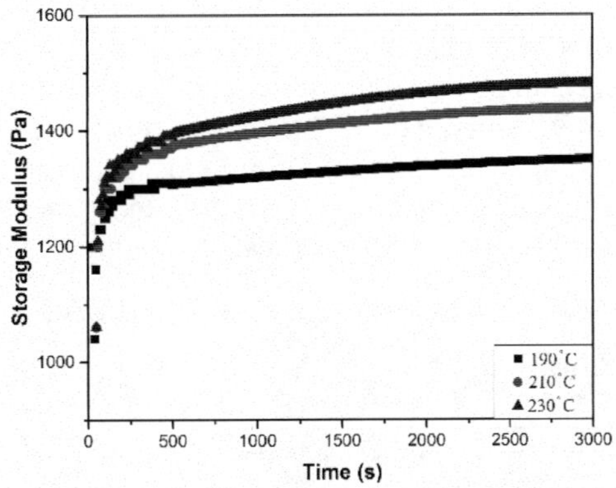

FIGURE 2. Storage modulus recovery of PP containing 2 wt % MWCNT as a function of temperature after shear rate of 0.5 s-1 for 2500 sec.

To confirm this claim, detailed concept of van der Waals forces were reviewed (5). The van der Waals interaction potential was calculated by

$$\phi = A_H f(a,r) \qquad (1)$$

where "r" and "a" were diameter and distance of particles respectively and A_H is Hamaker constant. Based on equation 3 temperature dependency of van der Waals forces was related to Hamakar constant. For calculation of Hamaker constant, the Lifshitz theory was utilized (5). Assuming the same absorption frequencies for all media and symmetric cases, Hamaker constant was calculated by

$$A_H = \frac{3}{4} k_B T \left(\frac{\varepsilon_A - \varepsilon_B}{\varepsilon_A + \varepsilon_B}\right)^2 + \frac{3h\upsilon_e}{16\sqrt{2}} \frac{(n_A^2 - n_B^2)^2}{(n_A^2 + n_B^2)^{3/2}} \qquad (2)$$

where ε_i, n_i, k_B, υ_e, h and T were static dielectric constant of materials, index of refraction of materials, Boltzman constant, main ultraviolet absorption frequencies, Plank constant and temperature respectively. Hamaker constant consists of two terms: First term was a zero-frequency energy of the van der Waals which includes the Keesom and Debye dipole contribution. Second part was dispersion energy and includes the London energy contribution. Based on equation 2 van der Waals force between two bodies in a medium was always attractive and Hamaker constant was positive. Temperature dependency of Hamaker constant was related to the first term of equation 2. Therefore due to temperature independency of Hamaker constant, second term (dispersion energy) should be dominant in Hamaker constant calculation (5).

Based on the findings of Larson dispersion energy contribution can be dominant when index of refraction of disperse phase and media are close to each other (5). In our case, index of refraction of polypropylene was 1.503 and for MWCNT was between 1.5-2.5 that were almost close to each other.

Hamaker constant calculation for our case showed with increasing temperature from 0 C to 100 C, A_H changed from $5.384*10^{-20}$ to $5.484*10^{-20}$ that could be considered almost constant. This calculation confirmed that dispersion energy contribution was dominant in Hamaker constant calculation and van der Waals force was independent of temperature.

The van der Waals interactions were short range forces because of the dispersion interactions-that was dominant in Hamaker constant calculation-were effective when dipoles were correlated with each other. The time lag between flactuation led to disruption of correlation and dispersion interaction began to decrease abruptly when molecules of surfaces became separated. Hamaker constant was dependent on distance at separations more that 5-10 nm. This effect was well-known as retardation effect (5). This means van der Waals forces could not affect the individual MWCNTs are longer than effective distance (5 nm) from initial aggregates. In this way, after initial aggregation (180 sec) two types of structure could be seen, initial MWNT aggregations and individual MWCNT that were in the longer distance. Figure 3 shows these two types of structure for a sample after pre-shearing for 2500 seconds and annealing at 190 C for 180 seconds.

Initial aggregation led to concentration depression in regions far away from aggregates. This process that was named "local dilution", causes concentration regime changes from concentrated to dilute. In dilute regime MWCNTs had no effect on the each other without any hydrodynamic interaction and could freely move and rotate due to high distance between particles. This type of motions prepared possibility of Brownian motions in the second stage of recovery. Figure 3 shows that initial aggregation and local dilution occurred in the first stage for sheared sample that was annealed for 180 seconds. Individual tubes that were far away from initial aggregates are showed in red circles.

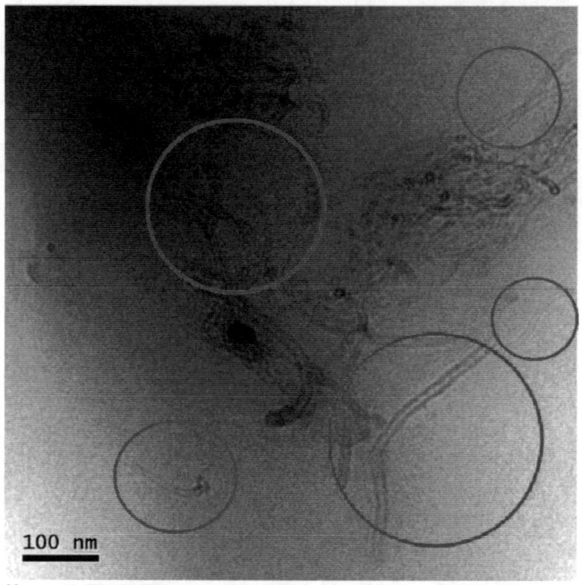

FIGURE 3. TEM images of different MWCNT arrangements in a polypropylene matrix: initial aggregates and individual MWCNTs.

Figure 2 shows storage modulus increases with temperature increment in the second stage recovery. Structural evolution acceleration with temperature in our case was in agreement with findings of some researchers (4). Dilute regime and temperature dependency of structural recovery in the absence of flow implied Brownian motions could be driving force for the second stage recovery.

To study Brownian motion possibility, rotary diffusion coefficient of a rigid rod in a dilute regime was showed in equation 3 (6)

$$D_r = \frac{3k_B T(Ln(L/_d)-0.8)}{\Pi\eta_s L^3} \qquad (3)$$

where k_B is Boltzman constant, T, L and d are temperature, length and diameter of nanoparticles and η_s is matrix viscosity. For an individual MWCNT, D_r was calculated $3.771*10^{-5}$ s^{-1}. Rotary diffusion time scale was proportional to inverse of rotary diffusion coefficient and is about 26500 sec. By considering equation 5 decreasing MWNCTs length due to extrusion in internal mixer -figure 3 confirmed this decrease (initial length of MWCNT is 10 μm)-led to intense increase in rotary diffusion coefficient and subsequently decrease rotary diffusion time scale (~5000 sec). This time scale was in agreement with the experimental time for second stage of recovery. Also the MWCNTs flexibility can also be an important parameter in shear response which helped rotational diffusion to form final structure due to possibility of MWCNTs locking in each other. Increasing the melt temperature leads to media viscosity decrement. Therefore rotary diffusion coefficient increases based on equation 5 and MWCNTs diffusion and motion toward initial aggregates would be easier. This stage was named "secondary aggregation". Furthermore decreasing rotary diffusion time scale led to acceleration of structural evolution occurred with secondary aggregation process at same time. Increasing the storage modulus was originated from particle-particle interactions that increased in higher temperature due to viscosity decrement (4). Thus, secondary aggregation due to increasing particle-particle interaction could be possible mechanism for evolution of the structure in the second stage recovery in long time. Figure 8 shows decrease of diagram slope with time in the second stage of recovery that was attributed to decrease number of diffusive MWCNTs and network enhancement in longer time.

CONCLUSION

The structural evolution of multi-walled carbon nanotube in polypropylene matrix by transient and oscillatory analysis coupled with TEM micrographs were investigated. Subsequent time sweep after stress growth experiments showed structural recovery is a two-stage mechanism that changes with MWCNT loadings and temperature. The first quick stage was temperature independent with remarkable role in storage modulus increment. This was attributed to van der Waals attractions between particles. Study of van der Waals concepts confirmed independency of temperature because dispersion energy contribution was dominant in Hamaker constant calculation. Calculation Hamaker constant for our case confirmed this claim. This first stage of recovery was named initial aggregation. Initial aggregation led to concentration decrement in distance far away from aggregates and change concentration regime. The Second stage that was accelerated with temperature originated from rotary diffusion of MWCNTs. Rotary diffusion time scale was in agreement with experimental time test (3000 sec) and endorsed this claim. This second stage of recovery was minor factor of storage modulus increment that was named secondary aggregation. This claim was confirmed by transmission electron micrograph.

REFERENCES

1. S. Abbasi, P. J. Carreau, and A. Derdouri, Polymer (Guildf)., vol. 51, no. 4, pp. 922–935, Feb. 2010.
2. M. J. Solomon, A. S. Almusallam, K. F. Seefeldt, A. Somwangthanaroj, and P. Varadan, Macromolecules, vol. 34, no. 6, pp. 1864–1872, Mar. 2001.
3. M. A. Treece and J. P. Oberhauser, Polymer (Guildf)., vol. 48, no. 4, pp. 1083–1095, Feb. 2007.
4. F. Khalkhal, P. J. Carreau, and G. Ausias, J. Rheol. (N. Y. N. Y)., vol. 55, no. 1, p. 153, 2011.
5. R. G. Larson, The structure and rheology of complex fluids, vol. 33. Oxford university press New York, 1999.
6. M. Doi and S. F. Edwards, The theory of polymer dynamics, vol. 73. oxford university press, 1988.

Numerical and Morphological Studies of Micro Injection Molded Plastic Micro Gears

Rabie El Otmani [a]*, Musa R. Kamal[b], Salim Derdouri[c] and M'Hamed Boutaous[d]

[a] *Chouaib Doukkali University of El Jadida, National School of Applied Sciences, Science Engineer Laboratory For Energy (LabSIPE), route nationale N°1 (route d'Azemmour), km 6 Azemmour EL HAOUZIA, BP : 1166 El Jadida Plateau 2400, Morocco.*

[b] *Department of Chemical Engineering, McGill University, Montreal, QC, Canada*

[c] *National Research Council, Boucherville, QC, Canada*

[d] *CETHIL, UMR 5008, CNRS, INSA Lyon, Universite Lyon 1, F-69621 Villeurbanne CEDEX, France*

Presenting author's email rabieelotmani@gmail.com
Reference author's email elotmani.r@ucd.ac.ma

Abstract. Interest in the micro-injection molding (μIM) process has grown as the demand for micro parts increased for various electronics, transportation, communications, biomedical and other applications. Recently, we have conducted an intensive research program to understand the details of this process and the material-process-property relationships for various polymers. In the present study, a micro-gear micro part was presented, in order to investigate the effects of processing conditions on microstructure, mechanical properties, and thermal properties, microtomed and examined using polarized light microscopy, and differential scanning calorimetry for investigation of morphology. Various microstructural features, such as morphological layer thickness and crystalline polymorphs, were observed and analyzed in light of the thermomechanical history. A skin-core region was observed, with spherulites predominating the core region, and highly oriented lamellae appeared in the skin layer. Numerical simulation of filling phase of micro gear gives us a detailed description of the thermal history and flow, in which contribute to better understanding on the origins differences in the morphologies among the layers across the part thicknesses in view of the prevailing process conditions.

Keywords: Microinjection, microgear,morphology,modeling.
PACS: 83.50.Uv; Material processing (extension, molding, etc.)

INTRODUCTION

Polymer micro gears are light in weight and can be produced using the micro-injection molding (μIM) process. They exhibit remarkable characteristics in terms of mechanical properties, including tooth stiffness, and flexural strength. Also, they have desirable tribological performance due to low friction and high resistance to wear. For all these reasons, plastic gears are mainly used in low and moderate power transmission systems.

Polyamides or nylons are among the earliest engineering plastics. Some of the key advantages of nylon for engineering plastics applications include thermal stability, toughness, good process ability and resistance to chemicals Kurokawa and Nagai [1, 2]. Senthilvelan, S. and Gnanamoorthy, R. [3] reported that injection molded high performance high cost short carbon fiber reinforced Nylon 66 gear is used in the highly stressed tooth region, while low cost unreinforced Nylon 66 is used in the hub region. There have been many attempts to study the micro gear injection molding process both experimentally and using simulation. Tom, A. M., Angelov, A. K. and Coulter, J. P. [4] reported an experimental investigationof the micro gears injection molding process to determine the optimum injection molding processing parameters for HDPE and POM semi-crystalline polymers as well as amorphous cyclic olefin copolymer (COC), focusing on mechanical properties. Barriere, T., Gelin, J. C., Michel, G., Sahli, M. and Quinard, C. [5] used injection molded polypropylene micro gears to investigate the geometry of the micro gears. Park, K. and Kim, Y. S. [6] evaluated poly(etheretherketone) (PEEK) Flexspline parts for a harmonic gear drive. They investigated the effect of mold temperature on crystallinity and mechanical properties using a tensile tests and differential scanning calorimetry (DSC).

There is currently limited literature to facilitate the fundamental understanding and characterization of polymer micro gear morphology. Only a few works report on the morphology of the micro gear. Generally, skin-core morphology has been observed in μIM. Kamal and co-workers [7] reported on a detailed evaluation of the relationships between the processing conditions and morphology development of polyoxymethylene (POM). A five-layers skin core morphology was observed in POM microinjected molding and injection speed and to a lesser extent low temperature were found to be the most important parameters influencing the microstructure features. The present work is part of a larger effort to elucidate the effects of thermal and flow history during the filling phase on the formation of microstructure of final micro injection molded parts. This extends earlier studies reported by the group [7], by coupling numerical simulation and experimental characterization in order to give some basic information about the various mechanisms produced during filling phase of the micro injection molding process. A detailed first-order prediction of flow and thermal history is used to explain the variation of morphological zones in the micro molded part. Polyamide-6 (PA-6) was chosen in this study because it is known to have excellent fatigue resistance and good melt thermal stability, up to about 300°C. We also use an industrial PA-6 micro gear supplied by Kleiss Gears Inc. Grantsburg USA (Figure 1).

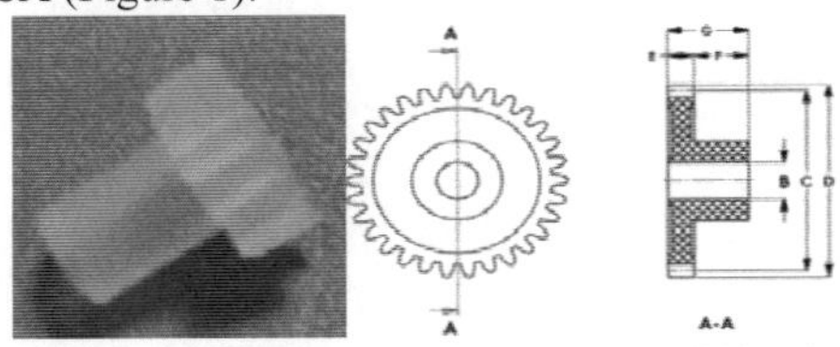

FIGURE 1. Geometry of micro gear 1 (B: bore (200 μm), C: pitch diameter (340μm), D: outside diameter (560 μm), E: face width (80 μm), F: hub width, G: total length (670μm), and number of teeth 10).

EXPERIMENTAL METHOD

Polarized light microscopy was used to examine the morphology of micro injection molded PA-6. For this purpose, 2-μm sections were obtained in both the flow and transverse directions across the micro gear, using a Leica Reichert Ultracut microtome at room temperature. The microtomed sections were viewed under transmitted polarized light using an Olympus BX50 optical microscope system, equipped with a Sony HD camera.

Differential scanning calorimetry (DSC) was used to assess the crystallinity developed locally. The samples were initially heated at a rate 10 °C/ min to 300°C, and cooled to 30°C at a rate of 20 °C/min. The peak crystallization temperature and the onset temperature were calculated by the supplied software.

ANALYSIS AND MODELLING OF POLYMER MELT FLOW AND HEAT TRANSFER DURING FILLING

We compare the experimental results with the predictions of numerical modelling. For the simulation of mold cavity filling phase, a custom finite element code developed by El-Otmani et al. [8] was used. For the rheological properties of the polymer, the 7-parameters Cross-WLF model [9], given in Equation (1) was employed.

$$\eta(T,\dot{\gamma},p)=\frac{\eta_0(T,p)}{1+\left[\frac{\eta_0\dot{\gamma}}{\tau^*}\right]^{1-n}} \tag{1}$$

$$\eta_0(T,p)=D_1\exp\left(-\frac{A_1(T-\bar{T})}{A_2+(T-\bar{T})}\right) \tag{2}$$

where n is the power-law index, τ^* is the threshold shear stress at which shear thinning starts, and $\eta_0(T, p)$ is the zero shear rate viscosity. Parameters A_1, A_2 and D_1 are material constants. The values of the Cross-WLF model constants used for PA-6 are taken from the work of Osorio et al. [9], as given in Table 1:

TABLE 1. Cross-WLF model constant [9]

Constant	D_1 (Pa)	$\bar{T}$ (K)	A_1	A_2	n	τ^*
Value	**$1.522*10^{12}$**	**323.15**	**27.826**	**51.6**	**0.4514**	**86,135**

The treatment assumes non-isothermal, viscous, incompressible, and laminar melt flow during the filling phase of the micro injection molding process. The governing equations are the incompressible Navier-Stokes Equation (3), the Continuity Equation (4) and the energy Equation (6) for the cavity and (7) for the mold:

$$\rho \frac{Du_i}{Dt} = \frac{\partial \left(\eta \left(\frac{\partial u_i}{\partial x_j} + \frac{\partial u_j}{\partial x_i} \right) \right)}{\partial x_j} - \frac{\partial p}{\partial x_i} + \rho g_i + \sigma k_i \frac{\partial \phi_i}{\partial x_j} \delta(\phi_i) \tag{3}$$

$$\frac{\partial u_i}{\partial x_i} = 0 \tag{4}$$

$$\dot{\gamma} = \sqrt{2 d_{ij} d_{ji}} \tag{5}$$

$$\rho c_p \frac{DT}{Dt} = \frac{\partial}{\partial x_j} \left(k \frac{\partial T}{\partial x_j} \right) + \eta (d_{ij} d_{ji}) \tag{6}$$

$$\rho_{mold} c_{p_{mold}} \frac{\partial T_{mold}}{\partial t} = \frac{\partial}{\partial x_j} \left(k_{mold} \frac{\partial T_{mold}}{\partial x_j} \right) \tag{7}$$

where D/Dt denotes the material derivative, σ is the surface tension taken here as 0.03 N/m, $\delta(\phi)$ is the Dirac function, k_i is the curvature. u_i are the components of the velocity vector, p is the pressure, η, ρ, c_p, k are respectively the viscosity, density, heat capacity and thermal conductivity of the fluid in each domain (polymer or air). Convection, transverse conduction, and viscous dissipation were considered in the energy equation, where the right hand terms represent the pure heat conduction in the polymer and the viscous dissipation heat source. The latter is linked to the strain tensor **d** defined as:

$$d_{ij} = \frac{1}{2} \left(\frac{\partial u_i}{\partial x_j} + \frac{\partial u_j}{\partial x_i} \right) \tag{8}$$

In this case, the classical level set method cited in the previous work [10] is used to follow the flow front during polymer filling of a 2D T shape cavity. The boundary and initial conditions used in this work are based on the work of El-Otmani *et. al.* [8]. On the other boundaries, a no-slip condition is imposed for the polymer flow, and a free slip condition is prescribed downstream at the polymer front for the air flow. The unfilled region in the mold cavity is represented by a fictitious fluid (air). The boundary conditions on the fictitious fluid should be chosen in such a way that they have little effect on the progression of the filled region. On the air-vent, only the fictitious fluid is allowed to leave the cavity.

For thermal boundary and initial conditions at the mold inlet, it is assumed that the polymer enters the cavity with a constant and homogeneous temperature Tinj (injection temperature). As an initial condition for the energy equation, a uniform temperature field over the entire domain is imposed: $T(x, y, 0) = T_0$

Within the mold material, pure heat conduction is assumed between the interior walls and an isothermal line corresponding to the cooling channels surface or representing the so-called thermal thickness of the mold.

RESULTS AND DISCUSSIONS

Morphology Along Flow and Transverse Directions

It is well known that crystalline morphology is sensitive to the melt temperature and cooling rate. For example, Figure 2 shows the morphology from the surface to the core region of our microtomed micro part. Three layers can be observed:

1. A non-spherultitic skin layer, having a high degree of chain orientation along the flow direction;
2. An intermediate layer, comprising a thin layer of twisted lamellae mixed with fine spherulites;
3. A random spherulitic core, with no preferred orientation.

This skin-core structure was only observed in the case of high injection speed. Increasing melt temperature or decreasing cooling rate tends to increase the size of spherulites. In micro injection molding processing conditions it produced a wide range of structures from embryonic spherulites to well-developed stacks of lamellar blocks.

FIGURE 2. Polarized transmitted light micrographs of microtomed sections of polyamide-6 in the flow direction at 4X magnification.

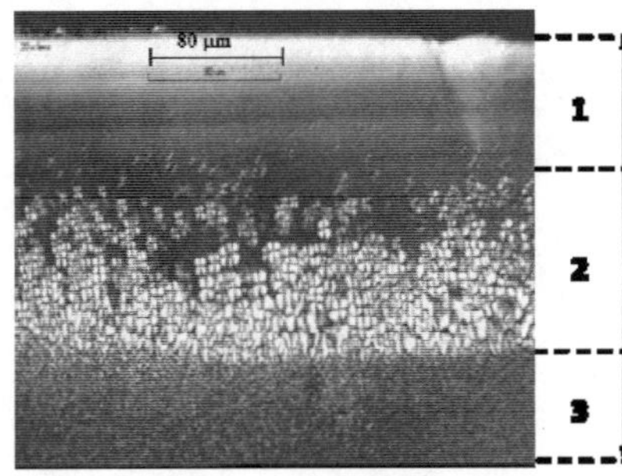

FIGURE 3. The three morphological zones in a micro injection molded nylon 6 of micro gear 3 viewed in the transverse cross section (magnification: 20X).

Along the transverse direction, the morphology of micro injection molded nylon 6 micro gear examined in the present study is in agreement with the results of the other reported studies. The nylon morphology reveals the following microstructures as shown in Figure 3:

1. A trans-crystalline skin with no visible spherulites.
2. A clear intermediate layer with random spherulites growing in the direction of the core
3. Very small spherulitic core region, which takes up the majority of the thickness. The spherulites are clearly seen at hight magnification, increasing in size from the outer region to the center.

Numerical Results

The evolution of viscosity of the polymer melt during the filling phase depicted by numerical simulation is shown in Figure 4, while the temperature evolution during the filling stage is presented in Figure 5.

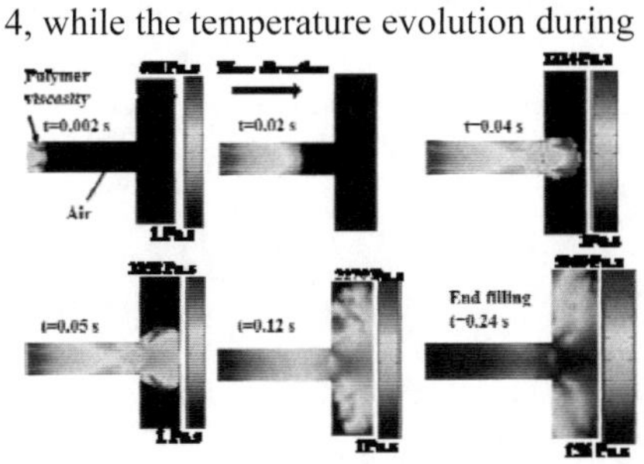

FIGURE 4. Evolution of the viscosity distribution during filling stage

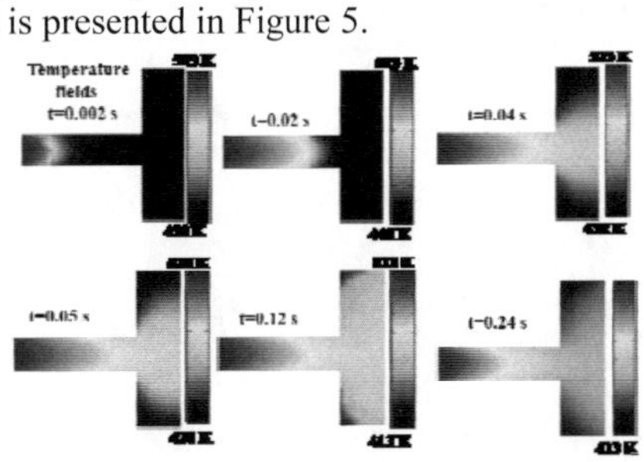

FIGURE 5. Temperature distribution during filling stage

The numerical simulation shows that the melt front viscosity is higher than the viscosity of the subsequent melt that follows it, and that the viscosity increases with time until the front reaches the border of the T shape. However, at the upper border, the viscosity reaches its maximum and remains the highest in the divergent region. Also, the viscosity distribution is not symmetrical. The explanation is that the melt front cools fast (cf. figure 5) and its temperature profile decrease drastically from the injected temperature to another temperature close to the crystallization temperature.

There is good agreement between the numerical simulation and experimental results regarding the presence of non-symmetrical distributions. It is interesting to note that the crystallization starts at a higher melt temperature and occurs over the whole micro gear thickness, which leads to low cooling rates. It is expected, therefore, that the core should produce larger spherulites at the end. The above arguments are in good agreement with the observed differences in spherulite sizes reported in this work. Therefore, the morphological differences are understandable in this context, since slow cooling/solidification rates that occur when the polymer melt impacts the border of the cavity end (T shape) are likely to yield the larger spherulites. Another factor that could play an important role is the high shear stress, which is higher in the micro injection molding process. It is known that stress enhances nucleation, which in turn, reinforces the influence of temperature (see Figure 5), by causing crystallization to occur earlier and at higher temperature. In the micro part, the boundary between the skin layer and the intermediate layer appears quite sharp, while the boundary between the intermediate layer and spherulitic core appears diffuse. We remark that the growth rate of the negative spherulites increase from the wall to the core of the T shape. Also we can see the transcrystallization close to the skin layer. The maximum radius measured in the microphotographs is around 20 microns. This observation is in accordance with the results in [11] for PA-66 which crystallizes at higher temperature, because it has a higher melting point.

We observed other structures like seeded crystallization or granular structure, which can be seen with processing at high mold temperature.

Three morphological zones are observed in the micro gear along the flow direction in the case of radial flow. The structure can be different depending on the mold temperature, injection speed and melt temperature. Different combinations produce varying distribution in the structure (size and shape) and thickness of the morphological zone.

CONCLUSION

The factors affecting the final mechanical properties of micro gear are many and complex. Despite the complexity of the situation, the following conclusions seem reasonable: Three morphological zones are depicted. The high viscosity distributions near the wall and in the melt flow front confirm the presence of all the observed structures in our morphological studies. Some flow lines are observed in the samples, which suggest the potential presence of zones of weakness or wear susceptibility in the micro gear.

ACKNOWLEDGEMENTS

The authors wish to thank NIPMMP, NSERC, McGill University and IMI/NRC Boucherville for supporting this research project.

REFERENCES

1. Kurokawa, M. and S. Nagai, *Reinforcement of polyamide 6 with thermotropic liquid crystalline polymer.* Polymer Engineering and Science, 1999. **39**(5): p. 872-880.

2. Kurokawa, M., Y. Uchiyama, and S. Nagai, *Performance of plastic gear made of carbon fiber reinforced polyether-ether-ketone: Part 2.* Tribology International, 2000. **33**(10): p. 715-721.

3. Senthilvelan, S. and R. Gnanamoorthy, *Selective carbon fiber reinforced Nylon 66 spur gears: Development and performance.* Applied Composite Materials, 2006. **13**(1): p. 43-56.

4. Tom, A.M., A.K. Angelov, and J.P. Coulter, *An experimental investigation of a micro injection molded mechanical device.* Manufacturing Engineering and Materials Handling, 2005 Pts A and B, 2005. **16**: p. 1157-1165.

5. Barriere, T., J.C. Gelin, G. Michel, M. Sahli, and C. Quinard, *Injection moulding of micro-parts: applications to micro-gears.* 10th ESAFORM Conference on Material Forming, Pts A and B, 2007. **907**: p. 671-676.

6. Park, K. and Y.S. Kim, *Effects of Mold Temperature on Mechanical Properties of an Injection-Molded Part with Microfeatures.* Journal of Polymer Engineering, 2009. **29**(1-3): p. 135-153.

7. Chu, J.S., M.R. Kamal, A. Derdouri, and A. Hrymak, *Morphology Development in the Gate Region of Microinjection-Molded Thermoplastics.* Polymer Engineering and Science, 2012. **52**(4): p. 787-794.

8. El Otmani, R., M. Zinet, M. Boutaous, and H. Benhadid, *Numerical Simulation and Thermal Analysis of the Filling Stage in the Injection Molding Process: Role of the Mold-Polymer Interface.* Journal of Applied Polymer Science, 2011. **121**(3): p. 1579-1592.

9. Osorio, A. and L.S. Turng, *Mathematical modeling and numerical simulation of cell growth in injection molding of microcellular plastics.* Polymer Engineering and Science, 2004. **44**(12): p. 2274-2287.

10. El Otmani, R., M. Boutaous, and H. Benhadid, *Simulation of Two-Phase Flow Using Conservative Level-Set Method without Re-Initialization Process.* Proceedings of the Asme Fluids Engineering Division Summer Conference - 2010 - Vol 1, Pts a-C, 2010: p. 1401-1410.

11. Boasson, E.H. and J.M. Woestenenk, *Some Aspects of the Crystallization of Nylon-66 (Polyhexamethylene-Adipamide).* Journal of Polymer Science, 1957. **24**(105): p. 57-61.

Thermo-optical Characterization of Polyester Fiber in Real Time via Quantitative Cross-polarized Optical Microscopy

Sebastião V. Canevarolo*, Luciana A. Bicalho and Jorge M. J. Silva

Departamento de Engenharia de Materiais, Universidade Federal de São Carlos
Rod.Washington Luis, km 235 - São Carlos - SP – Brazil

caneva@ufscar.br

Abstract. A real time quantitative thermo-optical characterization of a poly(ethylene terephthalate) (PET) fiber was performed using a modified polarized light optical microscope (PLOM) equipped with a coupled hot-stage CSS 450. A recently in-house developed optical module was installed in the PLOM in order to achieve quantitative optical measurements. That module is composed of two adjacent photocells, with one of them being covered by a polarizer filter. Therefore, the first photocell measures the changes in the transmitted light intensity which results from the scattered light whereas the second one (covered by the filter) quantify the variation in the cross-polarized light intensity. Since it is not possible to infer the evolution of the optical path difference from the cross-polarized light intensity unless the optical order is known, the images of the fibers were also recorded during the heating. A subsequent analysis of the changes in the fiber color helped to deduce about the evolution of OPD, which can be related to the birefringence and thus to the orientation of the fibers. The data of the thermo-optical analysis show a good agreement with the thermal transitions identified by differential scanning calorimetry (DSC): glass transition, cold crystallization and melting. Furthermore, the thermo-optical analysis provides a better understanding of the fiber orientation during a thermal treatment.

Keywords: Birefringence, polarized light microscopy, poly(ethylene terephthalate), thermo-optical analysis.
PACS: 81.70.Fy, 81.70.Pg, 78.20.Fm.

INTRODUCTION

The characterization of polymer microstructures is essential for various applications in industry. Among the many types of optical characterization procedures, the most common is the analysis between crossed polarizers [1]. However, the characterization of the morphology and microstructure of the polymeric materials it is not routinely performed in real time by optical microscopy with crossed polarizers. In a pioneer work in the 1960s, Magill studied in real time the crystallization kinetics by modifying a polarized light microscopy [2,3]. Later, in the 1990s, by using a camera and a detection device adapted to a polarized optical microscope with temperature control, Spruiell and his collaborators [4-7] have carried out experiments on the non-isothermal crystallization kinetics. Those modifications on microscope presented by Magill and Spruiell were designed to quantify in real time the changes in cross-polarized transmitted light intensity caused by the variation of the material birefringence.

Our group has been dedicated to the construction and application of optical devices to the real time investigation of polymers, blends, and composites during extrusion or thermal treatment. In this work, the thermo-optical behavior of a extruded fiber of poly(ethylene terephthalate) (PET) was investigated. Additional data are obtained about the relaxation and orientation of the fibers during heating, besides the traditional ones about the thermal transitions, obtained by differential scanning calorimetry. The device used in this study was originally used to quantitatively analyze the crystallization kinetics [8].

EXPERIMENTAL

Material

A commercial fiber of poly(ethylene terephthalate) (PET), supplied by Unifi textiles and containing 0.03% (w/w) of titanium dioxide (TiO_2), was used in this study.

Differential Scanning Calorimetry

A TA Instruments differential scanning calorimeter, Q2000, was used to measure the heat flow over time during non-isothermal conditions. The equipment was calibrate with indium (Tm=155.6°C, ΔH= 6.8calg^{-1}) at 10°Cmin^{-1}. Firstly, the sample was held at 280°C for 5min to remove the thermal history. Then, it was cooled to 25°C and heated again to 280°C at 10°Cmin^{-1}.

Quantitative Polarized Optical Microscopy

A schematic illustration of the experimental setup used for thermo-optical analysis is presented in **FIGURE 1** [8]. It consists of a polarized light optical microscopy with a coupled hot-stage and shearing system (CSS 450 Cambridge Shearing System). In order to measure the transmitted light intensity, a home-built detection module is positioned on the slot of the microscope, therefore avoiding the residual birefringence generated by the beam splitter of the microscope. This detection module is composed of two photocells (light dependent resistors, LDRs) and an analyzer, which is located just before one of the LDRs. A signal-handling box converts the LDR signals into voltages that are then transformed into digital signals by means of an analogic-digital interface (USB data acquisition device NI-DAQ 6812). Those digital signals are sent to a computer equipped with a software home-developed in LabView 8.6 platform (National Instruments); it was able to perform the data collection, real-time calculation, screen presentation and data saving. In addition, the software also controlled all parameters of the CSS 450 (4,5), it synchronously measured both light intensities with/without analyzer and it showed the images from the CCD camera (9).

The white light emitted by the lamp of the microscope is polarized before to reach the sample in the hot-stage. Thus, the photocells in the detection module allow to measure the transmitted intensity between cross polarizers (LDR with the analyzer) as well as to quantify the light intensity attenuation due to scattering (LDR without the analyzer). Also, a digital camera attached to the microscope permits to record simultaneously images of the samples.

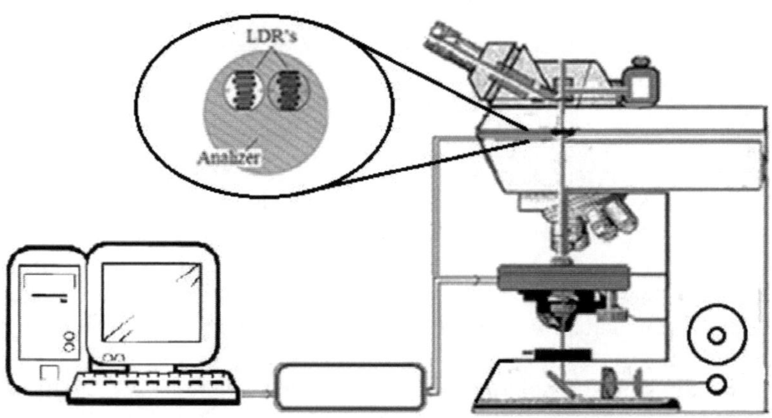

FIGURE 1. Schematic arrangement of the optical device highlighting the cross-polarized detection module fitted to the light polarized optical microscope.

The thermo-optical behavior of the fiber was analyzed in the temperature range 25-280°C; the procedure was similar to that one used in DSC experiments. The fiber was manually positioned on hot-stage aiming 45° from the crossed polarizers.

RESULTS AND DISCUSSION

The thermo-optical behavior of PET fiber was analyzed and compared with data obtained by DSC (**FIGURE 2)** In the DSC thermogram (**FIGURE 2a**), a change in the baseline, that can be assigned to the glass transition, is observed at about 75°C. At higher temperatures, between roughly 85 and 150°C, an exothermic peak indicates the occurrence of cold crystallization. A continuous increase of temperature leads to the melt of the fiber crystalline phase, which occurs between 225°C to 260°C, as it can be inferred from the endothermic peak in this temperature range.

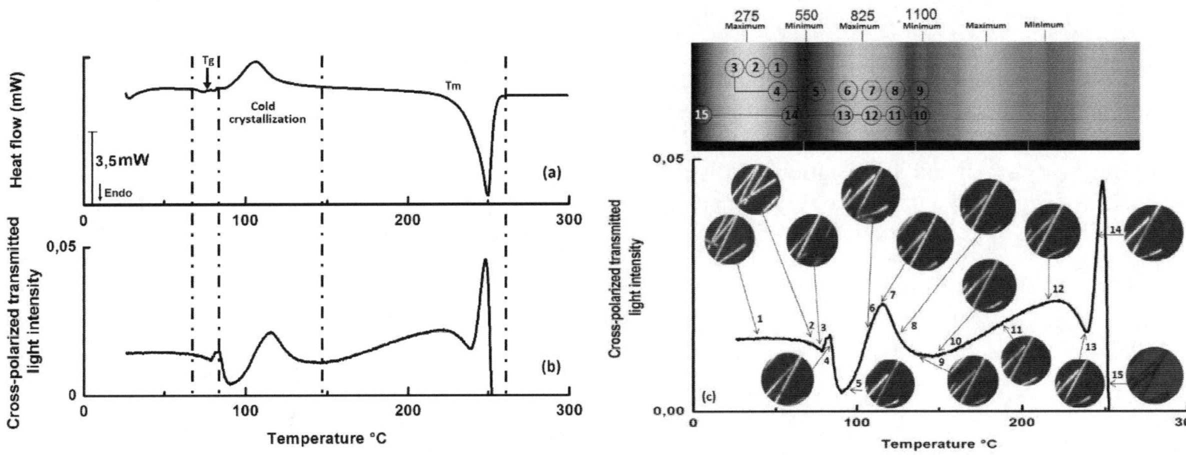

FIGURE 2. Thermo-optical behavior of PET fiber during the first heating: (a) DSC thermogram, (b) Cross-polarized transmitted light intensity in PLOM, (c) Evolution of OPD after combining the cross-polarized transmitted light intensity, the imagens of the camera and the Michel-Levy's chart for the PET fiber.

The cross-polarized transmitted light intensity measured during the first heating and at the same heating rate, shows several maxima and minima, when represented as a function of temperature (**FIGURE 2b**). The occurrence of these maxima and minima is not surprising; this just indicates that OPD go through several optical orders. Moreover, the maxima and minima occur in the same temperature ranges than those associated to thermal transitions in DSC. However, the cross-polarized transmitted light intensity curve alone did not allow to conclude about the evolution of OPD, since the optical order in which the fibers lie can not be inferred.

With the aim to gather information on the optical order, the images of the fibers, taken by the camera during the heating, were also analyzed. In fact, for white light, the passage through the different optical orders can be associated with various interference colors on the Michel-Levy's chart. Moreover, considering an average wavelength of 550nm, the maxima and minima of light intensity should occurs for multiple values of 275nm. Thus, by jointly analyzing the evolution in the color of the fibers and the cross-polarized transmitted light intensity (**FIGURE 2c**), it was possible to accurately track the progress of OPD. The OPD of the fibers at room temperature was measured by Berek's compensator and it was equal to 408nm (point 1 in **FIGURE 2c**). As the temperature increases to 70°C the intensity and, therefore the OPD, remains almost constant. After that, it decreases to a reach a minimum at a temperature slightly higher than Tg (75°C). An examination of the color of the fibers during heating shows that such minimum of intensity (point 3 in **FIGURE 2c**) corresponds to a minimum of OPD. The origin of this behavior will be explained later. After that temperature, the OPD increases continuously until a maximum of 1100 nm at 150°C (points 3 to 9 in **FIGURE 2c**). This increase of OPD coincides with the exothermic broad peak observed in DSC and therefore should be associated to cold crystallization phenomenon. Beyond 150°C the OPD decreases, vanishing at 255°C as indicated by the dark image. Furthermore, it can be observed a pronounced decrease of OPD between 225°C and 255°C, with two maxima and to minima occurring in this region (points 12 to 15 in **FIGURE 2c**). Such behavior of OPD should be attributed to the melting of crystalline regions as observed in DSC.

At temperatures around 75°C, there was a decrease in the cross-polarized transmitted light intensity due to the mechanical relaxation of the stretch fibers. This decrease can be associated with the approximation of Tg. The sine behavior in intensity was seen in temperatures from 85°C to 260°C and it occurred with maxima and minima of the cross-polarized transmitted light intensity. First maxima and minima occurred from 85°C to 140°C, in same region that took place the cold crystallization of PET fiber. From 235°C to 255°C, it was observed two maxima and two minima. This variation can be associated with the melting of PET fibers.

The morphological changes happening during the heating have also caused a variation in the transmitted light intensity (through the text the term turbidity will be also used; turbidity is defined as 1- *normalized transmitted light intensity*). While obtaining the curve shown in **FIGURE 2b**, the turbidity of the fibers was also measured In **FIGURE 3** it is possible to observe a pronounced decrease in the transmitted intensity between about 45 and 70°C. This increase of the turbidity can be explained by the relaxation of the stretch fibers, which leads to a decrease of its length and to an increase of its diameter. Notably, the mechanical relaxation seems to cease slightly below Tg as measured by DSC. In this temperature range, i.e., 45-70°C, the OPD is approximately constant (**FIGURE 2b**). If we recall that the birefringence is the ratio between OPD and the optical path, it becomes obvious that the birefringence decreases, in this range, during heating. Since the birefringence is a measure of the overall orientation of the fibers (orientation of crystalline phase plus the orientation of the amorphous one), it is not surprising to find a decrease of birefringence associated to the mechanical relaxation of the fibers. Between 70 and approximately 234°C, the transmitted intensity remains nearly constant, even though, according to DSC curve, the cold crystallization happens in a broad region below 150°C. A possible explanation for these results may be the crystals formed during this process were very small and it did not interact with light. A sharp increase of the transmitted light intensity between 234 and 265°C signals the melting of crystalline regions. Since the crystalline regions within the fiber are the main responsible for the light scattering, its melting eliminates almost completely the turbidity.

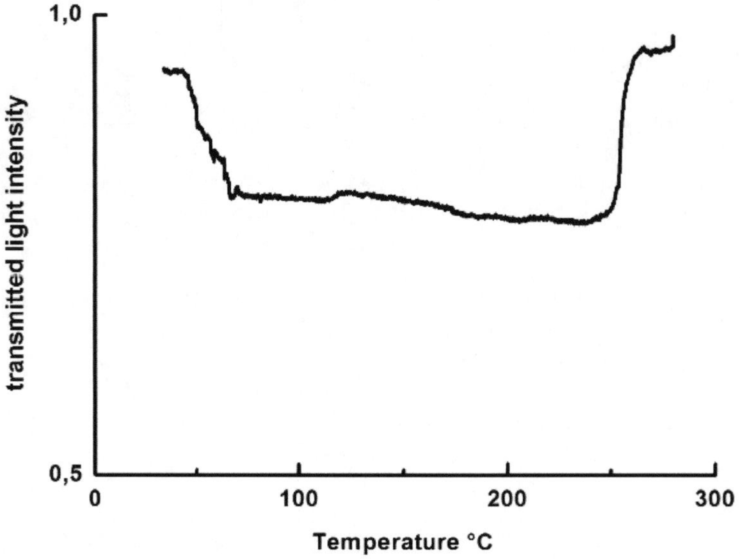

FIGURE 3. Transmitted light intensity (turbidity) of PET fibers measured in LPOM during the first heating.

CONCLUSIONS

In this work, a home-built detection module composed of two photocells and one analyzer was mounted in a polarized light optical microscope with a coupled hot-stage and shearing cell. This experimental setup allows to measure simultaneously the transmitted intensity (turbidity) and the transmitted intensity between cross polarizers. Since the evolution of the optical path difference (OPD) cannot be inferred from the transmitted intensity between cross polarizers unless the optical order is known, the images of the samples during the experiments were recorded by a camera installed in the microscope. By analyzing both the color evolution in light of Michel-Levy chart and the curve of transmitted intensity between cross polarizers it was possible to track the progression of the OPD of a PET fiber during heating. If the fiber diameter is constant, the variation of OPD relates directly with variation of the birefringence, which is a measure of overall orientation. In fact, the fiber diameter was assumed approximately constant during heating, except below T_g. In that temperature range the stretched fibers relax, increasing its diameter and causing a significant increase in turbidity. The birefringence reaches a maximum at about 150°C, which coincides with the end of the cold crystallization as measured by DSC, indicating that during this process the crystals grow in an oriented way. Moreover, the turbidity signal enabled to identify the beginning of melting. In conclusion, a positive correlation was found between optical analysis and thermal analysis (DSC) showing good reproducibility, and validating the thermo-optical method. Furthermore, the optical analyses have the advantage of providing additional information about the level of fiber orientation and stretching.

ACKNOWLEDGMENTS

The authors would like to acknowledge UNIFI for fiber donation, all the laboratory technicians for the support and the financial support from CAPES through scholarship to L. A. B. and Grant BJT 019/2012 to J. S. and from CNPq through a PQ scholarship311790/2013-5 to S. V. C..

REFERENCES

1. K. De Clerck, H. Rahier, B. Van Mele and P. Kiekens, *J. Appl. Polym. Sci*, **89**, 3840-3849 (2003).
2. J. H. A. MAGILL, *Polymer*, **2**, 221-233 (1961).
3. J. H. A. MAGILL, *Polymer*, **3**, 35-42 (1962).
4. Z. Ding and J. E. Spruiell, *Soc. Plast. E.*, 1485-1490 (1994).
5. Z. Ding and J. E. Spruiell, *J. Polym. Sci. Pol. Phys.*, **34**, 2783-2804 (1996).
6. P. Supaphol and J. E. Spruiell, *J. Polym. Sci. Pol. Phys.*, **36**, 681-692 (1998).
7. P. Supaphol and J. E. Spruiell, *J. Appl. Polym. Sci.*, **86**, 1009-1022 (2002)
8. J. H. Berl, A. M. C. Santos, A. A. Lucas and S. V. Canevarolo, *Conference Polymer Processing Society* PPS-Tel Aviv (2014).

Fatigue crack propagation behavior of simultaneously toughened and electrically conductive multi-functional epoxy composites

Gökhan Bakis[1)], Alexander Brückner[1)], Rico Zeiler[1)], Holger Deutges[1)], Volker Altstädt[1,a)]

[1]Lehrstühl für Polymere Werkstoffe, Universität Bayreuth, 95447 Bayreuth, Germany

[a)]Corresponding author: altstaedt@uni-bayreuth.de

Abstract. Simultaneously conductive and very tough resin formulations are desired for applications such as lightning strike protection or electrostatic dissipation during flight of commercial airplanes. Within the scope of this research, diglycidylether of bisphenol A (DGEBA) – 4,4'-diaminodiphenyl sulfone (4,4'-DDS) resin system was modified with fixed amount of carbon black (CB, 20 nm primary particle size, high structure, 1.5 wt%) in combination with various type of tougheners up to 10 wt%. As tougheners, core-shell and two different grades of functionalized tri-block copolymer of PMMA-b-PBA-b-PMMA were chosen. Following, fatigue crack propagation (*da/dN*) resistance of the multifunctional composites are characterized under cyclic tensional loading. Addition of various tougheners did not affect the T_g of the neat resin. Addition of all the tougheners increased dramatically the fatigue crack growth resistance of the DGEBA-4,4'DDS matrix by shifting the whole *da/dN* curve to higher ΔK ranges. Core shell particles increased the ΔK_{th}, stress intensity factor at initiation of crack propagation slightly higher compared to the tri-block copolymers. On the other hand, tri-block copolymer with higher soft-phase content (M52N) increased the stress intensity at the critical failure, ΔK_{cf}, at most, up to 95% by addition of 10 wt%.

1. INTRODUCTION

Fiber reinforced thermosets have been used increasingly in aviation and automotive industry as structural materials due to their spesific stiffness and strength, chemical resistance, etc. Recently, possible functionalities such as flame retardancy, thermal and electrical conductivity, simultaneous toughened and conductive formulations offer new application fields such as electric mobility (either automotive or aerospace) by promoting multifunctionality.

Mechanical performance of polymers and fiber reinforced polymer composites under cycling loading is a very important design parameter for structural applications. Although the mechanical properties of polymers and their fiber reinforced composites have been studied extensively under static loading since two decades, there is still lack of information in the field of dynamic-mechanical performance and underlying micro-mechanical mechanics under cyclic loading.

It is well known that thermoset matrix such as epoxy suffers from the inherent brittleness and low fatigue crack growth resistance. In literature, *inorganic additives* such as colloidal or fumed silica [1–3], aluminium, titanium oxides [4] and as well *organic additives* such as carboxyl terminated butadiene acrylonitrile rubber (CTBN) [5] and simultaneously both types of additives as hybrids [6] are used to enhance the fatigue crack growth resistance of epoxies. Briefly, the additive size, chemistry, interface strength, dispersion state as well as the size of the plastically deformed zone, cyclic loading frequency, applied stress intensity factor, etc. have enormous effect on the fatigue crack growth resistance of epoxies. Especially, the relationship between the particle or toughener domain size and plastically deformed zone size has a crucial importance affecting the micro-mechanical toughening mechanisms, in case of both organic and inorganic additives [2,7].

In the scope of this work, fatigue crack growth resistance of multifunctional epoxy composites having fixed amount of conductive additive (1.5 wt% carbon black) and an organic toughener with 2.5, 5, 7.5 and 10 wt% filler contents are characterized with fatigue crack growth rate (*da*/*d*N) test. As organic tougheners, core shell particles (2 µm average particle size), triblock copolymer M22N (promoting nano-phase seperation during curing, 30 nm average domain size, spherical) and tri-block copolymer M52N (promoting as well nanophase seperation during curing, <100 nm domain size, ellipsoidal) are chosen. The fatigue crack propagation resistance of the multifunctional composites were then characterized in close relation to the toughener morphology. The effect of the conductive additive network on the toughener distribution is as well studied.

2. EXPERIMENTAL METHODS

2.1. Materials

The epoxy resin used in this study is the diglycidylether of bisphenol A (DGEBA) resin, Baxxores 2200® (epoxy equivalent weight: 182 g mol^{-1} provided by BASF). As hardener, the 4,4'-diaminodiphenyl sulfone (4,4'-DDS) ORGANICA® (amine equivalent weight: 62.075 g mol^{-1} provided by Feinchemie GmbH) is chosen. The chemical structures of the DGEBA and 4,4'-DDS is shown below in figure 1.

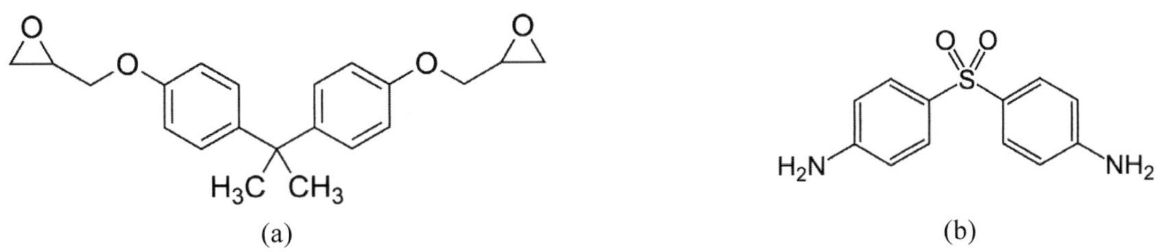

FIGURE 1. (a) DGEBA resin, (b) 4,4'- DDS hardener

Carbon black, Printex XE-2B (primary particle size: 20 nm, provided by Orion Engineered Carbons GmbH) with highly branched aggregate structure called *high structure* is used as conductive filler. As tougheners, core-shell particles, Albidur EP 2240-A (cross-linked polysiloxane core with stiff-epoxy compatible shell, SLS: 2 µm average particle size, provided as 40 wt% masterbatch in DGEBA by Evonik AG) and two different grades of functionalized tri-block copolymer of PMMA-b-PBA-b-PMMA (MAM), Nanostrength® (provided by ARKEMA) are used which are called as M52N and M22N. M22N has higher molecular weight, higher polarity but lower soft phase content compared to the M52N.

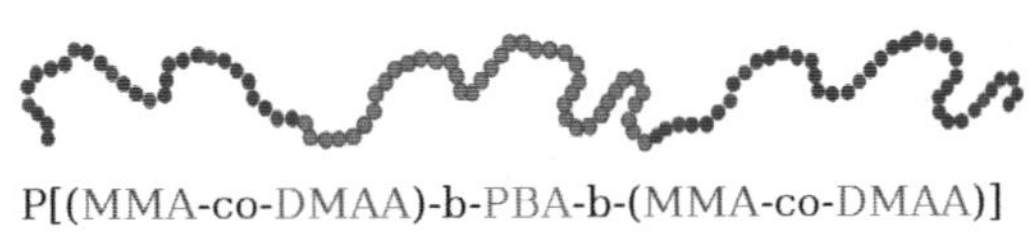

FIGURE 2. Molecular structure of the functionalized tri-block copolymers.

2.2. Characterization Methods

2.2.1. Dynamical Mechanical Thermal Analysis

Dynamical mechanical thermal analyses (DMTA) were performed according to DIN EN ISO 6721-7 using an Advanced Rheometric Expansion System (ARES RDAIII, Rheometrics Scientific) in torsional mode. An elastic deformation of 0.1 % with a frequency of 1 Hz was applied, from 25 to 250 °C at a heating rate of 3 K/min. According to the standard, rectangular samples were used (50-10-2 mm^3). The glass transition temperature (T_g) of the epoxy composites was determined as the maximum of the loss factor, *tan.delta*.

2.2.2. Fatigue Crack Growth Rate (da/dN) Test

Fatigue crack growth rate (*da/dN*) was measured as a function of the applied stress intensity factor ΔK during the test. Stress intensity factor K is calculated from the applied tensional force by using the *equation 1*, below. Tensional force was applied in sinusoidal waveform with the ratio of lowest (K_{min}) to highest (K_{max}) stress intensity factor R = 0.1. *da/d*N test conditions were set as room temperature (23 °C) with 50 % relative humidity. Universal servo-hydraulic testing machine (Hydropuls MHF, Schenck) was employed for the testing of all composites. Crack opening displacement (COD) was measured with a transducer (632.13F-20, MTS, Germany). CT type specimens were used and minimum 2 specimens per material were tested. A sharp crack with a length below 2 mm was introduced by kindly tapping the razor blade to the machined V-notch. Specimen thickness d was 4 mm and length from the loading point W was approx. 33 mm for all CT specimens.

$$\Delta K = \frac{\Delta F}{dW}\sqrt{\pi a}Y \qquad (1)$$

In the *equation 1* above, F, d, W, a and Y notations refer to the tensional sinusoidal force, sample thickness, length from the loading point, crack length and the geometrical factor respectively. During stable crack growth (called Paris regime), fatigue crack growth rate, *da/dN* shows a power law relationship to the stress intensity factor, ΔK which can be explained with the *equation 2*, below.

$$\frac{da}{dN} = C\Delta K^{m} \qquad (2)$$

In *equation 2* above, a, N, C and m notations refer to crack length, number of cycles, crack growth rate at 1 MPa√m and the slope at the Paris regime.

2.3. Composite Production

Within the scope of this work, except the neat resin, the amount of the carbon black (CB) is fixed for every composite which is 1.5 wt%. Following, the amount of the organic tougheners (core shell, M22N and M52N) are increased from 2.5 wt% up to 10 wt%. In case of tri-block copolymer M22N, it was only possible to produce samples without air bubbles only up to 5 wt% M22N content, since the degassing was not possible.

2.3.1. Production of carbon black – core shell epoxy composites

The calculated amount of DGEBA, carbon black and core shell – DGEBA masterbatch (40 wt% core shell) were weighted, and mixed at 90 °C in an oil bath with a laboratory mixer for 15 minutes. Following, the mixture was dispersed further with a three-roll-mill (EXAKT 120 EH-450, Germany). Mixing was done in one cycle with 15 µm and 5 µm gap distance in between rolls. Following the dispersion with three-roll mill, stoichiometric amount of 4,4'-DDS was weighted and mixed with laboratory mixer at 140 °C for 40 minutes to dissolve the hardener. The final mixture was then degassed at < 10 mbar vacuum environment for minimum 10 minutes. At last, degassed batch was cured in an aluminium mold at 180 °C for 150 minutes isothermally.

2.3.2. Production of carbon black – tri-block copolymer epoxy composites

Tri-block copolymers M52N and M22N are solid pellets which are soluble in DGEBA resin before curing reaction and form a phase seperated morphology during curing. Due to this fact, composite production with tri-block copolymers involves an additional step to dissolve the tougheners. First of all, calculated amounts of carbon black and DGEBA resin is dispersed with three-roll mill. Following, the tri-block copolymer is added to CB-DGEBA mixture and mixed at 140 °C for 20 minutes to dissolve the tri-block copolymer. The stoichiometric amount of 4,4'-DDS is weighted and added to the mixture at 140 °C and the final mixture is further mixed for another 40 minutes with laboratory mixer to dissolve the hardener. The final mixture was then degassed and cured as same as core shell-carbon black epoxy composites.

3. RESULTS AND DISCUSSIONS

At first, glass transition temperatures of the neat resin and multifunctional composites are determined by DMTA. *Tan.delta* (*tan.delta* = loss modulus (G^{II}) / storage modulus (G^{I})) is used to define the glass transition temperature of the composites. The results are shown below in fig. 3.

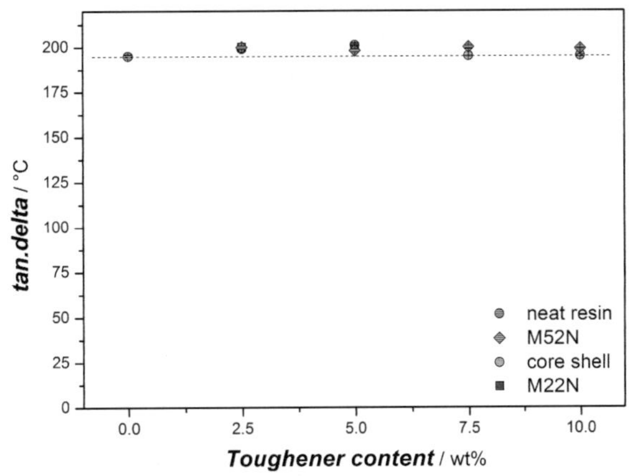

FIGURE 3. Glass transition temperatures of the neat epoxy and multifunctional composites with increasing toughener content. Except neat resin, all composites have 1.5 wt% carbon black, fixed.

Neat DGEBA – 4,4'-DDS system has T_g of 194.5 ± 0.7 °C. As it can be seen from fig. 3, the addition of organic tougheners with fixed CB content (1.5 wt%) did not significantly affect the T_g of the composites.
The fatigue crack growth rate (*da/dN*) of the multifunctional composites are characterized under cyclic tensional loading. The results are shown below in fig. 4 and 5.

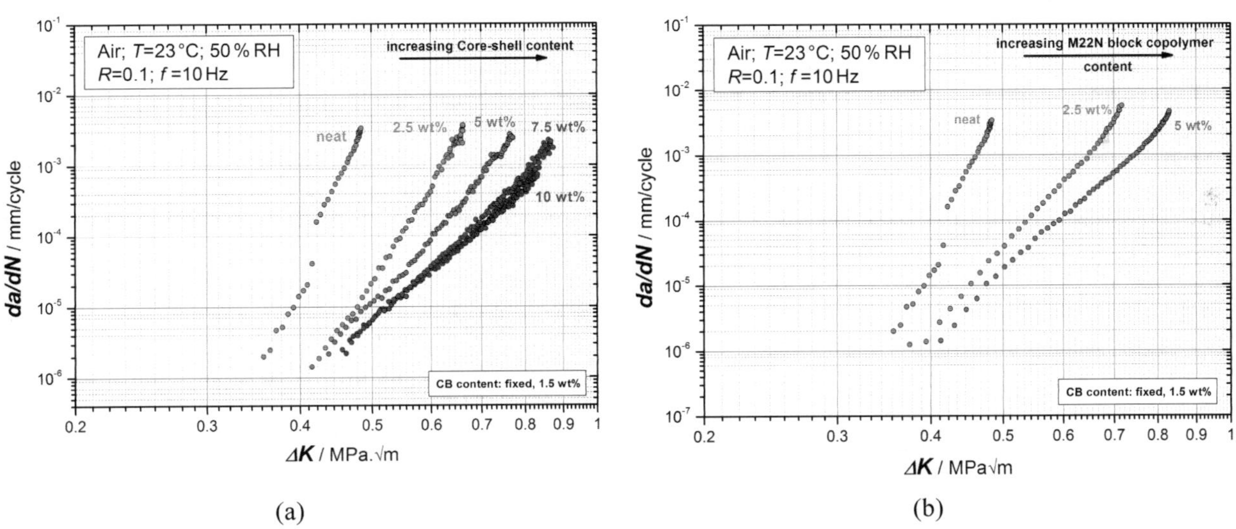

FIGURE 4. Fatigue crack growth rate (*da/dN*) of CB-CS (*a*) and CB-M22N (*b*) epoxy composites. Due to very high viscosity during degassing, composite with 1.5 wt% CB and only up to 5 wt% M22N toughener could be processed.

As it can be seen at fig. 4 and as well from fig. 5 (a), the addition of organic tougheners increased the crack propagation resistance dramatically even at 2.5 wt% toughener content. In addition, further increase of toughener content enhanced the fatigue crack growth resistance further by shifting the whole *da/dN* curve to higher ΔK ranges. By the addition of 10 wt% M52N toughener with 1.5 wt% CB content results in almost doubled stress intensity factor (0.94 MPa√m) at critical failure compared to neat epoxy (0.48 MPa√m).

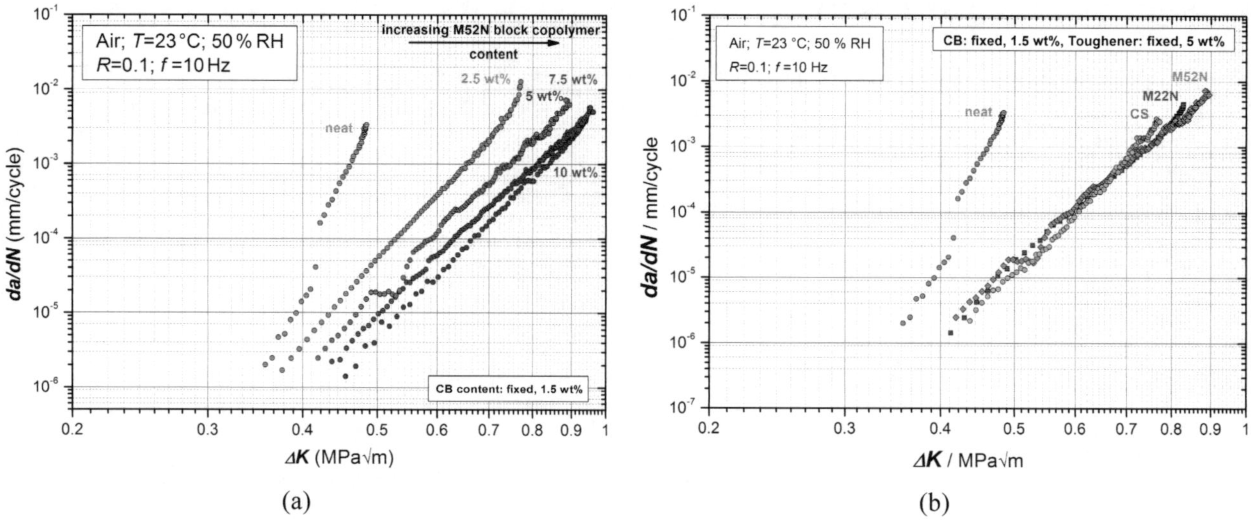

FIGURE 5. Fatigue crack growth rate (*da/dN*) of CB-M52N (*a*) and general comparison of all composites at 5 wt% (b).

In fig. 5 (b), *da/dN* results of composites having organic tougheners with 5 wt% concentration and 1.5 wt% CB are shown to compare the toughening performance. At the treshold region, addition of 5 wt% core shell resulted in a slightly better improvement of the crack initiation resistance showing 19% relative improvement of ΔK_{th} compared to 5 wt% of tri-block copolymers M52N and M22N with 17% and 15% relative improvement, respectively. On the other hand, at the critical failure region, the effect of the different tougheners becames clearer. The tri-block copolymer M52N with higher soft-phase content compared to M22N showed the highest improvement of the ΔK_{cf} by showing 95% higher stress intensity factor at critical failure compared to neat epoxy.

4. CONCLUSION

In this research, simultaneously toughened and electrically conductive epoxy formulations with fixed amount of conductive filler (1.5 wt%, CB) and increasing amount of organic tougheners (up to 10 wt%, CS, M22N and M52N) were characterized by the means of the fatigue crack growth resistance. The addition of the CB and various tougheners did not affect the T_g of the neat system negatively. The addition of core shell particles increased the resistance to the initiation of crack propagation slightly higher compared to the tri-block copolymers. However, tri-block copolymer with higher soft-phase content (M52N) resulted in the highest stress intensity factor at critical failure, ΔK_{cf} which is almost doubled compared to the neat epoxy.

5. REFERENCES

[1] B. R. K. Blackman, A. J. Kinloch, J. Sohn Lee, A. C. Taylor, R. Agarwal, G. Schueneman, and S. Sprenger, J. Mater. Sci. **42**, 7049 (2007).

[2] M. H. Kothmann, R. Zeiler, A. Rios De Anda, A. Brückner, and V. Altstädt, Polym. (United Kingdom) **60**, 157 (2015).

[3] M. Battistella, M. Cascione, B. Fiedler, M. H. G. Wichmann, M. Quaresimin, and K. Schulte, Compos. Part A Appl. Sci. Manuf. **39**, 1851 (2008).

[4] B. Wetzel, P. Rosso, F. Haupert, and K. Friedrich, Eng. Fract. Mech. **73**, 2375 (2006).

[5] H. R. Azimi, R. A. Pearson, and R. W. Hertzberg, J. Mater. Sci. **31**, 3777 (1996).

[6] H. R. Azimi, R. a Pearson, and R. W. Hertzberg, J. Appl. Polym. Sci. **58**, 449 (1995).

[7] J. F. Hwang, J. A. Manson, R. W. Hertzberg, G. A. Miller, and L. H. Sperling, Polym. Eng. Sci. **29**, 1477 (1989).

The Effect of Injection Moulding Temperature on PET Particles/Fibrils in Blends and MFCs

M. Kuzmanovic[a], L. Delva[a], L. Cardon[a], K. Ragaert[a]

[a] *CPMT Research Group, Department of Materials Science & Engineering, Faculty of Engineering and Architecture, Ghent University*

maja.kuzmanovic@ugent.be
kim.ragaert@ugent.be

Abstract. The microfibrillar composites of polypropylene (PP)/poly(ethylene terephthalate) (PET) have been prepared by twin-screw extrusion, followed by cold drawing. The employed stretch ratio was 4. Further processing was done by injection moulding at three different processing temperatures (210ºC, 230ºC, 280ºC) on PP/PET blends with wt% 70/30 Samples were subjected to extensive characterization in each step of MFC preparing. Fourier Transform Infrared (FTIR) spectroscopy was employed to determine the nature of the interaction between the polymers in the composites.. Thermogravimetric Analysis (TGA) were used to investigate degradation of polymers. The crystallization, melting behaviour and the crystallization morphology were investigated by Dynamic Scanning Calorimetry (DSC) and Polarized Optical Microscopy (POM). Influence of processing temperature on morphology was investigated by using Scanning Electron Microscopy (SEM). The observations from the fracture surfaces were discussed and compared with the mechanical properties, and the results have shown a significant influence of the injection moulding temperature on the morphology development and mechanical properties.

Keywords: microfibrillar reinforced composites, injection moulding, morphology, mechanical properties, polypropylene, poly(ethylene terephthalate).
PACS: 80.81

INTRODUCTION

The goal of this research is to investigate the effect of injection moulding temperature on the properties of the PP/PET injection moulding blends (IMBs) and microfibrillar composites (MFCs). It has already been shown that an improvement of mechanical properties can be obtained by the MFC concept [1, 2, 3], although not much attention has been dedicated to the specific influence of processing temperature during the injection moulding. However, some researchers have reported about the influence of screw speed [4], barrel temperatures [4], cooling time during moulding and the mould temperature [5]. Still, these researchers [3, 6, 7] have done studies on other compositions like PE/PET, PE/PEN blends and composites. Generally, in polymer processing it is of primary importance to understand the formation of dispersed phase, establish the relationships with morphology and with mechanical properties [2, 8, 9].

In present work, we represent the results of non-compatibilized blends and composites as we would like to explain how it affects the main morphology and mechanical properties made from two virgin polymers without additives, compatibilizers, fillers etc.

Methodology

In this work, PP was used as a matrix, and PET as a reinforcing element. PP/PET IMBs and MFCs were prepared in the weight ratio 70/30. The sample preparation was divided into IMB and MFC preparation, respectively. The IMBs were prepared by extrusion and injection moulding, while for the preparation of MFCs an additional stretching step was introduced between extrusion and injection moulding. In order to study effect of injection moulding temperature, the final processing was done at three different temperatures (210, 230 and 280°C). The relevant mechanical and thermal properties were investigated. Scanning Electron Microscopy (SEM) was used to investigate the morphology of the samples under the influence of different injection moulding temperatures.

Results

From the observation (Figure 1), it can be seen that discrete domains of minor phase are dispersed within continuous phase of major phase without adhesion between the different phases, which indicates completely immiscible blend [2, 10]. Furthermore, the appearance of some bigger ellipsoidal particles was found which probably come from the coalescence effect, (see the red circles on Figure 1 b and c).

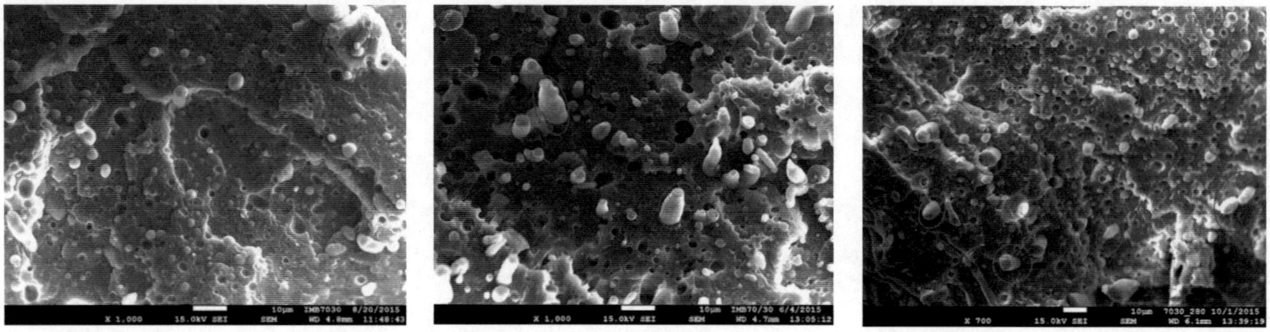

(a) (b) (c)

FIGURE 1. SEM micrographs of freeze-fracture surface under liquid nitrogen 70/30 IMB at (a) 210ºC, (b) 230ºC and (c) 280ºC

The effect of coalescence is the merging of two or more particles into one new larger particle which is not often in spherical shape. The coalescence effect is very dependent on the injection moulding temperature. It was found that at 210ºC the diameter of the particles were 1-2.5 μm, while the particle size at 280ºC can reach diameters up to 7 μm. Probably, the specimens processed at lower temperature of processing maintained the morphology formed by twin-screw extrusion, as the PET particles were in solid during processing at low T.

It can be seen a huge influence of processing temperature on PET particles. During processing at higher temperatures, there is a strong possibility that PET particles are melted again, as the phenomenon of coalescence was present in those samples. Similar conclusions have found other authors [3, 6].

REFERENCES

1. K. Friedrich, M. Evstatiev, S. Fakirov, O. Evstatieva, M. Ishiic and M. Harrassa, Microfibrillar reinforced composites from PET/PP blends: processing, morphology and mechanical properties, *Compos. Sci. Technol.* vol **65**, (1) 107–116, (2005).
2. L. Xu, G. J. Zhong, X. Ji, Z. M. Li, Crystallization behaviour and morphology of one-step reaction compatibilized microfibrillar reinforced isotactic polypropylene/poly(ethylene therephthalate) (iPP/PET) blends, *Chin. J. Polym. Sci.* Vol. **29**, No. 5, 540-551, (2011)
3. Z. M. Li, M. B. Yang, J. M. Feng, W. Yang, R. Huang, Morphology of in situ poly(ethylene terephthalate)/polyethylene microfiber reinforced composite formed via slit-die extrusion and hot-stretching, *Mater. Res. Bull.* **37**, 2185-2197, (2002)
4. K. Jayanarayanan, S. Thomas, K. Joseph, In situ microfibrillar blends and composites of polypropylene and poly (ethylene terephthalate): Morphology and thermal properties, *J Polym Res*, **18**, 1–11, (2011)
5. J. C. Viana, N. M. Alves, J. F. Mano, Morphology and Mechanical Properties of Injection Molded Poly(Ethylene Terephthalate), *Polym. Eng. Sci.*, Vol. **44**, No. 12, (2004)
6. Y. Lei, Q. Wua, Q. Zhang, Morphology and properties of microfibrillar composites based on recycled poly(ethylene terephthalate) and high density polyethylene, *Composites:* Part A **40**, 904–912, (2009)
7. K. Jayanarayanan, K. Joseph, S. Thomas, Microfibrils Reinforced Composites Based on PP and PET: Effect of Draw Ratio on Morphology, Static and Dynamic Mechanical Properties, Crystallization and Rheology, Charpter 15, book "Synthetic Polymer-Polymer Composites", 525-562, (2012)
8. C. Fuchs, D. Bhattacharyya, S. Fakirov, Microfibril reinforced polymer–polymer composites:Application of Tsai-Hill equation to PP/PET composites, *Compos. Sci. Technol* **66**, 3161–3171, (2006)
9. J. Andrzejewski, M. Szostak, M. Barczewski, J. Krasucki, T. Sterzynski, Fabrication of the Self-Reinforced Composites Using Co-Extrusion Technique, *J. Appl. Polym. Sci.* (2014)
10. Zhong-Ming Li, Wei Yang, Bang-Hu Xie, Kai-Zhi Shen, Rui Huang, Ming-Bo Yang, Morphology and Tensile Strength Prediction of in situ Microfibrillar Poly(ethylene terephthalate)/Polyethylene Blends Fabricated via Slit-Die Extrusion-Hot Stretching-Quenching, *Macromol. Mater. Eng.* **289**, 349–354 (2004)

Commercial Plastics: Tertiary Level Postgraduate Education in New Zealand

Neil R. Edmonds[1*], *John McKee*[2], *Peter N. Plimmer*[3]

[1]School of Chemical Sciences, The University of Auckland, New Zealand;
E-mail: nr.edmonds@auckland.ac.nz
[2]Zinniatech Ltd., Auckland, New Zealand
[3]P.N. Plimmer & Associates, Auckland, New Zealand

Summary: The Postgraduate Certificate in Engineering in Plastics (PGCertEng (Plastics)) is a joint initiative between The University of Auckland and Plastics New Zealand and was introduced to meet demand from industry for a high level programme focussed on and the chemistry, design and processing of plastics materials. This need arose from industries employing both engineering and science graduates who generally had little formal education in polymer materials or processing or knowledge of the commercial world of plastics. Four courses are delivered over a two year period with lectures delivered by both academic staff and industry experts.

Keywords: plastics, education, industry.

Introduction

The PGCertEng (Plastics) programme started in 2010 at the Plastics Centre of Excellence[1] at the University of Auckland and is run as week-long block courses to best accommodate the part-time study requirements of employers and their staff. In addition to the block course, students complete course and work-based assignments plus a final three hour exam. While full-time students can attend courses the highly commercial nature of the content means that some industry experience is required to fully benefit from the programme. Graduates have reported a very high level of satisfaction with the value of the programme to their employment.

The program grew from short courses run by Plastics New Zealand[2], the commercial body representing plastics manufacturers and associated industries in New Zealand. These courses provide an overview of the industry and are aimed at employees of manufacturing and supply companies. Course delivery is carefully structured to cater for the broad range of participants which may include engineering, processing, sales and marketing staff.

Four courses are delivered over a two-year period at a rate of one course each Semester, meaning that the full programme requires two years to complete. All courses are delivered as one-week blocks that include both theory and practical components, this is complemented by assignments and self-directed learning. Courses are

- Highly industry relevant
- Presented by industry experts
- Include intensive practical content
- Focus on processing, problem solving and product development
- Credits can be reassigned to Master of Engineering Studies

Course contents are given below:

Chemmat 740: Advanced Polymer Materials
Microstructure and morphology of semi-crystalline and amorphous polymers, including alloys and thermoplastic elastomers. The study of structure/property/processing inter-relationships for plastics materials. Characterisation of plastics materials, including spectroscopic, thermal and rheological analysis. Mechanical testing.

Chemmat 741 - Processing of Plastics
In-depth coverage of advanced processing techniques including associated rheology considerations. The study of additives, degradation processes and the prevention of degradation. Formulation of products and the mixing of materials.

Chemmat 742 - Plastics Materials Design
Heat transfer, drying. The application of commercial software for the design of products and tools. Material selection. Mold design and assembly, fabrication techniques. Environmental/lifetime considerations.

Chemmat 743 - Selected Topics in Plastics Processing
Reactive Compounding: polypropylene-based compatible alloys (Santoprene, Noryl, Alcryn), incompatible alloys (eg. toughened nylon), compatibilizers for multi-layer film (PE/PA, PE/EVOH). Overview of cast and blown film. Membranes. Fluoropolymers, biobased polymers, unusual polymers. Thermosets: elastomers, powder coating, reaction injection moulding, liquid moulding (siloxanes and urethanes). Sintering (eg. Teflon PTFE).

The PGCertEng (Plastics) program provides an in-depth treatment of key topics and is designed for both science and engineering graduates who are in full-time industry employment. Critical to this is providing necessary chemistry for engineers and mechanics for those with a science background where this is not part of undergraduate degrees.

'Polymeric Materials' (Gottfried W. Ehrenstein)[3] provides an excellent balance of both subjects at the appropriate level and is an ideal text for bridging this gap.

Practical Program

The following table illustrates a typical practical program for Advanced Polymer Materials. Note that the practical program includes field trips to companies involved in masterbatch manufacture, rotational moulding, tool production etc.

MONDAY/TUESDAY	WEDNESDAY/THURSDAY	FRIDAY
POLYMER SYNTHESIS PMMA, PA6/10 FTIR COLOUR DENSITY MOISTURE CONTENT MICROSCOPE/HOT STAGE	THERMAL ANALYSIS: DSC DMTA TGA THERMAL CONDUCTIVITY MFI CONE CALORIMETER GAS PERMEATION WEATHEROMETER	MECHANICAL TESTING TENSILE/ELONGATION MODULUS HARDNESS TOUGHNESS IMPACT STRENGTH

Assessment:
All courses require the completion of an open-book test plus two assignments. These three components make 45% of the final mark, the remaining 55% coming from a final three hour exam. Observations/Results from practical exercises must be recorded. While not marked, these form part of coursework requirements and may be assessed in the final exam[4].

Typical Open Book Test questions:

- What dominant attractive forces holds polyethylene, polycarbonate, polyamide (nylon), and PVC polymer chains together within their and amorphous and crystalline (if they have one) phases? Comment on the `special' attractive force which arises in the polyamide family.
- What useful information comes from 'stress/strain' testing?
- Explain, using morphological terminology and the concept of tacticity, why polypropylene is a semicrystalline polymer (with a Tg and a Tm) whilst PVC and PMMA are amorphous (only have a Tg).
- Explain, on the basis of polymer 'architecture' and morphology, why polycarbonate has a higher Tg than polystyrene. How are these Tg's measured? Why is the Tm of HDPE higher than that of LDPE and how are these Tm's measured?
- Define 'weight average molecular weight' (M_w), and explain how it impacts on polymer properties and processing.
- What are the benefits of glass reinforcement of polyamides or polyesters? Describe the simple laboratory test used to measure the heat resistance benefit.
- Using you knowledge of polymer structure and morphology, explain why high impact polystyrene, thermoplastic elastomers and thermoset elastomers are more impact resistant (tougher) than polymers from the nylon or polyester families? How is impact resistance measured?
- Compare and contrast the techniques used to manufacture graft, block and random copolymers. Why are these polymers useful?

Assignments: Assignments should be ~15 pages in lengths including relevant data sheets, tables etc.

1. A topic chosen by students relating the course content to a technical problem at their workplace.
2. Expanding on a topic selected from the course content. Some examples:
   - The use of analytical techniques in problem solving involving polymers.
   - The impact of heterogeneous catalyst (metallocene) on new polymer development.
   - Understanding the role of glass transition temperature manipulation in the development of (a) thermoplastic elastomers, (b) toughening of engineering polymers, (c) creep and dimensional instability, (d) shape- memory polymers and, (e) 'breathable' polyurethanes and polyesters.
   - Influence of molecular weight on polymer processing and end-use performance (how is it controlled?).
   - Discuss how three different polymerization techniques can be used to manufacture different "forms" of poly(methyl methacrylate), polyvinyl chloride), poly(styrene) and poly(tetrafluoroethylene).

Conclusions:
The programme has been well received by industry with 15-20 students attending each course. Typical student feedback includes:

- "From the first assignment, this course has been directly applicable to my work. I've gone from having no knowledge of certain aspects (e.g. UV degradation of polymers) to being referred to as the "expert" in our office. While this is personally satisfying, my manager and I both feel very strongly that the knowledge gained in these courses has been of great value to my company"
- "The value of this programme has been hugely important to my role as a product development engineer at [Company name] as it has given me a really good knowledge of polymer properties and their applications, as well as practical understanding of tools and polymer processing"
- "The value of this course has been immense. Gaining more product and processing knowledge has been invaluable in my role as I need to know a lot of different polymer products and the way they react when used in different processes"

The potential for delivery of the program to a broader audience is under consideration.

References

1. The Plastics Centre of Excellence: http://www.pcoe.org.nz
2. Plastics New Zealand: http://www.plastics.org.nz
3. Ehrenstein, Gottfried W., *Polymeric Materials*; Hanser Garder Publications, 2001.
4. Further details on the PGCertEng (Plastics) program can be found at http://www.engineering.auckland.ac.nz/uoa/pgcerteng-plastics

Structure construction of amorphous polymer at long time region and their mechanical properties.

Naoto Oda, Ryoko Nakano, Hiroshi Sekiguchi, Shigeru Yao*

Department of Chemical Engineering, Fukuoka University
8-19-1, Nanakuma, Jonan-ku, Fukuoka, Japan 814-0180

[Introduction]

Polymer is thought to be constructing a random structure in molten state. Especially, longer the longest relaxation time, polymer is in a relaxed state with lost the orientation history. Moreover, the amorphous polymer, cooled and solidified from such molten stage is thought to have no structural morphology and show stable mechanical property. However, in this time, we investigated holding time of molten stage dependence to tensile properties by using polystyrene virgin pellet and found that the tensile properties much depends on the holding time and became ductile. And also, we got the results that obvious structural construction were occurred by SAXS and WAXS measurements. These results shows that amorphous polymer construct ordered structure in the molten stage.

[Sample and Experiments]

The virgin polystyrene pellet was Toyo Styrene GPPS(G200C). Viscoelastic properties of melt state were measured by using rheometer. Hot pressed thin films (~ 50 μm) at various temperature (210 °C and 230 °C) and time (2, 6, and 10 min) were made and were used for tensile test, WAXS, and SAXS measurement.

[Results and Discussion]

Figure 1 shows viscoelastic properties of the sample at 210°C. From this result, the longest relaxation time is calculated about 0.03 sec.

Figure 2 shows results of tensile tests. From the figure, we found that the tensile elongation length and yield stress were increased with molding time. This result indicate that after the longest relaxation time, the mechanical properties of polystyrene is still changing.

Figure 3 shows the SAXA profile of these samples. From the profile of the 2min compressed thin film, there is no scattering in the large angle area, however with increasing the molding time, scattering appears due to the construction of small structures.

Figure 4 shows the WAXA profiles of these samples. Comparing these profiles, with increasing the molding time, new scattering peaks appearing in the small angle side.

[Summary]

With increasing the molding time, the mechanical properties changed to good. Also, some new structure will construct with the time. This results indicate that long molding time that is longer than the longest relaxation time, will change the inner structure of amorphous polymer and give a good mechanical property.

Keywords: **Morphology development, Longest relaxation time, Amorphous polymer**

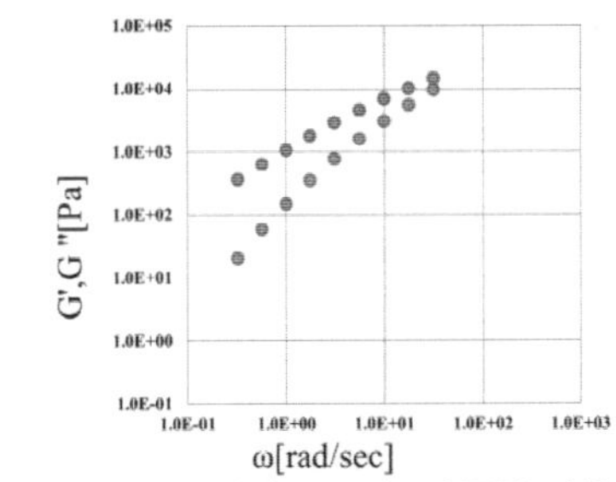

Fig.1 Viscoelastic properties of GPPS at 210 °C. G': blue circle, G": red circle.

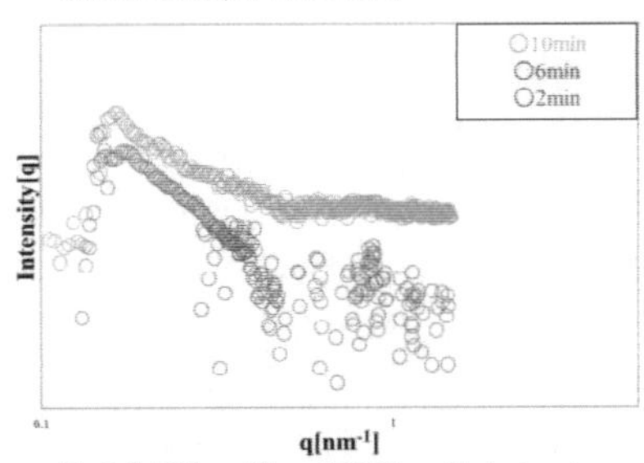

Fig.3 SAXS profiles of GPPS molded at various time.

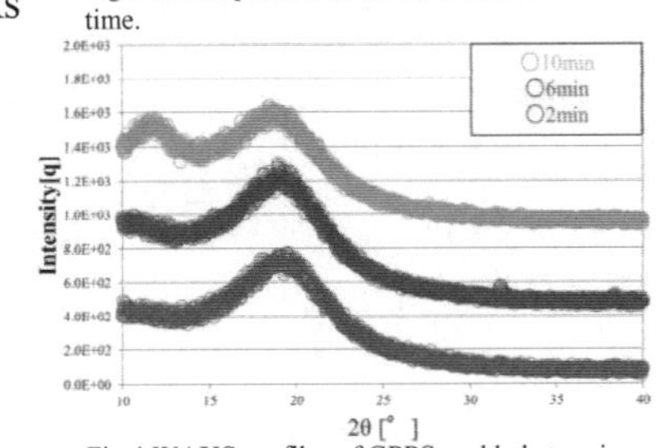

Fig.4 WAXS profiles of GPPS molded at various time.

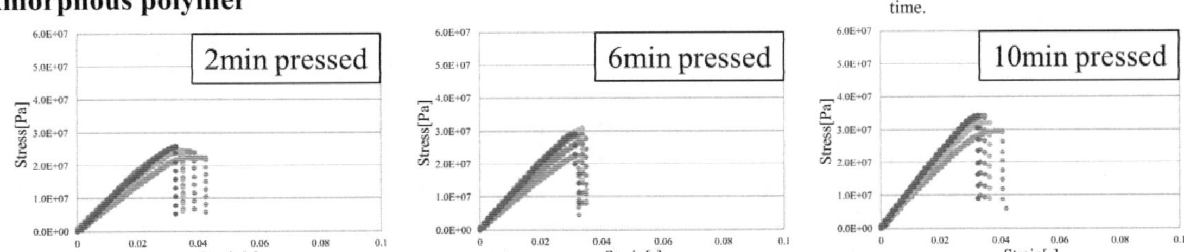

Fig.2 Tensile properties of GPPS molded at various time.

DMBS-INDUCED NUCLEATION OF ISOTACTIC POLYPROPYLENE UNDER HIGH PRESSURE IN ISOTHERMAL AND NON-ISOTHERMAL CONDITIONS

Przemyslaw Sowinski[a], Ewa Piorkowska[a], Severine A.E. Boyer[b], Jean-Marc Haudin[b],

[a] *Centre of Molecular and Macromolecular Studies, Polish Academy of Sciences, Sienkiewicza 112, 90 363 Lodz, Poland, przem_so@cbmm.lodz.pl, epiorkow@cbmm.lodz.pl*
[b] *MINES ParisTech, PSL Research University, Centre for Material Forming, UMR CNRS 7635, 1 Rue Claude Daunesse, 06904 Sophia Antipolis, France, severine.boyer@mines-paristech.fr, jean-marc.haudin@mines-paristech.fr*

Keywords
Crystallization, High Pressure, Nucleation, DMBS, Isotactic Polypropylene

Introduction
Among the numerous challenges faced in understanding polymer nucleation, one question, never reported until nowadays and even if the search for nucleating agents of PP is an old story since the pioneering works of Beck [1] and Binsbergen [2], is if nucleating agents could actually enhance the crystallization in a preferential-crystal under high pressure. The question stays of concern for isotactic polypropylene (PP) and more especially when studying the gamma-crystals. Usually, PP crystallizes in the monoclinic alpha-crystals under atmospheric pressure (P_{atm}), and also in the trigonal beta-crystals, when appropriate nucleating agents are used [3]. Under high pressure, the unique because of non-parallel chain alignment, orthorhombic gamma-crystals form, although it also requires high temperature [4-6].
Recently, we have studied the nucleation of the gamma-form in PP under high pressure. To nucleate crystallization of PP three nucleating agents were used, namely alpha-nucleants [Hyperform HPN-20E (Milliken Chemical), poly(tetrafluoroethylene) PTFE submicron particles (Polysciences Inc.) [7]] and a beta-nucleant [calcium pimelate]. New advance focuses on the nucleation of the gamma-phase by 1,3:2,4-bis(3,4-dimethylbenzylidene sorbitol) (NA) a clarifier for PP crystallizing under P_{atm} in the alpha form. Under P_{atm} this NA melts and dissolves in PP in temp range 180-200°C, and during cooling crystallizes in the form of fibrillar crystals, strongly nucleating crystallization of PP alpha form.
The study has focused on nucleation of the gamma form in PP under high pressure in isothermal conditions and also during cooling, because during processing crystallization is non-isothermal.

Materials
The polymer used was an isotactic PP (3250MR1, Atofina) grade for injection-molding. As NA 1,3:2,4-bis(3,4-dimethylbenzylidene sorbitol) - DMBS, Millad 3988i (Milliken Chemical), was applied. PP was mixed with 0.2, 0.4, 0.6 and 1wt% of NA in a Brabender batch mixer at 195°C, 60 rpm for 10 min (systems PP/M 0.2 to PP/M 1). PP without nucleating agent was processed in the same way to have a blank control sample.

The high-pressure crystallization was carried out using a custom-built hydrostatic cell, consisting of a barrel and a piston. It is described in detail elsewhere [8-10]. The hydrostatic pressure (P) and temperature (T) inside the cell were controlled with an accuracy of ± 0.5 MPa and 1 °C, respectively. Pressure during crystallization ranged from 1.3 to 300 MPa. Two kinds of experiments were conducted: nonisothermal crystallization during cooling (all materials) and isothermal at 200°C (only for PP and PP/M 0.2). For isothermal crystallization, samples were heated up to 250°C under a pressure of 1.3 MPa applied to ensure good thermal contacts, held for 5 min, cooled down to 200°C, pressurized and held under high P at 200°C: either 15 min under pressure from 1.3 to 300 MPa or for various dwell times from 2 min to 4 h under 200 MPa, similarly as in [8]. Then, samples were cooled down to 50°C, and depressurized. Procedure for nonisothermal crystallization [9] consisted of heating to 250°C under 1.3 MPa, pressurizing (50-300 MPa) and then cooling down. The cooling from 200°C to 100°C, during which crystallization could be expected was nearly linear with an average rate of 8°C min^{-1}.

The samples were compressed using an Instron tensile testing machine (Instron Corp., High Wycombe, UK). This equipment let us to measure cross-head movement, and to follow a volume change of samples, which was correlated with temperature, and allowed us to determine the temperature of maximum crystallization rate, taken as crystallization temperature, T_c. The samples were analyzed ex-situ: thermal properties by differential scanning calorimetry during heating (TA Instruments DSC 2920, New Castle, DE, USA) at 10°C min under P_{atm}, crystallinity and contents of crystallographic forms by wide-angle X-ray diffraction (WAXD) in the reflection mode in the 2θ range from 7 to 67° (Philips PW3830, Eindhoven, The Netherlands), spherulitic structure in 10-μm-thick microtomed sections by polarized light microscopy (PLM, PZO, Poland). For the identification of the crystallographic forms, $(130)_{\alpha}$ and $(117)_{\gamma}$ diffraction peaks were considered. Deconvolution of WAXS diffractograms was carried out by WAXSFIT program [11]. The contents of the alpha and gamma phases, K_{α} and K_{γ}, respectively, in the crystalline phase of PP were determined according to Turner-Jones et al. [12].

Selected Results & Discussion

Fig. 1a-c shows spherulitic structure, T_c and content of alpha and gamma forms in samples crystallized nonisothermally under elevated pressure.

Cooling in DSC evidenced NA caused an increase in crystallization peak temperature (T_c) during cooling (10°C/min) to 130°C from 115°C for pure PP, with NA content increasing up to 0.4 wt%, because of nucleation activity of the NA. For PP/M 0.2 elevation of T_c was observed only under P_{atm} and 50 MPa. However, increase of NA content from 0.2 wt% to 0.4 wt% resulted in an increase of T_c both under P_{atm} and under elevated pressure, up to 300 MPa. Under 50 MPa T_c of PP/M 0.2, 144°C, was below that of PP/M only by 3°C but the difference enlarged with increasing pressure up to 20°C at 300 MPa. Under P>50 MPa the effect of NA on T_c practically vanished in PP/M 0.2 and it behaved similarly to neat PP, as shown in Fig. 1b. Examination of spherulitic structure in thin sections demonstrated that fine grains crystallized in PP/M 0.2 under 50 MPa, evidencing strong nucleating activity. Under 200 MPa a structure formed consisting of larger spherulites surrounded by very small grains, which shows that NA acted when the larger ones were already growing. Similar features were visible in thin sections of PP/M 0.2 samples annealed in isothermal conditions under high pressure. Contrary to this, PP/M 0.4 revealed fine grain structure regardless of pressure, and its T_c always exceeded markedly that of neat PP. Further increase of NA content did have a significant influence on T_c or on the structure. The most possible reason is that crystallization temperature of NA in PP/M 0.2 under P > 50 MPa was below that of nucleation onset in neat PP, and that increase of NA content resulted in elevation of this temperature.

Content of crystallographic forms, calculated from WAXS diffractograms showed that under 1.3 MPa only pure PP crystallized exclusively in the alpha form, whereas in the other materials a small fraction (about 15%) of the gamma phase formed. The gamma content in all materials increased with increasing pressure. Under elevated pressure the presence of NA enhanced crystallization in the gamma form, less in PP/M 0.2 and more in the other materials, although the differences decreased with increasing pressure. Under 300 MPa pure gamma phase crystallized in PP/M samples, and only a small fraction of the alpha phase formed in neat PP. In the samples annealed in isothermal conditions at 200°C the content of γ-phase depended on dwell time and the γ content in PP/M 0.2 was higher than in neat PP, especially for short dwell times under 200 MPa and for dwell time of 4 h under 50-100 MPa.

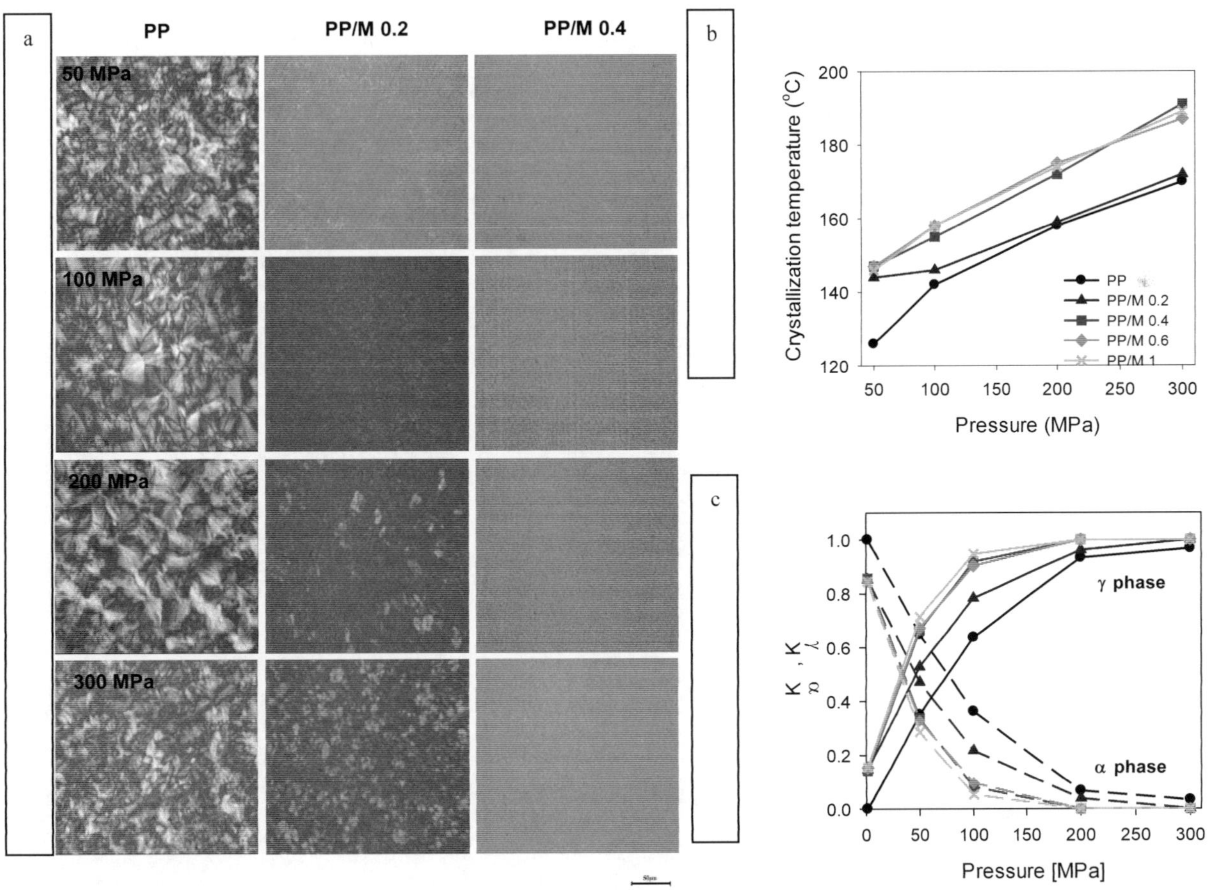

Figure 1. - a, PLM micrographs of thin sections of samples crystallized nonisothermally under elevated pressure. Nonisothermal crystallization - b, pressure dependence of crystallization temperature determined from volume change during cooling; - c, pressure dependence of alpha and gamma phase contents: K_α - dashed line, K_γ - solid line.

Conclusion

NA - Millad exhibited ability to nucleate the gamma form under elevated pressure, although, compared to the other nucleants, studied by us before [8], it showed a different behavior. The nucleation activity of this NA depended both on pressure and on its content in PP. At low content, NA markedly increased the temperature of nucleation onset only under moderate pressures, P <100 MPa, whereas under higher pressure this effect vanished, and nucleation started at the similar temperature as in neat PP. Analysis of spherulitic structure demonstrated that under P>100 MPa NA acted in temperatures lower than that of nucleation onset in neat PP, resulting in structures composed of larger spherulites surrounded by small-grains. Increase of NA content to 0.4 wt% caused an increase of nucleation temperature, and as a consequence, crystallization temperature, and also led to a drastic reduction of a grain size, regardless of pressure. The gamma content enlarged in all the materials with increasing pressure. The presence of NA increased the gamma content compared to neat PP, although less in PP/M 0.2 than in PP with higher NA content. The difference decreased with increasing pressure; under 300 MPa the pure gamma phase crystallized in nucleated PP regardless of NA content, and only a small fraction of the alpha phase formed in neat PP. The results show that to nucleate PP crystallization under high pressure in the gamma form higher contents of Millad are required than to nucleate PP under P_{atm}.

References

[1] Beck HN (1967) J Appl Polym Sci 11:673-685

[2] Binsbergen FL (1970) Polymer 11:253-267

[3] Varga J (1993) In: Karger-Kocsis J (ed) Polypropylene: structure, blends and composites, vol 1. Chapman & Hall, London, pp 56-115

[4] Bruckner S, Meille SV (1989) Nature 340:455-457

[5] Meille SV, Bruckner S, PorzioW (1990) Macromolecules 23:4114-4121

[6] Bruckner S, Meille SV, Sozzani P, Torri G (1990) Makromol Chem Rapid Commun 11:55-60

[7] Masirek R, Piorkowska E (2010) Eur Polym J 46:1436-1445

[8] Sowinski P, Piorkowska E, Boyer SAE, Haudin JM, Zapala K (2015) Colloid Polym Sci 293:665-675

[9] Sowinski P, Piorkowska E, Boyer SAE, Haudin JM (to be published)

[10] Psarski M, Galeski A, Piorkowska E (2000) Macromolecules 33:916-932

[11] Rabiej M, Rabiej S (2006), ATH, Bielsko-Biala

[12] Turner-Jones A, Aizlewood JM, Beckett DR (1964) Makromol Chem 75:134-158

Acknowledgement

CMMS PAN statutory funds and PAN/CNRS Collaborative Project 179990 High pressure crystallization and structure of original polymers and polymers based nanocomposites.

Structural Evolution of Polypropylene during Creep: an in-situ Study by Synchrotron Small Angle X-ray Scattering

Baobao Chang[a,b*], Konrad Schneider[a] and Gert Heinrich[a,b]

[a] *Leibniz-Institut für Polymerforschung Dresden (IPF), Germany*
[b] *Institut für Werkstoffwissenschaft, Technische Universität Dresden, Germany*
chang@ipfdd.de

Abstract. The objective of this study is to explore the structure evolution of isotactic polypropylene (iPP) during creep by in-situ synchrotron small angle X-ray scattering (SAXS). The results suggest that the structural evolution during creep differs in different well distinguished stages. During primary creep step, strain grows nonlinearly to a value smaller than 15 % in a short time. In this stage, the long period along drawing direction (L_a) increases with the time, whereas the long period perpendicular to drawing direction (L_p) decreases slightly. The change of long period is caused by elastic deformation of molecular chains in amorphous phase. In secondary creep step, strain increases linearly with time. Both L_a and L_p increase with time. What's more, the increase rate for L_a is larger than that for L_p. This can be explained by thickening of lamellae by cooperation motion of molecular chains together with their neighbors. The molecular chains in the amorphous region along drawing direction have a higher orientation degree in the normal of lamellae compared with that perpendicular to drawing direction, which enhances the probability for such cooperation motion and decreases the energy barrier, thus the increase rate for L_a is larger than that for L_p. In tertiary creep step, strain grows dramatically within a limited time, and the yielding of iPP happens in this stage. The reflections in 2D-SAXS pattern changes from light ellipse to a distinct ellipse firstly, and then grows into "two-spots" pattern. This significant transition indicates that the lamellae are tilted and rotated during the yielding process. After that, the tilted and rotated lamellae are disaggregated, and aligned along the drawing direction. The lamellar orientation degree calculated by Herman's orientation method is about 0.5. In addition, a streak perpendicular to the drawing direction appears in the SAXS patterns, which is a hint for the formation of highly oriented structure. Employing Ruland's streak method, it is calculated that the length of highly oriented structure is about 600 nm and grows to 850 nm finally.

Keywords: creep; polypropylene; in-situ small angle X-ray scattering
PACS: 82.35.Lr

INTRODUCTION

Isotactic polypropylene (iPP) is one kind of thermoplastic polymer which has been used in a wide variety of areas including packaging, textiles, laboratory equipment, automotive components, et. al. Considering its ever-increasing range of applications in our daily life and engineering field, it is of great significance to evaluation also the property response under permanent load conditions [1-3]. Different from uniaxial drawing, compression and impact test, creep test serves as an effective method for the long-term life time assessment of polymer materials.

Creep is generally defined as a time-dependent deformation of a material under constant load for a prolonged period of time at constant temperature. Creep behavior can be well elaborated within the framework of linear viscoelasticity [4]. Under this circumstance, the master curve of creep compliance can be constructed by using time-temperature superposition concept based on the William-Landel-Ferry equation [3, 5]. However, the prerequisite for linear viscoelasticity is that the creep compliance is assumed to be a function of time, beyond critical strain limit creep compliance will become a function of both time and stress [6]. As one kind of semi-crystalline polymer, it has been demonstrated that the crystalline morphology of iPP has a significant effect on its creep behavior. Sliding of tie chains and their detachment from lamellar blocks plays a key role in the time-dependent response of iPP [7]. Despite the fact that considerable attentions have been devoted, the study of the microstructure evolution during creep is not yet well understood.

In this study, the structural evolution of iPP during creep is characterized by in-situ synchrotron SAXS at PETRA III, Deutsches Elektronen-Synchrotron (DESY). The results revealed that the microstructural evolution differs in different creep stages, especially in secondary and tertiary creep. The evolution of long period, lamellae disaggregation, and highly oriented structure formation during creep is studied in detail.

EXPERIMENTAL

Materials and sample preparation

iPP used in this study is one kind of homopolymer, grade HD 120 MO, manufactured by Borealis (Linz, Austria). The weight-average and number-average molecular weights are 365 kg/mol and 67.6 kg/mol respectively [8], and the melt flow index is 8 g/10 min (at 230 °C and 2.16 kg).

In-situ synchrotron small angle X-ray scattering (SAXS) measurement

Creep tests were performed in force-controlled mode on a custom-made miniature tensile machine designed for on line study at the Synchrotron by Leibniz Institute of Polymer Research Dresden (IPF). To keep the beam in a fixed position during deformation, both grips were moved simultaneously in opposite direction throughout the measurement. In situ synchrotron SAXS measurements were performed at the MiNaXS beamline at Deutsches Elektronen Synchrotron (DESY), Hamburg, Germany [9]. The energy of the X-ray radiation was 11.5957 keV. The patterns are recorded using a Pilatus 1M detector. The distance between the specimen and the detector was 4961 mm. An exposure time of 0.5 s and time interval of 7.5 s is used to avoid burning of the specimen by X-ray during the test. All SAXS patterns were background corrected and normalized to unit sample thickness.

RESULTS AND DISCUSSION

2D-SAXS patterns during creep process

Fig. 1 shows some characteristic background corrected and normalized SAXS patterns of the iPP during creep according to the different states discussed below. The initial isotropic SAXS pattern (a) results from randomly oriented lamellae. Before region II (b), no obvious differences could be found. In region III (c), SAXS pattern grows from nearly isotropic one to slightly ellipse. The long axis of the ellipse is perpendicular to drawing direction. In region IV (d and e), slight ellipse changes into distinct ellipse. This significant transition indicates that lamellae are tilted and rotated. After that, the SAXS pattern transform from distinct ellipse into a "two-spot" pattern, meaning that lamellae are stacked in a relatively ordered organization, preferentially aligned along drawing direction. What's more, a streak appears in the equator, which provides a hint that highly oriented structure is formed in this region.

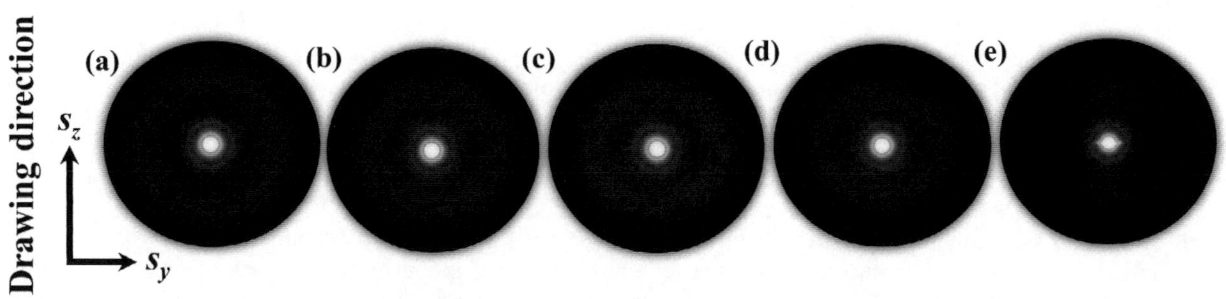

FIGURE 1: Characteristic 2D-SAXS patterns during different states of deformation

Evolution of long period

In order to get a comprehensive investigation on the evolution of crystalline structure during creep, slices in both horizontal (s_y) and vertical direction (s_z) are cut in 2D-SAXS patterns. The value of long period L (including one lamella and one inter lamellar amorphous phase) can be calculated from the peak position of 1-D SAXS curves based on Bragg's law: $L=1/s_{max}$, where s_{max} is the peak position, $s=2\sin\theta/\lambda$, s is scattering vector, λ is the incident wavelength. The evolution of long period along (L_a) and perpendicular (L_p) to drawing direction as a function of creep time are qualitatively shown in Fig. 2. It can be seen that the whole curve could be divided into 4 intervals for both L_a and L_p, which coincides with the evolution of macroscopic displacement.

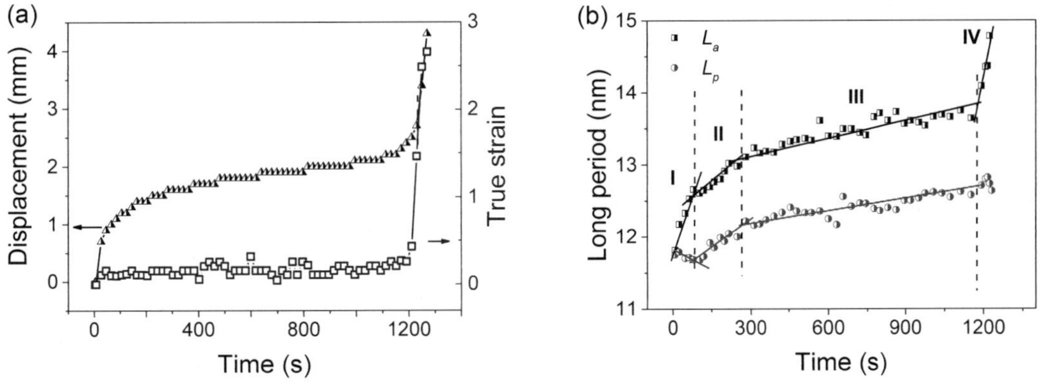

FIGURE 2: Creep curve of iPP under 8 MPa at 120 °C (a); evolution of long period L_a and L_p in and perpendicular to drawing direction during creep (b)

In region I, L_a increases linearly with the increase of creep time, whereas L_p decrease slightly. The deformation in region I belongs to elastic deformation of the amorphous phase. The amorphous regions with lower modulus compared to the crystalline regions were easier stretched. The increase of L_a suggests that the molecular chains along drawing direction are stretched [10], while the decrease of L_p means that the molecular chains are compressed in transversal direction. The compressing force comes from the crush of the expanded lamellae along drawing direction [11]. In region II and region III, the evolution of L_a and L_p show a very interesting trend. Generally, L_a increases as time increases, but the increase rate decreases from region II to region III. In the meantime, L_p increases as time increases, and the increase rate is higher in region II than that in region III. The increase of L_p in region II and III suggests that the compressing force shows up in region I has been overcome. In the work by Li [1], the elongation of micron-scale spherulites along the creep direction, accompanying the increase of nano-scale lamellar long spacing were also observed at creep strain below 17%, but the information about the evolution of microstructure perpendicular to drawing is missed unfortunately. Meanwhile, region III corresponding to secondary creep in macroscopic deformation, which is a steady state and occupies most of the creep process. So the evolution of L_a and L_p in region III, especially L_p is worthy of a more detailed study.

For thin polymer crystals a thermodynamic force capable of driving the thickening phenomena arises from the unequal free energies of the fold and lateral surfaces, the crystal moving in the direction of lowest free energy. A single molecular chain in the crystal can attain liquidlike mobility if it experiences a free energy fluctuation equal to the free energy of fusion (melting), which in the simplest approximation is given with only small error over a wide temperature range by

$$\Delta G = \Delta H(T_m^0 - T)/T_m^0 = \Delta H \Delta T / T_m^0 \quad (1)$$

where ΔH is the heat of fusion per mole of monomer and T_m^0 is the equilibrium melting point of the crystal. However, this isolated liquid like monomer unit might be unable to translate along the chain axis direction unless some of its neighbors also possess liquid like mobility; i. e., a certain degree of cooperation may be required for chain backbone motion through the crystal. If v is the minimum number of units required for this cooperative motion, then each unit must attain simultaneously the requisite free energy. The probability that v units will simultaneously suffer a free energy fluctuation of ΔG^* is proportional to the product of the individual probabilities, or $\exp(-v\Delta G/RT)$ [12],

$$\Delta G^* = \exp\left[v \Delta H \Delta T / T_m^0\right] \quad (2)$$

Obviously, the molecular chains in the amorphous region along drawing direction have a higher orientation degree in the normal of lamellae compared with that perpendicular to drawing direction, which enhances the probability for such cooperation and decreases the free energy fluctuation barrel, thus the increase rate for L_a is larger than that for L_p.

In region IV, a streak shows up in equator direction, which provides a hint that highly oriented structure may be formed in this region. In order to calculate the size of highly oriented structure, Ruland streak method [13] is used to analyze the streak in SAXS patterns. The results are given in Fig. 3.

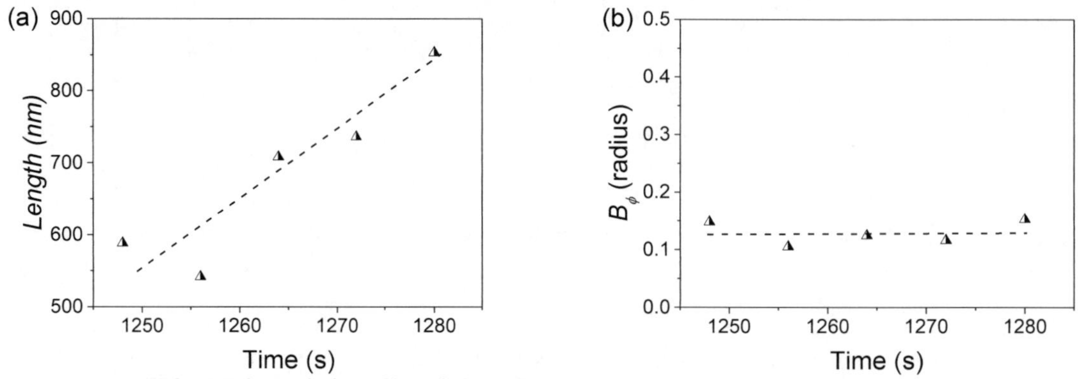

FIGURE 3: Evolution of length (a) and orientation degree B_ϕ (b) of highly oriented structure

It can be found in Fig. 3(a) that at the beginning, the length of the highly oriented structure is *ca*. 600 nm. With the increase of creep time, it grows to 870 nm finally. But the orientation degree of the highly oriented structure keeps constant, see Fig. 3(b). The formation of the highly oriented structure can be attributed to lamellae disaggregation induced molecular chains stretching along drawing direction during yielding [14].

ACKNOWLEDGMENTS

Baobao Chang thanks the Grant by China Scholarship Council. The authors acknowledge DESY, Hamburg, for beamtime and supports during the measurements.

REFERENCES

1. C. Jia, Q. Zhang, X. Liao, J. Zhu, L. Wu, K. Ni, Q. Yang, Z. An, and G. Li, *Polymer* **67**, 92-100 (2015).
2. Y. Lv, Y. Huang, M. Kong, J. Yang, Q. Yang, and G. Li, *Compos. Sci. Technol.* **96**, 71-79 (2014).
3. M. Ganß, B. K. Satapathy, M. Thunga, R. Weidisch, P. Pötschke, and A. Janke, *Macromol. Rapid. Comm.* **28**, 1624-1633 (2007).
4. K. S. Fancey, *J. Mater. Sci.* **40**, 4827-4831 (2005).
5. A. Genovese and R. A. Shanks, *Macromol. Mater. Eng.* **292**, 184-196 (2007).
6. J. Kolařík and A. Pegoretti, *Polymer* **47**, 346-356 (2006).
7. K.-H. Nitta and M. Takayanagi, *J. Polym. Sci. Polym. Phys.* **38**, 1037-1044 (2000).
8. J. Chen, K. Schneider, S. Gao, R. Vogel, and G. Heinrich, *Polymer* **76**, 182-190 (2015).
9. A. Buffet, A. Rothkirch, R. Dohrmann, V. Korstgens, M. M. Abul Kashem, J. Perlich, G. Herzog, M. Schwartzkopf, R. Gehrke, P. Muller-Buschbaum, and S. V. Roth, *J. Synchrotron. Radiat* **19**, 647-653 (2012).
10. N. S. Murthy and D. T. Grubb, *J. Polym. Sci. Polym. Phys.* **44**, 1277-1286 (2006).
11. T. Kida, T. Oku, Y. Hiejima, and K.-h. Nitta, *Polymer* **58**, 88-95 (2015).
12. I. C. Sanchez, J. P. Colson, and R. K. Eby, *J. Appl. Phys.* **44**, 4332-4339 (1973).
13. R. Perret and W. Ruland, *J. Appl. Crystallogr.* **3**, 525-532 (1970).
14. Y. Tian, C. Zhu, J. Gong, S. Yang, J. Ma, and J. Xu, *Polymer* **55**, 4299-4306 (2014).

The Effect of PDLA on the Crystallization and Morphology of PLLA

Wen-Bin Liau [a,b*], Tun-Hao Huang [a] and Shuang-Yuan Chang [b]

[a] *Institute of Polymer Science and Engineering, National Taiwan University, No. 1, Sec. 4, Roosevelt Rd., Taipei, Taiwan, R.O.C.*
[b] *Department of Materials Science and Engineering, National Taiwan University, No. 1, Sec. 4, Roosevelt Rd., Taipei, Taiwan, R.O.C.*

wbliau@ntu.edu.tw
wbliau@ccms.ntu.edu.tw

Abstract. The crystallization behaviors and morphology of PLLA with a few amount of PDLA (<10%) were studied. Besides the homo-crystallization of PLLA, PDLA and PLLA polymer chains can co-crystalliz together with 1:1 ratio to form a stereocomplex crystal (SC). The effect of SC on the crystallization behaviors and morphology of PLLA were investigated utilizing DSC, POM, and XRD. Interesting, the melting point of SC is about 50 °C higher than that of PLLA homo-crystal. Therefore, the SC was formed before the crystallization of PLLA. From the results of POM and XRD, it was found that the PLLA homo-crystals were able to be formed in three different sites: "inside of SC spherulite", "front edge of SC spherulite", and "outside of SC spherultes".

The crystallization behaviors in these sites were different. The induction time for the nucleation of PLLA in the site, "outside of SC spehrulites", was the longest among three sites. They were about the same in other sites, "inside of SC spherulite" and "front edge of SC spherulite". Obviously, SC acted as a nucleating agent and accelerated the nucleating of PLLA crystals. However, the growth rates of spherulite for three sites were about the same and slower than that of neat PLLA. Obviously, the amorphous of PLDA acted as impurity and reduced the growth rate of PLLA spherulites. Also, growth rate was not influenced by SC and resulted in the same growth rate for different sites.

Keywords: PLLA, PDLA, Sterepcplex crystal, morphology, crystallization behavior
PACS: 81

INTRODUCTION

Poly(L-lactic acid) (PLLA) has attracted much attention because of its biodergradability, biocompatibility, and good physical properties such as high strength, thermoplasticity, and fabricability.[1,2] Moreover, PLLA can be used for biomedical applications such as implants, bioresorbable sutures, tissue engineering scaffolds, and drug release systems as well as the application in packaging.[1,2] However, the hydrophobicity, stiffness, and brittleness of PLLA have limited the applications mentioned above.[3,4] To overcome these drawbacks, modifying PLLA by adding another polymer is a common way. During the modification, the biodegradability and biocompatibility must be retained. Hsu [5] reported that the incorporation of the triblock copolymer (PDLA-PEG-PDLA) into PLLA had produced a semicrystalline polymer of substantial modulus and strength. In 1987, Tsuji et al. blended the poly(D-lactic acid) (PDLA) with PLLA, and there came the result that the thermal property and crystallization behavior of the blend was different from the neat PLLA or PDLA.[6] The crystal of the blend was formed by the particular force with combining the L-form and D-form to turn into the double helix, and accumulating to become new stereocomplex crystal (SC).[7] Structure of SC is 3/1 helix with a = 0.916 nm, b = 0.916 nm, c = 0.870 nm, α = 109.28°, β = 109.28° and γ = 109.88°; besides, the characteristic peaks on the XRD are 12°, 21°, 24°. The melting point of the SC is 40℃ higher than neat PLLA or PDLA. Thus, the thermal stability could be improved. According to Tsuji et al., they pointed out that the particular force of the formation of SC was hydrogen bond, and the force (CH_3…O=C) was formed by the oxygen on the C=O and the hydrogen of CH_3 on the side group of the main chain.[8] Tsuji used the peaks of 1039cm^{-1} (C-CH_3 stretching) and 1749cm^{-1}(C=O stretching) varying with time, and he pointed out that the formation of hydrogen bond would be prior to the arrangement of SC to become order structure. The preparation of SC would be affected by the following factors: molecular weight of polymer, the blending percentage of L-form and D-form and the time of evaporating the solvent. Usually, the lower molecular weight, the blending percentage of L-form and D-form more closing to 1:1 and the longer time of evaporating the solvent, the easier the SC would be formed. In the reference, Tsuji et al. successfully allowed the high molecular weight of PLLA and PDLA to form SC by the solvent annealing. [9] It means that the particular force, which formed the SC, had already formed in the solvent state, and the particular force also needed enough time to be formed. Simultaneously, this means that SC was the most stable state on the thermodynamics. In the reference, the high molecular weight of PLA was more difficult to form the SC, and this might be due to the low mobility of PLA chain segment.[9] Same with PLLA and PDLA, SC also performed the traditional spherulites structure,[10] but both the nucleation ability and spherulite growth rate of SC were higher than neat PLLA and PDLA.[11] Under the same crystallization temperature, the spherulites density would be higher; spherulites would be smaller, and the spherulite growth rate of SC would be faster compared with neat PLLA or PDLA. In addition, SC could crystallize at higher temperature such as 160°C and higher. Under normal process, SC usually coexisted with both PLLA and PDLA crystals. There was the reference indicating that SC could act as the nucleating agent for the PLLA and PDLA to crystallize.[12] Also, Tsuji et al. added little PDLA in the PLLA, and reported that SC could act as the nucleating agent of the PLLA to accelerate the crystallization of PLLA. However, SC limiting the crystallization of PLLA also happened. [13]

The melting point of SC is higher than it of PLLA or PDLA crystals. Therefore, it is expected that SC will be the first one to be formed and PLLA or PDLA crystals will be formed lately in PLLA/PDLA blend. It is interesting to know that the PLLA or PDLA will be formed inside or outside of spherulite of SC. Also, it is interesting to know if the influence of SC on the crystallization of PLLA or PDLA depends on the sites. It is well known that the mechanical and thermal properties of blends depend on the morphology of blends. Therefore, the crystallization behaviors and morphology of PLLA/PDLA are important and need to be study more in detail. Therefore, we add little PDLA-PEG-PDLA block copolymer in PLLA and study the crystallization behavior and morphology of the blend.

EXPERIMENTAL

Synthesis of PDLA-PEG-PDLA

PDLA-PEG-PDLA block copolymer was synthesized by a ring opening polymerization. PEG (0.4 g) was added under nitrogen into a flame dried, 50 mL round-bottomed flask. The reaction flask was immersed in an oil bath and degassed under vacuum at 80 °C for 2 hours. Then, it was cooled down to room temperature and de-vacuumed. Under N_2 purging, monomer D-lactide (4 g), ethylhexanoate (Sn(Oct)2) (20 μL), and a magnetic stir bar were added into the reaction flask. Again, the reaction flask was immersed in an oil bath, degassed under vacuum at 120 °C, and sufficient time was allowed for the D-lactide to melt under the nitrogen. Finally, PDLA-PEG-PDLA block copolymer was polymerized at 120 °C under N_2 purging for 1 hour. The product was dissolved in chloroform and precipitated in a hexane/methanol (10:1) co-solvent. The solid precipitate, PDLA-PEG-PDLA block copolymer was then dried in vacuo at 80 °C for 48 h. The molecular weights of PDLA-PEG-PDLA copolymer obtained from GPC with THF as solvent, were Mn = 22000 and Mw = 28000 g/mol.

Preparation of PLLA/PDLA-PEG-PDLA Blends

Two kinds of blends with different weight ratio (PLLA/PLDA-PEG-PDLA : 95/5 and 90/10) were prepared utilizing solution casting. PLLA was used as purchased. PLLA and PLDA-PEG-PDLA were dissolved in chloroform. The film of blend was obtained after the evaporation of chloroform. The film was then dried further in vacuo at 80 °C for 2 dsys.

RESULT AND DISCUSSION

From DSC heating scan of casting film for blend (90:10), two melting points, 179 and 230 °C were observed. They are assigned as the melting of PLLA and SC respectively. Indeed, the melting point of SC is higher than it of PLLA crystal about 50 °C as reported in literature. Also, the existence of SC in casting film was able to reconfirmed using FTIR. Two characteristic peaks at 921 and 909 cm^{-1} were found. They are assigned as the characteristic peak for PLLA crystals and SC, respectively.

Table 1 shows DSC heating scan results of neat PLLA, PDLA-PEG-PDLA, and blend (90:10) after quenched from melt. T_g, T_{m1}, T_{m2}, and T_m^o are glass transition temperature, melting point of PLLA or PDLA crystal, melting point of SC, and equilibrium melting point of PLLA crystal, respectively. There is no phase separation in PDLA-

PEG-PDLA film since only one T_g was found. It indicates that PEG segments and PDLA segments are miscible. The glass transition of PDLA-PEG-PDLA is the lowest one due to the low molecular weight and low T_g of PEG segment. Also, only one T_g was observed in blend (90:10) that it means PLLA and PDLA-PEG-PDLA are miscible. Equilibrium melting point was calculated from Hoffman-Weeks plot. Obviously, SC do not have significant effect on the equilibrium melting point of PLLA.

TABLE 1. DSC heating scan results for neat PLLA, PDLA-PEG-PDLA, and blend (90:10)

	T_g (°C)	T_{m1} (°C)	T_{m2} (°C)	T_m^o (°C)
PLLA	61	180	N.A.	198
PDLA-PEG-PDLA	47	164	N.A.	N.A.
BLEND(90:10)	51	175	231	194

Isothermal crystallization of blend (90:10) was performed as following procedures. Film was heating to 250 °C and held for 3 min to release the thermal history during casting. Then, it was quenched to corresponding temperature and sufficient time was allowed to complete the crystallization. Figure 1 shows the XRD results for isothermal crystallization at 150 °C for various times. From the figure, it can be found that SC was formed already before 15 min while no or few PLLA crystals were formed. After 15 min, SC and PLLA crystal were formed simultaneously. However, the intensity increase of PLLA was faster than it of SC. The intensity of SC increased a little bit only. It reconfirmed that SC is much easier to be nucleated than PLLA at same crystallization temperature.

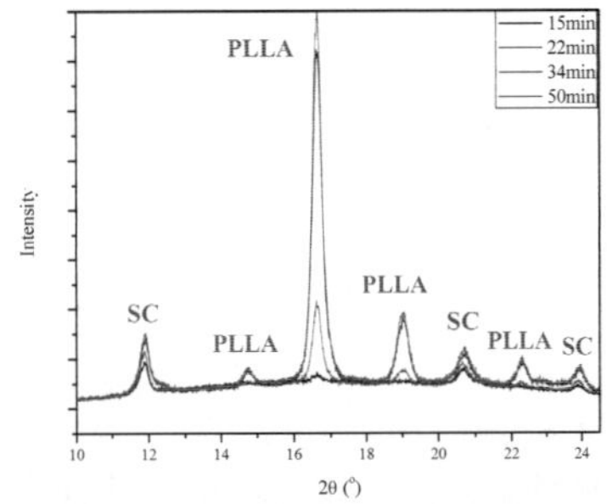

FIGURE 1. XRD results of isothermal crystallization at 150 °C for various times.

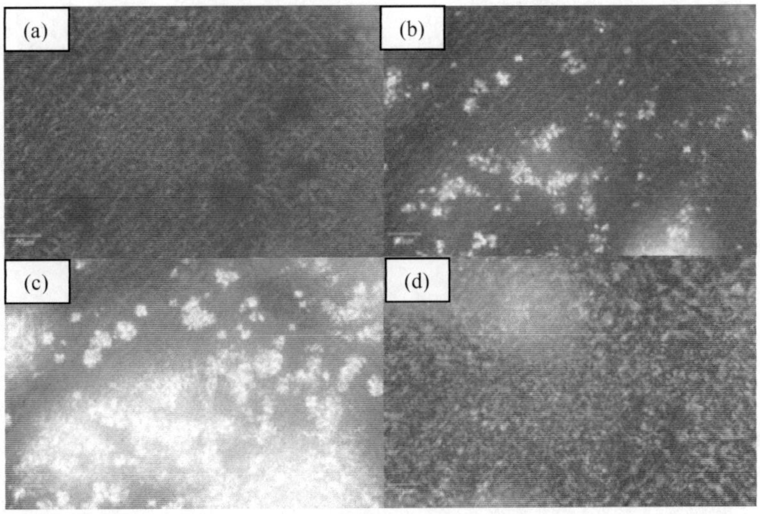

FIGURE 2. Polarized optical micrographs of blend (90:10) isothermal crystallization at 140 °C varying with time. (a) 4 min (b) 6 min (c) 7 min 30 s (d) 30 min.

Figure 2 shows the POM results of blend (90:10) isothermally crystallizing at 140 °C varying with time. From figure 2(a), it can be observed that the crystallization of SC was almost complete at 6 min and typical spherulites almost merged together. At From XRD result (not shown in here), there was very few PLLA crystal in this period. At 6 min (figure 2(b), in addition to SC spherulites, many light spots could be observed in picture. Those are the PLLA crystals. Light spots sprayed out everywhere with time. Comparing 2(a) and 2(d), it was found that the size of SC spherulites almost did not change. However, it is difficult to determine the crystallization behaviours of PLLA crystal form figure 2. Also, it is difficult to identify the sites that PLLA crystallization happens.

To understand the effect of SC crystal on PLLA crystallization in detail, the composition of PDLA-PEG-PDLA was reduced from 10 to 5 wt%. The procedures were modified as following. Isothermal crystallization was performed at 150 °C for 30 min after the releasing of thermal history at casting. In this step, the crystallization of SC was completed and some PLLA crystals were formed too. Then, the temperature was raised to 190 °C and held for 15 min. The PLLA crystal would completely melt while SC would not melt at all. Finally, blend was quenched to corresponding temperature for isothermal crystallization of PLLA. In this step, SC would not be formed any more. Figure 3 shows that PLLA could be crystallized at three different sites, outside, edge, and insdie of SC spherulites. It means PLLA crystals are formed in three different kinds of environment.

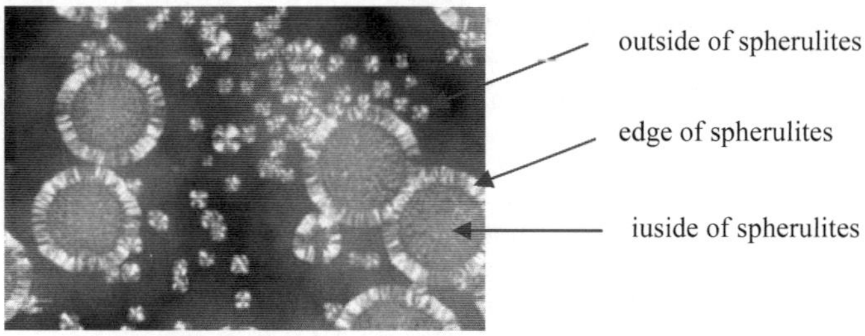

FIGURE 3. Polarized optical micrographs of blend (95:5). PLLA crystals in three different sites, outside, edge, and inside of SC spherulites

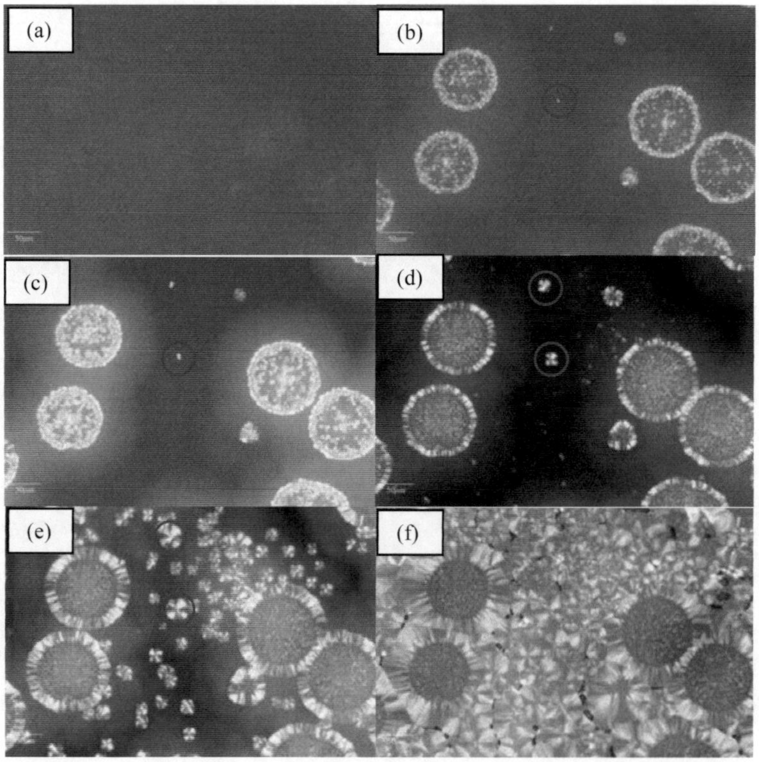

FIGURE 4. Polarized optical micrographs of blend (95:5) isothermal crystallization of PLLA at 130 °C varying with time. (a) 0 s (b) 50 s (c) 1 min 20 s (d) 3 min (e) 5 min (f) 18 min.

At the beginning of PLLA crystallization, only SC spherulites were observed (figure 4(a)). In 50 s, a lot of light spots in the edge and inside of SC spherulites were found. Only two light spots were formed outside of SC spherulites (red circle). Obviously, it shows that PLLA crystals in the edge of SC spherulites are formed earliest and easiest. The PLLA crystals outside of SC spherulites are formed latest and most difficult. The density of PLLA nuclei in the edge of SC spherulites is so high that the morphology of PLLA crystal likes tran-crystalline (columnar growing lamellae) at beginning (figure 4(d), 4(e)). Gradually it becomes typical spherulite with Maltese cross. On the other hand, the density of PLLA nuclei inside of SC spherulite is still high. However, the growth was constrained by the lamellae of SC and the size was small but distributed even inside of SC spherulite. The density of PLLA nuclei outside of SC spherulite is much low and it grows as a typical PLLA spherulite without SC.

RESULT AND DISCUSSION

In PLLA/PDLA-PEG-PDLA blend (90:10), SC is formed first and PLLA crystal is formed lately. The crystal growth is constrained by the SC lamellse and the size of PLLA is not able to be large. In PLLA/PDLA-PEG-PDLA blend (95:5), SC is able to be formed first and complete before the crystallization of PLLA in some condition. In such way, one can observe that SC likes the nuclei agent to accelerate the nucleation of PLLA. Formation of PLLA crystals occurs in three different sites. The nucleation of PLLA occurs earliest and easiest in the edge of SC spherulite. On the other hand, it occurs latest and most difficult outside of SC spherulite. There are two types of spherulites. One is typical spherulite that compose of PLLA lamellae. The other is core-shell type. The core composes of SC lamellae and even distributed PLLA crystal. The shell composes of PLLA lamellae.

ACKNOWLEDGMENTS

We gratefully acknowledge financial support from the National Science Council, Taiwan, ROC, through Grant No. MOST-103-2221-E-002-279.

REFERENCES

1. L.T. Lim, R. Auras, and M. Rubino, *Prog. Polym. Sci.* **33**, 820 (2008).
2. L.S. Nair and C.T. Laurencin, *Prog. Polym. Sci.* **32**, 762 (2007).
3. W. Tong, H. Yong, F. Zhongyong, W. Jia, and L. Suming, *Polym. Eng. Sci.*, **48**, 425 (2008).
4. Y. Hu, Y. S. Hu, V. Topolkaraev, A. Hiltner, and E. Baer *Polymer* **44**, 5711 (2003).
5. S. Rathi, X. Chen, E.B. Coughlin, S.L. Hsu, C.S. Golub, and M.J. Tzivanis *Polymer* **52**, 4184 (2011).
6. Y. Ikada, K. Jamshidi, H. Tsuji, and S. H. Hyon, *Macromolecules* **20**, 904 (1987).
7. H. Tsuji, *Macromol. Biosci.* **5**, 569 (2005).
8. J. Zhang, H. Sato, H. Tsuji, I. Noda, and Y. Ozaki, *Macromolecules* **38**, 1822 (2005).
9. H. Tsuji, S.H. Hyon, and Y. Ikada, *Macromolecules* **24**, 5651 (1991).
10. H. Tsuji and Y. Ikada, *Macromolecules* **26**, 6918 (1993).
11. H. Tsuji and Y. Tezuka, *Biomacromolecules* **5**, 1181 (2004).
12 H. Yamane and K. Sasai, *Polymer* **44**, 2569 (2003).
13. H. Tsuji, M. Katagiri, and J. Narita, *Macromol. Mater. Eng.* **296**, 887 (2011).

Decrosslinking of Peroxide Crosslinked HDPE in Ultrasonic Twin Screw Extruder: Simulation and Experiment

Avraam I. Isayev and Keyuan Huang

Department of Polymer Engineering, The Universaity of Akron, Akron, Ohio 44325-0301, USA

aisayev@uakron.edu

Abstract. A model for ultrasonic decrosslinking of peroxide crosslinked HDPE (XHDPE) is developed. The ultrasonic bubble cavitation in a viscoelastic solid containing small concentration of voids is considered. Ultrasonic power consumption and acoustic pressure in a bubble filled polymer are calculated using the linear acoustic analysis. The rupture of network is modeled by considering the breakage of main chains caused by bubble oscillations without assigned kinetic parameters. The nonisothermal flow of the decrosslinked XHDPE (DXHDPE) in the ultrasonic treatment zone in TSE is modeled based on a helical flow of a power-law fluid with parameters dependent on the gel fraction. The distribution of temperature, velocity, gel fraction, crosslink density and pressure in the ultrasonic treatment zone in TSE is calculated. The simulation results are compared with experimental data. A quantitative agreement for ultrasonic power consumption and a qualitative agreement on the dependence of gel fraction and crosslink density on ultrasonic amplitude are achieved.

Keywords: Extrusion, Recycling, Ultrasound, Polyethylene, Modeling, Experiment
PACS: 82. Physical chemistry and chemical physics

INTRODUCTION

Extensive experimental investigations of ultrasonic decrosslinking of XHDPE are carried out [1-4]. It is found that ultrasonic decrosslinking is a promising technology to obtain melt-processible materials exhibiting excellent mechanical performance. To optimize and scale up the ultrasonic decrosslinking of XHDPE, a science based technology is required. Thus, it is important to develop a model to describing the ultrasonic decrosslinking process of XHDPE.

Simulations of ultrasonic devulcanization of rubber vulcanizates under static [5-7] and continuous conditions [8-11] are carried out. The simulations allow one to calculate the distribution of pressure, velocity, temperature, viscosity, gel fraction and crosslink density in the ultrasonic treatment zone. Comparison of the simulation with the experiments shows a qualitative agreement for ultrasonic devulcanization of rubber under continuous condition [8-11] and quantitative agreement under static condition [7]. However, these simulations describe the ultrasonic devulcanization of vulcanizates as a kinetic process with kinetic parameters are either assumed or determined by the best fit to the experimental data. Since the dependence of the kinetic parameters on the ultrasonic amplitude is a priori unknown, the potential for the simulation of ultrasonic devulcanization is limited. The aim is to develop simulation model of ultrasonic decrosslinking of XHDPE in TSE without use of kinetic parameters and compare the simulation and experimental data.

THEORETICAL

The ultrasonic decrosslinking of XHDPE using TSE is described earlier [3, 4]. The polymer flows through the ultrasonic treatment zone consisting of the ultrasonic horn and co-rotating cylindrical elements with smooth surfaces. A curved surface of the ultrasonic horn provides a uniform gap between the surfaces. Due to the high aspect ratio of the circumferential and axial length to the thickness of ultrasonic treatment zone, the flow is approximated as a helical flow. The horn provides a longitudinal ultrasonic wave propagating in the direction perpendicular to the axial direction causing decrosslinking of XHDPE. Modeling of ultrasonic decrosslinking of XHDPE in TSE includes four aspects: (1) imposition of ultrasound induces the dynamics of bubble cavitation in the polymer; (2) propagation of ultrasound generates the energy dissipation; (3) induced bubble oscillations rupture the crosslinked network; (4) viscous and ultrasonic dissipations lead to nonisothermal flow with the viscosity of DXHDPE greatly affected by gel fraction and crosslink density.

Bubble Dynamics

Modeling of the bubble oscillation in a viscoelastic medium subjected to ultrasound was studied earlier [5] using a constitutive equation. Various constitutive equations including Lodge [12], Upper Convected Maxwell [13,14], K-BKZ [15] models are employed to predict the bubble dynamics in the medium. For the simplicity, in the present study, the viscoelasticity of medium was described by the Lodge model [12]. A gas bubble in a non-deformed polymer is at equilibrium. A force balance at the interface between the bubble and polymer is introduced and the equation of motion governing the dynamics of the bubble in the polymer is formulated with the adiabatic thermodynamic behavior. Due to the action of ultrasound, the pressure in the polymer is time dependent:

$$P_{\infty}(t) = P - P_A \sin \omega t \qquad (1)$$

where P and P_A are the hydrostatic and acoustic pressure, respectively, and ω is the ultrasonic frequency. The linear viscoelasticity of the polymer is approximated by a standard viscoelastic solid with the relaxation modulus:

$$G(t) = (G_u - G_N^0)e^{-\frac{t}{\lambda}} + G_N^0 \qquad (2)$$

where G_u is the elastic modulus in the glassy state, λ is the relaxation time and G_N^0 is the storage modulus of PE in the rubbery plateau. The λ is determined from the shear loss tangent at the ultrasonic frequency of 40 kHz. Substitution of $G(t)$ from Eq. (2) into equation of motion provides a system of the ordinary differential equations, governing the bubble dynamics using auxiliary functions [6].

Modeling of Ultrasonic Power Consumption

The ultrasonic power consumption during the propagation of ultrasound is due to the energy loss in a viscoelastic solid. In the present case, the propagation of ultrasound in a thin layer of XHDPE is considered, since the thickness of the ultrasonic treatment zone (2.5 mm) is small in comparison with the wavelength of the ultrasound. The sound impedance of the metal is by several orders of magnitude higher than that of the polymer. A strong reflection of ultrasound occurs at the interface. An earlier model for calculation of the ultrasonic power consumption in rubbers under static conditions was based on the propagation of ultrasound in a thin layer with small amount of bubbles [5].

Modeling of Decrosslinking

Due to the ultrasound induced bubble oscillations, the main chains in XHDPE around the bubble are deformed. For a material point, the distance from the center of a bubble ($R \geq R_0$) is determined [6]. The change of the local length caused by the bubble dynamics is characterized by the stretch tensor. Since the expansion and contraction motion of a bubble is equal-biaxial and XHDPE is incompressible, the explicit form of the stretch tensor is used [6]. The affine deformation of the main chains between the crosslinks around a pulsating bubble is assumed with the end-to-end vector in its equilibrium state. The end-to-end vector in the deformed state is evaluated. The breakage of the polymer chains between the crosslinks occurs when the magnitude of the end-to-end vector in the deformed state is larger than the contour length. The contour length between crosslinks is not uniquely defined, but obeys certain distribution. The fraction of the broken polymer chains at a distance from the center of an oscillating bubble is determined [6]. The probability distribution of the polymer chains between crosslinks is approximated by the Gaussian distribution. If the crosslinks remain intact and only breakage of main chains between crosslinks occurs, the normalized crosslink density of the gel after one ultrasonic cycle is found. It should be noted that XHDPE in TSE experiences not only ultrasonic decrosslinking, but also mechanical decrosslinking caused by the rotation of the screws [2-5]. Due to mechanical decrosslinking, the gel fraction of the material entering the ultrasonic treatment zone in TSE is lower than that of the material fed to TSE [3]. Therefore, the ratio of the crosslink density of DXHDPE in ultrasonic zone to that of entering the ultrasonic zone is modified to take into account the gel fraction of the material entering the ultrasonic treatment zone ζ_0:

$$\frac{\nu_{DXHDPE}}{\nu_0} = 1 - \zeta_0 \frac{8\pi}{3^{5/2}} \frac{x_{\max}^3 - x_{\min}^3}{1 + \frac{3N}{2}} \left| \bar{n} R_0^3 \ln(\bar{n} R_0^3) \right| \qquad (3)$$

Using Eq. (3) the normalized crosslink density of DXHDPE is calculated within one period of oscillation (t=2.5x10^{-5} s). The model is modified to incorporate the effect of number of cycles within the residence time. According to the assumption that only breakage of the main chains occurs during the ultrasonic decrosslinking of XHDPE, the gel fraction of DXHDPE at the end of ith period is calculated using the Horikx function describing the normalized crosslink density vs. the normalized gel fraction [16]:

$$1-\frac{\nu_{mech+ultra}(i)}{\nu_0}=1-\frac{(1-(1-\zeta_{mech+ultra}(i))^{1/2})^2}{(1-(1-\zeta_0)^{1/2})^2} \quad (4)$$

Thus, the gel fraction and crosslink density of DXHDPE is determined by using the Eqs. (3) to (4).

It should be noted that decrosslinking of crosslinked network assumes that only breakage of main chains occurs. From the experimental observations [1-4], this assumption is not necessary true. It is convenient to model the breakage of crosslinks based on the affine deformation. However, the model predicts that more than 90% crosslinks are broken within 10 cycles, significantly over-predicting the breakage of crosslinks and contradicting the experimental observations [1-4]. The reason of the over-prediction is the assumption of the affine deformation. Since the crosslinks in XHDPE are C-C bonds, the crosslinks are easily broken even under the imposition of small affine deformation. Thus, the criterion for the breakage of crosslinks has to be altered to not easily brake them. Unfortunately, from the classical polymer physics, no an explicit criterion is available. Thus, the modeling of the breakage of crosslinks is not addressed here.

Modeling of Nonisothermal Flow

For the flow simulation, a constitutive equation is needed. Lodge model is unable to predict the shear thinning of polymer under shear flow [17]. A strong shear thinning is observed in DXHDPE [2]. Therefore, the flow behavior of DXHDPE is described by a power-law fluid model with an Arrhenius temperature dependent viscosity:

$$\eta = K\dot{\gamma}^{n-1} = Ae^{T_b/T}\dot{\gamma}^{n-1} = Ae^{E/RT}\dot{\gamma}^{n-1} \quad (5)$$

where η is the shear viscosity, $\dot{\gamma}$ is the shear rate, K is the consistency, n is the power-law index, T_b is the material parameter, R is the gas constant and E is the activation energy of viscous flow. Parameters K, n and T_b are taken to be dependent on the gel fraction:

$$K = K(\zeta),\ n = n(\zeta),\ T_b = T_b(\zeta) \quad (6)$$

The explicit forms of Eq. (6) are determined by fitting to the experimentally measured viscosity. The flow in the ultrasonic treatment zone is a nonisothermal helical flow. The energy equation incorporates the rate of energy consumption due to the breakage of main chains and the rate of the ultrasonic energy dissipation. The concentration of the broken main chains in DXHDPE is equal to the decrease of the crosslink density.

Simulation of Ultrasonic Decrosslinking

Based on the above models, the simulation algorithms describing ultrasonic decrosslinking are developed. These algorithms include ultrasonic decrosslinking under static conditions without thermal degradation and ultrasonic decrosslinking in TSE without and with the thermal degradation. First, tthe ultrasonic decrosslinking of XHDPE under static conditions is considered. The difference between the ultrasonic decrosslinking in TSE and under static conditions is the absence of flow in the latter. The propagation of ultrasound in a polymer is a dynamic process accomplished by the local segmental motion. In contrast, the flow of a polymer melt occurs by the reptation of molecules which is a dynamic process of a larger scale. In fact, the measurement of sound velocity in HDPE melts at an ultrasonic frequency shows that the shear flow has an insignificant effect on the sound velocity and longitudinal loss tangent [18]. Accordingly, one can conclude that the ultrasonic decrosslinking under static conditions exhibits a great deal of similarities with that in TSE and may serve as a good starting point for simulation. Thus, simulation of the ultrasonic decrosslinking in TSE is approximated as a combination of many decrosslinking steps under static conditions.

EXPERIMENTAL

Materials

HDPE used was a rotational molding grade, 35 mesh powder (Paxon 7004, Exxon Mobil, Baytown, TX). Crosslinking agent was dicumyl peroxide in the pellet form (DC-40, Akrochem, Akron, OH). It contained 40 wt% peroxide and 60 wt% calcium carbonate as a carrier. Tetrakis [methylene (3,5-di-tert-butyl-4-hydroxyhydro cinnamate)] methane (Antioxidant 1010, Akrochem, Akron, OH) was used as a stabilizer during the swelling and rheology test. Xylene (mixture of isomers, ACS Reagent Grade, Sigma-Aldrich, WI) was used as a solvent in the swelling test. Details of preparation and decrosslinking of XHDPE were reported elsewhere [3,4].

RESULTS AND DISCUSSION

Figure 1 shows the predicted relative bubble radius as a function of dimensionless time at hydrostatic pressures of 0.5, 1, 2 and 3 MPa, an ultrasonic amplitude of 10 μm and a bubble volume fraction of 0.67%. It is seen from Figure 1 that the initial relative bubble radius decreases with the hydrostatic pressure. Also, after the bubble dynamics attains the periodic steady state both the maximum and minimum relative bubble radii decrease with the hydrostatic pressure. In order to quantify how the intensity of bubble dynamics vary with the hydrostatic pressure, the variation of bubble volume with respect to its initial volume, i.e. $x^3_{max} - x^3_{min}$, is considered. The values of $x^3_{max} - x^3_{min}$ at hydrostatic pressures of 0.5, 1, 2 and 3 MPa are 1.221, 1.1719, 1.0616 and 0.9405, respectively. Clearly, an increase of the hydrostatic pressure leads to an insignificant decrease of the intensity of bubble dynamics.

Figure 2 shows the calculated gel fraction (a) and crosslink density (b) of DXHDPE as a function of the treatment time at ultrasonic amplitudes of 5, 7.5 and 10 μm, a hydrostatic pressure of 1 MPa and a bubble volume fraction of 0.67%. It is seen from Figure 2 that the gel fraction and crosslink density of DXHDPE decreases with the treatment time and the ultrasonic amplitude.

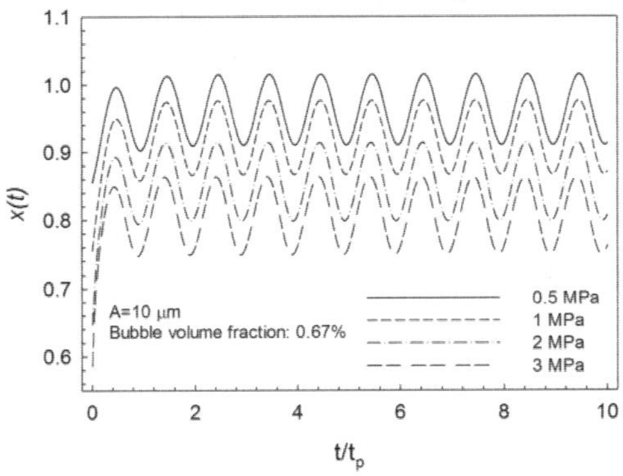

Figure 1. Predicted relative bubble radius as a function of dimensionless time at various hydrostatic pressures of 0.5, 1, 2 and 3 MPa at an ultrasonic amplitude of 10 μm and a bubble volume fraction of 0.67%.

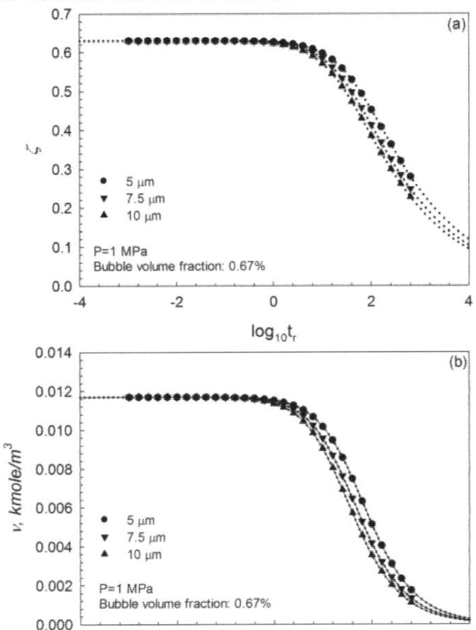

Figure 1. Predicted gel fraction (a) and crosslink the density (b) of ultrasonically decrosslinked XHDPE as a function of the treatment time at ultrasonic amplitudes of 5, 7.5 and 10 μm at a hydrostatic pressure of 1 MPa and a bubble volume fraction of 0.67%. Lines represent fittings by Eqs. (3) and (4).

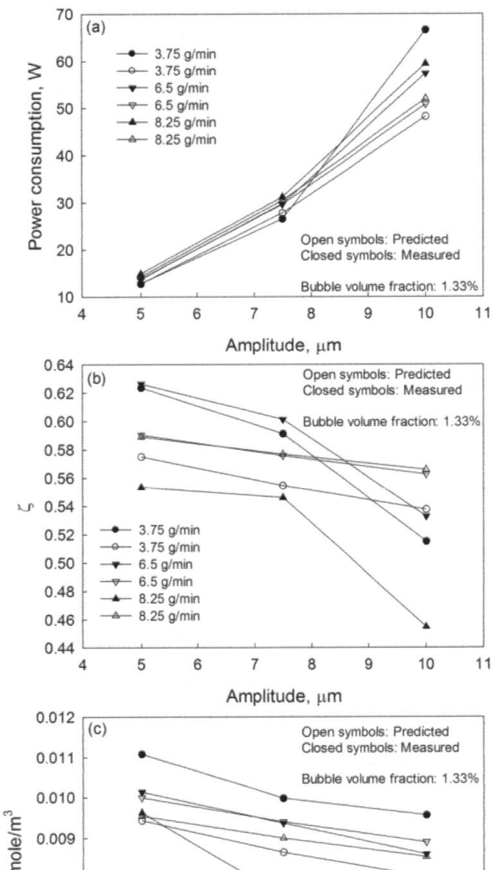

Figure 3. Measured and predicted ultrasonic power ultrasonically consumption during ultrasonic decrosslinking of XHDPE (a), gel fraction (b) and crosslinked density (c) of the ultrasonically decrosslinked XHDPE as function of the ultrasonic amplitude at flow rates of 3.75, 6.5 and8.25 g/min at a bubble volume Fractions of 1.33%.

Figure 3 shows the measured and predicted ultrasonic power consumption (a), gel fraction (b) and crosslink density (c) of DXHDPE at flow rates of 3.75, 6.5 and 8.25 g/min as functions of the ultrasonic amplitude. Figure 3a shows a fair agreement between the predicted and measured ultrasonic power consumption at all flow rates. However, the model cannot provide a quantitative agreement between the predicted and measured gel fraction and crosslink density of DXHDPE at flow rates of 3.75 and 8.25 g/min. It is seen from Figure 3c that the crosslink density of DXHDPE at a flow rate of 3.75 g/min is over-predicted, although the slope of the amplitude dependence is correct. The gel fraction of DXHDPE at a flow rate of 3.75 g/min is also over-predicted at amplitudes 5 and 7.5 μm, as seen from Figure 3b. It is seen from Figure 3c that the model correctly predicts the crosslink density of DXHDPE at a flow rate of 8.25 g/min and an amplitude of 5 μm, but under-predicts at amplitudes of 7.5 and 10 μm. Also, as seen from Figure 3b, the measured gel fraction of DXHDPE at a flow rate of 8.25 g/min is lower than the predicted one.

CONCLUSION

The numerical simulation and experimental study of ultrasonic decrosslinking of XHDPE under static condition and in TSE is carried out. The present model simulates the ultrasonic decrosslinking of XHDPE using only assigned bubble radius and volume bubble fraction. Comparison between the simulated and experimental results indicates a quantitative agreement for the ultrasonic power consumption, but a qualitative agreement with the gel fraction and crosslink density. A quantitative agreement on the gel fraction and crosslink density can be achieved by adjusting the bubble volume fraction. Also, incorporation of the thermal degradation into the simulation improves the prediction at high amplitudes. A future work should be directed toward modeling of ultrasonically induced bubble nucleation and crosslinks breakage.

ACKNOWLEDGEMENTS

Financial support for this work by the NSF Division of Engineering under Grant CMMI-1131342 is greatly appreciated. Authors also wish to thank the ExxonMobil Chemical Company for providing HDPE and the Akrochem Corporation for providing peroxide.

REFERENCES

1. A.I. Isayev, J. Jenkins, *SPE ANTEC Technical Papers*, 53, 109 (2008).
2. A.I. Isayev, K. Huang, *Poly. Eng. Sci.*, 54, 2715 (2014).
3. K. Huang, A.I. Isayev, J. Zhong, *J. Appl. Polym. Sci.*, 131, 40680 (2014).
4. K. Huang, A.I. Isayev, *RSC Adv.*, 4, 38877 (2014).
5. V.V. Yashin, A.I. Isayev, *Rubber Chem. Technol.*, 72, 741 (1999).
6. V. V. Yashin, A.I. Isayev, *Rubber Chem. Technol.*, 73, 325 (2000).
7. V. V. Yashin, C.K. Hong, A.I. Isayev, *Rubber Chem. Technol.*, 77, 50 (2004).
8. A.I. Isayev, S.P. Yushanov, J. Chen, *J. Appl. Polym. Sci.*, 59, 803 (1996).
9. A.I. Isayev, S.P. Yushanov, J. Chen, *J. Appl. Polym. Sci.*, 59, 815 (1996).
10. A.I. Isayev, S.P. Yushanov, S.H. Kim, V.Y. Levin, *Rheol. Acta.*, 35, 616 (1996).
11. A.I. Isayev, S.P. Yushanov, S.H. Kim, *Rubber Chem. Technol.*, 71, 168 (1998).
12. A.S. Lodge, *Trans. Faraday Society*, 52, 120 (1956).
13. J. Naude, F. Mendez, *J. Non-Newton. Fluid*, 155, 30 (2008).
14. A.C. Panpanastasiou, L.E. Scriven, C.W. Macosko, *J. Non-Newton. Fluid*, 16, 53 (1984).
15. M. Doi, S.F. Edward, The theory of polymer dynamics, Oxford University Press, New York (1986).
16. M.M. Horikx, *J. Polym. Sci.*, 19, 445 (1956).
17. R.G. Larson, Constitutive Equations for Polymer Melts and Solutions, Butterworths, MA (1988).
18. R. Gendron, J. Taibouet, J. Guevremont, M. Dumoulin, L. Piché, *Poly. Eng. Sci.*, 35, 79 (1995).

Manufacturing and Characterisation of Thermoplastic Composite of Hemp Shives and recycled Expanded Polystyrene

ALMUSAWI[a,b] A., LACHAT[a] R., ATCHOLI[a] K.E. and TIÉ BI S.[a]

[a] *Research Institute on Transport, Energy and Society ("IRTES-M3M"), University of Technology of Belfort-Montbéliard ("UTBM"), 90010 Belfort City, FRANCE.*

[b] *Ph.D. candidate, assistant teacher in Faculty of Civil Engineering, Al-Qadissyah University, IRAQ.*

Presenting author's email: aqil.almusawi@utbm.fr

Reference author's email: (remy.lachat ; kokou-esso.atcholi ; salima.tie-bi) @utbm.fr

Abstract. There are two main problems caused by the huge volume of expanded polystyrene (Styrofoam) waste: the large amount of space that would be needed if it were to be disposed of in landfills and the excessive transport cost if it were to be recycled or incinerated. Consequently, the solution to these problems is the execution of two processes: the first one is the process of recycling the polystyrene with acetone, which allows us to separate the polystyrene from the other wastes, thereby reducing the volume by more than 98%. Then the resulting fresh polystyrene paste is used in the second process as a binder with 60 wt % of hemp shive to produce lightweight (porous) biobased-thermoplastic composites. The hemp shive particles significantly reduce the time of acetone evaporation to be three hours instead of three weeks for drying the net polystyrene paste. As a result, the recycling and the mixing processes produce a non-thermoformed lightweight composite. While the thermoforming process has been adapted to improve the mechanical properties of this composite, this ductile composite has specific mechanical properties that are better than those of the brittle net recycled polystyrene.

Keywords: Polystyrene; Reinforced thermoplastics; Characterisation; Hemp Shive.

PACS: (62. 20.-x); (65. 60. +a)

1. INTRODUCTION

The greatest challenge of the recycling process is obtaining high quality recycled plastic. The high purity and the compatibility between the different types of plastic are forcing us to pass through many steps (collection, sorting, chipping, washing, and pelleting). [1][2]

The polystyrene waste represents the worst negative impact on the environment due to its low price (high commercial productions and few recycling companies) which will increase its huge quantity in municipal waste. A number of researchers have suggested dissolving the polystyrene by many organic solvents [3] and acetone is one of the best solvents [4], allowing us to eliminate most of the recycling steps, especially for expanded polystyrene (EPS), which has huge volume. Acetone has low health hazards, is quite solvent (reducing the volume of the EPS by 98.5%), and satisfies the economical collection step from the point of view of waste management [5]. This process is able to separate the polystyrene from the other waste: plastics, labels, oils, colors, etc., because the acetone breaks all the cross-links (Vander-Waals bonds) between the PS molecules. On the other hand, the acetone can be recycled due to its volatility and its immiscibility with water [6]. The major problem with this recycling process is the time of the acetone evaporation because the hardening of the exterior layer of polystyrene paste restricts and retards the acetone evaporation. Schmidt [4] investigated that the mechanical properties of dissolved polystyrene with acetone were not adequate due to the air bubbles and the brittleness characteristic. In our study, the mechanical properties were improved when the dissolved polystyrene (DisEPS) was reinforced with hemp shive particles. Moreover, the minimum fresh polystyrene paste percentage (40 wt %) was mixed with maximum hemp shive content (60 wt %) to produce biobased-thermoplastic composite (DisCOMP 60). This polystyrene proportion was less than that used by Poletto [7]. DisCOMP 60 has many advantages: hardening without the thermoforming process, low price, recyclability and partial biodegradability. In Borsoi's work [8], the biodegradability of polystyrene composite depended on the cotton load.

This paper explores how to produce lightweight biobased-thermoplastic composite. Accordingly, the hemp shive particle has been chosen because it contains about 78% of air voids [9]. This particle is very hydrophilic due to its nature [8] [9]. Consequently, it will absorb the acetone from the fresh polystyrene paste and reduce the hardening time. Therefore, the effect of heating and the thermoforming pressure was studied to investigate the specific mechanical properties, water-absorption and the thickness swelling of a composite of 60 weight % hemp shive.

2. MANUFACTURING PROCESS

2. 1. Materials

The **expanded polystyrene,** which had a density of (0.015 g/cm^3), was collected and then dissolved with acetone. The saturation percentage of acetone was 81 wt % (acetone/polystyrene), the result being a viscous liquid of polystyrene with a density of (0.96 g/cm^3). This means that the volume was reduced by (98.4 %). This commercial **acetone** was purchased from ARDEA France Ltd.

The **hemp shive particles** were purchased from CHANVRALIT France Ltd, having a bulk density of (0.121 g/cm^3). These coarse particles were washed with clean water and then air-dried at (100°C) for 24 hours. After that, they were ground by the universal cutting mill (PULVERISETTE 19) to obtain a particle size of (0.5 to 1.4 mm), this product having a bulk density of (0.171 g/cm^3). The grinding process was necessary to reduce the inter-particle voids by (30%) and the agglomeration of the particles during the mixing process.

2. 2. Compounding process

The composite of 60 wt % hemp shive with dissolved polystyrene (DisCOMP 60) was chosen because of this mixture has minimal acetone percentage. The other mix proportions had higher acetone percentages; for example, in the composites of more than 60 wt % hemp shive, a lot of additional acetone was needed to wet the hemp shive particles (150 wt % acetone/hemp shive). This technique was necessary to increase the workability of these mixtures during the mixing process and to prevent the fast setting of the paste. While, the composites of less than 60 wt % of hemp shive, the acetone percentages increased proportionally when increased DisEPS paste percentage in the mixture. Accordingly, DisCOMP 60 was mixed manually within 5 minutes and then pre-formed by cold mold pressure of 5 MPa for 5 minutes to reduce the inter-particle voids, and then the matrix between the particles was redistributed.

2. 3. Acetone evaporation and hardening time

The hemp shive particles significantly reduced the hardening time of this composite because the hardening time depends on the acetone evaporation from fresh DisEPS paste. In contrast, the hardening time of net DisEPS paste takes about 21 days [5]. Therefore, an approach study about the acetone evaporation was done to determine the acetone evaporation at different times for net DisEPS paste and DisCOMP 60, as shown in Figure (1).

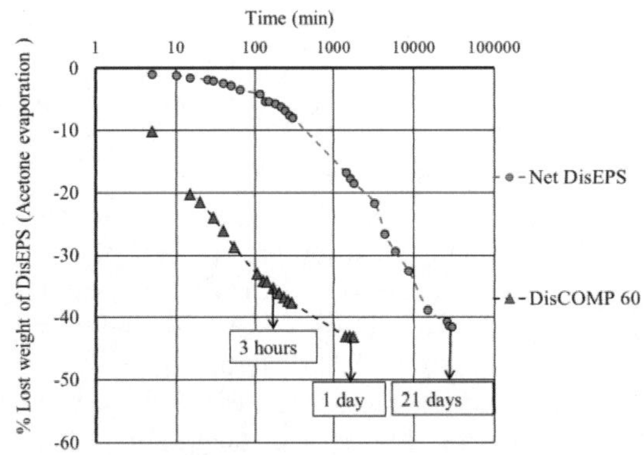

FIGURE 1. Acetone evaporation from fresh DisEPS paste versus the time.

We observed that after three hours, the fresh DisEPS paste of DisCOMP 60 lost about 35% of its weight (about 70% acetone) at the ambient temperature of 20±2°C. At this step, the composite has sufficient handling resistance and most of the acetone (≈95%) will evaporate after 24 hours, while the sample of net DisEPS (0 % hemp shive) evaporates the acetone very slowly compared with DisCOMP 60. It takes about 21 days, this result being in accordance with that obtained by Mwasha [5]. Interestingly, the acetone represents about 45% of net DisEPS.

3. PHYSICAL-MECHANICAL PROPERTIES

These biobased-thermoplastic composites of (50 x 50 x (2.75±0.25)) mm^3 were studied; firstly, without of thermoforming process and secondly when thermoformed them at 130°C for 30 minutes. Moreover, the thermoforming pressure was changed to be (0.5, 2.5, 5 and 20 MPa) in order to investigate the pressure effect on the physical and the mechanical properties.

3. 1. The Density and the Voids Content

The apparent composites' densities were measured by dividing their sample weights by their volumes, as seen in Figure (2.a). Then, the voids content of the composites were calculated according to ASTM (D 2734), for this reason the theoretical density of DisCOMP (without voids) was calculated by using the rule of mixture, it was about (1.308 g/cm3), considering that the absolute density of polystyrene and hemp fiber were (1.05 g/cm^3) [10] and (1.48 g/cm^3) [11], respectively.

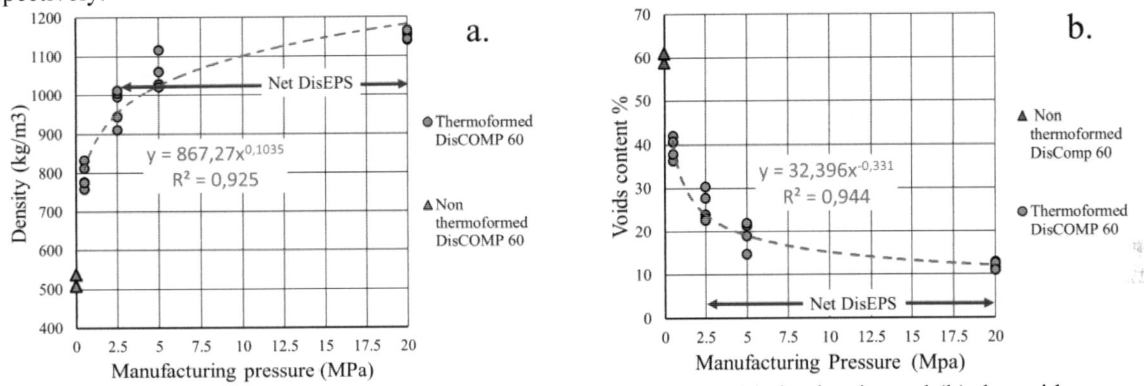

FIGURE 2. Effect of thermoforming and manufacturing pressure on (a) the density and (b) the voids content.

The non-thermoformed composite had a minimum density, then the density jumped due to temperature effects and light pressure of (0.5 MPa). This behavior can be explained by the collapse of the small gas bubbles inside of dried DisEPS, as shown in Figure (5.c). After that, the density increased more and more when the manufacturing pressure was increased because the hemp shive particles are very compressible materials. On the contrary, the voids content (intra-particle and inter-particle) decreased when this composite was thermoformed at (130°C) and when the manufacturing pressure was increased.

3. 2. Flexural Strength and Modulus of Elasticity

A flexural test of a sample of (50 x 10 x (2.5 to 3)) mm^3 at 3-point of load was achieved according to ASTM (D 790, ISO 178) to investigate the mechanical properties of these composites by using the universal tester (TA-XT2i) of cell load of (500N) and crosshead speed of (6 mm/min). The ultimate flexural strength and the modulus of elasticity were calculated and illustrated in Figure (3).

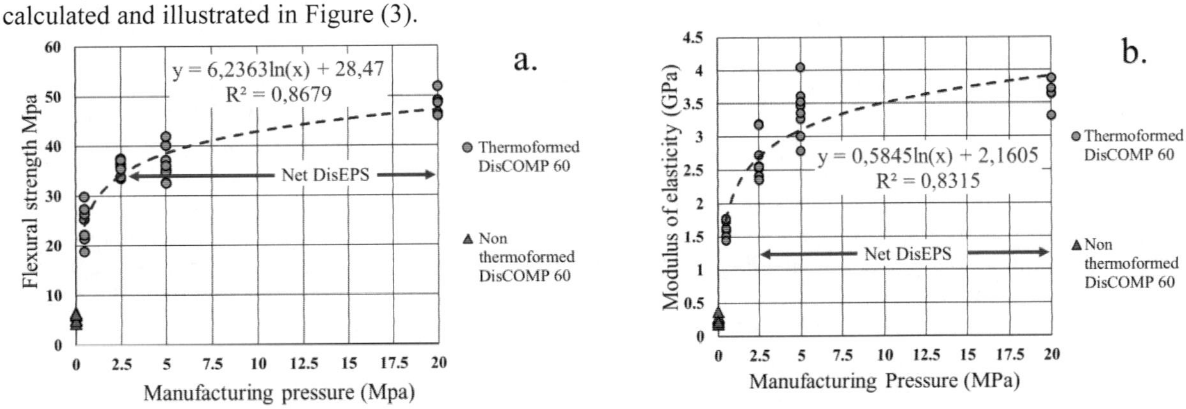

FIGURE 3. Effect of thermoforming process and manufacturing pressure on DisCOMP 60% (a) flexural strength and (b) modulus of elasticity.

The modulus of rupture (MOR) and the modulus of elasticity (MOE) jumped when this composite was thermoformed in accordance with the density. While the pressure represents the second principle factor to modify the mechanical properties when reducing the composite porosity. As a result, the composites (DisCOMP 60) were thermoformed by pressure of more than (2.5 MPa) had good specific mechanical properties, which were better than the net DisEPS as listed in Table (1).

3. 3. Water-Absorption and Dimensional Stability

All the composites of (50 x 50 x (2.5 to 3))mm^3 were immersed in water for 24 hours, according to ASTM (D 570-98, ISO 62) at a temperature of 20±2°C, then they were weighted to calculate the water-absorption, and their dimensions were measured to calculate the thickness swelling, as listed in Table (1).

For DisCOMP 60%, the water-absorption decreased when the composite was thermoformed and when the thermoforming pressure was increased; it was proportionally dependent on the voids content and the surface absorption. While the thickness swelling took on approximately the same value for all the composites. However, these composites lost their surface textures due to the particles swelling, whereas, the interior particles were either well enveloped by the matrix, or the interior voids content would represent a refuge space for the expanded particles.

TABLE 1. Physical and mechanical properties of net recycled expanded polystyrene and hemp shive-thermoplastic composites.

	Composite type	Density (kg/m^3)	Voids content (%)	Specific MOR (MPa)	Specific MOE (GPa)	Water absorption (%)	Thickness swelling (%)
	Net DisEPS	1017.75	3	33.63	1.263	0	0
DisCOMP 60	Non-thermoformed	523,09	60	10,73	0,518	40.48 ± 5.41	3.64 ± 1.82
	Thermoformed at 0.5 MPa	791,22	39	30,55	2,091	22.47 ± 5.47	3.04 ± 0.72
	Thermoformed at 2.5 MPa	973,73	25	36,57	2,758	13.88 ± 3.06	5.08 ± 1.32
	Thermoformed at 5 MPa	1057,14	19	34,32	3,199	9.18 ± 1.25	4.34 ± 0.44
	Thermoformed at 20 MPa	1154,10	11	41,91	3,151	7.83 ± 3.25	4.89 ± 1.20

3. 4. SEM Analysis

Scanned electron micrographs of EPS, hemp shive particles, and thermoformed DisEPS are represented in Figure (5). The very large air void explains the low density of expanded polystyrene, as shown in Figure (5.a), therefore the acetone reduced the volume by more than 98.4 %. In the other hand, the fresh DisEPS paste enveloped the hemp shive particles, and the particles tried to absorb the acetone during the mixing process, as shown in Figure (5.b). Equally important, the expanded gas bubble inside of the fresh DisEPS paste that caused by the acetone evaporation (at low pressure) pushed the thin layer of fresh DisEPS paste to contact the particle's surface, as shown in Figure (5.c). This phenomenon would modify the compounding and the mechanical interlocking between the particles and the PS matrix.

FIGURE 5. SEM pictures (a) EPS cells, (b) interface of hemp shive with fresh DisEPS paste and (c) hemp shive porosity and the interior gas bubble due to temperature effect on fresh DisEPS paste.

4. CONCLUSION

This experimental study succeeded in reducing the volume of recycled expanded polystyrene by more than 98.4%, separating the polystyrene from other wastes and cleaning it from the oils and the colors. As a result, we obtained a viscous liquid plastic without the use of a heating process. Then the polystyrene paste was used as a matrix in the second step to produce a biobased-thermoplastic composite of 60 wt % of hemp shive, and the results indicated that the hemp shive particles facilitated the acetone evaporation from the DisEPS paste. The non-thermoformed composite had 10.37 MPa and 518 MPa for specific flexural strength and specific modulus of elasticity, respectively. In the second steps, the temperature and the thermoforming pressure improved the mechanical properties of this composite to be better than those of the net recycled polystyrene. These results extended our knowledge to confirm that the hemp shive particles work as a reinforcement material. Further studies need to be carried out in order to validate the possibility of using this composite as a lightweight building material because it has good specific mechanical properties and is better than many construction materials such as Hempcrete.

ACKNOWLEDGMENTS

The authors would like to thank the Ministry of Higher Education and Scientific Research of Iraq for their financial support, and also want to thank ARCENS Emmanuelle for her scientific expertise in SEM.

REFERENCES

1- Chanda, M., & Roy, S. K. (2006). Plastics Technology Handbook *(Vol. 72). CRC Press.*
2- Aminudin, E., Din, M. F. M., Mohamad, Z., Noor, Z. Z., & Iwao, K. (2011). A Review on Recycled Expanded Polystyrene Waste as Potential Thermal Reduction in Building Materials. *In Proceedings of International Conference on Environment and Industrial Innovation (ICEII 2011).*
3- Gutiérrez, C., García, M. T., Gracia, I., de Lucas, A., & Rodríguez, J. F. (2012). Recycling of Extruded Polystyrene Wastes by Dissolution and Supercritical CO2 Technology. *Journal of Material Cycles and Waste Management*, *14*(4), 308-316.
4- Schmidt, P. N. S., Cioffi, M. O. H., Voorwald, H. J. C., & Silveira, J. L. (2011). Flexural Test on Recycled Polystyrene. *Procedural Engineering, 10, 930-935.*
5- Mwashaa, A., Armstrong-Richardson, A., & Wilsonc, W. (2013). Management of Polystyrene Wastes Using a Supercritical Solvent - Propanone, *The Journal of the Association of Professional Engineers of Trinidad and Tobago Vol.41, No.1, April/May 2013, pp.23-28.*
6- Márki, E., Lenti, B., Vatai, G., & Békássy-Molnár, E. (2001). Clean Technology for Acetone Absorption and Recovery. *Separation and Purification Technology, 22, 377-382.*
7- Poletto, M., Dettenborn, J., Zeni, M., & Zattera, A. J. (2011). Characterization of Composites Based on Expanded Polystyrene Wastes and Wood Flour. *Waste Management*, *31*(4), 779-784.
8- Borsoi, C., Berwig, K. H., Scienza, L. C., Zoppas, B. C., Brandalise, R. N., & Zattera, A. J. (2014). Behavior in Simulated Soil of Recycled Expanded Polystyrene/Waste Cotton Composites. *Materials Research, 17(1), 275-283.*
9- Cérézo, V. (2005). Propriétés Mécaniques, Thermiques et Acoustiques d'un Matériau à base de Particules Végétales: Approche Expérimentale et Modélisation Théorique (*Doctoral dissertation, Ecole Nationale des Travaux Publics de l'Etat*).
10- Gay, D. (2005). Matériaux Composites-*5eme éd. révisée (Lavoisier).*
11- Islam, M. S. (2008). The Influence of Fibre Processing and Treatments on Hemp Fibre/Epoxy and Hemp Fibre/PLA Composites. *University of Waikato, Faculty of Materials and Process Engineering, Hamilton, New Zealand.*

Efficiency Of Recycling Post-Consumer Plastic Packages

E.U. Thoden van Velzen[a], M. Jansen[b], M.T. Brouwer[a], A. Feil[b], K. Molenveld[a], and prof. Th. Pretz[b*] (reference author)

[a]*Food & Biobased Research, Wageningen-UR, Bornse Weilanden 9 6709 WG Wageningen The Netherlands*
[b]*Institut für Aufbereitung und Recycling, RWTH, Wüllnerstrasse 6, 52062 Aachen Germany*

ulphard.thodenvanvelzen@wur.nl
pretz@ifa.rwth-aachen.de

Abstract. The recycling of packaging waste is an important part of the EU circular economy package, with a political focus on raising the recycling targets for post-consumer plastic packaging waste (PPW). The recycling of PPW involves at least three steps; collection, sorting and mechanical recycling. In contrast to the first two steps, mechanical recycling is poorly documented, as it is considered a free market activity. In order to provide a complete chain description the mechanical recycling yields were determined. The recovery of mass was determined for the main plastic sorting products from both major collection systems: separate collection (SC) and mechanical recovery (MR) from municipal solid waste. This technical assessment was conducted with a laboratory set-up and standard mechanical recycling process. This analysis showed that there is a substantial sample-to-sample variation in polymeric composition between similar feedstocks and this variation is also observed in recovered masses. Next, the mechanical recycling of polyethylene feedstocks was studied more in depth. Six PE feedstocks with a gradual increasing level of complexity (from only transparent PE bottle bodies to the complete PE sorting product according to DKR 329), were prepared and mechanical recycled with the laboratory set-up. Since the polymeric composition of both the six feedstocks and the six floating milled goods were known, the net PE recycling yield could be calculated. The net PE yields are close to 100% for such a standard recycling process. Additionally, the compositional analysis revealed that contaminants are only partially removed by the standard mechanical recycling process.

Keywords: Recycling, post-consumer plastic packaging waste, washed milled goods, yields.
PACS: 89.20.Bb, 89.20.Kk

INTRODUCTION

The recycling of packaging waste is an important part of the EU circular economy package, with a common EU recycling target of 75% for all packaging wastes by 2030.[1] Plastic is the second-largest packaging material and achieving high recycling targets is relatively more challenging in comparison to the other materials, due to the compositional complexity of plastic packages. Hence, the political focus is on raising the recycling targets for especially post-consumer plastic packaging waste (PPW). This calls for a scientific technological description of the PPW recycling system. The recycling of post-consumer plastic packaging waste involves at least three different steps; collection, sorting and mechanical recycling. The first two steps are well-documented in several countries, since this a legal obligation and the financial remuneration depends on administrative compliance. The subsequent mechanical recycling step (the conversion of sorted plastic products into washed milled goods) is, however, considered a free market activity and therefore this recycling process step is not publically documented. Furthermore, the mechanical recycling step itself is only poorly described in the scientific literature, for limited combinations of collected and sorted plastic packaging feedstocks and processes.[2, 3]

In order to scientifically describe the Dutch recycling system for post-consumer plastic packaging waste, the mechanical recycling yields[1] had to be determined for all five sorted plastic products from both major collection systems: separate collection (SC) and mechanical recovery (MR) from municipal solid waste (MSW).[4] This was achieved by setting up a dedicated laboratory to mimic standard mechanical recycling processes. The standard mechanical recycling process is a simplification of the industrial process and is composed of a shredder, a washing vessel, a float-sink-separator, centrifuge and an oven. It yields two products (the floating fraction and the sinking fraction) as washed and dried milled goods, a waste-product (dried sludge in 3 size classes) and waste water. The recovery of mass is simply derived from the quotient of the product weights and the feedstock weight (equation 2).

[1] The relation between the weight of the feedstock and the weight of the products (washed milled goods).

The plastic yield can only be determined in case the concentration of plastic in the feedstock and the products are known (equation 3). However, in the industrial practice these concentrations are unknown. This is, nevertheless, relevant since the sorted plastic packaging products do contain different plastics and materials at two levels. At the object-level, the sorted plastic product contains besides the desired plastic packages also faulty sorted objects and some residual waste. At the packaging-level, almost all plastic packages are composed of multiple components made from different materials. For instance a typical small Dutch PET bottle is composed of a PET body (14.7 g), a PP label (0.4 g) and a PE cap with closure ring (2.0 g). In order to study the plastic yields, the composition of sorted PE feedstock was determined in great detail (object-level and component-level), mechanical recycling was conducted and the composition of the products was studied in great detail by manual NIR-assisted flake analysis. This was conducted in a systematic manner, starting with the composition of sorted PE feedstocks and a stepwise reduction in compositional complexity to the level of clean PE bottle bodies, see table 1.

METHODS

Samples of sorted plastic products (25-50 kg) were taken at sorting facilities and stored at 7°C. All plastic sorted products were sampled, hence PET (DKR[2] 328-1), PE (DKR 329), PP (DKR 324), Film (DKR 310) and Mixed Plastics (MP, DKR 350). The codes refer to the specifications to which these sorted plastic products need to comply.[5] The samples were manually object-wise sorted into 50 different categories of plastic packages and residual materials with a SIRO IOSYS near infrared (NIR) machine. The samples were reconstituted (with the exception of large metal and mineral objects to protect the equipment). Besides full samples also five partial reconstituted samples were prepared, see table 1. The reconstituted feedstock was milled with a WEIMA WLK-04 shredder and a 2 cm ø screen, yielding dirty milled flakes in the 1-5 cm size range. Samples were taken to determine the dry matter content (dm) and the net material content (nmc)[3]. 5 kg gross plastic was washed with 150 liter aqueous 0.01 M sodium hydroxide solution at 50°C for 5 minutes in a large metal stirring vessel. Only for plastic film material the amount of feedstock had to be reduced to 2.5 kg gross. The waste water was sieved over three different screens (2 mm, 1 mm and 0.5 mm), sampled and discarded. The three different types of collected sludge were dried and weighed. The moist washed plastic flakes were transferred into a 3 meter high conical shaped settling tank filled with 200 liter tap water to mimic industrial floatation separation. The floating plastic flakes were scooped up from the top of the tank after 2 minutes and the sinking plastic flakes were collected in the bottom sluice by opening the bottom valve. Both the low density plastic flakes and the high density plastic flakes were separately centrifuged and dried in an oven (85°C, overnight). Both dried products were weighted. The mass of dissolved matter ($m^{dm}_{dissolved}$) was estimated from the difference in the dry weight of the feedstock and the dried weights of the floating product, sinking product and sludges, see equation 1.

$$m^{dm}_{dissolved} = m^{gross}_{feedstock} * dm - m^{dm}_{floatingproduct} - m^{dm}_{sinkingproduct} - m^{dm}_{sludges} \tag{1}$$

The recovery of mass (R_m) of a product was calculated from quotient of the dry product weight and the dry feedstock weight, which in turn is derived from the product of the gross feedstock weight and the dry matter content.

$$R_m^{Floatingproduct} = \frac{m^{dm}_{floatingproduct}}{m^{gross}_{feedstock} * dm} \tag{2}$$

The PE yield (η^{PE}) for the six samples in table 1 is calculated from the mass of polyethylene flakes in the product divided by the PE composition of the feedstock expressed in dry weights, see equation 3.

$$\eta^{PE} = \frac{c^{PE}_{floatingproduct} * m^{dm}_{floatingproduct}}{c^{PE}_{feedstock} * m^{gross}_{feedstock} * nmc} \tag{3}$$

---

[2] Deutsche Gesellschaft für Kreislaufwirtschaft und Rohstoffe

[3] Dry matter content is determined by taking a gross sample of plastic flakes, weighing it, placing it an oven for 12 hours at 105°C and weighing it again; the ratio of the dry weight over the gross weight in the dry matter content. The net material content is determined by taking a gross sample of plastic flakes, weighing it, washing it and dry it in the same manner. The ratio of the net clean weight over the gross weight is the net material content.

TABLE (1). The six PE feedstocks tested, starting with the bare essence (the transparent PE bottle bodies) and gradually increasing the complexity to the PE sorting product according to DKR 329 by adding different contaminants.

Code	Feedstock composition
1	Only transparent PE bottles
2	PE bottles (all colors)
3	Only complete PE bottles and PE flasks including packaging components made from non-PE polymers such as labels, caps and closures
4	Only complete PE packages, hence including PE films
5	PE packages including faulty sorted objects from predominantly PP and PET
6 warm	SC DKR 329, sorted PE including faulty sorted objects and attached residual waste, washed with 50°C 0.01 M NaOH solution
6 cold	SC DKR 329, sorted PE including faulty sorted objects and attached residual waste, washed with cold 0.01 M NaOH solution

The material composition of the feedstocks were calculated from two datasets; sorting analysis of feedstocks on the level of packaging categories and average material compositions of packaging categories. This yielded a composition in terms of 6 polymer types (PE, PP, PET, PS, PVC and black) and 2 residual materials in percentages and hence also the concentration of PE in the feedstock ($c^{PE}_{feedstock}$). The material composition of the produced milled goods were determined for 500 gram samples by NIR-based flake sorting with a IOSYS-SIRO NIR analyser and yielded the concentration of PE in the product ($c^{PE}_{floatingproduct}$).

RESULTS

The measured recovered masses for several different sorted plastic products processed with the simple standard process are listed in table 2. The results for slightly more complex processes involving pre-screening and mixed plastics have been presented previously.[6] The average recovered mass of the main product (PE, PP, film: floating product; PET: sinking product; MP: both) varied between 83 and 94%. Although in general the recovered masses are higher for separately collected and sorted plastics (SC) in comparison to mechanical recovered and sorted plastics (MR), there also appears to be a substantial sample-to-sample variation. This is especially true for the ratio between the recovered masses of the floating and sinking product. For the polyolefinic feedstocks (PE, PP, Film) the sinking fraction is mostly composed from PET, PVC, PS packaging components and polyolefinic packages with large amounts of inorganic fillers. The use of these heavy plastic components and inorganic fillers will vary between packages and therefore, these recovered masses should preferably be used in an indicative manner.

TABLE (2). Recovery of mass for various sorted plastic products as feedstocks for a standard mechanical recycling process.

Feedstock	Dry matter content of the feedstock	Recovered mass of floating product	Recovered mass of sinking product	Sludge waste	Dissolved substances
PET SC	93%	11±1%	83±2%	3±2%	3±2%
PET MR	87%	10±2%	84±2%	3±2%	3±2%
PE SC	95%	94±2%	3±1%	0.8±0.3%	2±2%
PE MR	96%	93±2%	2±2%	1±2%	4±2%
PP SC	99%	90±1%	7±1%	0.7±0.6%	3±1%
PP MR	88%	83±2%	3±2%	6±2	8±2%
Film SC	97%	90±6%	4±4%	0.8±0.6%	6±3%
Film MR	87%	83±15%	4±2%	3±4%	10±9%
MP SC	98%	64±12%	32±11%	0.7±0.8%	5±1%
MP MR	87%	65±4%	28±1%	3.1±0.1%	4±3%

The recovered masses for the six different PE feedstocks with an increasing level of complexity are given in table 3 and the material composition of the feedstock and produced milled goods (floating products) are given in table 4.

TABLE (3). Recovery of mass (R_m) for the 6 different PE feedstocks of table 1.

Feedstock	Dry matter content of the feedstock	Recovered mass of floating product	Recovered mass of sinking product	Sludge waste	Dissolved substances
1	99.1%	98.1%	0.2%	0.5%	1±2%
2	98.5%	98.2%	0.0%	0.3%	1±2%
3	92.3%	97.9%	1.4%	0.9%	0±2%
4	91.0%	95.6%	1.4%	1.0%	2±2%
5	93.5%	94.4%	1.9%	0.9%	3±2%
6 warm	92.2%	96.8%	2.2%	0.9%	0±2%
6 cold	92.2%	95.1%	3.6%	0.9%	0±2%

TABLE (4). The composition of feedstocks (FS) and produced milled goods (MG) of the floating products for the 6 PE samples.

Material	1 & 2		3		4		5		6	
	FS	MG	FS	MG	FS	MG	FS	MG	FS	MG
PE	100.0%	100.0%	97.6%	97.2%	97.7%	95.8%	89.3%	89.9%	88.7%	90.0%
PP			0.1%	2.3%	0.1%	3.9%	6.8%	8.3%	6.7%	8.2%
PS			0.0%	0.1%	0.0%	0.0%	0.3%	0.0%	0.3%	0.2%
PVC			0.0%	0.0%	0.0%	0.0%	0.1%	0.1%	0.1%	0.3%
PET			0.0%	0.2%	0.0%	0.0%	0.8%	0.0%	0.8%	0.2%
Black plastics			0.1%	0.2%	0.1%	0.1%	0.9%	0.6%	0.9%	0.4%
Paper			2.2%	0.0%	2.1%	0.0%	1.9%	0.0%	2.2%	0.1%
Residual materials			0.1%	0.1%	0.1%	0.1%	0.1%	1.1%	0.4%	0.7%

Obviously, the highest recovered mass of the floating product is recorded for the samples that contained PE bottle bodies and the lower values for the samples that contained sorting faults and residual waste, but a poor correlation was found between the concentration PE of the feedstock and the recovered mass. This is most likely caused by the relative small differences in concentration (89-100%) and the relative large errors in concentration (±2%) and in recovered mass (±2%).

There is a moderate relationship between the recovered mass of the sinking product and the sum of the PS, PVC, PET and other material concentrations in the feedstocks. The moderate correlation is likely to be caused by the small differences in concentration (0-4%) as compared to the error in the concentration (±2%). Although the density separation process effectively removes a sinking product and hence heavy polymers such as PS, PVC and PET, still a substantial part of these heavy polymers is found in the floating product. This is agreement with previous DSC and ATR-IR analysis of comparable PE samples indicating 5-10% PP in the recycled PE.[7] This is also apparent from the good relationship between the concentration of PE in the feedstock and PE in the milled goods, see figure 1. Hence, although this simple standard recycling process is able to remove a part of the contaminants, it is not able to remove them all.

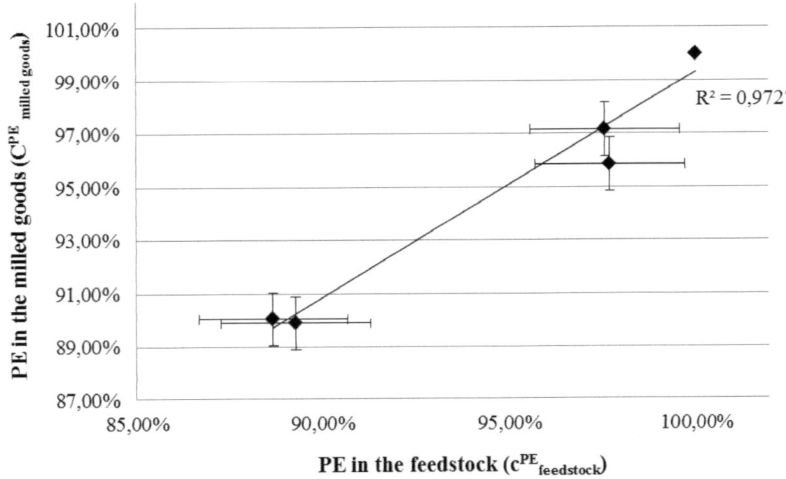

FIGURE 1. The relationship between the concentration of PE in the feedstock and the concentration PE in the milled goods for the six PE samples with a progressively more complex composition.

TABLE (5). Calculated net PE and PO (PE+PP) yields (ηPE and ηPO) for the 6 PE feedstocks.

	1	2	3	4	5	6 warm	6 cold
ηPE	94 ± 3%	101 ± 3%	98 ± 3%	95 ± 3%	97 ±3%	101 ±3%	99 ± 3%
ηPO	94 ± 2%	101 ± 2%	98% ± 2%	97% ± 2%	96% ± 2%	99 ± 2%	98 ± 2%

The net PE yields were calculated with equation 3 and the data in table 3 and table 4 and the net material contents. These net PE yields are listed in table 5. The error in the net PE yields could not be derived with error propagation laws due to the many assumptions made. Instead it was derived from a regression fit of the calculated net PE yields versus the PE concentration of the feedstock (see figure 2), which yielded a residual standard deviation of $\sigma = 2.7\%$ in the net PE yields. Also the net polyolefine yields (PE+PP) were calculated, see table 5. The residual standard deviation was estimated in a similar manner to be $\sigma = 2.4\%$. The net PE yields and the net PO yields are for all samples approximately equal to 100%.

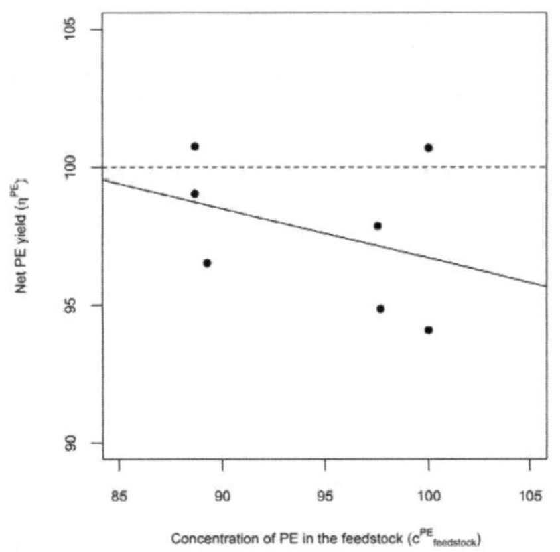

FIGURE 2. The poor relationship between the concentration polyethylene in the feedstock and the net PE yield was used to derive a residual standard deviation in the net PE yield. The dashed line refers to the optimal 100% yield.

A laboratory method for measuring the recovered masses and net polymer yields for the mechanical recycling of sorted plastic products into washed milled goods has been presented. There is a substantial sample-to-sample variation in polymeric composition between similar feedstocks and this variation is also observed in recovered masses. Polymeric contaminants are only partially removed by this standard mechanical recycling process. The net polymer yields are close to 100% for such a standard recycling process. This implies that the recovered mass of recycling can be estimated from a detailed compositional analysis (composition in categories, material composition per category and net material content). Nevertheless, the concentration of polymeric contaminants in recycled plastic still have to be determined by NIR assisted flake sorting.

ACKNOWLEDGMENTS

We are grateful for the help of Eric Boer with the statistical evaluation and application of the error propagation laws.

REFERENCES

1. EU "Closing the loop- an EU action plan for the circular economy", Brussels 2015 COM 614.
2. S. Yin et al., *Polym. Eng. Sci.* 2899-2909 **55**, (2015).
3. F. Perugini, M.L. Mastellone and U. Arena, Environmental Progress **24** 137-154 (2005).
4. C. Cimpan et al., *J. Environm. Managem.* **156**, 181-199 (2015).
5. The sorted PPW products need to comply to DKR specifications, website: http://www.gruener-punkt.de/en/download.html
6. M. Jansen, "Processing of plastic waste – from material following the DKR specifications to milled goods" EIMPACK Conference, Lisbon Portugal, 2012.
7. B. Luijsterburg and H. Goossens, *Resources Conservation and Recycling* **85**, 88-97 (2014).

The Effect of the Compatibilizer SEBS-g-GMA on the Blend PP-PET: Virgin and Recycled Materials

Karen Van Kets [a], Nicolas Van Damme [a], Laurens Delva [a] and Kim Ragaert [a]

[a]*Center for Polymer & Material Technologies, Department of Materials Science & Engineering, Faculty of Engineering & Architecture, Ghent University. Technologiepark 915, B-9052 Zwijnaarde*

Presenting author: ncvdamme.vandamme@ugent.be
Corresponding author: kim.ragaert@ugent.be

Abstract. In the carpet industry poly(ethylene terephthalate) (PET) and poly(propylene) (PP) are often used together within a single product. Mechanical recycling of these carpets results in a blend of PET and PP, which are immiscible. To enhance impact strength of this waste stream, the compatibilizer SEBS-g-GMA was used. More specific the transferability of earlier results with the compatibilizer, obtained on virgin PET-PP blends with amorphous PET (PETg), was assessed. Firstly, from these blends to blends with semi-crystalline PET (PETe) and secondly, from virgin to recycled materials. Two blends of virgin material were made containing 80 wt% PP and 20 wt% PETg or PETe. The effect of adding 2,5 wt% SEBS-g-GMA was assessed. Subsequently, post-industrial PP (r-PP) and post-consumer PETe (r-PETe) were blended and mechanical properties were measured for blends with and without compatibilizer. An increase in impact strength for the two virgin compatibilized blends (PP:PETg:SEBS-g-GMA and PP:PETe:SEBS-g-GMA) was expected and confirmed. A reduced effect of the compatibilizer on impact strength was observed for the recycled blends, due to the possible presence of contaminants. It was concluded that the results from virgin PETg-PP were directly transferable to virgin PETe-PP, but not entirely to recycled materials.

Keywords: recycling, compatibilizer, mechanical properties, morphology, polypropylene, poly(ethylene terephthalate)
PACS: 81.05.Lg, 81.20.Hy, 81.70.Bt, 83,80Tc

INTRODUCTION

When poly(ethylene terephthalate) (PET) and poly(propylene) (PP) are used for carpet manufacturing, the polymers are attached to one another and become physically inseparable for the purposes of mechanical recycling. Typically, such blends of polar and apolar polymers are inherently immiscible, which leads to reduced mechanical properties of the resulting product, toughness foremost among them [1]. In this It was the aim of this research to recover the blend's properties by increasing the miscibility via the use of a compatibilizer. there has been attempted to increase the toughness. A blend with a matrix of 80 wt% PP and a dispersed phase of 20 wt% PET was made. The main goal was to obtain a tougher material (and consequently also a higher elongation at break) and second to evaluate the influence of the compatibilizer on the yield strength.

The morphology and composition can give a prediction for the mechanical properties. If the amount of the dispersed phase is too high, the PET spheres enlarge and will present a bigger surface. The stress transfer at the interface between both polymers will decrease due to the weak interfacial adhesion and a wider interface. This results in reduced mechanical properties for the blend. On the other hand, if the amount of the dispersed phase is too small, the inter-particle distance becomes too high, which causes a more fracture sensitive material [2, 3].

Because these two polymers are immiscible in the melt state, the compatibilizer styrene-ethylene/butylene-styrene grafted with glycidyl methacrylate (SEBS-g-GMA) was added. This compatibilizer was selected by previous research [4], where it was found to sufficiently induce the brittle-to-ductile transition for an added percentage of 2,5 wt% to a mix of 80:20 PP with amorphous PET (PETg). SEBS-g-GMA increases the toughness and is made from a rubber component grafted with a functional group. It will act as an impact modifier and as coupling agent [5]. The different interactions between the compatibilizer and PET are an in-situ copolymerisation of epoxy ring-opening with the end-groups of PET. But also hydrogen bridges between the carboxylgroups, dipole-dipole interactions and Van Der Waals interactions can occur. PP is the matrix and only the Van Der Waals interactions are possible with the ethylene/butylene components of the rubber [1, 6, 7].

In this research the transferability of the compatibilizer's effect from PP:PETg blends to blends consisting of PP and semi-crystalline PET (PETe) was examined, as well as the transferability from these virgin to recycled industrial materials. The difference of completely amorphous PETg and semi-crystalline PETe is originated in a different backbone structure of the polymers. PETe is the common known condensation polymer from ethylene glycol and terephtalic acid. While the ethylene glycol in PETg is replaced by a diol with increasing free volume. This can be propylene glycol, but more often 1,4-bis(hydroxymethyl) cyclohexane is used [8, 9].

MATERIALS AND METHODS

Materials

All blends of PP and PET were mixed manually before adding them to the co-rotating twin-screw extruder (Coperion ZSK 18, 40 L/D) Compounding was performed at a temperature of 280°C, as preliminary research showed that melt compounding at increasing temperatures above the melt temperature improved the impact strength of the compatibilized blend [4]. The blends are named '*matrix : dispersed phase : compatibilizer*' (78 wt% :19,5 wt% :2,5 wt%). Wherein the matrix is PP or recycled PP (r-PP), the dispersed phase is PET where each time one of the following PET were mixed in the blend: PETg, PETe and recycled PETe (r-PETe). As compatibilizer SEBS-g-GMA was added by the side feeder. The virgin materials used in this project were PP (DOMO: PP Domolen 1040N, density: 0,91 g/cm³), PETg (EASTMAN: Eastar™ Copolyester 6763, density: 1,27 g/cm) and PETe (Equipolymers: PET Lighter C93, density: 1,34 g/cm³). The recycled materials were kindly donated by Beaulieu International Group (BINTG) located in Wielsbeke (Belgium). For all blends test bars were injection moulded (BOY 22S) at the same temperature of compounding and were left at room temperature for minimum five days prior to testing. These test bars include 'dog bone' specimens for tensile (ISO 527-2/1B) and rectangular bars for impact testing (ISO 179-1/1eA).

Methods

In the initial phase of the research, differential scanning calorimetry (DSC)-analysis (NETZSCH DSC 204 from Phoenix) was performed on all the used materials. The density was used to characterise the materials. The density of the melted PET-fibers and the granulates of the SEBS-g-GMA were gravimetrically determined by the immersion method according to ISO 1183. This was executed in propanol by means of a density kit and a Precisa balance. For the recycled materials there was a more profound analysis by Fourier transformation infrared (FTIR) (TENSOR 27 FTIR from Brüker). The different blends were tested on their mechanical properties. First the virgin blends, where the differences of amorphous to semi-crystalline PET were compared, and afterwards the recycled blends. The mechanical tests include a tensile and impact test. According to the Charpy method of ISO 179-1/1eA the rectangular test bars were notched 2 mm and tested for impact strength with a pendulum of 2 J. The tensile specimens were tested in accordance with ISO 527 on an Instron 5565 machine with a load cell of 2 kN. An extensometer with gauge length 25 mm was used for strain measurement in the initial part of the test where the Young's modulus is determined. There the cross-head rate of 10 mm/min was used. After removing the extensometer, a cross-head rate of 50 mm/min was applied on the tensile bars. The tensile offset yield at 0.2% (σ_y), strain at break (ε_b) and the Young's modulus (E) were determined (Bluehill2 software).. The reported values for the mechanical properties are the average of at least five measurements. All data was processed with SPSS Statistics 20 and screened for outliers. The outliers were deleted in order to work within the 95% confidence interval. Furthermore the morphology of the blends was examined by scanning electron microscopy (SEM) (Jeol JSM-7600F Field emission) of cryofractured surfaces and briefly the crystallinity was analysed using DSC.

RESULTS AND DISCUSSION

Properties of PP/PETg to PP/PETe (virgin)

It was found that the impact strength and strain at break were similar or increased when comparing the compatibilized blend to the PP matrix. The beneficial effect was even higher when compared to the uncompatibilized blends (Figure 1). The opposite was noted for Young's Modulus and the yield strength. The Young's modulus of the blends without compatibilizer was higher, since the PET had the effect of a filler [10], but their impact strength was weak. A proportional law-of-mixtures could not be used to predict the impact strength results, due to the phase separation effect [11]. It must be noted that the impact strength of PETg was higher than PETe.

The smaller dispersed spheres, the finer dispersion and the smaller inter particle distance, contribute all to a more ductile blend. This can be seen in the SEM-images (Figure 2). Therefore it can be said that the functionality of SEBS-g-GMA is transferable from amorphous PET to semi-crystalline PET.

As the blends with SEBS-g-GMA were compared to the blend without SEBS-g-GMA, the amount of crystallinity in PET decreases due to the copolymerisation between the compatibilizer and the PET during melt blending. This only relates to the semi-crystalline PET. It must be taken into account that SEBS-g-GMA has a high grafting ratio of 1,5 wt%. All the mechanical properties of the blends were compared to the matrix and the results are summarized in Table 1.

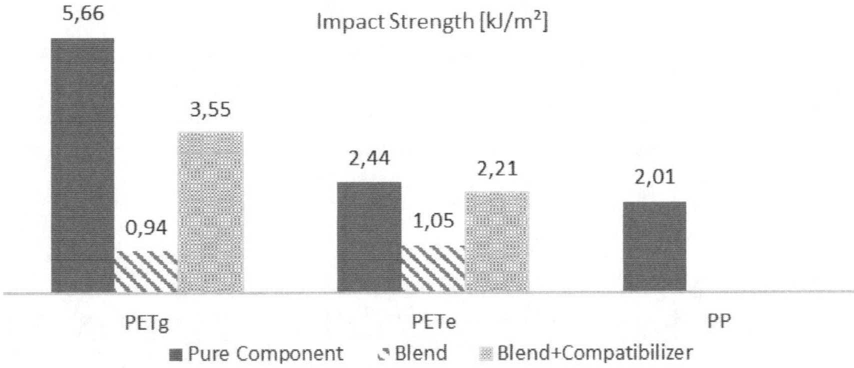

FIGURE 1. The impact strength of the virgin polymers and their blends

Properties from PETe to rPETe

SEBS-g-GMA was also added to a mixture of post-industrial PP (originated form carpet fibres) and post-consumer PET-fibres (origin unknown). The impact strength and strain at break increased with the addition of the compatibilizer, but not as significantly as with the virgin blends. The decrease in yield strength remains within the same trend, as does the Young's modulus but this difference is more pronounced (Table 1). When recycled materials were used, the amount of compatibilizer was insufficient to realize the brittle-to-ductile transition. Possible causes include unknown contamination and additives in the recycled polymers. This was also noticed by Kazemi et al. [12]. The contamination will, according to their interfacial tension, prefer to migrate in one of the polymers or they will prefer to be in the interface. This could cause interaction with the compatibilizer but also influence the migration of the dispersed phase in the matrix phase. Subsequently a different droplet size of the dispersed phase occurred compared to the blend of the virgin materials (Figure 2).

TABLE 1. A summarize of the results from the blends with and without compatibilizer

Material	E [MPa]	σ_y [MPa]	ε_b [%]	Impact strength [kJ/m²]
PP:PETg	1499 ± 116	18,49 ± 0,77	11,51 ± 0,64	0,94 ± 0,00
PP:PETg:SEBS-g-GMA	1134 ± 97	14,29 ± 0,23	34,41 ± 4,92	3,55 ± 0,26
PP:PETe	1628 ± 44	18,33 ± 1,64	16,14 ± 2,05	1,05 ± 0,09
PP:PETe:SEBS-g-GMA	1501 ± 61	14,00 ± 0,48	29,38 ± 0,59	2,21 ± 0,05
PP	1313 ± 100	16,70 ± 1,31	22,92 ± 2,84	2,01 ± 0,15
rPP	1673 ± 26	13,93 ± 0,29	28,48 ± 1,05	2,74 ± 0,18
rPP:rPETe	1926 ± 105	15,69 ± 0,19	12,66 ± 0,84	1,58 ± 0,01
rPP:rPETe:SEBS-g-GMA	1594 ± 38	13,98 ± 0,37	20,83 ± 4,73	1,70 ± 0,05

Morphology of the blends

Besides the smaller dispersed spheres, finer dispersion and smaller inter particle distance; there are other noteworthy remarks to the morphology of the compatibilized blends. These are the increased amount of droplets and a limited particle size distribution (Figure 2), which gives the blend a more uniform consistency than the non-compatibilized blends [13]. This can be correlated to the impact strength results, indicating a matrix with lowered crack propagation and less inducement to micro fractures [2].

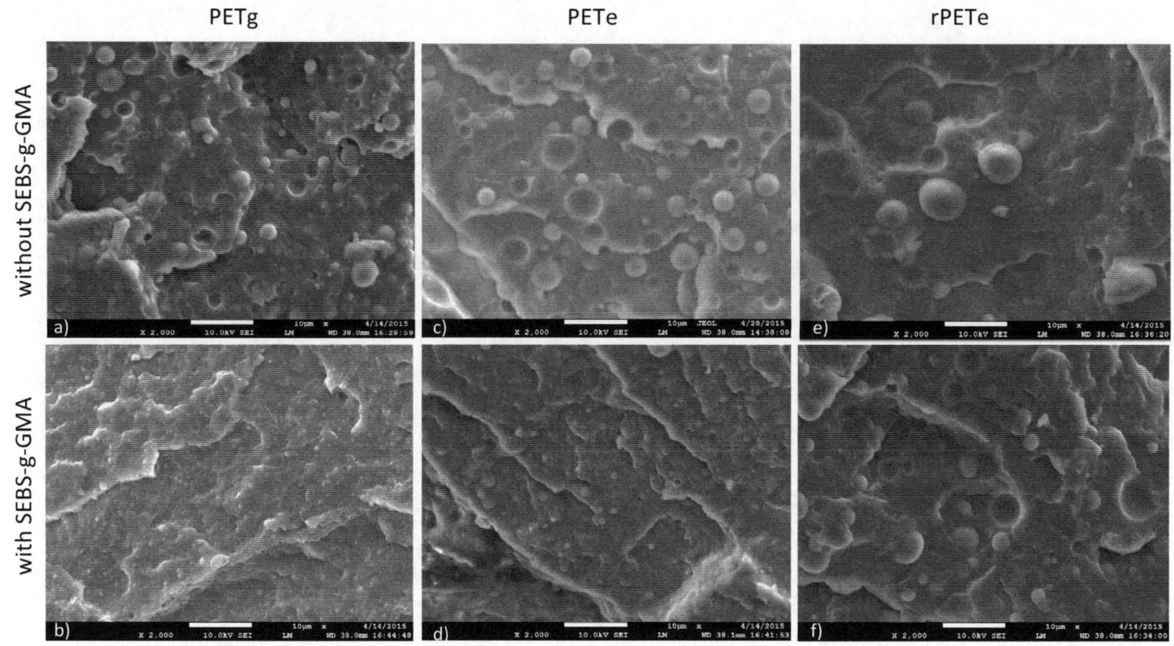

FIGURE 2. SEM-images of all the blends at x2000 magnification with a working distance of 38 mm; the blends without compatibilizer: (a) PP:PETg, (c) PP:PETe and (e) r-PP:r-PETe; the blends with compatibilizer (b) PP:PETg:SEBS-g-GMA, (d) PP:PETe:SEBS-g-GMA and (f) r-PP:r-PETe:SEBS-g-GMA

Conclusion

It can be concluded that the effect of SEBS-g-GMA is transferable from PETg to PETe in PP:PET blends of virgin material. Higher impact strength and larger elongation at break are observed. Due to the coupling effect of the compatibilizer, a better dispersion of PET in the PP matrix and a smaller inter particle distance are effectuated. However the values for impact and elongation at break are bigger for the blend with PETg. Resulting from its amorphous state and the better flow during processing, introducing a better mixture of the two phases.

The effect of the compatibilizer, when using post-industrial PP and post-consumer PET, is much smaller. Recycled material is unpredictable and the presence of contaminants is not exceptional. These impurities are a hindrance for a good dispersion and can interact with the SEBS-g-GMA, causing a decrease in amount of compatibilizer for the main components. Further research will decide the needed amount of compatibilizer for a sufficient brittle-to-ductile transition in r-PP:r-PET blends.

ACKNOWLEDGMENTS

This research was funded by the Faculty of Engineering and Architecture at Ghent University.

REFERENCES

1. Heino, M., J. Kirjava, and P. Hietaoja, *Compatibilization of polyethylene terephthalate/polypropylene blends with styrene–ethylene/butylene–styrene (SEBS) block copolymers*. Journal of applied polymer science, 1997. 65(2): p. 241 249.
2. Wu, Souheng, *Impact fracture mechanisms in polymer blends: Rubber-toughened nylon*. Journal of Polymer Science: Polymer Physics Edition, 1983. 21(5): p. 699-716.
3. Bucknall, C. and D. Paul, *Notched impact behaviour of polymer blends: Part 2: Dependence of critical particle size on rubber particle volume fraction*. Polymer, 2013. 54(1): p. 320-329.
4. Van Bruggen E., Koster R., Picken S. and Ragaert K., Influence of Processing Parameters and Composition on the Effective Compatibilization of Polypropylene - Poly(ethylene terephthalate) blends. In press in international Polymer Processing..
5. Bicerano, J., A, *Practical guide to polymeric compatibilizers for polymer blends, composites and laminates*. provided by http://www.specialchem4polymers. com, 2005. 22.
6. Koning, C., Van Duin, M., Pagnoulle, C., & Jerome, R., *Strategies for compatibilization of polymer blends*. Polymer Science, 1998. 23(4): p. 707-757.
7. Litmanovich, A.D., N.A. Platé, and Y.V. Kudryavtsev, *Reactions in polymer blends: interchain effects and theoretical problems*. Progress in polymer science, 2002. 27(5): p. 915-970.
8. Schwartz, S., et al., *FIBER PRODUCTION-PET modification with MPDiol glycol*. Chemical Fibers International, 2003. 53(6): p. 445-449.
9. Wu, J.H., et al., *Poly (ethylene terephthalate)/Poly (ethylene glycol-co-1, 3/1, 4-cyclohexanedimethanol terephthalate)/ Clay nanocomposites: Morphology and isothermal crystallization kinetics*. Journal of applied polymer science, 2013. 128(1): p. 487-497.
10. Bataille, P., S. Boisse, and H. Schreiber, *Mechanical properties and permeability of polypropylene and poly (ethylene terephthalate) mixtures*. Polymer Engineering & Science, 1987. 27(9): p. 622-626.
11. Hubo, S., Delva, L., Van Damme, N. &Ragaert, K., *Blending of recycled mixed polyolefins with recycled polypropylene:effect on physical and mechanical properties*. Polymer Processing Society 2015, Graz. Accepted for publication in AIP Conference Proceedings
12. Kazemi, Y., Ramezani Kakroodi, A., & Rodrigue, D. (2015). Compatibilization efficiency in post-consumer recycled polyethylene/polypropylene blends: Effect of contamination. *Polymer Engineering & Science*, 55(10), 2368-2376.
13. Inoya, H., Wei Leong, Y., Klinklai, W., Takai, Y., & Hamada, H., *Compatibilization of recycled poly (ethylene terephthalate) and polypropylene blends: Effect of compatibilization on blend toughness, dispersion of minor phase, and thermal stability*. Journal of Applied Polymer Science, 2012. 124(6): 5260-5269.

Adsorption Mechanisms of Emulsified Crude Oil Droplets onto Hydrophilic Open-Cell Polymer Foams

P. Cherukupally, [1] J. H. Hinestroza, [2] R. Farnood, [3] A.M. Bilton, [1, a)] and

C. B. Park [1, b)]

[1]*Department of Mechanical and Industrial Engineering, University of Toronto, 5 King's College Road, Toronto, Ontario M5S 3G8*
[2] *Fiber Science Program, Cornell, 37 Forest Home Drive, Ithaca, New York 14850*
[3]*Department of Chemical Engineering and Applied Chemistry, University of Toronto, 200 College Street, Toronto, Ontario M5S 3E5*

[a)]Corresponding author: bilton@mie.utoronto.ca
[b)]Corresponding author: park@mie.utoronto.ca

Abstract. Polyester/polyurethane open cell foam is examined for oil field water treatment. The oil droplet adsorption is influenced by foam-droplet bonding due to surface chemical interactions. Of the possible adsorption, acid-base, electrostatic (ES), and hydrophobicity interactions are among the most important. To determine the role of acid-base interactions on the adsorption, foam K_a and K_b values were measured. The foam is found to be a stronger base. To evaluate ES affect, adsorption experiments were performed for a series of emulsion pH levels using crude oil-in-water emulsions which were adjusted to acidic, neutral, and basic conditions. It was found that the turbidity removal rates (adsorption-separation quantification) were substantially increased during acidic conditions due to active interaction with the basic sites at the foam surface. Specifically, at pH5 the turbidity removal rates were faster perhaps due to optimal charge concentration and mobility of the ions where the electrostatic repulsion is minimal. At basic conditions, the turbidity removal rates were too slow could be due to electrostatic repulsive forces between hydroxyl ions-droplets and droplet-base foam. Based on the experimental results, to minimize the energy input, the separation process will be performed at pH5 where the turbidity rates are found to maximum.

INTRODUCTION

The petroleum industry is a critical component of the global energy sector. Unfortunately, petroleum extraction and processing results in environmentally harmful process waste containing difficult to remove emulsified organic compounds (OCs) [1]. Current treatment technologies are unable to economically and efficiently treat dilute aqueous solutions containing PM10 (Particulate Matter, PM10) oil droplets common in oil contaminated waste water [2-4]. This water contamination is a challenge for the Canadian oil sands industry, which generates 1.8 million cubic meters of toxic oil-contaminated tailing per day [6], and has failed to treat nearly 1 billion cubic meters of tailings to date due to currently available, ineffective treatment methods [5]. An economical and efficient water treatment technology for removal of dilute emulsified organics would make a positive impact on the petroleum industry.

Adsorption is an inexpensive, simple, and low waste method for contamination removal. However, current adsorbents, which are effective at treating dissolved and dispersed OCs, are ineffective for treatment emulsified OCs and have low adsorption capacity [5, 4]. A new class of materials, open cell foams containing high porosity micro-sized pores at are potential adsorbent materials for OCs [8, 8]. The authors' previous work has shown that high porosity, open cell polyester/polyurethane (PEPU) foam is capable of over 99% emulsified crude oil droplet removal from oil-in-water emulsions [8]. In this paper, we discuss the physicochemical principles such as acid-base and ionic electrostatic (IES) interactions with experimental evidence that may have contributed to the high removal rates. Furthermore, another variable of the adsorption process, effect of emulsions pH will be determined experimentally.

BACKGROUND

The governing mechanisms for solute deposition onto an adsorbent are chemisorption and physisorption. Chemisorption occurs due to Lewis acid-base, hydrogen, ionic, or covalent bonds. Meanwhile, in physisorption, the droplet adheres to the surface due to van der Waals attractive forces, surface roughness, or hydrophobicity. The deposition occurs on the surface in a monolayer or multi-layer form depending on the process conditions. In monolayer deposition, the solute deposits as a single layer of one molecule depth, whereas, in multilayer deposition, the solute deposits in multiple layers involving differing chemical bonds and physical forces at different layers [9, 10]. Multilayer deposition will result into higher adsorption mass uptakes, however, due to the number of bonds involved, the pores need to be large enough to accommodate bond lengths. Therefore, larger pores potentially aid adsorption due to accessible free volume. The adsorbate mass uptake by the porous media also depends on other forces such as capillary forces, electric field strength, and competitive sorption among species and water. In this paper, we restricted the scope to the multilayer depositions associated surface forces such as hydrogen bonding, electrostatic interactions, and Lewis acid-base interactions.

The authors' previous work showed the effective separation of oil-in-water (O/W) emulsions using open-cell foams by chemisorption and physisorption [8, 11]. Here we extend the study to droplet monolayer and multilayer deposition mechanisms. The foam is composed of negatively charged elements, such as Cl, C, N, and O at the surface. As a result, the polymer may possess negative point charges. In an acidic environment, the hydrogen ions could contribute to droplet coalescence (due to their negative surface charge), protonization of the foam surface, and multilayer deposition of the coalesced droplets onto the foam. The van der Waals forces, surface roughness, and capillary forces due to the micro-pore structure could all contribute to crude oil adsorption. However, this work focuses only on influence of acid-base and IES interactions on adsorption.

EXPERIMENTAL

In this paper, commercial polyester/polyurethane (PEPU) open cell foam is used as an adsorbent material. In previous work, the PEPU foam composition was characterized and found to have the following properties: surface elemental composition of C, O, N, and Cl; surface roughness of 68 μm; morphological features of pore size of 200-1100 μm, porosity of 96.80%, and throat size of 4 μm; and crosslinked polyurethane (PU) [8]. The foam Lewis acid-base profiles were obtained according to Gutmann theory using Inverse Gas Chromatography Surface Energy Analyzer, IGC-SEA (Surface measurement systems, Alperton, UK) and SEA analysis software. Approximately 10 mg of samples were tested at 50°C with helium gas at a flow rate of 5 sccm. For consistent adsorption experimental results, oil-in-water emulsions (O/W) to emulate oil-field water were developed using Texas raw crude and deionized water. The model emulsions were developed for 1% and 3% oil content by volume. The emulsion had an average droplet size of less than 10 μm [8] turbidity of 28300 NTU at 1% concentration, and turbidity of 61400 NTU at 3% concentration. The adsorption of the foam has been characterized for 1% and 3% oil concentrations for variable temperature and fluid circulation conditions. The details of the materials characterization and adsorption experiments can be found in [8].

The efficacy of crude oil removal is highly dependent on oil concentration, emulsion stability, and droplet size. To reduce the stability of emulsions, droplet coalescence was induced by physical destabilization at a low-mixing speed. The destabilization is also a function of ionic activity in electrolytic solutions. To determine the effect of ionic activity on coalescence and adsorption, the pH of the model emulsions was adjusted. The emulsions were doped with 1 mol% H_2SO_4 and 1 mol% NaOH for acidic and basic conditions, respectively. The pH of the emulsions was measured using an Oakton pH meter. The droplet adsorption onto foam was quantified by gravimetric analysis and change in the emulsion oil concentration or turbidity.

To determine the relationship between pH and adsorption, experiments were conducted with 1% and 3% oil concentration emulsions at pH 3, 5, 7, 9, and 11 at predesigned demulsification conditions of 40°C and 500 rpm. For each adsorption experiment, a foam sample of 40 mm x 40 mm x 15 mm with an approximate weight 750 mg was immersed in a 100 mL of model emulsions. At predefined time intervals foam samples were measured for mass uptake. The change in sample weight is expressed as a fractional mass uptake, M_f as expressed by.

$$Fractional\ mass\ uptake,\ M_f = \left[\frac{M_t - M_o}{M_o}\right] \tag{1}$$

where, M_t is the instantaneous adsorbent weight (g), and M_o is the initial weight of the adsorbent (g).

The process efficacy was quantified from change in initial and final turbidity values of the emulsions $\eta\%$,

$$Turbidity\ removal,\ \eta\% = \left[\frac{C_o - C}{C_o}\right] \tag{2}$$

where, C_0 and C are the initial and final turbidity of the emulsion (NTU). The turbidity of the emulsions was measured according to EPA (U.S. Environment Protection Agency) method 180.1 [12]. All the experiments were repeated at least three times to reduce the error in measurements.

RESULTS AND DISCUSSION

The PEPU foam K_a and K_b values (representative of material acidity and basicity) according to Guttman theory are 0.16 and 1.1, respectively. This indicates the foam is a stronger base and will have affinity to acids. The pH is a major influencing parameter of OCs adsorption onto polymeric adsorbents as the presence of charged ions can aid droplet coalescence and droplet attachment to the charged polymer surfaces [9, 10, 13]. The effect of initial emulsion pH on crude oil droplet adsorption was experimentally investigated for 1% and 3% oil concentration. The emulsions mass uptake, turbidity removal, and treatment times are listed in Table 1 and the trend plots are depicted in Figure 1. The pH had a strong influence on oil droplets adsorption for acidic conditions as expected. Results show over 99.9% turbidity removal was achieved at acidic and neutral conditions within the experimental time allowed. Conversely, the removal was slow at basic conditions and 99% removal was not achieved during the 24 hour experimental window.

TABLE 1. Crude oil-water separation by PEPU foam with the influence of pH

	1% oil concentration				3% oil concentration		
pH	**Turbidity removal (%)**	**Treatment time (hrs)**	**Mass uptake (g/g of adsor)**	**pH**	**Turbidity removal (%)**	**Treatment time (hrs)**	**Mass uptake (g/g of adsor)**
3	99.97	8	20.97	3	99.99	9	18.11
5	99.98	5.5	19.72	5	99.99	3	13.16
7	99.97	8	12.37	7	99.99	24	18.14
9	94.60	24	13. 17	9	93.78	24	17.92
11	85.31	24	13.66	11	23.10	24	10.84

Adsorption for 1% oil concentration, Figure 1a) shows the mass uptake verses time for 1% oil concentration. These experiments were halted when the turbidity removal exceeded 99% or the total experiment time reached 24 hours. At acidic conditions, as the pH increased from pH3 to pH5, there is a significant reduction in the treatment times from 8 hrs to 5 hrs. Despite this reduction in treatment time, the mass uptake remained the same at 20 g/g of adsorbent. At neutral pH7, the adsorption was quite similar to that of pH3 with respect to turbidity removal and treatment time, but there was an anomaly in mass uptake. The anomalous behavior could be due to absence of ES interactions at neutral conditions. Whereas at basic conditions, as the pH increases from pH9 to pH11, the turbidity removal and mass uptake are very slow. The results indicate that, at 1% oil concentration, for the same process conditions, pH5 delivers shortest treatment time, and therefore the smallest energy input for emulsion circulation and heating.

Adsorption for 3% oil concentration, Figure 1 (b) shows the treatment time and turbidity removal rates for 3% oil concentration. At acidic conditions, as the pH increases from pH3 to pH5, there is a substantial reduction in treatment times from 9 hours to 3 hours. In addition, with this pH increase, the mass uptake is reduced from 18.11 g/g to 13.16 g/g of adsorbent. At neutral pH7, the turbidity removal and mass uptake were quite similar to that of pH3 except drastic increase in treatment time. It could be due to absence of favorable ES interactions. However, for basic conditions, as the pH increased from pH9 to pH11, the oil adsorption and mass uptake are very slow and unable to reach 99% removal within the allowable 24 hour window. Moreover, at pH11 the turbidity removal was poor and only reach 23.10% within 24 hours. The results indicate that, at 3% oil concentration, pH5 is the preferred operating condition.

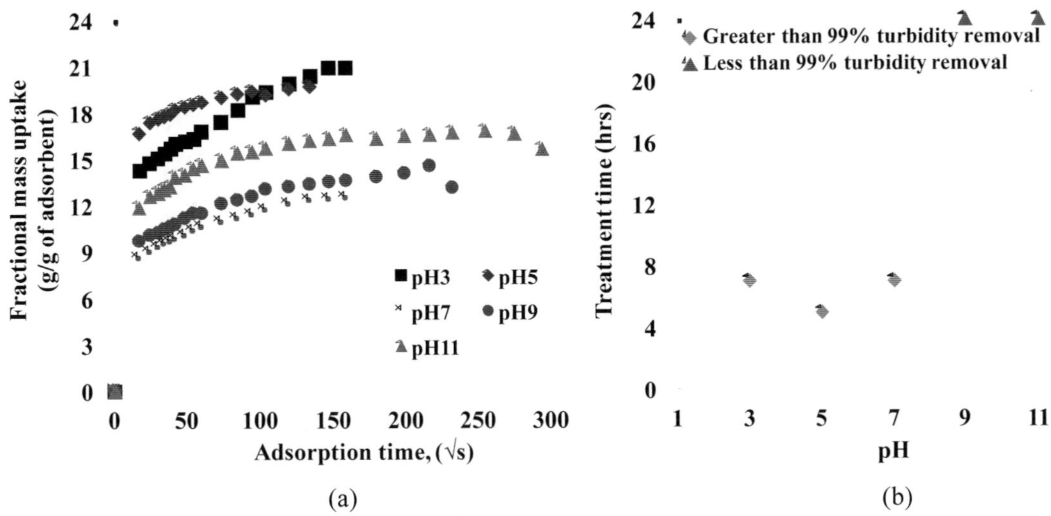

FIGURE 1. (a) Crude oil mass uptake with the effect of pH at 1% oil concentration, (b) Turbidity removal treatment time with the effect of initial pH at 1% oil concentration. Figure shows treatment times for over 99% turbidity removal at acidic and neutral conditions while at basic conditions (pH9 and pH11) the removal rates are less than 99% after 24 hrs (listed in Table 1)

Results show, for over 99% turbidity removal, the treatment times were substantially lower for acidic conditions. For basic conditions, the turbidity removal rates did not reach 99% after the allowed 24 hour of treatment time. Furthermore, at acidic conditions at pH5, the removal was found to be highest. This could be associated with a change in hydrogen and hydroxyl ion concentration, mobility, and interaction with droplets and foam surface. The oil droplets have negative surface charge, therefore, at acidic conditions, due to the opposite charge hydrogen ions, the droplet repulsion forces are decreased and coalescence occurs resulting in rapid demulsification. In addition, the hydrogen ions could protonate the basic foam surface, resulting droplets attachment. On the other hand, at pH9 and pH11, the coalescence and adsorption is lower due to intensified electrostatic repulsive forces from the OH^- and basic foam surface. At pH3, although H^+ are helpful to induce droplet coalescence and protonization, an excessive charge concentration might delay the coalescence and deposition due to ES repulsive forces between H^+. Meanwhile, at pH5, the H^+ concentration is just enough for ES forces act in favor of faster adsorption while at pH7 due to lower concentration the phenomenon is quite similar to pH3. To corroborate this hypothesis, additional foam surface characteristics such as surface energy, coverage, and heterogeneity will be presented in future publications.

CONCLUSIONS AND FUTURE WORK

The cost effective and energy efficient removal of emulsified OCs from dilute aqueous solutions typically seen in oil sands tailings is challenging. An inexpensive and reusable open cell PEPU foam adsorbent could overcome this issue. In this paper, the effect of pH on emulsified OC droplet adsorption was investigated. The oil adsorption mechanism and the forces of oil-water separation were hypothesized and effective pH operating conditions were experimentally determined. The oil adsorption rate was found to be highest at pH5 for both 1% and 3% oil concentration due to chemical potential, protonization, acid-base, and IES attractive forces. On the other hand, at pH11, the adsorption was poor due to IES repulsive forces. These results indicate that an operating at pH5 would result in the most energy efficient removal.

In future work, more accurate surface and bulk characterization of the foam and droplets will be presented for prediction of adsorption. Polymer foam surface energy profiles, zeta potentials, work of adhesion, and other properties will provide critical information about the material and spontaneity of the adsorption process. This would provide necessary information for new foam development, surface modification to improve adsorption properties, and to design energy efficient separation processes.

ACKNOWLEDGMENTS

The authors would like to acknowledge Lokesh Chokanathan for his assistance in adsorption experiments; and Daniel Burnett, Surface Measurement Systems (SMS) for the surface properties measurements. The authors would also like to acknowledge the financial support of Natural Sciences and Engineering Research Council of Canada (NSERC) and the Canada Foundation for Innovation (CFI).

REFERENCES

1. C.C. Small, S. Cho, Z. Hashisho, and A.C. Ulrich, J. Pet. Sci. Eng. **127**, 490-501 (2015).
2. United states environmental protection agency (EPA), "Particulate matter size effects on health," (online source: http://www3.epa.gov/pm/health.html)
3. E.W. Allen, J. Environ. Eng. Sci. **7**, 123-138 (2008).
4. A. Fakhru'l-Razi, A. Pendashteh, L.C. Abdullah, D.R.A. Biak, S.S. Madaeni, Z.Z. Abidin, J. Hazard. Mater. **170**, 530-551 (2009).
5. E.W. Allen, J. Env. Eng. Sci. **7**, 499-524 (2008).
6. J. Grant, S. Dyer, and D. Woynillowicz, "Fact or fiction: oil sands reclamation," (The Pembina Institute, 2008), pp. 30.
7. A. Rizvi, R.K.M. Chu, J.H. Lee, C.B. Park, ACS Appl Mat Inter. **2014**, 6, 21131-21140.
8. P.Cherukupally, J.P.Hinestroza, A.M.Bilton, and C.B. Park, "Emulsified oil-water separation using open cell polyester/polyurethane foam," *ACS Appl. Mater. Inter* (will be submitted).
9. A.W. Adamson and A. P. Gast "Long-Range Forces" in *Physical chemistry of surfaces* (John Wiley & Sons, Hoboken,1967) pp. 225-252.
10. H.J. Butt, K. Graf, and M. Kappl, "Adsorption" in *Physics and chemistry of interfaces* (John Wiley & Sons, Hoboken,1967) pp. 187-218.
11. P. Cherukupally, R.K.M. Chu, A.M. Bilton, and C.B. Park, "Oil diffusion properties of acoustic foams for oil contaminated water treatment," in Proceedings of the 29th Regional Polymer Processing Society Conference, Austria, September, 2015 (American Institute of Physics, 2016).
12. Methods for chemical analysis of water and wastes. (U.S. Environmental Protection Agency, Cincinnati, 1979), pp. 180.1-1.
13. M. Doğan, H. Abak, M. Alkan, J. Hazard. Mater., **164**, 177–181, (2009).
14. Tufenkji, N. and Elimelech, M., Environ. Sci. and Technol., **38**(2), 529-536, (2004).

Fabrication of advanced recycling process considered from the perspective of the physical degradation

Aya TOMINAGA[a], Hiroshi SEKIGUCHI[b], Ryoko NAKANO[b], Shigeru YAO[a,b,*], Eiichi TAKATORI[c]

[a] *Graduation School of Energy and Environment Systems, Fukuoka University*
[b] *Depertment of Chemical Engineering, Fukuoka University*
[c] *TOSOH A&R Center Co. Ltd.*
[a] *Tel: +81-92-871-6631, Fax: +81-92-865-6031, E-mail: td155502@cis.fukuoka-u.ac.jp, shyao@fukuoka-u.ac.jp*
**:corresponding author*

Abstract. To address the dual problems of the depletion of natural resources and the environmental impact of a growing volume of wastes, effective recycling of various products and materials has become an important challenge. Especially, the amount of plastic waste is extremely large, and thus it is very important to establish effective process of recycling them as soon as possible. However, the mechanical properties and durability properties of the products made from waste plastics are very inferior compared to those of products molded from virgin plastics, and thus their application area is presently limited. Usually, the reason of above phenomena has been thought to be caused by chemical degradation and there is no way to improve them. Recently, we found that recycled pre-consumer polypropylene (Pre-RPP), that was not chemically degraded and had almost same molecular properties of virgin polypropylene (VPP), shows very different inner structure and very poor mechanical properties compared to VPP. And by choosing suitable molding temperature, molding time, and cooling condition, the mechanical properties of Pre-RPP had changed very much and some of them shown almost same properties of VPP. By applying of this research results, we could dramatically improved the mechanical properties of waste containers and packaging recycling plastics by changing molding process.

Keywords: pre-consumer recycled polypropylene, physical degradation, mechanical property, molding condition, cooling condition, containers and packaging recycling plastics
PACS: 82.35.Lr Physical properties of polymers

1. Introduction

To address the dual problems of the depletion of natural resources and the environmental impact of a growing volume waste, effective recycling of various products and materials has become an important challenge. Especially, the amount of plastic waste become extremely large, and thus it is important to establish effective methods for recycling these materials rapidly.

There are essentially three ways to recycle plastics. In the thermal recycle, plastics are incinerated to produce heat. In the chemical recycle, monomers or organic chemicals are obtained from polymers, and they are reused. Finally, in the material recycle, recycled plastics are used for molding.

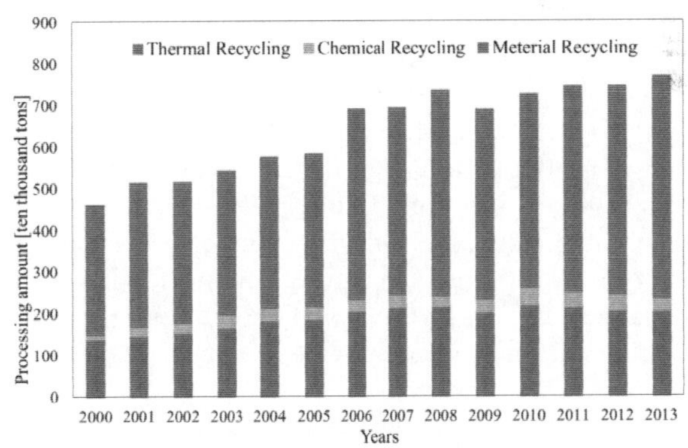

FIGURE 1. The ratio of thermal, chemical and material recycling for 13 years.

Among these methods, thermal recycle requires the least amount of energy. However, from an environmental perspective, this method should be considered to be a last resort. Chemical recycle requires large amounts of energy and very complicated processes, and even then it is not suitable for all macromolecular species. Therefore, ideally, material recycle should be repeated as many times as possible before the last thermal recycle. However, as shown in Figure 1, the proportion of material-recycled plastic has remained at about 30% for the past 13 years[1)]. And, the mechanical properties of the products of material recycled plastics are believed to be inferior to those of products molded from virgin plastics, and thus their application area is presently limited. Generally, the reason of the inferior mechanical properties were believed to be due to the chemical degradation of plastics, but there have been few detailed investigations on this subject. Especially, the physical properties of pre-consumer recycling plastic made from byproducts (for example, runners, etc.) and waste recycling plastics during molding have not been investigated much.

Recently, we investigated the mechanical properties of virgin and pre-consumer recycled polypropylene. At that time, transparent, injection-molding-grade virgin polypropylene (VPP) and pre-consumer recycled polypropylene (Pre-RPP), made from processed VPP, such as the runners generated during injection molding, were used. Various molecular properties, including the molecular weight, molecular weight distribution (Mw/Mn), density, and the melt flow ratio (MFR), of these two materials were measured and confirmed to be almost the same.

However, the mechanical properties of thin films (~100μm), the structure of the fractured surface, and the durability were quite different. Figure 2 shows stress-strain curves of VPP and Pre-RPP of thin films. The different colored lines subjected to the different tensile test.

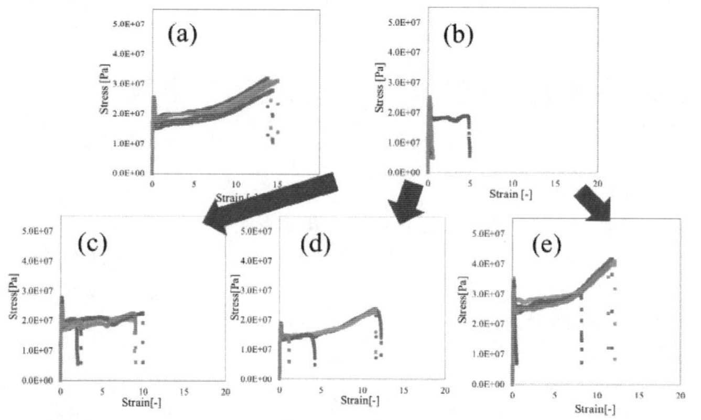

FIGURE 2. Stress-Strain curves after slow cooling and quench. (a) VPP 210°C, 2min, SC samples, (b) Pre-RPP 210°C, 2min, SC samples, (c) Pre-RPP 210°C, 10min, SC samples, (d) Pre-RPP 210°C, 2min, quench samples, (e) Pre-RPP 250°C, 10min, quench samples

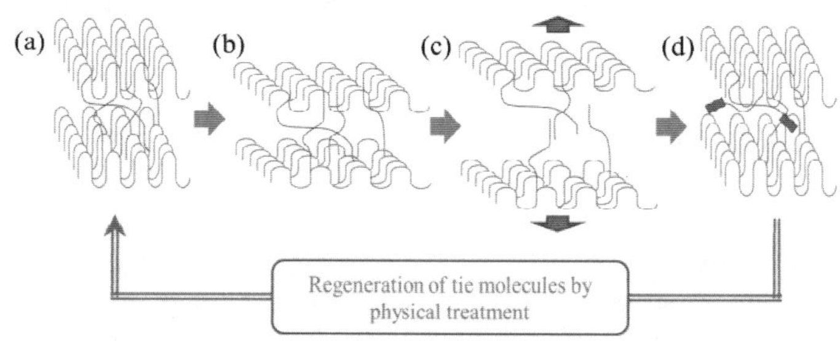

FIGURE 3. Schematic images of the influence of flow (shear) induced crystallization of inner structure of recycled crystalline polymer. (a) is original virgin crystal, (b) is the state of melt, (c) is the state of flow (shear) elongation, and (d) is recrystallized state.

The results of the tensile test of Figure 2(a), VPP (210°C, 2min, SC) (inside the brancket shows the compress molding condition such as (press temperature, press time, coolong condition). Coolong conditions are slow cooled at room temperature(SC), and rapidly cooled in ice bath(Quench). In this time, (210°C, 2min, SC) condition is set to be the standard compress condition.) are significantly ductile and variations are very small. However, as shown in Figure 2(b), almost all of the samples of Pre-RPP (210°C, 2min, SC) are very brittle. Figure 2(c) shows the stress-strain curves of Pre-RPP (210°C, 10min, SC). With changing the hot-press time, the stress-strain profiles changed to be ductile dramatically. Figure 2(d) shows stress-strain curves of Pre-RPP (210°C, 2min, Quench) samples. While the yield stresses decreased slightly, some films show ductile profiles that are similar to those of VPP. Figure 2(e) shows stress-strain curves of Pre-RPP (250°C, 10min, Quench). Almost all of these samples are ductile and the profiles are similar to that of VPP. The tensile fracture stress values of Pre-RPP (250°C, 10min, Quench) are higher than that of VPP.

From this study, we can improve the mechanical properties of Pre-RPP by selectiong long compress time, high compress temperature and rapidly cooling condition.

It is well known that a crystalline polymer will form flow (shear)-induced crystals when it had exposed to flow deformation. Thus, when Pre-RPP is exposed to high shear deformation, it is expected to have a quite different crystal structure than VPP.

Figure 3 shows a schematic diagram of the formation of the crystal structure of Pre-RPP. In the virgin state (Figure 3(a)), there are many tie-molecules between crystal lamella, and the mechanical properties and durability are very good. At the molding, temperature increases, and the crystal begins to melt and the thickness of lamella layer becomes thinner (Figure 3(b)). In this stage, the holding strength of the tie-molecules in lamella layer decreases. When the sample has subjected to elongation deformation, the tie-molecules dissociate from the lamella layer (Figure 3(c)). After this deformation and subsequent cooling and recrystallization, the dissociated molecule cannot return to the original lamella layer and becomes a dangling bond (Figure 3(d)). In this stage, the number of tie-molecules become very small and then the mechanical properties and durability become very poor. The situation shown in Figure 3(d) is the typical state of physical degradation. If we can restore the dangling bond to the lamella layer by using physical processes, the mechanical properties and durability should be regenerate and the values become closer to those of the virgin state.

We found that the mechanical properties of Pre-RPP improved by changing hot pressing and cooling conditions. However, when we make a compress film according to the method shown in Fig.4, the mechanical properties of

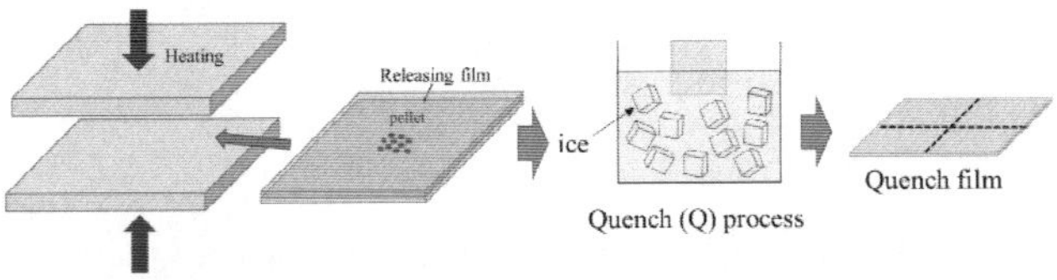

FIGURE 4(A). The step of producing quenching films.

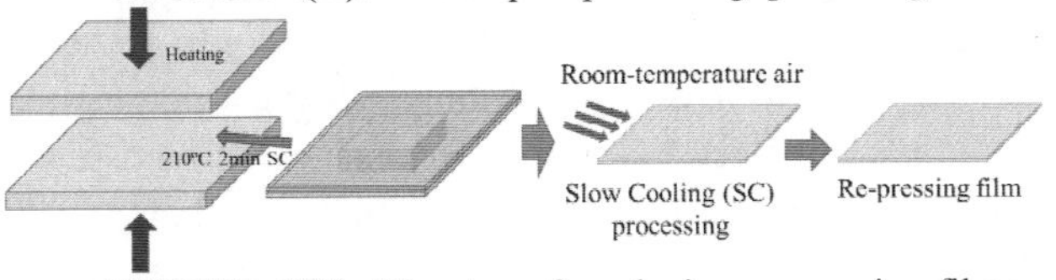

FIGURE 4(B). The step of producing re-pressing films.

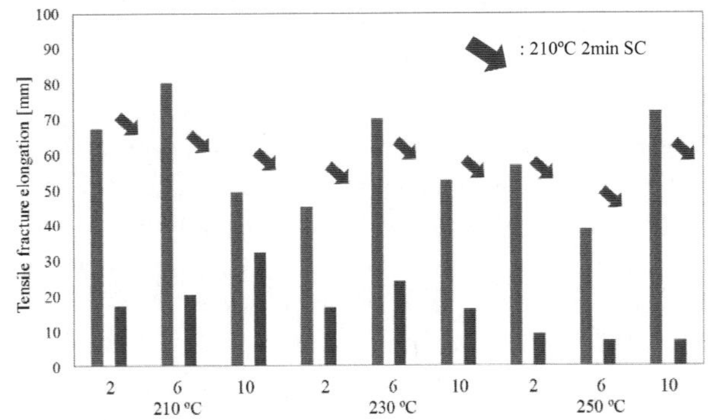

FIGURE 5. Results of tensile fracture elongation. Blue bar graph : Quench film, Red bar graph : Re-press (210°C, 2min, SC) film

improved films became poor once again by re-molding at 210°C 2min SC condition. Therefore, stabilization of the improved mechanical property is very important. It is well known that annealing is a method to stabilize the structure of the polymer. In this time, we investigated the effect of the annealing.

2. Experiments

2 − 1 Samples and molding conditions of test thin-films

Thin films were made from pellets of VPP and Pre-RPP by a press molding machine (MH-10; IMOTO Machinery Co. Ltd.). The compress condition (210°C 6min Quench) was selected for obtaining the mechanically improved thin film. To investigate the effect of annealing and re-molding, we used the press molding machine. Figure 6(A) shows how the annealing was carried out. A thin film was placed between release films (Polyimide film) and then between stainless steel plates. Figure 6(B) shows the re-molding process. The thin film was divided into 4 pieces and they were overlaid and press molded at 210 °C, 2 min, SC (standard molding condition).

Table 1 shows the molding and annealing conditions.

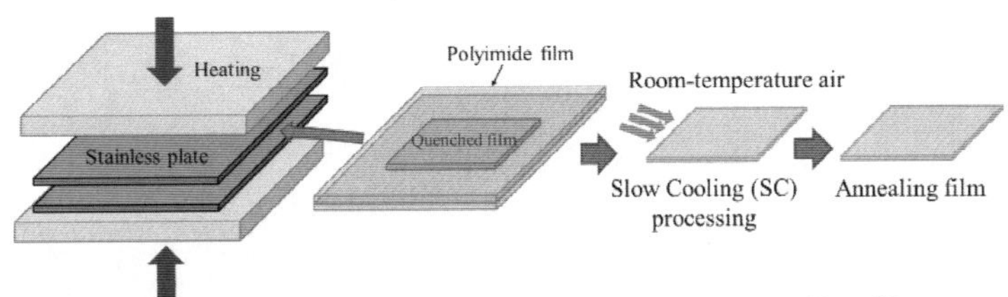

FIGURE 6(A). The step of producing annealing films.

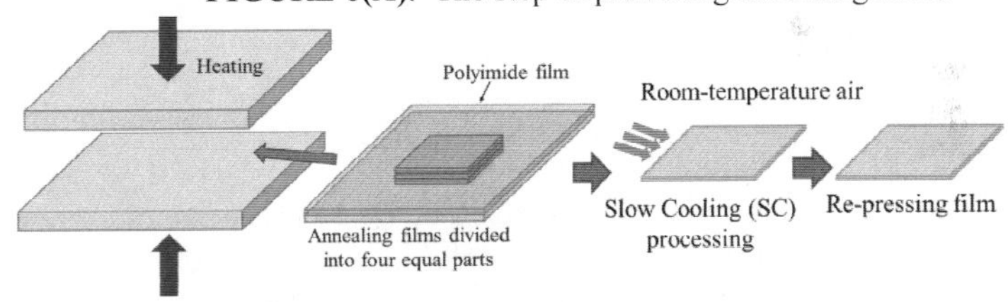

FIGURE 6(B). The step of producing re-pressing films after annealing process.

Table 1. The type of sample by molding and annealing conditions.

Sample	Molding and annealing conditions
Sample ①	Original molding condition (210°C 6min Quench)
Sample ②	Sample① → 210°C 2min SC
Sample ③	Sample① → 65°C 2 hour annealing
Sample ④	Sample① → 65°C 2 hour annealing → 210°C 2min SC
Sample ⑤	Sample① → 65°C 8 hour annealing
Sample ⑥	Sample① → 65°C 8 hour annealing → 210°C 2min SC

2 − 2 Measurement of tensile properties

The tensile test specimens were punched from the thin films fabricated under various compress mold conditions and annealing conditions. The shape of the thin films used in this study was according to JIS K 7113 2(1/3). The thickness of the thin films used in the tensile tests was about 60~100 μm. Five specimens were used for each tensile test. Tensile tests were carried out at 26.0 °C and 41.5% RH. The elongation rate was 5 mm/min. A screw-type uniaxial tester (LSC-02/30-2; Tokyo Testing Machine Company) was used for the evaluation. Toughness was evaluated from the area surrounded by the load-displacement curve by the initial cross-section of the specimen.

2 — 3 Measurement of small-angle X-ray scattering

To examine the inner structure of thin films produced in each conditions, SAXS was used. A thin film was cut into 10mm square, superimposed on about four pieces and irradiated with X-ray of 2.1~23keV. The equipment used beam line no.11 (BL11) of Kyushu Synchrotron Light Research Center.

3 . Results and Discussion

3 — 1 Tensile test results of annealing process

Figure 7 shows stress-strain curves of after annealing process. In Figure 7(a), Pre-RPP (210°C, 6min, Quench) samples are significantly ductile and variations are very small. Figure 7(b) shows the results of re-molded film(210°C, 6min, Quench→210°C, 2min, SC) and it shows that the mechanical properties became poor again. Figure 7(c) shows the results of annealed film(210°C, 6min, Q→65°C, 2 hours annealing) and it shows that the mechanical properties are almost remaind to the original values. Figure 7(d) shows the annealed-remold film (210°C, 6min, Q→65°C, 2 hours annealing→210°C, 2min, SC). This result indicates that annealing is effective for stabilizing the improved mechanical properties. Figure 7(e) shows the annealed film (210°C, 6min, Q→65°C, 8 hours annealing), and the results are brittle and variations became large. Figure 7(f) shows the annealed-remold film (210°C, 6min, Q→65°C, 8 hours annealing→210°C, 2min, SC), and they are very brittle and all of the sample aren't extended at all. From these results, we found that stabilization of the mechanical properties is achieved by annealing at 65°C, 2 hours. However, the annealing effects much depending on the annealing time and temperature.

Figure 8 shows summary of tensile fracture elongation test shown if Figure 7. As shown in this figure, the tensile fracture elongation was stabilized after performing annealing processing for 2 hours. However, it was not stabilized after performing annealing processing for 8 hours.

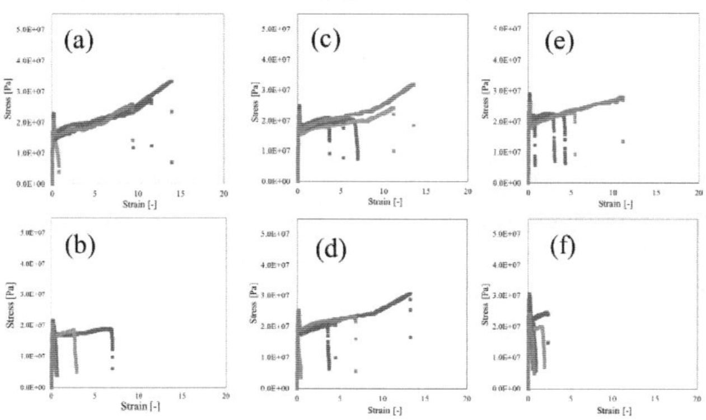

FIGURE 7. Stress-Strain curves after annealing process of Pre-RPP.
(a) sample 1, (b) sample 2, (c) sample 3, (d) sample 4, (e) sample 5, (f) sample 6

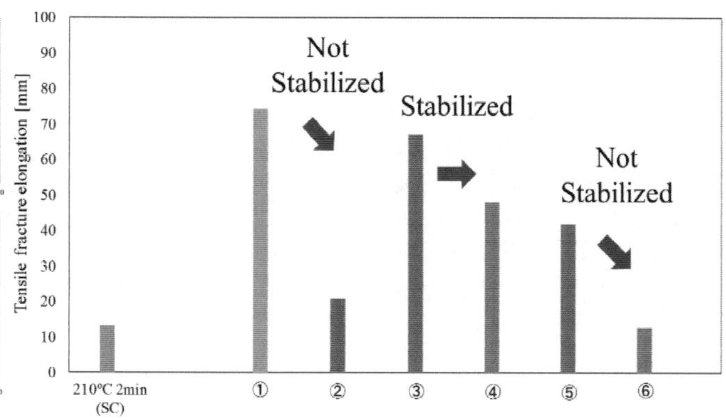

FIGURE 8. Summary of tensile fracture elongation of thin films produced by annealing process.

3 — 2 Results of small-angle X-ray scattering

Figure 9 shows results of small-angle X-ray scattering after annealing of Pre-RPP. In this figure, VPP (210°C, 2min, SC) and Pre-RPP (210°C, 2min, SC) are also shown in this graph. The SAXS profiles of SC samples are almost same and they show slightly shoulder in the wide angle region. The SAXS profiles of Quench samples show clear peak in the wider angle region compared to the SC samples. With increasing the annealing time, the peak becomes stronger. Even at that case, once compressed SC condition, the peak disappeared and the profile become usual SC profile.

From the results, we must consider the long period of amorphous and crystal layer by measuring degree of crystallinity and investigate the correlation

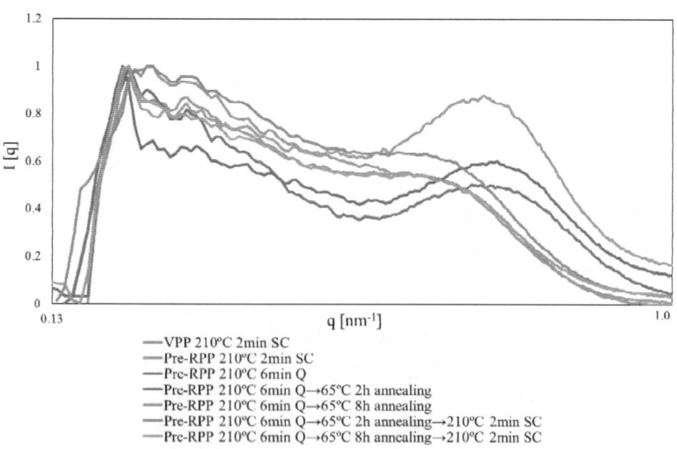

FIGURE 9. Results of small-angle X-ray scattering.

between the mechanical properties and the long period from those results.

4. Conclusion

In the previous report, we investigated the relation between tensile properties and inner structures of thin films of VPP and Pre-RPP under various hot-press molding conditions. Based on these results, we investigated stabilization method and found that suitable annealing would stabilize the improved mechanical properties by quench process.

In the small-angle X-ray scattering analysis, clear peak appears in the Quench process but it disappeared after SC remolding.

5. Acknowledgment

This research was supported by the Environment Research and Technology Development Fund (3K143013) of the Ministry of the Environment, Japan.

6. References

1. Home page of Plastic Waste Management Institute
2. Tominaga, A., Yao S., CONVERTECH & e-Print, 6(1), 52-57 (2016)
3. Tominaga A., Yao S., Sekiguchi H., Nakano R., Takatori E., Proceedings of PPS-30, AIP Conf. Proc. 1664, 150011-1 - 150011-4 (2015)
4. Yao S., Tominaga A., Y., Sekiguchi H., Takatori E., Nihon Reoroji Gakkaishi, Nihon Reoroji Gakkaishi (J. Soc. Rheol. JAPAN) 42(1), 61-64 (2014)
5. Takatori Eiichi, Shimura Toshitaka, Adachi Takayuki, Yao Shigeru and Shindou Yoshio, Nihon Reoroji Gakkaishi, Vol.42 (1), 45-49 (2014)
6. Takatori Eiichi, Shimura Toshitaka, Yao Shigeru and Shindou Yoshio, Nihon Reoroji Gakkaishi Vol. 42 (1), 39-43 (2014)
7. Amir Saffar et al., Polymer 55, 3156-3167 (2014)
8. Yao S., Tominaga A., Fujikawa Y., Sekiguchi H., Takatori E., Nihon Reoroji Gakkaishi, Nihon Reoroji Gakkaishi (J. Soc. Rheol. JAPAN) 41(3), 173-178 (2013)
9. Tominaga A., Yao S., Sekiguchi H., Nakano R., Takatori E., Kobunshi Ronbunshu Vol. 70, No.12, pp. 712-721 (2013)
10. Bai H. et al., Polymer 52, 2351-2360 (2011)
11. Ferrer-Balas D. et al., Polymer 42, 1697-1705 (2011)
12. Hedesiu C. et al., Polymer 48, 763-777 (2007)

Mechanical properties of virgin ABS/post-consumer ABS blends

Adriana M. Catelli de Souza[a*], Mayara G. Cucchiara[a]

[a] *Centro Universitário da FEI, Av. Humberto de Alencar Castelo Branco, 3972, CEP 09850-90, São Bernardo do Campo, SP, Brazil*

amcsouza@fei.edu.br

Abstract. Blends of virgin ABS (acrylonitrile butadiene styrene terpolymer) and post-consumer ABS coming from waste of electrical and electronic equipment (WEEE) was studied. Virgin ABS/post-consumed ABS blends were obtained by extrusion in five different proportions: 100 - 75/25 -50/50 - 25/75 -100 (wt%) using a co-rotational twin-screw extruder. Samples for flexural and impact tests were obtained by injection molding. After this procedure, some specimens were select for characterization and others were ground and injected again. Samples obtained by the first and second injection molding cycle were characterized by flexural and impact tests. Significant differences in mechanical properties were assessed by analysis of variance (ANOVA). vABS/pcABS blends, subjected to two injection molding cycles presented an increase in flexural modulus and flexural strength. Blends subjected to one or two injection molding cycles presented a slight increase in flexural modulus and flexural strength with increasing pcABS concentration. Impact strength of vABS/pcABS blends decreased with increasing pcABS concentration and with increasing the number of injection molding cycles.

Keywords: ABS, recycling, WEEE

PACS: 81

INTRODUCTION

Due to its good mechanical properties (toughness, rigidity and high level of strength) and ease processing, acrilonitrile-butadiene-styrene terpolymer (ABS) is used in many applications in electric and electronic equipment: computer casing, TV equipment, keyboards, mobile phone casing, printers casing, etc [1]. ABS consists of styrene – acrylonitrile (SAN) random copolymer mixed with and partially grafted to polybutadiene rubber. The polybutadiene component tends to form particles within a SAN matrix. Acrilonitrile provides chemical and heat resistance and toughness; butadiene acts as impact modifier [2].

The production of electric and electronic equipment is one of fasted growing areas of manufacturing industry [3]. As a result, huge amounts of plastic wastes from end-of-life electrical and electronic equipment (WEEE) are generated every year, creating serious environmental problems [4].

Mechanical recycling, where plastic waste is reused in manufacturing, is the preferred route for plastic waste management in terms of minimizing emission gases and energy use. Nevertheless, the mechanical properties of the recycled polymers are usually worse as compared to the virgin material due to degradation during lifetime (environmental degradation) and mainly during reprocessing (melting processing). An alternative to improve the performance of the reprocessed polymer is blending it with virgin polymer [5,6].

In the study of Scaffaro et al. [5], virgin ABS/ post-consumer ABS obtained in different compositions were recycled up to three times. The authors observed that the tensile properties decreased with the smallest concentration of post-consumer ABS during the first reprocessing. Flexural properties was unaffected by post-consumer ABS concentration and slight affected by reprocessing. Impact strength was the most affected property by post-consumer ABS concentration and number of reprocessing. Rahimi et al. [6] observed that the increase post-consumer ABS concentration in virgin ABS/ post-consumer ABS blends results in a decrease in the shrinkage.

The present work aims to study blends of virgin ABS and post-consumer ABS coming from electronic equipment housing in different proportions. The blends were characterized by mechanical properties (flexural and impact properties). Significant differences in mechanical properties were assessed by analysis of variance (ANOVA).

MATERIALS AND METHODS

Materials

Post-consumer ABS from electronic equipment housing (computers, printers, keyboards etc.) and virgin ABS with an injection-molding grade were used in this study. The post-consumer ABS were kindly provided by CEDIR (Centre for waste disposal of computers) belonging to São Paulo University (USP) - Brazil. The virgin ABS used was ABS Terluran GP 22 natural provided by Styrolution with melt index of 19g / 10min (220 ° C / 10kg), suitable for use in electrical and electronic equipment housings according to the manufacturer.

Methods

Post-consumer ABS was washed, cut and ground (particle diameter of 3mm approximately).

Virgin ABS/post-consumer ABS blends were prepared in a co-rotational twin-screw extruder Thermoscientific Haake Rheomex PTW 16 OS. The barrel temperature ranged from 210 to 190°C and the rotation speed was 150 rpm. TABLE 1 shows the blend compositions obtained in this study. The neat polymers suffered the same processing to ensure the same thermo-mechanical history.

Samples for flexural and impact tests were obtained by injection molding using Battenfeld injection molding machine HM 60/350 (injection pressure: 65 MPa, heating cylinder temperature profile of 190–200°C, mold temperature: 50°C). After this procedure, some specimens were select for characterization and others were ground and injected again.

Flexural tests were performed using an Instron tensile machine, according to standard conditions (ASTM D-790). Flexural modulus and flexural strength were determined. Charpy impact tests were performed using notched samples according to the ASTM D256-D (pendulum weight of 5.4J).

TABLE (1). Blend compositions obtained in this study.

	Virgin ABS (wt%)	Post-consumer ABS (wt%)
vABS	100	0
vABS/pcABS 75/25	75	25
vABS/pcABS 50/50	50	50
vABS/pcABS 25/75	25	75
pcABS	0	100

RESULTS AND DISCUSSION

Mechanical Properties

Figures 1 to 3 present the mechanical tests results (flexural and impact strength) of virgin ABS/post-consumer ABS blends (vABS/pcABS) subjected to one or two injection molding cycles.

The mechanical properties significant differences were evaluated by analysis of variance (Two-way ANOVA with replication), adopting a significance level p (t-student test) of 0.05. Two variables were considered: post-consumer ABS (pcABS) concentration and number of injection molding cycles (1 or 2). The results obtained are shown in Table 2.

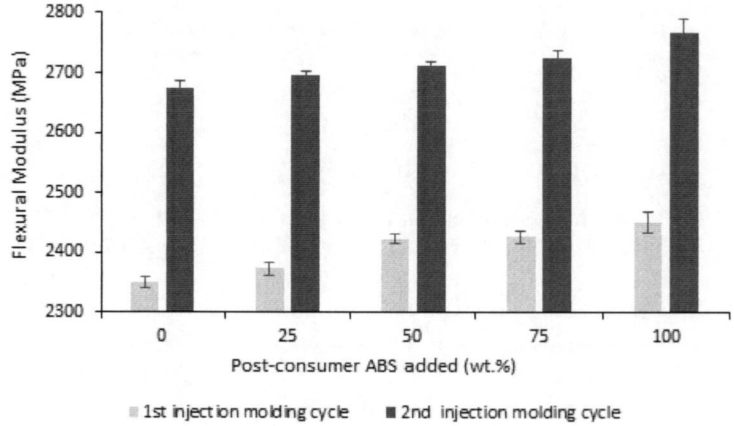

FIGURE 1. Flexural Modulus (MPa) of vABS/pcABS blends as a function of pcABS concentration.

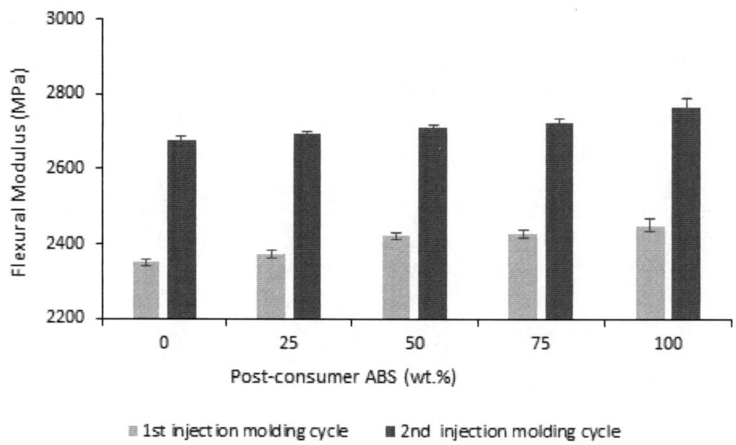

FIGURE 2. Flexural Strength (MPa) of vABS/pcABS blends as a function of pcABS concentration.

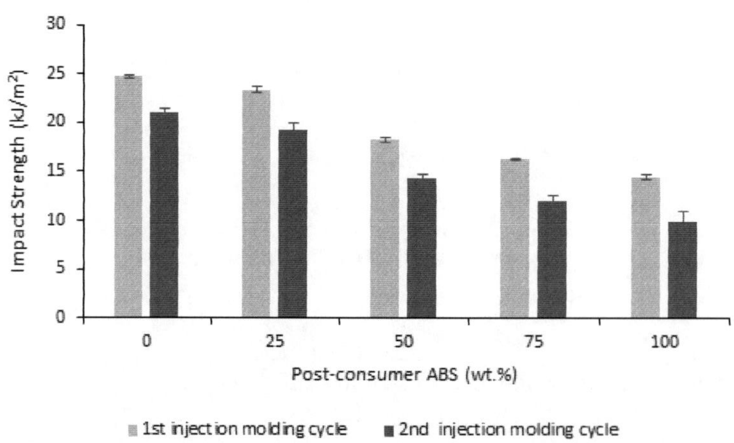

FIGURE 3. Impact Strength (MPa) of vABS/pcABS blends as a function of pcABS concentration.

TABLE (2). Analysis of variance of mechanical properties of vABS/pcABS blends.

Property	Source of variation	Sum of squares	Degree of freedom	Mean square	F	p-value
Flexural Modulus	pc ABS concentration	721894.4	1	721894.4	4568.0	**0.00**
	number of injection molding cycles	33699.7	4	8424.9	53.3	**0.00**
	Interaction	1530.3	4	382.6	2.4	**0.00**
Flexural Strength	pc ABS concentration	297.36	1	297.36	2193.19	**0.00**
	number of injection molding cycles	10.00	4	2.50	18.45	**0.00**
	Interaction	1.01	4	0.25	1.86	0.94
Impact Strength	pc ABS concentration	43.7	1	43.7	128.9	**0.00**
	number of injection molding cycles	866.9	4	216.7	638.9	**0.00**
	Interaction	5.8	4	1.4	4.3	**0.00**

vABS/pcABS blends, subjected to two injection molding cycles presented an increase in flexural modulus (about 12%) and in flexural strength (about 8% approximately) as compared to blends obtained by the first injection cycle. Also, vABS/pcABS blends, subjected to one or two injection molding cycles presented a slight increase in flexural modulus (1 to 4% - first injection cycle and 0.8 to 1.6% - second injection cycle) and in flexural strength (0.7 to 1.9% - first injection cycle and 0.6 to 2.8% - second injection cycle) with increasing pcABS concentration. The differences observed were statistically significant according to the ANOVA results shown in Table 2 ($p<0.05$ for source of variation: pcABS concentration and number of injection molding cycles). In this case, the number of injection molding cycles presented a more significant influence on the flexural properties.

Impact strength of vABS/pcABS blends decreased with increasing pcABS concentration (5 to 41% - first injection cycle and 7 to 53% - second injection cycle) and with increasing the number of injection molding cycles (15 to 31%). The differences observed were statistically significant according to the ANOVA results shown in Table 2.

Rahimi et al. [6] observed that the most affected properties of ABS, after being recycled for 5 times, were impact strength and shrinkage.

The observed mechanical properties can be related to possible ABS degradation during processing. Scaffaro et al. [5] concluded that the degradation of ABS is a complicated system because two opposite effects may occur during processing: formation of crosslinking structures in the PB phase and chain scission in the SAN matrix. At more severe degradation, breaking of butadiene/SAN grafts were observed.

The formation of crosslinking structures resulted in an increase in stiffness of the PB phase, reducing deformation and adhesion with the matrix. Consequently, the impact strength decreased since this property of ABS is dependent on PB phase. The increase in flexural modulus can also be related to increase in stiffness of the PB phase. In this case, formation of crosslinking structures in the PB phase prevailed over chain scission in the SAN matrix.

CONCLUSIONS

In this study, blends of virgin ABS and post-consumer ABS coming from electronic equipment housing were obtained in different proportions and characterized by flexural and impact properties. Significant differences in mechanical properties were assessed by analysis of variance (ANOVA).

vABS/pcABS blends, subjected to two injection molding cycles presented an increase in flexural modulus and flexural strength. The blends, subjected to one or two injection molding cycles presented a slight, but statistically significant, increase in flexural modulus and flexural strength with increasing pcABS concentration. Impact strength of vABS/pcABS blends decreased with increasing pcABS concentration and with increasing the number of injection molding cycles.

The mechanical properties behavior can be related to the formation of crosslinking structures in the PB phase leading to an increase in stiffness of the elastomeric phase resulting in a decrease in impact strength and an increase in flexural modulus. Probably, formation of crosslinking structures in the PB phase prevailed over chain scission in the SAN matrix during recycling.

ACKNOWLEDGMENTS

The authors would like to thank to CNPq and Centro Universitário FEI for financial support and to CEDIR for materials supply.

REFERENCES

1. V. Ramesh; M. Biswal;, S. Mohanty; S.K. Nayak Waste Manag. Res. **32**, 379-389 (2014).
2. P.A. Tarantilli; A.N. Mitsakaki; M.A. Petoussi Polym. Degrad. and Stab. **95**, 405-410 (2010).
3. J.C. Arnold; T. Watson; S. Alston; M. Carnie, C. Glover Polymer Testing **29**, 459-470 (2010).
4. L.B. Brennan; D. H. Isaac; J. C. Arnold J. Appl. Polym. Sci. **86**, 572-578 (2002)
5. R. Scaffaro; L. Botta; G. Di Benedetto Eur. Polym. J. **48**, 637-648 (2012).
6. M. Rahimi; M. Esfahanian; M. Moradi, J. Mater. Process. Tech. **214**, 2359-2365 (2014).

Effective Pelletizing Condition of Recycle Containers and Package Plastics

Nozomi Takenaka, Aya Tominaga, Ryoko Nakano, Hiroshi Sekiguchi, Shigeru Yao*

Department of Chemical Engineering, Fukuoka University
8-19-1, Nanakuma, Jonan-ku, Fukuoka, Japan 814-0180

[Introduction]

The mechanical properties of material-recycled containers and packaging plastics are very poor. This poor physical properties of the material-recycled plastics are believed to be due to the chemical degradation that results from the heat, ultraviolet light or oxidation in the recycling process. Such chemical degradation is irreversible because it is associated with the breaking of molecular chains. That is the one reason that there are few basic investigation about the material recycle.

However, we have recently suggested that the main reason of this phenomena is physical degradation. We also suggested the physical regeneration molding condition of those recycled plastics. In this time, we investigated pelletizing condition dependence of mechanical properties of recycle containers and package plastics. We found that the mechanical properties of injection molded samples also depends on the pelletizing condition that is estimated from the previous consideration.

[Sample and Experiments]

Recycled waste containers and packaging plastic pellet was offered from ECO FEEL Co. LTD. The pellet was not sorted and that was a mixture of polypropylene and polyethylene. Figure 1 is the photo of the pellet.

We re-pelletized the pellet by using twin crew extruder. The re-pelletize condition are listed in Table 1. Tensile and Izod test pieces are molded by injection molding.

Fig. 1 Pellets of recycled containers and packaging plastics.

[Results and Discussion]

Fig2 shows the tensile elongation and Izod impact strength of each samples. When we see the pelletizing temperature dependence, the averaged Izod impact strength and tensile elongation from sample 1 to 8 are rather smaller than sample 9 to 16. This indicates that higher pelletizing temperature is advantageous. Compering from sample 9 to 16, the sample 13 to 16 are very good. This indicates that low screw rotational speed is advantageous. Compering from sample 13 to 16, the sample 15 and 16 are better. This indicates that low taking speed is advantageous.

[Summary]

In this time, we investigated the recycled waste containers and packaging plastic pellet re-pelletize condition dependence of mechanical properties of injection molded test piece. From the results, we found that searching of the best the pelletize condition is very important for polymer recycling.

Keywords: Waste containers and packaging plastic, Material recycle, Pelletize

Table1 Pelletize conditions

Sample No.	Cooling Water Temperature (℃)	Pelletizing Temperature(℃)	Taking Speed (rpm)	Screw Rotational Speed (rpm)
1	16.3	200	90	200
2	9.8	200	90	200
3	18.4	200	45	200
4	10.1	200	45	200
5	14.6	200	90	100
6	2.7	200	90	100
7	16.2	200	45	100
8	6.3	200	45	100
9	17.1	230	90	200
10	7.9	230	90	200
11	16.3	230	45	200
12	13.1	230	45	200
13	14.3	230	90	100
14	4.9	230	90	100
15	15.9	230	45	100
16	6.4	230	45	100

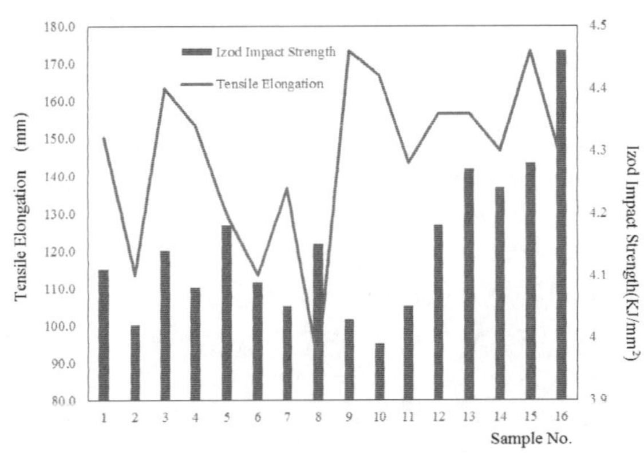

Fig.2 (a)Pelletize condition dependence elongation and Izod strength. (b)Comparison of the Izod strength of injection molded test piece and young modulus of press molded thin film.

DEVULCANIZATION OF WASTE TIRE RUBBER BY MICROWAVES

Fabiula D. B. de Sousa[1,3*], Carlos H. Scuracchio[2], Guo-Hua Hu[3] e Sandrine Hoppe[3]

[1]*Technology Development Center (CDTec), Universidade Federal de Pelotas, Pelotas, Rio Grande do Sul, Brazil*
[2] *Materials Engineering Department (DEMa), Universidade Federal da São Carlos (UFSCar), São Carlos, São Paulo, Brazil*
[3] *CNRS, Laboratoire Réactions et Génie des Procédés, LRGP, Nancy, França*
* *fabiuladesousa@gmail.com*

It is well known that the disposal of solid residues is an enormous problem, especially in the case of vulcanized rubbers like tires. The old practice of landfilling is out of using due to the rising land costs, environmental concerns, fire risks and the possibility of proliferation of mosquitoes and other disease-carrying vectors. Attempting find a solution, the devulcanization has been shown to be an effective technique, making possible to the rubber to flow and to be remolded. However, the process not only breaks sulfur links, but also creates new other bonds, modifying the structure of the material as a whole. In order to help the understanding of the structural modifications on the GTR as a result of devulcanization by microwaves, this work proposes an in-depth study of the chemical modifications as function of the exposure time of the GTR to microwaves. Some important conclusions were that the final temperature reached by the sample is the determining factor for the success of the process. So, the process must be well controlled to avoid degradation of the main chains.

Keywords: *GTR, recycling, devulcanization, microwaves.*

Introdução

The waste of tire rubber (*Ground Tire Rubber* - GTR) does not decompose easily due to its cross-linked structure (it is not biodegradable), presence of stabilizers and other additives [1, 2], being that the required time for it to degrade naturally is indeterminate [3]. It cannot be reprocessed as thermoplastic materials [4] by being infusible and insoluble in organic solvents due to the presence of three dimensional chemistry network [5].

Devulcanization of elastomers is defined as the process of total or partial break of the cross-linkings formed during the initial vulcanization [6-9]. This rubber is known as regenerated or reclaimed rubber [6], i.e., devulcanized rubber that recovered its viscous properties [10], as well as some features of the original compound [11].

The technique of microwave devulcanization uses volumetric heating of the material, promoting a more uniform heating then those achieved by more traditional methods of heating based on conduction and/or convection. As the interaction between the material and the microwave depends on the dielectric characteristics of the material, in mixtures with different types of materials it is possible the selectivity of certain regions, property that has been explored in thermoset processing with mineral fillers [12, 13]. This method is very useful because it provides an economic and environmentally friendly form of waste rubber reuse, making it possible in some cases its return to the same process and originally produced products, thus creating a similar product with equivalent physical properties [14]. The process has physical nature, i.e., does not involve chemicals during the process (is a kind of thermal devulcanization or controlled degradation [15]).

The main objective of this work is the study of the mechanism responsible for the GTR devulcanization by microwaves, through the results of extraction and swelling by solvent.

Experimental

Materials

Residues from the tread of truck tires (GTR) already sprayed and separated from non elastomeric components were used. This material was kindly donated by Pirelli Pneus Ltda.

Devulcanization of the GTR

GTR was devulcanized in a system composed by a microwave oven adapted with a stirrer with controlled speed. The process was carried out by using the maximum power of the oven (820 W). The GTR was exposed to microwaves for a period of 1 to 5 minutes, and 5.5 minutes. The temperature after the devulcanization process was measured with the aid of a rod thermometer.

Characterization

The GTRs devulcanized under all the different times of exposure to microwaves were extracted according to ASTM 2765-11 for the verification of the efficiency of the devulcanization process.

After extraction, gel fraction of desvulcanized GTRs were pressed in a hydraulic press at 170° C in a circular mold (Ø25x1 mm) for 5 min. Circular samples were immersed in 100 mL of toluene in controlled temperature of 25°C for 72 hours, according to ASTM D6814-2. The cross-linking density was calculated by using the Flory-Rehner equation.

Results and discussion

Figure 1 presents the graph of the exposure time of the GTR to microwaves *versus* gel fraction, cross-linking density and average temperature after treatment.

It can be clearly observed the dependence of the gel fraction and cross-linkings density of devulcanized rubber with the final temperature reached after its exposure to microwaves. In general, the higher the exposure time of the sample to the microwaves, the higher its final average temperature and the lower the gel fraction, proving the devulcanization. The cross-linking density did not present a trend because it is known that exposure to microwave changes the chemical structure of rubbers, and links are broken (cross-linkings) and others formed at the same time [16].

As a conclusion, the temperature rise is the mechanism responsible for devulcanization, and probably there are no additional mechanisms of interaction between the microwaves and the material.

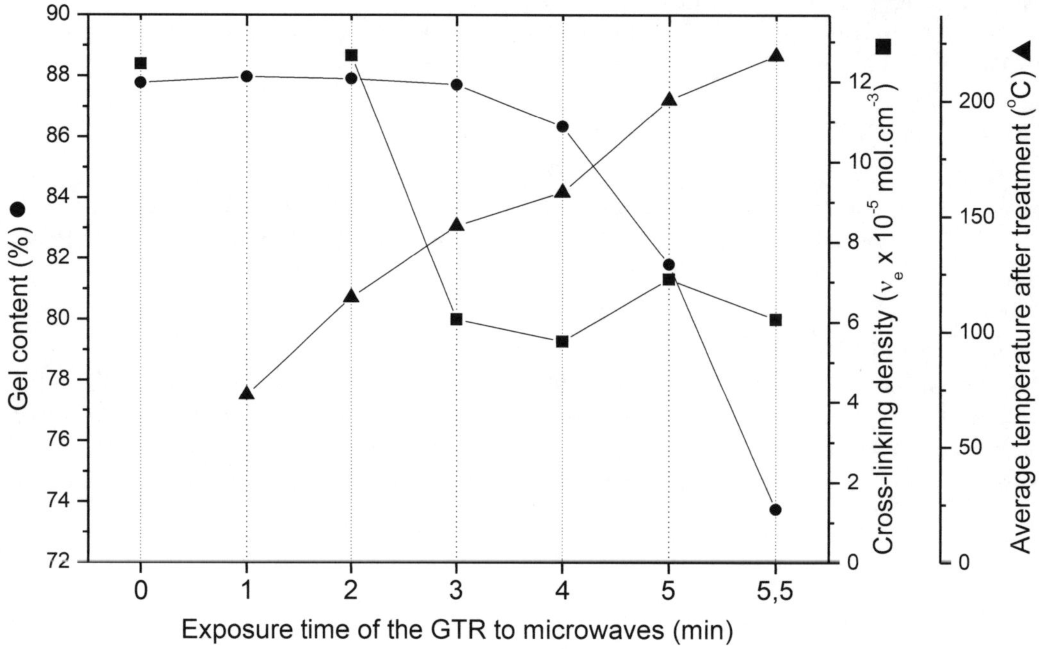

Figure 1 - Gel content, cross-linkings density and average temperature after treatment of GTRs desvulcanized under all the different times of exposure to microwaves.

Conclusions

The results presented show that the devulcanization technique is able to break the three-dimensional network composed by cross-linkings from initial vulcanization of the material by reducing the gel fraction values and the cross-linkings density, as increase the exposure time of the GTR to microwaves. The results show that the increase of the temperature of the samples due to microwaves devulcanization process is the mechanism responsible for its success.

Acknowledgments

The authos would like to thank Pirelli for the material donation, FAPESP (process number 2010/15799-6) and CNPq (process number 201891/2011-5) for the financial support.

References

1. B. Wu; M.H.Zhou *Waste Manage.* 2009, *29*, 355.
2. N. Roche; M.N. Ichchou; M. Salvia; A. Chettah *J. Elastomers Plast.* 2011, *43*, 317.
3. < http://www.inmetro.gov.br/painelsetorial/palestras/Zilda-Maria-Faria-Veloso-Ciclo-Vida-Pneus.pdf>, acesso em 26/04/2015.
4. J.M. Vergnaud; I.D. Rosca, *Methods of Recycling Waste Tire Rubber. Rubber curing and properties*, Taylor & Francis Group, New York, 2008.
5. K. Masaki; S.I. Ohkawara; T. Hirano; M. Seno; T. Sato *J. Appl. Polym. Sci.* 2004, *91*, 3342.
6. M. Sienkiewicz; J. Kucinska-Lipka; H. Janik; A. Balas *Waste Manage.* 2012, *32*, 1742.
7. J.S. Oh; A.I. Isayev *J. Appl. Polym. Sci.* 2004, *93*, 1166.

8. T. Kleps; M. Piaskiewicz; W. Parasiewicz *J. Therm. Anal. Calorim.* 2000, *60*, 271.
9. W.C. Warner *Rubber Chem. Technol.* 1994, *67*, 559.
10. Y.H. Li; S.H. Zhao; Y.Q. Wang *J. Polym. Environ.* 2012, *20*, 372.
11. < http://www.etrma.org/uploads/Modules/Documentsmanager/chemrisk--09-12-16-end-of-life-tyre.pdf>, acesso em 26/04/2015.
12. C.H. Scuracchio; D.A. Waki; R.E.S. Bretas *Polímeros: Ciência e Tecnologia* 2006; *16*, 46.
13. D.S. Novotny; R.I. Marsh; F.C. Masters; D.N. Tally. US Patent 4 104 205, 1978.
14. B. Adhikari; D. De; S. Maiti *Prog. Polym. Sci.* 2000, *25*, 909.
15. M. Myhre; D.A. MacKillop *Rubber Chem. Technol.* 2002, *75*, 429.
16. D. Hirayama; C. Saron *Ind. Eng. Chem. Res.* 2012, *51*, 3975.

Mechanical Joining of Metal Sheets and CFRTP by Using High Frequency Induction Heating

K. KUWABARA[a], T. YASUHARA[b]

[a]*Department of Environmental Symbiotic System, Nippon Institute of Technology*
[b]*Department of Products Engineering and Environmental Management, Nippon Institute of Technology*

1fuji2taka3nasuvi@gmail.com
yasut@nit.ac.jp

Abstract. Nowadays lightweight materials such as aluminum alloy or magnesium alloy, also High Tensile Strength Steel have been used for weight reduction of motor vehicles. On the other hand, Carbon Fiber Reinforced Plastics (CFRPs) have remarkable high specific strength, however they have been only adopted airplane or parts of high grade cars due to its cost. In these days, problems concerning of CFRP's applications for motor vehicles have decreased by various researches, multi material include CFRP could be used on motor vehicle designs. Generally it is still difficult to join different materials. Though there are a lot of reports about joining between CFRP and metals[1][2], the joining strength depends on the strength of rivets or the matrix material of CFRP. In this study, joining method between aluminum sheet and CFRTP (Carbon Fiber Reinforced Thermo-Plastic) sheet without rivets by using induction heating for carbon fiber is discussed, and joining strength were investigated.

Keywords: CFRTP, Mechanical Joining, High Frequency Induction Heating
PACS: 81.05.Qk Reinforced polymers and polymer-based composites

1. Introduction

CFRTPs are greatly expected for weight reduction of vehicles. However it is difficult to make whole part by CFRTP, several materials are generally used for motor vehicles. Joining method for different materials is usually using rivets or adhesion bond. In case of joining method between CFRPs.and metals, a lot of problems occur. Prepared holes for rivet make the strength weak because break of carbon fiber starting point of crack. The other hand, because adhesion bond joins interface between metal and matrix polymer of CFRP, the joining strength is almost same as the strength of matrix polymer. In this study, joining method between aluminum sheet and CFRTP sheet without rivet or adhesion bond are discussed.

2. Experimental

Induction heating apparatus was used for joining between an aluminum sheet and a CFRTP sheet as shown Fig. 1. The inner diameter of coil is 50mm, and frequency of electrical power supply is 50 kHz.

Firstly, heating characteristics of CFRTP by induction heating was measured. A CFRTP matrix polymer is PMMA, sheet of 10mm-wide, 30mm-long and 0.5mm-thick was set into the induction heating coil. Then the carbon fibers situated among CFRTP were heated by high frequency induction current and surface of the CFRTP was heated up to its softening point (about 170 degrees C). Relation between surface temperature measured by T type thermo-couple and voltage of power supply was shown in Fig.2. It is found that the surface temperature of CFRTP is affected by the thickness of CFRTP. Joining procedure of this study shows as below.

An aluminum sheet of 0.5mm-thick was fabricated by mechanical anchors (Fig.3) which were formed like a pin holder. The projection shape was about 1mm squared. The press mold with a punch made of ceramic material of 10mm-square was closed and aluminum sheet specimen and molten CFRTP were pressed and projection pins stuck into molten CFRTP. The compression force of about 2kN was applied.

FIGURE 1. Experimental apparatus

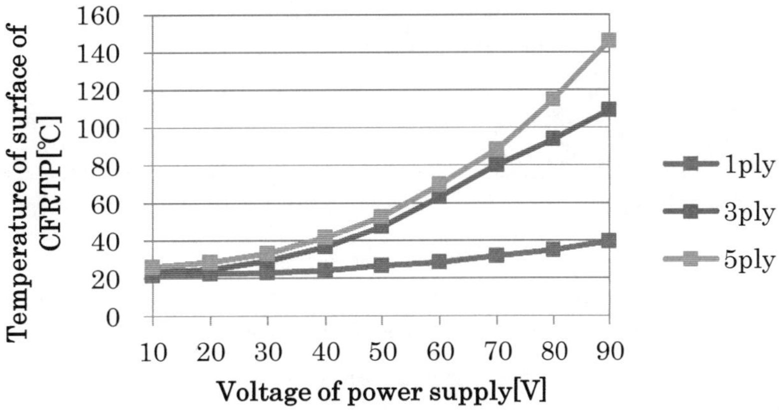

FIGURE 2. Relation between surface temperature of CFRTP and voltage of power supply applied high frequency induction heating

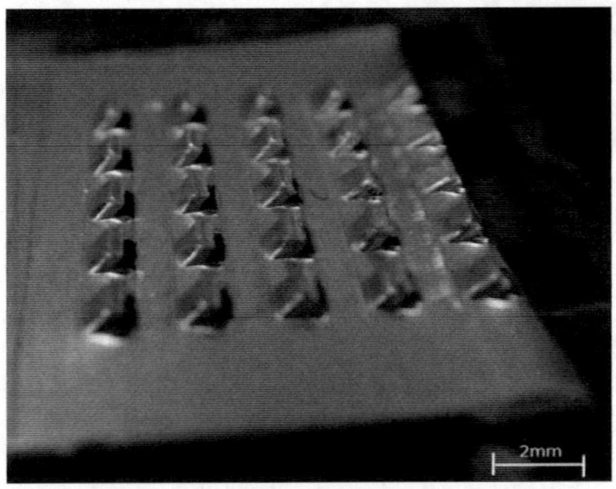

FIGURE 3. Anchors of aluminum plate fabricated by iron punch

Tensile shear test were performed and the specimen is shown in Fig.4. The cross-shape joint specimen was used for tensile peel test and the specimen is shown in Fig.5. The joining specimens were classified by direction of anchor shown in Fig 6 (a), (b). The direction was referred to as the Plus-anchor (Fig. 6 (a)), and another direction was referred to as the Minus-anchor (Fig.6 (b)).

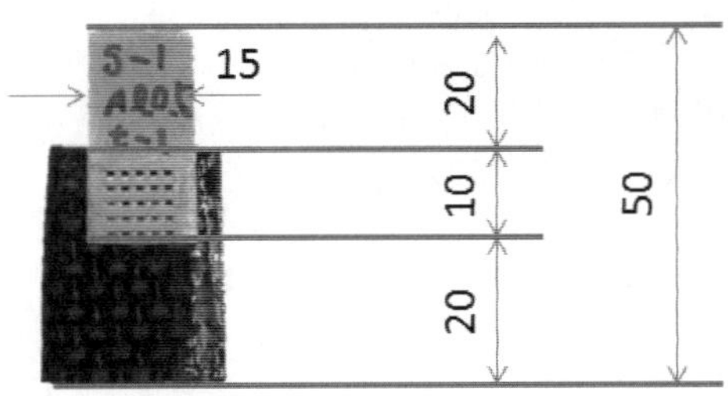

FIGURE 4. Joining specimens for tensile test

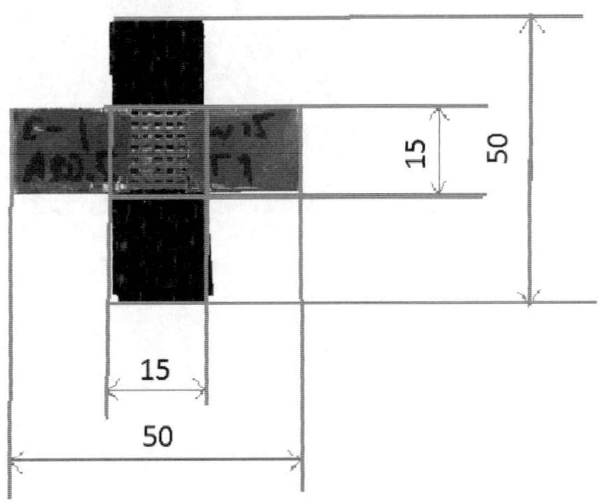

FIGURE 5. Joining specimens for tensile peel test

(a)Plus anchor (b) Minus anchor

FIGURE 6. Schematic diagram of Joining specimens classified by anchor's direction

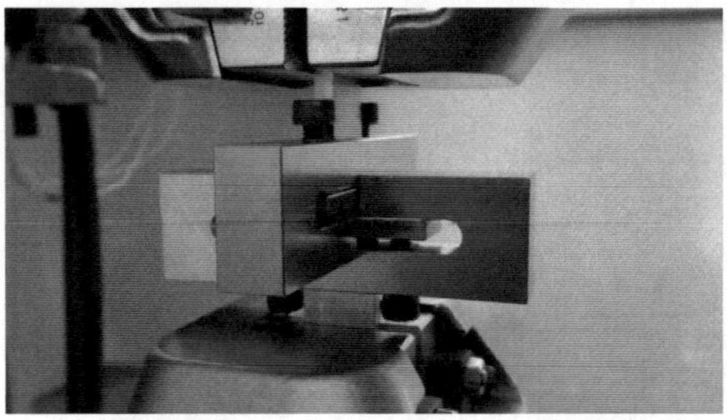

FIGURE 7. Jig for tensile peel test

3. Tensile shear test

Tensile shear test was performed. All specimens were broken after tensile test as shown in Fig. 8 Two typical curve were obtained in this experiment as shown in Fig.9. As a result shown in Fig. 9(a), the maximum tensile strength of Plus-anchor was about 50MPa. In case of Plus-anchor, maximum tensile strength was about 70MPa (Fig. 9(b)). This value was almost twice larger than the tensile strength of the pristine PMMA sheet. It can be say that carbon fibers of CFTTP contribute to the joining strength.

FIGURE 8 Broken specimen after tensile test

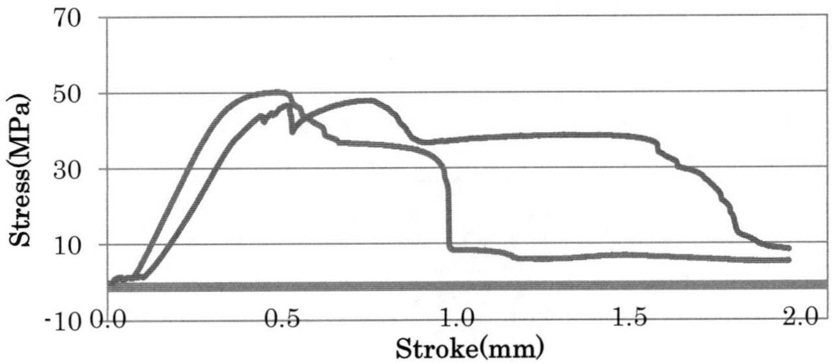

(a) Plus-anchor

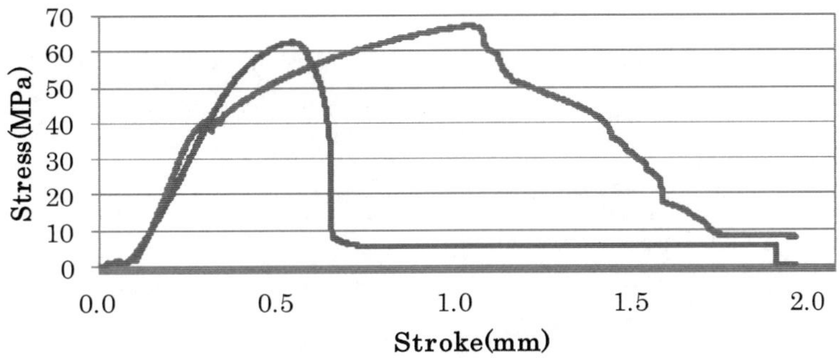

(b) Minus-anchor

FIGURE 9. Results of tensile shear test

4. Tensile peel test

Tensile peel tests were performed twice. As a result, maximum peel strength was about 15MPa as shown in Fig 10. All test piece peeled off at the interface. It is found that the projection pins don't contribute to the peeling strength.

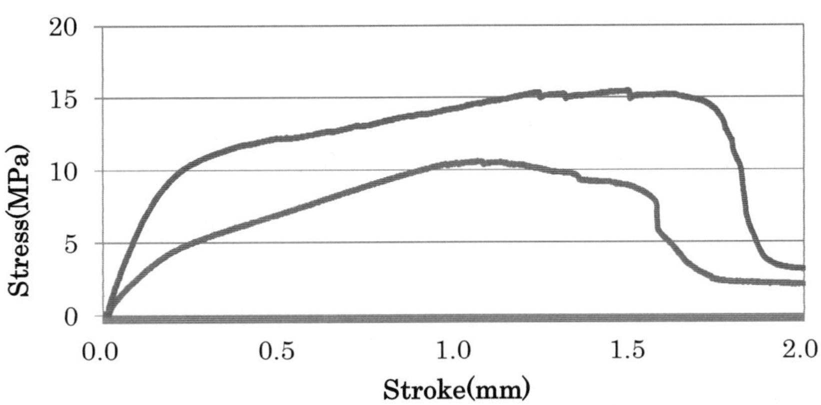

FIGURE 10. Result of tensile peel test

5. Observation of the cross-section of joinin specimen

The interface between carbon fiber and aluminum anchor was observed by SEM. Joining specimens were filled in embedding-resin, after curing specimen was cut along the cutting line as shown Fig.11. Fig.12.and Fig.13 shows to the image by SEM. From Fig. 13, some orthogonal fiber against anchor was broken. However, in case of parallel fiber against anchor, damage could particularly not be observed.

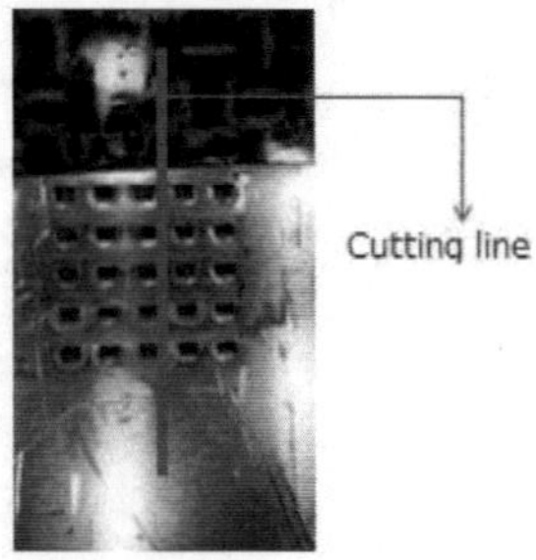

FIGURE 11. Cutting line for SEM observation of cross section between aluminum plate and CFRTP sheet

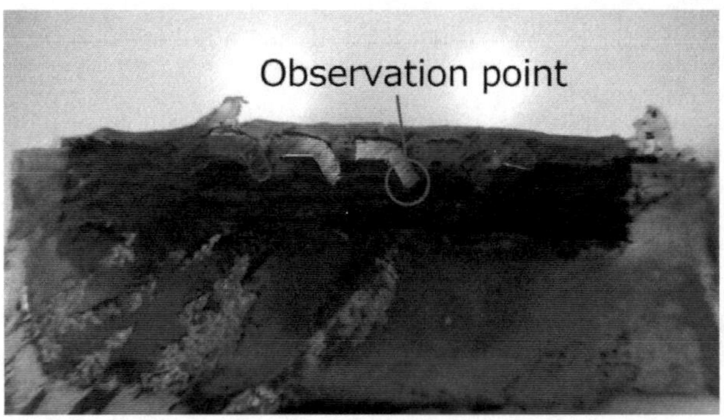

FIGURE 12. A cross section of joining specimen

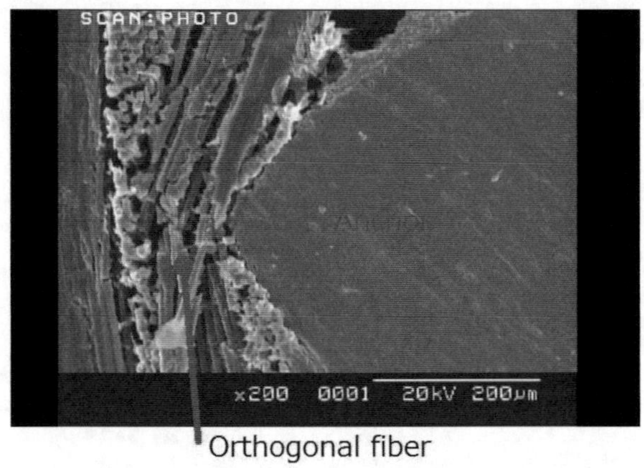

FIGURE 13. SEM observation of interface between carbon fibers and aluminum anchor

6. Summary

Joining method between CFRTP and aluminum sheet were discussed. CFRTP sheet was softened by induction heating and fabricated small aluminum anchors stuck into CFRTP. Then tensile shear test of specimens were carried out. This result indicates that the mechanical anchors of the aluminum sheet contribute to the enhancement of the joining strength.

Prevent method for galvanic corrosion of between CFRTP and metal and also improvement of peel strength should be considered in the future.

REFERENCES

[1] YAMAMOTO Tetsuya : Japan Welding Society 70(2), 262-265, 2001-03-15
[2] YAMAGUCHI Yasuhiro : Japan Welding Society 65(6), 463-468, 1996

Investigation of the Matrix Influence on the Laminate Properties of Epoxy- and Polyurethane-based CFRPs manufactured with HP-RTM-Process

F. Behnisch[a], P. Rosenberg[a*], K. A.Weidenmann[a,b], F. Henning[a]

[a]*Fraunhofer Institute for Chemical Technology, Josef-von-Fraunhofer-Str. 7, 76327 Pfinztal, Gemany*
[b]*Institute for Applied Materials, Engelbert-Arnold-Straße 4, 76131 Karlsruhe, Karlsruhe Institute of Technology (KIT), Germany*

Philipp.Rosenberg@ict.fraunhofer.de

Abstract. The high pressure resin transfer molding (HP-RTM) process has the potential for high-volume production of continuous fiber-reinforced components in the automotive industry. The development of robust equipment, new process variants and highly reactive matrix systems lead to significant reductions of the cycle time in recent years. The paper addresses the manufacturing of carbon fiber reinforced plastic (CFRP) laminates using different matrix systems. To evaluate the matrix influence on the material properties, matrix dominated test methods were selected for evaluation of the mechanical properties. The test plates were made with the HP-RTM process at constant process parameters using carbon fiber fabrics. Only the matrix-specific supplier instructions for processing of the matrix systems (mixing ratio, temperature of components in HP-RTM equipment and mold) were varied in the study. Three Polyurethane (PU) systems and one Epoxy (EP) system were used for the characterization of the matrix dominated properties. To identify the interlaminar shear properties, two test methods were selected and compared to each other: The Short-Beam Shear (SBS) Test and the Edge Shear Test (ESH). To obtain the damage tolerance under impact loading, the energy absorptions for each material combination during instrumented drop tower tests were investigated. The results show the impact of the different Epoxy and Polyurethane matrix systems on the laminate performance.

Keywords: CFRP, HP-RTM, ILSS, Short Beam Shear Test, Edge Shear Test, Impact test, Epoxy resin, Polyurethane
PACS: 81.05.Lg; 81.05.Qk; 81.05.Zx; 81.20.Hy; 81.70.Bt; 83.80.Ab; 89.20.Bb

INTRODUCTION

The HP-RTM process has been studied intensely in recent years [1–5]. During this time, different variants of the process were developed to fulfill the requirements of the industry for quick manufacturing of high-performance composites. The process variants, namely high-pressure injection resin transfer molding (HP-IRTM) and high-pressure compression transfer molding (HP-CRTM), were investigated with different epoxy resin systems. However, polyurethanes (PU) have gained high interest for the use in HP-RTM process due to a good adjustability of their processing behavior to the process and component requirements. PU resins furthermore promise comparable material properties with regard to epoxy resin systems. The co-authors investigated the quasi-static mechanical properties of epoxy and polyurethane based laminates. The results indicate that testing of quasi-static, fiber-dominated properties lead to no clear statement about the influence of different resin systems on the mechanical properties as the main load during testing is taken by the fiber [6]. The objective of this study is therefore to investigate matrix dominated test methods to gain a better understanding of matrix specific composite properties. The test plates were manufactured with constant and comparable process conditions to allow a direct comparison of the test results.

A widely used test method for determining the interlaminar shear strength is the Short Beam Shear (SBS) Test (figure 1a) according to the standard DIN EN ISO 14130 [7]. The test is similar to the three-point-bending test having a modified ratio of support distance L to sample thickness h (L/h) of 1/5 to obtain high shear stresses in the sample. The stresses are induced indirectly by the bending load and the resulting SBS values are therefore defined as the apparent interlaminar shear strength. The classical beam theory assumes a parabolic distribution of shear stresses in the cross section, having the maximum value in the middle of the sample thickness. At this point, the normal stresses theoretically disappear and a pure shear stress state is present. In the SBS test standard, an overview of approved types of failure and their interpretation is given. To obtain comparability between different samples a pure shear failure has to occur but even this does not guarantee a reliable value for the measured shear strength. Several publications indicate that the SBS theory does reflect the real stresses in the sample only partially and may lead to misinterpretations of the test results. The failure often occurred next to the loading or support rollers before a shear failure was observed [8, 9]. Finite Element Analysis (FEA) showed the appearance of shear and compressive stress concentrations near the force introduction and support areas which exceeded the theoretical stresses determined by the classical beam theory [10, 11].

Weidenmann et al. proposed the Edge Shear Test (ESH) as an alternative test method to obtain the interlaminar strength of fiber reinforced plastics which is schematically illustrated in figure 1b. This test uses a direct force introduction (F_p) on the laminate cross-section to apply shear load between two fabric layers of the sample. No overlap between the upper and lower force introducing areas should be present to transfer the load by interlaminar shearing to adjacent laminate layers. During the test, fixing jaws (F_{Fix}) support the sample to prevent tilting and thin specimens may be equipped with caps to achieve homogeneous force introduction and to avoid peeling of the specimens near the load. Weidenmann et al showed that the EST provides a reliable shear failure of the specimens [8]. The distribution of the load stresses in the sample can be predicted with a numerical analysis of the Compression Shear Test (CST) which has the identical force introduction principle. Comparing ESH and CST, the main difference is the sample geometry which is slightly curved for the CST test. The shear stress showed homogeneous distribution along the sample and small stress concentrations near the force application areas were identified [9].

The characterization of the damage behavior under dynamic impact load was carried out in accordance to the standard DIN EN ISO 6038 [12], which is also known as Compression After Impact (CAI) test. Firstly a sample is pre-damaged by an impactor in a drop tower (see also figure 1c). Subsequently, the residual compressive strength of the damaged sample is measured with a compression test rig. Having an instrumented drop tower, the absorbed energy in the specimen during the impact can be evaluated. This study is confined to the analysis of the energy absorption / time relation during the impact tests of specimens with different matrix systems at various energy levels. Delfosse and Poursartip evaluated an energy-based approach for investigating impact damage behavior. A full penetration of the sample by the impactor lead to approximately 30 % energy absorption by matrix damage and 70 % by fiber damage [13].

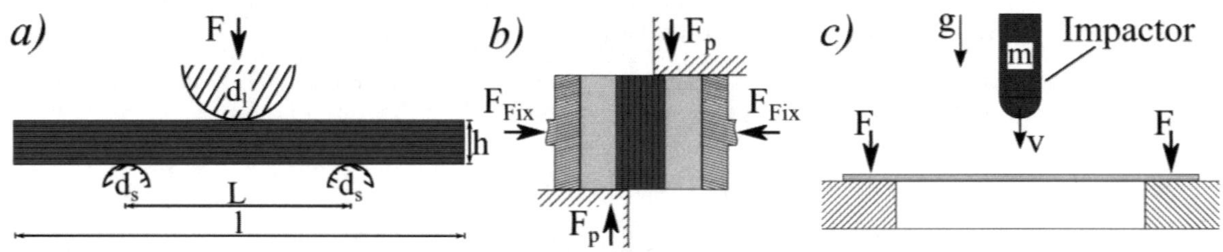

FIGURE 1. Test methods used in the study: a) Short Beam Shear Test; b) Edge Shear Test; c) Impact Test in drop tower

EXPERIMENTAL

The carbon fiber (CF) reinforcement used consists of 8 layers biaxial non-woven fabric from Zoltek (type MD300) with a symmetrical layup of $[0/90]_{4s}$. Each layer of the fabric is made of 150 g/m² carbon fibers in 0° direction and 150 g/m² in 90° direction. The biaxial layers are stitched with a polyester yarn having 4 g/m². The fabric is based on the 50 K heavy tow carbon fiber roving type Panex 35. Four different matrix systems were used for the investigations - one epoxy resin (EP1) and three polyurethane systems (PU1, PU2 and PU3). The resins were processed following the technical specifications of the respective supplier.

Manufacturing of the test samples with the HP-RTM process

The resin specific process parameters are given in table 1. The laminates were manufactured with the HP-RTM process using constant process parameters to ensure the direct comparability of the samples. A highly precise hydraulic press from Dieffenbacher type Compress Plus DCP G 3600/3200 AS with parallel holding system and a maximum press force of 36,000 kN was used. The oil tempered RTM plate mold has a central film gate, two vacuum blocks and a cavity size of 900 x 550 mm^2. Two HP-RTM machines from KraussMaffei were used for the infiltration, a KraussMaffei 8/3,2K for processing of the EP system and a KraussMaffei RimStar 8-4 for the processing of the PU systems. The machines are equipped with separate circulation units for the single components of the resins which are then mixed under high pressure of 120 bar in the mixing head during the injection step. The injection was carried out with constant process parameters for all matrix systems used. After placing the fabric layup in the cavity, the mold was closed by the press and vacuum was applied for 60 s. The press force for the injection was set to 500 kN and a resin amount of 710 g was injected with a constant flow rate of 40 g/s. After the mixing head had closed, the press force was increased with a press force increase rate of 750 kN/s to 5000 kN achieving the desired laminate thickness and to cure the resin. After demolding and post-curing of the plates the test samples were obtained by water-jet cutting. The epoxy resin system was processed with an additional component, an internal mold release agent (IMR) with a content of 2 % w/w. The manufactured CFRP laminates have a fiber volume content of 52.4 ± 3.7 Vol.-% and a thickness of 2.3 mm.

TABLE 1: Matrix specific process parameters

	EP1	PU1	PU2	PU3
Resin/Polyol temperature in equipment	80 °C	80 °C	70 °C	45 °C
Hardener/Isocyanate temperature in equipment	30 °C	35 °C	50 °C	45 °C
Mold temperature	120 °C	110 °C	90 °C	90 °C
Mixing ratio Resin:Hardener:IMR and Polyol:Isocyanate [% w/w]	100:24:2	100:132	100:121	100:178

Testing parameters

The rectangular specimens for the SBS test had a length (l) to thickness (h) ratio of l/h = 10 and a thickness (h) to width (w) ratio of w/h = 5. The diameter of the loading roller (d_l) was 5 mm and the support rollers had a diameter (d_s) of 2 mm. For each matrix system at least five tests at a test speed of 1 mm/min were conducted using a 50 kN testing machine type Inspect table 5 kN from Hegewald & Peschke and the shear strength was calculated at the maximum force following the test standard: $\tau_{SBS} = ¾\cdot F_{max}\cdot(w\cdot h)^{-1}$. The ESH specimens (length = 8 mm; width = 9 mm) were prepared with force introduction caps made of a 2 mm thick aluminum plate on each side to prevent the sample of pressure failure near the force introduction. At least eight samples of each matrix system showing pure shearing failure were considered for the evaluation of the ESH properties. The failure of the samples was well observed during the ESH tests in form of a significant drop of the force and the test was stopped subsequently. The interlaminar shear strength was then calculated using the maximum force during the test (F_{max}), relative to the shear area (l x w) of the sample: $\tau_{ESH} = F_{max}\cdot(l\cdot w)^{-1}$.

The energy absorption of the specimens during the instrumented impact was analyzed at nine impact energy (E_I) levels: 10, 15, 20, 24, 27, 32, 37, 42, 45 Joule. The impactor was a cylinder with a spherical head diameter of 19 mm and a mass of 4.84 kg. The sample size was 150 x 100 mm² (length x width). The tests were carried out on an Instron Dynatup® 9250HV drop tower at KIT. The force-time data and impact speed were recorded and the energy-time relation for each test was calculated by the integrated Instron Dynatup® Impulse Data Acquisition Software. Figure 2 shows a representative curve of the absorbed energy by time for an impact test. At maximum energy (E_{max}) the complete kinetic energy of the impactor is transferred to the sample having the maximum deflection at this point. The elastically stored energy (E_{el}) in the specimen is then returned to the impactor in form of kinetic energy. The remaining difference is the energy absorbed by damage mechanisms: $E_D = E_{max} - E_{el}$.

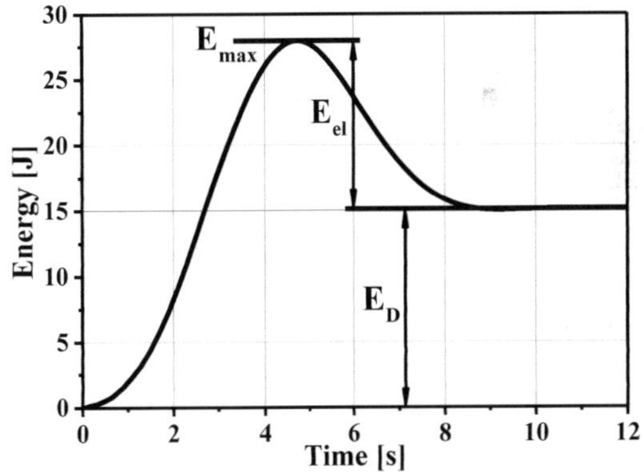

FIGURE 2. Schematic energy-time-plot for an impact test with maximum energy E_{max}, elastic energy E_{el} and damage energy E_D

RESULTS AND DISCUSSION

The interlaminar shear strength results (SBS and ESH) are given in figure 3a. The results of the SBS test show an average shear strength of > 60 MPa for each of the investigated materials whereas the PU1 has the highest average value of 69 MPa. The EP resin shows the highest standard deviation compared to the PU systems. In contrast the ESH test results are consistently lower compared to the SBS test results with an average deviation of 27 % (PU3) to 42 % (PU2). The measured average ESH shear strength values are in a range of 37 to 48 MPa. Comparing ESH and SBS test for the EP resin system the standard deviation is consistently higher compared the PU systems. The shear strength measured for PU2 in ESH test shows a decrease in its average value compared to the SBS test result and the standard deviation is significantly higher for the EHS.

Figure 3b shows the results of the impact tests at different impact energy levels. The ratio of energy absorbed by deformation (E_D) to the impact energy (E_I) is drawn as solid line and increases at lower impact energy levels. For the EP resin the maximum is reached at 27 J impact energy and for the PU resins the maximum appears in a range of 27 to 32 J. At impact energy levels > 27 J the EP resin samples drop significantly compared to the PU systems. The E_D energy has exceeded a critical value in failure behavior of the laminates (see also figure 4). The second relationship shown in figure 3b (dashed lines) is the ratio of the elastic energy (E_{el}) to the impact energy (E_I) which represents the elastic energy stored in deflection of the samples during the impact considering that the impactor will be rebounded. As the impact energy increases the elastic share reaches 0 % at 27 J for the EP samples which indicates that a critical value of a total material failure (see figure 4) is reached. The PU samples show a low amount of 5 to 12 % elastic energy absorption at 27 J which reaches the critical value of 0 % at 32 J impact energy.

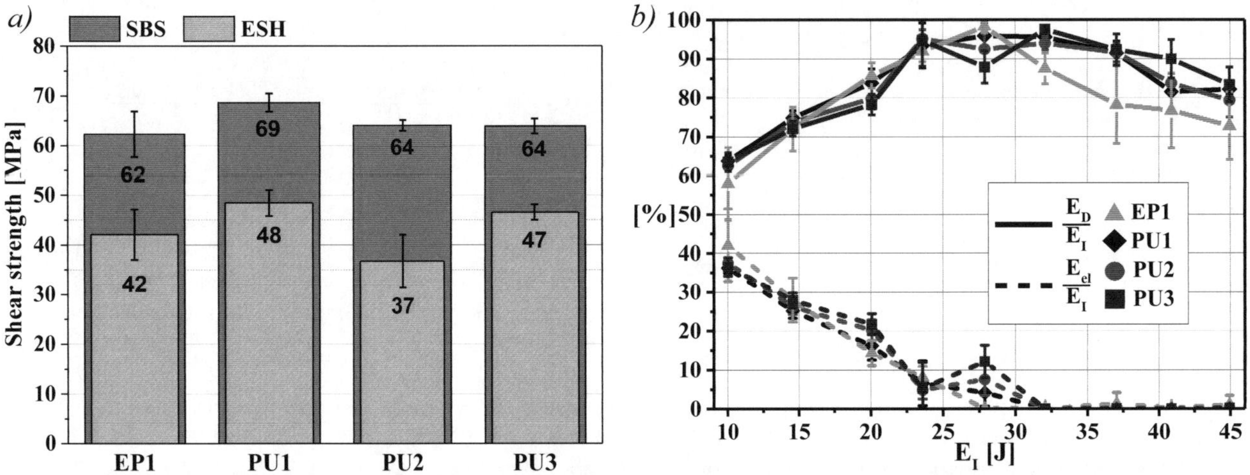

FIGURE 3. a) Comparison of interlaminar shear strength results of SBS and ESH test method; b) Result of the impact tests, ratio of the maximum energy E_{max} to the impact energy E_I at different impact energy levels

Comparing SBS to ESH test a significant drop of the shear strength can be observed in a range of 27 % (PU3) to 42 % (PU2). It has to be noted that a direct comparison of the results is only possible if a clear shear failure of the samples can be observed [7]. In this study, all SBS samples showed consistently complex failure modes. As already mentioned in the introduction, several publications have investigated this test method demonstrating, that a pure shear failure may not be realized with the SBS test. The indirect force introduction by the rollers causes significant high compression and shear stresses which lead to interferences of failure modes in the load introduction area [10, 11]. Thus, the results of the SBS test indicate that a realistic value of the interlaminar shear strength can hardly be obtained by using this method. In contrast the results of the ESH tests show a consistent interlaminar shear failure between two layers of the sample. The force-displacement-curves are characterized by a significant drop of the force during the test which is accompanied by an abrupt failure. The stress distribution and the influence of possible stress concentrations in the force introduction area are not precisely known and the determined values may not represent the real interlaminar shear strength. However, the reproducible sample failure in ESH test allows a more reliable comparative statement of interlaminar shear behavior with respect to the SBS test. Both test methods indicate that the PU1 specimens tend to have the highest interlaminar shear strength, although this statement has no absolute validity considering the standard variations. The drop of the PU2 shear strength in ESH test can be explained by the existence of many small micropores, which were identified in SEM. These voids are imperfections and act as starting points for sample failure which causes a decrease of the pure shear strength obtained by ESH test. Generally the PU based samples have a lower standard deviation than the EP and tend to have slightly higher average shear strength values with exception of PU2 in ESH test. The elongations at break of the PU materials (> 9 %) are higher

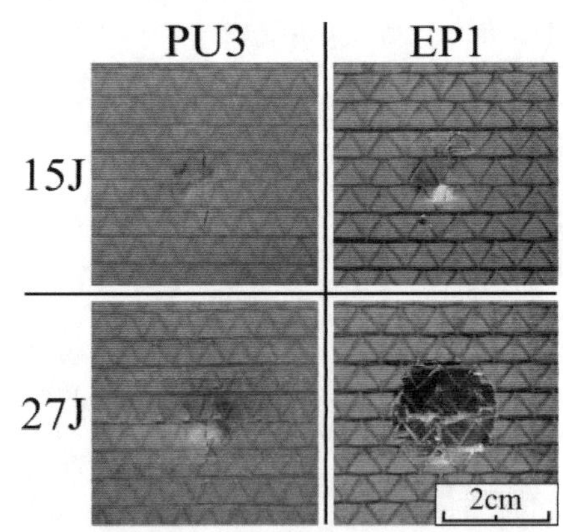

FIGURE 4. Comparison of the EP1 and PU3 samples at 15 and 27 J impact energy

compared to the EP system (< 6.6 %). This indicates that the higher PU ductility might lead to a better compensation of peak stresses on microscopic level.

The dynamic impact test curves at energies < 27 J indicate a quasi-linear and comparable behavior of the tested materials. Until this point, the impactor is prevented to punch trough the specimens of all materials and a significant content of elastic energy is present. The critical impact energy for the EP is reached at 27 J. At this energy the impactor punches through the samples and the elastic energy decreases to 0 % which indicates that the impactor is not rebounded anymore. This critical impact energy is higher for the PU materials (32 J). At energies > 32 J no elastic energy could be measured for all materials, the energy absorption caused by damage decreases and the impactor punches through the samples. It was clearly seen, that the PU based samples were able to absorb higher level of impact energy compared to the EP samples. Nevertheless, the real extend of damage could not be determined quantitatively. Possible delaminations, fiber and matrix cracks within the laminate layers are not visible on the specimen's surface. For a comprehensive assessment of the damage, additional testing methods like computer tomography have to be considered.

CONCLUSIONS

In this study laminates with different resin systems were manufactured using the HP-RTM process to investigate the matrix influence on the interlaminar shear strength and the materials behavior under dynamic load. Two test methods, SBS and ESH were used and compared to each other to obtain the shear strength. The results show that the SBS causes complex failure modes in contrast to the ESH test having a high reproducible interlaminar shear failure of all tested samples. The ESH values were consistently lower compared to the SBS and the negative influence of micropores could be observed with the ESH test method. The results indicate that the ESH test is more reliable than the SBS test to obtain comparable shear properties. More tests and simulations of the stresses in the sample are needed to evaluate and establish the ESH test method for composite characterization. The impacts tests indicate a higher energy absorption of PU materials compared to EP if the impact energy exceeds a critical value. Below that critical value the impact data is characterized by a comparable behavior but the visual analysis of the damage patterns shows that PU has a higher damage tolerance than EP based material. Further evaluations are necessary to gain a better understanding of the dynamic behavior regarding the investigated materials.

REFERENCES

[1] R. Chaudhari, P. Rosenberg, M. Karcher, S. Schmidhuber, P. Elsner, and F. Henning, "High Pressure RTM Process Variants for Manufacturing of Carbon Fiber Reinforced Composites," *19th International Conference on Composite Materials, 28 July-02 August 2013, Montreal, Canada*, 2013.

[2] R. Chaudhari, F. Viebahn, M. Karcher, P. Elsner, and F. Henning, "Characterization of high-performance composites manufactured by using high-pressure RTM process variants," *ACCE Conference Paper*, 2012.

[3] P. Rosenberg, R. Chaudhari, M. Karcher, F. Henning, and P. Elsner, "Investigating cavity pressure behavior in high-pressure RTM process variants," *Proceeding of the Polymer Processing Society 29th Annual Meeting*, 2013.

[4] R. Chaudhari, "Characterization of high-pressure resin transfer molding process variants for manufacturing high-performance composites," *PhD Thesis,* Fraunhofer Verlag, 2014.

[5] P. Rosenberg, R. Chaudhari, A. Pascal, M. Karcher, and F. Henning, "Effects of process parameters on cavity pressure and component performance in high-pressure RTM process variants," *ACCE Conference Paper*, 2014.

[6] P. Rosenberg, B. Thoma, and F. Henning, "Characterization of epoxy and polyurethane resin systems for manufacturing of high-performance composites in high-pressure RTM process," *ACCE Conference Paper*, 2015.

[7] *Faserverstärkte Kunststoffe - Bestimmung der scheinbaren interlaminaren Scherfestigkeit nach dem Dreipunktverfahren mit kurzem Balken*, DIN EN ISO 14130, 1998.

[8] K.A. Weidenmann, L. Baumgärtner, and B. Haspel, Eds, *The Edge Shear Test-an Alternative Testing Method for the Determination of the Interlaminar Shear Strength in Composite Materials*, 2015.

[9] F. Rosselli and M. H. Santare, "Comparison of the short beam shear (SBS) and interlaminar shear device (ISD) tests," *Composites Part A: Applied Science and Manufacturing*, vol. 28, no. 6, pp. 587–594, 1997.

[10] W. C. Cui and M. R. Wisnom, "Contact finite element analysis of three-and four-point short-beam bending of unidirectional composites," *Composites science and technology*, vol. 45, no. 4, pp. 323–334, 1992.

[11] M. Xie and D. F. Adams, "Contact finite element modeling of the short beam shear test for composite materials," *Computers & structures*, vol. 57, no. 2, pp. 183–191, 1995.

[12] *Luft- und Raumfahrt - Faserverstärkte Kunststoffe - Prüfverfahren - Bestimmung der Restdruckfestigkeit nach Schlagbeanspruchung*, DIN EN 6038, 2014.

[13] D. Delfosse and A. Poursartip, "Energy-based approach to impact damage in CFRP laminates," *Composites Part A: Applied Science and Manufacturing*, vol. 28, no. 7, pp. 647–655, 1997.

Compression Moulding of LFT Components with Increased Quality by Variothermal Mould Technology

Christian Hopmann, Christian Beste, Arne Boettcher*, Kai Fischer

Institute of Plastics Processing (IKV) in Industry and the Skilled Crafts at RWTH Aachen University
Seffenter Weg 201, D-52074 Aachen, Germany

arne.boettcher@ikv.rwth-aachen.de

Abstract. For the production of composite components made of long-fibre reinforced thermoplastics (LFT) the economic compression moulding process is well established. However, the low temperature of the compression mould during processing leads to a significant increase in both the material viscosity and in the compression force in particular for thin-walled components. In addition, low surface qualities result which prevent the application of such components in visible areas. To overcome these disadvantages, a variothermal compression moulding process has been developed at the Institute of Plastics Processing (IKV) at RWTH Aachen University. With the variothermal mould technology a reduction of the viscosity of the moulding compound during processing is achieved. This results in a reduction of compression force by about 60 %. Additionally, moulding compounds with higher fibre contents (up to 60 % by weight) can be processed and the surface quality of LFT compression moulding components can be significantly increased for the application in visible areas. This paper describes the functionality of the variothermal mould technology and its efficiency for the process of LFT compression moulding. Results of investigations regarding the influence of the mould temperature on the process properties and the component properties are presented and discussed.

Keywords: Composites, Compression Moulding, LFT, Variothermal, Temperature control
PACS: 61.41.+e; 81.05.Qk; 81.20.Hy

INTRODUCTION

Fibre reinforced plastics (FRP), also known as composites, are used in many industrial applications. The lightweight properties of FRP in the form of very high weight specific stiffness and strength are particularly decisive. For an economical production of FRP parts in large numbers low production costs and cycle times as well as high component complexities are required. In this context, thermoplastic composites are meltable and can be formed [4]. For further improvement of component quality, innovative solutions in the form of novel FRP manufacturing processes and materials are required [1, 2, 3].

For applications requiring a high number of components, the compression moulding of long-fibre reinforced thermoplastics (LFT) is well established. LFT lightweight components in medium and large series (10,000 to more than 100,000 parts/a) can be manufactured economically [5, 6]. During compression moulding the molten LFT charge is positioned into a mould, compression moulded and cooled under high pressure [7]. The short cycle times of less than one minute and the material recycling enable efficient and therefore cost-effective component manufacturing [6, 7, 8]. Compared to injection moulding of short fibre reinforced thermoplastics, a lower fibre damage during the plasticisation and compression moulding usually results in a higher average fibre length (> 4 mm) and higher mechanical properties in the component. Additionally shorter cycle times are feasible for large-area components [8, 9]. Therefore, the compression moulding of LFT is used for the production of large-area and ribbed components. Typical applications are especially instrument panel supports (IPS), technical front-ends and underbody structures [6].

A disadvantage is the inferior surface quality due to freezing effects in the cold compression mould during charge placement. Fibre imprints and freezing effects occur on the component's surface, which means that these components are not suitable in visual applications. Due to the necessary fluidity of the LFT compound in the mould, the fibre content and fibre length and thus the mechanical properties of the component are limited [11]. The low mould temperature leads to a significant increase in material viscosity and the required compression force during processing. Therefore, a cost-intensive compression technology is often used with high compression forces of up to 50,000 kN. Thus, a significant advantage in terms of process and component properties can be expected by optimising the temperature control of the compression mould. In this paper a new compression moulding process with a variothermal temperature control is presented and results of investigations regarding the influence of the mould temperature on the process properties as well as the component properties are presented and discussed.

APPROACH – VARIOTHERMAL TEMPERATURE CONTROL

Variothermal temperature control is well known for different applications in injection moulding in order to increase surface quality, flowability and to reduce flow lines [11, 12, 13]. To dynamically control the temperature of the compression mould additively manufactured cavity inserts with a three-dimensional network (Figure 1, left) of heating and cooling channels arranged directly behind the cavity surface are used. The prototype compression mould (400 x 200 mm²) was developed by Werkzeugbau Siegfried Hofmann GmbH, Lichtenfels, Germany. The additive manufacturing (LaserCusing®) of the mould inserts allows for high heating and cooling rates by an integration of netshaped temperature channels for complex component geometries, as well as a reduction of the masses of the cavity inserts. Depending on the inlet temperature of the used thermal oil, heating and cooling rates of up to 8.5 K/s can be achieved by this temperature control. Figure 1 right, shows an exemplary profile of the mould surface temperature, measured at the opened mould.

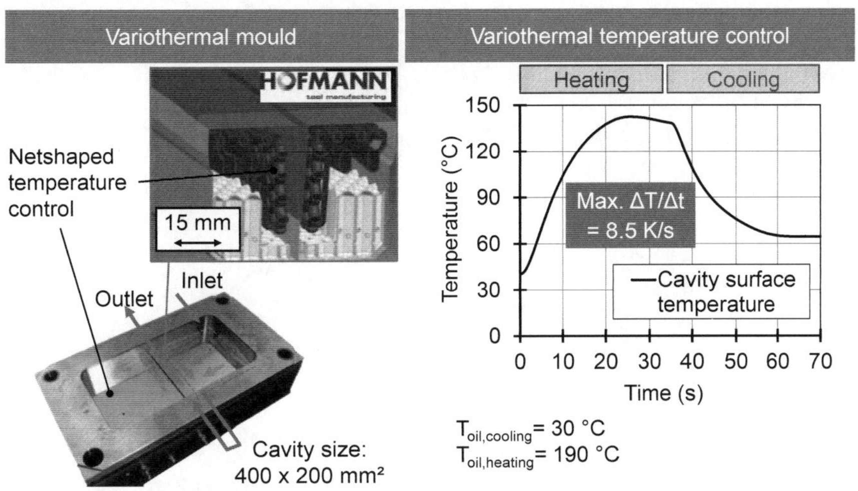

FIGURE 1. Variothermal mould for dynamic heating and cooling

EXPERIMENTAL

The objectives of the experimental investigations regarding the process properties are to reduce the required compression pressure and to process LFT with increased FWC at similar compression force. The objectives of the experimental investigations regarding the component properties are to increase the components surface quality as well as mechanical properties. To achieve these objectives the variothermal mould technology is used. As material long-fibre granulates (LFG) Celstran PP-GF40 and Celstran PP-GF60 from Ticona GmbH, Sulzbach, Germany, with a fibre weight content (FWC) of 40 wt.-% and 60 wt.-% and an initial fibre length of 10 mm are used. The material is extrusion based molten by a KMH 60 S of the company Kannegiesser GmbH, Vlotho, Germany. The LFT charge is positioned into the compression mould with a mass temperature of 220 °C. For components with a final thickness of 3 mm a mass of 293 g for the PP-GF40 and a mass of 353 g of the PP-GF60 is used. The closing speed of the compression mould is 10 mm/s.

For all investigations, a constant temperature of the mould surface is ensured during filling of the cavity by a long heating time of 8 min before placing the LFT charge in the mould. In Table 1 the constant mould surface temperature after 8 min heating depending on the heating circuit temperature of the oil $T_{oil,heating}$ is shown. A cooling circuit temperature of the oil $T_{oil,cooling} = 30$ °C is used for all investigations. For the variothermal process the temperature control of the mould is changed from the heating to the cooling circuit after the defined compression force is achieved. For the investigations a cooling time of 120 s is used. Reference components are produced with a constant mould temperature $T_{const.} = 60$ °C .

TABLE (1). Mould temperature T_{mould} after a heating time of 8 min depending on the heating circuit temperature $T_{oil,heating}$

$T_{oil,heating}$	T_{mould}
140 °C	126 °C
190 °C	169 °C

For the investigations of the required compression pressure depending on the mould temperature the charge (PP-GF40) is positioned at a defined position into the mould and compressed with a pressure of 10, 20, 30 and 50 bar relative to the area of the mould. As a result the filling of the mould is compared by images. For extending the compression moulding component properties investigations regarding the processing of LFT material with the high FWC of 60 wt.-% were done. Due to the high needed compression pressure LFT with this high FWC is not processable in common compression moulding process. To show the processability of this material the filling of the mould is stopped at defined compression pressure of 10, 20, 30 and 60 bar relative to the area of the mould and a heating circuit temperature of $T_{oil,heating}$ = 190 °C.

The potential of the mechanical properties of LFT components with a high FWC of 60 wt.-% is analysed by three point bending tests (DIN EN ISO 14125). Therefore, the mechanical properties of components produced with the PP-GF40 and the PP-GF60 LFT material with a compression pressure of 150 bar and $T_{oil,heating}$ = 190 °C are investigated. The surface topographies of test specimens (FWC = 40 wt.-%) is recorded by photographic images and 3d surface scans from confocal laser scanning microscope, Typ VK-X200, company Keyence Deutschland GmbH, Neu-Isenburg, Germany. The arithmetic mean roughness R_a (DIN EN ISO 4287) is analysed from the 3d surface scans of three components based on the altitude profile on two lines in each micrograph.

As the here presented investigations use long heating and cooling times for analysing the influence of the mould temperature on the process and component properties, the production of compression moulding components with using the benefits of the variothermal temperature control can be done in a cycle time of about 1 minute. Therefore, a heating circuit temperature of $T_{oil,heating}$ = 190 °C with a heating time of 22 s and a cooling circuit temperature $T_{oil,cooling}$ = 30 °C with a cooling time of 30 s is used.

RESULTS AND DISCUSSION

Process Properties

The results of the investigations regarding the required compression pressure at two heating circuit temperatures as well as a conventional isothermal mould temperature of $T_{const.}$ = 60 °C are shown in Figure 2. The filling of the mould increases with increasing compression pressure. By processing with the conventional mould temperature the cavity is not completely filled at a compression pressure of 50 bar. In contrast the mould is already filled at a compression pressure of only 30 bar at a heating circuit temperature $T_{oil,heating}$ = 140 °C and the required compression pressure is reduced to 50 %. A further increase of the heating circuit temperature to $T_{oil,heating}$ = 190 °C reduces the required compression pressure by 60 %. The results show, that an increased mould temperature potentially allows for a use of presses with significantly lower compression forces and lower investment costs in comparison to use a conventional isothermal mould temperature.

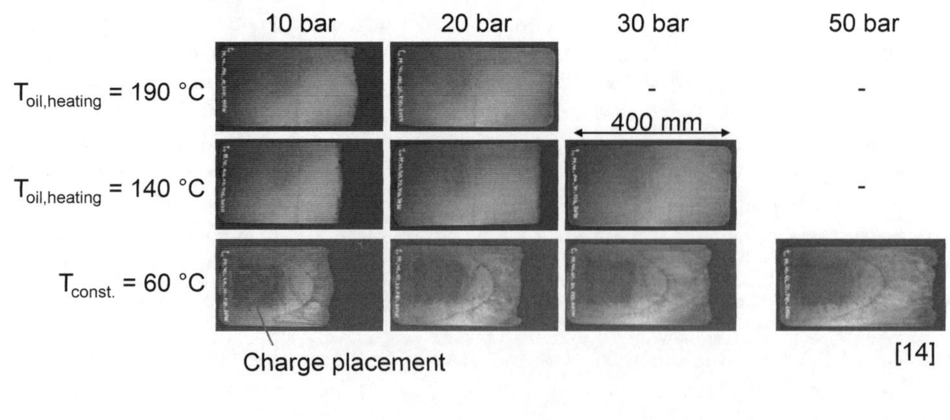

FIGURE 2. Flowability depending on the mould temperature (Exemplary: FWC = 40 wt.-%)

Figure 3 shows the filling of PP-GF60 in the cavity as a function of the compression pressure at a heating circuit temperature of $T_{oil,heating} = 190$ °C. For this temperature it is possible to process LFT with a FWC of 60 wt.-% by compression moulding with moderate pressures comparable to common compression moulding process with a FWC of 40 wt.-%. This allows for the processing of LFT with high FWC in compression moulding enabled by the variothermal process control.

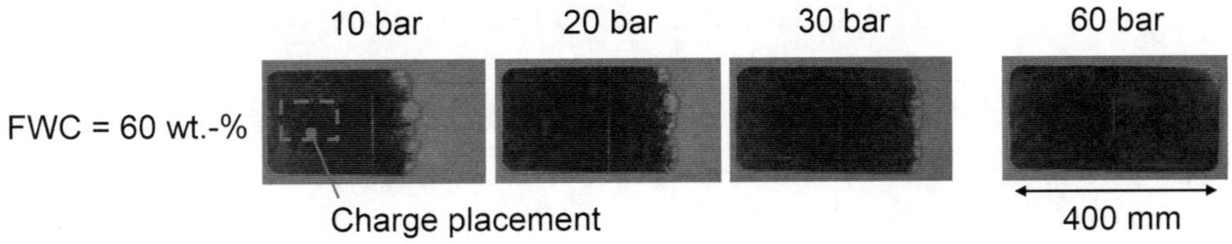

FIGURE 3. Processing of LFT with FWC of 60 wt.-% ($T_{oil,heating} = 190$ °C)

Component Properties

In Figure 4 the three-point bending properties of LFT samples with a FWC of 40 wt.-% and 60 wt.-% prepared from components processed with a heating circuit temperature of $T_{oil,heating} = 190$ °C are shown. The increase of the FWC results in an increase of the bending strength of 35 % and an increase of the bending modulus of 85 % both in flow direction and perpendicular to the flow direction. In conclusion the increased mechanical properties can lead to new applications of compression moulding components.

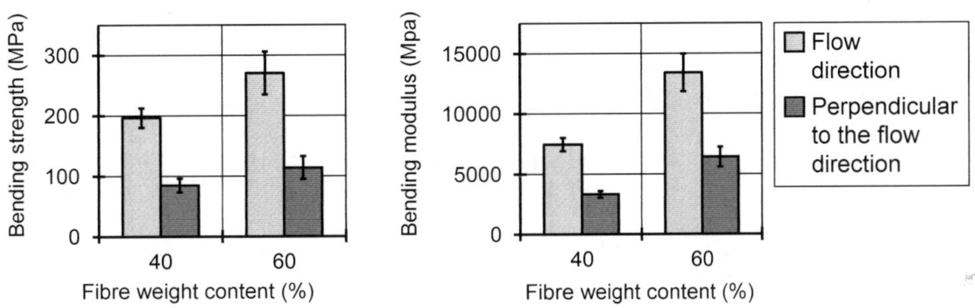

FIGURE 4. Three-point bending properties for different FWC

Figure 4 shows photographic images of components produced with conventional isothermal mould temperature ($T_{const.} = 60$ °C) in direct comparison with components produced at $T_{oil,heating} = 140$ °C. The differences in surface quality are obvious, particularly in the area of the charge placement. The surface topographies of test specimens, recorded by LSM, show the better surface properties as well (Figure 4, left).

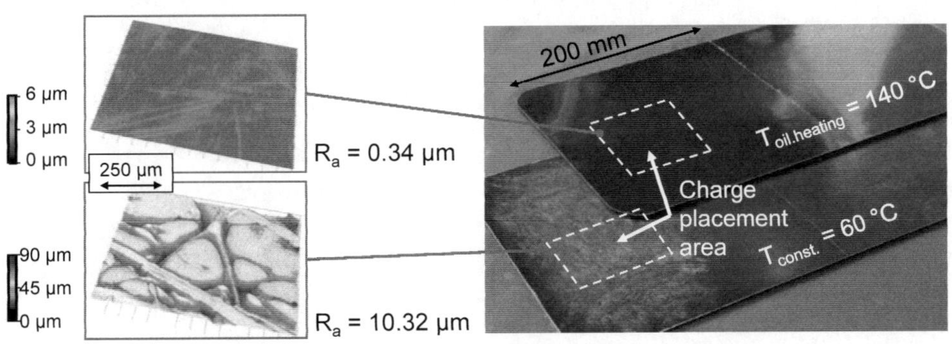

FIGURE 5. Surface qualities for different mould temperatures

The exemplified mean roughness of the cavity surface of R_a = 0.19 μm cannot be achieved by the test specimens produced at a heating circuit temperature of $T_{oil,heating}$ = 140 °C (R_a = 0.34 μm). The mean roughness of test specimens produced at $T_{const.}$ = 60 °C is R_a = 10.32 μm. The characteristic freezing effects on the surface can be prevented and significantly improved surface qualities are achieved by the use of variothermal temperature control in compression moulding.

CONCLUSION AND OUTLOOK

In this paper results of investigations of LFT compression moulding with high mould temperatures and a temperature control of the compression mould for the production of high-quality LFT components were shown. The high mould temperature avoid or limit the restrictive effects regarding compression force, processable FWC and surface quality of conventional compression moulding on the process and component properties. The results show that by increasing the mould temperature, the compression force required for the production of 3 mm thick components can reduced by 60 %. This reduction of the compression force allows for the reduction of investment costs for the press technology. Keeping the LFT viscosity on a low level with a high mould temperature enables the processing of LFT with a FWC of 60 wt.-%. Thus an increase of the mechanical properties can enlarge the field of applications while presses are the same. Components with high surface quality can be produced with $T_{oil,heating}$ = 140 °C in LFT compression moulding. In further investigations the production of LFT components by variothermal process control in short cycle times will be analysed as well as the filling behavior of more complex components.

ACKNOWLEDGMENTS

The research project (18078 N) of the Forschungsvereinigung Kunststoffverarbeitung has been sponsored as a part of the "industrielle Gemeinschaftsforschung und -entwicklung (IGF)" by the German Bundesministerium fuer Wirtschaft und Energie (BMWi) due to an enactment of the German Bundestag through the AiF. We would like to extend our thanks to all organisations mentioned.

REFERENCES

1 C. Dallner,"Werkstoffe fuer die Zukunft des Leichtbaus" in *Kunststoffe* 102, 2012, pp. 60-67

2 Ch. Hopmann; A. Böttcher and J. Fragner, "Funktionsintegrierte Kunststoffloesungen fuer die zukuenftige Mobilitaet" in *VDI Conference Kunststoffe im Automobilbau,* Mannheim, Germany, 2013

3 N.N., "An Assessment of Mass Reduction Opportunities for a 2017 – 2020 Model Year Vehicle Program" Market Analysis, Lotus Engineering Inc, Ann Arbor (MI), USA, 2010

4 M. Schuck, "Kunststoffe als Leichtbauwerkstoffe" in *Conference Metropolregion Nuernberg*, Ansbach, Germany, 2011

5 W. Krause: "Langfaserverstaerkte Thermoplaste im Automobilbau" Faserverbundwerkstoffe im Automobilbau und Transportwesen. Duebendorf, Germany, 2012

6 M. Schemme, "Langfaserverstaerkte Thermoplaste - Status und Perspektiven" in *Conference Thermoplastische Faserverbundkunststoffe*, Erlangen, Germany, 2014, pp. 1-29.

7 G. Ehrenstein, Faserverbund-Kunststoffe, Carl Hanser Verlag, Muenchen, Germany, 2006

8 S.T. Jespersen, "Methodology for evaluating new high volume composite manufacturing technologies", Ph.D. Thesis, Thèse École polytechnique fédérale de Lausanne, Switzerland, 2008

9 H. J. Wolf: "Zum Einfluß der Schneckenplastifizierung auf die Faserstruktur diskontinuierlich langglasfasergefuellter Thermoplaste" Ph.D. Thesis, TU Darmstadt, Germany, 1966

10 O. Herd, "Langglasfaser verstaerkte Kunststoffe und ihre Einsatzgebiete im Automobil" in *Conference Kunststoff Kolloquium*, Deggendorf, Germany, 2006

11 S. F. Bush; F. G.Torres; J. M. Methven: "Rheological characterisation of discrete long glass fibre (LGF) reinforced thermoplastic" in *Composites: Part A,* 31, (2000, pp. 1421-1431

12 P.-C. Chang and S.-J. Hwang, "Experimental Investigation Of Infrared Rapid Surface Heating For Injection Moulding" in *Proceedings of the 64th Annual Technical Conference (ANTEC) of The Society Of Plastics Engineers (SPE)*, Charlotte, NC, USA, 2006

13 M. Schoengart, „Variotherme Temperierung beim Spritzgießen – Verfahren und Moeglichkeiten zur gezielten Steigerung der Formteilqualitaet" *Conference Das Spritzgießwerkzeug – Kernstueck einer effizienten Produktion* Institute of Plastics Processing (IKV) at RWTH Aachen University, Aachen, Germany, 2012

14 Ch. Hopmann, C. Beste and A. Boettcher, "High Surface Quality, Half the Compression Force" in *Kunststoffe International*, 2016, pp. 29-31

Rheokinetic studies of polyamide for in-situ monitoring of T-RTM process: Rheology coupled to dielectric analysis and FTIR spectroscopy.

Mohamed Dkier a,b, Khalid Lamnawar a,b, Abderrahim Maazouz a,b,c,
a) Université de Lyon, F-69361 Lyon, France
b) CNRS, UMR 5223, INSA Lyon, Ingénierie des Matériaux Polymères, F-69621 Villeurbanne, France
c) Hassan II Academy of Science and Technology, 10 100 Rabat, Morocco

mohamed.dkier@insa-lyon.fr, Khalid.lamnawar@insa-lyon.fr, abderahim.maazouz@insa-lyon.fr

Abstract

The goal of this work is to develop study and characterize the structural evolution of thermoplastic matrix during the thermoplastic Resin Transfer Molding process. To attain this objective, the reaction kinetic and its properties for forming process of PA-matrix were investigated as an alternative for the production of thermoset systems. Through this work, various formulations of PA with anionic polymerization from caprolactam were studied. Reaction kinetic were determined depending on processing parameters and chemistry. Besides, we present some promising routes to in-situ monitor the process of thermoplastic composites with tailored mechanical and physical properties. The dielectric analysis technique and FTIR spectroscopy was coupled to rheology and used to monitor the reaction versus time, temperature and the type of catalyst. Indeed we characterize both the structure and chemorheological properties of PA formulations. Based on the obtained results, the T-RTM cycle processing window for good impregnation of the fiber reinforcement were purposed as well the injection time, the reaction temperature and composition.

Keywords: T-RTM, Anionic Polymerization, Rheokinetics, thermoplastic composites.

Introduction

Thermoplastic resin transfer molding (T-RTM) is a promising technology which could be used to mass produce lighter composites, suitable for large-scale use in automotive application. The obtained composite should be cheaper, safer and recyclable.
Components made using T-RTM are produced much quicker than thermosets due to the rapid and fast chemical reaction polymerization. Researchers are currently very active in the development of novel material systems for even faster polymerisation and improved processing ability [1–3]. A major difficulty in processing thermoplastic composites is to achieve cost-effective high-quality impregnation of the fiber reinforcement by the matrix resin. Unlike thermoset resins impregnation with thermoplastic resins is usually performed in their polymerized form, and difficulties with the impregnation arise from their high melt viscosity. A promising processing route for thermoplastic composites is to exploit very low monomer viscosity to achieve good fiber impregnation prior to *in situ* polymerization[2].
Different types of *in situ* polymerization have been studied in the literature. Up to now, polymerization of poly-(methyl methacrylate), linear polyurethane,[3] polystyrene,[4] and polyamide 6 (PA6) have been investigated [4-8]. Currently, the monitoring and modeling of each reaction during processing is a big challenge for scientist and industrial companies.
Through this work, Anionic ring-opening polymerization of ε-caprolactam to produce polyamide-6 are studied depending of the activator/catalyst type, temperature and reaction time. Consequently, the use

of the monomer for given thermoplastic matrix together with activator and catalyst can be an elegant solution according to the lower mixture viscsotiy required befor polymerization.

The main objective of this work is dealing with the determination of appropriate processing window for injection, polymerization and demolding during T-RTM cycle. By the way, the ultimate goal is to instrument the process with fast and adequate sensors to track this reaction. The process requirements for this one are the following:

1. *the initial viscosity of the reactants should be low to facilitate impregnation,*
2. *the reactants should be stable during storage,*
3. *the reaction should proceed at a cost-effective rate*
4. *the final material should have a modulus and strength high enough to obtain good mechanical properties.*

Figure 1 summarizes typical steps and the critical parameters to be identified. They are of extreme importance and necessity to give a better handle of this fast process and for its modeling.

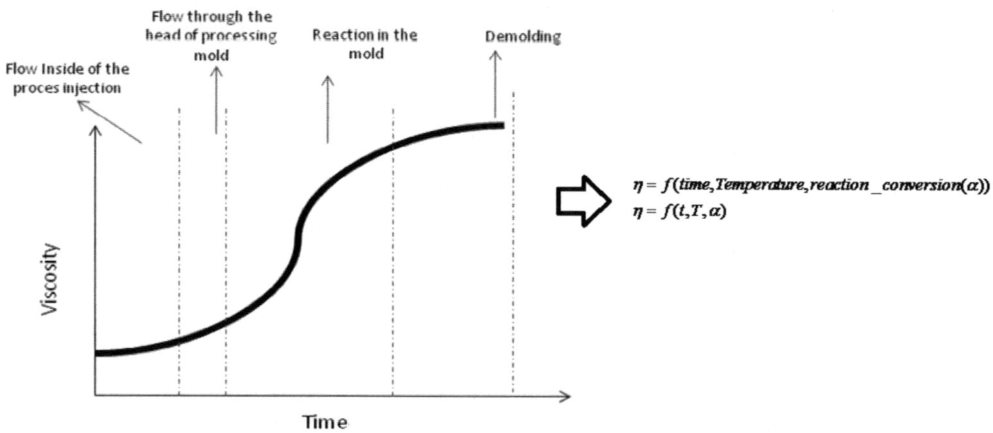

Figure 1: Typical cycle for T-RTM and physical parameters to be identified for modeling.

Experimental section and main results

The monomer used in this work was ε-caprolactam purchased from Brüggemann Chemical. It was The monomer presents a low moisture content (<100ppm) and displays a lower viscosity (5mPa.s at 100°C). it was stored at 50°C under vaccum to keep it dry without causing sublimation and recrystallisation. Different type of activator and catalyst was used to elaborate PA6 with high conversion rate. In this study both activator and catalyst were supplied by Brüggemann Chemical as reported in table1:

Reactif	Nomenclature	Chemical structure	Mw(g/mol)	Tm(°C)
Monomer	ε-caprolactame: **CL**	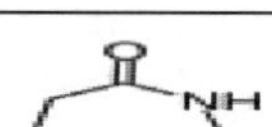	113	69
Activator	Caprolactamate de sodium: **C10**	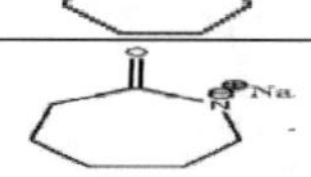	135	60
	Caprolactamate bromure magnésium: **C1**	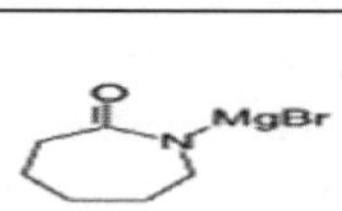	119	69
Catalyst	Dissocyanate : **C20P**	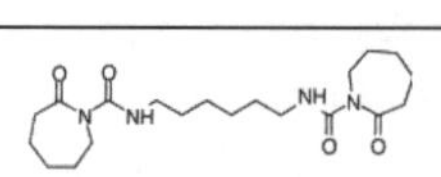	394	60

The reaction process of our system was investigated by means of dielectric analysis coupled to rheology as proposed by DRS "Dielectric Relaxation Spectroscopy" from TA instruments. FTIR coupled to SAOS measurements are also used in this study, a development device was employed (Rheonault system). In addition, In-situ dielectric measurements were conducted using a spectrometer coupled to a varicon sensor, at least to ensure a real-time control of T-RTM composite manufacturing applications. Hence, isothermal polymerizations between 190°C and 240°C of reactive systems are examined see example in Figure2. Upstream study was performed at different frequencies and it was demonstrated that at higher frequency lead to a direct relation between loss factor and conductivity. Each isothermal experiment was performed twice and the data showed good reproducibility.

The values of conductivity due to migrating charges as a function of reaction time, were derived from equation (1) and (2) and represented in figure2

$$\varepsilon' = \varepsilon'_{dip} + \varepsilon'_{ion} = \varepsilon_u + \frac{(\varepsilon_r - \varepsilon_u)}{1+(\omega\lambda_d)^2} \quad \textbf{(1)}$$

$$\varepsilon'' = \varepsilon''_{dip} + \varepsilon''_{ion} = \frac{(\varepsilon_r - \varepsilon_u)\omega\lambda_d}{1+(\omega\lambda_d)^2} + \frac{\sigma}{\omega\varepsilon_0} \quad \textbf{(2)}$$

$$\sigma(\omega) = \varepsilon_0 \omega \varepsilon'' \quad \textbf{(3)}$$

Where ε_u: unrelaxed permittivity, ε_r: relaxed permittivity, λ: dipolar relaxation time which was neglected in equation (3), σ : ionic conductivity.

The dielectric properties of the material can be related to the materiel transformation:

(i) the augmentation in the caprolactam conductivity correspond to melting of the sample and stabilization of temperature

(ii) the decrease of ionic conductivity after the point of maximum provides from the beginning of reaction

(iii) the plate of conductivity correlates well to the end of reaction

Under isothermal conditions, the dynamic viscosity modulus increases with reaction. In fact, the complex viscosity increases due to the increase of the molar masse. Indeed, we clearly determine the effect of temperature on the induction time and viscosity profile of a (100/3/3) formulation. As a result of low polymerization temperature (190°C), a slow increase in viscosity was observed, within the highest the temperature, the faster was the reaction.

The effect of activator/catalyst combinaition on PA complex viscosity can be also shown in sample in this example of figure2, for the combinations (CL-C1-C20), the polymerization is fastest and that the viscosity increases rapidly in time comparing to (CL-C10-C20) formulations, viscosity increase slowly, and it was shown to be influenced by crystallization at lower temperature.

Interestingly, the maximum pic of conductivity at the beginning of the reaction correspond nicely to the inflection time of viscosity especially at 230°C crystallization seems to hinder the mobility of specimens and there is a delay. DRS seem to be more suitable to detect a beginning of reaction then the apparent dynamic viscosity.

The present study highlight also that higher apparent conversion and molecular weight are obtained with C1/C20P as a activator/catalyst. Finally, this work demonstrates that the time of reaction is faster with C20P in association with C1. For processing, a curve calibration log (η) Vs log (σ) was plotted and validated for some formulation.

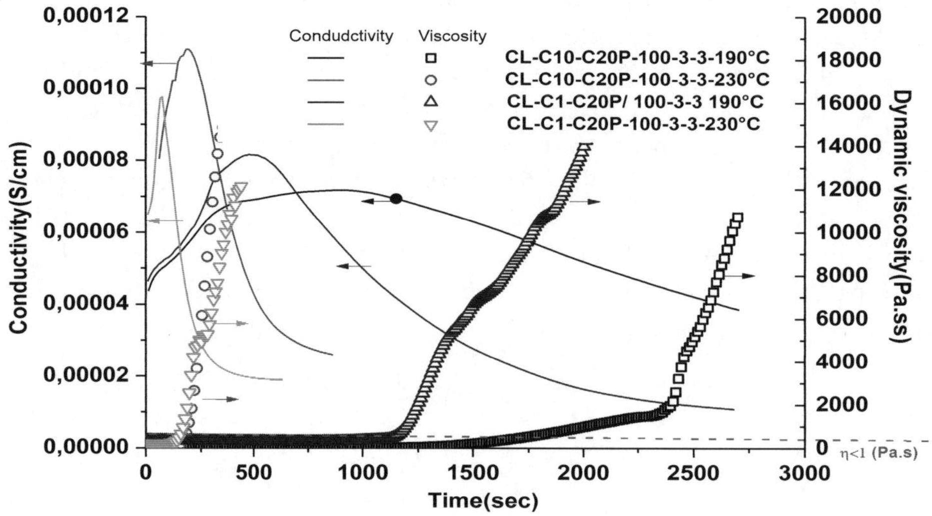

Figure 2: Example of results given evolution of dynamic viscosity modulus at the angular frequency of 1rad/s and ionic conductivity for two studied formulations (CL-C10-C2OP//CL-C1-C20P) frequency=2MHz; impact of type of catalyst at two different temperature (190°C and230°C)

Furthermore, reaction kinetic and its properties were also studied using FTIR spectroscopy coupled to rheology. A 3D FTIR spectrum were recorded for the different formulations (absorbance vs. time and wevenumbers) at iso-temperatures figure 3 shows an example of the obtained results. The peak area at (1740-1450 cm-1) was used for the quantitative analysis of the reaction rate. Indeed, the band at 1560cm-1 characteristic of C—N group is completely absent, then it starts growing as the reaction progress. In fact, the absorbance at 1560cm-1 corresponds to amide bonds in the PA produced. These bonds were monitored independently during the polymerization to follow the reaction kinetics.

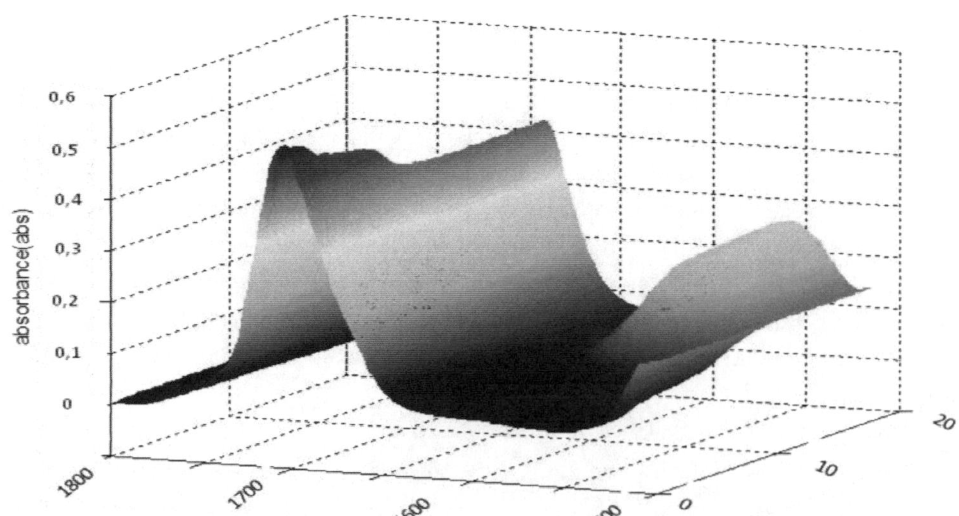

Figure3: 3D FTIR spectrum recorded at 230°C for system CL/C10/C20P- 100/4/4 in 1800cm-1/1450cm-1.

The extent of reaction during time was determined from evolution of amide area (peak area 1560cm-1). The normalized fractional conversion from FTIR absorption data was calculated based on equation (4).

$$\alpha = \frac{A(t_f) - A(t)}{A(t_f)} \quad \textbf{(4)}$$

In this equation, α is the fractional conversion, A(t) is the area of the peak during time, A(tf) in the area of the peak at the end of reaction.

Figure 4 portrays an example of results given the evolution of reaction conversion, which contains conversion determined from integration series of FTIR spectra taken during reaction times in comparison of conversion as established by in-situ dielectric measurement and molar masses measurements equations(5)

$$\alpha = \frac{\log\sigma - \log\sigma_0}{\log\sigma_\infty - \log\sigma_0} \quad \textbf{(5)}$$

Where: σ0, σ, σ∞ are respectively the ionic conductivities in the initial state before reaction, during the reaction and after complete reaction

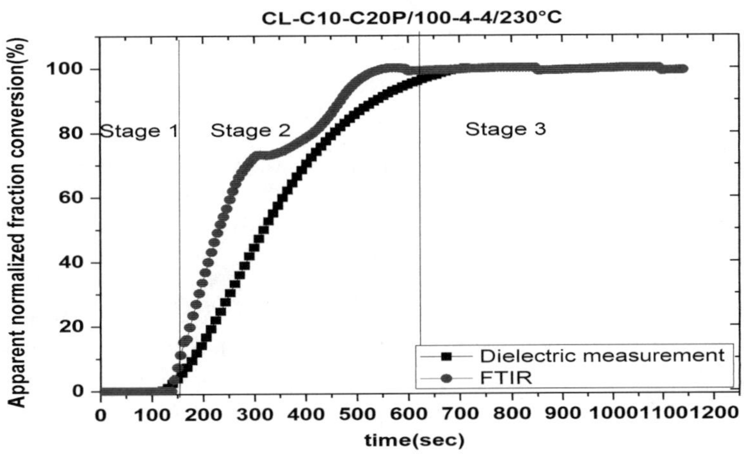

Figure 4: Example give the extent of reaction as a function of time at 230°C for CL/C10/C20P(100-4-4), comparison of conversion results obtained from FTIR spectroscopy, In-situ dielectric measurement.

The results show once again that the beginning of the reaction is very fast for CL/C10/C20P formulations at high temperature and the conversion reaches 80% after 4 min. Indeed, the experimental results given by in-situ dielectric measurement corroborate those obtained from the apparent normalized fraction conversion with FTIR monitoring. A final approximate conversion of 98% is found (i.e 230°C), it is comparable to the results obtained in the literature. Indeed, the progress of reaction can be separated to 3 part as illustrated in figure 4, before 200 second (i.e stage1) the conversion is close to zero and the viscosity modulus is lower than 1 Pa.s. In stage2, the conversion increase drastically recording the polymerization reaction. The viscosity increase up to 650 second, the reaction conversion reaches its maximums indicating the end of reaction (stage3).

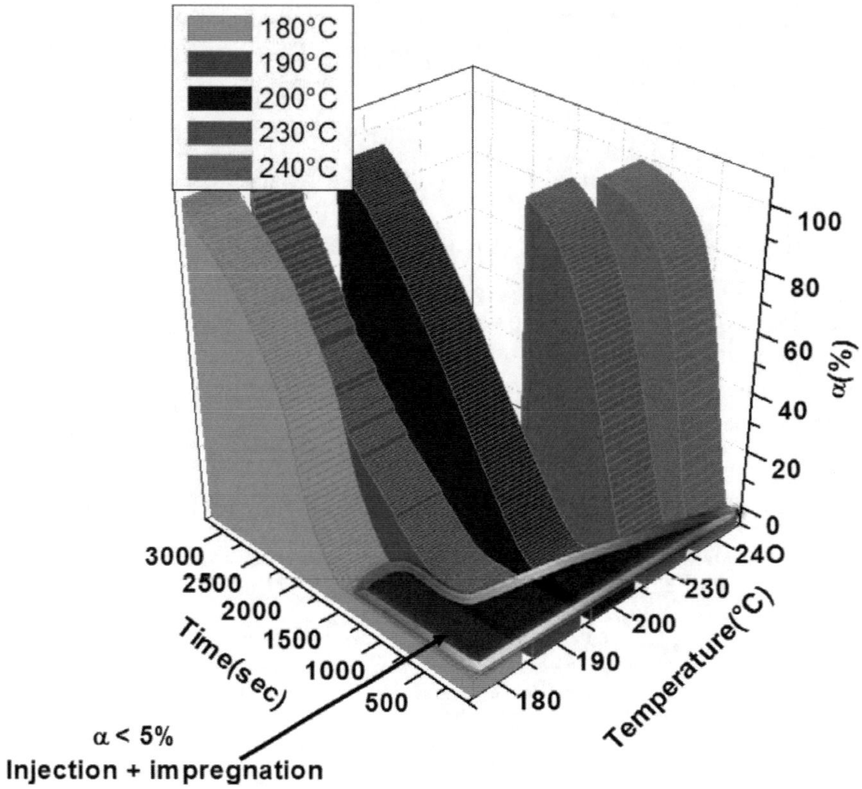

Figure5: Schematic Time-temperature-transformation (TTT) isothermal reaction diagram for a CL/C10/C20P-100/4/4 system, showing the critical injection and impregnation area.

Based on the obtained results, at a higher temperature were determined to facilitate impregnation. At lower temperature (190°C for example) crystallization can stop reaction and lower conversion is recorded after 650second. The limit viscosity is about 10 Pa.s at higher. Finally, a processing window has been defined to have a better handle of process. It seems to be quite narrow, leading to the necessity to carefully control of the processing parameters. Composites can be obtained at higher temperature and less reaction time. Moreover, higher molecular weight at optimized amount of activator/catalyst for PA matrix allowed us to tailor mechanical properties depending of final applications.

References

[1] Bourban PE. Liquid molding of thermoplastic composites. In: Kelly A, editor. Comprehensive composite materials, vol. 2, 2000.p. 965–77.

[2] Luisier A, Bourban P-E, Manson J-AE. In-situ polymerisation of polyamide 12 for thermoplastic composites. In ICCM 12, Paris, France; 5–9 July 1999.

[3] Luisier A, Bourban P-E, Manson J-AE. Reaction injection pultrusion of PA12 composites: process and modelling. Composites, Part A 2003;34:583–95.

[4] Zingraff L, Michaud V, Bourban PE, Månson JAE. Resin transfer moulding of anionically polymerised polyamide 12. Composites Part A: Applied Science and Manufacturing. 2005;36:1675-86.

[6] van Rijswijk K, Bersee HEN, Jager WF, Picken SJ. Optimisation of anionic polyamide-6 for vacuum infusion of thermoplastic composites: choice of activator and initiator. Composites Part A: Applied Science and Manufacturing. 2006;37:949-56.

[7] Zhang C-L, Feng L-F, Hu G-H. Anionic Polymerization of lactams: A Comparative Study on Various Methods of Measuring the Conversion of ε-Caprolactam to polyamide-6. Journal of Applied Polymer Science.101:1972-81.

Analysis of the State of Tension at the Bonding Area Between Components of a Full Thermoplastic Hybrid Composite

Ruggero Giusti[a], Giovanni Lucchetta[a]

[a] *Department of Industrial Engineering, University of Padova, via Venezia 1, 35131 Padova, Italy*

ruggero.giusti@dii.unipd.it

Abstract. The adhesion between two components of full plastic hybrid composite parts is ensured by the macromolecules interdiffusion across the interface. Technologies that are usually involved, such as injection overmolding an in-mold forming, require high temperature in order to promote this phenomenon. The aim of this work is to evaluate the warpage effect along the bonding area in relation to process setting and to evaluate the effect of an annealing treatment. A T-joint specimen was produced with polypropylene-matrix composites and it was made of a rectangular sheet base, which is glass woven fabric reinforced, and a short glass fiber reinforced stem. The molding process was numerically simulated and the results were post-processed with a structural code. The coupling of the software allowed to convert the warpage output in to a residual stress at the interface. The most significant process parameters effects on tensile strength were investigated in accordance to a design of experiments approach. Tensile test results of as-molded samples and after annealing samples were compared.

Keywords: In Mold Forming, Hybrid Composites, Warpage, State of Tension, Annealing, Hybrid Composite.
PACS: 81.05.Lg, 81.40.Ef, 83.50.Uv, 89.20.Bb

INTRODUCTION

Nowadays, environmental regulation and industrial standards push towards resources optimization. Especially the automotive industry is focuses on achieving increased recyclability [1], enhanced vehicle efficiency and minimizing engine emissions [2]. In order to meet these requirements the metal replacement is often taken into account [3], [4]. Thermoplastic hybrid composites are the best candidate to perform this change because of its high specific properties and larger design freedom [5], [6]. The in-mold forming or the back-injection molding of organo-sheet are recent technology which in-line couple thermoforming and injection molding without the necessity of under-cut [6]. An external thermoformed shell composes the final part, which is internal reinforced by an injected ribs system. In addition, features for assembly can be added in the same way. In both the cases, the adhesion between parts is ensured by a bonding mechanism which takes place at the matrix-matrix interface and it is based on the macromolecules inter-diffusion [7]-[12]. It is known that the weaker zone of an hybrid thermoplastic composite is the bonding area wherein reinforcement fibers are not present [13], [15]. Moreover, the breakage can take place at polymer-glass woven after a peeling of the external matrix layer. The aim of this work is to evaluate the process-induced effects of the in-mold forming technology on the welding resistance between parts of a full thermoplastic hybrid T-specimen. The effects of process parameter were evaluated in accordance to the design of experiment approach: melt temperature and packing pressure effects combined with a variothermal mold controlling system were investigated. Several full-factorial plan with two factor levels are scheduled in accordance to DOE method. The resistance of the welding was determined through tensile test and the UTS was assumed as response. Moreover, the residual stress distribution at the interface on the injected side, which is induced by the part warpage, was evaluated through CAE process simulation. On the basis of those results, a post-process annealing treatment was considered in order to reduce the pre-tensioning state [16], [17] and its effect was investigated through a second DOE based on annealing temperature and in-oven residence time. The results show that annealing treatment effectively acts as stress releasing providing a better performance of the interface.

EXPERIMENTAL

The hybrid T-joint specimen is made of a 2 mm thick composite laminate base with area of 22 mm x 42 mm and an overinjected stem with area of 4 mm x 20 mm x 50 mm. The interface has a nominal area of 4 mm x 20 mm. The material used for the base is the Tepex dynalite 104-RG600 which is a polypropylene matrix 50-wt% glass reinforced with balanced woven fabrics manufactured by Bond Laminate. The material used for the stem is the Celstran PP-GF30-0304 which is a polypropylene matrix 30-wt% long glass fiber reinforced manufactured by Celanese. All the samples were manufactured on a 100 ton electrical injection molding machine (Engel E-Motion 440/100).

A 2-cavity mold was designed in order to realize a T-welding joint at the end of the filling phase. The slots, which contain the laminates, are perpendicular to parting plane to allow the manual handling and the in-mold positioning after the laminate IR heating. Three 8 mm diameter channels were added close to the interface area in order to provide a better heat exchange efficiency during the thermoregulation phase.

The heating system is mainly composed by an IR lamp, which is manufactured by Krelus AG. The lamp is controlled by an electrical power unit and it was installed at the top of a tower frame with a lower net-vessel wherein the laminates were placed. The heating phase was calibrated by setting the most suitable combination of power heating supply and the residence time of the composite cut-outs. After having calibrated the emissivity of an IR pyrometer (PorTerBir by Seitron) it was set an emissivity of 0.94. The best combination provided an heating time of 60 seconds to reach the temperature of 200°C without matrix degradation. The laminates surfaces reached a temperature of 160°C after 10 s from the in mold placing. This parameter was set as constant value.

The tensile test of the T-specimen were performed on a MTS 322. The basis was not clamped but it was forced on a C70 steel plate with a central rectangular hole with sides 1 mm longer than the sides of the interface area. Each test was displacement controlled with a rate of 0.5 mm/min and a frequency acquisition of 20 Hz.

Experimental design

The process parameter effects on the ultimate tensile strength are evaluated in accordance to the design of experiment approach. More specifically, a full factorial design was selected with three parameters at two levels as scheduled in Table 1. The investigated parameters are the injection temperature, the holding pressure and the mold temperature. The mold temperature was regulated in two ways: a constant temperature was set at slower level, whereas the VARIOMOLD system (Wittmann-Battenfeld) was used in the upper one. The vario-thermal cycle was calibrated in order to heat the entire mold block and the transversal mold insert before the laminate placing so that the part cooling was contrasted. The thermal cycle is resumed in figure 1. Each experiment was replicated 7 time.

TABLE 1 Factors levels of the in mold forming process

Factor	Low level	High level
Melt temperature [°C]	220	260
Packing pressure [MPa]	10	20
Mold temperature [°C]	80	130/80 (Vario)

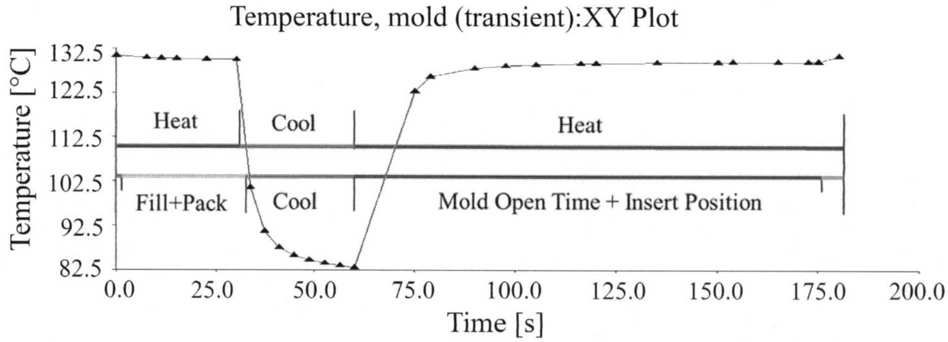

FIGURE 1. Time-Temperature plot of the injection molding cycle of the mold cavity under the laminates. The colored upper line indicates the thermoregulation phases, the lower one indicates the injection molding phases.

According to the previous experimental plan, each level was simulated in Moldflow environment. Cool FEM 3D analysis was set in order to consider both the vario-thermal cycle and the part-insert cooling after having in-mold-placed it at the heating temperature. The residual stress output was analysed by transferring the Moldflow results to

Autodesk Simulation Mechanical environment. An intermedium ANOVA was performed in order to evaluate the process parameter effect on the residual stress (based on the Von Mises average value on the stem-side interface) according to Table 1. Considering the most and the less performant settings, tensile test FEA analyses were carried out. This approach allowed to qualitatively evaluating the effect of residual stresses on the mechanical performance of the T-joint sample.

On the basis of the first ANOVA results, the parameter level combination which provided the higher and the lower tensile strength were involved in a deepening analysis based in a post-forming annealing treatment. The second experimental plan is based on the annealing temperature and on the in-oven residence time. Treatments scheduling are summarized in Table 2 and they are first carried out on the most performant sample. Both the parameters levels were chose according to [16] to allow relaxing of crystals with imperfect space conformation avoiding any change of the crystallinity phase-distribution. In this case, the tensile strength was qualitatively consider in order to evaluate the annealing effect. The annealing treatments were carried out in the MTS 322 climate chamber and a delay time was consider in order to allow the laminate sample reaching the room temperature before the time counting.

TABLE 2 Factors levels of the annealing treatment

Factor	Low level	High level
Annealing temperature [°C]	90	125
Holding time [min]	15	30

Finally, the relation between samples before and after annealing treatment was deepened following the third experimental plan, as it is resumed in Table 3. The annealing treatment (125°C for 30 min) was performed also on the less performant sample. A full factorial design with two factor at two level was performed.

TABLE 3 Factors levels for overall evaluation of as-molded and annealed T-samples

Factor	Low level	High level
Annealing	no	yes
Process parameters	worst (Tmelt-, Pp-, Tmold-)	best (Tmelt+, Pp+, Tmold-)

RESULT AND DISCUSSION

The analysis of variance (ANOVA) results (Table 4) of the screening plan on the tensile tested samples indicate that both the melt temperature and the packing pressure positively affect the bonding resistance being agree with literature results [13]-[15]. Moreover, also the interaction between melt temperature and mold temperature positively affects the experimental ultimate tensile stress.

TABLE 4 ANOVA results of the full factorial plan based on process parameters

Factor	SS	MS	F-Value	P-Value
T_melt	1.41E+13	1.41E+13	5.79	0.02
T_mold	3.91E+12	3.91E+12	1.61	0.211
p_post	1.70E+13	1.70E+13	6.98	0.011
T_melt*T_mold	1.01E+13	1.01E+13	4.15	0.047
T_melt*p_post	1.21E+12	1.21E+12	0.5	0.483
T_mold*p_post	2.39E+12	2.39E+12	0.98	0.326
T_melt*T_mold*p_post	2.64E+12	2.64E+12	1.09	0.302
Error	1.17E+14	2.43E+12		

By varying only the melt temperature from 220 to 260°C, the UTS increase from 9.11 MPa to 10.82 MPa. Conversely, varying the only packing pressure from 10 MPa up to 20 MPa, the increase was from 9.11 MPa to 9.65 MPa. The combined effect of the two parameters allows to reach the best performance of 11.65 MPa. With reference to the interaction plot (figure 2a), the mold temperature seem to affect negatively the UTS. It was supposed that this trend is related to an higher residual stress induced by a lower cooling of the part before the ejection.

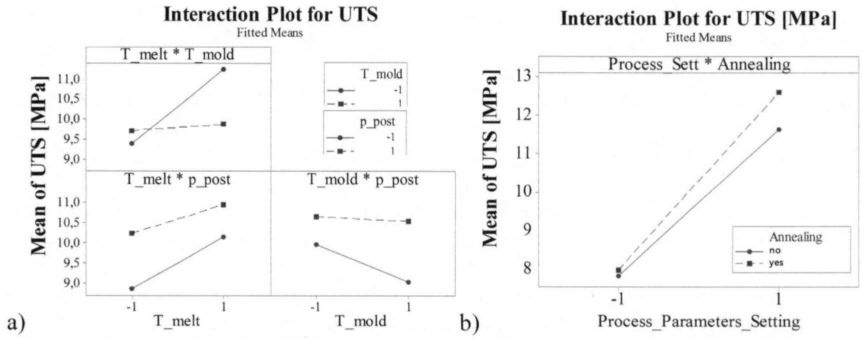

FIGURE 2. a) Interaction plot of the fist ANOVA between in-mold forming process parameters. b) Interaction plot of the third ANOVA between in-mold forming parameter setting and annealing treatment .

The first hypothesis was confirmed by the FEA analysis in Autodesk Simulation Mechanical, which converts the Moldflow outputs in residual stress at the end of the injection phase. The information transfer includes the "Initial Stresses" output, which is not directly available in the original environment. The influence of the process variables on the Von Mises Mean-value on the stem-side interface was investigated by using the same full factorial design reported in Table 1, which was performed in a simulation environment whit a consequent lack of replications.

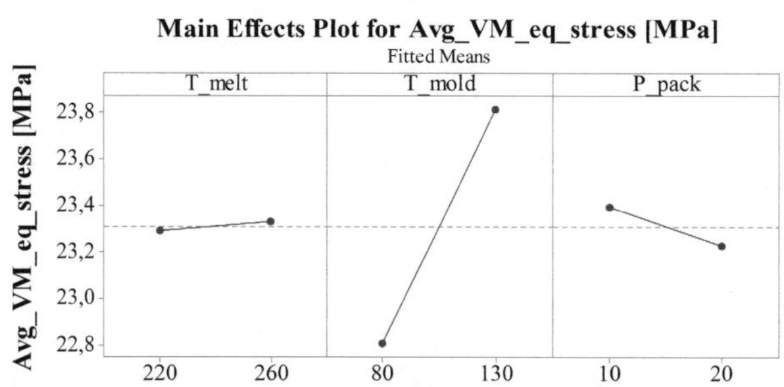

FIGURE 3. Main effects plot on the Average Von Mises equivalent stress on the stem-side interface on ASM results.

The simulation results (figure 3) show that all the process parameters that correspond to an increase in temperature of the part lead to an increase of the residual stresses. However, the most significant parameter that lead to the maximum residual stress is the mold temperature. The experimental and the simulated ANOVA results showed that melt temperature and holding pressure at their higher level can affect both the UTS and the Residual-Stresses. With reference to the reptation theory holding a part at higher temperature for extended time promotes the macromolecules interdiffusion across the interface providing a better healing of the interface [7]-[11]. However, as shown by simulation, the higher processing temperatures results in higher volumetric shrinkage and consequent generation of residual stresses (figure 3). Because of the last statement, it was supposed that an annealing treatment oriented to the stress release could improve the welding resistance. The annealing treatment was designed as a full two-levels factorial plane. The scheduled treatments was firstly carried out on the most performant samples (higher level of both the melt temperature and holding pressure). The ANOVA results of the second experimental screening plan (Table 5) on the ultimate tensile test indicate that both the annealing temperature and the holding time did not affected the bonding resistance.

TABLE 5 ANOVA results of the full factorial plan based on annealing parameters

Factor	SS	MS	F-Value	P-Value
Holding Time	0.0277	0.02771	0.01	0.925
Annealing Temperature	1.1428	1.14284	0.37	0.548
Time*Temperature	0.0057	0.00575	0	0.966
Error	74.0331	3.08471		

Nevertheless, the main effect of the annealing temperature leads an increase of the UTS. Performing the third ANOVA it was allowed to deep the slight relation between annealing and performances.

TABLE 6 ANOVA results of the full factorial plan based on process and annealing parameters

Factor	SS	MS	F-Value	P-Value
Process Parameters	124.324	124.324	34.97	0
Annealing	2.178	2.178	0.61	0.441
Process Parameters*Annealing	1.093	1.093	0.31	0.584
Error	85.326	3.555		

The last experiment indirectly confirmed the simulation results. Although higher melt temperature increases the residual stresses, performing an annealing treatment relaxes the pre-tensioning status and leads an improvement of the overall welding performances from 11.65 MPa to 12.60 MPa (figure 2b).

CONCLUSION

In this work, the process-induced effects of the in-mold forming technology on the welding resistance between parts of a full thermoplastic hybrid T-specimen were evaluated. Melt temperature and packing pressure combined with a variothermal mold conditioning system were investigated according to the design of experiment method.

The results of the experimental test showed that high melt temperature and high packing pressure markeledly affect on the ultimate tensile test of the hybrid T-specimen. By varying only the melt temperature from 220 to 260°C the UTS increase from 9.11 MPa to 10.82 MPa. On the other hand, varying the only packing pressure from 10 MPa up to 20 MPa, the increase is from 9.11 MPa to 9.65 MPa. The combined effect of the two parameters allows reaching the best performance of 11.65 MPa.

On the basis of FEA simulation, it was possible to understand the process-induced effect in terms of residual-stress. Since the higher process temperature are responsible of both a better welding, in accordance to the reptation theory, and an higher residual-stresses, the annealing treatment was considered. The effect was evaluated by supposing that it could act as pre-tensioning state reducer. Even if the screening ANOVA results showed that both the annealing temperature and the in-oven residence time are not significant, a slight increase of the UTS was appreciable.

In order to deepener the effect of the annealing treatment, the most effective combination of annealing parameter was applied also to the specimen manufactured at lower temperature and lower packing pressure. The process parameter setting which leads to higher temperature are slightly influenced by the annealing treatment and the UTS of the hybrid T-specimen increase from 11.65 MPa to 12.60 MPa.

REFERENCES

1. Regulation (EC) No 443/2009 of the European Parliament and of the Council
2. Directive 2000/53/EC on End-of-Life Vehicles
3. T. Müller, I. Kuehnert and T. Osswald "Injection Molding of High-Performance Hybrid Composites in Series-Production." 67th Society of Plastic Engineering Annual Technical Conference, Chicago, IL, 2009, pp. 1233-1237.
4. G. P. Holzinger, M. Wurtele, S. Schierl and S. Kruppa, "Efficient Injection Molding Processes for the Serial Production of Continuous Fiber Reinforced Parts" 2nd IMTC Conference Proceedings, Chemnitz, Germany, 2015, pp. 119-126.
5. M. Grujicic, *American Journal of Science and Technology*, **1**, pp. 168-182 (2014)
6. R. Giusti, A. Costa, A. Bicego and G. Lucchetta, "Re-Design of an Automotive Front-End in Full Thermoplastic Hybrid Composite" 2nd IMTC Conference Proceedings, Chemnitz, Germany, 2015, pp. 127-135.
7. R. P. Wool "Strength of Polymer Interfaces" Final Report Proposal No. 22966-MS, U.S. Army, (217) 222-2468 (1990).
8. L. J. Bastien and J. W. Gillespie, *Polymer Engineering and Science*, **31**, pp. 1720-1730 (1991).
9. C. J. G. Plummer et al, *Journal of Applied Polymer Science*, **87**, pp. 1267-1276 (2003).
10. F. Yang and R. Pitchumani, *Polymer Composites*, **24**, pp. 263-278 (2003).
11. V. Kagan, J. Mazza, I. Palley "Optimizing Mechanical Performance of Injection Molded Multiple Gated Rotating Thermoplastic Components: Part 2" SAE World Congress, Detroit, MI, 2001.
12. E. Lafranche, T. Renault and P. Krawczak, *Key Engineering Material*, **611-612**, pp. 821-828 (2014)
13. M. Fiorotto and G. Lucchetta, *International Journal of Material Forming*, **6**, pp. 179-185 (2013).
14. R. Giusti and G. Lucchetta, *Key Engineering Material*, **611-612**, pp. 915-921 (2014).
15. E. Moritzer et al, "Study of the bond strength of a combination consisting of composite sheet and short-fiber thermoplastic." Euro Hybrid Materials and Structures 2014, pp. 188-192 (2014).
16. W. Wang, G. Zhao, X. Wu and Z. Zhai, *Journal of Applied Polymer Science*, **132**, 42773 (2015).
17. S.Y. Kim, S.H. Kim, S.Y. Pak and J.R. Youn, *Journal of Applied Polymer Science*, **125**, 3029-3037 (2012).

Engineering Properties and Structure of Reactive Molded Polymers and Its Fiber Reinforced Composites

Hiroto Suenaga[a], Akira Ishigami[a], Kentaro Taki[b], Takashi Inoue[a], and Hiroshi Ito[a*]

[a] *Graduate School of Science and Engineering, Yamagata University, Japan*
[b] *Graduate School of Natural Science and Technology, Kanazawa University, Japan*

Presenting and Reference author's email address: ihiroshi@yz.yamagata-u.ac.jp

Abstract. Recently, fiber reinforced thermoplastics (FRTP) has attracted attention in the automotive and aviation fields. However, the thermoplastic resin is used as the matrix of FRTP is very difficult to the fiber bundles impregnate at a high melting temperature and high viscosity. Therefore, it has to be noted that in-situ polymerization using monomer with a low melting temperature and low viscosity is the one of the attractive alternative method. Here, we study the in-situ polymerization of ε-caprolactam as a monomer in order to fabricate polyamide 6 (PA6) and its composites by using reactive molding. The thermal property and the higher order-structure of polymerized PA6 and its composites were analyzed by differential scanning calorimeter (DSC). The engineering mechanical properties as bending and impact behaviors were evaluated. It found that increasing crystallinity of PA6 has resulted in enhance the flexural modulus and flexural strength in a linear relation. In the case of FRTP using PA6, the flexural modulus and flexural strength of Carbon and Glass FRTP showed high values, respectively, as compared with common PA6 composites materials.

Keywords: in-situ polymerization, polyamide 6, mechanical properties, polymer nano-alloy, fiber reinforced composites.
PACS: 81.05.Qk

1. INTRODUCTION

In recent years, epoxy thermoset plastic with the engineering plastic property and fibers reinforced plastic (FRP) as composite material with high strength, high modulus, and lightweight were interested in the place of the metal material. Carbon and glass fibers had been used in automotive part. However, the epoxy thermoset plastic/FRP composite has an environmental problem cause of difficult to recycle. Therefore, fibers reinforced thermoplastic (FRTP) with thermoplastics matrix was focused on the alternative material[1].

Polyamide6 (PA6) or Nylon6 was synthesized by the ring-opening polymerization of caprolactam monomer[2]. PA6 has a high tensile strength, high elasticity and luster. These properties are strongly dependent on the polymerization and plastic forming process.

One of the simple and comfortable of plastic forming is molding process. This process intensive to use for the FRTP resin transfer molding (RTM)[3]. In the RTM process, thermoplastic as a matrix was melted while applying pressure result to the reinforcing fibers bundle, was impregnated into the polymer matrix and polymer composite performs after cooling. However, the higher melting viscosity and long cooling time of the PA6 is difficult to fibers bundle impregnate. Therefore, using ε-caprolactam as a monomer in order to fabricate PA6 for impregnation process is an alternative method in the avoidance these problems. To optimize the polymerization of ε-caprolactam monomer, the in-situ polymerization had been essential to examined.

In this study, carried out in-situ polymerization using ε-caprolactam as a monomer by RTM the reactive molding process. The crystallinity, thermal property and mechanical property of molded article were investigated by DSC and bending test respectively.

2. MATERIALS AND METHOD

2.1 In-situ Polymerization

ε-caprolactam (ε-CL) monomer from Wako Pure Chemcal Industries, Ltd. and sodium hydride (NaH) as initiator from Aldrich were mixed at the melt temperature of 90, 100 and 110 °C. N-acetyl-ε-caprolactam (NA-εCL,) from Tokyo Chemical Industry, Ltd. as an activator was dropped into ε-CL and NaH mixture. After that poured the solution into the reactive mold by using hot press machine (Imoto Machinery Co.,Ltd, IMC-11 FA). Polymerization condition is 140 °C, 10 min and 10 MPa. Reinforce materials were used Glass Fiber (GF) and Carbon Fiber (CF). Reinforce materials Impregnate were 20 vol%. Pouring solution on top of reinforce materials.

FIGURE 1 presents the schematic of reactive molding for (a) non-reinforcement process, (b) reinforcement process. TABLE 1) presents the materials characteristics.

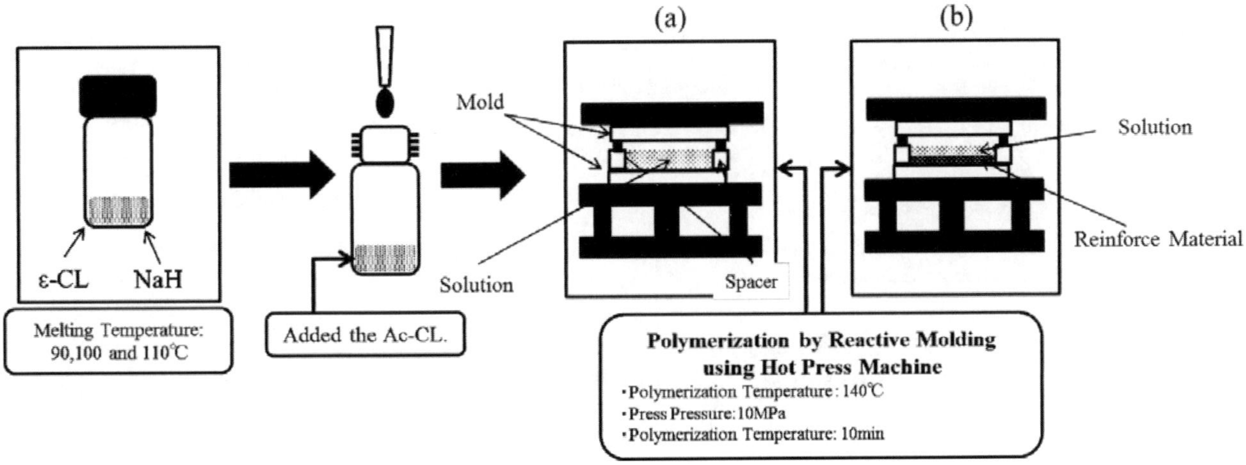

FIGURE 1. Schematic of reactive molding for (a) non-reinforcement process, (b) reinforcement process.

TABLE 1). Characteristics of materials.

Substance	Amount (g)	Content (wt-%)	Mole (mmol)	Content (mol-%)
ε-CL	14.00	97.6	1.2	95.9
NA-ε-CL	0.25	1.8	0.6	1.2
NaH	0.09	0.6	3.7	2.9

2.2 Experimental

Crystallinity of PA6 was investigated by thermal properties suing modulation differential scanning calorimetry (DSC, Q200; TA Instruments). The rate of temperature increase was 10 °C/min. The temperature range was 20 to 270 °C.

Mechanical was investigated by using bending test machine (Strograph VGS-1, Toyo Seiki Kogyo Ltd.) to determine the flexural strength and flexural modulus. Bending tests were conducted at a rate of 2 mm/min at 25°C.

3. RESULTS AND DISCUSSION

3.1 Thermal properties

Generally, increase the melting temperature can enhance the reactivity of monomer. However, it does not relate to the crystallinity and degree of molecular orientation change. Figure 2 and 3 illustrate DSC thermogram of PA6 after the reactive molding process at the various melting temperature of monomer/initiator mixture. The tendency of the crystal melting behavior begins at the temperature of 200 °C was observed.

Figure 3 shows the DSC thermogram of annealing PA6. After the annealing process at 200 °C and held for 1 h, it does not find a wide range of crystal melting behavior. It can be explained by PA6 crystal was melted by the annealing process and has effects on enhancing the crystal regularity of PA6.

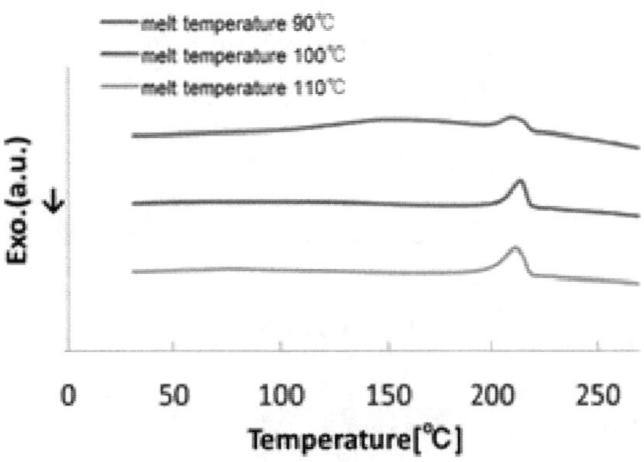

FIGURE 2. DSC curves of PA6 polymerized at various melting temperature.

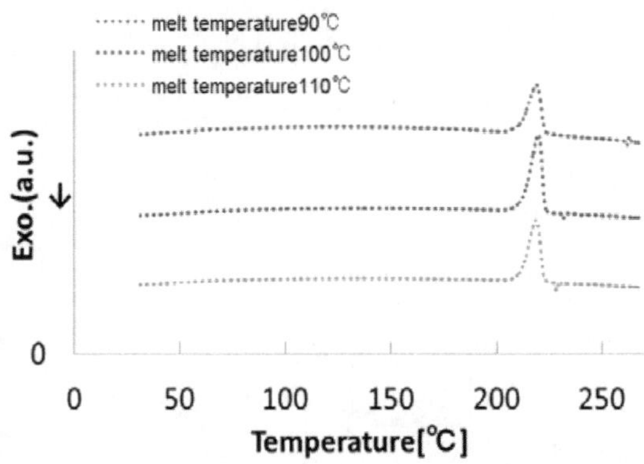

FIGURE 3. DSC curves of annealed PA6 polymerized at various melting temperature.

3.2 Relationship between crystallinity and physical properties

3.2.1 PA6 reactive molding

Figure 4 and 5 show the effect of melting temperature of monomer/initiator and the PA6 crystallinity on flexural modulus and flexural strength respectively. Increasing of crystallinity of PA6 results in flexural modulus and flexural strength increase. It can explain by the strength of PA6 crystal structure significant effect to its mechanical property. It can explain by annealing, micro crystals were melted and recrystallized, might be that the crystal size was uniform. Moreover, flexural strength and a flexural modulus of PA6 were increased by the annealing process.

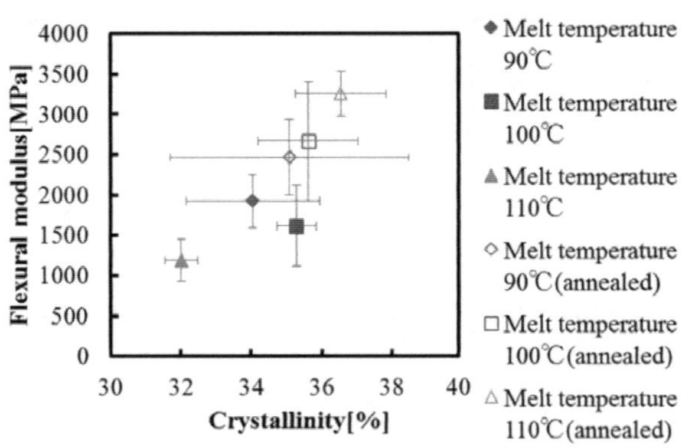

FIGURE 4. Relationship between degree of crystallinity and flexural modulus.

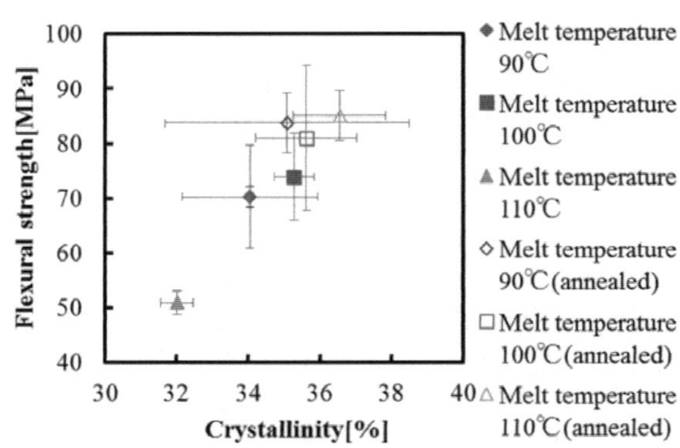

FIGURE 5. Relationship between degree of crystallinity and flexural strength.

3.2.2 Reinforcing material impregnated PA6 reactive molding

Figure.6 and 7 show the relationship between crystallinity and flexural modulus and flexural strength of GF/PA6. By increase crystallinity, the flexural modulus also the strength tended to increase. Furthermore, the flexural modulus and flexural strength were increased by annealing. It found that crystallinity doesn't change by annealing process.

Figure.8 and 9 show the relationship between crystallinity and flexural modulus and flexural strength of CF/PA6. Unilike the GF/PA, by increase crystallinity, the flexural modulus also the strength tended to decrease.

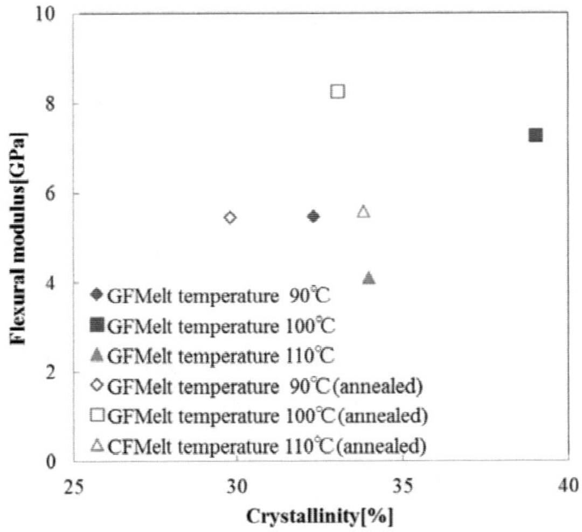

FIGURE 6. Relationship between degree of crystallinity and flexural modulus of GF/PA6.

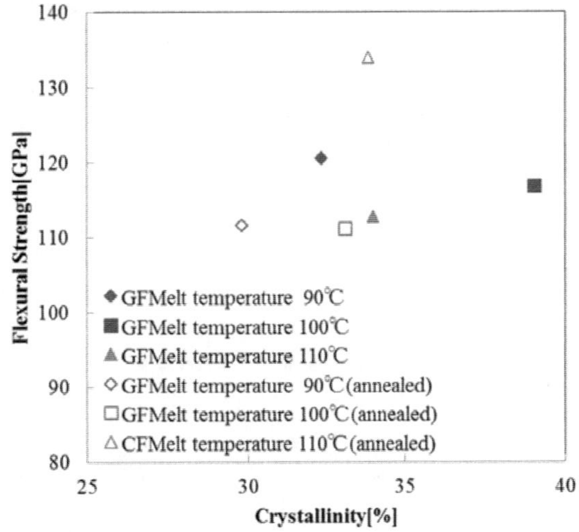

FIGURE 7. Relationship between degree of crystallinity and flexural strength of GF/PA6.

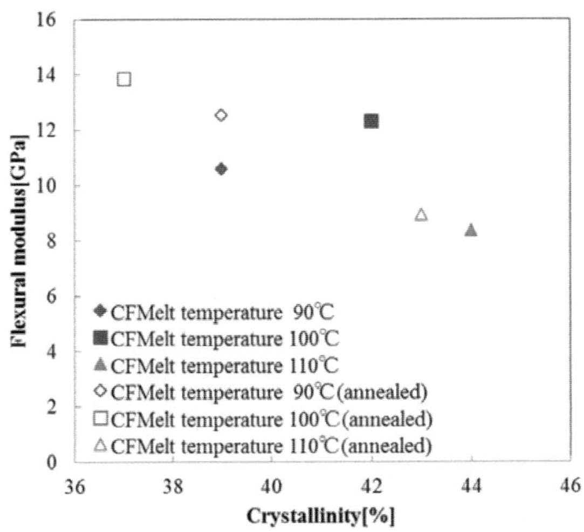

FIGURE 8. Relationship between degree of crystallinity and flexural modulus of CF/PA6.

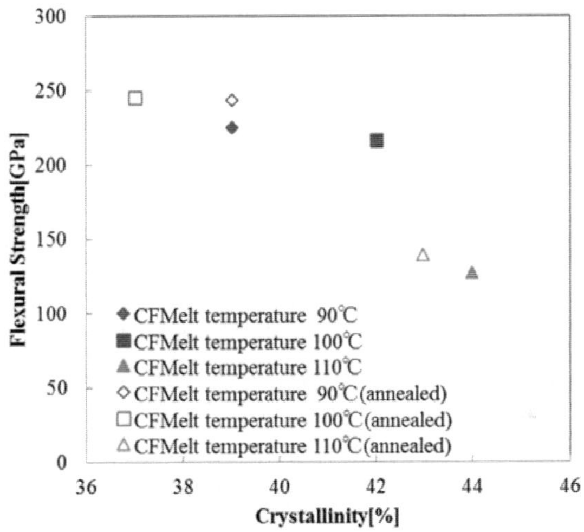

FIGURE 9. Relationship between degree of crystallinity and flexural strength of CF/PA6.

CONCLUSIONS

In-situ polymerization of PA6 using ε-caprolactam as monomer by reactive molding process was carried out. Increasing of the crystallinity of PA6 can enhance the mechanical property of PA6. Moreover, flexural strength and a flexural modulus of PA6 were increased by the annealing process.

Reinforcing material impregnated PA6 reactive molding and structure-property evaluated were carried out. By annealing, flexural strength and flexural modulus of the reinforcements impregnated moldings were increased. However, flexural strength of the GF-impregnated molding was reduced.

It shows that important to simultaneously control the crystallinity and reactivity of the monomer in-situ polymerization.

ACKNOWLEDGMENTS

This research was partially funded by Impulsing Paradigm Change Through disruptive Technologies (ImPACT) Program of Council for Science, Technology and Innovation (Cabinet Office, Government of Japan).

REFERENCES

1. K. van Rijswijk, H.E.N. Bersee, A. Beukers, S.J. Picken, A.A. van Greenen, *Optimisation of anionic polyamide-6 for vacuum infusion of thermoplastic composites: Influence of polymerisation temperature on matrix properties.* Polymer Testing, vol. 25, Issue 3, 2006, pp. 392-404.
2. Raj S. Dave, Robert L. Kruse, Lionel R. Stebbins, *Polyamides from lactams via anionic ring-opening polymerization: 1. Chemistry and some recent findings*, Polymer, Vol. 38, Issue 4, 1997, pp. 927-938.
3. D. Abraham, S. Matthews, R. McIlhagger, *A comparison of physical properties of glass fibre epoxy composites produced by wet lay-up with autoclave consolidation and resin transfer moulding*, Composites Part A: Applied Science and Manufacturing, Vol. 29, Issue 7, 1998, pp. 795-801.

Polyphenylene sulfide containing hydroxyl: synthesis and the effect on the micromechanical properties of composites Polyphenylene sulfide/carbon fiber

Hao-hao Ren [a], Gang Zhang [a]* and Jie Yang [a,b]*

[a]*Institute of Materials Science & Technology, Analysis and Testing Center, Sichuan University, Chengdu 610064, China*
[b]*State of Key Laboratory of Polymer Materials Engineering, Sichuan University, Chengdu 610064, 646149522@qq.com;*
gangzhang@scu.edu.cn.
ppsf@scu.edu.cn

Abstract: The aim of this study was synthesis of the polyphenylene sulfide (PPS) containing hydroxyl and investigation of the effect of modified PPS used as compatibilizer on the interfacial micromechanical properties of carbon fiber (CF) reinforced Polyphenylene sulfide composites. The high molecular weight and copoly(1,4-phenylene sulfide)-poly(2,5-phenylene sulfide phenol) (PPS-OH) was synthesized by the reaction of dihalogenated monomer and sodium sulfide via nucleophilic substitution polymerization. The results of FT-IR, X-ray diffraction, DSC, TGA and dissolvability, PPS-PPSOH was found to have analogous structure and improved hydrophilic property comparing with PPS. There was a good physical compatibility between the functional PPS and the pure PPS. However, the microbond test (apparent interfacial shear strength, τ_{app}) showed that the functional polyphenylene sulfide almost have no influence on the micro-mechanical properties between CF and PPS matrix.
Key words: Synthesis, Poly(phenylene sulfide), mechanical properties.
PACS: S17-523

INTRODUCTION

As is well known, Poly(1,4-phenylene sulfide) (PPS) is popular due to its heat resistance, excellent chemical resistance, good electronic properties, and low toxicity [1-2]. It is widely used in many areas, especially in mechanical engineering, chemicals and car industry [3-6]. But its applications have been somewhat limited because of the relative low mechanical strength of the pure PPS. In order to enhance the mechanical properties, fiber and particle fills reinforcing are usually used in industry or reported in the earlier literatures [7-8]. Recently, carbon fiber (CF) was largely used as reinforcing fiber materials in polymer-matrix composites, due to its superior reinforce-effect and excellent thermal properties [9]. For instance, PPS-SGS30 is a 15% CF reinforced composite in Sciengy Limited Company (Shandong, China), PPS-OC306 is a 30% CF reinforced composite with good electrical conductivity in Huapuhang Trading Limited Company (Guangzhou, China), and so on. Therefore, CF reinforced PPS composite was thought to be the outstanding composites in high performance engineering plastics [10-11]. However, the mechanical properties of PPS composites always are not excellent due to the poor interface interaction of the composites between fillers and matrix resin. Here, we have designed the compatibilizer (PPS-OH) expecting that it have excellent compatibility with the resin matrix and can interact with the surface of the fiber. The chemical structures, thermal properties, mechanical properties of these obtained copolymers were characterized. However, the modified PPS almost have no influence on the micro-mechanical properties between CF and PPS matrix. And the failure mechanism was analyzed.

EXPERIMENTAL

Materials. 1,4-chlorobenzene (DCB) (AR, Quzhou, Rainful Chemical Industry Company), 2,5-dichlorophenol (AR, Aladdin reagent), sodium sulfide ($Na_2S{\cdot}xH_2O$ $Na_2S\%{\approx}60\%$) (Nanfine Chemical Industry Group Co., Ltd), carbon fiber (7 μm diameter, Japan Toray); N-methyl-2-pyrrolidone (NMP) (JiangSu NanJing JinLong Chemical Industry Company).

Copolymers synthesis. The modified PPS was synthesized by a nucleophilic substitution reaction of DCB, active monomers and sodium sulfide in NMP at high pressure as depicted in **Scheme 1**. And the copolymer with optimal functionalization degree of ~5% -OH (named PPS-OH) was selected as the compatibilizer to reinforce the PPS/CF composite, due to its optimal balance between molecular weight, thermal stability and number of functional groups. Briefly, 250 ml of NMP, 65.0 g (0.50 mol) of sodium sulfide ($Na_2S{\cdot}xH_2O$ $Na_2S\%{\approx}60\%$) were added into the reactor. Then the reactor was heated to 200 °C and 26 ml liquid was removed during this course. Secondly, 69.87 g (0.475 mol) of DCB and 4.10 g (0.025 mol) of 2,5-dichlorophenol were added, then the reaction temperature was raised to 220 °C and maintained for 6 h. After reaction, the product was washed with hot water and acetone, then heated at 100 °C for 24 h under vacuum to yield 49.28g (yield: 90.1%) of a white powder.

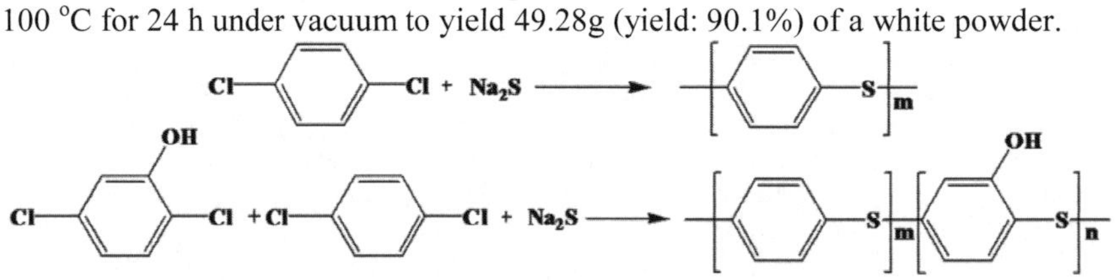

Scheme 1. Synthesis of copolymer PPS-OH, m:n=95:5.

Preparation of PPS-OH/PPS alloys. In this work, the PPS-OH/PPS alloys were prepared by the common mix-melting method at 310 °C in the internal mixer, with a screw speed of 150 rpm for 15 min. The PPS-OH/PPS alloys with PPS-OH contents of 3, 5, 8, 10 and 12wt% were named 3%PPS-OH/PPS, 5%PPS-OH/PPS, 8%PPS-OH/PPS, 10% PPS-OH/PPS and 12%PPS-OH/PPS respectively. After blending, the alloys were injection molded through an injection molding machine (CJ150NC, Zhengxiong Corporation, China) at the melt temperature of 300 °C.

Characterizations. Intrinsic viscosity of copolymers was obtained by α-chloronaphthalene at 208 °C with 0.400 g of polymer dissolved in 100 ml α-chloronaphthalene, using a Cannon-Ubbelodhe viscometer. The Intrinsic viscosity values were obtained by a one-point method (or Solomon–Ciuta equation) as follows:

$$\eta_{\text{int}} = \frac{\sqrt{2(\eta_{sp} - \ln \eta_r)}}{C}$$

where $\eta_r = \eta/\eta_0$, $\eta_{sp} = \eta/\eta_0 - 1$. And the eight-average molecular weight (Mw's) were calculated with the following formula [12]:

$$\eta_{sp} = 8.91\times10^{-5} M_w^{0.747}$$

Melt index was measured at 316 °C, 5 kg load. FT-IR spectroscopy measurements were determined by NEXUS670 FT-IR instrument. X-ray diffraction (XRD) (Philips X0pert Pro MPD) was used to study the polymers' crystalline characteristics. Universal Testing Machine Instron 5567 was used to study the mechanical strength of the samples at room temperature according to ISO 527 and ISO 178, respectively. Differential scanning calorimetry (DSC) was performed on a NETZSCH DSC 200 PC thermal analysis instrument with a heating rate of 10 °C/min. Thermogravimetric analysis (TGA) measurements were performed on a TGA Q500 V6.4 Build 193 thermal analysis instrument with a heating rate of 10 °C/min under a nitrogen atmosphere.

The microbond test was used to study the micromechanical property between modified PPS and

PPS/CF composite, the typical method in this work was reported in the previous literature [13], and the procedure was illustrated in Fig 1.

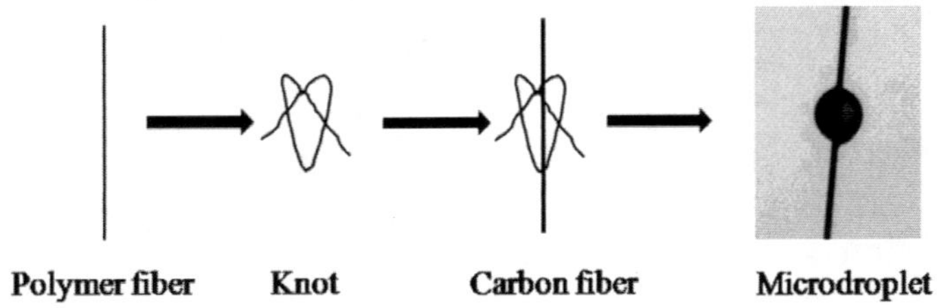

`**FIGURE 1.** Procedure to form thermoplastic resin on the single CF and the optical microscope photograph of one microdroplet.

Firstly, the pure PPS resin was melted and pulled into fiber, the resin fiber was knotted slightly around a carbon fiber. Then the sample was placed at a temperature of 320 °C in order to melt PPS completely. After melting completely and forming a testable droplet, the specimens were cooled to room temperature naturally. Similarly, the PPS blended regularly with different content of PPS-PPSOH or PPS-PPSA also was pulled and prepared the single fiber-microdroplet samples using the same method. The speed of microbond test is 0.02 μm/s. Meanwhile the apparent IFSS (τ_{app}) was calculated from the following equation [14]:

$$\tau_{\mathrm{app}} = \frac{F_{\max}}{\pi d_{\mathrm{f}} l_{\mathrm{e}}},$$

where F_{max} is the maximum pull-out force; d_f is the fiber diameter and l_e is the embedded length, both were measured using the optical microscope equipped with a high resolution color CCD camera (EC3001) and Studio Measure software (EC imagine). The τ_{app} of the samples were determined from at least 40 pull-out measurements in order to obtain a statistically significant average.

RESULT AND DISCUSSION

The intrinsic viscosity (η_{int}) and molecular weights of C-PPS

Table 1 Intrinsic viscosity, molecular weights, yields of PPS-OH

Product	η_{sp} (dl/g)	Mw ($\times 10^4$)	yields
PPS	0.31	5.72	95%
PPS-OH	0.33	6.03	96%

The intrinsic viscosity values, the molecular weights and MI of polymers were shown in Table 1. The results showed that the intrinsic viscosity of PPS-OH was 0.33 dl/g (PPS-OH), which was slightly higher than that of PPS (5.72×10^4).

The structure of the pure PPS and modified PPS

The structure of the pure PPS and PPS-OH were characterizied by FT-IR and XRD. The FI-IR spectra of polymers were shown in Fig 2, the results showed that PPS-OH had nearly the same characteristics with PPS. The absorption at 1180 cm^{-1} of copolymers could be assigned to the characteristic peaks of C-S bond. The new peak at 3420 cm^{-1} of PPS-OH was attributed to the stretching vibration of -OH. And the -C=O stretching vibration absorption at 1691 cm^{-1} appeared, which might be caused by the partial oxidation of 2,5-dichlorophenol during the reaction. Meanwhile, the stretching vibrations of bond C-Cl (715 cm^{-1}) became very weak after the polymerization. Thus, it can be concluded that polymerization take place during this reaction. The results of X-ray diffraction (Fig 4) showed that PPS-OH and PPS are crystalline. Besides, the crystallinity of PPS-PPSOH decreased slightly with the introduction of hydroxy, which could be explained well to be the decreasing regularity of the polymer chain.

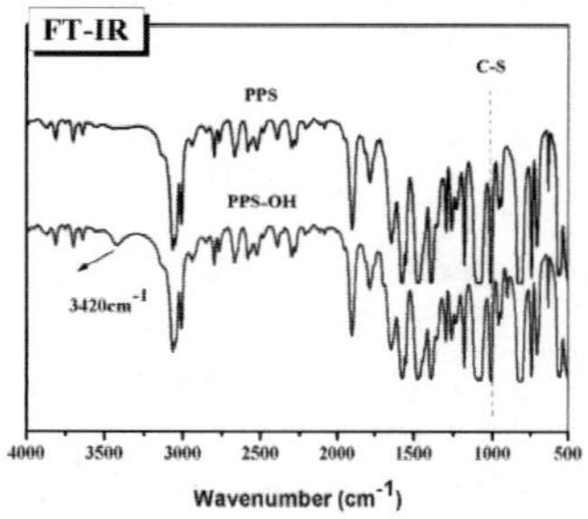

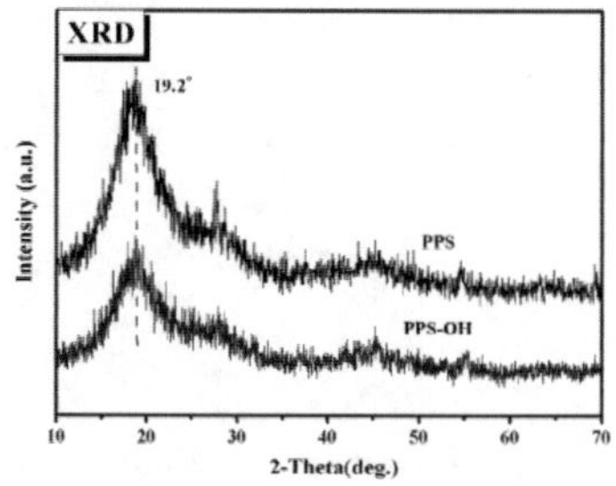

FIGURE 2. FI-IR spectra of PPS and PPS-OH. **FIGURE 3.** X-ray diffraction of PPS and PPS-OH.

Mechanical properties of copolymers

Table 2 Mechanical properties of PPS and PPS-OH.

Product	Tensile strength (MPa)	Elongation at break	Young's modulus (GPa)
PPS	66.4	4.65	1.76
PPS-OH	67.2	4.83	2.42

The average tensile strength, elongation at break and Young's modulus values of PPS and PPS-OH treated at 120 °C for 4 hours, were listed in Table 2. As shown in Table 2, the average tensile strength and Young's modulus of PPS-OH were 67.2 MPa and 2.42 GPa, respectively. Also it was found that the elongation at break of PPS-OH and its alloys was more than 4.5%, and the values were nearly to that of PPS.

Thermal properties of the prepared copolymers

Table 3 Thermal properties of PPS and PPS-OH.

Product	T_g (°C)	T_m (°C)	Δ H_r (J.g^{-1})	T_d (°C)	Char yield
PPS	86.2	282.3	39.13	480.9	43.9%
PPS-OH	88.2	284.9	31.86	477.3	44.5%

---, T_g had not been detected by DSC.

The thermal properties of copolymers were examined by DSC and TGA. The results were displayed in Table 3. From the table, we could find the compatibilizers PPS-OH exhibited higher T_m and T_g values than that of PPS, which might be caused by the formation of carbonyl. Meanwhile, the initial degradation temperatures of PPS-OH was 477.3 °C (the initial degradation temperatures of PPS was 480.9 °C), and provided about 44.5% char yied at 800 °C (the char yied of PPS was 43.9%). These results suggest that PPS-OH have good thermal property as well as PPS.

Interfacial shear strength between CF and PPS matrix

The τ_{app} between PPS and CF with or without PPS-OH were in Table 4. From Table 4, the maximum τ_{app} of PPS/CF composites which added 3%PPS-OH was about 36.6 MPa. Comparing with pure PPS/CF composite (36.1MPa), the τ_{app} of PPS/CF composites with 3%PPS-OH only increased 1.4%. As we know, there are large numbers of functional groups (e.g. epoxy groups, –C–OH, –C–O–C–) in the surface of CF [15-16]. So there was no physical or chemical reaction between the resin and CF, which might be caused by the oxidation of -OH during the reaction that influence the interaction between PPS-OH and CF.

Table 4 the τ_{app} between PPS and CF with or without PPS-OH.

Product	PPS	3% PPS-OH/PPS	5% PPS-OH/PPS	8% PPS-OH/PPS	12% PPS-OH/PPS
τ_{app} (MPa)	36.1	36.6	35.8	36.3	35.9

CONCLUSION

In this study, the modified PPS with higher content reactive groups (-OH) had been successfully synthesized. The resultant polymers were found to have moderate molecular weight, good thermal and mechanical properties. The copolymer was found to have similar structure with the pure PPS.

Through the microbond test, it was found that the functional PPS-OH had no obvious influence on the interfacial property of PPS/CF composite, which might be caused by the oxidation of hydroxyl.

REFERENCES

[1] Yan YG, Li YB, Yang Z, Zuo Y, Wei J, Cao X, Chen YR. Eur Polym J. 39: 411-416 (2003).
[2] Zhang DK, Yao HJ, Zhou DJ, Dai LB, Zhang J, Yuan SG. Polym Adv Technol. 25: 1590-1595 (2014).
[3] Li H, Lv GY, Zhang G, Ren HH, Fan XX, Yan YG. Polym Int. 63: 1707-1014 (2014).
[4] Zwettler N, Engbak JS, Lundsgaard R, Paranowska I, et al. React Funct Polym. 88:47-54 (2015).
[5] Tsuchida E, Yamamoto K, Miyatake K. React Funct Polym. 30: 269-277 (1996).
[6] Ferrara JA, Seferis JC, Sheppard CH. J Therm Anal. 42: 467-484 (1994).
[7] Xu HY, Feng ZZ, Chen J, Z H. Mater Sci and Eng A. 416: 66-73 (2006).
[8] Esteves JV, Goushegir SM, Santos JF, Canto LB, Jr EH, Amancio-Filho ST. Mater Des. 66:437-445 (2015).
[9] Kuram E, Tasci E, Altan AI, Medar MM, Yilmaz F, Ozcelik B. Mater Des. 49: 139-150 (2013).
[10] De Baere I, Van Paepegem W, Degrieck J. Polym Compos. 30: 1016-1026 (2009).
[11] Diez-Pascual AM, Naffakh M. Polymer. 53: 2369-2378 (2012).
[12] Stacy CJ. J Appl Polym Sci. 32: 3959-3969 (1986).
[13] Gonon L, Momtaz A, Van Hoyweghen D, Chabert B, Gérard J-F, Gaertner R. Polym Compos. 17:265–274 (1996)..
[14] Zhandarov S, Mäder E. Compos Sci Technol. 65: 149–160 (2005).
[15] Pittman jr CU, Wu Z, Jiang W, He GR, Wu B, Li W, et al. Carbon. 35: 929–943 (1997).
[16] Belfiore LA, Shah RJ, Cheng C. Polym Compos. 10: 122–133 (1987).

Mechanical Properties of 3-Dimesional FRP Prepared by VaRTM

Young–Ah Kang* and Ae Ri Kim

[a]*Department of Organic Material and Polymer Engineering, Dong-A University, Busan 604-714, Korea*

yakang@dau.ac.kr

Abstract. 3-dimensional fabric Reinforced Plastic was prepared by vacuum assisted resin transfer molding. The 3-dimensional fabric as a reinforcement composed of ultrahigh molecular weight polyethylene (UHMWPE), Aramid fibers and polyethylene terephthalate (PET) was knitted by Raschel Warp Knitting method, which can be applied in civil engineering and construction uses. The UHMWPE/Aramid hybrid fabric-reinforced plastic composites were prepared by vacuum-assisted resin transfer molding (VaRTM) apparatus with a newly designed mold. Unsaturated polyester resin was used as a matrix, which involved pre-curing for 24 h at room temperature and then post-curing at different temperatures from 80°C. For the obtained three-dimensional plastic composites, the weight ratio of fiber-to-matrix in the composite reached about 50-60 wt%, as estimated by TGA. Tensile and flexural properties are characterized for the plastic composite.

Keywords: VaRTM, FRP, 3-dimensional hybrid fabric.
PACS: 81.05.Qk

INTRODUCTION

Vacuum-assisted resin transfer molding (VaRTM) infuses resin in a rigid and closed molding. The fiber reinforcement is compressed by the pressure of atmosphere, and pressed tightly against the rigid surface of the mold. The VaRTM process is associated with low tooling cost and facilitates the production of high quality, defect-free structures. Using the VaRTM method, the vacuum expels air and VOCs from the freeform assembly and allows the resin to flow into the preform. A pressure of up to 1 atm provides the driving force for the resin to impregnate the dry preforms, and the compression force to compact the composite to the desired volume fraction of reinforcement-to-matrix. Accordingly, a high volume fraction of fiber reinforcement can be optimized to yield superior toughness and strength in a more flexible composite structure. In this study, the 3-dimensional FRP composite reinforced by a hybrid knitted fabric composed of ultrahigh molecular weight polyethylene (UHMWPE) and *para*-ARAMID was produced using an in-house prepared VaRTM apparatus.

METHODS

Unsaturated polyester resin (HF-123NHL, AEKYUNG CHEMICAL CO., LTD.) was used as a matrix material, and methyl ethyl ketone peroxide of 1.0 v% was used a catalyst which initiates crosslinking of unsaturated polyester resins in composites. The 3-dimensional hybrid fabric of UHMWPE/*p*-ARAMID/PET, which was kindly provided by DEARIM TEX CO., LTD.) was used as received without additional cleaning. The fabric was warp-knitted with UHMWPE, *p*-ARAMID, and PET filament by means of Raschel warp knitting machine. Schematic diagram of the VaRTM apparatus equipped with newly designed molds (300×300×30 mm in size) was used for the 3-dimensional composites (Fig. 1). Resin and initiator were mixed at a ratio of 100:1 by volume, and the mixture was stirred at room temperature. Two vacuum pumps were used, the pump 1 was to compact the preform by maintaining a constant vacuum and the pump 2 was to inject the resin into the mold. The resin mixture was injected into cavities in the mold at room temperature under constant vacuum. Residual air bubbles in the mold were extracted into a resin trap tank. The composite was cured for 2 h and 24 h under the constant vacuum and free vacuum at room temperature, respectively. The reinforcement and the obtained composites were shown in Fig. 2. Tensile tests of the 3-D FRP composites were measured using an Instron CRE testing machine, at a test speed of 2 mm/min.

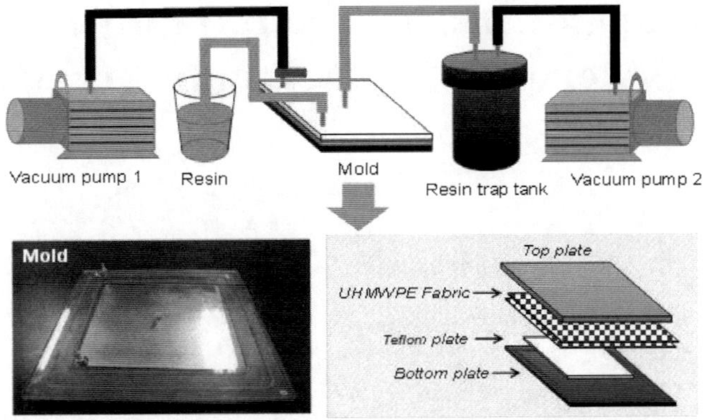

FIGURE 1. Schematics of the VaRTM apparatus for preparation of 3-dimensional composite.

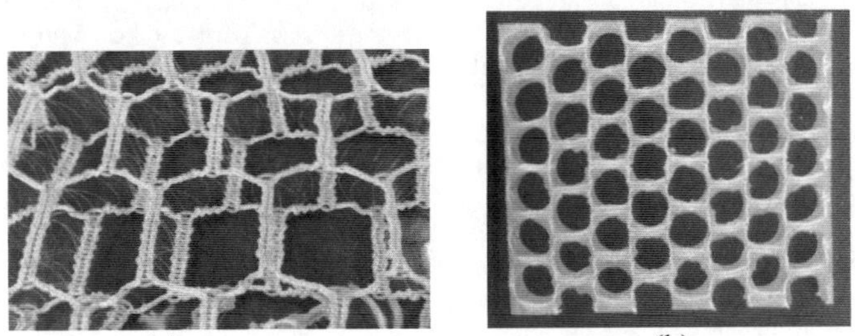

(a) (b)

FIGURE 2. Photographs of the reinforcement (a) and the 3-D FRP composites composed of circular holes

RESULTS AND DISCUSSION

Figure 3 shows tensile load-extension curves along longitudinal and transverse directions for the 3-D FRP composites, of which both directions were from the reinforcement knitted fabrics. The resistance to rupture, i.e., toughness, increased largely for the longitudinal samples while the initial slop, i.e., initial modulus, increased for the transverse samples. The increase of toughness for the longitudinal samples is due to a ductile deformation after the first rupture.

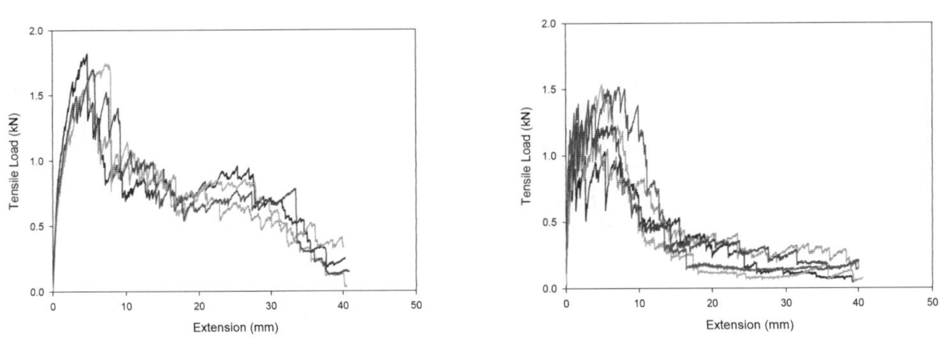

(a) (b)

FIGURE 3. Tensile behaviors of 3-D FRP composites along longitudinal (a) and transverse (b) directions.

ACKNOWLEDGMENTS

This research was supported by the Ministry of Trade, Industry & Energy(MOTIE), Korea Institute for Advancement of Technology(KIAT) through the Encouragement Program for The Industries of Economic Cooperation Region under Award No. R0002902.

Cross-Linking of Laser Sintered Polyamide 12 and Polyethylene Parts by Gamma Irradiation

Andreas Wegner[a,b*]

[a]University of Duisburg-Essen, Chair for Manufacturing Technology, Lotharstraße 1, 47057 Duisburg, Germany

[b]AM Polymer Research UG, Bismarckstraße 120, 47057 Duisburg, Germany

andreas.wegner@uni-due.de

Abstract. Laser sintering gets more and more importance as a small series production technique. However, there is only a little number of materials available for the process. In most cases parts are build up using polyamide 11 or polyamide 12. Therefore, parts produced with laser sintering sometimes do not have the required part properties for intended applications. Especially temperature stability or chemical resistance is sometimes not sufficient for parts made by polyamide 12.

Cross-linking by gamma irradiation is a well-established process for the post-processing of plastic parts in a cheap and fast single step treatment. It offers the possibility to improve temperature stability, the mechanical properties and the chemical resistance of parts. Therefore, it offers also great potential to improve part properties of laser sintered parts. However, there is only very limited knowledge on use of cross-linking for those parts.

Within the performed work cross-linking by gamma irradiation is used to post-process parts made by laser-sintering. Two different laser sintering materials are considered in the study. On the one hand a standard polyamide 12 from EOS (PA 2200) is used. This is modified with a cross-linking agent on two different ways in order to test that influence on the resulting part properties. On the other hand a new polyethylene material from ROWAK, which was developed in person, is used for the tests. This material does not require an additional cross-linking agent. The laser sintered parts are cross-linked using three different radiation levels. Afterwards material and part properties like temperature stability and mechanical properties are studied for the cross-linked parts. Additionally, these results are compared to those of parts which were not treated. Results show a significant influence of cross-linking on the performance of laser sintered parts with an improvement of some properties.

Keywords: Laser Sintering, Cross-Linking, Mechanical Properties, Polyamide 12, Polyethylene.
PACS: 62.20.de, 62.20fg, 62.20fk, 81.05.Lg, 81.20.Ev, 83.50.Uv, 83.80.Sg,

INTRODUCTION

Increasing competition, decreasing product life cycles, the wish for customized products and a shortage of resources cause the need for innovative manufacturing techniques for small series production [1]. Going beyond the stage of Rapid Prototyping on to Rapid Manufacturing, Additive Manufacturing offers possibilities for small series production of customized products and an increased freedom of design due to the lack of tools [2]. The laser sintering of plastic parts is, aside from beam melting of metal parts, one of only two AM-processes which have the capability to be used for Rapid Manufacturing in the near future [3]. In laser sintering parts are built up layer by layer. The machine produces the parts by repeating four stages for each layer: Firstly, the platform descends by the thickness of one layer. Secondly, powder is spread across the build platform by a leveling roller or coater. Thirdly, the layer is preheated to a temperature close to the material's melting point by a radiant heater. Then, a CO_2-laser beam melts the powder by tracing the actual cross section line after line using a scanner system. These steps are repeated until the parts are completed. [4]

Laser sintering has reached a high technical level within the past two decades. However, there is only a little number of materials available for the process. In about 90 to 95 % of all cases parts are build up using polyamide 12 (PA 12) or polyamide 11 [5, 6]. Alternative materials like polyethylene are seldom used. Reasons for that are high prices, a restricted availability, insufficient mechanical part properties, a more difficult processing or an insufficient understanding of the processing.

Cross-linking by gamma irradiation is a well-established process for the post-processing of plastic parts in a cheap and fast single step treatment. It offers the possibility to improve temperature stability, the mechanical properties and the chemical resistance of parts. Therefore, it offers also great potential to improve part properties of laser sintered parts. However, there is only very limited knowledge on use of cross-linking for those parts. The actual study should establish fundamental knowledge in the use of cross-linking as post processing of laser sintered parts. Therefore, the effects of cross-linking on the part properties of polyethylene (PE) and PA 12 parts manufactured by laser sintering are studied.

STATE OF THE ART

In the last years a huge number of scientific papers were published in the field of Additive Manufacturing. However, there is only one paper presenting results on the effects of cross-linking laser sintered parts by gamma radiation. Wendel describes in [7] a method to prepare laser sintered parts for a post treatment by cross-linking. She dips the laser sintered PA 12-parts in a cross linking agent and radiate them with a dose of 99 kGy gamma radiation in order to achieve a cross-linking effect. The study considers only the mechanical properties of parts oriented in x-direction. Results show an increase of 16 % for young's modulus, of 21 % for tensile strength and a reduction of 56 % for Elongation at break compared to unirradiated specimen produced with the same material. This data are being used as reference for the actual study.

However, cross-linking is a well-known process for the post treatment of plastic parts. Therefore, several publications can be found which study the effects of cross-linking on the materials properties of injection molded or extruded parts. In order to get some further reference values on the change of material properties by cross-linking, some papers especially on Polyethylene are evaluated. Cota describes in [8] the effect of gamma irradiation on PE using dose levels between 20 and 720 kGy. Results show an increase of yield stress by up to 30 % for compression test. Tensile tests show only a slight increase for low and very high dosage levels while for medium radiation levels even a decrease can be found. For modulus of elasticity at no dose level an increase was found. In a further study [9] the effects of irradiation doses between 0 and 220 kGy on the mechanical properties of PE were studied. Tensile tests show a maximum increase of tensile strength at ambient temperature by 5.6 % and of young's modulus by 36 %. Maximum increase of values is found for radiation levels between 99 and 198 kGy.

The evaluation of literature shows that there is only very limited knowledge on the use of cross-linking post processing by gamma radiation in laser sintering. Therefore, fundamental knowledge should be established in this field.

EXPERIMENTAL

Within this study two different materials were studied. One is a polyethylene form ROWAK named Rolaserit PEGR, which is a grey colored material. This was used due to the point that polyethylene does not need cross-linking agents. Specimens were produced with a standard parameter set which was established during material development in [10, 11]. Additionally, a standard laser sintering polyamide 12 type EOS PA 2200 was studied. In contrast to PE the PA 12 has to be modified with a cross-linking agent in order to improve part properties. Therefore, two different strategies were chosen while using Triallyl-Isocyanurat (TAIC). On the one hand, refreshed PA 12 powder (refresh rate 50 %) was mixed with 2 % TAIC firstly by a fast running mixer for 10 min and secondly by a slow mixer for another 30 min. After mixing material was sieved and stored for 24 h before processing. Afterwards, specimens were produced using an EOS Formiga P100 and an optimized parameter setting found in [12]. On the second hand, specimens were produced using refreshed PA 12 powder without modification. Parts were again produced with the optimized parameter set according to [12]. After the laser sintering process and after cleaning the parts, specimens were dipped in TAIC for 5 min and dried in standard climate afterwards. This method works for laser sintered parts due to their open surface porosity. Laser sintered parts show always a porosity of up to 5 %. Therefore, parts absorb the cross-linking agent in its open porosity. This way to modify the laser sintered parts with a crosslinking agent was also used in [7] and is studied in this content as a comparing method.

Cross-linking of the laser sintered specimen was performed by the company Beta-Gamma-Service by using three different radiation levels of gamma radiation. For PA 12 radiation levels were 66 kGy (2 x 33 kGy), 99 kGy (3 x 33 kGy) and 132 kGy (4 x 33 kGy). For PE the three radiation levels were chosen a little bit higher beginning at 99 kGy (3 x 33 kGy), over 132 kGy (4 x 33 kGy) and using 165 kGy (5 x 33 kGy) as highest level. For both materials some specimen was not irradiated being used for determination of reference values. For PA 12 besides unmodified specimen also those which were produced with PA 12 + 2 % TAIC are considered as reference.

In order to analyse the effects of cross-linking, tensile properties and vicat softening temperature were analyzed. Tensile tests were performed for the x- (x0) and the z-direction (z90) in compliance with DIN EN ISO 527-1 at a testing speed of 5 mm/min for both materials. PA 12 parts were stored according to ISO 1110 before testing. Additionally, the vicat B50 softening temperature was measured according to ISO 306 for both materials using a heating rate of 50 K/h and a weight of 50 N.

RESULTS AND DISCUSSION

After manufacturing of the specimen, these were cross-linked at different radiation levels. The irradiated parts were then tested on their mechanical properties and the softening temperatures. In the following chapter the measured results are presented and discussed.

For the standard polyamide 12 two different ways of material modification with a cross-linking agent were analyzed. The results for mechanical properties are compared to reference values of unmodified and unirradiated specimens. Figure 1 gives the values for young's modulus and elongation at break (EAB) as a function of radiation level in dependence of the part orientation. For the reference a young's modulus of 1618 N/mm² for the x-direction (x0) and 1586 N/mm² for the z-direction (z90) are measured, Figure 1a. Elongation at break is 24.4 % (x0) respectively 18.5 % (z90), Figure 1b. The processing of PA 2200 mixed with 2 % TAIC leads to a little drop of mechanical properties. A reason for that might be a reduced flowability of the powder due to the addition of TAIC. The Hausner ratio is reduced to 1.3 after addition of TAIC while the refreshed PA 12 shows a value of 1.21. Additionally, bulk density shows a slight decrease. Young's modulus is reduced to 1510 N/mm² for x0 and 1402 N/mm² for z90. Elongation at break for x0 falls to 21.1 %. In contrast to that EAB in z-direction increases to 19.3 %. When analyzing the results for the mixed and irradiated material an increase of young's modulus and a reduction of EAB is found. Young's modulus increases by circa 10 % compared to the reference independent from the orientation and the radiation level. Only at the lowest radiation level of 66 kGy the increase in x-direction is with 11.5 % a little bit higher than for the z-direction with 8.4 %. When results are compared to the values measured for the mixed but unirradiated PA 2200 the increase is much higher. The young's modulus rises by 17.5 to 19.5 % in x-direction and 22.6 to 25.4 % in z-direction. Therefore significant improvement of values is found for the young's modulus. At the same time EAB drops due to the cross-linking. This effect is well known from literature. However, drop in laser sintering is more significant compared to other manufacturing techniques. EAB in x- and z-direction is reduced by 31.5 to 50.5 % compared to the reference and by 26 to 52.5 % compared to the mixed and unirradiated material. In x-direction EAB values of 15.5 % are achieved for 66 kGy and 13 % for 99 as well as 132 kGy. Values in z-direction are between 9.2 and 12.7 %. In this case, EAB fell with an increase of the radiation level. However, no brittle behavior is found after the cross-linking. EAB values are even higher than some results found in the literature [13-15] for PA 12 without cross-linking.

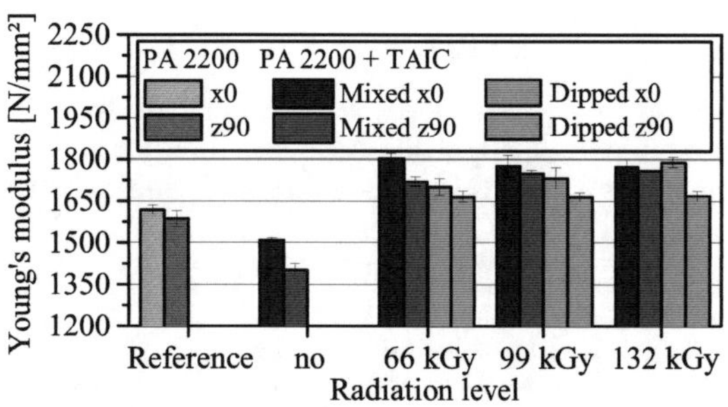

a. Young's modulus

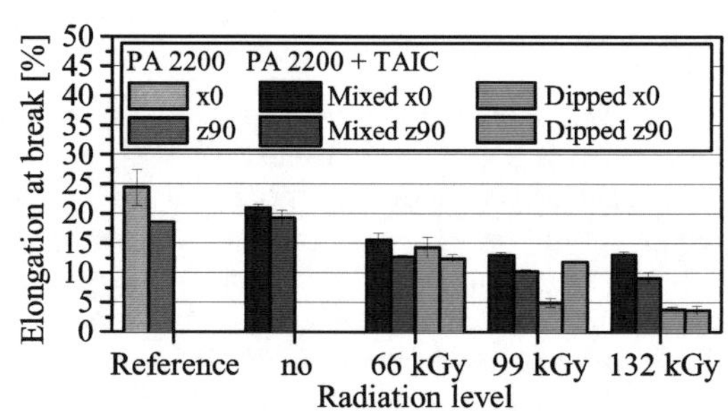

b. Elongation at break

FIGURE 1. Mechanical properties of cross-linked PA 12-parts for different build orientations as a function of radiation level

Additionally, the effect of another method of adding a cross-linking agent was studied. The parts manufactured with unmodified PA 2200 powder were dipped in TAIC after the process. The results for mechanical properties of the dipped and irradiated specimen are also shown in Figure 1. The young's modulus in x-direction rises compared to the reference by 5.1 to 10.6 % with an increase of the radiation level from 66 to 132 kGy. Only for the highest radiation level of 132 kGy a similar increase compared to the reference like for the mixed material is found. However, the improvement for the mixed is in reality much higher at 17.5 to 25.4 % due to the point that the addition of TAIC lead to a reduction of mechanical properties. Therefore, the dipping of specimen leads to no comparable improvement. In z-direction the improvement is even worse with a maximum increase of 5.2 % compared to the reference. Additionally, EAB is strongly influenced by the cross-linking when the specimen are dipped in TAIC. For a radiation level of 66 kGy reduction of EAB is 41.4 % for x0 respectivly 32.7 % for z90 which is similar to the results described before. For a radiation level of 99 kGy especially the EAB value in x-direction show a heavy drop by 79.4 % to a value of only 5 %. EAB in z-direction is not affected that way for the same radiation level. For a further increase of radiation level to 132 kGy EAB in both directions drops to under 4 %. Therefore, the material behaviour changes from ductile to brittle for the dipping method using high radiation levels. However, this change do not lead to superior values for young's modulus. Even tensile strength (results are not shown here due to limited space) show no singnificant improvement. For a radiation level of 66 kGy only an improvement of only 1 % is measured. For higher radiation level even a decrease of tensile strength by 6.5 to 16.7 % is found. In contrast to that, for the mixing method an increase of tensile strength by 6 to 8 % is achieved independent from the radiation level when values are compared to the unirradiated specimen which were build using the powder modified with TAIC. Even when compared to the reference, values show an improvement of 3 to 4 % in x-direction and circa 1.5 % in z-direction.

Besides the PA 12 material, polyethylene powder of the type Rolaserit PE was analysed. Figure 2 gives the values for young's modulus and elongation at break (EAB) as a function of radiation level dependent from the part orientation. In contrast to PA 12, PE does not need a modification with a cross-linking agent. The unirradiated specimen has a young's modulus of 1397 N/mm² for x0 and 1348 N/mm² for z90, Figure 2a. EAB is 4.9 % in x-direction and 2.9 % for the z-direction, Figure 2b. Basing on these results the improvement of mechanical properties due to cross-linking are studied. Young's modulus shows a rise of values when radiation limit is increased from 99 to 165 kGy. Highest values are achieved at a level of 165 % with 1502 N/mm² for x0 and 1435 N/mm² for z90. This correlates with an improvement by 7.5 % respectively 6.4 %. Improvement is here not as significant as found for PA 12 when powder is mixed with TAIC. Like for PA 12, also for PE EAB decreases when specimen are cross-linked. The reduction seems not to be influenced by the radiation level. For the x-direction EAB of irradiated specimen shows values between 3.6 and 3.9 % which is 20 to 26.5 % lower than the reference values. In z-direction EAB decreases by 17 up to 27.8 % when parts are cross-linked. EAB values vary between 2.1 and 2.4 %. Hereby, the percentaged decrease is lower as for PA 12. However, PE shows in general lower EAB values and a more brittle behavior. When considering tensile strength, values in x-direction are only slightly influenced by the radiation. The change of values varies between a decrease by 1.1 % and an increase by 1.7 %, while the increase is found for a radiation level of 165 kGy. Tensile strength in z-direction is reduced by 1.4 to 7.1 % when parts are cross-linked. For PE optimal radiation level is 165 kGy. For this setting the highest improvement of Young's modulus is achieved while decrease of EAB and tensile strength is not at the highest level. However, improvement of part properties by cross-linking is limited. One possible reason might be that part properties of PE are more significantly influenced by the inner structure of the parts especially the porosity as well as the surface roughness.

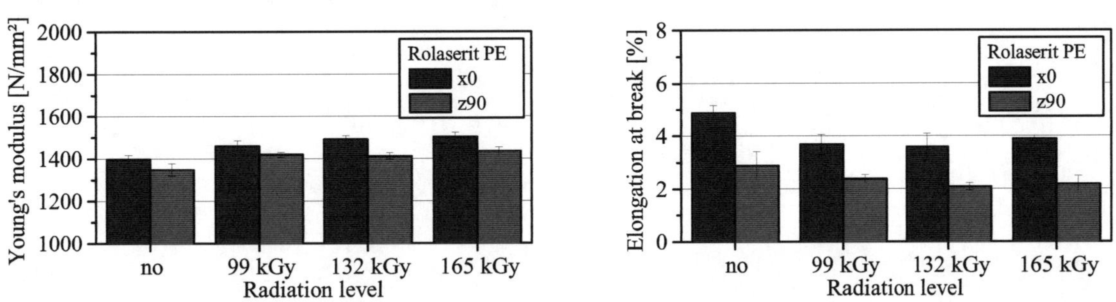

a. Young's modulus b. Elongation at break

FIGURE 2. Mechanical properties of cross-linked PE-parts for different build orientations as a function of radiation level

Additionally, vicat B50 softening temperature of the unirradiated and irradiated samples are measured in order to prove an increase of parts temperature resistance. The measured vicat B50 values are sumerized in dependancy of the rediation level in Figure 3a for the PA 12 samples and in Figure 3b for PE. The softening point of the unmodified reference sample lies at 161.3 °C while the value for the sample mixed with 2 % TAIC is at 159.9 °C. Softening temperature for the parts manufactured with the mixed powder increases only slightly by 1 to 1.7 % when sampels are irradiated with 66 to 132 kGy. In comparison to the mixed method, the effects using the dipping method are a little bit higher when using radiation levels of 99 or 132 kGy. Vicat softening point rises to circa 165 °C which results in an increase by more than 2 %. However, results show a standard deviation of 6 °C. Therefore the effects has no good reproducibilty. In contrast to PA 12, the change of softening point is significantly larger for PE, Figure 3b. The unirradiated sample shows a vicat B50 temperature of 74.4 °C. After cross-linking the softening temperature is between 7.1 % and 11 % higher than the reference. The biggest increase is achieved for a radiation level of 132 kGy. The softening temperature rises here to 82.6 °C. For higher radiation level a decrease of vicat B50 is found. Softening point for 165 kGy is only 80.8 °C. However, also this means a increase by 6.4 K compared to the reference.

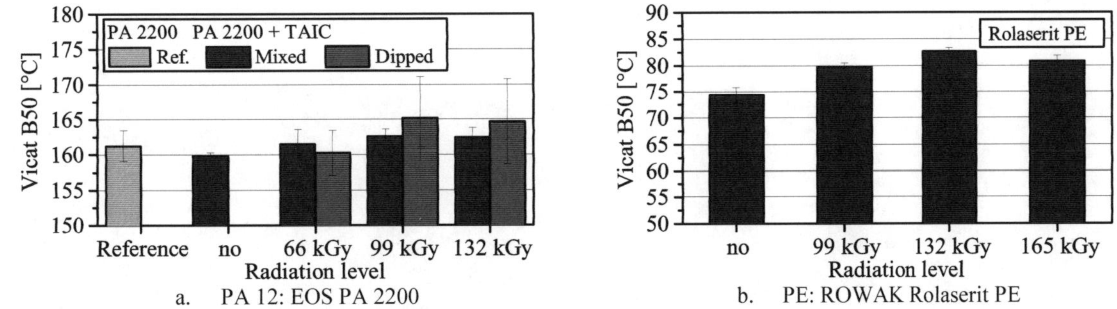

a. PA 12: EOS PA 2200 b. PE: ROWAK Rolaserit PE

FIGURE 3. Vicat B50 softening temperatures for different laser sintering materials as a function of radiation level

CONCLUSIONS

The study analyses the effects of using cross-linking by gamma irradiation on the part properties of laser sintered PA 12- and PE-parts. Results show, that for some properties a significant increase of values is possible. Optimal increase of properties are achieved for PA 12 at a dose level of 66 kGy when using PA 12 powder mixed with 2 % TAIC and for PE at a dose level of 165 kGy. However, the increase of young's modulus and softening temperature leads always also to a reduction of elongation at break which is the most critical part property in laser sintering. However, for PA 12 even after cross-linking a ductile part behavior can be achieved. The change of part properties is in most cases in a similar range as found in the literature. However, in some cases the effects are not as large as in the papers evaluated. Especially for the tensile strength of PA 12 and the young's modulus of PE the increase is on a lower level. In summary, results provide for the first time detailed information on the use of cross-linking by gamma irradiation as a post treatment process in laser sintering.

Future research should validate the found correlations and should establish further influences. Additionally, processing of the modified material should be improved in order to achieve superior properties. Results should be also transferred on other laser sintering materials in order to enhance knowledge in field of cross linking.

REFERENCES

1. E. Abele and G. Reinhart, *Zukunft der Produktion - Herausforderungen, Forschungsfelder, Chancen*, München: Carl Hanser Verlag, 2011.
2. R. Hague et al,: Proc. Inst. Mech. Eng., Part C: J. Mech. Eng. Sci., **217/1**, 25-30 (2004).
3. N. N., *VDI 3405 Additive Fertigungsverfahren – Grundlagen, Begriffe, Verfahrensbeschreibungen, 2014.*
4. A. Gebhardt, *Generative Fertigungsverfahren – Rapid Prototyping – Rapid Tooling – Rapid Manufacturing*, München: Carl Hanser Verlag, 2007.
5. R.D. Goodridge et al., *Prog. Mat. Sci.*, **57/2**, 229-267 (2012).
6. M. Schmid, *Selektives Lasersintern (SLS) mit Kunststoffen – Technologie, Prozesse und Werkstoffe*, München: Carl Hanser Verlag, 2015.
7. B. Wendel et al., "New Developments in Selective Laser Sintering of Polymers" in *Proceedings of the 5th LANE*, Universität Erlangen-Nürnberg, 2007, pp. 323-331.
8. S.S. Cota et al., *Braz. J. Chem. Eng.*, **24/02**, 259-265 (2007).
9. A. Mizera et al., *Int. J. Math. Comp. and Sim.*, **6/6**, 584-591 (2012).
10. A. Wegner, "New Materials for Laser Sintering: Processing conditions of Polyethylene and Polyoxymethylene" in *Proceedings of the Fraunhofer Direct Digital Manufacturing Conference 2016*, Berlin, 2016.
11. A. Wegner and G. Witt, "Optimale Energieeinträge für die Verarbeitung unterschiedlicher kommerzieller Polyamid 11 und 12-Pulver sowie eines neu entwickelten Polyethylen-Pulvers beim Laser-Sintern" in *Neue Entwicklungen in der Additiven Fertigung - Beiträge aus der wissenschaftlichen Tagung der Rapid.Tech 2015*, edited by G. Witt, A. Wegner and J. T. Sehrt, Berlin Heidelberg: Springer Vieweg Verlag, 2015, pp. 43-61.
12. A. Wegner, "Theorie über die Fortführung von Aufschmelzvorgängen als Grundvoraussetzung für eine robuste Prozessführung beim Laser-Sintern von Thermoplasten", Ph.D. Thesis, University of Duisburg-Essen, 2015.
13. T.L. Starr, T.J. Gornet, and J.S. Usher, Rapid Prototyping J., 17, 6 (2011), pp. 418-423.
14. U. Ajoku et al., *Proc. Inst. Mech. Eng., Part B: J. Eng. Man.*, **220**, 1077-1086 (2006).
15. T. J. Silverman et al., "Comparison of Material Properties and Microstructure of Specimens Built using the 3D Systems Vanguard HS and Vanguard HiQ+HS SLS Systems" in *Proceedings of the 18th International Solid Freeform Fabrication Symposium (SFF 2007)*, The University of Texas at Austin, 2007, pp. 392-401.

3D Printing Of Maize Protein By Fused Deposition Modeling

Laurent Chaunier[1, 2*], Eric Leroy[2], Guy Della Valle[1], Michèle Dalgalarrondo[1], Bénédicte Bakan[1], Didier Marion[1], Baptiste Madec[3] and Denis Lourdin[1]

[1]*BIA INRA-UR-1268, La Géraudière. 44316 Nantes Cedex 3, France*
[2]*GEPEA UMR-CNRS-6144, CRTT, 37 av. de l'université. 44606 St-Nazaire Cedex, France*
[3]*Ecole Centrale, Plateforme RFI, 1 rue de la Noë. 44321 Nantes Cedex 3, France*

Corresponding author: laurent.chaunier@nantes.inra.fr

Abstract. Additive Manufacturing (AM) opens new perspectives for biopolymers to obtain functional parts, like biomedical devices, by exploiting their biocompatibility and resorbability. Plasticized materials from zein, a storage protein from maize seed endosperm, display thermomechanical properties that could match with Fused Deposition Modeling (FDM). The objective of this work was to evaluate their thermo-rheological behavior and their structural modifications during processing. 20% glycerol was added to commercial zein, containing naturally about 4% lipids and 5% water. After storage at intermediate relative humidity (RH=59%), its glass transition temperature, measured by DSC, was Tg=42°C. The main mechanical relaxation, measured by DMA, was found at $T\alpha$=50°C, leading to a drop of the elastic modulus from E'=1.1GPa, at ambient temperature, to E'=0.6MPa at $T\alpha$+100°C. These values are in a similar range as those of standard polymers used for AM-FDM processing, such as PLA and ABS. The structure of zein was characterized at different scales by SDS-PAGE, reversed-phase HPLC, FTIR and WAXS, at each processing stage: (i) the initial formulation of the plasticized powdery material, (ii) after extrusion at 130°C for shaping printable filaments, and (iii) after deposition through the 3D printer nozzle ($Ø_{nozzle}$=0.5mm, $T_{printing}$=130°C, $v_{printing}\approx$10mm/s). The presence of disulfide bond cross-links was evidenced in extruded filaments and remains at the same level after printing. WAXS showed that, in these conditions, no molecular orientation was obtained in the deposited material. By tuning zein-based printable compositions, these results open the field of their processing as resorbable printed parts, with a controlled geometry and a designed tridimensional structure.

Keywords: 3D Printing, Additive Manufacturing, Biopolymer, Corn, FDM, Protein, Zein.

PACS: 81, 62, 65.

INTRODUCTION

Additive Manufacturing (AM) is a family of processes allowing the layer by layer production of functional finished plastic parts, from digital three-dimensional models [1, 2]. This allows now the production of functional finished parts. Applications can be found in various fields as for the production of prostheses or matrices for cells growth in the biomedical domain, complex mechanical parts for industry, or even edible printed foods [3-5]. The main advantage of AM is the fast transition from design to obtaining parts from a digital model providing desired properties [6].

For polymers, one of the common techniques is the Fused Deposition Modeling (FDM). It is based on the deposition layer-by-layer at high temperature of a molten thermoplastic polymer through a nozzle, on a solid substrate. Initially, the thermoplastic material is shaped as a filament, obtained by a previous extrusion step and which allows the feeding of the 3D printer (Fig.1). The deposited material must solidify quickly, by cooling below its glass transition temperature (Tg). The upper layers of the deposited material have to adhere to the previously deposed layer so as to obtain a finished solid part.

In the case of FDM applied to the pharmaceutical field, the manufacture of materials with specific porosity (*e.g.* variable deposition patterns for filling) could allow the production of innovative 3D printed tablets, for the controlled release of active constituents. As an example, in a printed part made from PVA, 100% fluorescein, taken as a model molecule, can be released under simulated physiological conditions in less than 6 hours with 10% density of the printed filling, while more than 10 hours are required for 50% or more [7].

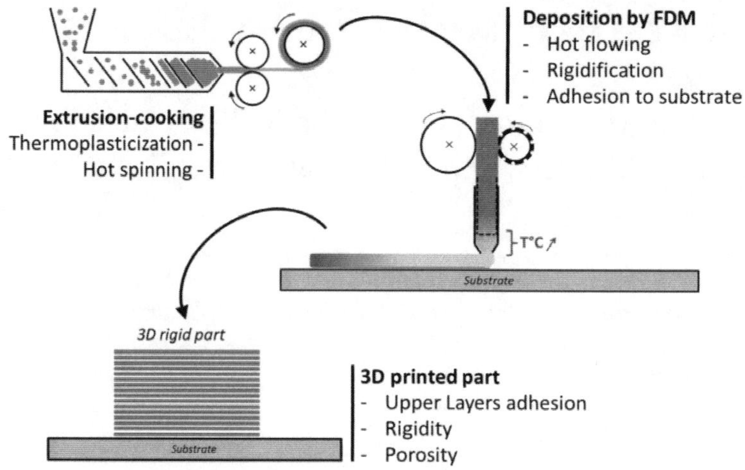

FIGURE 1. Scheme of the main steps and characteristics of FDM

Now, progresses should target the development of new 3D printable materials with adequate behavior for FDM like rheological properties, adhesion ability and mechanical properties [8].

Biopolymers are good candidates to obtain 3D printed parts with novel properties, such as biodegradability, or biocompatibility. Then the present work focuses on a plant protein : zein, a maize storage protein which represents about 5% of the grain dry weight and being a by-product of starch and bioethanol production. It is part of the prolamins family [9]. Considered for various applications, it is used as a water barrier in food or pharmaceutic industries and for the production of degradable plastics [10, 11]. Zeins are low charge proteins which amino acids sequences display both hydrophobic domains and hydrophilic ones (including glutamine). Accordingly, they constitute amphiphilic polymers that can be thermoplasticized after blending with polar or apolar plasticizers. As an example, in a previous study for a content of 22% glycerol (w/w), it presents a main mechanical relaxation linked to its glass transition temperature at about T_α=60°C, and a storage modulus (E') about 1GPa at ambient conditions [12, 13]. Plasticized zein, added with paracetamol as active ingredient, has been recently considered as extruded-molded caplets for controlling drug release [14]. Thus, this protein is a promising compound for processing as a thermoplastic biopolymer, in the case of Fused Deposition Modeling, to obtain tridimensional designed parts presenting a defined porosity. It could be used for the controlled release of bioactive molecules.

The present work proposes to investigate 3D printing abilities of plasticized zein and understand structure-properties relationships possibly leading to 3D printed parts with targeted functional properties.

MATERIALS AND METHODS

Raw Materials and Thermomechanical Processing

Zein from maize (Ref.Z3625) and glycerol were purchased from Sigma-Aldrich (F-38, Saint-Quentin Fallavier). The moisture content was determined by thermogravimetry after 2hr at 130°C (TA Instruments, F-78 Guyancourt).

Zein-based plasticized compositions were obtained after blending with glycerol. They were extruded using a SCAMIA single-screw device (SCAMEX, F-91 Crosne), with a 200mm barrel (screw L/D=10) and a cylindrical die ($Ø_{die}$=1.7mm). The screw speed was 20 rpm and temperature was T_{die}=130°C (Fig.2-a). In these conditions, the average flow rate is Q≈2g/min and the specific mechanical energy about SME≈300J.g^{-1}.

A 3D Printer with an operating volume close to 20^3cm^3 (ORDBot-Hadron, $Ø_{die}$=0.5mm) and standard synthetic polymers as feedstock filaments for 3D printing by FDM (*i.e.* Acrylonitrile Butadiene Styrene, ABS, and Polylactic Acid, PLA) were purchased from eMotion Tech (F-31 Toulouse), Figure 2-b. Printing trials with thermoplasticized zein were carried out at T_{nozzle}=130°C.

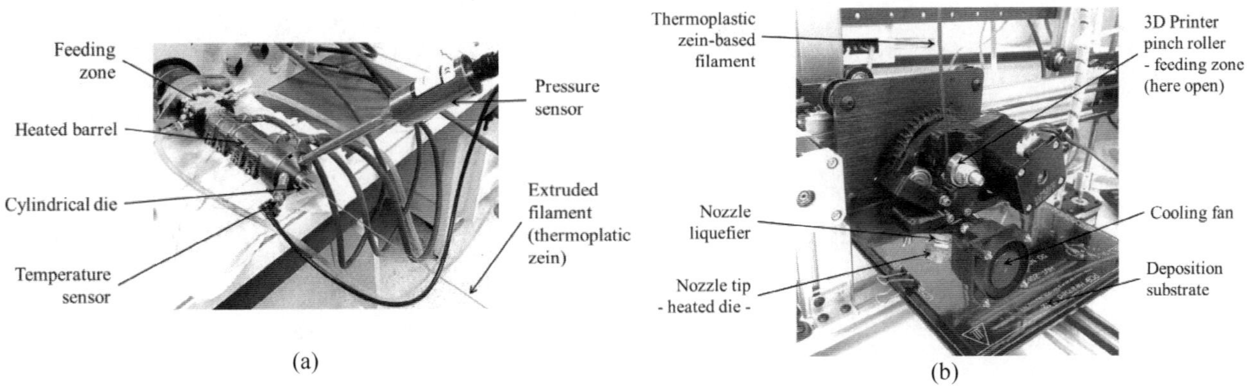

FIGURE 2. Extrusion of thermoplasticized zein as filament, at T_{die}=130°C (a). 3D Printer-FDM, ORDBot Hadron (20x20cm^2 solid deposition substrate, T_{nozzle}=130°C) : feeding zone with zein-based rigid filament with 20% glycerol content (Z20GLY) (b)

Glass Transition and Thermomechanical Properties of Zein-based materials

To determine the glass transition temperature, thermograms were recorded with an automated Differential Scanning Calorimeter DSC (Q100, TA Instruments, F-78 Guyancourt). Each specimen (10 mg) was placed into a sealed stainless-steel cell. Measurements were performed at 3 °C/min from 10 to 140 °C during a second scan, to cancel thermal events due to aging during storage. Extruded cylindrical samples (Ø≈1,75mm), cut in 20mm long specimen, were submitted to Dynamic Mechanical Analysis, DMA (DMA-50N-O1dB, Metravib, F-59 Lyon). The mechanical active length, between the grips, was 10 mm and extrudates were characterized in a tensile mode at 1 Hz, with a strain set at 0.1% and heating rate of 3 °C/min, up to 160°C.

Structural Characterization of Zein during Processing

The structure of the plasticized zein was evaluated at the different steps of the FDM processing, from the initial powdery composition, the extruded and the 3D-printed filament out of the nozzle of the 3D-printer. Each specimen was diluted to obtain a 2mg/ml solution in a Tris/HCl 50mM containing Sodium Dodecyl Sulfate 2% (SDS) buffer at pH=6.8 and submitted to Electrophoresis on a 12% Polyacrylamide Gels (PAGE). Samples were characterized by SDS-PAGE without and with 3% β-mercaptoethanol as reducing agent. After staining, migration bands were seen, and molecular weight was calculated by comparison with standard Mw markers (Euromedex, F-67 Souffelweyersheim) submitted to SDS-PAGE in the same conditions. The molecular orientation was assessed by direct beam X-ray scattering (Cu-$K_{\alpha 1}$=1.5405Å) on a Bruker D8 diffractometer (Madison, WI, US), after integration of the azimuthal angle (Φ from 0 to 360°) on diffractograms obtained for 2θ varying from 3° to 30°.

RESULTS AND DISCUSSIONS

Commercial zein presents a moisture content at MC=5% (total wet basis) and Tg=88°C. When plasticized with 20% glycerol, Tg decreases to 42°C which leads to decrease the storage modulus from E'=1.1 GPa at 20°C, to less than 0.6 MPa at 130°C, above the principal mechanical relaxation linked to its glass transition temperature (Fig.3). Such results are in the same range as those previously published in the literature [12].

Compositions containing 10% glycerol added, or without plasticizer, do not present such an important decrease of storage modulus, limiting their flowing abilities in the molten state above Tg, and then avoiding their application in 3D printing by FDM. In the case of glycerol content above 30%, lower mechanical properties are obtained at ambient conditions than with 20% (E'< 1GPa), but similar values at high temperature. Thus, extrudates based on zein with 20% glycerol, namely "Z20GLY", could be the most pertinent to carry out trials of 3D printing by FDM. Furthermore, despite different structures and relaxation mechanisms, the loss of the mechanical strength at high temperature allows to evaluate the printing temperature of Z20GLY at $T_{printing}$>120°C, whereas $T_{printing}$>180°C for PLA and $T_{printing}$>220°C for ABS.

In these conditions, each polymer is beyond its relaxation temperature, *i.e.* above Tg for amorphous Z20GLY (Tg=42°C, Fig.3) and ABS (Tg=100-120°C), or above its melting temperature as for PLA ($T_{melting}$=160°C). Even if the mechanical strength is below the one of standard polymers, like PLA and ABS which present moduli above 2GPa at 20°C, it is enough to launch feasibility trials with plasticized zein on a FDM 3D printer at 130°C (Fig.4-a).

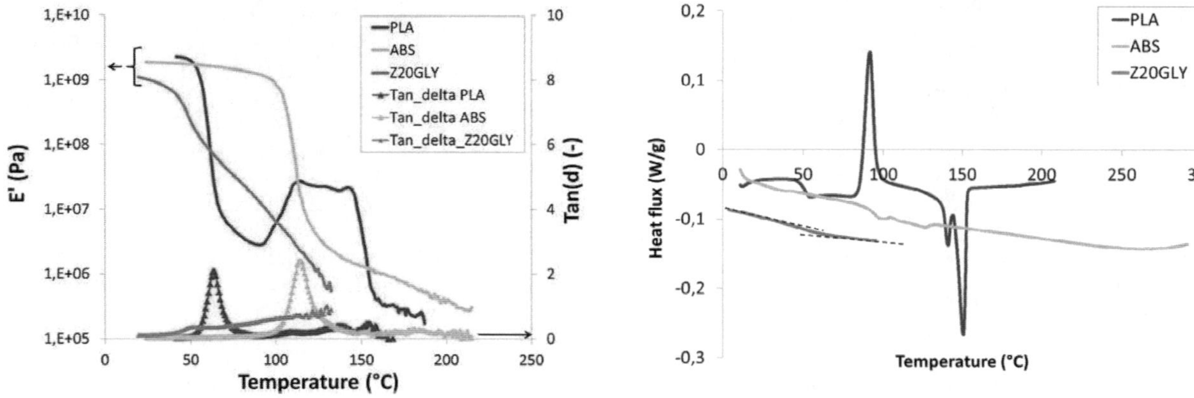

(a) (b)

FIGURE 3. Thermomechanical behavior of the plasticized zein with 20% glycerol added (Z20GLY), PLA and ABS filaments : evolution of the elastic modulus and the damping factor -tanδ- (a); characterization by DSC (b)

By direct beam X-ray characterization, no molecular orientation was obtained for the plasticized zein in these printing conditions ($Ø_{nozzle}$=0.5mm, $T_{printing}$=130°C, $v_{printing}$≈10mm/s) and its structure remained amorphous. As a raw material, before extrusion, zein is mainly composed of α-zein (Mw from 20 to 25kDa, Fig.4-b). The molecular characterization also showed that high temperature crosslinkings occurred during the extrusion of the thermoplasticized zein, by oxydation of cystein sulfhydryl groups (-SH) as disulfide bonds (S-S).

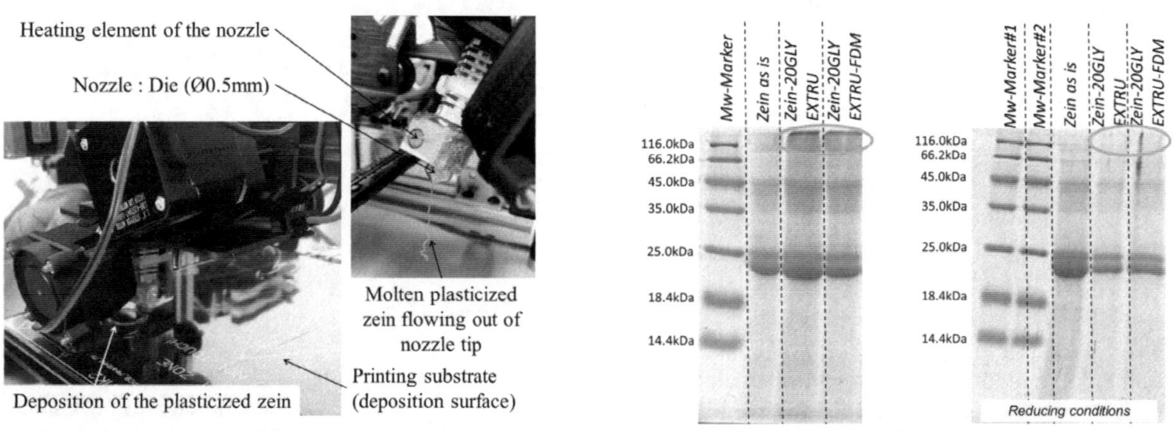

(a) (b)

FIGURE 4. 3D Printing trials of zein-based composition with 20% glycerol added : molten material flowing through the die at 130°C and its deposition (a). SDS-PAGE characterization of zein at the different steps of 3D printing (normal conditions on left hand and with reducing agent on right hand) : molecular weight markers "Mw-Marker", powdery raw material "zein as is", after adding 20% glycerol and extrusion "Zein-20GLY EXTRU" and finally after 3D-printing "Zein-20GLY EXTRU-FDM" (b)

The addition of β-mercaptoethanol leads to cancel such crosslinkings by a chemical reduction reaction (Fig.4-b, on right hand). Indeed, high molecular weight molecules (Mw>120kDa) are created by such thermomechanical treatment (SME≈300J/g, 130°C) and the crosslinked structure of the material remained in the 3D printed part.

CONCLUSIONS AND MAIN PROSPECTS

Our present data demonstrate that plasticized zein can be processed by 3D printing with Fused Deposition Modeling. Indeed, the thermomechanical properties of this plasticized biopolymer, as well as its hot melt flowing, match with process specifications. No molecular orientation was checked in the deposited filament at the outlet of the nozzle of the 3D printer and the amorphous structure of zein was shown as crosslinked after extrusion and remaining in the 3D printed part.

Such results will be used to study more accurately the rheological behavior of zein-based compositions in the molten state and their hot welding properties. Furthermore, the possible interactions of zein with different plasticizing compounds (*e.g.* polar, apolar, or ionic liquids) could lead to enhance their 3D printing abilities and to open the field of novel applications in the biomedical or pharmaceutical domains, for instance.

ACKNOWLEDGMENTS

The authors would like to thank Roselyne Désirest, André Lelion, Patrice Papineau and Bruno Pontoire (INRA, Nantes) for their helpful contributions to samples extrusion and characterization.

REFERENCES

1. ASTM. F2792-12a, Standard Terminology for Additive Manufacturing Technologies. *ASTM International,* West Conshohocken, PA (US), 2012.
2. D.T. Pham and R.S. Gault. *International Journal of Machine Tools & Manufacture* **38(10-11)**, 1257-1287 (1998).
3. A. Pfister, R. Landers, A. Laib, U. Hubner, R. Schmelzeisen and R. Mulhaupt. *Journal of Polymer Science Part a-Polymer Chemistry* **42(3)**, 624-638 (2004).
4. J.T. Belter and A.M. Dollar. *Plos One* **10(4)** (2015).
5. T.F. Wegrzyn, M. Golding and R.H. Archer. *Trends in Food Science & Technology* **27(2)**, 66-72 (2012).
6. A.W.L. Yao. *International Journal of Advanced Manufacturing Technology* **26(11-12)**, 1284-1288 (2005).
7. A. Goyanes, A.B.M. Buanz, A.W. Basit and S. Gaisford. *International Journal of Pharmaceutics* **476(1-2)**, 88-92 (2014).
8. B.N. Turner, R. Strong, and S. Gold. A. *Rapid Prototyping Journal* **20(3)**, 192-204 (2014).
9. T.J. Anderson and B.P. Lamsal. *Cereal Chemistry* **88(2)**, 159-173 (2011).
10. J.W. Lawton. *Cereal Chemistry* **79(1)**, 1-18 (2002).
11. E. Corradini, P.S. Curti, A.B. Meniqueti, A.F. Martins, A.F. Rubira and E.C. Muniz. *International Journal of Molecular Sciences* **15(12)**, 22438-22470 (2014).
12. E. Corradini, E.S. de Medeiros, A.J.F. Carvalho, A.A.S. Curvelo and L.H.C. Mattoso. *Journal of Applied Polymer Science* **101(6)**, 4133-4139 (2006).
13. G.W. Selling and K..D. Utt. *Polymer Degradation and Stability* **98(1)**, 184-189 (2013).
14. J. Bouman, P. Belton, P. Venema, E. van der Linden, R. de Vries and S. Qi. *Pharmaceutical Research* **32(8)**, 2775-2786 (2015).

Viscoelastic Modelling Of Gas Pore Collapse During Polymer Sintering

Florian Wohlgemuth[a] and Ingo Alig[a*]

[a]*Fraunhofer Institute for Structural Durability and System Reliability LBF Schlossgartenstraße 6, 64289 Darmstadt, Germany*

ingo.alig@lbf.fraunhofer.de

Abstract. Based on Frenkel's Newtonian viscous flow approach [1] and Mackenzie and Shuttleworth's [2] extension accounting for trapped gas, a model for the shrinkage of a gaseous pore in a viscoelastic polymer melt driven by surface tension was developed. The viscoelastic flow of polymeric materials was considered by using a simple viscoelastic model described by Bellehumeur [3]. In the simulation the viscoelastic relaxation leads to a pronounced increase of the characteristic time of shrinkage, but does not alter the final equilibrium gas sphere diameter. A first attempt at including gas diffusion into the polymer melt motivated by the works of Kontopoulou and Vlachopoulos [4] and Gogos [5] will be given. Numerical solutions of our simplified model show that gas diffusion leads to a complete collapse of the pore. The rate of diffusion can change the collapse mechanism and the time-dependence of pore radius and pressure tremendously.

Keywords: Viscoelasticity, Continuum Mechanics, Polymers, Additive Manufacturing, Selective Laser Sintering, Diffusion

PACS: 83.80.Sg, 83.60.Bc, 83.50.Jf, 83.10.Ff, 89.20.Kk, 81.05.Lg, 47.50.Cd, 47.57.Ng, 47.85.md

Introduction

The underlying mechanisms of additive manufacturing techniques of polymeric materials such as selective laser sintering are melting, coagulation and crystallisation (or vitrification) of the material. Understanding and modelling of particle coagulation and pore collapse are therefore key issues for process optimisation and material development. The early work by Frenkel [1] describing the early and late stage of metallic sintering as Newtonian flow driven by surface tension has often been applied to these sintering processes. Frenkel's early stage describes the coagulation of two spherical particles, the late stage the collapse of a pore inside a melt. For the early stage, various extensions, even to viscoelasticity [3], exist. The late stage has been extended to include gaseous pressure inside the pore [2] without taking the viscoelastic nature of polymeric materials into account. In this work, a viscoelastic extension of this late stage process is given.

Separate models for the diffusive dissolution of gaseous pores in polymer melts have been presented [4], [5]. These do however either model flow processes as purely viscous or neglect them completely. A first attempt towards the coupling of diffusion and viscoelastic flow using a constant transfer coefficient is made.

Viscoelastic Model

Frenkel [1], in his late stage model, assumed an infinite homogeneous incompressible viscous melt with a spherical pore trapped inside. The pore collapses due to the minimisation of surface energy. Mackenzie and Shuttleworth [2] extended the model by including an ideal gas trapped inside the pore. In this work, their model was further extended by assuming a viscoelastic melt. Due to volume conservation, the radial velocity profile $\dot{r}$, and thus the radial rate of strain $\frac{\partial \dot{r}}{\partial r}$, in the melt are given by:

$$\dot{r} = \frac{r_p^2 \dot{r}_p}{r^2} \Rightarrow \frac{\partial \dot{r}}{\partial r} = -2\,\frac{r_p^2 \dot{r}_p}{r^3} \tag{1}$$

$r_p(t)$ is the pore radius, $\dot{r}_p$ the velocity at the pore surface and r the radial coordinate in the melt.

To ensure volume conservation, $Tr(\boldsymbol{D}) = 0$, where $\boldsymbol{D}$ is the symmetric rate of strain tensor, needs to be fulfilled [6]. Therefore, two flows of half magnitude and opposite sign rectangular to the main flow need to be inserted. This is called the Eshelby correction [7] and common practice in later works [2],[3].

The viscoelastic equation of state to be used is the objective Maxwell model as used by Bellehumeur *et al.* in their extension of Frenkel's early stage model. This extension has been successfully applied to experimental results e.g. in [3],[8],[9],[10],[11]. The rheological equation has a relaxation time λ and a viscosity η:

$$\lambda\overset{\square}{\boldsymbol{\tau}} + \boldsymbol{\tau} = 2\eta\boldsymbol{D} \tag{2}$$

$\boldsymbol{\tau}$ is the stress tensor, $\boldsymbol{W}$ is the antisymmetric rate of strain tensor and the operator $^{\square}$ denotes a generalized invariant time derivative [3],[6],[12]:

$$\overset{\square}{\boldsymbol{\tau}} = \frac{D\boldsymbol{\tau}}{Dt} - \boldsymbol{W\tau} + \boldsymbol{\tau W} - \alpha(\boldsymbol{D\tau} + \boldsymbol{\tau D});\, \alpha \in \{0; 1; -1\} \tag{3}$$

Following Bellehumeur *et al.*, a steady flow approximation, i.e. $\frac{D\tau}{Dt} \approx 0$, is used. The flow work rate $\dot{W}_v$ can then be calculated to be [6]:

$$\dot{W}_v = \int dV \sum_{i,j=1}^{3} \tau_{ij} D_{ji} = \frac{8\pi}{3}\frac{\eta \dot{r_p} r_p^2}{\alpha\lambda} \ln\left(\frac{1 + \frac{4\alpha\lambda\dot{r_p}}{r_p}}{1 - \frac{2\alpha\lambda\dot{r_p}}{r_p}}\right) \tag{4}$$

By shrinking, the pore minimises its surface S and thus there is a change in surface energy per time:

$$\dot{W}_s = \sigma\dot{S} = 8\,\pi r_p \dot{r_p}\sigma \tag{5}$$

Here, σ is the surface tension.

Further, the gas trapped inside exerts a force against the shrinkage yielding a third energy rate.

$$\dot{W}_p = -pS\dot{r_p} \tag{6}$$

p is the gas pressure inside the pore which can be calculated from the ideal gas law. The initial value is p_0.

For energy conservation to hold, the three rates from equation (4), (5) and (6) need to add up to zero. Defining for convenience a function $G(r_p)$ as:

$$G(r_p) = \exp\left(-3\,\frac{\alpha\lambda}{\eta r_p}\left(\sigma - \frac{p_0 r_p(0)^3}{2r_p^2}\right)\right) \tag{7}$$

The differential equation can be written:

$$\dot{r_p} = \frac{r_p}{2\alpha\lambda}\frac{G(r_p) - 1}{2 + G(r_p)} \tag{8}$$

The differential equation can be written in the following reduced variables and parameters:

$$\begin{aligned} r_{red} &= \frac{r_p(t)}{r_p(0)} \\ t_{red} &= \frac{1}{2}\frac{\sigma}{r_p(0)\eta}t = \frac{t}{\tau_{Frenkel}} \\ r_{eq,red} &= \sqrt{\frac{p_0 r_p(0)}{2\sigma}} \\ \lambda_{red} &= -\frac{\alpha\lambda\sigma}{2r_p(0)\eta} = -\frac{\alpha\lambda}{\tau_{Frenkel}} \end{aligned} \tag{9}$$

In the reduced form, the solution to the original late stage Frenkel model reads:

$$r_{red}(t_{red}) = 1 - t_{red} \tag{10}$$

The differential equation of the Mackenzie and Shuttleworth model[1] has only one reduced parameter, the reduced equilibrium radius $r_{eq,red}$ at which the forces exerted by surface tension and by gas pressure are equal.

---

[1] The original work by Mackenzie and Shuttleworth further included a mean field approximation which is not included in this paper. The differential equation given here is thus their differential equation in the limit of $r_2 \rightarrow \infty$.

$$\frac{dr_{red}}{dt_{red}} = -1 + \frac{{r_{eq,red}}^2}{{r_{red}}^2} \tag{11}$$

The differential equation of our viscoelastic extension has the same reduced equilibrium radius and a further parameter λ_{red} which is basically the relaxation time rescaled onto the scale of the reduced time t_{red}.

$$\frac{dr_{red}}{dt_{red}} = \frac{r_{red}}{2\lambda_{red}} \frac{1 - \exp\left(6 \frac{\lambda_{red}}{r_{red}}\left(1 - \frac{{r_{eq,red}}^2}{{r_{red}}^2}\right)\right)}{2 + \exp\left(6 \frac{\lambda_{red}}{r_{red}}\left(1 - \frac{{r_{eq,red}}^2}{{r_{red}}^2}\right)\right)} \tag{12}$$

The Frenkel (F), the Mackenzie and Shuttleworth (M-S) and our viscoelastic (VE) model can be solved and compared in the chosen set of reduced variables and parameters. In figure 1, it can be seen that both the inclusion of gaseous pore pressure and of viscoelastic relaxation do slow down the rate of pore shrinkage separately. The equilibrium radius is found to be independent of the rheological equation used. It appears already in the viscous Mackenzie and Shuttleworth model.

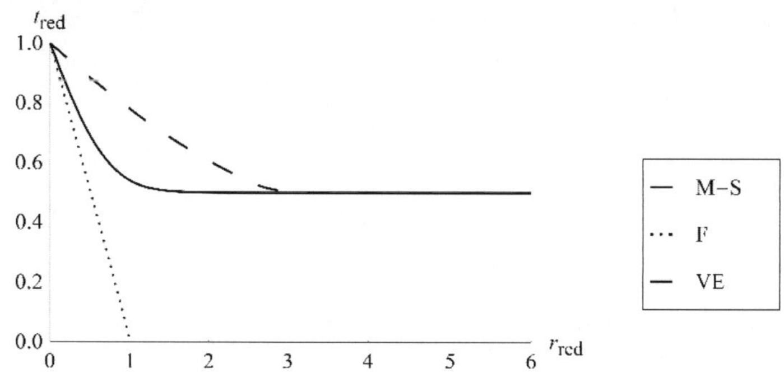

FIGURE 1. Plots of the numerical solutions for $r_{eq,red} = 0.5$ and $\lambda_{red} = 2$. The abbreviations in the legend read M-S = classical Mackenzie and Shuttleworth model, F = Frenkel's model and VE the viscoelastic model developed.

Figure 2 shows plots of the viscoelastic model for different reduced relaxation times. It can clearly be seen that higher relaxation times slow down the pore shrinkage more strongly which agrees with physical intuition.

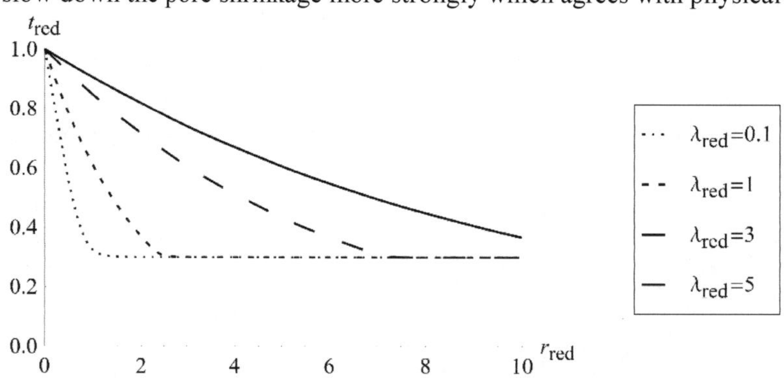

FIGURE 2. Plots of the viscoelastic modelling for $r_{eq,red} = 0.3$ and different values for λ_{red} as indicated by the legend.

Diffusion

A real melt will allow for diffusion of the gas into the melt. This can be taken into account by treating the pressure as an independent variable. This changes the original expression for the work rate due to pressure effects (equation (6)) to one with a time dependent pressure $p(t)$:

$$\dot{W}_p = -4\pi r_p^2 p(t) \dot{r_p} \tag{13}$$

For the pressure an additional differential equation can be obtained by considering mass conservation (see [4]). The total change of gas molecules inside the pore can – in the absence of chemical reactions – only be due to diffusion across the melt-gas boundary whose strength is controlled by the concentration gradient at the boundary:

$$\frac{4}{3}\pi\frac{d}{dt}\left(\frac{p(t)r_p^3}{RT}\right) = \left(4\pi r_p^2\right)D\left(\frac{\partial c(r,t)}{\partial r}\right)_{r=r_p}$$
$$\Rightarrow \frac{dp(t)}{dt} = \frac{1}{r_p}\left(3RTD\left(\frac{\partial c(r,t)}{\partial r}\right)_{r=r_p} - 3p(t)\dot{r_p}\right) \tag{14}$$

Here R is the universal gas constant, T the temperature, D the Fickian diffusion constant and $c(r,t)$ the gas concentration inside the polymer melt.

To obtain the concentration gradient, the concentration dynamics need to be modelled. The equations of Fickian diffusion in a moving medium read [13]:

$$\frac{\partial \tilde{c}}{\partial t_{red}} + \frac{dr_{red}}{dt_{red}}\frac{{r_{red}}^2}{\tilde{r}^2}\frac{\partial \tilde{c}}{\partial \tilde{r}} = \left(\frac{dt_{red}}{dt}\right)^{-1}\frac{D}{r_p(0)^2}\frac{1}{\tilde{r}^2}\frac{\partial}{\partial \tilde{r}}\left(\tilde{r}^2\frac{\partial \tilde{c}}{\partial \tilde{r}}\right) \tag{15}$$

$\tilde{r}$ is defined via $\tilde{r} = r/r_p(0)$ and $\tilde{c}$ as $\tilde{c} = c/c(r,0)$.

As boundary conditions, the concentration at the boundary between gas and melt is assumed to be given by Henry's law (see [4]) with reduced Henry's constant $\widetilde{K}_h$ and reduced pressure $\tilde{p} = p/p_0$. Furthermore, the initial concentration throughout the melt is taken as homogeneous:

$$\tilde{c}(r_{red}, t_{red}) = \widetilde{K}_h\tilde{p}$$
$$\tilde{c}(\tilde{r}, 0) = 1 \text{ for } \tilde{r} > r_{red} \tag{16}$$

As a first simplified approximation of the effects of gas diffusion, a constant gas transfer coefficient across the boundary is assumed. In this approximation the concentration gradient in the pressure differential equation (14) is replaced by a linear function in the pressure. This yields:

$$\frac{d\tilde{p}}{dt_{red}} = -\frac{\tilde{p}}{r_{red}}\left(3\frac{dr_{red}}{dt_{red}} - g\right) \tag{17}$$

The differential equation for r_{red} can again be obtained by requiring the work rates (4), (5) and (13) to add up to zero. The time evolution of the reduced pressure (17) and the reduced pore radius can be simulated numerically. An example is shown in figure 3 and 4. In figure 3, it can be seen that the pore first shrinks in a manner similar to the one in the viscoelastic model without gas diffusion but subsequently shows a complete collapse by diffusion. The pressure within the pore (figure 4) can actually increase through diffusion as the force exerted by surface tension grows with decreasing pore radius. A simulation with a much higher diffusion coefficient (not shown) does neither show this two regime process in the reduced radius nor the high increase in reduced pressure. This suggests that the respective relevance of diffusive and flow process depends strongly on the real parameters.

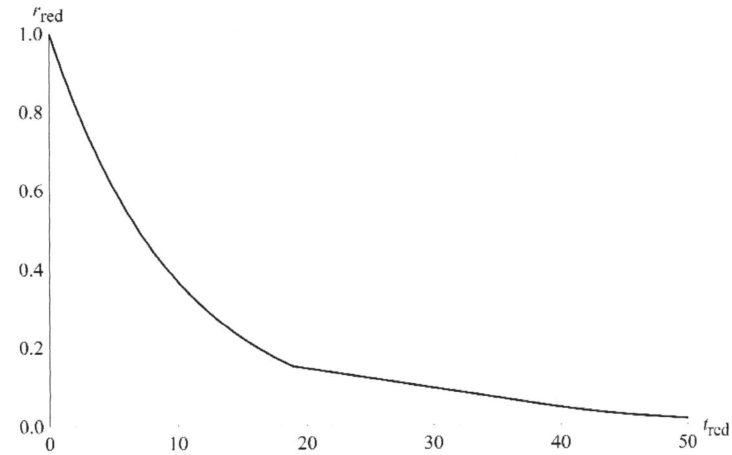

FIGURE 3. Reduced radius from the numerical simulation of the constant transfer coefficient model with $r_{eq,red} = 0.2$, $\lambda_{red} = 5$ and $g = -0.01$.

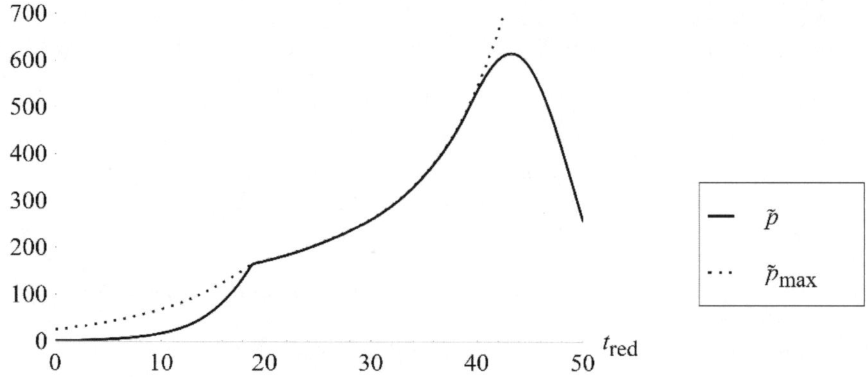

FIGURE 4. Reduced pressure and maximum possible reduced pressure from the numerical simulation of the constant transfer coefficient model with $r_{eq,red} = 0.2$, $\lambda_{red} = 5$ and $g = -0.01$.

Conclusion and Outline

Based on the ideas of Frenkel [1] and Mackenzie and Shuttleworth [2] a model for the shrinkage of a gaseous pore in a viscoelastic melt was developed using a simple but widely used viscoelastic model taken from Bellehumeur [3]. This model was solved numerically using *Mathematica*. Viscoelastic relaxation has a pronounced effect on the time scales of the shrinkage process in the simulation but does not alter the equilibrium pore radius. A first attempt at including gas diffusion was made motivated by the works of Kontopoulou and Vlachopoulos [4] and Gogos [5]. Numerical solutions of a simplified model showed that, depending on the diffusion process, diffusion can have a significant influence on viscoelastic polymer sintering and that the pressure inside the pore reaches high values without or with slow diffusion.

Future work will concentrate on a model incorporating Fickian diffusion as given by equation (15). Furthermore, implementation of a more elaborate viscoelastic rheological equation and comparison with experimental data are planned.

REFERENCES

1. J. Frenkel, *Journal of Physics* **IX**, 385-391 (1945).
2. J.K. Mackenzie and R. Shuttleworth, *Proc. Phys. Soc. B* **62**, 833-852 (1949).
3. C. T. Bellehumeur *et al.*, *Rheol Acta* **37**, 270-278 (1998).
4. M. Kontopoulou and J. Vlachopoulos, *Polymer Engineering and Science* **39**, 1189-1198 (1999).
5. G. Gogos, *Polymer Engineering and Science* **44**, 388-394 (2004).
6. P. Haupt, *Continuum Mechanics and Theory of Materials*, Berlin Heidelberg: Springer, 2000, p. 124,182,300,301.
7. A. J. Shaler, *Metals Transactions* **185**, 796-813 (1949).
8. J.D. Muller *et al.*, *J Mater Sci* **47**, 121–131 (2012).
9. M. Asgarpour *et al.*, *Journal of Applied Polymer Science* **119**, 2784–2792 (2011).
10. S. Dupin, "Etude fondamentale de la transformation du polyamide 12 par frittage laser : mécanismes physico-chimiques et relations microstructures/propriétés", Ph.D. Thesis, L'institut national des sciences appliquées de Lyon, 2012.
11. J.D. Muller and A. Maazouz, *Int J Mater Form* **Suppl 1**, 645 – 648 (2008).
12. D. D. Joseph, *Fluid Dynamics of Viscoelastic Liquids*, New York: Springer, 1990, p. 7.
13. R. B. Bird, W. E. Stewart and E. N. Lightfoot, *Transport Phenomena*, New York: John Wiley & Sons, Inc., 1960, p. 557/559.

Effect of Particle Size on the Properties of Highly-Filled Polymers for Fused Filament Fabrication

Christian Kukla [a*], Joamin Gonzalez-Gutierrez [b], Ivica Duretek [b], Stephan Schuschnigg [b] and Clemens Holzer [c]

[a]*Montanuniversitaet Leoben, Industrial Liaison Department, Peter Tunner Strasse 27, 8700 Leoben, Austria*

[b]*Montanuniversitaet Leoben, Department of Polymer Engineering and Science, Chair of Polymer Processing, Otto Gloeckel-Strasse 2, 8700 Leoben, Austria*

Presenting author's email: christian.kukla@unileoben.ac.at
Reference author's email: christian.kukla@unileoben.ac.at

Abstract. Fused Filament Fabrication (FFF) could replace injection molding as the shaping step in a process similar to powder injection molding (PIM). Herein after shaping by using a highly-filled polymer the part is debound and sintered to obtain a solid part of metal or ceramic. New feedstock materials have been developed that can be printed using conventional FFF equipment. And after debinding and sintering stainless steel parts can be obtained. However, there are many parameters that can affect the performance of the FFF feedstock materials. One important parameter is the particle size distribution of the filler particles. In this paper, feedstocks containing 316L steel powder with different particle size distributions were characterized in terms of viscosity and mechanical properties, and tested regarding the printability using a conventional FFF machine. It has been observed that the particle size significantly affects the properties of feedstock materials and thus their ability to be printed.

Keywords: additive manufacturing, fused filament fabrication, highly-filled polymer, suspension, viscosity, mechanical properties
PACS: 62.20.de, 62.20.fq, 66.20.Ej 81.05.Bx, 81.05.Lg, 81.05.Qk, 81.05.Rm, 81.20.Ev, 81.40.Jj, 81.70.Bt, 83.50.Uv, 83.60.Fg, 83.80.Hj, 83.80.Tc, 83.85.Jn

INTRODUCTION

Thermoplastic-based Fused Filament Fabrication (FFF) is one of the most widely used additive manufacturing processes in the world. The main reasons of its increasing popularity and use have been its reliability, safe and simple fabrication process, low cost, and the availability of a variety of thermoplastics for printing [1].

In FFF the printing material is heated until it melts or softens so far that it is able to flow and then it is extruded from a nozzle onto a substrate to build a structure in a layer-by-layer manner. The extrudate solidifies when its temperature decreases due to air convection [2].

As with other additive manufacturing techniques, FFF is able to produce parts with very intricate shape, for this reason it is believed that FFF could be used to replace injection molding machines when the number of parts needed is small or when the geometrical requirements are very demanding. This includes also Powder Injection Molding (PIM). In PIM a highly-filled polymeric system with metal or ceramic powders is shaped by injection molding, the polymeric system is then removed without affecting the shape of the molded part, and finally the part is sintered to obtain a solid part of metal or ceramic [3].

In order to successfully perform FFF, the quality of the filament is very important. This includes the material properties and the geometrical shape of the filament. In this paper, we investigated one important factor that affects the material properties strongly of highly filled systems: particle size distribution (PSD) of the filler powders [4, 5]. In short, this paper presents the effect of PSD on the modulus, elongation at break and apparent viscosity of highly-filled filaments with 316L stainless steel.

MATERIALS AND METHODS

The highly-filled filaments for FFF were composed of a multicomponent binder system and filler particles. The main binder component in the binder is a thermoplastic elastomer (Kraiburg TPE GmbH & Co. KG, Germany). The backbone was a grafted polyolefin (BYK Chemie GmbH, Germany). The filler particles in the filaments were 316L stainless steel of two particle size distribution: P_S (Smaller) (Epson Atmix Corporation, Japan) and P_L (Larger) (Carpenter Powder Products AB, Sweden). All filaments had a 55 vol-% content of metal powder.

The particle size distribution was measured by laser scattering. Scanning electron microscopy was used to observe the shape of the two different powders. The particle size distribution and the shape of the two types of powders are shown in Figures 1 and 2, respectively.

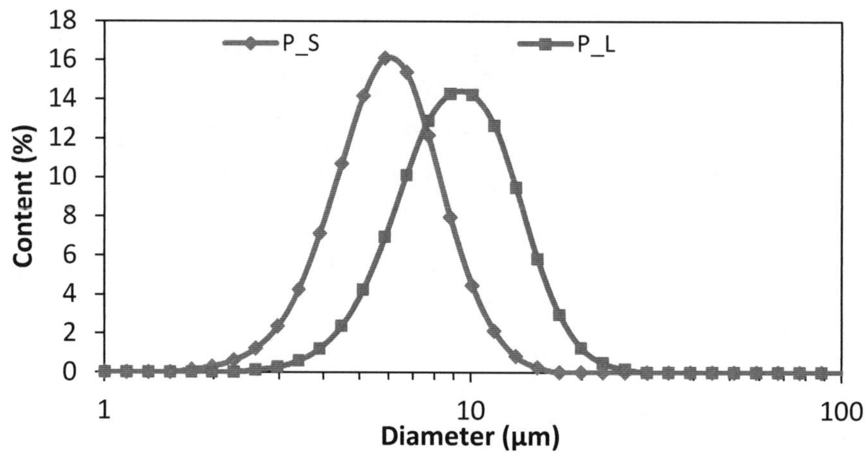

FIGURE 1. Particle size distributions of 316L stainless steel powders (P_S, small and P_L, large) measured by laser scattering.

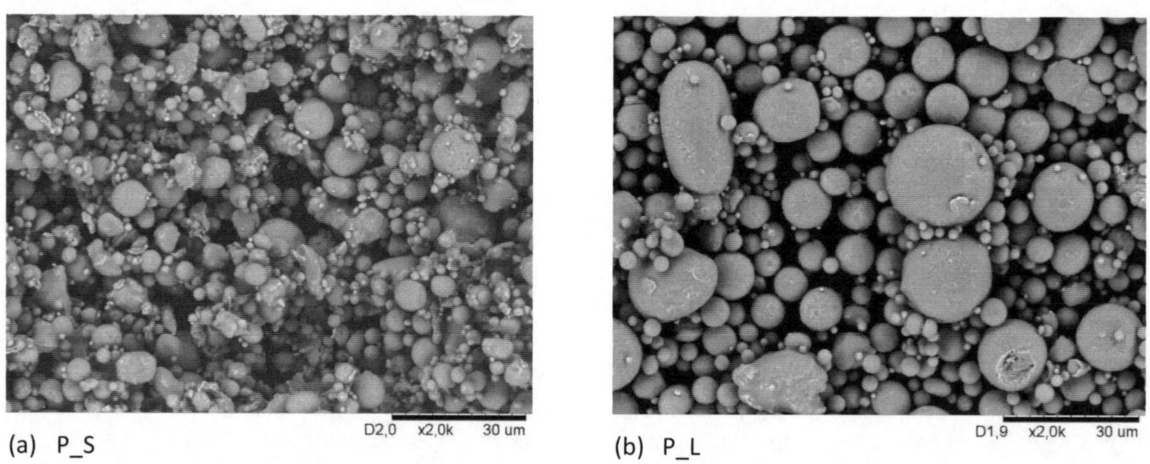

FIGURE 2. Scanning electron microscopy images of 316L powder (a) P_S, small and (b) P_L, larger particles.

P_S has a smaller average particle size of approximately D50 ≈ 5.5 μm, while P_L of approximately D50 ≈ 8.6 μm (Figure 1). Both powders have a normal distribution with a similar broadness; both powders have their standard deviations around 35 % from the mean value. Figure 2 shows that the majority of the particles in both types of powders are spherical. Thus it can be said that the only parameter changed is the average particle size

The feedstocks were compounded in a kneader with counter rotating rollers (Plasti-Corder PL2000, Brabender GmbH & Co. KG, Germany). The kneader temperature was set at 200 °C and the total kneading time was 30 minutes. After compounding, the feedstocks were grinded in a cutting mill fitted with a sieve with square perforations of 4 mm in length (Retsch SM200, Retsch GmbH, Germany). Two types of feedstocks were prepared: FT-P_S containing the smaller particles and FT-P_L containing the larger particles.

Filaments were prepared using a high pressure capillary rheometer as a ram extruder (Rheograph 2002, Göttfert Werkstoff-Prüfmaschinen GmbH, Germany). A barrel of 15 mm in diameter was used to melt the material before pushing it into a capillary. A round capillary with a diameter of 2 mm and length of 20 mm was used. Extrusion temperature was set at 185 °C and the piston speed was 0.5 mm/s. At the exit of the die a Teflon conveyor belt was placed to pull the filament as it was extruded at an approximately speed of 45 mm/s. At the same time when the filaments were produced the apparent viscosity was measured at the apparent shear rate of 113 s^{-1}, which corresponds to the piston speed and the geometry used during the measurement.

The mechanical properties of filaments were tested in a material tensile testing machine fitted with pneumatic grips (Zwick Z001 2.5 kN load cell, Zwick GmbH & Co. KG, Germany). The pneumatic pressure was 3 bar. Measurements were performed at 23 °C and 50 % relative humidity. Gauge length was set at 75 mm. In order to measure the apparent secant modulus an extension rate of 1 mm/min was used, afterwards the rate was increased to 10 mm/min until the sample broke. The apparent secant modulus was estimated between 0.05 % and 0.25 % extension.

After the filament characterization was made, printing trials were performed on a HAGE 3Dp-A2 fused filament deposition machine (HAGE Sondermaschinenbau GmbH & Co KG, Austria). Printing parameters were kept constant for both materials investigated.

RESULTS AND DISCUSSION

The properties for the materials investigated are summarized in Figure 3. Three properties are shown in Figure 3, the apparent viscosity measured during filament preparation (η), the secant modulus (E) and the strain at break (ε_{br}) for the filaments.

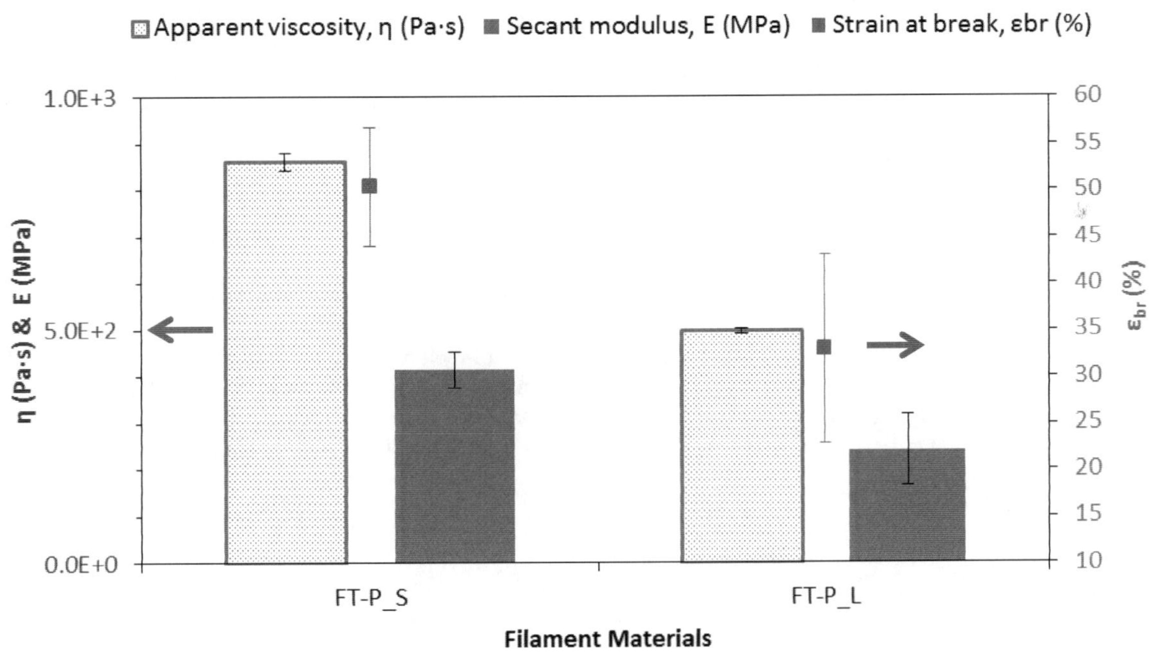

FIGURE 3. Apparent viscosity (η), secant modulus (E) and elongation at break (ε_{br}) for filaments with different particle sizes.

As it can be seen in Figure 3, increasing the average particle size of filler particles causes the apparent viscosity to decrease by ~42 %, the secant modulus decreases by ~42 % and the elongation at break decreases by ~35 %. A decrease in viscosity has also been reported in the literature when using coarser particles in PIM feedstocks [4]. Additionally, it has been reported in the literature of polymer-metal composites filled with particles that increasing the particle size decreases the modulus [5] and elongation at break [6] of the composite materials. Therefore all our results are in agreement with the literature. However, in order to determine if the filaments are useful or not, printing trials were performed.

Printing trials revealed that filaments FT-P_S can be printed and filaments FT-P_L cannot. Figure 4 shows some 3D printed parts produced with FT-P_S. Therefore it is clear that changing the average particle size of filler particles can lead to unprintable filaments. By looking at the results in Figure 3, one can speculate that decreasing the viscosity is not as crucial as maintaining a good level of modulus and elongation at break for the FFF printing process; nevertheless further investigation is need for conclusive evidence. But it can be concluded that in order to print feedstock materials with larger particles, adjustments to the binder formulation should be done in the future.

FIGURE 4. Parts produced by FFF using filaments with FT-P_S feedstock material (a) top view and (b) side view.

CONCLUSIONS

Highly filled composite materials (feedstock) have been prepared with 55 vol.-% of 316L stainless steel with two different particle size distributions. It was observed that increasing the average particle size from 5.5 to 8.6 μm decreases the apparent shear viscosity of feedstock, as well as the secant modulus and the elongation at break of filaments. The change in properties is significant since the filaments with smaller particles can be printed while the materials with larger particles cannot.

ACKNOWLEDGMENTS

This research has been performed as part of the REProMag (Resource Efficient Production of Magnets) project with the financial support of the European Commission in the Horizon 2020 FoF-framework under grant agreement 636881. We extend our gratitude to Ing. Thomas Schlauf (FOTEC Forschungs- und Technologietransfer GmbH) for performing the particle size distribution analysis and the scanning electron microscopy, as well as, Dr. Florian Arbeiter (Chair of Material Science and Testing of Polymers, Montanuniversitaet Leoben) for the mechanical characterization of filaments. We also thank DI. Stefan Hampel (Hage Sondermaschinenbau GmbH & Co KG) for performing the 3D printing trials.

REFERENCES

1. S. Masood and W. Song, *Materials & Design* **25**, 587–594 (2004)
2. Q. Sun, G.M. Rizvi, C.T. Bellehumeur and P. Gu, Free Form Fabrication Symposium, Austin, TX, 2003, pp. 313-323 http://sffsymposium.engr.utexas.edu/Manuscripts/2003/2003-29-Bellehumeur.pdf (accessed 09.03.2016)
3. J. Gonzalez-Gutierrez, G.B. Stringari, and I. Emri, "Powder injection moulding of metal and ceramic parts" in *Some Critical Issues for Injection Moulding*, edited by J. Wang, Rijeka, Croatia, InTech, 2012, pp. 65-88.
4. B. Hausnerova , T. Kitano , I. Kuritka , J. Prindis and L. Marcanikova, *International Journal of Polymer Analysis and Characterization*, **16**, 141-151 (2011).
5. W. Brostow, A. Buchman, E. Buchman and O. Olea-Mejia, *Polymer Engineering & Science*, **48**, 1977-1981 (2008).
6. Y.R. Uhm, J. Kim, K.J. Son, and C.S. Kim, *Research on Chemical Intermediates*, **40**, 2145 – 2153 (2014)

Influence of Temperature on the Flowability of Polymer Powders in Laser Sintering

Michael Van den Eynde, Leander Verbelen and Peter Van Puyvelde*

KU Leuven, Dept. Chemical Engineering, 3001 Leuven

michael.vandeneynde@cit.kuleuven.be
peter.vanpuyvelde@cit.kuleuven.be

Abstract. Powder flowability is a determining factor in many industrial applications. Multiple techniques exist for assessing whether a powder flows adequately or not. However, most of these techniques occur at ambient conditions, rather than the actual processing temperature. Especially for laser sintering (LS), in which polymer powders are spread at a strongly elevated temperature, thermal effects cannot be ignored. This work expands an existing set-up, the powder spreader, which is designed for analysing LS powder flowability, to include temperature control. The focus in this paper lies on the additions made to the set-up. Tests on the standard PA12 2200 of EOS illustrate the use of this expanded technique, as well as the level of temperature control attained.

Keywords: Powder Flowability, Laser Sintering, Temperature Effects, Polymers

PACS: 83.80.Fg, 81.05.Rm, 81.05.Lg

INTRODUCTION

Adequate powder flowability at elevated temperatures is a requirement for powders processed via laser sintering (LS). In LS, a form of additive manufacturing (AM), powder particles with a typical diameter of about 50 μm are pre-heated and spread into thin layers that are roughly two to three times the average particle diameter [1,2,3]. A laser then sinters the cross-section of the desired product in each of these layers. To reduce the product porosity and defects, the deposited powder layers need to be smooth, defect-free and preferably show a high packing density [1].

Various techniques and indices for assessing powder flow exist, such as the angle of repose, the Hausner ratio and the Hall flow meter. The choice of a right technique is crucial, as the obtained indices are not only material dependent, but are also strongly influenced by the applied flow field. This is a result of the limited understanding of the link between local grain interactions and macroscopic powder flow [4]. Consequently, different techniques can lead to different powder rankings [5,6]. As the flow field for laser sintering is rather specific, the analysis technique to predict powder flowability should ideally mimic this flow field as close as possible. Previous work proposes a new technique, namely the powder spreader, to assess flowability specifically for LS [7]. This technique fully mimics the spreading movement of the blade in an industrial sintering machine.

An issue with most current powder flowability measurements is the lack of temperature control. Most set-ups are only equipped to work at ambient conditions. However, as polymers tend to soften at elevated temperatures, temperature is expected to influence the inter-particle interactions noticeably. Literature on set-ups expanded for temperature control back this hypothesis. An influence of temperature on powder flowability is reported, with a notable transition around the glass transition temperature [8]. Therefore, this work expands the powder spreader to include heating of the powder bed. The goal is to thereby provide a technique that includes both the relevant flow field for LS as well as a relevant temperature range for this application. This expansion thus enables the assessment of whether the suggested temperature effects are also seen in this specific application.

EXPANSION OF THE POWDER SPREADER

FIGURE 1 shows the powder spreader set-up, as described in previous work. It is a modification of a commercial Elcometer 4340 Motorised Film Applicator. During the measurement, a spreading blade (Elcometer 3580 Casting Knife Film Applicator) pushes a heap of powder over the support plate, onto the measurement plate. The balance supporting the measurement plate then weighs the layer mass and thus provides a relevant density index for LS. The measurement plate is lowered with a precision of 10 μm, enabling the study of powder-on-powder deposition. Further details on the set-up as well as the protocol are found in the previous work [7].

Three expansions, shown in FIGURE 2, are made to this set-up. Firstly, infrared heating and a PID control loop enable temperature control. Secondly, a mechanical adjustment is made to the measurement plate, which improves stability and accounts for thermal expansion. Finally, the spreading blade is replaced by a thicker blade, which resembles more closely the types of blades used in industrial sintering machines.

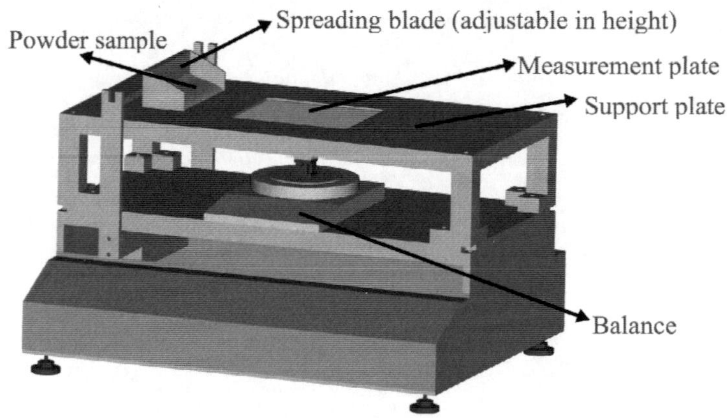

FIGURE 1: Powder spreader set-up as described in earlier work [7].

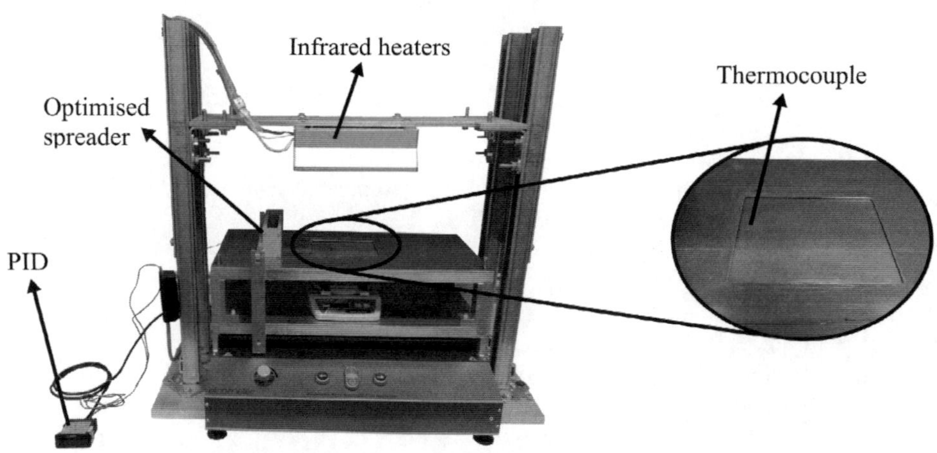

FIGURE 2. Expanded powder spreader set-up, containing the mechanically improved measurement platform, ceramic heaters and the optimised spreading blade.

Thermal Control

Infrared ceramic heaters are used as a heat input, similarly to the infrared heaters in industrial sintering machines. Two 650 W heaters are positioned above the measurement plate to provide homogenous heating. A PID control loop operates the heaters to ensure a relatively constant bed temperature. The input signal for this control loop is the temperature recorded by a thermocouple, positioned near the top layer of the deposited powder. To minimise convective heat loss to the environment, a glass casing is placed around the entire set-up. Fresh powder is loaded next to the building platform, where it receives some heat radiation. This way, it is pre-heated to a temperature below the temperature of the deposited layers, as is the case in an industrial sintering machine. An external thermocouple is used to quantify the temperature of the loaded powder during the experiment and validate the temperature control in the deposited powder layers.

Mechanical Expansions

The mayor concern when expanding the set-up to include temperature control is thermal expansion of the components. In the set-up described in previous work, the measurement plate expands, creating friction with the support plate and thus disabling an accurate weight measurement. Making the measurement plate smaller to account for expansion is not a viable solution, as at lower temperatures the gap between the measurement plate and the support plate becomes too large and excessive amounts of powder fall through. To solve the problem, several components were added to the measurement plate. FIGURE 3 illustrates these additions.

The old measurement plate is replaced by a rigid framework structure. The framework consists of a rectangular ring, in which a smaller measurement plate is added. This new measurement plate is lowered within the rectangular ring by a Newport vertical stage, thereby enabling again the study of multiple layers. The entire framework rests on the balance and is thus weighed. Therefore any friction between the measurement plate and the rectangular ring has no effect on the measurement. This allows for a close fit between the two parts so no powder falls through. As both the rectangular ring and the measurement plate expand with temperature, friction does not increase to values at which the platform gets stuck.

The only remaining problem is possible friction between the rectangular ring and the support plate. To solve this problem, a second addition is made. A thin plate with a hole of the same size as the hole in the rectangular ring is placed concentrically above the hole in the support plate. This way the gap between the rectangular ring and the support plate is covered. During powder deposition, the spreading blade pushes the thin plate down against the rectangular ring, thereby sealing the bed volume. This now allows for a larger gap between the rectangular ring and the support plate, eliminating any friction arising from mismatches in thermal expansion. A thin thermocouple is inserted between the thin plate and the support plate, ending in the powder bed. At its end it is slightly bend upwards, so it resides near the top layer of the powder bed.

Teflon layers are placed within the framework structure to slow down thermal conduction and thus shield the balance and micrometre from high temperatures. The support plate furthermore shields the balance from direct heat radiation. This way the balance and micrometre remain within their thermal operating range.

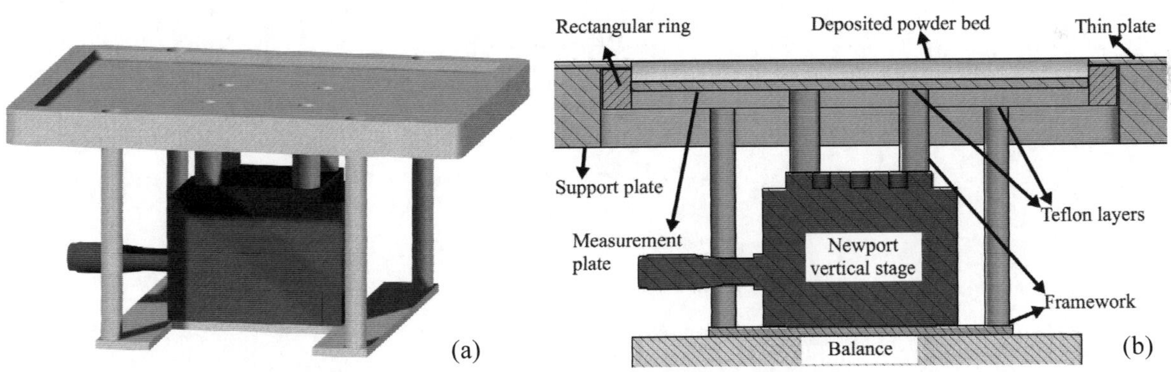

FIGURE 3. (a) Schematic illustration of the improved measurement platform (b) Cross-section of the improved measurement platform within the powder spreader.

Improved Spreading Blade

The Elcometer spreading knife is a tool originally intended for spreading liquid films. To get closer to the operation in an industrial sintering machine, it is replaced by a thicker blade, illustrated in FIGURE 4. The bottom of the blade is slightly inclined, by an angle of 7°, to aid in pushing down the powder. Adding a fourth wall, thereby closing the rectangle, further expands the blade. This way powder is loaded inside the new spreader, rather than pouring it in front of it. This enables a more constant powder load. To minimise the scraping effect of this extra wall, it is rounded at the bottom.

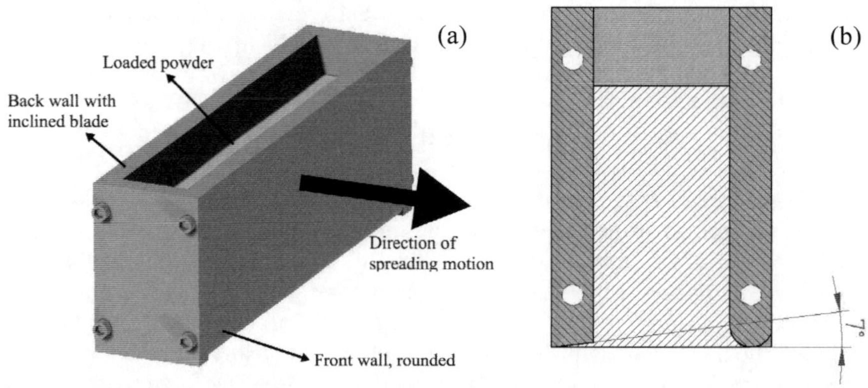

FIGURE 4. Schematic illustration (a) and cross-section (b) of the improved spreading blade, based on an industrial spreader.

ASSESSMENT OF THE SET-UP

To assess the validity of the temperature expansion, tests on the standard LS powder PA12 2200, distributed by EOS, are performed. Firstly, the ability to attain a constant bed temperature is assessed. An external thermocouple indicated the temperature at the top of the powder bed at various positions. Secondly, the temperature effect on the flowability of the PA12 2200 sample is evaluated at various temperatures from ambient conditions to 140 °C. The density of at least 20 subsequent layers is measured. To look solely at flowability and not at geometrical packing limitations, the packing ratio (PR), shown in equation 1, is defined in previous work [7]. It defines the ratio between the density achieved by the flow field of the application to the tapped density. This index is analogous to the inverse of the Hausner ratio (equation 2), which takes the ratio of the powder density resulting from free, aerated flow to the tapped density. In previous work, a tapped density of 520 kg/m^3 is found for PA12 2200 [7]. For the LS process the absolute value of the layer density has direct consequences. To obtain a dense part, it is beneficial to start from a dense powder bed. To compare the density of different polymers, this density is normalised by the material density, resulting in the packing density (ρ_{packing}). As shown in equation 3, this provides the inter-particle void fraction ϵ. Besides the quantitative layer density, the surface quality is still usable to assess flowability.

$$\text{PR} = \frac{\rho_{\text{layer}}}{\rho_{\text{tapped}}} \tag{1}$$

$$\frac{1}{\text{HR}} = \frac{\rho_{\text{bulk}}}{\rho_{\text{tapped}}} \tag{2}$$

$$\rho_{\text{packing}} = \frac{\rho_{\text{layer}}}{\rho_{\text{material}}} = 1 - \varepsilon \tag{3}$$

RESULTS

For a temperature set-point of 90 °C and 120 °C respectively, FIGURE 5 illustrates the measured surface temperature at various positions. It is seen that there is a slight fluctuation in the surface temperature depending on position. Due to the on-off action of the lamps, there is also a slight fluctuation of temperature in time. As measurement points are taken one by one, this time-effect also has some influence on FIGURE 5. The temperature fluctuations are however deemed sufficiently low to get a clear insight in the possible change in powder flow behaviour over a wide temperature range. If knowledge of changes in flow behaviour are required for very small temperature differences, more control needs to be added to the system. A possible refinement is control of the power output of the lamps, rather than an on-off mechanism. The powder loaded next to the bed reaches a temperature of roughly 70 °C and 80 °C respectively.

As a further remark, the temperature of the micrometre and the balance never exceeds 25 °C during the measurements. As this temperature is still well within the specifications of both components, the slight temperature increase has no negative effect on the measurement.

At room temperature a packing ratio of 0.87 was found for the PA12 2200 sample. This corresponds to a packing density of 43%. Smooth layers without any defects were deposited during the entire measurement. This is all in line with the results found in earlier work [7], showing that the mechanical alterations to the set-up did not negatively influence the measurement. A remark has to be made, namely that for more compressible powders the change of the spreading blade may still have a significant influence. As the new blade is thicker and can push the powder down, rather than scraping over it, this is expected to increase the packing ratio of powders that are sensitive to compression. Further tests on such powders are planned.

At higher temperatures up to the glass transition temperature (~50 °C), no change was noticeable in both the layer quality and the packing ratio. Above the glass transition the layer quality remained visually the same, but a slight increase in the packing ratio is seen. This may be due to softening of the polymer, which increases the compressibility by the spreading blade. However, as this increase is limited to a few percent, further studies on other polymers are advised.

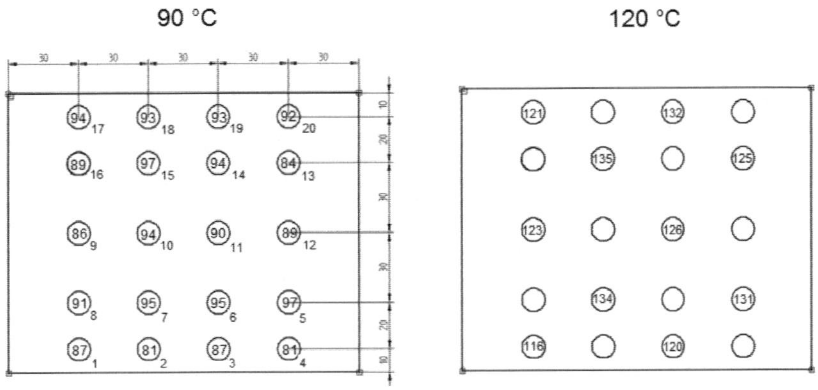

FIGURE 5: Measured surface temperature [°C] of the upper powder layer for an input value of 90 °C and 120 °C.

CONCLUSIONS

The expanded powder spreader set-up provides a viable testing method to assess LS powder flow at elevated temperatures. Tests on the LS standard PA (PA12 2200, EOS) prove that decent temperature control is attained and no friction problems arise during the measurement. As expected from its industrial use, smooth and defect-free layers were attained at all tested temperatures. Tests on other polymers are on-going to further investigate the temperature effects on LS powder flow.

ACKNOWLEDGMENTS

The authors would like to thank IWT and SIM-Flanders. Their funding of the research programme STREAM and its SBO POLYFORCE and ICON WAVE made the research of this paper possible.

REFERENCES

1. R. Goodridge, C. Tuck, R. Hague, *Prog. in Mater. Sci.* **57** 229-267 (2012)
2. G. Salmoria, J. Leite, C. Ahrens, A. Lago, A. Pires, *Polym. Test.* **26** 361-368 (2007)
3. J. Williams, A. Adewunmi, R. Schek, C. Flanagan, P. Krebsbach, S. Feinberg, S.Hollister, S. Das, *Biomater.* **26** 4817-4827 (2005)
4. A. Castellanos, J. Valverde, M. Quintanilla, *KONA Powder and Part. J.* **22** 66-81 (2004)
5. M. Krantz, H. Zhang, J. Zhu, *Powder Technol.* **194** 239-245 (2009)
6. A. Amado, M. Schmid, G. Levy, K. Wegener, Proceedings of Solid Freeform Fabrication, Austin, TX, 2011, pp.438-452
7. M. Van den Eynde, L. Verbelen, P. Van Puyvelde, *Powder Technol.* **286** 151-155 (2015)
8. A. Amado, M. Schmid, K.Wegener,Flowability of SLS powders at elevated temperature. Department of Mechanical and Process Engineering, Swiss Institute of Technology (2014)

Quality Improvement of FDM Parts by Parameter Optimization

F. Knoop[a, b*], A. Kloke[a, b], V. Schoeppner[a]

[a]*Kunststofftechnik Paderborn (KTP), Paderborn University, D-33098 Paderborn, Germany*
[b]*Direct Manufacturing Research Center (DMRC), Paderborn University, D-33100 Paderborn, Germany*

frederick.knoop@dmrc.de
volker.schoeppner@ktp.upb.de

Abstract. Fused Deposition Modeling (FDM) is an Additive Manufacturing (AM) technology which is used for prototypes, single-part-production and also small batch productions. For use as a final product, it is important that the parts have good mechanical properties, a high dimensional accuracy and smooth surfaces. The knowledge of the mechanical properties is very important for the design engineer when it comes to the component design. End-use parts out of the FDM process have to resist applied forces. In this paper, investigations were conducted with the polymer ABS-M30 from Stratasys Inc. To achieve a quality improvement of FDM parts, various toolpath parameters and orientations were used. Within the mechanical properties, the tensile, flexural and impact strength were evaluated. Furthermore, the tensile strength of FDM parts is compared to injection molded specimens. With optimized parameters, an increase of the tensile strength by up to 28 % and a doubling of the impact strength were possible.

Keywords: Additive Manufacturing, Fused Deposition Modeling, ABS-M30, Mechanical Properties, Quality Improvement

PACS: 81.40.Jj; 81.70.Bt; 81.05.Lg; 81.20.Hy; 83.80.Uv

INTRODUCTION

Fused Deposition Modeling (FDM) is an additive manufacturing process which has the huge benefit that no product-specific tool is needed to produce a thermoplastic component. Due to the layer by layer manufacturing, even very complex components can be built. The information for every layer is generated from a 3D CAD dataset. Therefore, only a few steps are necessary for the realization of a digital model to a physical component. That is one reason why these manufacturing processes have been proven very economical for individual production and small series production [1, 2, 3].

The FDM process is shown in Figure 1 a). The strand shaped raw material is heated up in the FDM head by two extrusion nozzles and then the molten material is deposited defined in X- and Y-direction on the build platform. After the completion of one layer, the building platform is lowered by one layer thickness and the next layer will be produced. The final part consists of model material and the support material is necessary to support overhangs and cavities. This support material can be removed in an additional post-processing step by breaking away or dissolving in an alkaline bath. The deposited material bonds with the layer beneath due to thermal fusion. This results in a strong and permanent bond between the layers [4].

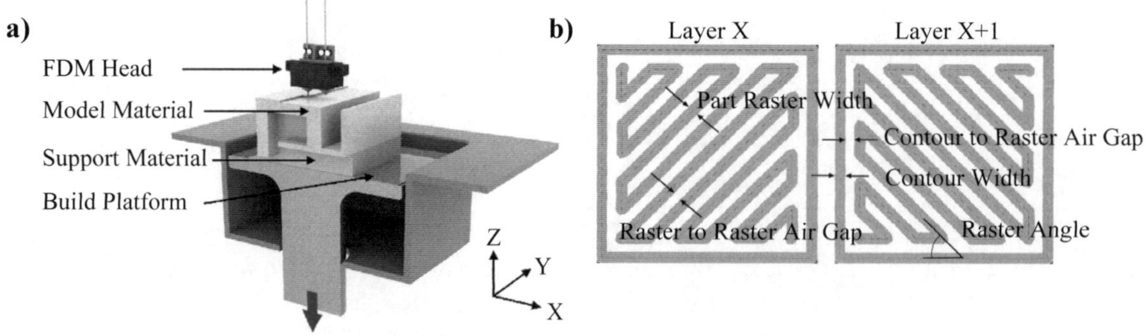

FIGURE 1. a) FDM Process b) Toolpath Parameter

First of all the contour of one layer is deposited and then the raster fill of the inner geometry is completed. By default the filling is done at an angle of 45° to the X-axis. In the next layer, the raster angle is changing by 90° so that there is an alternating raster fill for the parts (compare Figure 1 b). As known from previous publications, there is an anisotropic material behavior for FDM components [2, 5]. A filling with an angle of 45° should counteract an even stronger anisotropy. By using the Insight software from the manufacturer Stratasys, it is possible to change the filling strategy of every layer. The adjustable toolpath parameters include, but are not limited to: raster angle, filament width and air gap between two filaments. Every change of the toolpath parameters results in a change of the mechanical behavior [5, 6].

In this paper mechanical properties of FDM components manufactured with the polymer Acrylonitrile Butadiene Styrene (ABS-M30) are investigated. ABS-M30 was developed by Stratasys based on standard ABS and it should be distinguished by 25 to 70 % higher mechanical properties [7]. ABS is an amorphous mass plastic which is very suitable for the FDM process due to its good thermal properties. In addition to good strength and stiffness properties, ABS has a very good impact resistance at low temperatures. Furthermore, the characteristics dimensional stability, hardness and scratch resistance are to mention. Application areas include the industries of automotive, toys and housings for electrical appliances. A disadvantage is the low weather resistance so that UV radiation leads to yellowing and embrittlement [8]. All test specimens were manufactured with a Fortus 400mc from Stratasys. ABS-M30 can be processed with four different layer thicknesses, which are realized with different tip sizes. In this paper a T10 tip size (diameter of 0.01 inch) is used, thus a layer thickness of 0.127 mm (0.005 inch) is realized.

MECHANICAL PROPERTIES

The investigations of the mechanical properties (tensile, bending) were carried out with the universal testing system Instron 5569 with a load cell of 5 kN at an ambient temperature of 23 °C and a relative humidity of 50 %. The Charpy impact tests were conducted with the testing machine of the company Zwick "Type HIT 5-5P". All tests were performed according to European DIN standards. The test specimens for all tests were cuboids with the dimension 80 x 10 x 4 mm (L x W x H). The tests were first conducted on test specimens which were built up with standard parameters of Insight version 9.1. The test specimens are manufactured in three different build orientations: on its side (X), flat (Y) and upright (Z). Afterwards, a variation of the toolpath parameters was done. With the variation the parameters part raster width, air gap and raster angle were changed. The toolpath parameters can be seen in Table a). The parameter variation results in 12 experimental points for this layer thickness.

TABLE a). Absolute values of the parameter variation

Parameter	Values		
Part Raster Width	Thin: 0.33 – 0.35 mm		Thick: 0.50 – 0.52 mm
Filament Air Gap	0 mm	-0.017 mm (Thin)	-0.025 mm (Thick)
Raster Angle	0°	90°	0° / 90°

Tensile Strength

The tensile tests to determine the maximum tensile strength σ_{max} were conducted according to DIN EN ISO 527 and DIN EN ISO 2580 (ABS polymer). The velocity was 1 mm/min for the determination of the Young's Modulus and 50 mm/min for the determination of the elongation at break (ε_B). First of all, the tensile tests were conducted with specimens which are built with the standard FDM parameter (see Figure 3). The X-direction shows the highest tensile strength of $\sigma_{m,x}$ = 36.8 MPa with a standard deviation (SD) of 0.4 MPa and the highest elongation at break with an average of 6.2 % (SD = 1 %). The tensile strength of samples built in Y-direction is lower than the tensile strength in X-direction. Here, the average elongation at break is 4.3 % (SD = 1.3 %) and the tensile strength is $\sigma_{m,y}$ = 33.9 MPa with a standard deviation of 0.7 MPa. For the Z-direction, the lowest values are measured with an average of $\sigma_{m,z}$ = 25.5 MPa (SD = 2.1 MPa) and ε_B = 1.4 % (SD = 0.2 %). The Young's Modulus is nearly the same for the Y and Z-direction ($E_{m,y}$ = 2064 MPa and SD = 141.2 MPa) but here the X-direction has also a higher mean value with $E_{m,x}$ = 2203 MPa (SD = 144.3 MPa).

In the next step, the toolpath parameters are changed, hence there are six different curves for each direction (see Figure 2). The vertical axis shows the maximum tensile strength (MPa), while the horizontal axis indicates the air gap value (positive or negative). Figure 2 a) shows a significant increase for σ_{max} in X-direction by the use of a negative air gap. The highest gain is achieved for the unidirectional 90° and alternating 0°/90° direction. The filament width only has an influence for the unidirectional 90° direction. The highest strengths are recorded for the unidirectional 0° direction with a $\sigma_{x,max}$ = 43.7 MPa. This represents an increase of 18.7 % compared to the standard parameters in X-direction. The elongation at break for these specimens (unidirectional 0°, negative air gap) is around 20 %. However, the other parameters have still a very low value for ε_B.

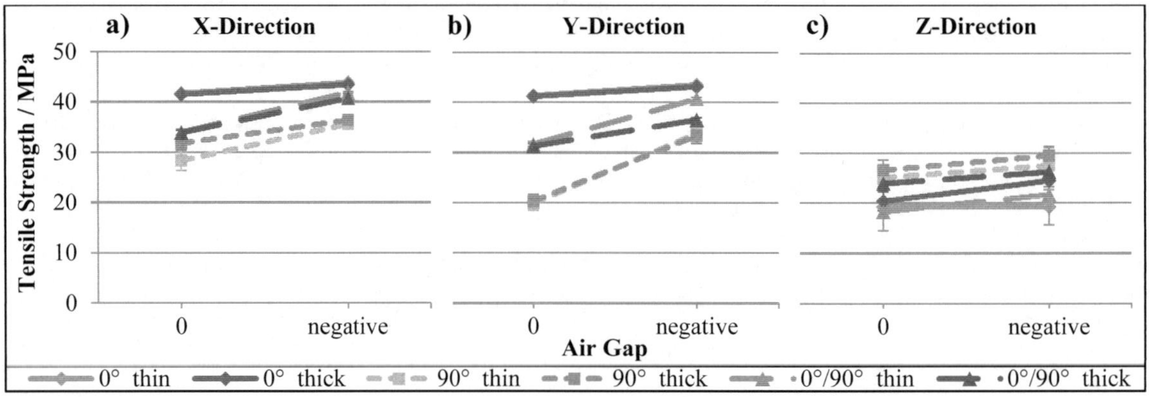

FIGURE 2. Results of the Tensile Tests as a Function of the Air Gap.

Figure 2 b) shows also an increase of the tensile strength with a negative air gap for specimens which are built in Y-direction. The behavior of the unidirectional 0° specimens are identical to the X orientated specimens. Therefore, the highest tensile strength is achieved for the 0°_thin and 0°_thick parameter with a negative air gap. The increase is around 28 % compared to the standard parameter. The reason for this is that the inner structure is the same for both orientations. An influence of the part raster width is only visible for the alternating 0°/90° specimens. For the Z-direction, (Figure 2 c), lower tensile strength properties are registered. Identical to the orientations described previously, σ_{max} increases with negative air gap. Specimens with a thick filament geometry achieve higher values than the thin filament geometry. However, the unidirectional 90° raster angle obtain the highest tensile strength in Z-direction ($\sigma_{z,max}$ = 29.5 MPa).

In Figure 3 the comparison between standard and optimized parameter is shown. Additionally, tensile tests with injection molded ABS-M30 specimens were conducted. The optimized X and Y parameters are unidirectional 0° filaments with negative air gap and thin part raster width. For the Z-direction the alternating 0°/90° raster angle with negative air gap and thick filament geometry achieved the best results. Therefore, we see that the tensile strength of the optimized FDM parameters (X- and Y-direction) are on one level with the injection molded specimens. Due to the inner structure (corners and edges lead to stress peaks), the elongation at break is not as high as in the injection molding process. The increase of the ε_B is around 230 % for the X-direction and 380 % for the Y-direction. With these optimized parameters it was not possible to achieve a higher elongation at break for upright manufactured specimens, but the maximum tensile strength was enhanced by 16 %. The Young's Modulus is around 2400 MPa for the optimized flat and sideways manufactured specimens and 2150 MPa for the upright direction.

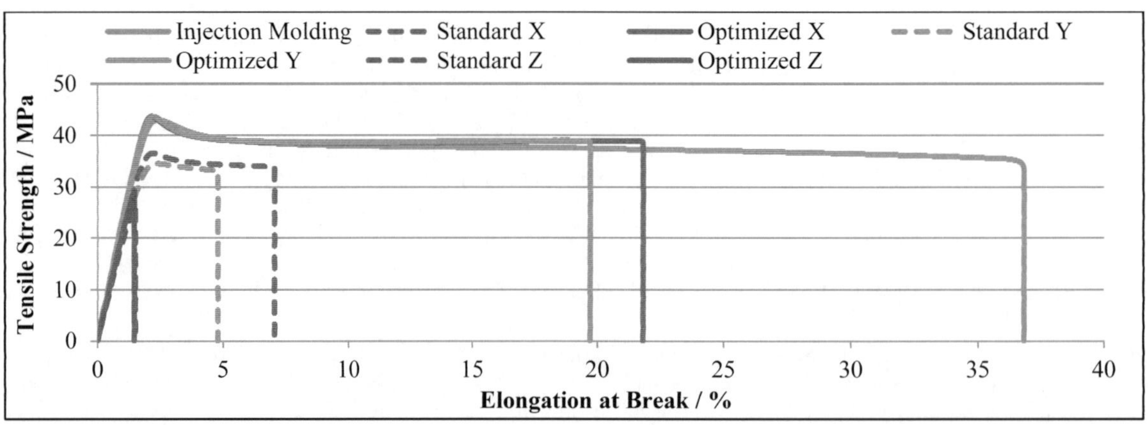

FIGURE 3. Stress-Strain Diagram of Standard and FDM Optimized Parameter in Comparison to Injection Molding Process.

Flexural Strength

The bending test is used to analyze the mechanical and deformation properties of specimens by applying a three point load. The tests were conducted according to DIN EN ISO 178. The velocity was 2 mm/min and a load was applied to the specimens until they broke. The span distance was 64 mm and the radius of the loading noses and the supports were 5 mm. In Figure 4 the results of the parameter variation are shown. In X-direction, the maximum flexural strength with standard parameters is 64.9 MPa (SD = 0.5 MPa) and the flexural modulus is around 2280 MPa (SD = 19.6 MPa). Only the unidirectional 0° orientation achieves higher values with an air gap of 0 mm (see Figure 4 a). A negative air gap leads to an increase of the flexural strength. The outcome of this is that all parameters (except unidirectional 90°) result in nearly one value of around 69.5 MPa (+ 7.1 %). The optimized flexural modulus for 0°_thin is about 2330 MPa.

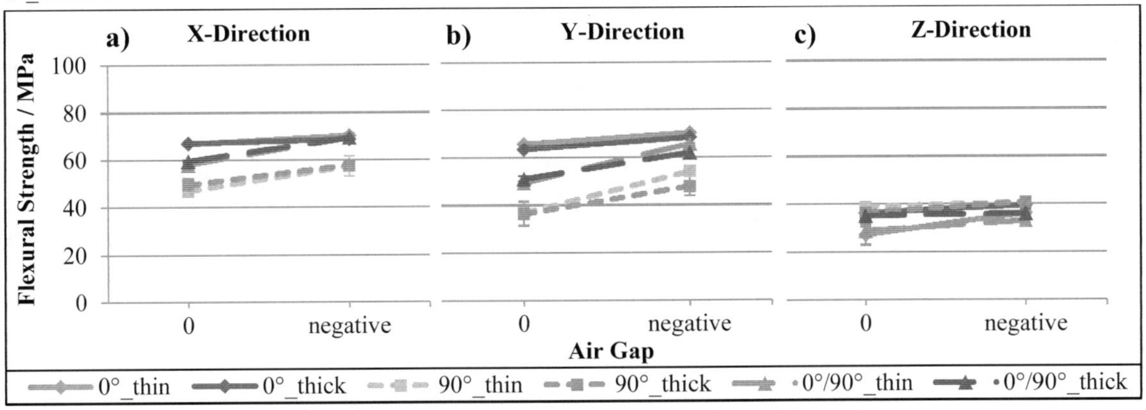

FIGURE 4. Results of the Bending Tests as a Function of the Air Gap.

For the flat lying specimens (Y-direction), which are built with standard parameters, the flexural strength and modulus are lower (σ_y = 58.1 MPa ± 0.9 MPa, E_y = 1955 MPa ± 23.2 MPa). Figure 4 b) shows that the negative air gap leads to higher values for the Y-direction as well. With an air gap of zero again only the unidirectional 0° raster angle has better values than the standard parameter. In Y-direction the thin filament geometries are above the thick filaments, if a negative air gap is used. The highest value is recorded for the 0°_thin parameter with $\sigma_{y,max}$ = 70.3 MPa, this represents an increase of 21 %.

The upright manufactured bending bars are, again, the weakest specimens. With standard parameter there is a maximum flexural strength of 35.7 MPa (SD = 2 MPa) and the modulus is just 1886 MPa (SD = 28.6 MPa). The use of negative air gaps achieves only a small gain for the flexural strength (compare Figure 4 c). It can also be seen that the thick filament geometry leads to higher values than the thin filament width. The maximum flexural strength and modulus is recorded for the unidirectional 90° raster angle, thick filament geometry and negative air gap ($\sigma_{z,max}$ = 40.9 MPa, $E_{z,max}$ = 2040 MPa). Here we have a growth of 14.6 %.

Impact Strength

The Charpy impact test is used to analyze material properties at a sudden and impulsive stress on a specimen. The specimen is fixed and breaks by a single swing of the pendulum (4 J). The Impact strength is not transferable on any specimens with other geometries and dimensions, so it is mostly used as quality assurance. The characteristic value here is the impact energy, which is determined on unnotched specimens according to the DIN EN ISO 179. For the standard parameter, the flat and sideways manufactured specimens achieve values of around 21 kJ/m². In contrast to this, the upright produced test specimens show very low values for the impact energy of around 3.7 kJ/m².

By varying the toolpath parameters it is possible to increase the impact energy. In detail this is possible for the 0° raster angle (see Figure 5 a, b). The highest data is recorded for the 0°_thin parameter with 51.1 kJ/m². In contrast to the mechanical tests before, here is no gain of the impact energy with a negative air gap for the 0° raster angle. The reason for this might be that the specimens with a negative air gap have a better filling degree and therefore less air inclusions between the strands. Due to the application of force there are cracks that move through the component. These cracks may terminate at the air inclusions and do not grow further. The growth of a crack leads to a component failure and to a lower impact energy.

In Y-direction (flat lying), the unidirectional 0° parameter with zero air gap has the highest values with around 79.2 kJ/m². This is an increase of over 360 % (see Figure 5 b). The higher values in Y-direction can be explained that the bonding between the layers are not in the power flow direction. The unidirectional 90° and alternating 0°/90° parameters have very low values for the impact energy for X- and Y-direction. The reason for this is that the force of the pendulum leads directly into the bonding between two layers and thus hits the weak point of the FDM process. This is also the reason for the low values in Z-direction (Figure 5 c). The values are between 4 and 6 kJ/m² and here is no influence of the air gap or the filament width recognizable.

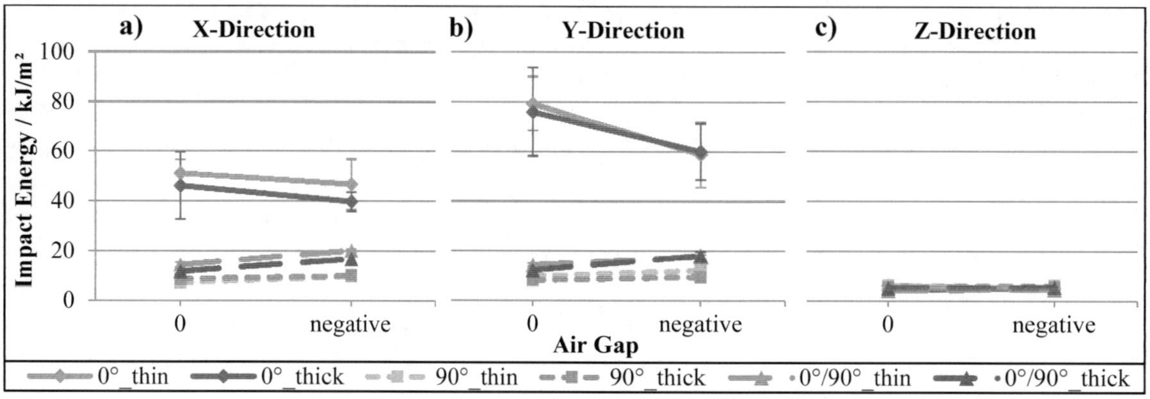

FIGURE 5. Results of the Charpy Impact Tests as a Function of the Air Gap.

CONCLUSIONS

In this paper the mechanical properties of test specimens manufactured with ABS-M30 were analyzed. Therefore, tensile, flexural and impact tests were conducted according to the European DIN standards. First of all, the initial state with standard parameters of Stratasy's Insight 9.1 and Fortus 400mc was recorded. Afterwards, the toolpath parameter filament width, air gap and raster angle were changed and the influence for each build direction was analyzed. The results show that the mechanical properties of ABS-M30 are dependent on the build direction as well as other materials mentioned in literature. The two major influence factors of the mechanical properties are the raster orientation and the air gap. With a unidirectional 0° filament orientation, it is possible to reach the same maximum tensile strength that is achievable with injection molded specimens. The elongation at break is relatively lower due to the inner structure. The filament width has shown an influence when test specimens are built up in the upright direction. The results of the flexural and impact test are similar to the tensile tests. These results show that optimized FDM parameters can lead to good mechanical properties even if they are low for the standard parameter. Almost all optimized parameters, which achieve the highest mechanical values, are built up with unidirectional 0° raster orientation and, therefore, the filaments are lying in the load direction. But if the applied force affects orthogonally to the 0° direction, it comes to lower mechanical properties. This leads to an anisotropy within the FDM component. This knowledge can be used to design certain component areas (in which the load is clearly known) with the optimized parameters.

REFERENCES

1. A. Gebhardt: "Generative Fertigungsverfahren / Additive Manufacturing", 4th Edition, Carl Hanser Verlag, München, 2013
2. A. Bagsik, V. Schoeppner: "Mechanical Properties of Fused Deposition Modeling Parts Manufactured with Ultem*9085", ANTEC 2011, Boston / Massachusetts, May 1th – 5th, 2011
3. T. Wohlers: "Wohlers Report 2015 – 3D Printing and Additive Manufacturing State of the Industry"; Annual Worldwide Progress Report, 2015
4. B. Wendel: "Prozessuntersuchung des Fused Deposition Modeling", Dissertation, Friedrich-Alexander-University of Erlangen-Nuernberg, 2009
5. S.H. Ahn, M. Montero, D. Odell et al.: "Anisotropic Material Properties of Fused Deposition Modeling ABS", Rapid Prototyping Journal, Vol. 8, No. 4, pp 248-257, 2002
6. O.S. Es-Said, J.Foyos, R. Noorani, M. Mendelson, R. Marloth: "Effect of Layer Orientation on Mechanical Properties of Rapid Prototyped Samples", Material and Manufacturing Processes, Vol. 15, No. 1, pp. 107-122, 2000
7. ABS-M30 Datasheet – Stratasys Inc. / Last Access 07.03.2016 http://usglobalimages.stratasys.com/Main/Files/Material_Spec_Sheets/MSS_FDM_ABSM30.pdf
8. H. Domininghaus: "Kunststoffe – Eigenschaften und Anwendungen", 8th Edition, Springer Verlag, Heidelberg, 2012

The Influence of PLA Composite Filaments on Molding Product made by Three-dimensional Printing

Kunihiro Araki[a], Goshi Hamabe[a], Keishi Date[a], and Tatsuya Tanaka[a]

[a] *Applied Materials Engineering Laboratory, Faculty of Science and Engineering, Doshisha University, 1-3, Tatara Miyakodani, Kyotanabe, Kyoto, 610-0321, Japan*

euo1501@mail4.doshisha.ac.jp

The objective of this study is to investigate mechanical and electrical properties of filler-reinforced Polylactic acid (PLA) composite product made by three-dimensional printing. The cellulose and the CNT were infused in PLA by a melt kneading method, and then the PLA composites were taken for processing the filaments. Then, three-dimensional printing was performed using the composite filaments. After that, the comparison was performed which was between pure PLA filament and PLA composite filament about the characteristics of the molding products. As the result of the tensile test of Cellulose-reinforced PLA composite, the tensile strength of about 60MPa has been found, although the reinforcing effect of adding cellulose was not found. However, by the CNT addition, the effect of reinforcing the PLA was observed. Also, the conductivity of CNT-reinforced PLA composite was expressed. As the FE-SEM observation, many holes pulled out cellulose from PLA were observed in the Cellulose-reinforced PLA composite. And also, measuring the areas of the aggregated CNT with each the form of pellets, filaments, and specimens, the areas of the aggregates of each condition were found that these tend to decrease.

Keywords: Additive manufacturing, Polylactic acid, Cellulose, Carbon nanotubes
PACS: 81.05.Pj

INTRODUCTION

Since about 2013, it had been experiencing the boom that is the manufacturing using the three-dimensional printers called "Additive Manufacturing". Even among them, "fused deposition modeling (FDM)" type of printers are inexpensive, so they are becoming popular in a general office and a home. PLA and Acrylonitrile-Butadiene-Styrene (ABS) are the famous resins used for FDM type printers. The characteristics of PLA are shown below: the shrinkage is small when cooling time, and it is the biodegradable resin. On the other hand, as compared with ABS, PLA has defects of the impact resistance and the heat resistance. The molding products made by FDM-type 3D printing using PLA filament have same demerits. So, we attempted to improve by making and using composite filaments composed of PLA and fillers. In this study, the PLA composite filaments included cellulose or carbon nanotube (CNT) were produced by a twin-screw extruder. Then, three-dimensional printing was performed using the composite filaments. After that, the comparison was performed which was between pure PLA filament and PLA composite filament about the characteristics of the molding products.

EXPERIMENTAL

Materials

The PLA (REVODE110 grade) used was purchased from Zhejiong Hisun Biomaterials Co., Ltd. (China). The density of the PLA was 1.25 g/cm^3. The PLA having the melting point was 160°C. The PLA was dried at least for 3h at 90°C in a vacuum oven before mixing step. The cellulose powder (KC FLOCK W-100GK grade) used as a biomass reinforcement was purchased from Nippon Paper Industries Co., Ltd. (Japan). The average diameter of cellulose powder was 37 μm (reference values). The cellulose powder was dried at least for 24h at 80°C in a hot-air oven before mixing step. The CNT (Nanocyl 7000) used as an electroconductive reinforcement was purchased from Nanocyl S.A. (Belgium). The CNT is thin multi-walled carbon nanotubes with a purity of 90% produced via the catalytic chemical vapor deposition process. The nanotubes have a typical diameter of 9.5 nm and an average length of 1.5 μm.

Composite preparation

The compounding process was performed using a co-rotating twin-screw extruder (ZSK18 MEGAlab, Coperion, Germany) having a screw outer diameter of 18mm and a barrel length of 720 mm (L/D=40). PLA pellets and the material (cellulose or CNT) were premixed before the mixing. The following filler weight fractions were used: 10 wt% cellulose and 5 wt% CNT. During melt-mixing, the premixed materials were fed simultaneously into the hopper by volumetric feeder.

The screw configuration is shown in Fig. 1. The configuration, a standard component of Coperion, included a kneading zone of one screw mixing element and five kneading elements with standard (45°) and neutral (90°) staggering angles. The screw rotational speed was kept constant at 150 rpm. The amount of supply was kept constant at 3.0 kg/h.

The set temperatures for barrel are shown in Table 1. On the mixing of PLA/Cellulose composite, the set temperatures continuously became lower toward the top end of the barrel in order to protect from thermal decomposition of cellulose. The resin temperature of PLA/Cellulose composite was about 185°C under the temperature condition. On the other hand, in the case where the mixing of PLA/CNT composite, the set temperatures having rising temperature were decided the previous research. The resin temperature of PLA/CNT composite was about 242°C under the temperature condition.

All composite materials were cooled down in a water bath and pelletized after the extrusion process. And then, the materials were dried at least for 24h at 80°C in a hot-air oven in order to remove the moisture by the cooling process.

TABLE 1 Temperature conditions for the mixing of PLA pellets and the fillers

	Heating zone [°C]						
	1	**2**	**3**	**4**	**5**	**6**	**7**
PLA/Cellulose	180	175	175	170	170	165	165
PLA/CNT	200	200	190	180	180	200	220
Flow direction	→						

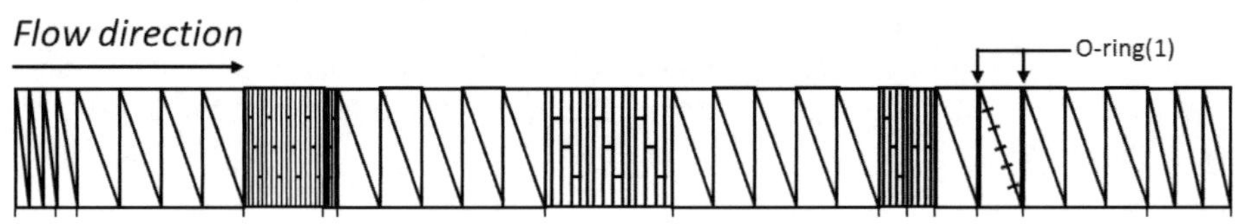

FIGURE 1. The screw configuration designed as standard component with conveying, kneading and mixing elements

Filament preparation

The composites were processed into a filament shape which can be used to three-dimensional printer. Two kinds of methods, using a filament extruder, and using a co-rotating twin-screw extruder, were taken for processing the filaments. PLA/CNT composite filament was made by the first method. PLA/CNT composite pellets were melted by a filament extruder (Noztek Pro Filament Extruder, Noztek, Britain), and wound in a rolled shape by a winder (Filament Winder, Noztek, Britain). A ribbon heater was located at the tip of the extrusion screw in order to prevent an overload of the motor caused by the melt insufficiency of the composites. The temperature conditions were as follows: the heater temperature of the extruder main body was 210°C, and the ribbon heater temperature was about 250°C. On the other hand, PLA/Cellulose composite filament was made by the second method. PLA/Cellulose composite pellets were melted by the co-rotating twin-screw extruder (the same one as before), and wound in a rolled shape by the winder which was made by oneself. The extrusion used the same conditions as the production of the PLA/Cellulose composite pellets, but the amount of supply was changed into 1.7 kg/h only. The reason for this value was suitable for winding filament under the extrusion condition. The composite filaments with diameters about 1.75 mm were made by the above method.

The composite filaments were made into specimens, which used for measuring tensile strength or volume resistivity, by FDM-type three-dimensional printer (SCOOVO X9, Opencube Inc., Japan). The molding conditions are shown in Table 2.

TABLE 2 Molding conditions for the making of specimens by FDM-type three-dimensional printer

	PLA/Cellulose	PLA/CNT
Layer height (First layer) [mm]	0.4 (0.35)	
Nozzle spped at perimeter (First layer) [mm/s]	15 (7.5)	
Nozzle spped at infill (First layer) [mm/s]	15 (7.5)	
Density	1	
Nozzle temperature [°C]	200	260
Bed temperature [°C]	80	135

Composite characterization

The tensile properties of the composites were measured according to Japanese Industrial Standards (JIS) K 7162. The tensile specimens were type A12 in the JIS K 7139. The tensile tests were carried out using a universal testing machine (Auto Graph AG-100kN, Shimadzu Co., Japan) at room temperature with crosshead speed of 1mm/min. The evaluation values set an average of the results for tests run on at least three specimens.

The volume resistivity of the composites was measured with two kinds of resistivity meters according to JIS K 6911. The size of specimens for measuring volume resistivity were 30 mm long, 20 mm wide, and 4 mm high. The specimens having a low volume resistivity were measured by the four-point probe method with a meter (Loresta-GP MCP-T610, Mitsubishi Chemical Analytech Co., Ltd, Japan). On the other hand, the specimens having a high volume resistivity were measured by the double ring electrode method with a meter (Hiresta-UP MCP-HT450, Mitsubishi Chemical Analytech Co., Ltd, Japan). In any case of the methods, the central part of the each specimen was measured only. The measurement was carried out for 5 seconds each. The evaluation values set an average of the results for tests run on at least five specimens.

Fracture surface of PLA/Cellulose composites was observed by a Field Emission Scanning Electron Microscope (FE-SEM) (JSM-7001F, JEOL Ltd., Japan). For this observation, the fracture surfaces after tensile breaking were cut and the surfaces were coated with Pt using sputter coating method for 5.0 nm to avoid the charging of the PLA during the observation. The observation was carried out at the acceleration voltage from 1.0 to 15.0 kV. When the low-magnification observation was carried out, the acceleration voltage was 1.0 kV, and when the high-magnification observation was carried out, the acceleration voltage was 15.0 kV. About PLA/CNT composite, the area of CNT agglomerates were measured in each production step: pellets, filaments, and specimens. This measurement was carried out in order to investigate whether the dispersion of CNT improve or not. At this time, total 10 agglomerates were selected optionally for each observation target. And the evaluation values set an average of the results.

RESULTS AND DISCUSSION

The results of tensile test are shown graphically in Figs. 2 (a) and (b). PLA/Ce (10) means the PLA composite including 10 wt% cellulose, and PLA/CNT (5) means the PLA composite including 5wt% CNT. Figure 2 (a) compares the tensile strength of specimen made by the undried filament (the specimen is called the before drying) with that made by the filament after drying (the specimen is called the after drying). As the result of the tensile test, the tensile strength of about 60MPa has been found, although the reinforcing effect of adding cellulose was not found. Also, the tensile strength of after drying was higher than that of before drying at either pure PLA or PLA/Ce (10). This result is considered to have an effect that water cooling when produced the filaments.

On the other hand, Fig. 2 (b) compares the tensile strength of pure PLA or PLA/CNT (5) specimen made at 200 °C with that made at 260 °C. Here, there was no data modeling at 200 ° C for PLA/CNT. This is because the PLA/CNT specimens could not made at 200 °C. Tensile strength of 260 °C of pure PLA was reduced than that of the 200 °C. This is considered to be the effect of resin degradation due to high temperature. However, by the CNT addition, the effect of reinforcing the PLA was observed.

The volume resistivity of the pure PLA or PLA/CNT (5) was measured in order to investigate the electrical conductivity. The result showed the volume resistivity of the PLA/CNT (5) was small value, about 7 ×100 Ω·cm. On the other hand, the pure PLA could not be measured even by using Hiresta UP, so the volume resistivity of pure PLA was presumed to be about 1013 Ω·cm or more. It can be seen that the electrical conductivity was given by the addition of CNT, and as a result the volume resistivity was lowered. Therefore it was also found by the three-dimensional printing it is possible to give conductivity to the molding parts.

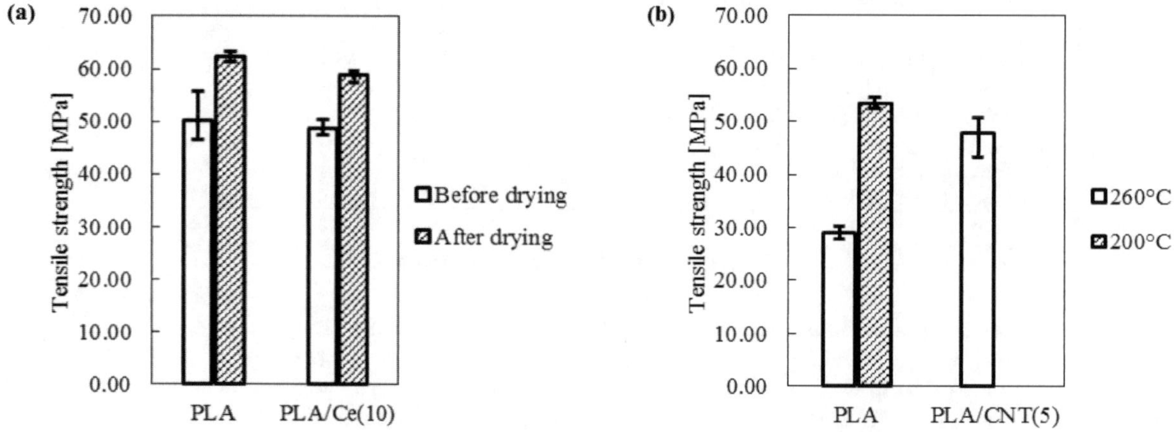

FIGURE 2. Tensile strength of (a) the pure PLA and the PLA with Cellulose content of 10 wt% and (b) the pure PLA and the PLA with CNT content of 5 wt%

Figures 3 (a) and (b) show the fracture cross-section of the PLA/Ce (10) specimen. Figure 3 (a) is a micrograph obtained by low-magnification observation, it can be seen that there is a difference in smoothness between the right hand and left hand. It was observed that fracture surface closer to the molding table is bumpy. The micrograph of bumpy cross-section of the same composite obtained by high-magnification observation as shown in Fig. 3 (b). Many holes pulled out cellulose from PLA were observed as shown in Fig. 3 (b). Therefore, a load transmission due to the filler is not performed successfully, it is believed to have not reached the tensile strength improvement. Figures 4 (a) and (b) show the fracture cross-section of the PLA/CNT (5) specimen. It could be seen that the good adhesion of PLA and the CNT was observed as shown in Fig. 4 (b). On the other hand, it could be seen that the CNT agglomerates having several-ten μm size were present in the composites as shown in Fig. 4 (a). The conductivity was expressed in this example the CNT was added 5 wt%, but the conductivity expression at a smaller amount of CNT will need to reduce the sizes of these aggregates. When preparing specimens of PLA/CNT using the three-dimensional printer, there were repeated a total of three times of melting: making of the composite pellets, making of the composite filaments, and making of the specimen. For this reason, as compared to the pellets condition, the dispersion of the aggregated CNT was expected to improve in the specimen. In order to research this assumption, the areas of the aggregated CNT were measured with each the form of pellets, filaments, and specimens. Figures 5 (a), (b), and (c) show the observation surface measured aggregates and an example of selected aggregates. The areas obtained from the measurement were followed: about 900 μm^2 in the pellet condition, about 750 μm^2 in the filament, and about 650 μm^2 in the specimen. Therefore, the areas of the aggregates of each condition were found that these tend to decrease.

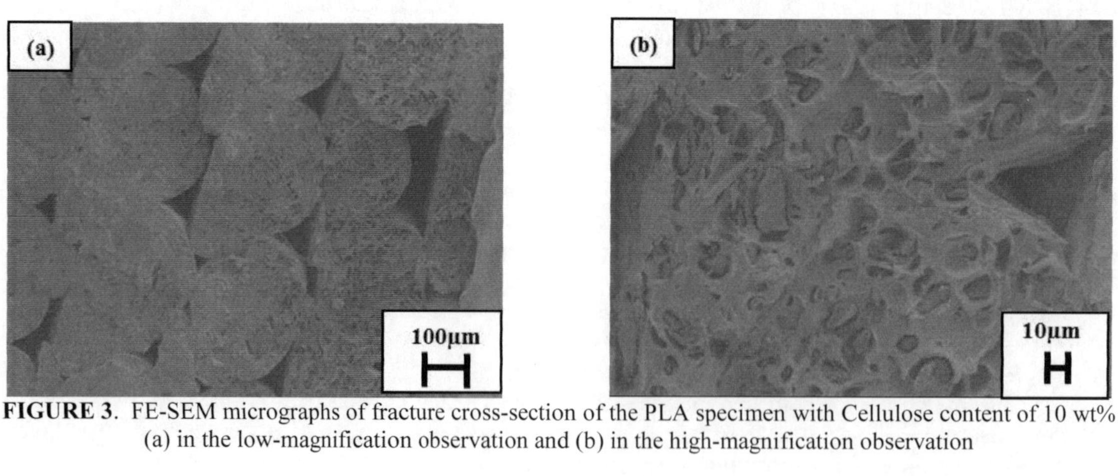

FIGURE 3. FE-SEM micrographs of fracture cross-section of the PLA specimen with Cellulose content of 10 wt% (a) in the low-magnification observation and (b) in the high-magnification observation

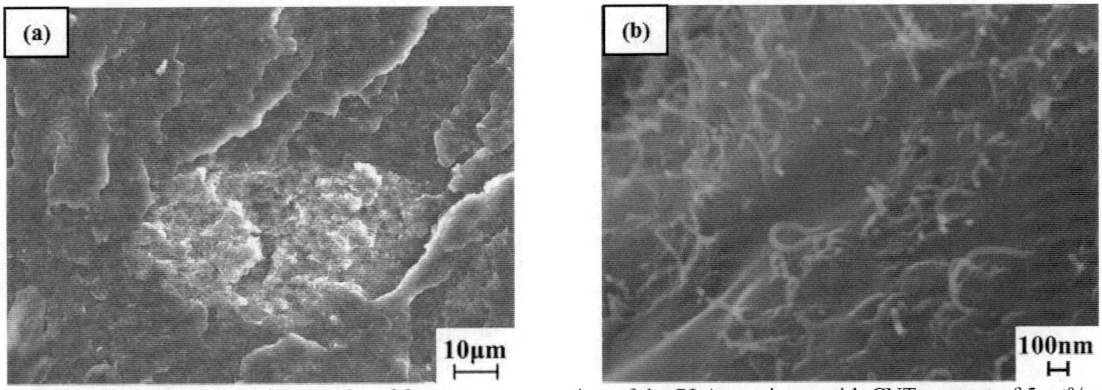

FIGURE 4. FE-SEM micrographs of fracture cross-section of the PLA specimen with CNT content of 5 wt%
(a) in the low-magnification observation and (b) in the high-magnification observation

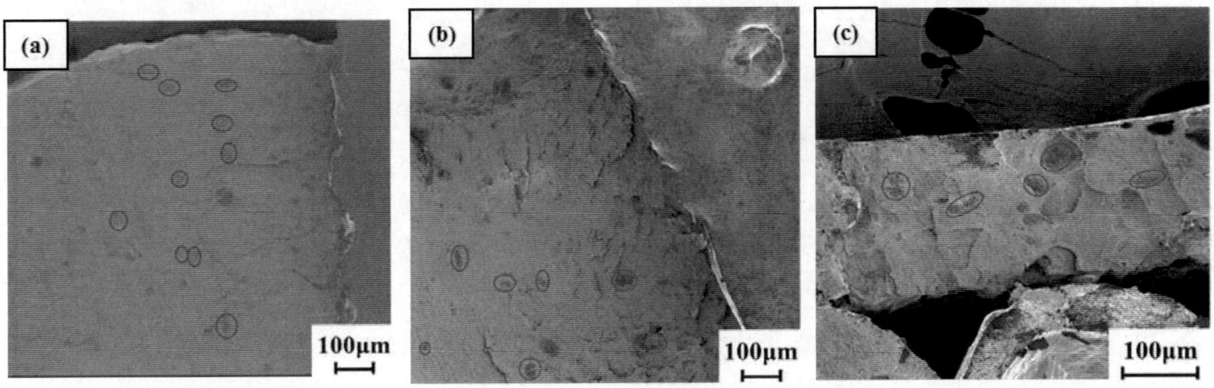

FIGURE 5. FE-SEM micrographs of fracture cross-section of the PLA with CNT content of 5 wt%
(a) in the pellet, (b) in the filament, and (c) in the specimen

CONCLUSION

In this study the cellulose and the CNT were infused in PLA by a melt kneading method, and then the PLA composites were taken for processing the each filament. As the result of the tensile test of Cellulose-reinforced PLA composite, the tensile strength of about 60MPa has been found, although the reinforcing effect of adding cellulose was not found. However, by the CNT addition, the effect of reinforcing the PLA was observed. Also, the conductivity of CNT-reinforced PLA composite was expressed. As the FE-SEM observation, many holes pulled out cellulose from PLA were observed in the Cellulose-reinforced PLA composite. And also, measuring the areas of the aggregated CNT with each the form of pellets, filaments, and specimens, the areas of the aggregates of each condition were found that these tend to decrease.

ACKNOWLEDGMENTS

This work was financially supported by a (advanced technological) research project on "Research and Development Center for Advanced Composite Materials" of Doshisha University and MEXT (the Ministry of Education, Culture, Sports, Science and Technology, Japan) -Supported Program for the Strategic Research Foundation at Private Universities, 2013-2017, the project S1311036.

REFERENCES

1. A. K. Sood, R. K. Ohdar, and S. S. Mahapatra, *Materials and Design*, **31**, 287-295 (2010)
2. B. Huang and S. Singamneni, *Journal of Composite Materials* 0, 1-21 (2014)
3. T. Villmow, P. Potschke, S. Pegel, L. Haussler, and B. Kretzschmar, *Polymer* **49**, 3500-3509 (2008)

32nd International Conference of the Polymer Processing Society Abstract

Authors: Timothy J. Coogan[1,2] and David Kazmer[1]

[1] *Plastics Engineering Department, University of Massachusetts Lowell (Lowell, MA)*

[2] *Saint-Gobain Northborough R&D Center (Northborough, MA)*

IF POSSIBLE, PLEASE USE THE FOLLOWING ABSTRACT FOR MY PRESENTATION. THANK YOU.

FDM Bond Strength: Experiments and Simulations Based on Healing Models

One of the major limitations of fused deposition modeling (FDM) is part strength of the products, due to the weak strength between fibers (bond strength). This presentation will describe the process parameters that affect bond strength, and a theoretical model is developed for predicting bond strength. Four-sided ABS boxes with a wall thickness equal to a single fiber width were printed according to a design of experiments. A set of tensile specimens were cut from the printed box and tested in the fiber and cross-fiber directions: smaller layer thicknesses, larger fiber widths, faster print speeds, and higher nozzle and platform temperatures increased interlayer bond strength.

A one-dimensional transient heat analysis provided an estimate of the temperature profile of the interface between layers. By transposing temperature-dependent diffusion coefficients onto the interface temperature profiles, the amount of polymer chain diffusion that occurred across the layers was calculated for all the tested processing conditions. Bond strength versus total diffusion was found to match well with polymer healing models, showing that the healing models developed for polymer welding can be applied to the non-isothermal FDM process.

The presented results should be used to accurately model and predict bond strength. Such models are useful for FDM part design and process control. The findings support that diffusion, wetting, and intimate contact all separately affect bond strength. Fundamental models for predicting FDM part strength should, therefore, consider all three of these factors. This presentation provides processing recommendations for producing the strongest FDM parts, which can be useful for companies producing FDM products as well as companies designing printers.

New approaches towards production of polymer powders for selective laser beam melting of polymers

Jochen Schmidt[a*], Marius Sachs[a], Stephanie Fanselow[a], Karl-Ernst Wirth[a] and Wolfgang Peukert[a]

[a]*Friedrich-Alexander-Universität Erlangen-Nürnberg, Institute of Particle Technology, Cauerstr. 4, D-91058 Erlangen, Germany*
jochen.schmidt@fau.de

Abstract. The opening of new fields of application of powder-based additive manufacturing methods like selective laser beam melting of polymers (LBM) currently is hindered by the very limited availability of materials showing good processability. So far, basically only polyamide-based materials are available as optimized powder materials for LBM. Two innovative methods for production of spherical polymer microparticles for LBM polymers are presented: wet grinding with subsequent rounding in a heated downer reactor and melt emulsification. Moreover, the possibilities of dry coating to tailor particle properties are illustrated. The influence of particle and powder properties, like bulk density, shape or flowability on the LBM processability and properties of devices obtained from the powders are outlined. The first approach, stirred media milling and rounding, is applicable for a variety of polymers with a quite high breakage elongation. The second approach, melt emulsification, allows for spherical polymer microparticles in a single process step. The dependency of powder properties on process parameters will be outlined exemplarily for the production of spherical polypropylene (PP) particles. PP powders of good flowability are obtained after spray drying of the product suspensions. The powders obtained by the two methods are characterized with respect to tensile strength, packing density and layer formation. Moreover the processabilty of the PP powder is proven by building single layers using a LBM device.

Keywords: laser beam melting of polymers, polybutylene terephthalate (PBT), polypropylene (PP), wet grinding, downer reactor, melt emulsification, dry coating
PACS: 81.05.Lg; 82.35.Lr; 81.20.Wk; 47.55.-t; 81.15.-z

INTRODUCTION

Additive manufacturing methods like selective laser beam melting of polymers (LBM) allow the tool-free and formless production of complex structures. With the advent of additive manufacturing, i.e. the application of generative methods in mass production, new requirements concerning the LBM process itself as well as the powder materials used have been defined [1-8]. Currently, the choice of commercially available LBM powders is very limited: about 95 % of the total market share is made up of polyamide. PA powders, for example, are typically produced by bottom-up approaches (c.f. emulsion polymerization) or precipitation processes are applied [2, 8]. Also top-down approaches like cryogenic grinding [9], wet grinding of polymers [10], co-extrusion [8] or melt emulsification [11] have been proposed to produce polymer microparticles. To widen the field of applications of LBM, new and cheap materials of good processability must be supplied. Thus, innovative processes being applicable to a wide variety of thermoplasts are preferable.

Within this contribution two novel approaches based on top-down methods to produce spherical polymer microparticles for LBM are discussed: The first approach is a process chain [6] which consists of a wet grinding step [10] followed by rounding of the ground polymer particles in a heated downer reactor. Finally, the flowability of he obtained spherical particles is increased by dry coating [12-14]. Powders for LBM can be produced by this method from any thermoplast that shows satisfactory breakage behaviour, i.e. which can be comminuted to the desired product particle size. The second approach is melt emulsification. In this process the polymer is melted in an appropriate continuous phase and the polymer melt is emulsified by shear and elongational stress. After cooling of the obtained melt emulsion a suspension of spherical polymer particles is obtained, i.e. size reduction and engineering of the particle shape are done in a single step [11]. Again, dry coating can be employed to increase the powder flowability. The processability [15] of the obtained powders in LBM is demonstrated by building single layer specimens: powders of good flowability and high bulk density are mandatory to obtain dense devices [14].

MATERIALS AND METHODS

Powder Materials

The first approach to produce LBM powders, i.e. the process chain mentioned in the introduction section (for details see [6]) is exemplified for polybutylene terephthalate (PBT). As raw material PBT 4520 Ultradur Highspeed (BASF) has been used. The following two PBT powders have been characterized and evaluated with respect to their processability:

- 'ground PBT': plate-like PBT micro particles produced by wet comminution in denatured ethanol (95%, VWR).
- 'rounded & dry coated PBT': powder produced by dry coating of rounded PBT powder with fumed silica. 'Ground PBT' was used as feed material for rounding.

The processes applied to obtain the PBT powders are addressed below.

The second approach, melt emulsification [11], is demonstrated for polypropylene (PP). PP 520P (Sabic) of a melting temperature of 140 °C to 170 °C was applied. The continuous phase was n-hexadecane (Alfa Aesar) and sodium dioctyl sulfosuccinate (AOT; Acros Organics) was the stabilizer.

For dry coating of the obtained spherical polymer microparticles, fumed silica (Evonik Industries) with a primary particle size of approx. 7 nm has been applied.

Characterization Methods

Particle Sizing

Particle size distributions were determined by laser diffraction particle sizing using a Mastersizer 2000 / Hydro 2000S (Malvern). In the case of wet grinding, the product suspensions have been diluted as appropriate prior to measurement and stabilized using sodium dodecyl sulphate solution (SDS, 98% (Merck)) Dry samples after rounding or melt emulsification and subsequent spray drying have been dispersed in aqueous SDS solution, too. Prior to particle sizing ultrasonication of the dispersions has been performed.

Scanning Electron Microscopy

Polymer particles have been characterized SEM using a Gemini Ultra 55 (Zeiss) device equipped with a through-the-hole detector. An acceleration voltage of 1 kV has been applied. SEM images were taken at appropriate magnifications.

Powder Flowability

Flowability of the products was assessed using a tensile strength tester [16]. For details on the measurement procedure see [6].

Image analysis & Quantification of powder deposition

To characterize the powder deposition behaviour of the PBT powders doctor blading (Quadruple Film Applicator Model 360 (Erichsen), gap size: 120 µm) was applied. A black-colored paper was used as substrate. The area covered with powder was quantified using the commercial program "Image J" for image analysis. The coverage of the substrate with powder in this model experiment allows to assess the LBM processability of the material during the deposition step (for details see [14]).

RESULTS AND DISCUSSION

Process Chain for Production of Spherical PBT Microparticles

The process chain consisting of wet grinding, rounding in a heated downer reactor and dry coating was applied to PBT: First, PBT 4520 Ultradur Highspeed (BASF) was pre-ground using a Pulverisette-14 (Fritsch) equipped with a 0.5 mm sieve ring operated at 20,000 min-1. The feed material was cooled with liquid nitrogen. A coarse PBT powder 'PBT < 0.5 mm' of size $x_{50,3}$ = 239 µm ($x_{10,3}$ = 79 µm, $x_{90,3}$ = 550 µm) was obtained. 'PBT < 0.5 mm' was used as feed material for wet grinding [10]. A mass fraction of about 18.5 % of PBT has been used and a stress energy of 1.91 mJ was chosen by setting the stirrer tip speed to 6.3 m s^{-1} equaling (800 min^{-1}) and using Yttria-stabilized ZrO_2 grinding beads (Tosoh) of 2 mm diameter (density: of 6,050 kg m^{-3}). The process temperature was (20 +/- 2) °C. After about 5.5 hours of grinding 'ground PBT' was obtained. The wet ground polymer consists of plate-like fragments and shows a rather broad size distribution with $x_{50,3}$ = 25.2 µm ($x_{10,3}$ = 6.6 µm, $x_{90,3}$ = 275.5 µm), see particle size distribution (volume sum, Q_3), depicted in Figure 1.

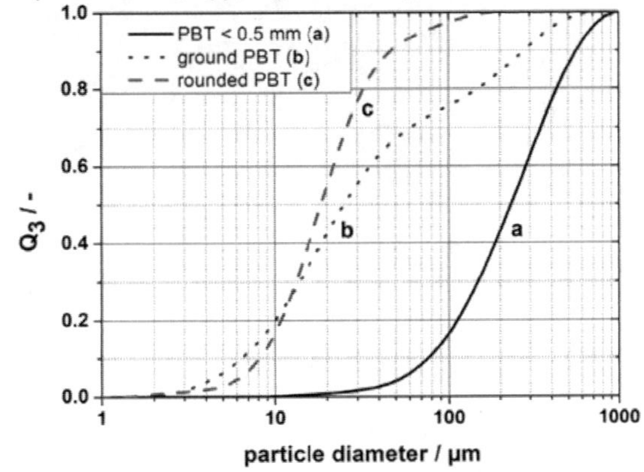

FIGURE 1. Particle size distributions (volume sum, Q_3) of PBT powders obtained by cryogenic grinding ('PBT < 0.5 mm'), wet grinding ('ground PBT') and after rounding in a downer reactor ('rounded PBT').

The rounding was performed in a heated downer reactor made of stainless steel with a diameter of 100 mm and a heated length of 4.5 m where the particles are molten, rounded and solidified. 'Ground PBT' was used as feed material. The heating is achieved using a nine stage heating system. Nitrogen is used as inter carrier gas. Powder dispersion is achieved using a PALAS RGB 1000 powder disperser. The aerosol enters the downer in the center of the reactor's cross section. A sheath gas flow (nitrogen), is applied and surrounds the aerosol in order to minimize the contact of polymer particles and reactor wall. To achieve a homogenous distribution of the sheath gas, this flow passes a sintered metal plate. Further details on the reactor setup as well as on the rounding process and the selection of proper process parameters can be found elsewhere [6]. Proper choice of temperature profile, residence time and particle concentration allows for production of almost perfect spherical particles, see Figure 2. The particle size distribution (volume sum, Q_3) of the rounded powder is shown in Figure 1. Some shift of the particle size distribution as compared to 'ground PBT' is notable which might be due e.g. improper aerosol generation or due to adhesion onto reactor components. 'Rounded PBT' already shows a better flowability as compared to 'ground PBT', however dry coating (T2F, Willy A. Bachofen AG) allows for further improvement. After 10 minutes of process time and applying a mass concentration of 0.1% silica a homogeneous nanoparticulate coating of the PBT particles is obtained. By the attachment of the nanoscale silica (guest) particles the surface roughness of the microparticles is increased, leading to a decrease of the adhesion force between them [17, 18]: the flowability of dry-coated powders increases, which is advantageous for the powder deposition step.

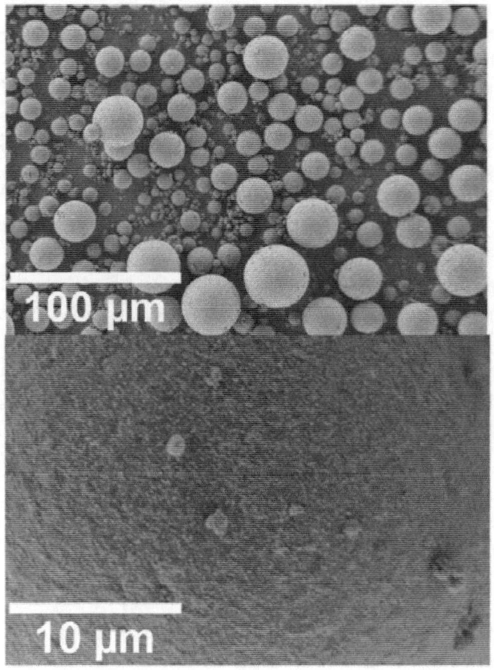

FIGURE 2. top: SEM image of 'rounded & dry coated PBT', bottom: SEM image of the composite particle surface.

Tensile Strength and Powder Deposition Behaviour of PBT Powders

The reduction of adhesion force by dry coating is also reflected by the tensile strength of the powders: whereas 'ground PBT' has a tensile strength of (15.7 +/- 0.8) Pa, which corresponds to a powder of bad flowability, (2.9 +/- 0.6) Pa tensile strength is observed for 'ground & dry coated PBT'. Also the bulk density roughly doubles (0.47 g/cm3) after changing shape and surface roughness. A quantitative evaluation of powder deposition by doctor blading and image analysis gives the possibility to assess the processability of the powders during the application of a new powder layer in LBM. Depending on the flowability only a certain area is covered with particles after the deposition step: For 'ground PBT' an area coverage of only (19.1 +/- 4.2) % was observed, whereas an almost complete surface coverage of (97.3 +/- 1.8) % is achieved for 'rounded & dry coated PBT'. S small tensile strength of powders (good flowability) correlates with homogeneous powder deposition.

Melt Emulsification of PP

Melt emulsification of PP was done in a pilot plant which is described in great detail elsewhere [11]. In short the device basically consists of a stirred tank were the pre-melt emulsion is formed, a gear pump for pumping of the pre-emulsion into a rotor-stator device (CD 1000, Cavitron), where the fine emulsification takes place, and a receiver tank, where the product emulsion is collected and cooled. The melt emulsification plant may be used at temperatures up to 250 °C and pressures up to 16 bar. PP particles of size $x_{50,3}$ = 22.1 µm have been obtained in a hexadecane-AOT system after 30 minutes at processing applying a polymer concentration of 5 wt%. The PP product particles have been collected by centrifugation and washed with hexane. After spray-drying of the suspension, PP powders (coarse fraction: $x_{50,3}$ = 29.8 µm, fine fraction: $x_{50,3}$ = 17.2 µm) are obtained which can be used in LBM directly or after dry coating to improve the powder flowability prior to LBM processing.

To investigate the processability, thin layers of PP were built in an EOS P380 device from material the fine fraction. A bed preheating temperature of 150 °C was adjusted. As powder bed and reference material commercially available polypropylene powder (Coathylene PD0580, DuPont) was used. For the construction of specimens a powder layer thickness of 300 µm thickness was deposited and an area of around 25 cm2 has been molten (exposure time 20 s thulium laser; λ = 1.94 µm, maximal power 120 W). Figure 3 shows sintered layers. In all cases, i.e. for rounded and rounded and dry coated material, dense layers are obtained. The PP powders produced at least show comparable LBM processability like the commercially available reference material.

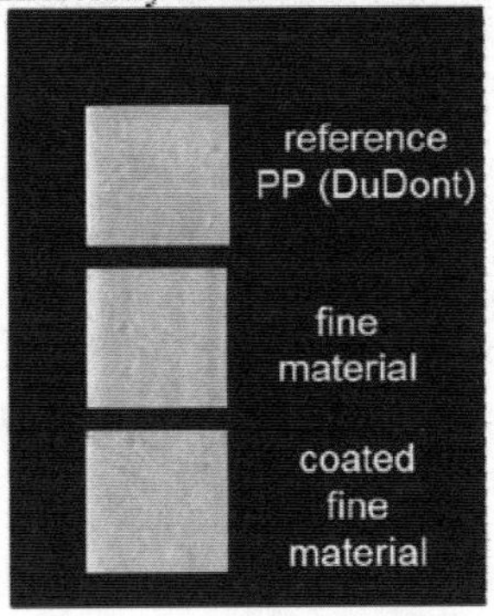

FIGURE 3. single layer specimens built from PP reference material as well as melt emulsified PP (as obtained and dry coated with silica).

CONCLUSIONS

Within this contribution two process routes, wet grinding / rounding and melt emulsification for the production of spherical polymer microparticles for LBM and dry coating to improve the powder flowability have been adressed: flowability is remarkably influenced by both, particle shape and surface roughness. Powders of good flowability are the prerequisite to obtain dense parts in LBM. Both approaches allow for powders of very good flowability and may be applied to a variety of polymers. Moreover, the processses are scalable, i.e. might be transferred to the plant scale, which makes them interesting for production of LBM powders.

ACKNOWLEDGMENTS

This study has been supported by Deutsche Forschungsgemeinschaft (DFG) in the framework of the collaborative research center CRC 814 "Additive Manufacturing" (projects A1, A2, A3). Financial support is gratefully acknowledged. The authors are grateful to Tobias Laumer and Michael Schmidt (Bayerisches Laserzentrum blz; Erlangen) for giving us the opportunity to use their LBM machine.

REFERENCES

1. J. P. Kruth, X. Wang, T. Laoui, L. Froyen, *Assembly Automation* **23** 357-371(2003).
2. T. Wohlers, Wohlers report 2014. 3D printing and additive manufacturing state of the industry annual worldwide progress report. Fort Collins, Col.: Wohlers Associates, 2014.
3. R. D. Goodridge, C. J. Tuck, R. J. M. Hague, *Progress in Materials Science* **57** 229–267 (2012).
4. B. Wendel, D. Rietzel, F. Kühnlein, R. Feulner, G. Hülder, Gerrit; E. Schmachtenberg, *Macromol. Mater. Eng.* **293** 799–809 (2008).
5. D. Drummer, D. Rietzel, F. Kühnlein, *Physics Procedia* **5** 533–542 (2010).
6. J. Schmidt, M. Sachs, C. Blümel, B. Winzer, F. Toni, K.-E. Wirth, W. Peukert, *Powder Technology* **261** 78–86 (2014).
7. G. Witt, Neue Entwicklungen in der Additiven Fertigung. Springer, Berlin, 2016.
8. M. Schmid, Selektives Lasersintern (SLS) mit Kunststoffen. Technologie, Prozesse und Werkstoffe. Hanser, München, 2015.
9. M. Wilczek, J. Bertling, D. Hintemann, *International Journal of Mineral Processing* **74** S425-S434 (2004).
10. J. Schmidt, M. Plata, S. Tröger, W. Peukert, *Powder Technology* **228** 84–90 (2012).
11. S. Fanselow, S. E. Emamjomeh, K.-E. Wirth, J. Schmidt, W. Peukert, *Chemical Engineering Science* **141** 282–292 (2016).
12. R. Pfeffer, R. N. Dave, D. Wei, M. Ramlakhan, *Powder Technology* **117** 40–67 (2001).
13. J. Yang, A. Sliva, A. Banerjee, R. N. Dave, R. Pfeffer, *Powder Technology* **158** 21–33 (2005).
14. C. Blümel, M. Sachs, T. Laumer, B. Winzer, J. Schmidt, M. Schmidt, W. Peukert, K.-E. Wirth, *Rapid Prototyping Journal* **21** 697–704 (2015).
15. I. Gibson, D. Shi, *Rapid Prototyping Journal* **3** 129–136 (1997).
16. A. Schweiger, I. Zimmermann, *Powder Technology* **101** 7–15 (1999).
17. H. Zhou, M. Götzinger, W. Peukert, *Powder Technology* **135-136**, 82–91 (2003).
18. M. Götzinger, W. Peukert, *Langmuir* **20** 5298–5303 (2004).

Insights into the Development of a Short-fiber Reinforced Polypropylene for Laser Sintering

R.G. Kleijnen[a,b*], J.P.W. Sesseg[b], M. Schmid[b] and K. Wegener[a]

[a]*Swiss Federal Institute of Technology, Department of Mechanical and Process Engineering, Leonhardstrasse 21, 8092 Zürich, Switzerland*
[b]*Inspire AG, Innovation Center for Additive Manufacturing Switzerland, Lerchenfeldstrasse 5, 9014 St. Gallen, Switzerland*

**Reference author: kleijnen@inspire.ethz.ch*

Abstract. Selective laser sintering (SLS) is a powder based additive manufacturing technology, and currently one of the most promising technologies in use for the manufacturing of medium sized series of functional parts. However, the availability of suitable materials is still very limited. Recently, a number of polypropylene materials has come onto the market as an addition to the SLS materials portfolio. The current study focuses on the improvement of mechanical properties of one of these materials by fiber reinforcement, a method well known in traditional polymer processing. Various amounts of wollastonite and glass fibers of different lengths (50 – 150 μm) were dry blended with two varieties of bonding agents and the polypropylene powder. The processability on an SLS machine of the produced blends was assessed, and the mechanical properties of produced parts were determined via tensile tests. Additionally, the nature of fracture was investigated with SEM. Through simple dry blending, homogeneous distributions of fibers could be achieved. It was found that both fiber length and amount play a critical role with respect to powder flowability, and therewith processability. The addition of the shortest glass fibers available with a length of 50 μm, led to the biggest improvement of tensile modulus. Addition of small quantities of longer glass fibers hinted towards even larger improvements, however processability issues prohibited these from being used in larger fractions.

Keywords: Selective Laser Sintering, fiber reinforced polypropylene, materials, Additive Manufacturing.
PACS: 81.05.Qk, 81.20.Ev

INTRODUCTION

The use of high strength and high elastic modulus fibers to reinforce a polymer material is well known. Whether in the form of impregnated mats (prepregs or sheet molding compounds) or melt blended directly in a polymer resin, fiber reinforced composites represent a great deal of what we consider state-of-the-art engineering materials today [1]. In an ideal case, composites combine the high strength and stiffness of the reinforcing fibers with the ductility and shape retention of a resin. In the portfolio for Selective Laser Sintering (SLS) materials, fiber reinforced polymers are also widespread. Mostly in the form of dry blends of fine polymer powder and short fibers, a great variety with respect to base material, fiber type, and fiber amount exists. An overview of the available materials is offered by Schmid [2].

Since the majority of the material consumption for SLS consists of polyamide-based polymers [3], it is not surprising that most of the existing reinforced compounds are based on the same material. Recently however, a new polypropylene material (trade name: iCoPP) was brought to the market [4]. This material exhibits excellent processing, mechanical, and surface properties, but suffers from a relatively low E-modulus. Where the modulus of conventional PP lies in the range of 1.1 – 2.0 GPa [5], the modulus of iCoPP is 0.7 GPa. This issue may be addressed by the addition of a reinforcing compound, such as glass or mineral fibers. In the current work, a variety of fibers of different lengths is dry blended in varying amounts with the base polypropylene powder and a bonding agent. The effect of the fiber presence on SLS processability and part mechanical properties is investigated, with the intention to discover the best compound and amount of reinforcing material, as well as to understand the mechanisms that lead to the observed properties, so that they may be improved upon in future work.

SLS Processing of Fiber Reinforced Composites

Selective Laser Sintering is a powder-based additive manufacturing technology. A typical commercial machine is shown in figure 1. To produce an object, layers of powder are laid upon each other. After deposition of each layer, a CO_2 -laser melts the cross-section of the object, thereby fusing the powder in the current layer with itself, as well as previously sintered layers. At the end of the process, excess, unmolten powder is removed and recycled, to leave only the intended object. In order for a material to be suitable for SLS, it must fulfill many requirements. A summary of these requirements is shown in the Ishikawa diagram in figure 2.The current investigations cover the entire process chain, and will focus on powder flowability, thermal properties, and prominently on the mechanical properties of the built parts.

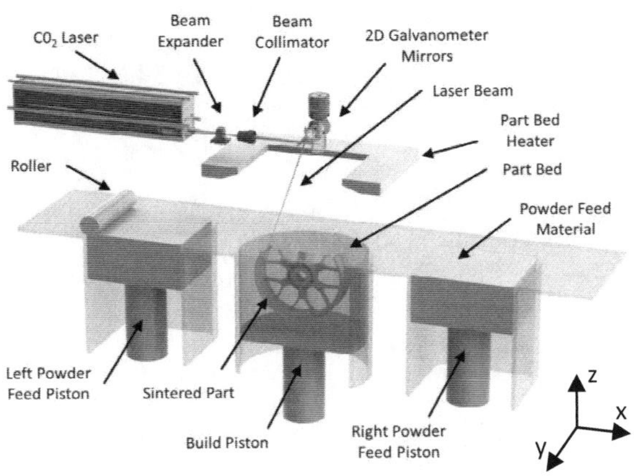

FIGURE 1. Schematic representation of a typical SLS machine.

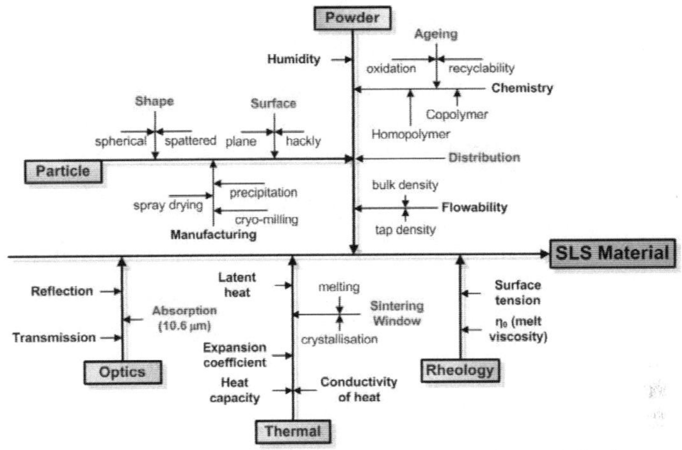

FIGURE 2. Influence factors on the processing and final properties of an SLS material.

The introduction of an additional compound in an SLS powder leads potentially to many challenges. A few examples of processing issues as they occurred over the course of the current investigation include complete inability to spread an even layer of powder, excessive curling, or sticking of the powder to the roller. Given an ideal particle size distribution for an SLS powder between 20 – 80 µm, it is assumed that the length of the reinforcing fibers must fall in the same order of magnitude, in order for the material to retain sufficient flowability.

Literature regarding the processability and part properties of fiber reinforced SLS materials is surprisingly sparse, with only a few authors having reported thorough investigations. A recent contribution from Jansson and Pejryd [6] deals with the characterization of carbon fiber–reinforced polyamide parts, with particular emphasis on part porosity. They found a drastic increase of porosity for the fiber-reinforced parts in comparison with neat polyamide parts. The reasons for this observation were sought in the low thermal expansion coefficient of the carbon fibers, as well as inefficient packing of the powder bed due to the presence of longer fibers.

Other authors predominantly covered polyamide materials filled with glass beads. For example, powder properties and pre-processing trials of polyamide 12 (PA12) blended with different amounts of glass beads were reported by Mousa et al. [7]. Negi et al. [8] subsequently performed an extensive investigation using response surface methodology (RSM) to find optimal processing parameters for a commercial PA12 filled with glass beads. A similar, yet statistically slightly less comprehensive study focusing on glass beads filled polyamide 11 (PA11) was carried out by Chung and Das [9]. In addition to optimizing process parameters, the authors proposed a heat transfer model, and were able to create functionally graded parts by locally varying the ratio of glass beads.

In the current investigation however, reinforcement of a polypropylene with glass and mineral fibers will play the central role. These systems have the potential to deliver a much desired mechanical properties set, which has been inaccessible with SLS as of yet.

Theoretical Tensile Properties of a Short-fiber Reinforced Polymer

Due to the necessity of sufficient powder flowability, short fibers with lengths between 50 to 150 µm are selected as reinforcing phase in the polymer matrix. This means the fibers are much shorter than the critical length required for complete stress transfer to the fiber up to its ultimate tensile strength. Unless a particularly strong adhesion between fiber and matrix can be achieved, a drastic increase of the tensile strength of the composite in comparison with the unfilled polymer is not expected. The presence of high tensile modulus fibers, no matter how short, however does lead to an increase in tensile modulus for the composite. The longitudinal modulus (along the fiber orientation) for a unidirectional short-fiber lamina can be derived from the Halpin-Tsai equations [10], and is given by:

$$E_{11} = \frac{1 + 2(l_f/d_f)\eta_L v_f}{1 - \eta_L v_f} E_m \tag{1}$$

Where:

$$\eta_L = \frac{(E_f/E_m) - 1}{(E_f/E_m) + 2(l_f/d_f)} \tag{2}$$

Where l_f and d_f are the length and diameter of the fibers respectively, v_f the volume fraction of fibers, E_m the tensile modulus of the matrix, and η_L a factor depending on the moduli of matrix and fibers, and the fiber aspect ratio (l_f/d_f). Additionally, the transversal modulus (perpendicular to the fiber orientation) of the lamina is given by:

$$E_{22} = \frac{1 + 2\eta_T v_f}{1 - \eta_T v_f} E_m \tag{3}$$

Where:

$$\eta_T = \frac{(E_f/E_m) - 1}{(E_f/E_m) + 2} \tag{4}$$

Finally, for a lamina with randomly oriented short fibers, the tensile modulus is given by a ratio between the longitudinal and transversal modulus:

$$E_{rand} = \frac{3}{8}E_{11} + \frac{5}{8}E_{22} \tag{5}$$

The above equations will be used as a reference for the experimental results. The theoretical values may be interpreted as the maximum achievable properties, since they are derived under the assumption of uniform fiber distribution, perfect adhesion between fibers and matrix, and a matrix free of voids. All three of these conditions are difficult to reproduce in practice.

MATERIALS AND METHODS

Materials. As laser sintering material, a poly(propylene-co-ethylene) random copolymer was used. The material consists of spherical particles with a narrow size distribution between 5–80 µm. Its melting point and crystallization onset are 125 °C and 97 °C respectively (DSC, 10 °C/min). Two different coupling agents were used, both based on polypropylene grafted with maleic anhydride. The first, available as Scona TPPP 9212 in the form of coarse powder, is especially designed for adhesion of the matrix with wollastonite fibers, the other, available as Polygroup MPP 2040 in the form of fine powder, is specifically intended to promote adhesion with glass fibers. As reinforcing additives, fibers of different materials and lengths were used, their properties are summarized in table 1 below.

TABLE 1. Selected properties of fibers used in this study

Trade name	Designation	Material	Tensile Modulus [GPa]	Length [µm]	Diameter [µm]
Fibertec 800U	W20	Wollastonite	303-530 [11]	20	3
Lanxess MF7982	G150	E-Glass	72.4 [1]	150	14
Lanxess MF7904	G50	E-Glass	72.4 [1]	50	14

Dry blend preparation. Dry blends of polymer powder, fibers, and coupling agent were prepared in a rotating tumbler mixer, in the presence of 15 stainless steel agitator balls (Ø = 7 mm) to promote mixing of the different density components. In a first step, fibers were blended together with the coupling agent for 2 hours, then the polymer powder was added, and further mixed for at least 6 hours, after which the agitator balls were sieved out. In each case, approximately 1 kg of powder blend was produced. The composition of the produced blends is listed in table 2.

SLS Processing. Selective laser sintering was performed on a DTM Sinterstation 2000, equipped with custom powder feed inserts to reduce the amount of required powder. Each build consisted of three separate stacks of Charpy impact and tensile bars (type ISO 527-1A), oriented both in the direction of application (X) and perpendicular to this direction (Y), as indicated in figure 1. Each stack was produced with a slightly different energy density to obtain an overview with regard to its influence on mechanical properties. For each blend, feed and part bed temperatures were adjusted and optimized in order to minimize processing issues such as curling or part bed fracture. All parts were sintered with a layer thickness of 100 μm, a scan spacing of 0.23 mm, and a scan speed of 5080 mm/s. Selected processing parameters for each of the blends are given in Table 2.

TABLE 2. Produced blends and selected process parameters for the produced test parts

Blend Designation	Fiber	Amount [wt. %]	Amount [vol. %]	Coupling Agent	Amount [wt. %]	T_{Feed} [°C]	$T_{Part\ Bed}$ [°C]	Laser Power [W]
iCoPP	None	-	-	-	-	50	122	15
10W20	W20	10	3	TPPP 9212	1	50	122	10-15-20
15W20	W20	15	5	TPPP 9212	1	50	122	10-15-20
10G150	G150	10	4	MPP 2040	1	59	115	12-14-16
10G50	G50	10	4	MPP 2040	1	61	116	12-14-16
20G50	G50	20	8	MPP 2040	2	64	115	12-16-20
30G50	G50	30	13	MPP 2040	3	64	115	16-20-24

Characterization. Tensile tests were performed on a Zwick Z100 universal testing machine with a load cell of 5 kN. Crosshead speed was 1 mm/min for the first 0.25% elongation, and was then increased to 20 mm/min. SEM images were obtained on a TESCAN VEGA3 SEM with an acceleration voltage of 5 kV. Samples were coated with a thin layer of carbon before analysis.

RESULTS AND DISCUSSION

SLS Processing

Parts could be built with the 10W20 and 15W20 blends with reasonable accuracy. Y-direction test parts were not built given the focus on glass fibers. The flowability of the 10G150 blend proved very sensitive to temperature, which explains the choice of a relatively low part bed temperature during the processing of this blend. A slightly higher part bed temperature of 120 °C in this case led to clumping of the powder, and ultimately a complete break-up of the powder bed. Another build with the 20G50 blend could be completed successfully with a part bed temperature as low as 109 °C. The mechanical properties of the manufactured parts however suffered considerably, especially for the parts sintered with lower laser power. A clear nucleating effect of the fibers could not be detected by DSC, which is positive for SLS, since additional nucleation narrows the processing window.

Mechanical Properties and Fracture Surface

The tensile moduli for virgin iCoPP and the composites are reported in Table 3. As a reference, the theoretical moduli calculated according to Eqs. (1) to (5) are listed as well. The predicted moduli are much higher than the ones measured. The reasons lie in the aforementioned assumptions in the Halpin-Tsai equations. For a start, in practice the fibers are not unidirectional, as can clearly be seen in the fracture surface in Figure 5. The theoretical longitudinal modulus E_{11} can therefore not be achieved. However, the observed tensile moduli are smaller than the modulus for a composite with randomly distributed fibers E_{rand}, and also even smaller than the transversal modulus for a unidirectional composite E_{22}. Poor adhesion of the fibers to the matrix despite addition of a coupling agent is an explanation for the witnessed behavior. This can also be seen from the fracture surface, where fiber pull-out has clearly taken place, but without any plastic deformation of the polypropylene matrix. The biggest improvement of modulus is achieved by adding 30 wt. % 50 μm glass fibers (30G50). However, this is also one of the composites with the poorest ultimate tensile strength. Again, the poor adhesion between fibers and matrix can be identified as main responsible. As can be seen in Figure 4, with increasing amount of fibers (going from 10G50 via 20G50 to 30G50), the ultimate tensile strength is reduced. This is a clear indication that the adhesive strength of fibers and matrix is much lower than the matrix tensile strength.

The presented results are the outcome of an exploratory study to investigate the effects of the addition of fibers to a polypropylene laser sintering material. Possibilities to improve therefore are ample. Increasing energy density or layer time to provide more time for flow and adhesion to take place may improve tensile properties for example.

TABLE 3. Theoretical and measured tensile moduli of produced parts sintered at 15 and 16W laser power

Blend	E_{11} [MPa]	E_{22} [MPa]	E_{Rand} [MPa]	E_{tx} [MPa]	E_{ty} [MPa]
iCoPP	-	-	-	703.90 ± 28.62	-
10W20	1335	1056	1161	709.7 ± 7.59	-
15W20	1532	1089	1255	519.8 ± 24.92	-
10G150	1503	1064	1229	782.09 ± 92.79	700.21 ± 22.40
10G0	1214	1064	1120	776.48 ± 9.79	636.52 ± 15.00
20G50	1480	1143	1269	893.40 ± 66.59	793.20 ± 10.51
30G50	1819	1245	1460	1117.59 ± 3.77	885.68 ± 8.51

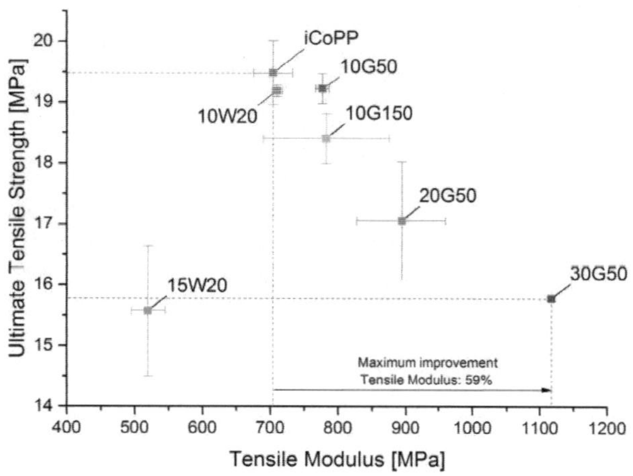

FIGURE 4. Ultimate tensile strength and tensile modulus of iCoPP and reinforced parts

FIGURE 5. Fracture surface of a 30G50 tensile bar built in X-direction (along application direction)

CONCLUSIONS

Various short reinforcing fibers were added in different amounts to a polypropylene laser sintering material in order to improve the tensile modulus. The tensile modulus could be improved by 59% by the addition of 30 wt. % 50 µm glass fibers, at the cost of ultimate tensile strength of the composite. The reason behind this behavior was the poor adhesion between fibers and matrix, as was revealed by investigation of the fracture surface of a tensile bar.

ACKNOWLEDGEMENTS

Caroline Durif of the Laboratory for High Performance Ceramics at Empa Dübendorf is gratefully acknowledged for the SEM image acquisition.

REFERENCES

1. Mallick, P. K. in *Fiber-reinforced composites: materials, manufacturing, and design* 1–30 (CRC press, 2007).
2. Schmid, M. in *Selektives Lasersintern (SLS) mit Kunststoffen* 131–166 (Hanser Verlag, 2015).
3. Wohlers, T. *Wohlers Report 2014.* (2014).
4. Schmid, M., Amado, F. & Levy, G. iCoPP - A new polyolefin for additive manufacturing (SLS). in *International Conference on Additive Manufacturing AM* (2011).
5. Mark, J. E. *Physical properties of polymers handbook.* (2007).
6. Jansson, A. & Pejryd, L. Characterisation of carbon fibre-reinforced polyamide manufactured by selective laser sintering. *Addit. Manuf.* **9,** 7–13 (2016).
7. Mousa, A. a., Pham, D. T. & Shwe, S. P. Pre-processing studies for selective laser sintering of glass beads-filled polyamide 12 composites. *Int. J. Rapid Manuf.* **4,** 28 (2014).
8. Negi, S., Dhiman, S. & Sharma, R. K. Determining the effect of sintering conditions on mechanical properties of laser sintered glass filled polyamide parts using RSM. *Measurement* **68,** 205–218 (2015).
9. Chung, H. & Das, S. Processing and properties of glass bead particulate-filled functionally graded Nylon-11 composites produced by selective laser sintering. *Mater. Sci. Eng. A* **437,** 226–234 (2006).
10. Mallick, P. K. in *Fiber-reinforced composites: materials, manufacturing, and design* 44–45 (CRC Press, 2010).
11. Nyco Minerals. Physical Properties of Wollastonite. 1–3 (2016). at <http://www.sandb.com/wp-content/uploads/Physical-Properties-Overview.pdf>

Highly filled self-assembled nanocomposites containing Layered Double Hydroxides (LDH) nanoparticles

Nelson M. Larocca and Luiz A. Pessan*

Federal University of São Carlos – Materials Engineering Department – São Carlos - Brazil
**Corresponding author: pessan@ufscar.br*

Abstract. Multilayered nanocomposites of polyelectrolytes and layered double hydroxides (LDH) were synthesized by a modified layer by layer (lbl) technique, called spin-spray layer by layer. This new technique allowed to produce films about 8 times faster and with much larger material usage efficiency than conventional dipping coating process. By using a hydrophobic substrate, we were also able to produce free standing nanocomposite films of about 15 μm. These films presented a high inorganic content (about 26 wt%) and yet a large transparency and flexibility. Electronic microscopy shows a structure with high spatial organization and orientation of layered inorganic nanoparticles and low concentration of LDH aggregates.

Keywords: Nanocomposites, Layered Double Hydroxides, layer by layer.
PACS: 81.07.Pr

INTRODUCTION

The layer-by-layer (LbL) assembly technique is a simple, inexpensive, and versatile process for the assembly of highly ordered multilayer film, allowing a precise control of the thickness of each layer at the nanoscale [1,2]. The lbl process can be used to assembly hybrid polymer-nanoparticles structures, like nanosheets and nanowires, thus producing nanocomposites which combine the properties of each type of material and have many potential applications, like photonic materials, sensors and membranes for ion transport [3]. However, despite these very attractive properties of lbl process, each cycle in the traditional lbl dipping method takes about 5 to 20 minutes to be produced, since it depends on times of diffusion and adsorption of polyelectrolytes molecules and colloidal particles in a substrate. Therefore, macroscopic nanocomposites films with micron-size thickness can take days to be prepared. In order to speed up the lbl assembly recently it was introduced the spin-spray lbl technique. In this method, the molecular layers are formed by spraying the polyelectrolytes or nanoparticles in a highly rotating substrate of a spin coater, in such a way that the adsorption of these specimens on the substrate is almost instantaneous [4,5]. This new lbl technique has been utilized in the present work to produce multilayered nanocomposite films of polyelectrolytes and Layered Double Hydroxides (LDH), a synthetic anionic clay. It is showed that the spin spray lbl technique allows the assembly of nanocomposites where the LDH nanoparticles are highly ordered, in a time assembly of about 10 times faster than the traditional dipping lbl process.

EXPERIMENTAL

Materials:

The following reagents were utilized for the LDH synthesis :$Al(NO_3)_3$ (Sigma-Aldrich); $Mg(NO_3)_2$ (J.T.Baker); $Na_2(CO_3)$ (J.T.Baker); NaOH (Mallinckrodt) and sodium acetate (Synth). The polyelectrolytes Poly(ethylene imine) (PEI) and Poly(acrylic acid) (PAA) were purchase from Sigma-Aldrich.

Methods :

The synthesis of the Layered Double Hydroxides of $Mg_3Al\text{-}CO_3$ (LDH-CO_3) was performed by the pH constant method, according to Yun and Pinnavaia [6] with subsequent hydrothermal treatment at 155 °C for 72 hours, in order to increase the size and crystallinity of the LDH- CO_3. Following this treatment, the LDH- CO_3 was anion exchanged first with Cl^- and after with acetate ânions, resulting in a LDH-acetate (LDH-ac) compounding, which is swellable in water and led to a stable aquous suspension [7].

The nanocomposites films were assembled by the spin-spray lbl technique, utilizing a home-built system described in Figure 1. Molecular layers were formed by spraying solutions of polyelectrolytes PEI, PAA solution and suspension of layered double hydroxides (LDH) nanoparticles. The concentrations of the polyelectrolytes solution and LDH-ac suspensions were 0.1% and 0.05%, respectively. Each tetralayer was produced with the following spraying sequence : PEI/PAA/LDH/PAA. This sequence was deposited in a hydrophobic substrate (polypropylene) on the spin coater, what allowed to detached the film from the substrate and achieve a free standing film after the spraying of 500 tetralayers.

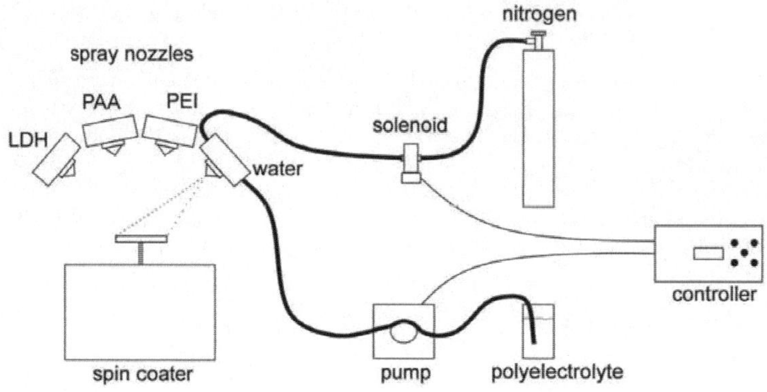

FIGURE 1. Schematics of the "spin-spray lbl" apparatus built by the author.

Characterization :

Atomic Force Microscopy (AFM) images were acquired with a Nanosurf easyscan 2Flex AFM instrument, in non-contact mode and transmission electron microscopy images were acquired with a CEM 120 (Philips) instrument, at 120 KV. Thermo-gravimetric analysis (TGA) were performed in a TGA Q50 (TA Instruments), from 25 °C to 1000 °C at 20 °C/min.

RESULTS AND DISCUSSIONS

Figure 2 show the AFM images of a hydrophobic substrate of polystyrene after the deposition of one and four tetralayers of PEI/PAA/LDH-ac/PAA. It can be seen that the substrate is not immediately covered with the first tetralayer. Instead, first are formed islands that grow until are coalesce and cover all the substrate. The more hydrophobic the substrate, the more difficult is to wet it and the more slow to cover all the surface [8].

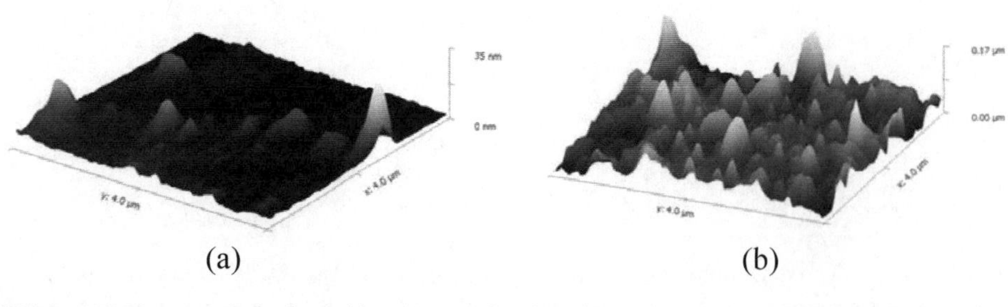

(a) (b)

FIGURE 2. AFM images of a hydrophobic substrate after deposition of tetralayers of PEI/PAA/LDH-ac/PAA : (a) 1 tetralayer; (b) 4 tetralayers.

Figure 3 shows the increase of mass of the lbl film with the tetralayers number. It can be seen a slow initial growth followed by a increase of growth rate that becomes nearly constant. From Figure 2 it can be inferred that the mass per tetralayer is about 0.4 mg. By take into account that the total mass of PEI, PAA and LDH-ac sprayed during a tetracycle is about 3 mg, then the material usage efficiency can be estimated in about 12 %. Even though this is still low efficiency, it is far more efficient than that found in conventional dipping lbl, i.e., about 1% [4].

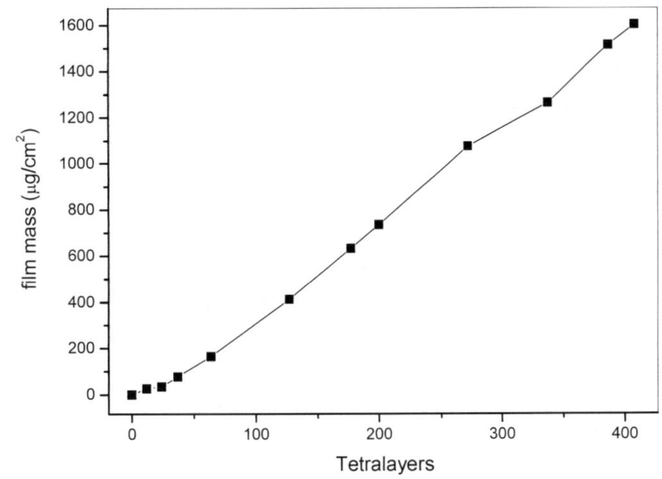

FIGURE 3. Growth of film mass of the PEI/PAA/LDH-ac/PAA system during the spin-spray lbl process.

Figure 4 shows the photo of a free standing film of PEI/PAA/LDH-ac/PAA nanocomposite assembled by spin-spray lbl, with 500 tetralayers and 15 μm thickness. This film has been achieved by detachment from the hydrophobic polypropylene substrate and shows high transparency. It took about 18 hours to assembly this film, while it is estimated that it would take about 160 hours to produce a film with 500 tetralayers by the dipping lbl process, i.e., the spin-spray lbl is at least about 8 times faster than the dipping counterpart. Thermo gravimetric analysis of this film showed that the LDH concentration of this film is 26 wt.%, or about 15 vol.%. It should be highlighted that even with this high inorganic concentration the film presented a high transparency, what indicates the absence of aggregates in the structure of the film.

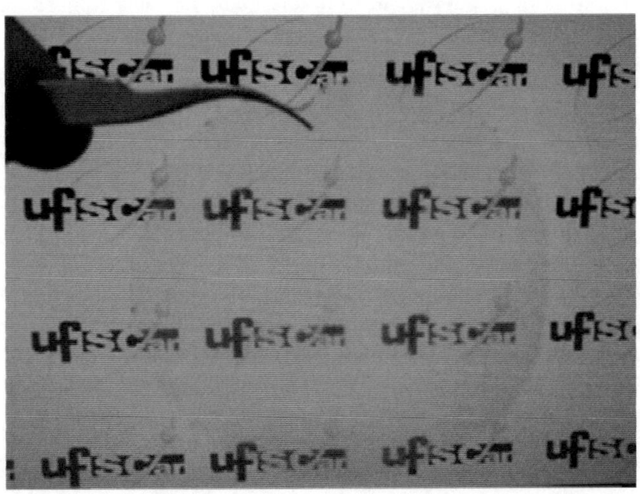

FIGURE 4. Free standing PEI/PAA/LDH-ac/PAA film with 500 tetralayers

Figure 5 shows the transmission electron microscopy image of the PEI/PAA/LDH-ac/PAA nanocomposite free standing film. It can be seen the presence of delaminated platelets, with about 2 nm thickness and length about 200 nm, which are aligned parallel to film surface. There are some aggregates or tactoids, but these are small and in a low concentration. This figure suggests an average regular separation distance among the platelets, what indicates a highly structured platelet disposition at nanoscale. TEM images of low magnification (not shown) unveil that this orientation and regular disposition is maintained throughout the film thickness.

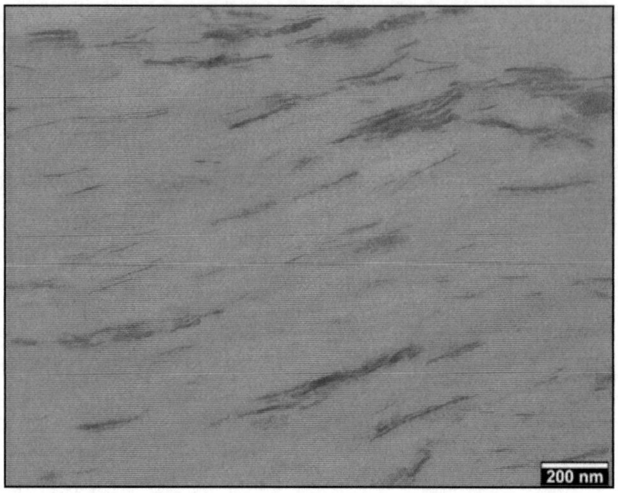

FIGURE 5. Transmission Electron Microscopy image of the PEI/PAA/LDH-ac/PAA film

CONCLUSIONS

It has been demonstrated that the spin-spray lbl process can successfully produce nanocomposites comprising multiples tetralayers of PEI/PAA/LDH-ac/PAA in a velocity about 8 times faster than the conventional dipping lbl and with material usage efficiency about 12 times larger. The nanocomposites synthesized by spin-spray lbl showed high spatial organization of the LDH platelets at nanoscale, what resulted in a macroscopic film which combine high transparency, flexibility and high inorganic content.

ACKNOWLEDGMENTS

The authors acknowledge support from FAPESP through project 09/50946-2 and CAPES.

REFERENCES

1. G. Decher, J.D. Hong, *Macromol. Chem. Macromol. Symp.* **46**,321(1991).
2. G. Decher *Science*, **277**,1232(1997).
3. S. Srivastava, N. Kotov. *Acc. Chem. Res*., **41**,1831(2008).
4. M.H. Merrill, C.T. Sun. *Nanotechnology*, **20**,075606(2009).
5. M. Salomäki, T. Peltonen, J. Kankare. *Thin Solid Films*,**520**,5550(2012).
6. S.K. Yun, T.J. Pinnavaia, *Chem. Mater*.,**7**,348(1995).
7. N. Iyi, Y. Ebina, T. Sasaki, *Langmuir* ,**24**,5591(2008).
8. D. Zimnitsky, V.V. Shevchenko, V.V. Tsukruk. *Langmuir*,**24**,5996(2008).

Process Monitoring by Thermography during Fused Deposition Modeling

Gerhard Traxler* and Thomas Köpplmayr

Profactor GmbH, Im Stadtgut A2, 4407 Steyr-Gleink, Austria

Gerhard.Traxler@profactor.at

Abstract. Infrared (IR) thermography can provide direct process metrology during Fused Deposition Modeling (FDM). In our study we investigate (i) the temperature conditions along the extruder, (ii) the melt temperature of the extruded material at the nozzle exit and (iii) the temperature change of the fabricated part during printing on a heated print bed. A mid-wavelength infrared (MWIR) camera was used at a distance of about half a meter from the printed object. The emissivity of the investigated parts was high enough to capture reliable data and reflected radiation from the print head could be deducted. During temperature measurement of the extruded material at the nozzle exit, microscope optics was used in order to adjust the spatial frequency of the sensor. We investigated the performance of different thermoplastic filaments during printing. Heat conduction to the print bed and temporal cooling of the part were analyzed in order to evaluate the potential of thermography for inline monitoring of the 3D printing process. Due to the low conductivity of the thermoplastic polymers heat conduction from the heated print bed to the part was low and rapid cooling by convection was observed, which indicates the importance of temperature control inside the printer to reduce thermal stresses.

Keywords: Fused Deposition Modeling (FDM), additive manufacturing (AM), thermography, inline inspection

PACS: 87.63.Hg, 81.05.Lg, 81.20.Hy, 83.50.Uv

INTRODUCTION

FDM is a 3D printing technique in which a thermoplastic filament is unwound from a coil and extruded through a heated nozzle to produce a part in a layerwise manner. In case of a poor quality of the joint layers, mechanical consistency of the final product will be also poor in consequence. Regarding process monitoring for documentation or possible inline process control, only very little work has been published up to now. Thermography of parts while they are being printed allows the detection of surface temperature and calculation of temperature gradients within the part, which can lead to thermal stresses and geometric distortions. Dinwiddie et al. [1] used an extended-range IR camera equipped with a lowpass cutoff filter to conduct temperature measurements of samples directly through a double pane soda lime glass window of a Fortus 900mc production printer. They also used a small lightweight uncooled longwave IR camera mounted on the liquefier head to monitor the thermoplastic as it was extruded.

Process monitoring by thermography is a common method for documenting process conditions and process results in laser-based manufacturing of metallic parts. Raghavan et al. [2] used a validated heat transfer and fluid flow laser welding model to examine how changes in processing parameters affect the relationships between top surface and subsurface temperatures and solidification parameters in Ti-6Al-4V. Krauss et al. [3] investigated process errors originating from insufficient heat dissipation as well as the limits for detecting pores and other irregularities by observation of the temperature distribution. Moylan et al. [4] discussed how imaged temperatures can be converted to true temperatures. In addition, they developed an experimental setup in which the existing viewport of the machine (window with IR-blocking glass) is replaced with a metal structure that protrudes into the build chamber a bit, allowing the thermal camera to be positioned closer to the working plane. In-situ process monitoring by thermography is also possible for electron-beam melting technology. Dinwiddie et al. [5] examined the focus size of the electron beam during preheating and melting, detecting molten and unmolten regions of the build, and searching for areas of unwanted porosity (defects). Similar studies employing IR cameras to measure part surface temperatures during the build process have been published by Price et al. [6], Gong et al. [7] or Rodriguez et al. [8].

In our study, we focused on process monitoring in FDM of plastic components using an open build chamber without the need of changing the design of the printer. We investigated both the melt temperature of the extruded material at the nozzle exit and the temperature change of the fabricated part during printing on a heated print bed.

MATERIALS AND EQUIPMENT

Fabrication of plastic parts out of black PLA and white ABS (3DMarkt.at, Klagenfurt, Austria) was performed on a HAGE3Dp-A2 printer (HAGE Sondermaschinenbau GmbH & CoKG, Obdach, Austria). For the output test a black PP filament (German RepRap GmbH, Feldkirchen, Germany) was used in addition. The printer nozzle is made out of brass with a 30 mm long melt channel (diameter 2 mm) and a 1 mm long hot-end (diameter 0.5 mm). The reflectivity of the nozzle was reduced by spray coating graphite before starting the measurement. During the output test a gcode command (G1 E50 F*XX*, where *XX* corresponds to the conveying speed in mm/min) was used to extrude 50 mm of filament through the preheated nozzle. The actual weight of the extruded filament was determined using an analytical balance (Sartorius AG, Göttingen, Germany). 50x50x5 mm^3 objects with 20% infill (45° grid) were printed with a layer height of 0.3 mm and 0.2 mm for the first and next layers, respectively. A thermal camera CEDIP-Silver450, 320x256 pixel, InSp-Chip, MWIR (3.6 µm…5.1 µm), CamLink, recording frequency 100 Hz, NEDT 25 mK, measurement range 95…240°C, exposure time 100 µs with optics f = 27 mm, NA = 3 and a microscope extension M 2:1 (60 µm/pixel) was used for temperature measurements. Calibration of the camera was performed using a black body in thermal equilibrium combined with interpolation following Planck's law for the defined temperature range. The emissivity was set to 100% for all measurements.

RESULTS AND DISCUSSION

Prior to monitoring the temperature of the printed material, we investigated the extruder nozzle itself. Figure Figure 1a shows the print head consisting of a brass nozzle, a heating element and a thermal barrier out of PEEK. At 230°C nominal extruder temperature, the measured temperature showed some oscillations with an amplitude of 5°C and a period of 30-35 s (Figure 1b). A phase shift between different positions along the nozzle could be observed. However, the temperature did not significantly decrease along the brass nozzle. At the lower position of the thermal barrier a temperature of 188°C was determined at a nominal extruder temperature of 200°C (Figure 1c).

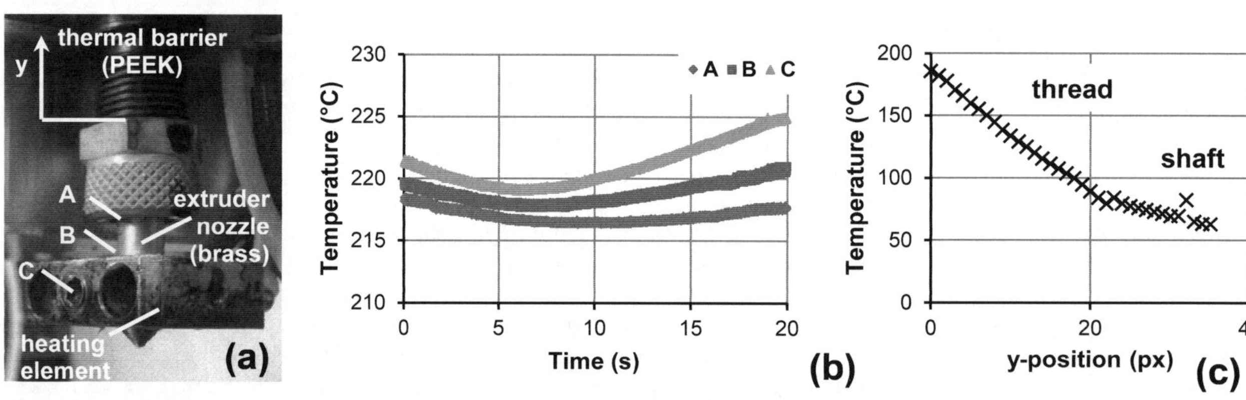

FIGURE 1. Extruder nozzle with heating element and thermal barrier (a), temperature at 3 pos. (A-C) over time with nominal extruder temperature of 230°C (b), temperature along the thermal barrier with nominal extruder temperature of 200°C (c).

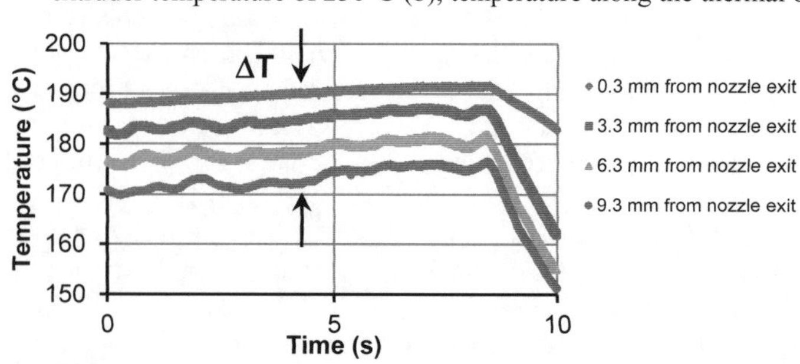

FIGURE 2. Temperature over time at four positions along an extruded PLA filament at 190°C with 2 mm/s. After 8 s the extruder was stopped.

TABLE (1). Temperature differences between 9 mm extruded filament at different conveying speeds for PLA, PP and ABS.

ΔT \ v	1 mm/s	2 mm/s	3 mm/s	4 mm/s
ABS	25-30°C	20°C	15°C	10°C
PLA	24°C	15-20°C	10-15°C	9°C
PP	35°C	30°C	20-25°C	15-20°C

Next, the temperature of the extruded filament (PLA with 2 mm/s at 190°C) was measured at four positions (one directly after the nozzle exit and three after 3 mm steps away from the nozzle, Figure 2). An evaluation of different materials and conveying speeds reveals that cooling of extruded PP filament occurs faster compared to PLA and ABS (Table 1). In order to avoid crack formation due to thermal differences, a closed build chamber might be useful when printing parts out of PP. Moreover, warpage is a critical issue due to the semicrystalline nature of the polymer, which might be reduced if the cooling was performed more uniformly.

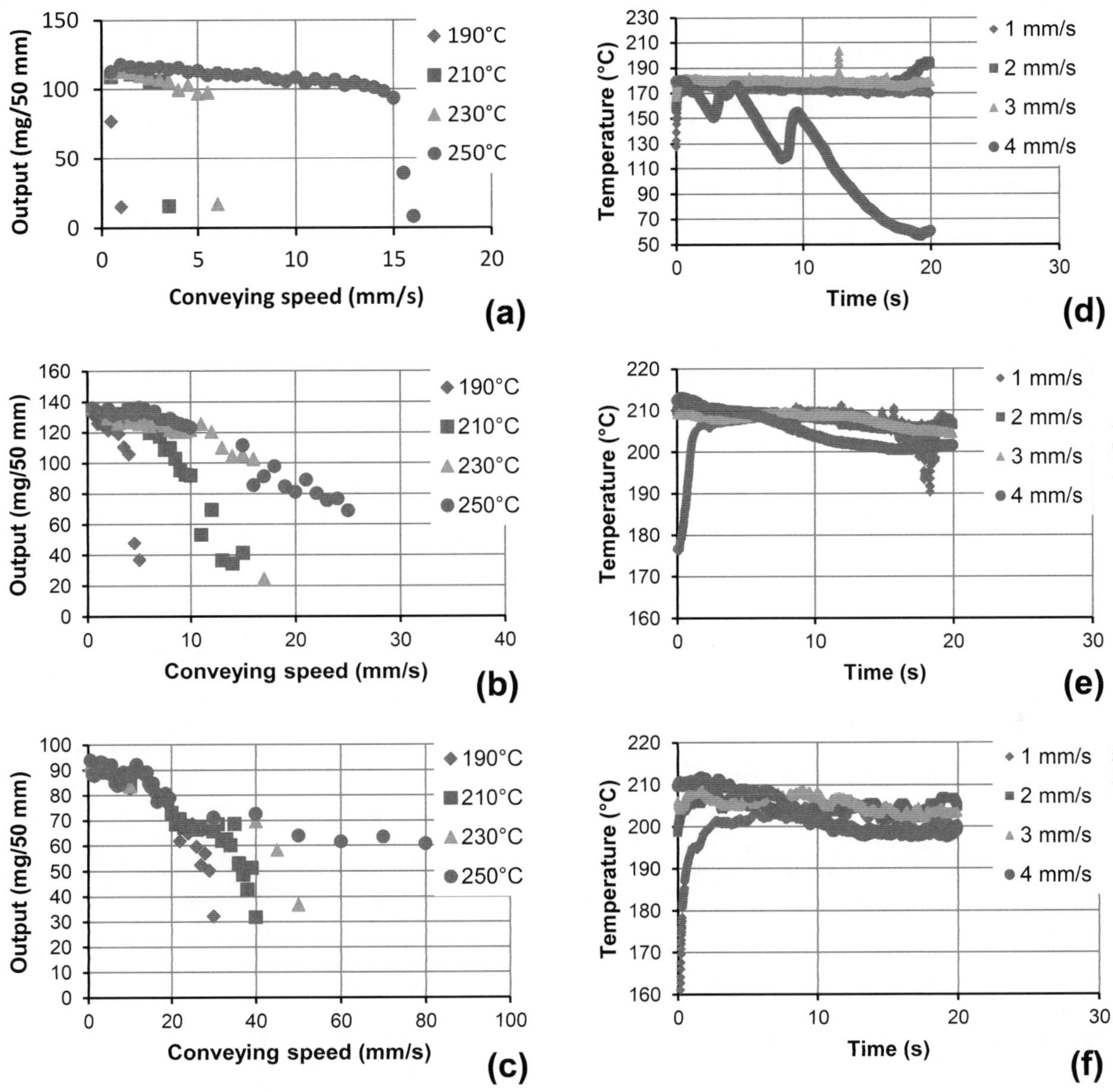

FIGURE 3. Output tests using ABS, PLA and PP (a-c, from top to bottom) by extruding 50 mm filament at defined nozzle temperature and conveying speed followed by weighting the sample; filament temperature measurements by thermography at nozzle exit for ABS, PLA and PP (d-f, from top to bottom) at 210°C nozzle temperature. The values for ABS deviate from the real temperature due to the lower emissivity of the material, which was set to 1 for all experiments.

In FDM three types of filament failure in the print head are known: (i) buckling of soft filaments, (ii) slip at the drive wheel due to high pressure generation in the nozzle and (iii) insufficient melting at higher conveying speeds. In order to investigate the performance of the printer, output test were performed by extruding 50 mm filament at defined nozzle temperatures and conveying speeds followed by weighting the sample. Buckling was observed for none of the investigated materials. In the case of ABS, temperature measurements at the nozzle exit reveal that the melt temperature is independent of the conveying speed as long as no failure occurs and slip at the drive wheel is the major type of failure (Figure 3). Due to the amorphous nature of the polymer, it softens without additional latent heat of fusion. In the case of PLA and PP, a temperature decrease of about 10°C after 11 s at 4 mm/s was observed. Failure occurs at higher velocities of 6 mm/s or 30 mm/s, respectively and is at least partly attributed to insufficient melting followed by high pressure generation due to the increasing viscosity of the material.

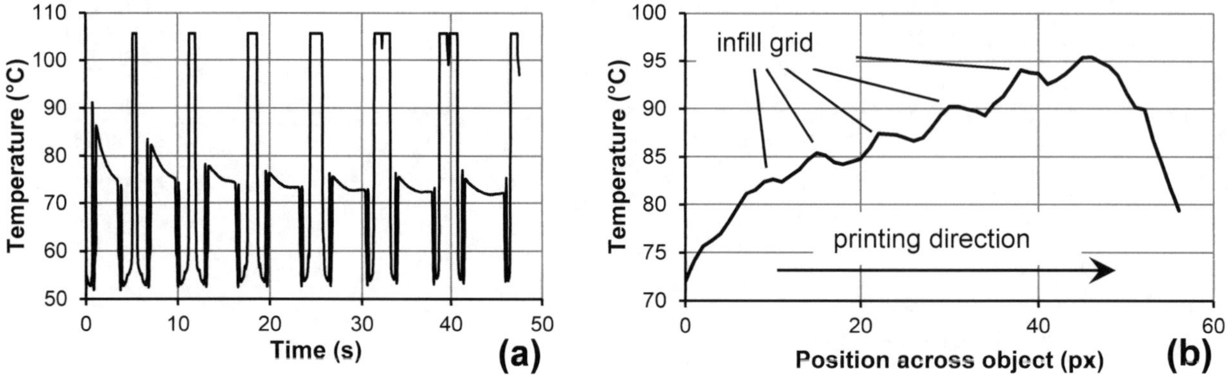

FIGURE 4. (a) Temperature evolution at fixed position in the first layer of a PLA printed part; (b) temperature distribution across the printed object in the top layer of the PLA part (first layer temperature 220°C, then 210°C, bed temperature 55°C).

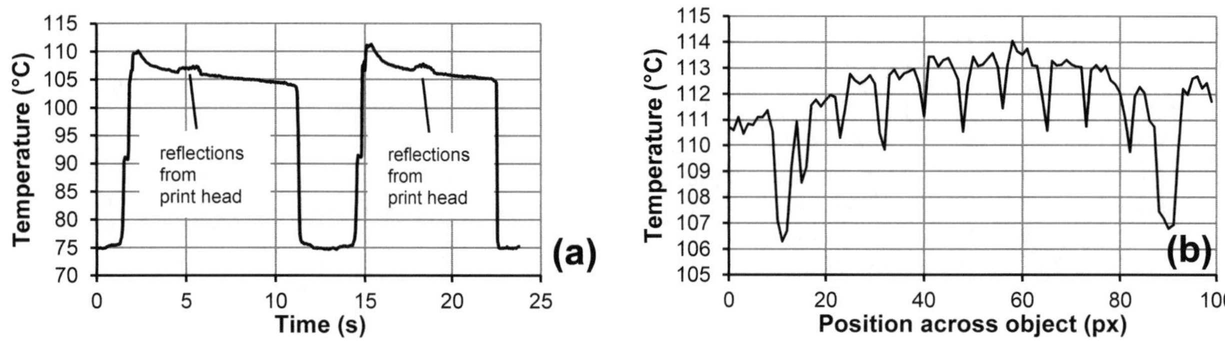

FIGURE 5. (a) Temperature evolution at fixed position in the first layer of a ABS printed part; (b) temperature distribution across the printed object in the top layer of the ABS part (first layer temperature 240°C, then 230°C, bed temperature 110°C).

Finally, we investigated the printing process itself by thermography. A measurement position was fixed in the first layer of a PLA print and the temperature was monitored during printing. The saturated temperature peaks in Figure 4a are caused the print head when it crosses the field of view. Printing is performed in 45° direction away from the measurement position by arranging lines one against the other. The sections between the peaks show the following characteristics: (i) The pulse duration decreases due to the increasing distance to the print head, (ii) the average amplitude decreases and (iii) the cooling occurs slower. After printing the 20% infill rectangular grid, the top layer was printed and the temperature was evaluated along a line across the printed object (Figure 4b). The lowest temperature could be measured at the edges of the object, which can cool faster due to the surrounding air at room temperature. The highest temperature was measured at the position, which was last crossed by the printhead. In between smaller peaks were observed, which are caused by the rectangular grid underneath. At these positions cooling was delayed due to the low thermal conductivity of the material.

ABS shows a similar behavior as illustrated in Figure 5. The outer frame of the object was printed and a fixed position was monitored over time. Due to the lower emissivity of ABS and the higher bed temperature, which is necessary to assure adhesion to the print bed, the reflections of the printhead have a stronger influence on the result compared to PLA. However, both the successive cooling of the first layer and the temperature differences in the top layer could be determined. Other materials such as PP, nylon or polycarbonate do not show sufficient adhesion to the print bed (glass) even at elevated print bed temperatures. Especially in the case of PP, sticky tapes such as packaging tapes can be used to enhance the printability due to partial welding of the printed material and the foil. Temperature monitoring has not been performed yet, but is expected to show similar results as mentioned above.

CONCLUSION AND OUTLOOK

The temperature conditions along the extruder, the melt temperature of the extruded material at the nozzle exit and the temperature change of the fabricated part during printing on a heated print bed could be measured by a thermal camera. Although heating of the 30 mm long extruder nozzle was homogeneous, the melt temperature decreased at conveying speeds of 4 mm/s for PLA and PP, whereas ABS caused slip at the drive wheel due to high pressure generation in the nozzle even at sufficiently high melt temperatures. During thermal monitoring of the printed parts, reflections by the printhead were observed, but still made the temperature measurements possible. Cooling occurred faster at the edges of the printed parts and at positions on the top layer with voids underneath. This makes thermal imaging an interesting method for quality control during printing. Therefore, a small lightweight longwave bolometer camera will be mounted on the print head and focused onto a close-up view of the heated extrusion tip (Figure 6). This view is expected to allow high resolution imaging and temperature measurements of the material as it is extruded and begins to cool in contact with the layer below.

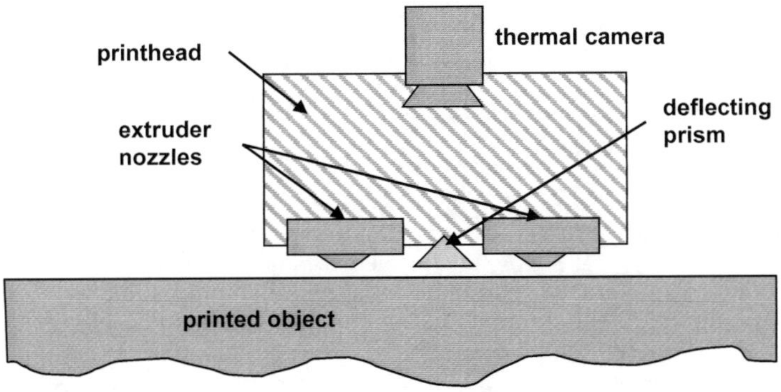

FIGURE 6. Setup for integrating a thermal camera into a FDM printer by using a deflecting prism to monitor both printheads simutaneously.

ACKNOWLEDGMENTS

Financial support by the Austrian Research Promotion Agency (FFG) within a national research project ("Next Generation 3D-Printing: Material- und Prozessentwicklung für die industrietaugliche Anwendung", project number 848624) is kindly appreciated. The authors would like to thank HAGE Sondermaschinenbau GmbH & CoKG for providing the FDM printer in the research project.

REFERENCES

1. R. B. Dinwiddie, L. J. Love and J. C. Rowe, "Real-time Process Monitoring and Temperature Mapping of a 3D Polymer Printing Process" in *Thermosense: Thermal Infrared Applications XXXV*, edited by G. R. Stockton and F. P. Colbert, Proc. of SPIE 8705, Bellingham, WA, 2013, pp. 87050L.
2. A. Raghavan, H. L. Wei, T. A. Palmer and T. DebRoy, *J. Laser Appl.* **25**, 052006 (2013).
3. H. Krauss, C. Eschey and M. F. Zaeh, "Thermography for Monitoring the Selective Laser Melting Process" in *Proceedings of the 23rd Annual International Solid Freeform Fabrication (SFF) Symposium*, The University of Texas, Austin, TX, 2012, pp. 999-1014.
4. S. Moylan, E. Whitenton, B. Lane and J. Slotwinski, "Infrared Thermography for Laser-Based Powder Bed Fusion Additive Manufacturing Processes" in *40th Annual Review of Progress in Quantitative Nondestructive Evaluation*, AIP Conference Proceedings 1581, American Institute of Physics, Melville, NY, 2014, pp. 1091-1196.
5. R. B. Dinwiddie, R. R. Dehoff, P. D. Lloyd, L. E. Lowe and J. B. Ulrich, "Thermographic in-situ process monitoring of the electron-beam melting technology used in additive manufacturing" in *Thermosense: Thermal Infrared Applications XXXV*, edited by G. R. Stockton and F. P. Colbert, Proc. of SPIE 8705, Bellingham, WA, 2013, pp. 87050K.
6. S. Price, K. Cooper and K. Chou, "Evaluations of Temperature Measurements by Near-Infrared Thermography in Powder-Based Electron-Beam Additive Manufacturing" in *Proceedings of the 23rd Annual International Solid Freeform Fabrication (SFF) Symposium*, The University of Texas, Austin, TX, 2012, pp. 761-773.
7. X. Gong, B. Cheng, S. Price and K. Chou, in *2013 ASME Early Career Technical Conference*, ASME Early Career Technical Journal, Birmingham, AL, 2013, pp. 59-66.
8. E. Rodriguez, J. Mireles, C. A. Terrazas, D. Espalin, M. A. Perez and R. B. Wicker, *Addit. Manuf.* **5**, 31-39 (2015).

Layer Strength of FDM Type 3D Printed Article for Nozzle Temperature and Filament Material

Min-Young Lyu [a,*], J. H. Park[b], S. J. Park[b], and K. H. Lee[a]

[a] *Department of Mechanical System Design Engineering, Seoul National University of Science and Technology*
[b] *Graduate School, Seoul National University of Science and Technology*

mylyu@seoultech.ac.kr
mylyu@seoultech.ac.kr

Abstract. 3D printing is getting attention in manufacturing industry recently. However there are still many supplement points for the application to industrial parts. There are several types of 3D printing method according to additive way, which are FDM, SLA, DLP, SLS, Polyjet, and so on. FDM type uses filament. Continuous filament with circular cross-section is fed into a barrel and is melted when it passes a nozzle which is heated by heating block. Melted filament is extruded through nozzle and it makes layer. Additional melted filament put on the previously made layer and this process makes three dimensional object. Layer strength in the 3D printed object is weaker than bulk material. Layer strength is dependent of tool path, nozzle temperature, nozzle moving speed, and so on. In this paper, layer strength has been examined according to nozzle temperature for PLA, PA6, and ABS filaments. Layer strengths for orthogonal direction to deposition are weaker than deposition direction by 27.9~50.3% according to filament material. Layer strengths for the deposition direction in 3D printed article have 50.2~95% of strength of bulk material. Layer strength for the orthogonal direction to deposition strongly depends on nozzle temperature and increases as nozzle temperature increases.

Keywords: Additive Manufacturing, 3D Printing, FDM (Fused Deposition Material), Layer Strength, Nozzle Temperature
PACS: 06

INTRODUCTION

Additive manufacturing (AM) is novel manufacturing process and is developed recently. AM is also called by 3D printing, rapid prototyping, rapid tooling, and etc. There are several type of 3D printing those are fused deposition modeling (FDM), selective laser sintering (SLS), stereo lithography apparatus (SLA), plastic sheet lamination (PSL), and so on[1-5]. FDM type 3D printing is widely used among them since it is easy to operation and various polymer materials can be used. The weak point of 3D printed product using FDM is low layer strength or deposition strength that is much lower than bulk strength of material [6-9]. Deposition strength is dependent of deposition direction and deposition temperature, and etc [10-12]. In this study deposition strengths of several materials have been investigated for various deposition temperatures.

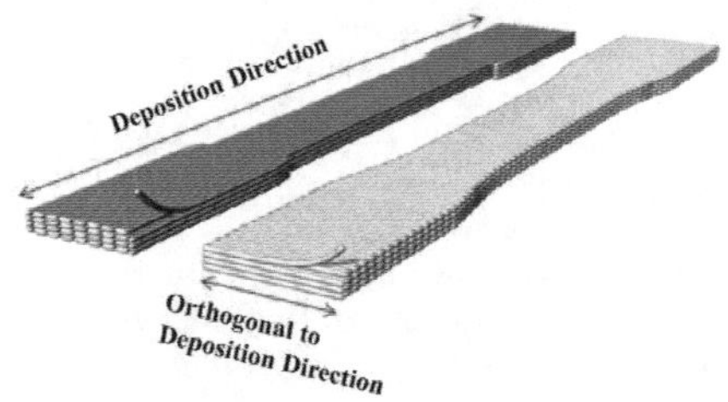

FIGURE 1. Tensile Test Specimen for Deposition Direction and Orthogonal to Deposition Direction in FDM Type 3D Printing.

EXPERIMENTAL

The filament used in this study for 3D printing with FDM type is summarized in Table 1.

TABLE (1). Filament Used in FDM Type 3D Printing.

	PLA	ABS	PA6
Test Method	ASTM D882	ASTM D638	ASTM D638
Tensile Stress(MPa)	103.4	46.1	73.5
Elongation at Break(%)	180	25	200

The operation condition of FDM type 3D printing is summarized in Table 2. The nozzle temperatures of FDM type 3D printer were in Table 2 for PLA, ABS, and PA6.

TABLE (2). Operation Condition of FDM Type 3D Printer

Item	Value
Layer Height	0.2mm
Nozzle Diameter	0.4mm
Nozzle Speed	80mm/s
Bed Temperature	90℃

TABLE (3). Nozzle Temperature of FDM Type 3D Printer

PLA	ABS	PA6
200℃	210℃	245℃
220℃	230℃	255℃
240℃	250℃	265℃

FDM type 3D printer used in this experiment was Edison Multi2 (ROKIT, S. Korea). Tensile tester for deposition strength was EZ20 (LLOYD INSTRUMENT).

RESULTS AND DISCUSSIONS

Test specimens after tensile test were shown in Figure 2 for PLA. Those specimens were manufactured by deposition direction as shown in left side of Figure 1.

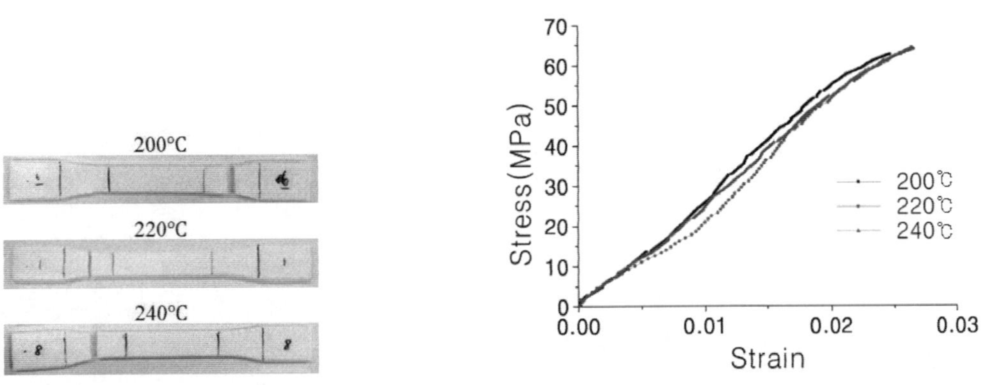

FIGURE 2. Tensile Specimens after Tensile Test and Tensile Behavior for the Specimen of Deposition Direction for PLA.

Higher deposition temperature shows higher tensile behavior. However the differences are small. Tensile Test specimens manufactured by orthogonal to deposition direction for PLA is shown in Figure 3. It also shows tensile behaviors. As deposition temperature increases tensile strength increases.

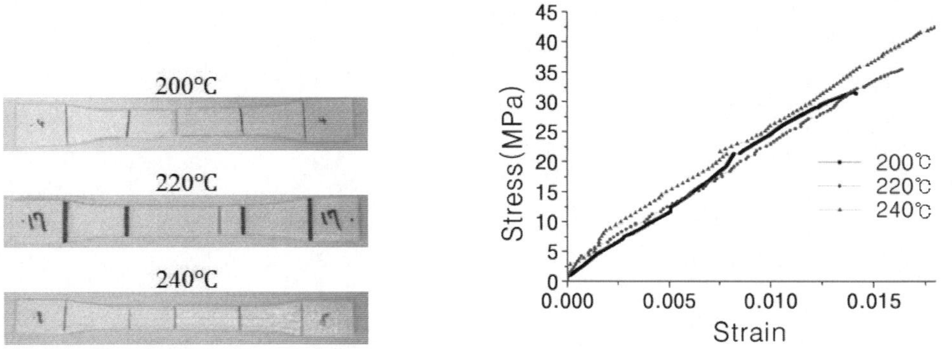

FIGURE 3. Tensile Specimens after Tensile Test and Tensile Behavior for the Specimen of Orthogonal to Deposition Direction for PLA.

Figure 4 shows tensile specimens for ASTM and rectangular shape. It also shows tensile behavior. ASTM specimen shows higher tensile property.

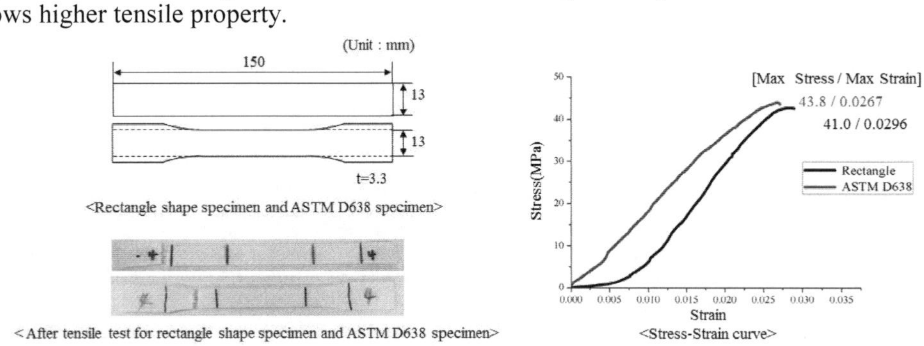

FIGURE 4. Comparisons of Tensile Test Specimens and Tensile Behavior for ABS.

For PLA, the tensile strength in deposition direction and orthogonal to deposition direction were about 63 MPa and 37 MPa respectively. The strength orthogonal to deposition direction has about 57% of deposition direction. For ABS, the tensile strength in deposition direction and orthogonal to deposition direction were about 43 MPa and 18 MPa respectively. The strength orthogonal to deposition direction has about 42% of deposition direction. For PA6, the tensile strength in deposition direction and orthogonal to deposition direction were about 33 MPa and 15 MPa respectively. The strength orthogonal to deposition direction has 46% of deposition direction.

CONCLUSIONS

Tensile behaviors of 3D printed specimens were investigated. Tensile strength of orthogonal to deposition direction showed less than half for the tensile strength of deposition direction. Tensile strength of orthogonal to deposition direction was highly dependent of deposition temperature.

ACKNOWLEDGMENTS

This work was supported by the Industrial Fundamental Technology Development Program funded by the Ministry of Trade, Industry and Energy (MOTIE) of Korea. (10051680, Development of high strength and environmental friendly polymer for 3D printing)

REFERENCES

1. D. G. Ahn, S. H. Lee, K. D. Kim and D. Y. Yang, *Transactions of Materials Processing,* **11**, 4 (2002).
2. G. D. Kim, J. Y. Kim, *Transactions of Korea Society of Machine Tool Engineers,* **15**, 4 (2006).
3. C. Park, M. H Kim, S. M. Hong, J. S. Go, B. S. Shin, *Journal of the Korean Society of Manufacturing Technology Engineers,* **24**, 3, (2015).
4. Raymond N. Chuk Vincent and J. Thomson, *Rapid Prototyping Journal*, **4**, 4, 185-196 (1998).
5. Kumar. S and Kruth J. P, *Material and Design,* **31**, 2, 850-856 (2010).
6. HS Ramanath, M Chandrasekaran, CK Chua, KF Leong and KD Shah, *Key Engineering Materials,* **334-335**, 1241-1244 (2007).
7. Weiyin Ma, Wing-Chung But and Peiren He, *Computer-Aied Design,* **36**, 1309-1325 (2004).
8. Michael Dawoud, Iman Taha and Samy J. Ebeid, *Journal of Manufacturing Processes,* **21**, 39-45 (2016).
9. Sung-Hoon Ahn, Michael Montero, Dan Odell, Shad Roundy and Paul K. Wright, *Rapid Prototyping Journal,* **8**, 4, 248-257 (2002).
10. C. W. Ziemian, D. E. Cipoletti, S. N. Ziemian, M. N. Okwara, K. V. Haile, *Annual International Solid Freeform Fabrication Symposium 2014,* 525-541 (2014).
11. O. S. Es-said, J. Foyos, R. Boorani, M. Mendelson and R. Marloth, *Materials and Manufacturing Process,* **5**, 1, 107-122 (2000).
12. Q. Sun, G. M. Rizvi, C. T. Bellehumeur and P. Gu, *Rapid Prototyping Journal,* **14**, 2, 72-80 (2008).

Fabrication and Characterization of Polymeric Origami Structure for Compression Properties

Mohamed Ali E. Kshad, Hani E. Naguib*

Department of Mechanical and Industrial Engineering
Department of Materials Science and Engineering
Institute of Biomaterials and Biomedical Engineering
University of Toronto, 5 King's College Road, Toronto, ON, Canada, M5S 3G8

naguib@mie.utoronto.ca

Abstract. Sandwich structures, which typically consist of two thin flat plates separated by a core, are commonly used in applications where the light weight considered as one of the influential design criteria. The core is usually designed in a certain way, allowing the structure to withstand the applied loads and to dissipate energy. The main function of the core in the sandwich structure is to redistribute the applied force by deformation and to dissipate the impact energy. The objective properties of sandwich cores are high stiffness to weight ratio, high bending stiffness, and high energy absorption ratio. The conventional designs of sandwich cores, such as foamed cores, honey-comb, and ribbed cores, have been used for many decades. Origami inspired sandwich cores have attracted attention during the last years, due to their high capacity for elastic deformation, force redistribution, bending and twisting resistance. In this paper, we present the results of a quasi-static compression test, and the energy absorbed by Muira origami folded cores, made by direct molding of polymeric blends. The material compositions used were polylactic acid (PLA) and thermoplastic polyurethane (TPU), and the weight percentage of these compositions were (100/0, 80/20, 65/35, 50/50, 20/80, and 0/100) (PLA/TPU). It has been found that cores with high PLA content have high stiffness values, while the cores with high TPU content have the high energy absorptions efficiency.

Keywords: Origami, metamaterials, polylactic acid (PLA), mechanical properties, energy absorption.
PACS: 81.05.Zx

INTRODUCTION

Different types of core structures designed and used to produce sandwich structures for lightweight applications to reduce density and to absorb applied energy. Using sandwich cores increases the moment of inertia of the panel, which allows the structure to resist bending loads and absorb the impact energy at low weight [1],[2]. The flat sheet of material can have enhanced mechanical properties by having local textures [3],[4]. Folding flat sheets to produce 3D core structure is another way to increase the moment of inertia and then enhance the mechanical properties of the structure [4]. The idea of origami comes from the Japanese art of paper folding, in which the two-dimensional flat paper folded to form a three-dimensional structure. The art of origami paper folding has stimulated engineers and researchers to develop structures to serve a vast range of applications, from aerospace systems, small scale devices, deployable equipment, self-assembling systems, medical devices, and in applications where, the lightweight is a critical design parameter [5],[6],[7]. Furthermore, it is applicable across various scales (Nano, Micro, and Macro) [5],[8],[9],[10]. Nowadays, origami cores are become more efficient sandwich interiors for such applications, compared to the classical sandwich cores such as honey-comb, foam, ribbed, corrugated, and other cores [6], [11]. There are limitless of 3D geometries that can be produced by the folding process, based on the design of the folding patterns. One of the common patterns of origami structures is Miura Origami. Miura origami core is a periodical unit cell tessellation; the unit cell consists of four parallelograms connected through straight lines called creases, the crease lines generating peak and valleys vertices [18],[19]. Presently, many research provided mathematical and analytical models for the origami metamaterials, including computer simulations and finite element analysis (FEA) [12],[13],[14]. Paimushin, et al. [15], introduced theoretical and experimental technique to determine the mechanical characteristics of folded cores in the form of Z-Crimp. Zhou et al. [16], provided a numerical study of Miura-origami folded cores under quasi-static loads using finite element method and it has been concluded that the folded core models with curved lines showed the best performance in compression and shear. J.M. Gattas [17], studied the Quasi-static impact of indented fold cores, using numerical simulation and experimental validation. An experimental study was presented by Dvoeglazov et al. [6], in which studied the geometrical parameters of Z-crimp origami pattern on the strength under transverse compression and longitudinal shear at the same volumetric density made of different materials. The current paper aims to present an experimental investigation, of Miura origami cores made of polymeric blends under compression loading. In the first step, the origami core was fabricated by direct molding, and then the compression test was conducted on the origami cores to find the compression behavior, and consequently the fracture modes of the origami cores. The material blends employed were PLA and TPU and the weight percentages of the compositions were (100/0, 80/20, 65/35, 50/50, 20/80, and 0/100) (PLA/TPU). The results demonstrated that cores with high percentage of PLA content have the higher stiffness values, while the cores with high TPU content have the higher energy absorptions efficiency.

EXPERIMENTAL WORK

Materials Processing: In this study, two polymers were used to produce the origami samples: Poly-Lactic Acid (PLA) (3052D), provided by Natureworks LLC (USA), and Thermoplastic polyurethane (TPU) (55D), provided by Lubrizol Engineering Polymers (USA). The twin screw micro-compounder (MICRO15, DSM; Geleen, Netherlands) was used to fabricate the material blends, in which the polymers melted at a temperature of 195 C (above the melting temperature of the polymers), and were then mixed at screw speed of 30 rpm for 10 minutes.

Sample Preparation: The samples were fabricated using a compression mold technique, for which a special mold containing the origami geometrical features used. The operating temperature was set above the melting temperature of the parent material, and then the mold was kept at a pressure of 3 tons for 30 minutes. Figure 1 shows the geometry details of Miura origami cores and the fabricated samples. The geometrical parameters of the origami core are listed in table 1.

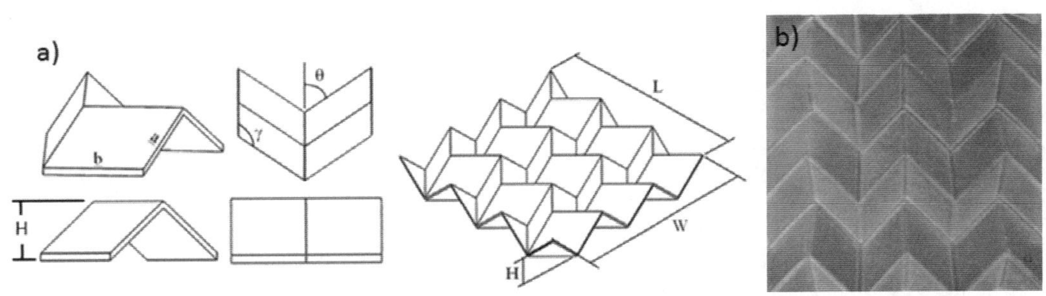

FIGURE 1. a) Origami unit cell and panel geometry b) fabricated origami cores (PLA sample)

Table (1). Miura Origami Geometrical Parameters

a (mm)	b (mm)	θ (°)	γ (°)	H (mm)	t (mm)	L (mm)	W (mm)
24	24	45	45	20	1.5	120	120

Quasi-static Compression Test: The compression test was conducted to determine the strength of origami cores in order to measures the capacity of origami cores to withstand compressive loads. The origami core samples were tested on a compression testing General Machine. Two thick steel plates were assembled with the testing machine, allowing the load to be uniformly distributed on the origami panel structure. The test was consistent with ASTM D1621/94 standard with 5 mm/min stroke; three specimens were tested from each material blend. Each sample has 9 periodical cells (see figure 1) with cell size of 40*40 mm.

Results & Discussion: The compression test curves of origami cores made of PLA/TPU blends and neat PLA and TPU show that the compression resistance stiffness is decreasing by the increase of the soft material (TPU) (figure 2). The highest compression stress is for neat PLA origami cores, while the lowest is for neat TPU core. It is observed that cores with high percentage of PLA are prone to fracture at the crease lines (figure 3). However, neat TPU core proceeds to high compressive strains without failure, but at the expense of very low compressive stiffness. These results make evident the possibility to tailor compression stiffness by controlling the amount of materials in the blends. Compression stiffness helps in tailoring other properties, like impact strength and relaxation modulus.

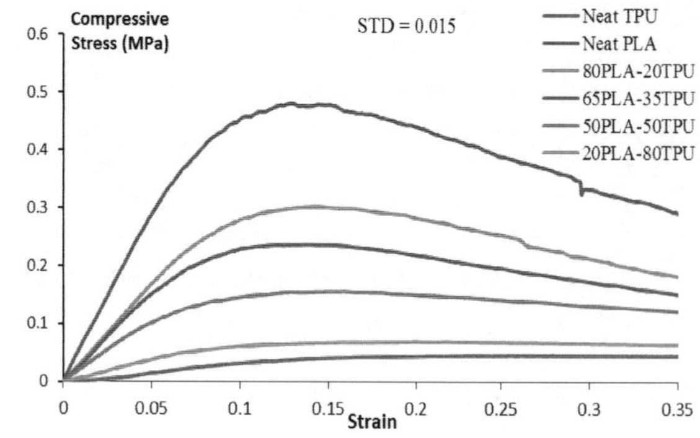

FIGURE 2. Compression test of origami core structures.

FIGURE 3. Fractured origami structures made of high percentage of PLA

The quasi-static compression is characterized by the energy-absorption diagram, which is obtained from the absorbed energy (W) as a function of the stress (equation 1), the area under the stress-strain curve [20].

$$W = \int_0^{\varepsilon} \sigma(c)dc \tag{1}$$

Other indicators used to characterize the energy absorbed of lightweight structures are the efficiency parameter (E_f) which is defined as the ratio of the absorbed energy up to a stress value divided by stress (equation (2)), and the ideality parameter (I), which is the ratio of the absorbed energy to the product of stress and strain (equation 3) [20].

$$E_f = \frac{\int_0^{e} \sigma(e)de}{\sigma} \tag{2}$$

$$I = \frac{\int_0^{e} \sigma(e)de}{\sigma e} \tag{3}$$

Figures (4-6) show the comparison of the above calculated indicators in which 80/20,65/35,and 50/50 PLA/TPU cores showed the highest energy absorption efficiency, specific energy absorption, and energy absorption ideality on the expense of reduced elastic modulus. The cores of neat PLA, and 80/20 PLA/TPU experienced fracture along crease lines; however, neat TPU, 20/80, and 50/50 PLA/TPU cores maintained behaved response with 100% recovery factor (figure 7). The 65/35 PLA/TPU core showed plastic deformation at crease lines, which prevented the full recovery after compression. The results suggest that there is a balance between compressive strength and energy absorption where both of them inversely affects the other. Origami cores with high percentage of PLA exhibit higher compressive strength at the expense of lower energy absorption and are prone to fracture at high compressive strains. On the other side, higher percentage of TPU reduces the compressive strength but offers higher energy absorption without damage even at higher compressive strains.

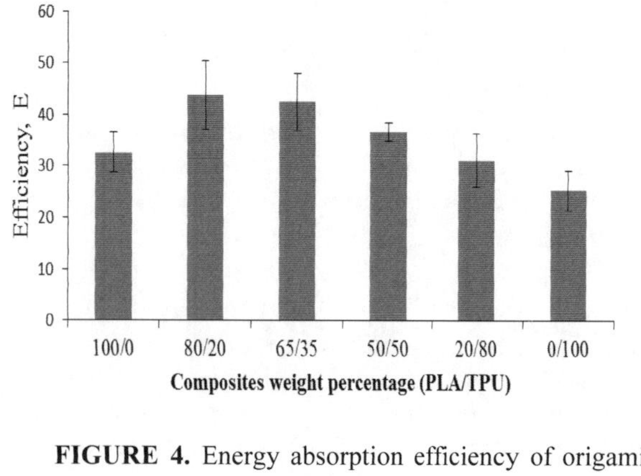

FIGURE 4. Energy absorption efficiency of origami cores

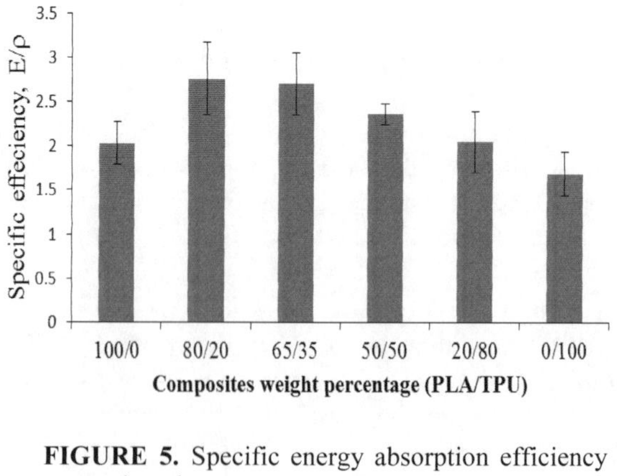

FIGURE 5. Specific energy absorption efficiency of origami cores

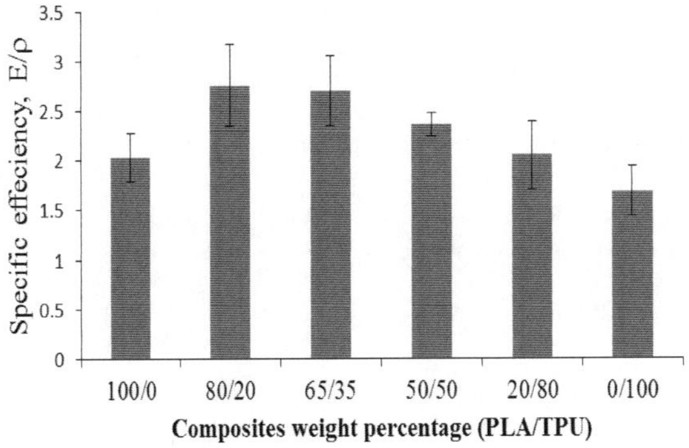

FIGURE 6. Energy absorption ideality of origami cores

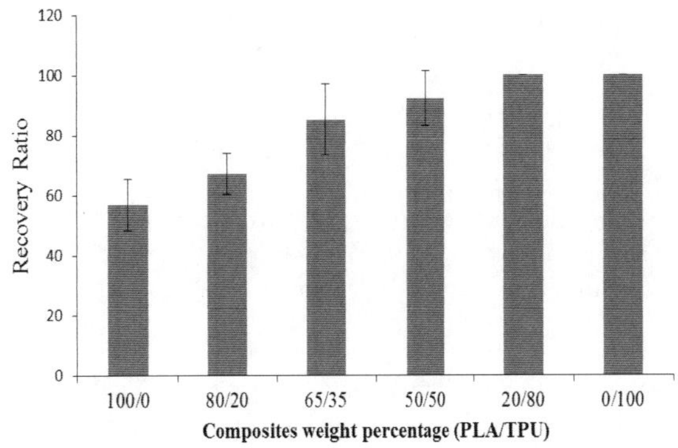

FIGURE 7. Recovery ratio of origami core

CONCLUSION

Several engineering applications can benefit from the potential advantages of origami cores, which are lightweight and offer high stiffness, and high energy absorption efficiency. In this work, Miura origami cores were fabricated from PLA and TPU blends using direct compression molding. The weight percentage of the material blends used were (100/0, 80/20, 65/35, 50/50, 20/80, and 0/100) (PLA/TPU). These different percentages of PLA and TPU reflect an increase in compressive strength or energy transfer, respectively. The optimal balance between PLA and TPU will depend on the desirable application. PLA origami cores offer the highest compressive modulus of 4.6 MPa. The high compressive modulus of PLA resulted in fracture at crease lines. The high modulus of PLA origami cores can be reduced to 2.79 MPa by adding 35% of TPU, which offers more elasticity and avoids failure at high compressive strains. This work provides numerical quantification of the compromise between compressive strength and energy absorption, which can be used to design origami structures with target values for both.

ACKNOWLDGEMENTS

The authors would like to acknowledge the following agencies for financial support: The Natural Science and Engineering Research Council (NSERC) of Canada, the Canada Research Chair Program, the Libyan Ministry of Higher Education and scientific research, Tripoli, and The Canadian Bureau for International Education (CBIE). And also the authors would like to extend their acknowledgements to Daniel Grozea, from Material Department for his technical support.

REFERENCES

1. A. Petras, "Design of Sandwich Structures," Proc. Est. Acad. Sci., pp. 4–8, 1998.
2. Ashby M. F., "Materials selection in mechanical design," 2011.
3. M. Schenk, "Folded Shell Structures," PhD. thesis, Clare College University of Cambridge, 2011.
4. M. Schenk and S. D. Guest, "Origami Folding: A Structural Engineering Approach," Origami 5 Fifth Int. Meet. Origami Sci. Math. Educ., pp. 1–16, 2011.
5. E. A. Peraza-Hernandez, D. J. Hartl, R. J. Malak Jr, and D. C. Lagoudas, "Origami-inspired active structures: a synthesis and review," Smart Mater. Struct., vol. 23, no. 9, p. 094001, 2014.
6. I. V Dvoeglazov and V. I. Khaliulin, "A Study of Z-Crimp Structural Parameters Impact on Strength under Transverse Compression and Longitudinal Shear," Russ. Aeronaut., vol. 56, no. 1, pp. 15–21, 2013.
7. S. S. Tolman, I. L. Delimont, L. L. Howell, and D. T. Fullwood, "Material selection for elastic energy absorption in origami-inspired compliant corrugations," Smart Mater. Struct., vol. 23, no. 9, p. 094010, 2014.
8. C. Cromvik and K. Eriksson, "Airbag Folding based on Origami Mathematics," Technology, 2006.
9. P. W. K. Rothemund, "Folding DNA to create nanoscale shapes and patterns.," Nature, vol. 440, no. 7082, pp. 297–302, 2006.
10. K. Miura, "Map fold a la Miura style, its physical characteristics and application to the space science," Research of Pattern Formation Edited by R Takaki KTK Scientific Publishers Tokyo. pp. 77–90, 1994.
11. S. Heimbs and P. Middendorf, "Aircraft Sandwich Structures with Folded Core Under Impact Load," 8th Int. Conf. Sandw. Struct. ICSS 8, pp. 369–380, 2008.

12. C. Lv, D. Krishnaraju, G. Konjevod, H. Yu, and H. Jiang, "Origami based Mechanical Metamaterials.," Sci. Rep., vol. 4, p. 5979, 2014.
13. Y. Xia and M. I. Friswell, "Equivalent models of corrugated laminates for morphing skins," in Proc. of SPIE Vol. 7977, 2011, vol. 7977, p. 79771I–79771I–10.
14. M. Eidini and G. H. Paulino, "Unraveling metamaterial properties in zigzag-base folded sheets," Sci. Adv., vol. 1, no. 8, pp. e1500224–e1500224, 2015.
15. V. N. Paimushin, I. M. Zakirov, and Y. A. Karpikov, "Theoretical and experimental technique of determining the mechanical characteristics of folded structure filler in the form of Z-crimp (shear of a filler in cross-sectional planes)," Russ. Aeronaut. (Iz VUZ), vol. 56, no. 3, pp. 234–246, 2013.
16. X. Zhou, H. Wang, and Z. You, "Mechanical properties of Miura-based folded cores under quasi-static loads," Thin-Walled Struct., vol. 82, pp. 296–310, 2014.
17. J. M. Gattas and Z. You, "Quasi-static impact of indented foldcores," Int. J. Impact Eng., vol. 73, pp. 15–29, 2014.
18. M. Schenk and S. D. Guest, "Geometry of Miura-folded metamaterials," Proc. Natl. Acad. Sci., vol. 110, no. 9, pp. 3276–3281, 2013.
19. Z. Y. Wei, Z. V. Guo, L. Dudte, H. Y. Liang, and L. Mahadevan, "Geometric mechanics of periodic pleated origami," Phys. Rev. Lett., vol. 110, no. 21, 2013.
20. M. Avalle, G. Belingardi, and R. Montanini, "Characterization of polymeric structural foams under compressive impact loading by means of energy-absorption diagram," Int. J. Impact Eng., vol. 25, no. 5, pp. 455–472, 2001.

Conducting Polymer Hydrogels (CPHs) For Electronic Skin Applications

Nazanin Khalili[a,b,c], Roy H. Kwon[a] and Hani E. Naguib[a,b,c*]

[a]Department of Mechanical and Industrial Engineering
[b]Institute of Biomaterials and Biomedical Engineering
[c]Department of Materials Science and Engineering
Toronto, Ontario, M5S 3G8, Canada, Email:Naguib@mie.utoronto.ca

nkhalili@mie.utoronto.ca
naguib@mie.utoronto.ca

Abstract. Conducting polymer based hydrogels exhibit physical properties analogous to those of human tissues. They are soft materials with high water content that can be made stimuli-responsive through altering their properties. Combining the unique features of hydrogels with the high electrical conductivity of conducting polymers, a mesh-like three dimensional network can be obtained in which the conjugated polymer acts as the single continuous conductive phase. These electrically conductive hydrogels respond to the outside pressure perturbations through a piezoresistive response which makes them promising candidates for resistive-type sensors with ultra-high sensitivity mimicking some major properties of human skin. In order to imitate natural skin, an electronic skin (E-skin) should be able to recognize both medium and low pressure perturbations. The former ranges from 10 to 100 kPa suitable for object manipulation while the latter is lower than 10 kPa for gentle touch. Herein, using a multiphase reaction, a polypyrrole (PPy) based hydrogel is fabricated as a pressure sensor for E-skin applications. Embedding different additives and formulating different microstructures, a highly sensitive pressure sensor with tailored elastic properties is produced. The fabricated PPy based hydrogel can elastically deform and recover upon applying a mechanically induced compression load within the range of human skin detection. The intrinsic elasticity of these hydrogels originates from the hollow micro-spherical structure of the polymerized PPy forming a cross-linked network within the gel.

Keywords: Piezoresistive sensor, PPy hydrogels, high sensitivity, electronic skin

INTRODUCTION

Human skin is the largest sensory organ in our body containing a multitude of receptors through which environmental stimuli are transduced into physiological signals to be interpreted by the brain. In order to mimic the complex and multifaceted properties of this barrier layer, a high sensitivity to low applied pressure along with a wide range of pressure sensing in the high-pressure regime is required. Fast response time in the range of milliseconds is another crucial design criteria for e-skins. Electronic skin (e-skin), also known as smart skin, is composed of a network of sensory elements that respond to outside stimuli. Accordingly, in order to ensure reliable human-robot interaction (HRI), intelligent sensing capabilities for robots in different applications including interactive wearable devices, artificial prosthetics, and minimally invasive surgeries are of critical importance.[1] This has boosted the demand of e-skins in application-oriented HRI research fields.[2] Mimicking major functionalities of human skin within one single electronic device entails integration of some of the major properties of human skin including high pressure sensitivity, flexibility, and conformability to complex shapes.

Conducting polymers (CPs) possess the physical and chemical properties analogous to those of organic polymers along with high electrical conductivity. They have high sensitivity, short response time, easy sensor element processing, and their applications are ensured at room temperature.[3] Polypyrrole (PPy) is one of the most widely utilized CPs due to its promising features that can be exploited in mechanical sensors as well as pressure sensors. However, the rigid conjugated-ring backbone of dense PPy results in a brittle and stiff structure which is not suitable to fulfill the flexibility and stretchability of an E-skin. Therefore, in order to have a PPy based sensor that can elastically deform and recover upon the application of a mechanically induced load, a hollow sphere structure can be employed.[4] In this work, a PPy based hydrogel using two different oxidants has been fabricated with high sensitivity to a wide range of applied pressure. The microspherical structure within the three dimensional network of the gel endows the sensor with high flexibility upon application of an external force.

FABRICATION PROCESS

The PPy hydrogels were fabricated using a bi-component multiphase reaction. Herein, the effect of two different oxidants on the performance of the PPy based hydrogels as piezoresistive sensors was investigated. The first set of samples was fabricated using ammonium persulfate (APS) as the initiator for the polymerization process of PPy. In order to prepare solution A, 0.736 ml of phytic acid (PA) acting as the cross-linker was added to 4.8 mmol of distilled pyrrole monomer and 2 ml of isopropyl alcohol. Solution B contained 4.8 mmol of APS dissolved in 2 ml of deionized water (DI).[4] Solutions were both agitated with a magnetic stirrer to obtain homogeneous solutions. The multiphase reaction was carried out at -4°C using a dry-ice bath to reduce the reaction rate. Solution B was added to solution A and rapidly mixed for 15 seconds. The gelation processwas then allowed to be well developed for 2 hours at room temperature and ambient air. The resulting hydrogel was alternatively immersed in deionized water and ethanol to remove and wash any unreacted monomer. The samples were dried at 70°C for 24 hours to reach their dehydrated state. The second set of samples were fabricated following a similar procedure as that of the PPy/APS where the oxidant was replaced by Iron(III) nitrate, $Fe(NO_3)_3$, which acts as both the oxidant and cross-linker in the multiphase reaction. The phytic acid was therefore removed from solution A for the fabrication process of the PPy/$Fe(NO_3)_3$ hydrogels. These conducting polymer hydrogels (CPHs) can then be used in the structure of a multilayer flexible electronic skin in which the sensors are interconnected through an arrayed structure as schematically demonstrated in Figure 1.

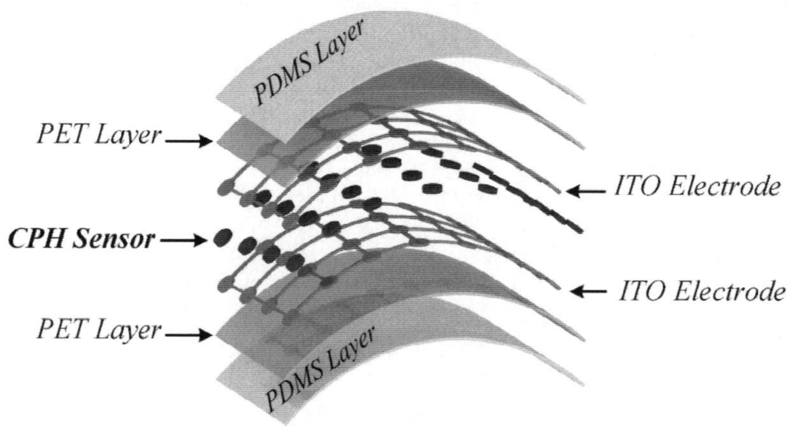

FIGURE 1. The surface profile of the cross section of the dehydrated PPy based hydrogels

RESULTS AND DISCUSSION

In order to gain an insight into the microstructure of the fabricated gel, the roughness of their cross sections in their dehydrated state along with its height profile was obtained using a surface roughness tester (Surftest SJ-210) as shown in Figure 2a and 2b. The roughness of the cross section of three samples for each set of gels was measured over a distance of 4000 μm. The average root mean squared roughness ($R_{q,avg}$) for the PPy/APS hydrogels was obtained as 11.82 μm whereas this value was 6.8 μm for the PPy/$Fe(NO_3)_3$ samples. This implies that the APS oxidant results in larger pores within the three-dimensional network of the gel structure. On the other hand, the Fe^{3+} ions play an important role in the ionic cross-linking process of the polymer while having an oxidizing capacity in the PPy/$Fe(NO_3)_3$ samples. It is worth mentioning that the polymerization process was carried out without any mechanical energy inserted into the system. The continuous and spontaneous occurrence of the gelation process contributes to a three-dimensional and continuous network with a low interfacial energy between the polymer network and the aqueous environment around it.[5] This static condition will also lead to expansion of the polymer chains along with an improvement in the $\pi - \pi$ stacking within the chains of the polymer.[6] The behavior of the as-synthesized PPy gels in terms of their shrinking-swelling properties was investigated through five cycles of DI water uptake and drying. The weight ratio of the gels ($\frac{w}{w_0}$) was measured after each cycle as depicted in Figure 2c and 2d. The initial state of the gels was the equilibrated sate of the as-prepared gels in DI water after a week. The swelling and shrinking states were respectively denoted by "sl_i" and "sr_i" where i ranges from 1 to 5. The swelling of the gels was induced by immersing the samples in DI water for duration of 1 min at room temperature. The swelled hydrogel was then dried in a vacuum oven and its weight was measured. For PPy/$Fe(NO_3)_3$ hydrogels the average swelling ratio was 54% and the shrinking ratio was nearly 13%. These values were 13% and 73% for the PPy/APS hydrogels. This indicates that the PPy based gels can be reversibly shrunk and swelled without collapsing in the shrinking state. Accordingly, these hydrogels can be utilized for chemical release, drug delivery, and artificial muscles applications.[7-9]

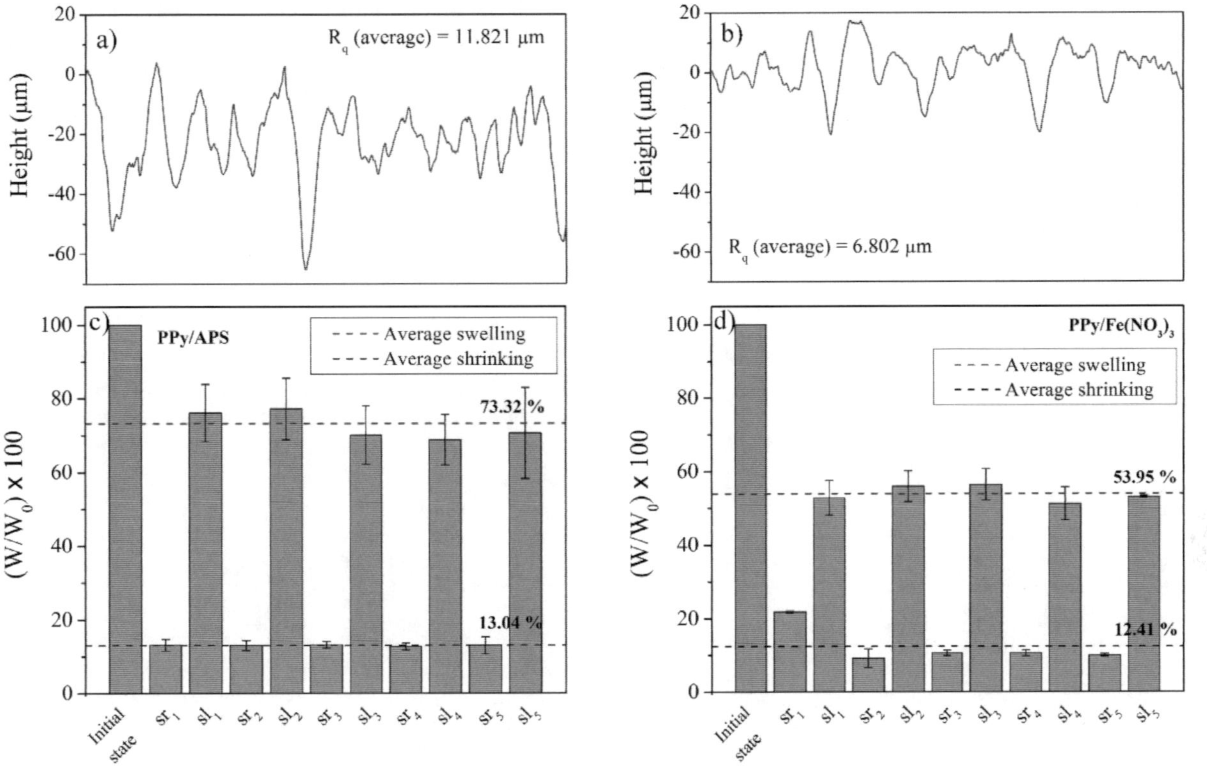

FIGURE 2. Surface profile of the cross section of the dehydrated PPy based hydrogels over a 4000 μm distance (a) PPy/APS, (b) PPy/$Fe(NO_3)$. Swelling-shrinking behavior of (c) PPy/APS, (d) PPy/$Fe(NO_3)$.

The piezoresistive response of the fabricated hydrogels was measured using a sandwiched assembly. The PPy hydrogels were placed between two sheets of copper foils and a compression load was applied in the range of 0.1 to 100 kPa while the change in the resistance of the sensor was being recorded simultaneously, as schematically depicted in Figure 3. The resistance change of the samples was obtained and the relative resistance change $(\frac{\Delta R}{R_0})$ along with the sensor's sensitivity corresponding to each measurement was also measured. The sensitivity of a piezoresistive sensor can be defined as the slope of the curve attained from the applied pressure vs. the resulting relative resistance change.

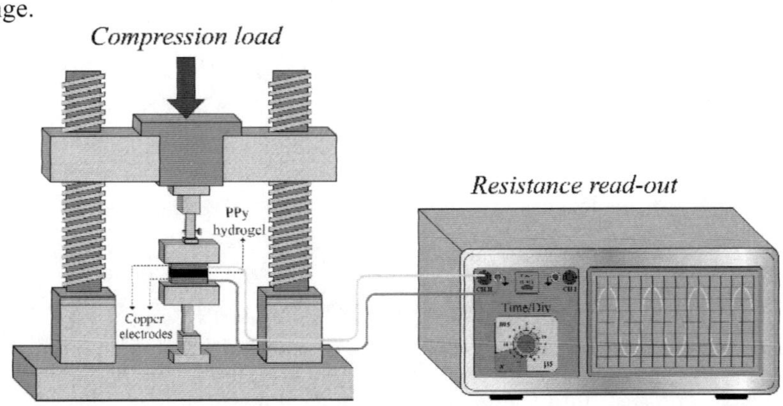

FIGURE 3. Schematic of the piezoresistive response testing of the CPHs.

Considering the fact that the major mechanism for the sensing response of these sensors is the change in the contact resistance between the electrode and the dried hydrogel, increasing the applied compression results in a larger real contact area and therefore, the contact resistance decreases. This increase in the real contact area between the electrode and the PPy based hydrogels stems from larger area and higher number of microcontacts as well as clusters. [10, 11] This results in more electrical current going through the sample and accordingly, the electrical conductivity of the sensors increases (Figure 3a and 3b). It was also observed that both PPy/APS and PPy/$Fe(NO_3)_3$ hydrogels exhibited higher sensitivity to the lower range of the applied load and as the pressure increases, the sensors become less sensitive to the load while they can detect a wider range of applied load (Figure 3c and 3d). This functionality of the PPy hydrogels makes them highly potential candidates for electronic skin applications where pressure sensitivity is a major characteristic to be mimicked. The properties of this stimuli-responsive layer can be fully tailored using different oxidant/initiators, conductive fillers, and fabrication processes. By changing the microstructure of the hydrogel along with its surface morphology, its response to the applied compression load can be controlled via developing various clusters of microcontacts as well as microstructures.

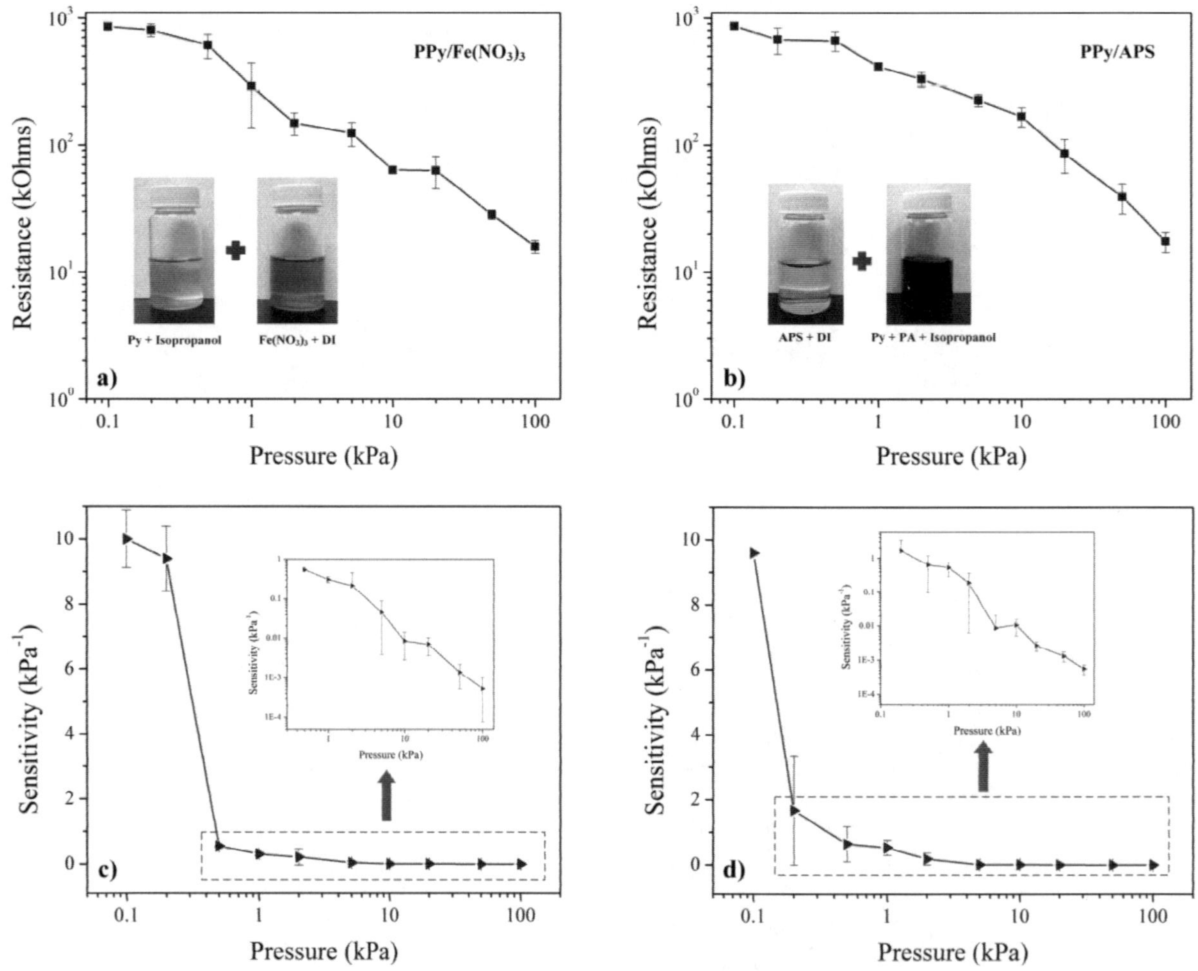

FIGURE 3. The piezoresistive response and sensitivity of (a) and (c) PPy/APS, and (b) and (c) PPy/$Fe(NO_3)_3$ hydrogels

CONCLUSIONS

PPy based hydrogels were fabricated using two different oxidants (APS and $Fe(NO_3)_3$). Using a continuous and spontaneous polymerization process through a bi-component multiphase reaction, a hydrogel with a three-dimensional network was obtained. The low interfacial energy between the phases of the product accounted for the continuous mesh like structure. Moreover, the Fe^{3+} ions acted as both the oxidant and cross linker in the gelation process of PPy/$Fe(NO_3)_3$. The results attained from the roughness measurement of their cross section indicated the PPy/APS gels posses larger pores compared to the other set of fabricated samples. The shrinking-swelling behavior of the gels was evaluated and it was shown that over more than 5 cycles of DI water uptake and drying, both sets of samples did not use their three-dimensional structure and did not collapse in their shrinking stage. Finally, it was demonstrated that as-synthesized PPy hydrogels can be used as piezoresistive sensors with higher sensitivity to applied loads within a lower range and a lower sensitivity to a wider range of higher applied loads. The change in the real contact area between the electrode and the dehydrated gels decreased the contact resistance and hence, increased the electrical conductivity of the samples. This stimuli-responsive characteristic of the PPy hydrogels makes endows them with a high potential in the electronic skin applications.

ACKNOWLEDGMENT

Financial support was provided by Natural Sciences and Engineering Research Council of Canada (NSERCCanada), Canada Research Chair Program (CRC), and the Canadian Foundation for Innovation (CFI).

REFERENCES

1. S. J. Benight, C. Wang, J. B. Tok and Z. Bao, *Progress in Polymer Science*, **38**, 2013, 1961-1977.
2. Y.-J. Yang, M.-Y. Cheng, W.-Y. Chang, L.-C. Tsao, S.-A. Yang, W.-P. Shih, F.-Y. Chang, S.-H. Chang and K.-C. Fan, *Sensors and Actuators A: Physical*, **143**, 2008, 143-153.
3. Wang, Lin-Xia, Xin-Gui Li, and Yu-Liang Yang. "Preparation, properties and applications of polypyrroles." Reactive and Functional Polymers 47.2 (2001): 125-139.
4. L.X. Wang, X.G. Li,. and Y.L. Yang, *Reactive and Functional Polymers*, **47**, 2001, 125-139.
5. L. Pan, A. Chortos, G. Yu, Y. Wang, S. Isaacson, R. Allen, Y. Shi, R. Dauskardt and Z. Bao, *Nature communications*, **5**, 2014.
6. T. Dai, X. Jiang, S. Hua, X. Wang, Y. Lu. *Chemical Communications*, **36**, 2008, 4279-4281
7. K. Lee, S. Cho, S.H. Park, A. J. Heeger, C.W. Lee, S.H. Lee, *Nature* **441**, 2006, 65-68.
8. D.J. Irvin, S.H. Goods, L.L. Whinnery, *Chemistry of materials*, **13**, 2001, 1143-1145.
9. J.D. Ehrick, S.K. Deo, T.W. Browning, L.G. Bachas, M.J. Madou, S. Daunert, *Nature materials*, **4**, 2005, 298-302.
10. N. Khalili, H.E. Naguib, and R.H. Kwon, *Soft Matter*, DOI: 10.1039/C6SM00204H, 2016.
11. N. Khalili, H.E. Naguib, and R.H. Kwon, Proceedings of SPIE, Behavior and Mechanics of Multifunctional Materials and Composites, March 2016, Las Vegas, NV.

Micro-FDM Process Capability and Comparison with Micro-Injection Moulding

Vito Basile[a], Claudia Pagano[b] and Irene Fassi[b]

[a] *Institute of Industrial Technologies and Automation - CNR, Via Lembo 38/F, 70124 Bari, Italy*
[b] *Institute of Industrial Technologies and Automation - CNR, Via Corti 12, 20133 Milan, Italy*

vito.basile@itia.cnr.it
irene.fassi@itia.cnr.it

Abstract. In recent years, fused deposition modelling technology (FDM) has become one of the most important additive manufacturing technology due to its capability to produce functional prototypes with complex shape in a cost effective way. Recently, the trend towards miniaturization invested also this technology, since the request of micro-component is rapidly growing due to the increasing number industrial sectors involved. Mechanical properties are fundamental in some sectors of high-quality micro-parts, so the knowledge of the influence of FDM process parameters on mechanical properties can be useful to extend its application and help the optimization of the parameter selection. In this context, the aim of this study is the analysis of the FDM capability and the room for improvement, through a comparison with a well-consolidated industrial process, such as the micro-injection moulding. Although FDM quality cannot compete with the technologies industrially used for final products, its low cost and short time are very attractive for some applications. Moreover, the comparison can be interesting since FDM is often used to manufacture prototypes eventually made with more performing industrial technologies, so that a measure of the quality and functionality of these prototypes can be extremely useful for product developers.

Keywords: Fused deposition modelling, micro-injection moulding, additive manufacturing, tensile test.
PACS: 81.05.Lg, 83.50.Uv, 81.70.Bt , 81.20.Hy, 89.20.Bb.

1. INTRODUCTION

The term additive manufacturing (AM) refers to several different processes, for a number of different materials, with the common principle of adding layer upon layer to build three-dimensional objects, opposed to subtractive manufacturing methodologies. AM, thus, includes groups of manufacturing technologies, such as 3D printing, rapid prototyping and stereolithography.

Formerly these technologies were used to fabricate preproduction models, but nowadays more and more products are made using these techniques, due to the great developments achieved.

The trend towards miniaturization invested also these technologies since the request of micro-component is rapidly growing due to the increasing applications in several industrial fields such as biomedical, electronics, optics and high-precision micro mechanical industry.

In recent years, fused deposition modelling technology (FDM) has become one of the most important AM technology due to its capability to produce functional prototypes with complex shape for industrial applications in a cost effective way 1. This technology found a wide diffusion since the physics behind the manufacturing process is relatively simple and machine and raw material costs are significantly low.

Mechanical properties are fundamental in some sectors of high-quality micro-parts, so the knowledge of the influence of FDM process parameters on mechanical properties can be useful to extend its application and help the optimization of the parameter selection [2, 3]. FDM quality cannot compete with the technologies industrially used for final products; however, the low cost and short time make it be very attractive for some applications.

The aim of this study, thus, is not a comparison between two manufacturing technologies, but the analysis of the FDM capability and the room for improvement. Nevertheless, the measurement of the mechanical properties of parts manufactured using an industrial process, namely micro-injection moulding (μIM), is presented and compared with the results of FDM specimens. This can be interesting also considering that FDM is often used to manufacture prototypes eventually made with more performing industrial technologies; therefore, the comparison can be a significant measure of the quality and functionality of these prototypes for product developers. Finally, for the sake of completeness a brief comparison of the costs and the time-to-market related to the two technologies has been added.

This study consists of two main sections. In the first section, the focus has been on FDM micro-part production capability and process parameter optimization; while in the second section, a comparison of the mechanical properties of the samples fabricated by μFDM and by μIM is presented.

2. MATERIAL AND METHODS

In order to define a miniaturized geometry for mechanical test specimens, several geometries have been proposed in recent years, down scaling the geometries proposed in the standards ISO 527 [4] and ASTM D638 [5]. Moreover, a few original geometries have been proposed in [6] and [7]. In this study a geometry obtained applying a scale factor of about 0.5 to the most important dimensions of geometry V of standards ASTM D638 is proposed and shown in **FIGURE 1a**.

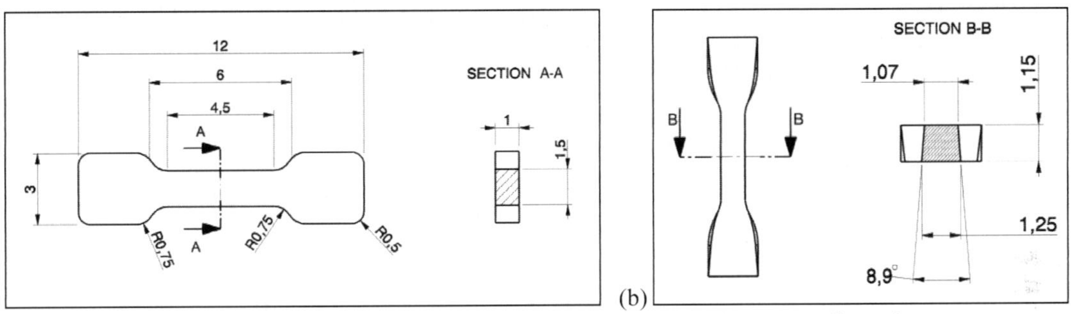

FIGURE 1. (a) Mini dog-bone geometry, (b) revisited model geometry (in mm).

This geometry, earlier proposed by Desma Tec is used in several research studies [7, 8, 9]. The dimensions of the proposed geometry is reduced enough to comply with the definition of micro-component reported by Giboz et al. [10], but, at the same time, big enough for the specimens to be used to perform mechanical tests.

The specimens are fabricated in Polylactic acid (PLA), a thermoplastic aliphatic polyester derived from renewable resources, biodegradable and biocompatible [11], which is becoming more and more attractive in biomedical applications such as orthopaedics, tissue engineering and micro-part fabrication [12, 13, 14].

The FDM samples have been manufactured using a ShareBot NextGeneration machine equipped with a double extrusion head, heated build platform, filament diameter of 1.75 mm, nozzle diameter of 0.35 mm and build volume of 250x200x200 mm. This FDM machine has a theoretical positioning resolution on axes X and Y of 0.06 mm and a positioning resolution of axis Z of 0.0025 mm.

Micro-injection moulding samples have been fabricated by a Desma Tec Formica 1K with a two cavities steel mould. The raw material has been obtained pelletizing the PLA filament. The granules have been dried for 6 hours at 75°C and then injected using a consolidated set of process parameters resulting from previous experiments: injection speed 150 mm/s, melt temperature 210 °C, injection pressure limit 1000 bar, holding pressure 1500 bar, holding time 1 s, cooling time 5 s and mould temperature 70 °C.

The tensile tests have been carried out using a Shimadzu EZ-S equipped with a 500 N load cell and crosshead speed 5 mm/min.

3. FDM MANUFACTURING CAPABILITY

The 3D model of the specimen in **FIGURE 1a** has been modelled using Solidworks SW and then imported to Slic3r FDM software to generate the extrusion head path layer by layer. Each layer consists of the contour and internal fill. Several values of the process parameters have been combined in order to obtain a variety of process conditions and identify the most influent parameters on the geometrical and dimensional properties of the specimens. A few parameters have been set invariable for all the tests, namely the raster to raster gap, the orientation, the raster angle, the temperature of the build platform and the filament cooling, which was always active. No gap has been inserted between the rasters and the orientation and raster angle have been kept at their optimal values for higher mechanical performance i.e. raster follows applied stress. Moreover, a contour approach has been chosen, so that the gauge length is completely filled by the contours. The other parameters, i.e. layer thickness, number of contours per layer, extruder speed, raster width and nozzle temperature have been the focus of this preliminary study.

For each of the 12 conditions ten miniaturized dog-bone specimens have been fabricated using the FDM machine. A sample, randomly selected among the ten, of each of the 12 runs has been cryogenically fractured and the sections obtained have been observed using an optical confocal microscope (Zeiss Discovery V.20) so that the most frequent defects have been identified. In **FIGURE 2** few images of the specimens are presented in order to show the most observed defects. The "stair step" effect is evident in all the pictures and in some conditions, it is even more pronounced. Several empty spaces can be seen inside the sample section and on its surface (especially in **FIGURE 2a** and **d**). The shape of the samples is never rectangular as the model, but trapezoidal with some conditions (**FIGURE 2c** and **d**) worse than others. The surface on the top is not planar and the separate filaments of the material are visible. The thickness of the samples varies extensively for different process conditions, producing very thin samples (i.e. **FIGURE 2b** and **c**) and thicker ones; however, it is always lower than the model thickness.

FIGURE 2. Sections of FDM samples.

The preliminary results have permitted to observe the main defects and identify the most influent process parameters. According to these results, a design of experiment (DOE) has been done with four factors and two levels of each; i.e. nozzle temperature (215-230 °C), layer thickness (0.20-0.33) extruder speed (12-30) and raster width (0.250-0.375). In order to reduce the number of experiment runs, an half factorial design has been carried out. Therefore, eight conditions have been generated and for each of them ten samples have been produced, which have been observed using the optical microscope. The width of the central part of the dog-bones has been measured from the images and the thickness using a calliper with resolution equal to 0.01 mm. Considering the possible trapezoidal shape of the specimens the measures have been done on the bottom side, i.e. where the sample is in contact with the substrate of the FDM machine (indicated as w_b) and on the opposite (top) side (indicated as w_t). An analysis of variance (ANOVA) has been done on the following variables: Δw_t (Δw_b), which is the difference between the measured w_t (w_b) and the nominal width (1.5 mm) and Δt, which is the difference between the measure of the thickness and the corresponding value of the model. The statistical results are presented in **FIGURE 3**.

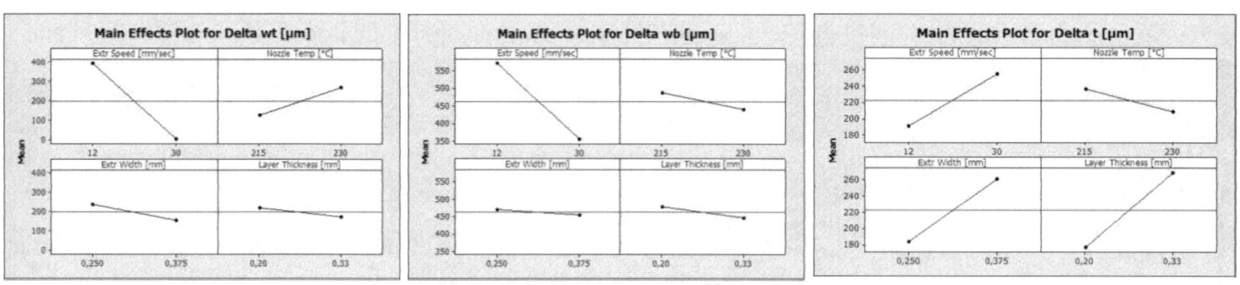

FIGURE 3. ANOVA main effects plots.

The extrusion speed is the most influent parameter on both sample widths, while the nozzle temperature is slightly influent on the top width. The thickness is affected by all the four factors with a slightly less influence from the nozzle temperature. The trend of the extrusion speed is opposite for the widths and the thickness. Therefore, in order to have a minimum deviation of the value of the widths the extrusion speed should be set at its highest value, but this would lead to a greater deviation of the thickness value. According to this qualitative consideration, the DOE optimization feature cannot find a process condition that allows a compensation of the deviation from the model. Hence, the defects peculiar to this manufacturing technology cannot be reduced acting only on the process parameters, probably enhanced by the reduced dimensions of the samples. In order to produce specimens with dimensions that meet the specification, a correction on the CAD model has been performed [15]. In **FIGURE 1b** the drawing of the revisited model is shown. The trapezoidal shape, observed in the experiments, has been compensated with a model geometry with a reversed trapezoidal section and dimensions derived from the mean of the measurements of the DOE.

Using the revisited model ten samples have been manufactured setting suitable process parameters (i.e. extruder speed: 12 mm/sec, nozzle temperature: 215 °C, extruder width: 0.25 mm, layer thickness: 0.20 mm) and measured as above described. In Table 1 the main dimensions of the samples of the DOE (FDM) and the revisited model (Rev. FDM) are presented. The values are calculated as the mean of the measurements (10 samples for each of the 8 conditions of the DOE and 10 samples for the revisited model) and its standard deviation.

Table 1. Main dimensions of the specimens

Model	w_t [mm]	w_b [mm]	t [mm]	Yield stress [MPa]	Elongation at break [%]	Density [mg/mm^3]
FDM	1.68±0.23	1.95±0.18	0.83±0.13	-	-	-
Rev. FDM	1.50±0.04	1.63±0.04	1.03±0.04	77 ± 4	13 ± 3	1.00 ± 0.02
μIM	1.42±0.01	1.51±0.01	1.00±0.01	93 ± 7	27 ± 7	1.13 ± 0.01

4. MICRO-FDM AND MICRO-INJECTION MOULDING COMPARISON

In order to compare the tensile test results of specimens made using FDM and μIM, ten samples of the dog-bones have been manufactured using a micro-injection moulding machine. As well-known, this technology has a very good replication capability and the dimensions of the samples, measured using the optical microscope, are very close to the nominal values (Table 1). The discrepancy on the top width is due to the draft on the side walls of the mould, necessary to facilitate the extraction of the part.

Stress-strain curves have been calculated from the force-displacement curves recorded and the section area and gauge length measured for each sample. The mean values and standard deviations of the yield stress and the elongation at break for each technology have been reported in Table 1. As expected, the μIM has better performances presenting higher values of both yield stress and the elongation at break. This behaviour is expected mainly because of the incomplete filling of the FDM samples. Indeed, although the raster-to-raster gap is set to zero, a gap between the adjacent filaments is likely to be present. On the contrary, the section of the μIM samples is expected to be fully compact with a density higher than in the FDM samples as confirmed by the measurements shown in Table 1. The density has been calculated from the weight of the samples, measured using a scale with accuracy of 1 mg, and the volume, calculated from the Zeiss microscope pictures and its proprietary software.

Nevertheless, the FDM samples shows good results considering that the difference between the yield stresses of the two technologies is only 17%. However, the elongation at break of the samples made using μIM is double the variable for FDM specimens, probably due to a more favourable cold drawing effect in the more uniform density of the μIM samples.

Finally, the two technologies have been briefly compared with respect to costs and time to market for the production of the micro-parts above described. Four main contributes can be considered: material, energy, machine, tools and the man power. The cost of the raw material is much higher for FDM which requires filament spool (> 30 €/Kg) rather than granules (~ 2 €/Kg) used for μIM. The technologies involve comparable amount of waste material (~ 50%), but considering small amount required to produce micro-parts, it has a very low impact on the process costs. The injection machine requires an electrical power of 2,6 kW while the FDM machine has a nominal power of 250 W, nevertheless the energy cost for each part (about 9 kJ) is comparable, taking into account the much shorter fabrication time per part for μIM (3,5 s) than for FDM (36 s). On the contrary, the costs of machine and tools are very different: considering the equipment used in this study the FDM Sharebot NG costs ~ 1500 € and requires no tools, whereas the micro-injection machine Desma Tec Formica 1K has a cost of ~ 200 k€ and requires a mould whose cost can be estimated ~ 5 k€. Due to the high cost of the machine, μIM would be much more favourable performed in outsourcing. Hence, this approach has been included in the comparison among the costs, whose results are reported in **Table 2**. The moulding cost, in case of outsourcing, is typically given per hour including the machine use, the energy and the man power, and it has been assumed 40 €/h. Finally, FDM necessitates post-processing actions done by an operator, who, however, does not require any specific technical skill or competence, which has, thus, been taken into account.

According with a linear model of the cost functions with fixed and variable costs, the total cost (C) versus the number of units (Q) for the two technologies is: $C^{FDM} = 1500 + 0.064\ Q$; $C^{\mu IM}_{OUTSOURCING} = 5000 + 0.039\ Q$ and

$C^{\mu IM}$ = 205000 + 0.00034 Q. For this study, the break-even point between FDM and μIM-outsourcing is 140000 units. Above this threshold, the μIM becomes more convenient than FDM.

Table 2. Estimated costs for micro-dog-bone production

Resources	Costs		
	FDM	**μIM (Outsourcing)**	**μIM**
Material	0.001590 €/piece	0.000086 €/piece	0.000086 €/piece
Energy	0.000250 €/piece	-	0.000250 €/piece
Manpower	0.062000 €/piece	-	-
Mould	-	5000.00 €	5000.00 €
Machine	1500.00 €	0.03900 €/piece	200000.00 €

The time to market is the main strength of 3D printing, since it does not require any specific tool or mould so that it can be only few days. On the contrary, the time to market of micro-injection moulding products can vary between 2 weeks and 2 months, depending on several parameters, including part geometry, mould structure and features, mould design process.

5. CONCLUSIONS

In this paper, fused deposition modelling (FDM) technology has been analysed in terms of process capability for micro-component manufacturing estimating the mechanical properties, the cost production and the time to market. The results have been compared to micro-injection-moulding (μIM) performance. The experimentations have been carried out on micro dog-bones samples, performing dimensional accuracy measurements, tensile tests and costs estimation. Although the FDM quality cannot compete with injection moulding, its low cost and short time to market are very attractive for some applications. The elongation at break is dramatically reduced for FDM micro samples (-52%) and their yield stress is 17% less than μIM samples, on contrary, their total unit cost is significantly cheaper than μIM-parts for small-scale production. Indeed, the total unit cost for μIM becomes more economic for lot size larger than 140000 pieces. For applications where small quantity is requested (up to few thousands of samples) and which do not require high quality, in terms of geometrical accuracy and mechanical properties, the FDM process has been proved to be the a valid choice.

6. ACKNOWLEDGEMENTS

The authors would like to thank Prof. F. Baldi of the University of Brescia for the cryogenic fractures. This work has been partially funded by the project Progetto Bandiera - Fabbrica intelligente *F@H for 3D plates*.

7. REFERENCES

1. N.S.A. Bakar et al., *Journal of Zheijang University Science* A, **11** (12): 972-977 (2010).
2. R.A. Giordano et al., *Journal of biomaterials science polymer edition*, **8** (1): 63-75 (2012).
3. M. Montero et al., *Proceedings of rapid prototyping and manufacturing conference*, SME, (2001).
4. E. Haberstroh et al, *Macromolecular Materials and Engineering*, **287** (12): 881-888(2002).
5. C.H. Wu et al, *Polymer Engineering and Science*, **45** (7), 1021-1030 (2005).
6. T. Koch et al., *Macromolecular Symposia* WILEY-VCH Verlag GmbH, **181** (1): 499-506 (2002).
7. L. Xie et al., *Microsystem Technologies*, **15** (6): 913-917 (2009).
8. F. Baldi et al, *Polymer Engineering and Science*, **54** (3): 512-521 (2014).
9. F. Costa et al., *Polymer Process Engineering '09 Conference*, 331-364 (2009).
10. J. Giboz et al., *Journal of Micromechanics and Microengineering* **17**: 96 (2007).
11. I.C. Celerino de Moraes Porto, Ailton De Souza Gomes, *InTech*, **436** (2012).
12. B. Gupta et al., *Progress in Polymer Science*, **32**: 455-482 (2007).
13. L.T. Lim et al., *Progress in Polymer Science*, **33**: 820-852 (2008).
14. A.P. Mathew et al., *Journal of Applied Polymer Science*, **97**: 2014-2025 (2005).
15. A. Noriega et al., *International J. of Advanced Manufacturing Technology*, **69** (9-12): 2301-2313 (2013).

Design and Numerical Simulation of a Piezoelectric Harvester using PVDF Polymer for Keyboard Application

YESSARI.Madiha[a,b*], HAJJAJI Abdelowahed[a] and ARBAOUI Abdezzahid[b]

[a] *Laboratoire des Sciences d'Ingénieur Pour l'Energie, Ecole Nationale des Sciences Appliquées, University Chouaib Doukkali - El Jadida - Morocco*
[b] *Laboratoire de Physique de la Matière Condensée, Faculté des Sciences, University Chouaib Doukkali - El Jadida – Morocco*

Presenting author: hajjaji.a@ucd.ac.ma

Reference author: yessari.madiha@gmail.com

Abstract. Finding new energy sources has received in recent years a particular interest among scientists. Our research focuses on the generation of micro-renewable energy. This energy is used to power various wireless devices, such as sensors, actuators and micro-electromechanical systems (MEMS). In this work, a new innovative form of piezoelectric harvester, using PVDF polymer, has been proposed, "piezo-spring". Their time deformation gives a time signal, useful for applications that need an instantaneous electrical signal. Materials and geometry were chosen such as to increase the stiffness of the spring and by this the piezoelectric conversion. A numerical simulation of the use of this device in a keyboard application was carried out using finite element calculation software.

Keywords: Piezoelectricity, Energy harvesting, Keyboard, PVDF polymer, Numerical simulation
PACS: ***77.84.Jd***

INTRODUCTION

With the growing use of electronics in recent years as well as barriers related to the use of traditional electrochemical batteries in terms of portability of the various miniaturized electrical and mechanical components; the use of self-powered electronic systems have become a main scientific interest. The micro-power generation and micro energy harvesting technologies are in high demand for reliable auto supply process, providing sufficient power for extended service life [1]. This can be achieved by using an electromechanical conversion of the energy lost from the human body by using piezoelectric materials [2-3].

Computers are one of the essential electronic devices used in our daily-life for an average of about 6 h each day [4-5] and the amount of energy used in key typing motions is simply wasted. The present work report the use of new micro-power harvesting device, recycling the continuous and repeating vertical typing force, wasted in one of the most popular input means used in data terminal equipment, the keyboard.

This paper presents a new design for the electronic keyboard using a flexible piezoelectric conversion mechanism of the mechanical energy from each single key stroke.

PIEZOELECTRIC KEYBOARD STRUCTURE

In data terminal keyboards, key displacement corresponding to the key character is converted into an electric signal, coded signal, and transmitted signal as a data communication code [6]

Fig. 1 shows the piezoelectric keyboard structure and signal flow. In this structure, the impact force is output from the key mechanism and is converted to an electrical signal directly by a piezoelectric element. As a result, the key input displacement is efficiently converted to electrical communication code.

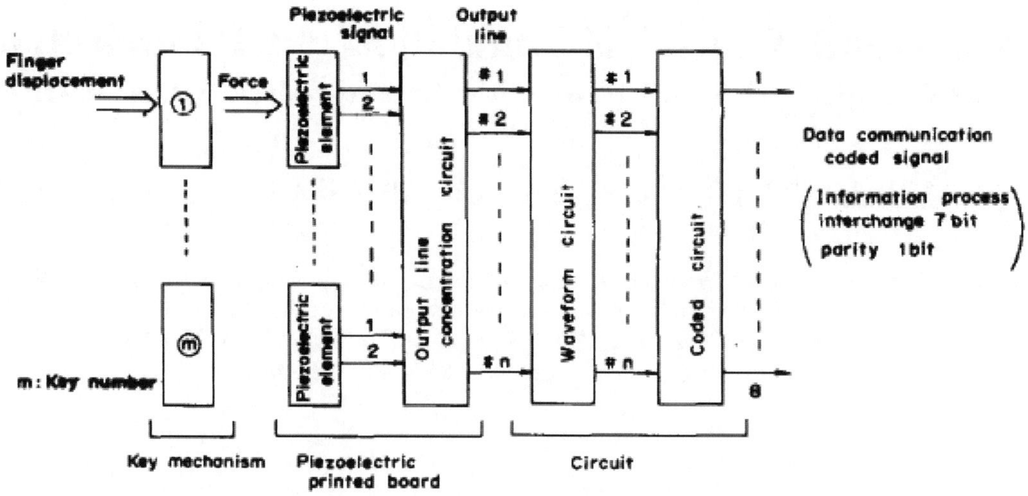

FIGURE 1. Piezoelectric keyboard structure and signal flow [6]

The piezoelectric device has the form of a compression spring, made mostly with PVDF polymer (Fig.2). The choice of dimensions and number of turns was done so as to increase the conversion and efficiency while respecting the volume allocated to a keyboard button.

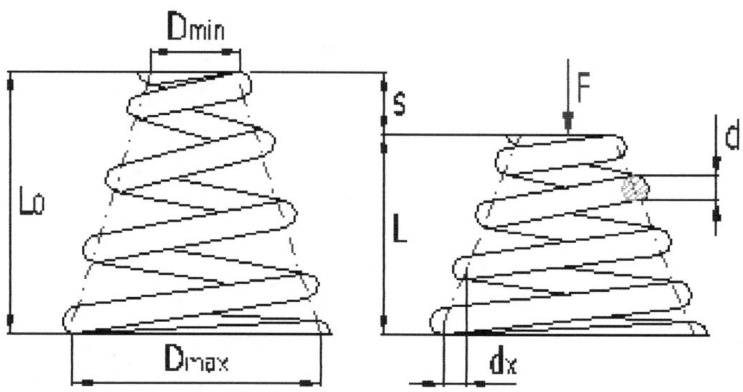

FIGURE 2. Conical compression spring

Springs are components known to be designed to maintain and accumulate the mechanical energy. Their functioning concerns the principle of the flexible deformation of the material. The deformation work performed in the spring during its compressive load represents the deformation energy to the tension or bending strength. This energy can be expressed as a function of the compression force and deflection (Eq.1):

$$W = \int_{0}^{s} F.ds \qquad (1)$$

Springs of conical shape made of helicoidal wire with a constant spacing between the active coils are able to absorb external forces applied in opposition to each other along the axis of the spring. They are usually used if the spring constant must increase progressively so that its compression.

The constant k of the spring being the intensity of the load (force or torque) which causes unit deformation (displacement or rotation) of the spring is expressed as:

$$k = \frac{dF}{ds} \qquad (2)$$

With:
k : spring constant [N / mm]
F : load on the spring [N]
S: deformation (compression) of the spring [mm]

Spring compression load depends on various shape parameters and material choices (Eq.3). This charge causes a deformation of the structure (Eq.4) and subsequently converting it by direct piezoelectric effect into an electric displacement (Eq.5). Thus, enhancing the compression spring load increases the amount of electrical energy converted by the piezoelectric effect.

$$F_c = \frac{(L_0 - L_s).M.d^4}{8.n.D^3_{\max}} \quad (3) \qquad s = \frac{2.F.n}{M.d^4}.DC \quad (4)$$

$$\{D\} = [e].\{S\} + [c^s]\{E\} \quad (5)$$

With:
M : elastic modulus of the material [MPa]
L_0 : unloaded spring length [mm]
L_S : full length [mm]
n: number of active coils
e : PVDF Coupling matrix [C/m²]
C^s : PVDF Elastic matrix for a constant charge [Pa]
E : Applied field [V/m]

The electric displacement is related to the electric current by the following equations:

$$\int_{\partial\Omega} D.nds = Q_0 \quad (6) \qquad \frac{\partial Q_0}{\partial t} = I_{cir} \quad (7)$$

After the electrode depositing on the two terminals of the PVDF spring and the polarization, the piezoelectric spring was wrapped with a circular section tube of aluminum and an insulating layer of Al2O3 between the two materials (Fig. 3). This choice of materials is so that remaining within the elastic deformation range of our polymer while increasing the load applied to the structure.

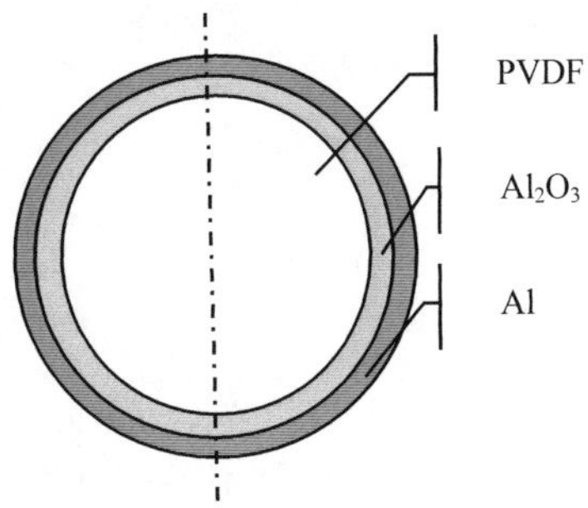

FIGURE 3. *Spring cross section*

SIMULATION RESULTS

The applied force depends on fingertips and can be estimated as shown in table. 1. For example the force applied by the middle finger on a button is estimated to 10N. This force displaces the button by about 5 mm and gives us a mechanical energy of 50 mJ to convert. A person with moderate skills in keyboard manipulation will do about 5 strokes per second, which gives us a mechanical energy of 0.25 w.

TABLE (1). The statical magnitude of force of each fingertip [7]

Finger	Thumb	Index	Middle	Ring	Pinkie
Load (N)	16	10	10	9	8

In our application piezoelectric conversion mechanism use the mechanical strain caused by the applied load to the spring put under each keyboard button and convert it into an electric energy (Fig 3). Typing movements are generally in a low and narrow frequency range, our simulation was limited to a frequency range of 3 to 10 Hz.

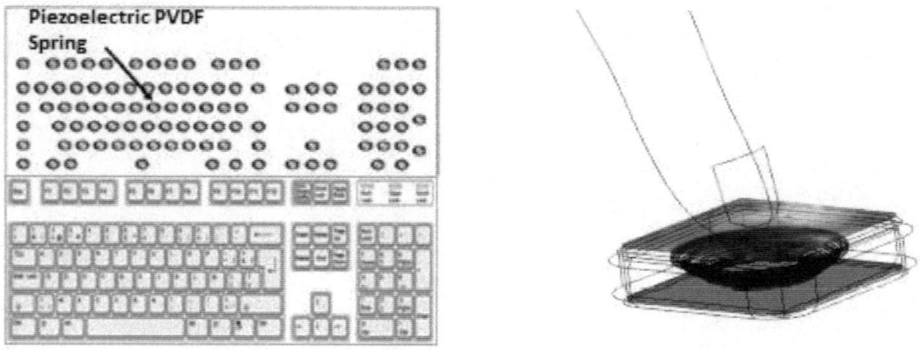

FIGURE 4. Keyboard structure using piezoelectric PVDF spring

A force of 8 N was applied to the button of a keyboard modeled by finite element calculation software inducing a displacement of 0.5 mm of the piezoelectric spring (fig 5).

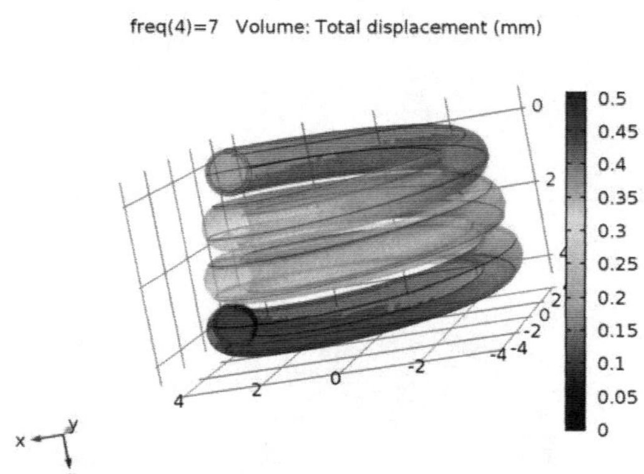

FIGURE 5. Piezoelectric PVDF spring displacement

Key displacement corresponding to the key character is converted into an electric displacement resulting from the piezoelectric conversion. The electric displacement was evaluated in a frequency range of 3 to 10 Hz (Fig. 6).

It can be noted that the frequency has no great influence on the electric displacement of the spring since the frequency variation domain is small and narrow.
An average value of $2.5X10^{-6}$ C / m² of the electric displacement had been harvested with a displacement of only 0.5 mm.
The spring forms may be a promising form in piezoelectric conversion from flexible structure.

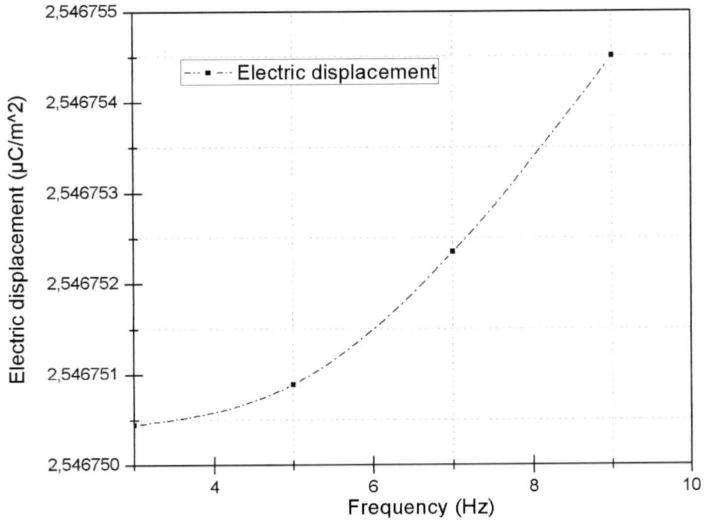

FIGURE 6. Electric displacement depending on frequency

CONCLUSION

A new flexible micro-power harvesting device was studied in the present work using a PVDF polymer. In order to convert efficiently the input displacement to an electrical communication code, a spring structure was proposed. The impact force is output from the key mechanism of a keyboard and converted to an electrical signal directly from the piezoelectric spring. The device provides a feasible method of reusing the wasted energy of typing motion and an optimization study needs to be done so that it can provide enough power to recharge batteries while the keyboard is in use.

REFERENCES

1. B. O Regan and M. Gratzel, *Nature Publishing Group* **353**, 737–740 (1991)
2. T. Starner, *IBM Syst. J.* **35** 618–629 (1996)
3. S. R. Pandian, *Proc. Int. Symp. on Technology and Society,* 54–61 (2004)
4. S. J. Clarke, J. C. Witt, B. E. Smith and A. L. Weaver, *J. Am. Acad. Neurol.* **56**, 1568–1570 (2001)
5. D. Marjolein, D. K. Heleen and M. B. Birgitte, *Int. J. Indust. Ergon.* **37**, 425–31 (2007)
6. P. Namiki, K. Sugiyama, T. Kitayama and T. Ueda, *IEEE Transactions on Components Hybrids and Manufacturing Technology* **4**, 0-310 (1981)
7. T. Wacharasindhu and J. W. Kwon, *J. Micromech. Microeng* ***18***,104016-104024 (2008)

Crystalline Supramolecular Interaction of crystalline side chain and its application

Ryoko Nakano, Ai Maeda, Hideaki Obuchi, Yuriko Tateishi, Yuri Kanazawa, Hiroshi Sekiguchi, Shigeru Yao*

Department of Chemical Engineering, Fukuoka University
8-19-1, Nanakuma, Jonan-ku, Fukuoka, Japan 814-0180

shyao@fukuoka-u.ac.jp

Abstract. Recently, we found that the side chain block co-polymer (SCCBC) constructed with long alkyl side chain monomers shows strong supramolecular interactions with polyethylene (PE). This supramolecular interaction was thought to be generated from the construction of pseudo-crystals with PE crystalline molecules and SCCBC's long alkyl side chains by arranging them on the crystalline potential curve of the PE crystal lattice. By selecting the chemical structure of another monomer of SCCBC, the surface properties of the PE film and inner properties of the PE porous membrane can be easily modified chemically with the help of this supramolecular interaction. In the present work, we have polymerized a new chemical structure of SCCBC, whose crystalline side chains have tetrafluoroethylene conformation. We investigated whether this fluorine-type SCCBC (F-SCCBC) shows supramolecular interactions with poly(tetrafluoroethylene) (PTFE). We also evaluated whether F-SCCBC can modify the surface properties of the PTFE film or the inner property of the PTFE porous membrane. The results showed that this F-SCCBC formed supramolecular interactions with PTFE and can modify the properties of both the PTFE film and PTFE porous membrane.

Keywords: poly(tetrafluoroethylene), supramolecular interaction, modification, side chain crystalline block co-polymer
PACS: 68.35.Np

INTRODUCTION

Poly(tetrafluoroethylene) (PTFE) is widely used in various fields because of its excellent chemical stability, high heat resistance, weather resistance, chemical resistance, and electrical insulation. However, because the solvent wettability of PTFE is very weak, it is extremely difficult to modify its surface property using a functional solvent or resin. To modify PTFE, alkali metal (Na) treatment, plasma treatment, etching, sputtering, and/or particulate spray have been proposed [1–7]. However, when surface properties are modified by these methods, the PTFE molecules break, and the mechanical strength become very weak; the surface structure also becomes rough. Moreover, these methods often do not result in the required properties. It is also impossible to modify the inner surfaces of porous membrane and shadow parts of body by these methods. Therefore, alternative methods need to be developed to modify PTFE. Although the use of polyethylene (PE) leads to the same problems as those of PTFE, we found that the side chain crystalline block co-polymer could modify the surface properties of PE by chemical interactions [8,9]. The side chain crystalline block copolymers that we constructed comprised two monomers: a monomer with a long alkyl side-chain (more than 10 carbon atoms) and another monomer with various functions (solvent affinity, hydrophilic, and absorption). These block copolymers have a low melting point and show side-chain crystallization. Thus, this block copolymer can crystallize because of its long alkyl side-chain (side-chain crystalline block copolymer: SCCBC). We found that this SCCBC is adsorbed on the PE crystal through supramolecular interactions. Using this interaction, we very easily modified not only the surface properties of PE but also the inner properties of the porous PE membrane. In the current study, we used 2,2,3,3,4,4,5,5,6,6,7,7-dodecafliorohepty lacrylate as a side chain crystalline monomer and fluorine-polymerized SCCBC (F-SCCBC). We estimated the PTFE surface modification ability of this F-SCCBC.

EXPERIMENT

Polymerization

The chemical structure of the side chain crystalline block co-polymer synthesized in this investigation is shown in Figure 1. We used 2,2,3,3,4,4,5,5,6,6,7,7-dodecaflioroheptylacrylate (DFA) as a side chain crystalline monomer and di(ethylene glycol) ethyl ether acrylate (DEEA) as a functional monomer unit. Living radical polymerization method (Nitroxide-Mediated radical Polymerization method: NMP method) was used for polymerization. 3,7-Dioxa-4-aza-6-phosphanonanoic acid, 4,5-bis(1,1-dimethylethyl)-6-ethoxy-2, 2-dimethyl-, 6-oxide (SG-1-MA; BlocBuilder©, ARKEMA) was used as the initiator. Polymerization was performed under a nitrogen atmosphere at 105 °C. The block copolymer was polymerized by sequential monomer addition. We first polymerized the DFA-unit. In this polymerization sequence, the sampled polymer did not solved; therefore, we could not measure the molecular weight of the polymer. After adding the DEEA-unit, the sampled polymer began dissolve in the solvent and was used for size exclusion chromatography.

TABLE i). Weight averaged molecular weight of F-SCCBC.

	All	DFA-unit	DEEA-unit
M_w	38,000	19,000	19,000
M_w/M_n		3.9	

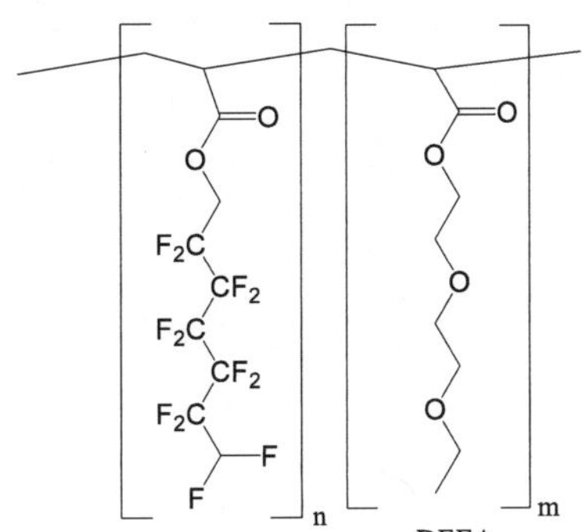

DFA DEEA

FIGURE 1. Chemical structure of F-SCCBC

PTFE film and PTFE porous membrane modification

The surface of PTFE film and inner surface of PTFE porous membrane were modified by dipping in and coating with the F-SCCBC butyl acetate solution. Figure 2 illustrates the modification methods. The concentration of this solution was 5 wt%. The porous PTFE membrane used was ADVANTEC membrane filter (T100A047A).

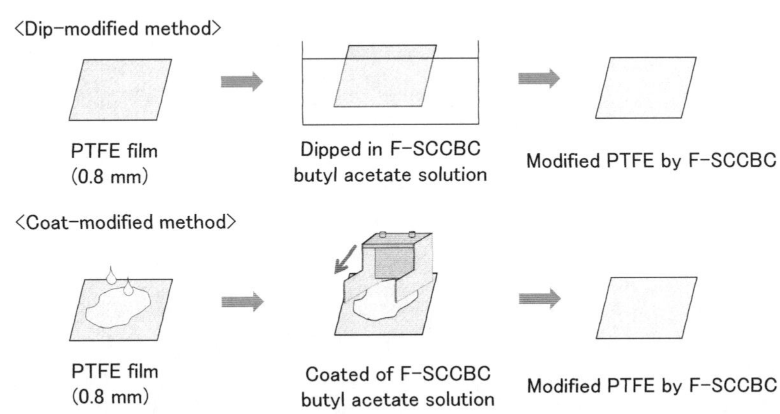

FIGURE 2. Illustration of modification methods.

PTFE film and PTFE porous membrane modification

To measure the surface modification ability of F-SCCBC, we measured the contact angle of pure water and peel strength (JIS K 6854-3 1/2). To measure the peel strength, we bonded original PTFE films, dip-modified PTFE films, and coat-modified PTFE films by epoxy resin and obtained the average peel strength of five samples.

RESULTS AND DISCUSSION

F-SCCBC

We first polymerized the DFA-unit. In this polymerization sequence, the sampled polymer did not solved, and hence, we could not measure the molecular weight. When a DEEA-unit is added, the sampled polymer began to dissolve the solvent used for size exclusion chromatography. Table 1 shows the estimated weight averaged molecular weight obtained from the monomer molar concentration ratio and polydispersity of the polymerized F-SCCBC. Figure 3 shows the polarizing microscopic image of the F-SCCBC. Almost all the F-SCCBC was amorphous, with some small crystalline domains.

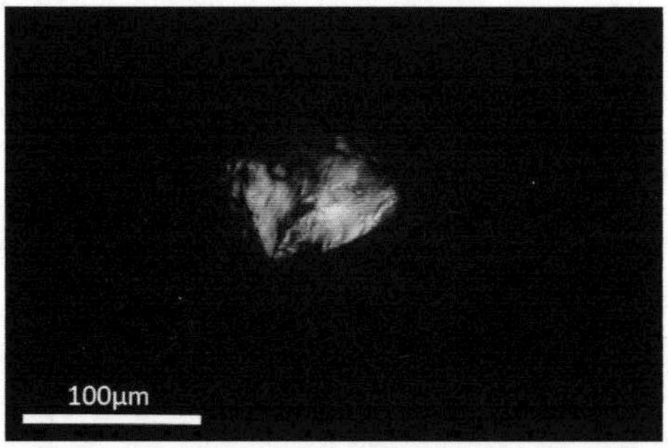

FIGURE 3. Polarizing microscope image of F-SCCBC.

Figure 4 shows wide angle X-ray scattering patterns of PTFE and F-SCCBC. Here, F-SCCBC does not show any crystalline scattering peak. However, the amorphous peak of F-SCCBC is similar to that of PTFE.

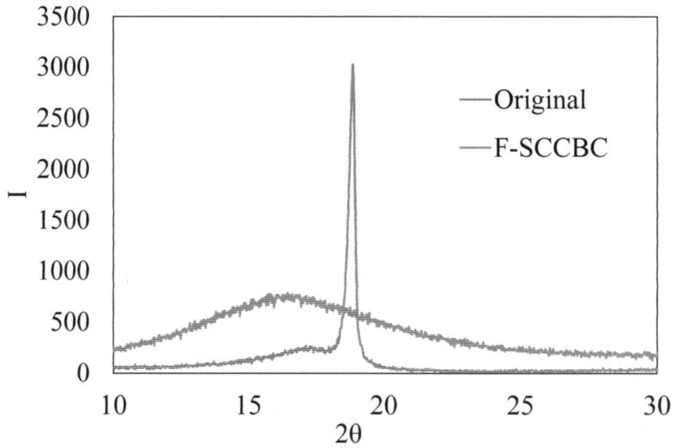

FIGURE 4. X-ray scattering patterns of PTFE(original) and F-SCCBC.

Surface morphology and Contact angle

Figure 5 shows a comparison of surface morphologies of original PTFE, dip-modified PTFE, and coat-modified PTFE. The original PTFE is very flat and smooth. In contrast, the dip-modified PTFE has a wavy surface. This structure is thought to result from the coagulation and self-organization of F-SCCBC while drying the F-SCCBC solution. For coat-modified PTFE, because the F-SCCBC solution was highly concentrated, the PTFE surface was uniformly coated with a very thick layer of the F-SCCBC solution.

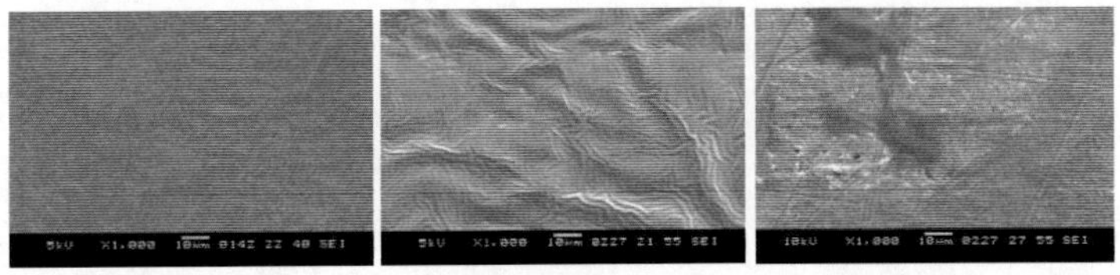

(a) (b) (c)

FIGURE 5. SEM images of the surface of original or modified PTFE film: (a) Original (b) dipped F-SCCBC acetate solution; (c) coated with F-SCCBC acetate solution.

Table 2 shows the average contact angles of each film, measured at five points. The contact angle of the original PTFE is very high, which implies that its surface is highly hydrophobic. However, the contact angles of dip-modified and coat-modified PTFE are about half of the original value. These results mean that when PTFE is modified by F-SCCBC, the PTFE surface becomes highly hydrophilic. The figures of Table 1 also show the modification effect.

TABLE iii). Contact angles and measured example photos.

	Original	**Dip-modified**	**Coat-modified**
Contact angle (degree)	101.2°	59.9°	54.8°
Measured example photos	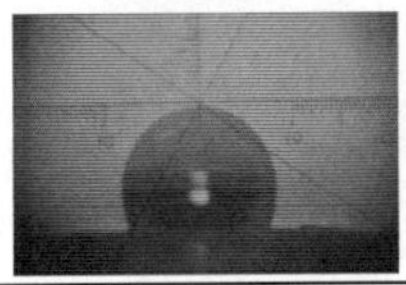	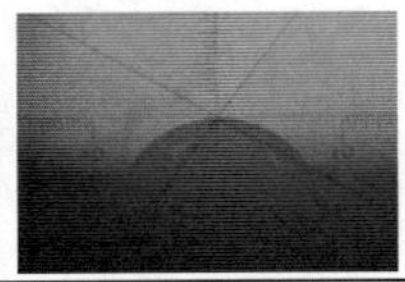	

Modification of inner pores of porous PTFE membrane

In our previous study [8, 9], we were able to easily modify the inner surface properties of PE porous membrane by dipping the membrane in the SCCBC solution or coating it with the solution. In the present work, we tried to modify the inner property of the porous PTFE membrane by dipping it in or coating it with the F-SCCBC solution. We then compared the shape of a water droplet on the original and modified porous membranes. Figure 6 compares the shape of a water droplet on the original PTFE porous membrane, dip-modified, and coat-modified membrane. In Figure 6(a), the water droplet keeps its original shape, which means that the surface of the porous PTFE membrane is highly hydrophobic. In contrast, in Figures 6(b) and (c), we can see the trace of sintering of the water droplet. It means when the PTFE membrane is dipped in or coated with the F-SCCBC solution, not only the surface but also the inner pores of the PTFE porous membrane very easily become hydrophilic. In the case of the dipped-modified case, in particular, the water droplet water expanded in the porous PTFE membrane. This means that the inner pores of the porous PTFE membrane became highly hydrophilic.

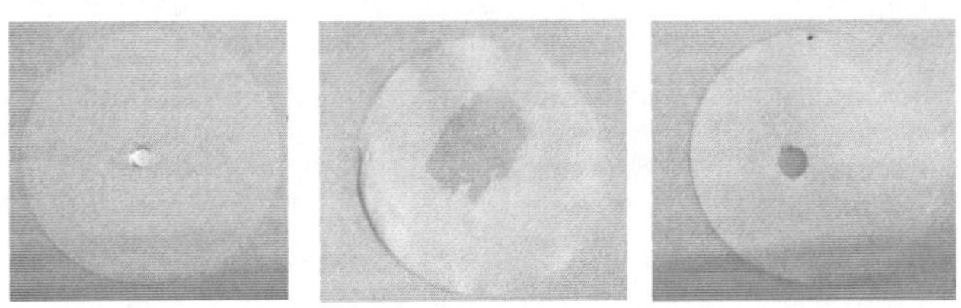

(a) (b) (c)

FIGURE 6. Shape comparison of a drop of water between the original PTFE porous membrane (a), dip-modified (b), and coat-modified (c).

Estimation of adsorption force by peel test

We measured the peel strength to estimate the adsorption force of the crystalline supramolecular interaction between F-SCCBC and PTFE. Figure 7 shows an example of the load-stroke relationship in the original PTFE, dip-modified PTFE, and coat-modified PTFE films. From this figure, it can be concluded that by modifying the PTFE surface with F-SCCBC, both the dip-modified and coat-modified PTFE films can be bonded by the epoxy resin.

Table 3 shows the results of the peel test. From this table, it can be concluded that F-SCCBC is adsorbed on PTFE by crystalline supramolecular interactions and this interaction is very strong.

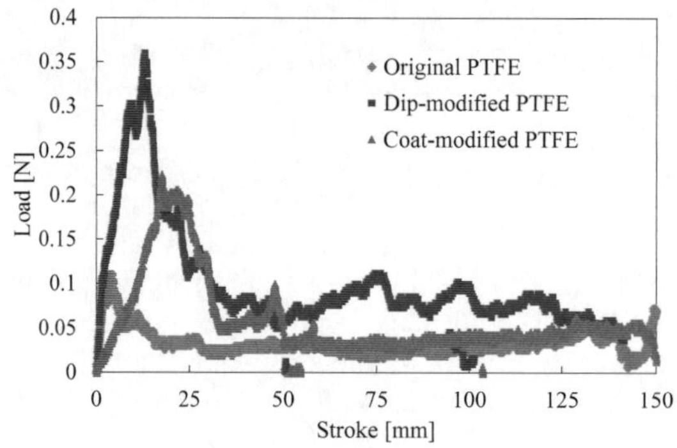

FIGURE 7. Load-stroke relationship of original PTFE, dip-modified PTFE, and coat-modified PTFE.

TABLE iiii). Results of the peel test.

	Load [N]	Peel Strength [N/m]
Original	0.11	2.30
Dip-modified	0.36	6.55
Coat-modified	0.22	3.90

CONCLUSION

In this work, we synthesized a new side chain crystalline block copolymer containing a long tetrafluoroethylene side chain unit (F-SCCBC). Similar to long alkyl side chain units, crystalline F-SCCBC copolymers showed supramolecular interactions with poly(tetrafluoroethylene) (PTFE) and very strong adsorption force on PTFE surface. By using the F-SCCBC, the PTFE surface can be modified from hydrophobic to hydrophilic. After the modification, two PTFE films can be very strongly connected by an epoxy resin. The inner properties of the PTFE porous membrane can be very easily modified by dipping the membrane in the dilute F-SCCBC solution or by coating it with a dilute F-SCCBC solution.

This investigation suggests that any SCCBC having similar structure of crystalline polymer has a crystalline supramolecular interaction with the crystalline polymer.

REFERENCES

1. I. Noh, K. Chittur, S.L. Goodman, J.A. Hubbell, *Polym. Chem.* **35**, 1499-1514 (1997).
2. B. Goknur, O. Guralp, *J. Adhes. Sci. Technol.* **21**, 883-898 (2007).
3. Y. Yamamoto, H. Akiyama, K. Ooka, K. Yamamura, Y. Oshikane, N. Zettsu, *Curr. Appl. Phys.* **12**, S63-S68 (2012).
4. J. Gao, J. Yu, Y. Ma, *Surf. Interface Anal.* **44**, 578–583 (2012).
5. A. Sarani, N. De Geyter, A. Yu. Nikiforov, R. Morent, C. Leys, J. Hubert, F. Reniers, *Surf. Coat. Technol.* **206**, 2226-2232 (2012).
6. M. Dübner, N.D. Spencer, C. Padeste, *Langmuir* **30**, 14971–14981 (2014).
7. Y.K. Cho, D. Park, H. Kim, H. Lee, H. Park, H.J. Kim, D. Jung, *Appl. Surf. Sci.* **296**, 79-85 (2014).
8. Y. Sano, R. Nakano, H. Sekiguchi, S. Yao, *Int. J. Mater. Sci. Appl.* **3/6**, 399-403 (2014).
9. R. Nakano, H. Sekiguchi, S. Yao, *Macromol. Symp.* **349/1**, 44–50 (2015).

Frictional Properties of Plants-Derived Polyamide against Surface Microstructures of Metal Counterpart Fabricated by Femtosecond Laser

Keisuke Nakamura[a], Yosuke Nishitani[b*] and Takeshi Kitano[c]

[a] *Department of Mechanical Engineering, Graduate school of Engineering, Kogakuin University, 2665-1 Nakano, Hachioji, Tokyo 1920015, Japan*
[b] *Department of Mechanical Engineering, Faculty of Engineering, Kogakuin University, 2665-1 Nakano, Hachioji, Tokyo 1920015, Japan*
[c] *Polymer Centre, Faculty of Technology, Tomas Bata University in Zlin, Czech Republic, T. G. M. 275, Zlin, 762 72 Czech Republic*

Reference author's email: at13152@ns.kogakuin.ac.jp

Abstract. The purpose of this study is to investigate the frictional properties of the plants-derived polyamide 11 (PA11) against microstructures on the metal counterpart fabricated by femtosecond laser in order to control the frictional properties of polymeric materials. This study discusses the frictional properties in terms of various factors: microstructure pattern, sliding direction, sliding speed and so on. The material used in this study was plants-derived polyamide (PA). Plants-derived PA was extruded by a twin screw extruder and injection-molded. Surface microstructures on SUS304 were fabricated by femtosecond laser. This SUS304 with microstructures using femtosecond laser was used as counterparts. Tribological properties were measured by a pin-on-plate type reciprocating-sliding wear tester under oil lubrication conditions. It was found that the frictional coefficient differs according to the type of the direction of microstructures on SUS304 and with or without microstructures. The frictional coefficient of PA11 against various microstructured SUS304 decreases in the following order: Textured (MD) > Non-textured > Textured (TD). The effect of sliding speed on the frictional coefficient decreases with increasing sliding speed. The frictional coefficient of PA11 against Textured (TD) is lower than that of Non-textured in wide sliding speed region. However, the wear loss of PA11 against Textured (TD) is heavier than that of Non-textured although the effect of sliding speed on the wear loss shows the complex behavior. These results may be attributed to the change in the mode of friction and wear mechanism by the microstructures.

Keywords: Microstructure, Femtosecond Laser, Frictional properties, Plants-derived Polyamide
PACS: 81.40.Pq, 81.16.Rf, 81.40.-z

INTRODUCTION

In recent years, laser surface texturing (LST) using femtosecond has been great attention due to its versatility and precision for surface texturing technique in order to control the tribological properties such as frictional and wear properties of various engineering materials[1)]. This is because the need to reduce friction and the amount of wear on machine element components involved in sliding contact is ever present[2)], and the surface texturing technology has been widely used to improve the functionality of materials surface wettability and tribological properties[3)]. Especially, in this field, the conduction heat transfer in materials treated under short time scales in the ablation process with femtosecond laser pulses is negligible compared with that generated using picosecond and nanosecond pulse lasers[4)]. Therefore, femtosecond lasers enhance precision and pure laser processing in surface texturing. Moreover, most frictional properties of surface texturing have focused on the modification of ceramics and metallic materials[5, 6)]. However, there are only but few published studies on the frictional properties of the polymeric materials against surface texturing using LST on metallic materials[7)]. On the other hand, the polymeric materials have replaced metals due to their superior performance, which is high specific strength, ease of processability, and low cost. Most polymeric materials are derived from petroleum supply resources, therefore there is a demand to replace them to biopolymer because of the problems of uncertainly of the petroleum supply and global warming[8)]. Some of the reasons are the renewability, sustainability and environmentally. In addition, supply of raw materials is unstable since many of biopolymers are made from edible biomass like as corn and starch. To solve these problems, the new engineering materials based on all inedible plants-derived materials such as polyamide 11 (PA11) and PA1010 are strongly required[9)]. In our previous works, we investigated the fabrication of microstructured surface of various polymeric materials and their frictional properties[10-13] and also the frictional properties of the polymeric materials against surface microstructures on the metal counterpart fabricated by femtosecond laser[7]. In particular, it was found that the frictional coefficients of polytetrafluoroethylene (PTFE) and polyoxymethylene (POM) against microstructures on SUS304 are lower than those of non-textured SUS304. However, the tribological properties of various polymeric materials against surface microstructures on the metal counterpart fabricated by femtosecond laser is not fully understand. It is therefore necessary to understand first those of plants-derived polymer, which is PA11, PA1010, and so on. The purpose of this study is to investigate the frictional properties of the plants-derived PA11 against surface microstructures on the metal counterpart fabricated by femtosecond laser in order to control the frictional properties of polymeric materials. This study discusses the frictional properties in terms of various factors: the relationship between microstructure pattern direction and sliding direction, and sliding speed.

EXPERIMENTAL

The material used in this study was plants-derived Polyamide 11 (PA11, Rilsan BMN O, Arkema Inc., Japan). PA11 was made from sebacic acid and decamethylendiamine, which are obtained from plants-derived castor oil. PA11 was dried in vacuum oven at 80°C for 12h. Subsequently PA11 was melt mixed by a twin screw extruder (TEX-30, Japan Steel Works, Ltd., Japan) and injection-molded (NS20-A, Nissei Plastic Industrial, Japan) to 30mm x 30mm x 3mm plates. The molding condition of PA11 was melt mixed and cylinder temperatures of 220°C and mold (cavity) temperature of 30°C. The molded plates of PA11 cut into pin shaped specimens (2mm x 2mm x 2mm) for tribological properties measurements. To keep the drying conditions of specimens for all measurements, they were kept in accordance with JIS K 6920-2 for at least 24h at 23°C in desiccators after injection molding and machinery cutting. Surface microstructures on the austenitic stainless steel (SUS304, 30mm x 30mm x 3mm) was fabricated by femtosecond laser (IFRIT, Cyber Laser Inc., Japan). This SUS304 with microstructures was used as counterpart of sliding wear test. The surface microstructures area in this study was a rectangle region of 3mm x 25mm. Fig. 1 illustrates the schematic diagram of microstructures area on the SUS304. Femtosecond laser was performed to generate linearly polarized laser pulses at a wave length of 800nm, and with pulse duration of 190fs. The processing parameters in this study were selected as laser power of 1W, laser speed of 600mm/min, pitch distance of 100μm and working distance, which is not equal to focal distance, between laser and SUS304 specimens of 46mm. The surface structures on the SUS304 by laser fabrication and the counterface of various microstructured SUS304 against PA11 were observed by scanning electron microscope (SEM, JSM6360LA, Jeol Ltd., Japan) and digital microscope (VHX-100, Keyence Co., Japan). Frictional properties were measured by a pin-on-plate reciprocating type sliding wear tester (HEIDON Type 38S, Shinto Scientific Co., Ltd., Japan) under oil lubrication sliding conditions, at room temperature. The test conditions were selected as sliding speed of 5, 10, 20, 50mm/sec, sliding distance of 20m and normal load of 5N. The silicone oil (KF-96-100cs, Shin-Etsu Chemical Co., Ltd.) was used as oil lubricant. The sliding surface on the SUS304 with microstructures after sliding testing against PA11 were observed by SEM. This surface of SUS304 after sliding test was cleaned in toluene to remove the extra silicone oil.

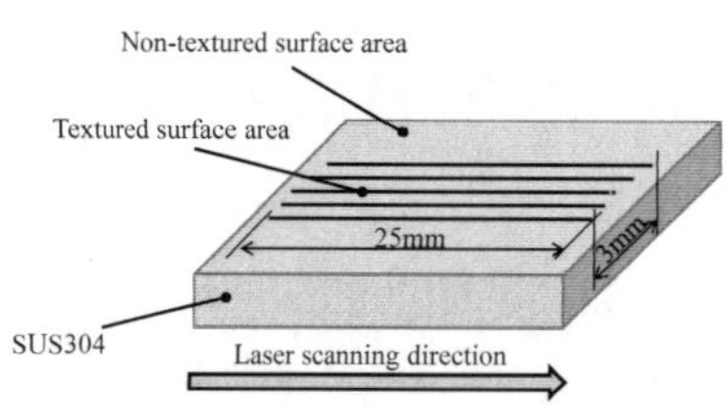

FIGURE 1. The schematic diagram of microstructures area on the SUS304[7].

RESULTS AND DISCUSSION

First, the surface microstructures on SUS304 fabricated by femtosecond laser are discussed. Fig. 2 shows the microscope photograph (Fig. 2(a), x50) and SEM photograph (Fig. 2(b), x10000) of the surface microstructures on the SUS304. In Fig. 2(a), the black central area (width 3mm) indicates the textured (laser processing) area, on the other hand, the gray areas in both sides indicate the non-textured areas. Fig. 2(b) is the enlarged the SEM photographs of the textured area on the SUS304 in Fig. 2(a). The laser-induced periodic surface structures, which have parallel structures to the laser scanning, are formed on the surface of SUS304. These microstructures are submicron scales periodic structures constituted of alternate crests and troughs, and have a periodicity much smaller than the laser wave length. The width of these crests is the average 250nm, and that of these troughs is the average 150nm, and the pitch distance is the average 400nm, which is a half of laser wave length. The formation of laser-induced periodic surface structures on SUS304 may be attributed to the spatial energy distribution[14]. According to the classical theory, the interference between the incident laser beam and scattered optical wave at the surface leads to these periodic distribution. However, the debate of origin and formation of laser-induced periodic surface structures is still ongoing and future investigations are needed to clarify this subject[15].

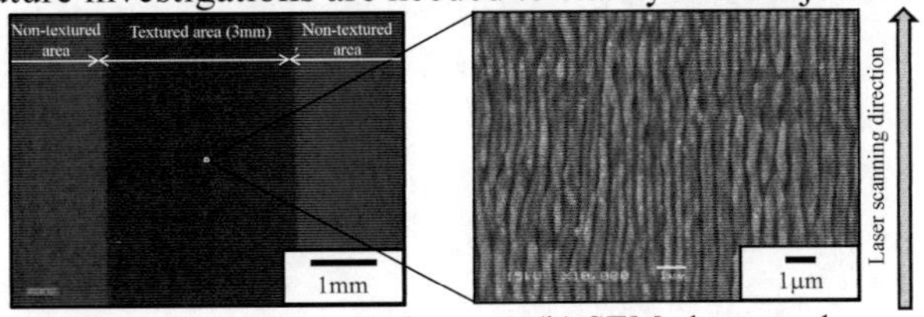

(a) Microscope photograph (b) SEM photograph

FIGURE 2. The microscope photograph and SEM photograph of microstructures area of SUS304[7].

Second, the frictional properties of the plants-derived PA11 (pin specimens) against various surface microstructures on the SUS304 by constant normal load and sliding speed testing using a pin-on-plate reciprocating type sliding wear tester under oil lubrication are discussed. In particular, the relationship between the direction of the microstructures and sliding direction of frictional properties is argued. Fig. 3 shows the schematic diagram and SEM photographs of relationship between sliding direction and microstructures. The sliding direction was fixed with a longitudinal direction of surface microstructures area on SUS304. The laser scanning direction, which in formed by the microstructures on SUS304, was used two different directions: first, parallel direction of microstructures sliding direction (longitudinal direction) (MD: Machine Direction, Fig. 3(b)), and the orthogonal direction of the microstructure to sliding direction (TD: Transverse Direction, Fig. 3(c)). These directions of microstructures compared to non-textured SUS304 were investigated in this study. Fig. 4 shows the relationship between frictional coefficient μ and sliding distance L (μ - L curves) of PA11 against various microstructured SUS304. μ - L curve of PA11 against non-textured shows sharp rise immediately after starting the sliding test, and is nearly straight up to 20m. On the other hand, those of PA11 against various textured SUS304 have a maximum peak immediately after starting the sliding test, and then abrupt decrease, and are almost linearly up to 20m. The average frictional coefficient μ, which is calculated value in the sliding distance 15 to 20m (steady state friction), differs according to the type of the direction of microstructures on SUS304 and with or without microstructures. Therefore, μ of PA11 against various SUS304 decreases in the following order: Textured (MD) > Non-textured > Textured (TD). These results may be attributed to the change in the mode of frictional mechanism caused by the type of the direction of microstructures on SUS304 and with or without microstructures. Because the frictional behaviors of polymeric material are highly influenced by their ability to form a transfer film on the counterface, it is essential to observe this factor for understanding the mechanisms of frictional behavior. Fig. 5 presents SEM photographs on the surface of SUS304 (counterface) after sliding wear testing by the pin-on-plate type sliding wear tester against PA11: Non-textured (Fig. 5(a)), Textured (MD) (Fig. 5(b)) and Textured (TD) (Fig. 5(c)), respectively. The morphology after sliding wear test of Non-textured SUS304 against PA11 (Fig. 5(a)) has bumpy thick transfer films with parallel frictional marks to sliding direction. On the other hand, that of Textured (MD) against PA11 (Fig. 5(b)) has patchy incoherent transfer films and is partially filled up with PA11 in the microstructures. To the contrary, that of Textured (TD) against PA11 (Fig. 5(c)) has a smooth microstructures. These differences in the formation of transfer film on the counterface are considered to change the frictional properties of PA11.

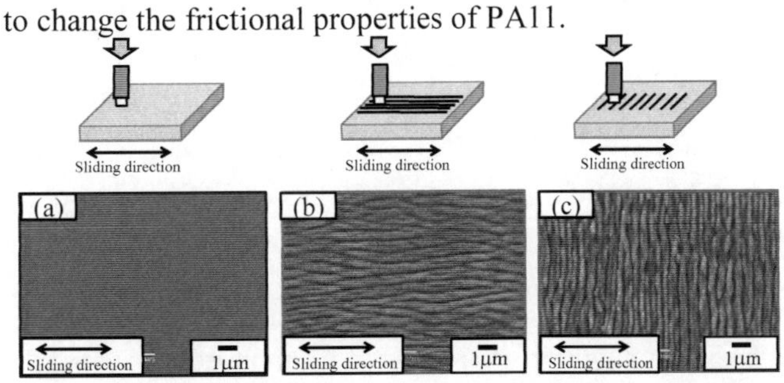

(a) Non-textured (b) Textured (MD) (c) Textured (TD)

FIGURE 3. The schematic diagram and SEM photographs of relationship between sliding direction and microstructures[7].

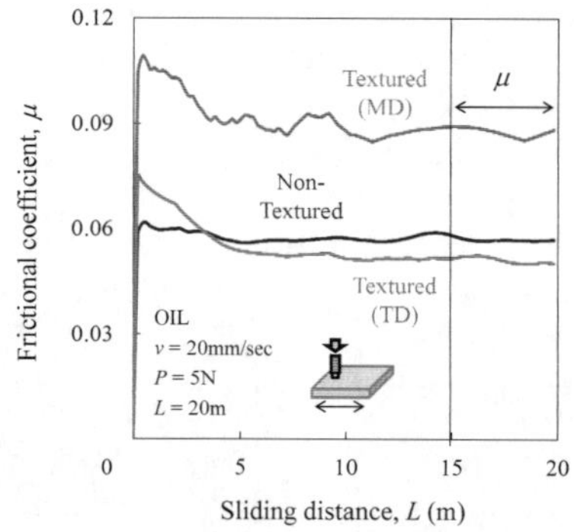

FIGURE 4. μ-L curves of plants-derived PA11 against various microstructured SUS304.

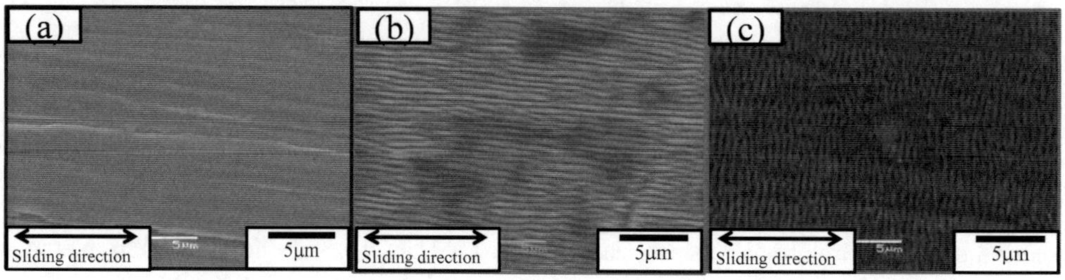

(a) Non-textured (b) Textured(MD) (c) Textured(TD)

FIGURE 5. SEM photographs of counterface of various microstructured SUS304 against plants-derived PA11.

Finally, the effect of sliding speed on the tribological properties of the plants-derived PA11 against microstructures on SUS304 is discussed. Fig. 6(a) and Fig. 6(b) shows the influence of sliding speed on the frictional properties μ and wear loss V of PA11 against SUS304 with or without microstructures. μ of both SUS304 with or without microstructures decrease with increasing the sliding speed. μ of PA11 against Textured (TD) is lower than that of Non-textured. To the contrary, V of PA11 against Textured (TD) is heavier than that of Non-textured although the influence of sliding speed on the V shows the complex behavior. These tendencies may be attributed to the change in the mode of wear mechanism by the sliding speed.

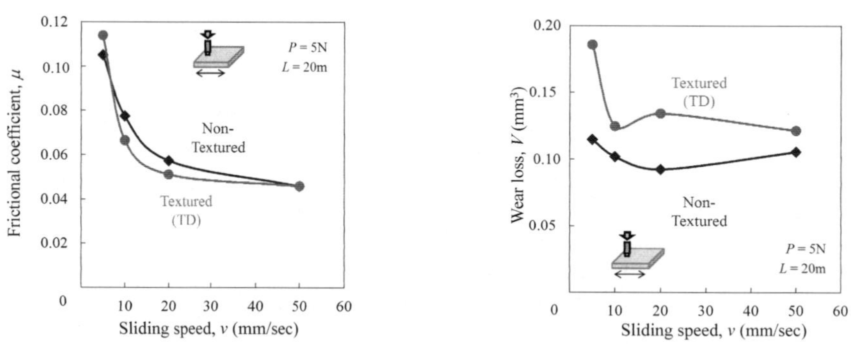

(a) Frictional coefficient (b) Wear loss

FIGURE 6. Influence of sliding speed on the tribological of plants-derived PA11 against various microstructured SUS304.

CONCLUSION

The purpose of this study was to investigate the frictional properties of the plants-derived polyamide 11 (PA11) against microstructures on the metal counterpart fabricated by femtosecond laser in order to control the frictional properties of polymeric materials. This study discussed the frictional properties in terms of various factors: microstructure pattern, sliding direction, sliding speed and so on. It was found that the frictional coefficient differs according to the type of the direction of microstructures on SUS304 and with or without microstructures. The frictional coefficient of PA11 against various microstructured SUS304 decreased in the following order: Textured (MD) > Non-textured > Textured (TD). The effect of sliding speed on the frictional coefficient decreased with increasing sliding speed. The frictional coefficient of PA11 against Textured (TD) was lower than that of Non-textured in wide sliding speed region. However, the wear loss of PA11 against Textured (TD) was heavier than that of Non-textured although the effect of sliding speed on the wear loss showed the complex behavior. These results may be attributed to the change in the mode of frictional and wear mechanism by microstructures.

ACKNOWLEDGMENTS

We would like to thank the Functional Micro-structured Surface Research Center of Kogakuin University (FMS, MEXT, Japan), and the Project Research of Research Institute for Science and Technology of Kogakuin University for funding this study and partial support by national budget of Czech Republic within the framework of the Centre of Polymer Systems project (reg. number: CZ. 1.05/2.1.00/ 03.0111).

REFERENCES

1. J. Bonse, R. Koter, M. Hartelt, D. Spaltmann, S. Pentzien, S. Hohm, A. Rosenfeld and J. Kruger, *Applied physics A* **117**, 103-110 (2014)
2. A. Kovalchenko, O. Ajayia, A. Erdemira, G. Fenskea and I. Etsion, *Tribology International* **38**, 219-225 (2005).
3. J. Bico, U. Thiele and D. Que´re´, *Colloids and Surfaces A: Physicochemical and Engineering Aspects* **206**, 41-46 (2002).
4. C. Y. Chen, C. J. Chung, B. H. Wu, W. L. Li, C. W. Chien, P. H. Wu and C. W. Cheng, *Applied physics A: Materials Science and Processing* **107**, 345-350 (2012).
5. G. Dumitru, V. Romano, Y. Gerbig, H. P. Weber and H. Haefke, *Applied physics A* **80**, 283-287 (2005).
6. X. Wang, K. Kato, K. Adachi, K. Aizawa, *Tribology International* **34**, 703–711 (2001).
7. Y. Nishitani, K. Nakamura and T. Kitano, *Journal of Material Testing Research Association* **61**, 1, 12-20 (2016)
8. L. T. Lim, R. Auras and M. Rubino, *Progress in Polymer Science* **33**, 820-852 (2008).
9. J. Mukaida, Y. Nishitani and T. Kitano," Influence of Types of Alkali Treatment on the Mechanical Properties of Hemp Fiber Reinforced Polyamide 1010 Composites" *The Proceeidngs of Regional Conference of Polymer Processing Society*, Graz 2015, Graz, Austria: Nat-104 (2015).
10. N. Naruse, Y. Nisitani and T. Kitano, "Fabrication of Micro-Structured Surface of Polymeric Materials using Femtosecond Laser," in *International Symposium on Fiber Science and Technology 2014*, ISF conference proceedings. (The society of fiber science and technology, Tokyo, Japan, 2014), PG1-44.
11. N. Naruse, K. Nakamura, Y. Nisitani and T. Kitano, "Fabrication of Micro-Structured Surface of PTFE using Femtosecond Laser and its Frictional Properties," in *2015 the Japan Society for Precision Engineering Spring Meeting*, JSPE conference proceedings. (The Japan Society for Precision Engineering, Tokyo, Japan, 2015), M61 (in Japanese).
12. K. Nakamura, N. Naruse, Y. Nishitani and T. Kitano, "Fabrication of Micro-Structured Surface of Biomass Polyamide Thermoplastic Elastomer using Femtosecond Laser and Their Frictional Properties," in *Tribology Conference Spring'15*, JAST conference proceedings. (Japanese Society of Tribologists, Kobe, Japan, 2015), D29 (in Japanese).
13. K. Nakamura, N. Naruse, Y. Nishitani and T. Kitano, "Fabrication of Micro-Structured Surface of Plants-Derived Polyamide using Femtosecond Laser and Their Frictional Properties," *The Proceeidngs of Regional Conference of Polymer Processing Society*, Graz 2015, Graz, Austria: Nan-105 (2015).
14. K. M. T. Ahmmed, C. Grambow and A.-M. Kietzig, *Micromachines* **5**, 1219-1253 (2014).
15. J. Bonse, J. Kruger, S. Hohm, A. Rosenfeld, *Journal of laser applications* **24**, 042006 (2012).

Tuning the Hydrolytic Degradation Rate of Poly-Lactic Acid (PLA) to more Durable Applications

Iozzino Valentina[a], Askanian Haroutioun[b],
Leroux Fabrice[b], Verney Vincent[b], Pantani Roberto[*a]

[a]*Department of Industrial Engineering. University of Salerno*
Via Giovanni Paolo II, 132, 84084 Fisciano (SA), ITALY
[b] *Institut de Chimie de Clermont Ferrand (ICCF), UMR 6296 Université Blaise Pascal, 24 Avenue des Landais, AUBIERE (cedex), France*
rpantani@unisa.it

Abstract. The disposal of polymeric waste is increasingly becoming an issue of international concern. The use of biodegradable polymers is a possible strategy to face most of the problems related to the disposal of the durable (non-biodegradable) polymers. Among biodegradable polymers, PLA (obtained from renewable sources) is a very attractive one, due to its good processability, biocompatibility, interesting physical properties. Hydrolysis is the major depolymerisation mechanism and the rate-controlling step of PLA biodegradation in compost. The propensity to degradation in the presence of water significantly limits specific industrial applications such as automobile, biomedical, electronic and electrical appliances, agriculture. Therefore the control of biodegradation rate is somewhat even more important than the characteristic of biodegradability itself. This is the reason why it is critical to find additives able to modulate the biodegradation rate of biodegradable polymers in relationship to the expected lifetime. Since the kinetics of hydrolysis strongly depend on the pH of the hydrolyzing medium, in this work some fillers able to control the pH of water when it diffuses inside the polymer were added to PLA. In particular, fumaric acid, a bio- and eco- friendly additive, was used. In order to control the dispersion of this molecule inside the polymer, it was intercalated in a biocompatible nanofiller: layered double hydroxides (LDH). These fillers were added to the material using melt-compounding technique, suitable for industrial application. The preliminary results obtained are encouraging toward the possibility of effectively controlling the degradation rate.

Keywords: PLA; bio-nanocomposites; hydrolysis; LDH.
PACS: 81.05.

INTRODUCTION

Over the years, polymers have found an increasingly important use in different sectors. The problem of the use of traditional polymers, beyond the necessity of using oil as a raw material, is their difficult to dispose after use. A very good alternative is the use of biodegradable polymers: their macromolecules, under exposition to atmospheric agents, break down into CO_2, H_2O and inorganic nontoxic material in a lapse of time of a few months. Biodegradation process consists in two different steps: hydrolysis, in which water molecules break down the polymer chains into simpler molecules, and effective biodegradation, in which microorganisms feed on the hydrolysis products of transforming them into compounds that return to be a part of the natural cycle (CO_2, CH_4, biomass).

PLA is one of the most interesting biodegradable polymers, it is an aliphatic polyester derived from 100% renewable sources (synthetic product from starch derived from maize) [1]. Several studies investigated peculiar aspects of this resin, going from crystallization kinetics [2] to thermo-mechanical degradation during injection molding [3] and melt compounding [4]. Some investigations have focused on the hydrolytic degradation of aliphatic polyesters, analyzing the factors which can influence such phenomenon [5-7]: it is generally accepted that, among other key factors like temperature, the hydrolysis of PLA depends on the media pH [8-9]. The propensity of PLA to degradation in the presence of water significantly limits specific industrial applications, particularly for durable products with long-term performance. The solution could be using some fillers that can influence the hydrolysis rate. As additives could be considered some nanoscale materials [10]: in particular, combination of two-dimensional layered material and intercalation technique offers a new area for developing nanohybrids with desired functionality [11]. Layered double hydroxides (LDHs) are also called anionic clays, mineral of this family is hydrotalcite

(Mg-Al-CO_3); molecules negatively charged can be incorporated between hydroxide layers as charge compensating anions through ion exchange.

MATERIALS

In this work the PLA 4032D produced by Natureworks was used. This PLA grade contains about 2% of D-lactide with a maximum degree of crystallinity of about 45% (Xc = 45%) and with molecular weights Mn = 119,000 g/mol, Mw = 207,000 g/mol [12-13].

Fumaric acid was acquired by Sigma Aldrich.

LDHs were obtained by co-precipitation according to the procedure described in the literature [14]. This filler was used both in nitrate form (LDH in the following), and intercalated with fumaric acid (LDH+fumaric acid).

EXPERIMENTAL PROCEDURES

Sample Production

The pellets of PLA and all the fillers were dried for 24 h under vacuum at a temperature of 60°C. The materials were melt compounded in a Minilab Haake Thermoscientific twin-screw mini-extruder, with a homogeneous temperature of 170°C and at 100 rpm, with a cycle time of 5 min. Several materials were obtained: pure PLA 4032D extruded; 4032D+3% (LDH); 4032D+3% (LDH+fumaric acid); 4032D+1% (fumaric acid). Percentage reported above are wt/wt; 1% fumaric acid corresponds about to the same quantity of organic acid intercalated in 3% (LDH+fumaric acid).

From all the extruded materials several amorphous films were obtained by compression molding (T=170°C) with an average thickness of 200-300 μm.

Hydrolysis at high temperature

In order to assess the significance of hydrolysis in the molten state, time sweep tests were performed. The equipment used to carry out the rheological measures was a rotary rheometer Haake Mars Thermo Scientific. Rheological tests were performed at 200°C, with a stress set to 500 Pa and a strain rate of 1 rad/s, maintaining all values constant for 3 hr. The materials were not dried before the test.

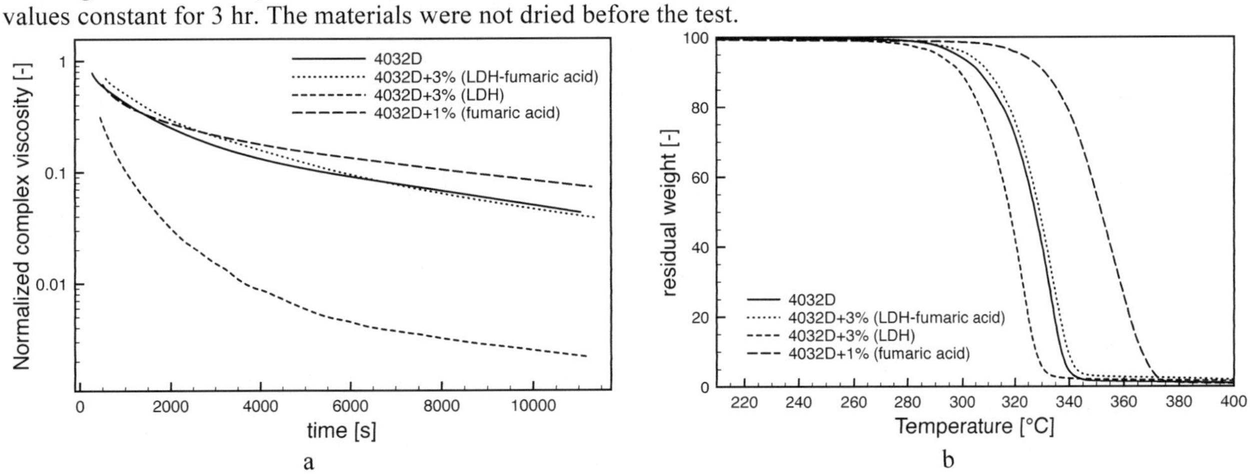

FIGURE 1. a) time evolution of viscosity during rheological tests at 1rad/s. The values were normalized with respect to starting values for each material; b) results of TGA measurements in nitrogen atmosphere at a heating rate of 10°C/min

As shown in Fig. 1a, all the materials present a decrease of viscosity during time, due to hydrolysis and thermal degradations. The presence of LDH enhances significantly the degradation, whereas fumaric acid alone seems to reduce the phenomenon. The intercalation of LDH with fumaric acid largely reduces the degradation, so that the compound with 3% (LDH+fumaric acid) behaves similary to 4032D alone. Similar results are found by TGA analysis: in the presence of fumaric acid the degradation temperature increases significantly (more than 20°C). On the contrary, LDH alone reduces the thermal stability. The intercalation of LDH with fumaric acid allows obtaining a compound with a thermal stability similar, and even slightly better, than PLA alone.

Hydrolysis in the solid state

Hydrolysis tests were conducted in distilled water at 58 °C whose pH was measured to be 6-7. This temperature was selected because it is the temperature adopted for biodegradation tests according to ASTM and ISO standards [15-16]. Several films (having a thickness of about 250 μm, weighing about 50 mg and dimensions 1.5x1.5 cm^2) of each of the samples were placed in a glass vessel containing water. The volume of water [ml] was fixed in the ratio 800:1 with the mass [g] of the dry samples, according to a previously defined protocol [12]. Water was replaced every day. At selected days of hydrolysis, the vessels containing the samples were emptied by sucking the liquid by the syringe, dried under vacuum at 60°C for 2.5 h and weighted. One of the films of each sample, after measuring its mass, was placed in a desiccator for further analysis. All the other vessels were refilled with water in the selected 800:1 ratio and put again in water at 58°C to continue the hydrolysis test. The hydrolysis of the samples was followed for about 50 days.

Weight loss

The results of hydrolysis in the solid state (at 58°C) are substantially different from what reported above concerning the degradation in the molten state. As clear from Fig. 2, PLA containing acid alone is the less resistant to hydrolysis process. The presence LDH intercalated with the same amount of acid instead improves significantly the stability of the material: the same weight loss than for the pure PLA is found when LDH+fumaric filler is dispersed by 3 %w/w but shifted at significant longer time (10 % weight loss is shifted from 39 to 46 days in presence of the filler) or at lower times (from 32 to 46 days with respect to the fumaric acid functionality intercalated or not).

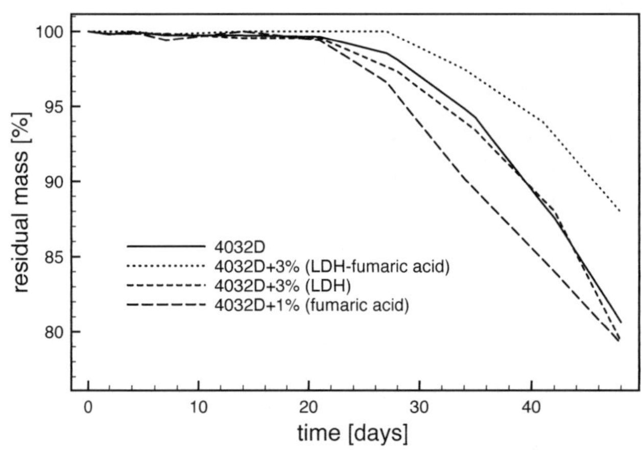

FIGURE 2. Weight loss during hydrolysis at 58°C

Gel Permeation Chromatography (GPC Analysis)

The equipment used to carry out the measures GPC is a high performance liquid chromatograph (HPLC) Waters, equipped with an auto-sampler. Analyzed samples, at different hydrolysis time, were dissolved in tetrahydrofuran, with a ratio (sample mass/solvent): 1/1[g/ml], at 50°C. The solutions obtained were then filtrated by a filter Chromafil PTFE 0,45 μm.

The results of molecular weight evolution during hydrolysis at 58°C substantially confirm the results of weight loss reported above: the loss of molecular weight is higher for the sample containing fumaric acid alone, the sample with 3% (LDH+fumaric acid), instead, show the lowest loss molecular weight. It is worth mentioning that the polydispersity index does not change significantly during the time of analysis.

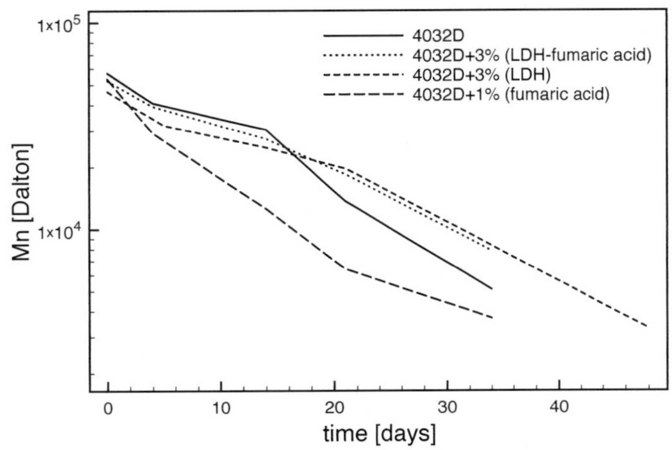

FIGURE 3. Number average molecular weight changes during hydrolysis at 58°C

Calorimetric analysis (DSC)

A differential scanning calorimetry (DSC) analysis was carried out on all the samples. The mass of the samples was about 6 mg. The tests were carried out by means of a DTA Mettler Toledo (DSC 822) under nitrogen atmosphere. The samples were heated from -10°C to 200°C at 10°C/min, held for 2 minutes at 200°C, cooled from 200°C to -10°C and then heated again from -10°C to 200°C. The thermal analysis was conducted to evaluate the glass transition temperature (Tg) and the crystalline degree (Xc) [17].

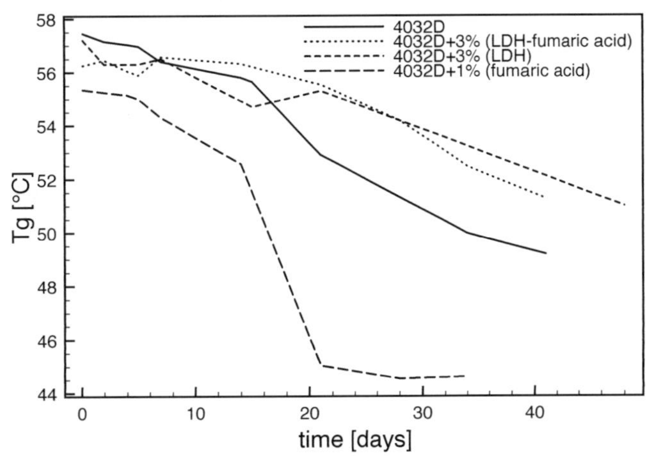

FIGURE 4. Changes of Tg during hydrolysis process, data collected during the cooling step at 10°C/min

Fig. 4 shows that Tg decreases with hydrolysis and the results are perfectly in agreement with variation of molecular weight: by adding a moderate percentage of LDH (with or without fumaric acid) allows a lower degradation rate in the solid state. The crystallinity degree, not reported here, was found to increase during the hydrolysis time, reaching a value of about 40% after 4 days, in agreement to what already reported in the literature [12].

CONCLUSIONS

In this work hydrolysis tests on a PLA grade were conducted. In order to control the rate of hydrolysis, fumaric acid and LDHs intercalated with fumaric acid was added as fillers. The hydrolysis process was followed, in function of time, both in the molten and in the solid state, by means of different techniques: rheology, TGA, variation of weight, variation of molecular weight and its distribution using GPC analysis, variation of crystallinity degree and glass transition temperature using DSC analysis.

From the analysis, it was observed that using 3%(LDH-fumaric acid) as filler makes PLA more resistant to the hydrolysis process in the solid state, without increasing the degradation rate in the molten state. The addiction of fumaric acid alone, on the contrary, allows a higher resistance to degradation at high temperatures, but induces a faster rate of hydrolysis in the solid state. Mineralization rates will have to be checked in standard conditions but, it appears now possible to increase or to decrease the time scale of biodegradability with these additives through the control of hydrolysis rate.

REFERENCES

1. D. Garlotta, *Journal of Polymers and the Environment* **9**, 63-84 (2001).
2. F. De Santis, R. Pantani, G. Titomanlio, *Thermochimica Acta* **522**, 128-134 (2011).
3. R. Pantani, F. De Santis, A. Sorrentino, F. De Maio, G. Titomanlio, *Polymer Degradation and Stability* **95**, 1148-1159 (2010).
4. F. De Santis and R. Pantani, *Journal of Polymer Research* **22**, 242 (2015). Doi: 10.1007/s10965-015-0885-1
5. R. Pantani, A. Sorrentino, *Polymer Degradation and Stability* **98**(5), 1089-1096 (2013).
6. M. R. Lostocco, S. J. Huang, *Polymer Degradation and Stability* **61**(2), 225-230 (1998).
7. D. E. Henton, P. Gruber, J. Lunt, J. Randall, *Natural fibers, biopolymers and biocomposities* ,CRC Press, 527-577 (2005).
8. S. De Jong, E. R. Arias, D. T. S. Rijkers, C. F. Van Nostrum, J. J. Kettenes-Van Den Bosch, W. E. Hennink, *Polymer* **42**(7), 2795-2802 (2001).
9. H. Tsuji, K. Nakahara, *Journal of Applied Polymer Science* **86**(1), 186-194 (2002).
10. G. Gorrasi, R. Pantani, M. Murariu, P. Dubois *Macromolecular Materials and Engineering* **299**, (2014). Doi: 10.1002/mame.201200424
11. S. Concilio, P. Iannelli, L. Sessa, R. Olivieri, A. Porta, F. De Santis, R. Pantani, S. Piotto, *Journal of Applied Polymer Science* **132**(33), 42357(1-8) (2015). Doi: 10.1002/app.42357
12. G. Gorrasi, R. Pantani, *Polymer Degradation and Stability* **98**, 1006-1014 (2013).
13. V. Speranza, A. De Meo, R. Pantani, *Polymer Degradation and Stability* **100**, 37-41 (2014).
14. L. Sisti, G. Totaro, M. Fiorini, A. Celli, C. Coelho, M Hennous, M., et al. *Journal of Applied Polymer Science* **130**(3), 1931-1940 (2013).
15. R. Pantani, F. De Santis, *AIP Conference Proceedings* **1695** (2015). Doi: 10.1063/1.4937344
16. R. Pantani, F. De Santis, F. Auriemma, C. De Rosa, R. Di Girolamo, *Polymer* (2016). Doi: 10.1016/j.polymer.2016.07.008
17. F. De Santis, G. Gorrasi, R. Pantani *Polymer Testing* **44**, (2015). Doi: 10.1016/j.polymertesting.2015.03.015

AUTHOR INDEX

AUTHOR INDEX

AUTHOR INDEX

AUTHOR INDEX

AUTHOR INDEX

AUTHOR INDEX

AUTHOR INDEX

AUTHOR INDEX

Polymer Processing Society
810 E. 10th St.
Lawrence, KS 66044
USA

ISBN 978-1-5108-4339-4
9 781510 843394

CURRAN ASSOCIATES INC.
proceedings.com

33rd Polymer Processing Society Annual Meeting 2017 (PPS-33)

Cancun, Mexico
10 - 14 December 2017

Volume 1 of 2